D1238059

International Geophysics
Volume 103

Ocean Circulation and Climate

A 21st Century Perspective

Edited by

Gerold Siedler
Helmholtz Centre for Ocean Research
Kiel, Germany

Stephen M. Griffies
NOAA Geophysical Fluid Dynamics Laboratory
Princeton, USA

John Gould
National Oceanography Centre
Southampton, UK

John A. Church
Centre for Australian Weather and Climate Research
A Partnership between CSIRO and the Bureau of
Meteorology, Hobart, Australia

AMSTERDAM • BOSTON • HEIDELBERG • LONDON
NEW YORK • OXFORD • PARIS • SAN DIEGO
SAN FRANCISCO • SINGAPORE • SYDNEY • TOKYO
Academic Press is an imprint of Elsevier

Academic Press is an imprint of Elsevier
The Boulevard, Langford Lane, Kidlington, Oxford OX5 1GB, UK
Radarweg 29, PO Box 211, 1000 AE Amsterdam, The Netherlands

British Library Cataloguing in Publication Data
A catalogue record for this book is available from the British Library

Library of Congress Cataloging-in-Publication Data
A catalog record for this book is available from the Library of Congress

ISBN: 978-0-12-391851-2
ISSN: 0074-6142

Printed and bound in China

13 14 15 16 17 10 9 8 7 6 5 4 3 2 1

Contents

Contributors

Numbers in parentheses indicate the Chapter on which the authors contributions begin.

Molly O. Baringer (Chapter 29), NOAA/Atlantic Oceanographic and Meteorological Laboratory, Miami, Florida, USA
molly.baringer@noaa.gov

Nicholas R. Bates (Chapter 30), Bermuda Institute of Ocean Sciences, Ferry Reach, Bermuda
nick.bates@bios.edu

Lisa Beal (Chapter 13), Rosenstiel School of Marine and Atmospheric Science, University of Miami, Miami, Florida, USA
lbeal@rsmas.miami.edu

Swadhin Behera (Chapter 15), Research Institute for Global Change, JAMSTEC, Yokohama, Japan
behera@jamstec.go.jp

Amy S. Bower (Chapter 13), Woods Hole Oceanographic Institution, Woods Hole, Massachusetts, USA
abower@whoi.edu

Tim P. Boyer (Chapter 28), National Oceanographic Data Center, NOAA, Silver Spring, Maryland, USA
tim.boyer@noaa.gov

Peter Brandt (Chapter 15), Helmholtz Centre for Ocean Research Kiel (GEOMAR), Kiel, Germany
pbrandt@geomar.de

F.O. Bryan (Chapter 8), National Center for Atmospheric Research, Boulder, Colorado, USA
bryan@ucar.edu

John L. Bullister (Chapter 10), National Oceanic and Atmospheric Administration—Pacific Marine Environmental Laboratory (NOAA-PMEL), Seattle, Washington, USA
John.L.Bullister@noaa.gov

Robert Burgman (Chapter 24), Florida International University, Miami, Florida, USA
rburgman@fiu.edu

Luca Centurioni (Chapter 12), Scripps Institution of Oceanography, University of California, San Diego, La Jolla, California, USA
lcenturioni@ucsd.edu

John A. Church (Chapter 27), Centre for Australian Weather and Climate Research, and CSIRO Marine and Atmospheric Research and Wealth from Oceans Flagship, GPO Box 1538, Hobart, Tasmania, Australia
John.Church@csiro.au

Vincent Combes (Chapter 14), College of Earth, Ocean and Atmospheric Sciences, Oregon State University, Corvallis, Oregon, USA
vcombes@coas.oregonstate.edu

Henk A. Dijkstra (Chapter 11), Institute for Marine Atmospheric research Utrecht, Department of Physics and Astronomy, Utrecht University, Utrecht, The Netherlands
H.A.Dijkstra@uu.nl

Catia M. Domingues (Chapter 27), Antarctic Climate and Ecosystems Cooperative Research Centre, University of Tasmania, Private Bag 80, Hobart, Tasmania, Australia
Catia.Domingues@csiro.au

Scott C. Doney (Chapter 31), Woods Hole Oceanographic Institution, Woods Hole, Massachusetts, USA
sdoney@whoi.edu

Sybren S. Drijfhout (Chapter 11), School of Ocean and Earth Science, University of Southampton, Southampton, United Kingdom, and Royal Netherlands Meteorological Institute, De Bilt, The Netherlands
sybren.drijfhout@knmi.nl

Paul J. Durack (Chapter 28), Program for Climate Model Diagnosis and Intercomparison (PCMDI), Lawrence Livermore National Laboratory (LLNL), Livermore, California, USA, and CSIRO Marine and Atmospheric Research, Hobart, Australia
durack1@llnl.gov

Rainer Feistel (Chapter 6), Leibniz Institute for Baltic Sea Research, Warnemuende, Germany
rainer.feistel@io-warnemuende.de

Georg Feulner (Chapter 2), Potsdam Institute for Climate Impact Research, Potsdam, Germany
georg.feulner@pik-potsdam.de

Gael Forget (Chapter 9), Department of Earth, Atmospheric and Planetary Sciences, Massachusetts Institute of Technology, Cambridge, Massachusetts, USA
gforget@mit.edu

B. Fox-Kemper (Chapter 8), Geological Sciences, Brown University, Providence, Rhode Island, USA
bfk@colorado.edu

Lee-Lueng Fu (Chapter 4), Jet Propulsion Laboratory, California Institute of Technology, Pasadena, California, USA
llf@jpl.nasa.gov

Alberto C. Naveira Garabato (Chapters 7 and 18), National Oceanography Centre, University of Southampton, Southampton, United Kingdom
acng@noc.soton.ac.uk

Marion Gehlen (Chapter 26), Laboratoire des Sciences du Climat et de l'Environnement (LSCE), UMR, CEA-CNRS, Gif-sur-Yvette, France
marion.gehlen@cea.fr

Peter R. Gent (Chapter 23), National Center for Atmospheric Research, Boulder, Colorado, USA
gent@ucar.edu

John Gould (Chapter 3), National Oceanography Centre, Southampton, United Kingdom
wjg@noc.soton.ac.uk

Stephen M. Griffies (Chapter 20), NOAA Geophysical Fluid Dynamics Laboratory, Princeton, New Jersey, USA
stephen.griffies@noaa.gov

Serge Gulev (Chapter 5), P.P. Shirshov Institute of Oceanology, Moscow, Russia
gul@sail.msk.ru

Patrick Heimbach (Chapter 21), Department of Earth, Atmospheric and Planetary Sciences, Massachusetts Institute of Technology, Cambridge, Massachusetts, USA
heimbach@mit.edu

Christoph Heinze (Chapter 26), Geophysical Institute, University of Bergen; Bjerknes Centre for Climate Research, and Uni Klima, Uni Research, Bergen, Norway
christoph.heinze@gfi.uib.no

David M. Holland (Chapter 16), Courant Institute of Mathematical Sciences, New York University, New York, USA
holland@cims.nyu.edu

Shiro Imawaki (Chapter 13), Japan Agency for Marine–Earth Science and Technology, Yokohama, Japan
imawaki@jamstec.go.jp

Simon A. Josey (Chapter 5), National Oceanography Centre, Southampton, United Kingdom
simon.josey@noc.soton.ac.uk

Ben P. Kirtman (Chapter 24), University of Miami—RSMAS, Miami, Florida, USA
bkirtman@rsmas.miami.edu

Arne Körtzinger (Chapter 30), Helmholtz Centre for Ocean Research Kiel (GEOMAR), Kiel, Germany
akoertzinger@geomar.de

Mojib Latif (Chapter 25), Helmholtz Centre for Ocean Research Kiel (GEOMAR), and Cluster of Excellence "The Future Ocean," Kiel University, Kiel, Germany
mlatif@geomar.de

Tong Lee (Chapter 22), Jet Propulsion Laboratory, California Institute of Technology, Pasadena, California, USA
Tong.Lee@jpl.nasa.gov

R. Lumpkin (Chapters 8 and 12), Physical Oceanography Division, Atlantic Oceanographic and Meteorological Laboratory, Miami, Florida, USA
Rick.Lumpkin@noaa.gov

Alison M. Macdonald (Chapter 29), Woods Hole Oceanographic Institution, Woods Hole, Massachusetts, USA
amacdonald@whoi.edu

Jennifer MacKinnon (Chapter 7), Scripps Institution of Oceanography, La Jolla, California, USA
jmackinn@ucsd.edu

David P. Marshall (Chapter 11), Department of Physics, University of Oxford, Oxford, United Kingdom
marshall@atm.ox.ac.uk

Shuhei Masuda (Chapter 22), Research Institute for Global Change, Japan Agency for Marine-Earth Science and Technology (JAMSTEC), Yokohama, Japan
smasuda@jamstec.go.jp

Cecilie Mauritzen (Chapters 10 and 17), CICERO Center for International Climate and Environmental Research Oslo, Oslo, Norway, and CICERO Center for Climate and Environmental Research—Oslo, Oslo, Norway
c.mauritzen@met.no

Nikolai Maximenko (Chapter 12), International Pacific Research Center, School of Ocean and Earth Science and Technology, University of Hawaii at Manoa, Honolulu, Hawaii, USA
maximenk@hawaii.edu

Trevor J. McDougall (Chapter 6), University of New South Wales, Sydney, New South Wales, Australia
trevor.mcdougall@unsw.edu.au

Herlé Mercier (Chapter 19), CNRS, Laboratoire de Physique des Océans, Plouzané, France
Herle.Mercier@ifremer.fr

Elaine R. Miles (Chapter 27), Bureau of Meteorology, GPO Box 1289, Melbourne, Victoria, Australia
e.miles@bom.gov.au

Didier P. Monselesan (Chapter 27), Centre for Australian Weather and Climate Research, and CSIRO Marine and Atmospheric Research and Wealth from Oceans Flagship, GPO Box 1538, Hobart, Tasmania, Australia
Didier.Monselesan@csiro.au

Rosemary Morrow (Chapter 4), Laboratoire des Etudes en Géophysique et Océanographie Spatiale, Observatoire Midi-Pyrénées, Toulouse, France
Rosemary.Morrow@cnes.fr

Rich Pawlowicz (Chapter 6), University of British Columbia, Vancouver, British Columbia, Canada
rich@eos.ubc.ca

Oscar Pizarro (Chapter 14), Department of Geophysics and Center for Oceanographic Research in the Eastern South Pacific (COPAS), University of Concepcion, Chile
opizarro@udec.cl

Bo Qiu (Chapter 13), School of Ocean and Earth Science and Technology, University of Hawaii, Honolulu, Hawaii, USA
bo@soest.hawaii.edu

Stefan Rahmstorf (Chapter 2), Potsdam Institute for Climate Impact Research, Potsdam, Germany
stefan.rahmstorf@pik-potsdam.de

Gilles Reverdin (Chapter 15), Laboratoire d'Océanographie et du Climate (LOCEAN), Paris, France
reve@locean-ipsl.upmc.fr

Monika Rhein (Chapter 10), Institute of Environmental Physics IUP, University of Bremen, Bremen, Germany
mrhein@physik.uni-bremen.de

Stephen R. Rintoul (Chapter 18), CSIRO Marine and Atmospheric Research, Antarctic Climate and Ecosystems Cooperative Research Centre, University of Tasmania, Centre for Australian Weather and Climate Research, Hobart, Tasmania, Australia
Steve.Rintoul@csiro.au

Bert Rudels (Chapter 17), Finnish Meteorological Institute, Erik Palménin aukio 1, P.O. Box 503, FI-00101 Helsinki, Finland
rudels@fmi.fi

Andreas Schiller (Chapter 22), Centre for Australian Weather and Climate Research, CSIRO Wealth from Oceans Flagship, Hobart, Australia
Andreas.Schiller@csiro.au

Frank A. Shillington (Chapter 14), Department of Oceanography and Nansen-Tutu Centre, University of Cape Town, Rondebosch, South Africa
frank.shillington@uct.ac.za

Gerold Siedler (Chapter 19), Helmholtz Centre for Ocean Research Kiel (GEOMAR), Kiel, Germany
gsiedler@geomar.de

Bernadette Sloyan (Chapter 3), CSIRO Marine and Atmospheric Research, GPO Box 1538, Hobart, Tasmania, Australia
Bernadette.Sloyan@csiro.au

Kevin Speer (Chapter 9), Department of Earth, Ocean, and Atmospheric Science, the Geophysical Fluid Dynamics Institute, Florida State University, Tallahassee, Florida, USA
kspeer@ocean.fsu.edu

Janet Sprintall (Chapter 19), Scripps Institution of Oceanography, U.C. San Diego, La Jolla, California, USA
jsprintall@ucsd.edu

Lou St Laurent (Chapter 7), Woods Hole Oceanographic Institution, Woods Hole, Massachusetts, USA
lstlaurent@whoi.edu

Tim Stockdale (Chapter 24), European Centre for Medium-Range Weather Forecasts, Shinfield Park, United Kingdom
Tim.Stockdale@ecmwf.int

Thomas F. Stocker (Chapter 1), Physics Institute, University of Bern, Bern, Switzerland
stocker@climate.unibe.ch

P. Ted Strub (Chapter 14), College of Earth, Ocean and Atmospheric Sciences, Oregon State University, Corvallis, Oregon, USA
tstrub@coas.oregonstate.edu

Toste Tanhua (Chapter 30), Helmholtz Centre for Ocean Research Kiel (GEOMAR), Kiel, Germany
ttanhua@geomar.de

John Toole (Chapter 17), Woods Hole Oceanographic Institution, Woods Hole, Massachusetts, USA
jtoole@whoi.edu

Anne Marie Treguier (Chapter 20), Laboratoire de Physique des Océans, LPO, Brest, France
treguier@ifremer.fr

Martin Visbeck (Chapter 3), Helmholtz Centre for Ocean Research Kiel (GEOMAR), Kiel, Germany
mvisbeck@geomar.de

Neil J. White (Chapter 27), Centre for Australian Weather and Climate Research, and CSIRO Marine and Atmospheric Research and Wealth from Oceans Flagship, GPO Box 1538, Hobart, Tasmania, Australia
Neil.White@csiro.au

Susan E. Wijffels (Chapter 28), CSIRO Marine and Atmospheric Research, Hobart, Australia
Susan.Wijffels@csiro.au

Carl Wunsch (Chapter 21), Department of Earth, Atmospheric and Planetary Sciences, Massachusetts Institute of Technology, Cambridge, Massachusetts, USA
cwunsch@mit.edu

Lisan Yu (Chapter 5), Woods Hole Oceanographic Institution, Woods Hole, Massachusetts, USA
lyu@whoi.edu

Acknowledgments

The editors want to thank all the mostly anonymous reviewers for their invaluable help in the preparation of the book and in checking and improving all the individual chapters. We also want to acknowledge the support of individual authors and institutions in preparing the figure set. G. S. wants to thank his colleagues at the Helmholtz Centre for Ocean Research Kiel (GEOMAR) for their manifold help and advice and his wife for her patience and support. S. M. G. thanks his colleagues at NOAA/GFDL for continued fruitful collaboration and inspiration over the past 20 years, resulting in a wonderful environment for conducting fundamental research into some of the hard-problems of climate science. He also thanks his family for patient support while completing this project. J. G. thanks the National Oceanography Centre for his position as a visiting scientist during the preparation of this book and his wife, Hilary, for her encouragement and support. J. A. C. thanks his family and colleagues for their generous support and was partly supported in the preparation of this publication by the Australian Government Department of Industry, Innovation, Climate Change, Science, Research and Tertiary Education, the Bureau of Meteorology and CSIRO through the Australian Climate Change Science Program.

Cover Graphics

Front cover images are views of the ocean developed since the first edition was published and were prepared at CSIRO Hobart, Australia.

Main image was prepared by Neil White and Lea Crosswell, CSIRO Hobart. Pacific: the trend of sea level from 1993 through 2012 from satellite altimeter missions, showing larger average rates of rise in the western Pacific/eastern Indian Oceans and smaller values in the eastern Pacific Ocean (see http://www.cmar.csiro.au/sealevel/sl_hist_last_15.html). Atlantic: $\log_{10}$ of the daily mean surface ocean speed in a 1/10th degree configuration of the Modular Ocean Model (MOM) used in a global coupled climate model developed at NOAA's Geophysical Fluid Dynamics Laboratory. Such ocean model resolution allows for a realistic representation of the ocean mesoscale eddy field. Indian: 1993–2002 mean zonal surface geostrophic velocity was calculated as described in Figure 12.3. It reveals a web of nearly ubiquitous jet-like features usually hidden by eddy noise.

Shadow image prepared by Bernadette Sloyan is a composite of the 3000+ floats of the CTD profiling Argo array and the lines along which the GO-SHIP project measures top to bottom ocean properties (see Chapter 3).

Preface

Recognition, by scientists and nonscientists alike, of the oceans' important role in our climate has grown enormously in the past 25 years. In part, this growth was driven by the World Ocean Circulation Experiment (WOCE), the planning for which started in the 1980s and on which the first edition of this book, published in 2001, was focused. Although the issues relating to climate variability and change were well known in 2001, they occupy a far more central position in oceanography and climate science in 2013. The increased focus on climate change and its socioeconomic consequences has arisen from the accumulation of unambiguous evidence of substantial changes in the climate system associated with human activities. The oceans are central to these changes as we document increases in the ocean heat and carbon content, sea-level rise, Arctic sea-ice loss, ocean acidification, and more. These changes are so important that they provide the impetus for many present-day ocean research programs. Hence it is no accident that a number of authors of the present volume also played key roles in the Fifth Assessment Report (AR5) of the Intergovernmental Panel on Climate Change (IPCC) completed in October 2013. The ocean science community is increasingly being asked to provide expertise to help chart a path toward sustainability in a world where significant change is inevitable. In parallel with this increased recognition, our understanding of the oceans and our capability to observe and model them has improved enormously. WOCE was a major driver of this improvement and progress has continued to be made by WOCE's successor projects in the World Climate Research Programme, by initiatives of the Global Climate Observing System and by many national programs.

We have been struck, as we trust our readers will be, by the enormous progress in ocean observations and modeling that has been made since the publication of the first edition of *Ocean Circulation and Climate*. In 2001, the first floats were deployed in the Argo profiling float array. Argo has since become a mainstay of our *in situ* ocean-observing system. In 2004, routine monitoring of the Atlantic Meridional Overturning Circulation started and continues to this day. Since the beginning of the twenty-first century, our observational capability, data delivery, and quality control systems have advanced to the point where ocean indices and gridded fields can be presented on a weekly or monthly basis, not only at the ocean surface but also for the subsurface ocean. We have a new and rigorous equation of the state of seawater and a redefinition of salinity published in 2010. We can now investigate interannual, and to some degree decadal, variability and change of sea level using more than 20 years of satellite altimetry measurements.

The first goal of WOCE was to develop ocean models for use in climate research. The significant improvement of coupled atmosphere-ocean general circulation models has in part been underpinned by progress made during and since WOCE, but has also been aided by the relentless increase in computing power. Models of the global ocean commonly, though not yet routinely, include mesoscale features (eddies, fronts, boundary currents) that were extremely rare in models circa 2001. Consequently, simulations, particularly those assimilating observations, play a central role in interpreting the current state of the ocean and in aiding the planning for new measurement programs. Progress in computer power has allowed not only more realistic representations of ocean physics but also the inclusion of ocean biogeochemistry. These models also no longer rely on artificial flux corrections to prevent model drift.

A deeper conceptual understanding of the role that the oceans play in climate has been prompted by the many improvements to the observational record and enhanced modeling capabilities. The particular role that the oceans play in long-term climate variations, trends, and in the predictability of these changes is greatly aided by idealized conceptual models, with significant progress made along these lines over the past decade.

This book aims at presenting ocean science and its relevance to climate more than a decade after the first edition of *Ocean Circulation and Climate*. When we were approached by Elsevier to consider editing a second edition, we were initially reticent, remembering the magnitude of the work involved in producing the first edition. However, our peers and younger colleagues reassured us of the value of the first edition as a unique resource documenting the state of knowledge of the oceans' role in climate following WOCE; a project in which we were all variously involved. We were encouraged to develop the second edition presenting "A 21st Century Perspective."

The achievements attained by the end of WOCE provided an almost entirely physically focused "snapshot" of the ocean as documented in the first edition. However, since 2001, there has been progress on a broader front. In part, this is due to the significant lengthening of the record of changing ocean properties with the WOCE era observations providing a "state of the ocean in the 1990s" benchmark. The greater maturity and comprehensiveness of the network of ocean observations was founded during WOCE and was subsequently stimulated by the 1999 and 2009 OceanObs conferences, with the second of these having a more interdisciplinary focus. For this new edition of the book, we aimed to have a strong involvement of scientists whose careers developed during and after WOCE. Our authors include a great proportion of younger scientists from the post-WOCE generation.

This second edition of *Ocean Circulation and Climate* leads the reader through the important areas of progress since the first edition. It is composed of six parts each prefaced by a short introduction. The reader is initiated in the ocean's role in climate by Part I. In Parts II – IV we discuss developments in ocean observations, processes in the ocean and elements of the global-scale circulation. Modeling of the ocean and the coupled climate system and its changes are discussed in Parts V and VI. While the book focuses on ocean physics, we have also included related paleoclimatic findings and descriptions of biogeochemical processes, marine ecosystems, and the carbon cycle. There are inevitable overlaps between chapters, and cross-references are inserted where appropriate. Production of the book ran in parallel with the preparation of the IPCC AR5 for which this book provides an ocean-focused complement addressing the underpinning scientific issues.

Despite the enormous progress made during and since the WOCE era, many challenges remain. Perhaps first and foremost is maintaining the *in situ* and satellite observational capability and filling the remaining gaps to provide a comprehensive knowledge and understanding of how and why the oceans are changing. Development of ocean and climate models will require continued and strengthened partnerships between ocean, atmospheric, and cryospheric modelers and with the observational community. The ocean–cryosphere interaction is an important and urgent challenge.

There are also new opportunities, particularly as the length of observational time series increases and as new and innovative techniques become established. For example, this edition has little focus on the various gravity missions that are beginning to bring new insights into ocean circulation and its variability. These missions are likely to have a much bigger impact on a subsequent edition of this book. More powerful computers and improved ocean model representation will allow greater resolution of ocean features and greater coupling of physical, cryospheric, biogeochemical, and ecosystem elements of the ocean.

With continued enthusiasm within the ocean science community to address the challenging issues of ocean climate, we are confident that the future for understanding the ocean circulation, its interaction with the atmosphere and the cryosphere, and the related biogeochemical and ecosystem sciences is bright.

Gerold Siedler
Helmholtz Centre for Ocean Research
Kiel, Germany

Stephen M. Griffies
NOAA Geophysical Fluid Dynamics Laboratory
Princeton, USA

John Gould
National Oceanography Centre
Southampton, UK

John A. Church
Centre for Australian Weather and Climate Research
A Partnership between CSIRO and the Bureau of
Meteorology, Hobart, Australia

Part I

The Ocean's Role in the Climate System

This book is concerned with the role of the ocean in the climate system, with a focus on physical processes relevant to scientifically describing and understanding that role. The two chapters in this Part I offer a context for the discussion extending over the nearly 30 chapters that follow. We are introduced in this Part to notions of the huge space and time scales over which ocean phenomena play a role in climate. Such phenomena extend from the millimeter and seconds relevant for irreversibly mixing heat, carbon, and other tracers, to the global scales of sea level changes, water mass variations, and air–sea interactions that comprise natural and anthropogenic climate fluctuations. We also encounter evidence seen for rather large fluctuations in the paleo record, and consider future climate scenarios suggested through computer simulations.

The ocean is often termed the flywheel of the climate system. This metaphor has relevance given that roughly 3 m of the ocean contains heat equivalent to the entire atmosphere. Furthermore, the ocean absorbs the relatively rapid synoptic atmospheric fluctuations in a way that provides a low pass filter to the climate system, and in so doing reflects the vast thermal inertia available in the ocean. Furthermore, the ocean transfers mechanical energy across space and time scales, thus adding both richness and complexity to ocean phenomena. These properties form part of the foundation for how the ocean interacts with other components of the earth climate system.

As oceanographers explore the ocean through theory, observations, and simulations, they continue to solidify the perspective that nearly all ocean phenomena play some role in climate. For example, climate scientists as recently as the early 1990s discounted the role of astronomical ocean tides, assuming they merely add a few harmonic wiggles to measurements yet have no major impact on climate. However, research on the role of tides and ocean mixing, and the associated role of ocean mixing on climate, provide striking evidence for the basic connection between tides and climate. Other examples can be identified where developing a fundamental understanding of ocean processes is prerequisite to establishing a deeper and more predictive description of climate. This theme forms part of the mandate for this book.

Science generally evolves through methodical additions to previous theories and observations, as well as through the somewhat chaotic paradigm shifts arrived at through new insights, data, and perspectives. Oceanography and climate science have seen their share of such evolutionary steps, with examples provided in the two chapters of this Part as well as others exhibited throughout this book. We trust that this Part will whet the appetite for readers as they further explore elements of how the ocean forms a fundamental component of the earth's climate system.

Chapter 1

The Ocean as a Component of the Climate System

Thomas F. Stocker
Physics Institute, University of Bern, Bern, Switzerland

Chapter Outline

1. SETTING THE SCENE

From certain points of view, the Earth appears as an almost perfect "aquaplanet" (Figure 1.1). This is a powerful visual testimony to the importance and key role of the ocean in the Earth's climate system. More than 70% of the entire Earth surface is covered by the ocean and therefore most of the lower boundary of the atmosphere is in contact with water or, in the high latitudes, with the seasonal sea ice cover.

The ocean constitutes a virtually unlimited reservoir of water for the atmosphere because it is by many orders of magnitude the largest body of readily accessible water on Earth. Water is the principal resource of all life and thus represents a global commons. A quantitative understanding of the world ocean must therefore be a top priority of Earth System science. Such an understanding rests on three pillars: (i) *in situ* and remote observations with worldwide coverage; (ii) theoretical understanding of processes within the ocean and at its boundaries to other Earth System components; and (iii) capability of simulating ocean and climate processes in the past, present, and future, using a hierarchy of physical–biogeochemical models. Each of these pillars are considered in Section 2, Sections 3 and 4, and Chapter 2, and Sections 5 and 6, respectively.

Due to its large spatial extent and as the principal water reservoir on Earth, the ocean supplies more than 80% of the water vapor for the atmosphere. When the mean atmospheric temperature is higher, the atmosphere contains more water vapor, which is drawn primarily from the ocean. In times of a colder atmosphere that water is returned to the ocean, or when the climate is significantly colder as during an ice age, a significant amount of water is transferred from the ocean to the large ice sheets on land. The ocean is thus the dominant source of the Earth's most important greenhouse gas, water vapor, which is primarily responsible for increasing the Earth's mean surface temperature by about 33 °C to its present value of about 15 °C. Therefore, the ocean is an essential component for habitability of the Earth.

The climate system can be usefully partitioned into seven *spheres* which are physically coupled through exchange fluxes of energy, momentum, and matter. This is schematically illustrated in Figure 1.2. The notion *sphere* should not suggest that the Earth System components are separate entities, they are intertwined. This is, for example, evident for the *hydrosphere* which is present throughout the Earth System. The ocean as the major part of the Earth's *hydrosphere* interacts with all components of the Earth System. It is also coupled to biogeochemical processes through exchange fluxes of substances such as carbon, nitrogen, and many others. Hence, the Earth System cannot be understood without detailed quantitative knowledge of the ocean, its physical properties and the various processes that determine its status and its response to forcings and perturbations.

The ocean is coupled to the *atmosphere* through exchanges of momentum, heat, water, and many substances such as oxygen, carbon dioxide, and other trace gases, and

Ocean Circulation and Climate, Vol. 103. http://dx.doi.org/10.1016/B978-0-12-391851-2.00001-5

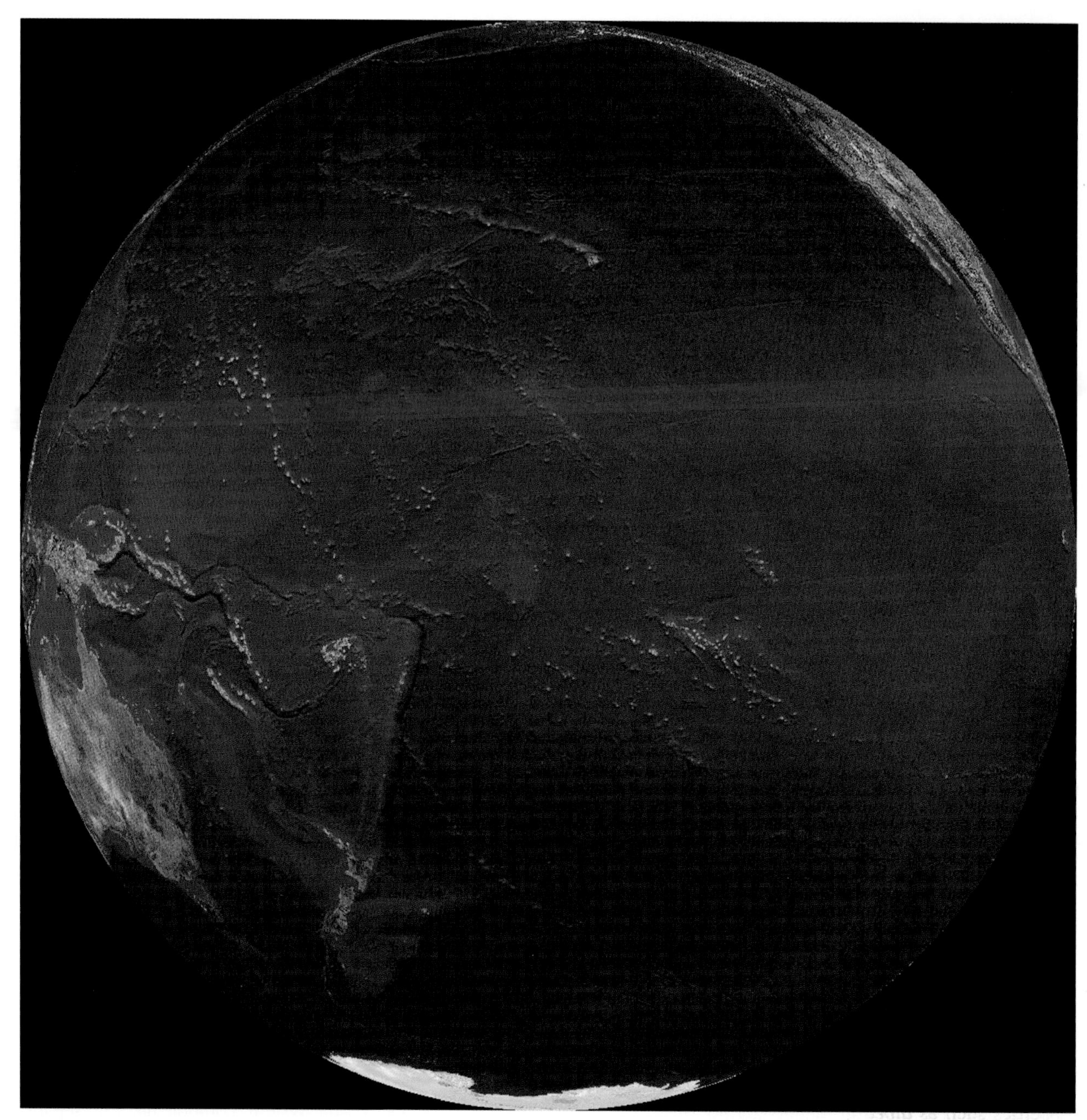

FIGURE 1.1 Ocean appearance of the Earth in an idealized cloud-free view constructed using Earth viewer (from 35,785 km above 10°S, 160°W).

minerals in the form of dust and suspended solids. The exchange of momentum, mediated by the action of wind systems, is the primary driver of the ocean circulation, but also fluxes of heat and freshwater, which influence the density of ocean water and its regional distribution, are important in determining the properties and the flow of water masses.

The *cryosphere* comprises the terrestrial ice sheets, glaciers, and sea ice, and these interact with the ocean directly and indirectly, both through the freshwater supply to the ocean and its effect on sea level (Chapter 27), and through the modification of the ocean–atmosphere heat exchange in the case of sea ice cover (Chapter 16). The ocean is also influenced by the terrestrial and marine *biosphere*, mainly through biogeochemical coupling via fluxes of carbon, oxygen, and nutrients. The *pedosphere*, that is, the land surface, directs river runoff and therefore the spatial distribution of freshwater delivery to the ocean, which modifies water mass properties and ocean circulation. The *lithosphere* comprises the solid Earth which supplies minerals through, for example, volcanism and weathering, and is responsible for processes that

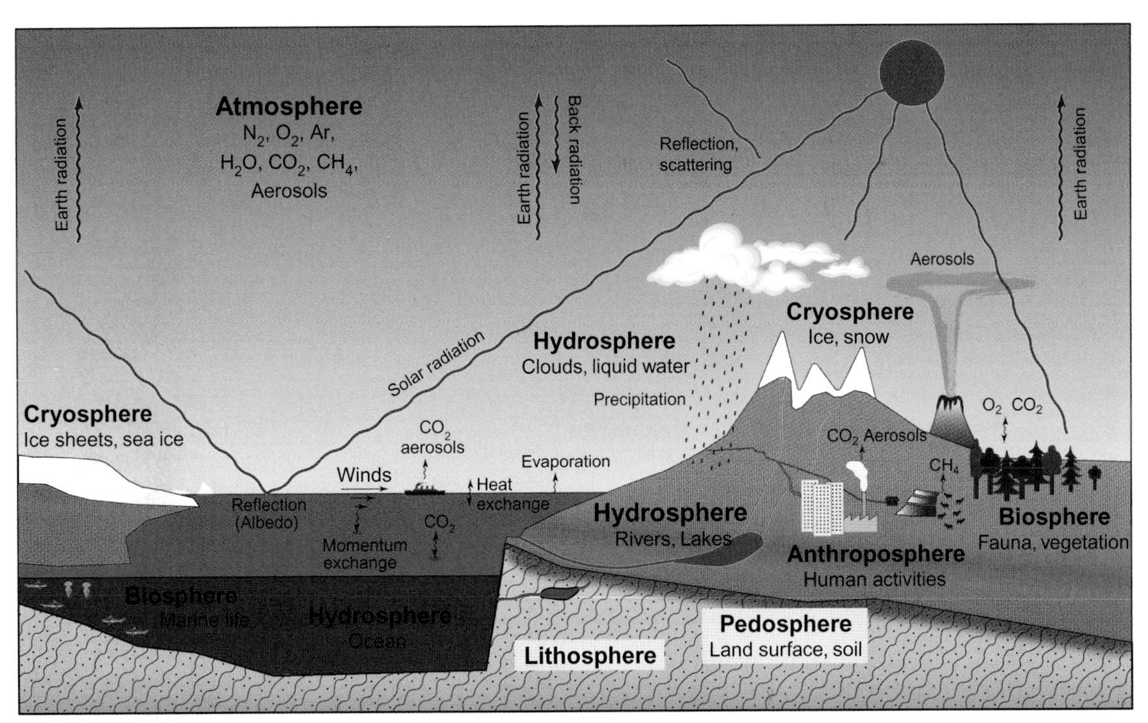

FIGURE 1.2 Illustration of the seven *spheres* of the Earth System, which are intertwined and physically coupled through exchange fluxes of energy, momentum, and matter, and biogeochemically coupled through fluxes of carbon and other substances.

influence the geochemistry of the climate system. It is a component that is relevant when Earth System processes on time scales of 10^5 years and longer are studied.

The latter part of the eighteenth century marks the beginning of the industrial use of coal, and therefore a seventh *sphere* has started to become important in the Earth System. It is the *anthroposphere*, indicated in Figure 1.2, which comprises all human activities, for example, emissions of greenhouse gases through the burning of fossil fuels, land use change, particularly deforestation, and the input of dust and chemical constituents into the various spheres. Further, anthropogenic land surface changes also impact physical properties such as *albedo*, surface roughness, and regional water and heat balances. Human activities have become important drivers of climate and environmental change, and we now live in an epoch in which we leave traces and imprints that will be detectable by our successors using today's classical analytical methods many centuries and millennia from now. It is thus appropriate to put this epoch into a longer-term context of geological epochs by naming it *Anthropocene*, as proposed by Crutzen and Stoermer (2000). Therefore, understanding of the *anthroposphere* and its influence on all the other Earth System spheres is an important prerequisite for assessing the future evolution of the Earth's spheres on a human time scale and thereby for a responsible stewardship of our only home.

2. THE OCEAN AS AN EXCHANGING EARTH SYSTEM RESERVOIR

The ocean covers about 71% of the Earth surface and has a mean depth of 3734 m (Talley et al., 2011) as estimated from the most recent geodetic data analysis (Becker et al., 2009). The ocean volume is about 1.34×10^{18} m^3 and thus contains more than 95% of the Earth's water that participates in the hydrological cycle. In addition, the ocean is a large reservoir of heat, and many substances that are cycled in the Earth System. Particularly, the ocean is the largest storage of carbon, apart from the lithosphere, which is not considered here.

The ocean is a large reservoir of heat with a strong seasonal cycle of temperature observed in the upper 250 m. In the northern and southern hemispheres, the peak-to-peak variations, that is, summer minus winter, of the heat content in the upper 250 m are about 1.4×10^{23} and 2×10^{23} J, respectively (Antonov et al., 2004). The seasonal heat fluxes of 9–13 PW (1 PW $= 10^{15}$ W) which effect these variations are mainly atmosphere–ocean heat fluxes, whereas the net meridional heat fluxes in the ocean are an order of magnitude smaller (see Section 3). The hemispheric asymmetry leads to a net seasonal cycle of the world ocean heat content with an amplitude of about 4×10^{22} J, which peaks in April and assumes a minimum in September

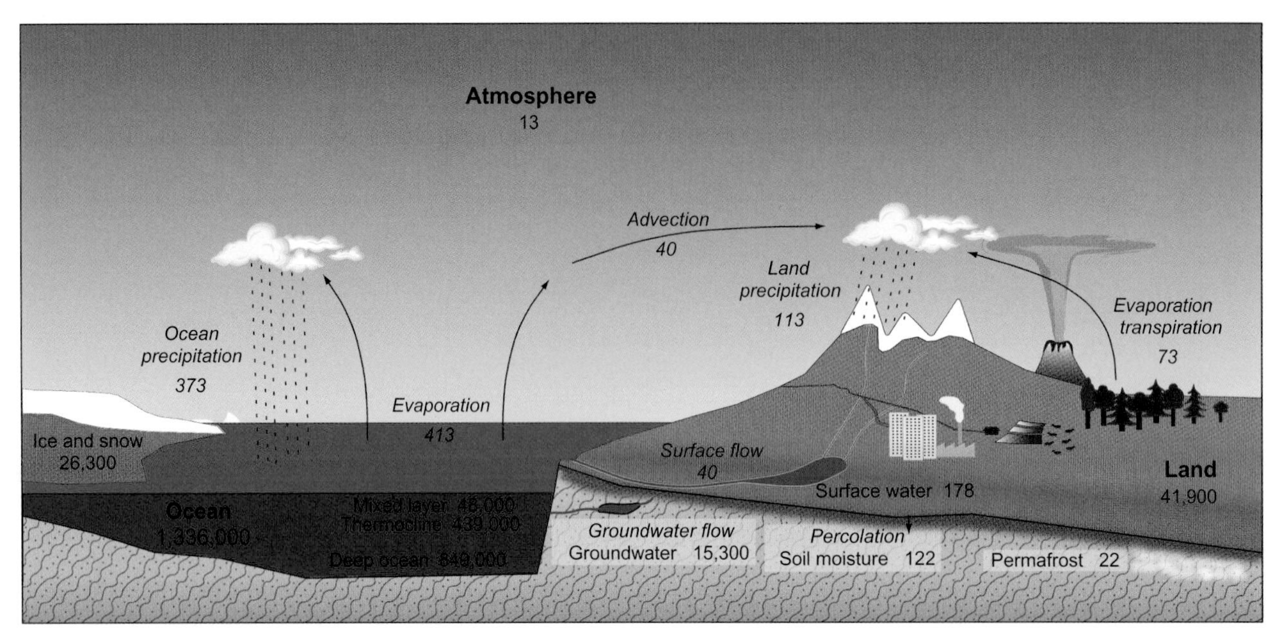

FIGURE 1.3 Inventories of water (in 10^{15} kg) and water fluxes (*italics*, in 10^{15} kg/yr) in the Earth System, based on estimates by Trenberth et al. (2007).

(Antonov et al., 2004; Fasullo and Trenberth, 2008). Seasonal variations of sea-surface temperature are around 2 °C in the tropics and the Southern Ocean, and exceed 10 °C between about 30°N and 50°N in the western parts of the North Atlantic and North Pacific Ocean basins. Together with the seasonal cycle in the wind stress (Risien and Chelton, 2008), this causes substantial changes in the depth of the mixed layer and the thermocline, which in turn determine the upwelling and the availability of nutrients for the marine biosphere.

Observations point to large seasonal variations in the water content of the atmosphere (Trenberth et al., 2007, 2011). Since the ocean is a fundamental component of the global water cycle, seasonal variations must also be present in the freshwater content of the ocean and are expressed most strongly near the surface because of air–sea coupling, continental runoff, and seasonal sea ice export from the polar regions. The summer-minus-winter differences of water content in the atmosphere are about 3×10^{15} kg in the northern, and 1.8×10^{15} kg in the southern hemisphere (Peixoto and Oort, 1992). There are large seasonal variations in the cross-equatorial water flux in the atmosphere, effected by the migration of the Intertropical Convergence Zone. In summer, there is water transport into the northern hemisphere of about 1.8 Sv, while in winter about 1.4 Sv is transported southward. This results in a net northward transfer of water in the atmosphere across the equator of about 0.4 Sv over the course of a year. Consequently, the northern hemisphere experiences an excess of precipitation over evaporation, which is then supplied to the ocean. The ocean closes the global water cycle by a net freshwater transport from the northern to the southern hemisphere of about 0.5 Sv (Wijffels, 2001). This is the net effect of the ocean circulation in all basins, and ocean-based observations are consistent with the estimate of atmospheric freshwater transport, within the uncertainties.

Annual-mean fluxes of water between different reservoirs in the climate system are schematically illustrated in Figure 1.3. They are given as mass fluxes in 10^{15} kg/year. More commonly in oceanography and hydrology, such fluxes are reported as volume fluxes. A widely used and by now standard unit of large-scale volume flux in oceanography is 1 Sv = 1 Sverdrup = 10^6 m^3/s.[1] The large-scale hydrological volume fluxes are more commonly reported in 10^3 km^3/year.[2] Global runoff is thus estimated at about 1.3 Sv (Labat et al., 2004; Trenberth et al., 2007).

Water is transported between the reservoirs by regionally varying evaporation and precipitation, and by transport in the atmosphere and on land (Chapter 5). On a global scale, almost 10 times more water is delivered to the ocean by precipitation than by runoff from the continents. The latter compensates the slight imbalance between evaporation and precipitation over the ocean. The residence

1. This unit was coined by the late Max Dunbar (1914–1995), Professor of Oceanography at McGill University, and is in honor of Harald U. Sverdrup (1888–1957), physical oceanographer, and author of one of the founding treatises in oceanography.

2. Max Dunbar also proposed the volume flux unit Bering, in honor of Vitus Bering (1681–1741), a Danish explorer, who discovered that Asia and America are two separate continents and who charted the west coast of Alaska. The volume flux 1 Be = 1 Bering = 10^3 km^3/yr is a convenient, yet not widely used, large-scale unit in hydrological sciences; 1 Sv ≈ 31.6 Be.

time τ of the water in the world ocean can be estimated by comparing the inventory V with the total influx F to the ocean, that is, $\tau = V/F$. For the ocean, a mean residence time of water is about 3200 years, whereas it is less than 10 days for the atmosphere. However, if one determines the age of a particular water mass in the ocean, the internal transport of water in the ocean, and in particular the vertical transport of water masses, cannot be ignored in this regard. Near-surface waters have a much shorter residence time due to vigorous atmosphere–ocean interaction on a seasonal time scale, for example, the mixed layer has a residence time of somewhat more than 100 years.

In order to appreciate the global significance of the ocean as a reservoir of water, heat, energy, and substances, particularly carbon, inventories, and gross fluxes of some quantities are summarized in Table 1.1. Large uncertainties exist for all quantities, specifically for those associated with the various forms of energy in the ocean (Ferrari and Wunsch, 2009). The total thermal energy of the ocean is about 3×10^{25} J and is given only for the purpose of an order of magnitude estimate of the time it would take for the shortwave radiative input from the Sun to bring a totally ice covered ocean, a condition suggested to have occurred during the "Snowball Earth" phase some 500–1000 million years ago (Pierrehumbert et al., 2011), to today's mean temperature: a surprisingly short time of 6 years (assuming an ice cover of 10 m and all ocean water at freezing point). The energy contained in the general circulation is about 2×10^{25} J (Wunsch and Ferrari, 2004), and therefore of the same order of magnitude, but this quantity is not physically relevant because most of it is potential energy that is not available for the circulation (Ferrari and Wunsch, 2010). The circulation in the ocean is driven by a total supply of energy from the atmosphere of about 2.6×10^{12} W which stems from the input of mechanical energy through the action of wind and from heat fluxes across the ocean surface to compensate for the dissipation of available potential energy (see Table 1.1, footnote h).

The ocean inventories are not constant through time but measurably affected by natural climate variability and anthropogenic climate change (Chapters 27, 28, and 30). The associated anthropogenic fluxes of mass of water, thermal energy, and carbon are also included in Table 1.1. Currently, further mass is added to the world ocean from the melting of the large ice sheets of Greenland and Antarctica, and of ice caps and glaciers around the world. Based on gravimetric satellite data, this amounts to about 3.6×10^{14} kg of water every year. Although only about 1% of global runoff, this additional mass flux constitutes more than 50% of the observed sea-level rise of the past decades, about 1.1 mm/year since 1972 (Church et al., 2011). The world ocean is also absorbing a large amount of thermal energy estimated at a rate of about 1.9×10^{14} W since 1972 (Church et al., 2011), and thus has a slowing effect on anthropogenic climate change, as it stores a substantial fraction of the heat, some of which would otherwise be observed in the atmosphere in response to the increase in greenhouse gas concentrations. This heat causes a warming of the upper ocean and an associated contribution to sea-level rise of about 0.8 mm/year since 1972. A significantly larger average energy flux into the ocean of about $2.3 \cdot 10^{14}$ W from 1993 to 2008 was reported by Lyman et al. (2010).

TABLE 1.1 Estimates of Inventories of Basic Ocean Quantities, Gross Fluxes Across the Ocean Surface, and Anthropogenic Perturbation Fluxes

Quantity	Ocean Inventory	Gross Flux	Anthropogenic Flux
Mass	1.3×10^{21} kg	4.1×10^{17} kg/year	$+3.6 \times 10^{14}$ kg/year[a]
Thermal energy	3×10^{25} J[b]	6×10^{16} W[c]	$+1.9 \times 10^{14}$ W[d]
Kinetic energy	3.8×10^{18} J[e]	1.2×10^{12} W[f]	?
Available potential energy	8×10^{20} J[g]	6×10^{16} W[h]	?
Carbon[i]	38,000 Gt	70 Gt/year	+2.2 Gt/year[j]

Fluxes into the ocean are positive.
[a] *From melting of glaciers and ice sheets from 2003 to 2009 (Riva et al., 2010).*
[b] *Referred to the freezing temperature of sea water (−1.8 °C), and assuming a global mean ocean temperature of 3.8 °C.*
[c] *Annual mean solar energy flux into the ocean (Stephens et al., 2012).*
[d] *From the change in the ocean heat storage from 1972 to 2008 (Church et al., 2011), and later corrected (J. Church, personal communications).*
[e] *Total kinetic energy as estimated by Wunsch (1998).*
[f] *Work done by wind (Munk and Wunsch, 1998).*
[g] *Referred to ocean bottom (Huang, 2010). A smaller estimate of 2×10^{20} J is by Ferrari and Wunsch (2010).*
[h] *Annual mean solar energy flux transferred to the oceans available potential energy. The net flux required to compensate dissipation is much smaller, ca. 1.4×10^{12} W (Oort et al., 1994) from thermal, and 1.2×10^{12} W from mechanical energy flux.*
[i] *In the global mean, 98.8% of dissolved inorganic carbon is in the form of HCO_3^- and CO_3^{2-}. 1 Gt = 10^{12} kg.*
[j] *Estimated mean ocean uptake of carbon 1990–2005 (Denman et al., 2007).*

The world ocean is also a large reservoir of carbon, which exchanges with the atmosphere on a very rapid time scale (Table 1.1, Chapter 30). Ocean carbon is present mainly in the form of dissolved bicarbonate and carbonate whose repartitioning is determined by the ocean's acidity–alkalinity balance. As a carbon reservoir, the ocean is over 60 times larger than the atmosphere and about 16 times larger than the terrestrial biosphere (Denman et al., 2007; Ciais et al., 2013). Carbon is transferred primarily between the atmosphere and the ocean through the gas exchange of CO_2. An associated carbon renewal time is estimated at about 540 years for the entire ocean but significantly faster, only about 13 years, for the carbon found in the surface ocean. The rapid gas exchange and the chemical equilibration between the different dissolved forms of carbon in the surface ocean generate an effective carbon buffering in the world ocean. Therefore, the ocean acts as an important storage of additional carbon from the atmosphere, which results from a variety of human activities, specifically the burning of fossil fuels and deforestation. The increase in the atmospheric carbon inventory, and hence the CO_2 concentration, would be about 70% larger than without the substantial storage effect of the world ocean. In consequence, the ocean plays an increasingly important role as a storage of anthropogenic "waste": the ocean takes up heat driven by changes in the Earth's energy balance, and it takes up carbon due to the increase of CO_2 in the atmosphere.

3. ATMOSPHERE–OCEAN FLUXES AND MERIDIONAL TRANSPORTS

The previous subsection has provided a global overview of the sizes of major inventories and fluxes associated with quantities important for the world ocean. This has merely set the scene, but more relevant information on the role of the ocean in the climate system is obtained from considering the major drivers, in particular their spatial structure at the ocean–atmosphere interface. A large input of mechanical energy is provided by the periodic variations of the differential gravitational attraction between the Earth and the Moon and, to a lesser extent, the Sun, which generate the tides around the globe. Together their supply of power to the ocean is estimated at about 3.5×10^{12} W (Munk and Wunsch, 1998). Tides have a crucial impact on ocean mixing in the interior of the ocean through the breaking of internal waves (Chapters 7 and 8), and dissipation around the continental boundaries and the ocean floor through turbulence, which is generated by the periodic tidal currents. Together with mixing effected by eddies in the mean flow, each with strongly regional patterns, mixing is important on a global scale as it ultimately determines the large-scale aspects of the internal distribution of water masses (Chapters 9 and 10). However, tides have little direct effect on the general circulation of the world ocean, because the tidal residual mean circulation is at least an order of magnitude smaller than the large-scale overturning circulation and smaller still than the large-scale horizontal flow (Bessières et al., 2008).

Instead, the general circulation of the world ocean is mainly driven by the atmosphere–ocean fluxes of three quantities that together supply mechanical, thermal, and available potential energy via the transfer of momentum, heat and freshwater to the ocean (Chapters 11 and 12). The annual mean values of these fluxes and their global distribution are depicted in Figure 1.4.

Momentum is imparted to the surface layers of the ocean through the action of the wind systems in the atmosphere and the associated horizontal stresses on the ocean surface. Surface wind stress is produced by the large-scale atmospheric circulation and partly influenced by air–sea fluxes of heat and water vapor. On a global scale, wind stress is oriented primarily zonally with meridional components that are much weaker. The strongest wind stresses are observed in the westerly wind belt of the Southern Ocean, where they force the Antarctic Circumpolar Current (Figure 1.4a). The wind stress causes most of the large-scale circulation, but only rather indirectly because its effect is strongly modified by the rotation of the Earth. This is achieved by Ekman transport of water over the top few tens of meters of the ocean. Ekman transport scales with the magnitude of the wind stress and is directed to the right (left) in the northern (southern) hemisphere, relative to the wind direction. Its spatial variations cause distortions of the ocean surface and interior layers, which in turn generate horizontal pressure gradients. It is these pressure gradients which drive the large-scale gyre circulations observed in all ocean basins (Chapters 11–14).

The net heat flux to the ocean is a result of the sum of shortwave solar radiative flux, longwave thermal radiative fluxes from the ocean surface (upwelling radiation) and the atmosphere (downwelling radiation), the sensible heat flux, and the latent heat flux. On the global scale, the ocean gains heat roughly between 20°S and 20°N and releases heat poleward of this area. This implies that in the global mean, the ocean must transport heat away from the equator. The world ocean is therefore not simply a passive reservoir but an active component participating in the global heat redistribution in the Earth's climate system (Chapter 29). Regions of strongest heat exchange are clearly identified in the global datasets presented in Figure 1.4b. They are spatially very limited and indicate particular ocean circulation regimes. Heat is taken up in excess of 100 W/m^2 in the eastern equatorial Pacific, where a major ocean upwelling system is located and which generates the most important and coherent mode of internal atmosphere–ocean variability, the El Niño–Southern Oscillation (ENSO) phenomenon. Strong heat release on the order of 150 W/m^2 and more is observed at the western boundaries of the ocean basins, and they coincide

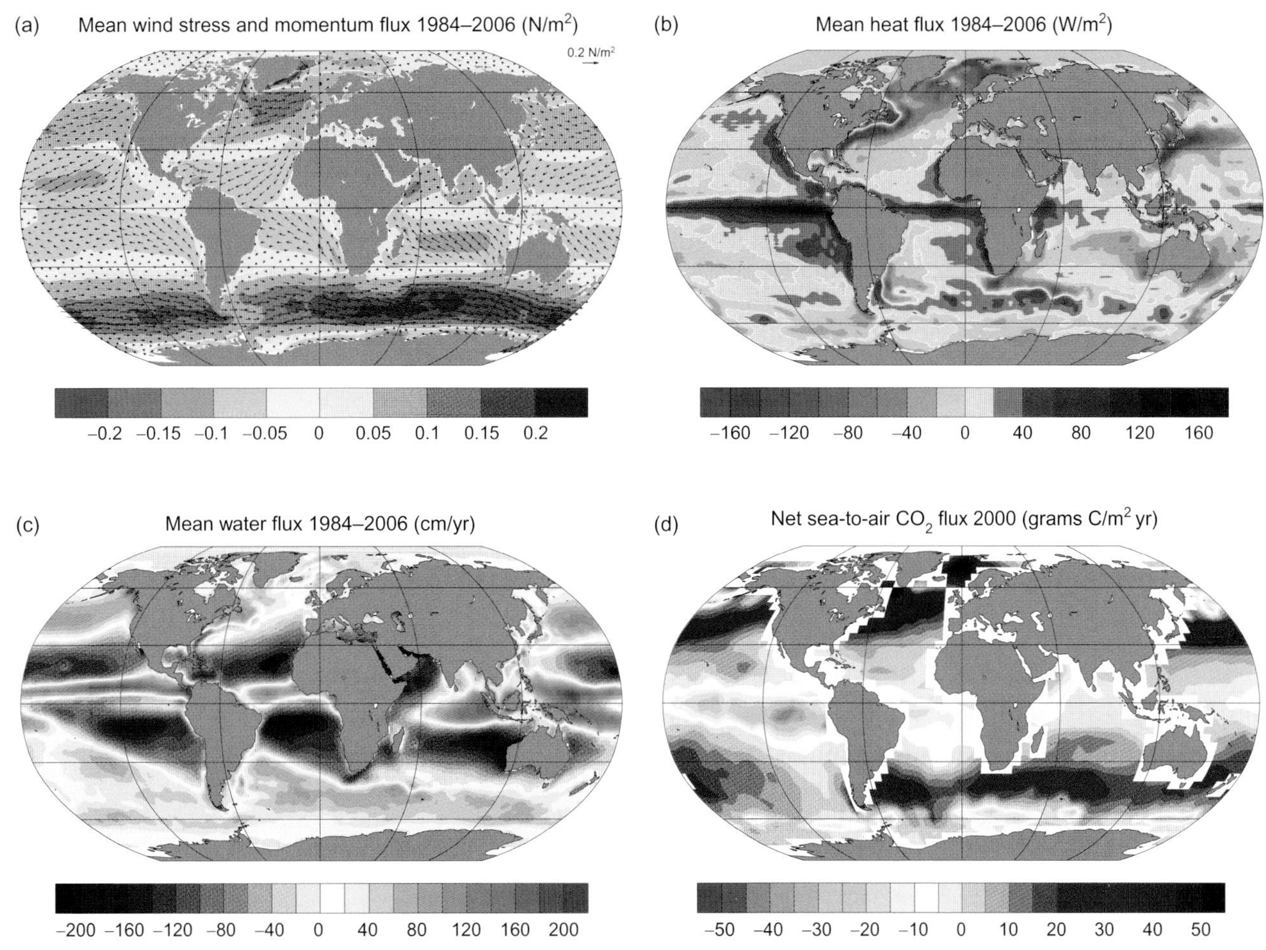

FIGURE 1.4 Major drivers of ocean processes and ocean circulation from various data sources. Shown are time-averaged quantities: (a) wind stress (arrows) and atmosphere-to-ocean momentum flux (colors); (b) net atmosphere-to-ocean heat flux; (c) net freshwater flux excluding river runoff, that is, precipitation minus evaporation, and (d) net atmosphere-to-ocean carbon flux in the year 2000, indicating a large imbalance caused by the uptake of anthropogenic carbon by the ocean. Momentum flux is everywhere into the ocean; positive (negative) values indicate that this flux is caused by a westerly (easterly) wind stress. In panels (b–d), fluxes are positive when they are from the atmosphere to the ocean. *Figures are redrawn based on data from Large and Yeager (2009), and for (d) from Takahashi et al. (2009).*

with the current systems of the Gulf Stream and the Kuroshio, and the Agulhas retroflection off South Africa. Large heat releases are also indicated in the Nordic Seas and the Arctic Ocean. The distinct heat gains and heat losses in the ocean at different latitudes imply meridional heat transports by the ocean circulation in each basin.

The gross flux of freshwater to the ocean is estimated at about 4.1×10^{17} kg/year (Table 1.1) and consists of precipitation (about 90%), river runoff (almost 10%), and ice-sheet melting (nearly 0.1%), the latter being partially compensated by net accumulation on the ice sheets. Imbalances in this freshwater cycle are caused by recent warming and ice-sheet melting. Although they are small (Table 1.1), they are measurable and have a large long-term impact through the rise in sea level. The freshwater balance is achieved mainly by precipitation and evaporation on the ocean's surface, each with a distinct spatial distribution that shows a largely zonal structure on a global scale (Figure 1.4c). This results in a net surface water balance which is characterized by freshwater gain in a narrow equatorial band on the order of 50–100 cm/year with a maximum in the Western Pacific warm pool, freshwater loss to the atmosphere in the subtropical dry zones, and net freshwater gain again in the higher latitudes of both hemispheres. On a global scale, the tropics and subtropics lose freshwater at a rate of about 1 Sv while gains are estimated north of 30°N at about 0.4 Sv, and south of 30°S at about 0.6 Sv (Talley, 2008). As inferred earlier for the heat fluxes, this also implies a meridional transport of freshwater by the ocean circulation in order to close the global water cycle.

A global view of the air-to-sea flux of carbon illustrates the large imprint of the natural carbon cycle at the ocean surface (Figure 1.4d). Carbon enters the ocean in the midlatitudes of the Pacific and Atlantic Oceans roughly in the areas of the Kuroshio and Gulf Stream, and the Nordic Seas. Also, carbon is taken up in a large band circling most of the

Southern Ocean. Major areas of carbon release are located in the tropical Pacific, the Arabian Sea and the northernmost Pacific Ocean, as well as around Antarctica (Takahashi et al., 2009). This latitudinal dependence of air–sea carbon fluxes is mainly temperature driven, as the solubility of CO_2 in warmer ocean water is lower than that in colder water. An imprint of the net uptake of anthropogenic carbon is contained in Figure 1.4d (Chapter 30), but this does not change much the dominant patterns of the gross fluxes.

In the time mean, atmosphere–ocean fluxes are indicators of convergent and divergent meridional fluxes in the ocean. With suitable boundary conditions, for example, no transport across the boundary of Antarctica, meridional fluxes can be calculated from the surface data (Large and Yeager, 2009). This is shown in Figure 1.5. Globally, the ocean transports heat at about 2×10^{15} W northward in the northern hemisphere but only about 0.5×10^{15} W southward in the southern hemisphere. The ocean transports freshwater southward at midlatitudes of the northern hemisphere and, equally, exports large amounts of freshwater from the Southern Ocean. The ocean therefore essentially supplies the freshwater that is lost to the atmosphere in the zones of excessive evaporation (Figure 1.4c). Atmosphere and ocean are therefore tightly coupled through the global water cycle. However, these estimates are plagued with large uncertainties as indicated by directly determined meridional heat fluxes on measurements along hydrographic sections, combined with dynamical constraints as calculated by Ganachaud and Wunsch (2003). They find larger ocean heat transports in the southern hemisphere. Uncertainties are larger still for estimates of the meridional freshwater fluxes in the ocean (Wijffels, 2001).

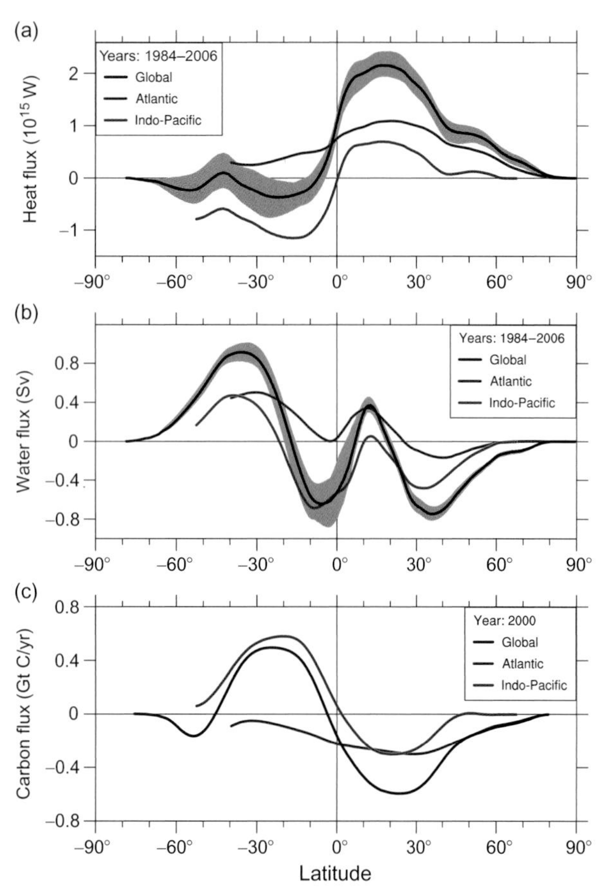

FIGURE 1.5 Global ocean and basin meridional fluxes 1984–2006 of (a) heat, (b) freshwater, and (c) carbon. Meridional fluxes are determined by integrating the data shown in Figure 1.4 in the Atlantic and Indo-Pacific ocean basins from north to south, assuming zero flux at the northern basin boundary and a globally uniform correction for each quantity to ensure zero flux at the southern end. Northward fluxes are positive. The range of interannual variability in global heat and freshwater fluxes is indicated by the turqouise band.

Although uncertainties and differences between the various approaches remain large, a robust picture emerges for the heat and freshwater in the different ocean basins. In the Atlantic ocean, heat is transported northward; in the Pacific and Indian Oceans combined, poleward in both hemispheres. This transport is effected mainly through the meridional overturning circulation (Ganachaud and Wunsch, 2003). The basin-wide meridional heat flux in the Pacific Ocean in the northern hemisphere is northward and carried by shallow overturning, but it amounts to less than half of that carried by the Atlantic Ocean (Talley, 2003).

In terms of freshwater fluxes, various processes need to be considered in addition to evaporation and precipitation over the ocean area: transfer of water from land to the ocean as river runoff, freshwater fluxes between ocean basins through straits, and freshwater transfer from one ocean basin to another via the atmosphere. Global river runoff is estimated at about $1.3\ \text{Sv} \approx 40 \times 10^{15}$ kg/yr (Figure 1.3). Flows through the Bering Strait (about 1 Sv, not considered in Figure 1.5) and the Indonesian Archipelago (about 10 Sv, Chapter 19) connect the Pacific Ocean with the Arctic basin and the Indian Ocean, respectively, and are important elements in the maintenance of the distinctly different salinity signatures of the ocean basins. Further, the Mediterranean Sea, as a marginal basin with an overall excess in evaporation over precipitation, and thus a negative freshwater balance, leaves a large-scale imprint on the salinity distribution of the mid-depth midlatitude North Atlantic Ocean. The global picture of the water cycle would not be complete without taking into account the inter-basin transports of freshwater through the atmosphere. The extra-tropical North Atlantic ocean loses about 0.1–0.3 Sv of freshwater through excess evaporation. This water vapor is then carried to the Pacific basin via the trade winds (Zaucker and Broecker, 1992). This is one of the mechanisms to maintain a significantly saltier North Atlantic Ocean (mean salinity of 35.75 at 200 m depth) than the North Pacific (34.50 at 200 m depth) (Levitus et al., 1994).

Meridional fluxes of carbon are depicted in Figure 1.5c estimated from integrating the atmosphere–ocean carbon fluxes of Takahashi et al. (2009). Consistent with Figure 1.4d, carbon is transported toward the equator on a global scale. In the Atlantic Ocean, carbon is transported southward at all latitudes. Note that these meridional fluxes are based on data from the modern ocean and therefore already contain a significant anthropogenic contribution. For example, under preindustrial conditions, all ocean basins are found to export carbon to the Southern Ocean. This transport has now changed direction in some areas due to the large uptake of anthropogenic carbon in the Southern Ocean (Figure 1.4d).

These large-scale air–sea fluxes and meridional fluxes of heat, freshwater, and substances are the manifestation of the tight coupling between the atmosphere and the ocean. Therefore, data as presented in Figures 1.4 and 1.5 are strong additional constraints for a global view of the ocean circulation.

4. GLOBAL-SCALE SURFACE AND DEEP OCEAN CIRCULATIONS

The global view of atmosphere–ocean fluxes and the implied meridional transports of heat, freshwater and carbon have pointed to the existence of a worldwide ocean current system. Its surface flow is characterized by basin-wide gyre circulations in the Atlantic, Pacific, and Indian Oceans (Chapters 11–14). The near-surface circulation can be divided roughly into five major phenomena: in the northern hemisphere, anticlockwise subpolar gyres (Chapter 17), clockwise subtropical gyres and mainly zonal equatorial current systems (Chapter 15); in the southern hemisphere, the anticlockwise subtropical gyres and the Antarctic Circumpolar Current (Chapter 18). A qualitative, global overview is given in Figure 1.6. For the strong subtropical gyres of the Pacific and Atlantic Oceans, transport estimates are at 42 Sv for the Kuroshio (Imawaki et al., 2001), and 30 Sv through the Strait of Florida (Lumpkin and Speer, 2003); about 134 Sv are carried by the Antarctic Circumpolar Current through the Drake Passage (Cunningham et al., 2003). This information, together with knowledge of the sea-surface temperature distribution, permits a rough estimate of the heat carried by the subtropical gyre in the North Atlantic Ocean. Assuming a volume transport of 30 Sv and a temperature difference of about 2 °C between the northward flowing warm water at the western boundary and colder water whose southward flow is spread over most of the Atlantic basin width east of the western boundary current, we obtain $F_H = \rho \cdot c \cdot \dot{V} \cdot \Delta T \approx 10^3 \times 4 \times 10^3 \times 30 \times 10^6 \times 2\mathrm{W} = 0.24 \times 10^{15}\mathrm{W}$ for the meridional heat flux by the subtropical gyre. However, estimates for the total meridional heat flux in the North Atlantic are about 1.3×10^{15} W (Ganachaud and Wunsch, 2000; Johns et al., 2011). This implies that there must exist another important type of circulation in the Atlantic ocean that carries the missing heat. This is the deep meridional overturning circulation of the Atlantic ocean, referred to as the Atlantic meridional overturning circulation (AMOC).

A global-scale circulation in the deep ocean (Chapter 10) has been suspected since the early days of ocean exploration, when it was realized that even in tropical latitudes, the deep ocean is very cold. Such waters could only be supplied from polar regions where the annual mean temperature is sufficiently cold. This also suggests that the deep waters derive from surface, or near-surface waters, of the high latitudes. The latest comprehensive international effort, in the framework of the World Ocean Circulation Experiment (WOCE, www-pord.ucsd.edu/whp_atlas), has measured the global distributions of temperature, salinity, and many tracers, and produced a comprehensive view of the global water masses and their physical and chemical characteristics. The view of the importance of a globe-encompassing deep circulation in the world ocean has been confirmed in great detail by this immense dataset.

Figure 1.7 shows a representative section through the Atlantic Ocean for temperature and salinity and indicates the three dominant water masses in the Atlantic. The coldest waters in the deep ocean can be traced back to regions around Antarctica where Antarctic Bottom Water (AABW) is formed. North Atlantic Deepwater (NADW) is less dense and hence is located between AABW and Antarctic Intermediate Water (AAIW). The presence and extent of these water masses suggest that deep water is being formed in the high latitudes in rather localized areas. In order to close the flow, the simplest possibility is that water upwells uniformly on a global scale as proposed by Munk (1966). He estimated a global mean upwelling rate of 0.7×10^{-7} m/s using mean vertical structures of temperature, salinity, and radiocarbon, from the central Pacific Ocean. This is consistent with a recent estimate of the global deep water production of 36 Sv which would need to be replenished by a global mean upwelling rate of 10^{-7} m/s (Ganachaud and Wunsch, 2000). Incidentally, this order of magnitude for global upwelling yields a renewal time for the entire ocean volume of about 1000 years, which is significantly faster than the earlier estimate based on Table 1.1. This time is also more consistent with estimates of the age of the oldest waters in the Pacific based on radiocarbon measurements and inverse calculations (Gebbie and Huybers, 2012). However, more detailed observations have uncovered a distinctly regional structure of upwelling, with most of it occurring in the Southern Ocean and equatorial regions (Döös et al., 2012; Marshall and Speer, 2012). In both areas, the upwelling is largely wind-driven by eddy-induced momentum transport and Ekman divergence.

Figure 1.7 also suggests that the deep circulation is essentially a meridional overturning circulation, which is characterized in the Atlantic Ocean by warm waters flowing northward and colder waters at depth flowing southward, each in western boundary currents. This view for the Atlantic Ocean was first proposed by Stommel (1957),

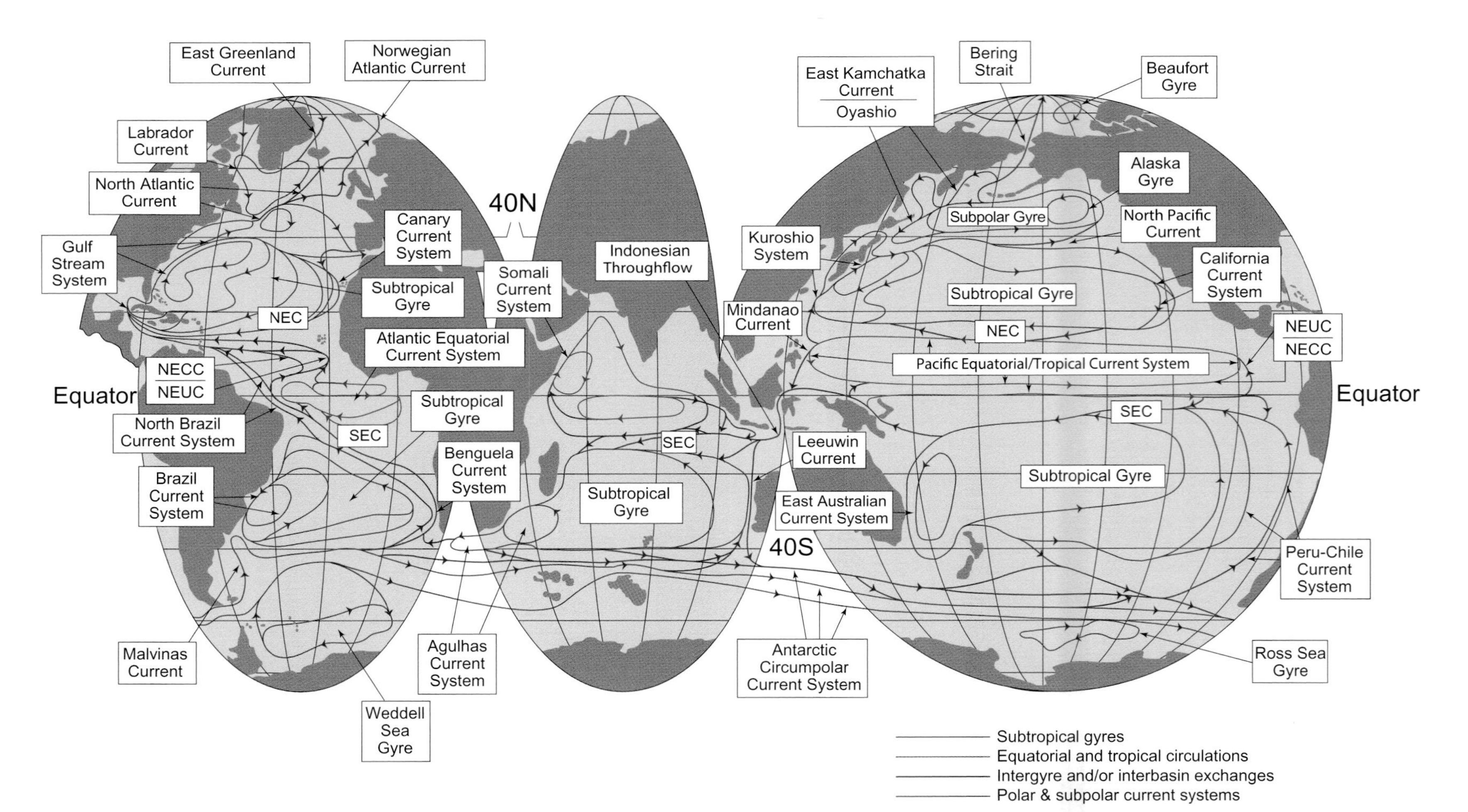

FIGURE 1.6 Illustration of surface circulation systems of the world ocean. NEC and SEC denote the Northern and Southern Equatorial Currents, NEUC and NECC are the Northern Equatorial Under and Counter Currents, respectively. *Figure from Talley et al. (2011), based on an earlier version of Schmitz (1996).*

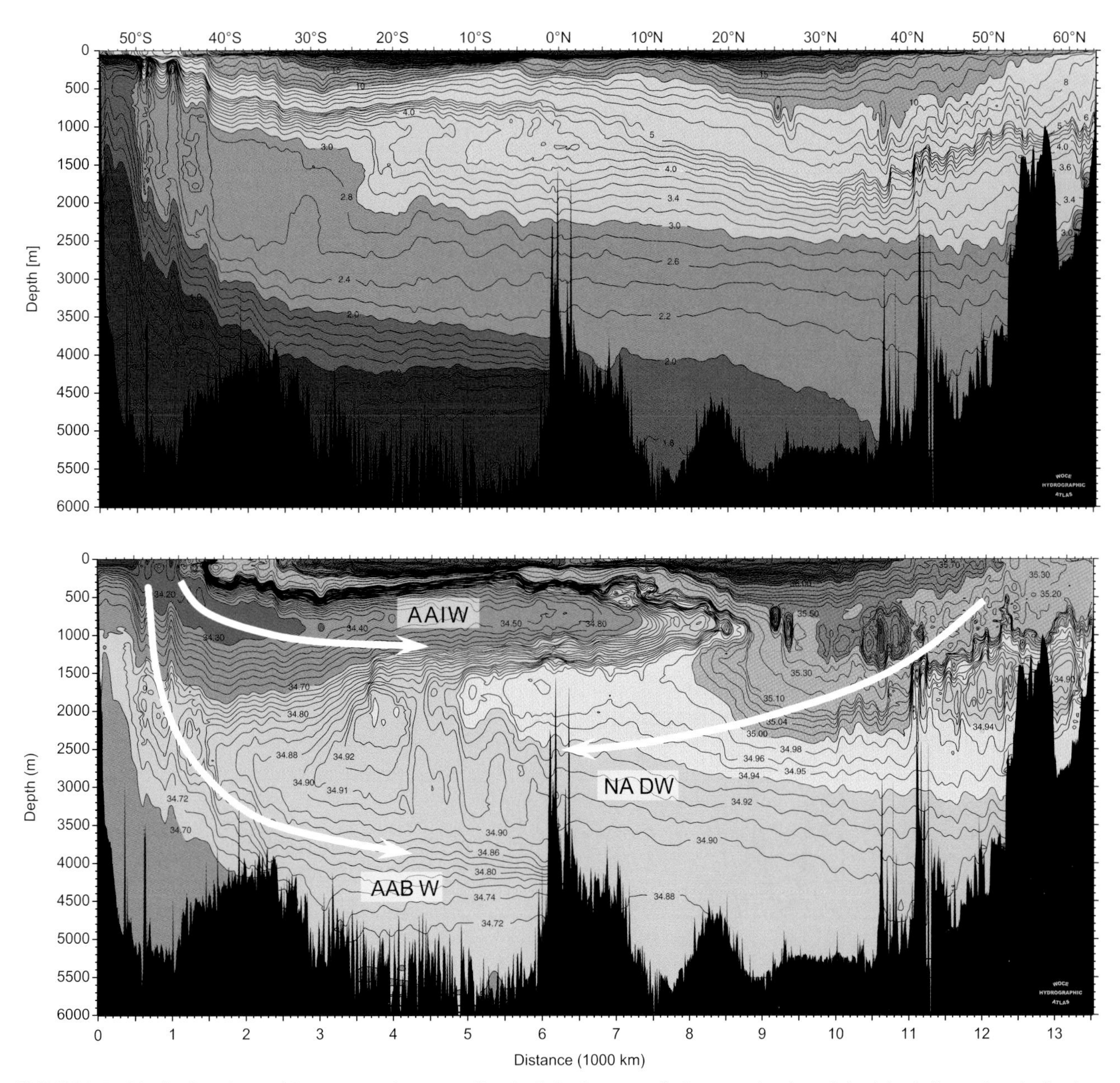

FIGURE 1.7 Distribution of potential temperature (upper panel) and salinity (lower panel) along a section through the Atlantic Ocean from the Southern Ocean to Iceland. The major water masses in this ocean basin are North Atlantic Deep Water (NADW), Antarctic Intermediate Water (AAIW), and Antarctic Bottom Water (AABW). They are characterized by specific ranges of temperature and salinity and thus visible in both quantities. Arrows schematically trace their pathways. *From the WOCE Hydrographic Atlas (Koltermann et al., 2011).*

based purely on dynamical considerations of a fluid on a rotating sphere, but inspired by earlier insight from direct measurements (Defant, 1941). Stommel (1958) extended this concept qualitatively to the entire ocean and provided estimates of the volume transport for each of the deep western boundary currents. It is worth noting that Stommel closed his landmark paper, presenting for the first time a global view of the deep circulation of the world ocean, by saying *One cannot pretend that it describes the abyssal circulation accurately in detail.* The quantitative and dynamically consistent analysis of the deep circulation was presented in a series of papers starting with Stommel and Arons (1960).

Stommel's Letter to the Editors (1958) prompted a series of iconic depictions of the global deep ocean circulation (Richardson, 2008); the most popular is the *Great Ocean Conveyor Belt* (Broecker, 1987b). Broecker and Peng (1982) introduced the term *large conveyor belt* and described the geochemical significance of this global circulation. A schematic illustration that indicates the flow

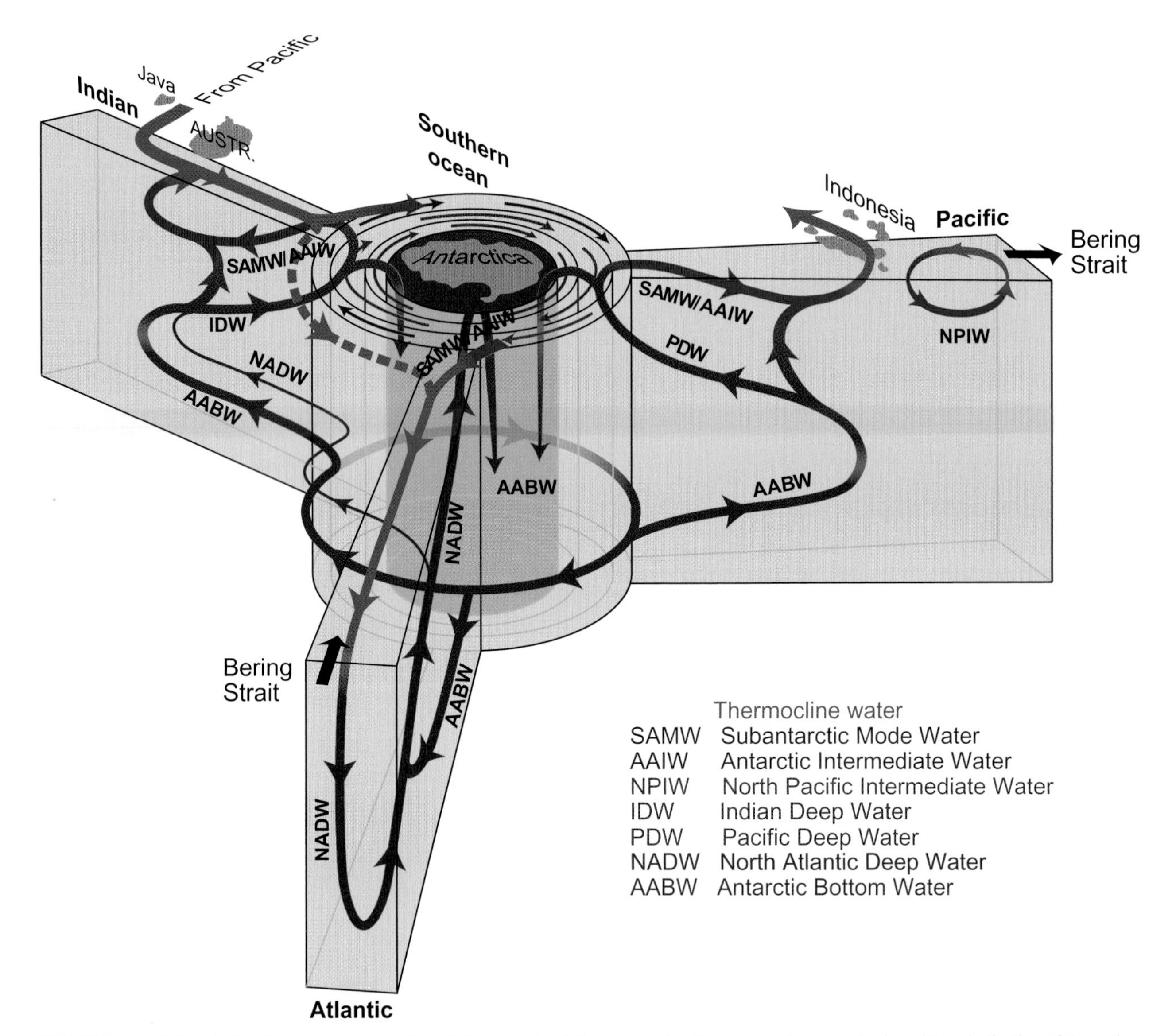

FIGURE 1.8 Highly idealized cartoon-type overview of the deep circulation connecting the three major ocean basins with an indication of the various water masses involved in this flow. *Figure adapted from Talley (2013), based on Schmitz (1996), after Gordon (1991).*

and the locations of the various water masses in the three major ocean basins is given in Figure 1.8. This concept has been useful in highlighting the importance of this circulation in the climate system and has also been inspiring for the understanding of abrupt climate changes in the past (Broecker and Denton, 1989; Stocker and Wright, 1991; Stocker, 2000; Clark et al., 2002). However, modern oceanography with the growing dataset of autonomous measurement devices deployed in the ocean (Chapter 3) (Roemmich and Gilson, 2009), dedicated arrays (Johns et al., 2011), a rich variety of satellite products (Chapter 4) (Hurlburt et al., 2009), and high-resolution ocean modeling (Chapters 20 and 22) (Maltrud and McClean, 2005) paints a complex picture of a turbulent ocean which, only in the multidecadal time, may bear some similarities with the cartoon-type view presented in Figure 1.8.

Turning briefly back to the question of the closure of the meridional heat transport in the ocean, which suggested the existence of a deep ocean circulation, we estimate the contribution of the overturning circulation in the Atlantic to the total meridional heat flux. Of the 36 Sv of deep water formed globally about 15 Sv are estimated to come from the North Atlantic (Ganachaud and Wunsch, 2000). As above, we can now calculate the heat that is carried in the North Atlantic by a northward flow of 15 Sv of warm surface waters at about 20 °C and an equal amount of cold NADW flowing southward at about 3 °C. These assumptions yield $F_H \approx 1.0 \times 10^{15}$W which is about four times the heat flux transported by the near-surface horizontal gyre circulation. The calculated estimates of the gyre and the meridional overturning circulation together are in good agreement with the observed value in the North Atlantic (see Figure 1.5a). In the North Pacific, on the other hand, overturning is only shallow and much less heat is carried northward (Talley, 2003).

5. LARGE-SCALE MODES OF VARIABILITY INVOLVING THE OCEAN

So far, attention has been focused on the global-scale, steady circulation patterns of the world ocean. However, it is well documented that atmosphere–ocean exchange fluxes of heat and freshwater vary also on interannual to multidecadal time scales and thus point to the ocean as an active component of global and regional climate variability. A comprehensive synthesis of nonseasonal, interannual to multidecadal sea-surface temperature variability observed worldwide is given by Deser et al. (2010).

The best studied atmosphere–ocean mode of variability is the ENSO phenomenon (Chapter 24, e.g., Cane, 2005; Deser et al., 2010; Sarachik and Cane, 2010). It is characterized by two anomalous ocean–atmosphere states in the equatorial Pacific Ocean (Figure 1.9). These anomalous states can last for several months to more than a year and recur irregularly on a typical time scale of 2–7 years: (i) El Niño with anomalous warm sea-surface temperatures in the eastern tropical Pacific, regionally reduced atmospheric pressure causing the trade winds to weaken there, and an equatorial rain band that extends from Indonesia into the central Pacific; (ii) La Niña with the opposite changes resulting in colder sea-surface water and higher atmospheric pressure in the eastern tropical Pacific, due to a shoaled thermocline and hence stronger coastal upwelling, in response to stronger trade winds. Paleoclimate reconstructions demonstrate the presence of this variability at least during the last millennium (Cobb et al., 2003), but possibly during the past 20,000 years (Rein et al., 2005). During the last 7000 years, no systematic trends in this variability could be detected (Cobb et al., 2013). This mode therefore appears to be a very stable feature of interannual variability in the climate system.

The important role the ocean plays in this mode of variability is evident in Figure 1.9. The vertical movements of the thermocline in the eastern part of the equatorial Pacific influence sea-surface temperatures and thus modify the strength of the trade winds. Atmosphere and ocean are coupled to produce a positive feedback, the *Bjerknes feedback* (Bjerknes, 1969), which is at the heart of the ENSO phenomenon (Cane, 2005). Stronger trades in the eastern equatorial Pacific cause the thermocline to rise through Ekman suction, which in turn produces colder sea-surface temperatures that tend to promote subsidence in the overlying atmosphere. An increase in the high pressure over the eastern equatorial Pacific follows, which further strengthens the trades and closes this positive feedback loop. But the Bjerknes feedback alone cannot explain the recurrence of ENSO. The oscillation involves wave propagation in the ocean: Kelvin waves travel toward the eastern boundary of the Pacific basin where, in turn, the perturbations generate Rossby waves, which then propagate westwards and into the basin off the equator. Together they modify the position and structure of the thermocline in such a way that the system is "recharged" after some years (Wang, 2001). Although basic aspects of this coupled atmosphere–ocean phenomenon are simulated by many comprehensive climate models, ENSO events in only a few models have occurrence frequencies similar to those observed (Guilyardi et al., 2009).

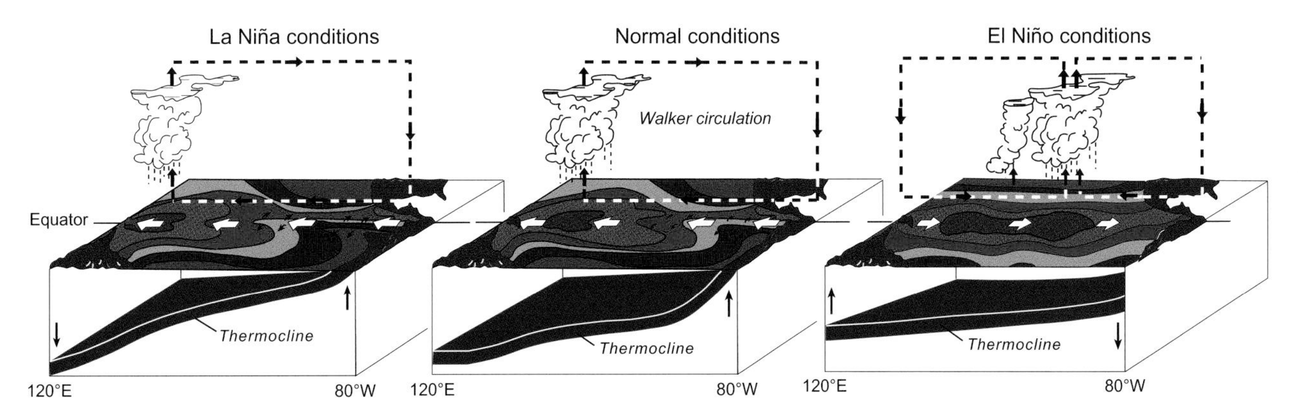

FIGURE 1.9 Schematic illustration of the normal conditions in the equatorial Pacific and two different phases of the El Niño–Southern Oscillation phenomenon. Normal conditions show an east–west gradient in sea-surface temperature (SST) in the equatorial Pacific with a western Pacific warm pool where convective rainfall occurs. The SST gradient is maintained by well-established trade winds with subsidence over the cold pool in the east and convection in the west as part of the Walker Circulation. The trade winds maintain the slope of the thermocline in the equatorial Pacific which, in turn, sets up the SST gradients. During an El Niño event, the trade winds slacken, SSTs warm across the equatorial Pacific, and the convective rain band extends and moves east. As the trade winds weaken, the thermocline relaxes downward. The opposite changes are observed during a La Niña event. El Niño conditions typically prevail from December through February. *Figures adapted from the PMEL TAO Project.*

While ENSO is generated and maintained by processes in the equatorial and tropical Pacific, teleconnections caused by ENSO have been documented around the world (Ropelewski and Halpert, 1987; Halpert and Ropelewski, 1992). The most notable phenomena during El Niño events are dry conditions over Indonesia, Australia, and southeastern Africa, wet conditions in Peru, Ecuador, and the Gulf of Mexico, and warm anomalies in northwestern North America up to Alaska. Effects of ENSO can also be identified to reach Europe (Brönnimann, 2007). Such teleconnections may be difficult to distinguish from the Pacific Decadal Oscillation (PDO; Deser et al., 2004), or other interannual to multidecadal variability that generate teleconnections (Liu and Alexander, 2007).

Also in the tropical Indian Ocean, a basin-wide mode of variability involving strong ocean–atmosphere interaction has been found in observations (Saji et al., 1999; Webster et al., 1999). It is now referred to as the Indian Ocean Dipole (IOD) and is a dominant feature of Indian Ocean variability (Schott et al., 2009). In its positive phase, sea-surface temperatures off the coast of Sumatra are colder, while they are anomalously warm in the western Indian Ocean. The colder temperatures west of Sumatra may be triggered by anomalous seasonal upwelling (Xie et al., 2002) which leads to atmospheric subsidence and stronger easterly winds along the equator. They reinforce the upwelling. This also constitutes the western branch of the Walker circulation in the atmosphere, which leads to convection over the warmer western Indian basin. The rising air enhances rainfall in the western part of the Indian Ocean and in eastern equatorial Africa. This is, again, the Bjerknes feedback, which sets up the prominent dipole pattern in sea-surface temperature in the tropical Indian Ocean. Overall, the physical processes in the tropical atmosphere and ocean that are at work for IOD are similar to those for ENSO.

IOD events are seasonally phase locked and tend to be significantly shorter than ENSO events. While many of the processes that operate during an IOD are also fundamental to ENSO, there are distinct differences that are primarily caused by the strong, thermally driven land–ocean interactions in the Indian Ocean (Li et al., 2003). The warming over the western Indian Ocean induces enhanced evaporation and cloud cover which dampens the warming. It also strengthens the Indian summer monsoon circulation with consequent stronger winds and enhanced mixing in the ocean surface (Webster et al., 1999). All this contributes to a more rapid demise of the IOD, which typically only lasts for a few months, with the strongest phase in October. In the past 40 years, IOD and ENSO did not occur synchronously (Saji et al., 1999) which suggests that they can operate independently. However, through the atmospheric Walker circulation in the tropical Indo-Pacific region, and further teleconnections in the atmosphere, it is not surprising that there are also phases of interaction between IOD and ENSO (Luo et al., 2010), for example, during the prominent IOD event in 1997, and the strong ENSO that unfolded in 1997/1998. IOD events thus may even trigger ENSO (Izumo et al., 2010).

IOD events were inferred from coral records extending over the past 6500 years (Abram et al., 2007). IOD events and associated changes therefore appear to be a persistent feature of tropical variability in the Indian Ocean, but the proxy records indicate that the duration of IOD events was longer during the middle Holocene resulting from a stronger Indian monsoon then. East African rainfall and drought changes found in paleoclimate records covering the past millennium also point to active IOD modes (Verschuuren et al., 2000; Mölg et al., 2006).

The PDO (Chapter 25) manifests itself as warm and cold anomalies of monthly North Pacific sea-surface temperature from November to March (Mantua and Hare, 2002; Deser et al., 2010). PDO variability is not periodic but shows power in two time windows of 15–25 years and 50–70 years. Paleoclimate reconstructions suggest that the PDO is a robust feature during the past 1000 years but the typical time scales of variations may not have been stable (MacDonald and Case, 2005; Shen et al., 2006). The mechanism of this atmosphere–ocean mode is not fully understood. While forcing due to ENSO teleconnections and reemergence of sea-surface temperature anomalies may play roles, stochastic forcing and modulation by the ocean appear to be dominant (Liu, 2012).

Multidecadal variations (Chapter 25) on time scales of 40–60 years and 50–90 years have been found both in coupled model simulations (Delworth et al., 1993) and in the analysis of observed surface temperature records (Schlesinger and Ramankutty, 1994; Sutton and Hodson, 2003). Both studies pointed to the North Atlantic as the center of action. This phenomenon is now referred to as Atlantic multidecadal oscillation (AMO) and is described by basin-wide, coherent sea-surface temperature anomalies in the Atlantic north of the equator (Deser et al., 2010). These anomalies are closely linked to the meridional overturning circulation in the Atlantic, and both temperature and salinity anomalies may be involved (Ou, 2012). When the overturning is stronger, warm anomalies are produced, and also saltier waters are advected northward along the western boundary. This provides a positive feedback for the overturning. Finally, a warm pool will develop and a geostrophic response results in an anticyclonic circulation which brings colder waters from the north to the mid-basin of the Atlantic Ocean, closing the cycle. This suggests that the typical time scale of the AMO may be determined by the volume of water that participates in this variability, which may explain the irregularity found in model simulations and paleoclimatic reconstructions (Gray et al., 2004). AMO might also influence the September sea ice extent in the Arctic (Day et al., 2012). Recent simulations over the twentieth century using coupled climate models show variability on this time scale but underestimate the amplitude and associated large-scale regional patterns of variation (Kavvada et al., 2013).

6. THE OCEAN'S ROLE IN PAST CLIMATE CHANGE

In order to understand and quantify the sensitivity of the ocean circulation to perturbations and its response to future changes in the major drivers, a combination of approaches, including multiproxy paleoclimate reconstructions and modeling, must be taken. Climate simulations using comprehensive coupled atmosphere–ocean models provide us with estimates of changes in ocean status and circulation for past climates (Braconnot et al., 2012). Understanding the ocean in the context of past climate change (Chapter 2) offers a complementary and increasingly quantitative insight into how the ocean has responded to rather large changes in the forcing fluxes and knowledge that is relevant to assess future changes in ocean status and circulation (see Section 7).

A few numbers demonstrate that the ocean played an important role in past climate change. The world ocean was the principal carbon storage during the course of the ice ages when atmospheric CO_2 concentrations changed between about 180 ppm during an ice age and about 280 ppm during interglacials (Lüthi et al., 2008). Global mean surface air temperatures during the Last Glacial Maximum (LGM, 26,000 to 20,000 years before the present) were about 4–7 °C colder than today (Jansen et al., 2007; Masson-Delmotte et al., 2013). Global mean cooling at the ocean surface during the LGM has been recently estimated at about 2 °C (MARGO Project Members, 2009), but for large areas with likely much larger cooling no reconstructions are available (Figure 1.10). Deep ocean temperatures were near the freezing point (Adkins et al., 2002). During the LGM, sea level was about 130 m lower than today (Clark et al., 2009; Church et al., 2013), whereas about 125,000 years ago, in the previous interglacial, it was 5.5–9 m higher than today (Dutton and Lambeck, 2012). These sea-level variations impact directly on the mean salinity of the world ocean with today's mean salinity being about 1.1 units lower than at the LGM.

Wind patterns during the last glacial period were significantly different from today due to the presence of the northern hemisphere ice sheets, more extended sea ice cover, and changed meridional temperature gradients. Although their

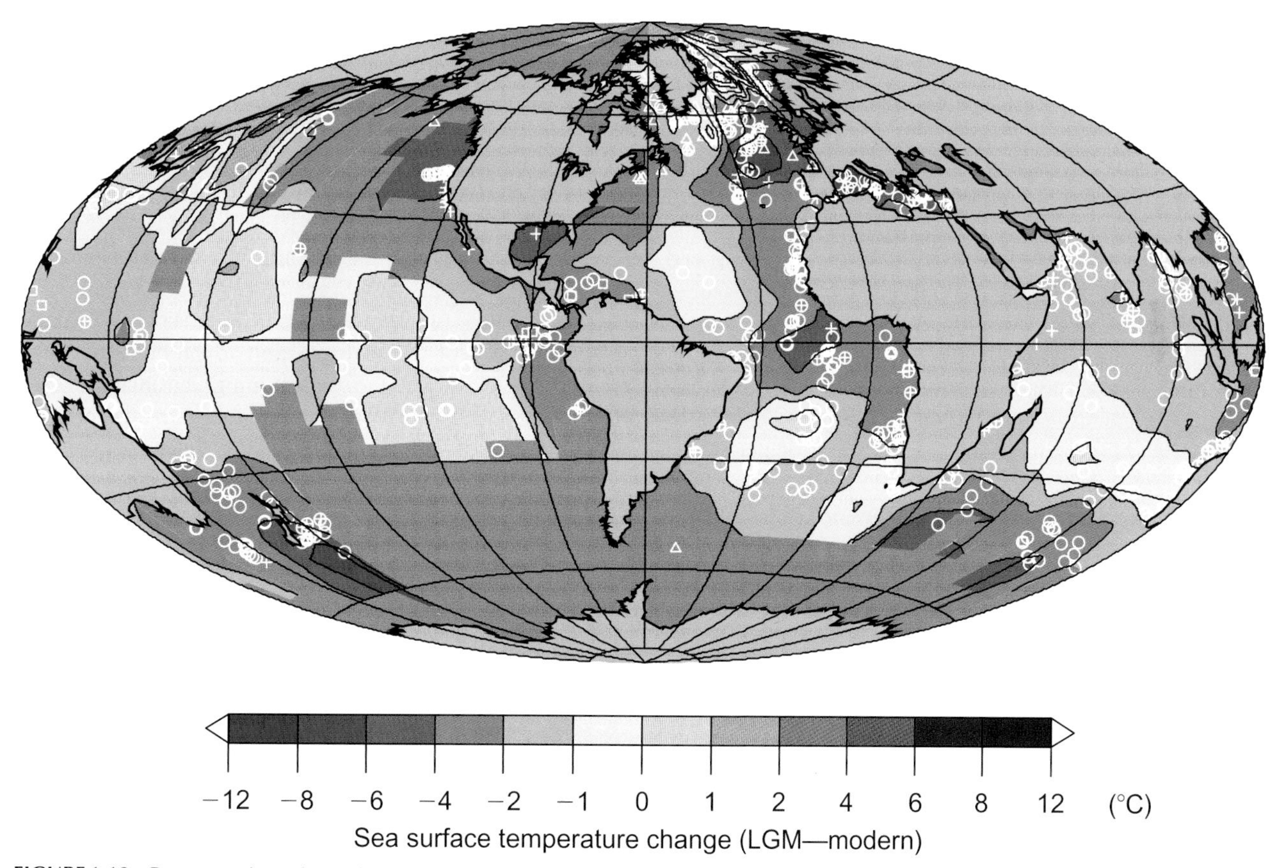

FIGURE 1.10 Reconstructions of annual mean sea-surface temperatures anomalies at the Last Glacial Maximum relative to today, based on a multiproxy approach using several paleothermometers. Uncertainties are large, and data coverage in large dynamically essential areas of the ocean, for example, the Southern Ocean, is poor or missing. *Figure from MARGO Project Members (2009).*

strength has likely changed, zonal wind belts still forced the surface gyre circulation, making it a robust feature of the ocean circulation throughout the ice age cycles.

Unfortunately, reconstructions of sea-surface salinity distributions in the past are still scarce and highly uncertain, but there is indication of a change in the east–west surface salinity gradient in the North Atlantic (Schäfer-Neth and Paul, 2003). During most of the last ice age, the Bering Strait was closed because sea level was below the depth of this shallow passage. With the Bering Strait closed, the Arctic salt balance and the Arctic–North Atlantic water exchange were very different from today. Together with modified air–sea fluxes and sea ice cover, it is therefore not unreasonable to expect major changes in the strength and structure of the deep circulation of the ocean in the past. Model simulations suggest that this closure also modified the sensitivity of the meridional overturning circulation to freshwater perturbations (Hu et al., 2012).

A major effort in paleoceanography therefore concerns the reconstruction of the ocean circulation in the past, in particular the ocean circulation during the LGM and its transient evolution to the circulation observed today. The rationale is that by documenting and understanding in detail how the ocean circulation operated during the last ice age, and during the transition to the Holocene leading up to the present, a wide variety of ocean states could be "sampled", which would allow us to capture the essential dynamic range of the ocean to perturbations and changes in the different forcing factors.

A palette of different physical and geochemical indicators measured on marine sediments has been developed to quantify changes in water mass characteristics of the ocean in the past (Lynch-Stieglitz, 2003). The stable isotope ratios of oxygen ($^{18}O/^{16}O$) and carbon ($^{13}C/^{12}C$), measured on the calcareous shells of different organisms, are widely used paleoceanographic quantities from which changes in temperature, water mass distribution, and sea level are inferred. Temperature information as in Figure 1.10 is also derived from Mg/Ca ratios and concentrations of alkenones; metal tracers such as Pa, Th, and Nd, and their isotopes, provide further constraints for changes in water masses (Rempfer et al., 2011).

With the growing number of paleoceanographic indicators and related datasets, it should therefore eventually be possible to assimilate the information into an ocean circulation model and reconstruct the past circulations employing the same approach of inverse modeling as for modern data. Such a study indicates that the deep circulation in the Atlantic Ocean of today is incompatible with the tracer distributions reconstructed for the last glacial (Marchal and Curry, 2008). Based on nutrient tracers, one can infer that the large-scale water mass distribution in the glacial Atlantic was dominated by a shallower water mass, Glacial North Atlantic Intermediate Water, lying over a more widespread AABW (Lynch-Stieglitz et al., 2007), and possibly reduced mixing between northern and southern water masses (Lund et al., 2011). However, various model simulations still give conflicting results for the Atlantic overturning strength during the LGM (Otto-Bliesner et al., 2007), and recent numerical model results suggest some threshold behavior of the overturning circulation with respect to the surface buoyancy flux and wind stress forcings (Oka et al., 2012).

An intriguing feature of past climate change, for which the ocean is thought to be a major player, is the sequence of 26 abrupt warmings as recorded in the Greenland ice core record (Stocker, 2000; North Greenland Ice Core Project members, 2004). The most recent and last in the series of these warmings occurred 11,650 years ago and marked the beginning of the current warm period, the Holocene. These events are commonly referred to as Dansgaard–Oeschger events in honor of the Danish and Swiss scientists *Willy Dansgaard* (1922–2011) and *Hans Oeschger* (1927–1998) who identified and interpreted them in the first deep ice cores from Greenland (Dansgaard et al., 1984) and noted their widespread climatic significance (Oeschger et al., 1984). The warmings in Greenland occur extremely rapidly, sometimes within a few years (Steffensen et al., 2008), and have amplitudes of 8° to 15 °C (Huber et al., 2006). Many independent indications from reconstructions of temperature and precipitation changes on land, sea-surface temperature changes, changes in water mass indicators in the ocean, and global-scale spatial correlations strengthen the view that the ocean, in particular the Atlantic overturning circulation, has played a fundamental role in these climate variations (Broecker and Denton, 1989; Stocker, 2000; Clark et al., 2002; Voelker and Workshop-Participants, 2002; Alley et al., 2003; Clement and Peterson, 2008; Fleitmann et al., 2009). This is because more than 75% of the heat transport in the North Atlantic Ocean depends on the overturning circulation (see Section 4), and variations of its strength are an efficient regulator of regional climate in this region. If this "heat pump" is switched off, a substantial impact on sea-surface temperatures and the atmosphere is expected.

A large body of climate model simulations, ranging from simplified conceptual models to comprehensive coupled climate models, demonstrates that the AMOC sensitively responds to changes in the surface buoyancy balance and the distribution of wind stress. Some studies argued that the overturning circulation has multiple equilibria (Stommel, 1961; Manabe and Stouffer, 1988; Stocker and Wright, 1991). Such a property would create hysteresis, with the surface freshwater balance in the North Atlantic being an important determinant which, depending on the climate state, could create the possibility of abrupt climate change in response to perturbations in the freshwater balance (Stocker and Marchal, 2000; Hu et al., 2012). The most effective perturbation of this kind during the last ice age was recurring massive iceberg discharges from the northern hemisphere ice sheets and Antarctica. Such an iceberg flow into the ocean would result in large freshwater lenses disrupting the deep water formation and hence reduce the oceanic meridional heat flux in the North Atlantic.

Direct identification of past changes in the AMOC remains a challenge as there are only indirect paleoceanographic indicators registering such changes. Metal tracer concentrations (Pa, Th, and Nd) and their isotopic ratios inform about water mass distributions and are, unlike the carbon isotopic ratios, less influenced by couplings with the carbon cycle. They therefore offer a way to quantitatively determine changes of the meridional overturning circulation in the past (Piotrowski et al., 2004). Model simulations suggest that such tracers may be able to quantify such changes in response to freshwater discharges from the north or the south, and discriminate between them (Rempfer et al., 2012). Time-resolved reconstruction of sea-surface temperature distributions also holds promises to quantify AMOC changes (Ritz et al., 2013).

Currently, the most convincing evidence for a large involvement of the meridional overturning circulation in the sequence of Dansgaard–Oeschger events during the last ice age comes from the remarkable interhemispheric coupling of these abrupt warmings. Today, the meridional heat flux in the Atlantic Ocean at 30°S is directed northward (Figure 1.5), and therefore this ocean imports heat from the Southern Ocean when the overturning circulation is operational. The paleoclimatic significance of this cooling effect of the Southern Ocean was recognized by Crowley (1992). Rapid changes of the Atlantic overturning circulation are therefore expected to produce climate signals of opposite sign in the South Atlantic Ocean to those in the North Atlantic Ocean, and by ocean–atmosphere coupling, in the overlying atmosphere. This led to the formulation of the "bipolar seesaw" hypothesis (Broecker, 1998; Stocker, 1998).

Ice cores from Greenland and Antarctica offer the unique possibility of producing highly time-resolved climate records that can be placed on one synchronized time scale by employing the rapid changes in the methane concentrations measured on these cores as a reliable global time marker. This approach revealed an asynchronous behavior between Greenland and Antarctica for the Antarctic warmings during the last ice age (Blunier et al., 1998; Capron et al., 2010). While abrupt warmings in Greenland are followed by millennial-scale coolings, temperature variations inferred from Antarctica isotope records were more gradual during the last ice age. A consistent pattern is that the Antarctic warming trends change into cooling trends at the time of the abrupt warming in the Greenland ice cores, that is, synchronous with the Dansgaard–Oeschger events. The fundamentally different temporal character of these events—abrupt in the north and gradual in the south—could be resolved by postulating a thermal damping influence of the large Southern Ocean heat reservoir (Stocker and Johnsen, 2003). Two testable predictions followed from this model of the "thermal bipolar seesaw": (i) all Dansgaard–Oeschger events should produce a warming in Antarctica and (ii) the duration of the cooling in Greenland is proportional to the amplitude of the warming in Antarctica.

The ice core from Dronning Maud Land in Antarctica produced a temperature record of significantly higher time resolution than previous Antarctic records. Moreover, the ice core is located in the South Atlantic sector of Antarctica and hence in an area which is expected to react particularly sensitively to the bipolar seesaw (EPICA Community Members, 2006). When synchronized onto the time scale of the Greenland ice core (North Greenland Ice Core Project members, 2004), each Dansgaard–Oeschger event could be associated with a corresponding Antarctic Isotope Maximum event (Figure 1.11). Further, the proportionality postulated by the thermal bipolar seesaw was also confirmed (EPICA Community Members, 2006; Capron et al., 2010). This remarkably coherent interhemispheric coupling persisted through the transition from the last ice age into the Holocene (Barker et al., 2009). Using the seesaw concept in an inverse approach, it can be suggested that Dansgaard–Oeschger events were a common feature of at least the past 800,000 years (Siddall et al., 2006; Barker et al., 2011). In summary, these results demonstrate the important global-scale role of changes in the ocean circulation for abrupt climate change and variations on the centennial-to-millennial time scale in the past.

Finally, the ocean is also the key climate system component to enable glacial–interglacial changes in atmospheric CO_2 (Fischer et al., 2010). Ice core records indicate CO_2 variations of about 100 ppm during the past eight ice age cycles (Lüthi et al., 2008). During the LGM, atmosphere and land together held about 400–800 Gt carbon less than today (Sigman and Boyle, 2000). This extra carbon must have been stored in the ocean and in sediments, and this required substantial physical, chemical, and biological changes in the ocean to permit the observed reduction in atmospheric CO_2. During the LGM, the sea-surface temperature was about 2 °C colder (Figure 1.10), which increases the solubility of CO_2 and reduces atmospheric CO_2 by about 20 ppm. As pointed out above, the uncertainty in this estimate is very large. The increase in salinity partially offsets this reduction by about 10 ppm. Other physical changes involve shifts in water mass distributions, larger sea ice cover, and increased stratification in the Southern Ocean, but these physical effects together are able to account for only about a third of the required reductions in atmospheric CO_2. Ocean chemistry was therefore also involved in the glacial–interglacial CO_2 changes. An increase in the ocean's alkalinity would

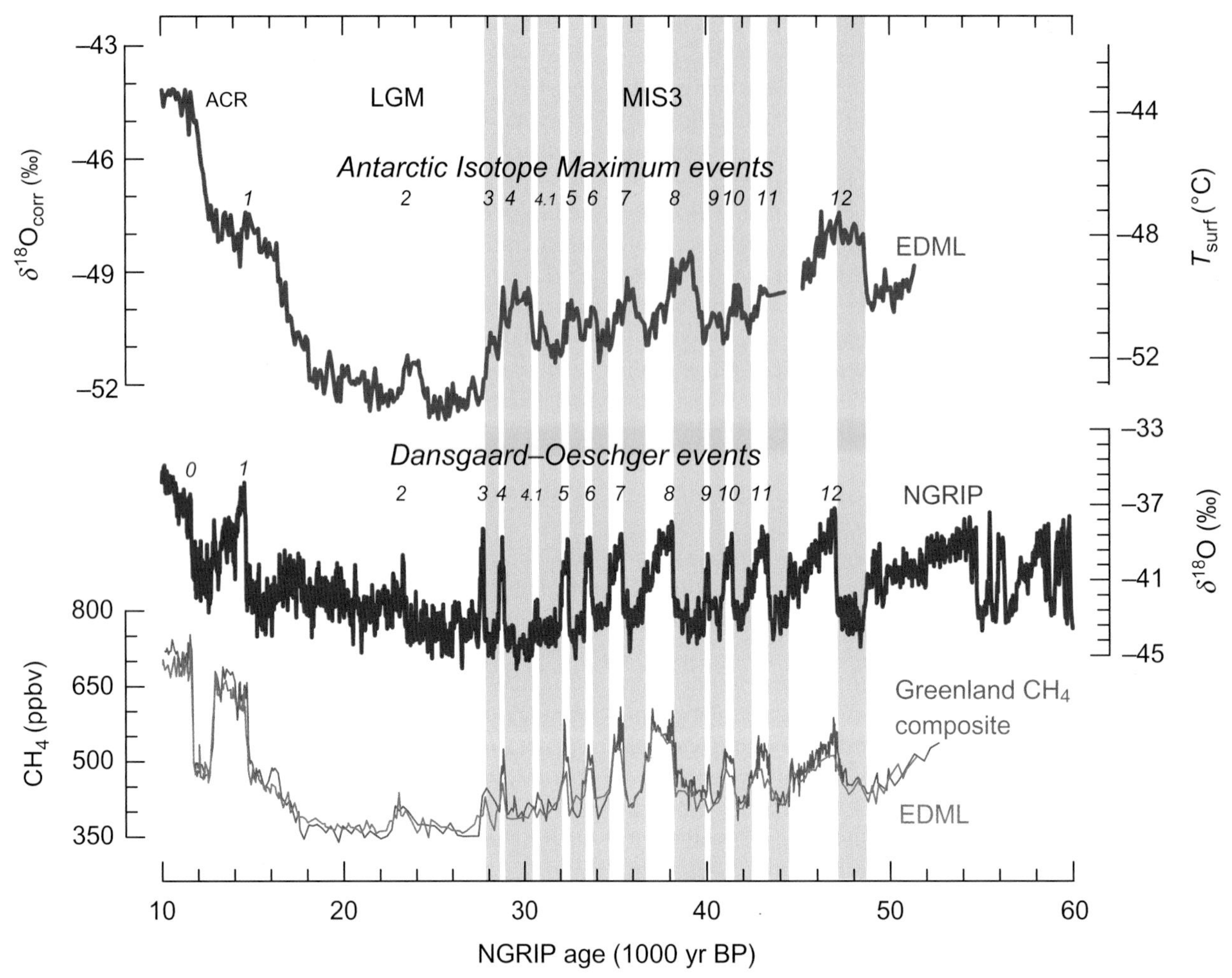

FIGURE 1.11 Sequence of abrupt climate changes during the Marine Isotope Stage 3 (MIS3) as recorded in ice cores from Antarctica (EDML, EPICA Dronning Maud Land) and Greenland (NGRIP, North Greenland Ice Core Project). The combination of these key paleoclimatic records provides convincing evidence for the role of the Atlantic meridional overturning circulation in determining the interhemispheric coupling during the last ice age. Gradual warmings and coolings in Antarctica (Antarctic Isotope Maximum events) occur synchronously with abrupt warmings in Greenland (Dansgaard–Oeschger events). This coherent interhemispheric coupling persists through much of the last ice age (Capron et al., 2010). During the Last Glacial Maximum (LGM), these events are absent or subdued; during the transition to the Holocene, the sequence reappears with the abrupt warming in Greenland at 14,500 years before the present (Dansgaard–Oeschger event 1), the Antarctic Cold Reversal (ACR), and Dansgaard–Oeschger event 0 being the last such abrupt event (Stenni et al., 2011). *Figure modified from EPICA Community Members (2006).*

also reduce atmospheric CO_2. This can be caused by changes in the calcium carbonate cycle on land (increase in weathering), in the ocean (reduced growth of coral reefs, a decrease in the production ratio of carbonate/organic carbon), or in the sediment (reduced carbonate burial).

Changes in the biological processes are also invoked to explain the glacial–interglacial CO_2 changes. More dust in the atmosphere during the ice age may have supplied more iron, a nutrient which could stimulate marine productivity and burial of organic matter.

None of these mechanisms, individually, is able to explain quantitatively the carbon cycle changes during a glacial–interglacial cycle (Kohfeld and Ridgwell, 2009). Therefore, a temporally coordinated succession of processes involving ocean physics, chemistry, and biology must have been at play to produce the observed tight relationship between reconstructed air temperatures and atmospheric CO_2 concentrations (Siegenthaler et al., 2005). Coupled climate–carbon cycle models begin to capture the complexity of these transient changes but are not yet able to quantitatively simulate ice age variations in a way that is fully consistent with the paleoclimatic and paleoceanographic evidence (Menviel et al., 2012).

7. THE OCEAN IN THE ANTHROPOCENE

The anthropocene is a new era during which human activities profoundly change the Earth System, by burning of fossil fuels, changing land use, and by deforestation. These impacts on all components of the climate system are now well documented (IPCC, 2007, 2013) and can, in many cases, be attributed to the direct and indirect effects of the increase of greenhouse gases in the atmosphere on global and regional scales (Stott et al., 2010; Bindoff et al., 2013). Although changes in the ocean are challenging to detect due to the large volume and the relative scarcity of observations, the ocean is an excellent integrator of changes owing to the long time scales involved in ocean processes. As already shown in Table 1.1, fundamental properties have been observed to change (Chapters 27, 28, and 30). Melting glaciers and ice sheets add mass to the ocean, thermal energy penetrates the ocean due to the widespread and long-term warming of the lower atmosphere observed in the twentieth and the beginning of the twenty-first century, a stronger atmospheric branch of the hydrological cycle causes a redistribution of salinity in the global ocean (Bindoff et al., 2007; Rhein et al., 2013), and the continuous rise of CO_2 in the atmosphere causes an ocean uptake of carbon (Rhein et al., 2013). All these changes have consequences for ocean circulation, distribution of water mass properties, and the chemical status of the ocean.

This is evident from the global zonal averages of the most recent data survey (Durack and Wijffels, 2010), and shown in Figure 1.12. A total of about 24×10^{22} J have entered the world ocean since 1955 (Levitus et al., 2012). The warming can be detected to be essentially global in the top 700 m of the ocean, and recent studies indicate that this is induced by human activity (Gleckler et al., 2012). In the Atlantic Ocean, the warming reaches deeper to about 2000 m because NADW transports the warmer waters efficiently to that depth (Figure 1.7). Since the 1990s, a small downward heat flux crossing 4000 m depth could be observed, which amounts to about 10^{12} W (Purkey and Johnson, 2010), thus less than 1% of the anthropogenic heat flux at the surface (Table 1.1). The time span of these observations is still too short to determine whether this is an anthropogenic signal or natural variability. The global extent of the signal and simple physical arguments, however, rather suggest this to be the early sign of whole-ocean warming.

The salinity changes in the top 500 m of the world ocean are significant and show a consistent pattern of freshening in the tropics and the high latitudes of both hemispheres, while the extra-tropics show increasing salinity (Durack and Wijffels, 2010). This is the signal that would be expected from an "accelerating" hydrological cycle: more water is evaporated in the dry zones of the subtropical gyres and more precipitation would occur in the high latitudes implying a stronger meridional moisture flux in the atmosphere (Held and Soden, 2006; Durack et al., 2012). The detection of this important change is facilitated by the integrating property of the ocean for changes in the atmospheric hydrological cycle. This is the response of the water cycle that comprehensive climate models consistently project for the twenty-first century (Meehl et al., 2007; Collins et al., 2013), and it appears that this trend has already started on a global scale.

Both the addition of freshwater from melting glaciers and ice sheets, and the heating from the surface reduce the density of the upper 500 m of the ocean. This strengthens the stratification and thus would likely reduce vertical mixing, which is an essential process in

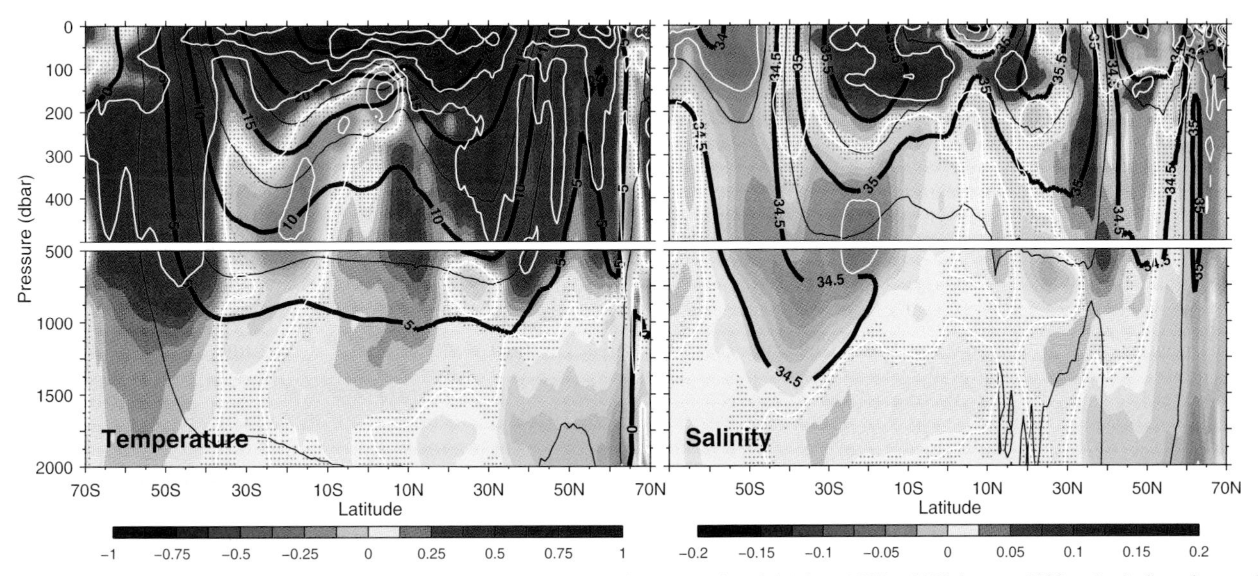

FIGURE 1.12 Global zonal mean 50-year trends of ocean temperature (left) and salinity (right) from 1950 to 2008 down to 2000 m depth. Superimposed black contours mark the zonal mean temperatures and salinities. Areas with trends that are not significant at the 90% confidence level are stippled. *Figure from Durack and Wijffels (2010) and P. Durack, personal communications.*

determining the ocean circulation and formation of water masses (see Section 4). Moreover, warming and freshening are amplified in the high latitudes where deep water is being formed. With both temperature and salinity changes conspiring to lower the densities of surface ocean waters, one would expect that the meridional overturning rates may reduce. Early coupled atmosphere–ocean model simulations (Chapter 23) have indicated the possibility that the ocean circulation may exhibit large-scale changes in deep water formation rates, particularly in the North Atlantic (Manabe and Stouffer, 1994). Most models, in fact, show a decrease of the AMOC in response to increases in greenhouse gas concentrations (Meehl et al., 2007; Collins et al., 2013), although the spread of this reduction is large, ranging from essentially no change to a reduction of 50% by the end of the twenty-first century. But the confidence in the representation of ocean processes determining deep water formation simulated by these models is only moderate.

Guided by insights from the paleoclimatic records, concerns about unpleasant surprises in a warming world were expressed by Broecker (1987a). Particularly, the warming could constitute a perturbation sufficiently strong for the meridional overturning circulation in the Atlantic Ocean to come to a halt. Some even speculated that the warming could trigger an abrupt shutdown of the "heat pump" with severe adverse implications for climate in Europe. However, there is no support from climate models for such extreme scenarios. In fact, as the changes in the AMOC are induced by the warming and the related strengthening of the water cycle, a slow-down of the circulation would only moderate the warming in the North Atlantic region but not lead to a net cooling.

If the AMOC indeed has multiple equilibrium states, as suggested by a number of models (Bryan, 1986; Stocker and Wright, 1991; Stocker and Marchal, 2000; Rahmstorf et al., 2005), the system could show a bifurcation with rising global temperatures. This behavior was suggested in a coupled climate model for the first time by Manabe and Stouffer (1993). In these simulations, the CO_2 concentrations were increased by 1%/yr up to double after 70 years, and quadruple levels after 140 years, and held constant thereafter, which resulted after 500 years of integration in a global mean warming of about 3.5–7 °C, respectively. The Atlantic overturning circulation decreased initially in both simulations, bifurcated then to a near shutdown in the $4\times CO_2$ experiment but recovered in the $2\times CO_2$ case. The slow-down occurred on the time scale of the forcing and was not abrupt. A central question was whether this, at that time comprehensive, model simulation would settle into this new equilibrium state. Later, continuation of these simulations showed that the Atlantic overturning recovered again after more than 1000 years of constant $4\times CO_2$ forcing (Stouffer and Manabe, 2003). This was attributed to a slow deep water warming, which gradually decreased the stability of the North Atlantic Ocean waters and promoted the deep water formation to return.

This early coupled climate model still employed so-called flux-corrections, and hence confidence in this result was limited. However, the most recent simulations with comprehensive models, which couple the different climate system components in a physically consistent way, confirm these early results obtained using simplified models. For high greenhouse concentrations, the occurrence of multiple equilibrium states in the meridional overturning circulation seems to be a robust result that does not depend on the model used (Mikolajewicz et al., 2007; Meehl et al., 2012), nor on its complexity (Stocker and Schmittner, 1997). Multiple equilibria are also found in complex coupled models applied in very different climatic conditions and configurations (Ferreira et al., 2011), suggesting that this is likely a general feature of the ocean in the Earth System.

Figure 1.13 presents an example of a simulation investigating the AMOC's response to an increase in the greenhouse gas concentrations since 1850 (Meehl et al., 2012), including an update (A. Hu, personal communication). The latest emission scenarios for climate model simulations (Moss et al., 2010) are used and extended to the year 2300, to achieve CO_2-stabilization at about 361, 543, 752, and 1962 ppm, which results in global mean warmings of 0.5, 2.2, 3.6, and 8.4 °C, respectively, relative to the mean of 1986–2005 (Figure 1.13a). The simulation using the highest emission scenario are carried further to year 2500 and reach a global mean warming of 9 °C. In these simulations, the response of the Atlantic overturning circulation shows a bifurcation between the warming of 3.6 and 8.4 °C. The evolution of the overturning indicates that the response of the AMOC changes fundamentally between the two highest scenarios. Whereas for the three lower scenarios, the circulation reduces but then recovers to almost the original value, for a higher forcing, the circulation spins down and almost vanishes by year 2200. That this seems to be a new stable state for at least several centuries is indicated by the continuation of this simulation up to year 2500. Therefore, a bifurcation point for the AMOC appears to exist in this model in which at a specific warming, and associated changes of the water cycle, a transition to a new equilibrium is triggered. The bifurcation already occurs between years 2020 and 2040 in these simulations, when the temperature differences between the scenarios are still rather small. The fact that a bifurcation is simulated demonstrates that there could be intrinsic irreversibility in the climate system. The Atlantic overturning circulation is not the only component in the climate system for which models have found such "points of no return."

The occurrence of bifurcation in the AMOC also depends on the history of the warming (Stocker and Schmittner, 1997). Thresholds are thus not absolute but

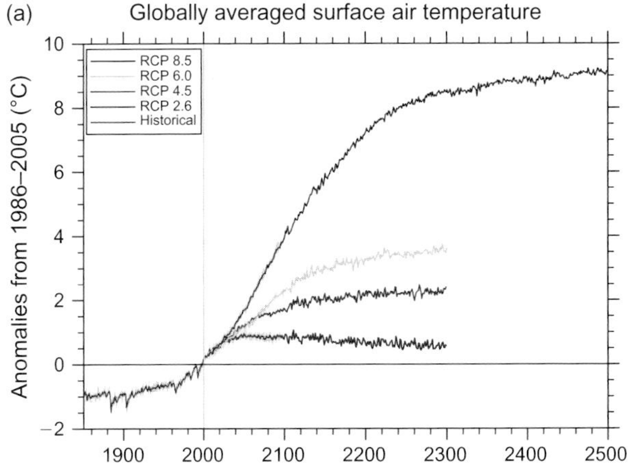

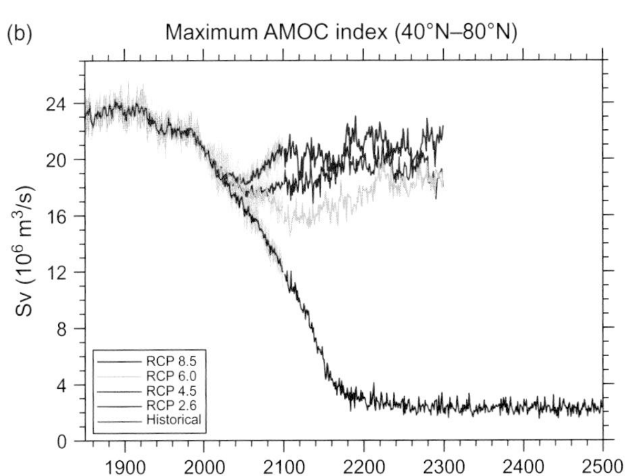

FIGURE 1.13 Projections of global mean temperature change (referenced to 1986–2005) (a) and Atlantic meridional overturning circulation (b) using a comprehensive coupled climate model (CCSM4) forced with concentrations from four greenhouse gas emission scenarios. *Figure based on Meehl et al. (2012) and updated by A. Hu (personal communication).*

may be avoided if different emission pathways are chosen. This is best illustrated in simple scenarios in which the atmospheric CO_2 concentration is increased at a given fraction per year until a value is reached and the concentration is held fixed thereafter (Manabe and Stouffer, 1993; Stocker and Schmittner, 1997). Figure 1.14 shows the critical CO_2 concentrations beyond which the AMOC assumes the new equilibrium state of shutdown. Results are obtained from many simulations using a coupled ocean–atmosphere model of reduced complexity that has two stable equilibrium states of the AMOC (Stocker et al., 1992; Stocker and Schmittner, 1997). It is evident that increased climate sensitivity and more rapid emissions both lower the CO_2 threshold. Therefore, a shutdown can be avoided by choosing a slower path to the same eventual CO_2 level. The reason for this rate-dependent stability is the fact that a faster increase of greenhouse gas concentrations in the atmosphere and the consequent rapid surface

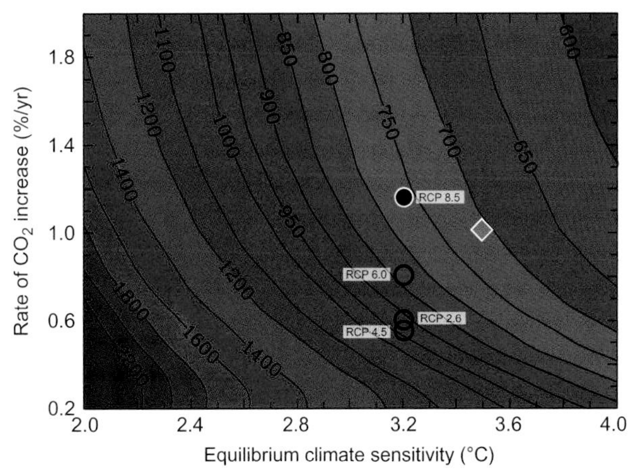

FIGURE 1.14 Contours of threshold CO_2 concentrations beyond which a bifurcation of the Atlantic meridional overturning circulation (AMOC) is simulated using a coupled model of reduced complexity that shows bistability of the AMOC. The threshold concentration depends on climate sensitivity and the rate of the CO_2 increase, and hence the warming. Circles indicate simulations presented in Figure 1.13 forced by the four emission scenarios RCP 2.6 to RCP 8.5. The AMOC recovers when the CO_2 concentrations remain below the threshold (open circles), but the AMOC shuts down if the CO_2 concentrations exceed the threshold (filled circle). The gray diamond indicates the early simulations by Manabe and Stouffer (1993) which showed an AMOC shutdown for more than 1000 years and a subsequent slow recovery over several centuries. *Figure modified from Stocker and Schmittner (1997).*

warming of the ocean enhance the stratification of the surface ocean, which reduces deep water formation and the overturning circulation. If the greenhouse gas increase proceeds more slowly, time is given for the heat to penetrate to the deeper layers of the ocean, which avoids the strong vertical density anomalies that would otherwise create an eventual shutdown. This suggests the rate at which the ocean takes up heat must be sufficiently slow to mix away the surface buoyancy increase created by the warming.

Results from some simulations using comprehensive coupled climate models are indicated in Figure 1.14 by symbols. Manabe and Stouffer (1993) report a climate sensitivity of their model of 3.5 °C. In a 1%/year CO_2 increase experiment, the threshold for a shutdown of the AMOC lies between 2 × and 4 × CO_2 concentration, consistent with the value of about 700 ppm obtained from the simplified model. The model used for the simulations presented in Figure 1.13 has a climate sensitivity of 3.2 °C (Meehl et al., 2012). Taking from every simulation the maximum of the time-varying rate of CO_2 increase as an indicator, the CO_2 thresholds are expected to be above 800 ppm for the three lower emission scenarios which have maximum CO_2 concentrations of about 361, 543, and 752 ppm, respectively. The simulation forced by the highest emission scenario RCP 8.5 reaches a CO_2 concentration of 1962 ppm and a maximum increase rate of 1.15%/year. The threshold predicted by the reduced complexity model is 770 ppm and therefore a bifurcation would be expected, as indeed simulated by the comprehensive model (Figure 1.13b). This lends strong support to the suggestion that the rate of buoyancy increase in the high-latitude surface ocean, caused by the combination of warming and enhanced meridional transport of freshwater, is therefore one of the determinants for an eventual AMOC shutdown. In order to avoid such a fundamental and long-lasting, if not irreversible, change of the circulation regime in the Atlantic Ocean, not only the level of warming, but also the speed of it, need to be limited. This insight provides a strong physical argument in favor of early and substantial greenhouse gas mitigation efforts.

An important question also concerns the response of the modes of natural variability to climate change. The ENSO phenomenon strongly influences regional climate in the equatorial Pacific region and even has implications on the global scale (see Section 5). Comprehensive climate models are capable of simulating ENSO variability, but many of them have difficulties in being realistic in the amplitude and frequency of occurrence and associated teleconnections. Early model simulations using a coupled climate model with sufficient resolution in the equatorial region showed an increase in the frequency of El Niño situations, and increased cooling during La Niña events, and they attributed this to the strengthening of the equatorial thermocline caused by the warming (Timmermann et al., 1999). However, the analysis of the evolution of ENSO characteristics (frequency and intensity) in both the past and the ongoing coordinated coupled model intercomparison projects does not show conclusive results (Collins et al., 2010; Stevenson, 2012). Some models show a strengthening, some a weakening of ENSO, and even when only those models are considered that exhibit the most realistic present-day ENSO simulations, the results remain, unfortunately, equivocal. Two robust results, however, are that: (i) ENSO continues to be a dominant feature of climate variability in the equatorial Pacific and (ii) none of the models indicates a lock-in into one of the ENSO states in response to the warming. The irregular sequence of warm El Niño and cold La Niña events is very likely to persist, possibly with altered relative frequencies and durations.

One of the longest-term and most certain commitments of the emissions of greenhouse gases is sea-level rise (Cazenave and Llovel, 2010; Church et al., 2013) (Chapter 27). The warming has direct and indirect effects on sea level. First, the well-observed uptake of heat by the world ocean to progressively deeper layers causes the so-called thermosteric contribution to sea-level rise due to the expansion of water as it warms. This contribution is rather well understood and quantified ranging from 0.1 to 0.41 m depending on the scenario by the end of the twenty-first century (Meehl et al., 2007). Second, melting glaciers and ice caps increase sea level but their contributions are projected to be significantly smaller (0.07–0.17 m). Third, the contribution from melting and possibly rapid responses of Greenland and Antarctica are potentially very large. A recent estimate for a dynamical response of the West Antarctic Ice-Sheet estimates an additional 3.3 m of which about 0.8 m may be realized by the end of the twenty-first century in a fast melt scenario (Bamber et al., 2009). However, such estimates and the possible location of thresholds remain highly uncertain (Joughin and Alley, 2011).

An important issue is the long-term commitment by the two contributions of ocean warming and ice sheet melting to global sea-level rise. The thermosteric contribution alone has a very long characteristic time scale, taking many centuries to equilibrate after greenhouse gases equilibrate (Plattner et al., 2008). This is due to the time required for the heat to penetrate to the deepest levels of the ocean. Likewise, the ice sheet melting and adjustment take more than 1000 years to come into equilibrium with the new climate. As before, there is indication that also the large ice sheets may exhibit bifurcation behavior, and thus thresholds for the warming that would preserve the ice sheets may exist (Gregory and Huybrechts, 2006; Robinson et al., 2012). With the recent insight and understanding that warming at the ocean–ice shelf interface is efficiently destabilizing ice streams (Holland et al., 2008), the ocean's crucial role in determining future sea level must be considered, as it influences the two dominant contributions to sea-level rise.

Currently, the ocean takes up carbon at the rate of about 2.2 Gt/year, which stems primarily from the combustion of fossil fuels and the effect of deforestation. Since industrialization, the ocean carbon reservoir has mitigated a substantial fraction of the increase of atmospheric CO_2. Estimates for the 1990s are that the ocean has taken up 118 Gt of the 244 Gt carbon emitted to the atmosphere by fossil fuel combustion (Denman et al., 2007), that is, today over half of the cumulative emissions are in the ocean. While this storage of anthropogenic carbon in the ocean prevents the associated radiative forcing and resulting warming in the atmosphere, it does have a measurable effect on the ocean's chemical status. The injection of CO_2 in the ocean increases the concentration of H^+-ions in the ocean because CO_2 reacts with H_2O to bicarbonate and a proton $\left(HCO_3^- + H^+\right)$. This makes the ocean more acidic, a process that is referred to as *ocean acidification*, and which is a direct effect of the increase of CO_2 concentrations in the atmosphere (Orr et al., 2005). The decrease of the ocean surface pH, a measure of the concentrations of H^+-ions: $pH = -\log_{10}(H^+)$, is well observed and amounts to about 0.1 pH-units since industrialization, that is, already 25% increase of the H^+-ion concentration has occurred (Feely et al., 2004). In the future, these changes are expected to increase significantly: depending on the emissions of CO_2, further acidification and reductions in the pH of 0.14–0.35 are projected (Orr et al., 2005), which will result in a 50% reduction of the carbonate ion CO_3^{2-}.

A more acid ocean has thus a significantly reduced CO_3^{2-} concentration and therefore, the volume of ocean in which $CaCO_3$, that is mainly shells of marine calcifying organisms, dissolves, becomes larger (Joos et al., 2011). Simulations with coupled climate–carbon cycle models demonstrate that annual-mean undersaturation is imminent in the Arctic Ocean by about 2030, to be followed by the Southern Ocean by about 2050 for high emission scenarios (Steinacher et al., 2009). These model simulations further suggest that undersaturation in the Arctic is reached during at least one month already at an atmospheric CO_2 concentration of 409 ppm, which is expected to be reached before 2020. Of great concern is this change in chemical status of the world ocean for marine calcifying organisms such as corals and plankton as it may become more difficult for them to maintain their calcareous skeletons in an increasingly corrosive environment (Hoegh-Guldberg et al., 2007; Hoegh-Guldberg and Bruno, 2010).

8. CONCLUDING THOUGHTS

The ocean is a vast reservoir of water, energy, carbon, and many other substances, and due to its large surface interacting with the atmosphere, it represents a key component in the Earth System. Human activities since the industrial revolution have caused a massive increase of the concentrations of the major greenhouse gases to levels unprecedented in the past 800,000 years. This increase has warmed the atmosphere and led to many changes worldwide. Such changes are now unequivocally observed in the climate system (IPCC, 2007, 2013). These include continuous melting of the large ice sheets of Greenland and Antarctica (Shepherd et al., 2012), Arctic sea ice retreat and global-scale ocean warming from the surface downward. The increased CO_2 concentration in the atmosphere also increases the ocean's acidity with potentially adverse effects on marine ecosystems (Chapter 31). Were it not for the ocean's ability to absorb substantial amounts of heat and carbon, the effects of worldwide anthropogenic climate change would be much larger. The ocean is therefore already an important mitigating element in the Earth System. Future research will show to what extent we can continue to rely on the mitigation power of the world ocean. We know already now that the uptake capacity of the ocean for carbon will decrease due to a shift in the chemical composition of the ocean's surface and the uptake efficiency of heat due to the increased stratification and reduced vertical mixing that can be anticipated in an ocean that warms from the surface.

The combination of paleoclimatic reconstructions from many sites, and the spatial correlation among them, reveals that the ocean has played an active role in climate history. In contrast to the ocean being merely a passive reservoir, the sequence of abrupt climate changes observed during the past ice ages suggests that the ocean has only limited stability to perturbations. Simulations indicate that the ocean's overturning circulation may undergo transitions to fundamentally different equilibrium states for sufficiently large perturbations. Bifurcation points initiating such irreversible paths may be closer than our current knowledge suggests.

While the understanding of the ocean, its internal processes and its interactions with the other components of the Earth System remain patchy and incomplete, the strong concern emerges that the world ocean, and the global web of ecosystems embedded in it, may be more vulnerable to human impact than ever assumed. For responsible stewardship of this planet, we therefore urgently need the implementation of the scientific knowledge into tangible and forward-looking decisions. This is the crucial contribution of science to international negotiations that aim to implement Article 2[3] of the United Nations Framework Convention on Climate Change (UNFCCC, 1992) with the goal to limit anthropogenic climate change (UNFCCC, 2010). This requires global mitigation efforts, which must be established without delay (Stocker, 2013), and in a sufficiently comprehensive manner (Steinacher et al., 2013), if global mean warming and related consequences on the Earth System, and in particular on the world ocean, are to be limited.

ACKNOWLEDGMENTS

I thank Beat Ihly for designing Figures 1.2 and 1.3, and redrawing Figures 1.4 and 1.5, and Paul Durack, Aixue Hu, and Lynne Talley for contributing figures, and John A. Church, Peter Gent, Beat Ihly, Kevin Trenberth, Christoph Raible, and two anonymous reviewers for thoughtful reviews. The efficient editorial work by John A. Church and Gerold Siedler is appreciated.

REFERENCES

Abram, N.J., Gagan, M.K., Liu, Z., Hantoro, W.S., McCulloch, M.T., Suwargadi, B.W., 2007. Seasonal characteristics of the Indian Ocean Dipole during the Holocene epoch. Nature 445, 299–302.

Adkins, J.F., McIntyre, K., Schrag, D.P., 2002. The salinity, temperature, and $\delta^{18}O$ of the glacial deep ocean. Science 298, 1769–1773.

Alley, R.B., et al., 2003. Abrupt climate change. Science 299, 2005–2010.

Antonov, J.I., Levitus, S., Boyer, T.P., 2004. Climatological annual cycle of ocean heat content. Geophys. Res. Lett. 31, L04304.

Bamber, J.L., Riva, R.E.M., Vermeersen, B.L.A., LeBrocq, A.M., 2009. Reassessment of the potential sea-level rise from a collapse of the West Antarctic Ice Sheet. Science 324, 901–903.

3. UNFCCC Article 2: The ultimate objective of this Convention [. . .] is to achieve [. . .] stabilization of greenhouse gas concentrations in the atmosphere at a level that would prevent dangerous anthropogenic interference with the climate system. [. . .]

Barker, S., Diz, P., Vautravers, M.J., Pike, J., Knorr, G., Hall, I.R., Broecker, W.S., 2009. Interhemispheric Atlantic seesaw response during the last deglaciation. Nature 457, 1097–1103.

Barker, S., Knorr, G., Edwards, R.L., Parrenin, F., Putnam, A.E., Skinner, L.C., Wolff, E., Ziegler, M., 2011. 800,000 years of abrupt climate variability. Science 334, 347–351.

Becker, J.J., et al., 2009. Global bathymetry and elevation data at 30 arc seconds resolution: SRTM30_PLUS. Mar. Geod. 32, 355–371.

Bessières, L., Madec, G., Lyard, F., 2008. Global tidal residual mean circulation: does it affect a climate OGCM? Geophys. Res. Lett. 35, L03609.

Bindoff, N.L., et al., 2007. Observations: oceanic climate change and sea level. In: Solomon, S. et al., (Eds.), Climate Change 2007: The Physical Science Basis. Contribution of Working Group I to the Fourth Assessment Report of the Intergovernmental Panel on Climate Change. Cambridge University Press, Cambridge, pp. 385–432.

Bindoff, N.L., et al., 2013. Detection and attribution of climate change: from global to regional. In: Stocker, T.F. et al., (Eds.), Climate Change 2013: The Physical Science Basis. Contribution of Working Group I to the Fifth Assessment Report of the Intergovernmental Panel on Climate Change. Cambridge University Press, Cambridge.

Bjerknes, J., 1969. Atmospheric teleconnections from the equatorial Pacific. Mon. Weather Rev. 97, 163–172.

Blunier, T., et al., 1998. Asynchrony of Antarctic and Greenland climate change during the last glacial period. Nature 394, 739–743.

Braconnot, P., Harrison, S.P., Kageyama, M., Bartlein, P.J., Masson-Delmotte, V., Abe-Ouchi, A., Otto-Bliesner, B., Zhao, Y., 2012. Evaluation of climate models using palaeoclimatic data. Nat. Clim. Chang. 2, 417–424.

Broecker, W.S., 1987a. Unpleasant surprises in the greenhouse? Nature 328, 123–126.

Broecker, W.S., 1987b. The biggest chill. Nat. Hist. 96, 74–82.

Broecker, W.S., 1998. Paleocean circulation during the last deglaciation: a bipolar seesaw? Paleoceanography 13, 119–121.

Broecker, W.S., Denton, G.H., 1989. The role of ocean-atmosphere reorganizations in glacial cycles. Geochim. Cosmochim. Acta 53, 2465–2501.

Broecker, W.S., Peng, T.-H., 1982. Tracers in the Sea. Eldigio Press, Palisades, New York.

Brönnimann, S., 2007. Impact of El Niño-Southern Oscillation on European climate. Rev. Geophys. 45, RG3003.

Bryan, F., 1986. High-latitude salinity effects and interhemispheric thermohaline circulations. Nature 323, 301–304.

Cane, M.A., 2005. The evolution of El Niño, past and future. Earth Planet. Sci. Lett. 230, 227–240.

Capron, E., et al., 2010. Millennial and sub-millennial scale climatic variations recorded in polar ice cores over the last glacial period. Clim. Past 6, 345–365.

Cazenave, A., Llovel, W., 2010. Contemporary sea level rise. Ann. Rev. Mar. Sci. 2, 145–173.

Church, J.A., et al., 2011. Revisiting the Earth's sea-level and energy budgets from 1961 to 2008. Geophys. Res. Lett. 38, L18601.

Church, J.A., et al., 2013. Sea level change. In: Stocker, T.F. et al., (Eds.), Climate Change 2013: The Physical Science Basis. Contribution of Working Group I to the Fifth Assessment Report of the Intergovernmental Panel on Climate Change. Cambridge University Press, Cambridge.

Ciais, P., et al., 2013. Carbon and other biogeochemical cycles. In: Stocker, T.F. et al., (Eds.), Climate Change 2013: The Physical Science Basis. Contribution of Working Group I to the Fifth Assessment Report of the Intergovernmental Panel on Climate Change. Cambridge University Press, Cambridge.

Clark, P.U., Pisias, N.G., Stocker, T.F., Weaver, A.J., 2002. The role of the thermohaline circulation in abrupt climate change. Nature 415, 863–869.

Clark, P.U., Dyke, A.S., Shakun, J.D., Carlson, A.E., Clark, J., Wohlfarth, B., Mitrovica, J.X., Hostetler, S.W., McCabe, A.M., 2009. The last glacial maximum. Science 325, 710–714.

Clement, A.C., Peterson, L.C., 2008. Mechanisms of abrupt climate change of the last glacial period. Rev. Geophys. 46, RG4002.

Cobb, K.M., Charles, C.D., Cheng, H., Edwards, L., 2003. El Niño/Southern Oscillation and tropical Pacific climate during the last millennium. Nature 424, 271–276.

Cobb, K.M., Westphal, N., Sayani, H.R., Watson, J.T., Di Lorenzo, E., Cheng, H., Edwards, R.L., Charles, C.D., 2013. Highly variable El Niño–Southern oscillation throughout the Holocene. Science 339, 67–70.

Collins, M., et al., 2010. The impact of global warming on the tropical Pacific Ocean and El Niño. Nat. Geosci. 3, 391–397.

Collins, M., et al., 2013. Long-term climate change: projections, commitments and irreversibility. In: Stocker, T.F. et al., (Eds.), Climate Change 2013: The Physical Science Basis. Contribution of Working Group I to the Fifth Assessment Report of the Intergovernmental Panel on Climate Change. Cambridge University Press, Cambridge.

Crowley, T.J., 1992. North Atlantic deep water cools the southern hemisphere. Paleoceanography 7, 489–497.

Crutzen, P.J., Stoermer, E.F., 2000. The "Anthropocene" IGBP Newslett. 41, 17–18.

Cunningham, S.A., Alderson, S.G., King, B.A., Brandon, M.A., 2003. Transport and variability of the Antarctic Circumpolar Current in Drake Passage. J. Geophys. Res. 108, 8084.

Dansgaard, W., Johnsen, S.J., Clausen, H.B., Dahl-Jensen, D., Gundestrup, N., Hammer, C.U., Oeschger, H., 1984. North Atlantic climatic oscillations revealed by deep Greenland ice cores. In: Hansen, J.E., Takahashi, T. (Eds.), Climate Processes and Climate Sensitivity, vol. 29. American Geophysical Union, Washington, pp. 288–298.

Day, J.J., Hargreaves, J.C., Annann, J.D., Abe-Ouchi, A., 2012. Sources of multi-decadal variability in Arctic sea ice extent. Environ. Res. Lett. 7, 034011.

Defant, A., 1941. Quantitative Untersuchungen zu Statik und Dynamik des Atlantischen Ozeans: Die absolute Topographie des physikalischen Meeresniveaus und der Druckflächen sowie die Wasserbewegungen im Raum des Atlantischen Ozeans. Wissenschaftliche Ergebnisse der Deutschen Atlantischen Expedition auf dem Forschungs-und Vermessungsschiff "Meteor" 1925-1927, vol. 6. Walter de Gruyter, Berlin, pp. 191–260.

Delworth, T., Manabe, S., Stouffer, R.J., 1993. Interdecadal variations of the thermohaline circulation in a coupled ocean-atmosphere model. J. Clim. 6, 1993–2011.

Denman, K.L., et al., 2007. Couplings between changes in the climate system and biogeochemistry. In: Solomon, S. et al., (Eds.), Climate Change 2007: The Physical Science Basis. Contribution of Working Group I to the Fourth Assessment Report of the Intergovernmental Panel on Climate Change. Cambridge University Press, Cambridge, pp. 499–587.

Deser, C., Phillips, A.S., Hurrell, J.W., 2004. Pacific interdecadal climate variability: linkages between the Tropics and North Pacific during boreal winter since 1900. J. Clim. 17, 3109–3124.

Deser, C., Alexander, M.A., Xie, S.-P., Phillips, A.S., 2010. Sea surface temperature variability: patterns and mechanisms. Ann. Rev. Mar. Sci. 2, 115–143.

Döös, K., Nilsson, J., Nycander, J., Brodeau, L., Ballarotta, M., 2012. The world ocean thermohaline circulation. J. Phys. Oceanogr. 42, 1445–1460.

Durack, P.J., Wijffels, S.E., 2010. Fifty-year trends in global ocean salinities and their relationship to broad-scale warming. J. Clim. 23, 4342–4362.

Durack, P.J., Wijffels, S.E., Matear, R.J., 2012. Ocean salinities reveal strong global water cycle intensification during 1950 to 2000. Science 336, 455–458.

Dutton, A., Lambeck, K., 2012. Ice volume and sea level during the last interglacial. Science 337, 216–219.

EPICA Community Members, 2006. One-to-one coupling of glacial climate variability in Greenland and Antarctica. Nature 444, 195–198.

Fasullo, J.T., Trenberth, K.E., 2008. The annual cycle of the energy budget. Part I: global mean and land-ocean exchanges. J. Clim. 21, 2297–2312.

Feely, R.A., Sabine, C.L., Lee, K., Berelson, W., Kleypas, J., Fabry, V.J., Millero, F.J., 2004. Impact of anthropogenic CO_2 on the $CaCO_3$ system in the oceans. Science 305, 362–366.

Ferrari, R., Wunsch, C., 2009. Ocean circulation kinetic energy: reservoirs, sources, and sinks. Annu. Rev. Fluid Mech. 41, 253–282.

Ferrari, R., Wunsch, C., 2010. The distribution of eddy kinetic and potential energies in the global ocean. Tellus 62A, 92–108.

Ferreira, D., Marshall, J., Rose, B., 2011. Climate determinism revisited: multiple equilibria in a complex climate model. J. Clim. 24, 992–1012.

Fischer, H., et al., 2010. The role of Southern Ocean processes in orbital and millennial CO_2 variations—a synthesis. Quat. Sci. Rev. 29, 193–205.

Fleitmann, D., et al., 2009. Timing and climatic impact of Greenland interstadials recorded in stalagmites from northern Turkey. Geophys. Res. Lett. 36, L19707.

Ganachaud, A., Wunsch, C., 2000. Improved estimates of global ocean circulation, heat transport and mixing from hydrographic data. Nature 408, 453–457.

Ganachaud, A., Wunsch, C., 2003. Large-scale ocean heat and freshwater transports during the World Ocean Circulation Experiment. J. Clim. 16, 696–705.

Gebbie, G., Huybers, P., 2012. The mean age of ocean waters inferred from radiocarbon observations: sensitivity to surface sources and accounting for mixing histories. J. Phys. Oceanogr. 42, 291–305.

Gleckler, P.J., et al., 2012. Human-induced global ocean warming on multidecadal timescales. Nat. Clim. Chang. 2, 524–529.

Gordon, A., 1991. The Role of Thermohaline Circulation in Global Climate Change. Lamont–Doherty Geological Observatory of Columbia University, Palisades, New York, pp. 44–51.

Gray, S.T., Graumlich, L.J., Betancourt, J.L., Pederson, G.T., 2004. A tree-ring based reconstruction of the Atlantic Multidecadal Oscillation since 1567 A.D. Geophys. Res. Lett. 31, L12205.

Gregory, J.M., Huybrechts, P., 2006. Ice-sheet contributions to future sea-level change. Philos. Trans. R. Soc. A 364, 1709–1731.

Guilyardi, E., Wittenberg, A., Fedorov, A., Collins, M., Wang, C., Capotondi, A., van Oldenborgh, G.J., Stockdale, T., 2009. Understanding El Niño in ocean-atmosphere general circulation models. Bull. Am. Meteorol. Soc. 90, 325–340.

Halpert, M.S., Ropelewski, C.F., 1992. Surface temperature patterns associated with the southern oscillation. J. Clim. 5, 577–593.

Held, I.M., Soden, B.J., 2006. Robust responses of the hydrological cycle to global warming. J. Clim. 19, 5686–5699.

Hoegh-Guldberg, O., Bruno, J.F., 2010. The impact of climate change on the world's marine ecosystems. Science 328, 1523–1528.

Hoegh-Guldberg, O., et al., 2007. Coral reefs under rapid climate change and ocean acidification. Science 318, 1737–1742.

Holland, D.M., Thomas, R.H., De Young, B., Ribbergaard, M.H., Lyberth, B., 2008. Acceleration of Jakobshavn Isbræ triggered by warm subsurface ocean waters. Nat. Geosci. 1, 659–664.

Hu, A., et al., 2012. Role of the Bering Strait on the hysteresis of the ocean conveyor belt circulation and glacial climate stability. Proc. Natl. Acad. Sci. U.S.A. 109, 6417–6422.

Huang, R.X., 2010. Ocean Circulation. Cambridge University Press, Cambridge, 791 pp.

Huber, C., Leuenberger, M., Spahni, R., Flückiger, J., Schwander, J., Stocker, T.F., Johnsen, S., Landais, A., Jouzel, J., 2006. Isotope calibrated Greenland temperature record over Marine Isotope Stage 3 and its relation to CH_4. Earth Planet. Sci. Lett. 243, 504–519.

Hurlburt, H.E., et al., 2009. High-resolution global and basin-scale ocean analyses and forecasts. Oceanography 22, 110–127.

Imawaki, S., Uchida, H., Ichikawa, H., Fukasawa, M., Umatani, S., ASUKA Group, 2001. Satellite altimeter monitoring the Kurosio transport south of Japan. Geophys. Res. Lett. 28, 17–20.

IPCC, 2007. Climate Change 2007: Solomon, S. et al., (Eds.), The Physical Science Basis. Contribution of Working Group I to the Fourth Assessment Report of the Intergovernmental Panel on Climate Change. Cambridge University Press, Cambridge.

IPCC, 2013. Climate Change 2013: The Physical Science Basis. In: Stocker, T.F. et al., (Eds.), Contribution of Working Group I to the Fifth Assessment Report of the Intergovernmental Panel on Climate Change. Cambridge University Press, Cambridge.

Izumo, T., Vialard, J., Lengaigne, M., de Boyer Montegut, C., Behera, S.K., Luo, J.-J., Cravatte, S., Masson, S., Yamagata, T., 2010. Influence of the state of the Indian Ocean Dipole on the following year's El Niño. Nat. Geosci. 3, 168–172.

Jansen, E., et al., 2007. Paleoclimate. In: Solomon, S. et al., (Eds.), Climate Change 2007: The Physical Science Basis. Contribution of Working Group I to the Fourth Assessment Report of the Intergovernmental Panel on Climate Change. Cambridge University Press, Cambridge, pp. 433–497.

Johns, W.E., et al., 2011. Continuous, array-based estimates of Atlantic Ocean heat transport at 26.58°N. J. Clim. 24, 2429–2449.

Joos, F., Frölicher, T.L., Steinacher, M., Plattner, G.-K., 2011. Impact of climate change mitigation on ocean acidification projections. In: Gattuso, J.-P., Hansson, L. (Eds.), Ocean Acidification. Oxford University Press, Oxford, pp. 272–290.

Joughin, I., Alley, R.B., 2011. Stability of the West Antarctic ice sheet in a warming world. Nat. Geosci. 4, 506–513.

Kavvada, A., Ruiz-Barradas, A., Nigam, S., 2013. AMO's structure and climate footprint in observations and IPCC AR5 climate simulations. Clim. Dyn. http://dx.doi.org/10.1007/s00382-013-1712-1.

Kohfeld, K.E., Ridgwell, A., 2009. Glacial-interglacial variability in atmospheric CO_2. In: Le Quéré, C., Saltzman, E.S. (Eds.), Surface Ocean—Lower Atmosphere Processes, vol. 187. American Geophysical Union, Washington, pp. 251–286.

Koltermann, K.P., Gouretski, V.V., Jancke, K., 2011. Hydrographic Atlas of the World Ocean Circulation Experiment (WOCE). In: Sparrow, M. et al., (Eds.), Atlantic Ocean, vol. 3. International WOCE Project Office, Southampton, UK.

Labat, D., Goddéris, Y., Probst, J.L., Guyot, J.L., 2004. Evidence for global runoff increase related to climate warming. Adv. Water Res. 27, 631–642.

Large, W.G., Yeager, S.G., 2009. The global climatology of an interannually varying air-sea flux data set. Clim. Dyn. 33, 341–364.

Levitus, S., Burgett, R., Boyer, T.P., 1994. In: NOAA Atlas NESDIS 3, World Ocean Atlas 1994. Salinity, vol. 3. NOAA, U.S. Department of Commerce.

Levitus, S., et al., 2012. World ocean heat content and thermosteric sea level change (0–2000 m), 1955–2010. Geophys. Res. Lett. 39, L10603.

Li, T., Wang, B., Chang, C.-P., Zhang, Y., 2003. A theory for the Indian Ocean Dipole-Zonal Mode. J. Atmos. Sci. 60, 2119–2135.

Liu, Z., 2012. Dynamics of interdecadal climate variability: a historical perspective. J. Clim. 25, 1963–1995.

Liu, Z., Alexander, M., 2007. Atmospheric bridge, oceanic tunnel, and global climatic teleconnections. Rev. Geophys. 45, RG2005.

Lumpkin, R., Speer, K., 2003. Large-scale vertical and horizontal circulation in the North Atlantic Ocean. J. Phys. Oceanogr. 33, 1902–1920.

Lund, D.C., Adkins, J.F., Ferrari, R., 2011. Abyssal Atlantic circulation during the Last Glacial Maximum: constraining the ratio between transport and vertical mixing. Paleoceanography 26, PA1213.

Luo, J.-J., Zhang, R., Behera, S.K., Masumoto, Y., Jin, F.-F., Lukas, R., Yamagata, T., 2010. Interaction between El Niño and extreme Indian Ocean Dipole. J. Clim. 23, 726–742.

Lüthi, D., et al., 2008. High-resolution carbon dioxide concentration record 650,000-800,000 years before present. Nature 453, 379–382.

Lyman, J.M., Good, S.A., Gouretski, V.V., Ishii, M., Johnson, G.C., Palmer, M.D., Smith, D.M., Willis, J.K., 2010. Robust warming of the global upper ocean. Nature 465, 334–337.

Lynch-Stieglitz, J., 2003. Tracers of past ocean circulation. In: Holland, H.D., Turekian, K.K. (Eds.), Treatise on Geochemistry, vol. 6. Elsevier, Amsterdam, pp. 433–451.

Lynch-Stieglitz, J., et al., 2007. Atlantic meridional overturning circulation during the Last Glacial Maximum. Science 316, 66–69.

MacDonald, G.M., Case, R.A., 2005. Variations in the Pacific Decadal Oscillation over the past millennium. Geophys. Res. Lett. 32, L08703.

Maltrud, M.E., McClean, J.L., 2005. An eddy resolving global 1/10° ocean simulation. Ocean Model. 8, 31–54.

Manabe, S., Stouffer, R.J., 1988. Two stable equilibria of a coupled ocean atmosphere model. J. Clim. 1, 841–866.

Manabe, S., Stouffer, R.J., 1993. Century-scale effects of increased atmospheric CO_2 on the ocean-atmosphere system. Nature 364, 215–218.

Manabe, S., Stouffer, R.J., 1994. Multiple-century response of a coupled ocean-atmosphere model to an increase of atmospheric carbon dioxide. J. Clim. 7, 5–23.

Mantua, N.J., Hare, S.R., 2002. The Pacific Decadal Oscillation. J. Oceanogr. 58, 35–44.

Marchal, O., Curry, W.B., 2008. On the abyssal circulation in the glacial Atlantic. J. Phys. Oceanogr. 38, 2014–2037.

MARGO Project Members, 2009. Constraints on the magnitude and patterns of ocean cooling at the Last Glacial Maximum. Nat. Geosci. 2, 127–132.

Marshall, J., Speer, K., 2012. Closure of the meridional overturning circulation through Southern Ocean upwelling. Nat. Geosci. 5, 171–180.

Masson-Delmotte, V., et al., 2013. Information from Paleoclimate Archives. In: Stocker, T.F. et al., (Eds.), Climate Change 2013: The Physical Science Basis. Contribution of Working Group I to the Fifth Assessment Report of the Intergovernmental Panel on Climate Change. Cambridge University Press, Cambridge.

Meehl, G.A., et al., 2007. Global Climate Projections. In: Solomon, S. et al., (Eds.), Climate Change 2007: The Physical Science Basis. Contribution of Working Group I to Fourth Assessment Report of the Intergovernmental Panel on Climate Change. Cambridge University Press, Cambridge, pp. 747–845.

Meehl, G.A., et al., 2012. Climate system response to external forcings and climate change projections in CCSM4. J. Clim. 25, 3661–3683.

Menviel, L., Joos, F., Ritz, S.P., 2012. Simulating atmospheric CO_2, ^{13}C and the marine carbon cycle during the Last Glacial-Interglacial cycle: possible role for a deepening of the mean remineralization depth and an increase in the oceanic nutrient inventory. Quat. Sci. Rev. 56, 46–68.

Mikolajewicz, U., Gröger, M., Maier-Reimer, E., Schurgers, G., Vizcaino, M., Winguth, A.M.E., 2007. Long-term effects of anthropogenic CO_2 emissions simulated with a complex earth system model. Clim. Dyn. 28, 599–633.

Mölg, T., Renold, M., Vuille, M., Cullen, N.J., Stocker, T.F., Kaser, G., 2006. Indian Ocean Zonal Mode activity in a multicentury-integration of a coupled AOGCM consistent with climate proxy data. Geophys. Res. Lett. 33, L18710.

Moss, R.H., et al., 2010. The next generation of scenarios for climate change research and assessment. Nature 463, 747–756.

Munk, W., 1966. Abyssal recipes. Deep Sea Res. 13, 707–730.

Munk, W., Wunsch, C., 1998. Abyssal recipes II, energetics of tidal and wind mixing. Deep Sea Res. 45, 1977–2010.

North Greenland Ice Core Project members, 2004. High-resolution climate record of the northern hemisphere back into the last interglacial period. Nature 431, 147–151.

Oeschger, H., Beer, J., Siegenthaler, U., Stauffer, B., Dansgaard, W., Langway, C.C., 1984. Late glacial climate history from ice cores. In: Hansen, J.E., Takahashi, T. (Eds.), Climate Processes and Climate Sensitivity, vol. 29. American Geophysical Union, Washington, pp. 299–306.

Oka, A., Hasumi, H., Abe-Ouchi, A., 2012. The thermal threshold of the Atlantic meridional overturning circulation and its control by wind stress forcing during glacial climate. Geophys. Res. Lett. 39, L09709.

Oort, A.H., Anderson, L.A., Peixoto, J.P., 1994. Estimates of the energy cycle of the oceans. J. Geophys. Res. 99, 7665–7688.

Orr, J.C., et al., 2005. Anthropogenic ocean acidification over the twenty-first century and its impact on calcifying organisms. Nature 437, 681–686.

Otto-Bliesner, B.L., Hewitt, C.D., Marchitto, T.M., Brady, E., Abe-Ouchi, A., Crucifix, M., Murakami, S., Weber, S.L., 2007. Last Glacial Maximum ocean thermohaline circulation: PMIP2 model intercomparisons and data constraints. Geophys. Res. Lett. 34, L12706.

Ou, H.-W., 2012. A minimal model of the Atlantic multidecadal variability: its genesis and predictability. Clim. Dyn. 38, 775–794.

Peixoto, J.P., Oort, A.H., 1992. Physics of Climate. American Institute of Physics, New York, USA, 520 pp.

Pierrehumbert, R.T., Abbot, D.S., Voigt, A., Koll, D., 2011. Climate of the Neoproterozoic. Annu. Rev. Earth Planet. Sci. 39, 417–460.

Piotrowski, A.M., Goldstein, S.L., Hemming, S.R., Fairbanks, R.G., 2004. Intensification and variability of ocean thermohaline circulation through the last deglaciation. Earth Planet. Sci. Lett. 225, 205–220.

Plattner, G.-K., et al., 2008. Long-term climate commitments projected with climate—carbon cycle models. J. Clim. 21, 2721–2751.

Purkey, S.G., Johnson, G.C., 2010. Warming of global abyssal and deep Southern Ocean Waters between the 1990s and 2000s: contributions to global heat and sea level rise budgets. J. Clim. 23, 6336–6351.

Rahmstorf, S., et al., 2005. Thermohaline circulation hysteresis: a model intercomparison. Geophys. Res. Lett. 32, L23605.

Rein, B., Lückge, A., Reinhardt, L., Sirocko, F., Wolf, A., Dullo, W.-C., 2005. El Niño variability off Peru during the last 20,000 years. Paleoceanography 20, PA4003.

Rempfer, J., Stocker, T.F., Joos, F., Dutay, J.-C., Siddall, M., 2011. Modelling isotopes of neodymium with a coarse resolution ocean circulation model: sensitivities to model parameters and source/sink distributions. Geochim. Cosmochim. Acta 75, 5927–5950.

Rempfer, J., Stocker, T.F., Joos, F., Dutay, J.-C., 2012. On the relationship between Nd isotopic composition and ocean overturning circulation in idealized freshwater discharge events. Paleoceanography 27, PA3211.

Rhein, M., et al., 2013. Observations: Ocean. In: Stocker, T.F. et al., (Eds.), Climate Change 2013: The Physical Science Basis. Contribution of Working Group I to the Fifth Assessment Report of the Intergovernmental Panel on Climate Change. Cambridge University Press, Cambridge.

Richardson, P.L., 2008. On the history of meridional overturning circulation schematic diagrams. Prog. Oceanogr. 76, 466–486.

Risien, C.M., Chelton, D.B., 2008. A global climatology of surface wind and wind stress fields from eight years of QuikSCAT scatterometer data. J. Phys. Oceanogr. 38, 2379–2413.

Ritz, S.P., Stocker, T.F., Grimalt, J.O., Menviel, L., Timmermann, A., 2013. Estimated strength of the Atlantic overturning circulation during the last deglaciation. Nat. Geosci. 6, 208–212.

Riva, R.E.M., Bamber, J.L., Lavallé, D.A., Wouters, B., 2010. Sea-level fingerprint of continental water and ice mass change from GRACE. Geophys. Res. Lett. 37, L19605.

Robinson, A., Calov, R., Ganopolski, A., 2012. Multistability and critical thresholds of the Greenland ice sheet. Nat. Clim. Chang. 2, 429–432.

Roemmich, D., Gilson, J., 2009. The 2004–2008 mean and annual cycle of temperature, salinity, and steric height in the global ocean from the Argo Program. Prog. Oceanogr. 82, 81–100.

Ropelewski, C.F., Halpert, M.S., 1987. Global and regional scale precipitation patterns associated with the El Niño/Southern Oscillation. Mon. Weather Rev. 115, 1606–1626.

Saji, N.H., Goswami, B.N., Vinayachandran, P.N., Yamagata, T., 1999. A dipole mode in the tropical Indian Ocean. Nature 401, 360–363.

Sarachik, E.S., Cane, M.A., 2010. The El Niño-Southern Oscillation Phenomenon. In: Cambridge University Press, Cambridge, 369 pp.

Schäfer-Neth, C., Paul, A., 2003. The Atlantic Ocean at the last glacial maximum: 1. Objective mapping of the GLAMAP sea-surface conditions. In: Wefer, G. et al., (Eds.), The South Atlantic in the Late Quaternary: Material Budget and Current Systems. Springer, Berlin, pp. 531–548.

Schlesinger, M.E., Ramankutty, N., 1994. An oscillation in the global climate system of period 65-70 years. Nature 367, 723–726.

Schmitz, W.J., 1996. On the world ocean circulation: Volume II. The Pacific and Indian Oceans/A global update. Technical Report WHOI-96-08, Woods Hole Oceanographic Institution, 237 pp.

Schott, F.A., Xie, S.-P., McCreary Jr., J.P., 2009. Indian ocean circulation and climate variability. Rev. Geophys. 47, RG1002.

Shen, C., Wang, W.-C., Gong, W., Hao, Z., 2006. A Pacific Decadal Oscillation record since 1470 AD reconstructed from proxy data of summer rainfall over eastern China. Geophys. Res. Lett. 33, L03702.

Shepherd, A., et al., 2012. A reconciled estimate of ice-sheet mass balance. Science 338, 1183–1189.

Siddall, M., Stocker, T.F., Blunier, T., Spahni, R., McManus, J., Bard, E., 2006. Using a maximum simplicity paleoclimate model to simulate millennial variability during the last four glacial periods. Quat. Sci. Rev. 25, 3185–3197.

Siegenthaler, U., et al., 2005. Stable carbon cycle-climate relationship during the Late Pleistocene. Science 310, 1313–1317.

Sigman, D.M., Boyle, E.A., 2000. Glacial/interglacial variations in atmospheric carbon dioxide. Nature 407, 859–869.

Steffensen, J.P., et al., 2008. High-resolution Greenland ice core data show abrupt climate change happens in few years. Science 321, 680–684.

Steinacher, M., Joos, F., Frölicher, T.L., Plattner, G.-K., Doney, S.C., 2009. Imminent ocean acidification in the Arctic projected with the NCAR global coupled carbon cycle-climate model. Biogeosciences 6, 515–533.

Steinacher, M., Joos, F., Stocker, T.F., 2013. Allowable carbon emissions lowered by multiple climate targets. Nature 499, 197–201.

Stenni, B., et al., 2011. Expression of the bipolar see-saw in Antarctic climate records during the last deglaciation. Nat. Geosci. 4, 46–49.

Stephens, G.L., et al., 2012. An update on Earth's energy balance in light of the latest global observations. Nat. Geosci. 5, 691–696.

Stevenson, S.L., 2012. Significant changes to ENSO strength and impacts in the twenty-first century: results from CMIP5. Geophys. Res. Lett. 39, L17703.

Stocker, T.F., 1998. The seesaw effect. Science 282, 61–62.

Stocker, T.F., 2000. Past and future reorganisations in the climate system. Quat. Sci. Rev. 19, 301–319.

Stocker, T.F., 2013. The closing door of climate targets. Science 339, 280–282.

Stocker, T.F., Johnsen, S.J., 2003. A minimum thermodynamic model for the bipolar seesaw. Paleoceanography 18, 1087.

Stocker, T.F., Marchal, O., 2000. Abrupt climate change in the computer: is it real? Proc. Natl. Acad. Sci. U.S.A. 97, 1362–1365.

Stocker, T.F., Schmittner, A., 1997. Influence of CO_2 emission rates on the stability of the thermohaline circulation. Nature 388, 862–865.

Stocker, T.F., Wright, D.G., 1991. Rapid transitions of the ocean's deep circulation induced by changes in surface water fluxes. Nature 351, 729–732.

Stocker, T.F., Wright, D.G., Mysak, L.A., 1992. A zonally averaged, coupled ocean-atmosphere model for paleoclimate studies. J. Clim. 5, 773–797.

Stommel, H., 1957. A survey of ocean current theory. Deep Sea Res. 4, 149–184.

Stommel, H., 1958. The abyssal circulation. Deep Sea Res. 5, 80–82.

Stommel, H., 1961. Thermohaline convection with two stable regimes of flow. Tellus 13, 224–230.

Stommel, H., Arons, A.B., 1960. On the abyssal circulation of the world ocean—I. Stationary planetary flow patterns on a sphere. Deep Sea Res. 6, 140–154.

Stott, P.A., Gillett, N.P., Hegerl, G.C., Karoly, D.J., Stone, D.A., Zhang, X., Zwiers, F., 2010. Detection and attribution of climate change: a regional perspective. Wiley Interdiscip. Rev. Clim. Change 1, 192–211.

Stouffer, R.J., Manabe, S., 2003. Equilibrium response of thermohaline circulation to large changes in atmospheric CO_2 concentration. Clim. Dyn. 20, 759–773.

Sutton, R.T., Hodson, D.L.R., 2003. Influence of the ocean on North Atlantic climate variability 1871-1999. J. Clim. 16, 3296–3313.

Takahashi, T., et al., 2009. Climatological mean and decadal change in surface ocean pCO_2, and net sea–air CO_2 flux over the global oceans. Deep Sea Res. II 56, 554–577.

Talley, L.D., 2003. Shallow, intermediate, and deep overturning components of the global heat budget. J. Phys. Oceanogr. 33, 530–560.

Talley, L.D., 2008. Freshwater transport estimates and the global overturning circulation: shallow, deep and throughflow components. Prog. Oceanogr. 78, 257–303.

Talley, L.D., 2013. Closure of the global overturning circulation through the Indian, Pacific and Southern Oceans: schematics and transports. Oceanography 26, 80–97.

Talley, L.D., Pickard, G.L., Emery, W.J., Swift, J.H., 2011. Descriptive Physical Oceanography. Elsevier, Amsterdam, 555 pp.

Timmermann, A., Oberhuber, J.M., Bacher, A., Esch, M., Latif, M., Roeckner, E., 1999. Increased El Niño frequency in a climate model forced by future greenhouse warming. Nature 398, 694–696.

Trenberth, K.E., Smith, L., Qian, T., Dai, A., Fasullo, J., 2007. Estimates of the global water budget and its annual cycle using observational and model data. J. Hydrometeorol. 8, 758–769.

Trenberth, K.E., Fasullo, J.T., Mackaro, J., 2011. Atmospheric moisture transports from ocean to land and global energy flows in reanalyses. J. Clim. 24, 4907–4924.

UNFCCC, 1992. United Nations Framework Convention on Climate Change (FCCC/INFORMAL/84 GE.05-62220 (E) 200705). United Nations, New York.

UNFCCC, 2010. The Cancun Agreements, FCCC/CP/2010/7/Add.1. United Nations Framework Convention on Climate Change.

Verschuuren, D., Laird, K.R., Cumming, B.F., 2000. Rainfall and drought in equatorial east Africa during the past 1,100 years. Nature 403, 410–414.

Voelker, A.H.L., Workshop-Participants, 2002. Global distribution of centennial-scale records for Marine Isotope Stage (MIS) 3: a database. Quat. Sci. Rev. 21, 1185–1212.

Wang, C., 2001. A unified oscillator model for the El Niño-Southern Oscillation. J. Clim. 14, 98–115.

Webster, P.J., Moore, A.M., Loschnigg, J.P., Leben, R.R., 1999. Coupled ocean-atmosphere dynamics in the Indian Ocean during 1997-98. Nature 401, 356–360.

Wijffels, S.E., 2001. Ocean transport of fresh water. In: Siedler, G. et al., (Eds.), Ocean Circulation & Climate: Observing and Modelling the Global Ocean. International Geophysics Series, vol. 77. Academic Press, pp. 475–488.

Wunsch, C., 1998. The work done by the wind on the oceanic general circulation. J. Phys. Oceanogr. 28, 2332–2340.

Wunsch, C., Ferrari, R., 2004. Vertical mixing, energy, and the general circulation of the oceans. Annu. Rev. Fluid Mech. 36, 281–314.

Xie, S.-P., Annamalai, H., Schott, F.A., McCreary, J.P., 2002. Structure and mechanisms of South Indian Ocean climate variability. J. Clim. 15, 864–878.

Zaucker, F., Broecker, W.S., 1992. The influence of atmospheric moisture transport on the fresh water balance of the Atlantic drainage basin: general circulation model simulations and observations. J. Geophys. Res. 97, 2765–2773.

Chapter 2

Paleoclimatic Ocean Circulation and Sea-Level Changes

Stefan Rahmstorf and Georg Feulner
Potsdam Institute for Climate Impact Research, Potsdam, Germany

Chapter Outline

1. INTRODUCTION

The oceans have been an important factor in shaping Earth's past climate. Their circulation has influenced the overlying atmosphere (see, e.g., reviews of Alley et al., 2002; Clark et al., 2002; Rahmstorf, 2002) and they have acted, as they still do, as a repository for heat and gases. One consequence of the ocean's response to climate change, which is acutely important for human society, is sea-level change (the book *Understanding sea-level rise and variability* edited by Church et al., 2010 gives an excellent overview of this topic; see Chapter 27). This chapter provides a brief introductory discussion of past ocean circulation and sea-level changes before the time of instrumental measurements.

We describe the methods used for reconstructing past ocean states using proxy data and models and give specific examples of ocean circulation and sea-level changes in Earth's history. This is a large field with a vast body of scientific literature, thanks to a great research community with many excellent scholars, and this chapter can certainly not provide a comprehensive review. Rather, we provide the reader with an introduction to some of the main methods, together with examples of the research questions being addressed. Inevitably, the choice of examples is to some extent subjective and dependent on the authors' areas of expertise. However, we hope to provide a basis of understanding this exciting and important topic in which many puzzles still wait to be solved.

The dominant drivers of paleoclimatic change depend on the timescale considered. Two important ones are tectonic and orbital timescales (Ruddiman, 2000). The former covers climatic changes on timescales of some millions of years up to the age of the Earth of 4.6 billion years. These changes are driven by tectonic processes associated with the solid Earth's internal heat, but also include the evolution of life which has altered the composition of Earth's atmosphere (Stanley, 2005). Plate tectonics are a key driver of Earth's long-term carbon cycle which controls atmospheric CO_2 concentration on multimillion year timescales by exchanging carbon with the Earth's crust. Plate tectonics also alter the Earth's geography with effects on climate, that

Ocean Circulation and Climate, Vol. 103. http://dx.doi.org/10.1016/B978-0-12-391851-2.00002-7

is, the position of continents and oceans, the formation of mountain ranges, and the opening or closing of ocean gateways.

On timescales of thousands to hundreds of thousands of years, the Earth's orbital cycles are an important (probably dominant) driver of climate changes. These Milankovitch cycles (Milankovitch, 1941) have main periods of 23,000 years (precession), 41,000 years (tilt of Earth's axis), and ~100,000 as well as ~400,000 years (eccentricity of Earth's orbit) and have a strong effect on the seasonal and latitudinal distribution of solar radiation (Ruddiman, 2000). These orbital cycles are pacemakers of the Quaternary glaciations (see Section 3).

In addition, climate has been changed by variability in the sun's radiation output and by internal processes in the climate system, such as through instabilities in ocean circulation or ice sheets. It is important to distinguish between changes in global-mean temperature (primarily dependent on Earth's global radiation balance but with small transient variations due to changes in ocean heat storage; Chapter 27) and those climate changes caused by redistribution of heat such as through changing oceanic or atmospheric heat transport (Chapter 29).

2. RECONSTRUCTING PAST OCEAN STATES

Past changes in ocean circulation and sea level have left a number of lasting records that can still be sampled today, for example, in the form of sediments on the ocean floor, uplifted marine terraces, or ancient corals. In order to interpret these proxy data and to provide quantitatively consistent scenarios of past ocean states and changes, a range of models is used. The combination of proxy data and models allows us to formulate and test hypotheses about mechanisms of past ocean and climate changes.

2.1. Proxies for Past Ocean Circulation

Although there are numerous sources of information about past ocean states in the sedimentary record, it is not easy to interpret this information in terms of specific changes in ocean circulation. In fact, this is an inverse problem where the products of a given ocean circulation (in terms of sediment deposition) are used to infer what the ocean circulation state may have been. Solving this inverse problem is complicated by the fact that what is found in the sedimentary record is usually the result of multiple influencing factors, the data have uncertainties both in dating and in the parameter values measured, and their time resolution is often seriously limited (e.g., by bioturbation of the sediment) so that they need to be interpreted as some kind of time average.

Tracers of past ocean circulation can be broadly grouped into three types: nutrient-type water mass tracers, conservative water mass tracers, and circulation rate tracers. In addition, special cases are the neodymium isotope ratios and nongeochemical tracers. An excellent, much more detailed review along these lines is found in Lynch-Stieglitz (2003); see also Chapter 26.

2.1.1. *Nutrient Water Mass Tracers*

Nutrient water mass tracers are those elements that are involved in biological activity and thus behave like nutrients, that is, like the key constituents of marine organic matter: carbon, nitrogen, and phosphorus. These are taken up by marine life during primary production near the ocean surface and hence tend to be depleted in surface waters. In many parts of the ocean, nitrogen or phosphorus are close to zero concentration at the surface since their availability is the limiting factor of primary production (these are *bio-limiting elements*). As dead organic matter sinks through the water column and decays there or on the seafloor, nutrients are returned to the water. Thus, water masses typically gain in nutrient concentration over time after they have left the surface ocean. In the present-day Atlantic, the main water masses of Antarctic Intermediate Water (AAIW), North Atlantic Deep Water (NADW), and Antarctic Bottom Water (AABW) reflect the initial nutrient content set by the respective surface value in the water mass formation region and by subsequent mixing between these water masses. In the deep Pacific Ocean, nutrient content reflects the initial value of the inflowing AABW, progressively increasing as the water mass ages along the pathway that it spreads.

The basic principle behind the use of nutrient-style watermass tracers is to use elements (or isotopes of elements) that behave like nutrients but that are preserved in sediments (typically in the calcium carbonate shells of bottom dwelling marine organisms such as foraminifera), so that their distribution during past epochs can be mapped using a large number of sediment cores from different ocean depths. Prominent among these tracers are carbon isotope ratios ($\delta^{13}C$) (Deuser and Hunt, 1969) and the cadmium/calcium ratio (Cd/Ca) (Boyle et al., 1976).

A basic precondition for using the composition of shells (or "tests") of marine organisms as water mass indicators is that the chemical composition of these shells faithfully reflects that of the seawater they grew in. This is illustrated for two species of foraminifera in Figure 2.1 for their carbon isotope ratios.

The choice of species is important: not all are as well suited as the two shown. Biological processes can lead to preferential uptake of the lighter ^{12}C as compared to ^{13}C, so that a systematic offset arises between $\delta^{13}C$ (a measure of relative ^{13}C content) in the shells relative to the ambient

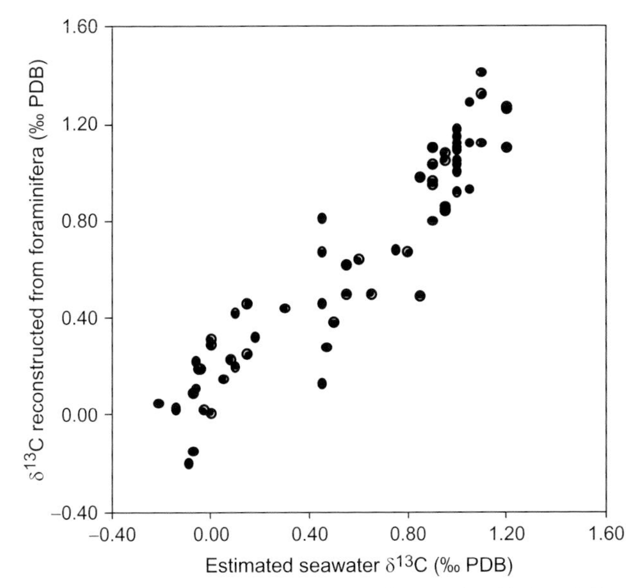

FIGURE 2.1 Carbon isotopic composition of core top benthic foraminifera in the genera *Cibicidoides* and *Planulina* versus estimated carbon isotopic composition of dissolved inorganic carbon in the water column above core location. *From Lynch-Stieglitz (2003) after Duplessy et al. (1984).*

water (Wefer and Berger, 1991). This is generally the case for planktonic foraminifera (i.e., those living in the water column), which thus do not record water mass properties as well as the benthic (i.e., bottom dwelling) species shown in Figure 2.1 (Spero and Lea, 1996). Hence, benthic foraminifera are commonly used to reconstruct past deepwater masses. Within benthic foraminifera, differences between species are thought to mainly arise because of their different choices of microhabitat (e.g., Tachikawa and Elderfield, 2002). Some (e.g., *Uvigerina*) live not on the surface of the sediments but within the pore water, which is depleted in $\delta^{13}C$ relative to the bottom water. This example highlights that much experience and a detailed understanding of the processes is required in order to arrive at robust conclusions from proxy data; it takes considerable time and detective work to develop proxies to the level where they can be properly interpreted.

The Cd/Ca ratio in shells of benthic foraminifera is another commonly used proxy for nutrient cycling in the ocean, because these elements are incorporated in the shells in proportion to their abundance in the water. However, there is a depth-dependent empirical factor that needs to be accounted for in reconstructing seawater cadmium concentrations from the shells (Boyle, 1992).

Cadmium in seawater is distributed much like major nutrients (e.g., phosphorus)—indeed it shows an almost linear relationship to phosphorus. Apparently, marine plankton metabolize cadmium like a nutrient, so it is depleted in warm surface waters where primary productivity occurs and enriched in deepwaters where organic matter decays (Boyle et al., 1976). A complication that needs to be considered in the interpretation is that the biological relation between cadmium and phosphorus is not exactly linear due to preferential uptake, so that a curved P versus Cd relation is expected along a water mass that is progressively nutrient-depleted by biological activity (Elderfield and Rickaby, 2000). In contrast, a straight mixing line would be obtained by progressive mixing of two distinct water masses with low and high nutrient content.

Further nutrient-style tracers include barium and zinc, measured as Ba/Ca and Zn/Ca ratios, respectively (Lynch-Stieglitz, 2003).

2.1.2. Conservative Water Mass Tracers

Conservative water mass tracers are those that (in contrast to nutrients) have no sources or sinks in the subsurface ocean, so that their properties are set at the ocean surface and the variations deeper in the water column reflect the transport and mixing of water masses (see Chapters 6–8 and 26). This is the case with temperature and salinity, the standard water mass tracers of modern physical oceanography. Such a conservative tracer is magnesium, as the Mg/Ca ratio in benthic foraminifera reflects the water temperature during calcification, that is, that in which the organisms lived (Rosenthal et al., 1997). The major challenge here is to determine the water temperature from Mg/Ca with sufficient accuracy in order to reconstruct the relatively small gradients in deepwater temperatures.

Another common conservative tracer is the oxygen isotope ratio, expressed as $\delta^{18}O$ (i.e., the relative deviation of ^{18}O content from a standard). The $\delta^{18}O$ value of ocean water is set at the surface depending on water exchange (e.g., by evaporation and precipitation), hence it is closely linked to salinity (Lynch-Stieglitz, 2003). Since ^{18}O is heavier than the common ^{16}O, it evaporates less easily. Therefore, continental ice (made from snow) is depleted in ^{18}O and the formation of large ice sheets leads to a sizeable increase in ^{18}O in ocean water. During the last glacial maximum (LGM), sea level was ~130 m lower than today; the missing water was ^{18}O depleted and stored on land in the form of ice. The remaining water therefore was proportionally ^{18}O enriched by about 1‰ in $\delta^{18}O$. In fact, this glacial ^{18}O peak can still be found today in the pore water of sediments formed at the time, now typically found at a depth of 20–60 m within the sediments, depending on location (Schrag et al., 2002).

The $\delta^{18}O$ in the calcite shells of foraminifera depends on that of the surrounding waters and on the temperature (there is a temperature-dependent biological fractionation) (e.g., Duplessy et al., 2002). Hence, if the $\delta^{18}O$ of the water or the salinity and ice volume effect is known (or negligible), $\delta^{18}O$ in these shells can be used as a proxy for local temperature (Figure 2.2). The slope shown there suggests that

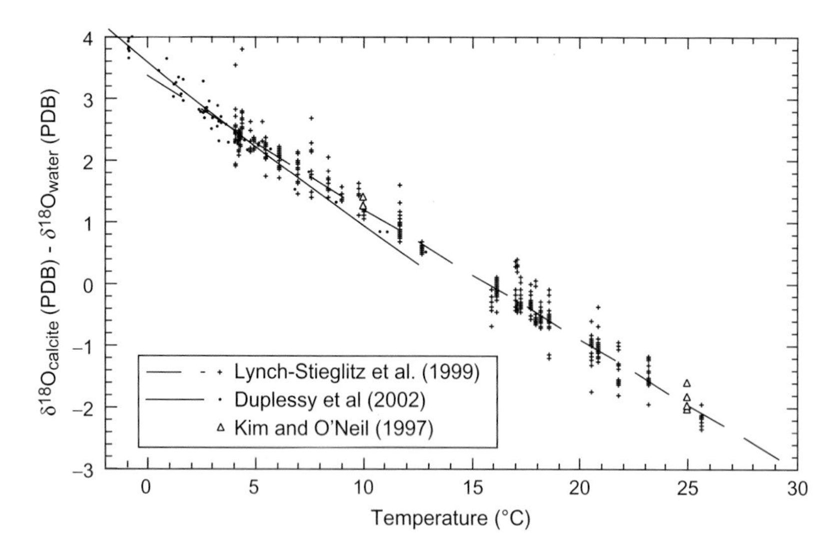

FIGURE 2.2 The isotopic fractionation, expressed as the difference between the $\delta^{18}O$ of foraminiferal calcite and the $\delta^{18}O$ of seawater versus temperature of calcification. Cibicidoides and Planulina from Lynch-Stieglitz et al. (1999) (—+) and Duplessy et al. (2002) (—•) are shown along with the inorganic precipitation experiments of Kim and O'Neil (1997) (Δ). *From Lynch-Stieglitz (2003).*

benthic $\delta^{18}O$ in calcite increases by ~0.25‰ per °C. However, $\delta^{18}O$ is useful as a water mass tracer even if the effects of temperature and salinity cannot be separated, since they are both conservative tracers (Lynch-Stieglitz, 2003).

2.1.3. *Circulation Rate Tracers*

The tracers discussed so far provide information on the water masses present in the past at locations for which we have sediment data, including on the relative contribution of waters from different source regions and thus relative renewal rates. But they do not reveal the rates of ocean circulation. Various methods have been applied to estimate rates.

One set of techniques uses radiocarbon (^{14}C), which is created in the atmosphere due to cosmic rays. While the surface ocean is close to equilibrium with atmospheric radiocarbon content, this decays below the surface according to the half-life of ^{14}C of 5730 years. In the modern ocean, this is a very useful measure of the age of water masses, that is, the time elapsed since they left the surface (Stuiver et al., 1983).

The problem with using ^{14}C in paleoclimatic studies is that it continues to decay after it is incorporated in calcite shells, so that the time recorded by ^{14}C ages of benthic foraminifera is the age of the deepwater plus the age of the foraminifera shell. To derive the deepwater age, an independent measure of the shell age is needed. This can be derived from the age of planktonic foraminifera collocated in the same sediment (Broecker et al., 1988). Or the method is applied to benthic corals, which can be dated independently using uranium dating (Adkins et al., 1998).

Another approach is measuring the ratio of protactinium to thorium, $^{231}Pa/^{230}Th$. These elements are created uniformly throughout the water column by uranium decay and rapidly removed into the sediments by reacting with sinking marine particles (Anderson et al., 1983). Since the efficiency with which both elements are removed is different, their ratio depends on how long this removal process has operated, for example, along the path of NADW flowing south in the Atlantic. This method has been used to infer NADW flow rates during the LGM (Yu et al., 1996). However, this ratio is also sensitive to particle fluxes and composition and interpretation is not straightforward.

Finally, the geostrophic method, a standard approach in physical oceanography, can also be applied to paleoclimate. It requires knowledge of density profiles in the past upper ocean, which can be estimated from $\delta^{18}O$ in benthic foraminifera (see Section 2.1.2). This method has been used to reconstruct past Gulf Stream flow in the Florida Straits (Lynch-Stieglitz et al., 1999). To reconstruct density profiles, benthic foraminifera are required from different depth levels and thus a range of cores from different water depths.

2.1.4. *Other Tracers*

Some other tracers exist, that neither behave like nutrients (i.e., with biological sources and sinks) nor are conservative. An example is neodymium isotopes, which have sources and sinks at the ocean–sediment interface, because neodymium precipitates in metallic crusts or is mobilized from detrital material. This gives water masses like NADW a distinctive neodymium isotope signature that can be used to assess the mixing of water masses of different origin (Rutberg et al., 2000).

2.2. Past Sea-Level Proxies

When studying sea level, we first need to distinguish relative from absolute sea-level changes. Absolute sea-level changes are changes in the sea surface height with respect to a fixed absolute reference frame, for example, the center of the Earth, as measured from a satellite. Relative sea-level change is what an observer (or tide gauge) at the coast would have experienced. It is measured locally, relative to the land and is thus the sum of absolute sea-level changes and vertical motions of the land. In the recent geological past (last ~4000 years), vertical land motion was the dominant factor of relative sea-level change on many coastlines.

Within absolute sea-level changes, it is useful to distinguish local from global changes. There are mechanisms (e.g., a change in wind regime) that can change sea level locally without having any effect on global-mean water volume and sea level. Changes in global-mean (or eustatic) sea level consist of changes in the volume of seawater (arising either from mean density changes or from addition of water, e.g., from melting continental ice) and of changes in the volume of the ocean basins that contain it (due to plate tectonics or isostatic adjustment).

Past sea-level changes are reconstructed using sea-level indicators (proxies) that have a specific (and known) relationship to sea level. A sea-level indicator is any biological, chemical, or physical proxy that can be reliably related to sea level. These include relic beaches, ancient corals, intertidal sediment, and historic human structures built at or close to contemporary sea level (e.g., fish ponds, ports, or coastal wells). The relationship between a proxy and sea level is established from modern, observable examples. When a fossil example of the sea-level indicator is located, it is dated and interpreted based on its modern counterparts. The resulting reconstruction estimates the unique position in time and space of former sea level as a sea-level index point. Compilations of index points allow patterns and trends in relative sea level to be described. Almost all of these record the local sea level relative to the land, and a major challenge in interpreting these data is to disentangle the land motions from true sea-level changes, and to obtain a reconstruction of global sea-level history that can be linked to climate. More detailed reviews are found, for example, in Lambeck et al. (2010) and Shennan et al. (2007).

2.2.1. Coastal Morphology and Corals

Waves, tides, sediment transport, or reef-building corals shape the coastline, and in some places, the resulting coastal land forms are found today at elevations far away from present-day sea level. Tidal notches that have been eroded out of coastal cliffs during extended periods of stable sea level can today be seen at different elevations (Pirazzoli and Evelpidou, 2013). Marine terraces form when the flat former beach front is uplifted or inundated. The sequence of reef terraces on Huon Peninsula (Papua New Guinea) is famous: the reef front from the last interglacial has been uplifted over 100 m above the present sea level due to seismic activity (Figure 2.3; McCulloch et al., 1999; Ota and Chappell, 1999). For large relative sea-level changes, these are useful indicators.

Since corals can grow tens of meters below the water surface but not above it, the envelope of (radiometrically dated) ancient corals can provide a lower limit to past relative sea level (e.g., Lambeck, 2002). Most useful are coral species that have a particularly close relation to sea level, such as Elkhorn coral (Acropora palmata). Coral microatolls are disc-shaped coral colonies that have been stopped from further upward growth by exposure at low tides, so that the center of the upper surface is dead and coral growth occurs laterally around the margin (Figure 2.4; Woodroffe and McLean, 1990). Because corals show annual growth bands, precise dating is possible, and microatolls can reveal interannual to decadal sea-level changes of the order of ±5 cm and millennial changes of the order of ±25 cm (Lambeck et al., 2010). However, since they

FIGURE 2.3 Uplifted coral reef terraces of the Huon Peninsula, Papua New Guinea. In this area, the land is moving upward at a rate of ~2 m/1000 years. Consequently, fringing coral reefs along the coast get uplifted. These ancient reefs now form a succession of "steps," or terraces, in the coastal landscape, with the youngest reefs closest to the coast, and the oldest reefs at higher elevation further back from the coast. The oldest reefs in this image are about 250,000 years old and are seen as terraces toward the top-left of the photo. *Photo: Sandy Tudhope.*

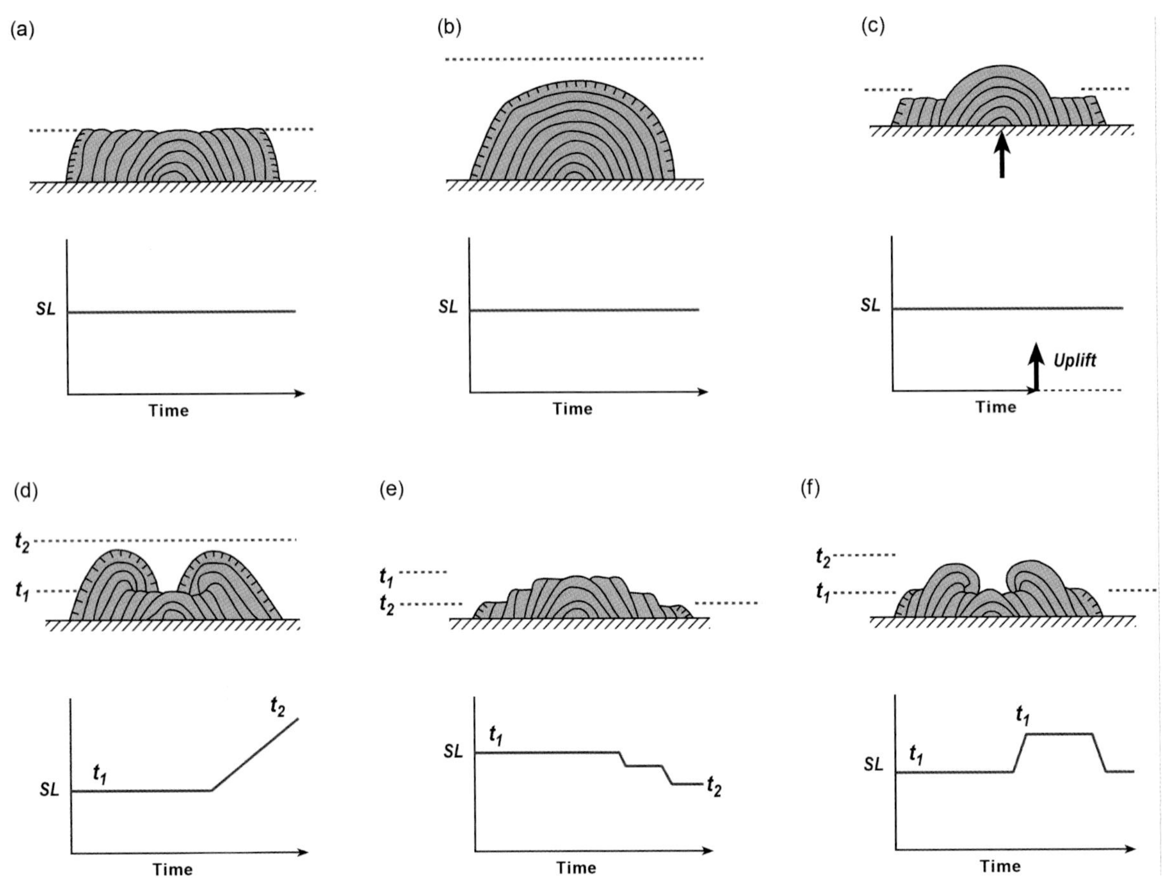

FIGURE 2.4 Schematic illustration of the response of coral, particularly the upper surface of microatolls, to changes of sea level or uplift of the land, as shown by annual banding within the coral skeleton. (a) If sea level (SL) remains constant from year to year, a massive coral that has grown up to sea level continues to grow outward but with its upper surface constrained at water level. The inner part of the upper surface is partly protected by the outer rim, which will usually be the only living part of the colony. (b) If the coral does not reach water level, it adopts a domed growth form and is not constrained by water level. Its upper surface could be several meters below sea level. (c) If the coast undergoes uplift (or sea level falls), a coral previously not limited by sea level may be raised above its growth limit and the exposed upper surface will die, but with continued lateral growth at a lower elevation. (d) If the water level increases, a coral previously constrained by exposure at low tides can resume vertical growth and begin to overgrow the formerly dead upper surface. (e) If water level falls episodically, then the microatoll adopts the form of a series of terracettes. (f) If there are fluctuations of water level with a periodicity of several years then the upper surface of the microatoll consists of a series of concentric undulations. Such a pattern can be seen on microatolls from reef flats in the central Pacific where El Niño results in interannual variations in sea level. *Reproduced from Lambeck et al. (2010).*

mainly record how low the water drops at low tide, highly localized events (such as storms) that change the shoreline and affect tidal flow and the formation of pools on reef flats can alter the results. As with most proxy data, it is the combination of data from different sites and/or using different methods that ultimately provides robust results.

2.2.2. Sediment Cores

Sheltered, low-energy coasts are often vegetated by salt marshes in temperate climate zones and mangroves in the tropics. The distribution of these ecosystems is fundamentally and intrinsically tied to tidal limits and sea level. Under regimes of rising relative sea level, organic and muddy sediment is deposited in these environments. Over time the sediment accumulates to form an important archive from which relative sea level can be reconstructed. The sediment is dated using the radiometric method (principally radiocarbon), while sea-level indicators preserved in the sediment such as identifiable plants, diatoms, and foraminifera are used to determine the vertical location of the marsh surface at any time relative to the tidal range (Scott and Medioli, 1978). Since the current vertical position of the sediment is used in reconstructing the past sea level, an important issue is that no compaction of the sediment has occurred after it has formed (Allen, 2000). The procedure is illustrated in Figure 2.5. This method is useful for high-resolution reconstruction of sea-level changes over the last few millennia, with a vertical precision of ±5–20 cm (e.g., Gehrels, 1994; Donnelly et al., 2004; Kemp et al., 2011).

A notable exception to the usual records of local relative sea level is the $\delta^{18}O$ in benthic foraminifera, which depends

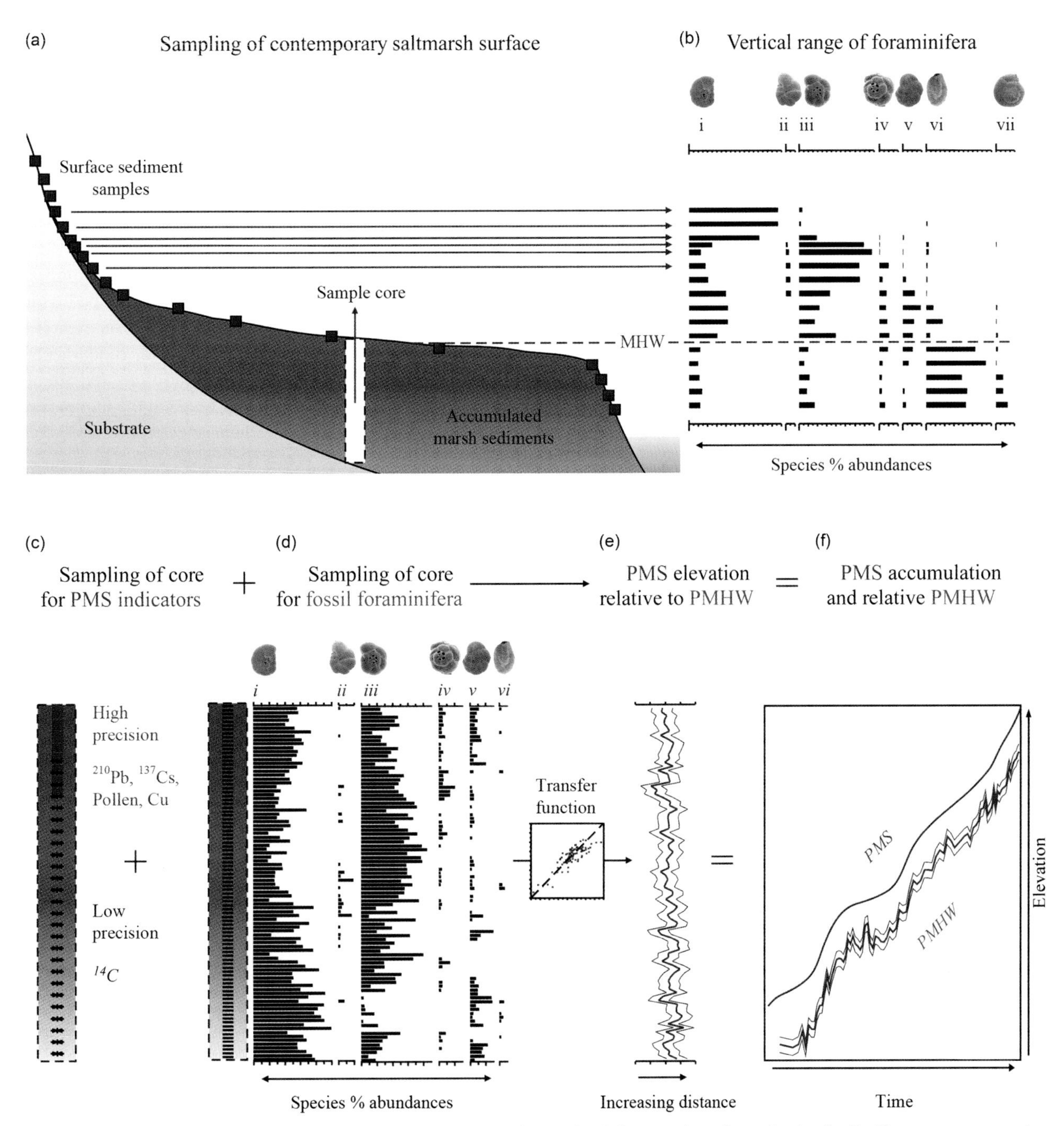

FIGURE 2.5 Schematic illustration of the steps involved in reconstructing sea-level changes using salt marsh microfossils. The contemporary surface distribution (a) of foraminifera species (b) is related to elevation of the marsh surface above a tidal datum and subsequently represented by a transfer function. Paleo-marsh surface (PMS) indicators are sampled in a sediment core (c), with dating control provided by radiometric techniques. The core is also analyzed for fossil foraminifera species abundances (d), which are interpreted in terms of PMS height (e) relative to paleo-mean high water (PMHW) using the transfer function arrived on the basis of steps (a) and (b). These combine (f) to reconstruct the PMS accumulation history and rate of relative sea-level rise. *Reproduced from Lambeck et al. (2010).*

on the total ice volume on land (see Section 2.1.2). Since ice volume changes are by far the dominant cause of global sea-level changes during glacial–interglacial cycles, δ^{18}O is a very useful global sea-level tracer on these timescales (Waelbroeck et al., 2002).

2.2.3. *Manmade Sea-Level Indicators*

Ancient buildings or artifacts and their relation to sea level can provide clues about past sea-level changes. The oldest example is the famous cave paintings of Cosquer cave in southern France from the last ice age (from 19 to 27 kyears BP); today the entrance of the then inhabited cave is 37 m below the sea surface (Lambeck and Bard, 2000). Ancient wells submerged off the coast of Israel have been dated to be between 8200- and 9500-years old (Sivan et al., 2004). Only divers can today visit the sunken city of Baia in Italy to marvel at the floor mosaics (Passaro et al., 2013).

More useful as a sea-level constraint are structures with a precise relation to sea level, such as ancient harbor walls or the Roman fish tanks which were connected to the sea with a series of canals and sluice gates for water exchange, which constrain the sea level at the time these ponds were built to within a narrow range (Lambeck et al., 2004). Remains of the Roman market at Pozzuoli include pillars that have borings by marine organisms up to 7 m above present-day sea level; the site must have been submerged up to that level and uplifted again between the time it was built and the present (due to volcanic activity). Remarkably, already in 1832, Lyell showed this in his *Principles of Geology* (Lyell, 1832). More recently, old paintings of Venice have been analyzed for the level up to which the palace walls are covered by brown algae (Camuffo and Sturaro, 2003; Figure 2.6). Canaletto and his students painted these palaces along the canals with great accuracy using a *camera obscura*. Results show a ~70-cm relative sea-level rise since the first half of the eighteenth century (mostly due to land subsidence) which is the main cause of the frequent floodings of Venice today (Carbognin et al., 2010).

2.3. Models

In principle, the full range of ocean models described elsewhere in this book can also be applied to paleoclimate. The major difficulties are the specification of paleoclimatic boundary conditions and computational cost. Computing a very different ocean circulation, like that of the LGM, requires millennia of model time until a new thermodynamic equilibrium of the circulation with the temperature and salinity fields is reached.

Ocean-only models require boundary conditions for the entire ocean surface, a demand that is very difficult to satisfy. Historically, ocean-only models have often been driven by prescribing surface temperatures and salinities using a relaxation boundary condition and prescribed wind forcing anomalies (e.g., Fichefet et al., 1994). But even if a global map of these quantities is available, fundamental problems with relaxation boundary conditions remain: if the surface temperature and salinity fields in the model perfectly match the data, then the heat and freshwater fluxes vanish, but if the fluxes are correct, then errors in temperature and salinity must exist. Hence relaxation boundary conditions cannot converge to the "true" solution but only approximate it in a first-order sense.

High-resolution (eddy-permitting) ocean models are starting to be applied to paleoclimatic studies (Ballarotta et al., 2013), but due to computational cost, these models typically can only be run for a limited time period of the order of decades, which means they remain close to the initial conditions for temperature and salinity in the water column and essentially diagnose a velocity field that is dynamically consistent with this initial distribution.

Coupled ocean–atmosphere models are best suited for paleoclimatic applications (Braconnot et al., 2007a,b) because they can simulate the surface climate together with atmospheric and oceanic circulations in a physically consistent

FIGURE 2.6 Left: A detail from B. Bellotto's painting S. Giovanni e Paolo (1741). The two arrows give the level of the algae belt in 1741 (lower) and today (upper) as derived from on-site observations. The painting shows that there were two front steps above the green belt. The displacement is 77 ± 10 cm. Right: The same door today. The picture was taken during low tide and the top step of the old front stairs is just visible (green arrow). The door was walled up with bricks in the first 70 cm above the front step to avoid water penetration. *From Camuffo and Sturaro (2003).*

way, including the air–sea fluxes which are crucial in driving the ocean circulation. The boundary conditions required are less demanding. For example, for a simulation of the LGM, besides the orbital parameters, one needs the atmospheric composition (greenhouse gases, dust load) and specification of the land surface (vegetation and ice cover). More comprehensive Earth system models require increasingly fewer external boundary conditions as more processes are included in the simulation. If an ice sheet model is included, then the ice sheet configuration need not be prescribed; if a vegetation model is included, then the vegetation cover is likewise predicted rather than prescribed, and with a closed carbon cycle the atmospheric CO_2 concentration is also predicted by the model. Only the latter approach can ultimately explain glacial cycles—as long as prescribed CO_2 is still included, the forcing already includes the same sawtooth-shaped 100-kyear cycles as the climate response, so even a simple linear model can produce reasonable glacial cycles. But the Milankovitch forcing—the ultimate external driver of the glacial cycles—does not resemble the climate response, so that obtaining realistic glacial cycles only from this forcing requires capturing the key nonlinearities in the climate system.

When going back deeper into Earth's geologic past, boundary conditions such as the atmospheric greenhouse gas content are increasingly poorly known, and the additional problem arises that the ocean's bottom topography also becomes more and more uncertain due to the action of plate tectonics.

Models of intermediate complexity (Claussen et al., 2002) are particularly suited for paleoclimate studies, not only because their computational speed allows for the long simulated time periods needed in paleoclimate experiments (e.g., related to the slow time scale of ice sheet formation), but also because modeling such a complex nonlinear system well outside the realm of experience (i.e., modern climate) is a process of learning by trial and error. Model runs are compared to proxy data, discrepancies are inevitably found, and their physical (or computational) reasons are investigated; on this basis, model improvements are made and the next round of experiments is performed, and so on—this learning process requires a large number of model experiments to reach a mature stage. Full-blown general circulation models, which explicitly simulate weather in the atmosphere with the associated short time steps, typically only allow a few model experiments and thus the first tentative steps in this development and learning process. Also, explicitly resolving synoptic timescales (i.e., weather) may not be needed for most paleoclimatic studies.

In terms of experimental design, historically the first approach was the simulation of time slices, that is, a "snapshot" of a climate state at a particular point (or period) in time, such as the mid-Holocene and the LGM (Braconnot et al., 2007a,b). In this case, a model is driven by boundary conditions that are unchanging in time, and it needs to be run until an equilibrium with these fixed boundary conditions has been reached. More advanced are transient simulations where boundary conditions are changing over time, for example, to simulate climate evolution over a full glacial cycle (Ganopolski et al., 2010). Particularly when boundary conditions are poorly known, sensitivity studies are a useful approach in which a whole range of possibilities is investigated in an ensemble of model runs.

Increasingly, models are used to simulate not just paleo-ocean circulation but also the transport, transformation, and sedimentation processes of particular tracers, so that the models effectively simulate the formation of a sedimentary sequence that can be directly compared to proxy data from sediment cores (Schmidt, 1999; Hesse et al., 2011). This is a promising alternative to comparing a modeled circulation state to one heuristically inferred backward from proxy data. One still would like to draw inferences about past ocean circulation, so the inverse problem remains, but it can be approached in a more quantitative manner with the help of such models, for example, by data assimilation techniques using proxy data.

Modeling sea-level changes in principle require models of all the processes that contribute to sea level. For the large global sea-level changes during glacial cycles, the problem essentially reduces to continental ice sheet modeling, the dominant contribution. The much smaller sea-level variations during the last few millennia, including the twentieth century rise, on the other hand, are caused by a more even mix of thermal expansion and ice sheet and glacier mass changes which are not easily modeled (see Chapter 16). As a complementary approach to these "bottom up" models of individual processes, semiempirical models have been developed, which link global sea level to global-mean temperature or radiative forcing with simple equations calibrated with empirical data (Rahmstorf, 2007; Grinsted et al., 2010). The equation used in a semiempirical model is typically a variation on the idea that the rate of sea-level rise is proportional to the amount of warming above a previous temperature level at which sea level was stable. In some cases (particularly for multicentury timescales), a time scale of the response is explicitly included.

To aid in the interpretation of sea-level proxies, an entirely different class of models is used: those that describe vertical land motions, for example, models of glacial isostatic adjustment (GIA; Argus and Peltier, 2010). This is important to derive absolute sea-level changes from proxy records of relative sea level, by subtracting the local vertical land movement.

3. THE OCEANS IN THE QUATERNARY

The Quaternary period covers roughly the last 2.5 million years, which are characterized by periodic glaciations. A prime target for reconstructions and models of the paleo-ocean has been the LGM about 20 kyears before present, because it is the most recent period with a massively different climate (so the signal is large). Interglacial climate, full glacial cycles, and millennial-scale events have also been targets of scientific interest, as well as of course the climate evolution of the most recent millennia preceding the twentieth century global warming.

3.1. The Last Glacial Maximum

The LGM is defined as the time period when the continental ice sheets reached their maximum total mass during the last ice age; Clark et al. identify the interval between 26.5 and 19 kyears BP with the LGM (Clark et al., 2009). Maximum ice sheet size coincides with a minimum in global sea level since continental ice mass is by far the dominant factor in glacial–interglacial sea-level changes. Figure 2.7 shows global sea-level history across the LGM based on four proxy records for relative sea level from far-field sites (i.e., sites that are remote from the location of the ice sheets) as well as an independent estimate of eustatic sea-level changes based on a large number of proxy estimates of the size of continental ice masses. Global sea level during the LGM was 120–135 m lower than at present (Peltier and Fairbanks, 2006).

The timing of the LGM coincides with a strong minimum in summer insolation in mid- to high northern latitudes due to orbital cycles, and it is thus naturally explained by the Milankovitch theory of glaciations. Between the LGM and the beginning of the Holocene ~10 kyears BP, peak northern summer insolation increased by some 40–50 W/m^2. This massive increase in solar heating was the driver of Northern Hemisphere deglaciation and consequent sea-level rise by ~130 m. Detailed analysis reveals some episodes of exceptionally rapid sea-level rise, known as meltwater pulses (Fairbanks, 1989). The most prominent is meltwater pulse 1A at 14.5 kyears BP, with an estimated sea-level rise of about 20 m at a rate reaching 4 m per century (Stanford et al., 2006; Deschamps et al., 2012). Clark et al. (2009) argue that meltwater pulse 1A included a major meltwater contribution from the West Antarctic Ice Sheet.

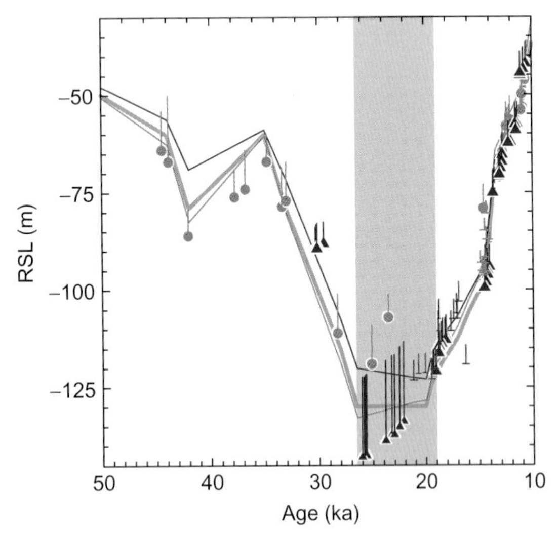

FIGURE 2.7 Sea-level reconstructions from four different sites (dots with depth uncertainty bars): New Guinea (blue dots), Sunda Shelf (blue half-pluses), Barbados (purple), Bonaparte Gulf (green). The blue and purple lines show sea-level predictions for New Guinea and Barbados, while the gray line shows eustatic sea level. The vertical gray bar indicates the time of the LGM. *From Clark et al. (2009).*

A comparison of glacial sea-level proxies with $\delta^{18}O$ from benthic foraminifera shells (the often-used Lisiecki–Raymo stack (Lisiecki and Raymo, 2005)) reveals some differences that can be explained by changes in deepwater temperature, since the calcite $\delta^{18}O$ depends on both ice volume and local temperature (Section 2.1.2). This difference suggests that deep ocean temperatures during the LGM must have been ~3 °C colder than today, and hence close to the freezing point of seawater.

Because of the relatively good availability of data, the large climate change signal, and relatively stable climate conditions over several millennia, the LGM was the target of the earliest proxy reconstructions of paleo-ocean circulation (Duplessy et al., 1988) and of a number of time slice climate model experiments with atmosphere models, with ocean models and with coupled ocean–atmosphere models, including systematic model intercomparison studies (Braconnot et al., 2007a,b). Figure 2.8 (left) shows the distribution of $\delta^{13}C$ in the modern and LGM Atlantic ocean, as a nutrient-like tracer of water masses (see Section 2.1.1). The $\delta^{13}C$ distribution of the modern ocean reflects the spread of the well-known water masses NADW and AABW. The LGM Atlantic likewise shows a low-nutrient water mass of northern origin and below it a high-nutrient water mass, but their boundary is higher up in the water column at about 2 km depth. The northern water mass is sometimes called glacial North Atlantic intermediate water (GNAIW); it can be interpreted as a shoaling of the flow of NADW which leaves room for a northward and upward extension of AABW during the LGM. This distribution of water masses was more recently confirmed by Cd/Ca ratios (Lynch-Stieglitz et al., 2007). More recent and detailed compilations of LGM proxy data also show a presence of AAIW in the Atlantic, the northward extent of which is subject to ongoing research. $\delta^{18}O$ data at 30 °S from the eastern and western side of the basin indicate a collapse of the east–west density gradient, which currently characterizes the outflow of NADW (see Section 2.1.3). While the protactinium/thorium ratio suggests a NADW renewal rate similar to today, these $\delta^{18}O$ data may indicate a much reduced NADW outflow into the Southern Ocean (Lynch-Stieglitz et al., 2007). Reconstructions of surface properties of the northern Atlantic at the same time suggest that NADW (or GNAIW) formation probably occurred to the south of the Greenland–Iceland–Scotland ridge during the LGM (Oppo and Lehman, 1993; Alley and Clark,

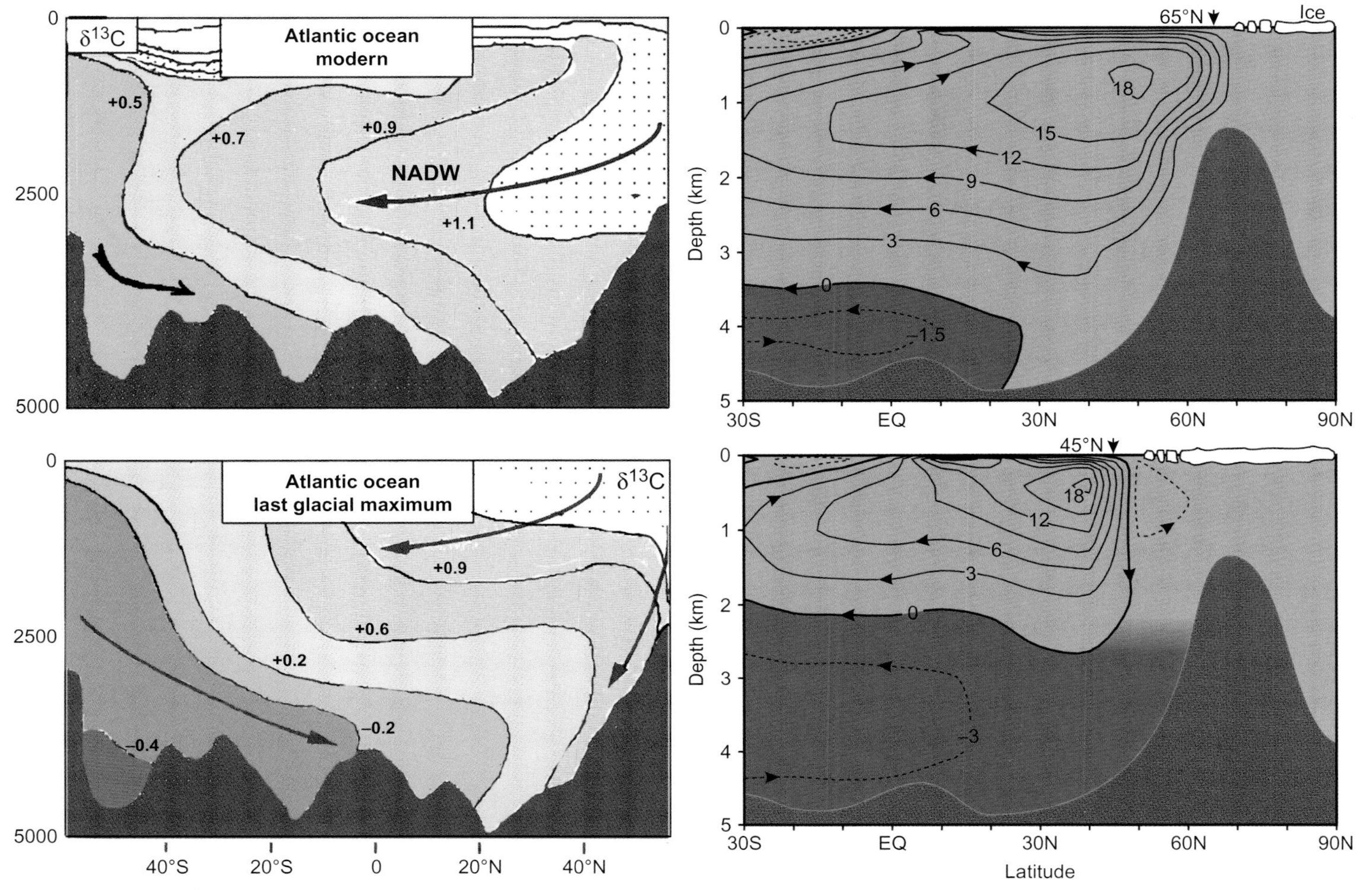

FIGURE 2.8 Left: Distribution of $\delta^{13}C$ in the modern and LGM Atlantic Ocean from Duplessy et al. (1988). Right: Stream function of Atlantic Ocean circulation for modern and LGM conditions in the climate model of Ganopolski et al. (1998). Dark blue shading indicates bottom water of Antarctic origin, brown the bottom topography.

1999), consistent with a southward extension of sea ice cover and in contrast to the modern ocean where it partly forms in the Nordic Seas and then overflows this ridge.

The first coupled ocean–atmosphere model simulation of LGM climate (Ganopolski et al., 1998), at the time still with prescribed continental ice sheets, produced an LGM circulation in the Atlantic broadly consistent with these proxy findings (see Figure 2.8). We can see a southward shift of deepwater formation, a shoaling of NADW flow and an upward and northward extension of AABW, and a similar NADW renewal rate as at present, combined with a near-breakdown of the outflow across 30 °S as the overturning cell recirculates within the Atlantic. This flow pattern is consistent with recent findings of a reversed gradient in the $^{231}Pa/^{230}Th$ ratio between north and south Atlantic during the LGM (Negre et al., 2010).

However, subsequent attempts at modeling LGM climate, including those for the Paleoclimate Modeling Intercomparison Project, have produced a variety of circulation patterns for the LGM Atlantic (Weber et al., 2007). This is perhaps not surprising since the stability properties of the thermohaline ocean circulation are highly nonlinear and dependent on a subtle density balance, where particularly the freshwater budget is difficult to get right in models even for the modern ocean (Hofmann and Rahmstorf, 2009). Systematic comparison of model results with the full suite of proxy data is needed to establish what range of circulation patterns is consistent with the data.

3.2. Abrupt Glacial Climate Changes

While the LGM was a period of frosty climate stability lasting for several 1000 years, the glacial time before the LGM as well as the period of deglaciation following the LGM were rather turbulent, punctuated by abrupt and massive, large-scale climate changes. An illustration is given in Figure 2.9, based on Greenland ice core data (often shown as standard because of their high resolution) and sediment data from the subtropical North Atlantic. The numbered warm events there are known as Dansgaard–Oeschger (DO) events while H1...H6 refers to Heinrich events, defined as episodes of massive continental ice discharge into the northern Atlantic as documented by ice-rafted debris in sediment cores. They could either result from internally or externally triggered ice sheet instability. Heinrich events do not stand out in Greenland temperature but tend to occur during cold periods preceding some strong DO events.

There is plentiful and strong evidence linking these abrupt climate events to changes in Atlantic Ocean circulation, reviewed, for example, in Alley (2007). From a mechanistic point of view, the perhaps most compelling piece of evidence for a major role of ocean heat transport changes is the "bipolar seesaw" (or "seasaw"): an antiphase behavior between the North Atlantic and the Southern Ocean/Antarctica (Stocker, 1998). Establishing the phase relationship required accurate relative dating between distant paleoclimatic records, which was achieved for Greenland and Antarctic ice cores using the globally synchronous variations in atmospheric methane that are recorded at both poles (Blunier and Brook, 2001; Figure 2.10). Methane changes must be synchronous since methane is a well-mixed greenhouse gas in the atmosphere, so they can be used to line up the records. Examination of the phase relation shows that during cold phases in Greenland (termed "stadials") Antarctic temperature increases, while at the time of abrupt warming in Greenland (the DO events) Antarctic temperature begins to drop and then continues to decline during Greenland warm phases (called interstadials).

The same pattern is found in climate model simulations in response to changes in the Atlantic meridional overturning circulation (Ganopolski and Rahmstorf, 2001; see Figure 2.11). This behavior can be nicely explained by a simple conceptual model consisting of changes in northward ocean heat transport coupled to a heat reservoir in the south, the Southern Ocean (Stocker and Johnsen, 2003).

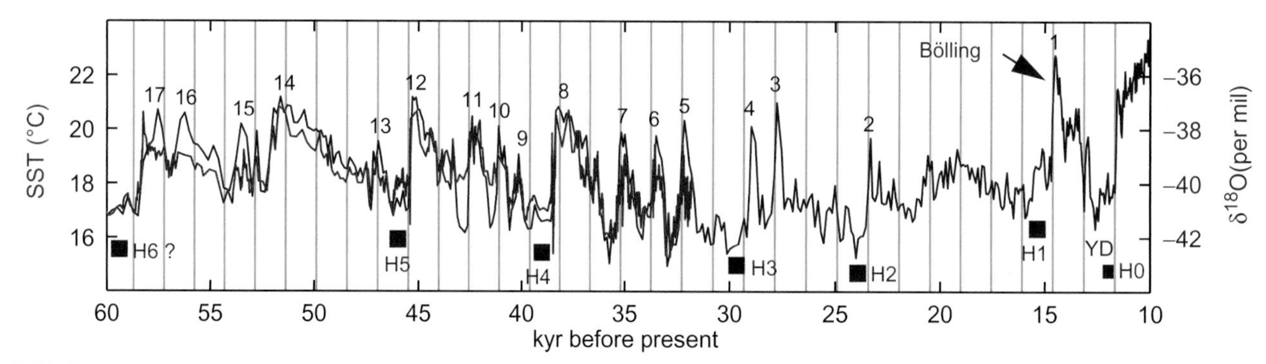

FIGURE 2.9 Proxy data from the subtropical Atlantic (green) (Sachs and Lehman, 1999) and from the Greenland ice core GISP2 (blue) (Grootes et al., 1993) show several Dansgaard–Oeschger (D/O) warm events (numbered). The timing of Heinrich events is marked in red. Gray lines at intervals of 1470 years illustrate the tendency of D/O events to occur with this spacing, or multiples thereof.

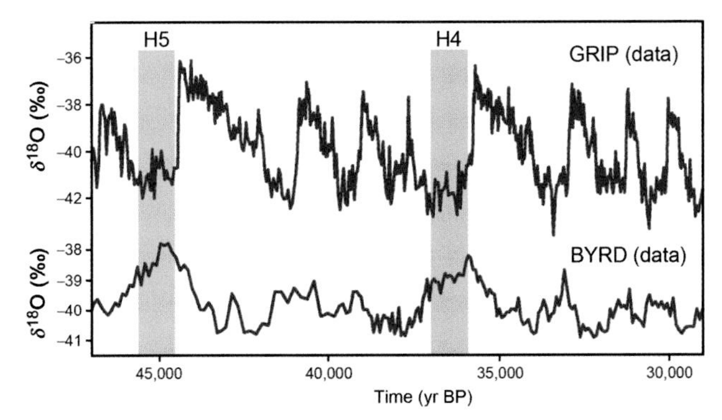

FIGURE 2.10 Oxygen isotopes as a temperature proxy for the Greenland ice core of GRIP and the Byrd ice core in Antarctica, synchronized on a common time scale. Graph by A. Ganopolski after (Blunier and Brook, 2001). Heinrich events 4 and 5 are marked as in Figure 2.9.

Subsequent more detailed data, including also millennial events from the previous glacial period, support this concept, showing a magnitude of Antarctic response that asymptotically approaches an equilibrium with increasing duration of North Atlantic stadials (Margari et al., 2010).

While compelling evidence thus points to ocean heat transport changes at the core of abrupt glacial climate events, the exact nature of these ocean circulation changes is harder to establish. The simple concept of a bistable AMOC, which is either turned "on" or "off," clearly does not have enough degrees of freedom to explain glacial and modern circulations as well as DO and H events. Neither does the concept of an AMOC that simply has different flow rates appear to explain the data. For example, why are DO warmings in Greenland so abrupt? Why do Heinrich events appear as prominent cooling at the Portuguese margin (Cacho et al., 1999) but not in Greenland (Figure 2.9)? And why are DO events associated with a large salinity increase near Iceland (Kreveld et al., 2000)?

One possible explanation for these features is a concept with three distinct circulation modes and transitions between them (see schematic Figure 2.12), which is based on time slice reconstructions using sediment cores (Sarnthein et al., 1994) as well as model experiments (Ganopolski and Rahmstorf, 2001). The central image shows a "cold mode" of the AMOC prevailing during the LGM and stadial periods. DO events occur when convection starts in the Nordic Seas (a situation that is stable in the Holocene but only metastable during glacial conditions, that is, in the latter case the circulation reverts spontaneously to the cold mode after some hundreds of years), which extends the AMOC northward, reduces sea ice cover there, and leads to strong warming over Greenland. This mechanism can explain the salinity increase during DO events (Figure 2.11b) and the shape and phasing of Greenland and Antarctic temperatures (Figure 2.11d and e). DO events can thus be viewed as a "flickering" between the Holocene and LGM modes of the AMOC, where the latter is the stable one during glacial

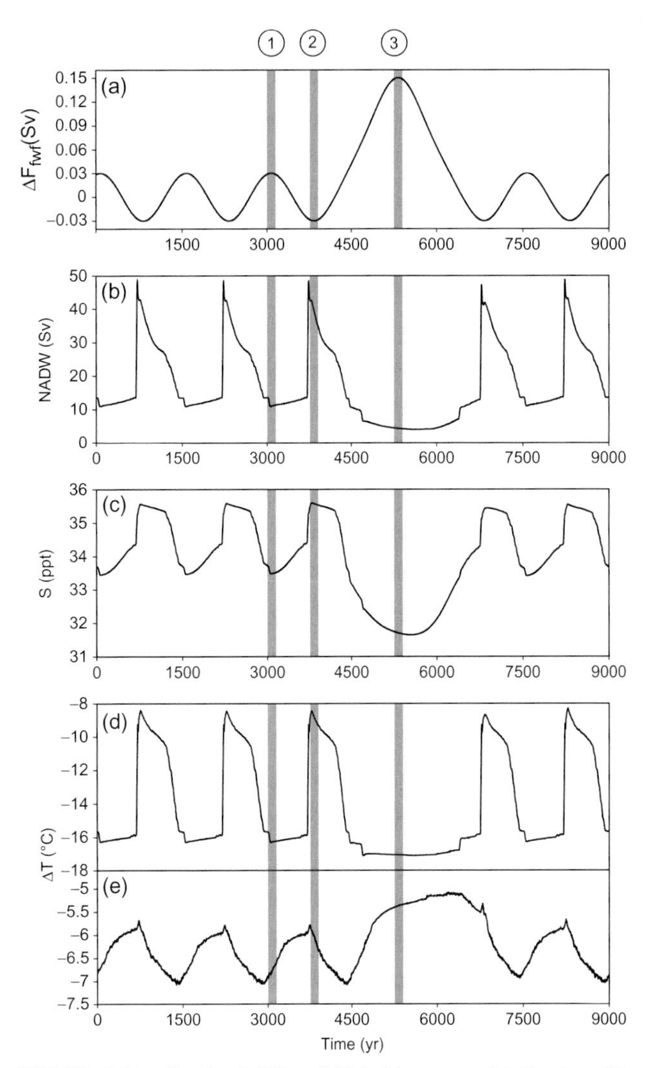

FIGURE 2.11 Simulated DO and Heinrich events. (a) Forcing, (b) Atlantic overturning, (c) Atlantic salinity at 60 °N, (d) air temperature in the northern North Atlantic sector (60–70 °N), and (e) temperature over Antarctica (temperature values are given as the difference from the present-day climate, ΔT). *From Ganopolski and Rahmstorf (2001).*

times. Other modeling attempts have been reviewed by Kageyama et al. (2010).

Heinrich events can be interpreted as a temporary shutdown of North Atlantic deepwater formation and flow, caused by a dilution of northern Atlantic surface waters due to meltwater release stemming from the iceberg discharge events (Otto-Bliesner and Brady, 2010). Such a shutdown would lead to cooling of the North Atlantic (but perhaps hardly affecting Greenland, since ocean heat transport already does not reach that far north in the cold mode) and warming in Antarctica, asymptotically approaching an equilibrium as seen in Figure 2.11e and in ice core data (Margari et al., 2010).

An interesting discussion has arisen about the timing of DO events, many of which tend to occur in intervals

FIGURE 2.12 Schematic of the three circulation modes explained in the text.

of 1500 years or multiples thereof (see the gray lines in Figure 2.9; Alley et al., 2001a; Schulz, 2002; Rahmstorf, 2003). A physical mechanism that can produce such a timing is stochastic resonance (Gammaitoni et al., 1998; Alley et al., 2001a,b; Ganopolski and Rahmstorf, 2002). Model simulations suggest that solar cycles could provide the weak regular trigger required by the stochastic resonance mechanism (Braun et al., 2005).

3.2.1. Deglaciation

Since the late 1990s and based on an increasingly large number of proxy records, the time history of deglaciation (the transition from the last ice age into the Holocene) has been interpreted as a globally near synchronous warming out of the ice age (synchronized in part by the global atmospheric CO_2 increase), superimposed with a north–south seesaw due to episodic changes in the Atlantic overturning circulation (Alley and Clark, 1999; Clark et al., 2002; Rahmstorf, 2002). During deglaciation, the melting of the ice sheets may have provided an irregular freshwater input disrupting the Atlantic Ocean circulation. The most recent data compilations have firmed up that interpretation and added much detail (Barker et al., 2009; Clark et al., 2012). In line with early model results for the response of the AMOC to freshwater forcing, they document an immediate antiphase response off South Africa and more gradual changes in Antarctica. Thus, a major role of AMOC changes in shaping the climate evolution during deglaciation can now be considered well established, thanks to the consistent picture that has emerged from many high-resolution proxy records as well as model simulations. Details of the sequence of events are still subject to active research.

3.3. Glacial Cycles

The prime characteristic of glacial cycles is the growth and decay of vast continental ice sheets, directly mirrored in the global ocean in the form of sea-level and salinity changes. Since the average depth of today's global ocean is 3790 m, the 130-m sea-level drop during the LGM amounts to ~3.5% of all ocean water being removed and stored on land, increasing the average salinity of the remaining ocean water by over 1 psu and increasing its average $\delta^{18}O$ content by 1.0‰±0.1‰ (Clark et al., 2009).

Figure 2.13 shows a reconstruction of eustatic sea-level changes over the last four glacial cycles based mainly on $\delta^{18}O$ and coral data. To first order, it shows a sawtooth pattern with a slow descent into full glacial conditions but comparatively rapid deglaciations, which can be explained by a fundamental asymmetry in ice sheet physics: continental ice sheets grow slowly by accumulating snow at their surface, but they can decay much more rapidly due to a combination of surface melting and ice flow (i.e., solid ice discharge into the ocean).

Figure 2.14 shows a recent attempt at modeling the last glacial cycle with an intermediate complexity coupled climate model driven by variations of the Earth's orbital parameters and atmospheric concentration of major greenhouse gases prescribed from ice core data (Ganopolski et al., 2010). The model contains a three-dimensional polythermal ice sheet model which successfully reproduces the history of ice sheet growth and decay and hence sea level, as shown in the figure, with some underestimation of the maximum ice sheet volume. The oscillations superimposed on the basic sawtooth shape result from the precession cycle in the orbital parameters which has a period of 23 kyears.

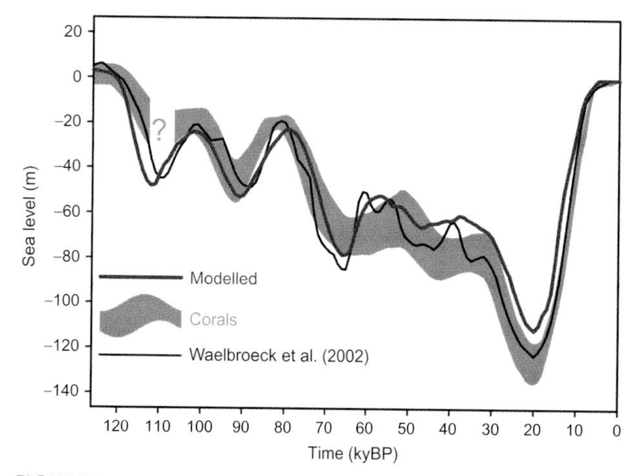

FIGURE 2.14 Simulation of the last glacial cycle with the CLIMBER-2 ocean–atmosphere ice sheet model compared to the data shown in Figure 2.13. *From Ganopolski et al. (2010).*

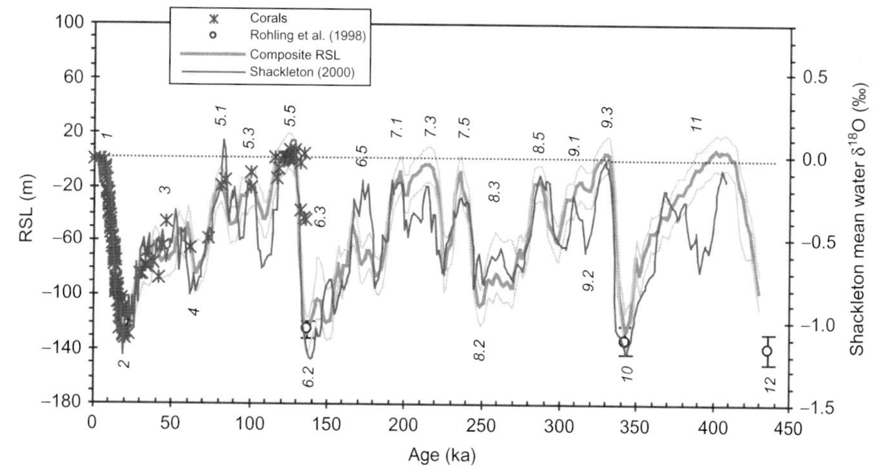

FIGURE 2.13 Left axis: Composite RSL curve (bold gray line) and associated confidence interval (thin gray lines). Crosses: coral reef relative sea-level data. Empty circles: Relative sea-level low stands estimated by Rohling et al. (1998). Right axis: variations in mean ocean water $\delta^{18}O$ derived by Shackleton (2000) from atmospheric $\delta^{18}O$ (black line). *Graph from Waelbroeck et al. (2002).*

A remaining challenge is to produce such simulations with predicted rather than prescribed greenhouse gas concentrations.

In the course of glacial cycles, ocean circulation changes have mainly been discussed with respect to abrupt events within the glacial and the specific sequence of events during the last deglaciation (see Section 3.2.1). However, if a weakening of the Atlantic overturning circulation is a general feature of the transitions from glacial to interglacial conditions, due to the massive northern meltwater input occurring at these times, then the bipolar seesaw may be part of the explanation of the time lag of atmospheric carbon dioxide concentration behind Antarctic temperature that is found in Antarctic ice cores (Ganopolski and Roche, 2009).

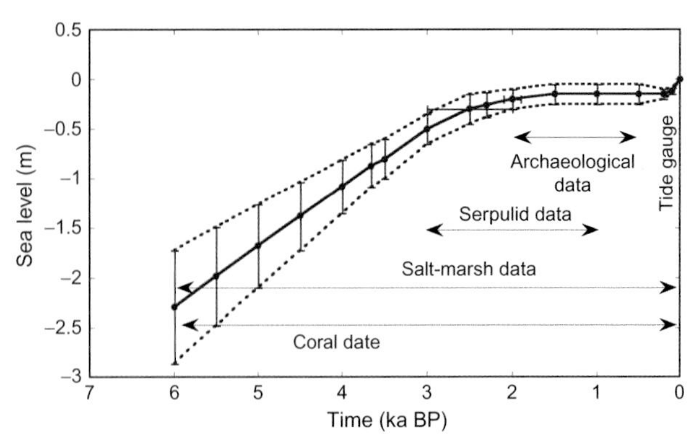

FIGURE 2.15 Current best estimates, including uncertainty estimates, of the eustatic sea-level change over the past 6000 years, as inferred from geological and archaeological data and from the tide gauge data for the past century. *From Lambeck et al. (2010).*

3.4. Interglacial Climates

Conditions during the current Holocene and previous interglacials do not differ as dramatically from modern climate as do glacial conditions, so it is more difficult to establish what changes in ocean circulation—if any—may have occurred. Global mean temperature likely was no more than 1.5 °C warmer than in the preindustrial time, if at all, although regional differences were larger (Turney and Jones, 2010; McKay et al., 2012). Lynch-Stieglitz et al. (2009) have attempted to reconstruct the flow of the Florida current over the past 8000 years from oxygen isotope data and found a small increase of 4 Sv (out of a total of ~30 Sv) over this period. Given this small change and the uncertainties of the proxy method, the main (and also interesting) conclusion probably is that the flow was rather stable over the last 8000 years.

Most discussion on interglacial climates has focused on sea-level changes, both over the Holocene and in previous interglacials. Global sea level during the Eemian interglacial, ~120,000 years BP, has been estimated as peaking at 5.5–9 m above present sea level (Kopp et al., 2009; Dutton and Lambeck, 2012). This is of considerable interest in the context of current global warming since it may provide clues about the response of ice sheets to warmer climate conditions. Data from the Eemian in combination with an ice sheet model ensemble have been used to constrain the stability threshold of the Greenland ice sheet (Robinson et al., 2011). This threshold could be crossed between 0.8 and 3.2 °C global warming above preindustrial conditions, with a best estimate of 1.6 °C (Robinson et al., 2012). Discussion on Eemian sea level continues, for example, about sea-level changes within the Eemian period and about the relative contributions of the Greenland and Antarctic ice sheets (Dahl-Jensen et al., 2013).

In the Holocene, sea level is characterized by the long tail of deglaciation due to the long time scale needed for melting continental ice. Different locations record different times when the relative postglacial sea-level rise ended, due to the interplay between eustatic sea-level rise and postglacial uplift (or in some places, subsidence). Lambeck et al. (2010) have constructed a eustatic sea-level curve for the past 6000 years, based on a multitude of relative sea-level data (Figure 2.15). Their best estimate shows how postglacial sea level rise came to an end between 2 and 3 kyears BP, after which sea level was approximately constant until the modern rise, registered by the tide gauges, started.

A compilation of relative sea-level records for the last two millennia from different parts of the world is shown in Figure 2.16. This shows some consistency but also large local deviations (e.g., records for Israel, Cook Islands). It is tempting to consider the consistent records as representative of eustatic sea level (the data have already been adjusted for GIA) while the deviating records are affected by local issues. More records need to be collected from different shores to build up a clearer picture.

Figure 2.17 shows proxy records from the US east coast (also shown in Figure 2.16) with an attempt to model the sea-level evolution with a semiempirical model as a function of global temperature. A successful fit is obtained for the last millennium but not the time before 1000 AD. Ongoing work suggests this may be due to the global temperature reconstruction that was used, which is warmer than others before 1000 AD and thus leads to sea-level rise in the model at a time when stable sea level is found in the proxy data.

4. THE DEEPER PAST

4.1. Challenges of Deep-Time Paleoceanography

Reconstructing the ocean circulation in the geological past becomes increasingly more difficult at earlier times. This is mainly due to the effects of plate tectonics which continuously

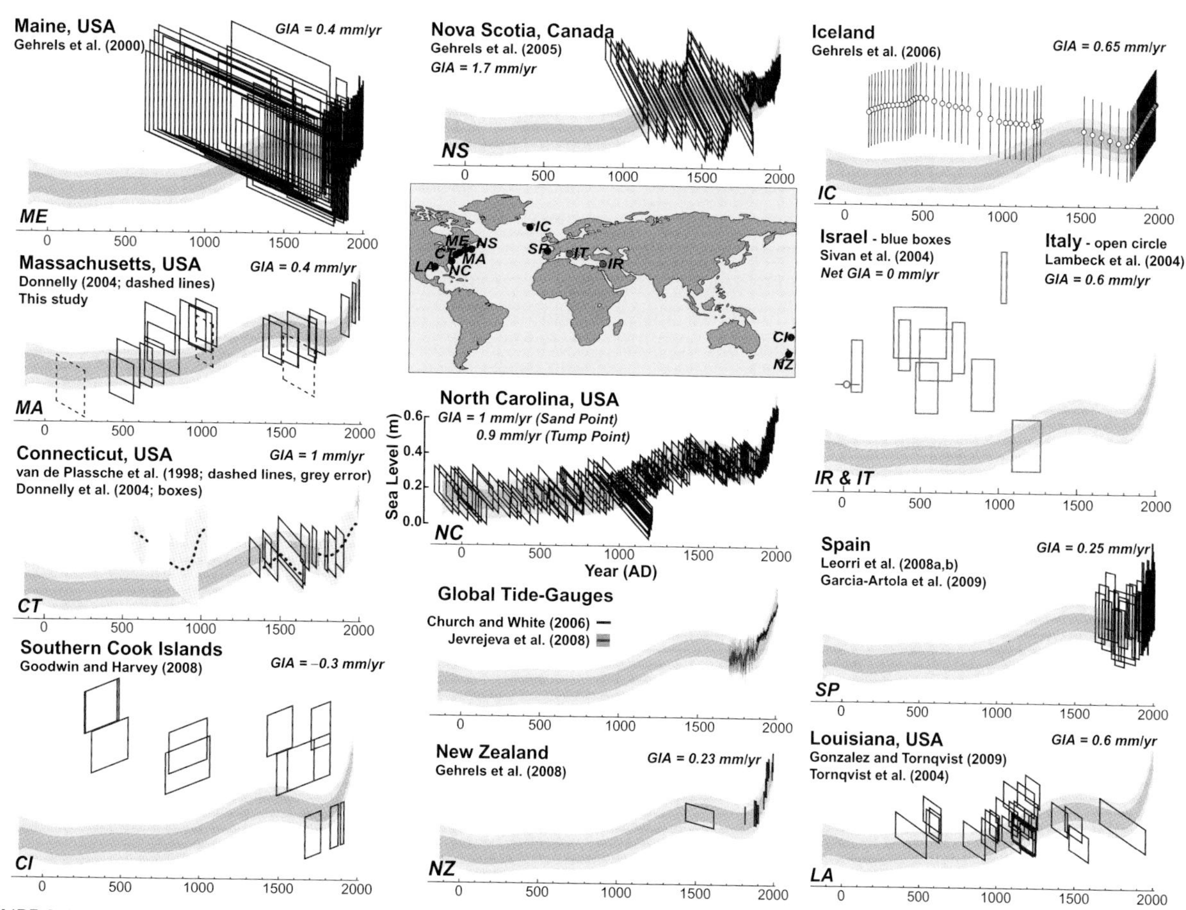

FIGURE 2.16 Late Holocene sea-level reconstructions after correction for GIA. Rate applied (listed) was taken from the original publication when possible. In Israel, land and ocean basin subsidence had a net effect of zero. Reconstructions from salt marshes are shown in blue, archaeological data in green, and coral microatolls in red. Tide gauge data expressed relative to AD 1950–2000 average. Vertical and horizontal scales for all datasets are the same and are shown for North Carolina. Datasets were vertically aligned for comparison with the summarized North Carolina reconstruction (pink). *From Kemp et al. (2011).*

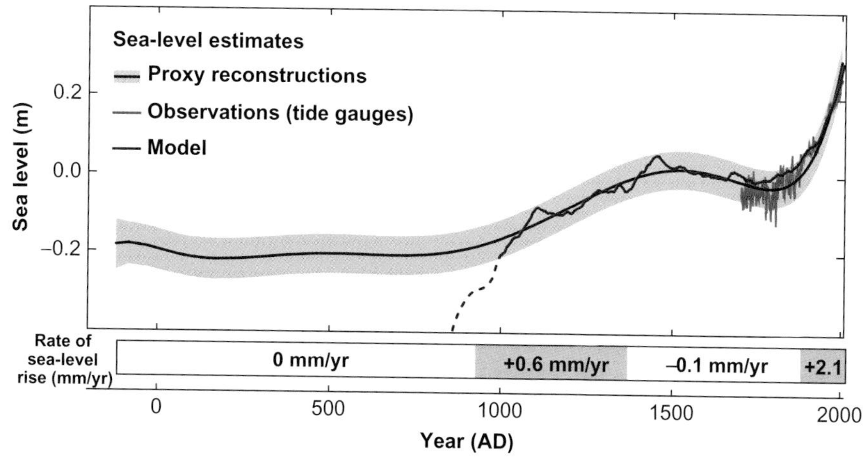

FIGURE 2.17 Sea-level reconstruction for North Carolina from salt marsh data (as in Figure 2.16, here shown blue) compared to tide gauge data (green) (Jevrejeva et al., 2006) and a semiempirical model (red). *From Kemp et al. (2011).*

change the position of continents and the topography of the ocean floor on geological timescales, both of which have a direct influence on ocean circulation. In contrast to the oceans in the Quaternary (Section 3), accurate paleogeographic reconstructions are therefore a prerequisite for understanding the ocean circulation during Earth's deeper past.

For the past 150 million years or so, paleogeographic reconstructions mostly rely on magnetic anomalies imprinted into the ocean floor and on the positions of the relatively stable mantle plumes. Unfortunately, these proxies are not available for earlier times, because oceanic crust is continuously formed at mid-ocean ridges and later returned to the mantle at subduction zones. In fact, the oldest parts of the ocean floor date back to the Jurassic period (200–146 million years ago), with the majority being much younger (Müller et al., 2008a; see Figure 2.18). For earlier times, paleogeographic reconstructions have to rely on paleomagnetic data (indicating the latitude and orientation of continents) and paleontological records (indicating the geographic distribution of species), making paleogeographic maps increasingly uncertain beyond the Cretaceous (146–66 million years ago) (Cocks and Torsvik, 2002; Torsvik and Cocks, 2004).

The continuous recycling of the oceanic crust also limits the sedimentary record and thus the availability of tracers of past ocean circulation (see Section 2.2), again making the Cretaceous the earliest time for which the physical state of the ocean can be investigated in detail. As a matter of fact, the Cretaceous with its warm greenhouse climate and small meridional temperature gradient is also a particularly interesting period in Earth's history. In particular, it has been suggested that the ocean circulation in the Cretaceous could have been radically different from today. The remainder of this section therefore concentrates on the climate during the Cretaceous and briefly reviews research on the ocean circulation during this time.

4.2. The Oceans During the Mid-Cretaceous Warm Period

Temperature reconstructions indicate that the global average surface air temperature was above modern levels for the entire Cretaceous, with particularly strong warming around 95 million years ago, when global surface air temperatures were about 20 °C warmer than today (Figure 2.19). In the following, we focus on ocean circulation during this mid-Cretaceous warm period, beginning with a look at the continental configuration at that time.

A reconstruction for the paleogeography during the mid-Cretaceous (90 million years ago) is shown in Figure 2.20. At that time, the break-up of the supercontinent Pangaea, which had begun in the Jurassic, has progressed to a point where continental landmasses known today had separated from each other, although in many cases only separated by narrow and very shallow seas. Note that the distribution of continents and the absence or presence of open ocean gateways can be of considerable importance for Earth's climate. The opening of the passages around Antarctica, for example, likely played a major role in the growth of its ice sheet and thus for the global and regional energy balance (Kennett, 1977).

One particularly striking feature of the geography during that period are epicontinental (or epeiric) seas, shallow seas covering large parts of what would later become continental North America, Europe, and Africa. This Cretaceous transgression is thought to be primarily caused by rapid seafloor spreading at mid-ocean ridges, which reduced the volume of ocean basins and thus led to a rise in eustatic sea level (Hays and Pitman, 1973). Estimates of mid-Cretaceous sea levels differ widely, however, roughly covering a range from 50 to 250 m above present day (Miller et al., 2005; Müller et al., 2008b).

Early model simulations indicated that the changes in paleogeography (as compared to present day) alone can account for about 5 °C warming during the mid-Cretaceous (Barron and Washington, 1984). Later studies, however, found only a minor contribution to the observed warming (Barron et al., 1995; Bice et al., 2000). Higher levels of atmospheric greenhouse gases (in particular carbon dioxide) are thought to explain the bulk of the warming during the Cretaceous.

Mid-Cretaceous atmospheric carbon dioxide levels can be estimated from proxy data and models of the global carbon cycle. Empirical estimates rely on $\delta^{13}C$ carbon isotope ratios in paleosols, alkenones, or planktonic foraminifera, distribution of stomatal pores in C3 plants, or $\delta^{11}B$ boron isotope ratios in planktonic foraminifera (Royer, 2006). As shown in Figure 2.21, atmospheric carbon dioxide concentrations during the middle Cretaceous were around 1000 ppm and thus significantly higher than today, albeit with a large uncertainty range (roughly 500–1500 ppm). A strong contribution of the higher carbon dioxide levels (possibly enhanced by methane) to the observed warming is therefore very likely. The geological record also indicates an ice-free world during the Jurassic and Cretaceous. These time periods are therefore an ideal testbed to study the internal dynamics of a climate system without polar icecaps.

Concerning geographic patterns, temperature proxy data for the mid-Cretaceous indicate tropical temperatures a few degrees warmer than today (possibly up to 40 °C), but significantly warmer polar regions. As for other warm periods in Earth's history, the mid-Cretaceous is therefore characterized by a significantly reduced meridional temperature gradient often referred to as an "equable" climate (Crowley and Zachos, 2000; Hay, 2008). For illustration, Figure 2.22 shows an example of a proxy-based latitudinal temperature distribution during the late Cretaceous as

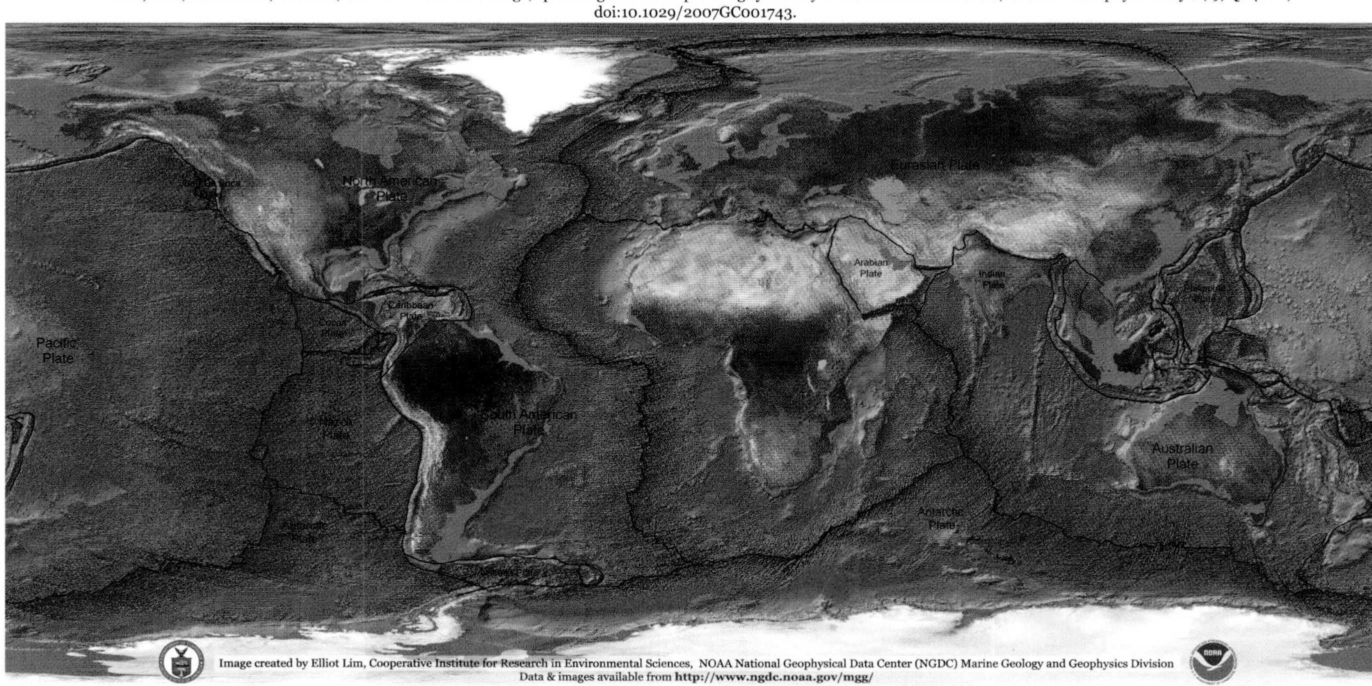

FIGURE 2.18 Age distribution of the oceanic crust on Earth. *Source: NOAA.*

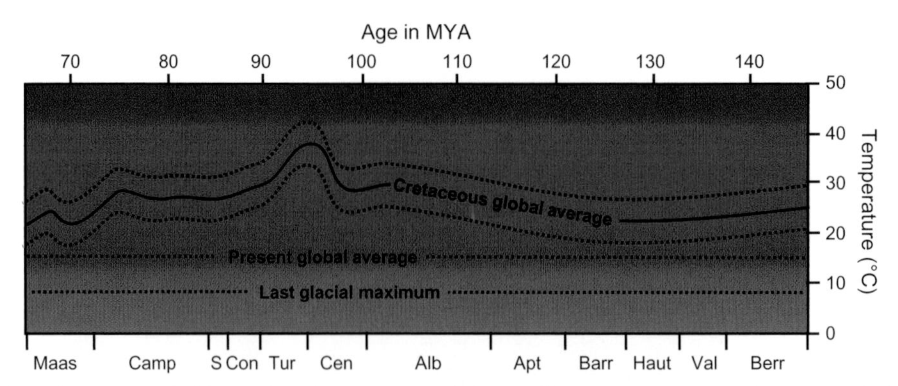

FIGURE 2.19 Schematic view of reconstructed global surface air temperatures during the Cretaceous. *Reproduced from Hay and Floegel (2012).*

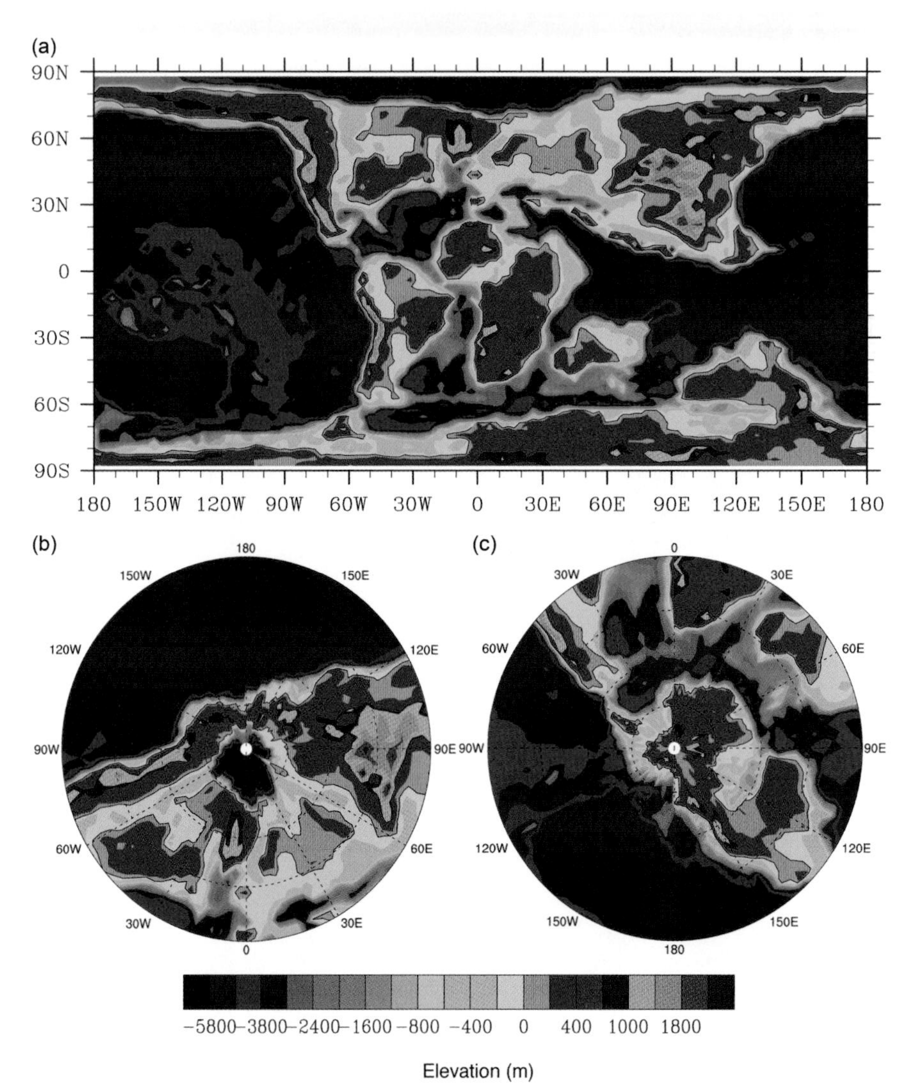

FIGURE 2.20 Paleogeography (elevation in meters) during the mid-Cretaceous (90 million years ago) in cylindrical (a), North polar (b), and South polar (c) projections (Sewall et al., 2007).

compared to the present-day climate. (Note that earlier studies had indicated an even flatter Cretaceous temperature gradient with "cool tropics," but this interpretation has now been shown to be biased by diagenesis, see, e.g., Pearson et al., 2001).

Equable climates during previous warm periods (and in particular during the Cretaceous) have received considerable attention because climate model experiments have generally had difficulties in reproducing the latitudinal temperature distribution inferred from proxy data

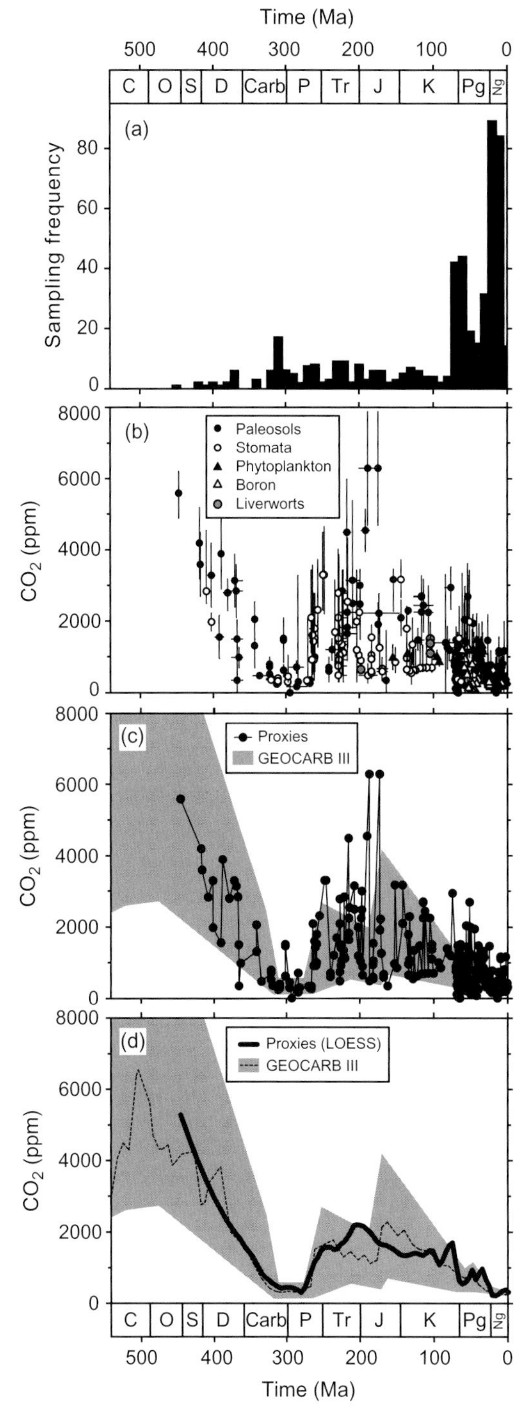

FIGURE 2.21 Sampling frequency (panel a) for proxy data for the evolution of atmospheric carbon dioxide concentrations during the Phanerozoic (the past 542 million years) as shown in panel b. Panel c: comparison of the proxy data (black) with results from carbon cycle modeling (gray, with uncertainty range). Panel d: comparison of best-guess prediction of the carbon cycle model (dashed line, range indicated by gray shading) with the smoothed proxy record (black). *Reproduced from Royer (2006).*

(Crowley and Zachos, 2000). It has often been suggested that an increased meridional heat transport in the ocean could have caused the reduced temperature gradient in the Cretaceous. One hypothesis postulates that—in contrast to the present-day situation —deepwater formation could have taken place at low latitudes, with warm saline water masses transported at depths toward the poles (Chamberlin, 1906). Model simulations for the Cretaceous do not support this hypothesis, however, and yield conflicting results concerning the open ocean sites of deepwater formation. Emanuel (2001) suggests that the higher intensity of tropical cyclones following an increase in tropical temperatures could lead to stronger vertical mixing in the tropics, resulting in an increased ocean heat transport toward the poles. Hotinski and Toggweiler (2003) propose that the presence of a circumglobal ocean passage at low latitudes could increase the meridional heat transport in the Cretaceous oceans. This hypothesis is another example of the more general importance of ocean gateways for ocean circulation and climate. Hays (2008) argues that a significant contribution from many sources of deepwater along the margins of the wide-spread shallow seas in the Cretaceous may be expected. This lack of a truly global overturning circulation could also help explain the evidence for ocean anoxia observed in the sedimentary record as frequent occurrences of black shales (Hay, 2008).

Clearly, much remains to be done to better understand the role of the ocean circulation for the equable climate problem. However, other factors could have contributed to the shallow temperature gradient. Forest expansion during in the Cretaceous has been shown to lead to high-latitude warming, for example (Otto-Bliesner and Upchurch, 1997; Zhou et al., 2012). Further, an increased meridional transport of latent heat in the atmosphere or changes in the radiative balance of mid and high latitudes have been suggested to contribute to the equable climate during the Cretaceous. For example, Abbot et al. (2009) suggest a high-latitude positive cloud feedback mechanism in which high-latitude warming leads to increased atmospheric convection, resulting in decreased cooling (or even warming) by clouds at high latitudes and thus a more equable climate. Further, coupled oceanic–atmospheric processes could be important. Rose and Ferreira (2013) propose that enhanced ocean heat transport could drive strong atmospheric convection in the mid-latitudes, leading to warming up to the poles due to higher humidity in the upper troposphere.

More detailed simulations with coupled Earth system models could shed further light on the processes responsible for the equable climate problem and on the role of the oceans in the Cretaceous climate system.

5. OUTLOOK

The proxy data and model results discussed here give an impression of the formidable detective work that is required to find and decipher information about past ocean states. Both proxy data and models are difficult and time-consuming to develop and understand, and their results are not always easy to interpret. Mistakes and blind alleys

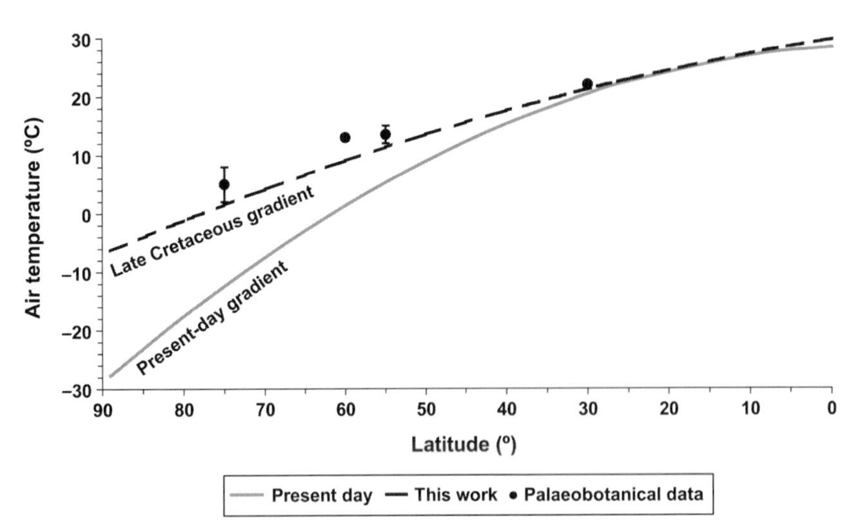

FIGURE 2.22 Example of a shallow temperature gradient during the late Cretaceous: temperature as a function of absolute paleolatitude based on $\delta^{18}O$ of fossil continental vertebrates (dashed line) and paleobotanical data (circles) as compared to the present-day gradient (gray line). *Reproduced from Amiot et al. (2004).*

are an inevitable part of this slow learning process. Yet, we hope we have been able to convince the readers that these efforts pay off and that slowly, more and more robust and precise information is emerging. This information about past climate and oceans is crucial for a better understanding of the rapid climate changes we are witnessing (and causing) at present.

The effort on paleoclimatic research is thus an effort well spent. While there are still more questions than answers, there is great promise that further painstaking work will be able to clarify many of the issues that still seem puzzling today. A better knowledge of the past is vital for an improved understanding of the dynamics of the Earth system, and it thus helps to provide the kind of knowledge humanity needs for a sustainable stewardship of our home planet.

ACKNOWLEDGMENTS

We thank Jean Lynch-Stieglitz and Kurt Lambeck for their excellent review articles, which greatly helped in the preparation of this book chapter. We also thank Andrew Kemp and Ben Horton for introducing us to sea-level proxies, and many colleagues, particularly Andrey Ganopolski, Michael Mann, Eric Steig, and Gavin Schmidt, for numerous discussions on paleoclimate.

REFERENCES

Abbot, D.S., Huber, M., Bousquet, G., Walker, C.C., 2009. High-CO2 cloud radiative forcing feedback over both land and ocean in a global climate model. Geophys. Res. Lett. 36, L05702. http://dx.doi.org/10.1029/2008GL036703.

Adkins, J.F., Cheng, H., Boyle, E.A., Druffel, E.R.M., Edwards, R.L., 1998. Deep-sea coral evidence for rapid change in ventilation of the deep North Atlantic 15,400 years ago. Science 280 (5364), 725–728. http://dx.doi.org/10.1126/science.280.5364.725.

Allen, J.R.L., 2000. Morphodynamics of Holocene salt marshes: a review sketch from the Atlantic and Southern North Sea coasts of Europe. Quat. Sci. Rev. 19 (12), 1155–1231. http://dx.doi.org/10.1016/s0277-3791(99)00034-7.

Alley, R.B., 2007. Wally was right: predictive ability of the North Atlantic "Conveyor belt" hypothesis for abrupt climate change. Annu. Rev. Earth Planet. Sci. 35, 241–272. http://dx.doi.org/10.1146/annurev.earth.35.081006.131524, Annual Review of Earth and Planetary Sciences. Annual Reviews, Palo Alto.

Alley, R.B., Clark, P.U., 1999. The deglaciation of the northern hemisphere: a global perspective. Annu. Rev. Earth Planet. Sci. 27, 149–182.

Alley, R.B., Anandakrishnan, S., Jung, P., 2001a. Stochastic resonance in the North Atlantic. Paleoceanography 16, 190–198.

Alley, R.B., Anandakrishnan, S., Jung, P., Clough, A., 2001b. Stochastic resonance in the North Atlantic: further insights. In: Seidov, D., Maslin, M., Haupt, B.J. (Eds.), The Oceans and Rapid Climate Change: Past, Present and Future, vol. 126. AGU, Washington, pp. 57–68, Geophysical Monograph.

Alley, R., et al., 2002. Abrupt Climate Change—Inevitable Surprises. National Academy Press, Washington, p. 230.

Amiot, R., Lécuyer, C., Buffetaut, E., Fluteau, F., Legendre, S., Martineau, F., 2004. Latitudinal temperature gradient during the Cretaceous upper Campanian-middle Maastrichtian: $\delta^{18}O$ record of continental vertebrates. Earth Planet. Sci. Lett. 226, 255–272. http://dx.doi.org/10.1016/j.epsl.2004.07.015.

Anderson, R.F., Bacon, M.P., Brewer, P.G., 1983. Removal of Th-230 and Pa-231 from the Open Ocean. Earth Planet. Sci. Lett. 62 (1), 7–23. http://dx.doi.org/10.1016/0012-821x(83)90067-5.

Argus, D.F., Peltier, W.R., 2010. Constraining models of postglacial rebound using space geodesy: a detailed assessment of model ICE-5G (VM2) and its relatives. Geophys. J. Int. 181 (2), 697–723. http://dx.doi.org/10.1111/j.1365-246X.2010.04562.x.

Ballarotta, M., Brodeau, L., Brandefelt, J., Lundberg, P., Döös, K., 2013. A Last Glacial Maximum world-ocean simulation at eddy-permitting resolution—part 1: experimental design and basic evaluation. Clim. Past Discuss. 9, 297–328.

Barker, S., Diz, P., Vautravers, M.J., Pike, J., Knorr, G., Hall, I.R., Broecker, W.S., 2009. Interhemispheric Atlantic seesaw response during the last deglaciation. Nature 457 (7233), 1097–1102. http://dx.doi.org/10.1038/nature07770.

Barron, E.J., Washington, W.M., 1984. The role of geographic variables in explaining paleoclimates: results from Cretaceous climate model sensitivity studies. J. Geophys. Res. 89, 1267–1279.

Barron, E.J., Fawcett, P.J., Peterson, W.H., Pollard, D., Thompson, S.L., 1995. A "simulation" of mid-Cretaceous climate. Paleoceanography 10, 953–962.

Bice, K.L., Scotese, C.R., Seidov, D., Barron, E.J., 2000. Quantifying the role of geographic change in Cenozoic ocean heat transport using uncoupled atmosphere and ocean models. Palaeogeogr. Palaeoclimatol. Palaeoecol. 161, 295–310.

Blunier, T., Brook, E.J., 2001. Timing of millennial-scale climate change in Antarctica and Greenland during the last glacial period. Science 291, 109–112.

Boyle, E.A., 1992. Cadmium and delta-C-13 paleochemical ocean distributions during the stage-2 glacial maximum. Annu. Rev. Earth Planet. Sci. 20, 245–287. http://dx.doi.org/10.1146/annurev.earth.20.1.245.

Boyle, E.A., Sclater, F., Edmond, J.M., 1976. Marine geochemistry of cadmium. Nature 263 (5572), 42–44. http://dx.doi.org/10.1038/263042a0.

Braconnot, P., Otto-Bliesner, B., Harrison, S., Joussaume, S., Peterchmitt, J.Y., Abe-Ouchi, A., Crucifix, M., Driesschaert, E., Fichefet, T., Hewitt, C.D., Kageyama, M., Kitoh, A., Laine, A., Loutre, M.F., Marti, O., Merkel, U., Ramstein, G., Valdes, P., Weber, S.L., Yu, Y., Zhao, Y., 2007a. Results of PMIP2 coupled simulations of the mid-Holocene and last glacial maximum—part 1: experiments and large-scale features. Clim. Past 3 (2), 261–277.

Braconnot, P., Otto-Bliesner, B., Harrison, S., Joussaume, S., Peterchmitt, J.Y., Abe-Ouchi, A., Crucifix, M., Driesschaert, E., Fichefet, T., Hewitt, C.D., Kageyama, M., Kitoh, A., Loutre, M.F., Marti, O., Merkel, U., Ramstein, G., Valdes, P., Weber, L., Yu, Y., Zhao, Y., 2007b. Results of PMIP2 coupled simulations of the Mid-Holocene and Last Glacial Maximum—part 2: feedbacks with emphasis on the location of the ITCZ and mid- and high latitudes heat budget. Clim. Past 3 (2), 279–296.

Braun, H., Christl, M., Rahmstorf, S., Ganopolski, A., Mangini, A., Kubatzki, C., Roth, K., Kromer, B., 2005. Solar forcing of abrupt glacial climate change in a coupled climate system model. Nature 438, 208–211.

Broecker, W.S., Andree, M., Bonani, G., Wolfli, W., Oeschger, H., Klas, M., Mix, A., Curry, W., 1988. Preliminary estimates for the radiocarbon age of deep water in the glacial ocean. Paleoceanography 3 (6), 659–669. http://dx.doi.org/10.1029/PA003i006p00659.

Cacho, I., Grimalt, J.O., Pelejero, C., Canals, M., Sierro, F.J., Flores, J.A., Shackleton, N.J., 1999. Dansgaard-Oeschger and Heinrich event imprints in the Alboran Sea paleotemperatures. Paleoceanography 14, 698–705.

Camuffo, D., Sturaro, G., 2003. Sixty-CM submersion of Venice discovered thanks to Canaletto's paintings. Clim. Change 58 (3), 333–343. http://dx.doi.org/10.1023/a:1023902120717.

Carbognin, L., Teatini, P., Tomasin, A., Tosi, L., 2010. Global change and relative sea level rise at Venice: what impact in term of flooding. Clim. Dyn. 35 (6), 1055–1063. http://dx.doi.org/10.1007/s00382-009-0617-5.

Chamberlin, T.C., 1906. On a possible reversal of deep-sea circulation and its influence on geologic climates. J. Geol. 14, 363–373.

Church, J.A., Woodworth, P.L., Aarup, T., Wilson, W., 2010. Understanding Sea-Level Rise and Variability. Blackwell, Oxford.

Clark, P.U., Pisias, N.G., Stocker, T.F., Weaver, A.J., 2002. The role of the thermohaline circulation in abrupt climate change. Nature 415, 863–869.

Clark, P.U., Dyke, A.S., Shakun, J.D., Carlson, A.E., Clark, J., Wohlfarth, B., Mitrovica, J.X., Hostetler, S.W., McCabe, A.M., 2009. The last glacial maximum. Science 325 (5941), 710–714. http://dx.doi.org/10.1126/science.1172873.

Clark, P.U., Shakun, J.D., Baker, P.A., Bartlein, P.J., Brewer, S., Brook, E., Carlson, A.E., Cheng, H., Kaufman, D.S., Liu, Z.Y., Marchitto, T.M., Mix, A.C., Morrill, C., Otto-Bliesner, B.L., Pahnke, K., Russell, J.M., Whitlock, C., Adkins, J.F., Blois, J.L., Clark, J., Colman, S.M., Curry, W.B., Flower, B.P., He, F., Johnson, T.C., Lynch-Stieglitz, J., Markgraf, V., McManus, J., Mitrovica, J.X., Moreno, P.I., Williams, J.W., 2012. Global climate evolution during the last deglaciation. Proc. Natl. Acad. Sci. U.S.A. 109 (19), E1134–E1142. http://dx.doi.org/10.1073/pnas.1116619109.

Claussen, M., et al., 2002. Earth system models of intermediate complexity: closing the gap in the spectrum of climate system models. Clim. Dyn. 18, 579–586.

Cocks, L.R.M., Torsvik, T.H., 2002. Earth geography from 500 to 400 million years ago: a faunal and palaeomagnetic review. J. Geol. Soc. 159, 631–644. http://dx.doi.org/10.1144/0016-764901-118.

Crowley, T.J., Zachos, J.C., 2000. Comparison of zonal temperature profiles for past warm time periods. In: Huber, B.T., MacLeod, K.G., Wing, S.L. (Eds.), Warm Climates in Earth History. Cambridge University Press, Cambridge, UK, pp. 50–76.

Dahl-Jensen, D., Albert, M.R., Aldahan, A., Azuma, N., Balslev-Clausen, D., Baumgartner, M., Berggren, A.M., Bigler, M., Binder, T., Blunier, T., Bourgeois, J.C., Brook, E.J., Buchardt, S.L., Buizert, C., Capron, E., Chappellaz, J., Chung, J., Clausen, H.B., Cvijanovic, I., Davies, S.M., Ditlevsen, P., Eicher, O., Fischer, H., Fisher, D.A., Fleet, L. G., Gfeller, G., Gkinis, V., Gogineni, S., Goto-Azuma, K., Grinsted, A., Gudlaugsdottir, H., Guillevic, M., Hansen, S.B., Hansson, M., Hirabayashi, M., Hong, S., Hur, S.D., Huybrechts, P., Hvidberg, C. S., Iizuka, Y., Jenk, T., Johnsen, S.J., Jones, T.R., Jouzel, J., Karlsson, N.B., Kawamura, K., Keegan, K., Kettner, E., Kipfstuhl, S., Kjaer, H. A., Koutnik, M., Kuramoto, T., Kohler, P., Laepple, T., Landais, A., Langen, P.L., Larsen, L.B., Leuenberger, D., Leuenberger, M., Leuschen, C., Li, J., Lipenkov, V., Martinerie, P., Maselli, O.J., Masson-Delmotte, V., McConnell, J.R., Miller, H., Mini, O., Miyamoto, A., Montagnat-Rentier, M., Mulvaney, R., Muscheler, R., Orsi, A.J., Paden, J., Panton, C., Pattyn, F., Petit, J.R., Pol, K., Popp, T., Possnert, G., Prie, F., Prokopiou, M., Quiquet, A., Rasmussen, S.O., Raynaud, D., Ren, J., Reutenauer, C., Ritz, C., Rockmann, T., Rosen, J.L., Rubino, M., Rybak, O., Samyn, D., Sapart, C.J., Schilt, A., Schmidt, A.M.Z., Schwander, J., Schupbach, S., Seierstad, I., Severinghaus, J.P., Sheldon, S., Simonsen, S.B., Sjolte, J., Solgaard, A.M., Sowers, T., Sperlich, P., Steen-Larsen, H.C., Steffen, K., Steffensen, J.P., Steinhage, D., Stocker, T.F., Stowasser, C., Sturevik, A.S., Sturges, W.T., Sveinbjornsdottir, A., Svensson, A., Tison, J.L., Uetake, J., Vallelonga, P., van de Wal, R.S.W., van der Wel, G., Vaughn, B.H., Vinther, B., Waddington, E., Wegner, A., Weikusat, I., White, J.W.C., Wilhelms, F., Winstrup, M., Witrant, E., Wolff, E.W., Xiao, C., Zheng, J., 2013. Eemian interglacial reconstructed from a Greenland folded ice core. Nature 493 (7433), 489–494. http://dx.doi.org/10.1038/nature11789.

Deschamps, P., Durand, N., Bard, E., Hamelin, B., Camoin, G., Thomas, A.L., Henderson, G.M., Okuno, J., Yokoyama, Y., 2012. Ice-sheet collapse and sea-level rise at the Bolling warming 14,600 years ago. Nature 483 (7391), 559–564. http://dx.doi.org/10.1038/nature10902.

Deuser, W.G., Hunt, J.M., 1969. Stable isotope ratios of dissolved inorganic carbon in Atlantic. Deep Sea Res. 16 (2), 221–225. http://dx.doi.org/10.1016/0011-7471(69)90078-3.

Donnelly, J.P., Cleary, P., Newby, P., Ettinger, R., 2004. Coupling instrumental and geological records of sea-level change: evidence from southern New England of an increase in the rate of sea-level rise in the late 19th century. Geophys. Res. Lett. 31, L05203. http://dx.doi.org/10.1029/2003gl018933.

Duplessy, J.C., Shackleton, N.J., Matthews, R.K., Prell, W., Ruddiman, W.F., Caralp, M., Hendy, C.H., 1984. C-13 record of benthic foraminifera in the last interglacial ocean—implications for the carbon-cycle and the global deep-water circulation. Quat. Res. 21 (2), 225–243. http://dx.doi.org/10.1016/0033-5894(84)90099-1.

Duplessy, J.-C., Shackleton, N.J., Fairbanks, R.B., Labeyrie, L., Oppo, D., Kallel, N., 1988. Deep water source variations during the last climatic cycle and their impact on the global deep water circulation. Paleoceanography 3, 343–360.

Duplessy, J.C., Labeyrie, L., Waelbroeck, C., 2002. Constraints on the ocean oxygen isotopic enrichment between the last glacial maximum and the Holocene: paleoceanographic implications. Quat. Sci. Rev. 21 (1–3), 315–330. http://dx.doi.org/10.1016/s0277-3791(01)00107-x.

Dutton, A., Lambeck, K., 2012. Ice volume and sea level during the last interglacial. Science 337 (6091), 216–219. http://dx.doi.org/10.1126/science.1205749.

Elderfield, H., Rickaby, R.E.M., 2000. Oceanic Cd/P ratio and nutrient utilization in the glacial Southern ocean. Nature 405 (6784), 305–310. http://dx.doi.org/10.1038/35012507.

Emanuel, K., 2001. Contribution of tropical cyclones to meridional heat transport by the oceans. J. Geophys. Res. 106, 14771–14781. http://dx.doi.org/10.1029/2000JD900641.

Fairbanks, R.G., 1989. A 17,000-year glacio-eustatic sea level record: influence of glacial melting rates on the younger Dryas event and deep-ocean circulation. Nature 342, 637–642.

Fichefet, T., Hovine, S., Duplessy, J.C., 1994. A model study of the Atlantic thermohaline circulation during the last glacial maximum. Nature 372, 252–255.

Gammaitoni, L., Hanggi, P., Jung, P., Marchesoni, F., 1998. Stochastic resonance. Rev. Mod. Phys. 70, 223–287.

Ganopolski, A., Rahmstorf, S., 2001. Rapid changes of glacial climate simulated in a coupled climate model. Nature 409, 153–158.

Ganopolski, A., Rahmstorf, S., 2002. Abrupt glacial climate changes due to stochastic resonance. Phys. Rev. Lett. 88 (3), 038501.

Ganopolski, A., Roche, D.M., 2009. On the nature of lead–lag relationships during glacial–interglacial climate transitions. Quat. Sci. Rev. 28, 3361–3378. http://dx.doi.org/10.1016/j.quascirev.2009.09.019.

Ganopolski, A., Rahmstorf, S., Petoukhov, V., Claussen, M., 1998. Simulation of modern and glacial climates with a coupled global model of intermediate complexity. Nature 391, 351–356.

Ganopolski, A., Calov, R., Claussen, M., 2010. Simulation of the last glacial cycle with a coupled climate ice-sheet model of intermediate complexity. Clim. Past 6 (2), 229–244.

Gehrels, W.R., 1994. Determining relative sea-level change from salt-marsh foraminifera and plant zones on the coast of Maine, USA. J. Coast. Res. 10 (4), 990–1009.

Grinsted, A., Moore, J.C., Jevrejeva, S., 2010. Reconstructing sea level from paleo and projected temperatures 200 to 2100 AD. Clim. Dyn. 34, 461–472. http://dx.doi.org/10.1007/s00382-008-0507-2.

Grootes, P.M., Stuiver, M., White, J.W.C., Johnsen, S., Jouzel, J., 1993. Comparison of oxygen isotope records from the GISP2 and GRIP Greenland ice cores. Nature 366, 552–554.

Hay, W.W., 2008. Evolving ideas about the Cretaceous climate and ocean circulation. Cretaceous Res. 29, 725–753. http://dx.doi.org/10.1016/j.cretres.2008.05.025.

Hay, W.W., Floegel, S., 2012. New thoughts about the Cretaceous climate and oceans. Earth Sci. Rev. 115, 262–272. http://dx.doi.org/10.1016/j.earscirev.2012.09.008.

Hays, J.D., Pitman, W.C., 1973. Lithospheric plate motion, sea level changes and climatic and ecological consequences. Nature 246, 18–22. http://dx.doi.org/10.1038/246018a0.

Hesse, T., Butzin, M., Bickert, T., Lohmann, G., 2011. A model-data comparison of delta C-13 in the glacial Atlantic Ocean. Paleoceanography 26, 16. http://dx.doi.org/10.1029/2010pa002085.

Hofmann, M., Rahmstorf, S., 2009. On the stability of the Atlantic meridional overturning circulation. Proc. Natl. Acad. Sci. U.S.A. 106 (49), 20584–20589. http://dx.doi.org/10.1073/pnas.0909146106.

Hotinski, R.M., Toggweiler, J.R., 2003. Impact of a Tethyan circumglobal passage on ocean heat transport and "equable" climates. Paleoceanography 18, 1007. http://dx.doi.org/10.1029/2001PA000730.

Jevrejeva, S., Grinsted, A., Moore, J.C., Holgate, S., 2006. Nonlinear trends and multiyear cycles in sea level records. J. Geophys. Res. 111, C09012. http://dx.doi.org/10.1029/2005JC003229.

Kageyama, M., Paul, A., Roche, D.M., Van Meerbeeck, C.J., 2010. Modelling glacial climatic millennial-scale variability related to changes in the Atlantic meridional overturning circulation: a review. Quat. Sci. Rev. 29 (21–22), 2931–2956. http://dx.doi.org/10.1016/j.quascirev.2010.05.029.

Kemp, A., Horton, B., Donnelly, J., Mann, M.E., Vermeer, M., Rahmstorf, S., 2011. Climate related sea-level variations over the past two millennia. Proc. Natl. Acad. Sci. U.S.A. 108 (27), 11017–11022. http://dx.doi.org/10.1073/pnas.1015619108.

Kennett, J.P., 1977. Cenozoic evolution of Antarctic glaciation, the circum-Antarctic ocean, and their impact on global paleoceanography. J. Geophys. Res. 82, 3843–3860. http://dx.doi.org/10.1029/JC082i027p03843.

Kim, S.T., O'Neil, J.R., 1997. Equilibrium and non-equilibrium oxygen isotope effects in synthetic carbonates. Geochim. Cosmochim. Acta 61, 3461–3475.

Kopp, R.E., Simons, F.J., Mitrovica, J.X., Maloof, A.C., Oppenheimer, M., 2009. Probabilistic assessment of sea level during the last interglacial stage. Nature 462 (7275), 863–867. http://dx.doi.org/10.1038/nature08686.

Kreveld, S.V., Sarnthein, M., Erlenkeuser, H., Grootes, P., Jung, S., Nadeau, M.J., Pflaumann, U., Voelker, A., 2000. Potential links between surging ice sheets, circulation changes, and the Dansgaard-Oeschger cycles in the Irminger Sea, 60-18 kyr. Paleoceanography 15, 425–442.

Lambeck, K., 2002. Sea-level change from mid-Holocene to recent time: an Australian example with global implications. In: Mitrovica, J.X.,

Vermeersen, L. (Eds.), Glacial Isostatic Adjustment and the Earth System. AGU, Washington, pp. 33–50.

Lambeck, K., Bard, E., 2000. Sea-level change along the French Mediterranean coast for the past 30 000 years. Earth Planet. Sci. Lett. 175 (3–4), 203–222. http://dx.doi.org/10.1016/s0012-821x(99)00289-7.

Lambeck, K., Anzidei, M., Antonioli, F., Benini, A., Espositol, A., 2004. Sea level in Roman times in the central Mediterranean and implications for recent change. Earth Planet. Sci. Lett. 224, 563–575.

Lambeck, K., Woodroffe, C., Antonioli, F., Anzidei, M., Gehrels, W.R., Laborel, J., Wright, A.J., 2010. Paeoenvironmental records, geophysical modeling, and reconstruction of sea-level trends and variability on centennial and longer time scales. In: Church, J.A., Woodworth, P.L., Aarup, T., Wilson, W. (Eds.), Understanding Sea-Level Rise and Variability. Blackwell, Oxford, pp. 61–121.

Lisiecki, L.E., Raymo, M.E., 2005. A Pliocene-Pleistocene stack of 57 globally distributed benthic delta O-18 records. Paleoceanography 20, Pa1003. http://dx.doi.org/10.1029/2004pa001071.

Lyell, C., 1832. Principles of Geology. Murray, London.

Lynch-Stieglitz, J., 2003. Tracers of past ocean circulation. In: Turkekian, K., Holland, H. (Eds.), Treatise on Geochemistry, vol. 6. Elsevier, pp. 433–451.

Lynch-Stieglitz, J., Curry, W.B., Slowey, N., 1999. A geostrophic transport estimate for the Florida current from the oxygen isotope composition of benthic foraminifera. Paleoceanography 14 (3), 360–373. http://dx.doi.org/10.1029/1999pa900001.

Lynch-Stieglitz, J., Adkins, J.F., Curry, W.B., Dokken, T., Hall, I.R., Herguera, J.C., Hirschi, J.J.M., Ivanova, E.V., Kissel, C., Marchal, O., Marchitto, T.M., McCave, I.N., McManus, J.F., Mulitza, S., Ninnemann, U., Peeters, F., Yu, E.F., Zahn, R., 2007. Atlantic meridional overturning circulation during the last glacial maximum. Science 316 (5821), 66–69. http://dx.doi.org/10.1126/science.1137127.

Lynch-Stieglitz, J., Curry, W.B., Lund, D.C., 2009. Florida Straits density structure and transport over the last 8000 years. Paleoceanography 24, 9. http://dx.doi.org/10.1029/2008pa001717.

Margari, V., Skinner, L.C., Tzedakis, P.C., Ganopolski, A., Vautravers, M., Shackleton, N.J., 2010. The nature of millennial-scale climate variability during the past two glacial periods. Nat. Geosci. 3 (2), 127–131. http://dx.doi.org/10.1038/ngeo740.

McCulloch, M.T., Tudhope, A.W., Esat, T.M., Mortimer, G.E., Chappell, J., Pillans, B., Chivas, A.R., Omura, A., 1999. Coral record of equatorial sea-surface temperatures during the penultimate deglaciation at Huon Peninsula. Science 283 (5399), 202–204. http://dx.doi.org/10.1126/science.283.5399.202.

McKay, N.P., Overpeck, J.T., Otto-Bliesner, B.L., 2012. The role of ocean thermal expansion in last interglacial sea level rise. Geophys. Res. Lett. 38, L14605. http://dx.doi.org/10.1029/2011gl048280.

Milankovitch, M., 1941. Kanon der Erdbestrahlung und seine Anwendung auf das Eiszeitenproblem. Königlich Serbische Akademie, Belgrad.

Miller, K.G., Kominz, M.A., Browning, J.V., Wright, J.D., Mountain, G.S., Katz, M.E., Sugarman, P.J., Cramer, B.S., Christie-Blick, N., Pekar, S.F., 2005. The phanerozoic record of global sea-level change. Science 310, 1293–1298. http://dx.doi.org/10.1126/science.1116412.

Müller, R.D., Sdrolias, M., Gaina, C., Roest, W.R., 2008a. Age, spreading rates, and spreading asymmetry of the world's ocean crust. Geochem. Geophys. Geosyst. 9, Q04006. http://dx.doi.org/10.1029/2007GC001743.

Müller, R.D., Sdrolias, M., Gaina, C., Steinberger, B., Heine, C., 2008b. Long-term sea-level fluctuations driven by ocean basin dynamics. Science 319, 1357–1362. http://dx.doi.org/10.1126/science.1151540.

Negre, C., Zahn, R., Thomas, A.L., Masque, P., Henderson, G.M., Martinez-Mendez, G., Hall, I.R., Mas, J.L., 2010. Reversed flow of Atlantic deep water during the last glacial maximum. Nature 468 (7320), 84. http://dx.doi.org/10.1038/nature09508.

Oppo, D., Lehman, S.J., 1993. Mid-depth circulation of the subpolar North Atlantic during the last glacial maximum. Science 259, 1148–1152.

Ota, Y., Chappell, J., 1999. Holocene sea-level rise and coral reef growth on a tectonically rising coast, Huon Peninsula, Papua New Guinea. Quat. Int. 55, 51–59. http://dx.doi.org/10.1016/s1040-6182(98)00024-x.

Otto-Bliesner, B.L., Brady, E.C., 2010. The sensitivity of the climate response to the magnitude and location of freshwater forcing: last glacial maximum experiments. Quat. Sci. Rev. 29 (1–2), 56–73. http://dx.doi.org/10.1016/j.quascirev.2009.07.004.

Otto-Bliesner, B.L., Upchurch Jr., G.R., 1997. Vegetation-induced warming of high-latitude regions during the late Cretaceous period. Nature 385, 804–807.

Passaro, S., Barra, M., Saggiomo, R., Di Giacomo, S., Leotta, A., Uhlen, H., Mazzola, S., 2013. Multi-resolution morpho-bathymetric survey results at the Pozzuoli-Baia underwater archaeological site (Naples, Italy). J. Archaeol. Sci. 40 (2), 1268–1278. http://dx.doi.org/10.1016/j.jas.2012.09.035.

Pearson, P.N., Ditchfield, P.W., Singano, J., Harcourt-Brown, K.G., Nicholas, C.J., Olsson, R.K., Shackleton, N.J., Hall, M.A., 2001. Warm tropical sea surface temperatures in the late Cretaceous and Eocene epochs. Nature 413, 481–487. http://dx.doi.org/10.1038/413481A0.

Peltier, W.R., Fairbanks, R.G., 2006. Global glacial ice volume and last glacial maximum duration from an extended Barbados sea level record. Quat. Sci. Rev. 25 (23–24), 3322–3337. http://dx.doi.org/10.1016/j.quascirev.2006.04.010.

Pirazzoli, P.A., Evelpidou, N., 2013. Tidal notches: a sea-level indicator of uncertain archival trustworthiness. Palaeogeogr. Palaeoclimatol. Palaeoecol. 369, 377–384. http://dx.doi.org/10.1016/j.palaeo.2012.11.004.

Rahmstorf, S., 2002. Ocean circulation and climate during the past 120,000 years. Nature 419, 207–214.

Rahmstorf, S., 2003. Timing of abrupt climate change: a precise clock. Geophys. Res. Lett. 30, 1510.

Rahmstorf, S., 2007. A semi-empirical approach to projecting future sea-level rise. Science 315 (5810), 368–370.

Robinson, A., Calov, R., Ganopolski, A., 2011. Greenland ice sheet model parameters constrained using simulations of the Eemian Interglacial. Clim. Past 7 (2), 381–396. http://dx.doi.org/10.5194/cp-7-381-2011.

Robinson, A., Calov, R., Ganopolski, A., 2012. Multistability and critical thresholds of the Greenland ice sheet. Nat. Clim. Chang. 2 (6), 429–432. http://dx.doi.org/10.1038/nclimate1449.

Rohling, E.J., Fenton, M., Jorissen, F.J., Bertrand, P., Ganssen, G., Caulet, J.P., 1998. Magnitudes of sea-level lowstands of the past 500,000 years. Nature 394, 162–165. http://dx.doi.org/10.1038/28134.

Rose, B.E.J., Ferreira, D., 2013. Ocean heat transport and water vapor greenhouse in a warm equable climate: a new look at the low gradient paradox. J. Clim. 26, 2117–2136. http://dx.doi.org/10.1175/JCLI-D-11-00547.1.

Rosenthal, Y., Boyle, E.A., Slowey, N., 1997. Temperature control on the incorporation of magnesium, strontium, fluorine, and cadmium into benthic foraminiferal shells from Little Bahama Bank: prospects for thermocline paleoceanography. Geochim. Cosmochim. Acta 61 (17), 3633–3643. http://dx.doi.org/10.1016/s0016-7037(97)00181-6.

Royer, D.L., 2006. CO2-forced climate thresholds during the Phanerozoic. Geochim. Cosmochim. Acta 70, 5665–5675. http://dx.doi.org/10.1016/j.gca.2005.11.031.

Ruddiman, W.F., 2000. Earths Climate: Past and Future. Freeman, New York, p. 465.

Rutberg, R.L., Hemming, S.R., Goldstein, S.L., 2000. Reduced North Atlantic deep water flux to the glacial Southern ocean inferred from neodymium isotope ratios. Nature 405 (6789), 935–938.

Sachs, J.P., Lehman, S.J., 1999. Subtropical North Atlantic temperatures 60,000 to 30,000 years ago. Science 286, 756–759.

Sarnthein, M., Winn, K., Jung, S.J.A., Duplessy, J.C., Labeyrie, L., Erlenkeuser, H., Ganssen, G., 1994. Changes in east Atlantic deep-water circulation over the last 30,000 years: eight time slice reconstructions. Paleoceanography 9, 209–267.

Schmidt, G.A., 1999. Forward modeling of carbonate proxy data from planktonic foraminifera using oxygen isotope tracers in a global ocean model. Paleoceanography 14 (4), 482–497. http://dx.doi.org/10.1029/1999pa900025.

Schrag, D.P., Adkins, J.F., McIntyre, K., Alexander, J.L., Hodell, D.A., Charles, C.D., McManus, J.F., 2002. The oxygen isotopic composition of seawater during the last glacial maximum. Quat. Sci. Rev. 21 (1–3), 331–342. http://dx.doi.org/10.1016/s0277-3791(01)00110-x.

Schulz, M., 2002. On the 1470-year pacing of Dansgaard-Oeschger warm events. Paleoceanography 17, 4-1–4-9.

Scott, D.S., Medioli, F.S., 1978. Vertical zonations of Marsh foraminifera as accurate indicators of former sea-levels. Nature 272 (5653), 528–531. http://dx.doi.org/10.1038/272528a0.

Sewall, J.O., van de Wal, R.S.W., van der Zwan, K., van Oosterhout, C., Dijkstra, H.A., Scotese, C.R., 2007. Climate model boundary conditions for four Cretaceous time slices. Clim. Past 3, 647–657. http://dx.doi.org/10.5194/cp-3-647-2007.

Shackleton, N.J., 2000. The 100,000-year ice-age cycle identified and found to lag temperature, carbon dioxide, and orbital eccentricity. Science 289, 1897–1902. http://dx.doi.org/10.1126/science.289.5486.1897.

Shennan, I., et al., 2007. Sea level studies. In: In: Elias, S.A. (Ed.), Encyclopedia of Quaternary Science, vol. 4. Elsevier, Amsterdam, pp. 2967–3095.

Sivan, D., Lambeck, K., Toueg, R., Raban, A., Porath, Y., Shirman, B., 2004. Ancient coastal wells of Caesarea Maritima, Israel, an indicator for relative sea level changes during the last 2000 years. Earth Planet. Sci. Lett. 222 (1), 315–330. http://dx.doi.org/10.1016/j.epsl.2004.02.007.

Spero, H.J., Lea, D.W., 1996. Experimental determination of stable isotope variability in Globigerina bulloides: implications for paleoceanographic reconstructions. Mar. Micropaleontol. 28 (3–4), 231–246. http://dx.doi.org/10.1016/0377-8398(96)00003-5.

Stanford, J.D., Rohling, E.J., Hunter, S.E., Roberts, A.P., Rasmussen, S.O., Bard, E., McManus, J., Fairbanks, R.G., 2006. Timing of meltwater pulse 1a and climate responses to meltwater injections. Paleoceanography 21, Pa4103. http://dx.doi.org/10.1029/2006pa001340.

Stanley, S., 2005. Earth System History. W.H. Freeman, New York, p. 567.

Stocker, T.F., 1998. The seesaw effect. Science 282, 61–62.

Stocker, T.F., Johnsen, S.J., 2003. A minimum thermodynamic model for the bipolar seesaw. Paleoceanography 18, art. no. 1087 http://onlinelibrary.wiley.com/http://dx.doi.org/10.1029/2003PA000920/abstract.

Stuiver, M., Quay, P.D., Ostlund, H.G., 1983. Abyssal water C-14 distribution and the age of the world oceans. Science 219 (4586), 849–851. http://dx.doi.org/10.1126/science.219.4586.849.

Tachikawa, K., Elderfield, H., 2002. Microhabitat effects on Cd/Ca and delta C-13 of benthic foraminifera. Earth Planet. Sci. Lett. 202 (3–4), 607–624. http://dx.doi.org/10.1016/s0012-821x(02)00796-3.

Torsvik, T.H., Cocks, L.R.M., 2004. Earth geography from 400 to 250 Ma: a palaeomagnetic, faunal and facies review. J. Geol. Soc. 161, 555–572. http://dx.doi.org/10.1144/0016-764903-098.

Turney, C.S.M., Jones, R.T., 2010. Does the Agulhas current amplify global temperatures during super-interglacials? J. Quat. Sci. 25 (6), 839–843. http://dx.doi.org/10.1002/jqs.1423.

Waelbroeck, C., Labeyrie, L., Michel, E., Duplessy, J.C., McManus, J.F., Lambeck, K., Balbon, E., Labracherie, M., 2002. Sea-level and deep water temperature changes derived from benthic foraminifera isotopic records. Quat. Sci. Rev. 21 (1–3), 295–305.

Weber, S.L., Drijfhout, S.S., Abe-Ouchi, A., Crucifix, M., Eby, M., Ganopolski, A., Murakami, S., Otto-Bliesner, B., Peltier, W.R., 2007. The modern and glacial overturning circulation in the Atlantic Ocean in PMIP coupled model simulations. Clim. Past 3 (1), 51–64.

Wefer, G., Berger, W.H., 1991. Isotope paleontology—growth and composition of extant Calcareous species. Mar. Geol. 100 (1–4), 207–248. http://dx.doi.org/10.1016/0025-3227(91)90234-u.

Woodroffe, C., McLean, R., 1990. Microatolls and recent sea-level change on coral atolls. Nature 344 (6266), 531–534. http://dx.doi.org/10.1038/344531a0.

Yu, E.-F., Francois, R., Bacon, M.P., 1996. Similar rates of modern and last-glacial ocean thermohaline circulation inferred from radiochemical data. Nature 379, 689–694.

Zhou, J., Poulsen, C.J., Rosenbloom, N., Shields, C., Briegleb, B., 2012. Vegetation-climate interactions in the warm mid-Cretaceous. Clim. Past 8, 565–576. http://dx.doi.org/10.5194/cp-8-565-2012.

Part II

Ocean Observations

Accurate observations of the physical and biogeochemical state of the ocean are an essential prerequisite for understanding how the oceans "work" both internally and through interaction with their enclosing basins and with the atmosphere and cryosphere. These latter two interactions are fundamental for defining the oceans' role in the earth's climate system. The remoteness and hostility of the oceans and their opaqueness to electromagnetic radiation together pose tremendous challenges to those who seek to observe the oceans. Our ability to address ocean and climate research questions is dependent on the availability of appropriate technologies to make observations, and the chapters in this Part II describe the development of physical measurements and demonstrate this dependence.

The use of water bottles and reversing thermometers continued virtually unchanged during the first half of the twentieth century but was complemented eventually by mechanical bathythermograph to study the upper-ocean variability. It is only since the 1970s that continuously profiling conductivity–temperature–depth probes, current meter moorings, floats and drifters together with expendable (XBT) probes, and the introduction of satellite navigation have brought fundamental changes in our observational capability. Satellite observations of sea-surface temperature, ocean color, and most importantly, since the 1990s, sea-surface height (SSH), now permit the mapping of the global ocean with high accuracy and resolution. The World Ocean Circulation Experiment (WOCE) in the 1990s combined state-of-the-art *in situ* measurements (including those from ships of opportunity) with satellite SSH and eventually led to the implementation of a global network of automatic Argo profilers, now a mainstay of our observing network.

WOCE also brought about changes in our approach to data handling leading to rapid reporting of data, internationally accepted quality control standards and the open exchange of observational data. These have allowed the analysis of the state of the global ocean and its changes. The volume and distribution of observations have changed enormously. We present examples of the coverage by *in situ* observations, in particular, Argo profiles, the network of XBT measurements on ships-of-opportunity, the global surface drifters, the OceanSITES time-series stations, the sea-level stations, and repeat hydrography sections.

Remote sensing using satellite altimeter measurements of SSH provide much improved knowledge of the oceanic circulation. The combination of altimeter data with *in situ* observations and with new geoid data for obtaining absolute geostrophic currents on small scales approaching 100 km have enabled the identification of a multitude of fronts and the determination of changes in the surface circulation. Such data are provided in almost real time as gridded SSH fields to the research community, making satellite altimetry a routine tool of marine research and applications. The relation of sea-surface temperature data from infrared or microwave radiation and ocean surface wind data from satellite scatterometers provides information on air–sea interaction on scales down to below 1000 km.

New instrumentation, both *in situ* and onboard satellites, new methods for providing data to the wider community, and new approaches in coordinating international observational programs are improving the knowledge on the oceanic circulation and exchange processes, and are providing constraints for modeling the ocean and climate.

Chapter 3

In Situ Ocean Observations: A Brief History, Present Status, and Future Directions

John Gould*, Bernadette Sloyan[†] and Martin Visbeck[‡]
*National Oceanography Centre, Southampton, United Kingdom
[†]CSIRO Marine and Atmospheric Research, GPO Box 1538, Hobart, Tasmania, Australia
[‡]Helmholtz Centre for Ocean Research Kiel (GEOMAR), Kiel, Germany

Chapter Outline

1. INTRODUCTION

Observations of the interior of the ocean are fundamental to understanding ocean dynamics and properties, monitoring changes in the oceans' state, (whether caused by natural or human influences), quantifying the forcing at the atmosphere–ocean (in some areas, atmosphere–ice–ocean) boundary, and determining the role and importance of the ocean in the climate system. *In situ* ocean observations also complement and provide ground truth for remotely sensed observations of the ocean from earth-observing satellites (Chapter 4). Both satellite and *in situ* observations are vital for ocean forecasting, ocean reanalysis and for assessing the fidelity of ocean and earth-system models and underpinning their future improvement (Chapters 21 and 22).

The technical and logistical challenges of making *in situ* ocean observations are legion; measurements often have to be made in areas far removed from land, in a corrosive liquid, at great pressure, and in a fluid that is effectively opaque to electromagnetic radiation. Capturing the oceans' variability requires repeated measurements over wide areas and yet with small spatial resolution. Detecting change demands measurements of high precision and stability over decadal and longer time scales. For these reasons, the history of scientifically focused, open ocean observations is relatively short: it may be said to have started with the voyage of *HMS Challenger* in the 1870s (Wyville Thomson and Murray, 1885; Figure 3.1).

Through the twentieth century, measurements became more accurate but remained relatively sparse and regionally focused until the 1990s. During that century, there were a number of initiatives and technical advances that, with hindsight, can be regarded as having been crucial steps in improving our ability to make systematic measurements within the global ocean. Often the driver for progress was the sequence of "*New observations lead to new understanding—new understanding points out the inadequacy of earlier observations—this understanding stimulates new technical development.*" The other major driver for progress was the application of advanced technologies

Ocean Circulation and Climate, Vol. 103. http://dx.doi.org/10.1016/B978-0-12-391851-2.00003-9

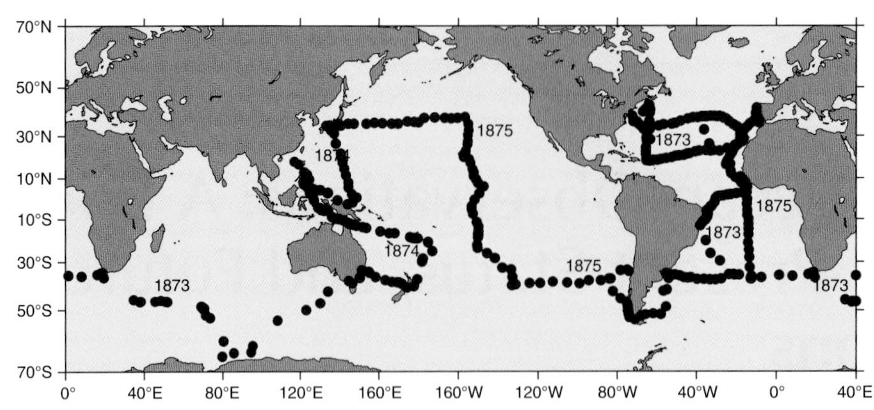

FIGURE 3.1 Track of *HMS Challenger*. This was the first major scientific exploration of the global ocean. The voyage lasted 4 years and covered more than 68,000 nautical miles.

to the oceans—solid state electronics in the 1960s and 1970s, miniaturized computing power, and satellite communication and navigation from the 1990s to the present day. Innovative exploitation of these advances has led to major advances in our observational capability. In the following, we briefly review the development of key observing technologies (Section 2) and their impact on the number and distribution of ocean observations (Section 3). Apart from the ever-present uncertainties of funding, the future for sustained ocean observations through the exploitation of emerging technologies (Section 3) and within the new Framework for Ocean Observations (Section 5) is bright.

2. DEVELOPMENT OF PRESENT OBSERVATIONAL CAPABILITY

Rather than considering the advances in observational capability on a parameter-by-parameter basis, we discuss the topic through a brief chronology of some of the most important technical developments that have enabled the establishment of the present multiparameter ocean observing systems. In recent years, progress has often been made in a number of key ocean parameters simultaneously through the mounting of major internationally coordinated observational programs and/or the adoption of new generic observing platforms or technologies.

2.1. Late Nineteenth to Mid-Twentieth Centuries

Improvements in navigation were the first drivers of systematic ocean observations. In the nineteenth century, following the introduction of Harrison's chronometer that enabled longitude to be determined, it became possible to estimate surface currents from a vessel's navigation and these were recorded in navigational logs. Such observations led to the compilation of surface currents in the Atlantic by (Rennell, 1832) posthumously by his daughter and more widely by Mathew Fontaine Maury (1855). The major motivation for Maury's work was not scientific but was to use knowledge of surface currents to shorten sea voyages and hence gain commercial advantage. This was also a driver for the measurement of ocean temperatures since it was recognized by Benjamin Franklin that temperatures changed across the Gulf Stream and that by navigating into water of the correct temperature ships could speed their voyages between America and Europe. Early ocean surface temperature measurements were made by dipping a simple mercury-in-glass thermometer into, first, wooden and later canvas buckets of water collected from the sea surface: a technique that remained in common use until the mid-late twentieth century. While wooden buckets were well insulated, canvas ones were less so and thus these temperature measurements are now known to be biased low due to evaporative cooling.

The ability to make subsurface observations of temperature and salinity developed substantially between the pioneering voyage of *HMS Challenger* (1872–1876), the Meteor Expedition to the Atlantic (1925–1927) (Wüst, 1935), and *Discovery* investigations in the Southern Ocean (starting in 1925) (Herdman, 1948) (Figure 3.2). Most temperature measurements on the *Challenger* were made with Six's maximum/minimum thermometers under the erroneous assumption of a monotonic decrease of temperature with depth. A small number of Negretti and Zambra reversing thermometers were also used on the Expedition (see Rice, 2001) and these subsequently became the standard method for determining subsurface temperatures until the 1970s. Carefully calibrated reversing thermometers could determine temperature to at best about 5 millidegrees. The difference between paired thermometers (one protected against pressure effects and the other unprotected) allowed the depth of the measurement to be determined to within 10 m.

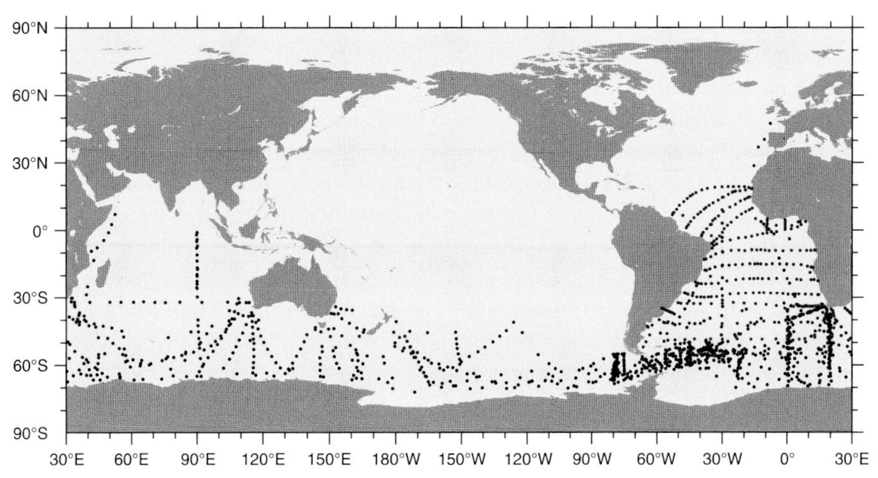

FIGURE 3.2 Stations worked during the Meteor Atlantic Expedition (red) and by the Discovery Investigations (black) (From NOAA WOD).

On *HMS Challenger*, salinity was determined by measuring density using a hydrometer and converting this value to a quantity of total dissolved solids. In 1902, the International Council for the Exploration of the Sea (ICES) was established to investigate the relationship between physical ocean properties and fisheries in the Northern Atlantic. One of ICES' earliest and most significant contributions to physical oceanography was the work of Martin Knudsen in standardizing the determination of salinity by titration against silver nitrate solution and building on the investigation of seawater chemistry by Dittmar (1884) and Forchhammer (1865). A Standard Seawater Service was established by ICES, the successor organizations of which continue to provide the internationally accepted standard (http://www.sea-technology.com/features/2011/0611/salinity.php).

Subsurface water samples were collected and thermometers deployed using strings of typically up to 12 water sampling bottles clamped to a wire and sequentially triggered to close the bottles and reverse the thermometers by weights (messengers) sliding down the wire. In deep water, multiple casts could be deployed at each station to sample the full-ocean depth but even so the deepest measurements were commonly separated by several hundred meters. Many designs of sampling bottle were used but by the 1920s, the Nansen bottle had become the generally used standard. Water samples were drawn from each bottle (occasionally with duplicates) and stored for analysis by titration (for salinity) either at sea or the end of the voyage. The Nansen bottle/reversing thermometer combination was used virtually unchanged until the 1960s/1970s when continuous profiling was introduced together with multisampler cassettes and plastic sample collection bottles (see Section 2.2).

Wüst (1935), using data from the Meteor Expedition that included dissolved oxygen (Winkler, 1888), developed the "core method" by which the spreading of subsurface water masses was inferred by tracing ocean property distributions. Velocity could not be determined in absolute terms below the ocean surface but vertical shear could be estimated by the dynamical method/geostrophy (Sandström and Helland-Hansen, 1903).

The next technological development that added substantially to the inventory of ocean observations was the mechanical bathythermograph invented by Athelstan Spilhaus in 1937 (Spilhaus, 1938) and developed further by Al Vine of the Woods Hole Oceanographic Institution during the early 1940s. This instrument enabled the thermal stratification of the upper 150 m of the ocean to be determined, first from submarines and later from underway surface ships, and made possible predictions of underwater sound propagation. The technology was simple, a bourdon tube for pressure, a bimetallic strip for temperature and the results scribed on a smoked glass slide. This instrument was eventually used systematically by the scientific community to make a major contribution to understanding the spatial variability, for example, of the Gulf Stream (Fuglister, 1963).

2.2. Second Half of Twentieth Century

The advent of solid state electronics and the use of "O"-rings to seal pressure cases had a profound effect on oceanographic instrumentation. It enabled small, battery-powered electronics to be fitted into modestly sized instruments. One of the first applications in the field of ocean physics was the development of Swallow's neutrally buoyant floats, with which he made the first, absolute measurements of deep currents[1] (Swallow, 1955) and then

1. Prior to 1955 measurements had been made by primitive recording current meters lowered from anchored ships. Such records were short and contaminated by navigational uncertainties. They could be summarized in a single one-page table (Bowden, 1954).

discovered (Crease, 1962) the first evidence of the existence of an energetic ocean dominated by mesoscale variability.

Technical development accelerated in the 1960s and early 1970s, leading to observational methods that we now regard as routine and greatly enhanced by improvements in navigation: LORAN and Decca Navigator in the 1960s followed by Transit satellite navigation in the 1970s. This meant that for the first time we knew (to an accuracy better than 1 km) where observations had been made and thus it was possible to make greatly improved estimates of surface currents. Chemical titrations for the determination of salinity were replaced by salinometers measuring electrical conductivity (Park and Burt, 1965a,b) and the traditional water bottle/reversing thermometer combination began to be replaced by continuous temperature and salinity profilers—the conductivity (salinity), temperature depth (C(S)TD) instrument. The first were inductive STDs, such as that marketed commercially by Bissett Berman, and these were followed in the early 1970s by the Neil Brown CTD that used a conductivity sensor with greater stability. Calibrations of measurements from these profilers still depended on mercury-in-glass thermometers and, in the absence of multiple samples in the vertical, were often limited to only a deep and shallow calibration point in each profile.

The limited measurements of deep ocean currents by Swallow's ship-tracked floats (lasting a few days and covering distances measured in tens of kilometers) were later enhanced by floats tracked over many months from fixed listening stations using low frequency sound propagation through the SOFAR channel. The CTD and SOFAR floats were developed specifically for use during the Mid-Ocean Dynamics Experiment conducted near Bermuda and allowed the first mapping of the ocean mesoscale velocity field (The MODE Group, 1978). Later the SOFAR float measurements were simplified by using fixed sound sources and receivers on drifting RAFOS (SOFAR spelled backwards) floats, (Rossby et al., 1986).

By the late 1960s, the mechanical bathythermograph began to be replaced by the expendable bathythermograph (XBT). The XBT measured temperature with a thermistor and estimated depth as a function of time using a fall-rate algorithm. The probe relayed its data to the deploying ship through a thin two-conductor copper wire spooled from reels on both the probe and the ship. The initial XBTs (model T4) reached approximately 450 m while later models reached greater depths. The majority of such probes were used by navies in antisubmarine operations but large-scale civilian use rapidly developed in experiments such as TRANSPAC. Such measurements greatly increased our knowledge of upper-ocean variability (e.g., Koblinsky et al., 1984; Talley and White, 1987). However, the reliance on fall-rate algorithms was later revealed to be problematic when attempts were made to merge these data with other sources in which pressure (depth) was measured directly (Wijffels et al., 2008).

Surface temperature measurements by ships were, by this time, being made using sensors inserted into the engine cooling water intake and recorded automatically. While generating more data, these measurements also introduced uncertainty due to the widely differing depths of these intakes on large and small ships, along with thermal contamination from the ships' machinery. This latter factor was later reduced by the use of hull contact sensors (Kent et al., 1991). The process of defining what is meant by "sea-surface temperature" (SST) is complex and was brought into sharp focus by the challenges of providing ground truth for satellite observations in which the measurement is of the skin temperature rather than a bulk interior value (Donlon et al., 2002). Following the advent of satellite SST measurements, these and *in situ* data were combined to produce global climatologies of which that by Reynolds (1988) was an early example. Today the GODAE high-resolution sea-surface temperature (GHRSST) consortium (www.ghrsst.org) seeks to produce and improve such climatologies.

The period of the 1960s and 1970s also saw significant advances in the measurement of subsurface ocean currents, using moored instruments. Microelectronics allowed the development of internally recording (on magnetic tape) current meters. Two current meter designs dominated the field in the west: the Aanderaa RCM4 and the Geodyne, while in the Soviet Union, the mechanical Alekseev instrument was used extensively. However, until the 1980s, individual records rarely exceeded 30 days and significant differences in instrument response depending on mooring type—with surface or subsurface buoyancy—were found. This led to the development of vector-averaging current meters that significantly reduced the wave-induced contamination of records from surface moorings. Together, these technologies allowed major experiments (Polygon (1970), Mid-Ocean Dynamics Experiment (MODE), and POLYMODE (1973–1978)) to explore and map the oceans' mesoscale variability (Freeland and Gould, 1976; The MODE Group, 1978; Kamenkovich et al., 1986).

During the 1970s first the NIMBUS and later the more accurate TIROS series of satellites started to provide global instrument tracking by measuring the Doppler shift of radio signals. The method was used to reveal the paths of surface drifters and ultimately developed into the Argos tracking system. Regional experiments used this technology: NORPAX in the North Pacific starting in 1975 (McNally et al., 1983), followed by a Gulf Stream Experiment in 1978 (Richardson, 1983) and culminating in the internationally coordinated deployment of 300 drifters in the Southern Ocean in 1978–1979 as a contribution to the First GARP (Global Atmospheric Research Project) Global Experiment (FGGE) (Garrett, 1980). While the FGGE buoys collected surface temperature and atmospheric pressure data, the quality of the near-surface velocity data suffered from the large size of the float bodies (windage), the poor performance of the "window-shade" drogues and the inability to detect without ambiguity if the drogue was still attached. (The technological developments during this era are described by Baker, 1981.)

Throughout the 1980s, various combinations of moored current meters and neutrally buoyant floats, CTD profilers, expendable probes, and surface drifters were used in a wide range of regional experiments. For example, the Tropical Ocean Global Atmosphere (TOGA) project was a concerted effort to collect data from the equatorial Pacific, combined with numerical modeling, aimed at understanding and predicting the evolution of the El Niño-Southern Ocean (ENSO) phenomenon. TOGA significantly enhanced the collection and distribution of *in situ* sea-level data, especially temperature profile data, from XBT probes, but most importantly led to the deployment of the Tropical Ocean Atmosphere (TAO, later TAO–TRITON) array of moorings measuring and reporting real-time upper ocean and atmospheric data (McPhaden et al., 1998) (Figure 3.3). The array has now expanded to cover the Atlantic and Indian Oceans (http://www.pmel.noaa.gov/tao/global/global.html). See Section 2.3.

The insights into ocean variability gained during the 1960s, 1970s, and 1980s highlighted the problems of interpreting sparsely sampled *in situ* data and of detecting long-term change in the oceans (Wunsch, 2001). It was the prospect of satellites carrying radar altimeters (as had been heralded by the brief 1978 SeaSat mission (see Chapter 4)) that led for the first time to the planning of an almost global-scale program—the World Ocean Circulation Experiment (WOCE)—that aimed to improve models of the oceans' role in climate by collecting comprehensive remotely sensed and *in situ* data (Thompson et al., 2001).

Improved understanding of the oceans' role in climate-related variability from WOCE-derived data has been well documented in the various chapters of Siedler et al. (2001) but it is worth noting that the strength of WOCE, in terms of producing a global-scale data set, is that the resulting data serves as a baseline against which past and future change may be assessed. Given this, considerable effort was made during WOCE to ensure high quality and internal consistency of the project's data sets. WOCE incorporated a number of subprograms, including ship-based hydrographic sections, western boundary current mooring arrays and XBT and surface drifter programs (the latter jointly with TOGA). In particular, the quality and comprehensiveness of temperature, salinity, and ocean chemistry data in the WOCE Hydrographic Programme (WHP) (Figure 3.4) in conjunction with the Joint Global Ocean Flux Study

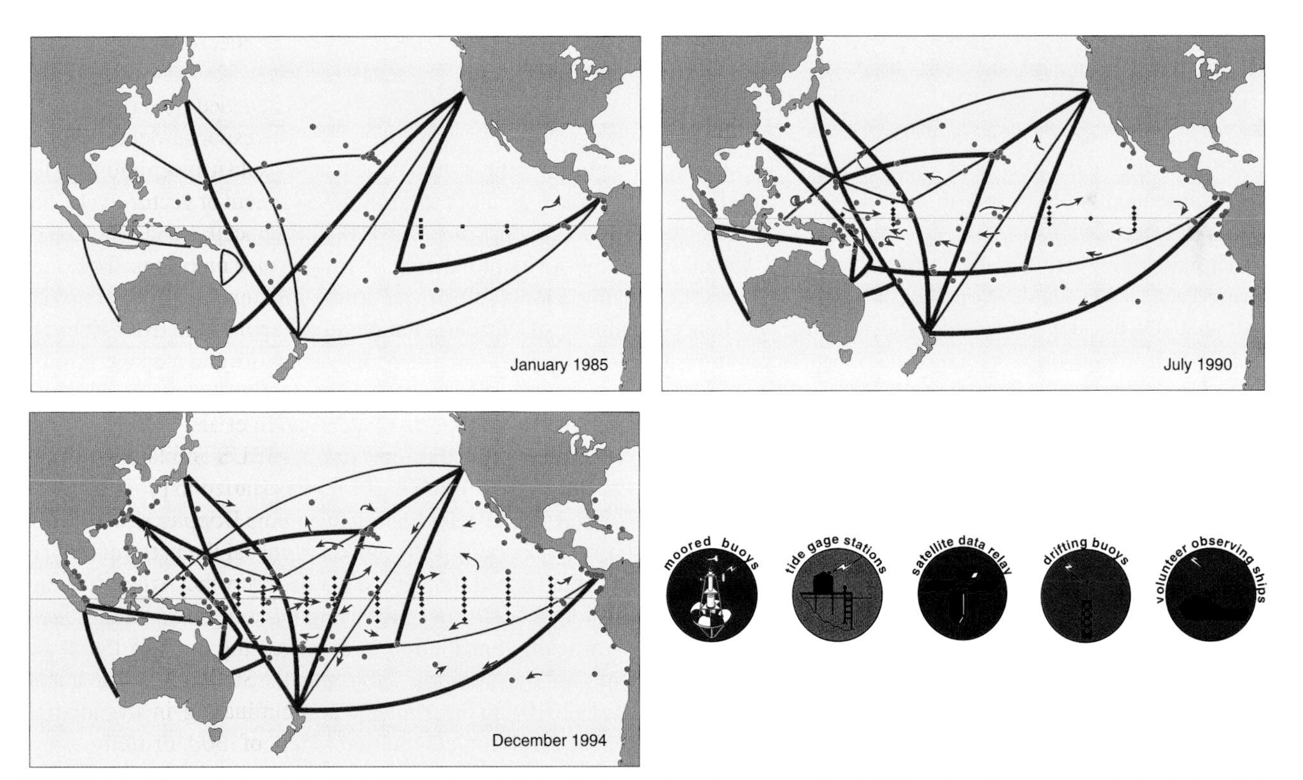

FIGURE 3.3 The evolution of the TAO–TRITON tropical Pacific Ocean observing system. (Top left) The observing system at the beginning of the TOGA program in 1985, (Top right) the evolving observing system in 1990, and (bottom) the sustained observing system that now comprises the TAO–TRITON observing system. Key: XBTs from volunteer observing ships (blue lines); Coastal tide gauges (yellow dots); Drifting buoy (curved arrows); Current meter, temperature, and salinity moorings and surface flux stations (red diamonds). *From McPhaden et al. (1998).*

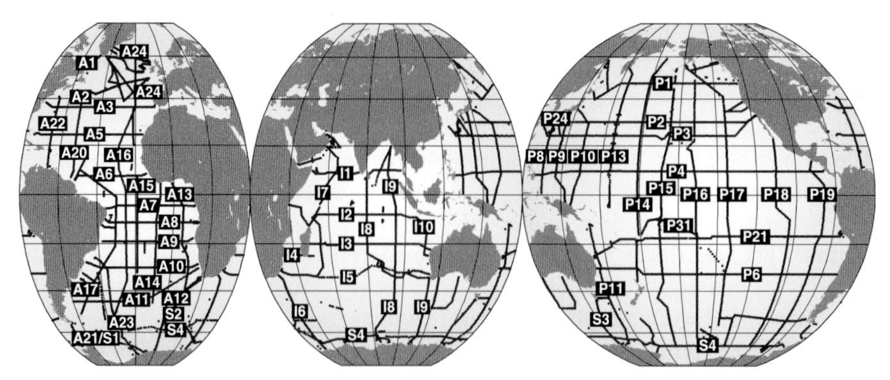

FIGURE 3.4 The WOCE one-time hydrographic sections. Each line consisted of full-depth stations at typically 100 km spacing measuring temperature, salinity, nutrients, and a suite of chemical tracers. WOCE International Project Office.

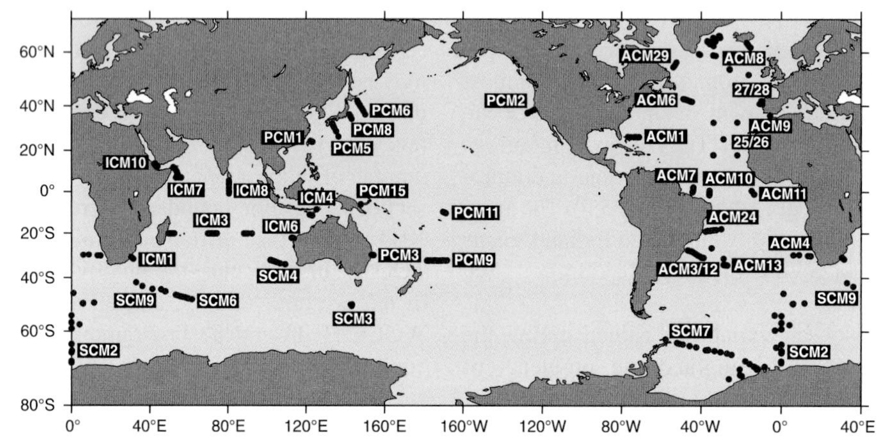

FIGURE 3.5 WOCE moored current meter arrays. Most mooring arrays were deployed for approximately 2 years. In many locations, these arrays provided the first time series of ocean velocity, temperature, and salinity. WOCE International Project Office.

(JGOFS) (King et al., 2001) was unprecedented. These data were subsequently described in a series of four atlases (Sparrow et al., 2012).

The most innovative and significant technical advance during WOCE was the development of a neutrally buoyant float that did not depend on acoustic tracking and hence could be deployed on a global scale. The Autonomous, LAgrangian Circulation Explorer (ALACE) floats (Davis et al., 1992) were ballasted to drift at depths around 1000 m and programed to surface at regular intervals by changing their buoyancy. Once at the surface, their positions were fixed by satellite (Argos tracking). Successive surfacing positions gave a measure of the time-averaged ocean currents for each drift segment. These floats allowed for the first time the collection of subsurface velocity data across entire ocean basins (Davis, 1998), (Davis and Zenk, 2001). Later in WOCE, the floats started the collection of, first, temperature and later temperature and salinity profiles, acquired as the floats rose to the surface. In 1998, the Profiling ALACE (PALACE) float was envisaged as the means of building a global array that would provide unprecedented observations of the upper 2000 m of the open oceans (Argo Science Team, 1998).

In the 1990s, the WOCE and TOGA programmes also provided an impetus for the systematic global-scale use of moorings. Since the 1980s, technology had matured to allow for deployments of full-ocean-depth moorings, carrying a range of self-recording instruments to observe variability of currents, temperature, salinity, and pressure. In WOCE, arrays of moorings across several “choke points” (geographic constrictions of important ocean flows) documented the transports of deep and shallow boundary currents and flow through passages and gaps as well as contributing to trans-basin transport estimates Figure 3.5. Retired ocean telephone cables complemented the choke point observations to observe changes in conductivity-weighted ocean transports. For example, more than 30 years of Florida Strait transport using this technique have made a significant contribution to documenting parts of the upper limb of the Atlantic subtropical gyre transport (Meinen et al., 2010). In the tropical Pacific, the TOGA moored array reached full implementation to record the state of the equatorial thermocline and provide real-time information of winds to support seasonal El Niño forecasting.

The late 1980s and early 1990s also saw the development of moored and ship-mounted Acoustic Doppler

Current Profilers (ADCPs) for the observation of ocean velocity profiles. The ADCP measures ocean currents using sound waves to detect the Doppler effect from small scattering particles in the ocean; as particles move toward or away from the sound source, the frequency of the return signal is either higher or lower. Assuming that the particles are advected by the ocean currents, the frequency shift is proportional to the speed of the current along the axis of the acoustic beam. Combining information from three or more beams allows derivation of the ocean velocity in all three coordinates. As the emitted sound travels through the water column, the ADCP measures the current at many different depths simultaneously.

The full exploitation of ship-mounted ADCPs was dependent on the arrival of another enabling technology, the Global Positioning System. This became increasingly available to civilian users during the 1990s and as well as enabling absolute positions to be determined instantaneously to meter accuracy, a ship's heading could also be determined with an accuracy far better than from previously used gyro-compasses. This provided the information needed to accurately determine the speed and direction of the ship over the ground and enabled ADCPs to be mounted on research vessels to provide underway upper-ocean velocity (to 800 m) observations. From the 1990s onward, ADCPs were also incorporated in CTD/multisampler packages, allowing velocity to be determined throughout the water column (Fischer and Visbeck, 1993; King et al., 2001).

During the 1990s, collaboration between WOCE and TOGA also led to an expansion of the collection of XBT data and significant improvements in the quality and quantity of data collected by surface drifters (see Chapter 12). As mentioned, the FGGE drifter extensively used in the 1970s did not provide high-quality surface ocean velocity data. A standardized WOCE/TOGA drifting buoy to suit observational requirements for meteorological and oceanographic applications was designed and deployed and is largely responsible for the improved data quality of Lagrangian surface measurements (Sybrandy and Niiler, 1991 and Chapter 12).

Many of the physical observations established by WOCE and TOGA were eventually subsumed into the framework of the much broader CLIVAR (Climate Variability and Predictability) project, established in 1995 as a component of the World Climate Research Program (WCRP). CLIVAR's focus on the coupled ocean-atmosphere system and its interest in a broad range of time scales (from seasonal to centennial) inevitably diluted the momentum in full-ocean-depth observations gained during WOCE. However, a much broader agenda of sustained ocean observations was developed under the auspices of GOOS (Global Ocean Observing System) responding to the need to understand the ocean's role in climate as identified by WCRP, the Global Climate Observing System (GCOS) established in 1992 and by the United Nations Framework Convention on Climate Change, UNFCC. Observations were also increasingly required to be delivered in near real time for use in operational ocean information products to guide and safeguard marine operations and deliver short-term ocean and weather forecasts. As the twentieth century closed, there was a growing recognition of the potential to build on the observational capabilities that had been established for limited-lifetime scientific experiments so as to provide a framework for an emerging sustained ocean observing system.

2.3. Twenty-First Century: Consolidation of Capabilities and Growth of Sustained Observations

The ocean observations collected during WOCE and TOGA were used to document the importance of the ocean in regional and global climate on short (days) to longer (decades and centuries) time scales (e.g., Siedler et al., 2001). Moreover, WOCE established the strong international collaborations among national research institutes and funding agencies that would be required if a coordinated ocean observing program were to be established, together with the systems needed to collate, quality control, and distribute data. Building on the success of WOCE, the aspirations of CLIVAR, and the requirements from GCOS and UNFCCC, and the continuing technological developments, in 1999 the ocean community held the first international conference solely focused on sustained ocean observations, the OceanObs'99 conference (Koblinsky and Smith, 2001). The Conference's goal was to provide the framework and to set feasible objectives for the establishment of the first decade of a sustained ocean observing system. The network of sustained ocean observations that emerged from this conference was organized primarily around observing platforms that had been developed during WOCE, TOGA, and CLIVAR.

The major observational programs that were delineated at OceanObs'99 were:

- A program called Argo that sought to establish and maintain a global-scale array of floats similar to the profiling ALACE float developed in WOCE. The array of 3000 instruments (roughly one every 300 × 300 km in the ice-free oceans deeper than 2000 m) would collect profile data (temperature and salinity) to 2000 m at nominal 10 day intervals. The data would be freely available in a real-time and in a climate-quality-controlled data set with a 6-month lag. Argo would also produce subsurface velocity estimates (The Argo Science Team, 2001; Figure 3.6).

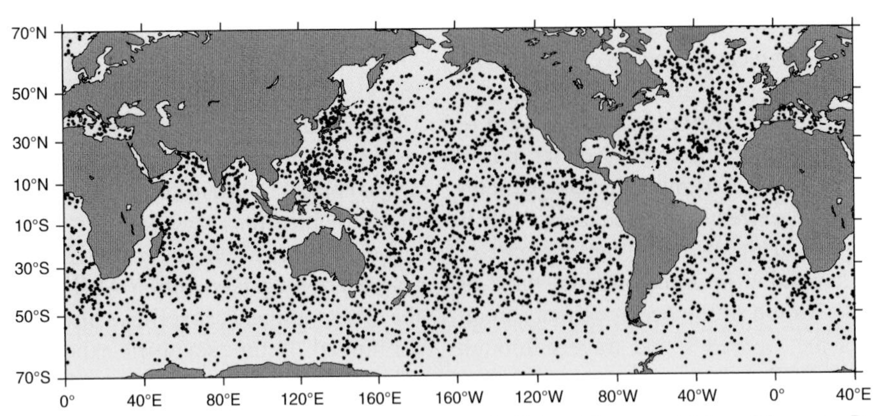

FIGURE 3.6 The array of Argo profiling floats that measure the temperature and salinity of the upper 2000 m of the ocean. Positions of floats that had delivered data within the past 30 days of March 2013. *Data from www-argo.ucsd.edu.*

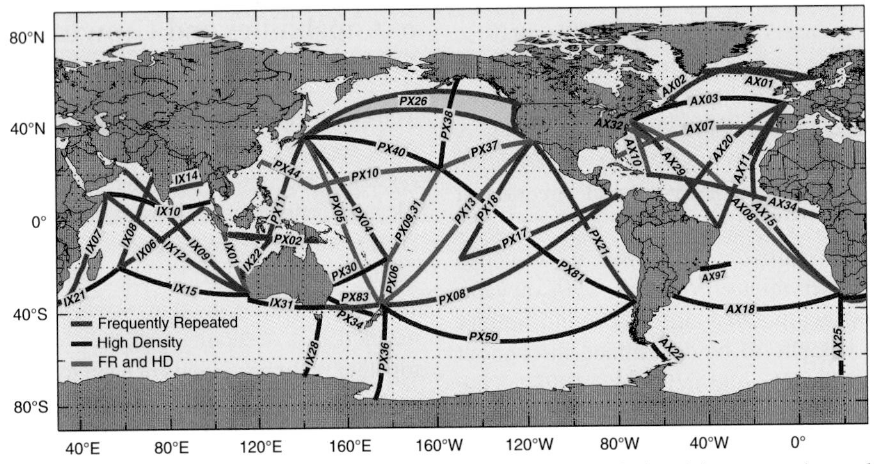

FIGURE 3.7 The present-day XBT network of transects across ocean basins. The XBTs are deployed from research vessels and ships of the Ship of Opportunity Program. The transects are sampled in two modes: HD and FR. Some transects include time series with more than 30 years of data (http://www.aoml.noaa.gov/phod/goos/xbt_network/index.php).

Argo reached its goal of 3000 operating floats by November 2007 and has since remained above that level (due in large part to the steady improvement in float operational lifetimes), thus providing continuous monitoring of the temperature, salinity, and velocity of the upper temperate and tropical oceans. The initial Argo design criterion of an array density of one active float per 300×300 km grid excluding regional seas and sea ice zones has not yet been fully achieved since some areas remain overpopulated while other regions are undersampled. The success of Argo lies not just in almost reaching design specification but in its data flow and quality control systems. Argo has successfully established a real-time data stream regardless of the national provider of the particular floats and a delayed-mode stream of climate-quality calibrated data.

- A global XBT network that would focus on high-resolution transects collecting temperature profiles in the upper 800–1000 m across ocean basins mostly using commercial vessels engaged in the Ship of Opportunity Program using semi-automatic XBT launchers. Although some XBT transects have been maintained for 30 years, the program was redesigned after the advent of Argo from a broad-scale sampling to a network with increased spatial and temporal resolution focusing on boundary and choke points currents or regions of high seasonal variability (tropical oceans) that would be complementary to the Argo global broad-scale array (Smith et al., 2001). Currently two modes of XBT transects are in operation: frequently repeated (FR), transects that are occupied 12–18 times per year with XBT deployment every 100–150 km, and high density (HD) transects occupied 4 times per year with XBT deployment every 25 km. (Figure 3.7)
- Following on from WOCE and TOGA, a Global Surface Drifter Program was established, which, in 2005, reached the design density provided by 1250 drifters (Figure 3.8). The drifters are needed to anchor satellite-based measurements of SST as a critical

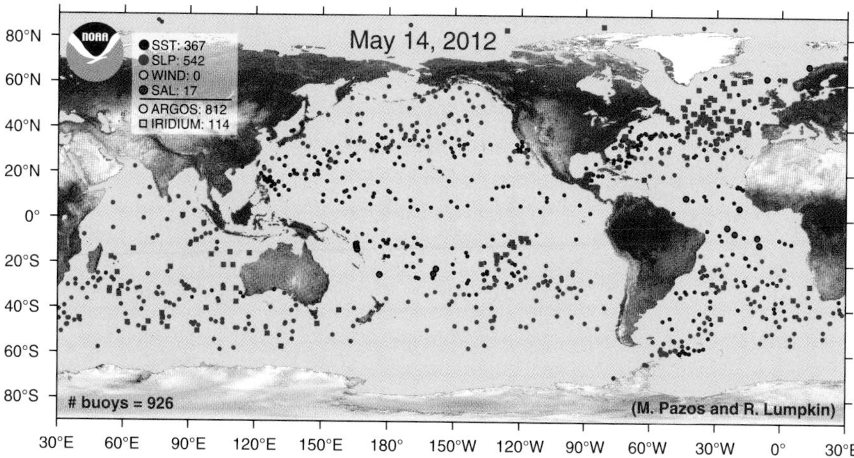

FIGURE 3.8 The global drifter array, as of August 20, 2012, that provides information of sea-surface velocity, temperature, and/or pressure. *Image from http://www.aoml.noaa.gov/phod/dac/index.php.*

component of the GHRSST (www.ghrsst.org). In addition, they are able to measure surface velocity. A subset of the drifters additionally observe atmospheric pressure and surface wind speed and direction. A dedicated data center assembles and provides uniform quality-controlled SST and surface velocity measurements. (This program and its contribution to surface current measurement are described in Chapter 12.)

– There are a small number of locations where regularly repeated observations over long time periods have given insights into physical and biogeochemical ocean variability and also into the processes that connect the upper ocean to its deep interior. Notable among these are Hydrostation S (since 1954) and the Bermuda Atlantic Timeseries (BATS) (since 1988) near Bermuda and the Hawaii Ocean Timeseries (HOT) (since 1988). Since 1980, a mooring (KIEL276) has been maintained by repeated deployments near 33°N 22°W and provides the longest record of deep currents and their variability (Siedler et al., 2005). More recent time series include the Cape Verde Ocean Observatory and the European Station for Timeseries in the Ocean since 1994 near the Canary Islands. In addition, some long-term observations were conducted from ocean weather ships that were instrumental in supporting early intercontinental air travel. The last weather ship "Mike" in the central Norwegian Sea was decommissioned in January 2010. Several of those sites have been continued with moored observatories, most notably ocean weather station "Bravo" in the central Labrador Sea and "Papa" in the Northeast Pacific. Building on these stations, and on the tropical surface mooring technology established during TOGA and the WOCE boundary current mooring arrays, OceanSITES (www.oceansites.org) (Figure 3.9) was initiated at OceanObs'99 as a network of full-depth and surface time series at key climate-relevant locations (Send et al., 2001). It has since developed into a global network of moorings at strategic locations in the ocean that measure a diverse range of ocean variables. This program now incorporates the
  - Tropical moored arrays (Pacific—TAO/TRITON, Atlantic—PIRATA, Indian-RAMA)
  - Arrays monitoring the North Atlantic overturning (MOVE, RAPID-WATCH, 16°N, 53°N, Denmark Strait, Faroe-Shetland Channel, and Fram Strait)
  - Sites documenting water mass property changes such as those mentioned in the previous paragraph and similar long-term mooring/time-series sites in the South Atlantic, Pacific, Indian, Arctic, and Southern Oceans.

  The time series from the observatories provide the means to develop accurate fields of watermass formation and transformation, as well as estimation of air–sea fluxes, and to allow quantification of the transports of major ocean current systems, and assessments of the variability of the vertical structure of the ocean and the role of eddy processes in the transport of heat and other properties.

– At OceanObs'99, Gould et al. (2001) articulated the need for continued systematic ship-based survey of the global ocean. This plan was supported by CLIVAR, GOOS, and the International Ocean Carbon Coordination Project and has since resulted in a program of hydrographic sections based on the WOCE lines and reoccupied at 5–10 year intervals. The Global Ocean Ship-based Hydrographic Investigations Program (GO-SHIP) (Hood et al., 2010) was formally established in 2007 to oversee the continuation of sustained ship-based hydrography and particularly addressed the needs of the community of scientists concerned with the ocean carbon and related chemical measurements. A global survey was completed in 2013 and

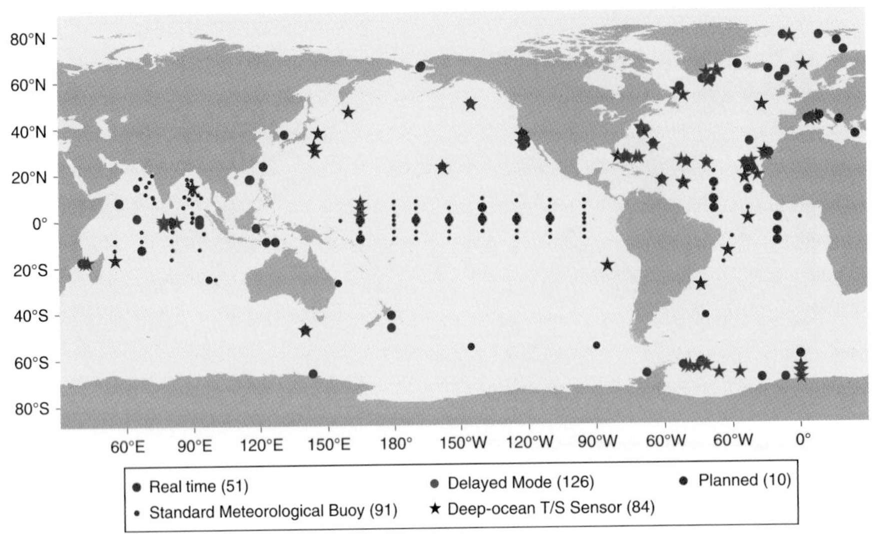

FIGURE 3.9 Map of OceanSITES time-series stations as of May 2013. OceanSITES is a worldwide system of long term, deep water reference stations measuring many variables and monitoring throughout the water column from air–sea interactions down to 5000 m. All green sites relay data in real time. The larger dots indicate high-quality air–sea flux measurements. Stars indicate sites measuring deep (below 2000 m and mostly near-sea-bed) temperature and salinity. *Image from http://www.oceansites.org/.*

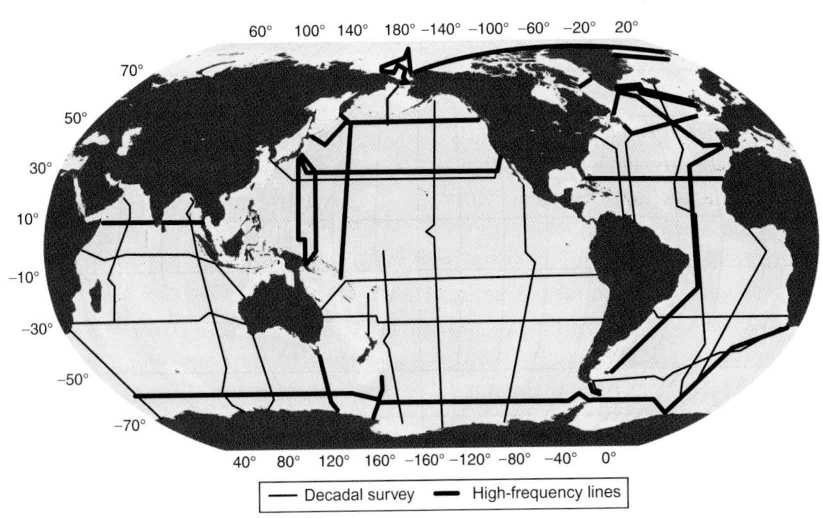

FIGURE 3.10 Repeat hydrographic sections of the GO-SHIP program. These sections are maintained by the cooperation of a number of countries with the objective of completing a global survey every 10 years. *Image from www.go-ship.org.*

reoccupation of sections are continuing. GO-SHIP (www.go-ship.org) (Figure 3.10) is developing formal international agreements for a sustained international repeat ship-based hydrography program, including an internationally agreed strategy and implementation plan; advocacy for national contributions to this strategy and participation in the global program; and providing a central forum for communication and coordination. GO-SHIP has brought up to date the manual of best practice for ship-based hydrography first produced by WOCE. The establishment of GO-SHIP recognizes that, despite numerous technological advances over the last several decades, ship-based hydrography remains the only method for obtaining the highest quality, high spatial and vertical resolution measurements of a suite of physical, chemical, and biological parameters over the full water column. Ship-based hydrography (and the sparse array of deep moorings) provide essential contributions toward documenting ocean changes throughout the water column. This is particularly important for the deep ocean below 2 km (52% of global ocean volume) that cannot yet be sampled by profiling floats although such floats are under development.

While these observing programs build on advances during the TOGA/WOCE era, two others have older roots.

- The systematic measurement of sea level extends back to the early nineteenth century and though such measurements were originally made in support of safe navigation, they now provide a valuable monitor of the consequences of changing ocean heat storage and are an essential element of the ENSO monitoring and forecasting system. These measurements have, since the early 1990s, been supplemented by systematic satellite altimetry but they remain an essential independent benchmark. (These topics are discussed in detail elsewhere in this book (Chapters 4, 15, and 27) Figure 3.11.)
- All these observing systems are focused on physical observations but ocean biogeochemistry also plays a key role in the climate system (Chapters 1, 26, 30, and 31). Though biological and chemical ocean observations have a long history, it is only in the past three decades through programs such as the JGOFS, IMBER, and SOLAS components of the International Geosphere–Biosphere Program, that these have been coordinated on a global scale and none yet match the density and global scale of physical measurements. The longest time series of observations are those made since the 1930s by the continuous plankton recorders towed by research and commercial vessels (http://www.globalcpr.org/). These data have documented the changing patterns of zooplankton distribution, many of which can be clearly linked to changes in ocean physics (Reid and Beaugrand, 2012).

In summary, over the first decade of the twenty-first century, progress in the *in situ* observing system (detailed in plenary and community white papers published from OceanObs'09) includes:

- Complete implementation of the core Argo mission, with more than 3500 (compared to a target of 3000) floats currently delivering high-quality data from the open and ice-free oceans. (It should be noted that there are a number of developments being made that enable profiling floats to collect deeper data and data from ice-covered regions—see Section 3).
- Improvement in global distribution and number of surface drifters for SST ground truthing. (1100 drifters, 470 measuring atmospheric pressure, and 700 measuring SST (March 2013)).
- Maintenance of Pacific Tropical moored array and extended coverage of the moored array into the Atlantic and Indian Oceans.
- Establishment of a growing network of multidisciplinary moored ocean observatories (OceanSITES).
- The reinvigoration of the science of SST estimation based on the synthesis of the multiple satellite platform data streams and *in situ* data; production of new and better SST products (with errors) via the Global High-Resolution SST project.
- Stabilization of high-quality sea-level observations under the Global Sea-Level Observing System (GLOSS) project.
- The transition of the global XBT network from broad-scale monitoring (taken over by Argo) to circulation monitoring via FR and HD lines with a global design.
- The success in internationally coordinated efforts to reoccupy a subset of hydrographic and tracer transects (GO-SHIP).

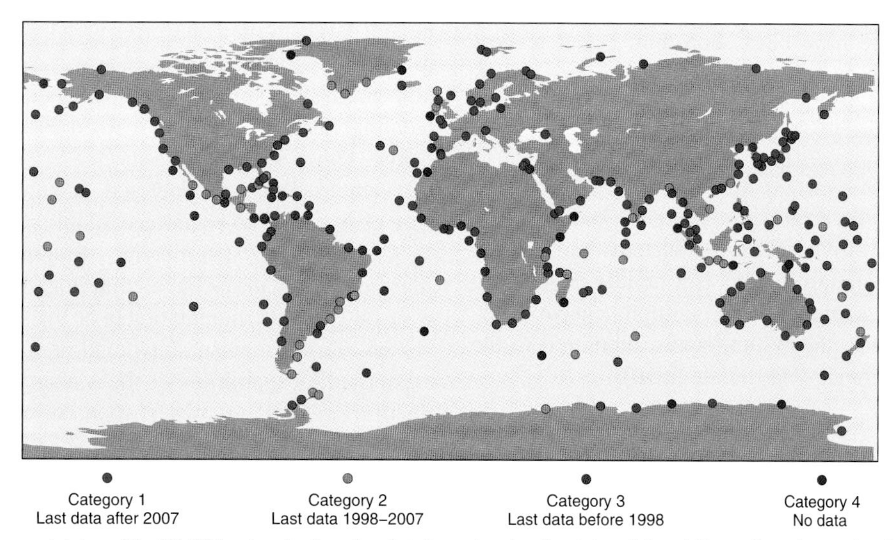

FIGURE 3.11 The present status of the GLOSS network of sea-level stations showing the status of data delivery from the sea-level sites that comprise the Global Core Network (see http://www.gloss-sealevel.org/).

3. EMERGING AND SPECIALIZED OCEAN OBSERVING TECHNOLOGIES

In the previous section, there was a focus on ocean observations within globally coordinated observing networks that can be sustained for many years. However, during WOCE, TOGA, and CLIVAR, oceanographers also developed and deployed other ocean observing systems, platforms and sensors, some of which may become part of sustained observing systems.

CLIVAR in particular identified three types of observing activities.

- It encouraged and contributed to the establishment of the aforementioned global observing networks in collaboration with GOOS, with the expectation that the data they provided would support ocean analyses and the assessment of ocean variability and change from seasons to decades. Moreover, these data would provide the basis for ocean and climate prediction.
- Second, CLIVAR sponsored several regional oceanographic experiments designed to test new observing capabilities in pilot mode. Examples are RAPID (monitoring the North Atlantic Meridional Overturning Circulation), TACE (in the tropical Atlantic), VOCALS (around South America), SPICE (Southeast Pacific), and KESS (Kuroshio extension). Each of these sites, operated for a period of 3–5 years in regions of particular interest to CLIVAR's research objectives, were instrumented with a mix of traditional and new technologies. The intention was to improve the understanding and representation of a key ocean process in models but also to develop and field-test innovative ocean observing systems that could eventually become part of the sustained ocean observing system. (RAPID is now in its nineth year of observations, for example, Johns et al., 2011, and www.noc.soton.ac.uk/rapidmoc/).
- In addition, there have been short-term ocean process experiments with ocean observations lasting from a few weeks to 2 years. These often employed observing technologies that were very likely to be too labor intensive or costly to become part of the sustained system.

Within these last two categories, there have been deployments of new, innovative, observing systems and sensors that we will now summarize. This is a rapidly developing field and the most appropriate and comprehensive references can be found in the White Papers submitted to the OceanObs'09 conference (Hall et al., 2010).

3.1. Advanced Observing Platforms

An increasing demand for ocean observations (in particular in remote areas, in winter and under extreme weather conditions), coupled with the high cost of research vessel operations have together stimulated an explosive growth of autonomous ocean observing platforms. Key climate processes occur in the Arctic and Southern Oceans; however, sustained ocean observations in multiyear sea-ice, the marginal sea ice region, and under floating ice-sheets have been both a logistic and technologic challenge. Modifications to and novel use of existing observation platforms are beginning to fill this data gap. Recent developments to Argo floats, both in software (ice detections algorithm and storage of multiple profiles) and hardware (rugged float bodies and antennae), are now extending the Argo program to the high latitude oceans (e.g., Wong and Riser, 2011). Moored profiling systems deployed on fast- and multiyear ice have also been developed to provide profiles of ocean properties under ice shelves and within the drifting ice pack (e.g., Timmermans et al., 2010). Novel use of miniature CTD systems deployed on animals, mainly seals, are now providing ocean observations in the high latitude oceans (Boehme et al., 2008).

Self-propelled ocean gliders, using the same buoyancy engines as Argo floats and, with their short wings and satellite navigation, capable of underwater navigation, are now used increasingly to routinely observe ocean property changes (e.g., http://www.ego-network.org). They can navigate the upper ocean (typically upper 1000 m) with an effective speed of less than 0.3 ms^{-1}. Most gliders have battery endurance from several weeks up to almost a year, depending on profiling depth and sensor suites. Gliders are expected soon to become part of the sustained ocean observing strategy. Other autonomous underwater vehicles (AUVs) include shallow and deep diving systems with propellers. A range of small vehicles with several hour endurance, well suited for coastal and near shore/ship applications, are available. Other systems can sample up to 6000 m and have been used to make measurements of bottom boundary layer mixing processes.

New near-surface platforms have been developed. These include large moored platforms with fast satellite communication and significant onboard power capable of supporting new multidisciplinary ocean observatories and complementing regional sea-floor cabled installations described below. Other systems use surface wave energy to propel the vehicle (Wave Glider) and unmanned sailing vessels may be expected to become available in the near future.

Consortia of commercial shipping companies and science organizations have come together to arrange for a wide range of instrumentation to be installed on commercial vessels (ferries, cruise ships, and container vessels). This expands the capabilities of the Voluntary Observing Ship fleets from previous observations of surface temperature and salinity, meteorological parameters and the deployment of expendable (XBT) probes to include surface layer chlorophyll (fluorescence), pCO_2, and upper-ocean velocities. A summary of issues relating to the use of commercial vessels can be found in the final report of OCEANSCOPE project (http://www.scor-int.org/Working_Groups/wg133.htm) cosponsored by SCOR and IAPSO.

New platforms are under development for use in the near-surface environment. Several designs for moored battery-powered winch systems are becoming available that allow sensitive sensor packages to be kept for most of the time below the euphotic and surface layers to reduce biofouling and mechanical stress and to profile to the surface when required.

A growing number of cabled sea-floor observatories will enable the deployment of multidisciplinary oceanographic sensor systems. The relatively high investment cost of sea-cables will limit the number of installations in the foreseeable future, but new sensors and the observation of changes in real time near the sea floor in particular will provide opportunities for ocean observations (see http://www.oceanobservatories.org/).

3.2. Specialized Observing Systems and Technologies

Some key ocean parameters (ocean turbulence (mixing) and biogeochemical processes) that are crucial for improved understanding of fundamental ocean dynamics and development of subgrid scale ocean parameterization in climate models cannot be adequately measured by techniques and observing platforms that are presently commonplace in the sustained observing system.

Ocean turbulence (mixing) occurs sporadically and on short time scales and small spatial scales and so require highly specialized observing platforms. Observations of turbulence near the ocean surface are difficult due to the influence of wave motions, while turbulence observations in the ocean interior are sporadic and weak (Thorpe, 2004). Nevertheless, ocean turbulence is an important parameter in both the surface and deep ocean for dispersal of phytoplankton and stratification, respectively. Direct observations of the turbulence are obtained using fast thermistors and shear probes. These instruments can be fitted to a number of different platforms: free falling (or loosely tethered) profilers; gliders; mooring; submarines; and AUV (Thorpe, 2004). Together with complementary observations of the ocean, stratification rates of diapycnal mixing can be estimated. The highly episodic nature of significant mixing events puts extra demands on the sampling and typically a large number of profiles need to be taken in a region to obtain reliable and representative estimates.

An alternative approach has been to analyze the vertical dispersion of trace elements in the ocean. A particularly successful method has been to inject an artificial inert tracer (SF_6 or SF_5CF_3) and follow its vertical and horizontal dispersion over many months to several years and large spatial scale (Ledwell et al., 1993). Several large-scale tracer release experiments have been conducted in the Subtropical North Atlantic (NATRE, SaltFinger, GUTRE), Greenland Sea, Deep Brazil Basin (BBTRE) and in the Southern Ocean (DIMES) and have provided estimates of the time and space integrated ocean mixing rates.

Because low frequency sound in the ocean can be detected over great distances and its speed is a function of ocean temperature and pressure, measuring acoustic travel times can be used to reveal the thermal structure of the ocean and its changes over time via a process known as acoustic tomography. The method uses cabled or moored arrays of sound sources and receivers to obtain tomographic images of ocean temperature distributions and their changes. High precision navigation of the moored source and receiver arrays, significant power consumption and potential impacts on marine mammals have so far limited the use of acoustic tomography in a sustained manner (Dushaw et al., 2010).

3.3. New Sensors

In general, ocean sensors are required to have low power consumption, to be compact, and to have low calibration drift. These requirements assume particular importance when deployment is made on autonomous or expendable platforms and much progress has been made in the past 20 years in improving these properties for temperature and salinity sensors. A diverse range of optical sensors are now available to measure ambient light, reflected light, and fluorescence and are also used to measure dissolved oxygen, particulate carbon, and chromophoric dissolved organic matter. A growing number of miniature wet chemistry analysis systems and even miniature mass spectrometers are becoming available.

Automatic image analysis and pattern detection will soon allow the routine identification of several types of plankton. Attempts to deploy gene-chips and highly specialized sensors will increase the range of multidisciplinary ocean observations in the coming decade. While the development of such novel biological and chemical sensors now appears unrelated to the mainstream of climate science, measurements by such sensors may eventually lead to a better understanding of the impacts of climate change on the health and productivity of the oceans.

Robotic platforms, low power electronics, and the possibility of small and capable sensors measuring a wide range of physical, chemical, and biological parameters will allow for significant enhancements of the present-day sustained ocean observations. At the same time, the improvement of ocean modeling capabilities and associated data assimilation techniques described in other chapters of this book will allow use of the full range of complementary *in situ* and remotely sensed data to provide interpolated ocean information to an increasing range of marine applications. This wider view of ocean observations was a major driver for the OceanObs'09 Conference.

4. CHANGES IN DATA VOLUME AND COVERAGE AND IMPLICATION FOR SYNTHESIS PRODUCTS

The diversity of *in situ* observations and the enormous changes over time in their quantity, spatial distribution, and precision create difficulties in producing the types of ocean syntheses that are required to assess global-scale change in the oceans and examining the significance of such changes for earth's climate.

Figure 3.12 from Durack (2011) shows the number and spatial distribution of ocean profile data (measuring salinity) in 5-year time windows from 1950 to 2010. The North Atlantic is the only ocean with reasonable data coverage throughout the entire period—a result of the proximity of many laboratories in Europe and North America with observational capability and in no small measure due to the long existence of the ICES (Went, 1972). Until the 1990s, the coverage was due in large part to the accumulation of observations made in short-term regional experiments. The WOCE survey of the ocean is clearly visible as long coast-to-coast hydrographic sections between 1990 and 2000 (the WOCE data were collected between 1990 and 1998). For many parts of the central South Pacific, Indian, and Southern Oceans, WOCE provided the first ever observations of the ocean interior. From 2000 until the present, we see the significant impact of Argo, in the first 4 years during its ramp-up phase that reached 1500 floats in 2004 and its design target of 3000 floats in 2007. Since 2007 no ice-free areas of the deep ocean have been devoid of measurements.

What the figure does not show is that there is also a distinct lack of wintertime observations, particularly at high latitude. This results from ship-based operations having been preferentially scheduled in the summer season to avoid weather-related disruptions. This is true even for the relatively well-sampled North Atlantic, where in an area bounded by 50–55°N and 35–45°W there are four times as many salinity profiles in NOAA's World Ocean Data base in April through September than in the winter months.

The evolution of the observing system has also resulted in significant changes in the depth distribution of observations (Figure 3.13, based on temperature profile measurements). The dominant data source in the early 1980s was the T4 (450 m) XBTs; these were then supplemented by 750 m (T7) probes and finally, post-2000, by Argo profiles.

While the full operation of the Argo program has led to an enormous increase in ocean observations from the sea-surface to 2000 m, the ocean below 2000 m is still poorly observed (Figure 3.14).

Here, we see that the number of T/S profiles of the upper ocean (here defined as profiles to 1000 m) has increased from typically 10,000 per decade to five times that number (and with better temporal and spatial distribution) thanks to Argo, while the situation for the deep ocean remains poor. With the exception of the decade of the 1990s when WOCE made a deliberate effort to sample the full water column, the number of TS profiles to 4000 m remains below 10,000 per decade (~3 per day!).

The fact that the data set of ocean observations is so heterogeneous (changing since the 1950s from discrete sampling levels, through continuous ship-based profiles to profiles from autonomous instruments and with large areas unsampled for long periods), with an ever-changing mix of data sources, presents enormous challenges to anyone trying to reconstruct the past state of the ocean and to unambiguously determine how that state has changed. For that reason much of what is known about ocean change has come from comparisons of reoccupations of trans-ocean sections (see, e.g., Wong et al., 2001; Bryden et al., 2005) or from the analysis of ocean time-series stations (e.g., Joyce and Robbins, 1996). Here, it is worth a note of caution. Each observing system element applies calibration and quality assurance before archival of the data. The critical analysis and interpretation of these quality assurance schemes is essential if analysis and interpretation are to be rigorous. The individual observing systems each strive to maximize data quantity and quality while delivering datasets as quickly and efficiently as is practical for each data type. System interoperability in data formats, metadata protocols, and modes of data delivery are clearly enormous advantages but they are not always achieved. There is also a need for data products, for example, gridded datasets with uncertainty estimates, in addition to the observational datasets. The documentation and characterization of products and datasets are essential, along with guidance on the suitability of datasets for a range of applications.

The synthesis and delivery of high-quality data and products for climate applications are major undertakings that have historically been underresourced. However, the production of these climate-quality observational products is vital for assessing global ocean change and variability, data assimilation model, initialization of climate and ocean only model and for the assessment of these models (see Chapters 21 and 22). As an example, calculations of the global ocean heat content and freshwater (salinity) change over the observational period (Chapters 27 and 28) are derived from the ocean temperature and salinity data. As we have noted, the temperature record is a synthesis of a number of observation platforms—ship bucket, ship-based CTD, XBT, and more recently Argo. Between 1967 and 2001, the ocean temperature profile data was mainly comprised of XBT data (56%) but since the advent in 2001 of Argo, these profiles have increasingly dominated the data record. The comparison of overlapping data sources allows the correction of instrumental shortcomings to be corrected, as, for example, with the fall rate of XBT probes (Cowley et al., 2012).

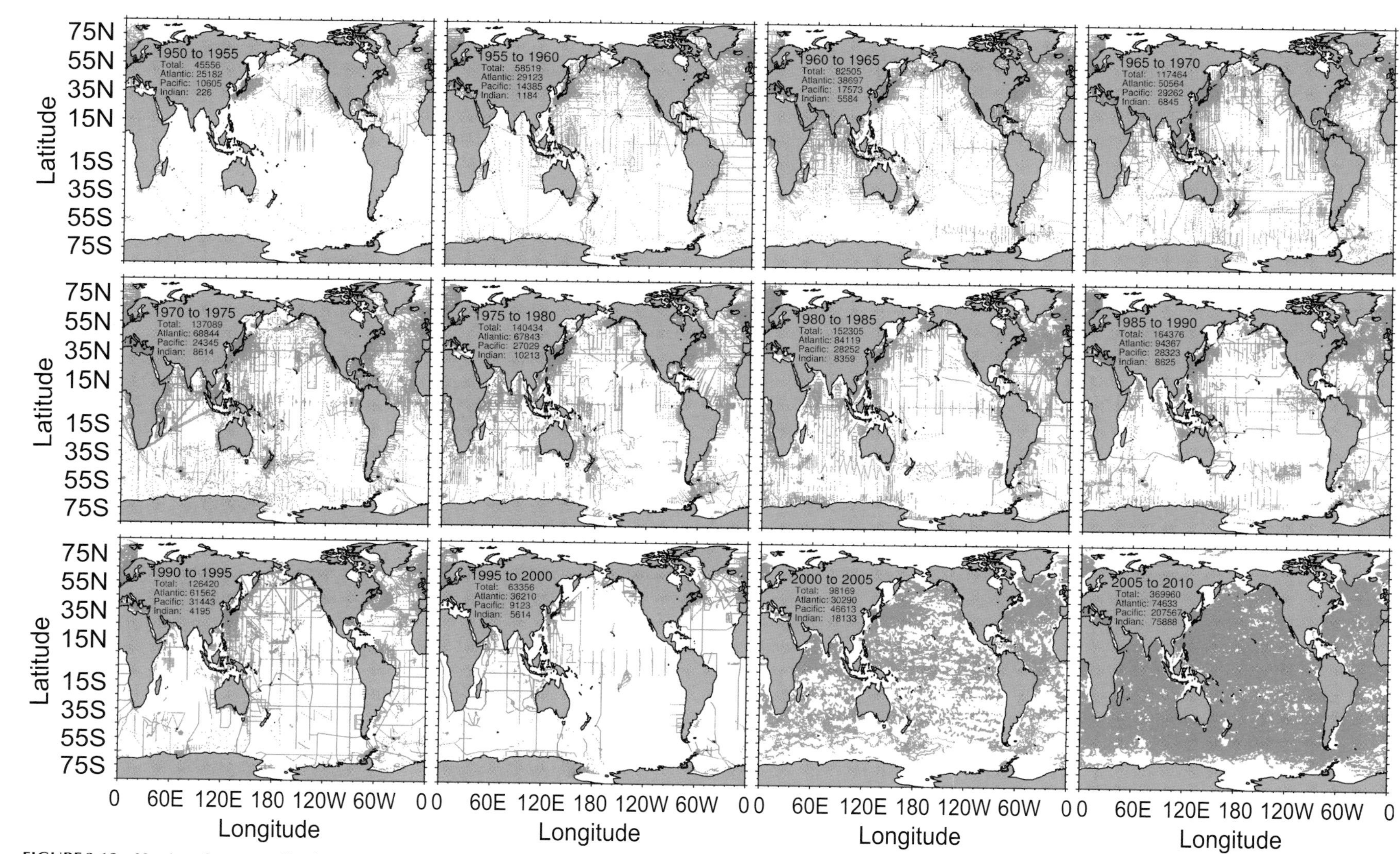

FIGURE 3.12 Number of ocean profiles from which data are available in 5-year temporal bins from 1950–1955 to 2005–2010. The increase in profiles in the past 10 years results from Argo. *From Durack (2011).*

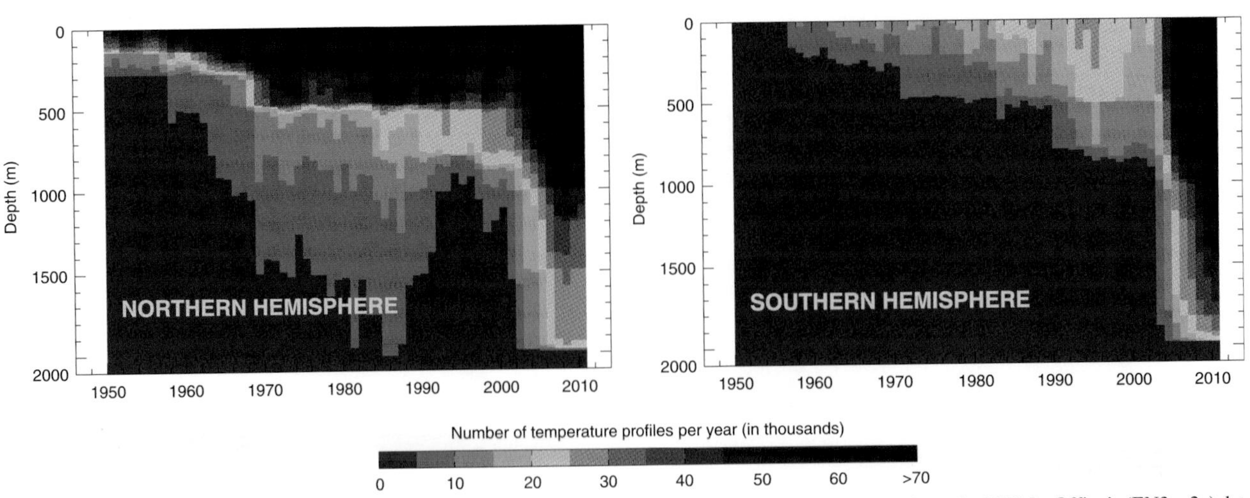

FIGURE 3.13 The global number of temperature observations per month as a function of depth, based on data from the UK Met Office's (EN3_v2a) data set (Ingleby and Huddleston, 2007).

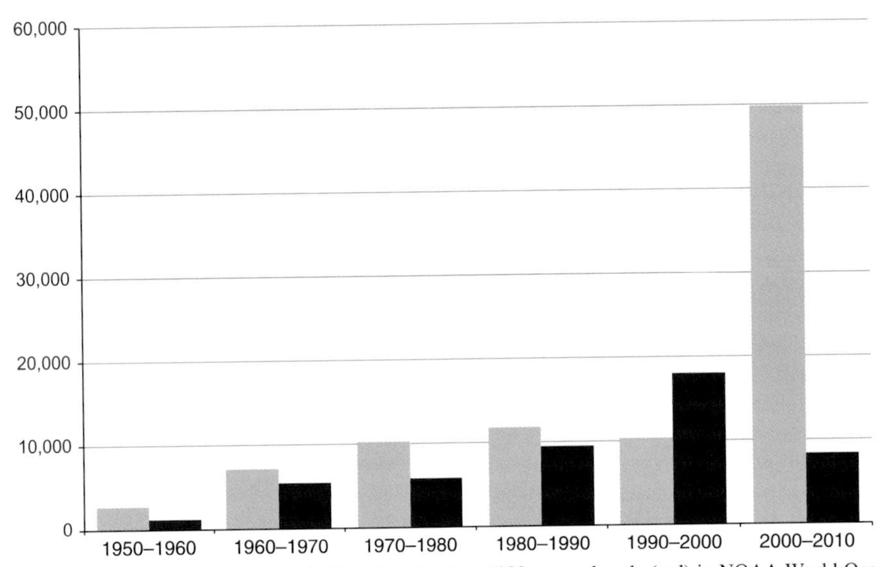

FIGURE 3.14 Number of salinity profiles per annum to 1000 m (blue) and to 4000 m per decade (red) in NOAA World Ocean Database.

Apart from the change in observation platforms, there has been an evolution of the temperature scale definitions dating back to the early 1900s (Preston-Thomas, 1990) requiring care when considering the smallest temperature changes. Similarly, the methods for the determination of salinity and the definition of salinity have evolved over the observational period (Chapter 6).

Ocean data products commonly use the World Ocean Database 09 (WOD09) (Boyer et al., 2009, http://www.nodc.noaa.gov/OC5/WOD09/pr_wod09.html) as their source of ocean observations. The WOD is an initiative of the National Oceanographic Data Center (NODC) and World Data Center for Oceanography (WDC) that collates ocean observations submitted to NODC/WDC by individual scientists and observation programs (i.e., Argo, global XBT, drifter, and GO-SHIP), and national and regional data centers. WOD09 provides a centralized quality-controlled database of all ocean observations and metadata from observational platforms. This database is used in the production of observation-based, objectively mapped ocean climatologies (e.g., World Ocean Atlas, 2009 (http://www.nodc.noaa.gov/OC5/WOA09/pr_woa09.html) and CSIRO Atlas for Regional Seas (http://www.marine.csiro.au/~dunn/cars2009/)). Ocean reanalysis models (e.g., SODA, ECCO-GECCO, and others) combine ocean observational data and an ocean general circulation model driven by known forcing (wind, surface forcing) to reconstruct an ocean climatology that is consistent with the observational record and dynamically balanced. Data assimilation methods recognize that the historical data are sparse and apply a general circulation model to reconstruct the time-evolving ocean properties and circulation. Ocean reanalyses are used for short term (seasonal) to decadal ocean forecasting efforts. Many of these products employ additional data quality control steps to further improve the reliability of the database (http://www.icdc.zmaw.de/wohp.html).

Such ocean climatologies are a great help in the documentation of temporal change since they provide baselines against which even sparse data sets may be compared. A recent example of such a comparison is that between the *Challenger* data from the 1870s and the recent Argo climatology (Roemmich et al., 2012).

5. THE FUTURE: OUTSTANDING ISSUES AND A NEW FRAMEWORK FOR GLOBAL OCEAN OBSERVING

We have explored the evolution of *in situ* ocean observations and have noted that the rapid advances of the past two decades have brought us to the point where a sustained observing system can be envisaged that largely meets the needs of a wide range of users. Earth's climate has entered a phase in which impacts are increasingly attributable to human activities (sometimes referred to as the anthropocene) and this has greatly increased the socioeconomic as well as scientific justification for ocean measurements. Ocean observations have shown that more than 90% of the extra heat energy stored by the Earth in the past 50 years is found in the oceans and that ocean salinity is a direct monitor of the global hydrological cycle (Solomon et al., 2007; Durack, 2011). It is therefore critically important that the observing system is capable of detecting and documenting global climate change so that policy makers (and the general public) can have access to climate observations and products to assess the present state of the ocean, cryosphere, atmosphere, and land and place them in context with the past. To be of both societal and scientific value, these observations need to be sustained over many decades and be of a quality that is adequate to address present-day concerns and those that may be of concern in the future.

Ocean observations are also needed to initialize and evaluate climate models and to improve predictions of climate change (Hurrell et al., 2010). Such assessments are essential for guiding national and international policies that relate to resources (such as fisheries, agriculture, and water supply) that are impacted by climate variability and change and efforts aimed at mitigating long-term climate change. The quality of the climate services provided to the public and policy makers is founded on advances in fundamental research for which comprehensive observations are also needed and these go beyond physics to include the mechanisms involved in ocean acidification, ocean deoxygenation, and the loss of marine biodiversity. All require globally consistent, regionally coordinated and systematic ocean observations of climate-relevant properties.

In order to guide the development and implementation of the ocean observing system, a framework is required that integrates the present rather independent networks described earlier—Argo, OceanSites, GO-SHIP, XBT, Global Drifter Program, Sea Level (GLOSS). Currently, these elements fall under the auspices of several programs, each with its own priorities and agendas. These are the Climate Variability and Predictability Experiment (CLIVAR) and the GOOS, part of the GCOS, which is in turn an element of the Global Earth-Observing System of Systems (GEOSS). All are fundamental elements of the Ocean Observing System (see Figure 3.15 for an artist's view).

5.1. Building on OceanObs'09

Since the OceanObs'99 conference, tremendous progress has been made toward deploying a truly global ocean observing system. Some of the elements of the *in situ* observing system have managed to reach their design specification (Argo and the Global Drifter Program), while others still have room for improvement. The status of the present ocean observing system (Figure 3.16) and community recommendations for its enhancement were reviewed by the OceanObs'09 Conference and related activities were set in place. (Hall et al., 2010, http://www.oceanobs09.net/proceedings).

Despite some significant inadequacies, the present system has allowed us to produce estimates of quantities that are essential for monitoring the oceans' role in climate (heat, freshwater and carbon storage, upper-ocean circulation and, with less confidence, changes in the ecological cycle). From these estimates and from our knowledge of the climate system, it is now possible to identify a set of essential ocean variables (EOVs) (Table 3.1). Some of them are already included in the list of the essential climate variables defined by the GCOS (World Meteorological Organisation, 2010), while others have to be confirmed and specified in detail. Once the community has agreed to a list, the decision will be for EOVs to be monitored in a sustained and consistent manner and with adequate precision for climate applications. The most cost effective network arrangements have to be determined and financial and governance arrangements have to be developed.

A key development between the two OceanObs conferences was an emphasis in 2009 on interdisciplinary ocean observations, a call for more effective international coordination, and tighter integration between observing activities and the data and information product delivery. In response to the new challenges, a systems approach to ocean observations (Figure 3.17) was developed and encapsulated in the Framework for Ocean Observing (FOO, http://www.oceanobs09.net/foo/). The FOO concept starts with societal drivers and the demands these generate for ocean observations and include:

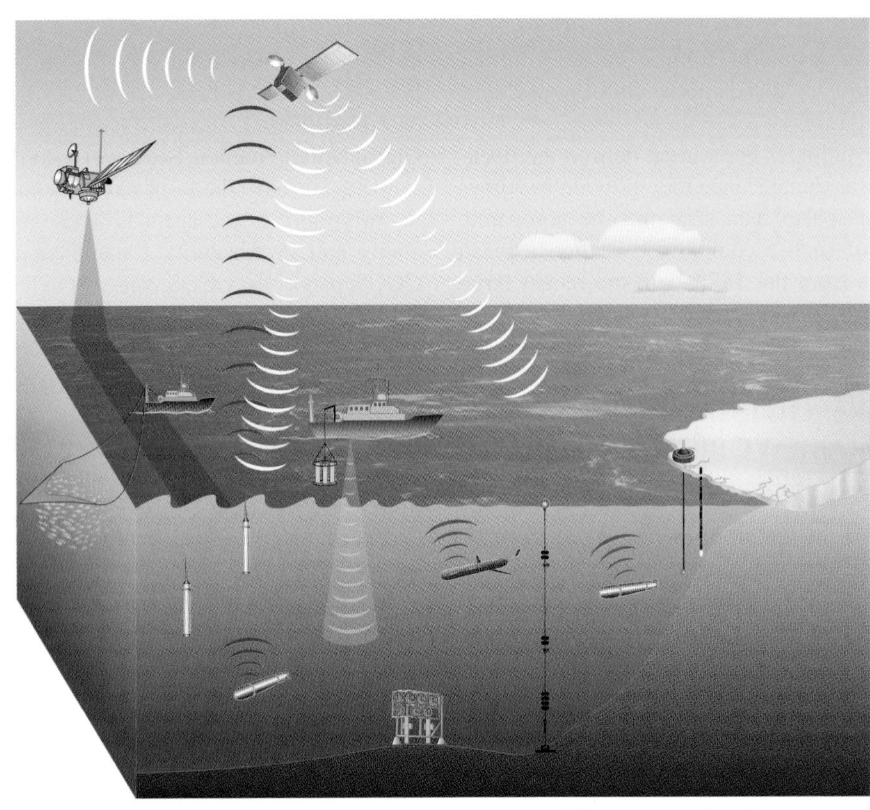

FIGURE 3.15 A picture of the integrated Ocean observing system incorporating many different observation platforms to provide the sustained high-quality routine observations of the ocean. *Image created by Louise Bell, CSIRO, Hobart, Australia.*

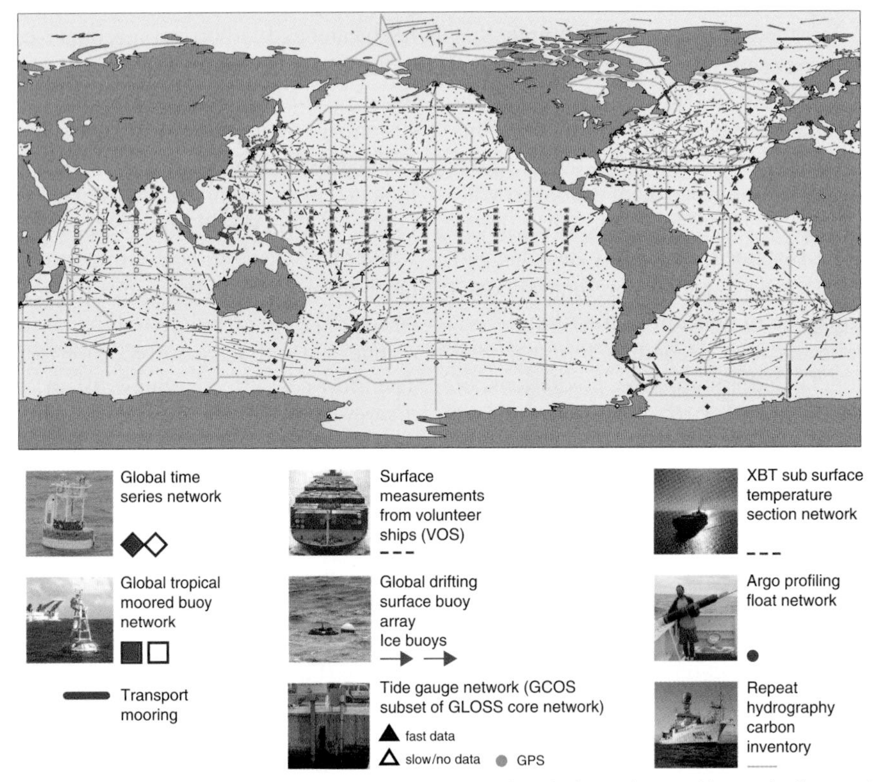

FIGURE 3.16 The global ocean observing system is comprised of numerous observing platforms that provide sustained ocean data from the sea surface to the abyssal ocean at varying temporal and spatial resolutions. This figure shows the observing system status on May 2013. JCOMM-GOOS.

TABLE 3.1 The Subset of the Essential Climate Variables That May Comprise the Essential Ocean Variables

Near-surface atmospheric variables	Ocean surface variables
Air temperature	Sea-surface temperature
Precipitation	Sea-surface salinity
Atmospheric pressure	Sea level
Surface radiation budget	Sea state
Wind speed/direction	Sea-ice coverage
Water vapor	Current speed/direction
	Ocean color (biological activity)
	Carbon dioxide (pCO_2)
	Ocean subsurface variables
	Temperature
	Salinity
	Current speed/direction
	Nutrients
	Carbon
	Ocean tracers
	Phytoplankton

Sustained observations of these variables are required for the generation of global climate products.

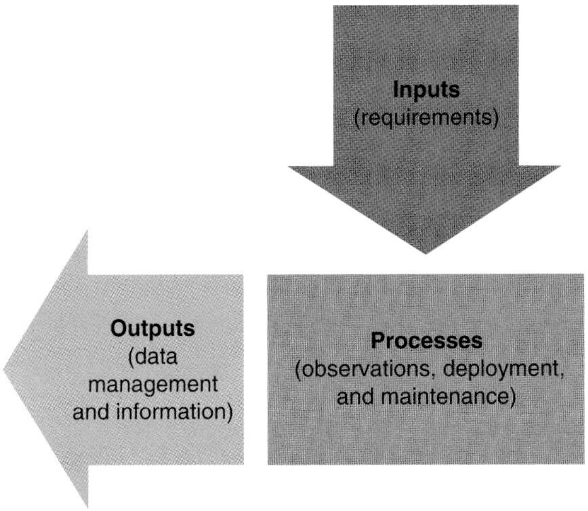

FIGURE 3.17 A systems approach to ocean observations, the guiding principle of the Framework for Ocean Observations.

- The need to document ocean change (measuring the responses to climate change, overfishing, and pollution).
- Initializing ocean models for climate predictions (e.g., El Niño, Tropical Atlantic Variability, Indian Ocean Dipole and their respective impacts on monsoon systems and decadal predictability).
- Initializing short-term ocean forecasts for marine operations (e.g., oil spill and pollution tracking, search-and-rescue).
- Regulatory matters of coastal states (e.g., Climate Change Convention, Convention of Biodiversity, Marine Spatial Planning, and associated demands).

The framework would guide the whole ocean observing community and be organized around a set of EOVs, an approach shown by GCOS to break down barriers to cooperation amongst funding agencies and observing networks.

Implementation would be guided by the level of "readiness" for immediate implementation of components that have already reached maturity, while encouraging innovation and capacity building for less mature observation streams and methods.

By taking a systems engineering approach, the FOO input requirements (observations) will be identified as the information needed to address a specific scientific problem or societal issue. The societal issues extend from short-timescale needs such as hazard warning to such long-timescale needs as knowledge of ecosystem limits appropriate to the sustainable exploitation of ocean resources. The mechanisms to deliver these observation elements will then be identified in terms of technologies and observing networks. The outputs (data and information products) will consist of the most appropriate syntheses of ocean observation streams to provide services, address scientific problems, or permit informed decisions on societal issues.

The vastness, remoteness, and harshness of the oceans mean that collecting any *in situ* observations is expensive. As a consequence, observing systems have been and will continue to be designed to measure as many variables as possible, so as to take full advantage of the limited number of observing platforms. These multiple sensors place demands on energy and thus a focus for FOO will be the avoidance of duplication between observing platforms and networks. However, the complementarity of observing networks (for instance between Argo and ship-based CTD observations) has enormous benefits in allowing intercalibration and eliminating systematic bias. Common standards for data collection and dissemination of EOV data will be adopted so as to maximize the utility of data.

The framework approach will be used to encourage partnerships between the research and operational communities so as to assess and improve the readiness levels of observation elements and data systems appropriate for each EOV. Similar partnerships will refine requirements. The framework should also enhance collaboration between developed and developing regions and promote the use of common standards and best practices.

In summary, the framework will promote a more consistent and integrated approach to the assessment of readiness, implementation, and setting standards for information sharing among the varied and largely autonomous observing elements. It should also lead to a well-defined set of requirements and goals, facilitate coordination between observing system elements and streamline implementation of sustained global-scale observations by applying a systems engineering approach and identifying best practices.

6. CONCLUSIONS

Since the early 1980s and 1990s, enormous progress has been made in gathering *in situ* ocean observations at global scales that complement information from ocean-focused earth-observing satellites. Building on global-scale research programs and stimulated by the impetus of the 1999 OceanObs Conference, the international community has implemented the first elements of what may become a sustained *in situ* global ocean observing system. Many ocean and climate analysis and forecasting centers are now dependent on measurements and derived products provided by the present observing system. This period of development has coincided with a growing awareness, not only in scientists but in some politicians and a large part of the general public of the need to monitor the oceans in order to quantify the impacts and progress of climate change and of the impacts of pollution and exploitation of the oceans' living resources.

The second OceanObs conference in 2009 built on this progress but also expanded the horizons of systematic ocean observing from its earlier largely physical and science-based perspective to one that recognized the lack of emphasis on systematic biological and biogeochemical observations and the need to underpin the case for sustained observations by reference to societal drivers.

The coming decade will see a more systematic approach to the sustained observing of the ocean than has been common in the past decades through the developing Framework for Ocean Observations. This framework has the potential to optimize the assembly of data from multiple and diverse platforms and eventually to optimize observing system design. If we continue to follow the example of the past two decades, innovative observation techniques made possible by experimental technologies will be incorporated into the mainstream of ocean observations.

Of course, any future development, expansion, and global-scale implementation of a sustained observing system will require commitment of funds from national and multinational sources and there will be competition for those funds within the ocean observing community and between the physical, biological, and solid-earth science communities. Success in competing for these funds will require a clear demonstration of the benefits, not just to the work of the ocean and climate science communities but also to societal issues such as better predicting the likely progress of sea-level rise and the oceans' role in predictions of seasonal climate.

While our ability to observe the oceans and to understand their role in the earth's climate system has advanced dramatically, national and international oversight structures and funding streams have changed much less—only a slowly growing number of governments or agencies are willing to acknowledge the need for a long-term (~decades) commitment of funds to observing programs and networks.

While these constraints may hinder progress, the dedication, persistence, and innovative nature of the observational marine science community that has already made such remarkable progress might be expected to overcome such obstacles.

The highest priority for the coming decade must be to sustain the present ocean observing system, while improving its coverage and data quality. The system can also be significantly enhanced by the following extensions of existing elements and by integration across elements:

- The sampling domain of autonomous platforms can become truly global through extensions to higher latitude, into marginal seas and the deep ocean, and through higher resolution observations in boundary current regions. Incremental technology developments and definition of new sampling requirements are needed for these extensions.
- Multidecadal ocean warming and ocean acidification have impacts on marine ecosystems with severe socio-economic consequences. Given the value of ocean ecosystems to human health and welfare, it is important to understand the links between ocean and climate variability and between marine chemical processes and their impact on marine ecosystems. Thus, there is an urgent need to *fully integrate biogeochemical and biological observations* into the ocean observing system.
- The global network measuring the physical state of the oceans provides a platform for multidisciplinary observations of biogeochemical and ecosystem impacts of climate change. Key requirements are further developments in low-power sensor accuracy and stability, and effective integration between autonomous and shipboard observational networks (e.g., definition of core variables; ensuring a sufficient quantity of reference-quality data for quality assurance of autonomous sensors).
- Improvements in the observation of the ocean surface layer and of air–sea exchanges require better utilization of research vessels and commercial shipping, improvements to automated measurement systems, better coordination across networks, and a review of sampling requirements for marine meteorology and ocean surface velocity.
- Strong commitment to preserve continuity, or in some cases, such as satellite measurements of the air–sea momentum flux from scatterometers and variations in the ocean mass field from gravity satellites, reinstate measurement missions. The principal challenge remains to advocate, plan, finance and press for executing the transition of the critical satellite sensors to sustained status.

A major effort is needed to ensure that data quality is maximized, that data access is simplified (including for data types extending across multiple observational networks), and that data products are useful and available. All measurements need to be documented, calibrated, and stored in internationally accessible data base systems. Most of the raw data from different observing networks will be combined using a range of techniques from simple statistical tools to full 4DVAR ocean data assimilation (Chapters 21 and 22) to produce ocean information products that are capable of addressing the user requirements outlined above. This information flow should be implemented by an integrated network of data information centers, data assimilating systems, and a number of routine and near real-time assessments of key ocean quantities with adequate time and space resolution.

REFERENCES

Argo Science Team, 1998. On the design and implementation of Argo: an initial plan for a global array of profiling floats. International CLIVAR Project Office Report 21, GODAE Report 5. GODAE International Project Office, Melbourne, Australia, 32pp.

Argo Science Team, 2001. Argo: The Global Array of Profiling Floats, Chapter 3.2 (pp. 248–258) in Koblinsky and Smith (2001).

Baker, D.J., 1981. Ocean Instruments and Experimental Design. In: Warren, B.A., Wunsch, C. (Eds.), Evolution of Physical Oceanography. The MIT Press, Cambridge, Massachusetts, USA, pp. 396–433 (Chapter 14).

Boehme, L., Meredith, M.P., Thorpe, S.E., Biuw, M., Fedak, M., 2008. Antarctic Circumpolar Current frontal system in the South Atlantic: Monitoring using merged Argo and animal-borne sensor data. J. Geophys. Res. Oceans 113, C09012. http://dx.doi.org/10.1029/2007JC004647.

Bowden, K.F., 1954. The direct measurement of subsurface currents in the oceans. Deep Sea Res. 2, 33–47.

Boyer, T.P., Antonov, J.I., Baranova, O.K., Garcia, H.E., Johnson, D.R., Locarnini, R.A., Mishonov, A.V., O'Brien, T.D., Seidov, D., Smolyar, I.V., Zweng, M.M., 2009. World Ocean Database 2009, Chapter 1: Introduction. In: Levitus, S. (Ed.), NOAA Atlas NESDIS 66, U.S. Gov. Printing Office, Washington, DC, 216 pp., DVD.

Bryden, H.L., Longworth, H.R., Cunningham, S., 2005. Slowing of the Atlantic meridional overturning circulation at 25°N. Nature 438, 655–657. http://dx.doi.org/10.1038/nature04385.

Cowley, R., Wijffels, S.E., Cheng, L., Boyer, T., Kizu, S., 2012. Biases in Expendable Bathythermograph Data: A New View Based on Historical Side-by-Side Comparisons. Journal of Atmospheric and Oceanic Technology 30 (6) (June 2013), 1195–1225. doi: http://dx.doi.org/10.1175/JTECH-D-12-00127.1.

Crease, J., 1962. Velocity measurements in the deep water of the western north Atlantic, summary. J. Geophys. Res. 67, 3173–3176.

Davis, R.E., 1998. Preliminary results from directly measuring mid-depth circulation in the tropical and South Pacific. J. Geophys. Res. 103, 24619–24639.

Davis, R.E., Webb, D.C., Regier, L.A., Dufour, J., 1992. The autonomous, Lagrangiam circulation explorer (ALACE). J. Atmos. Oceanic Technol. 9, 264–285.

Davis, R.E., Zenk, W., 2001. Subsurface lagrangian observations during the 1990s. In: Siedler, G., Church, J., Gould, J. (Eds.), Ocean Circulation and Climate, first ed. Academic Press, London, UK, pp. 123–139 (Chapter 3.2).

Dittmar, W., 1884. Report on the scientific results of the voyage of the H.M.S. Challenger during the years 1873–76, Sect. II, vol. 1, pt. 1–3.

Donlon, C.J., Minnett, P.J., Gentemann, C., et al., 2002. Toward improved validation of satellite sea surface skin temperature measurements for climate research. J. Clim. 15, 353–369.

Durack, P.J., 2011. Global Ocean Salinity: a climate change diagnostic. Doctor of Philosophy in Quantitative Marine Science (A Joint CSIRO and University of Tasmania PhD Program), University of Tasmania, 132pp.

Dushaw, B., et al., 2010. A global ocean acoustic observing network. In: Hall, J., Harrison, D.E., Stammer, D. (Eds.), Proceedings of OceanObs'09: Sustained Ocean Observations and Information for Society, vol. 2. Venice, Italy, 21–25 September 2009, ESA Publication WPP-306.

Fischer, J., Visbeck, M., 1993. Deep Velocity Profiling with Self-contained ADCPs. J. Atmos. Oceanic Technol. 10, 764–773.

Forchhammer, G., 1865. On the composition of sea-water in the different parts of the ocean. Philos. Trans. R. Soc. Lond. 155, 203–262.

Freeland, H.J., Gould, W.J., 1976. Objective analysis of mesoscale ocean circulation features. Deep Sea Res. 23, 915–923.

Fuglister, F.G., 1963. Gulf Stream at 60. Prog. Oceanogr. 1, 265–373.

Garrett, J.F., 1980. The availability of the FGGE drifting buoy system data set. Deep Sea Res. 27, 1083–1086.

Gould, W.J., Toole, J.M., et al., 2001. Investigating ocean climate variability: the need for systematic hydrographic observations within CLIVAR/GOOS. In: Koblinsky, C., Smith, N. (Eds.), Observing the Oceans in the 21st Century. Bureau of Meteorology, Melbourne, pp. 259–284.

Hall, J., Harrison, D.E., Stammer, D. (Eds.), 2010. Proceedings of OceanObs'09: Sustained Ocean Observations and Information for Society, Venice, Italy, 21-25 September 2009. http://dx.doi.org/10.5270/OceanObs09, ESA Publication WPP-306.

Herdman, H.F.P., 1948. Soundings taken during the Discovery investigations, 1932-39. Discov. Rep. 25, 39–106.

Hood, E.M., Sabine, C.L., Sloyan, B.M., (Eds.), 2010. The GO-SHIP Repeat Hydrography Manual: A Collection of Expert Reports and Guidelines. IOCCP Report Number 14, ICPO Publication Series Number 134. Available online at http://www.go-ship.org/HydroMan.html.

Hurrell, J., Delworth, T., Danabasoglu, G., Drange, H., Drinkwater, K., Griffies, S., Holbrook, N., Kirtman, B., Keenlyside, N., Latif, M., Marotzke, J., Murphy, J., Meehl, G., Palmer, T., Pohlmann, H., Rosati, T., Seager, R., Smith, D., Sutton, R., Timmermann, A., Trenberth, K., Tribbia, J., Visbeck, M., 2010. Decadal climate prediction: opportunities and challenges. In: Hall, J., Harrison, D.E., Stammer, D. (Eds.), Proceedings of OceanObs'09: Sustained Ocean Observations and Information for Society, vol. 2. Venice, Italy, 21–25 September 2009, ESA Publication WPP-306. http://dx.doi.org/10.5270/OceanObs09.cwp.45.

Ingleby, B., Huddleston, M., 2007. Quality control of ocean temperature and salinity profiles—historical and real-time data. J. Mar. Syst. 65, 158–175. http://dx.doi.org/10.1016/j.jmarsys.2005.11.019.

Johns, W.E., Baringer, M.O., Beal, L.M., Cunningham, S.A., Kanzow, T., Bryden, H.L., Hirschi, J.J.-M., Marotzke, J., Meinen, C.S., Shaw, B., Curry, R., 2011. Continuous, array-based estimates of Atlantic Ocean heat transport at 26.5N. J. Clim. 24 (10), 2429–2449. http://dx.doi.org/10.1175/2010JCLI3997.1.

Joyce, T.M., Robbins, P., 1996. The long-term hydrographic record at Bermuda. J. Clim. 9, 3121–3131.

Kamenkovich, V.M., Koshlyakov, M.N., Monin, A.S. (Eds.), 1986. Synoptic Eddies in the Ocean. D. Reidel, Dordrecht, Holland, 435 pp. ISBN 90-277-1925-X.

Kent, E.C., Truscott, B.S., Taylor, P.K., Hopkins, J.S., 1991. The Accuracy of ship's meteorological observations: results of the VSOP-NA. Geneva, Switzerland, World Meteorological Organisation, 86 pp (Marine Meteorology and Related Oceanographic Activities Report, (26 (WMO/TD-No. 455))).

King, B.A., Firing, E., Joyce, T.M., 2001. Shipboard observations during WOCE. In: Siedler, G., Church, J., Gould, J. (Eds.), Ocean Circulation and Climate, first ed. Academic Press, London, UK, pp. 99–122(Chapter 3.1).

Koblinsky, C.J., Smith, N.R. (Eds.), 2001. Observing the Oceans in the 21st Century—A Strategy for Global Ocean Observations. Australian Bureau of Meteorology, Melbourne, Australia, 604 pp. ISBN 0642 70618 2.

Koblinsky, C.J., Bernstein, R.L., Schmitz Jr., W.J., Niiler, P.P., 1984. Estimates of the geostrophic stream function in the Western North Pacific from XBT surveys. J. Geophys. Res. 89 (C6), 10,451–10,460. http://dx.doi.org/10.1029/JC089iC06p10451.

Ledwell, J.R., Watson, A.J., Laws, C.S., 1993. Evidence of slow mixing across the pycnocline from and open-ocean tracer-release experiment. Nature 364, 701–703.

Maury, M.F., 1855. The Physical Geography of the Sea. Harper and Brothers, New York, 280 pp.

McNally, G., Patzert, W., Kirwan, A.D., Vastano, A., 1983. The near-surface circulation of the North Pacific using satellite tracked drifting buoys. J. Geophys. Res. 0148-0227. 88 (C12). http://dx.doi.org/10.1029/0JGREA000088000C12007507000001.

McPhaden, M.J., et al., 1998. The Tropical Ocean-Global Atmosphere observing system: a decade of progress. J. Geophys. Res. 103, 14,169–14, 240.

Meinen, C.S., Baringer, M.O., Garcia, R.F., 2010. Florida Current transport variability: An analysis of annual and longer-period signals. Deep Sea Res. I 57 (7), 835–846. http://dx.doi.org/10.1016/j.dsr.2010.04.001.

Park, K., Burt, W.V., 1965a. Electrolytic conductance of sea water and the salinometer (Part1). J. Oceanogr. Soc. Jpn. 21 (2), 69–80.

Park, K., Burt, W.V., 1965b. Electrolytic conductance of sea water and the salinometer (Part2). J. Oceanogr. Soc. Jpn. 21 (3), 124–132.

Preston-Thomas, H., 1990. The International Temperature Scale of 1990 (ITS-90). Metrologia 27 (1), 3–10. http://dx.doi.org/10.1088/0026-1394/27/1/002, (2) p 107.

Reid, P.C., Beaugrand, G., 2012. Global synchrony of an accelerating risein sea surface temperature. J. Mar. Biol. Assoc. UK 92, 1435–1450.

Rennell, J., 1832. An Investigation of the Currents of the Atlantic Ocean. Rivington, London.

Reynolds, R.W., 1988. A real-time global sea surface temperature analysis. J. Clim. 1, 75–86.

Rice, A.L., 2001. The Challenger Expedition: the end of an era or a new beginning? In: Deacon, M., Rice, T., Summerhayes, C. (Eds.), Understanding the Oceans. UCL Press, London, UK, pp. 27–48, ISBN I-85728-706-1.

Richardson, P., 1983. Eddy kinetic energy in the North Atlantic from surface drifters. J. Geophys. Res. 88, 4355–4367.

Roemmich, D., Gould, W.J., Gilson, J., 2012. 135 years of global ocean warming between the Challenger expedition and the Argo Programme. Nat. Clim. Chang. 2, 425–428. http://dx.doi.org/10.1038/nclimate1461.

Rossby, T., Dorson, D., Fontaine, J., 1986. The RAFOS System. J. Atmos. Oceanic Technol. 3, 672–679.

Sandström, J.W., Helland-Hansen, B., 1903. Über die Berechnung von Meeresströmungen. Rep. Norw. Fish. Mar. Invest. 2 (4), 43.

Send, U., Weller, R., Cunningham, S., Eriksen, C., Dickey, T., Kawabe, M., Lukas, R., McCartney, M., Østerhus, S., 2001. Oceanographic timeseries observations. In: Koblinsky, C., Smith, N. (Eds.), Observing the Oceans in the 21st Century. Bureau of Meteorology, Melbourne, pp. 259–284.

Siedler, G., Church, J., Gould, J. (Eds.), 2001. Ocean Circulation & Climate—Observing and Modelling the Global Ocean. Academic Press, San Diego, International Geophysics Series Number 77. 715 pp.

Siedler, G., Armi, L., Müller, T.J., 2005. Meddies and decadal changes at the Azores Front from 1980 to 2000. Deep Sea Res. II 52 (3–4), 583–604.

Smith, N.R., Harrison, D.E., Bailey, R., Alves, O., Delcroix, T., Hanawa, K., Keeley, B., Meyers, G., Molinari, R., Roemmich, D., 2001. The upper ocean thermal network. In: Koblinsky, C., Smith, N. (Eds.), Observing the Oceans in the 21st Century. Bureau of Meteorology, Melbourne, pp. 259–284.

Solomon, S., Qin, D., Manning, M., Chen, Z., Marquis, M., Avery, K.B., Tignor, M., Miller, H.L. (Eds.), 2007. Contribution of Working Group I to the Fourth Assessment Report of the Intergovernmental Panel on Climate Change. Cambridge University Press, Cambridge, United Kingdom/New York, USA.

Sparrow, M., Chapman, P., Gould, J. (Eds.), 2012. In: The World Ocean Circulation Experiment (WOCE) Hydrographic Atlas Series, vol. 4. International WOCE Project Office, Southampton, UK, 2005–2011.

Spilhaus, A., 1938. A bathythermograph. J. Mar. Res. 1, 95–100.

Swallow, J.C., 1955. A neutral-buoyancy float for measuring deep currents. Deep Sea Res. 3, 74–81.

Sybrandy, A.L., Niiler, P.P., 1991. WOCE/TOGA Lagrangian Drifter Construction Manual. SIO Ref Series 91/6. WOCE International Project Office, WOCE Report No. 63/91, 58pp.

Talley, L.D., White, W.B., 1987. Estimates of Time and Space Scales at 300 Meters in the Midlatitude North Pacific from the TRANSPAC XBT Program. J. Phys. Oceanogr. 17, 2168–2188.

The MODE Group, 1978. The mid-Ocean dynamics experiment. Deep Sea Res. 25, 859–910.

Thompson, B.J., Crease, J., Gould, J., 2001. The origins, development and conduct of WOCE. In: Siedler, G., Church, J., Gould, J. (Eds.), Ocean Circulation and Climate. Academic Press, London, UK, pp. 31–43 (Chapter 1.3).

Thorpe, S.A., 2004. Recent developments in the study of ocean turbulence. Annu. Rev. Earth Planet. Sci. 32, 91–109. http://dx.doi.org/10.1146/annurev.earth.32.071603.152636.

Timmermans, M.-L., Krishfield, R., Laney, S., Toole, J., 2010. Ice-tethered profiler measurements of dissolved oxygen under permanent ice cover in the Arctic Ocean. J. Atmos. Oceanic Technol. 27, 1936–1949.

Went, A.E.J., 1972. Seventy Years Agrowing. A History of the International Council for the Exploration of the Sea, 1902–1972. ICES Rapp. et Proc. Verb. Réun, 165, 252.

Wijffels, S., Willis, J., Domingues, C.M., Barker, P., White, N.J., Gronell, A., Ridgway, K., Church, J.A., 2008. Changing expendable bathythermograph fall rates and their impact on estimates of thermosteric sea level rise. J. Clim. 21, 5657–5672. http://dx.doi.org/10.1175/2008JCLI2290.1.

Winkler, L., 1888. Die Bestimmung des in Wasser Gelösten Sauerstoffes. Ber. Dtsch. Chem. Ges. 21 (2), 2843–2855.

Wong, A.P.S., Riser, S.C., 2011. Profiling Float Observations of the Upper Ocean under Sea Ice off the Wilkes Land Coast of Antarctica. J. Phys. Oceanogr. 41, 1102–1115. http://dx.doi.org/10.1175/2011JPO4516.1.

Wong, A.P.S., Bindoff, N.L., Church, J.A., 2001. Freshwater and heat changes in the North and South Pacific Oceans between the 1960s and 1985–94. J. Clim. 14, 1613–1633.

World Meteorological Organisation, 2010. Implementation Plan for the Global Observing System for Climate in Support of the UNFCC (2010 Update). World Meteorological Organisation, Geneva, GCOS-138, 180 pp.

Wunsch, C., 2001. Global problems and global observations. In: Siedler, G., Church, J., Gould, J. (Eds.), Ocean Circulation and Climate. Academic Press, London, UK, (Chapter 2.1).

Wüst, G., 1935. Schichtung und Zirkulation des Atlantisches Ozeans. Das Bodenwasser und die Stratosphäre. Wissenshaftliche Ergebnisse der Deutschen Atlantik Expedition 'Meteor', 1925-1927 6, 288 pp, Berlin.

Wyville Thomson, C., Murray, J., 1885. The Voyage of H.M.S. Challenger 1873–1876. Narrative vol. I. First Part. Ch. III (Johnson Reprint Corporation, 1885); available at http://archimer.ifremer.fr/doc/00000/4751/.

Chapter 4

Remote Sensing of the Global Ocean Circulation

Lee-Lueng Fu* and Rosemary Morrow[†]

*Jet Propulsion Laboratory, California Institute of Technology, Pasadena, California, USA

[†]Laboratoire des Etudes en Géophysique et Océanographie Spatiale, Observatoire Midi-Pyrénées, Toulouse, France

Chapter Outline

1. INTRODUCTION

The planning of the World Ocean Circulation Experiment (WOCE) in the early 1980s was partly motivated by the feasibility of observing the world's oceans from space as demonstrated by Seasat. Satellite altimetry and scatterometry missions were developed as essential components of the observing system for the measurement of sea surface height (SSH) and wind, while sea surface temperature (SST) was routinely observed by various operational infrared and microwave satellite sensors. Satellite remote sensing was indeed a key contributor to the success of WOCE as reported in Siedler et al. (2001).

One of the legacies of WOCE has been the recognition of the importance of satellite remote sensing for monitoring and understanding ocean circulation and climate on a long-term basis. The past decade since the completion of WOCE has seen tremendous progress in ocean remote sensing. Precision altimetry demonstrated by TOPEX/Poseidon was succeeded by Jason-1, followed by Jason-2, and to be continued as operational missions in the future. The utility of multiple altimetry missions for enhanced spatial and temporal resolution has been exploited for revolutionary advances in the study of oceanic mesoscale eddies and operational applications. The first decade-long record of scanning scatterometry has revealed previously unknown small-scale features of the global wind field. Coupled with new microwave radiometer observations of SST, the new findings of the wind field have advanced our understanding of air–sea interaction of great importance to climate. A new technique of observing earth's gravity field and its variability from space was developed and demonstrated, leading to the capability of detecting the mass change in the ocean associated with ocean circulation as well as from melting ice on land. These advances in the study of ocean circulation and climate are reviewed in this chapter.

The topics discussed are arranged based on temporal and spatial scales in decreasing order. We begin with the ocean general circulation and we emphasize the contributions from remote sensing and its impact on the circulation at depth. The discussion of the variability of the large-scale ocean circulation is based on observations from space: SSH from altimetry, ocean mass and bottom pressure from gravity measurements, atmospheric forcing and air–sea interaction from scatterometry and radiometry. The progress made in global mean sea-level change benefits greatly

Ocean Circulation and Climate, Vol. 103. http://dx.doi.org/10.1016/B978-0-12-391851-2.00004-0

from a combination of altimetry and gravity measurements. Owing to the increasing number of altimetry missions, the advances in our understanding and processing of the data, coupled with increased sophistication in using multiple observations, the advance in the study of the oceanic mesoscale processes has been substantial over the past decade, and is discussed in detail in this review.

2. OCEAN GENERAL CIRCULATION

A major goal of oceanography is the determination of the global ocean general circulation, defined as time averaged currents of the world's oceans at all depths. It has long been recognized that satellite altimetry is a viable approach to the surface boundary condition by measuring the SSH that in combination with the knowledge of the geoid, determined from tracking satellite's motion in orbit, would lead to estimates of the surface geostrophic current velocity. There have been technical challenges to the approach: meeting the required accuracy of both the altimetry measurement and the knowledge of the geoid. Over the decade of the 1990s the TOPEX/Poseidon Mission demonstrated the centimetric accuracy of SSH measurement (Fu, 2001). However, the knowledge of the geoid was not sufficiently accurate for determining the surface circulation at scales shorter than 1000 km.

As part of WOCE, a large number of surface drifters were deployed in the global oceans to measure the surface current velocity (Niiler, 2001). The wind-driven Ekman component of the surface velocity was removed to obtain the surface geostrophic velocity (Niiler et al., 2003). Recognizing the relatively sparse coverage of the drifters, Niiler et al. (2003) used the dense coverage of the altimetry observations from multiple missions to correct for the sampling bias of the drifter observations. A mean dynamic topography of the global oceans was then constructed for estimating the surface geostrophic velocity. This result is unique in providing the details of the surface circulation at a resolution much higher than the geoid models would allow, but it might be less accurate at large scales owing to the lack of a geodetic reference. Using a technique that combined estimates from altimetry and geoid with *in situ* hydrographic and drifter observations, Rio et al. (2011) produced an estimate of global mean dynamic topography that is optimally constrained by observations at all scales (Figure 4.1). A similar approach was used by Maximenko et al. (2009) to obtain an estimate that is consistent with the dynamic topography derived from altimetry and the Gravity Recovery And Climate Experiment (GRACE) Geoid at large scales, while the information at small scales was dominated by drifter observations (the reader is referred to Chapter 12 for more discussions).

One of the objectives of geodetic satellite missions like GRACE and Gravity Field And Steady-State Ocean Circulation Explorer (GOCE) is to provide geoid models that can be used with altimetry observations for determining ocean general circulation. The geoid model derived from the GRACE mission has led to significant improvement in determining ocean circulation (e.g., Jayne, 2006) in comparison to previous results, but the improvement is limited to scales larger than 300 km (Tapley et al., 2005). The preliminary geoid models from the GOCE mission, which is more sensitive to small-scale gravity variability than GRACE, have probably improved the resolution to about 200 km (Rummel, 2013).

The mean dynamic topography produced from drifter and altimetry observations has provided a fruitful dataset for studying the ocean circulation dynamics. For example, Hughes (2005) discovered that the dynamic topography was

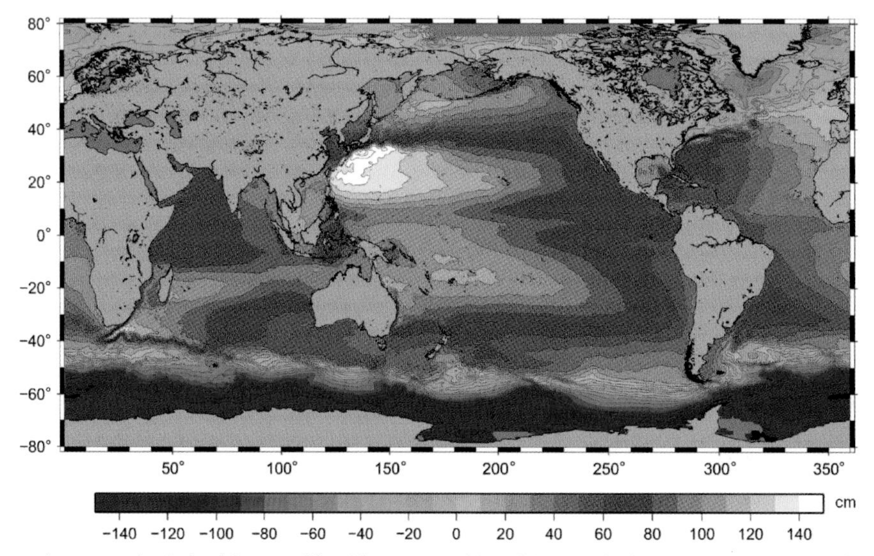

FIGURE 4.1 Mean dynamic topography derived from satellite altimetry, a geoid model, *in situ* hydrographic and surface drifter observations. *From Rio et al. (2011).*

sufficiently accurate to allow sensible calculations of highly differentiated quantities such as the relative vorticity advection. He performed analysis of the vorticity balance of the Antarctic Circumpolar Current (ACC) using the dataset and identified two modes of the mean current flow. One behaves like the stationary equivalent barotropic Rossby waves and the other exhibits divergence associated with topographic features.

A new feature of the ocean general circulation is visible only after a high-pass filter is applied to the mean dynamic topography, revealing jet-like features with small velocities of order 1 cm/s (Maximenko et al., 2008) (see Chapter 12). These features are also present in the upper ocean temperature observations as well as simulations by high-resolution ocean general circulation models (Richards et al., 2006). Multiple frontal features have also been found in the ACC as reported by Sokolov and Rintoul (2009a,b). By analyzing the gradient of altimetry-measured SSH, they were able to identify up to 12 distinct fronts across the ACC. Each front was basically associated with a given streamline associated with a contour of SSH along the entire circumpolar path. From the relationship between the ACC fronts and the SSH contours, the authors concluded that the ACC fronts had shifted southward by 60 km during the period from 1992 to 2007. By analyzing simulated virtual drifters advected by the surface velocity derived from 17 years' worth of satellite altimeter data, Thompson and Sallée (2012) found that cross-front transport of water is concentrated at the lee of major bathymetric features. They also stressed the effects of topography in breaking the circumpolar continuity of fronts and jets in the Southern Ocean.

The combination of satellite altimetry, gravity, and surface drifter observations has led to a mean dynamic topography that is probably the best that can be achieved with today's technology. Since the early 2000s, efforts like the OSCAR (Ocean Surface Current Analyses Real-time) project (Bonjean and Lagerloef, 2002) have produced routine estimates of the ocean surface current velocity. Dynamic topography based on hydrographic climatology was used with altimetry observations for estimating the surface geostrophic velocity, to which a wind-driven Ekman component was added using scatterometer wind to produce the total velocity. Recently, the new mean dynamic topography has been used for such efforts like the CTOH (Centre de Topographie des Océans et de l'Hydrosphère) project (Sudre and Morrow, 2008).

The advent of the global observations of the upper ocean density and mid-depth circulation by the Argo floats has provided an opportunity for constructing three-dimensional circulation of the upper ocean. However, the relatively sparse distribution of the floats presents a sampling issue similar to that for the surface drifter observations. Willis and Fu (2008) made an attempt to combine Argo and altimetry data to mitigate the sampling errors of the Argo observations and construct the three-dimensional geostrophic circulation of the North Atlantic Ocean north of 35° N. They applied linear regression analysis to the relationship between SSH and subsurface density and velocity fields. The dense and uniform coverage of altimetry data was exploited to make corrections to the Argo data for sampling errors caused by infrequent and irregular coverage of the floats. Shown in Figure 4.2 is the dynamic height at 1000 db averaged from 2004 to 2006. The small-scale details are striking when compared to previous results. This map is the first estimate of mid-depth dynamic topography based on well-sampled observations. The small-scale features are primarily resulting from the short duration of averaging not sufficient for removing the eddy variability. Applying this technique to the global ocean using a longer data set in the future will make a major advance in our knowledge of the upper ocean circulation.

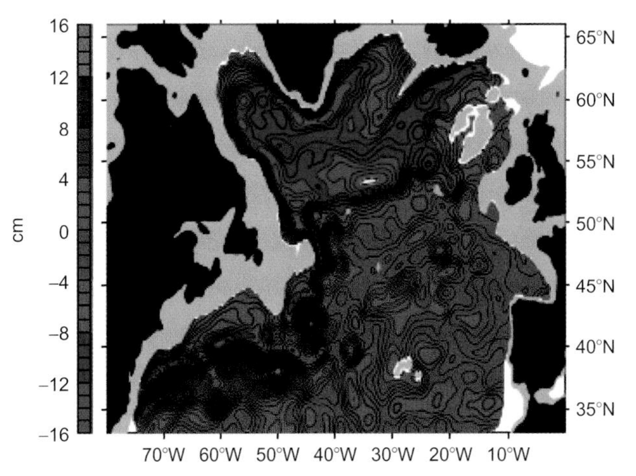

FIGURE 4.2 The dynamic height at 1000 dbar computed from Argo and altimetry observations. *From Willis and Fu (2008).*

A detailed analysis of the results of Willis and Fu (2008) by van Sebille et al. (2011) has revealed jet-like striations in the zonal current velocity at 1000 m depth as well as the vertical zonal velocity profiles in the upper ocean above 1000 m. This has further confirmed the existence of such features as noted previously in the mean dynamic topography from surface drifter and altimetry observations as well as ocean general circulation models. However, Schlax and Chelton (2008) suggested that these features might be merely the result of eddies that are not completely averaged out of the estimate of the mean circulation from a relatively short data record.

3. VARIABILITY OF THE LARGE-SCALE OCEAN CIRCULATION

The spatial and temporal variability of the large-scale ocean circulation, including the global mean sea level, has been studied from various observations: SSH from altimetry, ocean mass, and bottom pressure from gravity measurements, atmospheric forcing, and air–sea interaction from scatterometry and radiometry.

3.1. Sea Surface Height

Since the launch of TOPEX/Poseidon and ERS-1 in the early 1990s, there have been at least two satellite altimeters operating simultaneously for measuring the global SSH. These observations have provided the first opportunity to study the change of sea level and ocean circulation on time-scales longer than a decade. Displayed in Figure 4.3 is a map showing the trend of sea level estimated for the period of 1993–2009 (Willis et al., 2010, and many others referenced in that paper). This complex geographic pattern is perhaps one of the most significant recent findings in oceanography. Decadal changes in the ocean have been predicted by theories for a long time (e.g., Gill, 1982), but direct observations showing their geographic pattern rich in all scales have not been possible before.

Superimposed on the basin-scale variability expected of the decadal scales is small-scale noise exhibiting the residual effects of eddy variability, indicating that even a 17-year averaging is not sufficient to remove the eddy noise. Based on altimetric frequency spectrum of SSH, Hughes and Williams (2010) estimated the record length required to reduce the statistical errors from ocean eddies in studying low-frequency large-scale variability. Their estimates range from a decade in regions of low eddy energy to more than half a century in regions of high eddy energy. Nevertheless, new understanding of the decadal-scale change of the ocean has emerged from many studies motivated by the altimetry data.

While the globally averaged sea level measured by altimetry shows an overall rising trend at a rate about 3 mm/year (see Section 3.3), the regional trend has both positive and negative values. The global mean sea-level rise over the past 17 years is probably associated with global warming, but the regional variability to a large extent is due to natural causes. The largest regional sea-level rise takes place in the western tropical Pacific Ocean where the sea-level rise rate exceeds 10 mm/year. Much of this rise is related to the strengthening of the atmospheric Walker circulation that generates anticyclonic wind stress curls over the region. Using a simple reduced-gravity model, Timmermann et al. (2010) showed that the enhanced wind stress curl had caused Ekman flux convergence and led to the enhanced regional sea-level rise. This simple dynamic model was also adopted by Qiu and Chen (2006) in quantifying the decadal strengthening of the southern limb of the South Pacific subtropical gyre in connection with the surface wind forcing of the Southern Hemisphere westerlies. The study by Bromirski et al. (2011),

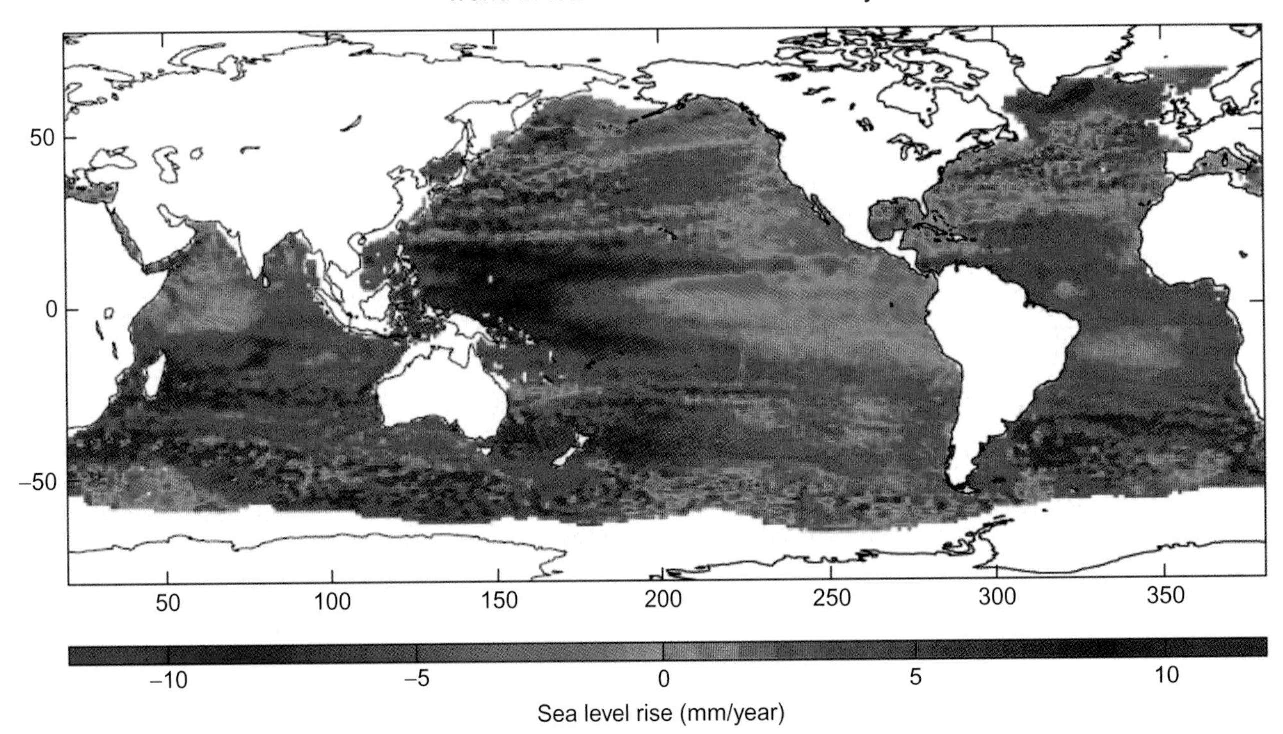

FIGURE 4.3 The linear trend of sea-level change from 1993 to 2009 estimated from satellite altimetry observations. *From Willis et al. (2010).*

based on analyzing sea level in relation to wind forcing over many decades, suggests that the minimal change along the Pacific coast of North America is caused by a long-term change of wind stress pattern.

It has been suggested that a super gyre of the Southern Ocean, connecting the subtropical gyre of the Indian Ocean with that of the South Pacific Ocean (Ridgway and Dunn, 2007; Roemmich, 2007), has been strengthening due to the increasing wind forcing from the effects of ozone depletion and global warming (Cai, 2006). This notion is consistent with Figure 4.3, showing a positive trend in the South Pacific and the Indian Ocean. The change in the South Pacific was studied in detail by Roemmich et al. (2007). The two maps shown in Figure 4.4a reveal the change in sea level over a period of nearly a decade. There is a large-scale rise of sea level. This change is apparently caused by the change in the wind forcing over the region, as shown in the map (Figure 4.4b) of the change of sea-level pressure. The increase in pressure caused a counterclockwise wind pattern that drives a convergence of surface water into the region east of New Zealand, leading to sea-level rise. This convergence is shown by the shade scale in terms of vertical motion. A negative value is associated with downward motion caused by the convergence.

The trend of sea level in the North Atlantic Ocean has been studied by Hakkinen and Rhines (2004, 2009). The North Atlantic subpolar gyre south of Greenland is characterized by a depression of SSH in association with a cyclonic (counterclockwise) circulation. This gyre is linked to the Atlantic meridional overturning circulation (AMOC) that controls the northward transport of heat in the Atlantic Ocean. The positive trend south of Greenland and the negative trend in the Gulf Stream region lead to a weakening of the North Atlantic subpolar gyre. However, the weakening trend has slowed down since 2004. Changing wind patterns are one apparent cause for the variability of the gyre. The ocean overturning circulation (i.e., the AMOC) also brings warm waters north from the subtopics, possibly contributing a negative warm-water feedback, leading to the weakening of the gyre. Together, atmospheric forcing and oceanic overturning circulation cycles can orchestrate large decadal changes in ocean temperatures and circulation, which contribute also to Atlantic Multidecadal Variability at the 50–100 year timescale.

Monitoring the AMOC has become a priority as climate simulations suggest a significant slowdown of the overturning during the next century as a result of global warming (Hu et al., 2009). Observing the AMOC is an extraordinary observational challenge because it requires measuring the net meridional transport integrated from east to west across the entire basin. Historically, full-depth hydrographic transects have been used (Bryden et al., 2005), but these provide only snapshots of the overturning.

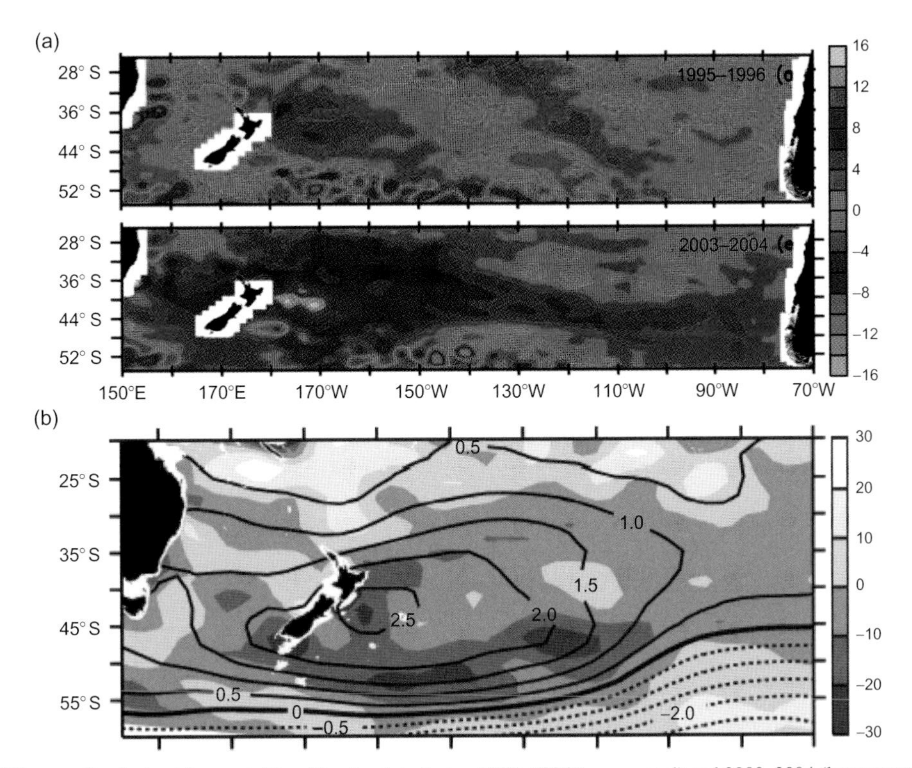

FIGURE 4.4 (a) SSH anomalies derived from satellite altimetry data during 1995–1996 (upper panel) and 2003–2004 (lower panel). (b) The change in sea-level pressure (contours, hPa) and the change in Ekman pumping (m/year) between an initial 5-year period centered on 1990 and another centered on 1999, from the NCEP–NCAR reanalysis. *From Roemmich et al. (2007).*

More recently, a dedicated mooring array has been deployed at 26.5° N, establishing a time series of overturning strength at that latitude (Cunningham et al., 2007). Efforts continue, however, to leverage the global ocean observing systems in order to supplement such dedicated measurements.

Willis (2010) showed that the AMOC could be measured using satellite altimeter observations in combination with temperature, salinity, and velocity data from the Argo array of profiling floats, for a narrow band of latitudes around 41° N. Willis showed that the overturning there experienced substantial seasonal to interannual variability, although less so than at 26.5° N. Despite predictions that the overturning will eventually slow due to global warming, Willis showed that there was no significant slowing at this latitude since 2002, and altimeter data suggest a small increase since 1992. However, in the subpolar gyre, combined altimetry and repeat hydrographic sections between Europe and Greenland show a net slowing of the upper ocean northward transport over the last 15 years (Gourcuff et al., 2011), which has an impact on the higher latitude overturning circulation. This work relied heavily on the high spatial and temporal resolution of the altimeter data to reduce the sampling error caused by small mesoscale features like eddies. Such analyses would suffer substantial increases in error without altimetry observations.

As noted earlier, a linear trend estimated from a record of limited duration is prone to statistical errors. A case in point is the region of the western tropical Pacific and Indian Oceans. If one computes trends over two different periods: 1993–2000 and 2000–2006, as reported in Lee and McPhaden (2008), one would find opposite signs in the two periods as illustrated in Figure 4.5. The wind stress measured by scatterometers (the upper two panels) provides explanations for the changes of SSH (the lower two panels). During 1993–2000, the negative trends of zonal wind stress in the Pacific imply stronger eastward trade winds, which forced higher SSH in the western Pacific. The positive trends of zonal wind stress in the Indian Ocean imply weaker trade winds, also leading to higher SSH. The reverse happened during 2000–2006. Such patterns of wind variability reflect the oscillation of the atmospheric Walker circulation as part of the large-scale air–sea interaction. The Ekman circulation resulting from the change of the wind led to more export of tropical surface water out of the western Pacific and less export out of the tropical Indian

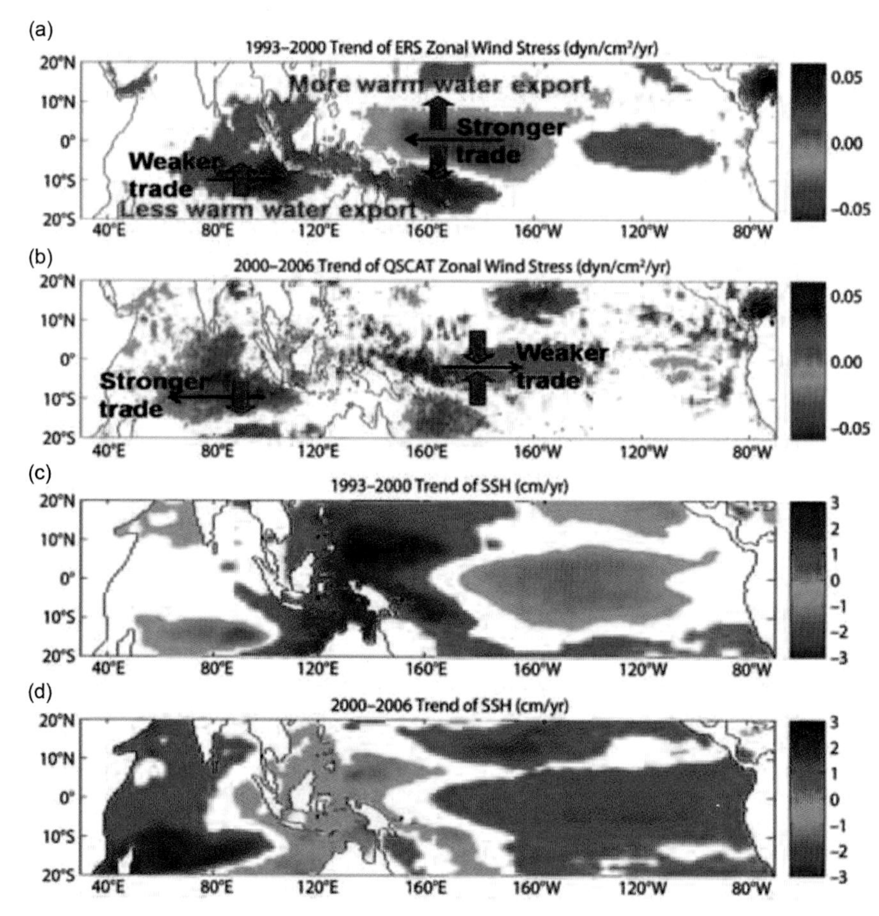

FIGURE 4.5 The linear trend of zonal wind stress (a) from 1993 to 2000 estimated from the ERS scatterometer data, (b) from 2000 to 2006 estimated from the QuikSCAT data. The linear trend of SSH estimated from satellite altimeter data (c) from 1993 to 2006, and (d) from 2000 to 2006. *From Lee and McPhaden (2008).*

Ocean during 1993–2000, with the opposite happening during 2000–2006. The combined observations from scatterometer and altimeter have provided a powerful database for understanding the mechanisms for the change of the ocean.

3.2. Ocean Mass and Bottom Pressure

A major surprise from altimetry observations in the 1990s was the large-scale rapid change of SSH in many regions of the open ocean. Through modeling studies, these SSH variations were found to be caused by the barotropic motions of the ocean (Chao and Fu, 1995; Fu and Smith, 1996; Fukumori et al., 1998). However, direct observations of such motions were difficult in part owing to their large scales. To confirm the altimetry observations of a 25-day oscillation in the Argentine Basin that were consistent with barotropic dynamics (Fu et al., 2001), three bottom pressure recorders were deployed to form a triangular array of dimension of about 400 km in the middle of the basin over a topographic ridge (Hughes et al., 2007). One of the recorders failed, but the remaining two provided interesting observations showing the existence of three forms of variability with one of them corresponding to the altimetry observations.

The advent of the GRACE Mission has opened a new door to the measurement of ocean bottom pressure by measuring the variability of earth's gravity field caused by the variations of ocean mass, which lead to variations of ocean bottom pressure (Tapley et al., 2004). Because of the limitation of spatial and temporal sampling, however, the scales resolved by GRACE are monthly and larger than about 500 km. Chambers (2006a,b) demonstrated the utility of the GRACE observations for estimating the ocean-mass-induced sea-level variations and estimated the root mean square (RMS) measurement errors to be about 2.5 cm in equivalent SSH at monthly timescales with 500-km smoothing (see also Chapter 27). Chambers and Willis (2008) showed consistency between GRACE-derived ocean bottom pressure and independent estimates from the differences between altimeter-measured SSH and Argo-derived steric component of SSH in the North Pacific Ocean on seasonal to interannual timescales. Bingham and Hughes (2006) demonstrated favorable comparisons of seasonal bottom pressure variations determined from GRACE and that from an ocean model in the North Pacific Ocean. Zlotnicki et al. (2007) and Ponte and Quinn (2009) also found favorable comparisons of the bottom pressure derived from GRACE with estimates from data-assimilative model calculations in the Southern Ocean. Ponte and Quinn (2009) also demonstrated the utility of the GRACE observations for studying mass exchange between ocean basins (also see Chambers and Willis, 2009).

The most significant impact of GRACE on the study of ocean mass is probably the determination of the contribution of ocean mass to the budget of the global mean sea-level variations to be discussed in Section 3.3. However, the synergy of GRACE and altimetry observations for ocean circulation studies is probably best demonstrated by a recent study of an anomalous warming event in the Southern Ocean (Boening et al., 2011). As reported by Lee et al. (2010), the SST over a large area (35°S–55°S, 110°W–160°W) of the Southern Ocean exceeded recorded maximum in the region by five standard deviations in late 2009. The nature of the ocean circulation during this event was revealed in Figure 4.6 (from Boening et al., 2011), showing coincidental positive anomalies in ocean bottom pressure derived from GRACE and SSH from altimetry. As shown in Figure 4.6, the SSH and ocean bottom pressure anomalies in late 2009 also reached their respective record high. The nearly same values of SSH and ocean bottom pressure indicate that the rise in SSH is primarily related to a convergence of ocean mass into the region leading to a high bottom pressure, and not caused by a warming of

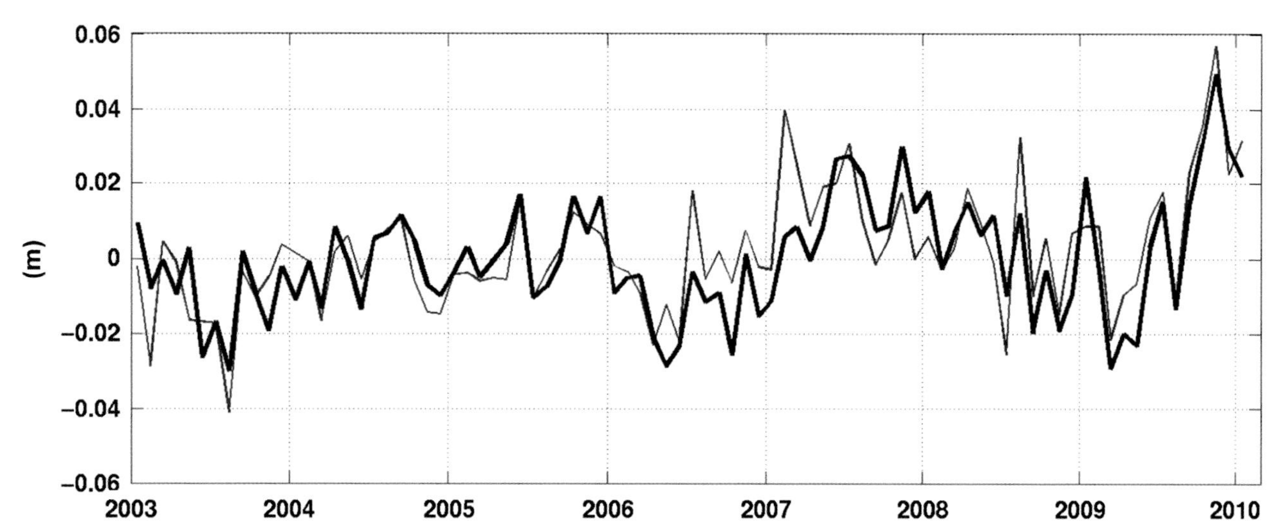

FIGURE 4.6 Time series of ocean bottom pressure (blue) and SSH (black) anomalies averaged over 90°W–140°W, 35°S–55°S. *From Boening et al. (2011).*

the water column. Simultaneous Argo float observations indicated that the SST anomalies were confined to the upper 50 m, consistent with the conclusions from the GRACE and altimetry observations.

The time series of ocean bottom pressure and altimetry SSH in Figure 4.6 indicate that the ocean variability in the region is dominated by barotropic dynamics in accordance to Chao and Fu (1995) and Fu and Smith (1996). The wind stress curl anomalies derived from scatterometer observations revealed a large-scale anticyclonic vortex over the region of the warming (not shown). Using the wind-driven barotropic vorticity equation, Boening et al. (2011) were able to simulate the observed ocean bottom pressure anomalies. The anticyclonic wind pattern was consistent with a high-pressure system, which was believed responsible for the SST anomalies due to reduced evaporative cooling and the advection of warm waters from the north to the region (Lee et al., 2010).

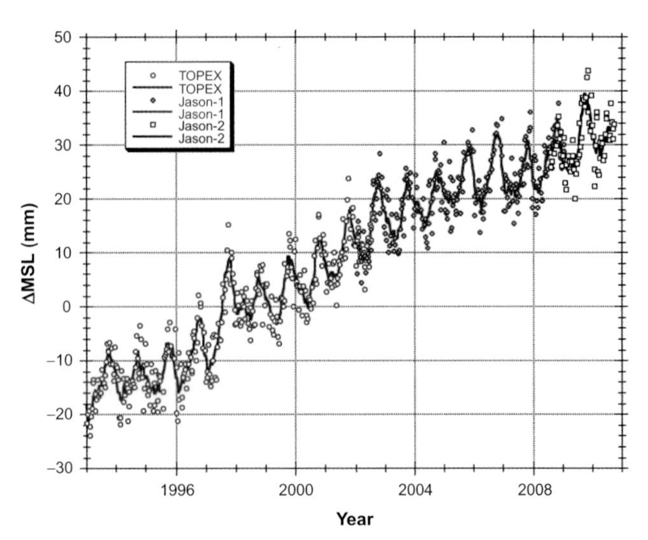

FIGURE 4.7 Global mean sea level computed from altimetry data every 10 days (dots). Blue, from TOPEX/Poseidon; red, from Jason-1; black, from Jason-2. The curves represent 60-day smoothing. *Adapted from Nerem et al. (2010).*

3.3. Global Mean Sea-Level Change (see also Chapter 27)

There has been substantial progress made in the past decade in monitoring and understanding global mean sea-level change. The capability of making this precise measurement from space has provided a unique global perspective for assessing the extent and consequence of climate change. TOPEX/POSEIDON provided the first demonstration of the feasibility of measuring the global mean sea-level change with an accuracy better than 1 mm/year from an optimally designed altimeter mission (Nerem and Mitchum, 2001). Jason-1 and OSTM/Jason-2, of a similar design, have extended the sea-level measurement from 1992 to present. Through extensive cross-calibration efforts conducted by an international science team, the time series of the global mean sea level exhibits smooth transitions over the overlap periods of adjacent missions (Figure 4.7).

The dominant signals are seasonal and interannual variations reflecting the global water cycle (Llovel et al., 2011) Note the amplification of the seasonal cycle during the 1997–1998 El Nino Southern Oscillation (ENSO) event, which also caused an anomalous increase in water vapor averaged over the global ocean derived from the microwave radiometer measurement onboard T/P (Keihm et al., 2009). These observations suggest that an ENSO event has a significant effect on the global hydrological cycle. The study by Boening et al. (2012) demonstrated that the recent slowdown of the global mean sea-level rise was caused by more water storage on land than ocean during a La Nina event. The salinity measurements from recently launched satellite missions such as SMOS (Soil Moisture and Ocean Salinity) and Aquarius hold promise to provide new information on the balance of precipitation and evaporation over the ocean in relation to the global hydrological cycle and mean sea-level change.

A clear trend of increasing global mean sea level is shown in Figure 4.7. The trend over the record from 1992 to 2010 is estimated to be 3.3 mm/year (Nerem et al., 2010) with an estimated uncertainty of 0.6 mm/year (Ablain et al., 2009). Over the past half century, the partition of the trend of sea-level rise is roughly equal between the effects of water density change and ocean mass change (Domingues et al., 2008). As the modern space and *in situ* observations become available to resolve the temporal variability, it has become clear that the rate of sea-level change and its causes vary on all timescales. Based on upper ocean temperature observations, Willis et al. (2004) estimated the rate of sea-level rise caused by ocean warming to be 1.6 mm/year from 1993 to 2003, accounting for about 2/3 of the total sea-level rise observed by altimetry over the same period.

The rate of upper ocean warming slowed down after 2004 (Lyman et al., 2010), leading to a smaller contribution of ocean warming to sea-level rise. At the same time, estimates from the GRACE observations of mass loss from melting ice sheets on Greenland and Antarctica suggest an increasing role of the addition of ocean mass in sea-level rise from 2002 to 2009 (Figure 4.8; from Velicogna, 2009). The rate of mass loss from the Greenland ice sheet increased from 137 Gt/year during 2002–2003 to 286 Gt/year during 2007–2009. A similar acceleration of melting was also observed in Antarctica. The total rate of mass loss from

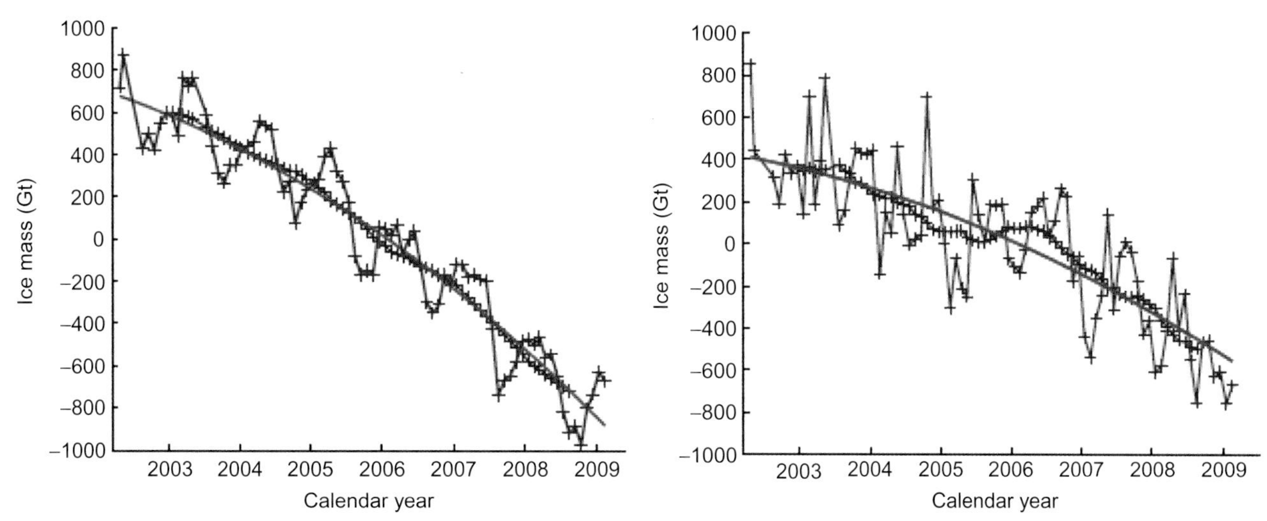

FIGURE 4.8 Time series (blue crosses and curves) of ice mass changes for the Greenland (left panel) and Antarctica (right panel) ice sheet estimated from the GRACE monthly observations from April 2002 to February 2009. The red crosses represent seasonal smoothing. The green line represents a quadratic fit. *From Velicogna (2009).*

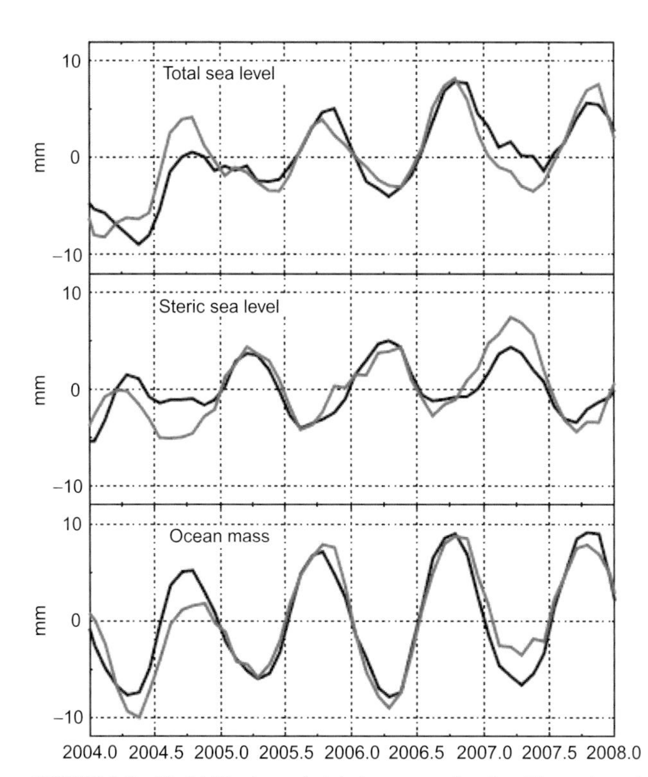

FIGURE 4.9 Variability in total global mean sea level and its steric and mass components. The black lines are the observed (top) total sea level from Jason-1, (middle) steric sea level from Argo, and (bottom) ocean mass from GRACE. The gray lines show the inferred variability computed from the other two observations, namely the total sea level computed from GRACE and Argo (top); the steric sea level computed from Jason-1 and GRACE (middle); ocean mass from Jason-1 and Argo (bottom). A 3-month boxcar smoothing is applied to each time series. *From Leuliette and Miller (2009).*

Greenland and Antarctica combined was estimated to be 532 Gt/year, or an equivalent rate of sea-level rise at 1.5 mm/year, almost half of the rate of total sea-level rise.

A major advance in understanding the variability of global mean sea level is through simultaneous analysis of altimetry, GRACE, and *in situ* observations. The global Argo float array reached its planned global coverage in 2007. The contribution of water density change in sea level has been reliably estimated globally since then. Subtracting the density contribution from the total sea-level observations from altimetry, one obtains an estimate of the contribution from the addition of ocean mass to sea-level rise (Figure 4.9; Leuliette and Miller, 2009; Willis et al., 2010).

3.4. Forcing by the Atmosphere and Air–Sea Interaction

For many decades oceanographers have relied on wind measurements from ships and buoys for studying the ocean's response to wind forcing and air–sea interaction processes. Such measurements suffer greatly from the poor sampling of the global ocean and cannot provide adequate information on the spatial and temporal variability of the ocean surface wind field. Radar scatterometry was first demonstrated by Seasat for global observations of ocean surface wind stress (Chelton et al., 1990). Similar scatterometers were flown on ERS-1, -2, and NSCAT (Quilfen et al., 1999). These were fan-beam scatterometers that provide global coverage in about 2 days for NSCAT and

5 days for ERS-1, 2 (Schlax et al., 2001). As a quick recovery of the loss of NSCAT after only 9 months' operation after launch, QuikSCAT was launched in 1999. Using a pencil beam scanning scatterometer design, the SeaWinds scatterometer on QuikSCAT has provided a decade's worth of high-quality observations of ocean surface wind with unprecedented spatial and temporal coverage (Chelton and Freilich, 2005).

Although scatterometry measures the ocean surface roughness that is directly related to wind stress, the lack of direct wind stress measurement in the ocean makes it difficult to translate the roughness measurement to wind stress. Instead, the roughness measurement is used to estimate the wind velocity at 10-m height under the condition of neutral stability in the atmospheric boundary layer above the ocean (Liu and Tang, 1996), using extensive *in situ* measurements of 10-m wind velocity and the contemporaneous stability of the boundary layer. This equivalent neutral stability wind can be readily converted to ocean surface wind stress using a typical bulk aerodynamic formula. The accuracy of QuikSCAT wind velocity observations is about 0.75 m/s in the along-wind direction and 1.50 m/s in the cross wind direction (Chelton and Freilich, 2005). The accuracy of wind direction improves with increasing wind speed. For wind speeds higher than 6 m/s, the direction accuracy is about 14° (Chelton and Freilich, 2005).

Using 8 years' worth of QuikSCAT data, Risien and Chelton (2008) created a global climatology of the ocean surface wind field. The wind observations were interpolated and smoothed onto a 0.25×0.25 grid with a resolution equivalent to 40-km block averaging. The gridded and smoothed wind stresses were vector averaged monthly to create seasonal cycle estimates using a harmonic analysis. Shown in Figure 4.10 are maps of January and July wind stress curl from the scatterometer climatology and the National Center for Environmental Prediction (NCEP) climatology (from Risien and Chelton, 2008). The seasonal comparison has revealed many small-scale features in the scatterometer maps that are missing from the NCEP maps. A striking feature is the correlation between wind stress curl and ocean currents. There are several factors causing this phenomenon.

The wind stress at the ocean surface is determined by the relative motion between the ocean and atmosphere at the

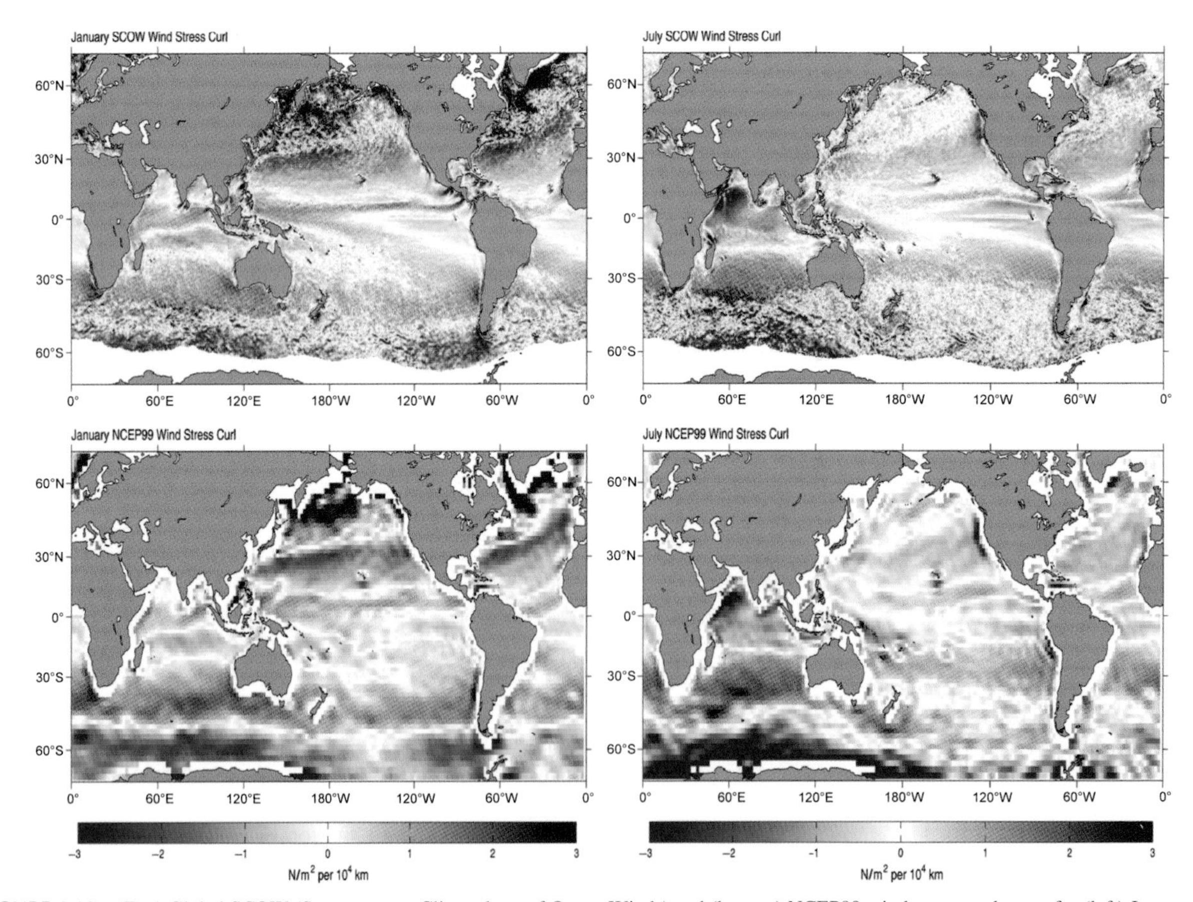

FIGURE 4.10 (Top) Global SCOW (Scatterometer Climatology of Ocean Winds) and (bottom) NCEP99 wind stress curl maps for (left) January and (right) July. The SCOW fields are plotted on a $0.25° \times 0.25°$ grid. The NCEP99 fields are plotted on a grid that has a zonally uniform spacing of 1.875° and a meridionally nonuniform spacing that varies from 1.89° at the poles to 2.1° near the equator. *From Risien and Chelton (2008).*

interface between the two fluids. The presence of strong currents like the Gulf Stream has appreciable effects on the wind stress, which is decreased in the along-current direction, creating a positive wind stress curl perturbation on the seaward side of the Gulf Stream and a negative perturbation on the landward side. This effect has been discussed in a number of studies (Cornillon and Park, 2001; Kelly et al., 2001; Chelton et al., 2004). Taking the effects of ocean currents into account of the wind work on the geostrophic circulation of the ocean, Hughes and Wilson (2008) estimated a reduction of the global energy input from the wind into ocean circulation by 20–35% from previous estimates.

Many of the small-scale features shown in Figure 4.10 are manifestations of complicated air–sea interactions that have been hypothesized (Wallace et al., 1989) and studied with sparse *in situ* observations (Hayes et al., 1989) and early generation of satellite observations with coarse resolution (Xie et al., 1998). The variability of SST, especially associated with a front across ocean currents and eddies, creates modulation of the stability of the marine atmospheric boundary layer. When wind blows over a warm ocean surface, the reduced stability and enhanced vertical mixing allow the upper level winds to exert more influence on surface wind and increase its speed. When wind blows over a cold ocean surface, the increased stability and reduced vertical mixing inhibit the upper level wind influence on the surface wind and lead to a decrease in surface wind speed. These mechanisms have recently been reviewed by Chelton and Xie (2010). Although there are more complicated processes involved in the interaction (e.g., Small et al., 2008; O'Neill et al., 2010), the simple scenario of an interplay between SST and surface wind provided by Chelton and Xie (2010) is quite useful. Divergence (convergence) of wind stress usually takes place when wind blows over a down wind gradient of increasing (decreasing) SST, while positive (negative) curl of wind stress usually takes place when wind blows over a cross-wind gradient of SST with the warmer water on the right (left) side of the wind direction.

A prominent feature in Figure 4.10 is the narrow zonal band of positive wind stress curl just north of the equator in the eastern Pacific as revealed by the QuikSCAT observations, but it is less well defined in the NCEP analysis. This narrow band of positive wind stress curl is caused by the northwestward trade winds blowing over the transition between the eastern equatorial Pacific cold tongue and the warm water to the north. The horizontal shear in the westward component of the wind creates a positive curl of the wind stress field. In fact, the interaction between SST and ocean surface wind in the tropical instability waves in the eastern tropical Pacific was one of the first foci of studies using a combination of the QuikSCAT observations with microwave SST observations from the TRMM Microwave Imager (TMI) (Liu et al., 2000; Chelton et al., 2001; Polito et al., 2001; Hashizume et al., 2002).

Another important feature in Figure 4.10 are the narrow bands of wind stress curl along the eastern boundary currents off the west coasts of South Africa, South America, Australia, and North America in the QuikSCAT observation, which are poorly resolved or completely missing in the NCEP analysis. The negative curl in the Southern Hemisphere is caused by the poleward along-shore wind blowing over the cold upwelled water. The same process in the Northern Hemisphere creates positive curl because the cold upwelled water is to the left of the wind as opposed to the right of the wind in the Southern Hemisphere. The band of negative wind stress curl around the southern tip of Greenland is also missing in the NCEP analysis. Another notable missing feature in the NCEP analysis is the dipolar curl associated with the strong winds through the mountain channels in Central America in winter. These winds are important energy source for the energetic ocean eddies propagating westward from the coast into the interior subtropical Pacific Ocean.

The study of the interaction between SST and ocean surface wind was primarily made possible by the advent of microwave SST observations. The prevalence of cloud cover over most part of the global oceans has prevented systematic analysis of the infrared SST observations with sufficient statistical basis. TMI, launched in 1997 as part of the payload of the Tropical Rainfall Measuring Mission (TRMM), was the first microwave SST sensor after the long hiatus since the Seasat Scanning Multichannel Microwave Radiometer (SMMR) that had demonstrated the feasibility of microwave SST measurement. However, the frequency channel of TMI used for SST estimation (10.7 GHz) was not optimized for SST. Furthermore the inclination of the TRMM orbit (35°) has also limited the utility of TMI for global studies. The flight of Advanced Microwave Scanning Radiometer for EOS (AMSR-E) onboard the Aqua satellite, launched in 2003, has provided the data base for the recent surge of studies of the relationship between SST and ocean surface wind reviewed by Chelton and Xie (2010). The polar orbit of Aqua and the more optimal frequency (6.9 GHz) of AMSR-E for SST estimation has remedied the two deficiencies of TMI mentioned earlier. Global all-weather observations of SST are available from AMSR-E every 2 days at a resolution of about 50 km, and accuracy of about 0.4 °C (Chelton and Wentz, 2005).

Figure 4.11 shows the comparison of AMSR-E SST observations with three other products from infrared observations and blended *in situ* observations in the Kuroshio Extension region of the western North Pacific Ocean. All available data were averaged over a 3-day period of 9–11 June 2003. The persistent cloud cover during the 3-day period has rendered the Advanced Very High Resolution Radiometer (AVHRR) image of limited

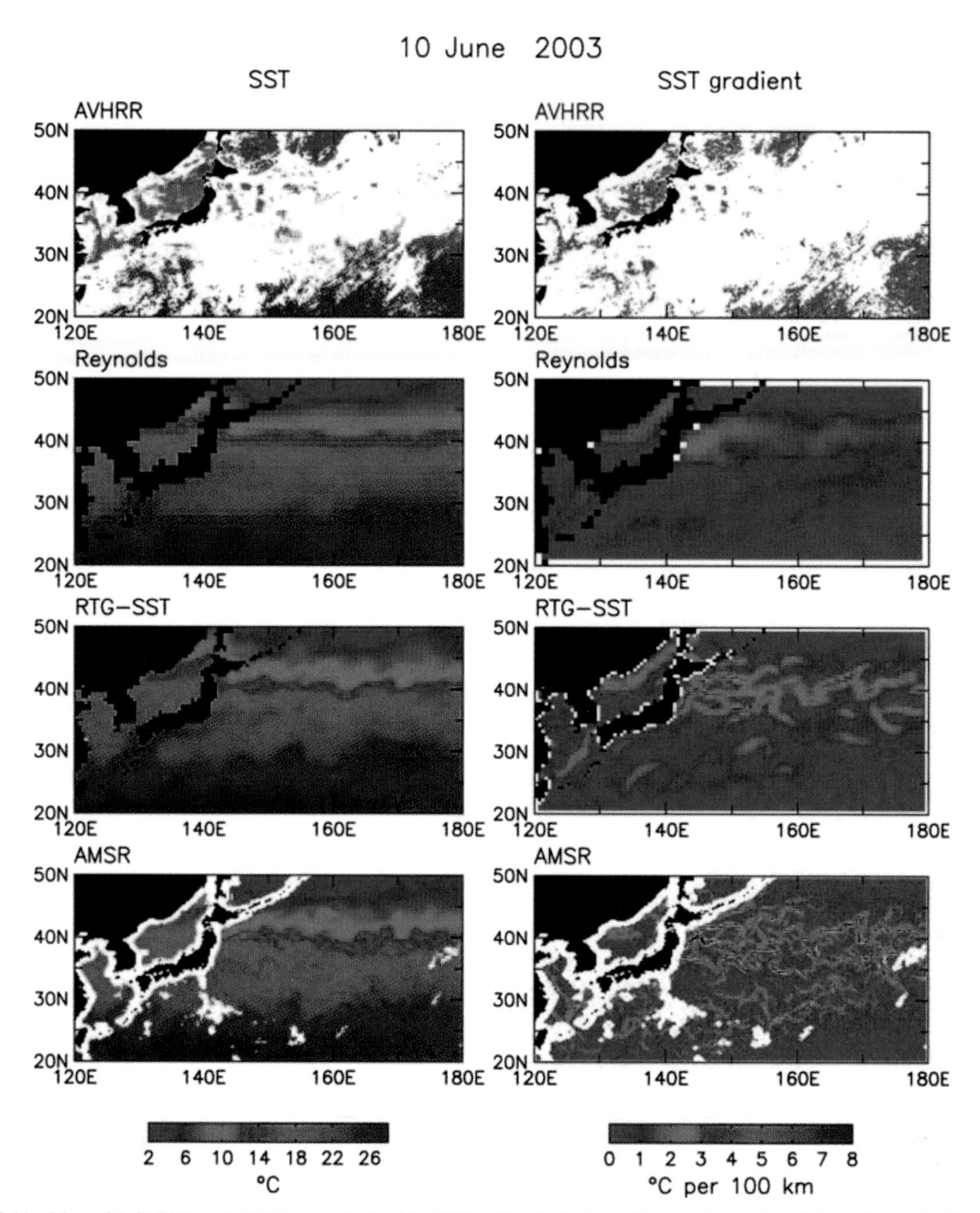

FIGURE 4.11 Maps of (left) SST and (right) the magnitude of the SST gradient in the Kuroshio extension region of the northwest Pacific constructed from (top to bottom) AVHRR, Reynolds, RTG-SST, and AMSR-E for the 3-day-averaging period June 9–11, 2003. The white areas in the top left are regions of persistent cloud cover over the 3-day-averaging period. The white areas in the bottom left correspond to regions of rain contamination. These white areas expand in the derivative fields in the top right and bottom right panels. *From Chelton and Wentz (2005).*

use. To get reasonable global coverage, *in situ* and satellite SST observations available over a period of 7 days have been merged to produce global analysis using an optimal interpolation technique (Reynolds et al., 2002). This product has been made available by National Oceanic and Atmospheric Administration (NOAA) since 1981. The Reynolds SST product, despite its low resolution, did provide sufficient information that allowed O'Neill et al. (2003) to study the SST effects on QuikSCAT wind observations in the Southern Ocean. Starting in 2001, NOAA began producing a new high-resolution product from daily averages of blended *in situ* and infrared SST observations on 0.5×0.5 grids using a variational analysis technique, referred to as RTG-SST analysis (Thiébaux et al., 2003). Although the RTG-SST product does reveal more detailed small-scale structure in the SST field, especially in the gradient field, the AMSR-E product is obviously superior to the other products.

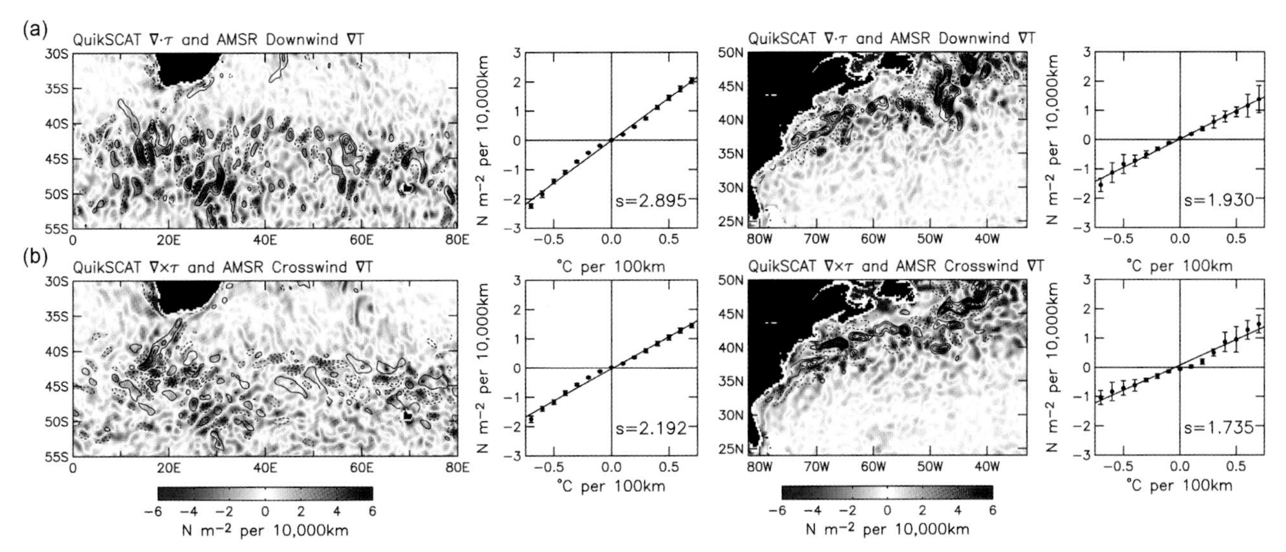

FIGURE 4.12 Maps and binned scatter plots for 2-month averages (January–February 2008) of spatially high-pass-filtered downwind and crosswind SST gradients overlaid, respectively, as contours on the associated wind stress divergence (a) and curl (b). The left panels are for the region of the Agulhas Current south of Africa. The right panels are for the Gulf Stream region. Winds and SST were obtained from QuikSCAT and AM SR-E, respectively. Positive and negative high-pass-filtered SST gradient components are shown as solid and dotted lines, respectively, with a contour interval of 0.5 °C per 100 km and with the zero contours omitted for clarity. The binned averages and standard deviations were computed over eight January–February time periods. *From Chelton and Xie (2010).*

Illustrated in Figure 4.12 are the relationships between SST and ocean surface wind stress reported in Chelton et al. (2004) and Chelton and Xie (2010), using the AMSR SST and QuikSCAT wind stress observations in two regions of energetic oceanic mesoscale variability: the Gulf Stream and the Agulhas Current. Both SST and wind stress observations were spatially high-pass-filtered to retain only scales shorter than approximately 1000 km. The positive correlation between SST and ocean surface wind stress is clearly demonstrated by the scatter plots of the SST gradient with wind stress curl and divergence. In sharp contrast to the negative correlation between SST and wind stress at scales larger than 1000 km, the positive correlation at smaller scales indicates that the air–sea interaction is driven by the spatial variations of SST. The SST-induced perturbations to wind stress are able to affect the upper ocean mixing and vertical velocity in upwelling and downwelling, which in turn affect the SST. A two-way interaction between SST and wind stress is established with significant effects on both the atmospheric and oceanic circulation. The reader is referred to Chelton and Xie (2010) for a review.

One of the best regions to illustrate the two-way interaction noted earlier is the Agulhas Current off the southern tip of South Africa. O'Neill et al. (2005) conducted the first study of this region using the AMSR-E SST and QuikSCAT wind stress observations. They showed that the SST spatial gradient estimated from the AMSR-E observations was a factor of 5 higher than that from the Reynolds SST analysis. Their analysis confirmed the relationship between SST gradient and wind stress curl and divergence first established by Chelton et al. (2001) in the eastern tropical Pacific Ocean. They also showed a positive correlation between SST and the cloud liquid water content measured by the AMSR-E, reflecting the effects of SST on vertical mixing in the atmosphere and the convergence/divergence of wind and moisture transport. The effects of SST on cloud formation, upper atmosphere wind, as well as rain rate, have been found in the region of the Gulf Stream (Minobe et al., 2008). Liu et al. (2007) also reported the effects of SST on the overlying atmosphere in the Agulhas Current region, revealing correlation between SST and cloud top temperature.

The effects of SST on low cloud cover are similar to those on ocean surface wind stress, namely, the transition from a negative correlation at scales larger than 1000 km to a positive correlation at smaller scales. The negative correlation between SST and low cloud cover at large scales causes a positive feedback to SST through the enhanced solar radiation at surface over warm SST and cloud-shielding over cold SST (Xie, 2004). On the other hand, the positive correlation between SST and low cloud cover at small scales leads to a negative cloud-radiative feedback on SST. Kelly et al. (2010) analyzed decade-long SSH and SST data to demonstrate the effects of the ocean's western boundary currents on SST and air–sea interaction. The coupling between SST and wind is thus important feedback to the forcing of the ocean currents. Proper representation of these feedback mechanisms in ocean–atmosphere coupled climate models is a key challenge in the study of climate change.

As noted in Chelton and Xie (2010), the SST boundary condition in the weather prediction model run by the European Center For Medium-Range Weather Forecast (ECMWF) changed from the low-resolution Reynolds SST analyses to the high-resolution RTG-SST analyses in 2001, leading to a dramatic increase in the intensity of wind speed variations in the model outputs at scales less than 1000 km. However, the strength of coupling between SST and ocean surface wind stress in the ECMWF model was much weaker than that estimated from the observations. A significant amount of effort has been made to improve the model's horizontal and vertical resolution as well as the parameterization scheme of vertical mixing, leading to much improved comparison to the observations. This is an example of the impact of the recent advancement in ocean remote sensing on the improvement of the understanding of air–sea interaction processes and the development of coupled ocean–atmosphere models.

Near-real-time satellite altimetry and SST data are also being used to improve hurricane and cyclone predictions. Hurricanes and cyclones will increase their intensity as they pass over regions of high SST, and that high SST will be maintained, despite the strong wind mixing, if the underlying heat content is large. In the Gulf of Mexico, the Loop Current regularly pinches off a warm, anticyclonic ring, which is associated with high sea level and SST, deepening isotherms in toward the centre and large heat content values. The NOAA/AOML group uses a simple two-layer model with altimetry and SST observations to determine the tropical cyclone heat potential in the tropical Atlantic and the Gulf of Mexico. Figure 4.13 shows that during Hurricane Katrina, the surface SST was greater than 28 °C everywhere in the Gulf of Mexico (upper panel), but Hurricane Katrina showed a sharp increase in intensity and a change in direction as it passed over the warm ring detached from the Loop Current (lower panel; Goni et al., 2009). Unfortunately for the residents of New Orleans, the warm ring was located close to the coast, and the Hurricane maintained its strong intensity until it struck land.

4. MESOSCALE EDDIES AND FRONTS

Since the start of the twenty-first century, there has been a renewed interest in the monitoring and understanding of mesoscale ocean processes and their impact on climate and operational oceanography. This has been enhanced by the improvement in ocean monitoring from satellite missions, by the improved *in situ* data coverage and collection, and from the increased resolution from global ocean models and assimilation schemes. If the 1990s were characterized by improved observations and theories on mesoscale eddy processes (e.g., Le Traon and Morrow, 2001), the decade from 2000 onward has been accompanied by the development of global operational oceanography applications steered by the GODAE project (GODAE, 2000). The operational requirements have led to an improvement not only in the quality and quantity of remote sensing data in near real time, but also in the long-term data processing chains, including reanalysis of products based on the complete time series, which have been invaluable for mesoscale circulation and climate studies.

4.1. Mapping the Eddy Field

Satellite altimetry has provided the mainstay for the long-term monitoring of the ocean mesoscale field over the last 20 years (Fu et al., 2010). However, the number of altimetric missions in flight varies over time, and the quality of the data is continually upgraded as the time series evolves. Today we are able to calculate more precise tidal

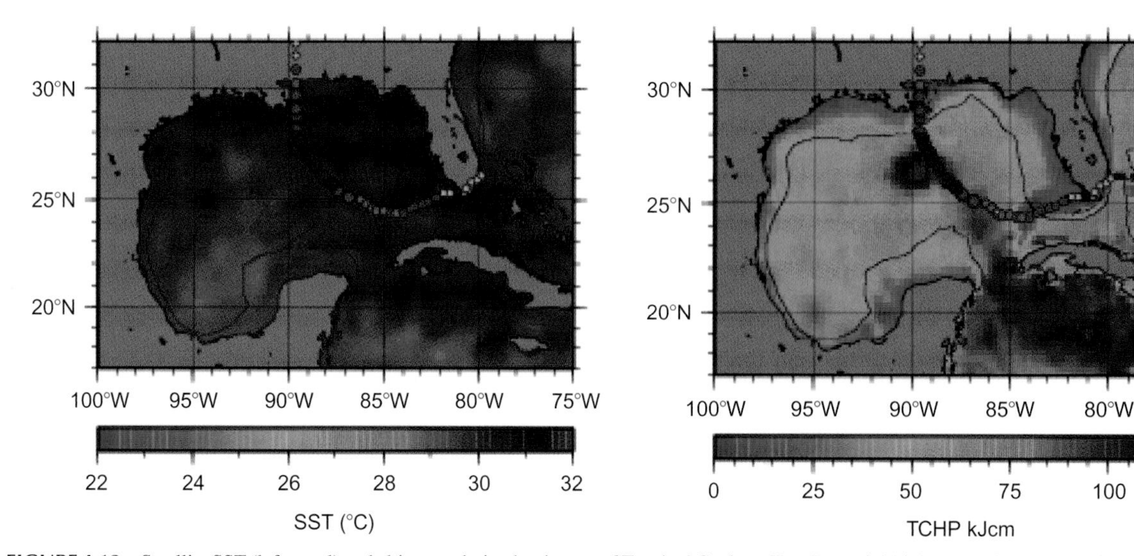

FIGURE 4.13 Satellite SST (left panel) and altimeter-derived estimates of Tropical Cyclone Heat Potential (right panel) in the Gulf of Mexico on August 28, 2005. The circles in different colors indicate the track and intensity of Hurricane Katrina. *Courtesy of G. Goni of NOAA/AOML.*

corrections, orbits, marine geoids, and mean sea surface and dynamic heights, which can be applied to earlier missions to improve their quality in re-analysis products (e.g., Dibarboure et al., 2011). Since 1992, the altimetric missions include the 10-day repeat TOPEX/Poseidon mission (followed by Jason-1 in 2001; and Jason-2 in 2008), the 35-day repeat ERS-1 and -2 and Envisat missions, and the 17-day repeat Geosat Follow-on mission, launched in 1999. In addition, the T/P and then Jason-1 missions were moved to a tandem "inter-leaved" orbit, after their successors (Jason-1 and Jason-2) were launched and validated, providing complementary sampling. The ensemble of these altimeter missions provides a complicated ground sampling pattern, and most scientists rely on the weekly gridded altimetric sea surface anomaly fields, which map the altimetric data using optimal interpolation, and are produced by AVISO (Le Traon and Dibarboure, 2002; Dibarboure et al., 2011).

Two versions of the weekly gridded altimeter maps exist. The first is the SSALTO/DUACS "Reference" time series, based on a two satellite configuration (either TOPEX/Poseidon & ERS or Jason-1, -2 & Envisat) which provides uniform sampling from October 1992 onward. This Reference time series is mainly used for long-term mesoscale climate variability studies, and particularly for interannual to decadal variability. The second version is called the "Updated" mapped time series, which includes all of the available altimetric missions. The Updated time series, with more observations each week after 1999, has a better representation of the amplitude and energy of the smaller mesoscale eddy field and is used extensively for short-term process studies and for operational analyses. Both the 2-altimeter "Reference" series and the multialtimeter "Updated" series provide a good representation of the larger mesoscale variability, greater than 150–200 km wavelength. The mesoscale mapping capabilities of the different altimetric data products are discussed in detail in Dibarboure et al. (2011), Chelton et al. (2011a,b), and Morrow and Le Traon (2011).

To illustrate the mapping capabilities of the "Reference" or "Updated" series, studies have compared the mapped altimetric data with simulations from a high-resolution ocean general circulation model. With two altimeters in the T/P-ERS "Reference" configuration, Le Traon and Dibarboure (2002) found that sea level could be mapped with an accuracy of better than 10% of the modeled signal variance, while velocity can only be mapped with an accuracy of 20–40% of the signal variance. Pascual et al. (2006) merged four altimeters flying from October 2002 to September 2005 (Jason-1, Envisat, T/P interleaved with Jason-1, GFO) to improve the estimation of the mesoscale surface circulation in the "Updated" series. Although similar spatial and temporal smoothing was applied, having more observations available between the standard tracks improved the mapping capabilities. In areas of intense variability, the RMS differences between SSH maps from two altimeters (the "Reference" series) and those from four altimeters (the "Updated" series) can reach up to 10 cm in SSH, and up to 400 $(cm/s)^2$ in eddy kinetic energy (EKE).

Inspired by the need to have a better resolution of narrow swift currents in coastal regions, a number of groups are reanalyzing the along-track altimetric data to improve their quality in the coastal zone. This means improving the quality of certain altimetric corrections, such as the tide correction, atmospheric high-frequency forcing correction, and the wet troposphere correction, so they are better adapted to the smaller space scales and rapid timescales of the coastal dynamics. Data editing is also modified for the coastal dynamics and the along-track filtering is refined (e.g., Roblou et al., 2011). This specific tuning has allowed monitoring of coastal currents of 20–40 km width in the NW Mediterranean Sea (e.g., the Liguro-Provencal current) (Bouffard et al., 2008; Birol et al., 2010), in the Bay of Biscay (Le Hénaff et al., 2010; Dussurget et al., 2011), or along the Indian coast (Durand et al., 2008).

Smaller-scale mesoscale eddies have also been revealed off Mallorca using combined altimetry and gliders (Bouffard et al., 2010), and in the Bay of Biscay from multi-altimeter data (Dussurget et al., 2011). In this latter study, the improved along-track altimetric data revealed a large number of low energy, small-scale eddies in the nearshore region with fewer, larger and more intense eddies offshore. The nearshore eddies with scales of 50–70 km were located close to bathymetric gradients in documented eddy-generation regions, and their evolution to fewer but larger-scale eddies offshore ($\sim$110 km) could be evidence of the upscale cascade of two-dimensional geostrophic turbulence. Altimetry and tide gauges have also been combined to give better sea-level coverage in coastal regions (e.g., Saraceno et al., 2008).

Efforts are currently underway to develop regional gridded altimetric products, using improved along-track data processing and less horizontal smoothing in the mapping process, for the period from 2001/2002 onward when 3–5 altimeters are in flight. These improved regional gridded products aim to provide a better representation of the smaller mesoscale dynamics revealed in the along-track data. Regional products, with latitudinal varying mapping scales, have been developed at higher 1/8° resolution, in the Mediterranean Sea (Pujol and Larnicol, 2005; Pascual et al., 2007), the Black Sea and in near-real time for the Mozambique channel.[1] Variable mapping scales are also being developed, which take into account the bathymetric constraint on eddies (with longer along-shelf scales, smaller cross-shelf scales), or varying nearshore–offshore scales (e.g., Dussurget et al., 2011).

1. http://www.aviso.oceanobs.com/en/data/products:sea-surface-height-products/regional/index.html

4.2. Wave Number Spectra and the Ocean Energy Cascade

The ocean's energy flux through different spatial scales, the energy cascade, is a fundamental problem for ocean dynamics, which has been studied extensively with numerical models but is difficult to observe on a global scale, except recently with satellite observations. Quasi-geostrophic turbulence theory predicts that baroclinic energy in the ocean tends to cascade from both small and large scales towards the deformation radius of the first baroclinic mode (10–100 km), where the energy is converted into barotropic energy (vertically uniform structure), which then cascades towards larger scales, the so-called inverse cascade.

This energy cascade can be observed using along-track altimetry, by examining the spectral energy of the ocean at different horizontal scales (i.e., in wave number, k, space), and at different locations. Ocean wave number spectra exhibit a power law like k^{-p}. For mesoscale eddy motions, the theory of quasi-geostrophic turbulence is thought to apply, leading to a k^{-3} spectrum for kinetic energy, or k^{-5} for SSH. Le Traon et al. (2008) using multiple altimeter data sets have suggested that the SSH spectrum in high eddy-energy regions is closer to $k^{-11/3}$ than to k^{-5} and consistent with the theory of surface quasi-geostrophic (SQG) turbulence that takes the effects of the surface boundary into consideration. However, Xu and Fu (2012) showed the wave number spectral slope in the wavelength range of 70–250 km varies globally, revealing slopes ranging from −4 in high eddy-energy regions to close to −2 in low eddy-energy regions outside the tropics (Figure 4.14). The findings indicate that the spectral slopes in the core regions of the major ocean current systems have values between the original geostrophic turbulence theory and

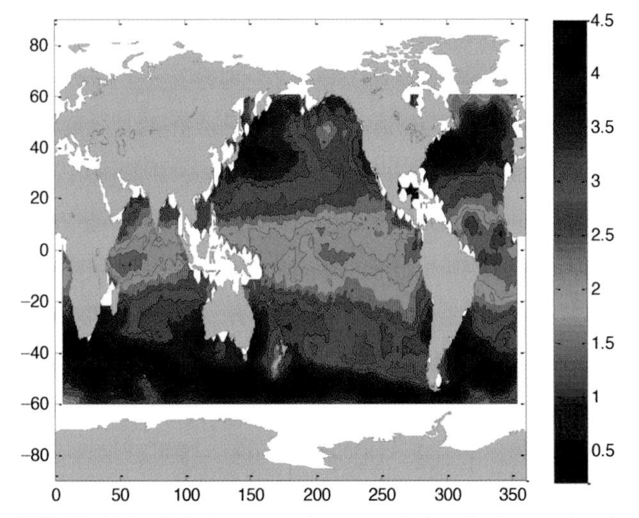

FIGURE 4.14 Global wave number spectral slope in the wavelength range of 70–250 km showing slopes ranging from −4 in high eddy-energy regions to close to −2 in low eddy-energy regions outside the tropics. *After Xu and Fu (2012).*

the SQG theory. The near k^{-4} spectrum suggests that the SSH variability at these wavelengths in the high eddy-energy regions might be governed by frontogenesis.

Scott and Wang (2005) made the first calculation of the energy cascade in wave number space based on two-dimensional observations using gridded satellite altimeter data. Their results revealed an inverse cascade at scales larger than the deformation radius. Because altimetry observations are primarily the perturbations of SSH associated with the motions of the first baroclinic mode, the observed inverse cascade is not consistent with the quasi-geostrophic turbulence theory, which allows an inverse cascade only for the barotropic modes. Scott and Arbic (2007) investigated the problem further and demonstrated that the inverse cascade is indeed dominated by baroclinic modes in a quasi-geostrophic model simulation. Later it was found that the theory of SQG turbulence could explain the inverse cascade of upper-ocean baroclinic energy (Capet et al., 2008).

Qiu et al. (2008) applied the technique of Scott and Wang (2005) to studying the seasonal energy exchange process between the mean flow and eddy field of the Subtropical Countercurrent in the South Pacific Ocean. They found that the instability of the mean flow generated meridionally elongated eddies that, through eddy–eddy interactions, transformed into zonally elongated eddies. The energy was thus transferred from eddy scales to larger zonal scales in an anisotropic inverse cascade process. These observations provide one of the first demonstrations of the detailed mechanism of energy exchange between mean flow and eddy variability and the transformation of scales.

4.3. Seasonal and Interannual Variations in Eddy Energy

The long altimetric time series has revealed how the ocean eddy energy is varying over time, and from one region to another. These observations of locally varying eddy fields have inspired a large number of investigations on the dynamics of this eddy adjustment in relation to larger-scale ocean variations. Long-lived ocean eddies may exist for 3 months to a few years, and during this time, their surface mixed layer is continually modified by seasonal forcing, although their eddy cores remain intact at depth. So amongst various sensing techniques, altimetry is particularly suited to studying the long-term evolution of mesoscale eddies, as it responds to the depth-integrated eddy structure.

For example, the North Pacific Subtropical Countercurrent (NPSTCC) around 20°N has been shown to have a well-defined seasonal modulation in its EKE field (Qiu, 1999). Using a 2½ layer model to represent the vertically sheared NPSTCC-North Equatorial Current system, they found that the observed seasonal modulation reflected the intensity of baroclinic instability of the sheared current system. In the South Pacific, Qiu and Chen (2004) detected high EKE bands with well-defined annual cycles along the eastward-flowing surface currents of both the South Tropical Countercurrent (STCC) between 21°S–29°S and the South Equatorial Countercurrent (SECC) centered near 9°S. Once again, the seasonal variation in the intensity of baroclinic instability was found to be responsible for the seasonal modulation of the STCC's EKE field around 25°S (Figure 4.15). However, the seasonal modulation of the SECC's EKE field around 10°S was mainly generated by the seasonal variation in the intensity of barotropic instability associated with seasonal changes in the strong horizontal shear of the SECC–SEC system. Recently, Qiu and Chen (2010) analyzed decade-long SSH data in the North Pacific and discovered interannual variability of the EKE of the Kuroshio Extension in relation to the effects of the Pacific Decadal Oscillation (PDO) on the stability of the Kuroshio Extension.

Brachet et al. (2004) have analyzed the temporal evolution of the monthly EKE over an 8-year period (from December 1992 to December 2000) and investigated the relationship with the wind stress in several regions. Their results follow previous studies that link seasonal EKE variations at high latitudes to seasonal wind stress variations, although other forcing mechanisms may also be important as discussed by Eden and Böning (2002). Clear EKE interannual variability was revealed. They hypothesized that a contraction of the subpolar and subtropical gyres due to the North Atlantic Oscillation (NAO) could explain a reduction of the eddy activity in the North Atlantic Current, in the Newfoundland Basin, and in the Azores Current. Penduff et al. (2004) also found that strong NAO events after 1994 were followed by gyre-scale EKE fluctuations with a 4–12-month lag, and then studied the complex and nonlinear adjustment processes with the aid of a numerical model. Altimetric EKE variations have also been studied in the Mediterranean Sea (Pujol and Larnicol, 2005) and in the Bay of

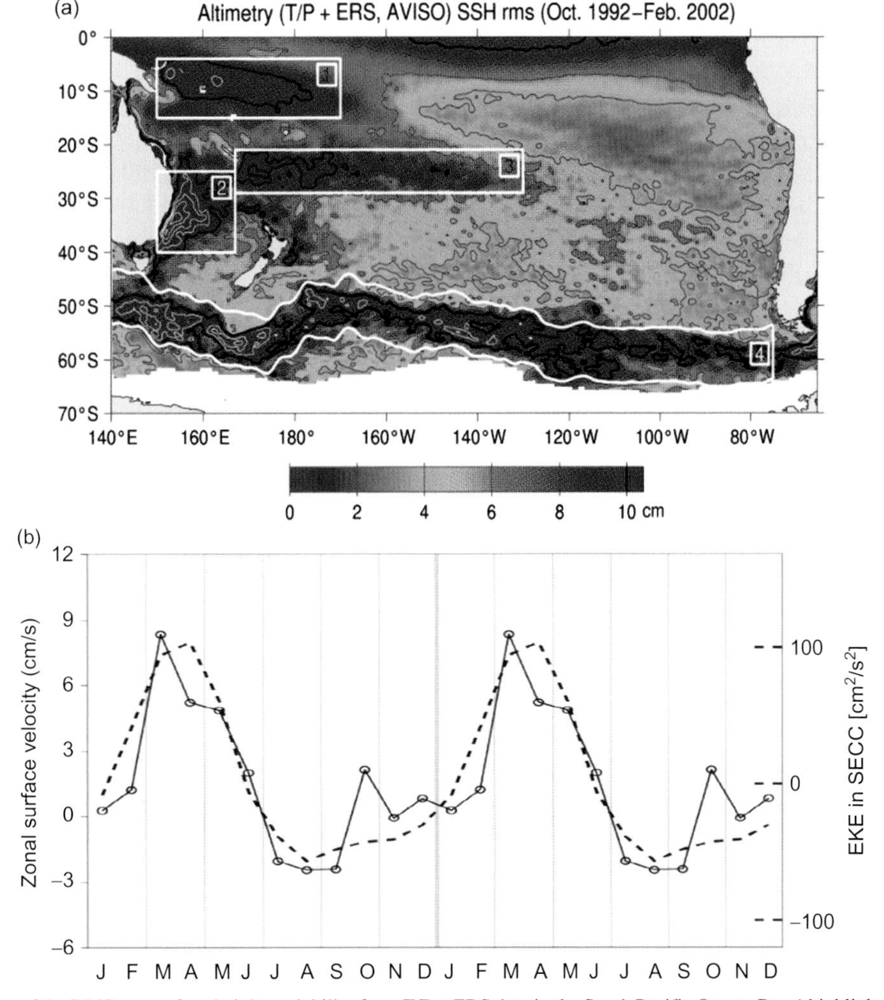

FIGURE 4.15 (a) Map of the RMS sea surface height variability from T/P + ERS data in the South Pacific Ocean. Box 1 highlights the SECC variability; Box 3, the STCC variability. (b) Zonal velocity shear between 0 and 600 m for the STCC–SEC from monthly climatological data (solid line) compared to altimeter EKE (dashed line). *After Qiu and Chen (2004).*

Biscay (Caballero et al., 2007), where the regional variations have been investigated in relation to different forcing mechanisms (varying wind stress, coastal currents, etc.).

In the Southern Ocean south of Tasmania, interannual variations in EKE can be linked to interannual changes in the position and strength of the Sub-Antarctic Front (Morrow et al., 2002). Over the entire Southern Ocean, the EKE variations also show a lagged response to the principal climate mode forcing from the Southern Annular Mode (Meredith and Hogg, 2006) and can also be modified regionally by ENSO (Morrow et al., 2010). The mechanisms that generate the observed 2 year lag in eddy energy have also been explored with model simulations (Hogg and Blundell, 2006; Morrow et al., 2010). The delay in EKE was caused by a positive feedback in the model; the increased potential energy creates favorable conditions for baroclinic instability, which increases eddy activity and the eddy momentum transfer to deeper layers. There, interaction with the bottom topography induces a stronger meridional deep circulation, which is intrinsically more unstable, increasing in turn the EKE. This feedback continues until the excess of large-scale potential energy is consumed. This example illustrates how the observed EKE variations from remote sensing can stimulate theoretical and modeling analyses, to give us an improved understanding of the interplay between the large-scale circulation changes and the eddy adjustment processes.

4.4. Tracking Individual Eddies

Although the ocean is a turbulent system, long-term altimetric monitoring of the global ocean mesoscale signal has shown that the instabilities propagate in an organized way. One fundamental property of the larger, long-duration ocean mesoscale instabilities is that they are in quasigeostrophic balance and can be influenced by the rotation of the Earth (beta-effect). Movies of ocean sea-level anomalies show a continual movement of mesoscale eddies toward the west at all latitude bands, except in regions of strong eastward currents or topographical obstacles.

A major discovery in the 1990s from the time series of multimission altimetry data was that the sea-level anomalies propagated westward with a wave speed up to twice as fast as that predicted from linear Rossby wave theory (Chelton and Schlax, 1996). This observational discovery led to a host of revised theories on Rossby wave propagation, which took into account the vertical shear in the background flow, bathymetric effects, etc. (for a review, see Fu and Chelton, 2001). Although the difference between these revised theories and the observed propagation speeds has been reduced, the observed propagation speeds remain faster than theory predicts.

Most of the difference occurs at mid-to-high latitude, and Chelton et al. (2007) raise the question of whether the observed propagation is due to Rossby waves at these latitudes, or whether the instabilities are associated with isolated nonlinear vortices. Nonlinear vortices also propagate westward (Cushman-Roisin et al., 1990) but interact differently with the background flow. Unlike Rossby waves, they also transport heat, salt, and tracers in their core waters, which can have consequences for the mid-latitude tracer budgets (see Section 4.6). In addition, many studies using *in situ* observations to determine the vertical structure of mid-latitude sea-level anomalies analyze a vertical "eddy" structure (Carton and Chao, 1999; de Ruijter et al., 1999; Morrow et al., 2003), although Roemmich and Gilson (2001) suggest that their subsurface temperature anomalies in the North Pacific mid-latitudes appear "wavelike."

A number of studies have tracked the propagation pathways of individual eddies at mid-to-high latitudes using the mapped multimission altimetric sea-level anomalies. Different automatic eddy tracking techniques have been developed and tested on altimetry data, based on the Okubo–Weiss parameter (e.g., Isern-Fontanet et al., 2003); the skewness of the relative vorticity (Niiler et al., 2003); criteria based on sea-level contours (Fang and Morrow, 2003; Chaigneau and Pizarro, 2005; Chelton et al., 2011a,b); wavelet decomposition of the sea-level anomalies (e.g., Lilly et al., 2003); or a geometric criterion using the winding angle approach (e.g., Souza et al., 2011). In the Agulhas retroflection region, the separation of warm-core Agulhas rings and their propagation into the South Atlantic provides the main mechanism of interocean exchange here (see de Ruijter et al., 1999 for a review). The number of Agulhas warm-core rings generated varies from year to year, and their pathways can vary in response to interannual changes in the South Atlantic gyre strength. However, depending on the tracking technique used, there can be differences in the number of eddies detected, their duration, and their propagation velocities (Chelton et al., 2011a,b; Souza et al., 2011).

A recent global analysis of eddy trajectories by Chelton et al. (2011a,b) confirms that eddies exist nearly everywhere in the world's ocean (Figure 4.16a). There are 6% more cyclones than anti-cyclones with lifetimes >4 months; however, eddies with the longest lifetimes and longest propagation tend to be anticyclonic. There are quiescent regions with very few eddies in the northeast and southeast Pacific Oceans. There may be weak eddies in the region, but with amplitudes or space scales that are smoothed out by the mapping process. The small number of eddies in the tropical band is attributed to the difficulty in tracking low-latitude eddies because of their fast propagation speeds and large spatial scales. However this region is also one where the linear Rossby wave signal dominates.

Eddies form nearly everywhere in the world oceans, but their formation rates are highest in the strong western boundary currents, the eastern boundary current systems,

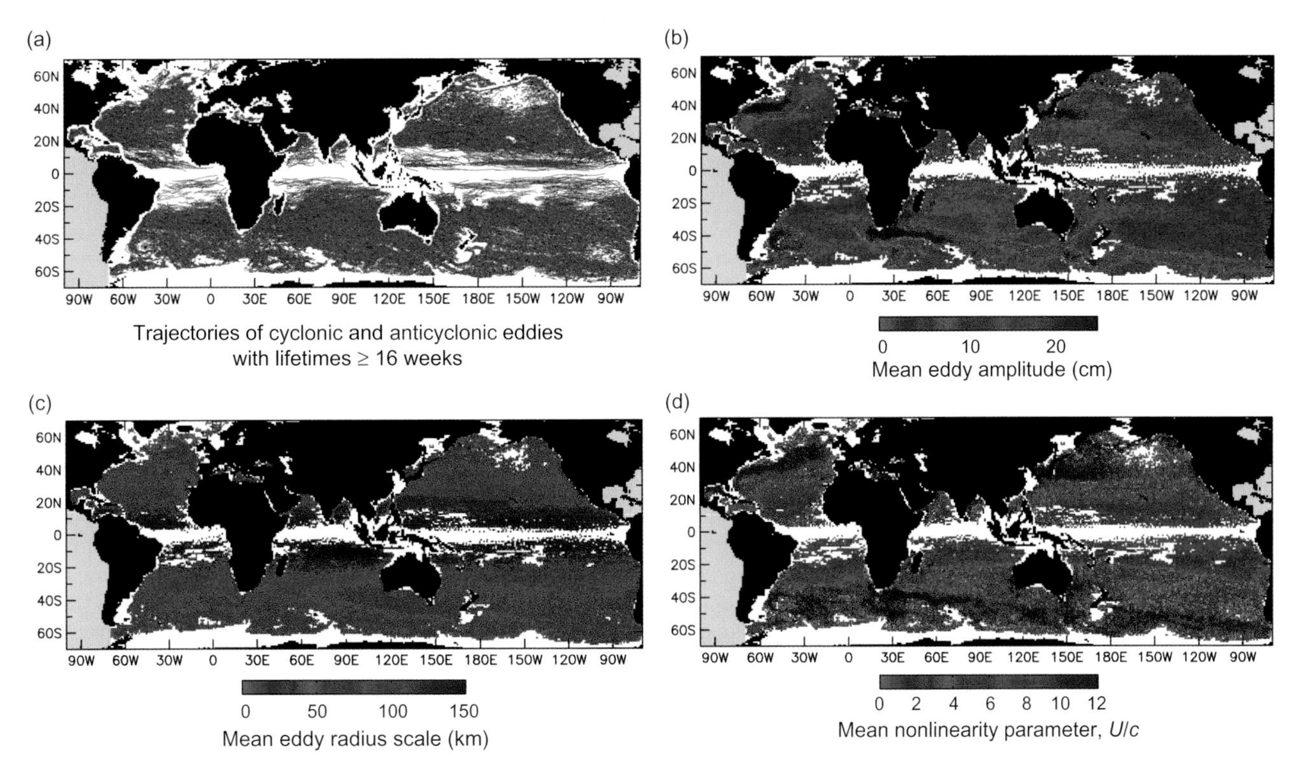

FIGURE 4.16 Global maps of the characteristics of eddies that were tracked for 16 weeks or longer in the first 16 years of the merged altimeter dataset (October 1992–December 2008): (a) The trajectories of cyclonic and anticyclonic eddies (blue and red lines, respectively); (b) the mean amplitude for each 1° square; (c) the mean scale for each 1° square, defined to be the effective radius at which the rotational speed averaged around an SSH contour is maximum; and (d) the mean nonlinearity parameter U/c (see text). *Adapted from Chelton et al. (2011a,b).*

the Circumpolar Current, or near bathymetric gradients (Chelton et al., 2011a,b). The global distribution of eddy formation supports the conclusions of previous studies that the generation mechanism is principally due to baroclinic instability of the vertically sheared flow, especially in regions where the flow has a non-zonal component (see Smith, 2007, and references therein).

The eddy amplitudes reflect the maps of global EKE, that is, individual eddies have larger amplitudes in the western boundary currents and along the ACC (Figure 4.16b). The mean eddy radius scales decrease from around 200 km near the equator to about 75 km at 60°S (Figure 4.16c). At mid-to-high latitudes, these scales are much larger than the baroclinic Rossby radius at which energy is input to the ocean from baroclinic instability. The larger eddy scales are consistent with the upscale transfer of energy that is expected from the SQG turbulence theory (e.g., Capet et al., 2008), if we are observing "mature" eddies. However, it is likely that we are missing part of the smaller-scale, newly formed eddies in these analyses, which are smoothed out by the mapping process at mid-to-high latitudes.

A key feature that distinguishes the observed mesoscale eddies from linear Rossby waves is their nonlinearity. This can be assessed from the ratio of the maximum rotational fluid velocity U within the eddy interior to its translation speed, c. When this nondimensional parameter U/c exceeds 1, there is trapped fluid within the eddy interior (Samelson and Wiggins, 2006). The eddy can then transport heat, salt, carbon, and biogeochemical properties such as nutrients and phytoplankton (see Section 4.6), and have an important influence on the global budgets for these tracers. The mean value of U/c exceeds 1 poleward of 20° latitude in both hemispheres, reaching values of 10–15 in eddy rich regions (Figure 4.16d; after Chelton et al., 2011a,b). So 99% of the extratropical eddies in this analysis were nonlinear by this U/c metric. The westward propagation speed of these nonlinear eddies is very close to the phase speed for nondispersive baroclinic Rossby waves.

In terms of their direction, most eddies propagating more than 1000 km had pathways differing by less than 15° from due west. However, consistent with the theory for the propagation of large, nonlinear eddies (Cushman-Roisin et al., 1990), there are small opposing meridional deflections of cyclones and anticyclones. As noted regionally by Morrow et al. (2004), and globally by Chelton et al. (2011a,b), long-lived cyclones tend to propagate poleward, and anticyclones tend to propagate equatorward. This implies a net equatorward heat flux from these long-lived eddies. The basic "rules" of westward or meridional propagation can be overturned in regions of strong mean currents or strong bathymetric steering. Eddies are advected by the mean circulation in the NW Pacific

(Isoguchi and Kawamura, 2003); in the ACC (Hughes et al., 1998; Fu, 2009), and in the North Atlantic (Brachet et al., 2004).

Eddy corridors also exist in a large number of regions, created by bathymetric gradients at eastern boundaries, such as the Leeuwin Current region (Fang and Morrow, 2003), or by perturbations of a current intersecting with an island, which occurs near the Canary Islands (Sangrà et al., 2009), or Hawaii (Yoshida et al., 2010). As an example, de Ruijter et al. (2004) analyzed eddies and dipoles south of Madagascar. They observed the regular formation of dipoles around south Madagascar, where the East Madagascar Current separates from the shelf. Two periods of enhanced dipole formation coincided with the negative phases of the Indian Ocean Dipole and the El Niño cycles, suggesting a connection between these tropical climate modes and the dipole train formation. A very regular train of dipoles started in December 1999 and continued through 2000. They stayed remarkably coherent and triggered an unusually early retroflection of the Agulhas Current in late 2000. Thus, interannual variability originating in the equatorial climate modes of the Indian Ocean can impact on the variability of the Agulhas retroflection and associated ring shedding.

4.5. Surface Currents from Multisensor Mapping

Weekly surface current products resolving mesoscale eddy fields have been developed based on altimetric geostrophic currents, Ekman currents from scatterometry-based ocean vector winds, and satellite SST. These products include the OSCAR (Bonjean and Lagerloef, 2002), Mercator SURCOUF (Larnicol et al., 2006), and the CTOH current product (Sudre and Morrow, 2008). The differences in these products depend on the algorithms used for deriving geostrophic currents at the equator, in the models of wind-driven turbulence, and the inclusion of an adjustment to the currents from SST gradients. These satellite-based products compare well with *in situ* surface drifters, particularly in the high eddy-energy regions. Their advantage over the *in situ* data is regular space-time sampling.

Mesoscale surface velocities have also been derived from a variety of satellite surface fields using feature tracking techniques. The maximum cross correlation technique tracks the displacements of small-scale features on high-resolution images of SST or ocean color, and attributes them to advection by ocean currents (e.g., Bowen et al., 2002). These fine-scale advected currents are available only in cloud-free regions, but can be blended with altimetric geostrophic currents for better coverage. Feature tracking of mesoscale structures has also been applied to surface height fields from altimetry (Fu, 2009), and from microwave SST maps using singularity or multifractal analyses (Turiel et al., 2008, 2009).

Very fine scale cross-track surface currents can also be derived from synthetic aperture radar (SAR) data (e.g., Johannessen et al., 2005; Liu and Hsu, 2009; Rouault et al., 2010). In addition, the two antennas on TerraSAR-X make the computation of cross track currents possible using along-track interferometry. These methods are very promising for regional applications where SAR data are collected. Although intensive processing is required to separate the surface current signature from surface waves, internal waves, and tide motions, this processing is becoming much more routine. Again the fine-scale currents are complementary to the altimeter-based geostrophic currents, and blended products are possible. Advanced Synthetic Aperture Radar (ASAR) surface current velocities have been used to monitor the position and intensity of the Agulhas Current, and have revealed the importance of nearshore processes and submesoscale eddies in the overall variability of the Agulhas Current system (Rouault et al., 2010).

4.6. Eddy Fluxes of Ocean Properties (see also Chapter 8)

Ocean eddies provide an important mechanism for transporting heat, salt, and nutrients in the ocean. Although eddy transport is small in the ocean interior, meridional eddy heat transport plays a dominant role in the heat balance of the strong western boundary currents, the ACC, and the tropical oceans (Jayne and Marotzke, 2002). If mesoscale eddies are important in the lateral transport of ocean tracers, their interaction also helps drive the submesoscale filamentation, which dominates the vertical transport (Ferrari, 2011). Global satellite observations provide a unique opportunity for estimating eddy transport and diffusion. This is an active field of research since eddy diffusivity impacts on ocean mixing.

The first global maps of observed eddy diffusion were based on geostrophic turbulence theory and global eddy statistics of altimetric eddy properties such as eddy kinetic energy, and eddy length and timescales (e.g., Stammer, 1998). Mean temperature and salinity gradients, obtained from climatological data, were then used to compute the mean eddy transport of heat and salt. Over the last decade, lateral eddy diffusivity has been derived using dispersion statistics, based on surface drifter observations or Lagrangian observations derived from altimetric currents (Sallée et al., 2008a, 2011), the altimetric advection of a tracer field (e.g., Marshall et al., 2006), or from Microcanonical multifractal formalism (MMF) applied to altimetric maps (Nieves and Turiel, 2009). In the Southern Ocean, the observed eddy diffusion in the ACC and western boundary currents is shown to play a key role in modifying the upper ocean heat budget, and limiting the regions where sub-Antarctic Mode waters can form in winter (Sallée et al., 2008b).

Our understanding of eddy heat transports has greatly benefited in the last decade by the availability of concurrent mesoscale altimetry and satellite SST fields, and vertical temperature profiles from Argo floats and XBT profiles. Qiu and Chen (2005) have estimated the basin-scale meridional eddy heat transport in the North Pacific. Their instantaneous calculations of eddy transports with a realistic vertical structure were somewhat less than the previous altimetric estimates based on climatological temperature gradients alone. The importance of subsurface information for determining eddy heat transport was also noted by Roemmich and Gilson (2001), who showed that the vertical tilt of the eddy core is essential in the eddy transport of heat. In the southeast Pacific, Chaigneau et al. (2011) showed that the cores of the anticyclonic and cyclonic eddies were located at different depths. The cyclones were formed from instabilities in the surface equatorward coastal currents, whereas the anticyclones with deeper cores were shed from the subsurface poleward Peru–Chile Undercurrent.

Time-varying heat budgets for the surface mixed-layer have also been established in different oceanic regions (Dong et al., 2007; Sallée et al., 2008b). More recently, combinations of altimetry with gliders have highlighted the fine-scale structure of the mesoscale processes. In the Labrador Sea, altimetry and gliders are being used to track eddies spawned from the west Greenland boundary current (Hátún et al., 2007). These eddies transport low-salinity upper waters and warm, saline Irminger waters at depth. Individual eddies can contribute to the rapid restratification of the Labrador sea interior after periods of deep winter convection, and may help explain why the deep convection region is so small, compared to the larger scale atmospheric forcing.

4.7. Submesoscale Dynamics

Although the altimeter constellation provides a good constraint on the larger mesoscale field, we still cannot adequately sample the smaller mesoscales or sub-mesoscales. As the name suggests, the sub-mesoscale concerns the small-scale features surrounding the mesoscale eddies, or around propagating planetary waves. These include the elongated filament structures associated with the stirring at the periphery of mesoscale eddies, but also coastal squirts and jets, and open-ocean fronts, and smaller vortices. These structures are ubiquitous in high-resolution radiometric SST and ocean color, observed under cloud-free conditions. We have only recently started to monitor the surface evolution of these structures using space observations, combined with either statistical techniques or models.

One solution currently in practice is to combine the high-resolution satellite altimetry with very high resolution SST products and/or regional numerical models (1–5 km resolution). The high-resolution models allow the sub-mesoscale features to develop, and the solution can be constrained towards the observed altimetric sea level, SST, and *in situ* observations. Modeling studies from the California Current (e.g., Capet et al., 2008) and the NE Atlantic (Paci et al., 2005; Klein et al., 2008) have shown that the smaller-scale structures are associated with active surface frontogenesis, and as a consequence, trigger most of the vertical velocity field in the upper 500 m depth. These smaller-scale structures also drive a large part of the horizontal and vertical momentum fluxes, and the inverse kinetic energy cascade, which is why they have a significant impact on the ocean circulation at larger scales. The vertical exchange in the filaments has consequences for the biogeochemical exchanges between the surface layers and at depth. It also has an impact on the subduction of mode waters, which do not occur along a theoretical north–south gradient as previously thought, but are subducted in the filament structures around the mesoscale eddies (Paci et al., 2005). The filaments greatly increase the depth range of these exchanges, and also increase the density of mode waters subducted.

Efforts are underway to derive these filament structures and upper ocean velocity fields from satellite data. The upper ocean three-dimensional velocity fields (u, v, w) can be derived from very high-resolution surface density observations (or satellite SST fields, as a proxy) combined with SQG theory (Lapeyre and Klein, 2006), from the base of the mixed layer down to 500 m depth. The validity of the SQG technique has been tested with high-resolution numerical models (Isern-Fontanet et al., 2008) in the North Atlantic. The results show that the SQG technique provides a reasonably good reconstruction of the upper ocean's velocity field, especially in the Gulf Stream and the North Atlantic Drift.

The temporal evolution of the surface lateral velocity field, derived from satellite altimetry or SST fields, is also being used with Lagrangian statistical techniques to derive the evolving smaller-scale dynamical fields. Although the surface geostrophic currents from altimetry are nondivergent, the temporal evolution of these mapped fields show eddies and meandering jets that interact, creating zones of divergence and convergence. Statistical techniques based on finite size or finite time Lyapunov exponents (FSLE or FTLE; d'Ovidio et al., 2009) can be used to derive the filament positions, associated with lateral transport "barriers" where larger vertical exchange occurs. These dynamical filament structures can be associated with strong SST or ocean color gradients, and can be used to delimit different ocean biomass groups (Figure 4.17; d'Ovidio et al., 2010) or identify feeding regimes for marine animals (Tew Kai et al., 2009; Cotté et al., 2011). Similarly, these filament positions can be tracked using MMF, based on

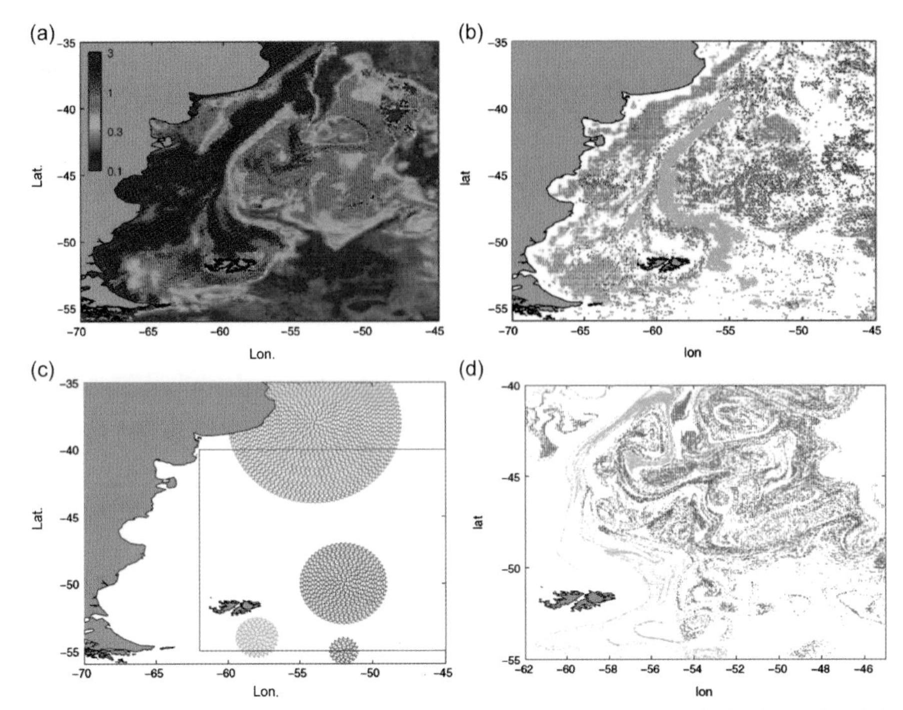

FIGURE 4.17 Ecological and physical sub-mesoscale structure from satellite data. (a) Total chlorophyll distribution (μg/L). (b) Dominant phytoplankton types identified from SeaWiFS composite by the PHYSAT algorithm during the spring bloom (November 24–December 1, 2001). Colors in (b) indicate diatoms (green), Prochlorococcus (red), Synechococcus (dark blue), nanoeukaryotes (yellow), Phaeocystis (magenta), and coccolithophorids (cyan). (c) Initial distribution of numerical tracers (August 20, 2001). (d) Distribution of these numerical tracers after 3 months of lateral advection driven by altimetry-derived geostrophic currents (November 28, 2001). Note the smaller region in (d). *After d'Ovidio et al. (2010).*

mesoscale SST products (Turiel et al., 2008) or altimeter data (Nieves and Turiel, 2009).

The horizontal velocity computed from surface gridded satellite products is also being used to stir large-scale tracer fields and in doing so, introduce smaller meso to sub-mesoscale structures. These techniques can help position regions where frontogenesis is dominated by lateral advection (Klein and Hua, 1990). In the subpolar gyre of the North Atlantic, Desprès et al. (2011) applied lateral stirring from altimetry to climatological sea surface salinity fields, and were able to improve the gradients and positions of surface salinity fronts in some regions.

Maps based on traditional nadir-looking altimetric configurations have allowed us to monitor and understand the larger oceanic mesoscale signals. However, we cannot observe the smaller mesoscale and sub-mesoscale signal between the tracks, and the background noise of the present generation of altimeters also masks the smaller scale signals in the along-track data. In the future, the technology proposed by the wide-swath Surface Water And Ocean Topography (SWOT) interferometric altimeter aims at improving the smaller space-timescales associated with mesoscale eddies and coastal zone dynamics (Fu and Ferrari, 2008). For SWOT, two SAR antennae measure simultaneously the radar signals reflected by the rough sea surface. The phase differences between the two signals and the range measurement allow interferometric calculation of the SSH. The intrinsic resolution of SAR is on the order of a few meters, allowing high spatial resolution of the measurements over two swaths of 60 km width. After smoothing, the random measurement noise can be reduced over the oceans to a level of 1 cm over 1 km × 1 km resolution cells. When these missions are finally flown, we expect to have a much better understanding of the important interactions between the mesoscale and sub-mesoscale dynamics.

4.8. Eddies and Biogeochemical Processes

Mesoscale eddies have a large influence on phytoplankton distribution, biogeochemical cycles, and pelagic ocean ecosystems, through their lateral stirring and mixing (e.g., Chelton et al., 2011a,b), and through the vertical advection of nutrients into the sunlit surface layer (where photosynthesis takes place) in and around the eddies (McGillicuddy et al., 1998; Garcon et al., 2001; Yoder et al., 2010).

The network of satellite ocean color observations provide one of the main tools for studying temporal sequences of phytoplankton chlorophyll *a* (Chl *a*) distribution over the global oceans, and have revealed the important impact of mesoscale eddies on the chlorophyll and primary production distribution. For example, by tracking individual eddies with altimetry and ocean color maps, composites can be made of the typical Chl *a* distributions over cyclonic and anticyclonic eddies. In regions with a strong northward gradient in Chl *a*, composites formed from thousands of eddies show positive Chl *a* anomalies to the southwest of eddy centers for cyclones, and to the southeast of the centers of anticyclones (Chelton et al., 2011a,b). These patterns are dominated by the lateral stirring of the surface chlorophyll distribution by the rotational effects of eddies (Killworth et al., 2004; Siegel et al., 2008; Chelton et al., 2011a,b). Other mechanisms can modify the "stirred" chlorophyll content, including the trapping and transport of marine organisms and nutrients from one biogeographical region to another within the eddy cores, and eddy "pumping" or nutrients into the euphotic zone from the shoaling of isopycnal surfaces during the intensification or decay of eddies (Siegel et al., 2008, and references therein).

Satellite ocean color observations have revealed the ubiquitous distribution of fine-scale, high Chl *a* filamentation patterns, related to sub-mesoscale dynamical processes. In the past few years, high-resolution model simulations have shown that this physical–biological coupling is strongly linked to the sub-mesoscale physics, through dynamical restratification, lateral stirring, and vertical advection (see Lévy, 2008 for a review). Indeed, frontogenesis triggers intense vertical velocities within submesoscale filaments, and thus potentially stimulates primary production, even under oligotrophic conditions (Lévy et al., 2001). During spring blooms when the nutrients are plentiful, submesoscale dynamics can restratify the surface mixed layer, thus accelerating the bloom over specific submesoscale features. As noted earlier, horizontal stirring is also important in shaping the ecosystem, and in enhancing or reducing vertical nutrient supplies (Martin et al., 2002; Lévy and Klein, 2004).

One of the difficulties in linking the physical and biogeochemical processes is that the physics is not instantaneously correlated to the biology; indeed, the phytoplankton distribution reflects the cumulative effects of lateral and vertical advection, nutrient supplies, and other biological processes. Techniques that follow the temporal evolution of the flow field provide better descriptions of the Chl *a* patterns. This can be provided by Lagrangian statistical techniques such as Lyapunov exponents, or through full coupling of OGCMs (ocean general circulation models) with biogeochemical models. For example, Lehahn et al. (2007) examined phytoplankton filaments with altimetric derived surface currents: the phytoplankton filaments were less correlated with instantaneous streamlines than with the time-evolving Lyapunov exponents.

In the future, the next generation of altimeters with finer spatial resolution, such as SWOT, Jason-CS, in conjunction with high-frequency, high-resolution ocean color data on geostationary orbits, will help enable a better monitoring and understanding of the phytoplankton distributions at the scale of these filaments.

5. SUMMARY AND OUTLOOK

This review has highlighted how much progress has been made in the past decade in our understanding of ocean circulation from space. One of the key outcomes concerns the multi-instrument analysis, which helps distinguish the different physical components in the ocean and climate system. This is clearly revealed in the combined altimetry–gravity mission analyses, which has allowed us to better understand the contributions to the global mean sea-level rise, and to separate the effects of mass changes from the melting glaciers or ice caps from steric changes due to a warming ocean. The synergy between remote sensing and *in situ* observations is best demonstrated by the analysis of the consistency of global sea-level budget though combined use of observations from altimetry, satellite gravity measurement, and Argo floats.

The advent of the satellite gravity measurement from the GRACE and GOCE missions has greatly improved the knowledge of the geoid at scales approaching 100 km. These improvements have made it possible to determine the ocean surface general circulation from truly global sampling from space, without any spatial bias resulting from uneven sampling from *in situ* observations. The dense altimetry observations are also used to minimize spatial sampling bias in *in situ* observations from surface drifters and Argo floats for achieving optimal estimates of ocean currents at the surface as well as at depths. The temporal variability of gravity measured by GRACE over the oceans revealed for the first time direct evidence of large-scale redistribution of mass in the ocean on seasonal and interannual timescales in response to atmospheric forcing.

The decade-long observations of the global ocean wind field from the QuikSCAT Mission have revealed the role of air–sea interaction in the measurement of ocean wind from space. The finding confirmed the understanding of the scatterometer measurement of radar backscatter from the ocean surface in terms of the relative motion between air and sea responsible for the wind stress acting on the ocean surface. Through the analysis of simultaneous scatterometer wind observations and SST observations from microwave radiometers, a positive correlation between SST and wind stress is discovered at scales shorter than 1000 km, indicating that the air–sea interaction is driven by the spatial variations of SST, in sharp contrast to the negative correlation at larger scales. This finding has led to important improvements in ocean–atmosphere coupled models.

A major advance in the usage of satellite altimetry data for ocean circulation studies is enabled by the advent of gridded SSH fields produced from merged observations from multiple missions. This data set is available on weekly basis on Mercator grids of 1/3° in latitude from 1993 to present. It has essentially revolutionized the study of ocean circulation, and satellite altimetry has become a routine tool for both research and applications. For example, the data have been used for tracking the movement of ocean eddies over the global oceans, leading to fundamental advances in the understanding of the mesoscale dynamics of the ocean. The combination of the SSH data with ocean color data gives us enormous insight into the biological coupling associated with mesoscale eddies and fronts.

Since altimetric sea-level rises and falls in response to surface and deep changes in the ocean, it provides the strongest constraint for the four-dimensional mesoscale ocean circulation estimation through data assimilation. Its unique monitoring capabilities are of utmost importance for the development of operational oceanography services (e.g., safe and efficient offshore activities, pollution monitoring, environmental management, security, and sustainable use of marine resources) and for ocean, ecosystem, and climate research (e.g., GODAE, 2000). However, a number of issues still need to be improved to correctly observe the full spectra of ocean processes. Coastal mesoscale data need to be improved to have better coverage and more accurate data available. Accurate mean dynamic topographies and bathymetry are also required for a better representation of eddy–mean interactions, and the role of bathymetric gradients in generating mesoscale instabilities. We are also missing accurate global observations of the smaller mesoscale signals, from 10 to 200 km, which are not well represented with the present altimetric coverage. Wide swath altimetry (e.g., SWOT) is also a very attractive possibility (Fu and Rodriguez, 2004) with the potential to sample the full spectra of ocean variability with a resolution better than 4–5 coordinated conventional altimeters.

ACKNOWLEDGMENTS

The research presented in the chapter was carried out in part (LLF) at the Jet Propulsion Laboratory, California Institute of Technology, under contract with the National Aeronautic and Space Administration, and at LEGOS (RM), supported by the Observatoire Midi Pyrénées and the French TèOSCA program. Support from the Jason-1 and OSTM/Jason-2 Projects is acknowledged.

REFERENCES

Ablain, M., Cazenave, A., Valladeau, G., Guinehut, S., 2009. A new assessment of the error budget of global mean sea level rate estimated by satellite altimetry over 1993–2008. Ocean Sci. 5, 193–201.

Bingham, R.J., Hughes, C.W., 2006. Observing seasonal bottom pressure variability in the North Pacific with GRACE. Geophys. Res. Lett. 33, L08607. http://dx.doi.org/10.1029/2005GL025489.

Birol, F., Cancet, M., Estournel, C., 2010. Aspects of the seasonal variability of the Northern Current (NW Mediterranean Sea) observed by altimetry. J. Mar. Syst. 81, 297–311. http://dx.doi.org/10.1016/j.jmarsys.2010.01.005.

Boening, C., Lee, T., Zlotnicki, V., 2011. A record high ocean bottom pressure in the South Pacific observed by GRACE. Geophys. Res. Lett. 38, L04602. http://dx.doi.org/10.1029/2010GL046013.

Boening, C., Willis, J.K., Landerer, F.W., Nerem, R.S., Fasullo, J., 2012. The 2011 La Niña: so strong, the oceans fell. Geophys. Res. Lett. 39, L19602. http://dx.doi.org/10.1029/2012GL053055.

Bonjean, F., Lagerloef, G., 2002. Diagnostic model and analysis of the surface currents in the tropical Pacific Ocean. J. Phys. Oceanogr. 32, 2938–2954.

Bouffard, J., Vignudelli, S., Herrmann, M., Lyard, F., Marsaleix, P., Ménard, Y., et al., 2008. Comparison of ocean dynamics with a regional circulation model and improved altimetry in the North-Western Mediterranean. Terr. Atmos. Oceanic Sci. 19, 117–133. http://dx.doi.org/10.3319/TAO.2008.19.1-2.117.

Bouffard, J., Pascual, A., Ruiz, S., Faugère, Y., Tintoré, J., 2010. Coastal and mesoscale dynamics characterization using altimetry and gliders: a case study in the Balearic Sea. J. Geophys. Res. 115, C10029. http://dx.doi.org/10.1029/2009JC006087.

Bowen, M., Emery, W.J., Wilkin, J.L., Tildesley, P., Barton, I.J., Knewtson, R., 2002. Extracting multiyear surface currents from sequential thermal imagery using the Maximum Cross Correlation technique. J. Atmos. Oceanic Technol. 19, 1665–1676.

Brachet, S., Le Traon, P.Y., Le Provost, C., 2004. Mesoscale variability from a high-resolution model and from altimeter data in the North Atlantic Ocean. J. Geophys. Res. 109, C12025. http://dx.doi.org/10.1029/2004JC002360.

Bromirski, P.D., Miller, A.J., Flick, R.E., Auad, G., 2011. Dynamical suppression of sea level rise along the Pacific coast of North America: indications for imminent acceleration. J. Geophys. Res. 116, C07005. http://dx.doi.org/10.1029/2010JC006759.

Bryden, H.L., Longworth, H.R., Cunningham, S.A., 2005. Slowing of the Atlantic meridional overturning circulation at 25° N. Nature 438, 655–657. http://dx.doi.org/10.1038/nature04385.

Caballero, A., Pascual, A., Dibarboure, G., Espino, M., 2007. Sea level and eddy kinetic energy variability in the Bay of Biscay inferred from satellite altimeter data. J. Mar. Syst. 72, 116–134.

Cai, W., 2006. Antarctic ozone depletion causes an intensification of the Southern Ocean super-gyre circulation. Geophys. Res. Lett. 33, L03712. http://dx.doi.org/10.1029/2005GL024911.

Capet, X., Klein, P., Hua, B.L., Lapeyre, G., McWilliams, J.C., 2008. Surface kinetic energy transfer in surface quasi-geostrophic flows. J. Fluid Mech. 604, 165–174.

Carton, J.A., Chao, Y., 1999. Carribean Sea eddies inferred from Topex/Poseidon altimetry and 1/6° Atlantic Ocean model simulation. J. Geophys. Res. 104 (C4), 7743–7752.

Chaigneau, A., Pizarro, O., 2005. Eddy characteristics in the eastern South Pacific. J. Geophys. Res. vol. 110. C06005. http://dx.doi.org/10.1029/2004JC002815.

Chaigneau, A., Le Texier, M., Eldin, G., Grados, C., Pizarro, Ó., 2011. Vertical structure of mesoscale eddies in the eastern South Pacific Ocean: a composite analysis from altimetry and Argo profiling floats. J. Geophys. Res. vol. 116. C11025. http://dx.doi.org/10.1029/2011JC007134.

Chambers, D.P., 2006a. Evaluation of new GRACE time-variable gravity data over the ocean. Geophys. Res. Lett. 33, L17603. http://dx.doi.org/10.1029/2006GL027296.

Chambers, D.P., 2006b. Observing seasonal steric sea level variations with GRACE and satellite altimetry. J. Geophys. Res. 111, C03010. http://dx.doi.org/10.1029/2005JC002914.

Chambers, D.P., Willis, J.K., 2008. Analysis of large-scale ocean bottom pressure variability in the North Pacific. J. Geophys. Res. 113, C11003. http://dx.doi.org/10.1029/2008JC004930.

Chambers, D.P., Willis, J.K., 2009. Low-frequency exchange of mass between ocean basins. J. Geophys. Res. 114, C11008. http://dx.doi.org/10.1029/2009JC005518.

Chao, Y., Fu, L.-L., 1995. A comparison between the TOPEX/POSEIDON data and a global ocean general circulation model during 1992–93. J. Geophys. Res. 100, 24965–24976.

Chelton, D.B., Freilich, M.H., 2005. Scatterometer-based assessment of 10-m wind analyses from the operational ECMWF and NCEP numerical weather prediction models. Mon. Weather Rev. 133, 409–429.

Chelton, D.B., Schlax, M.G., 1996. Global observations of oceanic Rossby waves. Science 272, 234–238.

Chelton, D.B., Wentz, F.J., 2005. Global microwave satellite observations of sea surface temperature for numerical weather prediction and climate research. Bull. Am. Meteorol. Soc. 86, 1097–1115.

Chelton, D.B., Xie, S.-P., 2010. Coupled ocean-atmosphere interaction at oceanic mesoscales. Oceanography 23 (4), 52–69.

Chelton, D.B., Mestas-Nuñez, A.M., Freilich, M.H., 1990. Global wind stress and Sverdrup circulation from the Seasat scatterometer. J. Phys. Oceanogr. 20, 1175–1205.

Chelton, D.B., Esbensen, S.K., Schlax, M.G., Thum, N., Freilich, M.H., Wentz, F.J., et al., 2001. Observations of coupling between surface wind stress and sea surface temperature in the eastern tropical Pacific. J. Clim. 14, 1479–1498.

Chelton, D.B., Schlax, M.G., Freilich, M.H., Milliff, R.F., 2004. Satellite radar measurements reveal short-scale features in the wind stress field over the world ocean. Science 303, 978–983.

Chelton, D.B., Schlax, M.G., Samelson, R.M., de Szoeke, R.A., 2007. Global observations of large oceanic eddies. Geophys. Res. Lett. 34, L15606. http://dx.doi.org/10.1029/2007GL030812.

Chelton, D.B., Schlax, M.G., Samelson, R.M., 2011a. Global observations of nonlinear mesoscale eddies. Prog. Oceanogr. 91, 167–216.

Chelton, D.B., Gaube, P., Schlax, M.G., Early, J.J., Samelson, R.M., 2011b. The influence of nonlinear mesoscale eddies on near-surface oceanic chlorophyll. Science 334, 328–332. http://dx.doi.org/10.1126/science.1208897.

Cornillon, P., Park, K.-A., 2001. Warm core ring velocities inferred from NSCAT. Geophys. Res. Lett. 28, 575–578.

Cotté, C., d'Ovidio, F., Chaigneau, A., Lévy, M., Taupier-Letage, I., Mate, B., et al., 2011. Scale-dependent interactions of resident Mediterranean whales with marine dynamics. Limnol. Oceanogr. 56 (1), 219–232.

Cunningham, S.A., et al., 2007. Temporal variability of the Atlantic meridional overturning circulation at 26.5 degrees N. Science 317 (5840), 935–938. http://dx.doi.org/10.1126/science.1141304.

Cushman-Roisin, B., Chassignet, E.P., Tang, B., 1990. Westward motion of mesoscale eddies. J. Phys. Oceanogr. 20, 97–113.

d'Ovidio, F., Isern-Fontanet, J., López, C., García-Ladona, E., Hernández-García, E., 2009. d'Comparison between Eulerian diagnostics and the finite-size Lyapunov exponent computed from altimetry in the Algerian Basin. Deep Sea Res. Part I 56, 15–31.

d'Ovidio, F., De Monte, S., Alvain, S., Danonneau, Y., Lévy, M., 2010. Fluid dynamical niches of phytoplankton types. Proc. Natl. Acad. Sci. U.S.A. 107, 18366–18370. http://dx.doi.org/10.1073/pnas.1004620107.

de Ruijter, W.P.M., Biastoch, A., Drijfhout, S.S., Lutjeharms, J.R.E., Matano, R.P., Pichevin, T., et al., 1999. Indian-Atlantic inter-ocean exchange: dynamics, estimation and impact. J. Geophys. Res. 104, 20,885–20,910.

de Ruijter, W.P.M., van Aken, H.M., Beier, E.J., Lutjeharms, J.R.E., Matano, R.P., Schouten, M.W., 2004. Eddies and dipoles around South Madagascar: formation, pathways and large-scale impact. Deep Sea Res. Part I 51, 383–400.

Després, A., Reverdin, G., d'Ovidio, F., 2011. Surface fronts and currents in the Irminger sea. Ocean Model. 39, 97–113.

Dibarboure, G., Pujol, M.-I., Briol, F., Le Traon, P.-Y., Larnicol, G., Picot, N., et al., 2011. Jason-2 in DUACS: first tandem results and impact on processing and products. Mar. Geod. 34, 214–241.

Domingues, C.M., Church, J., White, N., Gleckler, P., Wijffels, S., Barker, P., et al., 2008. Improved estimates of upper-ocean warming and multi-decadal sea-level rise. Nature 453 (7198), 1090–1093.

Dong, S., Gille, S.T., Sprintall, J., 2007. An assessment of the Southern Ocean mixed layer heat budget. J. Clim. 20, 4425–4442.

Durand, F., Shankar, D., Birol, F., Shenoi, S.S.C., 2008. Spatio-temporal structure of the East India Coastal Current from satellite altimetry. J. Geophys. Res. 114, C02013. http://dx.doi.org/10.1029/2008JC004807.

Dussurget, R., Birol, F., Morrow, R., Demey, P., 2011. Fine resolution altimetry data for a regional application in the Bay of Biscay. Mar. Geod. 34, 447–476.

Eden, C., Böning, C., 2002. Sources of eddy kinetic energy in the Labrador Sea. J. Phys. Oceanogr. 32, 3346–3363.

Fang, F., Morrow, R., 2003. Warm-core eddy propagation in the southeast Indian Ocean. Deep Sea Res. Part II 50, 2245–2261.

Ferrari, R., 2011. A frontal challenge for climate models. Science 332, 316. http://dx.doi.org/10.1126/science.1203632.

Fu, L.-L., 2001. Ocean circulation and variability from satellite altimetry. In: Siedler, G., Church, J., Gould, J. (Eds.), Ocean Circulation and Climate. Academic Press, San Diego, CA, pp. 141–172.

Fu, L.-L., 2009. Pattern and velocity of propagation of the global ocean eddy variability. J. Geophys. Res. 114, C11017. http://dx.doi.org/10.1029/2009JC005349.

Fu, L.-L., Chelton, D.B., 2001. Large-scale ocean circulation. In: Fu, L.-L., Cazenave, A. (Eds.), Satellite Altimetry and Earth Sciences. A Handbook of Techniques and Applications. Academic Press, San Diego, pp. 133–169.

Fu, L.-L., Ferrari, R., 2008. Observing oceanic submesoscale processes from space. Eos 89 (48), 488.

Fu, L.-L., Rodriguez, R., 2004. High-resolution measurement of ocean surface topography by radar interferometry for oceanographic and geophysical applications. In: Sparks, R.S.J., Hawkesworth, C.J. (Eds.), AGU Geophysical Monograph 150. State of the Planet: Frontiers and Challenges, vol. 19. IUGG, Washington, DC, pp. 209–224.

Fu, L.-L., Smith, R.D., 1996. Global ocean circulation from satellite altimetry and high-resolution computer simulation. Bull. Am. Meteorol. Soc. 77, 2625–2636.

Fu, L.-L., Cheng, B., Qiu, B., 2001. 25-Day period large-scale oscillations in the Argentine Basin revealed by the TOPEX/POSEIDON altimeter. J. Phys. Oceanogr. 31, 506–517.

Fu, L.-L., Chelton, D.B., Le Traon, P.-Y., Morrow, R., 2010. Eddy dynamics from satellite altimetry. Oceanography 23 (4), 14–25.

Fukumori, I., Raghunath, R., Fu, L.-L., 1998. The nature of global large-scale sea level variability in relation to atmospheric forcing: a modeling study. J. Geophys. Res. 103, 5493–5512.

Garcon, V.C., Oschlies, A., Doney, S.C., McGillicuddy, D.J., Waniek, J., 2001. The role of mesoscale variability on plankton dynamics in the North Atlantic. Deep Sea Res. Part II 48, 2199–2226.

Gill, A.E., 1982. Atmosphere-Ocean Dynamics. Academic Press, New York, p. 662.

GODAE Strategic Plan, 2000. Prepared by the International GODAE Steering Team. GODAE Report No. 6, December.

Goni, G.J., DeMaria, M., Knaff, J., Sampson, C., Ginis, I., Bringas, F., et al., 2009. Applications of satellite-derived ocean measurements to tropical cyclone intensity forecasting. Oceanography 22 (3), 176–183.

Gourcuff, C., Lherminier, P., Mercier, H., Le Traon, P.-Y., 2011. Altimetry combined with hydrography for ocean transport estimation. J. Atmos. Oceanic Technol. 28, 1324–1337. http://dx.doi.org/10.1175/2011JTECHO818.1.

Hakkinen, S., Rhines, P.B., 2004. Decline of subpolar North Atlantic circulation during the 1990s. Science 309, 555–559.

Hakkinen, S., Rhines, P.B., 2009. Shifting surface currents of the northern North Atlantic Ocean. J. Geophys. Res. 114, C04005. http://dx.doi.org/10.1029/2008JC004883.

Hashizume, H., Xie, S.-P., Fujiwara, M., Shiotani, M., Watanabe, T., Tanimoto, Y., et al., 2002. Direct observations of atmospheric boundary layer response to slow SST variations on the Pacific equatorial front. J. Clim. 15, 3379–3393.

Hátún, H., Eriksen, C.C., Rhines, P.B., 2007. Buoyant eddies entering the Labrador Sea observed with gliders and altimetry. J. Phys. Oceanogr. 37 (12), 2838–2854.

Hayes, S.P., McPhaden, M.J., Wallace, J.M., 1989. The influence of sea-surface temperature on surface wind in the eastern equatorial Pacific: weekly to monthly variability. J. Clim. 2, 1500–1506.

Hogg, A.M., Blundell, J.R., 2006. Interdecadal variability of the Southern Ocean. J. Phys. Oceanogr. 36, 1626–1644.

Hu, A., Meehl, G.A., Han, W., Yin, J., 2009. Transient response of the MOC and climate to potential melting of the Greenland Ice Sheet in the 21st century. Geophys. Res. Lett. 36, L10707. http://dx.doi.org/10.1029/2009GL037998.

Hughes, C.W., 2005. Nonlinear vorticity balance of the Antarctic Circumpolar Current. J. Geophys. Res. 110, C11008. http://dx.doi.org/10.1029/2004JC002753.

Hughes, C.W., Williams, S.D.P., 2010. The color of sea level: importance of spatial variations in spectral shape for assessing the significance of trends. J. Geophys. Res. 115, C10048. http://dx.doi.org/10.1029/2010JC006102.

Hughes, C.W., Wilson, C., 2008. Wind work on the geostrophic ocean circulation: an observational study of the effect of small scales in the wind stress. J. Geophys. Res. 113 (2), C02016. http://dx.doi.org/10.1029/2007JC004371, 10.

Hughes, C.W., Jones, M.S., Carnochan, S., 1998. Use of transient features to identify eastward currents in the Southern Ocean. J. Geophys. Res. 103 (C2), 2929–2943.

Hughes, C.W., Stepanov, V.N., Fu, L.-L., Barnier, B., Hargreaves, G.W., 2007. Three forms of variability in Argentine Basin Ocean bottom pressure. J. Geophys. Res. 112, C01011. http://dx.doi.org/10.1029/2006JC003679.

Isern-Fontanet, J., Garcia-Ladona, E., Font, J., 2003. Identification of marine eddies from altimeter maps. J. Atmos. Oceanic Technol. 20, 772–778.

Isern-Fontanet, J., Lapeyre, G., Klein, P., Chapron, B., Hecht, M.W., 2008. Three-dimensional reconstruction of oceanic mesoscale currents from surface information. J. Geophys. Res. 113, C09005. http://dx.doi.org/10.1029/2007JC004692.

Isoguchi, O., Kawamura, H., 2003. Eddies advected by time-dependent Sverdrup circulation in the western boundary of the subarctic North Pacific. Geophys. Res. Lett. 30 (15), 1794. http://dx.doi.org/10.1029/2003GL017652.

Jayne, S.R., 2006. The circulation of the North Atlantic Ocean from altimetry and the Gravity Recovery and Climate Experiment geoid. J. Geophys. Res. 111, C03005. http://dx.doi.org/10.1029/2005JC003128.

Jayne, S.R., Marotzke, J., 2002. The oceanic eddy heat transport. J. Phys. Oceanogr. 32, 3328–3345. http://dx.doi.org/10.1175/1520-0485(2002)032.

Johannessen, J.A., Kudryavtsev, V., Akimov, D., Eldevik, T., Winther, N., Chapron, B., 2005. On radar imaging of current features: 2. Mesoscale eddy and current front detection. J. Geophys. Res. 110 (C7), C07017, 34.

Keihm, S.S., Brown, J., Teixeira, S., Desai, W., Lu, E., Fetzer, C., et al., 2009. Ocean water vapor and cloud liquid water trends from 1992 to 2005 TOPEX Microwave Radiometer data. J. Geophys. Res. 114, D18101. http://dx.doi.org/10.1029/2009JD012145.

Kelly, K.A., Dickinson, S., McPhaden, M.J., Johnson, G.C., 2001. Ocean currents evident in satellite wind data. Geophys. Res. Lett. 28, 2469–2472.

Kelly, K.A., Small, R.J., Samelson, R., Qiu, B., Joyce, T.M., Cronin, M., et al., 2010. Western boundary currents and frontal air-sea interaction: Gulf Stream and Kuroshio Extension. J. Clim. 23, 5644–5667.

Killworth, P.D., Cipollini, P., Uz, B.M., Blundell, J.R., 2004. Physical and biological mechanisms for planetary waves observed in satellite-derived chlorophyll. J. Geophys. Res. 109, C07002. http://dx.doi.org/10.1029/2003JC001768.

Klein, P., Hua, B.L., 1990. The meso scale variability of the sea surface temperature: an analytical and numerical model. J. Mar. Res. 48, 729–763.

Klein, P., Hua, B.L., Lapeyre, G., Capet, X., LeGentil, S., Sasaki, H., 2008. Upper ocean turbulence from high 3-D resolution simulations. J. Phys. Oceanogr. 38, 1748–1763.

Lapeyre, G., Klein, P., 2006. Dynamics of the upper oceanic layers in terms of surface quasigeostrophy theory. J. Phys. Oceanogr. 36, 165–176.

Larnicol, G., Guinehut, S., Rio, M.-H., Drevillon, M., Faugere, Y., Nicolas, G., 2006. The Global Observed Ocean Products of the French Mercator project. In: Proceedings of 15 Years of Progress in Radar Altimetry Symposium. ESA Special Publication, Frascati, Italy, SP-614.

Le Hénaff, M., Roblou, L., Bouffard, J., 2010. Characterizing the Navidad current interannual variability using coastal altimetry. Ocean Dyn. 61, 425–437.

Le Traon, P.Y., Dibarboure, G., 2002. Velocity mapping capabilities of present and future altimeter missions: the role of high frequency signals. J. Atmos. Oceanic Technol. 19, 2077–2088.

Le Traon, P.-Y., Morrow, R.A., 2001. Ocean currents and mesoscale eddies. In: Fu, L.-L., Cazenave, A. (Eds.), Satellite Altimetry and Earth Sciences. A Handbook of Techniques and Applications. Academic Press, San Diego, pp. 171–215.

Le Traon, P.Y., Klein, P., Hua, B.L., Dibarboure, G., 2008. Do altimeter data agree with interior or surface quasi-geostrophic theory. J. Phys. Oceanogr. 38 (5), 1137–1142.

Lee, T., McPhaden, M.J., 2008. Decadal phase change in large-scale sea level and winds in the Indo-Pacific region at the end of the 20th century. Geophys. Res. Lett. 35, L01605. http://dx.doi.org/10.1029/2007GL032419.

Lee, T., Hobbs, W., Willis, J., Halkides, D., Fukumori, I., Armstrong, E., et al., 2010. Record warming in the South Pacific and western Antarctica associated with the strong central-Pacific El Nino in 2009–10. Geophys. Res. Lett. 37, L19704. http://dx.doi.org/10.1029/2010GL044865.

Lehahn, Y., d'Ovidio, F., Lévy, M., Heitzel, E., 2007. Stirring of the Northeast Atlantic spring bloom: a Lagrangian analysis based on multi-satellite data. J. Geophys. Res. 112, C08005. http://dx.doi.org/10.1029/2006JC003927.

Leuliette, E.W., Miller, L., 2009. Closing the sea level rise budget with altimetry, Argo, and GRACE. Geophys. Res. Lett. 36, L04608.

Lévy, M., 2008. The modulation of biological production by oceanic mesoscale turbulence. In: Weiss, J.B., Provenzale, A. (Eds.), Transport in Geophysical Flow: Ten Years After. Lecture Notes in Physics, vol. 744. Springer, Berlin, pp. 219–261. http://dx.doi.org/10.1007/978-3-540-75215-8_9.

Lévy, M., Klein, P., 2004. Does the low-frequency variability of mesoscale dynamics explain a part of the phytoplankton and zooplankton spectral variability? Proc. R. Soc. Lond. 460 (2046), 1673–1683. http://dx.doi.org/10.1098/rspa.2003.1219.

Lévy, M., Klein, P., Treguier, A.-M., 2001. Impacts of sub-mesoscale physics on phytoplankton production and subduction. J. Mar. Res. 59, 535–565. http://dx.doi.org/10.1357/002224001762842181.

Lilly, J.M., Rhines, P.B., Schott, F., Lavender, K., Lazier, J., Send, U., et al., 2003. Observations of the Labrador Sea eddy field. Prog. Oceanogr. 59 (1), 75–176. http://dx.doi.org/10.1016/j.pocean.2003.08.013.

Liu, A.K., Hsu, M.-K., 2009. Deriving ocean surface drift using multiple SAR sensors. Remote Sens. 2009 (1), 266–277. http://dx.doi.org/10.3390/rs1030266.

Liu, W.T., Tang, W.Q., 1996. Equivalent neutral wind. Jet Propulsion Laboratory Publications 96-19, Pasadena, CA, 8 pp (Available online at http://airsea-www.jpl.nasa.gov/data.html.

Liu, W.T., Xie, X., Polito, P.S., Xie, S.-P., Hashizume, H., 2000. Atmospheric manifestation of tropical instability waves observed by QuikSCAT and Tropical Rain Measuring Mission. Geophys. Res. Lett. 27, 2545–2548.

Liu, W.T., Xie, X., Niiler, P.P., 2007. Ocean–atmosphere interaction over Agulhas extension meanders. J. Clim. 20, 5784–5797. http://dx.doi.org/10.1175/2007JCLI1732.1.

Llovel, W., Becker, M., Cazenave, A., Jevrejeva, S., Alkama, R., Decharme, B., et al., 2011. Terrestrial waters and sea level variations on interannual time scale. Glob. Planet. Change 75 (1–2), 76–82. http://dx.doi.org/10.1016/j.gloplacha.2010.10.008.

Lyman, J.M., Good, S.A., Gouretski, V.V., Ishii, M., Johnson, G.C., Palmer, M.D., et al., 2010. Robust warming of the global upper ocean. Nature 465, 334–337.

Marshall, J., Shuckburgh, E., Jones, H., Hill, C., 2006. Estimates and implications of surface eddy diffusivity in the Southern Ocean derived from tracer transport. J. Phys. Oceanogr. 36, 1806–1821.

Martin, A.P., Richard, K.J., Bracco, A., Provenzale, A., 2002. Patchy productivity in the open ocean. Global Biogeochem. Cycles 16 (2), 1025. http://dx.doi.org/10.1029/2001GB001449, 9 pp.

Maximenko, N., Melnichenko, O.V., Niiler, P.P., Sasaki, H., 2008. Stationary mesoscale jet-like features in the ocean. Geophys. Res. Lett. 35, L08603. http://dx.doi.org/10.1029/2008GL033267.

Maximenko, N., Niiler, P., Centurioni, L., Rio, M.-H., Melnichenko, O., Chambers, D., et al., 2009. Mean dynamic topography of the ocean derived from satellite and drifting buoy data using three different techniques. J. Atmos. Oceanic Technol. 26 (9), 1910–1919.

McGillicuddy Jr., D.J., Robinson, A.R., Siegel, D.A., Jannasch, H.W., Johnson, R., Dickey, T.D., et al., 1998. Influence of mesoscale eddies on new production in the Sargasso Sea. Nature 394, 263–265.

Meredith, M.P., Hogg, A.M., 2006. Circumpolar response of Southern Ocean eddy activity to changes in the Southern Annular Mode. Geophys. Res. Lett. 33, L16608. http://dx.doi.org/10.1029/2006GL026499.

Minobe, S., Kuwano-Yoshida, A., Komori, N., Xie, S.-P., Small, R.J., 2008. Influence of the Gulf Stream on the troposphere. Nature 452, 206–209.

Morrow, R.A., Le Traon, P.Y., 2011. Recent advances in observing mesoscale ocean dynamics with satellite altimetry. Adv. Space Res. 50, 1062–1076.

Morrow, R.A., Brut, A., Chaigneau, A., 2002. Seasonal and interannual variations of the upper ocean energetics between Tasmania and Antarctica. Deep Sea Res. 50 (3), 339–356.

Morrow, R.A., Fang, F., Fieux, M., Molcard, R., 2003. Anatomy of three warm-core Leeuwin Current eddies. Deep Sea Res. Part II 50, 2229–2243.

Morrow, R.F., Birol, D., Griffin, J. Sudre, 2004. Divergent pathways of anticyclonic and cyclonic eddies. Geophys. Res. Lett. 31, L24311. http://dx.doi.org/10.1029/2004GL020974.

Morrow, R., Ward, M.L., Hogg, A.M., Pasquet, S., 2010. Eddy response to Southern Ocean climate modes. J. Geophys. Res. 115, C10030. http://dx.doi.org/10.1029/2009JC005894.

Nerem, S., Mitchum, G.T., 2001. Sea level change. In: Fu, L.-L., Cazenave, A. (Eds.), Satellite Altimetry and Earth Sciences: A Handbook for Techniques and Applications. Academic Press, San Diego, pp. 329–349, 423 pp.

Nerem, R., Chambers, D., Choe, C., Mitchum, G., 2010. Estimating mean sea level change from the TOPEX and Jason altimeter missions. Mar. Geod. 33 (1), 435–446.

Nieves, V., Turiel, A., 2009. Analysis of ocean turbulence using adaptive CVE on altimetric maps. J. Atmos. Oceanic Technol. 77 (4), 482–494. http://dx.doi.org/10.1016/j.jmarsys. 2008.12.001.

Niiler, P.P., 2001. The world ocean surface circulation. In: Siedler, G., Church, J., Gould, J. (Eds.), Ocean Circulation and Climate. International Geophysics Series, vol. 77. Academic Press, San Diego, pp. 193–204.

Niiler, P.P., Maximenko, N.A., Panteleev, G.G., Yamagata, T., Olson, D.B., 2003. Near-surface dynamical structure of the Kuroshio Extension. J. Geophys. Res. 108 (C6), 3193. http://dx.doi.org/10.1029/2002JC001461.

O'Neill, L.W., Chelton, D.B., Esbensen, S.K., 2003. Observations of SST-induced perturbations of the wind stress field over the Southern Ocean on seasonal timescales. J. Clim. 16, 2340–2354.

O'Neill, L.W., Chelton, D.B., Esbensen, S.K., Wentz, F.J., 2005. High-resolution satellite measurements of the atmospheric boundary layer response to SST variations along the Agulhas Return Current. J. Clim. 18, 2706–2723.

O'Neill, L.W., Esbensen, S.K., Thum, N., Samelson, R.M., Chelton, D.B., 2010. Dynamical analysis of the boundary layer and surface wind responses to mesoscale SST perturbations. J. Clim. 23 (3), 559–581.

Paci, A., Caniaux, G., Gavart, M., Giordani, H., Levy, M., Prieur, L., et al., 2005. A high-resolution simulation of the ocean during the POMME experiment: simulation results and comparison with observations. J. Geophys. Res. 110, C07S09. http://dx.doi.org/10.1029/2004JC002712.

Pascual, A., Faugere, Y., Larnicol, G., Le Traon, P.Y., 2006. Improved description of the ocean mesoscale variability by combining four satellite altimeters. Geophys. Res. Lett. 33 (2), Art. No. L02611.

Pascual, A., Pujol, M.I., Larnicol, G., Le Traon, P.Y., Rio, M.H., 2007. Mesoscale mapping capabilities of multisatellite altimeter missions: first results with real data in the Mediterranean Sea. J. Mar. Syst. 65, 190–211. http://dx.doi.org/10.1016/j.jmarsys.2004.12.004, 2005.

Penduff, T., Barnier, B., Dewar, W.K., O'Brien, J.J., 2004. Dynamical response of the oceanic eddy field to the North Atlantic Oscillation: a model-data comparison. J. Phys. Oceanogr. 34, 2615–2629.

Polito, P.S., Ryan, J.P., Liu, W.T., Chavez, F.P., 2001. Oceanic and atmospheric anomalies of tropical instability waves. Geophys. Res. Lett. 28, 2233–2236.

Ponte, R.M., Quinn, K.J., 2009. Bottom pressure changes around Antarctica and wind-driven meridional flows. Geophys. Res. Lett. 36, L13604. http://dx.doi.org/10.1029/2009GL039060.

Pujol, M.I., Larnicol, G., 2005. Mediterranean Sea eddy kinetic energy variability from 11 years of altimetric data. J. Mar. Syst. 58, 121–142.

Qiu, B., 1999. Seasonal eddy modulation of the North Pacific Subtropical Countercurrent: Topex/Poseidon observations and theory. J. Phys. Oceanogr. 29, 2471–2486.

Qiu, B., Chen, S., 2004. Seasonal modulations in the eddy field of the South Pacific Ocean. J. Phys. Oceanogr. 34, 1515–1527.

Qiu, B., Chen, S., 2005. Eddy-induced heat transport in the subtropical North Pacific from Argo, TMI and altimetry measurements. J. Phys. Oceanogr. 35, 458–473.

Qiu, B., Chen, S., 2006. Decadal variability in the large-scale sea surface height field of the South Pacific Ocean: observations and causes. J. Phys. Oceanogr. 36, 1751–1762.

Qiu, B., Chen, S., 2010. Eddy-mean flow interaction in the decadally modulating Kuroshio Extension system. Deep Sea Res. 57, 1097–1110.

Qiu, B., Scott, R., Chen, S., 2008. Length scales of eddy generation and nonlinear evolution of the seasonally-modulated South Pacific Subtropical Countercurrent. J. Phys. Oceanogr. 38, 1515–1528.

Quilfen, Y., Chapron, B., Bentamy, A., Gourrion, J., El Fouhaily, T., Vandemark, D., 1999. Global ERS 1 and 2 and NSCAT observations: upwind crosswind and upwind downwind measurements. J. Geophys. Res. 104, 11459–11469.

Reynolds, R.W., Rayner, N.A., Smith, T.M., Stokes, D.C., Wang, W., 2002. An improved in situ and satellite SST analysis for climate. J. Clim. 15, 1609–1625.

Richards, K.J., Maximenko, N.A., Bryan, F.O., Sasaki, H., 2006. Zonal jets in the Pacific Ocean. Geophys. Res. Lett. 33, L03605. http://dx.doi.org/10.1029/2005GL024645.

Ridgway, K., Dunn, J., 2007. Observational evidence for a Southern Hemisphere oceanic 'Supergyre'. Geophys. Res. Lett. 34, L13612. http://dx.doi.org/10.1029/2007GL030392.1.

Rio, M.H., Guinehut, S., Larnicol, G., 2011. New CNES-CLS09 global mean dynamic topography computed from the combination of GRACE data, altimetry, and in situ measurements. J. Geophys. Res. 116, C07018. http://dx.doi.org/10.1029/2010JC006505.

Risien, C.R., Chelton, D.B., 2008. A global climatology of surface wind and wind stress fields from eight years of QuikSCAT scatterometer data. J. Phys. Oceanogr. 38, 2379–2413.

Roblou, L., Lamouroux, J., Bouffard, J., Lyard, F., Le Hénaff, M., Lombard, A., et al., 2011. Post-processing altimeter data towards coastal applications and integration into coastal models. In: Vignudelli, S., Kostianoy, A.G., Cipollini, P., Benveniste, J. (Eds.), Coastal Altimetry. Springer, Berlin/Heidelberg, pp. 217–246.

Roemmich, D., 2007. Super spin in the southern seas. Nature 449, 34–35.

Roemmich, D., Gilson, J., 2001. Eddy transport of heat and thermocline waters in the North Pacific: a key to interannual/decadal climate variability? J. Phys. Oceanogr. 31, 675–688.

Roemmich, D., Gilson, J., Davis, R., Sutton, P., Wijffels, S., Riser, S., 2007. Decadal spin-up of the South Pacific subtropical gyre. J. Phys. Oceanogr. 37 (2), 162–173.

Rouault, M.J., Mouche, A., Collard, F., Johannessen, J.A., Chapron, B., 2010. Mapping the Agulhas Current from space: an assessment of ASAR surface current velocities. J. Geophys. Res. 115, C10026. http://dx.doi.org/10.1029/2009JC006050.

Rummel, R., 2013. Height unification using GOCE. J. Geodetic Sci. 2 (4), 355–362, Versita, ISSN 2081-9943, http://dx.doi.org/10.2478/v10156-011-0047-2.

Sallée, J.B., Speer, K., Morrow, R., Lumpkin, R., 2008a. An estimate of Lagragian eddy statistics and diffusion in the mixed layer of the Southern Ocean. J. Mar. Res. 66 (4), 441–446.

Sallée, J.B., Morrow, R., Speer, K., 2008b. Eddy heat diffusion and Subantarctic Mode Water formation. Geophys. Res. Lett. 35, L05607. http://dx.doi.org/10.1029/2007GL032827.

Sallée, J.B., Speer, K., Rintoul, S.R., 2011. Mean-flow and topography control on surface eddy-mixing in the Southern Ocean. J. Mar. Res. 69, 753–777.

Samelson, R.M., Wiggins, S., 2006. Lagrangian Transport in Geophysical Jets and Waves. Springer-Verlag, New York, 147 pp.

Sangrà, P., Pascual, A., Rodríguez-Santana, Á., Machín, F., Mason, E., McWilliams, J.C., et al., 2009. The Canary Eddy Corridor: a major pathway for long-lived eddies in the subtropical North Atlantic. Deep Sea Res. 56 (12), 2100–2114. http://dx.doi.org/10.1016/j.dsr.2009.08.008.

Saraceno, M., Strub, P.T., Kosro, P.M., 2008. Estimates of sea surface height and near-surface alongshore coastal currents from combinations of altimeters and tide gauges. J. Geophys. Res. 113, C11013. http://dx.doi.org/10.1029/2008JC004756.

Schlax, M.G., Chelton, D.B., 2008. The influence of mesoscale eddies on the detection of quasi-zonal jets in the ocean. Geophys. Res. Lett. 35, L24602.

Schlax, M.G., Chelton, D.B., Freilich, M.H., 2001. Sampling errors in wind fields constructed from single and tandem scatterometer datasets. J. Atmos. Oceanic Technol. 18, 1014–1036.

Scott, R.B., Arbic, B.K., 2007. Spectral energy fluxes in geostrophic turbulence: implications for ocean energetics. J. Phys. Oceanogr. 37 (3), 673–688.

Scott, R.B., Wang, F., 2005. Direct evidence of an oceanic inverse kinetic energy cascade from satellite altimetry. J. Phys. Oceanogr. 35 (9), 1650–1666.

Siedler, G., Church, J., Gould, J., 2001. Ocean Circulation & Climate. Academic Press, San Diego, 715 pp.

Siegel, D.A., Court, D.B., Menzies, D.W., Peterson, P., Maritorena, S., Nelson, N.B., 2008. Satellite and in situ observations of the bio-optical signatures of two mesoscale eddies in the Sargasso Sea. Deep Sea Res. Part II 55, 1218–1230.

Small, R.J., deSzoeke, S.P., Xie, S.-P., O'Neill, L., Seo, H., Song, Q., et al., 2008. Air-sea interaction over ocean fronts and eddies. Dyn. Atmos. Oceans 45, 274–319.

Smith, K.S., 2007. The geography of linear baroclinic instability in the Earth's oceans. J. Mar. Res. 65, 655–683.

Sokolov, S., Rintoul, S.R., 2009a. Circumpolar structure and distribution of the Antarctic Circumpolar Current fronts: 1. Mean circumpolar paths. J. Geophys. Res. 114, C11018. http://dx.doi.org/10.1029/2008JC005108.

Sokolov, S., Rintoul, S.R., 2009b. Circumpolar structure and distribution of the Antarctic Circumpolar Current fronts: 2. Variability and relationship to sea surface height. J. Geophys. Res. 114, C11019. http://dx.doi.org/10.1029/2008JC005248.

Souza, J.M.A.C., de Boyer Montegut, C., Le Traon, P.Y., 2011. Mesoscale eddies in the South Atlantic. Ocean Sci. Discuss. 8, 483–531. http://dx.doi.org/10.5194/osd-8-483-2011.

Stammer, D., 1998. On eddy characteristics, eddy transports and mean flow properties. J. Phys. Oceanogr. 28, 727–739.

Sudre, J., Morrow, R., 2008. Global surface currents: a high-resolution product for investigating ocean dynamics. Ocean Dyn. 58 (2), 101–118.

Tapley, B.D., Bettadpur, S., Ries, J.C., Thompson, P.F., Watkins, M., 2004. GRACE measurements of mass variability in the Earth system. Science 305, 503–505.

Tapley, B.D., Ries, J., Bettadpur, S., Chambers, D., Cheng, M., Condi, F., et al., 2005. GGM02—an improved Earth gravity field model from GRACE. J. Geod. 79 (8), 467–478. http://dx.doi.org/10.1007/s00190-005-0480-z.

Tew Kai, E., Rossi, V., Sudre, J., Weimerskirch, H., Lopez, C., Hernandez-Garcia, E., et al., 2009. Top marine predators track Lagrangian coherent structures. Proc. Natl. Acad. Sci. U.S.A. 106 (20), 8245–8250. http://dx.doi.org/10.1073/pnas.0811034106.

Thiébaux, J., Rogers, E., Wang, W., Katz, B., 2003. A new high-resolution blended real-time global sea surface temperature analysis. Bull. Am. Meteorol. Soc. 84, 645–656.

Thompson, A.F., Sallée, J.-B., 2012. Jets and topography: jet transitions and the impact on transport in the Antarctic Circumpolar Current. J. Phys. Oceanogr. 42, 956–972. http://dx.doi.org/10.1175/JPO-D-11-0135.1.

Timmermann, A., McGregor, S., Jin, F.-F., 2010. Wind effects on the past and future regional sea level trends in the Southern Indo-Pacific. J. Clim. 23, 4429–4437.

Turiel, A., Sol, J., Nieves, V., Ballabrera-Poy, J., Garcia-Ladona, E., 2008. Tracking oceanic currents by singularity analysis of Micro-Wave Sea Surface Temperature images. Remote Sens. Environ. 112, 2246–2260.

Turiel, A., Nieves, V., Garcia-Ladona, E., Font, J., Rio, M.-H., Larnicol, G., 2009. The multifractal structure of satellite sea surface temperature maps can be used to obtain global maps of streamlines. Ocean Sci. 5, 447–460.

van Sebille, E., Kamenkovich, I., Willis, J.K., 2011. Quasi-zonal jets in 3D Argo data of the northeast Atlantic. Geophys. Res. Lett. 38, L02606. http://dx.doi.org/10.1029/2010GL046267.

Velicogna, I., 2009. Increasing rates of ice mass loss from the Greenland and Antarctic ice sheets revealed by GRACE. Geophys. Res. Lett. 36, L19503. http://dx.doi.org/10.1029/2009GL040222.

Wallace, J.M., Mitchell, T.P., Deser, C., 1989. The influence of sea surface temperature on surface wind in the eastern equatorial Pacific: seasonal and interannual variability. J. Clim. 2, 1492–1499.

Willis, J.K., 2010. Can in situ floats and satellite altimeters detect long-term changes in Atlantic Ocean overturning? Geophys. Res. Lett. 37, L06602. http://dx.doi.org/10.1029/2010GL042372.

Willis, J.K., Fu, L.-L., 2008. Combining altimeter and subsurface float data to estimate the time-averaged circulation in the upper ocean. J. Geophys. Res. 113, C12017. http://dx.doi.org/10.1029/2007JC004690.

Willis, J.K., Roemmich, D., Cornuelle, B., 2004. Interannual variability in upper ocean heat content, temperature, and thermosteric expansion on global scales. J. Geophys. Res. 109, C12036. http://dx.doi.org/10.1029/2003JC002260.

Willis, J.K., Chambers, D.P., Kuo, C.-Y., Shum, C.K., 2010. Global sea level rise: recent progress and challenges for the decade to come. Oceanography 23 (4), 26–35.

Xie, S.-P., 2004. Satellite observations of cool ocean-atmosphere interaction. Bull. Am. Meteorol. Soc. 85, 195–208.

Xie, S.-P., Ishiwatari, M., Hashizume, H., Takeuchi, K., 1998. Coupled ocean-atmospheric waves on the equatorial front. Geophys. Res. Lett. 25, 3863–3866.

Xu, Y., Fu, L.-L., 2012. The effects of altimeter instrument noise on the estimation of the wavenumber spectrum of sea surface height. J. Phys. Oceanogr. 42, 2229–2233. http://dx.doi.org/10.1175/JPO-D-12-0106.1.

Yoder, J.A., Doney, S.C., Siegel, D.A., Wilson, C., 2010. Study of marine ecosystems and biogeochemistry now and in the future: examples of the unique contributions from space. Oceanography 23 (4), 104–117. http://dx.doi.org/10.5670/oceanog.2010.09.

Yoshida, S., Qiu, B., Hacker, P., 2010. Wind generated eddy characteristics in the lee of the island of Hawaii. J. Geophys. Res. 115, C03019. http://dx.doi.org/10.1029/2009JC005417.

Zlotnicki, V., Wahr, J., Fukumori, I., Song, Y.T., 2007. Antarctic circumpolar current transport variability during 2003–05 from GRACE. J. Phys. Oceanogr. 37, 230–244. http://dx.doi.org/10.1175/JPO3009.1.

Part III

Ocean Processes

The coupled ocean–atmosphere system is arguably the key component of earth's climate. The two media are intimately coupled through processes at the ocean surface where conditions cover the span from calm to the most tempestuous seas and from the hot, humid tropics to the frigid polar regions. The first chapter in this Part III describes what we know of the physical transfer processes between atmosphere and ocean. The key to recent progress has been the collection of high-quality data sets from ships, buoys, and satellites and the critical evaluation of these data followed by their combination into global flux data fields. The following chapters then cover the processes that determine how the changes that take place in the upper ocean as a result of varying air–sea fluxes then spread through the oceans' interior as a result of mixing, both lateral and across density surfaces.

Fundamental to all of these interior processes is the thermodynamic relationship between temperature, salinity, pressure, and density. Recent progress has for the first time allowed the equation of state that governs these relationships to be rigorously redefined in such a manner that the effects of regional variations in the composition of seawater can now be included. These advances have implications in particular for the formulation of ocean models. It should be noted that the redefinition is so recent that it has not been used in the preparation of the other chapters.

Turbulent mixing processes govern the manner in which properties set at the ocean surface penetrate deep into the ocean and determine the oceans' basic stratification. Measurements in recent decades have revealed the enormous geographical variability of this mixing and the broad range of processes involved. Lateral mixing, particularly that associated with the mesoscale variability that pervades the open ocean, also governs the distribution of properties and hence the structure of the oceans' circulation. Again an adequate understanding and parameterization of these processes is a fundamental requirement for the formulation of realistic models of the ocean, its circulation, and its role in climate.

This Part concludes with descriptions of the distribution and properties of surface-modified "mode" waters and of the formation and circulation of the deep water masses that occupy much of the oceans' volume. The advent of the Argo profiling float array has made it possible to observe unambiguously the changing properties of mode waters. Hence we can now monitor the impacts on the ocean of changing air–sea fluxes. This is a fundamental advance in our understanding of the oceans' role on climate. Despite the tremendous observational challenges involved, considerable progress has been made in observing changes in the formation and circulation of deepwater masses both in the deep ocean and at high latitudes. This progress is important to improve our understanding of the storage of carbon dioxide and heat resulting from human activity.

Chapter 5

Exchanges Through the Ocean Surface

Simon A. Josey*, Serge Gulev† and Lisan Yu‡
*National Oceanography Centre, Southampton, United Kingdom
†P.P. Shirshov Institute of Oceanology, Moscow, Russia
‡Woods Hole Oceanographic Institution, Woods Hole, Massachusetts, USA

Chapter Outline

1. INTRODUCTION

Exchanges of heat, water, and momentum across the air–sea interface play major roles in driving the circulation of the ocean and atmosphere on timescales ranging from less than a day to millennia. Obtaining reliable estimates of these exchanges remains a major challenge because of the complex nature of the transfer processes, their dependence on a broad range of variables, and the lack of observations of these variables at the desired sampling level across much of the global ocean.

Over the past decade, the air–sea flux research agenda has expanded, with the leading desire now being the identification of signals associated with anthropogenic climate change. In particular, basin-scale changes in evaporation and precipitation are expected as the result of a modified hydrological cycle (Held and Soden, 2006) and are likely to be responsible in part for observations of changing ocean salinity (Hosoda et al., 2009; Durack and Wijffels, 2010) (See Chapter 28). At the same time, the long-standing problem of obtaining closure of the ocean heat budget remains to be adequately resolved. A shift has also taken place in the development of air–sea flux datasets from products based solely on *in situ* observations (e.g., Josey et al., 1999) to harnessing of the combined power of satellite observations and atmospheric reanalyses to produce high resolution, more globally complete fields (Yu and Weller, 2007).

Here, an attempt is made to review some of the key developments in the subject area as well as to summarize the ocean–atmosphere exchange bulk formulae. A focal point running through much of this discussion is the ocean heat budget closure problem that remains a major unresolved issue in the field. This problem may be stated as follows: at multi-decadal timescales the global mean net heat flux into the ocean should be close to zero as observed changes in ocean heat content have placed a limit of about

Ocean Circulation and Climate, Vol. 103. http://dx.doi.org/10.1016/B978-0-12-391851-2.00005-2

0.5 W m^{-2} on the ocean warming signal in recent decades (Levitus et al., 2009). However, in many cases, available flux datasets have a global mean net heat flux in the range 10–30 W m^{-2}, a situation that is indicative of major unresolved biases in the heat flux estimates and one that has not improved with the new generation of atmospheric reanalyses (see in particular Section 3.3).

Space limitations prevent as full a discussion of many topics as we would have liked and we note that the report prepared by the Working Group on Air–Sea Fluxes (WGASF, 2000) remains an excellent resource for many air–sea interaction topics. Gas and dust exchanges, particularly the flux of CO_2, are treated elsewhere (see Chapters 30 and 31). Other topics of direct relevance include the ocean heat and freshwater transports that are covered in Chapter 29. Air–sea exchanges in the polar oceans are also a key topic given the major sea-ice reduction observed in the Arctic since 2007, and this region is covered in Chapters 16 and 17.

Within this chapter, the air–sea exchange formulae and the key properties of the climatological exchange fields are first summarized in Section 2. Section 3 follows with a discussion of measurement techniques and an overview of the wide range of flux datasets currently available with reference to the ocean heat budget closure problem. Variability of air–sea exchanges is a very broad research field given the many space- and time-scales over which it may be defined; three key topics in this area have been highlighted in Section 4. In Section 5, the ocean impacts of air–sea flux variability are considered with a focus on the framework of water mass transformation and how this may be employed to estimate surface-forced variability in the Atlantic overturning circulation. Finally, future prospects for improved flux datasets and enhanced observational constraints are discussed in Section 6 "Outlook and Conclusions."

2. AIR–SEA EXCHANGE FORMULAE AND CLIMATOLOGICAL FIELDS

2.1. Air–Sea Exchange Formulae

The basic processes governing the exchanges of heat, water, and momentum between the atmosphere and ocean have been discussed at length in the literature (see WGASF, 2000 for a full treatment and Gulev et al., 2010 for a recent update). Here we summarize the key formulae for estimating the fluxes that are parameterizations of the physical processes governing the exchanges. First, the net heat flux, Q_{Net}, is the sum of four components, two of which are turbulent in nature (the latent and sensible heat fluxes) and two radiative (the shortwave and longwave flux). The net freshwater flux ($E-P$) is simply the difference between evaporation (E) and precipitation (P). Q_{Net} and $E-P$ have some degree of mutual dependence as the evaporation term scales with the latent heat flux apart from a minor dependence of the latent heat of evaporation on sea surface temperature.

The turbulent terms tend to dominate net heat flux variability once any seasonal cycle has been removed, and they are treated first. Each is governed primarily by the wind speed and near-surface vertical gradients of humidity (for the latent term) and temperature (for the sensible heat). They have been widely estimated from measurements of the meteorological variables using the following 'bulk' formulae:

$$Q_{\mathrm{E}} = \rho L C_{\mathrm{e}} u (q_{\mathrm{s}} - q_{\mathrm{a}}) \tag{5.1}$$

$$Q_{\mathrm{H}} = \rho c_{\mathrm{p}} C_{\mathrm{h}} u (T_{\mathrm{s}} - (T_{\mathrm{a}} + \gamma z)) \tag{5.2}$$

where ρ is the density of air; L, the latent heat of vaporization; C_{e} and C_{h}, the stability and height-dependent transfer coefficients for latent and sensible heat, respectively; u, the wind speed; q_{s}, 98% of the saturation specific humidity at the sea surface temperature to allow for the salinity of sea water, and q_{a}, the atmospheric specific humidity; c_{p}, the specific heat capacity of air at constant pressure; T_{s}, the sea surface temperature; T_{a}, the surface air temperature with a correction for the adiabatic lapse rate, γ; and z, the measurement height for air temperature.

A third formula provides estimates of the wind stress on the ocean surface, which is equivalent to the momentum flux across the air–sea interface,

$$\begin{aligned} \tau_x &= \rho C_{\mathrm{D}} u_x \left(u_x^2 + u_y^2\right)^{1/2} \\ \tau_y &= \rho C_{\mathrm{D}} u_y \left(u_x^2 + u_y^2\right)^{1/2} \end{aligned} \tag{5.3}$$

where, τ_x, and τ_y are the zonal and meridional components of the wind stress; u_x and u_y are the corresponding components of the wind speed, respectively; and C_{D} is the drag coefficient. The transfer and drag coefficients depend upon wind speed, measurement heights, and atmospheric stability, with further potential influences from a range of processes, for example, wave characteristics for C_{D} (Taylor and Yelland, 2001) and sea spray for C_{e} (Andreas et al., 2008), see Section 3.1.1. A long series of field campaigns to obtain direct flux measurements, in particular through the eddy correlation method, has led to more accurate values for the transfer coefficients and the Coupled Ocean-Atmosphere Response Experiment (COARE) flux algorithm (Fairall et al., 2003, 2011; see Section 3.1.1). Significant uncertainties remain, however, particularly in the high wind speed regime.

Turning to the radiative heat fluxes, the net longwave flux is the residual of upward- and downward-directed components from the ocean and atmosphere, respectively. It has been determined empirically from *in situ* observations for various flux datasets using formulae that depend primarily on sea surface temperature, air temperature, humidity, and cloud amount. Josey et al. (2003) developed a new longwave formula from measurements made in the North Atlantic. The new formula expresses the combined effects of cloud cover and other parameters on atmospheric longwave in terms of an adjustment, ΔT_a, to the measured air temperature; where ΔT_a is the difference between the measured air temperature and the effective temperature of a blackbody that emits a radiative flux equivalent to the atmospheric longwave. The new formula was tested using independent measurements made on two more recent cruises and found to perform well, agreeing to within 2 W m^{-2} in the mean, at mid-high latitudes. The net shortwave flux may likewise be estimated via the empirical formula approach where the dependency is now primarily on cloud cover and solar elevation (e.g., Reed, 1977). For a more detailed discussion of empirical radiative flux formulae, see for example Josey (2011).

In atmospheric model reanalyses and satellite retrievals, a radiative transfer model-based approach is used to estimate the longwave and shortwave terms (e.g., Gupta et al., 1999; Zhang et al., 2004). Significant progress has been made with satellite-retrieved radiative fluxes over the past decade, see discussion in Section 3.2.2, which will likely provide the most reliable source of data on these fluxes into the future. However, pre-satellite era studies (i.e., before the mid-1980s) are forced to rely either on empirical estimates or output from reanalyses and there is the potential to significantly improve such estimates through inter-comparison with remote sensing measurements in the recent period where they overlap. Note, it is important that the longwave and shortwave fluxes are taken from the same product as they have a significant degree of mutual dependence resulting from the influence of cloud cover.

2.2. Climatological Fields

The climatological annual mean spatial variation of each heat flux component is shown in Figure 5.1 using two products: the combined satellite and reanalysis Objectively Analyzed Air-sea Fluxes (OAFlux) dataset for the turbulent fluxes (Yu and Weller, 2007) and the satellite-only International Satellite Cloud Climatology Project (ISCCP) fields for the radiative fluxes (Zhang et al., 2004). Estimates of these fields from other sources are qualitatively similar as regards the main features although significant quantitative differences do exist (see Section 3.2). The latent heat flux field shows major ocean heat losses, exceeding −200 W m^{-2}, associated with the Gulf Stream and Kuroshio, and broader, slightly weaker heat loss regions associated with the Trade Wind return flows. In contrast, the sensible heat flux tends to be an order of magnitude smaller than the latent, the exception being high latitude seas, which experience advection of very cold air originating over land.

The influence of cloud cover on the shortwave flux is evident in the near-equatorial band of reduced solar radiation at the ocean surface that is a response to the greater cloud amount associated with the Inter—Tropical Convergence Zone (ITCZ). Away from the Tropics, which exhibit peak values of 250 W m^{-2}, the shortwave flux declines toward the poles as expected, given its dependence on mean solar elevation. In comparison, the net longwave flux is a relatively uniform field with a small range (about −30 to −70 W m^{-2}). The spatial variations that are present arise from changes in the sea—air temperature difference, amount of water vapor, and cloud cover, with the ITCZ again evident.

The OAFlux and ISCCP component fields have been used to determine the net heat flux field shown in Figure 5.2a. Also shown, as examples, are net heat flux estimates from several other sources: the coupled ocean–atmosphere National Centers for Environmental Prediction (NCEP) Climate Forecast System Reanalysis (CFSR) (Saha et al., 2010), the Coordinated Ocean Research Experiments version 2 (COREv2) product (Large and Yeager, 2009), and the National Oceanography Centre version 1.1a (NOC1.1a) net heat flux field (Grist and Josey, 2003). Each of these products is discussed in more detail later. COREv2 has been developed to provide a globally balanced dataset for forcing ocean models. NOC1.1a is a balanced version of the earlier unbalanced NOC1.1 dataset (Josey et al., 1999) where linear inverse analysis with observed ocean heat transport constraints has been employed to provide closure of the ocean heat budget to within 2 W m^{-2}. The spatial variation for each of the four datasets is qualitatively similar, primarily reflecting the balance of the dominant latent and shortwave components. However, significant differences between the fields do exist, particularly in the boundary between net ocean heat loss and gain, and these reflect the continuing high level of uncertainty in this key ocean field. The OAFlux/ISCCP combination has a global mean net heat flux of 29 W m^{-2}, which is similar in magnitude to many earlier datasets, for example, University of Wisconsin-Milwaukee / Comprehensive Ocean-Atmosphere Data Set (UWM/COADS) (da Silva et al., 1994), NOC1.1, and unadjusted COREv2. The recent CFSR product has a smaller, but still unfeasibly large, global mean net heat flux bias of 15 W m^{-2} (note that observations of changes in ocean heat storage require closure of the surface budget to within 0.5 W m^{-2} at multidecadal timescales). The large imbalances in these and many other products reflect the global ocean heat budget closure problem (highlighted in the Introduction), which has still not been solved without resort to the use of heat transport constraints or selective adjustments of meteorological fields. Although such methods are capable of producing a balanced field, they remain unsatisfactory in the sense that closure has not been achieved on the basis of best knowledge of the exchange formulae alone. Future progress will likely see the position of the zero net heat flux boundary for CFSR and OAFlux/ ISCCP (and other products with large global imbalances) move toward those obtained with the two

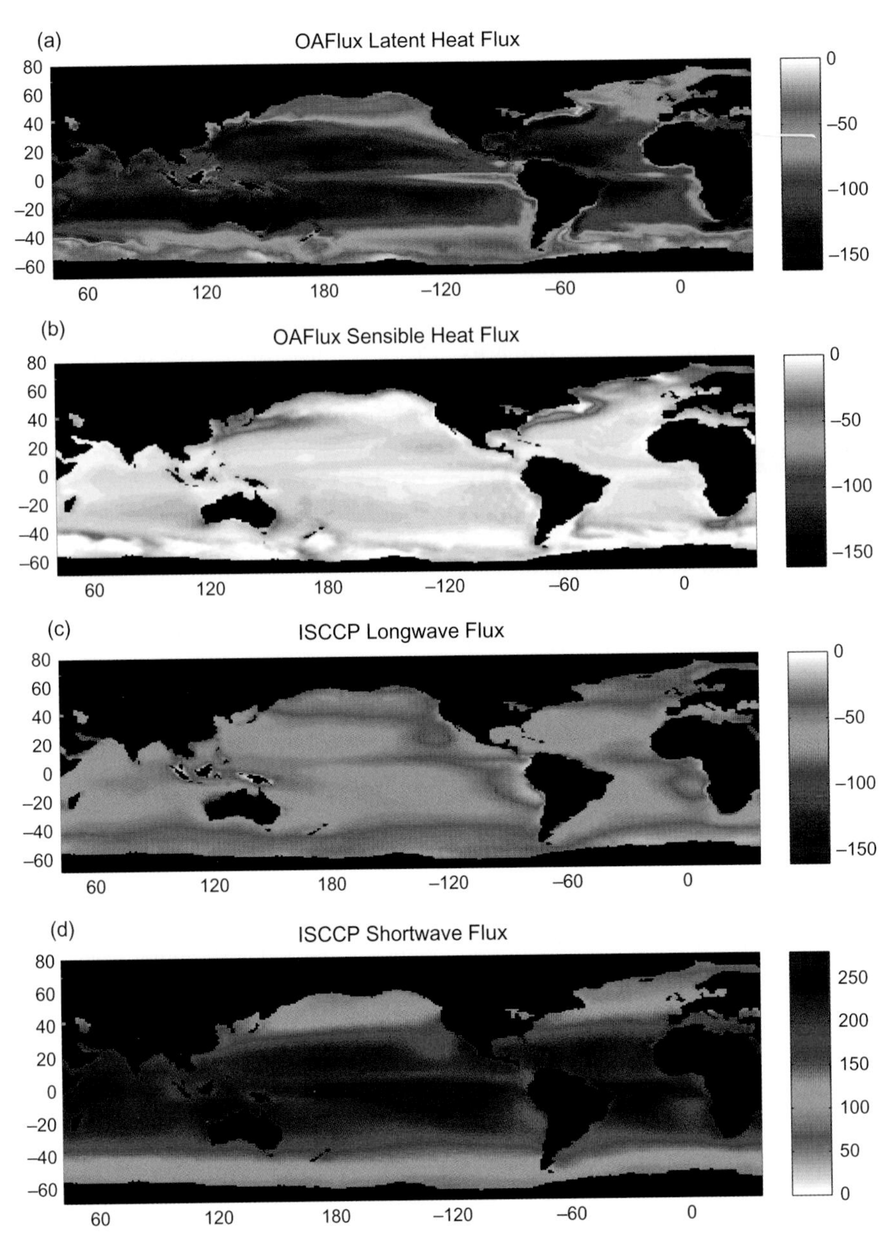

FIGURE 5.1 Components of the annual mean net heat flux (a) Latent Heat (OAFlux), (b) Sensible Heat (OAFlux), (c) Longwave Flux (ISCCP), and (d) Shortwave Flux (ISCCP). Units Wm^{-2}.

adjusted datasets COREv2 (global mean net heat flux 3 W m^{-2}) and NOC1.1a (−2 W m^{-2}).

It should also be noted that, in some regions, the globally unbalanced datasets such as NOC1.1 and OAFlux are in good local agreement (Josey, 2001; Yu et al., 2008) with high quality flux buoy measurements (e.g., in the Subduction region of the eastern subtropical North Atlantic). Forcing global closure by means of ocean heat transport constraints can introduce net heat flux biases of the order of 20 W m^{-2} in such regions where good agreement was previously obtained (Grist and Josey, 2003). Thus, closure of the global heat budget is likely to require large adjustments in some regions (e.g., toward Western boundaries) but not in others. Another element of the closure problem is the freshwater budget, which, although less strongly constrained than the heat budget, is potentially useful as the two budgets are mutually dependent because of the latent heat flux (equivalently evaporation) that appears in both.

Significant progress has been made over the past decade in the estimation of wind stress from space as a result of measurements made by passive microwave radiometers and scatterometers on various satellite missions. Yu and Jin (2012) have recently produced a new ¼° OAFlux vector wind analysis by synthesizing wind speed and direction retrievals from 12 sensors acquired during the satellite era from July 1987 onward (see their paper

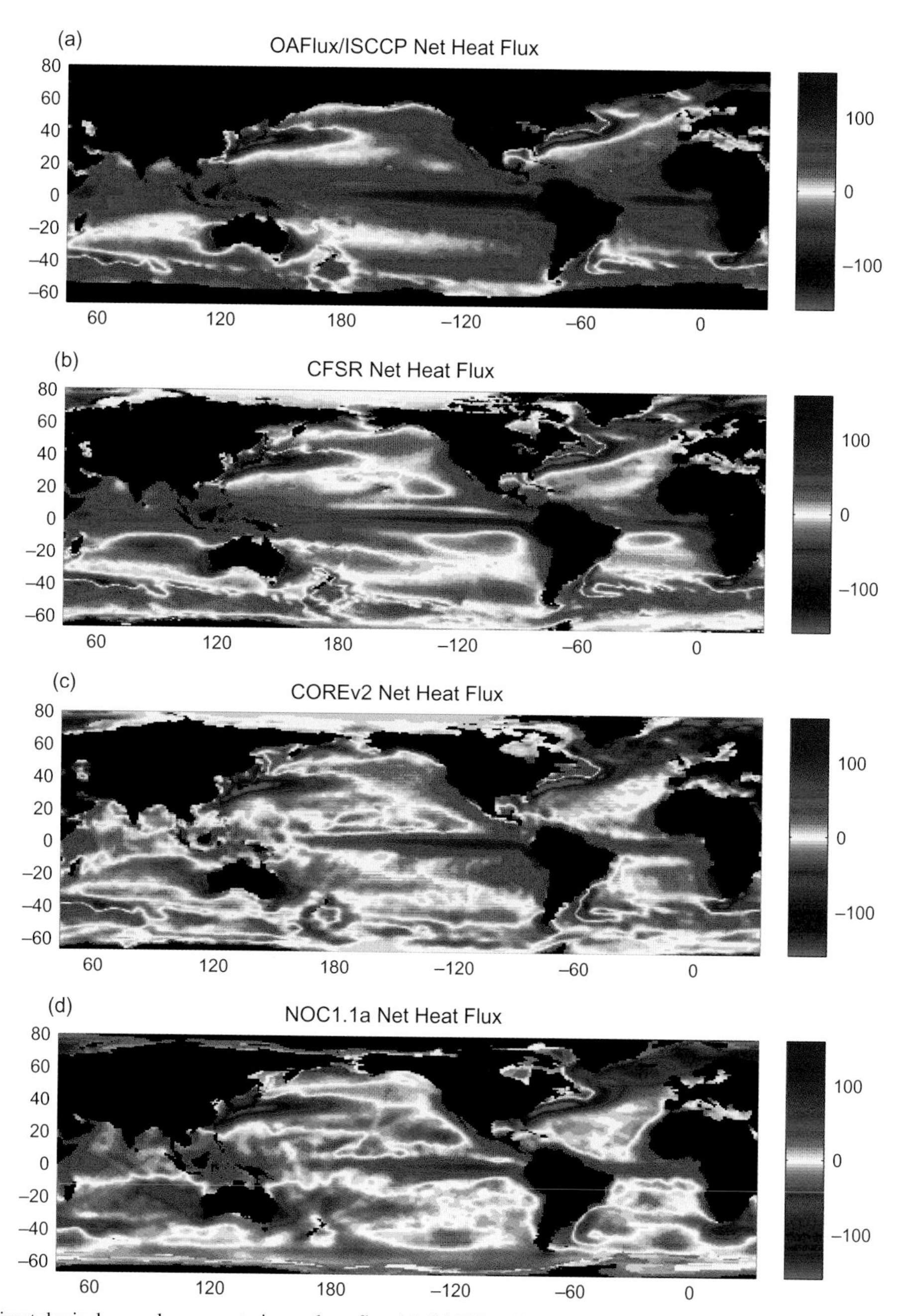

FIGURE 5.2 The climatological annual mean net air–sea heat flux (a) OAFLUX/ISCCP, (b) CFSR, (c) COREv2, and (d) NOC1.1a. Units Wm^{-2}.

for discussion of the different satellite missions and progress in this field). Their new product is used to illustrate the spatial variation of the annual mean surface wind stress (Figure 5.3; note that similar features are obtained from *in situ* observation-based datasets, for example, NOC1.1, Josey et al., 2002). The strongest feature is the intense westerly wind belt in the Southern Ocean which has strengthened in recent decades (e.g., Yang et al., 2007; Swart and Fyfe, 2012). The other main features are the subtropical and subpolar gyres; these also exhibit significant interdecadal variability (see, for example, Häkkinen et al., 2011 for the North Atlantic subpolar gyre). Ocean circulation fields derived from the wind stress include Ekman transport, upwelling and downwelling, and the Sverdrup transport (e.g., Josey et al., 2002; see Chapter 11 for a discussion on the wind-driven ocean circulation).

The climatological annual mean net freshwater flux field obtained using OAFlux for E and Global Precipitation Climatology Project version 2 (GPCPv2 Adler et al., 2003) for P is shown in Figure 5.4. The ocean gains freshwater primarily in the Tropics and mid-high latitudes, where P is dominant, and loses it to the atmosphere under the Trade

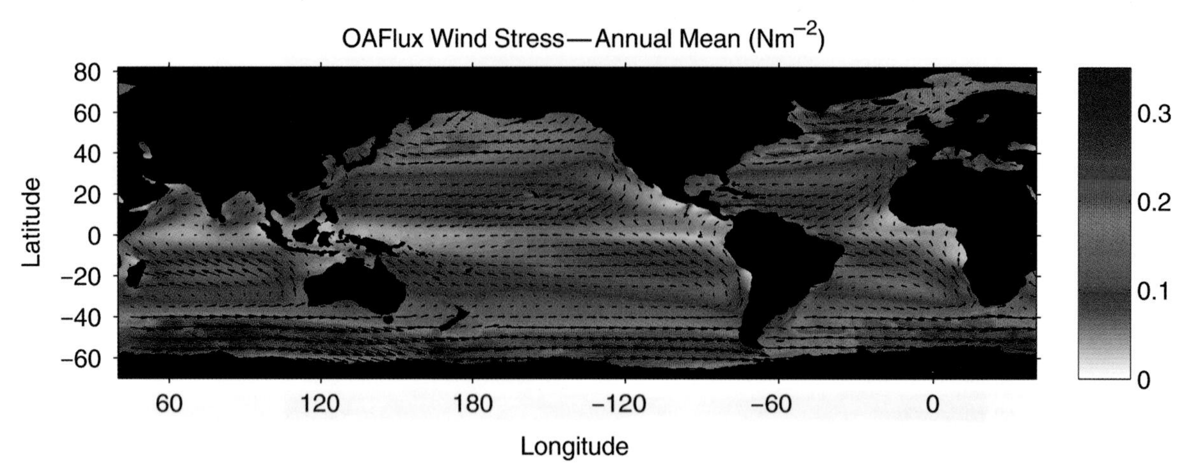

FIGURE 5.3 Global annual mean wind stress from OAFlux. Units N m^{-2}.

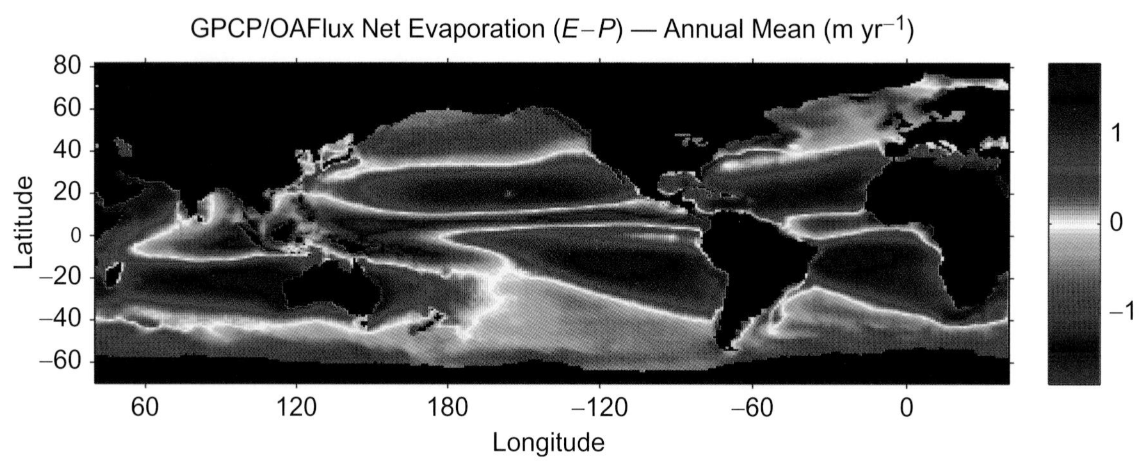

FIGURE 5.4 Global annual mean net evaporation ($E-P$) field from a combination of OAFlux (for E) and GPCPv2 (for P). Units m yr^{-1}.

Wind return flows where E is the major term. Significant differences remain between precipitation datasets; Béranger et al. (2006) compared 10 products and found reasonable qualitative but poor quantitative agreement, with strong regional variations. Schanze et al. (2010) evaluated eight evaporation (E) and seven precipitation (P) products from satellite-based global analyses and reanalyzed models using the ocean water budget constraint. They found that the ocean freshwater flux ($E-P-R$) is in balance to within the measurement uncertainties when E and P are from OAFlux and GPCP, but the budget is far from closed when E and P are from the reanalyses. River inflow forms about 10% of the freshwater flux to the oceans and uncertainty in this term has been significantly reduced through the analysis of Dai et al. (2009), who have produced a global monthly streamflow dataset that includes the 925 largest ocean-reaching rivers.

The relationship between ocean salinity and $E-P$ is of considerable interest, particularly given the potential of ocean salinity to register anthropogenic changes to the hydrological cycle (Section 4.2 and Chapter 28). Recent analyses reveal that over the past 40 years the surface ocean has tended to freshen in the Tropics and become more saline in the sub-tropics (Curry et al., 2003; Hosoda et al., 2009; Durack and Wijffels, 2010; Durack et al., 2012), which is consistent with expectations if an intensification of the hydrological cycle leads to an increase in the amplitude of the mean $E-P$ field (Held and Soden, 2006).

In addition to potential climate change-related signals, it is necessary to understand processes responsible for natural variability at multidecadal timescales and the seasonal cycle. Studies have shown that in some regions the $E-P$ flux can account for a significant component of multidecadal variability in surface salinity (Josey and Marsh, 2005; Cravatte et al., 2009; Mariotti, 2010), while in others, ocean advection and mixing processes have an important role in redistributing and dispersing the salinity anomalies away from the generation sites (Johnson et al., 2002; Mignot and Frankignoul, 2004; Grodsky et al., 2006; Reverdin et al., 2007; Ren and Riser, 2010).

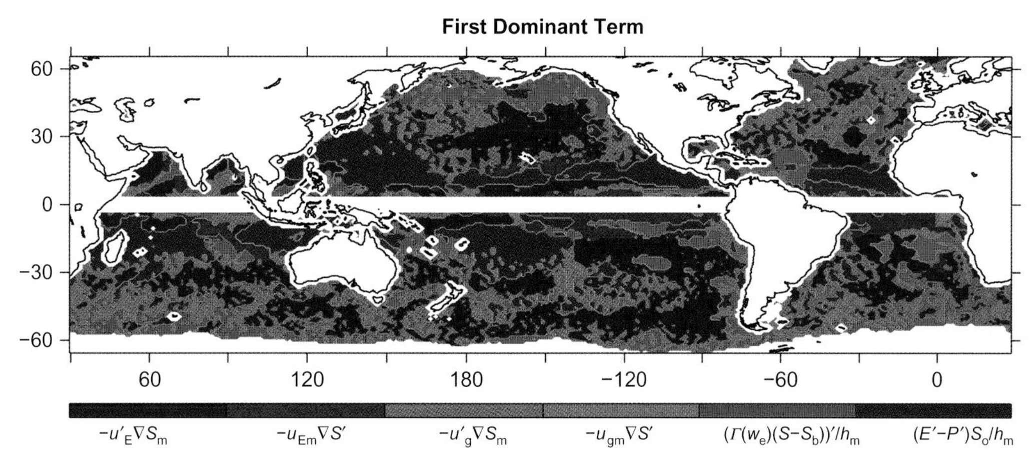

FIGURE 5.5 Map of the leading dominant process in governing seasonal changes of the mixed-layer salinity. *From Yu (2011).*

Yu (2011) has recently developed a better understanding of the relationship between seasonal changes in $E-P$ and salinity, which also requires a consideration of upper-ocean dynamics. A global map from her analysis that outlines the influence regime of key processes in governing seasonal changes of the mixed-layer salinity (MLS) is shown in Figure 5.5. The map is established from an analysis of MLS dynamics that allows key balance terms (i.e., $E-P$, the Ekman and geostrophic advection, vertical entrainment, and horizontal diffusion) to be computed from satellite-derived datasets and a salinity climatology. Major $E-P$ control on seasonal MLS variability (shown in red) is found in two regions: the tropical convergence zones featuring heavy rainfall and the western North Pacific and Atlantic under the influence of high evaporation. Within this regime, $E-P$ accounts for 40–70% of MLS variance with peak correlations occurring at a 2- to 4-month lead time. Outside of the tropics, the MLS variations are governed predominantly by the Ekman advection and, then, vertical entrainment. With satellite salinity measurements becoming available from SMOS and Aquarius, the relationship between surface salinity variability and $E-P$ is now being examined at various temporal- and spatial scales, for example in the Tropical Atlantic (Grodsky et al., 2012; Tzortzi et al., 2013).

3. MEASUREMENT TECHNIQUES AND REVIEW OF DATASETS

3.1. Flux Measurement and Estimation Techniques

3.1.1. *Advances in Parameterizations and* In Situ *Flux Measurements*

Direct surface measurements of air–sea turbulent and radiative fluxes are the basis for developing the air–sea flux parameterizations used in ocean and atmospheric general circulation models and the production of flux datasets. The past decade has been characterized not only by significant progress in quantification of the accuracy of flux-related variable measurements, but also by the development and standardization of observational techniques for research-quality meteorological observations from research vessels and buoys. Weller et al. (2008) provided a detailed overview of sensors, including those for gas and aerosol fluxes, and of the critical problem of their exploitation in different conditions. Furthermore, Bradley and Fairall (2007) developed a comprehensive guide for making accurate flux and meteorological measurements at sea.

Considering the turbulent fluxes of heat, moisture, and momentum, various field campaigns have been carried out and analyses of the data collected (e.g., Fairall et al., 2011) have demonstrated the accuracy of the COARE-3 algorithm (Fairall et al., 2003). For individual flux estimates, it has an uncertainty of about 5% for wind speeds in the range 0–10 ms^{-1}, increasing to 10% for the range 10–20 ms^{-1}. There remain major uncertainties at wind speeds >20 ms^{-1} and this high wind speed regime is a key area requiring further research. Significant differences also exist between transfer coefficients from different sources; in particular, the algorithms used for the NCEP/NCAR reanalysis (note NCAR is the National Center for Atmospheric Research and the NCEP/NCAR reanalysis is also commonly referred to as NCEP-1) and by Large and Yeager (2009) give higher values than the COARE-3 method (see Fig. 5 of Fairall et al., 2010), which is the one that is best supported by observations. In a recent analysis using aircraft measurements, Vickers et al. (2013) note the sensitivity of drag coefficient estimates to the method employed for averaging the surface stress and wind speed. In a recent analysis using aircraft measurements, Vickers et al. (2013) note the sensitivity of drag coefficient estimates to the method employed for averaging the surface stress and wind speed. The COARE algorithm has recently been refined to create COARE 3.5 (Edson et al., 2013).

Considerable effort has been devoted to determining the wind wave-induced component of wind stress (e.g., Taylor and Yelland, 2001; Oost et al., 2002). It is still unclear whether wave age-based parameterizations (e.g., Drennan et al., 2003) or scaling with wind wave steepness (e.g., Taylor and Yelland, 2001) provide the most accurate description of the effect. Drennan et al. (2005) argued that for well-developed seas, the steepness formulation is preferable, while for conditions dominated by young seas, the wave age-based formulation yields the best results. Neither formulation performs well in the presence of swell whose role in modifying the surface wind stress has remained uncertain since the 1990s (e.g., Dobson et al., 1994).

There is some evidence to suggest that wind stress stops growing or decreases under extremely high winds (Powell et al., 2003; Donelan et al., 2004). Further analysis by Makin (2005) and Andreas (2004) indicates that the behavior of the drag coefficient under very strong winds (observed in, e.g., tropical cyclones) is quite different from that predicted by routine parameterizations and does not imply a rapid growth in kinetic energy input into the ocean in storm conditions. Accurate estimation of wind stress additionally requires measurements or model diagnostics of surface currents, since they contribute to the tangential stress (e.g., Dawe and Thompson, 2006; Duhaut and Straub, 2006). Although the global effect is small, locally strong impacts on both wind stress and wind stress curl can occur in the equatorial oceans and western boundary current regions.

Processes associated with wind waves become critically important under strong winds when they affect air–sea heat and moisture exchanges by non-turbulent mechanisms associated with spray. Andreas et al. (2008) argue that spray-induced fluxes of heat and moisture make a significant contribution and suggest a formulation to account for this process. For winds stronger than 25–30 m s^{-1}, spray-induced fluxes become comparable to, or higher than, the interfacial components (Figure 5.6). Bao et al. (2011) proposed a formulation covering wind speeds up to 60 m s^{-1}. Their results are qualitatively comparable to those of Andreas et al. (2008) and demonstrate that the role of spray starts to be clearly visible from speeds of 30 m s^{-1}.

Accurate measurements of radiative fluxes are made at sea using longwave and shortwave radiometers and these have benefited from advances in the understanding of various sources of sensor bias, for example, heating of the sensor dome and shortwave transmission in the case of the longwave flux (Pascal and Josey, 2000). For a full discussion of sensor issues for both radiative and turbulent fluxes, see Weller et al. (2008).

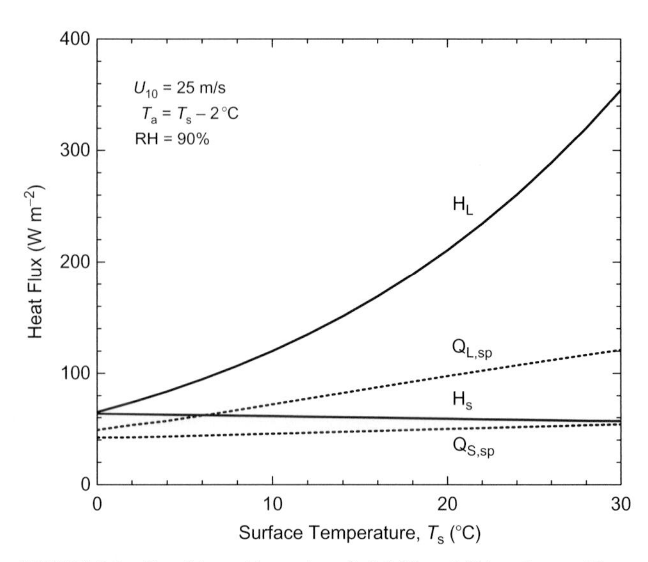

FIGURE 5.6 Sensible and latent interfacial (H_s and H_L) and spray ($Q_{S,sp}$ and $Q_{L,sp}$) fluxes as functions of the 10-m wind speed U_{10}. *After Andreas et al. (2008).*

3.1.2. *High Quality* In Situ *Surface Flux Datasets*

A major development over the past decade has been the increasing number of measurements from high-quality surface flux buoys. Together with research cruise measurements, buoy deployments now provide an increasingly valuable tool both for the analysis of flux related processes in the upper ocean and for evaluation of flux datasets. However, they do not yet constitute a global array as the buoys tend to be concentrated in the Tropics, with just a limited number of mid-latitude deployments and none at high latitudes. It should also be noted that in many cases the deployments have been for only a limited historical period (typically up to 1 year). The accuracy of the buoy measurements has improved with time; a recent example is the analysis by Colbo and Weller (2009) who find that the accuracy of annual mean measurements of each of the four heat flux components lies in the range 1.5–5 W m^{-2} for the WHOI Stratus buoy. Here are two examples illustrating the valuable new results at a regional scale that are possible with the advances in buoy data availability.

First, there has been considerable progress in the analysis of surface fluxes in western boundary current regions and their extensions, which are areas of the most intense air–sea interaction. In the Kuroshio extension region, measurements have been made using the KEO (Kuroshio Earth Observatory) and JAMSTEC-KEO (JKEO) buoys (Cronin et al., 2008). Bond and Cronin (2008) used KEO array-derived fluxes to determine the important role of SST in forming winter surface flux anomalies, while strong air–sea flux synoptic events were caused primarily by highly variable atmospheric conditions. Konda et al. (2010) shed further light on the mechanisms driving surface flux variability north and south of the Kuroshio during cold air outbreaks (strong winds orthogonal to the Kuroshio front) and monsoon conditions (winds aligned along the front). They found that during cold air outbreaks, surface fluxes north and south of the Kuroshio are clearly coordinated while in monsoon conditions, they exhibit independent behavior. This gives the first observational evidence in support of the earlier findings based on model and reanalysis data from Alexander and Scott (1997), Zolina and Gulev (2003), and Shaman et al. (2010).

Direct measurements of surface radiative fluxes in the past decade have contributed to a better understanding of the heat budget of the Eastern Tropical Pacific. Colbo and Weller (2007) present a detailed study of the heat budget of the upper ocean at the site of the WHOI Stratus buoy located under the persistent stratus decks to the west of Peru and Chile. de Szoeke et al. (2010) developed the Tropical Eastern Pacific Stratocumulus Synthesis dataset of surface turbulent and radiative flux measurements at 10-min resolution, spanning 7 years from cruise data, and the WHOI Stratus buoy measurements (Figure 5.7). This unique dataset allowed the accuracy of surface-gridded flux products (OAFlux and CORE turbulent fluxes and ISCCP-FR radiation) to be established and enabled the subsequent attribution of SST errors in coupled ocean–atmosphere climate models in this region. In particular, it was found that most coupled climate simulations have excessive insolation and thermal radiative cooling on the order of 20 W m^{-2}, which suggested that they have too few or radiatively ineffective clouds. This has revealed the need for further efforts to accurately observe tropical clouds and precipitation, which may significantly affect cloud properties (e.g., Zuidema et al., 2012).

3.2. Flux Datasets: Overview of Recent Products

The number of new heat flux datasets made available in the past 10 years has increased significantly compared to the preceding decade as is shown schematically in Figure 5.8 (note details of these datasets and expansions of acronyms not given previously are provided in the following sections). This increase has been driven by the advent of a wide range of higher resolution reanalysis products (e.g., CFSR, Modern Era Reanalysis for Research and Applications (MERRA)), the production of new satellite datasets the Hamburg Ocean Atmosphere Parameters and Fluxes from Satellite Data (HOAPS), and the combination of information from different sources to produce hybrid (CORE) and synthesis datasets (OAFLUX). Development of ship-based flux datasets (NOC1.1a, NOC2) remains limited by the inherent sampling problems that arise from the distribution of ship observations, with very limited amounts of information in many regions, particularly the southern hemisphere. No single product can be identified as the best available as each has its own strengths and weaknesses and the selection of dataset must be guided by the problem being tackled, for example, globally balanced products are needed for ocean model forcing.

3.2.1. Atmospheric Reanalyses

Surface flux products from atmospheric reanalyses offer globally complete, high temporal resolution (6 hourly typically; in some cases, 1 or 3 hourly) fields from numerical weather prediction models that assimilate a wide range of data types including direct (e.g., radiosonde) and indirect (e.g., satellite profiles) atmospheric observations in addition to ocean surface observations from ships. The inclusion of atmospheric data and the constraints provided by the model physics potentially allows reanalyses to provide estimates of surface fluxes in regions sparsely sampled by ships that are more accurate than are possible from surface observations alone. The NCEP (Kalnay et al., 1996) and European Centre for Medium range Weather Forecasting (ECMWF) (Uppala et al., 2005) reanalysis products spanning periods of 40–50 years have been widely used by the community and their utility demonstrated in many areas, for example, the characterization of atmospheric modes of variability and their impacts (Hurrell et al., 2003).

However, both the NCEP and ERA40 reanalysis fields have been found to exhibit significant biases in the surface fluxes under certain conditions and, as a consequence, must be used with caution. Specific examples include excessive

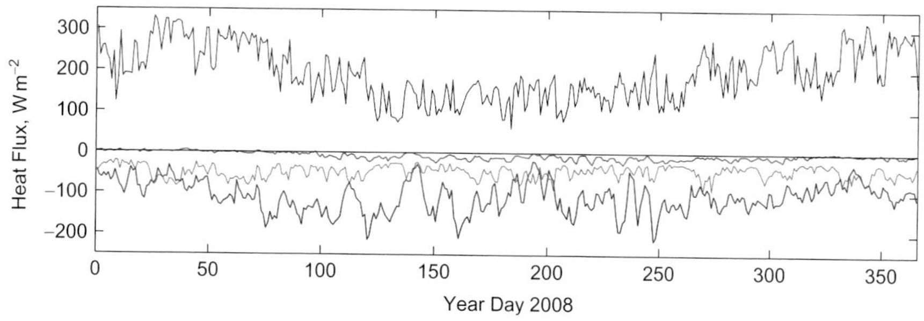

FIGURE 5.7 Time series of the heat flux components (shortwave-red, longwave-magenta, latent-blue, sensible-green, units Wm^{-2}) from the WHOI Stratus buoy for 2008 (Whelan et al., 2009), shown as an example of the reference data becoming available from surface flux moorings.

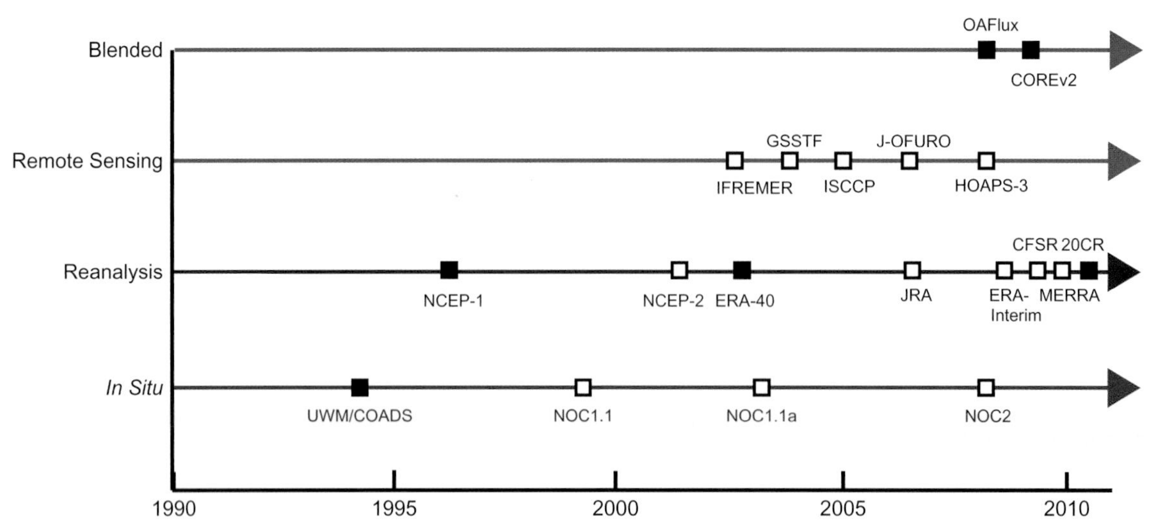

FIGURE 5.8 Schematic representation of various global heat flux datasets developed since the year 1990. The four parallel timelines show reanalysis, remote sensing, *in situ*, and blended products. The filled (unfilled) squares denote products covering greater (less) than 40-year periods.

estimates of the surface heat loss under conditions of high sea–air temperature difference (Renfrew et al., 2002) and under-representation of stratiform cloud that can result in overestimates of the shortwave input to the ocean. The existence of such biases has led to the hybrid approach discussed later in which reanalysis fields are adjusted at the level of the surface meteorological variables prior to estimating the fluxes (Large and Yeager, 2009; Brodeau et al., 2010). A further problem is the variation in data types assimilated by the atmospheric models, particularly during the transition to the satellite era in the 1980s, which can result in spurious variability in the reanalysis products and the blended datasets that make use of them (Brodeau et al., 2010).

The recent generation of new reanalysis products were developed with many of these problems in mind and they offer a higher spatial resolution than previously possible. They include the ECMWF Interim reanalysis (ERA-Interim, Dee et al., 2011), the Japanese 25-year reanalysis (JRA, Onogi et al., 2007), the NASA Modern Era Reanalysis for Research and Applications (MERRA, Rienecker et al., 2011), the NCEP Climate Forecast System Reanalysis (CFSR, Saha et al., 2010) and the NOAA-CIRES 20th Century Reanalysis V2 (20CR, Compo et al., 2011). At this stage it is not possible to give a critical assessment of the relevant strengths and weaknesses of the new generation reanalyses as they have not yet been fully evaluated. A particular concern that should be noted, however, is their representation of the different components of the hydrological cycle following work by Trenberth et al. (2011) who identified significant inconsistencies in many of the new products between the global average $E-P$ from the oceans and the corresponding atmospheric transport of water to land.

3.2.2. Satellite Observations

As for reanalyses, the past 10 years have seen an increasing number of surface flux products developed from satellite missions. Continuous Special Sensor Microwave Imager (SSM/I) measurements span the period from 1987 to date and have enabled the production of several global turbulent flux datasets with high spatial (1/4–1°) and temporal (daily) resolution (reviewed in Gulev et al., 2010). In these datasets, atmospheric humidity and air temperature are statistically retrieved from the brightness temperatures, although large uncertainties remain in these retrievals that severely limit the accuracy of the derived fluxes. SST is either derived from Advanced Very High Resolution Radiometer (AVHRR) radiances or from a mixture of microwave radiometer measurements. Wind speed is obtained from SSM/I measurements or from satellite scatterometers such as QuikSCAT.

The Hamburg Ocean Atmosphere Parameters and Fluxes from Satellite Data products (HOAPS, Andersson et al., 2010) provide turbulent heat flux (and precipitation) fields developed from observations at microwave and infrared wavelengths. However, in common with other satellite-based datasets, it spans only the relatively recent period starting in 1987 and thus cannot be used for studying long-term variability. In addition to HOAPS, there are several similar products that differ from each other by the retrieval algorithms and specific satellite missions used. These include the Japanese Ocean Flux Data sets with Use of Remote Sensing Observations (J-OFURO) product (Kubota et al., 2007), the Goddard Satellite-Based surface Turbulent Fluxes (GSSTF) data (Chou et al., 2003), and the Institut Français de Recherche pour l'Exploitation de la Mer (IFREMER) flux product (Bentamy et al., 2003, 2008).

A key limitation of remote sensing-based flux products is that it remains a major technical challenge to retrieve the near-surface atmospheric q_a and T_a terms required to accurately determine the turbulent fluxes (Curry et al., 2004; Andersson et al., 2010). All satellite flux products need to make particular assumptions for retrieval of q_a and T_a, which lead to further uncertainties in these variables, for example, the assumption of a globally constant relative humidity is often adopted to determine T_a from q_a (Yu, 2009). For satellite based products, q_a and T_a are inferred either from SSM/I integrated water vapor retrievals (e.g., Bentamy et al., 2003) or from a combination of SSM/I with SST and sounding profiles from Advanced Microwave Sounding Unit (AMSU) (Jackson et al., 2005; Roberts et al., 2010). Although the sounders do not directly provide near-surface measurements, they provide detailed profile information that can help to remove variability in total column measurements not associated with the surface. Satellite observations of precipitation have been determined using both microwave and infrared measurements, and these have been combined in the Global Precipitation Climatology Project (GPCP, Adler et al., 2003) that covers 1979 onward. Smith et al. (2009) have employed this dataset to reconstruct global ocean precipitation for 1900–2008 using a Canonical Correlation Analysis (CCA) with SST and SLP.

The surface radiative flux can be estimated indirectly using radiative transfer models initialized by satellite-derived cloud and aerosol properties and meteorological data from atmospheric reanalyses with top of the atmosphere (TOA) irradiance as inputs. Global cloud properties have been available from passive instruments on board operational weather satellites since 1983. Determining the physical properties of clouds from these measurements was a major objective of the International Satellite Cloud Climatology Project (ISCCP, Rossow and Schiffer, 1991). The ISCCP-derived cloud properties, together with temperature and humidity profiles, have been used by Zhang et al., (2004) to compute surface shortwave and longwave radiative fluxes. A parallel effort has been undertaken by the Global Energy and Water Cycle Experiment—Surface Radiation Budget (GEWEX-SRB) project (Stackhouse et al., 2004). The two sets of fluxes are in reasonably good agreement in the long-term mean (see Table 2 in Yu, 2009). However, when compared with ground-based observations, the uncertainty of these fluxes is about 10–15 W m^{-2}. Surface radiative fluxes have also been estimated through the Clouds and the Earth's Radiant Energy System (CERES) experiment using cloud property measurements from the Moderate-Resolution Imaging Spectroradiometer (MODIS). Despite the relative success of ISCCP, GEWEX-SRB, and CERES, surface radiation flux accuracy is constrained by limitations in retrieving the necessary cloud parameters, particularly the cloud base height, needed for the radiative transfer models (Trenberth et al., 2009).

The Cloud-Aerosol Lidar and Infrared Pathfinder Satellite Observation (CALIPSO) (Winker et al., 2010) and CloudSat (Stephens et al., 2008) missions combine active lidar measurements with passive infrared and visible images to probe the vertical structure and properties of clouds and aerosols. These vertical profiles significantly improve the CERES surface radiative flux estimates previously based only on MODIS-derived cloud properties (Kato et al., 2011). At present, CERES provides three monthly averaged data sets, including one termed the Energy Balanced and Filled (EBAF) flux product. EBAF is generated from TOA and surface clear-sky fluxes, all-sky fluxes, and the cloud radiative effect, where the TOA global mean net flux is constrained to lie within limits derived from observed ocean heat storage (about 0.5 W m^{-2}).

3.2.3. In Situ *Observations*

Ship-based flux datasets have a long history (e.g., Bunker, 1976) and continue to provide a valuable complementary source of information to both reanalysis and remote sensing products. However, their use remains limited by the severe under-sampling problems that arise from the strongly inhomogeneous distribution of ship observations. Note that these datasets also contain information from various classes of buoys, drifters, and platforms collected in the ICOADS dataset on which they are based (Woodruff et al., 2011) and the term *ship-based* is used here as, historically, voluntary observing ship data form the primary data source.

da Silva et al. (1994) developed the UWM/COADS flux dataset from a version of COADS spanning the period 1945–1989. They found that, in common with earlier ship-based datasets, UWM/COADS has a global mean net heat flux bias of some 30 W m^{-2} (see discussion in Section 2.2). An adjusted, globally balanced version of UWM/COADS was produced using linear inverse analysis with several ocean heat transport constraints. Likewise, the NOC1.1a flux dataset (Grist and Josey, 2003) was developed from the NOC1.1 (previously termed SOC) flux climatology using linear inverse analysis but with a wider range of constraints, primarily ocean heat transport estimates resulting from various WOCE sections. With these constraints, NOC1.1a has a global mean net heat flux of -2 W m^{-2}. Subsequent research has resulted in the NOC2 dataset (Berry and Kent, 2009) that uses optimal interpolation and features estimates for some of the sources of error in the ship-based flux fields, but still has a large global mean net heat flux bias of 24 W m^{-2}. The voluntary observing ship network has undergone a serious decline in coverage over the past decade (Kent et al., 2010), caused by a reduction in the number of participating vessels, that limits its ongoing use in producing ship-only flux datasets.

3.2.4. Blended Products

Blended flux products make use of data from different sources, principally reanalysis and satellite-based fields (e.g., Yu and Weller, 2007; Large and Yeager, 2009). The approaches used in these two studies are significantly different. The term *hybrid* may be used to refer to the approach of Large and Yeager (2009) and Brodeau et al. (2010) that employs plausible adjustments to individual variables based on a selection of datasets. In contrast, the term *synthesis* is a better description of the Yu and Weller (2007) methodology that uses the Gauss-Markov theorem to formulate a linear least-squares estimator to combine data from various sources and seeks the best fit to input data through minimization. The advantage of the synthesis approach is that the error information of each dataset is incorporated in the least-squares estimator so that the relative importance of satellite observations and atmospheric reanalyses is proportional to the accuracy of the input dataset.

The COREv2 turbulent fluxes, which cover 1948–2006, were developed by Large and Yeager (2009) who modified NCEP/NCAR reanalysis state variables prior to flux calculation using various adjustment techniques. When combined with satellite based radiative flux estimates, COREv2 provides a globally balanced net heat flux field for forcing ocean models. However, the adjustments employed were based on satellite and *in situ*-based observations spanning only limited periods (e.g., 1999–2004 for the wind speed adjustment) and the hybrid COREv2 product contains several fields (the two radiative flux components and precipitation) that are climatological means prior to 1984. Thus it is not clear to what extent it can be reliably used for studies of interdecadal variability. COREv2 features various plausible adjustments to the basic meteorological fields but in some cases, for example, q_a, these have been obtained with reference to earlier flux products that may also contain large errors. Thus, COREv2 should be regarded as one of many possible solutions to the problem of heat budget closure. Another solution has been put forward by Brodeau et al. (2010), who employ a hybrid approach comparable to that of Large and Yeager (2009) but using ERA40 instead of NCEP.

The Objectively Analysed Air–Sea heat flux (OAFlux) product covers the period from 1958 to the present with a 4-month delayed update and for the first time synthesizes state variables (sea surface temperature, air temperature and humidity, wind speed) from reanalyses and, where available, from satellite observations, prior to flux calculation (Yu and Weller, 2007). By combining these data sources, OAFlux avoids the severe spatial sampling problems that limit the usefulness of datasets based on ship observations and offers significant potential for studies of temporal variability. However, the balance of data sources used for OAFlux changed significantly in the mid-1980s with the advent of satellite data, and the consequences of this change need to be assessed. OAFlux has been developed using advanced statistical methods to improve all the basic meteorological fields (u, T_s, q_a and T_a) through an optimal synthesis of satellite- (SSM/I, SSMIS, AVHRR, AMSRE, TMI, QuikSCAT, ASCAT) and reanalysis (NCEP1, NCEP2, ERA40, and ERA-interim)-based estimates of these variables. Global latent and sensible heat fluxes are obtained from the synthesized fields using the bulk flux parameterization of Fairall et al. (2003). The synthesis framework requires an estimate of the errors in the chosen meteorological datasets and these have been obtained with reference to more than 120 active flux buoys over the global oceans (Yu et al., 2008). The OAFlux framework offers great potential for development of a flux product that makes best use of the advantages offered by the reanalysis and satellite data sources. However, it should be noted (see Section 2.2) that when combined with satellite radiative flux estimates from data such as ISCCP and SRB, it still has a significant (>25 W m^{-2}) global mean net heat flux bias and this problem remains to be resolved.

A further source of information on the net heat flux is the residual technique (e.g., Trenberth et al., 2001). In this case, the mean net heat flux field is determined as the residual of top-of-the-atmosphere satellite measurements of the radiative flux and atmospheric heat flux divergence estimates determined from atmospheric model reanalyses. This approach is attractive in that it avoids the need to estimate the surface flux field directly from sparse observations but is limited by the accuracy of the reanalyses employed for the calculation. Furthermore, it provides information on only the net heat flux and not the individual components.

3.3. Flux Datasets: Evaluation Techniques

The advent of high-quality measurements from the growing network of air–sea flux buoys has resulted in more sophisticated evaluations of new flux datasets that now go significantly beyond the old practice of intercomparison with earlier products, which have their own deficiencies.

Josey and Smith (2006) summarized evaluation techniques in a three-stage methodology with the following elements:

a. local evaluation of time-averaged fluxes and variables at specific grid locations with corresponding research quality data;
b. regional evaluation of either gridded flux product implied ocean transports or, preferably, area-averaged fluxes with corresponding research quality data from hydrographic sections;

c. global evaluation of gridded flux product area weighted mean fluxes through closure of the appropriate property budget within observational constraints.

For stage (a), research quality data typically consist of *in situ* point measurements from research buoys or ships at high frequency (typically averaged up to timescales from minutes to hours). This methodology has been adopted to some extent in subsequent studies (e.g., Yu et al., 2008; Praveen et al., 2011).

Studies using research buoy data in specific regions have already been discussed in Section 3.1.2. At a global scale, they are beginning to enable the identification of spatially dependent biases in particular flux products. In particular, Yu et al. (2008) used available buoy measurements (that tend to be concentrated in the Tropics) to evaluate OAFlux and three reanalyses (NCEP1, NCEP2, and ERA-40). Their results for OAFlux and NCEP1 are shown in Figure 5.9. OAFlux has consistently smaller differences with respect to the buoy measurements than NCEP-1 and this is also the case for the other two reanalyses (not shown). There have also been attempts to compare reanalyses and the other flux products to direct ship-based surface flux measurements (e.g., Smith et al., 2001; Brunke et al., 2011). They demonstrate that differences in turbulent fluxes may amount to several tens of W m^{-2} and that they grow significantly under strong winds. It should also be noted that there are conceptual problems in comparisons of gridded surface flux products to the buoy and ship measurements. These problems are related but not limited to the co-location of grid cell averages and point measurements and require further research (see Section 4.3).

Local comparisons against buoy measurements must be supplemented by regional and global evaluations; in particular the constraint provided by closure of the global ocean heat budget. The zonal mean net heat flux variation with latitude obtained from a selection of 10 flux datasets is shown in Figure 5.10. Each product considered has a qualitatively similar variation with net heat input to the oceans in the Tropics and release back to the atmosphere at mid-high

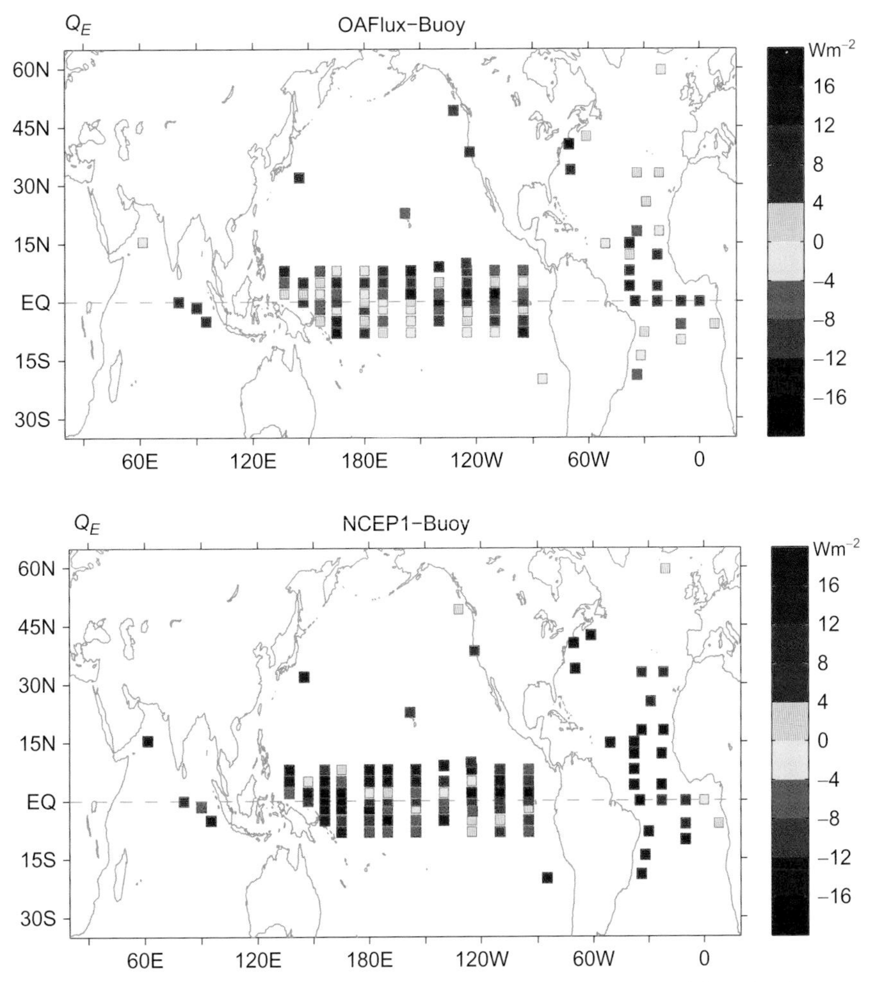

FIGURE 5.9 Mean latent heat flux difference for OAFlux-buoy (top) and NCEP1-buoy (bottom) at each buoy location, color coded according to the value of the difference. The mean difference at each location represents the average over the available measurement period. Units W m^{-2}. *Adapted from Yu et al. (2008).*

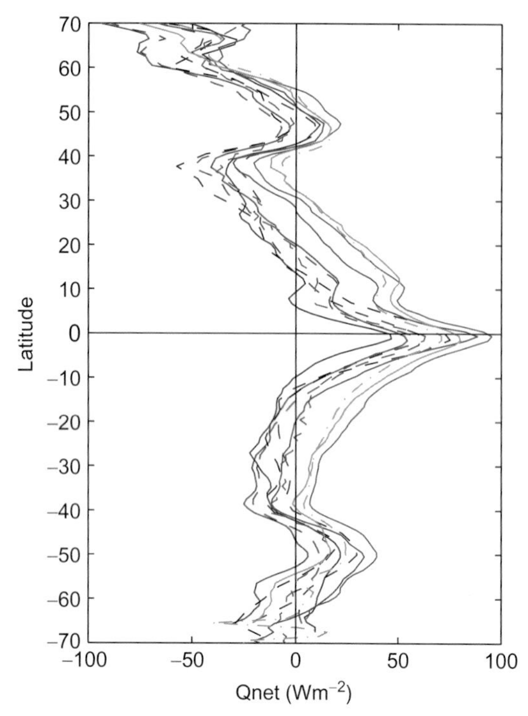

FIGURE 5.10 Zonal mean net air–sea heat flux from NCEP-1 (red dashed), ERA-40 (red solid), NCEP CFSR (blue solid), Trenberth residual (black dashed), NOC1.1a (green solid), NOC1.1 (gray solid), NOC2 (gray dash-dot), UWM/COADS (green dashed), OAFlux/ISCCP (purple solid), and COREv2 (purple dashed).

latitudes. Note the band of heat loss common to all products at 35–40°N, which includes the strong latent heat flux contributions from the Gulf Stream and Kuroshio. However, there are significant quantitative differences, with the spread at any given latitude typically of the order of 50 W m^{-2}; also, some products show stronger meridional gradients than others (note the relatively shallow gradients in NCEP-1 and ERA40 in the band 10–40°N). These differences reflect the ocean heat budget closure problem, with the NOC1.1, NOC2, and OAFlux/ISCCP products tending to have more heat gain/ less heat loss than the other datasets considered (COREv2, ERA40, NCEP-1, NOC1.1a, Trenberth residual, UWM/COADS). The recent NCEP–CFSR coupled reanalysis product lies between the two groupings.

Values for the global mean net heat flux from various datasets are listed in Table 5.1. There is a large spread in the net heat flux, which ranges from −2 to 33 W m^{-2}. The values highlight the global ocean heat budget problem defined in the Introduction: for most products, it is not possible to close the heat budget to within 10 W m^{-2} and in some cases the bias is of the order of 30 W m^{-2}. The values in the second column are for reanalysis-only products. One perplexing feature is that the first-generation reanalyses (ERA40, NCEP-1, NCEP-2) have a balance closer to zero (between 3 and 6 W m^{-2}), while the latest reanalyses (CFSR, ERA-interim, MERRA, 20CR) have a much larger residual

TABLE 5.1 Global Mean Net Heat Flux for Various Datasets (Units W m^{-2})

Product	Q_{Net}	Product	Q_{Net}
CFSR	15	COREv2 (unadjusted)	30
ERA-40	6	COREv2 (adjusted)	2
ERA-interim	11	NOC1.1	30
MERRA	21	NOC1.1a	−2
NCEP 1	3	NOC2	24
NCEP 2	3	OAFlux/ISCCP	29
20CR	13	OAFlux/SRB	33

Columns 1–2 detail reanalysis-only datasets; columns 3–4 detail ship-based and blended products. Note that the NOC and CORE net heat flux values are reproduced from the papers describing these products (Josey et al., 1999; Grist and Josey, 2003; Berry and Kent, 2009; Large and Yeager, 2009). For the reanalysis and OAFlux based datasets, the values have been obtained by interpolation onto a common grid for the period 1984–2007. It is not possible to repeat this procedure for all the products as in some cases they are based on a fixed period, for example, NOC1.1a was produced by applying inverse analysis to a climatology for the period 1980–1993.

(between 13 and 21 W m^{-2}). The cause of this difference remains to be established. The values in the fourth column are for ship-based and blended products; for these products, the global heat budget imbalance is particularly large (24–33 W m^{-2}) unless adjustments are made to achieve closure (COREv2 and NOC1.1a).

Despite the large mean differences between the various products, in well-sampled regions of the global ocean they can exhibit a striking degree of similarity in flux variability from 1 month to the next. This is illustrated for boxes in the North Atlantic and North Pacific in Figure 5.11. For the period of overlap between the four products considered, similar, month-to-month variations are obtained irrespective of the fact that some of the datasets considered have large global biases.

4. VARIABILITY AND EXTREMES

The subject of variability and extremes is a very broad topic given the many space and timescales over which variability may be defined and the wide variety of datasets that can be used for its investigation. A full review of this topic is not possible here; rather, three research areas that highlight some of the main issues are presented.

4.1. Impacts of Large-Scale Modes of Variability on Surface Fluxes

The impacts of large-scale modes of atmospheric variability are felt in many elements of the global climate system including the air–sea exchange fields. Indices of these modes have been provided by various techniques ranging

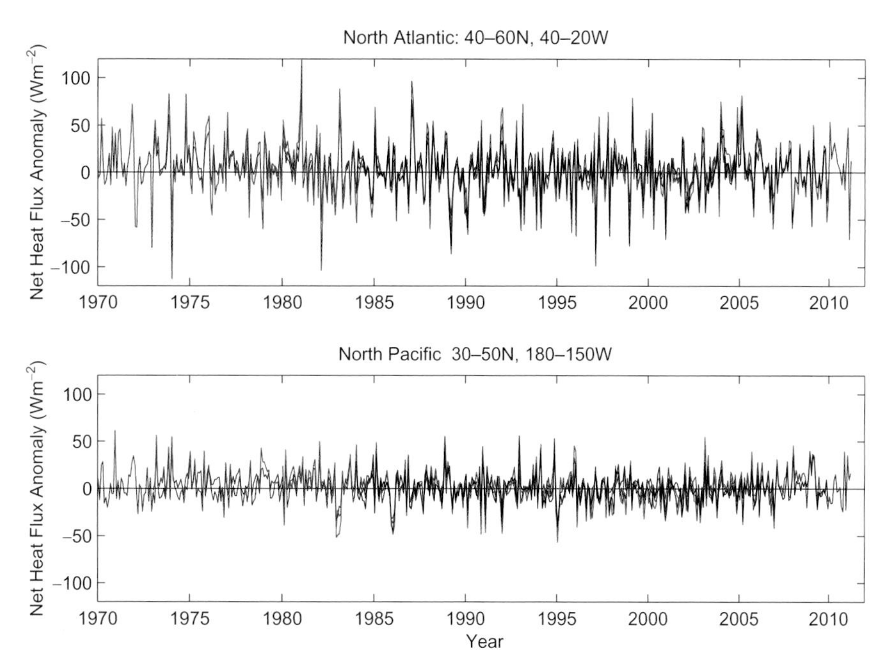

FIGURE 5.11 Monthly net heat flux anomaly for various products averaged over boxes in the North Atlantic and North Pacific. NCEP-1 (red), NOC1.1 (green), NOC2 (blue), and OAFlux/ISCCP (black). Units W m^{-2}.

from a simple two-point difference in sea level pressure (e.g., for the NAO, Hurrell, 1995) to principal component analysis (Barnston and Livezey, 1987). Variations in the strength of each mode lead to changes in the wind field, and the near-surface temperature and humidity; these in turn modify the latent and sensible heat loss. Impacts on surface exchange fields associated with several modes are considered here; many other modes exist (e.g., the Pacific Decadal Oscillation) and the treatment here is intended to be indicative rather than exhaustive.

Heat flux anomaly patterns for the featured modes composited on positive and negative states of each mode such that the patterns represent the anomaly for a unit positive value of the mode index are shown in Figure 5.12. The first of these modes to be recognised was the El Nino—Southern Oscillation (ENSO). It has a signal of strengthened heat loss in the eastern Tropical Pacific which is linked to the well-known change in the sea surface temperature in this region.

In the Atlantic, the North Atlantic Oscillation (NAO) and the East Atlantic Pattern (EAP) are the two major modes. The NAO is associated with heat flux anomalies over the Labrador Sea and the Gulf Stream (Josey et al., 2001) while the EAP is characterized by strong heat loss at 45–50°N. These modes also impact the freshwater flux (not shown) and, thus, the ocean surface salinity (see Josey and Marsh, 2005). In particular, the EAP has been linked to freshening of the eastern subpolar gyre from the 1960s to 1990s. It is also well known that NAO-related variations in surface buoyancy loss have an influence on whether the Greenland Sea or the Labrador Sea is the dominant location of North Atlantic dense water production (Dickson et al., 1996). The main focus of attention in the Southern Ocean has been the Southern Annular Mode (SAM), which has been associated with intensification of the main westerly wind belt in recent decades (e.g., Böning et al., 2008; Ciasto and Thompson, 2008). The annual net heat flux anomaly pattern for the SAM shows a tendency toward reduced heat loss at 40–55°S and enhanced heat loss further south.

4.2. Surface Flux Response to Anthropogenic Climate Change

In the context of the large level of uncertainty in the global mean net heat flux discussed earlier (Section 3.3), the change in this quantity expected as a result of the human influence on climate is small. Analysis of the changing heat budget in coupled climate models indicates that the anthropogenic signal in the global mean net air–sea heat flux is less than 1 W m^{-2} (Pierce et al., 2006). Observed changes in ocean heat storage provide a constraint on multidecadal variability in the surface net heat flux averaged over the global ocean. The observed warming of the ocean over the past 40 years is equivalent to an increase of 0.5 W m^{-2} in the net heat flux to the ocean (e.g., Levitus et al., 2009). Reliable detection of such a small ocean warming signal is not possible with the currently available heat flux datasets.

The strength of various components of the hydrological cycle is expected to change as a result of the ability of a warmer atmosphere to hold more moisture (e.g., Held and Soden, 2006). This in turn will have an impact on the transfers of water across the air–sea interface via

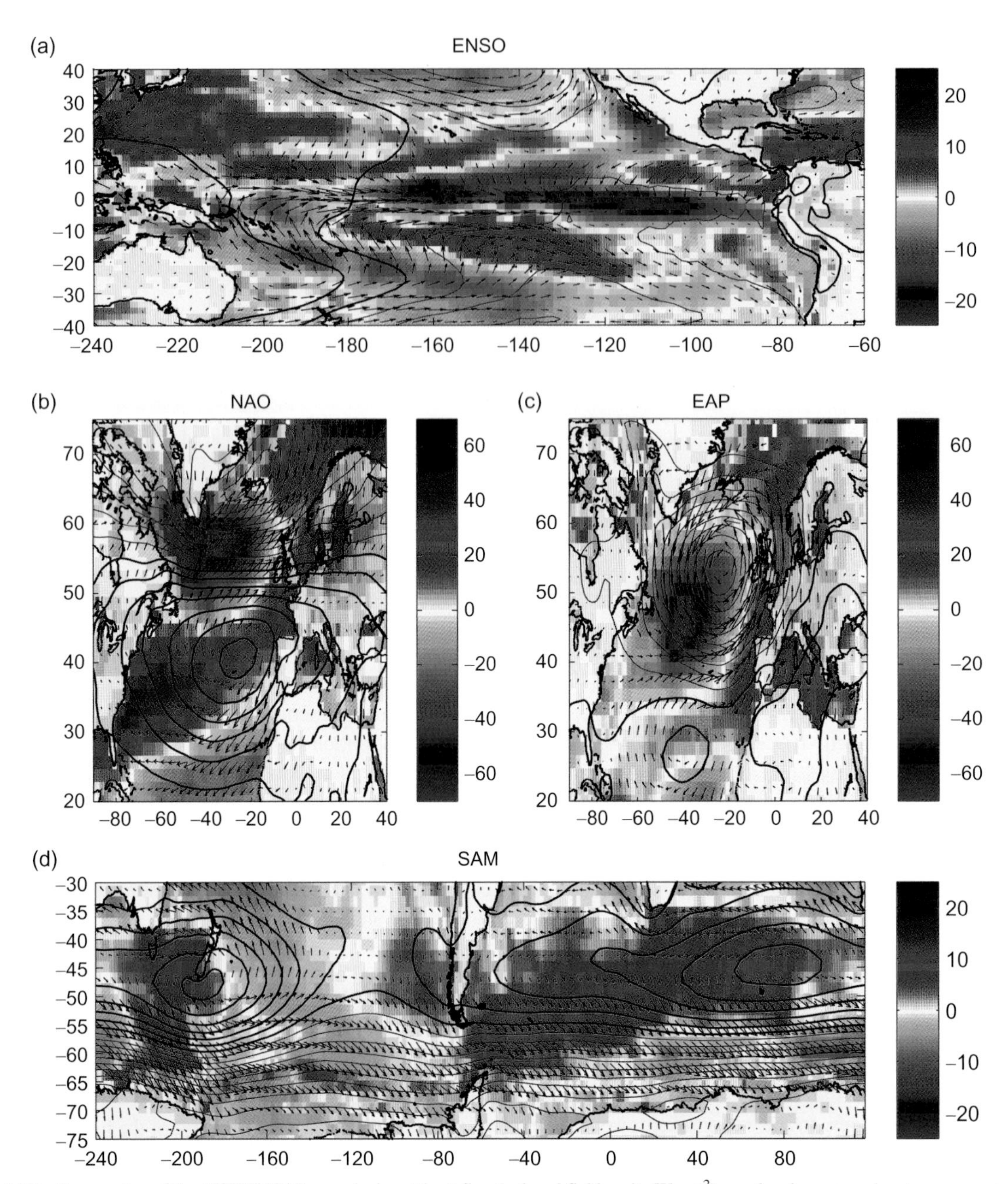

FIGURE 5.12 Composites of the NCEP/NCAR reanalysis net heat flux (colored field, units W m^{-2}), sea level pressure (contours, zero and positive values thick black, negative values thin black) and wind speed (arrows) for extreme states of ENSO (top panel), the NAO and EAP (middle panel), and SAM (bottom panel). Contour intervals are 0.5 (1) mb for ENSO, SAM (NAO, EAP). Composites have been produced for the period 1958–2006 using indices from the Climate Prediction Center for ENSO, NAO, and EAP, and the method of Marshall (2003) for the SAM. The ENSO and SAM anomalies are an annual average while the NAO and EAP anomalies are for Sep–Mar.

evaporation and precipitation. In a recent analysis, Schanze et al. (2010) have examined ocean mean net surface freshwater flux ($E-P$) using OAFlux for E and GPCPv2 for P. They note problems with the satellite-based datasets prior to 1987 caused by variations in data type. Given these problems and the known inconsistencies between the reanalyses, it is not yet possible to directly identify trends in ocean E and P from such datasets. Significant changes in salinity have been observed since about 1970 with the most saline ocean regions becoming saltier and the less saline regions fresher, which is consistent with an enhancement of regional ocean mean $E-P$ fields (Hosoda et al., 2009; Durack and Wijffels, 2010) (See Chapter 28). Salinity changes potentially provide a means to integrate the difficult-to-resolve $E-P$ signal (Yu, 2011), but it is not yet possible to use them to accurately infer changes in $E-P$ because of the complicating effect of the poleward migration of isopycnals (Durack and Wijffels, 2010).

Changes in the wind stress field at multidecadal timescales are strongly influenced by atmospheric modes of variability and this makes it difficult to identify anthropogenic change. The primary change in the global wind stress field in recent decades has occurred at southern high latitudes as is apparent in a range of reanalyses (Swart and Fyfe, 2012). Increasing Southern Ocean surface zonal wind stress is also evident in ECMWF Re-analysis (ERA40) data and supported by wind speed observations from SSM/I and an island station (Yang et al., 2007). The trend is linked with changes in the Southern Annular Mode (SAM) and has a seasonal dependence.

4.3. Transfers Under Extreme Conditions

We have already noted the level of uncertainty in turbulent flux parameterizations associated with extreme winds and this may contribute to differences between flux products. There is thus the need for a quantitative statistical analysis of the distributions of surface fluxes and related variables over the global ocean. Surface wind speed has been characterized by the Weibull distribution (e.g., Monahan, 2006a,b; Morrissey et al., 2010). Recently, Monahan (2007) analyzed non-Weibull behavior in surface winds and suggested using the maximum entropy distribution for the wind speed at sea. Comparisons of the probability distribution functions of surface wind speed in different products (Monahan, 2006b) show that differences in skewness between reanalysis and satellite winds are not negligible. Such differences are largest in the Southern Ocean where a strong interproduct scatter in higher quantiles is found. Smith et al. (2011) demonstrated large differences in various percentiles of surface turbulent fluxes from different datasets, even when the means and medians qualitatively agreed with each other.

Gulev and Belyaev (2012) using output from the NCEP/NCAR reanalysis analyzed statistical distributions of surface turbulent sensible and latent heat fluxes with the two-parameter modified Fisher–Tippett distribution, which effectively describes the statistical properties of turbulent fluxes and fits well with the observed probability of occurrence of fluxes in most areas. This enabled the accurate estimation of extreme sensible and latent fluxes that may locally amount to 1500–2000 W m^{-2} (for the 99th percentile) and can exceed 2000 W m^{-2} for the higher percentiles (see Figure 5.13). The strongest absolute extremes are observed in the subpolar latitudes for sensible heat and over the western boundary current regions and subtropics for latent heat. Looking to the future, analysis of surface flux probability distributions is likely to become an important component of flux dataset intercomparisons. However, this research is still at an early stage and the technique needs to be assessed using

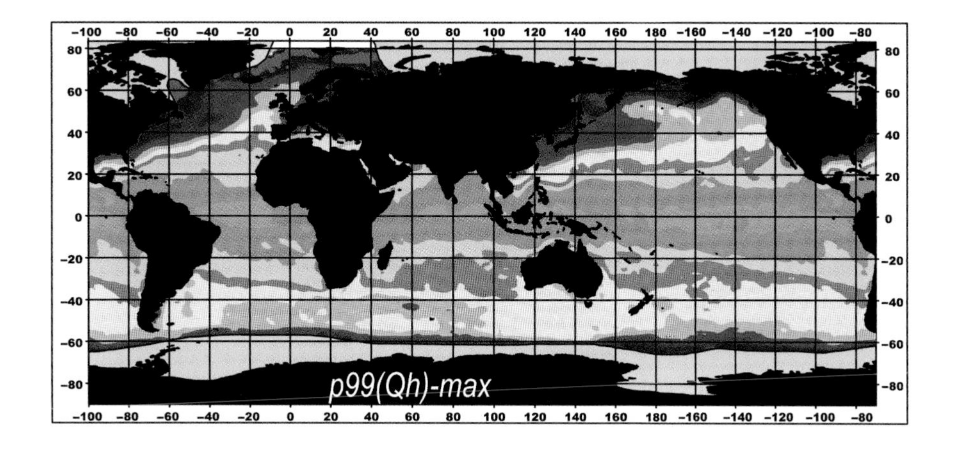

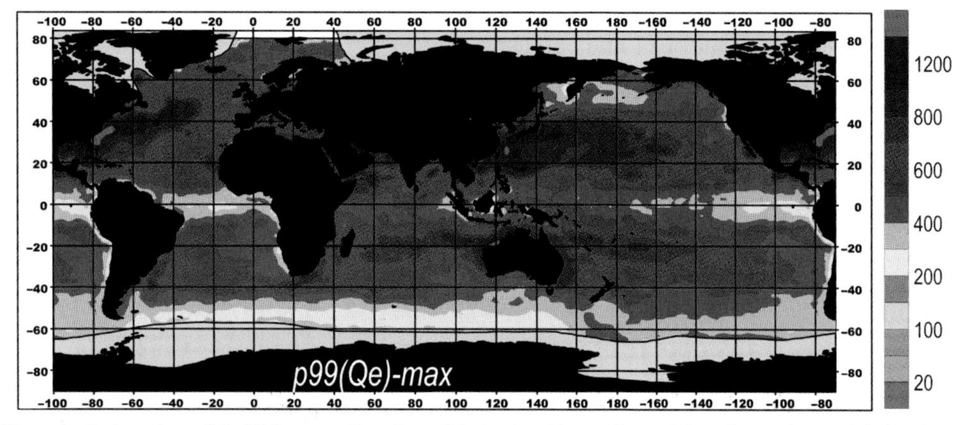

FIGURE 5.13 Climatological maxima of the 99th percentiles of sensible (top) and latent (bottom) heat fluxes observed during the period from 1948 to 2008. Units W m^{-2}. *Adopted from Gulev and Belyaev (2012).*

surface flux buoy observations in order to fully determine the accuracy of the flux estimates.

The existence of localized surface flux extrema that may not be adequately represented in coarser resolution flux products is one aspect of a more general scaling problem. Local extreme values of the sensible heat flux up to 1000 W m^{-2} and latent heat flux exceeding 2000 W m^{-2} have been documented in a number of observational studies (e.g., Renfrew et al., 2002; Tomita et al., 2010). Although strongly localized in space and short-lasting in time, these events may seriously affect the local ocean surface heat balances and significantly contribute to, for example, surface water mass transformation. Locally high winds and associated extremes in the wind stress may contribute to local mixing and surface circulation patterns (e.g., Pietri et al., 2013). The increasing spatial resolution of ocean GCMs (with ~1/12° models becoming standard for global simulations and higher resolutions used in regional simulations) requires the surface forcing function to be consistent in resolution with the model setting.

The present generation of ocean circulation models uses surface fluxes (or flux related variables) from reanalyses whose spatial resolution has increased in recent reanalyses to 0.5° or several tens of kilometers. However, the scaling problem raises some questions that remain open: What portion of mesoscale surface flux variability remains unexplained and unresolved in these products? To what extent may locally high fluxes change ocean model solutions? To answer these questions, a consolidated effort is required from the observational and modeling communities. The scaling problem is also of importance to the accurate evaluation of surface fluxes by direct local comparison to *in situ* ship and buoy measurements (Section 3.3), which requires a proper co-location of estimates in space and in time. When comparing a local point observation to a grid-cell average attributed to the forecast time step derived from reanalysis or operational analysis, one is also dealing with the uncertain fraction of surface flux variability that was integrated by the flux value diagnosed by the reanalysis model. This problem is less critical for co-location of single satellite retrievals; with *in situ* observations, however, it starts to be significant at the stage of the development of global satellite products synthesizing outputs from different satellites. This problem, first noted by Stoffelen (1998) for surface wind speed, now requires further attention.

5. OCEAN IMPACTS

5.1. Impacts on Near-Surface Ocean Layer Properties, Water Mass Transformation

The surface density flux, F_ρ, which scales with the buoyancy flux, expresses the contributions of the net heat flux and net evaporation to changes in density of the surface water layer,

$$F_\rho = -\rho\left(\alpha\frac{Q_{\mathrm{Net}}}{\rho c_{\mathrm{Pw}}} - \beta S\frac{E-P}{(1-S/1000)}\right) \quad (5.4)$$

Here, ρ is the density of water at the sea surface; c_{Pw}, the specific heat capacity of water; S, the sea surface salinity and α and β, the thermal expansion and haline contraction coefficients which are defined as follows,

$$\alpha = -\frac{1}{\rho}\frac{\partial\rho}{\partial T};\quad \beta = \frac{1}{\rho}\frac{\partial\rho}{\partial S} \quad (5.5)$$

Density fluxes are the basis for surface water mass transformation calculations (Walin, 1982; Large and Nurser, 2001; Marsh, 2000). The transformation rate measures the volume of water of density ρ, which is transformed, during a given period, into higher or lower densities according to its sign (See Chapter 9).

Over the past decade, density flux analyses have been carried out in a number of studies. Purely observation-based analyses are limited since accurate computation of the density flux requires knowledge of surface salinity at the same space and time resolution as surface temperature, which is still a problem. Howe and Czaja (2009) developed a climatology of surface density fluxes and estimates of surface water mass transformation using the NOC1.1a globally balanced surface fluxes (Grist and Josey, 2003) with temperatures and salinities from the World Ocean Atlas 2001 (Conkright et al., 2002). They estimated the maximum of the global transformation of surface water to lighter (heavier) density classes to be about 130 Sv (100 Sv) at low (high) densities. They also demonstrated that the haline contribution to surface water mass transformation is comparable to the thermal contribution at low densities (in the Tropics), while in mid- and subpolar latitudes the thermal contribution dominates. Recently Cerovecki et al. (2011) analyzed the impact of six different products (reanalyses, satellite based, blended) in estimation of surface density fluxes in the Southern Ocean where there is a large spread of surface flux estimates in different datasets. They demonstrated that, south of 45–50°S, the spread of estimates of surface density flux computed using the different datasets may vary by up to 100% of the mean values, which highlights the problems caused by the lack of observations in this region.

Diagnostic studies of model experiments by Ladd and Thompson (2001) and Gulev et al. (2003) identified the major characteristics of surface water mass transformation in the North Pacific and North Atlantic, respectively, and demonstrated the critical role of surface heat fluxes in water mass transformation. Gulev et al. (2003) found a close link between the NAO and the leading modes of surface water mass transformation in a coarse resolution ocean circulation model. The use of eddy-permitting models at resolutions of ¼–1/6° may shed more light on transformation but the representation of transformation in different experiments may be dependent to some extent on the model setting and parameterizations of lateral mixing. For instance, Gulev et al. (2007) showed that the 1/6° CLIPPER model demonstrates less skill in simulation of surface water mass transformation rates in the subpolar North Atlantic compared to the coarser resolution version of the same model. However, Marsh et al. (2005) in a ¼° model found realistic properties of surface water mass transformation in the North Atlantic.

5.2. Impacts of Surface Fluxes on Ocean Circulation

The air–sea exchanges of heat, freshwater, and momentum play key roles in determining both the mean state of, and variability in, the ocean circulation across a range of space and timescales. At large scales, variations in the wind stress in recent decades have had impacts on the strength of the main ocean gyres, for example, in the subpolar North Atlantic (Hakkinen et al., 2011). Variations in the heat flux have also played a significant role in the recent warming of the North Atlantic (Marsh et al., 2008; Grist et al., 2010). At regional scales, the link between strong buoyancy loss and dense water formation is well established, for example, in the Labrador and Nordic Seas (Dickson et al., 1996), and in the Mediterranean (Roether et al., 1996; Josey, 2003; Beuvier et al., 2010; Josey et al., 2011). Both the density change induced through buoyancy loss and the direct wind-driven response (i.e., doming of isopycnals) play a role in dense water formation and it is difficult to separate their effects (e.g., in the Norwegian sea, Gamiz-Fortis and Sutton, 2007; Grist et al., 2007, 2008).

High latitude variations in the surface buoyancy loss are expected to have an impact on the Atlantic Meridional Overturning Circulation (AMOC). By employing water mass transformation theory (Walin, 1982), it is possible to determine a surface-forced overturning stream function from the surface density flux field and use it to estimate this impact (Marsh, 2000). Grist et al. (2009, 2012) developed a variation of this method that included decadal averaging, to take account of the advective timescale of the surface-forced signal. Josey et al. (2009) employed this approach to estimate AMOC variability across a range of latitudes. Their analysis of the HadCM3 coupled climate model indicated that this method is capable of explaining about 40% of the AMOC variance in the range 35–65°N. They estimate surface-forced variability in the mid-high latitude North Atlantic over 1960–2008 using the method with NCEP/NCAR reanalysis surface fluxes as input (see Figure 5.14).

Interdecadal variability in the overturning circulation at the 1 to 2 Sv level is observed, but there is no indication of a trend in the period considered. In the HadCM3 analysis, the method breaks down south of about 35°N, which is consistent with a separate model analysis by Bingham et al., (2007) indicating decoupling of the AMOC variability north and south of this latitude. This method offers the potential for insights into AMOC variability at mid-high latitudes from surface flux information alone and may eventually provide a useful supplement to direct measurements from the Rapid mooring array at 26°N. However, other processes, including variations in ocean transports into the South Atlantic, are also likely to be important (Biastoch et al., 2008) (See also discussion in Chapters 9, 10 and 17).

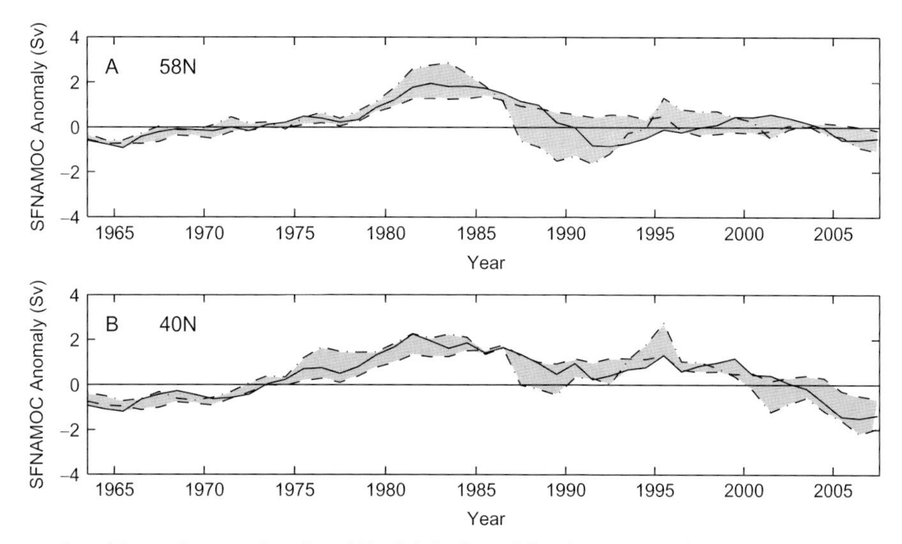

FIGURE 5.14 Reconstruction of the maximum surface-forced North Atlantic meridional overturning circulation anomaly (SFNAMOC, units Sverdrup, 1 Sv = 10^6 m^3 s^{-1}) at 40 and 58°N using density fluxes determined from the NCEP/NCAR reanalysis. Details of the method are given in Josey et al. (2009); the different lines are estimates based on surface flux fields integrated over 6 years (dash-dot line), 10 years (solid line), and 15 years (dashed line).

6. OUTLOOK AND CONCLUSIONS

6.1. Prospects for Improved Flux Datasets

Progress toward (a) resolution of the historical issues that have hampered air–sea flux analysis (principally the ocean heat budget closure problem) and (b) identification of the relatively small signals associated with anthropogenic climate change will require the use of data from satellite, *in situ*, and reanalysis sources to develop more accurate flux datasets. Satellite-based flux analysis faces the challenge of the type and variety of satellite observations being continually evolving, which creates substantial discontinuities in the satellite data record. Furthermore, they are limited to a relatively short time span (the past 20–30 years) and retrieval of key variables, principally air temperature and humidity, remains very difficult. Atmospheric reanalyses offer the hope of longer coverage; indeed, the recent 20CR product spans 1870–2009. However, they face ongoing problems including discontinuities caused by changes in the data type and volume assimilated (e.g., Krueger et al., 2013) and obtaining self-consistent budgets, for example, in the components of the hydrological cycle (Trenberth et al., 2011).

Blended products combining reanalysis and satellite data sources provide a potential advantage over either data type alone. For example, the objective synthesis approach employed for OAFlux has the advantage that it can integrate data from a range of sources and find an optimal solution in the least-squares sense. As it includes multiple data sources since the mid-1980s, objective synthesis tends to be less susceptible to observing system changes (e.g., satellite platform). However, it remains the case that prior to the satellite era, blended products to date are simply a combination of output from different reanalyses. The major change from *reanalysis only* to *reanalysis + satellite* as input data in the mid-1980s renders studies of multidecadal variability that span this period extremely problematic. Data from *in situ* observations, principally ship reports, provide a valuable third source of information. Including *in situ* observations within the synthesis process, that is, producing a synthesized *reanalysis + satellite + in situ* product is potentially a useful route forward. However, this creates the problem of lack of independent data for evaluation of the resulting product, so care must be taken in the selection of *in situ* data for incorporation in such an analysis.

Synthesis flux products, such as OAFlux, also offer the exciting prospect, in the coming decades, of a more accurate depiction of mesoscale to subsynoptic air–sea interaction than has previously been possible. As noted earlier, these scales are not resolved by even the latest atmospheric reanalyses although their spatial-temporal resolutions have increased considerably. One example is the frontal-scale air–sea coupling in the vicinity of the Gulf Stream (Figure 5.15). OAFlux shows a high level of coherence between the wind stress and the SST at small scales in this figure that is not captured by the reanalysis products. Further research into the synthesis of different types of products and development of very

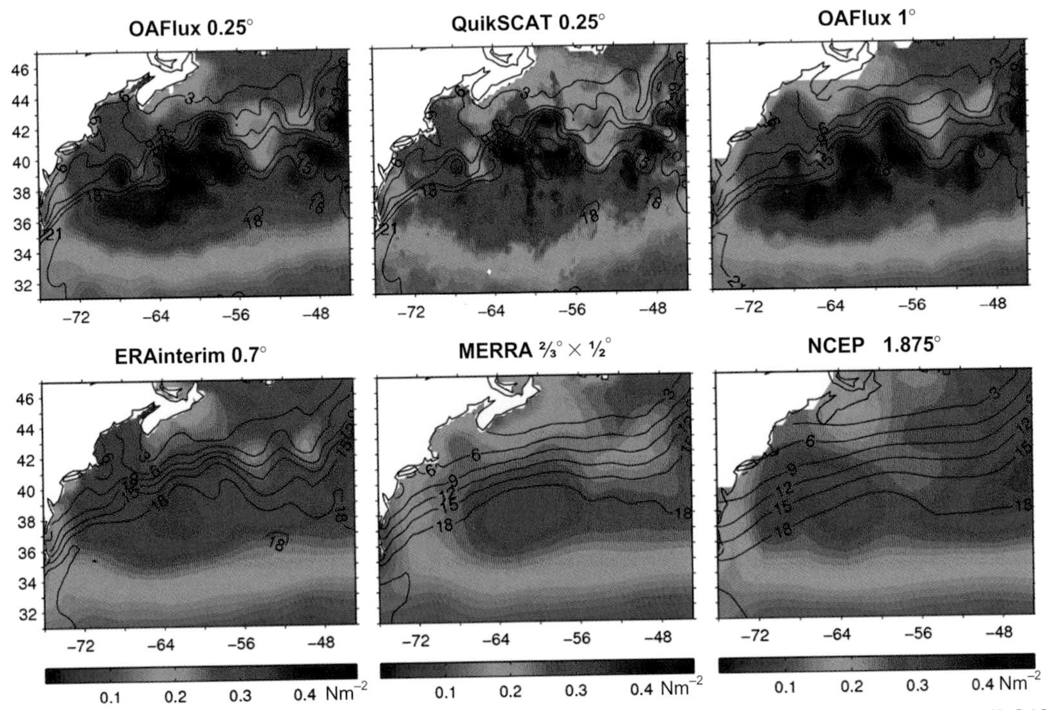

FIGURE 5.15 Frontal-scale relationship between wind stress (N m^{-2}, colors) and SST (°C) contours from the high-resolution ¼° OAFlux analysis, QuikSCAT, 1° OAFlux, ERAinterim, MERRA, and NCEP1. The fields show the mean for a cold surge event over Feb 1st 2007–Feb 14th 2007.

high resolution datasets is needed, with one major driver being the scaling problem (see Section 4.3).

6.2. Prospects for Enhanced Observational Constraints

Observational constraints that can be used to evaluate air–sea flux products continue to be provided primarily by surface flux buoy measurements at the local scale and hydrographic measurements of ocean heat and freshwater transports at the basin scale. The locations of past, existing, and planned surface flux moorings are shown in Figure 5.16. The number of surface flux buoys has increased significantly over the past decade to the extent that the tropics are relatively well covered, the sub-tropics sparsely covered, and a flux mooring has been deployed for the first time in the Southern Ocean, albeit at its northern margin close to Tasmania (Schulz et al., 2012). However, as yet there are no open ocean deployments poleward of 50° latitude in either hemisphere and this area represents a major source of uncertainty given the many key processes taking place at high latitudes, in particular dense water formation, that are strongly influenced by air–sea interaction. More generally, there is a pressing need to reduce the uncertainty of the bulk flux algorithms in high wind speed and sea spray regimes. The preceding discussion highlights the existence of uncertainties in both the spatial (e.g., mid-high latitude) and physical variable (e.g., high wind speed) domains. Such uncertainties are likely to be significant factors in the unresolved global ocean heat budget closure problem as well as an issue for the current generation of atmospheric reanalyses.

Heat budget analysis based on Argo profiling float subsurface observations is a potentially useful tool to check the fidelity of surface flux products. However, the uncertainty in ocean advection terms complicates the analysis of heat flux errors using this approach. One exception in which the advection terms can be eliminated is the warm pool region, where the warm waters bounded by a specific isotherm (e.g., 28 °C) form a bowl-shape. A heat budget equation can be formulated for the warm pool volume (Niiler and Stevenson 1982; Enfield and Lee 2004; Toole et al., 2004), linking the change of the surface heat flux to the change in volume and temperature of the pool. Likewise heat budget analysis for enclosed or semi-enclosed seas (such as the Mediterranean Sea, e.g., Sanchez-Gomez et al., 2011) is a potentially useful tool to assess the uncertainties in surface heat flux products although conclusions for such regions may not hold for the open ocean. At high latitudes, the combination of Argo data with observations from marine mammal-borne sensors is providing more accurate temperature fields (Grist et al., 2011) that may prove useful for constraining surface heat exchanges (See also Chapter 3).

6.3. Conclusions

To conclude, significant progress has been made in the field of air–sea exchange research over the past decade. The level of uncertainty in air–sea flux parameterizations has been reduced as a result of extensive field campaigns although significant questions remain, particularly in the high wind speed regime. A growing network of surface flux buoys is starting to provide the high quality observational basis for rigorous evaluation of the many new flux products now available. However, much fundamental research remains to be done. Identifying multi-decadal variability in the surface exchanges, particularly of freshwater, at the accuracy required for climate change analyses is going to

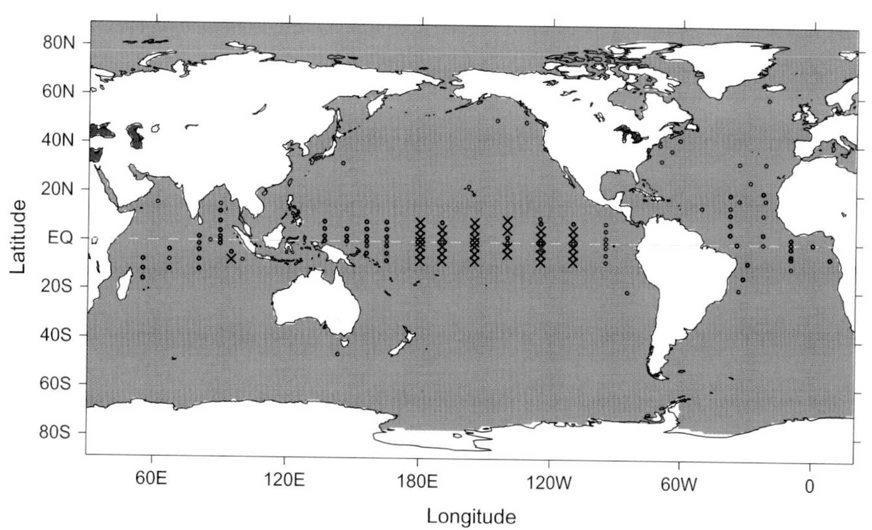

FIGURE 5.16 Global distribution of surface flux moorings. Symbols indicate the components of the heat exchange that are measured: all components (red circles). All except longwave (black circles) and all except longwave and shortwave (black crosses). *Adapted from Yu and Jin (2012).*

be a major challenge in the near future. Furthermore, a well-justified resolution of the ocean heat budget closure problem has yet to be obtained. The developing buoy network offers the possibility of this long-standing problem being finally resolved in the coming decade, but this will require the deployment of further surface flux moorings, particularly at mid-high latitudes. New techniques have been developed for generating air–sea flux products through synthesis of satellite observations with atmospheric reanalyses. Such products have the potential to overcome the sampling limitations that have been so problematic in the past and to yield new insights into the processes governing ocean–atmosphere interaction in the years ahead.

ACKNOWLEDGMENTS

The authors would like to thank Bernard Barnier and Robert Weller for many insightful comments on this chapter.

REFERENCES

Adler, R.F., et al., 2003. The Version-2 Global Precipitation Climatology Project (GPCP) monthly precipitation analysis (1979–present). J. Hydrometeorol. 4, 1147–1167.

Alexander, M.A., Scott, J.D., 1997. Surface flux variability over the North Pacific and North Atlantic Oceans. J. Clim. 10, 2963–2977.

Andreas, E.L., Persson, P.O.G., Hare, J.E., 2008. A bulk turbulent air-sea flux algorithm for high-wind, spray conditions. J. Phys. Oceanogr. 38, 1581–1596.

Andersson, A., Fennig, K., Klepp, C., Bakan, S., Graßl, H., Schulz, J., 2010. The Hamburg Ocean atmosphere parameters and fluxes from satellite data - HOAPS-3. Earth Syst. Sci. Data 2, 215–234. http://dx.doi.org/10.5194/essd-2-215-2010.

Andreas, E.L., 2004. Spray stress revisited. J. Phys. Oceanogr. 34, 1429–1440. http://dx.doi.org/10.1175/1520-0485.

Bao, J.W., Fairall, C.W., Michelson, S.A., Bianco, L., 2011. Parameterizations of Sea-Spray Impact on the Air-Sea Momentum and Heat Fluxes. Mon. Weather Rev. 139, 3781–3797. http://dx.doi.org/10.1175/MWR-D-11-00007.1.

Bentamy, A., Katsaros, K.B., Mestas-Nuñez, A.M., Drennan, W.M., Forde, E.B., Roquet, H., 2003. Satellite estimates of wind speed and latent heat flux over the global oceans. J. Clim. 16, 637–656.

Barnston, A.G., Livezey, R.E., 1987. Classification, seasonality and persistence of low-frequency atmospheric circulation patterns. Mon. Weather Rev. 115, 1083–1126.

Béranger, K., Barnier, B., Gulev, S., Crépon, M., 2006. Comparing twenty years of precipitation estimates from different sources over the world ocean. Ocean Dyn. 56–2, 104–138. http://dx.doi.org/10.1007/s10236-006-0065-2.

Berry, D.I., Kent, E.C., 2009. A new air-sea interaction gridded dataset from ICOADS with uncertainty estimates. Bull. Am. Meteorol. Soc. 90, 645–656. http://dx.doi.org/10.1175/2008BAMS2639.1.

Beuvier, J., Sevault, F., Herrmann, M., Kontoyiannis, K., Ludwig, W., Rixen, M., Stanev, E., Béranger, K., Somot, S., 2010. Modelling the mediterranean sea inter-annual variability during 1961-2000: focus on the eastern mediterranean transient. J. Geophys. Res. 115, C08017. http://dx.doi.org/10.1029/2009JC005950.

Biastoch, A., Böning, C.W., Lutjeharms, J.R.E., 2008. Agulhas leakage dynamics affects decadal variability in Atlantic overturning circulation. Nature 456, 489–492. http://dx.doi.org/10.1038/nature07426.

Bingham, R.J., Hughes, C.W., Roussenov, V., Williams, R.G., 2007. Meridional coherence of the North Atlantic meridional overturning circulation. Geophys. Res. Lett. 34, L23606. http://dx.doi.org/10.1029/2007GL031731.

Bond, N.A., Cronin, M.F., 2008. Regional weather patterns during anomalous air–sea fluxes at the Kuroshio Extension Observatory (KEO). J. Clim. 21, 1680–1697.

Böning, C.W., Dispert, A., Visbeck, M., Rintoul, S.R., Schwarzkopf, F.U., 2008. The response of the Antarctic circumpolar current to recent climate change. Nat. Geosci. 1, 864–869.

Bradley, F., Fairall, C., 2007. A guide to making climate quality meteorological and flux measurements at sea. NOAA Technical Memorandum OAR PSD-311, NOAA/ESRL/PSD, Boulder, CO, 108 pp.

Brodeau, L., Barnier, B., Penduff, T., Treguier, A.-M., Gulev, S., 2010. An ERA40 based atmospheric forcing for global ocean circulation models. Ocean Model. 31, 88–104.

Brunke, M.A., Wang, Z., Zeng, X., Bosilovich, M., Shie, C.-L., 2011. An Assessment of the uncertainties in Ocean surface turbulent fluxes in 11 reanalysis, satellite-derived, and combined global datasets. J. Clim. 24, 5469–5493. http://dx.doi.org/10.1175/2011JCLI4223.1.

Bunker, A.F., 1976. Computations of surface energy flux and annual air-sea interaction cycles of the North Atlantic Ocean. Mon. Weather Rev. 87, 329–340.

Cerovečki, Ivana, Talley, L.D., Mazloff, M.R., 2011. A comparison of Southern Ocean air–sea buoyancy flux from an Ocean state estimate with five other products. J. Clim. 24, 6283–6306. http://dx.doi.org/10.1175/2011JCLI3858.1.

Chou, S.-H., Nelkin, E., Ardizzone, J., Atlas, R.M., Shie, C.-L., 2003. Surface turbulent heat and momentum fluxes over global oceans based on the Goddard satellite retrievals, version 2 (GSSTF2). J. Clim. 16, 3256–3273.

Ciasto, L.M., Thompson, D.W.J., 2008. Observations of large-scale Ocean–Atmosphere interaction in the Southern hemisphere. J. Clim. 21, 1244–1259. http://dx.doi.org/10.1175/2007JCLI1809.1.

Colbo, K., Weller, R.A., 2007. The variability and heat budget of the upper ocean under the Chile-Peru stratus. J. Mar. Res. 65, 607–637.

Colbo, K., Weller, R.A., 2009. Accuracy of the IMET sensor package in the subtropics. J. Atmos. Oceanic Technol. 26, 1867–1890.

Compo, G.P., Whitaker, J.S., Sardeshmukh, P.D., Matsui, N., Allan, R.J., Yin, X., Gleason, B.E., Vose, R.S., Rutledge, G., Bessemoulin, P., Brönnimann, S., Brunet, M., Crouthamel, R.I., Grant, A.N., Groisman, P.Y., Jones, P.D., Kruk, M.C., Kruger, A.C., Marshall, G.J., Maugeri, M., Mok, H.Y., Nordli, Ø., Ross, T.F., Trigo, R.M., Wang, X.L., Woodruff, S.D., Worley, S.J., 2011. The twentieth century reanalysis project. Q. J. R. Meteorol. Soc. 137, 1–28. http://dx.doi.org/10.1002/qj.776.

Conkright, M.E., Locarnini, R.A., Garcia, H.E., O'Brien, T.D., Boyer, T.P., Stephens, C., Antonov, J.I., 2002. World Ocean Atlas 2001: Objective Analyses, Data Statistics, and Figures, CD-ROMDocumentation. National Oceanographic Data Center, Silver Spring, MD, 17 pp. Available at: http://www.nodc.noaa.gov/OC5/WOA01/docwoa01.html.

Cravatte, S., Delcoix, T., Zhang, D., McPhaden, M., LeLoup, J., 2009. Observed freshening and warming of the western Pacific Warm Pool. Clim. Dyn. 33, 565–589. http://dx.doi.org/10.1007/s00382-009-0526-7.

Cronin, M.F., Meinig, C., Sabine, C.L., Ichikawa, H., Tomita, H., 2008. Surface mooring network in the Kuroshio Extension. IEEE Syst. J. 2, 424–430.

Curry, R., Dickson, R., Yashayaev, I., 2003. Ocean evidence of a change in the fresh water balance of the Atlantic over the past four decades. Nature 426, 826–829.

Curry, J.A., et al., 2004. SEAFLUX. Bull. Am. Meteorol. Soc. 85, 409–424.

Dai, A., Qian, T., Trenberth, K.E., Milliman, J.D., 2009. Changes in continental freshwater discharge from 1948 to 2004. J. Clim. 22, 2773–2792.

Dawe, J.T., Thompson, L., 2006. Effect of ocean surface currents on wind stress, heat flux, and wind power input to the ocean. Geophys. Res. Lett. 33. http://dx.doi.org/10.1029/2006GL025784, L09604.

Dee, D.P., et al., 2011. The ERA-Interim reanalysis: configuration and performance of the data assimilation system. Q. J. R. Meteorol. Soc. 137, 553–597. http://dx.doi.org/10.1002/qj.828.

de Szoeke, S.P., Fairall, C.W., Wolfe, D.E., Bariteau, L., Zuidema, P., 2010. Surface flux observations on the southeastern tropical Pacific Ocean and attribution of SST errors in coupled ocean–atmosphere models. J. Clim. 23, 4152–4174. http://dx.doi.org/10.1175/2010JCLI3411.1.

Dickson, R., Lazier, J., Meincke, J., Rhines, P., 1996. Long-term coordinated changes in the convective activity of the North Atlantic. In: Willebrand, D.A.J. (Ed.), Decadal Climate Variability: Dynamics and Predictability. Springer-Verlag, Berlin, pp. 211–262.

Dobson, F.W., Smith, S.D., Anderson, R.J., 1994. Measuring the relationship between wind stress and sea state in the open ocean in the presence of swell. Atmos. Ocean 32, 61–81.

Donelan, M.A., Haus, B.K., Reul, N., Plane, W.J., Stiassnie, M., Grabber, H.C., Brown, O.B., Saltzman, E.S., 2004. On the limiting aerodynamics roughness on the ocean in very strong winds. Geophys. Res. Lett. 31. http://dx.doi.org/10.1029/2004GL019460, L18306.

Drennan, W., Graber, H.C., Hauser, D., Quentin, C., 2003. On the wave age dependence of wind stress over pure wind seas. J. Geophys. Res. 108, 8062. http://dx.doi.org/10.1029/2000JC000715.

Drennan, W.M., Taylor, P.K., Yelland, M.J., 2005. Parameterizing the sea surface roughness. J. Phys. Oceanogr. 35, 835–848.

Duhaut, T.H.A., Straub, D.N., 2006. Wind stress dependence on ocean surface velocity: implications for mechanical energy input to ocean circulation. J. Phys. Oceanogr. 36 (2), 202–211.

Durack, P.J., Wijffels, S.E., 2010. Fifty-year trends in global ocean salinities and their relationship to broad-scale warming. J. Clim. 23, 4342–4362.

Durack, P.J., Wijffels, S.E., Matear, R.J., 2012. Ocean salinities reveal strong global water cycle intensification during 1950–2000. Science 336, 455–458. http://dx.doi.org/10.1126/science.1212222.

Edson, J.B., Jampana, V., Weller, R.A., Bigorre, S.P., Plueddemann, A.J., Fairall, C.W., Miller, S.D., Mahrt, L., Vickers, D., Hersbach, H., 2013. On the Exchange of Momentum over the Open Ocean. J. Phys. Oceanogr. 43, 1589–1610. http://dx.doi.org/10.1175/JPO-D-12-0173.1.

Enfield, D.B., Lee, S., 2004. The heat balance of the western hemisphere warm pool. J. Clim. 18, 2662–2681.

Fairall, C.W., Bradley, E.F., Hare, J.E., Grachev, A.A., Edson, J.B., 2003. Bulk parameterization of air-sea fluxes: updates and verification for the COARE algorithm. J. Clim. 16, 571–591.

Fairall, C., Barnier, B., Berry, D.I., Bourassa, M.A., Bradley, E.F., Clayson, C.A., de Leeuw, G., Drennan, W.M., Gille, S., Gulev, S., Kent, E.C., McGillis, W.R., Quartly, G.D., Ryabinin, V., Smith, S.R., Weller, R.A., Yelland, M.J., Zhang, H.-M., 2010. Observations to quantify air-sea fluxes and their role in climate variability and predictability. In: Community White Paper in Proceedings of the "OceanObs'09: Sustained Ocean Observations and Information for Society" Conference, Venice, Italy, 21-25 September 2009. ESA Publication, Paris, WPP-306, 15 pp.

Fairall, C.W., Yang, M., Bariteau, L., Edson, J.B., Helmig, D., McGillis, W., Pezoa, S., Hare, J.E., Huebert, B., Blomquist, B., 2011. Implementation of the coupled ocean-atmosphere response experiment flux algorithm with CO_2, dimethyl sulfide, and O_3. J. Geophys. Res. 116, C00F09. http://dx.doi.org/10.1029/2010JC006884.

Grist, J.P., Josey, S.A., 2003. Inverse analysis of the SOC air-sea flux climatology using ocean heat transport constraints. J. Clim. 16 (20), 3274–3295.

Grist, J.P., Josey, S.A., Sinha, B., 2007. Impact on the ocean of extreme greenland sea heat loss in the HadCM3 coupled ocean-atmosphere model. J. Geophys. Res. 112. http://dx.doi.org/10.1029/2006JC003629, C04014.

Grist, J.P., Josey, S.A., Sinha, B., Blaker, A.T., 2008. Response of the Denmark Strait overflow to Nordic Seas heat loss. J. Geophys. Res. 113. http://dx.doi.org/10.1029/2007JC004625, C09019.

Grist, J.P., Marsh, R.A., Josey, S.A., 2009. On the relationship between the North Atlantic meridional overturning circulation and the surface-forced overturning stream function. J. Clim. 22 (19), 4989–5002. http://dx.doi.org/10.1175/2009JCLI2574.1.

Grist, J.P., Josey, S.A., Marsh, R., Good, S.A., Coward, A.C., deCuevas, B.A., Alderson, S.G., New, A.L., Madec, G., 2010. The roles of surface heat flux and ocean heat transport convergence in determining Atlantic Ocean temperature variability. Ocean Dyn. 60 (4), 771–790. http://dx.doi.org/10.1007/s10236-010-0292-4.

Grist, J.P., Josey, S.A., Boehme, L., Meredith, M.P., Davidson, F.J.M., Stenson, G., Hammill, M., 2011. Temperature signature of high latitude Atlantic boundary currents revealed by marine mammal-borne sensor and Argo data. Geophys. Res. Lett. 38. http://dx.doi.org/10.1029/2011GL048204, L15601.

Grist, J.P., Josey, S.A., Marsh, R.A., 2012. Surface estimates of the Atlantic overturning in density space in an eddy-permitting ocean model. J. Geophys. Res. 117. http://dx.doi.org/10.1029/2011JC007752, C06012.

Grodsky, S.A., Carton, J.A., Bingham, F.M., 2006. Low frequency variation of sea surface salinity in the tropical Atlantic. Geophys. Res. Lett. 33. http://dx.doi.org/10.1029/2006GL026426, L14604.

Grodsky, S.A., Reul, N., Lagerloef, G., Reverdin, G., Carton, J.A., Chapron, B., Quilfen, Y., Kudryavtsev, V.A., Kao, H.-Y., 2012. Haline hurricane wake in the Amazon/Orinoco plume: AQUARIUS/SACD and SMOS observations. Geophys. Res. Lett. 39, L20603. http://dx.doi.org/10.1029/2012GL053335.

Gulev, S.K., Belyaev, K.P., 2012. Probability distribution characteristics for surface air-sea turbulent heat fluxes over the global ocean. J. Clim. 25, 184–206. http://dx.doi.org/10.1175/2011JCLI4211.1.

Gulev, S.K., Barnier, B., Kocher, H., Molines, J.-M., Cottet, M., 2003. Water mass transformation in the North Atlantic and its impact on the meridional circulation: insights from on ocean model forced by NCEP/NCAR reanalysis surface fluxes. J. Clim. 16, 3085–3110.

Gulev, S.K., Barnier, B., Molines, J.-M., Penduff, T., Chanut, J., 2007. Impact of spatial resolution of simulated surface water mass transformation in the Atlantic. Ocean Model. 19, 138–160.

Gulev, S., Josey, S.A., Bourassa, M., Breivik, L.-A., Cronin, M.F., Fairall, C., Gille, S., Kent, E.C., Lee, C.M., McPhaden, M.J., Monteiro, P.M.S., Schuster, U., Smith, S.R., Trenberth, K.E., Wallace, D., Woodruf, S.D., 2010. Surface energy and CO_2 fluxes in the global ocean-atmosphere-ice system. In: Plenary White Paper in Proceedings of the "OceanObs'09: Sustained Ocean Observations and Information for Society" Conference, Venice, Italy, 21-25 September 2009. ESA Publication, Paris, WPP-306, 20pp.

Gupta, S.K., Ritchey, N.A., Wilber, A.C., Whitlock, C.H., Gibson, G.G., Stackhouse Jr., P.W., 1999. A climatology of surface radiation budget derived from satellite data. J. Clim. 12, 2691–2710.

Häkkinen, S., Rhines, P.B., Worthen, D.L., 2011. Atmospheric blocking and Atlantic multidecadal ocean variability. Science 334, 655–659.

Held, I.M., Soden, B.J., 2006. Robust responses of the hydrological cycle to global warming. J. Clim. 19, 5686–5699. http://dx.doi.org/10.1175/JCLI3990.1.

Hosoda, S., Suga, T., Shikama, N., Mizuno, K., 2009. Global surface layer salinity change detected by Argo and its implication for hydrological cycle intensification. J. Oceanogr. 65 (4), 579–586.

Howe, N., Czaja, A., 2009. A new climatology of air-sea density fluxes and surface water mass transformation rates constrained by WOCE. J. Phys. Oceanogr. 39, 1432–1447. ISSN:0022-3670.

Hurrell, J.W., 1995. Decadal trends in the North Atlantic oscillation regional temperatures and precipitation. Science 269, 676–679.

Hurrell, J.W., Kushnir, Y., Visbeck, M., Ottersen, G., 2003. An overview of the North Atlantic Oscillation. In: Hurrell, J.W., Kushnir, Y., Ottersen, G., Visbeck, M. (Eds.), The North Atlantic Oscillation: Climate Significance and Environmental Impact. Geophysical Monograph Series, vol. 134. p. 279.

Jackson, D.L., Wick, G.A., Bates, J.J., 2005. Near-surface retrieval of air temperature and specific humidity using multisensor microwave satellite observations. J. Geophys. Res. 111, D10306.

Johnson, E.S., Lagerloef, G.S.E., Gunn, J.T., Bonjean, F., 2002. Surface salinity advection in the tropical oceans compared with atmospheric freshwater forcing: a trial balance. J. Geophys. Res. 107, 8014. http://dx.doi.org/10.1029/2001JC001122.

Josey, S.A., 2001. A comparison of ECMWF, NCEP–NCAR, and SOC surface heat fluxes with moored buoy measurements in the subduction region of the northeast Atlantic. J. Clim. 14, 1780–1789.

Josey, S.A., 2003. Changes in the heat and freshwater forcing of the Eastern mediterranean and their influence on deep water formation. J. Geophys. Res. 108 (C7), 3237. http://dx.doi.org/10.1029/2003JC001778.

Josey, S.A., 2011. Air-sea fluxes of heat, freshwater and momentum. In: Schiller, A., Brasington, G.B. (Eds.), Operational Oceanography in the 21st Century. Springer, Berlin, pp. 155–184. http://dx.doi.org/10.1007/978-94-007-0332-2_6.

Josey, S.A., Marsh, R., 2005. Surface freshwater flux variability and recent freshening of the North Atlantic in the eastern subpolar gyre. J. Geophys. Res. 110. http://dx.doi.org/10.1029/2004JC002521, C05008.

Josey, S.A., Smith, S.R., 2006. Guidelines for evaluation of air-sea heat, freshwater and momentum flux datasets, CLIVAR Global Synthesis and Observations Panel (GSOP) White Paper, July 2006, pp. 14. Available at: http://www.clivar.org/sites/default/files/gsopfg.pdf.

Josey, S.A., Kent, E.C., Taylor, P.K., 1999. New insights into the ocean heat budget closure problem from analysis of the SOC air-sea flux climatology. J. Clim. 12 (9), 2856–2880.

Josey, S.A., Kent, E.C., Sinha, B., 2001. Can a state of the art atmospheric general circulation model reproduce recent NAO related variability at the air-sea interface? Geophys. Res. Lett. 28 (24), 4543–4546.

Josey, S.A., Kent, E.C., Taylor, P.K., 2002. On the wind stress forcing of the ocean in the SOC climatology: comparisons with the NCEP/NCAR, ECMWF, UWM/COADS and Hellerman and Rosenstein Datasets. J. Phys. Oceanogr. 32 (7), 1993–2019.

Josey, S.A., Pascal, R.W., Taylor, P.K., Yelland, M.J., 2003. A new formula for determining the atmospheric longwave flux at the ocean surface at mid-high latitudes. J. Geophys. Res. 108 (C4). http://dx.doi.org/10.1029/2002JC001418.

Josey, S.A., Grist, J.P., Marsh, R.A., 2009. Estimates of meridional overturning circulation variability in the North Atlantic from surface density flux fields. J. Geophys. Res. 114. http://dx.doi.org/10.1029/2008JC005230, C09022.

Josey, S.A., Somot, S., Tsimplis, M., 2011. Impacts Of atmospheric modes of variability on Mediterranean Sea surface heat exchange. J. Geophys. Res. 116. http://dx.doi.org/10.1029/2010JC006685, C02032.

Kalnay, E., et al., 1996. The NMC/NCAR 40-year reanalysis project. Bull. Am. Meteorol. Soc. 77, 437–471.

Kato, S., Rose, F.G., Sun-Mack, S., Miller, W.F., Chen, Y., Rutan, D.A., Stephens, G.L., Loeb, N.G., Minnis, P., Wielicki, B.A., Winker, D.M., Charlock, T.P., Stackhouse, P.W., Xu, K.-M., Collins, W., 2011. Improvements of top-of-atmosphere and surface irradiance computations with CALIPSO, CloudSat, and MODIS derived cloud and aerosol properties. J. Geophys. Res. 116. http://dx.doi.org/10.1029/2011JD16050, D19209.

Kent, E.C., et al., 2010. The Voluntary Observing Ship Scheme. In: Community White Paper in Proceedings of the "OceanObs'09: Sustained Ocean Observations and Information for Society" Conference, Venice, Italy, 21-25 September 2009. ESA Publication, Paris, WPP-306.

Konda, M., Ichikawa, H., Tomita, H., Cronin, M.F., 2010. Surface heat flux variations across the Kuroshio Extension as observed by surface flux buoys. J. Clim. 23, 5206–5221.

Krueger, O., Schenk, F., Feser, F., Weisse, R., 2013. Inconsistencies between long-term trends in storminess derived from the 20CR reanalysis and observations. J. Clim. 26, 868–874.

Ladd, C., Thompson, L., 2001. Water mass formation in an isopycnal model of the North Pacific. J. Phys. Oceanogr. 31, 1517–1537.

Large, W.G., Nurser, A.J.G., 2001. Ocean surface water mass transformation. In: Siedler, G., Church, J., Gould, J. (Eds.), Ocean Circulation and Climate: Observing and Modelling the Global Ocean. International Geophysics Series 77. USA, Academic Press, San Fransisco CA, pp. 317–336.

Large, W.G., Yeager, S.G., 2009. The global climatology of an interannually varying air-sea flux data set. Clim. Dyn. 33, 341–364.

Levitus, S., Antonov, J.I., Boyer, T.P., Locarnini, R.A., Garcia, H.E., Mishonov, V.A., 2009. Global ocean heat content 1955–2008 in light of recently revealed instrumentation problems. Geophys. Res. Lett. 36. http://dx.doi.org/10.1029/2008GL037155, L07608.

Makin, V.K., 2005. A note on the drag of the sea surface at hurricane winds. Bound. Layer Meteorol. 115, 169–176.

Mariotti, A., 2010. Recent changes in the Mediterranean water cycle: a pathway toward long-term regional hydroclimatic change? J. Clim. 23, 1513–1525. http://dx.doi.org/10.1175/2009JCLI3251.1.

Marsh, R., 2000. Recent variability of the North Atlantic thermohaline circulation inferred from surface heat and freshwater fluxes. J. Clim. 13 (18), 3239–3260.

Marsh, R., Josey, S.A., Nurser, A.J.G., de Cuevas, B.A., Coward, A.C., 2005. Water mass transformation in the North Atlantic over 1985-2002 simulated in an eddy-permitting model. Ocean Sci. 1, 127–144, SRef-ID: 1812-0792/os/2005-1-127.

Marsh, R., Josey, S.A., de Cuevas, B.A., Redbourn, L.J., Quartly, G.D., 2008. Mechanisms for recent warming of the North Atlantic: insights gained with an eddy-permitting model. J. Geophys. Res. 113. http://dx.doi.org/10.1029/2007JC004096, C04031.

Marshall, G.J., 2003. Trends in the Southern Annular Mode from observations and reanalyses. J. Clim. 16, 4134–4143.

Mignot, J., Frankignoul, C., 2004. Interannual to interdecadal variability of sea surface salinity in the Atlantic and its link to the atmosphere in a coupled model. J. Geophys. Res. 109. http://dx.doi.org/10.1029/2003JC002005, C04005.

Monahan, A.H., 2006a. The probability distribution of sea surface wind speeds. Part I: theory and sea winds observations. J. Clim. 19, 497–520.

Monahan, A.H., 2006b. The probability distribution of sea surface wind speeds. Part II: dataset intercomparison and seasonal variability. J. Clim. 19, 521–534.

Monahan, A.H., 2007. Empirical models of the probability distribution of sea surface wind speeds. J. Clim. 20, 5798–5814.

Morrissey, M.L., Albers, A., Greene, J.S., Postawko, S., 2010. An isofactorial change-of-scale model for the wind speed probability density function. J. Atmos. Oceanic Technol. 27, 257–273.

Niiler, P.P., Stevenson, J., 1982. On the heat budget of tropical warm water pools. J. Mar. Res. 40 (Suppl.), 465–480.

Onogi, K., Tsutsui, J., Koide, H., Sakamoto, M., Kobayashi, S., Hatsushika, H., Matsumoto, T., Yamazaki, N., Kamahori, H., Takahashi, K., Kadokura, S., Wada, K., Kato, K., Oyama, R., Ose, T., Mannoji, N., Taira, R., 2007. The JRA-25 reanalysis. J. Meteorol. Soc. Jpn. 85, 369–432.

Oost, W.A., Komen, G.J., Jacobs, C.M.J., van Oort, C., 2002. New evidence for a relation between wind stress and wave age from measurements during ASGAMAGE. Bound. Layer Meteorol. 103, 409–438.

Pascal, R.W., Josey, S.A., 2000. Accurate radiometric measurement of the atmospheric longwave flux at the sea surface. J. Atmos. Oceanic Technol. 17 (9), 1271–1282.

Pierce, D.W., Barnett, T.P., AchutaRao, K.M., Gleckler, P.J., Gregory, J.M., Washington, W.M., 2006. Anthropogenic warming of the oceans: observations and model results. J. Clim. 19, 1873–1900.

Pietri, A., Testor, P., Echevin, V., Chaigneau, A., Mortier, L., Eldin, G., Grados, C., 2013. Fine scale vertical structure of the upwelling system of Southern Peru as observed from glider data. J. Phys. Oceanogr. 43, 631–646. http://dx.doi.org/10.1175/JPO-D-12-035.1.

Powell, M.D., Vickery, P.J., Reinhold, T.A., 2003. Reduced drag coefficient for high wind speeds in tropical cyclones. Nature 422, 279–283.

Praveen, K.B., Vialard, J., Lengaigne, M., Murty, V.S.N., McPhaden, M.J., 2011. TropFlux: air-sea fluxes for the global tropical oceans—description and evaluation. Clim. Dyn. 38, 1521–1543. http://dx.doi.org/10.1007/s00382-011-1115-0.

Reed, R.K., 1977. On estimating insolation over the ocean. J. Phys. Oceanogr. 7, 482–485.

Ren, L., Riser, S., 2010. Observations of decadal-scale salinity changes in the thermocline of the North Pacific Ocean. Deep -Sea Res. II 57, 1161–1170.

Renfrew, I.A., Moore, G.W.K., Guest, P.S., Bumke, K., 2002. A comparison of surface-layer and surface turbulent-flux observations over the Labrador Sea with ECMWF analyses and NCEP reanalyses. J. Phys. Oceanogr. 32, 383–400.

Reverdin, G., Kestenare, E., Frankignoul, C., Delcroix, T., 2007. Surface salinity in the Atlantic Ocean (30S-50N). Prog. Oceanogr. 73, 311–340. http://dx.doi.org/10.1016/j.pocean.2006.11. 004.

Rienecker, M.M., Suarez, M.J., Gelaro, R., Todling, R., Bacmeister, J., Liu, E., Bosilovich, M.G., Schubert, S.D., Takacs, L., Kim, G.-K., Bloom, S., Chen, J., Collins, D., Conaty, A., da Silva, A., et al., 2011. MERRA—NASA's Modern-Era retrospective analysis for research and applications. J. Clim. 24, 3624–3648. http://dx.doi.org/10.1175/JCLI-D-11-00015.1.

Roberts, B., Clayson, C.A., Robertson, F.R., Jackson, D.L., 2010. Predicting near-surface characteristics from SSM/I using neural networks with a first guess approach. J. Geophys. Res. 115, D19. http://dx.doi.org/10.1029/2009JD013099.

Rossow, W.B., Schiffer, R.A., 1991. ISCCP cloud data products. Bull. Am. Meteorol. Soc. 72, 2–20. http://dx.doi.org/10.1175/1520-0477(1991)072<0002:ICDP>2.0.CO;2.

Roether, W., Manca, B.B., Klein, B., Bregant, D., Georgopolous, D., Beitzel, V., Kovacevic, V., Luchetta, A., 1996. Recent changes in eastern Mediterranean deep waters. Science 271 (5247), 333–335.

Sanchez-Gomez, E., Somot, S., Josey, S.A., Elguindi, N., Déqué, M., 2011. Evaluation of the mediterranean sea water and heat budgets simulated by an ensemble of high resolution regional climate models. Clim. Dyn. 37, 2067–2086. http://dx.doi.org/10.1007/s00382-011-1012-6.

Saha, S., et al., 2010. The NCEP climate forecast system reanalysis. Bull. Am. Meteorol. Soc. 91, 1015–1057.

Schanze, J.J., Schmitt, R.W., Yu, L., 2010. The global oceanic freshwater cycle: a state-of-the-art quantification. J. Mar. Res. 68, 569–595.

Schulz, E., Josey, S.A., Verein, R., 2012. First air-sea flux mooring measurements in the Southern Ocean. Geophys. Res. Lett. 39. http://dx.doi.org/10.1029/2012GL052290, L16606.

Shaman, J., Samelson, R.M., Skyllingstad, E., 2010. Air-sea fluxes over the Gulf stream region: atmospheric controls and trends. J. Clim. 23, 2651–2670.

da Silva, A.M., Young, C.C., Levitus, S., 1994. Algorithms and Procedures, vol. 1, Atlas of Surface Marine Data. NOAA Atlas NESDIS 6, 74 pp.

Smith, S.R., Legler, D.M., Verzone, K.V., 2001. Quantifying uncertainties in NCEP reanalyses using high-quality research vessel observations. J. Clim. 14, 4062–4072.

Smith, T.M., Arkin, P.A., Sapiano, M.R.P., 2009. Reconstruction of near-global annual precipitation using correlations with sea surface temperature and sea level pressure. J. Geophys. Res. Atmos. 114, D12107.

Smith, S.R., Hughes, P.J., Bourassa, M.A., 2011. Acomparison of nine monthly air–sea flux products. Int. J. Climatol. 31, 1002–1027. http://dx.doi.org/10.1002/joc.2225.

Stackhouse Jr., P.W., Gupta, S.K., Cox, S.J., Mikovitz, J.C., Zhang, T., Chiacchio, M., 2004. 12-year surface radiation budget dataset. GEWEX News 14, 10–12.

Stephens, G.L., et al., 2008. CloudSat mission: performance and early science after the first year of operation. J. Geophys. Res. 113. http://dx.doi.org/10.1029/2008JD009982, D00A18.

Stoffelen, A., 1998. Toward the true near-surface wind speed: error modeling and calibration using triple collocation. J. Geophys. Res. 103, 7755–7766.

Swart, N.C., Fyfe, J.C., 2012. Observed and simulated changes in the Southern Hemisphere surface westerly wind-stress. Geophys. Res. Lett. 39. http://dx.doi.org/10.1029/2012GL052810, L16711.

Taylor, P.K., Yelland, M.J., 2001. The dependence of sea surface roughness on the height and steepness of the waves. J. Phys. Oceanogr. 31, 572–590.

Tomita, H., Kubota, M., Cronin, M.F., Iwasaki, S., Konda, M., Ichikawa, H., 2010. An assessment of surface heat fluxes from J-OFURO2 at the KEO and JKEO sites. J. Geophys. Res. 115. http://dx.doi.org/10.1029/2009JC005545, C03018.

Toole, J.M., Zhang, H., Caruso, M.J., 2004. Time-dependent internal energy budgets of the Tropical warm water pools. J. Clim. 17, 1398–1410.

Trenberth, K.E., Caron, J.M., Stepaniak, D.P., 2001. The atmospheric energy budget and implications for surface fluxes and ocean heat transports. Clim. Dyn. 17, 259–276.

Trenberth, K.E., Fasullo, J.T., Kiehl, J., 2009. Earth's global energy budget. Bull. Am. Meteorol. Soc. 90, 311–324. http://dx.doi.org/10.1175/2008BAMS2634.1.

Trenberth, K.E., Fasullo, J.T., Mackaro, J., 2011. Atmospheric moisture transports from ocean to land and global energy flows in reanalyses. J. Clim. 24, 4907–4924.

Tzortzi, E., Josey, S.A., Srokosz, M.A., Gommenginger, C., 2013. Tropical Atlantic salinity variability: new insights from SMOS. Geophys. Res. Lett. 40, 2143–2147. http://dx.doi.org/10.1002/grl.50225.

Uppala, S.M., et al., 2005. The ERA-40 reanalysis. Q. J. R. Meteorol. Soc. 131, 2961–3012.

Vickers, D., Mahrt, L., Andreas, E.L., 2013. Estimates of the 10-m neutral sea surface drag coefficient from aircraft eddy-covariance measurements. J. Phys. Oceanogr. 43, 301–310.

Walin, G., 1982. On the relation between sea-surface heat-flow and thermal circulation in the ocean. Tellus 34, 187–195.

Weller, R.A., Bradley, E.F., Edson, J.B., Fairall, C.W., Brooks, I., Yelland, M.J., Pascal, R.W., 2008. Sensors for physical fluxes at the sea surface: energy, heat, water, salt. Ocean Sci. 4, 247–263.

WGASF, 2000. Intercomparison and validation of ocean–atmosphere energy flux fields—final report of the Joint WCRP/SCOR Working Group on Air–Sea Fluxes(WGASF). In: Taylor, P.K. (Ed.), WCRP-112, WMO/TD-1036, 306 pp.

Whelan, S.P., Lord, J., Galbraith, N.R., Weller, R.A., Farrar, J.T., Grant, D., Grados, C., de Szoeke, S.P., Moffat, C.F., Zappa, C.J., Yang, M., Straneo, F., Fairall, C.W., Zuidema, P., Wolfe, D., Miller, M., Covert, D., 2009. Stratus 9/VOCALS ninth setting of the Stratus Ocean Reference Station & VOCALS Regional Experiment, WHOI-2009-03, Technical Report. 127pp.

Winker, D.M., et al., 2010. The CALIPSO mission: a global 3D view of aerosols and clouds. Bull. Am. Meteorol. Soc. 91, 1211–1229. http://dx.doi.org/10.1175/2010BAMS3009.1.

Woodruff, S.D., Worley, S.J., Lubker, S.J., Ji, Z., Freeman, J.E., Berry, D.I., Brohan, P., Kent, E.C., Reynolds, R.W., Smith, S.R., Wilkinson, C., 2011. ICOADS Release 2.5: extensions and enhancements to the surface marine meteorological archive. Int. J. Climatol. 31, 951–967. http://dx.doi.org/10.1002/joc.2103, CLIMAR-III Special Issue.

Yang, X.Y., Huang, R.X., Wang, D.X., 2007. Decadal changes of wind stress over the Southern Ocean associated with Antarctic ozone depletion. J. Clim. 20, 3395–3410.

Yu, L., 2009. Sea surface exchanges of momentum, heat, and freshwater determined by satellite remote sensing. In: Steele, J., Thorpe, S., Turekian, K. (Eds.), Encyclopedia of Ocean Sciences 2e. Academic Press, London, UK, pp. 202–211.

Yu, L., 2011. A global relationship between the ocean water cycle and near-surface salinity. J. Geophys. Res. 116. http://dx.doi.org/10.1029/2010JC006937, C10.

Yu, L., Jin, X., 2012. Buoy perspective of a high-resolution global ocean vector wind analysis constructed from passive radiometers and active scatterometers (1987–present). J. Geophys. Res. 117. http://dx.doi.org/10.1029/2012JC008069, C11013.

Yu, L., Weller, R.A., 2007. Objectively analyzed air-sea heat fluxes for the global ice-free oceans (1981–2005). Bull. Am. Meteorol. Soc. 88, 527–539.

Yu, L., Jin, X., Weller, R.A., 2008. Multidecade global flux datasets from the Objectively Analyzed Air-sea Fluxes (OAFlux) project: latent and sensible heat fluxes, ocean evaporation, and related surface meteorological variables. Woods Hole Oceanographic Institution, OAFlux Project Technical Report. OA-2008-01, 64 pp. Woods Hole, Massachusetts.

Zhang, Y.-C., Rossow, W.B., Lacis, A.A., Oinas, V., Mishchenko, M.I., 2004. Calculation of radiative fluxes from the surface to top of atmosphere based on ISCCP and other global data sets: refinements of the radiative transfer model and the input data. J. Geophys. Res. 109, D19105. http://dx.doi.org/10.1029/2003JD004457.

Zolina, O., Gulev, S.K., 2003. Synoptic variability of ocean-atmosphere turbulent fluxes associated with atmospheric cyclones. J. Clim. 16, 3023–3041.

Zuidema, P., Li, Z., Hill, R.J., Bariteau, L., Rilling, B., Fairall, C., Brewer, A., Albrecht, B., Hare, J., 2012. On trade-wind cumulus cold pools. J. Atmos. Sci. 69, 258–280. http://dx.doi.org/10.1175/JAS-D-11-0143.1.

Chapter 6

Thermodynamics of Seawater

Trevor J. McDougall*, Rainer Feistel[†] and Rich Pawlowicz[‡]
*University of New South Wales, Sydney, New South Wales, Australia
[†]Leibniz Institute for Baltic Sea Research, Warnemuende, Germany
[‡]University of British Columbia, Vancouver, British Columbia, Canada

Chapter Outline

1. INTRODUCTION

The thermodynamic properties of seawater include variables such as density (or specific volume), sound speed, specific heat capacity, the adiabatic lapse rate, and the freezing temperature. In order to undertake physical oceanographic studies, expressions are needed for these and other thermodynamic quantities in terms of properties that can be measured at sea, namely the salinity, temperature, and pressure of seawater samples. There is a long history of the measurement of salinity in the ocean and a significant milestone was reached in 1980 with the adoption of the Practical Salinity Scale (UNESCO, 1981; Fofonoff, 1985) that allows the calculation of Practical Salinity from measurements of temperature, pressure, and the electrical conductivity of seawater.

Accurate knowledge of the specific volume of seawater in terms of measured quantities is needed so that the "thermal wind" and geostrophic relationships can be accurately calculated from ocean measurements. The algorithm expressing specific volume as a function of salinity, temperature, and pressure is often called the "equation of state" of seawater. The horizontal gradients of specific volume and of Practical Salinity at constant pressure in the ocean are usually very small, and measuring them places strong demands on ocean instrument engineering and on ocean instrument calibration laboratories. Our present ability to measure Practical Salinity with an accuracy of $0.003\,\mathrm{g\,kg^{-1}}$ at sea is testament to the innovation and persistence of oceanographers and engineers over more than a century.

Some of the variation of *in situ* temperature that is observed in the ocean is simply caused by changes in pressure, and if seawater properties such as temperature and salinity are to be used to distinguish between the origins of water parcels and to detect the influence of mixing processes, this "adiabatic" variation in *in situ* temperature must be removed. This dependence of *in situ* temperature on pressure when no exchange of heat or salt takes place is called "the adiabatic lapse rate" and it can be evaluated given algorithms for both the specific volume and the specific heat capacity of seawater. The temperature corrected for this passive effect of pressure is called "potential temperature" and this temperature variable has been used to good effect in physical oceanography for a century. The calculation of potential temperature is also central to the ability to correct *in situ* density for much of the passive effects of the compression of seawater due to increases in pressure; this correction leads from *in situ* density to potential density. Subsequent adjustments to *in situ* density have led to the concept of "neutral surfaces" along which fluid parcels can be exchanged without encountering vertical buoyant restoring forces.

The variables and algorithms described in the above paragraphs include almost all of the thermodynamic concepts that have found practical use in the field of oceanography

Ocean Circulation and Climate, Vol. 103. http://dx.doi.org/10.1016/B978-0-12-391851-2.00006-4

during the twentieth century. We have used Practical Salinity together with algorithms for specific volume and the adiabatic lapse rate to calculate *in situ* density, potential temperature, and potential density, and we have treated Practical Salinity and potential temperature as conservative variables since we have described, tracked, and mixed water masses on the Practical Salinity—potential temperature diagram. Moreover, we have treated potential temperature as being proportional to the "heat content" of seawater in that (a) we have calculated the meridional transport of potential temperature in the ocean and compared this to the area-integrated heat flux crossing the air–sea interface (using a constant specific heat capacity), and (b) in ocean models we have interpreted the air–sea heat flux as being an air–sea flux of potential temperature (using a constant specific heat capacity).

We have known for some time that these practices regarding Practical Salinity and potential temperature were less than perfect, but until quite recently we had no practical alternatives. Regarding the use of potential temperature, Fofonoff (1962) demonstrated that potential temperature is not a conservative variable by analyzing the mixing of two seawater parcels of contrasting potential temperatures and Absolute Salinities. While Fofonoff showed that potential temperature was not an ideal variable in this respect, it was not clear how to quantify the importance of this issue in oceanographic practice, or how oceanographic practice could be improved. The idea of using potential enthalpy in place of potential temperature to represent the "heat content" per unit mass of seawater emerged a decade ago (McDougall, 2003), and this idea became a practical solution to this issue because of the availability of a Gibbs function for seawater (Feistel, 2003, 2008). The beauty of describing the thermodynamic properties of seawater using a thermodynamic potential function (such as a Gibbs function) is that this enables the evaluation of entropy, internal energy, and enthalpy. These thermodynamic variables could not be evaluated from the 1980 equation of state since it is not possible to construct these quantities from the specific heat capacity alone, but knowledge of these variables is necessary in order to apply the First Law of Thermodynamics to the ocean.

The limitations of Practical Salinity were better known to the oceanographic community than the deficiencies of potential temperature. For example, even before Practical Salinity was endorsed for international use, Brewer and Bradshaw (1975) pointed out that the conductivity of seawater does not account for the influence of non-conducting species on the density of seawater and they provided a theoretically estimated expression for a correction to density based on additional data on the concentrations of nutrients and quantities associated with carbon chemistry, an expression that is in close agreement with the recent and more exhaustive model-based approach of Pawlowicz et al. (2011) (see also Wright et al., 2011).

These deficiencies in our use of Practical Salinity and potential temperature have now been addressed with the thermodynamic properties of seawater being updated in 2010 to what is known as TEOS-10, being shorthand for "The International Thermodynamic Equation Of Seawater—2010." TEOS-10 defines thermodynamic potentials for pure water, seawater, ice and humid air, thus allowing all the thermodynamic properties of liquid pure water, water vapor, ice, seawater and moist air to be evaluated in an internally self-consistent manner.

This chapter summarizes the three new oceanographic variables that have been introduced with the adoption of TEOS-10, namely Absolute Salinity, Preformed Salinity, and Conservative Temperature. The ways in which these new variables should be used in oceanographic practice is discussed. A good place to start this chapter is to revise exactly what was the international standard for the thermodynamic properties of seawater from 1980 to 2009, namely EOS-80.

EOS-80 is the acronym of the International Equation of State of Seawater—1980 (UNESCO, 1981) which is the equation of state of seawater that had been used in oceanography for the past thirty years, from 1980 to 2009. EOS-80 consists of five separate algorithms, namely (1) the Practical Salinity Scale, PSS-78, (2) the expression for the density of seawater as a function of Practical Salinity, temperature, and pressure, (3) the Millero et al. (1973) algorithm for the specific heat capacity of seawater at constant pressure, (4) the Chen and Millero (1977) expression for the sound speed of seawater and (5) the Millero and Leung (1976) formula for the freezing temperature of seawater. Because these five algorithms were defined with separate polynomials, the thermodynamic properties are not totally self-consistent. Three other algorithms supported under the auspices of the Joint Panel on Oceanographic Tables and Standards (JPOTS) concerned the conversion between hydrostatic pressure and depth, the calculation of the adiabatic lapse rate, and the calculation of potential temperature. The expressions for the adiabatic lapse rate and for potential temperature could in principle have been derived from the other algorithms of the EOS-80 set, but in fact they were based on the formulas of Bryden (1973). The scientific rationale behind the adoption of EOS-80 was summarized in Fofonoff (1985). Importantly, we note that the First Law of Thermodynamics is concerned with the evolution of the thermodynamic quantities entropy, internal energy, and enthalpy, and EOS-80 did not provide for the calculation of these quantities.

The TEOS-10 properties of seawater are based on a Gibbs function (or Gibbs potential) of seawater which is a function of Absolute Salinity S_A (rather than of Practical Salinity S_P), temperature, and pressure. Absolute Salinity is traditionally defined as the mass fraction of dissolved material in seawater. The use of Absolute Salinity as the salinity argument for the Gibbs function and for all other thermodynamic functions (such as specific volume and specific enthalpy) is a major departure from previous practice (EOS-80). Absolute Salinity is preferred over Practical Salinity because the thermodynamic properties of seawater are directly influenced by the mass of dissolved constituents whereas Practical Salinity depends only on electrical conductivity. Consider, for example, exchanging a small amount of pure water with the same mass of silicate in an otherwise isolated seawater sample at constant temperature and pressure. Since silicate is predominantly non-ionic, the conductivity (and therefore Practical Salinity S_P) is almost unchanged but the Absolute Salinity is increased, as is the density. Similarly, if a small mass of say NaCl is added and the same mass of silicate is taken out of a seawater sample, the mass fraction absolute salinity will not have changed (and so the density should be almost unchanged) but the Practical Salinity will have increased.

While Absolute Salinity is the salinity variable that is to be used in journals to describe the salinity of seawater, the salinity variable that is to be stored in national databases must remain Practical Salinity. Note that the practice of storing one type of salinity in national data bases (Practical Salinity) but using a different type of salinity in publications (Absolute Salinity) is exactly analogous to our present practice with temperature; *in situ* temperature t is stored in data bases (since it is the measured quantity) but the temperature variable that is used in publications is a calculated quantity, being potential temperature θ under EOS-80 and is now Conservative Temperature Θ under TEOS-10.

Some advantages of TEOS-10 compared to EOS-80 are

- The Gibbs function approach allows the calculation of internal energy, entropy, enthalpy, potential enthalpy and the chemical potentials of seawater as well as the freezing temperature, and the latent heats of melting and of evaporation. These quantities were not available from the International Equation of State 1980 (EOS-80) but are essential for the accurate accounting of "heat" in the ocean and for the consistent and accurate treatment of air–sea and ice–sea heat fluxes. For example, the new TEOS-10 temperature variable, Conservative Temperature, Θ, is defined to be proportional to potential enthalpy and is a very accurate measure of the "heat" content per unit mass of seawater; Θ is two orders of magnitude more conservative than potential temperature θ.
- For the first time, the influence of the spatially varying composition of seawater is taken into account through the use of Absolute Salinity. In the open ocean, this has a non-trivial effect on the horizontal density gradient computed from the equation of state, and thereby on the ocean velocities and heat transports calculated via the "thermal wind" relation. The density differences caused by the non-constant composition of seawater are up to ten times as large as the uncertainty in density caused by the precision of Practical Salinity measurements at sea (and are much larger in coastal waters).
- The thermodynamic quantities available from this Gibbs function-based approach are totally consistent with each other. These consistent and accurate thermodynamic quantities include those of seawater, ice, and humid air, so that, for example, we now have accurate and thermodynamically consistent algorithms for the entropy and enthalpy of ice crystals high in the atmosphere.
- The new salinity variable, Absolute Salinity, is measured in SI units. Moreover, the treatment of freshwater fluxes in ocean models will be consistent with the use of Absolute Salinity but is only approximately so for Practical Salinity.
- A single algorithm for seawater density (the 48-term computationally-efficient expression $\hat{\rho}(S_A, \Theta, p)$) can now be used for ocean modeling, for observational oceanography, and for theoretical studies. By contrast, for the past 30 years, we have used different algorithms for density in ocean modeling and in observational oceanography.
- The Reference Composition of standard seawater supports marine physicochemical studies such as the solubility of sea salt constituents, the alkalinity, the pH, and ocean acidification through rising concentrations of atmospheric CO_2.

The decision by IOC (the Intergovernmental Oceanographic Commission) to adopt TEOS-10 in June 2009 and the earlier such recommendations of SCOR (Scientific Committee on Oceanic Research) and IAPSO (International Association for the Physical Sciences of the Oceans) were based on the results of the SCOR/IAPSO Working Group 127 (WG127) on the "Thermodynamics and Equation of State of Seawater." The approach taken by WG127 has been to develop thermodynamic potentials (also known as fundamental equations of state, such as Gibbs functions or Helmholtz functions, see Alberty (2001) for a technical review) for pure fluid (i.e., liquid and gaseous) water, seawater, ice, and humid air from which all their thermodynamic properties can be derived by purely mathematical manipulations (such as differentiation). This approach ensures that the various thermodynamic properties are complete, independent, and self-consistent (in that they obey the Maxwell cross-differentiation relations). Gibbs functions are expressed in terms of temperature and pressure, while Helmholtz functions require temperature and density as independent variables. For a given substance, Gibbs and Helmholtz functions (as well as other options) are mathematically fully equivalent but differ in their convenience of practical use. A short historical account of the work of SCOR/IAPSO Working Group 127 and the evolution of the scientific thinking leading to TEOS-10 can be found in Pawlowicz et al. (2012).

2. ABSOLUTE SALINITY S_A AND PREFORMED SALINITY S_*

The "raw" physical oceanographic data collected from ships and from autonomous platforms (e.g., Argo floats) are stored in national oceanographic data bases as

- Practical Salinity (S_P, unitless, PSS-78) and
- *in situ* temperature (t, °C, ITS-90) as functions of
- sea pressure (p, dbar), at a series of
- longitudes, latitudes, and observation times.

Under TEOS-10 the same "raw" physical oceanographic data are to be stored in national databases. This is done to maintain continuity in the archived salinity variable, and because Practical Salinity is virtually the measured variable related to the electrical conductivity of seawater, after correction for temperature and pressure effects. However, all the thermodynamic properties, including an accurate estimate of "heat content," are functions of the mass of solute in seawater (denoted as its absolute salinity), rather than its electrical conductivity. The precisely defined measure of absolute salinity used in TEOS-10 to determine the thermodynamic properties of seawater is denoted Absolute Salinity S_A.

The first step in processing oceanographic data is therefore to calculate Absolute Salinity. This step is the most apparent change relative to past practices involving the International Equation of State 1980 (EOS-80). The simplest way to calculate Absolute Salinity is to first estimate a Reference Salinity and then to add a small correction that takes into account the spatial variations in the composition of seawater.

As discussed in Pawlowicz (2010), Wright et al. (2011) and IOC et al. (2010), there are actually several contenders for the title of the "absolute salinity" of seawater, namely "Solution Salinity," "Added-Mass Salinity," and "Density Salinity." The paper of Wright et al. (2011) presents a clear and readable account of this difficult subject; however, the nuances surrounding these different definitions of absolute salinity need not concern most physical oceanographers. Under TEOS-10, the words Absolute Salinity and symbol S_A are reserved for "Density Salinity" such as can be deduced using laboratory measurements with a vibrating beam densimeter.

The TEOS-10 international standard was adopted while recognizing that the techniques for estimating Absolute Salinity, and for standardizing its measurements against the SI, will likely improve over the coming decades. Algorithms for evaluating Absolute Salinity in terms of Practical Salinity, longitude, latitude, and pressure or other relevant quantities will be updated from time to time, after relevant appropriately peer-reviewed publications have appeared. Oceanographers should always state in their published work which method or version of the available software was used to calculate Absolute Salinity.

2.1. Reference-Composition Salinity S_R

Millero et al. (2008) defined a Reference-Composition Seawater as a table of exact mole fractions of the main chemical constituents of seawater. This Reference Composition is a fundamental underpinning component of TEOS-10 and it forms a precisely defined baseline for all future discussions of the properties of seawater. It was introduced by Millero et al. (2008) as their best estimate of the composition of Standard Seawater, which is seawater from the surface waters of a certain region of the North Atlantic that has been processed and bottled as a calibration standard. The Reference-Composition Salinity (this is often shortened to "Reference Salinity") is the mass of dissolved material in Reference-Composition Seawater, as calculated from the table of exact mole fractions. Standard Seawater can thus be considered the best physical realization of Reference-Composition Seawater, and the measurements of properties of Standard Seawater are then the best estimates of the properties of Reference-Composition Seawater.

For the range of salinities where Practical Salinity S_P is defined (that is, in the range $2 < S_P < 42$) Millero et al. (2008) show that

$$S_R \approx u_{PS} S_P \quad \text{where } u_{PS} \equiv (35.16504/35)\,\text{g kg}^{-1}. \qquad (6.1)$$

In the range $2 < S_P < 42$, this equation expresses the Reference-Composition Salinity S_R of a seawater sample on the Reference-Composition Salinity Scale (Millero et al., 2008). While the TEOS-10 Gibbs function itself is valid for all salinities between pure water and the precipitation point of calcium carbonate ($\sim$90 g kg^{-1}), no suitable conductivity equation is currently available for Practical Salinities greater than 42. An extension of Practical Salinity into the region $0 < S_P < 2$, while not formally part of PSS-78, is available in the work of Hill et al. (1986) and this has been adopted in the GSW Oceanographic Toolbox of TEOS-10 (McDougall and Barker, 2011).

For practical purposes, relationship (6.1) can be taken to be an equality since the approximate nature of this relation only reflects the extent to which Practical Salinity, as determined from measurements of conductivity ratio, temperature, and pressure, varies when a seawater sample is heated, cooled, or subjected to a change in pressure but without exchange of mass with its surroundings other than by dilution or evaporation of pure water. The Practical Salinity Scale of 1978 was designed to satisfy this property as accurately as possible within the constraints of the polynomial approximations used to determine Chlorinity (and hence Practical Salinity) in terms of the measured conductivity ratio.

From Equation (6.1), a seawater sample of Reference Composition whose Practical Salinity S_P is 35 has a Reference Salinity S_R of $35.16504\,\mathrm{g\,kg^{-1}}$. Millero et al. (2008) estimate that the absolute uncertainty in this value is $\pm 0.007\,\mathrm{g\,kg^{-1}}$. The difference between the numerical values of Reference and Practical Salinities can be traced back to the original practice of determining salinity by evaporation of water from seawater and weighing the remaining solid material. This process also evaporated some volatile components and most of the $0.16504\,\mathrm{g\,kg^{-1}}$ salinity difference is due to this effect.

2.2. Absolute Salinity S_A

For Standard Seawater, the Reference-Composition Salinity is the best estimate of the Absolute Salinity. However, for real seawater, a correction must be added that takes into account the effects of varying relative composition. This is because the relationship between electrical conductivity and the mass of solute is biased by the addition of relatively non-conductive material.

In the Gibbs SeaWater (GSW) library of TEOS-10 computer algorithms, this calculation is accomplished by the function *gsw_SA_from_SP*. Hence this function is perhaps the most fundamental of the TEOS-10 algorithms as it is the gateway leading from oceanographic measurements to all the thermodynamic properties of seawater. A call to this function can be avoided only if one is willing to ignore the influence of the spatial variations in the composition of seawater on seawater properties (such as density and specific volume). If this is indeed the intention, then the remaining TEOS-10 functions must be called with the salinity argument being Reference Salinity S_R, and most definitely not with Practical Salinity S_P. When this is done, it should be clearly stated that Reference Salinity is being used, not Absolute Salinity.

The *gsw_SA_from_SP* function which calculates Absolute Salinity S_A first interpolates the global Absolute Salinity Anomaly Ratio (R^δ) data set to the given pressure, latitude, and longitude. The Absolute Salinity Anomaly Ratio $R^\delta \equiv \delta S_A^{\mathrm{atlas}}/S_R^{\mathrm{atlas}}$ is the ratio of Absolute Salinity Anomaly $\delta S_A \equiv S_A - S_R$ and the corresponding Reference-Composition Salinity S_R of a stored atlas data set. The values of R^δ are interpolated onto the latitude, longitude, and pressure of an oceanographic observation and R^δ takes values no larger than 0.001 in the global ocean.

This interpolated value of R^δ is then used to calculate Absolute Salinity S_A according to

$$S_A = u_{PS} S_P \left(1 + R^\delta\right) = S_R \left(1 + R^\delta\right) \quad \text{Non-Baltic.} \tag{6.2}$$

In this expression $u_{PS} S_P = (35.16504\,\mathrm{g\,kg^{-1}}/35) S_P$ is the Reference-Composition Salinity S_R, which is the best estimate of Absolute Salinity of a Standard Seawater sample (see Equation 6.1). In this way we see that Absolute Salinity is given by

$$S_A = S_R + \delta S_A, \tag{6.3}$$

and in the ocean interior (but not in the Baltic Sea) the Absolute Salinity Anomaly is given by $\delta S_A = S_R R^\delta = \delta S_A^{\mathrm{atlas}} S_R / S_R^{\mathrm{atlas}}$. Plots of the Absolute Salinity Anomaly are shown in Figure 6.1; notice that the ocean is deliberately extended over land near coastlines so that any location that is genuinely in the ocean does not get excluded.

In the Baltic Sea, Absolute Salinity is related to Reference-Composition Salinity by $S_A - S_R = 0.087\,\mathrm{g\,kg^{-1}} \times (1 - S_P/35)$ (from Equation A.5.16 of IOC et al., 2010; following Feistel et al., 2010a), so that S_A is given by

$$S_A = \frac{(35.16504 - 0.087)\,\mathrm{g\,kg^{-1}}}{35} S_P + 0.087\,\mathrm{g\,kg^{-1}} \quad \text{Baltic Sea.} \tag{6.4}$$

The differences in the formulae for the open ocean and the Baltic Sea arise from the effects of river salts. Since river salts and their effects will vary in different marginal seas, similar but not identical formulae will eventually be developed for other regions. However, at present, the effects of river salts on the physical properties of seawater have not been well studied. We note that the empirical coefficient $0.087\,\mathrm{g\,kg^{-1}}$ in Equation (6.4) for the Baltic Sea has varied significantly over recent decades (Feistel et al., 2010a).

The approach described above for estimating Absolute Salinity is based on laboratory measurements of the density of seawater samples collected from around the world's oceans as described in McDougall et al. (2012). However, if open ocean measurements are available of the Total Alkalinity (TA), Dissolved Inorganic Carbon (DIC), and the nitrate and silicate concentrations, but not of density anomalies, then an alternative formula is available to calculate Absolute Salinity. Pawlowicz et al. (2011) have used a chemical model of conductivity and density to arrive at the expression

$$(S_A - S_R)/\left(\mathrm{g\,kg^{-1}}\right) = \left(55.6\Delta\mathrm{TA} + 4.7\Delta\mathrm{DIC} + 38.9\mathrm{NO_3^-} + 50.7\mathrm{Si(OH)_4}\right)/\left(\mathrm{mol\,kg^{-1}}\right)$$

in terms of the values of the nitrate and silicate concentrations in the seawater sample (measured in $\mathrm{mol\,kg^{-1}}$), the differences ΔTA and ΔDIC, between the TA and DIC of the sample and the corresponding values of our best estimates of TA and DIC in Standard Seawater. For Standard Seawater our best estimates of TA and DIC are $0.0023\,(S_P/35)\,\mathrm{mol\,kg^{-1}}$ and $0.00208(S_P/35)\,\mathrm{mol\,kg^{-1}}$ respectively (see the discussion in Wright et al., 2011).

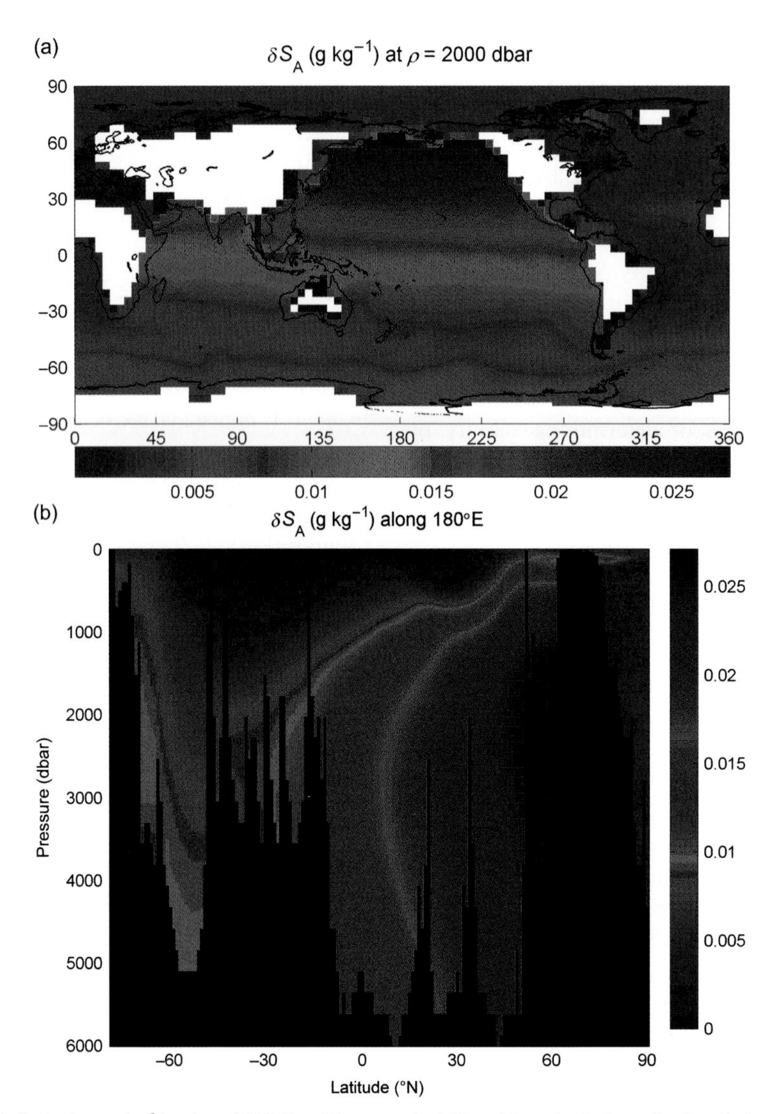

FIGURE 6.1 (a) Absolute Salinity Anomaly δS_A at $p = 2000$ dbar. The ocean is deliberately extended over land so that any location that is genuinely in the ocean does not get excluded from the algorithm. (b) A vertical section of Absolute Salinity Anomaly δS_A along 180°E in the Pacific Ocean.

The largest influence of the variable seawater composition in the open ocean occurs in the northern North Pacific where $S_A - S_R = \delta S_A$ is as large as $0.027\,\mathrm{g\,kg^{-1}}$ (see Figure 6.1), this being the difference between Absolute Salinity and the estimate of Absolute Salinity which can be made on the basis of Practical Salinity alone. This increment of salinity equates to an increment of density of approximately $0.020\,\mathrm{kg\,m^{-3}}$.

In order to gauge the importance of the spatial variation of seawater composition, the northward gradient of density at constant pressure is shown in Figure 6.2 for the data in a world ocean hydrographic atlas deeper than 1000 m. The vertical axis in this figure is the magnitude of the difference between the northward density gradient at constant pressure when the TEOS-10 algorithm for density is called with S_A (as it should be) compared with calling the same TEOS-10 density algorithm with S_R as the salinity argument. Figure 6.2 shows that the "thermal wind" is misestimated by more than 2% for 58% of the data in the world ocean below a depth of 1000 m if the effects of the variable seawater composition are ignored. When this same comparison is done for only the North Pacific, it is found that 60% of the data deeper than 1000 m has "thermal wind" misestimated by more than 10% if S_R is used in place of S_A. These improvements in evaluating the "thermal wind" in the ocean are about six times as large as the extra accuracy that TEOS-10 brings to the calculation of the "thermal wind" compared with using EOS-80 when the compositional variations are ignored (see Appendix A.5 of the TEOS-10 Manual, IOC et al., 2010). This demonstrates the importance of compositional variations to the calculation of the "thermal wind" and to ocean dynamics.

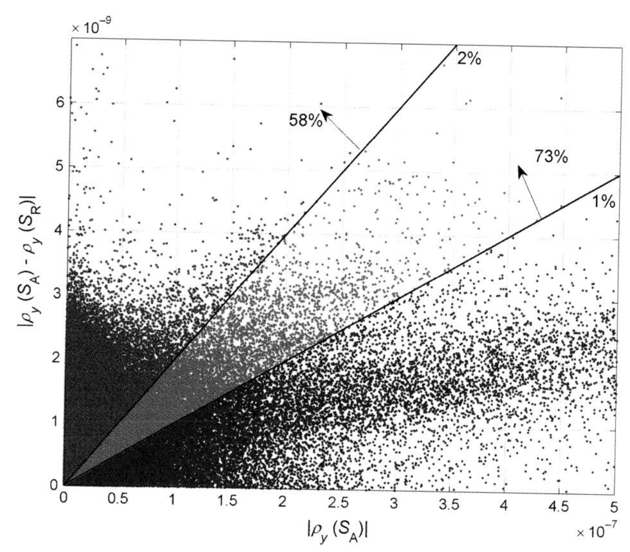

FIGURE 6.2 The northward density gradient at constant pressure (the horizontal axis) for data in the global ocean atlas of Gouretski and Koltermann (2004) for $p > 1000$ dbar. The vertical axis is the magnitude of the difference between evaluating the density gradient using S_A versus S_R as the salinity argument in the TEOS-10 expression for density.

2.3. Preformed Salinity S_*

Absolute Salinity S_A, Reference Salinity S_R and Practical Salinity S_P are all conservative salinity variables under the processes of (i) adiabatic pressure changes, and (ii) turbulent mixing, but none of these three salinity variables are conservative in the presence of (iii) biogeochemical processes. Preformed Salinity S_* is designed to be a conservative salinity variable which is unaffected by biogeochemical activity in the ocean; it is defined as Absolute Salinity less the contributions of biogeochemical processes to Absolute Salinity.

Based on the work of Pawlowicz et al. (2011) the difference between Absolute Salinity and Preformed Salinity is approximately proportional to the difference between Absolute Salinity and Reference Salinity, with the proportionality constant being 1.35. That is, $S_A - S_* \approx 1.35\,(S_A - S_R)$ so that Preformed Salinity is related to Practical Salinity by

$$S_* = \frac{35.16504\,\mathrm{g\,kg^{-1}}}{35} S_P\left(1 - 0.35R^{\delta}\right) \quad \text{Non-Baltic.} \quad (6.5)$$

Note that $(35.16504\,\mathrm{g\,kg^{-1}}/35)S_P$ is Reference Salinity S_R,which is the best estimate of Absolute Salinity for a Standard Seawater sample. The simplicity of this relationship arises from empirical correlations between the observed biogeochemically determined changes in different chemical constituents in the ocean. However, the correction is needed because changes in density and conductivity that arise from changes in these biogeochemical parameters

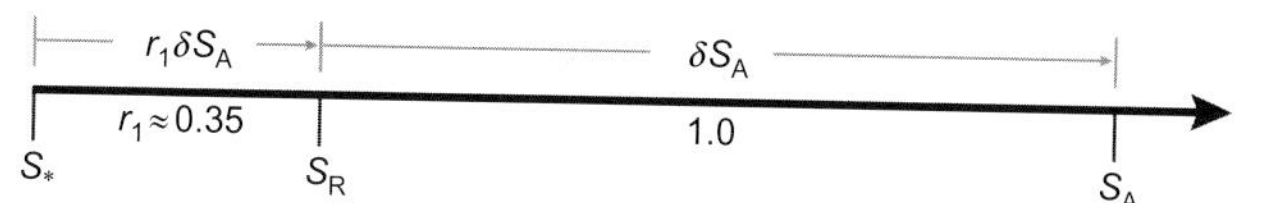

FIGURE 6.3 Number line of salinity, illustrating the differences between Preformed Salinity S_*, Reference Salinity S_R, and Absolute Salinity S_A for seawater whose composition differs from that of Standard Seawater.

are not the same as those that arise from additions or subtractions of an equal mass of "sea salt." Figure 6.3 illustrates the relationships between Preformed Salinity S_*, Reference Salinity S_R, and Absolute Salinity S_A.

In the Baltic Sea, the deviations of Absolute Salinity from Reference Salinity are not due to non-conservative biogeochemical processes but are rather due to the anomalous composition entering the Baltic from rivers. Since these anomalous constituents are conservative, Preformed Salinity S_* in the Baltic Sea is Absolute Salinity S_A so that S_* is evaluated using Equation (6.4).

The largest influence of the variable seawater composition in the open ocean occurs in the northern North Pacific where $S_R - S_*$ is almost $0.01\,\mathrm{g\,kg^{-1}}$, and the difference between Absolute Salinity and the conservative Preformed Salinity, $S_A - S_*$, is as large as $0.036\,\mathrm{g\,kg^{-1}}$, equivalent to an increment of density of approximately $0.028\,\mathrm{kg\,m^{-3}}$.

What then is the appropriate use of Preformed Salinity S_*? This salinity variable is the one that can be treated as being conservative. Hence, in contrast to the evolution equation of Absolute Salinity S_A, the evolution equation for Preformed Salinity S_* does not contain non-conservative source terms caused by biogeochemistry (see Appendix A.20 of IOC et al., 2010). This means that Preformed Salinity S_* is ideal for use as

- **i.** the salinity variable that is advected and diffused in forward ocean models,
- **ii.** the salinity variable that is advected and diffused in inverse ocean models, and
- **iii.** the salinity variable that is averaged when forming a hydrographic atlas.

In these applications, the salinity difference $S_A - S_*$ is added to the averaged atlas or model salinity variable to form Absolute Salinity S_A before other thermodynamic properties (such as density etc.) are calculated. The use of Preformed Salinity S_* in ocean modeling is discussed in Section 7 below and in Appendix A.20 of IOC et al. (2010).

3. THE GIBBS-FUNCTION APPROACH TO EVALUATING THERMODYNAMIC PROPERTIES

The Gibbs function of seawater $g(S_A,t,p)$ is defined as the specific Gibbs energy expressed in terms of the independent variables Absolute Salinity, *in situ* temperature, and pressure, and is given as the sum of the Gibbs function for pure water $g^{W}(t,p)$ and the saline part of the Gibbs function $g^{S}(S_A,t,p)$ so that

$$g(S_A,t,p) = g^{W}(t,p) + g^{S}(S_A,t,p), \tag{6.6}$$

and at zero Absolute Salinity the thermodynamic properties of seawater are forced to be equal to those of pure water. This consistency is also maintained with respect to the Gibbs function of ice and the Helmholtz function of humid air (Feistel et al., 2010b; IAPWS, 2010) so that the properties along the phase-equilibrium curves can be accurately determined (such as the freezing temperature as a function of Absolute Salinity and pressure). The careful alignment of the thermodynamic potentials of pure water, ice Ih and seawater is described in Feistel et al. (2008). Ice Ih is the naturally abundant form of ice, having hexagonal crystals.

The internationally accepted thermodynamic description of the properties of pure water (IAPWS-95) is the official pure-water basis upon which the Gibbs function of seawater is built according to Equation (6.6). Derived from the original IAPWS-95 Helmholtz function, the Gibbs function of liquid water $g^{W}(t,p)$ is valid over extended ranges of temperature and pressure from the freezing point to the critical point ($-22\,^{\circ}\mathrm{C} < t < 374\,^{\circ}\mathrm{C}$ and $600\ \mathrm{Pa} < (p + P_0) < 1000\ \mathrm{MPa}$); however, it is a computationally expensive algorithm. Part of the reason for this computational intensity is that the IAPWS-95 formulation is in terms of a Helmholtz function, which has the pressure as a function of temperature and density, so that an iterative procedure is needed to form the Gibbs function $g^{W}(t,p)$ (see, e.g., Feistel et al., 2008).

For practical oceanographic use in the oceanographic ranges of temperature and pressure, from less than the freezing temperature of seawater (at any pressure), up to 40 °C and in the pressure range $0 < p < 10^4$ dbar we also recommend the use of the pure water part of the Gibbs function of Feistel (2003) which has been approved by IAPWS as the Supplementary Release, IAPWS-09 (IAPWS, 2009b). This Gibbs function fits the underlying thermodynamic potential of IAPWS-95 extremely well, with the rms misfit between IAPWS-09 and IAPWS-95 for the variables density, thermal expansion coefficient, and specific heat capacity being a factor of between 20 and 100 less than the corresponding error in the laboratory data to which IAPWS-95 was fitted. Hence, in the oceanographic range of parameters, IAPWS-09 and IAPWS-95 may be regarded as equally accurate thermodynamic descriptions of pure liquid water. The computer code based on IAPWS-09 is approximately a factor of 65 faster than that based on IAPWS-95.

Most of the experimental seawater data that were already used for the construction of EOS-80 were exploited again for the IAPWS-08 (IAPWS, 2008) formulation after their careful adjustment to the new temperature and salinity scales and use of the improved pure-water reference IAPWS-95. Additionally, the TEOS-10 seawater Gibbs function was significantly improved (compared to EOS-80) by making use of theoretical relations such as the ideal-solution law and the Debye–Hückel limiting law, as well as by incorporating additional accurate measurements such as the temperatures of maximum density, sound speed, vapor pressures, and mixing heats, and implicitly by the enormous background data set which underlies IAPWS-95 (Wagner and Pruß, 2002; Feistel, 2003, 2008). The accuracy of high-pressure seawater densities has increased and the known sound-speed inconsistency of EOS-80 for deep-ocean waters of 0.5 m s^{-1} has been resolved through the use of IAPWS-95 for the pure-water part.

The TEOS-10 potential functions describing ice Ih (IAPWS-06) and humid air (IAPWS-10) similarly incorporate all the available accurate thermodynamic data, and these functions have been constructed to be exactly compatible with each other and with the Gibbs function of seawater (Feistel et al., 2008). Because of this exact compatibility, these potential functions can be used to calculate the properties at thermodynamic equilibrium between the different phases, such as between ice and seawater. For example, freezing occurs at the temperature t_f at which the chemical potential of water in seawater μ^{W} equals the chemical potential of ice μ^{Ih}. Thus, t_f is found by solving the implicit equation

$$\mu^{W}(S_A,t_f,p) = \mu^{Ih}(t_f,p) \tag{6.7}$$

or equivalently, in terms of the two Gibbs functions (subscript means partial derivative),

$$g(S_A,t_f,p) - S_A g_{S_A}(S_A,t_f,p) = g^{Ih}(t_f,p). \tag{6.8}$$

The Gibbs function for ice Ih, $g^{Ih}(t,p)$, is defined by IAPWS-06 (IAPWS, 2009a) and Feistel and Wagner (2006). Similar relations hold for the equilibrium between seawater and humid air (Feistel et al., 2010b). For example, evaporation from the sea surface ceases at relative humidities less than 100% because of the lowered vapor pressure of seawater compared to freshwater. These and similar thermal effects of freezing and evaporation are properly accounted for automatically in the consistent TEOS-10 formulation and in the software libraries available for the evaluation of those properties.

The term "sea ice" is used to describe the composite system of ice in equilibrium with trapped pockets of liquid seawater. A Gibbs function can be constructed for this composite system, dependent on the mass fraction of trapped seawater. As described in Section 5.4 of Feistel et al. (2010c), this composite Gibbs function of sea ice can be used to find the latent heat of melting of sea ice, its thermal expansion coefficient, and the adiabatic lapse rate of sea ice.

4. THE FIRST LAW OF THERMODYNAMICS AND CONSERVATIVE TEMPERATURE Θ

The fundamental thermodynamic relation for a system composed of a solvent (water) and a solute (sea salt) relates the total differentials of thermodynamic quantities for the case where the transitions between equilibrium states are reversible. This restriction is satisfied for infinitesimally small changes of an infinitesimally small seawater parcel. The fundamental thermodynamic relation is

$$\mathrm{d}h - v\mathrm{d}P = (T_0 + t)\mathrm{d}\eta + \mu \mathrm{d}S_\mathrm{A}. \qquad (6.9)$$

The left-hand side of Equation (6.9) is often written as $\mathrm{d}u + (p + P_0)\mathrm{d}v$ where $(p + P_0) = P$ is the absolute pressure. Here h is the specific enthalpy (i.e. enthalpy per unit mass of seawater), u is the specific internal energy, $v = \rho^{-1}$ is the specific volume, $(T_0 + t) = T$ is the absolute temperature, η is the specific entropy and μ is the chemical potential of seawater. In terms of the Gibbs function of seawater $g(S_\mathrm{A}, t, p)$ and its partial derivatives (in terms of its natural variables S_A, t and p, written as subscripts of g) these thermodynamic quantities are given by

$$h = g - (T_0 + t)g_\mathrm{T}, \quad u = g - (T_0 + t)g_\mathrm{T} - (P_0 + p)g_\mathrm{P} \qquad (6.10)$$

$$v = g_\mathrm{P}, \; \eta = -g_\mathrm{T}, \text{ and } \mu = g_{S\mathrm{A}}. \qquad (6.11)$$

In fluid dynamics we usually deal with material derivatives, $\mathrm{d}/\mathrm{d}t$, that is, derivatives defined following the fluid motion, $\mathrm{d}/\mathrm{d}t = \partial/\partial t + \mathbf{u}\cdot\nabla$ where $\mathbf{u}$ is the fluid velocity. In terms of this type of derivative, and assuming local thermodynamic equilibrium (i.e., that local thermodynamic equilibrium is maintained during the temporal change), the fundamental thermodynamic relation is

$$\frac{\mathrm{d}h}{\mathrm{d}t} - \frac{1}{\rho}\frac{\mathrm{d}P}{\mathrm{d}t} = (T_0 + t)\frac{\mathrm{d}\eta}{\mathrm{d}t} + \mu\frac{\mathrm{d}S_\mathrm{A}}{\mathrm{d}t}. \qquad (6.12)$$

Note that the same symbol t is used in $(T_0 + t)$ for the *in situ* temperature on the Celsius temperature scale as is used for time in the expression for the material derivatives.

The First Law of Thermodynamics is carefully derived in Appendix B of the TEOS-10 Manual (IOC et al., 2010) and is

$$\rho\left(\frac{\mathrm{d}h}{\mathrm{d}t} - \frac{1}{\rho}\frac{\mathrm{d}P}{\mathrm{d}t}\right) = -\nabla\cdot\mathbf{F}^\mathrm{R} - \nabla\cdot\mathbf{F}^\mathrm{Q} + \rho\varepsilon + h_{S_\mathrm{A}}\rho S^{S_\mathrm{A}}, \qquad (6.13)$$

where $\mathbf{F}^\mathrm{R}$ is the sum of the boundary and radiative heat fluxes and $\mathbf{F}^\mathrm{Q}$ is the sum of all the molecular diffusive fluxes of heat, being the usual molecular heat flux directed down the temperature gradient plus a term proportional to the molecular flux of salt (the Dufour effect). Lastly, ε is the rate of dissipation of kinetic energy per unit mass, and $h_{S_\mathrm{A}}\rho S^{S_\mathrm{A}}$ is the rate of increase of enthalpy due to the interior source term of Absolute Salinity caused by remineralization. This last term has been shown to be negligible (see Appendix A.21 of IOC et al., 2010) and will be ignored here.

Following Fofonoff (1962) we note that an important consequence of Equation (6.13) is that when two finite-sized parcels of seawater are mixed at constant pressure and under ideal conditions, the total amount of enthalpy (rather than entropy) is conserved. To see this, one combines Equation (6.13) with the continuity equation $\partial\rho/\partial t + \nabla\cdot(\rho\mathbf{u}) = 0$ to find the following divergence form of the First Law of Thermodynamics,

$$\partial(\rho h)/\partial t + \nabla\cdot(\rho\mathbf{u}h) - \frac{\mathrm{d}P}{\mathrm{d}t} = -\nabla\cdot\mathbf{F}^\mathrm{R} - \nabla\cdot\mathbf{F}^\mathrm{Q} + \rho\varepsilon. \qquad (6.14)$$

One then integrates over the volume that encompasses both fluid parcels while assuming there to be no radiative, boundary, or molecular fluxes across the boundary of the control volume. This control volume may change with time as the fluid moves (at constant pressure), mixes, and contracts. The dissipation of kinetic energy by viscous friction is commonly ignored during such mixing processes but in fact the dissipation term does cause a small increase in the enthalpy of the mixture with respect to that of the two original parcels. Apart from this non-conservative source term, under these assumptions Equation (6.14) reduces to the statement that the volume-integrated amount of ρh is the same for the two initial fluid parcels as for the final mixed parcel; that is, the total amount of enthalpy is unchanged.

This result of non-equilibrium thermodynamics is of the utmost importance in oceanography. The fact that enthalpy is conserved when fluid parcels mix at constant pressure is the central result upon which all of our understanding of "heat fluxes" and of "heat content" in the ocean rests. The importance of this result cannot be overemphasized; it must form part of all our introductory courses on oceanography and climate dynamics. As important as this result is, it does not however follow that enthalpy is the best variable to represent "heat content" in the ocean. Enthalpy is actually a very poor representation of "heat content" because it does not possess the "potential" property; that is, enthalpy does not remain unchanged when a parcel's pressure is changed in an adiabatic and isohaline manner.

It will become apparent that potential enthalpy h^0 (referenced to zero sea pressure) is the best thermodynamic variable to represent "heat content" in the ocean. Potential enthalpy h^0 is related to the *in situ* enthalpy h through the pressure integral of specific volume according to (see Section 3.2 of the TEOS-10 Manual, IOC et al., 2010)

$$
\begin{aligned}
h^0(S_A,t,p) &= h(S_A,t,p) - \int_{P_0}^{P} v(S_A,\theta[S_A,t,p,p'],p')\mathrm{d}P' \\
&= h(S_A,t,p) - \int_{P_0}^{P} \hat{v}(S_A,\Theta,p')\mathrm{d}P'.
\end{aligned} \tag{6.15}
$$

In the first line of this equation, specific volume is considered to be a function of *in situ* temperature (which is equal to the potential temperature $\theta[S_A,t,p,p']$ with respect to reference pressure p' as the pressure is changed during the integration) while in the second line it is regarded as being a function of Conservative Temperature Θ which is constant during the adiabatic and isohaline pressure change. It proves convenient to define a new temperature variable called Conservative Temperature Θ as being proportional to potential enthalpy, so that

$$\Theta \equiv h^0/c_p^0, \tag{6.16}$$

where c_p^0 is the fixed constant 3991.867957119 63 J kg^{-1}K^{-1}. This constant is arbitrary and was chosen to make Conservative Temperature as similar as possible to potential temperature at the sea surface. Truncation to say six significant figures would have been entirely appropriate, but the above number is now enshrined in the TEOS-10 computer code and cannot easily be changed; it is part of the TEOS-10 definition of seawater as endorsed by the Intergovernmental Oceanographic Commission.

When two fluid parcels of mass m_1 and m_2 undergo irreversible and complete mixing at constant pressure, the thermodynamic quantities that are conserved during the mixing process are mass, Absolute Salinity, and enthalpy, while potential enthalpy and Conservative Temperature will not be exactly conserved. During the mixing process, the non-conservative production of Conservative Temperature, $\delta\Theta$, is given by (with $m = m_1 + m_2$ being the mass of mixed fluid)

$$m_1\Theta_1 + m_2\Theta_2 + m\delta\Theta = m\Theta. \tag{6.17}$$

Enthalpy in the functional form $h = \hat{h}(S_A,\Theta,p)$ is expanded in a Taylor series of S_A and Θ about the values S_A and Θ of the mixed fluid, retaining terms to second order in $[S_{A2} - S_{A1}] = \Delta S_A$ and in $[\Theta_2 - \Theta_1] = \Delta\Theta$. Then h_1 and h_2 are evaluated and the fact that enthalpy is conserved is used to find that

$$\delta\Theta = \frac{1}{2}\frac{m_1 m_2}{m^2}\left\{\frac{\hat{h}_{\Theta\Theta}}{\hat{h}_\Theta}(\Delta\Theta)^2 + 2\frac{\hat{h}_{\Theta S_A}}{\hat{h}_\Theta}\Delta\Theta\Delta S_A + \frac{\hat{h}_{S_A S_A}}{\hat{h}_\Theta}(\Delta S_A)^2\right\}. \tag{6.18}$$

The reason why the mixing is taken to occur at constant pressure is simply that in order to mix, the two fluid parcels must be at the same physical location. Turbulent diffusion can be considered the result of mixing of fluid parcels of equal mass so that $\frac{1}{2}m_1 m_2 m^{-2}$ is $\frac{1}{8}$. This result of non-equilibrium thermodynamics, Equation (6.18), shows how the non-conservative production of Conservative Temperature caused by mixing across an ocean front might be estimated. Moving on from this parcel mixing argument, we now proceed to develop an evolution equation for Conservative Temperature in a turbulent ocean, since it is this type of evolution equation that is incorporated into forward ocean models.

The First Law of Thermodynamics, Equation (6.13), can be written in terms of Conservative Temperature Θ, as

$$\rho\left(\hat{h}_\Theta\frac{\mathrm{d}\Theta}{\mathrm{d}t} + \hat{h}_{S_A}\frac{\mathrm{d}S_A}{\mathrm{d}t}\right) = -\nabla\cdot\mathbf{F}^R - \nabla\cdot\mathbf{F}^Q + \rho\varepsilon, \tag{6.19}$$

where the partial derivatives of specific enthalpy in the functional form $h = \hat{h}(S_A,\Theta,p)$ are (from McDougall, 2003)

$$\hat{h}_\Theta \equiv \left.\frac{\partial h}{\partial\Theta}\right|_{S_A,p} = \frac{(T_0+t)}{(T_0+\theta)}c_p^0, \tag{6.20}$$

$$\hat{h}_{S_A} \equiv \left.\frac{\partial h}{\partial S_A}\right|_{\Theta,p} = \mu(S_A,t,p) - \frac{(T_0+t)}{(T_0+\theta)}\mu(S_A,\theta,0), \tag{6.21}$$

and

$$\hat{h}_P \equiv \left.\frac{\partial h}{\partial P}\right|_{S_A,\Theta} = v = \frac{1}{\rho}. \tag{6.22}$$

When the First Law of Thermodynamics Equation (6.19) is carefully averaged in a turbulent ocean, the same combination of non-conservative production terms in Equation (6.18) also appears in the turbulent evolution equation for Conservative Temperature, in both the epineutral (along neutral tangent planes) and vertical diffusion terms as follows (Graham and McDougall, 2013)

$$
\begin{aligned}
\frac{\mathrm{d}\hat{\Theta}}{\mathrm{d}t} &= \left.\frac{\partial\hat{\Theta}}{\partial t}\right|_n + \hat{\mathbf{v}}\cdot\nabla_n\hat{\Theta} + \tilde{e}\frac{\partial\hat{\Theta}}{\partial z} = \gamma_z\nabla_n\cdot\left(\gamma_z^{-1}K\nabla_n\hat{\Theta}\right) + \left(D\hat{\Theta}_z\right)_z + \varepsilon/\hat{h}_\Theta \\
&+ K\left(\frac{\hat{h}_{\Theta\Theta}}{\hat{h}_\Theta}\nabla_n\hat{\Theta}\cdot\nabla_n\hat{\Theta} + 2\frac{\hat{h}_{\Theta S_A}}{\hat{h}_\Theta}\nabla_n\hat{\Theta}\cdot\nabla_n\hat{S}_A + \frac{\hat{h}_{S_A S_A}}{\hat{h}_\Theta}\nabla_n\hat{S}_A\cdot\nabla_n\hat{S}_A\right) \\
&+ D\left(\frac{\hat{h}_{\Theta\Theta}}{\hat{h}_\Theta}\hat{\Theta}_z^2 + 2\frac{\hat{h}_{\Theta S_A}}{\hat{h}_\Theta}\hat{\Theta}_z\hat{S}_{A_z} + \frac{\hat{h}_{S_A S_A}}{\hat{h}_\Theta}\left(\hat{S}_{A_z}\right)^2\right).
\end{aligned} \tag{6.23}
$$

Here the left-hand side is the material derivative of the thickness-weighted Conservative Temperature $\hat{\Theta}$ of density coordinate averaging. The advection of the material derivative is performed by the thickness-weighted horizontal velocity $\hat{\mathbf{v}}$ and the temporally averaged dianeutral velocity $\tilde{e}$ of density coordinates, while K and D are the epineutral and vertical diffusion coefficients respectively. The two-dimensional gradient of properties along the neutral tangent plane is designated ∇_n in the above equation and γ_z^{-1} is the average of the reciprocal of the vertical gradient of Neutral Density or of the locally-referenced potential density, while γ_z is the reciprocal of γ_z^{-1}. Graham and McDougall (2013) have shown that

- the non-conservative production terms in this equation (the last two lines), when vertically integrated from the ocean floor to the surface and expressed in terms of a surface heat flux, are typically no larger than 1 mW m^{-2}, which is a factor of 100 less than the corresponding non-conservative production of potential temperature, θ,
- the dominant non-conservative terms in Equation (6.23) are those due to $\hat{h}_{\Theta\Theta}$,
- the non-conservative terms in Equation (6.23) are caused only by the "cabbeling" nonlinearities of the equation of state, namely by $\hat{v}_{\Theta\Theta}$, $\hat{v}_{\Theta S_A}$ and $\hat{v}_{S_A S_A}$ (since $\hat{h}_{\Theta\Theta}$, $\hat{h}_{\Theta S_A}$ and $\hat{h}_{S_A S_A}$ are the pressure integrals of $\hat{v}_{\Theta\Theta}$, $\hat{v}_{\Theta S_A}$ and $\hat{v}_{S_A S_A}$, see Equation 6.22), and
- the non-conservative production of Conservative Temperature caused by turbulent mixing processes is always positive for seawater. This non-conservation is more than a factor of 10 less than that due to the dissipation of kinetic energy and a factor of 1000 less than that due to the non-conservative production of entropy.

Because Conservative Temperature also possesses the "potential" property, it is a very accurate representation of the "heat content" of seawater. The difference $\theta - \Theta$ between potential temperature θ and Conservative Temperature Θ at the sea surface is shown in Figure 6.4 (after McDougall, 2003). If an ocean model is written with potential temperature as the prognostic temperature variable rather than Conservative Temperature, and is run with the same constant value of the isobaric specific heat capacity c_p^0, the neglect of the non-conservative source terms that should appear in the prognostic equation for θ means that such an ocean model incurs errors in the model output. These errors will depend on the nature of the surface boundary condition; for flux boundary conditions, the errors are as shown in Figure 6.4.

The difference between potential temperature and Conservative Temperature (see Figure 6.5) can be as large as $\theta - \Theta = -1.4\,^\circ$C but more typically the difference lies in a range that is 0.2 °C wide (see Figure 6.4). To put a temperature difference of 0.2 °C in context, this is the typical difference between *in situ* and potential temperatures for a pressure difference of 2000 dbar, and it is approximately 80 times as large as the typical differences between t_{90} and t_{68} in the ocean (these being the temperatures measured on the International Temperature Scales of 1968 and 1990).

If Figure 6.5 were to be used to quantify the errors in oceanographic practice incurred by assuming that θ is a conservative variable, one might select property contrasts that were typical of a prominent oceanic front and decide that because the non-conservative production of potential temperature $\delta\theta$ at this one front is small, that the issue can be ignored. But the observed properties in the ocean result from a large and indeterminate number of such prior mixing events and the non-conservative production of θ accumulates during each of these mixing events. How can we possibly estimate the error that is made by treating potential temperature as a conservative variable during all of these unknowably many past individual mixing events?

This seemingly difficult issue is partially resolved by considering what is actually done in ocean models today. These models carry a temperature conservation equation

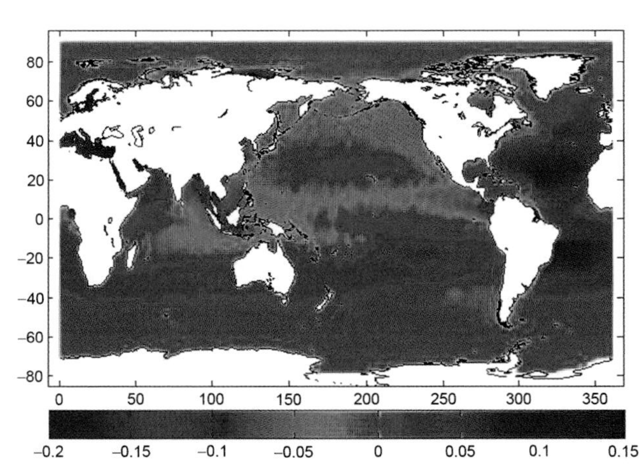

FIGURE 6.4 Contours (in °C) of the difference $\theta - \Theta$ between potential temperature θ and Conservative Temperature Θ at the sea surface of the annually averaged atlas of Gouretski and Koltermann (2004).

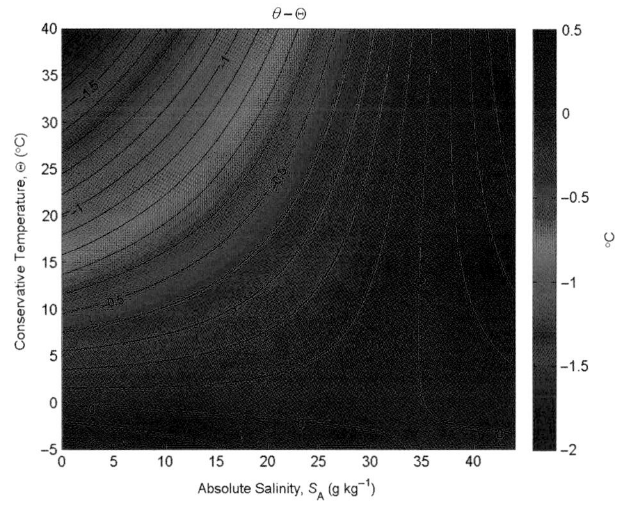

FIGURE 6.5 Contours (in °C) of the difference between potential temperature and Conservative Temperature, $\theta - \Theta$. This plot illustrates the non-conservative production of potential temperature θ in the ocean.

that does not have non-conservative source terms, so that the model's temperature variable can be interpreted as being Θ. This being the case, the temperature difference contoured in Figure 6.5 illustrates the error that is made by interpreting the model temperature as being θ. That is, the values contoured in Figure 6.5 are representative of the error, expressed in °C, caused by assuming that θ is a conservative variable. The contoured values of temperature difference encapsulate the accumulated non-conservative production that has occurred during all the uncountably many mixing processes that have led to the ocean's present state.

This section has largely demonstrated the benefits of potential enthalpy and Conservative Temperature from the viewpoint of conservation equations, but the benefits can also be proved by the following parcel-based argument. First, the air–sea heat flux needs to be recognized as a flux of potential enthalpy, which is exactly c_p^0 times the flux of Conservative Temperature. Second, we note (from above) that while it is the *in situ* enthalpy that is conserved when parcels mix, a negligible error is made when potential enthalpy is assumed to be conserved during mixing at any depth. Third, note that the ocean circulation can be regarded as a series of adiabatic and isohaline movements during which Θ is absolutely unchanged (because of its "potential" nature) followed by a series of turbulent mixing events during which Θ is almost totally conserved. Hence it is clear that Θ is the quantity that is advected and diffused in an almost conservative fashion and whose surface flux is exactly proportional to the air–sea heat flux. Hence we may identify $c_p^0\Theta$ as the "heat content" per unit mass of seawater.

It can also be shown (see Figure 6.6, from the TEOS-10 Manual, IOC et al., 2010) that the parameterized lateral diffusion of "heat" along neutral tangent planes can be more than 1% different when such lateral diffusive heat fluxes are estimated using gradients of potential temperature rather than gradients of Conservative Temperature.

Note that it is the existence of the seawater Gibbs function that has enabled the calculation of enthalpy and potential enthalpy, since it is not possible to find potential enthalpy from simply integrating the specific heat capacity c_p at $p=0$ dbar (such as from the EOS-80 algorithm for c_p) because this integration throws up an unknown non-linear function of salinity. Similarly, McDougall (2003) has shown that the product of the specific heat capacity at $p=0$ dbar and potential temperature, $c_p(S_A,\theta,p=0)\theta$, is no better at approximating potential enthalpy than is simply using potential temperature with a constant specific heat capacity; this has been the standard oceanographic practice for the calculation of meridional heat fluxes and for calculating the tendency of potential temperature at the sea surface caused by air–sea heat fluxes.

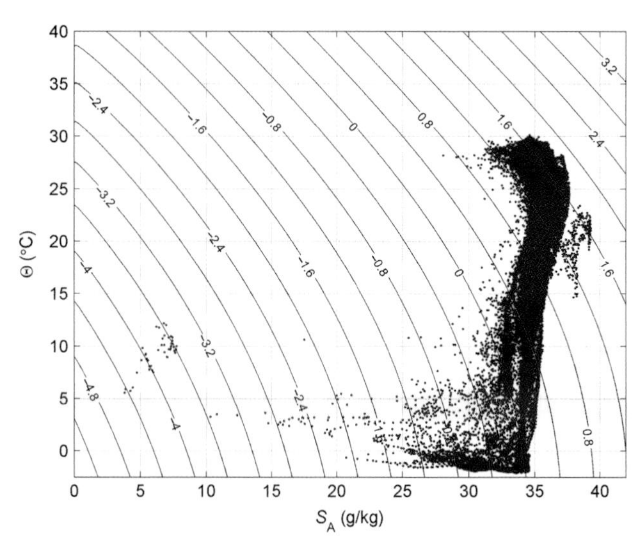

FIGURE 6.6 Contours of $(|\nabla_n\theta|/|\nabla_n\Theta|-1)\times 100\%$ at $p=0$, showing the percentage difference between the epineutral gradients of θ and Θ. The red dots are from the global ocean atlas of Gouretski and Koltermann (2004) at $p=0$.

5. THE 48-TERM EXPRESSION FOR SPECIFIC VOLUME

While the TEOS-10 Gibbs function is a function of *in situ* temperature, there is a need in oceanography to express specific volume and other variables as functions of Conservative Temperature, since this temperature variable has the "potential" property and is an accurate approximation of the "heat content" per unit mass of a seawater parcel. Hence we have formed a computationally efficient 48-term rational function expression for density, $\hat{\rho}(S_A,\Theta,p)$, described in Appendix A.30 and Appendix K of the TEOS-10 Manual (IOC et al., 2010). Seawater density data was fitted in a "funnel" of data points in (S_A,Θ,p) space that extends to a pressure of 8000 dbar. At the sea surface the "funnel" covers the full range of temperature and salinity while for pressures greater than 6500 dbar, the maximum Conservative Temperature of the fitted data is 10 °C and the minimum Absolute Salinity is $30\,\mathrm{g\,kg^{-1}}$. That is, the 48-term fit has been performed over a region of parameter space that includes water that is approximately 10 °C warmer and $5\,\mathrm{g\,kg^{-1}}$ fresher in the deep ocean than the seawater that exists in the present ocean.

The rms error of this 48-term approximation to the TEOS-10 density over the oceanographic "funnel" is $0.00046\,\mathrm{kg\,m^{-3}}$; this can be compared with the rms uncertainty of $0.004\,\mathrm{kg\,m^{-3}}$ of the underlying laboratory density data to which the TEOS-10 Gibbs function was fitted. Similarly, the appropriate thermal expansion coefficient,

$$\alpha^{\Theta} = -\frac{1}{\rho}\frac{\partial\rho}{\partial\Theta}\bigg|_{S_A,p}, \tag{6.24}$$

of the 48-term equation of state is different from the same thermal expansion coefficient evaluated directly from the sum of the IAPWS-09 and IAPWS-08 Gibbs functions with an rms error in the "funnel" of $0.069 \times 10^{-6}\,\text{K}^{-1}$; this is to be compared with the rms error of the thermal expansion coefficient of the laboratory data to which the Feistel (2008) Gibbs function was fitted of $0.73 \times 10^{-6}\,\text{K}^{-1}$. In terms of the evaluation of density gradients, the haline contraction coefficient evaluated from the 48-term equation is many times more accurate than the thermal expansion coefficient.

In dynamical oceanography, it is the thermal expansion and haline contraction coefficients α^{Θ} and β^{Θ} which are the most important aspects of the equation of state since the "thermal wind" is proportional to $\alpha^{\Theta}\nabla_p\Theta - \beta^{\Theta}\nabla_p S_A$ and the vertical static stability is given in terms of the buoyancy frequency N by $g^{-1}N^2 = \alpha^{\Theta}\Theta_z - \beta^{\Theta}(S_A)_z$. Hence for dynamical oceanography the 48-term rational function expression for density retains essentially the full accuracy of TEOS-10. The use of the 48-term expression for density has several advantages over using the exact formulation, namely,

- it is a function of Conservative Temperature, so eliminating the need to be continually converting between Conservative and *in situ* temperatures in order to evaluate density,
- it is computationally faster (by a factor of 6.5) to use the 48-term expression rather than directly using the Gibbs functions of pure water IAPWS-09 and of sea salt IAPWS-08 (which in turn is a further multiplicative factor of 65 faster than using IAPWS-95 for the pure water part),
- ocean models will be able to use this 48-term equation of state, and it is advantageous for the fields of observational and theoretical oceanography to use the same equation of state as ocean models.

6. CHANGES TO OCEANOGRAPHIC PRACTICE UNDER TEOS-10

For the past thirty years, we have taken the "raw" data of Practical Salinity S_P (PSS-78), *in situ* temperature t (now ITS-90) and pressure p and we have used an algorithm to calculate potential temperature θ in order to analyze and publish water-mass characteristics on the $S_P - \theta$ diagram. On this $S_P - \theta$ diagram, we have drawn curved contours of potential density at constant pressure using EOS-80, under the assumption that the relative chemical composition of seawater is constant, and used straight lines as approximations to the products of mixing between two water parcels in the absence of external sources of heat, under the assumption that salinity and temperature-dependent variations in heat capacity are negligible. However, the errors incurred in making these assumptions are much greater than the precision of our measurements. In addition, neglect of changes in the relative composition can make significant differences in large-scale density gradients at constant pressure, and hence in geostrophic transports.

Under TEOS-10, this practice has now changed: density and potential density (and all types of geostrophic streamfunctions including dynamic height anomaly) are now not functions of Practical Salinity S_P but rather are functions of Absolute Salinity S_A. In addition, although potential temperature θ is still available, a new variable called Conservative Temperature Θ is recommended. The use of S_A accounts for changes in the relative composition of seawater, and the use of Θ provides a better measure of the "heat content," corrected for the effects of pressure.

Under TEOS-10 it is not possible to draw isolines of potential density on a $S_P - \theta$ diagram. Rather, because of the spatial variations of seawater composition, a given value of potential density defines an area on the $S_P - \theta$ diagram, not a curved line. Hence, for the analysis and publication of ocean data under TEOS-10, we need to change from using the $S_P - \theta$ diagram, which was appropriate under EOS-80, to using the $S_A - \Theta$ diagram. It is on this $S_A - \Theta$ diagram that the isolines of potential density can be drawn under TEOS-10.

The various oceanographic properties that rely on the equation of state have been written in terms of S_A and Θ in a suite of software (the GSW Oceanographic Toolbox), and all of the oceanographic variables in common use (including several different geostrophic streamfunctions) have been written using a reduced 48-term expression for density to ensure consistency between ocean models, observational studies and theoretical work. Under TEOS-10, the observed variables (S_A,t,p), together with longitude and latitude, are used to first form Absolute Salinity S_A using a function *gsw_SA_from_SP*, and then Conservative Temperature Θ is calculated using a second function *gsw_CT_from_t*. Oceanographic water masses are then analyzed on the $S_A - \Theta$ diagram, on which true potential density contours can be drawn. In low-accuracy situations where corrections for differing relative composition are not desired, an intermediate variable called the Reference Salinity S_R may be used as an estimate for the Absolute Salinity.

Conservative Temperature Θ has the advantage over potential temperature θ of more accurately representing the "heat content" of seawater, and is also much closer (by a factor of a hundred) to being a conservative variable than is potential temperature. Heat is exchanged between the ocean and its atmosphere and ice boundaries as a flux of potential enthalpy, which is exactly $c_p^0 \equiv 3991.867\,957\,119\,63\ \text{J kg}^{-1}\,\text{K}^{-1}$ times the density times the corresponding flux of Θ. The transport of potential enthalpy $c_p^0\Theta$ in the ocean, and in particular across ocean sections, can be regarded as the transport of "heat" irrespective of whether there are non-zero fluxes of mass and/or of salt across such ocean sections (IOC et al., 2010).

Absolute Salinity S_A has the advantage over S_P of more accurately representing the mass of solute in seawater, and is much closer (by at least an order of magnitude) to being a conservative variable under conditions of isobaric mixing than is Practical Salinity or Reference Salinity. Fluxes and exchanges of mass and freshwater can now be considered in units of mass flux.

The recommended nomenclature, symbols and units of thermodynamic quantities to be used by oceanographers are listed in Appendix L of the TEOS-10 Manual (IOC et al., 2010). In particular, in order to avoid confusion while the use of Practical Salinity in scientific publications is phased out, salinity should be specifically identified as being either Practical Salinity with the symbol S_P or Absolute Salinity with the symbol S_A. When describing the use of TEOS-10, it is the TEOS-10 Manual that should be referenced.

The TEOS-10 salinity variable to appear in publications is Absolute Salinity S_A. TEOS-10 allows for different methods and measurement techniques that can be used to estimate S_A, and the technology underlying these methods may develop further in the future. Therefore the method and/or version number of the software that is used to convert Reference Salinity S_R into Absolute Salinity S_A should always be stated in publications.

7. OCEAN MODELING USING TEOS-10

Ocean models treat their salinity and temperature variables as being conservative, with the choice of variables to date being Practical Salinity and potential temperature. Converting ocean models to be TEOS-10 compatible requires several changes. The model's temperature variable needs to

i. accurately represent the "heat content" per unit mass of seawater and

ii. be as conservative as possible under ocean mixing processes.

Conservative Temperature Θ has these properties, whereas potential temperature θ does not. Fortunately it is relatively easy to change ocean models to have Conservative Temperature as their temperature variable. With the expression for density being cast in terms of Absolute Salinity S_A and Conservative Temperature Θ as $\hat{\rho}(S_A, \Theta, p)$, the interior of an ocean model can be written totally in terms of this one temperature variable, Θ. In the air–sea interaction module of an ocean model, the sea-surface-temperature (SST) needs to be evaluated for use in bulk air–sea flux formulae. This conversion from Θ to SST needs to be done just at the sea surface in the air–sea interaction module.

The current practice in numerical models is to treat salinity as a perfectly conserved quantity in the interior of the ocean. In order to continue this practice, the appropriate model salinity variable is Preformed Salinity S_*. Preformed Salinity S_* and Absolute Salinity S_A are related to Reference Salinity S_R by Equations (A.20.1) and (A.20.2) of the TEOS-10 Manual, repeated here

$$S_* = S_R\left(1 - r_1 R^\delta\right), \tag{6.25}$$

$$S_A = S_*\left(1 + F^\delta\right), \tag{6.26}$$

where

$$R^\delta \equiv \frac{\delta S_A^{\text{atlas}}}{S_R^{\text{atlas}}} \tag{6.27a}$$

and

$$F^\delta = \frac{[1 + r_1]R^\delta}{\left(1 - r_1 R^\delta\right)}. \tag{6.27b}$$

Recall that the Absolute Salinity Anomaly Ratio, $R^\delta \equiv \delta S_A^{\text{atlas}}/S_R^{\text{atlas}}$, is the ratio of the values of Absolute Salinity Anomaly and Reference-Composition Salinity from a stored atlas. The constant r_1 is taken to be 0.35.

Because Preformed Salinity S_* is designed to be a conservative salinity variable, blind to the effects of biogeochemical processes, its evolution equation is in the conservative form (see Appendix A.21 of IOC et al., 2010),

$$\frac{\mathrm{d}\hat{S}_*}{\mathrm{d}t} = \gamma_z \nabla_n \cdot \left(\gamma_z^{-1} K \nabla_n \hat{S}_*\right) + \left(D \frac{\partial \hat{S}_*}{\partial z}\right)_z. \tag{6.28}$$

Here the over-tilde of $\hat{S}_*$ indicates that this variable is the thickness-weighted average Preformed Salinity, having been averaged between a pair of closely-spaced neutral tangent planes. The material derivative on the left-hand side of Equation (6.28) is with respect to the sum of the Eulerian and quasi-Stokes velocities of height coordinates (equivalent to the description in Appendix A.21 of IOC et al. (2010) in terms of the thickness-weighted mean horizontal velocity and the mean dianeutral velocity), while the right-hand side of this equation is the standard notation indicating that $\hat{S}_*$ is being diffused along neutral tangent planes with the diffusivity K and in the vertical direction with the diapycnal diffusivity D. The model is initialized with values of Preformed Salinity using Equation (6.25) based on observations of Practical Salinity and on the interpolated global observed data base of R^δ (this can be achieved by a call to *gsw_Sstar_from_SP* in the GSW library of computer algorithms).

In order to evaluate density during the running of an ocean model, Absolute Salinity must be evaluated based on the model's primary salinity variable, Preformed Salinity, and Equation (6.26). This can be done by carrying the following evolution equation for F^δ

$$\frac{\mathrm{d}F^\delta}{\mathrm{d}t} = \gamma_z \nabla_n \cdot \left(\gamma_z^{-1} K \nabla_n F^\delta\right) + \left(D \frac{\partial F^\delta}{\partial z}\right)_z + \tau^{-1}\left(F^{\delta\,\text{obs}} - F^\delta\right). \tag{6.29}$$

The model variable F^δ (note that $F^\delta = S_A/S_* - 1$) is initialized based on observations of $R^\delta \equiv \delta S_A^{atlas}/S_R^{atlas}$ and the use of Equation (6.27b) (this is best done by calling *gsw_Fdelta* in the GSW library). From Equation (6.29) we see that F^δ is advected and diffused like any other tracer, but in addition, there is a non-conservative source term $\tau^{-1}(F^{\delta obs} - F^\delta)$ which serves to restore the model variable F^δ towards the observed value (found previously using *gsw_Fdelta*) with a restoring time τ that can be chosen to suit particular modeling needs. At the time of writing, the choice of this restoring time remains a research task (see the discussion in Appendix A.20 of the TEOS-10 Manual, IOC et al., 2010).

In summary, the approach for handling salinity in ocean models suggested in IOC et al. (2010) carries the evolution Equations (6.28) and (6.29) for $\hat{S}_*$ and F^δ, while $\hat{S}_A$ is calculated from these two model variables at each time step according to

$$\hat{S}_A = \hat{S}_*\left(1 + F^\delta\right). \tag{6.30}$$

It is this salinity, $\hat{S}_A$, which is used as the argument for the model's expression for density at each time step of the model.

The Baltic Sea is somewhat of an exception because its compositional variations are not due to biogeochemistry but to anomalous riverine input of dissolved salts that behave conservatively. Preformed Salinity S_* in the Baltic is equal to Absolute Salinity S_A, which implies that $r_1 = -1$ and $F^\delta = 0$ in the Baltic Sea. Hence in the Baltic, an ocean model simply puts $S_A = S_*$ and the value of Absolute Salinity Anomaly δS_A is immaterial during the running of the model. Of course the values of δS_A in the Baltic are important for relating Absolute Salinity and Preformed Salinity to measured values of Practical Salinity there. The discharges (mass fluxes) of river water and of Absolute Salinity should both appear as source terms at the edges of the Baltic Sea in the model.

The use of an existing climatology for F^δ and the introduction of a rather arbitrary relaxation time τ are less than desirable features of this way of treating salinity in ocean models. An alternative strategy is available in an ocean model that includes biogeochemical processes and carries evolution equations for TA and DIC, as well as nitrate and silicate concentrations. Having these quantities available during the running of an ocean model allows the use of the following equation (from Pawlowicz et al., 2011) to evaluate Absolute Salinity

$$\left(S_A - S_*\right)/\left(\mathrm{g\,kg^{-1}}\right) = \left(73.7\Delta\mathrm{TA} + 11.8\Delta\mathrm{DIC} + 81.9\mathrm{NO_3^-} + 50.6\mathrm{Si(OH)_4}\right)/\left(\mathrm{mol\,kg^{-1}}\right).$$

Under this approach, Preformed Salinity would be carried as the model's conservative prognostic salinity variable as in Equation (6.28), and the above equation for $S_A - S_*$ in terms of the biogeochemical variables would be used to evaluate Absolute Salinity for use in the model's expression for specific volume.

If an ocean model is to be run for only a short time (less than a century), then it may be sufficiently accurate to carry only one salinity variable, namely Absolute Salinity S_A without including a non-conservative source term for Absolute Salinity. For longer integrations, the neglect of the non-conservative biogeochemical source term means that the model's salinity variable S_A will depart from reality. A more detailed discussion of these points is available in Appendix A.20 of the TEOS-10 Manual (IOC et al., 2010).

In summary, the changes needed to make ocean models TEOS-10 compatible are

- **i.** use an equation of state in terms of S_A and Θ, $\hat{\rho}(S_A, \Theta, p)$, such as the TEOS-10 48-term expression of IOC et al. (2010),
- **ii.** have Conservative Temperature Θ as the model's temperature variable (note that SST needs to be evaluated in the model's air–sea flux module at the sea surface only),
- **iii.** incorporate the effects of the spatially variable seawater composition, using the techniques of Appendix A.20 of IOC et al. (2010) as summarized above,
- **iv.** restoring boundary conditions for ocean-only models can be imposed on the model variables S_* and Θ,
- **v.** model output salinities and temperatures are best made as Absolute Salinity S_A and Conservative Temperature Θ, consistent with the variables that will be published in oceanographic journals.

8. SUMMARY

In accordance with resolution XXV-7 of the Intergovernmental Oceanographic Commission (IOC) at its 25th Assembly in June 2009, and the several Releases and Guidelines of the International Association for the Properties of Water and Steam, the TEOS-10 thermodynamic description of seawater, of ice and of moist air has been adopted for use by oceanographers in place of the International Equation Of State—1980 (EOS-80). The computer software to implement this change is available at the web site www.TEOS-10.org.

Perhaps the most apparent changes compared with the International Equation of State of Seawater (EOS-80) are (i) the adoption of Absolute Salinity S_A instead of Practical Salinity S_P (PSS-78) as the salinity argument for the thermodynamic properties of seawater, and (ii) the use of Conservative Temperature Θ in place of potential temperature θ. Importantly, Practical Salinity is retained as the salinity variable that is stored in national data bases because Practical Salinity is virtually the measured variable (whereas Absolute Salinity is a calculated variable) and so that national data bases do not become corrupted with incorrectly labeled and stored salinity data.

Absolute Salinity is calculated from the computer algorithm of McDougall et al. (2012) or by other means, as the sum of Reference Salinity and the Absolute Salinity Anomaly. There are subtle issues in defining what is exactly meant by "absolute salinity" and at least four different definitions are possible when compositional anomalies are present. The definition adopted in TEOS-10 is the one that yields the most accurate estimates of seawater density since the ocean circulation is sensitive to rather small horizontal gradients of density. The algorithm that estimates Absolute Salinity Anomaly represents the state of the art as at 2010, but this area of oceanography received only marginal attention before TEOS-10 and is still relatively immature. It is likely that the accuracy of this algorithm will improve as more seawater samples from around the world ocean have their density accurately measured. After such future work is published and the results distilled into a revised algorithm, such an algorithm will be served from www.TEOS-10.org.

Conservative Temperature Θ is in some respects quite similar to potential temperature θ in that the same thought experiment is involved with their definitions. In both cases, one takes a seawater sample at an arbitrary pressure in the ocean and one imagines decreasing the pressure on the seawater parcel in an adiabatic and isohaline manner until the sea pressure $p = 0\,\text{dbar}$ is reached. The temperature of the fluid parcel at the end of this thought experiment is defined to be the potential temperature θ. Similarly, the enthalpy at the end of this thought experiment is defined to be the potential enthalpy h^0, and Conservative Temperature Θ is simply potential enthalpy divided by the fixed "specific heat capacity" $c_p^0 \equiv 3991.867\,957\,119\,63\ \text{J kg}^{-1}\ \text{K}^{-1}$.

Potential temperature θ has been used in oceanography as though it is a conservative variable, and yet the specific heat of seawater varies by 5% at the sea surface, and potential temperature is not conserved when seawater parcels mix. The First Law of Thermodynamics can be very accurately regarded as the statement that potential enthalpy h^0 and Conservative Temperature Θ are conservative variables in the ocean. This, together with the knowledge that the air–sea heat flux is exactly the air–sea flux of potential enthalpy (i.e., the air-sea flux of $c_p^0\Theta$) means that potential enthalpy can be treated as the "heat content" per unit mass of seawater, and fluxes of potential enthalpy in the ocean can be treated as "heat fluxes." Just as it is perfectly valid to talk of the flux of salinity anomaly ($S_A -$ constant) across an ocean section even when the mass flux across the section is non-zero, so it is perfectly valid to treat the flux of $c_p^0\Theta$ across an ocean section as the "heat flux" even when the fluxes of mass and of salt across the section are non-zero; this is discussed at length in McDougall (2003).

Conservative Temperature Θ can be readily evaluated from *in situ* temperature t, and the difference between potential temperature θ and Conservative Temperature can be as large as $\theta - \Theta = -1.4\,°\text{C}$ but is more typically smaller, with 95% of the surface $\theta - \Theta$ data lying within a range of $0.2\,°\text{C}$. In terms of density, a θ range of $0.2\,°\text{C}$ corresponds to a density range of $0.04\,\text{kg}\,\text{m}^{-3}$. This density difference is almost twice as large as that caused by ignoring the maximum value of $S_A - S_R$ of $0.027\ \text{g kg}^{-1}$ in the deep North Pacific.

We might summarize the magnitudes of the main differences between TEOS-10 and EOS-80 as follows. In the near-surface ocean the difference between potential temperature and Conservative Temperature is the largest change to be encountered in the switch from EOS-80 to TEOS-10, with typical differences of $0.2\,°\text{C}$ and the largest difference being $1.4\,°\text{C}$. By contrast, away from the sea surface at pressures greater than 1000 dbar, it is the new definition of Absolute Salinity that causes the largest differences between TEOS-10 and EOS-80, particularly in the Pacific Ocean. The new salinity variable leads to improvements in the isobaric density gradient (i.e., the "thermal wind") of between 2% and 10% in the global ocean at pressures greater than 1000 dbar.

The temperature variable in ocean models has been taken to be potential temperature θ, but to date the non-conservative source terms that are present in the evolution equation of potential temperature have not been included. To be TEOS-10 compatible, ocean models need to treat their temperature variable as Conservative Temperature Θ. Ocean models should be initialized with Θ rather than θ, the output temperature must be compared to observed Θ data rather than to θ data, and during the model run, any air–sea fluxes that depend on the sea-surface temperature (SST) must be calculated at each model time step using an algorithm for potential temperature in terms of Conservative Temperature, $\theta = \hat{\theta}(S_A, \Theta)$. In the GSW computer software library, this algorithm has the name *gsw_pt_from_CT*. This calculation of SST need only be done at the sea surface in the module that calculates the air–sea heat flux using bulk formulae.

The recommended treatment of salinity in ocean models is to carry evolution equations for both Preformed Salinity S_* and another variable, F^δ, which is related to the Absolute Salinity Anomaly, so that Absolute Salinity can be calculated at each time step of the model and used to accurately evaluate density (as discussed in Section 7 above and in Appendix A.20 of the TEOS-10 Manual, IOC et al., 2010).

Under TEOS-10, the observed variables (S_P,t,p), together with longitude and latitude, are first used to form Absolute Salinity S_A, and then Conservative Temperature Θ is evaluated. Oceanographic water masses are then analyzed on the $S_A - \Theta$ diagram, and potential density ρ^Θ contours can also be drawn on this $S_A - \Theta$ diagram. The computationally efficient 48-term expression for the density of seawater (of Appendix K in the TEOS-10 Manual) is a convenient and accurate equation of state for observational and theoretical studies and for ocean modeling. Also, it is this 48-term expression for density that will be the basis for updated TEOS-10 compliant algorithms for Neutral Density γ^n (Jackett and McDougall, 1997) and for ω-surfaces (Klocker et al., 2009). In this regard, it should be noted that because of the nonlinear nature of the equation of the state of seawater, planetary potential vorticity is not proportional to fN^2, not even in a lake (see Section 3.20 of the TEOS-10 Manual, IOC et al., 2010).

In summary, under EOS-80, we used the observed variables (S_P,t,p) to first form potential temperature θ and then we analyzed water masses on the $S_P-\theta$ diagram, and we were able to draw curved contours of potential density on this same $S_P-\theta$ diagram. Under TEOS-10, the observed variables (S_P,t,p), together with longitude and latitude, are used to first form Absolute Salinity S_A using *gsw_SA_from_SP*, and then Conservative Temperature Θ is calculated using *gsw_CT_from_t*. Oceanographic water masses are then analyzed on the $S_A-\Theta$ diagram, and potential density contours can be drawn on this $S_A-\Theta$ diagram.

In addition to the present chapter, two other introductory articles about TEOS-10, namely "Getting started with TEOS-10 and the Gibbs Seawater (GSW) Oceanographic Toolbox" (McDougall and Barker, 2011), and "What every oceanographer needs to know about TEOS-10 (The TEOS-10 Primer)" (Pawlowicz, 2010), are also available from www.TEOS-10.org. The key concepts of TEOS-10 are also described in an introductory set of lecture slides that are publicly available (McDougall, 2012).

ACKNOWLEDGMENTS

This chapter summarizes much of the work of SCOR/IAPSO Working Group 127 and the remaining members of WG127 are thanked for their valuable contributions over the 6-year life of the working group from 2005 to 2011. In particular, Dr Dan Wright of Canada was a key member of this group who contributed greatly over the full range of our activity, and his untimely death in 2010 was a terrible loss to the oceanographic community. This chapter is dedicated to Dan's memory. Dr Paul Barker is thanked for the excellent job he has done with the TEOS-10 web site and the GSW computer software library. This chapter has benefited from the valuable comments of the two reviewers and from the editor Stephen M. Griffies.

REFERENCES

Alberty, R.A., 2001. Use of Legendre transforms in chemical thermodynamics. Pure Appl. Chem. 73, 1349–1380.

Brewer, P.G., Bradshaw, A., 1975. The effect of non-ideal composition of seawater on salinity and density. J. Mar. Res. 33, 157–175.

Bryden, H.L., 1973. New polynomials for thermal expansion, adiabatic temperature gradient and potential temperature of seawater. Deep Sea Res. 20, 401–408.

Chen, C.-T., Millero, F.J., 1977. Sound speed of seawater at high pressures. J. Acoust. Soc. Am. 62, 1129–1135.

Feistel, R., 2003. A new extended Gibbs thermodynamic potential of seawater. Prog. Oceanogr. 58, 43–114.

Feistel, R., 2008. A Gibbs function for seawater thermodynamics for −6 to 80 °C and salinity up to 120 g kg^{-1}. Deep Sea Res. Part I 55, 1639–1671.

Feistel, R., Wagner, W., 2006. A new equation of state for H_2O Ice Ih. J. Phys. Chem. Ref. Data 35 (2), 1021–1047.

Feistel, R., Wright, D.G., Miyagawa, K., Harvey, A.H., Hruby, J., Jackett, D.R., McDougall, T.J., Wagner, W., 2008. Mutually consistent thermodynamic potentials for fluid water, ice and seawater: a new standard for oceanography. Ocean Sci. 4, 275–291. http://www.ocean-sci.net/4/275/2008/os-4-275-2008.pdf.

Feistel, R., Weinreben, S., Wolf, H., Seitz, S., Spitzer, P., Adel, B., Nausch, G., Schneider, B., Wright, D.G., 2010a. Density and absolute salinity of the Baltic Sea 2006–2009. Ocean Sci. 6, 3–24. http://www.ocean-sci.net/6/3/2010/os-6-3-2010.pdf.

Feistel, R., Wright, D.G., Kretzschmar, H.-J., Hagen, E., Herrmann, S., Span, R., 2010b. Thermodynamic properties of sea air. Ocean Sci. 6, 91–141. http://www.ocean-sci.net/6/91/2010/os-6-91-2010.pdf.

Feistel, R., Wright, D.G., Jackett, D.R., Miyagawa, K., Reissmann, J.H., Wagner, W., Overhoff, U., Guder, C., Feistel, A., Marion, G.M., 2010c. Numerical implementation and oceanographic application of the thermodynamic potentials of liquid water, water vapour, ice, seawater and humid air—part 1: background and equations. Ocean Sci. 6, 633–677. http://www.ocean-sci.net/6/633/2010/os-6-633-2010.pdf and http://www.ocean-sci.net/6/633/2010/os-6-633-2010-supplement.pdf.

Fofonoff, N.P., 1962. Physical properties of seawater. In: Hill, M.N. (Ed.), The Sea, vol. 1. Wiley-Interscience, New York, pp. 3–30.

Fofonoff, N.P., 1985. Physical properties of seawater: a new salinity scale and equation of state for seawater. J. Geophys. Res. 90, 3332–3342.

Gouretski, V.V., Koltermann, K.P., 2004. WOCE global hydrographic climatology. Berichte des Bundesamtes für Seeschifffahrt und Hydrographie Tech. Rep. 35/2004, 49 pp.

Graham, F.S., McDougall, T.J., 2013. Quantifying the non-conservative production of Conservative Temperature, potential temperature, and entropy. J. Phys. Oceanogr. 43, 838–862.

Hill, K.D., Dauphinee, T.M., Woods, D.J., 1986. The extension of the Practical Salinity Scale 1978 to low salinities. IEEE J. Oceanic Eng. 11, 109–112.

IAPWS, 2008. Release on the IAPWS Formulation 2008 for the Thermodynamic Properties of Seawater. The International Association for the Properties of Water and Steam. Berlin, Germany, September 2008, available from www.iapws.org. This release is referred to in the text as IAPWS-08.

IAPWS, 2009a. Revised Release on the Equation of State 2006 for H_2O Ice Ih. The International Association for the Properties of Water and Steam. Doorwerth, The Netherlands, September 2009, available from http://www.iapws.org. This release is referred to in the text as IAPWS-06.

IAPWS, 2009b. Supplementary Release on a Computationally Efficient Thermodynamic Formulation for Liquid Water for Oceanographic Use. The International Association for the Properties of Water and Steam. Doorwerth, The Netherlands, September 2009, available from http://www.iapws.org. This release is referred to in the text as IAPWS-09.

IAPWS, 2010. Guideline on an Equation of State for Humid Air in Contact with Seawater and Ice, Consistent with the IAPWS Formulation 2008 for the Thermodynamic Properties of Seawater. The International Association for the Properties of Water and Steam. Niagara Falls, Canada, July 2010, available from http://www.iapws.org. This guideline is referred to in the text as IAPWS-10.

IOC, SCOR, IAPSO, 2010. The International Thermodynamic Equation of Seawater——2010: Calculation and Use of Thermodynamic Properties. Intergovernmental Oceanographic Commission, Manuals and Guides No. 56, UNESCO (English), 196 pp. Available from http://www.TEOS-10.org.

Jackett, D.R., McDougall, T.J., 1997. A neutral density variable for the world's oceans. J. Phys. Oceanogr. 27, 237–263.

Klocker, A., McDougall, T.J., Jackett, D.R., 2009. A new method for forming approximately neutral surfaces. Ocean Sci. 5, 155–172.

McDougall, T.J., 2003. Potential enthalpy: a conservative oceanic variable for evaluating heat content and heat fluxes. J. Phys. Oceanogr. 33, 945–963.

McDougall, T.J., 2012. The International Thermodynamic Equation of Seawater- 2010; Introductory Lecture Slides, available from www.TEOS-10.org.

McDougall, T.J., Barker, P.M., 2011. Getting Started with TEOS-10 and the Gibbs Seawater (GSW) Oceanographic Toolbox. 28pp. SCOR/IAPSO WG127, ISBN 978-0-646-55621-5, available from www.TEOS-10.org.

McDougall, T.J., Jackett, D.R., Millero, F.J., Pawlowicz, R., Barker, P.M., 2012. A global algorithm for estimating Absolute Salinity. Ocean Sci. 8, 1123–1134. http://www.ocean-sci.net/8/1123/2012/os-8-1123-2012.pdf. The computer software is available from www.TEOS-10.org.

Millero, F.J., Leung, W.H., 1976. The thermodynamics of seawater at one atmosphere. Am. J. Sci. 276, 1035–1077.

Millero, F.J., Perron, G., Desnoyers, J.F., 1973. Heat capacity of seawater solutions from 5 to 35 °C and 0.05 to 22‰ chlorinity. J. Geophys. Res. 78, 4499–4506.

Millero, F.J., Feistel, R., Wright, D.G., McDougall, T.J., 2008. The composition of Standard Seawater and the definition of the Reference-Composition Salinity Scale. Deep Sea Res. Part I 55, 50–72.

Pawlowicz, R., 2010. What every oceanographer needs to know about TEOS-10 (The TEOS-10 Primer), unpublished manuscript, available from www.TEOS-10.org.

Pawlowicz, R., McDougall, T., Feistel, R., Tailleux, R., 2012. An historical perspective on the development of the Thermodynamic Equation of Seawater—2010. Ocean Sci. 8, 161–174. http://www.ocean-sci.net/8/161/2012/os-8-161-2012.pdf.

Pawlowicz, R., Wright, D.G., Millero, F.J., 2011. The effects of biogeochemical processes on oceanic conductivity/salinity/density relationships and the characterization of real seawater. Ocean Sci. 7, 363–387. http://www.ocean-sci.net/7/363/2011/os-7-363-2011.pdf.

UNESCO, 1981. The practical salinity scale 1978 and the international equation of state of seawater 1980. UNESCO technical papers in marine science 36, 25 pp.

Wagner, W., Pruß, A., 2002. The IAPWS formulation 1995 for the thermodynamic properties of ordinary water substance for general and scientific use. J. Phys. Chem. Ref. Data 31, 387–535.

Wright, D.G., Pawlowicz, R., McDougall, T.J., Feistel, R., Marion, G.M., Sci, Ocean, 2011. Absolute salinity, "density salinity" and the reference-composition salinity scale: present and future use in the seawater standard TEOS-10. Ocean Sci. 7, 1–26. http://www.ocean-sci.net/7/1/2011/os-7-1-2011.pdf.

Chapter 7

Diapycnal Mixing Processes in the Ocean Interior

Jennifer MacKinnon*, Lou St Laurent† and Alberto C. Naveira Garabato‡

*Scripps Institution of Oceanography, La Jolla, California, USA

†Woods Hole Oceanographic Institution, Woods Hole, Massachusetts, USA

‡National Oceanography Centre, University of Southampton, Southampton, United Kingdom

Chapter Outline

1. INTRODUCTION

Physical processes in the ocean span a vast range of spatial and temporal scales. The winds, tides, and atmospheric buoyancy forcing of the ocean are processes that occur over horizontal scales of the order of 100–1000 km, driving basin-scale gyres, the meridional overturning circulation, and wave motions such as Kelvin, Rossby, and internal waves. A range of dynamical processes ultimately lead to viscous dissipation at small scales, both at ocean boundaries and in the interior of the deep ocean. The resultant turbulent mixing plays a primary role in the thermodynamic balance of the ocean.

Much of the past half century of research has revolved around attempts to reconcile global estimates of how much turbulent mixing is needed to explain observed water property distributions, or to close global energy budgets, with mixing rates inferred from small-scale observations. In a simplistic two-dimensional view, the meridional overturning circulation (MOC) consists of cold water sinking at the poles and upwelling at lower latitudes as it is slowly warmed by heat turbulently diffusing down from the surface. Matching the rate of diapycnal upwelling to rates of deep water production at high latitudes gives the canonical average required diapycnal eddy diffusivity of $1 \times 10^{-4}\ \mathrm{m^2\ s^{-1}}$ below the main thermocline (Munk, 1966; Munk and Wunsch, 1998). In a related calculation, Munk and Wunsch (1998) and Wunsch and Ferrari (2004) argue that approximately 2 TW of power is required to replenish potential energy at the rate it is released by the MOC itself. St Laurent and Simmons (2006) demonstrate that there is a rough equivalence between the total oceanic dissipation by turbulence and the power required estimates for the ocean interior. The main candidates for external energy sources are the wind and the tides, which together approximately supply the needed power channeled largely but not entirely through the internal wave field (Section 4.1).

The original Munk paper touched off an observational search for the canonical $10^{-4}\ \mathrm{m^2\ s^{-1}}$ diffusivity that has

Ocean Circulation and Climate, Vol. 103. http://dx.doi.org/10.1016/B978-0-12-391851-2.00007-6

lasted decades. Initial reports of diffusivities a factor of 10 lower than the Munk value (Gregg, 1987; Ledwell et al., 1998) gave rise to the impression that we were "missing" mixing, a notion that is still pervasive. As one offshoot of the supposed discrepancy, a vein of reasoning was developed that much of the MOC could be closed adiabatically through wind-driven Ekman suction and upwelling in the Southern Ocean (Toggweiler and Samuels, 1998; Marshall and Radko, 2006; Wolfe and Cessi, 2010; Nikurashin and Vallis, 2011). In this view, turbulent mixing was needed only at the deepest levels of the ocean, with an associated much lower power requirement.

However, more recent evidence suggests the quest for missing mixing may be a red herring, for several reasons. First, there is often a depth mismatch in the discussion. The original Munk calculation for water was deeper than 1 km, below the main thermocline, while the observations by Gregg (1987) and Ledwell et al. (1998) were in or above the main thermocline. The bulk of microstructure observations of mixing taken below the main thermocline in fact do show values on the order of $10^{-4}\,m^2\,s^{-1}$ or larger (St Laurent and Simmons, 2006; Waterhouse et al., 2013; Section 5.2). Second, a preponderance of inverse models and related calculations demonstrate that average diffusivities of $1-10\times10^{-4}\,m^2\,s^{-1}$ are required at all depths below the main thermocline to close mass or tracer budgets in individual ocean basins (e.g., Ganachaud and Wunsch, 2000; Talley et al., 2003; Lumpkin and Speer, 2007). Many of these estimates are for domains that do not include the Southern Ocean.

Finally, the two-dimensional, zonally averaged view of the MOC can hide essential pathways of water transformation. In a zonally averaged view, it appears that there are two somewhat distinct overturning cells, one involving North Atlantic Deep Water (NADW) sinking in the North Atlantic and upwelling in the Southern Ocean, and the second involving Antarctic Bottom Water (AABW) sinking near Antarctica and upwelling slowly into the bottom of the upper cell (see Chapter 10). However, the three-dimensional circulation appears to be more like a mobius strip, in which the majority of NADW does not directly return to the North Atlantic after upwelling in the Southern Ocean, but cools and sinks as AABW. The majority of the AABW diabatically upwells into intermediate water in the Indian and Pacific Oceans, and then returns to the surface through a combination of diabetic and adiabatic processes (Lumpkin and Speer, 2007; Talley et al., 2011; Talley, 2013). The process is illustrated schematically in Figure 7.1. From this Lagrangian perspective, most water parcels on the so-called conveyor belt gain buoyancy through diapycnal mixing below the main thermocline at some point during their journey.

There currently appears to be rough agreement between the power required to drive the overturning circulation, the power available to the internal wave field, and the global sum of observed mixing rates, with all estimates around 2–3 TW (Wunsch and Ferrari, 2004; St Laurent and Simmons, 2006; Ferrari and Wunsch, 2009). Given the sparse nature of microstructure observations, narrowing the comparison down further is a daunting task. Instead, a major emphasis over the past decade has been on process studies targeted at understanding specific dynamical regimes, with the hope that such understanding could then be extrapolated globally (Section 5.2). At the same time, the combination of increasing sophistication and decreasing

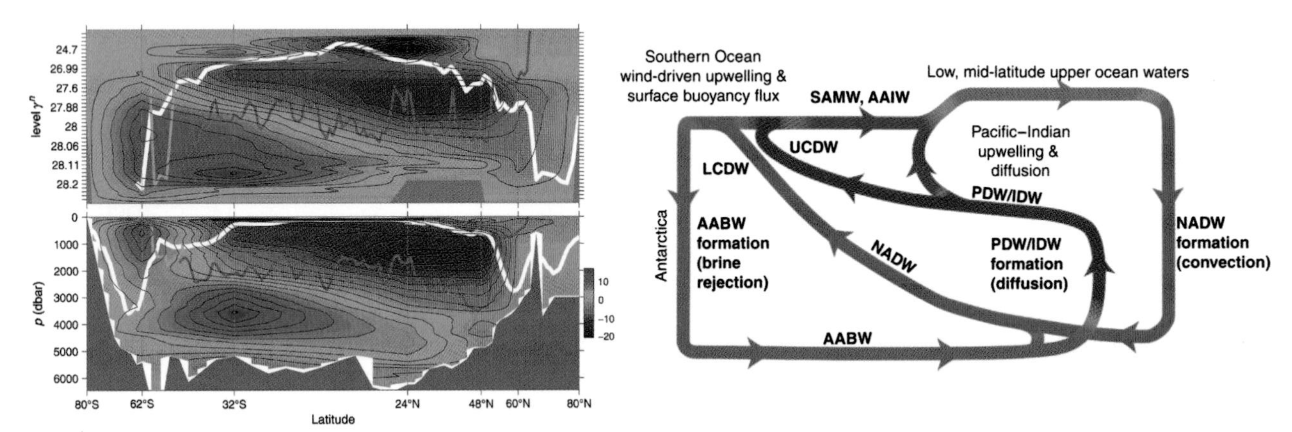

FIGURE 7.1 Left panels: zonally averaged stream functions for the global meridional overturning circulation in density (top) and pressure (bottom) coordinates, every 2 Sv contoured. Typical winter mixed-layer densities/depths (white), the mean depth of ocean ridge crests (dark gray), and the depth of the Scotia Arc east of Drake Passage (light gray) are also shown. Here, it appears that the North Atlantic Deep Water (NADW) and Antarctic Bottom Water (AABW) form distinct overturning cells. Right panel: schematic of three-dimensional global overturning circulation. Here, it is clear that diapycnal mixing, particularly, in the Pacific and Indian Oceans, plays a fundamental role in the upwelling that ultimately returns deep and bottom waters toward the surface. Additional labeled water masses include South Atlantic Mode Water (SAMW), Antarctic Intermediate Water (AAIW), Lower and Upper Circumpolar Deep Water (LCDW and UCDW), and Pacific and Indian Deep Waters (PDW and IDW). *Left panels are reproduced from Lumpkin and Speer (2007). Right panel is reproduced from Talley et al. (2011).*

spurious diffusion in large-scale numerical models has shown that circulation and tracer distributions are extremely sensitive not just to the average value of diapycnal diffusivity but to its detailed geographical distribution (Section 5.3; Griffies et al., 2010). Because the global energy budget of ocean mixing has been well reviewed elsewhere (Munk and Wunsch, 1998; Wunsch and Ferrari, 2004; St Laurent and Simmons, 2006; Kuhlbrodt et al., 2007; Ferrari and Wunsch, 2009), the bulk of this chapter is dedicated to describing the wide range of dynamic processes responsible for creating turbulence in the ocean interior and the associated complex geography of diapycnal mixing.

Though much of the global energetics discussion revolves around deep and abyssal mixing rates, there has also been increasing interest in diapycnal mixing in the top few hundred meters of the ocean. Elevated upper ocean diffusivity is especially important in tropical regions, where it can significantly influence mixed-layer heat content and associated air–sea heat fluxes. So before moving on to the "zoo" of deep ocean processes, we first briefly review recent developments in our understanding of upper ocean turbulence.

Our review is far from comprehensive. We neglect a number of major areas relevant to a discussion of ocean mixing that have recently been reviewed elsewhere. This includes the topic of double diffusion (Schmitt, 2012), which is known to be important in many regions, such as the tropical Atlantic (Schmitt et al., 2005) and the Arctic (Timmermans et al., 2003). We also neglect the area of biogenic turbulence, concerning turbulence levels generated by the kinetic activities of marine animals (Dewar et al., 2006; Young, 2012). The classic turbulence problem of the bottom boundary layer is also ignored, despite recent interest especially in coastal oceanography (e.g., Perlin et al., 2005). In addition, the role of nonlinearities in the equation of state for seawater is not discussed here (Klocker and McDougall, 2010). Finally, turbulent entrainment in deep overflows, essential for setting the water mass characteristics of all deep and bottom waters of the world's oceans, is nicely reviewed by Legg (2012) and discussed in Chapter 10.

2. MIXING BASICS

Throughout this chapter, we are primarily concerned with mixing resulting from small-scale three-dimensional turbulence. Molecular diffusion of heat and salt is a slow process, characterized by diffusivities of the order 1.4×10^{-7} and $1.5 \times 10^{-9}\ \mathrm{m^2\ s^{-1}}$, respectively (Gill, 1982). Turbulent stirring of fluid acts to dramatically increase gradients and accelerate the rate of irreversible mixing. Such mixing increases the potential energy of stratified water. The associated turbulent fluxes are often characterized by an effective or turbulent diffusivity. A commonly used formulation relates the diapycnal diffusivity to the turbulent dissipation rate, ϵ, a common measure of the strength of turbulence, through an assumed mixing efficiency,

$$\kappa_\rho = \Gamma \frac{\epsilon}{N^2} \tag{7.1}$$

where Γ is the mixing efficiency and N is the buoyancy frequency, a measure of stratification strength. In practice, the mixing efficiency is often taken to be $\Gamma = 0.2$ (Osborn, 1980). In reality, Γ is likely to vary with the background stratification and the strength of the turbulence (Shih et al., 2005; Ivey et al., 2008), but a discussion of these issues is beyond the purview of this chapter. Interested readers are referred to recent reviews by Staquet and Sommeria (2002), Ivey et al. (2008), Moum and Rippeth (2009), and Hughes et al. (2009). Here we specify the strength of observed mixing rates using both the turbulent dissipation rate (typical values of 10^{-10} to $10^{-9}\ \mathrm{W\ kg^{-1}}$ in the deep ocean) and the turbulent diapycnal diffusivity (often just referred to as the diffusivity, with typical values of 10^{-5} to $10^{-4}\ \mathrm{m^2\ s^{-1}}$), depending on the metric used in the original studies being referenced.

Turbulence in the ocean may be produced through a range of instabilities, details of which are reviewed by Smyth and Moum (2001), Wunsch and Ferrari (2004), Thorpe (2005), and Ivey et al. (2008). Away from the direct influence of surface fluxes, turbulence is often related to vertical shear or convective instabilities (Alford and Pinkel, 2000; Smyth and Moum, 2012). The tendency for a fluid to undergo shear instability is controlled by the gradient Richardson number, $\mathrm{Ri} = N^2/S^2$, which reflects the counteracting stabilizing and destabilizing effects of stratification and shear $[S^2 = (\mathrm{d}u/\mathrm{d}z)^2 + (\mathrm{d}v/\mathrm{d}z)^2]$. Many models include a turbulent diffusivity that is a function of the Richardson number, but it is only applied to vertically coarse and generally low-frequency shear features that are well resolved by the model (Section 5.3). Observations and process studies described below demonstrate that most turbulence in the ocean interior is produced by small-scale or high-frequency motions that are not generally resolved by global or even regional scale models, and are unlikely to be resolved in the foreseeable future. Hence there is a premium on understanding the nature and geography of the dynamics driving turbulent mixing so that it may be properly parameterized. Our focus is not on the details of the turbulence *per se*, but on the larger-scale dynamics that set the stage and supply the energy for turbulent mixing, largely by moving energy to small vertical scales where shear is large.

Observationally, the most direct way to measure turbulent mixing is through purposeful dye release (Ledwell et al., 1993, 2000, 2011), a complicated endeavor. The turbulent dissipation rate may be estimated using microstructure instruments that measure either velocity or scalar fluctuations within the inertial subrange that characterizes turbulence on the smallest scales, typically centimeters or smaller (Moum et al., 1995; Gregg, 1998; Lueck et al., 2002). These estimates are accurate, but the instruments are specialized and difficult to operate. The turbulent dissipation rate may also be estimated from measurements of the outer scales of turbulent overturns, typically meters, which may be made from a variety of instruments. Moving to even larger scales, recent techniques have allowed the strength of turbulence to be inferred from measurements of internal waves, whose breaking is presumed to produce the turbulence. Such finescale methods are described in Section 5.1.

3. TURBULENCE IN AND BELOW THE SURFACE MIXED LAYER

Diapycnal mixing processes in the upper ocean are an important component of the coupled climate system. The ocean surface boundary layer is the subject of vigorous interactions with the overlying atmosphere and cryosphere. Heat, fresh water, and momentum are exchanged across the ocean surface at large rates, and upper ocean water mass properties are modified by turbulent processes. A striking feature of the upper ocean (and one that is defining of ocean circulation) is the glaring contrast between the energy flow across the ocean surface and that through the base of the mixed layer into the stratified ocean below. Of the approximately 65 TW of power that has been estimated to be imparted to the ocean surface by the wind, less than 5% is thought to be transmitted to the ocean interior (Huang, 1998; Wunsch and Ferrari, 2004; Ferrari and Wunsch, 2009). Since surface fluxes are well reviewed elsewhere (Large and Nurser (2001) and Chapter 5), here we focus on the processes through which surface fluxes may be transmitted into the ocean interior through turbulent mixing at and below the mixed-layer base. Just below the mixed layer a strongly stratified "transition layer" mediates the transfer of heat, nutrients, and dissolved gasses to the deeper ocean (Johnston and Rudnick, 2009). Turbulence in the transition layer is driven primarily by a combination of shear extending down below the mixed layer, penetrative convection, and breaking high-frequency internal waves. Mixed-layer deepening by direct wind forcing and convection are represented in existing bulk formula (Price et al., 1986; Large et al., 1994; Moum and Smyth, 2001). Here we describe a few processes that are areas of active research and are not represented in most mixed-layer parameterizations.

3.1. Langmuir Turbulence

Langmuir turbulence occurs when a surface boundary layer is forced by wind in the presence of surface waves. The combination drives elongated counterrotating vortices organized into the wave direction (McWilliams et al., 1997), generally thought to be caused by an instability arising from the interaction of the Stokes drift induced by the surface waves and the shear of the upper ocean flow (Craik and Leibovich, 1976; McWilliams et al., 2004; Sullivan et al., 2004). Even though the Stokes drift is confined to a shallow vortex layer on the order of the significant wave height deep, downwelling jets originating within that layer penetrate well beyond it (Polton and Belcher, 2007). Because of these jets, the influence of surface waves may be felt, via Langmuir turbulence, throughout the depth of the mixed layer. It has been suggested, in fact, that Langmuir turbulence may be more effective than shear-driven turbulence in deepening the mixed layer (Skyllingstad and Denbo, 1995).

The importance of Langmuir turbulence and its significance in inducing diapycnal mixing in the upper ocean are commonly assessed by reference to the turbulent Langmuir number $\mathrm{La_t} = (u_*/u_{s0})^{1/2}$, where u_* is the surface friction velocity in the water and u_{s0} is the surface Stokes drift (McWilliams et al., 1997; Li et al., 2005). Langmuir turbulence becomes a dominant process when $\mathrm{La_t} \leq 0.3-0.5$, whereas shear-driven turbulence is dominant for significantly higher $\mathrm{La_t}$ (e.g., Li et al., 2005; Grant and Belcher, 2009). Climatological estimates of $\mathrm{La_t}$ suggest that mixing in the ocean surface boundary layer is often dominated by Langmuir turbulence (Li et al., 2005; Belcher et al., 2012). Crucially, the mixing associated with Langmuir cells cannot be parameterized by local closure schemes and requires an alternate approach. More details of Langmuir cell dynamics and parameterization are nicely reviewed by Sullivan and McWilliams (2010).

3.2. Inertial Motions

Another mechanism via which wind forcing may destabilize a one-dimensional upper ocean is the generation of inertial motions. As a natural resonant frequency of the system, inertial motions in the surface mixed layer are efficiently forced by time-variable wind stresses, in particular those with strong inertially rotating components such as due to passing mid-latitude storms. The generation process is often modeled using the damped-slab model (Pollard and Millard, 1970; D'Asaro, 1985), which describes the temporal evolution of inertial oscillations in a one-dimensional mixed layer of constant depth as a balance between wind forcing and a parameterized (linear) damping. Integrated estimates of power going into near-inertial motions derived from global wind products range from 0.3 to 1.2 TW (Alford, 2001; Watanabe and Hibiya, 2002; Jiang et al., 2005), and are highly sensitive to the specific wind products used as well as assumptions about the damping rate of near-surface oscillations (Plueddemann and Farrar, 2006). Physically, the damping rate represents energy loss both through shear instabilities at the base of the mixed layer (Crawford and Large, 1996; Skyllingstad et al., 2000) and radiation of near-inertial internal waves (Section 4.1.2). The damped-slab model has been tuned to produce estimates of upper ocean inertial velocity that are in good qualitative correspondence with observations, given the values of the damping coefficient in the range 2–10 days (D'Asaro et al., 1995). The geography of near-inertial motion generation varies seasonally, but generally follows storm tracks. Near-inertial waves and associated mixing are a key component of the ocean response to hurricanes (Price et al., 2008). The global patterns of power input into inertial motions calculated using a slab model approach bear many similarities to the geography of near-surface near-inertial kinetic energy measured from surface drifter tracks (Figure 7.3). Convergences and divergences of inertial motions pump energy into the ocean interior in the form of internal waves, which are considered further in Section 4.1.2.

3.3. An Equatorial Example

Much of the recent observational progress in understanding turbulent mixing in the transition layer comes from equatorial studies, where dynamics are further complicated by the presence of the strongly sheared equatorial undercurrent. Regional and global ocean models show that coupled air–sea phenomena such as ENSO (see Chapters 15 and 24) are quite sensitive to mixing in the transition layer, as the rate of downward heat flux out of the mixed layer affects SST (Harrison and Hallberg, 2008). An example of the rich field of turbulence present beneath the equatorial mixed layer is shown in Figure 7.2. The turbulent dissipation rate (bottom panel) shows bursts of turbulence extending well below the mixed layer (upper black line) on most nights, with separate patches of turbulence at times present at the upper edge of the equatorial undercurrent (lower black line). Bursts of high-frequency oscillations penetrating below the mixed layer have been well documented at the equator (e.g., visible in the middle panel in Figure 7.2), though they likely occur elsewhere as well (Lien et al., 2002). A variety of theories have been proposed to explain observed high-frequency motions in this depth range, from generation by shear instabilities acting on the upper edge of the equatorial undercurrent (Moum et al., 2011; Smyth et al., 2011) to internal waves triggered by nocturnal convection bursts impinging on the stratified mixed-layer base (Gregg et al., 1985; Wijesekera and Dillon, 1991), to the obstacle effect as Langmuir cells or other undulations of the mixed-layer base are advected by mixed-layer currents over the stratified layer below (Polton et al., 2008). The spate of recent equatorial mixing observations also highlights the compounding effects of processes with very different timescales. For example, while the bursts of turbulence visible in Figure 7.2 clearly have a diurnal pattern, Moum et al. (2009) demonstrate that slow modulation by passing tropical instability waves is enough to nudge the underlying undercurrent shear past the threshold for shear instability, with resultant turbulent mixing large enough to be an order-one player in local vertical heat budgets. While questions remain as to the dynamics driving transition layer mixing at the equator (or elsewhere), it is becoming increasingly clear that large-scale-modeled upper ocean budgets and air–sea fluxes are sensitive to how these effects are parameterized (Section 5.3).

3.4. Fronts and Other Lateral Processes

While mixing near the surface of the ocean is often discussed and parameterized as a one-dimensional process, the phenomenology of mixing is greatly enriched in the

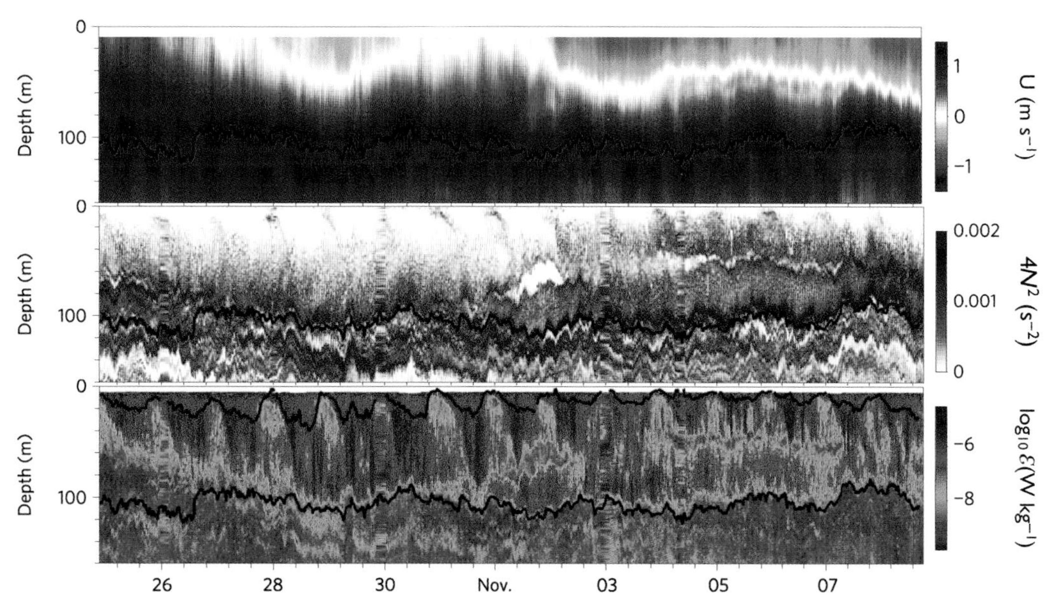

FIGURE 7.2 The rich structure of near-surface turbulence. Profiling time series at 0° E, 140° W in boreal autumn 2008. Zonal velocity: the core (eastward velocity maximum) of the eastward-flowing EUC is shown as a black line (top); $4N^2$ (middle); turbulence dissipation rate, ϵ (lower). The mixed layer is defined by the upper black line in ϵ as the depth at which ρ deviates by 0.01 kg m^{-3} from its surface value. *Reproduced from Moum et al. (2009).*

presence of significant lateral density gradients. The widespread occurrence of such gradients on horizontal scales down to 1–2 km throughout the ocean extratropics has been demonstrated by observations with towed instrumentation (Rudnick and Ferrari, 1999; Hosegood et al., 2006; Capet et al., 2008) and has also been suggested by numerical simulations of the ocean's submesoscale (e.g., Lapeyre et al., 2006; Capet et al., 2008; Fox-Kemper et al., 2008). The results of such simulations are consistent with an interpretation of the observed submesoscale lateral density structure as arising from fast-growth (growth rate of $\sim f$), small-scale (lateral scale on the order of 1 km) ageostrophic baroclinic instabilities (e.g., Stone, 1966, 1970; Molemaker et al., 2005; Boccaletti et al., 2007) of mixed-layer fronts, which result in an adiabatic restratification of the upper ocean. The mixed-layer fronts are thought to be formed by strong convective mixing induced by atmospheric forcing (followed by a rapid relaxation of lateral density gradients to geostrophy through Rossby adjustment; e.g., Tandon and Garrett, 1995) or, more commonly, by the straining of the large-scale horizontal density distribution by the mesoscale eddy field (Treguier et al., 1997; Ferrari et al., 2008, 2010; Fox-Kemper et al., 2008). At any rate, it is now recognized that the upper ocean density field is fundamentally three dimensional and that this can qualitatively alter the effects of wind and buoyancy forcing on turbulent dissipation and diapycnal mixing in the upper ocean.

Evidence for the latter statement has proliferated in recent years. For example, numerical simulations of buoyancy-driven convection at a mixed-layer front (e.g., Taylor and Ferrari, 2010) reveal that, while classical vertical convection (in which the turbulent buoyancy flux supplies the production of turbulent kinetic energy) occurs in a relatively shallow upper layer, the bulk of the depth range affected by the forcing experiences symmetric instability, a gravitational centrifugal form of instability undergone by gravitationally stable fluid with negative potential vorticity in which shear production (by perturbations growing along isopycnals) is the primary source of turbulent kinetic energy (e.g., Thorpe and Rotunno, 1989; Haine and Marshall, 1998). Similarly, wind forcing of a mixed-layer front has been shown to have qualitatively distinct impacts on the upper ocean relative to the one-dimensional mixed-layer scenario that is traditionally considered. Upfront wind forcing (i.e., oriented in the direction opposite to the quasi-geostrophic flow of the surface ocean) induces an Ekman flow directed toward the dense side of the front and thereby restratifies the upper ocean (e.g. Thomas, 2005; Thomas and Ferrari, 2008; Mahadevan et al., 2010).

Resonant wind forcing by upfront winds has been found to provide a particularly rapid and effective mechanism for injecting near-inertial internal waves into the stratified upper ocean (Forryan et al., 2013). Downfront wind forcing, in turn, destratifies the upper ocean and promotes turbulent dissipation and diapycnal mixing. The means by which it does so centrally involves symmetric instability (Taylor and Ferrari, 2010; Thomas and Taylor, 2010). Since kinetic energy is extracted by the instability from the geostrophic flow and ultimately dissipated in small-scale turbulence, wind-forced symmetric instability leads to a significant reduction of the wind work that is available for increasing the kinetic energy of the ocean's general circulation (Thomas and Taylor, 2010), and can produce rates of upper ocean turbulent dissipation and diapycnal mixing that greatly exceed expectations from wind-forced boundary layer theory (D'Asaro et al., 2011).

In regions of strong frontogenesis, ageostrophic secondary circulation may also strongly interact with internal gravity waves, transferring energy through them toward turbulent dissipation at small scales (Thomas, 2012). Finally, it has been argued that turbulent dissipation and diapycnal flow in the upper ocean may be induced by the interaction of the mesoscale eddy field with the ocean surface, which forces eddy buoyancy fluxes to be directed horizontally (i.e., diapycnally) in the mixed layer (Treguier et al., 1997; Radko and Marshall, 2004; Ferrari et al., 2008). The coupling of mesoscale eddy stirring with air–sea heat fluxes can also act to reduce variance in near-surface tracer fields (Shuckburgh et al., 2011).

4. MIXING IN THE OCEAN INTERIOR

Diapycnal mixing in the ocean interior fluxes heat, salt, and dissolved gasses across density classes and influences the slowly evolving meridional overturning circulation. Away from direct influence of the surface or bottom boundaries, power to supply turbulent mixing must be imported into the ocean interior, largely through the propagating internal waves that produce most observed shear. Mixing may also occur near the ocean floor and the resultant mixed fluid can spread out along isopycnals into the interior. Here, we provide a basic review of some of the processes acting to drive mixing well below the mixed layer. This is followed by a discussion of processes driving mixing in the Southern Ocean (Section 4.4), which have a flavor all of their own.

4.1. Internal Wave Breaking

Internal wave breaking has long been argued to be the dominant source of turbulent mixing away from boundary layers. Energy is added into the internal wave field primarily by the tides and winds, producing internal tides and near-inertial internal waves, respectively. In regions of very strong near-bottom flows, internal waves may also be created by a lee-wave mechanism, an example of which will be presented in Section 4.4. In both tide- and wind-forced internal waves, power input into the ocean is often dominated by energy flux into relatively large-scale internal waves, with vertical wavelengths from hundreds to thousands of meters (Gill, 1984; St Laurent and Garrett, 2002) (Figure 7.3). However, dissipation results from the breaking of small-scale waves (generally tens of meters or less) through shear or convective instabilities (Alford and Pinkel, 2000; Staquet and Sommeria, 2002). The geography of internal wave mixing thus is controlled by the combination of the generation geography, wave propagation and refraction, and processes that move energy toward smaller, more dissipative scales of motion.

4.1.1. Dissipation Near Internal Tide Generation Sites

Internal tides are generated in areas where the barotropic tide (diurnal or semidiurnal) sloshes over rough or steep topography, the process of which is reviewed by St Laurent and Garrett (2002) and Garrett and Kunze (2007). Global patterns of internal tide generation reflect the product of barotropic tidal strength and topographic roughness, and resemble maps of deep-sea energy loss from the barotropic tide (Egbert and Ray, 2001; Jayne and St Laurent, 2001; Simmons et al., 2004b). Globally, there is approximately 1 TW of power going into internal tides (Wunsch and Ferrari, 2004). Recent work has conceptually divided the problem into nearfield and farfield components representing, respectively, the fates of comparatively higher-mode waves that break near the generation region and low-mode waves that propagate away.

The nearfield part of tidal dissipation has been well studied observationally in the past two decades. For example, both the Hawaii Ocean Mixing Experiment (HOME) (Klymak et al., 2006) and the Brazil Basin Experiment (Polzin et al., 1997) find turbulence elevated by orders of magnitude within several horizontal wavelengths (hundreds of kilometers or less) of internal tide generation sites. There are a range of processes responsible. Strong flow over steep slopes can produce very strong mixing driven by a combination of convective instabilities, breaking internal lee waves, and hydraulic jumps. Recent work in the very energetic

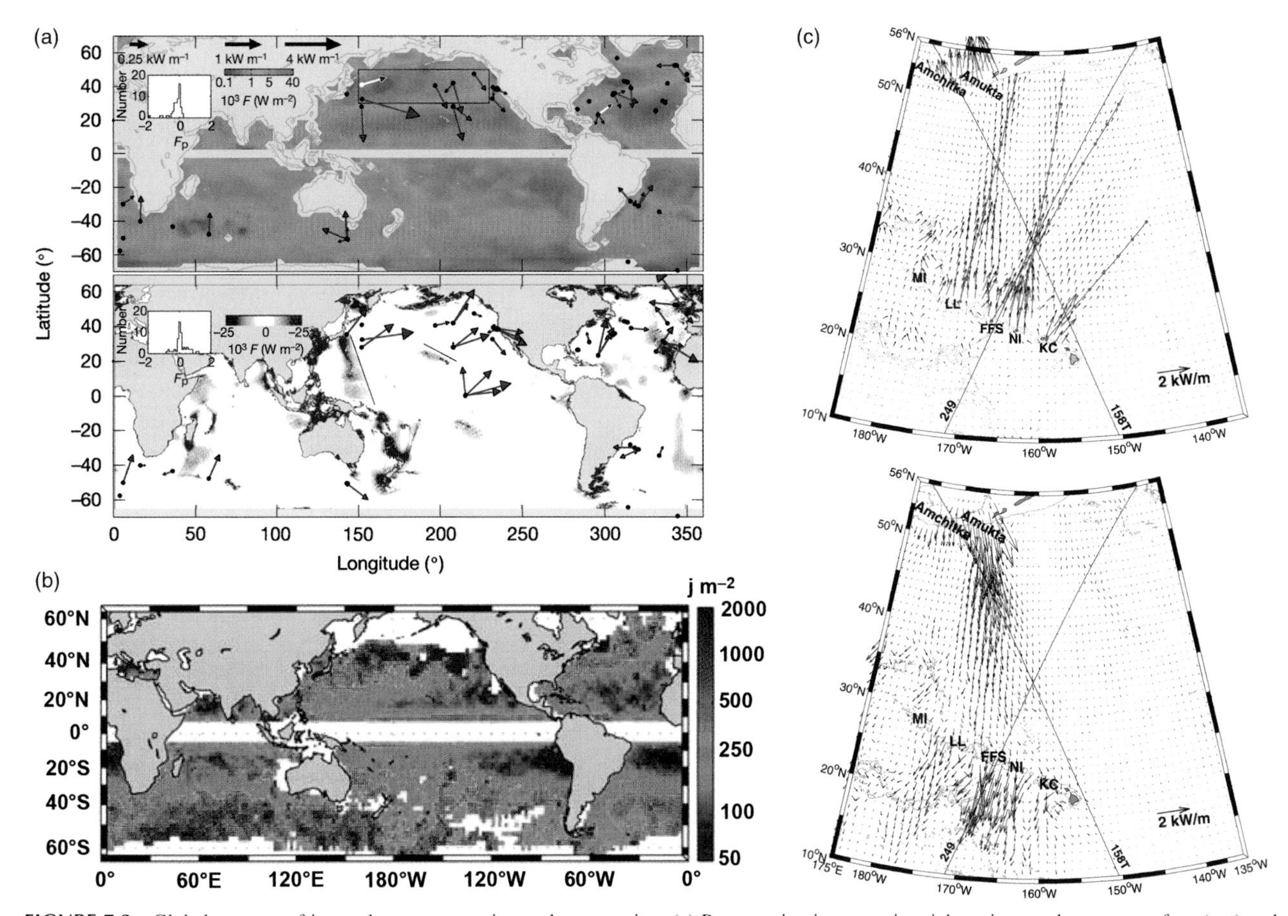

FIGURE 7.3 Global patterns of internal wave generation and propagation. (a) Power going into near-inertial motions at the ocean surface (top) and internal tides (bottom). Arrows represent near-inertial and internal tide energy fluxes as measured at available historical moorings (Alford, 2003). (b) Depth-integrated near-inertial kinetic energy in the surface mixed layer as estimated from surface drifter tracks (Chaigneau et al., 2008). (c) Mode-1 internal tide energy fluxes in the North Pacific estimated from satellite altimetry, decomposed into northward and southward components (Zhao and Alford, 2009).

Luzon Strait showcases the range of phenomenology possible (Alford et al., 2011; Klymak et al., 2012; Pinkel et al., 2012). Slightly further aloft, higher-mode wave-like motions may achieve critical Froude numbers leading to strong breaking, a process observed in HOME (Klymak et al., 2008) and the subject of recent modeling and theoretical parameterization attempts (Legg and Klymak, 2008; Klymak et al., 2010). Dissipation rates are also elevated in high-mode beam-like structures seen in some observations (Lien and Gregg, 2001; Martin et al., 2006; Cole et al., 2009; Johnston et al., 2011), models (Gerkema et al., 2006), and laboratory experiments (Peacock et al., 2008).

Turbulence may be elevated near rough topography even apart from influence of such direct tidally forced nonlinearities. For example, over the deep rough topography of the eastern Brazil Basin, turbulent dissipation is elevated for several kilometers above the bottom, with a magnitude that steadily decreases with increasing height (Polzin et al., 1997). Polzin (2009) suggests that the pattern is set by the rate at which an upward propagating quasi-linear internal tide steadily loses energy through weakly nonlinear wave–wave interactions to higher-mode waves, which in turn have higher shear and dissipate locally. The hypothesis is roughly consistent with the empirical formulation proposed by St Laurent et al. (2002) and the analytical formulation of Polzin (2004) now being implemented in large-scale models (Section 5.3). In this case, and perhaps in similar environments of deep rough topography, measurements are consistent with most of the energy going into the internal tide being dissipated locally, that is within a few hundred kilometers (Polzin, 2004; Waterhouse et al., 2013).

Yet in other locations, particularly over tall steep topography, the majority of generated internal tide energy escapes to propagate up to thousands of kilometers across ocean basins in the form of low-mode internal waves. Evidence for these propagating waves can be seen in satellite altimetry (Zhao and Alford, 2009), *in situ* flux measurements (Alford, 2003; Althaus et al., 2003; Rudnick et al., 2003; Alford et al., 2007; Zhao et al., 2010), and high-resolution models (Simmons, 2008) (Figure 7.3c). The ultimate fate of this energy is less clear: some is bled into the ambient internal wave field through nonlinear wave–wave interactions producing the "background" diffusivity of the main thermocline and abyss (Section 4.1.3), one subset of which is Parametric Subharmonic Instability (PSI), some may become trapped in mesoscale shear or vorticity, and the remainder likely scatters over deep ocean topography or breaks on distant continental slopes (Section 4.1.4).

4.1.2. Dissipation Near-Inertial Wave Generation Sites

Though near-inertial internal waves provide one of the most prominent peaks in any oceanic energy spectrum, comprising half the kinetic energy in the internal wave field and a larger percentage of the shear, remarkably little is known about their generation, evolution, or decay (Ferrari and Wunsch, 2009). Generation mechanisms include wind forcing at the ocean surface, loss of balance of mesoscale or submesoscale features (Section 4.3), or nonlinear interactions between other internal waves (McComas and Müller, 1981). At the ocean surface, time-variable wind stresses force inertial motions in the surface mixed layer, since the inertial frequency is the natural "ringing" frequency of any fluid on a rotating planet (Section 3.2). Horizontal convergences and divergences of this moving mixed-layer water at the edges of the forced region create vertical velocities with an inertial period at the mixed-layer base, which in turn force near-inertial internal waves in the stratified region below. The horizontal wavelength of the propagating waves is heavily influenced by the beta effect, namely that for a patch large enough to feel the latitudinal change of inertial frequency, motions at the northern end of the patch gradually get out of phase with those near the southern end. Inertial motions are also sensitive to mesoscale vorticity, which can add or detract from the planetary vorticity to change the effective inertial frequency felt by these motions. Detailed dynamics of the generation and initial propagation are described by D'Asaro (1985), D'Asaro et al. (1995), Young and Ben-Jelloul (1997), and Moehlis and Llewellyn-Smith (2001). Simple models predict that most of the energy goes into low-mode internal waves (Gill, 1984; Zervakis and Levine, 1995). However, the higher, shear-containing modes are also of interest as they may provide a more direct pathway to mixing.

As with the internal tide, some of the wind-generated energy probably dissipates in the upper ocean, particularly that of higher-mode waves (Alford and Gregg, 2001), leading to maps of estimated upper ocean diffusivities (e.g., Figure 7.7b) that mirror patterns of mixed-layer near-inertial energy in Figure 7.3. Such upper ocean dissipation likely has a seasonal cycle (Jing and Wu, 2010; Whalen et al., 2012). High-mode near-inertial internal waves are particularly sensitive to interactions with mesoscale vorticity, making associated patterns of mixing very sensitive to ambient conditions (Rainville and Pinkel, 2004; Danioux et al., 2008; Elipot et al., 2010).

Yet substantial wind-generated near-inertial energy is also clearly reaching the deep sea, as evidenced by a deep seasonal cycle of near-inertial energy (Mihaly et al., 1998; Alford and Whitmont, 2007; van Haren, 2007; Silverthorne and Toole, 2009) and direct observations of downward energy flux. Alford et al. (2012) observe up to a third of near-inertial power input into the surface mixed layer of the North Pacific transiting vertically through 800-m depth. Low-mode near-inertial waves can also be observed propagating equatorward from the mid-latitude storm track

(Alford, 2003; Figure 7.3). The ultimate fate of these waves, like that of propagating low-mode internal tides, is unknown. Several candidates for dissipation of low-mode internal waves are discussed in the following subsections.

One of the persistent mysteries regarding near-inertial internal waves is the presence of large numbers of waves with properties suggesting upward propagating group velocities (e.g., Alford, 2010). Most of the shear resides in waves with small vertical wavelengths and associated very slow vertical propagation speeds, making bottom reflection of surface-generated waves of this wavelength an unlikely possibility. Some types of wave–wave interactions (Section 4.1.3) and loss of mesoscale balance (Section 4.3) may also produce upward propagating near-inertial waves, although the rates and pathways are somewhat unclear. Given that turbulent mixing is intricately related to small-scale shear and that near-inertial motions are the largest source of small-scale shear, improving our understanding of these processes seems essential.

4.1.3. *Wave–Wave Interactions*

A leading candidate for energy loss from propagating internal waves is nonlinear interactions with an ambient internal wave field. Wave–wave interactions redistribute energy to a variety of frequencies and wavenumbers. The broadband internal wave continuum displays a remarkably narrow range of spectral shapes (Polzin and Lvov, 2011), presumably reflecting attractive states for the underlying nonlinear dynamics (Müller et al., 1986; Lvov et al., 2010). Within the continuum, there is a general tendency for energy to flow toward higher vertical wavenumbers, where it is likely to lead to wave breaking and dissipation (McComas and Müller, 1981; Lvov et al., 2010) (Section 5.1).

Propagating low-mode tidal or near-inertial internal waves may bleed energy through the continuum, creating essentially a smeared out wake of dissipation along wave propagation paths. In turn, the energy lost from low-mode waves likely supplies and maintains the continuum, with regional differences in continuum shape reflecting features of the primary forcing wave (Polzin and Lvov, 2011). While documenting the relationship between continuum energy levels and mixing rates is an important step, the ability to predict the resultant global patterns of turbulent mixing requires a prognostic relationship between available tidal or near-inertial internal wave energy, the continuum it supplies, and the eventual rate of small-scale wave breaking (Polzin, 2009).

PSI is a specific type of wave–wave interaction that transfers energy from a low-mode wave to two high-mode waves near half the frequency. The interaction can occur anywhere where the subharmonic waves are within the internal wave band ($f \leq \omega \leq N$), where f is the local inertial frequency. The interaction may be particularly efficient where the subharmonic of the tidal frequency is equal to the local inertial frequency (Hibiya et al., 2002; MacKinnon and Winters, 2003, 2005; Furuichi et al., 2005; Nikurashin and Legg, 2011). For the semidiurnal tide, this occurs near a latitude of 29°N/S. Numerical studies predict significantly elevated mixing at these latitudes (MacKinnon and Winters, 2005; Simmons, 2008) with some suggestively corroborating observational evidence (Hibiya and Nagasawa, 2004). However, a major field campaign to track the internal tide northward from Hawaii found only a moderate effect near 29°N, with diffusivity in the upper ocean elevated by a factor of 2–4 over background levels. The discrepancy between observations and the more catastrophic numerical results is likely due to the complex and time-variable nature of the internal tide in that region (Alford et al., 2007; Hazewinkel and Winters, 2011; MacKinnon et al., 2013). Other studies have found evidence of PSI of strong internal tides at lower latitudes (Carter and Gregg, 2006), of the diurnal internal tide near 14°N (Alford, 2008), and of equatorward-propagating near-inertial internal waves between 5° and 15°N (Nagasawa et al., 2000).

4.1.4. *Distant Graveyards*

The remainder (perhaps the majority) of propagating low-mode internal wave energy scatters to higher modes and dissipates due to interaction with topography, often in distant locations that are far from the generation site. Some of this interaction may occur with mid-basin topographic features (Gilbert and Garrett, 1989; Lueck and Mudge, 1997; Toole et al., 1997; Müller and Liu, 2000; Johnston et al., 2003). Altimetric evidence suggests that in basins such as the North Pacific, with relatively smooth bottom topography, the majority of low-mode internal tide energy propagates virtually unscathed for thousands of kilometers (Ray and Mitchum, 1996, 1997; Zhao and Alford, 2009). Care must be taken in interpreting patterns of tidal fluxes such as Figure 7.3c, as some of the apparent decay in altimetric fluxes may be due to loss of coherence as waves refract in an evolving mesoscale (rendering them invisible to this detection technique) rather than genuine dissipation. Comparable observations of long-range propagating low-mode near-inertial wave patterns are not available. Much of this low-mode energy may dissipate where waves scatter and reflect off continental slopes and shelves. Simple theory predicts that an internal wave hitting a near-critical slope (one with the same angle from vertical as the ray path of the incident wave) will reflect into smaller-scale waves that are more likely to break (Eriksen, 1985, 1998; Ivey et al., 2000; McPhee-Shaw and Kunze, 2002; Nash et al., 2004; Kelly and Nash, 2010). The details of such processes are poorly constrained by existing observations but likely are sensitive to both the steepness and the three-dimensional roughness of the slope in question. For example, Klymak et al. (2011) observed that a substantial portion of mode-one tidal energy impinging on a supercritical slope in the South China Sea reflects back as low-mode waves, while the remainder propagates onshore as a dissipative bore. In contrast, incoming internal tides hitting the rough Oregon slope were observed to shoal, refract, and break (Martini et al., 2011; Kelly et al., 2012). Canyons littered along continental slopes may also focus incoming waves (Kunze et al., 2002; Gregg et al., 2011), and canyon mixing may contribute significantly to basin-wide averages (Kunze et al., 2012). Open questions here include what percentage of the total available internal tide energy dissipates where waves hit the continental slope, how that mixing is distributed both laterally along the global coastline and vertically, and how efficiently such boundary mixing communicates with the interior. As an example of the later issue, Garrett (1991, 2001) points out that the reduction of mixing efficiency near the bottom (essentially because you are mixing already mixed water) renders extrapolation of boundary mixing values to basin-wide averages a somewhat tricky endeavor.

4.2. Mixing in Fracture Zones

Measurements of very strong rates of turbulent mixing in deep fracture zones (FZ) place them on a short list for globally significant mixing hotspots. One of the most well-studied sites is the Romanche Fracture Zone in the equatorial Atlantic, where velocity, hydrographic, and microstructure measurements have been taken (e.g., Polzin et al., 1996a; Ferron et al., 1998). There, some of the largest abyssal turbulence levels ever observed were measured ($\epsilon > 10^{-5}\,\mathrm{W\,kg^{-1}}$), resulting from the strong northward flow of Antarctic Bottom Water descending over a series of sills (see also Chapter 19 for discussion of flows through abyssal channels). Spatial patterns in the turbulence associated with flow through the Romanche FZ suggest that a significant portion of the observed energy dissipation is associated with hydraulic jumps occurring at several sills. While the total area of elevated mixing is small, Polzin et al. (1996b) argue the net impact in terms of diapycnal buoyancy flux is equivalent to a diffusivity of $10^{-5}\,\mathrm{m^2\,s^{-1}}$ acting over a $1.5 \times 10^6\,\mathrm{km^2}$ region. The Romanche FZ thus stands as perhaps the single most important water mass conversion pathway for AABW in the Atlantic (Bryden and Nurser, 2003). Elevated levels of mixing have also been inferred at the Samoan Passage, an analogous "choke point" for bottom water entering the Pacific (Roemmich et al., 1996). Exceptionally strong turbulent mixing associated with accelerated sill flows within the Samoan Passage was recently confirmed by direct microstructure observation (M. Alford, G. Carter and J. Girton, personal communication 2012).

Similarly strong mixing was observed in the Atlantis II FZ, which is the main passageway for northward flow of Lower Circumpolar Deep Water and Antarctic Bottom Water across the Southwest Indian Ridge into the main Indian Ocean Basin (Donohue and Toole, 2003). MacKinnon et al. (2008) measure a net 3 Sv northward transport of deep and bottom water. They estimate dissipation rates and diffusivities up to and above $10^{-3}\,\mathrm{m^2\,s^{-1}}$ in the bottom 2 km of the water column, in and below the main northward jet (Figure 7.4). Unlike the Romanche measurements, the MacKinnon et al. (2008) measurements were taken well downstream of the entrance sill and hence the observed mixing is likely not due to hydraulic effects (although other data suggest hydraulic features are likely present upstream). Instead, the pattern of increasing diffusivity with depth was consistent with an observed increase in finescale shear with depth (likely due to internal waves), above the background level expected by the Garrett and Munk canonical spectrum. They hypothesize that the shear associated with the mean flow may provide critical layers to enhance internal wave breaking.

Turbulence and mixing levels at several other smaller FZs have also been studied. This includes an unnamed FZ along Mid-Atlantic Ridge of the Brazil Basin (St Laurent et al., 2001), where high spatio-temporal measurements indicated evidence for mixing by externally forced turbulence (i.e., with an energy source other than the near-bottom flow alone). In the case of the Brazil Basin, the depth-integrated dissipation rates were observed to modulate with the spring–neap cycle (St Laurent et al., 2001). This modulation was not observed in the near-bottom turbulence levels alone, ruling out that the signal was caused by frictional processes or mixing near sills, but instead pointing to the internal tide as a mechanism for providing a source of turbulence into the abyssal interior. Thurnherr et al. (2005) examined the details of the turbulent events observed in the Brazil Basin FZs and found that within the canyons, the largest dissipation levels were likely related to flows that had accelerated over sills (Figure 7.4). Their analysis suggests that sill-related mixing contributes at least as much to the diapycnal buoyancy flux in the canyon as tidally forced internal wave breaking.

Other mid-ocean ridge features, such as those associated with rift valleys, have also been found to be characterized by strong near-bottom flows and elevated turbulence levels. The Lucky Strike site of the Mid-Atlantic Ridge (St Laurent and Thurnherr, 2007) is a narrow passage feature within the rift valley. There, the largest near-bottom dissipation rates were again associated with hydraulically controlled flow over sills. As was the case in the Romanche FZ, dissipation rates as large as $\epsilon = 10^{-5}\,\mathrm{W\,kg^{-1}}$ were found to characterize the area just downstream of the sill, with associated diffusivities of $3 \times 10^{-2}\,\mathrm{m^2\,s^{-1}}$. But unlike the Romanche FZ, which sits at nearly 5000-m depth, the Lucky Strike site is at 2000 m, allowing the elevated turbulence levels there

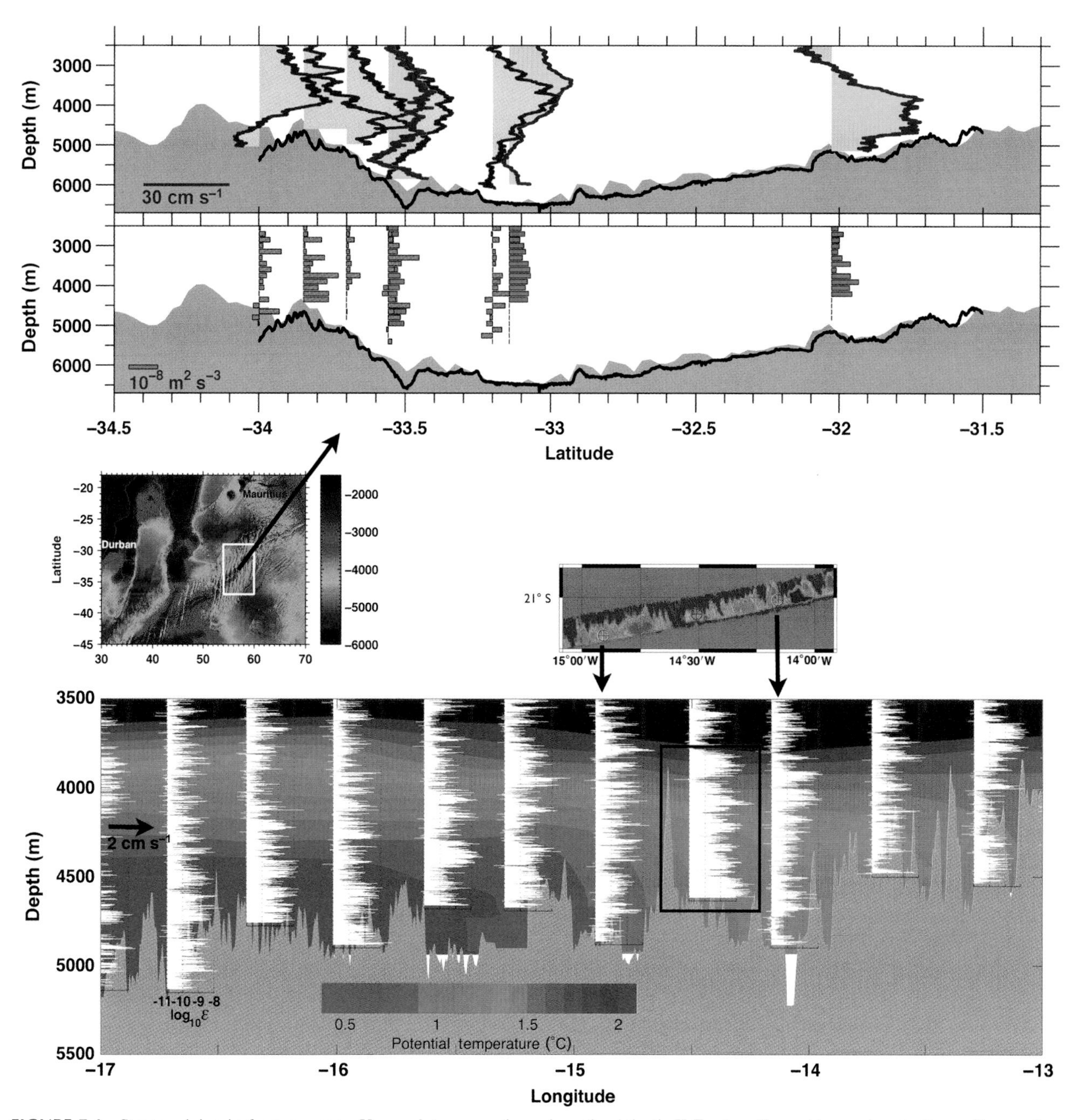

FIGURE 7.4 Strong mixing in fracture zones. Upper plots are sections along the Atlantis II Fracture Zone. Along-axis velocity profiles are from MacKinnon et al. (2008) and a CLIVAR 2009 repeat transect (J. Swift and G. Johnson, personal communication) shown in the upper subpanel in blue and red, respectively, and the corresponding dissipation rate estimates in the lower subpanel. Lower panel: multibeam bathymetry of the unnamed FZ canyon studied by St Laurent et al. (2001) and Thurnherr et al. (2005). The locations of three microstructure profiles are shown upstream, just downstream, and further downstream of a prominent sill (boxed area). Potential temperature layers are indicated by the color shading in the background. Turbulent dissipation rates (W/kg) are shown, with reference axis indicated on the profile on the left.

to provide mixing at the level of the North Atlantic Deep Water (NADW). Given the exceptional levels of turbulence and mixing at Lucky Strike, St Laurent and Thurnherr (2007) suggest that in bulk such sites may be significant contributors to basin averaged diapycnal mixing of mid-depth water masses such as NADW. Based on an analysis of bathymetric data for the region between 20° and 60°N, they speculate that flow through rift valley passages may contribute up to 50% of the mixing along the 2000-m-depth level in this region. Given the relative dearth of measurements in such "commonplace" fracture zones, their relative contribution to total mixing for deep isopycnals, and the ultimate power source for this mixing, remain important open questions.

4.3. Mesoscale Dissipation as a Source of Turbulent Mixing

The global mesoscale energy budget is surprisingly uncertain. Energy is input into large-scale motions primarily by the winds, with a global integral around 1 TW (Ferrari and Wunsch, 2009). Resultant large-scale currents decay through baroclinic or other instabilities into mesoscale eddies. Wunsch and Ferrari (2004) and Ferrari and Wunsch (2009) discuss several candidate mechanisms that may dissipate mesoscale energy, including bottom drag, loss of balance into ageostrophic motions, generation of internal lee waves, and suppression by wind work. Each of these mechanisms could potentially dissipate between 0.1 and 1 TW. The relative importance of each of these processes, and the associated amount of turbulent mixing produced by each, is as yet unclear. Some evidence points to enhanced dissipation in western boundary currents (Zhai et al., 2010). Several examples of ageostrophic dissipation processes in the upper ocean are discussed in Section 3.4, while two mechanisms relevant to the deep ocean are described below.

Internal lee-wave generation provides a complementary mechanism to internal tide generation to explain observed enhanced turbulent dissipation rates over rough topography. The impingement of the ocean's quasi-geostrophic circulation on small-scale (with characteristic horizontal scales of 1–10 km) topographic roughness may represent a globally significant source of internal waves. Scott et al. (2011) and Nikurashin and Ferrari (2011) provide independent estimates of the global rate of internal wave generation by quasi-geostrophic flow over topography and find it to be in the range of 0.2–0.5 TW, that is, between approximately 25% and 60% of the wind work on the ocean's general circulation and the rate of mesoscale eddy generation by baroclinic instability (Wunsch and Ferrari, 2004; Ferrari and Wunsch, 2009). However, some recent evidence suggests these techniques may be significant overestimates (K. Polzin, personal communication). The Southern Ocean is the area of strongest predicted lee-wave generation, as will be discussed in Section 4.4. Related work by Dewar and Hogg (2010) demonstrate that, when geostropically balanced features interact with topography, a suite of unbalanced motions can result, many being associated with enhanced turbulent dissipation.

Direct interaction between mesoscale eddies and internal waves in the stratified interior of the ocean may represent another important pathway from the mesoscale to turbulent mixing. Yet there are few theories and even fewer observations of the process. Common types of interaction discussed in the literature include an internal wave propagating in a sheared flow that encounters a critical layer (Winters and D'Asaro, 1994; Kunze et al., 1995) or internal waves trapped in the lowered effective vorticity of a horizontally sheared flow (Kunze, 1985; Rainville and Pinkel, 2004). Both of these phenomena involve wave refraction in relatively simple symmetric mean flows without substantial transfer of energy from the mesoscale flow to internal waves. In contrast, recent works by Buhler and McIntyre (2005) and Polzin (2010) consider the more general case of a fully three-dimensional mesoscale flow. They show that in a mesoscale flow with significant horizontal strain, internal waves can strongly refract and become trapped, a situation Buhler and McIntyre (2005) dub "wave capture." Conceptually, a wavepacket of almost any initial wavenumber orientation will rotate to align with the axis of maximum strain, and phase lines will be squeezed together in a similar way to contours of a passive scalar. Both horizontal and vertical wavenumbers exponentially grow until the wave presumably breaks. Polzin (2010) shows suggestive observational evidence in support of this theory and describes the effective viscous drag on the mesoscale as characterized by a horizontal viscosity of 50 $m^2 s^{-1}$. Ferrari and Wunsch (2009) globally extrapolate Polzin's results to produce a net mesoscale power loss of 0.35 TW, with most of the resultant diapycnal mixing presumably taking place in the upper kilometer of the ocean in regions of high mesoscale kinetic energy, consistent with some of the findings of Whalen et al. (2012). Thomas (2012) point out that, in environments with order 1 Rossby number (e.g., frontal regions), secondary circulations arise, making the nature of the nonlinear interaction more complex.

4.4. In-Depth Example: Southern Ocean Mixing (see also Chapter 18)

Both Scott et al. (2011) and Nikurashin and Ferrari (2011) highlight the Southern Ocean as the most prominent region of internal lee-wave generation in the world ocean (Section 4.3). There, the deep-reaching, multi-jet Antarctic Circumpolar Current (ACC) flows over small-scale topographic roughness associated with numerous ridges and plateaus in each of the major ocean basins. The consequent internal wave activity in the form of lee waves has been inferred in measurements of velocity finestructure downstream of the Kerguelen Plateau (Polzin and Firing, 1997) and in the Scotia Sea (Naveira Garabato et al., 2004). At each of these sites, finestructure evidence suggested the turbulence levels were substantially enhanced due to the breaking of internal lee waves. High rates of diapycnal mixing for the Southern Ocean as a whole have been suggested by inverse studies of the circulation (Heywood et al., 2002; Lumpkin and Speer, 2007; Zika et al., 2009), with deep ocean average diffusivity levels of $\kappa_\rho \sim 10^{-4}$ $m^2 s^{-1}$. However, these estimates are indirect, being based mainly on mass balance or the structure of the large-scale thermohaline fields.

More targeted studies have examined the Kerguelen Plateau and the Southeast Pacific/Southwest Atlantic sectors of the Southern Ocean, where the ACC flows around and over complex topography, with direct measurements of internal waves and turbulence. Waterman et al. (2013a) present microstructure measurements in the ACC, in the context of the Kerguelen Plateau's northern flank. They find a systematic enhancement of dissipation rates above background levels ($\epsilon \sim 10^{-9}\,\mathrm{W\,kg^{-1}}$) in the upper 1000–1500 m of the water column, and elevated dissipation and mixing rates ($\epsilon \sim 10^{-9}\,\mathrm{W\,kg^{-1}}$, $\kappa \sim 10^{-4}\,\mathrm{m^2\,s^{-1}}$) in deep ocean sites where the ACC jets impinge on complex small-scale topography. In several of these cases, there is a noticeable discrepancy between microstructure and finescale parameterizations of dissipation that may be related to the dynamics of wave-mean-flow interaction (Section 5.1).

Finestructure estimates in the Drake Passage include those described by Naveira Garabato et al. (2004), who inferred very large dissipation and diffusivity levels throughout the Drake Passage. However, other studies using the same parameterization as part of larger-scale Southern Ocean (Sloyan, 2005) and global ocean (Kunze et al., 2006) examinations inferred somewhat more modest Drake Passage mixing levels, with local enhancements generally confined to depths below 1500 m. Some of the discrepancy can be attributed to difference in implementation of the finescale parameterization (Section 5.1). In the case of Sloyan (2005), the enhanced mixing levels were concentrated into the frontal zones. Another Drake Passage study, Thompson et al. (2007), focused on the upper 1000 m of the water column and examined vertical overturns implied by inversions in temperature and density profile data from expendable instruments (XCTDs and XBTs). That work reported diffusivity levels implied by Thorpe Scales (Dillon, 1982) reaching $\kappa = 10^{-3}\,\mathrm{m^2\,s^{-1}}$, and a strong seasonal cycle to the mixing.

The first direct measurements of turbulence levels in the Drake Passage were made in 2010 as part of the Diapycnal and Isopycnal Mixing Experiment in the Southern Ocean (DIMES). DIMES is a joint tracer release and microstructure sampling experiment, focused on examining the spatial variation in mixing levels as a mid-depth tracer cloud evolves as it passes from the Pacific through the Drake Passage into the Scotia Sea (Ledwell et al., 2011). During the first year of the experiment, the tracer injected on the 27.9-kg $\mathrm{m^{-3}}$ neutral density surface evolved from a very small highly concentrated patch to an O(1000 km) cloud in the Pacific sector just upstream of Drake Passage. The vertical (diapycnal) diffusivity acting in the cloud was $\kappa = 1.3 \times 10^{-5}\,\mathrm{m^2\,s^{-1}}$, consistent a mean dissipation level of $10^{-10}\,\mathrm{W\,kg^{-1}}$ measured by microstructure sampling. These mixing levels are the same order of magnitude as background mixing in the mid-latitude thermocline and seem to suggest elevated mixing in the Southern Ocean is not as widespread as some previous studies have predicted.

Within Drake Passage, measurements were also made near the Phoenix Ridge. This mid-ocean ridge site is the first of a series of significant topographic regions that the ACC passes on its eastward path into the Atlantic. Measurements described by St Laurent et al. (2012) show enhanced levels of turbulence in the frontal zones (Figure 7.5a).

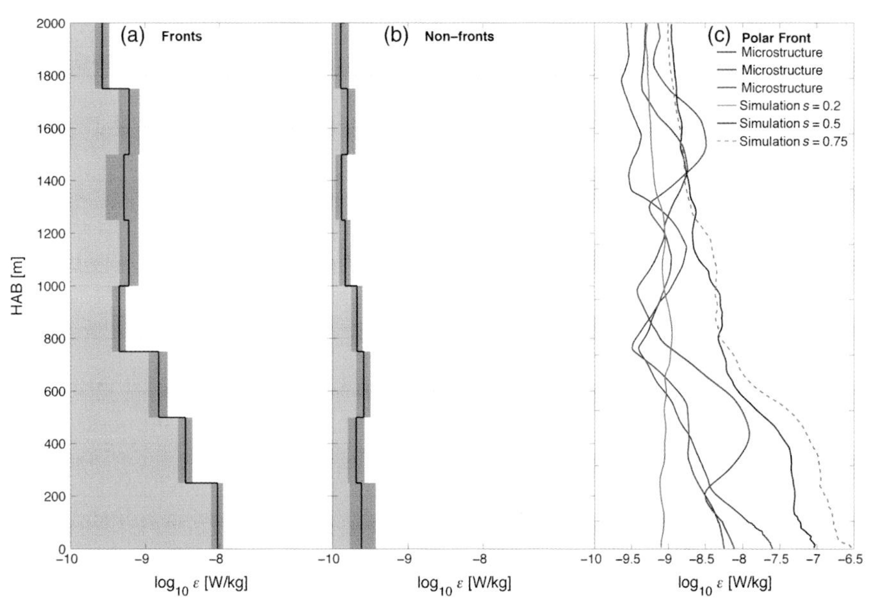

FIGURE 7.5 Average height-above-bottom (HAB) profiles of turbulent kinetic energy dissipation rates of frontal (a) and nonfrontal regions (b) as estimated from the microstructure data from Drake Passage surveys done at and just upstream of the Phoenix Ridge (St Laurent et al., 2012). These represent ensemble means using 250-m HAB bins from approximately 12 profiles each. In each bin, the line denotes the mean, and the shading about the mean indicates the 95% confidence interval. Lighter gray shading fills the gap between the estimates and the oceanic background dissipation rate level of $10^{-10}\,\mathrm{W\,kg^{-1}}$. Individual turbulent dissipation rate profiles (colored curves) and numerical simulations (Nikurashin and Ferrari, 2010; gray curves) from the Polar Front above the Phoenix Ridge are also shown (c). Measured profiles span the Polar Front from south (blue) to north (red). The numerical simulations were done using several values of a steepness parameter(s) found to characterize the finescale steepness of the ridge topography, and hence the spectral character of the generated lee waves.

Turbulent dissipation rates exceed 1×10^{-8} W kg^{-1} at heights-above-bottom (HAB) reaching 1000 m, supporting diffusivities from 1×10^{-4} to 1×10^{-3} m^2 s^{-1}. These elevated mixing rates decay to background rates above 2000 m, suggesting that the energy in the deep internal wave field is locally driving turbulence only to mid-depth. Outside the frontal zones, turbulence levels show no enhancement (Figure 7.5b), indicating that without a deep-reaching current, there is no mechanism to generate lee waves. In the specific case of the Polar Front at the Phoenix Ridge, observed turbulent dissipation rates and depth structure show some similarity to the numerical simulations of Nikurashin and Ferrari (2010) (Figure 7.5c). They found that near-bottom inertial oscillations accompany the generation of lee waves, leading to instability and enhanced dissipation near the bottom. Dissipation profiles from the simulations (black and gray curves, shown for simulations according to internal wave/topographic steepness ratios (s)) show similar depth-dependent structures to measured profiles in frontal zones (left panel), with the general decay of dissipation levels with height. However, the magnitude of the Nikurashin and Ferrari (2010) prediction appears to be a significant overestimate (Figure 7.5 and Waterhouse et al., 2013).

While most turbulent dissipation measurements are limited to the Austral summer season, inferred diffusivities calculated from EM-APEX floats (Ledwell et al., 2011) and repeat hydrography (Thompson et al., 2007) show mixing rates are somewhat elevated in the winter. Given that the strength of the near-bottom flow incident on bathymetric features appears to be the most critical indicator of mixing intensity via lee-wave processes, eddy variability of the ACC may dictate the temporal variability of mixing in the deep Southern Ocean rather than seasonal forcing. Yearlong moorings were recently deployed as part of the DIMES experiment—analysis of that data over the next few years will hopefully shed some light on the dynamics and variability of energetic mixing in the Drake Passage specifically and the Southern Ocean more generally.

5. DISCUSSION

5.1. Finescale Parameterizations of Turbulent Mixing

Over the past two decades, a "finescale parameterization" of turbulent dissipation and diffusivity has been developed that combines observations of internal wave energy levels with theoretical models of turbulence as controlled by wave–wave interaction rates (Section 4.1.3; Polzin et al., 1995; Gregg et al., 2003; Kunze et al., 2006; Polzin et al., 2013). As this method has gained increasingly widespread use in recent years, we feel it deserves a few comments here. The basic idea is that in a steady state the rate of downscale energy transfer through a broadband internal wave continuum by wave–wave interactions can be equated to the dissipation rate at small scales. The rate of downscale energy transfer can be estimated using properties of internal wave strain or shear measured at vertical scales of the order 10–100 m, raising the tantalizing possibility that mixing in the ocean could be observed (however crudely) using a much larger variety of instruments than specialized microstructure sensors.

Most formulations are based on the empirically derived Garrett–Munk (GM) vertical wavenumber spectra of internal wave shear and strain, both of which are nearly white (flat) at larger scales and then drop off with a -1 slope beyond a cutoff wavenumber (k_c in Figure 7.6) (Gregg and Kunze, 1991). Physically, motions at scales larger than the cutoff (smaller wavenumbers) are interpreted as weakly nonlinear internal waves, while motions at smaller scales become more strongly nonlinear, eventually leading to wave breaking (D'Asaro and Lien, 2000). For the empirically derived GM spectrum, the transition occurs at a wavelength of $2\pi/k_c = 10$ m. For other observations, the cutoff appears to move toward lower wavenumbers with higher spectral energy levels (Gargett, 1990). Polzin et al. (1995) interpret this as the internal wave field maintaining a constant Richardson number, which is related to the wave-turbulence transition point suggested by D'Asaro and Lien (2000).

The rate of downscale energy transfer through the weakly nonlinear range, and thus the dissipation rate, tends to scale quadratically with the spectral level ($\hat{E}$), a scaling consistent between theory (McComas and Müller, 1981; Henyey et al., 1986; Müller et al., 1986; Lvov et al., 2004), observations (Gregg, 1989; Polzin et al., 1995; Gregg et al., 2003), and numerical simulations (Winters and D'Asaro, 1997). Henyey et al. (1986) physically interpret this transfer rate as the rate at which small-scale waves are being refracted toward dissipative scales by interaction with larger-scale shear. Following Gregg et al. (2003), Kunze et al. (2006), and Polzin et al. (2013), the dissipation rate can be written as

$$\epsilon = \epsilon_0 \left(\frac{N}{N_0}\right)^2 \hat{E}^2 L(R_w,\theta) \qquad (7.2)$$

where $\hat{E}$ is a measure of the observed internal wave spectral level integrated out to k_c, R_w is the shear-to-strain ratio, which provides a measure of the average frequency content of a wavefield, and θ is latitude. $\hat{E}$ is typically calculated from vertical profiles of either shear or strain, which give estimates of kinetic and potential energy, respectively. The $L(R_w, \theta)$ term includes the theoretical dependence on downscale energy transfer rate on both average wavefield frequency content (through R_w) and latitude (Polzin et al., 1995; Gregg et al., 2003).

In ideal circumstances, both shear and strain are measured at vertical resolution comparable to the cutoff

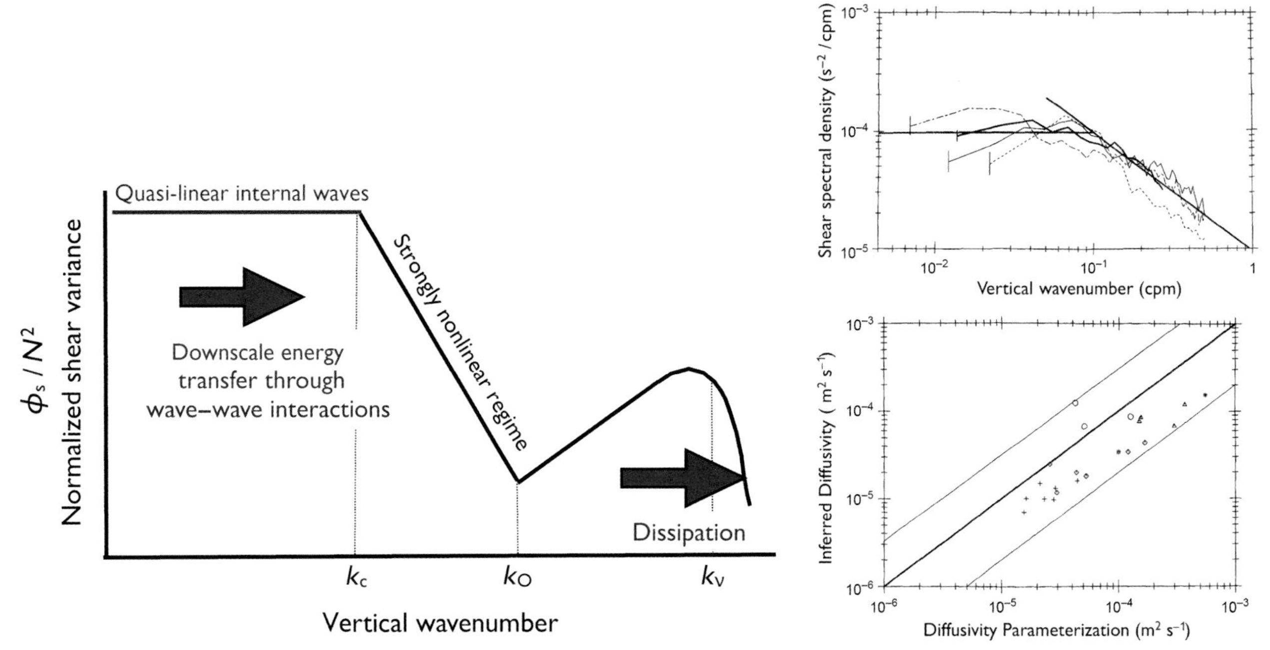

FIGURE 7.6 Left: sketch of idealized vertical wavenumber spectra of stratification normalized shear showing steady-state spectral shapes for the internal wave regime (low wavenumbers/large vertical scales), the transition range, and the turbulent subrange at high wavenumbers/small vertical scales. Wavenumbers indicated on the x-axis correspond to the Kolmogorov scale (k_v), the Ozmidov scale (k_O), and the edge of the quasi-linear internal wave regime (k_c). The blue arrows schematically indicate the direction of energy transfer from large to dissipative scales. Right: several observed vertical wavenumber spectra (upper) and comparison of diffusivity inferred from the finescale method to that inferred from microstructure data (lower). *Right panels are reproduced from Polzin et al. (1995).*

wavelength, approximately 10 m. Realistically, many observations are limited in one way or another and a modified version of Equation (7.2) is used. For example, the Lowered ADCP data used by Kunze et al. (2006) is noisy at scales smaller than about 50 m. They thus calculate $\hat{E}$ by integrating out measured spectra to the highest non-noisy wavenumber, typically much lower than the "real" k_c would be, which can yield biased results. Other studies attempt to apply the method using measurements of either shear or strain alone, with an assumed valued of R_w (e.g., Wijesekera et al., 1993), which can also bias results in regions where wave frequency varies.

In addition to measurement limitations, finescale parameterizations may be inappropriate where the underlying physics is not as assumed by theory. For example, the type of directly breaking internal tides observed by both Klymak et al. (2008) and Alford et al. (2011) require no spectral cascade and are not best described by a finescale model. Recent observations in the Southern Ocean also show a systematic elevation of finescale mixing estimates compared to those from a microstructure profiler (St Laurent et al., 2012; Waterman et al., 2013a). Waterman et al. (2013b) suggest that wave-mean flow interactions may produce a different relationship between spectral variance levels and the rate of downscale energy transfer than predicted by the underlying theory. The method also fails in more subtle ways where downscale energy transfer is influenced by scattering from topography (Kunze et al., 2002) or in shallow water (MacKinnon and Gregg, 2003).

Nevertheless, the method often produces values (Figure 7.6) and patterns that are qualitatively consistent with the mixing dynamics described above. Some examples of large-scale mixing patterns estimated using the finescale technique are shown in Figure 7.7 and discussed in Section 5.2. As the "cottage industry" of finescale measurements and methods continues to grow, further detailed comparison with microstructure measurements will help further constrain the method's accuracy in different environments.

5.2. Global Values and Patterns

The plethora of diapycnal mixing processes described above suggest a complex and evolving geography for mixing in the global ocean, of which microstructure observations sample an incredibly small portion. Most early microstructure measurements were limited to the upper ocean (St Laurent and Simmons, 2006). Deep profiling, particularly to depths greater than 2000 m, did not become common practice until the 1990s (Toole et al., 1994). Waterhouse et al. (2013) have compiled a significant percentage of available microstructure in the open ocean, the locations of which are shown in Figure 7.7a by red diamonds. Though the average microstructure-measured diffusivity below the main thermocline is on the order of $10^{-4}\ m^2\ s^{-1}$ (St Laurent and

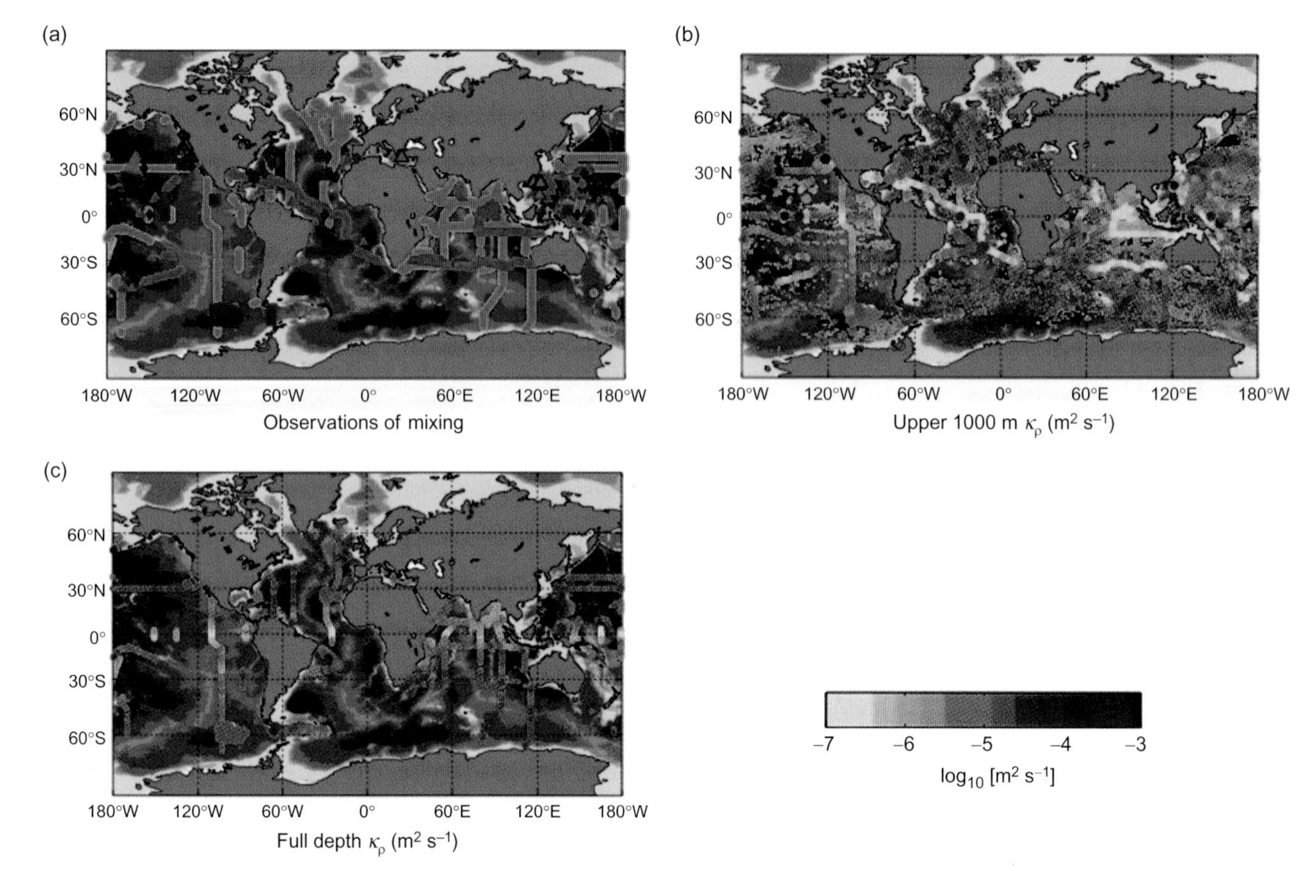

FIGURE 7.7 (a) Compiled observations of direct and indirect mixing measurements with red squares denoting microstructure measurements. Red ◇ are historical published microstructure measurements, green △ represent diffusivities calculated from ship board shear, yellow △ are inferred diffusivities from LADCP/CTD profiles of Kunze et al. (2006), and magenta △ are diffusivities calculated from overturns of density profiles from moored profilers. Depth averaged diffusivity, κ_ρ, plotted as $\log_{10}$ [m^2 s^{-1}] from (b) the upper ocean (down to 1000 m) and (c) from the full-water column. Background diffusivity map in (b) comes from the strain-based inferences of diffusivity of Whalen et al. (2012) from Argo floats. *Reproduced from Waterhouse et al. (2013).*

Simmons, 2006; Waterhouse et al., 2013), estimates from different locations vary considerably reflecting the different dominant processes. Though the data are sparse, Waterhouse et al. (2013) compare available depth-integrated dissipation rates from microstructure to the global map of power input into the internal wave field and conclude that the sampling locations, taken as a set, are not especially biased toward energetic or quiet locations—in other words, that the average observed 10^{-4} m^2 s^{-1} value is a reasonable one.

When mixing estimates from the finescale parameterization are included, a clearer pattern of mixing begins to emerge in both the upper (Figure 7.7b) and deep ocean (Figure 7.7c). Most of the data in the upper ocean come from Whalen et al. (2012), who apply the finescale strain parameterization to 3 years of Argo profiles. They find strong correlations between elevated dissipation rates and topographic roughness (Section 4.1.1), mixed-layer inertial energy (Section 4.1.2), and mesoscale eddy kinetic energy (Section 4.3), and demonstrate a strong seasonal cycle to mixing in regions with energetic near-inertial motions (Section 3.2). Data are sparse in the deep ocean—much of what are shown in Figure 7.7c comes from Kunze et al. (2006). They also find systematically elevated mixing above rough topography. Though the finescale parameterization comes with significant uncertainty and possibly systematic biases (Section 5.1), the coverage afforded provides geographical guidance for both future fieldwork planning and preliminary attempts to incorporate mixing patterns into large-scale models (Section 5.3).

5.3. Representing Patchy Mixing in Large-Scale Models: Progress and Consequences

General circulation models used for climate research parameterize the impact of subgridscale processes, including turbulent mixing, because of their necessarily limited resolution. The current implementations generally do not include most of the spatial variability of mixing patterns described above (see Simmons et al., 2004a; Jayne, 2009; Griffies et al., 2010, for discussion of recent work). Below the surface mixed layer, climate models used in the IPCC-AR4 assessment (2007) employ a combination of a simple Richardson number parameterization for diffusivity and a horizontally uniform background diffusivity profile such as that suggested by Bryan and Lewis (1979) which crudely replicates the observed increase of diffusivity with depth. The Richardson number dependent components (Pacanowski and Philander, 1981; Large et al., 1994) are necessary for reasonable representation of large-scale shear flows such as the Equatorial Undercurrent, but rely on the resolved Richardson number to predict mixing. For almost all the processes described above, regional or global scale models will never be able to explicitly represent the motions with critical Richardson numbers. Often the limiting factor is horizontal rather than vertical resolution, as the nonlinear motions that directly lead to turbulence, such as breaking internal waves, have horizontal scales of kilometers or less.

Development of parameterizations that represent the full geography of diapycnal mixing is essential, as evidence is accumulating that patchy mixing can have significant consequences for global circulation patterns. Since both the magnitude and distribution of diapycnal mixing are likely to change in a future climate (as, e.g., wind stress patterns evolve), accurate prediction of future or past climate requires development of parameterizations of turbulent mixing that are based on appropriate physics. In the past decade, a spate of studies have shown that many features of global ocean circulation are sensitive to the distribution of diapycnal mixing, thorough reviews of which can be found in Jayne (2009) and Friedrich et al. (2011).

The U.S. Climate Variability and Predictability (CLIVAR) Program recently established a series of Climate Process Teams (CPTs) to develop and implement parameterizations for unresolved processes in climate models. One of these CPTs focused on mixing and entrainment in overflows, with results described in Legg et al. (2009) and Danabasoglu et al. (2010). A combination of data analysis, theory, and idealized numerical modeling led to, among other things, the development of an improved formulation for mixing related to shear instability that allowed for vertical transport of turbulence over a larger region than just that of unstable local Richardson number (Jackson et al., 2008). Implementation of this new scheme in global models improved representations of both deep overflows and other strongly sheared flows such as the Pacific Equatorial Undercurrent (Legg et al., 2009).

The first substantial attempt to parameterize internal wave-driven mixing was to represent the nearfield part of internal tide dissipation, which is the portion of generated internal waves that break near rough topography at which they are created. The result is a global map of dissipation that mirrors that of internal tide generation spots, with most of the elevated mixing at depth (St Laurent et al., 2002). The parameterization is essentially given by the product of power going into internal tides, itself a function of topographic roughness, barotropic tidal strength and deep stratification, and an empirically derived vertical decay scale to represent the observed enhancement of turbulent mixing at depth. This scheme is now implemented in the Community Climate System Model (CCSM4) of NCAR (Jayne, 2009; Gent et al., 2011), the Modular Ocean Model of GFDL (Simmons et al., 2004a), and GFDL's Generalized Ocean Layer Dynamics model (GOLD). The parameterization appears to significantly modify circulation patterns, particularly by enhancing the deep limb of the MOC (Figure 7.8; Jayne, 2009). An updated version with a vertical structure function based on wave–wave interaction dynamics as described in Polzin (2009) is now being tested at GFDL with promising results (Melet et al., 2013).

More generally, heterogeneous diapycnal fluxes in the abyssal ocean rule out the Stommel–Arons conceptual

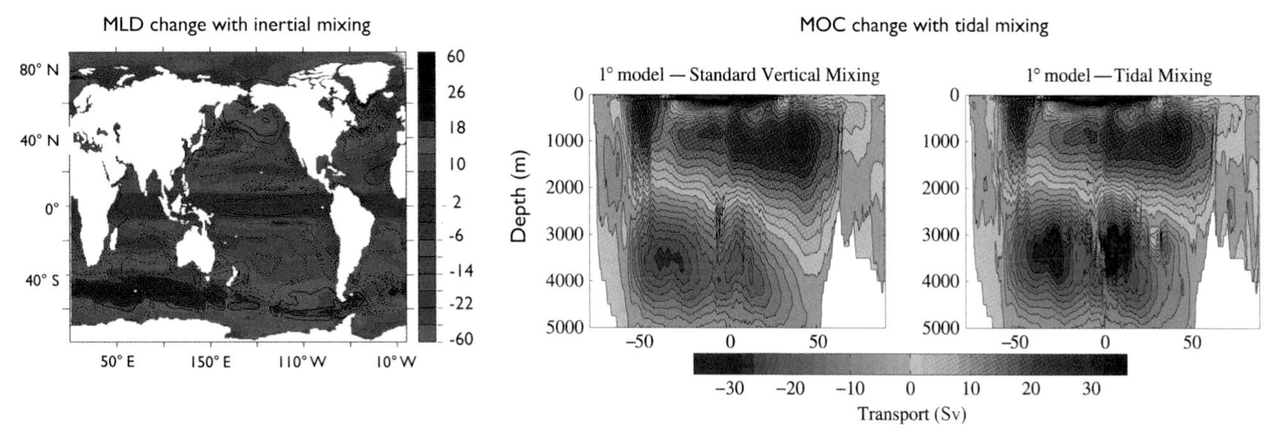

FIGURE 7.8 Examples of the sensitivity of large-scale circulation patterns to diapycnal mixing parameterizations in the upper and deep ocean. Left: change in mixed-layer depths, in meters, when near-inertial motions are explicitly represented in a global model. Right: snapshots of the modeled MOC using a standard parameterization for diapycnal diffusivity and one that includes elevated mixing over rough topography. *Left panel is reproduced from Jochum et al. (2013). Right panel is reproduced from Jayne (2009).*

picture, in which deep diapycnal upwelling stretches water columns and leads to uniform poleward flow to conserve potential vorticity (Stommel and Arons, 1960). Instead, isolated mixing hotspots should lead to limited regions of meridional flow (Samelson, 1998; Huang and Jin, 2002; Katsman, 2006; Emile-Geay and Madec, 2009), as indicated in observations (Davis, 1998; Hogg and Owens, 1999; St Laurent et al., 2001). Furthermore, certain profiles of bottom-enhanced turbulent buoyancy fluxes can actually lead to local diapycnal downwelling, with consequent substantial changes in the abyssal circulation (Simmons et al., 2004a; Saenko and Merryfield, 2005). A variety of recent modeling studies show that everything from the strength of the MOC to the deep ocean stratification to the distribution of passive tracers respond to changing patterns of imposed deep diffusivity (Hasumi and Suginohara, 1999; Scott and Marotzke, 2002; Gnanadesikan et al., 2004; Simmons et al., 2004a; Saenko and Merryfield, 2005; Palmer et al., 2007; Friedrich et al., 2011). Though the nature of the circulation changes is not entirely consistent between models, the sensitivity to mixing patterns is a persistent feature of such experiments, making an improved understanding and parameterization of the processes described in this chapter essential.

The geography of upper ocean mixing also has significant impact on circulation, water properties, fluxes of heat, dissolved greenhouse gasses, and biologically essential nutrients (Harrison and Hallberg, 2008). For example, an initial attempt at representing mixing in the upper ocean from near-inertial motions is described by Jochum et al. (2013). Their first step was to increase the frequency of ocean-atmosphere coupling to every 2 h, which allowed generation of an energetic field of near-inertial motions in the mixed layer. Their second step was to parameterize mixing due to unresolved vertically propagating near-inertial internal waves using an *ad hoc* vertical decay scale. The results suggest that NIWs lead to a 20–50% deeper ocean mixed layer under the storm tracks and the trade winds, largely from inertial shear at the mixed-layer base (Figure 7.8b). Of particular note, the tropical deepening leads to a cooler SST and a substantial shift in global precipitation, sea level pressure, and the resulting surface winds. Upper ocean mixing is also crucial for supplying nutrients to the euphotic zone, and changes in the rate and pattern of diapycnal nutrient fluxes may have significant effects on primary production rates (Gnanadesikan et al., 2002).

While these attempts are a good start, only a fraction of the internal wave energy available for mixing is represented. So far, unaccounted for are about 2/3 of a TW in the low-mode internal tides, most of the power in near-inertial waves, high-frequency breaking waves in the upper ocean, and lee waves in the Southern Ocean. Attempts are underway as part of a new CPT to develop and refine related parameterizations for diapycnal mixing related to internal wave breaking. However, for many of the mixing processes described in this chapter, the relevant physics is not yet well-enough understood.

6. SUMMARY AND FUTURE DIRECTIONS

The search for observations confirming the expected levels of mixing in the ocean interior has revealed enormous geographical variability and an incredibly rich range of turbulent processes. Overall there appears to be approximately the same power available to turbulent mixing as is required to drive the deep overturning circulation. Over the past decade, emphasis has moved toward an appreciation of the complex patterns of mixing in space and time and the diverse range of associated dynamics. The next step is to tease apart how the specific geography of mixing (Figure 7.7) is compatible with the details of water mass transformation in individual basins. For example, Huussen et al. (2012) look at the energy budget for the Indian Ocean equatorward of 32°S. They find that while there is overall consistency between the power available in the internal wave field and that required for water mass transformation by inverse models, there is a discrepancy in depths. In particular, they find that mixing at mid-depths inferred by a finescale parameterization (1000–3000 m) is not sufficient to produce the required water mass transformations. Further efforts to approach such problems from all angles are clearly needed. Though there is no shortage of pressing open questions, we find the following one particularly intriguing:

- Where does the low-mode internal wave energy seen crossing ocean basins dissipate? What percentage of the energy dissipates steadily as waves propagate and what percentage dissipates when waves hit continental slopes? If the latter is large, what are the implications for regional and basin-wide circulation patterns?
- What percentage of mesoscale energy is dissipated through irreversible mixing in the stratified ocean interior? What are the dominant processes? Is the dissipation primarily in the deep or upper ocean?
- What is the role of strong, often hydraulically controlled mixing near fracture zones or other deep rough topography? Can the resultant mixed fluid be exported to form a significant percentage of diapycnal mass transport across certain isopycnals?

Moving into the next decade, we expect that new observations will continue to be a primary driver of progress. Microstructure sensors are moving beyond the traditional vertical profilers onto a range of platforms, including fixed-point moorings (Moum and Nash, 2009), autonomous underwater vehicles (AUVs) and gliders (Wolk et al., 2009), horizontally towed vehicles, and even onto the CTD rosette (J. Nash, personal communication). In situations where dissipation rates are set by the rate of energy cascade from large to small scales, mixing may be inferred from finescale measurements of the internal wave field or direct measurements of turbulent overturns (Sections 2 and 5.1), both of which can be made by a growing variety of ship-based and autonomous instruments such as Argo floats. Finally, increasingly high-resolution models are proving invaluable for looking at everything from the details of turbulent instabilities at the smallest scales (Smyth and Moum, 2012; Venayagamoorthy and Fringer, 2012) to global patterns of internal wave propagation and destruction (Simmons, 2008; Arbic et al., 2012). We have every expectation that the next decade of ocean mixing research will bring new surprises.

ACKNOWLEDGMENTS

We are very appreciative for the detailed and thoughtful comments of nine reviewers, which have considerably improved the perspective and precision of the text. J. A. M and L. St L. are participants in an NSF- and NOAA-funded Climate Process Team, within which many relevant discussions have occurred.

REFERENCES

Alford, M.H., 2001. Internal swell generation: the spatial distribution of energy flux from the wind to mixed layer near-inertial motions. J. Phys. Oceanogr. 31, 2359–2368.

Alford, M.H., 2003. Energy available for ocean mixing redistributed though long-range propagation of internal waves. Nature 423, 159–163.

Alford, M.H., 2008. Observations of parametric subharmonic instability of the diurnal internal tide in the South China Sea. Geophys. Res. Lett. 35, L15602. http://dx.doi.org/10.1029/2008GLO34,720.

Alford, M.H., 2010. Sustained, full-water-column observations of internal waves and mixing near Mendocino Escarpment. J. Phys. Oceanogr. 40, 2643–2660.

Alford, M., Gregg, M., 2001. Near-intertial mixing: modulation of shear, strain and microstructure at low latitude. J. Geophys. Res. 106 (C8), 16947–16968.

Alford, M., Pinkel, R., 2000. Observations of overturning in the thermocline: the context of ocean mixing. J. Phys. Oceanogr. 30 (5), 805–832.

Alford, M.H., Whitmont, M., 2007. Seasonal and spatial variability of near-inertial kinetic energy from historical moored velocity records. J. Phys. Oceanogr. 37 (8), 2022–2037.

Alford, M.H., MacKinnon, J.A., Zhao, Z., Pinkel, R., Klymak, J., Peacock, T., 2007. Internal waves across the Pacific. Geophys. Res. Lett. 34. http://dx.doi.org/10.1029/2007GL031566.

Alford, M.H., et al., 2011. Energy flux and dissipation in Luzon Strait: two tales of two ridges. J. Phys. Oceanogr. 41 (11), 2211–2222.

Alford, M.H., Cronin, M.F., Klymak, J.M., 2012. Annual cycle and depth penetration of wind-generated near-inertial internal waves at Ocean Station Papa in the Northeast Pacific. J. Phys. Oceanogr. 42 (6), 889–909.

Althaus, A., Kunze, E., Sanford, T., 2003. Internal tide radiation from Mendocino Escarpment. J. Phys. Oceanogr. 33, 1510–1527.

Arbic, B.K., Richman, J.G., Shriver, J.F., Timko, P.G., Metzger, E.J., Wallcraft, A.J., 2012. Global modeling of internal tides within an eddying ocean general circulation model. Oceanography 25 (2), 20–29.

Belcher, S.E., et al., 2012. A global perspective on mixing in the ocean surface boundary layer. Geophys. Res. Lett. 39 (18). http://dx.doi.org/10.1029/2012GL052932.

Boccaletti, G., Ferrari, R., Fox-Kemper, B., 2007. Mixed layer instabilities and restratification. J. Phys. Oceanogr. 37 (9), 2228–2250.

Bryan, K., Lewis, L., 1979. A water mass model of the world ocean. J. Geophys. Res. 84, 2503–2517.

Bryden, H.L., Nurser, A.J.G., 2003. Effects of strait mixing on ocean stratification. J. Phys. Oceanogr. 33 (8), 1870–1872.

Buhler, O., McIntyre, M., 2005. Wave capture and wave-vortex duality. J. Fluid Mech. 534, 67–95.

Capet, X., McWilliams, J.C., Molemaker, M.J., Shchepetkin, A.F., 2008. Mesoscale to submesoscale transition in the California Current System. Part ii: Frontal processes. J. Phys. Oceanogr. 38 (1), 44–64.

Carter, G.S., Gregg, M.C., 2006. Persistent near-diurnal internal waves observed above a site of M_2 barotropic-to-baroclinic conversion. J. Phys. Oceanogr. 36 (6), 1136–1147.

Chaigneau, A., Pizarro, O., Rojas, W., 2008. Global climatology of near-inertial current characteristics from Lagrangian observations. Geophys. Res. Lett. 35, L13603. http://dx.doi.org/10.1029/2008GL034060.

Cole, S.T., Rudnick, D.L., Hodges, B.A., 2009. Observations of tidal internal wave beams at Kauai Channel, Hawaii. J. Phys. Oceanogr. 39, 421–436.

Craik, A., Leibovich, S., 1976. A rational model for Langmuir circulations. J. Fluid Mech. 73, 401–426.

Crawford, G.B., Large, W.G., 1996. A numerical investigation of resonant inertial response of the ocean to wind forcing. J. Phys. Oceanogr. 26, 873–891.

Danabasoglu, G., Large, W.G., Briegleb, B.B., 2010. Climate impacts of parameterized Nordic Sea overflows. J. Geophys. Res. 115. http://dx.doi.org/10.1029/2010JC006243.

Danioux, E., Klein, P., Riviere, P., 2008. Propagation of wind energy into the deep ocean through a fully turbulent mesoscale eddy field. J. Phys. Oceanogr. 38, 2224–2241.

D'Asaro, E., 1985. The energy flux from the wind to near-inertial motions in the mixed layer. J. Phys. Oceanogr. 15, 943–959.

D'Asaro, E.A., Lien, R.-C., 2000. The wave-turbulence transition for stratified flows. J. Phys. Oceanogr. 30 (7), 1669–1678.

D'Asaro, E., Eriksen, C., Levine, M., Niller, P., Paulson, C., van Meurs, P., 1995. Upper-ocean inertial currents forced by a strong storm. Part 1: data and comparisons with linear theory. J. Phys. Oceanogr. 25, 2909–2936.

D'Asaro, E.A., Lee, C., Rainville, L., Harcourt, R., Thomas, L., 2011. Enhanced turbulence and energy dissipation at ocean fronts. Science 332, 318–322.

Davis, R.E., 1998. Preliminary results from directly measuring mid-depth circulation in the tropical and South Pacific. J. Geophys. Res. 103, 24619–24639.

Dewar, W.K., Hogg, A.M., 2010. Topographic inviscid dissipation of balanced flow. Ocean Model. 32, 1–13.

Dewar, W., Bingham, R., Iverson, R., 2006. Does the marine biosphere mix the ocean? J. Mar. Res. 64 (4), 541–561.

Dillon, T.M., 1982. Vertical overturns: a comparison of Thorpe and Ozmidov length scales. J. Geophys. Res. 87, 9601–9613.

Donohue, K., Toole, J., 2003. A near-synoptic survey of the Southwest Indian Ocean. Deep Sea Res. Part II 50, 1893–1931.

Egbert, G.D., Ray, R.D., 2001. Estimates of M_2 tidal energy dissipation from TOPEX/Poseidon altimeter data. J. Geophys. Res. 106 (C10), 22475–22502.

Elipot, S., Lumpkin, R., Prieto, G., 2010. Intertial oscillations modification by mesoscale vorticity. J. Geophys. Res. 114 (C06003). http://dx.doi.org/10.1029/2008JC005,170.

Emile-Geay, J., Madec, C., 2009. Geothermal heating, diapycnal mixing and the geothermal heating, diapycnal mixing and the abyssal circulation. Ocean Sci. 5, 203–217.

Eriksen, C.C., 1985. Implications of ocean bottom reflection for internal wave spectra and mixing. J. Phys. Oceanogr. 15, 1145–1156.

Eriksen, C., 1998. Internal wave reflection and mixing at Fieberling Guyot. J. Geophys. Res. 103, 2977–2994.

Ferrari, R., Wunsch, C., 2009. Ocean circulation kinetic energy: reservoirs, sources, and sinks. Annu. Rev. Fluid Mech. 41, 253–282.

Ferrari, R., McWilliams, J., Canuto, V., Dubovikov, M., 2008. Parameterization of eddy fluxes near oceanic boundaries. J. Clim. 21, 2770–2789.

Ferrari, R., Griffies, S.M., Nurser, A., Vallis, G., 2010. A boundary-value problem for the parameterized mesoscale eddy transport. Ocean Model. 32, 143–156.

Ferron, B., Mercier, H., Speer, K., Gargett, A., Polzin, K., 1998. Mixing in the Romanche fracture zone. J. Phys. Oceanogr. 28 (10), 1929–1945.

Forryan, A., Garabato, A.C.N., Polzin, K.L., Waterman, S.N., 2013. Rapid injection of near-inertial kinetic energy into the stratified upper ocean at an Antarctic Circumpolar Current front. J. Geophys. Res. Oceans (submitted).

Fox-Kemper, B., Ferrari, R., Hallberg, R., 2008. Parameterization of mixed layer eddies. Part i: theory and diagnosis. J. Phys. Oceanogr. 38 (6), 1145–1165.

Friedrich, T., Timmermann, A., Decloedt, T., Luther, D., Mouchet, A., 2011. The effect of topography-enhanced diapycnal mixing on ocean and atmospheric circulation and marine biogeochemistry. Ocean Model. 39, 262–274.

Furuichi, N., Hibiya, T., Niwa, Y., 2005. Bispectral analysis of energy transfer within the two-dimensional oceanic internal wave field. J. Phys. Oceanogr. 35, 2104–2109.

Ganachaud, A., Wunsch, C., 2000. Improved estimates of global ocean circulation, heat transport and mixing from hydrographic data. Nature 408, 453–457.

Gargett, A.E., 1990. Do we really know how to scale the turbulent kinetic energy dissipation rate ϵ due to breaking of oceanic internal waves? J. Geophys. Res. 95 (C9), 15971–15974.

Garrett, C., 1991. Marginal mixing theories. Atmos. Ocean 29 (2), 313–339.

Garrett, C., 2001. An isopycnal view of near-boundary mixing and associated flows. J. Phys. Oceanogr. 31 (1), 138–142.

Garrett, C., Kunze, E., 2007. Internal tide generation in the deep ocean. Annu. Rev. Fluid Mech. 39, 57–87. http://dx.doi.org/10.1146/annurev.fluid.39.050,905.110,227.

Gent, P.R., et al., 2011. The community climate system model version 4. J. Clim. 24 (19), 4973–4991.

Gerkema, T., Staquet, C., Bouruet-Aubertot, P., 2006. Non-linear effects in internal-tide beams, and mixing. Ocean Model. 12 (3–4), 302–318.

Gilbert, D., Garrett, C., 1989. Implications for ocean mixing of internal wave scattering off irregular topography. J. Phys. Oceanogr. 19 (11), 1716–1729.

Gill, A.E., 1982. Atmosphere-Ocean Dynamics. Academic Press, San Diego, (662 pp).

Gill, A., 1984. On the behavior of internal waves in the wakes of storms. J. Phys. Oceanogr. 14 (7), 1129–1151.

Gnanadesikan, A., Slater, R., Gruber, N., Sarmiento, J., 2002. Oceanic vertical exchange and new production: a comparison between models and observations. Deep Sea Res. Part II 49 (1–3), 363–401.

Gnanadesikan, A., Dunne, J., Key, R., Matsumoto, K., Sarmiento, J., Slater, R., 2004. Oceanic ventilation and biogeochemical cycling: understanding the physical mechanisms that produce realistic distributions of tracers and productivity. Global Biogeochem. Cycles 18 (4), GB4010.

Grant, A.L., Belcher, S.E., 2009. Characteristics of Langmuir turbulence in the ocean mixed layer. J. Phys. Oceanogr. 39 (8), 1871–1887.

Gregg, M., 1987. Diapycnal mixing in the thermocline. J. Geophys. Res. 92 (C5), 5249–5286.

Gregg, M.C., 1989. Scaling turbulent dissipation in the thermocline. J. Geophys. Res. 94 (C7), 9686–9698.

Gregg, M., 1998. Estimation and geography of diapycnal mixing in the stratified ocean. In: Imberger, J. (Ed.), Physical Processes in Lakes and Oceans. American Geophysical Union, pp. 305–338. Coastal and Estuarine Studies.

Gregg, M.C., Kunze, E., 1991. Shear and strain in Santa Monica Basin. J. Geophys. Res. 96 (C9), 16709–16719.

Gregg, M., Peters, H., Wesson, J., Oakey, N.S., Shay, T., 1985. Intensive measurements of turbulence and shear in the equatorial undercurrent. Nature 318, 140–144.

Gregg, M.C., Sanford, T.B., Winkel, D.P., 2003. Reduced mixing from the breaking of internal waves in equatorial waters. Nature 422, 513–515.

Gregg, M.C., Hall, R.A., Carter, G.S., Alford, M.H., Lien, R., Winkel, D.P., Wain, D.J., 2011. Flow and mixing in Ascension, a steep, narrow canyon. J. Geophys. Res. 116, C07016. http://dx.doi.org/10.1029/2010JC006610.

Griffies, S., et al., 2010. Problems and prospects in large-scale ocean circulation models. In: Hall, J., Harrison, D., Stammer, D. (Eds.), Proceedings of OceanObs'09: Sustained Ocean Observations and Information for Society, Venice, Italy, 21–25 September 2009, vol. 2. ESA Publication. http://dx.doi.org/10.5270/OceanObs09.cwp.38, WPP-306.

Haine, T.W.N., Marshall, J., 1998. Gravitational, symmetric, and baroclinic instability of the ocean mixed layer. J. Phys. Oceanogr. 28, 634–658.

Harrison, M., Hallberg, R., 2008. Pacific subtropical cell response to reduced equatorial dissipation. J. Phys. Oceanogr. 38 (9), 1894–1912.

Hasumi, H., Suginohara, N., 1999. Effects of locally enhanced vertical diffusivity over rough bathymetry on the world ocean circulation. J. Geophys. Res. 104, 23367–23374.

Hazewinkel, J., Winters, K., 2011. Psi of the internal tide on a β-plane: flux divergence and near-inertial wave propagation. J. Phys. Oceanogr. 41 (9), 1673–1682.

Henyey, F.S., Wright, J., Flatté, S.M., 1986. Energy and action flow through the internal wave field. J. Geophys. Res. 91, 8487–8495.

Heywood, K., Naveira Garabato, A., Stevens, D., 2002. High mixing rates in the abyssal Southern Ocean. Nature 415, 1011–1013.

Hibiya, T., Nagasawa, M., 2004. Latitudinal dependence of diapycnal diffusivity in the thermocline estimated using a finescale parameterization. Geophys. Res. Lett. 31, L01301.

Hibiya, T., Nagasawa, M., Niwa, Y., 2002. Nonlinear energy transfer within the oceanic internal wave spectrum at mid and high latitudes. J. Geophys. Res. 107 (C11).

Hogg, N.G., Owens, W.B., 1999. Direct measurements of the deep circulation with the Brazil Basin. Deep Sea Res. Part II 46, 335–353.

Hosegood, P., Gregg, M.C., Alford, M.H., 2006. Sub-mesoscale lateral density structure in the oceanic surface mixed layer. Geophys. Res. Lett. 33 (22), L22604. http://dx.doi.org/10.1029/2006GL026797.

Huang, R.X., 1998. On the balance of energy in the oceanic general circulation. Sci. Atmos. Sin. 22 (4), 562–574.

Huang, R., Jin, X., 2002. Deep circulation in the South Atlantic induced by bottom-intensified mixing over the midocean ridge. J. Phys. Oceanogr. 32, 1150–1164.

Hughes, G.O., Hogg, A.M.C., Griffiths, R.W., 2009. Available potential energy and irreversible mixing in the meridional overturning circulation. J. Phys. Oceanogr. 39 (12), 3130–3146.

Huussen, T.N., Naveira-Garabato, A.C., Bryden, H.L., McDonagh, E.L., 2012. Is the deep Indian Ocean MOC sustained by breaking internal waves? J. Geophys. Res. 117 (C8), C08024.

Ivey, G.N., Winters, K.B., DeSilva, I.P.D., 2000. Turbulent mixing in a sloping benthic boundary layer energized by internal waves. J. Fluid Mech. 418, 59–76.

Ivey, G., Winters, K., Koseff, J., 2008. Density stratification, turbulence, but how much mixing? Annu. Rev. Fluid Mech. 40, 169–184.

Jackson, L., Hallberg, R., Legg, S., 2008. A parameterization of shear-driven turbulence for ocean climate models. J. Phys. Oceanogr. 38 (5), 1033–1053.

Jayne, S.R., 2009. The impact of abyssal mixing parameterizations in an ocean general circulation model. J. Phys. Oceanogr. 39, 1756–1775.

Jayne, S., St Laurent, L.C., 2001. Parameterizing tidal dissipation over rough topography. Geophys. Res. Lett. 28, 811–814.

Jiang, J., Lu, Y., Perrie, W., 2005. Estimating the energy flux from the wind to ocean inertial motions: the sensitivity to surface wind fields. Geophys. Res. Lett. 32, L15610. http://dx.doi.org/10.1029/2005GL023289.

Jing, Z., Wu, L., 2010. Seasonal variation of turbulent diapycnal mixing in the northwestern Pacific stirred by wind stress. Geophys. Res. Lett. 37. http://dx.doi.org/10.1029/2010GL045418.

Jochum, M., Briegleb, B.P., Danabasoglu, G., Large, W.G., Norton, N.J., Jayne, S.R., Alford, M.H., Bryan, F.O., 2013. The impact of oceanic near-inertial waves on climate. J. Clim. 26 (9), 2833–2844. http://dx.doi.org/10.1175/JCLI-D-12-00181.1.

Johnston, T.M.S., Rudnick, D.L., 2009. Observations of the transition layer. J. Phys. Oceanogr. 39, 780–797.

Johnston, T.M.S., Merrifield, M.A., Holloway, P.E., 2003. Internal tide scattering at the Line Islands Ridge. J. Geophys. Res. 108, 3365. http://dx.doi.org/10.1029/2003JC001844.

Johnston, T.S., Rudnick, D.L., Carter, G.S., 2011. Internal tidal beams and mixing near Monterey Bay. J. Geophys. Res. 116. http://dx.doi.org/10.1029/2010JC006592.

Katsman, C.A., 2006. Impacts of localized mixing and topography on the stationary abyssal circulation. J. Phys. Oceanogr. 36, 1660–1671.

Kelly, S.M., Nash, J.D., 2010. Internal-tide generation and destruction by shoaling internal tides. Geophys. Res. Lett. 37 (23).

Kelly, S., Nash, J., Martini, K., Alford, M., Kunze, E., 2012. The cascade of tidal energy from low to high modes on a continental slope. J. Phys. Oceanogr. 42 (7), 1217–1232.

Klocker, A., McDougall, T., 2010. Influence of the nonlinear equation of state on global estimates of dianeutral advection and diffusion. J. Phys. Oceanogr. 40 (8), 1690–1709.

Klymak, J.M., Moum, J.N., Nash, J.D., Kunze, E., Girton, J.B., Carter, G.S., Lee, C.M., Sanford, T.B., Gregg, M.C., 2006. An estimate of tidal energy lost to turbulence at the Hawaiian Ridge. J. Phys. Oceanogr. 36, 1148–1164.

Klymak, J.M., Pinkel, R., Rainville, L., 2008. Direct breaking of the internal tide near topography: Kaena Ridge, Hawaii. J. Phys. Oceanogr. 38, 380–399.

Klymak, J.M., Legg, S., Pinkel, R., 2010. A simple parameterization of turbulent tidal mixing near supercritical topography. J. Phys. Oceanogr. 40 (9), 2059–2074.

Klymak, J.M., Alford, M.H., Pinkel, R., Lien, R.C., Yang, Y.J., 2011. The breaking and scattering of the internal tide on a continental slope. J. Phys. Oceanogr. 41, 926–945. http://dx.doi.org/10.1175/2010JPO4500.1.

Klymak, J.M., Legg, S., Alford, M.H., Buijsman, M., Pinkel, R., Nash, J.D., 2012. The direct breaking of internal waves at steep topography. Oceanography 25 (2), 150–159.

Kuhlbrodt, T., Griesel, A., Montoya, M., Levermann, A., Hofmann, M., Rahmstorf, S., 2007. On the driving processes of the Atlantic meridional overturning circulation. Rev. Geophys. 45. http://dx.doi.org/10.1029/2004RG000166.

Kunze, E., 1985. Near-inertial wave propagation in geostrophic shear. J. Phys. Oceanogr. 15 (5), 544–565.

Kunze, E., Schmitt, R.W., Toole, J.M., 1995. The energy balance in a warm-core ring's near-inertial critical layer. J. Phys. Oceanogr. 25, 942–957.

Kunze, E., Rosenfeld, L.K., Carter, G.S., Gregg, M.C., 2002. Internal waves in monterey submarine canyon. J. Phys. Oceanogr. 32 (6), 1890–1913.

Kunze, E., Firing, E., Hummon, J.M., Chereskin, T.K., Thurnherr, A.M., 2006. Global abyssal mixing inferred from lowered ADCP shear and CTD strain profiles. J. Phys. Oceanogr. 36 (8), 1553–1576.

Kunze, E., MacKay, C., McPhee-Shaw, E.E., Morrice, K., Girton, J.B., Terker, S.R., 2012. Turbulent mixing and exchange with interior waters on sloping boundaries. J. Phys. Oceanogr. 42 (6), 910–927. http://dx.doi.org/10.1175/JPO-D-11-075.1.

Lapeyre, G., Klein, P., Hua, B.L., 2006. Oceanic restratification forced by surface frontogenesis. J. Phys. Oceanogr. 36 (8), 1577–1590.

Large, W.G., Nurser, A.J.G., 2001. Ocean surface water mass transformation. In: Siedler, G., Church, J., Gould, J. (Eds.), 2001, Ocean Circulation and Climate–Observing and Modelling the Global Ocean. International Geophysics Series, vol. 77. Academic Press, pp. 317–336. ISBN 0-12-641351-7 (Chapter 5.1).

Large, W., McWilliams, J., Doney, S., 1994. Oceanic vertical mixing—a review and a model with a nonlocal boundary-layer parameterization. Rev. Geophys. 32 (4), 363–403.

Ledwell, J., Watson, A., Law, C., 1993. Evidence for slow mixing across the pycnocline from an open-ocean tracer-release experiment. Nature 364, 701–703.

Ledwell, J.R., Watson, A.J., Law, C.S., 1998. Mixing of a tracer in the pycnocline. J. Geophys. Res. 103 (C10), 21499–21529.

Ledwell, J., Montgomery, E., Polzin, K., St Laurent, L., Schmitt, R., Toole, J., 2000. Evidence for enhanced mixing over rough topography in the abyssal ocean. Nature 403 (6766), 179–182.

Ledwell, J.R., St Laurent, L.C., Girton, J.B., Toole, J.M., 2011. Diapycnal mixing in the Antarctic Circumpolar Current. J. Phys. Oceanogr. 41, 241–246.

Legg, S., 2012. Overflows and convectively driven flows. In: Chassignet, E., Cenedese, C., Verron, J. (Eds.), Buoyancy-Driven Flows. Cambridge Univ. Press, Cambridge, UK, pp. 203–239.

Legg, S., Klymak, J.M., 2008. Internal hydraulic jumps and overturning generated by tidal flow over a steep ridge. J. Phys. Oceanogr. 38, 1949–1964.

Legg, S., et al., 2009. Improving oceanic overflow representation in climate models: the gravity current entrainment climate process team. Bull. Am. Meteorol. Soc. 90 (5), 657–670.

Li, M., Garrett, C., Skyllingstad, E., 2005. A regime diagram for classifying turbulence large eddies in the upper ocean. Deep Sea Res. Part I 52 (2), 259–278.

Lien, R.-C., Gregg, M., 2001. Observations of turbulence in a tidal beam and across a coastal ridge. J. Geophys. Res. 106, 4575–4591.

Lien, R.-C., D'Asaro, E.A., McPhaden, M., 2002. Internal waves and turbulence in the Upper Central Equatorial Pacific: Lagrangian and eulerian observations. J. Phys. Oceanogr. 32 (9), 2619–2639.

Lueck, R.G., Mudge, T.D., 1997. Topographically induced mixing around a shallow seamount. Science 276, 1831–1833.

Lueck, R., Wolk, F., Yamazaki, H., 2002. Oceanic velocity microstructure measurements in the 20th century. J. Oceanogr. 58 (1), 153–174.

Lumpkin, R., Speer, K., 2007. Global ocean meridional overturning. J. Phys. Oceanogr. 37, 2550–2562.

Lvov, Y.V., Polzin, K.L., Tabak, E.G., 2004. Energy spectra of the ocean's internal wave field: theory and observations. Phys. Rev. Lett. 92 (12), 128501-1–128501-4.

Lvov, Y.V., Polzin, K.L., Tabak, E.G., Yokoyama, N., 2010. Oceanic internal wavefield: theory of scale-invariant spectra. J. Phys. Oceanogr. 40, 2605–2623.

MacKinnon, J., Gregg, M., 2003. Mixing on the late-summer New England shelf—solibores, shear and stratification. J. Phys. Oceanogr. 33 (7), 1476–1492.

MacKinnon, J.A., Winters, K.B., 2003. Spectral evolution of bottom-forced internal waves. In: Muller, P., Henderson, D. (Eds.), Near-Boundary Processes and Their Parameterization, Proceedings of the 13th 'Aha Huliko'a Hawaiian winter workshop, University of Hawaii, USA, pp. 73–83.

MacKinnon, J.A., Winters, K., 2005. Subtropical catastrophe: significant loss of low-mode tidal energy at 28.9 degrees. Geophys. Res. Lett. 32 (15). http://dx.doi.org/10.1029/2005GL023376.

MacKinnon, J.A., Johnston, T.M.S., Pinkel, R., 2008. Strong transport and mixing of deep water through the Southwest Indian Ridge. Nat. Geosci. 1, 755–758.

MacKinnon, J.A., Alford, M.H., Sun, O., Pinkel, R., Zhao, Z., Klymak, J., 2013. Parametric subharmonic instability of the internal tide at 29N. J. Phys. Oceanogr. 43 (1), 17–28.

Mahadevan, A., Tandon, A., Ferrari, R., 2010. Rapid changes in mixed layer stratification driven by submesoscale instabilities and winds. J. Geophys. Res. 115, C03017. http://dx.doi.org/10.1029/2008JC005,203.

Marshall, J., Radko, T., 2006. A model of the upper branch of the meridional overturning of the Southern Ocean. Prog. Oceanogr. 70, 331–345.

Martin, J.P., Rudnick, D.L., Pinkel, R., 2006. Spatially broad observations of internal waves in the upper ocean at the Hawaiian ridge. J. Phys. Oceanogr. 36 (6), 1085–1103.

Martini, K.I., Alford, M.H., Kunze, E., Kelly, S.H., Nash, J.D., 2011. Observations of internal tides on the Oregon continental slope. J. Phys. Oceanogr. 41 (9), 1772–1794.

McComas, C.H., Müller, P., 1981. The dynamic balance of internal waves. J. Phys. Oceanogr. 11, 970–986.

McPhee-Shaw, E., Kunze, E., 2002. Boundary layer intrusions from a sloping bottom: a mechanism for generating intermediate nepheloid layers. J. Geophys. Res. Oceans 107 (C6). http://dx.doi.org/10.1029/2001JC000801.

McWilliams, J., Sullivan, P., Moeng, C., 1997. Langmuir turbulence in the ocean. J. Fluid Mech. 334, 1–30.

McWilliams, J., Restrepo, J., Lane, E., 2004. An asymptotic theory for the interaction of waves and currents in coastal waters. J. Fluid Mech. 511, 135–178.

Melet, A., Hallberg, R., Legg, S., Polzin, K., 2013. Sensitivity of the Pacific Ocean state to the vertical distribution of internal-tide driven mixing. J. Phys. Oceanogr. 43 (3), 602–615. http://dx.doi.org/10.1175/JPO-D-12-055.1.

Mihaly, S.F., Thomson, R., Rabinovich, A.B., 1998. Evidence for nonlinear interaction between internal waves of inertial and semidiurnal frequency. Geophys. Res. Lett. 25 (8), 1205–1208.

Moehlis, J., Llewellyn-Smith, S.G., 2001. Radiation of mixed layer near-inertial oscillations into the ocean interior. J. Phys. Oceanogr. 31, 1550–1560.

Molemaker, M.J., McWilliams, J.C., Yavhen, I., 2005. Baroclinic instability and loss of balance. J. Phys. Oceanogr. 35 (9), 1505–1517.

Moum, J.N., Nash, J.D., 2009. Mixing measurements on an equatorial ocean mooring. J. Atmos. Oceanic. Technol. 26, 317–336.

Moum, J.N., Rippeth, T.P., 2009. Do observations adequately resolve the natural variability of oceanic turbulence? J. Mar. Syst. 77, 409–417.

Moum, J., Smyth, W., 2001. Encyclopedia of Ocean Sciences. Academic Press, pp. 3093–3100 (Chap. Upper ocean mixing).

Moum, J., Gregg, M.C., Lien, R.-C., Carr, M., 1995. Comparison of turbulence kinetic energy dissipation rate estimates from two ocean microstructure profilers. J. Atmos. Oceanic. Technol. 12, 346–366.

Moum, J.N., Lien, R.-C., Perlin, A., Nash, J.D., Gregg, M.C., Wiles, P.J., 2009. Sea surface cooling at the equator by subsurface mixing in tropical instability waves. Nat. Geosci. 2, 761–765.

Moum, J.N., Nash, J.D., Smyth, W.D., 2011. Narrowband oscillations in the upper equatorial ocean. Part i: interpretation as shear instabilities. J. Phys. Oceanogr. 41 (3), 397–411.

Müller, P., Liu, X., 2000. Scattering of internal waves at finite topography in two dimensions. Part I: theory and case studies. J. Phys. Oceanogr. 30, 532–549.

Müller, P., Holloway, G., Henyey, F., Pomphrey, N., 1986. Nonlinear interactions among internal gravity waves. Rev. Geophys. 24 (3), 493–536.

Munk, W.H., 1966. Abyssal recipes. Deep Sea Res. 13, 707–730.

Munk, W., Wunsch, C., 1998. Abyssal recipes II: energetics of tidal and wind mixing. Deep Sea Res. Part I 45, 1977–2010.

Nagasawa, M., Yoshihiro, N., Hibiya, T., 2000. Spatial and temporal distribution of the wind-induced internal wave energy available for deep water mixing in the north pacific. J. Geophys. Res. 105 (C6), 13933–13943.

Nash, J.D., Kunze, E., Toole, J.M., Schmitt, R.W., 2004. Internal tide reflection and turbulent mixing on the continental slope. J. Phys. Oceanogr. 34, 1117–1134.

Naveira Garabato, A., Polzin, K., King, B., Heywood, K., Visbeck, M., 2004. Widespread intense turbulent mixing in the Southern Ocean. Science 303, 210–213.

Nikurashin, M., Ferrari, R., 2010. Radiation an dissipation of internal waves generated by geostrophic motions impinging on small-scale topography: application to the Southern Ocean. J. Phys. Oceanogr. 40, 2025–2042.

Nikurashin, M., Ferrari, R., 2011. Global energy conversion rate from geostrophic flows into internal lee waves in the deep ocean. Geophys. Res. Lett. 38. http://dx.doi.org/10.1029/2011GL046576.

Nikurashin, M., Legg, S., 2011. A mechanism for local dissipation of internal tides generated at rough topography. J. Phys. Oceanogr. 41, 378–395.

Nikurashin, M., Vallis, G., 2011. A theory of deep stratification and overturning circulation in the ocean. J. Phys. Oceanogr. 41, 485–502.

Osborn, T.R., 1980. Estimates of the local rate of vertical diffusion from dissipation measurements. J. Phys. Oceanogr. 10, 83–89.

Pacanowski, R., Philander, S., 1981. Parameterization of vertical mixing in numerical models of tropical oceans. J. Phys. Oceanogr. 11 (11), 1443–1451.

Palmer, M.D., Garabato, A.C.N., Stark, J.D., Hirschi, J.J.-M., Marotzke, J., 2007. The influence of diapycnal mixing on quasi-steady overturning states in the Indian Ocean. J. Phys. Oceanogr. 37, 2290–2304.

Peacock, T., Echeverri, P., Balmforth, N.J., 2008. An experimental investigation of internal tide generation by two-dimensional topography. J. Phys. Oceanogr. 38, 235–242.

Perlin, A., Moum, J.N., Klymak, J.M., 2005. Response of the bottom boundary layer over a sloping shelf to variations in alongshore wind. J. Geophys. Res. 110. http://dx.doi.org/10.1029/2004JC002500.

Pinkel, R., Buijsman, M., Klymak, J.M., 2012. Breaking Topographic Lee Waves in a Tidal Channel in Luzon Strait. Oceanography 25 (2), 160–165.

Plueddemann, A.J., Farrar, J.T., 2006. Observations and models of the energy flux from the wind to mixed-layer inertial currents. Deep Sea Res. 53, 5–30.

Pollard, R., Millard, R., 1970. Comparison between observed and simulated wind-generated inertial oscillations. Deep Sea Res. 17, 153–175.

Polton, J., Belcher, S., 2007. Langmuir turbulence and deeply penetrating jets in an unstratified mixed layer. J. Geophys. Res. 112. http://dx.doi.org/10.1029/2007JC004205.

Polton, J.A., Smith, J.A., MacKinnon, J.A., Tejada-Martinez, A., 2008. Rapid generation of high-frequency internal waves beneath a wind and wave forced oceanic surface mixed layer. Geophys. Res. Lett. 35, L13602. http://dx.doi.org/10.1029/2008GL033856.

Polzin, K., 2004. Idealized solutions for the energy balance of the finescale internal wave field. J. Phys. Oceanogr. 34, 231–246.

Polzin, K.L., 2009. An abyssal recipe. Ocean Model. 30, 298–309.

Polzin, K.L., 2010. Mesoscale eddy—internal wave coupling. Part II: energetics and results from PolyMode. J. Phys. Oceanogr. 40, 789–801.

Polzin, K., Firing, E., 1997. Estimates of diapycnal mixing using LADCP and CTD data from I8S. Int. WOCE Newslett. 29, 39–42.

Polzin, K.L., Lvov, Y.V., 2011. Toward regional characterizations of the oceanic internal wavefield. Rev. Geophys. 49. http://dx.doi.org/10.1029/2010RG000,329.

Polzin, K.L., Toole, J.M., Schmitt, R.W., 1995. Finescale parameterizations of turbulent dissipation. J. Phys. Oceanogr. 25, 306–328.

Polzin, K.L., Oakey, N.S., Toole, J.M., Schmitt, R.W., 1996a. Fine structure and microstructure characteristics across the northwest Atlantic Subtropical Front. J. Geophys. Res. 101 (C6), 14111–14121.

Polzin, K.L., Speer, K.G., Toole, J.M., Schmitt, R.W., 1996b. Intense mixing of Antarctic Bottom Water in the equatorial Atlantic Ocean. Nature 380, 54–57.

Polzin, K., Toole, J., Ledwell, J., Schmitt, R., 1997. Spatial variability of turbulent mixing in the abyssal ocean. Science 276, 93–96.

Polzin, K.L., Garabato, A.C.N., Huussen, T.N., Sloyan, B.M., Waterman, S.N., 2013. Finescale parameterizations of turbulent dissipation by internal wave breaking. J. Geophys. Res. (submitted).

Price, J.F., Weller, R.A., Pinkel, R., 1986. Diurnal cycling: observations and models of the upper ocean response to diurnal heating, cooling, and wind mixing. J. Geophys. Res. 91, 8411–8427.

Price, J., Morzel, J., Niiler, P., 2008. Warming of SST in the cool wake of a moving hurricane. J. Geophys. Res. 113, C07010.

Radko, T., Marshall, J., 2004. The leaky thermocline. J. Phys. Oceanogr. 34, 1648–1662.

Rainville, L., Pinkel, R., 2004. Observations of energetic high-wavenumber internal waves in the Kuroshio. J. Phys. Oceanogr. 34, 1495–1505.

Ray, R., Mitchum, G., 1996. Surface manifestation of internal tides generated near Hawaii. Geophys. Res. Lett. 23 (16), 2101–2104.

Ray, R., Mitchum, G., 1997. Surface manifestation of internal tides in the deep ocean: observations from altimetry and island gauges. Prog. Oceanogr. 40 (1), 135–162.

Roemmich, D., Hautala, S., Rudnick, D., 1996. Northward abyssal transport through the Samoan Passage and adjacent regions. J. Geophys. Res. 101, 14039–14055.

Rudnick, D.L., Ferrari, R., 1999. Compensation of horizontal temperature and salinity gradients in the ocean mixed layer. Science 283, 526–529.

Rudnick, D., et al., 2003. From tides to mixing along the Hawaiian Ridge. Science 301, 355–357.

Saenko, O., Merryfield, W., 2005. On the effect of topographically enhanced mixing on the global ocean circulation. J. Phys. Oceanogr. 35, 826–834.

Samelson, R.M., 1998. Large scale circulation with locally enhanced vertical mixing. J. Phys. Oceanogr. 28 (4), 712–726.

Schmitt, R.W., 2012. Finger puzzles. J. Fluid Mech. 692, 1–4.

Schmitt, R., Ledwell, J., Montgomery, E., Polzin, K., Toole, J., 2005. Enhanced diapycnal mixing by salt fingers in the thermocline of the tropical Atlantic. Science 308, 685–688.

Scott, J.R., Marotzke, J., 2002. The location of diapycnal mixing and the Meridional Overturning Circulation. J. Phys. Oceanogr. 32, 3578–3595.

Scott, R.B., Goff, J.A., Garabato, A.C.N., Nurser, A.J.G., 2011. Global rate and spectral characteristics of internal gravity wave generation by geostrophic flow over topography. J. Geophys. Res. 116. http://dx.doi.org/10.1029/2011JC007005.

Shih, L.H., Koseff, J.R., Ivey, G.N., Ferziger, J.H., 2005. Parameterization of turbulent fluxes and scales using homogeneous sheared stably stratified turbulence simulations. J. Fluid Mech. 525, 193–214.

Shuckburgh, E., Maze, G., Ferreira, D., Marshall, J., Jones, H., Hall, C., 2011. Mixed layer lateral eddy fluxes mediated by air-sea interaction. J. Phys. Oceanogr. 41, 130–144.

Silverthorne, K.E., Toole, J.M., 2009. Seasonal kinetic energy variability of near-inertial motions. J. Phys. Oceanogr. 39 (4), 1035–1049.

Simmons, H.L., 2008. Spectral modification and geographic redistribution of the semi-diurnal internal tide. Ocean Model. 21, 126–138.

Simmons, H., Jayne, S., Laurent, L.S., Weaver, A., 2004a. Tidally driven mixing in a numerical model of the ocean general circulation. Ocean Model. 6, 245–263.

Simmons, H.L., Hallberg, R.W., Arbic, B.K., 2004b. Internal wave generation in a global baroclinic tide model. Deep Sea Res. Part II 51, 3043–3068.

Skyllingstad, E.D., Denbo, E.D., 1995. An ocean large-eddy simulation of Langmuir circulations and convection in the surface mixed layer. J. Geophys. Res. 100, 8501–8522.

Skyllingstad, E.D., Smyth, W.D., Crawford, G.B., 2000. Resonant wind-driven mixing in the ocean boundary layer. J. Phys. Oceanogr. 30, 1866–1890.

Sloyan, B.M., 2005. Spatial variability of mixing in the Southern Ocean. Geophys. Res. Lett. 32, L18603. http://dx.doi.org/10.1029/2005GL023,568.

Smyth, W.D., Moum, J.N., 2001. Three-dimensional turbulence. In: Encyclopedia of Ocean Sciences, vol. 6. Academic Press, pp. 2947–2955.

Smyth, W.D., Moum, J.N., 2012. Ocean mixing by Kelvin-Helmholtz instability. Oceanography 25 (2), 140–149.

Smyth, W.D., Moum, J.N., Nash, J.D., 2011. Narrowband oscillations in the upper equatorial ocean. Part ii: properties of shear instabilities. J. Phys. Oceanogr. 41 (3), 412–428.

Staquet, C., Sommeria, J., 2002. Internal gravity waves: from instabilities to turbulence. Annu. Rev. Fluid Mech. 34, 559–593.

St Laurent, L., Garrett, C., 2002. The role of internal tides in mixing the deep ocean. J. Phys. Oceanogr. 32, 2882–2899.

St Laurent, L.C., Simmons, H., 2006. Estimates of power consumed by mixing in the ocean interior. J. Clim. 19, 4877–4890.

St Laurent, L.C., Thurnherr, A.M., 2007. Intense mixing of lower thermocline water on the crest of the Mid-Atlantic Ridge. Nature 448, 680–683.

St Laurent, L.C., Toole, J.M., Schmitt, R.W., 2001. Buoyancy forcing by turbulence above rough topography in the abyssal Brazil Basin. J. Phys. Oceanogr. 31, 3476–3495.

St Laurent, L., Simmons, H., Jayne, S., 2002. Estimating tidally driven mixing in the deep ocean. Geophys. Res. Lett. 29 (23). http://dx.doi.org/10.1029/2002GL015633.e.

St Laurent, L.C., Naveira Garabato, A.C., Ledwell, J.R., Thurnherr, A.M., Toole, J.M., Watson, A.J., 2012. Turbulence and diapycnal mixing in Drake Passage. J. Phys. Oceanogr. 42 (12), 2143–2152. http://dx.doi.org/10.1175/JPO-D-12-027.1.

Stommel, H., Arons, A., 1960. On the abyssal circulation of the world ocean—ii. An idealized model of the circulation pattern and amplitude in oceanic basins. Deep Sea Res. 6, 217–233.

Stone, P.H., 1966. On non-geostrophic baroclinic stability. J. Atmos. Sci. 23 (4), 390.

Stone, P.H., 1970. On non-geostrophic baroclinic stability 2. J. Atmos. Sci. 27 (5), 721.

Sullivan, P., McWilliams, J., 2010. Dynamics of winds and currents coupled to surface waves. Annu. Rev. Fluid Mech. 42, 19–42.

Sullivan, P., McWilliams, J., Melville, W., 2004. The oceanic boundary layer driven by wave breaking with stochastic variability. Part 1. direct numerical simulations. J. Fluid Mech. 507, 143–174.

Talley, L.D., 2013. Closure of the global overturning circulation through the Indian, Pacific, and Southern Oceans: schematics and transports. Oceanography 26 (1), 80–97.

Talley, L.D., Reid, J.L., Robbins, P.E., 2003. Data-based meridional overturning streamfunctions for the global ocean. J. Clim. 16 (10), 3213–3226.

Talley, L.D., Pickard, G.L., Emery, W.J., Swift, J.H., 2011. Descriptive Physical Oceanography: An Introduction, sixth ed. Elsevier, Boston.

Tandon, A., Garrett, C., 1995. Geostrophic adjustment and restratification of a mixed-layer with horizontal gradients above a stratified layer. J. Phys. Oceanogr. 25 (10), 2229–2241.

Taylor, J.R., Ferrari, R., 2010. Buoyancy and wind-driven convection at mixed layer density fronts. J. Phys. Oceanogr. 40, 1222–1242.

Thomas, L.N., 2005. Destruction of potential vorticity by winds. J. Phys. Oceanogr. 35, 2457–2466.

Thomas, L.N., 2012. On the effects of strain on symmetric instability and near-inertial motions. J. Fluid Mech. 711, 620–640.

Thomas, L.N., Ferrari, R., 2008. Friction, frontogenesis, and the stratification of the surface mixed layer. J. Phys. Oceanogr. 38, 2501–2518.

Thomas, L.N., Taylor, J.R., 2010. Reduction of the usable wind-work on the general circulation by forced symmetric instability. Geophys. Res. Lett. 37, L18606. http://dx.doi.org/10.1029/2010GL044,680.

Thompson, A.F., Gille, S.T., MacKinnon, J.A., Sprintall, J., 2007. Spatial and temporal patterns of small-scale mixing in Drake Passage. J. Phys. Oceanogr. 37, 572–592.

Thorpe, S., 2005. The Turbulent Ocean. Cambridge Univ. Press, Cambridge.

Thorpe, A.J., Rotunno, R., 1989. Nonlinear aspects of symmetric instability. J. Atmos. Sci. 46, 1285–1299.

Thurnherr, A.M., St Laurent, L., Speer, K., Toole, J., Ledwell, J., 2005. Mixing associated with sills in a canyon in the midocean ridge flank. J. Phys. Oceanogr. 35, 1370–1381.

Timmermans, M., Garrett, C., Carmack, E., 2003. The thermohaline structure and evolution of the deep waters in the Canada Basin, Arctic Ocean. Deep Sea Res. Part I 50, 1305–1321.

Toggweiler, J., Samuels, B., 1998. On the ocean's large-scale circulation near the limit of no vertical mixing. J. Phys. Oceanogr. 28 (9), 1832–1852.

Toole, J.M., Polzin, K.L., Schmitt, R.W., 1994. Estimates of diapycnal mixing in the abyssal ocean. Science 264, 1120–1123.

Toole, J.M., Schmitt, R.W., Polzin, K.L., Kunze, E., 1997. Near-boundary mixing above the flanks of a midlatitude seamount. J. Geophys. Res. 102 (C1), 947–959.

Treguier, A.M., Held, I.M., Larichev, V.D., 1997. On the parameterization of quasi-geostrophic eddies in primitive equation ocean models. J. Phys. Oceanogr. 27, 567–580.

van Haren, H., 2007. Longitudinal and topographic variations in North Atlantic tidal and intertial energy around latitudes 30 $\pm$ 10 N. J. Geophys. Res. 112. http://dx.doi.org/10.1029/2007JC004193.

Venayagamoorthy, S.K., Fringer, O.B., 2012. Examining Breaking Internal Waves on a Shelf Slope Using Numerical Simulations. Oceanography 25 (2), 132–139.

Watanabe, M., Hibiya, T., 2002. Global estimates of the wind-induced energy flux to inertial motions in the surface mixed layer. Geophys. Res. Lett. 29 (8). http://dx.doi.org/10.1029/2001GL014422.

Waterhouse, A.F., et al., 2013. Global patterns of diapycnal mixing from measurements of the turbulent dissipation rate. J. Phys. Oceanogr. (submitted).

Waterman, S., Garabato, A.C.N., Polzin, K.L., 2013a. Internal waves and turbulence in the Antarctic Circumpolar Current. J. Phys. Oceanogr. 43 (2), 259–282. http://dx.doi.org/10.1175/JPO-D-11-0194.1.

Waterman, S., Polzin, K.L., Garabato, A.C.N., Sheen, K.L., 2013b. Suppression of internal wave breaking by lee wave—mean flow interactions in the Antarctic Circumpolar Current. J. Phys. Oceanogr. (submitted).

Whalen, C.B., Talley, L.D., MacKinnon, J.A., 2012. Spatial and temporal variability of global ocean mixing inferred from Argo profiles. Geophys. Res. Lett. 39, L18612. http://dx.doi.org/10.1029/2012GL053196.

Wijesekera, J., Dillon, T., 1991. Internal waves and mixing in the upper Equatorial Pacific Ocean. J. Geophys. Res. 96, 7115–7125.

Wijesekera, H., Padman, L., Dillon, T., Levine, M., Paulson, C., 1993. The application of internal-wave dissipation models to a region of strong mixing. J. Phys. Oceanogr. 23, 269–286.

Winters, K., D'Asaro, E., 1994. Three-dimensional wave instability near a critical level. J. Fluid Mech. 272, 255–284.

Winters, K.B., D'Asaro, E.A., 1997. Direct simulation of internal wave energy transfer. J. Phys. Oceanogr. 27, 1937–1945.

Wolfe, C.L., Cessi, P., 2010. What sets the strength of the middepth stratification and overturning circulation in eddying ocean models? J. Phys. Oceanogr. 40 (7), 1520–1538.

Wolk, F., Lueck, R.G., St Laurent, L., 2009. Turbulence measurements from a glider. In: OCEANS 2009, MTS/IEEE Biloxi—Marine Technology for Our Future: Global and Local Challenges, pp. 1–6.

Wunsch, C., Ferrari, R., 2004. Vertical mixing, energy and the general circulation of the oceans. Annu. Rev. Fluid Mech. 36, 281–412.

Young, K., 2012. Biogenic inputs to ocean mixing. J. Exp. Biol. 215, 1040–1049.

Young, W., Ben-Jelloul, M., 1997. Propagation of near-inertial oscillations through a geostrophic flow. J. Mar. Res. 55, 735–766.

Zervakis, V., Levine, M., 1995. Near-inertial energy propagation from the mixed layer: theoretical considerations. J. Phys. Oceanogr. 25, 2872–2889.

Zhai, X., Johnson, H.L., Marshall, D.P., 2010. Significant sink of ocean-eddy energy near western boundaries. Nat. Geosci. 3, 608–612.

Zhao, Z., Alford, M.H., 2009. New altimetric estimates of mode-one M2 internal tides in the Central North Pacific Ocean. J. Phys. Oceanogr. 39, 1669–1684.

Zhao, Z., Alford, M.H., MacKinnon, J.A., Pinkel, R., 2010. Long-range propagation of the semidiurnal internal tide from the Hawaiian Ridge. J. Phys. Oceanogr. 40 (4), 713–736.

Zika, J.D., Sloyan, B., McDougall, T.J., 2009. Diagnosing the Southern Ocean overturning from tracer fields. J. Phys. Oceanogr. 39 (11), 2926–2940.

Chapter 8

Lateral Transport in the Ocean Interior

B. Fox-Kemper*, R. Lumpkin[‡] and F.O. Bryan[§]

*Geological Sciences, Brown University, Providence, Rhode Island, USA

[‡]Physical Oceanography Division, Atlantic Oceanographic and Meteorological Laboratory, Miami, Florida, USA

[§]National Center for Atmospheric Research, Boulder, Colorado, USA

Chapter Outline

1. INTRODUCTION

The motions of the oceans span a vast range of length and time scales, consistent with the broad range of forcing from small and fast (wind gusts), large and fast (tides, diurnal cycle) to small and slow (coastal erosion), to large and slow (Milankovich and tectonic change). However, even forcing over a narrow band of space and time scales does not result in a narrow banded response, as nonlinear processes connect different scales and hydrodynamic instabilities produce new scales that differ from those of the forcing.

High-resolution models, satellites, and drifter observations generally agree about the surface signature of the mean and mesoscale motions of the surface ocean. Even coarse-resolution models approximate many nonlinear effects, such as the formation of the advective thermocline, meridional overturning circulations, and western boundary currents. However, the instabilities and small-scale processes that produce transient mesoscale and smaller motions are still not resolved routinely in global scale simulations, which is likely to be true for a few decades to come. Continued study of mesoscale processes and their parameterization will provide increasingly accurate understanding and models. This chapter focuses on the present understanding and remaining questions about the effect of these processes on the larger, steadier oceanic "general circulation."

Precise usage of "general circulation" is rarely exercised; it usually refers to a circulation governed by steady or simplified equations or observations that somehow reduce variability. Before computers, general circulation usually meant assuming steady, linear, strongly diffusive equations. Since computers, we refer to "general circulation models" or GCMs, that are too coarse to resolve some phenomena. Whatever is not resolved, apparently, is not part of the "general circulation." Now that we can afford to simulate all of the globe, GCM is sometimes taken as "global circulation model" or "global climate model" where

 http://dx.doi.org/10.1016/B978-0-12-391851-2.00008-8

"circulation" and "climate" apparently are that which is resolved and everything else is not.

General circulation models have parameterizations to approximate all phenomena smaller or faster than the model can resolve directly. Which phenomena need to be parameterized depends on the model resolution and an assessment of the phenomena that dominate that scale. The emphasis here is on the present global ocean models, which have a resolution too coarse to directly resolve features that are a few hundred kilometers in the horizontal or smaller. Model resolution has increased steadily with computing power and so the definition of general circulation grows to include more phenomena but direct solution of the Navier–Stokes equations for the global ocean is still centuries away, based on current trends. So, parameterizations of some unresolved ocean phenomena will remain part of oceanography for the foreseeable future. Mesoscale and smaller processes transport tracers, momentum, and energy, and these transports constitute a non-negligible contribution to the global ocean's general circulation and stratification.

This chapter reviews principles behind the processes that dominate the lateral transport of fluid properties: the general circulation and mesoscale eddies. Away from the surface, both general circulation and eddies are thought to be nearly conservative in many properties. The conservation principles, and how they are formulated into guiding principles for the development of parameterizations, constitutes the first few sections. A brief discussion of the interactions between the unresolved and resolved processes follows. Later sections present high-resolution numerical and observational evidence to illustrate some parameterization principles in present practice and suggest future improvements.

2. THEORY OF MASS, TRACER, AND VECTOR TRANSPORT

This section contains an introduction to the relevant equations of fluid motion in the forms most useful for the discussion of transport processes to come in later sections. The compressible equations with a generic equation of state are a useful starting point. They are contrasted against the "primitive" equations usually simulated in ocean models, both in depth-coordinate and density-coordinate versions. Care is taken with the connections between energy and buoyant restoring force. The importance of the vertical coordinate, averaging operations, and relevant "eddy" correlations are also discussed in preparation for later sections.

2.1. Fundamental Equations

The general equations of motion for a rotating, stratified fluid are the rotating Navier–Stokes equations, combined with the first and second laws of thermodynamics. With appropriate initial and boundary conditions, density (ρ), absolute pressure (P), three velocity components $\left(\vec{u}\right)$, specific internal energy (e), specific entropy (s), salinity (S) and passive tracer (c) in mass fraction units may be determined using these equations and standard thermodynamic relations (e.g., Vallis, 2006).

The fundamental continuity and tracer conservation equations for a compressible fluid is used later and so are given here:

$$\frac{\partial \rho}{\partial t}+\nabla_i(\rho u_i)=0, \quad \frac{\partial \rho c}{\partial t}+\nabla_i(\rho u_i c)=\rho \dot{c}. \tag{8.1}$$

For consistency with later sections where tensor notation is more convenient, Cartesian tensors are used throughout, as also Einstein summation (repeated indices indicate summation over all coordinates). For example, $\nabla_i(\rho u_i) \equiv \frac{\partial \rho u}{\partial x}+\frac{\partial \rho v}{\partial y}+\frac{\partial \rho w}{\partial z}$, and $u_i u_i \equiv u_j u_j \equiv u^2+v^2+w^2$. The ∇ operator with a subscript denotes partial differentiation in that direction. Griffies (2004) has a thorough discussion on using tensor notation for ocean modeling. The dot above c indicates rate of change of concentration due to non-reversible or non-conservative effects such as sources, sinks, and diffusion. Cartesian coordinates and flux form equations (where each tendency may be balanced against the divergence of a flux) are used. The flux form helps to compare transport among different equation sets to come. The equations can easily be converted to curvilinear or spherical coordinates with the metric tensor formalism described in Griffies (2004); some care is required to maintain the flux-conservation principles. For motions that span only a fraction of the earth's circumference, such as a single model grid point where a parameterization may apply, these equations are sufficiently accurate to describe the tangent plane to the spherical earth at that location.

2.1.1. *Primitive Equations*

Present large-scale ocean models do not solve the Navier–Stokes and thermodynamic equations, which have unwanted complexity, such as a time step limited by the speed of sound. Instead, the Boussinesq, hydrostatic, traditional, and geoid simplification approximations are typical (Griffies and Adcroft, 2008; Young, 2010; Chapter 20) although exceptions exist (e.g., McDougall et al., 2002; Mahadevan, 2006). The hydrostatic approximation $\nabla_z P = -\rho g$ is appropriate for large-aspect-ratio flow, and the background pressure ($P_0 - \rho_0 g z$) is hydrostatic, but the hydrostatic approximation will not be made to the vertical momentum equation so that the symmetries are more apparent.

$$\nabla_i u_i = 0, \quad b = \widetilde{b}(S, \Theta, P_0 - \rho_0 g z), \tag{8.2}$$

$$\frac{\partial S}{\partial t}+\nabla_i(u_i S)=\dot{S}, \quad \frac{\partial c}{\partial t}+\nabla_i(u_i c)=\dot{c}, \quad \frac{\partial}{\partial t}\Theta+\nabla_i(u_i \Theta)=\dot{\Theta}, \tag{8.3}$$

$$\frac{\partial u_j}{\partial t} + \int_{jzk} f u_k + \nabla_j (u_j u_j) + \nabla_j p - b\delta_{zj} = \dot{u}_j, \quad (8.4)$$

where the buoyancy is $b = |g|(\rho_0 - \rho)/\rho_0$,[1] the dynamic pressure is $p = (P - P_0)/\rho_0 + gz$, and the constant background values are ρ_0 for density and $P_0(x, y, t)$ for pressure variations due to the sea surface height or atmosphere. When the coordinate names x, y, z appear as indices, for example, ϵ_{jzk}, only that component in the x (zonal), y (meridional), or z (vertical) direction is intended. For example, the vertical velocity is $w = u_z$. Tildes denote thermodynamic relations $\left(\text{e.g., } \tilde{b}(S, \Theta, P_0 - \rho_0 g z)\right)^2$ as opposed to values (e.g., b). Conservative forces (gravity and centrifugal) are represented by an averaged value of gravitational acceleration (g) and deviations from the local geoid (z). Nonconservative forces ($\dot{u}$), heating, and irreversible and diffusive processes ($\dot{S}$, $\dot{\Theta}$, $\dot{c}$) are included but not specified. The traditional approximation reduces the directionality of the Coriolis force based on the axis of planetary rotation (Ω) to the local vertical component f. The Kronecker δ and Levi–Civita totally antisymmetric symbol (ϵ) are needed to provide the direction of the Coriolis and buoyancy forces.[3]

The temperature variable Θ can be ordinary temperature, potential temperature, or conservative temperature to have (Equation 8.3) represent seawater thermodynamics with increasing accuracy (McDougall, 2003; Nycander, 2011; Young, 2010; Chapter 6). Near-surface forcing by the sun and infrared radiation can be accounted for via $\dot{\Theta} = \dot{e}/c_p^0$. Latent and sensible heat exchange with the atmosphere, evaporation and precipitation, and turbulent boundary layer mixing also contribute to the right sides ($\dot{u}$, $\dot{S}$, $\dot{c}$, $\dot{\Theta}$). Away from the boundaries, the right sides of Equation (8.3) are generally very small. Thus, water mass analysis can detect where waters were "formed" after decades or centuries of advection and weak diffusion. The mixing processes that do contribute to nonzero right sides of Equation (8.3), along with their sources and rates, are reviewed in Chapter 7.

Boussinesq models have no conversion of internal energy to mechanical energy, but do allow conversion between potential and kinetic energy via sinking of dense water or rising of light water via wb (Young, 2010). The Boussinesq energy equations are

$$\frac{\partial}{\partial t}\frac{u_j u_j}{2} + \nabla_i u_i \left(\frac{u_j u_j}{2} + p\right) - wb = u_j \dot{u}_j, \quad \frac{\partial}{\partial t} h^{\ddagger} + \nabla_i u_i h^{\ddagger} + wb = \frac{\partial \tilde{h}^{\ddagger}}{\partial \Theta}\dot{\Theta} + \frac{\partial \tilde{h}^{\ddagger}}{\partial S}\dot{S}, \quad (8.5)$$

$$\text{where } \tilde{h}^{\ddagger} \equiv \int_Z^0 \tilde{b}(S, \Theta, P_0 - \rho_0 g z')\,dz'. \quad (8.6)$$

The enthalpy (and the Boussinesq dynamic enthalpy $h^{\ddagger}$) is a key thermodynamic variable because it is very nearly conserved during mixing (Young, 2010; chapter 6). In a stably-stratified ocean, wb keeps water parcels near a fixed location. The restoring force for internal waves (where $wb < 0$) is a good example. A second important role for wb is as the source of energy for baroclinic instabilities, which convert mean potential energy to eddy energy by correlating eddy motions with water buoyancy (thus $wb > 0$ on average). Likewise, unstable density profiles convect with $wb > 0$.

2.1.2. Minimal-Disturbance Planes and Slopes

In a stratified ocean, arbitrary adiabatic displacement of a water parcel typically results in a buoyancy anomaly and force that act to restore or destabilize the parcel. One direction will generally maximize this effect and motions in a plane perpendicular to this direction will minimize it. Potential density or buoyancy b, neutral density (McDougall, 1987), and the Boussinesq dynamic enthalpy, $\tilde{h}^{\ddagger}$ in Equation (8.5) above (Young, 2010; Nycander, 2011) are thermodynamic variables that can be used to estimate the direction in which displacements create maximal disturbances of stratification or energy. The "diapycnal" direction ($\mathcal{D}_i$), the "dianeutral" direction ($\mathcal{N}_i$), and the "P vector" ($\mathcal{P}_i$) are all maximal-disturbance directions given by

$$\mathcal{D}_i = \nabla_i b, \quad \mathcal{N}_i = \left(\frac{\partial \tilde{b}}{\partial \Theta}\right)_{S,z} \nabla_i \Theta + \left(\frac{\partial \tilde{b}}{\partial S}\right)_{\Theta,z} \nabla_i S,$$

$$\mathcal{P}_i = \left(\frac{\partial \tilde{h}^{\ddagger}}{\partial \Theta}\right)_{S,z} \nabla_i \Theta + \left(\frac{\partial \tilde{h}^{\ddagger}}{\partial S}\right)_{\Theta,z} \nabla_i S,$$

Each estimates a direction for maximum effect by displacement or mixing. Displacements in the perpendicular plane result in a minimal disturbance. Large scale oceanic motions are generally believed to be dominantly oriented along a minimal-disturbance plane. At different levels of thermodynamic accuracy, these planes are generally called "isopycnals" (with potential density implied), "neutral planes," or "P planes." They are not equal or exactly parallel, and only the simplest, least accurate case $\mathcal{D}_i$ can be thought of as determined by a global surface (isopycnals, where potential density and b are constant). No unique surface or thermodynamic variable connects all the local neutral planes

1. Young (2010) uses a different form that is slightly simpler thermodynamically but asymptotically identical.

2. As noted by Vallis (2006) and Young (2010), $P_0 - \rho_0 g z$ is the approximation to thermodynamic pressure appropriate for energetically consistent Boussinesq equations. So this thermodynamic relation gives buoyancy as a function of salinity, temperature, and pressure.

3. The traditional approximation is best justified when accompanied by the hydrostatic approximation. It can be relaxed, with some added complications (Sheremet, 2004).

or local P planes into global surfaces. However, a minimal-disturbance slope can always be calculated locally using $\mathcal{S}_\gamma = -\mathcal{D}_\gamma/\mathcal{D}_z$, $\mathcal{S}_\gamma = -\mathcal{N}_\gamma/\mathcal{N}_z$, $\mathcal{S}_\gamma = -\mathcal{P}_\gamma/\mathcal{P}_z$, or similar approximations with corresponding levels of thermodynamic sophistication. Greek indices, for example, γ, indicate throughout that the index is to vary only over the horizontal directions, that is, x and y. Depending on this level of sophistication, different terms are used to describe the fluxes in different directions: skew, isopycnal, and diapycnal (buoyancy); adiabatic and diabatic (energy); isentropic and irreversible (entropy); epineutral and dianeutral (neutral planes). Here, we use minimal-disturbance plane and maximal-disturbance direction to apply generally.

2.1.3. *Density-Coordinate Continuity and Tracer Equations*

Since lateral motions are thought to be oriented along the minimal-disturbance plane, a density-like variable that approximates this slope globally—usually potential density or buoyancy—is often used as a vertical coordinate in models. If these surfaces are not too steep, then the hydrostatic approximation is valid and the conservation laws for volume and tracer may be written for this density-coordinate model as (McDougall and Dewar, 1998; Hallberg, 2000):

$$\frac{\partial h}{\partial t} + \nabla_\gamma h u_\gamma = -\frac{\partial}{\partial \varrho}\left(h\frac{D\varrho}{Dt}\right) \equiv -\frac{\partial w_\mathrm{e}}{\partial \varrho}, \quad \frac{\partial}{\partial t}(hc) + \nabla_\gamma h u_\gamma c = \frac{\partial F_\mathrm{c}}{\partial \varrho} - \frac{\partial w_\mathrm{e} c}{\partial \varrho}. \tag{8.7}$$

The change in height with ϱ is $h \equiv -\rho_0 z_\varrho$. Note that the precise meaning of "horizontal" and "vertical" holds some complexity in density-coordinate models; Young (2012) offers much to clarify.

If ϱ is materially conserved, then the entrainment velocity (w_e) vanishes, and these equations are formally identical to the *compressible* mass conservation and tracer equations (Equation 8.1), with "layer thickness" (h) in place of fluid density (ρ). A crucial element of modern eddy parameterizations results from idealizing their transport as flowing along 2D compressible, minimal-disturbance surfaces. They are thus assumed to inhabit only a subset of the possible motions in the 3D nearly-incompressible flow governed by the Boussinesq equations. The formal connection between Equations (8.7) and (8.1) is useful when considering how to parameterize the effects of eddy stirring (Dukowicz and Smith, 1997).

2.2. Steady, Conservative Equations

The oceanic "general circulation" is often intended to imply a purely steady $\left(\frac{\partial}{\partial t}\cdot = 0\right)$ or time-mean $\left(\frac{\partial}{\partial t}\bar{\cdot} \approx 0\right)$ solution to the primitive equations or approximations thereof. For steady, conservative flow, time-derivatives and the non-conservative terms on the right side of Equations (8.3)–(8.5) are neglected. A total specific energy, or Bernoulli function, can be found by eliminating wb between the equations in Equation (8.5). These simplifications yield properties that are conserved in the direction of flow.

$$u_i \nabla_i S = 0, \quad u_i \nabla_i \Theta = 0, \quad u_i \nabla_i \left(\frac{u_j u_j}{2} + p + h^\ddagger\right) = 0, \tag{8.8}$$

$$u_i \nabla_i Q(\vartheta) \equiv u_i \nabla_i \left(\left[\epsilon_{jkl} \nabla_k u_l + 2\Omega_j\right] \nabla_j \vartheta\right) = \frac{\partial \vartheta}{\partial x}\frac{\partial b}{\partial y} - \frac{\partial \vartheta}{\partial y}\frac{\partial b}{\partial x}. \tag{8.9}$$

In Equation (8.8), gradients of salinity, conservative temperature, and Bernoulli function must all be perpendicular to the direction of steady motion. The Ertel potential vorticity $Q(\vartheta)$, based on a materially conserved tracer ϑ, will itself be materially conserved if the buoyancy is zero or ϑ and b are functionally related at each depth. In either case, the right side of Equation (8.9) vanishes. For example, if $\vartheta = \Theta$ and S is constant at that level, then $Q(\Theta)$ is conserved. Most often, $Q(b)$ is used, as b is a conserved tracer if the thermodynamic equation of state can be approximated and combined using Equation (8.8) (e.g., by a local linearization of $\widetilde{b}$).

According to Equation (8.8), conservative, steady flow proceeds in such a way that salinity, conservative temperature, buoyancy, potential vorticity, and Bernoulli function are all constant along streamlines. Many successful theories for steady oceanic flows result (e.g., Sverdrup, 1947; Charney, 1955; Welander, 1959; Stommel and Arons, 1960; Stommel and Schott, 1977; Luyten et al., 1983; Rhines, 1986). These steady solutions are the quantitative basis of understanding for the oceanic gyres, meridional overturning circulation, and generally the oceanic conveyor (Broecker, 1987), as well as air mass analysis (reviewed by Hoskins et al., 1985), and water mass analysis (e.g., Talley and McCartney, 1982; Levitus et al., 1993). These conservative, steady methods can often be adapted to weak diffusion (Welander, 1971; Rhines and Young, 1982; Haynes and McIntyre, 1987; Samelson and Vallis, 1997) and identification of related water masses over long distances, respecting the nonlinear equation of state (McDougall and Jackett, 2007; McDougall and Klocker, 2010).

Now we turn to another aspect of an understanding of the general circulation—how the large-scale, time—mean flow differs from the steady flow, which involves the averaged effects of mesoscale eddies and other smaller variability on the general circulation.

2.3. Reynolds-Averaged Equations

The ocean is not steady, nor is it adiabatic, isentropic, or inviscid. The remainder of this chapter focuses on the unsteady behavior, in particular the contribution from unsteady $O(100\text{ km}, 10\text{ day})$ mesoscale ocean eddies. Diabatic and irreversible effects and mixing are discussed in Chapter 7. Here, all unsteady mesoscale motions are called eddies.

The study of eddy effects on the general circulation began soon after the Mid-Ocean Dynamics Experiment (MODE) (e.g., Holland and Lin, 1975; McWilliams and Flierl, 1976) and fundamental understanding soon followed (Rhines and Holland, 1979; Holland and Rhines, 1980; Rhines and Young, 1982). Eddies play a leading order role in the overturning of the Antarctic Circumpolar Current (Johnson and Bryden, 1989; Henning and Vallis, 2005; Radko and Marshall, 2006), while others have explained the effects of eddies on the gyres (Scott and Straub, 1998; Berloff and McWilliams, 1999; Fox-Kemper and Pedlosky, 2004; Henning and Vallis, 2004; Radko and Marshall, 2004; Fox-Kemper, 2005; Fox-Kemper and Ferrari, 2009). At depth, eddy-induced tracer transport usually exceeds the Eulerian mean transport (Rhines and Holland, 1979; Lozier, 1997, 2010). In western boundary currents, two-way interaction is possible between the time–mean flow and the eddies (Pedlosky, 1984; Edwards and Pedlosky, 1998; Jochum and Malanotte-Rizzoli, 2003; Fox-Kemper, 2004; Fox-Kemper and Ferrari, 2009; Grooms et al., 2011), and the recirculation gyres nearby (Nurser, 1988; Fox-Kemper and Pedlosky, 2004; Kravtsov et al., 2006; Waterman and Jayne, 2011). Eddies can connect basins where mean currents cannot (Gordon et al., 1992; Hallberg and Gnanadesikan, 2006). Numerical studies of oceanic eddies are numerous, and recently even coupled climate models have been run with partially resolved eddies (McClean et al., 2011; Delworth et al., 2012). However, it is likely that parameterization of eddy effects will continue for some decades in centennial-scale climate simulations and high-complexity (e.g., biogeochemical) models. Analytic studies, in particular, benefit from good approximations of eddy effects on the general circulation (Radko and Marshall, 2006; Fox-Kemper and Ferrari, 2009; Smith and Marshall, 2009; Grooms et al., 2011).

Early inclusion of eddies took the form of "eddy viscosity," where the viscosities are increased until nonconservative terms enter the dominant momentum balance (e.g., Munk, 1950; Parsons, 1969). Explicit physical, rather than numerical, discussion of eddy viscosity in coarse-resolution models are rare (Smith and McWilliams, 2003), although results are sensitive to the viscosity chosen (Jochum et al., 2008). Sometimes more complex effects of eddies can be treated as a viscosity (Johnson and Bryden, 1989; Fox-Kemper and Ferrari, 2009). However, all coarse-resolution models use either explicit eddy viscosity or numerical schemes that amount to the same. In eddy resolving models, choosing diffusivity and viscosity carefully allows cascades of energy and enstrophy and aids accuracy in boundary current separation (Smagorinsky, 1963; Leith, 1996; Griffies and Hallberg, 2000; Chassignet and Garraffo, 2001; Arbic et al., 2007; Bryan et al., 2007; Fox-Kemper and Menemenlis, 2008).

In modern coarse-resolution models, a combination of "eddy diffusivity" and "eddy-induced velocity" is typically used (Redi, 1982; Gent and McWilliams, 1990). One reason is that mesoscale eddies—at least baroclinic mesoscale eddies—appear in the large-scale momentum equation at lower order through redistribution of buoyancy and potential vorticity rather than directly through Reynolds stresses (Andrews and McIntyre, 1978b; Greatbatch and Lamb, 1990; Gent and McWilliams, 1996; Wardle and Marshall, 2000; Eden, 2010b; Grooms et al., 2011; Marshall et al., 2012). Further, eddy stresses that are not connected to buoyancy and potential vorticity transport are more difficult to parameterize and sometimes result in "negative viscosities" (e.g., Berloff, 2005), as the steady Bernoulli conservation law in Equation (8.8) does not result in a local time–mean balance.

Thus, it is the eddy transport of active and passive scalar tracers that is the primary focus of present coarse-resolution modeling effort and the remainder of this chapter. The time–mean, coarse-resolution equations can be written, returning to Cartesian coordinates, as

$$\nabla_i\left(\bar{u}_i\bar{S}+\overline{u_i'S'}\right)=\bar{\dot{S}},\ \ \nabla_i\left(\bar{u}_i\bar{\Theta}+\overline{u_i'\Theta'}\right)=\bar{\dot{\Theta}},$$
$$\nabla_i\left(\bar{u}_i\overline{Q(b)}+\overline{u_i'Q(b)'}\right)\approx 0,\ \ \nabla_i\left(\bar{u}_i\bar{c}+\overline{u_i'c'}\right)=\bar{\dot{c}}. \tag{8.10}$$

Here, overbars denote averaging to a coarse-resolution and slowly-varying in time, and primes denote mesoscale and submesoscale deviations from that mean.[4]

The fluxes appear entirely inside a divergence in Equation (8.10), which has led some authors to treat rotational eddy fluxes as less important (e.g., Marshall and Shutts, 1981; Bryan et al., 1999; Eden et al., 2007b). Often, it is suggested that removing or adjusting a rotational contribution will lead to fluxes that are more aligned with their corresponding gradient or more likely to yield positive diffusivities. The rotational change may be chosen broadly, as in a bounded domain rotational fluxes are not uniquely defined (Fox-Kemper et al., 2003). However, here eddy fluxes will be diagnosed both from drifters using observed trajectory anomalies from the time-mean flow and from a realistic model transporting passive tracers. These fluxes and diffusivities can be kinematically related to the theory of diffusion by continuous movements of Taylor (1921)—if their rotational parts are left unchanged. Changing the rotational fluxes violates this connection to fundamental fluid processes. Here fluxes will be

4. The averaging procedure is assumed to have the properties of a Reynolds-average, or $\overline{\overline{(\cdot)}}=\overline{(\cdot)}$, $\overline{(\cdot)'}=0$, and $\overline{\overline{(*)}(\cdot)'}=\overline{(*)}\,\overline{(\cdot)'}=0$.

related directly to an anisotropic diffusivity, which implies positive diffusivity nearly everywhere, natural boundary conditions, but at the cost of fluxes that are not necessarily down their mean gradient.

Even when the non-conservative terms on the right of Equation (8.10) are negligibly small in comparison to the advection terms on the left so that the flow is still "conservative," the eddy correlations on the left may still be as large as the mean tracer transport. Thus, the unsteady, conservative ocean solutions are potentially quite different from the steady, conservative oceanic solutions. Approaching the infinite Péclet limit requires some care, as even weak non-conservative terms may strongly affect the conservative fluxes (e.g., Jones and Young, 1994; Eden et al., 2007a), but the primary discussion of non-conservative mixing in this volume is found in Chapter 7.

Averaging the density-coordinate equations (Equation 8.7) is also useful. Often the averaging will be taken at a fixed density, which follows the displacement of the density surfaces and can make fewer explicit eddy correlations appear (de Szoeke and Bennett, 1993; Young, 2012). Weighting each average by thickness or treating hu as a unit (thickness-weighting) in the density-coordinate equations is equivalent to treating ρu_i as a unit (Hesselberg or Favre averaging) in the compressible fluid equations (Greatbatch and McDougall, 2003).

In the remainder of this chapter, estimates of the eddy properties, eddy fluxes, and eddy correlations in the Equation (8.10) will be presented. From data, the Lagrangian displacements of fluid parcels will be approximated from surface drifter trajectories. Assuming a scale separation between the background tracer gradients and the decorrelation length and assuming the properties of the fluid parcels are conserved over the decorrelation time, the tracer fluxes may be related to the parcel displacements. Some of the theory of Taylor (1921) is reviewed so that these results may be understood in relation to eddy diffusivity. Direct analysis of multiple tracers in high-resolution global ocean models is also presented. This method has been used before in simpler contexts to estimate the Lagrangian transports (Plumb and Mahlman, 1987; Bratseth, 1998). Finally, studies using satellite and *in situ* data together with turbulence or hydrodynamic stability theories are also refer to, to show that different approaches find similar results.

From Figure 8.1 one should expect fair consistency between the estimates. Shown are the mean and eddy kinetic energies of the drifters, model, and AVISO multi-satellite reconstruction. The model used is an improved version of the Maltrud and McClean (2005) global 0.1° POP model, with climatological "normal-year" forcing (Large and Yeager, 2004). Consistent with the conclusions of McClean et al. (2006), the model agrees with the AVISO altimetry. However, the drifter eddy kinetic energy is quite a bit larger in the eastern side of the subtropical and subpolar gyres than either the satellite geostrophic velocity or model total velocity indicates.[5] The higher drifter kinetic energy likely results from smaller mesoscale and submesoscale features that are not resolved *spatially* by either the satellite or model (Fratantoni, 2001; Capet et al., 2008; Lévy et al., 2010; Lumpkin and Elipot, 2010; Fox-Kemper et al., 2011).

2.4. Diffusion by Continuous Movements

Taylor (1921) successfully quantifies the effects of small-scale discrete and continuous motions of a fluid. A result of particular interest here is that while the root-mean-square (rms) of parcel displacements increases on integration over a time τ (first linearly, then as the square root), the covariance of parcel displacements and parcel velocities increases at first and then saturates after the decorrelation timescale, T, when the velocity autocorrelation following a parcel displacement goes to zero. After saturation, Lagrangian displacements Y' and Lagrangian velocities V' obey:

$$\overline{Y'(\tau)Y'(\tau)} \approx 2\overline{u'^2}\tau T, \quad \overline{Y'(\tau)V'(\tau)} \approx 2\overline{u'^2}T. \tag{8.11}$$

The constant velocity-displacement covariance can be interpreted as a form of eddy diffusivity for stationary, homogeneous turbulent flow.

This general approach can be extended in a stationary, homogeneous, incompressible, 3D flow (Batchelor, 1949; Monin et al., 2007, p. 542), by using the fluid parcel displacement covariance:

$$D_{ij}(\tau) = \overline{Y_i'(\tau)Y_j'(\tau)} = \int_{t_0}^{t_0+\tau}\int_{t_0}^{t_0+\tau} \overline{V_i'(x,t_1)V_j'(x,t_2)}\mathrm{d}t_1\mathrm{d}t_2. \tag{8.12}$$

The average (overbar) is an ensemble average over many displaced fluid parcels. For sufficiently long times $\tau \gg T_{ij}$, the covariance becomes linear in time (τ):

$$D_{ij}(\tau) \approx \sqrt{\overline{u_i'^2 u_j'^2}}T_{ij}\tau,$$
$$T_{ij} = \frac{\int_0^\infty \left(\overline{V_i'(x,s)V_j'(x,s)} + \overline{V_j'(x,s)V_i'(x,s)}\right)\mathrm{d}s}{\sqrt{\overline{u_i'^2 u_j'^2}}}. \tag{8.13}$$

No summation is implied on the repeated indices of D_{ij} and T_{ij}. Note that the Eulerian velocity scales u are used to estimate the size of the Lagrangian velocity correlations, consistent with homogeneous, stationary turbulence where these scales are closely related.

5. The drifter data is filtered to remove inertial motions (5-day lowpass), and the timescale of the filtering is similar to that used for the altimetry (7 days).

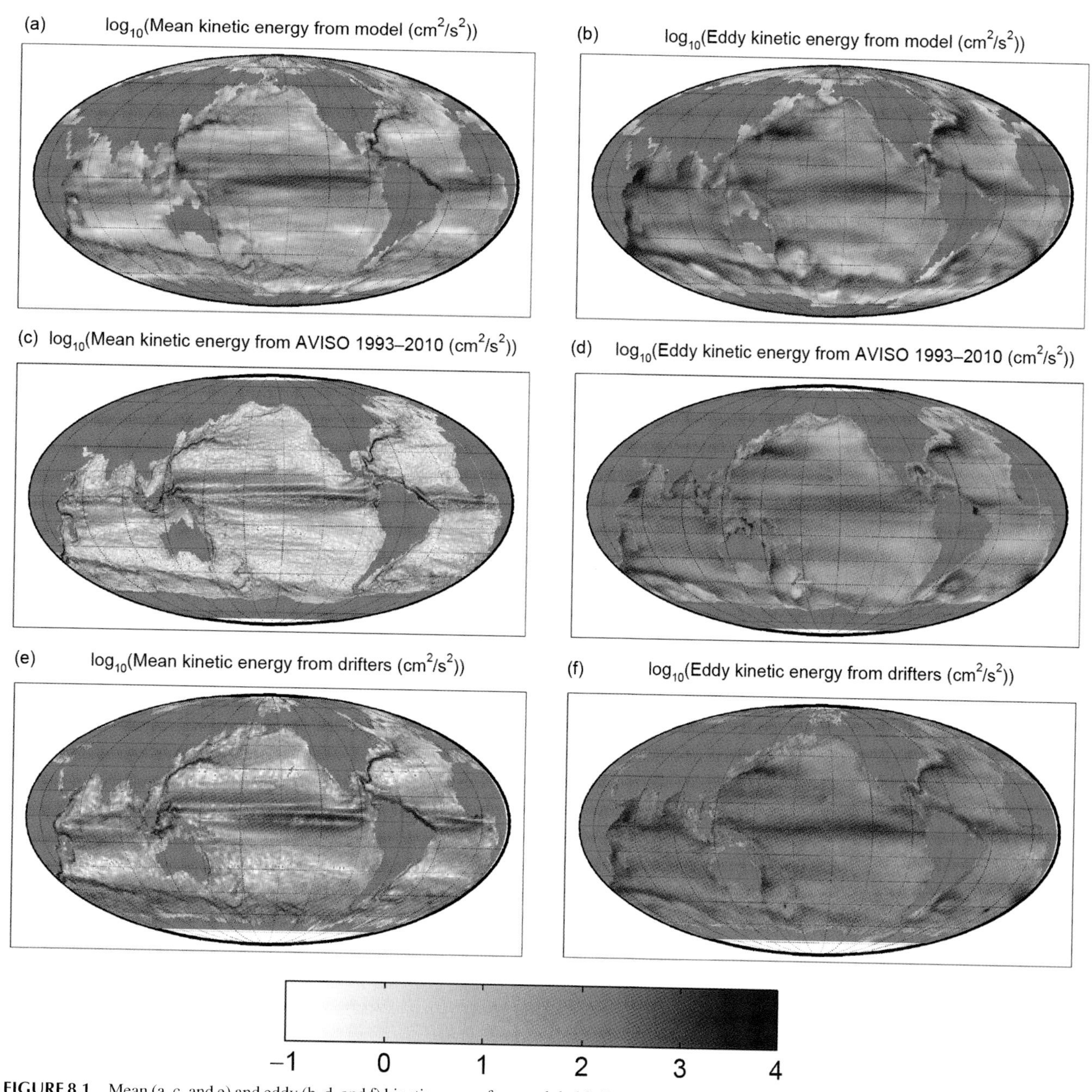

FIGURE 8.1 Mean (a, c, and e) and eddy (b, d, and f) kinetic energy from a global 0.1° model (see text for description), SSALTO/DUACS multi-satellite Maps of Absolute Dynamic Topography (MADT) product distributed by AVISO (c and d) which is based on weekly data, and from surface drifters (e and f; Lumpkin and Garraffo, 2005; Lumpkin and Garzoli, 2005), which are filtered to remove variability on timescales less than 5 days. Drifter mean kinetic energy is the energy in the time averaged flow while eddy kinetic energy is energy in motion with 5–7 day or longer timescales. Model mean kinetic energy is the energy of time-averaged flow, while eddy kinetic energy is the deviation from time-mean.

The proportionality with time of the displacement covariance for long times (τ) may be associated with an eddy diffusivity tensor, K_{ij}:

$$\begin{aligned} K_{ij} &\approx \frac{1}{2}\sqrt{\overline{u_i'^2 u_j'^2}}\,T_{ij} = \frac{1}{2}\frac{\mathrm{d}D_{ij}(\tau)}{\mathrm{d}\tau} \\ &= \int_0^{\infty} \frac{1}{2}\left(\overline{V_i'(x,s)V_j'(x,s)} + \overline{V_j'(x,s)V_i'(x,s)}\right)\mathrm{d}s. \quad (8.14) \end{aligned}$$

K_{ij} is symmetric by (8.14) and depends solely on the correlations of Lagrangian velocity displacements and velocity magnitudes in different directions. These velocities are likely to differ in each direction when symmetries are broken, for example by gravity, rotation, or other body forces. Similarly, the components of the decorrelation timescale T_{ij} are likely to vary if eddies tend to remain more coherent in one direction versus another, as is the case in an "eddy street" or turbulent wake. If a tracer is nearly conserved over the timescale T_{ij}, then every displaced fluid parcel carries its tracer with it, and the diffusivity K_{ij} may be used to diffuse the average tracer concentration as $K_{ji}\nabla_i\bar{c} \approx -\overline{u'_j c'}$, or

$$\frac{\partial \bar{c}}{\partial t} + \nabla_i(\bar{c}\bar{u}_i) = \nabla_j(K_{ji}\nabla_i\bar{c}) + \bar{\dot{c}}, \tag{8.15}$$

where all averages are at a fixed depth (Eulerian).

2.4.1. Diagnosing Eigenvectors, Eigenvalues, and Principal Axes of Diffusivities

Any symmetric tensor can be fully described by its real eigenvalues and orthogonal eigenvectors. The equation that is satisfied by the ith eigenvalue $\lambda(i)$ and ith eigenvector $v(i)$ of a tensor M is:

$$M_{kl}v_l(i) = \lambda(i)v_k(i). \tag{8.16}$$

Thus, for that particular eigenvector $v(i)$, the action of matrix multiplication by the tensor is the same as multiplication by the *scalar* eigenvalue $\lambda(i)$. However, vectors not aligned with $v(i)$ acted upon by the tensor will *not* be likely to behave as though multiplied by a scalar. A 3×3 tensor has three eigenvectors and eigenvalues, and they are real for symmetric tensors. For real and symmetric tensors, the eigenvectors may always be chosen to be an orthonormal set, often called the *principal axes* of the tensor.

The (non-negative) eigenvalues of the diffusivity K_{ij} represent a typical value of the diffusivity in each of three different eigenvector directions. In the absence of a background flow, a tracer is diffused when its gradient is oriented partly along a principal axis. The component of the gradient will be reduced *down-gradient along that direction alone* by the associated eigenvalue (diffusivity). Thus, for a tracer whose gradient is oriented along an eigenvector, the diffusivity is effectively a scalar by Equation (8.16). However, it is uncommon for an arbitrary tracer gradient to align with a principal diffusivity axis. In this case, the tracer is diffused *anisotropically*, that is, at different rates along each principal axis simultaneously. Anisotropic diffusion results from a symmetric K with distinct eigenvalues.

Only isotropic, homogeneous turbulence with no background flow is likely to be represented well by a scalar diffusivity for arbitrary tracers (i.e., three equal eigenvalues). In this case, any gradient direction can be treated as an eigenvector direction and a scalar diffusivity can be used in place of K_{ij}. In a typical oceanic flow, the rms velocities and decorrelation times in Equation (8.14) differ in each-direction. Even in isotropic, homogeneous turbulence, a background flow often leads to an anisotropic effective diffusivity (Taylor, 1953; Ferrari and Nikurashin, 2010).

Much confusion has arisen from apparently negative components of the diffusivity tensor in a realistic flow. Frenkiel (1952) shows examples of autocorrelation functions and the corresponding displacement covariances, and finds that rarely does the autocorrelation change sign frequently and vigorously, so the diagonal elements of D_{ij} are generally positive and only weakly related to the details of the autocorrelation shape. To put it another way, the diffusivity eigenvalues are generally likely to be positive. The trace of a tensor is not affected by rotation, including rotation into and out of the principal axis orientation. Thus, the sum of the diagonal elements tends to be positive in all coordinate systems. Indeed, the displacement covariance is most naturally measured in the principal axis coordinate system if known, as the tensor will be diagonal in that basis. In other coordinate systems, the elements may be positive or negative, which is not necessarily an indication of anti-diffusive behavior and negative eigenvalues, just an indication that the coordinate system is not aligned with the principal axes.

Eddy diffusivities must be positive in a sense found from the tracer variance equation, derived from Equations (8.3) and (8.10), to be

$$\frac{\partial \overline{c'^2}}{\partial t} + \nabla_i\left(\overline{u_i}\,\overline{c'^2}\right) + \nabla_i\left(\overline{u'_i c'^2}\right) + 2\overline{u'_i c'}\,\nabla_i\bar{c} = \overline{\dot{c}'c'}. \tag{8.17}$$

The first two terms do not create variance, they just transport it materially with the mean flow. If the third, triple-correlation term is neglected (often done but rarely justified), then the balance is

$$2\overline{u'_i c'}\,\nabla_i\bar{c} \approx -2(\nabla_j\bar{c})K_{ji}(\nabla_i\bar{c}) \approx \overline{\dot{c}'c'}. \tag{8.18}$$

For most forms of dissipation (e.g., molecular diffusivity), $\overline{\dot{c}'c'}$ is negative definite. When $-K$ is projected into the tracer gradient on both of its indices, $-(\nabla_j\bar{c})K_{ji}(\nabla_i\bar{c})$, the result is negative definite for arbitrary tracer gradients if the symmetric part of K *has all positive eigenvalues*. It is not necessary for all elements of K to be positive. Relatedly, a numerical model with negative diffusivities is usually numerically unstable. However, a diffusivity tensor that has negative elements in a few coordinate systems but has positive eigenvalues is not a concern.

Attempting to infer eddy diffusivity on scales where the transport is not diffusive (e.g., $t < T_{ij}$) can be misleading (e.g., Stommel, 1949). Similarly, attempting to infer a scalar diffusivity from an anisotropic diffusive process may be misleading. A 2D example is given by rotating a diagonal K by an angle θ:

$$\begin{bmatrix} \overline{u'c'} \\ \overline{v'c'} \end{bmatrix} = -\begin{bmatrix} \kappa_{xx}\cos^2\theta + \kappa_{yy}\sin^2\theta & -\kappa_{xx}\cos\theta\sin\theta + \kappa_{yy}\cos\theta\sin\theta \\ -\kappa_{xx}\cos\theta\sin\theta + \kappa_{yy}\cos\theta\sin\theta & \kappa_{yy}\cos^2\theta + \kappa_{xx}\sin^2\theta \end{bmatrix} \begin{bmatrix} \frac{\partial \bar{c}}{\partial x} \\ \frac{\partial \bar{c}}{\partial y} \end{bmatrix}. \tag{8.19}$$

If $\kappa_{xx} \gg \kappa_{yy}$ and $\frac{\partial \bar{c}}{\partial x} \sim \frac{\partial \bar{c}}{\partial y}$, then a diagnosis of the scalar "diffusivity" k by $\overline{u'c'} = -k\frac{\partial \bar{c}}{\partial x}$ would yield a negative "diffusivity" k for all $\pi/4 < \theta < 3\pi/4$.

2.5. Sources of Anisotropy in Oceanic Diffusion

The best-studied form of anisotropic diffusion results from diffusion in the presence of a shear flow. The resulting spreading, or shear dispersion, has a strongly anisotropic diffusion operator. In Taylor (1953, 1954), there are early quantifications of this effect, where it is demonstrated that in laminar pipe flow, the *effective diffusion* in the along-pipe direction (D_{eff}) greatly exceeds the molecular or eddy diffusion across the pipe (D). Indeed, the virtual along-pipe diffusion is inversely proportional to the small-scale (molecular) diffusion, $D_{\text{eff}}=D+U^2a^2/(48D)$, where U and a are the flow speed and pipe radius (i.e., maximum eddy scale). In oceanographic application, Okubo (1967) and Young et al. (1982) describe a related effect from oscillatory shear, such as from internal waves, interacting with small-scale turbulence. Young and Jones (1991); Jones and Young (1994) review shear dispersion work up to that date, and extend solutions toward shorter timescales and anomalous and chaotic diffusion regimes where Taylor's scaling does not apply. The jet-like structures found by Treguier et al. (2003) and Maximenko et al. (2005) or the Antarctic Circumpolar Current (Ferrari and Nikurashin, 2010) are examples of flows that should result in strong shear dispersion. Of particular relevance is Smith (2005), who develops scalings for passive tracer transport along and across quasi-geostrophic jets, and shows that the along-jet scaling is inversely proportional to the isotropic diffusivity and depends on eddy scale and flow rate much as in Taylor's theory.

Shear dispersion results from an interaction of the steady and isotropic, homogeneous eddying flows. However, other phenomena are also expected to result in anisotropic diffusion. Oceanic turbulence is quite heterogeneous, as shown in Figure 8.1. Horizontal velocities exceed vertical ones in the ocean, but it also seems likely that the two components of horizontal velocity will vary, as well as their correlations. From Equation (8.14), these effects also impact the diffusivity. Furthermore, jets and potential vorticity gradients are thought to suppress diffusivity in the cross-jet and PV-gradient direction (Rogerson et al., 1999; Ferrari and Nikurashin, 2010). Coherent structures and vortices have a strongly anisotropic direction of propagation (Griffa et al., 2008; Chelton et al., 2011). Finally, instabilities are often quite anisotropic; such "noodle" instabilities produce strong anisotropy (Berloff et al., 2009). Both "noodling" and coherent eddy drift also tend to align according to potential vorticity gradients—from the flow or from the planetary beta effect. At present it is unclear which of these effects are dominant in the ocean. Indeed, even the degree to which they are distinct from each other is presently unclear.

Two studies attempting to measure rotational effects on anisotropy found sensitive dependence on rotational fluxes (Eden and Greatbatch, 2009; Eden, 2010a). However, their method assumes a limited form for K (the symmetric part has two equal eigenvalues, which is not anisotropic diffusion in the sense described here). Furthermore, they subtract rotational contributions (Eden et al., 2007a), rendering their diagnostic framework too different from the assumptions of Taylor (1921, 1953) to compare directly. Riha and Eden (2011) find strong contrast between along- and across-front transports in agreement with other studies (Berloff et al., 2009; Ferrari and Nikurashin, 2010).

2.6. The Veronis Effect

Gravity breaks the symmetry of stratified turbulent flow, leading to anisotropic diffusion. If we assume uncorrelated velocities in each direction and reduced vertical velocities, then an anisotropic eddy diffusivity is a natural form to assume for the result of Equation (8.14):

$$K_{ij}=\begin{bmatrix}\kappa_h & 0 & 0\\ 0 & \kappa_h & 0\\ 0 & 0 & \kappa_v\end{bmatrix}_{ij} \tag{8.20}$$

This assumed diffusivity tensor is diagonal, so the eigenvalues are the diagonal elements and the eigenvectors can be taken to be the x, y, and z directions. The eigenvalues are chosen such that the diffusion in the x- and y-directions is the same, while the diffusion in the vertical direction is smaller.

Suppose that, with great effort, an oceanography cruise measures the eddy flux of temperature along a section, providing $\overline{u'\Theta'}$, $\overline{w'\Theta'}$, and $\bar{\Theta}(x,z)$. Delighted, the intrepid crew inverts to find the 2D eddy diffusivity components,

$$\overline{u_i'\Theta'}=-\begin{bmatrix}\kappa_h & 0\\ 0 & \kappa_v\end{bmatrix}_{ij}\nabla_j\bar{\Theta}, \rightarrow \kappa_h(\Theta)=\frac{-\overline{u'\Theta'}}{\nabla_x\bar{\Theta}}, \kappa_v(\Theta)=\frac{-\overline{w'\Theta'}}{\nabla_z\bar{\Theta}}. \tag{8.21}$$

And since the number of degrees of freedom equals the number of equations to satisfy, the result exactly satisfies the observational evidence (assuming $\nabla\bar{\Theta}\neq 0$).

The next year, another cruise measures both salinity and temperature. The temperature results prove nearly the same, aside from interannual variability, but the salinity results are confusing. The salinity fluxes are exactly predicted by diffusion formulae

$$\kappa_h(S)=\frac{-\overline{u'S'}}{\nabla_x\bar{S}}, \kappa_v(S)=\frac{-\overline{w'S'}}{\nabla_z\bar{S}}, \tag{8.22}$$

but the diffusivities disagree with those diagnosed from Θ: $\kappa_h(\Theta)\neq\kappa_h(S)$, $\kappa_v(\Theta)\neq\kappa_v(S)$. How is this possible? Do salinity and temperature have different eddy diffusivities? Not necessarily.

If the salinity and temperature gradients do not align, then they will sample different combinations of diffusivity eigenvalues, and therefore be diffused differently. But, is not all that taken into account using the tensor K_{ij}? Yes, if indeed the correct form for K_{ij} was assumed in Equation (8.20), different gradient directions would be correctly treated.[6] Just as mean flows have preferred directions, so too eddy fluxes occur fastest along dynamically selected directions, which may not be horizontal and vertical. The assumed form (Equation 8.20) for K_{ij} does not allow this to occur.

Consider instead simultaneous equations on salinity and temperature,

$$\overline{u_i'\Theta'} = -R_{ij}\nabla_j\bar{\Theta}, \ \overline{u_i'S'} = -R_{ij}\nabla_j\bar{S}. \tag{8.23}$$

Now for 2D observations, there are four equations, but only 2 degrees of freedom can be matched by Equation (8.20). However, a 2×2 matrix for R_{ij} will do nicely, and indeed we expect it will have a symmetric part representing diffusion with its eigenvectors aligned and perpendicular to the principal axis of maximal disturbance, regardless of whether that direction is exactly vertical or somewhat tilted.

If an incorrect form for R is used, it may still be possible to tune to reproduce the transport for one tracer as it was in Equation (8.20). However, other tracers will be diffused incorrectly. One expects largest eigenvectors to fall along the minimal-disturbance plane, so a mistaken diffusion ansatz such as Equation (8.20)—even as a numerical artifact (Griffies et al., 2000)—results in spurious diabatic or dianeutral mixing of different water masses called the *Veronis effect* (Veronis, 1975).

What if there are more than two tracers, or more than three in three dimensions? Well, since Equation (8.3) is linear in tracer concentration, the equations for tracer concentration and fluxes can be added and subtracted as desired. If all of the tracers have only a large-scale gradient, then only three whose gradients are not aligned are needed to map the diffusivity tensor and any more can be related to a linear combination of these three. However, most tracers not only have a large-scale gradient, but also variations on larger and smaller scales that differ from other tracers. In this case, a least-squares or otherwise optimized tensor can average appropriately over variations in the local gradient (Bachman and Fox-Kemper, 2013).

2.7. Streamfunction and Diffusivity

If the experiment measuring temperature and salinity fluxes were conducted in a region where small-scale mixing in the maximal-disturbance direction was weak, then present theory would predict the result to be nearly (Griffies, 1998)

$$\overline{u'c'} = -\begin{bmatrix} \kappa_h & 0 \\ 2\kappa_h\mathcal{S} & \mathcal{S}\kappa_h\mathcal{S} \end{bmatrix}\nabla\bar{c}. \tag{8.24}$$

The minimal-disturbance slope appears ($\mathcal{S}$ from Section 2.1.2), and indeed the eigenvectors are aligned along it, but the tensor is not symmetric! It is perhaps easier to contemplate this as the combination of two distinct eddy effects: diffusion, the symmetric part of this tensor, and advection, represented by a streamfunction (Redi, 1982; Gent and McWilliams, 1990; Griffies, 1998; Bachman and Fox-Kemper, 2013):

$$\overline{u'c'} = -\left(\begin{bmatrix} \kappa_h & \kappa_h\mathcal{S} \\ \kappa_h\mathcal{S} & \mathcal{S}\kappa_h\mathcal{S} \end{bmatrix} + \begin{bmatrix} 0 & -\kappa_h\mathcal{S} \\ \kappa_h\mathcal{S} & 0 \end{bmatrix}\right)\nabla\bar{c}. \tag{8.25}$$

In the Taylor (1921, 1953) theory, the 3D diffusion tensor in Equation (8.14) was explicitly symmetric. This symmetry is a consequence of the fluid having constant density in that case.[7] If, instead, displaced parcels are correlated with variations in density, then an advection-like term arises. That is, if $\overline{V'\rho'}\neq 0$, then mass will tend to move in the direction the denser parcels tend to go on average, even if there is no mean velocity $\overline{V'}=0$.

An elegant theory that helps explain this effect is that of Dukowicz and Smith (1997) (an extension of Monin and Yaglom, 1971, to compressible cases and stratified flow). In their theory, they consider stochastic relocation of fluid parcels, and the evolution of the probability distribution of the location of those parcels. The probability that a parcel originally located at $\boldsymbol{x}$ at time t will later be at z at time $t+\Delta t$ is $p(z, t+\Delta t|\mathbf{x}, t)$. Two important quantities can be derived from this distribution, the motion of the center of the parcel displacement and the evolving correlation of displacements. These are

$$v_i(x,t) = \lim_{\Delta t\to 0}\frac{1}{\Delta t}\int d^3z(z_i - x_i)p(z,t+\Delta t|x,t), \tag{8.26}$$

$$K_{ij}(x,t) = \frac{1}{2}\lim_{\Delta t\to 0}\frac{1}{\Delta t}\int d^3z(z_i - x_i)(z_j - x_j)p(z,t+\Delta t|x,t). \tag{8.27}$$

Comparing the second relation with Equation (8.13), if the integral over all possible relocations weighted by their probability can be associated with the average used by Taylor (1921), then $(z_i - x_i)\sim X_i'$ and this K_{ij} is the equivalent to K_{ij} in Taylor's theory.

However, the centroid displacement velocity v in Equation (8.26) does not have an analog in that theory. Dukowicz and Smith (1997) show that probability-weighted

6. Although nonlocal, incomplete scale-separation effects might still be an issue.

7. Boussinesq fluids are not sufficiently constant in density to qualify, for reasons that will be made clear.

averaging over all relocations of parcels in the 3D compressible equations (Equation 8.1) yields

$$\frac{\partial \bar{\rho}}{\partial t}+\nabla_i\left(\bar{\rho}v_i-\nabla_j K_{ji}\bar{\rho}\right)=\nabla_j K_{ji}\,\nabla_i\bar{\rho},$$
$$\frac{\partial \overline{\rho c}}{\partial t}+\nabla_i\left(\overline{\rho c}v_i-\frac{\overline{\rho c}}{\bar{\rho}}\nabla_j K_{ji}\bar{\rho}\right)=\nabla_j K_{ji}\nabla_i\overline{\rho c}+\overline{\rho\dot{c}}. \quad (8.28)$$

In an incompressible fluid, then the simplest way to satisfy the former is to set $\bar{u}_i=v_i-\nabla_j K_{ji}$, since incompressibility already ensures $\nabla_i\bar{u}_i=0$.[8] This assumption relates the relocation of the probability center to the Eulerian mean velocity $\bar{u}_i$, and recovers (Equation 8.15) from the second equation above. Note that when K_{ij} varies spatially, this implies that v_i *can be divergent even in a nondivergent flow*! This property is generally true of Lagrangian velocities (Andrews and McIntyre, 1978a), and here results from the relatively few constraints placed on the stochastic parcel displacements—only the combination of diffusion and Lagrangian advection is required to conserve volume, not each independently.

If the fluid is compressible, then Equation (8.28) can still be reduced, and Dukowicz and Smith (1997) argue that the same assumption for v is the simplest. Then,

$$\frac{\partial \bar{\rho}}{\partial t}+\nabla_i[\bar{\rho}\bar{u}_i]=\nabla_j\left[K_{ji}\nabla_i\bar{\rho}\right],\ \frac{\partial \overline{\rho c}}{\partial t}+\nabla_i\left[\overline{\rho c}\,\bar{u}_i-\frac{1}{\bar{\rho}}K_{ij}\nabla_j\bar{\rho}\right]$$
$$=\nabla_i\left(K_{ij}\nabla_j\overline{\rho c}\right)+\overline{\rho\dot{c}}. \quad (8.29)$$

The effect of compressibility is therefore *advective* (bracketed) *and diffusive* (right side). Note that the right side is guaranteed to be diffusive, as K_{ij} is symmetric. Combining together the advective parts into U yields

$$\frac{\partial \bar{\rho}}{\partial t}+\nabla_i[\bar{\rho}U_i]=0,\ \frac{\partial \overline{\rho c}}{\partial t}+\nabla_i[\overline{\rho c}\,U_i]$$
$$=\nabla_i\left(K_{ij}\nabla_j\overline{\rho c}\right)+\overline{\rho\dot{c}},\ U_i\equiv\bar{u}_i-\frac{1}{\bar{\rho}}K_{ij}\nabla_j\bar{\rho}.$$

Dukowicz and Smith (1997) also exploit the similarity of the compressible equations (Equation 8.1) to the density-coordinate equations (Equation 8.7). Considering stochastic eddy displacements *within the ϱ surfaces* results in

$$\frac{\partial \bar{h}^{\varrho}}{\partial t}+\nabla_\alpha\left[\bar{h}^{\varrho}\left(\overline{u_\alpha}^{\varrho}-\frac{1}{\bar{h}^{\varrho}}K_{\alpha\beta}\nabla_\beta\bar{h}^{\varrho}\right)\right]=-\frac{\partial \bar{e}^{\varrho}}{\partial\varrho}, \quad (8.30)$$

$$\frac{\partial \overline{hc}^{\varrho}}{\partial t}+\nabla_\alpha\left[\overline{hc}^{\varrho}\left(\overline{u_\alpha}^{\varrho}-\frac{1}{\bar{h}^{\varrho}}K_{\alpha\beta}\nabla_\beta\bar{h}^{\varrho}\right)\right]$$
$$=\nabla_\alpha\left[K_{\alpha\beta}\nabla_\beta\overline{hc}^{\varrho}\right]+\overline{z_\rho\dot{c}}^{\varrho}-\frac{\partial\overline{ec}^{\varrho}}{\partial\varrho}+\frac{\partial\overline{F_c}^{\varrho}}{\partial\varrho}. \quad (8.31)$$

Here, it has been assumed that the entrainment velocity and the stochastic along-ϱ displacements are statistically independent. The form of Equations (8.30) and (8.31) can be applied directly in density-coordinate models.

The additional velocity $u_i^*=-\frac{1}{\bar{h}^{\varrho}}K_{\alpha\beta}\nabla_\beta\bar{h}^{\varrho}$, is sometimes called the *eddy-induced velocity*, and is closely related to the quasi-Stokes and bolus velocities (McDougall and McIntosh, 2001). It is a result of the fact that as the tracer anomalies flow along the minimal layer, they may correlate with a bolus of excess water or a thinning of the layer. Just as the Stokes drift results from a correlation of larger velocities when a wave crest is present, the quasi-Stokes velocity results from a correlation between tracer anomaly and bolus presence.

The net effect of the stochastic process in Equations (8.30) and (8.31) is to move thickness h along the ϱ surfaces, horizontally and potentially anisotropically in the two horizontal directions. The eddy-induced velocity u_i^* alone is required in the thickness Equation (8.30), and both u_i^* and diffusion appear in the tracer Equation (8.31). It is incorrect to call any of the terms in Equation (8.30) a diffusion (Gent, 2011), even though the diffusivity $K_{\alpha\beta}$ appears in u_i^*.

Note that only one (four-component) diffusivity $K_{\alpha\beta}$ appears (in contrast to frequent modeling practice), which is related via Equation (8.27) to the likely correlations of displacements within the ϱ layer, just as the generalization of Taylor (1921) in Equation (8.14) relates diffusivity to the displacements of fluid parcels. These kinematic relations do not depend directly on whether c is active or passive, or if there is more than one tracer, which one is presently under consideration. Only the assumption of stochastic displacements that conserve h and ch during their displacement is required. To contrast against Eden et al. (2007a) and others who find different diffusivities for different tracers (e.g., Lee et al., 1997; Smith and Marshall, 2009), two crucial differences should be considered: (1) Was the diagnosed diffusivity sufficiently general to capture heterogeneity or anisotropy in the flow? and (2) Were the different tracers similar in their rate of non-conservative properties? Bachman and Fox-Kemper (2013) demonstrate that even when non-conservative processes are much smaller than conservative eddy transports, changing the non-conservative rates may affect the conservative rate estimates.

Transforming Equation (8.30) to depth-coordinates yields the continuity equation for the eddy-induced velocity,

$$\nabla_i\left(\bar{u}_i+u_i^*\right)=0,\ \nabla_i\bar{u}_i=0\ \rightarrow\nabla_i u_i^*=0. \quad (8.32)$$

The u^* velocity as found by Dukowicz and Smith (1997) is similar, but not identical to, that utilized or diagnosed elsewhere (Andrews and McIntyre, 1978a; Plumb, 1979; Plumb and Mahlman, 1987; Gent and McWilliams, 1990; McDougall and McIntosh, 1996, 2001; Eden et al., 2007a; Eden and Greatbatch, 2009; Eden, 2010a; Bachman and Fox-Kemper, 2013). The Dukowicz and Smith (1997) velocity is distinguished by its

8. Different choices have also been explored (Dukowicz and Smith, 1997; Dukowicz and Greatbatch, 1999; Smith, 1999).

incompressibility in three dimensions and its association to the stochastic minimal-disturbance displacements of thickness and tracer. To ensure incompressibility is satisfied, often a vector streamfunction is used to generate u^*. Thus,

$$\epsilon_{ijk}\nabla_j\Psi^*_k = u^*_i. \tag{8.33}$$

For small slopes the streamfunction generates the horizontal component of the eddy-induced velocity via

$$u^*_\alpha = K_{\alpha\beta}\nabla_z\mathcal{S}_\beta = \epsilon_{\alpha jk}\nabla_j\Psi^*_k. \tag{8.34}$$

In component form,

$$u^*_x = -\nabla_z\Psi^*_y = K_{x\beta}\nabla_z\mathcal{S}_\beta,\ u^*_y = \nabla_z\Psi^*_x = K_{y\beta}\nabla_z\mathcal{S}_\beta,$$
$$u^*_z = \nabla_x\Psi^*_y - \nabla_y\Psi^*_x.$$

For arbitrary spatial variations in K, determining the streamfunction Ψ^* from u^* or K and $\mathcal{S}$ involves solving the two differential equations in z. Gent and McWilliams (1990) and Smith and Gent (2004) propose a slightly different form, which is identical if the $K_{\alpha\beta}$ are not a function of z:

$$\begin{aligned} u^{gm}_\alpha &= -\nabla_z(K_{\alpha\beta}\mathcal{S}_\beta) = \epsilon_{\alpha jk}\nabla_j\Psi^{gm}_k,\ u^{gm}_x = -\nabla_z\Psi^{gm}_y \\ &= \nabla_z(K_{x\beta}\mathcal{S}_\beta),\ u^{gm}_y = \nabla_z\Psi^{gm}_x = \nabla_z(K_{y\beta}\mathcal{S}_\beta),\ u^{gm}_z \\ &= \nabla_x\Psi^{gm}_y - \nabla_y\Psi^{gm}_x = -\nabla_\alpha(K_{\alpha\beta}\mathcal{S}_\beta). \end{aligned} \tag{8.35}$$

This form does not require solving any differential equations, except perhaps in implementing boundaries (Ferrari et al., 2010).

Another way of considering the effects of the streamfunction is as a "skew" flux of tracers along minimum-disturbance planes. To make this clear, note that the divergence of the cross-product of Ψ^* with the tracer gradient is the same as the divergence of advection of the tracer by u^*,

$$\nabla_i\left(\Psi^*_j\epsilon_{ijk}\nabla_k c\right) = \nabla_i\left(c\epsilon_{ijk}\nabla_j\Psi^*_k\right) = \nabla_i\left(cu^*_i\right). \tag{8.36}$$

The relation (8.36) depends only on the symmetry of the derivatives and the antisymmetry of ϵ, so it applies to Ψ^{gm} as well.

Ψ^{gm} has special properties if the tracer being advected is buoyancy,

$$\nabla_i\left(bu^*_i\right) = \nabla_\alpha\left(-K_{\alpha\beta}\nabla_\beta b\right) + \nabla_z\left[(\nabla_\alpha b)K_{\alpha\beta}(\nabla_\beta b)\right]. \tag{8.37}$$

Thus, the action of the Ψ^{gm} has two parts: a horizontal diffusion by the $K_{\alpha\beta}$ and the final term is an upward buoyancy transport if $K_{\alpha\beta}$ has positive eigenvalues. This vertical flux ensures consistent extraction of potential energy from the mean in Equation (8.5) by $\overline{w'b'} > 0$.

Some authors have suggested that potential vorticity conservation is more fundamental (Killworth, 1997; Dukowicz and Greatbatch, 1999; Treguier, 1999). In practice, these differences amount to relatively small corrections compared to other assumptions, if the typical displacement distance is less than a substantial range of latitude (Smith, 1999; Zhao and Vallis, 2008; Fox-Kemper and Ferrari, 2009; Grooms et al., 2011).

Griffies (1998) uses a tensor form to make the relationships above clear upon transformation to z coordinates. For an arbitrary Ψ^*, from Equation (8.34) or Equation (8.35), the divergence of tracer flux is

$$\nabla_i\overline{u'_ic'} \approx -\nabla_i\left\{\left(\begin{bmatrix} 0 & 0 & \Psi^*_y \\ 0 & 0 & -\Psi^*_x \\ -\Psi^*_y & \Psi^*_x & 0 \end{bmatrix} + \begin{bmatrix} K_{xx} & K_{xy} & K_{x\alpha}\mathcal{S}_\alpha \\ K_{xy} & K_{yy} & K_{y\alpha}\mathcal{S}_\alpha \\ K_{x\alpha}\mathcal{S}_\alpha & K_{y\alpha}\mathcal{S}_\alpha & \mathcal{S}_\alpha K_{\alpha\beta}\mathcal{S}_\beta \end{bmatrix}\right)_{ij}\nabla_j\bar{c}.\right\} \tag{8.38}$$

If the Gent and McWilliams (1990) form (Equation 8.35) for Ψ^{gm} is used, this simplifies to

$$\nabla_i\overline{u'_ic'} \approx -\nabla_i\left\{\begin{bmatrix} K_{xx} & K_{xy} & 0 \\ K_{xy} & K_{yy} & 0 \\ 2K_{x\alpha}\mathcal{S}_\alpha & 2K_{y\alpha}\mathcal{S}_\alpha & \mathcal{S}_\alpha K_{\alpha\beta}\mathcal{S}_\beta \end{bmatrix}_{ij}\nabla_j\bar{c}\right\}. \tag{8.39}$$

Now, comparing Equation (8.39) with Equations (8.24) and (8.25) connnects this theory to the hypothetical cruise data (Equation 8.24).

Eden and colleagues (Eden et al., 2007a; Eden and Greatbatch, 2008; Eden, 2010a) allow the possibility of a different eddy transport for every tracer. By doing so, they are able to simplify each tracer variance Equation (8.17), so that eddy diffusivity occurs only when non-conservative effects are present as in Equation (8.18). Their treatment is best-suited to problems where different tracers have radically different rates of variance dissipation $\overline{\dot{c}'c'}$. Their diagnosis is limited in the number of degrees of freedom it can measure: expressing Equation (8.2) from both Eden et al. (2007a) and Eden et al. (2007b) in terms of R from (Equation 8.14) yields

$$\begin{aligned} \overline{u'_ib'} &= -K^{eg}_{ij}\nabla_j\bar{b} \\ &= -\begin{bmatrix} K_{\text{dia}} & -B_z & B_y \\ B_z & K_{\text{dia}} & -B_x \\ -B_y & B_x & K_{\text{dia}} \end{bmatrix}_{ij}\nabla_j\bar{b} + \text{rotational flux}, \end{aligned} \tag{8.40}$$

which applies throughout the nearly adiabatic ocean interior. The authors are able, by choosing specialized rotational fluxes, to reduce the number of free parameters from the four here (K_{dia}, B_x, B_y, B_z) to three that can be determined by the buoyancy flux. However, the tensor in Equation (8.38) potentially has 9 degrees of freedom, and at least 6 of them are distinct measures of turbulence statistics representing the symmetric tensor (Equation 8.14). Many of the Transformed Eulerian Mean papers propose similar diagnoses of specialized forms of R based on buoyancy fluxes alone (e.g., Andrews and McIntyre, 1976; Ferrari and Plumb, 2003). However, there are many more degrees of freedom in R than can be prescribed by the three components of a single flux, which is the root cause of the appearance of the uncertain parameters in those analyses. It is also the reason why the diagnoses below can proceed and find reasonable diffusivities and eddy-induced advection without recourse to rotational flux manipulations.

3. OBSERVATIONS AND MODELS OF SPATIAL VARIATIONS OF EDDY STATISTICS

The isoneutral diffusivity tensor K should vary in space as the rms velocities certainly vary from location to location, as does the decorrelation timescale tensor:

$$K_{\alpha\beta} \approx \frac{1}{2}\sqrt{\overline{u'^2_\alpha u'^2_\beta}}\,T_{\alpha\beta} = \frac{1}{2}\frac{\mathrm{d}D_{\alpha\beta}(\tau)}{\mathrm{d}\tau}. \tag{8.41}$$

Further, shear dispersion effects (Taylor, 1953; Smith, 2005) and potential vorticity barrier effects (Ferrari and Nikurashin, 2010) will vary spatially as the mean flow and potential vorticity gradients vary. Measurements of the components of the K tensor are rare, even at the surface. In this section, we compare 2D measurements from surface drifters of $K_{\alpha\beta}$ with a model estimate, and then describe the variations with depth and 3D structure of the model estimate.

To begin with, consider the global distribution of horizontal eddy kinetic energy, closely related to the trace of the K tensor. Figure 8.1 shows the logarithm of the mean and eddy kinetic energy, as estimated from a high-resolution model (Grooms et al., 2011; Fox-Kemper et al., 2013), from surface drifters, and from SSALTO/DUACS multi-satellite altimetry (SSALTO/DUACS Team, 2013). Note that the products are in rough agreement, despite slightly different averaging procedures and the fact that the altimetry provides only the geostrophic velocity.[9] Eddy kinetic energy is largest surrounding the regions where the mean kinetic energy is largest, such as western boundary currents, in the tropical jets, and in the core of the Antarctic Circumpolar Current. Regions of high eddy energy are broader than the tightly concentrated mean kinetic energy, which is consistent with the formation and dispersal of eddies from instabilities of the energetic mean features and asymptotics (Grooms et al., 2011).

The mean kinetic energy is highly anisotropic, with narrow features that extend for thousands of kilometers in many cases. It seems unlikely that this structure would lead to eddies with isotropic decorrelation timescales and isotropic kinetic energy. The eddy kinetic energy is more isotropic (not shown), with typically less than a factor of 2 difference between the zonal and meridional rms velocities—much less than the ratio of the diffusion eigenvalues. These conditions suggest shear dispersion is expected to be an important source of anisotropy.

In the analysis of drifter data, it is commonly the goal to eliminate the effects of shear dispersion and measure the cross-stream diffusivity. Following Davis (1991), the first stage of calculating a diffusivity from drifter observations is correlating the displacements with the drifter velocity, after removing a time–mean velocity for each analysis location.

$$K_{\alpha\beta}(\boldsymbol{x},t) = -\langle V'_\alpha(t_0|\boldsymbol{x},t_0)X'_\beta(t_0-t|\boldsymbol{x},t_0)\rangle, \tag{8.42}$$

where $V'_\alpha(t|x,t_0)$ represents the value of the α component of V' at time t that passes through location $\boldsymbol{x}$ at time t_0. In practice, this quantity is calculated for lag $t-t_0$ along the trajectories of the drifters, and then averaged into spatial bins to approximate an Eulerian field of K. Comparing Equation (8.42) to Equations (8.14) and (8.27) shows the close relationship of this diffusivity estimate to the preceding ones, apart from the specific averaging operator. One of the major concerns when mapping diffusivities and time scales from drifter or float data is this use of finite bins rather than fixed positions, which makes shear dispersion magnify the resulting diffusivity (c.f., Bauer et al., 1998). In simulation experiments with isotropic eddy energy in a flow with mean shear, Oh and Zhurbas (2000) showed that shear dispersion can be minimized by eigenanalysis of the diffusivity tensor (Equation 8.41) and taking the smaller eigenvalue to represent an approximate isotropic lateral diffusivity. Smith (2005) arrives at a similar result in an idealized model. The effect of shear dispersion is then reflected almost entirely in the larger horizontal eigenvalue. Yet, in terms of parameterizations for coarse-resolution ocean models, the goal is *not* to eliminate shear dispersion, but to predict and include it correctly.

The Oh and Zhurbas (2000) approach was exploited to map diffusivities in the Pacific and Atlantic Oceans from drifter data through 1999 (Zhurbas et al., 2003; Zhurbas and Oh, 2004). The global distribution of the minor eigenvalue of K in Figure 8.2c and ratio of minor to major eigenvalue is shown in Figure 8.3, calculated from drifter data through 2010. The values were calculated using lag times of 5, 10, 15, and 20 days. The figure depicts the average of the 10, 15, and 20 day values, since these lag times exceed the integral timescale and so estimate the diffusive limit of Taylor (1921). Thus, this figure updates the results of Zhurbas and Oh (2004) and extends them into the Indian Ocean.

Anisotropic diffusion should be included in coarse-resolution model parameterizations so that the modeled general circulation matches (Equation 8.10). For this reason, both large and small drifter eigenvalues are shown

9. In the model, a mean velocity over 6 years was used as the mean, and all perturbations from that mean were taken as eddies. The figure shows two degree averages of mean KE and eddy KE. Drifter KE was calculated from drogued surface drifter trajectories. First, the trajectories were binned into trajectories originating in one degree boxes, and the mean velocity in each bin was removed. Then variability on timescales faster than 5 days was removed by lowpass filtering (Lumpkin and Garraffo, 2005; http://www.aoml.noaa.gov/phod/dac/drifter_climatology.html). The remaining variability is associated with eddy velocities u'. Horizontal MKE and EKE were then calculated for each one degree bin. The 7-day SSALTO/DUACS absolute geostrophic velocity fields were broken down into the time-mean over 1993–1999 and the perturbation KE (SSALTO/DUACS Team, 2013). Thus, all variability faster than 7 days was reduced in the averaging process as was done deliberately for the drifters. The ME and EKE from the satellites is shown on the grid of this analysis. In all datasets, the eddy kinetic energy varies by more than 2 orders of magnitude in the horizontal.

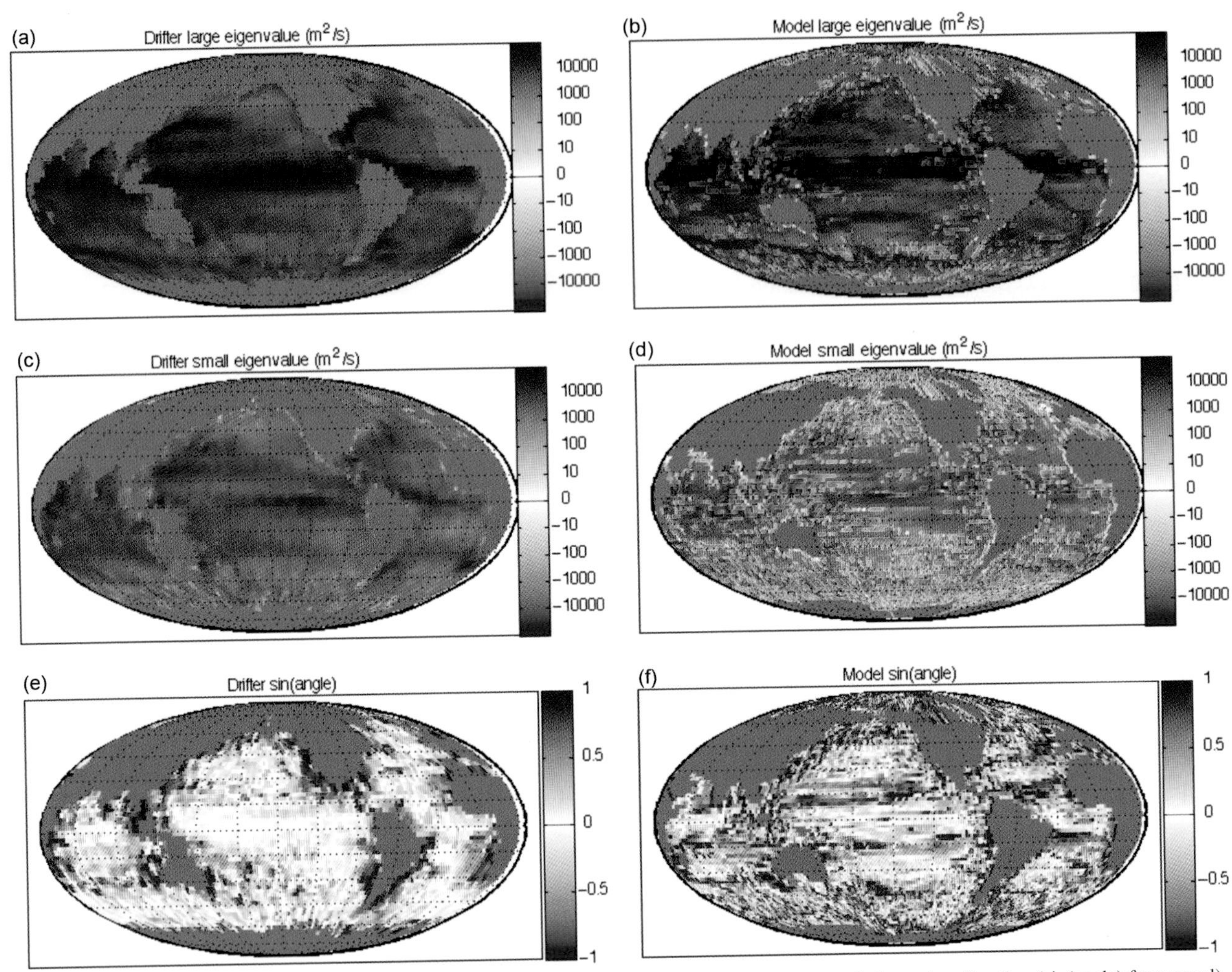

FIGURE 8.2 Surface drifter (a, c, and e) and model (b, d, and f) horizontal diffusivity eigenvalues and eigenvalue direction (sin(angle) from zonal).

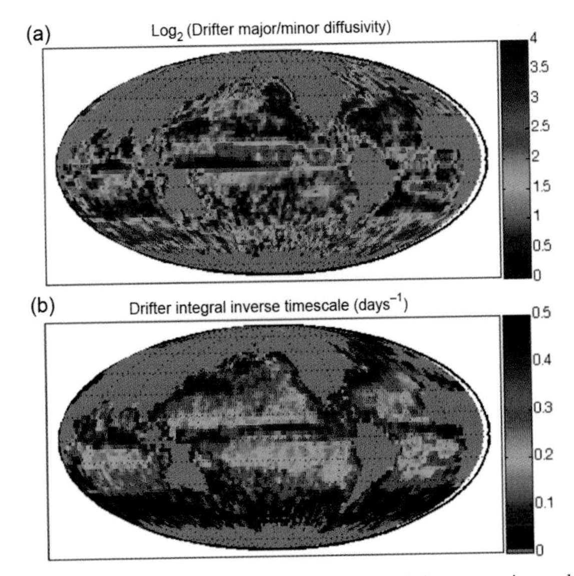

FIGURE 8.3 (a) Log of the major to minor diffusivity tensor eigenvalues from drifters. (b) Drifter integral inverse timescale (days^{-1}).

in Figure 8.2. A complementary diagnostic approach is used (Fox-Kemper et al., 2013) to produce model estimates of $K_{\alpha\beta}$ and of the right side of Equation (8.38). Nine passive tracers (the tracers are oriented in three directions—vertical, zonal, meridional—and restored to this initial distribution with three different timescales—no restoring, 1 year restoring, and 3 year restoring) are simulated in the model already described. A least-square method (Bratseth, 1998; Bachman and Fox-Kemper, 2013) is used to invert the relationship

$$\overline{u_i'c'} = -R_{ij}\nabla_j\bar{c} + \xi. \tag{8.43}$$

and solve for the components of R and reduce the magnitude of the scatter ξ. By Equation (8.38), we expect R to be closely related to Ψ and K. The larger eigenvalues of the R tensor for the surface model gridpoint, which are associated with nearly horizontal transport, are what is shown in Figure 8.2. The inversion for R becomes indeterminate if one of the tracer gradients vanishes. Figure 8.2 shows many regions where evaluating this tensor is compromised by the small vertical gradients in the mixed layer. However, direct comparison with the surface drifters requires the surface value (diagnosed diffusivities quickly decrease with depth). Both the minor and major eigenvalues agree between the drifters and model where the model diagnoses are not contaminated by the weak vertical tracer gradients. Despite the different methods, assumptions, models and measurement biases, and despite difficulties with exceptionally long integral timescales for the major eigenvalue, the pattern and magnitude of the eigenvalues typically agree within a factor of 2. Bachman and Fox-Kemper (2013) estimate R to within a few percent, but only in an idealized simulation.

The bottom two panels show the direction of the eigenvectors, that is, if the large horizontal eigenvalue is associated with zonal diffusion, this value is zero. While the agreement is far from perfect, away from the coastlines and boundary currents, the dominant eigenvector tends to be zonal. At the equator, this is expected from shear dispersion. It is presently unclear whether shear dispersion or additional eddy anisotropy is required to produce this effect in other regions. Figure 8.3a shows the ratio of the minor eigenvalue to major eigenvalue in the drifter data. In the regions of strong shear, such as the ACC and western boundary currents, these eigenvalues are separated by a factor of ten or more. In the gyre interior, these ratios decrease to between one and two, with the larger eigenvalue being nearly zonal.

The Lagrangian integral time and length scales are estimated by

$$T = -\frac{K_2}{\langle u_2' u_2' \rangle}, \quad L = -\frac{K_2}{\sqrt{\langle u_2' u_2' \rangle}}. \tag{8.44}$$

where K_2 is the minor horizontal eigenvalue of $K_{\alpha\beta}$ and u_2 is the minor eigenvalue of the Lagrangian velocity covariance matrix (Zhurbas et al., 2003). The global distribution of these scales is shown in Figure 8.3b. These values are qualitatively in agreement with earlier regional drifter-based studies (c.f., Krauss and Boning, 1987; Poulain and Niiler, 1989; Swenson and Niiler, 1996; Lumpkin et al., 2002). This result is consistent with Lumpkin et al. (2002), who showed that Lagrangian time scales calculated by four different methods—integrating to the first zero crossing, integrating the squared autocorrelation function to a large fixed lag (Richman et al., 1977; Lumpkin et al., 2002, use 120 days), fitting a "yardstick" to derive the integral length scale (Rupolo et al., 1996) and a parametric approach (Griffa et al., 1995)—all give qualitatively the same distribution of time scales in the North Atlantic.

The surface drifters and the model agree at the surface, but at depth the only information available is from the model. Figure 8.4 shows all components of the R tensor from the model at 318 m depth. Below the mixed layer, the tracer gradients tend not to vanish, so the model inversion is more robust. The horizontal components of the tensor (indicated by K and related to the isoneutral diffusivity) are dominantly symmetric, consistent with the Taylor (1921) and Dukowicz and Smith (1997) theory and the results of Eden (2010a). The outer row and column of R should be reduced by a factor of S, and indeed they are

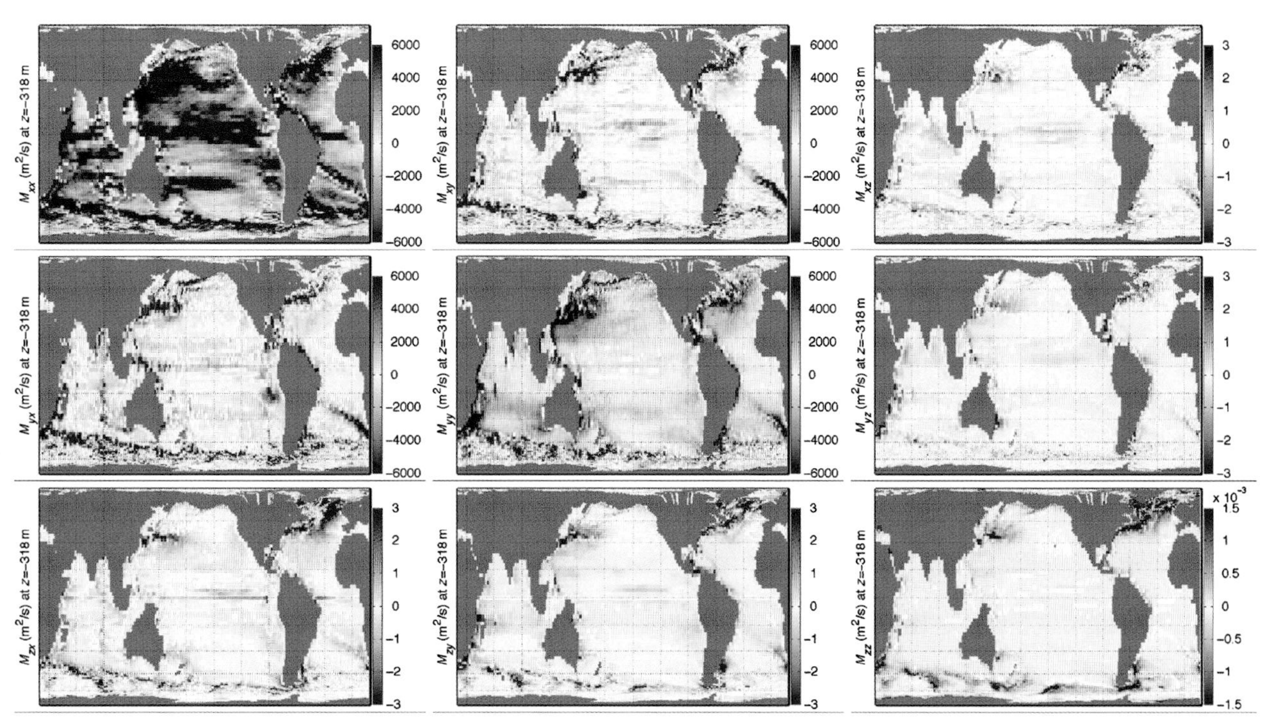

FIGURE 8.4 Components of the R tensor at 318 m depth, with the $K_{\alpha\beta}$ part in the upper left four panels.

about 1000 times smaller than the $K_{\alpha\beta}$ components. The R_{zz} component is smaller by another factor of 1000, consistent with the squared slope in Equations (8.38) and (8.39).

The outer row and column of the R tensor are not symmetric. Instead, these portions of the tensor combine the antisymmetric part related to the streamfunction and the symmetric part related to the projection of the minimal-disturbance diffusion of tracers into the vertical direction in Equation (8.38). Indeed, if the K_{ij} for these two different processes is identical and the Gent and McWilliams (1990) parameterization is correct, then as in Equation (8.39) the R_{xz} and R_{yz} elements of R should be zero. They are not exactly zero, but they are a great deal smaller than R_{zx} and R_{zy}. If the estimated uncertainty in the tracer gradient inversion above, along with uncertainty due to limited simulation length and other diagnostic issues, is taken into account, then the R_{xz} and R_{yz} elements are indistinguishable from zero in most locations (Fox-Kemper et al., 2013). Bachman and Fox-Kemper (2013) show close agreement with Equation (8.39) in an idealized model. Thus, the horizontal mixing $K_{\alpha\beta}$ that generates the symmetric and antisymmetric parts of R in the Dukowicz and Smith (1997) theory should be the same, that is, the same $K_{\alpha\beta}$ in the Gent and McWilliams (1990) parameterization as in the Redi (1982) parameterization. It is possible that small, but nonzero R_{xz} and R_{yz} elements arise from the distinction between Ψ^*, which is diagnosed in R, and Ψ^{gm} which cancels exactly in Equation (8.39). Unfortunately, the decay with depth in these fields is too weak and uncertainty too great to distinguish between Ψ^{gm} from Equation (8.35) and Ψ^* from Equation (8.34).

Figure 8.5 shows the eigenvectors and eigenvalues of the R tensor. As R is not symmetric, these eigenvalues can be imaginary, but typically they are not. Instead, the eigenvalues remain diffusion-like with positive eigenvalues, but the eigenvectors are not orthogonal (as they are for a symmetric tensor). The first eigenvalue is quite similar to the R_{xx} zonal component, and the associated eigenvectors are typically zonal. Thus, the horizontal direction of typical rapid diffusion along the neutral plane is the zonal direction. The second eigenvalue is weaker, and is often related to the R_{yy} meridional component of the R tensor. The third eigenvalue is much smaller and is related to the vertical eddy transport, the R_{zz} component of the Redi tensor. The horizontal components of the R tensor are nearly symmetric, as it should be to be consistent with the Taylor displacement correlation tensor (in agreement with the lack of such antisymmetry found by Eden, 2010a).

Fox-Kemper et al. (2013) go further in evaluating the eigenvector directions and consider their projection onto the horizontal gradient of potential vorticity. Generally, the largest direction of diffusion is perpendicular to the potential vorticity gradient, while the second largest diffusion is in the potential vorticity gradient direction, as in Ferrari and Nikurashin (2010). Where the eigenvectors are not aligned with the zonal and meridional directions is precisely where there are strong currents, consistent with both shear dispersion and transport barriers. However, it is

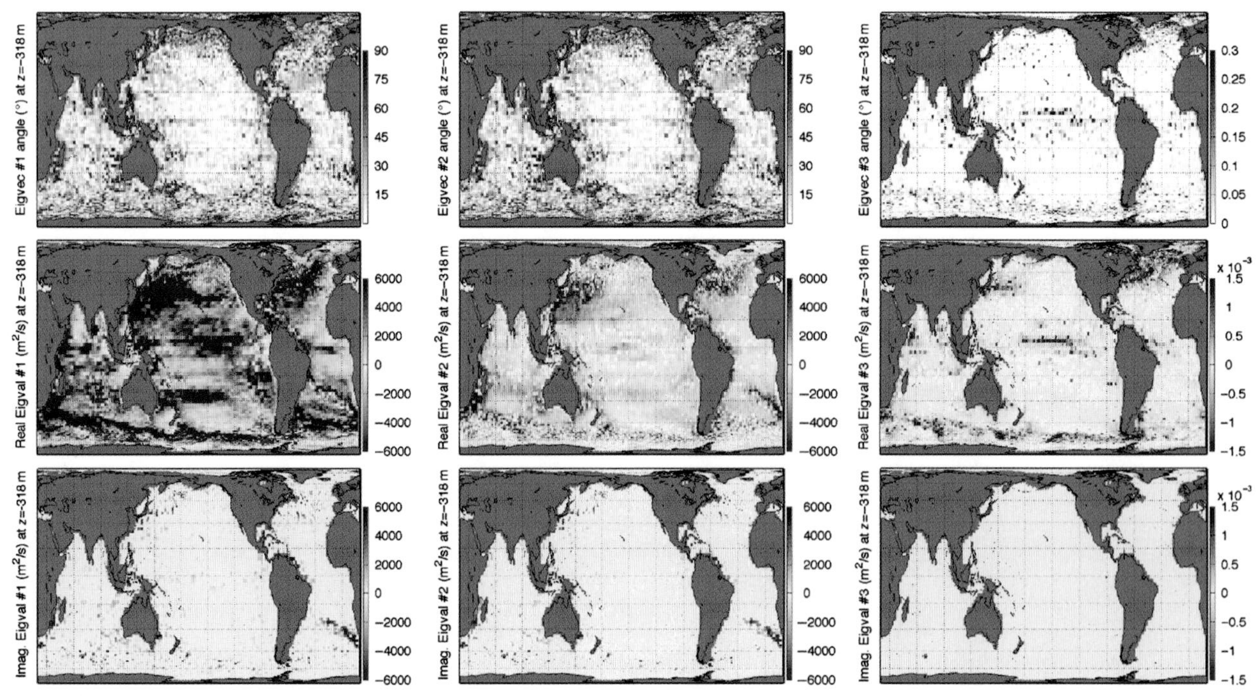

FIGURE 8.5 Eigenanalysis of the R tensor. (Top) Eigenvector direction of symmetric part of R, $(R_{ij}+R_{ji})/2$, as angle in degrees of deviation from expectation. Largest magnitude eigenvalue direction versus zonal direction (top, left), middle magnitude eigenvalue direction versus meridional direction (top, center), smallest magnitude eigenvalue direction versus vertical direction (top, right). Real (middle) and imaginary (bottom) parts of the largest eigenvalue (left), second eigenvalue (center), and smallest eigenvalue (right) of R. All values shown are at 318 m depth.

unclear if the zonal transport preference is a result of shear dispersion in zonal jets, a preference for zonal propagation of coherent eddies, or a feature of typical instabilities.

Other realistic diagnostic studies (Jayne and Marotzke, 2002; Ferreira et al., 2005; Marshall et al., 2006; Abernathey et al., 2010; Griesel et al., 2010) generally agree with the magnitude and pattern of diffusivities here. However, these studies all use a scalar diffusivity and a limited number of tracers or single tracer for diagnosis, and thus are prone to fail when the tracer gradient is not aligned with a principal diffusion axis. The Nakamura (1996) diagnosis method also reveals only a single scalar diffusivity, which is a complicated average of eigenvalues depending on contour shape. Figure 8.5 indicates that the principal axis direction is not trivially predicted, especially in the region of strong currents. Other authors have attempted to use uncertain gauge choices or other diagnostic techniques such as rotational-divergent flux decompositions to increase the ability of a scalar diffusivity to represent the data (Eden et al., 2007a; Griesel et al., 2010). However, such approaches typically rely on unmeasurable assumptions (Fox-Kemper et al., 2003). The results here indicate that the K_{ij} tensor is likely too anisotropic for such approaches to work well, in agreement with surface drifter analysis (Zhurbas and Oh, 2004).

4. MESOSCALE ISONEUTRAL DIFFUSIVITY VARIATION PARAMETERIZATIONS

Many studies have shown that the methods of parameterizing lateral transport by mesoscale and smaller features affect the general circulation (Veronis, 1975; Danabasoglu and McWilliams, 1995; Gent et al., 1995; McDougall et al., 1996; Hirst and McDougall, 1998, etc.). Increasing resolution to finer than 0.1° continues to affect the general circulation (Oschlies, 2002; Lévy et al., 2010) due to submesoscale processes (Fox-Kemper et al., 2011).

The majority of current eddy parameterization schemes are based on the work of Redi (1982) and Gent and McWilliams (1990). As described above, Griffies (1998) shows how these two important papers provide the basic flux-gradient relationship needed to represent a Dukowicz and Smith (1997) closure in a z-coordinate model. Density-coordinate models require a different form of closure Equations (8.30) and (8.31). However, all of these approaches require a prediction of the isoneutral diffusivity $K_{\alpha\beta}$, which is spatio-temporally variable and anisotropic as shown above. Cleverly, Gent and McWilliams (1990) note that this tensor needs prediction, but decline to propose how.

How to model the spatio-temporal and flow-dependent variations of $K_{\alpha\beta}$? The first conceptually successful theory specifically for this purpose is that of Visbeck et al. (1997). Their form is isotropic and based on a timescale from linear instability theory (Eady, 1949). It is

$$K_{xx} = K_{yy} = \frac{0.0015f}{\sqrt{\mathrm{Ri}}} l_z^2 \tag{8.45}$$

where l_z is intended to be a baroclinic zone width inspired by Green (1970). However, in typical implementations, the gridscale or Rhines scale is used for l_z. Similar scalings for geophysical horizontal diffusivities based on linear instability, deformation radius, or other physical scales are also commonly discussed (Green, 1970; Stone, 1972; Larichev and Held, 1995; Held and Larichev, 1996).

Some parameterizations of the spatial variation of K are based on diagnostic analyses rather than theoretical considerations (Danabasoglu and Marshall, 2007). This approach has the advantage of getting the right pattern of diffusivity, but maybe for the wrong reasons and with the wrong sensitivity to changes in the flow.

More recently, parameterizations based on a production–dissipation of eddy kinetic energy (Eden and Greatbatch, 2008) have been tested in a simplified form (Eden et al., 2009). Under additional assumptions and simplifications, this balance can be related to a parameter similar to the Eady timescale, but one that varies with depth, so it is different from the Visbeck et al. (1997) parameterization. While this approach has the advantage over linear-instability-based approaches of being valid at finite amplitude, many aspects of the finite amplitude dynamics such as the inverse cascade, anisotropy, barotropic production, and realistic eddy kinetic energy dissipation are presently neglected. Many authors have shown that these effects contribute to the eddy fluxes of tracer (Smith et al., 2002; Arbic and Flierl, 2004; Arbic et al., 2007; Thompson and Young, 2007), so production–dissipation balances as in Eden and Greatbatch (2008) will not be fully explored until they have been extended to include such effects. Even so, Eden et al. (2009) compare the present simplified parameterizations of $K_{\alpha\beta}$, their bias from climatology data, and the level of impact on model results.

4.1. Parameterizations Versus Diagnosed K

Figure 8.6 compares the diffusivities that resulted from the different parameterizations tested by Eden et al. (2009). All of the $K_{\alpha\beta}$ they tested were horizontally isotropic, so $K_{\alpha\beta} = \kappa$, which according to Oh and Zhurbas (2000) may be compared to the smaller horizontal eigenvalue of the tensor K. While the diffusivities vary widely, the diagnosis-based (Danabasoglu and Marshall, 2007) and production–dissipation-based (Eden and Greatbatch, 2008) estimates agree with each other more than they do with any of the other estimates. Using these two parameterizations in a simulation resulted in similar modeled salinity and temperature climatologies and biases, while using a constant diffusivity and the Visbeck et al. (1997) parameterization produced radically different and less realistic simulation climatologies (Eden et al., 2009). The pattern

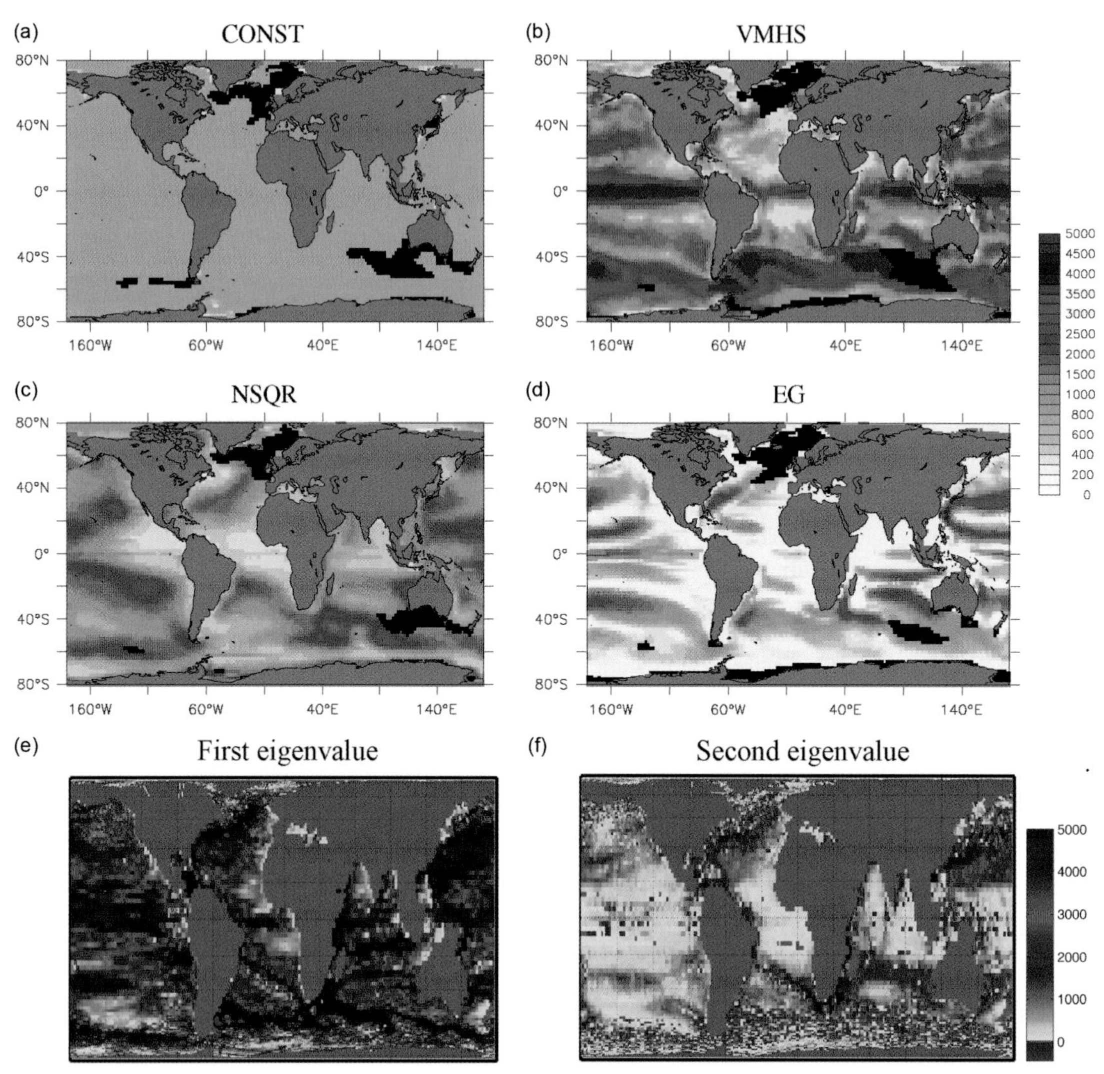

FIGURE 8.6 Comparison of diffusivity variations realized using different common parameterizations of spatial variation of (isotropic) diffusivity $K_{\alpha\beta}=\kappa$. (a) Constant diffusivity, (b) Visbeck et al. (1997), (c) Danabasoglu and Marshall (2007), (d) Eden and Greatbatch (2008), and (e and f) first and second eigenvalue from Figure 8.5. Black areas result from landmarks on the sphere remapping onto this projection. *Figures (a–d) are taken from Eden et al. (2009).*

of parameterization diffusivities (Figure 8.6c and d) is similar to the weaker horizontal eigenvalue diagnosed from the model here (Figures 8.5 (center) and 8.6f), but the parameterization magnitudes are too small. No isotropic parameterization predicts diffusivities nearly as large as the major eigenvalue in the tracer-diagnosed model (Figures 8.5 and 8.2) or drifter estimate (Figure 8.2), but a parameterization of shear dispersion might.

4.1.1. Eddy Scales Versus Instability Scale

Estimates of the characteristics of eddies short of prescribing a $K_{\alpha\beta}$ estimate are also useful in grounding the mixing estimates. Satellite estimates of the statistics of oceanic mesoscale variability have been invaluable in setting model parameters at realistic values and validating theory (e.g., Wunsch and Stammer, 1995; Stammer, 1997, 1998; Kushner and Held, 1998; Scott and Wang, 2005; Arbic and Scott, 2008). Other studies using data-assimilating models have been used to estimate eddy statistics (Abernathey et al., 2010; Mazloff et al., 2010; Tulloch et al., 2011). A successful parameterization will be in agreement with such estimates in terms of eddy kinetic energy (Figure 8.1), eddy integral timescale (Figure 8.3), and eddy lengthscale (related to timescale and kinetic energy). Predicting these scalars accurately allows prediction of the minor eigenvalue of $K_{\alpha\beta}$. The anisotropy of $K_{\alpha\beta}$ associated with the major eigenvalue is an additional effect, which may be associated with the longer separation distance of parcels in a sheared mean flow (Taylor, 1953), sustained eddy correlation timescales in a particular direction, or other effects.

Many argue that the integral length and timescales should be closely related to the fastest-growing or energetically favored modes from linear instability theory. Recent

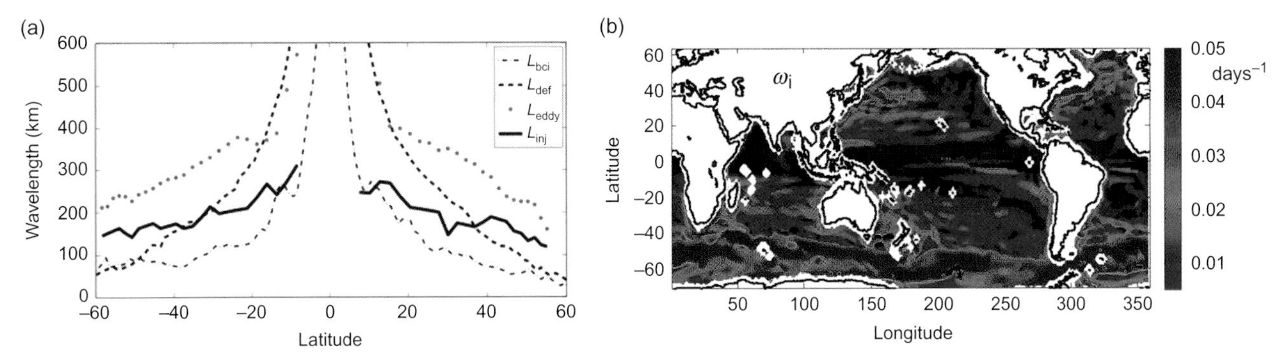

FIGURE 8.7 (a) Typical wavelength of baroclinic instability (L_{bci}), Rossby deformation radius (L_{def}), and energy-containing scale (L_{eddy}), and instability energy injection scale (L_{inj}) as estimated by Tulloch et al. (2011) based on ocean reanalysis (Forget, 2010) and AVISO altimetry. (b) Baroclinic instability inverse timescale (Tulloch et al., 2011).

studies have examined the structure, growth rates, and length scales of linear instability throughout the world, based on realistic stratification and shear profiles (Smith, 2007; Tulloch et al., 2011). The most recent and comprehensive of these studies is Tulloch et al. (2011). Figure 8.7 reproduces the most relevant results from that study—the eddy length and time scales—which may be compared to the drifter integral inverse timescale in Figure 8.3.

4.1.2. Eddy Versus Instability Spatial Scale

Even when production–dissipation balance is assumed, an additional prediction of length or time scale is needed to arrive at the units of diffusivity as in Equation (8.41). Part of the spatial-variation of K parameterization may be based on an *eddy mixing length* conceptual framework (Held and Larichev, 1996; Spall, 2000; Thompson and Young, 2007). In Eden et al. (2009) a simplified eddy lengthscale (minimum of the Rhines scale and deformation radius) is used. Prandtl (1925) first proposed a mixing length to describe momentum transfer across a boundary layer (eddy viscosity) by a mass of moving turbulent fluid. Prandtl's and Taylor's theory share a conceptual similarity to the thermodynamic mean free path. In some subgrid models, a prognostic model to predict the turbulent lengthscale is used (see Burchard and Petersen, 1999; Pope, 2000). The eddy length scale in Figure 8.7a shows that different measures of scalar eddy length scale vary by perhaps a factor of 4, which is an estimate of the magnitude of the inverse cascade. Diffusivity would be in error by the square of the error in the estimate of eddy integral lengthscale. Thus, if the (Eden et al., 2009) lengthscale were increased to reflect that the energy-containing eddy scale is 1–2 times larger than the assumed deformation radius scale, the parameterization diffusivity would be 1–4 times stronger and very close to the diagnosed K in Figure 8.6.

4.1.3. Eddy Versus Instability Time Scale

While there are some similarities in pattern, the instability timescale (Figure 8.7b) is longer than the drifter integral timescale (3b) by about a factor of 10, and the instability timescale increases as the equator is approached, consistent with $1/|f| \to \infty$. The drifter timescale is *shorter* near the equator, by contrast. Tulloch et al. (2011) argue that a particular weighting of the Eady (1949) timescale does a good job of predicting observed eddy statistics. It remains to be seen if this better linear instability timescale can improve the diffusivity in a parameterization like Visbeck et al. (1997); the disagreement between the instability and decorrelation timescales here is not encouraging.

4.2. New Parameterization Approaches and Future Developments

As is clear from the results of Eden et al. (2009), the eddy parameterization details strongly affect simulations. The Gent and McWilliams (1990) parameterization applies in the ocean interior, and treatment of how the parameterization is adjusted as a boundary is approached has recently been advanced. Early approaches used unphysical tapering schemes (Gnanadesikan et al., 2007), but recently two novel approaches have made the boundary treatment more physical. The first approach transitions from the neutral, along-ϱ surface mixing plus velocity described above to a purely horizontal diffusion in the surface layer, where diabatic effects render the approach less meaningful (Treguier et al., 1997; Danabasoglu et al., 2008; Ferrari et al., 2008). The deliberate transition from eddy transport along neutral planes to across occurs over the depth where diabatic processes occur sporadically (the transition layer) across a gridcell and is complete across the depths where diabatic processes occur consistently (the mixed layer). The second approach is that of Ferrari et al. (2010), which raises the order of Equation (8.35) to include a vertical smoothing of the streamfunction compared to the neutral slope, which they argue is related to the dominance of low-mode vertical structures in the production of the u^*. Based on the differential order of the smoothing operator, they gain extra boundary conditions that may be used to satisfy surface and bottom boundary conditions. Conceivably, a similar approach might adopt the differential form of Equation (8.34) instead of

Equation (8.35). Danabasoglu et al. (2008) and Ferrari et al. (2010) produce smoother profiles of eddy fluxes as the boundaries are approached, consistent with high-resolution models. Significantly, both novel boundary methods are independent of other specifications of the horizontal and vertical variations of $K_{\alpha\beta}$, so they may continue to be used with new parameterizations of the spatial variations.

While not a lateral transport in the ocean interior, the mixed layer eddy parameterization for submesoscale eddy restratification of Fox-Kemper et al. (2008) and Fox-Kemper et al. (2011) shares many features with the Gent and McWilliams (1990) parameterization. Bachman and Fox-Kemper (2013) show that a matching symmetric parameterization, as in Redi (1982), should be used with the Fox-Kemper et al. (2008) antisymmetric parameterization. However, the stratification and shear of the mixed layer are simpler than in the full ocean depth, so Fox-Kemper et al. (2008) were able to handle the surface and mixed layer base boundary problem conditions and the flow-dependent spatial variation of mixed layer eddies all at once. Continued development of the parameterization approaches in the previous section, guided by the diagnostic results from eddy-resolving models and data in the preceding section, may soon allow the mesoscale problem to be similarly complete.

Eddies and their transport across the Antarctic Circumpolar Current (ACC) are of present interest due to the recent DIMES experiment (Gille et al., 2007), as well as the potentially significant impact of changing winds over the ACC (Lovenduski et al., 2008). Some high-resolution models have shown that eddy sensitivity to wind may be difficult to reproduce in parameterizations (Hallberg and Gnanadesikan, 2006), but some results are more encouraging, that coarse-resolution models may be doing a reasonable job (Gent and Danabasoglu, 2011) or may be improved with new scalings (Abernathey et al., 2011). Along the way, properties of the ACC eddy transport have been discovered (Smith and Marshall, 2009; Abernathey et al., 2010; Ferrari and Nikurashin, 2010).

Currently, a number of groups are developing and using eddy-permitting models on a large or global scale (e.g., McClean et al., 2011; Delworth et al., 2012). As computer speed increases, these approaches will become standard and the coarse-resolution, eddy-free models using the mesoscale parameterizations in the preceding section will only be used in contexts such as millennial-scale and longer paleoclimate simulations, high complexity Earth system modeling, large ensemble projects, etc. This new generation of eddy-permitting models will require a different kind of parameterization, more akin to the Large Eddy Simulation subgrid models in use for engineering and boundary layer turbulence applications. Some early progress in this direction has been made (Smagorinsky, 1963; Leith, 1996; Roberts and Marshall, 1998; Griffies and Hallberg, 2000; Fox-Kemper and Menemenlis, 2008; Hecht et al., 2008; Chen et al., 2011; San et al., 2011, 2013; Graham and Ringler, 2013).

5. CONCLUSIONS AND REMAINING QUESTIONS

The processes that dominate the lateral transport in the ocean interior, both the mean flow and mesoscale eddy transport, are presently observed, diagnosed from high-resolution simulations, and parameterized in coarse resolution models with modest success. These estimates do not all agree, and many *ad hoc* assumptions are yet to be understood in a larger context of appropriate theory.

The agreement between the modeled eddy tracer transport and the anisotropic diffusivity in the surface drifter diagnosis suggests that the effects of anisotropy and shear dispersion, while perhaps not strictly part of the "eddy diffusivity," are nonetheless missing from a coarse resolution model and should be approximated with subgrid eddy closures.

It is useful to summarize here the aspects of these lateral transport processes that are clear and consistent. The transport in the ocean interior is largely along directions that minimize disturbances of energy or stratification, at a rate closely related to the mean and rms horizontal velocities and eddy correlation scales. The eddy kinetic energy, eddy length scale, and eddy time scale are not spatially homogeneous, nor are the mean flow statistics. Heterogeneity in the eddy statistics, as well as interactions with the spatially variable mean flow, lead to large variations in the rate of eddy tracer transport. There is a relationship between linear instability scales and those of the finite amplitude eddies, but distinctions between the two roughly rationalizes the disagreement in magnitude between extant parameterizations and diagnosed diffusivity eigenvalues. The horizontal eddy diffusivity is properly a tensor based on its connection to displacement covariances, and strong anisotropy is common, although continued study of mechanisms is needed to understand and properly parameterize anisotropy. Finally, observations of tracers worldwide have led to good maps of tracer concentrations against which models may be validated, but even more observations and modeling are needed to account for all tracers and correlations of interest, including the biological and chemical.

Model errors in the transport of tracers result in significant model biases, but improvement has occurred and will continue. As a general rule, parameterizations should be as simple as possible and diagnoses as general as possible. Otherwise, errors cannot be diagnosed and excessive tuning may make our solutions look good for the wrong reasons.

ACKNOWLEDGMENT

The authors thank the reviewers and editor for patient, hard work and useful suggestions that greatly improved this chapter. The altimeter products were produced by the SSALTO/DUACS Team (2013) and distributed by AVISO, with support from CNES (http://www.aviso.oceanobs.com/duacs/). B. F.-K. was supported by NSF OCE 0825614. R. L. was supported by NOAA's Office of Climate

Observations and the Atlantic Oceanographic and Meteorological Laboratory. F. O. B. was supported by NSF through its sponsorship of NCAR. The model simulations here were run by John Dennis, with resources from IBM and the NCAR CISL HPC Advisory Panel (CHAP).

REFERENCES

Abernathey, R., Marshall, J., Mazloff, M., Shuckburgh, E., 2010. Enhancement of mesoscale eddy stirring at steering levels in the southern ocean. J. Phys. Oceanogr. 40, 170–184.

Abernathey, R., Marshall, J., Ferreira, D., 2011. Dependence of Southern Ocean meridional overturning on wind stress. J. Phys. Oceanogr. 41, 2261–2278.

Andrews, D.G., McIntyre, M.E., 1976. Planetary waves in horizontal and vertical shear; asymptotic theory for equatorial waves in weak shear. J. Atmos. Sci. 33, 2049–2053.

Andrews, D.G., McIntyre, M.E., 1978a. An exact theory of nonlinear waves on a Lagrangian-mean flow. J. Fluid Mech. 89, 609–646.

Andrews, D.G., McIntyre, M.E., 1978b. Generalized Eliassen-Palm and Charney-Drazin theorems for waves on axisymmetric flows in compressible atmospheres. J. Atmos. Sci. 35, 175–185.

Arbic, B.K., Flierl, G.R., 2004. Baroclinically unstable geostrophic turbulence in the limits of strong and weak bottom Ekman friction: application to midocean eddies. J. Phys. Oceanogr. 34, 2257–2273.

Arbic, B.K., Scott, R.B., 2008. On quadratic bottom drag, geostrophic turbulence, and oceanic mesoscale eddies. J. Phys. Oceanogr. 38 (1), 84–103.

Arbic, B.K., Flierl, G.R., Scott, R.B., 2007. Cascade inequalities for forced—dissipated geostrophic turbulence. J. Phys. Oceanogr. 37, 1470–1487.

Bachman, S., Fox-Kemper, B., 2013. Eddy parameterization challenge suite. I: eady spindown. Ocean Model. 64, 12–28.

Batchelor, G.K., 1949. Diffusion in a field of homogeneous turbulence. 1. Eulerian analysis. Aust. J. Sci. Res. A Phys. Sci. 2 (4), 437–450.

Bauer, S., Swenson, M., Griffa, A., Mariano, A., Owens, K., 1998. Eddy mean flow decomposition and eddy-diffusivity estimates in the tropical Pacific Ocean 1. Methodology. J. Geophys. Res. Oceans 103 (C13), 30855–30871.

Berloff, P., 2005. On dynamically consistent eddy fluxes. Dyn. Atmos. Oceans 38, 123–146.

Berloff, P.S., McWilliams, J.C., 1999. Large-scale, low-frequency variability in wind-driven ocean gyres. J. Phys. Oceanogr. 29, 1925–1949.

Berloff, P., Kamenkovich, I., Pedlosky, J., 2009. A model of multiple zonal jets in the oceans: dynamical and kinematical analysis. J. Phys. Oceanogr. 39 (11), 2711–2734.

Bratseth, A.M., 1998. On the estimation of transport characteristics of atmospheric data sets. Tellus 50A, 451–467.

Broecker, W., 1987. The biggest chill. Nat. Hist. 96, 74–82.

Bryan, K., Dukowicz, J.K., Smith, R.D., 1999. On the mixing coefficient in the parameterization of bolus velocity. J. Phys. Oceanogr. 29 (9), 2442–2456.

Bryan, F.O., Hecht, M.W., Smith, R.D., 2007. Resolution convergence and sensitivity studies with North Atlantic circulation models. Part I: the western boundary current system. Ocean Model. 16, 141–159.

Burchard, H., Petersen, O., 1999. Models of turbulence in the marine environment—a comparative study of two-equation turbulence models. J. Mar. Syst. 21, 29–53.

Capet, X., Mcwilliams, J.C., Mokemaker, M.J., Shchepetkin, A.F., 2008. Mesoscale to submesoscale transition in the California current system. Part I: flow structure, eddy flux, and observational tests. J. Phys. Oceanogr. 38, 29–43.

Charney, J.G., 1955. The Gulf Stream as an inertial boundary layer. Proc. Natl. Acad. Sci. U.S.A. 41 (10), 731–740.

Chassignet, E.P., Garraffo, Z.D., 2001. Viscosity parameterization and the Gulf Stream separation. In: Muller, P., Henderson, D. (Eds.), From Stirring to Mixing in a Stratified Ocean. Proceedings of the 12th 'Aha Huliko'a Hawaiian Winter Workshop. University of Hawaii, pp. 37–41.

Chelton, D.B., Schlax, M.G., Samelson, R.M., 2011. Global observations of nonlinear mesoscale eddies. Prog. Oceanogr. 91 (2), 167–216.

Chen, Q., Gunzburger, M., Ringler, T., 2011. A scale-invariant formulation of the anticipated potential vorticity method. Mon. Weather Rev. 139, 2614–2629.

Danabasoglu, G., Marshall, J., 2007. Effects of vertical variations of thickness diffusivity in an ocean general circulation model. Ocean Model. 18, 122–141.

Danabasoglu, G., McWilliams, J., 1995. Sensitivity of the global ocean circulation to parameterizations of mesoscale tracer transports. J. Clim. 8 (12), 2967–2987.

Danabasoglu, G., Ferrari, R., McWilliams, J.C., 2008. Sensitivity of an ocean general circulation model to a parameterization of near-surface eddy fluxes. J. Clim. 21 (6), 1192–1208.

Davis, R., 1991. Observing the general circulation with floats. Deep Sea Res. Part A 38 (Supplement 1), S531–S557.

Delworth, T., Rosati, A., Anderson, W., Adcroft, A., Balaji, V., Benson, R., Dixon, K., Griffies, S., Lee, H., Pacanowski, R., et al., 2012. Simulated climate and climate change in the GFDL CM2. 5 High-resolution coupled climate model. J. Clim. 25, 2755–2781.

de Szoeke, R.A., Bennett, A.F., 1993. Microstructure fluxes across density surfaces. J. Phys. Oceanogr. 23 (10), 2254–2264.

Dukowicz, J., Greatbatch, R., 1999. The bolus velocity in the stochastic theory of ocean turbulent tracer transport. J. Phys. Oceanogr. 29, 2232–2239.

Dukowicz, J., Smith, R., 1997. Stochastic theory of compressible turbulent fluid transport. Phys. Fluids 9, 3523–3529.

Eady, E.T., 1949. Long waves and cyclone waves. Tellus 1, 33–52.

Eden, C., 2010a. Anisotropic rotational and isotropic residual isopycnal mesoscale eddy fluxes. J. Phys. Oceanogr. 40 (11), 2511–2524.

Eden, C., 2010b. Parameterising meso-scale eddy momentum fluxes based on potential vorticity mixing and a gauge term. Ocean Model. 32, 58–71.

Eden, C., Greatbatch, R.J., 2008. Diapycnal mixing by meso-scale eddies. Ocean Model. 23 (3–4), 113–120.

Eden, C., Greatbatch, R.J., 2009. A diagnosis of isopycnal mixing by mesoscale eddies. Ocean Model. 27 (1–2), 98–106.

Eden, C., Greatbatch, R.J., Olbers, D., 2007a. Interpreting eddy fluxes. J. Phys. Oceanogr. 37, 1282–1296.

Eden, C., Greatbatch, R.J., Willebrand, J., 2007b. A diagnosis of thickness fluxes in an eddy-resolving model. J. Phys. Oceanogr. 37, 727–742.

Eden, C., Jochum, M., Danabasoglu, G., 2009. Effects of different closures for thickness diffusivity. Ocean Model. 26 (1–2), 47–59.

Edwards, C.A., Pedlosky, J., 1998. Dynamics of nonlinear cross-equatorial flow. Part i: potential vorticity transformation. J. Phys. Oceanogr. 28 (12), 2382–2406.

Ferrari, R., Nikurashin, M., 2010. Suppression of eddy diffusivity across jets in the Southern Ocean. J. Phys. Oceanogr. 40 (7), 1501–1519.

Ferrari, R., Plumb, R.A., 2003. Residual circulation in the ocean. In: Proceedings of the 13th 'Aha Huliko'a Hawaiian Winter Workshop 13, pp. 219–228.

Ferrari, R., McWilliams, J.C., Canuto, V.M., Dubovikov, M., 2008. Parameterization of eddy fluxes near oceanic boundaries. J. Clim. 21 (12), 2770–2789.

Ferrari, R., Griffies, S.M., Nurser, A.J.G., Vallis, G.K., 2010. A boundary-value problem for the parameterized mesoscale eddy transport. Ocean Model. 32 (3–4), 143–156.

Ferreira, D., Marshall, J., Heimbach, P., 2005. Estimating eddy stresses by fitting dynamics to observations using a residual-mean ocean circulation model and its adjoint. J. Phys. Oceanogr. 35, 1891–1910.

Forget, G., 2010. Mapping ocean observations in a dynamical framework: a 2004–06 ocean atlas. J. Phys. Oceanogr. 40, 1201–1221.

Fox-Kemper, B., 2004. Wind-driven barotropic gyre II: effects of eddies and low interior viscosity. J. Mar. Res. 62 (2), 195–232.

Fox-Kemper, B., 2005. Reevaluating the roles of eddies in multiple barotropic wind-driven gyres. J. Phys. Oceanogr. 35 (7), 1263–1278.

Fox-Kemper, B., Ferrari, R., 2009. An eddifying Parsons model. J. Phys. Oceanogr. 39 (12), 3216–3227.

Fox-Kemper, B., Menemenlis, D., 2008. Can large eddy simulation techniques improve mesoscale-rich ocean models? In: Hecht, M., Hasumi, H. (Eds.), Ocean Modeling in an Eddying Regime. AGU Geophysical Monograph Series, vol. 177. Wiley, pp. 319–338.

Fox-Kemper, B., Pedlosky, J., 2004. Wind-driven barotropic gyre I: circulation control by eddy vorticity fluxes to an enhanced removal region. J. Mar. Res. 62 (2), 169–193.

Fox-Kemper, B., Ferrari, R., Pedlosky, J., 2003. On the indeterminacy of rotational and divergent eddy fluxes. J. Phys. Oceanogr. 33 (2), 478–483.

Fox-Kemper, B., Ferrari, R., Hallberg, R., 2008. Parameterization of mixed layer eddies. Part I: theory and diagnosis. J. Phys. Oceanogr. 38 (6), 1145–1165.

Fox-Kemper, B., Danabasoglu, G., Ferrari, R., Griffies, S.M., Hallberg, R.W., Holland, M.M., Maltrud, M.E., Peacock, S., Samuels, B.L., 2011. Parameterization of mixed layer eddies. III: implementation and impact in global ocean climate simulations. Ocean Model. 39, 61–78.

Fox-Kemper, B., Bryan, F.O., Dennis, J., 2013. Global diagnosis of the mesoscale eddy flux tracer gradient relationship. Ocean Model. in preparation.

Fratantoni, D.M., 2001. North atlantic surface circulation during the 1990's observed with satellite-tracked drifters. J. Geophys. Res. 106 (C10), 22,067–22,093.

Frenkiel, F.N., 1952. On the statistical theory of turbulent diffusion. Proc. Natl. Acad. Sci. U.S.A. 38 (6), 509–515.

Gent, P.R., 2011. The Gent–Mcwilliams parameterization: 20/20 hindsight. Ocean Model. 39, 2–9.

Gent, P.R., Danabasoglu, G., 2011. Response to increasing Southern Hemisphere winds in CCSM4. J. Clim. 24, 4992–4998.

Gent, P.R., McWilliams, J.C., 1990. Isopycnal mixing in ocean circulation models. J. Phys. Oceanogr. 20, 150–155.

Gent, P.R., McWilliams, J.C., 1996. Eliassen-palm fluxes and the momentum equation in non-eddy-resolving ocean circulation models. J. Phys. Oceanogr. 26, 2539–2546.

Gent, P.R., Willebrand, J., McDougall, T.J., McWilliams, J.C., 1995. Parameterizing eddy-induced tracer transports in ocean circulation models. J. Phys. Oceanogr. 25, 463–474.

Gille, S.T., Speer, K., Ledwell, J.R., Naveira Garabato, A.C., 2007. Mixing and stirring in the Southern Ocean. Eos 88, 39.

Gnanadesikan, A., Griffies, S.M., Samuels, B.L., 2007. Effects in a climate model of slope tapering in neutral physics schemes. Ocean Model. 16 (1–2), 1–16.

Gordon, A.L., Weiss, R., Smethie Jr., W., Warner, M., 1992. Thermocline and intermediate water communication. J. Geophys. Res. 97 (C5), 7223–7240.

Graham, J.P., Ringler, T., 2013. A framework for the evaluation of turbulence closures used in mesoscale ocean large-eddy simulations. Ocean Model. 65, 25–39.

Greatbatch, R.J., Lamb, K.G., 1990. On parameterizing vertical mixing of momentum in non-eddy-resolving ocean models. J. Phys. Oceanogr. 20, 1634–1637.

Greatbatch, R.J., McDougall, T.J., 2003. The non-Boussinesq temporal residual mean. J. Phys. Oceanogr. 33 (6), 1231–1239.

Green, J.S.A., 1970. Transfer properties of the large-scale eddies and the general circulation of the atmosphere. Q. J. R. Meteorol. Soc. 96, 157–185.

Griesel, A., Gille, S.T., Sprintall, J., McClean, J.L., LaCasce, J.H., Maltrud, M.E., 2010. Isopycnal diffusivities in the Antarctic Circumpolar Current inferred from Lagrangian floats in an eddying model. J. Geophys. Res. Oceans 115, C06006.

Griffa, A., Owens, K., Piterbarg, L., Rozovskii, B., 1995. Estimates of turbulence parameters from Lagrangian data using a stochastic particle model. J. Mar. Res. 53 (3), 371–401.

Griffa, A., Lumpkin, R., Veneziani, M., 2008. Cyclonic and anticyclonic motion in the upper ocean. Geophys. Res. Lett. 35 (1), L01608.

Griffies, S.M., 1998. The Gent-McWilliams skew flux. J. Phys. Oceanogr. 28 (5), 831–841.

Griffies, S.M., 2004. Fundamentals of Ocean Climate Models. Princeton University Press, Princeton, NJ.

Griffies, S.M., Adcroft, A.J., 2008. Formulating the equations for ocean models. In: Hecht, M., Hasumi, H. (Eds.), Ocean Modeling in an Eddying Regime. AGU Geophysical Monograph Series, vol. 177. Wiley, pp. 281–317.

Griffies, S., Hallberg, R., 2000. Biharmonic friction with a smagorinsky-like viscosity for use in large-scale eddy-permitting ocean models. Mon. Weather Rev. 128 (8), 2935–2946.

Griffies, S., Pacanowski, R., Hallberg, R., 2000. Spurious diapycnal mixing associated with advection in a z-coordinate ocean model. Mon. Weather Rev. 128 (3), 538–564.

Grooms, I., Julien, K., Fox-Kemper, B., 2011. On the interactions between planetary geostrophy and mesoscale eddies. Dyn. Atmos. Oceans 51, 109–136.

Hallberg, R., 2000. Time integration of diapycnal diffusion and richardson number-dependent mixing in isopycnal coordinate ocean models. Mon. Weather Rev. 128 (5), 1402–1419.

Hallberg, R., Gnanadesikan, A., 2006. The role of eddies in determining the structure and response of the wind-driven southern hemisphere overturning: results from the Modeling Eddies in the Southern Ocean (MESO) project. J. Phys. Oceanogr. 36 (12), 2232–2252.

Haynes, P., McIntyre, M.E., 1987. On the evolution of vorticity and potential vorticity in the presence of diabatic heating and frictional or other forces. J. Atmos. Sci. 44, 828–841.

Hecht, M.W., Holm, D.D., Petersen, M.R., Wingate, B.A., 2008. Implementation of the LANS-alpha turbulence model in a primitive equation ocean model. J. Comput. Phys. 227 (11), 5691–5716.

Held, I.M., Larichev, V.D., 1996. A scaling theory for horizontally homogeneous baroclinically unstable flow on a beta plane. J. Atmos. Sci. 53, 946–952.

Henning, C.C., Vallis, G.K., 2004. The effects of mesoscale eddies on the main subtropical thermocline. J. Phys. Oceanogr. 34 (11), 2428–2443.

Henning, C.C., Vallis, G.K., 2005. The effects of mesoscale eddies on the stratification and transport of an ocean with a circumpolar channel. J. Phys. Oceanogr. 35 (5), 880–896.

Hirst, A., McDougall, T., 1998. Meridional overturning and dianeutral transport in a z-coordinate ocean model including eddy-induced advection. J. Phys. Oceanogr. 28 (6), 1205–1223.

Holland, W.R., Lin, L.B., 1975. On the origin of mesoscale eddies and their contribution to the general circulation of the ocean. I. A preliminary numerical experiment. J. Phys. Oceanogr. 5, 642–657.

Holland, W.R., Rhines, P.B., 1980. An example of eddy-induced ocean circulation. J. Phys. Oceanogr. 10, 1010–1031.

Hoskins, B.J., McIntyre, M.E., Robertson, A.W., 1985. On the use and significance of isentropic potential vorticity maps. Q. J. R. Meteorol. Soc. 111 (470).

Jayne, S., Marotzke, J., 2002. The oceanic eddy heat transport. J. Phys. Oceanogr. 32 (12), 3328–3345.

Jochum, M., Malanotte-Rizzoli, P., 2003. On the generation of North Brazil Current rings. J. Mar. Res. 61 (2), 147–173.

Jochum, M., Danabasoglu, G., Holland, M., Kwon, Y.O., Large, W.G., 2008. Ocean viscosity and climate. J. Geophys. Res. Oceans 113 (C6).

Johnson, G.C., Bryden, H.L., 1989. On the size of the Antarctic Circumpolar Current. Deep Sea Res. Part A 36 (1), 39–53.

Jones, S.W., Young, W.R., 1994. Shear dispersion and anomalous diffusion by chaotic advection. J. Fluid Mech. 280, 149–172.

Killworth, P.D., 1997. On the parameterization of eddy transfer. Part I: theory. J. Mar. Res. 55, 1171–1197.

Krauss, W., Boning, C.W., 1987. Lagrangian properties of eddy fields in the northern North Atlantic as deduced from satellite-tracked buoys. J. Mar. Res. 45 (259–291).

Kravtsov, S., Berloff, P., Dewar, W.K., Ghil, M., McWilliams, J.C., 2006. Dynamical origin of low-frequency variability in a highly nonlinear midlatitude coupled model. J. Clim. 19, 6391–6408.

Kushner, P.J., Held, I.M., 1998. A test, using atmospheric data, of a method for estimating oceanic eddy diffusivity. Geophys. Res. Lett. 25, 4213–4216.

Large, W.G., Yeager, S.G., 2004. Diurnal to decadal global forcing for ocean and sea-ice models: the data sets and flux climatologies. NCAR Technical Note, NCAR/TN-460+STR.

Larichev, V.D., Held, I.M., 1995. Eddy amplitudes and fluxes in a homogeneous model of fully developed baroclinic instability. J. Phys. Oceanogr. 25, 2285–2297.

Lee, M.-M., Marshall, D.P., Williams, R.G., 1997. On the eddy transfer of tracers: advective or diffusive? J. Mar. Res. 55, 483–505.

Leith, C.E., 1996. Stochastic models of chaotic systems. Phys. D 98, 481–491.

Levitus, S., Conkright, M.E., Reid, J.L., Najjar, R.G., Mantyla, A., 1993. Distribution of nitrate, phosphate and silicate in the world oceans. Prog. Oceanogr. 31 (3), 245–273.

Lévy, M., Klein, P., Tréguier, A.-M., Iovino, D., Madec, G., Masson, S., Takahashi, K., 2010. Modifications of gyre circulation by submesoscale physics. Ocean Model. 34, 1–15.

Lovenduski, N.S., Gruber, N., Doney, S.C., 2008. Toward a mechanistic understanding of the decadal trends in the Southern Ocean carbon sink. Global Biogeochem. Cycles 22 (3).

Lozier, M.S., 1997. Evidence for large-scale eddy-driven gyres in the North Atlantic. Science 277 (5324), 361–364.

Lozier, M.S., 2010. Deconstructing the conveyor belt. Science 328 (5985), 1507–1511.

Lumpkin, R., Elipot, S., 2010. Surface drifter pair spreading in the North Atlantic. J. Geophys. Res. Oceans 115, C12017.

Lumpkin, R., Garraffo, Z., 2005. Evaluating the decomposition of tropical Atlantic drifter observations. J. Atmos. Oceanic. Technol. 22 (9), 1403–1415.

Lumpkin, R., Garzoli, S.L., 2005. Near-surface circulation in the tropical Atlantic Ocean. Deep Sea Res. Part I 52 (3), 495–518.

Lumpkin, R., Treguier, A., Speer, K., 2002. Lagrangian eddy scales in the northern Atlantic Ocean. J. Phys. Oceanogr. 32, 2425–2440.

Luyten, J.R., Pedlosky, J., Stommel, H., 1983. The ventilated thermocline. J. Phys. Oceanogr. 13, 292–309.

Mahadevan, A., 2006. Modeling vertical motion at ocean fronts: are nonhydrostatic effects relevant at submesoscales? Ocean Model. 14, 222–240.

Maltrud, M.E., McClean, J.L., 2005. An eddy resolving global 1/10° ocean simulation. Ocean Model. 8, 31–54.

Marshall, J.C., Shutts, G., 1981. A note on rotational and divergent eddy fluxes. J. Phys. Oceanogr. 11, 1677–1680.

Marshall, J., Shuckburgh, E., Jones, H., Hill, C., 2006. Estimates and implications of surface eddy diffusivity in the Southern Ocean derived from tracer transport. J. Phys. Oceanogr. 36 (9), 1806–1821.

Marshall, D.P., Maddison, J.R., Berloff, P.S., 2012. A framework for parameterizing eddy potential vorticity fluxes. J. Phys. Oceanogr. 42, 539–557.

Maximenko, N.A., Bang, B., Sasaki, H., 2005. Observational evidence of alternating zonal jets in the world ocean. Geophys. Res. Lett. 32 (L12607). http://dx.doi.org/10.1029/2005GL022728.

Mazloff, M.R., Heimbach, P., Wunsch, C., 2010. An eddy-permitting Southern Ocean state estimate. J. Phys. Oceanogr. 40 (5), 880–899.

McClean, J.L., Maltrud, M.E., Bryan, F.O., 2006. Measures of the fidelity of eddying ocean models. Oceanography 19, 104–117.

McClean, J.L., Bader, D.C., Bryan, F.O., Maltrud, M.E., Dennis, J.M., Mirin, A.A., Jones, P.W., Kim, Y.Y., Ivanova, D.P., Vertenstein, M., Boyle, J.S., Jacob, R.L., Norton, N., Craig, A., Worley, P.H., 2011. A prototype two-decade fully-coupled fine-resolution CCSM simulation. Ocean Model. 39, 10–30.

McDougall, T.J., 1987. Neutral surfaces. J. Phys. Oceanogr. 17, 1950–1964.

McDougall, T., 2003. Potential enthalpy: a conservative oceanic variable for evaluating heat content and heat fluxes. J. Phys. Oceanogr. 33 (5), 945–963.

McDougall, T., Dewar, W., 1998. Vertical mixing and cabbeling in layered models. J. Phys. Oceanogr. 28 (7), 1458–1480.

McDougall, T.J., Jackett, D.R., 2007. The thinness of the ocean in s-theta-p space and the implications for mean diapycnal advection. J. Phys. Oceanogr. 37, 1714–1732.

McDougall, T.J., Klocker, A., 2010. An approximate geostrophic streamfunction for use in density surfaces. Ocean Model. 32 (3–4), 105–117.

McDougall, T.J., McIntosh, P.C., 1996. The temporal-residual mean velocity. Part I: derivation and the scalar conservation equations. J. Phys. Oceanogr. 26, 2653–2665.

McDougall, T.J., McIntosh, P.C., 2001. The temporal-residual-mean velocity. Part II: isopycnal interpretation and the tracer and momentum equations. J. Phys. Oceanogr. 31, 1222–1246.

McDougall, T., Hirst, A., England, M., McIntosh, P., 1996. Implications of a new eddy parameterization for ocean models. Geophys. Res. Lett. 23 (16), 2085–2088.

McDougall, T., Greatbatch, R., Lu, Y., 2002. On conservation equations in oceanography: how accurate are boussinesq ocean models? J. Phys. Oceanogr. 32 (5), 1574–1584.

McWilliams, J.C., Flierl, G.R., 1976. Optimal, quasi-geostrophic wave analyses of mode array data. Deep Sea Res. 23, 285–300.

Monin, A.S., Yaglom, A.M., 1971. Statistical Fluid Mechanics: Mechanics of Turbulence. MIT Press, Cambridge, MA, English, updated Edition.

Monin, A.S., Yaglom, A.M., Lumley, J.L., 2007. Statistical Fluid Mechanics: Mechanics of Turbulence. Dover Publications, Mineola, NY.

Munk, W.H., 1950. On the wind-driven ocean circulation. J. Meteorol. 7 (2), 79–93.

Nakamura, N., 1996. Two-dimensional mixing, edge formation, and permeability diagnosed in an area coordinate. J. Atmos. Sci. 53, 1524–1537.

Nurser, A., 1988. The distortion of a baroclinic Fofonoff gyre by wind forcing. J. Phys. Oceanogr. 18, 243–257.

Nycander, J., 2011. Energy conversion, mixing energy, and neutral surfaces with a nonlinear equation of state. J. Phys. Oceanogr. 41 (1), 28–41.

Oh, I., Zhurbas, V., 2000. Study of spatial spectra of horizontal turbulence in the ocean using drifter data. J. Phys. Oceanogr. 30 (7), 1790–1801.

Okubo, A., 1967. Effect of shear in an oscillatory current on horizontal diffusion from an instantaneous source. Int. J. Oceanol. Limnol. 1 (3), 194–204.

Oschlies, A., 2002. Improved representation of upper-ocean dynamics and mixed layer depths in a model of the North Atlantic on switching from eddy-permitting to eddy-resolving grid resolution. J. Phys. Oceanogr. 32, 2277–2298.

Parsons, A.T., 1969. A two-layer model of Gulf Stream separation. J. Fluid Mech. 39, 511–528.

Pedlosky, J., 1984. The equations for geostrophic motion in the ocean. J. Phys. Oceanogr. 14 (2), 448–455.

Plumb, R.A., 1979. Eddy fluxes of conserved quantities by small-amplitude waves. J. Atmos. Sci. 36, 1699–1704.

Plumb, R.A., Mahlman, J.D., 1987. The zonally averaged transport characteristics of the GFDL general circulation/transport model. J. Atmos. Sci. 44, 298–327.

Pope, S.B., 2000. Turbulent Flows. Cambridge University Press, Cambridge.

Poulain, P.M., Niiler, P.P., 1989. Statistical analysis of the surface circulation in the California Current System using satellite-tracked drifters. J. Phys. Oceanogr. 19, 1588–1603.

Prandtl, L., 1925. Bericht uber die entstehung der turbulenz. Z. Angew. Math. Mech. 5, 136–139.

Radko, T., Marshall, J., 2004. The leaky thermocline. J. Phys. Oceanogr. 34, 1648–1662.

Radko, T., Marshall, J., 2006. The Antarctic Circumpolar Current in three dimensions. J. Phys. Oceanogr. 36, 651–669.

Redi, M.H., 1982. Oceanic isopycnal mixing by coordinate rotation. J. Phys. Oceanogr. 12, 1154–1158.

Rhines, P.B., 1986. Vorticity dynamics of the oceanic general circulation. Annu. Rev. Fluid Mech. 18, 433–497.

Rhines, P.B., Holland, W.R., 1979. A theoretical discussion of eddy-driven mean flows. Dyn. Atmos. Oceans 3, 289–325.

Rhines, P.B., Young, W.R., 1982. Homogenization of potential vorticity in planetary gyres. J. Fluid Mech. 122, 347–367.

Richman, J., Wunsch, C., Hogg, N., 1977. Space and time scales of mesoscale motion in the western North Atlantic. Rev. Geophys. Space Phys. 15, 385–420.

Riha, S., Eden, C., 2011. Lagrangian and Eulerian lateral diffusivities in zonal jets. Ocean Model. 39 (1–2), 114–124.

Roberts, M.J., Marshall, D.P., 1998. Do we require adiabatic dissipation schemes in eddy-resolving models? J. Phys. Oceanogr. 28, 2050–2063.

Rogerson, A.M., Miller, P.D., Pratt, L.J., Jones, C.K.R.T., 1999. Lagrangian motion and fluid exchange in a barotropic meandering jet. J. Phys. Oceanogr. 29, 2635–2655.

Rupolo, V., Hua, B.L., Provenzale, A., Artale, V., 1996. Lagrangian velocity spectra at 700 m in the western North Atlantic. J. Phys. Oceanogr. 26, 1591–1607.

Samelson, R.M., Vallis, G.K., 1997. Large-scale circulation with small diapycnal diffusion: the two thermocline limit. J. Mar. Res. 55, 223–275.

San, O., Staples, A.E., Wang, Z., Iliescu, T., 2011. Approximate deconvolution large eddy simulation of a barotropic ocean circulation model. Ocean Model. 40, 120–132.

San, O., Staples, A.E., Iliescu, T., 2013. Approximate deconvolution large eddy simulation of a stratified two-layer quasigeostrophic ocean model. Ocean Model. 63, 1–20.

Scott, R.B., Straub, D.N., 1998. Small viscosity behavior of a homogeneous, quasigeostrophic, ocean circulation model. J. Mar. Res. 56, 1225–1258.

Scott, R.B., Wang, F., 2005. Direct evidence of an oceanic inverse kinetic energy cascade from satellite altimetry. J. Phys. Oceanogr. 35, 1650–1666.

Sheremet, V.A., 2004. Laboratory experiments with tilted convective plumes on a centrifuge: a finite angle between the buoyancy force and the axis of rotation. J. Fluid Mech. 506, 217–244.

Smagorinsky, J., 1963. General circulation experiments with the primitive equations I: the basic experiment. Mon. Weather Rev. 91 (3), 99–164.

Smith, R.D., 1999. The primitive equations in the stochastic theory of adiabatic stratified turbulence. J. Phys. Oceanogr. 29, 1865–1880.

Smith, K., 2005. Tracer transport along and across coherent jets in two-dimensional turbulent flow. J. Fluid Mech. 544, 133–142.

Smith, K.S., 2007. The geography of linear baroclinic instability in earth's oceans. J. Mar. Res. 65 (5), 655–683.

Smith, R., Gent, P., 2004. Anisotropic Gent-McWilliams parameterization for ocean models. J. Phys. Oceanogr. 34, 2541–2564.

Smith, K.S., Marshall, J., 2009. Evidence for enhanced eddy mixing at middepth in the Southern Ocean. J. Phys. Oceanogr. 39 (1), 50–69.

Smith, R., McWilliams, J., 2003. Anisotropic horizontal viscosity for ocean models. Ocean Model. 5, 129–156.

Smith, K.S., Boccaletti, G., Henning, C.C., Marinov, I., Tam, C.Y., Held, I.M., Vallis, G.K., 2002. Turbulent diffusion in the geostrophic inverse cascade. J. Fluid Mech. 469, 13–48.

Spall, M.A., 2000. Generation of strong mesoscale eddies by weak ocean gyres. J. Mar. Res. 58 (1), 97–116.

SSALTO/DUACS Team, 2013. SSALTO/DUACS user handbook: (M) SLA and (M)ADT near-real time and delayed time products. Tech. Rep. CLS-DOS-NT-06-034, SALP-MU-P-EA-21065-CLS, Centre National d'Etudes Spatiales.

Stammer, D., 1997. Global characteristics of ocean variability estimated from regional TOPEX/Poseidon altimeter measurements. J. Phys. Oceanogr. 27, 1743–1769.

Stammer, D., 1998. On eddy characteristics, eddy transports, and mean flow properties. J. Phys. Oceanogr. 28, 727–739.

Stommel, H.M., 1949. Horizontal diffusion due to oceanic turbulence. J. Mar. Res. 8, 199–225.

Stommel, H., Arons, A.B., 1960. On the abyssal circulation of the world ocean—II. An idealized model of the circulation pattern and amplitude in oceanic basins. Deep Sea Res. 6, 217–233.

Stommel, H., Schott, F., 1977. Beta spiral and determination of absolute velocity-field from hydrographic station data. Deep Sea Res. 24, 325–329.

Stone, P.H., 1972. A simplified radiative-dynamical model for the static stability of rotating atmospheres. J. Atmos. Sci. 29 (3).

Sverdrup, H.U., 1947. Wind-driven currents in a baroclinic ocean; with appplication to the equatorial currents of the eastern Pacific. Proc. Natl. Acad. Sci. U.S.A. 33, 318–326.

Swenson, M.S., Niiler, P.P., 1996. Statistical analysis of the surface circulation. J. Geophys. Res. Oceans 101 (C10), 22631–22646.

Talley, L.D., McCartney, M.S., 1982. Distribution and circulation of Labrador Sea-water. J. Phys. Oceanogr. 12 (11), 1189–1205.

Taylor, G.I., 1921. Diffusion by continuous movements. Proc. Lond. Math. Soc. 20, 196–212.

Taylor, G.I., 1953. Dispersion of soluble matter in solvent flowing slowly through a tube. Proc. R. Soc. Lond. A Math. Phys. Sci. 219 (1137), 186–203.

Taylor, G.I., 1954. The dispersion of matter in turbulent flow through a pipe. Proc. R. Soc. Lond. A Math. Phys. Sci. 223 (1155), 446–468.

Thompson, A.F., Young, W.R., 2007. Two-layer baroclinic eddy heat fluxes: zonal flows and energy balance. J. Atmos. Sci. 64, 3214–3231.

Treguier, A., 1999. Evaluating eddy mixing coefficients from eddy-resolving ocean models: a case study. J. Mar. Res. 57, 89–108.

Treguier, A.M., Held, I.M., Larichev, V.D., 1997. Parameterization of quasigeostrophic eddies in primitive equation ocean models. J. Phys. Oceanogr. 27, 567–580.

Treguier, A., Hogg, N., Maltrud, M., Speer, K., Thierry, V., 2003. The origin of deep zonal flows in the brazil basin. J. Phys. Oceanogr. 33, 580–599.

Tulloch, R., Marshall, J., Hill, C., Smith, K.S., 2011. Scales, growth rates, and spectral fluxes of baroclinic instability in the ocean. J. Phys. Oceanogr. 41 (6), 1057–1076.

Vallis, G.K., 2006. Atmospheric and Oceanic Fluid Dynamics: Fundamentals and Large-Scale Circulation. Cambridge University Press, Cambridge.

Veronis, G., 1975. The role of models in tracer studies. In: Numerical Models of Ocean Circulation: Proceedings of a Symposium Held at Durham, New Hampshire, October 17–20, 1972. National Academy of Sciences, pp. 133–146.

Visbeck, M., Marshall, J.C., Haine, T., Spall, M., 1997. Specification of eddy transfer coefficients in coarse resolution ocean circulation models. J. Phys. Oceanogr. 27, 381–402.

Wardle, R., Marshall, J., 2000. Representation of eddies in primitive equation models by a PV flux. J. Phys. Oceanogr. 30, 2481–2503.

Waterman, S., Jayne, S., 2011. Eddy-driven recirculations from a localized, transient forcing. J. Phys. Oceanogr. 42, 430–447.

Welander, P., 1959. An advective model of the ocean thermocline. Tellus 11 (3), 309–318.

Welander, P., 1971. The thermocline problem. Philos. Trans. R. Soc. Lond. A Math. Phys. Sci. 270, 415–421.

Wunsch, C., Stammer, D., 1995. The global frequency-wavenumber spectrum of oceanic variability estimated from TOPEX/POSEIDON altimetric measurements. J. Geophys. Res. 100 (C12), 24,895–24,910.

Young, W.R., 2010. Dynamic enthalpy, conservative temperature, and the seawater boussinesq approximation. J. Phys. Oceanogr. 40 (2), 394–400.

Young, W.R., 2012. An exact thickness-weighted average formulation of the boussinesq equations. J. Phys. Oceanogr. 42, 692–707.

Young, W.R., Jones, S., 1991. Shear dispersion. Phys. Fluids A Fluid Dynam. 3, 1087–1101.

Young, W.R., Rhines, P.B., Garrett, C.J.R., 1982. Shear-flow dispersion, internal waves and horizontal mixing in the ocean. J. Phys. Oceanogr. 12, 515–527.

Zhao, R., Vallis, G.K., 2008. Parameterizing mesoscale eddies with residual and Eulerian schemes, and a comparison with eddy-permitting models. Ocean Model. 23, 1–12.

Zhurbas, V., Oh, I., 2004. Drifter-derived maps of lateral diffusivity in the pacific and atlantic oceans in relation to surface circulation patterns. J. Geophys. Res. Oceans 109 (C5), C05015.

Zhurbas, V., Oh, I., Pyzhevich, M., 2003. Maps of horizontal diffusivity and Lagrangian scales in the Pacific Ocean obtained from drifter data. Oceanology 43 (5), 622–631.

Chapter 9

Global Distribution and Formation of Mode Waters

Kevin Speer* and Gael Forget[†]
*Department of Earth, Ocean, and Atmospheric Science, the Geophysical Fluid Dynamics Institute, Florida State University, Tallahassee, Florida, USA
[†]Department of Earth, Atmospheric and Planetary Sciences, Massachusetts Institute of Technology, Cambridge, Massachusetts, USA

Chapter Outline

Of all the problems in ocean circulation, that of the amount and manner of water-mass formation is the least tractable.

Worthington, 1976

1. MODE WATER OBSERVATIONS

Mode waters are water masses readily identifiable by their relatively uniform properties over large volumes, expanding both in depth and area of the ocean. They are a result of ocean convection, driven by air–sea interaction, but also spread under their own dynamics. A schematic depiction is provided to illustrate the interplay of diabatic and adiabatic processes leading to mode waters in the ocean (Figure 9.1). The mode water layer occupies a large fraction of the upper ocean water column. For the majority of the year, the mode water is isolated from the atmosphere by the seasonal thermocline; it may recirculate adiabatically at depth, but it can also be affected by diapycnal mixing along the way. In winter, the superficial stratification is eroded as vigorous air–sea fluxes extract heat from the ocean. As a result, warm water is transformed into colder water. This process propagates downward as convective mixing is activated and deepens the mixed layer. Once the mode water layer itself outcrops, the atmosphere can impact the entire layer directly. When the atmospheric forcing reverts to net heating, the near surface water warms and caps the mode water layer. Note that the warming does not penetrate as deep as cooling, since convective mixing is suppressed in summer. This asymmetry provides the basic process for net mode water formation over the full cycle. The fate of the newly formed mode water can vary: it can be exported and subduct into the main thermocline or it can recirculate and be reentrained into the mixed layer. But more generally mode waters and their geography have to be examined as part of the global, three-dimensional, ocean circulation.

Although our basic understanding of the large-scale formation and distribution of mode water is well established, studies have progressed toward finer scale descriptions, and more quantitative assessments of production and dissipation by various processes. The observation of mode waters has been based on traditional shipboard hydrographic measurements, but the arrival of autonomous hydrography has redefined mode water observations by resolving more of the structure and property variations generated by the convective processes responsible for mode water formation.

Hanawa and Talley (2001) give an excellent summary of the geographical distribution of mode waters, their definition in terms of vertical profiles of temperature, salinity, and stratification, and their historical background. With a focus on studies of the North Pacific Ocean, Oka and Qiu (2012) provide a comprehensive discussion of mode water research for this region including dynamical and eddy effects. Sallée et al. (2010) summarized mode water

Ocean Circulation and Climate, Vol. 103. http://dx.doi.org/10.1016/B978-0-12-391851-2.00009-X

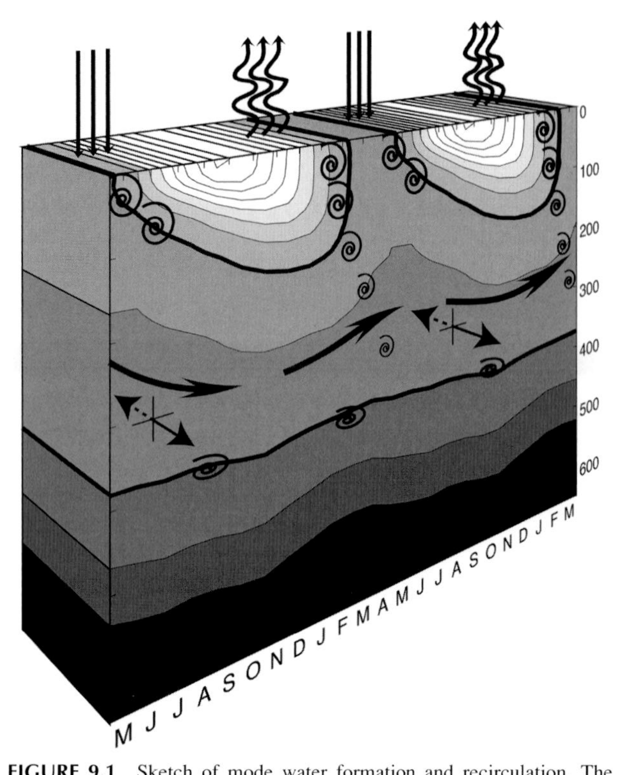

FIGURE 9.1 Sketch of mode water formation and recirculation. The mode water layer is delineated by the two thick black contours. The contours are of temperature versus depth (m) over two seasonal cycles as measured by Argo profilers in the pool of "Eighteen Degree Water." Thick arrows in the ocean interior denote adiabatic mode water recirculation, whereas spirals represent diabatic processes that can form or consume mode water. The lower axis is labeled with months of the year. The deep winter mixed layer (February) is capped by heating (buoyancy gain; straight arrows) as summer approaches. This cap or seasonal thermocline is subsequently eroded by cooling or buoyancy loss (squiggly arrows) in fall to early winter (October–January) leading up to the next deep cycle. *Adapted from Forget et al. (2011).*

distributions in the Southern Ocean and calculated subduction rates, thus determining the net exchange between surface layers and sub-surface mode waters taking place regionally and in circumpolar average. Recent work on the North Atlantic Eighteen Degree Water (CLIMODE, discussed below), studies of eddy fluxes and mode waters in numerical models of increasing resolution, and the growing body of literature on results derived from model-data syntheses further demonstrate the continued interest in mode water formation and propagation and the insight mode waters provide about the nature of air–sea interaction, ocean dynamics, and turbulent mixing processes. Along with higher resolution models, the increased space and time observational sampling due to the Argo hydrographic program[1] is a common theme of recent work, as is the use of state estimation.

The historical context might be said to go back as far as Parr (1938) who, as noted by Tsuchiya (1968), used the rate of change of thickness (vertical distance between isopycnal surfaces) with density in his circulation studies. His intention was for reference level guidance in a study of circulation off New England; however, while his plots provided a clear illustration of thick Eighteen Degree Water (see also Worthington, 1959) in a hydrographic section from Cape Hatteras to Bermuda, mode water was not the object of study. The analysis of circulation using thermostads on carefully chosen surfaces related to density, as defined by Tsuchiya (1968) and subsequent work, was a giant step toward the type of analysis that followed under the general heading of mode water research.

We will not review the important and necessary background of the development of isentropic analysis and the various density parameters that derived from this over the years; this is well covered by the discussions in McCartney (1977, 1982), Worthington (1981) and references therein. Hanawa and Talley (2001) provide some interesting background to the development of mode water studies from the theoretical perspective as well, the evolution of thermocline theories, and the notion of subduction. Oka and Qiu (2012) reviewed progress in mode water studies dealing with the North Pacific Ocean over the past decade. Their thorough discussion ranges over the large body of literature devoted to this region, revealing mode water generation and spreading, air–sea interaction, dissipation, and climate implications. Our primary aim here is to provide a new volumetric accounting of the global upper ocean and in the process update Worthington's 1981 census, at least for the part of the ocean shallower than 2000 m, using observations and methods acquired and deployed since the time of his study. We also summarize the concepts of water mass formation and recent practice.

Winter buoyancy loss mixes the upper ocean down to a depth determined by both the air–sea fluxes and the underlying stratification (Figure 9.1). Atmospheric circulation patterns, storm tracks, thermal wind near fronts and other processes produce spatial variability in both fluxes and stratification, and the deepest mixed layers (producing the largest volumetric contributions or modes, hence the name mode waters) tend to be recreated near western boundary currents. The numerous factors involved suggest that many varieties of mode water ought to exist and indeed one of the dominant themes of recent research is the description of new types of mode water.

Not only are new types being found but the distribution of well-known varieties is being extended. The Argo program, releasing profiling temperature, salinity, and depth instruments over the world ocean, has enabled mode water studies to progress with much improved sampling, especially in winter, during formation periods and areas of commonly rough weather, where shipboard sampling

1. http://www.argo.ucsd.edu; http://argo.jcommops.org; The Argo Program is part of the Global Ocean Observing System.

is sparse. Argo is by design a large-scale observational system, hence it does not provide the sampling (except by chance) of small-scale features of potential importance to mode water formation and evolution, particularly near boundaries or fronts. Such sampling requires ship-based techniques or autonomous platforms such as gliders. Also, the high latitude ice-covered regions still suffer from relatively poor sampling due to the special difficulties in deploying, recovering, and telemetering data from these regions.

Tracer chemistry and its application to mode waters has proceeded in parallel to physical oceanographic studies and informed us about the circulation, formation and ventilation, and processing and reprocessing of mode waters. Atmospheric gases, such as CFCs, enter the ocean and have been used with great skill to infer ventilation and circulation rates (see review by Jenkins (2003); for a recent example see Waugh, et al., 2013). In this survey, we maintain an emphasis on the global air–sea formation rates and the temperature, salinity, density, and stratification or potential vorticity as descriptive and quantitative tools for representing mode waters. Formation mechanisms are an interesting domain in themselves, linking air–sea interaction to small-scale convection, stirring, and mixing; these phenomena will be bypassed here by using net rates of water mass convergence in density and *T–S* classes, skipping the details of processes involved in this link. Similarly, subduction processes, whereby recently formed water masses enter the interior of the ocean, are not reviewed here.

To begin our survey, we return to the basic question of the definition of mode waters on the global scale. This brings us to the notion of a water mass census, or volumetric calculation in temperature and salinity classes.

2. GLOBAL WATER MASS CENSUS OF THE UPPER OCEAN

Worthington's 1981 global water mass census still stands as a monument to water mass description, including mode waters, and he discussed their contribution to the overall volumetric measure of ocean water types (defined by temperature–salinity (*T–S*) categories). Worthington (1981) focused on deep water and he noted the large areas of little or no full-depth sampling, hence the incompleteness of the estimates; nevertheless, he was able to draw a broad picture of the ocean's full stratification in terms of water types and link it to formation processes. Here, we are concerned with the upper layers of the ocean accessible to present-day profiling floats. These are the layers that contain mode waters, and which even at the time of Worthington's 1981 census were broadly sampled, if at coarse spatial resolution (he used 5×5 degree blocks), and infrequently during the year. Missing the wintertime formation cycle of mode waters (aside from occasional sampling) is a serious gap, and the advent of the Argo program roughly 25 years later brings access to the entire seasonal cycle at greater areal coverage, and, often, higher spatial resolution within that area.

To illustrate the global setting of mode waters, we have constructed a new water mass census using hydrographic data from the Argo program. Each ocean was split into subregions, and the Argo *T/S* probability density was computed for each subregion. These probability densities were converted into a volumetric census by multiplying by the total volume of the corresponding subregion. Finally, all subregions were added together to obtain the displayed volumetric census estimate.

The volumetric diagram presented here puts mode water distributions into a global context—mode waters are to a large extent responsible for the shape of the census since they contribute extraordinarily to the volume in *T–S* classes (Figure 9.2a) lying near the surface. An annotated version of this graph is provided (Figure 9.2b) as a reference for subsequent discussion.

It is important to recall that water below 2000 m is not sampled in the Argo program hence is missing from this diagram, which emphasizes the upper ocean. Water below this depth occupies the largest *T–S* volume classes in the world ocean (and is prominent in Worthington's census). Regions that have not been densely sampled by Argo (Arctic Ocean, marginal seas, and the Southern Ocean south of 60° S) are also omitted in this volumetric estimate. In this presentation, the northern limits to the Pacific and Atlantic Oceans are set to the Aleutian Arc, Davis Strait, and the Greenland–Scotland ridge. A division of 0.025 psu (henceforth salinity units are dropped) and 0.1 °C has been used to construct the volume elements presented here, but finer divisions can be used to investigate individual basins and particular mode waters.

The three primary branches or prongs (as Worthington called them) on the diagram are ridges of elevated volume following the *T–S* relation of the major ocean basins. They are derived from the different freshwater fluxes and marginal sea influences in these basins, in order from fresh (left) to salty (right), are the fresher North Pacific Central Waters, the Southern Hemisphere waters in the central prong encompassing all the major basins, and the saltier North Atlantic Central Waters. The influence of the Mediterranean Sea outflow shows as a "wing" of water masses extending toward higher salinity near 10 °C.

Each prong contains the mode waters within these regions; the central prong, for example, contains central waters of all three ocean basins, as well as Subantarctic Mode Water (SAMW) from the Southern Ocean. As SAMW evolves north of the Subantarctic Front it ranges from near 14 °C, 35.4 to cooler, fresher, denser values near 5–6 °C and 34.4–34.5 in the South Pacific. Worthington's (1981) census showed a weak isolated peak near 5–6 °C,

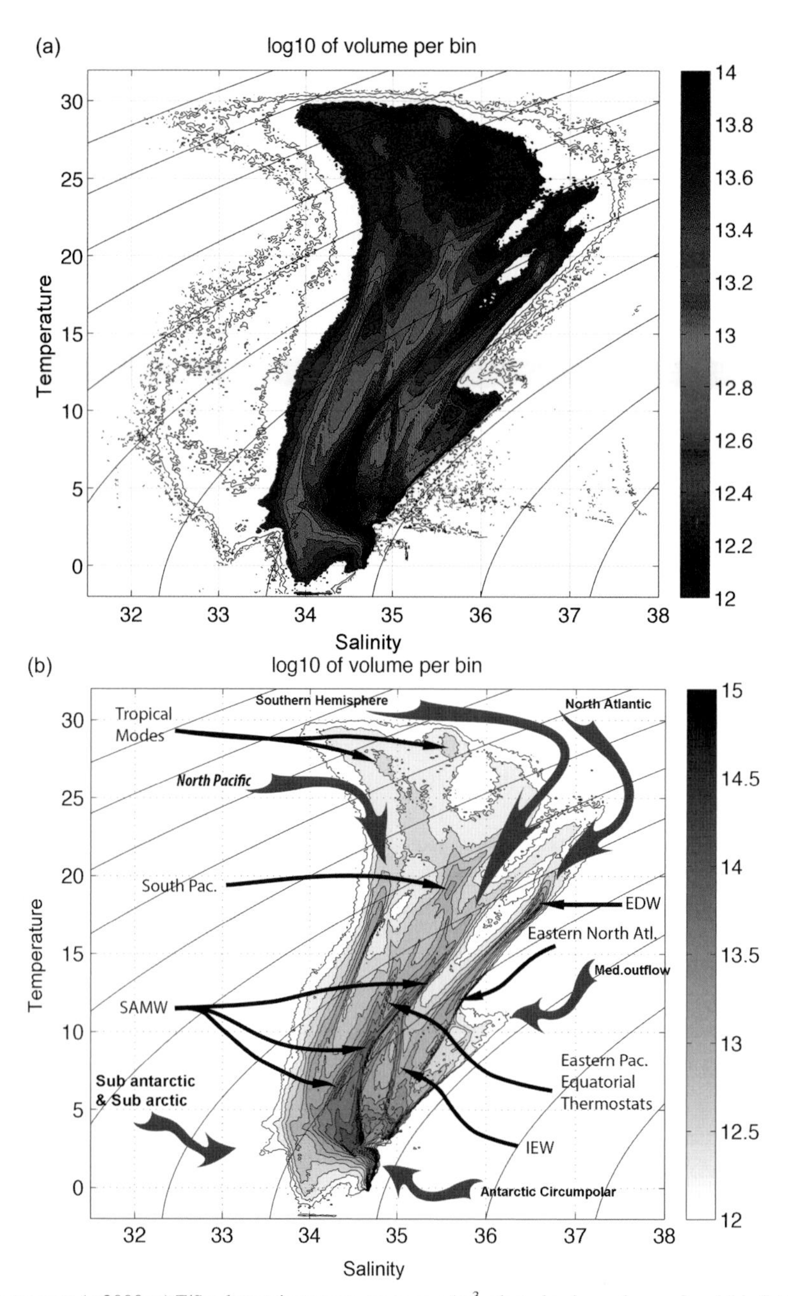

FIGURE 9.2 (a) Global upper ocean (<2000 m) *T*/*S* volumetric water mass census (m^3, plotted as log volume of each bin 0.1 °C and 0.025 psu) estimated from Argo profiles collected over the Atlantic, Pacific and Indian Oceans including their Southern Ocean sectors. Three primary warm prongs are due to waters (from left to right) of the North Pacific, Southern Hemisphere (north of 60°S), and Atlantic Oceans. Complex branching occurs, some of which is due to the superposition of water masses from widely separated basins. (b) Reproduction of figure (a) with selected water masses labeled (thin arrows) and an indication of ocean basins (thick arrows).

34.4 not distinct here, but this census does show a well-defined volumetric ridge linking SAMW across this entire range and across three-quarters of the globe.

In the global census, numerous isolated peaks are present. However, what dominates are the *ridges* of volumetric *T–S* relationships that are the natural result of a *T–S* relation imprinted on the basic ocean stratification. Deviations of thickness that locally define mode waters exist together with this basic state. At temperatures less than 10–15 °C, that is, below the range roughly associated with the main thermocline, the *T–S* relation and ridges are naturally linked to subpolar mode water generation. Subtropical mode waters sit, more or less, on top of the main thermocline and have more localized volumetric contributions. While it may be possible to define regional stratification and examine volume anomalies with respect to that, on the global scale this is difficult. The logarithmic scale is a simpler way to account for the basic stratification.

We find the familiar Eighteen Degree Water, which has salinity near 36.5, emerging as a volumetric peak. Interestingly, the cooler eastern North Atlantic mode waters near 10–12 °C show strongly on this diagram. North Pacific mode water is also apparent. Less familiar volumetric peaks are present as well; for instance, the peak near 19 °C and 35.6, an extension of South Pacific subtropical mode water, also appeared on Worthington's (1981) census, but was thought to be spurious due to insufficient sampling in the South Pacific. A search for other possible sources near these *T–S* values would show some overlap with other basins.

As noted earlier, mode waters themselves are found to be composed of various water types. The subpolar mode waters are an obvious example of inhomogeneity but even the usually very distinctive subtropical mode waters are not perfectly uniform; they are also spread to some extent along a ridge in this diagram and the width of the ridge may itself be an indication of the variable sources from air–sea fluxes and vertical and lateral mixing. Other processes, such as double diffusion, tend to restrain the excursions from a *T–S* relation. Observational studies are mapping these discrete water types out with greater and greater precision.

A fascinating degree of structure exists in the warm waters greater than 20 °C, which was not available to Worthington (1981) to discuss in the global context. These are tropical modes of a sort, not what is usually termed mode water, but which nevertheless form volumetrically coherent bodies of water. As noted by Hanawa and Talley (2001), some of these have been described under Worthington's (1976) term "Subtropical Underwater," for the Atlantic Ocean. At temperatures near 25 °C, warm, high salinity volumetric maxima overlap each other in contributions from distinct ocean basins. Volumetric peaks in this part of the diagram are partly due to this overlapping of water types from different basins.

Deeper equatorial water masses appear at cooler temperatures; for instance, Tsuchiya's (1968) eastern Pacific equatorial thermostads lie along the central branch near 10–13 °C and 34.7–34.9. A prominent ridge lying roughly along a curve from 5 to 10 °C with salinity remaining close to 35 is Indian Equatorial Water (IEW on Figure 9.2b).

On the cold side of the diagram, very fresh subpolar North Pacific waters extend to lower salinity near 5 °C, at relatively small volumes. These waters stand out in remarkable contrast to the salty Mediterranean and Red Sea contributions, forming "wings" of water mass populations on the volumetric diagram. These structures naturally arise from global freshwater fluxes, due to excess evaporation in the warm Atlantic and Indian Oceans (and marginal seas) and excess precipitation in the colder Pacific Ocean. At the lower left part of the diagram, volumetric contributions appear from water fresher than 34 and colder than 2 °C, coming from the upper layers of northern and southern polar regions. A polar version of the typical heat flux driven convective process arises from the seasonal sea–ice freezing and melting cycle, producing mixed layers that tend, on the diagram, to lower salinities.

3. GLOBAL DISTRIBUTION OF MODE WATER

Global maps of mode water have been presented by McCartney (1982) and Talley (1999), and reviewed by Hanawa and Talley (2001). Recent descriptions of mode waters emphasize spatial and time dependent structure and previously unrecognized sites of formation (Roemmich and Cornuelle, 1992; Tsuchiya and Talley, 1998; McCarthy and Talley, 1999; Oka and Qiu, 2012). The notion of multiple "vintages" of mode water has been described in several studies; for instance, Harvey and Arhan (1988) provide a rather complete description of the existence of mode water varieties and their separation by fronts in the North Atlantic Current. Oka et al. (2012) describe mode water variations related to the state of the Kuroshio Extension. Mode water varieties have been identified in the South Atlantic as well (Provost et al., 1999), and this richness is probably typical.

A multitude of mode waters revealed by detailed observations in the North Pacific is discussed in Oka and Qiu's (2012) review (see also Oka et al., 2011) and they pay particular attention to dissipation in mode water, or what can be termed mode water consumption. The very process that generates mode waters forms not one but many neighboring water masses of nearly equal *T*, *S*, and density. The differences are enough to restratify the underlying mode water slightly in the presence of eddy stirring as the water masses relax to equilibrium. Moreover, as the mode water itself ages, it is subject to vertical mixing (e.g., Qiu et al., 2006) and double-diffusive processes that slowly modify its *T* and *S*, so that when it recirculates in the gyre it is modified from the original and encounters younger mode water of different characteristics.

In addition, Oka and Qiu (2012) emphasize the dynamical role North Pacific mode water plays, as the various neighboring layers are stacked at the southern side of the gyre to contribute to the maintenance of the Subtropical Countercurrent. Marchese and Gordon (1996) discussed a dynamical front separating fresher, more homogenous water in the recirculation south of the Gulf Stream from saltier water to the east, extending into the subtropical convergence zone. Presumably similar processes may be inferred in other gyres, but the consequences are not yet clear. This is a valuable area of research.

Figure 9.3 shows the typical vertical density gradient near 300 m depth, mapped from the dataset of Argo profiles. These data were complemented by animal-borne CTD data (or "seal-based") from high southern latitudes, a new sampling method available to oceanographers to study the extensive regions of seasonal sea–ice cover in

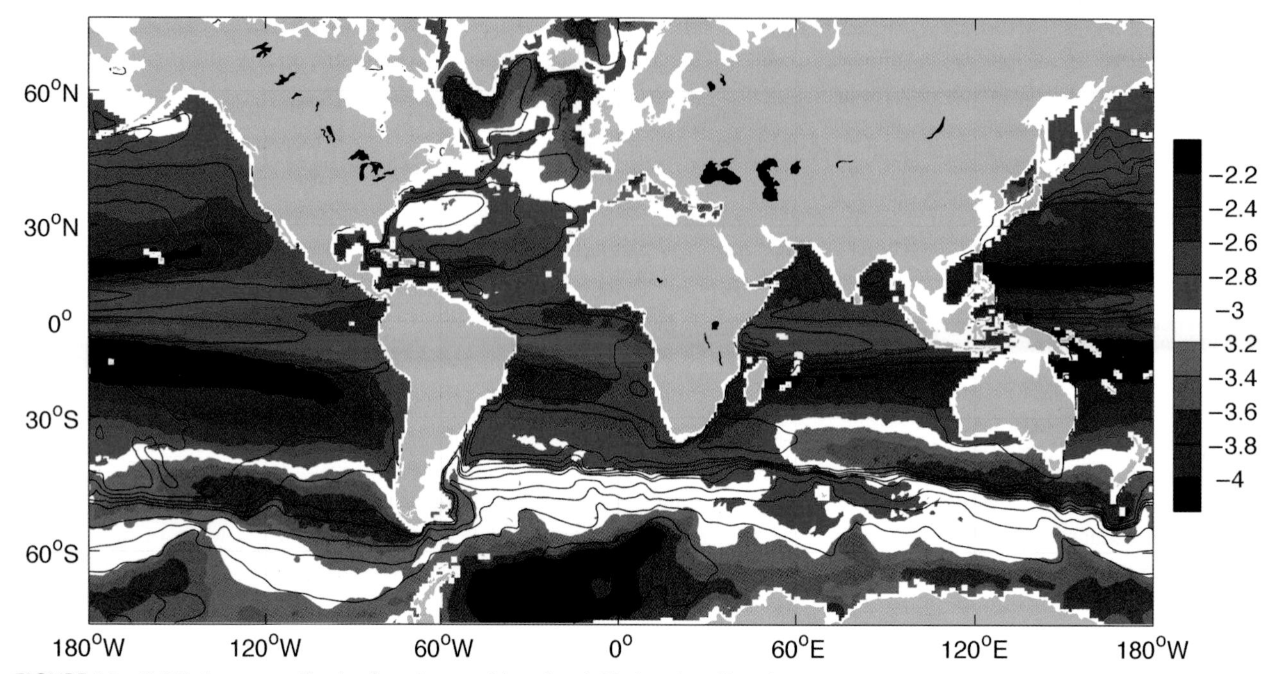

FIGURE 9.3 Gridded mean stratification from Argo, seal-based, and ship-based profiles of temperature and salinity, at 300 m depth. The color shading is, in kg/m^4, the logarithm of the vertical derivative of potential density referenced to 300 m depth. Stronger stratification is red, weaker is blue. Streamlines of the top to bottom volume transport as estimated in ECCO (v. 4) for 1992–2010 are displayed (black contours). Subtropical stratification is strong while high latitude stratification (poleward of about 60°) is very weak. Variation on this basic pattern is dominated by weakly stratified subtropical and subpolar mode waters in the North Atlantic and North Pacific, and by SAMW (blue shaded waters near 45°S) in the Southern Ocean. Lower gradients in the tropics are associated with tropical thermostads.

the Southern Ocean (e.g., Charrassin et al., 2008; Roquet et al., 2011). Such a slice through the global ocean at one shallow depth is dominated by the weak stratification at polar latitudes and the strong stratification in the lower latitudes. Weakly stratified mode waters also appear as low values of the vertical density gradient. This particular choice of depth captures the primary mode waters but does not necessarily illustrate their relative intensity well, since their cores are found at different depths and densities. In order to distinguish individual mode waters, subjective choices of property values or contouring is usually required.

In the map of stratification (Figure 9.3), Subantarctic Mode Water (SAMW) dominates the southern mid to high latitudes, and North Atlantic Subtropical Mode Water (STMW) and Subpolar Mode Water (SPMW) dominate the high northern latitudes, with North Pacific STMW and tropical thermostads somewhat weaker on this scale.

Each of the subtropical gyres of the Southern Hemisphere also shows its own version of mode waters, and again, numerous types have been identified (see e.g., Hanawa and Talley, 2001). To some extent, these differences are a result of particular definitions of mode water and the great variety of weaker mode waters in the southern hemisphere. The dominance of the Eighteen Degree Water in the North Atlantic may be the result of the geometry of the gyre in this basin, greater "pooling" of water in the strong recirculation, and stronger air–sea fluxes directed on this mode water as a result of adjacent continental and polar influences. Conversely, and consistent with this, is the smallness of the STMW area in the southern hemisphere (Hanawa and Talley, 2001), apparently due, in part, to the arrangement of the continents and less dry continental air masses to drive intense air–sea fluxes.

McCartney (1977) has described mode waters in the Southern Ocean in his pioneering work on what he termed Subantarctic Mode Water (SAMW). Due to the geometry of the Southern Ocean, the mode waters are spread along the northern flank of the Antarctic Circumpolar Current (ACC) from the tip of Africa all the way around and back into the Atlantic Ocean through Drake Passage.

In the region where SAMW is initially formed, at lowest densities, just downstream of the Agulhas Retroflection near 40°S, the Crozet Plateau appears to exert some control on the branching of the ACC, with consequences for mode waters. Farther downstream, the current bifurcates again around the Kerguelen Plateau. The merging of ACC and Agulhas waters upstream of the two plateaus gives rise to strong lateral mixing, influencing, in turn, the generation of nascent SAMW (Sallée et al., 2006). Past the Kerguelen Plateau, near 80°E, the main fronts of the ACC diverge; the STF moves northward, whereas the main branch of the SAF follows the northern flank of the Southeast Indian Ridge (Sandwell and Zhang, 1989). This is the region of deeper winter mixed layers described by Talley (1999), and McCarthy and Talley (1999).

Fine (1993) divided the Indian Ocean SAMW into three density ranges: 26.65–26.7 kg/m^3 (1000 kg/m^3 has been subtracted from the potential density to produce "sigma" units), which dominates in the southwestern region, 26.7–26.8 kg/m^3, which dominates in the central region, and 26.8–26.85 kg/m^3, which dominates the south-eastern region. These divisions are a result of the large-scale wind-driven circulation, bathymetry, and to some extent the dynamics of mode water itself, and divisions such as hers fit into the notion of mode water as a collection of distinct neighboring water masses.

The northward spreading of mode water into the Indian subtropical gyre is mainly due to northward advection, itself a result not only of wind stress forcing but also the density gradients induced by the fluxes that create mode water. Stramma (1992) showed that southeast of Africa the Subtropical Front (STF) is associated with a geostrophic transport of some 60 Sv (1 Sv = 106 m^3/s) and that this transport is reduced to less than 10 Sv (and the frontal system itself is less well defined) as Australia is approached (Schodlock et al., 1997). The surface water between the subtropical gyre and SAF becomes progressively cooler and denser as it moves east; this allows the development of northward thermal wind, thus gradually carrying away water from the northern side of the ACC and into the subtropical gyre (e.g., Koch-Larrouy et al., 2010).

A schematic summary of the mode water distribution (and subduction) in the Southern Ocean is shown in Figure 9.4 from Sallée et al. (2010). The progression of mode waters from least dense in the Indian Ocean to greater densities is illustrated. Some of these mode waters recirculate in the major subtropical gyres and some are carried east within the ACC. Ultimately, SAMW approaches the density of Antarctic Intermediate Water (AAIW). Recently, water in surface mixed layers within this density class has been found in the southwestern Pacific, starting south of the Campbell Plateau, and well west of the traditional site of formation in the southeast Pacific Ocean (Sallée et al., 2008). As a result of increased Argo sampling, our view of mode water distribution and density classes has changed in important ways in the Southern Ocean and elsewhere.

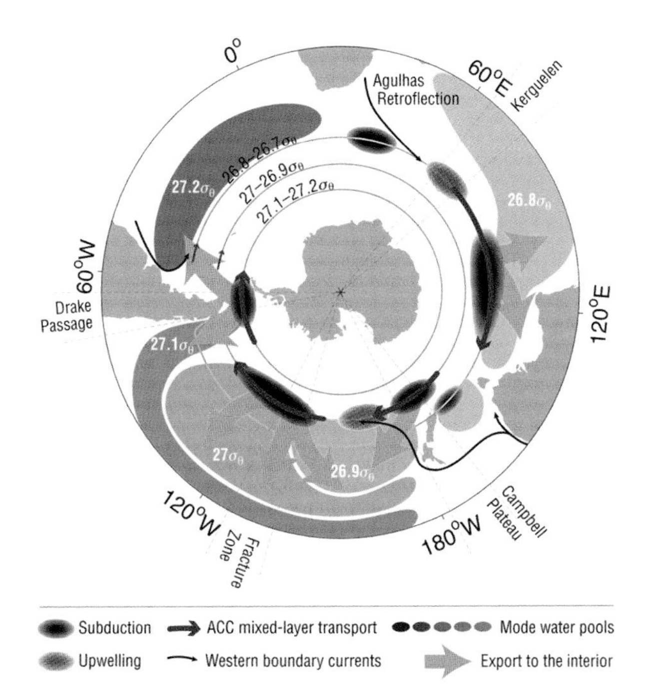

FIGURE 9.4 Subpolar Southern Ocean mode waters, transport processes, and key sites of subduction and export. *From Sallée et al. (2010).*

The renewal rate of mode waters in Southern Ocean sectors or basins is similar to that in other oceans (Hartin et al., 2011). Eddy compensation of Ekman and geostrophic advection is apparent in the observational estimates of net mass convergence into mode water. An important point is that the circumpolar averaged structure often used to summarize Southern Ocean formation and subduction hides strong regional variability and is not representative of the local balances. Net meridional transports are of interest, though, and several studies of net convergence in mode waters summarized in Sallée et al. (2010) suggest that the primary recipient of net northward transport in the upper ocean is SAMW.

Partly as a result of the large increase in information available from the Argo hydrography, state estimation is another theme of recent mode water research. As the results of such estimates become accessible to a larger body of observational investigators, both distributions and property fluxes become part of the description. The gap between descriptive and quantitative studies of water mass distributions has diminished as a result. Notable ocean state-estimates used in mode water studies include the global Ocean Comprehensible Atlas (OCCA; Forget, 2010) and the Southern Ocean State Estimate (SOSE; Mazloff et al., 2010), both part of the ECCO project. Here, we use these results as a tool to discuss formation in a globally consistent manner. As model resolution improves, the discussion of mode water formation mechanisms has progressed into the eddy-resolving domain.

State estimates have provided more complete integration of data and forcing for Eighteen Degree Water (Forget et al., 2011) using OCCA and for SAMW and AAIW using SOSE (Sallée and Rintoul, 2011; Cerovecki et al., 2013). In effect, circulation and eddy fluxes are becoming part of the observational context for *in situ* experiments. For the purpose of this review, air–sea fluxes (Figure 9.5) are taken from the latest ECCO state estimate (ECCO v.4) that was designed and produced by one of the authors (G.F.). A summary of the new features that were added in ECCO v.4, and a list of its data constraints can be found in Wunsch and Heimbach (2013).

The Argo and seal-based profile data used here to derive volumetric and stratification estimates are also included as constraints in ECCO v.4. It should be stressed, however, that the volumetric and stratification estimates presented rely on the *in situ* data alone, rather than on ECCO v.4. In either case, when uncertainty estimates are omitted (e.g., in Figures 9.2 and 9.6), one should assume that they might be sizable.

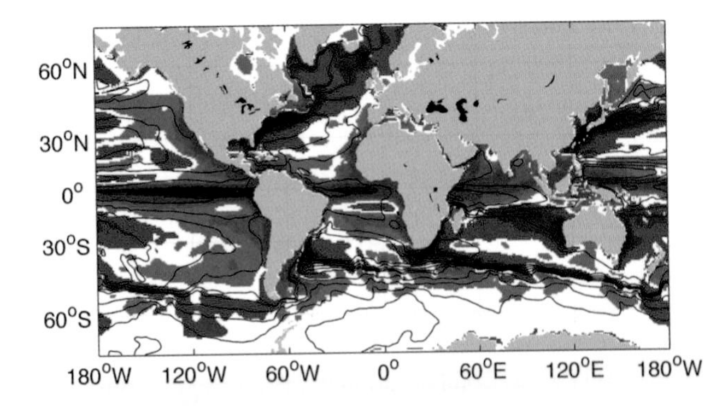

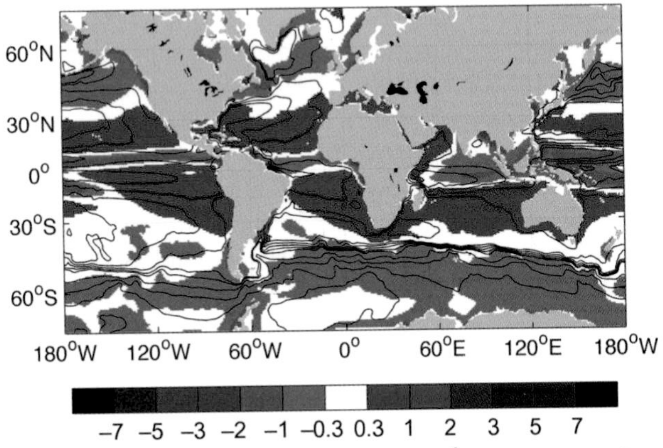

FIGURE 9.5 Annual mean buoyancy flux (mg/m^2/s) to the ocean surface, due to air–sea heat flux (top) and freshwater flux (bottom), as estimated in ECCO. Negative values indicate buoyancy gain (e.g., in the tropics). Streamlines of the top to bottom volume transport (black contours) overlaid.

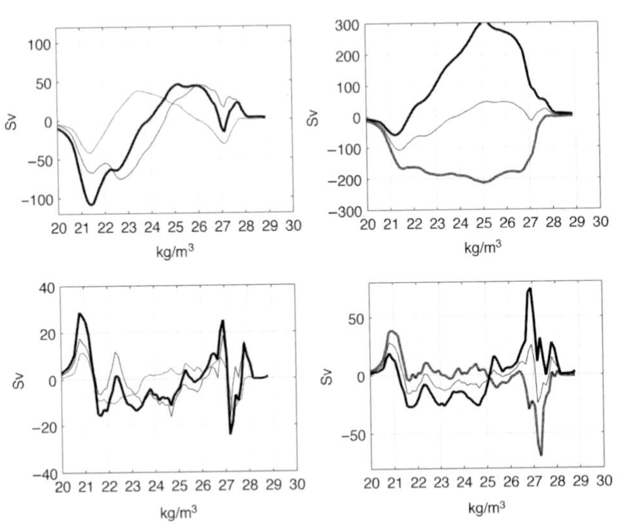

FIGURE 9.6 Annual transformation (top) and formation (bottom) versus density by air–sea heat and freshwater fluxes over all oceans as estimated by ECCO, for the period average 1992–2010. Time mean transformation rate (**F**, top left, thick solid curve), and seasonal terms (top right). Time mean transformation rate contributions from heat flux (thin solid curve) and freshwater flux (thin dashed curve) are also displayed. For seasonal terms (top right), fall–winter mean (thick solid curve), and spring–summer mean (thick dashed curve) are shown, together with the time mean (thin solid curve). Bottom panels show the corresponding formation rate in bins of width 0.2 kg/m^3. Negative values are buoyancy gain (warming and rainfall) and positive values are buoyancy loss (cooling and evaporation).

4. FORMATION OF MODE WATER

The most straightforward way to derive the formation rate of a water mass would be to find a water mass that disappears and reappears each year and measure its volume change. Madeira Mode Water is such an example and this extreme cycle was exploited by Siedler et al. (1987) to estimate its annual renewal rate strictly from hydrography. Similarly, Roemmich and Cornuelle (1992) examined the seasonal cycle of subtropical mode water in the South Pacific Ocean, with quantitative results. In general, we expect that not all of a given water mass is renewed and we wish to know the relation between air–sea fluxes and other processes that renew and destroy the particular characteristics defining the water mass.

Walin's (1982) concept and methodology, a version of which was derived independently by Tziperman (1986), has inspired many studies of water mass formation and in particular, the formation rates of mode waters due to air–sea fluxes.

Large and Nurser (2001) provide an overview of water mass formation calculations. Numerous authors have adapted the technique to more general situations and applied the methods to model studies and time varying problems.

For a global water mass volume between two density (ρ) surfaces one can write:

$$dV/dt = d\mathbf{F}/d\rho + d^2D/d\rho^2$$

where V is the water mass volume, **F** is the transformation due to air–sea fluxes at the surface density outcrop, and D is diabatic flux from any source. Here, t is time and ρ is density. Thus, the balance provides an estimate of globally averaged mixing if storage (term on the left) and air–sea fluxes (first term on the right) are known to a sufficient precision. In regional analyses, water mass import–export rate must also be taken into account. The air–sea flux term (first on the right) is, however, often analyzed alone for estimates of water mass formation. This term is based on surface air–sea fluxes, but recent analyses also take into account the vertical penetration of light into the stratified ocean (e.g., Bozec et al., 2008).

Large-scale transformation of water from one density to another occurs globally because of heating at low latitudes and cooling at high latitudes (Figure 9.5). Temperatures at the ocean surface fluctuate with the seasonal cycle, and in the presence of mixing, surface heating and cooling do not exactly cancel out over the surface temperature outcrop. Overall, mixing balances the net transformation to higher and lower temperatures due to air–sea fluxes by returning water to intermediate temperatures and densities (e.g., Tziperman and Speer, 1994; Nurser et al., 1999).

The dominant forcing mechanism of mode waters is thought to be the strong fluxes that result from cold dry continental air masses that spread over warm western boundary currents and neighboring mode waters. Ekman transports can catalyze mode water formation, especially in subpolar regions, as water is carried from the light to the dense side of the gyre and mixes down. This effect was found to be key to the flux balance of mode water in the southern Indian Ocean (Sallée et al., 2006).

Annual transformation for the global ocean was computed from the ECCO (v. 4) estimates of air–sea fluxes and sea-surface density (Figure 9.6). The fall–winter average rates are here defined as the October to March average for the northern hemisphere, and the April to September average for the southern hemisphere. Spring–summer average rates were similarly calculated. The results show the general pattern of warming (buoyancy gain) at low densities in the tropics and cooling (buoyancy loss) at high latitudes where surface densities are greater. A deviation occurs at high density (sigma near 28 kg/m^3) due to the region of net buoyancy gain in the Southern Ocean. This average is the sum of much larger seasonal components driving water to higher (winter) and lower (summer) sea-surface density.

The lower panels of Figure 9.6 show the formation rate (convergence of transformation). On the cold side, water converges into mode waters, while on the warm side water converges into the warm pools of the tropics. The largest formation rates are associated with the generation of warm pools in the tropics (sigma near 21 kg/m^3), and the positive and negative peaks associated with the formation of SAMW (near 27 kg/m^3). The middle-range, from 22 to 26 kg/m^3, shows net loss, which must be supplied by diapycnal mixing between the warm and cold water (see also Tziperman and Speer, 1994). Water mass convergence may be calculated from air–sea heat and freshwater fluxes separately over ocean SST and SSS (Speer et al., 1995) showing the tendency for formation to occur on well-defined water types. The notion of convergence into particular water masses as opposed to widespread forcing across water types is a strong diagnostic statement about coupled air–sea interaction on long time scales.

The volumetric census may be used to estimate seasonal changes in volume, and be compared to the corresponding air–sea flux transformation (Forget et al., 2011). Volumes were estimated as in Figure 9.2, but for each calendar month separately, and for the distribution of observed potential density (Figure 9.7; sigma units, referenced to the surface). The volume seasonality was then computed as the difference between the February to April average and the July to September average for the Northern Hemisphere (and similarly for the Southern Hemisphere).

Errors for volumetric estimates come from several sources, including the smaller scales that Argo does not resolve, and the statistical model used to convert local observations to volumes. The resolution error contribution

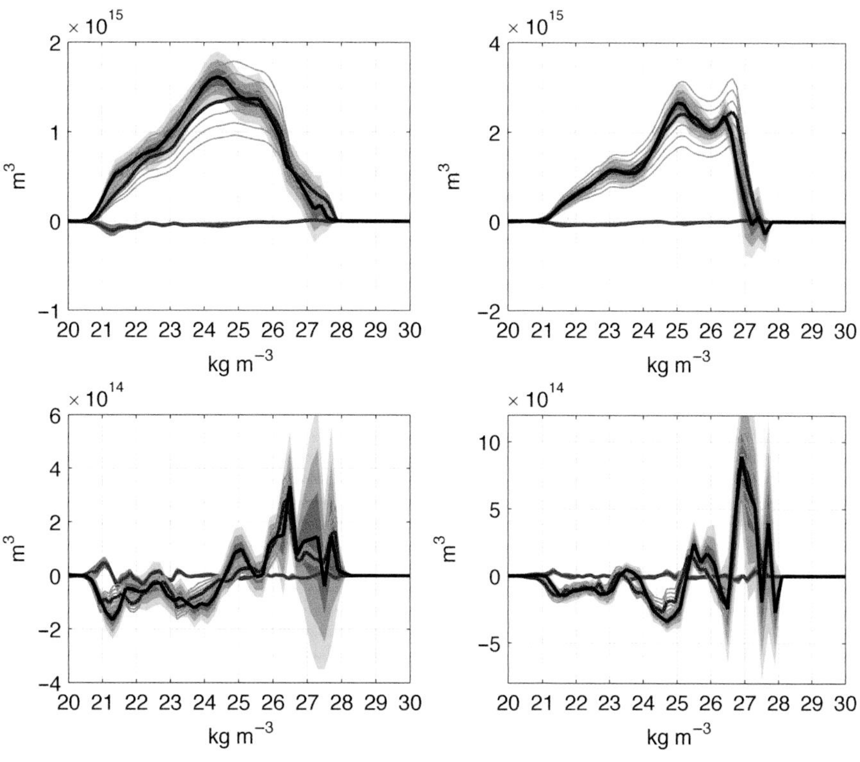

FIGURE 9.7 Seasonal volumetric changes, cumulative from lower to higher density (thick lines, top panels) and in 0.2 kg/m^3 density bins (bottom panels), computed for the Northern Hemisphere (left) and Southern Hemisphere (right; shading shows the first through third standard error due to the limited space and time sampling and analysis). Air–sea heat (red) and freshwater (green) flux contribution to seasonal water mass transformation (top panels) and formation (bottom panels) are shown. For comparison, ranges (10%, 20%, 30%) around the climatological seasonal cycles are also displayed.

was estimated from the differences between the Argo-based result and the analogous result obtained by subsampling the ECCO state estimate (that does not include eddies) at the Argo profile locations. The model error contribution was estimated from the differences between subsampled ECCO and full ECCO estimates. The shading in Figure 9.7 shows 1, 2, and 3 standard errors.

Corresponding seasonal contributions of the transformation rates were computed as the time integral of the fall–winter anomalies. As in Figure 9.6, the fall–winter average covers October to March for the Northern Hemisphere, and April to September for the Southern Hemisphere. However, the integrals were restricted to the ocean basins covered by the volumetric estimates. Quantifying errors in (trans)formation rate estimates remains an elusive goal. In Figure 9.7 we simply report a range of estimates around the climatological seasonal cycles, to which we added or removed a constant 10%, 20%, or 30% of the seasonal amplitude. Systematic errors of comparable magnitude were previously assessed in North Atlantic estimates (Forget et al., 2011).

The power of the methodology becomes apparent from the comparison of the observed seasonal volume changes to the corresponding (ECCO) air–sea flux driven transformation (Figure 9.7). The global estimates are in good agreement, indicating that the dominant balance over the seasonal cycle is between air–sea fluxes—predominantly heat flux—and storage, or volume change. A similar conclusion was drawn previously for North Atlantic Eighteen Degree Water (Forget et al., 2011).

Mode waters are a dominant part of this widespread seasonal cycle. They are the major positive peaks on the dense side of the seasonal formation graphs (bottom panels). Indeed, the largest signal in global formation is the formation of water in the Southern Hemisphere (the lower right panel of Figure 9.7) on a density range close to 27 kg/m^3, which is the SAMW. This may be visualized by mapping the regional contributions to the formation rate at this density (Figure 9.8) as explained by Maze and Marshall (2011); Maze et al., (2009). The broad structure dictated by the global seasonality is apparent. However, regional extrema (e.g., in the South Pacific) are also evident where air–sea fluxes predominantly act to transform water masses. Such intense regional water mass transformation is thought to be a key ingredient to the mode water's geography (Figure 9.3). The chosen density class happens to capture the formation of a portion of the subpolar mode water in the North Atlantic as well. In fact, there is an analogy between the subpolar mode water and subantarctic mode water, as they both experience an evolution of their temperature, salinity, density, thickness, etc., along an advective path, following the subpolar gyre in one case and the Antarctic Circumpolar Current in the other case.

Figure 9.9 depicts in more detail the seasonal progression of global upper ocean volumetric changes. These calculations illustrate the wintertime build-up of volume and summertime decrease in the upper ocean in each hemisphere in the upper ocean, lighter than a given density. The volumetric cycle within the primary mode water density layers of each hemisphere (Figure 9.9, lower panels) shows the more skewed and narrow distribution of volume change within layers owing to the limited seasonal (late winter) outcrop window. Results show that within fairly wide error bounds, the air–sea fluxes produce the right amount of water in the correct density classes to explain the volumetric cycle.

State estimates, as noted above, offer the prospect of determining all terms in the formation equation and the realistic detailed balances responsible for maintaining mode waters. Thus, Forget et al. (2011) analyzed the OCCA estimate for the full budget of Eighteen Degree Water in the North Atlantic, including not only volumetric changes and formation by air–sea fluxes (i.e., the two terms shown in Figure 9.6) but also contributions by diapycnal mixing and ocean transports. Similarly, Cerovecki et al. (2013) produced estimates of Subantarctic Mode Water formation and transports across 30°S and between basins, showing that

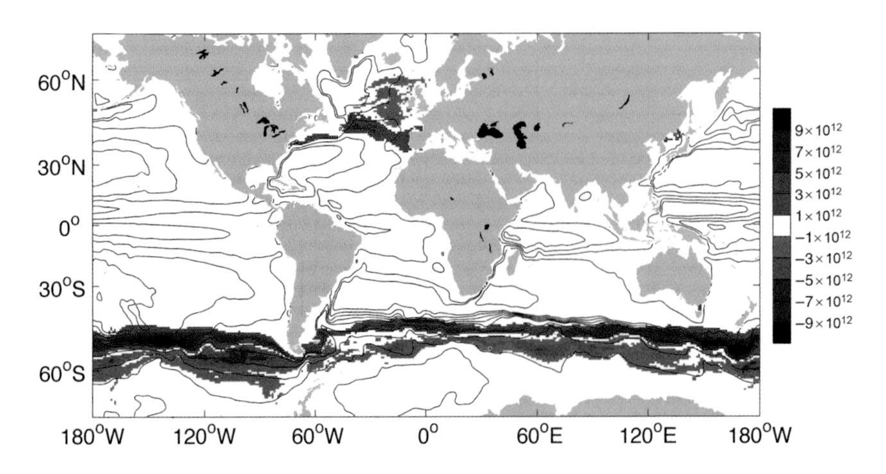

FIGURE 9.8 Seasonal fall–winter anomaly in formation rate due to air–sea heat flux, for the density classes 26.875–27.000 kg/m^3, in units of Sv/m^2. Streamlines of the top to bottom volume transport overlaid (black). Note that formation is not the same as subduction, which involves lateral transport and convergence of mode water.

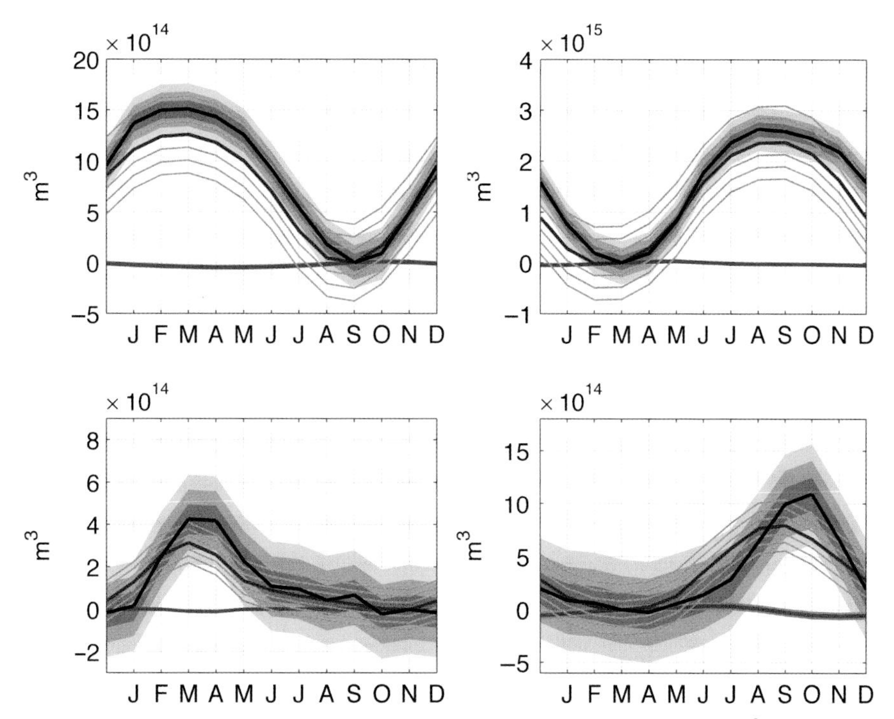

FIGURE 9.9 Seasonal cycle of the Argo volumetric census (black curves, top left: waters lighter than 26.5 kg/m^3 in the Northern Hemisphere; top right: waters lighter than 26.9 kg/m^3 in the Southern Hemisphere). Seasonal cycle of the isopycnal layer 26.5 $\pm$ 0.1 kg/m^3 (bottom left, Northern Hem.), and the layer 26.9 $\pm$ 0.1 kg/m^3 (bottom right, Southern Hemisphere). The red curves show the air–sea heat flux contribution to seasonal water mass transformation (top) and formation (bottom). Green curves show fresh water flux contribution.

air–sea driven transformation, diapycnal fluxes, export, and storage terms were all of roughly similar importance in the SOSE. It is possible that the relatively short integration period of OCCA and SOSE led to unrepresentative balances, but nevertheless the method provides insightful diagnostics of simulations.

Although our focus is on the time–mean and seasonal global cycle, it is important to recognize that this masks the rich array of local processes that come into play on the regional level. These can be viewed as elaborations on the diapycnal effects that are required to balance air–sea forcing, or that emerge as the system evolves in time. In particular, diapycnal fluxes are essential, but determining the nature of these fluxes, whether they be small-scale vertical mixing or meso-scale eddies near the surface, and where and when they operate, is difficult. Great progress has been made in distinguishing diapycnal processes operating on mode waters in recent years, including horizontal eddy fluxes (Sallée et al., 2008; Davis et al., 2013), vertical mixing (Sloyan et al., 2010), and cabbelling (using inverse methods, e.g., Yun and Talley, 2003; and from a numerical simulation, e.g., Urakawa and Hiroyasu, 2012).

It is possible to examine the dynamical implications of air–sea interaction that emphasizes the large-scale forcing of ocean circulation by considering the PV fluxes at the surface. In the next section we summarize recent results about the way air–sea interaction can be viewed as driving PV into and out of water masses, thus linking the thermodynamics to a dynamical framework.

5. PV FRAMEWORK

We note briefly the development of the idea of PV boundary sources for mode waters in the context of the PV description. The explosion of PV interpretations of mode waters in the early 1980s due, largely, to the descriptive work of McCartney (1982) and that of McDowell et al. (1982), resulted in a new paradigm for mode water generation and propagation. The distribution and spreading constraints described by McCartney (1982) and McCartney and Talley (1982) (see also Brambilla and Talley, 2008; Brambilla et al., 2008) showed how powerful this idea was in unifying the approach to mode water research.

The compelling nature of the water mass formation concepts and their consequences for mode water, including the intimate link to mixing, led to the formulation of a large project to study the formation and maintenance of Eighteen Degree Water (EDW) in the North Atlantic, called CLIMODE (Marshall et al., 2009, from: Climate Variability and Predictability (CLIVAR) Mode Water Dynamic Experiment). This project was designed to address the role of lateral and diapycnal mixing in the upper ocean in comparison to air–sea fluxes, and to assess where EDW forms and where it circulates after formation.

Joyce et al. (2009) suggest that Ertel PV ideas may be key to understanding EDW formation near the strong Gulf Stream front. The large vertical and horizontal shears of such a front create circumstances of low slantwise (along angular momentum surfaces) stability and net exchange or mixing can take place. Key to these ideas is the notion that Lagrangian mean flow may be present even if cross-front circulation appears to average out in the along-front direction (Thomas and Joyce, 2010).

Maze and Marshall (2011) suggest that the EDW may be viewed as a system that loses PV to surface boundary fluxes along the northern edge, following the Gulf Stream, and gains PV via mixing processes. These mixing processes can act over a longer time in the slower recirculation of mode water south of the Gulf Stream, whereas the loss occurs in strong seasonal fluxes along the frontal boundary.

PV mixing was also a key conclusion of Qiu et al. (2007) for the North Pacific subtropical mode water during the Kuroshio Extension System Study (KESS). Essentially, high-PV cold-core eddies from the poleward side of the front carry their high PV into the body of the mode water, where it eventually dissipates. This process of cold-core ring generation is modulated over inter-annual time scales by the large-scale winds over the Pacific.

The PV interpretation of Maze and Marshall (2011) is coarse-scale and only the time-averaged result is illustrated in Figure 9.10. Wintertime convection is central to the flux of PV at the boundary and the effects of stress. Czaja and Hausmann (2009) also investigate diabatic and mechanical forcing at the sea surface and their interpretation of PV fluxes emphasizes buoyancy fluxes at high latitude and mechanical fluxes (e.g., Ekman advection of surface density) at low latitudes.

6. MODE WATER AND CLIMATE

The best direct evidence of mode water stability through time, at least in the North Atlantic, is that between the year 1873, when the Challenger Expedition measured hydrography between the Virgin Islands and Bermuda, and 1958 (Schroeder et al., 1959), EDW remained close to 18 °C. Evidently this relative stability continues through the present, (e.g., Forget et al., 2011) or for more than 140 years (though some variations are evident). This does not mean that the formation rate of EDW is constant. Billheimer and Talley (2013), for instance, suggest that formation ceased in a recent winter. Warren's (1972) deduction that mode water is insensitive to meteorological fluctuations is, to some extent, supported by the centennial persistence; however, a 100-year time scale is much longer than the roughly 5-year renewal time scale of EDW that was the basis of the year-to-year insensitivity. More fundamental controls seem to be at play, though what these are remains unclear. It may be as simple as the fact that 18 °C is close to the average temperature of the atmosphere near the latitude of the western boundary current separation, hence the recirculating pool of mode water tends to this value, similar to the control on the mean temperature in the ground.

Multicentury simulations (e.g., Langehaug et al., 2012) further suggest that large-scale water mass transformations are relatively stable over time and related to the large-scale gyre circulation and sea–ice distributions. These phenomena tend to put the key air–sea interaction (heat loss) for the transformation of northward flowing water on the eastern side of the subpolar gyre, hence may relate more to the stability of the subpolar mode waters than the subtropical mode waters.

Some important variations of mode water properties have been observed with time-series stations and the

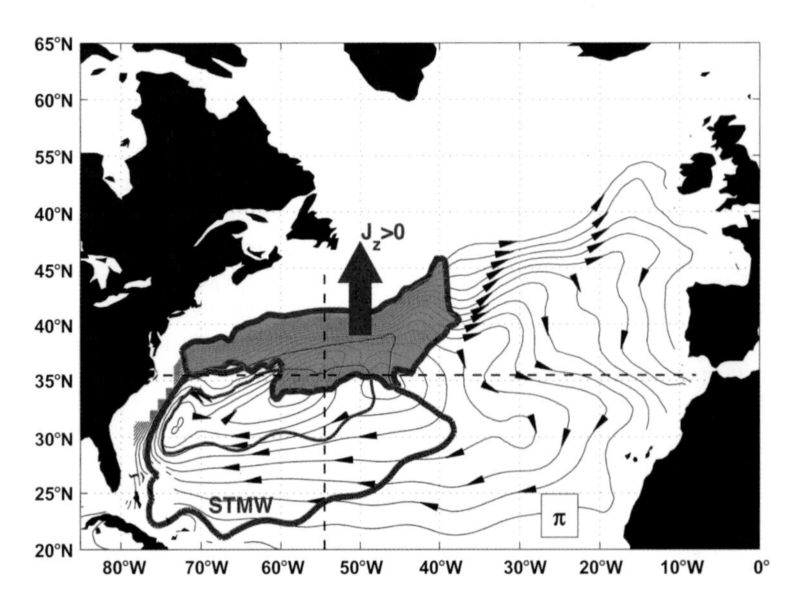

FIGURE 9.10 The removal of PV from the EDW STMW by air–sea fluxes is shown (blue area; this is the vertical component of the PV flux, including stress) and the horizontal flux of PV within the gyre back to mode water. Annual mean Bernoulli function π on the sigma 26.4 kg/m^3 surface is displayed (black lines with arrows). PV is replenished by mixing. *From Maze and Marshall (2011).*

modern extensive coverage of the Argo program. Durack and Wjiffels (2010) described the salinity changes occurring globally in the upper ocean and suggest that mode waters around the globe appear to be freshening. The mechanism appears to be due to the migration of outcrops into fresher areas, the migration itself being driven by surface warming.

Recent work links mode water variations in the North Pacific Ocean to climatic shifts in the winds (Davis et al., 2011). The suggestion has been made that eastern South Pacific mode waters are linked to ENSO (Wong and Johnson, 2003). At yet longer time scales, Sagawa et al. (2011) suggest that formation of STMW in the North Pacific was enhanced during the last glacial maximum, based on the oxygen isotope record of reduced intermediate-depth temperatures. This kind of study, together with simulations (e.g., Wainer et al., 2012) linking past climate to modern conditions, is of great value for understanding the future of climate change, and records that additionally enable us to deduce stratification from paleo-proxy records would be especially valuable.

Time series in the Atlantic also suggest links to fluctuations in atmospheric patterns such as the NAO. Talley and Raymer (1982) discussed variability (17.1 to $\sim$18.5 °C over 1954–1978), but also brought up the important point that the core properties are expected to be somewhat different at different locations within the water mass, closer and farther from formation areas. This notion of vintages or structure within mode waters, as noted earlier, makes single point observations problematic for climate interpretations, since advection of *T–S* gradients introduces time variation. Larger scale climatic links between mode water, PV, Gulf Stream position, and the subpolar gyre were investigated by Joyce et al. (2000). Their work brings together several dynamical elements, including PV gradients and transport, into the interpretation of low-frequency variability of mode water and boundary currents, possibly coupled with the atmosphere.

Large-scale variations of air–sea fluxes and variability in the Kuroshio Extension are linked to variations in the North Pacific STMW (Oka et al., 2012). Though Durack and Wjiffels (2010) note freshening trends, less clear is the nature of any feedback between these changes and the atmosphere. While the basic mechanism of climate memory may be linked to mode waters (as deep mixed layers), the detailed mechanisms of feedback between mode waters *per se* and the atmosphere are yet to be understood.

Downes et al. (2011) estimated water mass transformation rates from both surface buoyancy fluxes and interior diapycnal fluxes in the region south of 30°S, using the ECCO model-based state estimation as "ground-truth" and three free-running coupled climate models. The main effect of the surface buoyancy fluxes in the model's Southern Ocean was to convert Upper Circumpolar Deep Water and AAIW into lighter Subantarctic Mode Water. This might also suggest that mode water "captures" climate variability imprinted by surface fluxes over the Southern Ocean. This notion is consistent over the wider range of coupled climate models used in CMIP5 (Sallée et al., 2013).

Given the apparent stability of mode waters (e.g., Eighteen Degree Water), what role does mode water play in climate change? As shown by Wong et al. (1999), Aoki et al. (2003, 2005), Banks et al. (2002), Banks and Bindoff (2003), they are the receptacle of climate anomalies that are driven into them by air–sea fluxes. Thus, although the mode water itself might remain fairly well defined, there are variations within that develop over time and are significant. Recall that in rough volumetric terms, mode waters define ridges and peaks; fine-volume structure on top of this basic state experiences annihilation and recreation in a complicated dependence on circulation and air–sea interaction. Whether the *T–S* volumetric approach is a fruitful way to analyze climate change remains to be seen—and probably depends on the continued support for the large-scale observational programs producing the data needed to construct such diagrams. But promising results from climate model analyses of fluxes in *T–S* space suggest that this methodology will be useful (Döos et al., 2012; Zika et al., 2012).

We should finally reemphasize that understanding the basic state and stratification of climate models is fundamental to trying to understand how the model responds to changes such as increasing CO_2, and mode waters are a key component of the ocean's state. Thus, properly representing mode waters and diagnosing their balance in models is a useful target. The nature of the mode water response to the carbon cycle changes will likely become a key question for future climate. Recent work with coupled models (Downes et al., 2011; Sallée et al., 2013) points to the enduring usefulness of the water mass formation approach.

7. CONCLUSIONS

As sampling resolution increases, the distribution and characteristics of mode waters are revealed to be more varied and complex, reflecting better the diverse processes that form it, cause it to spread, and consume it. Surface fluxes of heat, freshwater, and stress combine, together with the particular flow and physical conditions of the surface layer, to create volumes of water that act to absorb climate anomalies and maintain coherence over decades and longer.

Descriptive studies of mode water have produced some of the most exciting and useful foundations for our understanding of the large-scale upper ocean circulation. Quantitative water mass studies link their distribution to formation rates, and from these some insight about air–sea interaction, ocean mixing and other diabatic processes may be obtained.

The description of mode water in terms of PV led to advances in large-scale ocean dynamics. Mode water is one of the fundamental components of the ocean, and the structure of mode water—its fine-scale varieties—can be a significant contributor to large-scale dynamics. The global stress, heat, and freshwater forcing of the ocean are directly linked to PV fluxes; however, it is still not clear how to exploit this relationship for the coupled air–sea problem. The question of PV sources at the boundary, whether it is the sea-surface or seafloor, links the large-scale water mass formation problem to small-scale physical processes, and constitutes an area of active current research.

ACKNOWLEDGMENTS

We would like to thank the editors for their uncommon patience, and reviewers for their rapid assessment and advice to help improve the manuscript. K. S. was supported by NSF OCE-0622670 and OCE-0927583. G. F. was supported in part by NOAA grant NA10OAR4310135, NOPP/NASA grant NNX08AV89G, and through the ECCO project. Major support for the ECCO project is provided by the NASA Physical Oceanography Program.

REFERENCES

Aoki, S., Yoritaka, M., Masuyama, A., 2003. Multidecadal warming of subsurface temperature in the Indian sector of the Southern Ocean. J. Geophys. Res. 108 (C4), 8081. http://dx.doi.org/10.1029/2000JC000307.

Aoki, S., Bindoff, N.L., Church, J.A., 2005. Interdecadal water mass changes in the Southern Ocean between 30°E and 160°E. Geophys. Res. Lett. 32. http://dx.doi.org/10.1029/2004GL022220, L07607.

Banks, H.T., Bindoff, N.L., 2003. Comparison of observed temperature and salinity changes in the Indo-Pacific with results from the coupled climate model HadCM3: processes and mechanisms. J. Clim. 16, 156–166.

Banks, H., Wood, R., Gregory, J., 2002. Changes to Indian Ocean Subantarctic Mode Water in a coupled climate model as CO2 forcing increases. J. Phys. Oceanogr. 32, 2816–2827.

Billheimer, S., Talley, L.D., 2013. Near-cessation of Eighteen Degree Water renewal in the western North Atlantic in the warm winter of 2011-2012. J. Geophys. Res. Oceans, Submitted.

Bozec, A., Bouruet-Aubertot, P., Iudicone, D., Crépon, M., 2008. Impact of penetrative solar radiation on the diagnosis of water mass transformation in the Mediterranean Sea. J. Geophys. Res. 113. http://dx.doi.org/10.1029/2007JC004606, C06012.

Brambilla, E., Talley, L., 2008. Subpolar mode water in the northern Atlantic: 1. averaged properties and mean circulation. J. Geophys. Res. 113, C04025.

Brambilla, E., Talley, L.D., Robbins, P.E., 2008. Subpolar Mode Water in the northeastern Atlantic: 2. Origin and transformation. J. Geophys. Res. 113. http://dx.doi.org/10.1029/2006JC004063, C04026.

Cerovečki, I., Talley, L.D., Mazloff, M.R., Maze, G., 2013. Subantarctic mode water formation, destruction, and export in the eddy-permitting Southern Ocean state estimate. J. Phys. Oceanogr. 43, 1485–1511. doi: http://dx.doi.org/10.1175/JPO-D-12-0121.1.

Charrassin, J.B., Hindell, M., Rintoul, S.R., Roquet, F., Sokolov, S., Biuw, M., Costa, D., Boehme, L., Lovell, P., Coleman, R., Timmermann, R., Meijers, A., Meredith, M., Park, Y.H., Bailleul, F., Goebel, M., Tremblay, Y., Bost, C.A., McMahon, C.R., Field, I.C., Fedak, M.A., Guinet, C., 2008. Southern Ocean frontal structure and sea-ice formation rates revealed by elephant seals. Proc. Natl. Acad. Sci. U.S.A. 105, 11634–11639. http://dx.doi.org/10.1073/pnas.0800790105.

Czaja, A., Hausmann, U., 2009. Observations of entry and exit of potential vorticity at the sea surface. J. Phys. Oceanogr. 39, 2280–2294.

Davis, X., Rothstein, L., Dewar, W., Menemenlis, D., 2011. Numerical investigations of seasonal and interannual variability of North Pacific Subtropical Mode Water and its implication for Pacific climate variability. J. Clim. 24, 2648–2665.

Davis, X.J., Straneo, F., Kwon, Y.-O., Kelly, K.K., Toole, J.M., 2013. Evolution and formation of North Atlantic Eighteen Degree Water in the Sargasso Sea from moored data. Deep Sea Res. Part II 91, http://dx.doi.org/10.1016/j.dsr2.2013.02.024.

Döos, K., Nilsson, J., Nycander, J., Brodeau, L., Ballarotta, M., 2012. The World ocean thermohaline circulation. J. Phys. Oceanogr. 42, 1445–1460.

Downes, S.M., Gnanadesikan, A., Griffies, S.M., Sarmiento, J., 2011. Water mass exchange in the Southern Ocean in coupled climate models. J. Phys. Oceanogr. 41, 1756–1771. http://dx.doi.org/10.1175/2011JPO4586.1.

Durack, P.J., Wijffels, S.E., 2010. Fifty-year trends in global ocean salinities and their relationship to broad-scale warming. J. Clim. 23, 4342–4362. doi: http://dx.doi.org/10.1175/2010JCLI3377.1.

Fine, R.A., 1993. Circulation of Antarctic Intermediate Water in the South Indian Ocean. Deep Sea Res. Part I 40, 2021–2042.

Forget, G., 2010. Mapping ocean observations in a dynamical framework: a 2004–06 Ocean Atlas. J. Phys. Oceanogr. 40, 1201–1221.

Forget, G., Maze, G., Buckley, M., Marshall, J., 2011. Estimated seasonal cycle of North Atlantic Eighteen Degree Water Volume. J. Phys. Oceanogr. 41, 269–286.

Hanawa, K., Talley, L., 2001. Mode waters. In: Siedler, G., Church, J. (Eds.), Ocean Circulation and Climate. International Geophysics Series, Academic Press, London, pp. 373–386.

Hartin, C.A., Fine, R.A., Sloyan, B.M., Talley, L.D., Chereskin, T.K., Happel, J., 2011. Formation Rates of Subantarctic Mode Water and Antarctic Intermediate Water within the South Pacific. Deep Sea Res. Part I 58 (2011), 524–534.

Harvey, J., Arhan, M., 1988. The water masses of the central North Atlantic in 1983-84. J. Phys. Oceanogr. 18, 1855–1875. http://dx.doi.org/10.1175/1520-0485(1988)018<1855:TWMOTC>2.0.CO;2.

Jenkins, W.J., 2003. Tracers of ocean mixing. In: Elderfield, H. (Ed.), Treatise on Geochemistry, vol. 6. Elsevier, Oxford, pp. 223–246.

Joyce, T.M., Deser, C., Spall, M.A., 2000. The relation between decadal variability of subtropical mode water and the North Atlantic Oscillation. J. Clim. 13, 2550–2569. http://dx.doi.org/10.1175/1520-0442(2000)013<2550:TRBDVO>2.0.CO;2.

Joyce, T.M., Thomas, L., Bahr, F., 2009. Wintertime observations of Subtropical Mode Water formation within the Gulf Stream. Geophys. Res. Lett. 36. http://dx.doi.org/10.1029/2008GL035918, L02607.

Koch-Larrouy, A., Morrow, R., Penduff, T., Juza, M., 2010. Origin and mechanism of Subantarctic Mode Water formation and transformation in the Southern Indian Ocean. Ocean Dyn. 60, 563–583. http://dx.doi.org/10.1007/s10236-010-0276-4.

Langehaug, H.R., Rhines, P.B., Eldevik, T., Mignot, J., Lohmann, K., 2012. Water mass transformation and the North Atlantic Current in three multicentury climate model simulations. J. Geophys. Res. 117, http://dx.doi.org/10.1029/2012JC0080, C11001.

Large, W.G., Nurser, A.J.G., 2001. Ocean surface water mass transformation. In: Ocean Circulation and Climate. Academic Press, New York, pp. 317–336.

Marchese, P., Gordon, A.L., 1996. The eastern boundary of the Gulf Stream recirculation. J. Mar. Res. 54, 521–540.

Marshall, J., et al., 2009. The CLIMODE field campaign: observing the cycle of convection and restratification over the Gulf Stream. Bull. Am. Meteorol. Soc. 90, 1337–1350.

Maze, G., Marshall, J., 2011. Diagnosing the observed seasonal cycle of Atlantic subtropical mode water using potential vorticity and its attendant theorems. J. Phys. Oceanogr. 41, 1986–1999. http://dx.doi.org/10.1175/2011JPO4576.1.

Maze, G., Forget, G., Buckley, M., Marshall, J., Cerovecki, I., 2009. Using transformation and formation maps to study the role of air-sea heat fluxes in North Atlantic Eighteen Degree Water formation. J. Phys. Oceanogr. 8, 1818–1835.

Mazloff, M., Heimbach, P., Wunsch, C., 2010. An eddy-permitting Southern Ocean State Estimate. J. Phys. Oceanogr. 40 (5), 880–899. http://dx.doi.org/10.1175/2009JPO4236.1.

McCarthy, M.C., Talley, L.D., 1999. Three-dimensional potential vorticity structure in the Indian Ocean. J. Geophys. Res. 104, 13251–13267.

McCartney, M., 1977. Sub-antarctic mode water. In: A Voyage of Discovery: George Deacon 70th Anniversary, Deep Sea Research, Suppl, pp. 103–119.

McCartney, M.S., 1982. The subtropical recirculation of mode waters. J. Mar. Res. 40 (Suppl.), 427–464.

McCartney, M., Talley, L.D., 1982. The subpolar mode water of the North Atlantic Ocean. J. Phys. Oceanogr. 12, 1169–1188.

McDowell, S., Rhines, P.B., Keffer, T., 1982. North Atlantic potential vorticity and its relation to the general circulation. J. Phys. Oceanogr. 12, 1417–1436.

Nurser, A.J.G., Marsh, R., Williams, R.G., 1999. Diagnosing water mass formation from air–sea fluxes and surface mixing. J. Phys. Oceanogr. 29, 1468–1487. http://dx.doi.org/10.1175/1520-0485(1999)029<1468:DWMFFA>2.0.CO;2.

Oka, E., Qiu, B., 2012. Progress of North Pacific mode water research in the past decade. J. Oceanogr. 68, 5–20.

Oka, E., Kouketsu, S., Toyama, K., Uehara, K., Kobayashi, T., Hosoda, S., Suga, T., 2011. Formation and subduction of Central Mode Water based on profiling float data, 2003–08. J. Phys. Oceanogr. 41, 113–129.

Oka, E., Qiu, B., Kouketsu, S., Uehara, K., Suga, T., 2012. Decadal seesaw of the Central and Subtropical Mode Water formation associated with the Kuroshio Extension variability. J. Oceanogr. 68, 355–360. http://dx.doi.org/10.1007/s10872-011-0098-0.

Parr, A.E., 1938. Analysis of current profiles by a study of pycnometric distortion and identifying properties. J. Mar. Res. 1, 4, 1937–1938, pp. 269–290.

Provost, C., Escoffier, C., Maamaatuaiahutapu, K., Kartavtseff, A., Garcon, V., 1999. Subtropical mode waters in the South Atlantic Ocean. J. Geophys. Res. 104 (C9), 21,033–21,049.

Qiu, B., Hacker, P., Chen, S., Donohue, K.A., Watts, D.R., Mitsudera, H., Hogg, N.G., Jayne, S.R., 2006. Observations of the subtropical mode water evolution from the Kuroshio extension system study. J. Phys. Oceanogr. 36, 457–473. http://dx.doi.org/10.1175/JPO2849.1.

Qiu, B., Chen, S., Hacker, P., 2007. Effect of mesoscale eddies on Subtropical Mode Water variability from the Kuroshio Extension System Study (KESS). J. Phys. Oceanogr. 37, 982–1000. http://dx.doi.org/10.1175/JPO3097.1.

Roemmich, D., Cornuelle, B., 1992. The subtropical mode waters of the South Pacific Ocean. J. Phys. Oceanogr. 22, 1178–1187. http://dx.doi.org/10.1175/1520-0485(1992)022<1178:TSMWOT>2.0.CO;2.

Roquet, F., Charrassin, J.-B., Marchand, S., Boehme, L., Fedak, M., Reverdin, G., Guinet, C., 2011. Delayed-mode calibration of hydrographic data obtained from animal-borne satellite relay data loggers. J. Atmos. Oceanic Technol. 28, 787–801. http://dx.doi.org/10.1175/2010JTECHO801.1.

Sagawa, T., Yokoyama, Y., Ikehara, M., Kuwae, M., 2011. Vertical thermal structure history in the western subtropical North Pacific since the Last Glacial Maximum. Geophys. Res. Lett. 38, L00F02. http://dx.doi.org/10.1029/ 2010GL045827.

Sallée, J.B., Rintoul, S., 2011. Parameterization of eddy-induced subduction in the Southern Ocean surface-layer. Ocean Model. 39 (1–2), 146–153.

Sallée, J.-B., Wienders, N., Speer, K., Morrow, R., 2006. Formation of subantarctic mode water in the southeastern Indian Ocean. Ocean Dyn. 1–18. http://dx.doi.org/10.1007/s10236-005-0054-x.

Sallée, J.-B., Morrow, R., Speer, K., 2008. Eddy heat diffusion and Subantarctic Mode Water formation. Geophys. Res. Lett. 35. http://dx.doi.org/10.1029/2007GL032827, L05607.

Sallée, J.B., Speer, K., Rintoul, S., Wijffels, S., 2010. Southern ocean thermocline ventilation. J. Phys. Oceanogr. 40, 509–529. http://dx.doi.org/10.1175/2009JPO4291.1.

Sallée, J.B., Shuckburgh, E., Bruneau, N., Meijers, A.J.S., Bracegirdle, T.J., Wang, Z., Roy, T., 2013. Assessment of the Southern Ocean water mass circulation and characteristics in CMIP5 models: historical bias and forcing response. J. Geophys. Res. 118, 1–15.

Sandwell, D.T., Zhang, B., 1989. Global mesoscale variability from the Geosat Exact Repeat Mission: correlation with ocean depth. J. Geophys. Res. 94, 17971–17984.

Schodlock, M.P., Tomczak, M., White, N., 1997. Deep sections through the South Australian Basin and across the Australian–Antarctic Discordance. Geophys. Res. Lett. 24, 2785–2788.

Schroeder, E., Stommel, H., Menzel, D., Suttcliff Jr., W., 1959. Climatic stability of 18 C water at Bermuda. J. Geophys. Res. 64, 363–366.

Siedler, G., Kuhl, A., Zenk, W., 1987. The Madeira mode water. J. Phys. Oceanogr. 17, 1561–1570.

Sloyan, B.M., Talley, L.D., Chereskin, T.K., Fine, R., Holte, J., 2010. Antarctic Intermediate Water and subantarctic mode water formation in the Southeast Pacific: the role of turbulent mixing. J. Phys. Oceanogr. 40, 1558–1574. http://dx.doi.org/10.1175/2010JPO4114.1.

Speer, K.G., Isemer, H.J., Biastoch, A., 1995. Water mass formation from revised COADS data. J. Phys. Oceanogr. 25, 2444–2457.

Stramma, L., 1992. The South Indian Ocean Current. J. Phys. Oceanogr. 22, 421–430.

Talley, L.D., 1999. Antarctic Intermediate Water in the South Atlantic. In: Wefer, G., Berger, W.H., Siedler, G., Webb, D. (Eds.), The South Atlantic: Present and Past Circulation. Springer-Verlag, pp. 219–238.

Talley, L.D., Raymer, M.E., 1982. Eighteen degree water variability. J. Mar. Res. 40 (Suppl.), 757–775.

Thomas, L.N., Joyce, T.M., 2010. Subduction on the northern and southern flanks of the Gulf Stream. J. Phys. Oceanogr. 40, 429–438. http://dx.doi.org/10.1175/2009JPO4187.1.

Tsuchiya, M., 1968. Thermostads and circulation in the upper layer of the Atlantic Ocean. Prog. Oceanogr. 16, 235–237.

Tsuchiya, M., Talley, L.D., 1998. A Pacific hydrographic section at 88°W: water-property distribution. J. Geophys. Res. 103, 12,899–12,918.

Tziperman, E., 1986. On the role of interior mixing and air–sea fluxes in determining the stratification and circulation of the oceans. J. Phys. Oceanogr. 16, 680–693.

Tziperman, E., Speer, K.G., 1994. A study of water mass transformation in the Mediterranean Sea: analysis of climatological data and a simple 3-box model. Dyn. Atmos. Oceans 21, 53–82.

Urakawa, L.S., Hiroyasu, H., 2012. Eddy-resolving model estimate of the cabbeling effect on the water mass transformation in the Southern Ocean. J. Phys. Oceanogr. 42, 1288–1302. http://dx.doi.org/10.1175/JPO-D-11-0173.1.

Wainer, I., Goes, M., Murphy, L.N., Brady, E., 2012. Changes in the intermediate water mass formation rates in the global ocean for the Last Glacial Maximum, mid-Holocene and pre-industrial climates. Paleoceanography 27. http://dx.doi.org/10.1029/2012PA00229, PA3101.

Walin, G., 1982. On the relation between sea-surface heat flow and thermal circulation in the ocean. Tellus 34, 187–195. http://dx.doi.org/10.1111/j.2153- 3490.1982.tb01806.x.

Warren, B.A., 1972. Insensitivity of subtropical mode water characteristics to meteorological fluctuations. Deep Sea Res. 19, 1–19.

Waugh, D.W., Primeau, F., DeVries, T., Holzer, M., 2013. Recent changes in the ventilation of the Southern Oceans. Science 339 (6119), 568–570. http://dx.doi.org/10.1126/science.1225411.

Wong, A.P.S., Johnson, G.C., 2003. South Pacific eastern subtropical mode water. J. Phys. Oceanogr. 33 (7), 1493–1509.

Wong, A.P.S., Bindoff, N.L., Church, J.A., 1999. Large-scale freshening of intermediate waters in the Pacific and Indian oceans. Nature 400, 440–443.

Worthington, L.V., 1959. The 18° Water in the Sargasso Sea. Deep Sea Res. 5, 297–305.

Worthington, L.V., 1976. On the North Atlantic circulation. In: Oceanographic Studies, The John Hopkins University, Baltimore, MD, 110 pp.

Worthington, L.V., 1981. The water masses of the world ocean: some results of a fine-scale census. In: Warren, B.A., Wunsch, C. (Eds.), Evolution of Physical Oceanography, Scientific Surveys in Honor of Henry Stommel. MIT Press, Cambridge, MA, pp. 42–69, copyright 1981.

Wunsch, C., Heimbach, P., 2013. Two decades of the Atlantic meridional overturning circulation: anatomy, variations, extremes, prediction, and overcoming its limitations. J. Clim. 26, 7167–7186. doi: http://dx.doi.org/10.1175/JCLI-D-12-00478.1.

Yun, J.-Y., Talley, L.D., 2003. Cabbeling and the density of the North Pacific Intermediate Water quantified by an inverse method. J. Geophys. Res. 108 (C4), 3118. http://dx.doi.org/10.1029/2002JC001482.

Zika, J.D., England, M.H., Sijp, W.P., 2012. The ocean circulation in thermohaline coordinates. J. Phys. Oceanogr. 2, 708–724. http://dx.doi.org/10.1175/JPO-D-11-0139.1.

Chapter 10

Deepwater Formation

John L. Bullister*, Monika Rhein† and Cecilie Mauritzen‡

*National Oceanic and Atmospheric Administration—Pacific Marine Environmental Laboratory (NOAA-PMEL), Seattle, Washington, USA

†Institute of Environmental Physics IUP, University of Bremen, Bremen, Germany

‡CICERO Center for International Climate and Environmental Research Oslo, Oslo, Norway

Chapter Outline

1. INTRODUCTION

The very limited regions of the World Oceans where deepwater formation takes place are critically important as "windows" to the vast volume of water in the deep ocean. The formation and subduction of water masses—also known as the "physical pump"—largely determine the capacity of the ocean to store heat, freshwater and carbon. During the past 40 years about 93% of the global heat increase has been stored in the world's oceans (Solomon et al., 2007). Since the beginning of the industrial era, the ocean has also acted as a significant sink for anthropogenic CO_2 emissions (Sabine et al., 2004; Le Quéré et al., 2009; also see Chapter 30). Deepwater formation and the deep ocean circulation thus play an important role in the Earth's climate through the uptake and storage of heat and anthropogenic CO_2 (Sigman and Boyle, 2000; Russell et al., 2006). The capacity of the deep ocean to take up and store anthropogenic CO_2 will continue to have a major impact on the CO_2 content of the atmosphere far into the future. Paleooceanographic studies indicate that there have been large changes in the rates of deepwater formation processes and in deep circulation patterns in the past, associated with rapid climate change (e.g., Clark et al., 2002; Lynch-Stieglitz et al., 2007). Long (25 kyr) radiocarbon records from deep sea corals show that periods of relative isolation of the deep ocean water masses followed by subsequent enhanced ventilation events (via upwelling and degassing) in the Southern Ocean are related to large changes in atmospheric CO_2 levels between glacial and interglacial periods (Burke and Robinson, 2012; see also Chapter 2).

Observational programs during the past several decades have detected significant changes in the properties of deepwaters, both near the formation regions and along the circulation pathways far into the interior of the deep ocean. These changes are related in part to decadal variability and partly to long-term climate change.

In this chapter we discuss recent developments in the understanding of formation and propagation of the densest water masses of the world ocean, and seek evidence for change in properties, formation rates and circulation strength during recent decades, as these changes will affect (in poorly understood ways) the uptake and storage rates of

Ocean Circulation and Climate, Vol. 103. http://dx.doi.org/10.1016/B978-0-12-391851-2.00010-6

heat and CO_2, as well as the contribution of the deep ocean to global sea level change.

1.1. Circulation and Distribution of NADW and AABW

Two key water masses contribute to the renewal of the deepest waters of the World Ocean, those found beneath 2000–3000 m, namely North Atlantic Deep Water (NADW), produced in the northern North Atlantic and the Nordic Seas, and Antarctic Bottom Water (AABW), formed at high southern latitudes at a number of locations around Antarctica. In those regions high buoyancy losses, primarily due to cooling but also due to increased salinity in sea-ice formation regions, create waters of such high density that they can sink to deep (>2000 m) and abyssal (>4000 m) depths. The pathways of NADW and AABW can be traced throughout the deep and abyssal ocean (Figure 10.1) by following their hydrographic and chemical signatures, specific for each water mass as they are gradually modified by mixing with surrounding waters and through a variety of biogeochemical processes. With timescales of decades to centuries or longer, deepwaters rise to shallower depths and eventually return to the formation regions to start the cycle over again. Based on property distributions, some of the main pathways of the deepwaters in the Atlantic Ocean were mapped out early in the twentieth century, with increasing level of details added to the global picture by the many oceanographers that have followed. An excellent and thorough discussion of the history of physical oceanography and early pioneering studies in the field is presented in Chapter S1 of Talley et al. (2011).

In the northern North Atlantic, the densest components of NADW are formed from overflow of dense waters from the Nordic (Greenland–Iceland–Norwegian) seas southward across sills between Greenland and Scotland. The pioneering oceanographers that mapped out ocean circulation a century ago did not recognize the Greenland–Scotland Overflows as significant contributors to the Atlantic circulation. Instead they considered the Labrador and Irminger Seas to be the northernmost sources of dense water. It was not until the 1950s that the importance of the Greenland–Scotland overflows was recognized (Cooper, 1955). These dense, cold overflow waters originate from well-ventilated, mid-depth waters in the Nordic seas. They encounter less dense water after crossing the sills, and entrain considerable amounts of ambient water during descent to the bottom.

Both the AABW and NADW formation processes involve surface waters and, as a result, newly formed AABW and NADW are recognized in the deep ocean as water of relatively young age (i.e., the time since the water lost contact with the surface ocean), with elevated levels of dissolved

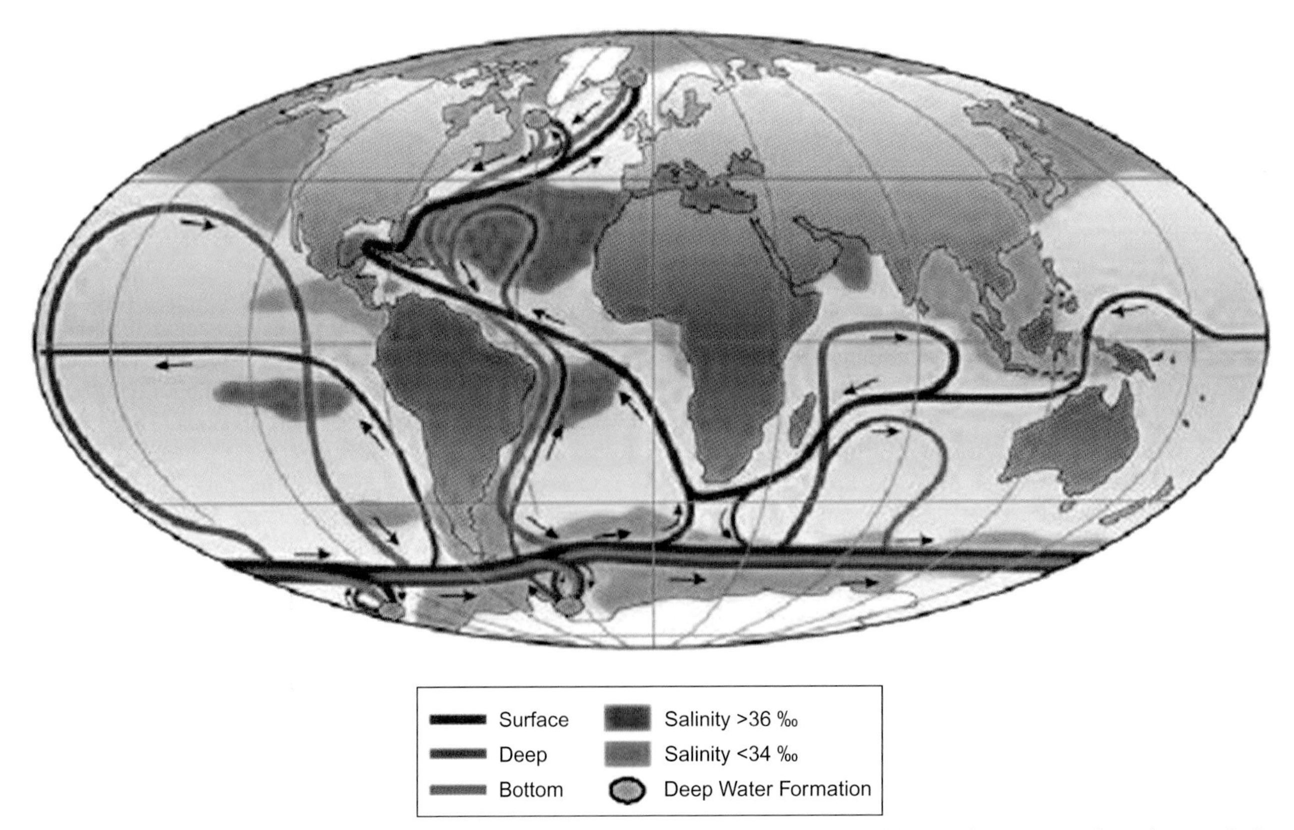

FIGURE 10.1 Schematic of the large-scale thermohaline circulation of the global ocean. Key regions of deep- and bottom-water formation occur in the northern North Atlantic and in regions around the Antarctic continent. *From Rahmstorf (2002).*

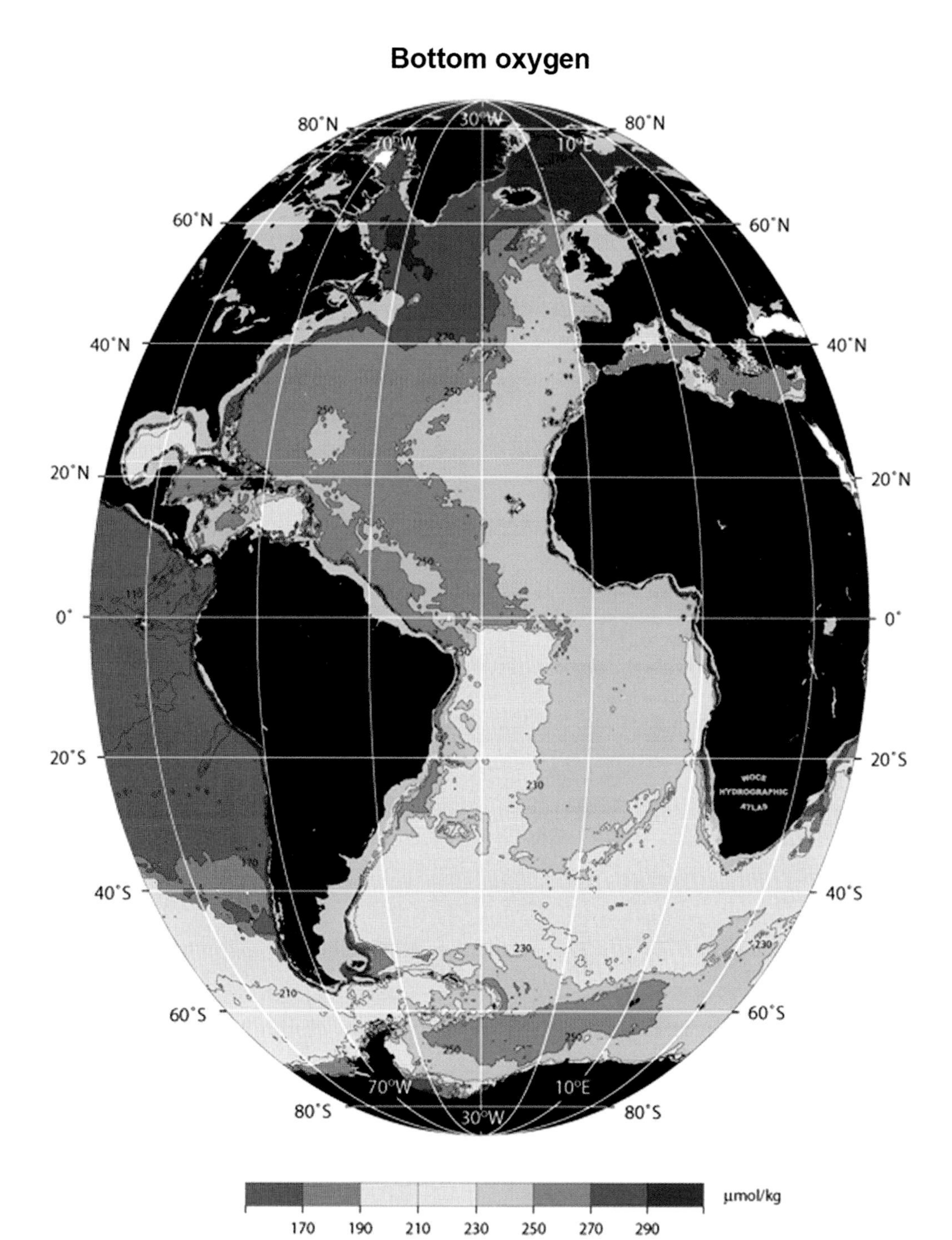

FIGURE 10.2 The distribution of dissolved oxygen (μmol kg^{-1}) near the bottom in the deep Atlantic Ocean. *From Koltermann et al. (2011).*

oxygen and transient tracers such as the chlorofluorocarbons (CFCs). In the Atlantic Ocean, high levels of dissolved oxygen are present in newly formed NADW and the southward propagation of this signal can be traced from the northern source regions along the western basin of the North Atlantic (Figure 10.2). Newly formed AABW also contains elevated levels of oxygen and is identified by the high dissolved oxygen signal at abyssal depths in the South Atlantic (Figure 10.2). In the Atlantic Ocean where northern (NADW) and southern (AABW) source waters meet, the AABW flows below the NADW. AABW is colder, fresher, and denser (at the relevant depth) than NADW.

A canonical view of the World Ocean circulation as seen from the Southern Ocean was presented by Gordon (1991) and Schmitz (1996a,b), who discussed the pathways and inter-basin exchanges of the global-scale thermohaline circulation. This figure was updated by Lumpkin and Speer (2007) (see Figure 10.3), who applied inverse modeling to a set of World

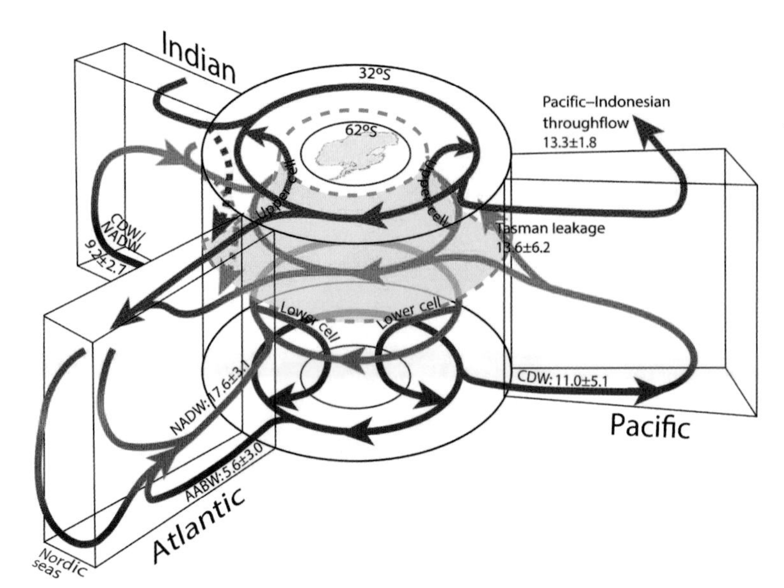

FIGURE 10.3 Schematic of the major circulation pathways of the global ocean, with estimates of mean flows (in Sverdrups (Sv), where 1 Sv = 10^6 m^3 s^{-1}). *From Lumpkin and Speer (2007).*

Ocean Circulation Experiment (WOCE) hydrographic sections occupied in the 1990s, adding air–sea fluxes of heat and freshwater as well as direct current measurements, to obtain quantitative estimates of circulation strength.

In this schematic, NADW formed in the North Atlantic sinks and flows southward, where it encounters and mixes with the underlying layer of dense AABW, which is undergoing an upward overturning loop in the Atlantic Ocean (Figure 10.3). As it reaches the Southern Ocean, the modified NADW upwells in the circumpolar region, with some of the denser components continuing southward and contributing to the formation and sinking of new AABW in the marginal seas around Antarctica. Other components of this cell (referred to in Figure 10.3 as AABW in the deep Atlantic, Circumpolar Deep Water (CDW) in the Pacific and CDW/NADW in the Indian Oceans) are exported northward from the circumpolar region across 32°S. These deepwaters gradually become lighter by mixing, and eventually contribute to the upper cells of global circulation. According to the schematic, the combined export of NADW in the South Atlantic (increased to a total of ~17 Sv by entrainment of underlying AABW) is roughly three times as large as the upward overturning of AABW in the Atlantic. In contrast to the North Atlantic, there is no deepwater formation in the North Pacific and North Indian Ocean. In the two other basins, Indian and Pacific, the upward CDW overturning is of roughly comparable size (~10 Sv).

The volumetric distribution of AABW-derived versus NADW-derived waters in the global ocean has been calculated by Johnson (2008) (Figure 10.4), using data from the WOCE Global Hydrographic Climatology (Gouretski and Koltermann, 2004; collected primarily in the 1990s). Using carefully chosen end-member water mass properties in a least-squares analysis, Johnson (2008) estimates the distribution of the depth integral of the fraction of NADW (Figure 10.4 upper panel) and AABW (Figure 10.4, lower panel) in the global ocean. NADW dominates in the North Atlantic, with equivalent layer thicknesses of more than 2000 m present in entire the North Atlantic and as far south as about 30°S in South Atlantic. The relative contribution of NADW-derived waters is reduced farther south but extends as a circumpolar band in the Southern Ocean and into the Indian and Pacific Oceans as a layer of thickness 250–500 m. Very thick layers of AABW exist as a circumpolar band in the southern ocean (Figure 10.4) and can be traced as far as 40°N in the Atlantic. AABW forms very thick layers throughout much of the Indian and Pacific Basins. The volume of the global ocean occupied by AABW is estimated to be about 36% versus about 21% for NADW. It is important to note, as Johnson (2008) comments, that due to the long timescales involved in the global deep overturning circulation, these AABW and NADW distributions do not necessarily reflect current rates of AABW and NADW formation, but rather a long-term average of processes contributing to the accumulation of these waters over decade-to-century timescales.

1.2. Observed Heat Content Changes in AABW

Until recently, most of the observed ocean heat content increase has been found in the upper 700 m of the World Ocean (see Chapter 27). At mid-depths there is little evidence of any net change in temperature, whereas near the bottom a significant signal of warming is now beginning to emerge (Purkey and Johnson, 2010). As part of the CLIVAR and the Global Ocean Ship-based Hydrographic Investigations Program (GO-SHIP) Repeat Hydrography surveys and other

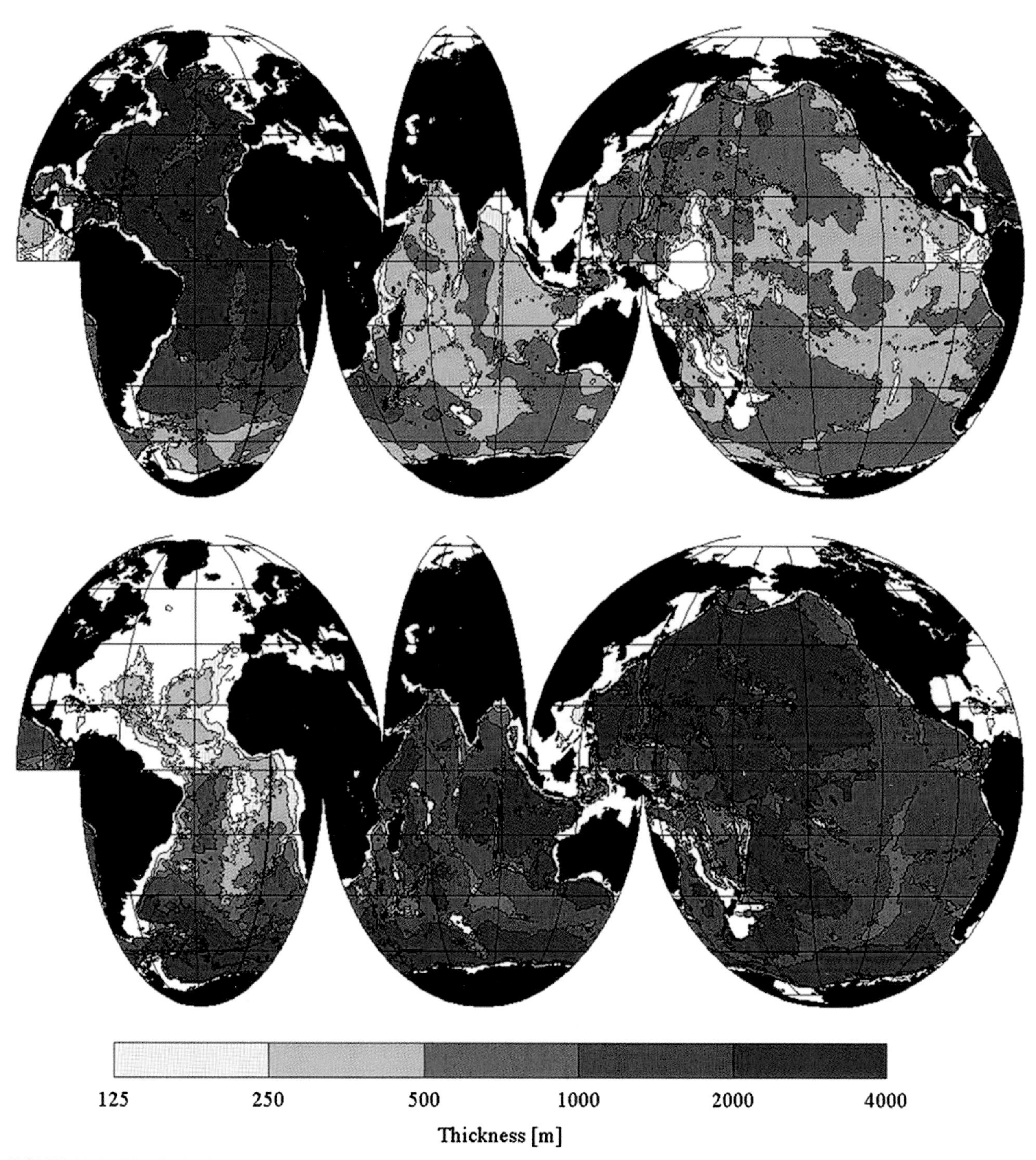

FIGURE 10.4 The distribution of the depth integral of the fraction of NADW (upper panel) and AABW (lower panel) in the global ocean. NADW-derived waters dominate in the North Atlantic, while AABW-derived waters dominate in the Antarctic Circumpolar Current (ACC) and in the Indian and Pacific basins. *From Johnson (2008).*

efforts, key hydrographic sections occupied as part of the WOCE program are being reoccupied on roughly decadal intervals to try to detect changes in water column properties along the sections. In the Southern Ocean, these repeat sections have allowed detection of a warming signal in AABW, with the largest change observed close to the AABW formation areas (Figure 10.5; Purkey and Johnson, 2010). The upper ocean warming is transferred into the deep ocean in the formation areas of AABW and the deep ocean warming signal can be seen to advect along the spreading pathways of AABW: it has been observed throughout the South and North Pacific (Figure 10.5;

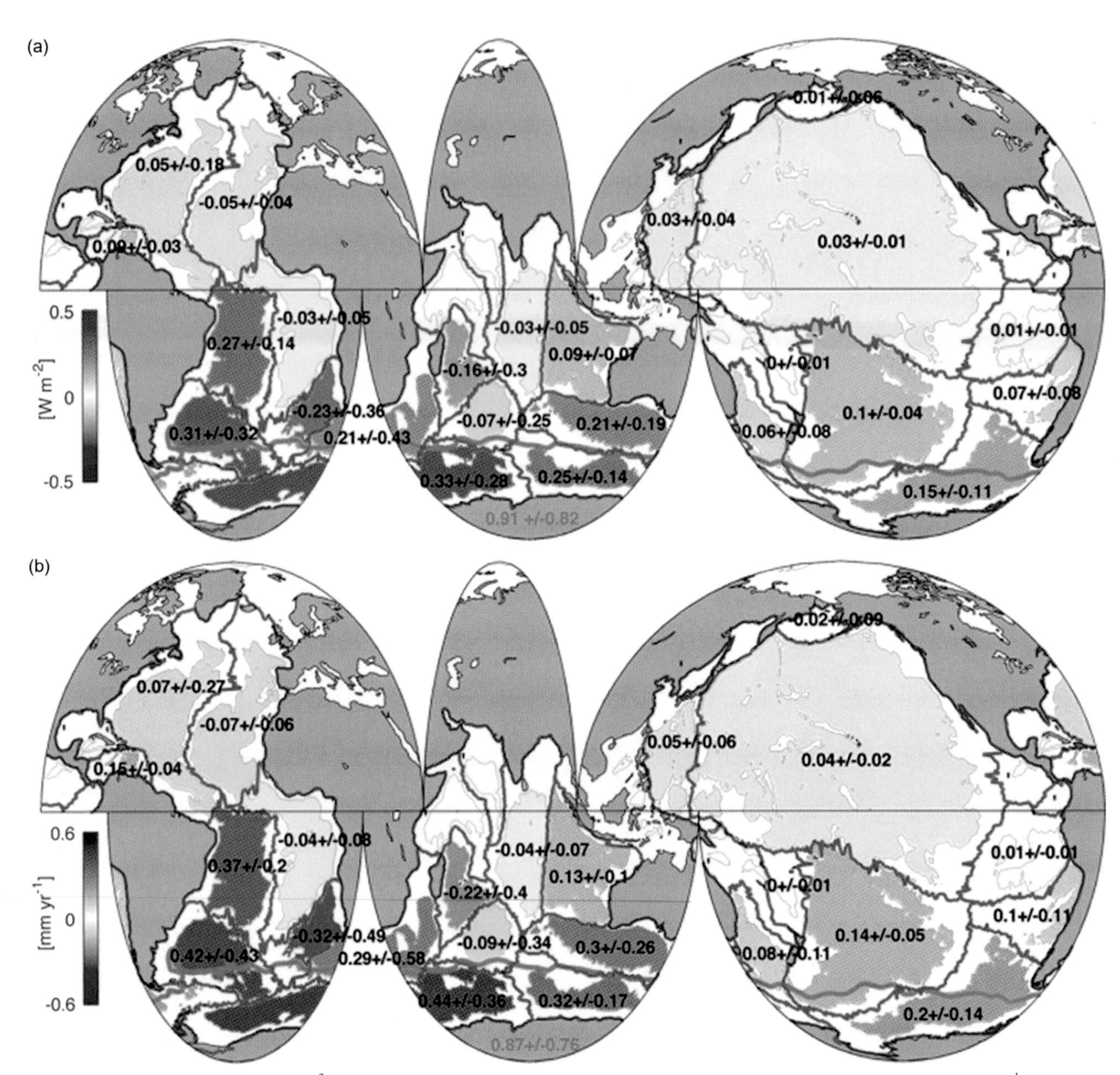

FIGURE 10.5 (a) Local heat flux (W m^2) from the 1900s to the 2000s implied by local warming below 4000 m. (b) sea level rise (mm yr^{-1}) from 1900s to 2000s due to thermal expansion of water below 4000 m (Purkey and Johnson, 2010).

Fukasawa et al., 2004; Kawano et al., 2006; Johnson et al., 2007), in the western basins of the South and North Atlantic (Johnson and Doney, 2006; Zenk and Morozov, 2007; Johnson et al., 2008b), and the eastern Indian Ocean (Johnson et al., 2008a). Much further downstream in the North Pacific, the warming cannot be caused directly by advection (which would be too slow a process) but might propagate to this region via internal waves within four decades (Masuda et al., 2010).

The net result of these observed changes in AABW to global heat uptake and sea level rise (due to thermal expansion of seawater) is that global abyssal ocean and deep Southern Ocean warming during the 1990s and 2000s contributed about 16% to the total ocean heat uptake during this period and that warming of these waters contributed about 9% to the sea level rise due to thermal expansion (Purkey and Johnson, 2010).

1.3. Observed Heat Content Changes in Upper and Lower NADW

Though globally there has been little net change in heat content at mid-depth, as mentioned above, the North Atlantic Ocean has seen significant changes in ocean heat content at all depths. In a recent pentadal analysis of changes in heat and salinity content in the North Atlantic north of 20°N during the past 50 years, Mauritzen et al. (2012) distinguish between the changes in various water masses including Upper NADW (UNADW), Lower NADW (LNADW) and AABW, and find that each developed differently. In the Subtropical Gyre, AABW has lost heat (Figure 10.6a, dotted line), not because its average temperature has dropped, but because its volume has shrunk, consistent with the findings in the previous section. LNADW has also lost heat (Figure 10.6a, dashed–dotted line), but in this case the loss is due to a temperature drop, not a change in volume. UNADW saw a warm pulse coming through from the Subpolar Gyre in the 1970s (Figure 10.6a and b, dashed line), and presently UNADW in the Subtropical Gyre carries comparable heat content to that at the beginning of the time series. As for LNADW, the change in heat content for UNADW is primarily due to changes in temperature. Key conclusions from this study are that heat uptake of the UNADW in the Subpolar Gyre is closely related to the North Atlantic Oscillation (NAO) (see Figure 10.6b), and that the continual cooling of LNADW is most likely a readjustment from a well-documented early twentieth century warm period in the North Atlantic and the Arctic. These are excellent examples of how internal variability of the climate system still can fully overwhelm the anthropogenic drivers regionally.

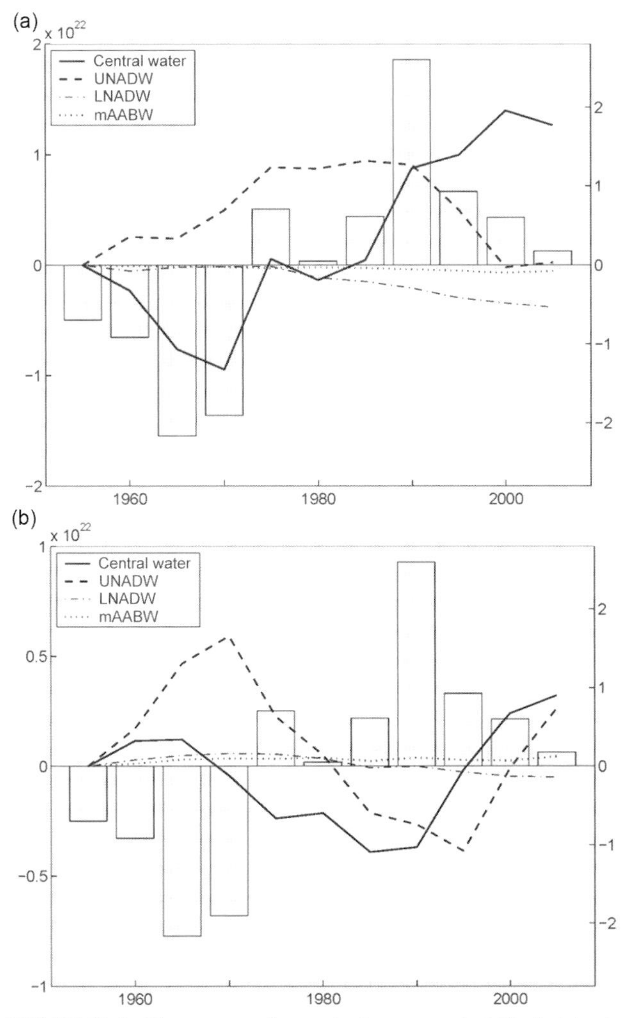

FIGURE 10.6 Heat content changes in the extratropical North Atlantic between 1955 and 2005. (a) In the Subtropical Gyre, between 20°N and 50°N; (b) in the Subpolar Gyre, between 50°N and the Greenland–Scotland Ridge. The water masses are defined as follows: (1) the thermocline waters, or Central Waters (CW): $\sigma_1 < 32.15$; (2) the Upper North Atlantic Deep Water (UNADW): $\sigma_1 > 32.15$ and $\sigma_2 < 37.0$; (3) the Lower North Atlantic Deep Water (LNADW): $\sigma_2 > 37.0$ and $\sigma_4 < 45.9$; and (4) toward the bottom: modified Antarctic Bottom Water (mAABW). *From Mauritzen et al. (2012).* Bars indicate the wintertime NAO index. *From Hurrell et al. (2013).*

2. PROCESSES OF DEEPWATER FORMATION

There are a number of processes that can contribute to dense water formation—here we emphasize those that we consider of specific importance for this chapter. In general, dense water is formed by convection as a result of buoyancy loss. The buoyancy loss may be due to air–sea fluxes, or sea–ice fluxes, or due to nonlinear effects of the equation of state, which may cause two water masses (which by themselves are light) to become denser when they mix. The deepest convection is most likely to occur in geographically localized regions, such as the Gulf of Lyons in the Mediterranean Sea and in the Greenland and Labrador Seas, where bowl-like topography and cyclonic winds cause strong cyclonic currents and therefore doming of isopycnals. Atmospheric cooling can then make the surface waters extremely dense. Deep convection will be discussed in Section 2.1.

The two former examples, the Gulf of Lyons and the Greenland Sea, are located within marginal seas with shallow sills. The densest waters produced in these regions can therefore not directly enter the adjacent large ocean basins. Waters crossing the Strait of Gibraltar and the Greenland–Scotland Ridge originate instead from intermediate depths in the Mediterranean and in the Nordic Seas, respectively. The overflowing waters entrain upper-ocean waters in the North Atlantic as they descend, and this entrainment may be as influential for the resulting water mass properties as the variability at the original locations of formation upstream of the sills. This process will be discussed in Section 2.2.

In a number of locations along the continental margins of Antarctica, rapid sea-ice formation and export can increase the salinity of the underlying waters. Additional buoyancy losses due to cooling and in some cases, circulation of water beneath large floating ice shelves, and addition of glacial water by melting below ice shelves, can lead to the formation of extremely dense shelf waters. These dense shelf waters cascade downward along the continental slope, and entrain surrounding water to form new AABW. Details of these processes are discussed in Section 2.3.

2.1. Deep Convection: The Example of Formation of Upper North Atlantic Deep Water

UNADW is primarily formed in the Labrador Sea and is therefore just as often called Labrador Sea Water (LSW). LSW is found with a wide range of properties, and it is likely that only lighter, shallower (down to ~1500 m) LSW (also called upper LSW-uLSW) is formed in milder winters and that dense LSW (DLSW), reaching as deep as 2400 m, is formed only after a series of harsher winters. LSW is characterized by its high oxygen and CFC levels, indicative of recent ventilation; low potential vorticity, indicative of the deep convection; and low salinity. LSW is found in the western basin as a thick layer of water with temperature just below 3 °C. LSW spreads southward mainly with the Deep Western Boundary Current (DWBC) but also along interior pathways in the Newfoundland basin from the subpolar into the subtropical gyre (Bower et al., 2009, 2011; Kieke et al., 2009). Further south, the DWBC is the main pathway (Fine et al., 2002; LeBel et al., 2008). Away from its source, the upper LSW loses its salinity signature through mixing with salty water of Mediterranean origin and by encountering Antarctic Intermediate Water that is high in silica and low in salinity. In the global overturning circulation, LSW contributes on average roughly 10–12 Sv (Bryden et al., 2005; LeBel et al., 2008).

LSW is formed by deep convection (Marshall and Schott, 1999) in late winter (February/March), where winter time buoyancy loss increases the density of surface water in the central Labrador Sea sufficiently to erase the density stratification in the upper 1000–2500 m. The maximum convection depth depends on the magnitude of the net buoyancy flux and the initial stratification in the previous autumn. After the preconditioning during the winter months (spatial scale ~100 km), active deep convection occurs in localized plumes (scale ~1 km), vigorously mixing the water column from the surface to the convection depth, and producing a vertically homogenized water column with a minimum in planetary potential vorticity (e.g., Talley and McCartney, 1982). Lateral exchange between the convection site and the surroundings (10 km) covers the surface with less dense water within about a week, and spreads the newly formed LSW first into the Labrador Sea, and then preferably along the main circulation pathways into the Irminger Sea, across the mid-Atlantic Ridge into the Northeast Atlantic, and with the Deep Western Boundary Current into the subtropics (e.g., Talley and McCartney, 1982; Sy et al., 1997; Rhein et al., 2002; Figure 10.7). An excellent summary of the LSW formation process is presented by Lazier et al. (2001)

2.2. Entrainment: The Example of the Formation of the Lower North Atlantic Deep Water

Formation of Greenland–Scotland Overflow Waters (Denmark Strait Overflow Water—DSOW, and Iceland–

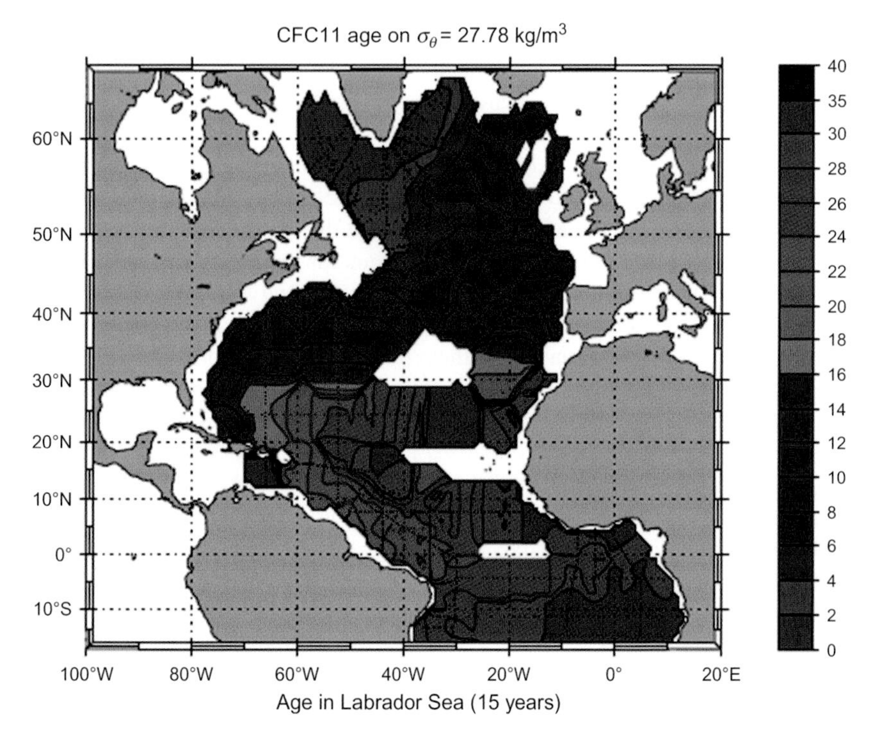

FIGURE 10.7 CFC-11 age on density surface σ_θ=27.78, highlighting the spreading of LSW into the Atlantic Ocean. CFC-11 ages shown are "effective" ages, or ages since the water mass left the source region, and have been calculated by subtracting a "relic" age of 15 years from the measured pCFC11 age of the water. *From Fine et al. (2002).*

Scotland Overflow Water—ISOW) is discussed in Chapter 17, and we direct the interested reader to that chapter. ISOW and DSOW originate from intermediate, well-ventilated layers of the Nordic Seas, and their density is very high relative to waters found at the same depth in the North Atlantic. DSOW is the coldest; it exits through the 640 m deep Denmark Strait. The ISOW is warmer; it exits primarily through the 840 m deep Faroe-Bank Channel. Together these overflows are now estimated at roughly 6 Sv, with 3 Sv in the Denmark Strait (Ross, 1984; Girton et al., 2001; and Macrander et al., 2005; Jochumsen et al., 2012; although when expanding the temperature range up to 2 °C, Dickson et al., 2008 estimate the DSOW transport to be 4 Sv), 1 Sv across the Iceland–Faroe Ridge (Hansen and Østerhus, 2000) and 2.2 Sv in the Faroe-Bank Channel, colder than 2 degrees (Hansen and Osterhus, 2007). After overflowing the sills, the overflows are surrounded by lighter water and descend to the bottom. During the descent, they are significantly modified by entrainment of ambient water that more than doubles the transport (Dickson and Brown, 1994; Mauritzen et al., 2005). The pathways of ISOW and DSOW are guided by topography. The main part of the Iceland–Scotland Overflow Water turns westward mainly through the Charlie–Gibbs Fracture Zone (Saunders, 1994), and some flows through other gaps in the Reykjanes Ridge (Bower et al., 2002). In the Irminger Sea, the ISOW joins the Denmark Overflow Waters continuing southward as part of the Deep Western Boundary Current (Figure 10.8). During time periods when the strong North Atlantic Current happens to be located above the Gibbs Fracture Zone, the whole water column flows eastward and the main western pathway for ISOW is blocked (e.g., Schott et al., 1999). Part of the ISOW remains in the Eastern Atlantic and flows south along the mid-Atlantic Ridge (Fleischmann et al., 2001).

Although ISOW and DSOW can be separated by their hydrographic and chemical signature in the North Atlantic, both are referred to as Lower North Atlantic Deep Water in the North Atlantic, recognizable by high oxygen and high salinity. LNADW is found between UNADW and AABW and contribute, on average, 10–15 Sv to the global overturning circulation (Bryden et al., 2005; Le Bel et al., 2008; see also Figure 10.3).

There have been several process experiments at the sills to describe and quantify the descent, mixing and entrainment as these plumes of dense water spill into the deep North Atlantic. Entrainment involves significant sinking of upper ocean waters in the subpolar North Atlantic (Saunders, 2001). Girton and Sanford (2003) discuss the descent of the Denmark Strait Overflow Water down to 2500 m at Cape Farewell, showing evidence to support the hypothesis that the rate of descent is friction-controlled. According to their measurements, the entrainment increases markedly 125 km downstream of the sill, where the slope increases and the density-difference to the surrounding waters decreases markedly. A similar increase in vertical mixing and entrainment by overlying waters was found in the Iceland–Scotland overflow 100 km downstream of the Faroe Bank Channel (Mauritzen et al., 2005; Fer et al., 2010). The entrained

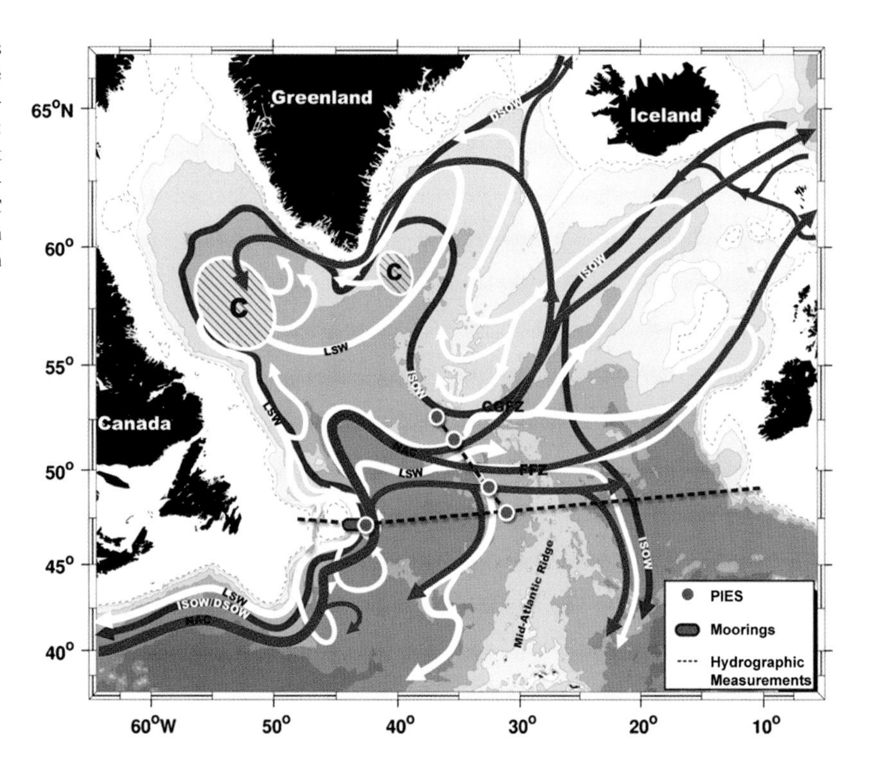

FIGURE 10.8 Circulation of main water masses in the subpolar North Atlantic. White: LSW, white hatched areas named C: convection areas where formation of UNADW (LSW) could occur, blue: ISOW and DSOW, red: North Atlantic Current NAC (upper 500 m). PIES are inverted echo-sounders with bottom pressure sensors. The NAC is included since its strong currents could reach down to the bottom and carry deep and bottom water. *After Rhein et al. (2011).*

waters into the Faroe-Bank plume are much warmer and lighter than those downstream of the Denmark Strait, and cause the ISOW to appear significantly warmer than the DSOW in the North Atlantic (Figure 10.8). In fact, already 100 km downstream of the sill, the main temperature mode of transportation of the plume is 3 °C, and significant velocities occurs for temperatures up to 6 °C (Mauritzen et al., 2005). Further downstream of the Faroe-Bank Channel mixing is enhanced by rough topography (ridges and canyons) along the descent (Wåhlin et al., 2008). The overflow plume has been monitored further downstream using bottom-following floats by Prater and Rossby (2005) who found variability in the overflows even downstream of the sill on timescales of 1–4 days. Short-term variability is common in both overflows, and is likely to be connected with baroclinic instability (Kida et al., 2009 and references therein), in which case the timescale is set by the latitude of the overflows. The mechanisms of entrainment are further elaborated on by Kida et al. (2009), who describe the upper ocean circulation response as a cyclonic topographic beta plume with transport much larger than that required by the downward entrainment itself.

2.3. Shelf and Under-Ice Processes: The Example of Formation of AABW

Quantities of very dense AABW are formed at a number of locations along the margins of the Antarctic continent (Figure 10.9). Near-surface and under-ice processes involved in the formation are described and discussed by Foster and Carmack (1976) and Foldvik et al. (1985). Salt rejection following sea ice formation forms dense surface water, which after flowing below the ice shelves (thereby cooling and incorporating basal meltwater) sinks to the bottom at the continental shelves. While descending along the continental slope, these dense, near-freezing shelf waters produced on the continental shelves entrain a significant amount of warmer and more saline intermediate waters (Gill, 1973; Carmack and Foster, 1975; Foster and Carmack, 1976; Foldvik et al., 1985). The intermediate depth waters that intrude onto the continental shelves and are involved in this entrainment process have characteristic properties, which are derived in part from mixing of Antarctic Surface Waters (ASW) and Circumpolar Deep Water. Therefore, the properties of newly formed varieties of dense AABWs depend not only on the properties of the very cold, high-density shelf-derived components, but also very significantly on the amounts and characteristics of the surrounding waters entrained during the mixing and sinking process. For example, in the main AABW formation region (Weddell Sea), the new bottom water contains 35–50% Warm Deep Water (WDW), a local variety of the Circumpolar Deep Water (Huhn et al., 2008).

Both in the southern Weddell and Ross Seas (Figure 10.9), the broad continental shelves and bathymetric features facilitate the formation and accumulation of dense shelf waters. The salinity of these shelf waters can vary depending on the extent of brine rejection during sea-ice formation and export, and on the pathways of the shelf waters. As an example, in the far southern Weddell

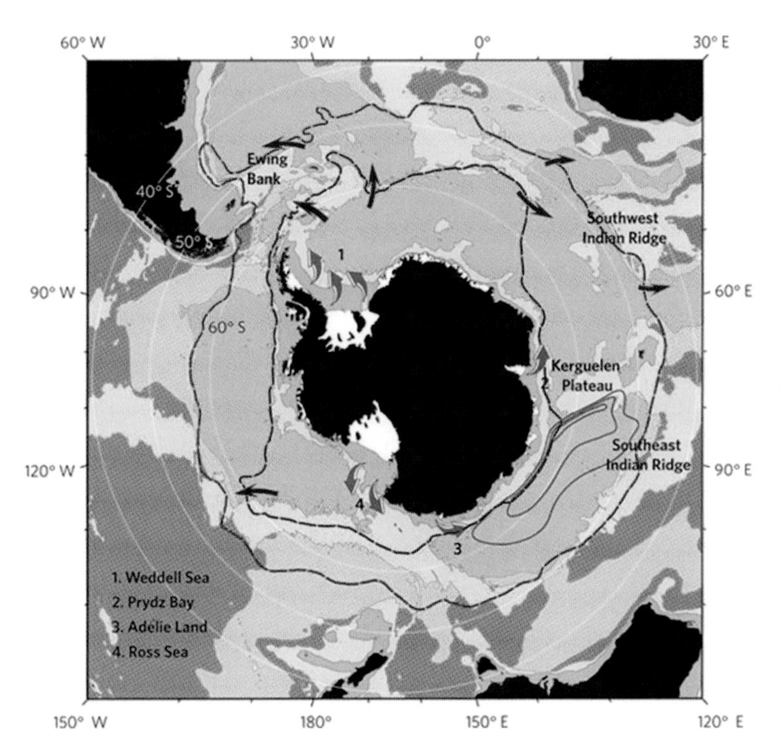

FIGURE 10.9 Major AABW formation regions and outflow pathways from the Southern Ocean. AABW (green) is generated from sinking of dense shelf waters and entrainment of surrounding waters at a number of locations including the Weddell Sea, Prydz Bay, Adélie Land, and Ross Sea, where it sinks to depth (purple arrows). On moving northward across the bounds of the Antarctic Circumpolar Current (red dashed lines), AABW mixes with intermediate water to form Antarctic Circumpolar Current bottom water (light pink). Areas shaded in dark pink indicate bottom water less dense than Antarctic Circumpolar Current bottom water. Relatively undiluted AABW has been shown to leave the Southern Ocean along Ewing Bank and through gaps in the Southwest Indian Ridge (blue arrows). *From Orsi (2010).*

Sea along the Filchner–Ronne Ice Shelf, several varieties of very dense shelf waters have been described including High-Salinity Shelf Water (HSSW) (Foster and Carmack, 1976), which is formed by brine rejection during sea-ice formation processes on the broad shelf region with temperatures near the surface freezing point. New Weddell Sea Bottom Water (WSBW) is formed as this HSSW descends along the slope and mixes with surrounding modified Circumpolar Deep Water (mCDW), ultimately derived in part from the Antarctic Circumpolar Current (ACC). Foldvik et al. (1985) describe a process in the southern Weddell Sea where HSSW can flow beneath the large floating ice-shelves in this region to produce extremely cold, slightly less saline Ice Shelf Waters (ISW) with temperatures below the surface freezing point. As seawater circulates beneath large floating ice shelves, it induces melting along the base of the ice-shelf and incorporates small amounts of subglacial meltwater. As this melting of the lower ice shelf occurs, small bubbles of air trapped in the glacial ice are released and dissolve in the surrounding waters, causing anomalies in the noble gas (e.g., helium and neon) content of the water. The contribution of this fresh glacial meltwater to the dense shelf waters and to new WSBW can be detected and quantified using dissolved helium and neon measurements (e.g., Schlosser et al., 1990; Huhn et al., 2008) and other tracers.

Observations (e.g., Foldvik et al., 2004) have detected the northward outflow of relatively narrow plumes of dense ISW from the Filchner depression northward and downslope into the Weddell Sea. Current meter measurements give flows in the range of ~ 1.6 Sv (at $\theta < -1.9$). Mixing and entrainment of surrounding warmer, deeper waters increase the volume flow to about 4.3 Sv (at $\theta < -0.8$, characteristic of WSBW). Additional mixing with surrounding water also leads to the production of slightly warmer ($-0.7 < \theta < 0$), less dense Weddell Sea Deep Water. A general schematic of the various processes and components involved in the formation of new AABW is shown in Figure 10.10.

The flow of deep and bottom waters through the western and northwestern Weddell Sea has been discussed by Gordon et al. (1993), Fahrbach et al. (2001) and Huhn et al. (2008) (see also Figure 10.9). In addition to the shelf regions in the southwest Weddell Sea, there is evidence for an additional source of deep and bottom-water formation in the Western Weddell Sea (Huhn et al., 2008) along the Larsen Ice Shelf, separate from the processes along Filchner–Ronne Ice shelves in the southern Weddell Sea. As in the southern Weddell Sea, both HSSW and ISW are involved in the deep- and bottom-water formation process, with the contribution of glacial meltwater estimated from noble gas anomalies (Huhn et al., 2008).

Another significant source of dense AABW formation is the Ross Sea (e.g., Jacobs et al., 1970; Whitworth and Orsi, 2006; Gordon et al., 2009). As with the Weddell Sea, dense HSSW and ISW waters are produced in this region. The outflow of these dense shelf waters can be observed cascading downward along the slope in a series of near-bottom gravity currents (Gordon et al., 2009) and leading to the production of new AABW.

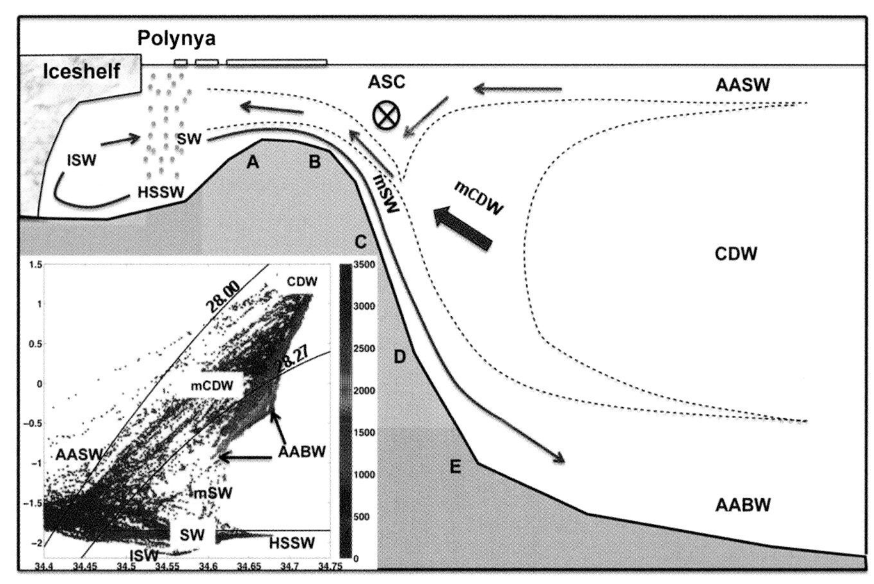

FIGURE 10.10 Schematic of formation and export of dense shelf waters and downslope formation of new AABW along the Adélie and George V Coast. ASC is the Antarctic Shelf Current; SW is Shelf Water; other key water masses are labeled and are defined in the text. Inset shows plot of θ–S from CTD data collected during the 2000/2001 expedition along the Adélie and George V Coast. Color bar is shaded for depth. Black lines are neutral density (γ^n) isolines. *From Williams et al. (2010).*

A third significant source of AABW formation is along the Adélie Land Cost in East Antarctica (140–149°E) (e.g., Gordon and Tchernia, 1972; Rintoul, 1998; Williams et al., 2008, 2010). Williams et al. (2010) describe the export of dense shelf waters from two independent sources in this region and the downward spreading of these along the continental slope to produce new AABW. The primary source is from the Adélie Depression, which exports a higher salinity shelf water via the Adélie sill. The formation of dense shelf waters in this region has been linked to the presence of a coastal polynya and enhanced sea ice formation and brine rejection, along with the storage capacity of the continental shelf in this region (see also Chapter 16).

Ohshima et al. (2013) discuss an additional process of AABW formation in a polynya region near Cape Darnley (~69°E) to the west of Prydz Bay. AABW formation in the Cape Darnley Polyna site does not appear to require the presence of a large ice shelf or storage reservoir but instead is primarily the result of the salt flux and density increase due to extremely high rates of sea ice formation and export in this region. This is in contrast to the other major sites identified for AABW formation along the Antarctic continent margin, including the Weddell and Ross Seas, where large continental ice shelves and embayments are involved in dense shelf-water formation, and along the Adélie Coast, where dense shelf waters are the product of high sea-ice production in a coastal polyna, combined with the presence of sufficient storage capacity on the continental shelf. In the Ohshima et al. (2013) study, elephant seals instrumented with CTD sensors detected the presence of dense shelf waters in the Cape Darnley polyna region and moorings detected downslope currents of cascading new AABW. Ohshima et al. (2013) estimate a formation rate of dense shelf waters of about 0.3–0.8 Sv in the Cape Darnley region, corresponding to 6–13% of total AABW production. Based on their findings in the Cape Darnley region that AABW can be produced from sea-ice formation in polynas along a narrow section of continental margin with limited dense water storage capacity, Ohshima et al. (2013) speculate that other similar polynas along the Antarctic continental margin may also be capable of producing significant quantities of new AABW.

In summary, the processes by which dense waters are formed and sink around the Antarctic continent to produce new ABBW are complex, involve mixing of components of differing composition and occur in a number of locations which can be difficult to monitor. In many cases, the source regions are extraordinarily difficult to reach and in some cases inaccessible. The processes that form deepwater often occur under ice shelves, and the outflows and cascading gravity plumes can be steered by a complex set of multiple, small, local bathymetric features (channels). The overall spreading of the various components of AABW into the World Ocean is indicated in Figure 10.9.

2.3.1. Formation Rates and Spreading of AABW

Orsi et al. (1999) define AABW as waters with neutral density (γ^{η}) > 28.27, where $\gamma^{\eta} = 28.27$ is the density found at the depth of the deepest sill at Drake Passage, a chokepoint between the southern tip of South America and the Antarctic peninsula. Therefore these densest forms of AABW cannot communicate directly with each other across Drake Passage. The spreading of this dense water mass in pure form to adjacent basins is limited by deep sills. Figure 10.11 shows the mean CFC-11 concentration in the densest AABW layer ($\gamma^{\eta} > 28.27$) around Antarctica, based on data collected along hydrographic sections prior to and during the WOCE program (~1984–1996).

In the Atlantic sector of the Southern Ocean, extremely high CFC-11 concentrations (> 3.5 pmol kg^{-1}) are present in newly formed WSBW in the southwestern Weddell Sea (~ 60°W) (see Figure 10.11). In the Pacific sector of the Southern Ocean, CFC-11 concentrations of ~2 pmol kg^{-1} are present in RSBW in the northwestern Ross Sea (~180°E), and CFC-11 concentrations >3 pmol kg^{-1} are found in ALBW in the Pacific/Indian sector of the Southern Ocean along the Adélie Land Coast south of Australia (~150°E). Relatively high CFC concentrations are carried along as dense AABW spreads in adjacent deep basins, and extends northward (north of ~50°S) into the deep basins in the southwest Atlantic Ocean and southwest Indian Ocean (see Figure 10.9).

Estimates of the production rates of this dense form of AABW based on CFC inventory techniques were made in Orsi et al. (1999). Since most of the CFC-11 data in the Orsi et al. (1999) study used to calculate inventories for the Atlantic sector of the Southern Ocean were collected during the period 1984–1990 and most of the CFC-11 data for the Indian and Pacific sectors were collected during the period 1991–1996, inventory-based calculations for formation rates of AABW were performed separately for the two regions. The total calculated AABW formation rate for the two sectors was about 8.1 Sv, with the Atlantic sector (primarily generated in the Weddell Sea) production rate of 4.9 Sv, somewhat larger than the combined Indian–Pacific sector rate (primarily from the Ross Sea and Adélie Land sources) of ~3.2 Sv. The Orsi et al. (1999) AABW formation rate in the Atlantic sector is compatible to results inferred from current meter observations (3–4 Sv; Fahrbach et al., 1991), inverse modeling approaches (5–6 Sv; Sloyan and Rintoul., 2001; Lumpkin and Speer, 2007), as well as other hydrography- or tracer-based estimates of 4–5 Sv (Gordon et al., 1993; Gordon, 1998; Huhn et al., 2008).

Mixing of these very dense AABW with waters of the Antarctic Circumpolar Current produces somewhat less dense varieties of AABW. It is mostly these modified varieties of "AABW," from the lower portion of the Circumpolar Deep Waters (Mantyla and Reid, 1983), which are

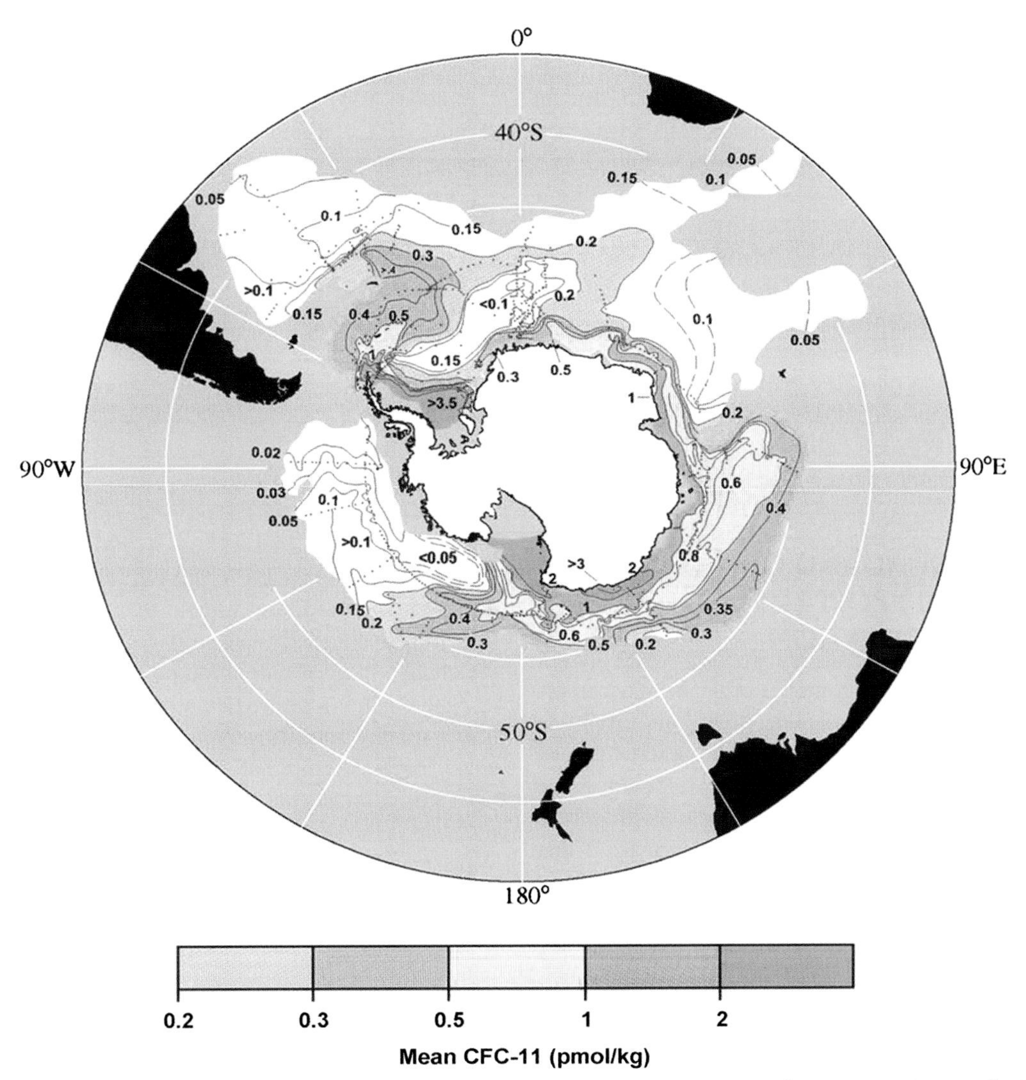

FIGURE 10.11 Distribution of the mean (depth-averaged) concentration of dissolved CFC-11 (in pmol kg^{-1}) in the layer of dense ($\gamma^{\eta} > 28.27$) AABW. *From Orsi et al. (1999).*

exported at the northern edges of the ACC and flow as deep western boundary currents to fill much of the volume of the abyssal South Atlantic, Indian, and Pacific Basins as "AABW." The cores of these northward-flowing deep western boundary currents in the south Atlantic, Indian, and Pacific Oceans can be clearly identified by characteristic properties (e.g., low temperatures, salinity and elevated levels of silicate, dissolved oxygen, and CFCs).

Orsi et al. (2002) also considered the somewhat less dense (densities $\gamma^{\eta} < 28.27$) CDW exported out of the ACC along with "true" AABW (densities $\gamma^{\eta} > 28.27$) to estimate the total Southern Ocean input to the deep ocean. When these additional waters were considered, the total production rate of dense waters of southern origin was ~21 Sv, of comparable magnitude to the estimates of NADW production based on CFC inventories (Smethie and Fine, 2001; LeBel et al., 2008). Lumpkin and Speer (2007) estimate a total northward transport of about 20 Sv out of the Southern Ocean in the bottom limb of the global meridional circulation, in good agreement with the Orsi et al. (2002) estimate. A number of additional estimates of northward transport of AABW into the individual basins have also been made using inverse models (e.g., Ganachaud and Wunsch, 2000; Sloyan and Rintoul, 2001) with values of roughly 7–11 Sv each for the Pacific and Indian Oceans, and about one half that for the Atlantic.

Van Sebille et al. (2013) discuss the processes and pathways by which the varieties of new AABW formed in various regions around the Antarctic continent cross the Southern Ocean and flow into the subtropical abyssal basins of the Atlantic, Indian, and Pacific Oceans. Using the analysis of trajectories in the Southern Ocean State Estimate (SOSE) model, they find that there is a large degree of mixing of the major sources of AABW by the time they reach 30°S in the abyssal subtropical basins. The ACC plays a major role in this amalgamation process and the majority of particles (~70%) from the major AABW source regions in the model undergo at least one complete circumpolar loop before they reach the subtropics. With the strong advection of the ACC, the homogenization process means that changes that occur near individual source regions of new AABW can be conveyed to all of the different basins. In the model, the major pathways from the Southern Ocean to the subtropical basins in the model are in relatively narrow and strong boundary currents (Orsi et al., 1999; Fukamachi et al., 2010), with about 60–80% of the particles ending up in the Pacific, 20–30% in the Indian and 5–10% in the Atlantic. AABW exported from the Weddell Sea is the major source and is the dominant contributor to each basin.

3. INTERANNUAL AND DECADAL VARIABILITY IN PROPERTIES, FORMATION RATE, AND CIRCULATION

In the long-term mean, the properties of the deep ocean are determined by a balance between the downward mixing in the warm surface regions through isopycnal and diapycnal diffusion and the formation and sinking of cold dense waters at high latitudes (Munk and Wunsch, 1998). There are two aspects of change in deepwater formation that can perturb this equilibrium state: (i) changes in the *formation rate* and (ii) changes in the *properties* (temperature, salinity, biogeochemical properties) of the newly formed water (Bindoff and McDougall, 1994). We will here look for changes in both of these aspects.

3.1. Labrador Sea Water: Variability in Properties and Formation Rate

During the past several decades, the temperature and salinity in the upper 2000 m of the central Labrador Sea (Figure 10.12; Yashayaev and Loder, 2009) have undergone strong multiyear fluctuations. The most intensive convection—caused by a succession of very strong buoyancy loss in winter—took place between 1987 and 1994, forming a deep-reaching (about 2500 m) anomalous cold and fresh LSW. Since then, in accord with relatively mild winters, LSW gradually warmed, became more saline and less dense, with convection depths shallower than 1300 m.

An exception was the winter of 2007/2008, when a localized large heat loss over the convection area cooled and freshened the upper 1600 m of the water column, and produced a denser mode of LSW than the previous years. The next winter was again milder, with a warmer and less dense LSW. The atmospheric variability is frequently expressed as the North Atlantic Oscillation (NAO) index (e.g., Hurrell, 1995), with a high NAO index favoring the transport of cold continental air masses to the central Labrador Sea, and thus initiating large heat loss (Figure 10.13; Rhein et al., 2011), leading to deep and intense convection (e.g., Dickson et al., 1996).

Two different methods to identify the various modes of LSW are reported in the literature. Yashayaev (2007) defined LSW classes by their T/S properties and σ_2 density, and tagged the LSW with the time period when it was formed (see Figure 10.12): the coldest and densest LSW class $LSW_{1987-1994}$, and the cold but less dense LSW_{2000}. Other authors define two LSW modes, the upper LSW (uLSW) with σ_Θ between 27.68 and 27.74, and LSW [also named classical LSW or deep LSW (dLSW)] between 27.74 $<\sigma_\Theta>$ 27.80 (Stramma et al., 2004; Kieke et al., 2006), the uLSW definition following earlier work of Pickart (1992) and Pickart et al. (1996). The dLSW corresponds roughly to the $LSW_{1987-1994}$. The definition of fixed density classes is more suitable to calculate the change in the CFC inventories between two different years and thus the change in LSW formation rates (Rhein et al., 2002).

The changing properties of the LSW can be used as a tracer for the spreading pathways of the water mass. As an example, the normalized planetary potential vorticity (PPV) time series (Pena-Molino et al., 2011) from moored profilers in the DWBC at 39°N southeast of Cape Cod (Line W) is presented in Figure 10.14. The time series 2001–2008 reflect the changes observed in the Labrador Sea, but shifted by roughly a pentad. In the first half of the record, the PPV minima was found at about 1500 m, that is, showing the cold and dense dLSW formed between 1987 and 1994, while in the second half the PPV minimum shifts to 700 m into the density range of uLSW, reflecting the Labrador Sea conditions after 1994 with the formation of uLSW.

Estimation of a deepwater formation rate requires an indirect calculation—to measure the strength of the process directly is not possible yet. Formally, formation rate is defined as the net annual volume of water across the material surface bounding the water mass in the formation area (Rhein et al., 2002; Haine et al., 2008), following Nurser and Marshall (1991). The formation rate is not per se equivalent to the ventilation rate. A volume of water in the density range of LSW could get ventilated in the next winter; but since it has already been LSW, it is not counted as a newly formed volume. For example, consider the upper bound of LSW to be located at 300 m depth. Buoyancy loss increases the density of the upper 300 m to be denser than the upper bound of LSW, and deep convection starts, homogenizing the water column from the surface to the convection depth. The upper 300 m of the water column counts as a newly formed LSW, while the water column from the surface to the convection depth was ventilated. Clearly the ventilated volume could be much larger than the newly formed volume.

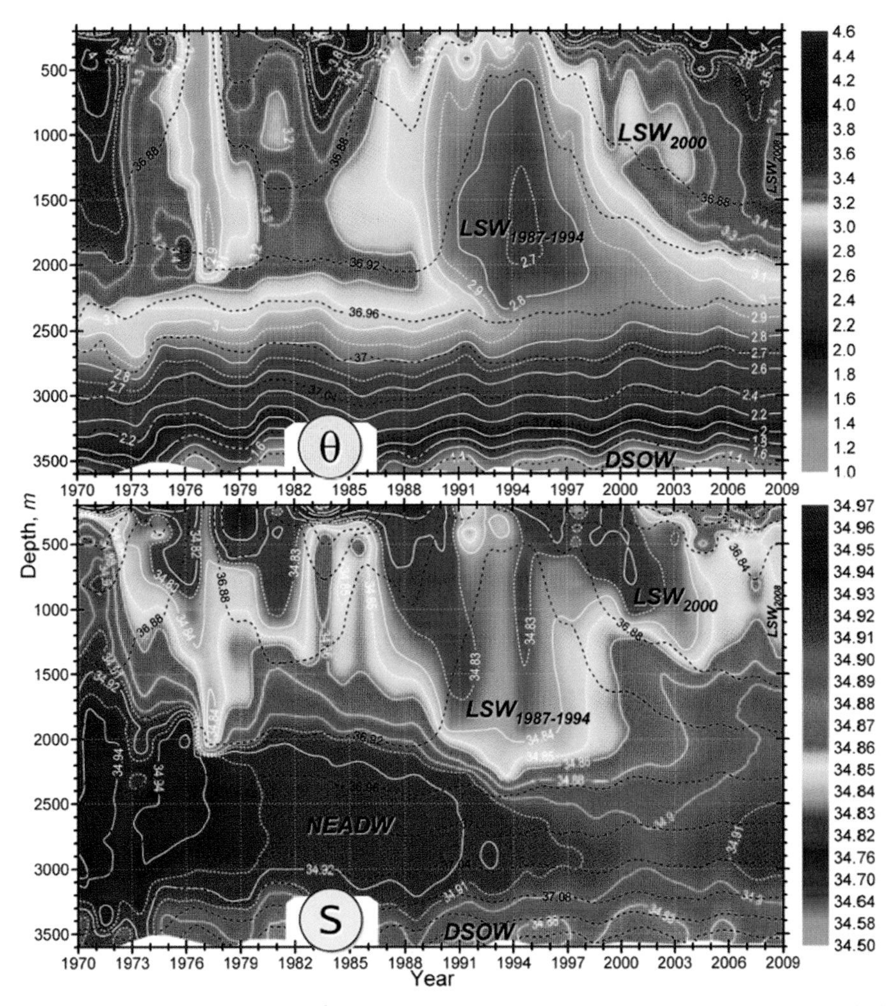

FIGURE 10.12 Time evolution of potential temperature (θ) (°C) and salinity (S) in the central Labrador Sea. Annual vertical property profiles were constructed by applying isopycnal averaging to all the measurements in water depths greater than 3250 m and within 150 km of the AR7W line, following Yashayaev (2007). The potential density anomaly (ref. 2000 dbar) isolines (dashed) reflect changes in the density of various LSW classes. *From Yashayaev and Loder (2009).*

In the literature, deep convection is usually associated with large volumes of newly formed deepwater, albeit in a qualitative sense. In the early 1990s, estimates of mean LSW formation rates between 2 and 8.6 Sv have been inferred mainly from geostrophy combined with mass, heat, or salt budget considerations. The individual results are nicely summarized in Haine et al. (2008). Climatological air/sea fluxes and sea surface data were used first by Speer and Tziperman (1992). Several LSW formation rate time series using this method have been published since then, ranging from about 2 to 10 Sv in the high NAO phase between 1984 and 1997 (e.g., Marsh, 2000; Khatiwala et al., 2002; Myers and Donnelly, 2008). Myers and Donnelly (2008) combined both time series of surface hydrographic data and of modified NCEP/NCAR reanalysis fluxes in the Labrador Sea for the time period 1960–1999. They estimated dLSW formation rates of up to 6 Sv between 1987 and 1994, and less than 2 Sv after 1995, while the formation of uLSW increased up to 8 Sv in 1997. The time series and the NAO index are correlated (coefficient = 0.45, significant at the 99% level). The LSW formation rates calculated by different authors and from different reanalysis products differ by about a factor of 4 (e.g., Haine et al., 2008). A significant problem with this approach is the uncertainties in the air–sea flux calculations (Myers and Donnelly, 2008).

Since 1997, repeat basin-wide chlorofluorocarbon (CFCs) distributions have became available, and have emerged as an important tool to quantify formation rates of LSW (Rhein et al., 2002) as well as their variability (e.g., Kieke et al., 2006, 2007; LeBel et al., 2008; Rhein et al., 2011). The method was first applied to AABW (Orsi et al., 1999), and together with a vigorous error analysis, used for LSW (Rhein et al., 2002). Changes in large-scale CFC inventories within the LSW density range were also used to infer changes in LSW formation rates (Kieke et al., 2006, 2007). An increase of the CFC inventory

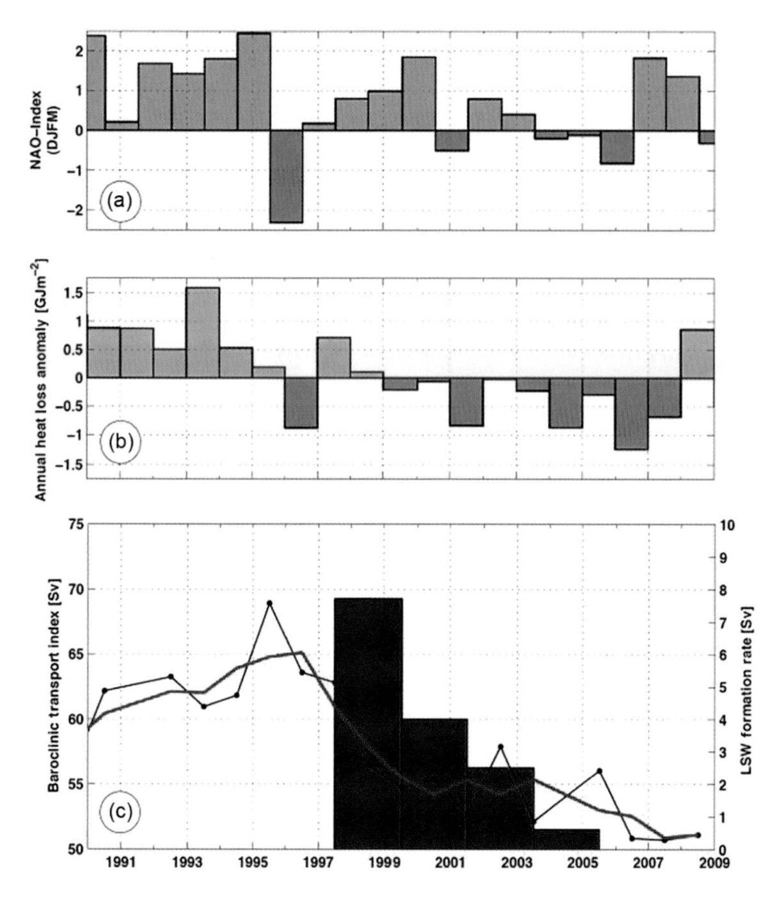

FIGURE 10.13 (a) Annual wintertime index of the North Atlantic Oscillation (NAO) for the years 1990–2009 following Jones et al. (1997). (b) Annual heat loss anomaly (GJ m^{-2}), central Labrador Sea. The mean heat loss from 1974 to 2009 has been subtracted. (c) Baroclinic volume transport (black line) between Bermuda and the Labrador Sea for the upper 2000 m following Curry and McCartney (2001) for the same period. Three-year low-pass filtered data of this time series are added as a green line. The scale on the right indicates the formation rate of LSW (in Sv) shown as red bars. *(a) From Rhein et al. (2011); (b) surface flux data are from NCEP.*

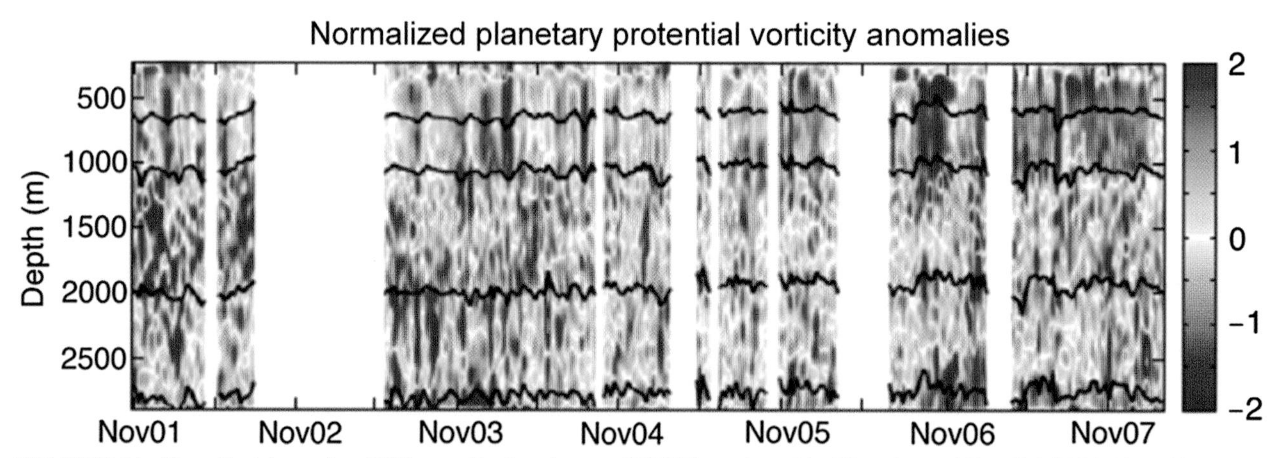

FIGURE 10.14 Normalized time series of PPV anomalies from the central DWBC mooring at Line W southeast of Cape Cod (39°N). Neutral isopycnals separating the water masses uLSW, dLSW, and overflow waters are drawn in black. *From Pena-Molino et al. (2011).*

in the LSW is related to the CFC content of the water volume being transferred into the LSW density classes, that is, to the formation rate. In the ventilated LSW volume, the CFCs became homogenized, but that does not lead to an additional CFC increase.

Uncertainties in the calculation of the CFC inventories come from the incomplete or too coarse data coverage of the regions considered and the extrapolation methods used (e.g., Kieke et al., 2006, 2007). To calculate formation rates from CFC inventory changes, an assumption about the CFC saturation degree of the water volume that is transferred to LSW is needed, and an assessment of the CFC sinks: export of CFC-bearing water masses and removal of LSW (and thus of CFCs) into other density classes, for instance through entrainment into the overflow water masses (Rhein et al., 2002; LeBel et al., 2008). The LSW formation rates published differ by less than a factor of 2 for the time period 1970–1997 (Smethie and Fine, 2001; Rhein et al., 2002; LeBel et al., 2008), and this is mainly caused by different assumptions about the saturation degree of the CFC concentrations in water that is transformed into LSW.

From 1997 to 2005, LSW formation rate has been calculated for uLSW and dLSW, with dLSW roughly equivalent to $LSW_{1984-1994}$. No dLSW was formed in that time period, and the production of uLSW decreased from about 8 Sv in 1997–1999 to 2 Sv in 2003–2005 (Rhein et al., 2011). Rhein et al. (2002) and Kieke et al. (2006) demonstrated a linear correlation between high CFC concentrations and low salinities in the LSW layers. Due to a coarser spatial coverage of tracer data in 2005 compared to earlier years, tracer concentrations were partly reconstructed from such linear correlations between CFC-12 and salinity, also including salinities from Argo profiles. This approach was tested successfully on the inventories from previous years, which had a much higher spatial CFC coverage.

Yashayaev and Loder (2009) concluded from shipboard CTD data and Argo profiles that the convection in the Labrador Sea in winter 2007 was weak, and convection depths were shallower than 700 m. For the first time since 1994, the convective overturning extended to 1600 m depth in winter 2008, and reached dLSW. The intense convection activity was caused by increased oceanic buoyancy loss, associated with below normal air temperatures in this region (Figures 10.12 and 10.13; see also Yashayaev and Loder, 2009). In winter 2009, the Labrador Sea fell back into a state of shallower convection depths. Compared to the significant changes observed in former years, the effect of the deep convection observed in winter 2008 on the layer thickness of the two LSW modes was minor (Rhein et al., 2011).

3.2. Greenland–Scotland Ridge Overflow Water: Variability in Properties and Overflow Rate

The overflow waters have also changed significantly during the past five decades, partly as a consequence of changes in the water masses in the Nordic Seas and partly as a consequence of changes in the water masses that the overflow waters entrain and mix with in the North Atlantic. In contrast, there does not appear to be a significant, long-term change in volume transport across the Greenland–Scotland Ridge, neither across the Denmark Strait nor across the Iceland–Scotland Ridge (Hansen and Osterhus, 2007; Olsen et al., 2008; Yashayaev and Dickson, 2008).

The densest waters overflowing the Iceland–Scotland Ridge exit through the Faroe Bank Channel. This overflow is remarkably persistent; as Hansen and Osterhus (2007) observe, the dense and cold waters were present on each of the more than 3200 days of their 10 year monitoring program (1995–2005). The Marine Laboratory (Aberdeen, Scotland) has maintained two standard hydrographic sections directly upstream, in the Faroe Shetland Channel, for more than a hundred years. These display significant variability both in temperature and salinity, but only the latter exhibits a long-term trend, namely a freshening trend from ca. 1970 until the end of the century (Turrell et al., 1999), at which point it halted. The overflow's salinity has remained unaltered for the past decade (Yashayaev and Dickson, 2008). However, the Subpolar Mode waters that the overflows entrain have become saltier each year since 1996, after decades of freshening. As a result, the overflow waters downstream of the ridge have become more saline in the past decade, a signal that can be followed into the western Labrador Sea (Figure 10.15), despite the lack of change in salinity at the sill (Yashayaev and Dickson, 2008). These authors also point out that as the Iceland–Scotland Overflow plume crosses the Reykjanes Ridge and enters the Irminger Basin, it mixes with Labrador Sea Water as well, and becomes imprinted with the time variability of that water mass as well, such that there is little resemblance of the Faroe Bank sill, both in property and variability, by the time the plume reaches the Deep Western Boundary Current.

The Denmark Strait Overflow waters have a much more direct route to the Deep Western Boundary Current, and a much steeper descent, than the Iceland–Scotland Overflow Waters. The properties of the DSOW in the Labrador Sea are directly related to those of the overflow (Figure 10.16). As for the ISOW, there has been no long-term trend in temperature, only significant interannual variability. And as for the ISOW, there has been a long-term freshening of DSOW since the 1960s (Dickson et al., 2002), which appears to have flattened out in the past decade (Yashayaev and Dickson, 2008).

To summarize, through diverse and circuitous routes, large hydrographic variability in the upper ocean in the Nordic Seas and Subpolar Gyre is carried into the deep North Atlantic. The time evolution of SPMW, LSW, ISOW, and DSOW is interconnected. Of the various deep components of NADW, only LSW has measurable indications in variability in formation rate.

3.3. Relationship Between Formation Rates of NADW and Changes in the AMOC

It is a relatively robust result of the future scenarios run in the Assessment Report 4 (AR4) of the IPCC [Meehl et al., 2007] that the strength of the Atlantic Meridional Overturning Circulation (AMOC) will decrease during the twenty-first century. The meridional transport of heat through the overturning circulation is of the same order of magnitude as the heat transport in the atmosphere; thus

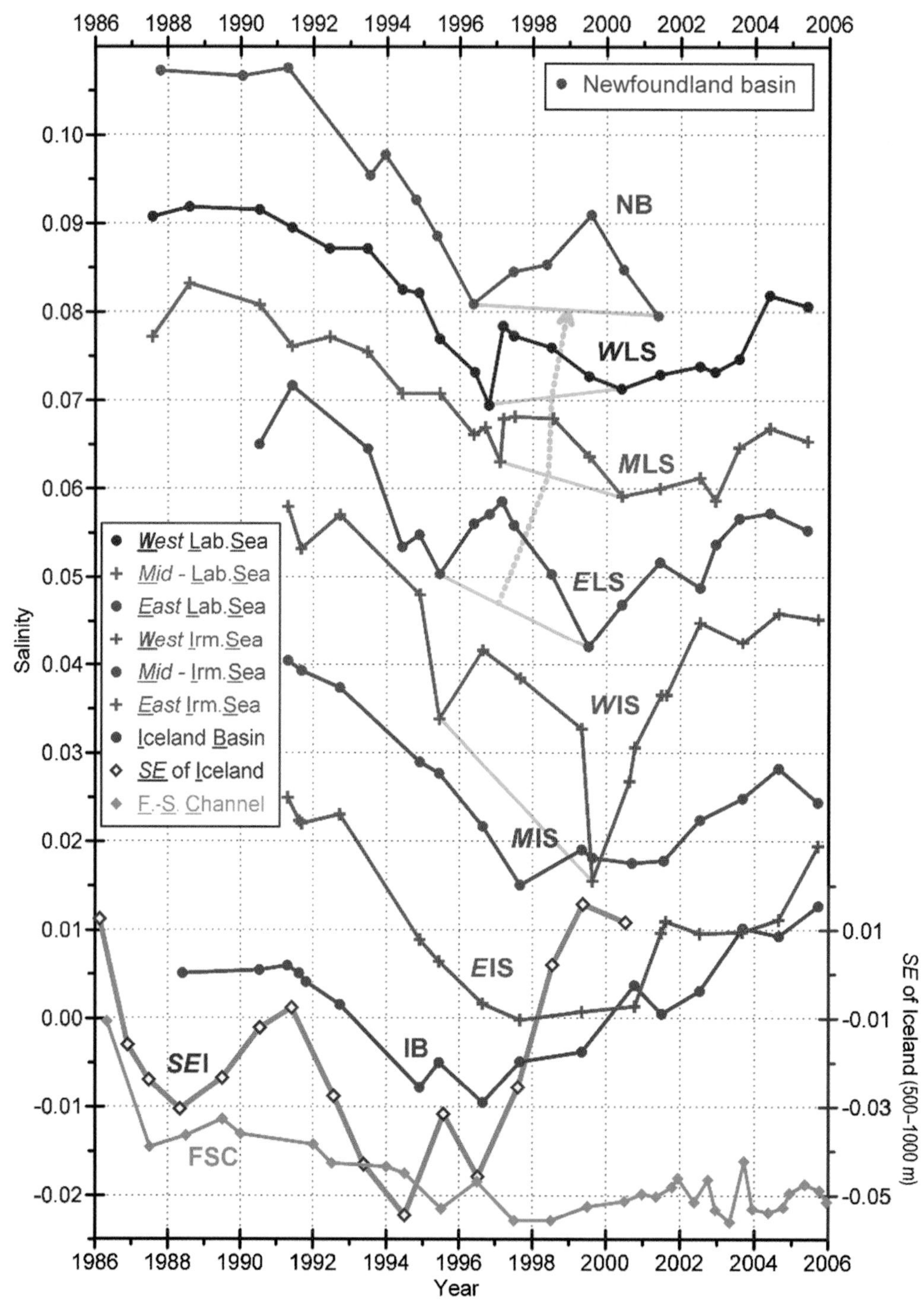

FIGURE 10.15 Time series of salinity anomalies (relative to the 1987–2005 mean salinity profile; see Yashayaev and Dickson, 2008) for ISOW, becoming NEADW, at selected locations along its spreading pathway from the sill of the Faroe–Shetland Channel to the Labrador Sea. The density range occupied by this salinity-maximum layer is defined as $37.00 < \sigma_2 < 37.06$ in the Iceland basin and as $36.98 < \sigma_2 < 37.04$ west of the mid-Atlantic Ridge.; the salinity range of the Subpolar Mode Water that is entrained by the NEADW at the head of the Icelandic Basin (South East of Iceland) is also shown.

changes in the MOC volume and heat transport would affect climate and regional sea level. Since it is currently prohibitively expensive to monitor the ocean's large-scale circulation everywhere directly, and since there are only a few places where time series of more than a few years exist, several indirect methods are commonly employed to attempt to quantify changes in ocean circulation that may already be occurring. These methods include merging

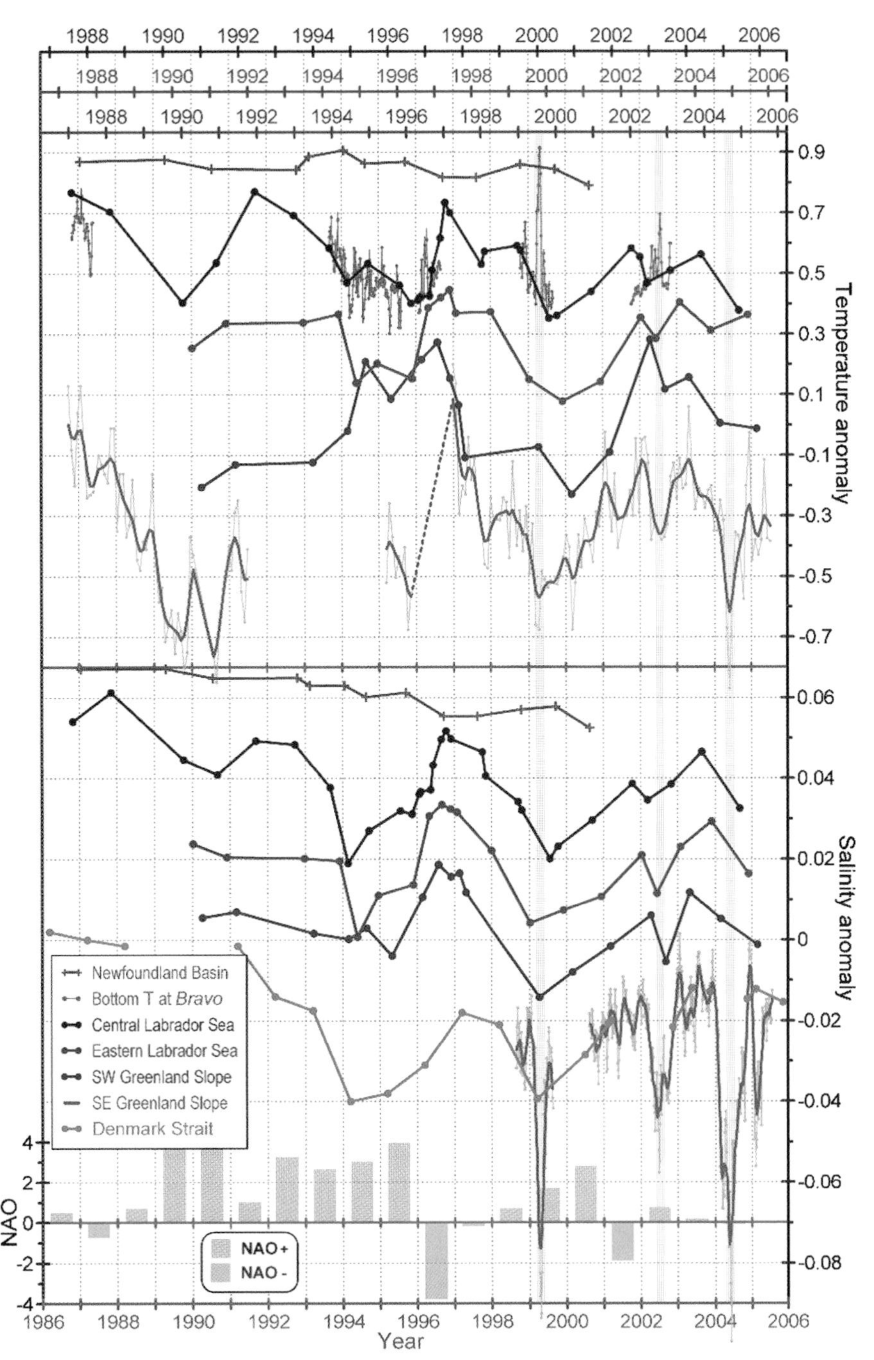

FIGURE 10.16 Time series of the salinity and temperature of the Denmark Strait Overflow Water from the sill to the Labrador Sea and Newfoundland Basin since 1986. Note that the time-axes for series in the Labrador Basin are displaced in time to reflect the advective time-lag. Note that the Newfoundland Basin time series is given the same time-axis as the Central Labrador Sea. In general, there is a 1-year lag between changes at the sill and those in the abyssal Labrador Sea. A bar graph of the NAO index is also shown for comparison. *From Yashayaev and Dickson (2008).*

ocean observations with numerical ocean simulations, inferring circulation strength from surface buoyancy fluxes and density structure, and inferring them from changes in sea surface height and subsurface hydrographic fields.

Figure 10.17 synthesizes many recent estimates of AMOC strength at various latitudes in the North Atlantic, using a variety of methods, covering different time periods. These estimates are not fully independent, but they do nevertheless yield some insights into the potential tendencies for a long-term change. All of the estimates seem to agree that the span in interannual to decadal variability ranges around 2–3 Sv. According to modeling studies (Knight et al., 2005), this is more than large enough to have a significant climate impact (and yet small enough to

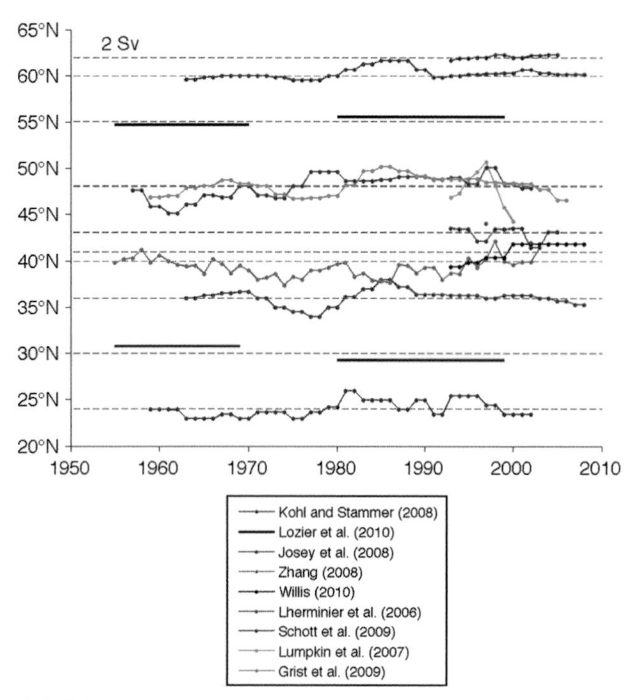

FIGURE 10.17 Estimates of AMOC transport at various latitudes in the North Atlantic from observations and model assimilations.

considerably challenge any monitoring array). However, the various estimates do not paint a coherent picture of the long-term trends, neither in the subtropical nor in the subpolar gyre. This may partly reflect the fact, pointed out by many authors, that the AMOC is not one, coherent structure, but contains many shortcuts that will tend to de-correlate variability. There is evidence that the northward flow across the boundary from the Subtropical to the Subpolar Gyre has been increasing since the early 1990s (Figure 10.17; Zhang, 2008; Willis, 2010), especially since that finding would be consistent with observed flushing of warm and saline waters into the Subpolar Gyre during that period (Holliday et al., 2008). However, more likely than not, these results demonstrate that there has been no clear long-term trend in the AMOC in the past four decades.

What then about the changes in dense water formation rate in the Subpolar Gyre? Correlations between observational time series and model studies have shown that the intensity of deepwater formation is linked to the local buoyancy fluxes (e.g., Dickson et al., 1996; Häkkinen, 1999; Eden and Willebrand, 2001), a fact that was used to infer formation rates from buoyancy fluxes (e.g., Speer and Tziperman, 1992; Marsh, 2000; Myers and Donnelly, 2008). There is not necessarily a direct link between changes in formation rate and changes in overturning strength, due to the buffering effect of water mass volume changes (Mauritzen and Häkkinen, 1999). Nevertheless, many modeling simulations suggest a strong relationship between anomalies in deepwater formation rates and the strength of the MOC (e.g., Gregory et al., 2005; Stouffer et al., 2006; Biastoch et al., 2008). In the latter study, positive AMOC anomalies follow roughly 2–3 years after the onset of intense LSW production. The AMOC signal spreads rapidly from the subpolar North Atlantic to the subtropics, thereby reducing in strength. However, the AMOC time series is dominated by high-frequency fluctuations due to local wind forcing and (in high-resolution models) eddy variability, obscuring the decadal-scale and longer term changes associated with deepwater formation (Biastoch et al., 2008; Böning et al., 2008). Grist et al. (2009) examined the role of surface thermohaline forcing on the AMOC variability in three IPCC coupled climate models. Similar to the results by Biastoch et al. (2008), a lag of 2–4 years was found between a maximum in surface forcing in the subpolar North Atlantic and the maximum AMOC strength at 48°N. The authors showed that the model AMOC variability can be simulated solely from the 10-year averaged surface forced stream function, which in turn is correlated to the deepwater formation.

In a reanalysis study of the variability of the AMOC by the GECCO (German partner for Estimating the Circulation and Climate of the Ocean) consortium, Köhl and Stammer (2008) find, in contrast to Böning et al. (2008) and Biastoch et al. (2008), that the variability in LSW formation does not affect the AMOC strength, neither at 48°N nor in the subtropics. Instead, subpolar density fluctuations caused by DSOW anomalies explain about a third of the AMOC variability at 25°N (Köhl and Stammer, 2008) (but, as discussed in Section 3.2, there is currently no observational evidence of such DSOW anomalies).

Most observational time series of deepwater transports are restricted to the western boundary current, and continuous measurements of the basin wide transports of the lower limb of the AMOC are only carried out at 16°N. The time series of the NADW transport in the western basin from 2000 to 2009 exhibits a weakening by about 3 Sv/decade, mostly focused in the LSW layer (Send et al., 2011). The trend is overlain by shorter term fluctuations of more than 10 Sv. It is possible that this observed trend is related to the reduced LSW formation since the mid 1990s; however, the mechanisms behind, and observational evidence for, a relationship between AMOC strength and dense water formation rate remains obscure.

3.4. Antarctic Bottom Water: Variability in Properties and Formation Rate

During the past decade, a number of studies have looked at variability in properties and export rates of dense waters around Antarctica. Changes in the AABW could reflect changes in the properties, volume transport or mixing ratios of the sources of the dense waters contributing to the formation of AABW. Studies have revealed a long-term freshening of shelf waters in the Weddell Sea (Hellmer et al., 2011) and Ross Seas (Jacobs et al., 2002; Jacobs and Giulivi, 2010). There have also been observations of trends in temperature in mid depth waters in the southern ocean (Gordon, 1982) involved in the AABW formation process, including warming during recent decades (Gille, 2002, 2008; Robertson et al., 2002; Smedsrud, 2005), with a possible interruption in this trend in the Weddell Sea in the late 1990s (Fahrbach et al., 2011). There is evidence that such changes in the source water components may impact the properties and amounts of newly formed bottom waters exported from these source regions on relatively short timescales. Observed changes in deep-and bottom-water formation rates and properties have included an ~20% reduction of the contribution of WSBW into the deep Weddell Sea Basin near the prime meridian and near the tip of the Antarctic Peninsula (Kerr et al., 2009) in the 1980s and 1990s, a cooling and freshening of bottom waters in the Australian–Antarctic basin from about 1994–2003 (Aoki et al., 2005) and an acceleration of the freshening of AABW formed in the Pacific and Indian Oceans (Rintoul, 2007) during the period between 1995 and 2005. Shimada et al. (2012) discuss possible causes for the observed freshening and warming of AABW in the Australian–Antarctic basin between the 1990s and 2000s. They hypothesize that the freshening may be due to freshening of the dense shelf waters in the Ross Sea and Adélie/George V Land region, but that the warming may be due to both an overall decrease in the transport of RSBW and an enhanced vertical diffusion of heat from the overlying ambient water, due to reduced vertical stratification. Therefore, the combination of changes in both the properties and transport of RSBW may contribute to the warming and freshening of AABW in this region. Couldrey et al. (2013) discuss repeat hydrographic surveys from the mid-1990s to the late 2000s in the eastern Weddell Gyre at 30°E, which show warming and density decreases of AABW entering from the Indian sector of the Southern Ocean. They associate these changes with increased entrainment of warmer CDW into shelf water plumes in the Cape Darnley region to the east and suggest that the change in entrainment may be related to a southward migration of the southern boundary of the ACC closer to the coast due to changes in atmospheric forcing (Sallée et al., 2008). They also suggest that this process of southward migration of the ACC could also lead to warming of AABW in other regions of Antarctica where the flow of the ACC is close to the coast.

Using current meter arrays, a number of studies have measured the rates and variability of dense shelf-water formation in the Weddell and Ross Seas and along the Adélie Land coast. These include measurements on the shelf and slope near the source region as well as the outflow of the waters at considerable distances away from the source regions. Current meter moorings have documented the transport of dense waters northward from the northwest Weddell Sea and Falkland Islands (Whitworth et al., 1991; Fahrbach et al., 1994, 2001).

Gordon et al. (2010) discuss variability in the export of WSBW from the Weddell Sea, based on 8 years of observations of abyssal waters near the tip of the Antarctic Peninsula. Using temperature and salinity measurements from moorings in this region, they observe a strong seasonal cycle in bottom temperatures, with a cold pulse typically in the May/June time frame and a warm pulse in Oct/Nov, with significant year-to-year variability in the timing and amplitude of the pulses. The cold pulses are thought to reflect the release of shelf waters from the southwestern Weddell Sea during the previous Austral summer. They propose that seasonal fluctuation and variability are produced by seasonal cycles in regional winds over the western Weddell Sea. The significant interannual fluctuations over the observational period are hypothesized to be linked to variability of the wind driven circulation in the Weddell gyre and ultimately to large-scale features including the Southern Annular Mode (SAM) (Hall and Visbeck, 2002; Gordon et al., 2007) and the El Nino/Southern Oscillation. Gordon et al. (2010) hypothesize that a positive SAM (associated with stronger westerly winds) produces conditions favorable to increased export of WSBW. They also suggest that longer time series of AABW outflows are required to understand the response of the southern ocean meridional circulation to long-term warming and other climate change-related phenomena.

Programs have monitored the outflow of AABW from other important regions. Fukamachi et al. (2010) discuss the export of AABW in the region to the east of the Kerguelen plateau (see Figure 10.3). Using an array of current meters over a 2-year period, they were able to estimate the transport and structure of this deep western boundary current and determine that it is a major pathway for AABW transport from the southern ocean to abyssal waters in the eastern Indian and Pacific oceans, perhaps comparable in magnitude to the export of AABW from the Weddell Sea. The contributions of this and other DWBCs to southern ocean overturning can be difficult to assess due to complex recirculation gyres, interaction of the DWBC with other circulation features, and mixing and entrainment with surrounding waters. As with the Gordon et al. (2010) study in the Weddell Sea, the Fukamachi et al. (2010) results indicate that in order to gain a better understanding of the magnitude of the circulation and its response to changes in forcing, sustained coherent observations are needed at a number of locations in the southern ocean.

Purkey and Johnson (2012) discuss statistically significant large-scale warming and contraction of AABW volumes between the 1980s and 2000s, based on data collected along a large set of repeated hydrographic sections crossing most of the deep ocean basins. The contractions of AABW are largest in the Southern Ocean and adjacent deep basins, but smaller decreases of AABW could also be detected in most of the major outflow routes for AABW into the Atlantic, Pacific, and Indian Oceans. The budgets and contraction are considered consistent with a slowdown of the bottom, southern limb of the meridional circulation. Purkey and Johnson (2012) discuss the possible relationship of this slowdown with observed freshening of shelf waters in the formation regions, possibly due to enhanced glacial melting, changes in sea ice extent and increased precipitation. Such increases in freshwater content would increase the stability of the water column on the shelves and could lead to an overall reduction in the overturning circulation (Stouffer, et al., 2007). The implied decrease between the 1980s and 2000s in AABW formation rates (~−8 Sv) in Purkey and Johnson (2012) is large when compared to estimates of total AABW formation rate (~8 Sv) derived in Orsi et al. (1999) from CFC inventories (which inherently integrate any variability in the formation rate over the previous ~40 year period from the 1950s to the 1990s) and from inverse models (see Section 1). This suggests a possibility that AABW formation rates had already undergone a slowdown by the 1980s and may have been significantly higher in previous decades.

As already mentioned other processes, however, might also contribute to changes in the AABW properties. The ventilation of AABW in the Weddell Sea was studied by Huhn et al. (2013) using 26 years (1984–2011) of CFC measurements along the Prime Meridian and across the Weddell Sea. They found that during that time period the deep and bottom water became older (i.e., less ventilated). These results, however, indicate that the reduced AABW ventilation was most likely transferred to the AABW during the entrainment process near the ice shelves. During the descent, Circumpolar Deep Water, called Warm Deep Water in the Weddell Sea, is entrained into the deep and bottom waters, and CDW/WDW aged even more in these 26 years (Figure 10.18). It might well be that at least part of the AABW warming signal is caused by entrainment of older and also warmer CDW/WDW, and not by a change in the near-surface processes (Couldrey et al., 2013; Huhn et al., 2013).

The observed changes in the AABW may be related to changes in the Southern Annular Mode, which has trended positive in recent decades due to past ozone depletion and may continue to so do under global warming scenarios (Thompson et al., 2011). This positive trend in the SAM is associated with stronger westerly winds moving poleward, along with a southward migration of the ACC. Such a shift may lead to warmer temperatures, increased glacial melting and precipitation. In addition to changes in the cold, high-density shelf components involved in formation of new AABW, observed warming of CDW (Böning et al., 2008) and changes in entrainment rates of

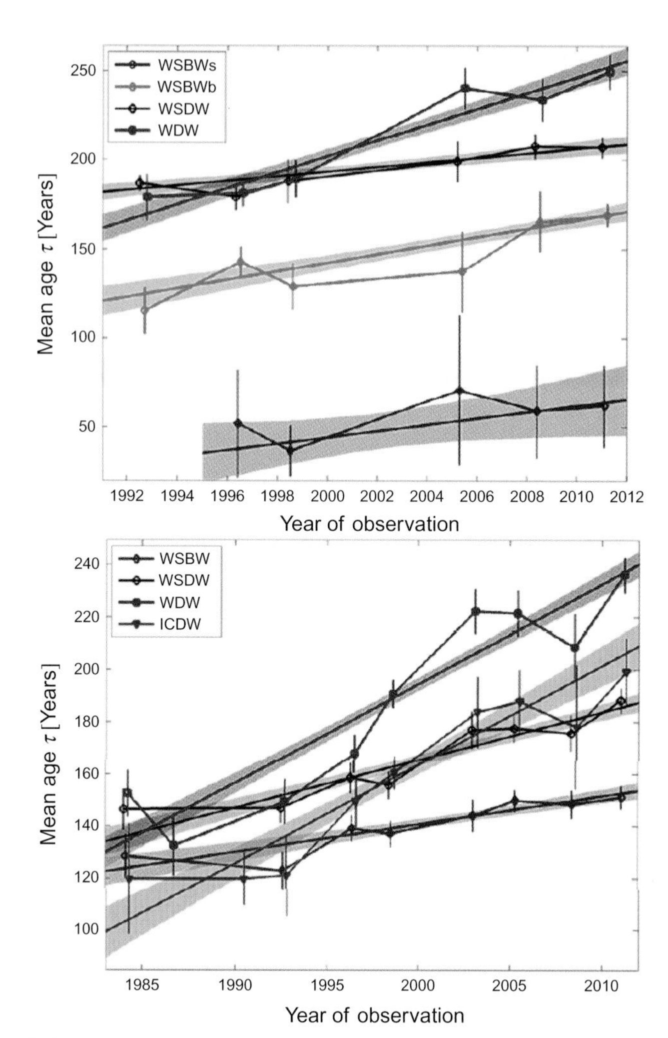

FIGURE 10.18 Age trends of deepwater masses from 1984 to 2011 at a section from southeast to northwest across the Weddell Sea Gyre (top) and along the Prime Meridian in the Weddell Sea (bottom). WDW: warm Deep Water, the Weddell Sea variation of Circumpolar Deep Water (CDW), WSDW and WSBW are Weddell Sea Deep and Bottom Water, the Weddell Sea precursors to AABW. ICDW is Lower Circumpolar Deep Water, found at the northern part of the Prime Meridian section. *From Huhn et al. (2013).*

this water mass could affect the properties of new AABW (Couldrey et al., 2013; Huhn et al., 2013).

A number of studies have been made to monitor the flow of the ACC at "chokepoint" sections in the Southern Ocean, where the flow is constricted most tightly between the major land masses (the South American, African and Australian continents) and the Antarctic continent. Monitoring techniques have included sea level gauges, current meters, bottom pressure recorders, and repeat hydrographic sections. A variety of monitoring programs at Drake Passage, which is the narrowest of the chokepoint sections, have revealed that the ACC transport varies on timescales from days to years in response to variability of winds over the Southern Ocean, but the mean transport of the ACC there (~135 Sv) has varied relatively little during the past several decades (Böning et al., 2008) despite overall increase in winds during this period. A detailed discussion of sustained monitoring programs at Drake Passage, as well as justification for the need for continue observations to understand the response of the ACC to changes in climate forcing, is given in Meredith et al. (2011). Such sustained monitoring programs at Southern Ocean chokepoint sections, in addition to monitoring at specific outflow regions at bottom-water formation sites around the Antarctic continent, will be essential in detecting and quantifying long-term changes in volume and properties of new AABW.

4. CONCLUSIONS AND OUTLOOK

Current models project a wide range of scenarios for the future uptake of heat (Boe et al., 2009) and CO_2 (see Chapter 30) in the ocean. Possible changes in the deep circulation of the ocean as a consequence of anthropogenic climate change and global warming (Solomon et al., 2007) include a possible slowdown in the MOC, significant reduction in NADW formation and changes in the ACC. For example, Srokosz et al. (2012) review evidence from both observational and modeling studies that link variability in the AMOC with climate patterns, with possible impacts on air temperatures, precipitation, and sea level in the bordering regions.

Much progress has been made during the past several decades in understanding the changes in deepwater properties and deepwater formation rates. A number of techniques have been developed to estimate the rates of these processes and their variability. Future studies will require sustained monitoring of these processes near the source regions and subsequent transport along the main pathways in both hemispheres, using a variety of techniques (e.g., repeat hydrographic surveys, moored current arrays, deep-reaching autonomous profiling Argo floats) as well as the continuing development of models to better understand the relationships between changes in deepwater formation and the overall strength of the MOC.

REFERENCES

Aoki, S., Rintoul, S.R., Ushio, S., Watanabe, S., Bindoff, N.L., 2005. Freshening of the Adélie Land Bottom Water near 140°E. Geophys. Res. Lett. 32. http://dx.doi.org/10.1029/2005GL024246, L23601.

Biastoch, A., Böning, C.W., Getzlaff, J., Molines, J.M., Madec, G., 2008. Causes of interannual-decadal variability in the meridional overturning circulation of the midlatitude north Atlantic Ocean. J. Clim. 21 (24), 6599–6615. http://dx.doi.org/10.1175/2008JCLI2404.1.

Bindoff, N.L., Mcdougall, T.J., 1994. Diagnosing climate change and ocean ventilation using hydrographic data. J. Phys. Oceanogr. 24 (6), 1137–1152.

Boe, J., Hall, A., Qu, X., 2009. Deep ocean heat uptake as a major source of spread in transient climate change simulations. Geophys. Res. Lett. 36. http://dx.doi.org/10.1029/2009GL040845, L22701.

Böning, C.W., Dispert, A., Visbeck, M., Rintoul, S.R., Schwarzkopf, F.U., 2008. The response of the Antarctic Circumpolar Current to recent climate change. Nat. Geosci. 1, 864–869.

Bower, A.S., Le Cann, B., Rossby, T., Zenk, W., Gould, J., Speer, K., Richardson, P.L., Prater, M.D., Zhang, H.M., 2002. Directly measured mid-depth circulation in the northeastern North Atlantic Ocean. Nature 419 (6907), 603–607.

Bower, A.S., Lozier, M.S., Gary, S.F., Boning, C.W., 2009. Interior pathways of the North Atlantic meridional overturning circulation. Nature 459. http://dx.doi.org/10.1038/nature07979.

Bower, A.S., Lozier, M.S., Gary, S.F., 2011. Export of Labrador sea water from the subpolar North Atlantic: a Lagrangian perspective. Deep Sea Res. Part II 58 (17–18), 1798–1818.

Bryden, H.L., Longworth, H.R., Cunningham, S.A., 2005. Slowing of the Atlantic meridional overturning circulation at 25°N. Nature 438, 655–657. http://dx.doi.org/10.1038/nature04385.

Burke, A., Robinson, L.F., 2012. The Southern Ocean's role in carbon exchange during the last deglaciation. Science 335 (6068), 557–561.

Carmack, E.C., Foster, T.D., 1975. On the flow of water out of the Weddell Sea. Deep Sea Res. 22, 711–724.

Clark, P.U., Pisias, N.G., Stocker, T.F., Weaver, A.J., 2002. The role of the thermohaline circulation in abrupt climate change. Nature 415, 863–869.

Cooper, L.H.N., 1955. Deep water movements in the North Atlantic as a link between climate changes around Iceland and biological productivity of the English Channnel and Celtic Sea. J. Mar. Res. 14 (4), 347–362.

Couldrey, M.P., Jullion, L., Naveira Garabato, A.C., Rye, C., Herráiz-Borreguero, L., Brown, P.J., Meredith, M.P., Speer, K.L., 2013. Remotely induced warming of Antarctic Bottom Water in the eastern Weddell gyre. Geophys. Res. Lett. http://dx.doi.org/10.1002/grl.50526.

Curry, R.G., McCartney, M.S., 2001. Ocean gyre circulation changes associated with the North Atlantic Oscillation. J. Phys. Oceanogr. 31, 3374–3400.

Dickson, R.R., Brown, J., 1994. The production of North Atlantic deep water: sources, rates and pathways. J. Geophys. Res. C 99, 12319–12341.

Dickson, R., Lazier, J., Meincke, J., Rhines, P., Swift, J., 1996. Long-term coordinated changes in the convective activity of the North Atlantic. Prog. Oceanogr. 38, 241–295.

Dickson, R.R., Yashayaev, I., Meincke, J., Turrell, W.R., Dye, S.R., Holfort, J., 2002. Rapid freshening of the deep North Atlantic Ocean over the past four decades. Nature 416, 832–837.

Dickson, R.R., Dye, S., Jónsson, S., Köhl, A., Macrander, A., Marnela, M., Meincke, J., Olsen, S., Rudels, B., Valdimarsson, H., Voet, G., 2008. The overflow flux west of Iceland. Variability, origins and forcing. In: Dickson, R.R., Meincke, J., Rhines, P. (Eds.), Arctic-Subarctic Ocean Fluxes. Defining the Role of the Northern Seas in Climate. Springer, Dordrecht, pp. 443–474.

Eden, C., Willebrand, J., 2001. Mechanism of interannual to decadal variability of the North Atlantic circulation. J. Clim. 14, 2266–2280.

Fahrbach, E., Knoche, M., Rohardt, G., 1991. An estimate of water mass transformation in the southern Weddell Sea. Mar. chem. 35 (1), 25–44.

Fahrbach, E., Rohardt, G., Schroder, M., Strass, V., 1994. Transport and structure of the Weddell Gyre. Ann. Geophys. 12, 840–855.

Fahrbach, E., Harms, S., Rohardt, G., Schröder, M., Woodgate, R.A., 2001. Flow of bottom water in the northwestern Weddell Sea. J. Geophys. Res. 106, 2761–2778.

Fahrbach, E., Hoppema, M., Rohardt, G., Boebel, O., Klatt, O., Wisotzki, A., 2011. Warming of deep and abyssal water masses along the Greenwich meridian on decadal-time scales: the Weddell gyre as a heat buffer. Deep Sea Res. Part II 58, 2508–2523. http://dx.doi.org/10.1016/j.dsr2.2011.06.007.

Fer, I., Voet, G., Seim, K.S., Rudels, B., Latarius, K., 2010. Intense mixing of the Faroe Bank Channel overflow. Geophys. Res. Lett. 37. http://dx.doi.org/10.1029/2009GL041924, L02604.

Fine, R.A., Rhein, M., Andrié, C., 2002. Using a CFC effective age to estimate propagation and storage of climate anomalies in the Deep Western North Atlantic Ocean. Geophys. Res. Lett. 29 (24), 2227. http://dx.doi.org/10.1029/2002 GL015618.

Fleischmann, U., Hildebrandt, H., Putzka, A., Bayer, R., 2001. Transport of newly ventilated deep water from the Iceland Basin to the Westeuropean Basin. Deep Sea Res. Part I 48, 1793–1819.

Foldvik, A., Gammelsrød, T., Torresen, T., 1985. Circulation and water masses on the Southern Weddell Sea shelf. In: Jacobs, S.S., Weiss, R.F. (Eds.), Oceanology of the Antarctic Continental Shelf. Antarctic Research Series, vol. 43. AGU, Washington DC, pp. 5–20.

Foldvik, A., Gammelsrød, T., Østerhus, S., Fahrbach, E., Rohardt, G., Schröder, M., Nicholls, K.W., Padman, L., Woodgate, R.A., 2004. Ice Shelf Water overflow and bottom water formation in the southern Weddell Sea. J. Geophys. Res. 109, C02015.

Foster, T.D., Carmack, E.C., 1976. Frontal zone mixing and Antarctic Bottom Water formation in the southern Weddell Sea. Deep Sea Res. 23, 301–317.

Fukamachi, Y., Rintoul, S.R., Church, J.A., Aoki, S., Sokolov, S., Rosenberg, M.A., Wakatsuchi, M., 2010. Strong export of Antarctic Bottom Water east of the Kerguelen plateau. Nat. Geosci. 3. http://dx.doi.org/10.1038/NGEO842.

Fukasawa, M., Freeland, H., Perkin, R., Watanabe, T., Uchida, H., Nishima, A., 2004. Bottom water warming in the North Pacific Ocean. Nature 427, 825–827.

Ganachaud, A., Wunsch, C., 2000. Improved estimates of global ocean circulation, heat transport and mixing from hydrographic data. Nature 408, 453–457.

Gill, A.E., 1973. Circulation and bottom water formation in the Weddell Sea. Deep Sea Res. 20, 111–140.

Gille, S.T., 2002. Warming of the Southern Ocean since the 1950s. Science 295, 1275–1277.

Gille, S.T., 2008. Decadal-scale temperature trends in the Southern Hemisphere ocean. J. Clim. 21, 4749–4765.

Girton, J.B., Sanford, T.B., 2003. Descent and modification of the overflow Plume in the Denmark Strait. J. Phys. Oceanogr. 33, 1351–1364.

Girton, J.B., Sanford, T.B., Käse, R.H., 2001. Synoptic sections of the Denmark Strait Overflow. Geophys. Res. Lett. 28, 1619–1622.

Gordon, A.L., 1982. Weddell deep water variability. J. Mar. Res. 40 (Suppl.), 199–217.

Gordon, A.L., 1991. The role of thermohaline circulation in global climate change. In: Lamont-Doherty Geological Observatory Biennial Report. Columbia University, Palisades, NY, pp. 44–51.

Gordon, A.L., 1998. Western Weddell sea thermohaline stratification. In: Ocean, Ice and Atmosphere: Interactions at the Antarctic Continental Margin, Antarctic Research Series, vol. 75. pp. 215–240.

Gordon, A.L., Tchernia, P.L., 1972. Waters of the continental margin off Adélie coast, Antarctica. Antarct. Res. Ser. 19, 59–69.

Gordon, A.L., Huber, B.A., Hellmer, H.H., Ffield, A., 1993. Deep and bottom water of the Weddell Sea's western rim. Science 262, 95–97.

Gordon, A.L., Visbeck, M., Comiso, J.C., 2007. A possible link between the Weddell Polynya and the Southern annular mode. J. Clim. 20, 2558–2571.

Gordon, A.L., Orsi, A.H., Muench, R., Huber, B.A., Zambianchi, E., Visbeck, M., 2009. Western Ross Sea continental slope gravity currents. Deep Sea Res. Part II 56 (13–14), 796–817.

Gordon, A.L., Huber, B., McKee, D., Visbeck, M., 2010. A seasonal cycle in the export of bottom water from the Weddell Sea. Nat. Geosci. 3. http://dx.doi.org/10.1038/NGEO916.

Gouretski, V.V., Koltermann, K.P., 2004. WOCE global hydrographic climatology. Ber. Bundesamtes Seeschifffahrt Hydrogr. 35, 52 pp. +2 CD-ROMs.

Gregory, J.M., Dixon, K.W., Stouffer, R.J., Weaver, A.J., Driesschaert, E., Eby, M., Fichefet, T., Hasumi, H., Hu, A., Jungclaus, J.H., Kamenkovich, I.V., Levermann, A., Montoya, M., Murakami, S., Nawrath, S., Oka, A., Sokolov, A.P., Thorpe, R.B., 2005. A model intercomparison of changes in the Atlantic thermohaline circulation in response to increasing atmospheric CO_2 concentration. Geophys. Res. Lett. 32 (12). http://dx.doi.org/10.1029/2005GL023209, L12703.

Grist, J.P., Marsh, R., Josey, S.A., 2009. On the Relationship between the North Atlantic Meridional Overturning Circulation and the Surface-Forced Overturning Streamfunction. J. Clim. 22 (19), 4989–5002.

Haine, T., Böning, C.W., Brandt, P., Fischer, J., Funk, A., Kieke, D., Kvaleberg, E., Rhein, M., 2008. North Atlantic Deep Water Transformation in the Labrador Sea, Recirculation through the Subpolar Gyre, and Discharge to the Subtropics. In: Dickson, R.R., Meincke, J., Rhines, P. (Eds.), Arctic–Subarctic Ocean Fluxes—Defining the Role of the Northern Seas in Climate. Springer. ISBN 978-1-4020-6773-0.

Häkkinen, S., 1999. Variability of the simulated meridional heat transport in the North Atlantic for the period 1951–1993. J. Geophys. Res. 104, 10991–11007.

Hall, A., Visbeck, M., 2002. Synchronous variability in the Southern Hemisphere atmosphere, sea ice, and ocean resulting from the annular mode. J. Clim. 15 (21), 3043–3057.

Hansen, B., Osterhus, S., 2007. Faroe Bank channel overflow 1995–2005. Prog. Oceanogr. 75 (4), 817–856.

Hansen, B., Østerhus, S., 2000. North Atlantic-Nordic seas exchanges. Prog. Oceanogr. 45, 109–208.

Hellmer, H.H., Huhn, O., Gomis, D., Timmermann, R., 2011. On the freshening of the northwestern Weddell Sea continental shelf. Ocean Sci. 7, 305–316. http://dx.doi.org/10.5194/os-7-305-2011.

Holliday, N., Hughes, S., Bacon, S., Beszczynska-Moller, A., Hansen, B., Lavin, A., Loeng, H., Mork, K., Osterhus, S., Sherwin, T., Walczowski, W., 2008. Reversal of the 1960s to 1990s freshening trend in the northeast North Atlantic and Nordic Seas. Geophys. Res. Lett. 35 (3). http://dx.doi.org/10.1029/2007GL032675, 510.

Huhn, O., Hellmer, H.H., Rhein, M., Rodehacke, C., Roether, W., Schodlok, M.P., Schröder, M., 2008. Evidence of deep- and bottom-water formation in the Western Weddell Sea. Deep Sea Res. Part II 55, 1098–1116.

Huhn, O., Rhein, M., Hoppema, M., van Heuven, S., 2013. Decline of deep and bottom water ventilation and slowing down of anthropogenic carbon storage in the Weddell Sea, 1984–2011. Deep Sea Res. Part I 76, 66–84.

Hurrell, J.W., 1995. Decadal trends in the North-Atlantic oscillation—regional temperatures and precipitation. Science 269 (5224), 676–679. http://dx.doi.org/10.1126/science.269.5224.676.

Hurrell, J., National Center for Atmospheric Research Staff (Eds.), Last modified 16 Sep 2013. The Climate Data Guide: Hurrell North Atlantic Oscillation (NAO) Index (PC-based). Retrieved from https://climatedataguide.ucar.edu/climate-data/hurrell-north-atlantic-oscillation-nao-index-pc-based.

Jacobs, S.S., Giulivi, C.F., 2010. Large multidecadal salinity trends near the Pacific–Antarctic continental margin. J. Clim. 23, 4508–4524. http://dx.doi.org/10.1175/2010JCLI3284.1.

Jacobs, S., Amos, A.F., Bruchhausen, P.M., 1970. Ross Sea oceanography and Antarctic Bottom Water formation. Deep Sea Res. 17, 935–962.

Jacobs, S.S., Guilivi, C.F., Mele, P., 2002. Freshening of the Ross Sea during the late 20th century. Science 297 (386–389), 2002.

Jochumsen, K., Quadfasel, D., Valsimarsson, H., Jonsson, S., 2012. Variability of the Denmark Strait Overflow: moored time series from 1996–2011. J. Geophys. Res. 117. http://dx.doi.org/10.1029/2012JC008244.

Johnson, G.C., 2008. Quantifying Antarctic Bottom Water and North Atlantic Deep Water volumes. J. Geophys. Res. 113. http://dx.doi.org/10.1029/2007JC004477, C05027.

Johnson, G.C., Doney, S.C., 2006. Recent western South Atlantic bottom water warming. Geophys. Res. Lett. 33. http://dx.doi.org/10.1029/2006GL026769, L14614.

Johnson, G.C., Mecking, S., Sloyan, B.M., Wijffels, S.E., 2007. Recent bottom water warming in the Pacific Ocean. J. Clim. 20, 5365–5375.

Johnson, G.C., Purkey, S.G., Bullister, J.L., 2008a. Warming and freshening in the abyssal southeastern Indian Ocean. J. Clim. 21, 5351–5363. http://dx.doi.org/10.1175/2008JCLI2384.1.

Johnson, G.C., Purkey, S.G., Toole, J.M., 2008b. Reduced Antarctic meridional overturning circulation reaches the North Atlantic Ocean. Geophys. Res. Lett. 35. http://dx.doi.org/10.1029/2008GL035619, L22601.

Jones, P.D., Jónsson, T., Wheeler, D., 1997. Extension to the North Atlantic Oscillation using early instrumental pressure observations from Gibraltar and South-West Iceland. Int. J. Climatol. 17, 1433–1450.

Kawano, T., Fukawasa, M., Kouketsu, S., Uchida, H., Doi, T., Kaneko, I., Aoyama, M., Schneider, W., 2006. Bottom water warming along the pathways of lower circumpolar deep water in the Pacific Ocean. Geophys. Res. Lett. 33. http://dx.doi.org/10.1029/2006GL027933, L23613.

Kerr, R., Mata, M.M., Garcia, C.A.E., 2009. On the temporal variability of the Weddell Sea Deep Water Masses. Antarct. Sci. 21 (4), 383–400. http://dx.doi.org/10.1017/S0954102009001990.

Khatiwala, S., Schlosser, P., Visbeck, M., 2002. Rates and mechanisms of water mass transformation in the Labrador Sea as inferred from tracer observations. J. Phys. Oceanogr. 32 (2), 666–686.

Kida, S., Yang, J.Y., Price, J.F., 2009. Marginal sea overflows and the upper-ocean interaction. J. Phys. Oceanogr. 39, 387–403.

Kieke, D., Rhein, M., Stramma, L., Smethie, W.M., LeBel, D.A., Zenk, W., 2006. Changes in the CFC inventories and formation rates of Upper Labrador Sea Water, 1997–2001. J. Phys. Oceanogr. 36 (1), 64–86.

Kieke, D., Rhein, M., Stramma, L., Smethie, W.M., Bullister, J.L., LeBel, D.A., 2007. Changes in the pool of Labrador Sea Water in the subpolar North Atlantic. Geophys. Res. Lett. 34. http://dx.doi.org/10.1029/2006GL028959, L06605.

Kieke, D., Klein, B., Stramma, L., Rhein, M., Koltermann, K.P., 2009. Variability and propagation of Labrador Sea Water in the southern subpolar North Atlantic. Deep Sea Res. Part I 56 (10), 1656–1674.

Knight, J.R., Allan, R.J., Folland, C.K., Vellinga, M., Mann, M.E., 2005. A signature of persistent natural thermohaline circulation cycles in observed climate. Geophys. Res. Lett. 32. http://dx.doi.org/10.1029/2005GL024233.

Köhl, A., Stammer, D., 2008. Variability of the Meridional Overturning in the North Atlantic from the 50 years GECCO State Estimation. J. Phys. Oceanogr. 38, 1913–1930.

Koltermann, K.P., Gouretski, V.V., Jancke, K., 2011. In: Sparrow, M., Chapman, P., Gould, J. (Eds.), Hydrographic Atlas of the World Ocean Circulation Experiment (WOCE). Volume 3: Atlantic Ocean. International WOCE Project Office, Southampton, UK. ISBN 090417557X.

Lazier, J., Pickart, R., Rhines, P., 2001. Deep convection. In: Siedler, G., Church, J., Gould, J. (Eds.), Ocean Circulation and Climate—Observing and Modelling the Global Ocean. International Geophysics Series, vol. 77. Academic Press. ISBN 0-12-641351-7.

Le Bel, D.A., Smethie, W.M., Rhein, M., Kieke, D., Fine, R.A., Bullister, J.L., Min, D.-H., Roether, W., Weiss, R.F., Andrie, C., Smythe-Wright, D., Jones, E.P., 2008. The formation rate of North Atlantic Deep Water and Eighteen Degree Water calculated from CFC-11 inventories observed during WOCE. Deep Sea Res. Part I 55, 891–910.

Le Quéré, C., et al., 2009. Trends in the sources and sinks of carbon dioxide. Nat. Geosci. 2, 831–836. http://dx.doi.org/10.1038/ngeo689.

Lumpkin, R., Speer, K., 2007. Global Ocean Meridional circulation. J. Phys. Oceanogr. 37, 2550–2562. http://dx.doi.org/10.1175/JPO3130.1.

Lynch-Stieglitz, J., Adkins, J.F., Curry, W.B., Dokken, T., Hall, I.A., Herguera, J.C., Hirschi, J.J.-M., Ivanova, E.V., Kissel, C., Marchal, O., Marchitto, T.M., McCave, I.N., McManus, J.F., Mulitza, S., Ninnemann, U., Peeters, F., Yu, E.-F., Zahn, R., 2007. Atlantic meridional overturning circulation during the Last Glacial Maximum. Science 316, 66–69. http://dx.doi.org/10.1126/science.1137127.

Macrander, A., Send, U., Valdimarsson, H., Jonsson, S., Käse, R.H., 2005. Interannual changes in the overflow from the Nordic Seas into the Atlantic Ocean through Denmark Strait. Geophys. Res. Lett. 32, http://dx.doi.org/10.1029/2004GL021463, L06606.

Mantyla, A.W., Reid, J.L., 1983. Abyssal characteristics of the World Ocean waters. Deep Sea Res. 30, 805–833.

Marsh, R., 2000. Recent variability of the North Atlantic thermohaline circulation inferred from surface heat and freshwater fluxes. J. Clim. 13, 3239–3260.

Marshall, J., Schott, F., 1999. Open-ocean convection: observations, theory, and models. Rev. Geophys. 37 (1), 1–64. http://dx.doi.org/10.1029/98RG02739.

Masuda, S., Awaji, T., Sugiura, N., Matthews, J.P., Toyoda, T., Kawai, Y., Doi, T., Kouketsu, S., Igarashi, H., Katsumata, K., Uchida, H., Kawano, T., Fukasawa, M., 2010. Simulated rapid warming of abyssal North Pacific. Science 329, 319–322. http://dx.doi.org/10.1126/science.1188703.

Mauritzen, C., Häkkinen, S., 1999. On the relationship between dense water formation and the Meridional Overturning Cell in the North Atlantic Ocean. Deep Sea Res. Part I 46, 877–894.

Mauritzen, C., Price, J., Sanford, T., Torres, D., 2005. Circulation and mixing in the Faroese Channels. Deep Sea Res. Part I 52, 883–913.

Mauritzen, C., Melsom, A., Sutton, R.T., 2012. Importance of density-compensated temperature change for deep North Atlantic Ocean heat uptake. Nat. Geosci. 5, 905–910. http://dx.doi.org/10.1038/ngeo1639.

Meehl, G.A., et al., 2007. Global climate projections. In: Solomon, S., Qin, D., Manning, M., Chen, Z., Marquis, M., Averyt, K.B., Tignor, M., Miller, H.L. (Eds.), Climate Change 2007: The Physical

Science Basis. Contribution of Working Group I to the Fourth Assessment Report of the Intergovernmental Panel on Climate Change. Cambridge University Press, Cambridge, UK.

Meredith, M.P., Woodworth, P.L., Chereskin, T.K., Marshall, D.P., Allison, L.C., Bigg, G.R., Donohue, K., Heywood, K.J., Hughes, C.W., Hibbert, A., Hogg, A.M., Johnson, H.L., Jullion, L., King, B.A., Leach, H., Lenn, Y.-D., Morales Maqueda, M.A., Munday, D.R., Naveira Garabato, A.C., Provost, C., Sallée, J.-B., Sprintall, J., 2011. Sustained monitoring of the Southern Ocean at Drake Passage: past achievements and future priorities. Rev. Geophys. 49. http://dx.doi.org/10.1029/2010RG000348, RG4005.

Munk, W., Wunsch, C., 1998. Abyssal recipes II: energetics of tidal and wind mixing. Deep Sea Res. Part I 45 (12), 1977–2010.

Myers, P.G., Donnelly, C., 2008. Water mass transformation and formation in the Labrador sea. J. Clim. 21 (7), 1622–1638. http://dx.doi.org/10.1175/2007JCLI1722.1.

Nurser, A.J.G., Marshall, J.C., 1991. On the relationship between subduction rates and diabetic forcing of the mixed layer. J. Phys. Oceanogr. 21 (12), 1793–1802.

Ohshima, K.I., Fukamachi, Y., Williams, G.D., Nihashi, S., Roquet, F., Kitade, Y., Tamura, T., Hirano, D., Herraiz-Borreguero, L., Field, I., Hindell, M., Aoki, S., Wakatsuchi, M., 2013. Antarctic Bottom Water production by intense sea-ice formation in the Cape Darnley polynya. Nat. Geosci. http://dx.doi.org/10.1038/ngeo1738.

Olsen, S.M., Hansen, B., Quadfasel, D., Osterhus, S., 2008. Observed and modelled stability of overflow across the Greenland-Scotland ridge. Nature 455 (7212), 519–522.

Orsi, A.H., 2010. Recycling bottom waters. Nat. Geosci. 3, 307–309. http://dx.doi.org/10.1038/ngeo854.

Orsi, A.H., Johnson, G.C., Bullister, J.L., 1999. Circulation, mixing and production of Antarctic Bottom Water. Prog. Oceanogr. 43, 55–109.

Orsi, A.H., Smethie, W.M., Bullister, J.L., 2002. On the total input of Antarctic Waters to the deep ocean: a preliminary estimate from chlorofluorocarbon measurements. J. Geophys. Res. 107. http://dx.doi.org/10.1029/2001JC000976.

Pena-Molino, B., Joyce, T.M., Toole, J.M., 2011. Recent changes in the Labrador Sea Water within the Deep Western Boundary Current southeast of Cape Cod. Deep Sea Res. Part I 58 (10), 1019–1030. http://dx.doi.org/10.1016/j.dsr.2011.07.006.

Pickart, R.S., 1992. Water mass components of the North Atlantic deep western boundary current. Deep Sea Res. 9, 1553–1572.

Pickart, R.S., Smethie, W.M., Lazier, J.R.N., Jones, E.P., Jenkins, W.J., 1996. Eddies of newly formed upper Labrador Sea water. J. Geophys. Res. Oceans 101 (C9), 20711–20726. http://dx.doi.org/10.1029/96JC01453.

Prater, M.D., Rossby, T., 2005. Observations of the Faroe Bank Channel overflow using bottom-following RAFOS floats. Deep Sea Res. Part II 52, 481–494.

Purkey, S.G., Johnson, G.C., 2010. Warming of global abyssal and deep Southern Ocean waters between the 1990s and 2000s: contributions to global heat and sea level rise budgets. J. Clim. 23, 6336–6351. http://dx.doi.org/10.1175/2010JCLI3682.

Purkey, S.G., Johnson, G.C., 2012. Global Contraction of Antarctic Bottom Water between the 1980s and 2000s. J. Climate. 25, 5830–5844. http://dx.doi.org/10.1175/JCLI-D-11-00612.1.

Rahmstorf, S., 2002. Ocean circulation and climate during the past 120,000 years. Nature 419, 207–214.

Rhein, M., Fischer, J., Smethie, W.M., Smythe-Wright, D., Weiss, R.F., Mertens, C., Min, D.-H., Fleischmann, U., Putzka, A., 2002. Labrador sea water: pathways, CFC inventory, and formation rates. J. Phys. Oceanogr. 32 (2), 648–665.

Rhein, M., Kieke, D., Hüttl-Kabus, S., Rößler, A., Mertens, C., Meißner, R., Klein, B., Böning, C.W., Yashayaev, I., 2011. Deep-water formation, the subpolar gyre, and the meridional overturning circulation in the subpolar North Atlantic. Deep Sea Res. Part II 58 (17–18), 1819–1832.

Rintoul, S.R., 1998. On the origin and influence of Adélie Land Bottom Water. In: Jacobs, S., Weiss, R. (Eds.), Ocean, Ice, and Atmosphere: Interactions at the Antarctic Continental Margin. Antarctic Research Series, vol. 75. American Geophysical Union, pp. 151–171.

Rintoul, S., 2007. Rapid freshening of Antarctic Bottom Water formed in the Indian and Pacific Oceans. Geophys. Res. Lett. 34. http://dx.doi.org/10.1029/2006GL028550, L06606.

Robertson, R., Visbeck, M., Gordon, A.L., Fahrbach, E., 2002. Long-term temperature trends in the deep waters of the Weddell Sea. Deep Sea Res. Part II 49, 4791–4806.

Ross, C.K., 1984. Temperature-salinity characteristics of the "overflow" water in Denmark Strait during "Overflow '73". Rapp. P. V. Reun. Cons. Int. Explor. Mer. 185, 111–119.

Russell, J.L., Dixon, K.W., Gnanadesikan, A., Stouffer, R.J., Toggweiler, J.R., 2006. The Southern Hemisphere Westerlies in a warming world: propping open the door to the deep ocean. J. Clim. 19, 6382–6390.

Sabine, C.L., Feely, R.A., Gruber, N., Key, R.M., Lee, K., Bullister, J.L., Wanninkhof, R., Wong, C.S., Wallace, D.W.R., Tilbrook, B., Millero, F.J., Peng, T.-H., Kozyr, A., Ono, T., Rios, A.F., 2004. The oceanic sink for anthropogenic CO_2. Science 305 (5682), 367–371.

Sallée, J.B., Speer, K., Morrow, R., 2008. Southern Ocean fronts and their variability to climate modes. J. Clim. 21 (12), 3020–3039.

Saunders, P.M., 1994. The flux of overflow water through the Charlie-Gibbs fracture zone. J. Geophys. Res. 99, 12343–12355.

Saunders, P.M., 2001. The dense northern outflows. In: Siedler, G., Church, J., Gould, J. (Eds.), Ocean Circulation and Climate—Observing and Modelling the Global Ocean. International Geophysics Series, vol. 77. Academic Press. ISBN 0-12-641351-7.

Schlosser, P., Bayer, R., Foldvik, A., Gammelsrød, T., Rohardt, G., Münnich, K.O., 1990. Oxygen 18 and helium as tracers of ice shelf water and water/ice interaction in the Weddell Sea. J. Geophys. Res. 95 (C3), 3253–3263.

Schmitz Jr., W.J., 1996a. On the World Ocean Circulation. Volume 1. Some Global Features/North Atlantic Circulation. Woods Hole Oceanographic Institution, Woods Hole, MA, No. WHOI-96-03-VOL-1.

Schmitz Jr., W.J., 1996b. On the World Ocean Circulation: Volume 2 The Pacific and Indian Oceans/A Global Update. Woods Hole Oceanographic Institution, Woods Hole, MA, No. WHOI-96-08.

Schott, F., Stramma, L., Fischer, J., 1999. Interaction of the North Atlantic Current with the deep Charlie Gibbs Fracture Zone throughflow. Geophys. Res. Lett. 26, 369–372.

Send, U., Lankhorst, M., Kanzow, T., 2011. Observation of decadal change in the Atlantic meridional overturning circulation using 10 years of continuous transport data. Geophys. Res. Lett. 38. http://dx.doi.org/10.1029/2011GL049801, L24606.

Shimada, K., Aoki, S., Ohshima, K.I., Rintoul, S.R., 2012. Influence of Ross Sea Bottom Water changes on the warming and freshening of the Antarctic Bottom Water in the Australian-Antarctic Basin. Ocean Sci. 8, http://dx.doi.org/10.5194/os-8-419-2012.

Sigman, D.M., Boyle, E.A., 2000. Glacial/interglacial variations in atmospheric carbon dioxide. Nature 407, 859–869.

Sloyan, B.M., Rintoul, S.R., 2001. The Southern Ocean limb of the global deep overturning circulation. J. Phys. Oceanogr. 31, 143–173.

Smedsrud, L.H., 2005. Warming of the deep water in the Weddell Sea along the Greenwich meridian: 1977–2001. Deep Sea Res. Part I 52, 241–258.

Smethie Jr., W.M., Fine, R.A., 2001. Rates of North Atlantic Deep Water formation calculated from chlorofluorocarbon inventories. Deep Sea Res. Part I 48, 189–215.

Solomon, S., Qin, D., Manning, M., Chen, Z., Marquis, M., Avery, K.B., Tignor, M., Miller, H.L. (Eds.), 2007. Climate Change 2007: The Physical Basis. Contribution of Working Group I to the Fourth Assessment Report of the Intergovernmental Panel on Climate Change. Cambridge University Press, Cambridge, UK.

Speer, K., Tziperman, E., 1992. Rates of water mass formation in the North-Atlantic Ocean. J. Phys. Oceanogr. 22 (1), 93–104.

Srokosz, M., Baringer, M., Bryden, H., Cunningham, S., Delworth, T., Lozier, S., Marotzke, J., Sutton, R., 2012. Past, present, and future changes in the Atlantic Meridional Overturning Circulation. Bull. Am. Meteorol. Soc. 93, 1663–1676. http://dx.doi.org/10.1175/BAMS-D-11-00151.1.

Stouffer, R.J., Russell, J., Spelman, M.J., 2006. Importance of oceanic heat uptake in transient climate change. Geophys. Res. Lett. 33 (17). http://dx.doi.org/10.1029/2006GL027242, L17704.

Stouffer, R.J., Seidov, D., Haupt, B.J., 2007. Climate response to external sources of freshwater: North Atlantic vs. the Southern Ocean. J. Clim. 20, 436–448.

Stramma, L., Kieke, D., Rhein, M., Schott, F., Yashayaev, I., Koltermann, K.P., 2004. Deep water changes at the western boundary of the subpolar North Atlantic during 1996 to 2001. Deep Sea Res. Part I 41, 1033–1056.

Sy, A., Rhein, M., Lazier, J.R.N., Koltermann, K.P., Meincke, J., Putzka, A., Bersch, M., 1997. Surprisingly rapid spreading of newly formed intermediate waters across the North Atlantic Ocean. Nature 386, 675–679.

Talley, L.D., McCartney, M.S., 1982. Distribution and circulation of Labrador sea-water. J. Phys. Oceanogr. 12 (11), 1189–1205.

Talley, L.D., Pickard, G.L., Emery, W.J., Swift, J.H., 2011. Descriptive physical oceanography: an introduction. Academic Press.

Thompson, D.W.J., Solomon, S., Kushner, J., England, M.H., Grise, K.M., Karoly, D.J., 2011. Signatures of the Antarctic ozone hole in Southern Hemisphere surface climate change. Nat. Geosci. 4, 741–749. http://dx.doi.org/10.1038/NGEO1296.

Turrell, W.R., Slesser, G., Adams, R.D., Payne, R., Gillibrand, P.A., 1999. Decadal variability in the composition of Faroe-Shetland Channel bottom water. Deep Sea Res. Part I 46, 1–25.

van Sebille, E., Spence, P., Mazloff, M.R., England, M.H., Rintoul, S.R., Saenko, O.A., 2013. Abyssal connections of Antarctic Bottom Water in a Southern Ocean State Estimate. Geophys. Res. Lett. http://dx.doi.org/10.1002/grl.50483.

Wåhlin, A.K., Darelius, E., Cenedese, C., Lane-Serff, G., 2008. Laboratory observations of enhanced entrainment in dense overflows in the presence of submarine canyons and ridges. Deep Sea Res. Part I. http://dx.doi.org/10.1016/j.dsr.2008.02.007.

Whitworth III, T., Orsi, A.H., 2006. Antarctic bottom water production and export by tides in the Ross Sea. Geophys. Res. Lett. 33. http://dx.doi.org/10.1029/2006GL026357.

Whitworth, T., Nowlin, W.D., Pillsbury, R.D., Moore, M.I., Weiss, R.F., 1991. Observations of the Antarctic Circumpolar Current and deep boundary current in the southwest Atlantic. J. Geophys. Res. 96, 15105–15118. http://dx.doi.org/10.1029/91JC01319.

Williams, G.D., Bindoff, N.L., Marsland, S.J., Rintoul, S.R., 2008. Formation and export of dense shelf water from the Adélie Depression, East Antarctica. J. Geophys. Res. 113. http://dx.doi.org/10.1029/2007JC004346, C04039.

Williams, G.D., Aoki, S., Jacobs, S.S., Rintoul, S.R., Tamura, T., Bindoff, N.L., 2010. Antarctic Bottom Water from the Adélie and George V land coast, East Antarctica (140–149°E). J. Geophys. Res. 115. http://dx.doi.org/10.1029/2009JC005812, C04027.

Willis, J.K., 2010. Can in situ floats and satellite altimeters detect long-term changes in Atlantic Ocean overturning? Geophys. Res. Lett. 37. http://dx.doi.org/10.1029/2010GL042372, L06602.

Yashayaev, I., 2007. Hydrographic changes in the Labrador Sea, 1960–2005. Prog. Oceanogr. 73, 242–276.

Yashayaev, I., Dickson, R.R., 2008. Transformation and fate of overflows in the northern North Atlantic. In: Dickson, R.R., Meincke, J., Rhines, P. (Eds.), Arctic-Subarctic Ocean Fluxes: Defining the Role of the Northern Seas in Climate. Springer, Dordrecht, The Netherlands, pp. 505–526.

Yashayaev, I., Loder, J.W., 2009. Enhanced production of Labrador Sea Water in 2008. Geophys. Res. Lett. 36. http://dx.doi.org/10.1029/2008GL036162, L01606.

Zenk, W., Morozov, E., 2007. Decadal warming of the coldest Antarctic Bottom Water flow through the Vema Channel. Geophys. Res. Lett. 34. http://dx.doi.org/10.1029/2007GJ030340, L14607.

Zhang, R., 2008. Coherent surface-subsurface fingerprint of the Atlantic meridional overturning circulation. Geophys. Res. Lett. 35 (20). http://dx.doi.org/10.1029/2008GL035463.

Part IV

Ocean Circulation and Water Masses

Though this book has a focus on the global-scale ocean and its circulation, there are features that, though regional in character, have common underlying dynamics and processes. In this Part IV, we explore the commonality that links these often geographically widespread features. The knowledge of water-mass properties has been used from the earliest days of ocean science to describe the pathways of oceanic circulation. We now know that changes in the circulation are closely involved in climate variability. Observations provide information on the distribution of water masses, from short scales of oceanic fronts and mesoscale eddies to ocean-wide and global scales. Temporal and spatial changes let us recognize features of the large-scale circulation of the ocean as well as smaller-scale processes from internal waves, eddies, and Rossby waves to decadal variability. One can identify source regions of specific water masses, learn about mixing processes, and obtain a description of the global ocean circulation. This knowledge base is used for initializing and checking ocean and climate models.

Basic processes are best understood by conceptual models in which conditions are simplified. The state of the development of such models of the wind-driven and the thermohaline circulation are discussed in the first chapter of this Part. This discussion is followed by a series of chapters emphasizing observations and their interpretation by modeling. We start with the global surface circulation where major advances in the knowledge of spatial structures and their changes have been achieved through the combination of satellite data sets with *in situ* data from drifters, hydrography, and moored instruments. One noteworthy result is the zonally banded structure found in many parts of the oceans.

We follow with chapters describing the principal features of the oceans' western and eastern boundary currents and the complex circulation system and climatic variability of the tropics. High-latitude oceans where the influence of the frozen regions plays a unique role are then discussed finally as the passages or "choke points" through which ocean basins are linked and which serve as key locations for monitoring ocean circulation changes.

A simplified look at global water mass transports reveals the overturning circulation with the key sinking region in the Subarctic and Arctic North Atlantic. Deep-ocean flow brings water to the south, being modified by intrusions from adjacent seas, followed by transports around the Antarctic continent and upwelling and upward mixing in various regions. An upper-ocean return flow brings warmer water back through the choke points of the Drake Passage, the Indonesian Archipelago and around southern Africa, and in the meridional western boundary currents. The processes and recent findings in the western boundary regions are discussed, with a comparison to the different features of eastern boundary regimes. These meridional flows are coupled by the mostly zonal tropical currents, including the ENSO region in the Pacific, and through the processes and currents in the subpolar and polar regions, including the major transport band of the Antarctic Circumpolar Current. The deep flows are largely controlled by topography, and critical regions and the signals for climatic changes in the deep ocean are discussed in this Part.

Chapter 11

Conceptual Models of the Wind-Driven and Thermohaline Circulation

Sybren S. Drijfhout*,†, David P. Marshall‡ and Henk A. Dijkstra§

*School of Ocean and Earth Science, University of Southampton, Southampton, United Kingdom

†Royal Netherlands Meteorological Institute, De Bilt, The Netherlands

‡Department of Physics, University of Oxford, Oxford, United Kingdom

§Institute for Marine Atmospheric research Utrecht, Department of Physics and Astronomy, Utrecht University, Utrecht, The Netherlands

Chapter Outline

1. INTRODUCTION

The field of physical oceanography has been transformed over the past two decades by the emergence of the modern global ocean observing system and of global eddy-permitting ocean circulation models. These observations and models have revealed the complexity of the ocean circulation across the spectrum of spatial and temporal scales. Nevertheless, if one is to make sense of all this complexity, it is essential to have a portfolio of simple conceptual models that can provide dynamic understanding. Conceptual models provide the vocabulary with which to articulate complex dynamic balances and circulation patterns in observations and numerical models. Even conceptual models that fail can have value in directing the community to seek alternative mechanisms: "all models are wrong, some models are useful" (usually attributed to George Box). The aim of this chapter is to present an overview of the conceptual models that articulate our current understanding of the wind-driven and thermohaline circulation (THC). In doing so, we also aim to set the theoretical stage for the chapters that follow.

2. WIND-DRIVEN CIRCULATION

Momentum is transferred from the atmosphere to the ocean by the viscous shear stress at the surface and the form stress across the surface waves. The precise relation between wind speed and stress remains uncertain but is a function of sea-state, the surface waves representing an integrated response to remote wind forcing. Moreover, Pacanowski (1987), and more recently, Duhaut and Straub (2006) have highlighted that the surface stress also depends on the surface ocean current speed; this provides a surprisingly large reduction in the energy input to the ocean from the atmosphere (Hughes and Wilson, 2008; Scott and Xu, 2009; Zhai et al., 2012). The effect is particularly important in driving vertical motion in the ocean interior through the damping of ocean eddies. More interesting, albeit smaller, corrections arise from considering the wave-driven circulation

Ocean Circulation and Climate, Vol. 103. http://dx.doi.org/10.1016/B978-0-12-391851-2.00011-8

(McWilliams and Restrepo, 1999). In the following, we set aside such considerations and focus on the response of the ocean to a prescribed surface wind stress.

2.1. Ekman Layer and Ekman Overturning Cells

The direct effect of the surface wind stress is felt over a relatively thin surface layer of a few tens of meters, known as the Ekman layer. Under an assumption of a small Rossby number, classical Ekman theory predicts a decaying spiral in the velocity vector with depth, from a surface velocity directed at 45° to the surface wind stress, to the geostrophic value beneath (Ekman, 1905). While Ekman spirals are observed if data is averaged over several months (Ralph and Niiler, 1999), the detailed structure of the spiral differs from classical theory due to the turbulent nature of the stress and the role of inertial oscillations. Fortunately, the crucial quantity for the large-scale ocean circulation is the integrated momentum balance over the Ekman layer. Assuming a three-way balance between surface wind stress, Coriolis force, and pressure gradient force, the depth-integrated flow of the Ekman layer consists of a geostrophic component and an Ekman component to the right of the wind stress direction in the northern hemisphere, and to the left of the wind stress direction in the southern hemisphere. Assuming the ocean is incompressible to leading order results in Ekman's remarkable formula for the vertical upwelling velocity at the base of the Ekman layer:

$$w_{\mathrm{Ek}} = \mathbf{k} \cdot \nabla \times \frac{\tau_{\mathrm{S}}}{\rho_0 f} \tag{11.1}$$

where τ_{S} is the surface wind stress vector, f is the Coriolis parameter, ρ_0 is a reference value of the density of seawater and $\mathbf{k}$ is a unit vertical vector.

Ekman upwelling explains the origin of many anomalous features in the sea surface temperature field, such as the cold tongue in the eastern equatorial Pacific, and the cold coastal upwelling zones on the eastern margins of ocean basins. Note that the former is a region in which Equation (11.1) breaks down due to the vanishing of the Coriolis parameter on the equator, yet the integrated upwelling remains robust as long as the integral extends to a few degrees either side of the equator. There is a remarkable coincidence of the primary areas of biological activity, as revealed, for example, by the SEAWIFS satellite observations of surface color, and areas of Ekman upwelling due to the cold, upwelled water fluxing nutrients to the surface euphotic zone.

Surface Ekman currents are mostly compensated by geostrophic currents within the upper few hundred meters of the ocean, the latter supported by basin wide pressure gradients. The one exception is in the Southern Ocean where the absence of continental barriers at the latitude of Drake Passage requires that the poleward geostrophic transport (to compensate for the equatorward Ekman transport) occurs in abyssal layers, supported by a pressure gradient across the Drake Passage ridge; note, however, that the net meridional overturning circulation (MOC) is very different in the Southern Ocean when one takes into account the eddy-induced circulation (see Section 3.2).

The patterns of surface Ekman and interior geostrophic currents are closed by Ekman upwelling and downwelling and give rise to Ekman overturning cells. These Ekman overturning cells in turn mechanically pump warm, buoyant water into the ocean interior and raise cold dense water from below, generating large-scale horizontal buoyancy gradients and thus, by the thermal wind equation,

$$f \frac{\partial \mathbf{u}_{\mathrm{g}}}{\partial z} = -\frac{g}{\rho_0} \mathbf{k} \times \nabla \rho, \tag{11.2}$$

the large-scale geostrophic circulation of the ocean interior. Here $\mathbf{u}_{\mathrm{g}}$ is the geostrophic velocity, z is height, g is the gravitational acceleration and ρ is density.

2.2. Sverdrup Balance

Sverdrup (1947) derived a further remarkable relation for strength and structure of the depth-integrated circulation. Assuming a sufficiently low Rossby number such that relative vorticity can be neglected, and assuming the magnitude of the vertical velocity at the sea floor to be much smaller than that of the Ekman upwelling velocity, the dominant depth-integrated vorticity balance is between the input of vorticity from the wind-stress curl and the advection of planetary vorticity:

$$\beta \int v \mathrm{d}z = \frac{1}{\rho_0} \mathbf{k} \cdot \nabla \times \tau_{\mathrm{S}}, \tag{11.3}$$

where v is the meridional velocity and β is the variation of the vertical component of the planetary vorticity with latitude. The depth-integrated meridional transports predicted by Sverdrup balance do not generally integrate to zero across an ocean basin. However, Sverdrup additionally noted that the depth-integrated circulation is closed by intense western boundary currents within which relative vorticity and topographic interactions, among other processes, are not negligible. Assuming that the depth-integrated circulation vanishes at the eastern boundary and using Equation (11.3) to infer the depth-integrated zonal transport required to satisfy continuity gives a complete theory for the strength and structure of the depth-integrated circulation outside the western boundary currents (Figure 11.1).

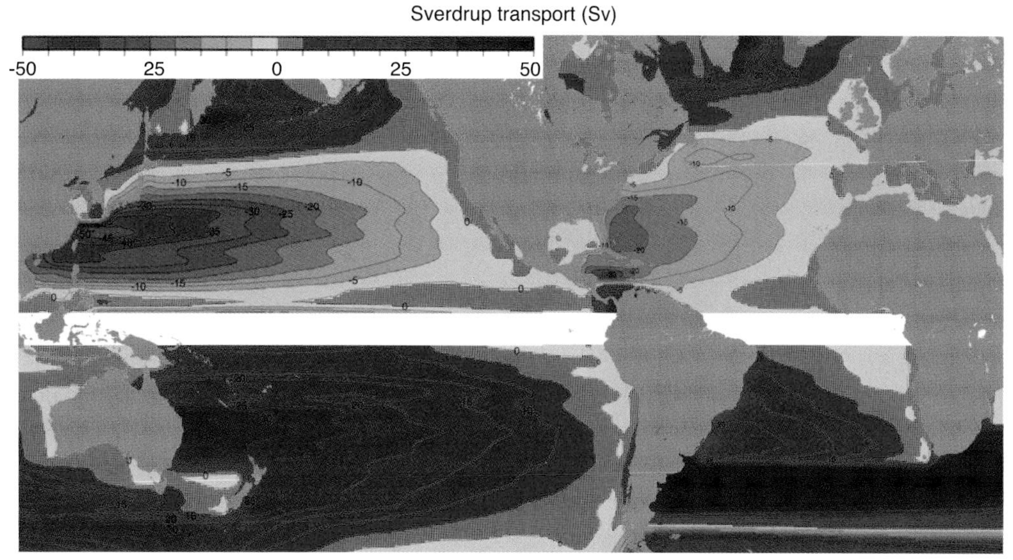

FIGURE 11.1 Estimate of the depth-integrated circulation (Sv) predicted by the Sverdrup balance, where positive values (yellow–red) correspond to anticlockwise circulation. *Adapted from Talley et al. (2011) and based on data from the NCEP reanalysis (Kalnay et al., 1996).*

While there have been a number of attempts to validate Sverdrup balance from observations, this turns out to be impossible without an accurate estimate of the absolute circulation at some reference level. Recently Wunsch (2011) investigated the validity of Sverdrup balance over the upper part of the ocean. While it is unlikely that Sverdrup balance holds in a local sense, for example due to topographic forcing, its successful explanation of structure and strength of the subtropical gyres (and to a lesser extent the subpolar gyres where topographic interactions are of greater importance) is one of the great successes of geophysical fluid dynamics.

2.3. Western Boundary Currents and Inertial Recirculation

The Sverdrup circulation is closed by intense boundary currents at the western margin of ocean basins. Our theoretical understanding of western boundary currents has mostly been developed with *homogenous models* in which density variations are neglected and the circulation is assumed independent of depth (aside from any Ekman layers). The first such western boundary current solution was obtained by Stommel (1948) through the addition of linear drag, providing a mechanism for removing the vorticity input by the surface wind stress curl; this solution was extended to lateral friction and a no-slip boundary condition by Munk (1950). However, even excessively large frictional coefficients give excessively narrow and intense western boundary currents within which relative vorticity is far from negligible, making these solutions somewhat inconsistent. An alternative paradigm was explored by Fofonoff (1954) and Charney (1955), of an inertial boundary current in which potential vorticity is conserved to leading order within the western boundary current. However, these purely inertial boundary current solutions are unable to reconnect smoothly to the Sverdrup solution in the interior due to the absence of any vorticity sink to balance the vorticity source from the wind stress curl. Both paradigms were unified through the numerical calculations of Bryan (1963) and Veronis (1966), who showed that the western boundary currents become increasingly inertial as the frictional damping is reduced. The circulation develops a north–south asymmetry, with an intense inertial recirculation developing close to the point at which the boundary current leaves the western boundary.

A notable feature of both the Bryan and Veronis models is that the transport exceeds the theoretical prediction of Sverdrup balance in the weakly-dissipative limit; this is known as *inertial recirculation*. In particular, in the Veronis solution (with free-slip boundary conditions), this inertial recirculation unrealistically swamps the Sverdrup gyre. A neat theoretical explanation for the inertial regime was provided by Niiler (1966), by integrating the steady vorticity equation over the area enclosed by a streamline, equivalent to the circulation budget along that streamline. Since inertia acts only to redistribute vorticity, dissipation is required along each streamline to balance the circulation input by the surface wind stress. In the Stommel and Munk models this is achieved by the streamlines each passing through the dissipative western boundary current. However, in the limit of weak dissipation, either the western boundary current must extend into the ocean interior, increasing the distance over which dissipation is active, or the gyre transport must increase, increasing the magnitude of the frictional sink of circulation. The latter serves a simple explanation for the presence of inertial recirculation. More modern treatments understand that transient eddies play a major role in maintaining the time-mean circulation budget (e.g., Marshall, 1981).

While the form and magnitude of friction, lateral boundary condition, inertia, eddies, and other processes such as bottom topography play a role in setting the detailed structure of western boundary currents, the ultimate cause of the western intensification of the circulation can be attributed to the latitudinal variation of the Coriolis parameter—the so-called *β effect*—as proposed in Stommel (1948). This can be clearly seen in the analysis of Marshall and Pillar (2011) in which the individual terms in the vorticity equation are converted into *rotational forces*: the rotational Coriolis force acts unambiguously to accelerate the western boundary currents, opposed by inertial and frictional forces. This acceleration of the western boundary currents is related to the tendency of the Coriolis force to propagate features westward via the Rossby wave mechanism. The role of bottom topography is discussed in Section 2.5.

2.4. Vertical Structure of the Wind-Driven Circulation

We now turn to the vertical structure of the wind-driven circulation. The geostrophic streamlines in the North Pacific (Figure 11.2) show that the circulation of the subtropical gyre both weakens and contracts toward its northwestern corner with depth. It is worth noting that the modern theoretical understanding of this vertical structure arose not from the advent of numerical circulation models, but rather from a simplification of the dynamic framework from that of a continuously stratified fluid to a stacked series of vertical layer of different densities. Through this simplification, the resultant problem became mathematically and conceptually tractable. Here we attempt to sketch the essential physical

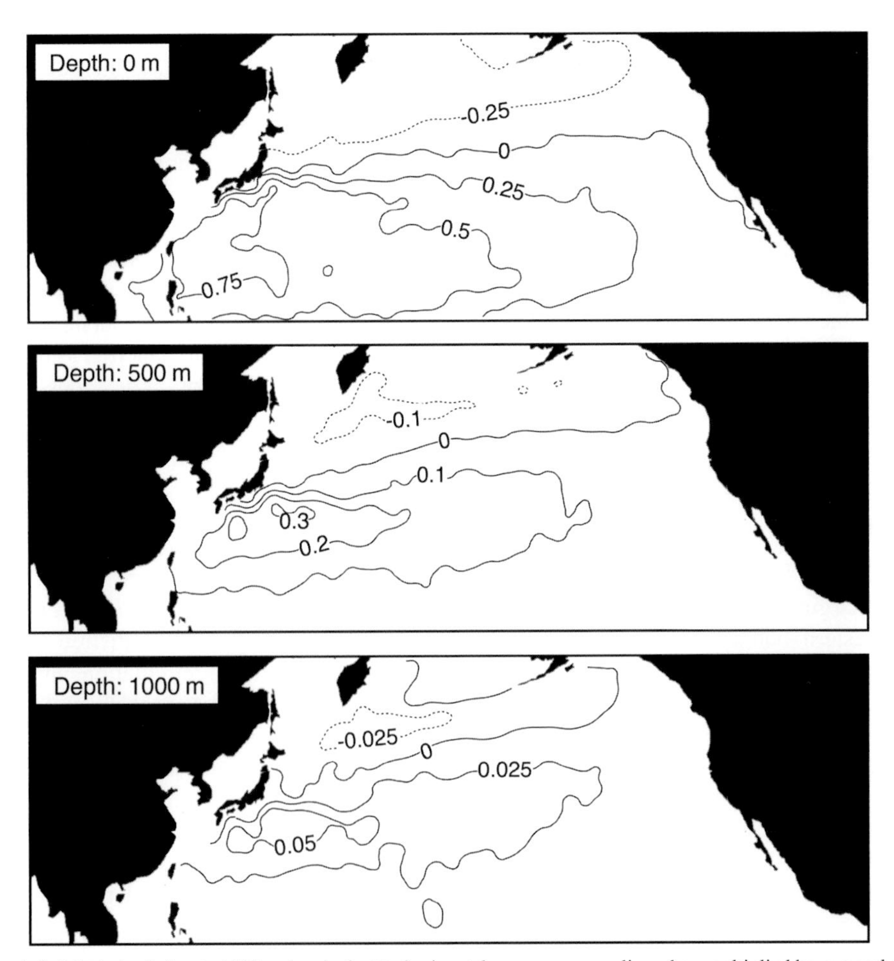

FIGURE 11.2 Dynamic height (m) relative to 1500 m (equivalent to horizontal pressure anomalies when multiplied by $\rho_0 g$ and assuming a level of no motion at 1500 m). The contours are streamlines for the geostrophic flow and show the subtropical gyre weakening and contracting to the northwestern corner with depth. *Based on data from the WOCE Global Hydrographic Climatology (Gouretski and Koltermann, 2004).*

ideas from a heuristic perspective; for detailed derivations of the solutions, the reader is referred to Rhines and Young (1982b), Luyten et al. (1983) and standard texts such as Pedlosky (1996), Vallis (2006) and Olbers et al. (2012).

The key dynamical concept underlying these conceptual models is that fluid parcels materially conserve their potential vorticity, f/h, in the absence of forcing and dissipation in the ocean interior, where h is the thickness of a layer bounded by two density surfaces. Following Rhines and Young (1982b), suppose that the subtropical gyre can be represented through three layers, and initially suppose that the Sverdrup circulation is confined to the upper layer that is exposed directly to vorticity forcing through a prescribed Ekman downwelling (Figure 11.3a). Rhines and Young's first major insight was that the flow in the upper layer deforms the interface with the layer below through thermal wind balance, and hence also deforms the potential vorticity in the second layer (Figure 11.3b).

Now the potential vorticity contains blocked contours that intersect the eastern boundary, along which potential-vorticity conserving flow is impossible. However, there are also unblocked contours that do not intersect the eastern boundary, along which potential vorticity conserving flow is possible, assuming that a western boundary current solution can be found to close the circulation. Finally, the strength of the circulation along the unblocked contours remains to be determined. Rhines and Young (1982a) noted that eddies tend to homogenize the potential vorticity field within closed contours in numerical models, an application of the Prandtl–Batchelor theorem, from which the flow can be determined (Figure 11.3c and d). Note that the presence of flow in layer two in turn deforms the potential vorticity field and opens the possibility of flow in layer three, and so on.

An alternative, but complementary, *ventilated thermocline* model of the vertical structure was subsequently put forward by Luyten et al. (1983). The key idea is that surface buoyancy forcing means that density layers outcrop at the sea surface, directly exposing each outcropped layer to surface forcing at some latitude. Now the possibility arises of potential vorticity contours threading down from the surface mixed layer, where fluid parcels are *subducted* into the ocean interior, providing in essence a mapping of the two-dimensional surface density field onto a three-dimensional interior stratification. These ventilated contours can coexist with the *homogenized pool* along the unblocked potential vorticity contours, and the *shadow zone* of no motion along blocked potential vorticity contours (Figure 11.4). Guided by the overall structure of the Rhines and Young (1982b) and Luyten et al. (1983) solutions, extensions to continuous stratification duly followed (Huang, 1989).

A further important point is that the water mass properties of the ocean interior map onto the extreme winter, rather than annual-mean conditions at the sea surface (Iselin, 1939). This was explained by Stommel's elegant *mixed-layer demon* (Stommel, 1979) and demonstrated afterward in a clean numerical calculation by Williams

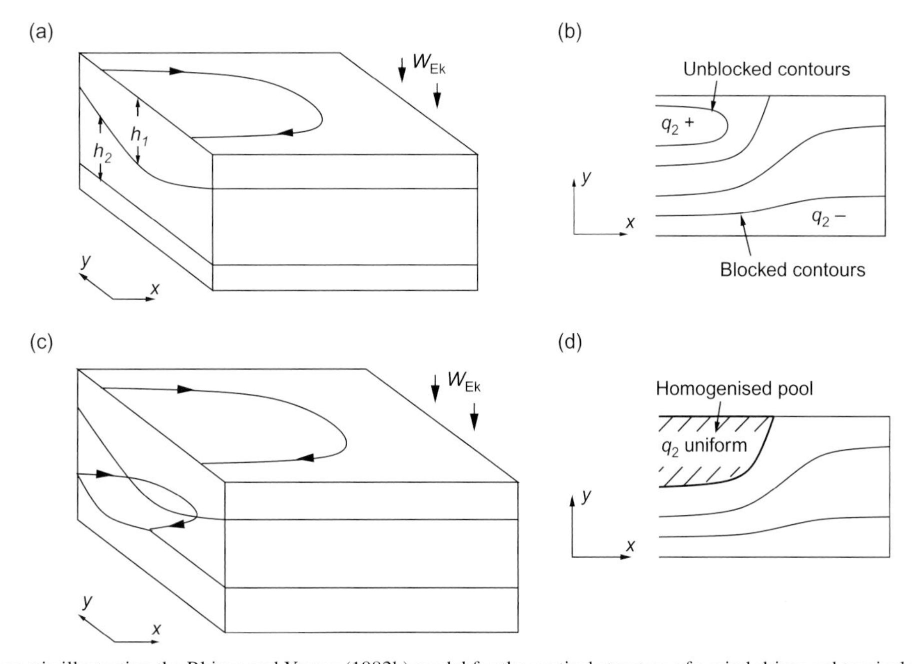

FIGURE 11.3 Schematic illustrating the Rhines and Young (1982b) model for the vertical structure of a wind-driven subtropical gyre. Panel (a) shows the three-dimensional structure of the circulation and layer thickness and panel (b) the potential vorticity in layer 2, assuming the flow is confined to the upper layer. Panels (c) and (d) show the same after flow has been incorporated in layer 2 by homogenizing the potential vorticity field within the unblocked potential vorticity contours. See the text for more details.

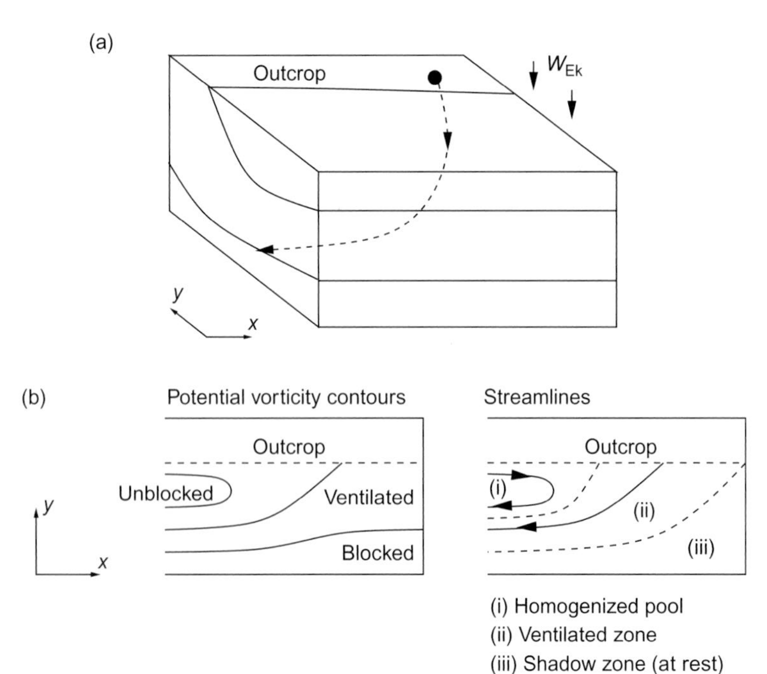

FIGURE 11.4 Schematic illustrating the ventilated thermocline model (Luyten et al., 1983). Panel (a) shows a fluid parcel in layer 2 being subducted beneath the surface outcrop, after which it conserves its potential vorticity since it is shielded from the surface Ekman pumping. Panel (b) shows plan views of the potential vorticity contours (left) and streamlines (right) in layer 2. See the text for further details.

et al. (1995). The simplest explanation is obtained by noting that the annual poleward and equatorward migration of surface density outcrops greatly exceeds the distance a fluid parcel moves in latitudinal direction over a year within the subtropical gyre. Thus, most fluid parcels are subducted from the mixed layer into the ocean interior. They are subsequently re-entrained into the surface mixed layer when the surface outcrop sweeps equatorward the following winter. It is only fluid parcels subducted in a narrow window around late March in the northern hemisphere (or late September in the southern hemisphere) that escape permanently into the ocean interior and hence that set the water mass properties of the ocean interior (Figure 11.5).

While the ideal-fluid thermocline models of Rhines and Young (1982b) and Luyten et al. (1983) have been successful at explaining aspects of the large-scale structure of wind-driven subtropical gyres and, at the very least, have provided an extremely useful conceptual paradigm for discussing more complex structures observed in numerical models, there is no suggestion that these models capture all aspects of the gyre interior. For example, geostrophic eddies almost certainly modify the structure of the ventilated thermocline model (Cessi and Fantini, 2004).

Moreover, the conceptual models remain open at the western boundary and the homogenous model has taught us that it is far from obvious that the western boundary current can be considered a passive agent as far as the ocean interior is concerned. Marshall (2000) and Polton and Marshall (2003) have derived an integral condition that

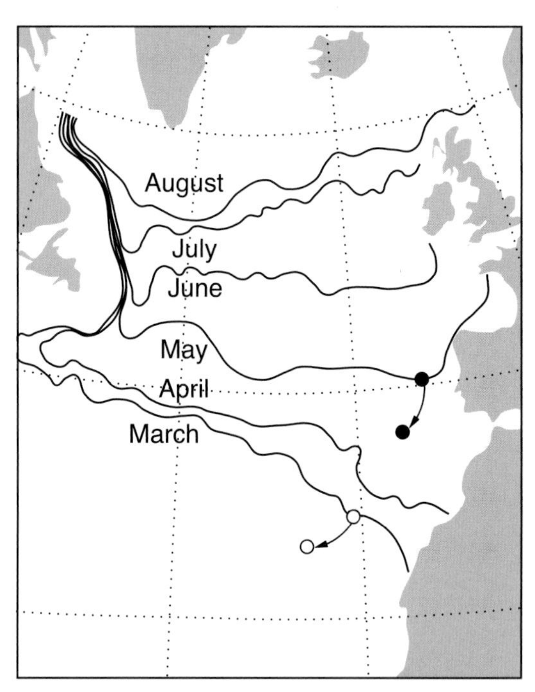

FIGURE 11.5 Schematic illustrating of the Stommel demon mechanism. The figure shows estimates of the location of the outcrop of the $\sigma_t = 26.5$ density surface in the North Atlantic between March and August. Superimposed are notional displacements of two fluid parcels in the subtropical gyre over an annual cycle. The solid black parcel is subducted in May but re-entrained the following winter when the outcrop sweeps equatorward. In contrast, the white parcel is subducted in March and permanently escapes into the ocean interior (Stommel, 1979). *Redrawn from Woods (1985).*

closely resembles the *diffusive thermocline* theories of Robinson and Stommel (1959) and Salmon (1990) in which vertical advection is balanced by vertical diffusion. Consistent with the suggestion of Welander (1971) and numerical results of Samelson and Vallis (1997), the integral constraint suggests that a diffusive thermocline forms as an internal boundary layer at the base of the quasi-adiabatic ventilated thermocline. These results are closely related to the *leaky thermocline* model of Radko and Marshall (2004).

2.5. Role of Bottom Topography

Each of the above conceptual models neglects any effects of variable bottom topography. There is no inconsistency with the Rhines and Young (1982b) and Luyten et al. (1983) models because of this neglect, since the flow vanishes at depth in these models and thus there can be no interaction with the bottom topography. However, abyssal currents do not vanish in general. A prevalent explanation of how the depth-averaged circulation interacts with variable bottom topography is due to the joint effects of baroclinicity and relief, or *JEBAR* (for an overview, see Sarkisyan (2006)) in which a term representing the cross product of the gradients of potential energy anomaly and inverse depth arises in the depth-averaged momentum equation. However, as noted by Cane et al. (1998), JEBAR fails to give a true measure of the effect of topography on the circulation, clearly seen in the limit that the abyssal circulation vanishes (Figure 11.6a) but the JEBAR "forcing" remains finite.

Instead, it is necessary to fully account for the dynamical feedbacks between the abyssal and depth-integrated circulations. Marshall (1995a) and Marshall (1995b) developed a conceptual model in which the circulation is in linear vorticity balance and fluid parcels conserve their potential vorticity (assumed uniform on density surfaces). Crucially, the requirement that fluid parcels at the sea floor materially conserve their density as they navigate the variable bottom topography provides a dynamical feedback between the depth-integrated and bottom flow: as fluid parcels rise over topography, the density surfaces are heaved up, generating meridional thermal wind velocities throughout the fluid column that dominate the former in the linear vorticity balance of the fluid column. Whereas unforced flow follows f/H contours, where H is the ocean depth, in the weakly-stratified limit and in the presence of a strong, surface-intensified stratification and weak bottom currents, both unforced surface and depth-integrated flow are only weakly steered by the bottom topography—a process Veronis (1981) termed *baroclinic compensation* (Figure 11.6b).

A more meaningful measure of the influence of bottom topography on the circulation is provided by the vertical velocity at the sea floor, equivalent to the bottom pressure torque if the flow is geostrophic:

$$w_b = -\mathbf{u}_b \cdot \nabla H \approx \frac{1}{\rho_0 f} \mathbf{k} \times \nabla p_b \cdot \nabla H, \qquad (11.4)$$

where $\mathbf{u}_b$ and w_b are the horizontal and vertical velocity components at the sea floor and p_b is the bottom pressure. At the sea floor, the horizontal circulation is mostly around, rather than over the topography. Nevertheless the generalization of the Sverdrup relation (11.3) to allow for variable bottom topography is:

$$\beta \int v \mathrm{d}z = \frac{1}{\rho_0} \mathbf{k} \cdot \nabla \times \tau_S - f w_b \qquad (11.5)$$

The second term on the right-hand side represents the compression and stretching of vortex tubes as they pass up and down the bottom slope, and vanishes when integrated over any area enclosed by a contour of constant depth H. This helps explain why the traditional Sverdrup relation works so well over much of the ocean interior,

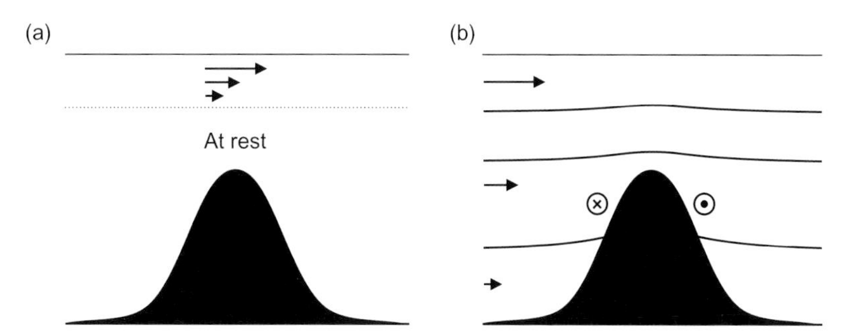

FIGURE 11.6 (a) Schematic diagram illustrating the shortcoming of the JEBAR model of flow-topography interactions (following Cane et al., 1998). Motion is confined to a surface layer such that there is no interaction of the flow with the bottom topography, yet the JEBAR forcing term is finite. (b) Illustration of the physical mechanism behind "baroclinic compensation." As the bottom flow passes over a seamount, it generates a bottom vertical velocity w_b, and raises isopycnals throughout the fluid column, deflecting the flow around the seamount by thermal wind balance. Thus, in the depth-integrated linear vorticity budget, the bottom topography contributes both to vortex stretching, through w_b, and advection of planetary vorticity, through $\int \beta v \mathrm{d}z$. For realistic ocean parameters, it turns out that the latter process is dominant, that is, the bottom flow is mostly deflected *around* the seamount but the surface flow is largely unaffected by the topography (Marshall, 1995a).

despite the neglect of topographic interactions (see, e.g., Wunsch, 2011). Nevertheless, the bottom pressure torque clearly does play a role in shaping the structure of western boundary currents. For further discussion, see Salmon and Ford (1995), Hughes and de Cuevas (2001) and Eden and Olbers (2010).

3. THERMOHALINE CIRCULATION

The thermohaline circulation is usually defined as that resulting from a change to the temperature or salinity by heat and freshwater exchange with the atmosphere. The *MOC*, defined as the zonal integral of the meridional and vertical circulation in each ocean basin, is often used as a surrogate for the THC. This is because the MOC is easier to calculate in models and inferred from observations (e.g., through end point monitoring; Rayner et al., 2011), and because of the conceptual simplicity of projecting the circulation onto two dimensions. However, the MOC also incorporates wind-driven Ekman overturning cells, and other flow features that may have nothing to do with thermohaline forcing (e.g., homogeneous flow over variable topography can project onto the MOC). Some of these shortcomings can be mitigated by calculating the MOC in neutral density coordinates (e.g., Döös and Webb, 1994). A recent and valuable extension has been to calculate an overturning stream-function in temperature-salinity coordinates (Döös et al., 2012; Zika et al., 2012), which more closely resembles the classical concept of a THC.

3.1. Energetics and Global Perspective

In contrast to the atmosphere, the ocean is both heated and cooled at its surface. This is an inefficient way of driving a circulation: consider the analogy of boiling a pan of water by placing it under the grill. In such an ocean, with weak diapycnal mixing, the resultant circulation can be expected to be weak and, along with the stratification, confined to a thin surface boundary layer (Sandström, 1916). Thus the deep stratification and circulation observed in the ocean must be maintained by mechanical forcing as well (Munk and Wunsch, 1998). A nice physical interpretation invoking Bjerknes' circulation theorem (Bjerknes, 1898) is given by Olbers et al. (2012): if there is no mechanical forcing, the baroclinic production of circulation can only balance the dissipative sink of circulation if the heating takes places at higher pressure, that is, deeper in the water, than the cooling.

A simple heuristic model of the global overturning circulation and stratification has been developed by Gnanadesikan (1999), where stratification is represented by two homogeneous layers separated by a pycnocline (Figure 11.7a). The model is cast in terms of the depth of the pycnocline, h, considered uniform, except in the Southern Ocean and northern high latitudes where the pycnocline outcrops. An equation is posited for the volume budget of the surface layer. Gnanadesikan identifies four sources of water for the surface layer: two that deepen the pycnocline—Southern Ocean winds which drive a northward Ekman transport, T_{Ek}, and turbulent diapycnal mixing which drives a diapycnal upwelling transport, T_{U}—and two that shallow the pycnocline—Southern Ocean eddies which give a poleward eddy bolus transport in the Southern Ocean, T_{Eddy}, and North Atlantic Deep Water (NADW) formation, T_{NADW}. Each of T_{U}, T_{Eddy} and T_{NADW} are parameterized as functions of h; see Gnanadesikan (1999) for further details. The Gnanadesikan model applies to the overturning cell associated with NADW, but a similar model can be constructed for the Antarctic Bottom Water (AABW) cell (Nikurashin and Vallis, 2011) with the notion that this overturning cell must be diffusive and diabatic (Figure 11.7b); Nikurashin and Vallis (2012) combine elements of each of these models. The plot of neutral density from the WOCE A16 section in Figure 11.7c nicely motivates the choice of density structure assumed in the conceptual models.

To the extent that the pycnocline is a proxy for the potential energy of the ocean, this balance is indicative of the mechanical energy budget of the ocean. The dominant mechanical energy sources for the ocean are Southern Ocean winds and tides (Munk and Wunsch, 1998; Wunsch and Ferrari, 2004), the latter because the most important process that gives rise to turbulent mixing is the breaking of internal waves (St Laurent and Garrett, 2002). These waves are generated by either winds at the surface, or the interaction of the tidal flow with bottom topography and the interaction of the eddy field with the bottom topography. In contrast, convection in high-latitude water mass formation sites is a mechanical energy sink (Marotzke and Scott, 1999), as is baroclinic instability in the Southern Ocean.

An alternative point of view is taken by Hughes et al. (2009) and Tailleux (2009). They argue that *available potential energy* is crucial in understanding the energy budget of the THC. They illustrate this point through a simple thought experiment, sketched in Figure 11.8. In both cases (a) and (b) the potential energy is the same, yet the available potential energy is very different: a full-depth convective overturning cell will develop in case (b), whereas no flow develops in case (a). However, note that case (b) is convectively unstable and thus cannot be maintained in equilibrium. Case (c) differs only slightly from case (b) in that we have modified the abyssal density in order to maintain convective equilibrium and the available potential energy is again small. Nevertheless, the point that buoyancy forcing can modify the available potential energy by modifying the background potential energy (Winters et al., 1995; Hughes et al., 2009; Tailleux, 2009; Saenz et al., 2012) is important.

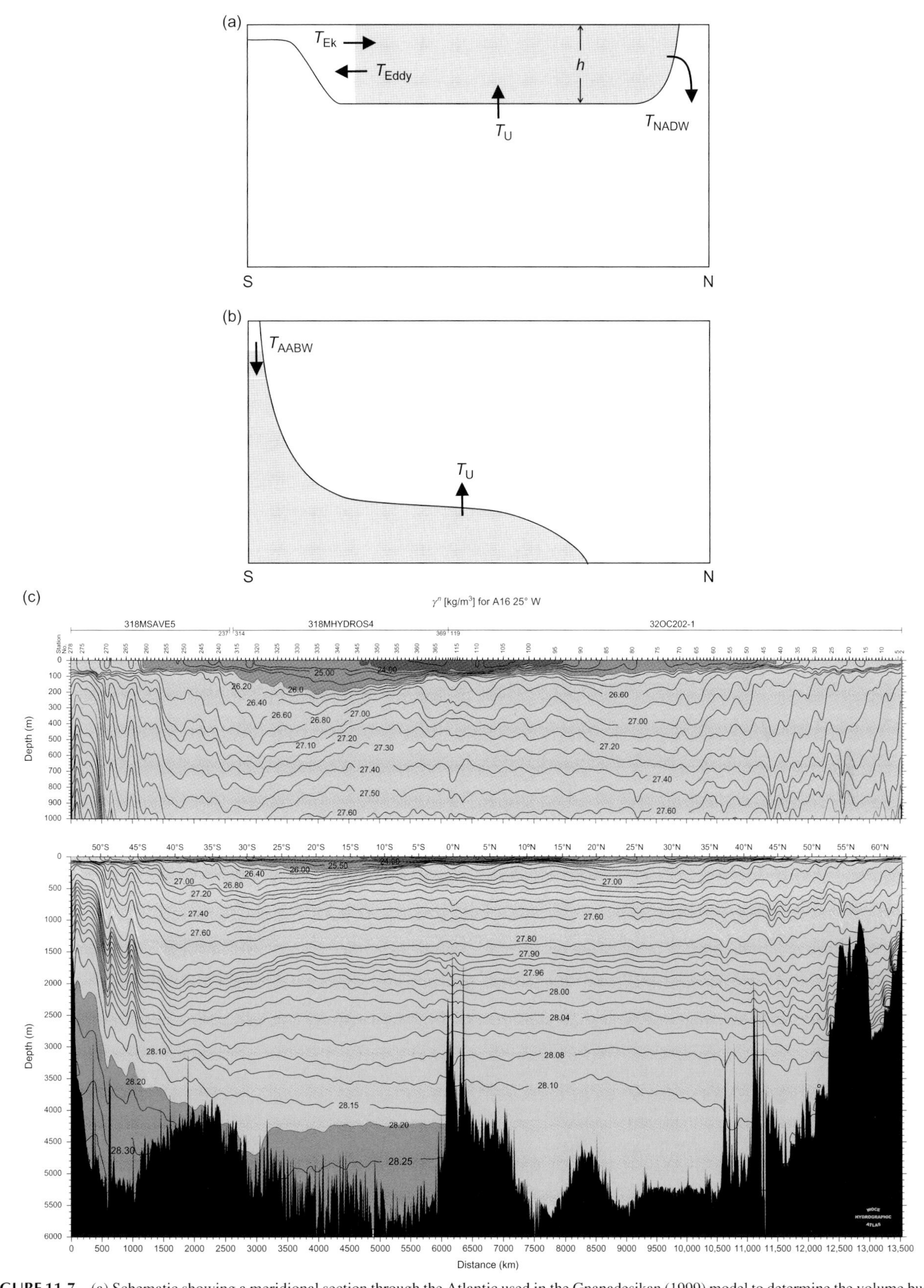

FIGURE 11.7 (a) Schematic showing a meridional section through the Atlantic used in the Gnanadesikan (1999) model to determine the volume budget of the surface pycnocline layer (shaded) and overturning circulation associated with the NADW cell; the transports and layer thickness are as defined in the text. (b) Similar models can be constructed for other water masses such as AABW (Nikurashin and Vallis, 2011). (c) Neutral density γ^n along the WOCE A16 section. *Panel (c) source: WOCE atlas, see http://www-pord.ucsd.edu/whp_atlas/.*

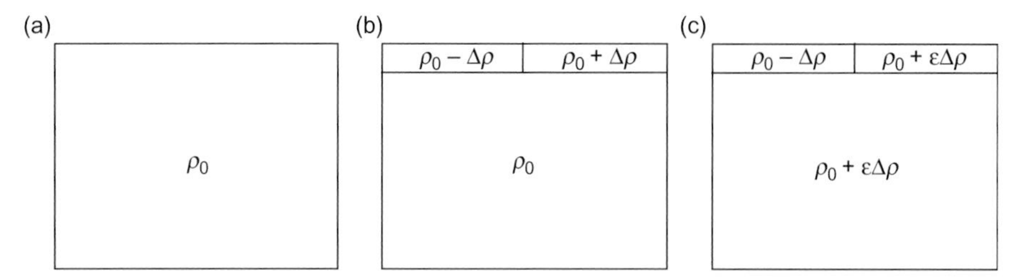

FIGURE 11.8 Schematic diagram illustrating three oceans with the same total potential energy but different available potential energy. (a) Uniform density, ρ_0; (b) two small volumes at the surface are made less and more dense by an amount $\Delta\rho$; (c) one of the small surface volumes is made less dense by an amount $\Delta\rho$ and the remainder of the ocean is made denser by an amount $\varepsilon\Delta\rho$ where $\varepsilon = h/(2H - h)$ where h and H are the depth of the surface box and total ocean depth. In (a) there is no available potential energy, in (b) the available potential energy is large, and in (c) the available potential energy is small. Note, however, that (b) is convectively unstable and not sustainable in equilibrium. *Adapted from Hughes et al. (2009).*

3.2. Role of the Southern Ocean and Relation to the Antarctic Circumpolar Current

A success of the Gnanadesikan model is that it captures both the classical diffusive limit of Munk (1966), in which deep water formation is balanced by diffusive upwelling as sketched in Figure 11.9a and the quasi-adiabatic pole-to-pole circulation limit (Klinger et al., 2003; Wolfe and Cessi, 2009) sketched in Figure 11.9b, in which Southern Ocean winds exert a strong control on the strength of the THC (Toggweiler and Samuels, 1995; Tsujino and Suginohara, 1999; Klinger et al., 2003). The latter is perhaps more realistic given the low levels of diapycnal mixing observed in the ocean interior (Ledwell et al., 1993).

Within the Southern Ocean, the quasi-adiabatic circulation is aligned along the isopycnals, at odds with the Eulerian-mean *Deacon cell* that results from closing the equatorward Ekman transport with a poleward geostrophic flow in abyssal layers (Section 2.1). The resolution of this paradox involves recognizing that water masses are advected by the sum of the Eulerian-mean and eddy-induced circulation (Danabasoglu et al., 1994), the latter arising through correlations in anomalies in the velocity and isopycnal layer thickness (Gent et al., 1995). The residual overturning circulation is quasi-adiabatic in the ocean interior, but crosses isopycnals within the surface mixed layer, at a rate controlled by the surface buoyancy forcing (Marshall, 1997; Marshall and Radko, 2003).

Marshall and Radko (2003) described the upper and deep Southern Ocean overturning circulation using residual mean theory. This view was extended by Wolfe and Cessi (2011) to model the (adiabatic) pole-to-pole MOC. In this case the MOC is solely forced by buoyancy fluxes at the surface, in combination with up- and downward Ekman pumping. Wolfe and Cessi (2011) obtained good agreement with eddy-resolving simulations. Ito and Marshall (2008) applied the same theory to the deep MOC associated with AABW formation and export. They derived a scaling for the diapycnal mixing rate needed to allow a transport of AABW across isopycnals that outcrop from the ocean floor.

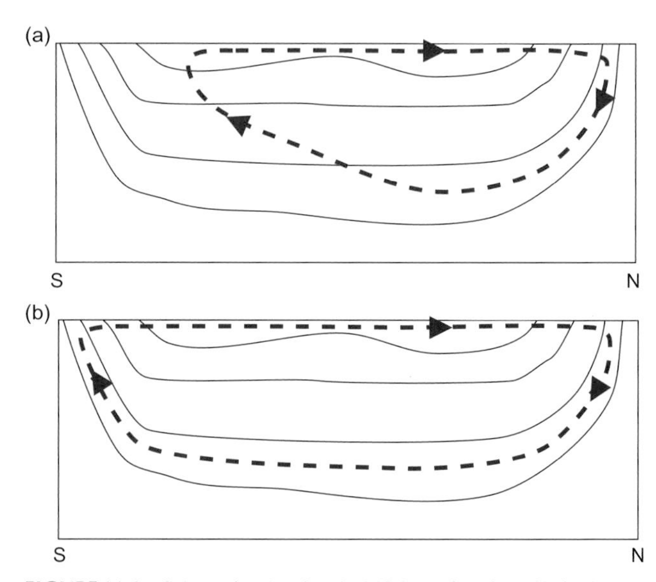

FIGURE 11.9 Schematics showing the MOC as a function of latitude and depth in the (a) diffusive and (b) adiabatic regimes. The solid lines are neutral density surfaces. See the main text for further details.

This scaling was combined with eddy-permitting simulations to predict an increase in AABW export in response to strengthening of the mid-latitude westerlies (Saenko et al., 2012). Nikurashin and Vallis (2012) extended the model of the diffusively driven deep overturning associated with AABW export to include a quasi-adiabatic upper cell associated with NADW export. Their model agreed qualitatively with coarse-resolution simulations using parameterized eddy transports. A further step has been taken by Stewart and Thompson (2013) who employed residual mean theory as a model of both overturning cells, now including an idealized Antarctic continental shelf, in concert with idealized eddy-resolving simulations of a sector of the Southern Ocean. In this way, they connected the MOC across the Antarctic Circumpolar Current (ACC) with the overturning across the Antarctic Slope Front, a weaker westward current that flows along the continental slope around most of Antarctica. By connecting the overturning in the ACC with that over the Antarctic continental shelf, the influence of shelf processes on the global MOC could be analyzed.

To the extent that most isopycnals outcrop in the Southern Ocean, the Gnanadesikan model also provides a useful pedagogical model of the ACC, since the depth of the global pycnocline is related to the strength of the ACC through thermal wind balance (Gnanadesikan and Hallberg, 2000; Allison et al., 2010). Thus, contrary to the traditional view that the ACC is controlled by *local* wind, buoyancy forcing and eddies (Rintoul et al., 2001), the ACC volume transport can also be increased by decreasing the rate of NADW formation (Fučkar and Vallis, 2007; Saenz et al., 2012) and by increasing the global diapycnal mixing (Munday et al., 2011). Finally, we note that the equilibrium of both ACC and MOC in the Southern Ocean exhibit far less sensitivity to the Southern Ocean wind stress in models with explicit, rather than parameterized, eddies (e.g., Hallberg and Gnanadesikan, 2006; Farneti et al., 2010; Abernathey et al., 2011; Munday et al., 2013), consistent with the idea of Straub (1993) that the Southern Ocean is close to a critical threshold for baroclinic instability.

3.3. Water Mass Formation

The deep ocean is filled with waters that sink along the Antarctic continental slope and overflow sills in the Nordic Seas (Mauritzen, 1996; Orsi et al., 2001). The only place where open ocean convection may directly feed the deeper ocean is in the Labrador Sea, but Labrador Sea convection is now thought to contribute only a small amount to the deep branch of the THC (Pickart and Spall, 2007). Both the Nordic Seas and Labrador Sea can be associated with an interior convective region, surrounded by a cyclonic boundary current. Deep water is provided by densification of the boundary current, where most of the sinking occurs, instead of in the interior convective region (Spall, 2004; Iovino et al., 2008). A useful conceptual model of these processes has been developed by Straneo (2006). The boundary current is made denser by surface buoyancy loss and exchanges with the interior by (sub)mesoscale eddies (Figure 11.10). So, eddy exchanges densify the boundary current and restratify the interior convective column (Gelderloos et al., 2011). An important consequence of decoupling the locations of sinking and densification (Spall and Pickart, 2001) is that the overturning streamfunctions in depth and density coordinates are very different at subpolar latitudes.

In the Southern Ocean, bottom water is formed on the shelves by surface cooling in polynyas (semi-permanent areas of open water in sea ice), brine rejection by freezing sea ice and basal melting and freezing under ice shelves and glacier tongues (Jacobs, 2004). Bottom water formation is likely enhanced by tidal stirring (Whitworth and Orsi, 2006) and upwelling of Circumpolar Deep Water driven by cascading bottom water in submarine canyons that connect the shelves and slopes around Antarctica (Kämpf, 2005). Dense water formation on the Arctic shelves may play a similar role as on the Antarctic shelves (Hansen and Østerhus, 2000), but the relation between Arctic bottom water and the overflow across the Greenland

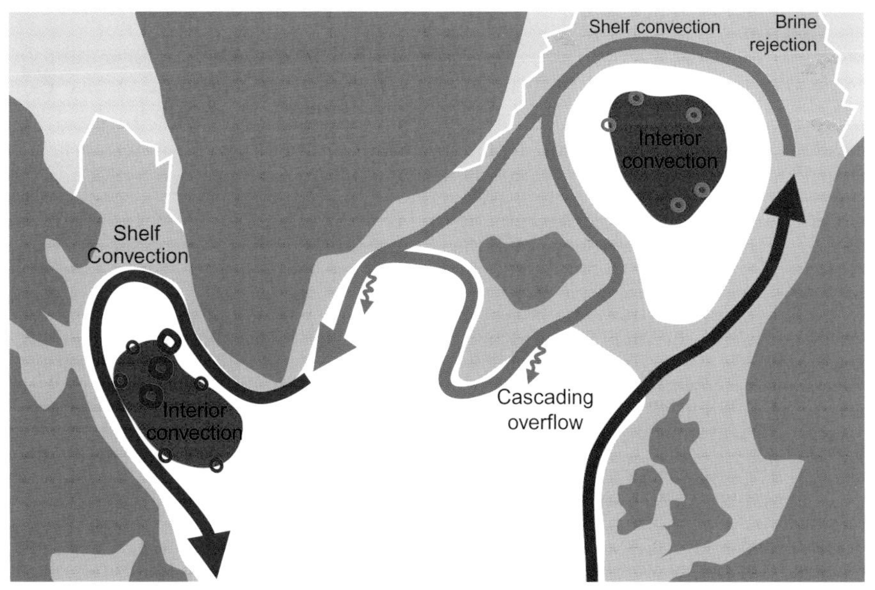

FIGURE 11.10 Schematic showing the myriad of processes that increase the density of the boundary current as it navigates around the margin of the subpolar North Atlantic and Greenland-Norwegian Sea. Sea ice is indicated by the cyan shading and shallow bathymetry by gray shading. See the main text for further details of the processes involved.

Scotland ridge appears to be weak (Eldevik et al., 2009). After sinking along the continental slopes and cascading from overflow sills, deep and bottom water is carried by slope currents which entrain overlying, less dense water and increase their volume flux significantly (Price and Baringer, 1994). Eddy resolving models have made great progress in explicitly simulating many of the processes involved (see Figure 11.10 for a schematic of these processes), but the treatment of diapycnal mixing still remains a major challenge (Legg et al., 2008). Continued development of conceptual models is required to better understand these water-mass formation processes and their impact on the large-scale circulation.

3.4. Three-Dimensional Structure of the THC

The classical theory for the lateral structure of the abyssal circulation is due to Stommel and Arons (Stommel, 1958; Stommel and Arons, 1961a,b). This theory is based on the following assumptions: (i) deep water is supplied by convection at a few, highly-localized locations; (ii) the abyssal ocean can be modeled as a slab of uniform thickness; (iii) abyssal water upwells at a known rate to balance the localized deep water formation; (iv) the interior abyssal circulation is in geostrophic balance. The resultant linear vorticity balance and vortex stretching mean that the meridional velocity is everywhere poleward and so, to resolve the paradox of fluid flowing toward its source, Deep Western Boundary Currents (DWBCs) are required to close the circulation.

There are however many limitations to the Stommel and Arons model. Elements neglected include non-uniform upwelling (Marotzke, 1997; Polzin et al., 1997), a layered structure that allows for the distinction between different water masses, geographically varying bottom topography (Stephens and Marshall, 2000), stratification that distorts the β effect, and forcing from eddies and waves. Nevertheless, while the structure of the poleward interior flows might be questioned, the idea that DWBCs are required and, indeed, dominate the abyssal circulation is the most valuable and robust result.

The predicted DWBC in the North Atlantic was shortly thereafter "discovered" using neutral floats (Swallow and Worthington, 1961), although the measurement was highly fortuitous due to the turbulent nature of the flow (Wunsch, 1996)! Nevertheless the existence of DWBCs is not in doubt, due mainly to transient tracer observations. The DWBC is the only example of a major ocean current having been predicted on theoretical grounds before it was observed.

In an attempt to allow for more complexity, Kawase (1987) developed a model that included coupling between the strength of the upwelling and the abyssal flow, and introduced baroclinicity as observed by the opposing flow of water from northern and southern polar sources. This model shows that the baroclinic dynamics of the large-scale Rossby waves are important in understanding the abyssal circulation. Another aspect of more complex dynamics being important to the abyssal flow is the existence of large-scale recirculation gyres (McCartney, 1992; Schott et al., 2004; McCartney, 2007). These recirculation gyres can be driven by eddy fluxes and cross-isopycnal mixing (Spall, 2000), and by pressure torques associated with bottom topography (Straub and Rhines, 1990).

A problem of particular interest with regard to the spreading of the abyssal flow is how the deep flow crosses the equator and how the potential vorticity budget is affected when the Coriolis parameter changes sign. Edwards and Pedlosky (1998) proposed mean flow instability and eddy generation near the equator as the main mechanisms. There, deep equatorial jets are found to interact with the DWBC (Richardson and Fratantoni, 1999). Instability of the DWBCs occurs in many other places, most noticeably at 8°S in the Atlantic, where the whole DWBC breaks up in eastward propagating eddies (Dengler et al., 2004). However, the role of dissipation, inertia, and mean flow instabilities in shaping the cross-equatorial deep flow is still not well established (Stephens and Marshall, 2000).

The upwelling branch of the THC is difficult to observe because it is spread out over all ocean basins, associated with weak upward motion everywhere. A Lagrangian calculation of upwelling of what originally was NADW in an eddy permitting ocean model (Döös and Coward, 1997) emphasizes upwelling on the poleward edge of the ACC. An upwelling fluid parcel traces a helical path, progressing upward and poleward while circuiting around Antarctica, until it reaches the surface Ekman layer where it is transported equatorward by the Ekman flow (Drijfhout et al., 2003). Most of the return flow is then transformed to Subantarctic and Intermediate Mode Water (Sloyan and Rintoul, 2001), forming the Southern Ocean limb of the NADW overturning cell. The densest part follows the Tasmanian route (Speich et al., 2002), after which it crosses the Atlantic equator (Drijfhout, 2005). Lighter waters are further transformed at the surface into subtropical thermocline water. The same type of diagnosis also suggests that both thermocline and intermediate waters enter the Atlantic via Agulhas leakage (Drijfhout et al., 2003), with a minor part following the direct path of the cold water route (Rintoul, 1991).

The wind-driven flow has a strong imprint on the pathways of the upper branch return flow. An illustration of the coupling between wind-driven and thermohaline flow is given by an arbitrary trajectory involved in the NADW overturning cell (Figure 11.11). This can be

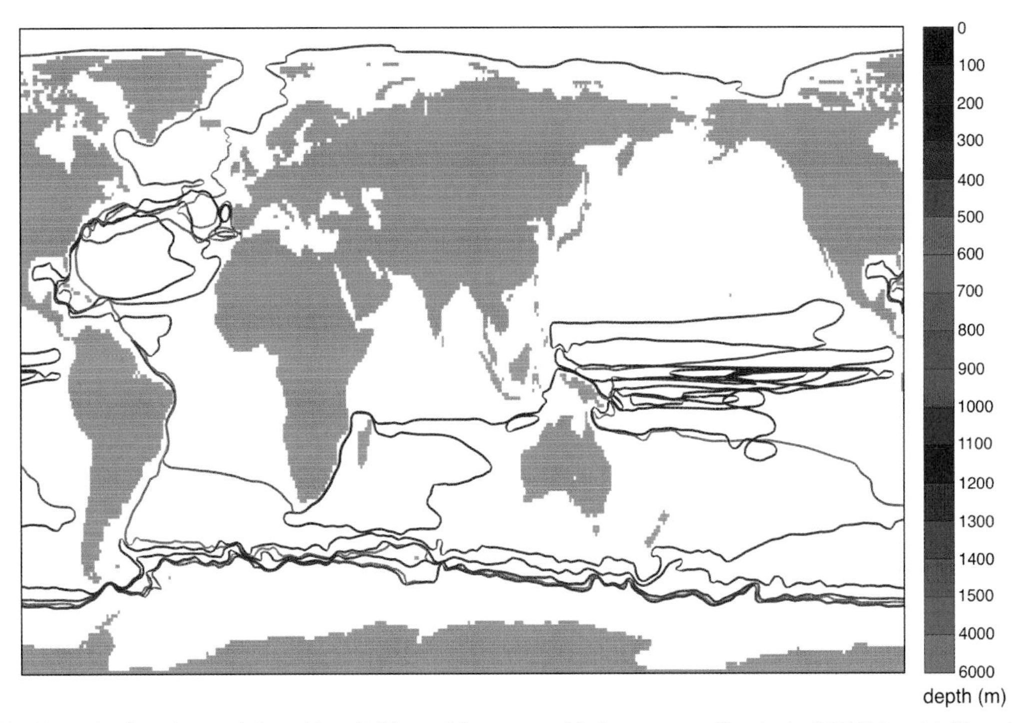

FIGURE 11.11 Example of a trajectory followed by a fluid parcel that starts as Mediterranean outflow in the OCCAM model. The color indicates the depth (m) of the fluid parcel at each point along the trajectory. See the main text for further details (Drijfhout et al., 2003).

contrasted with the traditional "conveyor-belt" picture from Gordon (1986) and Broecker (1991). The trajectory also underscores the role of the ACC in the THC. The isopycnal slope across the ACC determines the Southern Ocean eddy transport. A strong mean flow overturning recirculation (Deacon Cell) is canceled by transient eddies, tilting gyres, and stationary meanders in the ACC (Döös and Webb, 1994; Drijfhout, 2005). A residual diabatic Deacon Cell results (Speer et al., 2000), which is the Southern Ocean limb of the meridional overturning cell.

The overturning cell associated with AABW upwells in the interior in a deep to intermediate depth poleward return flow in all three ocean basins (Schmitz, 1995). Part of the overturning cell associated with NADW also upwells in the interior of all ocean basins but the majority is now thought to upwell to the surface in the Southern Ocean.

3.5. Feedbacks and Multiple Equilibria

Stommel (1961) developed a box model in which the opposing influences of thermal and freshwater boundary conditions result in multiple equilibrium states. Subsequently it was recognized that this model might be relevant to the THC, potentially explaining the abrupt transitions in climate and ocean circulation suggested by palaeodata (Clark et al., 2002; Alley et al., 2003). Investigating whether the Stommel mechanism is relevant for the global ocean stimulated Rooth (1982) to extend Stommel's box model, and inspired Bryan (1986) to confirm that multiple equilibria can be obtained in an ocean general circulation model.

The Stommel box model consists of well-mixed boxes with a circulation between the boxes that scales linearly with their density difference, representing the concept that THC scales with the meridional pressure gradient (e.g., Griesel and Morales Maqueda, 2006; de Boer et al., 2010). As noted by Straub (1996), this assumption is dynamically inconsistent with the Stommel and Arons model discussed in Section 3.4; indeed geostrophic balance suggests that the THC should scale with the zonal, rather than meridional, pressure gradient. The precise relation between the THC and meridional density gradients remains a topic of ongoing research, but recent work suggests that it is necessary to consider separately the dynamics of the western boundary layer and basin interior/eastern boundary (Wright et al., 1995; Brüggemann et al., 2011; Callies and Marotzke, 2012); also see the discussion in Marshall and Pillar (2011). Note that irrespective of the preceding issue, Stommel's model is mechanically-driven in the sense that mixing is required to maintain the meridional density and pressure gradients.

Vallis (2000); Wolfe and Cessi (2010); Cimatoribus (2013) have attempted to reconcile geostrophy with the north–south density difference scaling. The strength of the deep branch of the AMOC is determined by the east–west density difference, through thermal wind balance, with the densest subducted water flowing in the DWBC, and the least dense water at the eastern boundary. If the basin is connected to a periodic channel mimicking the Southern Ocean then, for adiabatic flow, the density range of subducted water maps onto the density range of water upwelling in the channel, with the lightest water upwelling at the northern tip of the channel. The water that upwells near the tip of Drake Passage is then the same water flowing along the eastern boundary of the deep branch of the AMOC. As a result, the east–west density difference across the deep branch of the AMOC must be closely related to the density difference along the west coast of the Atlantic basin, that is, the density difference between the water near Newfoundland and the water just north of Cape Horn. There are subtleties involved in taking the correct vertical average at those locations, and extending this argument to multiple basins, but with reasonable metrics one will always find a close correspondence between the two density differences (Cimatoribus, 2013).

This argument also reflects the competition between AAIW and NADW, because AAIW subducts just north of where the least dense NADW class upwells. If AAIW becomes denser, the density of the least dense class of NADW flowing southward in the North Atlantic has to adjust (increase), decreasing the east–west density difference across the deep branch of the AMOC. The reason is that all water taking part in NADW that is lighter than the water upwelling near Drake Passage has to upwell diffusively within the Atlantic basin and can no longer be part of the lower branch of the AMOC. If AAIW becomes as dense as the densest class of NADW, the east–west density difference across the deep branch of the AMOC becomes zero and the AMOC effectively collapses. It should be noted that the AMOC responds immediately to changes in east–west density difference, while changes in the north–south density difference involve a century-scale adjustment time. However, the north–south density difference across the whole Atlantic eventually sets the east–west density difference across the deep branch in the northern hemisphere, underscoring the essential role for Southern Ocean processes in affecting the AMOC.

A simpler, more pedagogical formulation of the Stommel's model (Figure 11.12) was developed by Marotzke (2000). He prescribed fixed temperatures and fixed freshwater fluxes for the two boxes to enable analytical solutions. For moderate freshwater forcing, there are two stable and one unstable equilibria. The first stable equilibrium solution corresponds to a fast circulation in which freshwater forcing has insufficient time to create a significant salinity gradient, and hence the density gradient is dominated by the prescribed temperature gradient, with sinking in the high-latitude box. In contrast, the second stable equilibrium solution corresponds to a slow circulation in which freshwater forcing has sufficient time to create a substantial salinity gradient that overcomes the prescribed temperature gradient such that sinking occurs in the low-latitude box.

In a climate change scenario in which freshwater forcing increases, a critical threshold occurs, beyond which the high-latitude sinking equilibrium solution ceases to exist through a saddle-point bifurcation. The mechanism by which the flow makes transitions from one to another equilibrium is the salt advection feedback. This positive feedback consists of advection of mean salinity gradients by the perturbation flow. A freshwater anomaly in the polar box leads to decreased flow through a reduction in the pressure difference, and hence to reduced salt advection from the equatorial box such that the initial anomaly is amplified.

Since the 1990s, many authors have proposed extensions to the original Stommel model, for example, focusing on the coupling to atmospheric feedbacks (Nakamura et al., 1994; Tziperman and Gildor, 2002), sea-ice feedbacks

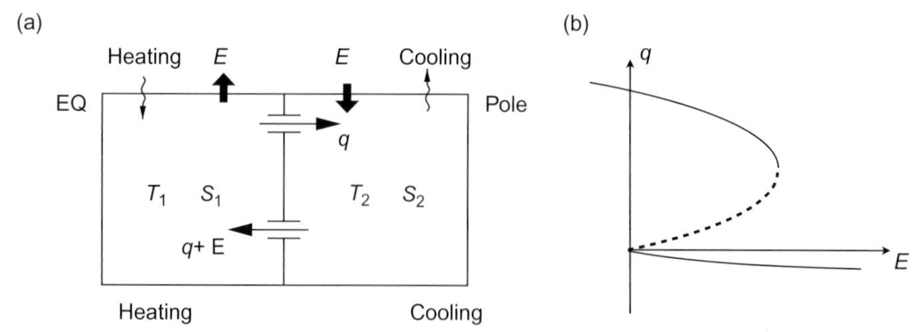

FIGURE 11.12 Schematic diagram showing the two-box model of Stommel (1961). (a) The boxes are well-mixed with temperatures and salinities T_1, T_2 and S_1, S_2. A circulation, strength q, flows through pipes connecting the boxes. Buoyancy forcing is through restoring thermal boundary conditions and a prescribed freshwater transport (E) through the sea surface. (b) Bifurcation diagram for the stable (solid lines) and unstable (dashed line) equilibrium solutions for circulation strength, q, as a function of freshwater transport, E.

(Jayne and Marotzke, 1999), wind feedbacks (Longworth et al., 2005) and energy constraints on mixing (Guan and Huang, 2008). Some feedbacks are stabilizing (wind, temperature-precipitation), some destabilizing (sea-ice, atmospheric eddy transports), but in all cases multiple equilibria remain possible for a wide range of parameters.

The results from box models on multiple equilibria have been extended to zonally-averaged models (Marotzke et al., 1988; Wright and Stocker, 1991) and three-dimensional models of different degrees of idealization (Bryan, 1986; Manabe and Stouffer, 1988; Marotzke and Willebrand, 1991; Rahmstorf, 1995). Dijkstra and Weijer (2003) showed that multiple equilibria of the THC in a hierarchy of models are connected, ranging from box models to coarse resolution general circulation models. In all cases the salt advection feedback associated with the perturbation flow is the main actor in creating the multiple equilibrium regime (Huisman et al., 2010).

To reconcile the Gnanadesikan model with the existence of multiple equilibria associated with the different boundary condition for temperature and salinity, Johnson et al. (2007) incorporated the Stommel (1961) equations into the Gnanadesikan (1999) model. In this formulation the THC remains mechanically driven, but thermohaline feedbacks allow for multiple equilibria (Fürst and Levermann, 2012). Cimatoribus et al. (2013) extended the model further and found two regimes for the thermally-driven branch of the THC: one that is primarily driven by Southern Ocean winds and one that is primarily driven by meridional density gradients.

In global ocean models, the Atlantic overturning also reverses when the solution switches from the thermal to the salinity driven circulation (Dijkstra, 2007). The reversal in circulation implies that AAIW has become denser and flows deeper than NADW (Sijp and England, 2006). This result is in line with Klinger and Marotzke (1999), who show that small density differences between the northern and southern hemispheres determine strong asymmetries in global overturning. The competition between northern and southern hemispheres is essentially between AAIW and NADW, with NADW being associated with the dominant overturning cell in the present climate. AABW and NADW overturning cells coexist with almost equal strength, implying another type of competition between those two overturning cells: if one cell increases in strength the other cell decreases (Cox, 1989; England, 1993), but their interrelation seems more linear than the competition between NADW and AAIW. There is no evidence that multiple equilibria are associated with the AABW overturning cell (Trevena et al., 2008). This is probably due to the fact that temperature decreases and salinity increases with latitude southward. In the Southern Ocean salinity is determined more by sea ice formation than by precipitation. As a result, both the thermal and haline components act together.

Despite the presence of multiple equilibria across a wide hierarchy of models, it is important to stress that multiple equilibria have yet to be found in a state-of-the-art, eddy-permitting ocean general circulation model. In part this may be due to biases in the salt-advection feedback in such models, as discussed in the next section.

3.6. Does the South Atlantic Determine the Stability of the THC?

We conclude with a discussion of some recent results suggesting a pivotal role for the South Atlantic in controlling THC stability. While a clear consensus on this topic is yet to emerge, it nicely illustrates the value of conceptual models in two respects: (i) new ideas can lead to surprising conclusions that force us to revise our views as to how the ocean works; (ii) most state-of-the-art climate models seem to misrepresent the pertinent South Atlantic feedback, potentially biasing these models toward an overly stable THC.

Rahmstorf (1996) studied a simple box model in which stability of the THC turns out to be controlled by the freshwater transport from the "Southern Ocean" into the "Atlantic." de Vries and Weber (2005) extended this idea to a multi-basin configuration, making a distinction by overturning and gyre transport of freshwater. They argued that net evaporation from the Atlantic to the Pacific has to be compensated by freshwater import by overturning and gyre circulation across the southern boundary of the Atlantic (Figure 11.13). Connecting the freshwater budget to Stommel's advective salt feedback implies that multiple equilibria can only exist when the overturning circulation exports freshwater from the Atlantic. The implication is that multiple equilibria exist only when the upper branch of the overturning is saltier than the NADW that leaves the Atlantic. It must be dominated by thermocline water from the Indian Ocean. When fresh, intermediate water becomes more important, the THC is monostable and the advective salt feedback is damping.

de Vries and Weber (2005) assumed that enhancing the freshwater transport by the gyre, M_{az}, would cause an equally large decrease in freshwater transport by the overturning circulation, M_{ov}. This is the case since the overall evaporation over the Atlantic remains unaffected. They determined two almost similar present-day climate states but with a different sign of M_{ov}. After a large freshwater perturbation, the THC with northward salt transport remains collapsed, while the THC with southward salt transport recovered. The different behavior of the THC can be associated with a salinity anomaly that develops at the southern boundary of the Atlantic (Weber et al., 2007), the sign of which is determined by M_{ov}. This salinity

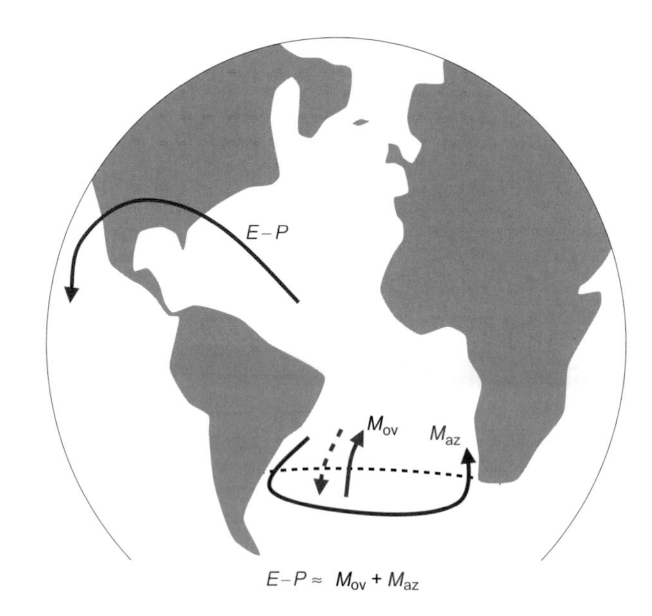

FIGURE 11.13 Schematic diagram illustrating the freshwater budget of the Atlantic. In equilibrium, the net freshwater transport into the South Atlantic by the gyre, M_{az}, and overturning circulation, M_{ov}, must be balanced by the net evaporation minus precipitation, $E-P$ (the latter including any run-off from rivers). As discussed in the text, M_{ov} may play a central role in controlling the stability of the THC.

anomaly spreads northward and restarts convection if it is positive, but not if it is negative.

Dijkstra (2007) has computed bifurcation diagrams for an implicit ocean model and found that the first bifurcation point that marks the boundary of the multiple equilibrium regime almost coincides with a sign change in M_{ov}; furthermore, M_{ov} decreases monotonically along the thermally driven branch until the point where the branch terminates. Huisman et al. (2010) and Cimatoribus et al. (2013) have found further proof for the crucial role of M_{ov} in THC stability.

To underscore the role of M_{ov}, Cimatoribus et al. (2012) performed an experiment in which M_{ov} was gradually decreased by applying a slowly increasing freshwater dipole in the southern South Atlantic. Freshwater import by the gyre slowly increased, freshwater transport by the overturning decreased. They found that the THC indeed can collapse without freshwater input in the northern North Atlantic, because the basin-averaged salt advection feedback becomes so strong that the highly variable atmospheric forcing provides enough noise to cross the threshold between the two states, challenging the traditional North Atlantic view of the THC and its stability.

A number of coupled climate models were analyzed with respect to this feedback in Weber et al. (2007) and Drijfhout et al. (2010). They found consistent positive M_{ov}, implying these models would not possess a stable haline driven state, while estimates based on observations point to a negative M_{ov} (Weijer et al., 1999; Bryden et al., 2011). This suggests that a bias exists in many coupled climate models with respect to the stability of the THC: most models possess a South Atlantic thermocline that is overly fresh, with southward salt transport by the overturning circulation, whereas observations suggest northward salt transport. For the newest generation of climate models the bias appears to be less (Weaver et al., 2012).

Another point is that atmospheric feedbacks may counteract the oceanic salt advection feedback, possibly degrading the role of M_{ov} as an indicator of the stability of the THC. Yin and Stouffer (2007) discussed two versions of the GFDL climate model, with the older R30 version showing a stable haline-driven state, but not the newer CM2.1 version. They attributed the difference to the more pronounced atmospheric feedbacks in the CM2.1 model. On the other hand, the R30 model featured negative M_{ov}, while the CM2.1 model featured positive M_{ov}. To study the impact of atmospheric feedbacks on the stability of the THC more systematically, Cimatoribus et al. (2011) diagnosed these by linear regressions to local and hemisphere-wide SST variations in a coupled climate model. The regression functions were used to isolate the impact of each atmospheric feedback on the bifurcation structure of the THC. den Toom et al. (2012) found that atmospheric feedbacks indeed reduce the width of the multiple equilibrium regime but cannot eliminate it. A cautionary note has to be made, however, since the coupled model used by Cimatoribus et al. (2011) is relatively simple and may underestimate changes in the Intertropical Convergence Zone and El Niño/La Niña cycles.

Hysteresis experiments in coupled models sometimes indicate a more complicated relation between M_{ov} and the existence of multiple equilibria. Hawkins et al. (2011) showed that bistability already occurred for positive M_{ov}. They did confirm, however, the decrease of M_{ov} along the thermally driven branch, leading to decreased stability. Consistent with this feature, Liu et al. (2012) argued that a better indicator would be $\partial\Delta M_{ov}/\partial\Psi$, with Ψ being the overturning strength and ΔM_{ov} being the difference in M_{ov} at 35°S and at 80°N. While this indicator better resembles a true feedback term, it is also more complicated to diagnose. A similar feedback term was proposed in Sijp et al. (2012), where the derivative of M_{ov} to salinity was used. They found that the Atlantic either resides in a salty or in a fresh state, with intermediate salinities being unstable. Crossing the salinity-threshold leads to a transition from a salty ocean and thermally driven overturning to a fresh ocean and a haline-driven overturning, and vice-versa.

4. TRANSIENT BEHAVIOUR OF THE WIND-DRIVEN AND THERMOHALINE CIRCULATION

In the previous two sections, we focused on the (statistical) mean properties of the ocean circulation. However, the ocean responds both to transient surface forcing and generates its own intrinsic variability and hence displays strong variability on subannual to centennial time scales. The null hypothesis for all this variability is that it is a response of the slower ocean system to fast atmospheric variability (Hasselmann, 1976). However, processes such as ocean adjustment to forcing and specific instabilities may cause high energy into certain frequency bands of variability.

The ocean responds to temporally changing wind patterns through adjustment. On time scales from months to years, this adjustment is associated with Rossby waves. The classical picture of adjustment to wind stress changes at midlatitudes was developed by Anderson and Gill (1975) using a multi-layer quasi-geostrophic model. The basin crossing times of long westward traveling Rossby waves determine the time scale of the adjustment. When these waves hit the western boundary, they reflect into short Rossby waves which participate in building up the western boundary layer.

Satellite measurements of sea surface height have detected westward phase propagation consistent with baroclinic Rossby waves (Chelton and Schlax, 1996). The phase speeds found, however, are faster than in classical theory by roughly a factor of two at mid-latitudes and even more at high latitudes. Explanations of this discrepancy have been sought in Rossby wave phase speed enhancement by mean flow advection (Killworth et al., 1997), potential vorticity homogenization (Dewar, 1998), bottom topography (Tailleux and McWilliams, 2001) and lateral boundaries (LaCasce, 2000).

In a closed basin, the response to a changing wind forcing can be decomposed into classical Rossby basin modes (Pedlosky, 1987). In LaCasce (2000) and Cessi and Primeau (2001), low-frequency basin modes were found that are only weakly affected by dissipation. These modes involve interactions between Rossby waves in the interior and eastern boundary pressure oscillations. Rather than having small horizontal scales (as occurs in the absence of boundary oscillations), the modes have basin-wide scales. The periods are determined by the long wave Rossby wave transit time and hence are in the order of decades for midlatitude basins.

The ability of these modes to be resonantly excited depends on the efficiency with which energy fluxed onto the western boundary can be transmitted back to the eastern boundary (Primeau, 2002a). The boundary pressure feedback on the interior flow forms a mechanism through which the interior circulation at one latitude induces flow in a remote (unforced) region, such as another hemisphere or another ocean basin (Cessi and Otheguy, 2003; Cessi, 2004). The basin modes can be strongly deformed by the mean flow and for example, strong gyres lead to higher frequencies and spatial localization of variability (Ben Jelloul and Huck, 2005).

Johnson and Marshall (2002) use a reduced gravity model to study the adjustment of the upper, warm limb of the THC to anomalous forcing (Figure 11.14). Changes

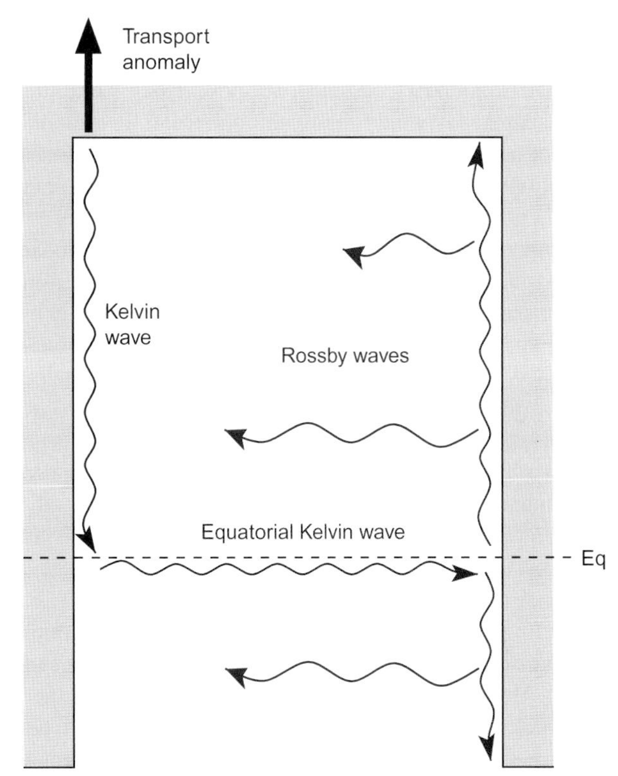

FIGURE 11.14 Schematic diagram illustrating the adjustment of the upper ocean to a deep water formation transport anomaly as described by Johnson and Marshall (2002). Initially a Kelvin wave propagates along the western margin of the basin (blue), reaching the equator within a few months. The signal then continues as an equatorial Kelvin wave and an eastern boundary Kelvin wave (red). The latter radiates westward-propagating Rossby waves into the basin interior (green). Further details are given in the text.

in deep convection initiate boundary Kelvin waves that propagate in a few months along the western boundary to the equator. Kelvin waves along the equator and eastern boundary take the signal eastward and northward, respectively. Eastern boundary deformations of the pycnocline lead to an interior adjustment by Rossby waves, similar to the processes determining the Rossby basin modes. Only a small fraction of the change in THC strength is communicated across the equator to the Southern Hemisphere. The equator hence acts as a low-pass filter (equatorial buffer) to MOC anomalies and MOC changes on time scales up to decadal are confined to the hemispheric basin in which they were generated (Johnson and Marshall, 2004).

In addition to basic adjustment processes, ocean flows generate internal variability through instability processes. Zonal currents can become unstable to barotropic and baroclinic instability (Pedlosky, 1987) giving rise to eddies with typical time scales of a few months. Rossby basin modes can also become unstable (LaCasce and Pedlosky, 2004) and poleward of a certain latitude, baroclinic wave components will break down into small-scale eddies with a barotropic component. These features also propagate westward but at faster speeds than their baroclinic counterparts (Isachsen et al., 2007).

Western boundary currents can show strong variability, even under steady wind forcing, through mean flow modification of barotropic and baroclinic instabilities and nonlinear interactions of unstable modes. The so-called double-gyre flow (Cessi and Ierley, 1995; Jiang et al., 1995) serves as a prototype to understand the variability in western boundary currents due to nonlinear processes. Multiple equilibria (Dijkstra and Katsman, 1997; Primeau, 2002b) may cause regime switches (Sura et al., 2001). Gyre modes (Simonnet and Dijkstra, 2002) may introduce oscillatory low-frequency (decadal time scale) variability in these flows and are, for example, thought to be responsible for path variations in the Kuroshio in the Pacific (Pierini et al., 2009). Strong nonlinear interaction between baroclinic eddies can also lead to low-frequency variability in a mechanism termed the turbulent oscillator (Berloff et al., 2007).

Apart from the possibility of multiple equilibria discussed in Section 3, the Atlantic MOC is also susceptible to large-scale oscillatory instabilities (Greatbatch and Zhang, 1995; Chen and Ghil, 1996; Colin de Verdiére and Huck, 1999; Huck and Vallis, 2001; te Raa and Dijkstra, 2002). te Raa and Dijkstra (2003) show that in an idealized single-hemisphere three-dimensional basin, multidecadal and centennial time scale modes are present. Growth of perturbations is caused by a feedback involving spatial correlations between density anomalies and MOC anomalies. The propagation of density perturbations is westward for the multidecadal modes and the time scale is set by the basin crossing time. The time scale of centennial modes arises through a propagation of a buoyancy anomaly along the mean overturning loop. For the global thermohaline ocean circulation, the time scale of these modes becomes millennial (Weijer and Dijkstra, 2003).

The ocean circulation is forced by atmospheric wind stresses and buoyancy fluxes which, in addition to a mean and specific forcing frequency (e.g., the seasonal cycle), have also a strong noise component. In the North Atlantic, spatial correlations of this forcing are related to large-scale atmospheric variability patterns such as the North Atlantic Oscillation (Hurrell, 1995). The temporal characteristics of the noise forcing are considered to be white (Hurrell and Deser, 2010).

As the ocean has a high thermal inertia, it integrates (Hasselmann, 1976) the high-frequency noise components of the atmospheric forcing to give a red noise response (with the properties of a first-order autoregressive process, AR(1)). In fact, such a response is found in observed sea-surface temperature variability (Frankignoul and Hasselmann, 1977; Dommenget and Latif, 2002). Saravanan and McWilliams (1998) considered the effect of ocean memory through advection and showed that enhanced decadal variability can occur due to spatial resonance, that is, the impact of spatially coherent atmospheric forcing on an advective ocean.

The stochastic forcing of an ocean without a memory provided by advection (Figure 11.15a) is considered the "null-hypothesis" of interannual to multidecadal climate

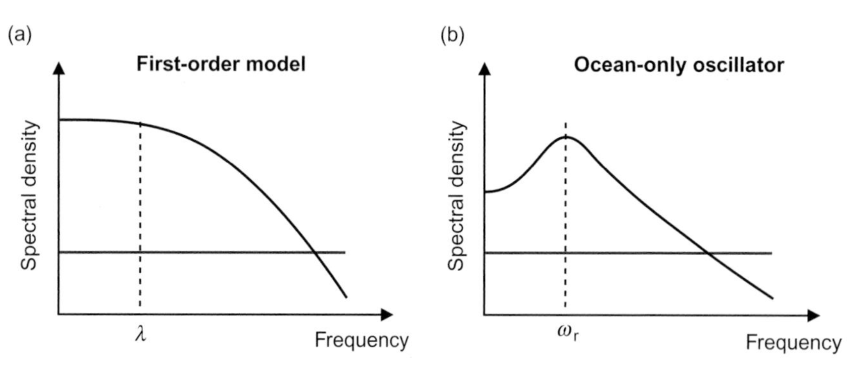

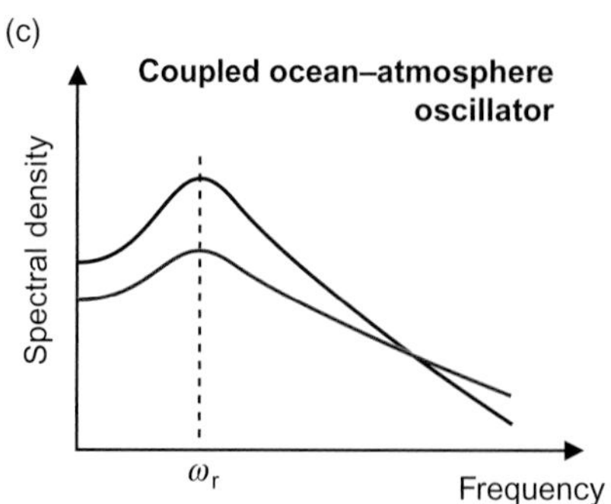

FIGURE 11.15 Spectral density of oceanic (black) and atmospheric (red) variability. (a) Null-hypothesis: a white atmospheric forcing induces a red-noise ocean response where λ is the damping time scale. (b) Presence of ocean memory (through spatial resonance or oscillatory modes) creates enhanced energy at a frequency of ω_r. (c) Coupled interactions between ocean and atmosphere lead to enhanced energy at a frequency of ω_r. *Adapted from a figure by Mojib Latif.*

variability (Latif, 1998). Both spatial resonance and oscillatory modes (such as the Rossby basin modes, gyre modes, multidecadal and centennial modes) may contribute to this variability and lead to specific enhancement of energy at particular frequency bands (Figure 11.15b). The only case known in which ocean-atmospheric interaction is responsible for a mode of variability is the El Niño/Southern Oscillation phenomenon (Figure 11.15c)—see Chapters 15 and 23.

While for the Pacific Decadal Oscillation (Mantua et al., 1997) the null-hypothesis cannot be rejected, oscillatory modes may be involved to create enhanced multidecadal variability in the North Atlantic (Griffies and Tziperman, 1995). Multidecadal oscillatory modes may be effectively excited by atmospheric noise by rectification mechanisms (Frankcombe et al., 2009) and by non-normal growth (Zanna et al., 2010; Sevellec and Fedorov, 2013). Aspects of these oscillatory modes (such as the westward propagation) can be found in general circulation models (Delworth et al., 1993; Dijkstra et al., 2006). However, the response of state-of-the-art models is quite diverse (Danabasoglu, 2008) and it is not clear whether the excited oscillatory mode mechanism is dominant in such models—see Chapter 25.

5. DISCUSSION AND PERSPECTIVE

In this chapter, we have reviewed a number of simple conceptual models of the steady and time-varying wind-driven circulation and THC. Such models can never fully describe the detail and complexity of the ocean but they are invaluable in attempting to describe and understand more complex circulation patterns observed in nature and in state-of-the-art ocean–atmosphere models. As stated in the introduction, conceptual models provide the vocabulary with which we are able to describe complex dynamical balances and circulation patterns in observations and numerical circulation models.

The theory of the wind-driven circulation is at a mature stage of development, with good understanding of what sets the strength and structure of the major wind-driven circulation patterns, as reviewed in Section 2. Nevertheless, some challenges remain, in particular concerning the interactions between western boundary currents and the vertical structure of the wind-driven gyres, as discussed in Section 2.4. Moreover, we have only a basic understanding of the impact of variable bottom topography on the circulation, where we feel that the prevalent JEBAR conceptual model is misleading as discussed in Section 2.5. While it is clear that bottom topography plays a major role in the dynamics of western boundary currents, again we have only the most basic understanding. In particular, we need to move away from the paradigm of dissipation occurring in lateral boundary layers as in classical models of western boundary currents, and move toward the paradigm of dissipation occurring in sloping bottom boundary layers. A related issue that we have not discussed here, but remains poorly understood despite research over several decades, is the dynamics of boundary layer separation (see Chassignet and Marshall, 2008, for a review).

In contrast, while significant progress has been made in understanding some of the basic dynamics of the THC, as reviewed in Section 3, many major challenges remain. For example, we have poor understanding of the processes that set the overall strength of the circulation and when it comes to the possibility of multiple equilibria, there are competing views with no clear consensus—see Chapters 1 and 23 for further discussion. For instance, in the Southern Ocean both the Eulerian mean and eddy-driven overturning cells appear to be largely driven by local winds, but whether the residual overturning is driven by buoyancy forcing in the North Atlantic, or by local (wind, mixing) processes is still under debate. Using the concept of a residual overturning circulation, however, may be the way forward, as well as using energy requirements. In coupled climate models, as well as in higher-resolution ocean-only models (Griffies et al., 2009; Drijfhout et al., 2010), simulation of the THC remains problematic and possibly associated with large biases. For instance, global ocean-sea ice models with an atmospheric forcing based on bulk formulations of the air-sea fluxes exhibit spurious trends in the AMOC. In general, such models have to use sea surface salinity restoring to constrain spurious AMOC trends (Griffies et al., 2009). Sensitivity experiments with small variations in precipitation and sea surface salinity restoring produce a wide range of AMOC transports, ranging from upward drifts to more than 22 Sv to nearly collapsed states with less than 7 Sv (Behrens et al., 2013). The message here is that the AMOC is too sensitive to the advective salt feedback because the counteracting effect of thermal damping is absent in ocean-only models. A way forward could be the inclusion of thermal feedbacks associated with large-scale air/sea interaction (Rahmstorf and Willebrand, 1995; Luksch et al., 1990). This holds for both conceptual models as well as realistic ocean-only models forced with reanalysis products. Such conceptual models of large-scale air/sea interaction could be extended to statistical models based on observed relations between AMOC and weather regimes and atmospheric patterns like the North Atlantic Oscillation. In general, we believe that the use of conceptual models of the coupled ocean/atmosphere, which are widely used for, for example, ENSO, will become increasingly important for other climate phenomena.

Another reason for the relatively poor state of understanding of the THC is that so many of the key dynamical ingredients that form the THC—convection, overflows, deep-western boundary currents, boundary waves, geostrophic eddies, diapycnal mixing, to name a few—occur on spatial scales that cannot, and will not for the foreseeable future, be routinely resolved by numerical ocean circulation models. Thus, conceptual models have a role not only in articulating a detailed dynamical description of the THC, but also in figuring out which processes are most critical to capture in numerical ocean circulation models, and to develop adequate physical parameterizations of those processes that cannot be explicitly resolved.

Rapid progress has been made over the past two decades in understanding the transient behavior of the wind-driven circulation and THC, as reviewed in Section 4. Modes of variability are best investigated in coupled climate models, to allow for two-way feedbacks. At present, coupled climate models show a bewildering variety of modes and timescales (Srokosz et al., 2012). Sometimes, dominant modes of AMOC variability change between two versions of the same model (Danabasoglu, 2008; Danabasoglu et al., 2012), or even between multi-century time slices of the same model (Kwon and Frankignoul, 2011). Apparently, a variety of ocean modes can exist, and which one is the dominant mode sensitively depends on details of the background state. At present, a theory of how AMOC variability depends on ocean characteristics is lacking, and some progress might be expected by theoretical investigations in this direction. Also, decadal prediction attempts, comparing free evolving models with models using data-assimilation, and careful comparison between models and observations, in particular the RAPID-array and the Argo program, will be helpful in not only improving our models, but also increasing our insight in the processes that are instrumental in driving AMOC variability. Because the AABW overturning cell is largely shielded from the atmosphere, more diffusively-driven, and associated with small amounts of heat transport, the variability of this overturning cell has not been studied well, and adequate observations of benchmark models in this respect are lacking. Also, we believe the way forward must come through a combination of theory development and conceptual understanding, in combination with increased realism in coupled and uncoupled ocean modeling. There exist significant model sensitivities with respect to the overturning circulation and its variability, and there is a critical need for improving our understanding of the mechanisms responsible for these.

A major caveat with most conceptual models developed for ocean circulation is that they either assume a laminar ocean, or adopt an overly simplistic parameterization for the turbulent geostrophic eddy field. Even the highest-resolution numerical ocean circulation models lie in a very different dynamical regime to the ocean. For example, the Reynolds number, defined by taking the transport of the Gulf Stream divided by a scale depth and lateral viscosity, is typically O(1) in a 1° resolution ocean model, O(10^3) in a 1/12° resolution ocean model, and O(10^{10}) in the ocean. Recent progress suggests that the global stratification and circulation of the oceans, and its adjustment, are acutely sensitive to the magnitude of Southern Ocean eddy fluxes (e.g., Allison et al., 2011; Nikurashin and Vallis, 2012). Moreover, the sensitivity of the circulation and stratification to changes in wind forcing is very different in models with explicit, rather than parameterized, eddies (e.g., Hallberg and Gnanadesikan, 2006; Farneti et al., 2010; Abernathey et al., 2011; Munday et al., 2013). This is likely to remain a major stumbling block unless a breakthrough occurs in our understanding of, and ability to parameterize, the geostrophic eddy field and its dependence on the background flow. One idea that shows significant promise is the development of stochastic parameterizations of geostrophic eddies (e.g., Grooms and Majda, 2013), following the successful application of similar ideas to the atmosphere (for a review see Palmer and Williams, 2010). A caveat, however, is that such closures, stochastic or otherwise, should remain consistent with fundamental physical constraints such as conservation of angular momentum and energy (Marshall et al., 2012) and adiabaticity (Gent and McWilliams, 1990). Nevertheless, we believe that ocean models will eventually be non-deterministic, as is already the case for their atmospheric counterparts.

Finally, it is worth remembering that the major conceptual advances result from one or two individuals thinking creatively and laterally about a problem. For example, the rapid progress in our understanding of the vertical structure of the wind-driven circulation in the 1980s (Rhines and Young, 1982a; Luyten et al., 1983; see Section 2.5) was made possible by realizing that the three-dimensional problem could be reduced to a series of coupled two-dimensional problems through the use of a layered model, rather than the rapid development of numerical ocean circulation models. Similarly, the rapid progress in our understanding of the role of eddies in the Southern Ocean circulation was facilitated by the key insight of Gent and McWilliams (1990); Gent et al. (1995) that an geostrophic eddy parameterization should be adiabatic. As such it is always difficult to predict where the next conceptual breakthrough will occur, although observations often play a pivotal guiding role, as do developments in related fields such as meteorology. What is clear, as numerical ocean circulation models become ever more complex, is that the simple conceptual models remain as vital a tool as ever and one that future generations of physical oceanographers and climate scientists must learn to embrace and be trained to use with skill.

ACKNOWLEDGMENTS

We thank Carsten Eden, Stephen M. Griffies and an anonymous reviewer for helpful suggestions.

REFERENCES

Abernathey, R., Marshall, J., Ferreira, D., 2011. The dependence of Southern Ocean meridional overturning on wind stress. J. Phys. Oceanogr. 41, 2261–2278.

Alley, R.B., et al., 2003. Abrupt climate change. Nature 299, 2005–2010.

Allison, L., Johnson, H., Marshall, D., Munday, D., 2010. Where do winds drive the Antarctic Circumpolar Current? Geophys. Res. Lett. 37, http://dx.doi.org/10.1029/2010GL043355, L12605.

Allison, L., Johnson, H., Marshall, D., 2011. Spin-up and adjustment of the Antarctic Circumpolar Current and global pycnocline. J. Mar. Res. 69, 167–189.

Anderson, D.L.T., Gill, A.E., 1975. Spin-up of a stratified oceaan, with applications to upwellings. Deep Sea Res. 22, 583–596.

Behrens, E., Biastoch, A., Böning, C., 2013. Spurious AMOC trends in global ocean sea-ice models related to subarctic freshwater forcing. Ocean Model. 69, 39–49.

Ben Jelloul, M., Huck, T., 2005. Low-frequency basin modes in a two-layer quasi-geostrophic model in the presence of a mean gyre flow. J. Phys. Oceanogr. 35, 2167–2186.

Berloff, P.S., Hogg, A.M., Dewar, W.K., 2007. The turbulent oscillator: a mechanism of low-frequency variability of the wind-driven ocean gyres. J. Phys. Oceanogr. 37, 2362–2386.

Bjerknes, V.W., 1898. über einen hydrodynamischen fundamentaksatz und seine anwendung, besonders auf die mechanik der atmosphäre und des weltmeeres. K. Svenska Vet. Akad. Handlingar 31, 1–35.

Broecker, W.S., 1991. The great ocean conveyor. Oceanography 4, 79–89.

Brüggemann, N., Eden, C., Olbers, D., 2011. A dynamically consistent closure for zonally averaged ocean models. J. Phys. Oceanogr. 41, 2242–2258.

Bryan, K., 1963. A numerical investigation of a nonlinear model of a wind-driven ocean. J. Atmos. Sci. 20, 594–606.

Bryan, F., 1986. High-latitude salinity effects and interhemispheric thermohaline circulations. Nature 323, 301–304.

Bryden, H.L., King, B.A., McCarthy, G.D., 2011. South Atlantic overturning circulation at 24S? J. Mar. Res. 69, 38–55.

Callies, J., Marotzke, J., 2012. A simple and self-consistent geostrophic-force-balance model of the thermohaline circulation with boundary mixing. Ocean Sci. 8, 49–63.

Cane, M.A., Kamenkovich, V.M., Krupitsky, A., 1998. On the utility and disutility of JEBAR. J. Phys. Oceanogr. 28, 519–526.

Cessi, P., 2004. Global seiching of thermocline waters between the Atlantic and the indian-Pacific ocean basins. Geophys. Res. Lett. 31, L04302.

Cessi, P., Fantini, M., 2004. The eddy-driven thermocline. J. Phys. Oceanogr. 34, 2642–2658.

Cessi, P., Ierley, G.R., 1995. Symmetry-breaking multiple equilibria in quasi-geostrophic, wind-driven flows. J. Phys. Oceanogr. 25, 1196–1205.

Cessi, P., Otheguy, P., 2003. Oceanic teleconnections: remote response to decadal wind forcing. J. Phys. Oceanogr. 33, 1604–1617.

Cessi, P., Primeau, F., 2001. Dissipative selection of low-frequency modes in a reduced gravity basin. J. Phys. Oceanogr. 31, 127–137.

Charney, J.G., 1955. The Gulf Stream as an inertial boundary layer. Proc. Natl. Acad. Sci. U.S.A. 41, 731.

Chassignet, E.P., Marshall, D.P., 2008. Gulf stream separation in numerical ocean models. In: Hecht, M., Hasumi, H. (Eds.), Ocean Modeling in an Eddying Regime. Geophysical Monograph Series, American Geophysical Union, Washington, D.C., pp. 39–61.

Chelton, D.B., Schlax, M., 1996. Global observations of oceanic Rossby waves. Science 272, 234–238.

Chen, F., Ghil, M., 1996. Interdecadal variability in an hybrid coupled ocean–atmosphere model. J. Phys. Oceanogr. 26, 1561–1578.

Cimatoribus, A., 2013. Impact of atmospheric and oceanic feedbacks on the stability of the meridional overturning circulation. Ph.D. thesis, Universiteit Utrecht.

Cimatoribus, A.A., Drijfhout, S.S., Dijkstra, H.A., 2011. A global hybrid coupled model based on atmosphere-SST feedbacks. Clim. Dyn. 38, 745–760.

Cimatoribus, A.A., Drijfhout, S.S., den Toom, M., Dijkstra, H.A., 2012. Sensitivity of the Atlantic meridional overturning circulation to South Atlantic freshwater anomalies. Clim. Dyn. 39, 2291–2306.

Cimatoribus, A.A., Drijfhout, S.S., Dijkstra, H.A., 2013. Meridional overturning circulation: stability and ocean feedbacks in a box model. Clim. Dyn. http://dx.doi.org/10.1007/s00382-012-1576-9 (published online).

Clark, P.U., Pisias, N.G., Stocker, T.F., Weaver, A.J., 2002. The role of the thermohaline circulation in abrupt climate change. Nature 415, 863–869.

Colin de Verdiére, A., Huck, T., 1999. Baroclinic instability: an oceanic wavemaker for interdecadal variability. J. Phys. Oceanogr. 29, 893–910.

Cox, M.D., 1989. An idealized model of the world ocean. Part I: the global-scale water masses. J. Phys. Oceanogr. 19, 1730–1752.

Danabasoglu, G., 2008. On multidecadal variability of the Atlantic meridional overturning circulation in the Community Climate System Model version 3. J. Clim. 21, 5524–5544.

Danabasoglu, G., McWilliams, J.C., Gent, P.R., 1994. The role of mesoscale tracer transport in the global ocean circulation. Science 264, 1123–1126.

Danabasoglu, G., Yeager, S.G., Kwon, Y.-O., Tribbia, J.J., Phillips, A.S., Hurrell, J.W., 2012. Variability of the Atlantic Meridional Overturning Circulation in CCSM4. J. Clim. 25, 5153–5172.

de Boer, A., Gnanadesikan, A., Edwards, N., Watson, A., 2010. Meridional density gradients do not control the Atlantic overturning circulation. J. Phys. Oceanogr. 40, 368–380.

Delworth, T.L., Manabe, S., Stouffer, R.J., 1993. Interdecadal variations of the thermohaline circulation in a coupled ocean–atmosphere model. J. Clim. 6, 1993–2011.

Dengler, M., Schott, F.A., Eden, C., Brandt, P., Fischer, J., Zantopp, R.J., 2004. Break-up of the Atlantic deep western boundary current into eddies at 8S. Nature 432, 1018–1020.

den Toom, M., Dijkstra, H.A., Cimatoribus, A.A., Drijfhout, S.S., 2012. Effect of atmospheric feedbacks on the stability of the Atlantic meridional overturning circulation. J. Clim. 25, 4081–4096.

de Vries, P., Weber, S.L., 2005. The Atlantic freshwater budget as a diagnostic for the existence of a stable shut down of the meridional overturning circulation. Geophys. Res. Lett. 32, L09606.

Dewar, W.K., 1998. On "too fast" baroclinic planetary waves in the general circulation. J. Phys. Oceanogr. 28, 1739–1758.

Dijkstra, H.A., 2007. Characterization of the multiple equilibria regime in a global ocean model. Tellus A 59A, 695–705.

Dijkstra, H.A., Katsman, C.A., 1997. Temporal variability of the wind-driven quasi-geostrophic double gyre ocean circulation: basic bifurcation diagrams. Geophys. Astrophys. Fluid Dyn. 85, 195–232.

Dijkstra, H.A., Weijer, W., 2003. Stability of the global ocean circulation: the connection of equilibria in a hierarchy of models. J. Mar. Res. 61, 725–743.

Dijkstra, H.A., te Raa, L.A., Schmeits, M., 2006. On the physics of the Atlantic Multidecadal Oscillation. Ocean Dyn. 56, 36–50.

Dommenget, D., Latif, M., 2002. Analysis of observed and simulated sst spectra in the midlatitude. Clim. Dyn. 19, 277–288.

Döös, K., Coward, A.C., 1997. The Southern Ocean as the major upwelling zone of north Atlantic deep water. WOCE Newslett. 27, 3–4.

Döös, K., Webb, D.J., 1994. The Deacon cell and the other meridional cells of the Southern Ocean. J. Phys. Oceanogr. 24, 429–442.

Döös, K., Nilsson, J., Nycander, J., Brodeau, L., Ballarotta, M., 2012. The world ocean thermohaline circulation. J. Phys. Oceanogr. 42, 1445–1460.

Drijfhout, S.S., 2005. What sets the surface eddy mass flux in the Southern Ocean? J. Phys. Oceanogr. 35, 2152–2166.

Drijfhout, S.S., de Vries, P., Döös, K., Coward, A.C., 2003. Impact of eddy-induced transport on the lagrangian structure of the upper branch of the thermohaline circulation. J. Phys. Oceanogr. 33, 2114–2155.

Drijfhout, S.S., Weber, S.L., Swaluw, E., 2010. The stability of the MOC as diagnosed from model projections for pre-industrial, present and future climates. Clim. Dyn. 37, 1575–1586.

Duhaut, T.H.A., Straub, D.N., 2006. Wind stress dependence on ocean surface velocity: implications for mechanical energy input to ocean circulation. J. Phys. Oceanogr. 36, 202–211.

Eden, C., Olbers, D., 2010. Why western boundary currents are diffusive: a link between bottom pressure torque and bolus velocity. Ocean Model. 32, 14–24.

Edwards, C.A., Pedlosky, J., 1998. Dynamics of nonlinear cross-equatorial flow. Part I: potential vorticity transformation. J. Phys. Oceanogr. 28, 2382–2406.

Ekman, V.W., 1905. On the influence of the earth's rotation on ocean currents. Ark. Mat. Astron. Fys. 2, 1–53.

Eldevik, T., Nilsen, J., Iovino, D., Olsson, K.A., Sandø, S.A., Drange, H., 2009. Observed sources and variability of Nordic Seas overflows. Nat. Geosci. 2, 406–410.

England, M.H., 1993. Representing the global-scale water masses in ocean general circulations models. J. Phys. Oceanogr. 23, 1523–1552.

Farneti, R., Delworth, T.L., Rosati, A., Griffies, S.M., Zeng, F., 2010. The role of mesoscale eddies in the rectification of the Southern Ocean response to climate change. J. Phys. Oceanogr. 40, 1539–1557.

Fofonoff, N.P., 1954. Steady flow in a frictionless homogeneous ocean. J. Mar. Res. 13, 254–262.

Frankcombe, L.M., Dijkstra, H.A., von der Heydt, A., 2009. Noise-induced multidecadal variability in the North Atlantic: excitation of normal modes. J. Phys. Oceanogr. 39, 220–233.

Frankignoul, C., Hasselmann, K., 1977. Stochastic climate models. II: application to sea-surface temperature anomalies and thermocline variability. Tellus 29, 289–305.

Fučkar, N.S., Vallis, G.K., 2007. Interhemispheric influence of surface buoyancy conditions on a circumpolar current. Geophys. Res. Lett. 34. http://dx.doi.org/10.1029/2007GL030379, L14605.

Fürst, J., Levermann, A., 2012. A minimal model for wind- and mixing-driven overturning: threshold behavior for both driving mechanisms. Clim. Dyn. 38, 239–260.

Gelderloos, R., Katsman, C.A., Drijfhout, S.S., 2011. Assessing the roles of three eddy types in restratifying the Labrador Sea after deep convection. J. Phys. Oceanogr. 41, 2102–2119.

Gent, P.R., McWilliams, J.C., 1990. Isopycnal mixing in ocean circulation models. J. Phys. Oceanogr. 20, 150–155.

Gent, P.R., Willebrand, J., McDougall, T., McWilliams, J.C., 1995. Parameterizing eddy-induced tracer transports in ocean circulation models. J. Phys. Oceanogr. 25, 463–474.

Gnanadesikan, A., 1999. A simple predictive model of the structure of the oceanic pycnocline. Science 283, 2077–2081.

Gnanadesikan, A., Hallberg, R.W., 2000. On the relationship of the circumpolar current to southern hemisphere winds in coarse-resolution ocean models. J. Phys. Oceanogr. 30, 2013–2034.

Gordon, A.L., 1986. Interocean exchange of thermocline water. J. Geophys. Res. 91, 5037–5046.

Gouretski, V.V., Koltermann, K. P., 2004. WOCE global hydrographic climatology. *Ber. Bundesamte Seeschifffahrt Hydrogr. Report*, **35**.

Greatbatch, R.J., Zhang, S., 1995. An interdecadal oscillation in an idealized ocean basin forced by constant heat flux. J. Clim. 8, 82–91.

Griesel, A., Morales Maqueda, M.A., 2006. The relation of meridional pressure gradients to North Atlantic deep water volume transport in an ocean general circulation model. Clim. Dyn. 26, 781–799. http://dx.doi.org/10.1007/s00382-006-0122-z.

Griffies, S.M., Tziperman, E., 1995. A linear thermohaline oscillator driven by stochastic atmospheric forcing. J. Clim. 8, 2440–2453.

Griffies, S.M., et al., 2009. Coordinated Ocean-ice Reference Experiments (COREs). Ocean Model. 26, 1–46.

Grooms, I., Majda, A.J., 2013. Efficient stochastic superparameterization for geophysical turbulence. Proc. Natl. Acad. Sci. U.S.A. 110, 4464–4469.

Guan, Y.P., Huang, R.X., 2008. Stommel's box model of the thermohaline circulation revisited—the role of mechanical energy supporting mixing and the wind-driven gyration. J. Phys. Oceanogr. 38, 909–917.

Hallberg, R.W., Gnanadesikan, A., 2006. The role of eddies in determining the structure and response of the wind-driven southern hemisphere overturning: results from the Modeling Eddies in the Southern Ocean (MESO) project. J. Phys. Oceanogr. 36, 2232–2252.

Hansen, B., Østerhus, S., 2000. North Atlantic—Nordic seas exchanges. Prog. Oceanogr. 45, 109–208.

Hasselmann, K., 1976. Stochastic climate models. I: theory. Tellus 28, 473–485.

Hawkins, E., Smith, R.S., Allison, L.C., Gregory, J.M., Woollings, T.J., Pohlmann, H., De Cuevas, B., 2011. Bistability of the Atlantic overturning circulation in a global climate model and links to ocean freshwater transport. Geophys. Res. Lett. 38, L10605.

Huang, R.X., 1989. The generalized eastern boundary conditions and the three-dimensional structure of the ideal fluid thermocline. J. Geophys. Res. 94, 4855–4865.

Huck, T., Vallis, G., 2001. Linear stability analysis of the three-dimensional thermally-driven ocean circulation: application to interdecadal oscillations. Tellus A 53A, 526–545.

Hughes, C.W., de Cuevas, B.A., 2001. Why western boundary currents in realistic oceans are inviscid: a link between form stress and bottom pressure torques. J. Phys. Oceanogr. 31, 2871–2885.

Hughes, C.W., Wilson, C., 2008. Wind work on the geostrophic ocean circulation: an observational study on the effect of small scales in the wind stress. J. Geophys. Res. 113. http://dx.doi.org/10.1029/2007JC004371, C02016.

Hughes, G.O., Hogg, A.M., Griffiths, R.W., 2009. Available potential energy and irreversible mixing in the meridional overturning circulation. J. Phys. Oceanogr. 39, 3130–3146.

Huisman, S.E., den Toom, M., Dijkstra, H.A., Drijfhout, S., 2010. An indicator of the multiple equilibria regime of the Atlantic meridional overturning circulation. J. Phys. Oceanogr. 40, 551–567.

Hurrell, J.W., 1995. Decadal trends in the North Atlantic Oscillation: regional temperatures and precipitation. Science 269, 676–680.

Hurrell, J.W., Deser, C., 2010. North Atlantic climate variability: the role of the North Atlantic Oscillation. J. Mar. Syst. 79, 231–244.

Iovino, D., Straneo, F., Spall, M.A., 2008. On the effect of a sill on dense water formation in a marginal sea. Deep Sea Res. 66, 325–345.

Isachsen, P.E., LaCasce, J.H., Pedlosky, J., 2007. Rossby wave instability and apparent phase speeds in large ocean basins. J. Phys. Oceanogr. 37, 1177–1191.

Iselin, C., 1939. The influence of vertical and lateral turbulence on the characteristics of the waters at mid-depths. Trans. Am. Geophys. Union 20, 414–417.

Ito, T., Marshall, J.C., 2008. Control of lower limb circulation in the Southern Ocean by diapycnal mixing and mesoscale eddy transfer. J. Phys. Oceanogr. 38, 2832–2845.

Jacobs, S.S., 2004. Bottom water production and its links with the thermohaline circulation. Antarct. Sci. 16, 427–437.

Jayne, S.R., Marotzke, J., 1999. A destabilizing thermohaline circulation-atmosphere-sea ice feedback. J. Clim. 12, 642–651.

Jiang, S., Jin, F.F., Ghil, M., 1995. Multiple equilibria, periodic, and aperiodic solutions in a wind-driven, double-gyre, shallow-water model. J. Phys. Oceanogr. 25, 764–786.

Johnson, H.L., Marshall, D.P., 2002. A theory for the surface Atlantic response to thermohaline variability. J. Phys. Oceanogr. 32, 1121–1132.

Johnson, H.L., Marshall, D.P., 2004. Global teleconnections of meridional overturning circulation anomalies. J. Phys. Oceanogr. 34, 1702–1722.

Johnson, H.L., Marshall, D.P., Sproson, D.A.J., 2007. Reconciling theories of a mechanically driven meridional overturning circulation with thermohaline forcing and multiple equilibria. Clim. Dyn. 29, 821–836.

Kalnay, E., et al., 1996. The NCEP/NCAR 40-year reanalysis project. Bull. Am. Meteorol. Soc. 77, 437–471.

Kämpf, J., 2005. Cascading-driven upwelling in submarine canyons at high latitudes. J. Geophys. Res. 110. http://dx.doi.org/10.1029/2004JC002554.

Kawase, M., 1987. Establishment of deep ocean circulation driven by deep-water production. J. Phys. Oceanogr. 17, 2294–2317.

Killworth, P.D., Chelton, D.B., de Szoeke, R.A., 1997. The speed of observed and theoretical long extratropical planetary waves. J. Phys. Oceanogr. 27, 1946–1966.

Klinger, B.A., Marotzke, J., 1999. Behavior of double-hemispheric thermohaline flows in a single basin. J. Phys. Oceanogr. 29, 382–399.

Klinger, B.A., Drijfhout, S., Marotzke, J., Scott, J.R., 2003. Sensitivity of basin-wide meridional overturning to diapycnal difffusion and remote wind forcing in an idealized Atllantic-Southern Ocean geometry. J. Phys. Oceanogr. 33, 249–266.

Kwon, Y.-O., Frankignoul, C., 2011. Stochastically-driven multidecadal variability of the Atlantic meridional overturning circulation in CCSM3. Clim. Dyn. 38, 859–876.

LaCasce, J.H., 2000. Baroclinic Rossby waves in a square basin. J. Phys. Oceanogr. 30, 3161–3178.

LaCasce, J.H., Pedlosky, J., 2004. The instability of Rossby basin modes and the oceanic eddy field. J. Phys. Oceanogr. 34, 2027–2041.

Latif, M., 1998. Dynamics of interdecadal variability in coupled ocean–atmosphere models. J. Clim. 11, 602–624.

Ledwell, J.R., Watson, A.J., Law, C.S., 1993. Evidence for slow mixing across the pycnocline from an open ocean tracer release experiment. Nature 364, 701703.

Legg, S., Jackson, L., Hallberg, R.W., 2008. Eddy-resolving modeling of overflows. In: Hecht, M.W., Hasumi, H. (Eds.), Ocean Modeling in an Eddying Regime. American Geophysical Union, Washington, D.C., pp. 63–82.

Liu, Z., et al., 2012. Younger Dryas cooling and the Greenland climate response to CO_2. Proc. Natl. Acad. Sci. U.S.A. 109, 11 101–11 104.

Longworth, H., Marotzke, J., Stocker, T.F., 2005. Ocean gyres and abrupt change in the thermohaline circulation: a conceptual analysis. J. Clim. 18, 2403–2416.

Luksch, U., von Storch, H., Maier-Reimer, E., 1990. Modeling North Pacific SST anomalies as a response to anomalous atmospheric forcing. J. Marine Sys. 1, 155–168.

Luyten, J.R., Pedlosky, J., Stommel, H., 1983. The ventilated thermocline. J. Phys. Oceanogr. 13, 292–309.

Manabe, S., Stouffer, R.J., 1988. Two stable equilibria of a coupled ocean–atmosphere model. J. Clim. 1, 841–866.

Mantua, N.J., Hare, S., Zhang, Y., Wallace, J.M., Francis, R.C., 1997. A Pacific interdecadal climate oscillation with impacts on salmon production. Bull. Am. Meteorol. Soc. 78, 1069–1079.

Marotzke, J., 1997. Boundary mixing and the dynamics of three-dimensional thermohaline circulations. J. Phys. Oceanogr. 27, 1713–1728.

Marotzke, J., 2000. Abrupt climate change and thermohaline circulation: mechanisms and predictability. Proc. Natl. Acad. Sci. U.S.A. 97, 1347–1350.

Marotzke, J., Scott, J.R., 1999. Convective mixing and the thermohaline circulation. J. Phys. Oceanogr. 29, 2962–2970.

Marotzke, J., Willebrand, J.U., 1991. Multiple equilibria of the global thermohaline circulation. J. Phys. Oceanogr. 21, 1372–1385.

Marotzke, J., Welander, P., Willebrand, J., 1988. Instability and multiple steady states in a meridional-plane model of thermohaline circulation. Tellus 40, 162–172.

Marshall, J.C., 1981. On the parameterization of geostrophic eddies in the ocean. J. Phys. Oceanogr. 11, 257–271.

Marshall, D., 1995a. Influence of topography on the large-scale ocean circulation. J. Phys. Oceanogr. 25, 1622–1635.

Marshall, D., 1995b. Topographic steering of the Antarctic Circumpolar Current. J. Phys. Oceanogr. 25, 1636–1650.

Marshall, D., 1997. Subduction of water masses in an eddying ocean. J. Mar. Res. 55, 201–222.

Marshall, D.P., 2000. Vertical fluxes of potential vorticity and the structure of the thermocline. J. Phys. Oceanogr. 30, 3102–3112.

Marshall, D.P., Pillar, H.R., 2011. Momentum balance of the wind-driven and meridional overturning circulation. J. Phys. Oceanogr. 41, 960–978.

Marshall, J., Radko, T., 2003. Residual-mean solutions for the Antarctic Circumpolar Current and its associated overturning circulation. J. Phys. Oceanogr. 33, 2341–2354.

Marshall, D.P., Maddison, J.R., Berloff, P.S., 2012. A framework for parameterizing eddy potential vorticity fluxes. J. Phys. Oceanogr. 42, 539–557.

Mauritzen, C., 1996. Production of dense overflow waters feeding the North Atlantic across the Greenland-Scotland ridge. Part 1: evidence for a revised circulation scheme. Deep Sea Res. 43, 769–806.

McCartney, M.S., 1992. Recirculating components to the deep boundary current of the northern North Atlantic. Prog. Oceanogr. 29, 283–383.

McCartney, M.S., 2007. A cyclonic gyre in the Atlantic Australian Basin. Prog. Oceanogr. 75, 675–750.

McWilliams, J.C., Restrepo, J.M., 1999. The wave-driven ocean circulation. J. Phys. Oceanogr. 29, 2523–2540.

Munday, D., Allison, L., Johnson, H., Marshall, D., 2011. Remote forcing of the Antarctic Circumpolar Current by diapycnal mixing. Geophys. Res. Lett. 38. http://dx.doi.org/10.1029/2011GL046849, L08609.

Munday, D.R., Johnson, H.L., Marshall, D.P., 2013. Eddy saturation of equilibrated circumpolar currents. J. Phys. Oceanogr. 43, 507–532.

Munk, W., 1950. On the wind-driven ocean circulation. J. Meteorol. 7, 79–93.

Munk, W., 1966. Abyssal recipes. Deep Sea Res. 12, 707–730.

Munk, W., Wunsch, C., 1998. Abyssal recipes ii: energetics of tidal and wind mixing. Deep Sea Res. 45, 1977–2010.

Nakamura, M., Stone, P.H., Marotzke, J., 1994. Destabilization of the thermohaline circulation by atmospheric eddy transports. J. Clim. 7, 1870–1882.

Niiler, P.P., 1966. On the theory of wind-driven ocean circulation. Deep Sea Res. Oceanogr. Abstr. 13, 597–606.

Nikurashin, M., Vallis, G., 2011. A theory of deep stratification and overturning circulation in the ocean. J. Phys. Oceanogr. 41, 485–502.

Nikurashin, M., Vallis, G., 2012. A theory of the interhemispheric meridional overturning circulation and associated stratification. J. Phys. Oceanogr. 42, 1652–1667.

Olbers, D., Willebrand, J., Eden, C., 2012. Ocean Dynamics Springer, Berlin Heidelberg.

Orsi, A.H., Jacobs, S.S., Gordon, A.L., Visbeck, M., 2001. Cooling and ventilating the abyssal ocean. J. Phys. Oceanogr. 28, 2923–2926.

Pacanowski, R.C., 1987. Effect of equatorial currents on surface stress. J. Phys. Oceanogr. 17, 833–838.

Palmer, T., Williams, P., 2010. Stochastic Physics and Climate Modelling. Cambridge University Press, Cambridge, UK.

Pedlosky, J., 1987. Geophysical Fluid Dynamics. Springer-Verlag, New York.

Pedlosky, J., 1996. Ocean Circulation Theory. Springer, New York.

Pickart, R.S., Spall, M.A., 2007. Impact of Labrador Sea convection on the North Atlantic meridional overturning circulation. J. Phys. Oceanogr. 37, 2207–2227.

Pierini, S., Dijkstra, H.A., Riccio, A., 2009. A nonlinear theory of the Kuroshio Extension bimodality. J. Phys. Oceanogr. 39, 2212–2229.

Polton, J.A., Marshall, D.P., 2003. Understanding the structure of the subtropical thermocline. J. Phys. Oceanogr. 33, 1240–1249.

Polzin, K.L., Toole, J.M., Ledwell, J.R., 1997. Spatial variability of turbulent mixing in the abyssal ocean. Science 276, 93–96.

Price, J.F., Baringer, M.O., 1994. Outflows and deep water production by marginal seas. J. Geophys. Res. 33, 161–200.

Primeau, F., 2002a. Long Rossby wave basin-crossing time and the resonance of low-frequency basin modes. J. Phys. Oceanogr. 32, 2652–2665.

Primeau, F.W., 2002b. Multiple equilibria and low-frequency variability of the wind-driven ocean circulation. J. Phys. Oceanogr. 32, 2236–2256.

Radko, T., Marshall, J., 2004. The leaky thermocline. J. Phys. Oceanogr. 34, 1648–1662.

Rahmstorf, S., 1995. Bifurcations of the Atlantic thermohaline circulation in response to changes in the hydrological cycle. Nature 378, 145–149.

Rahmstorf, S., Willebrand, J., 1995. The role of temperature feedback in stabilizing the thermohaline circulation. J. Phys. Oceanogr. 25, 787–805.

Rahmstorf, S., 1996. On the freshwater forcing and transport of the Atlantic thermohaline circulation. Clim. Dyn. 12, 799–811.

Ralph, E.A., Niiler, P.P., 1999. Wind-driven currents in the tropical pacific. J. Phys. Oceanogr. 29, 2121–2129.

Rayner, D., et al., 2011. Monitoring the Atlantic meridional overturning circulation. Deep Sea Res. Part II 58, 1744–1753.

Rhines, P.B., Young, W.R., 1982a. Homogenization of potential vorticity in planetary gyres. J. Fluid Mech. 122, 347–367.

Rhines, P.B., Young, W.R., 1982b. A theory of the wind-driven circulation. I. Mid-ocean gyres. J. Mar. Res. 40, 560–596.

Richardson, P.L., Fratantoni, D.M., 1999. Float trajectories in the deep western boundary current and deep equatorial jets of the tropical atlantic. Deep Sea Res. 33, 305–333.

Rintoul, S.R., 1991. South Atlantic interbasin exchange. J. Geophys. Res. 96, 2675–2692.

Rintoul, S., Hughes, C., Olbers, D., 2001. The Antarctic circumpolar current system. In: Siedler, S., Church, J., Gould, J. (Eds.), Ocean Circulation and Climate. 1 Academic Press, San Diego, pp. 271–302.

Robinson, A.R., Stommel, H., 1959. The oceanic thermocline and associated thermohaline circulation. Tellus 11, 295–308.

Rooth, C., 1982. Hydrology and ocean circulation. Prog. Oceanogr. 11, 131–149.

Saenko, O.A., Gupta, A.S., Spence, P., 2012. On Challenges in Predicting Bottom Water Transport in the Southern Ocean. J. Clim. 25, 1349–1356.

Saenz, J.A., Hogg, A.M., Hughes, G.O., Griffiths, R.W., 2012. On the mechanical power input to the oceans from buoyancy and wind. Geophys. Res. Lett. 39, L21609.

Salmon, R., 1990. The thermocline as an internal boundary layer. J. Mar. Res. 48, 437–469.

Salmon, R., Ford, R., 1995. A simple model of the joint effect of baroclinicity and relief on ocean circulation. J. Mar. Res. 53, 211–230.

Samelson, R.M., Vallis, G.K., 1997. Large-scale circulation with small diapycnal diffusion: the two-thermocline limit. J. Mar. Res. 55, 223–275.

Sandström, J.W., 1916. Meteorologische studien—à schedischen hochgebirge. Göteborgs K. Vetensk. Vitt. Handkl. 10, 2–42.

Saravanan, R., McWilliams, J.C., 1998. Advective ocean–atmosphere interaction: an analytical stochastic model with implications for decadal variability. J. Clim. 11, 165–188.

Sarkisyan, A.S., 2006. Forty years of JEBAR—the finding of the joint effect of baroclinicity and bottom relief for the modeling of ocean climatic characteristics. Izv. Atmos. Oceanic Phys. 42, 534–554.

Schmitz, W.J., 1995. On the interbasin-scale thermohaline circulation. Rev. Geophys. 33, 151–173.

Schott, F.A., Zantopp, R., Stramma, L., Dengler, M., Fischer, J., Wibaux, M., 2004. Circulation and deep-water export at the western exit of the subpolar North Atlantic. J. Phys. Oceanogr. 34, 817–843.

Scott, R.B., Xu, Y., 2009. An update on the wind power input to the surface geostrophic flow of the world ocean. Deep Sea Res. 56, 295–304.

Sevellec, F., Fedorov, A.V., 2013. The leading, interdecadal eigenmode of the Atlantic meridional overturning circulation in a realistic ocean model. J. Clim. 26, 2160–2183.

Sijp, W.P., England, M.H., 2006. Sensitivity of the Atlantic thermohaline circulation and its stability to basin-scale variations in vertical mixing. J. Clim. 19, 5467–5478.

Sijp, W.P., Gregory, J.M., Tailleux, R., Spence, P., 2012. The key role of the western boundary in linking the AMOC strength to the north–south pressure gradients. J. Phys. Oceanogr. 42, 628–643.

Simonnet, E., Dijkstra, H.A., 2002. Spontaneous generation of low-frequency modes of variability in the wind-driven ocean circulation. J. Phys. Oceanogr. 32, 1747–1762.

Sloyan, B.M., Rintoul, S.R., 2001. Circulation, renewal, and modification of antarctic mode and intermediate water. J. Phys. Oceanogr. 31, 1005–1030.

Spall, M.A., 2000. Buoyancy-forced circulations around islands and ridges. J. Mar. Res. 58, 957–982.

Spall, M.A., 2004. Boundary currents and watermass transformation in marginal seas. J. Phys. Oceanogr. 34, 1197–1213.

Spall, M., Pickart, R.S., 2001. Where does dense water sink? a subpolar gyre example. J. Phys. Oceanogr. 31, 810–826.

Speer, K., Rintoul, S.R., Sloyan, B., 2000. The diabatic Deacon cell. J. Phys. Oceanogr. 30, 3212–3222.

Speich, S., Blanke, B., de Vries, P., Drijfhout, S.S., Döös, K., Ganachaud, A., Marsh, R., 2002. Tasman leakage: a new route in the global ocean conveyor belt. Geophys. Res. Lett. 29. http://dx.doi.org/10.1029/2001GL01458.

Srokosz, M., Baringer, M., Bryden, H., Cunningham, S., Delworth, T., Lozier, S., Marotzke, J., Sutton, R., 2012. Past, Present, and Future Changes in the Atlantic Meridional Overturning Circulation. Bull. Am. Meteorol. Soc. 93, 1663–1676.

Stephens, J.C., Marshall, D.P., 2000. Dynamical pathways of Antarctic Bottom Water in the atlantic. J. Phys. Oceanogr. 30, 622–640.

Stewart, A.L., Thompson, A.F., 2013. Connecting antarctic cross-slope exchange with Southern Ocean overturning. J. Phys. Oceanogr. 43, 1453–1471. http://dx.doi.org/10.1175/JPO-D-12-0205.1.

St Laurent, L., Garrett, C., 2002. The role of internal tides in mixing the deep ocean. J. Phys. Oceanogr. 32, 2882–2899.

Stommel, H., 1948. The westward intensification of wind-driven ocean currents. Trans. Am. Geophys. Union 29, 202–206.

Stommel, H., 1958. The abyssal circulation. Deep Sea Res. 5, 80–82.

Stommel, H., 1961. Thermohaline convection with two stable regimes of flow. Tellus 13, 224–230.

Stommel, H., 1979. Determination of water mass properties of water pumped down from the Ekman layer to the geostrophic flow below. Proc. Natl. Acad. Sci. U.S.A. 76, 3051–3055.

Stommel, H., Arons, A.B., 1961a. On the abyssal circulation of the world ocean—I. Stationary planetary flow patterns on a sphere. Deep Sea Res. 6, 140–154.

Stommel, H., Arons, A.B., 1961b. On the abyssal circulation of the world ocean, II: an idealized model of the circulation pattern and amplitude in oceanic basins. Deep Sea Res. 6, 217–233.

Straneo, F., 2006. On the connection between dense water formation, overturning, and poleward heat transport in a convective basin. J. Phys. Oceanogr. 36, 1822–1840.

Straub, D., 1993. On the transport and angular momentum balance of channel models of the Antarctic Circumpolar Current. J. Phys. Oceanogr. 23, 776–782.

Straub, D., 1996. An inconsistency between two classical models of the ocean buoyancy driven circulation. Tellus 48A, 477–481.

Straub, D., Rhines, P.B., 1990. Effects of large-scale topography on abyssal circulation. J. Mar. Res. 48, 223–253.

Sura, P., Fraedrich, K., Lunkeit, F., 2001. Regime transitions in a stochastically forced double-gyre model. J. Phys. Oceanogr. 31, 411–426.

Sverdrup, H.U., 1947. Wind-driven currents in a baroclinic ocean with application to the equatorial current in the eastern pacific. Proc. Natl. Acad. Sci. U.S.A. 33, 318–326.

Swallow, J.C., Worthington, L.V., 1961. An observation of a deep countercurrent in the western North Atlantic. Deep Sea Res. 8, 1–19.

Tailleux, R., 2009. On the energetics of stratified turbulent mixing, irreversible thermodynamics, Boussinesq models and the ocean heat engine controversy. J. Fluid Mech. 638, 339–386.

Tailleux, R., McWilliams, J.C., 2001. The effect of bottom pressure decoupling on the speed of extratropical, baroclinic Rossby waves. J. Phys. Oceanogr. 31, 1461–1476.

Talley, L., Pickard, G., Emery, W., Swift, J., 2011. Descriptive Physical Oceanography, sixth ed. Elsevier, Amsterdam, The Netherlands.

te Raa, L.A., Dijkstra, H.A., 2002. Instability of the thermohaline ocean circulation on interdecadal timescales. J. Phys. Oceanogr. 32, 138–160.

te Raa, L., Dijkstra, H.A., 2003. Modes of internal thermohaline variability in a single-hemispheric ocean basin. J. Mar. Res. 61, 491–516.

Toggweiler, J.R., Samuels, B., 1995. Effect of Drake Passage on the global thermohaline circulation. Deep Sea Res. 42, 477–500.

Trevena, J., Sijp, W.P., England, M.H., 2008. Stability of Antarctic Bottom Water formation to freshwater fluxes and implications for global climate. J. Clim. 21, 3310–3326.

Tsujino, H., Suginohara, N., 1999. Thermohaline circulation enhanced by wind forcing. J. Phys. Oceanogr. 29, 1506–1516.

Tziperman, E., Gildor, H., 2002. The stabilization of the thermohaline circulation by the temperature-precipitation feedback. J. Phys. Oceanogr. 32, 2707–2714.

Vallis, G.K., 2000. Large-scale circulation and production of stratification: effects of wind, geometry and diffusion. J. Phys. Oceanogr. 30, 933–954.

Vallis, G.K., 2006. Atmospheric and Oceanic Fluid Dynamics. Cambridge University Press, Cambridge, UK.

Veronis, G., 1966. Wind-driven ocean circulation. II: numerical solution of the non-linear problem. Deep Sea Res. 13, 31–55.

Veronis, G., 1981. Dynamics of the large-scale ocean circulation. In: Warren, B., Wunsch, C. (Eds.), Evolution of Physical Ocenaography. MIT Press, Cambridge, pp. 140–183.

Weaver, A.J., et al., 2012. Stability of the Atlantic meridional overturning circulation: a model intercomparison. Geophys. Res. Lett. 39. http://dx.doi.org/10.1029/2012GL053763.

Weber, S.L., et al., 2007. The modern and glacial overturning circulation in the Atlantic ocean in PMIP coupled model simulations. Clim. Past Discuss. 3, 51–64.

Weijer, W., Dijkstra, H.A., 2003. Multiple oscillatory modes of the global ocean circulation. J. Phys. Oceanogr. 33, 2197–2213.

Weijer, W., de Ruijter, W.P.M., Dijkstra, H.A., Van Leeuwen, P.J., 1999. Impact of interbasin exchange on the Atlantic overturning circulation. J. Phys. Oceanogr. 29, 2266–2284.

Welander, P., 1971. The thermocline problem. Philos. Trans. R. Soc. A 270, 415–421.

Whitworth, T., Orsi, A.H., 2006. Antarctic Bottom water production and export by tides in the Ross Sea. Geophys. Res. Lett. 33. http://dx.doi.org/10.1029/2006GL026357.

Williams, R.G., Spall, M.A., Marshall, J.C., 1995. Does Stommel's mixed layer "demon" work? J. Phys. Oceanogr. 25, 3089–3102.

Winters, K.B., Lombard, P.N., Riley, J.J., dAsaro, E.A., 1995. Available potential energy and mixing in density-stratified fluids. J. Fluid Mech. 289, 115–228.

Wolfe, C.L., Cessi, P., 2009. Overturning circulation in an eddy-resolving model: the effect of the pole-to-pole temperature gradient. J. Phys. Oceanogr. 39, 125–142.

Wolfe, C.L., Cessi, P., 2010. What sets the strength of the middepth stratification and overturning circulation in eddying ocean models? J. Phys. Oceanogr. 40, 1520–1538.

Wolfe, C.L., Cessi, P., 2011. The Adiabatic Pole-to-Pole Overturning Circulation. J. Phys. Oceanogr. 41, 1795–1810.

Woods, J.D., 1985. The physics of thermocline ventilation. In: Nihoul, J.C.J. (Ed.), Coupled Ocean–Atmosphere Models. Elsevier, Woods: Amsterdam, pp. 543–590.

Wright, D.G., Stocker, T.F., 1991. A zonally averaged model for the thermohaline circulation, part i: model development and flow dynamics. J. Phys. Oceanogr. 21, 1713–1724.

Wright, D.G., Vreugdenhil, C.B., Hughes, T.M.C., 1995. Vorticity dynamics and zonally averaged ocean circulation models. J. Phys. Oceanogr. 25, 2141–2154.

Wunsch, C., 1996. The Ocean Circulation Inverse Problem. Cambridge University Press, Cambridge, UK.

Wunsch, C., 2011. The decadal mean ocean circulation and Sverdrup balance. J. Mar. Res. 69, 417–434.

Wunsch, C., Ferrari, R., 2004. Vertical mixing, energy, and the general circulation of the oceans. Annu. Rev. Fluid Mech. 36, 281–314.

Yin, J., Stouffer, R.J., 2007. Comparison of the stability of the Atlantic thermohaline circulation in two coupled atmosphere ocean general circulation models. J. Clim. 20, 4293–4315.

Zanna, L., Heimbach, P., Moore, A.M., Tziperman, E., 2010. Optimal excitation of interannual Atlantic meridional overturning circulation variability. J. Clim. 40, 413–427.

Zhai, X., Johnson, H.L., Marshall, D.P., Wunsch, C., 2012. On the wind power input to the ocean general circulation. J. Phys. Oceanogr. 42, 1357–1365.

Zika, J.D., England, M.H., Sijp, W.P., 2012. The ocean circulation in thermohaline coordinates. J. Phys. Oceanogr. 42, 708–724.

Chapter 12

Ocean Surface Circulation

Nikolai Maximenko*, Rick Lumpkin[†] and Luca Centurioni[‡]
*International Pacific Research Center, School of Ocean and Earth Science and Technology, University of Hawaii at Manoa, Honolulu, Hawaii, USA
[†]Atlantic Oceanographic and Meteorological Laboratory, NOAA, Miami, Florida, USA
[‡]Scripps Institution of Oceanography, University of California, San Diego, La Jolla, California, USA

This chapter is dedicated to the memory of Peter Niiler.

Chapter Outline

1. OBSERVED NEAR-SURFACE CURRENTS

Ocean currents connect local basins into the single Global Ocean. Together with atmospheric circulation, they serve to maintain the planetary balance of heat and freshwater. Currents are hard to measure. It required decades of international effort and generations of scientists and engineers before the observational network was built that monitors today's surface currents globally and in near real time.

1.1. Global Drifter Program and History of Lagrangian Observations

For many centuries, the large-scale pattern of ocean currents was inferred from the downstream fate of flotsam and debris of identifiable origin (c.f., Sverdrup et al., 1942). This information could be combined with *in situ* current estimates from ship drift—derived from the difference between the absolute motion of a ship and its relative motion through the water—to create maps of ocean currents and transport pathways. However, the speed of ocean surface currents can only be poorly estimated from this approach. Without robust knowledge of travel times and exact pathways, flotsam does not provide this information; ship drift in the pre-Global Positioning System (pre-GPS) period could suffer from relatively large navigational errors, resulting in surface current errors of O (20 cm s^{-1}) (Richardson and McKee, 1984). In addition, the force of the wind on a ship is not negligible, creating systematic biases in regions of large-scale, persistent winds such as the Trades or Westerlies (Richardson and Walsh, 1986; Reverdin et al., 1994; Lumpkin and Garzoli, 2005).

To reduce the effect of wind forcing on surface current measurements, investigators developed Lagrangian drifting buoys (or drifters) that had a small drag area above the waterline compared to the drag area below. They did this by attaching a sea anchor, or drogue, to the floating surface buoy. In the presatellite era, the drifter's position was determined by triangulation from a fixed point (land or an anchored ship). Observations of drogued drifters were collected as early as the mid-1700s off the US northeast coast (Franklin, 1785; Davis, 1991) and worldwide in the *Challenger* oceanographic survey of 1872–1876 (Thomson, 1877; Niiler, 2001).

Ocean Circulation and Climate, Vol. 103. http://dx.doi.org/10.1016/B978-0-12-391851-2.00012-X

With the introduction of radio, visual observations were no longer necessary to track drifters. Drifters with radio transmissions could be tracked via triangulation from shore receivers or from aircraft, allowing for larger arrays and observations in harsher conditions. This technology was used in the US Coastal Ocean Dynamics Experiment (CODE) of 1981–1982, in which 164 radio-tracked drifters were deployed off the US west coast (Davis, 1985a). These CODE drifters (also known as "Davis" drifters) had four cruciform-shaped drogue panels centered at a depth of ~60 cm beneath the surface, attached to a central vertical tube that contained the transmitter, batteries, and antenna (Davis, 1985a). Four floats at the upper, outer corners of each drogue panel provided additional buoyancy.

In the 1970s, satellites opened the possibility of worldwide drifter measurements. In the pre-GPS (mid-1990s) era, drifter positions were determined from the Doppler shift of the drifter's transmissions as measured by the satellite; the most prominent example of this technique is the Argos system created by the US agencies National Oceanic and Atmospheric Administration (NOAA) and National Aeronautics and Space Administration (NASA) and the French space agency Centre National d'Etudes Spatiales (CNES) and operated by Collecte Localisation Satellites (CLS) in Toulouse, France. Most of the modern drifters are tracked by the Argos system. In recent years, a growing number of data buoys are taking advantage of the Iridium satellite network to relay data globally, with GPS for positioning.

Several independent groups developed and deployed Argos-tracked drifters starting in the 1970s, leading to a number of competing designs for the instrument. In 1975, as part of the North Pacific Experiment (NORPAX), 165 drifters were released which were 3 m tall, 38 cm-diameter cylinders, most drogued at 30 m with a 9 m-diameter parachute (McNally et al., 1983). In 1976–1978, an array of 35 drifters, drogued at 200 m with either a 25 kg weight or a "window shade" drogue (a flat panel, usually weighted at the bottom), was deployed in the Gulf Stream region (Richardson, 1980). These drifters included a tether strain sensor to indicate drogue presence, but this sensor often failed shortly after deployment (Richardson, 1980). A large array of over 300 drifters was deployed as part of the Global Atmosphere Research Program (GARP) First Global GARP Experiment (FGGE) in 1979–1980 in the Southern Ocean (Garrett, 1980). FGGE drifters were not all identical, but the majority had a 3.4-m surface float with 100 m line acting as a drogue, weighted at the end with 29.5 kg of chain (Pazan and Niiler, 2001). From 1981–1984, 113 "HERMES" drifters, drogued with a window shade at 100 m depth, were deployed in the eastern and northern North Atlantic (Krauss and Böning, 1987). In 1983–1985, 53 "TIROS" drifters were deployed in the tropical Atlantic as part of the programs Seasonal Response of the Equatorial Atlantic (SEQUAL) and Programme Français Océan et Climat dans l'Atlantique Équatorial (FOCAL). These drifters had a 20-m^2 window shade drogue at 20 m depth, or (for the "mini-TIROS") a 2.2-m^2 window shade drogue at 5 m depth (Richardson and Reverdin, 1987) and did not include a sensor to indicate drogue presence. In 1987–1988, two arrays totaling 77 Argos-tracked drifters were deployed off Point Arena, California, with rigid "tristar" drogues resembling a radar reflector centered at 15 m depth (Brink et al., 1991). An array of 49 of these drifters was also deployed off the coast of British Columbia in 1987 (Paduan and Niiler, 1993). In addition to these designs, the CODE drifter mentioned above was converted to satellite positioning and used in the Gulf of Mexico Surface Current Lagrangian Program (SCULP) deployments of 1993–1998 (LaCasce and Ohlmann, 2003).

It was clear by the early 1980s that a global array of drifters would be invaluable for oceanographic and climate research, but data from existing regional deployments could not be easily combined due to the disparate water-following characteristics of the various drifter designs (WCRP, 1988; Niiler, 2001). Logistical demands for creating and maintaining a global array also called for a design that was relatively inexpensive to manufacture and transport and easy to deploy. In 1982, the World Climate Research Program (WCRP) declared that a standardized drifter should be developed to meet these constraints (WCRP, 1988; Niiler, 2001). This development took place under the Surface Velocity Program (SVP) of the Tropical Ocean Global Atmosphere (TOGA) experiment and the World Ocean Circulation Experiment (WOCE). NOAA's Atlantic Oceanographic and Meteorological Laboratory (AOML; P.I. Donald Hansen), MIT's Draper Laboratory (P.I. John Dahlen), and the Scripps Institution of Oceanography (SIO; P.I. Peter Niiler) proposed competing designs for what was to become the Global Drifter Program (GDP) drifter. In 1985–1989, the water-following characteristics of a number of drogue designs were evaluated by attaching vector-measuring current meters to the tops and bottoms of the drogues (Niiler et al., 1987, 1995). Significant problems were identified with several of the drogue types: window shade drogues could twist at an angle and create a sailing force across a current; parachute drogues could collapse and subsequently provide very little drag; line-and-chain drogues (such as used in the FGGE drifters) provided too little drag area compared to the surface float, resulting in significant wind- and wave-driven slip with respect to the current at the drogue depth (Niiler et al., 1987, 1995; Niiler and Paduan, 1995; Pazan and Niiler, 2001). The radar reflector-shaped tristar drogue developed at SIO had somewhat lower slip than the holey-sock drogue developed at AOML (Niiler et al., 1995), but cost more and was more difficult to ship and deploy. By 1993, a design for the GDP drifter had emerged which combined the holey-sock drogue of the AOML drifter, centered at a depth of 15 m beneath the ocean surface, with the reinforced tether ends and spherical surface float of the SIO drifter. This design (Sybrandy and Niiler, 1991) became the blueprint for future GDP drifter development.

The 1991 GDP drifter design was extremely robust, but was relatively expensive and heavy, with a surface float of 40 cm in diameter and a drogue of 6.44 m length, 92 cm in diameter and a total weight of 45 kg (Sybrandy and Niiler, 1991). A smaller, less-expensive redesign (Sybrandy et al., 1992) was introduced to reduce manufacturing and shipping costs, while retaining the same water-following characteristics. This 20 kg "mini" drifter has approximately one-third the manufacturing cost of the original design, has a smaller surface float, thinner tether, and smaller drogue (61 cm in diameter), while retaining the ratio of the drag area (cross-sectional area) of the drogue to all other components. This "drag area ratio" is 40:1 for a GDP drifter (original or "mini"), which results in 0.7 cm s^{-1} of downwind slip in 10 m s^{-1} winds when the drogue is attached (Niiler et al., 1995); for comparison, a standard FGGE-type drifter had a drag area ratio of 10:1–12:1 and a downwind slip of 8 cm s^{-1} in 10 m s^{-1} winds (Niiler and Paduan, 1995; Pazan and Niiler, 2001). Since the collection of Argos fixes is irregular in time and depends on latitude, the drifters' velocity data are interpolated to 6 h intervals with a kriging algorithm (Hansen and Poulain, 1996). The slip of a drogued GDP drifter (original or "mini" version) has not been measured at wind speeds greater than 10 m s^{-1}, but comparisons with altimetric observations suggest that it may exceed 0.07 cm s^{-1} per 1 m s^{-1} wind in these conditions (Niiler et al., 2003b).

When a drifter loses its drogue, the downwind slip increases by ~8 cm s^{-1} per 10 m s^{-1} wind (Pazan and Niiler, 2001), a result that can be exploited to recover low-frequency currents from undrogued drifter data (Pazan and Niiler, 2004). The presence of the drogue was originally detected via a submergence sensor on the surface float of the GDP drifter, as the float is frequently pulled underwater by the drogue. However, this sensor could fail or provide spurious results. In the late 2000s, the submergence sensor was replaced by the more reliable tether strain sensor at the base of the surface float. During the period of faulty submergence sensors in the early to mid-2000s, many undiagnosed drogue losses contaminated the velocity observations of supposedly drogued drifters in the drifter data base (Grodsky et al., 2011), explaining the time-varying behavior of drifter motion noted by Rio et al. (2011a). Subsequently, Rio (2012) developed a method to segregate drogue-off data, although this methodology tended to be overly conservative (i.e., it tended to discard known drogue-on data from some drifters). Lumpkin et al. (2013) adapted Rio's methodology to estimate as closely as possible the date of drogue loss for each drifter in the dataset, and applied this methodology to demonstrate that spurious low-frequency variations in current systems such as the Antarctic Circumpolar Current (ACC) subsequently disappear. Lumpkin et al. (2013) also demonstrated that the existing submergence records could be reinterpreted in a number of cases, and showed that transmission frequency anomalies could also be used to indicate drogue loss. These results have now been exploited in a systematic reevaluation of drogue presence for all drifters in the dataset since October 1992. These multiple criteria (tether strain, anomalous downwind motion, and frequency anomalies) are now used operationally to evaluate drogue presence.

Most of the global array of drifters is managed by the GDP, a component of NOAA's contribution to the Global Ocean Observing System (GOOS) and Global Climate Observing System (GCOS) and a scientific project of the Data Buoy Cooperation Panel (DBCP) of the World Meteorological Organization (Needler et al., 1999). The DBCP constitutes the data buoy component of the Joint WMO-IOC (Intergovernmental Oceanographic Commission) Technical Commission for Oceanography and Marine Meteorology (JCOMM). A full explanation of the structure and operating principles of the DBCP can be found at http://www.jcommops.org/dbcp/. The GDP maintains approximately 90% of the global drifter array, and important contributions are provided by several weather centers by way of barometer upgrades of GDP drifters, purchases of drifters for specific regional programs such as the Surface Marine observation program of the European Meteorological Services Network (E-SURFMAR) and deployment opportunities.

The scientific objectives of the GDP are to provide operational, near-real time surface velocity, sea surface temperature, and sea-level pressure observations for numerical weather forecasting, research, and *in situ* calibration and validation of satellite observations. The GDP is managed in close cooperation between NOAA/AOML in Miami, Florida, SIO in La Jolla, California, and commercial drifter manufacturers. AOML arranges and conducts drifter deployments with numerous national and international partners, processes the data, supervises data quality control, maintains files that describe each drifter, and hosts the GDP website (www.aoml.noaa.gov/phod/dac). SIO supervises the manufacturing industry, acquires most of the drifters, upgrades the technology, and develops new sensors. The GDP is funded by NOAA's Climate Program Office, and has considerable synergy with the Office of Naval Research, which supports instrument development at SIO and the deployment of additional drifters for regional circulation studies (c.f., Sections 4.1–4.4). Drifter data are made available in near-real time on the Global Telecommunications System for weather forecasting efforts, and in delayed mode (approximately three months, after quality control and interpolation) at the GDP web page and (with an additional six months' delay) at the data archive at Canada's Integrated Science Data Management

(ISDM; formerly MEDS), a Responsible National Oceanographic Data Center (RNODC) for drifting buoy data.

The modern dataset of GDP drifters includes all drifters that had a holey-sock drogue centered at 15 m depth, with a drag area ratio of ~40:1. AOML drifters with spar-shaped surface floats and holey-sock drogues were first deployed in February 1979 as part of the TOGA/Equatorial Pacific Ocean Circulation Experiment (EPOCS) (Hansen and Poulain, 1996). Large-scale deployments of the first modern GDP drifters took place in 1988 (WCRP, 1988) in the tropical Pacific Ocean. This effort was expanded to global scale as part of WOCE and the Atlantic Climate Change Program. The first GDP drifter deployments in the tropical Indian Ocean were conducted in 1985, and sustained North Atlantic deployments began in 1989. Sustained South Atlantic deployments began in 1993, at which point the global array had grown to over 500 drifters. Sustained Southern and Indian Ocean deployments began in 1994 (Niiler, 2001), and tropical Atlantic deployments in 1997 (Lumpkin and Garzoli, 2005). In October 1999, when the global array had reached >700 drifters, the OceanObs'99 conference was held to define research and operational oceanography needs, and translate these into goals for various components of the GOOS and GCOS (Needler et al., 1999). The GOOS/GCOS requirement for drifters was determined to be a global array of 1250 drifters at a nominal resolution of $5° \times 5°$, enough to provide sufficiently dense drifter SST observations to keep the globally averaged maximum potential SST satellite bias error smaller than 0.5 °C (Zhang et al., 2009). In September 2005, the global array reached the goal size of 1250 instruments, and the size of the array has subsequently fluctuated between 875 and ~1350 drifters with an annually averaged size near 1250. The density of quality-controlled observations as of July 2011 is shown in Figure 12.1.

1.2. Mean Surface Circulation

Currently, drifting buoys provide the most reliable measurements of ocean currents at nearly global coverage. Trajectories of even a few drifters can outline the structure of surface circulation in a most graphical manner (McNally et al., 1983). Observations with moored current meters, with ship-borne or lowered acoustic Doppler current profilers (ADCP) or with high-frequency radars (HFR) can provide accurate velocity measurements at particular locations, along lines or near coast lines, but their global distribution is very heterogeneous and these observations will not be addressed in this chapter.

Mean streamlines, computed from the trajectories of nearly 15,000 SVP/GDP drifters, ensemble-averaged in 1/4° bins, shown in Figure 12.2a, delineate the main features of the large-scale surface ocean circulation. These features range from large converging anticyclonic vortices in all five subtropical oceans to narrow jets, some exceeding 1 m s^{-1} in average velocity and 2 m s^{-1} in data of individual drifters. Western boundary currents (WBC) are apparent in Figure 12.2a along eastern shelves of all continents and the dynamics, inducing these currents, is described in Chapter 13 of this book. Chapter 18 addresses the complex system of currents, existing in the Southern Ocean, Chapter 14 describes eastern boundary currents, and Chapter 15 discusses the structure and dynamics of tropical currents.

Perhaps the most striking feature of Figure 12.2a is the complexity of the structure of many frontal zones emerging from the ensemble of high-quality observations, even though these data have been collected quasi-randomly in space and in time. The typical ocean jet is unstable, varying in time, and surrounded by an area of elevated eddy kinetic

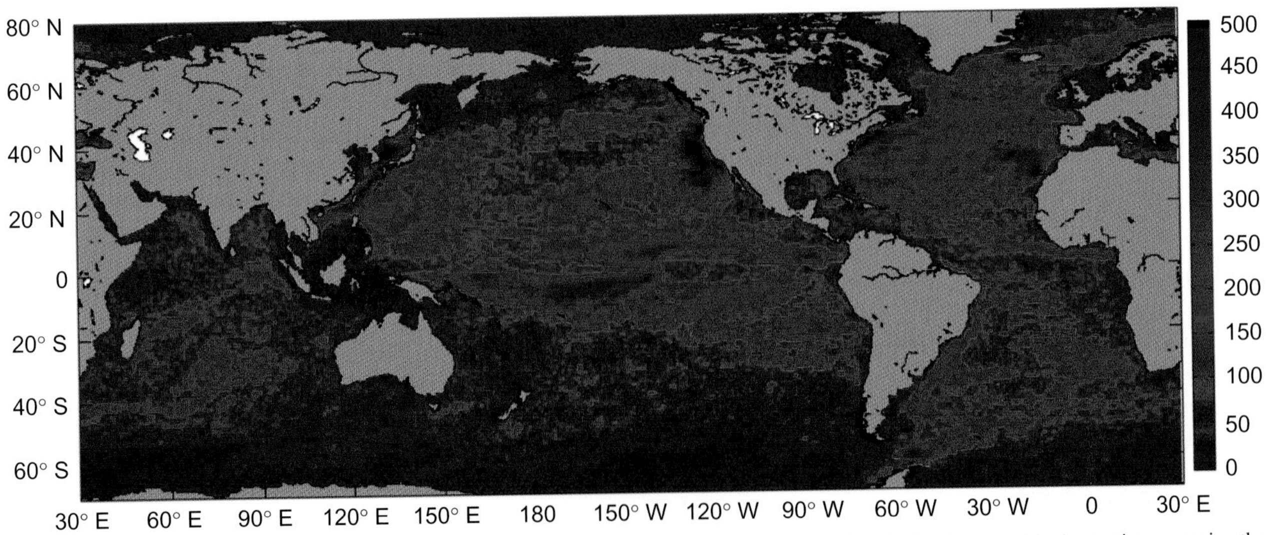

FIGURE 12.1 Density of drifter observations (drifter days per square degree) in the quality-controlled GDP database, with observations spanning the period February 15, 1979–July 1, 2011. Black indicates >500 drifter days per square degree.

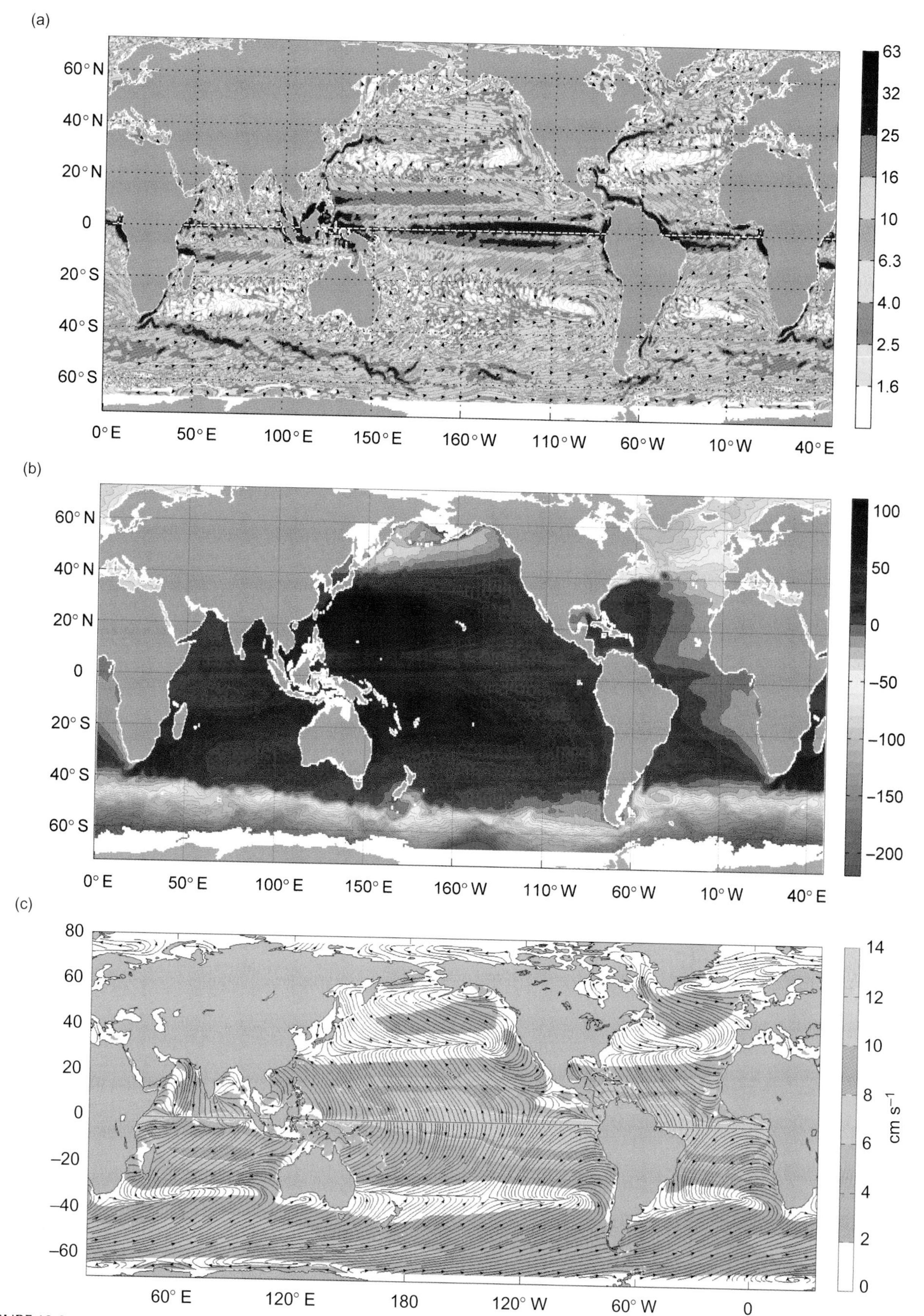

FIGURE 12.2 Ensemble-mean drifter streamlines (a), mean dynamic topography (b), and mean Ekman currents at 15 m depth (c). Units are cm s^{-1} (a and c) and cm (b). *Reproduced from Maximenko et al. (2008).*

energy (EKE) (see Chapter 8). Eddies are known to effectively mix properties of the ocean. One would expect that this mixing (and, correspondingly, averaging data in time) would reduce all contrasts, would broaden fronts and jets, and would eliminate small-scale details from their systems. Yet, many jets remain narrow in the multiyear ensemble average. The Agulhas return current demonstrates as many as seven meanders in the long-time mean (Pazan and Niiler, 2004), while at various times the number of the meanders may vary (Boebel et al., 2003). Centurioni et al. (2008) report four permanent meanders of the California Current, existing despite strong ocean variability in the region. The Kuroshio Extension and Subarctic Front form a system of meandering and branching fronts, visible in time or ensemble averages (e.g., Niiler et al., 2003a). Although the Azores Current exhibits tremendously complex structure on snapshots, its signature in the multiyear average extends all the way across the North Atlantic and nearly reaches the American coast in the west (Klein and Siedler, 1989; Niiler et al., 2003b). Multiple fronts/jets in the ACC, anchored to features on the bottom topography, create tremendously complex flow structure in the Southern Ocean (Hughes, 2005).

The robustness of these mesoscale structures in the time mean can be partly explained by their attachment to bathymetric features. The stability of the extensions may be provided by association of ocean jets with fronts. Baroclinic fronts at the ocean surface are expressions of the three-dimensional interface, separating large water masses. The role of fronts as barriers was studied in the North Pacific by Niiler et al. (2003a) and in the North Atlantic by Brambilla and Talley (2006).

Even the statistics of "perfect" drifters are subject to a number of biases (Davis, 1991), associated with heterogeneous distribution of buoys in space and in time (Figure 12.1). Some heterogeneity results from the deployment scheme of the drifters and induces biases common to all kinds of *in situ* observations (high density of observations in some regions and sparse coverage in other, peak of observations in one season or year followed by background activity at other time periods, etc.). Other biases are essentially due to the Lagrangian nature of drifter observations; in these cases, the sampling scheme is not independent from the sampled velocity field. These effects include dispersion of buoys from areas of high drifter density to areas of lower density, stronger dispersion in the direction of stronger mixing (Freeland et al., 1975), and the effects of surface convergences and divergence that attract or repel drifters from certain areas (e.g., see the pattern of drifter density in the equatorial Pacific in Figure 12.1).

Additional biases may be introduced when Lagrangian and Eulerian statistics are combined. For example, Uchida et al. (1998) found that with a perfect correspondence between drifter- and altimetry-derived velocities collocated along the ASUKA (The Affiliated Surveys of the Kuroshio off Ashizuri–Misaki) line, drifters preferentially sampled locations/periods of larger velocities. As a result, mean drifter currents were up to 30% stronger than the Eulerian mean. Maximenko (2004) showed that this bias appears only in the statistics of Lagrangian particles crossing an Eulerian line (more drifters cross the line when the current is stronger, and no drifter can cross the line at zero velocity) and is eliminated if bin averages are used or data are interpolated onto a space–time Eulerian grid. In some cases, satellite observations, calibrated or referenced using collocated *in situ* measurements, can help to mitigate biases by providing nearly uniform Eulerian datasets in broad space and timescale ranges. Techniques used to assess mean geostrophic and Ekman currents are exemplified in the following sections.

2. GEOSTROPHIC SURFACE CIRCULATION

In most cases advection of properties by ocean currents makes dynamic systems nonlinear and, therefore, complex. However, there are a few regimes in the range of observed ocean parameters which offer simple balances (see Chapter 11). An example of such a regime is geostrophy, in which the Coriolis force acting on a moving water parcel is compensated by the horizontal pressure gradient. For slow, low-frequency currents at the sea surface away from the equator, for which temporal changes, nonlinearity, and viscosity can be neglected, this balance can be simplified to

$$\left(\vec{k}\cdot f\right) \times \vec{V}_g = -g\cdot\nabla h, \tag{12.1}$$

where $\vec{k}$ is the vertical unit vector, f is the Coriolis parameter, g is the acceleration due to gravity, $\vec{V}_g$ is the geostrophic velocity vector, h is the sea level, and ∇ is the horizontal gradient operator. Equation (12.1) is often used to assess currents using sea level from satellite altimetry or using dynamic topography derived from hydrographic vertical profiles under the assumption that currents become negligibly small at some depth.

2.1. High-Resolution Mean Dynamic Topography

The straightforward way to calculate the mean absolute dynamic ocean topography (MDOT), the signal in sea level due to the ocean circulation, is through determining the deviation of the mean sea surface (MSS), describing the absolute shape of the ocean surface, from the geoid, the equipotential surface defined by rotation and gravity of the earth with its complex distribution of mass. Theoretically, the geoid coincides with the ocean surface in the absence of currents. Both the MSS and the geoid models have severe limitations. The MSS products are derived by combining many satellite altimetry missions, from repeat-track to geodetic, providing very different horizontal resolution and covering different time periods. On the planetary scale, variations of the geoid due to gravity exceed 150 m, while the range of the sea level due to the ocean dynamics is only about 3 m. The GRACE (Gravity Recovery and Climate Experiment; Tapley et al., 2004) mission has improved the model of the geoid, allowing derivation of the MDOT with 400–500 km resolution. The GOCE (Gravity Field and Steady-State Ocean Circulation Explorer; Johannessen et al., 2003) satellite further advanced the geoid model, allowing MDOT products with 150–250 km resolution (c.f., Le Traon et al., 2001; Rio et al., 2011b; Haines et al., 2011; Knudsen et al., 2011). In addition to accurate measurements of gravity, satellite missions can also measure changes in the gravity field caused by movement of large water masses. These variations are beyond the accuracy of and are not accounted for in the modern MDOT products.

To advance MDOT products to the smaller scales resolved by a constellation of satellite altimeters, a few groups around the world developed techniques that allow assessment of the MDOT based on the synthesis of satellite and *in situ* observations. The idea is largely based on the use of Equation (12.1). The data from an altimeter, flying along repeat tracks, can be freed from the unknown geoid by subtracting the time-mean. The absolute reference for the time-varying part (sea-level anomaly gradient) can then be derived from *in situ* velocity measurements, for example, provided by drifters. Imawaki et al. (2001) successfully applied this technique to study variations of transport of the Kuroshio Current (KC) along the ASUKA line south of Japan.

Two-dimensional velocity maps (Ducet et al., 2000) are computed and distributed by AVISO (Archiving, Validation, and Interpretation of Satellite Oceanographic Data, 1996) as weekly maps on a global 1/3° or 1/4° grid. On large scales, away from strong jets, the biggest error in Equation (12.1) originates from Ekman currents (see Section 3.1). Rio and Hernandez (2003, 2004) removed Ekman currents from drifter velocities using correlations with the local low-frequency wind in 5° boxes. Maximenko et al. (2009) calculated latitude-dependent regression coefficients, using mean NCEP reanalysis wind and the GRACE model of the geoid. In both techniques, coarse-resolution MDOT, derived with help of the geoid model, was refined on mesoscale, using sea-level gradients, estimated from Equation (12.1). To further improve spatial coverage of the *in situ* observations, Rio and Hernandez (2004) added historical hydrographic profiles. Recently, Rio et al. (2011a) updated the Rio'05 MDOT, using accumulated data and a more advanced Ekman model.

The MDOT map of Maximenko et al. (2009) is shown in Figure 12.2b. Major fronts, jets, and currents, described in Section 1.2 are even more sharply delineated in this figure than in the ensemble average of drifter velocities in Figure 12.2a. This is because the "eddy noise," significant in the random ensemble of drifters, has been efficiently suppressed by the technique, using (quasi-) continuous data provided by satellite altimetry.

Outside of strong currents, streamlines of geostrophic velocity differ significantly from streamlines of full (or drifter) velocity. These differences are particularly remarkable in the subtropics and tropics and are defined by the mean Ekman currents, shown in Figure 12.2c. Easterly winds, dominant in the equatorial regions in the Pacific and Atlantic, induce poleward Ekman currents on both sides of the equator. This divergence in the upper ocean is compensated by the upwelling waters fed by the subsurface geostrophic flows, directed toward the equator as seen in Figure 12.2b.

Mean surface circulations in Pacific and Atlantic subtropical gyres provide an excellent illustration of the complex underlying dynamics and kinematics. Convergence of Ekman currents along approximately 30° latitude indicates production by the wind stress of anticyclonic vorticity. According to Sverdrup (1947), the steady state under such forcing can be achieved through the equatorward flow, advecting planetary vorticity (see Chapter 11). Such flows are seen in Figure 12.2b throughout all four subtropical gyres with dynamic topography monotonously increasing from east to west.

2.2. Striated Patterns

Improved MDOT not only allows enhancement of the details of known currents, but also reveals new structures. Figure 12.3, showing the map of zonal geostrophic velocity calculated from such a MDOT and high-pass filtered in two-dimensional (latitude–longitude) space, highlights a web of nearly ubiquitous jet-like features (Maximenko et al., 2008). These features (striations) also exist in long-time ensemble averages of subsurface data of XBT (expendable bathythermograph) and velocities of near-surface drifters. However, in the face of eddy noise that is typically an order of magnitude stronger than striations, the striations are only significant in time averages of quasicontinuous datasets such as satellite altimetry. XBT profiles suggest that the subsurface striations extend to at least 400–800 m depth without noticeable tilt. Numerical models suggest that striations may remain coherent vertically throughout the entire water column (Maximenko et al., 2008).

The dynamics of these striations is not understood yet. Their zonal orientation and characteristics along with their ubiquity lead Nakano and Hasumi (2005), Galperin et al. (2004), and Richards et al. (2006) to conclusions on the relevance of the "Rhines mechanism." This mechanism (Rhines, 1975) is based on the theory of two-dimensional turbulence, whose free development induces zonal jets

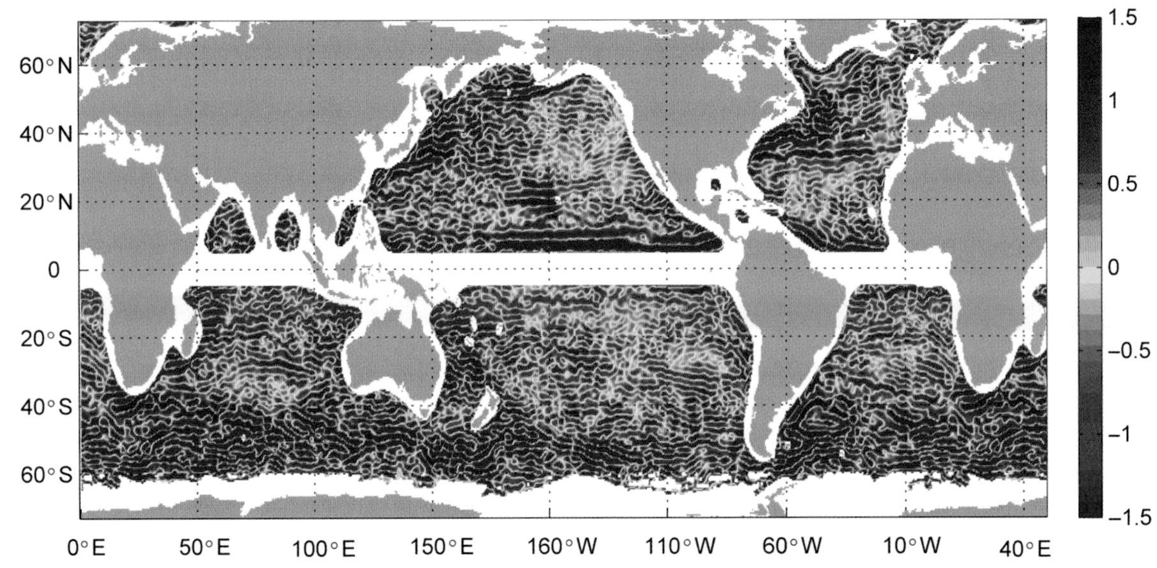

FIGURE 12.3 1993–2002 mean zonal surface geostrophic velocity calculated from the MDOT of Maximenko and Niiler (2005) high-pass filtered with a two-dimensional Hanning filter of 4° half-width. Units are cm s^{-1}. *Reproduced from Maximenko et al. (2008).*

through inverse energy cascade toward larger scales, which becomes limited in the meridional direction on the β-plane by the Rhines scale $L_R = \sqrt{2u'/\beta}$, where u' is the r.m.s. eddy velocity and β the meridional gradient of the Coriolis parameter. Baldwin et al. (2007) demonstrated how zonal jets can be generated by breaking Rossby waves, homogenizing potential vorticity within zonal bands.

Careful analysis of the characteristics of time-mean (or stationary) striations reveals deviations from the predictions of these theories. Most importantly, many of these striations do not appear to be inertial jets. This is particularly clear in the eastern parts of subtropical oceans, where mean geostrophic flows are directed toward the equator (Figure 12.2b). When moving across the striations, water parcels only slightly deviate toward the east or west along the axes of striations. In this sense, stationary striations in the east demonstrate properties of waves rather than jets. In addition, these striations appear to be not exactly zonal but have their axes systematically tilted in the zonal direction at angles, consistent with the dynamic of linear Rossby waves trapped at their eastern edges.

Such dynamics can be described as the β-plume (Pedlosky, 1996), generated by a local source of vorticity. Classical β-plumes form a set of nearly zonal currents and extend infinitely to the west from the forcing areas. β-plume dynamics have been suggested to be responsible for the formation of the Azores Current (Kida et al., 2007) and the Hawaiian Lee Countercurrent (Xie et al., 2001). Recently, Centurioni et al. (2008) have found the link between the striated pattern of zonal current velocity off the California coast and areas of upwelling and downwelling, induced by nonlinear interaction of wind-forced Ekman currents and permanent meanders of the California Current. In support of the β-plume mechanism, analysis of striations in Figure 12.3 shows that many of them have at their eastern tips features (such as an island or sea mount or cape in the coast line, etc.) that may anchor vorticity production by wind stress curl or through a more complex dynamical process.

While averaging data in time to suppress the noise from eddies is important (Huang et al., 2007), the very same averaging may produce artificial striations by smearing moving eddies (Schlax and Chelton, 2008; Scott et al., 2008). Huang et al. (2007) and Scott et al. (2008) show that without applying time averaging, the anisotropy of velocity data is very low and increases with the increase of the averaging period. Schlax and Chelton (2008) used a simple statistical model to demonstrate that the signal from the strongest eddies may remain significant even in very long time averages. At the same time, Scott et al. (2008) showed that the distribution of eddies in space is not random and that in some regions the eddies are moving along preferred paths. Although not included into the model of Schlax and Chelton (2008), this heterogeneity is seen in eddy trajectories, shown in their Figure 1.

Buckingham and Cornillon (2013) used data of Chelton et al. (2011) to isolate contribution of eddies to striations. They showed that time-averaged signal from correctly positioned eddies correlates highly with striations in MDOT. At the same time, they found that, contrary to Schlax and Chelton (2008), contributions from moderate-amplitude eddies is important and amplitudes of striations significantly exceed expectations from eddies.

Maximenko et al. (2005) discussed the role of time averaging for a different kind of striation—quasizonal jet-like features seen in the satellite sea-level anomaly as crests and troughs propagating in the subtropical latitudes toward the equator. They showed that eddy "streaks" in time averages, corresponding to the distances traveled by individual eddies during the average time, are significantly shorter than the length of the striations. In a wide range of periods, individual streaks interact without canceling each other; this is only possible if eddies of the same sense of rotation are aligned along the axes of striations.

The mechanisms that set these alignments or preferred areas of eddy formation or preferred paths of their propagation are not understood yet, but they point to the existence of higher organization of the eddy field than suggested by today's theory. While eddy terms in the vorticity balance of stationary striations are significant, their contribution to the balance differs between the striations (Melnichenko et al., 2010). Striations, having thermal signature at the sea surface, seem to be able to sustain their structure through the locally induced wind stress (Sasaki and Nonaka, 2006). A number of groups are currently working to understand the correspondence between the striated patterns of ocean currents, linear waves, nonlinear eddies, instability of the large-scale flow, and various kinds of external forcing.

2.3. Variability and Trends

EKE derived from AVISO geostrophic currents and from drifter observations is shown in Figure 1 of Chapter 8. For the drifter observations, only drogued drifters were used; velocities were first low-pass filtered at 5 days to remove tides and inertial oscillations, then the time-mean currents—calculated in 1° bins—were removed. As noted by previous researchers (Fratantoni, 2001), the drifter observations have higher EKE throughout most of the world oceans, due to the presence of ageostrophic motion (see Section 3) and geostrophic motion not resolved in the smoothed AVISO fields.

Long-term variations in surface EKE have been observed in a number of regions. A southward shift in the position of the Gulf Stream extension has been inferred from altimeter-derived EKE for the period 1993–2008 (Lumpkin et al., 2009). The Kuroshio Extension has been observed to vary in its stability on decadal timescales, transitioning from a relatively stable phase with lower EKE to an unstable phase with higher EKE, an oscillation perhaps driven by the Pacific Decadal Oscillation (Qiu and Chen, 2005). Over the time period 1993–2000, the confluence of the Brazil and Malvinas Currents in the southwest Atlantic Ocean shifted southward (Lumpkin and Garzoli, 2010), with a corresponding shift in the EKE maximum of the confluence front. This trend may have been part of a multidecadal oscillation forced by SST anomalies imported from the Indian Ocean along the Agulhas–Benguela pathway (Lumpkin and Garzoli, 2010).

The spatial pattern of the trend in the sea-level rise, observed by satellite altimeters between 1993 and 2009 (Willis et al., 2010), has a complex structure and reflects low-frequency changes in surface circulation. More on seasonal and interannual variability can be found in Chapter 24 and decadal variations in Chapter 25.

3. AGEOSTROPHIC CURRENTS

While geostrophy, discussed in the previous section, offers a simplified description of an important class of currents, most of the processes responsible for generation, equilibration, and damping of the currents are more complex. The following subsections will discuss important examples of ageostrophic currents.

3.1. Motion Driven by Wind

The most energetic ageostrophic currents near the ocean surface are directly wind driven. The frequency of the ocean surface's inertial (resonant) response to large-scale wind variations is set by the Coriolis effect, with potential modification by the background vorticity (Kunze, 1985). The resulting near-inertial oscillations are a prominent feature of Eulerian (c.f., Alford and Whitmont, 2007) and Lagrangian (c.f., Elipot and Lumpkin, 2008) spectra, and may represent a significant portion of the energy input needed to maintain the abyssal stratification via interior diapycnal mixing (Munk and Wunsch, 1998). The distribution of near-inertial variance in the mixed layer has been mapped globally using surface drifter observations (Chaigneau et al., 2008); compared to the planetary vorticity set by the Coriolis parameter f, the peak frequency of near-inertial oscillations is blue-shifted equatorward of 30° N/S latitude and is demonstrably affected by the background geostrophic vorticity (Elipot et al., 2010). It is uncertain how much of this energy leaves the upper ocean in trapping regions (c.f., Polzin, 2008) to propagate into the ocean interior.

At lower frequencies, the ocean response to wind forcing was first described by Ekman (1905). With the assumption of a constant diffusivity in an upper boundary layer, Ekman showed that this response was 45° to the right (left) of the wind at the surface in the northern (southern) hemisphere, with the exponentially decaying currents rotating further to the right (left) with increasing depth within the turbulent boundary layer, and a net transport at 90° to the right (left) of the wind. Subsequent research has questioned the assumption of a constant diffusivity and the resulting specific details of the Ekman spiral, although some observations are quantitatively consistent with Ekman's spiral and net transport (c.f., Chereskin, 1995). Using 1503 drogued GDP drifter observations in the period 1988–1996, Ralph and Niiler (1999) demonstrated the validity of Ekman theory across the tropical Pacific basin. Ralph and Niiler (1999) first removed time-mean geostrophic currents using hydrography-based climatologies, along with the wind-driven downwind slip (0.07 cm s^{-1} per 1 m s^{-1} wind; Niiler et al., 1995), and low-pass filtered the residual drifter currents at 5 days to remove near-inertial oscillations and high-frequency tides. They then considered a number of models for regressing the residuals onto the winds in 2° (meridional) by 5° (zonal) bins. They found that the best-fit model was close to $u_{Ek} = Au_*/\sqrt{|f|}$, where u_{Ek} is the magnitude of the Ekman current, $u_* = \sqrt{\tau/\rho}$ is the friction velocity due to the surface wind stress τ, ρ is the water density, and the best-fit coefficient $A = 0.065 \pm 0.002\ \mathrm{s}^{-0.5}$. The Ekman current was directed to the right (left) of the wind in the northern (southern) hemisphere, with an off-wind angle varying according to the ratio of the drogue depth to the Ekman scale depth $H_* \propto u_*/\sqrt{|f|}$ in a manner consistent with a rotating Ekman spiral (Figure 5 of Ralph and Niiler (1999); Figure 4.1.9 of Niiler (2001)). Niiler (2001) updated this study using drifter data through 1999, and found a similar result with $A = 0.081 \pm 0.013\ \mathrm{s}^{-0.5}$ for NCEP reanalysis winds.

Rio and Hernandez (2004) expanded upon Ralph and Niiler's (1999) study by removing the time-varying geostrophic motion using altimetry. Rio and Hernandez (2004) modeled the Ekman response as $\vec{u}_{\mathrm{Ek}} = b\,\vec{\tau}\,\exp(i\theta)/\sqrt{|f|}$ (i.e., proportional to wind stress, not wind speed as in Ralph and Niiler, 1999) and allowed the coefficients b and angle off the wind θ to vary spatially and seasonally on a 5° grid. They found the angle θ varied in a manner consistent with spatial and seasonal changes in upper-ocean stratification and consequent stretching/compressing of the Ekman spiral compared to the 15 m drogue depth of the drifters. An update of this Ekman parameterization was used to derive the 2009 CLS mean dynamic topography (Rio et al., 2011a). Figure 12.2c shows the mean streamlines of Ekman currents computed using the NCEP reanalysis wind and data from drifters, altimetry, and GRACE.

Although drifter slip has been measured at wind speeds up to 8 m s^{-1}, little is known of their water-following characteristics at higher wind speeds (Niiler et al., 2003a). Wind models with spatially varying coefficients, such as that of Rio and Hernandez (2004), indicate a larger downwind motion at higher latitudes than suggested by the Ralph and Niiler (1999) model with its constant coefficient. This may indicate enhanced downwind slip at high wind and wave states. A number of candidates could account for this: it is possible that wave-driven Stokes drift is generating significant downwind motion in the extreme wave states found, for example, in the ACC (Niiler et al., 2003b). Alternatively, the drifters may be trapped in Langmuir jets at the convergence of the wind- and wave-driven cells (c.f., Thorpe, 2005), and thus not exhibiting slip with respect to currents at 15 m but rather biased sampling compared to an Eulerian average. It is also possible that in extreme wave states the drogue is jerked upward such that its mean depth is significantly above 15 m. Direct measurements of pressure and currents at the top and bottom of the drogue are obviously needed in these conditions to better understand the measurements.

3.2. Centrifugal Effects

Eddy kinetic energy from altimetry is often compared to that from drifters, with the difference attributed to wind-driven motion and smaller scales not resolved by altimetry. Another source of discrepancy, when altimetric currents are derived from pure geostrophy, is the centrifugal effect of the nonlinear term in the momentum equation. This can play a nonzero role in the mean momentum budget in a region if vortices of one sign dominate, for example the cyclonic cold-core rings south of the mean Gulf Stream jet. To see the effect of this term, consider an azimuthally-symmetric vortex governed by the momentum equation

$$v^2/r + fv = g\partial_r \eta, \tag{12.2}$$

where r is the distance from the center of the vortex, v the speed of the vortex, g is gravity, f the Coriolis parameter, η is sea surface height, and $\partial_r \eta$ is its derivative along r. Suppose altimetry gives $\eta = fa(r^2 - R^2)/2g$ to radius R and $\eta = 0$ for $r > R$, with a constant, and geostrophy is assumed (the first term in the momentum equation is dropped). Then $v_{geo} = ar$ within the solid-body vortex. However, when the centrifugal term is included, a Taylor series expansion for small Rossby number a/f gives $v = ar(1 - a/f + \ldots)$. For a cyclonic vortex ($a/f > 0$), the actual velocity is less than the estimate from pure geostrophy; for an anticyclonic vortex, the magnitude of the velocity is higher. For the Kuroshio south of Japan, Uchida et al. (1998) showed that centrifugal effects due to the curvature of the current axis explain well the 10% difference between drifter and geostrophic velocities.

Fratantoni (2001) compared EKE from altimeter-derived geostrophic velocities and from 2-day low-passed drifter velocities in the North Atlantic basin, and found that drifter EKE was generally higher by O (100 cm^2 s^{-2}). He attributed this to the directly wind-forced motion in the drifter velocities. However, he found a fascinating pattern in the drifter minus altimeter EKE (his Plate 8) in the immediate region of the Gulf Stream front: altimeter EKE exceeded drifter EKE by >250 cm^2 s^{-2} immediately south of the mean Gulf Stream extension, while the opposite was true immediately to the north. Fratantoni (2001) attributed this to differences in the Gulf Stream location in the two datasets due to differences in resolution and smoothing between the data. However, this pattern is exactly what one would expect if the centrifugal term is playing a significant role in the surface momentum budget, as noted by Niiler (2003a) who found that the centrifugal term can alter the magnitude of currents on either side of the Kuroshio jet by as much as 25%. Consistent with this interpretation, Niiler (2003a) found maximum values of altimeter-derived geostrophic EKE in the Kuroshio extension region shifted south of the maximum indicated by drifters (their Figure 7). Maximenko and Niiler (2006) mapped dominant eddies using the skewness of the probability density function of sea-level anomaly $\langle h'^3 \rangle / \langle h'^2 \rangle^{3/2}$, derived from satellite altimetry, and energy-weighted angular velocity $\omega_{\mathrm{E}} = \left\langle \vec{V} \times \mathrm{d}\vec{V}/\mathrm{d}t \right\rangle / \left\langle \vec{V}^2 \right\rangle$, computed from drifter trajectories. The two maps demonstrate remarkable resemblance and reflect the prevalence of anticyclonic eddies north of the Gulf Stream and cyclones on the southern side.

3.3. Nonlinear Interactions with Baroclinic Features

The ageostrophic component of ocean currents also plays a major role in the three-dimensional upper-ocean circulation, for which much of our knowledge derives from theoretical and idealized, process-modeling, numerical studies.

The secondary circulation created by a wind blowing parallel over a geostrophically balanced current was studied by Lee et al. (1994) who concluded that, after the near-inertial fluctuations dissipate, the Ekman vertical velocity is enhanced by the horizontal shear of a geostrophic flow when a nonlinear model is used. In this case, a downwelling/upwelling pattern emerges in the mixed layer. For a wind blowing in the direction of the current, the upwelling area is narrow and confined in the region of positive relative vorticity and the downwelling region is broad and on the side of negative relative vorticity patch. The situation is reversed if the wind and the jet flow in opposite directions. While this mechanism could potentially play a role in determining the magnitude of the vertical fluxes of heat, momentum, and nutrients, the observational evidence is still scant.

When a wind stress is applied to a subtropical anticyclonic eddy (Lee and Niiler, 1998), cold water is advected over warm water, which induces vertical mixing and downwelling. A nonlinear numerical model solution shows that a downwelling region results, which is larger and more intense than the upwelling area located on the opposite side of the eddy. The residual secondary circulation induced by the nonlinear interaction between the wind-induced Ekman flow and the eddy resembles a jet that cuts across the eddy core. The ageostrophic velocity computed from a Regional Ocean Modeling System (ROMS) simulation of the California Current System (CCS) (Marchesiello et al., 2003) shows a pattern that is consistent with the effect described by Lee and Niiler (1998) when a wind parallel to the coast is blown over the permanent meanders of the CCS (Centurioni et al., 2008), which are essentially a series of alternating cyclonic and anticyclonic mesoscale features.

4. REGIONAL SURFACE OCEAN DYNAMICS

The true complexity of the dynamics of surface currents can be illustrated through more careful look at particular regions, highlighted in the following sections.

4.1. Drifter Studies in the California Current System

The CCS is a broad and generally southward eastern boundary current of the subtropical North Pacific (Chapter 14). Upwelling in the CCS region occurs in summer (i.e., when northwesterly winds occur) between 35° N–38° N and 46° N, and downwelling primarily occurs in winter in the same region (i.e., when the winds shift to the SE). South of 35° N–38° N upwelling occurs throughout the year (Huyer, 1983). In the past four decades, Lagrangian drifters were used extensively to map the structure of the surface circulation of the CCS and of the near-shore surface flow. Investigations of the CCS with Lagrangian drifters include regional studies (e.g., McNally et al., 1983; Davis, 1985b; Poulain and Niiler, 1989; Swenson et al., 1992; Swenson and Niiler, 1996; Winant et al., 1999; Austin and Barth, 2002) as well as studies leveraging on the large historical dataset accumulated through nearly three decades of deployments (Centurioni et al., 2008) of GDP drifters (Niiler, 2001).

The drifter-derived, 15-m deep geostrophic velocity is often approximated by subtracting the Ekman velocity (e.g., Ralph and Niiler, 1999; Centurioni et al., 2009) from the measured drifter velocity. Such an approximation holds for most large-scale flows, including when currents are not in cyclostrophic balance (Equation 12.2), when the flow is weakly nonlinear, when drifters are not inside Langmuir cells and when the Stokes' drift is not a significant fraction of the total velocity. Additionally, the tidal and near-inertial components of the flow are filtered out along the drifter track. Because of their characteristic space and timescales, the ageostrophic, non-Ekman currents, are in most cases filtered out by averaging all of the available observations within rather large spatial bins. An interesting aspect of the CCS is that the vector correlation between the drifter-derived geostrophic velocity residual, computed with respect to the mean, and the corresponding quantity obtained from satellite altimetry data is large offshore of the US west coast (>0.8) and drops to low or negative values in a near-shore strip about 200 km wide, suggesting that the near-shore flow is either nongeostrophic or has time and spatial scales, which are not adequately resolved by the weekly, 1/3°, satellite altimetry maps (Centurioni et al., 2008). Vigorous highly nonlinear eddies 600–800 m deep (Chereskin et al., 2000) as well as cold water filaments (Strub et al., 1991) have been observed in the upwelling regions of the US west coast and the timescales of those coherent structure range from 2 to 7 days near the coast (Swenson et al., 1992; Chereskin et al., 2000), and are of the order of 30 days and longer as the eddies move further offshore (Swenson et al., 1992).

The combined analysis of SVP/GDP drifter and satellite altimetry data has revealed the existence of at least four permanent meanders of the CCS, which are connected to a set of slanted bands of primarily eastward zonal flow (Figure 12.4; see also Centurioni et al., 2008). The spatial structure of the CCS meanders is seasonal (Figure 12.5), with the strongest near-surface currents observed in summer (JAS) and the most pronounced meanders occurring in fall (OND).

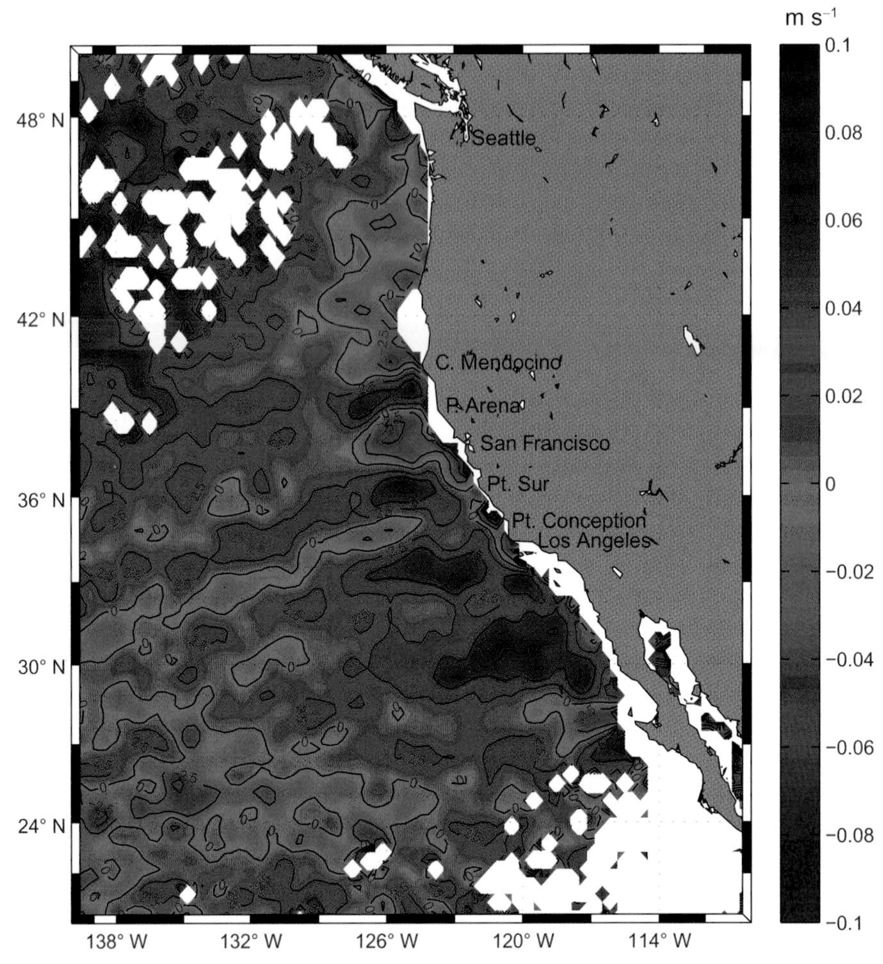

FIGURE 12.4 Near-surface unbiased geostrophic zonal flow in the eastern North Pacific. Year-long ensemble-average between July 1992 and June 2011. Note the four bands of eastward velocity whose easternmost expression is colocated with the CCS meanders. Refer to Centurioni et al. (2008) for the methodology used to compute the velocity field.

An interesting scientific question is which physical processes are responsible for the meandering structure of the CCS system. The acceleration of the near-surface water particles computed from drogued drifters is significantly larger in the near-shore strip and where the meanders occur than further offshore in the proximity of the "eddy desert" region (Cornuelle et al., 2000). The implication of this finding is that the departure from a pure geostrophic balance is larger in the CCS meanders than elsewhere. Eddy and mean vorticity fluxes are likely to act as one of the forcing mechanisms that dictate the shape of the CCS (Centurioni et al., 2008).

Direct velocity observations from Lagrangian GDP drifters are therefore particularly important in regions where the departure from geostrophy is largest or where the time and space scales of the flow are not adequately sampled by the ongoing satellite altimetry missions. The residence time of the drifters in the CCS is short especially near the coast because the surface flow there is highly divergent. Long-term and sustainable deployment programs are therefore important to further our understanding of the dynamics of the near-shore part of the CCS.

4.2. Drifter Studies off Senegal

Another surface divergent region where drifters' deployments have begun in recent years is the tropical North Atlantic off Senegal (Chapter 14). The Senegalese coast has perhaps the largest seasonal and interannual variation of oceanographic conditions of the entire west coast of Africa. It is also the African coastal region where the Pacific El Niño's influence is felt the strongest (Bhatt, 1989; Roy and Reason, 2001). On a seasonal and interannual basis, the surface waters from the equatorial region moving northward are separated by a strong frontal region perpendicular to the coast from the waters with origins in the subtropical gyre (Wooster et al., 1976) that move generally southward (Lumpkin and Garzoli, 2005). This frontal region moves south to north along the Senegal coast on both seasonal and interannual timescales (Wooster et al., 1976).

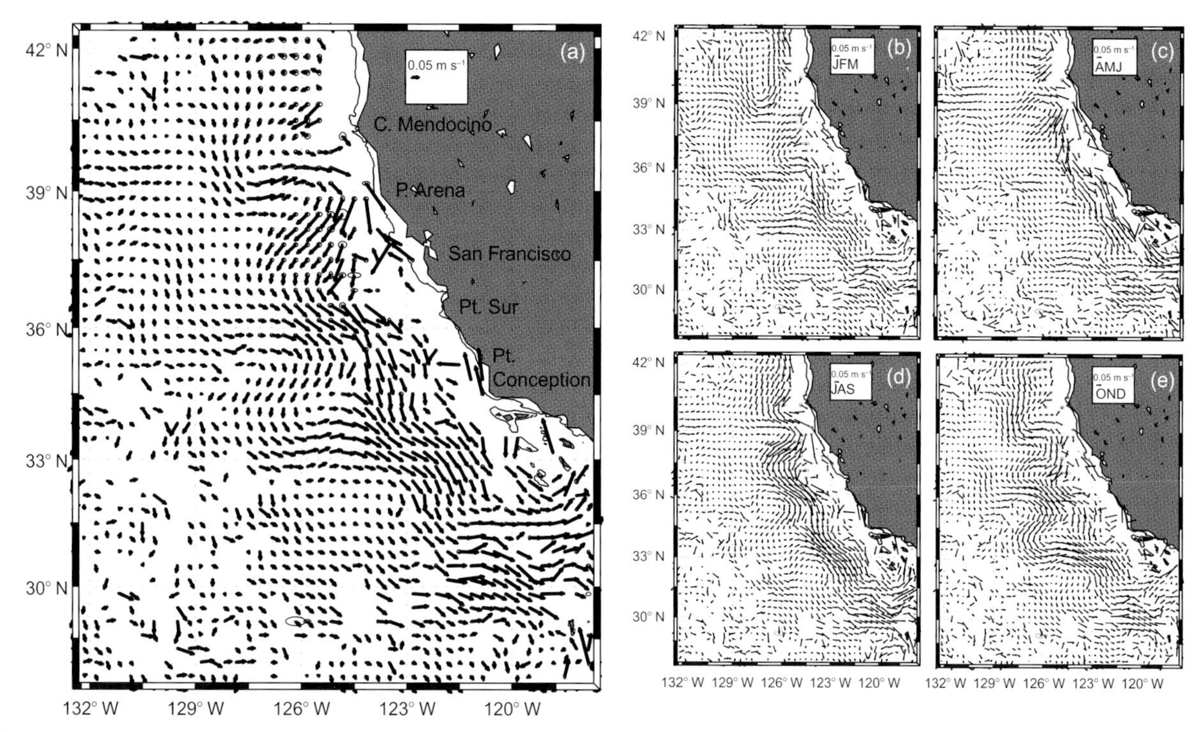

FIGURE 12.5 Near-surface unbiased geostrophic velocity field in the California Current System. Year-long (a) and seasonal (b–e) ensemble-average between July 1992 and June 2011. Note the increase of the intensity of the currents in summer (JAS) and the larger extent of the meanders in fall (OND).

Time-averaged confluence front can be seen in Figure 12.2a around 15–20°N. Although the drifter observations are sparser in this area than in any other part of the North Atlantic (Figure 12.1), a time-mean geostrophic, near-surface flow to the north, is indicated to exist by the data (Figure 12.2a).

The strongest upwelling system appears to be associated with the southward wind-driven circulation that is set up north of Cape Verde during La Niña years, generally in the early spring season (Roy and Reason, 2001). Under these conditions, a strong jet of upwelled water flows southward and leaves the coast at the tip of Cape Verde, and appears to set up a quasi-zonal frontal system nearly 10° of longitude into the eastern North Atlantic (Wooster et al., 1976). The flow patterns off the coast of Senegal express both upwelling and downwelling conditions and are further complicated by seasonal fresh water inflow from the Senegal, Gambia, and Casamance rivers (Bhatt, 1989).

Lagrangian GDP drifters are currently being used to improve the description of the near-surface circulation off the coast of Senegal. Lagrangian methods, when joined together with satellite data, are very powerful in determining the patterns of flow, especially when repeated over several years and seasons. But these data can be very misleading when small numbers over short timescales are used to make inferences. The emerging picture from more than two years of targeted drifter deployment off Senegal is that of a year-round, offshore directed near-surface flow north of Cap Vert peninsula, the westernmost point of Africa. In the ongoing project, the time-dependent sketches obtained by combining the unbiased geostrophic velocity field derived with the technique of Centurioni et al. (2008), the time-dependent, altimetry-derived geostrophic currents, and the Ekman flow were computed at weekly intervals and have confirmed the existence of broad filaments of intense (in excess of 15–20 cm s^{-1}) currents and rings off the coast of Senegal.

4.3. Interaction of the Kuroshio with the South China Sea

The KC, a major WBC of the North Pacific (Chapter 13), is about 140 km wide east of the islands of Luzon and Taiwan (Figure 12.6) and flows essentially northeastward with a speed of 1–1.5 m s^{-1} (Centurioni et al., 2009). The Kuroshio emerges out of a region dominated by eddy-like flow just north of where the North Equatorial Current ends in the proximity of the Philippines coast (Centurioni et al., 2004; Rudnick et al., 2011). The volume transport of the KC is between 23 and 32 Sv (1 Sv = 10^6 m^3s^{-1}) at 18.5° N (Yaremchuk and Qu, 2004) and averages 21.5 Sv through the East Taiwan Channel (Johns et al., 2001). Part of the KC seems to recirculate southward east of the Ilan Ridge and northeastward east of the Ryukyu Islands (Nitani, 1972), as also suggested by the synthesis of drifter and satellite altimetry data (Imawaki et al., 2003).

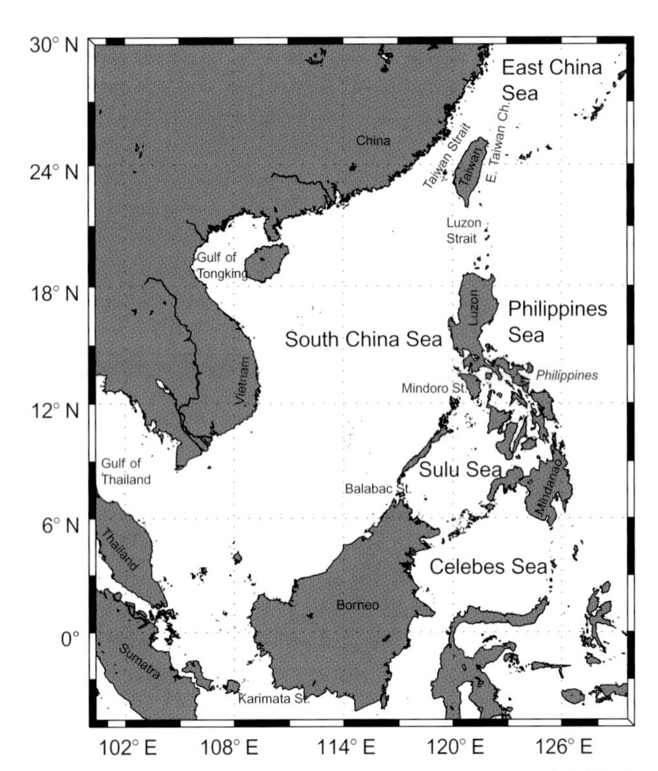

FIGURE 12.6 Map of marginal seas in the southwestern part of the North Pacific. Interaction of the Kuroshio with the South China Sea is described in Section 4.3 and with the East China Sea in Section 4.4.

The interaction of the KC with the South China Sea (SCS, Figure 12.6) occurs primarily through the Luzon Strait (LS), which has a sill depth generally above 2400 m and a bathymetry complicated by numerous islands and channels. The KC usually flows straight north across the LS, but occasionally it intrudes into the SCS, sometimes forming a loop current (Hu et al., 2000). A somewhat similar phenomenon occurs in the Gulf of Mexico where the Gulf Stream leaps from Yucatan to Florida but occasionally intrudes as a deeply penetrating loop current that sheds eddies into the Gulf of Mexico (Hofmann and Worley, 1986; Romanou et al., 2004).

Direct velocity observations from SVP/GDP drifters between November 1986 and May 2002 have provided further evidence of the occurrence of Philippine Sea water (PSW) intrusions into the interior of the SCS between October and January (Centurioni et al., 2004, 2009), with 15-m depth westward currents of the order of 1 m s^{-1} within the LS. Such strong current cannot be solely explained by Ekman currents generated by the northeast monsoon. The 15-m depth Ekman velocity that would be produced by monsoon winds is at most 0.25 m s^{-1} (Centurioni et al., 2009), and the stronger surface currents observed by the drifters indicate that a deeper current system to the west must be present (Centurioni et al., 2004, 2009). The associated westward Ekman transport through the LS computed from satellite Special Sensor Microwave Imager (SSM/I) winds reaches a maximum of 0.6 Sv in December (Centurioni et al., 2004). Drifter velocity data also confirm the existence of a fall through winter anticyclonic circulation (loop) southwest of Taiwan (Centurioni et al., 2004), agreeing with previous hydrographic observations (Nitani, 1972).

Many of the drifters that enter the SCS through the LS during the northeast monsoon do not move back to the Philippine Sea (PS) within the KC loop, but move rapidly into a current along the western margin of the deep SCS basin and continue to travel southward along the coast of Vietnam, with speeds in excess of 1 m s^{-1} (Hu et al., 2000; Centurioni et al., 2004, 2009). During the Asian southwest monsoon, which occurs in boreal spring and summer months, drifters deployed in the SCS have shown that the exchange of surface water between the SCS and the PS is reversed, with drifters now exiting the SCS through the LS and the Taiwan Strait (Centurioni et al., 2007).

The swift jet off Vietnam, whose surface signature is clearly revealed by the drifters, is mainly concentrated inshore of the 200 m isobaths and has speeds two to three times larger than previous estimates (Liu et al., 2004). The downwelling region offshore of Vietnam that stretches parallel to the coast is probably responsible for maintaining the pressure gradient balanced by the Coriolis force acting on the southward flowing jet (Gill, 1982). New subsurface observations are needed there to elucidate the vertical structure of this boundary current. At about 11° N, offshore southeast Vietnam, the current veers onto the continental shelf probably as a consequence of the approximately westward Ekman transport that develops during the northeast monsoon and associated downwelling and an extended region of Ekman pumping (Liu et al., 2004; Centurioni et al., 2009). Because of the strong northeasterly monsoon wind blowing in the direction of the sheared jet, a relatively strong secondary circulation with upwelling concentrated to the east and southeast of the jet should arise (Lee et al., 1994), thus enhancing the slopes of the isopycnals that maintain the dynamical balance of this current system (see also Section 3.3).

The results mentioned earlier strongly suggest that a net westward volume transport of PSW into the SCS occurs through the LS in the top few hundred meters between October and January. One interesting and still unanswered scientific question is where does the compensating SCS outflow occur? All the other straits that connect the SCS with the adjacent basins from which the fall/winter PSW inflow could be balanced by outflow are much shallower than the LS: the Taiwan Strait is on average 60 m deep, the Malacca, Gaspar and Karimata Straits have sills shallower than 50 m, and the Balabac and the Mindoro Straits connect the SCS to the Sulu Sea with passages that are about 100 and 450 m deep, respectively. It is still unclear how the volume balance of the SCS can be closed only by way of outflow from shallow passages (or deep and narrow in the case of the Mindoro Strait). Qu (2000) used hydrographic data to compute a geostrophic volume transport from 0 to 400 m of 5.3 Sv westward in January through February. Drifter releases in the north SCS suggest the existence of a current system that connects the LS region with the southwest equatorial SCS. A 5 Sv outflow from the SCS through the Karimata Strait (approximately 200 km wide and 30 m deep) would require a persistent average current through fall and winter of 0.8 m s^{-1}. It is not known whether or not such flow exists. Outflow from the SCS could also occur through the LS below the inflowing PSW layer, which requires water mass conversion that transforms the warm and salty PSW surface water into the cooler and fresher SCS deeper water. The concept of SCS water outflow below the PSW layer from the LS is not new since SCS waters characterized by a salinity minimum in the potential density/salinity space, which are normally found between 350 and 1350 m depth (Chen and Huang, 1996), were tracked as far north as Japan and are thought to have escaped through the LS (Chen, 2005). Furthermore, using the distribution of dissolved oxygen, Li and Qu (2006) showed the evidence for a sandwiched vertical structure in transports through the LS, with outflows in the intermediate layer (700–1500 m) and inflows above and below.

Quantifying the net inflow of PS/KC water into the SCS is an important step to better document the meridional heat fluxes in the north Pacific. New drifter deployments in the southern SCS will provide useful data to better understand the structure of the SCS near-surface circulation and therefore to address some of the science questions mentioned earlier.

4.4. Interaction of the Kuroshio with the East China Sea

The KC is known to interact with the East China Sea (ECS, Figure 12.6) by way of intrusions of the latter on the continental shelf northeast of Taiwan (Gawarkiewicz et al., 2011). SVP/GDP drifters from the historical data archive together with repeat releases designed to map the intrusion events suggest that deep excursions of the KC on the ECS continental shelf have a meandering nature and occur all year round, perhaps with a weak seasonal modulation (Vélez-Belchí et al., 2013).

The dynamics of the intrusion is still unclear, but there is evidence of a link between cyclonic mesoscale eddies coalescing with, or being advected by the KC east of Taiwan, low transport of the KC through the East Taiwan Channel, and ECS shelf intrusions of the KC (Vélez-Belchí et al., 2013). The time variations of the northward transport of the KC through the East Taiwan Channel on the I-Lan Ridge were measured with an intensive current array PCM-1 during WOCE (Johns et al., 2001). Remarkably, this transport varies with a period near 100 days, and no significant seasonal cycle was detected by these direct measurements. The variations of this transport are correlated with the presence of mesoscale eddies directly east of Taiwan in the northern PS (Zhang et al., 2001). A presence of a cyclonic eddy is accompanied with low KC transport and anticyclonic eddy with a high KC transport. The correlation between the WOCE PCM-1 KC transport and the AVISO gridded weekly sea level anomaly (SLA) data is maximum (0.83) at 23.9° N, 123.2° E. This suggests that the SLA at that location is a reasonable proxy for the KC transport variability through the East Taiwan Channel and can be used as a zero order predictor for the KC–ECS interaction events (Vélez-Belchí et al., 2013).

5. APPLICATIONS

Knowing surface currents is important for many applications. Since 1992, a constellation of satellite altimeters monitors the shape of the ocean surface with an accuracy and resolution sufficient to resolve large oceanic eddies (Pascual et al., 2006) and major ocean currents. Sea-level data are used in circulation and climate studies and are assimilated in many ocean models, including operational current forecast systems designed during the Global Ocean Data Assimilation Experiment (GODAE). For the latter models, other direct current datasets are also used for assimilation, such as high-frequency radar (HFR) and Autonomous Doppler Current Profiler (ADCP) observations (Davidson et al., 2009). Velocity data from Lagrangian drifters, however, are not suitable for operational oceanography because the drogue detection algorithm and sensors do not currently work in real time. Drifter velocity data, on the other hand, are one of the few independent datasets available to validate outputs from ocean forecasts (Hurlburt et al., 2009). For example, they are used to validate the Mercator Océan NEMO global model (Lellouche et al., 2013). In delayed mode, drifter velocities are also important for an accurate sea state estimate (Wunsch et al., 2009) facilitating the use of satellite-measured anomalies and, in the end, increasing forecast quality.

Operational surface current products, available from a number of teams, are actively used for tracking oil spills (c.f., Hackett et al., 2009), search and rescue, and ship routing (Davidson et al., 2009), maritime safety and iceberg drift, larval dispersal and harmful algal blooms (De Mey et al., 2009) and by the Navy for maritime operations (Jacobs et al., 2009).

One emerging application is to study the dynamics of marine debris. The trajectory of an object floating on the surface of the ocean is set by currents, waves, and wind, and depends on the geometry and mass distribution of the object. Light flotsam such as balloons are mainly blown by wind, while heavy objects, such as the GDP drifters, derelict fishing gear, and ship wrecks are mainly carried by currents. This latter group also includes microplastic, tiny pieces, produced by breaking old plastic, which can survive in water for many decades. This class of debris represents the most common type of marine debris, which poses the greatest risks for navigational safety and the health of the marine ecosystem. Because vertical excursions of flotsam are suppressed by their buoyancy, floating debris accumulates in open-ocean and coastal regions of downwelling.

The GDP dataset provides rich information about the pathways of marine debris. The main pathways are well represented by the mean streamlines, shown in Figure 12.2a. To eliminate the bias existing in drifter statistics due to highly heterogeneous density of the drifter ensemble, Maximenko et al. (2012) developed a statistical model, based on probability of observed drifter excursion between different boxes. Lumpkin et al. (2011) developed a technique to assess the probability that a drifter will run aground on different parts of the coastline. This model, accounting for probabilistic displacements of particles and their chance to end on the coast, was used for a numerical experiment with an initially homogeneous density of marine debris (Lumpkin et al., 2011). Figure 12.7 shows that after 10 years of wandering around the ocean, most of the debris (nearly 70% of the initial mass) is still floating, collected in five subtropical "garbage patches." While the North Pacific (Moore et al., 2001) and North Atlantic (Law et al., 2010) garbage patches are well known, three others have been confirmed only recently (Eriksen et al., 2013) and are being currently documented by researchers. Vertical bars on Figure 12.7 show the global pattern of "clean" and "dirty" shores, correlating well with the direction of dominant winds.

Another experiment with the same model can simulate the motion of debris generated on the northeastern coast of Japan. This helps to statistically predict the fate of debris originating from the tsunami of March 11, 2011. Figure 12.8 illustrates that debris will likely disperse over large area and the center of mass of the debris field will first drift toward east and later will recirculate into the garbage patch. Two and a half years after tsunami, scarce reports from sea generally confirm predictions of the model. Their analysis emphasizes importance of wind force, acting on high-windage objects, which are the most visible members of the debris ensemble at rough sea.

6. FUTURE DIRECTIONS

Important advances in our ability to observe surface currents have been made in recent decades. With the help of the GOOS, including the network of the GDP drifters and satellite altimeters, surface currents are monitored on space scales larger than 100 km and timescales longer than one week. These observations are complemented by satellite-borne measurements of the geoid and of surface winds, particularly via scatterometers, to separately resolve the geostrophic and ageostrophic components of surface currents. These observations provide invaluable information for basic and applied research on the ocean and climate dynamics. Yet, most operational activity requires resolution of a few kilometers delivered to the user within hours. The importance of smaller scales is not only in better description of details of ocean currents. The submesoscale contains most vertical motions, sustaining life in the upper

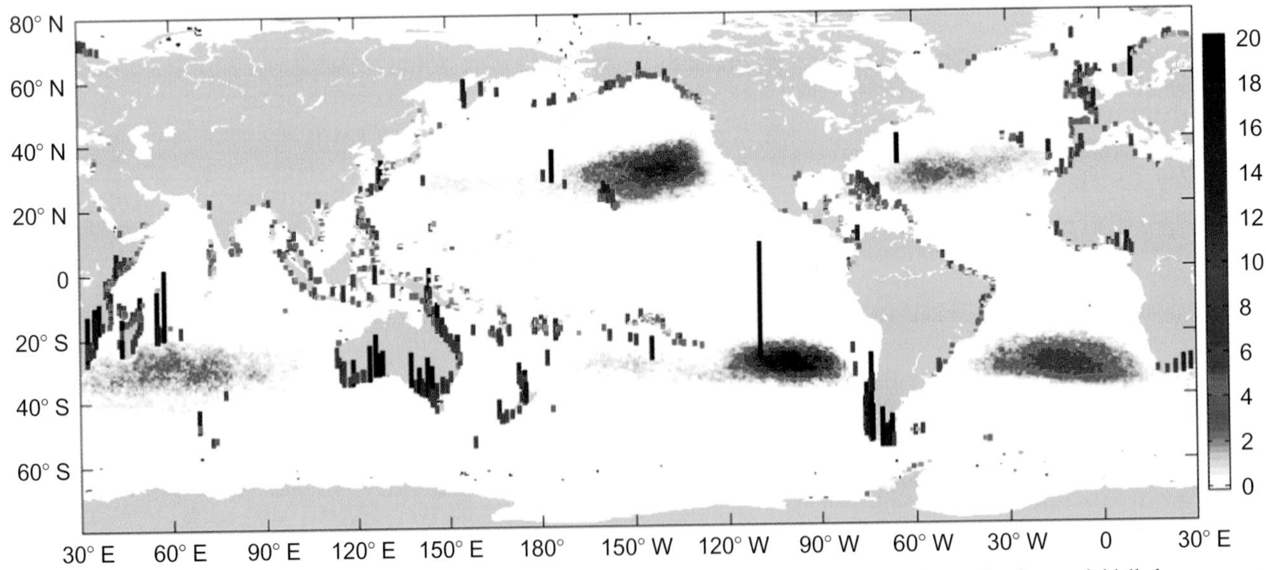

FIGURE 12.7 Distribution of the concentration of floating marine debris in arbitrary units, after 10 years of integration from an initially homogeneous distribution of concentration unity. Vertical bars indicate the concentration of material that has washed ashore, with color corresponding to 10 times the value in the color bar. *Reproduced from Lumpkin et al. (2011).*

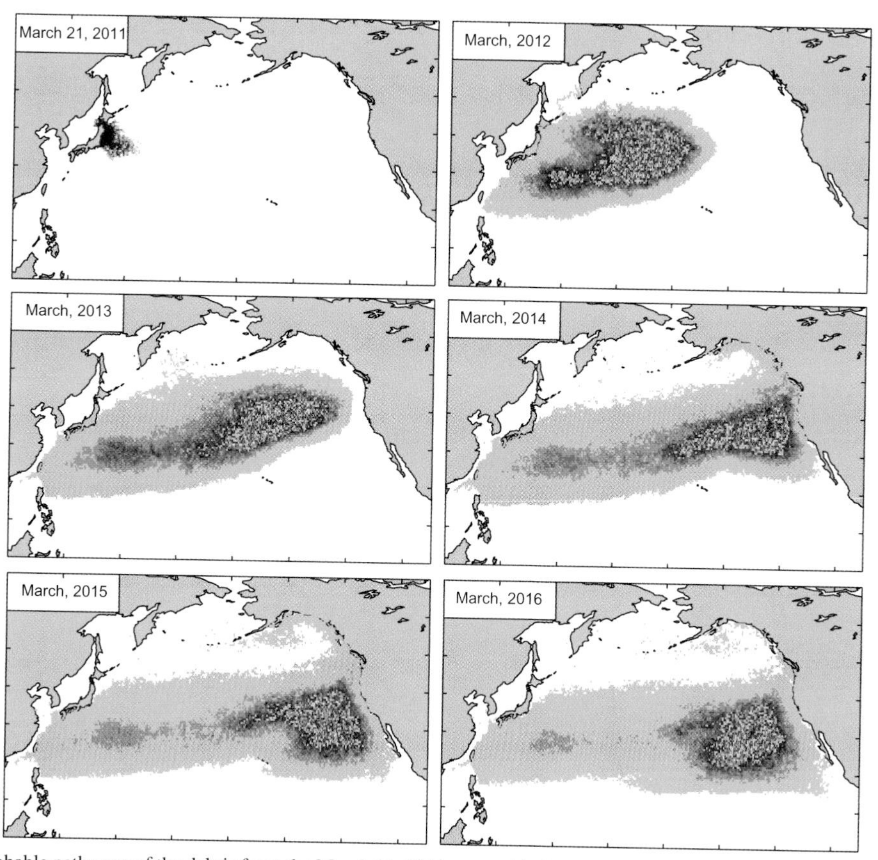

FIGURE 12.8 The probable pathways of the debris from the March 11, 2011 tsunami in Japan as estimated from historical trajectories of drifting buoys. *Reproduced from Maximenko et al. (2011).*

ocean (c.f., Klein et al., 2008; Lapeyre, 2009). Upwelling in local fronts commonly induces chlorophyll blooms, in which complex ecosystems quickly build up. Downwellings, on the contrary, are generated by surface convergences and collect marine debris and other floating pollutants. Experiments with ultra-high-resolution models also indicate that the vorticity of ocean currents, whose balance plays a principal role in the dynamics of ocean circulation, varies on scales of kilometers or less. Fine resolution will also allow expanding observations closer to the shore, the area of the greatest interest of most users. NASA's Surface Water Ocean Topography (SWOT) mission will approach these parameters using a wide-swath interferometry between a pair of radar altimeters, installed on a single satellite. The launch of SWOT is planed after 2020. Drifters, measuring full velocity, will be essential for future studies of submesoscale processes. At the same time, while increased resolution and accuracy along the trajectory are feasible thanks to new satellite systems (e.g., Iridium and GPS), implementing and maintaining a drifter array with submesoscale coverage is still a task for the future.

Conversion of the sea-level signal into velocity is not straightforward on small scales, where geostrophic Equation (12.1) is not a good approximation. New techniques (such as surface quasi-geostrophy (SQG) (Klein et al., 2008)) will need to be developed and implemented that will require better understanding of the momentum balance in the upper ocean. Velocities, measured with the array of GDP drifters (after some corrections for the wind-forced slip), are interpreted as ocean currents at a nominal depth of 15 m. Neither the motion of the drogue in rough seas where the wave height is a significant fraction of the drogue's depth nor much details of the upper-ocean velocity shear are well documented. No model or theory existing today approaches the full complexity of surface currents, and the very definition of the "sea surface" in the presence of breaking waves is far from trivial. For example, Kudryavtsev et al. (2008) point out inconsistencies of modern models of the mixed layer and suggest the wind force is not just applied at the sea surface, but momentum may be "injected" into deeper layers by breaking wind waves. Validation of this idea and progress in understanding the details of the momentum balance will require implementation in large numbers of existing velocity profilers and development of new, more efficient technologies.

New technologies are also required to monitor near-surface currents covered by ice. Implementation of engineering ideas (Freeman et al., 2010), allowing remote sensing of the ocean surface velocity vector with satellite, would further revolutionize the science of the upper ocean.

Gaps in observations of ocean currents can be filled using models of different complexity, synthesizing other datasets, from diagnostic models combining in a simple way wind- and pressure-driven velocities (e.g., Uchida and Imawaki (2003); OSCAR (Bonjean and Lagerloef, 2002); SCUD (Maximenko and Hafner, 2010)) to data assimilating systems (e.g., ECCO (Wunsch et al., 2009); HYCOM (Chassignet et al., 2009)). Enhancement of the observational dataset of surface velocity will help to validate and improve these models.

ACKNOWLEDGMENTS

N. M. was partly supported by NASA Ocean Surface Topography Science Team through Grants NNX08AR49G and NNX13AK35G, and also by JAMSTEC, by NASA through Grant NNX07AG53G, and by NOAA through Grant NA11NMF4320128 through their sponsorship of the International Pacific Research Center. R. L. was supported by NOAA's Office of Climate Observations and the Atlantic Oceanographic and Meteorological Laboratory. L. C. was supported by NOAA's Office of Climate Observations and by the Office of Naval Research. IPRC/SOEST Publication 962/8896.

REFERENCES

Alford, M.H., Whitmont, M., 2007. Seasonal and spatial variability of near-inertial kinetic energy from historical moored velocity records. J. Phys. Oceanogr. 37 (8), 2022–2037.

Archiving, Validation, and Interpretation of Satellite Oceanographic Data (Aviso), 1996. Aviso Handbook for Merged TOPEX/Poseidon Products, third ed. Aviso, Toulouse, France.

Austin, J.A., Barth, J.A., 2002. Drifter behavior on the Oregon-Washington shelf during downwelling-favorable winds. J. Phys. Oceanogr. 32 (11), 3132–3144.

Baldwin, M.P., Rhines, P.B., McIntyre, M.E., 2007. The jet-stream conundrum. Science 315, 467–468.

Bhatt, U.S., 1989. Circulation regimes of rainfall anomalies in the African South Asian monsoon belt. J. Clim. 2 (10), 1133–1144.

Boebel, O., Rossby, T., Lutjeharms, J., Zenk, W., Barron, C., 2003. Path and variability of the Agulhas return current. Deep Sea Res. Part II 50, 35–56.

Bonjean, F., Lagerloef, G., 2002. Diagnostic model and analysis of the surface currents in the tropical Pacific Ocean. J. Phys. Oceanogr. 32, 2938–2954.

Brambilla, E., Talley, L., 2006. Surface drifter exchange between North Atlantic subtropical and subpolar gyres. J. Geophys. Res. 111, C07026. http://dx.doi.org/10.1029/2005JC003146.

Brink, K.H., Beardsley, R.C., Niiler, P.P., Abbott, M., Huyer, A., Ramp, S., Stanton, T., Stuart, D., 1991. Statistical properties of near-surface flow in the California coastal transition zone. J. Geophys. Res. 96, 14693–14706.

Buckingham, C.E., Cornillon, P.C., 2013. The contribution of eddies to striations in absolute dynamic topography. J. Geophys. Res. Oceans 118, 448–461. http://dx.doi.org/10.1029/2012JC008231.

Centurioni, L.R., Niiler, P.P., Lee, D.K., 2004. Observations of inflow of Philippine Sea surface water into the South China Sea through the Luzon Strait. J. Phys. Oceanogr. 34 (1), 113–121.

Centurioni, L.R., Niiler, P.P., Kim, Y.Y., Sheremet, V.A., Lee, D.K., 2007. Near surface dispersion of particles in the South China Sea. In: Griffa, A.D.K.A., Mariano, A.J., Özgökmen, T., Rossby, T. (Eds.), Lagrangian Analysis and Prediction of Coastal and Ocean Dynamics. Cambridge University Press, pp. 73–75.

Centurioni, L.R., Ohlmann, J.C., Niiler, P.P., 2008. Permanent meanders in the California current system. J. Phys. Oceanogr. 38 (8), 1690–1710.

Centurioni, L.R., Niiler, P.P., Lee, D.K., 2009. Near-surface circulation in the South China Sea during the winter monsoon. Geophys. Res. Lett. 36, L06605. http://dx.doi.org/10.1029/02008GL037076.

Chaigneau, A., Pizarro, O., Rojas, W., 2008. Global climatology of near-inertial current characteristics from Lagrangian observations. Geophys. Res. Lett. 35, L13603. http://dx.doi.org/10.1029/2008GL034060.

Chassignet, E.P., Hurlburt, H.E., Metzger, E.J., Smedstad, O.M., Cummings, J., Halliwell, G.R., Bleck, R., Baraille, R., Wallcraft, A.J., Lozano, C., Tolman, H.L., Srinivasan, A., Hankin, S., Cornillon, P., Weisberg, R., Barth, A., He, R., Werner, F., Wilkin, J., 2009. U.S. GODAE: global ocean prediction with the HYbrid Coordinate Ocean Model (HYCOM). Oceanography 22 (2), 64–75.

Chelton, D.B., Schlax, M.G., Samelson, R.M., 2011. Global observations of nonlinear mesoscale eddies. Prog. Oceanog. 91, 167–216. http://dx.doi.org/10.1016/j.pocean.2011.01.002.

Chen, C.T.A., 2005. Tracing tropical and intermediate waters from the South China Sea to the Okinawa Trough and beyond. J. Geophys. Res. Oceans 110 (C5), C05012. http://dx.doi.org/10.1029/2004JC002494.

Chen, C.-T.A., Huang, M.-H., 1996. A mid-depth front separating the South China Sea water and the Philippine Sea water. J. Oceanogr. 52 (1), 17–25.

Chereskin, T.K., 1995. Evidence for an Ekman balance in the California Current. J. Geophys. Res. 100, 12727–12748.

Chereskin, T.K., Morris, M.Y., Niiler, P.P., Kosro, P.M., Smith, R.L., Ramp, S.R., Collins, C.A., Musgrave, D.L., 2000. Spatial and temporal characteristics of the mesoscale circulation of the California Current from eddy-resolving moored and shipboard measurements. J. Geophys. Res. Oceans 105 (C1), 1245–1269.

Cornuelle, B.D., Chereskin, T.K., Niiler, P.P., Morris, M.Y., Musgrave, D. L., 2000. Observations and modeling of a California undercurrent eddy. J. Geophys. Res. 105 (C1), 1227–1243.

Davidson, F.J.M., Allen, A., Brassington, G.B., Breivik, Ø., Daniel, P., Kamachi, M., Sato, S., King, B., Lefevre, F., Sutton, M., Kaneko, H., 2009. Applications of GODAE ocean current forecasts to search and rescue and ship routing. Oceanography 22 (3), 176–181. http://dx.doi.org/10.5670/oceanog.2009.76.

Davis, R.E., 1985a. Drifter observations of coastal surface currents during CODE—the method and descriptive view. J. Geophys. Res. 90 (C3), 4741–4755.

Davis, R.E., 1985b. Drifter observations of coastal surface currents during CODE: the statistical and dynamical views. J. Geophys. Res. 90 (C3), 4756–4772.

Davis, R., 1991. Observing the general circulation with floats. Deep Sea Res. 38, S531–S571.

De Mey, P., Craig, P., Davidson, F., Edwards, C.A., Ishikawa, Y., Kindle, J.C., Proctor, R., Thompson, K.R., Zhu, J., the GODAE Coastal and Shelf Seas Working Group (CSSWG) Community, 2009. Applications in coastal modeling and forecasting. Oceanography 22 (3), 98–205. http://dx.doi.org/10.5670/oceanog.2009.79.

Ducet, N., Le Traon, P.Y., Reverdin, G., 2000. Global high-resolution mapping of ocean circulation from TOPEX/Poseidon and ERS-1 and -2. J. Geophys. Res. 105 (C8), 19477–19489.

Ekman, V.W., 1905. On the influence of the earth's rotation on ocean currents. Arch. Math. Astron. Phys. 2 (11), 1–52.

Elipot, S., Lumpkin, R., 2008. Spectral description of oceanic near-surface variability. Geophys. Res. Lett. 35, L05606. http://dx.doi.org/10.1029/2007GL032874.

Elipot, S., Lumpkin, R., Prieto, G., 2010. Modification of inertial oscillations by the mesoscale eddy field. J. Geophys. Res. 115, C09010. http://dx.doi.org/10.1029/2009JC005679.

Eriksen, M., Maximenko, N.A., Thiel, M., Cummins, A., Lattin, G., Wilson, S., Hafner, J., Zellers, A., Rifman, S., 2013. Plastic pollution in the South Pacific subtropical gyre. Mar. Pollut. Bull. 68, 71–76. http://dx.doi.org/10.1016/j.marpolbul.2012.12.021.

Franklin, B., 1785. Sundry marine observations. Trans. Am. Philos. Soc. Ser. 1 2, 294–329.

Fratantoni, D.M., 2001. North Atlantic surface circulation during the 1990's observed with satellite-tracked drifters. J. Geophys. Res. 106, 22067–22093. http://dx.doi.org/10.1029/2000JC000730.

Freeland, H.J., Rhines, P., Rossby, H.T., 1975. Statistical observations of trajectories of neutrally buoyant floats in the North Atlantic. J. Mar. Res. 33, 383–404.

Freeman, A., Zlotnicki, V., Liu, T., Fu, L.-L., Holt, B., Kwok, R., Yueh, S., Fukumori, I., Vazquez, J., Siegel, D., Lagerloef, G., 2010. Ocean measurements from space in 2025. Oceanography 23 (4), 144–161.http://dx.doi.org/10.5670/oceanog.2010.12.

Galperin, B., Nakano, H., Huang, H.-P., Sukoriansky, S., 2004. The ubiquitous zonal jets in the atmospheres of giant planets and Earth's oceans. Geophys. Res. Lett. 31, L13303. http://dx.doi.org/10.1029/2004GL019691.

Garrett, J.F., 1980. The availability of the FGGE drifting buoy system data set. Deep Sea Res. 27, 1083–1086.

Gawarkiewicz, G., Jan, S., Lermusiaux, P.F.J., McClean, J.L., Centurioni, L., Taylor, K., Cornuelle, B., Duda, T.F., Wang, J., Yang, Y.J., Sanford, T., Lien, R.-C., Lee, C., Lee, M.-A., Leslie, W., Haley Jr., P.J., Niiler, P.P., Gopalakrishnan, G., Velez-Belchi, P., Lee, D.-K., Kim, Y.Y., 2011. Circulation and intrusions Northeast of Taiwan: chasing and predicting uncertainty in the cold dome. Oceanography 24 (4), 110–121. http://dx.doi.org/10.5670/oceanog.2011.99.

Gill, A., 1982. Atmosphere-Ocean Dynamics. Elsevier, 662 pp.

Grodsky, S.A., Lumpkin, R., Carton, J.A., 2011. Spurious trends in global surface drifter currents. Geophys. Res. Lett. 38, L10606. http://dx.doi.org/10.1029/2011GL047393.

Hackett, B., Comerma, E., Daniel, P., Ichikawa, H., 2009. Marine oil pollution prediction. Oceanography 22 (3), 168–175. http://dx.doi.org/10.5670/oceanog.2009.75.

Haines, K., Johannessen, J.A., Knudsen, P., Lea, D., Rio, M.-H., Bertino, L., Davidson, F., Hernandez, F., 2011. An ocean modelling and assimilation guide to using GOCE geoid products. Ocean Sci. 7, 151–164. http://dx.doi.org/10.5194/os-7-151-2011.

Hansen, D.V., Poulain, P.-M., 1996. Quality control and interpolations of WOCE-TOGA drifter data. J. Atmos. Oceanic Technol. 13, 900–909.

Hofmann, E.E., Worley, S.J., 1986. An investigation of the circulation of the Gulf of Mexico. J. Geophys. Res. 91 (C12), 14221–14236.

Hu, J., Kawamura, H., Hong, H., Qi, Y., 2000. A review on the Currents in the South China Sea: seasonal circulation South China sea warm current and Kuroshio intrusion. J. Oceanogr. 56 (6), 607–624.

Huang, H.-P., Kaplan, A., Curchitser, E.N., Maximenko, N.A., 2007. The degree of anisotropy for mid-ocean currents from satellite observations and an eddy-resolving model simulation. J. Geophys. Res. Oceans 112, C09005. http://dx.doi.org/10.1029/2007JC004105.

Hughes, C.W., 2005. Nonlinear vorticity balance of the Antarctic Circumpolar Current. J. Geophys. Res. 110, C11008. http://dx.doi.org/10.1029/2004JC002753.

Hurlburt, H.E., Brassington, G.B., Drillet, Y., Kamachi, M., Benkiran, M., Bourdallé-Badie, R., Chassignet, E.P., Jacobs, G.A., Le Galloudec, O., Lellouche, J.-M., Metzger, E.J., Oke, P.R., Pugh, T.F., Schiller, A., Smedstad, O.M., Tranchant, B., Tsujino, H., Usui, N., Wallcraft, A.J., 2009. High-resolution global and basin-scale ocean analyses and forecasts. Oceanography 22 (3), 110–127. http://dx.doi.org/10.5670/oceanog.2009.70.

Huyer, A., 1983. Coastal upwelling in the California current system. Prog. Oceanogr. 12 (3), 259–284.

Imawaki, S., Uchida, H., Ichikawa, H., Fukasawa, M., Umatani, S., ASUKA Group, 2001. Satellite altimeter monitoring the Kuroshio transport South of Japan. Geophys. Res. Lett. 28 (1), 17–20.

Imawaki, S., Uchida, H., Ichikawa, K., Ambe, D., 2003. Estimating the high-resolution mean sea-surface velocity field by combined use of altimeter and drifter data for geoid model improvement. Space Sci. Rev. 108, 195–204.

Jacobs, G.A., Woodham, R., Jourdan, D., Braithwaite, J., 2009. GODAE applications useful to navies throughout the world. Oceanography 22 (3), 182–189. http://dx.doi.org/10.5670/oceanog.2009.77.

Johannessen, J.A., Balmino, G., Le Provost, C., Rummel, R., Sabadini, R., Sunkel, H., Tscherning, C.C., Visser, P., Woodworth, P., Hughes, C. W., LeGrand, P., Sneeuw, N., Perosanz, F., Aguirre-Martinez, M., Rebhan, H., Drinkwater, M., 2003. The European gravity field and steady-state ocean circulation explorer satellite mission: impact in geophysics. Surv. Geophys. 24, 339–386.

Johns, W.E., Lee, T.N., Zhang, D.X., Zantopp, R., Liu, C.T., Yang, Y., 2001. The Kuroshio east of Taiwan: moored transport observations from the WOCE PCM-1 array. J. Phys. Oceanogr. 31 (4), 1031–1053.

Kida, S., Price, J.F., Jiayan, Y., 2007. The upper-oceanic response to overflows: a mechanism for the Azores Current. J. Phys. Oceanogr. 38, 880–895.

Klein, B., Siedler, G., 1989. On the origin of the Azores Current. J. Geophys. Res. 94, 6159–6168.

Klein, P., Hua, B.L., Lapeyre, G., Capet, X., Le Gentil, S., Sasaki, H., 2008. Upper ocean turbulence from high-resolution 3D simulations. J. Phys. Oceanogr. 38, 1748–1763.

Knudsen, P., Bingham, R., Andersen, O.B., Rio, M.-H., 2011. Enhanced mean dynamic topography and ocean circulation estimation using GOCE preliminary models. In: Presentataion at the 4th International GOCE User Workshop, 31 March–01 April 2011, Munich, Germany, Available at: https://earth.esa.int/download/goce/4th_Int_GOCE_User_Wkshp_2011/Enhanced_Mean_Dynamic_Topography_P.Knudsen.pdf.

Krauss, W., Böning, C., 1987. Lagrangian properties of eddy fields in the northern North Atlantic as deduced from satellite-tracked buoys. J. Mar. Res. 45, 259–291.

Kudryavtsev, V., Shrira, V., Dulov, V., Malinovsky, V., 2008. On the vertical structure of wind-driven sea currents. J. Phys. Oceanogr. 38, 2121–2144.

Kunze, E., 1985. Near-inertial wave propagation in geostrophic shear. J. Phys. Oceanogr. 15 (5), 544–565.

LaCasce, J.H., Ohlmann, C., 2003. Relative dispersion at the surface of the Gulf of Mexico. J. Mar. Res. 61 (3), 285–312.

Lapeyre, G., 2009. What vertical mode does the altimeter reflect? On the decomposition in baroclinic modes and on a surface-trapped mode. J. Phys. Oceanogr. 39, 2857–2874.

Law, K.L., Morét-Ferguson, S., Maximenko, N.A., Proskurowski, G., Peacock, E.E., Hafner, J., Reddy, C.M., 2010. Plastic accumulation in the North Atlantic Subtropical Gyre. Science 329, 1185–1188.

Le Traon, P.-Y., Schaeffer, P., Guinehut, S., Rio, M.-H., Hernandez, F., Larnicol, G., Lemoine, J.M., 2001. Mean ocean dynamic topography from GOCE and altimetry. In: Proceedings of the 1st International GOCE Workshop, April 23–24. ESTEC, Noordwijk, The Netherlands, Available at: http://earth.esa.int/goce04/first_igw/papers/Letraon_etal.pdf.

Lee, D.-K., Niiler, P.P., 1998. The inertial chimney: the near-inertial energy drainage from the ocean surface to the deep layer. J. Geophys. Res. 103 (C4), 7579–7591.

Lee, D.K., Niiler, P., Warnvarnas, A., Piacsek, S., 1994. Wind-driven secondary circulation in ocean mesoscale. J. Mar. Res. 52 (3), 371–396.

Lellouche, J.-M., Le Galloudec, O., Drévillon, M., Régnier, C., Greiner, E., Garric, G., Ferry, N., Desportes, C., Testut, C.-E., Bricaud, C., Bourdallé-Badie, R., Tranchant, B., Benkiran, M., Drillet, Y., Daudin, A., De Nicola, C., 2013. Evaluation of global monitoring and forecasting systems at Mercator Océan. Ocean Sci. 9, 57–81. http://dx.doi.org/10.5194/os-9-57-2013.

Li, L., Qu, T., 2006. Thermohaline circulation in the deep South China Sea basin inferred from oxygen distributions. J. Geophys. Res. 111, C05017. http://dx.doi.org/10.1029/2005JC003164.

Liu, Q.Y., Jiang, X., Xie, S.P., Liu, W.T., 2004. A gap in the Indo-Pacific warm pool over the South China Sea in boreal winter: seasonal development and interannual variability. J. Geophys. Res. Oceans 109, C07012.

Lumpkin, R., Garzoli, S.L., 2005. Near-surface circulation in the tropical Atlantic Ocean. Deep Sea Res. Part I 52, 495–518. http://dx.doi.org/10.1016/j.dsr.2004.09.001.

Lumpkin, R., Garzoli, S., 2010. Interannual to decadal changes in the Southwestern Atlantic's surface circulation. J. Geophys. Res. Oceans 116, C01014. http://dx.doi.org/10.1029/2010JC006285.

Lumpkin, R., Goni, G., Dohan, K., 2009. State of the ocean in 2008: surface currents. In: Peterson, T., Baringer, M. (Eds.), State of the Climate in 2008, vol. 90. Bulletin of the American Meteorological Society, pp. S12–S15.

Lumpkin, R., Maximenko, N., Pazos, M., 2011. Evaluating where and why drifters die. J. Atmos. Oceanic Technol. 29, 300–308. http://dx.doi.org/10.1175/JTECH-D-11-00100.1.

Lumpkin, R., Grodsky, S., Rio, M.-H., Centurioni, L., Carton, J., Lee, D., 2013. Removing spurious low-frequency variability in surface drifter velocities. J. Atmos. Oceanic Technol. 30 (2), 353–360. http://dx.doi.org/10.1175/JTECH-D-12-00139.1.

Marchesiello, P., McWilliams, J.C., Shchepetkin, A., 2003. Equilibrium structure and dynamics of the California Current System. J. Phys. Oceanogr. 33 (4), 753–783.

Maximenko, N., 2004. Correspondence between Lagrangian and Eulerian velocity statistics at the ASUKA line. J. Oceanogr. 60 (4), 681–687.

Maximenko, N., Hafner, J., 2010. SCUD: Surface CUrrents from Diagnostic model. IPRC Technical Note, No. 5, February 16, 2010, 17 p.

Maximenko, N.A., Niiler, P.P., 2005. Hybrid decade-mean global sea level with mesoscale resolution. In: Saxena, N.K. (Ed.), Recent Advances in Marine Science and Technology 2004. PACON International, Honolulu, Hawaii, pp. 55–59.

Maximenko, N.A., Niiler, P.P., 2006. Mean surface circulation of the global ocean inferred from satellite altimeter and drifter data. In: 15 Years of Progress in Radar Altimetry, ESA Publication SP-614, July 2006.

Maximenko, N.A., Bang, B., Sasaki, H., 2005. Observational evidence of alternating zonal jets in the world ocean. Geophys. Res. Lett. 32, L12607. http://dx.doi.org/10.1029/2005GL022728.

Maximenko, N.A., Melnichenko, O.V., Niiler, P.P., Sasaki, H., 2008. Stationary mesoscale jet-like features in the ocean. Geophys. Res. Lett. 35, L08603. http://dx.doi.org/10.1029/2008GL033267.

Maximenko, N., Niiler, P., Centurioni, L., Rio, M.-H., Melnichenko, O., Chambers, D., Zlotnicki, V., Galperin, B., 2009. Mean dynamic topography of the ocean derived from satellite and drifting buoy data using three different techniques. J. Atmos. Oceanic Technol. 26 (9), 1910–1919.

Maximenko, N., Hafner, J., Lumpkin, R., 2011. Modeling distribution of marine debris before and after tsunami of March 11, 2011. In: Presentation at the Ocean Surface Topography Science Team Meeting, San Diego, California, October 19–21. http://www.aviso.oceanobs.com/fileadmin/documents/OSTST/2011/oral/.

Maximenko, N., Hafner, J., Niiler, P., 2012. Pathways of marine debris from trajectories of Lagrangian drifters. Mar. Pollut. Bull. 65, 51–62. http://dx.doi.org/10.1016/j.marpolbul.2011.04.016.

McNally, G.J., Patzert, W.C., Kirwan, J.A.D., Vastano, A.C., 1983. The near-surface circulation of the North Pacific using satellite tracked drifting buoys. J. Geophys. Res. 88, 7634–7640.

Melnichenko, O.V., Maximenko, N.A., Schneider, N., Sasaki, H., 2010. Quasi-stationary striations in basin-scale oceanic circulation: vorticity balance from observations and eddy-resolving model. Ocean Dyn. 60 (3), 653–666.

Moore, C., Moore, S.L., Leecaster, M.K., Weisberg, S.B., 2001. A comparison of plastic and plankton in the north pacific central gyre. Mar. Pollut. Bull. 42, 1297–1300. http://dx.doi.org/10.1016/S0025-326X(01)00114-X, 2001-12-01.

Munk, W.H., Wunsch, C., 1998. Abyssal recipes II: energetics of tidal and wind mixing. Deep Sea Res. Part I 45, 1977–2010.

Nakano, H., Hasumi, H., 2005. A series of zonal jets embedded in the broad zonal flows in the Pacific obtained in eddy-permitting ocean general circulation models. J. Phys. Oceanogr. 35, 474–488.

Needler, G., Smith, N., Villwock, A., 1999. The Action Plan for GOOS/GCOS and Sustained Observations for CLIVAR. Available at: http://www.oceanobs09.net/work/oo99/docs/Needler.pdf.

Niiler, P., 2001. The world ocean surface circulation. In: Siedler, G., Church, J., Gould, J. (Eds.), Ocean Circulation and Climate. International Geophysics Series, vol. 77. Academic Press, pp. 193–204.

Niiler, P.P., 2003. A brief history of drifter technology. In: Autonomous and Lagrangian Platforms and Sensors Workshop. Scripps Institution of Oceanography, La Jolla, CA.

Niiler, P.P., Paduan, J.D., 1995. Wind-driven motions in the northeast Pacific as measured by Lagrangian drifters. J. Phys. Oceanogr. 25, 2819–2830.

Niiler, P.P., Davis, R., White, H., 1987. Water-following characteristics of a mixed-layer drifter. Deep Sea Res. 34, 1867–1882.

Niiler, P.P., Sybrandy, A., Bi, K., Poulain, P., Bitterman, D., 1995. Measurements of the water-following capability of holey-sock and TRISTAR drifters. Deep Sea Res. 42, 1951–1964.

Niiler, P.P., Maximenko, N.A., Panteleev, G.G., Yamagata, T., Olson, D. B., 2003a. Near-surface dynamical structure of the Kuroshio

Extension. J. Geophys. Res. 108, 3193. http://dx.doi.org/10.1029/2002JC001461.

Niiler, P.P., Maximenko, N.A., McWilliams, J.C., 2003b. Dynamically balanced absolute sea level of the global ocean derived from near-surface velocity observations. Geophys. Res. Lett. 30, 2164. http://dx.doi.org/10.1029/2003GL018628.

Nitani, H., 1972. Beginning of the Kuroshio. In: Stommel, H., Yoshida, K. (Eds.), Kuroshio: Its Physical Aspects. University of Tokyo Press, pp. 129–163.

Paduan, J.D., Niiler, P.P., 1993. Structure of velocity and temperature in the Northeast Pacific as measured with Lagrangian drifters in Fall 1987. J. Phys. Oceanogr. 23, 585–600.

Pascual, A., Faugère, Y., Larnicol, G., Le Traon, P.-Y., 2006. Improved description of the ocean mesoscale variability by combining four satellite altimeters. Geophys. Res. Lett. 33, L02611. http://dx.doi.org/10.1029/2005GL024633.

Pazan, S.E., Niiler, P.P., 2001. Recovery of near-surface velocity from undrogued drifters. J. Atmos. Oceanic Technol. 18, 476–489.

Pazan, S.E., Niiler, P.P., 2004. New global drifter data set available. Eos 85, 17.

Pedlosky, J., 1996. Ocean Circulation Theory. Springer-Verlag, Berlin, pp. 405–409.

Polzin, K., 2008. Mesoscale eddy–internal wave coupling. Part I: symmetry, wave capture and results from the mid-ocean dynamics experiment. J. Phys. Oceanogr. 38 (11), 2556–2574.

Poulain, P.M., Niiler, P.P., 1989. Statistical-analysis of the surface circulation in the California current system using satellite-tracked drifters. J. Phys. Oceanogr. 19 (10), 1588–1603.

Qiu, B., Chen, S., 2005. Variability of the Kuroshio Extension Jet, recirculation gyre, and mesoscale eddies on decadal time scales. J. Phys. Oceanogr. 35, 2090–2103.

Qu, T.D., 2000. Upper-layer circulation in the South China Sea. J. Phys. Oceanogr. 30 (6), 1450–1460.

Ralph, E.A., Niiler, P.P., 1999. Wind-driven currents in the tropical Pacific. J. Phys. Oceanogr. 29 (9), 2121–2129.

Reverdin, G., Frankignoul, C., Kestenare, E., McPhaden, M.J., 1994. Seasonal variability in the surface currents of the equatorial Pacific. J. Geophys. Res. 99, 20323–20344.

Rhines, P.B., 1975. Waves and turbulence on a beta-plane. J. Fluid Mech. 69, 417–443.

Richards, K.J., Maximenko, N.A., Bryan, F.O., Sasaki, H., 2006. Zonal jets in the Pacific Ocean. Geophys. Res. Lett. 33, L03605. http://dx.doi.org/10.1029/2005GL024645.

Richardson, P., 1980. Gulf Stream ring trajectories. J. Phys. Oceanogr. 10, 90–104.

Richardson, P.L., McKee, T.K., 1984. Average seasonal variation of the Atlantic equatorial currents from historical ship drift. J. Phys. Oceanogr. 14, 1226–1238.

Richardson, P., Reverdin, G., 1987. Seasonal cycle of velocity in the Atlantic North Equatorial Countercurrent as measured by surface drifters, current meters, and ship drifts. J. Geophys. Res. 92, 3691–3708.

Richardson, P., Walsh, D., 1986. Mapping climatological seasonal variations of surface currents in the Tropical Atlantic using ship drifts. J. Geophys. Res. 91, 10537–10550.

Rio, M.H., 2012. Use of Altimeter and Wind Data to Detect the Anomalous Loss of SVP-Type Drifter's Drogue. J. Atmos. Oceanic Technol. 29, 1663–1674. http://dx.doi.org/10.1175/JTECH-D-12-00008.1.

Rio, M.H., Hernandez, F., 2003. High-frequency response of wind-driven currents measured by drifting buoys and altimetry over the world ocean. J. Geophys. Res. 108 (C8), 3283. http://dx.doi.org/10.1029/2002JC001655.

Rio, M.H., Hernandez, F., 2004. A mean dynamic topography computed over the world ocean from altimetry, in situ measurements, and a geoid model. J. Geophys. Res. 109, C12032. http://dx.doi.org/10.1029/2003JC002226.

Rio, M.H., Guinehut, S., Larnicol, G., 2011a. New CNES-CLS09 global mean dynamic topography computed from the combination of GRACE data, altimetry, and in situ measurements. J. Geophys. Res. 116, C07018. http://dx.doi.org/10.1029/2010JC006505.

Rio, M.-H., Mulet, S., Guinehut, S., Lambin, J., Bronner, E., 2011b. High-resolution mean dynamic topography in the Kerguelen area from altimetry, GOCE data, and oceanographic in situ measurements. In: Presentation at the OSTST meeting, October 20, 2011, San Diego, USA, Available at: http://www.aviso.oceanobs.com/fileadmin/documents/OSTST/2011/oral/02_Thursday/Splinter%202%20GEO/03%20Helene%20%20Rio%20rev%20MDT_KEOPS_rio.pdf.

Romanou, A., Chassignet, E.P., Sturges, W., 2004. Gulf of Mexico circulation within a high-resolution numerical simulation of the North Atlantic Ocean. J. Geophys. Res. Oceans 109, C01003. http://dx.doi.org/10.1029/2003JC001770.

Roy, C., Reason, C., 2001. ENSO related modulation of coastal upwelling in the eastern Atlantic. Prog. Oceanogr. 49 (1–4), 245–255.

Rudnick, D.L., Jan, S., Centurioni, L., Lee, C.M., Lien, R.-C., Wang, J., Lee, D.-K., Tseng, R.-S., Kim, Y.Y., Chern, C.-S., 2011. Seasonal and mesoscale variability of the Kuroshio near its origin. Oceanography 24 (4), 52–63.

Sasaki, H., Nonaka, M., 2006. Far-reaching Hawaiian Lee Countercurrent driven by wind-stress curl induced by warm SST band along the current. Geophys. Res. Lett. 33, L13602. http://dx.doi.org/10.1029/2006GL026540.

Schlax, M.G., Chelton, D.B., 2008. The influence of mesoscale eddies on the detection of quasi-zonal jets in the ocean. Geophys. Res. Lett. 35, L24602. http://dx.doi.org/10.1029/2008GL035998.

Scott, R.B., Arbic, B.K., Holland, C.L., Sen, A., Qiu, B., 2008. Zonal versus meridional velocity variance in satellite observations and realistic and idealized ocean circulation models. Ocean Model. 23, 102–112.

Strub, P.T., Kosro, P.M., Huyer, A., 1991. The nature of the cold filaments in the California Current System. J. Geophys. Res. Oceans 96 (C8), 14743–14768.

Sverdrup, H.U., 1947. Wind-driven currents in a baroclinic ocean; with application to the equatorial currents of the eastern Pacific. Proc. Natl. Acad. Sci. U.S.A. 33, 318–326.

Sverdrup, H.U., Johnson, M.W., Fleming, R.H., 1942. The Oceans: Their Physics, Chemistry and General Biology. Prentice Hall, Englewood Cliffs, NJ, 1087 pp.

Swenson, M.S., Niiler, P.P., 1996. Statistical analysis of the surface circulation of the California Current. J. Geophys. Res. Oceans 101 (C10), 22631–22645.

Swenson, M.S., Niiler, P.P., Brink, K.H., Abbott, M.R., 1992. Drifter observations of a cold filament off point Arena, California, in July 1988. J. Geophys. Res. 97 (C3), 3593–3610.

Sybrandy, A.L., Niiler, P.P., 1991. WOCE/TOGA Lagrangian drifter construction manual. WOCE Rep. 63, SIO Ref. 91/6. Scripps Institution of Oceanography, La Jolla, CA, 58 pp.

Sybrandy, A.L., Niiler, P.P., Martin, C., Scuba, W., Charpentier, E., Meldrum, D.T., 1992. Global Drifter Program Barometer Drifter Design Reference. Data Buoy Cooperation Panel Report 4, Revision 1.3 (December 2003).

Tapley, B.D., Bettadpur, S., Watkins, M., Reigber, C., 2004. The gravity recovery and climate experiment: mission overview and early results. Geophys. Res. Lett. 31, L09607. http://dx.doi.org/10.1029/2004GL019920.

Thomson, C.W., 1877. A Preliminary Account of the General Results of the Voyage of the HMS Challenger. MacMillan, London.

Thorpe, S.A., 2005. The Turbulent Ocean. Section 9.4. Langmuir Circulation. In: Cambridge University Press, pp. 251–257.

Uchida, U., Imawaki, S., 2003. Eulerian mean surface velocity field derived by combining drifter and satellite altimeter data. Geophys. Res. Lett. 30, 1229. http://dx.doi.org/10.1029/2002GL016445.

Uchida, H., Imawaki, S., Hu, J.-H., 1998. Comparison of Kuroshio surface velocities derived from satellite altimeter and drifting buoy data. J. Oceanogr. 54, 115–122.

Vélez-Belchí, P., Centurioni, L.R., Lee, D.-K., Jan, S., Niiler, P.P., 2013. Eddy induced Kuroshio intrusions onto the continental shelf of the East China Sea. J. Mar. Res. 71, 83–107.

Willis, J.K., Chambers, D.P., Kuo, C.-Y., Shum, C.K., 2010. Global sea level rise: recent progress and challenges for the decade to come. Oceanography 23 (4), 26–35. http://dx.doi.org/10.5670/oceanog.2010.03.

Winant, C.D., Alden, D.J., Dever, E.P., Edwards, K.A., Hendershott, M.C., 1999. Near-surface trajectories off central and southern California. J. Geophys. Res. 104 (C7), 15713–15726.

Wooster, W.S., Bakun, A., Mclain, D.R., 1976. Seasonal upwelling cycle along Eastern Boundary of North-Atlantic. J. Mar. Res. 34 (2), 131–141.

World Climate Research Program, 1988. WOCE surface velocity program planning committee report of first meeting: SVP-1 and TOGA pan-Pacific surface current study. WMO/TD-No.323, WCRP-26. World Meteorological Organization, Wormley, 33 pp.

Wunsch, C., Heimbach, P., Ponte, R.M., Fukumori, I., the ECCO-GODAE Consortium Members, 2009. The global general circulation of the ocean estimated by the ECCO-Consortium. Oceanography 22 (2), 88–103. http://dx.doi.org/10.5670/oceanog.2009.41.

Xie, S.-P., Liu, W.T., Liu, Q., Nonaka, M., 2001. Far-reaching effects of the Hawaiian islands on the Pacific ocean-atmosphere system. Science 292, 2057–2060.

Yaremchuk, M., Qu, T.D., 2004. Seasonal variability of the large-scale currents near the coast of the Philippines. J. Phys. Oceanogr. 34 (4), 844–855.

Zhang, D.X., Lee, T.N., Johns, W.E., Liu, C.T., Zantopp, R., 2001. The Kuroshio east of Taiwan: modes of variability and relationship to interior ocean mesoscale eddies. J. Phys. Oceanogr. 31 (4), 1054–1074.

Zhang, H.-M., Reynolds, R.W., Lumpkin, R., Molinari, R., Arzayus, K., Johnson, M., Smith, T.M., 2009. An integrated global observing system for sea surface temperature using satellites and in situ data: research to operations. Bull. Am. Meteorol. Soc. 90 (1), 31–38.

Chapter 13

Western Boundary Currents

Shiro Imawaki*, Amy S. Bower[†], Lisa Beal[‡] and Bo Qiu[§]

*Japan Agency for Marine–Earth Science and Technology, Yokohama, Japan

[†]Woods Hole Oceanographic Institution, Woods Hole, Massachusetts, USA

[‡]Rosenstiel School of Marine and Atmospheric Science, University of Miami, Miami, Florida, USA

[§]School of Ocean and Earth Science and Technology, University of Hawaii, Honolulu, Hawaii, USA

Chapter Outline

1. GENERAL FEATURES

1.1. Introduction

Strong, persistent currents along the western boundaries of the world's major ocean basins are some of the most prominent features of ocean circulation. They are called "western boundary currents," hereafter abbreviated as WBCs. WBCs have aided humans traveling over long distances by ship, but have also claimed many lives due to their strong currents and associated extreme weather phenomena. They have been a major research area for many decades; Stommel (1965) wrote a textbook entitled *The Gulf Stream: A Physical and Dynamical Description*, and Stommel and Yoshida (1972) edited a comprehensive volume entitled *Kuroshio: Its Physical Aspects*, both milestones of WBC

Ocean Circulation and Climate, Vol. 103. http://dx.doi.org/10.1016/B978-0-12-391851-2.00013-1

study. This chapter is devoted to describing the structure and dynamics of WBCs, as well as their roles in basin-scale circulation, regional variability, and their influence on atmosphere and climate. Deep WBCs are described only in relation to the upper-ocean WBCs.

A schematic global summary of major currents in the upper-ocean (Schmitz, 1996; Talley et al., 2011), spanning the depth interval from the sea surface through the main thermocline down to about 1000 m, is shown in Figure 1.6 in Chapter 1. Major WBCs are labeled as well as other currents. More detailed schematics of each WBC are shown in the following sections on individual oceans.

1.2. Wind-Driven and Thermohaline Circulations

Anticyclonic subtropical gyres (red flow lines in Figure 1.6) dominate the circulation at midlatitudes in each of the five ocean basins. These gyres are primarily wind-driven, where the equatorward Sverdrup transport in the interior of each ocean, induced by the curl of the wind stress at the sea surface, is compensated by a strong poleward current at the western boundary (Stommel, 1948). Readers are referred to Huang (2010) and Chapter 11 for details on the physics of the wind-driven circulation, including WBCs. The poleward WBCs of these subtropical gyres are the Gulf Stream, Brazil Current, Agulhas Current, Kuroshio, and East Australian Current (EAC). These subtropical WBCs carry warm waters from low to high latitudes, thereby contributing to global meridional heat transport and moderation of Earth's climate. According to linear wind-driven ocean circulation theory, WBCs separate from the western boundary at the latitude where the zonal integral of wind stress curl over the entire basin is zero. In fact, the dynamics of the separation process are very subtle, and actual separation latitudes are considerably lower than the latitude of zero wind stress curl, due to various details discussed in the following sections. The reproduction of WBC separation has been a benchmark of numerical models of general ocean circulation. After separation, the WBCs feed into the interior as meandering jets called WBC extensions.

Some WBCs also carry waters as part of the thermohaline circulation, involving inter-gyre and inter-basin exchanges as shown by green flow lines in Figure 1.6. For example, there is leakage via the Agulhas Current around the southern tip of Africa into the South Atlantic, the North Brazil Current affects cross-equatorial exchange from the South Atlantic into the North Atlantic, and the Gulf Stream and North Atlantic Current carry warm waters northward up into the Nordic Seas. Readers are referred to Chapter 11 for the thermohaline circulation and meridional overturning circulation (MOC), and Chapter 19 for inter-ocean and inter-basin water exchanges.

1.3. Transport

WBCs typically have widths of about 100 km, speeds of order 100 cm s^{-1}, and volume transports between 30 and 100 Sv (1 Sv $= 10^6\ m^3\ s^{-1}$). Their volume transport can be estimated as the compensation, at the western boundary, of the Sverdrup transport calculated from wind stress curl over the interior ocean. However, the local volume transport is usually larger than predicted by Sverdrup theory, due to a thermohaline component and/or lateral recirculations adjacent to the WBC.

Volume transports of most WBCs have an annual signal, which through Sverdrup theory corresponds to the annual cycle of wind stress curl over the interior ocean. However, the observed signal is considerably weaker than estimated from simple theory. This is thought to be due to the blocking of fast barotropic adjustment by ridge topography, while the baroclinic signal is too slow to transmit an annual cycle to the western boundary. A unique seasonality is observed in the volume transport of the Somali Current in the northern Indian Ocean, where the flow reverses annually with the reversal of the Asian monsoon winds. The Somali Current could not be classified as part of the subtropical gyre, but will be described in detail in the following sections, because of this uniqueness and its behavior extending into the subtropics.

1.4. Variability

Intrinsic baroclinic and barotropic instabilities of the WBCs result in meanders and ring shedding, and consequently, eddy kinetic energy (EKE) levels are elevated in WBC regions. Figure 13.1 shows the global distribution of climatological mean EKE (Ducet et al., 2000), estimated from almost 20 years of sea surface height (SSH) obtained by satellite altimeters, assuming geostrophic balance. The figure shows clearly that the EKE of WBCs and their extensions is much higher than in the interior. Especially, extensions of the Gulf Stream, Kuroshio, and Agulhas Current show very high EKE.

The EKE is also high in the transition from the Agulhas Current to its extension, located south of Africa. Another western boundary region of high eddy activity is located between Africa and Madagascar, caused by the Mozambique eddies, which replace the more standard continuous WBC there. EKE is enhanced at the western boundary of the northern Indian Ocean, due to the unique seasonal reversal of the Somali Current. See their details in the Indian Ocean section.

1.5. Structure of WBCs

WBCs have a baroclinic structure. This is illustrated for the Kuroshio south of Japan in Figure 13.2, which shows the vertical section of 2-year Eulerian-mean temperature and velocity during the World Ocean Circulation Experiment (WOCE). As in other WBCs, the flow is the strongest near

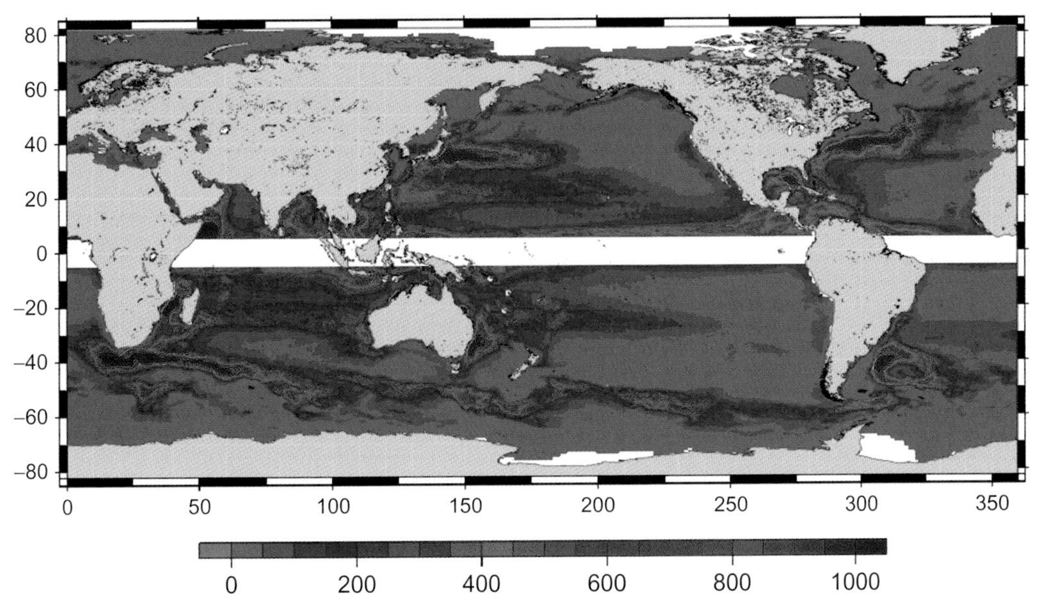

FIGURE 13.1 Global distribution of the climatological mean EKE (in cm^2 s^{-2}) at the sea surface derived from satellite altimetry data obtained during 1993–2011. The equatorial regions are blank because the Coriolis parameter is too small for geostrophic velocities to be estimated accurately from altimetric SSH. *From Ducet et al. (2000) and Dibarboure et al. (2011).*

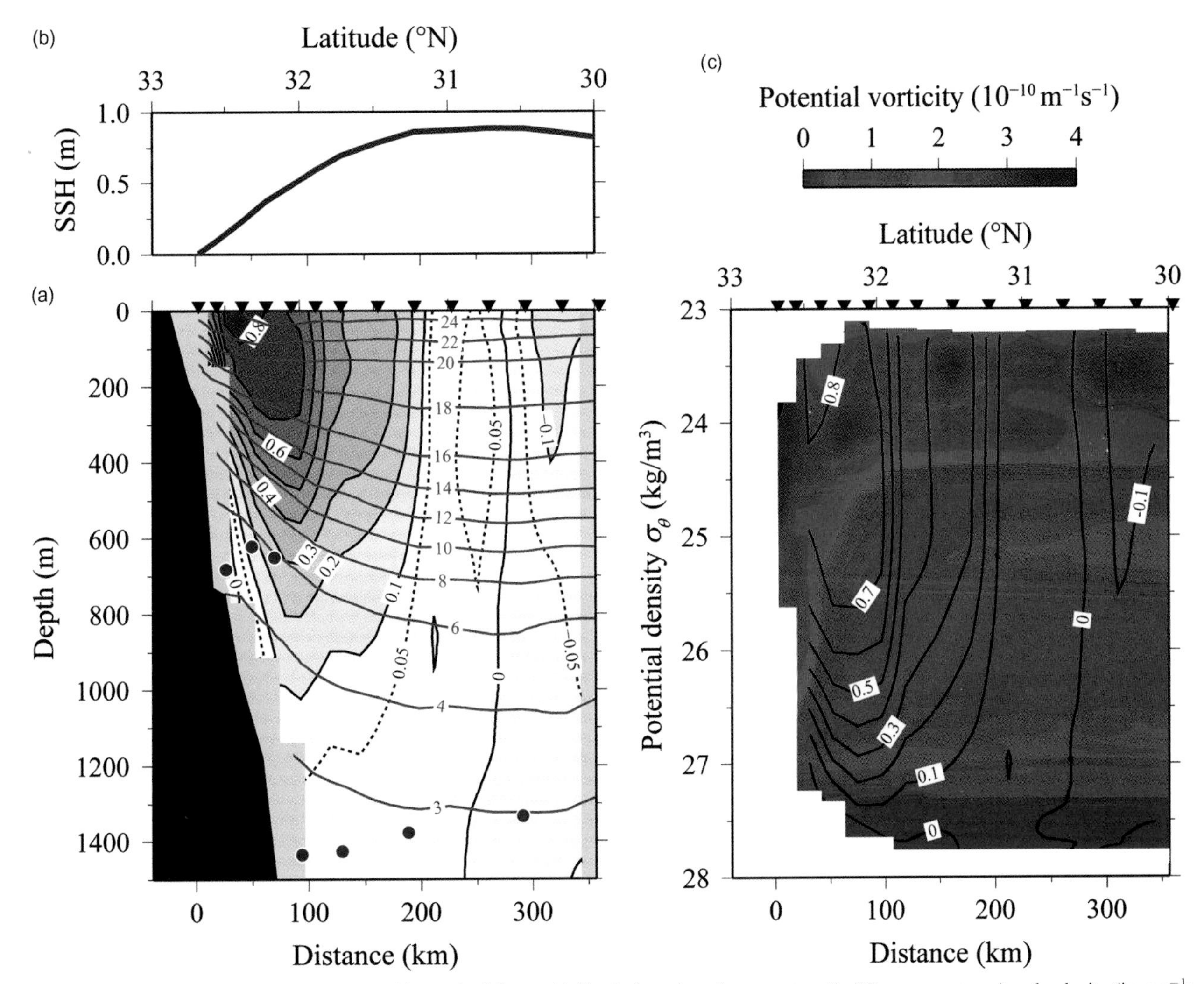

FIGURE 13.2 Vertical structure of the Kuroshio south of Japan. (a) Vertical section of temperature (in °C; green contours) and velocity (in m s^{-1}; positive, eastnortheastward; color shading with black contours), averaged over 2 years from October 1993 through November 1995. During that period, the Affiliated Surveys of the Kuroshio off Cape Ashizuri were carried out intensively (Uchida and Imawaki, 2008). Velocity is estimated from hydrographic data assuming geostrophy, being referred to observed velocities at locations shown by blue dots. Distance is directed offshore. (b) SSH profile relative to the coastal station, estimated from the surface velocity assuming geostrophy. (c) Section of potential vorticity (in m^{-1} s^{-1}; color shading; Beal et al., 2006) plotted in potential density σ_θ space. Overlaid are velocity contours (black) same as in (a); contours associated with the strong shear near the coast are omitted for the sake of visibility. *Courtesy of Dr. Hiroshi Uchida.*

the sea surface and decreases with depth. The velocity core, defined as the strongest along-stream flow at a given depth on a cross-stream vertical section, shifts offshore with increasing depth. Vertical and horizontal shears of the WBC are the strongest on the coastal (cyclonic) side, accompanying a strong gradient of sea surface temperature (SST). Geostrophic balance results in a SSH difference across a WBC of order 100 cm, with SSH higher on the offshore side (Figure 13.2b). The horizontal pressure gradient associated with this SSH difference is compensated by the baroclinic cross-stream pressure gradient associated with the main thermocline, which deepens by several hundred meters moving offshore across the current. As a result, the pressure gradient and velocity weaken with depth.

The SSH difference across a WBC has been found to be well correlated with its total volume transport, because the vertical structure is relatively stable and hence an increase (decrease) of the total transport results from a proportional increase (decrease) of transport of each layer of the WBC (Imawaki et al., 2001). This relationship has been used to estimate a time series of Kuroshio transport from satellite altimetry data.

Despite high lateral velocity shears, WBCs inhibit cross-frontal mixing owing to the strong potential vorticity front across their flow axis and to kinematic steering (Bower et al., 1985; Beal et al., 2006). Figure 13.2c shows the potential vorticity front and its location relative to the velocity core in the case of the Kuroshio. The potential vorticity front is related to strongly sloping isopycnals and dramatic changes in layer depth across the current. Steering or trapping of particles results when the speed of the WBC is greater than the meander or eddy phase speeds. As a result, water masses at the same density can remain distinct within a WBC down to intermediate depths.

1.6. Air–Sea Fluxes

Midlatitude WBCs, and particularly their extensions, are regions of strong air–sea interaction, and therefore are important to Earth's climate (see Chapter 5). Figure 13.3a

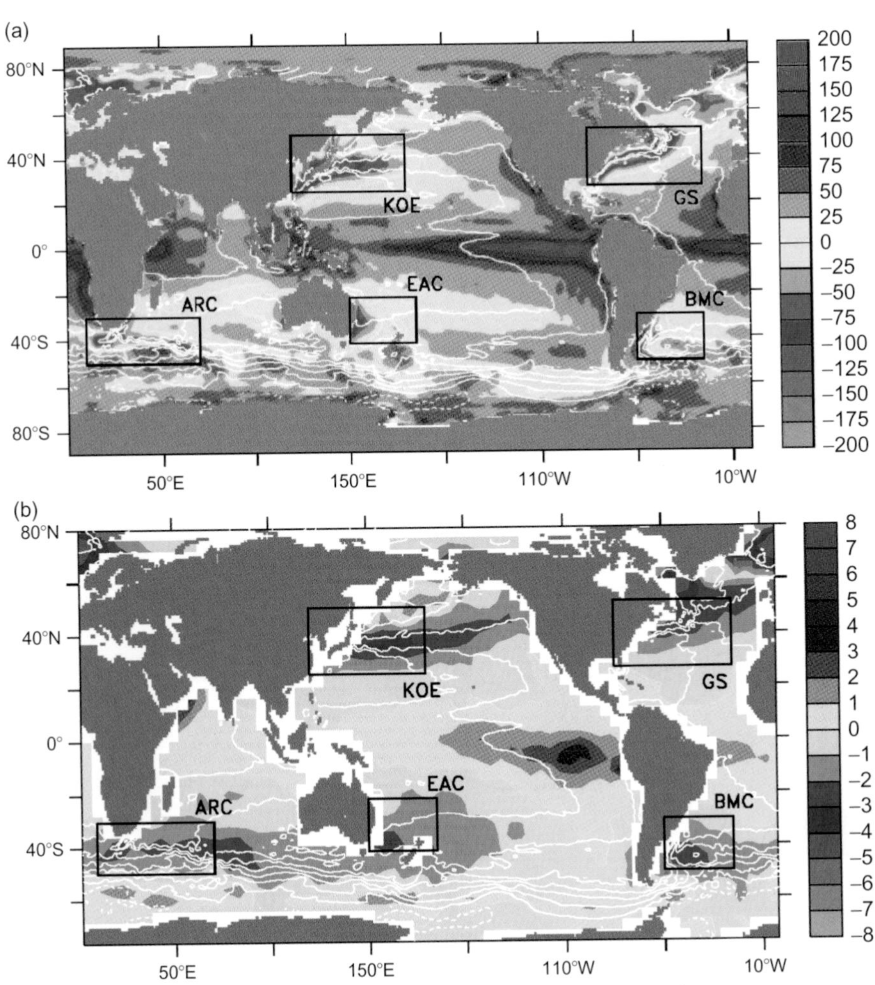

FIGURE 13.3 Global distribution of the climatological mean (a) latent plus sensible heat flux (in W m^{-2}; positive, atmosphere to ocean; Yu and Weller, 2007) and (b) CO_2 flux (in mol m^{-2} year^{-1}; positive, ocean to atmosphere; Takahashi et al., 2009) at the sea surface; the latter is for the reference year 2000 (non-El Niño conditions). White contours indicate mean sea surface dynamic height (Rio and Hernandez, 2004). ARC, Agulhas Return Current; KOE, Kuroshio–Oyashio Extension; EAC, East Australian Current; GS, Gulf Stream; and BMC, Brazil/Malvinas Current. *From Cronin et al. (2010).*

shows the global distribution of climatological mean net heat flux at the sea surface (Yu and Weller, 2007). The net heat flux is clearly the largest over the midlatitude WBCs, because warm water transported by the poleward WBCs from low to mid latitudes is cooled and evaporated by cold, dry continental air masses carried over the WBC regions by prevailing westerly winds. These large heat fluxes, together with moisture fluxes to the atmosphere and sharp SST fronts, contribute to the development of atmospheric disturbances. Recent studies show that storm tracks are found preferentially along WBCs and their extensions (e.g., Hoskins and Hodges, 2002; Nakamura and Shimpo, 2004; Nakamura et al., 2004, 2012), and effects of the sharp SST fronts can be detected even in the upper troposphere (e.g., Minobe et al., 2008).

Figure 13.3b shows the global distribution of the climatological mean flux of carbon dioxide (CO_2) from the ocean to the atmosphere (Takahashi et al., 2009). WBCs and their extension regions absorb large amounts of CO_2, because large wintertime heat loss leads to the formation of dense water, which is subducted into the interior ocean as a subsurface or intermediate mode water, carrying CO_2 away from the surface (Cronin et al., 2010). This is called a "physical pump." The "biological pump" associated with spring blooms also plays an important role in the very large uptake of CO_2 in WBC regions (Ducklow et al., 2001).

1.7. Observations

Since WBCs are characterized by relatively small scales, high velocities, often large vertical extent, and energetic variability, observing and monitoring them is a challenging task. This challenge has attracted many inspired scientists and resourceful engineers to tackle the measurement of these highly energetic signals. However, none of the currently available technologies and methods can satisfy all the requirements for an observing system of high spatial and temporal resolutions (see Chapters 3 and 4). Therefore, merged hybrid approaches are required, using sets of coastal and offshore end-point moorings, reference moorings for surface flux, inverted echo sounders with pressure gauges, submarine cables, research vessels, ships of opportunity, neutrally buoyant floats, underwater gliders, etc. (Cronin et al., 2010; Send et al., 2010). On the basis of those observational data, numerical model studies, including data assimilation, contribute to further understanding of WBCs. For climate studies, long time series, including mooring arrays maintained for longer than 10 years, are needed, as well as sustained satellite observations of vector winds, SSH (altimetry), SST, and sea surface salinity.

1.8. WBCs of Individual Ocean Basins

In the following sections, the features of major WBCs in different oceans are described in detail; we focus mostly on subtropical WBCs. They are the Gulf Stream System in the North Atlantic, the Brazil and Malvinas Currents in the South Atlantic, the Somali and Agulhas Currents in the Indian Ocean, the Kuroshio System in the North Pacific, and the EAC in the South Pacific (Figure 1.6).

2. NORTH ATLANTIC

2.1. Introduction

The series of WBCs in the North Atlantic, collectively referred to here as the Gulf Stream System, have helped shape human history in the Western Hemisphere. Their swift surface currents influenced the expansion of European civilization toward North America, and the advection of warm subtropical waters to high northern latitudes profoundly impacts climate on both sides of the Atlantic. The Gulf Stream System is also important on a global scale, being the primary conduit for the delivery of warm, saline waters to the Nordic and Labrador Seas, and therefore a central component of the global thermohaline circulation.

Several earlier works and reviews cover the classical ideas about WBC theory and observations in general, and the Gulf Stream specifically. Among these are Stommel's (1965) and Worthington's (1976) monographs, and Fofonoff's (1981) article in *Evolution of Physical Oceanography*. Schmitz and McCartney (1993) and Hogg and Johns (1995) summarize the observations up to the mid-1990s. Focus here will be mainly on recent (post-1995) advances in observation and understanding of low-frequency variability in the Gulf Stream System and its connection to the atmosphere, with some attention to earlier seminal contributions and some work that was not described in previous review articles.

The main components of the Gulf Stream System are shown in Figure 13.4. The first is the Florida Current, which originates where the Gulf of Mexico's Loop Current enters the Florida Straits. After leaving the confines of the Straits, the current is referred to as the Gulf Stream and continues northward along the continental shelf break of the eastern United States to the latitude of Cape Hatteras (35°N, 75°W). The current separates from the continental shelf near Cape Hatteras, flowing northeastward over the slope and into deepwater as a single free jet, the Gulf Stream Extension (GSE). Large-amplitude, propagating meanders develop along the GSE path—some of these meanders pinch off to form Warm (Cold) Core Rings north (south) of the mean path. The GSE is flanked by the cyclonic Northern Recirculation Gyre and anticyclonic Southern Recirculation Gyre (NRG and SRG). Near 40°N, 50°W, where the Grand Banks and Southeast Newfoundland Ridge extend southward into the abyssal plain, the GSE separates into several branches, including the recirculation return flows, the Azores Current, and the North Atlantic Current. The latter meanders northward off the eastern flanks of the Grand Banks and Flemish Cap to the so-called Northwest Corner near 52°N, where it turns abruptly eastward as a multibranched meandering flow toward the mid-Atlantic ridge.

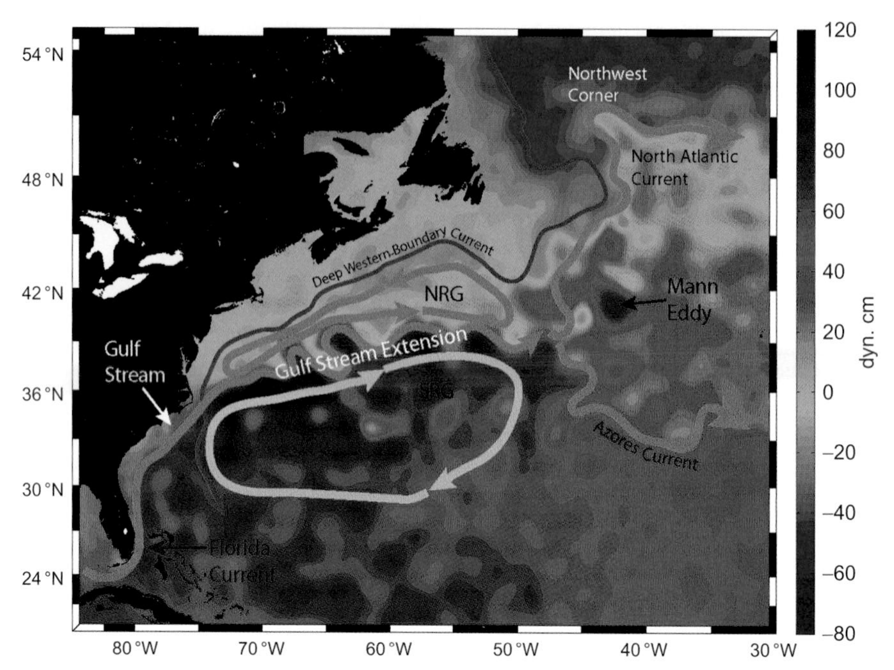

FIGURE 13.4 Map of Absolute Dynamic Topography (in dynamic cm; color shading) on September 21, 2011 for the western North Atlantic from AVISO (Archiving, Validation, and Interpretation of Satellite Oceanographic Data) Web site (http://www.aviso.oceanobs.com/), with schematic of currents in the Gulf Stream System, including the Northern and Southern Recirculation gyres (NRG and SRG).

2.2. Florida Current

The Florida Current is perhaps the most well-documented current in all the ocean basins. Its proximity to land and confinement within the Florida Straits (90 km wide, 700 m deep) have allowed frequent observations of its velocity structure and volume transport using a variety of measurement techniques. By the early 1990s, the mean transport (about 32 Sv), a small but significant annual cycle in transport (with a summer maximum and a winter minimum and peak-to-peak range of 4–6 Sv), and the water mass composition (13 Sv of South Atlantic origin) were already well established (Niiler and Richardson, 1973; Larsen and Sanford, 1985; Lee et al., 1985; Molinari et al., 1985; Leaman et al., 1987; Schmitz and Richardson, 1991). The velocity structure is laterally asymmetric, with the mean surface velocity maximum of about 180 cm s^{-1} pressed up against the western boundary (Leaman et al., 1987; Beal et al., 2008). The current extends to the bottom of the channel with a mean velocity of about 10 cm s^{-1}. Most observational and model results point to local and regional wind stress variability as the cause of the annual cycle in transport (Anderson and Corry, 1985; Schott and Zantopp, 1985).

Of particular significance in the study of Florida Current variability has been the nearly continuous measurement of the daily transport at 27°N since 1982, based on the voltage difference across a succession of abandoned underwater telephone cables (Larsen, 1992). From the first 16 years of cable-derived transports, Baringer and Larsen (2001) confirmed earlier estimates of the mean transport as well as the amplitude and phasing of the annual cycle. But they also showed that the annual cycle weakened in the second half of the record. They further found that interannual transport variability was inversely correlated with the North Atlantic Oscillation (NAO) index (Hurrell, 1995), with the NAO leading Florida Current transport by about 18 months. This suggested a connection between wind stress variability over the North Atlantic subtropical gyre and transport in the Florida Straits.

Meinen et al. (2010) combine the cable-derived transport time series with other *in situ* transport estimates to produce a 40-year time series. They argued for caution when attempting to explain changes in the amplitude or phasing of the annual cycle, pointing out that the dominance of subannual transport variability (containing 70% of the total variance and caused by the frictional effect of fluctuating along-channel winds; Schott et al., 1988) can contaminate the annual cycle, which contains only about 10% of the total variance. On longer timescales, Meinen et al. (2010) showed that lagged correlation between Florida Current transport and the NAO index was statistically significant during 1982–1998, but not before or after that time period.

DiNezio et al. (2009) use 25 years of cable-derived Florida Current transport and wind fields from the NCEP (National Centers for Environmental Prediction)/NCAR (National Center for Atmospheric Research) Reanalysis Project to examine the importance of wind stress curl variability over the North Atlantic to interannual transport variability in the Florida Current. They found that positive NAO index is associated with positive wind stress curl anomalies (contrary to what might be expected from a simple strengthening of the westerlies during the positive phase of the NAO), and a weakening of the southward Sverdrup circulation in the interior. This variability is correlated well with Florida Current transport at a lag of about half that predicted by classical baroclinic Rossby wave theory, and accounts for about 50% of the transport variability in the Florida Current. The faster baroclinic response time is consistent with some recent observational and theoretical studies that suggest that changes to the background potential vorticity distribution imposed by topography or the mean baroclinic circulation, or interaction with the atmosphere may speed up the westward propagation of long Rossby waves (see DiNezio et al., 2009 and references therein). DiNezio et al. (2009) also pointed out that other sources of Florida Current transport variability may lie upstream since nearly half of the transport is of South Atlantic origin.

2.3. Gulf Stream Separation

At Cape Hatteras, the Gulf Stream separates from the edge of the continental shelf and flows obliquely across the slope into deepwater (Figure 13.4). Unlike some of the other major WBCs, the Gulf Stream separation latitude varies by less than ± 50 km (Auer, 1987; Lee and Cornillon, 1995). The pioneering work on WBC separation based on idealized linear and nonlinear theory was focused primarily on the impact of wind stress patterns (Stommel, 1948; Munk, 1950; Munk et al., 1950; Charney, 1955). Since then, a number of other factors have been found to be important, including topography and adjacent currents; see Dengg et al. (1996) and Tansley and Marshall (2000) for reviews. For example, evidence is increasing that interaction of the GSE with the Deep Western Boundary Current (DWBC) and/or the NRG plays an important role. Several observational studies based on multiyear hydrography, velocity, and remote sensing records have revealed a correlation between north–south shifts in the GSE path northeast of Cape Hatteras and the strength of the southwestward flow in the NRG (Rossby and Benway, 2000; Rossby et al., 2005; Peña-Molino and Joyce, 2008). Some idealized modeling studies have also demonstrated how variability in the strength of the southwestward currents north of the GSE can impact the separation latitude (Thompson and Schmitz, 1989; Ezer and Mellor, 1992; Spall, 1996a,b; Joyce et al., 2000; Zhang and Vallis, 2007). A common feature of the model Gulf Stream–DWBC crossover studies is that the DWBC apparently alters the background potential vorticity field and effectively isolates the overlying Gulf Stream from the sloping topography, thus allowing it to cross the continental slope with a minimum of vortex stretching.

Accurate simulation of Gulf Stream separation has been a benchmark for assessing the performance of numerical model simulations of North Atlantic general circulation; see review by Hecht and Smith (2008). Significant improvement was only achieved when computing resources were sufficient to resolve the radius of deformation for the first baroclinic mode with several model grid points (grid spacing of about 10 km or less) (Smith et al., 2000; Chassignet and Garraffo, 2001). However, Bryan et al. (2007) argued that a grid spacing less than 10 km is not sufficient to represent Gulf Stream separation and northward penetration of the North Atlantic Current east of the Grand Banks (Figure 13.4), an even more elusive feature in North Atlantic simulations. They showed that lower subgrid scale dissipation is also necessary, as this allows for a more energetic DWBC.

2.4. Gulf Stream Extension

After the Gulf Stream has left the constraint of the continental slope, it develops meanders in its path that grow to maximum amplitude around 65°W (e.g., Lee and Cornillon, 1995). This downstream widening of the meander envelope leads to some of the highest oceanic EKE levels, reaching maximum values near 3000 $cm^2\ s^{-2}$ at the sea surface (e.g., Fratantoni, 2001). Lee and Cornillon (1995, 1996a,b) provide a comprehensive description of the frequency–wave number spectrum of GSE meanders based on 8 years of Advanced Very High Resolution Radiometer (AVHRR) imagery. For more discussion of the dynamics of the subannual variability in the GSE path, including a review of intrinsic baroclinic and barotropic instability of the current, the reader is referred to Hogg and Johns (1995), Cronin and Watts (1996), and references therein.

Trajectories of hundreds of freely drifting, long-range subsurface floats (SOFAR: Sound Fixing And Ranging, and RAFOS: SOFAR spelled backward) have extended the remotely sensed view of the GSE to the thermocline level and deeper; see Davis and Zenk (2001) for a general review of Lagrangian techniques and observations in the ocean. Ocean eddies naturally disperse floats over large areas, making it possible (with a sufficient number of trajectories) to map the horizontal structure of mean subsurface velocity and its variability over entire basins. This technique was used in several studies to reveal the horizontal structure of mean velocity of the GSE, the North Atlantic Current, and their adjacent recirculations (e.g., Owens, 1991; Carr and Rossby, 2000; Zhang et al., 2001b; Bower et al., 2002). Such direct measurement of the structure of the large-scale subsurface circulation is not readily achieved by any other means.

The GSE maintains a remarkably rigid baroclinic velocity structure even as its path undergoes large-amplitude meanders. This was first demonstrated by Halkin and Rossby (1985) at 73°W based on 16 sections of *absolute* velocity collected over 3 years with the Pegasus velocity profiler. After aligning all the sections in a stream-wise coordinate system (with the origin at the jet core), they found that two-thirds of the variability in the Eulerian frame was due to meandering of the GSE and not changes in the jet's velocity structure itself. Subsequent observational studies showed that the baroclinic velocity structure is more or less maintained as far east as 55°W, as shown in Figure 13.5 (Hogg, 1992; Johns et al., 1995; Sato and Rossby, 1995; Bower and Hogg, 1996). The constancy of the GSE's upper-ocean velocity structure has been further demonstrated recently by Rossby et al. (2010) based on a 17-year time series of weekly GSE crossings at 70°W by a container vessel, the *MV Oleander*, equipped with a hull-mounted acoustic Doppler current profiler (ADCP).

The inherently Lagrangian nature of float trajectories has been exploited to make inferences about the kinematics and dynamics of the GSE and North Atlantic Current. For example, Shaw and Rossby (1984) diagnosed the presence of significant vertical motions in the GSE based on the temperature change along the trajectories of 700 m SOFAR floats. Using isopycnal RAFOS floats, it was found that this vertical motion, as well as associated cross-stream exchange, is highly structured around GSE meanders, with

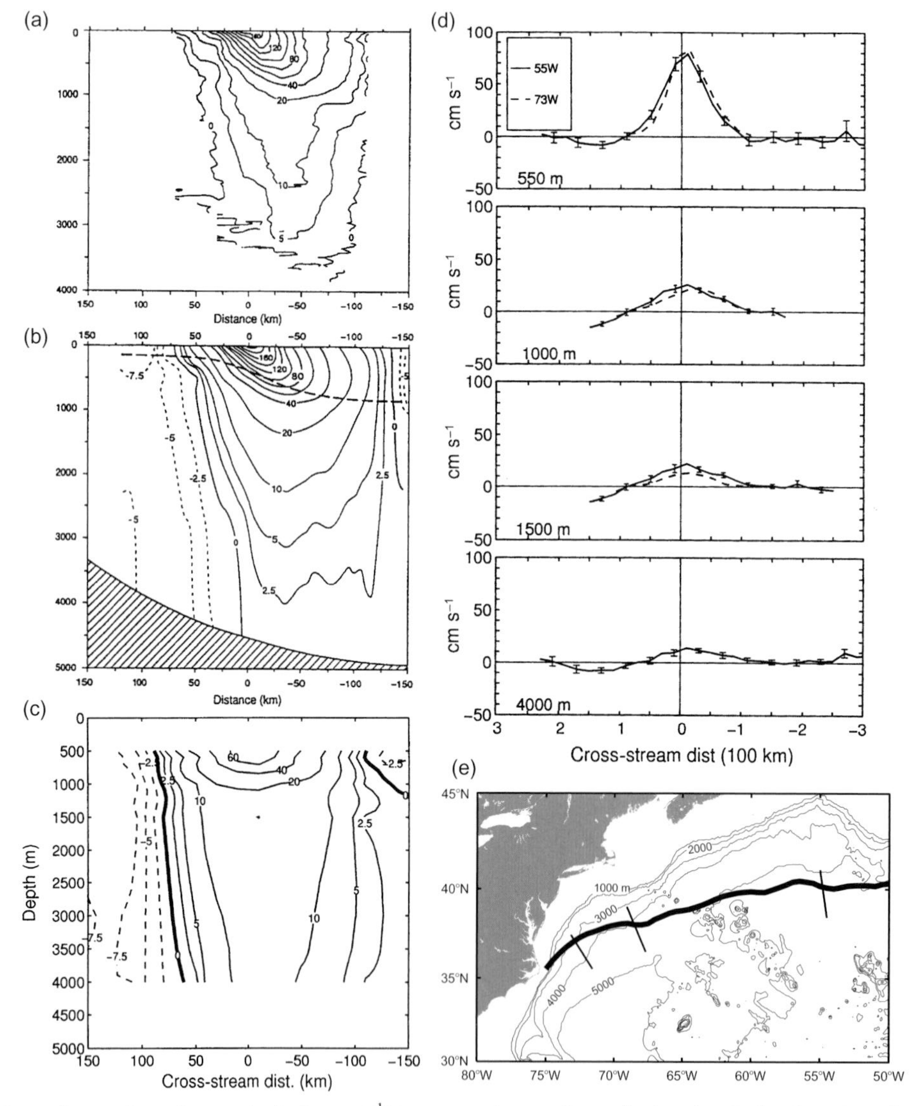

FIGURE 13.5 Sections of mean along-stream velocity (in cm s^{-1}) in stream-wise coordinates for three longitudes along the path of the GSE: (a) 73°W, (b) 68°W, and (c) 55°W. Downstream velocities are contoured with solid lines. Negative cross-stream distance is directed offshore. (d) Direct comparison of along-stream velocity (in cm s^{-1}) at 73°W and 55°W for four depths, showing similarity of peak speeds and cross-stream structure. Error bars show 95% confidence levels for the mean at 55°W. (e) Map showing locations of the three sections depicted in (a)–(c). The mean GSE path is drawn as a wide black line. *Panels (a) through (d): from Bower and Hogg (1996).*

upwelling in the thermocline approaching anticyclonic meander crests and vice versa moving toward cyclonic meander troughs (Bower and Rossby, 1989; Song and Rossby, 1995). This work led to a view of the GSE in the region of propagating meanders in which many fluid particles are constantly being expelled and replaced by others (Bower, 1991; Bower and Lozier, 1994; Lozier et al., 1996). A number of theoretical, numerical, and observational studies of fluid particle behavior in time-dependent jets followed (e.g., Samelson, 1992; Cushman-Roisin, 1993; Pratt et al., 1995; Duan and Wiggins, 1996; Lozier et al., 1997; Rypina et al., 2011).

The NRG and SRG are largely barotropic and swell the mean stream-wise transport of the GSE from 88 Sv just downstream of separation at 73°W (Halkin and Rossby, 1985), to 115 Sv at 68°W (Johns et al., 1995), and to a maximum of 150 Sv at 60°W (Hogg, 1992).

Several studies have used remote sensing observations to show that the mean path of the GSE is displaced 20–65 km farther north in fall compared to spring (Auer, 1987; Lee and Cornillon, 1995; Kelly et al., 1999). Based on 130 historical hydrographic sections across the GSE between Cape Hatteras and the New England Seamount Chain, Sato and Rossby (1995) found that the baroclinic transport in the upper 300 m also peaked in fall, when the path is at its northern extreme. However, baroclinic transport relative to 2000 m peaked in early summer and had peak-to-peak amplitude of 8 ± 3 Sv. They pointed out that the phasing of the annual cycle in the 0–2000 m transport is consistent with Worthington's (1976) hypothesis that winter convection in the SRG would deepen the thermocline and result in maximum transport in spring or early summer. However, they also showed that the downward displacement of isotherms occurred at depths below the depth of winter convection.

Kelly et al. (1999), using more than 4 years of altimetry-derived observations of the SSH difference across the GSE, found, like Sato and Rossby (1995), that the northerly fall position of the GSE was associated with an annual peak in surface geostrophic transport. They showed using historical hydrographic data that the seasonal change in surface transport was due to seasonal heating and was limited to the upper 250 m. The 17-year *MV Oleander* time series of upper-ocean transport shows a weak, surface-intensified annual cycle in layer transport with a maximum in mid-September, having amplitude of 4.3% of the mean at 55 m, and 1.5% at 205 m, compared to an average scatter around 1-year means of 15% (Rossby et al., 2010).

Some studies have shown that interannual-to-decadal variability in the GSE path is larger than the annual cycle, and is correlated with the NAO index. For example, Joyce et al. (2000) constructed a long time series of GSE position by using historical bathythermograph (BT) data over 35 years (1954–1989). They found significant correlation between the mean latitude of the GSE path and the NAO index, with the GSE lagging by 1 year or less. Frankignoul et al. (2001) extended the scope of this study by analyzing 6 years of TOPEX/Poseidon altimetric data and 45 years of BT observations of subsurface temperature. They reported that the GSE was very far north during the TOPEX/Poseidon years due to an extended period of high positive NAO index. They concluded that the GSE responds passively to the NAO with a delay of 1 year or so, and that this relatively rapid response time is associated with NAO-related buoyancy fluxes over the recirculation gyres.

2.5. Air–Sea Interaction

A recent large, multiinstitutional program, called Climate Variability and Predictability (CLIVAR) Mode Water Dynamics Experiment (CLIMODE), has made considerable progress toward a better understanding of the influence of the Gulf Stream System on climate variability. Recent reviews cover the regional (Kelly et al., 2010) and basin-scale (Kwon et al., 2010) interactions between the atmosphere and the Gulf Stream System. Here a few of the major features are highlighted; the reader is referred to these review articles and the references therein for a more thorough discussion.

The Gulf Stream System, like most other WBCs, is a region of strong heat loss to the atmosphere. This is due in large part to advection of warm water to midlatitudes, where cold, dry continental air masses carried over the warm water by prevailing westerly winds generate elevated latent and sensible heat fluxes. The annual average net heat flux over the GSE reaches a maximum of nearly 200 W m^{-2} out of the ocean, the highest of any of the major WBCs (Yu and Weller, 2007; Figure 13.3a). On synoptic timescales, values of turbulent heat flux to the atmosphere as high as 1000 W m^{-2} have been recently observed using the direct covariance method (The Climode Group, 2009). This transfer of heat from the ocean to the atmosphere leads to a sharp drop in northward heat transport by the ocean at the latitude of Gulf Stream separation (see, e.g., Trenberth and Caron, 2001).

These air–sea heat fluxes and the large SST gradients associated with the Gulf Stream System contribute to localized development of extratropical disturbances, leading to a storm track that is anchored to the current's path (see, e.g., Hoskins and Hodges, 2002; Nakamura et al., 2004). Joyce et al. (2009) used 22 years (1983–2004) of daily air–sea fluxes and combined reanalysis/scatterometer wind fields along with subsurface temperature observations to show that regions of maximum (2–8-day period) variance in latent and sensible heat flux, as well as meridional wind and wind divergence, all shifted in phase with north–south shifts in the GSE path.

While the impact of the Gulf Stream System on regional, near-surface atmospheric variability is becoming more clear (Kelly et al., 2010), the importance of WBCs in general, and the GSE specifically, to large-scale climate variability has been more difficult to unravel. A major step toward a better understanding was made by Minobe et al. (2008), who showed that the effects of the sharp SST gradient at the GSE can be detected in the upper troposphere. The annual climatology of upward motion from the European Center for Medium-Range Weather Forecasting (ECMWF) analyses is the strongest over the warm core of the GSE and extends into the upper troposphere (Figure 13.6a). Minobe et al. (2008) showed that this upward motion is associated with strong wind convergence at the sea surface (Figure 13.6a), and that the upper tropospheric divergence also tracks the path of the Gulf Stream and GSE in a climatological sense (Figure 13.6b). They went on to point out that the occurrence of cold cloud tops, indicative of high altitude, was elevated over the mean path of the Gulf Stream and GSE (Figure 13.6c). The full implications of such a connection between the lower and upper atmosphere over the Gulf Stream System on the large-scale atmospheric circulation are as yet unknown (Kwon et al., 2010).

2.6. North Atlantic Current

The North Atlantic Current extends to the highest latitude of any of the world's *subtropical* WBCs, about 52°N (Figure 13.4). As such, it represents the continuation of northward heat transport that is part of the thermohaline circulation, and therefore, it is important to include in this review. Its velocity structure is similar to that of the GSE near 55°W; namely, it has significant baroclinic and barotropic velocity structure, although peak velocities in the upper-ocean are about half. The synoptic and time-mean North Atlantic Current both extend to the ~4000 m deep sea floor (Meinen and Watts, 2000; Fischer and Schott, 2002; Schott et al., 2004). As might be expected for such a deep-reaching current, the total mean northward volume transport by the North Atlantic Current is large, 140–150 Sv at 43°N (Meinen and Watts, 2000; Schott et al., 2004). Meinen and Watts (2000) argued that 50–60 Sv of this transport recirculates in the quasi-permanent anticyclonic Mann Eddy located at the offshore edge of the current (Mann, 1967), 86–96 Sv recirculates in a larger loop around the Newfoundland Basin, and only 30 Sv exits the basin to the east (Schmitz and McCartney, 1993). There is some evidence for one or more branches leaving the North Atlantic Current at various latitudes along the flanks of the Grand Banks and Flemish Cap, and flowing eastward toward the mid-Atlantic ridge (Krauss et al., 1987); however, other studies show that most or all of the upper-ocean transport continues northward to the Northwest Corner near 52°W before turning eastward (Lazier, 1994; Pérez-Brunius et al., 2004a,b; Woityra and Rossby, 2008).

Historical hydrographic data, surface drifters, and subsurface floats have revealed that the North Atlantic Current generally follows the 4000 m isobath (Rossby, 1996; Kearns and Rossby, 1998; Carr and Rossby, 2000; Fratantoni, 2001; Zhang et al., 2001b; Bower et al., 2002; Orvik and Niiler, 2002). Unlike the propagating GSE meanders, North Atlantic Current meanders are largely locked to topographic features, including the Southeast Newfoundland Ridge, the Newfoundland Seamounts, and Flemish Cap (Carr and Rossby, 2000).

The penetration of the North Atlantic Current along the western boundary to the latitude of the Northwest Corner has been even more difficult to reproduce in ocean general

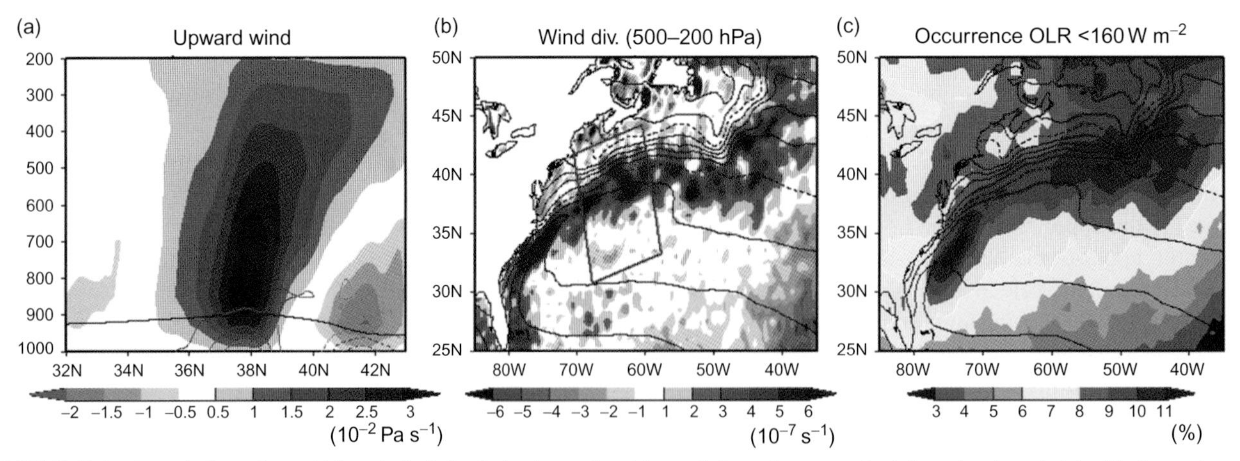

FIGURE 13.6 Annual climatology of (a) vertical wind velocity (upward positive; color), marine–atmospheric boundary layer height (black curve), and wind convergence (contours for $\pm 1, 2, 3 \times 10^{-6}\ s^{-1}$) averaged in the along-front direction in the green box in (b), based on the ECMWF analysis; (b) upper tropospheric wind divergence averaged between 200 and 500 hPa (color); (c) occurrence frequency of daytime satellite-derived outgoing long-wave radiation levels lower than 160 W m^{-2} (color). Contours in (b) and (c) are for mean SST, with 2 °C contour interval and dashed contours for 10° and 20 °C. *From Minobe et al. (2008).*

circulation models than the Gulf Stream separation; some success has been achieved with model grid separation that resolves the first baroclinic Rossby radius at all latitudes (<10 km) and sufficiently low subgrid scale dissipation (Smith et al., 2000; Bryan et al., 2007). Lower resolution and/or higher viscosity suppress advection of warm water along the eastern flank of the Grand Banks, resulting in large SST and air–sea heat flux errors when compared to observations (Bryan et al., 2007). Even when the North Atlantic Current path is represented well, its volume transport in models is still too low by a factor of 2 (Bryan et al., 2007).

3. SOUTH ATLANTIC

3.1. Introduction

The major WBCs of the South Atlantic include the northward-flowing North Brazil Current (referred to as the North Brazil Undercurrent south of about 5°S; Stramma et al., 1995), the southward-flowing Brazil Current, and northward-flowing Malvinas Current (Figure 13.7). Although the North Brazil Current is considered to be the principle conduit for the flow of warm water into the North Atlantic in the upper arm of MOC, the focus in this chapter is on subtropical mostly and additionally subpolar, not tropical, WBCs. Comprehensive discussions of the western tropical South Atlantic circulation can be found in Schott et al. (1998) and Johns et al. (1998).

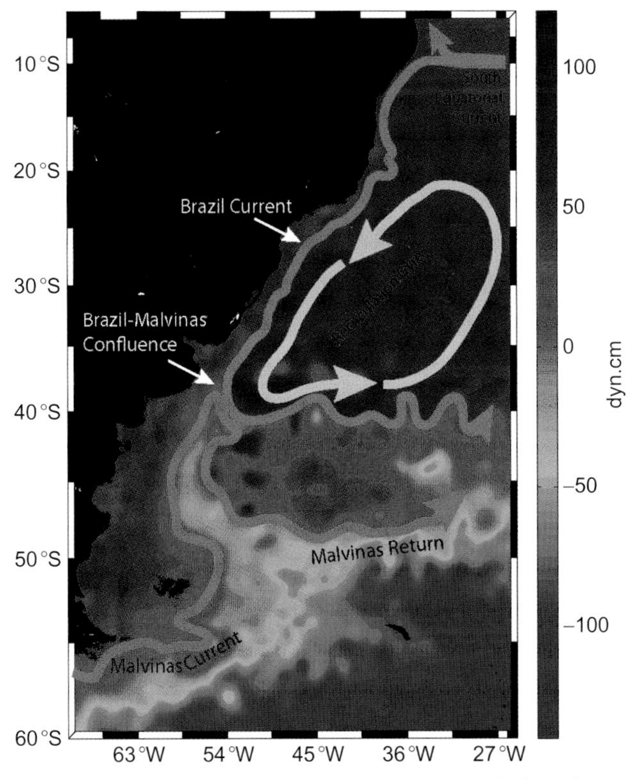

FIGURE 13.7 Map of Absolute Dynamic Topography (in dynamic cm; color shading) on December 22, 2010 for the western South Atlantic from AVISO Web site, with schematic of currents in the South Atlantic WBC system.

3.2. Brazil Current

The origins of the Brazil Current are in the South Equatorial Current (SEC), the northern limb of the South Atlantic subtropical gyre (Figure 13.7). According to Peterson and Stramma (1991) and references therein, the SEC has two main branches: transport in the northern branch feeds the North Brazil Current and equatorial countercurrents, while the southern branch (~16 Sv) bifurcates at the western boundary, with most transport (12 Sv) supplementing the North Brazil Current and a smaller fraction (4 Sv) turning southward as the Brazil Current (Stramma et al., 1990). This bifurcation at the western boundary is typically located south of 10°S.

Knowledge of the volume transport of the Brazil Current and its low-frequency variability has suffered significantly from the lack of long-term, direct velocity observations, and by the fact that on the order of half of the total Brazil Current transport is over the continental shelf, where estimating currents from hydrography is less reliable (Peterson and Stramma, 1991). Geostrophic transport estimates of the Brazil Current from 12° to 25°S, relative to various intermediate levels of no motion, are all less than 11 Sv (Peterson and Stramma, 1991). The only transport estimate based on direct velocity measurements, made using the Pegasus profiler at 23°S, is 11 Sv southwestward, of which 5 Sv was estimated to be flowing over the shelf (Evans and Signorini, 1985). The Pegasus velocity profiles revealed a three-layer current structure, with the southward-flowing Brazil Current confined to the upper 400 m, overlying an intermediate northward flow with Antarctic Intermediate Water (AAIW) characteristics and a deep southward flow carrying North Atlantic Deep Water (NADW).

As the Brazil Current flows southward, it continues to hug the continental shelf break. Garfield (1990) used infrared imagery in the latitude range 21–35°S to show that the inshore edge of the current lies over the 200 m isobath on average, and is always inshore of the 2000 m isobath. South of 24°S, the Brazil Current geostrophic transport, defined as the southward flow of warm subtropical waters above about 400 m, increases to 20 Sv due to the influence of an anticyclonic recirculation cell adjacent to the Brazil Current (Garzoli, 1993). The 20 Sv is considerably less than the estimates of the interior northward Sverdrup transport, which vary from 30 to 60 Sv (Veronis, 1973, 1978; Hellerman and Rosenstein, 1983). Gordon and Greengrove (1986) suggested that the deficit in southward transport by the Brazil Current relative to the northward interior Sverdrup transport might be compensated for by the southward flow of NADW in the DWBC. Some studies have suggested, based on water mass characteristics, that AAIW flows southward, rather than northward, under the Brazil Current in this latitude range, leading to total geostrophic transports around 70–76 Sv at 37°S (McCartney and Zemba, 1988; Zemba and McCartney, 1988; Peterson, 1990).

3.3. Brazil Current Separation and the Brazil–Malvinas Confluence

The Brazil Current separates from the western boundary where it meets the northward-flowing Malvinas (Falkland) Current, the subpolar WBC of the South Atlantic (Gordon and Greengrove, 1986; Olson et al., 1988). After colliding over the continental slope, both currents turn offshore and develop large-amplitude meanders and eddies (Figure 13.7). This highly energetic region is called the Brazil–Malvinas Confluence (hereafter the Confluence). Multiyear records of the Confluence latitude based on remote sensing observations show excursions of the WBC separation point along the western boundary as large as 900 km, with a mean latitude of separation at about 36–38°S (Olson et al., 1988; Peterson and Stramma, 1991; Goni and Wainer, 2001). This contrasts sharply with the more stable separation latitudes of the Gulf Stream and Kuroshio in the Northern Hemisphere (Olson et al., 1988). The mean separation/Confluence latitude is well north of the latitude of zero wind stress curl in the South Atlantic, 47–48°S (Hellerman and Rosenstein, 1983). Veronis (1973) speculated that the premature separation was related to the northward-flowing Malvinas Current, and Matano (1993) found support for this idea using analytical and numerical models.

Some of the separated Brazil Current flows generally southeastward, alongside the Malvinas Return Current, transporting relatively warm subtropical waters poleward to about 46°S at 53°W before turning back northeastward (Figure 13.7). This anticyclonic meander occasionally pinches off warm, saline eddies into the subantarctic region (see, e.g., Gordon, 1989).

3.4. Malvinas Current

The Malvinas Current, which originates as a branch of the Antarctic Circumpolar Current, transports relatively cold, fresh subantarctic water northward along the 1000–1500 m isobaths of the Patagonian slope to the Confluence near 38°S (Figure 13.7). Spadone and Provost (2009) estimated a mean volume transport of 34.3 Sv based on 14 years of altimetric data "calibrated" with two independent periods of current-meter observations. It has been shown that relatively cold, fresh subpolar waters injected into the South Atlantic via the Malvinas Current can make their way to the Benguela Current system in the eastern South Atlantic, at times making up 50% of the waters transported northward in the upper limb of the MOC (Garzoli et al., 1997; Garzoli and Matano, 2011).

3.5. Annual and Interannual Variability

Significant annual cycles have been observed in Brazil and Malvinas Current transport and the latitude of the Confluence. Olson et al. (1988) used a multiyear record of AVHRR imagery and Geodetic Satellite (GEOSAT) altimetric data in the Confluence region to document an increase in Malvinas Current transport and a northward shift of the Confluence latitude during austral winter, and vice versa in summer. Witter and Gordon (1999) and Goni and Wainer (2001) used altimetric data to show that the annual and semi-annual signals account for most (up to 75%) of the observed variability in the position of the Confluence. Using 9 years of AVHRR images, Saraceno et al. (2004) argued that the latitude where the two currents collide is quite stable near 39.5°S, 53.5°W, but the orientation of the merged front swings from northeastward in winter to southeastward in summer, leading to a distinct seasonal cycle where the Confluence crosses the 1000 m isobath. Goni and Wainer (2001) further argued that the latitude of the Confluence is most sensitive to Brazil Current transport, and only correlated with Malvinas Current transport when Brazil Current transport is low.

An annual signal was also observed in Malvinas Current transport; but its amplitude exhibits strong interannual modulation. Spadone and Provost (2009) found very little energy at the annual period from 1993 to 2000; but after 2000 there was significant energy at the semi-annual and annual periods. Monthly mean transports for the whole record showed an annual peak of 37 Sv in July/August (austral winter).

Regarding variability at longer timescales, Witter and Gordon (1999) computed empirical orthogonal functions from 4 years of TOPEX/Poseidon altimetric data and found significant interannual variability in the gyre scale circulation, characterized by zonal shifts in the center of the subtropical gyre and associated variations in the strength of the Brazil Current. The interannual changes in the subtropical gyre circulation were found to be correlated well with variations in the large-scale wind stress curl over the South Atlantic. Using 15 years of altimetric and surface drifter observations, Lumpkin and Garzoli (2011) documented a multiyear southward shift of the Confluence latitude at a rate of 0.6–0.9°decade^{-1} from 1992 to 2007. A comparable shift of the latitude of maximum wind stress curl averaged across the basin led the authors to conclude, like Garzoli and Giulivi (1994) and Witter and Gordon (1999), that the separation of the Brazil and Malvinas Currents from the western boundary is coupled to the basin-wide wind stress pattern on interannual-to-decadal, as well as annual, timescales. Using the latitude of maximum wind stress curl as a proxy for the Confluence latitude prior to 1992, Lumpkin and Garzoli (2011) went on to report a weak northward shift in the Confluence latitude from 1979 to 1992, suggesting that the shift observed from 1992 to 2007 may be part of a multidecadal oscillation.

4. INDIAN OCEAN

4.1. Somali Current

4.1.1. *Introduction*

The Somali Current could be classified as part of the tropical gyre and therefore not a major WBC. However, because the Indian Ocean is cut off to the north by the Asian continent and the monsoon winds are so strong, the Somali Current extends into the subtropics and is worthy of inclusion here. In comparison to other WBC systems, there are sparse measurements of the Somali Current, particularly considering that its flow reverses seasonally with the reversal of the Asian monsoon winds. Only three coordinated occupations have been undertaken over the past 50 years. The International Indian Ocean Experiment during 1964–1966 resulted in Wyrtki's (1971) hydrographic atlas and a first look at the monsoon variability of the Somali Current (Swallow and Bruce, 1966). The Indian Ocean Experiment (INDEX) included a study of the Somali Current during the onset of the 1979 southwest monsoon (Leetmaa et al., 1982; Swallow et al., 1983). And in 1995 there were several crossings of the current during WOCE and Joint Global Ocean Flux Study (e.g., Beal and Chereskin, 2003). In addition, there have been moored arrays on the equator and off the Horn of Africa and in the Socotra Passage (Schott et al., 1990, 1997). The circulation during the southwest monsoon is better measured and understood than that during the northeast monsoon. Little progress has been made in understanding the Somali Current over the past 10–15 years, mainly due to the dangers of Somali piracy, which continue to preclude *in situ* observations.

4.1.2. *Origins and Source Waters*

During the southwest monsoon, which peaks from July to September, the Somali Current flows northward from the equator up to the tip of the Arabian Basin at 25°N (Figures 13.8 and 13.9a). In contrast, during the northeast monsoon of December through February, the Somali Current flows (less strongly) southward between about 10°N and the equator (Figure 13.9b).

Throughout the southwest monsoon, the waters of the Somali Current largely originate in the SEC, flowing across the equator via the East African Coastal Current (EACC) (Schott et al., 1990). Therefore, water properties of the Somali Current include influence from the Indonesian Throughflow (Warren et al., 1966) and tropical surface waters, and are cooler and fresher than the interior of the Arabian Sea (Wyrtki, 1971). Surface water properties are also strongly influenced by evaporation and upwelling along the path of the monsoon jet, which follows the Somali coast and extends offshore from the Horn of Africa (Findlater, 1969). During the northeast monsoon, when the Somali Current flows to the south, it is partially fed by the North Monsoon Current, which flows westward across the Arabian Sea from the Bay of Bengal (Figure 13.9b). These waters are again fresher than the interior of the Arabian Sea. The annual mean Somali Current flows to the north since the southwest monsoon is stronger than its counterpart (Figure 13.9c).

4.1.3. *Velocity and Transport*

The velocity structure of the summer Somali Current is understood to develop and deepen over the course of each southwest monsoon (Schott and McCreary, 2001), but there is an extreme degree of variability in both its strength and path on intraseasonal and interannual timescales (Luther, 1999; Wirth et al., 2002; Beal and Donohue, 2013). Weak northward flow is established in April, well before the onset of the southwest monsoon, by the arrival of annual Rossby waves at the western boundary (Brandt et al., 2002; Beal and Donohue, 2013). At the beginning of the monsoon, the Somali Current is weak and shallow, and overlies a southward undercurrent (Quadfasel and Schott, 1982). At this time, the Somali Current is largely Ekman-driven, balanced by southward Ekman transport in the interior of the Arabian Sea, while the geostrophic flow is undeveloped (Hastenrath and Greischar, 1991; Beal et al., 2003). By the end of the season, the Somali Current can reach speeds of 350 cm s^{-1} at the surface and deepen to over 2000 m (Swallow and Bruce, 1966), resulting in a V-shaped structure similar to other WBCs (Beal and Chereskin, 2003). The transport has been measured to be as much as 70 Sv in late summer (Fischer et al., 1996) and is by then largely balanced by southward geostrophic transport in the interior (Beal et al., 2003). Less is understood about the development of the Somali Current in response to the northeast monsoon, although it is clearly weaker and shallower and almost purely Ekman-driven (Figure 13.9b; Hastenrath and Greischar, 1991). Volume transports are about 5–10 Sv (Quadfasel and Schott, 1983). A southward undercurrent appears to persist throughout the winter, connected to eastward undercurrents along the equator (Jensen, 1991). This undercurrent carries the Red Sea Water away from the Gulf of Aden (Schott and Fischer, 2000).

4.1.4. *Separation from the Western Boundary*

Unlike other steady WBCs, the Somali Current is associated with two quasi-stationary eddies that have coastal

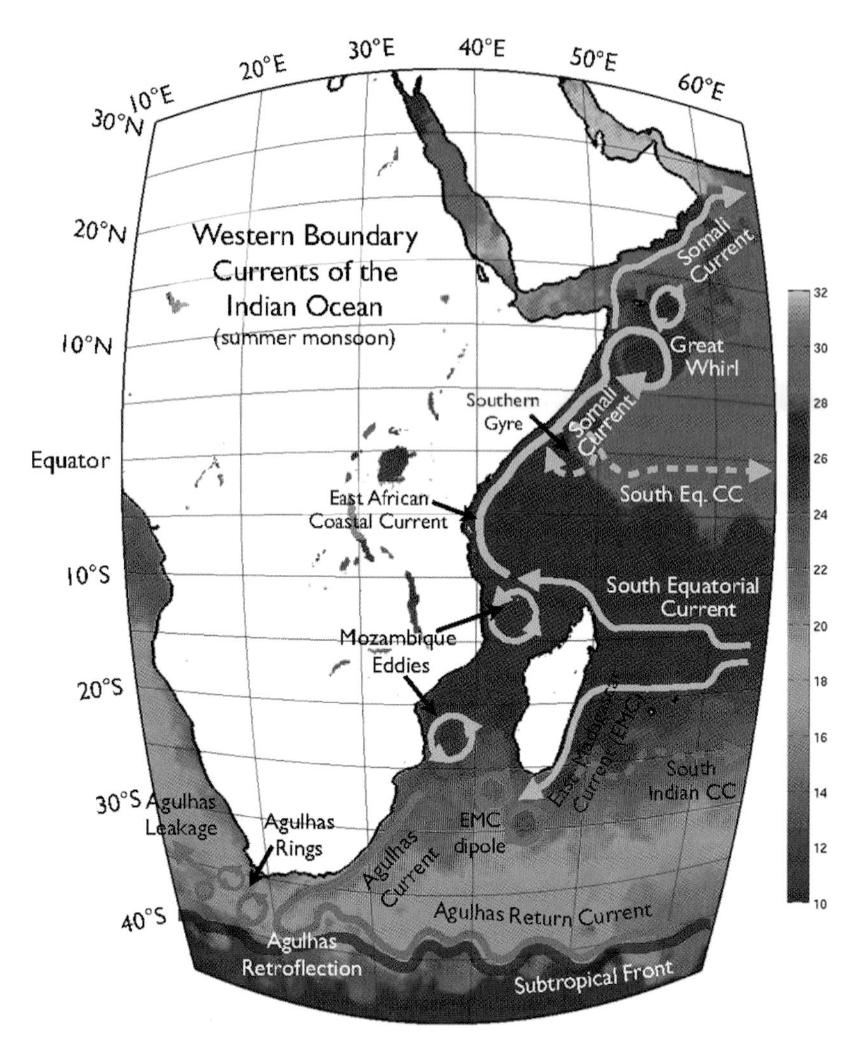

FIGURE 13.8 Schematic of the WBC system of the Indian Ocean, showing the Agulhas and Somali Currents, their sources and associated features, and the leakage of Agulhas waters into the Atlantic. SST (in °C; color shading) is for June 29, 2009 from the NAVOCEANO (United States Naval Oceanographic Office) Level 4 analysis, produced by interpolation of infrared and passive microwave observations, made available through the GHRSST (Group for High-Resolution Sea Surface Temperature) project (Donlon et al., 2007). CC: Countercurrent.

separations: the Southern Gyre and Great Whirl. The existence of these quasi-stationary eddies, rather than a continuous boundary current, has been shown to be the result of alongshore southwesterly winds and the slanted angle of the coastline, the latter arresting their northward migration (Cox, 1979; McCreary and Kundu, 1988).

The shallow (100 m deep) Southern Gyre is formed by the Somali Current separating from the coast at about 3°N and looping back across the equator (Düing et al., 1980; Jensen, 1991; Schott and McCreary, 2001). This circulation may result from an inertial overshoot of the EACC (Anderson and Moore, 1979), or from the local wind stress close to the equator which drives offshore currents (Cane, 1980). Monthly means from the global drifter climatology show that the Southern Gyre is a relatively short-lived feature and, on the decadal average, is an open loop that feeds into the South Equatorial Counter Current (SECC) (not shown).

Swallow et al. (1983) found that as the winds strengthen northward in June, the Great Whirl spins up between 5° and 10°N (Schott and McCreary, 2001). Using a 3.5 layer model, Jensen (1991) suggests that the Great Whirl is formed by barotropic instability where the kinetic energy in the Somali Current is at a maximum and the gradient in relative vorticity is the largest. Its northern edge typically lies close to the axis of the monsoon jet, that is, the latitude of zero wind stress curl. Altimeter and drifter data show that there is weak anticyclonic flow, a precursor to the Great Whirl, as early as April, due to remote forcing, and that the Great Whirl typically remains until the beginning of November, more than a month after the monsoon winds are gone (Beal and Donohue, 2013).

A portion of the Somali Current continues northward through the Socotra Passage (off the Horn of Africa), across the mouth of the Gulf of Aden, and along the coast of Oman (Quadfasel and Schott, 1982; Schott et al., 1997) (Figure 13.9a), before a final, broad separation from the coast. Crossing the Gulf of Aden, the current can trigger or interact with eddies, which subsequently propagate westward toward the mouth of the Red Sea (Fratantoni et al., 2006; Al Saafani et al., 2007).

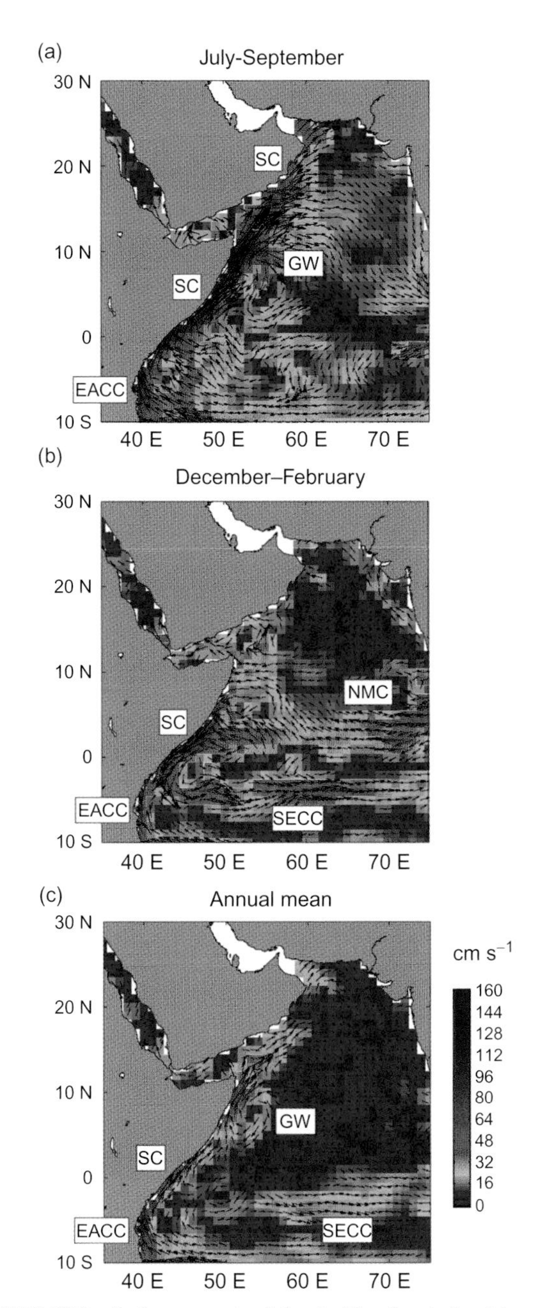

FIGURE 13.9 Surface currents of the Arabian Sea during (a) summer monsoon, (b) winter monsoon, and (c) annual mean, from the global drifter climatology (1993–2010). Color shading shows current speed (in cm s^{-1}), and arrows, current directions. Features are the Somali Current (SC), Great Whirl (GW), East African Coastal Current (EACC), North Monsoon Current (NMC), and South Equatorial Counter Current (SECC). *Data from http://www.aoml.noaa.gov/phod/dac/dac_meanvel.php (Lumpkin and Garraffo, 2005).*

In the past, numerical experiments have suggested that the Southern Gyre migrates northward to coalesce with the Great Whirl toward the end of the southwest monsoon, and the circulation and Somali Current collapse. However, more recent higher resolution models (Wirth et al., 2002), and 18 years of satellite observations (Beal and Donohue, 2013) show that the Great Whirl is ringed by smaller cyclones 70% of the time and hence its variability and stability are dominated by mutual eddy advection. There appears to be no sequence of events attributed to the collapse of the circulation, other than an initiation by the change in wind forcing.

During the northeast monsoon, the Somali Current flows to the south and separates from the coast just south of the equator, where there is a confluence with the EACC. Both currents then feed into the SECC (Figure 13.9b).

4.1.5. WBC Extension

The Somali Current does not have a recognized extension. During the southwest monsoon, the northern Somali Current, which flows through the Socotra Passage and along the coast of Oman, feeds gradually into the interior of the Arabian Sea between 15° and 25°N (Figure 13.9a). The curl of the wind stress vanishes at about 9°N, coincident with the northern arm of the Great Whirl.

4.1.6. Air–Sea Interaction and Implications for Climate

Much research on coupled modes over the Somali Current system relates to the effect of coastal upwelling cells inshore of the current to rainfall and wind anomalies. The relationship between the Somali Current and these upwelling cells is largely unexplored, although both will be weaker when monsoon winds are weaker.

A decrease in coastal upwelling strengthens monsoon rainfall over India by increasing SST and thus local evaporation and water vapor transport (Shukla, 1975; Izumo et al., 2008). Such a decrease has been related to weaker alongshore winds during monsoon onset, which are often related to El Niño conditions. In addition, coastal upwelling inshore of the Somali Current creates significant SST variability over small scales, which couples with variability in the monsoon jet, such that cool SSTs slow down local winds. This causes a feedback via local Ekman suction (pumping) downwind (upwind) of the SST anomaly, which tends to move the Ekman suction downwind (Halpern and Woiceshyn, 1999; Vecchi et al., 2004; Seo et al., 2008). Hence, air–sea coupling can feedback on the oceanic circulation.

On longer timescales, the Simple Ocean Data Assimilation reanalysis (Carton et al., 2000) shows that there may be a weakening trend in the Somali Current during 1950–1991, due to a decreasing cross-equatorial transport related to a trend in the reanalyzed winds (Schoenefeldt and Schott, 2006).

4.2. Agulhas Current

4.2.1. Introduction

The Agulhas Current is the WBC of the southern Indian Ocean subtropical gyre (Lutjeharms, 2006) and flows southwestward along the east coast of southern Africa between about 27° and 37°S (Figure 13.8). Its mean transport is 70 Sv at 32°S, making it the strongest WBC in either hemisphere at this latitude (Bryden et al., 2005). Once the Agulhas Current reaches the African cape, it separates and loops anticlockwise south of the continent to feed into the eastward Agulhas Return Current (Figure 13.8). This loop, known as the Agulhas Retroflection, sheds rings, eddies, and filaments of Agulhas waters into the Atlantic down to depths of more than 2000 m (Gordon et al., 1992; Boebel et al., 2003; Van Aken et al., 2003). Estimates of this "Agulhas leakage" are highly uncertain, ranging from 2 to 15 Sv, with about four to six Agulhas Rings shed annually (de Ruijter et al., 1999; Dencausse et al., 2010). Together with a leakage of waters south of Tasmania from the East Australia Current, which is described in the South Pacific section, Agulhas leakage forms the so-called Southern Hemisphere Supergyre, which links the subtropical gyres of the Pacific, Indian, and Atlantic Oceans (see Chapter 19).

More needs to be learned about the variability of the Agulhas Current and Retroflection, and especially about changes in leakage. On subseasonal timescales, variability of the current is dominated by four to five southward-propagating, solitary meanders per year (Grundlingh, 1979; Lutjeharms and Roberts, 1988; Bryden et al., 2005) (Figure 13.10). There is no consensus on seasonality (Ffield et al., 1997; Matano et al., 2002; Dencausse et al., 2010), but variations in retroflection and ring-shedding have been related to El Niño/Southern Oscillation on interannual timescales (de Ruijter et al., 2004). For example, during an anomalous upstream retroflection coincident with La Niña (2000–2001), no Agulhas rings were shed for 5 months. On climate timescales, peaks in Agulhas leakage have been linked to glacial terminations (Peeters et al., 2004) and to the resumption of a stronger Atlantic MOC (Knorr and Lohmann, 2003) (Figure 13.11). A simulation of the twentieth century ocean suggests that Agulhas leakage is currently increasing under the influence of global climate change (Biastoch et al., 2009).

4.2.2. Origins and Source Waters

Waters of the Agulhas Current originate in the marginal seas of the northern Indian Ocean, in the Pacific, in the Southern Ocean, and within the subtropical gyre itself. To the north of the Agulhas Current, where the island of Madagascar shades the western boundary from the interior of the gyre, the poleward boundary flow is split into two: a direct route via the East Madagascar Current and a route via eddies advected through the Mozambique Channel (Figure 13.8). Long-term moorings show four or five large (350 km) anticyclonic eddies per year in the Mozambique Channel, carrying a mean southward transport of 17 Sv (Ridderinkhof et al., 2010). Relatively fresh waters from the Indonesian Throughflow and those formed in the high rainfall region along the equator (Tropical Surface Water), as well as salty waters at intermediate depth from the Red and Arabian Seas (Red Sea Water and Arabian Sea Low Oxygen Water), feed into the Agulhas Current mainly via these Mozambique eddies (Beal et al., 2006). Salty Subtropical Surface Water and waters subducted seasonally in the southeastern region of the gyre (South East Indian Subantarctic Mode Water; Hanawa and Talley, 2001) feed into the Agulhas Current mainly via the East Madagascar Current. This current is less well measured than the Channel flow; at 20°S the geostrophic transport is estimated at 20 Sv (Donohue and Toole, 2003), while at the tip of Madagascar it is 35 Sv (Nauw et al., 2008). In addition to these boundary flow sources, a strong southwestern subgyre recirculates waters into the Agulhas Current (Stramma and Lutjeharms, 1997), including AAIW, which enters the Indian Ocean from the Southern Ocean at about 60°E (Fine, 1993). Finally, NADW is found everywhere below 2000 m within the Agulhas Current system, with 2 Sv flowing northeastward within the (leaky) Agulhas Undercurrent (Beal, 2009) and another 9 Sv flowing eastward with the Agulhas Return Current (Arhan et al., 2003).

4.2.3. Velocity and Vorticity Structure

The surface core of the Agulhas Current has maximum velocities over 200 cm s^{-1} and typically sits above the continental slope in over 1000 m water depth (Grundlingh, 1983). The vertical velocity structure is V-shaped, with the core of the current progressing offshore with depth such that the cross-stream scale of the flow (and geostrophic balance) is preserved (Figure 13.10b; Beal and Bryden, 1999). The Agulhas Current is more barotropic than the Gulf Stream, typically penetrating to the foot of the continental slope at 3000 m depth or more. Below about 1000 m, between the deep core of the Agulhas Current and the continental slope, the Agulhas Undercurrent flows in the opposite direction with speeds of 20–50 cm s^{-1} (Beal, 2009). Vertical and horizontal shears are at a maximum on the cyclonic, inshore side of the Agulhas Current, except within the undercurrent core, where shears are small. Comparisons of direct and geostrophic velocities have shown that the along-stream flow field (cross-stream momentum balance) is essentially geostrophic below 200 m (Beal and Bryden, 1999).

The velocity field of the Agulhas Current is highly variable, with a decorrelation timescale of 10 days in the along-stream component at 32°S (Bryden et al., 2005). The meander mode having a 50–70-day timescale dominates

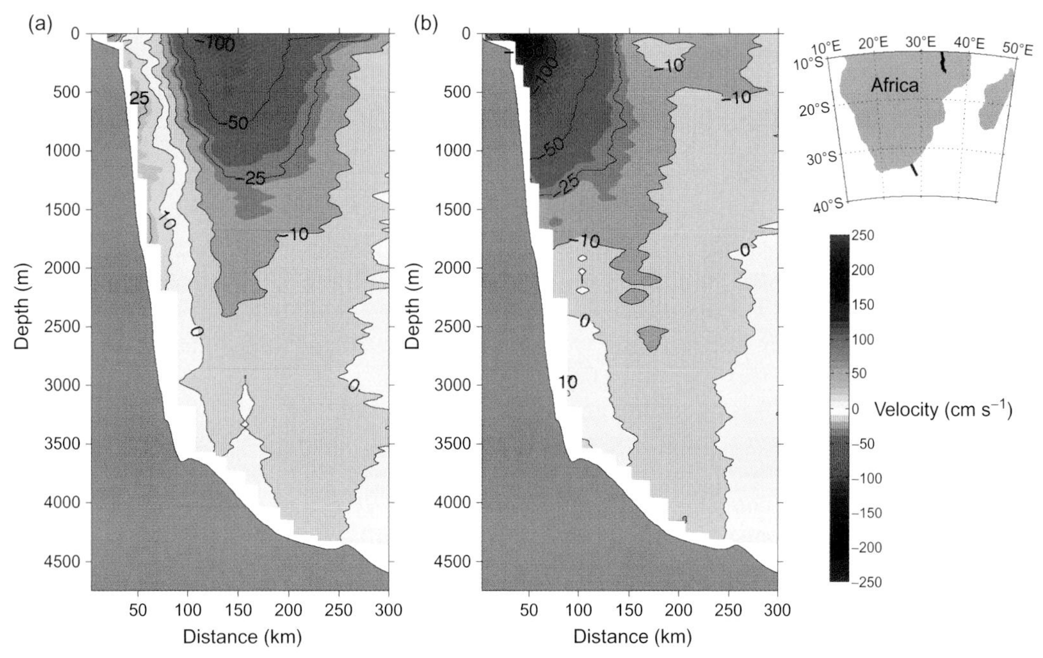

FIGURE 13.10 Velocity structure of the Agulhas Current near 34°S in (a) April 2010 during a solitary meander, when the current is located in offshore deepwater, and (b) November 2011, when the current is attached to the continental slope. Index map shows the position of the section. Velocity component (in cm s^{-1}; positive, eastnortheastward) perpendicular to the section is shown. Velocities were obtained from Lowered ADCP, during the Agulhas Current Time-series Experiment. *From Beal and Bryden (1999).*

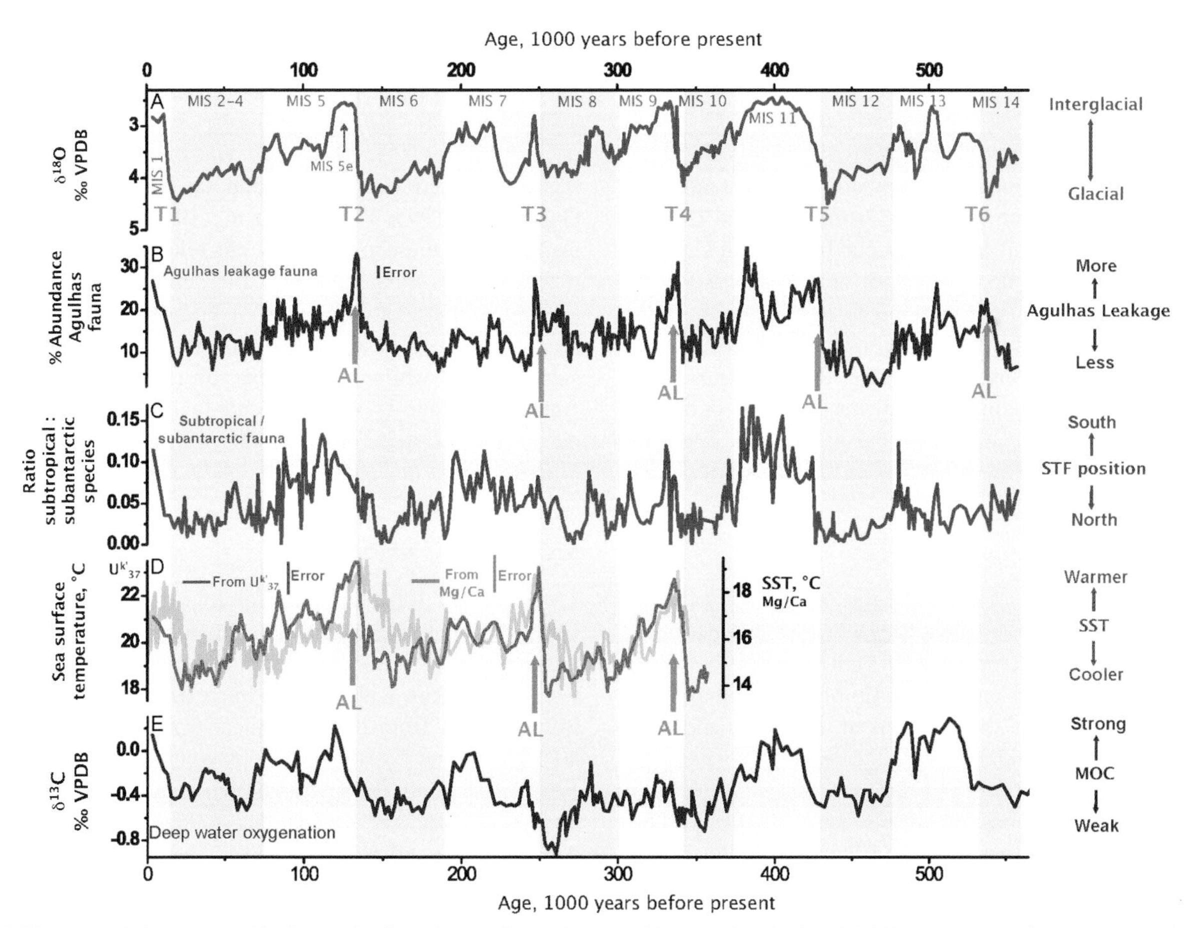

FIGURE 13.11 Paleoceanographic time series from the Agulhas leakage corridor spanning the last 570,000 year, adapted from Beal et al (2011). (a) Stable oxygen isotope profile, a proxy for glacial–interglacial variations in global climate. Marine isotope stages are labeled and highlighted by vertical blue/red shading. T1–T6 mark terminations of the past six glacial periods. (b) Abundance of tropical planktonic foraminiferal marker species, indicating maximum Agulhas leakage (AL) during glacial terminations. (c) Ratio of subtropical to subantarctic species, which are related to north–south migrations of the subtropical front. (d) SST derived from temperature-sensitive biomarkers $U^{k'}$ (brown line), and Mg/Ca ratios (gray line). Both reconstructions show maximum SST during glacial terminations, coinciding with Agulhas leakage events. (e) Benthic $\delta^{13}C$ from deep Pacific, thought to be linked to ocean ventilation and the strength of the Atlantic overturning circulation. Overturning strength appears to increase at each glacial termination, leading to the hypothesis that Agulhas leakage may trigger changes. *From Beal et al. (2011).*

in both the Agulhas Current and Undercurrent velocity fields (Beal, 2009), and results from the growth of barotropic instabilities generated when anticyclones from the Mozambique Channel or dipoles from the East Madagascar Current interact with the mean flow field (Schouten et al., 2002; Tsugawa and Hasumi, 2010). A cross-section of an Agulhas Current meander is shown in Figure 13.10a.

The strongest potential vorticity gradients ($>1.5\ m^{-1}\ s^{-1}\ km^{-1}$) in the Agulhas Current appear within the thermocline and just inshore of its velocity core. Here, relative vorticity contributes to the potential vorticity front, but layer depth changes dominate its structure (Beal and Bryden, 1999). The gradient of layer depth with offshore distance changes sign below the neutral density of 27.2 and this leads to weak potential vorticity gradients in the intermediate and deep layers. Hence, strong, cross-stream water-property gradients at these depths are largely due to kinematic steering (Bower et al., 1985; Beal et al., 2006), which maintains a separation between Tropical Surface Water and Red Sea Water inshore of the dynamical front, and Subtropical Surface Water and AAIW offshore.

4.2.4. Separation, Retroflection, and Leakage

The Agulhas Current separates from the African continent well before the latitude of zero wind stress curl, and subsequent to separation, there is retroflection and leakage. Early separation and leakage occur because the African cape lies north of the latitude of zero wind stress curl (and subtropical front, Figure 13.8), and hence there is a gap in the boundary through which Indo-Pacific waters can leak westward into the Atlantic. Retroflection occurs because the longitudinal slant of the African continental slope is westward, rather than eastward. This gives rise to southwestward flow at separation and the current must subsequently loop, or retroflect, back eastward to rejoin the Sverdrup gyre, as governed by the large-scale wind stress curl.

It is difficult to establish the mean geographical separation point of the Agulhas Current, since its path does not significantly diverge from that of the African continental slope until the latter ends at the tip of the Agulhas Bank. Theory suggests that the positions of separation and retroflection are linked and that they affect leakage. For example, separation will be farther to the northeast when Agulhas Current transport is greater, because isopycnal outcropping along the concave coastline will occur sooner (Ou and de Ruijter, 1986). In this case, the separated Agulhas Current has a more southward trajectory and greater inertia, and can attach more easily to the Agulhas Return Current with less leakage (van Sebille et al., 2009). Hence, in the absence of other far-field changes, a stronger Agulhas Current leads to less leakage and a more easterly (early) retroflection. Over the 20-year satellite record, the position of the retroflection has not varied greatly (Dencausse et al., 2010), perhaps because it is steered by the Agulhas Plateau, a region of shallow topography southeast of the African cape (Speich et al., 2006). However, this inertial theory, together with variations in the wind field, is able to explain many of the paleoclimate observations of Agulhas leakage variability (Beal et al., 2011).

Retroflection of a WBC after separation is intrinsically unsteady and leads to the shedding of rings (Nof and Pichevin, 1996; Pichevin et al., 1999; van Leeuwen and de Ruijter, 2009). The spatial scale of Agulhas Rings (200–300 km) is much larger than mesoscale eddies (Schouten et al., 2000; van Aken et al., 2003), because they result from an unsteady flow (not an unstable flow) and their scale is governed by the flow-force, or momentum flux, of the outgoing Agulhas Return Current (Pichevin et al., 1999). These rings appear to carry most of the leakage of Agulhas waters into the Atlantic, with smaller cyclones, patches, and filaments carrying the rest (Richardson, 2007). The timing and frequency of Agulhas Rings have been related to various upstream processes, including the interaction of currents with Madagascar (Penven et al., 2006), the radiation of Rossby waves from the eastern boundary (Schouten et al., 2002), and the downstream propagation of instabilities (meanders and transport pulses) in the Agulhas Current (Lutjeharms and van Ballegooyen, 1988; Goni et al., 1997; Pichevin et al., 1999). However, it is unclear how these parameters are related to the strength of the Agulhas leakage, if at all. In a simulation of the twentieth century ocean with a nested, eddy-resolving Agulhas region, Biastoch et al. (2009) find that leakage increases significantly, while the number of rings is unchanged. Agulhas leakage is very difficult to measure in the real ocean because it is fundamentally a Lagrangian transport (van Sebille et al., 2010).

4.2.5. WBC Extension

The extension of the Agulhas Current is the Agulhas Return Current (Figure 13.8), which flows eastward from the Agulhas Retroflection as a strongly barotropic current of width 60–80 km, with distinct water masses and a marked front separate from the subtropical front at least as far as 40°E (Read and Pollard, 1993). Its volume transport is over 100 Sv (including 9 Sv of NADW), reducing to about one quarter this strength upon reaching 76°E (Lutjeharms and Ansorge, 2001; Arhan et al., 2003). It is strongly meandering, with three quasi-stationary troughs (loops toward the equator) at the Agulhas Plateau, at 33°E, and at 39°E, with decreasing amplitude toward the east (Quartly and Srokosz, 1993; Boebel et al., 2003). Cyclones are frequently shed from these troughs and propagate westward, sometimes to be reabsorbed by the adjacent trough.

4.2.6. *Air–Sea Interaction*

Latent and sensible heat fluxes increase three to five times over the warm waters of the Agulhas Current system and there is a deepening of the marine–atmospheric boundary layer, and increased formation of convective clouds (Jury and Walker, 1988; Lee-Thorp et al., 1998; Rouault et al., 2000). Over the Agulhas Return Current, the response of surface winds and sensible heat flux to SST fronts are almost twice as strong during austral winter than during summer (O'Neill et al., 2005). The Agulhas Current system influences storm track positions and storm development, as well as regional atmospheric circulation patterns (Reason, 2001; Nakamura and Shimpo, 2004), and has been linked to extreme rainfall events and tornadoes over southern Africa (Rouault et al., 2002).

Uniquely among WBCs, the Agulhas Current system is thought to be an important source of continental moisture (Gimeno et al., 2010). Rainfall over Africa is correlated with SST anomalies over the larger Agulhas Current system, which are associated with Indian Ocean Dipole and El Niño/Southern Oscillation cycles. Overall warming of the system since the 1970s may have increased the sensitivity of African rainfall to these cycles (Behera and Yamagata, 2001; Zinke et al., 2004).

4.2.7. *Implications for Climate*

Paleoceanographic records and models have suggested links between Agulhas leakage strength and past climate change (Figure 13.11; Beal et al., 2011). In particular, an assemblage of planktonic foraminifera characteristic of modern-day Agulhas waters found in marine sediment records show that dramatic increases in Agulhas leakage have occurred at the onset of each glacial termination over the last 550,000 years (Peeters et al., 2004). Weaker Agulhas leakage is associated with glacial climate and appears to be correlated with a more northerly position of the subtropical front and a weaker Atlantic overturning circulation (Figure 13.11). Moreover, during the last deglaciation, the delay in and then abrupt warming of the North Atlantic (Bølling warm event) have been attributed to changes in Agulhas leakage through its influence on Atlantic overturning (Knorr and Lohmann, 2003; Chiessi et al., 2008).

Ocean and coupled model studies corroborate these climate data, showing that Agulhas leakage variability can impact Atlantic overturning on a number of timescales. Planetary waves associated with Agulhas Rings can cause small decadal oscillations in the overturning (Biastoch et al., 2008), and buoyancy forcing associated with the advection of saline Agulhas waters into the North Atlantic enhances deepwater formation (Weijer et al., 2002), strengthening the MOC 15–30 years after an increase in leakage. The Agulhas leakage strength is affected by changes in the strength and position of the southeast trade winds and/or Southern Hemisphere westerlies (de Ruijter, 1982; Biastoch et al., 2009; Sijp and England, 2009). In a warming climate, the westerlies shift poleward, increasing the gap between the African continent and the subtropical front, thereby increasing leakage (Beal et al., 2011). This ties with inertial theory as discussed previously (de Ruijter et al., 1999). A simulation of the twentieth century ocean (with a nested, eddy-resolving Agulhas region) shows that Agulhas leakage may be increasing now, under anthropogenic climate change (Biastoch et al., 2009), which could strengthen Atlantic overturning at a time when warming and fresh meltwater input in the North Atlantic are predicted to weaken it.

5. NORTH PACIFIC

5.1. Upstream Kuroshio

The Kuroshio is the WBC of the wind-driven subtropical gyre in the North Pacific. Its origin can be traced back to the Philippine coast, where the westward-flowing North Equatorial Current (NEC) bifurcates (around 15°N) and has its northern limb feeding into the nascent Kuroshio (Nitani, 1972; see Figure 13.12a). This bifurcation, and hence the Kuroshio, tends to shift northward with increasing depth (Reid, 1997, see his figure 5), due to the ventilation of the wind-driven, baroclinic subtropical gyre (Pedlosky, 1996). On seasonal timescales, the Kuroshio east of the Philippine coast tends to migrate northward and have a smaller volume transport in winter (November/December), and to shift southward and have a larger transport in summer (June/July). On interannual timescales, the Kuroshio begins at a more northern latitude and has a weaker volume transport during El Niño years (e.g., Qiu and Lukas, 1996; Kim et al., 2004; Kashino et al., 2009; Qiu and Chen, 2010).

The Kuroshio becomes a more coherent and identifiable boundary jet downstream of the Luzon Strait at 22–24°N, east of Taiwan (e.g., Centurioni et al., 2004). This is in part due to the addition of mass from the interior, wind-driven Sverdrup gyre. Moored current-meter observations show that the Kuroshio has a mean volume transport of 21.5 Sv east of Taiwan (Johns et al., 2001; Lee et al., 2001). The Kuroshio path and transport in the latitude band from 18° to 24°N are highly variable due to westward-propagating, energetic mesoscale eddies from the interior ocean (Zhang et al., 2001a; Gilson and Roemmich, 2002; see Figure 13.1). These impinging eddies have a dominant period of $\sim$100 days and are generated along the North Pacific Subtropical Countercurrent (STCC) as a result of baroclinic instability between the surface eastward-flowing STCC and the subsurface westward-flowing NEC (Qiu,

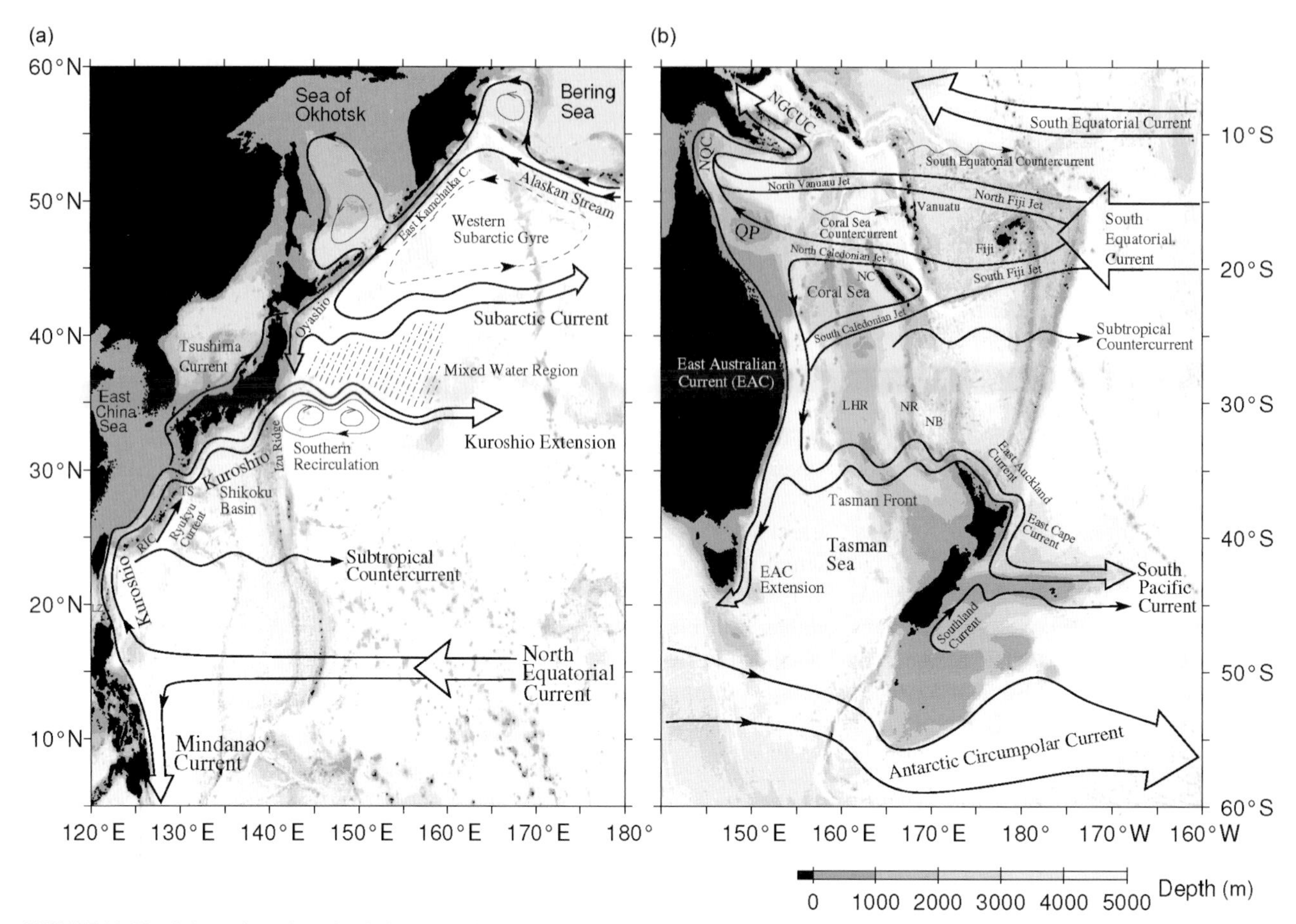

FIGURE 13.12 Schematic surface circulation pattern in (a) the western North Pacific and (b) the western South Pacific. Gray shading shows depth (in m). Abbreviations in (a) are: LZ, Luzon Strait; TS, Tokara Strait; and RIC, Ryukyu Island Chain, and in (b) are: NGCUC, New Guinea Coastal Undercurrent; NQC, North Queensland Current; QP, Queensland Plateau; NC, New Caledonia; LHR, Lord Howe Rise; NR, Norfolk Ridge; and NB, Norfolk Basin.

1999; Roemmich and Gilson, 2001). Perturbations induced by these impinging eddies force part of the northward-flowing Kuroshio to divert to the east of the Ryukyu Island Chain from 24°N, 124°E to 28°N, 130°E, contributing to the formation of the Ryukyu Current (Ichikawa et al., 2004; Andres et al., 2008).

North of 24°N, the main body of the Kuroshio enters the East China Sea where the Kuroshio path is topographically steered by the steep continental slope and approximately follows the 200 m isobaths (e.g., Lie et al., 1998). From repeat hydrography, the mean Kuroshio transport across the PN section (PN stands for Pollution Nagasaki; nominally from 27.5°N, 128.25°E to 29°N, 126°E) is estimated at 23.7–25.0 Sv (Ichikawa and Beardsley, 1993; Kawabe, 1995). With the time-mean Sverdrup transport across 28°N estimated at ~45 Sv (Risien and Chelton, 2008), this suggests that only 53–55% of Sverdrup return flow is carried poleward by the Kuroshio inside the East China Sea. The remaining ~20 Sv are likely carried northward by the offshore Ryukyu Current, although this is yet to be confirmed observationally.

Shielded to the east by the Ryukyu Island Chain, the Kuroshio inside the East China Sea avoids the direct impact from the westward-propagating interior eddy perturbations. Instead, the Kuroshio variability along the continental shelf break here is dominated by frontal meanders that tend to originate northeast of Taiwan and grow rapidly in amplitude while propagating downstream. The frontal meanders have typical wavelengths of 100–350 km, wave periods of 10–20 days, and downstream phase speeds of 10–25 cm s^{-1} (Sugimoto et al., 1988; Qiu et al., 1990; Ichikawa and Beardsley, 1993; James et al., 1999). When reaching the Tokara Strait at 29°N, 130°E, the fully developed frontal meanders can result in lateral Kuroshio path fluctuations as large as 100 km (e.g., Kawabe, 1988; Feng et al., 2000). Based on tide gauge measurements across the Tokara Strait, the Kuroshio transport has been inferred to reach a seasonal maximum in spring/summer and a minimum in fall. Interannually, the Kuroshio transport at the Tokara Strait is inferred to increase in the year preceding El Niño events and to drop significantly during the El Niño years (Kawabe, 1988).

5.2. Kuroshio South of Japan

Exiting from the Tokara Strait, the Kuroshio enters the deep Shikoku Basin, and its mean eastward volume transport increases to 52–57 Sv (Qiu and Joyce, 1992; Imawaki et al., 2001). This transport increase is due to both the confluence of the northward-flowing Ryukyu Current and the excitation of a southern recirculation gyre. Subtracting the contribution from the recirculation reduces the net eastward mean transport of the Kuroshio south of Japan to 34–42 Sv. Seasonally, the Kuroshio transport south of Japan varies by about 10 Sv, much smaller than the 40 Sv inferred from wind-driven Sverdrup theory (Isobe and Imawaki, 2002). Near 139°E, the Kuroshio encounters the meridionally oriented Izu Ridge that parallels 140°E south of Japan. Its presence restricts the Kuroshio from exiting the Shikoku Basin either near 34°N, where a deep passage exists, or south of 33°N, where the ridge height drops.

On interannual timescales, the Kuroshio in the Shikoku Basin is known for its bimodal path fluctuations between straight and meandering paths. In its "straight path," the Kuroshio flows along the Japanese coast, while a "large meander path" signifies a curving, offshore path (Kawabe, 1995). In addition to these two paths, the Kuroshio also inhabits a third, relatively stable path that loops southward over the Izu Ridge. It is interesting to note that while the large meander path persisted for several years in the 1970s and 1980s, since the 1990s it has occurred only once in mid-2004 for a period of about 1 year. During the past two decades, the Kuroshio path south of Japan largely vacillated between the straight path and the third path, detouring over the Izu Ridge (e.g., Usui et al., 2008). Theoretical and modeling studies attempting to explain the multiple path state of the Kuroshio south of Japan have a long history. Relevant reviews and references can be found in Qiu and Miao (2000) and Tsujino et al. (2006). In addition to be important for fisheries south of Japan, the bimodal Kuroshio path fluctuations have recently been shown to impact on development and tracks of wintertime extratropical cyclones that pass over south of Japan (Nakamura et al., 2012).

5.3. Kuroshio Extension

After separating from the Japanese coast at 36°N, 141°E, the Kuroshio enters the open basin of the North Pacific, where it becomes the Kuroshio Extension (KE). The Kuroshio separation latitude is located to the south of the zero Sverdrup transport stream-function line at 40°N in the North Pacific (Risien and Chelton, 2008). This southerly separation of the Kuroshio is due to the combined effect of the coastal geometry of Japan and the inertial nature of the Kuroshio/KE jet (Hurlburt et al., 1996). Free from the constraint of coastal boundaries, the KE has been observed to be an eastward-flowing inertial jet accompanied by large-amplitude meanders and energetic pinched-off eddies (e.g., Mizuno and White, 1983; Yasuda et al., 1992). Compared to its upstream counterpart south of Japan, the KE is accompanied by a stronger southern recirculation gyre. A lowered-ADCP survey across the KE southeast of Japan revealed that the eastward volume transport reached 130 Sv, which is more than twice the maximum Sverdrup transport in the subtropical North Pacific (Wijffels et al., 1998). Recent profiling float and moored current-meter observations have further revealed the existence of a recirculation north of the KE jet with a transport of about 25 Sv (Qiu et al., 2008; Jayne et al., 2009).

In addition to the high level of mesoscale eddy variability, an important feature emerging from recent satellite altimeter measurements and eddy-resolving ocean model simulations is that the KE system exhibits clearly defined decadal modulations between a stable and an unstable dynamic state (Vivier et al., 2002; Qiu and Chen, 2005; Taguchi et al., 2007). Figure 13.13 shows that the KE paths were relatively stable in 1993–1995, 2002–2005, and 2010. In contrast, spatially convoluted paths prevailed in 1996–2001 and 2006–2009. These changes in path stability are merely one manifestation of the decadally modulating KE system. When the KE jet is in a stable dynamic state, available satellite altimeter data further reveal that its eastward transport and latitudinal position tend to be greater and more northerly, its southern recirculation gyre tends to strengthen, and the regional EKE level tends to decrease. The reverse is true when the KE jet switches to an unstable dynamic state.

Transitions between the two dynamic states of KE are caused by the basin-scale wind stress curl forcing in the eastern North Pacific related to the Pacific decadal oscillations (PDOs) (Qiu and Chen, 2005; Taguchi et al., 2007). Specifically, when the central North Pacific wind stress curl anomalies are positive (i.e., positive PDO phase; see Figure 13.14), enhanced Ekman flux divergence generates negative local SSH anomalies. As these wind-induced negative SSH anomalies propagate westward into the KE region after a delay of 3–4 years, they weaken the zonal KE jet, leading to an unstable state of the KE system with a reduced recirculation gyre and an active EKE field. The negative, anomalous wind stress curl forcing during the negative PDO phase, on the other hand, generates positive SSH anomalies through the Ekman flux convergence. After propagating into the KE region in the west, these anomalies stabilize the KE system by increasing the KE transport and by shifting its position northward.

Decadal modulations in the dynamic state of KE can exert a significant impact on regional water mass formation and transformation processes. During the unstable state of the KE system, for example, the elevated eddy variability brings upper-ocean high potential vorticity water of the Mixed Water Region southward, creating a stratified upper-ocean condition in the southern recirculation gyre

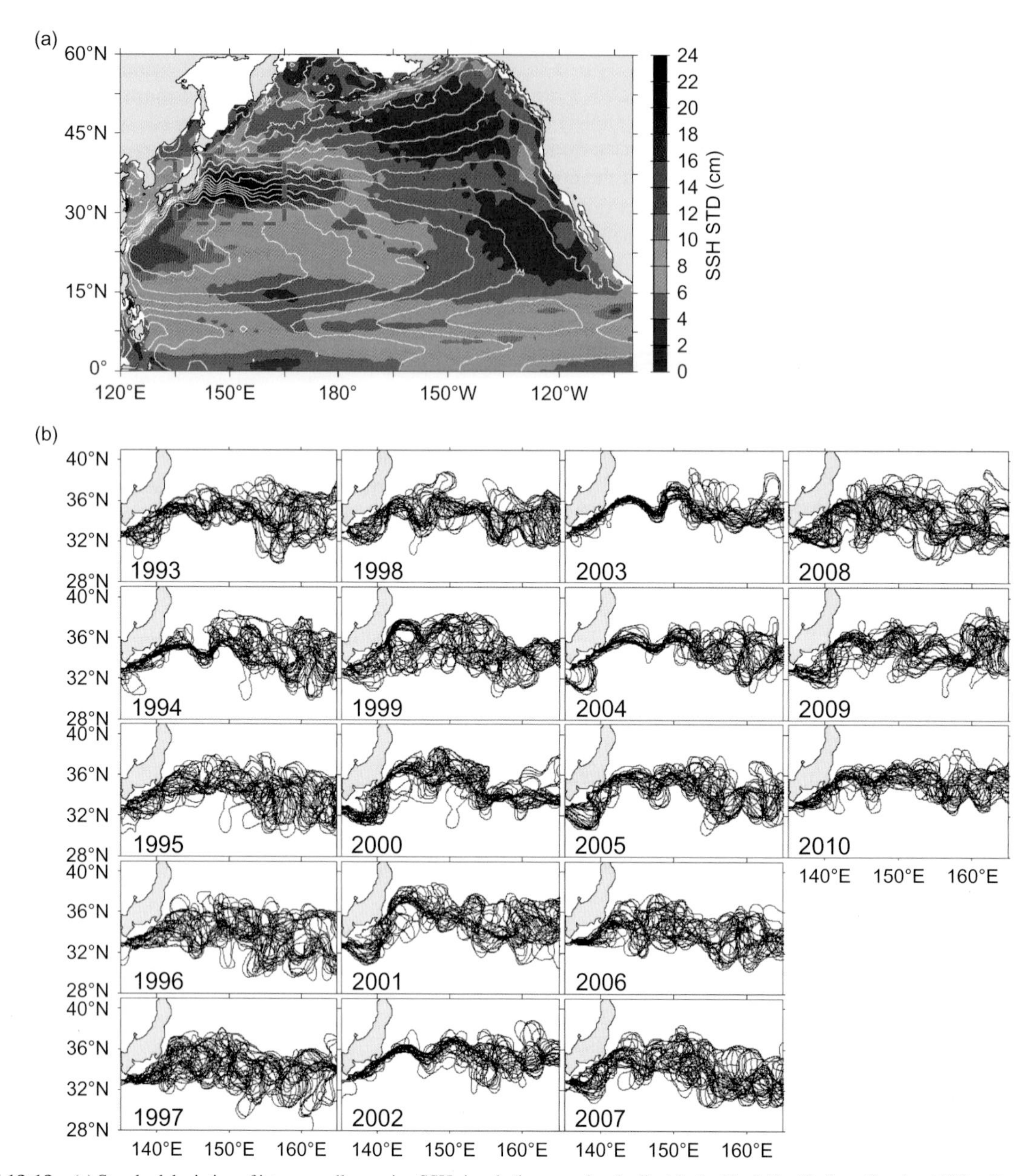

FIGURE 13.13 (a) Standard deviation of interannually varying SSH signals (in cm; color shading) in the North Pacific from October 1992 to December 2010. White contours denote the mean SSH field with contour intervals at 0.1 m. (b) Yearly paths of the Kuroshio and KE defined by the 1.7 m contours in the weekly SSH fields. Paths are plotted every 14 days. *Adapted from Qiu and Chen (2005).*

region, which is unfavorable for the wintertime deep convection and Subtropical Mode Water (STMW) formation (Qiu et al., 2007a; Sugimoto and Hanawa, 2010). In addition, changes in the dynamic state of KE are also important for the evolution of formed STMW. While it tends to remain trapped within the recirculation gyre during the unstable state of the KE jet, STMW tends to be carried away from its formation region during the stable state of KE (Oka, 2009; Oka et al., 2011).

By transporting warmer tropical water to the midlatitude ocean, the expansive KE jet provides a significant source of heat and moisture for the North Pacific midlatitude atmospheric storm tracks (Nakamura et al., 2004). By modifying the path and intensity of the wintertime overlying storm tracks, changes in the dynamic state of KE can alter not only the stability and pressure gradient within the local atmospheric boundary layer, but also the basin-scale wind stress pattern (Frankignoul and Sennéchael, 2007; Kwon et al., 2010). Specifically, a dynamically stable (unstable) KE tends to generate a positive (negative) wind stress curl in the eastern North Pacific basin, resulting in negative (positive) local SSH anomalies through Ekman divergence (convergence). This impact on wind stress induces a delayed negative feedback with a preferred period of about 10 years and is likely the cause for the enhanced decadal variance observed in the midlatitude North Pacific (Qiu et al., 2007b).

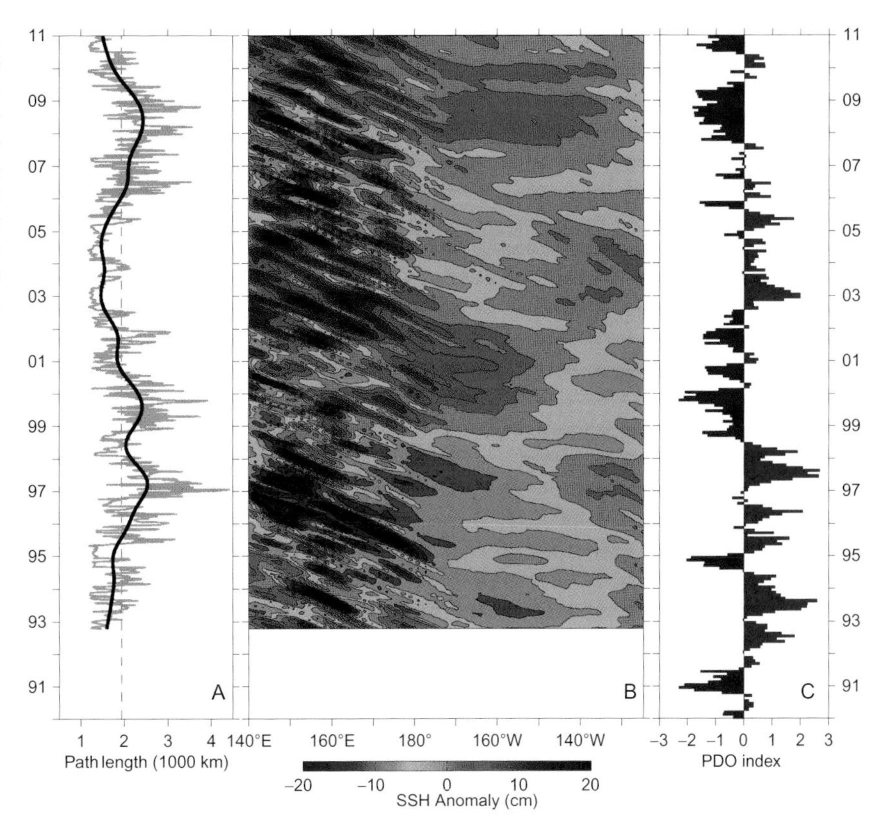

FIGURE 13.14 (a) Time series of the KE jet path length (in km) integrated from 141° to 153°E. A small value indicates a stable KE jet and a large value, a convoluted and dynamically unstable KE jet (see Figure 13.13). Gray line shows the weekly time series and black line shows the low-pass filtered time series. (b) SSH anomalies (in cm; color shading) versus time along the zonal band of 32–34°N from the AVISO satellite altimeter data. (c) PDO index versus time from http://jisao.washington.edu/pdo/PDO.latest.

6. SOUTH PACIFIC

6.1. Upstream EAC

Mirroring the NEC bifurcation off the Philippine coast, the wind-driven, westward-flowing SEC splits upon reaching the Australian coast, feeding into the northward-flowing North Queensland Current and southward-flowing EAC (Ridgway and Dunn, 2003; see Figure 13.12b). Unlike its counterpart in the Northern Hemisphere, however, the SEC in the western South Pacific is heavily affected by complex topography. The presence of the island ridges of Fiji (near 18°S and 178°E), Vanuatu (near 15°S and 167°E), and New Caledonia (near 22°S and 165°E) fractures the SEC, channeling it into localized zonal jets known as the North and South Fiji Jets, the North and South Caledonian Jets, and the North Vanuatu Jet (Webb, 2000; Stanton et al., 2001; Gourdeau et al., 2008; Qiu et al., 2009). In addition, the existence of the shallow Queensland Plateau just south of the SEC bifurcation near 18°S causes the EAC to begin as a doubled boundary jet system straddling the Queensland Plateau.

Constrained by the basin-scale surface wind forcing, the transport of the SEC entering the Coral Sea between New Caledonia and the Solomon Islands (near 9°S and 160°E) is about 22 Sv. This SEC volume transport has a seasonal maximum in October–December and a minimum in April–June (Holbrook and Bindoff, 1999; Kessler and Gourdeau, 2007). Concurrent with its seasonal transport increase, the SEC bifurcation tends to shift equatorward in October–December, and is accompanied by a summer transport increase in EAC along the eastern coast of Australia. The amplitude of seasonal change of the EAC transport has been estimated at 4–6 Sv (Ridgway and Godfrey, 1997; Roemmich et al., 2005; Kessler and Gourdeau, 2007).

Compared to the interior Sverdrup transport of ~35 Sv along 30°S (Risien and Chelton, 2008), the observed poleward transport of the EAC is about 20–22 Sv (Ridgway and Godfrey, 1994; Mata et al., 2000). This discrepancy is largely due to the presence of an open western boundary in the equatorial Pacific, through which part of the SEC inflow is lost to the Indian Ocean via the Indonesian Throughflow, which is shown schematically northwest of Australia in Figure 1.6 (Godfrey, 1989).

6.2. East Australian Current

After the SEC's bifurcation near 18°S, the poleward-flowing EAC evolves into a narrow, swift boundary jet with strong vertical shear over the upper 1000 m. The EAC has short-term transport variations with a dominant timescale of 90–180 days (Mata et al., 2000; Bowen et al., 2005), likely caused by intrinsic nonlinear variability of the EAC (Bowen et al., 2005; Mata et al., 2006) or forced by eddy signals propagating into the EAC (Nilsson and Cresswell, 1981; Cresswell and Legeckis, 1986). Similar to the subtropical North Pacific, a high EKE band exists at latitudes 20–30°S in the western subtropical South Pacific (Figure 13.1). Dynamically, this high EKE band is caused by baroclinic instability of the surface, eastward-flowing Subtropical Countercurrent, and the underlying, westward-flowing SEC (Qiu and Chen, 2004).

The main flow of the EAC detaches from the Australian coast at 30–34°S and crosses the northern Tasman Sea. Like the Kuroshio in the Northern Hemisphere, the latitude of the EAC separation is located equatorward of the zero Sverdrup transport stream-function line in the South Pacific (along about 50°S; Risien and Chelton, 2008). The presence of New Zealand and the inertial nature of the EAC jet have been found to be responsible for this equatorward separation latitude of the EAC (Tillburg et al., 2001). Offshore, it separates into an eastward branch, known as the Tasman Front, and a northeastward branch that connects to the eastward-flowing Subtropical Countercurrent in the 20–30°S band. The path of the Tasman Front is influenced by the meridionally aligned Lord Howe Rise (along about 162°E) and Norfolk Ridge (along about 168°E), which it must negotiate. Over the Lord Howe Rise, the isotherms of the Tasman Front tend to detour southward before turning northward to wrap around the southern edge of the Norfolk Ridge and into the Norfolk Basin (Uddstrom and Oien, 1999). Along its path, the Tasman Front is highly variable and is often accompanied by wave-like disturbances that propagate westward against the direction of the background mean flow (Nilsson and Cresswell, 1981). After impinging upon the Australian coast, many of these disturbances develop into isolated cyclonic eddies, migrating poleward into the southern Tasman Sea. The cyclonic eddy detachment has a frequency of about three eddies per year.

After reaching the northern tip of New Zealand and joined by flows feeding in from the east, a portion of the Tasman Front turns southeastward, forming the East Auckland Current (EAUC) along the northeast coastline of the North Island of New Zealand. The southeastward transport of the EAUC is highly variable, with a mean value of about 9 Sv (Stanton, 2001; Stanton and Sutton, 2003). The EAUC is renamed the East Cape Current after it flows around East Cape, the easternmost point of the North Island. Three topographically constrained, quasi-permanent, cyclonic eddies, known as the North Cape Eddy, the East Cape Eddy, and the Wairarapa Eddy, are observed along the EAUC and East Cape Current paths (Roemmich and Sutton, 1998). The subtropical-origin East Cape Current continues southward along the east coast of New Zealand until it turns eastward near 43°S to rejoin the interior Sverdrup circulation (Sutton, 2001), merging with the Southland Current to become the eastward-traveling South Pacific Current.

6.3. EAC Extension

While the main portion of the EAC separates from the Australian coast near 34°S, the remainder continues southward along the Australian coast to as far south as Tasmania and is known as the EAC Extension. *In situ* observations at the Maria Island coast station off the east coast of Tasmania reveal that both the temperature and salinity have increased steadily over the past 60 years, consistent with a southward expansion of the EAC Extension (Ridgway, 2007; Hill et al., 2008; see Figure 13.15a and b). Given temperature and salinity trends of 2.28 °C century^{-1} and 0.34 psu century^{-1}, observed at the Maria Island station, the EAC Extension is estimated to have expanded southward by about 350 km from 1944 to 2002. Over the same period, an increase in the net volume transport through the Tasman Sea is estimated at $\sim$10 Sv (Figure 13.15d).

The poleward expansion of the EAC Extension occurs at the expense of the Tasman Front. In other words, the strengths of the EAC Extension and the Tasman Front are anticorrelated (Hill et al., 2011). Concurrent with the multi-decadal poleward expansion of the EAC Extension into the southern Tasman Sea, a significant thermocline cooling has been detected in the northern EAC region from 1975 to 1990, reflecting the weakening of the upstream EAC (Ridgway and Godfrey, 1996).

The long-term intensification and southward expansion of the EAC Extension are caused by changes in the basin-scale surface wind field. Specifically, the strengthening and southward migration of the Southern Hemisphere westerlies enhance and expand the downward Ekman pumping in the subtropical South Pacific north of 50°S. These changes induce a spin-up and southward expansion of both the interior subtropical gyre and the EAC Extension, with the latter having a delay of adjustment of several years (Bowen et al., 2006; Qiu and Chen, 2006; Roemmich et al., 2007; Hill et al., 2010; see Figure 13.15).

Note that rather than a confined change in the South Pacific, spin-up of the wind-driven subtropical gyre also occurs in the South Atlantic and Indian Oceans and is connected to the upward trend in the Southern Annular Mode signals of the Southern Hemisphere atmospheric circulation (Cai, 2006; Roemmich, 2007). On the western side of the Tasman Sea, some of the EAC Extension turns west south of Tasmania, and connects to the southern Indian Ocean subtropical circulation, forming the so-called Southern Hemisphere Supergyre (Speich et al., 2002; Ridgway and Dunn, 2007). This inter-ocean exchange is known as Tasman leakage. As the wind-driven South Pacific subtropical gyre intensifies and shifts southward, the "outflow" from the South Pacific to southern Indian Ocean likely intensifies. It is important for future studies to clarify how this outflow intensification can modify the global overturning circulation.

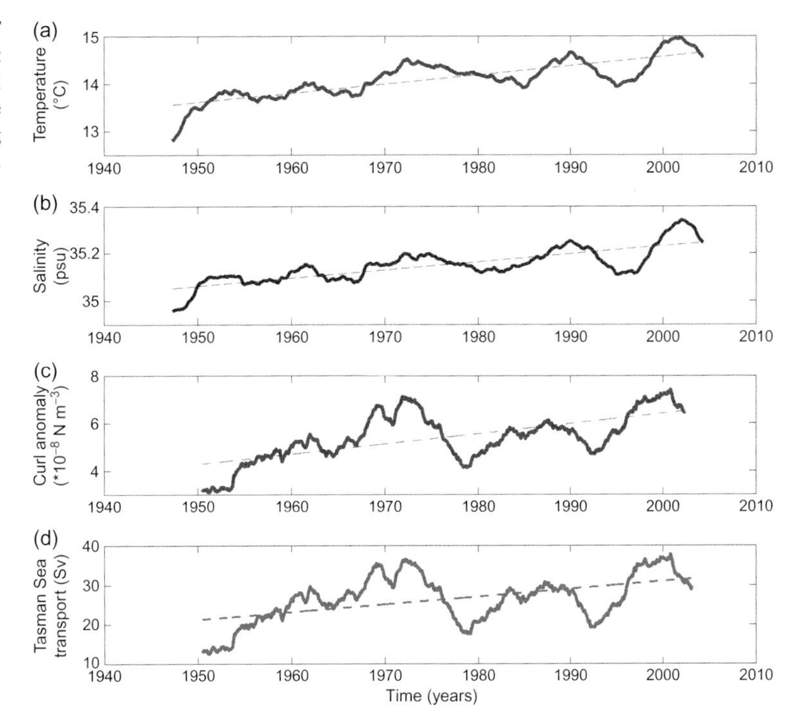

FIGURE 13.15 Low-pass filtered time series of (a) SST (in °C) and (b) salinity (in psu) at the Maria Island coast station, (c) South Pacific regional mean wind stress curl (in N m^{-3}; 20–50°S, 180–80°W), and (d) net volume transport (in Sv) through the Tasman Sea, calculated using Godfrey's Island Rule. Dashed lines show the linear trend. *Adapted from Hill et al. (2008).*

7. CONCLUDING REMARKS

7.1. Separation from the Western Boundary

The latitude of separation of WBCs from their continental boundaries has an important impact on ocean circulation, air–sea fluxes, and even climate. Separation dynamics are subtle and this chapter has shown that different controls dominate the latitude of separation of each WBC. However, the latitude of separation is *always* lower than that inferred from linear wind-driven circulation theory.

In the North Atlantic, the separation latitude of the Gulf Stream is fairly stable, and interaction with the DWBC below the Gulf Stream and/or the NRG located north of the GSE seems to play an important role. In the South Atlantic, the separation latitude of the Brazil Current is less stable, probably because it collides with the fairly strong, northward-flowing Malvinas Current well before reaching the latitude of zero wind stress curl. It separates from the boundary with the Malvinas Current, as the Brazil–Malvinas Confluence.

In the Indian Ocean, the African continent disappears before the Agulhas Current reaches the latitude of zero wind stress curl. This fact results in leakage of Indian Ocean water into the South Atlantic, and retroflection of the Agulhas Current eastward to rejoin the Indian Ocean subtropical gyre. Theory suggests that the positions of separation and retroflection are linked and affect leakage.

In the North Pacific, the separation latitude of the Kuroshio is likely governed by the combined effects of the coastal geometry of Japan and the inertial nature of the Kuroshio and KE jet. In the South Pacific, the separation latitude of the EAC is quite variable because of the presence of New Zealand offshore and the inertial nature of the EAC jet. There is leakage of South Pacific water into the Indian Ocean as the EAC Extension turns west south of Tasmania.

7.2. Northern and Southern Hemispheres

The difference in land mass distribution between the Northern and Southern Hemispheres leads to fundamental differences among WBCs and the circulations they feed into. The Agulhas Current and EAC leak waters into the Atlantic and Indian Oceans, respectively, because they run out of western boundary well before the latitude of zero wind stress curl. This creates a Southern Hemisphere "super-gyre," which connects the subtropical gyres of the South Pacific, Indian, and South Atlantic Oceans. The Southern Hemisphere WBCs interact strongly with the Antarctic Circumpolar Current, especially in the Indian and Atlantic sectors, where heat is transported toward the pole via eddies associated with the WBC extensions. Heat loss over WBC extensions of the Northern Hemisphere tends to be stronger than over those of the Southern Hemisphere, because adjacent larger continental land masses on the west provide colder, dryer air masses over them.

The most conspicuous difference between the Northern and Southern Hemispheres is that there are far fewer observations of WBCs in the Southern Hemisphere. This is particularly acute for the Brazil and Agulhas Currents, where long-term observations are needed.

7.3. Recent and Future Studies

In the Gulf Stream, research on mesoscale variability and pathway prediction has decreased over the past two decades. The latest emphasis is primarily on the role of the Gulf Stream in the MOC and climate, and its variability on seasonal and longer timescales. Little attention has been given to obtaining new *in situ* observations of the Brazil Current and its confluence with the Malvinas Current historically, with the most recent studies relying heavily on the analysis of remote sensing observations (e.g., Lumpkin and Garzoli, 2011).

Research into the Agulhas Current has accelerated over the last 3–4 years, with several international observation and modeling programs. However, many observations are coming to an end, and an international group of scientists is cooperating to establish a sustained array in the near future. Biastoch et al. (2009) developed a realistic simulation of the Agulhas Current system using a high-resolution, regional nest in a global ocean model. Climate modelers are becoming more interested in Agulhas Current research after the very recent advent of coupled climate models with eddy-resolving ocean models. For the Somali Current, there is growing interest in utilizing autonomous observing platforms, including Argo floats, underwater gliders, and surface drifters, in order to overcome the piracy problem in that region.

For the Kuroshio south of Japan, the research pace has been somewhat slow after the WOCE program, partly because the prominent large meander was absent. The KE research is strong, with recent interest focused on the KE jet and its recirculation gyre dynamics, carrying out intensive observations using a large set of inverted echo sounders equipped with bottom pressure gauges and current-meters. The role of KE front on the midlatitude atmospheric circulation is also targeted by an international group of oceanographers and meteorologists. For the EAC, observations using current-meter moorings, subsurface gliders, and hydrography are being pursued as a part of sustained marine observing system.

For future studies, we suggest four topics to be prioritized: air–sea interaction, boundary separation, submesoscale dissipation, and interaction with the deep ocean. The large heat and carbon fluxes between ocean and atmosphere associated with WBCs and their extensions are important topics in climate science. A better understanding of transfer processes and air–sea coupling on multiple timescales is needed. The separation latitudes of WBCs, as well as subsequently the mean latitude of WBC extensions, have a significant impact on regional meteorology and climate variability. Research into the dynamics of separation has exposed more possible mechanisms, but not yet identified which processes are most important. Our understanding of the mesoscale eddy variability of WBCs has improved due to an accumulation of *in situ* and satellite observations, and eddy-resolving numerical models. However, the dissipation of mesoscale variability is largely via submesoscale processes, which are largely unobserved and unresolved in general circulation models. Those processes can be addressed soon by remote sensing with the advent of the Surface Water and Ocean Topography satellite mission, which will measure SSH with a spatial resolution of less than 10 km. Finally, our description of the pathways of deep WBCs and our understanding of the interaction between upper- and deep-ocean WBCs are still evolving beyond the seminal contributions of Stommel and Arons (1960a,b).

ACKNOWLEDGMENTS

We thank two anonymous reviewers and the editors (G. Siedler and J. Gould) for constructive and valuable comments, which have improved the manuscript considerably. Lynne Talley, Pierre-Yves Le Traon, Hiroshi Uchida, and Heather Furey helped us prepare the figures.

REFERENCES

Al Saafani, M.A., Shenoi, S.S.C., Shankar, D., Aparna, M., Kurian, J., Durand, F., Vinayachandran, P.N., 2007. Westward movement of eddies into the Gulf of Aden from the Arabian Sea. J. Geophys. Res. 112, C11004. http://dx.doi.org/10.1029/2006JC004020.

Anderson, D.L.T., Corry, R.A., 1985. Seasonal transport variations in the Florida Straits: a model study. J. Phys. Oceanogr. 15, 773–786.

Anderson, D.L.T., Moore, D.W., 1979. Cross-equatorial inertial jets with special relevance to the very remote forcing of the Somali Current. Deep Sea Res. 26, 1–22.

Andres, M., Park, J.-H., Wimbush, M., Zhu, X.-H., Chang, K.-I., Ichikawa, H., 2008. Study of the Kuroshio/Ryukyu Current system based on satellite-altimeter and *in situ* measurements. J. Oceanogr. 64, 937–950.

Arhan, M., Mercier, H., Park, Y.-H., 2003. On the deep water circulation of the eastern South Atlantic Ocean. Deep Sea Res. Part I 50, 889–916.

Auer, S.J., 1987. Five-year climatological survey of the Gulf Stream System and its associated rings. J. Geophys. Res. 92, 11,709–11,726.

Baringer, M., Larsen, J., 2001. Sixteen years of Florida Current transport at 27°N. Geophys. Res. Lett. 28, 3179–3182.

Beal, L.M., 2009. A time series of Agulhas Undercurrent transport. J. Phys. Oceanogr. 39, 2436–2450.

Beal, L.M., Bryden, H.L., 1999. The velocity and vorticity structure of the Agulhas Current at 32°S. J. Geophys. Res. 104, 5151–5176.

Beal, L.M., Chereskin, T.K., 2003. The volume transport of the Somali Current during the 1995 southwest monsoon. Deep Sea Res. Part II 50, 2077–2089.

Beal, L.M., Chereskin, T.K., Bryden, H.L., Ffield, A., 2003. Variability of water properties, heat and salt fluxes in the Arabian Sea, between the onset and wane of the 1995 southwest monsoon. Deep Sea Res. Part II 50, 2049–2075.

Beal, L.M., Chereskin, T.K., Lenn, Y.D., Elipot, S., 2006. The sources and mixing characteristics of the Agulhas Current. J. Phys. Oceanogr. 36, 2060–2074.

Beal, L.M., de Ruijter, W.P.M., Biastoch, A., Zahn, R., SCOR/WCRP/IAPSO Working Group 136, 2011. On the role of the Agulhas system in ocean circulation and climate. Nature 472, 429–436. http://dx.doi.org/10.1038/nature09983.

Beal, L.M., Donohue, K.A., 2013. The Great Whirl: observations of its seasonal development and interannual variability. J. Geophys. Res. 118, 1–13. http://dx.doi.org/10.1029/2012JC008198.

Beal, L.M., Hummon, J.M., Williams, E., Brown, O.B., Baringer, W., Kearns, E.J., 2008. Five years of Florida Current structure and transport from the Royal Caribbean Cruise Ship *Explorer of the Seas*. J. Geophys. Res. 113, C06001. http://dx.doi.org/10.1029/2007JC004154.

Behera, S.K., Yamagata, T., 2001. Subtropical SST dipole events in the southern Indian Ocean. Geophys. Res. Lett. 28, 327–330.

Biastoch, A., Böning, C.W., Lutjeharms, J.R.E., 2008. Agulhas leakage dynamics affects decadal variability in Atlantic overturning circulation. Nature 456, 489–492.

Biastoch, A., Böning, C.W., Schwarzkopf, F.U., Lutjeharms, J.R.E., 2009. Increase in Agulhas leakage due to poleward shift of the Southern Hemisphere westerlies. Nature 462, 495–498.

Boebel, O., Lutjeharms, J.R.E., Schmid, C., Zenk, W., Rossby, T., Barron, C.N., 2003. The Cape Cauldron, a regime of turbulent inter-ocean exchange. Deep Sea Res. Part II 50, 57–86.

Bowen, M.M., Sutton, P.J.H., Roemmich, D., 2006. Wind-driven and steric fluctuations of sea surface height in the southwest Pacific. Geophys. Res. Lett. 33, L14617. http://dx.doi.org/10.1029/2006GL026160.

Bowen, M.M., Wilkin, J.L., Emery, W.J., 2005. Variability and forcing of the East Australian Current. J. Geophys. Res. 110, C03019. http://dx.doi.org/10.1029/2004JC0222533.

Bower, A.S., 1991. A simple kinematic mechanism for mixing fluid parcels across a meandering jet. J. Phys. Oceanogr. 21, 173–180.

Bower, A.S., Hogg, N.G., 1996. Structure of the Gulf Stream and its recirculations at 55°W. J. Phys. Oceanogr. 26, 1002–1022. http://dx.doi.org/10.1175/1520-0485(1996)026<1002:SOTGSA>2.0.CO;2.

Bower, A.S., Le Cann, B., Rossby, H.T., Zenk, W., Gould, J., Speer, K., Richardson, P.L., Prater, M.D., Zhang, H.-M., 2002. Directly measured mid-depth circulation in the northeastern North Atlantic Ocean. Nature 419, 603–607.

Bower, A.S., Lozier, M.S., 1994. A closer look at particle exchange in the Gulf Stream. J. Phys. Oceanogr. 24, 1399–1418.

Bower, A.S., Rossby, T., 1989. Evidence of cross-frontal exchange processes in the Gulf Stream based on isopycnal RAFOS float data. J. Phys. Oceanogr. 19, 1177–1190.

Bower, A.S., Rossby, H.T., Lillibridge, J.L., 1985. The Gulf Stream—barrier or blender? J. Phys. Oceanogr. 15, 24–32.

Brandt, P., Stramma, L., Schott, F., Fischer, J., Dengler, M., Quadfasel, D., 2002. Annual Rossby waves in the Arabian Sea from TOPEX/Poseidon altimeter and in situ data. Deep Sea Res. 50, 1197–1210.

Bryan, F.O., Hecht, M.W., Smith, R.D., 2007. Resolution convergence and sensitivity studies with North Atlantic circulation models. Part I: the western boundary current system. Ocean Model. 16, 141–159.

Bryden, H.L., Beal, L.M., Duncan, L.M., 2005. Structure and transport of the Agulhas Current and its temporal variability. J. Oceanogr. 61, 479–492.

Cai, W., 2006. Antarctic ozone depletion causes an intensification of the Southern Ocean super-gyre circulation. Geophys. Res. Lett. 33, L03712. http://dx.doi.org/10.1029/2005GL024911.

Cane, M., 1980. On the dynamics of equatorial currents, with application to the Indian Ocean. Deep Sea Res. 27A, 525–544.

Carr, M.-E., Rossby, H.T., 2000. Pathways of the North Atlantic Current from surface drifters and subsurface floats. J. Geophys. Res. 106, 4405–4419.

Carton, J.A., Chepurin, G., Cao, X., Giese, B.S., 2000. A Simple Ocean Data Assimilation analysis of the global upper ocean 1950–1995, Part 1: methodology. J. Phys. Oceanogr. 30, 294–309.

Centurioni, L.R., Niiler, P.P., Lee, D.-K., 2004. Observations of inflow of Philippine Sea water into the South China Sea through the Luzon Strait. J. Phys. Oceanogr. 34, 113–121.

Charney, J.G., 1955. The Gulf Stream as an inertial boundary layer. Proc. Natl. Acad. Sci. U.S.A. 41, 731–740.

Chassignet, E.P., Garraffo, Z.D., 2001. Viscosity parameterization and the Gulf Stream separation. In: From Stirring to Mixing in a Stratified Ocean: Proceedings of the 12th 'Aha Huliko'a Hawaiian Winter Workshop, 2001. University of Hawaii at Manoa, Honolulu, pp. 39–43.

Chiessi, C.M., et al., 2008. South Atlantic interocean exchange as the trigger for the Bølling warm event. Geology 36, 919–922.

Cox, M.D., 1979. A numerical study of Somali Current eddies. J. Phys. Oceanogr. 9, 311–326.

Cresswell, G.R., Legeckis, R., 1986. Eddies off southeastern Australia. Deep Sea Res. 22, 1527–1562.

Cronin, M., et al., 2010. Monitoring ocean–atmosphere interactions in western boundary current extensions. In: Hall, J., Harrison, D.E., Stammer, D. (Eds.), Proceedings of OceanObs'09: Sustained Ocean Observations and Information for Society, Venice, Italy, September 21–25, 2009, vol. 2. http://dx.doi.org/10.5270/OceanObs09.cwp.20. ESA Publication WPP-306.

Cronin, M.F., Watts, D.R., 1996. Eddy–mean flow interaction in the Gulf Stream at 68°W. Part 1: Eddy energetics. J. Phys. Oceanogr. 26, 2107–2131.

Cushman-Roisin, B., 1993. Trajectories in Gulf Stream meanders. J. Geophys. Res. 98, 2543–2554.

Davis, R., Zenk, W., 2001. Subsurface Lagrangian observations during WOCE. In: Siedler, G., Church, J., Gould, J. (Eds.), Ocean Circulation and Climate. International Geophysics Series, vol. 77. Academy Press, New York, pp. 123–139.

Dencausse, G., Arhan, M., Speich, S., 2010. Spatio-temporal characteristics of the Agulhas Current retroflection. Deep Sea Res. Part I 57, 1392–1405.

Dengg, J., Beckmann, A., Gerdes, R., 1996. The Gulf Stream separation problem. In: Krauss, W. (Ed.), The Warmwatersphere of the North Atlantic Ocean. Gebruder Borntrager, Stuttgart, pp. 253–290.

de Ruijter, W.P.M., 1982. Asymptotic analysis of the Agulhas and Brazil Current systems. J. Phys. Oceanogr. 12, 361–373.

de Ruijter, W.P.M., Biastoch, A., Drijfhout, S.S., Lutjeharms, J.R.E., Matano, R.P., Pichevin, T., van Leeuwen, P.J., Weijer, W., 1999. Indian–Atlantic inter-ocean exchange: dynamics, estimation, and impact. J. Geophys. Res. 104, 20,885–20,910.

de Ruijter, W.P.M., van Aken, H.M., Beier, E.J., Lutjeharms, J.R.E., Matano, R.P., Schouten, M.W., 2004. Eddies and dipoles around South Madagascar: formation, pathways and large-scale impact. Deep Sea Res. 51, 383–400.

Dibarboure, G., Pujol, M.-I., Briol, F., Le Traon, P.Y., Larnicol, G., Picot, N., Mertz, F., Ablain, M., 2011. Jason-2 in DUACS: updated system description, first tandem results and impact on processing and products. Mar. Geod. 34, 214–241.

DiNezio, P.N., Gramer, L.J., Johns, W.E., Meinen, C.S., Baringer, M.O., 2009. Observed interannual variability of the Florida Current: wind forcing and the North Atlantic Oscillation. J. Phys. Oceanogr. 39, 721–736. http://dx.doi.org/10.1175/2008JPO4001.1.

Donlon, C., et al., 2007. The Global Ocean Data Assimilation Experiment High-resolution Sea Surface Temperature Pilot Project. Bull. Am. Meteorol. Soc. 88, 1197–1213. http://dx.doi.org/10.1175/BAMS-88-8-1197.

Donohue, K.A., Toole, J.M., 2003. A near-synoptic survey of the southwest Indian Ocean. Deep Sea Res. Part II 50, 1893–1931.

Duan, J., Wiggins, S., 1996. Fluid exchange across a meandering jet with quasiperiodic variability. J. Phys. Oceanogr. 26, 1176–1188.

Ducet, N., Le Traon, P.Y., Reverdin, G., 2000. Global high-resolution mapping of ocean circulation from TOPEX/Poseidon and ERS-1 and -2. J. Geophys. Res. 105, 19,477–19,498.

Ducklow, H.W., Steinberg, D.K., Buesseler, K.O., 2001. Upper ocean carbon export and the biological pump. Oceanography 14, 50–58. http://dx.doi.org/10.5670/oceanog.2001.06.

Düing, W., Molinari, R.L., Swallow, J.C., 1980. Somali Current: evolution of surface flow. Science 209, 588–590.

Evans, D.L., Signorini, S.S., 1985. Vertical structure of the Brazil Current. Nature 315, 48–50.

Ezer, T., Mellor, G.L., 1992. A numerical study of the variability and the separation of the Gulf Stream induced by surface atmospheric forcing and lateral boundary flows. J. Phys. Oceanogr. 22, 660–682.

Feng, M., Mitsudera, H., Yoshikawa, Y., 2000. Structure and variability of the Kuroshio Current in Tokara Strait. J. Phys. Oceanogr. 30, 2257–2276.

Ffield, A., Toole, J., Wilson, D., 1997. Seasonal circulation in the south Indian Ocean. Geophys. Res. Lett. 24, 2773–2776.

Findlater, J., 1969. A major low-level air current near the Indian Ocean during the northern summer. Q. J. R. Meteorol. Soc. 95, 280–362.

Fine, R.A., 1993. Circulation of Antarctic Intermediate Water in the South Indian Ocean. Deep Sea Res. Part I 40, 2021–2042.

Fischer, J., Schott, F.A., 2002. Labrador Sea Water tracked by profiling floats—from the boundary current into the open North Atlantic. J. Phys. Oceanogr. 32, 573–584.

Fischer, J., Schott, F., Stramma, L., 1996. Current transports of the Great Whirl–Socotra Gyre system during summer monsoon, August 1993. J. Geophys. Res. 101, 3573–3687.

Fofonoff, N.P., 1981. The Gulf Stream system. In: Warren, B.A., Wunsch, C. (Eds.), Evolution of Physical Oceanography. MIT Press, Cambridge, MA, pp. 112–139.

Frankignoul, C., de Coetlogon, G., Joyce, T., Dong, S., 2001. Gulf Stream variability and ocean–atmosphere interactions. J. Phys. Oceanogr. 31, 3516–3529. http://dx.doi.org/10.1175/1520-0485(2002)031<3516:GSVAOA>2.0.CO;2.

Frankignoul, C., Sennéchael, N., 2007. Observed influence of North Pacific SST anomalies on the atmospheric circulation. J. Clim. 20, 592–606.

Fratantoni, D., 2001. North Atlantic surface circulation during the 1990's observed with satellite-tracked drifters. J. Geophys. Res. 106, 22,067–22,093. http://dx.doi.org/10.1029/2000JC000730.

Fratantoni, D.M., Bower, A.S., Johns, W.E., Peters, H., 2006. Somali Current rings in the eastern Gulf of Aden. J. Geophys. Res. 111, C09039. http://dx.doi.org/10.1029/2005JC003338.

Garfield, N., 1990. The Brazil Current at subtropical latitudes. PhD Thesis, University of Rhode Island, 122 pp.

Garzoli, S.L., 1993. Geostrophic velocity and transport variability in the Brazil–Malvinas Confluence. Deep Sea Res. 40, 1379–1403.

Garzoli, S.L., Giulivi, C., 1994. What forces the variability of the southwestern Atlantic boundary currents? Deep Sea Res. Part I 41, 1527–1550.

Garzoli, S.L., Goni, G.J., Mariano, A., Olson, D., 1997. Monitoring the upper southeastern Atlantic transport using altimeter data. J. Mar. Res. 55, 453–481.

Garzoli, S.L., Matano, R., 2011. The South Atlantic and the Atlantic Meridional Overturning Circulation. Deep Sea Res. Part II 58, 1837–1847.

Gilson, J., Roemmich, D., 2002. Mean and temporal variability in the Kuroshio geostrophic transport south of Taiwan (1993–2001). J. Oceanogr. 58, 183–195.

Gimeno, L., Drumond, A., Nieto, R., Trigo, R.M., Stohl, A., 2010. On the origin of continental precipitation. Geophys. Res. Lett. 37, L13804. http://dx.doi.org/10.1029/2010GL043712.

Godfrey, J.S., 1989. A Sverdrup model of the depth-integrated flow for the world ocean allowing for island circulations. Geophys. Astrophys. Fluid Dyn. 45, 89–112.

Goni, G.J., Garzoli, S.L., Roubicek, A.J., Olson, D.B., Brown, O.B., 1997. Agulhas ring dynamics from TOPEX/POSEIDON satellite altimeter data. J. Mar. Res. 55, 861–883.

Goni, G.J., Wainer, I., 2001. Investigation of the Brazil Current front variability from altimeter data. J. Geophys. Res. 106, 31,117–31,128. http://dx.doi.org/10.1029/2000JC000396.

Gordon, A.L., 1989. Brazil–Malvinas Confluence—1984. Deep Sea Res. 36, 359–384.

Gordon, A.L., Greengrove, C.L., 1986. Geostrophic circulation of the Brazil–Falkland confluence. Deep Sea Res. 33, 573–585.

Gordon, A.L., Weiss, R.F., Smethie, W.M., Warner, M.J., 1992. Thermocline and Intermediate Water communication between the South Atlantic and Indian Oceans. J. Geophys. Res. 97, 7223–7240.

Gourdeau, L., Kessler, W.S., Davis, R.E., Sherman, J., Maes, C., Kestenare, E., 2008. Zonal jets entering the Coral Sea. J. Phys. Oceanogr. 38, 715–725.

Grundlingh, M., 1979. Observation of a large meander in the Agulhas Current. J. Geophys. Res. 84, 3776–3778.

Grundlingh, M., 1983. On the course of the Agulhas Current. S. Afr. Geogr. J. 65, 49–57.

Halkin, D., Rossby, T., 1985. The structure and transport of the Gulf Stream at 73°W. J. Phys. Oceanogr. 15, 1439–1452. http://dx.doi.org/10.1175/1520-0485(1985)015<1439:TSATOT>2.0.CO;2.

Halpern, D., Woiceshyn, P.M., 1999. Onset of the Somali Jet in the Arabian Sea during June 1997. J. Geophys. Res. 104, 18, 041–18, 046.

Hanawa, K., Talley, L.D., 2001. Mode waters. In: Siedler, G., Church, J., Gould, J. (Eds.), Ocean Circulation and Climate. International Geophysics Series, vol. 77. Academy Press, New York, pp. 373–386.

Hastenrath, S., Greischar, L., 1991. The Monsoonal current regimes of the tropical Indian Ocean: observed surface flow fields and their geostrophic and wind-driven components. J. Geophys. Res. 96, 12,619–12,633.

Hecht, M.W., Smith, R.D., 2008. Towards a physical understanding of the North Atlantic: a review of model studies in an eddying regime. In: Hecht, M.W., Hasumi, H. (Eds.), Ocean Modeling in an Eddying Regime. Geophysical Monograph Series, vol. 177. American Geophysical Union, Washington, DC, pp. 213–239. http://dx.doi.org/10.1029/177GM15.

Hellerman, S., Rosenstein, M., 1983. Normal monthly wind stress over the world ocean with error estimate. J. Phys. Oceanogr. 13, 1093–1104.

Hill, K.L., Rintoul, S.R., Coleman, R., Ridgway, K.R., 2008. Wind forced low frequency variability of the East Australia Current. Geophys. Res. Lett. 35, L08602. http://dx.doi.org/10.1029/2007GL032912.

Hill, K.L., Rintoul, S.R., Oke, P.R., Ridgway, K.R., 2010. Rapid response of the East Australian Current to remote wind forcing: the role of barotropic–baroclinic interactions. J. Mar. Res. 68, 413–431.

Hill, K.L., Rintoul, S.R., Ridgway, K.R., Oke, P.R., 2011. Decadal changes in the South Pacific western boundary current system revealed in observations and ocean state estimates. J. Geophys. Res. 116, C01009. http://dx.doi.org/10.1029/2009JC005926.

Hogg, N.G., 1992. On the transport of the Gulf Stream between Cape Hatteras and the Grand Banks. Deep Sea Res. 39, 1231–1246. http://dx.doi.org/10.1016/0198-0149(92)90066-3.

Hogg, N.G., Johns, W.E., 1995. Western boundary currents. Rev. Geophys. 33, 1311–1334. http://dx.doi.org/10.1029/95RG00491.

Holbrook, N.J., Bindoff, N.L., 1999. Seasonal temperature variability in the upper southwest Pacific Ocean. J. Phys. Oceanogr. 29, 366–381.

Hoskins, B.J., Hodges, K.I., 2002. New perspectives on the Northern Hemisphere winter storm tracks. J. Atmos. Sci. 59, 1041–1061.

Huang, R.X., 2010. Ocean Circulation: Wind-driven and Thermohaline Processes. Cambridge University Press, Cambridge, 791 pp.

Hurlburt, H.E., Wallcraft, A.J., Schmitz Jr., W.J., Hogan, P.J., Metzger, E.J., 1996. Dynamics of the Kuroshio/Oyashio current system using eddy-resolving models of the North Pacific Ocean. J. Geophys. Res. 101, 941–976.

Hurrell, J., 1995. Decadal trends in the North Atlantic Oscillation: regional temperatures and precipitation. Science 269, 676–679.

Ichikawa, H., Beardsley, R.C., 1993. Temporal and spatial variability of volume transport of the Kuroshio in the East China Sea. Deep Sea Res. 40, 583–605.

Ichikawa, H., Nakamura, H., Nishina, A., Higashi, M., 2004. Variability of north-eastward current southeast of northern Ryukyu Islands. J. Oceanogr. 60, 351–363.

Imawaki, S., Uchida, H., Ichikawa, H., Fukasawa, M., Umatani, S., the ASUKA Group, 2001. Satellite altimeter monitoring the Kuroshio transport south of Japan. Geophys. Res. Lett. 28, 17–20.

Isobe, A., Imawaki, S., 2002. Annual variation of the Kuroshio transport in a two-layer numerical model with a ridge. J. Phys. Oceanogr. 32, 994–1009.

Izumo, T., de Boyer Montégut, C., Luo, J.-J., Behera, S.K., Masson, S., Yamagata, T., 2008. The role of the western Arabian Sea upwelling in Indian monsoon rainfall variability. J. Clim. 21, 5603–5623.

James, C., Wimbush, M., Ichikawa, H., 1999. Kuroshio meanders in the East China Sea. J. Phys. Oceanogr. 29, 259–272.

Jayne, S., Hogg, N., Waterman, S., Rainville, L., Donahue, K., Watts, D., Tracey, K., McClean, J., Maltrud, M., Qiu, B., Chen, S., Hacker, P., 2009. The Kuroshio Extension and its recirculation gyres. Deep Sea Res. 56, 2088–2099.

Jensen, T.G., 1991. Modeling the seasonal undercurrents in the Somali Current system. J. Geophys. Res. 96, 22,151–22,167.

Johns, W.E., Lee, T.N., Beardsley, R.C., Candela, J., Limeburner, R., Castro, B., 1998. Annual cycle and variability of the North Brazil Current. J. Phys. Oceanogr. 28, 103–128.

Johns, W.E., Lee, T.N., Zhang, D., Zantopp, R., Liu, C.-T., Yang, Y., 2001. The Kuroshio east of Taiwan: moored transport observations from the WOCE PCM-1 array. J. Phys. Oceanogr. 31, 1031–1053.

Johns, W.E., Shay, T.J., Bane, J.M., Watts, D.R., 1995. Gulf Stream structure, transport, and recirculation near 68°W. J. Geophys. Res. 100, 817–838. http://dx.doi.org/10.1029/94JC02497.

Joyce, T.M., Deser, C., Spall, M.A., 2000. The relation between decadal variability of subtropical mode water and the North Atlantic Oscillation. J. Clim. 13, 2550–2569.

Joyce, T., Kwon, Y.-O., Yu, L., 2009. On the relationship between synoptic wintertime atmospheric variability and path shifts in the Gulf Stream and the Kuroshio Extension. J. Clim. 22, 3177–3192.

Jury, M., Walker, N., 1988. Marine boundary layer modification across the edge of the Agulhas Current. J. Geophys. Res. 93, 647–654.

Kashino, Y., Espana, N., Syamsudin, F., Richards, K.J., Jensen, T., Dutrieux, P., Ishida, A., 2009. Observations of the North Equatorial Current, Mindanao Current, and the Kuroshio Current system during the 2006/07 El Niño and 2007/08 La Niña. J. Oceanogr. 65, 325–333.

Kawabe, M., 1988. Variability of Kuroshio velocity assessed from the sea level difference between Naze and Nishinoomote. J. Oceanogr. Soc. Jpn. 44, 293–304.

Kawabe, M., 1995. Variations of current path, velocity, and volume transport of the Kuroshio in relation with the large meander. J. Phys. Oceanogr. 25, 3103–3117.

Kearns, E.J., Rossby, H.T., 1998. Historical position of the North Atlantic Current. J. Geophys. Res. 103, 15,509–15,524.

Kelly, K.A., Singh, S., Huang, R.X., 1999. Seasonal variations of sea surface height in the Gulf Stream region. J. Phys. Oceanogr. 29, 313–327.

Kelly, K.A., Small, R.J., Samelson, R.M., Qiu, B., Joyce, T.M., Kwon, Y., Cronin, M.F., 2010. Western boundary currents and frontal air-sea interaction: Gulf Stream and Kuroshio Extension. J. Clim. 23, 5644–5667. http://dx.doi.org/10.1175/2010JCLI3346.1.

Kessler, W.S., Gourdeau, L., 2007. The annual cycle of circulation of the southwest subtropical Pacific, analyzed in an ocean GCM. J. Phys. Oceanogr. 37, 1610–1627.

Kim, Y., Qu, T., Jensen, T., Miyama, T., Mitsudera, H., Kang, H., Ishida, A., 2004. Seasonal and interannual variations of the North Equatorial Current bifurcation in a high-resolution OGCM. J. Geophys. Res. 109, C03040. http://dx.doi.org/10.1029/2003JC002013.

Knorr, G., Lohmann, G., 2003. Southern Ocean origin for the resumption of Atlantic thermohaline circulation during deglaciation. Nature 424, 532–536.

Krauss, W., Fahrbach, E., Aitsam, A., Elken, J., Koske, P., 1987. The North Atlantic Current and its associated eddy field southeast of Flemish Cap. Deep Sea Res. 34, 1163–1185.

Kwon, Y., Alexander, M.A., Bond, N.A., Frankignoul, C., Nakamura, H., Qiu, B., Thompson, L., 2010. Role of the Gulf Stream and Kuroshio–

Oyashio systems in large-scale atmosphere–ocean interaction: a review. J. Clim. 23, 3249–3281. http://dx.doi.org/10.1175/2010JCLI3343.1.

Larsen, J.C., 1992. Transport and heat flux of the Florida Current at 27°N derived from cross-stream voltages and profiling data: theory and observation. Philos. Trans. R. Soc. Lond. 338, 169–236.

Larsen, J.C., Sanford, T.B., 1985. Florida Current volume transports from voltage measurements. Science 227, 302–304.

Lazier, J.R.N., 1994. Observations in the Northwest Corner of the North Atlantic Current. J. Phys. Oceanogr. 24, 1449–1463.

Leaman, K.D., Molinari, R.L., Vertes, P.S., 1987. Structure and variability of the Florida Current at 27°N: April 1982–July 1984. J. Phys. Oceanogr. 17, 565–583.

Lee, T., Cornillon, P., 1995. Temporal variation of meandering intensity and domain-wide lateral oscillations of the Gulf Stream. J. Geophys. Res. 100, 13,603–13,613.

Lee, T., Cornillon, P., 1996a. Propagation of Gulf Stream meanders between 74° and 70°W. J. Phys. Oceanogr. 26, 205–224.

Lee, T., Cornillon, P., 1996b. Propagation and growth of Gulf Stream meanders between 75° and 45°W. J. Phys. Oceanogr. 26, 225–241.

Lee, T.N., Johns, W.E., Liu, C.-T., Zhang, D., Zantopp, R., Yang, Y., 2001. Mean transport and seasonal cycle of the Kuroshio east of Taiwan with comparison to the Florida Current. J. Geophys. Res. 106, 22,143–22,158.

Lee, T.N., Schott, F.A., Zantopp, R.J., 1985. Florida Current: low-frequency variability as observed with moored current meters during April 1982 to June 1983. Science 227, 298–302.

Lee-Thorp, A.M., Rouault, M., Lutjeharms, J.R.E., 1998. Cumulus cloud formation above the Agulhas Current. S. Afr. J. Sci. 94, 351–354.

Leetmaa, A., Quadfasel, D.R., Wilson, D., 1982. Development of the flow field during the onset of the Somali Current, 1979. J. Phys. Oceanogr. 12, 1325–1342.

Lie, H.-J., Cho, C.-H., Lee, J.-H., Niiler, P.P., Hu, J.-H., 1998. Separation of the Kuroshio water and its penetration onto the continental shelf west of Kyushu. J. Geophys. Res. 103, 2963–2976.

Lozier, M.S., Bold, T.J., Bower, A.S., 1996. The influence of propagating waves on cross-stream excursions. J. Phys. Oceanogr. 26, 1915–1923.

Lozier, M.S., Pratt, L.J., Rogerson, A.M., Miller, P.D., 1997. Exchange geometry revealed by float trajectories in the Gulf Stream. J. Phys. Oceanogr. 27, 2327–2341.

Lumpkin, R., Garraffo, Z., 2005. Evaluating the decomposition of tropical Atlantic drifter observations. J. Atmos. Oceanic Technol. 22, 1403–1415.

Lumpkin, R., Garzoli, S., 2011. Interannual to decadal changes in the western South Atlantic's surface circulation. J. Geophys. Res. 116, C01014. http://dx.doi.org/10.1029/2010JC006285.

Luther, M.E., 1999. Interannual variability in the Somali Current 1954–1976. Nonlinear Anal. 35, 59–83.

Lutjeharms, J.R.E., 2006. The Agulhas Current. Springer, Berlin, 330 pp.

Lutjeharms, J.R.E., Ansorge, I.J., 2001. The Agulhas return current. J. Mar. Syst. 30, 115–138.

Lutjeharms, J.R.E., Roberts, H.R., 1988. The Natal Pulse: an extreme transient on the Agulhas Current. J. Geophys. Res. 93, 631–645.

Lutjeharms, J.R.E., van Ballegooyen, R.C., 1988. The retroflection of the Agulhas Current. J. Phys. Oceanogr. 18, 1570–1583.

Mann, C.R., 1967. The termination of the Gulf Stream and the beginning of North Atlantic Current. Deep Sea Res. 14, 337–359. http://dx.doi.org/10.1016/0011-7471(67)90077-0.

Mata, M.M., Tomczak, M., Wijffels, S., Church, J.A., 2000. East Australian Current volume transports at 30°S: estimates from the World Ocean Circulation Experiment hydrographic sections PR11/P6 and the PCM3 current meter array. J. Geophys. Res. 105, 28,509–28,526.

Mata, M.M., Wijffels, S.E., Church, J.A., Tomczak, M., 2006. Eddy shedding and energy conversions in the East Australian Current. J. Geophys. Res. 111, C09034. http://dx.doi.org/10.1029/2006JC003592.

Matano, R.P., 1993. On the separation of the Brazil Current from the coast. J. Phys. Oceanogr. 23, 79–90.

Matano, R.P., Beier, E.J., Strub, P.T., Tokmakian, R., 2002. Large-scale forcing of the Agulhas variability: the seasonal cycle. J. Phys. Oceanogr. 32, 1228–1241.

McCartney, M., Zemba, J., 1988. Thermocline, intermediate and deep circulation in the southwestern South Atlantic. In: SAARI Meeting Report, May 24–26, 1988. Lamont-Doherty Geological Observatory of Columbia University, Palisades, NY, pp. 28–29 (Abstract).

McCreary, J.P., Kundu, P.K., 1988. A numerical investigation of the Somali Current during the southwest monsoon. J. Mar. Res. 46, 25–58.

Meinen, C.S., Baringer, M.O., Garcia, R.F., 2010. Florida Current transport variability: an analysis of annual and longer-period signals. Deep Sea Res. Part I 57, 835–846. http://dx.doi.org/10.1016/j.dsr.2010.04.001.

Meinen, C.S., Watts, D.R., 2000. Vertical structure and transport on a transect across the North Atlantic Current near 42°N: time series and mean. J. Geophys. Res. 105, 21,869–21,891.

Minobe, S., Kuwano-Yoshida, A., Komori, N., Xie, S., Small, R.J., 2008. Influence of the Gulf Stream on the troposphere. Nature 452, 206–209.

Mizuno, K., White, W.B., 1983. Annual and interannual variability in the Kuroshio Current system. J. Phys. Oceanogr. 13, 1847–1867.

Molinari, R.L., Wilson, W.D., Leaman, K., 1985. Volume and heat transports of the Florida Current: April 1982 through August 1983. Science 227, 295–297.

Munk, W.H., 1950. On the wind-driven ocean circulation. J. Meteorol. 7, 79–93.

Munk, W.H., Gross, G.W., Carrier, G.F., 1950. Note on the dynamics of the Gulf Stream. J. Mar. Res. 9, 218–238.

Nakamura, H., Nishina, A., Minobe, S., 2012. Response of storm tracks to bimodal Kuroshio path states south of Japan. J. Clim. 25, 7772–7779. http://dx.doi.org/10.1175/JCLI-D-12-00326.1.

Nakamura, H., Sampe, T., Tanimoto, Y., Shimpo, A., 2004. Observed associations among storm tracks, jet streams, and midlatitude oceanic fronts. In: Wang, C., Xie, S.-P., Carton, J.A. (Eds.), Earth's Climate: The Ocean-Atmosphere Interaction. Geophysical Monograph Series, vol. 147. American Geophysical Union, Washington, DC, pp. 329–346.

Nakamura, H., Shimpo, A., 2004. Seasonal variations in the Southern Hemisphere storm tracks and jet streams as revealed in reanalysis datasets. J. Clim. 17, 1828–1844.

Nauw, J.J., van Aken, H.M., Webb, A., Lutjeharms, J.R.E., de Ruijter, W.P.M., 2008. Observations of the southern East Madagascar Current and undercurrent and countercurrent system. J. Geophys. Res. 113, C08006. http://dx.doi.org/10.1029/2007JC004639.

Niiler, P.P., Richardson, W.S., 1973. Seasonal variability in the Florida Current. J. Mar. Res. 21, 144–167.

Nilsson, C.S., Cresswell, G.R., 1981. The formation and evolution of East Australian Current warm-core eddies. Prog. Oceanogr. 9, 133–183.

Nitani, H., 1972. Beginning of the Kuroshio. In: Stommel, H., Yoshida, K. (Eds.), Kuroshio: Its Physical Aspects. University of Tokyo Press, Tokyo, pp. 129–163.

Nof, D., Pichevin, T., 1996. The retroflection paradox. J. Phys. Oceanogr. 26, 2344–2358.

Oka, E., 2009. Seasonal and interannual variation of North Pacific Subtropical Mode Water in 2003–2006. J. Oceanogr. 65, 151–164.

Oka, E., Suga, T., Sukigara, C., Toyama, K., Shimada, K., Yoshida, J., 2011. "Eddy-resolving" observation of the North Pacific Subtropical Mode Water. J. Phys. Oceanogr. 41, 666–681.

Olson, D., Podesta, G., Evans, R., Brown, O., 1988. Temporal variations in the separation of the Brazil and Malvinas Currents. Deep Sea Res. 35, 1971–1990.

O'Neill, L.W., Chelton, D.B., Esbensen, S.K., Wentz, F.J., 2005. High-resolution satellite measurements of the atmospheric boundary layer response to SST variations along the Agulhas Return Current. J. Clim. 18, 2706–2723.

Orvik, K.A., Niiler, P., 2002. Major pathways of Atlantic water in the northern North Atlantic and Nordic Seas toward Arctic. Geophys. Res. Lett. 29, 1896. http://dx.doi.org/10.1029/2002GL015002.

Ou, H.W., de Ruijter, W.P.M., 1986. Separation of an internal boundary current from a curved coast line. J. Phys. Oceanogr. 16, 280–289.

Owens, W.B., 1991. A statistical description of the mean circulation and eddy variability in the northwestern North Atlantic using SOFAR floats. Prog. Oceanogr. 28, 257–303.

Pedlosky, J., 1996. Ocean Circulation Theory. Springer-Verlag, Berlin, 453 pp.

Peeters, F.J.C., et al., 2004. Vigorous exchange between the Indian and Atlantic oceans at the end of the past five glacial periods. Nature 430, 661–665.

Peña-Molino, B., Joyce, T.M., 2008. Variability in the slope water and its relation to the Gulf Stream path. Geophys. Res. Lett. 35, L03606. http://dx.doi.org/10.1029/2007GL032183.

Penven, P., Lutjeharms, J.R.E., Florenchie, P., 2006. Madagascar: a pacemaker for the Agulhas Current system? Geophys. Res. Lett. 33, L17609. http://dx.doi.org/10.1029/2006GL026854.

Pérez-Brunius, H., Rossby, T., Watts, D.R., 2004a. A method for obtaining the mean transports of ocean currents by combining isopycnal float data with historical hydrography. J. Atmos. Oceanic Technol. 21, 298–316.

Pérez-Brunius, H., Rossby, T., Watts, D.R., 2004b. Absolute transports of mass and temperature for the North Atlantic Current—subpolar front system. J. Phys. Oceanogr. 34, 1870–1883.

Peterson, R.G., 1990. On the volume transport in the southwestern South Atlantic Ocean. Eos 71, 542, (Abstract).

Peterson, R.G., Stramma, L., 1991. Upper-level circulation in the South Atlantic Ocean. Prog. Oceanogr. 26, 1–73.

Pichevin, T., Nof, D., Lutjeharms, J.R.E., 1999. Why are there Agulhas rings? J. Phys. Oceanogr. 29, 693–707.

Pratt, L.J., Lozier, M.S., Beliakova, N., 1995. Parcel trajectories in quasigeostrophic jets: neutral modes. J. Phys. Oceanogr. 25, 1451–1466.

Qiu, B., 1999. Seasonal eddy field modulation of the North Pacific Subtropical Countercurrent: TOPEX/POSEIDON observations and theory. J. Phys. Oceanogr. 29, 2471–2486.

Qiu, B., Chen, S., 2004. Seasonal modulations in the eddy field of the South Pacific Ocean. J. Phys. Oceanogr. 34, 1515–1527.

Qiu, B., Chen, S., 2005. Variability of the Kuroshio Extension jet, recirculation gyre and mesoscale eddies on decadal timescales. J. Phys. Oceanogr. 35, 2090–2103.

Qiu, B., Chen, S., 2006. Decadal variability in the large-scale sea surface height field of the South Pacific Ocean: observations and causes. J. Phys. Oceanogr. 36, 1751–1762.

Qiu, B., Chen, S., 2010. Interannual-to-decadal variability in the bifurcation of the North Equatorial Current off the Philippines. J. Phys. Oceanogr. 40, 2525–2538.

Qiu, B., Chen, S., Hacker, P., 2007a. Effect of mesoscale eddies on Subtropical Mode Water variability from the Kuroshio Extension System Study (KESS). J. Phys. Oceanogr. 37, 982–1000.

Qiu, B., Chen, S., Hacker, P., Hogg, N., Jayne, S., Sasaki, H., 2008. The Kuroshio Extension northern recirculation gyre: profiling float measurements and forcing mechanism. J. Phys. Oceanogr. 38, 1764–1779.

Qiu, B., Chen, S., Kessler, W.S., 2009. Source of the 70-day mesoscale eddy variability in the Coral Sea and the North Fiji Basin. J. Phys. Oceanogr. 39, 404–420.

Qiu, B., Joyce, T.M., 1992. Interannual variability in the mid- and low-latitude western North Pacific. J. Phys. Oceanogr. 22, 1062–1079.

Qiu, B., Lukas, R., 1996. Seasonal and interannual variability of the North Equatorial Current, the Mindanao Current and the Kuroshio along the Pacific western boundary. J. Geophys. Res. 101, 12,315–12,330.

Qiu, B., Miao, W., 2000. Kuroshio path variations south of Japan: bimodality as a self-sustained internal oscillation. J. Phys. Oceanogr. 30, 2124–2137.

Qiu, B., Schneider, N., Chen, S., 2007b. Coupled decadal variability in the North Pacific: an observationally-constrained idealized model. J. Clim. 20, 3602–3620.

Qiu, B., Toda, T., Imasato, N., 1990. On Kuroshio front fluctuations in the East China Sea using satellite and in-situ observational data. J. Geophys. Res. 95, 18,191–18,204.

Quadfasel, D., Schott, F., 1982. Water-mass distributions at intermediate layers off the Somali coast during the onset of the southwest monsoon. J. Phys. Oceanogr. 12, 1358–1372.

Quadfasel, D., Schott, F., 1983. Southward subsurface flow below the Somali Current. J. Geophys. Res. 33, 1307–1312.

Quartly, G.D., Srokosz, M.A., 1993. Seasonal variations in the region of the Agulhas Retroflection: studies with Geosat and FRAM. J. Phys. Oceanogr. 23, 2107–2124.

Read, J.F., Pollard, R.T., 1993. Structure and transport of the Antarctic Circumpolar Current and Agulhas Return Current at 40°E. J. Geophys. Res. 98, 12,281–12,295.

Reason, C.J.C., 2001. Evidence for the influence of the Agulhas Current on regional atmospheric circulation patterns. J. Clim. 14, 2769–2778.

Reid, J.L., 1997. On the total geostrophic circulation of the Pacific Ocean: flow pattern, tracers, and transports. Prog. Oceanogr. 39, 263–352.

Richardson, P.L., 2007. Agulhas leakage into the Atlantic estimated with subsurface floats and surface drifters. Deep Sea Res. Part I 54, 1361–1389.

Ridderinkhof, H., van der Werf, P.M., Ullgren, J.E., van Aken, H.M., van Leeuwen, P.J., de Ruijter, W.P.M., 2010. Seasonal and interannual variability in the Mozambique Channel from moored current observations. J. Geophys. Res. 115, C06010. http://dx.doi.org/10.1029/2009JC005619.

Ridgway, K.R., 2007. Long-term trend and decadal variability of the southward penetration of the East Australian Current. Geophys. Res. Lett. 34, L13613. http://dx.doi.org/10.1029/2007GL030393.

Ridgway, K.R., Dunn, J.R., 2003. Mesoscale structure of the mean East Australian Current system and its relationship with topography. Prog. Oceanogr. 56, 189–222.

Ridgway, K.R., Dunn, J.R., 2007. Observational evidence for a Southern Hemisphere oceanic supergyre. Geophys. Res. Lett. 34, L13612. http://dx.doi.org/10.1029/2007GL030392.

Ridgway, K.R., Godfrey, J.S., 1994. Mass and heat budgets in the East Australian Current: a direct approach. J. Geophys. Res. 99, 3231–3248.

Ridgway, K.R., Godfrey, J.S., 1996. Long-term temperature and circulation changes off eastern Australia. J. Geophys. Res. 101, 3615–3627.

Ridgway, K.R., Godfrey, J.S., 1997. Seasonal cycle of the East Australian Current. J. Geophys. Res. 102, 22,921–22,936.

Rio, M.-H., Hernandez, F., 2004. A mean dynamic topography computed over the world ocean from altimetry, in situ measurements, and a geoid model. J. Geophys. Res. 109, C12032. http://dx.doi.org/10.1029/2003JC002226.

Risien, C.M., Chelton, D.B., 2008. A global climatology of surface wind and wind stress fields from eight years of QuikSCAT scatterometer data. J. Phys. Oceanogr. 38, 2379–2413.

Roemmich, D., 2007. Super spin in the southern seas. Nature 449, 34–35.

Roemmich, D., Gilson, J., 2001. Eddy transport of heat and thermocline waters in the North Pacific: a key to interannual/decadal climate variability? J. Phys. Oceanogr. 31, 675–687.

Roemmich, D., Gilson, J., Davis, R., Sutton, P., Wijffels, S., 2007. Decadal spin-up of the South Pacific subtropical gyre. J. Phys. Oceanogr. 37, 162–173.

Roemmich, D., Gilson, J., Willis, J., Sutton, P., Ridgway, K., 2005. Closing the time-varying mass and heat budgets for large ocean areas: the Tasman box. J. Clim. 18, 2330–2343.

Roemmich, D., Sutton, P., 1998. The mean and variability of ocean circulation past northern New Zealand: determining the representativeness of hydrographic climatologies. J. Geophys. Res. 193, 13,041–13,054.

Rossby, T., 1996. The North Atlantic Current and surrounding waters: at the crossroads. Rev. Geophys. 34, 463–481. http://dx.doi.org/10.1029/96RG02214.

Rossby, T., Benway, R.L., 2000. Slow variations in mean path of the Gulf Stream east of Cape Hatteras. Geophys. Res. Lett. 27, 117–120.

Rossby, T., Flagg, C.N., Donohue, K., 2005. Interannual variations in upper-ocean transport by the Gulf Stream and adjacent waters between New Jersey and Bermuda. J. Mar. Res. 63, 203–226.

Rossby, T., Flagg, C., Donohue, K., 2010. On the variability of Gulf Stream transport from seasonal to decadal timescales. J. Mar. Res. 68, 503–522. http://dx.doi.org/10.1357/002224010794657128.

Rouault, M., Lee-Thorp, A.M., Lutjeharms, J.R.E., 2000. The atmospheric boundary layer above the Agulhas Current during along-current winds. J. Phys. Oceanogr. 30, 40–50.

Rouault, M., White, S.A., Reason, C.J.C., Lutjeharms, J.R.E., Jobard, I., 2002. Ocean–atmosphere interaction in the Agulhas Current region and a South African extreme weather event. Weather Forecast 17, 655–669.

Rypina, I.I., Pratt, L.J., Lozier, M.S., 2011. Near-surface transport pathways in the North Atlantic Ocean: looking for throughput from the subtropical to the subpolar gyre. J. Phys. Oceanogr. 41, 911–925.

Samelson, R.M., 1992. Fluid exchange across a meandering jet. J. Phys. Oceanogr. 22, 431–444.

Saraceno, M., Provost, C., Piola, A.R., Bava, J., Gagliardini, A., 2004. Brazil Malvinas Frontal System as seen from 9 years of advanced very high resolution radiometer data. J. Geophys. Res. 109, C05027. http://dx.doi.org/10.1029/2003JC002127.

Sato, O.T., Rossby, T., 1995. Seasonal and low frequency variability in dynamic height anomaly and transport of the Gulf Stream. Deep Sea Res. 42, 149–164.

Schmitz, W.J., 1996. On the world ocean circulation: volume I: some global features/North Atlantic circulation. Woods Hole Oceanographic Institution Technical Report, WHOI-96-03, Woods Hole, MA, 141 pp.

Schmitz, W.J., McCartney, M.S., 1993. On the North Atlantic circulation. Rev. Geophys. 31, 29–49. http://dx.doi.org/10.1029/92RG02583.

Schmitz, W.J., Richardson, P.L., 1991. On the sources of the Florida Current. Deep Sea Res. 38, S379–S409.

Schoenefeldt, R., Schott, F.A., 2006. Decadal variability of the Indian Ocean cross-equatorial exchange in SODA. Geophys. Res. Lett. 33, L08602. http://dx.doi.org/10.1029/2006GL025891.

Schott, F.A., Fischer, J., 2000. Winter monsoon circulation of the northern Arabian Sea and Somali Current. J. Geophys. Res. 105, 6359–6376.

Schott, F., Fischer, J., Gartenicht, U., Quadfasel, D., 1997. Summer monsoon response of the northern Somali Current, 1995. Geophys. Res. Lett. 24, 2565–2568.

Schott, F., Fischer, J., Stramma, L., 1998. Transports and pathways of the upper-layer circulation in the western tropical Atlantic. J. Phys. Oceanogr. 28, 1904–1928.

Schott, F.A., Lee, T.N., Zantopp, R., 1988. Variability of structure and transport of the Florida Current in the period range of days to seasonal. J. Phys. Oceanogr. 18, 1209–1230.

Schott, F.A., McCreary, J.P., 2001. The monsoon circulation of the Indian Ocean. Prog. Oceanogr. 51, 1–123.

Schott, F., Swallow, J.C., Fieux, M., 1990. The Somali Current at the equator: annual cycle of currents and transports in the upper 1000 m and connection to neighbouring latitudes. Deep Sea Res. 37, 1825–1848.

Schott, F.A., Zantopp, R.J., 1985. Florida Current: seasonal and interannual variability. Science 227, 308–311.

Schott, F., Zantopp, R., Stramma, L., Dengler, M., Fischer, J., Wibaux, M., 2004. Circulation and deep water export at the western exit of the subpolar North Atlantic. J. Phys. Oceanogr. 34, 817–843.

Schouten, M.W., de Ruijter, W.P.M., van Leeuwen, P.J., 2002. Upstream control of Agulhas Ring shedding. J. Geophys. Res. 107, 3109. http://dx.doi.org/10.1029/2001JC000804.

Schouten, M.W., de Ruijter, W.P.M., van Leeuwen, P.J., Lutjeharms, J.R.E., 2000. Translation, decay and splitting of Agulhas rings in the southeastern Atlantic Ocean. J. Geophys. Res. 105, 21,913–21,925.

Send, U., et al., 2010. A global boundary current circulation observing network. In: Hall, J., Harrison, D.E., Stammer, D. (Eds.), Proceedings of OceanObs'09: Sustained Ocean Observations and Information for Society, Venice, Italy, September 21–25, 2009, vol. 2. http://dx.doi.org/10.5270/OceanObs09.cwp.78. ESA Publication WPP-306.

Seo, H., Murtugudde, R., Jochum, M., Miller, A.J., 2008. Modeling of mesoscale coupled ocean–atmosphere interaction and its feedback to ocean in the western Arabian Sea. Ocean Model. 25, 120–131.

Shaw, P.-T., Rossby, H.T., 1984. Towards a Lagrangian description of the Gulf Stream. J. Phys. Oceanogr. 14, 528–540.

Shukla, J., 1975. Effect of Arabian Sea-surface temperature anomaly on Indian summer monsoon: a numerical experiment with the GFDL model. J. Atmos. Sci. 32, 503–511.

Sijp, W.P., England, M.H., 2009. Southern Hemisphere westerly wind control over the ocean's thermohaline circulation. J. Clim. 22, 1277–1286.

Smith, R.D., Maltrud, M.E., Bryan, F.O., Hecht, M.W., 2000. Numerical simulation of the North Atlantic Ocean at 1/10°. J. Phys. Oceanogr. 30, 1532–1561.

Song, T., Rossby, T., 1995. Lagrangian studies of fluid exchange between the Gulf Stream and surrounding patterns. J. Phys. Oceanogr. 25, 46–63.

Spadone, A., Provost, C., 2009. Variations in the Malvinas Current volume transport since 1992. J. Geophys. Res. 114, C02002. http://dx.doi.org/10.1029/2008JC004882.

Spall, M.A., 1996a. Dynamics of the Gulf Stream/Deep Western Boundary Current crossover, Part I: entrainment and recirculation. J. Phys. Oceanogr. 26, 2152–2168.

Spall, M.A., 1996b. Dynamics of the Gulf Stream/Deep Western Boundary Current crossover, Part II: low frequency internal oscillations. J. Phys. Oceanogr. 26, 2169–2182.

Speich, S., Blanke, B., de Vries, P., Drijfhout, S., Döös, K., Ganachaud, A., Marsh, R., 2002. Tasman leakage: a new route in the global ocean conveyor belt. Geophys. Res. Lett. 29, 1416. http://dx.doi.org/10.1029/2001GL014586.

Speich, S., Lutjeharms, J.R.E., Penven, P., Blanke, B., 2006. Role of bathymetry in Agulhas Current configuration and behavior. Geophys. Res. Lett. 33, L23611. http://dx.doi.org/10.1029/2006GL027157.

Stanton, B.R., 2001. Estimating the East Auckland Current transport from model winds and the Island Rule. N.Z. J. Mar. Freshwater Res. 35, 531–540.

Stanton, B., Roemmich, D., Kosro, M., 2001. A shallow zonal jet south of Fiji. J. Phys. Oceanogr. 31, 3127–3130.

Stanton, B., Sutton, P., 2003. Velocity measurements in the East Auckland Current north-east of North Cape, New Zealand. N.Z. J. Mar. Freshwater Res. 37, 195–204.

Stommel, H., 1948. The westward intensification of wind-driven ocean currents. Trans. Am. Geophys. Union 29, 202–206.

Stommel, H., 1965. The Gulf Stream: A Physical and Dynamical Description, second ed. University of California Press, Berkeley, 248 pp.

Stommel, H.M., Arons, A., 1960a. On the abyssal circulation of the World Ocean—I. Stationary planetary flow patterns on a sphere. Deep Sea Res. 6, 140–154.

Stommel, H.M., Arons, A., 1960b. On the abyssal circulation of the World Ocean—II. An idealized model of the circulation pattern and amplitude in oceanic basins. Deep Sea Res. 6, 217–233.

Stommel, H., Yoshida, K. (Eds.), 1972. Kuroshio: Its Physical Aspects. University of Tokyo Press, Tokyo, 517 pp.

Stramma, L., Fischer, J., Reppin, J., 1995. The North Brazil Undercurrent. Deep Sea Res. 42, 773–795.

Stramma, L., Ikeda, Y., Peterson, R.G., 1990. Geostrophic transport in the Brazil Current region north of 20°S. Deep Sea Res. 37, 1875–1886.

Stramma, L., Lutjeharms, J.R.E., 1997. The flow field of the subtropical gyre of the South Indian Ocean. J. Geophys. Res. 102, 5513–5530.

Sugimoto, S., Hanawa, K., 2010. Impact of Aleutian Low activity on the STMW formation in the Kuroshio recirculation gyre region. Geophys. Res. Lett. 37. http://dx.doi.org/10.1029/2009GL041795, L03606.

Sugimoto, T., Kimura, S., Miyaji, K., 1988. Meander of the Kuroshio front and current variability in the East China Sea. J. Oceanogr. Soc. Jpn. 44, 125–135.

Sutton, P., 2001. Detailed structure of the Subtropical Front over Chatham Rise, east of New Zealand. J. Geophys. Res. 106, 31,045–31,056.

Swallow, J.C., Bruce, J.G., 1966. Current measurements off the Somali coast during the southwest monsoon of 1964. Deep Sea Res. 13, 861–888.

Swallow, J.C., Molinari, R.L., Bruce, J.G., Brown, O.B., Evans, R.H., 1983. Development of near-surface flow pattern and water mass distribution in the Somali Basin in response to the southwest monsoon of 1979. J. Phys. Oceanogr. 13, 1398–1415.

Taguchi, B., Xie, S.-P., Schneider, N., Nonaka, M., Sasaki, H., Sasai, Y., 2007. Decadal variability of the Kuroshio Extension: observations and an eddy-resolving model hindcast. J. Clim. 20, 2357–2377.

Takahashi, T., et al., 2009. Climatological mean and decadal changes in surface ocean pCO_2, and net sea-air CO_2 flux over the global oceans. Deep Sea Res. Part II 56, 554–577. http://dx.doi.org/10.1016/j.dsr2.2008.12.009.

Talley, L.D., Pickard, G.L., Emery, W.J., Swift, J.H., 2011. Descriptive Physical Oceanography: An Introduction, sixth ed. Elsevier, London, 555 pp.

Tansley, C.E., Marshall, D.P., 2000. On the influence of bottom topography and the Deep Western Boundary Current on Gulf Stream separation. J. Mar. Res. 58, 297–325.

The Climode Group, 2009. The Climode Field Campaign: observing the cycle of convection and restratification over the Gulf Stream. Bull. Am. Meteorol. Soc. 90, 1337–1350.

Thompson, J.D., Schmitz, W.J., 1989. A limited-area model of the Gulf Stream: design, initial experiments, and model-data intercomparison. J. Phys. Oceanogr. 19, 791–814.

Tillburg, C.E., Hurlburt, H.E., O'Brien, J.J., Shriver, J.F., 2001. The dynamics of the East Australian Current System: the Tasman Front, the East Auckland Current, and the East Cape Current. J. Phys. Oceanogr. 31, 2917–2943.

Trenberth, K.E., Caron, J.M., 2001. Estimates of meridional atmosphere and ocean heat transports. J. Clim. 14, 3433–3443.

Tsugawa, M., Hasumi, H., 2010. Generation and growth mechanism of the Natal Pulse. J. Phys. Oceanogr. 40, 1597–1612.

Tsujino, H., Usui, N., Nakano, H., 2006. Dynamics of Kuroshio path variations in a high-resolution GCM. J. Geophys. Res. 111, C11001. http://dx.doi.org/10.1029/2005JC003118.

Uchida, H., Imawaki, S., 2008. Estimation of the sea level trend south of Japan by combining satellite altimeter data with in situ hydrographic data. J. Geophys. Res. 113, C09035. http://dx.doi.org/10.1029/2008JC004796.

Uddstrom, M.J., Oien, N.A., 1999. On the use of high-resolution satellite data to describe the spatial and temporal variability of sea surface temperatures in the New Zealand region. J. Geophys. Res. 104, 20,729–20,751.

Usui, N., Tsujino, H., Nakano, H., Fujii, Y., 2008. Formation process of the Kuroshio large meander in 2004. J. Geophys. Res. 113, C08047. http://dx.doi.org/10.1029/2007JC004675.

van Aken, H.M., van Veldhoven, A.K., Veth, C., de Ruijter, W.P.M., van Leeuwen, P.J., Drijfhout, S.S., Whittle, C.P., Rouault, M., 2003. Observations of a young Agulhas ring, Astrid, during MARE in March 2000. Deep Sea Res. Part II 50, 167–195.

van Leeuwen, P.J., de Ruijter, W.P.M., 2009. On the steadiness of separating meandering currents. J. Phys. Oceanogr. 39, 437–448.

van Sebille, E., Biastoch, A., van Leeuwen, P.J., de Ruijter, W.P.M., 2009. A weaker Agulhas Current leads to more Agulhas leakage. Geophys. Res. Lett. 36, L03601. http://dx.doi.org/10.1029/2008GL036614.

van Sebille, E., van Leeuwen, P.J., Biastoch, A., de Ruijter, W.P.M., 2010. Flux comparison of Eulerian and Lagrangian estimates of Agulhas leakage: a case study using a numerical model. Deep Sea Res. Part I 57, 319–327.

Vecchi, G.A., Sahng-Ping, X., Fischer, A.S., 2004. Ocean–atmosphere covariability in the western Arabian Sea. J. Clim. 17, 1213–1224.

Veronis, G., 1973. Model of world ocean circulation: I. Wind-driven, two-layer. J. Mar. Res. 31, 228–288.

Veronis, G., 1978. Model of world ocean circulation: III. Thermally and wind driven. J. Mar. Res. 36, 1–44.

Vivier, F., Kelly, K.A., Thompson, L., 2002. Heat budget in the Kuroshio Extension region, 1993–1999. J. Phys. Oceanogr. 32, 3436–3454.

Warren, B., Stommel, H., Swallow, J.C., 1966. Water masses and patterns of flow in the Somali Basin during the southwest monsoon of 1964. Deep Sea Res. 13, 825–860.

Webb, D.J., 2000. Evidence for shallow zonal jets in the South Equatorial Current region of the southwest Pacific. J. Phys. Oceanogr. 30, 706–720.

Weijer, W., de Ruijter, W.P.M., Sterl, A., Drijfhout, S.S., 2002. Response of the Atlantic overturning circulation to South Atlantic sources of buoyancy. Glob. Planet. Change 34, 293–311.

Wijffels, S.E., Hall, M.M., Joyce, T., Torres, D.J., Hacker, P., Firing, E., 1998. Multiple deep gyres of the western North Pacific: a WOCE section along 149°E. J. Geophys. Res. 103, 12,985–13,009.

Wirth, A., Willebrand, J., Schott, F., 2002. Variability of the Great Whirl from observations and models. Deep Sea Res. Part II 49, 1279–1295.

Witter, D.L., Gordon, A.L., 1999. Interannual variability of South Atlantic circulation from four years of TOPEX/POSEIDON satellite altimeter observations. J. Geophys. Res. 104, 20, 927–20, 948.

Woityra, W., Rossby, T., 2008. Current broadening as a mechanism for anticyclogenesis at the Northwest Corner of the North Atlantic Current. Geophys. Res. Lett. 35, L05609. http://dx.doi.org/10.1029/2007GL033063.

Worthington, L.V., 1976. On the North Atlantic circulation. The Johns Hopkins Oceanographic Studies, vol. 6. Johns Hopkins University Press, Baltimore, MD, 110 pp.

Wyrtki, K., 1971. Oceanographic atlas of the International Indian Ocean Expedition. National Science Foundation, Washington, DC, 531 pp.

Yasuda, I., Okuda, O., Hirai, M., 1992. Evolution of a Kuroshio warm-core ring—variability of the hydrographic structure. Deep Sea Res. 39, 131–161.

Yu, L., Weller, R.A., 2007. Objectively analyzed air-sea heat fluxes for the global ice-free oceans (1981–2005). Bull. Am. Meteorol. Soc. 88, 527–539.

Zemba, J.C., McCartney, M.S., 1988. Transport of the Brazil Current: it's bigger than we thought. Eos 69, 1237, (Abstract).

Zhang, D., Lee, T.N., Johns, W.E., Liu, C.-T., Zantopp, R., 2001a. The Kuroshio east of Taiwan: modes of variability and relationship to interior ocean mesoscale eddies. J. Phys. Oceanogr. 31, 1054–1074.

Zhang, H.-M., Prater, M.D., Rossby, T., 2001b. Isopycnal Lagrangian statistics from the North Atlantic Current RAFOS float observations. J. Geophys. Res. 106, 13,817–13,836.

Zhang, R., Vallis, G.K., 2007. The role of bottom vortex stretching on the path of the North Atlantic western boundary current and on the northern recirculation gyre. J. Phys. Oceanogr. 37, 2053–2080.

Zinke, J., Dullo, W.-C., Heiss, G.A., Eisenhauer, A., 2004. ENSO and Indian Ocean subtropical dipole variability is recorded in a coral record off southwest Madagascar for the period 1659 to 1995. Earth Planet. Sci. Lett. 228, 177–194.

Chapter 14

Currents and Processes along the Eastern Boundaries

P. Ted Strub*, Vincent Combes*, Frank A. Shillington[†] and Oscar Pizarro[‡]

*College of Earth, Ocean and Atmospheric Sciences, Oregon State University, Corvallis, Oregon, USA

[†]Department of Oceanography and Nansen-Tutu Centre, University of Cape Town, Rondebosch, South Africa

[‡]Department of Geophysics and Center for Oceanographic Research in the Eastern South Pacific (COPAS), University of Concepcion, Chile

Chapter Outline

1. INTRODUCTION AND GENERAL BACKGROUND

The processes that shape the ocean circulation along the eastern margins of the major ocean basins include both external forcing (at the surface and lateral boundaries) and internal dynamical responses within the fluid volume. For the current systems along the eastern boundaries of the major oceans, the questions are: "What physical processes are at work? How do they differ from each other in the present ocean? How do they vary on seasonal and longer timescales?" An increasingly important question is, "How will they be affected by climate change," a question which we cannot answer very well at this time. Accordingly, this chapter mostly describes the existing processes, as we understand them now.

Interest in eastern boundary currents (EBCs) is often motivated by the economically important fisheries of the

Ocean Circulation and Climate, Vol. 103. http://dx.doi.org/10.1016/B978-0-12-391851-2.00014-3

eastern boundary upwelling systems (Mackas et al., 2006; Chavez and Messie, 2009; Checkley et al., 2009; Freon et al., 2009), especially the extremely large catches of small pelagic fish off Peru and Chile (Chavez et al., 2003; Montecino et al., 2006; Bakun and Weeks, 2008; Bakun et al., 2010). Given their economic importance, it is natural to ask how the ecosystems of the EBCs might change due to climate change and global warming. The coupled (atmosphere–ocean–land) climate models that are being used to estimate future changes in regional climate are particularly poor at simulating the present conditions in the ocean's eastern boundaries (Gent et al., 2011). Errors in the modeled surface temperatures are of order 5 °C in the midlatitude EBCs, which implies errors in the processes involved in upwelling, vertical mixing and horizontal advection. The same processes that affect temperature also affect the distribution of nutrients and other aspects of habitat, raising concern that even good models of marine ecosystems and fisheries will give poor results if changes in the physical processes are poorly simulated. Increases in the horizontal resolution of the atmospheric components of the climate models provide some improvements (Gent et al., 2010; see also Chapter 23) and there are parallel efforts underway to improve the resolution of the ocean models. Improvements in these models need to be evaluated in terms of the processes at work in the eastern boundary oceanic systems, the focus of this chapter.

In Volumes 11 and 14 of *The Sea* (Robinson and Brink, 1998b, 2006) are found summaries of the physical characteristics of individual EBCs (Volume 11) and their biophysical interactions (Volume 14). Freon et al. (2009, and other papers in the same volume) provide additional descriptions of the midlatitude eastern boundary upwelling systems (EBUS, also called eastern boundary upwelling ecosystems, EBUEs, or simply midlatitude EBCs), which are outlined in white in Figure 14.1. Despite their characterization as "midlatitude" currents, the systems extend into the tropics in the NE Atlantic and SE Pacific. Syntheses of EBC physical characteristics are found in Hill et al. (1998), while Mackas et al. (2006), Freon et al. (2009) and Checkley et al. (2009) summarize the ecosystem attributes and fisheries of the same systems.

Rather than organizing our discussion by describing each specific EBC, in the present review we focus on the common physical processes that affect all EBCs, dividing the systems into low-, mid and high-latitude domains. In Figure 14.1, the low-latitude regions occupy the eastern boundaries between the outlined midlatitude systems, while the high-latitude regions are those poleward of the outlined EBUS. The systems also group naturally into Northern versus Southern Hemisphere Systems, due to systematic differences in the wind fields and the coastline geometry. Descriptions of the common processes are given for examples from each of the systems, although there are inevitably more examples from the systems with which the authors are most familiar. We apologize for this bias.

We begin our review of EBC circulation in the low latitudes. The tropical eastern boundaries are of interest both for their own importance and because they directly affect the boundary circulation at midlatitudes (Section 3) and

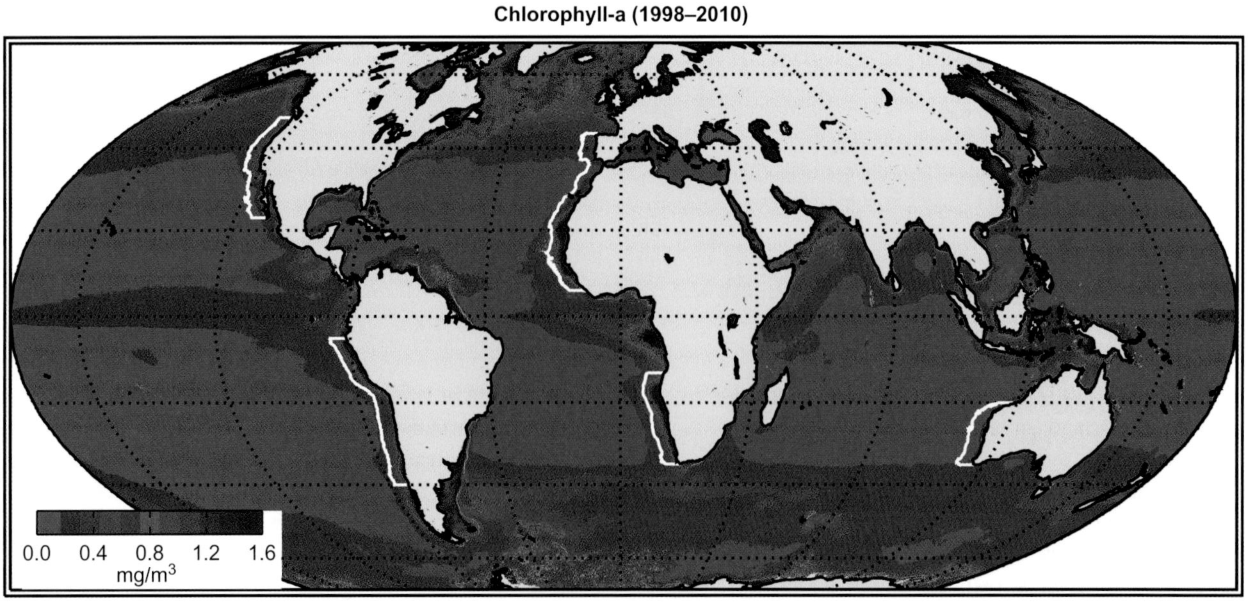

FIGURE 14.1 Mean surface pigment concentrations from the SeaWiFS ocean color sensor, averaged over its full mission (September 1997–December 2010), with the subtropical ("midlatitude") eastern boundary current systems (EBCs) outlined in white.

even the highest latitudes (Section 4). Within each latitude band, we describe the processes that affect the mean circulation and changes in that circulation on intra-seasonal, seasonal and longer timescales. These processes primarily involve responses to forcing by surface winds and fluxes of heat and freshwater, along with inflows along the boundaries of the regions. EBC regions are strongly affected by climate variability on all timescales. At the end of this chapter we provide a brief discussion of the ways in which EBC processes may interact with climatic variability and change, but leave a consideration of global climate change to other chapters in this book (Part V).

1.1. Dominant Processes

The processes of importance in EBCs are the same as in most coastal regions, described briefly below. Chapters in Volumes 10 and 13 of *The Sea* (Robinson and Brink, 1998a, 2005; Brink, 2005) provide more detailed examinations of coastal ocean processes.

1.1.1. Surface Forcing on Intraseasonal, Seasonal and Interannual Timescales

Surface wind stress: In most regions, strong alongshore and/or cross-shore winds are found, associated with synoptic weather systems, seasonal expansion, and contraction of the midlatitude high- and low-pressure systems, interannual and decadal-scale atmospheric climate modes. The coastal winds are also affected by the geometry of the coastline (capes and bays), the orography of the land next to the coast, and its interaction with the structure of the atmospheric marine boundary layer (Beardsley et al., 1987). In particular, coastal mountains that are high enough to intersect the top of the marine boundary layer serve as a "coastline" for the atmosphere, allowing perturbations in the height of the marine boundary layer to propagate poleward as coastal trapped waves (CTWs) (Reason and Steyn, 1990; Rogers et al., 1995, 1998; Garreaud and Rutllant, 2003).

Sea breeze/monsoonal winds: Since the surface of the land heats and cools more rapidly than the surface of the ocean, convective cells are established over land on a wide range of timescales, resulting in an onshore component of the wind when the land is warmer. Examples include diurnal sea breeze effects and seasonal monsoons (Bielli et al., 2002). In addition, thermal low-pressure systems over land in summer combine with the high-pressure systems over the eastern ocean basins to strengthen the quasi-geostrophic equatorward surface winds over the coastal ocean (Bakun and Nelson, 1991).

Wind stress curl (WSC): Broad regions of WSC result from large-scale wind patterns, while mesoscale features of WSC accompany strong wind jets. Risien and Chelton (2008) present global and regional maps of wind stress, wind stress divergence, and curl (annual, January and July means) from eight years (1999–2007) of QuikSCAT scatterometer data that provide a perspective for the ocean's eastern boundaries. Our regional wind fields are similar to those of Risien and Chelton but use the full 10-year data set.

Surface heat flux: Changes in surface heating and cooling result from changes in latitude, cloud cover, wind speed, humidity, and SST. With their low SST values, EBCs tend to be regions where heat enters the ocean, even though cloud cover may reduce the input of heat due to solar radiation. Low SST values (due to upwelling) reduce losses of heat due to infrared emission, latent, and sensible heat fluxes, resulting in the net downward heat flux into the ocean from the atmosphere (Da Silva et al., 1994; Roske, 2006).

Surface and coastal fresh water (buoyancy) flux: Direct precipitation follows global patterns: high values in the tropics, especially around the Intertropical Convergence Zones (ITCZ), low values in the subtropics, then increased precipitation with latitude. Large rivers have outfalls along some EBCs, mostly at low or midlatitude (the Columbia River off the western U.S., the Bio Bio River off Chile, the Congo and Niger Rivers off tropical West Africa, etc.). The higher latitude coasts (latitudes of 40–65°), especially those with high mountain ranges, tend to have large rates of precipitation and multiple short rivers that combine to act as line sources of fresh water, as described along the coast of Alaska (Royer, 1982; Stabeno et al., 2004; Weingartner et al., 2005) and the Iberian Peninsula (Relvas et al., 2007).

1.1.2. Incoming/Outgoing Currents and Geophysical Waves

Surface currents: These include low-latitude currents such as the North Equatorial Countercurrent (NECC) and midlatitude currents such as the West Wind Drift currents in each of the basins. Alongshore winds that drive upwelling also create alongshore sea surface height (SSH) and pressure gradients that drive onshore geostrophic currents at the western boundaries of the EBCs; although the magnitude of the currents are small, the transports can contribute significantly to poleward currents in the EBCs. This is especially true in the Leeuwin Current (Godfrey and Ridgway, 1985) and has been noted in the Canary Current (Aristegui et al., 2009).

Subsurface currents: Poleward undercurrents (PUCs) are commonly found in midlatitude EBC systems (Fonseca, 1989; Neshyba et al., 1989; Hill et al., 1998), although their dynamics are still under investigation. Subsurface onshore flow is also a necessary component of upwelling, sometimes coming from the PUC. The properties of the upwelling "source water" are important in determining the biochemical response to upwelling (phytoplankton blooms and subsequent hypoxic decay).

Geophysical waves: Equator-trapped Kelvin and Yanai waves propagate to the east, where some of their energy reflects as Rossby waves (Philander, 2000) and some excites CTWs. CTWs propagate poleward (Allen, 1975; Smith, 1978; Brink, 1991) with a large range of frequencies, horizontal, and vertical structures. As they propagate poleward, some of their energy moves offshore to the west in the form of off-equatorial Rossby waves (Moore and Philander, 1977; Philander, 1978; McCreary and Chao, 1985; Clarke and Van Gorder, 1994; Pizarro et al., 2001) and eddies (Chelton et al., 2011a).

1.1.3. Internal Responses of the Coastal Ocean and Larger-Scale Boundary Currents

Coastal upwelling: The process of coastal upwelling is well understood and described (Smith, 1968; Allen and Smith, 1981; Richards, 1981; Huyer, 1983, 1990; Brink, 1991; Lentz, 1992; Lentz and Fewings, 2012). Equatorward, alongshore winds cause offshore surface Ekman transport, subsurface onshore flow and upwelling of water that is rich in nutrients, low in temperature, oxygen, and pH. The onshore flow may occur at mid-depth over the shelf or in a bottom boundary layer (Smith, 1981; Lentz and Trowbridge, 1994). Even if the vertical structure of the currents differs from the theoretical Ekman spiral, the surface Ekman transport is close to the theoretical value (Lentz, 1992), allowing estimates of the volume of upwelled water from measurements of alongshore winds. The nutrient concentrations in the upwelled water depend on whether the water comes from above or below the nutricline–pycnocline depth, which can be modulated by equatorial Kelvin Waves and CTWs from differing origins and with different timescales (Barber and Chavez, 1983; Chavez, 2005). Changes in the properties of the source water for the upwelling can amplify some of the harmful effects of upwelling, such as hypoxia and acidity (Chan et al., 2008). Increased attention to CO_2 and carbon budgets, oxygen, and acidity has mirrored the increased concern for the climate-related issues in the global ocean (Hauri et al., 2009; Evans et al., 2011; Hales et al., 2012).

Ekman pumping (upwelling) due to WSC: Because integrated surface Ekman transports are perpendicular to the surface wind stress, the curl of the surface wind stress creates divergence of the surface transports and upwelling (vertical velocity) (Bakun and Nelson, 1991). If there were no horizontal advection or propagation of SSH features, sustained upwelling in a region would bring denser water underneath it, creating lower SSH beneath the areas with positive WSC (in the Northern Hemisphere). Horizontal advection and propagation distort this simple relationship between WSC and SSH in many regions, but it is still useful as a diagnostic relationship.

Alongshore pressure gradients: Persistent alongshore winds move water in the downwind direction along the coast, creating alongshore height and pressure gradients that oppose the winds (Hickey and Pola, 1983). In a steady state, these may balance the alongshore wind stress and also drive small onshore–offshore geostrophic transports. Relaxations in the winds allow the pressure gradients to briefly accelerate the water toward the lower SSH and eliminate the pressure gradients. This occurs over the period of time needed for the first several modes of CTWs to propagate alongshore through the system (several hours to a few days). Alongshore pressure gradients can also be created by spatial gradients in density and steric height, as happens in the Leeuwin Current, off western Australia (Godfrey and Ridgway, 1985; Thompson, 1987; Ridgway and Condie, 2004).

Buoyancy currents: Where freshwater flows from land into the coastal ocean, it creates a rise in SSH and localized pressure gradients around the fresher water. The general response in the Earth's rotating frame of reference is to create a current that carries the plume along the coast to the right (left) of the outflow (facing seaward) in the Northern (Southern) Hemisphere (Garvine, 1982, 1999; Hill, 1998; Peliz et al., 2002, 2003).

Topographic steering: Both coastal geometry (capes and bays) and bathymetric changes cause perturbations in alongshore and onshore–offshore circulation patterns (Trowbridge et al., 1998; Castelao and Barth, 2007). Capes and bays also interact with winds in the atmospheric boundary layer (ABL) to cause hydraulic jumps (decelerations of the winds and deepening of the ABL) upstream of capes and expansion fans (acceleration of the winds and thinning of the ABL as winds expand into the wider region next to the coast) downstream of capes (Beardsley et al., 1987; Winant et al., 1988; Haack et al., 2008). These changes in the wind stress create patterns of WSC and consequent convergences and divergences of the surface flow, even without bottom topography. Bottom topographic features, in the form of subsurface canyons and banks, steer the flow due to the tendency to conserve potential vorticity (Hickey, 1989). Horizontal currents overshoot changes in topography that are too abrupt (smaller than a Rossby Radius), creating regions of flow separation from the coast. All of these perturbations in the flow field can propagate as CTW.

Instabilities, eddies and filaments: In the midlatitude EBUS, the presence of a surface equatorward flow over a PUC, with slanting ispycnals (horizontal density gradients) can create baroclinically unstable conditions (Allen et al., 1991; Barth et al., 2000). Thus, the seasonal jet that develops over the upwelling front strengthens, moves offshore, and develops unstable meanders and eddies (Strub et al., 1991; Kelly et al., 1998; Strub and James, 2000; Aristegui et al., 2009). These instabilities occur with no need for variability in the coastal geometry or bottom bathymetry, but those features can act to provide preferred locations for the instabilities to form (Narimousa and Maxworthy, 1989). The most rapidly growing instabilities have alongshore spatial scales of order several hundred kilometers (Allen et al., 1991; Pierce et al., 1991), which are the observed scales for mesoscale meanders and the largest eddies in the midlatitude EBUS. The larger eddies move offshore and may serve as the main mechanism for mixing coastal water into the deep offshore region (Crawford, 2005; Chelton et al., 2011a). Subsurface eddies are also generated by instabilities of the PUC, leading to subsurface, anticyclonic "inter-thermocline eddies," with water mass characteristics of the undercurrent (Huyer et al., 1998; Hormazabal et al., 2004).

Geophysical waves: Poleward propagating CTWs and westward propagating Rossby waves are internal responses to forcing within EBCs, as well as arriving from outside of their boundaries (see references above).

1.2. Data and Model Fields

To illustrate the surface forcing and ocean response, we present fields from satellites and a global numerical ocean circulation model: Wind and WSC come from 10 years of satellite scatterometer (QuikSCAT) data. AVISO (Archiving, Validation, and Interpretation of Satellite Oceanography)-gridded sea surface height anomalies (SSHAs) are available for 19 years of data from multiple altimeters. The analyses we show below use data from the same 10 years as used for the scatterometer wind data. Although we have recalculated the results with the full 19 years of data and from 17 years that exclude the 1997–1998 El Niño years, the results are virtually the same. We use the same periods for winds and SSHA for consistency. The mean SSH at each point is removed to eliminate the poorly known marine geoid. Gridded (0.1°) surface velocity fields come from the Japanese Ocean–GCM for the Earth Simulator (OFES) model (Masumoto et al., 2004; Sasaki et al., 2004, 2006; Ohfuchi et al., 2007), driven by 9 years of QuikSCAT winds (2000–2008). "Textbook" schematic surface circulation patterns (primarily from Talley et al., 2011) are overlaid on the model current vectors. These fields require two warnings. First, the lengths of these time series are too short to avoid biases by decadal-scale climate variability. However, their regular sampling in space and time allows them to represent their particular period of time without biases and provides a consistent frame of reference for our discussion. Second, the model fields do not always reproduce the expected schematic circulation patterns, due to biases in the models and inaccurate model forcing. Likewise, the "textbook" schematic representations of currents may include subjective biases and differ from those drawn by other authors. These differences serve as a reminder that details of the ocean's circulation and its variability are still uncertain in many regions.

2. LOW-LATITUDE EBCs

The tropical ocean current systems along the western boundary of our domains of interest in Figure 14.2 are relatively simple, resembling the zonal currents in the middle of the tropical oceans (reviewed in detail in Chapter 15). In contrast, circulation patterns in the far eastern tropics are more complex, due to the seasonally migrating Trade Winds and Intertropical Convergence Zone (ITCZ), which is the convergence of Southeast and Northeast Trade Winds (with low wind speeds, strong convection and precipitation). The major low-latitude external processes include surface forcing by wind stress, WSC and fresh water. Especially important are equatorial signals that arrive from the west and move poleward or reflect back to the west.

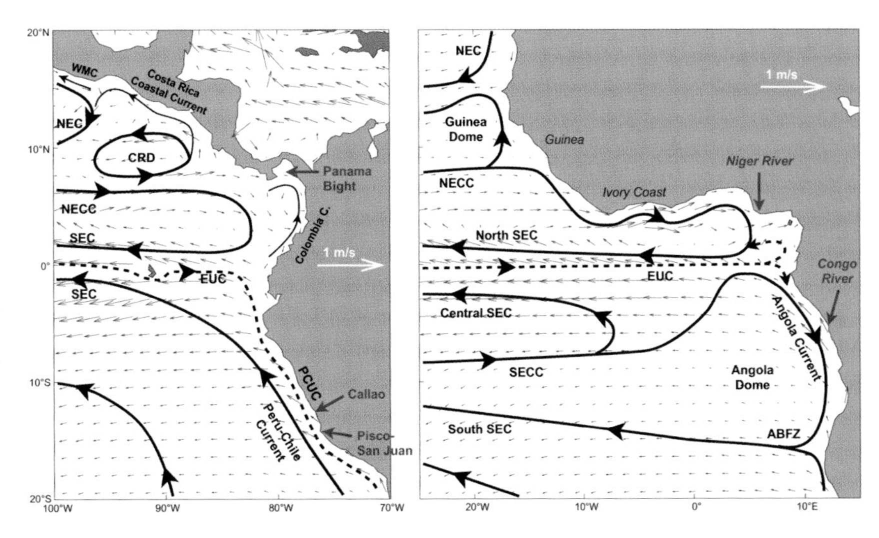

FIGURE 14.2 Schematics of the current systems in the eastern tropical Pacific and Atlantic Oceans, based on Talley et al. (2011), overlain on the long-term mean surface currents from the Ocean–GCM For the Earth Simulator (OFES) model (gridded vectors, Ohfuchi et al., 2007). Schematic surface currents are depicted by solid lines, subsurface currents by dashed lines. NEC: North Equatorial Current; CRD: Costa Rica Dome; NECC: North Equatorial Counter Current; SEC: South Equatorial Current; EUC: Equatorial Undercurrent; ABFZ: Angola-Benguela Frontal Zone.

Each basin has a different mix of these external processes. Buoyancy forcing is important in some regions, due to major rivers and migrating bands of high precipitation associated with the ITCZ. For example, when the ITCZ moves south in boreal winter in the eastern Pacific, heavy precipitation and runoff create a line source of fresh water along the coasts of Colombia and Ecuador (Enfield, 1976; Cucalon, 1987). In the eastern tropical Atlantic, fresh water inputs from the Niger and Congo Rivers contribute to the strong and shallow (10–20 m deep) pycnoclines found in the area (Ajao and Houghton, 1998). Internal responses of the ocean in these regions include upwelling (coastal and open-ocean Ekman pumping), excitation of geophysical waves (CTW and Rossby), surface and subsurface currents, and the generation of mesoscale eddies. Thorough reviews of the circulation in the eastern tropical Pacific include Badon-Dangon (1998), Fiedler (2002), Herguera (2006), Kessler (2006), and Fiedler and Talley (2006). For the eastern tropical Atlantic see Picaut (1983, 1985), Ajao and Houghton (1998), Stramma and Schott (1999), Lumpkin and Garzoli (2005, 2011), Roy (2006), and Jouanno et al. (2011).

2.1. The Mean Circulation in the Eastern Tropics

In Figure 14.2, the mean flow in the eastern tropics is simpler in the southeast Pacific, where the westward South Equatorial Current (SEC) is found both north and south of the equator. The southern branch of the SEC connects back to the equatorward Peru–Chile Current (also called the Humboldt Current). Currents in the southeast Atlantic Ocean are more complex, with the addition of the eastward South Equatorial Countercurrent (SECC, between 5°S and 10°S) that flows into the poleward Angola Current next to the African coast south of the equator. The Angola Current continues south until it converges with the Benguela Current at the Angola–Benguela Frontal Zone (ABFZ), near 17°S, where it turns offshore along the southern flank of the cyclonic Angola Dome (which is centered between 5°E and 10°E, at ~10°S). The Angola Current and Dome do not appear in the global OFES model field, but this is thought to be a problem with the global model, since these features are evident in regional models and observational data, most clearly in the analysis of 14 years of surface drifter trajectories by Lumpkin and Garzoli (2005). In the Pacific, a similar SECC and a poleward Peru–Chile Countercurrent are sometimes described offshore of the coastal branch of the equatorward Peru–Chile Current (Strub et al., 1995), although their existence as permanent features is a matter of debate (they are not included in the Figure 14.2 schematic). North of the SECC in the Atlantic, branches of the SEC flow to the west both north and south of the Equator, as they do in the Pacific. These are sometimes named the North and Central SEC, to distinguish them from the South SEC, which flows to the west south of the SECC.

In the Northern Hemisphere, the tropical current systems in the two basins are more similar to each other. The NECCs are strong features of the circulation in both basins. These flow to the east and separate the SECs from the westward North Equatorial Currents (NECs). In the Pacific, one branch of the NECC turns clockwise toward the south and contributes to the westward SEC north of the Equator. Farther east, in the Panama Bight, the Colombia Current is a strong, shallow northeastward flow along the northern coast of South America, strongest in boreal summer (Stevenson, 1970; Cornejo–Rodriguez and Enfield, 1987). Kessler (2006) does not name the Colombia Current but confirms the small cyclonic gyre in the Panama Bight. Another branch of the NECC flows north around the cyclonic Costa Rica Dome and joins the poleward Costa Rica Coastal Current (Kessler, 2006). Kessler also includes a separate poleward flow along the coast of Mexico between 16°N and 20°N, called the West Mexican Current. Lavin et al. (2006) observed the West Mexican Current during two summers to have surface speeds of 0.15–0.35 ms^{-1} and volume transports of 1.5–5.4 Sv to the north (a Sverdrup, Sv, is $10^6\ m^3\ s^{-1}$). Water properties found by Lavin et al. are similar to California Current water to the north, rather than Costa Rica Coastal Current water, supporting Kessler's finding that the West Mexican Current is not continuous with the Costa Rica Coastal Current.

In the Atlantic, the NECC also splits into two branches: one branch of the NECC flows into the Gulf of Guinea along its northern coast (the Ivory Coast), where it is called the Guinea Current, before turning south to join the North SEC. The other branch of the NECC flows north and contributes to the cyclonic circulation around the Guinea Dome, as documented in both observations and a regional model by Siedler et al. (1992). The Costa Rica and Guinea Domes are represented only weakly in the global model current fields in Figure 14.2.

The basin-scale circulation pattern derived by Munk (1950) from idealized winds representing the North Pacific included a narrow zonal band of cyclonic flow with a NECC along 5°N and poleward flow along the eastern boundary between 5°N and 15°N. Consistent with this picture, the mean coastal currents in the eastern tropics in Figure 14.2 are often poleward: In the North Pacific between the Equator and 20°N; in the North Atlantic, inshore of the Guinea Dome (10–15°N); and in the South Atlantic inshore of the Angola Dome (0–15°S). One exception in the North Atlantic is the eastward flow of the Guinea Current along the zonal Ivory Coast and southward along the coast to the Equator. This pattern, however, is not that different from the path of the Pacific NECC and its clockwise turn into the SEC west of the Panama Bight. The greatest exception to the dominance of poleward surface flow is in the South Pacific along the coast of Peru (5–18°S), where upwelling and equatorward currents occur all year. This picture changes when the subsurface currents are considered.

Looking beneath the surface, PUCs are often found in the midlatitude EBCs (Neshyba et al., 1989). In the SE tropical Pacific, a PUC has been well documented over the upper slope and shelf between 100 and 300 m depth next to the coast off Peru at 10°S (Huyer et al., 1987). Chemical tracers link this PUC back to the Equatorial Undercurrent (Tsuchiya, 1985; Lukas, 1986) and as far as 40°S along the Chilean coast (Silva and Neshyba, 1979). In the eastern tropical Atlantic, PUCs have been reported under a very shallow (15 m) equatorward flow off Gabon at 3–4°S and beneath the Guinea Current (flowing to the west along the Guinea Coast), with increasing salinity toward the east in the undercurrent along the Guinea Coast (Ajao and Houghton, 1998). Verstraete (1992) also identifies the water properties in the undercurrents off Gabon and the Guinea Coast as South Atlantic Central Water, typical of the Equatorial Undercurrent (EUC). These observations support the continuation of the Atlantic EUC into PUCs in the low-latitude eastern boundaries. There is little evidence of a continuous PUC in the NE tropical Pacific next to Central America, due to the more complex nature of the circulation caused by the presence of the NECC, the Costa Rica Dome and the eddies created by the strong wind jets through the mountain gaps (see below).

2.2. Seasonal Changes in Surface Wind Forcing, Surface Currents, and SSH Anomalies

Given that the boundary currents in low-latitude EBCs are affected by the NECC and SECs, what controls the tropical current systems? As introduced under "Ekman Pumping" (above), positive (negative) values of surface WSC in the Northern (Southern) Hemisphere create divergences of the surface Ekman transports, requiring upwelling and subsurface convergences of denser water, which create lower values of SSH. The slopes of the SSH then drive geostrophic currents. A classic example of this occurs in the central equatorial Pacific (Figure 14.3), where a zonal band of lower dynamic height is found between ~5°N and 15°N along north–south transects in the mid-Pacific (150–160°W, Wyrtki and Kilonsky, 1984). This is also the latitude occupied by a zonal band of positive WSC as determined by QuikSCAT data (Risien and Chelton, 2008). Along the southern half of the dynamic height "valley" (5–10°N), where the height slopes downward toward the pole in Figure 14.3, the shuttle data recorded the eastward NECC in the upper 150 m, a simple case of geostrophic motion. Elsewhere, the dynamic height surfaces slope down toward the Equator and currents in the upper 200 m are westward (the NEC between 9°N and 18°N and the SEC south of 3°N). The exceptions are at depth: the EUC within a couple of degrees of the equator in the upper 200 m and the off-equatorial Northern and Southern Subsurface Countercurrents (or Tsuchiya jets)

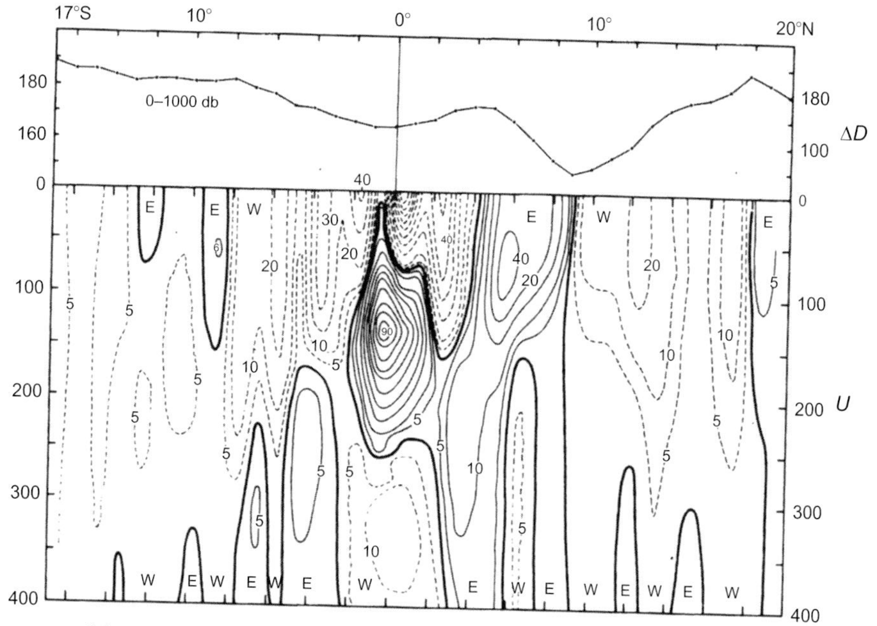

FIGURE 14.3 One-year mean of dynamic topography (top) and eastward current velocity (bottom, contours of velocity in cm s^{-1}) from the Tahiti to Hawaii shuttle transects. *From Wyrtki and Kilonsky (1984).*

below 200 m. There is also a hint of eastward SECC flow near 10°S, at this central Pacific location. Except for this SECC, the zonal currents in Figure 14.3 are similar to those expected at the western boundary (100°W) of the Pacific region in Figure 14.2.

Climatological fields of winter and summer surface wind stress and WSC in the eastern equatorial Pacific and Atlantic Oceans are presented in Figure 14.4, for comparison to the schematic circulation patterns in Figure 14.2 and altimeter SSH anomaly (SSHA) fields in Figure 14.5. The wind fields reveal that a major cause for the differences between the Northern and Southern Hemispheres in both basins is the location of the ITCZ. The decrease in the wind speed near the ITCZ creates positive (negative) WSC to the north (south) of the ITCZ. Positive curl is also caused by the curvature of the wind stress streamlines, sometimes spreading the positive WSC south of the ITCZ. Nevertheless, Figure 14.4 shows a band of strongly positive WSC that stays north of the equator, while migrating toward the north and south during boreal summer and winter, respectively. For example, along the western boundary of the Pacific regions in Figure 14.4 (120°W), the ITCZ and center of the positive WSC band is located near 7°N in boreal winter, moving to 13°N in boreal summer, with an average position similar to the region of low dynamic height farther west in the central Pacific in Figure 14.3. Given the lower SSHA values expected under the bands of positive WSC, the eastward NECC and its seasonally shifting location are direct consequences of the winds associated with the ITCZ and its movement. In the western tropical Atlantic, the ITCZ also stays north of the equator, moving from near the equator in boreal winter to around 7°N in boreal summer.

Because the location of the ITCZ is north of the equator, winds approaching it from the south are deflected to the right after crossing the equator, with curvature that contributes to the negative WSC south of the ITCZ. This is most evident in boreal summer (Figure 14.4, bottom), when the ITCZ is farthest north. One consequence of this includes the intensification of the Colombia Current (noted above) due to downwelling favorable winds in boreal summer; another is the strengthening of the upwelling-favorable component to the monsoonal winds along the northeast coast of the Gulf of Guinea. Since winds are weaker near the ITCZ, another general result of its southward displacement in boreal winter is a decrease in the strength of the Southeast Trade Winds along the Equator, reducing equatorial upwelling and the westward wind stress that

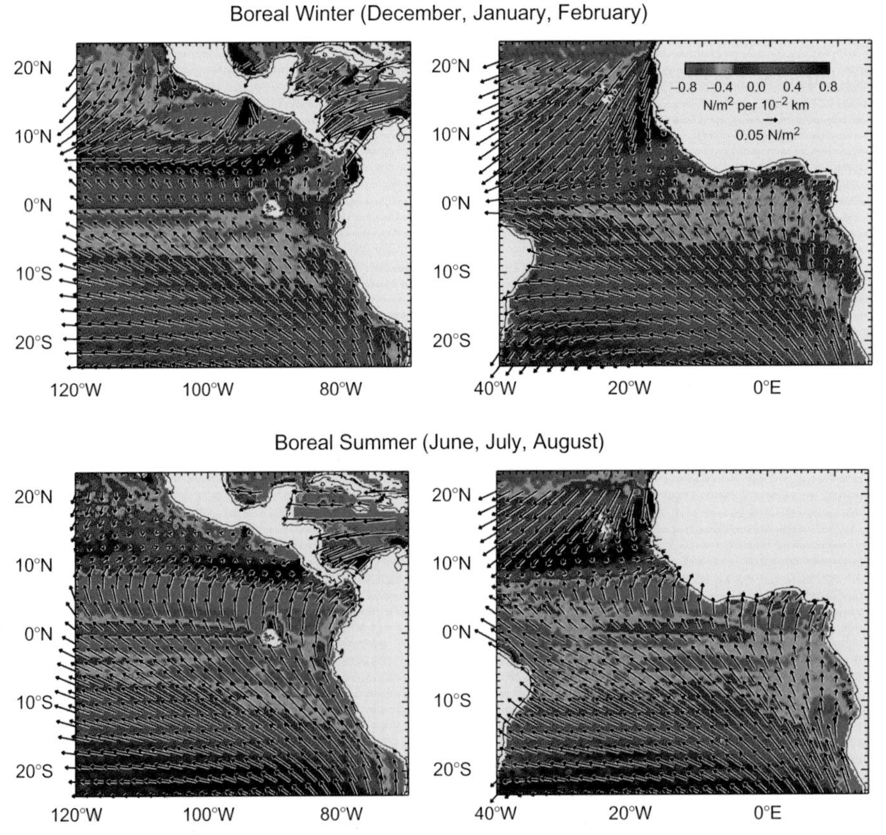

FIGURE 14.4 Surface wind stress (vectors) over wind stress curl (colors) for boreal winter (top) and boreal summer (bottom), from 10 years of QuikSCAT data. Left: Pacific; right: Atlantic.

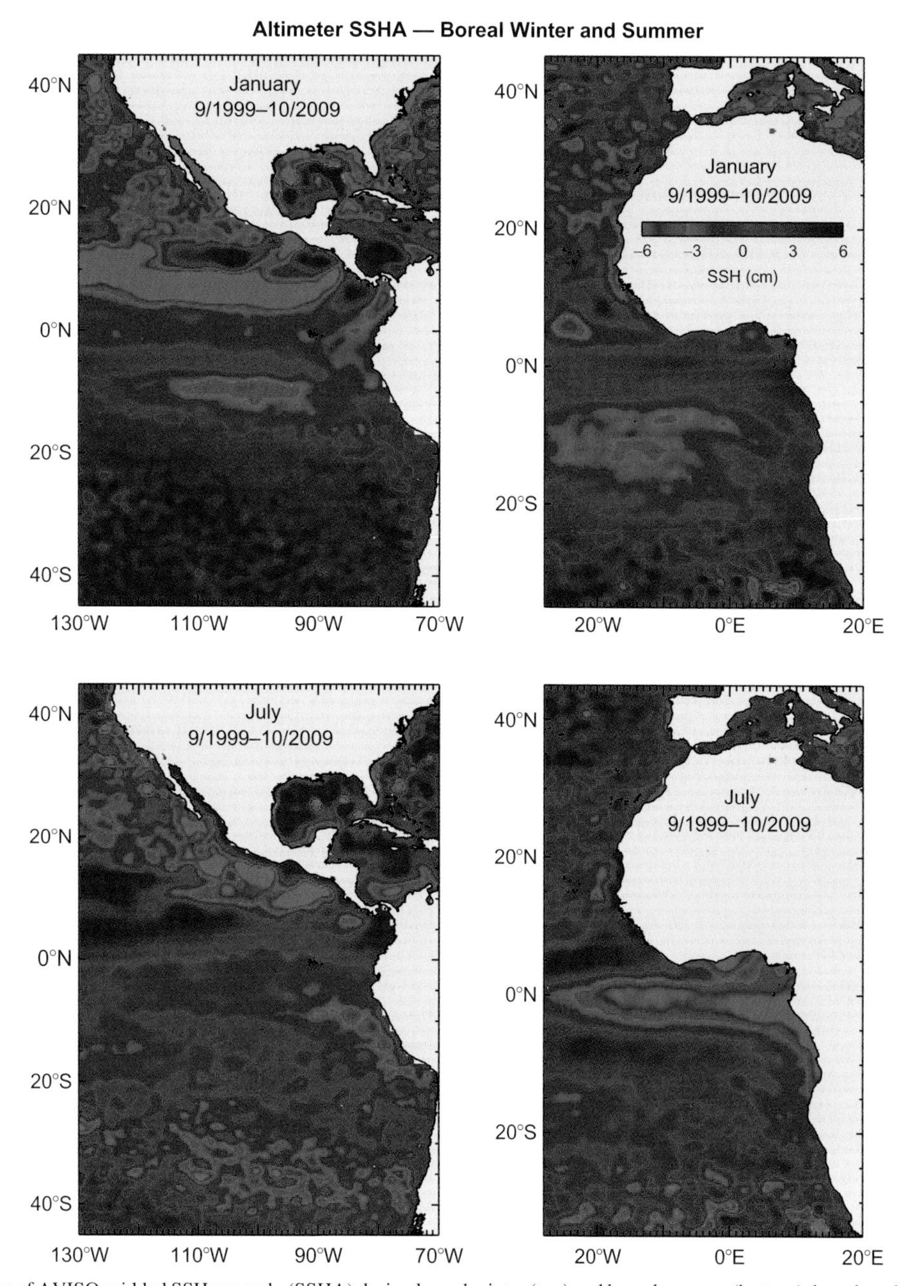

FIGURE 14.5 Maps of AVISO gridded SSH anomaly (SSHA) during boreal winter (top) and boreal summer (bottom), based on the same 10-year period as the scatterometer data in Figure 14.4. Left: Pacific; right: Atlantic.

maintains the lower sea levels in the eastern equatorial regions of both the Pacific and Atlantic.

The relation between surface wind forcing and sea level can be qualitatively appraised by comparing the wind stress and curl in Figure 14.4 with the summer and winter fields of altimeter SSHA in Figure 14.5. In comparing these fields, there are a number of caveats. First, in forming the SSHA fields, the long-term mean SSH pattern is removed during the process of eliminating the unknown marine geoid, which has not been done for the wind stress. This leaves only the temporal variability in the SSHA fields and hides any "permanent" features in the SSH and circulation. More importantly, the Ekman pumping relationship between WSC and SSHA assumes a steady state, rather than the seasonally varying patterns we are examining. In the real ocean, it takes time to establish the Ekman pumping relationship and then that pattern is altered by the propagation of geophysical waves, which carry SSHA signals into and out of a given region. Thus our comparison of wind stress and WSC to SSHA is qualitative and we return to the propagating signals below.

With these limitations in mind, what circulation features can we see in the two fields? Starting in the Northern

Hemisphere Pacific in the boreal winter, a band of positive WSC is found between 5°N and 11°N, except where it bends to the north near Central America. This is roughly collocated with a band of low SSHA between 7°N and 14°N. The NECC flows along the southern half of the low, near 6°N, turning cyclonically as it approaches Central America. In boreal summer, the bands of positive WSC and low SSHA are both farther north and the stronger north–south SSHA gradient east of 115°W implies a stronger NECC. This places the NECC approximately at 10°N, extending to the coast of Central America, where the high SSHA connects to a band of positive SSH that stretches to the north along the coast to central California. The other strong features in the eastern Pacific are the alternating regions of positive and negative WSC next to Central America in boreal winter, created by intense wind jets through the mountain gaps (Barton et al., 1993; Chelton et al., 2000a,b). These correspond closely to the areas of negative and positive SSHA, as expected for Ekman pumping. The strong mountain-gap winds and oceanic response dominate the regional ocean dynamics next to the coast of Central America, making it difficult to compare to the schematic of branching NECC currents in Figure 14.2. In boreal summer, the continuity of the band of high SSHA next to the coast from Central America to central California provides evidence that the poleward Inshore Countercurrent identified by Lynn and Simpson (1987) along the coast of central and southern California during late summer may be part of a much larger eastern Pacific circulation pattern.

In the northeast tropical Atlantic, the band of positive WSC in boreal winter stretches from 0°N to 10°N near the Brazil coast to the African coast between 10° and 20°N, weakest in the middle of the basin. The low SSHA under this band is weaker than in the Pacific, even reversing sign in the middle of the basin where the WSC is weakest. The low SSHA next to the African coast is most likely a result of direct coastal upwelling (Smith, 1981; Mittelstaedt, 1983), with an extra boost from the WSC. At approximately 3°N, along the northern edge of the band of higher SSHA, the NECC extends nearly continuously across the basin to equatorial Africa, representing the schematic current along the Guinea Coast in Figure 14.2. Moving to the far eastern Atlantic, the tongue of high SSHA next to the equatorial African coast in boreal winter corresponds to a poleward Angola Current, from the equator to 10°S, and a current flowing counter clockwise around the Gulf of Guinea. Moving away from the coast to consider a north–south transect around the zero-meridian, one finds alternating high and low SSHA bands between 20°S and the equator, producing alternating eastward and westward geostrophic currents. Although the locations of the bands do not correspond to the arrows in the Figure 14.2 schematic, there is agreement in the sense that the eastern South Atlantic is more complex than the eastern South Pacific, with multiple branches of an SEC. We also note that the SSHA field in the southeast Pacific should produce a SECC and poleward countercurrent offshore of Peru and Chile. As mentioned above, some authors include these as components of the southeast Pacific circulation (Codispoti et al., 1989), but their existence and behavior are open research questions.

Satellite data demonstrate the role of geophysical waves in these seasonal cycles of SSHA, which was missing in our earlier comparison between winds and surface heights. Animations of the changes in SSHA with 5-day time increments show a great deal of variability during the course of the year—the fields from boreal winter do not simply reverse to create the fields from boreal summer. The signals are easier to follow in the Atlantic, where the distance and travel times are shorter and the signals are more coherent. Even there, however, the signals are due to a mix of local Ekman pumping and propagation of distantly forced waves. The SSHA signals cross the tropical Atlantic toward the east along the equator as Kelvin waves in approximately a month; at 3–10° away from the equator they travel westward as Rossby waves, taking 3 months or longer to cross the basin. Schouten et al. (2005) use Hovmuller diagrams (time and distance plots of SSH contours) to track the seasonal cycle of altimeter SSHA signals along several selected pathways across the Atlantic.

Starting in the mid-basin equatorial Atlantic in January–March, positive SSHA (Figure 14.5) is due to the seasonal decrease in the trade winds and upwelling, resulting from the southward shift of the ITCZ. The positive SSHA spreads east along the equator to the coast and north and south along the African coast to 4°, then propagates slowly to the west during April–July. After reaching the coast of Brazil north and south of the equator, the high SSHA signals move to the equator, merge and propagate back across the basin to the east during August. While the high SSHA signals are moving westward north and south of the equator, low SSHA signals develop at both ends of the equator in April as the trade winds strengthen, merge along the equator in May–June and concentrate eastward into the July low tongue visible in Figure 14.5. The high SSHA signal described above converges on the western equatorial coast and crosses the basin eastward along the equator in August–September, reversing the strong July low. By October, the eastern equatorial Atlantic is filled with positive values of SSHA, although some of this is the general steric rise of SSH due to heating during the summer. A weaker low develops along the equatorial eastern Atlantic and Guinea Coast in December that does not obviously propagate along the equator from the west and is quickly replaced by high SSHA by mid-January, starting the process over.

Schouten et al. (2005) also follow the SSHA signals along the eastern boundaries of the tropical Atlantic. The low (high) SSHA signals that develop in July (January) next to equatorial Africa both propagate to approximately 10°N and 20°S, with speeds of 15 cm s^{-1} in the north and 75 cm s^{-1} in the south. The strong positive SSHA signal that arrives along the eastern equator in October propagates only to about 8°S and 5°N (the Guinea Coast), before initiating westward Rossby waves and dissipating. The weak low of short duration that develops in December in the Gulf of Guinea appears to originate and dissipate in place.

There is a long history of investigations into the two periods of upwelling and cooler water observed along the Guinea Coast in summer (major upwelling) and December (minor upwelling) (Picaut, 1983, 1985). These studies were initially motivated by the fact that the local winds are not strong enough to account for the amount of observed summer upwelling (as indicated in Figure 14.5 by the low SSHA values). In the summary by McCreary (1984) of the observational, theoretical, and modeling studies carried out in the 1970s and early 1980s, another puzzle was the fact that the upwelling signal arrives earlier at depth than at the surface. Modeling and analysis of data collected during 1983 and 1984 support the hypothesis that strongly varying seasonal winds in the western tropical Atlantic excite equatorial Kelvin waves that travel along the equator and then poleward in each hemisphere (Picaut, 1983, 1985; McCreary, 1984; Verstraete, 1992; Carton and Zhou, 1997). The combination of distant and local forcing produces the stronger upwelling in summer with a secondary upwelling in winter, along with vertically propagating signals that explain the earlier arrival of signals at depth. A more recent modeling study by Jouanno et al. (2011) provides a different emphasis, crediting turbulent vertical mixing with the upward transport of cold water and nutrients into the mixed layer, although still requiring vertical movement of the nutricline to allow mixing to accomplish the final transport. Putting aside differences about the specific mechanisms, the general conclusion is that the upwelling along the Ivory Coast is strongly affected by distantly forced signals, which travel along the equatorial and coastal wave guides to reach the Gulf of Guinea. The altimeter SSHA fields described above support this conclusion for the strong summer upwelling (July, Figure 14.5). The altimeter also sees low SSHA values along the eastern tropical Atlantic and Guinea Coast in December but is less conclusive about its origin (Schouten et al., 2005).

3. MIDLATITUDE EBCs: THE EBUS

The coastal geometries of the large-scale subtropical EBCs outlined in white in Figure 14.1 are more similar in appearance to each other than are the low-latitude EBCs. The relatively straight coasts are oriented approximately in a north–south direction, with capes, bays and "bights" of various sizes in each (Figure 14.6). This description does not apply to the Atlantic coasts north of the Iberian Peninsula, where the complex bottom bathymetry and continental geometry make it difficult to include in a discussion of typical EBC processes. The reader is directed to individual reviews of these regions in Robinson and Brink (1998b, 2006). The other less typical EBC is the Leeuwin Current, along the west coast of Australia. Geometrically, the coast is similar to SW Africa south of 22°S. Between the Equator and 20°S, however, there are the Indonesian Islands, open ocean and the westward SEC.

In the Pacific and Atlantic Oceans, these are the well-studied Eastern Boundary Upwelling Systems. Wind stress and WSC are the dominant external forces, although buoyancy forcing is also a factor on regional scales near the outflows of rivers, which may be large, individual rivers such as the Columbia River in the California Current (Hickey et al., 2009, 2010), the Bio Bio River in the Peru–Chile Current (Montecino et al., 2006), the Orange and Cunene Rivers off SW Africa (Shillington, 1998). In other regions, the fresh water comes from combinations of smaller rivers, as found along the Iberian Peninsula (Relvas et al., 2007). A special situation is created by the outflow of warm, salty, and dense water from the Mediterranean Sea, splitting the Canary Current into two subsystems. Internal processes include upwelling, downwelling, barotropic and baroclinic instabilities in coastal jets (equatorward and poleward) and the generation of Rossby waves and CTWs. In the Leeuwin Current (and perhaps others), large-scale north–south density gradients create differences in steric height and pressure, another example of buoyancy forcing that is capable of driving currents against the prevailing winds (Godfrey and Ridgway, 1985; Church et al., 1989).

3.1. Mean and Seasonal Circulation

3.1.1. Large-Scale Currents

One view of the EBUS (Figure 14.6) is based on long-term means of the circulation, characterizing them as broad, shallow, and slow surface flows which transport cool and relatively low-salinity waters equatorward through the midlatitudes, forming the eastern branches of the anticyclonic subtropical gyres (Hill et al., 1998; Mackas et al., 2006; Freon et al., 2009). Even offshore of the poleward Leeuwin Current, the large-scale circulation is usually described as an equatorward flow needed to complete the subtropical gyre and connect the South Indian Current to the SEC (Domingues et al., 2007). The equatorward currents are driven by equatorward winds in the semi-permanent atmospheric high-pressure systems, which expand and contract in summer and winter (respectively). Poleward of around 40° latitude, the disappearance of the high-pressure system allows winter storms to create seasonal alternations between upwelling and downwelling conditions. Water

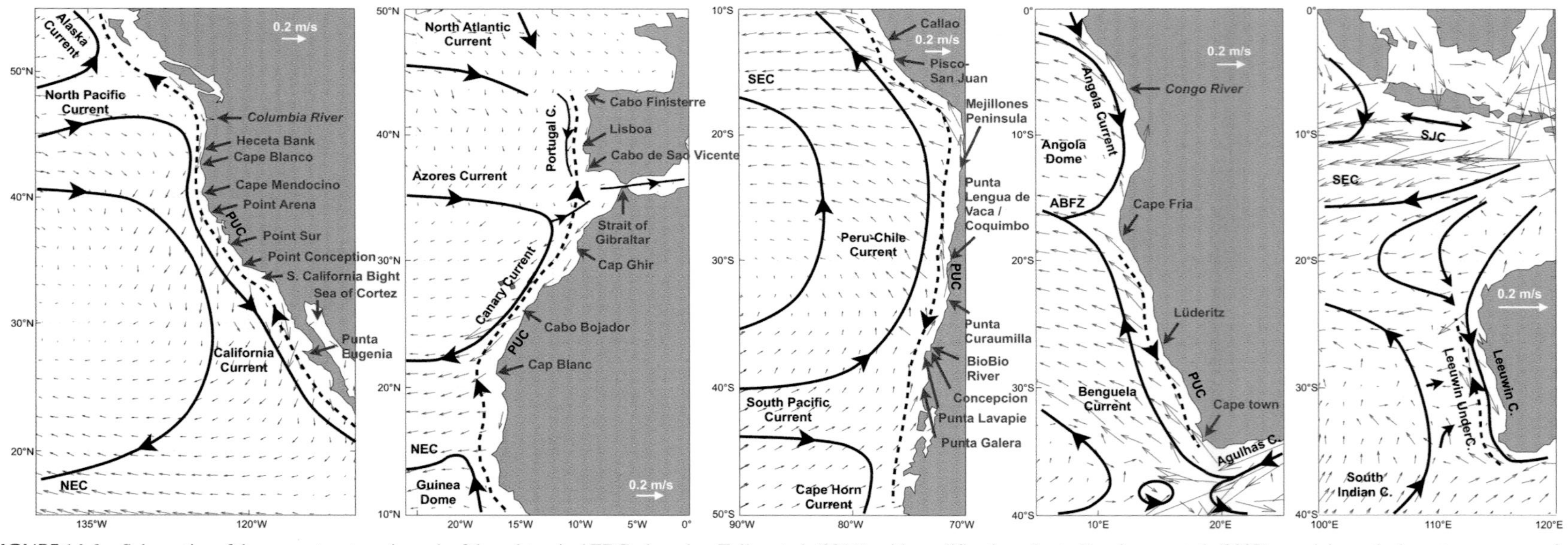

FIGURE 14.6 Schematics of the current systems in each of the subtropical EBCs, based on Talley et al. (2011), with modifications due to Domingues et al. (2007), overlain on the long-term mean surface currents of the OFES general circulation model (as in Figure 14.2). Note the different latitude limits of the figures, although each figure covers 40° of latitude.

flows into the higher-latitude regions of the EBUS in the Pacific and North Atlantic systems from the west in the West Wind Drift (WWD) Currents, individually named for their locations (the North and South Pacific or North Atlantic Currents). Inflow to the poleward Leeuwin Current north of Australia is in the form of tropical water that arrives from the Indian Ocean through the South Java Current and from the Pacific Ocean via the Indonesian Throughflow and the SEC. Beneath the surface flow in each of the EBUS are undercurrents of opposite direction to the surface, poleward in the Pacific and Atlantic Systems and equatorward under the Leeuwin Current.

An alternate view of the same regions consists of complex mesoscale circulation patterns, seen in high-resolution satellite-derived "snapshots" of SST and surface chlorophyll-a concentrations. In the early 1980's, improvements in satellite imagery and in-water mapping techniques revealed the three-dimensional, rapidly varying nature of these structures and changed our fundamental view of upwelling systems (and other coastal regions). At the same time, however, temporal averages of the in-water and satellite fields in these regions over complete (multiple) seasons recovered the classical depiction of smooth and slowly varying fields, albeit with much greater variances than would be found if the mean currents were also the instantaneous currents (Kelly, 1985; Kosro and Huyer, 1986; Kosro, 1987; Field and Shillington, 2006; Mackas et al., 2006; Montecino et al., 2006; Relvas et al., 2007). In Section 3.2 we return to the more complex, temporally varying fields. Here we continue the description of the smoother, temporally averaged mean and seasonally varying fields, represented by the schematic patterns in Figure 14.6 and the altimeter SSHA fields in Figure 14.5.

Surface wind forcing is shown in Figure 14.7. The top two rows present the seasonal changes between winter and summer winds stress (vectors) and WSC (colors) fields for the Pacific and Atlantic EBCs, continuing the tropical fields from Figure 14.4. Similar fields calculated from merchant ship data by Bakun and Nelson (1991) provide a relatively good representation of the seasonal changes in large-scale patterns of coastal wind stress and a qualitative view of the WSC. The systematic and high-resolution sampling of the scatterometer provide a more detailed view of the spatial variations in the wind fields:

- Winds are predominantly equatorward, with offshore wind stress maxima usually within ~100 km from the coast, creating bands of cyclonic WSC next to the coast;
- The strongest seasonal upwelling in winter occurs at the lower latitudes (usually <20°);
- In summer, both the offshore regions of anticyclonic WSC and the cyclonic bands next to the coast intensify and move poleward, creating a narrow (100 km) region of intense upwelling forcing (wind stress and curl) next to the coast;
- In winter, complete reversals to vector mean poleward (downwelling-favorable) wind stress occur only at the highest latitudes: in the California Current north of ~40°N, in the Peru–Chile Current south of ~38°S, and over the most southern portion (42–44°S) of the Benguela Current.
- Even where the climatological wind stresses remain equatorward or neutral, the higher latitudes are forced by synoptic storms with strong poleward winds and high precipitation, which are not well represented by the vector mean wind stresses in Figure 14.7.

The atmospheric forcing divides the EBUS into higher latitude regions of "seasonal" upwelling, with winter downwelling and storms, and lower latitude regions of "perennial upwelling," with winter upwelling maxima at the lowest latitudes. Precipitation at the higher latitudes contributes to buoyancy forcing of the coastal circulation in the NE Atlantic, NE and SE Pacific. The difference between winter downwelling and summer upwelling appears the strongest (in terms of wind forcing) for the northern California Current. However, the circulation along the Iberian Peninsula in winter is dominated by poleward flow typical of downwelling systems, opposing the weak mean winds (Figure 14.5; Haynes and Barton, 1990; Torres and Barton, 2006; Relvas et al., 2007). Based on the currents, we characterize the large northern sections of both the California and Canary Currents in winter as downwelling systems. In that sense, the separation between (northern) "seasonal" and (southern) "perennial" upwelling system is the clearest in the Canary Current, where a physical division is enhanced by the interruption of the circulation created by the Gulf of Cadiz and the Mediterranean outflow (Aristegui et al., 2006, 2009). The Southern Hemisphere EBUS have smaller high-latitude regions of seasonal alternations between upwelling and downwelling.

In Figure 14.6, the two Pacific EBCs mirror each other more closely than do the Atlantic systems, with a continental boundary that continues along higher latitudes, allowing the inflowing North and South Pacific Currents to move their areas of bifurcation north and south with the seasons. The North Atlantic Current also moves seasonally and flows eastward along the northern coast of the Iberian Peninsula in winter, while the circulation along the west coast of Iberia is poleward, joining the eastward flow into the Bay of Biscay at Cap Finisterre. The summer circulation in the North Atlantic is as portrayed in Figure 14.6, with the North Atlantic Current flowing into the equatorward current along the Iberian Peninsula.

In the South Atlantic, the Benguela Current lies offshore of the Benguela upwelling area and has been defined as the EBC of the anticyclonic South Atlantic subtropical gyre (Peterson and Stramma, 1991). Transport of the Benguela Current is supplied by the southern limb of the South Atlantic subtropical gyre, by south Indian Ocean waters

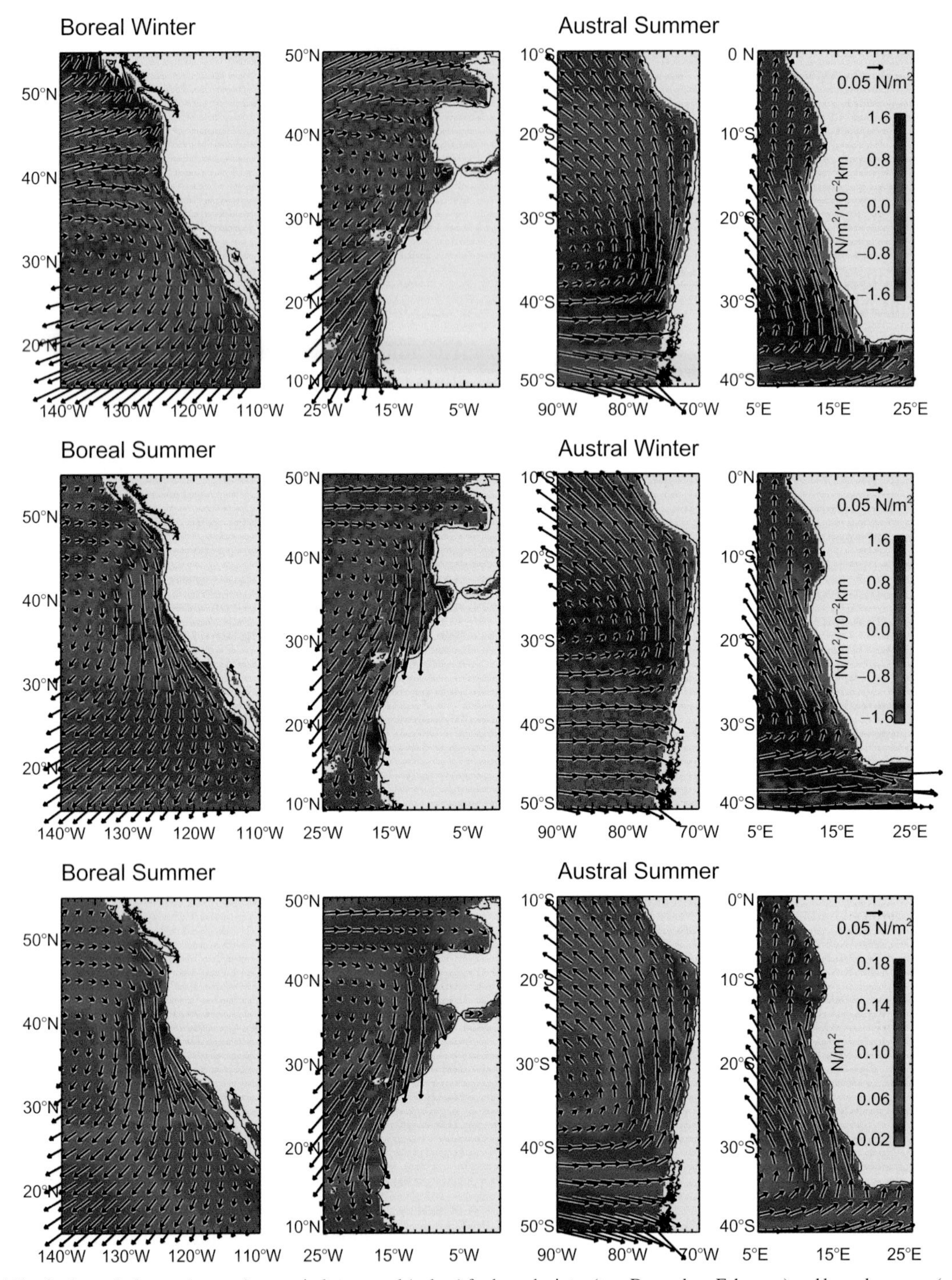

FIGURE 14.7 Surface wind stress (vectors) over wind stress curl (colors) for boreal winter (top, December–February) and boreal summer (middle, June–August), from 10 years of QuikSCAT data. (Bottom) Surface wind stress (vectors) over wind stress magnitude (colors) for summer in each hemisphere. Vectors are subsampled every 2°; curl and wind stress magnitude show the full 0.25° resolution. From left to right: North Pacific; North Atlantic; South Pacific; South Atlantic.

via the Agulhas Current (Figure 14.6), and can be influenced by Subantarctic Surface Water via perturbations in the Subtropical Front. Geostrophic transports derived from observational data by Gordon et al. (1987) suggest that as much as 10 Sv of a total flow of 15 Sv into the upper 1500 m of the South Atlantic is fed by the Agulhas Current, with the rest (5 Sv) being fed by the South Atlantic Current. Transport into the Atlantic from the Agulhas is mostly in the form of warm anticylonic Agulhas rings and cyclonic eddies generated by the retroflecting Agulhas Current near the western tip of the Agulhas Bank (Lutjeharms, 2006) and represents only 10% of the Agulhas transport, the other 90% retroflecting back toward the east, south of Africa. See Chapter 19 for a more detailed discussion of the large-scale exchange between the Indian and South Atlantic Oceans. The relative contributions of Indian and Pacific Ocean waters to northward transports in the South Atlantic are a matter of debate. However, the net effect of the inflow from the Agulhas is that the Benguela Current System has both a warm temperate (~26 °C) northern boundary off Angola and a warm temperate (~24 °C) southern boundary, unlike the cold poleward boundaries found in the other EBUS. The impact of this warm, salty water is largely on the southern Benguela upwelling system, although anticyclonic rings from the retroflection region occasionally move to the north and interact strongly with the coastal system, drawing long filaments of upwelled water offshore. A well-documented example of this occurred in 1989 (Duncombe Rae et al., 1989, 1992; Lutjeharms et al., 1991).

The boundaries between the midlatitude and low-latitude areas are the location of dramatic confluences and frontal regions in the Atlantic, less so in the Pacific. The analysis of altimeter data by Schouten et al. (2005) identifies tropical signals that travel as far as approximately 10°N and 20°S along the eastern boundary of the Atlantic Ocean, the locations of confluences between equatorward flow in the midlatitude EBCs and poleward flow at lower latitudes. The confluence along NW Africa has been well studied and described by Mittelstaedt (1983, 1991) and Barton (1987, 1998): summer downwelling along the region between 15°N and 20°N (represented by high coastal SSHA in Figure 14.5) creates poleward flow that merges with the NECC to meet the equatorward currents from midlatitude summer upwelling in an energetic confluence near 20°N, with subsurface intrusions. In winter, upwelling between 15°N and 20°N extends the northern perennial upwelling to the south and creates low SSHA and equatorward flow next to the coast that moves the confluence to the south.

Another energetic confluence creates the ABFZ at ~15–17°S, where the narrow and warm coastal Angola Current (Ajao and Houghton, 1998) flows south and meets the cool Benguela surface water (Shannon et al., 1987; Field and Shillington, 2006; Veitch et al., 2006). The strongest SST gradients in the ABFZ occur in April–May, when the Angola current travels farthest south; the weakest SST gradients occur in August –September (Hardman-Mountford et al., 2003). Seasonal changes in intensity of the ABFZ have been related to fluctuations in the upwelling-favorable wind stress, while the north–south migration of the front is related to changes in the poleward flow in the northern Benguela Current, which is (in turn) related to the curl of the wind stress (Colberg and Reason, 2006).

Connections between the midlatitude and low-latitude systems in the Pacific lack the strong surface convergences of the Atlantic. There is no surface poleward surface current in the low-latitude section of the Peru–Chile Current and the sparse observations of the poleward currents next to Central America (the Costa Rica and West Mexican Coastal Currents) characterize them as local and disconnected (Lavin et al., 2006). However, the lack of surface convergences does not represent a lack strong connections between the mid- and low-latitude systems. Despite the shallow equatorward flow at the surface, the strong undercurrent off Peru is directly connected to the equator and extends to southern Chile (Silva and Neshyba, 1979; Lukas, 1986). In the California current, the direct pathway back to the equator is long and crosses a number of complex geographical regions and wind-driven current features. Nevertheless, the tropical water properties of the undercurrent off California demonstrate a connection to the tropical system. The fields of monthly SSHA from the equator to the Gulf of Alaska presented by Strub and James (2002a) show a progression of high coastal SSHA signals (associated with poleward flow) that start off Mexico between 15°N and 20°N in July–August and move poleward to reach Vancouver Island by November–December. Figure 14.5 demonstrates a similar pattern, with a continuous band of high SSHA in July stretching from Central America to the tip of Baja California, extending to central California in August (not shown). This is consistent with observations during the late summer period of a poleward Inshore Countercurrent along southern and central California (Lynn and Simpson, 1987). Thus, poleward inflows at the equatorward ends of the all of the midlatitude EBCs interact strongly with the midlatitude equatorward surface circulation that is usually portrayed as their dominant characteristic.

The EBC that is the exception compared to the other midlatitude EBCs is the Leeuwin Current, next to western Australia (Figure 14.6). Although winds in the region are equatorward throughout the year (not shown), the Leeuwin Current flows poleward at the surface, forced by a large-scale meridional pressure gradient (Thompson, 1987) that is created by differences in steric heights (Cresswell and Golding, 1980; Godfrey and Ridgway, 1985; Godfrey and Weaver, 1991; Smith et al., 1991; Feng et al., 2003). Water properties responsible for the high steric heights north of Australia are advected both from the Indian Ocean in the South Java Current and from the Pacific Ocean through the Indonesian Throughflow, resulting in steric height differences between northwest and southwest Australia of approximately 0.3 meters (Reid, 2003). The winds do influence the seasonal variability of the Leeuwin Current: maximum poleward transport occurs in austral winter (June–July), when the equatorward winds are weakest (Feng et al., 2003, 2005).

The specific path taken by water flowing into the Leeuwin Current from the north is not well known. Domingues et al. (2007, their Figures 1 and 11) use five years of Lagrangian trajectories from a global ocean circulation model and find that both the South Java Current and Indonesian Throughflow feed into the westward SEC between 10°S and 15°S, which then connects to the Leeuwin Current through several counterclockwise meanders between ~95°E and 112°E. The southern branches of these meanders join and flow eastward as the East Gyral Current (not named in Figure 14.6) between 15°S and 20°S. We note that the OFES surface currents in Figure 14.6 show no obvious pathway for water to enter the northern Leeuwin Current (another model deficiency and subject for continued research). Figure 1 of Domingues et al. includes a broad, equatorward West Australia Current offshore of the Leeuwin Current, as also described by Talley et al. (2011). However, the figures shown by Talley et al. restrict the subtropical gyre of the South Indian Ocean to the western half of the basin, emphasizing eastward flow into the East Gyral Current and into the western boundary of the Leeuwin Current, based on the dynamic topography of Reid (2003) and the results of Schott and McCreary (2001). Figure 11 of Domingues et al. (2007) also emphasizes the flow of subtropical Indian Ocean water into the Leeuwin Current from the west along the length of its poleward journey. We have included eastward flow into the Leeuwin Current in Figure 14.6's schematic circulation, since it is consistent with a conceptual model of the mechanics of the poleward flow, known as the Joint Effect of Baroclinicity and Relief (JEBAR, see below). Additional descriptions of the Leeuwin Current can be found in Church and Craig (1998) and Condie and Harris (2006) and in the special volume introduced by Waite et al. (2007).

Although the poleward flow against prevailing winds is most extreme in the Leeuwin Current, it is also found in winter in the northern Canary Current. As noted above, scatterometer wind fields next to the Iberian Peninsula in Figure 14.7 show slightly equatorward winds during winter, while poleward currents are systematically observed in winter along the Iberian coast (Barton, 1998; Aristegui et al., 2006, 2009; Relvas et al., 2007). The inflow of Mediterranean water along the southern end of the Iberian Peninsula may provide the analog for the tropical water at the northern end of Australia. In the northern region of the California Current in winter, both currents and winds are observed to be poleward and the dynamics are thought to be more directly wind driven. These poleward currents are represented in Figure 14.5 by positive SSHA values in bands next to the coasts of Iberia and the U.S. west coast in January. We will continue the discussion of poleward currents after completing the description of the wind-driven upwelling systems.

3.1.2. Coastal Currents and Undercurrents

Returning to the more "typical" summertime EBUS in the Atlantic and Pacific, the effects of coastal upwelling must be added to the large-scale equatorward currents in Figure 14.6. The equatorward winds in Figure 14.7 cause surface Ekman transports away from the coast and upwelling, as described in Section 1.1.3. The cross-shelf circulation creates denser water with lower surface heights (0.1–0.2 m) next to the coast and a sharp horizontal density front between the upwelled and offshore water. The front extends along the coast and a surface jet flows equatorward along the front, driven by the density and height (i.e., pressure) differences across the front, in approximate geostrophic balance. The offshore scale of these phenomena is initially given by the internal Rossby radius of deformation (of order 10 km in most EBCs), although other processes move the fronts and jets farther offshore. An additional complexity is provided by the effects of WSC. Ekman pumping and additional upwelling are created away from the coast by the strong coastal bands of cyclonic WSC seen in Figure 14.7. In systems where significant amounts of fresh water enters next to the coast during the upwelling season, a buoyancy-driven poleward surface current is found next to the coast, pushing the upwelled water, front and frontal jet farther offshore. This is typical of the Iberian coast, as described and pictured in Aristegui et al. (2009, their Figure 2).

As already described, alongshore PUCs are observed in all EBUS, usually located over the upper continental slope and/or outer shelf bottom (Neshyba et al., 1989). In the Pacific, the chemical signature of the Peru–Chile Undercurrent has been used to trace it from its connection to the Equatorial Undercurrent at ~5°S (Tsuchiya, 1985; Lukas, 1986) to southern Chile at ~42°S (Silva and Neshyba, 1979). Off the U.S. West Coast, the poleward velocities in the PUC have been observed by Pierce et al. (2000) to extend continuously from Southern California to Canada (33–51°N), using shipboard acoustic Doppler current profiler data. In the Atlantic, the PUC under the Benguela Current is observed at a number of locations and described by Shillington (1998), Field and Shillington (2006), Shillington et al. (2006), and Hutchings et al. (2009). Separate PUCs in the Canary Current are found off NW Africa and Iberia (Mittelstaedt, 1983; Barton, 1998; Aristegui et al., 2006). The PUC along NW Africa has been traced from the confluence of the NECC and Canary Current (15–20°S) as far as 28°N off NW Africa using T–S relationships. Whether there is continuity between the PUCs off NW Africa and Iberia is still a research question (Barton, 1998; Aristegui et al., 2009).

Direct measurements of the PUCs with current meters have been made in the Peru–Chile Current System off Peru and Chile (Huyer et al., 1987; Shaffer et al., 1997; Pizarro et al., 2002). The longest records report mean velocities of 13 cm s^{-1} for the period 1991–2010 in the core of the undercurrent at 30°S off Chile. This current is modulated by strong intra-seasonal fluctuations that are related to coastally trapped waves of equatorial origin. Two seasonal maxima are recorded in the poleward flow, during spring and fall, in response to a combination of local wind forcing and remotely forced tropical disturbances (Pizarro et al., 2002; Ramos et al., 2006). The estimated mean transport of the Peru–Chile Undercurrent is about 1 Sv (Huyer et al., 1987; Shaffer et al., 1999). Similar velocities and transports have been reported in the Canary Current PUC off NW Africa at several sites between 20°N and –26°N (Barton, 1998) and in the California Current PUC (Halpern, 1976; Freitag and Halpern, 1981; Pierce et al., 2000). The best estimates of the mean PUC off southern California come from a decade of ship-born ADCP measurements. Gay and Chereskin (2009) use this systematic sampling to show a mean PUC with velocities of 7 cm s^{-1} and a transport of 1.7 Sv, centered at ~200 m depth just north of the Southern California Bight. In the Benguela Current, the mean velocity in the PUC is a somewhat lower 5 cm s^{-1}, after filtering out signals from CTWs (Shillington, 1998).

In most of the EBUS, the undercurrents are cited as a source of the upwelled water and a factor in determining the properties of that water, especially the nutrients, oxygen, CO_2, and acidity. Codispoti et al. (1989) describe a "nutrient trap," in which water upwelled from the PUC enriches the surface ecosystem, resulting in an increase of biomass that moves equatorward with the surface flow until the biological waste products sink into the PUC and travel back to the original location of the upwelling, decaying to produce regenerated nutrients and lower oxygen concentrations, only to be upwelled and repeat the cycle again. Codispoti et al. described this process along the South American coast, where the waters under the surface layer are hypoxic in the extreme. As oxygen is consumed, the concentrations of CO_2 and carbonic acid increase, decreasing the pH. If the central gyres generally increase in acidity (in the future), these processes may drive coastal pH values even lower than would otherwise occur. Recent increases in hypoxia and acidity along the Oregon coast have been interpreted as a warning of larger-scale increases in acidity (Chan et al., 2008).

Summarizing the coastal currents, the mean (temporally averaged) upwelling circulation pattern that should be added to the large-scale midlatitude EBC pattern consists of offshore Ekman transport at the surface and subsurface onshore return flow. The upwelled and denser water fills the shelf inshore of a front between the upwelled and offshore water, with an equatorward jet along the front. A PUC flows next to the bottom, over the upper slope and outer shelf. As with the large-scale circulation, this simple description of the two-dimensional upwelling circulation is an idealization. Temporally averaged fields may resemble this picture, but an instantaneous field of currents is much more complex and three dimensional. The schematic figure of Hill et al. (1998) depicts these currents in a three-dimensional sense. Figure 2 of Aristegui et al. (2009) provides variations on the general arrangement of alongshore currents, including wider and narrower shelves and the inner poleward (buoyancy-driven) current that occurs when there is a strong influx of fresh water at the coast.

3.1.3. Dynamical Models of Equatorward and Poleward Currents

What dynamical processes drive the large-scale, temporally averaged circulation patterns in the EBUS? We have used the process of Ekman pumping to relate surface forcing by the WSC to upwelling velocity and SSHA. A more dynamically correct description relates the upwelling velocities to the vertically integrated meridional (N–S) velocity through conservation of vorticity, resulting in the Sverdrup relation at a point. This states that the vertically integrated meridional (N–S) velocity is determined by the WSC and the meridional derivative of the Coriolis parameter, $\beta = \partial f / \partial y$

$$v = \frac{\text{curl}\,\tau}{\rho\beta}$$

Although this relationship is difficult to verify at a point, it is consistent with drifter and satellite observations in the central California Current (Kelly et al., 1998) and also with ADCP measurements off southern California (Peter Gay, personal communication, 2013). More importantly, it very successfully predicts the basin-scale two-dimensional vertically integrated transport patterns, by integrating the WSC from east to west (Munk, 1950). The physical reason for the east–west integration is that Rossby waves propagate westward, bringing to any location the net effect of wind forcing from all points east of that location (Anderson and Gill, 1975; Philander and Yoon, 1982). As stated above, Munk (1950) applied this method to an idealized version of the North Pacific basin and wind field (midlatitude westerlies, tropical easterlies). The steady state circulation that results from these "Svedrup dynamics" consists of a large anticyclonic gyre at midlatitudes, with slow equatorward flow along the eastern boundary.

In Figure 14.7, the large-scale regions of anticyclonic forcing are revealed by negative WSC (blue–green colors) offshore in the Northern Hemisphere and positive WSC (yellow and red) in the Southern Hemisphere. In a band within several hundred kilometers of the coast, however, interactions with the land slow the equatorward winds and create bands of cyclonic WSC (red or blue in the Northern or Southern Hemispheres, respectively). Using the Sverdrup relation and starting at the coast, westward integration of the WSC produces poleward integrated transports under the narrow bands of cyclonic WSC next to the coast, with equatorward flow farther offshore. The cyclonic bands of WSC were known and discussed by Munk (1950), who pointed out the agreement of the theory with poleward coastal transports observed during a summer cruise off central and southern California.

The discussion of Sverdrup transport supplies one model for the mean equatorward and poleward transports: If the cyclonic WSC within ~100 km of the coast drives vertically averaged poleward flows, combined with an equatorward surface current along an upwelling jet, the result is an equatorward surface current and a PUC. To test this Sverdrup model, a number of papers estimate vertically averaged transports from WSC fields using the Sverdrup model and find agreement with direct estimates of the average transports from measured currents or results from numerical models (e.g., Marchesiello et al., 2003; Penven et al., 2005; Veitch et al., 2010). In an early test of the relationship at a point, Halpern (1976) used current meter measurements and winds from meteorological buoys off Oregon and found better correlations between alongshore currents and estimates of the WSC than between the currents and wind stress. A more recent result calculated 16–18 Sv of mean equatorward transport in the eastern South Pacific between the coast (~71°W) and the South Pacific Rise (~113°W) across the WOCE P06 line at 32.5°S, within and above the main thermocline (Shaffer et al., 2004). This is consistent with estimates of the transport, based on the Sverdrup relation and satellite estimates of WSC. More qualitatively, climatological averages of geostrophic transports next to central California by Chelton (1984) and Lynn and Simpson (1987) confirm their general poleward nature where the WSC is cyclonic, as do numerous observations off the Iberian Peninsula (Haynes and Barton, 1990; Aristegui et al., 2006, 2009). However, PUCs develop in analytic (Pedlosky, 1974) and numerical (McCreary and Chao, 1985) models driven by winds with and without WSC, leading to a consideration of other dynamics.

Two related models help to explain the poleward flow in these systems and may provide clues for the undercurrents. The first is more general, consisting of the "arrested topographic wave" of Csanady (1978). Hickey and Pola (1983) successfully apply this model to explain the seasonal changes in the alongshore pressure gradient in the California Current. In this steady state, linear model, assuming barotropic flow and linear friction, with an offshore sloping bottom (shelf), the across-shelf momentum balance is simply the geostrophic balance for the alongshore current, v. The alongshore momentum balance adds the wind stress and bottom friction to the Coriolis force and pressure gradient. Combining these with conservation of mass, Csanady derives a vorticity equation that is an analog for heat diffusion, in which SSH is the analog for temperature and the alongshore distance is the analog for time. The wind stress takes the place of heat flux and its influence on the height fields travels only in the direction of CTWs (poleward in EBCs). Assuming no alongshore slope in the surface height at some distance offshore, Csanady first imposes a constant alongshore wind stress. The solution in the far offshore region is simply offshore Ekman transport. At the coast, the balance is between wind stress and friction, so the currents are in phase with the wind. As one moves alongshore, the response of the ocean spreads offshore. For a wind stress that is periodic in the alongshore direction, the minimum in sea level at the coast occurs poleward of the maximum upwelling-favorable wind stress, as observed off Oregon (Halliwell and Allen, 1987).

Although the model is steady state, Hickey and Pola (1983) apply it to each month of the seasonal cycle of alongshore winds from the U.S. west coast. This works because the ocean responds quickly to imposed forcing, adjusting the surface heights, pressures, and currents through the alongshore propagation of CTWs (CTW), the first modes of which travel the length of the California Current in several days to a week, allowing a steady state to be reached easily in monthly time steps. Hickey and Pola find good agreement between seasonally changing model SSH values at the coast and tide gauge observations of coastal sea levels (adjusted by adding a long-term mean steric height), better agreement than provided by a purely local model of wind-driven upwelling. The model produces a pattern of currents in the same direction as the wind, with offshore changes in the current patterns governed by the slope of the bottom topography in relation to the wind scale, friction, etc. The patterns of currents and surface heights are consistent with geophysical fluid dynamics for a barotropic coastal ocean but have nothing to say (so far) about poleward currents flowing against the wind (surface or subsurface).

In Csanady's original 1978 paper, he extends the model to include an offshore height or pressure gradient (applied in the alongshore direction), such as provided by offshore current systems or an imposed density gradient. In this form, removing the wind stress and adding an imposed density and pressure gradient, it becomes the conceptual model of Huthnance (1984), the Joint Effect of Baroclinicity and Relief (JEBAR). This model produces a flow from high to low pressure that is trapped over the sloping shelf by conservation of vorticity. But the imposed alongshore pressure gradients also induce an onshore geostrophic transport, such as drawn schematically flowing into the Leeuwin Current in Figure 14.6. This transport is small, but it accumulates along the current and the conservation of mass requires that the speed of the current increase in the alongshore direction, resulting in an increase in the cross-shore height difference that balances the alongshore flow. This "model" (as summarized by Hill, 1998), is limited in that the velocity cannot increase without bound; if the alongshore distance is too great, other forces (turbulence and friction) will come into play when velocities exceed some limit.

The JEBAR steady state dynamics are consistent with the flow in the Leeuwin Current, with an additional seasonal wind-driven component that explains the maximum poleward flow when the equatorward winds are minimum. Relvas et al. (2007) report that the JEBAR model has also been compared well to the observations of the poleward currents in the Canary Current off the Iberian Peninsula in winter. There the alongshore density gradient is represented by an alongshore rise (to the north) in the 27.0 isopycnal of 150 m over 6 degrees of latitude, consistent with an onshore geostrophic velocity of 0.02 m s^{-1}. This small current produces approximately 2 Sv of onshore transport along the coast, some of which feeds the poleward current.

These steady state models explore the roles of winds, density and pressure gradients, friction, vorticity, and sloping bottom topography as driving forces for poleward currents, countercurrents, and undercurrents. The two cases described, one with wind stress and no offshore pressure gradient and the other with an imposed density and pressure gradient, represent end members of experiments with different mixes of the two forcing functions. It would be instructive to apply the mixed model to each of the EBC regions in the extremes of the seasons, to evaluate the importance of the two types of forcing in each case. More ambitious studies are underway using more complete, high-resolution coastal circulation models in each of the EBCs. As the results of those calculations are analyzed, it would be useful to cast those results in terms of the balances of forces embodied by simpler models such as the Sverdrup, arrested topographic wave and JEBAR models.

3.2. Higher Frequency Mesoscale Variability

3.2.1. Wind Forcing of "Upwelling Centers"

During the upwelling-favorable wind forcing, the intensity of wind stress and WSC is concentrated around geographic "upwelling centers," often associated with capes or other changes in the coastline geometry. This is illustrated in the bottom row of Figure 14.7, which presents the same mean wind stress vectors as above, but over the scalar means of the wind stress magnitudes (color), for summer in each region. The fields of mean wind stress magnitude in summer have the full resolution of 0.25° and show the localized maxima in wind stress more clearly than the subsampled vectors. The locations identified in the figure are: the California Current near Capes Mendocino and Blanco (40°N, 43°N) and (weaker) Pta. Eugenia (27°N); the Canary Current at Cap Blanc (21°N), Cabo Bojador (26°N), Cap Ghir (31°N), Cabo de Sao Vicente (37°N), Lisboa (39°N) and Cap Finisterre (43°N); the Peru–Chile Current near Concepcion (37°S) and Pta. Galera (40°S); and the Benguela near Cape Town (34°S), Luderitz (27°S) and Cape Fria (17°S). During winter (not shown), the strongest equatorward winds move to lower latitudes, creating new seasonal upwelling centers (off Chile near Coquimbo (30°S)) or strengthening the lower-latitude summer centers (Cape Fria off SW Africa). Winds at a few of the upwelling centers in the middle of these regions remain strong in winter and summer, most notably at Luderitz off SW Africa and Cap Ghir off NW Africa (a somewhat weaker persistent center exists at Punta Eugenia along Baja California).

The stronger winds in some of the upwelling centers take the form of equatorward, low-level jet-like structures centered near the top of the atmospheric marine boundary layer, which is usually only 200–500 m thick. The reported offshore distances of these wind jets range from around 20 to over 100 km, depending on several processes and the details of the observational and modeling studies used to investigate the jets. Rogers et al. (1998) and Munoz and Garreaud (2005) provide literature reviews. Along a relatively straight section of northern California coast, Beardsley et al. (1987) describe a wind maximum 20–30 km offshore, located under the downward sloping (toward the coast) inversion layer at the top of the marine boundary layer. The addition of variable coastal geometry and topography creates more complex wind patterns. Wind speeds may become critical and capes may cause inshore minima in wind speed and offshore maxima. Wind maxima that form close to the coast can extend 100–200 km downstream and offshore (Rogers et al., 1998; Winant et al., 1988; Burke et al., 1999; Edwards et al., 2001; Koračin et al., 2004; Garreaud and Munoz, 2005; Dever et al., 2006). In regions where upwelling produces a wide band of cold surface temperatures next the coast, wind stress is decreased over the cold water, moving the core of the wind jet farther offshore (Perlin et al., 2007; Haack et al., 2008). By adding diurnal heating over land, Bielli et al. (2002) create diurnal temporal variability of the low-level wind jet's three-dimensional structure, while the jet stays within 30–50 km of the coast. Thus, the upwelling-favorable winds typical of all EBUS can take on a variety of temporally and spatially varying three-dimension structures, due to interactions with topography and marine boundary layer dynamics.

Synoptic variability in wind forcing, with timescales of 2–3 days, is created at the higher latitudes (mostly) in winter by cyclonic storms. The strongest winds are poleward but the alternating direction of storm winds can result in vector means of monthly winter winds that are weak, even though the wind speeds at any instant are high. Another source of synoptic atmospheric variability that is more noticeable at the low- and midlatitudes is the alongshore propagation of atmospheric low-pressure cells called coastal lows. They have a relatively shallow warm core with alongshore and cross-shore scales of 1000 and 500 km, respectively (Dorman, 1985; Reason and Jury, 1990; Garreaud et al., 2002; Dorman and Koracin, 2008). The coastal lows have been interpreted as midlatitude atmospheric waves that propagate poleward in the marine boundary layer, trapped and guided by the coastal mountains (atmospheric CTWs, Gill, 1982). According to Garreaud et al. (2002), they are more common during fall to spring off Chile, but can occur during all seasons. Their movement past a coastal location creates a relaxation of the upwelling winds or even a reversal to downwelling events, a typical occurrence during the upwelling season in most of the systems.

3.2.2. The Coastal Ocean's Role in Creating Upwelling Centers

The above discussion of intensified wind forcing at "upwelling centers" ignores the ocean's role. Upwelling centers are also created or intensified by internal ocean dynamics related to coastal geometry and changes in bathymetry associated with capes and bays. Early analytic and numerical models and observations considered only changes in the local winds (Yoshida, 1967; Smith, 1968; and McCreary et al., 1991). The effects of bottom topography (canyons and ridges) were investigated in analytic models by Killworth (1978), while Peffley and O'Brien (1976) used early numerical models to quantify the effects of capes. One of the results of these analyses is the hypothesis that upwelling is strongest along the equatorward (lee) side of capes, during equatorward winds. Following the motion of a water parcel, Arthur (1965) came to the same conclusion based on a Lagrangian vorticity balance. Using scaling arguments, the two terms that often contribute most to the changes in vertical velocity are the advection of planetary vorticity (βv) and the Lagrangian rate of change of relative vorticity ($\frac{D\xi}{Dt}$). For equatorward flow, these oppose each other on the upstream portion of capes and reinforce each other downstream of the most western extension of the capes, leading to the hypothesized preference for upwelling centers on the equatorward side of capes.

Capes are also the locations where coastal jets often separate from the coast. Observational studies off Oregon (Barth et al., 2000) and modeling studies off SW Africa (Veitch et al., 2010) and south–central Chile (Mesias et al., 2003; Aguirre et al., 2012) have shown that the capes are key locations where coastal jets separate from the coast, due to both enhancements of the wind forcing and internal ocean dynamics. At the northern limit of the Canary Current, an extreme example of cape-induced separation occurs at Cap Finisterre, where the coast makes a right angle (Relvas et al., 2007). Likewise, submarine canyons (sometimes associated with bays) affect internal dynamics and may serve as conduits for upwelled and downwelled water (Hickey, 1997). Gulfs and bays may also act as "upwelling shadow" zones that play an important ecological role (Graham and Largier, 1997; Marin et al., 2001; Marín and Moreno, 2002 and other papers in the same volume). In the most complex cases, river plumes and surface wind forcing interact with bays and subsurface canyons or ridges to create regions of highly productive ecosystems (Hickey and Banas, 2008).

Lists of upwelling centers have been produced for each of the EBUS. In the Peru–Chile Current south of 10°S, the most recognized are Callao (12°S), Pisco-San Juan (15°S; Brink, 1980), Mejillones Peninsula (~23°S; e.g., Sobarzo and Figueroa, 2001), Punta Lengua de Vaca (~30°S; Rutllant and Montecino, 2002), Punta Curaumilla (~33°S; Johnson et al., 1980), and Punta Lavapie (~37°S; Arcos and Salamanca, 1984; Mesias et al., 2003; Aguirre et al., 2012). Although there is less of a tradition in identifying an agreed-upon set of upwelling centers in the California Current, organized observational programs and individual efforts have concentrated around "persistent" upwelling "hot spots" such as over the Heceta Bank complex (44°N, Venegas et al., 2008) and coastal promontories such as Cape Blanco (42.7°N, Moum et al., 1988; Barth et al., 2000); Cape Mendocino (40°N, Kelly, 1985), Point Arena (39°N, Dever et al., 2006; Pringle and Dever, 2009; Largier et al., 2006), Point Sur (36°N, Breaker and Mooers, 1986), Point Conception (35°N, Brink and Muench, 1986), and Punta Eugenia (27–28°N, Espinosa-Carreon et al., 2004, 2012). In like fashion, in-water data and satellite-derived SST and surface chlorophyll-a concentrations in the Canary Current off northwest Africa have been used to identify the dominant upwelling centers: Cabo Vert (15°N), Cabo Blanc (21–22°N), Dakhla (24°N), Cabo Bojador (26°N), Cabo Yubi (27–28°N) and the region between Cabo Ghir and Cabo Beddouzza (30–33°N) (Mittelstaedt, 1991; Barton, 1998; Aristegui et al., 2006; Nieto et al., 2012). An unusual feature of the NW Africa system is the Canary Island archipelago, which interacts with the upwelling filaments in the transition zone 100–500 km offshore of the coast (Barton, 1998; Barton et al., 1998). In the Canary Current off the Iberian Peninsula, cold and chlorophyll-rich filaments identify the 5–6 dominant upwelling centers during the summer upwelling season, most associated with prominent capes (Haynes et al., 1993; Barton, 1998): Cap Finisterre (43°N) or slightly south of the cape (Barton et al., 2001), Cabo Roca (39°N) and Cabo sao Vicente (37°N). The greatest number of filaments were observed at Cabo Roca, followed by Cabo sao Vicente and Cap Finisterre (Haynes et al., 1993).

In the Benguela Current Systems, the process of identifying and characterizing these upwelling centers has been put to use to provide a check on the realism of regional ocean circulation models. For example, seven centers of enhanced upwelling have been identified in the Benguela Current System (Nelson and Hutchings, 1983; Lutjeharms and Meeuwis, 1987; Lutjeharms and Stockton, 1987), associated with areas of locally enhanced WSC and changes of coastline orientation (Shannon and Nelson, 1996). These have all been reproduced in models of the Benguela equilibrium state by Veitch et al. (2006, 2010), providing confidence in the model. The model also reproduces the qualitative difference between perennial upwelling in the north (with a slight maximum in winter) and the seasonal variability in the south (strongest in summer), created by the seasonal movement of the atmospheric South Atlantic Anticyclone along a northwestward axis (Preston-Whyte and Tyson, 1993). The Lüderitz upwelling cell at 27.5°S separates the "perennial" and "seasonal" upwelling domains, which also display differences in the nature of the thermal front that marks the seaward extent of the upwelled water. South of Lüderitz, the upwelling front is well defined and tends to follow the shelf edge, whereas it is more diffuse in the north (Shannon and Nelson, 1996). Thus, the observed characteristics of the upwelling systems provide metrics for ocean models that are increasingly being used for operational applications.

An aspect of the EBUS that most models do not yet include is the coupled nature of the atmosphere–ocean–land systems (Perlin et al., 2007). Present models usually describe the ocean as responding to the wind, ignoring the winds' response to the ocean SST in a coupled fashion. Some of the present models include the more well-known dynamics of sea–breezes, in which the diurnal cycle of land temperatures change the sign of the land–sea temperature difference to create alternating onshore and offshore winds during the daily cycle (Gille et al., 2005). However, the winds also respond directly to the ocean's local SST, with decreasing wind stress over colder water (Chelton et al., 2007). Since the ocean SST is directly affected by wind-driven upwelling and turbulent mixing in the upper ocean, the result is a fully coupled system: upwelling decreases SST next to the coast, which decreases the wind stress next to the coast, increasing the offshore WSC and moving the upwelling farther offshore than in the uncoupled response (Perlin et al., 2007; Haack et al., 2008). To capture these effects, future operational and climate models of the coastal ocean–atmosphere–land systems must ultimately be coupled. This is especially needed (and difficult) for the climate models, since the ocean–land temperature difference is the central component of one hypothesized consequence of global warming: an increase in upwelling strength (Bakun, 1990).

3.2.3. CTWs: Integration of Lower-Latitude Forcing

Coastal margins serve as a guide for CTWs, with periods of several days to weeks, bringing the influence of wind forcing and other disturbances to a local coastal region from lower latitudes. CTW behavior and dynamics are well understood as a result of extensive studies between 1965 and 1990 (Allen, 1975, 1980; Allen and Smith, 1981; Chapman and Brink, 1987; Huyer, 1990; Brink, 1991, 1998; and McCreary et al., 1991). In a manner analogous to the way in which Rossby waves carry wind-driven ocean responses to more westward longitudes in the open ocean, CTWs carry the accumulated effects of wind forcing from lower to higher latitudes along eastern ocean boundaries (Denbo and Allen, 1987; Halliwell and Allen, 1987). Chapman (1987) makes use of these "long wave models" to predict local changes in alongshore ocean currents at a point, intregrating the effects of winds and sea levels along a coastal pathway leading to that point from lower latitudes.

Off northern Chile ($\sim$20°S) in the large, sheltered region of the coast where wind forcing is weak, CTWs with periods of $\sim$50 days take on a special role. Due to weak wind-forced upwelling, CTWs become the main source of vertical displacements of the nutricline needed to sustain primary production. The propagation of waves of this period has compared well to theory and the waves have been traced back to the equator in the mid-Pacific, generated by Madden–Julian oscillations in atmospheric winds (Pizarro et al., 2001).

On the largest time and space scales, CTW dynamics connect the EBUS with the energetic equatorial regions. A significant fraction of the variability in sea level, alongshore currents and eddy kinetic energy (EKE) in EBUS can be directly traced to equatorial origins (e.g., Enfield and Allen, 1980; Chelton and Davis, 1982; Pares-Sierra and O'Brien, 1989; Pizarro et al., 2001; Hardman-Mountford et al., 2003). This is particularly dramatic for the Pacific Ocean, where large equatorial fluctuations of the thermocline and zonal currents related to ENSO propagate poleward in both hemispheres after reaching the South American coast. Periods of these larger-scale disturbances range from months to years (Clarke and Van Gorder, 1994).

3.2.4. Separation of the Jet from the Shelf: The Coastal Transition Zone

The relatively smooth and low-energy description of the long-term average circulation in the EBUS presented in Section 3.1 stands in contrast to the complex circulation patterns one finds in "instantaneous" snapshots of surface temperature and chlorophyll concentrations from the same regions. The first view of these patterns came from high resolution ($\sim$1 km) images of SST and surface chlorophyll-a pigment concentrations provided by the Advanced Very-High-Resolution Radiometer and the Coastal Zone Color Scanner sensors (respectively), beginning in 1979–1980 (Breaker and Gilliland, 1981; Traganza et al., 1983; Nihoul et al., 1997). At the same time, the three-dimensional, subsurface fields associated with these features were captured by newly developed techniques for rapid in-water surveys of velocities, turbulence and water properties (Kosro and Huyer, 1986; Moum et al., 1988; Barth et al., 2000, 2005a). The combination of satellite images and underwater measurements off the U.S. west coast, demonstrated that these structures transported significant amounts of water and plankton biomass (Bernstein et al., 1977; Abbott and Zion, 1985; Flament et al., 1985; Rienecker et al., 1985; Kosro and Huyer, 1986; Breaker and Mooers, 1986; Strub et al., 1991). Similar filaments were identified in each of the EBUS: the Canary Current (Fiuza, 1983; Mittelstaedt, 1983, 1991; Barton, 1987, 1998); the

Benguela Current (Shillington, 1998; Strub et al., 1998a; Lutjeharms et al., 1991) and the Humboldt Current (Fonseca and Farias, 1987; Strub et al., 1998b; Sobarzo and Figueroa, 2001; Hormazabal et al., 2004). These observations led to numerical model experiments that explain much of the structure as arising from intrinsic baroclinic and barotropic instabilities of coastal upwelling jets, requiring no similar spatial or temporal variability in the wind forcing (Ikeda and Emery, 1984; McCreary and Chao, 1985; McCreary et al., 1987, 1991; Marchesiello et al., 2003; Veitch et al., 2010). Thus, the mesoscale circulation features are essential components of the circulation and their effect can be seen in the increase in mean EKE over the EBCs, most especially in the California and Leeuwin Currents (Kelly et al., 1998; Ducet et al., 2000; Strub and James, 2000).

During the spring and summer periods of persistent upwelling-favorable winds, the upwelling fronts and jets eventually venture past the shelf break and enter the deep ocean domain, due to several processes. The accumulation of simple Ekman transports at the surface and onshore return flow at depth can fill the region inshore of the jet with upwelled water and push the jets offshore (deSzoeke and Richman, 1984). This has been observed to occur quickly in regions with the strongest wind forcing, such as the northern California region during the 1981–1982 Coastal Ocean Dynamics Experiment (Beardsley and Lentz, 1987; and other papers in the same volume) and the 1987–1988 Coastal Transition Zone (CTZ) experiments (Brink and Cowles, 1991; and other papers in the same volume). Meanders in the coastline or shelf-break can also cause the jet to simply overshoot the shelf and move into the deep ocean (Barth et al., 2000, 2005b; Mesias et al., 2003; Veitch et al., 2010). Barotropic and baroclinic instabilities have already been mentioned. At low frequencies, fluid particles that experience alongshore displacements from higher to lower latitudes respond to changes in planetary vorticity (the decreased Coriolis parameter) by changing their relative vorticity and inducing meanders that excite Rossby waves, which propagate westward into the ocean interior (Clarke and Shi, 1991).

The curl of the wind stress is also a factor in controlling the offshore movement of the jet. A modeling study by Aguirre et al. (2012) addresses the question: What are the relative roles of alongshore wind stress versus WSC in controlling the offshore movement of the upwelling front jet and the expansion of the mesoscale structure over the deep ocean? Because they are highly correlated, it is difficult to separate these aspects of wind forcing from field measurements. However, Aguirre et al. (2012) are able to separate them in a detailed modeling study of the wind-driven circulation off south–central Chile. The model is forced by climatological, seasonally varying scatterometer wind stress fields, in one case using the complete wind field. In the other case, only the meridional component of the wind stress (approximately alongshore) is used, spatially averaged and uniform in the east–west direction. This produces fields of wind stress with magnitudes similar to the actual winds, but with zero WSC. The results show that a realistic coastal upwelling jet is created by the alongshore wind stress, even when the WSC is zero. However, the jet does not separate from the coast without the WSC. Use of the full wind fields (with the WSC) causes offshore upwelling through Ekman pumping that creates an offshore density field consistent with an offshore jet. This allows the coastal jet to leave the coast in a manner similar to observations in altimeter SSH fields and hydrography. This demonstrates that the realistic WSC field is needed in the 100 km next to the coast to simulate the observed nature of the coastal circulation, a process that the coarse grids used in climate models do not include.

Once it leaves the shelf and evolves into a free jet, flow instabilities create the complex patterns of meandering currents, filaments and closed, westward propagating eddies in the deep ocean seen in the satellite images. The increase in the mesoscale variability and EKE creates a "CTZ", several hundred kilometers in width (Brink and Cowles, 1991; Strub et al., 1991; Barton et al., 1998; Hormazabal et al., 2004; Chaigneau and Pizarro, 2005; Colas et al., 2011). In general, eddies are ubiquitous in the global ocean, capable of persisting for several months to several years, propagating hundreds of kilometers in a predominantly westward direction with speeds similar to Rossby waves (Kelly et al., 1998; Chelton et al., 2011a). Strub and James (2000) used wave-number spectra along altimeter tracks that parallel the coast of California to show the offshore movement of coastally generated EKE from summer to winter. Wavelengths typical of coastal jet meanders are 200–300 km, longer than the dominant wavelengths of the ambient offshore eddy field (100–150 km). Energy at the longer wavelengths moves 500 km offshore in the California Current between summer and winter, approximately 3 cm s^{-1}.

The development of strong mesoscale eddy fields is not limited to equatorward jets, as demonstrated by the poleward Leeuwin Current, the winter poleward flow in the Canary Current off the Iberian Peninsula and the Alaska Current (Section 4) (Crawford, 2005; Relvas et al., 2007; Waite et al., 2007). As with equatorward jets, meanders in poleward jets pinch off eddies that carry water with coastal characteristics westward into the deep ocean (Crawford, 2005; Waite et al., 2007; and other papers in the same volume; Gaube et al., 2013). The global paths of eddies presented by Chelton et al. (2011a) include all of the eastern boundaries as sources of westward traveling eddies of both signs. As an example, Relvas et al. (2007) describe the development of eddies in the winter current along the west coast of the Iberian Peninsula, which flows poleward over the shelf break, next to the narrow shelf.

The current develops instabilities, meanders and cut off eddies, in a process characterized by four phases: "adjustment," starting with a tongue of lighter water; "eddy development" as the tongue separates from the shelf, initiated by interactions with topographic features; "eddy interaction," during which mostly anticyclonic eddies develop from meanders with wavelengths of 150 km and generate dipoles that migrate offshore; and a final "decay" phase, in which the eddies travel westward and decrease in strength. This description of eddy development and decay applies to most instances of eddy generation and decay in EBCs, for either poleward or equatorward jets.

Although the above description attributes the eddy field to instabilities of a jet that originates next to the coast or over the shelf break, offshore eddies originating elsewhere may impinge the coast and entrain coastal water. This occurs occasionally in the Benguela Current (Lutjeharms et al., 1991; Duncombe Rae et al., 1992; Shillington, 1998), where Agulhas rings may travel to the northeast and interact with the coastal circulation, creating exceptionally long filaments. It also occurs regularly in the Gulf of Alaska (Section 4), where eddies created in the northeast off Yakutat move westward in the Alaska Current, drawing water off of the wide shelf in the northern Gulf as they travel (Mackas et al., 2006).

One question about the mesoscale fields of circulation regards their role in the cross-shelf transport of water properties, biomass and whole ecosystems. Water in a coastal jet that meanders offshore may return to the coastal shelf; but it does not necessarily return with the same properties. Nutrients in the cold, upwelled water that are carried offshore of the shelf support the growth of phytoplankton and zooplankton (Mackas et al., 1991), which are consumed by higher trophic levels or simply die and are transformed into detritus that falls out of the water column over the deep ocean before the water returns to the shelf. Eddies that detach from the jet trap water in their centers and carry that water for months or years (Chelton et al., 2011a,b). Near eastern boundaries, even those that are predominantly downwelling, the water trapped within mesoscale eddies is continental in nature and enriched in nutrients, including trace nutrients such as iron. These allow the eddy to remain productive over great distances, as well documented in the Alaska Gyre (Batchelder et al., 2005, and other papers in the same volume; Crawford et al., 2000; Crawford, 2002, 2005). Another region of enhanced eddy transport is found off Northwest Africa, where offshore flow from the Canary Current passes the Canary Islands and generates long-lived eddies of both signs, but with a preference for anticyclones (Aristegui et al., 1994). Sangra et al. (2009) provide a quantitative description of the eddies and define five "eddy corridors" off NW Africa, of which the Canary Eddy Corridor produces the most eddies. This corridor is a region of high EKE and transport of water with coastal biogeochemical properties (including biomass) through the eddy corridor and into the central oligotrophic Atlantic, as far west as 132°W. Thus, on the local and regional scales, the mesoscale circulation fields determine the characteristics of the habitat and ecosystem dynamics within the EBCs; on the larger scale, through the mediation of eddy transports, the productive eastern boundaries contribute to the momentum, heat, salt and biogeochemical balances of the adjacent open ocean.

An example of the mesoscale features revealed by satellite-derived SST and surface chlorophyll-a concentration is presented in Figure 14.8, representing the region off northern California and Oregon at the end of summer. The center panel shows the bathymetry of the shelf, including the wider Heceta Bank region between 44.0°N and 44.6°N. In the SST field (left panel), the coldest water stays inshore of an equatorward jet that flows along the outer edge of the shelf between 45°N and 44°N. When the shelf narrows abruptly along the southern flank of Heceta Bank at 44°N, the jet turns cyclonically back toward the coast, narrowing the band of cold water next to the coast but often overshooting the bank (Barth et al., 2005b). Over time, the meander expands westward, creating a larger cyclonic meander that is eventually cut off, leaving a cyclonic eddy that continues to move to the west. Similar meanders and eddies are generated where the flow is deflected around capes such as Cape Blanco (42.7°N). Currents in the larger meanders spin off their own, smaller eddies, creating a multitude of eddies on a range of spatial scales. Some of the same meanders and eddies are apparent in the surface chlorophyll field (right panel), indicating that the upwelled nutrients are supporting phytoplankton blooms that form the base of the food chain in this system and that the entire enriched habitat is being carried offshore by the advective field and closed eddies. The fact that the cold, rich filaments seem to end in the offshore regions may be caused by subduction of the returning branches of the meanders, as well as by warming of the water and consumption of the phytoplankton. Similar patterns are found in all of the EBCs.

Within the mesoscale field, there are multiple frontal features, which also increase in number and move offshore with a seasonal cycle. Figure 14.9 shows the field of frontal probabilities, calculated from automatic frontal detection algorithms using hourly SST fields collected by NOAA geostationary satellites during a four year period (Castelao et al., 2006). For each pixel in the study domain, the percent of the images in which it was visible (not cloudy) and was flagged as part of a front is plotted for each 2-month calendar period. By the end of the upwelling period (September–October) in Figure 14.9, fronts are present for 7–12% of the time at locations extending 300–500 km offshore of the central California Current system. One of the most frequent locations for fronts is along the edge of

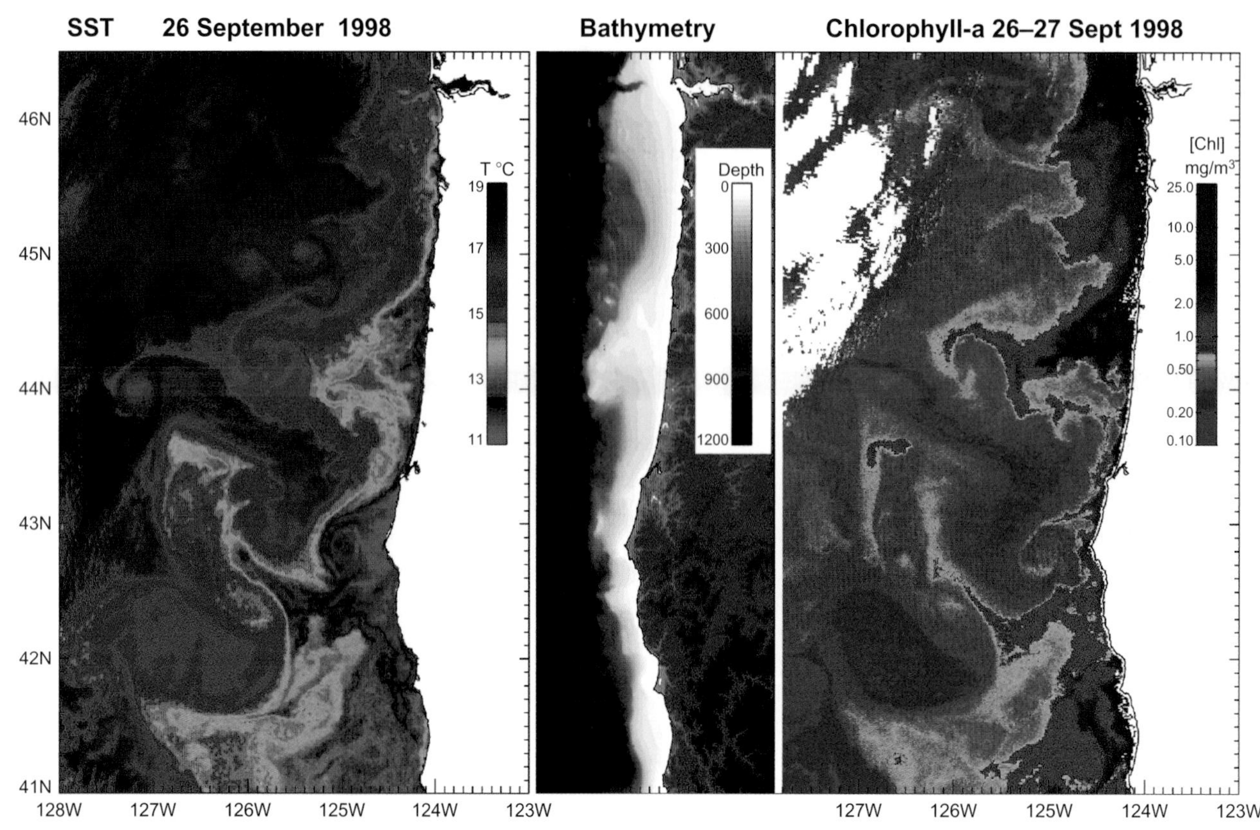

FIGURE 14.8 Left: SST (AVHRR); middle: bottom depth (m); and (right): Surface Chlorophyll-a Concentrations (SeaWiFS). Snapshots in autumn off Oregon. A southward meandering jet flows along the offshore edge of the cold, upwelled water (blues and greens in the left panel) that is closer to the coast. The upwelled water is also rich in chlorophyll (reds and yellows in right panel). The bottom topography around the Heceta Bank (44–45°N) and coastal geometry around Cape Blanco (42.7°N) deflect the jet offshore, adding topographic control to nonlinear instabilities that create the myriad jets and eddies seen in the images of SST and pigment concentration.

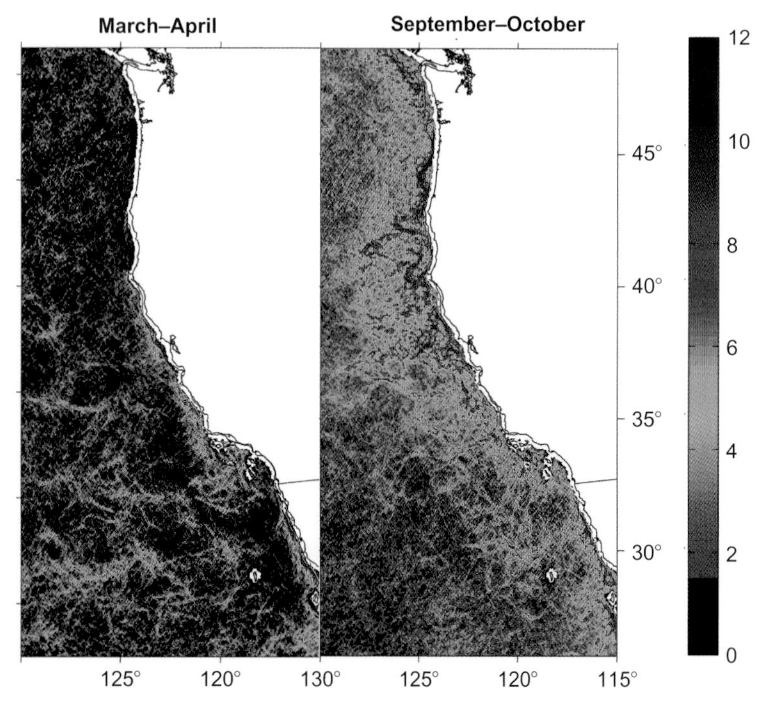

FIGURE 14.9 Probabilities (0–12%) of finding an SST front at a given pixel during the months of March–April and September–October in the California Current System, based on 4 years of hourly geostationary SST fields (4-km resolution). An automated frontal detection algorithm is used, which creates the narrow land mask (blank white region) next to the coast. The 200 m isobaths is drawn with a solid black line, most visible in the right hand panel. *Modified from Castelao et al. (2006).*

the shelf (the black 200 m contour) that borders the Heceta Bank off Oregon (~44.3°N), demonstrating the degree of topographic steering exerted by this submarine bank. By the end of winter (March–April in Figure 14.9), the frontal probabilities drop below 4% over most of this region, except next to the coast south of 40°N, where a new season of upwelling has begun. The region with frontal probabilities reaching 10–12% expands northward and offshore in May–July, reaching the peak coverage shown in Figure 14.9 at the end of summer (September–October). Since frontal regions are also the location of convergences of plankton, the seasonal development of the frontal structures associated with the mesoscale circulation features represents a seasonal expansion of the habitat favorable for foraging by higher trophic levels, helping to sustain the ecologically and economically important populations of shellfish, fish, marine birds and mammals.

Figures similar to Figure 14.8 have been shown by many authors. Frontal probabilities, similar to Figure 14.9, have been calculated and presented for the Canary Current in summer between 34°N and 43°N by Relvas et al. (2007). The frontal distributions reveal locations (39.5°N off Cape Carvoeiro, 36.5°N south of Cape Sao Vicente), where topography controls the location of the coastal fronts and jets to such a degree that the fronts coalesce to create a single, thick line, similar to the appearance of the jet around the Heceta Bank of Oregon off 44°N in Figure 14.9. There are also regions devoid of fronts, presumably due to topographic steering of the jets. Elsewhere the frontal distributions show a relatively even distribution over the 100–200 km next to the coast, similar to Figure 14.9.

3.2.5. *Synthesis of the Subtropical EBCs*

Figure 14.10 portrays a schematic of the seasonal development of the California Current System (Strub and James, 2000; see also Hickey, 1979, 1998; Mackas, 2006). How well does this represent the other subtropical EBCs? The winter poleward currents in the CCS extend from 30°N to 50°N, as also represented by high SSHA next to the coast during January in Figure 14.5. Currents that alternate between poleward in winter and equatorward in summer are also typical in the Canary Current off the Iberian Peninsula (37–43°N), along a southern section of the Peru–Chile Current (35–42°S) and next to the most southern part of the Benguela Current (30–34°S). The simplest description of the dynamics of the poleward current in the winter California Current (also called the Davidson Current) is primarily wind driven, with poleward winds, onshore Ekman transports and an upward slope of SSH next to the coast, which drives the poleward geostrophic currents next to the coast (Allen and Newberger, 1996). Hickey and Pola (1983) describe the winds and alongshore pressure gradients in the California Current (from tide gauge sea levels) as consistent with an "arrested topographic wave" model (Csanady, 1978). Along the Iberian coast, the winter poleward flow is opposed by weakly equatorward vector mean winds and Relvas et al. (2007) give at least partial credit for the poleward currents to the influence of the alongshore density gradient and sloping topography, that is, the JEBAR mechanism of Huthnance (1984, see also Hill, 1998). At the lowest latitudes of two of these systems (off Ecuador and Peru between 0°S and 10°S and off NW Africa between 10°N and 15°N) wind-driven coastal upwelling and equatorward currents occur in winter, due to the seasonal shifts in position of wind around the ITCZ.

The transition to upwelling in the California Current can be rapid (Huyer et al., 1979), in some years occurring over several days and affecting currents between southern California and Canada (Lentz, 1987; Strub et al., 1987). This is sometimes driven by the passage of a final winter storm and the onset of a "blocking" high-pressure system that produces a long period of large-scale equatorward winds (Strub and James, 1988). In other years, the transition occurs sporadically, with alternating storms and increasingly upwelling-favorable winds. An example of a late, "sequential" transition occurred in 2005, as documented by Kosro et al. (2006) and Barth et al. (2007), with negative consequences for the marine ecosystem (Sydeman et al., 2006; Thomas and Brickley, 2006). This shows that timing and spatial extent of the transition is more a response to the characteristics of the regional wind forcing rather than an intrinsic behavior of the ocean.

In the other EBC regions, the altimeter fields do not show similar, rapid transitions from winter downwelling to sustained summer upwelling on as large a scale. In the Canary Current, low SSHA values that represent winter upwelling seen south of 20°S in January continue to be visible through April, without spreading north. High SSHA replaces it in May and then low coastal sea levels appear simultaneously along the coast of NW Africa north of 20°N in June, along with low SSHA next to the Iberian coast. Along the Chilean coast, the lower values of coastal SSHA seen in January between 35°S and 42°S develop in place in October–December and persist through March, with no connection to other regions along the coast. Likewise in the Benguela Current, high SSHA values along the coast are replaced by low SSHA from ~18°S to 34°S in December and continue to strengthen and weaken through March–May. This is in contrast to the apparent propagation of SSHA signals in the lower latitudes of the Atlantic, described by Schouten et al. (2005).

The development of jets along upwelling and downwelling fronts, with unstable meanders that develop into filaments and eddies, appears to be a universal characteristic of the EBCs, with similar behavior in systems with

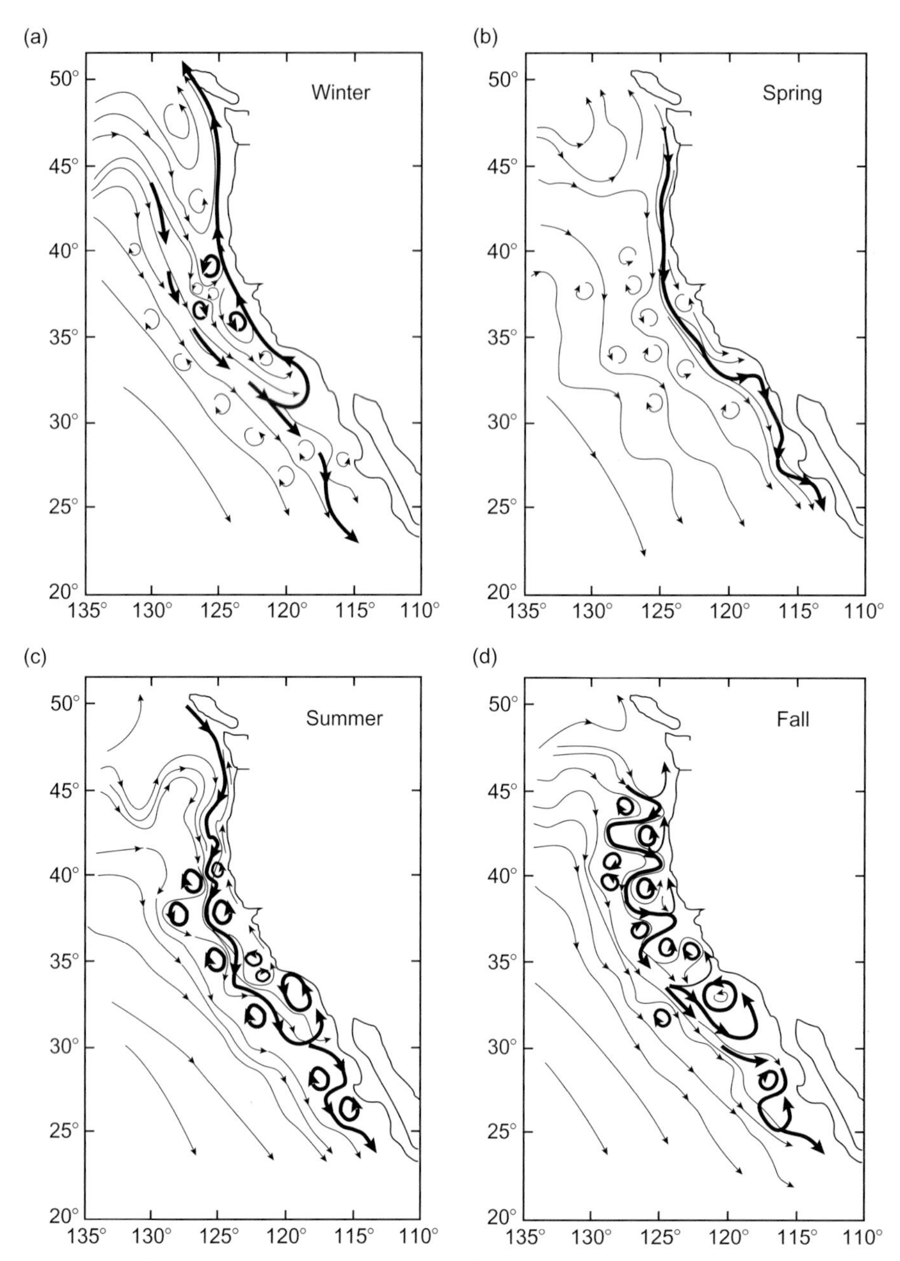

FIGURE 14.10 Schematic diagram of the conceptual seasonal development of the circulation patterns in the California Current System. *From Strub and James (2000).*

equatorward and poleward jets. This is not to say that the mechanisms that govern the instabilities and eddy formation are the same in all regions. Field programs and modeling studies will continue to clarify the dynamics of each system. In the upwelling systems, the eddy field that develops next to the coast in spring and moves offshore in summer continues to move offshore as the winter wind forcing returns and reverses conditions next to the coast. As shown by Chelton et al. (2011a) and in more detail in the Gulf of Alaska (below), some of these eddies continue to propagate for multiple years, carrying water with coastal properties into the adjacent deep ocean over moderately large distances.

4. HIGH-LATITUDE EBCs

The three high-latitude EBC regions are the Alaska Current in the NE Pacific, the Cape Horn Current in the SE Pacific, and the Norwegian Coastal Current in the NE Atlantic. There is no high latitude EBC in the SE Atlantic. The upstream currents for each of the high-latitude EBCs are the regional west wind drift currents in each basin, flowing to the east under strong westerly winds. In each case the west wind drift currents split into poleward and equatorward branches. The poleward current is the EBC of interest, flowing next to shelves that include complex outer islands, inland seas, and glacially carved fjords. After a poleward journey of 10–15°, the similarity between the systems ends. The Alaska Current continues around the closed Gulf of Alaska, returning westward in a counterclockwise direction. The Cape Horn Current continues past the end of the South American continent and joins the eastward Antarctic Circumpolar Current (some water flows through the Straights of Magellan to the coast of Argentina). The Norwegian Coastal Current continues poleward into the Barents Sea. During the poleward transit, freshwater enters each system from the coast, due to heavy precipitation and melting snow and ice, creating baroclinic buoyancy forcing. To this is added strong poleward wind forcing by winter storms. General descriptions of all three systems are found in Talley et al. (2011).

The physical processes at work along the high-latitude EBCs are conceptually straightforward. Poleward winds drive downwelling and increased coastal sea levels, which are augmented by the influx of fresh water at the coast. Poleward geostrophic currents balance the onshore–offshore gradients in SSH and pressure, along boundaries that are convoluted by fjords and islands. Perhaps the most important (and more complex) process is the creation of mesoscale eddies by instabilities in the boundary current, as it interacts with the convoluted topography. These eddies are key elements in the transport of enriched coastal water and plankton populations into the offshore waters, which are classified as "high-nitrate, low-chlorophyll" (HNLC) oligotrophic systems. For this reason, the transport of micronutrients such as iron from the coastal waters to the offshore deepwaters is an important component of the offshore ecosystem. Details of the eddy generation depend on local coastal geometry, currents and stratification, as examined for one region in the Gulf of Alaska (Di Lorenzo et al., 2005), and so should differ within and between systems, a topic open for future research.

Here we focus primarily on the Alaska Current, as an example of an eddy-shedding poleward EBC, which also exchanges water along the ocean margin with its subtropical EBUS neighbor to the south, the California Current. Briefly looking at the other two high-latitude EBCs, the Norwegian Coastal Current is located at the highest latitudes (58–71°N) of the three systems, and is separated from the nearest subtropical EBC of interest to us (off the Iberian Peninsula south of 44°N) by the large and complex North Atlantic Shelf Seas. For reviews of the Norwegian Coastal Currents see Hansen et al. (1998) and Gaard et al. (2006). Reviews of the North Atlantic regions between 44°N and 58°N can be found in Simpson (1998), Sharples and Holligan (2006), Rodhe (1998), and Rodhe et al. (2006).

In the SE Pacific, the poleward Cape Horn Current may be the most similar to the Alaska Current, in that direct exchanges occur between it and the subtropical EBC to its north, the Peru–Chile Current. Meridional displacements of the inflowing WWD current can affect both the subtropical and high-latitude EBCs, as is also true in the NE Pacific. However, most of the research in this region has been dedicated to the study of the Chilean inland seas, motivated by the use of the area for salmon farming. This activity has the potential for the creation of eutrophic and anoxic conditions, due to the introduction of large amounts of organic "fish food." The region is also subject to episodes of red-tide blooms, which are affected by tidal mixing, freshwater input and circulation patterns. Results from this research are summarized in Silva and Palma (2008, and other papers in the same volume) and Pantoja et al. (2011, and other papers in the same volume). Studies of specific fjords are also reported in Cáceres and Valle-Levinson (2004), Cáceres et al. (2003, 2006), Valle-Levinson et al. (2003), Valle-Levinson and Moraga-Opazo (2006). See also Montecino et al. (2006).

4.1. The Gulf of Alaska Circulation

The Gulf of Alaska is a semi-enclosed basin delimited by the Canadian and Alaskan mountains on the east and north, by the Aleutian Peninsula on the west and open to the south (Figure 14.11). The ocean circulation is subject to the combined effects of atmospheric forcing, discharges of fresh water, coastline irregularity and complex bottom topography (Royer, 1998; Stabeno et al., 2004). The complex bathymetry includes the presence of relatively large islands (Kodiak Island, Queen Charlotte Island, Vancouver Island), numerous inlets, and deep canyons intersecting a continental shelf that is wide along much of the coast. Along the Aleutian Island archipelago, the 7500 m deep Aleutian Trench also influences the path of the ocean currents along the continental margin.

4.1.1. *Atmospheric Forcing*

Located at the terminus of the Pacific storm track, the atmospheric forcing in the Gulf of Alaska spans a wide range of frequencies from high-frequency local winds to interannual and decadal scales of variations. On average, the wind stress pattern (Figure 14.11) exhibits a cyclonic circulation related to the Aleutian low pressure center over the Aleutian Peninsula. These winds result in Ekman pumping and upwelling in the central Gulf of Alaska and coastal downwelling in the northern and eastern regions. The upwelling in the central gyre provides high concentrations of macronutrients such as nitrate but a lack of iron keeps productivity low (it is a HNLC region). From May to September, the cyclonic winds over the northern Gulf of Alaska are weaker

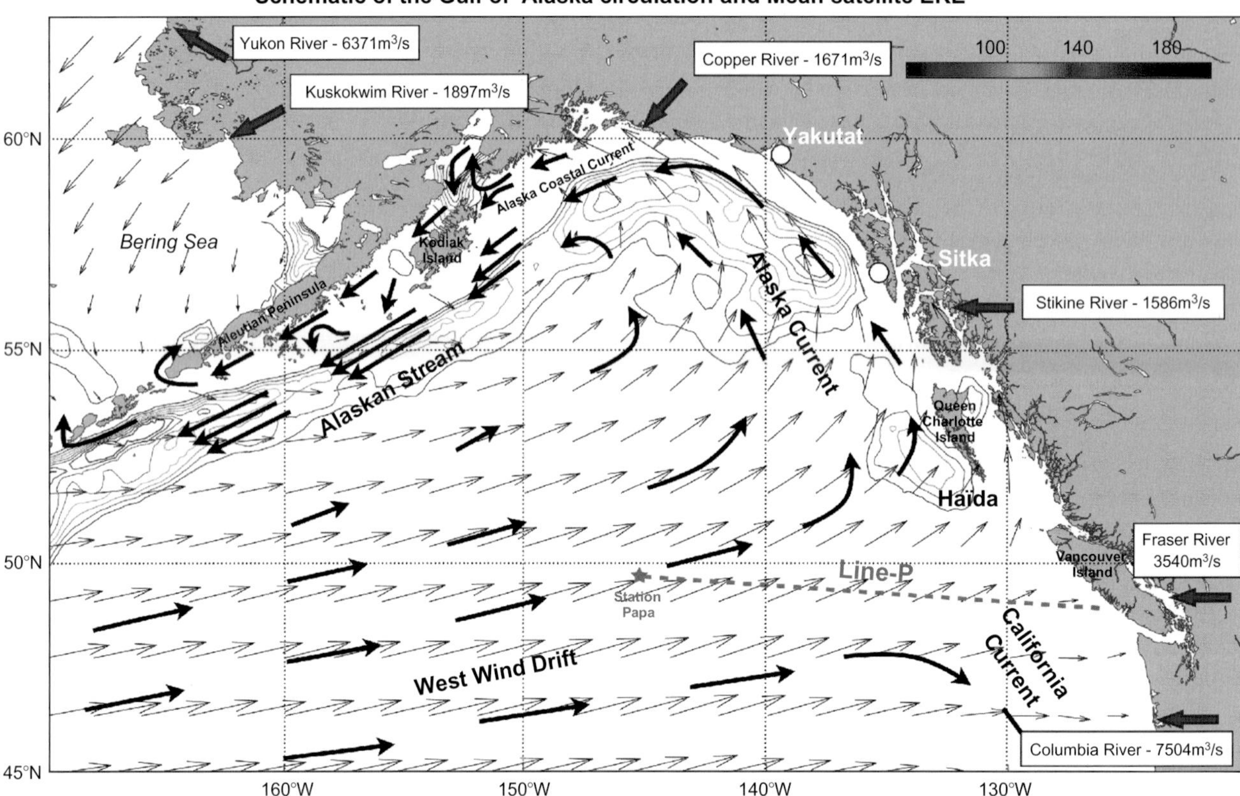

FIGURE 14.11 Annual wind stress over the Gulf of Alaska from the QuikSCAT scatterometer (light gray vectors). Schematic of the surface circulation (dark black vectors). Mean eddy kinetic energy derived from the AVISO gridded altimeter data set ($cm^2\ s^{-2}$, colored contours). Mean river discharge for the major rivers around the basin.

with periods of coastal upwelling (Stabeno et al., 2004). Analyses of the historical fish catch and estimates of stock size (in particular for Alaskan salmon) have revealed coincidental changes in the physical system and ecosystem in the North Pacific, which have been associated with interannual and interdecadal atmospheric variability, such as the Pacific Decadal Oscillation (Francis and Hare, 1994; Mantua et al., 1997; Ingraham et al., 1998).

4.1.2. Fresh Water Input

Fresh water input from precipitation and river discharges along the coast of the Gulf of Alaska, Washington, and Oregon plays an important role in the baroclinic structure of the coastal currents. The Columbia River ($\sim$7500 $m^3\ s^{-1}$, at the mouth), the Frasier River ($\sim$3500 $m^3\ s^{-1}$), Stikine River ($\sim$1600 $m^3\ s^{-1}$), and Copper River ($\sim$1700 $m^3\ s^{-1}$) are the main rivers along the coastline and inject fresh and iron-rich waters into the coastal circulation, noting that the poleward flow along Oregon and Washington coasts in winter continues into the Gulf of Alaska. Nevertheless, it is the numerous small rivers and streams that contribute the most to the net fresh water discharges. For the state of Alaska, Royer (1982) estimates that precipitation from the Pacific Storm track produces a water discharge from the coastal mountains of around 23,000 $m^3\ s^{-1}$. This discharge is spread over the Alaskan coastline (as a line source) and is comparable to the discharge of the Mississippi River (with an average of 17,000 $m^3\ s^{-1}$). Although the strong seasonal maximum (minimum) in precipitation occurs from September to November (June to July), the precipitation is in the form of snow during winter, creating a delay in its contribution to river discharges. The river discharges, therefore, have a maximum of 40,000 $m^3\ s^{-1}$ in October, and a minimum of 10,000 $m^3\ s^{-1}$ in February (Stabeno et al., 2004). The importance of the contribution of freshwater discharge from frozen sources is highlighted in a study examining the timing and volume of watersheds along the Gulf of Alaska (GOA). Neal et al. (2010) estimate that "discharge from glaciers and ice fields accounts for 47% of total freshwater discharge, with 10% coming from glacier volume loss associated with rapid thinning and retreat of glaciers along the GOA." Future near-shore and coastal ocean circulation can be expected to change if there are alterations in the timing and volume of buoyant water inflow to the coastal Gulf of Alaska, such as hypothesized due to future loss of glaciers in a warming climate.

4.1.3. Oceanic Circulation

The circulation in the Gulf of Alaska is dominated by the cyclonic subarctic gyre and the Alaska Coastal Current next to the northern continental shelf (Figure 14.11). As its source, Freeland (2006) estimates that approximately 60% of the North Pacific Current flows into the Gulf of Alaska. The Alaska Current is a broad eastern current rich in meanders and eddies, such as the Haïda and Sitka eddies. The geostrophic component of the Alaska Current is maintained by the poleward surface wind stress, downwelling and influx of buoyant fresh water along the coast. As the Alaska Current flows along the Aleutian Peninsula it becomes the Alaskan Stream, which has a more developed and richer eddy field associated with the strong instability of the mean currents (Melsom et al., 1999). The Alaskan Stream, which flows along the steep shelf break, has been associated with current speeds as high as 120 cm s^{-1} (Reed and Stabeno, 1999). The baroclinic Alaska Coastal Current (over the broad shelf) is principally driven by winds and freshwater runoff, with a strong seasonal cycle. The downwelling winds intensify the Alaska Coastal Current during winter, but it is during autumn that the Alaska Coastal Current has a maximum baroclinic structure (due to maximum freshwater runoff). Surprisingly, Miller et al. (1994) suggest that the Alaska Current and eddies generated in the eastern GOA remain mostly unchanged during the strengthening of the Aleutian low after the 1976–1977 climate shift, despite an intensification of the Alaskan Stream.

The Gulf of Alaska supports a rich ecosystem with abundant fisheries. How does it do this with in a system characterized by low-nitrate freshwater runoff, the prominent coastal downwelling and a High-nitrate Low Chlorophyll center? Ladd et al. (2005) hypothesize that three mechanisms contribute to cross-shelf exchanges of nutrients that support the highly productive ecosystem. (1) Bathymetric steering along canyons and other across-shelf topographic features, combined with strong tidal mixing, moves enriched, coastal water to the edge of the wide shelf in the northern Gulf. (2) Episodes of downwelling relaxation produce offshore surface flow and an onshore transport of deeper, nutrient rich waters over the shelf. (3) Finally, eddy transport plays an important role in moving water from the shelf edge into the deep interior.

Eddies in the Gulf of Alaska have been studied for several decades using various in situ (drifters) and satellite observations (Tabata, 1982). In the eastern Gulf of Alaska, three major groups of eddies have been identified according to the location of their formation: the Yakutat eddies, the Sitka eddies, and the Haïda eddies. These three eddy groups are strong and persistent enough to leave a footprint in the satellite mean EKE (Strub and James, 2002a), sometimes surviving for over three years (MacFarlane and McKinnell, 2005, and other papers in the same volume; Miller et al., 2005, and other papers in the same volume). Most are associated with an anticyclonic rotation, containing a warmer and less saline core (Crawford, 2002).

The (most northern) Yakutat eddies travel west with the Alaska Current, next to the edge of the wide shelf along the northern boundary of the Gulf of Alaska (Mackas et al., 2006), then enter the Alaskan Stream, where they grow in diameter and height (Crawford et al., 2000). The Sitka and Haïda eddies usually propagate offshore to the west across the open ocean. Haïda eddies can be generated by the simple advection of buoyant water around the Queen Charlotte Island (Di Lorenzo et al., 2005), although the main eddy generation mechanism is the baroclinic instability of the coastal currents. Eddies generally form in winter and detach from the continental margin in late winter and spring. Interannual changes in surface winds also modulate the seasonal development of the Haïda and Sitka eddies (Combes and Di Lorenzo, 2007; Combes et al., 2009). Specifically, during coastal downwelling conditions, relaxations of winds produce positive SSHAs that are converted into a field of large-amplitude eddies, due to the fast growth rates of instabilities at high latitudes. Coastally trapped waves, originating at lower latitudes (as far away as the equator during strong El Niño events), are credited with contributing to higher SSHAs along the coast, destabilizing the Alaska Current and promoting the development of strong anticyclonic eddies (Melsom et al., 1999). Eddies along the Aleutian Peninsula however do not seem to be correlated with any large-scale forcing and appear to be intrinsic (Combes and Di Lorenzo, 2007). The asymmetry in the eddy generating mechanisms between eddies along the Alaska Current (forced regime) and eddies along the Alaskan Stream (intrinsic regime) has important implications in predicting the response of the Gulf of Alaska to different atmospheric scenarios, and hence to climate change.

Eddies play an important role in the transport of coastal waters offshore. For example, Crawford (2005; and other papers in the same volume) estimates that around 35–60% of the northward heat transport during winter is carried offshore by the Haïda eddies and that 15% of the total river freshwater north of the Columbia River is transported by these eddies. Not only do eddies along the eastern basin transport heat and freshwater, they also transport chlorophyll in their outer rings and dominate the chlorophyll distribution offshore (Crawford et al., 2007; Ladd et al., 2007, 2009). To sustain the phytoplankton biomass during the eddies' multiyear lives, primary production must be enhanced by iron, since the Gulf is a known HNLC region. One mechanism that may maintain the highly productive region inside the anticyclonic eddies is the relaxation of the depressed isopycnals as the eddies spin down and decay, bringing macronutrients into the photic zone. It is, however, the transport of iron in the coastal water that is trapped in the eddies that maintains the primary productivity and high-chlorophyll concentrations. The hypothesis that the center of the Gulf of Alaska is iron limited has been supported by field studies such as the Subarctic Ecosystem Response to Iron Experiment Study (SERIES), which stimulated a significant phytoplankton bloom by injecting dissolved iron into the waters at Station Papa (Boyd et al., 2004). At higher trophic levels, Batten and Crawford (2005) use data from plankton recorders between 2000 and 2001 to confirm the presence of higher concentrations of coastal zooplankton species within the cores of the productive Alaskan eddies. Thus, the eddies generated in the richer coastal waters along the eastern boundary of the Alaska Gyre perform a well-documented function in transporting enriched coastal water and ecosystems into the more oligotrophic central gyre, sustaining their continued productivity by several processes during multiyear journeys.

5. CLIMATE VARIABILITY AND THE OCEAN'S EASTERN BOUNDARIES

Separate chapters in this book (Part V) discuss the ocean's role in climate variability from the global perspective. Our discussion here is primarily from the EBC perspective; but even in this more limited topical domain, the body of literature is vast. Rather than a summary of these many efforts, we provide only an overview of the topic, with a few examples and references to the literature. A good start on past reviews is provided by the EBC overviews in volumes 11 and 14 of *The Sea* (Robinson and Brink, 1998a,b, 2006), along with Freon et al. (2009, and other papers in the same volume).

The natural questions to ask are easy to state but complex to answer:

a. What are the most important of the EBC processes for climate variability?
b. How are EBC processes affected by climate variability?
c. How do EBC processes affect the larger-scale climate variability?
d. Are these processes well simulated in climate models?
e. What changes do the models predict in EBC processes due to global climate change over the next 50–100 years?

We briefly discuss (a)–(d). Several approaches can be taken in addressing these questions, focusing on: changes in the large-scale climate modes; variability in the dominant individual processes; statistical relationships between climate modes and regional processes; and downscaling global climate models to regional domains, with better resolution of the processes. Question (e) is the topic addressed by the ongoing series of IPCC (Intergovernmental Panel On Climate Change) reports. At present, the IPCC models do a poor job of representing the coastal upwelling systems along eastern ocean boundaries (Gent et al., 2011; see Chapter 23).

5.1. The Dominant Processes

All of the processes listed in Section 1.2 affect, or are affected by, the large-scale modes of climate variability. These interactions with the climate system occur on multiple spatial scales and all timescales: diurnal, synoptic (2–10 day), CTW (10–30 day), Madden–Julian (30–60 day), annual, interannual (ENSO: 3–7 year) and decadal, and multi-decadal (North Atlantic, Arctic and Pacific Decadal Oscillations, 10–60 years). Which are the most important in the different regions?

At low latitudes, changes in the equatorial waves and currents that affect the low- and midlatitude EBCs are of primary importance and have received a great deal of attention in the Pacific. The above discussion of the annual cycles in SSHA also highlights the importance of tropical geophysical waves in the Atlantic. These waves and other signals in tropical regions respond quickly and strongly to changes in the location of the ITCZ and Trade Wind systems, including the fields of WSC. In the eastern tropical Pacific, the wind jets through the mountain gaps of Central America are a dominant forcing process, while the large-scale temperature differences between the ocean and African land mass are more important for the wind forcing in the eastern tropical Atlantic.

The eastern ocean boundaries in the midlatitudes are connected to the tropics through the coastal wave guide, as described in numerous studies. Inflow of tropical water from low- to midlatitudes occurs in both subsurface PUCs and surface currents. The subsurface connection back to the equator has been made most clearly for the SE Pacific (Silva and Neshyba, 1979; Lukas, 1986). The PUCs in each of the other EBCs have been traced along their midlatitude regions to their boundaries with the low latitudes, but not as far as the equator (Barton, 1998; Shillington, 1998; Pierce et al., 2000). At the surface, Figure 14.5 makes it clear that SSHA fields in the Southern Hemisphere EBCs are connected to equatorial systems over a short and direct pathway. The pathway is longer and more complex between the Northern Hemisphere EBCs and the equator, but is shorter and more direct to the NECC. In the Atlantic EBCs, poleward surface currents along the low-latitude boundaries meet the equatorward midlatitude currents in strong confluences and frontal regions (Barton, 1998; Field and Shillington, 2006). Studies of the ABFZ show that the strength and position of the SST front depends on both the strength of the incoming currents and the local wind forcing, making them subject to climatological changes in the strength of the currents and winds in both the low- and midlatitude systems. Although there are no similar fronts in the Pacific, changes in the PUCs bring tropical signals to the midlatitudes and are sensitive to climatic changes along the equator.

At the opposite ends of the midlatitude systems, inflow from the west wind drift currents and along the coast from the higher latitudes affects the two Pacific and the North Atlantic systems. Properties of that inflow in some cases determine the initial properties of the upwelled water, with consequences for the ecosystems. As an example, upwelled water in the northern portion of the California Current in 2002 was colder and richer than usual, creating a larger bloom of phytoplankton that subsequently sank and decayed, contributing to hypoxic bottom waters over the shelf. This had not been commonly observed previously and had negative effects on the economically important crab and bottom fish populations. Various analyses led to the conclusion that the source water for upwelling in 2002 originated in the central Alaska Gyre, brought to the outer flank of the upwelling region by the large-scale circulation (Freeland et al., 2003; Huyer, 2003, and other papers in the same volume). Since 2002, hypoxic conditions have continued to develop each summer, although the causes are thought to involve more factors than simply the advection of different source waters from the Alaska Gyre (Chan et al., 2008), including changes in strength and timing of summer upwelling. The lesson that can be learned from this experience is that EBC properties (including the abundance of higher trophic level species) are sensitive to relatively moderate changes in the large-scale circulation and regional forcing, such as expected from climate variability and change. Further discussions of economic effects of regional climate variability are found in Carr and Broad (2000).

As one moves to higher latitudes, changes in the strength and timing of wind forcing become more dominant as sources of variability in the ocean's boundaries. The fields of wind stress and WSC are controlled, in turn, by the large-scale atmospheric pressure systems, with connections to both lower and higher latitudes. Thus, the forces that control inflow and surface forcing in the midlatitudes are all subject to changes in the large-scale climate. The same is true of the high-latitude EBCs, which receive their strongest forcing from synoptic storms. Changes in the position of the atmospheric jet streams affect the storm tracks, with direct consequences on the forcing by wind and fresh water buoyancy flux. For the high-latitude EBCs, the variability in precipitation becomes as important as forcing by the wind (Royer, 1998; Stabeno et al., 2004).

5.2. Climate Modes

One way to reduce the complexity of large-scale climate variability is to describe that variability as a combination of a small number of global or basin-scale modes. Seasonal variability (see also Chapter 24) is one type of mode that we usually characterize as composed of fixed or variable cycles, which we examine and remove. The remaining nonseasonal variability includes intraannual, interannual and longer signals (see also Chapter 25), which are usually determined from the leading patterns of statistical variability, using variations of "factor analysis," "empirical orthogonal functions," etc. (Richman, 1986). These analyses extend the pioneering correlation analysis of Walker (1924) and Walker and Bliss (1930, 1937). The longest historical climate-related time series are fields of SLP and SST, whose statistical analyses reveal atmospheric, oceanic and coupled climate modes. Atmospheric examples include the North Pacific Oscillation (NPO, Walker and Bliss, 1930; Rogers, 1981), the Southern Oscillation (SO, Walker, 1924), the Pacific/North American (PNA) pattern of atmospheric connections (Namias et al., 1988), the Aleutian Low (AL, Trenberth and Hurrell, 1995), and the North Atlantic Oscillation (Hurrell et al., 2003). Oceanic or coupled examples include the El Niño, usually coupled with the Southern Oscillation as ENSO, with indexes such as the MEI (the Multivariate ENSO Index, Wolter and Timlin, 1998). In the Atlantic, there is a warming of the waters off SW Africa similar to the El Niño, called a Benguela Niño (Florenchie et al., 2003; Gammelsrød et al., 1998; Rouault et al., 2007, 2009). When the analyses of SST and altimeter SSH are restricted to the North Pacific Ocean (north of 20°N), the leading modes become the longer-period Pacific Decadal Oscillation (Mantua et al., 1997) and the North Pacific Gyre Oscillation (NPGO, Di Lorenzo et al., 2008). Discussions of the primary global climate modes as calculated from long SST data sets ($\sim$150 years) can be found in Enfield and Mestas-Nunez (1999), with application to the eastern Pacific by Mestas-Nunez and Enfield (2001). Deser et al. (2011) relate the SST modes to surface fluxes of heat determined by the atmospheric climate modes, coupled ocean–atmosphere modes and modes intrinsic to ocean dynamics. Di Lorenzo et al. (2013) offer a hypothetical synthesis of the interactions between interannual and interdecadal modes in the North Pacific, which could increase our ability to predict changes in these modes, if verified by further testing.

5.2.1. *Use of Modes as Predictors*

How do basin-scale or global climate modes affect the ocean circulation in the eastern boundaries. A simple technique is to correlate the variables of interest with the time series of each mode, "projecting" the local data set onto the mode. Another methodology combines the climate modes with the leading modes (EOFs) of the local (eastern boundary) data set of interest, treats the climate mode as the predictor and quantifies the statistical Principal Estimator

Patterns (PEPs, Davis, 1977, 1978), similar to Canonical Correlation Analysis (Preisendorfer, 1988). This quantifies the covariability between basin-scale climate modes (ENSO, SO, PDO, North Atlantic Oscillation, Arctic Oscillation, Antarctic Oscillation, etc.) and EBC variability. Thomas et al. (2012) use both approaches (correlations with time series of the leading modes, PEPs) to explore the relationships between satellite-derived surface chlorophyll concentrations in the Pacific EBUS (the predicted fields) and various forcing fields and climate indexes over subregions of the Pacific Ocean. The results identify ENSO as the strongest mode that affects the pigment field, not surprising given previous results (next section). The PDO is the next best predictor. The single variable that covaries most closely with the pigment fields is the altimeter SSHA, which is interpreted as a proxy for mixed layer depth. That result is ambiguous, since positive correlations of pigment concentrations and deeper mixed layers may mean that the total chlorophyll amount is lower, that the chlorophyll amount is the same but diluted over a deeper layer or that the chlorophyll is concentrated in a deep chlorophyll maximum, beneath the mixed layer (and thus hidden from the satellite sensor).

5.2.2. Interannual and Interdecadal Effects on Pacific EBCs

In the low-latitude regions of the Pacific, described in Section 2, the dominant mode of variability is the well-known ENSO mode, with timescales of 3–7 years (Philander, 2000). The effects of El Niño on the eastern boundaries include the deepening of the mixed layer due to the eastward propagation of equatorial Kelvin waves and the poleward propagation of the deepened isopycnals along the coastal wave guide. The propagation speeds of the ENSO signals reported by Enfield and Allen (1980) and Chelton and Davis (1982) are 40–100 km day^{-1}, slower than expected for CTWs (200–500 km day^{-1}, Huyer, 1990), but similar to the phase speeds predicted by Clarke and van Gorder (1994) for much slower modes (still coastally trapped). As the signal moves poleward, it excites Rossby waves, which travel westward and reduce the energy in the propagating coastal signal. The higher SST values along the Equator also affect the tropical atmospheric convection and the mid- and high-latitude atmospheric pressure systems (through the Hadley and Walker circulations). Changes in the pressure systems alter the storm tracks and patterns of midlatitude winds and precipitation. The opposite effects are created by the cold phase of ENSO (the La Niña). Although early discussions of El Niño described a weakening of upwelling-favorable winds during El Niño at low- and midlatitudes, observations indicate a strengthening of those winds off Peru (Halpern, 2002) and elsewhere (Chavez, 1996). The collapse of fisheries that was blamed on weakened winds is usually caused by the deepening of the pycnocline/nutricline, with the consequence that upwelled water comes from above the nutricline, eliminating the injection of nutrients into the mixed layer from below (Barber and Chavez, 1983; Chavez et al., 2003; Chavez, 2005).

The result of the interactions described above is that the effects of El Niño arrive at mid and high latitudes through both the oceanic and atmospheric pathways (Chelton and Davis, 1982; Simpson, 1984; Emery and Hamilton, 1985; Mysak, 1985), with an increased dominance of the atmospheric contribution as one moves to higher latitudes. Off the U.S. West Coast, El Niño periods include stronger winter storms off California and slightly drier conditions off Oregon and Washington. In the Gulf of Alaska, the effects include positive anomalous air temperature, precipitation and along-shore winds during winter (Stabeno et al., 2004). Although propagation of equatorial signals has been documented in the Gulf of Alaska's coastal variability (Meyers and Basu, 1999), the separation of the atmospheric and oceanic contributions is difficult.

The 1997–1998 El Niño was thoroughly analyzed with respect to its effect on the Pacific EBCs. Altimeter fields of SSHA clearly show the movement of the high SSHA bands along the eastern boundaries of both hemispheres (Strub and James, 2002c). In that event, two pulses of high SSH arrived at the coast of South America, having traveled from the west along the Equator. The first arrived in May and propagated quickly down the coast of South America, where autumn and winter winds helped its passage. Along the North American coast, the poleward movement of that first pulse was opposed by spring and summer winds and currents. When the second pulse arrived in November, however, the seasons had reversed and the signal traveled more easily up the North American coast. At the same time, winds along each EBC are also affected by the ENSO conditions at the Equator. In an effort to separate the effects due to the oceanic pathway from those due to the atmospheric pathway during 1997–1998, Strub and James (2002b) report correlations between SSHA south of 30°N and SSHA along the rest of the North American coast. Correlations to the southern SSHA values decrease with increasing latitude, while correlations with local winds increase as one moves from 30°N to 50°N. A similar mix of low-latitude oceanic connections and mid-to-high-latitude atmospheric connections might be expected for longer period climate change processes in both basins, if the climate changes originate in the tropics.

On interdecadal scales, the leading mode of atmospheric variability is related to the NPO, AL and PNA, although they are all related to the strength and position of the Aleutian Low (Namias et al., 1988). By impacting the ocean's heat budget and hence the SST, interdecadal fluctuations of the atmospheric PNA closely parallel the leading mode of the North Pacific SST variability, the Pacific Decadal Oscillation (PDO). In particular, during the 1976–1977 winter, the atmosphere and climate systems were observed to shift abruptly, with an intensification of the Aleutian low and hence storm intensity and downwelling-favorable winds (Miller et al., 1994). In the ocean, the 1976–1977 climate shift was linked to a cooling of the sea surface temperature in the central Pacific and a warming off the coast of western North America, both consistent with wind stress and WSC associated with the intensified Aleutian Low.

A problem is created by the long periods of the interdecadal climate modes: The number of years in the time series of instrumental measurements is not long enough to allow us to statistically characterize these modes. On the one hand, their long period makes it tempting to use them as predictors. On the other hand, we have only seen one to several realizations of their behavior. This is illustrated by the PDO. After the 1997–1998 El Niño, the PDO changed from positive to negative. Since the existing record prior to 1997 shows the PDO to remain in a given state for at least 10 years, once it has switched and held its sign for several years, predictions were made that it would remain in the negative state for a decade or more, with the associated characteristics of that state (cooler temperatures around the boundaries of the Pacific, better returns of salmon off Oregon and Washington, etc.). Fortunately, the predictions were made with appropriate caveats about the uncertainty of its conclusions, since the PDO switched sign again a year later and has continued to fluctuate with higher frequencies than seen during most of the twentieth century (Schwing et al., 2005). This is the practical meaning of the statement that we do not have a long enough historical record to define the statistical characteristics of the PDO with any certainty. If the period is 60 years, we would require 300 years to observe five independent realizations of the phenomenon (5 degrees of freedom).

5.3. Changes in Processes

5.3.1. Changes in Process Strength: The Example of Upwelling-Favorable Winds

One of the best known predictions concerning EBCs and climate applies to even longer timescales: Bakun (1990) has identified the strength of wind-driven upwelling as a prime candidate for change during global warming, hypothesizing that overall warming of the land will intensify the thermal lows over the continents in summer, with stronger equatorward (upwelling-favorable) winds due to the increased land–sea pressure gradient. Analyses by Schwing and Mendelssohn (1997, 2002) have found some evidence for an increase in upwelling-favorable wind stress in the Comprehensive Ocean-Atmosphere Data Set (COADS, merchant ship) data over the past 50 years off Central California, consistent with the prediction. We note that use of the COADS data set for analysis of trends must always address artificial trends in the data caused by changes in the methods used by merchant vessels to collect the environmental data (Ramage, 1987; Cardone et al., 1990). Bakun et al. (2010) provide an additional statistical analysis to support the hypothesis that increases in upwelling-favorable winds will accompany increased greenhouse gas concentrations. To do this, they correlate present winds with atmospheric water vapor content, since water vapor is a stronger greenhouse gas than CO_2 and is predicted to increase in concentration as the atmosphere warms. They conclude that upwelling forcing increases with increased concentrations of atmospheric water vapor (specific humidity) off Peru. They discuss a number of caveats regarding the interpretation of these results.

Looking for more direct evidence, studies in the Canary Current are supportive of a decrease in upwelling over the past several decades, rather than an increase, as summarized by Aristegui et al. (2009) and Barton et al. (2013). Lemos and Pires (2004) reported decreases in measured upwelling-favorable wind stress along the Iberian coast from 1940 to 2000, while COADS data showed a slight SST warming. Alvarez–Salgado et al. (2008) also found a decrease in upwelling forcing using pressure-derived winds from atmospheric models along the Iberian coast between 1966 and 2006. Lemos and Sanso (2006) used in-water data to show warming of the upper 500 m next to the Iberian coast. Aristegui et al. (2009) show trends in satellite SST between 1998 and 2007, which are positive over the Canary Current off NW Africa. Barton et al. (2013) review a wide range of wind and SST data in the Canary Current from 15°N to 43°N and find a general increase in SST throughout the system, with no change or a decrease (Iberia) in upwelling-favorable wind stress. These findings are inconsistent with increased upwelling and contradict inferences by McGregor et al. (2007) of decreasing SST from sediment records 25 km offshore of Cape Ghir ($\sim$31°N). The sediment proxy for SST is based on changes in the chemical composition of cell walls in coccolithophorid phytoplankton, which covary with changes in the temperature of the water surrounding the living coccolithophorid cells. Barton et al. (2013) suggest that the coccolithophorids may have redistributed to greater depths (colder water) as the surface temperature and stratification increased over the last 100 years.

Changes in SST are somewhat inconclusive in proving increasing or decreasing upwelling. This is demonstrated off Peru during El Niño conditions, during which upwelling continues or even increases, but upwelled water is drawn from above the thermocline, with little effect on the SST (Barber and Chavez, 1983; Chavez, 2005). A similar situation can occur in the midlatitude EBUS, where signals from various types of wind events at lower latitudes can propagate into the region and deepen the thermocline, causing continued upwelling to occur entirely above the thermocline, with no effect on SST (Chavez, 1996; Rutllant and Montecino, 2002; Montecino et al., 2006). Considering basic processes, however, the Bakun hypothesis predicts lower pressures over the land, greater pressure differences between land and ocean and stronger upwelling winds. This hypothesis remains open for testing (Mote and Mantua, 2002). It would seem a good candidate for targeted studies with both new in situ data (collected over the last 20–30 years) and with coupled models, using both idealized models and downscaled boundary conditions from climate models.

Some of the other processes discussed above should also be examined for changes in the data sets that are newly available. Location and strength of wind stress and WSC fields around the ITCZ might be tested with reanalysis wind fields. Changes in the position and strength of SST fronts associated with various features discussed could be examined with reprocessed satellite SST fields, extending the climatology of Hardman–Mountford et al. (2003) and model studies of Colberg and Reason (2006). Alongshore pressure gradients could use the nearly 20 years of altimeter data. Other variables and processes that could be investigated with in water data sets (World Ocean Atlas) include the depth and stratification of the mixed layer, strength and water properties of the PUCs and other source waters for upwelling.

5.3.2. *Phenology: Changes in Characteristics of Seasonal Cycles*

In addition to affecting the mean state of processes, changes in climate may also take the form of changes in the nature of the higher frequency variability, a topic often called "phenology," especially when applied to biological seasonal cycles (Bograd et al., 2009; Henson et al., 2009; Platt et al., 2009; Racault et al., 2012). For example, in the California Current System, there are several decades of papers that attempt to quantify long-term changes in the annual date of the "Spring Transition"—a rapid switch from downwelling to upwelling conditions (Huyer et al., 1979; Strub and James, 1988; Bilbao, 1999; Pierce et al., 2006; Barth et al., 2007; Henson and Thomas, 2007). More generally, one can investigate changes in the length of seasons, the strength of seasonal peaks, the timing and speed of the transitions, etc. (Bograd et al., 2009). These seasonal timescales, in turn, provide the context for quantification of higher-frequency within-season variability—the length and intensity of winter storms and storm surges (Bromirski et al., 2003), the frequency of upwelling events and relaxations, daily changes in the depth of the marine boundary layer, strength of the sea breeze, cloud cover, and photosynthetically available radiation, etc. Differences between the timing of physical and biological cycles (Kosro et al., 2006) are also important, as they can lead to a mismatch (Sinclair, 1988) between the beginning of biological cycles (the end of zooplankton winter hibernation; migration, breeding and birth of fish; entrance into the coastal ocean by anadromous fish) and the physical cycles of environmental conditions needed to sustain biological populations during critical periods in their life histories (first feeding vs. the onset of upwelling, nutrient supply and turbulent mixing; phytoplankton and zooplankton biomass; the presence of predators).

As a case study, a series of papers analyzed the consequences of a well-observed delay in the onset of upwelling off Oregon in 2005 (Hickey et al., 2006; Kosro et al., 2006), which serves to illustrate the kind of changes one might find due to altered atmospheric patterns. Although upwelling began off southern and central California in spring, much as usual, upwelling off northern California, Oregon, and Washington was delayed and developed episodically, finally becoming persistent in July. The unusual surface wind conditions are attributed to the position of the atmospheric jet stream, which stayed farther south than usual in spring, delivering storms to the Oregon coast that repeatedly returned the winds to southerly and downwelling favorable (Barth et al., 2007). Late storms disrupted the breeding of marine birds off Northern California (Sydeman et al., 2006) and the delay in upwelling meant that the surface waters were more stratified than usual in the northern California Current, resisting mixing when northerly winds became more persistent. The net result was that the lower portion of the water column over the shelf felt the effects of upwelling winds farther to the south, communicated through CTW dynamics (Hickey et al., 2006), while the surface water remained warm and stratified until persistent upwelling-favorable winds arrived in July (Kosro et al., 2006). As upwelling progresses, it transforms vertical density gradients next to the coast into horizontal density gradients across the front. Thus a delay in upwelling that creates stronger vertical stratification leads to a stronger front and more energetic frontal jet, resulting in changes in the eddy field and the transport of coastal water properties from the shelf to the deep ocean. This case study does not represent "the expected change" due to a warmer climate but rather the type of changes in regional coastal ocean conditions that can result from perturbations of the larger-scale climate (jet stream, storm tracks, timing of seasonal transitions). It also makes the point that the coastal ocean responds strongly to changes in local winds, overriding the expected "normal" seasonal development of currents and water properties. Climatic changes in surface forcing will be directly felt in the coastal waters of EBCs.

5.4. Relating Modes to Models

A type of analysis that is increasing in use places nested regional models within the earth-system models used to predict the global response to long-term changes in greenhouse gases. This is referred to as "downscaling" the climate models. At the beginning of this chapter, we referred to comparisons that show the present coupled climate models to predict SST values in the eastern boundaries under present climatic conditions that are warmer by 5 °C or more than the observed values (Gent et al., 2010). Increasing the horizontal resolution of the atmospheric model reduced those errors but did not eliminate them (Gent et al., 2011). If much higher-resolution coupled models are to be nested within a selection of the global models, can we use the models' representation of the twentieth century climate modes to choose the more appropriate global models for specific regions?

As an example, Furtado et al. (2010) calculate the first two leading modes of variability of SLP and SST in the North Pacific Ocean, from the output fields of the 24 coupled climate models used in the IPCC Fourth Assessment Report. They find generally good agreement for the first modes of SLP and SST with the atmospheric AL and oceanic PDO modes. The second modes, however, do not represent as well the NPO and NPGO empirical patterns of variability. They also examine the interannual variability of the models and find most to lack the observed connection between the ENSO and AL–PDO modes. Concentrating on the tropics, the observed differences between the western and central ENSO modes are not well represented. One general conclusion of the study is the importance of the atmospheric NPO mode in all of the models, with the suggestion that this should be a key test of the models. For other basins, different models might be better. The point is that efforts to downscale specific models should begin with an assessment of the global model's ability to represent the climate modes that are most important for the region of interest.

5.5. Effects of EBCs on Climate

So far we have emphasized the response of EBCs to changing climate. Interactions between EBCs and the surrounding ocean and atmosphere also affect the regional climate. Upwelling maintains the cooler temperatures that allow these regions to cool the atmosphere. Thus, in studies of the global heat balance, EBCs are regions of net ocean heating. As with heat that enters the ocean in the tropical regions, the mechanisms that move heat from the EBCs to the higher latitudes (its ultimate fate) remain to be explored.

Upwelling also brings water with high CO_2 concentrations to the surface, releasing CO_2 to the atmosphere, while primary productivity reduces pCO_2 and allows the uptake of CO_2 in EBCs. The balance between these two processes varies greatly with fluctuating wind forcing over short timescales (Torres et al., 1999, 2002; Evans et al., 2011). This makes it difficult to determine even the correct sign for air–sea CO_2 fluxes in EBCs. There may be some hope to combine regional statistical relationships between carbon chemistry and satellite estimates of SST, chlorophyll pigment and winds in order to quantify regional net fluxes of CO_2 (Hales et al., 2012). This could also help in monitoring ocean acidity and hypoxia, which are both closely tied to CO_2 chemistry.

6. SUMMARY

Each of the eastern boundary systems is unique in at least one aspect, while other attributes are shared by pairs of systems, inviting comparative studies. The California Current System is subject to the strongest and largest region of seasonal wind reversals (upwelling–downwelling), perhaps explaining its strong seasonal transitions. In the Peru–Chile Current System, the region of strong upwelling-favorable winds extends to lower latitudes than in the other systems. It is also the only system without a well-defined, low-latitude poleward coastal surface current, although the Peru–Chile Countercurrent is observed at different times and places and the PUC is strong and continuous from the equator to southern Chile. In the SE Atlantic, the termination of the African continent allows the Indian Ocean's western boundary current (the Agulhas Current) to penetrate into the Atlantic and create large warm-core rings that interact with the Benguela Current. The Benguela is thus the only EBC bounded by warm water on the north and south. In the NE Atlantic, the Canary Current is bisected by the Mediterranean Outflow, an interruption in its geometry, wind forcing, and water properties. This has more effect on the Canary Current along the Iberian Peninsula than along the coast of NE Africa.

The two Northern Hemisphere systems are more complex in the tropics, due to the fact that the seasonally migrating ITCZ stays north of the Equator. In addition, the tropical coastal circulation in the North Pacific is complicated by interactions with the wind jets through the mountain gaps of Central America; while the tropical coastal currents in the North Atlantic must negotiate the zonal Guinea Coast and its monsoonal winds. The midlatitude EBCs in the Northern Hemisphere are also separated from the equatorial influences by greater distances, along with the more complex tropical systems. The two Southern Hemisphere systems have simpler and more direct pathways between the Equator and the midlatitudes along the coastal wave guides. However, the Northern Hemisphere systems are more directly connected to the NECC. The fundamentally different midlatitude EBC is the Leeuwin Current along western Australia, with poleward surface currents that oppose the winds. This is attributed to an along-shore pressure gradient, set up by water properties at its northern end.

The physical, chemical, and biological conditions in EBCs are sensitive to many factors that can be changed by climatic fluctuations in their surface forcing and the water that enters their ecosystems through their boundaries. Changes in the mean, seasonal, and higher-frequency attributes of the wind, heat, and freshwater fluxes are known to respond to the present modes of climate variability. Changes in the timing of seasonal transitions are also documented. Climate variability is communicated from the lower latitudes along oceanic pathways due to CTW dynamics and through atmospheric pathways such as the Hadley and Walker circulation cells. Likewise, the water entering through the WWD currents from the central gyres has been observed to change in properties such as oxygen concentrations and these changes have been amplified by biochemical processes in the EBCs. Measurements of ocean acidity also show changes that have direct effects on the natural ecosystem and on commercial aquaculture of shellfish (Barton et al., 2012).

Despite the sensitivity of the EBCs to climate variations on a range of scales, the models used to assess possible consequences of anthropogenic climate change do not reproduce present conditions in the EBCs well. Moreover, only a subset of those models reproduce the known modes of climate variability that affect the eastern boundaries of the ocean basins, and even those models only reproduce the strongest first modes. A specific improvement needed in the models is the inclusion of coupled ocean–atmosphere dynamics in coastal regions, especially in upwelling systems typical of eastern ocean boundaries. As the models are improved, their ability to reproduce the known responses of EBCs to climate fluctuations should be one of the criteria used in their evaluation.

ACKNOWLEDGMENTS

The authors are indebted to E. D. Barton for his thorough review of the manuscript, which resulted in a much stronger and more comprehensive treatment of the full range of eastern boundary currents. Other improvements resulted from many suggestions from an anonymous reviewer and editors J. Gould and G. Siedler, an editor of seemingly limitless patience and positive encouragement. Support for P. T. S. was provided by NASA grants NNX08AR40G and NNX10A092G for the altimeter and scatterometer analyses, as well as by the NOAA–OSU Cooperative Institute for Oceanographic Satellite Studies. Dr. C. James processed altimeter and scatterometer data and created the figures that use those data. OP thanks the support from PFB-31/2007 COPAS-Sur Austral.

REFERENCES

Abbott, M.R., Zion, P.M., 1985. Satellite observations of phytoplankton variability during an upwelling event. Cont. Shelf Res. 4, 661–680.

Aguirre, C., Pizarro, Ó., Strub, P.T., Garreaud, R., Barth, J.A., 2012. Seasonal dynamics of the near-surface alongshore flow off central Chile. J. Geophys. Res. 117. http://dx.doi.org/10.1029/2011JC007379, C01006.

Ajao, E.A., Houghton, R.W., 1998. Coastal ocean off Equatorial West Africa from 10°N-10°S. In: Robinson, A.R., Brink, K.H. (Eds.), The Sea, Vol. 11. The Global Coastal Ocean—Regional Studies and Syntheses. John Wiley and Sons, New York, pp. 605–631.

Allen, J.S., 1975. Coastal trapped waves in a stratified ocean. J. Phys. Oceanogr. 5, 300–325.

Allen, J.S., 1980. Models of wind-driven currents on the continental shelf. Annu. Rev. Fluid Mech. 12, 389–433.

Allen, J.S., Newberger, P.A., 1996. Downwelling on the Oregon continental shelf. Part I: response to idealized forcing. J. Phys. Oceanogr. 26, 2011–2035.

Allen, J.S., Smith, R.L., 1981. On the dynamics of wind-driven shelf currents. Philos. Trans. R. Soc. Lond. A 302, 617–634.

Allen, J.S., Walstad, L.J., Newberger, P.A., 1991. Dynamics of the coastal transition zone jet, 2, Nonlinear finite amplitude behavior. J. Geophys. Res. 96, 14, 995–14, 15016.

Alvarez-Salgado, X.A., Labarta, U., Fernandez-Reiriz, M.J., Figueiras, F.G., Roson, G., Piedracoba, S., Filgueira, R., Cabanas, J.M., 2008. Renewal time and the impact of harmful algal blooms on the extensive mussel raft culture of the Iberian coastal upwelling system (SW Europe). Harmful Algae 7, 849–855.

Anderson, D.L.T., Gill, A., 1975. Spin up of a stratified ocean, with applications to upwelling. Deep Sea Res. 22, 583–596.

Arcos, D., Salamanca, M., 1984. Distribucion de clorofila y condiciones oceanograficas superficiales frente a Chile central (latitudes 32°S-38°S, Febrero 1982). Biol. Pesquera 13, 5–14.

Aristegui, J., Sangra, P., Hernandez-Leon, S., Canton, M., Hernandez-Guerra, A., Kerling, J.L., 1994. Island-induced eddies in the Canary islands. Deep Sea Res. Part I 41 (10), 150901525. http://dx.doi.org/10.1016/0967-0637(94)90058-2.

Aristegui, J., Alvarez-Salgado, X.A., Barton, E.D., Figueiras, F.G., Hernandez-Leon, S., Roy, C., Santos, A.M.P., 2006. Oceanography and fisheries of the Canary Current/Iberian region of the eastern North Atlantic. In: Robinson, A.R., Brink, K.H. (Eds.), The Sea, Vol. 14. The Global Coastal Ocean, Interdisciplinary Regional Studies and Syntheses. Harvard University Press, pp. 879–933.

Aristegui, J., Barton, E.D., Alvarez-Salgado, X.A., Santos, A.M.P., Figueiras, F.G., Kifani, S., Hernandez-Leon, S., Mason, E., Machu, E., Demarq, H., 2009. Sub-regional ecosystem variability in the Canary Current upwelling. Prog. Oceanogr. 83, 33–48. http://dx.doi.org/10.1016/j.pocean.2009.07.031.

Arthur, R.S., 1965. On the calculation of vertical and horizontal motion in eastern boundary currents from the determination of horizontal motion. J. Geophys. Res. 70, 2799–2804.

Badon-Dangon, A., 1998. Coastal Circulation from the Galapagos to the Gulf of California. In: Robinson, A.R., Brink, K.H. (Eds.), The Sea, Vol. 11. The Global Coastal Ocean: Regional Studies and Syntheses. John Wiley and Sons, NY, pp. 315–343.

Bakun, A., 1990. Global climate change and intensification of coastal upwelling. Science 247, 198–201.

Bakun, A., Nelson, C.S., 1991. The seasonal cycle of wind stress curl in sub-tropical eastern boundry current regions. J. Phys. Oceanogr. 21, 1815–1834.

Bakun, A., Weeks, S.J., 2008. The marine ecosystem off Peru: What are the secrets of its fishery productivity and what might its future hold? Prog. Oceanogr. 79, 290–299. http://dx.doi.org/10.1016/j.pocean.2008.10.027.

Bakun, A., Field, D.B., Redondo-Rodriguez, A., Weeks, S., 2010. Greenhouse gas, upwelling-favorable winds, and the future of coastal ocean upwelling ecosystems. Glob. Chang. Biol. 16 (4), 1213–1228. http://dx.doi.org/10.1111/j.1365-2486.2009.02094.x.

Barber, R.T., Chavez, F.P., 1983. Biological consequences of El Niño. Science 222, 1203–1210.

Barth, J.A., Pierce, S.D., Smith, R.L., 2000. A separating coastal upwelling jet at Cape Blanco, Oregon and its connection to the California Current System. Deep Sea Res. Part II 47, 783–810.

Barth, J.A., Pierce, S.D., Cowles, T.J., 2005a. Mesoscale structure and its seasonal evolution in the northern California Current system. Deep Sea Res. Part II 52, 5–28. http://dx.doi.org/10.1016/j.dsr2.2004.09.026.

Barth, J.A., Pierce, S.D., Castelao, R.M., 2005b. Time-dependent, wind-driven flow over a shallow midshelf submarine bank. J. Geophys. Res. 110. http://dx.doi.org/10.1029/2004JC002761, C10S05.

Barth, J.A., Menge, B.A., Lubchenco, J., Chan, F., Bane, J.M., Kirincich, A.R., McManus, M.A., Nielsen, K.J., Pierce, S.D., Washburn, L., 2007. Delayed upwelling alters nearshore coastal ocean ecosystems in the Northern California Current. Proc. Natl. Acad. Sci. U.S.A. 104, 3719–3724.

Barton, E.D., 1987. Meanders, eddies and intrusions in the thermohaline front off northwest Africa. Deep Sea Res. 24, 1–6.

Barton, E.D., 1998. Eastern boundary of the North Atlantic: Northwest Africa and Iberia. In: Robinson, A.R., Brink, K.H. (Eds.), The Sea, Vol. 11. Global Coastal Ocean: Regional Studies and Syntheses. John Wiley and Sons, NY, pp. 633–657.

Barton, E.D., et al., 1993. Supersquirt: dynamics of the gulf of tehuantepec, Mexico. Oceanography 6, 23–30.

Barton, E.D., Aristegui, J., Tett, P., Canton, M., Garcia-Braun, J., Hernandez-Leon, S., Nykjaer, L., Almeida, C., Almunia, J., Ballesteros, S., Basterretxea, G., Escanez, J., Garcia-Weill, L., Hernandez-Guerra, A., Lopez-Laatzen, F., Molina, R., Montero, M.F., Navarro-Perez, E., Rodriguez, J.M., van Lenning, K., Velez, H., Wild, K., 1998. The transition zone of the Canary Current upwelling region. Prog. Oceanogr. 41, 455–504.

Barton, E.D., Inall, M.E., Sherwin, T.J., Torres, R., 2001. Vertical structure, turbulent mixing and fluxes during Lagrangian observations of an upwelling filament system off Northwest Iberia. Prog. Oceanogr. 51, 249–267.

Barton, A., Hales, B., Waldbusser, G., Langdon, C., Feely, R., 2012. The Pacific oyster, Crassostrea gigas, shows negative correlation to naturally elevated carbon dioxide levels: implications for near-term ocean acidification impacts. Limnol. Oceanogr. 57, 698–710. http://dx.doi.org/10.4319/lo.2012.57.3.0698.

Barton, E.D., Field, D.B., Roy, C., 2013. Canary Current upwelling: more or less? Prog. Oceanogr. 116, 167–178. http://dx.doi.org/10.1016/j.pocean.2013.07.007.

Batchelder, H.P., Strub, P.T., Lessard, E.J., Weingartner, T.J., 2005. U.S. GLOBEC biological and physical studies of plankton, fish and higher trophic level production, distribution, and variability in the northeast Pacific. Deep Sea Res. Part II 52, 1–4.

Batten, S.D., Crawford, W.R., 2005. The influence of coastal origin eddies on oceanic plankton distributions in the eastern Gulf of Alaska. Deep Sea Res. Part II 52 (1–2). http://dx.doi.org/10.1016/j.dsr2.2005.02.009.

Beardsley, R.C., Lentz, S.J., 1987. The Coastal Ocean Dynamics Experiment collection: an introduction. J. Geophys. Res. 92, 1455–1463. http://dx.doi.org/10.1029/JC092iC02p01455.

Beardsley, R.C., Dorman, C.E., Friehe, C.A., Rosenfeld, L.K., Winant, C.D., 1987. Local atmospheric forcing during the Coastal Ocean Dynamics Experiment 1. A description of the marine boundary layer and atmospheric conditions over a northern California upwelling region. J. Geophys. Res. 92, 1467–1488.

Bernstein, R.L., Breaker, L., Whritner, R., 1977. California Current eddy formation: ship, air and satellite results. Science 195, 353–359.

Bielli, S., Barbour, P., Samelson, R., Skyllingstad, E., Wilczak, J., 2002. Mon. Weather Rev. 130, 992–1008.

Bilbao, P.A., 1999. Interanual and interdecadal variability in the timing and strength of the spring transitions along the United States west coast. M.S. Thesis. Oregon State University, Corvallis, Oregon, USA, 115 pp.

Bograd, S.J., Schroeder, I., Sarkar, M., Qiu, X., Sydeman, W.J., Schwing, F.B., 2009. Phenology of coastal upwelling in the California Current. Geophys. Res. Lett. 36. http://dx.doi.org/10.1029/2008GL035933.

Boyd, P.W., Law, C.S., Wong, C.S., et al., 2004. The decline and fate of an iron-induced subarctic phytoplankton bloom. Nature 428, 549–553.

Breaker, L., Gilliland, R.P., 1981. A satellite sequence of upwelling along the California coast. In: Richards, F.A. (Ed.), Coastal Upwelling. American Geophysical Union, Washington, D.C, pp. 87–94.

Breaker, L.C., Mooers, C.N.K., 1986. Oceanic variability off the central California coast. Prog. Oceanogr. 17, 61–135.

Brink, K.H., 1980. Propagation of barotropic continental shelf waves over irregular bottom topography. J. Phys. Oceanogr. 10 (5), 765–778.

Brink, K.H., 1991. Coastal-trapped waves and wind-driven currents over the continental shelf. Annu. Rev. Fluid Mech. 23, 389–412.

Brink, K.H., 1998. Wind-driven currents over the continental shelf. In: Robinson, A.R., Brink, K.H. (Eds.), The Sea, Vol. 10. The Global Coastal Ocean—Processes and Methods. John Wiley and Sons, New York, pp. 3–20.

Brink, K.H., 2005. Coastal physical processes overview. In: Robinson, A.R., Brink, K.H. (Eds.), The Sea, Vol. 13. The Global Coastal Ocean—Multiscale Interdisciplinary Processes. Harvard University Press, Cambridge, MA & London, pp. 37–59.

Brink, K.H., Cowles, T.J., 1991. The coastal transition zone program. J. Geophys. Res. 96, 14637–14647.

Brink, K.H., Muench, R.D., 1986. Circulation in the Point Conception–Santa Barbara Channel region. J. Geophys. Res. 91, 877–895.

Bromirski, P.D., Flick, R.E., Cayan, D.R., 2003. Storminess Variability along the California Coast: 1858-2000. J. Clim. 16, 982–993.

Burke, S.D., Haack, T., Samelson, R.M., 1999. Mesoscale simulation of supercritical, subcritical, and transcritical flow along coastal topography. J. Atmos. Sci. 56, 2780–2795.

Cáceres, M., Valle-Levinson, A., Atkinson, L., 2003. Observations of cross-channel structure of flow in an energetic tidal channel. J. Geophys. Res. 108, 1517–1539.

Cáceres, M., Valle-Levinson, A., 2004. Transverse variability of flow on both sides of a sill/contraction combination in a fjord-like inlet of southern Chile. Estuar. Coast. Shelf Sci. 60, 325–338.

Cáceres, M., Valle-Levinson, A., Molinet, C., Castillo, M., Bello, M., Moreno, C., 2006. Lateral variability of flow over a sill in a channel of southern Chile. Ocean Dynam. http://dx.doi.org/10.1007/s10236-006-0077-y.

Cardone, V.J., Greenwood, J.G., Cane, M.A., 1990. On trends in historical wind data. J. Clim. 3, 113–127.

Carr, M.-E., Broad, K., 2000. Satellites, society, and the Peruvian fisheries during the 1997–1998 El Niño. In: Halpern, D. (Ed.), Satellites. Oceanography and Society. Elsevier, Amsterdam, pp. 171–191.

Carton, J.A., Zhou, Z., 1997. Annual cycle of sea surface temperature in the tropical Atlantic Ocean. J. Geophys. Res. 102, 27, 813–27, 824.

Castelao, R., Barth, J.A., 2007. The role of wind stress curl in jet separation at a cape. J. Phys. Oceanogr. 37, 2652–2670. http://dx.doi.org/10.1175/2007JPO3679.1.

Castelao, R.M., Mavor, T.P., Barth, J.A., Breaker, L.C., 2006. Sea surface temperature fronts in the California Current System from geostationary satellite observations. J. Geophys. Res. 111. http://dx.doi.org/10.1029/2006JC003541, C09026.

Chaigneau, A., Pizarro, O., 2005. Eddy characteristics in the eastern South Pacific. J. Geophys. Res. 110. http://dx.doi.org/10.1029/2004JC002815.

Chan, F., Barth, J.A., et al., 2008. Emergence of Anoxia in the California Current Large Marine Ecosystem. Science 319 (5865), 920. http://dx.doi.org/10.1126/science.1149016.

Chapman, D.C., 1987. Application of wind-forced long coastal-trapped wave theory along the California coast. J. Geophys. Res. 92, 1798–1816.

Chapman, D.C., Brink, K.H., 1987. Shelf and slope circulation induced by fluctuating offshore forcing. J. Geophys. Res. 0148-0227 92 (C11). http://dx.doi.org/10.1029/0JGREA000092000C11011741000001.

Chavez, F.P., 1996. Forcing and biological impact of onset of the 1992 El Niño in central California. Geophys. Res. Lett. 23, 265–268.

Chavez, F.P., 2005. Biological consequences of interannual to multidecadal variability. In: Robinson, A.R., Brink, K.H. (Eds.), The Sea, Vol. 13. The Global Coastal Ocean—Multiscale Interdisciplinary Processes. Harvard University Press, Cambridge, MA & London, pp. 643–679.

Chavez, F.P., Messie, M., 2009. A comparison of eastern boundary upwelling systems. Prog. Oceanogr. 83 (1–4), 80–96. http://dx.doi.org/10.1016/j.pocean.2009.07.032.

Chavez, F.P., Ryan, J., Lluch-Cota, S.E., Niquen, M.C., 2003. From anchovies to sardines and back: multidecadal change in the Pacific Ocean. Science 299, 217–221.

Checkley, D.M., Alheit, J., Oozeki, Y., Roy, C., 2009. Climate Change and Small Pelagic Fish. Cambridge University Press, Cambridge, 408 pp.

Chelton, D.B., 1984. Seasonal variability of alongshelf geostrophic velocity off central California. J. Phys. Oceanogr. 12, 757–784.

Chelton, D.B., Davis, R.E., 1982. Monthly mean sea-level variability along the west coast of North America. J. Phys. Oceanogr. 12, 757–784.

Chelton, D.B., Freilich, M.H., Esbensen, S.K., 2000a. Satellite observations of the wind jets off the Pacific coast of Central America. Part I: case studies and statistical characteristics. Mon. Weather Rev. 128, 1993–2018.

Chelton, D.B., Freilich, M.H., Esbensen, S.K., 2000b. Satellite observations of the wind jets off the Pacific coast of Central America. Part II: regional relationships and dynamical considerations. Mon. Weather Rev. 128, 2019–2043.

Chelton, D.B., Schlax, M.G., Samelson, R.M., 2007. Summertime coupling between sea surface temperature and wind stress in the California current system. J. Phys. Oceanogr. 37, 495–517.

Chelton, D.B., Schlax, M.G., Samelson, R.M., 2011a. Global observations of nonlinear mesoscale eddies. Prog. Oceanogr. 91, 167–216. http://dx.doi.org/10.1016/j.pocean.2011.01.002.

Chelton, D.B., Gaube, P., Schlax, M.G., Early, J.J., Samelson, R.M., 2011b. The influence of nonlinear mesoscale eddies on near-surface oceanic chlorophyll. Science 334, 328–332. http://dx.doi.org/10.1126/science.1208897.

Church, J.A., Craig, P.D., 1998. Australia's Shelf Seas: diversity and complexity. In: Robinson, A.R., Brink, K.H. (Eds.), The Sea, Vol. 11. The Global Coastal Ocean: Regional Studies and Syntheses. John Wiley and Sons, NY, pp. 933–964.

Church, J.A., Cresswell, G.R., Godfrey, J.S., 1989. The Leeuwin current. In: Neshyba, S.J., Mooers, C.N.K., Smith, R.L., Barber, R.T. (Eds.), Poleward Flows along Eastern Ocean Boundaries. Coastal and Estuarine Studies, vol. 34. Springer-Verlag, New York, pp. 230–252.

Clarke, A.J., Shi, C., 1991. Critical frequencies at ocean boundaries. J. Geophys. Res. 96, 10,731–10,738.

Clarke, A.J., Van Gorder, S., 1994. On ENSO coastal currents and sea level. J. Phys. Oceanogr. 24, 661–680.

Codispoti, L.A., Barber, R.T., Friederich, G.E., 1989. Do nitrogen transformations in the poleward undercurrent off Peru and Chile have a globally significant influence? In: Neshyba, S.J., Mooers, C.N.K., Smith, R.L., Barber, R.T. (Eds.), Poleward Flows Along Eastern Ocean Boundaries. Coastal and Estuarine Studies, Vol. 34. Springer-Verlag, New York, pp. 281–314.

Colas, F., McWilliams, J.C., Capet, X., Kurian, J., 2011. Heat balance and eddies in the Peru-Chile current system. Clim. Dyn. http://dx.doi.org/10.1007/s00382-011-1170-6.

Colberg, F., Reason, C.J.C., 2006. A model study of the Angola Benguela Frontal Zone: sensitivity to atmospheric forcing. Geophys. Res. Lett. 33. http://dx.doi.org/10.1029/2006GL027463, L19608.

Combes, V., Di Lorenzo, E., 2007. Intrinsic and forced interannual variability of the Gulf of Alaska mesoscale circulation. Prog. Oceanogr. 75, 266–286.

Combes, V., Di Lorenzo, E., et al., 2009. Interannual and Decadal variations in cross-shelf transport in the Gulf of Alaska. J. Phys. Oceanogr. 39 (4), 1050–1059.

Condie, S.A., Harris, P.T., 2006. Interactions between physical, chemical, biological, and sedimentological processes in Australia's shelf seas. In: Robinson, A.R., Brink, K.H. (Eds.), The Sea, Vol. 14. The Global Coastal Ocean, Interdisciplinary Regional Studies and Syntheses. Harvard University Press, pp. 1413–1449.

Cornejo-Rodriguez, M.P., Enfield, D.B., 1987. Propagation and forcing of high-frequency sea level variability along the west coast of South America. J. Geophys. Res. 92, 14,323–14,334.

Crawford, W.R., 2002. Physical characteristics of Haida Eddies. J. Oceanogr. 58, 703–713.

Crawford, W.R., 2005. Heat and fresh water transport by eddies into the Gulf of Alaska. Deep Sea Res. Part II 52, 893–908.

Crawford, W.R., Cherniawsky, J.Y., Foreman, M.G.G., 2000. Multi-year meanders and eddies in Alaskan Stream as observed by TOPEX/Poseidon altimeter. Geophys. Res. Lett. 27, 1025–1028.

Crawford, W.R., Brickley, P.J., Thomas, A.C., 2007. Mesoscale eddies dominate surface phytoplankton in northern Gulf of Alaska. Prog. Oceanogr. 75 (2), 287–303.

Cresswell, G.R., Golding, T.J., 1980. Observations of a south-flowing current in the southeastern Indian Ocean. Deep Sea Res. 27, 449–466.

Csanady, G.T., 1978. The arrested topographic wave. J. Phys. Oceanogr. 8, 47–62.

Cucalon, E., 1987. Oceanographic variability off Ecuador associated with an El Niño event in 1982-1983. J. Geophys. Res. 92, 14309–14322.

Da Silva, A.M., Young, C.C., Levitus, S., 1994. Atlas of Surface Marine Data, vol. 1–5. 6–10, NOAA Atlas NESDIS.

Davis, R.E., 1977. Techniques for statistical analysis and prediction of geophysical fluid systems. Geophys. Astrophys. Fluid Dyn. 8, 245–277.

Davis, R.E., 1978. Predictability of sea level pressure anomalies over the North Pacific Ocean. J. Phys. Oceanogr. 94 (8), 233–246.

Denbo, D.W., Allen, J.S., 1987. Large-scale response to atmospheric forcing of shelf currents and coastal sea level off the west coast of North America: May-July 1981 and 1982. J. Geophys. Res. 92, 1757–1782.

Deser, C., Alexander, M.A., Xie, S.-P., Phillips, A.S., 2011. Sea surface temperature variability: patterns and mechanisms. Ann. Rev. Mar. Sci. 2, 115. http://dx.doi.org/10.1146/annurev-marine-120408-151453.

DeSzoeke, R.A., Richman, J.G., 1984. On wind-driven mixed layers with strong horizontal gradient: a theory with applications to coastal upwelling. J. Phys. Oceanogr. 14, 364–377.

Dever, E.P., Dorman, C.E., Largier, J.L., 2006. Surface boundary layer variability off northern California, USA, during upwelling. Deep Sea Res. Part II 53, 2887–2905.

Di Lorenzo, E., Foreman, M.G.G., Crawford, W.R., 2005. Modelling the generation of Haida Eddies. Deep Sea Res. Part II 52, 853–873.

Di Lorenzo, E., Schneider, N., Cobb, K.M., Franks, P.J.S., Chhak, K., Miller, A.J., McWilliams, J.C., Bograd, S.J., Arango, H., Curchitser, E., Powell, T.M., Riviere, P., 2008. North Pacific Gyre Oscillation links ocean climate and ecosystem change. Geophys. Res. Lett. 35, L08607. http://dx.doi.org/10.1029/2007GL032838.

Di Lorenzo, E., Combes, C., Keister, J., Strub, T., Thomas, A., Franks, P.J.S., Ohman, M., Bracco, A., Furtado, J., Bograd, S., Peterson, W., Schwing, F., Chiba, S., Taguchi, B., Hormazabal, S., Parada, C., 2013. Mechanisms of Pacific Ocean climate and ecosystem variability. Oceanography, Submitted.

Domingues, C.M., Maltrud, M.E., Wijffels, S.E., Church, J.A., Tomczak, M., 2007. Simulated Lagrangian pathways between the Leeuwin Current System and the upper-ocean circulation of the southeast Indian Ocean. Deep Sea Res. Part II 54, 797–817.

Dorman, C.E., 1985. Evidence of Kelvin waves in California's marine layer and related eddy generation. Mon. Weather Rev. 113, 827–839.

Dorman, C.E., Koracin, D., 2008. Interaction of the summer marine layer with an extreme California coastal bend. Mon. Weather Rev. 136, 2894–2922.

Ducet, N., Le Traon, P.-Y., Reverdin, G., 2000. Global high-resolution mapping of ocean circulation from TOPEX/Poseidon and ERS-1 and -2. J. Geophys. Res. 105, 19477–19488.

Duncombe Rae, C.M., Shannon, L.V., Shillington, F.A., 1989. An Agulhas ring in the South Atlantic Ocean. S. Afr. J. Sci. 85, 747–748.

Duncombe Rae, C.M., Shillington, F.A., Agenbag, J.J., Taunton-Clark, J., Grundlingh, M.L., 1992. An Agulhas ring in the South Atlantic Ocean and its interaction with the Benguela upwelling frontal system. Deep Sea Res. 39, 2009–2027.

Edwards, K.A., Rogerson, A.M., Winant, C.D., Rogers, D.P., 2001. Adjustment of the marine atmospheric boundary layer to a coastal cape. J. Atmos. Sci. 58, 1511–1528.

Emery, W.J., Hamilton, K., 1985. Atmospheric forcing of interanual variability in the northeast Pacific ocean: connections with El Niño. J. Geophys. Res. 90, 857–868.

Enfield, D.B., 1976. Oceanography of the region north of the equatorial front, physical aspects. FAO Fish. Rep. 185, 299–334.

Enfield, D.B., Allen, J.S., 1980. On the structure and dynamics of monthly mean sea level anomalies along the Pacific coast of North and South America. J. Phys. Oceanogr. 10, 557–578.

Enfield, D.B., Mestas-Nunez, A.M., 1999. Multiscale variabilities in global sea surface temperatures and their relationships with tropospheric climate patterns. J. Clim. 12, 2719–2733.

Espinosa-Carreon, T.L., Strub, T., Beier, E., Ocampo-Torres, F., Gaxiola-Castro, G., 2004. Seasonal and interannual variability of satellite-derived chlorophyll pigment, surface height, and temperature off Baja California. J. Geophys. Res. 109. http://dx.doi.org/10.1029/2003JC002105, C03039.

Espinosa-Carreón, T.L., Gaxiola-Castro, G., Beier, E., Strub, P.T., Kurczyn, J.A., 2012. Effects of mesoscale processes on phytoplankton chlorophyll off Baja California. J. Geophys. Res. 117. http://dx.doi.org/10.1029/2011JC007604, C04005.

Evans, W., Hales, B., Strutton, P., 2011. The seasonal cycle of surface ocean pCO2 on the Oregon shelf. J. Geophys. Res. Oceans 116. http://dx.doi.org/10.1029/2010JC006625, C05012.

Feng, M., Meyers, G., Pearce, A., Wijffels, S., 2003. Annual and interannual variations of the Leeuwin Current at 32°S. J. Geophys. Res. 108 (C11), 3355. http://dx.doi.org/10.1029/2002JC001763.

Feng, M., Wijffels, S., Godfrey, S., Meyers, G., 2005. Do eddies play a role in the momentum balance of the Leeuwin Current? J. Phys. Oceanogr. 35, 964–975.

Fiedler, P.C., 2002. The annual cycle and biological effects of the Costa Rica Dome. Deep Sea Res. Part I 49, 321–338.

Fiedler, P.C., Talley, L.D., 2006. Hydrography of the eastern tropical Pacific: a review. Prog. Oceanogr. 69, 143–180. http://dx.doi.org/10.1016/j.pocean.2006.03.008.

Field, J.G., Shillington, F.A., 2006. Variability of the Benguela Current System. In: Robinson, A.R., Brink, K.H. (Eds.), The Sea, Vol. 14. The Global Coastal Ocean, Interdisciplinary Regional Studies and Syntheses. Harvard University Press, pp. 835–863.

Fiuza, A.F.G., 1983. Upwelling patterns off Portugal. In: Suess, E., Thiede, J. (Eds.), Coastal Upwelling: Its Sediment Record, A. Plenum Press, New York, pp. 85–98.

Flament, P., Armi, L., Washburn, L., 1985. The evolving structure of an upwelling element. J. Geophys. Res. 90, 11, 765–11, 778.

Florenchie, P., Lutjeharms, J.R.E., Reason, C.J.C., Masson, S., Rouault, M., 2003. The source of Benguela Niños in the South Atlantic Ocean. Geophys. Res. Lett. 30, 1505. http://dx.doi.org/10.1029/2003GL017172.

Fonseca, T., 1989. An overview of the Poleward Undercurrent and upwelling along the Chilean coast. In: Neshyba, S.J., Mooers, C.N.K., Smith, R.L., Barber, R.T. (Eds.), Poleward flows along eastern ocean boundaries. Coastal and Estuarine Studies, Vol. 34. Springer-Verlag, New York, pp. 203–228.

Fonseca, T.R., Farias, M., 1987. Estudia del proceso de surgencia en la costa chilena utilizando percepcion remota. Invest. Pesqu. (Chile) 34, 33–46.

Francis, R.C., Hare, S.R., 1994. Decadal-scale regime shifts in the large marine ecosystems of the Northeast Pacific: a case for historical science. Fish. Oceanogr. 3, 279–291.

Freeland, H.J., 2006. What proportion of the North Pacific Current finds its way into the Gulf of Alaska? Atmos. Ocean 44 (4).

Freeland, H.J., Gatien, G., Huyer, A., Smith, R.L., 2003. Cold halocline in the northern California Current: an invasion of subarctic water. Geophys. Res. Lett. 30 (3), 1141. http://dx.doi.org/10.1029/2002GL016663.

Freitag, H.P., Halpern, D., 1981. Hydrographic observations off northern California during May 1977. J. Geophys. Res. 86, 4248–4252.

Freon, P., Barange, M., Aristegui, J., 2009. Eastern boundary upwelling ecosystems: integrative and comparative approaches. Prog. Oceanogr. 83 (1–4), 1–14. http://dx.doi.org/10.1016/j.pocean.2009.08.001.

Furtado, J., Di Lorenzo, E., Schneider, E., Bond, N., 2010. North Pacific decadal variability and climate change in the IPCC AR4 models. J. Clim. http://dx.doi.org/10.1175/2010JCLI3584.1.

Gaard, E., Gislason, A., Melle, W., 2006. Iceland, Faroe and Norwegian coasts. In: Robinson, A.R., Brink, K.H. (Eds.), The Sea, Vol. 14. The Global Coastal Ocean, Interdisciplinary Regional Studies and Syntheses. Harvard University Press, pp. 1073–1105.

Gammelsrød, T., Bartholomae, C.H., Boyer, D.C., Filipe, V.L.L., O'Toole, M.J., 1998. Intrusion of warm surface water along the Angolan–Namibian coast in February–March 1995: the 1995 Benguela Niño. S. Afr. J. Mar. Sci. 19, 41–56.

Garreaud, R.D., Munoz, R.C., 2005. The low level jet off the west coast of subtropical South America: structure and variability. Mon. Weather Rev. 133, 2246–2261. http://dx.doi.org/10.1175/MWR2972.1.

Garreaud, R.D., Rutllant, J., 2003. Coastal lows along the subtropical west coast of South America: numerical simulation of a typical case. Mon. Weather Rev. 131, 891–908.

Garreaud, R.D., Rutllant, J., Fuenzalida, H., 2002. Coastal lows in north-central Chile: mean structure and evolution. Mon. Weather Rev. 130, 75–88.

Garvine, R.W., 1982. A steady state model for buoyant surface plume hydrodynamics in coastal waters. Tellus 34, 293–306.

Garvine, R.W., 1999. Penetration of buoyant coastal discharge onto the continental shelf: a numerical model experiment. J. Phys. Oceanogr. 29, 1892–1909.

Gaube, P., Chelton, D.B., Strutton, P.G., Behrenfeld, M.J., 2013. Satellite Observations of Chlorophyll, Phytoplankton Biomass and Ekman Pumping in Nonlinear Mesoscale Eddies. J. Geophys. Res., Submitted.

Gay, P.S., Chereskin, T.K., 2009. Mean structure and seasonal variability of the poleward undercurrent off southern California. J. Geophys. Res. Oceans 114. http://dx.doi.org/10.1029/2008jc004886, C02007.

Gent, P.R., Yeager, S.G., Neale, R.B., Levis, S., Bailey, D.A., 2010. Improvements in a half degree atmosphere/land version of the CCSM. Clim. Dyn. 34, 819–833. http://dx.doi.org/10.1007/s00382-009-0614-8.

Gent, P.R., Danabasoglu, G., Donner, L.J., Holland, M.M., Hunke, E.C., Jayne, S.R., Lawrence, D.M., Neale, R.B., Rasch, P.J., Vertenstein, M., Worley, P.H., Yang, Z.-L., Zhang, M., 2011. The Community Climate System Model version 4. J. Clim. 24, 4973–4991. http://dx.doi.org/10.1175/2011JCLI4083.1.

Gill, A.E., 1982. Atmosphere-Ocean Dynamics. Academic Press, New York, NY, 662 pp.

Gille, S.T., Llewellyn Smith, S.G., Statom, N.M., 2005. Global observations of the landbreeze. Geophys. Res. Lett. 32. http://dx.doi.org/10.1029/2004GL022139, L05605.

Godfrey, J.S., Ridgway, K.R., 1985. The large-scale environment of the poleward-flowing Leeuwin Current, Western Australia: longshore steric height gradients, wind stresses and geostrophic flow. J. Phys. Oceanogr. 15, 418–495.

Godfrey, J.S., Weaver, A.J., 1991. Is the Leeuwin Current driven by Pacific heating and winds? Prog. Oceanogr. 27, 25–272.

Gordon, A.L., Lutjeharms, J.R.E., Grundlingh, M.L., 1987. Stratification and circulation at the Agulhas retroflection. Deep Sea Res. 34 (4A), 565–599.

Graham, W.M., Largier, J.L., 1997. Upwelling shadows as nearshore retention sites: the example of Monterrey Bay. Cont. Shelf Res. 17, 509–532.

Haack, T., Chelton, D., Pullen, J., Doyle, J.D., Schlax, M., 2008. Summertimeinfluence of SST on surface wind stress off the U.S. West Coast from the U.S. Navy COAMPS Model. J. Phys. Oceanogr. 38, 2414–2437.

Hales, B., Strutton, P., Saraceno, M., Letelier, R., Takahashi, T., Feely, R., Sabine, C., Chavez, F., 2012. Satellite-based prediction of pCO2 in coastal waters. Prog. Oceanogr. http://dx.doi.org/10.1016/j.pocean.2012.03.001.

Halliwell Jr., G.R., Allen, J.S., 1987. Wave number-frequency domain properties of coastal sea level response to alongshore wind stress along the west coast of North America, 1980-1984. J. Geophys. Res. 92 (C11), 11761–11788.

Halpern, D., 1976. Measurements of a near-surface wind stress over an upwelling region near the Oregon coast. J. Phys. Oceanogr. 6, 108–112.

Halpern, D., 2002. Offshore Ekman transport and Ekman pumping off Perú during the 1997–1998 El Niño. J. Geophys. Res. Lett. 20, 19-1–19.4.

Hansen, B., Stefansson, U., Svendsen, E., 1998. Iceland, Faroe and Norwegian Coasts. In: Robinson, A.R., Brink, K.H. (Eds.), The Sea, Vol. 10. The Global Coastal Ocean—Processes and Methods. John Wiley and Sons, New York, pp. 733–758.

Hardman-Mountford, N.J., Richardson, A.J., Agenbag, J.J., Hagen, E., Nykjaer, L., Shillington, F.A., Villacastin, C., 2003. Ocean climate of the South East Atlantic observed from satellite data and wind models. Prog. Oceanogr. 59, 181–221.

Hauri, C., Gruber, N., Alin, S., Fabry, V.J., Feely, R.A., Hales, B., Plattner, G.-K., Wheeler, P., 2009. Ocean acidification in the California Current system. Oceanography 22, 60–71.

Haynes, R., Barton, E., 1990. A poleward flow along the Atlantic coast of the Iberian Peninsula. J. Geophys. Res. 95, 11425–11441.

Haynes, R., Barton, E., Pilling, J., 1993. Development, persistence, and variability of upwelling filaments off the Atlantic coast of the Iberian Peninsula. J. Geophys. Res. 98 (C12), 22681–22692.

Henson, S., Thomas, A.C., 2007. Interannual variability in timing of bloom initiation in the California Current System. J. Geophys. Res. 112, http://dx.doi.org/10.1029/2006/2006JC003960, C08007.

Henson, S., Dunne, J., Sarmiento, J., 2009. Decadal variability in North Atlantic phytoplankton blooms. J. Geophys. Res. 114. http://dx.doi.org/10.1029/2008JC005139, C04013.

Herguera, J.C., 2006. Coastal biogeochemical and ecological process from the eastern tropical North Pacific to the Gulf of California. In: Robinson, A.R., Brink, K.H. (Eds.), The Sea, Vol. 14. The Global Coastal Ocean, Interdisciplinary Regional Studies and Syntheses. Harvard University Press, pp. 391–439.

Hickey, B., 1979. The California current system: hypotheses and facts. Prog. Oceanogr. 8, 191–279.

Hickey, B., 1989. Patterns and processes of circulation over the shelf and slope. In: Landry, M.L., Hickey, B. (Eds.), Coastal Oceanography of Washington and Oregon. Elsevier, Amsterdam, pp. 41–55.

Hickey, B.M., 1997. Response of a narrow submarine canyon to strong wind forcing. J. Phys. Oceanogr. 27 (5), 697–726.

Hickey, B., 1998. Coastal oceanography of western North America from the tip of Baja California to Vancouver Island. In: Robinson, A.R., Brink, K.H. (Eds.), The Sea, Vol. 11. The Global Coastal Ocean—Processes and Methods. John Wiley and Sons, New York, pp. 345–393.

Hickey, B.M., Banas, N.S., 2008. Why is the northern end of the California Current System so productive? Oceanography 21 (4), 90–107.

Hickey, B., Pola, N., 1983. The seasonal alongshore pressure gradient on the west coast of the United States. J. Geophys. Res. 8, 7623–7633.

Hickey, B., MacFadyen, A., Cochlan, W., Kudela, R., Bruland, K., Trick, C., 2006. Evolution of chemical, biological, and physical water properties in the northern California Current in 2005: remote or local wind forcing? Geophys. Res. Lett. 33. http://dx.doi.org/10.1029/2006GL026782, L22S02.

Hickey, B.M., McCabe, R., Geier, S., Dever, E., Kachel, N., 2009. Three interacting freshwater plumes in the Northern California Current System. J. Geophys. Res. 114. http://dx.doi.org/10.1029/2008JC004907, C00B03.

Hickey, B.M., Kudela, R.M., Nash, J.D., et al., 2010. River influence on the shelf ecosystems: introduction and synthesis. J. Geophys. Res. 115. http://dx.doi.org/10.1029/2009JC005452, COOB17.

Hill, A.E., 1998. Buoyancy effects in coastal and shelf seas. In: Robinson, A.R., Brink, K.H. (Eds.), The Sea, Vol. 10. The Global Coastal Ocean—Processes and Methods. John Wiley and Sons, New York, pp. 21–62.

Hill, A.E., Hickey, B.M., Shillington, F.A., Strub, P.T., Brink, K.H., Barton, E.D., Thomas, A.C., 1998. Eastern Ocean boundaries. In: Robinson, A.R., Brink, K.H. (Eds.), The Sea, Vol. 11. Global Coastal Ocean: Regional Studies and Syntheses the Sea. John Wiley and Sons, New York, pp. 29–67.

Hormazabal, S., Shaffer, G., Leth, O., 2004. Coastal transition zone off Chile. J. Geophys. Res. 109. http://dx.doi.org/10.1029/2003JC001956, C01021.

Hurrell, J.W., Kushnir, Y., Ottersen, G., Visbeck, M., 2003. An overview of the North Atlantic Oscillation. In: Hurrell, J.W., Kushnir, Y., Ottersen, G., Visbeck, M. (Eds.), The North Atlantic Oscillation: Climate Significance and Environmental Impacts, Geophysical Monography Series, vol. 134.

Hutchings, L., Van Der Lingen, C.D., Shannon, L.J., Crawford, R.J.M., Verheye, H.M.S., Bartholomae, C.H., Van Der Plas, A.K., Louw, D., Kreiner, A., Ostrowski, M., Fidel, Q., Barlow, R.G., Lamont, T., Coetzee, J., Shillington, F., Veitch, J., Currie, J.C., Monteiro, P.M.S., 2009. The Benguela Current: an ecosystem of four components. Prog. Oceanogr. 83, 15–32. http://dx.doi.org/10.1016/j.pocean.2009.07.046.

Huthnance, J., 1984. Slope currents and JEBAR. J. Phys. Oceanogr. 14, 795–810.

Huyer, A., 1983. Coastal upwelling in the California Current System. Prog. Oceanogr. 12, 259–284.

Huyer, A., 1990. Shelf circulation. In: Le Mehaute, B., Hanes, D.M. (Eds.), The Sea, Vol 9: Ocean Engineering Science. Wiley, pp. 423–466.

Huyer, A., 2003. Preface to special section on enhanced Subarctic influence in the California Current, 2002. Geophys. Res. Lett. 30 (15), 8019. http://dx.doi.org/10.1029/2003GL017724.

Huyer, A., Sobey, E.J.C., Smith, R.L., 1979. The spring transition in currents over the Oregon continental shelf. J. Geophys. Res. 84 (C11), 6995–7011.

Huyer, A., Smith, R.L., Paluszkiewicz, T., 1987. The Peru Undercurrent: a study in variability. Deep Sea Res. 38 (Suppl. 1), S247–S271.

Huyer, A., Barth, J.A., Kosro, P.M., Shearman, R.K., Smith, R.L., 1998. Upper-ocean water mass characteristics of the California Current, summer 1993. Deep Sea Res. Part II 45, 1411–1442.

Ikeda, M., Emery, W.J., 1984. Satellite observations and modeling of meanders in the California Current System off Oregon and northern California. J. Phys. Oceanogr. 14, 1434–1450.

Ingraham Jr., W.J., Ebbesmeyer, C.C., Hinrichsen, R.A., 1998. Imminent climate and circulation shift in North Pacific Ocean could have major impact on marine resources. Eos 79, 197–201.

Johnson, D., Fonseca, T., Sievers, H., 1980. Upwelling in the Humboldt Coastal Current Near Valparaiso, Chile. J. Mar. Res. 38, 1–16.

Jouanno, J., Marin, F., du Penhoat, Y., Molines, J.M., Sheinbaum, J., 2011. Seasonal modes of surface cooling in the Gulf of Guinea. J. Phys. Oceanogr. 41, 1408–1416. http://dx.doi.org/10.1175/JPO-D-11-031.1.

Kelly, K.A., 1985. The influence of winds and topography on the sea surface temperature patterns over the northern California slope. J. Geophys. Res. 90, 11783–11798.

Kelly, K.A., Beardsley, R.C., Limeburner, R., Brink, K.H., Paduan, J.D., Chereskin, T.K., 1998. Variability of the near-surface eddy kinetic energy in the California Current based on altimetric, drifter, and moored current data. J. Geophys. Res. 103, 13067–13083.

Kessler, W.S., 2006. The circulation of the eastern tropical Pacific: a review. Prog. Oceanogr. 69, 181–217. http://dx.doi.org/10.1016/j.pocean.2006.03.009.

Killworth, P.D., 1978. Coastal upwelling and Kelvin waves with small longshore topography. J. Phys. Oceanogr. 8, 188–205.

Koračin, D., Dorman, C.E., Dever, E.P., 2004. Coastal perturbations of marine layer winds, wind stress, and wind stress curl along California and Baja California in June 1999. J. Phys. Oceanogr. 34, 1152–1173.

Kosro, P.M., 1987. Structure of the coastal current field off northern California during the Coastal Ocean Dynamics Experiment. J. Geophys. Res. 92, 1637–1654.

Kosro, P.M., Huyer, A., 1986. CTD and velocity surveys of seaward jets off northern California, July 1981 and 1982. J. Geophys. Res. 91, 7680–7690.

Kosro, P.M., Peterson, W.T., Hickey, B.M., Shearman, R.K., Pierce, S.D., 2006. The physical vs. the biological spring transition: 2005. Geophys. Res. Lett. 33. http://dx.doi.org/10.1029/2006GL027027, L22S03.

Ladd, C., Stabeno, P., Cokelet, E.D., 2005. A note on cross-shelf exchange in the northern Gulf of Alaska. Deep Sea Res. Part II 52, 667–679.

Ladd, C., Mordy, C.W., Kachel, N.B., Stabeno, P.J., 2007. Northern Gulf of Alaska eddies and associated anomalies. Deep Sea Res. Part I 54, 487–509. http://dx.doi.org/10.1015/j. dsr.2007.01.006.

Ladd, C., Crawford, W.R., Harpold, C.E., Johnson, W.K., Kachel, N.B., Stabeno, P.J., Whitney, F., 2009. Deep Sea Res. Part II 56, 2460–2473. http://dx.doi.org/10.1016/j.dsr2.2009.02.007.

Largier, J.L., et al., 2006. WEST: a northern California study of the role of wind-driven transport in the productivity of coastal plankton communities. Deep Sea Res. Part II 53, 2833–2849.

Lavin, M.F., Beier, E., Gomez-Valdes, J., Godinez, V.M., Garcia, J., 2006. On the summer poleward coastal current off SW Mexico. Geophys. Res. Lett. 33. http://dx.doi.org/10.1029/2005GL024686.

Lemos, R.T., Pires, H.O., 2004. The upwelling regime off the West Portuguese coast, 1941-2000. Int. J. Climatol. 24, 511–524.

Lemos, R.T., Sanso, B., 2006. Spatio-temporal variability of ocean temperature in the Portugal Current System. J. Geophys. Res. 111, http://dx.doi.org/10.1029/2005JC003051, C04010.

Lentz, S.J., 1987. A description of the 1981 and 1982 spring transitions over the northern California shelf. J. Geophys. Res. 92, 1545–1567.

Lentz, S.J., 1992. The surface boundary layer in coastal upwelling regions. J. Phys. Oceanogr. 22, 1517–1539.

Lentz, S.J., Fewings, M.R., 2012. The wind- and wave-driven inner-shelf circulation. Ann. Rev. Mar. Sci. 4, 317–343. http://dx.doi.org/10.1146/annurev-marine-120709-142745.

Lentz, S.J., Trowbridge, J.H., 1994. The bottom boundary layer over the northern California shelf. J. Phys. Oceanogr. 24, 1186–1201.

Lukas, R., 1986. The termination of the Equatorial Undercurrent in the eastern Pacific. Prog. Oceanogr. 16, 63–90.

Lumpkin, R., Garzoli, S.L., 2005. Near-surface circulation in the Tropical Atlantic Ocean. Deep Sea Res. Part I 52, 495–518. http://dx.doi.org/10.1016/j.dsr.2004.09.001.

Lumpkin, R., Garzoli, S., 2011. Interannual to decadal changes in the western South Atlantics surface circulation. J. Geophys. Res. 116, http://dx.doi.org/10.1029/2010JC006285, C01014.

Lutjeharms, J.R.E., 2006. The Agulhas Current. Springer, Berlin, 329 pp.

Lutjeharms, J.R.E., Meeuwis, J.M., 1987. The extent and variability of South-East Atlantic upwelling. S. Am. J. Mar. Sci. 5, 51–62.

Lutjeharms, J.R.E., Stockton, P., 1987. Kinematics of the upwelling front off southern Africa. S. Afr. J. Mar. Sci. 5, 35–49.

Lutjeharms, J.R.E., Shillington, F.A., Duncombe Rae, C.M., 1991. Observations of extreme filaments in the South East Atlantic Ocean. Science 253, 774–776.

Lynn, R.S., Simpson, J.J., 1987. The California Current System: the seasonal variability of its physical characteristics. J. Geophys. Res. 92 (C12), 12,947–12,966.

MacFarlane, G.A., McKinnell, S.M., 2005. Linking the open ocean to coastal ecosystems. Deep Sea Res. Part II 52, 665–666. http://dx.doi.org/10.1016/j.dsr2.2004.10.002.

Mackas, D., 2006. Interdisciplinary Oceanography of the Western North American Continental Margin: Vancouver Island to the tip of Baja California. In: Robinson, A.R., Brink, K.H. (Eds.), The Sea, Vol. 14. The Global Coastal Ocean, Interdisciplinary Regional Studies and Syntheses. Harvard University Press, pp. 441–501.

Mackas, D.L., Washburn, L., Smith, S.L., 1991. Zooplankton community patterns associated with a California Current cold filament. J. Geophys. Res. 96, 14781–14797.

Mackas, D.L., Strub, P.T., Thomas, A.C., Montecino, V., 2006. Eastern ocean boundaries—Pan regional overview. In: Robinson, A.R., Brink, K.H. (Eds.), The Sea, Vol. 14. The Global Coastal Ocean, Interdisciplinary Regional Studies and Syntheses. Harvard University Press, pp. 21–60.

Mantua, N.J., Hare, S.R., Zhang, Y., Wallace, J.M., Francis, R.C., 1997. A pacific interdecadal climate oscillation with impacts on salmon production. Bull. Am. Meteorol. Soc. 78, 1069–1079.

Marchesiello, P., McWilliams, J.C., Shchepetkin, A., 2003. Equilibrium structure and dynamics of the California Current System. J. Phys. Oceanogr. 33, 753–783.

Marín, V.H., Moreno, C.A., 2002. Wind-driven circulation and larval dispersal: a review of its consequences in coastal benthic recruitment. In: Castilla, J.C., Largier, J.L. (Eds.), The Oceanography and Ecology of the Nearshore and Bays in Chile. Proceedings of the International Symposium on Linkages and Dynamics of Coastal Systems: Open Coasts and Embayments. Ediciones Universidad Católica de Chile, Santiago, Chile, pp. 47–63.

Marin, V., Escribano, R., Delgado, L., Olivares, G., Hidalgo, P., 2001. Nearshore circulation in a coastal upwelling site off the northern Humboldt Current System. Cont. Shelf Res. 21, 1317–1329.

Masumoto, Y., Sasaki, H., Kagimoto, T., Komori, N., Ishida, A., Sasai, Y., Miyana, T., Motoi, T., Mitsudera, H., Takahashi, K., Sakuma, H., Yamagata, T., 2004. A fifty-year eddy-resolving simulation of the world ocean—preliminary outcomes of OFES (OGCM for the Earth Simulator). J. Earth Simul. 1, 31–52.

McCreary, J.P., 1984. Equatorial Beams. J. Mar. Res. 42, 395–430.

McCreary, J.P., Chao, S.Y., 1985. Three-dimensional shelf circulation along an eastern ocean boundary. J. Mar. Res. 43, 13–36.

McCreary, J.P., Kundu, P.K., Chao, S.-Y., 1987. On the dynamics of the California Current System. J. Mar. Res. 45, 1–32.

McCreary, J.P., Fukamachi, Y., Kundu, P.K., 1991. A numerical investigation of jets and eddies near an eastern ocean boundary. J. Geophys. Res. 96, 2515–2534.

McGregor, H., Dima, M., Fischer, H., Mulitza, S., 2007. Rapid 20th-century increase in coastal upwelling off northwest Africa. Science 315, 637–639. http://dx.doi.org/10.1126/science.1134839.

Melsom, A., Meyers, S.D., Hurlburt, H.E., Metzger, E.J., O'Brien, J.J., 1999. ENSO effects on Gulf of Alaska eddies. Earth Interact. 3 (1), Available at http://earthinteractions.org/.

Mesias, J.M., Matano, R.P., Strub, P.T., 2003. Dynamical analysis of the upwelling circulation off central Chile, 2003. J. Geophys. Res. 108 (C3). http://dx.doi.org/10.1029/2001/JC001135.

Mestas-Nunez, A.M., Enfield, D.B., 2001. Eastern equatorial Pacific SST variability: ENSO and Non-ENSO components and their climate associations. J. Clim. 14, 391–402.

Meyers, S.D., Basu, S., 1999. Eddies in the eastern Gulf of Alaska from TOPEX POSEIDON altimetry. J. Geophys. Res. 104, 13333–13343.

Miller, A.J., Cayan, D.R., Barnett, T.D., Graham, N.E., Oberhuber, J.M., 1994. The 1976–77 climate shift of the Pacific Ocean. Oceanography 7 (1), 21–26.

Miller, L.A., Robert, M., Crawford, W.R., 2005. The large westward propagating Haida Eddies of the Pacific eastern boundary, Editorial. Deep Sea Res. Part II 52, 845–851.

Mittelstaedt, E., 1983. The upwelling area off Northwest Africa: a description of phenomena related to coastal upwelling. Prog. Oceanogr. 12, 307–331.

Mittelstaedt, E., 1991. The ocean boundary along the northwest African coast: circulation and oceanographic properties at the sea surface. Prog. Oceanogr. 26, 307–355.

Montecino, V., Strub, P.T., Chavez, F., Thomas, A., Tarazona, J., Baumgartner, T., 2006. Bio-physical interactions off western South America. In: Robinson, A.R., Brink, K.H. (Eds.), The Sea, Vol. 14. The Global Coastal Ocean, Interdisciplinary Regional Studies and Syntheses. Harvard University Press, pp. 329–390.

Moore, D.W., Philander, S.G.H., 1977. Modelling of the tropical oceanicic circulation. In: Goldberg, E.D., McCave, I.N., O'Brien, J.J., Steele, J.H. (Eds.), The Sea, Vol. 6. Wiley (Interscience), New York, pp. 319–361.

Mote, P.W., Mantua, N.J., 2002. Coastal upwelling in a warmer future. Geophys. Res. Lett. 29, http://dx.doi.org/10.1029/2002GL016086.

Moum, J.M., Caldwell, D.R., Stabeno, P.J., 1988. Mixing and intrusions in a rotating cold-core feature off Cape Blanco, Oregon. J. Phys. Oceanogr. 18, 823–833.

Munk, W.H., 1950. On the wind-driven circulation. J. Meteorol. 7 (2), 79–93.

Munoz, R., Garreaud, R., 2005. Dynamics of the low level jet off the west coast of subtropical South America. Mon. Weather Rev. 133, 3661–3677. http://dx.doi.org/10.1175/MWR3074.1.

Mysak, L.A., 1985. On the interannual variability of eddies in the northeast Pacific Ocean. In: Wooster, W.S. (Ed.), El Niño Effects in the Eastern Subarctic Pacific Ocean. University of Washington, Seattle, pp. 97–106.

Namias, J., Yuan, X., Cayan, D.R., 1988. Persistence of North Pacific sea surface temperature and atmospheric flow patterns. J. Clim. 1 (7), 682–703.

Narimousa, S., Maxworthy, T., 1989. Application of a laboratory model to the interpretation of satellite and field observations of coastal upwelling. Dyn. Atmos. Oceans 13, 1–46. http://dx.doi.org/10.1029/2010GL042385.

Neal, E.G., Hood, E., Smikrud, K., 2010. Contribution of glacier runoff to freshwater discharge into the Gulf of Alaska. Geophys. Res. Lett. 37, L06404.

Nelson, G., Hutchings, L., 1983. The Benguela upwelling area. Prog. Oceanogr. 12 (3), 333–356.

Neshyba, S.J., Mooers, C.N.K., Smith, R.L., Barber, R.T. (Eds.), 1989. Poleward Flows Along Eastern Ocean Boundaries. Coastal and Estuarine Studies, Vol. 34. Springer-Verlag, New York.

Nieto, K., Demarcq, H., McClatchie, S., 2012. Mesoscale frontal structures in the Canary Current Upwelling System: new front and filament detection algorithms applied to spatial and temporal patterns. Remote Sens. Environ. 123, 339–346.

Nihoul, J.C.J., Strub, P.T., La Violette, P.E., 1997. Remote sensing. In: Robinson, A.R., Brink, K.H. (Eds.), The Sea, Vol. 10. The Global Coastal Ocean—Processes and Methods. John Wiley and Sons, New York, pp. 329–357.

Ohfuchi, W., Sasaki, H., Masumoto, Y., Nakamura, H., 2007. "Virtual" atmospheric and oceanic circulation in the Earth Simulator. Bull. Am. Meteorol. Soc. 88 (6), 861–866.

Pantoja, S., Iriarte, J.L., Daneri, G., 2011. Oceanography of the Chilean Patagonia. Cont. Shelf Res. 31, 149–153. http://dx.doi.org/10.1016/j.csr.2010.10.013.

Pares-Sierra, A., O'Brien, J.J., 1989. The seasonal and interannual variability of the California Current System: a numerical model. J. Geophys. Res. 94, 3159–3180.

Pedlosky, J., 1974. Longshore currents and the onset of upwelling over bottom slope. J. Phys. Oceanogr. 4, 310–320.

Peffley, M.B., O'Brien, J.J., 1976. A three-dimensional simulation of coastal upwelling off Oregon. J. Phys. Oceanogr. 6, 164–180.

Peliz, A., Rosa, T., Santos, A.M.P., Pissarra, J.L., 2002. Jets, eddies and counterflows in the Western Iberian upwelling system. J. Mar. Syst. 35, 61–77.

Peliz, A., Dubert, J., Haidvogel, D., 2003. Subinertial response of a density driven Eastern Boundary Poleward Current to wind forcing. J. Phys. Oceanogr. 33, 1633–1650.

Penven, P., Echevin, V., Pasapera, J., Colas, F., Tam, J., 2005. Average circulation, seasonal cycle, and mesoscale dynamics of the Peru Current System: a modeling approach. J. Geophys. Res. 110. http://dx.doi.org/10.1029/2005JC002945, C10021.

Perlin, N., Skyllingstad, E.D., Samelson, R.M., Barbour, P.L., 2007. Numerical simulation of air–sea coupling during coastal upwelling. J. Phys. Oceanogr. 37, 2081–2093.

Peterson, R.G., Stramma, L., 1991. Upper-level circulation in the South Atlantic Ocean. Prog. Oceaongr. 26, 1–73.

Philander, S.G.H., 1978. Forced oceanic waves. Rev. Geophys. Space Phys. 16 (1), 15–46.

Philander, S.G.H., 2000. El Niño, La Niña, and the Southern Oscillation. Academic Press, San Diego, CA, 289 pp.

Philander, S.G.H., Yoon, J.-H., 1982. Eastern boundary current upwelling. J. Phys. Oceanogr. 12, 862–879.

Picaut, J., 1983. Propagation of the seasonal upwelling in the eastern equatorial Atlantic. J. Phys. Oceanogr. 13, 18–37.

Picaut, J., 1985. Major dynamics affecting the eastern tropical Atlantic and Pacific Oceans. CalCOFI Rep. XXXVI, 41–50.

Pierce, S.D., Allen, J.S., Walstad, L.J., 1991. Dynamics of the coastal transition zone jet, 1, Linear stability analysis. J. Geophys. Res. 96, 14979–14993.

Pierce, S.D., Smith, R.L., Kosro, P.M., Barth, J.A., Wilson, C.D., 2000. Continuity of the poleward undercurrent along the eastern boundary of the mid-latitude north Pacific. Deep Sea Res. Part II 47, 811–829.

Pierce, S.D., Barth, J.A., Thomas, R.E., Fleischer, G.W., 2006. Anomalously warm July 2005 in the northern California Current: historical context and the significance of cumulative wind stress. Geophys. Res. Lett. 33. http://dx.doi.org/10.1029/2006GL027149, L22S04.

Pizarro, O., Clarke, A.J., Van Gorder, S., 2001. El Niño sea level and currents along the South American coast: comparison of observations with theory. J. Phys. Oceanogr. 31, 1891–1903.

Pizarro, O., Shaffer, G., Dewitte, B., Ramos, M., 2002. Dynamics of seasonal and interannual variability of the Peru-Chile Undercurrent. Geophys. Res. Lett. 29 (12). http://dx.doi.org/10.1029/2002GL014790.

Platt, T., White, G.N., Zhai, L., Sathyendranath, S., Roy, S., 2009. The phenology of phytoplankton blooms: ecosystem indicators from remote sensing. Ecol. Model. 220. http://dx.doi.org/10.1016/j.ecolmodel.2008.11.022.

Preisendorfer, R.W., 1988. Principal Component Analysis in Meterorology and Oceanography. Elsevier, New York, 425 pp.

Preston-Whyte, R.A., Tyson, P.D., 1993. The Atmosphere and Weather of Southern Africa. Oxford University Press, Capetown, SA, 374 pp.

Pringle, J.M., Dever, E.P., 2009. Dynamics of wind-driven upwelling and relaxation between Monterey Bay and Point Arena: local-, regional-, and gyre-scale controls. J. Geophys. Res. 114. http://dx.doi.org/10.1029/2008JC005016, C07003.

Racault, M.-F., Le Quere, C., Buitenhuis, E., Sathyendranath, S., Platt, T., 2012. Phytoplankton phenology in the global ocean. Ecol. Indic. 14, 152–163.

Ramage, C.S., 1987. Secular change in reported wind speeds over the ocean. J. Clim. Appl. Meteorol. 26, 525–528.

Ramos, M., Pizarro, O., Bravo, L., Dewitte, B., 2006. Seasonal variability of the permanent thermocline off northern Chile. Geophys. Res. Lett. 33. http://dx.doi.org/10.1029/2006GL025882, L09608.

Reason, C.J.C., Jury, M.R., 1990. On the generation and propagation of the southern African coastal low. Q. J. R. Meteorol. Soc. 116, 1133–1151.

Reason, C.J.C., Steyn, D.G., 1990. Coastally trapped disturbances in the lower atmosphere: dynamic commonalities and geographic diversity. Prog. Phys. Geogr. 14, 178–198.

Reed, R.K., Stabeno, P.J., 1999. A recent full-depth survey of the Alaskan Stream. J. Oceanogr. 55 (1), 79–85.

Reid, J.L., 2003. On the total geostrophic circulation of the Indian Ocean: flow patterns, tracers and transports. Prog. Oceanogr. 56, 137–186.

Relvas, P., Barton, E.D., Dubert, J., Oliveira, P.B., Peliz, A., da Silva, J.C.B., Santos, A.M.P., 2007. Physical oceanography of the western Iberia ecosystem: latest views and challenges. Prog. Oceanogr. 74, 149–173. http://dx.doi.org/10.1016/j.pocean.2007.04.021.

Richards, F.A., 1981. Coastal Upwelling. American Geophysical Union, Washington, DC.

Richman, M.B., 1986. Rotation of principal components. Int. J. Climatol. 6, 293–335.

Ridgway, K.R., Condie, S.A., 2004. The 5500-km long boundary flow off western and southern Australia. J. Geophys. Res. 109. http://dx.doi.org/10.1029/2003JC001921, C04017.

Rienecker, M.M., Mooers, C.N.K., Hagan, D.E., Robinson, A.R., 1985. A cool anomaly off northern California: an investigation using IR imagery and in situ data. J. Geophys. Res. 90, 4807–4818.

Risien, C.M., Chelton, D.B., 2008. A global climatology of surface wind and wind stress fields from eight years of QuikSCAT scatterometer data. J. Phys. Oceanogr. 38, 2379–2413. http://dx.doi.org/10.1175/2008JPO3881.1.

Robinson, A.R., Brink, K.H., 1998a. The Sea, Vol. 10. The Global Coastal Ocean: Processes and Methods. Wiley, New York.

Robinson, A.R., Brink, K.H., 1998b. The Sea, Vol. 11. The Global Coastal Ocean: Regional Studies and Syntheses. Wiley, New York.

Robinson, A.R., Brink, K.H., 2005. The Sea, Vol. 13. Multiscale Interdisciplinary Processes. Harvard University Press, Cambridge, MA.

Robinson, A.R., Brink, K.H., 2006. The Sea, Vol. 14A & 14B. Global Coastal Ocean, Interdisciplinary Studies and Syntheses. Harvard University Press, Cambridge, MA & London.

Rodhe, J., 1998. The Baltic and North Seas: a process-oriented review of the physical oceanography. In: Robinson, A.R., Brink, K.H. (Eds.), The Sea, Vol. 11. John Wiley and Son, Inc, pp. 699–732.

Rodhe, J., Tett, P., Wulff, F., 2006. The Baltic and North Seas: a regional review of some important physical-chemical-biological interaction processes. In: The Sea, Vol. 14B. The Global Coastal Ocean: Regional Studies and Syntheses. Wiley, New York, pp. 39–57.

Rogers, J.C., 1981. The North Pacific Oscillation. Int. J. Climatol. 1, 39–57.

Rogers, D.P., Johnson, D.W., Friehe, C.A., 1995. The stable internal boundary layer over a coastal sea. Part II: gravity waves and ageostrophic wind forcing. J. Atmos. Sci. 52, 684–696.

Rogers, D.P., Dorman, C.E., Edwards, K.A., et al., 1998. Highlights of Coastal Waves 1996. Bull. Am. Meteorol. Soc. 79, 1307–1326.

Roske, F., 2006. A global heat and freshwater forcing dataset for ocean models. Ocean Model. 11, 235–237.

Rouault, M., Illig, S., Bartholomae, C., Reason, C.J.C., Bentamy, A., 2007. Propagation and origin of warm anomalies in the Angola Benguela upwelling system in 200. J. Mar. Syst. 68 (3–4), 473–488. http://dx.doi.org/10.1016/j.jmarsys.2006.11.010.

Rouault, M., Penven, P., Pohl, B., 2009. Warming in the Agulhas Current system since the 1980's. Geophys. Res. Lett. 36. http://dx.doi.org/10.1029/2009GL037987, L12602.

Roy, C., 2006. A note on coastal upwellings and fisheries in the Gulf of Guinea. In: Robinson, A.R., Brink, K.H. (Eds.), The Sea, Vol. 14B. The Global Coastal Ocean, Interdisciplinary Regional Studies and Syntheses. Harvard University Press, pp. 865–877.

Royer, T.C., 1982. Coastal fresh water discharge in the northeast Pacific. J. Geophys. Res. 87, 2017–2021.

Royer, T.C., 1998. Coastal processes in the northern North Pacific. In: Robinson, A.R., Brink, K.H. (Eds.), The Sea, Vol. 11. The Global Coastal Ocean: Regional Studies and Syntheses. John Wiley and Sons, New York, pp. 395–414.

Rutllant, J., Montecino, V., 2002. Multiscale upwelling forcing cycles and biological response off north-central Chile. Rev. Chil. Hist. Nat. 75, 217–231.

Sangra, P., Pascual, A., Rodriguez-Santana, A., Machin, F., Mason, E., McWilliams, J.C., Pelegri, J.L., Dong, C., Rubio, A., Aristegui, J., Marrero-Diaz, A., Hernandez-Guerra, A., Martinez-Marrero, A., Auladell, M., 2009. The Canary Current Corridor: a major pathway for long-lived eddies in the subtropical North Atlantic. Deep Sea Res. Part I 56, 2100–2114. http://dx.doi.org/10.1016/j.dsr.2009.08.008.

Sasaki, H., Sasai, Y., Kawahara, S., Furuichi, M., Araki, F., Ishida, A., Yamanaka, Y., Masumoto, Y., Sakuma, H., 2004. A series of eddy-resolving ocean simulations in the world ocean. OFES (OGCM for the Earth Simulator) Project, OCEAN-04, vol. 3. 1535–1541.

Sasaki, H., Nonaka, M., Masumoto, Y., Sasai, Y., Uehara, H., Sakuma, H., 2006. An eddy-resolving hindcast simulation of the quasi-global ocean from 1950 to 2003 on the Earth Simulator. In: Ohfuchi, W., Hamilton, K. (Eds.), High Resolution Numerical Modeling of the Atmosphere and Ocean. Springer, New York.

Schott, F.A., McCreary Jr., J., 2001. The monsoon circulation of the Indian Ocean. Prog. Oceanogr. 51, 1–123.

Schouten, J.W., Matano, R.P., Strub, P.T., 2005. A description of the seasonal cycle of the equatorial Atlantic from altimeter data. Deep Sea Res. Part I 52, 477–493. http://dx.doi.org/10.1016/j.dsr.2004.10.007.

Schwing, F.B., Mendelssohn, R., 1997. Increased coastal upwelling in the California Current System. J. Geophys. Res. 102, 3421–3438.

Schwing, F.B., Mendelssohn, R., 2002. Common and uncommon trends in SST and wind stress in the California and Peru–Chile current systems. Prog. Oceanogr. 53, 141–162.

Schwing, F.B., Batchelder, H., Crawford, W., Mantua, N., Overland, J., Polovina, J., Zhao, J.-P., 2005. Decadal-scale climate events. In: King, J.R. (Ed.), Report of the Study Group on Fisheries and Ecosystem Responses to Recent Regime Shifts: PICES Scientific Report No. 28. Institute of Ocean Sciences, Sidney, Canada.

Shaffer, G., Pizarro, O., Djurfeldt, L., Salinas, S., Rutllant, J., 1997. Circulation and low-frequency variability near the Chile coast: remotely forced fluctuations during the 1991–1992 El Niño. J. Phys. Oceanogr. 27, 217–235. http://dx.doi.org/10.1175/1520-0485(1997)027<0217:CALFVN>2.0CO;2.

Shaffer, G., Hormazábal, S., Pizarro, O., Djurfeldt, L., Salinas, S., 1999. Seasonal and interannual variability of currents and temperature over the slope off central Chile. J. Geophys. Res. 104, 29951–29961. http://dx.doi.org/10.1029/1999JC900253.

Shaffer, G., Hormazábal, S., Pizarro, O., Ramos, M., 2004. Circulation and variability in the Chile Basin. Deep Sea Res. Part I 51, 1367–1386. http://dx.doi.org/10.1016/j.dsr.2004.05.006.

Shannon, L.V., Nelson, G., 1996. Wefer, G. (Ed.), The Benguela: large-scale features and processes and system variability. The South Atlantic: Present and Past Circulation, pp. 163–210.

Shannon, L.V., Agenbag, J.J., Buys, M.E.L., 1987. Large- and meso-scale features of the Angola-Benguela Front. In: Payne, A.I.L., Gulland, J.A., Brink, K.H. (Eds.), The Benguela and Comparable Ecosystems, S. Afr. J. Mar. Sci., vol. 5. pp. 11–34.

Sharples, J., Holligan, P.M., 2006. Interdisciplinary studies in the Celtic Seas. In: Robinson, A.R., Brink, K.H. (Eds.), The Sea, Vol. 14B. The Global Coastal Ocean, Interdisciplinary Regional Studies and Syntheses. Harvard University Press, pp. 1003–1031.

Shillington, F.A., 1998. Benguela upwelling system off southwestern Africa, coastal segment (16,E). In: Robinson, A.R., Brink, K.H. (Eds.), The Sea, Vol. 11. The Global Coastal Ocean—Regional Studies and Syntheses. John Wiley and Sons, New York, pp. 583–604.

Shillington, F.A., Reason, C.J.C., Rae, D., Florenchie, C.M.P., Penven, P., 2006. Large scale physical variability of the Benguela Current Large Marine Ecosystem (BCLME). In: Shannon, L.V., Hempel, G., Malanotte-Rizzoli, P., Moloney, C., Woods, J. (Eds.), The Benguela:

Predicting a Large Marine Ecosystem. Large Marine Ecosystems, vol. 14. Elsevier B.V./Ltd, pp. 49–70.

Siedler, G., Zangenberg, N., Onken, R., Morliere, A., 1992. Seasonal changes in the tropical Atlantic circulation: observation and simulation of the Guinea Dome. J. Geophys. Res. 97, 703–715.

Silva, N., Neshyba, S., 1979. On the southernmost extension of the Peru-Chile Undercurrent. Deep Sea Res. 26A, 1387–1393.

Silva, N., Palma, S., 2008. The CIMAR Program in the austral Chilean channels and fjords. In: Silva, N., Palma, S. (Eds.), Progress in the Oceanographic Knowledge of Chilean Interior Waters, from Puerto Montt to Cape Horn. Comite Oceanographfico Nacional—Pontificia Universidad Catolica de Valparaiso, Valparaiso, Chile, pp. 11–15.

Simpson, J.J., 1984. El Niño-induced onshore transport in the California Current during 1982-1983. Geophys. Res. Lett. 11, 233–236.

Simpson, J.H., 1998. The Celtic Seas. In: Robinson, A.R., Brink, K.H. (Eds.), The Sea, Vol. 11. The Global Coastal Ocean—Regional Studies and Syntheses. John Wiley and Sons, New York, pp. 659–698.

Sinclair, M., 1988. Marine Populations: An Essay on Population Regulation and Speciation. Washington Sea Grant, 252 pp.

Smith, R.L., 1968. Upwelling. Oceanogr. Mar. Biol. Annu. Rev. 6, 11–46.

Smith, R.L., 1978. Poleward propagating perturbations in currents and sea level along the Peru coast. J. Geophys. Res. 83, 6083–6092.

Smith, R.L., 1981. A comparison of the structure and variability of the flow field in three coastal upwelling regions: Oregon, Northwest Africa and Peru. In: Richards, F.A. (Ed.), Coastal Upwelling. American Geophysical Union, Washington, D.C, pp. 107–118.

Smith, R.L., Huyer, A., Godfrey, J.S., Church, J.A., 1991. The Leeuwin Current off Western Australia, 1986-1987. J. Phys. Oceanogr. 21, 323–345.

Sobarzo, M., Figueroa, D., 2001. The physical structure of a cold filament in a Chilean upwelling zone (Peninsula de Mejillones, Chile, 23°S). Deep Sea Res. 48, 2699–2726.

Stabeno, P.J., Bond, N.A., Hermann, A.J., Kachel, N.B., Mordy, C.W., Overland, J.E., 2004. Meteorology and oceanography of the Northern Gulf of Alaska. Cont. Shelf Res. 24, 859–897.

Stevenson, M., 1970. Circulation in the Panama Bight. J. Geophys. Res. 75, 659–672.

Stramma, L., Schott, F., 1999. The mean flow field of the tropical Atlantic Ocean. Deep Sea Res. Part II 46, 279–303.

Strub, P.T., James, C., 1988. Atmospheric conditions during the spring and fall transitions in the coastal ocean off western United States. J. Geophys. Res. 93 (C12), 15,561–15,584.

Strub, P.T., James, C., 2000. Altimeter-derived variability of surface velocities in the California Current System: 1. Seasonal circulation and eddy statistics. Deep Sea Res. Part II 47, 831–870.

Strub, P.T., James, C., 2002a. Altimeter-derived surface circulation in the large-scale NE Pacific gyres. Part 1: annual variability. Prog. Oceanogr. 53, 163–183.

Strub, P.T., James, C., 2002b. Altimeter-derived surface circulation in the large-scale NE Pacific gyres. Part 2: 1997–1998 El Niño anomalies. Prog. Oceanogr. 53, 185–214.

Strub, P.T., James, C., 2002c. The 1997–1998 El Niño signal along the southeast and northeast Pacific boundaries—an altimetric view. Prog. Oceanogr. 54, 439–458.

Strub, P.T., Allen, J.S., Huyer, A., Smith, R.L., 1987. Large-scale structure of the spring transition in the coastal ocean off western North America. J. Geophys. Res. 92, 1527–1544.

Strub, P.T., Kosro, P.M., Huyer, A., Collaborators, C.T.Z., 1991. The nature of the cold filaments in the California Current System. J. Geophys. Res. 96 (C8), 14,743–14,768.

Strub, P.T., Mesias, J.M., James, C., 1995. Satellite observations of the Peru-Chile countercurrent. Geophys. Res. Lett. 22, 211–214.

Strub, P.T., Shillington, F.A., James, C., Weeks, S.J., 1998a. Satellite comparison of the seasonal circulation in the Benguela and California Current Systems. S. Afr. J. Mar. Sci. 19, 99–112.

Strub, P.T., Mesias, J.M., Montecino, V., Ruttlant, J., Salinas, S., 1998b. Coastal ocean circulation of western South America. In: Robinson, A.R., Brink, K.H. (Eds.), The Sea, Vol. 11. The Global Coastal Ocean—Regional Studies and Syntheses. John Wiley and Sons, New York, pp. 273–313.

Sydeman, W.J., Bradley, R.W., Warzybok, P., Abraham, C.L., Jahncke, J., Hyrenbach, K.D., Kousky, V., Hipfner, J.M., Ohman, M.D., 2006. Planktiverous auklet Ptychoramphus aleuticus responses to the anomaly of 2005 in the California Current. Geophys. Res. Lett. 33. http://dx.doi.org/10.1029/2006GL026736, L22S09.

Tabata, S., 1982. The anticyclonic, baroclinic Eddy Off Sitka, Alaska, in the Northeast Pacific-Ocean. J. Phys. Oceanogr. 12, 1260–1282.

Talley, L.D., Pickard, G.L., Emery, W.J., Swift, J.H., 2011. Descriptive Physical Oceanography, an Introduction. Elsevier, Amsterdam, NL, 555 pp. plus color plates.

Thomas, A.C., Brickley, P., 2006. Satellite measurements of chlorophyll distribution during spring 2005 in the California Current. Geophys. Res. Lett. 33. http://dx.doi.org/10.1029/2006GL026588.

Thomas, A.C., Strub, P.T., Weatherbee, R.A., James, C., 2012. Satellite views of Pacific chlorophyll variability: comparisons to physical variability, local versus nonlocal influences and links to climate indices. Deep Sea Res. Part II. http://dx.doi.org/10.1016/j.dsr2.2012.04.008.

Thompson, R.O.R.Y., 1987. Continental-shelf-scale model of the Leeuwin Current. J. Mar. Res. 45, 813–827.

Torres, R., Barton, E., 2006. Onset and development of the Iberian poleward flow along the Galician coast. Cont. Shelf Res. 26, 1134–1153.

Torres, R., Turner, D., Silva, N., Rutllant, J., 1999. High short-term variability of CO2 fluxes during an upwelling event off the Chilean coast at 30°S. Deep Sea Res. Part I 46, 1161–1179.

Torres, R., Turner, D., Rutllant, J., Sobarzo, M., Antezana, T., González, H., 2002. CO2 outgassing off central Chile (31-30°S) and northern Chile (24-23°S) during austral summer 1997: the effect of wind intensity on the upwelling and ventilation of CO2 rich waters. Deep Sea Res. Part I 49, 1413–1429.

Traganza, E.D., Silva, V.M., Austin, D.M., Hanson, W.E., Bronsink, S.H., 1983. Nutrient mapping and recurrence of coastal upwelling centers by satellite remote sensing: its implication to primary production and the sediment record. In: Suess, E., Thiede, J. (Eds.), Coastal Upwelling: its Sediment Record. Plenum, New York, pp. 61–83.

Trenberth, K.E., Hurrell, J.W., 1995. Decadal climate variations in the Pacific. In: National Research Council, Natural Climate Variability on Decade-to-Century Time Scales. National Academy Press, Washington, D.C, pp. 472–481.

Trowbridge, J.H., Chapman, D.C., Candela, J., 1998. Topographic effects, straights and the bottom boundary layer. In: Robinson, A.R., Brink, K.H. (Eds.), The Sea, Vol. 10. The Global Coastal Ocean—Processes and Methods. John Wiley and Sons, New York, pp. 21–62.

Tsuchiya, M., 1985. The subthermocline phosphate distribution and circulation in the far eastern Equatorial Pacific Ocean. Deep Sea Res. 32, 299–313.

Valle-Levinson, A., Moraga-Opazo, J., 2006. Observations of bipolar residual circulation in two equatorward-facing semiarid bays. Cont. Shelf Res. 26, 179–193.

Valle-Levinson, A., Atkinson, L.P., Figueroa, D., Castro, L., 2003. Flow induced by upwelling winds in an equatorward facing bay: Gulf of Arauco, Chile. J. Geophys. Res. 108. http://dx.doi.org/10.1029/2001JC001272.

Veitch, J.A., Florenchie, P., Shillington, F.A., 2006. Seasonal and interannual fluctuations of the Angola Benguela Frontal Zone (ABFZ) using 4.5 km resolution satellite imagery from 1982 to 1999. Int. J. Remote Sens. 27 (5), 987–998.

Veitch, J., Penven, P., Shillington, F., 2010. Modeling Equilibrium Dynamics of the Benguela Current System. J. Phys. Oceanogr. 40, 1942–1964.

Venegas, R.M., Strub, P.T., Beier, E., Letelier, R., Cowles, T., Thomas, A.C., James, C., Soto-Mardones, L., Cabrera, C., 2008. Satellite-derived variability in chlorophyll pigments, wind stress, sea surface height and temperature in the northern California Current System. J. Geophys. Res. 113. http://dx.doi.org/10.1029/2007JC004481, C03015.

Verstraete, J.M., 1992. The seasonal upwelling in the Gulf of Guinea. Prog. Oceanogr. 29, 1–60.

Waite, A.M., Thompson, P.A., Pesant, S., Feng, M., Beckley, L.E., Domingues, C.M., Gaughan, D., Hanson, C.E., Holl, C.M., Koslow, T., Meuleners, M., Montoya, J.P., Moore, T., Muhling, B.A., Paterson, H., Rennie, S., Strzelecki, J., Twomey, L., 2007. The Leeuwin Current and its eddies: an introductory overview. Deep Sea Res. Part II 54, 789–796.

Walker, G.T., 1924. Correlation in seasonal variations of weather, IX. A further study of world weather. Memo. India Meteorol. Dep. 24 (9), 275–333.

Walker, G.T., Bliss, E.W., 1930. World Weather, IV. Memo. R. Meteorol. Soc. 3 (24), 81–95.

Walker, G.T., Bliss, E.W., 1937. World Weather, VI. Memo. R. Meteorol. Soc. 4 (239), 119–139.

Weingartner, T.J., Danielson, S., Royer, T.C., 2005. Freshwater Variability and Predictability in the Alaska Coastal Current. Deep Sea Res. 52, 169–192.

Winant, C.D., Dorman, C.E., Friehe, C.A., Beardsley, R.C., 1988. The marine layer off Northern California: an example of supercritical channel flow. J. Atmos. Sci. 45, 3588–3605.

Wolter, K., Timlin, M.S., 1998. Measuring the strength of ENSO—how does 1997-8 rank? Weather 53, 315–324.

Wyrtki, K., Kilonsky, B., 1984. Mean water and current structure during the Hawaii-to-Tahiti shuttle experiment. J. Phys. Oceanogr. 14, 242–254.

Yoshida, K., 1967. Circulation in the eastern tropical oceans with special reference to upwelling and undercurrents. Jpn. J. Geophys. 4 (2), 1–75.

Chapter 15

The Tropical Ocean Circulation and Dynamics

Swadhin Behera*, Peter Brandt† and Gilles Reverdin‡
*Research Institute for Global Change, JAMSTEC, Yokohama, Japan
†Helmholtz Centre for Ocean Research Kiel (GEOMAR), Kiel, Germany
‡Laboratoire d'Océanographie et du Climate (LOCEAN), Paris, France

Chapter Outline

1. INTRODUCTION

The tropical ocean circulations are fundamental to the Earth's climate system and its variations. The large flux of solar radiation in the tropics calls for strong poleward heat transport by ocean and atmosphere to maintain the global energy balance. This basic requirement, augmented by the rotation of the earth and other regional processes, creates and drives the major features that constitute the global ocean circulations. In a chapter in the first edition of this book, Godfrey et al. (2001) covered some of the recognized circulations particularly as observed during the World Ocean Circulation Experiment (WOCE). In the decade since the publication of that chapter, there has been considerable progress in our understanding of the ocean's role in climate. The arrival of new ocean observation techniques during this time has given deeper insights into important processes. In addition, the modeling of the ocean as well as of the coupled ocean–atmosphere system has become more realistic. Hence, this follow-up chapter attempts to cover some of those newly identified ocean and climate processes.

Part of the large-scale climate system is the energetic current system of the tropics, particularly the narrow east- and westward current bands that dominate the interior tropics. The strongest current of the central Pacific and Atlantic is the eastward-flowing Equatorial Undercurrent (EUC). Forced by the easterly winds at the equator, it supplies thermocline waters to the upwelling regions of the eastern part of both oceans (see Chapter 14). Remarkable instabilities of the zonal current system of the tropics lead to the generation of tropical instability waves (TIWs) that reshape pronounced temperature and salinity fronts in the equatorial cold tongue regions and redistribute heat and salt vertically and horizontally in the upper ocean. The evolution of the equatorial cold tongue in the eastern Pacific as well as Atlantic is strongly linked to the intermittent occurrence of vigorous diapycnal mixing in the vertical shear zones of mean and fluctuating currents resulting in a downward heat transport below the mixed layer. Intrinsic ocean dynamics are similarly identified as the cause of interannual sea surface temperature (SST) variability and may themselves be able to force variability in the atmosphere.

Ocean Circulation and Climate, Vol. 103. http://dx.doi.org/10.1016/B978-0-12-391851-2.00015-5

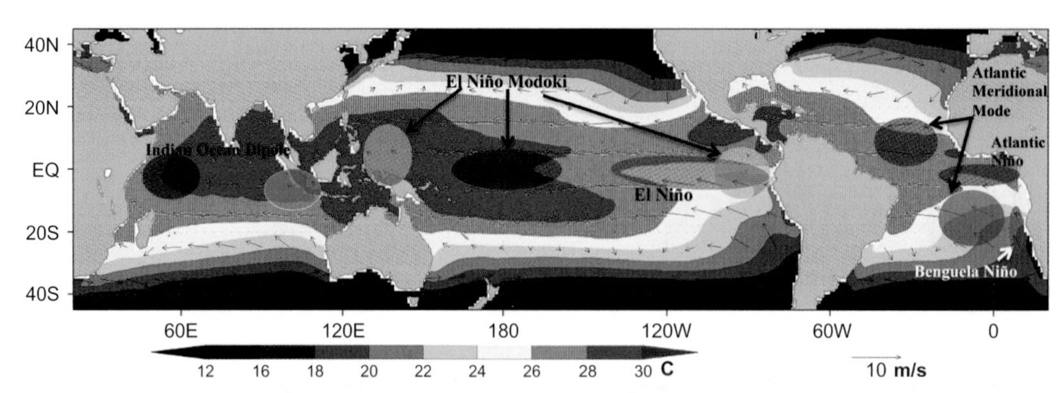

FIGURE 15.1 Modes of climate variations in the tropical and subtropical oceans overlaid on the annual mean SST (Rayner et al., 2003) and surface winds (Kalnay et al., 1996).

Due to the presence of a strong meridional overturning circulation (MOC, see Chapter 11), the Atlantic plays a particular role in the climate system of the earth. During the past two decades, much effort has been put into the quantification of the Atlantic MOC, which is suspected to change under ongoing anthropogenic forcing. The mean state of the tropical Atlantic Ocean largely results from the presence of a strong MOC, and it will likely respond to a changing MOC. In addition, the SST in the eastern equatorial Atlantic shows interesting variations. Coupled ocean–atmosphere and intrinsic ocean dynamics that control this SST variability are known to play an important role in affecting the climates of the surrounding continents. However, these are the regions where state-of-the-art ocean–atmosphere coupled general circulation models (CGCMs) have the largest model errors, thus limiting local climate predictions.

The circulation of the Indian Ocean is more complex. As compared to the other two basins, a somewhat different circulation regime is observed, particularly north of the equator, because of the effects of seasonal changes in monsoon winds. The North Equatorial Current in the Indian Ocean flows eastward during boreal summer and westward during boreal winter. This reversal helps the water exchange between the low-salinity of the Bay of Bengal and the high-salinity of the Arabian Sea. In addition, the currents along the Somali coast reverse direction from southward in winter to northward in summer (see Chapter 13). During the latter season, the Somali Current system is marked by several energetic eddies. The one known as the Great Whirl recirculates the water around it 300 km off the coast. This coastal region also experiences intense upwelling during the summer, and the circulation regime penetrates deeper, affecting the movement of water masses below the thermocline and thus aiding the energy and mass exchanges between the two hemispheres.

The semiannual eastward winds of the tropical Indian Ocean drive the swift Yoshida–Wyrtki jets (Wyrtki, 1973) on the equator during the Indian monsoon transition months of April–May and October–November. In contrast to the Atlantic and Pacific, the eastward surface jet brings warm water to the eastern Indian Ocean and helps to maintain the Indo-Pacific warm pool. Interannual variations in the eastward heat transports together with the equatorial ocean dynamics give rise to the ocean–atmosphere coupled variations in the basin (Figure 15.1), known as the Indian Ocean dipole (IOD).

The tropics are regions where the ocean dynamics largely influences the SST. Oceanic processes such as mean and eddy advection as well as diapycnal mixing play key roles in the tropical heat balance. Furthermore, evaporation is large in the warm and windy tropical oceans (Figure 15.1). The associated regional distributions of rainfall bands and the diabatic heating in the tropics are largely responsible for driving the global atmospheric circulation, which in turn drives the oceanic circulation. These are part of ocean and atmosphere feedback processes. It is because of this that the tropical oceans have long been recognized as the most important regions for large-scale ocean–atmosphere interactions responsible for coupled climate variations on a wide range of spatial and temporal scales (see Chapter 24). For instance, the ocean–atmosphere interactions in the tropical Pacific are associated with the most dominant mode of climate variation (Figure 15.1), the El Niño/Southern Oscillation (ENSO) (see Chapter 24). During the warm phase of ENSO, the eastern equatorial Pacific becomes warmer than normal, giving rise to a basin-scale climate perturbation that has large impacts on the climate and, hence, on socioeconomic conditions of many parts of the world. In the Indian Ocean, the coupled phenomenon is recognized as the IOD (Figure 15.1). A similar climate mode exists in the Atlantic Ocean, known as the Atlantic Niño, though it is less intense than that in the Pacific (Figure 15.1). Since these climate modes have large impacts on human well-being, it is important for us to develop reliable prediction models in an attempt to mitigate the impacts of disasters related to climate variations.

Over recent decades, considerable progress has been made in the ENSO prediction. While our ability to predict ENSO has improved dramatically, models sometimes have problems in predicting the triggering and the seasonal evolution of ENSO. Efforts to improve our understanding of ENSO dynamics and to refine the ENSO prediction system started during the World Climate Research Programme (WCRP)'s Tropical Ocean Global Atmosphere (TOGA) experiment (McPhaden et al., 2010) in the 1980s and continued in the early years of WCRP's CLImate VARiability and Predictability programme (CLIVAR). The scope and dimensions of this research have also expanded considerably in recent years to explore the role of other climate phenomena such as IOD, Atlantic Niño, and other regional subtropical modes. These recent studies clearly indicate the need to improve our understanding of those climate phenomena to extend our success in ENSO-based seasonal climate prediction to other domains.

2. TROPICAL PACIFIC VARIABILITY

In the tropical Pacific, an east-west slope in the thermocline is maintained by the prevailing easterlies. The thermocline slope adjusts to Sverdrup balance at low frequencies. In the eastern Pacific, the steady winds and shallow thermocline are not conducive to a complex vertical structure in the ocean. But in the warm pool regions of the western Pacific, the upper layer is very thick and the winds vary greatly leading to an elaborate vertical structure (Figure 15.2a).

2.1. Western Pacific Warm Pool

The tropical western Pacific Ocean is characterized by a deep thermocline that extends partly through the Indonesian Throughflow to parts of the eastern Indian Ocean (see also Chapter 19). The structure and oceanographic variability (Figure 15.2b) in this region are strongly related with the Intertropical Convergence Zone (ITCZ)/South Pacific Convergence Zone (SPCZ) structure and their variability on seasonal (monsoons), interannual, and interdecadal timescales. The associated heat balance creates a near-surface warm pool, and the strong rainfall often forms a fresh pool. It is also through this area that a large part of the thermocline water formed in the tropics reaches the equatorial band where it is later upwelled, mostly in the equatorial and eastern Pacific upwelling systems. There are also exchanges with the Indian Ocean through the Indonesian Throughflow (see Chapter 19), clearly modulated on seasonal and interannual timescales (Gordon, 2005). Although mass balance in the equatorial Pacific (Meinen and McPhaden, 2001; Izumo, 2005; Meinen, 2005) implies that the equatorial discharge/recharge on ENSO timescales happens largely through changes in the interior transports,

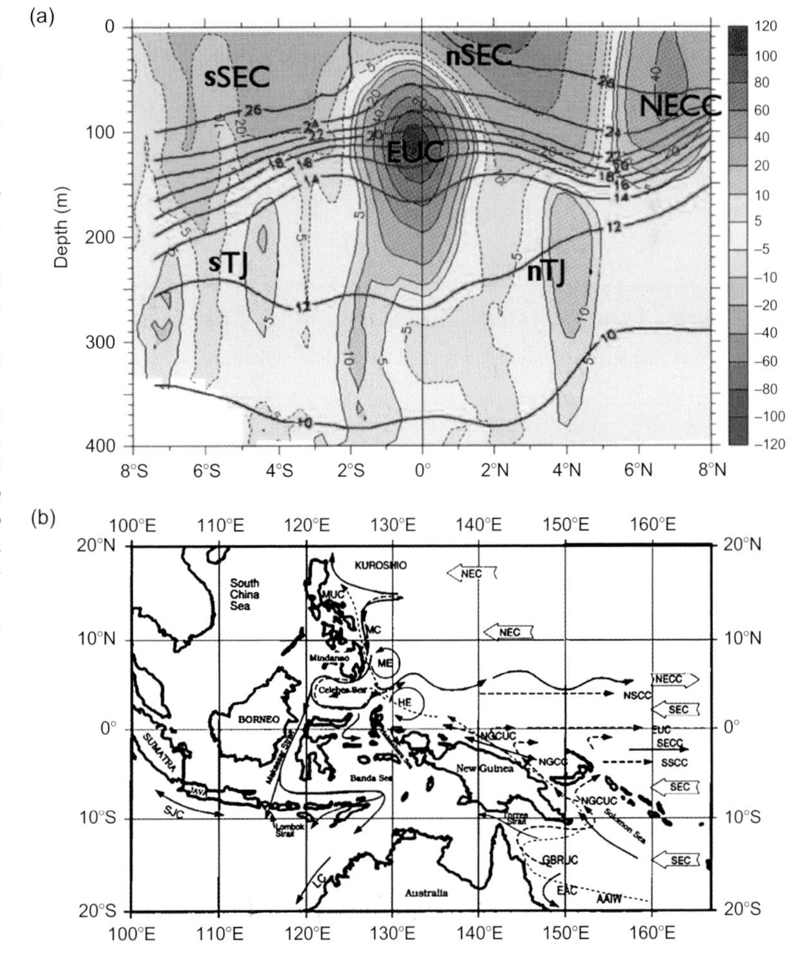

FIGURE 15.2 (a) The structure of the upper tropical Pacific zonal current at 140°W from acoustic Doppler current profiler (ADCP) measurements of Johnson et al. (2002). The currents are Equatorial Under Current (EUC), South Equatorial Current (SEC) with its Southern (sSEC) and Northern branch (nSEC), Northern Equatorial Countercurrent (NECC) and the subsurface countercurrents or Tsuchiya Jets with its Southern (sTJ)/ South Subsurface Countercurrent (SSCC) and Northern branch (nTJ)/North Subsurface Countercurrent (NSCC). The shading interval is cm/s and the green contour lines mark temperature in degree Celsius. The figure is adopted from W. S. Kessler (personal communications, 2012). (b) The complex circulation regime in the western tropical Pacific (adapted from Fine et al. 1994). NEC, North Equatorial Current; NECC, North Equatorial Countercurrent; SEC, South Equatorial Current; EUC, Equatorial Undercurrent; NSCC and SSCC, North and South Subsurface Countercurrent; MC, Mindanao Current; MUC, Mindanao Undercurrent; ME, Mindanao Eddy; HE, Halmahera Eddy; NGCC, New Guinea Coastal Current; NGCUC, New Guinea Coastal Undercurrent; GBRUC, Great Barrier Reef Undercurrent; EAC, East Australian Current; LC, Leeuwin Current; AAIW, Antarctic Intermediate Water.

the situation is not as clear on longer timescales, for which investigations have been less comprehensive (see Chapter 25). These equatorward thermocline transports in the western Pacific have likely changed on decadal timescales (McPhaden and Zhang, 2004), as well as the properties that they transport (e.g., "spiciness" anomalies, Giese et al., 2002) and their link to the subtropical cells (STCs).

In some areas, such as the western part of the SPCZ, heat content variability is very large on decadal timescales and the typical ENSO timescales are not dominant. This is seen in both surface variability (driven by latent heat exchanges) and thermocline structure (as a result of variations in the wind stress curl (Chang et al., 2001; Luo and Yamagata, 2001; Cibot et al., 2005)). This southern portion of the warm pool is also the site of considerable decadal variability in freshwater as a result of precipitation changes related to the changes in the SPCZ position, and to some extent associated with the Pacific decadal variability (Delcroix et al., 2007). The whole western Pacific warm pool has experienced fairly regular surface warming (outside of ENSO events) in recent decades (average 0.29 °C in 50 years) and associated surface freshening (average 0.34 PSU in 50 years), with a pattern (Figure 15.3) noticeably distinct from the one of the Pacific Decadal Oscillation (PDO) and much more equatorially symmetric (Cravatte et al., 2009). Changes in the ENSO characteristics (in particular, frequency of ENSO Modoki/central Pacific ENSO events as explained in the subsection 2.2) have only moderately (30%) contributed to these changes (Singh et al., 2011), which have also been attributed to anthropogenically induced hydrological changes (Terray et al., 2012).

The freshwater pool is another important aspect of the western Pacific. How the waters of the fresh pool are renewed and exchanged with the higher latitudes and how the

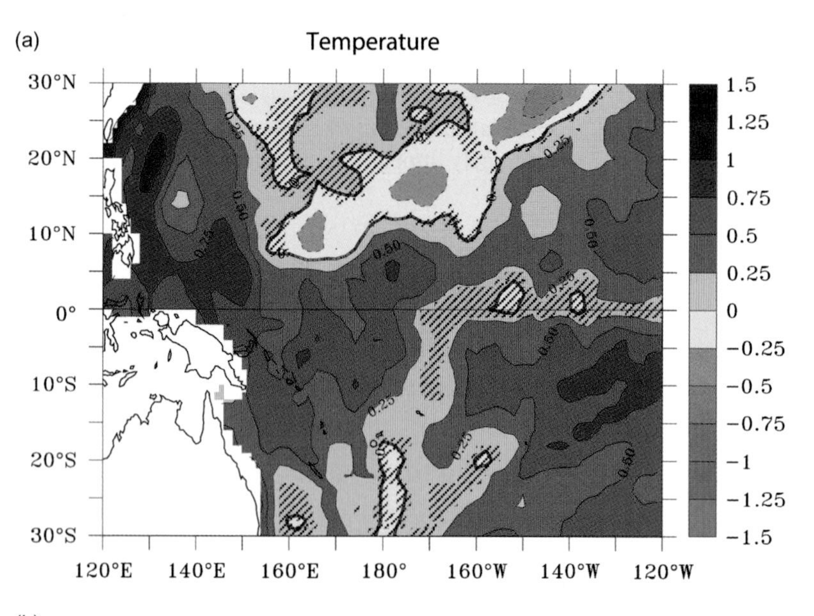

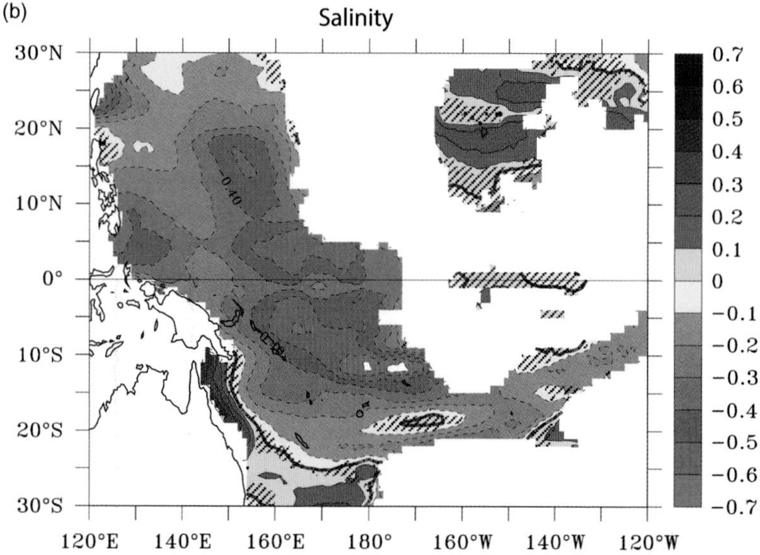

FIGURE 15.3 Linear trends in (a) SST and (b) SSS (from Cravatte et al., 2009). Units are °C/50 years and pss/50 years.

thermocline water feeds the equatorial band are the foci of renewed interest. Those were rather schematically investigated in the past (WEPOCS program in the early 1980s: Lindstrom et al., 1987), in particular in the Southern Hemisphere, where complex bathymetry and narrow passages connect the deep basins, inducing complex and intense western boundary currents. Model studies have suggested that a substantial fraction of the transports toward the equator are through these western boundary current systems (Blanke and Raynaud, 1997; Izumo et al., 2002; Fukumori et al., 2004). The transports in the southwestern Pacific also contribute to the bulk of the Indonesian Throughflow (Blanke et al., 2002; Talley and Sprintall, 2005). The details of this circulation and transports are now being investigated in the international Southwest Pacific Ocean Circulation and Climate Experiment (SPICE) program, which reveals a rather complicated situation in the Solomon and northern Coral Seas with multiple western boundary currents, zonal jets, and island effects (Webb, 2000; Ridgway and Dunn, 2003; Cravatte et al., 2011). The connection to the regions further north is through different straits (primarily, but not uniquely, Vitiaz Strait and Saint-George Channels). The variability of the respective transports in these passages is still being actively investigated. In a model study, Kessler and Gourdeau (2007) point to different paths in the Bismarck Sea between direct inflow toward the equator (in particular, in the thermocline to the EUC) and the flow through the New Guinea Coastal Undercurrent toward the western Pacific, where it connects with the flow from the Northern Hemisphere, and feeds into both the Indonesian Throughflow and the equatorial Pacific circulation (Qu and Lindstrom, 2002; Kawabe et al., 2008) with large seasonal variability and probably interannual, decadal, or even variability on longer timescales.

In the western Pacific, the circulation north of the equator is as complex as the one south of it (Figure 15.2b, see also Chapter 19) because of the presence of large island archipelagos and various straits and passages. It has been the focus of moderate activity in the past 10 years, but there has been renewed interest in recent years and planning for a focused experiment is underway. Some of the recent studies have focused on western boundary currents off Mindanao, the Mindanao Current that extends over at least 100 km from the coast and whose core is at 100 m (Kashino et al., 2005) below surface, and the Mindanao Undercurrent, the presence and intermittency of which are still debated. Indications are that southward transport extends at depth with total average transport probably on the order of 35 Sv (Wijffels et al., 1995) and with large seasonal (Qu et al., 2008) to interannual variability (Wang and Hu, 2006; Kashino et al., 2009). Being closely related to the annual fluctuation of the Mindanao Dome (e.g., Masumoto and Yamagata, 1991; Tozuka et al., 2002), the Mindanao Current transport is maximum in early summer and minimum in fall. Apparently, no significant seasonal variation is seen in the Mindanao Undercurrent. On the interannual timescale, the Mindanao Current was shown to correspond well with ENSO (e.g., Kim et al., 2004; Kashino et al., 2009). The connection of this current system with the Celebes and Halmahera Seas and the role of transport through the South China Sea and its variability still remain as important questions to be investigated.

As we commented, the western Pacific warm pool area lies on one of the major pathways of thermocline waters of the shallow MOC that links the subtropical subduction and equatorial upwelling in the STC (Liu et al., 1994; McCreary and Lu, 1994). Although covered in Chapter 24, it is worth mentioning here that the decadal modulation of ENSO that relies on slow oceanic processes is an area of active study. This involves both numerical and conceptual modeling of the ocean and ocean-atmosphere system and observations (e.g., Chang et al., 2006b). One field of investigation has been with the advection of subducted anomalies by the mean circulation (Deser et al., 1996). One difficulty has been to find evidence that the signals actually reach the equatorial band (Schneider et al., 1999; Giese et al., 2002). Some evidence from observations now indicate that it might actually be the case (e.g., Sasaki et al., 2010), although longer timeseries data will be necessary to clarify what those signals imply and their effectiveness in coupling ENSO dynamics with subduction in the subtropical regions. Less clear results from coarser data are described by Kessler (1999). Another view suggests that the changing meridional circulation also contributes to changes in meridional transports (the "V" hypothesis); indeed, significant changes of the interior STC circulation are seen on decadal timescales (Zhang and McPhaden, 2006). The contribution of interior circulation could be compensated by the contribution arising from western boundary currents. However, this fact is yet to be confirmed either from observations or model results that in any case would be highly model dependent.

On the eastern Pacific, one of the major features is the TIWs commonly observed in the SST patterns, distorting the SST front into $\sim$ 1000 km-long cusp shapes usually rounded to the south and pointed to the north. In the eastern Pacific, the winds are the strongest in June–December, SEC, and NECC are the strongest in boreal fall and so is the equatorial upwelling. All these lead to the strongest meridional shear and sharpest temperature front, both of which are conducive to the TIWs. Hence, we find most of the TIWs to appear in June–July and persist through February–March.

2.2. Climate Variations: ENSO and ENSO Modoki

The ENSO is the joint variability of ocean and atmosphere in the tropical Pacific explained by a simple mechanism proposed by Bjerknes (1969). The positive Bjerknes-type ocean–atmosphere feedback amplifies, for example, initial warm perturbations in the eastern Pacific into large anomalies through ocean–atmosphere interactions and eventually develops an El Niño event. ENSO events typically last 12–18 months and occur every 2–7 years. ENSO variability is explained by equatorial wave dynamics, which play a crucial role in giving rise to the quasi-oscillatory nature of ENSO. Many studies have investigated the phenomenon by simulating the coupled tropical ocean–atmosphere system with models of varying complexity (e.g., Philander et al., 1984; Cane and Zebiak, 1985; Yamagata, 1985; Hirst, 1986; Philander, 1990; McCreary and Anderson, 1991; Neelin et al., 1998; Chang et al., 2006a,b). Oceanic Kelvin and Rossby waves help to propagate the energy and momentum received by the ocean from the wind stress (e.g., Hansen and Paul, 1984). Since the propagation speeds of similar atmospheric waves are far greater than those of their oceanic counterparts, the timescale of the equatorial oceanic waves adjusts the scale of the ocean–atmosphere interactions rather than the atmospheric response to SST, which is shorter than a week. In turn, oceanic Kelvin and Rossby waves can be strongly modified by the air–sea coupling. The Bjerknes feedback can destabilize these waves, giving rise to unstable coupled modes that resemble the slow westward propagating oceanic Rossby wave mode and the eastward propagating oceanic Kelvin wave mode. In fact, in model experiments, a breed of modes is generated through dynamical adjustment of the ocean depending on the strength and the nature of the air–sea coupling.

In the present climate, El Niño and La Niña exhibit significant asymmetry in their spatial structure and duration (e.g., Okumura et al., 2011). Moreover, ENSO shows substantial variations on timescales of decades to centuries. Most of the long-term variations in ENSO activity and characteristics have been found to depend on the state of the tropical Pacific climate system. The mean state is expected to change in the twenty-first century under the influence of global warming. However, the response of ENSO to increased greenhouse gases is still a topic of considerable research (e.g., Deser et al., 2010), and based on the published results, it is not possible to rule out any of the possibilities: an increased, decreased, or unchanged level of ENSO activity (e.g., Vecchi and Wittenberg, 2010).

Individual ENSO events differ in their evolutionary cycle and such differences are often referred to as different "flavors" of ENSO (e.g., Trenberth and Smith, 2006). The characteristic of this difference is evident in the last three decades. This has given rise to diversity in the tropical Pacific variability though adding to the complexity of the understanding of ENSO variability. Since the 1980s, a new type of El Niño known as the El Niño Modoki (Ashok et al., 2007) is frequently observed in the tropical Pacific and is attributed to global climate change (Yeh et al., 2009).

The El Niño Modokis are different from the canonical El Niños and are characterized by maximum warming in the central Pacific flanked by the cooling on both sides of the basin. Such an event occurred in 2004 (Ashok et al., 2007) with associated double Walker cells (Figure 15.4). Sometimes such events are also termed as "Dateline El Niño" (Larkin and Harrison, 2005), "Warm Pool El Niño" (Kug et al., 2009), or "Central Pacific El Niño" (Kao and Yu, 2009). However, it is not clear if those definitions have been introduced to categorize the same or different El Niño flavors. It is also unclear whether observational and ocean datasets are consistent in reproducing the various aspects of ENSO diversity with a statistically significant characterization of the different flavors.

The link between the El Niño and El Niño Modoki is not clearly evident in the published articles. It appears that the decadal signal might be at the heart of the process. The recent decadal condition in the eastern Pacific does not favor development of warm anomalies in the eastern Pacific. This might be a reason why the El Niño Modoki has become the new face of tropical Pacific variability (Larkin and Harrison, 2005; Ashok et al., 2007; Ashok and Yamagata, 2009; Kao and Yu, 2009; Kug et al., 2009) in recent decades. Unlike the pre-1980s, when the central Pacific played host to only a trans-Niño signal (Trenberth and Stepaniak, 2001) for a few months, El Niño Modoki events last from boreal summer through boreal winter and also cause global teleconnections that differ from those of the canonical El Niños (Ashok et al., 2007; Weng et al., 2008, 2009a,b; Wang and Hendon, 2007; Cai and Cowan, 2009; Taschetto and England, 2009; Pradhan et al., 2011; Kim et al., 2012; Ratnam et al., 2012).

The decadal variations associated with ENSO Modoki are also evident in the decadal sea-level variation. In the recent decades, it is manifested by higher than normal sea level in the central Pacific flanked by lower than normal sea level on either side of the basin during the early part of this century. The abnormal condition is evidently aided by frequent occurrences of El Niño Modoki events and associated wind convergence to the dateline during 2000–2004 (Behera and Yamagata, 2010). The sea-level rise in the central Pacific succeeded a phase of lower than normal sea level associated with La Niña Modoki events toward the end of last century. The influence can even be seen in remote regions such as the coasts of California and Mauritius through atmospheric teleconnections. Hence, ENSO Modoki is another factor for the sea-level variations besides the variations that arise from ENSO (e.g., Chambers et al., 2002; Cazenave et al., 2004; Church et al., 2006; Llovel et al., 2009) and IOD.

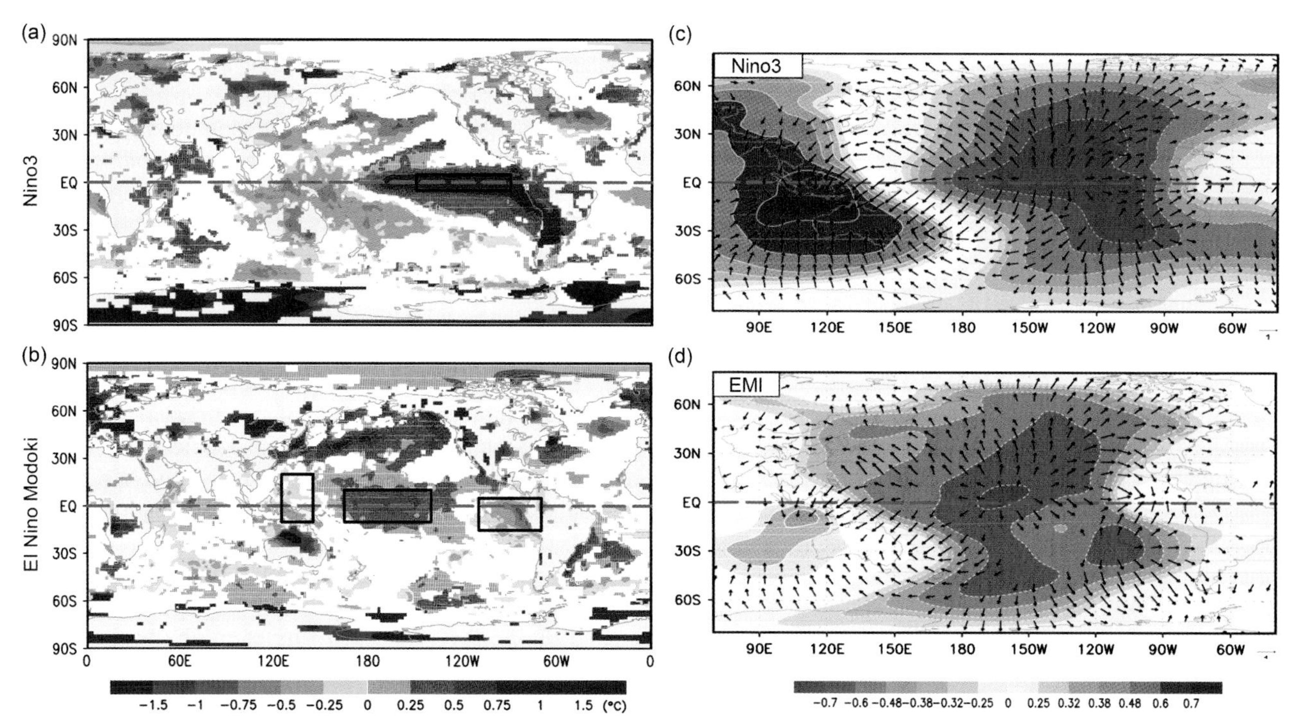

FIGURE 15.4 Composites of summer SSTA over oceans and skin temperature anomalies over land for the three largest events of (a) El Niño (1982, 1987, and 1997) and (b) El Niño Modoki (1994, 2002, and 2004). Anomalies that are not significant at the 80% level are suppressed. Partial correlation patterns of the 200 hPa divergent winds (arrow) and velocity potential anomalies (shading) with (c) Niño3 and (d) EMI. The wind vectors with correlation coefficients of either zonal or meridional component that are not significant at the 90% level are omitted. *Derived from Weng et al. (2008).*

3. TROPICAL ATLANTIC VARIABILITY

3.1. MOC and Western Boundary Circulation in the Tropical Atlantic

The Atlantic Ocean plays a particular role in the climate system of the earth due to the presence of a strong MOC, resulting in a net northward ocean heat transport in both hemispheres. Within this overturning cell, warm-water masses of Southern Hemisphere origin and cold-water masses produced in the North Atlantic pass through the tropical belt. The strength of the MOC is thought to affect circulation, water mass distribution, and stratification in the tropics with an impact on surface winds and precipitation, as suggested by dedicated model experiments (Zhang and Delworth, 2005; Chang et al., 2008). The mechanisms of MOC variability vary among different climate models, but often the variability is found to be related to the North Atlantic Oscillation (e.g., Delworth and Mann, 2000). However, changes in the tropical Atlantic, such as changes in the hydrological cycle and associated upper-ocean freshwater content, may also affect northward transported water masses and thus impact deepwater formation in the North Atlantic (Vellinga and Wu, 2004; Hazeleger and Drijfhout, 2006; Menary et al., 2012).

The water masses involved in the MOC mostly pass through the western boundary region of the tropical Atlantic (Schott et al., 2003). In addition to the MOC flow, the tropical wind field shapes the circulation pattern at the western boundary. The near-surface wind-driven circulation is characterized by a strong seasonal cycle associated with the seasonal migration of the ITCZ and corresponding seasonal intensification of the equatorial easterlies during the period of farthest northern ITCZ position in late boreal summer. The circulation is composed of the South Equatorial Current (SEC) and part of the Southern Hemisphere subtropical gyre with three main branches into the western boundary regime. The southern SEC (sSEC) bifurcates at around 10–14°S (Rodrigues et al., 2007) into the southward-flowing Brazil Current and the northward flowing North Brazil Under Current (NBUC). The central SEC (cSEC) with core latitude at about 4°S results in the loss of the undercurrent character of the northward western boundary current. Further downstream, this is known as the North Brazil Current (NBC). North of the cSEC, the surface flow is generally westward, forming another westward current maximum just north of the equator at about 2°N and called as the northern SEC (nSEC).

The NBC retroflection that is produced by the separation of the NBC from the western boundary at about 7°N supplies the eastward flow within the EUC and the North Equatorial Countercurrent (NECC)/North EUC (NEUC). The northward transport of the NBUC was estimated by Schott et al. (2005) from repeated ship sections at 5°S and 11°S as well as from a moored boundary current array at 11°S to be about 22 Sv in the upper 1100 m of the water column. The total transports include

1. a deeper part consisting of lower South Atlantic Central Water, Antarctic Intermediate Water, and Upper Circumpolar Deep Water of about 7 Sv at 11°S reducing toward the north to about 4 Sv at 5°S;
2. the main part of the NBUC of about 14 Sv at both latitudes that mainly supplies the EUC; and
3. the flow near the surface (above $\sigma_\theta = 24.5$ kg m^{-3}) that increases from 1 Sv at 11°S to 4 Sv at 5°S (Schott et al., 2005).

During the 4 years of moored observations, interannual transport variations were remarkably low and varied only by ±1.2 Sv (Schott et al., 2005). Calculation of the geostrophic NBC transport using hydrographic data from five decades of observations draws a different picture of interannual to multidecadal transport variability. The total transport of the NBC was found to roughly double in strength from weak flow periods in the early 1970s and 1980s to strong flow periods in the 1960s, and late 1980s to early 1990s (Zhang et al., 2011). A proof of such large NBC current fluctuations by direct current observations (derived from several repeat ship sections or a few years of moored measurements) is still outstanding. Forced ocean models show in general much smaller NBC transport variations of only a few Sv (Hüttl and Böning, 2006; Biastoch et al., 2008). Low-frequency NBC (and associated STC) variability was identified by Hüttl and Böning (2006) in response to remote changes in the North Atlantic MOC, associated with a heat flux-related variability in LSW formation. Variability of the MOC originating from the Agulhas leakage dynamics is similarly found to be associated with NBC variability (Biastoch et al., 2008).

The equator-crossing, northward transport of water masses from the Southern Hemisphere as part of the MOC becomes evident in the tropical water mass distribution. The hemispheric origin of intermediate and central water masses can be well identified from their T/S structure, Southern Hemisphere water masses being lower in salinity compared to their Northern Hemisphere counterparts. Kirchner et al. (2009) found by analyzing historical hydrographic and Argo data that in the thermocline and central water layers, the South Atlantic water dominates with a proportion of 80–90% in the region south of 10°N. Intermediate water masses are found to penetrate farther to the north compared to water masses above them due to the generally weaker currents and a northward excursion of the North Equatorial Current in this layer.

The retroflection of the NBC at the western boundary into the NECC/NEUC and the EUC connects the western boundary regime with the eastern tropical Atlantic (Schott et al., 1998; Fratantoni et al., 2000; Jochum and Malanotte-Rizzoli, 2003) contributing as well to the coastal upwellings off northwestern Africa (Glessmer et al., 2009) and southwestern Africa (Lass and Mohrholz, 2008). The EUC transport that is almost exclusively composed of South Atlantic water supplies to the equatorial upwellings of the central and eastern Atlantic. The upwelled water masses are then redistributed in both hemispheres via poleward Ekman transport, resulting in a net northward flow of South Atlantic water (Fratantoni et al., 2000; Hazeleger and Drijfhout, 2006).

The pathway of South Atlantic water along the western boundary is thought to consist of the transport within the frictional western boundary current and the transport carried by NBC rings that are regularly shed from the NBC retroflection. From numerical simulations, Fratantoni et al. (2000) estimated both pathways to contribute about 43% and 21%, respectively, to the total simulated MOC of 14 Sv. In the simulations, about 3–4 NBC rings are generated during the year; each ring carrying about 1 Sv of South Atlantic water northward. The analysis of altimeter data revealed a larger number of NBC rings (3–7) varying on interannual timescales (Goni and Johns, 2003). Using shipboard and mooring data from NBC Ring Experiment (1998–2000), Johns et al. (2003) described an annual generation of 8–9 NBC rings. Out of these rings, a few were associated with weak or no surface signals, which do not allow them to be observed by remote sensing techniques. The total transport by NBC rings estimated from the *in situ* observations was 9.3 Sv per year (Johns et al., 2003) and represents about half of the AMOC transport of about 18 Sv as estimated from zonal hydrographic sections across the Atlantic (Lumpkin and Speer, 2007; Talley et al., 2003).

As part of the MOC, the Deep Western Boundary Current (DWBC) transports North Atlantic Deep Water (NADW) into the South Atlantic. The Atlantic DWBC forms along the continental slope east of Greenland and carries dense and cold-water masses from the Denmark Strait overflow (lower North Atlantic Deep Water, lNADW), the Iceland–Scotland sills (middle NADW), and the Labrador Sea (upper NADW) to the south. In the equatorial zone, NADW pathways are complicated by the change of sign in planetary vorticity (Edwards and Pedlosky, 1998) and involve large and complicated zonal excursions into the interior (Richardson and Fratantoni, 1999; Rhein and Stramma, 2005). It is generally accepted that the DWBC splits into a shallow (<2500 m) and a deep (>2500 m) branch due to topographical constraints at about 8°N. Moored observations at 44°W on the equator show the shallow branch of the DWBC to be attached to the continental slope with a core at 2000 m depth, carrying about 13 Sv of NADW (Fischer and Schott, 1997).

At 35°W, the mean zonal velocity currents as revealed by 13 cross-equatorial shipboard current profiling sections show two deep zonal currents carrying uNADW, one

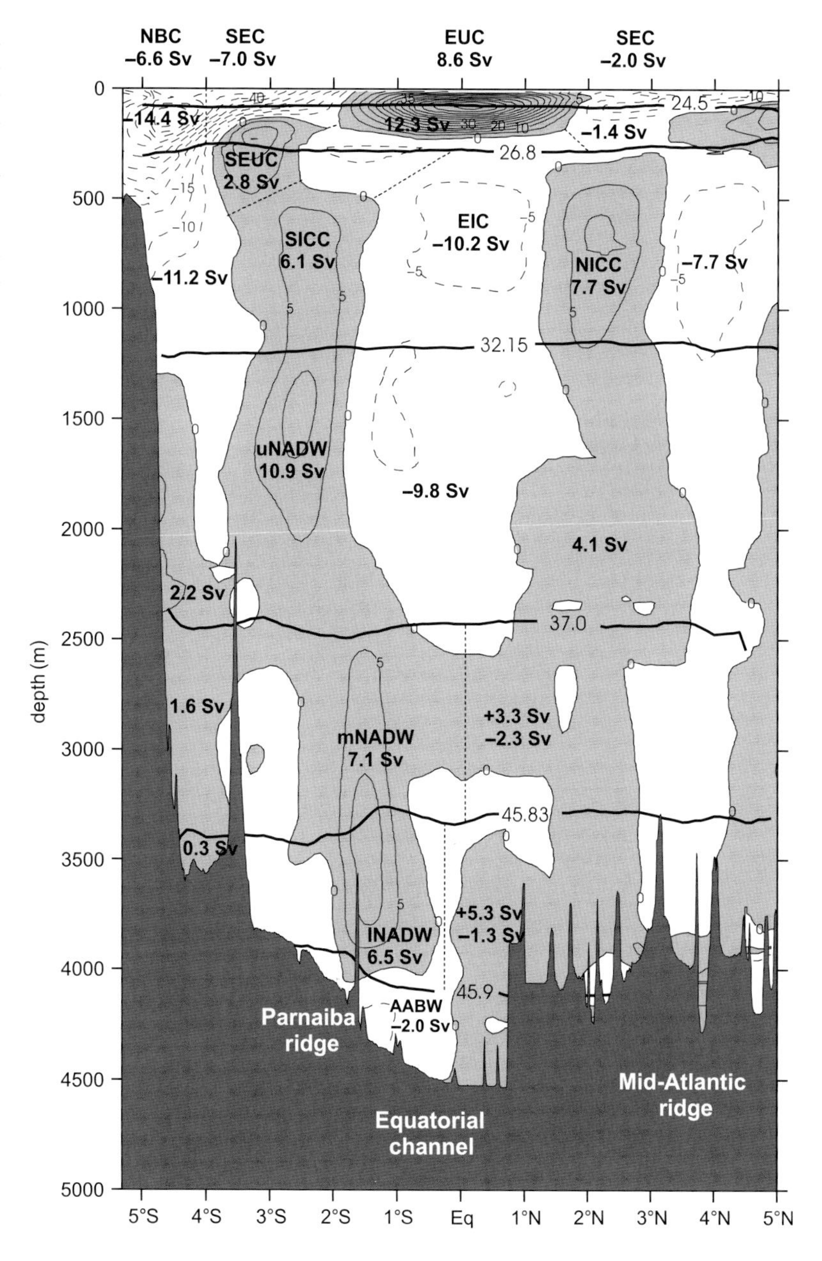

FIGURE 15.5 Mean zonal current distribution across 35°W from the 13 sections, with transports (in Sv $=10^6$ m^3/s) of the different current branches marked. Upper ocean currents are North Brazil Current (NBC), South Equatorial Current (SEC), Equatorial Under Current (EUC), South Equatorial Undercurrent (SEUC), South, Equatorial, and North Intermediate Countercurrents (SICC, EIC, and NICC); deep and bottom water masses are upper, middle, and lower North Atlantic Deep Water (uNADW, mNADW, and lNADW) and Antarctic Bottom Water (AABW). *From Schott et al. (2003).*

situated on the continental slope, while a second core is found detached from topography between 2°S and 3°S transporting a total of 13.1 Sv (Schott et al., 2003). The deep branch of the DWBC at 35°W is found to be attached to the Parnaiba Ridge slightly north of 2°S carrying 7.1 Sv of mNADW and 6.5 Sv of lNADW (Figure 15.5).

A large part, but not all, of the NADW arriving at 35°W appears to continue southward at the western boundary: Repeated shipboard current profiling observations indicate that the DWBC is reestablished at 5°S having a velocity core at about 2000 m depth (Schott et al., 2005); downstream at about 8°S the DWBC disintegrates into series of deep ocean eddies carrying NADW farther southward (Dengler et al., 2004). As also indicated by moorings at 11°S southward transport at 5°S and 11°S is confined to the uNADW and mNADW layers and amounts to about 20 Sv, which is comparable to the eastward transports within the same two layers at 35°W. There is, however, no net southward transport in the lNADW layer at the western boundary in the tropical Atlantic (Dengler et al., 2004; Schott et al., 2005) south of the equator. From an inverse study, Lux et al. (2001) also determined intense water mass transformations in the order of 10 Sv from lower and middle NADW into middle and upper NADW. However, this process would require intense abyssal mixing that is generally thought of as being low at the equator (e.g., Gregg et al., 2003). The fate of lNADW remains an unsolved issue of the mean deep equatorial circulation.

3.2. Climate Variability

The SST variability in the equatorial Atlantic is characterized by a pronounced seasonal cycle (see Chapter 24). During boreal spring, when the ITCZ is at its southernmost position, the equatorial Atlantic is uniformly warm at about 27–29°C. With the northward migration of the ITCZ during boreal spring, equatorial easterlies strengthen and force a shallowing of the thermocline in the eastern equatorial Atlantic. The strongest SST cooling in the equatorial region can be observed during early boreal summer reaching minimum temperatures at the sea surface of about 22–24°C in July and August forming the eastern Atlantic cold tongue (Figure 15.6). Studies of the mixed layer heat budget of the cold tongue suggest that the dominant cooling term during boreal summer is diapycnal mixing as evidenced from the dominant residual of the heat budget of Foltz et al. (2003) or Wade et al. (2011). Direct observation of diapycnal mixing by means of microstructure measurements confirms this assertion showing that the diapycnal mixing is mainly associated with the vertical shear between westward surface flow and EUC below; stratification plays additionally an important role in shaping the temporal and regional distribution of large downward heat flux (Hummels et al., 2013). Due to the intensification of the equatorial easterlies, the westward surface currents accelerate and the enhanced horizontal shear between the westward near-surface flow and the eastward flow of the EUC results in the generation of TIWs (Foltz et al., 2004; Jochum and Murtugudde, 2006; von Schuckmann et al., 2008). Together with wind-forced mixed Rossby-gravity waves that are more important in the region east of 10°W, the TIWs are the main source of intraseasonal SST variability (Athie and Marin, 2008). The meridional velocity fluctuations of intraseasonal waves result in a meridional displacement of the strong SST gradient that is the strongest north of the equatorial cold tongue. The vertical sheer of intraseasonal waves is also found to contribute to increase in the diapycnal mixing during boreal summer (Hummels et al., 2013), while their associated lateral mixing tends to oppose these effects by lateral eddy fluxes into the equatorial cold tongue region (Jochum and Murtugudde, 2006; Peter et al., 2006).

The seasonal variation in SST dominates over its interannual variation in the equatorial Atlantic with the interannual variations often remaining remarkably low at 1–2°C (Brandt et al., 2011a; Caniaux et al., 2011). The ratio of interannual to seasonal SST variations in the equatorial Atlantic considerably differs from those in the Pacific, where interannual variations dominate. Nevertheless, the interannual SST variations characterize distinct patterns of coupled ocean–atmosphere variability, collectively referred to as tropical Atlantic variability (TAV), which partly gains its importance from the associated modification of rainfall over heavily populated regions of Africa and South America.

TAV is dominated by two modes of behavior that are tightly phase-locked to the pronounced seasonal cycle (Chang et al., 2006a; Kushnir et al., 2006). The thermodynamic meridional mode, which peaks during boreal spring, is characterized by a north–south SST gradient that drives cross-equatorial wind anomalies from the cold hemisphere

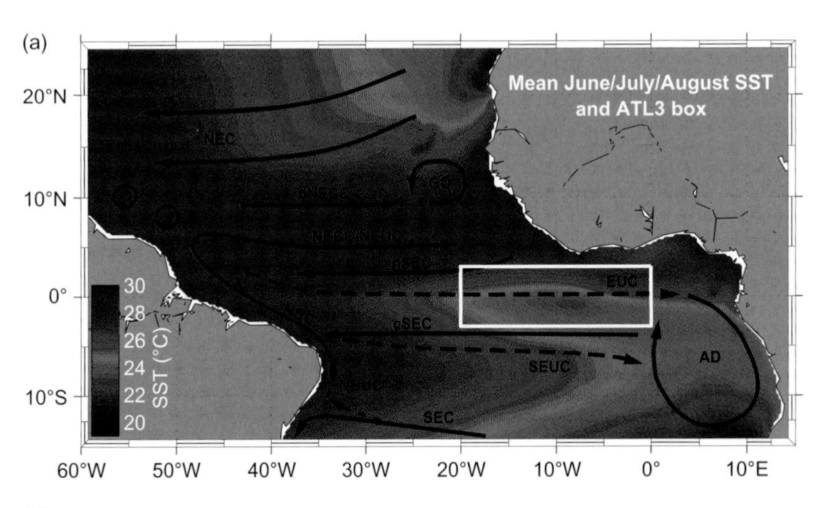

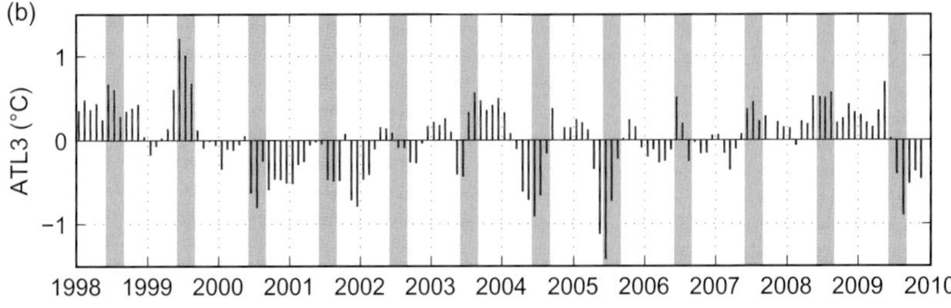

FIGURE 15.6 Mean June–August SSTs (averaged from 1998 to 2009) in the tropical Atlantic (a). Also included are the main surface (solid) and thermocline (dashed lines) current bands: North Equatorial Current (NEC), North Equatorial Undercurrent (NEUC), North Equatorial Countercurrent (NECC), its northern branch (nNECC), Equatorial Undercurrent (EUC), South Equatorial Current (SEC), its northern and central branches (nSEC, cSEC), South Equatorial Undercurrent (SEUC), North Brazil Current and Undercurrent (NBC, NBUC), as well as the cyclonic circulations around the Guinea and Angola Domes (GD, AD). The white box in (a) defines the region for the ATL3 SST index shown in (b). Monthly SST residuals are calculated by subtracting the mean seasonal cycle; gray bars highlight boreal summer (June–August). The microwave optimally interpolated (OI) SST dataset used here is available at www.remss.com. *From Brandt et al. (2011a).*

to the warm (Carton et al., 1996; Chang et al., 1997). The increase of initial SST anomalies is in agreement with the wind-evaporation-SST feedback mechanism, which enhances temperatures in the warm hemisphere due to the local reduction of the trade winds and associated reduction in evaporation (e.g., Joyce et al., 2004). Dommenget and Latif (2000) suggested that the variability between Northern and Southern Hemispheres is uncorrelated; likely forcing being the NAO or the variability of the Southern Hemisphere subtropical high (Czaja et al., 2002; Mélice and Servain, 2003). The dynamic zonal mode is characterized by an east-west SST gradient along the equator, which is most pronounced during boreal summer when the Atlantic cold tongue is well developed (Zebiak, 1993; Carton and Huang, 1994). The zonal mode is often referred to as the Atlantic counterpart to the Pacific El Niño. Like the Pacific El Niño, the Atlantic Niño is the result of the Bjerknes feedback mechanism, which, however, is weaker than that in the Pacific (Keenlyside and Latif, 2007). The strength of this feedback mechanism peaks in boreal summer with a secondary maximum found during boreal winter, characterized by a second intensification of equatorial easterly winds and associated increased upwelling and thermocline slope. Interannual zonal mode variability in the equatorial Atlantic during boreal winter was found to be dominantly independent from those observed during boreal summer, that is, the Atlantic Niño (Okumura and Xie, 2006).

In general, a strong impact of the Pacific El Niño on tropical Atlantic variability can be expected. However, such an impact is found to be complicated by the fact that the response of the tropical Atlantic to strong Pacific El Niño includes warm as well as cold boreal summer anomalies in the equatorial cold tongue region. Chang et al. (2006a) suggested that the dynamic ocean–atmosphere interaction in the equatorial Atlantic plays an important role in TAV and might overcome the more direct atmospheric response to an El Niño forcing. This makes the relationship between Pacific El Niño and Atlantic Niño ambiguous and inconsistent. In this context the state of the meridional mode was found to be important. Lübbecke and McPhaden (2012) concluded that the state of the meridional mode sets the sign of the wind curl forcing north of the equator. The absence of downwelling Rossby waves and reflected equatorial Kelvin waves during years with cold northern hemisphere SST anomalies and correspondingly weak or absent boreal spring northern tropical Atlantic warming in response to a Pacific El Niño forcing, perhaps, results in a continuation or amplification of the initial cooling in the eastern Atlantic in response to El Niño. Downwelling Rossby waves and reflected equatorial Kelvin waves, on the other hand, might overcome the initial cooling in the eastern equatorial Atlantic during years of warm Northern Hemisphere SST anomalies and corresponding strong warming response to El Niño during spring.

Intrinsic ocean dynamics might also play a role in modifying SST in the equatorial Atlantic besides the large-scale ocean–atmosphere interaction. For example, the instability of the mean and seasonally varying energetic near-surface flow field that generates the TIWs were identified in idealized simulations to generate stochastic SST variability on interannual timescales (Jochum et al., 2004). Also regular oscillations, which are the result of intrinsic ocean dynamics, are observed on interannual timescales in the equatorial Atlantic. High baroclinic mode variability, the so-called equatorial deep jets, oscillates at a period of 4.5 years in the deep equatorial Atlantic (Johnson and Zhang, 2003; Bunge et al., 2008; Brandt et al., 2011b). Their regular oscillations are in general agreement with an equatorial basin mode characterized by east- and westward propagation of high baroclinic mode equatorial Kelvin and Rossby waves (Greatbatch et al., 2012). Interestingly, these equatorial deep jets are found to propagate their energy from great depths toward the surface, thus having an impact on surface flow, SST, and atmospheric parameter (Brandt et al., 2011b). A simulation of equatorial deep jets in high-resolution ocean simulations is until now limited to simulations with very idealized forcing (d'Orgeville et al., 2007; Ménesguen et al., 2009) or result in an inadequate description of deep jet characteristics in more realistic simulations (Eden and Dengler, 2008).

Similar to the effect of the ENSO on the eastern boundary upwelling systems of the Pacific Ocean, the eastern boundary upwelling system of the tropical South Atlantic undergoes climate variations with warmer than normal oceanic events during the so-called Benguela Niños (see Chapter 14). These events that occur on interannual timescales, and are linked to Atlantic Niños during some but not all years, have a strong impact on the strength of primary productivity, fisheries, hypoxia, and rainfall of the region (Florenchie et al., 2003; Rouault et al., 2007). Two main effects, which force this kind of interannual climate variability, are discussed in the literature: (1) A remote effect mainly driven by the zonal winds in the western equatorial Atlantic (e.g., Florenchie et al., 2003; Lübbecke et al., 2010) and (2) a local forcing due to alongshore wind variability associated with the subtropical anticyclone (e.g., Richter et al., 2010). The intermittent feedback between remote and local processes appears to be responsible for the formation of Benguela Niños.

4. TROPICAL INDIAN OCEAN VARIABILITY

4.1. Monsoon Ocean Circulations and Upwelling Regimes

Unlike the other two tropical basins, the Indian Ocean is marked by its complex geometry, monsoon winds, and intraseasonal oscillations (ISOs) besides the interannual climate variations. A couple of review articles (Schott and McCreary, 2001; Shankar et al., 2002) provide a good overview of the present understanding of the tropical Indian Ocean circulations, particularly related to the monsoon regimes. The upper ocean currents are mostly Ekman driven forced by the monsoon winds (e.g., Hastenrath and Greischar, 1991). However, the geostrophic flows associated with the monsoon currents are not weak compared to the strong surface Ekman flows. Strong geostrophic flows and transports associated with the monsoon currents are reported in some of the recent studies based on the hydrographic data (Murty et al., 1992; Bruce et al., 1994; Donguy and Meyers, 1995; Gopalakrishna et al., 1996; Bruce et al., 1998; Vinayachandran et al., 1999; Murty et al., 2000). The monsoon winds and the associated circulations cause a southward transport of heat in the upper layers. The southwest monsoon winds dominate the annual mean (Figure 15.1) to give rise to mean southward Ekman transports in both hemispheres and a southward heat transport. The boreal summer winds produce strong upwelling along the Somali coast of the western Indian Ocean (Figure 15.7). The Somali current flows northward and several energetic eddies such as the Great Whirl (Bruce, 1968) embedded in the current recirculate the coastal water and carry nutrients far off the shore. The current regimes around the equator are less dramatic during this season, except for the fact that the NEC in the Indian Ocean flows eastward during the summer monsoon season (Figure 15.7).

Coastal currents along other coasts (Figure 15.7) are also influenced by the monsoon winds and reverse direction seasonally: These include the East India Coastal Current (EICC) (McCreary et al., 1993; McCreary et al., 1996; Schott and McCreary, 2001; Shankar et al., 1996; Shetye et al., 1991a, 1993, 1996; Vinayachandran et al., 1996) and the West India Coastal Current (WICC) (McCreary et al., 1993; Shetye et al., 1990). The EICC flows northeastward from February until September with a peak in March–April and southwestward from October to January with a peak in November. The surface circulation in the interior of the Bay of Bengal is best organized from February to May, when the EICC acts as a western boundary current to form the basin-wide anticyclonic gyre. Along the west coast of India, the currents are mostly equatorward at the surface with a weak (5–10 cm s^{-1}) poleward coastal undercurrent. Ship-drift observations suggest that the southward flow appears along the west coast in March, reaches peak strength in July, and disappears in October (Cutler and Swallow, 1984; Shetye and Shenoi, 1988). Syntheses from drifters that refine the view from the ship drifts are available in Miami AOML GDP DAC (Molinari et al., 1990). The WICC flows northward during the winter monsoon months of December–February (Shetye et al., 1991b). The ship-drift currents indicate that the northward WICC is at its strongest in the year, though the coastal winds were very light or absent at that time. This property lends strong support for the importance of remote forcing (Behera and Salvekar, 1998).

Among the other coastal currents that are extensively studied is the Leeuwin Current (Figure 15.7) that flows southward along the West Australian coast (see Chapter 14). It is about 50 km wide, extends to depths of about 250 m, and has an average speed of about 30 cm s^{-1} with a

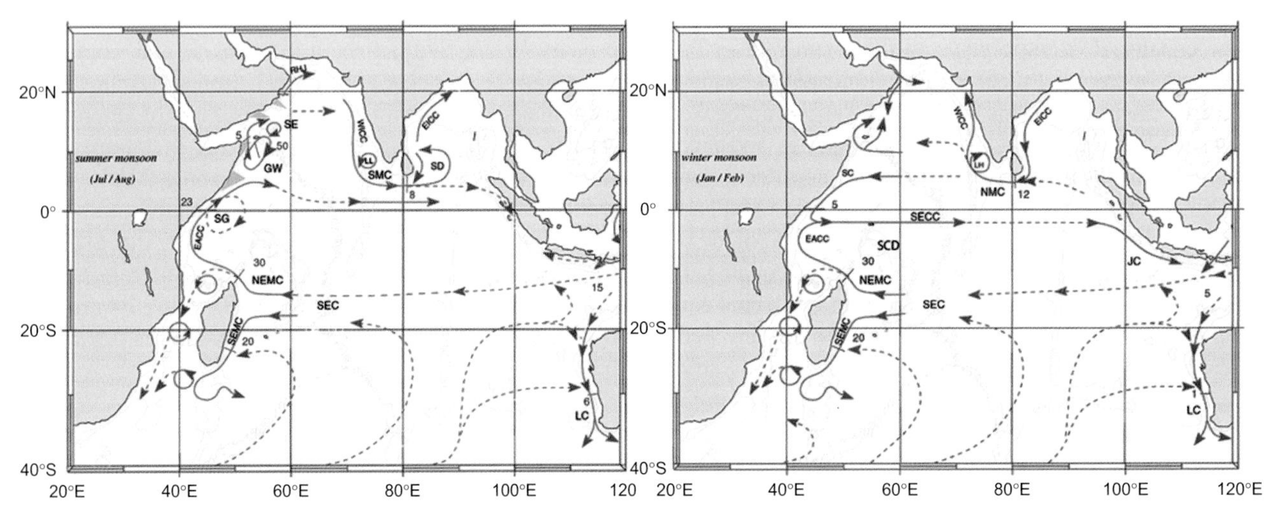

FIGURE 15.7 A schematic representation of identified current branches during the summer (left) and winter (right) monsoons adapted from Schott and McCreary (2001). Current branches indicated are the South Equatorial Current (SEC), South Equatorial Countercurrent (SECC), Northeast and Southeast Madagascar Current (NEMC and SEMC), East African Coast Current (EACC), Somali Current (SC), Southern Gyre (SG) and Great Whirl (GW) and associated upwelling wedges, Socotra Eddy (SE), Ras al Hadd Jet (RHJ) and upwelling wedges off Oman, West Indian Coast Current (WICC), Laccadive High and Low (LH and LL), East Indian Coast Current (EICC), Southwest and Northeast Monsoon Current (SMC and NMC), South Java Current (JC), Leeuwin Current (LC), Sri Lanka Dome (SD) and Seychelles Dome (SCD).

seasonal maximum of 60 cm s^{-1} (Smith et al., 1991). It follows the coast around Cape Leeuwin at the southwest corner of Australia to beyond 120°E (Cresswell and Petersen, 1993). An undercurrent known as the Leeuwin Undercurrent is also observed off the shelf break below the surface currents. A number of modeling studies have proposed a few processes such as the Indonesian Throughflow (Godfrey and Weaver, 1991; Schiller et al., 1998; Schneider, 1998), the meridional pressure-gradient field off the west coast of Australia (McCreary et al., 1986), and the local alongshore winds to explain the Leeuwin Current dynamics. Recent studies have linked the interannual variability in the current system and the associated changes in the heat content to the tropical climate variability in the Pacific (Feng et al., 2008).

The equatorial Kelvin and Rossby waves, and the coastal Kelvin wave in addition to the Ekman drift can in general explain the major surface circulations in the Arabian Sea and the Bay of Bengal. Numerical simulations even with simple reduced-gravity models (e.g., McCreary et al., 1993; Potemra et al., 1991; Yu et al., 1991; Perigaud and Delecluse, 1992; Behera and Salvekar, 1995) could simulate most features of the monsoon currents. Unlike the Pacific Ocean, where the equatorial Kelvin wave generated signals get lost to the poles along the eastern boundaries, the equatorial signals in the Indian Ocean get into the Bay of Bengal as coastal waves and influence the local circulations. From the sensitivity and model analysis, it is found that Ekman pumping and remote forcing from the equatorial Indian Ocean could in general explain the Bay of Bengal circulation. In the Arabian Sea, the continuity of the monsoon current along west coast of India to south of Sri Lanka is primarily the result of remote forcing by the winds along the east coasts of India and Sri Lanka. The winds help to generate the Lakshadweep High off southwest India and significantly force the monsoon currents in the eastern Arabian Sea. In the rest of the Arabian Sea, Ekman pumping together with Rossby waves radiated from the west coast of India produces the curving flows associated with the geostrophic monsoon currents (Brandt et al., 2002).

The equatorial wind in the basin has a strong semiannual cycle. Hence, the equatorial currents in the Indian Ocean are quite different in character to those in the other two tropical oceans. During the monsoon transition seasons of spring and fall, the otherwise weak equatorial winds become stronger and generate the strong equatorial currents (like Yoshida-jet, Yoshida, 1959)), known as the Yoshida–Wyrtki jet (Wyrtki, 1973; O'Brien and Hurlburt, 1974; Reverdin, 1987; Han et al., 1999), which transports the warm waters to the east. This helps to maintain the eastern Indian Ocean warm pool, part of the Indo-Pacific warm pool that is a major driver of the global climate system. Observational (Reppin et al., 1999) and modeling studies suggest that the seasonal jets are actually composed of spells with timescales of a month or less. Barrier layers seem to play an important role in the phase and amplitude of the seasonal jet (Masson et al., 2003). As the zonal pressure gradient on the equator adjusts to the transitional winds, semiannual Kelvin waves are generated. Anomalous events evolve, however, as in the tropical Pacific, sometimes owing to an imbalance between the interannual equatorial winds and the east-west pressure gradient. This leads to the development of IODs as discussed in the next subsection 4.2.

Shallow and deep overturning cells are observed in the basin associated with the major circulation and upwelling regimes. The pathways and the mechanisms are reasonably well understood for the former since it is coupled strongly with the monsoon winds. The shallow-cross-equatorial cell connects the upwelling regimes of the northwestern Arabian Sea with the subduction regimes in the Southern Hemisphere (Schott et al., 2002, 2004; Miyama et al., 2003). The northward cross-equatorial flow ensues at thermocline depth as a part of the Somali Current, whereas the southward upper-layer return flow is carried by Ekman transport in the ocean interior. Subducted water masses are then carried westward within the SEC. The SEC bifurcates at the Madagascar coast, a part turning southward and the rest eventually joining the East African Coastal Current to flow to the equator with a fraction also entering the Mozambique Channel. In the southern Indian Ocean, strong upwelling at 3–12°S drives a hemispheric STC. The deep overturning cell is composed of inflow of Circumpolar Deep Water through either the Crozet–Madagascar basins in the west or the Australian Basin in the east (see Chapter 19).

In addition to the seasonal upwelling regimes off the Somali–Oman coasts, several upwelling domes are observed in the basin (Figure 15.7). In the southwestern Bay of Bengal, strong positive wind-stress curl east of Sri Lanka gives rise to the dome known as the Sri Lanka Dome (Vinayachandran and Yamagata, 1998). The Sri Lanka Dome appears first east of Sri Lanka during May, matures in July, and begins to decay in September. Those are well captured even in simple reduced-gravity model solutions (McCreary et al., 1993; Behera et al., 1998). West of Sri Lanka, there exists another upwelling dome that arises in response to local wind-stress curl as well as the gap winds (Luis and Kawamura, 2002). In the southern tropical Indian Ocean, such a thermocline dome is called the Seychelles Dome (Hermes and Reason, 2008; Vialard et al., 2008; Yokoi et al., 2008; Tozuka et al., 2010).

The thermocline ridge in the southwestern tropical Indian Ocean has appeared in some of the earlier studies in various contexts (e.g., McCreary et al., 1993; Reverdin and Fieux, 1987; Woodberry et al., 1989; Masumoto and Meyers, 1998; Behera et al., 2000; Xie et al., 2002; Rao and Behera, 2005; Izumo et al., 2008). Though the dome is captured in most coupled model simulations (Yokoi et al., 2009), large variations in space and time are seen in the simulations owing to model biases (Nagura et al., 2012). The local winds, which are easterly in the southern Indian Ocean and westerly in the equatorial region, generate Ekman upwelling and contribute to the formation of the dome. In addition, the dome and the heat content of the region are affected by interannual Rossby waves propagating from the east. Due to the shallow thermocline, the SSTs over these domes are significantly influenced by vertical entrainment and diffusion and hence are important for both local as well as regional climate variations.

In addition to the thermocline domes, barrier layers formed by the freshwater fluxes from rain and rivers also induce ocean and climate variations. In particular, the barrier layers in eastern Indian Ocean near Java-Sumatra, Myanmar, east coast of India, and the southwestern tip of the Indian peninsula are discussed in detail in the recent literatures (e.g., Masson et al., 2002). The barrier layer of the southeast Arabian Sea off the Indian peninsula is shown to enhance the spring SST warming and hence influences the monsoon onset over India (Masson et al., 2004).

In the southern part of the tropical Indian Ocean, south of about 10°S, the directions of the currents remain approximately unchanged from season to season, and steady-state ocean Sverdrup circulation, which takes the Indonesian Throughflow into account (Godfrey and Golding, 1981), explains much of the structure of those currents. The Agulhas Current, the Mozambique Current eddies, and the East Madagascar Current (EMC) form a complex western boundary current system along the East African coast. Just outside of the southern subtropics, a countercurrent known as the subtropical South Indian Ocean Countercurrent (SICC) has been recently identified from the altimeter-derived absolute geostrophic surface velocities (Siedler et al., 2006). It is a narrow, eastward-flowing current between 22°S and 26°S confined to the Rossby wave trains. The narrow branch of the SEC approaches Madagascar near 18°S and feeds the southern EMC, which continues westward around the southern tip of Madagascar. It then partially retroflects and nourishes the SICC. In recent times, warming in the Agulhas Current system has also been detected as a part of the Indian Ocean warming. The SST of the tropical Indian Ocean has warmed up recently and hence rainfall has increased. This has increased the pressure gradient between the equator and the subtropics, leading to increased trade winds in the South Indian Ocean (Rouault and Behera, personal communications), an increase in wind-stress curl in the subtropics, an increase in sea level of the southern Indian Ocean, and ultimately an intensification of the Agulhas Current and associated warming in the Agulhas Current system (Backeberg et al., 2012) south of South Africa.

4.2. The IOD

Recent studies show that anomalous events evolve in the tropical Indian Ocean as an ocean–atmosphere coupled phenomenon known as the IOD (Saji et al., 1999; Webster et al., 1999; Yamagata et al., 2003a, 2004). The stronger than normal seasonal southeasterly winds along the Sumatra coast during positive IOD events cause SST cooling (Figure 15.8) by coastal upwelling (Vinayachandran et al., 1999, 2002; Murtugudde et al., 2000) and evaporation (Behera et al., 1999). Equatorial winds reverse direction from westerlies to easterlies during the peak phase of the positive IOD events and establish the characteristic dipole in the SST anomalies. Hence, the IOD index is defined as the difference between SST anomalies of the western (50–70°E, 10°S–10°N) and eastern (90–110°E, 10°S–Eq) tropical Indian Ocean as indicated by the boxes in Figure 15.8c. The dipole in SST anomalies is accompanied by abundant rainfall over the western Indian Ocean–East Africa and scarce rainfall over eastern Indian Ocean–Indonesia. This is similar to the Bjerknes-type of air–sea interaction in the tropical Pacific (Bjerknes, 1969).

The oceanic adjustments in response to the equatorial winds give rise to a subsurface dipole, which provides a kind of delayed oscillator mechanism (cf. Schopf and Suarez, 1988) required to reverse the phase of the surface dipole in the following year (Rao et al., 2002). The associated mechanism is related to the propagation of oceanic Rossby–Kelvin waves seen in observed data (Rao et al., 2002; Feng and Meyers, 2003) and coupled model results (Gualdi et al., 2003; Yamagata et al., 2004). Xie et al. (2002) suggested that the Rossby waves, coupled to the wind anomalies, are dominantly forced by ENSO, whereas Rao and Behera (2005) have distinguished regions that are influenced by the IOD: Wind-stress curl associated with the IOD forces the westward propagating downwelling long Rossby waves north of 10°S. In contrast, the ENSO influence dominates over the upwelling dome south of 10°S.

The processes that trigger an IOD event are not completely understood yet. Studies show roles of off-Sumatra anomalous southeasterlies (e.g., Behera et al., 1999; Saji et al., 1999), atmospheric pressure variability in the eastern Indian Ocean (e.g., Gualdi et al., 2003; Li et al., 2003), favorable changes in winds in relation to the Pacific ENSO and the Indian monsoon (e.g., Annamalai et al., 2003), oceanic conditions of the Arabian Sea related to the Indian monsoon (Prasad and McClean, 2004), influences from the southern extratropical region (e.g., Lau and Nath, 2003), and trade winds of the southern Indian Ocean (Hastenrath and Polzin, 2004). As found in recent observations, wind and subsurface temperature hold signals that lead the SST variations associated with IOD. Using moored buoy data from the eastern Indian Ocean, Horii et al. (2008) detected a cold temperature anomaly at the depth of the

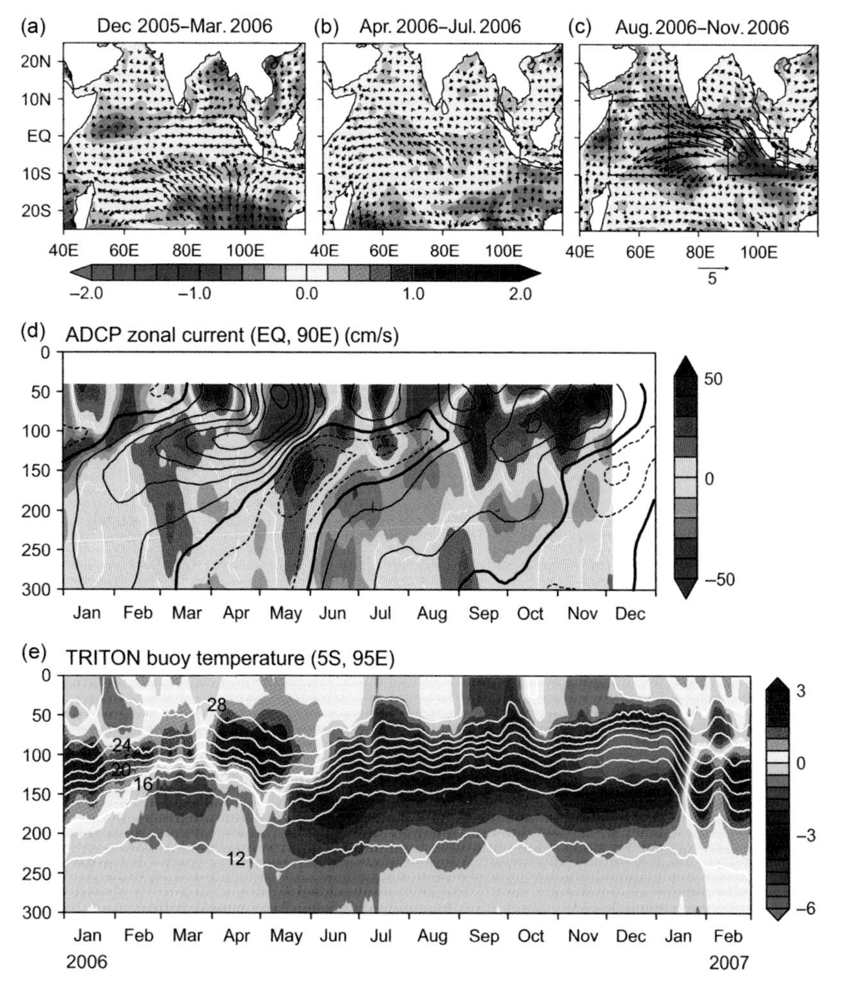

FIGURE 15.8 Observed SST and surface wind anomalies averaged for (a) December 2005–March 2006, (b) April–July 2006, and (c) August–November of 2006 positive IOD event depicting the inception, development, and mature stages. (d) and (e) show the corresponding development of the IOD event as observed by the equatorial ADCP and near-equatorial TRITON buoy. *Adapted from Horii et al. (2008).*

thermocline in May 2006, 3 months prior to the development of the positive IOD in that year (Figure 15.8). These cold anomalies are generated by vigorous monsoon winds in the southern Indian Ocean (e.g., Behera et al., 2006) and propagate to the East African coast as Rossby waves and then reflect to the Sumatra coast as equatorial Kelvin waves.

The ISOs–Madden Julian Oscillations (MJOs) originating from the tropical Indian Ocean (as discussed in the following subsection) are shown to play a significant role in the IOD evolution. In recent studies, Rao and Yamagata (2004) and Rao et al. (2007) have examined the possible link between the ISO activity and the IOD termination using multiple datasets. They observed strong, 30–60 day, oscillations in equatorial zonal winds prior to the termination of all IOD events, except for the event of 1997. This may be a reason why the 1997 IOD event was sustained until early February 1998 instead of usual termination around December. Thus the strong westerlies associated with the ISO excite anomalous downwelling Kelvin waves that terminate the coupled processes in the eastern Indian Ocean by deepening the thermocline in the east, as discussed by Fischer et al. (2005) for the 1994 IOD event. Gualdi et al. (2003) suggested that the anomalously high ISO activity in the northern summer might explain the aborted IOD event of 1974.

Like ENSO, the IOD can exert its influence on various parts of the globe via atmospheric teleconnection (Saji and Yamagata, 2003) and by interacting with other modes of climate variability. Through the changes in the atmospheric circulation, IOD influences the rainfall of the Indian summer monsoon (Behera et al., 1999; Ashok et al., 2001; Cherchi et al., 2007), of East Asia (Guan and Yamagata, 2003; Guan et al., 2003), of the East Africa Short Rains (Behera et al., 2003b, 2005; Black et al., 2003; Clark et al., 2003; Manatsa et al., 2008; Manatsa and Behera, 2013), of Sri Lanka (Zubair et al., 2003), of Australia (Ashok et al., 2003; Ummenhofer et al., 2008, 2013), and of Brazil (Chan et al., 2008). Recent studies have shown the relatively stronger impacts of IOD on the stream flows in the western part of Indonesia (Sahu et al., 2011) and extreme events of Malaysia (Tangang et al., 2008). The precipitations over the northern part of India, the Bay of Bengal, Indochina, and the southern part of China were enhanced during the 1994 positive IOD event (Behera et al., 1999; Guan and Yamagata, 2003; Saji and Yamagata, 2003).

Positive IOD events give rise to warm and dry summers in East Asia as is seen during 1961 and 1994 (Guan and Yamagata, 2003; Yamagata et al., 2003b, 2004). The mechanism called the "Silk Road process" may contribute to strengthening the equivalent barotropic Bonin High in East Asia (Enomoto et al., 2003). Recently, Behera et al. (2013) have shown that IOD-induced circum-global Rossby wave trains influence the summer conditions, particularly in late July and early August over Europe: Some of the extreme summers of Western Europe are associated with the IODs.

4.3. MJO with Indian Ocean Focus

The MJO is a prime manifestation of intraseasonal variability (periods of 30–100 days) in the tropics and has been of continuous interest since the 1990s, and the TOGA–COARE experiment in particular because of its potential interaction with ENSO. Its strongest amplitudes in the atmosphere are found over the warm pools of the Indian Ocean and the western Pacific, and are associated with a strong modulation of atmospheric deep convection, rainfall, and surface zonal winds. The zonal wind anomalies in response to the atmospheric convection are somewhat different in the Indian Ocean sector in the sense that the strongest westerly anomalies are found west of the convection. In contrast, they tend to be found under the convection region in the western Pacific. The MJO needs to be clearly distinguished from the westerly wind bursts, which are usually of shorter timescales and are embedded in the active phase of the MJO. Recent developments in atmospheric modeling or coupled modeling and the various examples of scale interactions, which have generated considerable interest in the recent decade, are covered in review papers by Zhang (2005) and in the dedicated volume of Lau and Waliser (2005), and its recently updated version (2012). Here, the main emphasis will be on ocean processes, and mostly what are new in the last 10 years and which are mostly focused on the Indian Ocean.

The tropical Indian Ocean has been the site for major experiments, starting from MISMO and CIRENE in 2006–2007 (Vialard et al., 2009a,b) to CINDY–DYNAMO in 2011–2012. The region has some of the largest signals in SST, in particular in austral summer, when the warm pool extends far into the western Indian Ocean south of the equator (Duvel and Vialard, 2007). MJO events in the Indian Ocean and (less systematically) in the western Pacific tend to propagate eastward. The active convective phases are followed to their west with significantly cooler waters. An issue not completely solved for the Indian Ocean is how this SST signal impacts the MJO and whether it is part of the dynamics (Duvel, 2012, and references therein). It is also a region of usually shallow seasonal thermocline (the Seychelles–Chagos Dome (Figure 15.7)) with a minimum thermocline depth in January. For instance, ocean heat budgets in the areas of maximum SST signal related to the MJO (Waliser et al., 2003; Duvel and Vialard, 2007) suggest that the SST changes are either related to the heat loss associated in particular with the increased winds west of the convective regions or to vertical processes, depending on the particular region. The large SST increase during the inactive periods is related not only to the weak winds and low cloudiness (Jayakumar et al., 2011) but also to the occasional large diurnal SST cycles (Bellenger et al., 2010). These very warm mid-day SSTs could contribute after a while to mid-tropospheric shallow or intermediate convection and thus could play a role in the initiation of the deep convection and MJO events in this region (Bellenger et al., 2010), although this is still a topic under discussion.

The winds associated with MJO generate planetary waves in the ocean through Ekman pumping or induction of zonal equatorial momentum. The waves contribute to propagate ocean signals to other regions (e.g., Vialard et al., 2009a,b) in the basin. In the boreal summer season, the MJO also has signatures north of the equator, which are associated with the active/break cycle of the monsoon, which propagate northward (Goswami, 2005) and are traced in surface temperature patterns (Vialard et al., 2012).

4.4. IOD, ENSO, and Monsoon Interactions

The importance of the remote influence of the Pacific ENSO on the Indian Ocean has long been recognized. Indeed a basin-wide SST anomaly of almost uniform polarity, which appears as the dominant EOF mode, is highly correlated with ENSO in the Pacific (Cadet, 1985; Klein et al., 1999) and known as the Indian Ocean basin mode (IOBM). Tozuka et al. (2007) have also found that the first EOF mode of the decadal (9–35 years) SST anomaly represents a basin-wide uniform mode that has close connection with the Pacific ENSO-like decadal variability. The El Niño-induced warming of the Indian Ocean may act as a source of heating as most parts of the basin remain anomalously warm for several months after the peak phase of El Niño. Therefore, the warm phase of IOBM can, in turn, feed back into the atmosphere, inducing a delayed influence on regional climate. Xie et al. (2009) equated the warm IOBM with a "capacitor" in the sense that the warming introduced by El Niño stayed as a charge in the capacitor and exerts a delayed effect by discharging the charge on the northwestern Pacific climate after the El Niño itself is terminated in the Pacific.

The basin-wide anomaly is often first established in the west and spreads eastward as the ENSO event matures. However, the IOD does appear as a seesaw mode dramatically in a physical space during typical event years (Meyers et al., 2007) irrespective of the phase of ENSO

(Behera et al., 2008), proving that it is an inherent mode of Indian Ocean variations.

While the ENSO-related basin-wide warming in the Indian Ocean is easier to understand and predict, the ENSO and IOD interaction cannot be explained easily. Because of simultaneous occurrences of IOD and ENSO in certain event years, sometimes it is assumed that the IOD is ENSO induced (e.g., Allan et al., 2001; Dommenget and Latif, 2002; Baquero-Bernal et al., 2002). However, the independent evolution of the IOD is demonstrated in observational data (Behera et al., 2003a) and model experiments. Behera et al. (2006) reported that ocean–atmosphere conditions related to the IOD are realistically simulated in the absence of Pacific Ocean–atmosphere coupling (noENSO) in the SINTEX-F CGCM experiment. This demonstrates that intrinsic processes within the basin are capable of generating IOD. Most of the other model studies (Iizuka et al., 2000; Yu et al., 2002; Gualdi et al., 2003; Lau and Nath, 2003; Yamagata et al., 2004; Cai et al., 2005; Fischer et al., 2005; Behera et al., 2006) have also reported an independent IOD mode except for the model study of Baquero-Bernal et al. (2002). The interactions between IOD and ENSO not only affect their amplitudes but also the periodicity of their inherent variations. In the absence of ENSO variability, in the noENSO experiment explained above, Behera et al. (2006) found that the interannual IOD variability is predominantly biennial.

The independent evolutions of IOD do not exclude the possibility of nonlinear interactions between the anomalous Walker cells of the Indian and Pacific Oceans associated with IOD and ENSO when they co-occur. Anomalous latent heat flux and vertical heat convergence associated with the modified Walker circulation in ENSO co-occurring years are shown to contribute to the alteration of western pole anomalies (Behera et al., 2006). Conversely, based on a statistical analysis, Behera and Yamagata (2003) showed that IOD modulates the Darwin pressure variability, that is, one pole of the Southern Oscillation and could influence ENSO (Annamalai et al., 2005; Izumo et al., 2008). The other mechanism is related to oceanic processes and the passage of ENSO signals through the Indonesian Throughflow. It is understood that the mature ENSO signal in the western Pacific intrudes into the eastern Indian Ocean through the coastal waveguide around the Australian continent (Clarke and Liu, 1994; Meyers, 1996; Schiller et al., 1998; Wijffels and Meyers, 2004), and is known as the Clarke–Meyers effect (Yamagata et al., 2004).

It has been found that the frequent occurrence of the IOD in the last couple of decades has weakened the ENSO–monsoon relationship (Ashok et al., 2001; Behera and Yamagata, 2003). Based on a 150-year-long coral record, Abram et al. (2008) demonstrated an intensification of the IOD magnitude and frequency since the mid-twentieth century, and explained this shift as an enhancement of upwelling in the eastern Indian Ocean. It was suggested that IOD-induced upwelling was a result of a weakening of Walker circulation in the tropical Pacific due to greenhouse warming. In another study, Nakamura et al. (2009) confirmed the recent shift in IOD variability and revealed the changes in ENSO–monsoon and IOD–monsoon relationships in the context of East African short rains (EASR) (Behera et al., 2005). In the period frequented by the IOD (e.g., in the 1990s), when the relationship between the ENSO and Indian summer monsoon rainfall (ISMR) is weakened, the correlation of the coral IOD index with the ISMR is strong, in addition to that with the EASR that follow the Indian summer monsoon.

5. PROGRESSES IN TROPICAL CLIMATE PREDICTIONS

Climate prediction systems now increasingly rely on the CGCMs (see Chapter 23). The skill of these CGCM-based dynamical predictions, particularly for the Indo-Pacific sector, has improved greatly in the recent decades and has established user confidence as compared to the simple coupled models (e.g., Cane and Zebiak, 1985; Cane et al., 1986) and statistical prediction systems. This has been achieved through improvements in model physics as well as through the development of good data assimilation systems. In dynamical seasonal prediction, errors in the predicted mean state may hinder the model's capability to produce accurate forecasts (cf. Lee et al., 2010). For example, Sperber and Palmer (1996) showed that models with a more realistic mean state tend to perform better for the interannual variability of all-Indian monsoon rainfall related to ENSO. Therefore, it is important to address these systematic biases by either removing or correcting them for better climate predictions. While modelers continue to work on the improvements of the mean state, availability of a greater number of coupled prediction models is helping the effort to make predictions based on several model results. These multimodel ensemble predictions are shown to have better skills (Figure 15.9) than any individual model (e.g., Wang et al., 2009).

Seasonal to interannual climate predictions based on CGCMs are now produced by several operational and research centers as a part of the climate services to the society. Some of the models have shown excellent skills in predicting tropical climate variations such as ENSO and IOD at very long-lead times. For example, the SINTEX-F model of JAMSTEC has predicted ENSO events at lead times of up to 1.5–2 years (e.g., Luo et al., 2008). The predicted magnitudes at long-lead times ($\geq$12 months), however, are much weaker than the observations and with a large phase delay in predicting the onset of some of the strong El Niño events. This was the case for

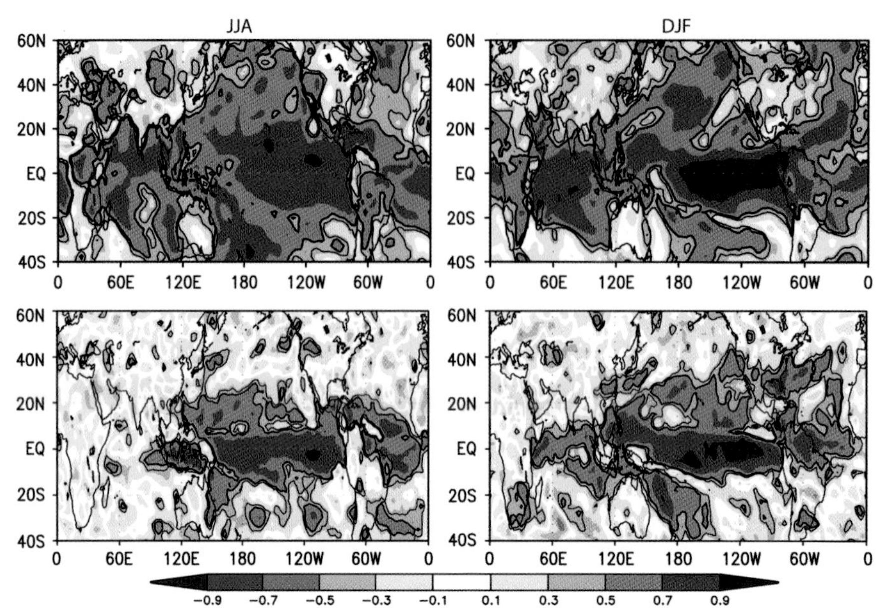

FIGURE 15.9 Multimodel-ensemble skills for 2-m air temperature (upper panels) and precipitation (lower panels) for June–August (left panels) and December–February (right panels) based on the anomaly correlations between observation and 1-month lead seasonal prediction for 1981–2003 obtained from 14 CliPAS models' MME. *After Wang et al. (2009).*

the 1997–1998 event. One of the factors that hinders the long-lead prediction of ENSO is the difficulty that CGCMs have in representing the spatial and temporal characteristics of westerly wind bursts in the western equatorial Pacific. Such was the case during late 1996–mid-1997. The La Niña events on the other hand are better predicted at longer lead times, since they are not much influenced by the wind bursts. The 1984–1986, 1999–2001, 2007–2009, and other events are predicted well up to 1.5–2 years lead time.

As in the case of ENSO, CGCMs are proving to be useful in predictability experiments of the IOD. Using the NASA Seasonal-to-Interannual Prediction Project-coupled model system, Wajsowicz (2005) has shown a remarkable predictability of the IOD at 3 months lead time for the decade 1993. Higher skills are reported by Luo et al. (2007) at a lead of 4 months. In the presence of chaotic and energetic ISOs in the Indian Ocean, it is understood that a large number of ensemble members of model predictions could improve the long-range forecasts of IOD. Nevertheless, substantial amounts of effort are required to improve the performances of both atmospheric and oceanic GCMs in simulating the tropical Indian Ocean climate. The flat zonal thermocline, a common bias found in several prediction models, influences the ocean–atmosphere coupling and the atmospheric convections. This in turn would affect the probability density function of IOD predictions.

6. OUTLOOKS

The tropical regions are the major driver of climate variations in the world besides being responsible for initiating some of the large-scale ocean circulations. Though better sampled compared to the extratropics, vast parts of the tropical oceans are sparsely observed. Considering the importance of the oceans and to monitor long-term changes in regions of inadequate data coverage, a major worldwide observational experiment was conducted in the decade 1990–2000. The WOCE had covered the world ocean with a network of *in situ* measurement stations extending from the surface to the ocean bottom (see Chapter 3). In recent times, we have seen the arrival of Argo floats—a new mode of observations that revolutionized the process studies in the data sparse oceans. In addition to the great increase in the *in situ* observation of temperature, observation of salinity has become a key added component of this system. The near-surface salinity observations are also boosted by recent satellite observations (see Chapter 4). Both Aquarius (http://aquarius.nasa.gov) and Soil Moisture and Ocean Salinity (SMOS; http://www.smos-sos.org/) satellites are now usable tools to describe large-scale variability of sea surface salinity, in the tropics, which represent a critical component to understand the ocean freshwater link to the land hydrological cycle. Dedicated experiments (e.g., Salinity Processes in the Upper Ocean Regional Study-2 (http://spurs.jpl.nasa.gov/SPURS)) in the near future are expected to find the link among hydrological cycle, upper ocean salinity, diapycnal mixing, and ocean circulation at least on subseasonal to seasonal timescales.

The circulations in the tropical Pacific are dominated by trade winds on seasonal timescales and climate variability such as ENSO on the interannual timescales. These seasonal and interannual variations in ocean circulations and

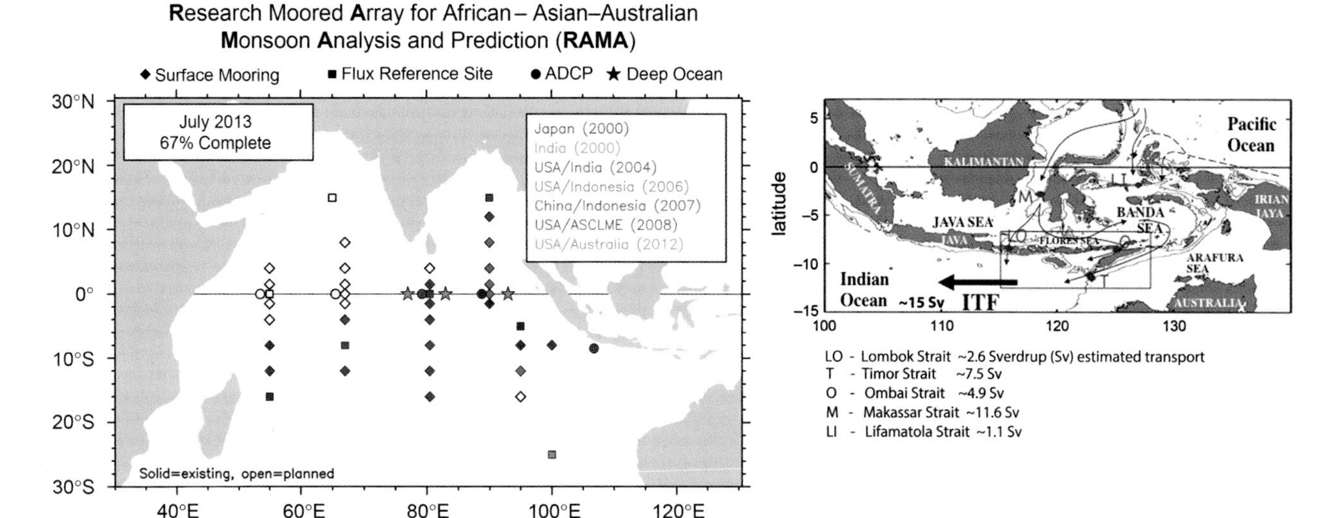

FIGURE 15.10 (Left) The RAMA moored buoy array of the Indian Ocean Observing System, which is based on a constellation of Earth observing satellites complemented by a variety of *in situ* measurement arrays. (Right) The observed transports of the Indoneseian Throughflow. *(Left) NOAA PMEL Web site. (Right) Based on Sprintall et al. (2009).*

climate have influenced and in turn are being influenced by extratropical oceans and associated processes. For example, some of the low-frequency variations in ENSO are shown to be originating from the subtropical northern and southern Pacific. Some of the variations also affect the interbasin exchanges of mass, heat, and freshwater between the Pacific and the Indian Oceans through the Indonesian Throughflow: It is a critical element in the global climate system because the heat and buoyancy it carries from the tropical Pacific to the Indian Ocean affect not only the short-term regional climate variations such as the monsoon but also the long-term global climate change as it is a major component of the global conveyer belt. Therefore, long-term monitoring (such as the International Nusantara Stratification and Transport (INSTANT) program) of the heat and freshwater transports through the Indonesian channels is very important to understand the extent to which ocean currents influence global climate.

In earlier decades, particularly during the TOGA period, much of the focus of tropical ocean research was on understanding El Niño-related processes and El Niño prediction. With the end of TOGA and the beginning of global CLIVAR program under the WCRP, the scope of climate variability and predictability studies has expanded from the tropical Pacific and ENSO-centric basis to the global domain.

Considering the recent progress in our understanding of the Indian Ocean climate variability and in particular IOD, development of a systematic observing system in the tropical Indian Ocean is a necessity not only for the understanding of the ocean and the climate processes but also to improve predictability of the tropical climate variations. The observation system in the basin is well behind those in the tropical Pacific and Atlantic Oceans. Some progress has been made in recent times (Figure 15.10), particularly the extension of the TAO/TRITON array to the eastern Indian Ocean with the deployment of two JAMSTEC TRITON buoys (http://www.jamstec.go.jp/iorgc/iomics/index.html) at (1.5°S, 90°E) and (5°S, 95°E) followed by the deployment of several RAMA buoys (http://www.pmel.noaa.gov/tao/rama/data.html) by NOAA/PMEL (McPhaden et al., 2009). Near coastal mooring-buoys and upward looking ADCPs are also deployed under the National Data Buoy Program of India (review article: Masumoto et al., 2010). The Indonesian Throughflow that carries the water from the western Pacific to the eastern Indian Ocean should also be resolved in a more precise way in next-generation coupled models aided by the observations and understanding gained through the INSTANT program (Figure 15.10, right panel). INSTANT is a coordinated international effort to measure the flow of water between the Pacific and Indian Oceans via the Seas of Indonesia (Gordon et al., 2003). In addition, the monitoring of subsurface conditions has become easier with the availability of reasonably populated Argo floats. It is possible now to detect early signals of ENSO and IOD at least a month ahead. The situation is expected to further improve with the increase of ocean observations in these areas. Current international efforts by the WCRP/CLIVAR and the Earth Observing System/Global Earth Observation System of Systems/Global Ocean Observing System (EOS/GEOSS/GOOS) to establish a long-term monitoring system in the tropical Indian Ocean (similar to its counterpart the Tropical Atmosphere Ocean/Triangle Trans-Ocean Buoy Network (TAO/TRITON) in the Pacific) will increase the forecast skills of IOD by providing better initial conditions, particularly related to the mass and heat transports from the western Pacific to the Indian Ocean and such exchanges between the Arabian Sea and the Bay of Bengal.

The Atlantic MOC and the western boundary currents that are largely associated with the MOC are particularly important for the tropical climate variations. Warm-water masses of Southern Hemisphere origin are exchanged with the cold-water masses produced in the North Atlantic within this large overturning cell, through the tropical belt with its energetic zonal current bands. The strength of the MOC is thought to affect circulation, water mass distribution, and stratification in the tropics with an impact on surface winds and precipitation in areas bordering the North Atlantic, thus affecting a large part of the human population as noted in the last two reports of the Intergovernmental Panel on Climate Change (IPCC). Observations and models also indicate that MOC variability is tightly linked to the tropical North Atlantic climate change through oceanic teleconnections. For example, the subsurface tropical North Atlantic experiences strong warming by a slowdown of MOC, though there is an atmospheric-induced surface cooling above (e.g., Schmidt et al., 2012). Therefore, monitoring of the Atlantic MOC is of great importance for tropical Atlantic research. SAMOC (http://www.aoml.noaa.gov/phod/research/moc/samoc/) program in the South Atlantic has complemented well the long-term observations of the RAPID MOC (http://www.noc.soton.ac.uk/rapidmoc/) program strengthening the observations in both hemispheres. This will help to understand the coherence of MOC signals across the tropical ocean that might result from subpolar North Atlantic and/or Agulhas leakage variability (Böning et al., 2006; Biastoch et al., 2008).

In addition to the advances in our knowledge of the earlier climate processes, several new modes of variations are identified and discussed in the decades following TOGA and WOCE. Besides the IOD of the Indian Ocean, there is a second coupled-ocean–atmosphere mode called the El Niño Modoki in the tropical Pacific. Both these modes have significant impacts on regional climate variations worldwide. The IOD influences on the interbasin variations, particularly related to ENSO and its predictability, have been extensively studied recently (Behera et al., 2006; Luo et al., 2010). The Indo-Pacific interbasin coupling is crucial to the evolution of both El Niño and extreme IOD and their predictions at long-lead times. The predictability is seen to improve by making ensembles of multiple model results. Perturbing the initial state with random noises is another approach, which has shown promise for the short-term climate predictions and intraseasonal oscillations. In any case, it is found to be important to resolve and predict the other associated or nonassociated climate modes in order to improve the ENSO predictability on longer lead periods.

It is important to reduce the model biases so as to precisely predict the equatorial SST, which is crucial for a proper representation of ENSO in the model predictions under the WCRP–CLIVAR initiatives. It will also be important for the future goals of IPCC. Biases in the tropical Atlantic are the worst among model simulations hampering predictions and the understanding of recent trends in the Atlantic sector. Different oceanic and atmospheric processes might contribute to eastern tropical Atlantic SST errors in the simulated mean climate state, possibly amplified by the ocean–atmosphere feedbacks of the coupled models (Toniazzo and Woolnough, 2013). Process studies using high-resolution ocean and atmosphere models and dedicated observational studies might help to identify sources of model errors that are hypothesized to originate from misrepresentations of diverse processes such as atmospheric convection, cloud processes, and aerosols. In the ocean, those could be the connection between equatorial and coastal upwelling regions (including advection and wave processes along the equatorial and coastal waveguide) and local wind forcing in the coastal upwelling regions (Xu et al., 2013).

The large-scale changes in the basic state of our climate system will also be affecting the ocean circulations and modes of climate variations. Frequent occurrences of extreme IOD and El Niño Modoki in recent decades (Ashok and Yamagata, 2009) perhaps are associated with the weakened Walker circulation in response to anthropogenic forcing (Vecchi et al., 2006). It is conceivable that the intensified IOD activity (Abram et al., 2008; Behera et al., 2008) will play a more important role in El Niño evolution under the present global warming trend. This may have implications for our future projection of ENSO in a warmer world. The large-scale ocean circulations, particularly both deep and shallow MOCs, are also important for the low-frequency climate variations. It is, therefore, important to monitor these ocean processes for the better understanding of climate variations and to improve their predictability.

ACKNOWLEDGMENTS

We thank anonymous reviewers for their very constructive comments that helped to improve the quality of the chapter.

REFERENCES

Abram, N.J., Gagan, M.K., Cole, J.E., Hantoro, W.S., Mudelsee, M., 2008. Recent intensification of tropical climate variability in the Indian Ocean. Nat. Geosci. 1, 849–853.

Allan, R., et al., 2001. Is there an Indian Ocean dipole, and is it independent of the El Niño–Southern Oscillation? CLIVAR Exchanges, vol. 6, No. 3. International CLIVAR Project Office, Southampton, United Kingdom, pp. 18–22.

Annamalai, H., Murtugudde, R., Potemra, J., Xie, S.P., Liu, P., Wang, B., 2003. Coupled dynamics over the Indian Ocean: spring initiation of the zonal mode. Deep Sea Res. Part II 50, 2305–2330.

Annamalai, H., Xie, S.P., McCreary, J.P., Murtugudde, R., 2005. Impact of Indian Ocean sea surface temperature on developing El Niño. J. Clim. 18, 302–319.

Ashok, K., Yamagata, T., 2009. Climate change: the El Niño with a difference. Nature 461, 481–484.

Ashok, K., Guan, Z., Yamagata, T., 2001. Impact of the Indian Ocean Dipole on the decadal relationship between the Indian monsoon rainfall and ENSO. Geophys. Res. Lett. 28, 4499–4502.

Ashok, K., Guan, Z., Yamagata, T., 2003. Influence of the Indian Ocean Dipole on the Australian winter rainfall. Geophys. Res. Lett. 30, 1821. http://dx.doi.org/10.1029/2003GL017926.

Ashok, K., Behera, S.K., Rao, S.A., Weng, H., Yamagata, T., 2007. El Niño Modoki andits possible teleconnection. J. Geophys. Res. 112, C11007. http://dx.doi.org/10.1029/2006JC003798.

Athie, G., Marin, F., 2008. Cross equatorial structure and temporal modulation of intraseasonal variability at the surface of the Tropical Atlantic Ocean. J. Geophys. Res. 113, C08020.

Backeberg, B.C., Penven, P., Rouault, M., 2012. Impact of intensified Indian Ocean winds on mesoscale variability in the Agulhas system. Nat. Clim. Chang. 2, 608–612. http://dx.doi.org/10.1038/nclimate1587.

Baquero-Bernal, A., Latif, M., Legutke, S., 2002. On dipole-like variability in the tropical Indian Ocean. J. Clim. 15, 1358–1368.

Behera, S.K., Salvekar, P.S., 1995. Numerical study of the interannual variability in Indian Ocean. Mausam 46, 409–422.

Behera, S.K., Salvekar, P.S., 1998. Numerical investigation of coastal circulation around India. Mausam 49, 345–360.

Behera, S.K., Yamagata, T., 2003. Influence of the Indian Ocean Dipole on the Southern Oscillation. J. Meteorol. Soc. Jpn. 81, 169–177.

Behera, S., Yamagata, T., 2010. Imprint of the El Niño Modoki on decadal sea level changes. Geophys. Res. Lett. 37, L23702. http://dx.doi.org/10.1029/2010GL045936.

Behera, S.K., Salvekar, P.S., Ganer, D.W., Deo, A.A., 1998. Interannual variability in simulated east coast circulation. Indian J. Mar. Res. 27, 115–120.

Behera, S.K., Krishnan, R., Yamagata, T., 1999. Unusual ocean-atmosphere conditions in the tropical Indian Ocean during 1994. Geophys. Res. Lett. 26, 3001–3004.

Behera, S.K., Salvekar, P.S., Yamagata, T., 2000. Simulation of interannual SST variability in the tropical Indian Ocean. J. Clim. 13, 3487–3499.

Behera, S.K., Rao, S.A., Saji, H.N., Yamagata, T., 2003a. Comments on "A cautionary note on the interpretation of EOFs" J. Clim. 16, 1087–1093.

Behera, S.K., Luo, J.-J., Masson, S., Yamagata, T., Delecluse, P., Gualdi, S., Navarra, A., 2003b. Impact of the Indian Ocean Dipole on the East African short rains: a CGCM study. CLIVAR Exchanges 27, 43–45.

Behera, S.K., Luo, J.-J., Masson, S., Delecluse, P., Gualdi, S., Navarra, A., Yamagata, T., 2005. Paramount impact of the Indian Ocean Dipole on the East African short rain: a CGCM study. J. Clim. 18, 4514–4530.

Behera, S.K., Luo, J.-J., Masson, S., Rao, S.A., Sakuma, H., Yamagata, T., 2006. A CGCM study on the interaction between IOD and ENSO. J. Clim. 19, 1688–1705.

Behera, S.K., Luo, J.-J., Yamagata, T., 2008. The Unusual IOD Event of 2007. Geophys. Res. Lett. 35, L14S11. http://dx.doi.org/10.1029/2008GL034122.

Behera, S.K., Ratnam, J.V., Masumoto, Y., Yamagata, T., 2013. Origin of extreme summers in Europe—the Indo-Pacific connection. Clim. Dyn. 41, 663–676. http://dx.doi.org/10.1007/s00382-012-1524-8.

Bellenger, H., Takayuba, Y.N., Ushiyama, T., Yoneyama, K., 2010. Role of diurnal warm layers in the diurnal cycle of convection over the tropical Indian Ocean during MISMO. Mon. Weather Rev. 23, 2426–2433.

Biastoch, A., Böning, C.W., Lutjeharms, J.R.E., 2008. Agulhas leakage dynamics affects decadal variability in Atlantic overturning circulation. Nature 456, 489–492.

Bjerknes, J., 1969. Atmospheric teleconnections from the equatorial Pacific. Mon. Weather Rev. 97, 163–172.

Black, E., Slingo, J., Sperber, K.R., 2003. An observational study of the relationship between excessively strong short rains in coastal East Africa and Indian Ocean SST. Mon. Weather Rev. 31, 74–94.

Blanke, B., Raynaud, S., 1997. Kinematics of the Pacific equatorial undercurrent: an Eulerian and Lagrangian approach from GCM results. J. Phys. Oceanogr. 27, 1038–1053.

Blanke, B., Speich, S., Madec, G., Maugé, R., 2002. A global diagnostic of interior ocean ventilation. Geophys. Res. Lett. 29, 1267. http://dx.doi.org/10.1029/2001GL013727.

Böning, C.W., Scheinert, M., Dengg, J., Biastoch, A., Funk, A., 2006. Decadal variability of subpolar gyre transport and its reverberation in the North Atlantic overturning. Geophys. Res. Lett. 33, L21S01. http://dx.doi.org/10.1029/2006GL026906.

Brandt, P., Stramma, L., Schott, F., Fischer, J., Dengler, M., Quadfasel, D., 2002. Annual Rossby waves in the Arabian Sea from TOPEX/POSEIDON altimeter and in-situ data. Deep Sea Res. Part II 49, 1197–1210.

Brandt, P., Caniaux, G., Bourlès, B., Lazar, A., Dengler, M., Funk, A., Hormann, V., Giordani, H., Marin, F., 2011a. Equatorial upper-ocean dynamics and their interaction with the West African monsoon. Atmos. Sci. Lett. 12, 24–30.

Brandt, P., Funk, A., Hormann, V., Dengler, M., Greatbatch, R.J., Toole, J.M., 2011b. Interannual atmospheric variability forced by the deep equatorial Atlantic Ocean. Nature 473, 497–500. http://dx.doi.org/10.1038/nature10013.

Bruce, J.G., 1968. Comparison of near surface dynamic topography during the two monsoons in the western Indian Ocean. Deep Sea Res. 15, 665–667.

Bruce, J.G., Johnson, D.R., Kindle, J.C., 1994. Evidence for eddy formation in the eastern Arabian Sea during the northeast monsoon. J. Geophys. Res. 99, 7651–7664.

Bruce, J.G., Kindle, J.C., Kantha, L.H., Kerling, J.L., Bailey, J.F., 1998. Recent observations and modeling in the Arabian Sea Laccadive high region. J. Geophys. Res. 103, 7593–7600.

Bunge, L., Provost, C., Hua, B.L., Kartavtseff, A., 2008. Variability at intermediate depths at the equator in the Atlantic Ocean in 2000-06: annual cycle, equatorial deep jets, and intraseasonal meridional velocity fluctuations. J. Phys. Oceanogr. 38 (8), 1794–1806. http://dx.doi.org/10.1175/2008JPO3781.1.

Cadet, D.L., 1985. The Southern Oscillation over the Indian Ocean. J. Climatol. 5, 189–212.

Cai, W., Cowan, T., 2009. La Niña Modoki impacts Australia autumn rainfall variability. Geophys. Res. Lett. 36, L12805. http://dx.doi.org/10.1029/12009GL037885.

Cai, W., Hendon, H., Meyers, G., 2005. Indian Ocean dipole-like variability in the CSIRO Mark 3 coupled climate model. J. Clim. 18, 1449–1468.

Cane, M.A., Zebiak, S.E., 1985. A theory for El Niño and the Southern Oscillation. Science 228, 1085–1087.

Cane, M.A., Zebiak, S.E., Dolan, S.C., 1986. Experimental forecasts of El Niño. Nature 321, 827–832.

Caniaux, G., Giordani, H., Redelsperger, J.-L., Guichard, F., Key, E., Wade, M., 2011. Coupling between the Atlantic cold tongue and the

West African monsoon in boreal spring and summer. J. Geophys. Res. 116, C04003.

Carton, J.A., Huang, B., 1994. Warm events in the tropical Atlantic. J. Phys. Oceanogr. 24, 888–903.

Carton, J.A., Cao, X., Giese, B.S., da Silva, A.M., 1996. Decadal and interannual SST variability in the tropical Atlantic Ocean. J. Phys. Oceanogr. 26, 1165–1175.

Cazenave, A., Minh, K.D., Gennero, M.C., 2004. Present-day sea level rise: From satellite and in-situ observations to physical causes, In: Hwang, C., Shum, C., Li, J. (Eds.), Satellite Altimetry for Geodesy, Geophysics and Oceanography, Int. Assoc. Geod. Symp., vol. 126, Springer, New York, pp. 23–31.

Chambers, D.P., Mehlhaff, C.A., Urban, T.J., Fujii, D., Nerem, R.S., 2002. Low-frequency variations in global mean sea level: 1950–2000. J. Geophys. Res. 107 (C4), 3026. http://dx.doi.org/10.1029/2001JC001089.

Chan, S., Behera, S., Yamagata, T., 2008. Indian Ocean Dipole influence on South American rainfall. Geophys. Res. Lett. 35, L14S12. http://dx.doi.org/10.1029/2008GL034204.

Chang, P., Ji, L., Li, H., 1997. A decadal climate variation in the tropical Atlantic Ocean from thermodynamic air-sea interactions. Nature 385, 516–518.

Chang, P., Geise, B., Ji, L., Seidel, H., Wang, F., 2001. Decadal change in the south tropical Pacific in a global assimilation analysis. Geophys. Res. Lett. 28, 3461–3464.

Chang, P., Fang, Y., Saravanan, R., Ji, L., Seidel, H., 2006a. The cause of the fragile relationship between the Pacific El Niño and the Atlantic Niño. Nature 443 (7109), 324. http://dx.doi.org/10.1038/nature05053.

Chang, P., Yamagata, T., Schopf, P., Behera, S.K., Carton, J., Kessler, W.S., Meyers, G., Qu, T., Schott, F., Shetye, S., Xie, S.-P., 2006b. Climate fluctuations of tropical coupled system—the role of ocean dynamics. J. Clim. 19, 5122–5174.

Chang, P., Zhang, R., Hazeleger, W., Wen, C., Wan, X., Ji, L., Haarsma, R.J., Breugem, W.-P., Seidel, H., 2008. Oceanic link between abrupt changes in the North Atlantic Ocean and the African monsoon. Nat. Geosci. 1 (7), 444–448.

Cherchi, A., Gualdi, S., Behera, S., Luo, J.J., Masson, S., Yamagata, T., Navarra, A., 2007. The influence of Tropical Indian Ocean SST on the Indian summer monsoon. J. Clim. 20, 3083–3105.

Church, J.A., White, N.J., Hunter, J.R., 2006. Sea-level rise at tropical Pacific and Indian Ocean islands. Glob. Planet. Change 53, 155–168. http://dx.doi.org/10.1016/j.gloplacha.2006.04.001.

Cibot, C., Maisonnave, E., Terray, L., Dewitte, B., 2005. Mechanisms of tropical Pacific interannual-to-decadal variability in the ARPEGE/ORCA global coupled model. Clim. Dyn. 24, 823–842. http://dx.doi.org/10.1007/s00382-004-0513-y.

Clark, C.O., Webster, P.J., Cole, J.E., 2003. Interdecadal variability of the relationship between the Indian Ocean zonal mode and East African coastal rainfall anomalies. J. Clim. 16, 548–554.

Clarke, A.J., Liu, X., 1994. Interannual sea level in the northern and eastern Indian Ocean. J. Phys. Oceanogr. 24, 1224–1235.

Cravatte, S., Delcroix, T., Zhang, D., McPhaden, M., Leloup, J., 2009. Observed freshening of the warming Western Tropical Pacific and extension of the Warm/Fresh Pool in recent decades. Clim. Dyn. 33, 565–589. http://dx.doi.org/10.1007/s00382-009-0526-7.

Cravatte, S., Ganachaud, A., Duong, Q.P., Kessler, W.S., Eldin, G., Dutrieux, P., 2011. Observed circulation in the Solomon Sea from SADCP data. Prog. Oceanogr. 88, 116–130.

Cresswell, G.R., Petersen, J.L., 1993. The Leeuwin Current south of western Australia. Aust. J. Mar. Freshw. Res. 44, 285–303.

Cutler, A.N., Swallow, J.C., 1984. Surface currents of the Indian Ocean (to 25S, 100E): Compiled from historical data archived by the Meteorological Office, Bracknell, UK. Report 187, Institute of Oceanographic Sciences, Wormley, England, 36 charts, 8pp.

Czaja, A., van der Vaart, P., Marshall, J., 2002. A diagnostic study of the role of remote forcing in tropical Atlantic variability. J. Clim. 15, 3280–3290.

d'Orgeville, M., Hua, B.L., Sasaki, H., 2007. Equatorial deep jets triggered by a large vertical scale variability within the western boundary layer. J. Mar. Res. 65, 1–25.

Delcroix, T., Cravatte, S., McPhaden, M., 2007. Decadal variations and trends in tropical Pacific sea surface salinity since 1970. J. Geophys. Res. 112, C03012. http://dx.doi.org/10.1029/2006JC003801.

Delworth, T.L., Mann, M.E., 2000. Observed and simulated multidecadal variability in the Northern Hemisphere. Clim. Dyn. 16, 661–676. http://dx.doi.org/10.1007/s003820000075.

Dengler, M., Schott, F.A., Eden, C., Brandt, P., Fischer, J., Zantopp, R., 2004. Break-up of the Atlantic deep western boundary current into eddies at 8°S. Nature 432, 1018–1020.

Deser, C., Alexander, M.A., Timlin, M.S., 1996. Upper-ocean thermal variations in the North Pacific during 1970-91. J. Clim. 9, 1840–1855.

Deser, C., Phillips, A.S., Alexander, M.A., 2010. Twentieth century tropical sea surface temperature trends revisited. Geophys. Res. Lett. 37, L10701. http://dx.doi.org/10.1029/2010GL043321.

Dommenget, D., Latif, M., 2000. Interannual to decadal variability in the tropical. Atlantic. J. Clim. 13, 777–792.

Dommenget, D., Latif, M., 2002. A cautionary note on the interpretation of EOFs. J. Clim. 15, 216–225.

Donguy, J.R., Meyers, G., 1995. Observations of geostrophic transport variability in the western tropical Indian Ocean. Deep Sea Res. Part I 42, 1007–1028.

Duvel, J.-P., 2012. New topics and advances, oceans and air-sea interaction, intraseasonal variability in the atmosphere-ocean climate system, In: Lau, W.K.M., Waliser, D.E. (Eds.), Intraseasonal Variability in the Atmosphere-Ocean Climate System, second ed., Springer, New York, pp. 513–530.

Duvel, J.-P., Vialard, J., 2007. Indo-Pacific sea surface temperature perturbations associated with intraseasonal oscillations of the tropical convection. J. Clim. 20, 3056–3082.

Eden, C., Dengler, M., 2008. Stacked jets in the deep equatorial Atlantic Ocean. J. Geophys. Res. 113, C04003.

Edwards, C.A., Pedlosky, J., 1998. Dynamics of nonlinear cross-equatorial flow. Part I: potential vorticity transformation. J. Phys. Oceanogr. 28, 2382–2406.

Enomoto, T., Hoskins, B.J., Matsuda, Y., 2003. The formation of the Bonin high in August. Q. J. R. Meteorol. Soc. 587, 157–178.

Feng, M., Meyers, G., 2003. Interannual variability in the tropical Indian Ocean: a two-year time scale of IOD. Deep Sea Res. Part II 50 (12/13), 2263–2284.

Feng, M., Biastoch, A., Böning, C., Caputi, N., Meyers, G., 2008. Seasonal and interannual variations of upper ocean heat balance off the west coast of Australia. J. Geophys. Res. 113, C12025. http://dx.doi.org/10.1029/2008JC004908.

Fine, R.A.R., Lukas, F.M., Bingham, M.J. Warner, Gammon, R.H., 1994. The Western Equatorial Pacific: a water mass crossroads. J. Geophys. Res. 99, 25063–25080.

Fischer, J., Schott, F., 1997. Seasonal transport variability of the deep western boundary current in the Atlantic. J. Geophys. Res. 102 (C13), 27751–27769.

Fischer, A.S., Terray, P., Delecluse, P., Gualdi, S., Guilyardi, E., 2005. Triggers for tropical Indian Ocean variability and links to ENSO in a constrained coupled climate model. J. Clim. 18, 3428–3449.

Florenchie, P., Lutjeharms, J.R.E., Reason, C.J.C., Masson, S., Rouault, M., 2003. The source of Benguela Niños in the South Atlantic Ocean. Geophys. Res. Lett. 30 (10), 1505. http://dx.doi.org/10.1029/2003GL017172.

Foltz, G.R., Grodsky, S.A., Carton, J.A., McPhaden, M.J., 2003. Seasonal mixed layer heat budget of the tropical Atlantic Ocean. J. Geophys. Res. 108, 3146.

Foltz, G.R., Carton, J.A., Chassignet, E.P., 2004. Tropical instability vortices in the Atlantic Ocean. J. Geophys. Res. 109, C03029. http://dx.doi.org/10.1029/2003JC001942.

Fratantoni, D., Johns, W., Townsend, T., Hurlburt, H., 2000. Low latitude circulation and mass transport pathways in a model of the tropical Atlantic Ocean. J. Phys. Oceanogr. 30, 1944–1966.

Fukumori, I., Lee, T., Cheng, B., Menemenlis, D., 2004. The origin, pathway, and destination of Niño3 water estimated by a simulated passive tracer and its adjoint. J. Phys. Oceanogr. 34, 582–604.

Giese, B.S., Urizar, S.C., Fuckar, N., 2002. Southern hemisphere origins of the 1976 climate shift. Geophys. Res. Lett. 29, 1014. http://dx.doi.org/10.1029/2001GL013268.

Glessmer, M.S., Eden, C., Oschlies, A., 2009. Contribution of oxygen minimum zone waters to the coastal upwelling off Mauritania. Prog. Oceanogr. 83, 143–150.

Godfrey, J.S., Golding, T.J., 1981. The Sverdrup relation in the Indian Ocean, and the effect of Pacific-Indian Ocean Throughflow on Indian Ocean circulation and on the East Australian Current. J. Phys. Oceanogr. 11, 771–779.

Godfrey, J.S., Weaver, A.J., 1991. Is the Leeuwin Current driven by Pacific heating and winds? Prog. Oceanogr. 27, 225–272.

Godfrey, J.S., Johnson, G.C., McPhaden, M.J., Reverdin, G., Wijffels, S.E., 2001. The tropical ocean circulation. In: Siedler, G., Church, J., Gould, J., Griffies, S. (Eds.), Ocean Circulation & Climate: Observing and Modelling the Global Ocean. Academic Press, London, pp. 215–246.

Goni, G.J., Johns, W., 2003. Synoptic study of warm rings in the North Brazil Current retroflection region using satellite altimetry. In: Goni, G.J., Malanotte-Rizzoli, P. (Eds.), Interhemispheric Water Exchange in the Atlantic Ocean. Elsevier Oceanogr. Ser., vol. 68. Elsevier, Amsterdam, pp. 335–356, Chapter 13.

Gopalakrishna, V.V., Pednekar, S.M., Murty, V.S.N., 1996. T-S variability and volume transport in the central Bay of Bengal during southwest monsoon. Indian J. Mar. Sci. 25, 50–55.

Gordon, A.L., 2005. Oceanography of the Indonesian Seas and their throughflow. Oceanography 18 (4), 14–27.

Gordon, A.L., Susanto, R.D., Vranes, K., 2003. Cool Indonesian Throughflow as a consequence of restricted surface layer flow. Nature 425, 825–828.

Goswami, B.N., 2005. South Asian monsoon. In: Lau, W.K.M., Waliser, D.E. (Eds.), Intraseasonal Variability in the Atmosphere-Ocean Climate System. Praxis Springer, Berlin, pp. 19–55.

Greatbatch, R.J., Brandt, P., Claus, M., Didwischus, S.-H., Fu, Y., 2012. On the width of the equatorial deep jets. J. Phys. Oceanogr. 42, 1729–1740.

Gregg, M.C., Sanford, T.B., Winkel, D.P., 2003. Reduced mixing from the breaking of internal waves in equatorial waters. Nature 422, 513–515.

Gualdi, S., Guilyardi, E., Navarra, A., Masina, S., Delecluse, P., 2003. The interannual variability in the tropical Indian Ocean as simulated by a CGCM. Clim. Dyn. 20, 567–582.

Guan, Z., Yamagata, T., 2003. The unusual summer of 1994 in East Asia: IOD Teleconnections. Geophys. Res. Lett. 30, 1544–1547. http://dx.doi.org/10.1029/2002GL016831.

Guan, Z., Ashok, K., Yamagata, T., 2003. Summer-time response of the tropical atmosphere to the Indian Ocean dipole sea surface temperature anomalies. J. Meteorol. Soc. Jpn. 81, 531–561.

Han, W., McCreary, J.P., Anderson, D.L.T., Mariano, A.J., 1999. On the dynamics of the eastward surface jets in the equatorial Indian Ocean. J. Phys. Oceanogr. 29, 2191–2209.

Hansen, D., Paul, C.A., 1984. Genesis and effects of long waves in the equatorial Pacific. J. Geophys. Res. 89, 10431–10440.

Hastenrath, S., Greischar, L.L., 1991. The monsoonal current regimes of the tropical Indian Ocean. Observed surface flow fields and their geostrophic and wind-driven components. J. Geophys. Res. 96, 12619–12633.

Hastenrath, S., Polzin, D., 2004. Dynamics of the surface wind field over the equatorial Indian Ocean. Q. J. R. Meteorol. Soc. 130, 503–517.

Hazeleger, W., Drijfhout, S., 2006. Subtropical cells and meridional overturning circulation pathways in the tropical Atlantic. J. Geophys. Res. 111, 3013–3025.

Hermes, J.C., Reason, C.J.C., 2008. Annual cycle of the South Indian Ocean (Seychelles-Chagos) thermocline ridge in a regional ocean model. J. Geophys. Res. 113, C04035. http://dx.doi.org/10.1029/2007JC004363.

Hirst, A.C., 1986. Unstable and damped equatorial modes in simple coupled ocean-atmosphere models. J. Atmos. Sci. 45, 830–852.

Horii, T., Hase, H., Ueki, I., Masumoto, Y., 2008. Oceanic precondition and evolution of the 2006 Indian Ocean dipole. Geophys. Res. Lett. 35, L03607. http://dx.doi.org/10.1029/2007GL032464.

Hummels, R., Dengler, M., Bourlès, B., 2013. Seasonal and regional variability of upper ocean diapycnal heat flux in the Atlantic cold tongue. Progr. Oceanogr. 111, 52–74. http://dx.doi.org/10.1016/j.pocean.2012.11.001.

Hüttl, S., Böning, C.W., 2006. Mechanisms of decadal variability in the shallow subtropical–tropical circulation of the Atlantic Ocean: a model study. J. Geophys. Res. 111, C07011. http://dx.doi.org/10.1029/2005JC003414.

Iizuka, S., Matsuura, T., Yamagata, T., 2000. The Indian Ocean SST dipole simulated in a coupled general circulation model. Geophys. Res. Lett. 27, 3369–3372.

Izumo, T., 2005. The equatorial undercurrent, meridional overturning circulation, and their roles in mass and heat exchanges during El Niño events in the tropical Pacific ocean. Ocean Dyn. 55, 110–123.

Izumo, T., Picaut, J., Blanke, B., 2002. Tropical pathway, equatorial undercurrent variability and the 1998 La Niña. Geophys. Res. Lett. 29, 2080–2083.

Izumo, T., Montegut, C.D.B., Luo, J.-J., Behera, S.K., Masson, S., Yamagata, T., 2008. The role of the western Arabian Sea upwelling in Indian Monsoon rainfall variability. J. Clim. 21, 5603–5623.

Jayakumar, A., Vialard, J., Lengaigne, M., Gnanaseelan, C., McCreary, J.P., Praveen Kumar, B., 2011. Processes controlling the surface temperature signature of the Madden-Julian Oscillation in the thermocline ridge of the Indian Ocean. Clim. Dyn. 37, 2217–2234.

Jochum, M., Malanotte-Rizzoli, 2003. On the generation of North Brazil Current rings. J. Mar. Res. 61, 147–173.

Jochum, M., Murtugudde, R., 2006. Temperature advection by tropical instability waves. J. Phys. Oceanogr. 36, 592–605.

Jochum, M., Murtugudde, R., Malanotte-Rizzoli, P., Busalacchi, A., 2004. Earth climate: the ocean-atmosphere interaction. In: Wang, C., Xie, S.-P., Carton, J.A. (Eds.), Geophys. Monogr., vol. 147. American Geophysical Union, Washington, DC, pp. 181–188.

Johns, W.E., Zantopp, R., Goni, G., 2003. Cross-gyre transport by North Brazil Current rings. In: Goni, G.J., Malanotte-Rizzoli, P. (Eds.), Interhemispheric Water Exchange in the Atlantic Ocean. Elsevier Oceanogr. Ser., vol. 68. Elsevier, Amsterdam, pp. 411–442, Chapter 16.

Johnson, G.C., Zhang, D., 2003. Structure of the Atlantic Ocean equatorial deep jets. J. Phys. Oceanogr. 33 (3), 600–609. http://dx.doi.org/10.1175/1520-0485(2003)033h0600:SOTAOEi2.0.CO;2.

Johnson, G.C., Kunze, E., McTaggart, K.E., Moore, D.W., 2002. Temporal and spatial structure of the equatorial deep jets in the Pacific Ocean. J. Phys. Oceanogr. 32, 3396–3407.

Joyce, T.M., Frankignoul, C., Yang, Y., Phillips, H.E., 2004. Ocean response and feedback to the SST dipole in the tropical Atlantic. J. Phys. Oceanogr. 34, 2525–2540.

Kalnay, E., et al., 1996. The NCEP/NCAR 40 year Reanalysis Project. Bull. Am. Meteorol. Soc. 77, 437–471.

Kao, H.-Y., Yu, J.-Y., 2009. Contrasting eastern-Pacific and central-Pacific types of ENSO. J. Clim. 22, 615–632.

Kashino, Y., Ishida, A., Kuroda, Y., 2005. Varaibaility of the Mindanao current: mooring observation results. Geophys. Res. Lett. 32, L18611. http://dx.doi.org/10.1029/2005GL023880.

Kashino, Y., Espana, N., Syamsudin, F., 2009. Observations of the North Equatorial Current, Mindanao Current, and the Kuroshio Current system during the 2006/7 El Niño and 2007/8 La Niña. J. Oceanogr. 65, 325–333.

Kawabe, M., Kashino, Y., Kuroda, Y., 2008. Variability and linkages of New Guinea coastal undercurrent and lower equatorial intermediate current. J. Phys. Oceanogr. 38, 1780–1793.

Keenlyside, N., Latif, M., 2007. Understanding equatorial Atlantic interannual variability. J. Clim. 20 (1), 131–142. http://dx.doi.org/10.1175/JCLI3992.1.

Kessler, W.S., 1999. Interannual variability in the subsurface high-salinity tongue south of the equator at 165°E. J. Phys. Oceanogr. 29, 2038–2049.

Kessler, W.S., Gourdeau, L., 2007. The annual cycle of circulation of the Southwest subtropical Pacific, analyzed in an ocean GCM. J. Phys. Oceanogr. 37, 1610–1627.

Kim, Y.Y., Qu, T., Jensen, T., Miyama, T., Mitsudera, H., Kang, H.-W., Ishida, A., 2004. Seasonal and interannual variations of the NEC bifurcation in a high-resolution OGCM. J. Geophys. Res. 109, C03040. http://dx.doi.org/10.1029/2003JC002013.

Kim, J.-S., Kim, K.-Y., Yeh, S.-W., 2012. Statistical evidence for the natural variation of the central Pacific El Niño. J. Geophys. Res. 117, C06014. http://dx.doi.org/10.1029/2012JC008003.

Kirchner, K., Rhein, M., Hüttl-Kabus, S., Böning, C.W., 2009. On the spreading of South Atlantic Water into the northern hemisphere. J. Geophys. Res. C05019. http://dx.doi.org/10.1029/2008JC005165.

Klein, S.A., Soden, B.J., Lau, N.C., 1999. Remote sea surface temperature variations during ENSO: Evidence for a tropical atmospheric bridge. J. Clim. 12, 917–932.

Kug, J.-S., Jin, F.-F., An, S.-I., 2009. Two types of El Niño events: cold tongue El Niño and warm pool El Niño. J. Clim. 22, 1499–1515.

Kushnir, Y., Robinson, W.A., Chang, P., Robertson, A.W., 2006. The physical basis for predicting Atlantic sector seasonal-to-interannual climate variability. J. Clim. 19, 5949–5970.

Larkin, N.K., Harrison, D.E., 2005. On the definition of El Niño and associated seasonal average U.S. weather anomalies. Geophys. Res. Lett. 32, L13705. http://dx.doi.org/10.1029/2005GL022738.

Lass, H.U., Mohrholz, V., 2008. On the interaction between the subtropical gyre and the subtropical cell on the shelf of the SE Atlantic. J. Mar. Syst. 74, 1–43.

Lau, N.-C., Nath, M.J., 2003. Atmosphere–ocean variations in the Indo-Pacific sector during ENSO episodes. J. Clim. 16, 3–20.

Lau, W.K.M., Waliser, D.E. (Eds.), 2005. Intraseasonal Variability of the Atmosphere-Ocean Climate System. Springer, Heidelberg, Germany, p. 474.

Lee, J.-Y., et al., 2010. How accurately do coupled climate models predict the leading modes of Asian-Australian monsoon interannual variability? Clim. Dyn. 35, 267–283. http://dx.doi.org/10.1007/s00382-010-0857-4.

Li, T., Wang, B., Chang, C.P., Zhang, Y., 2003. A theory for the Indian Ocean Dipole–Zonal Mode. J. Atmos. Sci. 60, 2119–2135.

Lindstrom, E., Lukas, R., Fine, R., Firing, E., Gdfrey, S., Meyers, G., Tsuchiya, M., 1987. The western Equtorial Pacific Circulation Study. Nature 330, 533–537.

Liu, Z., Philander, G.S., Pacanowski, R.C., 1994. A GCM study of the tropical-subtropical upper-ocean circulation. J. Phys. Oceanogr. 24, 2606–2623.

Llovel, W., Cazenave, A., Rogel, P., Berge-Nguyen, M., 2009. 2D reconstruction of past sea level (1950–2003) using tide gauge records and spatial patterns from a general ocean circulation model. Clim. Past Discuss. 5, 1109–1132. http://dx.doi.org/10.5194/cpd-5-1109-2009.

Lübbecke, J.F., McPhaden, M.J., 2012. On the inconsistent relationship between Pacific and Atlantic Niños. J. Clim. 25, 4294–4303. http://dx.doi.org/10.1175/JCLI-D-11-00553.1.

Lübbecke, J.F., Boning, C.W., Keenlyside, N.S., Xie, S.-P., 2010. On the connection between Benguela and equatorial Niños and the role of the South Atlantic Anticyclone. J. Geophys. Res. 115, C09015. http://dx.doi.org/10.1029/2009JC005964.

Luis, A.J., Kawamura, H., 2002. Dynamics and mechanism for sea surface cooling near the Indian tip during winter monsoon. J. Geophys. Res. 107, 3187. http://dx.doi.org/10.1029/2000JC000455.

Lumpkin, R., Speer, K., 2007. Global ocean meridional overturning. J. Phys. Oceanogr. 37, 2550–2562.

Luo, J.J., Yamagata, T., 2001. Long-term El Niño-Southern Oscillation (ENSO)-like variation with special emphasis on the South Pacific. J. Geophys. Res. 106, 22211–22227.

Luo, J.-J., Masson, S., Behera, S., Yamagata, T., 2007. Experimental forecasts of Indian Ocean dipole using a coupled OAGCM. J. Clim. 20 (10), 2178–2190.

Luo, J.-J., Masson, S., Behera, S.K., Yamagata, T., 2008. Extended ENSO predictions using a fully coupled ocean–atmosphere model. J. Clim. 21, 84–93.

Luo, J.-J., Zhang, R., Behera, S., Masumoto, Y., Jin, F.-F., Lukas, R., Yamagata, T., 2010. Interaction between El Niño and Extreme Indian Ocean Dipole. J. Clim. 23, 726–742.

Lux, M., Mercier, H., Arhan, M., 2001. Interhemispheric exchanges of mass and heat in the Atlantic Ocean in January–March 1993. Deep Sea Res. 48A, 605–638.

Manatsa, D., Behera, S.K., 2013. On the epochal strengthening in the relationship between rainfall of east africa and IOD. J. Clim. 26, 5655–5673.

Manatsa, D., Chingombe, W., Matarira, C.H., 2008. The impact of the positive Indian Ocean dipole on Zimbabwe droughts. Int. J. Clim. 28, 2011–2029. http://dx.doi.org/10.1002/joc.1695.

Masson, S., Delecluse, P., Boulanger, J.P., Menkes, C., 2002. A model study of the seasonal variability and formation mechanisms of barrier layer in the eastern equatorial Indian Ocean. J. Geophys. Res. 107 (C12), 8017. http://dx.doi.org/10.1029/2001JC000832.

Masson, S., Delecluse, P., Boulanger, J.-P., 2003. Impacts of salinity on the eastern Indian Ocean during the termination of the fall Wyrtki Jet. J. Geophys. Res. 108, 3067. http://dx.doi.org/10.1029/2001JC000833, C3.

Masson, S., Luo, J.-J., Madec, G., Vialard, J., Durand, F., Gualdi, S., Guilyardi, E., Behera, S.K., Delecluse, P., Navarra, A., Yamagata, T., 2004. Impact of barrier layer on winter-spring variability of the southeastern Arabian Sea. Geophys. Res. Lett. 32, L07703. http://dx.doi.org/10.1029/2004GL021980.

Masumoto, Y., Meyers, G., 1998. Forced Rossby waves in the southern tropical Indian Ocean. J. Geophys. Res. 103, 27589–27602.

Masumoto, Y., Yamagata, T., 1991. Response of the western tropical Pacific to the Asian winter monsoon: the generation of the Mindanao Dome. J. Phys. Oceanogr. 21 (9), 1386–1398. http://dx.doi.org/10.1175/1520-0485.

Masumoto, Y., Yu, W., Meyers, G., D'Adamo, N., Beal, L., de Ruijter, W. P.M., Dyoulgerov, M., Hermes, J., Lee, T., Lutjeharms, J.R.E., McCreary Jr., J.P., McPhaden, M.J., Murty, V.S.N., Obura, D., Pattiaratchi, C.B., Ravichandran, M., Reason, C., Syamsudin, F., Vecchi, G., Vialard, J., Yu, L., 2010. Observing systems in the Indian Ocean. In: Hall, J., Harrison, D.E., Stammer, D. (Eds.), Proceedings of the "OceanObs'09: Sustained Ocean Observations and Information for Society" Conference (vol. 2), Venice, Italy, 21–25 September 2009, ESA Publication WPP-306.

McCreary, J.P., Anderson, D.L.T., 1991. An overview of coupled models of El Niño and the Southern Oscillation. J. Geophys. Res. 96, 3125–3150.

McCreary, J.P., Lu, P., 1994. On the interaction between the subtropical and the equatorial ocean circulation: the subtropical cell. J. Phys. Oceanogr. 24, 466–497.

McCreary, J.P., Shetye, S.R., Kundu, P.K., 1986. Thermohaline forcing of eastern boundary currents: with application to the circulation off the west coast of Australia. J. Mar. Res. 44, 71–92.

McCreary Jr., J.P., Kundu, P.K., Molinari, R.L., 1993. A numerical investigation of dynamics, thermodynamics and mixedlayer processes in the Indian Ocean. Prog. Oceanogr. 31, 181–244.

McCreary, J.P., Han, W., Shankar, D., Shetye, S.R., 1996. Dynamics of the East India Coastal Current: 2. Numerical solutions. J. Geophys. Res. 101, 13993–14010.

McPhaden, M.J., Zhang, D., 2004. Pacific Ocean circulation rebounds. Geophys. Res. Lett. 31, L18301. http://dx.doi.org/10.1029/2004GL020727.

McPhaden, M.J., Meyers, G., Ando, K., Masumoto, Y., Murty, V.S.N., Ravichandran, M., Syamsudin, F., Vialard, J., Yu, L., Yu, W., 2009. RAMA: The Research Moored Array for African-Asian-Australian Monsoon Analysis and Prediction. Bull. Am. Meteorol. Soc. 90, 459–480.

McPhaden, M.J., Busalacchi, A.J., Anderson, D.L.T., 2010. A TOGA retrospective. Oceanography 23 (3), 86–103.

Meinen, C.S., 2005. Meridional extent and interannual variability of the Pacific Ocean tropical subtropical warm water exchange. J. Phys. Oceanogr. 35, 323–335.

Meinen, C.S., McPhaden, M.J., 2001. Interannual variability in warm water volume transports in the equatorial Pacific during 1993-99. J. Phys. Oceanogr. 31, 1324–1345.

Mélice, J.L., Servain, J., 2003. The tropical Atlantic meridional SST gradient index and its relationships with the SOI, NAO and Southern Ocean. Clim. Dyn. 20, 447–464.

Menary, M., Park, W., Lohmann, K., Vellinga, M., Palmer, M., Latif, M., Jungclaus, J., 2012. A multimodel comparison of centennial Atlantic meridional overturning circulation variability. Clim. Dyn. 38, 2377–2388. http://dx.doi.org/10.1007/s00382-011-1172-4.

Ménesguen, C., Hua, B., Fruman, M., Schopp, R., 2009. Dynamics of the combined extra-equatorial and equatorial deep jets in the Atlantic. J. Mar. Res. 67 (3), 323–346. http://dx.doi.org/10.1357/002224009789954766.

Meyers, G., 1996. Variation of Indonesian Throughflow and El Niño-Southern Oscillation. J. Geophys. Res. 101, 12255–12263.

Meyers, G., McIntosh, P., Pigot, L., Pook, M., 2007. The years of El Niño, La Niña and interactions with the tropical Indian Ocean. J. Clim. 20, 2872–2880.

Miyama, T., McCreary Jr., J.P., Jensen, T.G., Loschnigg, J., Godfrey, S., Ishida, A., 2003. Structure and dynamics of the Indian Ocean cross-equatorial cell. Deep Sea Res. Part II 50, 2023–2047.

Molinari, R.L., Olson, D., Reverdin, G., 1990. Surface current distributions in the tropical Indian Ocean derived from compilations of surface buoy trajectories. J. Geophys. Res. 0148-022795, 7217–7238. http://dx.doi.org/10.1029/89JC03749.

Murtugudde, R.G., McCreary, J.P., Busalacchi, A.J., 2000. Oceanic processes associated with anomalous events in the Indian Ocean with relevance to 1997-1998. J. Geophys. Res. 105, 3295–3306.

Murty, V.S.N., Sarma, Y.V.B., Rao, D.P., Murty, C.S., 1992. Water characteristics, mixing and circulation in the Bay of Bengal during southwest monsoon. J. Mar. Res. 50, 207–228.

Murty, V.S.N., Sarma, M.S.S., Lambata, B.P., Gopalakrishna, V.V., Pednekar, S.M., Rao, A.S., Luis, A.J., Kaka, A.R., Rao, L.V.G., 2000. Seasonal variability of upper-layer geostrophic transport in the tropical Indian Ocean during 1992–1996 along TOGA-I XBT tracklines. Deep Sea Res. Part I 47, 1569–1582.

Nagura, M., Sasaki, W., Tozuka, T., Luo, J.-J., Behera, S.K., Yamagata, T. 2012. Longitudinal biases in the Seychelles Dome simulated by 35 ocean-atmosphere coupled general circulation models. J. Geophys. Res. http://dx/doi.org/10.1029/2012JC008352, in press.

Nakamura, N., Kayanne, H., Iijima, H., McClanahan, T.R., Behera, S., Yamagata, T., 2009. The Indian Ocean Mode shift in the Indian Ocean climate under global warming stress. Geophys. Res. Lett. 36, L23708. http://dx.doi.org/10.1029/2009GL040590.

Neelin, J.D., Battisti, D.S., Hirst, A.C., Jin, F., Wakata, Y., Yamagata, T., Zebiak, S.E., 1998. ENSO theory. J. Geophys. Res. 103, 14261–14290.

O'Brien, J.J., Hurlburt, H.E., 1974. Equatorial jet in the Indian Ocean: theory. Science 184, 1075–1077.

Okumura, Y., Xie, S.-P., 2006. Some overlooked features of tropical Atlantic climate leading to a new Niño-like phenomenon. J. Clim. 19, 5859–5874.

Okumura, Y.M., Ohba, M., Deser, C., 2011. A proposed mechanism for the asymmetric duration of El Niño and La Niña. J. Clim. 24, 3822–3829. http://dx.doi.org/10.1175/2011JCLI3999.1.

Perigaud, C., Delecluse, P., 1992. Annual sea level variations in the southern tropical Indian Ocean from Geosat and shallow-water simulations. J. Geophys. Res. 97, 20169–20178.

Peter, A., Henaff, M.L., duPenhoat, Y., Menkes, C., Marin, F., Vialard, J., Caniaux, G., Lazar, A., 2006. A model study of the seasonal mixed layer heat budget in the equatorial Atlantic. J. Geophys. Res. 111, C06014.

Philander, S.G.H., 1990. El Niño, La Niña, and the Southern Oscillation. Academic, San Diego, California, 293 pp.

Philander, S.G.H., Yamagata, T., Pacanowski, R.C., 1984. Unstable air-sea interaction in the tropics. J. Atmos. Sci. 41 (4), 604–613.

Potemra, J.T., Luther, M.E., O'Brien, J.J., 1991. The seasonal circulation of the upper ocean in the Bay of Bengal. J. Geophys. Res. 96, 12667–12683.

Pradhan, K.P., Ashok, K., Preethi, B., Krishnan, R., Sahai, A.K., 2011. Modoki, IOD and Western North Pacific typhoons: possible implications for extreme events. J. Geophys. Res. 116, D18108. http://dx.doi.org/10.1029/2011JD015666.

Prasad, T.G., McClean, J.L., 2004. Mechanisms for anomalous warming in the western Indian Ocean during dipole mode events. J. Geophys. Res. Oceans 109. http://dx.doi.org/10.1029/2003jc001872.

Qu, T., Lindstrom, E., 2002. A climatological interpretation of the circulation in the western south Pacific. J. Phys. Oceanogr. 32, 2492–2508.

Qu, T., Gan, J., Ishida, A., et al., 2008. Semiannual variation in the western tropical Pacific Ocean. Geophys. Res. Lett. 35, L16602. http://dx.doi.org/10.1029/2008GL035058.

Rao, S.A., Behera, S.K., 2005. Subsurface influence on SST in the tropical Indian Ocean structure and interannual variabilities. Dyn. Atmos. Oceans 39, 103–135.

Rao, S.A., Yamagata, T., 2004. Abrupt termination of Indian Ocean dipole events in response to intra-seasonal oscillations. Geophys. Res. Lett. 31, L19306. http://dx.doi.org/10.1029/2004GL020842.

Rao, S.A., Behera, S.K., Masumoto, Y., Yamagata, T., 2002. Interannual variability in the subsurface tropical Indian Ocean with a special emphasis on the Indian Ocean Dipole. Deep Sea Res. Part II 49, 1549–1572.

Rao, S.A., Masson, S., Luo, J.-J., Behera, S.K., Yamagata, T., 2007. Termination of Indian Ocean dipole events in a general circulation model. J. Clim. 20, 3018–3035. http://dx.doi.org/10.1175/JCLI4164.1.

Ratnam, J.V., Behera, S.K., Masumoto, Y., Takahashi, K., Yamagata, T., 2012. Anomalous climatic conditions associated with the El Niño Modoki during boreal winter of 2009. Clim. Dyn. 39, 227–238. http://dx.doi.org/10.1007/s00382-011-1108-z.

Rayner, N.A., Parker, D.E., Horton, E.B., Folland, C.K., Alexander, L.V., Rowell, D.P., Kent, E.C., Kaplan, A., 2003. Global analyses of sea surface temperature, sea ice, and night marine air temperature since the late nineteenth century. J. Geophys. Res. 108 (D14), 4407. http://dx.doi.org/10.1029/2002JD002670.

Reppin, J., Schott, F.A., Fischer, J., Quadfasel, D., 1999. Equatorial currents and trans- ports in the upper central Indian Ocean: annual cycle and interannual variability. J. Geophys. Res. 104, 15495–15514.

Reverdin, G., 1987. The upper equatorial Indian Ocean: the climatological seasonal cycle. J. Phys. Oceanogr. 17, 903–927.

Reverdin, G., Fieux, M., 1987. Sections in the western Indian Ocean—variability in the temperature structure. Deep Sea Res. 34, 601–626. http://dx.doi.org/10.1016/0198-0149(87)90007-0.

Rhein, M., Stramma, L., 2005. Seasonal variability in the Deep Western Boundary Current around the eastern tip of Brazil. Deep Sea Res. 52A, 1414–1428.

Richardson, P.L., Fratantoni, D.M., 1999. Float trajectories in the deep-western boundary current and deep equatorial jets of the tropical Atlantic. Deep Sea Res. Part II 46, 305–333.

Richter, I., Behera, S.K., Masumoto, Y., Taguchi, B., Komori, N., Yamagata, T., 2010. On the triggering of Benguela Niños: remote equatorial versus local influences. Geophys. Res. Lett. 37, L20604. http://dx.doi.org/10.1029/2010GL044461.

Ridgway, K., Dunn, J., 2003. Mesoscale structure of the East Australian Current system and its relationship with topography. Prog. Oceanogr. 56, 189–192.

Rodrigues, R.R., Rothstein, L.M., Wimbush, M., 2007. Seasonal variability of the south equatorial current bifurcation in the Atlantic Ocean: a numerical study. J. Phys. Oceanogr. 37, 16–30.

Rouault, M., Illig, S., Bartholomae, C., Reason, C.J.C., Bentamy, A., 2007. Propagation and origin of warm anomalies in the Angola Benguela upwelling system in 2001. J. Mar. Syst. 68, 473–488. http://dx.doi.org/10.1016/j.jmarsys.2006.11.010.

Sahu, N., Behera, S., Yamashiki, Y., Takara, K., Yamagata, T., 2011. IOD and ENSO impacts on the extreme stream-flows of Citarum river in Indonesia. Clim. Dyn. 39, 1673–1680. http://dx.doi.org/10.1007/s00382-011-1158-2.

Saji, N.H., Yamagata, T., 2003. Possible impacts of Indian Ocean Dipole Mode events on global climate. Clim. Res. 25, 151–169.

Saji, N.H., Goswami, B.N., Vinayachandran, P.N., Yamagata, T., 1999. A dipole mode in the tropical Indian Ocean. Nature 401, 360–363.

Sasaki, Y.N., Schneider, N., Maximenko, N., Lebedev, K., 2010. Observational evidence for propagation of decadal spiciness anomalies in the North Pacific. Geophys. Res. Lett. 37, L07708. http://dx.doi.org/10.1029/2010GL042716.

Schiller, A., Godfrey, J.S., McIntosh, P.C., Meyers, G., Wijffels, S.E., 1998. Seasonal near-surface dynamics and thermodynamics of the Indian Ocean and Indonesian Throughflow in a global ocean general circulation model. J. Phys. Oceanogr. 28, 2288–2312.

Schmidt, M.W., Chang, P., Hertzberg, J.E., Them, T.R., Link, J., Otto-Bliesner, B.L., 2012. Impact of abrupt deglacial climate change on tropical Atlantic subsurface temperatures. Proc. Natl. Acad. Sci. U.S.A. 109, 14348–14352. http://dx.doi.org/10.1073/pnas.1207806109.

Schneider, N., 1998. The Indonesian Throughflow and the global climate system. J. Clim. 11, 676–689.

Schneider, N., Miller, A.J., Alexander, M.A., et al., 1999. Subduction of decadal North Pacific temperature anomalies: observations and dynamics. J. Phys. Oceanogr. 29, 10156–10170.

Schopf, P.S., Suarez, M.J., 1988. Vacillations in a coupled ocean-atmosphere model. J. Atmos. Sci. 45, 549–566.

Schott, F., McCreary, J., 2001. The monsoon circulation of the Indian Ocean. Prog. Oceanogr. 51, 1–123.

Schott, F., Fischer, J., Stramma, L., 1998. Transports and pathways of the upper-layer circulation in the western tropical Atlantic. J. Phys. Oceanogr. 28, 1904–1928.

Schott, F., Dengler, M., Schoenefeldt, R., 2002. The shallow thermohaline circulation of the Indian Ocean. Progr. Oceanogr. 53, 57–103.

Schott, F.A., Dengler, M., Brandt, P., Affler, K., Fischer, J., Bourlès, B., Gouriou, Y., Molinari, R.L., Rhein, M., 2003. The zonal currents and transports at 35°W in the tropical Atlantic. Geophys. Res. Lett. 30, 1349. http://dx.doi.org/10.1029/2002GL016849.

Schott, F.A., McCreary Jr., J.P., Johnson, G., 2004. Shallow overturning circulations of the tropical-subtropical oceans. In: Wang, C., Xie, S.-P., Carton, J.A. (Eds.), Earth Climate: The Ocean-Atmosphere Interaction. American Geophysical Union: Geophysical Monograph, Washington, DC, pp. 261–304.

Schott, F., Dengler, M., Zantopp, R.J., Stramma, L., Fischer, J., Brandt, P., 2005. The shallow and deep western boundary circulation of the South Atlantic at 5°-11°S. J. Phys. Oceanogr. 35, 2031–2053.

Shankar, D., McCreary, J.P., Han, W., Shetye, S.R., 1996. Dynamics of the East India Coastal Current: 1. Analytic solutions forced by interior Ekman pumping and local alongshore winds. J. Geophys. Res. 101, 13975–13991.

Shankar, D., Vinayachandran, P.N., Unnikrishnan, A.S., 2002. The monsoon currents in the north Indian Ocean. Prog. Oceanogr. 52, 63–120.

Shetye, S.R., Shenoi, S.S.C., 1988. The seasonal cycle of surface circulation in the coastal North Indian Ocean. Proc. Indian Acad. Sci. 97, 53–62.

Shetye, S.R., Gouveia, A.D., Shenoi, S.S.C., Sundar, D., Michael, G.S., Almeida, A.M., Santanam, K., 1990. Hydrography and circulation off the west coast of India during the southwest monsoon 1987. J. Mar. Res. 48, 359–378.

Shetye, S.R., Shenoi, S.S.C., Gouveia, A.D., Michael, G.S., Sundar, D., Nampoothiri, G., 1991a. Wind-driven coastal upwelling along the western boundary of the Bay of Bengal during the southwest monsoon. Cont. Shelf Res. 11, 1397–1408.

Shetye, S.R., Gouveia, A.D., Shenoi, S.S.C., Michael, G.S., Sundar, D., Almeida, A.M., Santanam, K., 1991b. The coastal current off western India during the northeast monsoon. Deep Sea Res. 38, 1517–1529.

Shetye, S.R., Gouveia, A.D., Shenoi, S.S.C., Sundar, D., Michael, G.S., Nampoothiri, G., 1993. The western boundary current of the seasonal subtropical gyre in the Bay of Bengal. J. Geophys. Res. 98, 945–954.

Shetye, S.R., Gouveia, A.D., Shankar, D., Shenoi, S.S.C., Vinayachandran, P.N., Sundar, D., Michael, G.S., Nampoothiri, G., 1996. Hydrography and circulation in the western Bay of Bengal during the northeast monsoon. J. Geophys. Res. 101, 14011–14025.

Siedler, G., Rouault, M., Lutjeharms, J.R.E., 2006. Structure and origin of the subtropical South Indian Ocean Countercurrent. Geophys. Res. Lett. 33 (24), L24609. http://dx.doi.org/10.1029/2006GL027399.

Singh, A., Delcroix, T., Cravatte, S., 2011. Contrasting the flavors of El Niño-Southern Oscillation using sea surface salinity observations. J. Geophys. Res. 116, C06016. http://dx.doi.org/10.1029/2010JC006862.

Smith, R.L., Huyer, A., Godfrey, J.S., Church, J.A., 1991. The Leeuwin Current off western Australia, 1986–1987. J. Phys. Oceanogr. 21, 322–345.

Sperber, K.R., Palmer, T.N., 1996. Interannual tropical rainfall variability in general circulation model simulations associated with the atmospheric model intercomparison project. J. Clim. 9, 2727–2750.

Sprintall, J., Wijffels, S.E., Molcard, R., Jaya, I., 2009. Direct estimates of the Indonesian Throughflow entering the Indian Ocean: 2004–2006. J. Geophys. Res. 114, C07001. http://dx.doi.org/10.1029/2008JC005257.

Talley, L., Sprintall, J., 2005. Deep expression of the Indonesian Throughflow: Indonesian intermediate water in the South Equatorial Current. J. Geophys. Res. 110, C10009. http://dx.doi.org/10.1029/2004JC002826.

Talley, L.D., Reid, J.L., Robbins, P.E., 2003. Data-based meridional overturning streamfunctions for the global ocean. J. Clim. 16, 3213–3226.

Tangang, F., Juneng, L., Salimun, E., Vinayachandran, P.N., Seng, Y.K., Reason, C.J.C., Behera, S.K., Yasunari, T., 2008. On the roles of the northeast cold surge, the Borneo vortex, the Madden-Julian Oscillation, and the Indian Ocean Dipole during the worst 2006/2007 flood in southern Peninsular Malaysia. Geophys. Res. Lett. 35, L14S07. http://dx.doi.org/10.1029/2008GL033429.

Taschetto, A.S., England, M.H., 2009. El Niinop Modoki impacts on Australian rainfall. J. Clim. 22, 3167–3174.

Terray, L., Corre, L., Cravatte, S., Delcroix, T., Reverdin, G., Ribes, A., 2012. Near-surface salinity as Nature's rain gauge to detect human influence on the water cycle. J. Clim. 25, 958–977.

Toniazzo, T., Woolnough, S., 2013. Development of warm SST errors in the southern tropical Atlantic in CMIP5 decadal hindcasts. Clim. Dyn. http://dx.doi.org/10.1007/s00382-013-1691-2, in press.

Tozuka, T., Kagimoto, T., Masumoto, Y., Yamagata, T., 2002. Simulated multiscale variations in the western tropical Pacific: the Mindanao Dome revisited. J. Phys. Oceanogr. 32 (5), 1338–1359. http://dx.doi.org/10.1175/ 1520-0485.

Tozuka, T., Luo, J.-J., Masson, S., Yamagata, T., 2007. Decadal Indian Ocean dipole simulated in an ocean-atmosphere coupled model. J. Clim. 20, 2881–2894.

Tozuka, T., Yokoi, T., Yamagata, T., 2010. A modeling study of interannual variations of the Seychelles Dome. J. Geophys. Res. 115, C04005. http://dx.doi.org/10.1029/2009JC005547.

Trenberth, K.E., Smith, L., 2006. The vertical structure of temperature in the tropics: different flavors of El Nino. J. Clim. 19, 4956–4973.

Trenberth, K.E., Stepaniak, D.P., 2001. Indices of El Niño Evolution. J. Clim. 14, 1697–1701.

Ummenhofer, C.C., England, M.H., McIntosh, P.C., Meyers, G.A., Pook, M.J., Risbey, J.S., Sen Gupta, A., Taschetto, A.S., 2008. What causes southeast Australia's worst droughts? Geophys. Res. Lett. 36, L04706. http://dx.doi.org/10.1029/2008GL036801.

Ummenhofer, C., McIntosh, P., Pook, M., Risbey, J., 2013. Impact of surface forcing on southern hemisphere atmospheric blocking in the Australia-New Zealand sector. J. Clim. http://dx.doi.org/10.1175/JCLI-D-12-00860.1, in press.

Vecchi, G.A., Wittenberg, A.T., 2010. El Niño and our future climate: where do we stand? WIREs Clim Change 1, 260–270. http://dx.doi.org/10.1002/wcc.33.

Vecchi, G.A., Soden, B., Wittenberg, A.T., Held, I.M., Leetmaa, A., Harrison, M.J., 2006. Weakening of tropical Pacific atmospheric circulation due to anthropogenic forcing. Nature 441, 73–76.

Vellinga, M., Wu, P., 2004. Low-latitude freshwater influences on centennial variability of the Atlantic thermohaline circulation. J. Clim. 17, 4498–4511.

Vialard, J., Foltz, G., McPhaden, M., Duvel, J.-P., de Boyer Montégut, C., 2008. Strong Indian Ocean sea surface temperature signals associated with the Madden-Julian Oscillation in late 2007 and early 2008. Geophys. Res. Lett. 35, L19608. http://dx.doi.org/10.1029/2008GL035238.

Vialard, J., Duvel, J.-P., McPhaden, M., BouruetAubertot, P., Ward, B., Key, E., Bourras, D., Weller, R., Minnett, P., Weill, A., Cassou, C., Eymard, L., Fristedt, T., Basdevant, C., Dandonneau, Y., Duteil, O., Izumo, T., De Boyer-Montégut, C., Masson, S., 2009a. Cirene: air sea interactions in the Seychelles-Chagos thermocline ridge region. Bull. Am. Meteorol. Soc. 90, 45–61.

Vialard, J., Shenoi, S.S.C., McCreary, J.P., Shankar, D., Durand, F., Fernando, V., Shetye, S.R., 2009b. Intraseasonal response of the northern Indian Ocean coastal waveguide to the Madden-Julian Oscillation. Geophys. Res. Lett. 36, L14606. http://dx.doi.org/10.1029/2009GL038450.

Vialard, J., Jayakumar, A., Gnanaseelan, C., Lengaigne, M., Sengupta, D., Goswami, B.N., 2012. Processes of 30-90 days sea surface temperature variability in the northern Indian Ocean during boreal summer. Clim. Dyn. 38, 1901–1916. http://dx.doi.org/10.1007/s00382-011-1015-3.

Vinayachandran, P.N., Shetye, S.R., Sengupta, D., Gadgil, S., 1996. Forcing mechanisms of the Bay of Bengal circulation. Curr. Sci. 71, 753–763.

Vinayachandran, P.N., Yamagata, T., 1998. Monsoon response of the Sea around Sri Lanka: generation of thermal domes and anticyclonic vortices. J. Phys. Oceanogr. 28, 1946–1960.

Vinayachandran, P.N., Saji, N.H., Yamagata, T., 1999. Response of the equatorial Indian Ocean to an anomalous wind event during 1994. Geophys. Res. Lett. 26, 1613–1616.

Vinayachandran, P.N., Iizuka, S., Yamagata, T., 2002. Indian Ocean Dipole mode events in an ocean general circulation model. Deep Sea Res. Part II 49, 1573–1596.

von Schuckmann, K., Brandt, P., Eden, C., 2008. Generation of tropical instability waves in the Atlantic Ocean. J. Geophys. Res. 113, C08,034.

Wade, M., Caniaux, G., duPenhoat, Y., 2011. Variability of the mixed layer heat budget in the eastern equatorial Atlantic during 2005-2007 as inferred using Argo floats. J. Geophys. Res. 116, C08006.

Wajsowicz, R.C., 2005. Potential predictability of tropical Indian Ocean SST anomalies. Geophys. Res. Lett. 32, L24702. http://dx.doi.org/10.1029/2005GL024169.

Waliser, D.E., Murtugudde, R., Lucas, L., 2003. Indo-Pacific Ocean response to atmospheric intraseasonal variability. Part I: Austral summer and the Madden-Julian Oscillation. J. Geophys. Res. 108, 3160. http://dx.doi.org/10.1029/2002JC001620.

Wang, G., Hendon, H.H., 2007. Sensitivity of Australian rainfall to Inter–El Niño variations. J. Clim. 20, 4211–4226.

Wang, Q., Hu, D., 2006. Bifurcation of the North Equatorial Current derived from altimetry in the Pacific Ocean. J. Hydrodyn. 18, 620–626.

Wang, B., Lee, J.-Y., et al., 2009. Advance and Prospect of Seasonal Prediction: assessment of the APCC/CliPAS 14-model ensemble retroperspective seasonal prediction (1980-2004). Clim. Dyn. 33, 93–117. http://dx.doi.org/10.1007/s00382-008-0460-0.

Webb, D.J., 2000. Evidence of shallow zonal jets in the South Equatorial Current region of the Southwest Pacific. J. Phys. Oceanogr. 30, 706–720.

Webster, P.J., Moore, A., Loschnigg, J., Leban, M., 1999. Coupled ocean-atmosphere dynamics in the Indian Ocean during 1997-98. Nature 40, 356–360.

Weng, H., Ashok, K., Behera, S.K., Rao, S.A., Yamagata, T., 2008. Impacts of recent El Niño Modoki on dry/wet conditions in the Pacific Rim during boreal summer. Clim. Dyn. 29, 113–129.

Weng, H., Behera, S.K., Yamagata, T., 2009a. Anomalous winter climate conditions in the Pacific rim during recent El Niño Modoki and El Niño events. Clim. Dyn. 32, 663–674.

Weng, H., Wu, G., Liu, Y., Behera, S.K., Yamagata, T., 2009b. Anomalous summer climate in China influenced by the tropical Indo-Pacific Oceans. Clim. Dyn. 36, 769–782. http://dx.doi.org/10.1007/s00382-009-0658-9.

Wijffels, S., Firing, E., Toole, J., 1995. The mean structure and variability of the Mindanao Current at 8°N. J. Geophys. Res. 100, 18421–18435.

Wijffels, S., Meyers, G., 2004. An intersection of oceanic waveguides: variability in the Indonesian throughflow region. J. Phys. Oceanogr. 34, 1232–1253.

Woodberry, K.E., Luther, M.E., O'Brien, J.J., 1989. The Wind-Driven seasonal circulation in the southern tropical Indian Ocean. J. Geophys. Res. 94 (C12), 17985–18002.

Wyrtki, K., 1973. An equatorial jet in the Indian Ocean. Science 181, 262–264.

Xie, S.-P., Annamalai, H., Schott, F., McCreary, J.P., 2002. Structure and Mechanisms of South Indian Ocean climate variability. J. Clim. 15, 864–878.

Xie, S.-P., Hu, K., Hafner, J., Tokinaga, H., Du, Y., Huang, G., Sampe, T., 2009. Indian Ocean capacitor effect on Indo-western Pacific climate during the summer following El Nino. J. Clim. 22 (3), 730–747. http://dx.doi.org/10.1175/2008JCLI2544.1.

Xu, Z., Li, M., Patricola, C.M., Chang, P., 2013. Oceanic origin of southeast tropical Atlantic biases. Clim. Dyn. Online publication date: 20-Aug-2013.

Yamagata, T., 1985. Stability of a simple air-sea coupled model in the tropics. Coupled ocean-atmosphere models. In: Nihoul, J.C.J. (Ed.), Elsevier Oceanogr. Series, pp. 637–657.

Yamagata, T., Behera, S.K., Rao, S.A., Guan, Z., Ashok, K., Saji, H.N., 2003a. Comments on "Dipoles, Temperature Gradient, and Tropical Climate Anomalies". Bull. Am. Meteorol. Soc. 84, 1418–1422.

Yamagata, T., Behera, S.K., Guan, Z., 2003b. The role of the Indian Ocean in climate forecasting with a particular emphasis on summer conditions in East Asia. In: ECMWF Workshop proceedings, 102–114. pp.

Yamagata, T., Behera, S.K., Luo, J.-J., Masson, S., Jury, M.R., Rao, S.A., 2004. The coupled ocean-atmosphere variability in the tropical Indian Ocean. Earth's climate: The ocean-atmosphere interaction. Geophys. Monogr. 147, 189–211, AGU.

Yeh, S.-W., Kug, J., Dewitte, B., Kwon, M.-H., Kirtman, Ben P., Jin, F.F., 2009. El Niño in a changing climate. Nature 461, 511–514. http://dx.doi.org/10.1038/nature08316.

Yokoi, T., Tozuka, T., Yamagata, T., 2008. Seasonal variation of the Seychelles Dome. J. Clim. 21, 3740–3754.

Yokoi, T., Tozuka, T., Yamagata, T., 2009. Seasonal variations of the Seychelles Dome simulated in the CMIP3 models. J. Clim. 39, 449–457.

Yoshida, K., 1959. A theory of the Cromwell Current (the equatorial under current) and of equatorial upwelling. J. Oceanogr. Soc. Jpn. 15, 159–170.

Yu, L., O'Brien, J.J., Yang, J., 1991. On the remote forcing of the circulation in the Bay of Bengal. J. Geophys. Res. 96, 20440–20454.

Yu, J.-Y., Mechoso, C., McWilliams, J., Arakawa, A., 2002. Impacts of the Indian Ocean on the ENSO cycle. Geophys. Res. Lett. 29 (8). http://dx.doi.org/10.1029/2001GL014098.

Zebiak, S.E., 1993. Air–sea interaction in the equatorial Atlantic region. J. Clim. 6, 1567–1586.

Zhang, C., 2005. Madden-Julian oscillation. Rev. Geophys. 43, 1–36, RG2003/2005.

Zhang, R., Delworth, T.L., 2005. Simulated tropical response to a substantial weakening of the Atlantic thermohaline circulation. J. Clim. 18, 1853–1860.

Zhang, D., McPhaden, M.J., 2006. Decadl variability of the shallow Pacific meridional overturning circulation: relation to tropical sea surface temperatures in observations and climate changemodels. Ocean Model. 15, 250–273.

Zhang, D., Msadek, R., McPhaden, M.J., Delworth, T., 2011. Multidecadal variability of the North Brazil Current and its connection to the Atlantic meridional overturning circulation. J. Geophys. Res. 116, C04012. http://dx.doi.org/10.1029/2010JC006812.

Zubair, L., Rao, S.A., Yamagata, T., 2003. Modulation of Sri Lankan Maha rainfall by the Indian Ocean Dipole. Geophys. Res. Lett. 30, 1063. http://dx.doi.org/10.1029/2002GL015639, 2.

Chapter 16

The Marine Cryosphere

David M. Holland
Courant Institute of Mathematical Sciences, New York University, New York, USA

Chapter Outline

1. INTRODUCTION

1.1. Marine Cryosphere

The cryosphere is loosely defined as the component of the earth's climate system consisting of frozen water. In one way or another, all components of the cryosphere interact with and influence the ocean. The cryosphere can be considered to consist of sea ice, land ice, and atmospheric ice. While atmospheric ice, essentially consisting of frozen water clouds, has an indirect impact on the ocean through precipitation, including snowfall onto sea ice and land ice, and atmospheric radiation influences, we discuss it no further here as it is more of an atmospheric science phenomenon. We focus on ice that is more or less in direct contact with the ocean (see Figure 16.1). Land ice, often referred to as a glacier, is in direct contact with the ocean wherever it flows, under the action of gravity, into the ocean to form floating ice shelves along the periphery of ice sheets and ice caps. A particular type of land ice, marine permafrost, is also found beneath the seafloor, where it holds a potential for interaction with the ocean. Sea ice, by its very nature, is formed at the surface of the ocean and is perhaps the most well understood and studied aspect of ice–ocean interaction. Depending upon the time of year, up to 5% of the global ocean surface is covered by ice (NSIDC, 2007). Collectively, we refer to the sea ice, glacier, and permafrost elements as the marine cryosphere.

The interactions of sea ice, land ice, and marine permafrost with the oceans are largely confined to the polar oceans, with the only exception being some marine terminating glaciers found south of the Arctic Circle in, for instance, Alaska (Meier et al., 1980), and north of the Antarctic Circle in Chile (Warren, 1993). The bathymetric layouts of the far northern and southern oceanographic basins are essentially morphological opposites. Looking to the north, the Arctic basin is largely landlocked, the only exchanges with the rest of the global ocean occurring with the North Atlantic Ocean through the Fram Strait and Barents Sea passages, as well as the modest connectivity

Ocean Circulation and Climate, Vol. 103. http://dx.doi.org/10.1016/B978-0-12-391851-2.00016-7

FIGURE 16.1 Photographs of typical (a) sea ice and (b) iceberg and ice-shelf areas of the polar oceans. Sea ice is marine in origin originating from freezing of sea water. Icebergs are meteoric ice, born from ice shelves, originating with evaporation of sea water that falls as snow onto the great ice sheets. *http://nsidc.org/cryosphere/quickfacts/seaice.html; http://summitvoice.files.wordpress.com/2011/07/ice.jpg*.

to the Pacific Ocean through Bering Strait, and some exchange through the Canadian Arctic Archipelago. This geographic restriction makes it somewhat easier to observe the water mass exchanges in the north as compared with the south (see Chapter 17). The bathymetry of the Southern Ocean is completely open to the global ocean, with no restriction to exchange with the neighboring South Atlantic, South Pacific, and Indian Oceans (see Chapter 18).

The difference in land configuration in the north and south is a key factor in the distinction between the sea-ice and land-ice cover of the two regions, discussed at length in Sections 2 and 3, respectively. The Arctic Ocean, consisting principally of the area north of 80°N, is surrounded by land that is largely nonglaciated and has river runoff at the surface in the boreal summer that remains largely trapped in the Arctic basin leading to a relatively stratified ocean and thick sea-ice cover. The Southern Ocean, essentially in an unconfined basin and with almost no surface runoff of freshwater, is more weakly stratified and has a much thinner sea-ice cover compared with the Arctic. In other words, there is not really an "Antarctic Ocean" analog to the Arctic Ocean; instead, there exists the Southern Ocean, much of which is subpolar and cannot be compared in any real sense to the truly polar Arctic Ocean.

There are a number of distinctions to be drawn between the coastal and the continental shelf zones of the two polar regions. The Arctic continental shelves cover approximately 10% of the area of the Arctic Ocean and have an average depth of little more than 100 m (Jakobsson, 2002). The situation in the south is fundamentally different with the continental shelf around Antarctica occupying a much smaller percentage area of the Southern Ocean and the average depth of the continental shelf being far greater, on average 600 m (Anderson et al., 1980). The global mean continental shelf depth is estimated to be 150 m (Bott, 1971), and thus, the Antarctic shelf is often referred to as being overdeepened. The deepness is a result of cumulative, erosive activities of a grounded ice sheet (SCAR, 1997). This distinction in average depth is of fundamental importance to the manner in which warm subpolar waters can or cannot go aboard the continental shelves and reach the periphery of the major ice sheets. Another distinct feature of the Antarctic continental shelf is that it is foredeepened, that is, the shelf depth becomes deeper as one travels from the shelf break in toward the margin of the ice sheet (Pope and Anderson, 1992).

The modern marine cryosphere of the north consists of perennial Arctic sea-ice cover, the Greenland Ice Sheet, and marine permafrost; in the south, there is the seasonal sea-ice cover and the more massive Antarctic Ice Sheet. On longer timescales, back through the Pleistocene, the northern cryosphere has undergone larger areal changes as ice sheets occupied the Canadian (Broecker et al., 1989) and Scandinavian (Rinterknecht et al., 2006) land masses more often than not during that epoch.

The character of the seasonal and perennial sea-ice cover of the polar seas has been well noted by voyagers dating back over the past millennium, certainly as early as the time of the Vikings in the north (Ogilvie et al., 2000). In more recent times, over the past century or so, explorers and oceanographers have set sail deep into both the Arctic Ocean (Nansen, 1902) and the Southern Ocean to the coast of the Antarctic continent (Ross, 1847). The mere existence of sea ice, aside from its physical implications for ocean circulation in the high latitudes, has in itself greatly hampered the exploration and understanding of polar ocean dynamics. The difficulty of cutting through a meter or more of frozen ocean surface and the associated risks of being trapped in the "pack" for a year or more have brought many an expedition to a tragic end (e.g., the Franklin expedition; Berton, 1988) or at least a difficult ending (e.g., Amundsen's travel through the Northwest Passage; Berton, 1988). In the 1890s, Nansen deliberately set his ship into the Arctic ice pack on the Siberian coast, and riding with the transpolar drift, he crossed close to the North Pole, ultimately to emerge into the Fram Strait area, thereby demonstrating the basic circulation pattern of Arctic sea ice (Nansen, 1902). In a somewhat analogous, albeit not so planned adventure, Ernest Shackleton was set into the pack of the Weddell Sea, and through the trajectory of his ship and travels, the circulation pattern of the ice of the Weddell Sea was better established (Shackleton, 1919).

While sea ice is an evident participant in the marine cryosphere, perhaps not so obvious is the role of land ice. Both the continent of Antarctica (Drewry et al., 1983) and the island of Greenland (Bamber et al., 2001) are almost everywhere covered by an ice sheet several kilometers in thickness. In some locations, these ice sheets actually sit upon bedrock that is below modern-day sea level. Such areas, referred to as marine ice sheets (Mercer, 1978), have direct interaction with the ocean. In some instances, the marine ice sheets terminate at the ocean in vast extensive floating ice shelves, such as in the Ross (Shabtaie and Bentley, 1987) and Weddell seas (Swithinbank et al., 1988), while in other cases, the termination can be rather more abrupt with the glacier only presenting a rather modest-sized interface to the ocean, as in the case of many outlet glaciers in Greenland (Csathó et al., 1999). Large portions of the West Antarctic Ice Sheet were discovered to be well below sea level during expeditions undertaken in the International Geophysical Year (Bentley and Ostenso, 1961). Such marine-based glaciated areas can be thought of as being an extension of the global ocean. There is some evidence from seafloor sediment records that the ocean has occupied these areas from time to time during past interglacial periods (Scherer et al., 1998). It is the waxing and waning of the marine ice sheets that holds the greatest present interest for potential rapid global sea-level change in the current century and beyond (Alley et al., 2005).

1.2. Ice Physics

To better understand the manner in which ice interacts with the ocean, we first review a few of the fundamental physical properties of ice and freezing ocean water. The water molecule, H_2O, has more solid phases than any other known substance. The most common form of solid ice takes on a hexagonal pattern, termed hexagonal ice, or ice Ih, or simply ice. Such ice, created from water at temperatures below 0 °C, settles into a regular crystalline structure with hexagonal symmetry (Hobbs, 1971). At even lower temperatures, ice crystallizes into structures with cubic symmetry, but we are only concerned in an oceanographic context with the hexagonal form. One exception is clathrates, discussed in Section 4 in the context of marine permafrost.

A phase diagram for water, showing the occurrence of the liquid, solid, and gas phases as functions of ambient temperature and pressure, reveals that the melting line has a negative slope (Figure 16.2). This means that as the pressure on ice is increased it tends to move from the solid toward the liquid phase. This is a consequence of ice expanding as it solidifies, giving rise to the remarkable fact that the water solid phase (Ih) is less dense than its liquid phase, so that ice shelves, icebergs, and sea ice all float upon the ocean, rather than resting on the seafloor, as would be the case if the ocean were composed of almost any other liquid substance.

The mechanical strength of ice varies depending upon the scale under consideration. At the smallest scale, that is of an individual floe of sea ice or a small iceberg, ice is a rigid crystalline solid, gaining its strength from hydrogen bonding, and it fractures in a brittle fashion (Petrovic, 2003). By contrast, on the large scale, sea ice (Mellor, 1983) and ice sheets (Weertman, 1983) follow distinctly different flow regimes than at the small scale. We will further investigate these large-scale dynamical regimes in Sections 4.2 and 4.3 as they are of great consequence to understanding the impact of ice upon the ocean. Deeper insight into the small- versus larger-scale behavior of ice comes from recognizing that, at the level of the crystal structure, individual molecules are linked together in a single crystal, but at the larger scale, ice is a polycrystalline material and its strength is dependent on that structure (Hooke, 1981). Ice has both an elastic behavior, arising from small displacements from their equilibrium position of atoms in a crystal, and a plastic behavior which emerges from the motion of dislocations that are the defects in the otherwise perfect crystal structure. An in-depth treatment of the molecular and aggregate strength properties of ice is found in the textbook by Petrenko and Whitworth (1998).

The thermodynamical properties of ice govern its growth and decay. One has to consider the density, specific heat, latent heat, and related thermodynamical properties of ice to model its evolution in a given thermodynamic regime. For sea ice, which is marine in origin, the thermodynamics can be complicated by the presence of significant amounts of brine locked into the ice (Cox and Weeks, 1982), something not found in glacier ice which is generally meteoric in origin. Whereas freshwater reaches maximum density at a temperature of approximately 4 °C, the impact of salinity in typical seawater is to move the temperature of maximum density to the freezing point (Fujino et al., 1974). A thorough review of ice thermodynamics is provided in the textbook by Lock (1990), which delves into the thermodynamical Gibbs function, which allows one to develop the phase diagram for ice, and thus to understand the manner in which water switches between its phases of liquid, gas, and solid (see also Chapter 6).

The freezing point of water is dependent upon both pressure and salinity (Fujino et al., 1974). The details of these dependencies have a fundamental impact on ice formation and decay in the ocean. The near-ubiquitous

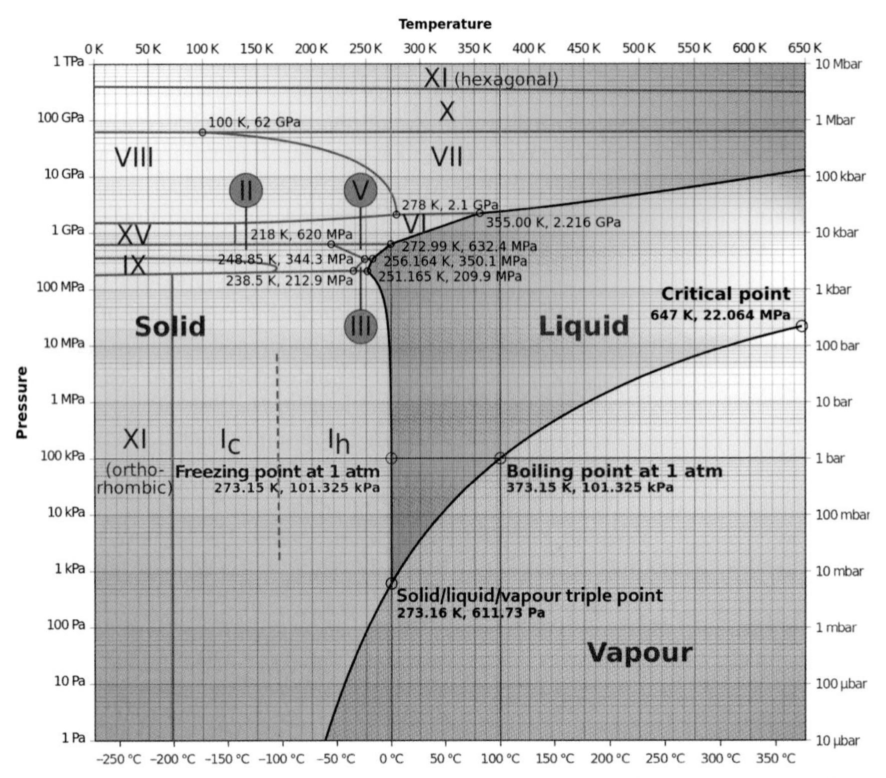

FIGURE 16.2 Phase diagram for water shown logarithmic in pressure and linear in temperature. The roman numerals indicate various ice phases. The most common form of ice, Ih, and that most relevant to understanding ice–ocean interaction are located near the center of the diagram. *http://en.wikipedia.org/wiki/Phase_diagram.*

presence of salt in the ocean lowers the freezing point below the nominal 0 °C mark of pure water, reaching down to −2 °C for typical ocean salinity. The deeper a piece of ice is found in the ocean, the greater the pressure and the lower the freezing point yet again. Combined, the effects of salinity and pressure can reduce the freezing point for some deep glaciers to below −3 °C. The effect of the depth dependence on the freezing temperature can result in more rapid melting for the deeper parts of a glacier than for its shallower parts.

Salinity plays a crucial role in the melting of ice for yet another reason. The molecular diffusivity of salt is approximately one hundred times smaller than that of heat, and as a result, the transfer of heat at the interface where ice and water meet is much more efficient than that of salt. This is because turbulence is suppressed in the upper few millimeters of the boundary layer so that exchanges between ice and ocean occur only by molecular diffusion (Josberger and Martin, 1981). This diffusivity contrast has the effect of making the interface less salty than the ambient fluid during melting, leading to rising of the freezing point, and ultimately slowing of the rate of ice melt compared to the rate that would occur if such an effect were absent (Holland and Jenkins, 1999). In the case of freezing, the situation is further complicated by the presence of convection at the interface due to the unstable fluid stratification that occurs when ocean water freezes and salt is rejected, by and large, from the newly forming ice.

1.3. Ocean Impacts

The presence of an ice cover, whether sea ice or ice shelf, has a profound effect on the underlying upper ocean. Perhaps most notable is the transformation of a low-albedo surface (typical ocean value of 0.10) to a high-albedo surface (typical snow cover surface value of 0.90). The abrupt decrease in transmission and increase in reflection of shortwave radiation as an open-ocean area is converted to a sea-ice-covered area completely alter the surface energy balance (Allison et al., 1993). This gives rise to the so-called ice-albedo effect whereby an increase in sea-ice cover leads to less surface absorption of heat energy, which in turn promotes further growth of sea ice and further decrease in surface energy absorption (Robock, 1980). In a paleoclimate context, such an effect has been widely implicated in the "snowball earth" phenomenon (Hoffman and Schrag, 2002) whereby almost the entire global ocean surface is covered by ice. The white surfaced planet earth absorbs little surface radiation in such a scenario, thereby locking the planet into an icy state. The converse, a planet with no ice cover absorbs an enormous amount of radiation, allowing it to stay in an ice-free state. Likely, a bifurcation point exists in the earth's climate system allowing the earth to toggle between ice-covered and near-ice-free states, like the present (North et al., 1981).

Aside from its impact as a radiative barrier, the sea-ice cover is a highly effective barrier to gas exchange between ice and ocean. A key greenhouse gas, carbon dioxide, is continuously exchanged between the ocean and atmosphere, the rate of exchange being controlled by the partial pressure of gas in the atmosphere and the wind speed, and the temperature and salinity of the ocean, which controls the solubility of the gas in seawater (Wanninkhof, 1992) (see also Chapter 30). It is known from observational data that this exchange is highly seasonal dependent, demonstrating the key moderating role played of the ice cover (Golubev et al., 2006). In ice-shelf-covered portions of the ocean, there is no atmospheric-ocean exchange of gas, but there is input of gas into the ocean where the base of an ice shelf melts, as gases dissolved in the ice are released into the ocean, providing a glacial source of dissolved oxygen (Rodehacke et al., 2007).

Throughout the global ocean, the cycle of surface evaporation and precipitation, driven by the atmospheric state and ocean surface conditions, ice-covered or otherwise, helps set the salinity of the ocean surface. Salinity is the determining factor on ocean density in cold waters in part because the thermal expansion coefficient of water decreases with decreasing temperature, hence the importance of surface freshwater fluxes at high latitudes in controlling the downwelling branches of the global thermohaline circulation (THC), also known as the global conveyor belt (Manabe and Stouffer, 1995). The formation of sea ice at high latitudes results in the injection of salt into the surface ocean, often resulting in unstable stratification leading to deep or bottom convection (Aagaard and Carmack, 1989), enhancing the conveyor belt. On the flip side, the melting of sea ice along the sea-ice margin can lead to an input of freshwater that can suppress or even shut down convection, thereby slowing down the conveyor-belt transport of heat from low latitudes to high latitudes.

Particular to the Arctic is the input of substantial river runoff mentioned earlier ($\sim$3500 km^3/year, Lammers et al., 2001); in fact, some 10% of the global river runoff goes into the Arctic basin (Peterson et al., 2002). The arrangement of basin layout and river input in the Arctic leads to the strongest halocline to be found anywhere in the global ocean (Nguyen et al., 2012).

The strong halocline effectively insulates the Arctic sea-ice cover from the vast, latent heat supply of the Atlantic layer below and thereby allows a substantial thickness, at least historically, of sea-ice cover to persist in the Arctic Ocean (Steele and Boyd, 1998). By contrast, the Southern Ocean has almost no surface freshwater runoff from the Antarctic continent, and the water column is at best marginally stable and thus more easily allows for the upwelling of deep, warm water (Foldvik and Gammelsrød, 1988) that can contribute to a thinner sea-ice cover. The melting at the base of the ice shelves and calving of icebergs cannot really be thought of as a kind of "runoff," as the melting related to those processes ($\sim$2400 km^3/year; Rignot et al., 2013) occurs largely at depth within the ocean. The resulting, diluted meltwater contributions are evidently not large enough in quantity to enforce substantive stratification in the surrounding ocean. Massive, persistent convective events, such as the Weddell Polynya event of the mid 1970s, are indicative of this borderline stability in the south (Holland, 2001).

Throughout most of the global ocean, the input of turbulent kinetic energy into the surface ocean leads to a downward mixing of surface properties over a depth scale referred to as the mixed-layer depth (de Boyer Montegut et al., 2004). This depth represents a balance between buoyancy input at the surface ocean, which tends to thin the mixed layer, and mechanical stirring, which tends to deepen the layer. Over a sea-ice-covered ocean, as compared to the open ocean, this balance is altered both in terms of buoyancy and mechanical input details, but the same fundamental principles apply (Lemke et al., 1990).

The mixed layer beneath an ice shelf is another distinct environment for mixed-layer physics (Holland and Jenkins, 2001). Whereas sea ice does move with respect to the ocean surface, an ice-shelf base does not, and consequently, there is much less mechanical input into the mixed layer at the base of an ice shelf. That notwithstanding, the base of an ice shelf tends to be inclined somewhat and thus an effective gravity current can flow upward along the basal slope, leading to a unique source of mechanical mixing, different from that beneath sea ice, which tends to be relatively flat on such scales. A one-dimensional numerical model of a plume flowing up the base of an ice shelf has been provided by Jenkins (1991).

The final impact of ice on the ocean that we mention is that of global sea-level change. Ice floating upon the ocean, such as sea ice or ice shelves, makes only a minor contribution to sea-level change as the ice upon melting freshens the relatively salty ocean, thereby raising the sea level through modest mass contributions (Jenkins and Holland, 2007). More impressive, land ice grounded on bedrock below current-day sea level has a volume above flotation (i.e., the volume of grounded ice above and beyond that which is simply displacing ocean water) that can contribute substantially to global sea-level change. Marine-based West Antarctica has been estimated to hold more than 3 m of potential global sea-level change (Bamber et al., 2009).

1.4. Relation to Other Chapters

This chapter, while dealing primarily with the physics and phenomenology of sea ice and land ice, implicitly draws in material that is discussed elsewhere in this book, such as mixed-layer dynamics in relation to sea-ice cover and bottom water formation in relation to sea-ice growth and land-ice basal melting. The state of sea ice and land ice in the polar regions is greatly influenced by oceanic heat transport from the subpolar oceans into the polar (see Chapter 11). The freshwater produced by melting sea ice and land ice may exert influence on the overturning circulation in both the Nordic Seas (see Chapter 17) and the southern boundary of the Southern Ocean (see Chapter 18), particularly in the deep-water formation regions of the Ross and Weddell Seas.

The outline of the remainder of this chapter is as follows. Sea ice is discussed in Section 2, followed by land ice in Section 3, and a brief mention of marine permafrost in Section 4. Recent advances in observations and modeling capabilities both for sea ice and land ice, as well as of the waters of the polar oceans, are treated in Section 5, which covers emerging capabilities. Finally, a review of documented, recent changes in the marine cryosphere and an outlook for the twenty-first century are given in Section 6.

2. SEA ICE

The surfaces of the polar oceans have a net energy loss through the course of an annual cycle. An important consequence is that the surface of the ocean freezes over, forming sea ice, and as solid ice is less dense than liquid water, the ice floats upon the ocean surface, turning it from dark blue to bright white. With the swing of the seasons alternating between the northern and the southern hemisphere, the sea-ice cover of the Arctic and Antarctic grow and retreat out of phase with one another, producing one of the largest seasonal phenomena on the earth's surface (see Figure 16.3). The Arctic sea-ice cover varies from a wintertime maximum of $15 \times 10^6\ \mathrm{km}^2$ to a summertime minimum of 7, whereas the Antarctic swings between 20 and 4 (Gloersen et al., 1992).

The detailed process by which sea ice forms depends on the state of the ocean's surface, as a more turbulent ocean will lead to a somewhat different ice cover than a calm ocean. In either case, atmospheric cooling of the ocean results in its surface being slightly supercooled in the first instance, which in turn initiates nucleation of ice crystals and frazil ice. The frazil ice platelets coalesce, leading to a type of ice known as nilas in relatively calm seas and to pancake ice in more turbulent seas, where distinct floes are discernible (ASPECT, 2012). Floes can be pushed to ride up upon one another, thus thickening the cover. Once a sufficient cover exists, additional growth can occur at the underside of the ice, whereby loss of heat by conduction through the cover leads to formation of congelation ice directly to the underside.

Growth of ice can also occur at the ice surface. This happens when snow cover accumulates on the ice surface to the extent that the weight of the snow is sufficient to depress the freeboard of the ice below sea level, causing flooding of the ice surface and conversion of wet snow to ice, the resulting ice being referred to as snow ice (Jeffries et al., 1997).

Ice that forms from the sea, that is, ice of marine origin, has a significant salt content, at least initially. The salinity of the ocean is approximately 35 psu, and when frazil ice forms, this salt is rejected. However, the rapid formation of sea ice leads to some salt being trapped into pockets within the larger sea-ice conglomerate, giving the ice a bulk salinity of approximately 3–4 psu (Niedrauer and Martin, 1979). These distributed pockets become very concentrated in salt and do not freeze. Over time, however, they tend to drain down through the ice under the action of gravity. Melt water washing through the ice during summertime as the result of surface melt pond formation is another factor that gradually reduces the salt content of sea ice (Fetterer and Untersteiner, 1998). Glacier ice, by contrast, is generally of meteoric origin and has very limited salt content, which is largely aeolian in origin.

A fully developed ice cover is usually relatively mobile, blown around by the winds and dragged by the ocean currents. Most of the Arctic and Antarctica sea-ice cover is mobile, but there are areas near the coast where the ice is stagnant and referred to as landfast ice (Mahoney et al., 2007). The main characteristics of an ice cover are its thickness and concentration. The thickness of Arctic sea ice ranges from thin, that is, 0.1 m, to thick, that is, up to 6 m, whereas Antarctic sea ice tends to be relatively thin overall, generally spanning the range 0.1–1 m. In some locations, the sea ice can pile up substantially upon itself, forming ridges that can reach up to 20 m in thickness. The term "concentration" is used to describe the fractional areal coverage of the ocean surface by sea ice and ranges between 0% and 100%. While in the case of landfast ice the coverage is full, 100%, the majority of the ice cover has leads, or fracture zones, in which the ocean surface is directly exposed to the polar atmosphere. Even though the leads occupy only a few percent of the sea-ice region, it is because of the great efficiency with which they can exchange heat and moisture with the atmosphere that they are an important aspect of the ice cover.

A major distinction between sea ice of the north and that of the south is the age of the ice. Arctic ice can have a significant multiyear fraction, meaning that the ice has survived several seasons—largely the result of the enclosed nature of the Arctic basin. By contrast, Antarctic ice is almost exclusively first-year ice as each summer season the ice cover is largely melted away. A similarity of Arctic and Antarctic ice is that they both have marginal ice zones, that is, transition zones between the interior, densely concentrated pack ice and the ice-free open ocean. The marginal ice zone migrates outward during winter and

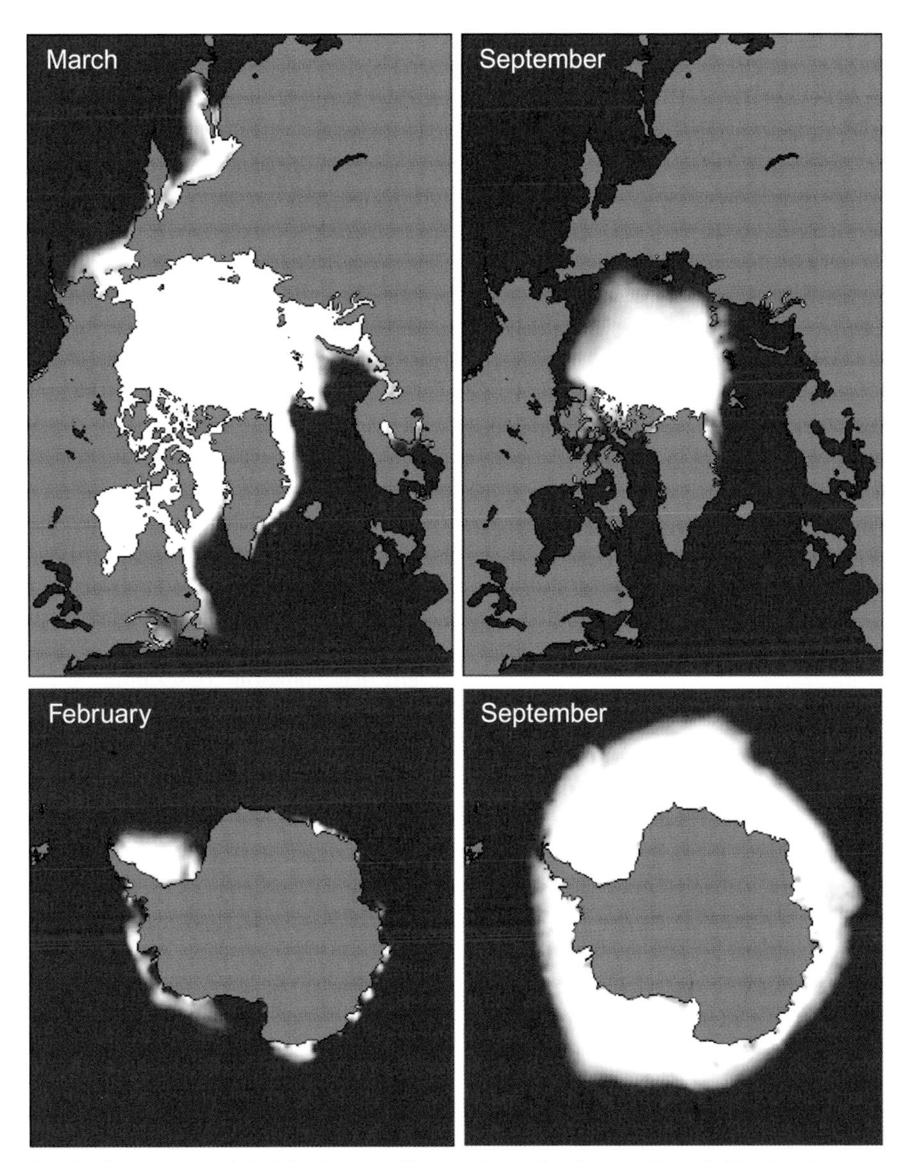

FIGURE 16.3 Seasonal cycle of sea-ice cover in both hemispheres. Top panels: Arctic winter maximum in March and summer minimum in September. Bottom panels: Antarctic summer minimum in March and winter maximum in September. Blue areas are open ocean, gray areas are continental land mass, and white is sea ice. *http://nsidc.org/cryosphere/sotc/sea_ice.html.*

retreats inward during summer, giving rise to an area of coverage known as the seasonal ice zone.

2.1. Observations

The traditional observational strategy for sea ice has been to make observations from ships traveling near or through the pack. Much of this data is dependent upon the subjective evaluations of an onboard ice observer, leading to some concerns over data quality. An effort has been made to standardize the observations, leading to a more quantitative global data set (ASPECT, 2012). From a ship, estimates of sea-ice concentration, often quoted in "egg-chart" fashion as fractions of eight, and sea-ice thickness are feasible. While a single ship-based cruise provides a relatively small areal data sample, taken cumulatively over many cruises, the global data set is of value to sea-ice researchers.

Complementing the ship-based data set is a submarine observational data set, based primarily on upward looking sonar that can give an estimate of ice areal coverage and draft (Moritz and Wensnahan, 2011). The draft can be converted to ice thickness, subject to assumptions about the ice density and snow cover. The data can be used to evaluate decadal and multidecadal trends in sea-ice thickness, spanning back to 1958 (Kwok and Rothrock, 2009). Presently, this type of data only exists for the Arctic region.

Another approach used to collect sea-ice data along a track has been the development and deployment of buoys placed on the surface of the ice, both in the Arctic (IABP, 2012) and the Antarctic (IPAB, 2012). The original version of these buoys collected surface air temperature and sea-level pressure data, and reported their position, all highly valuable as input to weather forecasting models and subsequent atmospheric reanalysis products. The buoys provided a very clear view of the circulation pattern of Arctic and Antarctic sea ice and the relation of that circulation to the sea-level pressure fields, and hence wind patterns. More recent development in sea-ice buoys now has them placed on the ice surface but with instrumentation extending all the way through the ice into the top few meters of the ocean. The buoys now have the ability to measure the temperature through the sea ice as well as the ice thickness and the entire mass balance of an ice floe. Such buoys have been developed that can not only be deployed in ice but also float on the ocean in the instance that the ice floe melts entirely (SIMB, 2012).

The largest leap forward in sea-ice observations occurred in the 1970s with the deployment of microwave satellite remote sensing instrumentation (Gloersen et al., 1992). The data provided the first global view of the entire Arctic and Antarctic sea-ice fields, their seasonal variability, and their interannual trends. Data sets were limited initially to primarily sea-ice concentration and used a variety of techniques to convert brightness radiance to concentration, producing broadly similar data products. More recently, developments using particle tracking, or image correlation techniques, with data sets from visible or active microwave sensors have allowed sea-ice velocity fields to be assessed, giving an impressive view of the large gyres that exist with the ice packs (Vesecky et al., 1988). Most recently, laser altimetry has provided the first glimpses of sea-ice thickness, based on measurements of the freeboard of ice (Hvidegaard and Forsberg, 2002).

2.2. Modeling

To understand the physics of large-scale sea-ice behavior, a model is needed, theoretical in nature and numerical in implementation. From a pragmatic view point, a numerical model once developed can be used to make projections of sea-ice behavior. Sea-ice physics, on the large scale, is complex, even more so than the ocean in at least one respect—while the continuum hypothesis upon which ocean modeling is based has some reasonable foundation, it is not the case for sea ice. Sea ice is composed of individual floes at scales ranging from 1 km down to 1 m, making the description of the sea-ice fluid exceptionally challenging. That notwithstanding, parameterizations of sea ice allow one to describe in a model the aggregate behavior of the underresolved smaller-scale features. Based on comparisons of model simulations with observational data, it can be claimed that there is indeed some accuracy and skill in the current generation of large-scale sea-ice models (Fichefet et al., 2003). The early-generation sea-ice models were "stand-alone" in the sense that the atmospheric and oceanic boundary conditions were fed to the sea-ice model equations as fixed boundary conditions (Flato and Hibler, 1992) as opposed to the present generation where the boundary conditions are dynamical, coming from atmosphere and ocean models to which the sea-ice model is coupled. Subsequent development of coupling strategies has now led to the point where sea-ice models are fully coupled with ocean and atmospheric general circulation models used in climate-change simulation scenarios (IPCC, 2007). We next discuss the major conceptual components of a sea-ice model: ice thermodynamics (growth/decay) and dynamics (motion).

Sea-ice thermodynamical equations describe the flow of heat through the sea-ice medium (Maykut and Untersteiner, 1971). As the aspect ratio of ice is such that ice is thin, the flow direction of heat is largely vertical. Brine pockets aside, the flow of heat is governed by a one-dimensional (vertical) equation describing diffusion of heat by molecular processes, aided by the fact that the thermal conductivity of sea ice is a well-established quantity (Ebert and Curry, 1993). The addition of a snow cover requires only a step change in the thermal conductivity to a smaller value (as snow is more insulating than ice) (Grenfell, 1991). The main boundary conditions are that of the flux of heat applied at the base of the sea ice and at its upper surface. Here we describe the surface fluxes and defer the basal fluxes to the mixed-layer discussion below.

At the surface of the ice (or snow), there are two fluxes to consider, the turbulent and the radiative. The turbulent fluxes in turn divide into two categories: the latent heat flux associated with evaporation from leads or sublimation from the ice (snow) surface and the sensible heat flux due to convective heat transfer over the leads or ice (snow) surface. The turbulent fluxes are calculated using bulk parameterizations that seek to capture the unresolved role of turbulence in moisture and heat transfer though empirical relations based on the time-averaged (i.e., bulk) air flow properties. The radiative fluxes also split into two subcategories: the longwave and the shortwave radiation. The simulated upwelling and downwelling components of the shortwave radiation are critically dependent on the description of the surface albedo. The sum of these four fluxes leads to a surface heat source or sink for the sea ice, helping to drive either ice melt at the surface or ice growth at the base (Oberhuber et al., 1993).

A unique aspect of sea-ice modeling, as compared with land-ice or even ocean modeling, is the treatment of the mass-conservation equation. While ice is considered incompressible from a density point of view, similar to the approximation made in ocean modeling, the three-dimensional volume of ice is actually split into two separate equations: an equation for ice thickness (i.e., the vertical average over an area) and an equation for the areal fraction, the area actually covered by sea ice (with the other fraction denoted as lead area). Taking the combined product of the ice thickness and areal concentration equation allows one to reconstitute the original three-dimensional mass (volume) conservation equation. This mathematical splitting of the volume (mass) conservation equation into separate equations for thickness and concentration is done as it is natural to model sea-ice thickness and concentration as separate variables and also to ease comparisons with observations. While this is a rather clever split, one complication that arises is how to apply heat sources and sinks to ice growth and decay. That is, just as the volume equation has been split into two equations, the heat source/sink terms need also to be split, but it is not at all obvious how to partition the heat source/sink. Specifically, given a unit of heat loss from the ocean, should that heat be used to grow the ice vertically and extend the thickness or horizontally and extend the areal coverage? The inverse question can be asked in the case of a unit of heat gain. Current-generation models have a relatively *ad hoc* formulation of how to split this heat flux, albeit one that generates reasonable model simulations. Further research is required to better understand and parameterize this heat flux split.

The sea-ice dynamics equations describe how the pack ice moves under the influence of various forces: ocean surface tilt, Coriolis effect, wind, currents, and most interestingly, internal ice dynamics, often referred to as ice rheology. The main discussion in theoretical developments of sea ice centers on an appropriate rheology—or flow law—that accurately describes how ice floes interact with one another and how they exchange momentum. Ocean waters are treated as a Newtonian fluid whereby stress is linearly related to strain through a fixed viscous coefficient. By contrast, sea ice is thought of as an elastic–plastic fluid (Rothrock, 1975). At small strain, the stress is described as elastic and is reversible. At larger strain, the stress is finite and plastic, meaning the ice fails under tension, compression, or shear stress. An appropriate description, one that is both physically accurate and numerically efficient, has been a target of sea-ice dynamicists for decades. Progress has been made, with the current generation of models simulating ice quite reasonably, starting notably from the early work of Hibler (1979) in which the observed sea-ice thickness pattern of the Arctic Ocean, with thick 6-m ice on the Canadian side and thin 1-m ice on the Russian side, was reasonably simulated for the first time.

2.3. Ocean Mixed-Layer Interaction

The field of oceanography took a major theoretical step forward with the discovery of the phenomenon of the Ekman layer (Ekman, 1905). During his trans-Arctic cruise, Nansen noticed that sea ice did not flow in the direction of the prevailing wind but drifted somewhat to the right. Nansen had this observation passed to Ekman who subsequently worked on the mathematical details of how sea ice and the upper ocean interact. Ekman's mathematics described a layer beneath the sea ice that progressively veered the water current to the right in a spiral fashion, with deeper water moving slower due to viscous forces. The underlying dynamics is a simple balance between Coriolis force and vertical stress. The resulting spiral structure has been directly observed beneath sea ice (Hunkins, 1966), as well as at the open ocean surface (Lenn and Chereskin, 2009) and at the seafloor (Kundu, 1976). The Ekman layer is of significant dynamical consequence as it drives a net transport of water to the right (left) of the surface ocean stress in the northern (southern) hemisphere. In the context of the marine cryosphere, this has an important role to play in the upwelling or downwelling of waters along the margins of the Arctic and Antarctic (England et al., 1993).

Away from the sea-ice-covered portions of the global ocean, the surface water layer is mixed down to some depth, generally in the range 10–100 m. In the resulting surface, mixed-layer water properties are almost homogeneous in the vertical (de Boyer Montegut et al., 2004). The key principle underpinning this mixed-layer depth is the balance between turbulent kinetic energy input and its ability to overcome the potential energy of the ambient stratification. The turbulent input comes via surface wind stress, ocean current shear, and wave breaking, and the opposing potential energy barrier is simply due to the ambient stratification (Large et al., 1994). In the Arctic Ocean, one of the most stratified in the world, the top 100–200 m is well stratified with respect to the deeper layers of water, making exchange between the upper, cold halocline and the deeper, warmer Atlantic layer quite difficult from an energy point of view (Aagaard and Carmack, 1989). By contrast, the Southern Ocean is only weakly stratified and a modest input of turbulent energy can overcome the weak background stratification (Martinson, 1990).

Vigorous exchange in the upper ocean, driven by turbulence, rapidly homogenizes scalar properties. In the case of a sea-ice cover, the exchange between the liquid ocean and solid ice surface, specifically at the liquid–solid interface, occurs via molecular transfer processes. The underlying principle is that turbulent fluid exchange is dampened at a solid interface as the scales of turbulence decay as one approaches the solid interface (Kader and Yaglom, 1972). Extremely close to the interface, the final exchange of scale properties occurs via molecular (diffusive) processes. At the ice–ocean interface, both salinity and temperature play a central role in the melt (growth) rates at the interface. Furthermore, as salt diffusion is approximately 100 times slower than that of temperature, there is an interesting

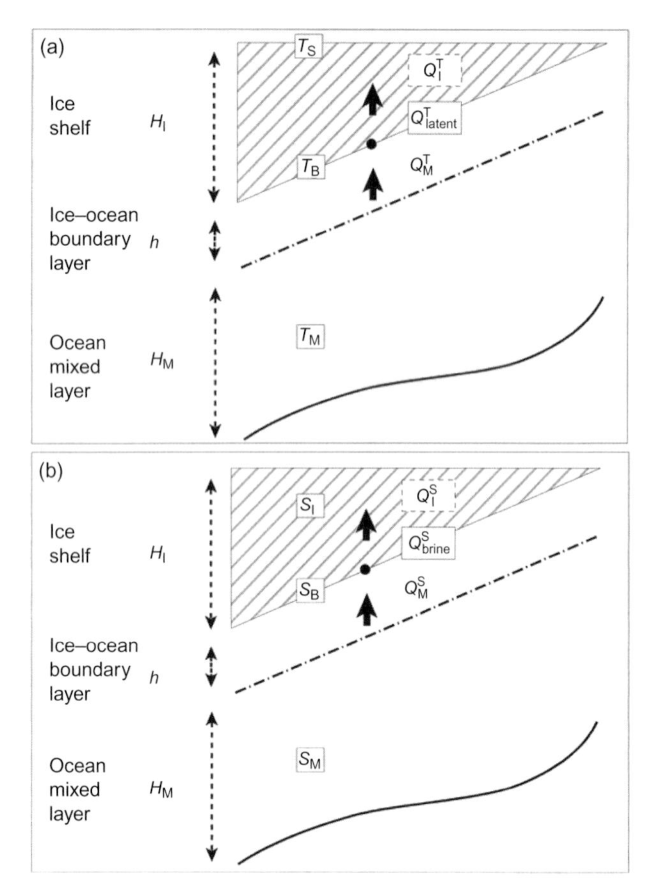

FIGURE 16.4 Schematic representation of the viscous-sublayer model of ice–ocean interaction showing (a) the heat and (b) the salt balance at the base of sea ice (or equivalently at an ice-shelf base). The thickness of ice H_I and ocean mixed-layer H_M are of the order of meters, while the ice–ocean boundary layer (i.e., viscous sublayer) h is of the order of millimeters. Temperatures (salinities) in the ice interior T_S (S_I), at the base T_B (S_B), and ocean mixed-layer T_M (S_M) are shown. The heat (salt) fluxes through the ice Q_I^T (Q_I^S), at the interface Q_{latent}^T (Q_{brine}^S), and in the ocean Q_M^T (Q_M^S) are shown. The key control is that the fluxes of heat and salt balance precisely at the interface. *From Holland and Jenkins (1999).*

competitive dynamic at the interface. The end result is that the ice–ocean interface melt (growth) rate is controlled by the relative rates at which heat and salt can diffuse into the boundary layer. At the interface, there are three physical relations to be satisfied simultaneously: the empirical pressure and salinity-dependent freezing point of sea water, the balance of heat flux into and out of the interface, and the balance between the salt flux into the interface and the freshening effect of melt water from the interface. An illustration of the various scalars and fluxes is presented in Figure 16.4. This set of constraints gives rise to the "three-equation" model of ice–ocean interaction (McPhee, 2008).

2.4. Polynyas

Derived from the Russian word meaning "a large opening in an otherwise sea-ice covered area," polynyas occur at various locations throughout the wintertime sea-ice cover. Their existence is at first glance surprising, as the surface energy balance in polar winter is strongly negative such that the ocean is losing a vast amount of heat, and one would expect it to be ice covered. Two physical processes that allow for the occurrence of polynya are, first, the existence of solid boundaries such as coastline which allow ice to be blown away from the coast and slowly replenished by new ice growth (Pease, 1987), and second, vertical convective processes that can raise warm water from depth (Holland, 2001). In either case, adequate heat is brought into the mixed layer or ice transported away from the mixed layer, such that sustained open water areas are possible, even in wintertime conditions.

From the perspective of the climate modeling and physical oceanographic communities, recurrent polynyas are worthy of study (Willmott et al., 2007) because they are sites where (1) water mass transformation takes place through the combined effects of cooling and frazil ice formation, (2) large ocean-to-atmosphere heat (several hundred Watts per square meter) and moisture fluxes occur, and (3) atmospheric CO_2 is sequestered into the ocean by physical–chemical processes and biological activity.

In coastal polynyas, where the water is shallow, the ocean quickly cools down to the freezing point at all depths (Lemke, 2001). The heat supplied to the atmosphere then originates only from the latent heat of fusion due to the continuous production of sea ice. Where offshore winds occur, sea ice is constantly formed and blown away from the coast, allowing yet more sea ice to be formed, in effect creating a type of sea-ice factory. During sea-ice formation, oceanic salt is released into the water because it is not incorporated into the ice crystals. This salt increases the local density of the ocean water, in some cases to the point where it can sink to the deep ocean. Coastal polynyas therefore represent important sources of dense deep and bottom water, which ventilate the abyss and drive the oceanic global conveyor belt.

In open ocean polynyas, the supply of oceanic heat is less restricted (Lemke, 2001). Therefore, large amounts of sensible heat from deeper ocean layers are lost to the atmosphere and little ice is created. The substantial cooling of the ocean water can lead to deep convection, which homogenizes the ocean waters to great depths and produces deep water in the open ocean. Consequently, both types of polynyas have a large influence on global ocean circulation and on the earth's climate.

2.5. Impact on Water Masses, and Circulation

The mere existence of a sea-ice cover provides a distinctive surface boundary forcing compared to other regions of the global ocean. From an oceanographic point of view, the most important implication of the sea-ice cover is the large input of salt that sea-ice growth inputs to the ocean surface during winter formation. On the flip side, the summer melt of the sea ice provides an enormous freshwater input at the surface. Taken together, these two processes result in an effective salt pump that moves salt from the near surface by convectively driving it down into the deeper ocean (Broecker and Peng, 1987).

A second impact of the ice cover is a modification to the input of wind stress (Joffre, 1982). Well known is that the curl of wind stress is the major input to the Sverdrup balance governing large-scale ocean circulation. Where there is extensive sea-ice cover, particularly near coastal regions, the exchange of momentum between atmosphere and ocean is fundamentally altered by the existence of the sea-ice cover. In the case of a lengthy sea-ice margin, the curl of the wind stress on the ocean can be radically different from anywhere else in the ocean as there is a sharp line where the curl abruptly changes (Quadfasel et al., 1987).

2.6. Biogeochemical Ramifications

Despite the low temperatures of the upper polar oceans, the presence of sea ice provides a platform that allows one of the largest and most important components of the earth's ecosystem to exist. Despite the low light levels, high salinities, and poor surface gas exchange, microorganisms thrive in this environment, particularly algae (Lizotte, 2001). As the regulation of carbon dioxide between the ocean and the atmosphere is particularly important to the marine biosphere, and sea ice plays a key role in modulating gas exchange and allowing organisms grow, once again the existence of a sea-ice cover is seen to be important in how the global ocean interacts with the atmosphere and ultimately the amount of carbon dioxide stored in the atmosphere, and hence global air temperatures (Rysgaard et al., 2011).

3. LAND ICE

Land ice, built up through the process of evaporation over the global ocean and subsequent precipitation over the great ice sheets of Greenland and Antarctica, has traditionally been thought to be one of the most passive components of the climate system and thus of not much consequence to ocean circulation, particularly on timescales of centuries or less (Figure 16.5). The pace of interaction was considered to be "glacial." Since the early 1990s, however, an unprecedented rapid change in the margins of the ice sheets has been observed through the eyes of satellites (Nick et al., 2009), and the cause for change has been largely deduced to be the global ocean, via the import of warm waters to the periphery of the ice sheets (Pritchard et al., 2012). Suddenly, land ice has become a major player in rapid global sea-level change and water mass modification processes in the polar oceans (Alley et al., 2005).

3.1. Observations

The global ocean only directly interacts with the ice sheets where the periphery of the ice terminates in the ocean, either as an ice shelf, which extends out onto the ocean surface as a floating tongue, or a tidewater glacier which has an abrupt transition from land ice to ocean with no floating tongue (Meier and Post, 1987). It was once thought that the disintegration of an ice shelf, which by definition floats on the ocean, would have no impact on the flow into the ocean of the inland ice that fed it (Van der Veen, 1985). That viewpoint has been disproved during the past decade as ice shelves, such as the Larsen B Ice Shelf, rapidly disintegrated and the flow toward the ocean of the inland ice behind accelerated manyfold (Scambos et al., 2004).

This knowledge comes from an unprecedented suite of observational equipment placed *in situ* on the periphery of the ice sheets, and airborne and remote sensing instrumentation (Mohr et al., 1998). Remote sensing techniques have shown a doubling in speed of the Jakobshavn glacier in west Greenland, following the breakup of the ice tongue in front of that glacier (Alley et al., 2008). That particular breakup is believed to have been caused by the sudden arrival of warm waters into the fjord connecting to the glacier (Holland et al., 2008). Laser altimetry has captured a drop in surface elevation over the Greenland outlet glaciers, and their analogs in the Antarctic, most notably in the Pine Island Glacier region, located in the Amundsen Sea (Pritchard et al., 2009). At the same time, correlation techniques applied to imagery have demonstrated that the velocity of these outlet glaciers has increased, coincident with the drop in surface elevations (Joughin et al., 2004). Furthermore, the occurrence of simultaneous change between neighboring glaciers suggests the ocean to be the main driver of this change (Shepherd et al., 2004).

3.2. Modeling

The physical basis for modeling land ice holds some similarity with sea ice, described in the previous section, but also has significant differences. A main difference is that glacial ice is treated as a three-dimensional continuum (Blatter et al., 2011), with no analog for the description of concentration (or open space) as is the case for sea ice. Glacial ice does indeed fracture, but such fractures occupy a small portion of the total areal coverage and are not yet widely considered in the overall integrity or strength of an ice sheet or shelf. There are some efforts to correct this modeling deficiency by incorporating the concept of "damage mechanics" into the ice equations (Pralong et al., 2003).

The thermodynamics of glacial ice is analogous to sea ice in almost every respect; that is, boundary conditions apply to the ice surface and base, and molecular diffusion

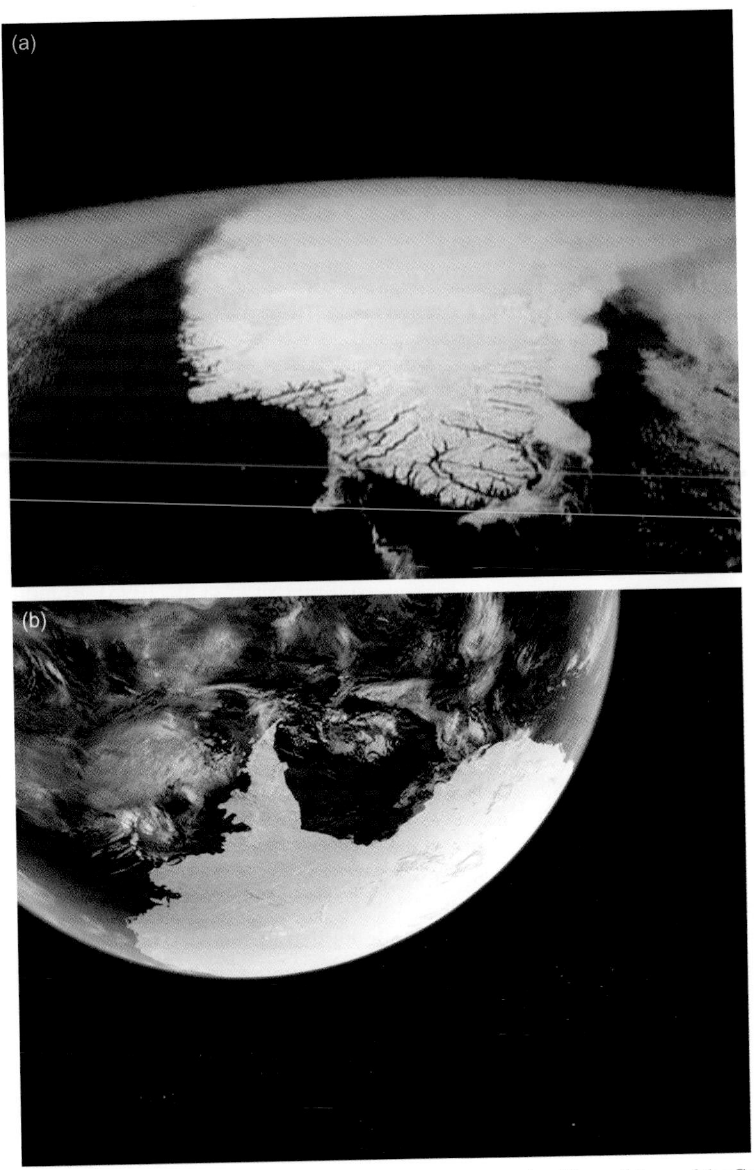

FIGURE 16.5 High-altitude views of the two major ice sheets (a) Greenland and (b) Antarctica. The periphery of the Greenland Ice Sheet interacts with the surrounding ocean through relatively narrow, deep fjords, some of which have floating tongues and others which terminate abruptly as a vertical face at the ice–ocean interface. By contrast, the interaction of the Antarctic Ice Sheet with the surrounding ocean is through ice shelves ranging in size from modest to massive, with the three largest being the Ross, Weddell, and Amery. *http://www.climatepedia.org/about-ice-sheets; http://news.discovery.com/earth/zooms/ocean-warming-melting-antarctica-ice-sheets-120426.html.*

of heat is the dominant process in the ice interior (Hutter, 1982). Given the thickness of glacial ice, tending to be greater than 1 km, it responds to thermal anomalies on the timescale of hundreds or thousands of years. Sea ice, on the other hand, is typically of order 1 m thick and responds to anomalies at the ice–ocean or atmosphere–ocean boundary rapidly, certainly well less than 1 year.

The dynamics of glacial ice bear little resemblance to those of sea ice, not only because of the absence of fracture or void space in the treatment of glacial ice but also because, unlike sea ice, glacial ice is in general not currently modeled to fracture. This is quite surprising, given the fact that one of the most dramatic natural events to occur is the calving, or fracturing, of large pieces of glacial ice into the coastal ocean at the margins of the great ice sheets. From the coast, the icebergs are transported by the mean ocean currents and winds and subsequently melt into the ocean, providing a distributed source of freshwater into the ocean. In order for this process to be adequately captured in global climate models, significant progress is going to be required in the physical description of ice calving, and hence of ice rheology. This advance is in turn dependent upon the acquisition of adequate observational data to better understand the mechanism underlying ice calving.

3.3. Ocean Mixed-Layer Interaction

From an oceanographic perspective, the thermodynamic interaction of the base of an ice shelf with the ocean is crucial for two reasons: firstly, the melting process modifies water masses such as Antarctic Bottom Water (AABW) (Foster and Carmack, 1976), and second, the disintegration of an ice shelf could eventually contribute to a change in sea level (Alley et al., 2005). Here we discuss the current treatment of thermodynamic interaction. The physical description is almost identical to that of the analogous phenomena at the base of sea ice, as discussed in Section 2. This is reasonable since the two environments have many features in common.

One important distinction is that the base of an ice shelf can be located at considerably higher pressure than that of sea ice, which is nominally at atmospheric pressure. At depths of 2 km, such as at the grounding zone of the Filchner-Ronne Ice Shelf in West Antarctica, the pressure increase causes a lowering of the freezing point to approximately 1.5 °C below the value at the ocean surface (Fujino et al., 1974). This has the interesting and important consequence that if water at the surface freezing point (−1.9 °C), the coldest water that can be formed by direct interaction with the atmosphere, reaches ice at the deep grounding zones, it finds itself least 1.5 °C warmer than the local freezing point and can therefore cause high rates of basal melting.

This thermodynamical effect leads to an ocean dynamical circulation creating the ice pump mechanism (Figure 16.6). Effectively deep ice is melted because of the pressure dependence of freezing temperature. The resulting melt water rises along the base of the ice shelf

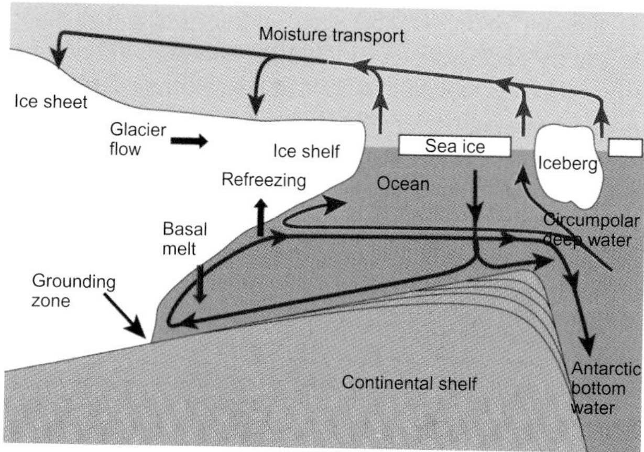

FIGURE 16.6 Schematic representation of the two-dimensional circulation under an ice shelf. Salt rejected by winter sea-ice growth forms dense, high-salinity water, which sinks and flows beneath the ice shelf. This causes melt when it comes into contact with deep ice at the grounding zone. The freshened plume rises under the base of the shelf and can either refreeze as marine ice or mix with warm, salty Circumpolar Deep Water (CDW) to form Antarctic Bottom Water (AABW). The circulation is often referred to as an "ice pump." *From AMISOR (2013).*

under the action of gravity because the meltwater is always more buoyant than the ambient—a consequence of the equation of state for seawater at low temperatures. Depending upon the ambient conditions, the rising water may become supercooled as the pressure lessens and the freezing point rises, leading to the possibility that frazil ice may form and be deposited at the base of the ice shelf (Lewis and Perkins, 1986). This entire cycle in which ice is melted at depth and then refrozen at shallower ice-shelf drafts is thought of as an ice pump circulation—moving floating ice from deep to shallow elevations at the ice-shelf base.

In the discussion of sea-ice basal melting in Section 2.3, the concept of the viscous-sublayer model has been introduced (see again Figure 16.4). This model has also been adopted for use at the base of an ice shelf (Holland and Jenkins, 1999), despite not yet being thoroughly validated for use in such a regime. Other than the greater depth, another feature distinguishing a sub-ice-shelf base from a sea-ice base is that an ice-shelf base tends to be inclined, typically by 0.1% slope, but steeper in some locations, particularly near the grounding zone. Field campaigns are only now being carried out to make direct measurements of the turbulent boundary layer at the base of a few Antarctic ice shelves, with validation and improvement of parameterizations of transport of fluxes through the ice shelf–ocean boundary layer likely to occur within the next few years. The first field results suggest that, at least in the instance of the Pine Island Ice Shelf in West Antarctica, the slope of the interface and the stratification that results from the melting along the sloping base form an intense feedback strengthening the melt rates. This feedback likely contributes to the large-scale channelization of the ice-shelf base (Stanton et al., 2013).

We have thus far mentioned facets of the ocean mixed layers beneath a sea-ice cover and also beneath an ice shelf. Along the front of an actively calving glacier, there often exists a sea-ice cover with icebergs embedded, sprinkled through the sea ice. The resulting mélange of sea ice and iceberg makes for a truly complex ocean mixed-layer environment, one in which the usual turbulent kinetic energy input is also influenced by the stirring due to deep iceberg keels as well as buoyancy input due to freshwater release as the bergs melt at depth (Burton et al., 2012). This is an area of ocean mixed-layer modeling that has not been much addressed to date and one that is in need of *in situ* observations upon which a solid theoretical foundation can be laid.

While sea ice is an obvious form of ice of marine origin, there is also a form that develops beneath an ice shelf and can reach impressive thickness. For instance, in the middle of the Filchner-Ronne Ice Shelf, the thickness of the bottom-accumulated marine ice layer is estimated to be almost half the total ice-shelf thickness (Oerter et al.,

1992). This marine ice is formed in the water column as part of the ice pump mechanism, mentioned above. The presence of significant thickness of marine ice at the base of an ice shelf could have significant ramifications for the strength of the ice shelf as a whole (Holland et al., 2009) and, if the marine ice survives to the ice front without being remelted, could play an important role in the manner in which an ice-shelf front calves.

3.4. Impacts on Water Masses

The continental shelves of Antarctica, particularly the western edges of the Ross and Weddell Seas, are well-known locations for formation of AABW (Foster and Carmack, 1976). This formation is the result of wintertime production of sea ice which causes the surface salinities to reach the highest values found anywhere in the Southern Ocean. As salinity controls density for waters near the freezing point, these cold salty surface waters sink under vertical convection and flow northward down over the continental shelf break and fill the abyssal global ocean. Approximately half of the global ocean volume is constituted by this water mass. But as the continental shelves of Antarctica are foredeepened, some fraction of the dense water formed over the continental shelves first goes southward and into the sub-ice-shelf cavity (see again Figure 16.6). The water that goes beneath the ice shelf generally travels inland all the way back to the grounding zone where, because of the dependence of the melting point of ice on pressure, the newly arrived water causes ice-shelf basal melting. The resulting melt water, known as Ice Shelf Water (ISW), is buoyant and rises along the ice-shelf base as mentioned earlier. In some areas, this ISW plays an important role in AABW production. The production of much of the AABW over the south western Weddell Sea continental shelf is regarded as being a result of ISW mixing with ambient warm, deep water on its way down the continental slope (Seabrooke et al., 1971). In the Ross Sea, some significant proportion of AABW is created by a similar mechanism (Jacobs et al., 1970). In the eastern Weddell Sea, and in many other parts of Antarctica, the melt from ice shelves freshens the continental shelf enough to inhibit the production of high-salinity shelf waters, which in turn inhibits the production of AABW. Turning to sea-ice cover impacts of ISW, it has been shown in numerical experiments that the mere existence of ISW in the Weddell Sea results in a fresher ocean surface over the Weddell Sea and consequently there is an impact on increased sea-ice cover development (Hellmer et al., 2006).

While dense water formation does occur in the Greenland Sea, this is truly open-ocean convection and is not forming along the coast of Greenland where glacier-ocean interaction might have an influence. Instead, it is the melting of the Greenland outlet glaciers by warm waters, from sources distinct from those feeding the convective deep-water formation zones that hold the potential to impact the convective zones (Aagaard and Carmack, 1989). The formation of deep water in the Greenland Sea is a major component of the global conveyor belt, and such formation can in principle be inhibited by the melt water coming from the Greenland ice sheet (Rahmstorf, 1999). In the transition between glacial and interglacial periods, it is thought that the massive freshwater pulse forming from either the Laurentide or the Greenland Ice Sheets feeds freshwater that caps the convective zones (Broecker and Denton, 1990).

3.5. Geochemical Tracers

Understanding the basal melting and water mass modifications beneath an ice shelf has been hampered by the difficulty of accessing the cavity. Setting up sustained monitoring of the ice-shelf cavity openings has also been problematic. Aside from traditional CTD (conductivity, temperature, and depth) observations that can be made at such locations either through the ice shelf (Nicholls et al., 1991) or along the ice front (Jacobs et al., 2002), one can also measure the concentration of various geochemical tracers and, to some degree, deduce the amount of glacial input to the ocean arising from basal melt (Smethie and Jacobs, 2005). Additionally, by using measurements of geochemical tracer concentrations to supplement the more common observations of temperature and salinity, one can better constrain the calculation of the amount of mixing between various water masses in the cavity (Schlosser et al., 1990). Another rationale for using geochemical tracers is that temperature and salinity are "nonconservative" when involved in the melting of ice in the sense that latent heat and freshwater exchange occur, transforming the two properties. The use of additional, independent geochemical tracers can remove ambiguity in water mass mixing calculations in such environments.

From an oceanographic point of view, geochemical tracers can be either naturally occurring isotopes of, for instance, oxygen, helium, or neon, or those of anthropogenic origin such as CFCs (chlorofluorocarbons). Over the past half century, CFCs have been entering the atmosphere and therefore the interior of the ocean through gas exchange at its surface (Beining and Roether, 1996). For the purpose of water mass analysis and mixing ratios, if one knows the amount of CFC in parent water masses, then one can better estimate the amount of relative mixing between two or more such water masses based on CFC measurements in the mixed-water mass. In the instance of an ice-shelf base, knowing CFC concentration in the high salinity shelf water (HSSW) that forms in the open ocean surface adjacent to the ice front and knowing the CFC concentration in ISW that subsequently forms when HSSW interacts with the base of an ice shelf, one is able to deduce the amount of input meltwater. One complicating factor in the case of the ice-shelf regime is that the presence of a variable sea-ice cover in the formation region means that equilibration between atmosphere and ocean is not always established, making the overall calculations subject to greater uncertainty (Smethie and Jacobs, 2005).

Other than CFCs, oxygen, helium, and neon have also been used to study glacial melt into the ocean. Oxygen, helium, and neon gas are present in glacial ice trapped in air bubbles or as clathrates (see Section 4). These gases are released from the ice and dissolve in seawater as a result of ice-shelf basal melting. The low solubility of helium and neon in seawater results in concentrations well above solubility equilibrium with the atmosphere, producing a glacial melt signal that can be readily traced from an ice front, across the continental shelf and into the abyssal ocean (Schlosser et al., 1990).

Geochemical tracers aside, another powerful constraint on calculating water mass mixing and formation in the polar regions is that relating to the constrained ratio which temperature and salinity change must follow as a water mass gives up heat to melt ice (Gade, 1979). As ice melts, the source water doing the melting is cooled and freshened in a specific ratio depending on physical parameters such as the latent heat and specific heat capacities of water and ice. For instance, for the Ross Ice Shelf cavity, an analysis of the temperature–salinity diagram of the various water masses in the cavity, coupled with the melting–freezing (M–F) constraint mentioned above, has led to an improved quantification of the water mass transformations occurring there (Figure 16.7).

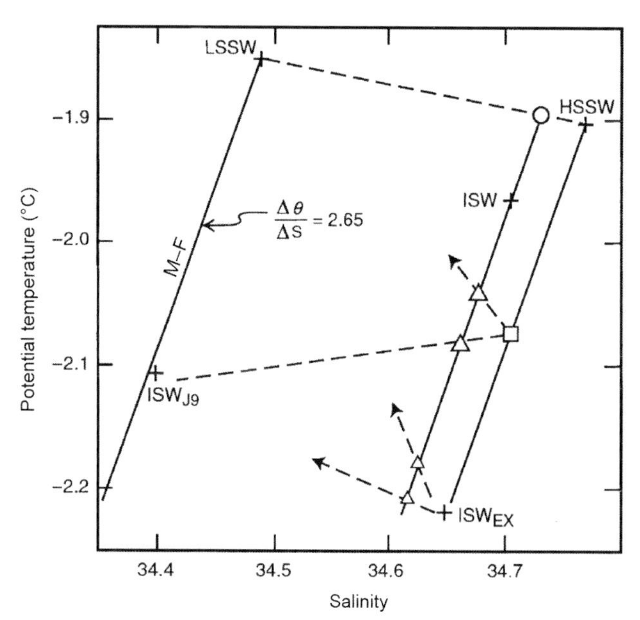

FIGURE 16.7 A potential temperature–salinity diagram illustrating melting, freezing, and mixing processes that can occur beneath an ice shelf. Such diagrams can be useful in identifying the production of Ice Shelf Water (ISW) and thus inferring the amount of glacial basal melt into the ocean. By melting and freezing (M–F) at the base of the ice shelf, the properties of the ocean mixed layer at the base move only along the solid lines, which have potential temperature to salinity ($\Delta\theta/\Delta S$) slopes of 2.65. The "+" symbols represent observed water masses. The dashed lines indicate possible mixing between observed varieties of ISW and the high- and low-salinity shelf waters (HSSW and LSSW). This diagram is for the Ross Ice Shelf cavity. *From Smethie and Jacobs (2005).*

3.6. Sea-Level Change

On the timescale of the earth's glacial cycles, the volume of the global ocean changes substantially, being approximately 100 m or more lower in global sea level during a glacial period than the sea level of the present interglacial (Siddall et al., 2003; Chapter 27). This glacial cycle change in global sea level is thought to be forced by the amount of solar radiation reaching the earth, the periodicity of which is known as the Milankovitch cycles, which are dependent upon the precession, obliquity, and eccentricity of the earth–sun system (Imbrie et al., 1992). Remarkably, in these regular cycles of continental ice-sheet growth and decay, there are periods in which the sea-level changes abruptly. A remarkable example is the event known as meltwater-pulse 1A that occurred approximately 14,000 years ago during the last deglaciation (Stanford et al., 2006). Over a period of a few hundred years, the global sea level at that time rose by at least 5 m. The implication from this is that future global sea level could also change as rapidly.

The mechanism by which such rapid past change could have occurred remains obscure. One theory is that sudden import of warm ocean waters to the periphery of the ice shelves, particularly the marine-based portions, leads to a rapid wasting of the ice that triggers major change over a short time period. The change in ocean circulation would have to be driven by changes in atmospheric circulation, and the mechanism for that change is equally obscure. Continued collection of observational data and development of coupled global models will likely unravel this mystery in the coming years.

4. MARINE PERMAFROST

In the previous two sections (Section 2 and 3), we have discussed the interaction of ice that floats on the surface of the ocean with the upper ocean. Ice is also found on the bottom of the ocean, embedded in the sediment. There are two types of such ice, one being a pure ice matrix and the other involving trapped methane.

4.1. Pure Ice

During glacial periods, large portions of the Siberian Arctic continental shelves have been covered by ice, and the subsequent flooding of these shelves by ocean waters in interglacial times has resulted in large areas of what was originally land permafrost being transferred to the marine permafrost category. In general, marine (or subsea) permafrost is formed either by such inundations or by areas of seafloor with mean annual ocean temperatures below 0 °C. A more technical definition of marine permafrost calls for the seafloor ice structure to exist for at least a 2-year period (PAGE21, 2013), a condition easily met along the Siberian Arctic shelves. The thickness of the marine permafrost layer has been assessed using direct coring and reaches to 100 m in some areas. Electromagnetic techniques have also been employed based on the fact that sediments in an ice-bonded state have a different electrical resistivity than in the unbonded state (Corwin, 1983). Seafloor permafrost has been found only in the Arctic.

4.2. Methane Clathrates

There is a peculiar water-molecule-based structure, formed under modest pressure (depths greater than 300 m of ocean) and low temperature (below 0 °C), that results in the trapping of a gas such as methane in a cage surrounded by water molecules (Max, 2003). The solid is similar to ice (Figure 16.8a), except for the trapped foreign molecule. These clathrate hydrates collectively form a component of the marine permafrost layer. The clathrate can exist at depths in the marine sediments where the sediment is colder than the freezing point of the clathrate, itself a function of pressure (Figure 16.8b). Clathrates are suspected to be located along the Arctic coastline beneath the marine permafrost that has formed as a result of glacial advances, mentioned above. However, they are not restricted to such locations. Oceanic deposits seem to be widespread on the continental shelf and can occur at depth within the sediments or close to the sediment–water interface. They may cap even larger deposits of gaseous methane (Kvenvolden, 1995). It has been conservatively estimated that the amount of carbon in the clathrate marine form is more than twice that found in all fossil fuels on the earth (USGS, 2012).

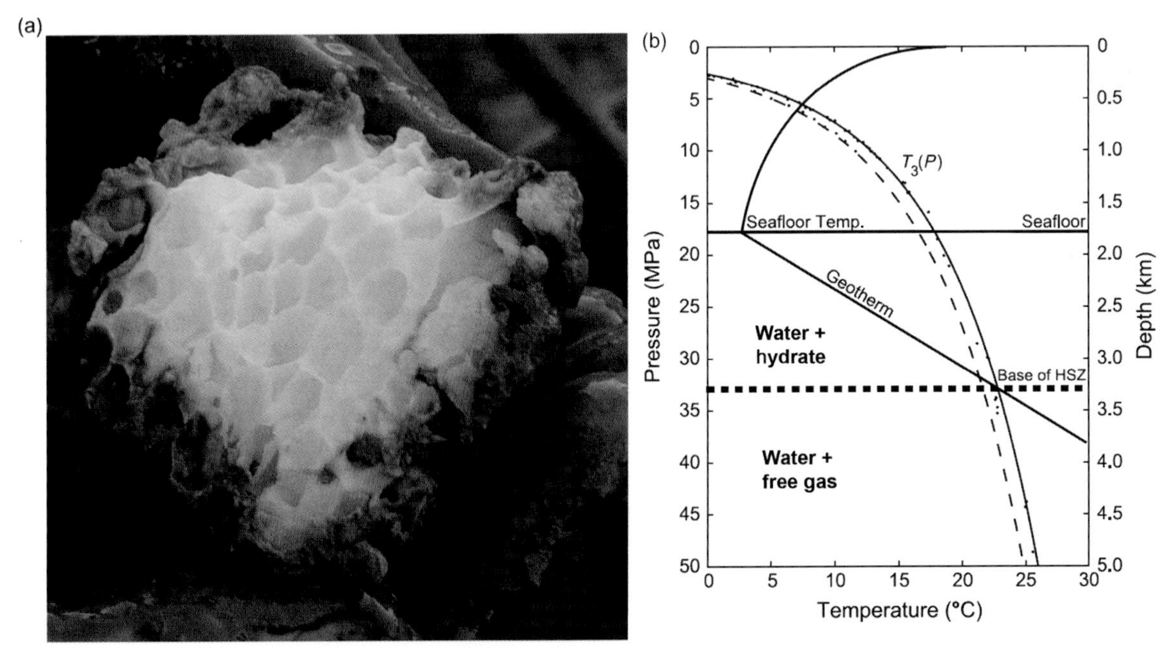

FIGURE 16.8 (a) A methane clathrate block found in a depth of about 1200 m of water in the upper meter of the sediment. (b) Schematic profile of temperature through the ocean and uppermost marine sediments. The temperature for clathrate stability, $T_3(P)$, increases with pressure (or depth). Experimental data for $T_3(P)$ in pure water (dashed line) and seawater (solid line) are shown. The base of the hydrate stability zone (HSZ) is defined by the intersection of the Geotherm with $T_3(P)$. Warming the ocean temperature, and hence the Geotherm, will shallow the HSZ. *Panel (a) from http://en.wikipedia.org/wiki/File:Gashydrat_mit_Struktur.jpg; panel (b) from Buffet and Archer (2004).*

5. EMERGING CAPABILITIES

In this section, we discuss the emerging capabilities in observing sea ice and the ocean beneath sea ice and ice shelves, some of the most difficult parts of the ocean to study. Improvements and future directions in sea-ice and land-ice modeling are also given; advances in ocean modeling are provided elsewhere in Chapter 20.

5.1. Ice-Capable Observations

Starting in the 1970s, the observation of sea ice was revolutionized by satellite remote sensing using passive microwave instrumentation, leading to unprecedented sea-ice concentration detail (Gloersen et al., 1992). Subsequent developments have led to sea-ice velocity fields determined from visual or active radar imagery (Emery et al., 1997). Now emerging is the possibility of determining the sea-ice thickness field, the most difficult of the sea-ice-related basic fields to measure by remote sensing (Laxon et al., 2003). The relevant instruments are the CryoSat radar altimeter (Wingham et al., 2006) and the ICESat laser altimeter (Schutz et al., 2005), which are able to give relatively precise observations of sea-ice thickness, and thus thickness change. This new ability to measure thickness is a key missing piece needed to more fully observe future Arctic sea-ice change, as it now allows the possibility, combined with the sea-ice concentration observations and sea-ice velocity, to make estimates of sea-ice volume and mass transport in the coming years.

A number of ice shelves in West Antarctica are undergoing significant mass loss due to what is assumed to be increasing rates of basal melt (Shepherd et al., 2004). There has been some efforts to determine basal melt rates using a combination of remote sensing data sets, basically surface elevation and surface velocity to infer ice flux divergence, and thus to evaluate basal melt (Joughin and Padman, 2003). As new satellite instruments come on board over the next few years, it is expected that observations of melt rate beneath all ice shelves will improve. However, these methods do have difficulties when the ice shelf is not in a steady-state configuration. Satellite altimetric methods can give an estimate of surface elevation change, which can then be used for corrections, but these in turn rely on snow compaction models.

Another important process of ice-shelf mass loss is that of calving. This is a process that can be well observed from space, given the number of satellites now in orbit producing high-resolution imagery (Herried et al., 2011). It will be possible to form an inventory of icebergs, starting from the birth to their ultimate demise. Such iceberg tracking will also provide a data set of freshwater melt input to the ocean from the icebergs in both polar seas (Gladstone and Bigg, 2002).

Aside from just taking an inventory of icebergs, there is still the nagging question of the physical processes that lead to calving in the first place (Bassis, 2011). To gain insight, it will be necessary to make measurements of both the strain and stress fields, in high spatial and high temporal resolution, dictating the need to make such observations *in situ*, either on the ice surface or with rapid, repeat airborne surveys. On the ice, the placement of newly developed passive seismic instrumentation may turn out to be useful to map out the stress field before, during, and after a major calving event (MacAyeal et al., 2009).

5.2. Ocean-Capable Observations

There are a significant number of emerging technologies to tackle the notoriously difficult problem of measuring ocean properties either beneath sea ice or in a sub-ice-shelf cavity. The subsurface polar ocean is one area where remote sensing by satellite has not been able to contribute much as the ice cover provides a surface obstacle to sensing. This implies that *in situ* observations are needed throughout the polar oceans. Perhaps the major breakthrough in this area has been the successful development and deployment of so-called ITP (Ice-Tethered-Profilers) (Krishfield et al., 2008; ITP, 2012). These instruments are profiling CTD devices that travel up and down a cable passed through access hole drilled through either sea ice (see Figure 16.9a) or an ice shelf. The data are retrieved via satellite in almost real time and have enormously increased the observational hydrographic data set in the Arctic basin (Toole et al., 2011). While the bulk have been deployed in the Arctic basin to date, their usage is now spilling over into the Antarctic. The Antarctic, because of its more seasonal sea-ice cover, is a more challenging environment for ITP deployment.

These ITP instruments are in some ways analogous to the Argo float program. Argo floats, of course, drift freely with ocean currents and surface from time to time to transmit their data (Send et al., 2009), but obviously this strategy does not work well, or at all, in an ice-covered environment (Klatt et al., 2007). Traditional Argo floats are now being ice hardened, so they can, if the opportunity arises, pop up through leads in sea ice (see Figure 16.9b). Another issue is that Argo floats determine their position by reaching the ocean surface and using a GPS receiver but, again, cannot do so in an ice-covered environment. If they are to operate in a region where they cannot surface often enough to satisfactorily fix the position of the data they are measuring, they need to use acoustics to navigate (Howe and Miller, 2004): arrays of subsurface acoustic beacons are currently being deployed to provide the necessary acoustic infrastructure.

A shortcoming of both the freely floating ITP and Argo buoy is that they are not able to be navigated to specific locations, for example, to perform repeat track surveys.

The use of AUV (Autonomous Underwater Vehicles) has been successful in exploring beneath sea-ice cover and ice shelves for short periods of time (McPhail et al., 2009). Such vehicles are now being engineered to have long range and long duration, including the capability to "hibernate" on the seafloor for extended periods of time. Although this is a hazardous enterprise, steps can be taken to minimize the risk (Jalving et al., 2008). For more sustained missions, lasting several months, and possibly up to a year in the near future, the development of the

FIGURE 16.9 Dominant, emerging observational technologies used to observe the polar ocean beneath the sea-ice and ice-shelf covers. (a) Ice Tethered Profiler (ITP) anchored to ice, (b) Argo float modified for under-ice deployment,

(Continued)

FIGURE 16.9—CONT'D (c) Glider deployed in sea-ice-covered sea, (d) Satellite Relay Data Logger (SRDL) marine-mammal tags attached to a seal, (e) Distributed Temperature Sensing (DTS) fiber-optic cable prepared for deployment through an Antarctic ice shelf. *http://www.polardiscovery.whoi.edu/expedition1/topic-itp.html; http://www.euro-argo.eu/content/download/21814/314255/file/0830%20Klatt-euroargo-18062010.pdf (Glider photo by Denise Holland); http://science.nationalgeographic.com/science/photos/antarctic-science-revealed/ (DTS photo by David Holland).*

underwater glider (see Figure 16.9c) perhaps offers the greatest hope of obtaining hydrographic sections in ice-covered oceans (Eichhorn, 2009). Like the Argo buoy for under-ice missions, the gliders do need navigational assistance from subsurface sound beacons. As an example, such a glider setup has been recently deployed for Davis Strait in Baffin Bay (Lee et al., 2004).

The ubiquitous presence of marine mammals in the polar oceans, such as seals, has led to the development of the SRDL (Satellite Relay Data Logger). This is a miniature CTD, coupled with a small Argos transmitter, that is attached to the back of the neck of a seal (Fedak et al., 2002). The seal is captured in a net, held for approximately 15 min during which a fast-acting glue, attaches the SRDL to the seal (see Figure 16.9d) (Teilmann et al., 2000). Such data sets have the limitation that they only provide data for approximately 1 year, as the SRDL falls off the seal during the annual molt. Another issue is that the data are only collected wherever the seal decides to travel. However, as the behavioral patterns of various seals species is now beginning to be learned, the SRDL are becoming more useful for data collection in selected regions of the polar oceans (Campagna et al., 1999).

A surprising property of a fiber-optical cable is that laser light shot along the length of the cable can be used to determine the temperature at approximately every 1 m for more than a 10-km stretch (Bao et al., 1993). This is possible because the scattered light can be analyzed by a detector, and based on physical principles of scattering, along with in-field calibration of the cable, one can obtain a time series of temperature along the cable at frequent time and space intervals (Selker et al., 2006). This technique is known as DTS (Distributed Temperature Sensing) (see Figure 16.9e for a typical deployment). Such a cable was recently deployed through the Ross Ice Shelf and a year-long data set of ice-shelf and sub-ice-shelf temperatures was collected (Stern et al., 2013).

5.3. Ice-Capable Modeling

Traditionally, sea-ice models have been developed based on finite difference discretizations of the governing equations and solved on either a regular Cartesian or a latitude–longitude grid (Holland et al., 1993). There are, of course, much more advanced discretizations and gridding schemes now developed in other areas of science. Ice researchers have begun to apply such techniques, for example, finite element modeling (Timmermann et al., 2009) which allows for very irregular grids (see Figure 16.10a). Such grids can be refined to very high resolution in narrow passageways and channels such as in the Canadian Arctic Archipelago (Lietaer et al., 2008).

While discretization and gridding dictate the spatial accuracy of the numerical solutions, time-stepping efficiency is also a major issue. Global-scale climate models seek to make each component submodel as efficient as possible. For the most part, sea-ice models have used explicit time-integration schemes (CICE, 2012), but more recent efforts have sought to bring implicit, matrix-based, methods to bear on the problem. For instance, a matrix-free Newton–Krylov technique (Lemieux et al., 2012) has been applied to the sea-ice equations and shows promise for allowing a significant speed up in integration time.

Numerics aside, aspects of the basic physics of sea ice still need to be improved. Parameterizations of a host of subgrid-scale phenomena continue to be tackled by researchers; for instance, summer melt ponds and their drainage, brine inclusions, and surface albedo (Holland et al., 2001). On the dynamics side, there is still uncertainty regarding the underlying rheology with the earliest rheology focusing on isotropic behavior and more recent studies looking into the anisotropic behavior of sea ice (Feltham, 2008). The most common rheology currently in use is a combination of elastic, viscous, and plastic behaviors (Hunke and Dukowicz, 1997). One feature still missing in all these rheologies is that of a tensile strength, appropriate to the description of landfast ice, as seen along parts of the coast of the Arctic and Antarctic (König Beatty and Holland, 2010). Current rheologies have not incorporated this phenomenon, but it is needed as the presence of landfast ice radically alters the formation of coastal waters, feeding the deep ocean (Barber and Massom, 2007).

The modeling of land ice, including that of ice shelves, has lagged significantly behind that of sea ice, but recent efforts, bolstered by the need for land-ice modeling in IPCC class models, are beginning to change that situation (CISM, 2012; see Figure 16.10b). The task is significant as it requires major advances in both the numerical modeling and the understanding of the physics of land ice and ice shelves, and also in the coupling of the ice models to ocean general circulation models. To date, there has not been a plausible coupling in which the ocean volume and ice volume swap physical space with one another when either the ocean melts the ice and invades its physical space or vice versa (Lipscomb et al., 2009).

6. CRYOSPHERIC CHANGE

In this final section, we look back on recent changes in the cryosphere and look ahead to projections for sea ice, land ice, and marine permafrost in the present century and beyond.

6.1. Observed Sea-Ice Change

The radical reduction in Arctic sea-ice areal extent over the past few decades has caught the world by surprise. It was not so long ago, in the early 1990s, when the Arctic sea ice was thought to be a stable, robust feature of the earth's

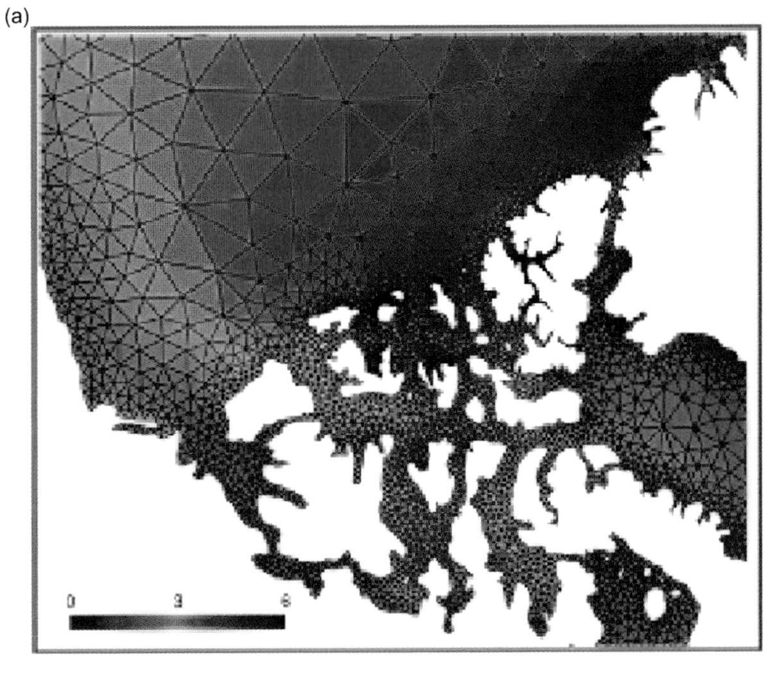

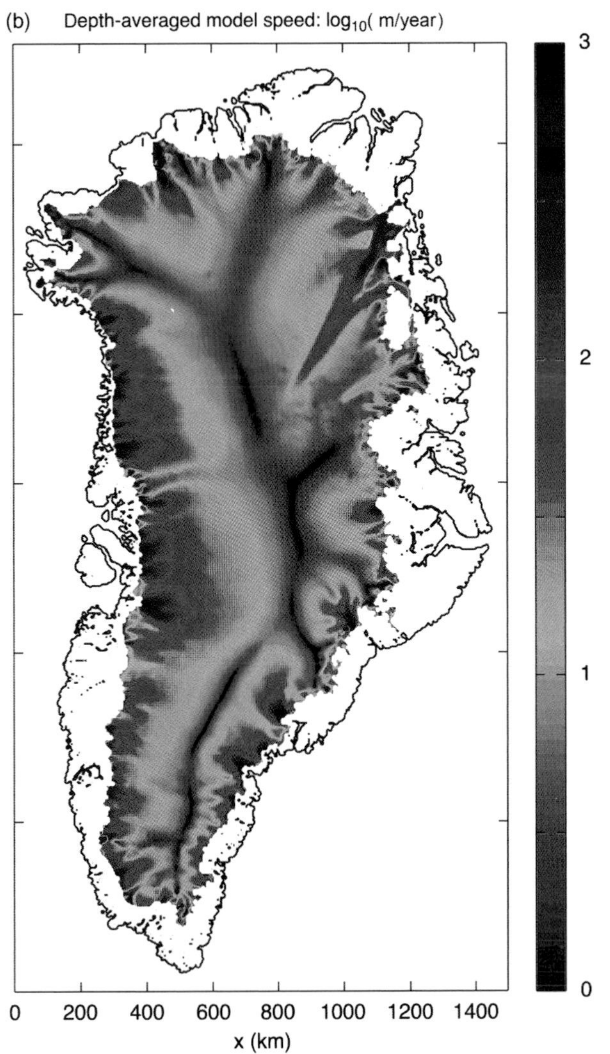

FIGURE 16.10 Examples of emerging technologies in sea-ice and ice-sheet (shelf) modeling. (a) Sea-ice-thickness modeling based on highly variable grid resolution capable of resolving flow through narrow straits. (b) High-resolution model of the flow of the entire Greenland Ice Sheet. Note the bright red areas showing the speedup of the ice sheet as it emerges toward the ocean through narrow outlet fjords. *https://www.uclouvain.be/en-376557.html; http://public.lanl.gov/sprice/images/GIS_vels.jpg.*

modern climate system, at least on the short timescale of years to decades. However, satellite remote sensing has unequivocally shown a nearly 50% decrease in the extent of the Arctic pack ice during the summer minimum (Figure 16.11, blue curve). In fact, the past year (2012) has set an all-time record low (NSIDC, 2012). The cause of this decline is tied up with the occurrence of global warming over the past century or so (Stroeve et al., 2007), as the ice-albedo feedback effect, once it takes hold, contributes to the accelerated melting. There is of course natural variability in the Arctic sea-ice cover, as driven, for example, by the Arctic Oscillation, a pattern of winds that can either reinforce ice export out of the Artic or hold it within the basin (NOAA, 2012). Coupled climate models have been successful in hindcasting Arctic sea-ice decline in general, but none of the models to date have been able to accurately hindcast the actual extreme decline, nor the extreme minima in past years, 2007 and 2012 (Rampal et al., 2011). This indicates that the coupled models are not yet fully able to capture extreme events, suggesting more research is needed either in the individual model components or in the model couplings to advance this capability.

The story with the Antarctic sea ice is just the opposite, with a possible small trend upward in extent (Figure 16.11, red curve), perplexing sea-ice researchers, and global climate modelers. While it is thought that the ice-albedo is playing a major role for Arctic sea ice, it is confounding that no such mechanism appears to be at play in the Antarctic. Of course, a major distinction between the Arctic and Antarctic is that in the former the sea ice is present in the summer in appreciable quantity but the Antarctic sea ice largely disappears, thus possibly reducing its susceptibility to the ice-albedo feedback effect. Additionally, the Arctic is effectively landlocked by the surrounding continents, whereas in the Antarctic, the inverse is the case and this may play a role. There is some evidence that wind patterns in the Antarctic have controlled the Antarctic sea ice (Holland and Kwok, 2012). Still, the apparent lack of an ice-albedo feedback remains a major question for understanding global climate change and the cryosphere.

6.2. Sea-Ice Projections

On the short timescale, looking ahead just a few months, an international initiative called the "Sea Ice Outlook" (SEARCH, 2012) has joined together a large number of sea-ice modeling efforts so as to provide a summary of the expected September Arctic sea-ice minimum. Monthly reports are released through the summer in an attempt to provide the scientific community and stakeholders with the best available information.

On the longer decadal and centennial timescales, sea-ice hindcasts and projections have been organized under the IPCC rubric, which has put together simulations based on ensemble runs of those global climate models that contain some form of a sea-ice component. The simulations captured to some degree the climatological annual mean, seasonal cycle, and temporal trends of sea-ice area during the modern satellite era, although there was considerable

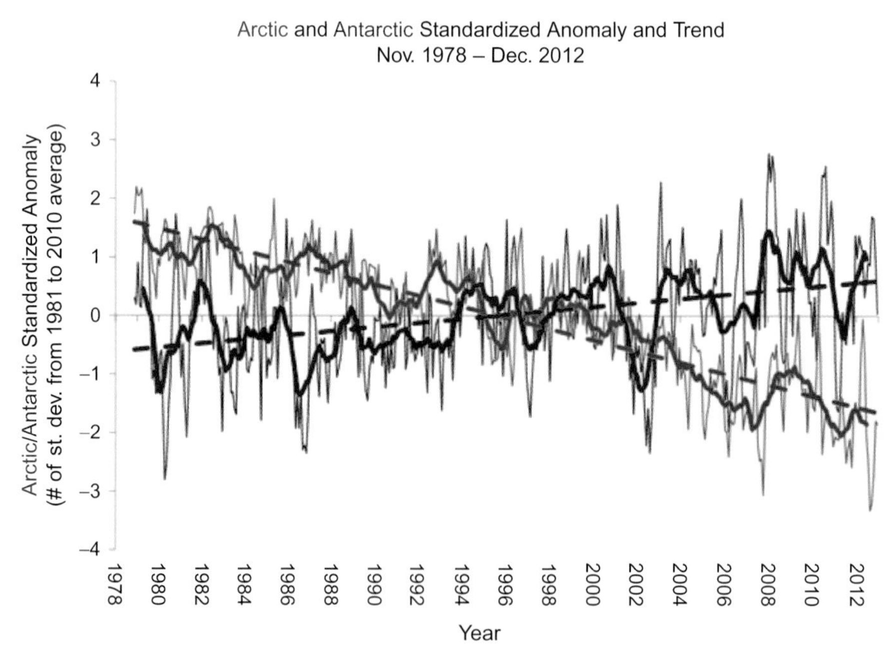

FIGURE 16.11 Arctic (blue curve) and Antarctic (red curve) standardized anomalies and trends of sea-ice cover over the period 1978–2012. Thick lines indicate 12-month running means, and thin lines indicate monthly anomalies. The Arctic shows a definite trend downward, while the Antarctic shows a slight increase. *http://nsidc.org/cryosphere/sotc/sea_ice.html.*

scatter among the models. Each member of the ensemble performed differently, and the interannual variability within each member model is not yet fully understood. An interesting aspect of the simulations is that multimodel ensemble means show promising estimates close to observations for the late twentieth century (Zhang and Walsh, 2006). With regard to sea-ice thickness distributions, considerable deficiencies in sea-ice properties in many model results have been seen. Most of the models have difficulty in reproducing the spatial sea-ice thickness distribution of the past, particularly in summer (Gerdes and Köberle, 2007). Taken collectively, the models indicate a significant reduction in Arctic sea-ice cover over the coming decades. However, the large scatter between individual model simulations at present leads to much uncertainty as to when a seasonally ice-free Arctic Ocean might be realized (Serreze et al., 2007). Of real concern is the fact that the models failed to capture the extreme minimum that occurred in 2007 (prior to the issuance of the IPCC report). An important criterion for future development and improvement of sea-ice models will be their ability to capture such extreme events.

Projections from IPCC Antarctic sea-ice models show there is larger uncertainty in model performance in the Southern Hemisphere (Arzel et al., 2006) with a much larger disagreement between the model ensemble members for the Southern Ocean sea-ice simulations. Many of the models have an annual sea-ice extent cycle that differs markedly from that observed over the past 30 years (Turner et al., 2013). In contrast to the satellite data, which exhibit a slight increase in extent, the mean extent of the models over the period 1979–2005 shows a decrease (Turner et al., 2013). Several studies have shown that the response of the Southern Ocean to an increase in atmospheric greenhouse gas concentrations is delayed compared with other regions (Goosse and Renssen, 2005), mainly because of the large thermal inertia of the Southern Ocean. This highlights the earlier discussion of the bathymetric contrasts between the Arctic and Southern oceans, the fact that the former is largely disconnected from the World Ocean while the latter is not. Aside from an inability to properly capture the recent ice extent trends, the coupled models have also not been able to capture the regional spatial variability of sea-ice cover over the Southern Hemisphere in the past few decades, during which time the Ross Sea has been gaining cover while the Amundsen has been losing it (Stammerjohn et al., 2008). Again, this suggests need for further research to improve sea-ice models, and their coupling with the ocean and atmosphere.

6.3. Observed Land-Ice Change

The recent breakup of ice shelves in both Greenland (e.g., Jakobshavn) and Antarctica (e.g., Larsen B) has also caught researchers by surprise. The Jakobshavn Ice Shelf has been undergoing periodic retreat over the past 100 years or so (Sohn et al., 1998), but with a dramatic thinning and retreat occurring starting in 1997 (Joughin et al., 2004). No less dramatic was the breakup of Larsen B, occurring in just a few weeks in early 2002. From an analysis of the marine sediments located beneath the former Larsen B Ice Shelf, it has been demonstrated that the recent collapse has been unprecedented in the Holocene (Domack et al., 2005). There is some evidence to suggest that the Greenland breakup was forced by the arrival of warm waters (Holland et al., 2008) and that the Antarctic case was driven by the arrival of warm air over the Antarctic Peninsula (Marshall et al., 2006). In the case of both Jakobshavn and the Larsen B ice shelves, the inland glaciers connected to the shelves have sped up dramatically, with a doubling or betterment in their speeds (Rignot et al., 2004). The breakup of these ice shelves has led to a refocus toward both the sensitivity of these ice shelves to climate change and the implications of ice-shelf decay for the stability of grounded ice (De Angelis and Skvarca, 2003).

A far more disconcerting example of recent glacial change is the observed thinning of several of the ice shelves fringing the Amundsen Sea in West Antarctica (Pritchard et al., 2012). As those ice shelves serve as a front to the massive marine-based West Antarctic Ice Sheet, any negative mass balance signal in that area could hold serious consequences for global sea-level rise. Using satellite laser altimetry combined with modeling of the firn layer, it has been deduced that ice shelves in this sector of Antarctica are thinning through increased basal melting (Pritchard et al., 2012). In fact, the greatest thinning occurs where warm, deep water can gain access to the continental shelf, and ultimately the grounding zone of the glaciers, via deep troughs in the shelves. From ocean modeling studies, it appears that slowly changing wind stress patterns over the Amundsen Sea are responsible for the changes in warm water volume being brought aboard the continental shelf (Thoma et al., 2008). Understanding what controls the variability of the large-scale wind patterns in the polar regions and indeed on the global scale is an area of active research and holds the ultimate key to how warm water can arrive at the base of an ice shelf (Kerr, 2008). Both Greenland and Antarctica are showing an overall accelerating mass loss (Figure 16.12).

6.4. Land-Ice Projections

The projections of land ice change, and the implicit role that climate change and warm ocean waters could have on breaking up of ice shelves and of causing acceleration of inland ice into the ocean, have been lacking. This shortcoming is implicitly acknowledged by IPCC which failed to provide an estimate of the potential dynamical contribution of glaciers to sea-level rise over the present century

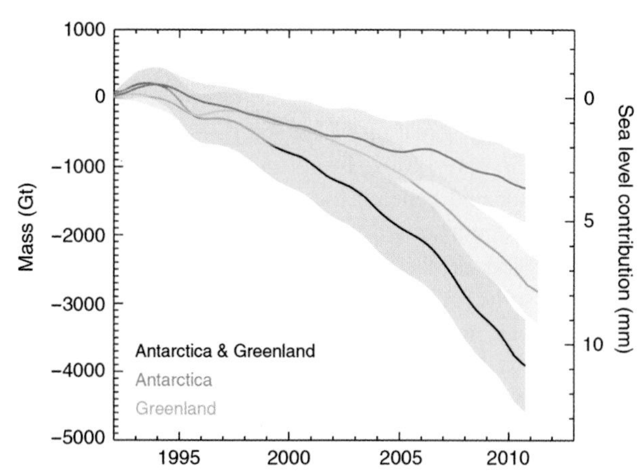

FIGURE 16.12 Change in the Antarctic Ice Sheet and Greenland Ice Sheet during 1992–2010 determined from a reconciliation of measurements acquired by various satellite instrumentation. Also shown is the equivalent global sea-level contribution (right axis), calculated assuming that 360 Gt of ice corresponds to 1 mm of sea-level rise. *From Shepherd et al. (2012).*

(IPCC, 2007). This apparent shortcoming serves to underscore the difficulty of this task, which the research community has only relatively recently addressed in earnest (IPCC, 2010). The projections of the contribution to sea level of the Antarctic Ice Sheet to sea-level rise are uncertain (Alley et al., 2005) as they range from a few centimeters rise over the next century to a meter or more, the latter implying devastating consequences for coastal habitats and infrastructure (Nicholls and Cazenave, 2010). Part of the difficulty in producing credible projections is the lack of observational evidence upon which to base such model (Alley et al., 2005). That situation is changing with the emerging ice and ocean observational technology as discussed in Section 5, but it will take decades to build up meaningful observational records in these difficult-to-access regions. Progress in model coupling is likely to proceed at a faster pace (Lipscomb et al., 2009), but will have to await the availability of basic observation data, again highlighting the need for the science community to have a sustainable ocean and ice observational system in place around the periphery of the great ice sheets.

6.5. Marine Permafrost

Concern has been raised that the considerable deposits of methane, in the form of sub-seafloor clathrates or gas trapped beneath other permafrost, could be released under a scenario of global warming leading to a runaway feedback on global warming. Methane is a powerful greenhouse gas, having a global warming potential of more than 60 times that of carbon dioxide. Over the Arctic continental shelf, methane gas release has indeed been detected (Kort et al., 2012). The observed distribution of dissolved methane and possible mechanisms of gas release in connection with observed dynamics of coastal ocean environments suggest that areas of the Arctic shelves are important natural sources of methane to the atmosphere and that such areas tend to be affected by ongoing global change (Shakhova et al., 2007), possibly being susceptible to ongoing and future changes in sea-ice and ocean circulation.

The speculative scenario of a release of vast amounts of methane or its clathrate form is commonly referred to as the clathrate gun hypothesis (Kennett et al., 2003), the main idea being that the seafloor permafrost would itself initially melt under conditions of a warming global ocean thereby triggering the release of methane. In its original form, the hypothesis proposed that the clathrate gun could cause abrupt runaway warming on a timescale of less than a human lifetime (Kennett et al., 2003) and might have been responsible for warming events in and at the end of the last ice age (Behl, 2000), but more recent research now suggests this to be unlikely (Sowers, 2006). Past suspected releases during the Permian–Triassic extinction event and the Paleocene–Eocene Thermal Maximum have been linked to such gas release and past climate change (Benton and Twitchett, 2003).

7. SUMMARY

In this chapter, we have looked into aspects of the interaction of the World Ocean with the marine cryosphere, focusing on the Arctic and Antarctic sea-ice covers, the outlet glaciers of the Greenland and Antarctic ice sheets, and to a lesser extent, the marine permafrost beneath the seafloor. We have discussed the fundamental chemical and physical properties of freezing water and ice that organize the manner in which ice forms or melts in the ocean, be it sea ice or land ice, and have seen that heat exchange at the ice–ocean interface is unique in ocean dynamics because of the fundamental role played by molecular exchange. The description of these fine-scale processes and their encapsulation as parameterizations for use in global climate models is a maturing subject.

On the large scale, the transport of heat to the polar ocean is also relatively well understood, but what is not understood is the manner in which that heat transport changes through natural variability or anthropogenic forcing. The global ocean impacts the marine cryosphere by transporting warm water to the base of sea ice or the base of an ice shelf, and the ensuing melt impacts back onto the global ocean through the conveyor belt by both influencing surface convection and bottom water formation. The ramifications of these large-scale ice–ocean interactions and potential feedbacks on the climate system are still an area of active research.

Remote-sensing observational technologies have been delivering data sets that provide an unprecedented global view into the marine cryosphere. Perhaps the largest advance, and surprise, of the past decades has been the witnessing of the enormous change occurring in both sea ice and land ice. This change has largely reshaped a viewpoint in which the marine cryosphere was thought to be a relatively fixed feature of the climate system on the timescale of decades or a century. We have now begun to see that viewpoint unravel and be replaced by one where the new norm is Arctic sea ice being halved in summer extent and a number of outlet glaciers doubling their seaward velocity. Such change shows the sensitivity of the marine cryosphere to climate change, whether natural or anthropogenic, and the pressing need for further sub-ice, polar ocean observational data sets and numerical modeling to provide a credible and robust ability to project the future of the marine cryosphere.

REFERENCES

Aagaard, K., Carmack, E.C., 1989. The role of sea ice and other fresh water in the Arctic circulation. J. Geophys. Res. Oceans 94 (C10), 14485–14498.

Alley, R.B., Clark, P.U., Huybrechts, P., Joughin, I., 2005. Ice-sheet and sea-level changes. Science 310 (5747), 456–460.

Alley, R.B., Fahnestock, M., Joughin, I., 2008. Understanding glacier flow in changing times. Science 322 (5904), 1061–1062.

Allison, I., Brandt, R.E., Warren, S.G., 1993. East Antarctic sea ice: Albedo, thickness distribution, and snow cover. J. Geophys. Res. 98 (C7), 12417–12429.

AMISOR, 2013. The Amery Ice Shelf Ocean Research project. http://www.antarctica.gov.au/.

Anderson, J.B., Kurtz, D.D., Domack, E.W., Balshaw, K.M., 1980. Glacial and glacial marine sediments of the Antarctic continental shelf. J. Geol. 88 (4), 399–414.

Arzel, O., Fichefet, T., Goosse, H., 2006. Sea ice evolution over the 20th and 21st centuries as simulated by current AOGCMs. Ocean Model. 12 (3), 401–415.

ASPECT, 2012. Antarctic sea ice processes and climate. http://aspect.antarctica.gov.au/.

Bamber, J.L., Layberry, R.L., Gogineni, S.P., 2001. A new ice thickness and bed data set for the Greenland ice sheet: 1. Measurement, data reduction, and errors. J. Geophys. Res. Atmos. 106 (D24). http://dx.doi.org/10.1029/2001JD900054.

Bamber, J.L., Riva, R.E.M., Vermeersen, B.L.A., LeBroq, A.M., 2009. Reassessment of the potential sea-level rise from a collapse of the West Antarctic Ice Sheet. Science 324 (5929), 901–903. http://dx.doi.org/10.1126/science.1169335.

Bao, X., Webb, D.J., Jackson, D.A., 1993. 32-km distributed temperature sensor based on Brillouin loss in an optical fiber. Opt. Lett. 18 (18), 1561–1563.

Barber, D.G., Massom, R.A., 2007. The role of sea ice in Arctic and Antarctic polynyas. Elsevier Oceanography Series 74, 1–54.

Bassis, J.N., 2011. The statistical physics of iceberg calving and the emergence of universal calving laws. J. Glaciol. 57 (201), 3–16.

Behl, R.J., 2000. Carbon isotopic evidence for methane hydrate instability during quaternary interstadials. Science 288 (5463), 128–133. http://dx.doi.org/10.1126/science.288.5463.128.

Beining, P., Roether, W., 1996. Temporal evolution of CFC 11 and CFC 12 concentrations in the ocean interior. J. Geophys. Res. 101 (C7), 16455–16464.

Bentley, C.R., Ostenso, N.A., 1961. Glacial and subglacial topography of West Antarctica. J. Glaciol. 3 (29), 882–912.

Benton, M.J., Twitchett, R.J., 2003. How to kill (almost) all life: the end-Permian extinction event. Trends Ecol. Evol. 18 (7), 358–365. http://dx.doi.org/10.1016/S0169-5347(03)00093-4.

Berton, P., 1988. The Arctic Grail: The Quest for the Northwest Passage and The North Pole, 1818–1909. McLelland & Stewart, Toronto. ISBN 0-7710-1266-7.

Blatter, H., Greve, R., Abe-Ouchi, A., 2011. Present state and prospects of ice sheet and glacier modelling. Surv. Geophys. 32 (4–5), 555–583.

Bott, M.H.P., 1971. Evolution of young continental margins and formation of shelf basins. Tectonophysics 0040-195111 (5), 319–327. http://dx.doi.org/10.1016/0040-1951(71)90024-2.

Broecker, W., Denton, G., 1990. What drives glacial cycles? Sci. Am. 262 (1), 49–56.

Broecker, W.S., Peng, T.H., 1987. The oceanic salt pump: does it contribute to the glacial-interglacial difference in atmospheric CO2 content? Global Biogeochem. Cycles 1 (3), 251–259.

Broecker, W.S., Kennett, J.P., Flower, B.P., Teller, J.T., Trumbore, S., Bonani, G., Wolfli, W., 1989. Routing of meltwater from the Laurentide Ice Sheet during the Younger Dryas cold episode. Nature 341, 318–321. http://dx.doi.org/10.1038/341318a0.

Buffett, B., Archer, D., 2004. Global inventory of methane clathrate: sensitivity to changes in the deep ocean. Earth Planet. Sci. Lett. 227 (3), 185–199.

Burton, J.C., Amundson, J.M., Abbot, D.S., Boghosian, A., Cathles, L.M., Correa-Legisos, S., Darnell, K.N., Guttenberg, N., Holland, D.M., MacAyeal, D.R., 2012. Laboratory investigations of iceberg capsize dynamics, energy dissipation and tsunamigenesis. J. Geophys. Res. Earth Surf. (2003–2012) 117 (F1). http://dx.doi.org/10.1029/2011JF002055.

Campagna, C., Fedak, M.A., McConnell, B.J., 1999. Post-breeding distribution and diving behavior of adult male southern elephant seals from Patagonia. J. Mammal. 80, 1341–1352.

CICE, 2012. The Los Alamos Sea Ice Model. http://climate.lanl.gov/Models/CICE/.

CISM, 2012. Community Ice Sheet Model. http://oceans11.lanl.gov/trac/CISM.

Corwin, R.F., 1983. Marine permafrost detection using galvanic electrical resistivity methods. In: Offshore Technology Conference, 2–5 May, Houston, Texas. ISBN 978-1-61399-075-9.

Cox, G., Weeks, W.F., 1982. Equations for Determining the Gas and Brine Volumes in Sea Ice Samples. US Army Corps of Engineers, Cold Regions Research & Engineering Laboratory.

Csathó, B.M., Bolzan, J.F., Van Der Veen, C.J., Schenk, A.F., Lee, D.C., 1999. Surface velocities of a Greenland outlet glacier from high-resolution visible satellite imagery 1. Polar Geogr. 23 (1), 71–82.

De Angelis, H., Skvarca, P., 2003. Glacier surge after ice shelf collapse. Science 299, 1560–1562.

de Boyer Montegut, C., Madec, G., Fischer, A., Lazar, A., Iudicone, D., 2004. Mixed layer depth over the global ocean: an examination of profile data and a profile-based climatology. J. Geophys. Res. Oceans 109 (C12).

Domack, E., Duran, D., Leventer, A., Ishman, S., Doane, S., McCallum, S., Prentice, M., 2005. Stability of the Larsen B ice shelf on the Antarctic Peninsula during the Holocene epoch. Nature 436 (7051), 681–685.

Drewry, D.J., Jordan, S.R., Jankowski, E., 1983. Measured properties of the Antarctic ice sheet: surface configuration, ice thickness, volume and bedrock characteristics. Ann. Glaciol. 3, 83–91.

Ebert, E.E., Curry, J.A., 1993. An intermediate one-dimensional thermodynamic sea ice model for investigating ice-atmosphere interactions. J. Geophys. Res. Oceans 98 (C6), 10085–10109.

Eichhorn, M., 2009. A new concept for an obstacle avoidance system for the AUV "SLOCUM Glider" operation under ice. In: Oceans 2009-Europe. IEEE, pp. 1–8.

Ekman, V.W., 1905. On the influence of the earth's rotation on ocean currents. Ark. Mat. Astron. Fys. 2 (11), 1–52.

Emery, W.J., Fowler, C.W., Maslanik, J.A., 1997. Satellite-derived maps of Arctic and Antarctic sea ice motion: 1988 to 1994. Geophys. Res. Lett. 24 (8), 897–900.

England, M.H., Godfrey, J.S., Hirst, A.C., Tomczak, M., 1993. The mechanism for Antarctic Intermediate Water renewal in a world ocean model. J. Phys. Oceanogr. 23 (7), 1553–1560.

Fedak, M., Lovell, P., McConnell, B., Hunter, C., 2002. Overcoming the constraints of long range radio telemetry from animals: getting more useful data from smaller packages. Integr. Comp. Biol. 42 (1), 3–10.

Feltham, D.L., 2008. Sea ice rheology. Annu. Rev. Fluid Mech. 40, 91–112.

Fetterer, F., Untersteiner, N., 1998. Observations of melt ponds on Arctic sea ice. J. Geophys. Res. 103 (C11), 24821–24824.

Fichefet, T., Goosse, H., Maqueda, M.A.M., 2003. A hindcast simulation of Arctic and Antarctic sea ice variability, 1955–2001. Polar Res. 22 (1), 91–98.

Flato, G.M., Hibler III., W.D., 1992. Modeling pack ice as a cavitating fluid. J. Phys. Oceanogr. 22 (6), 626–651.

Foldvik, A., Gammelsrød, T., 1988. Notes on Southern Ocean hydrography, sea-ice and bottom water formation. Palaeogeogr. Palaeoclimatol. Palaeoecol. 0031-018267 (1–2), 3–17. http://dx.doi.org/10.1016/0031-0182(88)90119-8.

Foster, T.D., Carmack, E.C., 1976. Frontal zone mixing and Antarctic Bottom Water formation in the southern Weddell Sea. Deep Sea Res. 23 (4), 301–317.

Fujino, K., Lewis, E.L., Perkin, R.G., 1974. The freezing point of seawater at pressures up to 100 bars. J. Geophys. Res. 79, 1792–1797.

Gade, H.G., 1979. Melting of ice in sea water: a primitive model with application to the antarctic ice shelf and icebergs. J. Phys. Oceanogr. 9, 189–198.

Gerdes, R., Köberle, C., 2007. Comparison of Arctic sea ice thickness variability in IPCC Climate of the 20th Century experiments and in ocean–sea ice hindcasts. J. Geophys. Res. 112, C04S13. http://dx.doi.org/10.1029/2006JC003616.

Gladstone, R., Bigg, G.R., 2002. Satellite tracking of icebergs in the Weddell Sea. Antarct. Sci. 14 (3), 278–287.

Gloersen, P., Campbell, W.J., Cavalieri, D.J., Comiso, J.C., Parkinson, C.L., Zwally, H.J., 1992. Arctic and Antarctic Sea Ice, 1978–1987: Satellite Passive-Microwave Observations and Analysis. Scientific and Technical Information Program, National Aeronautics and Space Administration, Washington, DC.

Golubev, V.N., Sokratov, S.A., Shishkov, A.V., 2006. Effect of sea ice extent on atmosphere–ocean CO2 exchange. Geophys. Res. Abstr. 8, 03989, SRef-ID: 1607-7962/gra/EGU06-A-03989.

Goosse, H., Renssen, H., 2005. A simulated reduction in Antarctic sea-ice area since 1750: implications of the long memory of the ocean. Int. J. Climatol. 25, 569–579. http://dx.doi.org/10.1002/joc.1139.

Grenfell, T.C., 1991. A radiative transfer model for sea ice with vertical structure variations. J. Geophys. Res. Oceans 96 (C9), 16991–17001.

Hellmer, H., et al., 2006. Sea ice feedbacks observed in western Weddell Sea. Eos 87 (18), 173. http://dx.doi.org/10.1029/2006EO180001.

Herried, B., Porter, C.C., Morin, P.J., Howat, I.M., 2011. Rapidice viewer: a web application to observe near real-time changes in Polar ice sheets and Glaciers with a multi-sensor multi-temporal approach. In: AGU Fall Meeting Abstracts, vol. 1. p. 0455.

Hibler, W.D., 1979. A dynamic thermodynamic sea ice model. J. Phys. Oceanogr. 9, 815–846. http://dx.doi.org/10.1175/1520-0485(1979)009<0815:ADTSIM>2.0.CO;2.

Hobbs, P., 1971. Ice Physics. Oxford Classic Texts.

Hoffman, P.F., Schrag, D.P., 2002. The snowball Earth hypothesis: testing the limits of global change. Terra Nova 14 (3), 129–155.

Holland, D.M., 2001. Explaining the Weddell Polynya—a large ocean eddy shed at Maud Rise. Science 292 (5522), 1697–1700.

Holland, D.M., Jenkins, A., 1999. Modeling thermodynamic ice-ocean interactions at the base of an ice shelf. J. Phys. Oceanogr. 29.8, 1787–1800.

Holland, D.M., Jenkins, A., 2001. Adaptation of an isopycnic coordinate ocean model for the study of circulation beneath ice shelves. Mon. Weather Rev. 129 (8), 1905–1927.

Holland, P.R., Kwok, R., 2012. Wind-driven trends in Antarctic sea-ice drift. Nat. Geosci. 5, 872–875. http://dx.doi.org/10.1038/ngeo1627.

Holland, D.M., Mysak, L.A., Manak, D.K., Oberhuber, J.M., 1993. Sensitivity study of a dynamic thermodynamic sea ice model. J. Geophys. Res. Oceans (1978–2012) 98 (C2), 2561–2586.

Holland, M.M., Bitz, C., Weaver, A.J., 2001. The influence of sea ice physics on simulations of climate change. J. Geophys. Res. Oceans (1978–2012) 106 (C9), 19639–19655.

Holland, D.M., Thomas, R.H., deYoung, B., Ribergaard, M.H., Lyberth, B., 2008. Acceleration of Jakobshavn Isbrae triggered by warm subsurface ocean waters. Nat. Geosci. 1, 659–664. http://dx.doi.org/10.1038/ngeo316.

Holland, P.R., Corr, H.F., Vaughan, D.G., Jenkins, A., Skvarca, P., 2009. Marine ice in Larsen Ice Shelf. Geophys. Res. Lett. 36 (11). http://dx.doi.org/10.1029/2009GL038162.

Hooke, R. LeB, 1981. Flow law for polycrystalline ice in glaciers' comparison of theoretical predictions, laboratory data, and field. Rev. Geophys. Space Phys. 19 (4), 664–672.

Howe, B.M., Miller, J.H., 2004. Acoustic sensing for ocean research. Mar. Technol. Soc. J. 38 (2), 144–154.

Hunke, E.C., Dukowicz, J.K., 1997. An elastic-viscous-plastic model for sea ice dynamics. J. Phys. Oceanogr. 27 (9), 1849–1867.

Hunkins, K., 1966. Ekman drift currents in the Arctic Ocean. Deep Sea Res. 13, 607–620.

Hutter, K., 1982. A mathematical model of polythermal glaciers and ice sheets. Geophys. Astrophys. Fluid Dyn. 21 (3–4), 201–224.

Hvidegaard, S.M., Forsberg, R., 2002. Sea-ice thickness from airborne laser altimetry over the Arctic Ocean north of Greenland. Geophys. Res. Lett. 29 (20), 1952.

IABP, 2012. International Arctic Buoy Programme. http://iabp.apl.washington.edu/index.html.

Imbrie, J., Boyle, E.A., Clemens, S.C., Duffy, A., Howard, W.R., Kukla, G., Toggweiler, J.R., 1992. On the structure and origin of major glaciation cycles 1. Linear responses to Milankovitch forcing. Paleoceanography 7 (6), 701–738.

IPAB, 2012. International Program for Antarctic Buoys. http://ipab.aq/.

IPCC, 2007. Climate change 2007: the physical science basis. In: Solomon, S., Qin, D., Manning, M., Chen, Z., Marquis, M., Averyt, K.B., Tignor, M., Mille, H.L. (Eds.), Contribution of Working Group I to the Fourth Assessment Report of the Intergovernmental Panel on Climate Change. Cambridge University Press, Cambridge, UK.

IPCC, 2010. Stocker, T.F., Qin, D., Plattner, G.-K., Tignor, M., Allen, S., Midgley, P.M. (Eds.), Workshop Report of the Intergovernmental Panel on Climate Change Workshop on Sea Level Rise and Ice Sheet Instabilities. IPCC Working Group I Technical Support Unit, University of Bern, Bern, Switzerland, p. 227. https://www.ipcc-wg1.unibe.ch/publications/supportingmaterial/SLW_WorkshopReport.pdf.

ITP, 2012. Ice Tethered Profiler. http://www.whoi.edu/page.do?pid=20756.

Jacobs, S.S., Amos, A.F., Bruchhausen, P.M., 1970. Ross Sea oceanography and Antarctic bottom water formation. Deep Sea Res. 17 (6).

Jacobs, S.S., Giulivi, C.F., Mele, P.A., 2002. Freshening of the Ross Sea during the late 20th century. Science 297 (5580), 386–389.

Jakobsson, M., 2002. Hypsometry and volume of the Arctic Ocean and its constituent seas. Geochem. Geophys. Geosyst. 3 (1). http://dx.doi.org/10.1029/2001GC000302.

Jalving, B., Faugstadmo, J.E., Vestgard, K., Hegrenaes, O., Engelhardtsen, O., Hyland, B., 2008. Payload sensors, navigation and risk reduction for AUV under ice surveys. In: 2008 IEEE/OES Autonomous Underwater Vehicles. IEEE, pp. 1–8.

Jeffries, M.O., Worby, A.P., Morris, K., Weeks, W.F., 1997. Seasonal variations in the properties and structural composition of sea ice and snow cover in the Bellingshausen and Amundsen Seas, Antarctica. J. Glaciol. 43 (143), 138–151.

Jenkins, A., 1991. A one-dimensional model of ice shelf-ocean interaction. J. Geophys. Res. 96 (C11), 20,671. http://dx.doi.org/10.1029/91JC01842.

Jenkins, A., Holland, D.M., 2007. Melting of floating ice and sea level rise. Geophys. Res. Lett. 34, L16609. http://dx.doi.org/10.1029/2007GL030784.

Joffre, S.M., 1982. Momentum and heat transfers in the surface layer over a frozen sea. Boundary Layer Meteorol. 24 (2), 211–229.

Josberger, E.G., Martin, S., 1981. A laboratory and theoretical study of the boundary layer adjacent to a vertical melting ice wall in salt water. J. Fluid Mech. 111, 439–473.

Joughin, I., Padman, L., 2003. Melting and freezing beneath Filchner-Ronne Ice Shelf, Antarctica. Geophys. Res. Lett. 30, 1477. http://dx.doi.org/10.1029/2003GL016941.

Joughin, I., Abdalati, W., Fahnestock, M., 2004. Large fluctuations in speed on Greenland's Jakobshavn Isbrae glacier. Nature 432 (7017), 608–610. http://dx.doi.org/10.1038/nature03130.

Kader, B.A., Yaglom, A.M., 1972. Heat and mass transfer laws for fully turbulent wall flows. Int. J. Heat Mass Transf. 15 (12), 2329–2351.

Kennett, J.P.K.G., Cannariato, I.L., Hendy, R.J. Behl, 2003. Methane Hydrates in Quaternary Climate Change: The Clathrate Gun Hypothesis. American Geophysical Union, Washington, DC. ISBN 0-87590-296-0.

Kerr, R.A., 2008. Winds, not just global warming, eating away at the ice sheets. Science 322 (5898), 33.

Klatt, O., Boebel, O., Fahrbach, E., 2007. A profiling float's sense of ice. J. Atmos. Oceanic Technol. 24 (7), 1301–1308.

König Beatty, C., Holland, D.M., 2010. Modeling landfast sea ice by adding tensile strength. J. Phys. Oceanogr. 40 (1), 185–198.

Kort, E.A., Wofsy, S.C., Daube, B.C., Diao, M., Elkins, J.W., Gao, R.S., Zondlo, M.A., 2012. Atmospheric observations of Arctic Ocean methane emissions up to 82 [deg] north. Nat. Geosci. 5 (5), 318–321.

Krishfield, R., Toole, J.H., Proshutinsky, A., Timmermans, M.L., 2008. Automated ice-tethered profilers for seawater observations under pack ice in all seasons. J. Atmos. Oceanic Technol. 25 (11), 2091–2105.

Kundu, P.K., 1976. Ekman veering observed near the ocean bottom. J. Phys. Oceanogr. 6, 238–242.

Kvenvolden, K., 1995. A review of the geochemistry of methane in natural gas hydrate. Org. Geochem. 23 (11–12), 997–1008. http://dx.doi.org/10.1016/0146-6380(96)00002-2.

Kwok, R., Rothrock, D.A., 2009. Decline in Arctic sea ice thickness from submarine and ICESat records: 1958–2008. Geophys. Res. Lett. 36 (15). http://dx.doi.org/10.1029/2009GL039035.

Lammers, R.B., Shiklomanov, A.I., Vörösmarty, C.J., Fekete, B.M., Peterson, B.J., 2001. Assessment of contemporary Arctic river runoff based on observational discharge records. J. Geophys. Res. Atmos. 106 (D4), 3321–3334.

Large, W.G., McWilliams, J.C., Doney, S.C., 1994. Oceanic vertical mixing: a review and a model with a nonlocal boundary layer parameterization. Rev. Geophys. 32 (4), 363–403.

Laxon, S., Peacock, N., Smith, D., 2003. High interannual variability of sea ice thickness in the Arctic region. Nature 425 (6961), 947–950.

Lee, C.M., Petrie, B., Gobat, J.I., Soukhovtsev, V., Abriel, J., Van Thiel, K., Scotney, M., 2004. An observational array for high-resolution, year-round measurements of volume, freshwater, and ice flux variability in Davis Strait. Cruise Report for R/V Knorr 179-05, 22 September-4 October 2004 (No. APL-UW-TR-0408).

Lemieux, J.-F., Knoll, D., Tremblay, B., Holland, D.M., Losch, M., 2012. A comparison of the Jacobian-free Newton-Krylov method and the EVP model for solving the sea ice momentum equation with a viscous-plastic formulation: a serial algorithm study. J. Comp. Phys. 231 (17), 5926–5944.

Lemke, P., 2001. Open windows to the polar oceans. Science 292 (5522), 1670–1671.

Lemke, P., Owens, W.B., Hibler, W.D., 1990. A coupled sea ice-mixed layer-pycnocline model for the Weddell Sea. J. Geophys. Res. Oceans 95 (C6), 9513–9525.

Lenn, Y., Chereskin, T.K., 2009. Observation of Ekman Currents in the Southern Ocean. J. Phys. Oceanogr. 39, 768–779.

Lewis, E.L., Perkin, R.G., 1986. Ice pumps and their rates. J. Geophys. Res. Oceans 91 (C10), 11756–11762.

Lietaer, O., Fichefet, T., Legat, V., 2008. The effects of resolving the Canadian Arctic Archipelago in a finite element sea ice model. Ocean Model. 24 (3), 140–152.

Lipscomb, W., Bindschadler, R., Bueler, E., Holland, D., Johnson, J., Price, S., 2009. A community ice sheet model for sea level prediction: building a next-generation community ice sheet model. Los Alamos, New Mexico, 18–20 August 2008. Eos 90 (3), 23. http://dx.doi.org/10.1029/2009EO030004.

Lizotte, M.P., 2001. The contributions of sea ice algae to Antarctic marine primary production. Am. Zool. 41 (1), 57–73.

Lock, G.S.H., 1990. Growth and Decay of Ice. Cambridge University Press.

MacAyeal, D.R., Okal, E.A., Aster, R.C., Bassis, J.N., 2009. Seismic observations of glaciogenic ocean waves (micro-tsunamis) on icebergs and ice shelves. J. Glaciol. 55 (190), 193–206.

Mahoney, A., Eicken, H., Gaylord, A.G., Shapiro, L., 2007. Alaska landfast sea ice: links with bathymetry and atmospheric circulation. J. Geophys. Res. Oceans 112 (C2). http://dx.doi.org/10.1029/2006JC003559.

Manabe, S., Stouffer, R.J., 1995. Simulation of abrupt climate change induced by freshwater input to the North Atlantic Ocean. Nature 378 (6553), 165–167. http://dx.doi.org/10.1038/378165a0.

Marshall, G.J., et al., 2006. The impact of a changing Southern Hemisphere Annular Mode on Antarctic Peninsula summer temperatures. J. Clim. 19, 5388–5404.

Martinson, D.G., 1990. Evolution of the southern ocean winter mixed layer and sea ice: Open ocean deepwater formation and ventilation. J. Geophys. Res. 5 (C7), 11,641. http://dx.doi.org/10.1029/JC095iC07p11641.

Max, M.D., 2003. Natural Gas Hydrate in Oceanic and Permafrost Environments. Kluwer Academic Publishers. ISBN 0-7923-6606-9.

Maykut, G.A., Untersteiner, N., 1971. Some results from a time-dependent, thermodynamic model of sea ice. J. Geophys. Res. 76, 1550–1575.

McPhail, S.D., Furlong, M.E., Pebody, M., Perrett, J.R., Stevenson, P., Webb, A., White, D., 2009. Exploring beneath the PIG Ice Shelf with the Autosub3 AUV. In: Oceans 2009-Europe. IEEE, pp. 1–8.

McPhee, M.G., 2008. Air-Ice-Ocean Interaction. Springer.

Meier, M.F., Post, A., 1987. Fast tidewater glaciers. J. Geophys. Res. Solid Earth (1978–2012) 92 (B9), 9051–9058.

Meier, M.F., Post, A., Rasmussen, L.A., Sikonia, W.G., Mayo, L.R., 1980. Retreat of Columbia Glacier, Alaska—a preliminary prediction. 80-10. USGS Open-File Report.

Mellor, M., 1983. Mechanical behavior of sea ice. No. CRREL-83-1, New Hampshire.

Mercer, J.H., 1978. West Antarctic ice sheet and CO2 greenhouse effect: a threat of disaster. Ohio State University, Institute of Polar Studies.

Mohr, J.J., Reeh, N., Madsen, S.N., 1998. Three-dimensional glacial flow and surface elevation measured with radar interferometry. Nature 391 (6664), 273–276.

Moritz, R., Wensnahan, M., 2011. Sea-ice Thickness and Draft Statistics from Submarine ULS, Moored ULS, and a Coupled Model. National Snow and Ice Data Center. Digital media, Boulder, CO.

Nansen, F., 1902. The Oceanography of the North Polar Basin. Longmans, Green, and Company.

Nguyen, A.T., Kwok, R., Menemenlis, D., 2012. Source and pathway of the Western Arctic upper halocline in a data-constrained coupled ocean and sea ice model. J. Phys. Oceanogr. 42, 802–823.

Nicholls, R.J., Cazenave, A., 2010. Sea-level rise and its impact on coastal zones. Science 328 (5985), 1517–1520.

Nicholls, K.W., Makinson, K., Robinson, A.V., 1991. Ocean circulation beneath the Ronne ice shelf. Nature 354 (6350), 221–223.

Nick, F.M., Vieli, A., Howat, I.M., Joughin, I., 2009. Large-scale changes in Greenland outlet glacier dynamics triggered at the terminus. Nat. Geosci. 2 (2), 110–114.

Niedrauer, T.M., Martin, S., 1979. An experimental study of brine drainage and convection in young sea ice. J. Geophys. Res. Oceans 84 (C3), 1176–1186.

NOAA, 2012. National oceanographic and atmosphere administration, climate indicators, Arctic oscillation. http://www.arctic.noaa.gov/detect/climate-ao.shtml.

North, G.R., Cahalan, R.F., Coakley, J.A., 1981. Energy balance climate models. Rev. Geophys. 19 (1), 91–121.

NSIDC, 2007. National Snow and Ice Data Center, Bootstrap Sea Ice Concentrations from Nimbus-7 SMMR and DMSP SSM/I. http://nsidc.org/api/metadata?id=nsidc-0079.

NSIDC, 2012. National Snow and Ice Data Center, Arctic sea ice extent settles at record seasonal minimum. http://nsidc.org/arcticseaicenews/2012/09/arctic-sea-ice-extent-settles-at-record-seasonal-minimum/.

Oberhuber, J.M., Holland, D.M., Mysak, L.A., 1993. A thermodynamic-dynamic snow sea-ice model. In: Ice in the Climate System. Springer, Berlin, pp. 653–673.

Oerter, H., Kipfstuhl, J., Determann, J., Miller, H., Wagenbach, D., Minikin, A., Graf, W., 1992. Evidence for basal marine ice in the Filchner-Ronne ice shelf. Nature 358, 399–401.

Ogilvie, A.E.J., Barlow, L.K., Jennings, A.E., 2000. North Atlantic climate c.ad 1000: Millennial reflections on the Viking discoveries of Iceland, Greenland and North America. Weather 55 (2), 34–45. http://dx.doi.org/10.1002/j.1477-8696.2000.tb04028.x.

PAGE21, 2013. Changing permafrost in the Arctic and its global effects in the 21st century. http://page21.org/.

Pease, C.H., 1987. The size of wind-driven coastal polynyas. J. Geophys. Res. 92 (C7), 7049–7059.

Peterson, B.J., Holmes, R.M., McClelland, J.W., Vörösmarty, C.J., Lammers, R.B., Shiklomanov, A.I., Rahmstorf, S., 2002. Increasing river discharge to the Arctic Ocean. Science 298 (5601), 2171–2173.

Petrenko, V.F., Whitworth, R.W., 1998. The Physics of Ice. Oxford Press.

Petrovic, J.J., 2003. Review Mechanical properties of ice and snow. J. Mater. Sci. 38 (1), 1–6. http://dx.doi.org/10.1023/A:1021134128038.

Pope, P.G., Anderson, J.B., 1992. Late quaternary glacial history of the Northern Antarctic Peninsula's Western Continental Shelf: evidence from the marine record. In: Elliot, D.H. (Ed.), Contributions to Antarctic Research III. American Geophysical Union, Washington, DC. http://dx.doi.org/10.1029/AR057p0063.

Pralong, A., Funk, M., Luthi, M.P., 2003. A description of crevasse formation using continuum damage mechanics. Ann. Glaciol. 37 (1), 77–82.

Pritchard, H.D., Arthern, R.J., Vaughan, D.G., Edwards, L.A., 2009. Extensive dynamic thinning on the margins of the Greenland and Antarctic ice sheets. Nature 461 (7266), 971–975.

Pritchard, H.D., Ligtenberg, S.R.M., Fricker, H.A., Vaughan, D.G., Van den Broeke, M.R., Padman, L., 2012. Antarctic ice-sheet loss driven by basal melting of ice shelves. Nature 484 (7395), 502–505.

Quadfasel, D., Gascard, J.C., Koltermann, K.P., 1987. Large-scale oceanography in Fram Strait during the 1984 Marginal Ice Zone Experiment. J. Geophys. Res. Oceans 92 (C7), 6719–6728.

Rahmstorf, S., 1999. Shifting seas in the greenhouse? Nature 399 (6736), 523–524.

Rampal, P., Weiss, J., Dubois, C., Campin, J.-M., 2011. IPCC climate models do not capture Arctic sea ice drift acceleration: consequences in terms of projected sea ice thinning and decline. J. Geophys. Res. 116, C00D07. http://dx.doi.org/10.1029/2011JC007110.

Rignot, E., Casassa, G., Gogineni, P., Krabill, W., Rivera, A., Thomas, R., 2004. Accelerated ice discharge from the Antarctic Peninsula following the collapse of Larsen B ice shelf. Geophys. Res. Lett. 31 (18), L18401.

Rignot, E., Jacobs, S., Mouginot, J., Scheuchl, B., 2013. Ice-shelf melting around Antarctica. Science 341 (6143), 266–270. http://dx.doi.org/10.1126/science.1235798.

Rinterknecht, V.R., Clark, P.U., Raisbeck, G.M., Yiou, F., Bitinas, A., Brook, E.J., Marks, L., Zelčs, V., Lunkka, J.-P., Pavlovskaya, I.E., Piotrowski, J.A., Raukas, A., 2006. The last deglaciation of the Southeastern sector of the scandinavian ice sheet. Science 311 (5766), 1449–1452. http://dx.doi.org/10.1126/science.1120702.

Robock, A., 1980. The seasonal cycle of snow cover, sea ice and surface albedo. Mon. Weather Rev. 108.3, 267–285.

Rodehacke, C.B., Hellmer, H.H., Huhn, O., Beckmann, A., 2007. Ocean/ice shelf interaction in the southern Weddell Sea: results of a regional numerical helium/neon simulation. Ocean Dyn. 57 (1), 1–11.

Ross, J.C., 1847. A Voyage of Discovery and Research in the Southern and Antarctic Regions V1: During the Years 1839–43. Albemarle Street, London.

Rothrock, D.A., 1975. The energetics of the plastic deformation of pack ice by ridging. J. Geophys. Res. 80 (33), 4514–4519.

Rysgaard, S., Bendtsen, J., Delille, B., Dieckmann, G.S., Glud, R.N., Kennedy, H., Tison, J.L., 2011. Sea ice contribution to the air–sea CO2 exchange in the Arctic and Southern Oceans. Tellus B 63 (5), 823–830.

Scambos, T.A., Bohlander, J.A., Shuman, C.A., Skvarca, P., 2004. Glacier acceleration and thinning after ice shelf collapse in the Larsen B embayment, Antarctica. Geophys. Res. Lett. 31 (18). http://dx.doi.org/10.1029/2004GL020670.

SCAR, 1997. Webb, P.N., Cooper, A.K. (Eds.), Report of a Workshop on Antarctic Late Phanerozoic Earth System Science Hobart, Australia. http://www.scar.org/publications/reports/.

Scherer, R.P., Aldahan, A., Tulaczyk, S., Possnert, G., Engelhardt, H., Kamb, B., 1998. Pleistocene collapse of the West Antarctic ice sheet. Science 281 (5373), 82–85.

Schlosser, P., Bayer, R., Foldvik, A., Gammelsrød, T., Rohardt, G., Münnich, K.O., 1990. Oxygen 18 and helium as tracers of ice shelf water and water/ice interaction in the Weddell Sea. J. Geophys. Res. 95 (C3), 3253–3263. http://dx.doi.org/10.1029/JC095iC03p03253.

Schutz, B.E., Zwally, H.J., Shuman, C.A., Hancock, D., DiMarzio, J.P., 2005. Overview of the ICESat mission. Geophys. Res. Lett. 32 (21), L21S01002E.

Seabrooke, J.M., Hufford, G.L., Elder, R.B., 1971. Formation of Antarctic bottom water in the Weddell Sea. J. Geophys. Res. 76 (9), 2164–2178. http://dx.doi.org/10.1029/JC076i009p02164.

SEARCH, 2012. Study of Arctic environmental change, "Sea Ice Outlook" http://www.arcus.org/search/seaiceoutlook/index.php.

Selker, J.S., Thévenaz, L., Huwald, H., Mallet, A., Luxemburg, W., Van De Giesen, N., Parlange, M.B., 2006. Distributed fiber-optic temperature sensing for hydrologic systems. Water Resour. Res. 42 (12). http://dx.doi.org/10.1029/2006WR005326.

Send, U., Burkill, P., Gruber, N., Johnson, G.C., Körtzinger, A., Koslow, T., Wijffels, S., 2009. Towards an integrated global observing system: in-situ observations. In: Proceedings of OceanObs '09: Sustained Ocean Observations and Information for Society, Venice, Italy, 21–25.

Serreze, M.C., Holland, M.M., Stroeve, J., 2007. Perspectives on the Arctic's shrinking sea-ice cover. Science 315 (5818), 1533–1536.

Shabtaie, S., Bentley, C.R., 1987. West Antarctic ice streams draining into the Ross Ice Shelf: configuration and mass balance. J. Geophys. Res. 92 (B2), 1311–1336. http://dx.doi.org/10.1029/JB092iB02p01311.

Shackleton, E., 1919. South: The Story of Shackleton's 1914–17 Expedition. Century Publishing, London. ISBN 0-7126-0111-2.

Shakhova, N., et al., 2007. Methane release on the Arctic East Siberian shelf. Geophys. Res. Abstr. 9, 01071.

Shepherd, A., Wingham, D., Rignot, E., 2004. Warm ocean is eroding West Antarctic ice sheet. Geophys. Res. Lett. 31 (23). http://dx.doi.org/10.1029/2004GL021106.

Shepherd, A., et al., 2012. A reconciled estimate of ice-sheet mass balance. Science 338 (6111), 1183–1189. http://dx.doi.org/10.1126/science.1228102.

Siddall, M., Rohling, E.J., Almogi-Labin, A., Hemleben, C., Meischner, D., Schmelzer, I., Smeed, D.A., 2003. Sea-level fluctuations during the last glacial cycle. Nature 423 (6942), 853–858.

SIMB, 2012. Seasonal ice mass balance buoy. http://imb.crrel.usace.army.mil/SeasonalIBinst.htm.

Smethie Jr., W.M., Jacobs, S.S., 2005. Circulation and melting under the Ross Ice Shelf: estimates from evolving CFC, salinity and temperature fields in the Ross Sea. Deep Sea Res. Part I 52 (6), 959–978.

Sohn, H.-G., Jezek, K.C., van der Veen, C.J., 1998. Jakobshavn Glacier, West Greenland: 30 years of spaceborne observations. Geophys. Res. Lett. 25 (14), 2699–2702. http://dx.doi.org/10.1029/98GL01973.

Sowers, T., 2006. Late quaternary atmospheric CH 4 isotope record suggests marine clathrates are stable. Science 311 (5762), 838–840. http://dx.doi.org/10.1126/science.1121235, PMID 16469923.

Stammerjohn, S.E., Martinson, D.G., Smith, R.C., Yuan, X., Rind, D., 2008. Trends in Antarctic annual sea ice retreat and advance and their relation to El Niño–Southern Oscillation and Southern Annular Mode variability. J. Geophys. Res. Oceans 113 (C3). http://dx.doi.org/10.1029/2007JC004269.

Stanford, J.D., Rohling, E.J., Hunter, S.E., Roberts, A.P., Rasmussen, S.O., Bard, E., Fairbanks, R.G., 2006. Timing of meltwater pulse 1a and climate responses to meltwater injections. Paleoceanography 21 (4). http://dx.doi.org/10.1029/2006PA001340.

Stanton, T.P., Shaw, W.J., Truffer, M., Corr, H.F.J., Peters, L.E., Riverman, K.L., Bindschadler, R., Holland, D.M., Anandakrishnan, S., 2013. Channelized ice melting in the ocean boundary layer beneath Pine Island Glacier, Antarctica. Science 1236–1239. http://dx.doi.org/10.1126/science.1239373.

Steele, M., Boyd, T., 1998. Retreat of the cold halocline layer in the Arctic Ocean. J. Geophys. Res. Oceans 103 (C5), 10419–10435.

Stern, A., Dinniman, M.S., Zagorodnov, V., Tyler, S.W., Holland, D.M., 2013. Intrusion of warm surface water beneath the McMurdo Ice Shelf, Antarctica. J. Geophys. Res. Oceans, (in press).

Stroeve, J., Holland, M.M., Meier, W., Scambos, T., Serreze, M., 2007. Arctic sea ice decline: faster than forecast. Geophys. Res. Lett. 34, L09501. http://dx.doi.org/10.1029/2007GL029703.

Swithinbank, C., Brunk, K., Sievers, J., 1988. A glaciological map of Filchner-Ronne ice shelf, Antarctica. Ann. Glaciol. 11, 150–155.

Teilmann, J., Born, E.W., Acquarone, M., 2000. Behaviour of ringed seals tagged with satellite transmitters in the North Water polynya during fast-ice formation. Can. J. Zool. 77 (12), 1934–1946.

Thoma, M., Jenkins, A., Holland, D., Jacobs, S., 2008. Modelling Circumpolar Deep Water intrusions on the Amundsen Sea continental shelf, Antarctica. Geophys. Res. Lett. 35, L18602.

Timmermann, R., Danilov, S., Schröter, J., Böning, C., Sidorenko, C., Rollenhagen, K., 2009. Ocean circulation and sea ice distribution in a finite element global sea ice–ocean model. Ocean Model. 27 (3), 114–129.

Toole, J.M., Krishfield, R.A., Timmermans, M.L., Proshutinsky, A., 2011. The ice-tethered profiler: Argo of the Arctic. Oceanography 24 (3), 126–135. http://dx.doi.org/10.5670/oceanog.2011.64.

Turner, J., Bracegirdle, T.J., Phillips, T., Marshall, G.J., Hosking, J.S., 2013. An initial assessment of Antarctic sea ice extent in the CMIP5 models. J. Clim. 26 (5), 1473–1484.

USGS, 2012. Gas (methane) hydrates—a new frontier. USGS fact sheet.

Van der Veen, C.J., 1985. Response of a marine ice sheet to changes at the grounding line. Quat. Res. 24 (3), 257–267.

Vesecky, J.F., Samadani, R.A., Smith, M.P., Daida, J.M., Bracewell, R.N., 1988. Observation of sea-ice dynamics using synthetic aperture radar images: automated analysis. IEEE Trans. Geosci. Remote Sens. 26 (1), 38–48.

Wanninkhof, R., 1992. Relationship between wind speed and gas exchange over the ocean. J. Geophys. Res. Oceans 97 (C5), 7373–7382.

Warren, C.R., 1993. Rapid recent fluctuations of the calving San Rafael Glacier, Chilean Patagonia: climatic or non-climatic? Geogr. Ann. 75A (3), 111–125.

Weertman, J., 1983. Creep deformation of ice. Annu. Rev. Earth Planet. Sci. 11, 215.

Willmott, A.J., Holland, D.M., Maqueda, M.A.M., 2007. Polynya modelling. Elsevier Oceanography Series 74, 87–125.

Wingham, D.J., Francis, C.R., Baker, S., Bouzinac, C., Brockley, D., Cullen, R., Wallis, D.W., 2006. CryoSat: a mission to determine the fluctuations in Earth's land and marine ice fields. Adv. Space Res. 37 (4), 841–871.

Zhang, X., Walsh, J.E., 2006. Toward a seasonally ice-covered Arctic Ocean: scenarios from the IPCC AR4 model simulations. J. Clim. 19, 1730–1747.

Chapter 17

The Arctic and Subarctic Oceans/Seas

Cecilie Mauritzen*, Bert Rudels† and John Toole‡
*CICERO Center for Climate and Environmental Research—Oslo, Oslo, Norway
†Finnish Meteorological Institute, Erik Palménin aukio 1, P.O. Box 503, FI-00101 Helsinki, Finland
‡Woods Hole Oceanographic Institution, Woods Hole, Massachusetts, USA

Chapter Outline

1. INTRODUCTION

The Arctic can, despite its small size (3% of Earth's surface), affect global climate in many ways. The most well known is the positive albedo-climate feedback, in which melting ice reduces Earth's albedo, thus allowing more solar heating of the ocean that in turn induces the melting of more ice (Brooks, 1949; Donn and Ewing, 1968; Budkyo, 1969; Kellogg, 1975; Holland and Bitz, 2003). Lian and Cess (1977) estimated that this feedback mechanism amplifies climate sensitivity to greenhouse gas forcing by roughly 25%. In a warming climate, storm tracks over the North Atlantic are projected to move further north (Yin, 2005): the associated eddy heat flux providing another positive feedback. On the other hand, recent years of reduced sea ice extent have been found to correlate with a weaker jet stream and more southerly storm paths over the North Atlantic (Francis et al., 2009; Francis and Vavrus, 2012; Liu et al., 2012), elucidating that regional climate change is not simple and linear, but a tug-of-war between many opposing forces. The Arctic/Subarctic is particularly exposed in this respect; large interannual variability dominates most time series.

Oceanographically, the most important Arctic Ocean feedbacks to global climate relate first to its function as a *transformer of sea water*: The Arctic Ocean receives warm water from the North Pacific and North Atlantic and exports fresher and colder waters to the North Atlantic, partly as buoyant surface waters and partly as dense overflows. This process contributes to the particularly efficient ocean uptake of anthropogenic carbon in this region (see Chapter 30). Variations in the properties and strengths of the return flows can affect the world's overturning circulation (see Chapters 10, 11, 19). Another feedback to the global climate relate to the Arctic Ocean's potential function as a *melter of ice*: The warm-water inflows from the North Pacific and North Atlantic can, if brought in contact with, or close proximity to, sea ice, cause it to melt and thereby influence Earth's albedo (see Chapter 16). The warm-water inflows can also contribute to the release of methane from thawing subsea permafrost (see Chapter 26). These oceanographic feedbacks are key issues in this chapter. For earlier, comprehensive reviews see for instance

Ocean Circulation and Climate, Vol. 103. http://dx.doi.org/10.1016/B978-0-12-391851-2.00017-9

Coachman and Aagaard (1974), Carmack (2000), and Rudels et al. (2012a).

During the past decade, a small revolution has occurred with respect to observations of the High North, not least because of the technological advances in autonomous ocean monitoring (see, e.g., the Community White Papers resulting from the *OceanObs'09 conference*; www.OceanObs09.net), but also because of the International Polar Year 2007–2008 (IPY), which resulted in the largest international effort ever, both scientifically and financially, to investigate the polar regions (Ortiz et al., 2011). The World Climate Research Programme (WCRP) was involved in the planning of IPY from the start. During its 24th session, the Joint Scientific Committee of the WCRP (Reading, March 2003) asked the Cryosphere and Climate project (CliC) to organize preparations for the IPY within WCRP. Generally, WCRP has directed its efforts toward the Arctic through its core projects, the Arctic Climate System Study (ACSYS, 1994–2003) and CliC (2000–present).

Legacies of the 2007–2008 IPY include continuation of inter- and trans-disciplinary science, coordinated monitoring systems, coordinated data management and public outreach. Public interest was further peaked by the extremes in Arctic sea ice extent observed in those years: a record minimum in summer sea ice extent was documented by satellites in 2007, a value only surpassed by the late-summer 2012 extent (see, e.g., www.nsidc.org). At the same time, important scientific insights into the Arctic system emerged; many of these are reported here.

1.1. Geography

We define the region discussed in this chapter, the Arctic and Subarctic Seas, as those waters bounded by Bering Strait, the Greenland–Scotland Ridge, and by the 60 °N line between Greenland and the tip of Labrador in Canada (Figure 17.1). The Arctic Ocean itself consists of 50% shallow shelf seas and 50% deep basins. North of Eurasia the shelves form shallow, broad (~500 km wide) shelf seas, separated zonally by small island groups into the Barents Sea, Kara Sea, Laptev Sea, East Siberian Sea, and Chukchi Sea. Except for the roughly 200 m-deep Barents Sea, the mean depths of these seas are around 50 m or less. North of the American continent, the Alaskan, and Beaufort Sea shelves are by comparison narrow (~<100 km).

The Eurasian Basin is divided by the Gakkel Ridge into the Nansen Basin (4000 m) and the Amundsen Basin (4500 m). The Lomonosov Ridge, with sill depth of 1870 m, separates the Eurasian Basin from the Amerasian Basin consisting of the Makarov Basin and the large Canada Basin, both about 4000 m deep and separated by the Mendeleyev and the Alpha ridges.

The Arctic Ocean opens to the North Pacific through the narrow, 50 km, and shallow, 50 m, Bering Strait. There are shallow, ~200 m deep, direct communication pathways between the Arctic Ocean and the North Atlantic via the Canadian Arctic Archipelago and Baffin Bay. The three main Canadian Archipelago gateways are Nares Strait (220 m), Lancaster Sound (125 m), and Jones Sound (including both Cardigan Strait (210 m) and Hell Gate (150 m)) that channels the flow through to the northern Baffin Bay. The 2400-m deep Baffin Bay connects in the south to the Labrador Sea through the 640-m deep Davis Strait. The Labrador Sea also receives water from Hudson Bay. The mean depth of Hudson Bay is only 100 m, but it receives a substantial fraction of the runoff from northern North America and is, with an area of more than 1 million km^2, the second largest bay in the world.

The Arctic Ocean connects to the Nordic Seas by the 500 km wide and 2600-m deep Fram Strait and by the wider but shallower (200–300 m deep) Barents Sea. The Nordic Seas, also partitioned by bathymetric ridges, consist of the Greenland Sea (3500 m), Lofoten Basin (3000 m), Norwegian Basin (3500 m), and the Iceland Sea (2000 m). The Nordic Seas communicate with the North Atlantic over the Greenland–Scotland Ridge, which has a sill depth of 840 m in the Faroe Bank Channel and 640 m in Denmark Strait.

2. EXCHANGES WITH THE SUBPOLAR OCEANS AND BEYOND

The major water masses of the Arctic and Subarctic Seas derive from inflows originating in the North Atlantic and Pacific Oceans. The mean sea level of the Pacific Ocean is higher than in the Atlantic Ocean. Older estimates place the difference at roughly 20 cm (Shtokman, 1957; Coachman and Aagaard, 1966), whereas a more recent estimate (see Chapter 21) yields roughly 40 cm. The sea level difference between the Pacific and the Atlantic is generally invoked, within a semigeostrophic framework, to account for the transport of Pacific Water down the pressure gradient from the Pacific through the narrow Bering Strait to the Arctic Ocean (Stigebrandt, 1984; Wijffels et al., 1992). Observations support this framework: despite the fact that there are three types of Pacific Water in the Bering Strait—Anadyr Water, Bering Shelf Water, and Alaska Current Water (Woodgate and Aagaard, 2005)—recent direct velocity measurements with a very dense array of instruments show the northward flow to be strongly coherent across the strait (Woodgate et al., 2005a; Beszczynska-Möller et al., 2011). The low-salinity Pacific water entering through Bering Strait accumulates in the upper layers of the Canada Basin, greatly influencing the vertical stratification in the Arctic Ocean (Figure 17.2a).

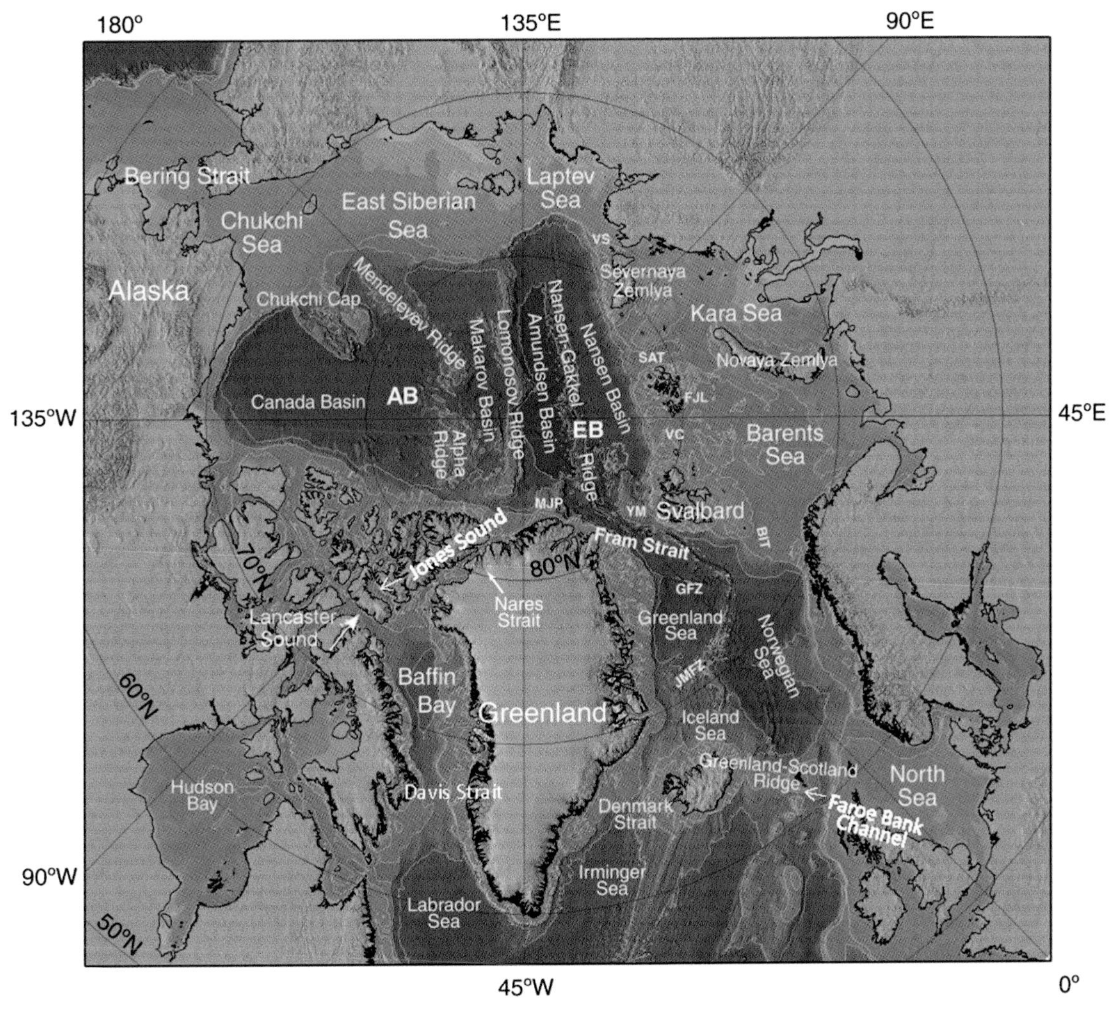

FIGURE 17.1 Bathymetry and place names in the Arctic/Subarctic seas. The bathymetry is from IBCAO updated data base (Jakobsson et al., 2008). The 200, 500, 2000, and 4000 m isobaths are drawn. Place names are in the Arctic/Subarctic seas. Acronyms are defined as follows: AB, Amerasian Basin; BIT, Bear Island Trough; EB, Eurasian Basin; FJL, Franz Josef Land; GFZ, Greenland Fracture Zone; JMFZ, Jan Mayen Fracture Zone; MJP, Morris Jessup Plateau; SAT, St. Anna Trough; YM, Yermak Plateau; VC, Victoria Channel; and VS, Vilkiltskij Strait. Lambert Equal Area projection. *Adapted from Rudels et al. (2012a).*

The Pacific inflow is compensated by export flows to the subpolar North Atlantic through the Canadian Archipelago and across the Greenland–Scotland Ridge.

At the Greenland–Scotland Ridge, northward currents carry warm and saline Atlantic water into the Subarctic and Arctic Seas where it undergoes tremendous transformations and eventually returns to the North Atlantic both as dense overflows and as colder, fresher, upper-ocean waters (Aagaard and Carmack, 1989; Mauritzen, 1996a; Carmack, 2000; Tsubouchi et al., 2012). The northernmost limb of this overturning circulation extends throughout the Arctic Ocean, manifested by a subsurface temperature maximum at 150–400 m depth, termed here the Arctic Atlantic Water (Figure 17.2a).

The straits in the Canadian Archipelago are deeper and, because of the stratification, dynamically wider than the Bering Strait. The currents in these openings appear more analogous to the exchange circulation about the Greenland–Scotland Ridge, with flows in both directions, partitioned by depth and cross-strait position (Beszczynska-Möller et al., 2011). These passages export polar water from the Arctic Ocean to the North Atlantic.

2.1. Volume Transports

At the Greenland–Scotland Ridge, the warm Atlantic Water flows poleward, primarily along either side of the Faroe Islands. The branch entering west of the Faroes has been monitored since 1997 at a site just inside the Norwegian Sea. The estimated mean volume transport of waters warmer than 4 °C was 4 Sv (10^6 m^3/s) for the period 1997–2001 (Østerhus et al., 2005), and 3.5 Sv for the period 1997–2008 (Hansen et al., 2010). The branch east of the Faroes, in the Faroe–Shetland Channel, has been monitored

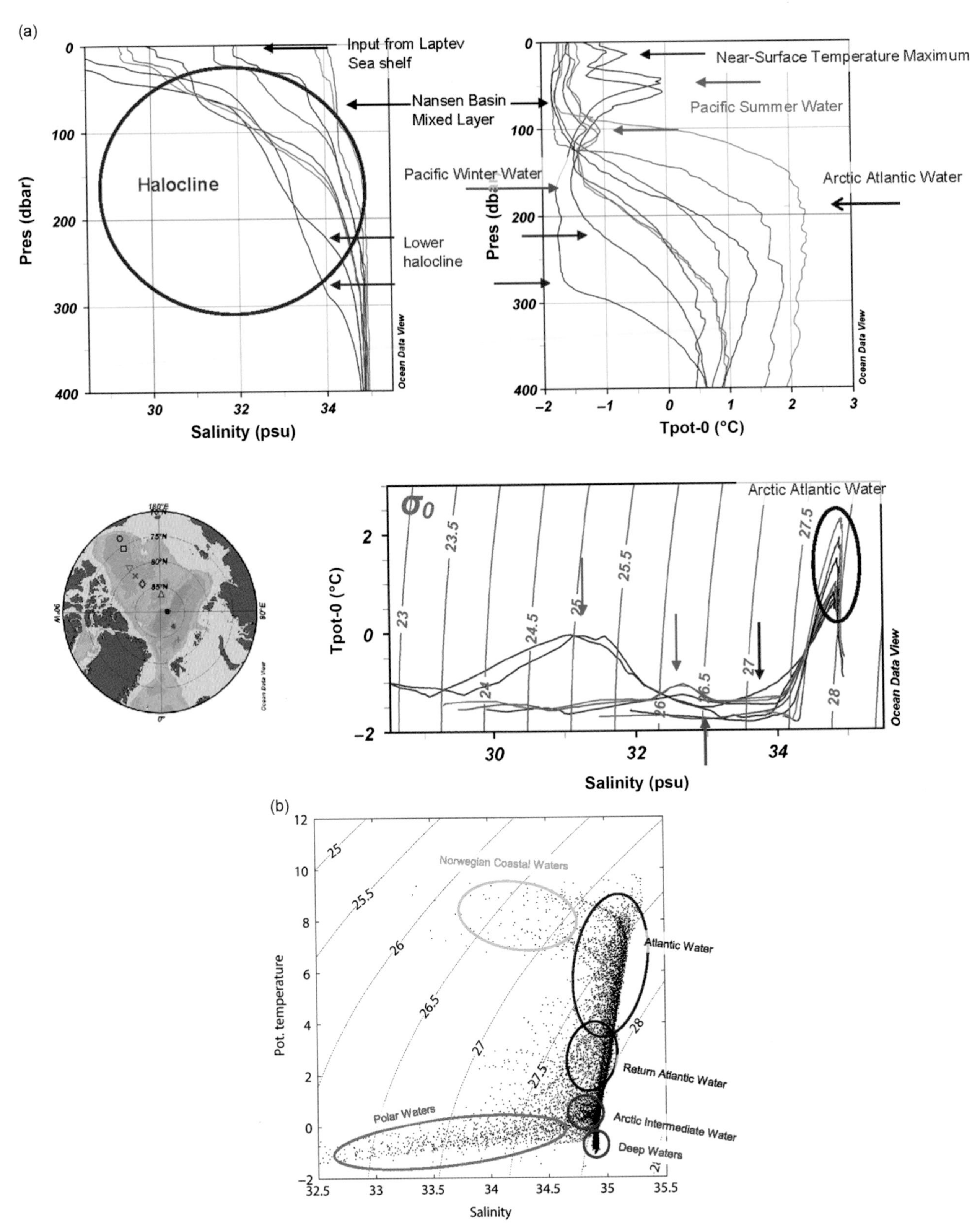

FIGURE 17.2 Salinity–potential temperature (°C) diagram for (a) the upper Arctic Ocean showing also temperature and salinity profiles from a trans-Arctic section and (b) the Nordic Seas. Lines of constant potential density are plotted with thin lines: note that at low temperatures these are more vertical, indicating that salinity has a greater influence on density. *(a) Adapted from Rudels (2012). In preparing this figure and figure 17.13 Ocean Data View (Schlitzer, 2013) has been used.*

since 1994; the average transport for 1994–2011 was estimated to be 2.7 Sv (Berx et al., 2013). The third inflow branch, in Denmark Strait between Iceland and Greenland, transports roughly 0.8 Sv of Atlantic Water (Jónsson and Valdimarsson, 2012). Together, these three inflow branches carry roughly 7 Sv of warm Atlantic Water into the Nordic Seas (Berx et al., 2013), a number very near the 6.8 Sv estimated by Mauritzen (1996b) based on data from the 1970s and 1980s.

The equatorward return flows across the Greenland–Scotland Ridge consist of both light, low-salinity waters, primarily the surface waters of the East Greenland Current, and dense overflow waters which exit (1) west of Iceland, in the Denmark Strait (estimated at roughly 3 Sv by a number of investigators including Ross, 1984; Girton et al., 2001; Macrander et al., 2005, although when expanding the temperature range up to 2 °C, Dickson et al., 2008 estimate the transport to be 4 Sv); (2) east of Iceland, through the Faroe Channels (estimated at 2.2 Sv, colder than 2 °C by Hansen and Østerhus, 2007, using measurements from 1995 to 2005); and (3) across the Iceland-Faroe Ridge (an intermittent flow averaging roughly 1 Sv; Hansen and Østerhus, 2000).

The strength of the Barents Sea inflow to the Arctic Ocean has been monitored since 1997 (Ingvaldsen et al., 2004). It consists of part Atlantic Water (transport estimated at roughly 2 Sv by Skagseth et al., 2008) and part fresher Norwegian Coastal Current waters (1.8 Sv; Skagseth et al., 2011). The Atlantic Water, having high salinity, is a freshwater sink relative to a reference salinity of 34.8, but the Norwegian Coastal Current adds freshwater to the Arctic Ocean (Dickson et al., 2007; Skagseth et al., 2008). The main temperature mode of the Atlantic Water flowing into the Barents Sea is 6–6.5 °C.

The Fram Strait opening has been continuously monitored, with increasingly improved resolution, since 1997 (Schauer and Beszczynska-Möller, 2009; Beszczynska-Möller et al., 2011). The exchanges, both northward and southward through the passage, are large—the strength of the monthly averaged north and south transports is around 10 Sv—with a net transport around 2 Sv southward (Rudels et al., 2008; Schauer et al., 2008). The northward transport consists of two branches: the almost-barotropic current over the continental slope off Spitsbergen and the western baroclinic branch that follows the Mohn and Knipovich ridges to Fram Strait (Walczowski et al., 2005). Both branches continue into the Arctic Ocean but recirculating limbs within the Strait return perhaps half of the northward transport back into the Nordic Seas. These recirculating waters have nearly the same characteristics as the inflowing water, indicating that little transformation occurs in the Strait (Rudels et al., 2008). Indeed, by calculating volume transport in temperature classes, so that the recirculation cancels, a fairly clear picture emerges of the net exchanges through Fram Strait between the Nordic Seas and the Arctic Ocean proper. In this framework, the net northward transport (primarily in the West Spitsbergen Current) is 3.2 Sv, all warmer than 2 °C (the main temperature mode is in the range 3–3.5 °C). The net southward transport (primarily in the East Greenland Current) is 5.2 Sv, all colder than 2 °C, the main temperature being in the range from −1 to −0.5 °C (Mauritzen et al., 2011).

At Bering Strait, year-round moorings have been deployed almost continuously since 1980 (Beszczynska-Möller et al., 2011). Based on these data, the annual mean inflow of Pacific Water has been estimated to be about 0.8 Sv (Roach et al., 1995; Woodgate et al., 2006), with a strong seasonal cycle; transport is greatest in summer at1.2 Sv and falls to 0.4 Sv in winter. Salinity of the northward flow varies from 32.5 in summer to above 33 in winter when sea ice also drifts from the Bering Sea northward into the Chukchi Sea. The inflow is warm in summer, perhaps 4 °C, but near the freezing point in winter. Arguably the greatest impact of the Pacific inflow on the Arctic is on the freshwater budget, accounting for 30% of the freshwater input to the Arctic Ocean (Serreze et al., 2006; Woodgate et al., 2006; Dickson et al., 2007). Recent observations have shown that the low-salinity Alaskan Coastal Current may make the Pacific freshwater input to the Arctic larger than previously thought (Woodgate et al., 2006).

The Canadian Arctic Archipelago is an outflow passage from the Arctic, primarily for Pacific Waters, but also for some denser Atlantic-derived waters (Jones et al., 2003). The transports through the straits are difficult to determine by direct measurement because of the sea ice environment but also because of (until recently) close proximity to the magnetic North Pole, which complicates compass determination of the flow direction. Before 2000, only a short period, up to a year, direct current measurements had been collected (Sadler, 1976; Tang et al., 2004) and transports were mostly estimated by geostrophic computation giving a total estimated transport through the archipelago of about 2 Sv (Muench, 1971). In the last 10 years, new techniques have been developed and direct current measurements have been undertaken in all passages (Melling et al., 2008). The transports in Barrow Channel (Lancaster Sound) have now been monitored for several years (Prinsenberg et al., 2009), as has the flow through Hell Gate and Cardigan Strait (Melling et al., 2008). The flow through Nares Strait has been measured with an extensive array in the Kennedy Channel (Melling et al., 2008; Münchow and Melling, 2008; Rabe et al., 2010). These observations indicate an average volume transport of 0.7 Sv through Lancaster Sound (Prinsenberg et al., 2009) and 0.7 Sv through Nares Strait (Münchow and Melling, 2008). The mean flow through Hell Gate and Cardigan Strait has been determined to be 0.3 Sv (Melling et al., 2008). The total transport through the Archipelago is thus estimated to be roughly 1.5–2 Sv (Beszczynska-Möller et al., 2011). Farther south, these waters continue equatorward through Davis Strait, where the net southward transport has been determined by direct current measurements to be around 2.5 Sv (Cuny et al., 2005; Curry et al., 2011). In the last few years, several glider sections have been obtained across Davis Strait, adding detailed hydrographic data to the direct current measurements (Azetsu-Scott et al., 2012).

3. CURRENTS AND WATER MASS TRANSFORMATIONS IN THE ARCTIC/SUBARCTIC

At very low temperatures, such as in the Arctic Ocean, the vertical stability of the water column is primarily determined by salinity, not by temperature (Figure 17.2a). The large freshwater input to the Arctic Ocean, by river runoff and net precipitation, therefore creates strong vertical stability in the upper Arctic Ocean. This stability inhibits vertical mixing and the water column is therefore characterized by advective water masses. Traditionally, five different water masses or layers are identified (Figure 17.2a): (1) the Polar Mixed Layer, the surface waters that are homogenized by surface stress and haline-driven convection during winter; (2) the halocline, a water mass located below the Polar Mixed Layer where the salinity increases strongly with depth whereas the temperature is uniformly close to freezing or is incised by warm and cold intrusions created by additional water masses: Pacific Summer Water and Pacific Winter Water layers (Steele et al., 2004; Shimada et al., 2005) and in summer, a third temperature extrema, the Near-Surface Temperature Maximum Layer (Perovich and Maykut, 1990; Jackson et al., 2010); (3) the warm Atlantic water inflowing from the Nordic Seas, commonly defined as a subsurface layer with temperatures above 0 °C and denoted Arctic Atlantic Water (Mauritzen, 1996a) to distinguish it from the Atlantic Water found in the Nordic Seas; (4) the intermediate water, located below the Arctic Atlantic Water but communicating freely between the Eurasian and Amerasian basins across the Lomonosov Ridge, also known as upper Polar Deep Water (Rudels et al., 1994); and (5) the various deep and bottom waters found in the Canadian, Makarov, Amundsen, and Nansen basins.

In the Nordic Seas, the vertical stability is much weaker than in the central Arctic and convection readily extends to 1000 m in the Lofoten Basin and 2000 m in the Greenland Sea. One tends to differentiate between (Figure 17.2b) (1) Atlantic Water (high-salinity water which enters the Nordic Seas with a temperature of 7–10 °C that subsequently cools to well below 5 °C before it enters the Arctic Ocean); (2) recirculating Atlantic Water (typically 2–3 °C; entering the East Greenland Current in the Fram Strait beneath polar waters and returning southwards); (3) Arctic Intermediate Water (low-salinity water mass at roughly −0.7 to 0 °C; created in the Greenland and Iceland Seas); (4) Greenland Sea Deep Water and 5) Norwegian Sea Deep Water. A comprehensive water mass classification for the Arctic and Subarctic Seas is given in Rudels et al. (2012a).

We organize this subsection as a tour of the region, following the pathway of the Atlantic Water inflow, beginning in the Nordic Seas.

3.1. The Norwegian Atlantic Current

The Atlantic Waters that enter the Nordic Seas west and east of the Faroe Islands are clearly depicted in surface drifter analyses within the Nordic Seas (Andersson et al., 2011; Koszalka et al., 2011; see Figure 17.3). The Atlantic Water occupies a much larger region in the Nordic Seas than the two current cores (Figure 17.3a), implying a long residence time of Atlantic Water in this sea. Flow variability is greatest near the cores of the Norwegian Atlantic Current (NAC) (Figure 17.3b), consistent with those cores generating eddies via instability.

RAFOS floats deployed in the subsurface NAC in the western branch north of the Faroes (Rossby et al., 2009) clearly show that despite the two-core structure of the current, these branches are not distinct in a Lagrangian sense; nearly all floats did at some point visit the eastern branch and continue northwards. Thus there is strong exchange between the branches, as evidenced by the large eddy kinetic energy in the current (Figure 17.3b) and also in the broad extent of the Atlantic water mass (Figure 17.3a).

The Lofoten Basin is unique as the region of highest eddy kinetic energy (Figure 17.3b). Eddies spread from the NAC and fill the entire basin with Atlantic Water. The root mean square velocities here are around 20 cm/s, comparable to the instantaneous flow speeds in the NAC cores (Andersson et al., 2011; Koszalka et al., 2011). Atlantic Water spends more time in this basin than further south or north along its circulation path. Drifter experiments indicate that the residence time for Atlantic Water is 1–3 years in the Lofoten Basin (Gascard and Mork, 2008).

The long residence time for Atlantic Water in the Nordic Seas (500 days vs. 60 days if the water were to follow the main branches only; Koszalka et al., 2012) gives the water ample time to cool during transit. The transformation of Atlantic Water, from 7 to 10 °C at the Greenland–Scotland Ridge to well below 5 °C in the Fram Strait, is the largest water mass transformation occurring within the Arctic/Subarctic Seas.

3.2. Arctic Ocean

3.2.1. Inflows and Boundary Currents in the Arctic Ocean

3.2.1.1. Barents Sea and Fram Strait

A considerable fraction of the NAC and most of the Norwegian Coastal Current enter the Arctic Ocean through

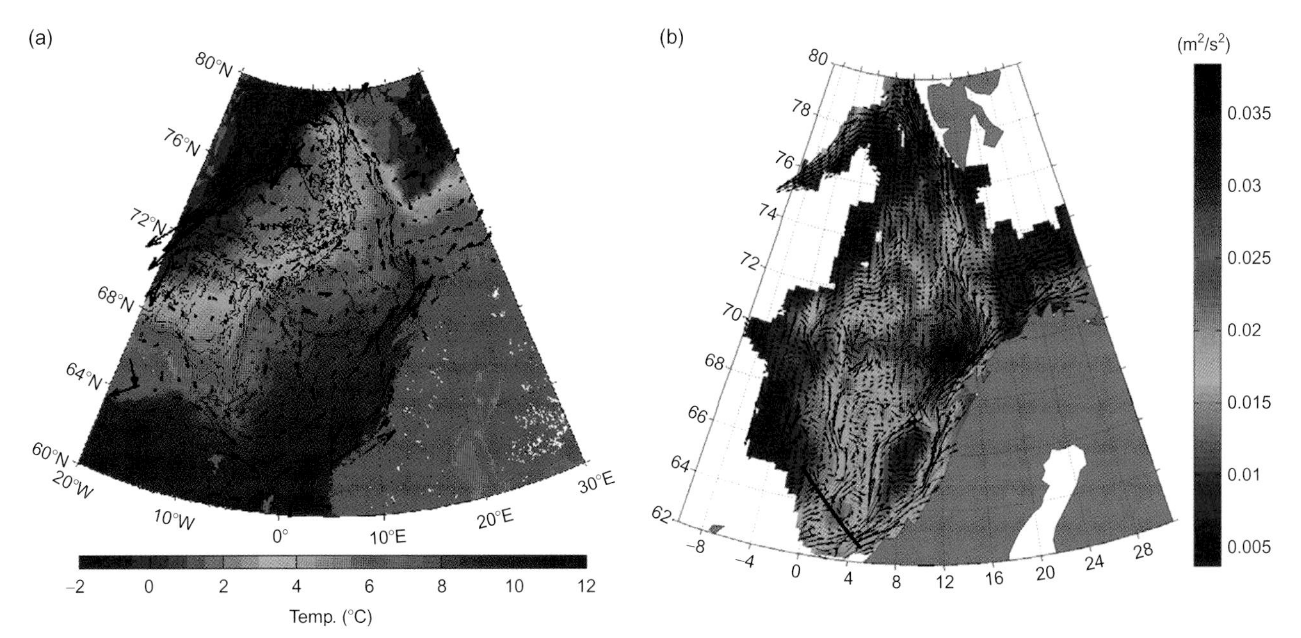

FIGURE 17.3 Surface drifters in the Nordic Seas. (a) Color field: mean surface temperature from the TOPAZ Reanalysis for the year 2003, vectors: mean velocities (cm/s), using clustering method by Koszalka and LaCasce (2010), colored in dark/light gray for speeds larger/less than 12 cm/s. (b) color field: Eddy kinetic energy (m^2/s^2), vectors: mean velocities (cm/s), using a binning method (Andersson et al., 2011).

the Barents Sea opening (Figure 17.4). The Barents Sea is a shallow shelf sea which experiences comparatively little river runoff and thus salinity changes are related mainly to the seasonal freezing and melting cycle. In winter, the presence of coasts and islands contributes to the formation of lee polynyas and extensive sea ice over the shallow banks west of Novaya Zemlya and around Franz Josef Land and in Storfjorden in southern Svalbard (Midttun, 1985). Some of the inflow water, mainly the water of the Norwegian Coastal Current, passes through Kara Gate south of Novaya Zemlya and then receives the runoff from Ob and Yenisey in the Kara Sea and from Lena in the Laptev Sea to form the low-salinity water encountered on the Siberian shelves. The other, mainly Atlantic, part of the Barents Sea inflow moves north of Novaya Zemlya. The southern part of the Barents Sea is largely ice-free year-round and the bulk of the Atlantic inflow becomes denser by air–sea exchange (cooling). The transformed waters continue to the Arctic Ocean down the St. Anna Trough (Figure 17.4) as a stratified column.

The Atlantic Water that enters the Arctic Ocean west of Svalbard interacts with sea ice in the vicinity of Svalbard and a less saline upper layer is created by ice melt. North of Svalbard, this upper part is advected eastward together with the main Atlantic inflow as a boundary current along the Eurasian continental slope. The upper layer becomes stratified by ice melt in summer and in winter it is again homogenized by cooling, freezing, and brine rejection down to the thermocline above the Atlantic layer.

The coalescing of the two inflow branches in and east of St. Anna Trough causes the Fram Strait inflow to be displaced farther offshore. The circulation of the Atlantic waters around the Arctic Ocean was denoted the Arctic Circumpolar Boundary Current and discussed by Rudels et al. (1999). As the combined boundary current evolves, the less saline and cooler Barents Sea branch waters interleave with the warmer, more saline Atlantic waters of the Fram Strait branch, resulting in both an Atlantic Water temperature maximum centered at 300–400 m depth and also an initial deep salinity minimum below the temperature maximum. North of the Laptev Sea, both branches are overrun by the low-salinity shelf water coming from the Laptev Sea and the connection between the Atlantic water and the sea ice and the atmosphere via the upper winter mixed layers is broken, isolating the warm Atlantic water as well as the winter mixed layers found in the Nansen Basin and in the Barents/Kara seas from the surface forcing and the sea ice. These submerged upper layers eventually evolve into halocline waters (Rudels et al., 1996, 2004).

3.2.1.2. Kara and Laptev Seas

The largest transformation of the boundary current occurs between the eastern Kara Sea and the eastern Laptev Sea (Schauer et al., 1997). Here the temperature and salinity maxima of the Atlantic core are reduced by 1 °C and 0.05, respectively, and the smooth profiles of the Fram Strait inflow branches develop strong inversions in temperature and salinity. The cause for such strong changes could be local heat loss to the atmosphere (Polyakov et al., 2011) or the result of mixing between the two inflow branches (Rudels, 2012). Most of the shelf water exported from the

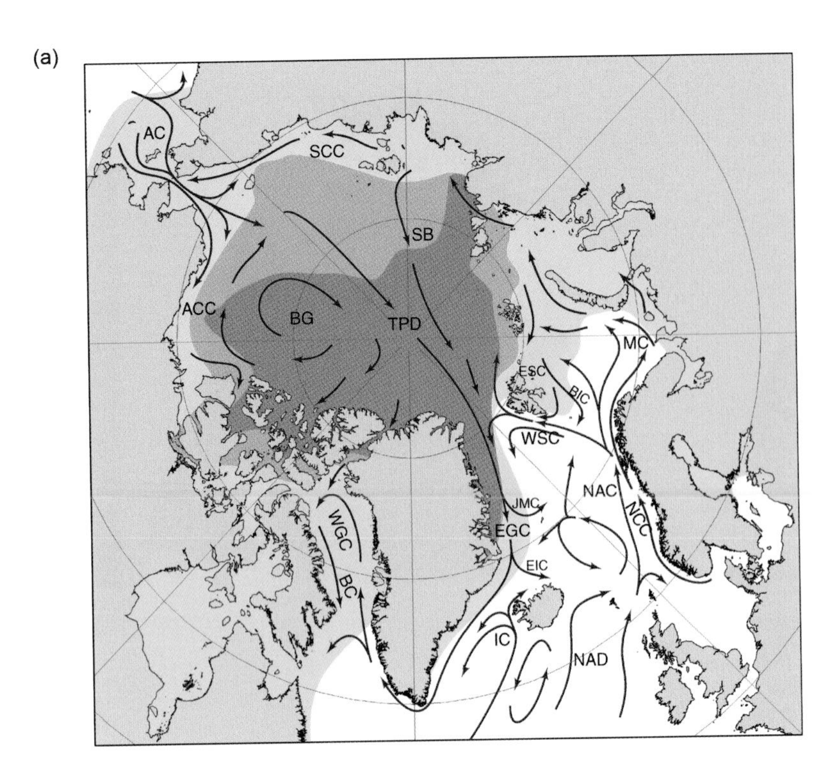

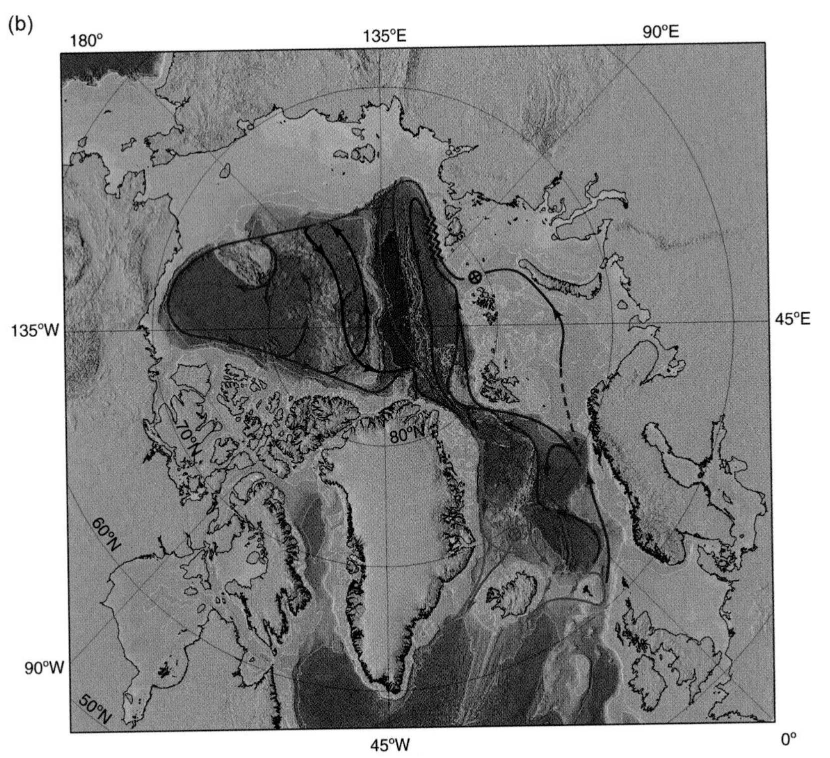

FIGURE 17.4 Circulation schematic for the Arctic/Subarctic Seas. (a) The upper ocean. Warm Atlantic currents are indicated by red arrows, cold less saline polar and arctic currents by blue arrows. Low-salinity transformed currents are shown by green arrows. The maximum ice extent is shown in blue and the minimum ice extent in red. The 2007 sea ice minimum is shown in dark red. AC, Anadyr Current; ACC, Alaskan Coastal Current; BC, Baffin Current; BIC, Bear Island Current; BG, Beaufort Gyre; EGS, East Greenland Current; EIC, East Iceland Current; ESC, East Spitsbergen Current; IC, Irminger Current; JMC, Jan Mayen Current; MC, Murman Current; NAD, North Atlantic Drift; NAC, Norwegian Atlantic Current; NCC, Norwegian Coastal Current; SB, Siberian branch (of the Transpolar Drift); SCC, Siberian Coastal Current; TPD, Transpolar Drift; WGC, West Greenland Current; and WSC, West Spitsbergen Current. (b) Intermediate depth ocean. The interactions between the Barents Sea and Fram Strait inflow branches north of the Kara Sea as well as the recirculation and different inflow streams in Fram Strait and the overflows across the Greenland–Scotland Ridge are indicated. *Adapted from Rudels et al. (2012a).*

Laptev Sea and the East Siberian Sea is not dense enough to become halocline water. Instead, it overflows and covers the denser winter mixed layers offshore (initially formed from ice melting on top of the Atlantic water of the Fram Strait and Barents Sea branches) (Figure 17.2a).

The interior shelves are, by definition, more isolated and their flushing depends more strongly on river input and on the exchanges across the shelf break to provide the saline end-member for the shelf waters. These cross-shelf-break exchanges are frequently episodic in nature, with wind-driven upwelling events responsible for bringing saline Atlantic Water onto the shelves and eddy formation by instability and flow disturbances related to shelf-break canyons causing export of shelf waters to the deep basins (Pickart, 2004; Kinney et al., 2008; Pickart et al., 2009; Watanabe, 2011).

3.2.1.3. Deep-Basin Ridges

The boundary current splits and partly separates from the continental slope at bathymetric features, allowing Atlantic and intermediate waters to be carried into the interior, into the Nansen and Amundsen basins along the Gakkel and Lomonosov Ridges, respectively. The part that crosses the Lomonosov Ridge into the Amerasian Basin subsequently enters the Makarov Basin at the Medeleyev Ridge (Rudels et al., 1994) and the Canada Basin at the Chukchi Cap (Smith et al., 1999; Smethie et al., 2000). Some of this water forms a loop in the northern Canada Basin while the rest moves around the Chukchi Cap (Shimada et al., 2004). The question of how the Atlantic water enters the southern Canada Basin is not fully resolved. Shimada et al. (2004) discuss two pathways, one along the slope between the Chukchi Cap and the Chukchi continental slope and one passing north of and around the Chukchi. Another possibility is that at least the upper part of the Atlantic water circulates anti-cyclonically around the southern Canada Basin (the Beaufort Sea) as suggested by Coachman and Barnes (1963) and Newton and Coachman (1974) and revived by McLaughlin et al (2009).

3.2.1.4. Chukchi Sea

In the shallow Chukchi Sea, the entire water column is homogenized across the entire shelf in winter and the water masses created contribute, together with the Bering Strait winter water, to the temperature minimum and the nutrient maximum observed, especially in the Canada Basin, around 150 m depth: the Pacific Winter Water (Jones and Anderson, 1986). As in the Barents Sea, most of the sea ice formed in the Chukchi Sea melts there and little is exported to the Arctic Ocean.

Similar to the situation on the Eurasian side, branching, modification, merging, and subsequent interleaving of inflowing Pacific waters occur in the Chukchi Sea (Woodgate et al., 2005b; Aagaard et al., 2006; Pickart et al., 2009). As on the Atlantic side, upon reaching the shelf break, vorticity constraints initially limit the off-shelf extent of these Chukchi Sea currents; the inflows deflect to the right and form eastward-flowing boundary currents about the shelf break with the various branches coalescing to form one flow around Barrow Canyon. A high-resolution mooring program located just east of Barrow Canyon, complemented by numerical modeling studies, has elucidated some of the water mass transformation processes affecting these Pacific Waters—time dependence appears central (Weingartner et al., 1998; Pickart, 2004; Nikolopoulos et al., 2009). Observations show a significant loss of transport between Bering Strait and the shelf-break boundary current around Barrow Canyon. Wind stress (perhaps now more effective due to thinner ice cover) has been suggested to intermittently divert waters from the shelf-break boundary current into the interior (Shimada et al., 2006; Nikolopoulos et al., 2009). A related process involves wind-driven upwelling of Atlantic Water onto the shelf in fall and early winter, where it may be influenced by convective mixing and brine rejection to create a denser water mass (Pickart et al., 2009). Relaxation or reversal of the along-shore winds can then drive these modified waters back off the shelf. This process would be less active in winter and early spring when the extensive sea ice cover impedes wind stress acting on the ocean.

The interior shelves must in some places, at some times, also produce dense water by freezing and brine rejection. The high salinity observed in the Canada Basin deep and bottom waters indicates that the Chukchi Sea and the Beaufort Sea shelves must occasionally form water dense enough to ventilate the deep Canada Basin (Aagaard et al., 1985).

3.2.2. Upper interior Arctic Ocean

3.2.2.1. Horizontal Transformation

Transport of boundary current waters from the continental slopes into the interior of the Arctic Ocean is facilitated by at least two key mechanisms. First, instabilities that develop in the boundary currents cast-off lenses of modified shelf and/or boundary current waters. These lenses may be carried and/or self-advect throughout the Arctic interior (e.g., Muench et al., 2000; Pickart et al., 2005). These eddy features can be surprisingly intense with azimuthal flow speeds of order 10 cm/s but with horizontal diameters only around 10 km (D'Asaro, 1988; Krishfield et al., 2002). Upper-ocean eddies are also generated by instabilities of mixed layer fronts in the ocean interior (Timmermans et al., 2008), while eddies of Arctic Atlantic Water have been observed about its core depth (Aagaard et al., 2006; Walsh et al., 2007; Dmitrenko et al., 2008). Ultimately these eddies decay (the central Arctic is not fully populated by eddies) but the decay mechanisms are not understood.

Another mechanism for lateral spreading of water property anomalies into the Arctic interior is by thermohaline intrusions (Carmack et al., 1998; Walsh and Carmack, 2003; Spall et al., 2008; McLaughlin et al., 2009). These features are particularly notable at the depth of the Atlantic Water core, extending from boundary currents. Owing perhaps to the weak flows of the Arctic interior, individual intrusion features appear coherent over distances of thousands of kilometers (Carmack et al., 1998). Lateral pressure gradients created by double-diffusively driven vertical buoyancy flux divergences have been invoked as the driving mechanism for these intrusions (e.g., Stern, 1967; Toole and Georgi, 1981). If so, many of the realizations of these features in observations may be of fossil or near-fossilized intrusions as the vertical salinity profile through them is nearly monotonic (and so providing little potential energy store for double diffusive instabilities to tap).

3.2.2.2. Vertical Transformation

As at lower latitudes, the dominant water mass transformations that occur in the Arctic Ocean interior are driven by surface exchange processes, but in the Arctic, the exchanges involve both air–sea and ice–sea fluxes. The latter frequently result in significant buoyancy fluxes at the ocean surface related to salt rejection during freezing and freshwater release during melting, both heavily influenced by the dominant dependence of sea water density on salinity at low temperatures. However, sea ice and snow also temper these buoyancy-forcing mechanisms due to their insulating nature (limiting winter heat loss) and high albedo (limiting solar warming in summer). The net result has been a fragile equilibrium that sustains an expansive perennial sea ice cover, but that state may now be changing (Richter-Menge and Jeffries, 2011) (see also Section 4).

3.2.2.3. Melting of Sea Ice

One of the long-standing scientific issues in Arctic oceanography has been the ocean's role in the sea ice balance (Maykut and Untersteiner, 1971; Carmack and Melling, 2011). It has long been known that there is sufficient heat stored in the subsurface cores of Atlantic and Pacific Summer Water layers to melt a significant fraction if not all of the Arctic sea ice, were that heat somehow transported upward to the ice–ocean interface. Inhibiting such vertical transport is the huge vertical density gradient associated with the Arctic salinity stratification: the halocline (Figure 17.2a). The ice–ocean balance question has thus focused on possible mechanisms to upwell and/or turbulently mix upward these subsurface ocean heat anomalies where they might directly influence the ice–ocean boundary layer (McPhee et al., 2005; Yang, 2006; Fer, 2009; Turner, 2010).

A principal mixing mechanism believed active at lower latitudes—turbulence driven by instability and breaking of the internal wave field—appears remarkably weak in the Arctic. In comparison with the canonical background wave field (e.g., Garrett and Munk, 1972 and subsequent updates), the super-inertial energy in the deep Arctic is an order of magnitude less energetic (Levine et al., 1985, 1987; Halle and Pinkel, 2003). Turbulent diapycnal diffusivity in the ocean interior is correspondingly small (Rainville and Winsor, 2008; Lenn et al., 2009). Possible reasons for the weak wave field include absence of resonant near-inertial forcing by wind (Alford, 2003) and reduction of wind forces on the ocean by internal ice stresses (Rainville and Woodgate, 2009). As these latter authors note, reduction of sea ice cover due to global warming raises the interesting prospect of more efficient wind forcing in the future, possibly leading to more internal-wave-driven diapycnal mixing. Exceptions to the weak wave field are regions of abrupt, shallow bathymetry such as the Yermak Plateau, where internal waves are generated in response to the barotropic tide incident on the bathymetry, break locally and result in local "hot spots" of turbulent mixing (e.g., Padman and Dillon, 1991; D'Asaro and Morison, 1992).

The depth interval between the Atlantic Water core at 300–400 m depth and the base of the isothermal halocline or Pacific Winter Water at 100–200 m has a thermohaline stratification capable of supporting the diffusive form of double diffusion (cold freshwater overlying warmer saltier water). High-vertical-resolution profile data through this segment of the water column frequently show a thermohaline staircase consisting of homogenized layers around 1 m thick, separated by thin interfaces with enhanced vertical gradients (Neal et al., 1969; Padman and Dillon, 1987). Upward vertical heat flux across these interfaces is believed to drive convective motions that maintain the layers, but estimates of the flux based on molecular diffusion through interfaces sampled with microstructure probes and theoretical predictions for a double diffusive staircase are quite small, suggesting only weak vertical transport and water mass modification by this mechanism (Timmermans et al., 2008; Lenn et al., 2009; Turner, 2010). Although the vertical heat flux in the ocean interior appears weak on average, modeling studies suggest that even a small increase in the flux acting over time might result in significant thinning of the sea ice cover (e.g., Polyakov et al., 2010).

However, when considering ocean influences on Arctic sea ice, the key focus must be the ice–ocean boundary layer and the surface mixing zone (Dewey et al., 1999; McPhee, 2008). Indeed, all heat that is exchanged between the ocean and sea ice must pass through these critical layers. The large, stabilizing vertical salinity gradient that characterizes the upper Arctic Ocean limits winter mixed layer depths to around 50 m or less (McPhee, 2008; Toole et al., 2010) and greatly impedes turbulent transport of sub-mixed-layer heat anomalies to the ice–ocean interface. In addition, submesoscale instabilities within the surface layer may be continually acting to restratify the surface layer via lateral processes, further impeding vertical fluxes from the stratified interior to the ice–ocean interface (Timmermans et al., 2012). Consequently, current thinking is that much of the ocean heat that causes basal sea ice melting derives locally from solar ocean warming through ice leads (McPhee, 2008; Perovich et al., 2008; Steele et al., 2010;

Jackson et al., 2012). In a twist to the canonical ice–albedo feedback mechanism, Stanton et al. (2012) suggest that the area of a lead is dictated more by lateral mechanical ice divergence than lateral melting, though the susceptibility of sea ice to convergent rafting and creation of leads must depend on ice thickness, which in turn is strongly dictated by the basal freeze–melt balance. In those broadening Arctic areas that are seasonally ice free for extended periods, solar radiation is causing significant ocean warming that, in combination with wind-driven lateral subduction, has led to a great reservoir of ocean heat just below the winter mixed layer base (Steele et al., 2004; McPhee, 2008; Toole et al., 2010). The efficacy of the vertical buoyancy gradient in continuing to insulate the surface layer and sea ice from subsurface heat has become a key question.

3.2.3. *The Intermediate and Deep Arctic Ocean*

3.2.3.1. Intermediate Waters

There are two external sources for the intermediate waters in the Arctic Ocean. Arctic Intermediate Water is created in the Greenland Sea (Figure 17.2b) and joins the West Spitsbergen Current and enters the Arctic Ocean through Fram Strait. The Arctic Intermediate Water can be detected as a deep, 1500 m, salinity minimum in the Nansen Basin around the Yermak Plateau. It cannot be identified from the temperature–salinity properties alone east of St. Anna Trough; further, it lies too deep to cross the Lomonosov Ridge.

The second source is the Barents Sea branch of the Arctic inflow, which contributes a less saline, $S \sim 34.85$, water with temperature $\theta \sim -0.5$ °C. This water stays on the slope, following roughly the 600 m isobath to the eastern Kara Sea, where it is deflected into the Fram Strait branch below the Atlantic layer, between 800 and 1200 m and above the Arctic Intermediate Water contribution (Schauer et al., 1997). This low-salinity intermediate water dominates the water column and the boundary current north of the Laptev Sea and also in the Amundsen Basin and along the Lomonosov Ridge, implying that the Barents Sea branch located at 600 m on the continental slope has circulated along and partly left the slope north of the Laptev Sea and entered the Eurasian Basin, moving toward Fram Strait.

Part of the boundary current continues along the continental slope and crosses the Lomonosov Ridge into the Amerasian Basin. Here no external advective input of intermediate water takes place and the intermediate layers are ventilated by dense waters created by freezing and brine rejection on the shelves and then sinking down the slope, entraining ambient water until they reach and enter their appropriate density levels (Rudels, 1986). This causes a restratification of the water column. The more saline and dense the water is initially, the deeper it sinks. Less saline water supplies the halocline and cools and freshens the Atlantic layer, while more saline water entrains warm Atlantic water and brings that heat and salt downward into the intermediate layer. The Atlantic water is then losing heat downward rather than upward and the low-salinity intermediate water from the Barents Sea branch crossing the Lomonosov Ridge becomes warmer and more saline and the intermediate depth part of the TS curve changes from being concave upward into a linear shape with salinity increasing and temperature decreasing with depth. This TS structure, stable in both temperature and salinity, is characteristic of the upper Polar Deep Water (Rudels et al., 1994).

3.2.3.2. The Deep Waters

The deep waters in the Arctic Ocean were historically thought to derive from the Greenland Sea (Nansen, 1915). However, the salinities in the Arctic Ocean deep waters are too high: they indicate that processes within the Arctic Ocean are capable of ventilating even the deepest layers (Aagaard, 1981; Aagaard et al., 1985; Rudels, 1986; Anderson et al., 1999). An inflow of deep water from the Nordic Seas does occur, but like the Arctic Intermediate Water, it is only clearly distinguished in temperature and salinity in the vicinity of Fram Strait and around the Yermak Plateau (Rudels et al., 2000a). Farther into the Eurasian Basin, the temperature and salinity of the deep and bottom waters are considerably higher than those of the Nordic Seas deep water, indicating that highly saline water sinks down the continental slope from the shelves, entraining ambient water and eventually, when it reaches its neutral isopycnal, turns into a boundary current and enters the interior. This has been observed in the Storfjorden outflow in Fram Strait (Quadfasel et al., 1988) and north of Severnaya Zemlya (Rudels et al., 2000b). In the Amerasian Basin, the advective input from the boundary current lacks the deepest, coldest component because of the sill depth of the Lomonosov Ridge. The deep waters in the Amerasian Basin thus are warmer and more saline than the Eurasian Basin Deep Water. The deep water created and/or modified in the Arctic Ocean returns to and modifies the deep waters of the Nordic Seas. The circulation of the deep waters in the Arctic Mediterranean are thus closely connected (Aagaard et al., 1985; Rudels, 1986). We refer to Rudels et al. (2000b), Timmermans and Garrett (2006), and Carmack and McLaughlin (2011) for further reading.

3.3. Canadian Archipelago and Baffin Bay

Arctic outflow through the Canadian Archipelago drains from the upper 200 m in the Arctic Ocean. In the Baffin Bay, the upper 200 m of the water column is cold and fresh, but more saline than the Polar and Pacific water that comprise the bulk of the outflow through the Archipelago.

This shows that the equatorward-flowing waters are transformed by local surface processes in the Baffin Bay, such as cooling, ice formation and brine rejection. Below the surface layer, a warmer, ~1 °C, and more saline, ~34.5, water mass is encountered (Riis Carstensen, 1936). This warm "Atlantic" layer originates from the West Greenland Current that enters Baffin Bay through Davis Strait from the south. The West Greenland Current also brings some of the Polar Water of the East Greenland Current into Baffin Bay, which contributes to the low-salinity upper layer. The colder deep and bottom waters (~ −0.5 °C, 34.45) by contrast derive from the Arctic Ocean through Nares Strait (Bailey, 1956), more specifically it is drawn from the Arctic Ocean lower halocline water that may ultimately be supplied by the Barents Sea inflow branch (Rudels et al., 2004).

The circulation in Baffin Bay is cyclonic, with the West Greenland Current moving northward along the Greenland coast, entering Nares Strait and Lancaster Sound but not the Arctic Ocean. The inflow in the West Greenland Current is estimated between 0.4 Sv (Curry et al., 2011) and 0.8 Sv (Cuny et al., 2005), consisting of low-salinity Polar Water but also Atlantic Water that was carried northward into Baffin Bay. On the western side, the Baffin Current carries the bulk of the low Arctic salinity water south through Davis Strait into the Labrador Sea.

3.4. East Greenland Current

The other outflow from the Arctic Ocean occurs in the western Fram Strait with the East Greenland Current. The East Greenland Current carries a variety of water masses out of the Arctic: sea ice and low-salinity Polar Water in the surface, various flavors of Arctic Atlantic Water that have traced one or more of the gyres in the different basins, and finally upper Polar Deep Water. In addition, the East Greenland Current carries a substantial amount of return Atlantic Water on its route toward the North Atlantic.

North of Fram Strait, the differences between the Arctic Atlantic Water and upper Polar Deep Water from the Eurasian Basin and from the Amerasian basins are clearly seen, but mixing in the strait between the outflow streams and the recirculating waters—Return Atlantic Water and Arctic Intermediate Water—usually removes these distinctions (Rudels et al., 2005).

The core of the low-salinity Polar outflow in the East Greenland Current is located over the continental slope. However, low-salinity water is also transported over the Greenland shelf. The main part of the low-salinity outflow derives from the Polar Mixed Layer and the halocline waters of Atlantic origin but Pacific water also exits through Fram Strait. The Pacific outflow varies strongly in time and may occasionally be absent (Falck et al., 2005).

The low-salinity upper part of the East Greenland Current continues southward over the slope and shelf. The upper layer becomes deeper, more homogenized and more saline as it moves south (Rudels et al., 2005) but the freshwater transport appears to remain constant (Nilsson et al., 2008). The properties of the East Greenland Current thus change as it passes by the Greenland Sea. The intermediate water becomes colder and less saline, indicative of exchanges along the front. This inflow is balanced by a loss of Atlantic Water as well as deep water to the Greenland Sea (Rudels et al., 2005).

4. EVIDENCE OF LONG-TERM CHANGES IN THE ARCTIC/SUBARCTIC

4.1. Introduction

Paleoclimate records show that the Arctic has experienced a general cooling trend for the past 2000 years. This cooling trend is not seen in time series of the northern hemisphere as a whole, and can be attributed primarily to a reduction in summer insolation which was enhanced at higher latitudes (Kaufman et al., 2009). Though the reduction in solar insolation has continued through the twentieth century, the cooling trend reversed abruptly in the late nineteenth/early twentieth century (Figure 17.5), coinciding with the rise in global temperatures (Kaufman et al., 2009). During the twentieth century, the annual average Arctic air temperature rose by roughly 2 °C (Figure 17.6), more than twice the global average, consistent with the known amplification

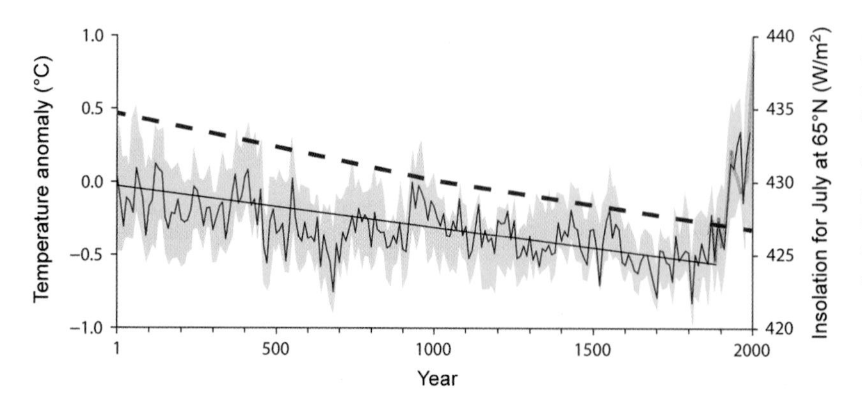

FIGURE 17.5 Proxy of Arctic summer surface air temperature anomaly relative to the 1961–1990 reference period, for the past 2000 years. A linear trend is overlaid. The red line is the observed 10-year mean Arctic temperature through 2008. The gray shadows encompass ±2 standard errors of the proxy values as evaluated for each 10-year interval. Also shown (blue dashed line) is the summer insolation at 65°N. *Adapted from Kaufman et al. (2009).*

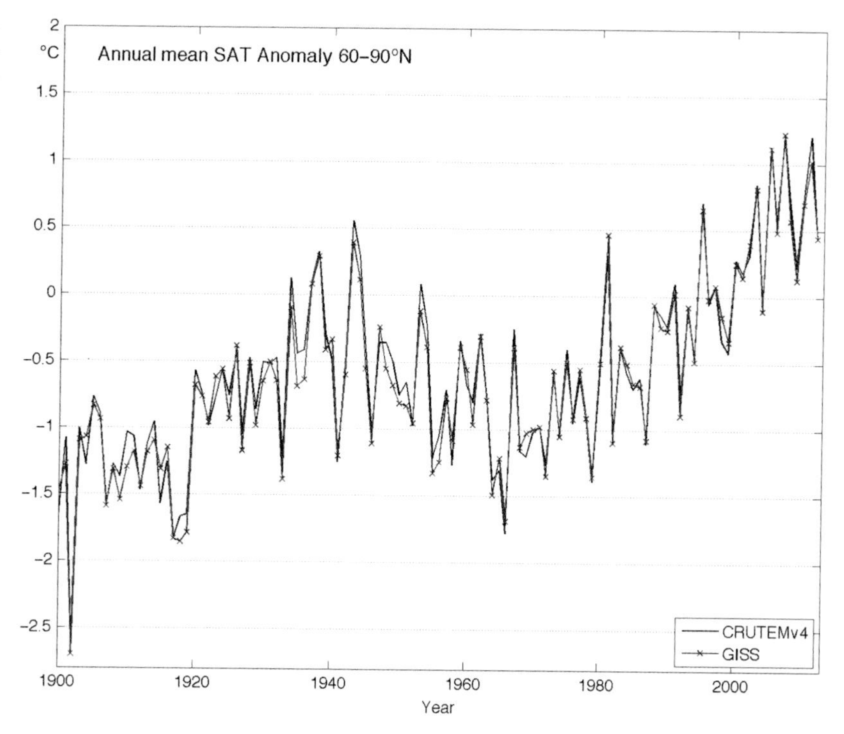

FIGURE 17.6 Annual average surface air temperatures (SAT) in the latitude range 60–90°N for the period 1900–2011 relative to the 1981–2010 mean value, based on land stations north of 60°N. *Data are from the CRUTEM4v dataset at www.cru.uea.ac.uk/cru/data/temperature/ and from the GISTEMP dataset at http://data.giss.nasa.gov/gistemp/.*

of climate change in the Arctic (Serreze and Francis, 2006). The warming reached a peak around 1950, resulting in the retreat of glaciers and melting of permafrost and sea ice. Thereafter, a cooling period was experienced until 1970, followed by warming up to the present. Superimposed on these long-term variations, the time series are characterized by centennial (Figure 17.5), decadal (Figures 17.5 and 17.6) and interannual (Figure 17.6) variability.

Due to all the natural variability in the Arctic—caused by internal couplings between atmosphere, oceans, cryosphere and biosphere but also by changes in solar insolation and irradiance—it is only after ~1990 that we can begin to recognize anthropogenic emissions (greenhouse gases, aerosols, and ozone-reducing substances) as a dominant cause of Arctic change. Using a formal "detection and attribution" method, Gillett et al. (2008) have shown that anthropogenic influence on Arctic air temperature is now detectable and distinguishable from the influence of natural forcings.

In recent decades, the most compelling record of Arctic change has been in the reduction in horizontal extent of the sea ice field. Satellite remote sensing has documented a near 50% reduction in late-summer areal extent from the measurements started in the late 1970s to the present with September 2012 and 2007 being the latest record minima (Figure 17.7; Jeffries et al., 2012). While the late-summer change has been most dramatic, a reduction in ice extent is seen in all seasons (Figure 17.7).

In addition to the decrease in extent, the Arctic sea ice has also thinned. By piecing together satellite observations and drifting buoy records to track ice parcels over several years, a record of sea ice age since the early 1980s has been computed, showing extensive loss in recent years of the older (multiyear) ice classes (Maslanik et al., 2007, 2011; Tschudi et al., 2010). A similar result is found when comparing direct sea ice thickness measurement made by submarines to satellite observations of sea ice thickness in the 2000s (Kwok and Rothrock, 2009). First- and second-year ice now dominates relative to multiyear floes (Rigor and Wallace, 2004). And the largest 1-year reduction in sea ice thickness was reported to follow the record sea ice extent minimum summer of 2007 (Giles et al., 2008).

The direct mechanisms leading to reductions in sea ice volume are presently vigorously debated; in particular, the ocean's role is in question, as will be discussed below. Indirectly, though, we know that human emissions now are a significant factor; Min et al. (2008) have shown that the changes in sea ice extent in recent years could not have taken place without anthropogenic contributions.

4.2. Evidence for Change in the Arctic Ocean

4.2.1. Atlantic Inflow

There exists only one century-long oceanic time series from the Arctic, namely the standard hydrographic section in the Barents Sea that runs along the 33°30′E meridian from the Kola Bay northward to 75°N, commonly known as the Kola section. The section crosses both the low-salinity Coastal Current and most of the Atlantic flow through the Barents

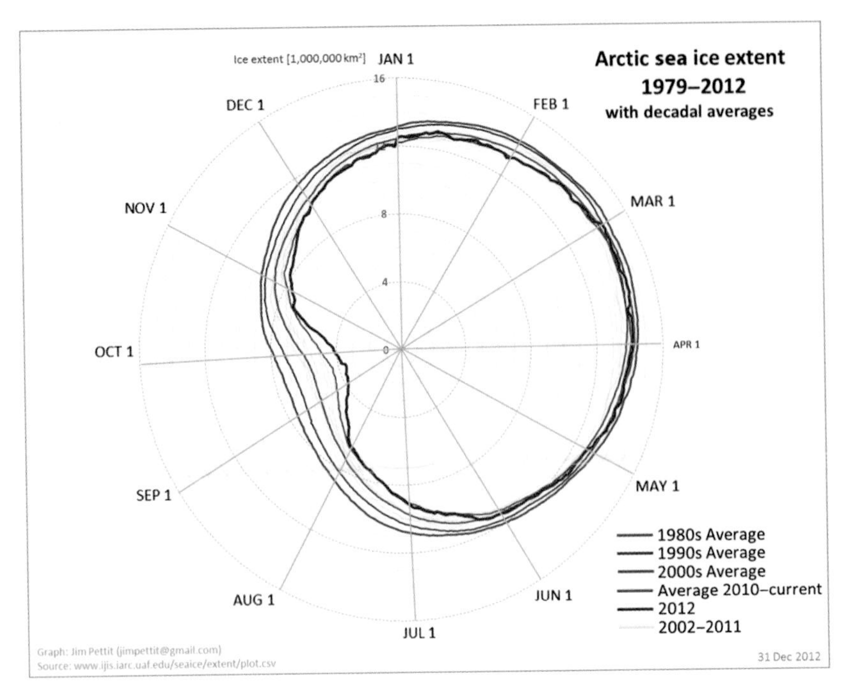

FIGURE 17.7 Time history of the areal extent of the Arctic sea ice coverage based on satellite remote-sensing data. In this depiction, azimuthal position indicates time of year while radial distance gives the ice extent. Averages over selected time periods, including the single year 2012, are shown. Note that sea ice extent in 2012 was lower in all seasons relative to decadal averages for the 1980s and 1990s. *Figure obtained from Jim Pettit, Cryosphere Today Arctic Sea Ice Area 1979-2012. Digital image. Pettit Climate Graphs. Web. 21 December, 2012.*

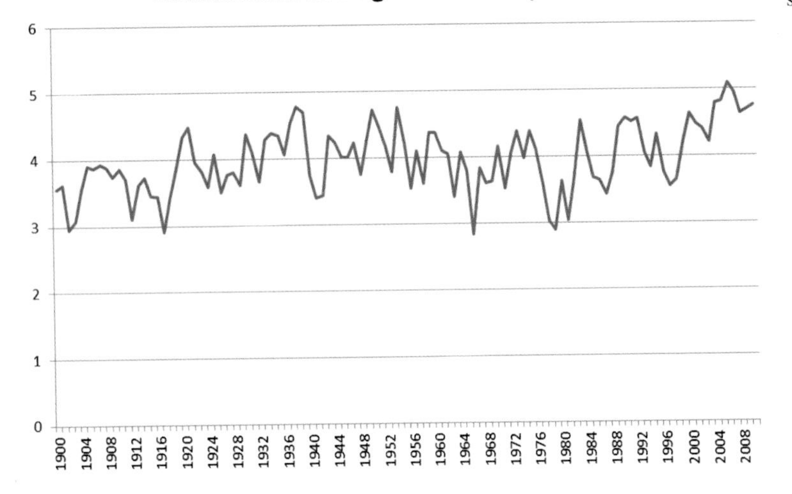

FIGURE 17.8 Average ocean temperature at the Kola section for the period 1900–2010 (Karsakov, 2009).

Sea and was occupied for the first time in May 1900.[1] The low-frequency changes in average temperature of the Kola section (Figure 17.8; Karsakov, 2009) largely mirror the average surface air temperature of the Arctic (Figure 17.6), not unexpected for a shallow and ice-free region. However, contrary to the air temperatures, the ocean temperature at Kola at present is not higher than the average ocean temperatures in the 1930s and 1940s (compare Figures 17.6 and 17.8).

1. For more information on the historical background on the Kola section, as well as on the PINRO Institute that runs the section and is responsible for the data, see www.pinro.ru/labs/indexhid_e.htm?top=hid/kolsec1_e.htm.

Observations from the many standard hydrographic sections in the northern North Atlantic and the Nordic Seas are compiled in the annual ICES Reports on Ocean Climate (www.ices.dk/marineworld/oceanclimate.asp). They show a large-scale pattern of upper-ocean warming from the 1990s to the 2000s (Figure 17.9, left panel), coherent with the pattern at the Kola section (Figure 17.8). They also mark a reversal of the 30-year long period of low-salinity conditions in the upper North Atlantic (Holliday et al., 2008; Figure 17.9, right panel).

On shorter time scales, the interannual temperature variability of the Kola section can be seen as a propagating signal along the path of the NAC. In fact, time dependence

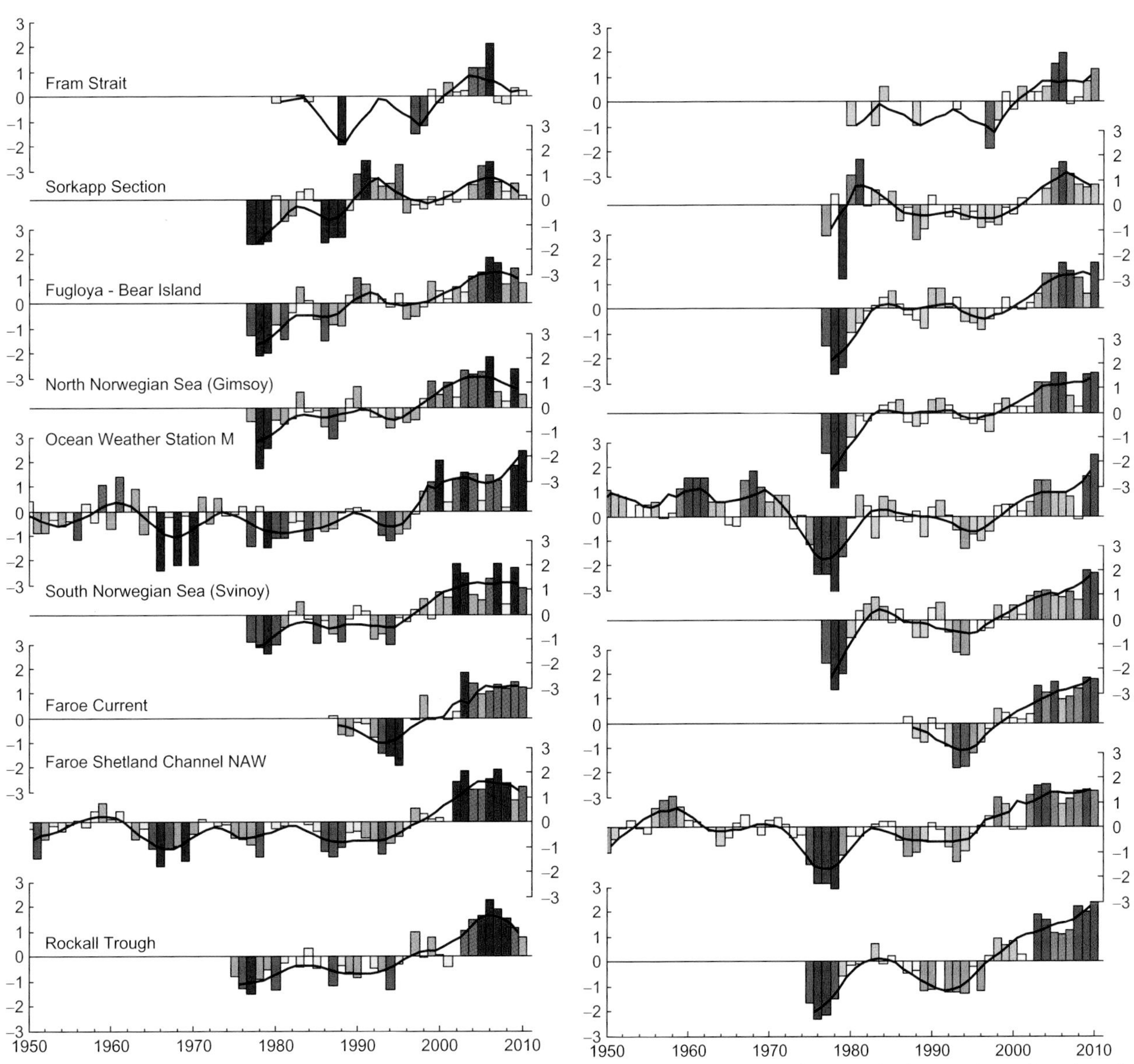

FIGURE 17.9 Time series of (left) upper-ocean temperature anomalies and (right) salinity anomalies from sustained ocean observations along the pathways of Atlantic Inflow from (bottom) the Rockall Trough to (top) the Fram Strait. Data are presented as normalized anomalies from the long-term mean (1988–2006 for Faroe Current and Fram Strait, 1978–2006 for all others). *Updated from Holliday et al. (2008). Courtesy Sarah Hughes.*

in water mass characteristics has allowed for tracing the boundary current in the Arctic. The conjectured boundary layer circulation by Rudels et al. (1994) (Figure 17.4), which was based on the water mass properties observed in the different basins, has largely been supported by the tracking of a warm pulse of Atlantic water entering the Arctic Ocean in the late 1980s or early 1990s (Quadfasel et al., 1991). The warm pulse could be followed into the Amundsen Basin in 1996 (Rudels et al., 2004) and along the Lomonosov Ridge in 1994 (Swift et al., 1997), moving toward Fram Strait. It was observed to enter the Makarov Basin in 1993 (Carmack et al., 1995), recirculate around the basin and then move along the Lomonosov Ridge from Greenland toward Siberia (Morison et al., 2006). A second pulse has been traced along the NAC and around the Eurasian Basin in 2003–2005 (Dmitrenko et al., 2008; Holliday et al., 2008; Schauer et al., 2008), growing warmer through 2007 (Polyakov et al., 2011) and subsiding by 2009 (Polyakov et al., 2012; Figure 17.10).

Note that the horizontal scale and temperature scale vary from one cascaded section to another. Warming in the Eurasian Basin is associated with the warm Atlantic Water pulse, which entered the polar basin in the 2000s; in

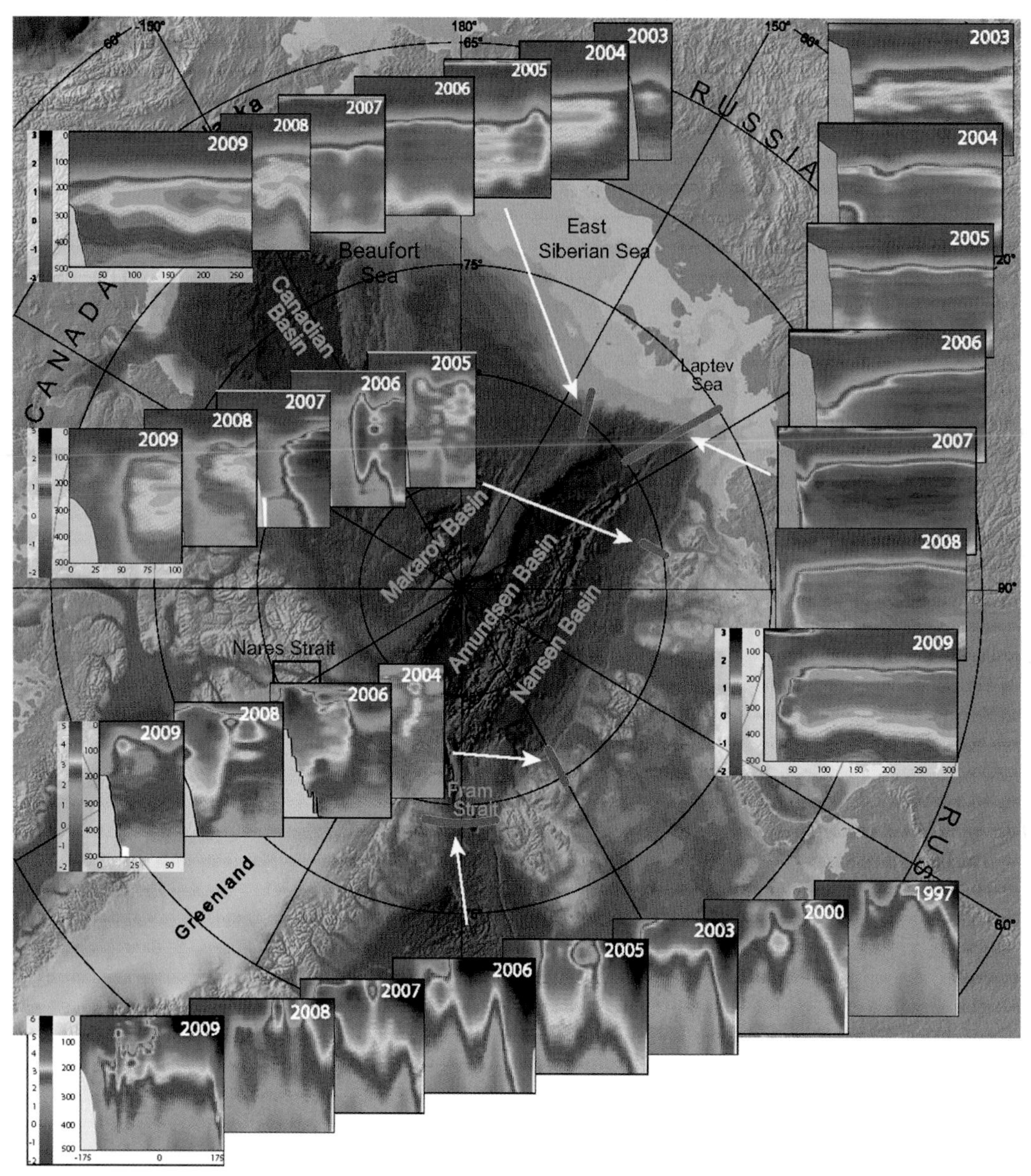

FIGURE 17.10 Vertical cross-sections of *in situ* ocean temperature (°C) from the Arctic Ocean. The five series of cascaded plots show temperatures measured at the five locations shown by yellow lines on the map. In each section, the horizontal axis shows distance from the southern end of the section (km) and the vertical axis shows depth (m).

contrast, the warm anomaly in the Canada Basin is related to an earlier pulse of warm water, which entered the Arctic Ocean interior through Fram Strait in the early 1990s. Adapted from Polyakov et al. (2011).

4.2.2. Upper Arctic Ocean

In a global climate perspective, the Arctic/Subarctic is the recipient of increased precipitation due to an acceleration of the hydrological cycle in a warming world. The runoff from Arctic rivers has been found to have clearly increased over the last few decades despite large interannual variability (Shiklomanov and Lammers, 2009) The largest increase has been on the European side, which has seen an upward trend (2.9 ± 0.5 km^3/year) since 1936. Reflecting these changes, there is ample evidence that the freshwater content of the upper Arctic Ocean has been increasing in recent decades. Rabe et al. (2011) considered salinity change in the upper Arctic Ocean (above the 34.0-isohaline) and found that between the periods 1992–1999 and 2006–2008 the freshwater content increased by 8400 ± 2000 km^3. They attributed the change to enhanced

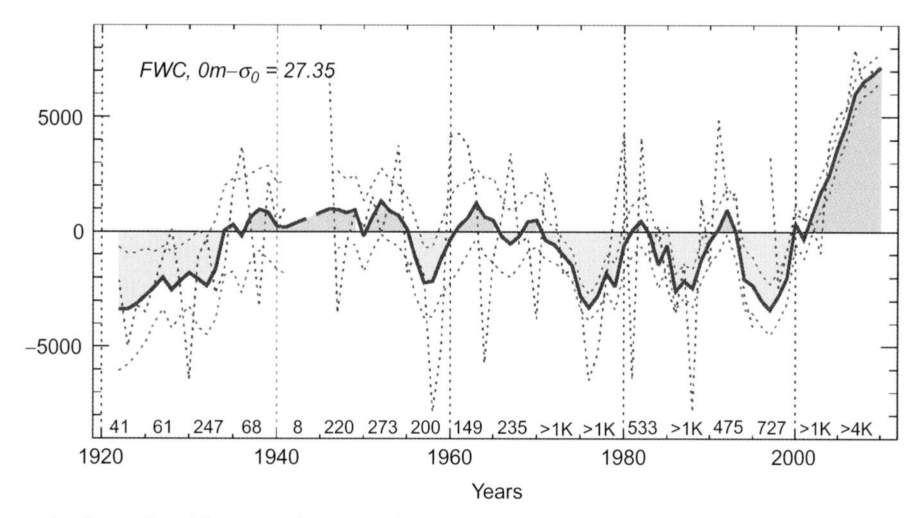

FIGURE 17.11 Composite time series of 7-year running mean of upper ocean (between the surface and the isopycnal $\sigma_0=27.35$) Arctic Freshwater Content (km^3). *Updated from Polyakov et al. (2008).*

river runoff and sea ice melt. McPhee et al. (2009) compared data from 2008 to a climatology of the second half of the twentieth century (the Polar Hydrographic Climatology; Steele et al., 2001) and found a similar result, namely an increase in freshwater content of 8500 km^3, primarily on the Pacific side of the Lomonosov Ridge. Both these studies used a reference salinity of 35. In an update to Polyakov et al. (2008), using the local climatological mean as the reference salinity, an increase in freshwater content from the 1990s to the 2000s of similar magnitude is found (Figure 17.11). This time series, which is pieced together from the highly unevenly distributed station data set in the Arctic in the twentieth century, shows large variability in ocean freshwater content throughout the century. Nevertheless, the rapid and large rise in freshwater from the 1990s to the 2000s is unprecedented in the time series.

We will therefore return to the discussion of changes in salinity on circulation in Section 4.2.4, since density is primarily controlled by salinity at these latitudes (Section 3/ Figure 17.2a).

Upper-ocean temperatures in the Arctic have increased in many regions over the last 30 years. Part of this signal is a reflection of increased heat flux into the Arctic from subpolar latitudes. For example, Shimada et al. (2006) argue that increased Bering Strait flow is related to sea ice reduction in the adjacent Arctic region through a feedback mechanism in which wind stress driving of the relatively warm Pacific Water becomes more effective as the ice cover thins. Subsequently, Woodgate et al. (2010) documented a doubling of the heat input by the Bering Strait flow in 2007 relative to 2001, due both to increased flow *and* increased ocean temperature.

Upper-ocean warming may also be associated with enhanced absorption of solar short-wave radiation in regions of reduced sea ice coverage and/or concentration—the classical albedo feedback mechanism. Evidence from the Canada Basin suggests there are at least two manifestations of this mechanism. The temperature of the Pacific Summer Water layer has increased significantly since the 1970s (McPhee, 2008; Toole et al., 2010). Steele et al. (2010) show that 80% of ocean warming and ice melt in the Canada Basin in summer 2007 derived from solar short-wave radiation. Those waters warmed at the surface in the ice-free regions are subducted below the cold, fresh surface layer of the central basin by the cyclonic wind forcing that characterizes the region forming the Polar Mixed Layer. More open water results in more ocean warming and higher Polar Mixed Layer temperature. Leads within the ice pack can also be sites of enhanced ocean warming by short-wave radiation. Sea ice floes adjacent to leads act to hold the ocean surface temperatures close to the freezing point even in summer, but the buoyancy source associated with ice melting causes the upper ocean to stratify. Solar energy that penetrates and warms the upper ocean is thus insulated from the surface cooling by the ice and the Near-Surface Temperature Maximum is formed (Perovich and Maykut, 1990). The decrease in sea ice concentration in recent years has resulted in a more prominent expression of the Near-Surface Temperature Maximum layer (Jackson et al., 2010; Toole et al., 2010). Lying within the depth range of the winter surface mixed layer, the heat contained in the Near-Surface Temperature Maximum will directly impact the amount of sea ice growth in the subsequent winter. Lying below a strong halocline, Pacific Summer

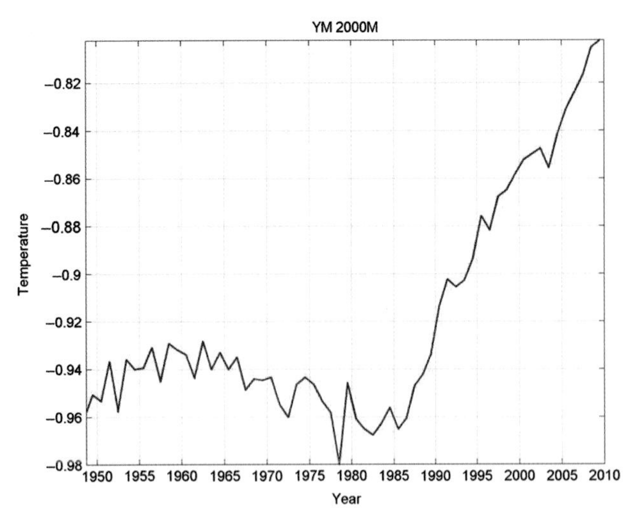

FIGURE 17.12 Temperature time series (1948–2009) at 2000 m (near the sea floor) at Weather Station Mike at 66°N, 2°E in the Norwegian Sea. *Courtesy S. Østerhus (see http://web.gfi.uib.no/forskning/mike/oceanweather/).*

Water heat is less easily entrained into the surface mixing layer to inhibit winter ice growth. But as Toole et al. (2010) note, this Pacific Summer Water heat will not simply vanish and may eventually be transported up to the ice–ocean interface to limit sea ice growth.

4.2.3. Abyss

After the 1980s, deep and bottom water formation in the Greenland Sea practically ceased. As a consequence, the warmer and more saline Arctic Ocean deep waters have started to replace the colder, less saline Greenland Sea Deep Water. The salinity maximum in the Greenland Sea at 2500 m, deriving from the Eurasian Basin Deep Water, has increased in strength, and further up in the water column a mid-depth temperature maximum has been created by the inflow of Amerasian Basin Deep Water (Meincke et al., 1997). The convection in the Greenland Sea is currently limited to the level of the Arctic Intermediate Water, above the temperature maximum, and the doming of the isopycnals, previously characterizing the Greenland Sea water column, has changed into a depression of the isopycnals in the center of the basin (Budéus et al., 1998; Budéus and Ronski, 2009).

This change in the Greenland Sea, and the extension of the influence of the Arctic Ocean deep waters, have had an impact on the intermediate and deep waters in the Nordic Seas (Rudels et al., 2012b): In the Fram Strait during the last 30 years, the densest water found at the sill has changed from the less saline Greenland Sea Deep Water to the more saline and warmer Eurasian Basin Deep Water. And in the Norwegian Sea, the 60-year long time series of ocean hydrography at Ocean Weather Station Mike (at 66°N, 2°E; see Dinsmore, 1996) shows a sudden 0.1 °C increase in deep ocean temperatures around 1985, from which it has never recovered (Figure 17.12).

The abyssal Arctic Ocean has also seen increases in temperature. Comparison of potential temperature and salinity sections across the Eurasian Basin occupied in 1996 and 2007 shows that the temperature in the lower 1500 m increased by about 0.015 °C over this decade (Figure 17.13). Later observations, for example, from the Polarstern cruise in 2011, documented a continued increase in the deep water temperature (Rudels et al., 2013).

4.2.4. Evidence for Changes in Circulation

There is ample evidence of short-term variability in the currents of the Nordic Seas. Analysis of 15 years of current meter measurements at the Svinøy section and at the Barents Sea opening (Anonymous, 2012; Figure 17.14) show variability at a multitude of frequencies. Much of the variability appears related to high-frequency (interannual and higher) atmospheric forcing. For instance, the Norwegian Coastal Current in the Barents Sea likely varies in accordance with wind-generated coastally-trapped waves (Skagseth et al., 2011). Furevik (1998) and Ingvaldsen et al. (2004) report that the variability of the transport of Atlantic Water at the Barents Sea opening and in the Fram Strait is governed by local wind forcing (in addition to upstream changes in flow position or strength). The observations also indicate that there is a compensation effect: if the transport in one limb of the circulation is large, the other is typically small (Lien et al., 2013). Others have presented evidence that these upper-ocean flows vary in response to the seasonal atmospheric forcing and passing of low pressure systems (Orvik and Skagseth, 2003; Ingvaldsen et al., 2004; Skagseth et al., 2004; Olsen et al., 2008; Richter et al., 2009). The largest seasonal signals are found within the closed geostrophic contours of the Greenland, Norwegian, and Lofoten Basins. This flow variability is primarily barotropic in nature and driven by the wind (Isachsen et al., 2003).

The propagation of hydrographic anomalies discussed in Section 4.2.1 may actually be an indication of changes in circulation strength. It has been argued, quite convincingly, that since such propagating signals are anomalies from local conditions, not the property of a particular water type, and since the propagation speeds vary, the signals are actually an indication of a strengthening/weakening of the current strength as opposed to a passive advection of anomalously warm/cold water mass (Sundby and Drinkwater, 2007). The direct evidence is inconclusive due to the shortness of the time series (consider, for instance, Figure 17.14), but current meter results from the Fram Strait do show that the annual average volume transport with water warmer than 4 °C was indeed 80%

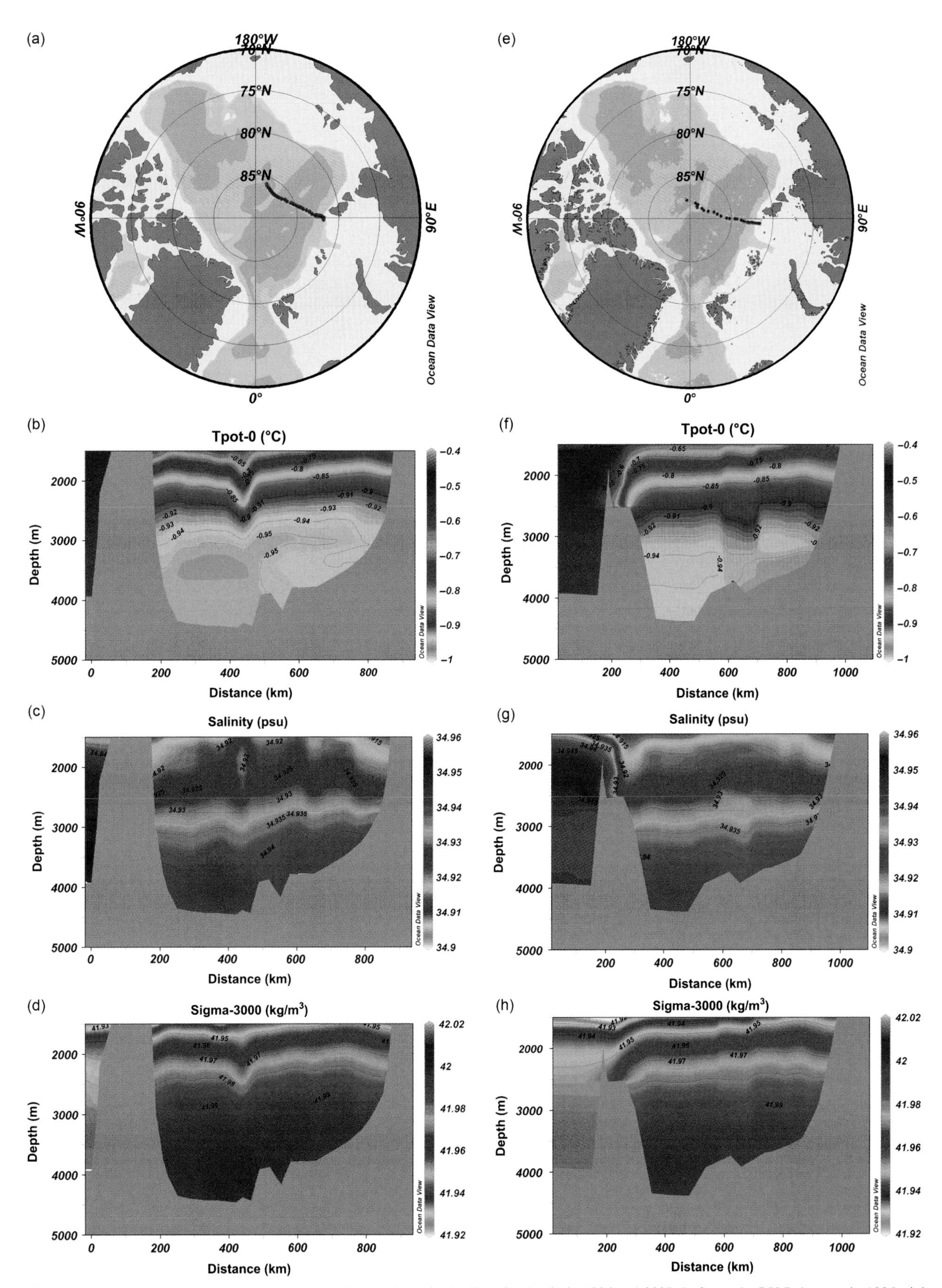

FIGURE 17.13 Comparison of hydrographic conditions at depth in the Eurasian Basin in 1996 and 2007. Left panels: RV Polarstern in 1996; right panels: RV Polarstern in 2007. Top row (a) and (e): Maps of cruise tracks, which cross the Eurasian Basin from the eastern Kara Sea and across the Lomonosov Ridge into the Makarov Basin; second row (b) and (f): potential temperature; third row (c) and (g): salinity; bottom row (d) and (h): σ_3 (potential density referenced to 3000 m). *Adapted from Rudels et al. (2013).*

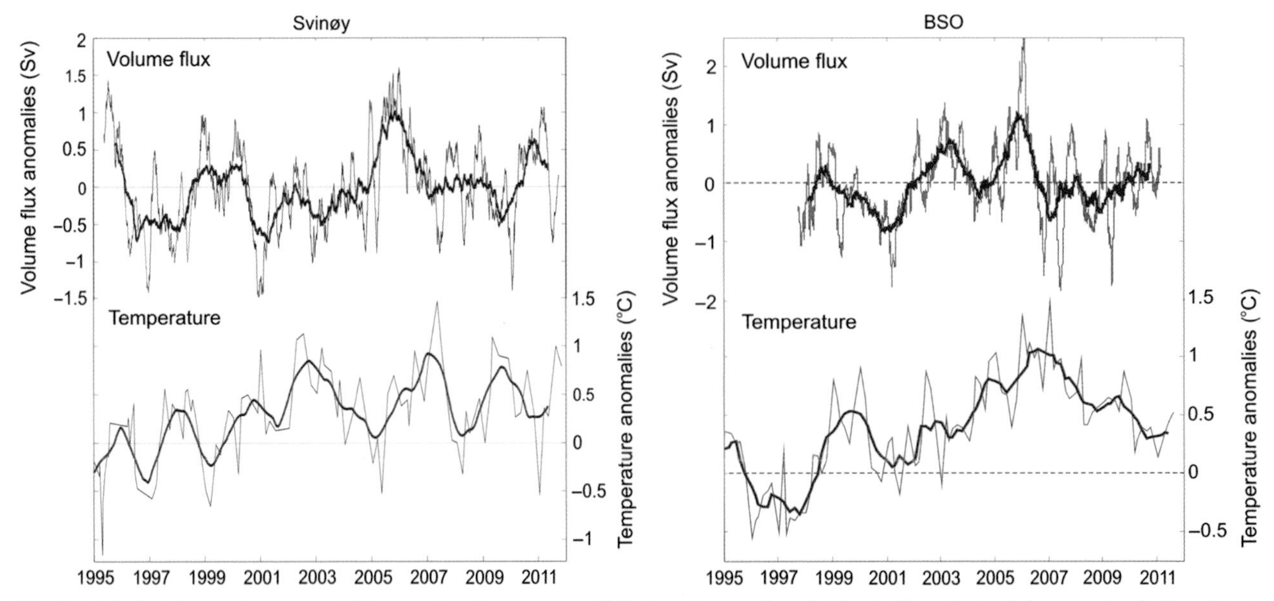

FIGURE 17.14 Measured transports ((Sv), upper) and temperature ((°C), lower) anomalies of Atlantic Water in the Svinøy section (left) and in the Barents Sea Opening (right). In the Svinøy section, the data are from the mooring S1 at 62°48′N, 4°15′E, using the instruments at depths of 100 and 300 m. In the Barents Sea Opening the transports are based on available current meters between 71°30′ and 73°30′N, and restricted to water with temperature above 3 °C. The presented data are low-pass filtered applying a Hanning window of effective length 1 year (Anonymous, 2012).

higher in 2006 than the 2002–2008 mean (Polyakov et al., 2011), that is, just before the largest temperature anomaly yet observed entered the Arctic Ocean (Figure 17.10).

There is, as yet, little evidence to suggest that there has been any long-term change in the volume exchange rates between the Arctic/Subarctic and the surrounding ocean basins (Atlantic and Pacific). Indeed, many pieces of evidence point to the opposite conclusion, namely that it is likely that there has been no observable long-term change in the exchange. First of all, most of the long-term direct current measurements referenced in Section 2.1 were acquired during the past 10 to 15 years. Nevertheless, the transport numbers are practically identical to those presented by Mauritzen (1996b), based on data from the 1970s and 1980s. Second, during the more than 15 years of monitoring in the Faroe–Shetland Channel, there has been no apparent long-term change in the inflow volume transport, though the Atlantic inflow has become warmer and more saline since 1994 (Berx et al., 2013). Similarly, measurements made at the Svinøy section, monitoring the eastern branch of the NAC since 1995, and of the Atlantic Water at the Barents Sea opening, monitored since 1998, show no detectable long-term trend in transport (Figure 17.14) though an upward trend in water temperature was observed at both locations over this time, with an absolute maximum in 2007 at both Svinøy section and the Barents Sea opening (Figure 17.14). Thirdly, there is no evidence of long-term change in the dense overflows to the North Atlantic. The 10-year time series in the Faroe Bank channel (Olsen et al., 2008) does not show a long-term trend in transport. Measurements in the Denmark Strait overflow show significant interannual variations (Macrander et al., 2005; Jochumsen et al., 2012), but the time series is too short to evaluate any long-term change in dense water volume transport.

In the Arctic Ocean, however, there is evidence of long-term as well as short-term change in circulation. North of Alaska, for example, moored observations over the continental shelf and shelf break have documented significant high-frequency variability, including flow reversals in response to changes in both local and remote wind forcing (Pickart et al., 2009, 2011). Ocean circulation changes on decadal time scales in the Arctic interior have been inferred from variations in wind stress patterns (e.g., Proshutinsky and Johnson, 1997), in particular, related to the spinup and spindown of the Beaufort Gyre. As mentioned in Section 4.2, circulation changes in the Arctic are closely connected to changes in salinity through the equation of state. Therefore, the buildup of freshwater in the Arctic since the 1990s, which was discussed in Section 4.2.2 (Figure 17.11), is likely a signature of circulation changes as well. To evaluate the causes of recent changes in freshwater content, Giles et al. (2012) and Morison et al. (2012) used satellite measurements of sea surface elevation to deduce changes in freshwater content as well as in geostrophic circulation. The former team considers the anticyclonic winds over the Beaufort Sea—the Beaufort High—as a cause of freshwater accumulation due to Ekman convergence (Proshutinsky et al., 2002) and they find that the winds have indeed strengthened over the 15 years of their

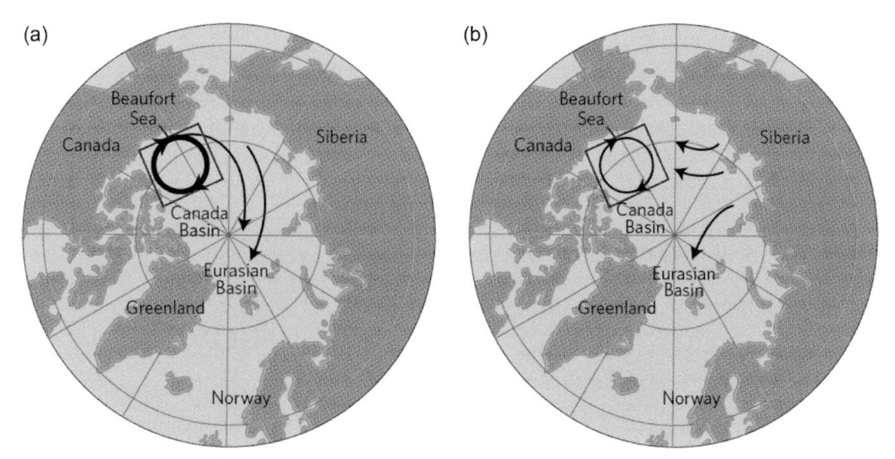

FIGURE 17.15 Schematic of Arctic Ocean circulation changes. (a) During the negative phase of the Arctic Oscillation, river runoff from Siberia is quickly swept out of the Arctic Ocean toward the Atlantic and the Beaufort High tends to be strong, leading to strong cyclonic circulation and enhanced freshwater storage in the Beaufort Gyre. (b) During the positive phase of the Arctic Oscillation, freshwater from the Russian rivers is brought toward the Canadian Basin whereas the Beaufort Gyre itself tends to relax. *From Mauritzen (2012).*

study (Giles et al., 2012). Morison et al. (2012) find that Russian river output has been steered in a more easterly direction during those years due to the recent sustained high phase period of the Arctic Oscillation, another mechanism that would cause accumulation of freshwater in the Canadian Basin. The Beaufort High is typically strong when the Arctic Oscillation is weak, such that the two mechanisms together may explain the large accumulation of freshwater in the Arctic Ocean in the past decade and a half. And the two circulation regimes of Proshutinsky and Johnson (1997) can be reformulated into these two scenarios: the negative Arctic Oscillation phase, when the Transpolar Drift is carrying Russian river runoff quickly out of the Arctic toward the Fram Strait and the positive Arctic Oscillation phase, when the runoff is partly directed toward the Canadian Basin (Figure 17.15).

5. CONCLUSIONS

We introduced this chapter by alluding to two ways in which the Arctic can influence the Earth's climate: as transformer of sea water and as melter of ice. As a consequence of the first mechanism—the transformation of sea water—cold and fresh waters are exported to the World Oceans, both in the surface ocean and in the deep ocean (the dense overflows that contribute to the Meridional Overturning Circulation in the Atlantic). Both observational and modeling studies through the past 50 years suggest that the Meridional Overturning Circulation has changed in strength and perhaps even in sign during different climatic conditions, and that it may be particularly sensitive to changes in salinity. A natural test of this sensitivity was performed in the mid-1960s, when an excess of freshwater was exported from the high north, and resulted in an accumulation of 19,000 km^3 of freshwater in the subpolar North Atlantic between 1965 and 1995 (Curry and Mauritzen, 2005). Roughly half that amount (10,000 km^3) accumulated in the late 1960s at an approximate rate of 2000 km^3/yr. But as large as that anomaly was, there is not sufficient evidence to suggest beyond reasonable doubt that there was an associated long-term change in the Atlantic Meridional Overturning Circulation during those years (see Chapters 9 and 10). Neither was that anomaly larger than the characteristic internal variability of that basin, so it can therefore not be formally attributed to human activities (Stott et al., 2008). The excess freshwater storage in the Arctic presently is roughly of the same magnitude (Section 4.2.2). Thus, only if the accumulation of freshwater in the Canadian Basin continues for many more years before collapsing will a significant change in the North Atlantic overturning circulation be likely.

The second category—as melter of ice from below—is normally disregarded due to the thick layer of fresh and cold water that lies on top of the warmer Pacific and Atlantic inflows. However, Arctic surface waters have warmed, likely from changes in albedo, from 1993 to 2007 and are likely contributing to melting of sea ice (Jackson et al., 2010). And as both the Pacific and Atlantic inflow temperatures have increased, so have their potential for influencing the sea ice above. The Pacific inflow is closest to the sea surface, and as discussed in Section 4.2.2 it has likely played a significant part in contributing to anomalous sea ice melt in recent years (e.g., Shimada et al., 2006; Woodgate et al., 2010). The Atlantic layer lies several hundred meters below the sea surface in most of the Arctic, but the layer has shoaled by 75–90 m in recent decades, and model results suggest that it might already be affecting melting of sea ice (Polyakov et al., 2010).

Upwelling of warmer Atlantic water onto the continental shelves may be having a more immediate impact.

Atlantic Water may even be melting glacier ice. Unusually warm Atlantic water carried in the boundary current around the southern tip of Greenland and drawn into the fjords of Greenland has been proposed as an explanation for the accelerated retreat of the Jakobshavn Glacier since 1997 (Holland et al., 2008). Subsurface intrusions of relatively warm waters appear to be entering fiords on the Greenland east coast as well and may be inducing glacier basal melting and accelerated ice flow and calving there too (Straneo et al., 2012; Sutherland and Straneo, 2012; Sciascia et al., 2013). Wind driven upwelling at the shelf break may play a role in transporting the warm waters landward.

Thus it appears that the oceans are contributing to changes in ice cover in northern latitudes. Changes in the cryosphere, in turn, have a profound impact on the air–sea heat and momentum exchanges throughout the Arctic, and on the atmospheric and oceanic stratification and circulation throughout the world.

The most immediate consequences of Arctic/Subarctic ocean change are seen in the high north itself. The warm NAC is key to traditional livelihood along the coast of Norway and northwestern Russia. Variations in oceanic conditions, in particular in ocean temperature, have been linked to variations in marine ecosystems in the North Sea, the Skagerak, along the coast of Norway and in the Barents Sea (see, e.g., Alheit and Hagen, 1997; Loeng and Drinkwater, 2007). Already at the beginning of the twentieth century, the Norwegian oceanographer and Nobel peace prize recipient Fridtjof Nansen voiced the opinion in Norwegian newspapers that variations in surface temperature in the Norwegian Sea could be a source of predictability for the Norwegian economy (its fisheries, forestry, etc.) with a horizon of several years (Jølle, 2009). And indeed, in synthesizing advances made in marine ecosystem research in the following 100 years, Dickson and Østerhus (2007) point to several cases of enormous swings in the ecosystem related to ocean temperature change, in particular the rise and spread of West Greenland cod fishery in the middle decades of the twentieth century and the dislocation of the traditional herring migration pattern in the 1960s to 1990s. With the warming of the Barents Sea in the 2000s (Figure 17.8), cod is again spawning there (Sundby and Nakken, 2008), for the first time in 40 years.

The Arctic is presently undergoing extreme changes in its physical environment. Here we document changes observed in the ocean, relating primarily to temperature and salinity, but also to circulation. The contribution of manmade activities to these changes is still largely unknown, although advances are made continually as techniques and time series are improving. For instance, the reversal toward higher salinities in the Atlantic water flowing from the Atlantic into the Nordic Seas and the Arctic Ocean presently (Figure 17.9) is partly explainable by human influences (Stott et al., 2008). It is therefore essential that the state of the Arctic/Subarctic be observed continually, using all the recent advances made in high-latitude monitoring and in Earth System understanding. The signature of manmade climate change in the Arctic/Subarctic Seas may possibly begin to emerge above the high level of natural variability within the next decade, if investigations of the Arctic atmosphere and sea ice are to be used as guides.

ACKNOWLEDGMENTS

The authors have received invaluable guidance and direct assistance from a great number of colleagues, and we would like to express our gratitude to these colleagues, including Agnieszka Beszczynska-Möller, Alexey Karsakov, Benjamin Rabe, Bob Dickson, Darrell Kaufman, Detlef Quadfasel, Eberhard Fahrbach, Eddy Carmack, Igor Polyakov, Igor Yashayaev, Ilker Fer, Inga Koszalka, James Overland, Jim Swift, Jonathan Overpeck, Leo Timokhov, Maria Andersson, Muyin Wang, Penny Holliday, Randi Ingvaldsen, Sarah Hughes, Pål Erik Isachsen, Seung-Ki Min, Svein Østerhus, Ursula Schauer, and Waldemar Walczowski. The authors would like to pay a special tribute to Tim Boyd, Kathrine Giles, Seymour Laxon, and Eberhard Fahrbach, whose untimely deaths were great losses for the community.

B. R. has received funding from the European Union 7th Framework Programme (FP7 2007–2013), under Grant agreements no. 212643 (THOR) & n.308299 (NACLIM). J. T.'s contribution to this work was supported by grants from the U.S. National Science Foundation. C. M.'s contribution was made possible through generous grants from the Norwegian Meteorological Institute and CICERO Center for International Climate and Environmental Research—Oslo.

REFERENCES

Aagaard, K., 1981. On the deep circulation in the Arctic Ocean. Deep Sea Res. Part A 28 (3), 251–268.

Aagaard, K., Carmack, E.C., 1989. The role of sea ice and other fresh-water in the Arctic circulation. J. Geophys. Res. 94, 14485–14498.

Aagaard, K., Swift, J.H., Carmack, E.C., 1985. Thermohaline circulation in the Arctic Mediterranean seas. J. Geophys. Res. 90 (C3), 4833–4846.

Aagaard, K., Weingartner, T.J., Danielson, S.L., Woodgate, R.A., Johnson, G.C., Whitledge, T.E., 2006. Some controls on flow and salinity in Bering Strait. Geophys. Res. Lett. 33, L19602. http://dx.doi.org/10.1029/2006GL026612.

Alford, M.H., 2003. Improved global maps and 54-year history of windwork on ocean inertial motions. Geophys. Res. Lett. 30, 1424–1427.

Alheit, J., Hagen, E., 1997. Long-term climate forcing of European herring and sardine populations. Fish. Oceanogr. 6, 130–139. http://dx.doi.org/10.1046/j.1365-2419.1997.00035.x.

Anderson, L.G., Jones, E.P., Rudels, B., 1999. Ventilation of the Arctic Ocean estimated by a plume entrainment model constrained by CFCs. J. Geophys. Res. 104 (C6), 13423–13429.

Andersson, M., Orvik, K.A., Koszalka, I., LaCasce, J.H., Mauritzen, C., 2011. Variability of the Norwegian Atlantic Current and associated eddy field from surface drifters. J. Geophys. Res. 116. http://dx.doi.org/10.1029/2011JC007078.

Anonymous, 2012. Havforskningsrapporten 2012. Fisken og Havet særnummer 1-2012, Havforskningsinstituttet, Bergen. ISSN 0802 0620.

Azetsu-Scott, K., Petrie, B., Yeats, P., Lee, C., 2012. Composition and fluxes of freshwater through Davis Strait using multiple chemical tracers. J. Geophys. Res. 117, C12011, http://dx.doi.org/10.1029/2012JC008172 DOI:10.1029/2012JC008172#Link to external resource: 10.1029/2012JC008172.

Bailey, W.B., 1956. On the origin of deep Baffin Bay water. J. Fish. Board Can. 13 (3), 303–308.

Berx, B., Hansen, B., Østerhus, S., Larsen, K.M., Sherwin, T., Jochumsen, K., 2013. Combining in-situ measurements and altimetry to estimate volume, heat and salt transport variability through the Faroe Shetland Channel. Ocean Sci. Discuss. 10, 153–195. http://dx.doi.org/10.5194/osd-10-153-2013.

Beszczynska-Möller, A., Woodgate, R.A., Lee, C.M., Melling, H., Karcher, M., 2011. A synthesis of exchanges through the main oceanic gateways to the Arctic Ocean. Oceanography 24 (3), 76–93.

Brooks, C.E.P., 1949. Causes of climatic fluctuations. Q. J. R. Meteorol. Soc. 75 (324), 172–185.

Budéus, G., Ronski, S., 2009. An integral view of the hydrographic development in the Greenland Sea over a decade. Open Oceanogr. J. 1874–2521. http://dx.doi.org/10.2174/1874252100903010008.

Budéus, G., Schneider, W., Krause, G., 1998. Winter convective events and bottom water warming in the Greenland Sea. J. Geophys. Res. 103 (C9), 18513–18527.

Budkyo, M., 1969. The polar ice and climate (interrelationships between Arctic ice cap and climate, and between ice cover and industrial plant heat production). Tr. Naunchn. Issled. Inst. Gidrometrol. Priborostr (Leningrad) 2–36.

Carmack, E.C., 2000. The Arctic Ocean's freshwater budget: sources, storage and export. Nato Sci. Ser. 2 Environ. Secur. 70, 91–126.

Carmack, E., McLaughlin, F., 2011. Towards recognition of physical and geochemical change in Subarctic and Arctic Seas. Prog. Oceanogr. 90 (1), 90–104.

Carmack, E.C., Melling, H., 2011. Cryosphere: warmth from the deep? Nat. Geosci. 4 (1), 7–8. http://dx.doi.org/10.1038/ngeo1044.

Carmack, E.C., Macdonald, R.W., Perkin, R.G., McLaughlin, F.A., 1995. Evidence for warming of Atlantic water in the southern Canadian Basin. Geophys. Res. Lett. 22, 1961–1964.

Carmack, E.C., Aagaard, K., Swift, J.H., Perkin, R.G., McLaughlin, F., Macdonald, R.W., Jones, E.P., 1998. Thermohaline transitions. In: Imberger, J. (Ed.), Physical Processes in Lakes and Oceans. Coastal and Estuarine Studies, vol. 54. American Geophysical Union, Washington, DC, pp. 179–186.

Coachman, L.K., Aagaard, K., 1966. On the water exchange through Bering Strait. Limnol. Oceanogr. 44–59.

Coachman, L.K., Aagaard, K., 1974. Physical oceanography of the arctic and subarctic seas. In: Herman, Y. (Ed.), Marine Geology and Oceanography of the Arctic Seas. Springer, New York, pp. 1–72.

Coachman, L.K., Barnes, C.A., 1963. The movement of Atlantic water in the Arctic Ocean. Arctic 16 (1), 8–16.

Cuny, J., Rhines, P.B., Kwok, R., 2005. Davis Strait volume, freshwater and heat fluxes. Deep Sea Res. Part I 52 (3), 519–542. http://dx.doi.org/10.1016/j.dsr.2004.10.006.

Curry, R., Mauritzen, C., 2005. Dilution of the northern North Atlantic Ocean in recent decades. Science 308 (5729), 1772–1774.

Curry, B., Lee, C.M., Petrie, B., 2011. Volume, freshwater, and heat fluxes through Davis Strait, 2004–05. J. Phys. Oceanogr. 41, 429–436. http://dx.doi.org/10.1175/2010JPO4536.1.

D'Asaro, E.A., Morison, J.H., 1992. Internal waves and mixing in the Arctic Ocean. Deep Sea Res. Part A 39, S459–S484. http://dx.doi.org/10.1016/ S0198-0149(06)80016-6.

D'Asaro, E.A., 1988. Observations of small eddies in the Beaufort Sea. J. Geophys. Res. 93 (C6), 6669–6684.

Dewey, R., Muench, R., Gunn, J., 1999. Mixing and vertical heat flux estimates in the Arctic Eurasian Basin. J. Mar. Syst. 21, 199–205. http://dx.doi.org/10.1016/S0924-7963(99)00014-7.

Dickson, B., Østerhus, S., 2007. One hundred years in the Norwegian Sea. Nor. Geogr. Tidsskr. Nor. J. Geogr. 61 (2), 56–75.

Dickson, R.R., Rudels, B., Dye, S., Karcher, M., Meincke, J., Yashayaev, I., 2007. Current estimates of freshwater flux through Arctic and subarctic seas. Prog. Oceanogr. 73 (3–4), 210–230.

Dickson, R.R., Dye, S., Jónsson, S., Köhl, A., Macrander, A., Marnela, M., Meincke, J., Olsen, S., Rudels, B., Valdimarsson, H., Voet, G., 2008. The overflow flux west of iceland: variability, origins and forcing. In: Dickson, R.R., Meincke, J., Rhines, P. (Eds.), Arctic-Subarctic Ocean Fluxes. Springer, Dordrecht, pp. 443–474.

Dinsmore, R.P., 1996. Alpha, Bravo, Charlie... Ocean Weather Ships 1940-1980. OCEANUS 39 (02), 9–11.

Dmitrenko, I.A., Polyakov, I.V., Kirillov, S.A., Timokhov, L.A., Frolov, I.E., Sokolov, V.T., Simmons, H.L., Ivanov, V.V., Walsh, D., 2008. Toward a warmer Arctic Ocean: spreading of the early 21st century Atlantic Water warm anomaly along the Eurasian Basin margins. J. Geophys. Res. 113 (C5), C05023.

Donn, W.L., Ewing, M., 1968. The theory of an ice-free Arctic Ocean. Meteorol. Monogr. 8 (30), 100–105.

Falck, E., Kattner, G., Budeus, G., 2005. Disappearance of Pacific Water in the northwestern Fram Strait. Geophys. Res. Lett. 32, L14619. http://dx.doi.org/10.1029/2005GL023400.

Fer, I., 2009. Weak vertical diffusion allows maintenance of cold halocline in the central Arctic. Atmos. Oceanic Sci. Lett. 3, 148–152.

Francis, J.A., Vavrus, S.J., 2012. Evidence linking Arctic amplification to extreme weather in mid-latitudes. Geophys. Res. Lett. 39, L06801. http://dx.doi.org/10.1029/2012GL051000.

Francis, J.A., Chan, W., Leathers, D.J., Miller, J.R., Veron, D.E., 2009. Winter Northern Hemisphere weather patterns remember summer Arctic sea-ice extent. Geophys. Res. Lett. 36, L07503. http://dx.doi.org/10.1029/2009GL037274.

Furevik, T.,1998. On the Atlantic inflow in the Nordic Seas: bifurcation and variability (Dr. scient thesis). University of Bergen, Bergen.

Garrett, C., Munk, W.H., 1972. Space-time scales of internal waves. Geophys. Fluid Dyn. 2, 225–264.

Gascard, J.-C., Mork, K.A., 2008. Climatic importance of large-scale and mesoscale circulation in the Lofoten Basin deduced from Lagrangian observations. In: Dickson, R.R., Meincke, J., Rhines, P. (Eds.), Arctic-Subarctic Ocean Fluxes. Springer, Dordrecht, pp. 131–143.

Giles, K.A., Laxon, S.W., Ridout, A.L., 2008. Circumpolar thinning of Arctic sea ice following the 2007 record ice extent minimum. Geophys. Res. Lett. 35 (22), L22502.

Giles, K.A., Laxon, S.W., Ridout, A.L., Wingham, D.J., Bacon, S., 2012. Western Arctic Ocean freshwater storage increased by wind-driven spin-up of the Beaufort Gyre. Nat. Geosci. http://dx.doi.org/10.1038/ngeo1379.

Gillett, N.P., Stone, D.A., Stott, P.A., Nozawa, T., Karpechko, A.Y., Hegerl, G.C., Wehner, M.F., Jones, P.D., 2008. Attribution of polar warming to human influence. Nat. Geosci. 1, 750–754. http://dx.doi.org/10.1038/ngeo338.

Girton, J.B., Sanford, T.B., Käse, R.H., 2001. Synoptic sections of the Denmark Strait Overflow. Geophys. Res. Lett. 28 (8), 1619–1622. http://dx.doi.org/10.1029/2000GL011970.

Halle, C., Pinkel, R., 2003. Internal wave variability in the Beaufort Sea during the winter of 1993/1994. J. Geophys. Res. 108 (C7), 3210.

Hansen, B., Østerhus, S., 2000. North Atlantic–Nordic Seas exchanges. Prog. Oceanogr. 45, 109–208.

Hansen, B., Østerhus, S., 2007. Faroe Bank Channel overflow 1995–2005. Prog. Oceanogr. 75, 817–856.

Hansen, B., Hátún, H., Kristiansen, R., Olsen, S.M., Østerhus, S., 2010. Stability and forcing of the Iceland-Faroe inflow of water, heat, and salt to the Arctic. Ocean Sci. 6 (4), 1013–1026.

Holland, M.M., Bitz, C.M., 2003. Polar amplification of climate change in coupled models. Clim. Dyn. 21, 221–232.

Holland, D.M., Thomas, R.H., De Young, B., Ribergaard, M.H., Lyberth, B., 2008. Acceleration of Jakobshavn Isbrae triggered by warm subsurface ocean waters. Nat. Geosci. 1 (10), 659–664.

Holliday, N.P., Hughes, S.L., Bacon, S., Beszczynska-Möller, A., Hansen, B., Lavin, A., Loeng, L., Mork, K.A., Østerhus, S., Sherwin, T., Walczowski, W., 2008. Reversal of the 1960s to 1990s freshening trend in the northeast North Atlantic and Nordic Seas. Geophys. Res. Lett. 35, L03614. http://dx.doi.org/10.1029/2007GL032675.

Ingvaldsen, R.B., Asplin, L., Loeng, H., 2004. Velocity field of the western entrance to the Barents Sea. J. Geophys. Res. 109, C03021. http://dx.doi.org/10.1029/2003JC001811.

Isachsen, P.E., LaCasce, J.H., Mauritzen, C., Häkkinen, S., 2003. Wind-driven variability of the large-scale recirculating flow in the Nordic Seas and Arctic Ocean. J. Phys. Oceanogr. 33, 2534–2550.

Jackson, J.M., Carmack, E.C., McLaughlin, F.A., Allen, S.E., Ingram, R.G., 2010. Identification, characterization, and change of the near-surface temperature maximum in the Canada Basin, 1993–2008. J. Geophys. Res. 115, C05021.

Jackson, J.M., Williams, W.J., Carmack, E.C., 2012. Winter sea-ice melt in the Canada Basin, Arctic Ocean. Geophys. Res. Lett. 39 (3).

Jakobsson, M., Macnab, R., Mayer, L., Anderson, R., Edwards, M., Hatzky, J., Schenke, H.W., Johnson, P., 2008. An improved bathymetric portrayal of the Arctic Ocean: implications for ocean modeling and geological, geophysical and oceanographic analyses. Geophys. Res. Lett. 35, L07602. http://dx.doi.org/10.1029/2008GL033520.

Jeffries, M.O., Richter-Menge, J.A., Overland, J.E. (Eds.), 2012. Arctic Report Card 2012. http://www.arctic.noaa.gov/reportcard.

Jochumsen, K., Quadfasel, D., Valdimarsson, H., Jónsson, S., 2012. Variability of the Denmark Strait overflow: moored time series from 1996–2011. J. Geophys. Res. 117, C12003. http://dx.doi.org/10.1029/2012JC008244.

Jølle, H.D., 2009. Polar prestisje og vitenskapelig ære. Fridtjof Nansens kamp for havforskning ved universitetet. Hist. Tidsskr. 4, 611–637.

Jones, E.P., Anderson, L.G., 1986. On the origin of the chemical properties of the Arctic Ocean halocline. J. Geophys. Res. 91 (C9), 10759–10767.

Jones, E.P., Swift, J.H., Anderson, L.G., Lipizer, M., Civitarese, G., Falkner, K.K., Kattner, G., McLaughlin, F.A., 2003. Tracing Pacific water in the North Atlantic Ocean. J. Geophys. Res. 108 (C4), 3116. http://dx.doi.org/10.1029/2001JC001141.

Jónsson, S., Valdimarsson, H., 2012. Water mass transport variability to the North Icelandic shelf, 1994–2010. ICES J. Mar. Sci. 69 (5), 809–815.

Karsakov, A.L., 2009. 110 year Oceanographic Researches on the Barents Sea on the Kola Section. PINRO Press, Murmansk, 138.

Kaufman, D.S., Schneider, D.P., McKay, N.P., Ammann, C.M., Bradley, R.S., Briffa, K.R., Miller, G.H., Otto-Bliesner, B.L., Overpeck, J.T., Vinther, B.M., Arctic Lakes 2k Project Members, 2009. Recent warming reverses long-term Arctic cooling. Science 325 (5945), 1236–1239. http://dx.doi.org/10.1126/science.1173983.

Kellogg, W.W., 1975. Climate change and the influence of man's activities on the global environment. In: Singer, S.F., Reidel, D. (Eds.), The Changing Global Environment. Springer, Netherlands, pp. 13–23.

Kinney, J.C., Maslowski, W., Okkonen, S., 2008. On the processes controlling shelf-basin exchange and outer shelf dynamics in the Bering sea. Deep Sea Res. Part II 56 (17), 1351–1362. http://dx.doi.org/10.1016/j.dsr2.2008.10.023.

Koszalka, I., LaCasce, J.H., 2010. Lagrangian analysis by clustering. Ocean Dyn. 60 (4), 957–972. http://dx.doi.org/10.1007/s10236-010-0306-2.

Koszalka, I., LaCasce, J.H., Andersson, M., Orvik, K.A., Mauritzen, C., 2011. Surface circulation in the Nordic Seas from clustered drifters. Deep Sea Res. Part I 58 (4), 468–485.

Koszalka, I., LaCasce, J.H., Mauritzen, C., 2012. In pursuit of anomalies—analyzing the poleward transport of Atlantic Water with surface drifters. Deep Sea Res. Part II 85, 96–108. http://dx.doi.org/10.1016/j.dsr2.2012.07.035.

Krishfield, R.A., Plueddemann, A.J., Honjo, S., 2002. Eddies in the Arctic Ocean from IOEB ADCP data. WHOI Report No. WHOI-2002-09. Woods Hole Oceanographic Institution, MA.

Kwok, R., Rothrock, D.A., 2009. Decline in Arctic sea ice thickness from submarine and ICESat records: 1958–2008. Geophys. Res. Lett. 36, L15501. http://dx.doi.org/10.1029/2009GL039035.

Lenn, Y.D., Wiles, P.J., Torres-Valdes, S., Abrahamsen, E.P., Rippeth, T.P., Simpson, J.H., Bacon, S., Laxon, S.W., Polyakov, I., Ivanov, V., Kirillov, S., 2009. Vertical mixing at intermediate depths in the Arctic boundary current. Geophys. Res. Lett. 36, L05601. http://dx.doi.org/10.1029/2008GL036792.

Levine, M.D., Paulson, C.A., Morison, J.H., 1985. Internal waves in the Arctic Ocean: observations and comparison with lower latitude climatology. J. Phys. Oceanogr. 15, 800–809.

Levine, M.D., Paulson, C.A., Morison, J.H., 1987. Observations of internal gravity waves under the arctic ice pack. J. Geophys. Res. 92 (C1), 779–782.

Lian, M.S., Cess, R.D., 1977. Energy balance climate models: a reappraisal of ice-albedo feedback. J. Atmos. Sci. 34 (7), 1058–1062.

Lien, V.S., Vikebø, F.B., Skagseth, Ø., 2013. One mechanism contributing to co-variability of Atlantic inflow branches to the Arctic. Nat. Commun. 4, 1488. http://dx.doi.org/10.1038/ncomms2505.

Liu, J., Curry, J.A., Wang, H., Song, M., Horton, R., 2012. Impact of declining Arctic sea ice on winter snowfall. Proc. Natl. Acad. Sci. U.S.A. 109, 4074–4079.

Loeng, H., Drinkwater, K., 2007. An overview of the ecosystems of the Barents and Norwegian Seas and their response to climate variability. Deep Sea Res. Part II 54 (23–26), 2478–2500.

Macrander, A., Send, U., Valdimarsson, H., Jonsson, S., Kase, R.H., 2005. Interannual changes in the overflow from the Nordic Seas into the Atlantic Ocean through Denmark Strait. Geophys. Res. Lett. 32, L06606. http://dx.doi.org/10.1029/2004GL021463.

Maslanik, J.A., Fowler, C., Stroeve, J., Drobot, S., Zwally, J., Yi, D., Emery, W., 2007. A younger, thinner Arctic ice cover: increased potential for rapid, extensive sea-ice loss. Geophys. Res. Lett. 34 (24), L24501.

Maslanik, J., Stroeve, J., Fowler, C., Emery, W., 2011. Distribution and trends in Arctic sea ice age through spring 2011. Geophys. Res. Lett. 38 (13), L13502.

Mauritzen, C., 1996a. Production of dense overflow waters feeding the North Atlantic across the Greenland Sea-Scotland Ridge. Part 1: evidence for a revised circulation scheme. Deep Sea Res. 43, 769–806.

Mauritzen, C., 1996b. Production of dense overflow waters feeding the North Atlantic across the Greenland Sea-Scotland Ridge. Part 2: an inverse model. Deep Sea Res. 43, 807–835.

Mauritzen, C., 2012. Oceanography: Arctic freshwater. Nat. Geosci. 5 (3), 162–164.

Mauritzen, C., Hansen, E., Andersson, M., Berx, B., Beszczynska-Möller, A., Burud, I., Christensen, K.H., Debernard, J., de Steur, L., Dodd, P., Gerland, S., Godøy, Ø., Hansen, B., Hudson, S., Høydalsvik, F., Ingvaldsen, R., Isachsen, P.E., Kasajima, Y., Koszalka, I., Kovacs, K.M., Køltzow, M., LaCasce, J.H., Lee, C.M., Lavergne, T., Lydersen, C., Nicolaus, M., Nilsen, F., Nøst, O.A., Orvik, K.A., Reigstad, M., Schyberg, H., Seuthe, L., Skagseth, Ø., Skarðhamar, J., Skogseth, R., Sperrevik, A., Svensen, C., Søiland, H., Teigen, S.H., Tverberg, V., Wexels Riser, C., 2011. Closing the loop—approaches to monitoring the state of the Arctic Mediterranean during the International Polar Year 2007–2008. Prog. Oceanogr. 90 (1), 62–89. http://dx.doi.org/10.1016/j.pocean.2011.02.010.

Maykut, G., Untersteiner, N., 1971. Some results from a time-dependent thermodynamic model of Arctic sea ice. J. Geophys. Res. 76 (6), 1550–1575.

McLaughlin, F.A., Carmack, E.C., Williams, W.J., Zimmermann, S., Shimada, K., Itoh, M., 2009. Joint effects of boundary currents and thermohaline intrusions on the warming of Atlantic water in the Canada Basin, 1993–2007. J. Geophys. Res. 114 (C1), C00A12.

McPhee, M.G., 2008. Air-ice-ocean interaction: turbulent ocean boundary layer exchange process. Springer Science + Business Media B.V., New York, 215 pp.

McPhee, M.G., Kwok, R., Robins, R., Coon, M., 2005. Upwelling of Arctic pycnocline associated with shear motion of sea ice. Geophys. Res. Lett. 32, L10616. http://dx.doi.org/10.1029/2004GL021819.

McPhee, M.G., Proshutinsky, A., Morison, J.H., Steele, M., Alkire, M.B., 2009. Rapid change in freshwater content of the Arctic Ocean. Geophys. Res. Lett. 36, L10602. http://dx.doi.org/10.1029/2009GL037525.

Meincke, J., Rudels, B., Friedrich, H.J., 1997. Water-mass formation and distribution in the Nordic Seas during the 1990s. ICES J. Mar. Sci. 61 (5), 846–863.

Melling, H., Agnew, T.A., Falkner, K.K., Greenberg, D.A., Craig, M.L., Münchow, A., Petrie, B., Prinsenberg, S.J., Samelson, R.M., Woodgate, R.A., 2008. Freshwater fluxes via Pacific and Arctic outflows across the Canadian polar shelf. In: Dickson, R.R., Meincke, J., Rhines, P. (Eds.), Arctic-Subarctic Ocean Fluxes. Springer, Dordrecht, pp. 193–247.

Midttun, L., 1985. Formation of dense bottom water in the Barents Sea. Deep Sea Res. Part A 32 (10), 1233–1241.

Min, S.-K., Zhang, X., Zwiers, F.W., Agnew, T., 2008. Human influence on Arctic sea ice detectable from early 1990s onwards. Geophys. Res. Lett. 35, L21701. http://dx.doi.org/10.1029/2008GL035725.

Morison, J., Steele, M., Kikuchi, T., Falkner, K., Smethie, W., 2006. Relaxation of central Arctic Ocean hydrography to pre-1990s climatology. Geophys. Res. Lett. 33, L17604. http://dx.doi.org/10.1029/2006GL026826.

Morison, J., Kwok, R., Peralta-Ferriz, C., Alkire, M., Rigor, I., Andersen, R., Steele, M., 2012. Changing Arctic Ocean freshwater pathways. Nature 481 (7379), 66–70.

Muench, R.D., 1971. The physical oceanography of the northern Baffin Bay region. North Water Project. Science Report 1. Arctic Institute of North America, University of Calgary, Calgary, Alberta, Canada.

Muench, R.D., Gunn, J.T., Whitledge, T.E., Schlosser, P., Smethie Jr., W., 2000. An Arctic Ocean cold core eddy. J. Geophys. Res. 105 (C10), 23997–24006.

Münchow, A., Melling, H., 2008. Ocean current observations from Nares Strait to the west of Greenland: interannual to tidal variability and forcing. J. Mar. Res. 66 (6), 801–833.

Nansen, F., 1915. Spitsbergen Waters. Videnskabs-selskabets skrifter I. Matematisk-Naturvidenskabelig klasse, vol. I(3), Christiania, Norway, 145 pp.

Neal, V.T., Neshyba, S., Denner, W., 1969. Thermal stratification in the Arctic Ocean. Science 166, 373–374.

Newton, J.L., Coachman, L.K., 1974. Atlantic water circulation in the Canada Basin. Arctic 27, 297–303.

Nikolopoulos, A., Pickart, R.S., Fratantoni, P.S., Shimada, K., Torres, D.J., Jones, E.P., 2009. The western Arctic boundary current at 152°W: structure, variability, and transport. Deep Sea Res. Part II 56, 1164–1181.

Nilsson, J., Björk, G., Rudels, B., Winsor, P., Torres, D., 2008. Liquid freshwater transport and Polar Surface Water characteristics in the East Greenland Current during the AO-02 Oden expedition. Prog. Oceanogr. 0079-661178 (1), 45–57. http://dx.doi.org/10.1016/j.pocean.2007.06.002.

Olsen, S.M., Hansen, B., Quadfasel, D., Østerhus, S., 2008. Observed and modelled stability of overflow across the Greenland–Scotland ridge. Nature 455, 519–523.

Ortiz, J.D., Falkner, K.K., Matrai, P.A., Woodgate, R.A., 2011. The Changing Arctic Ocean: Special Issue on the International Polar Year (2007–2009). Oceanography 24 (3), 14–16. http://www.tos.org/oceanography/archive/24-3.html.

Orvik, K.A., Skagseth, Ø., 2003. The impact of the wind stress curl in the North Atlantic on the Atlantic inflow to the Norwegian Sea toward the Arctic. Geophys. Res. Lett. 30, 1884. http://dx.doi.org/10.1029/2003GL017932.

Østerhus, S., Turrell, W.R., Jonsson, S., Hansen, B., 2005. Measured volume, heat, and salt fluxes from the Atlantic to the Arctic Mediterranean. Geophys. Res. Lett. 32, L07603. http://dx.doi.org/10.1029/2004GL022188.

Padman, L., Dillon, T.M., 1987. Vertical fluxes through the Beaufort Sea thermohaline staircase. J. Geophys. Res. 92, 799–806.

Padman, L., Dillon, T.M., 1991. Turbulent mixing near the Yermak Plateau during the Coordinated Eastern Arctic Experiment. J. Geophys. Res. 96, 4769–4782. http://dx.doi.org/10.1029/90JC02260.

Perovich, D.K., Maykut, G.A., 1990. Solar heating of a stratified ocean in the presence of a static ice cover. J. Geophys. Res. 95 (C10), 18233–18245.

Perovich, D.K., Richter-Menge, J.A., Jones, K.F., Light, B., 2008. Sunlight, water, and ice: extreme Arctic sea ice melt during the summer of 2007. Geophys. Res. Lett. 35, L11501. http://dx.doi.org/10.1029/2008GL034007.

Pickart, R.S., 2004. Shelfbreak circulation in the Alaskan Beaufort Sea: mean structure and variability. J. Geophys. Res. 109, C04024. http://dx.doi.org/10.1029/2003JC001912.

Pickart, R.S., Weingartner, T.J., Zimmermann, S., Torres, D.J., Pratt, L.J., 2005. Flow of winter-transformed Pacific water into the western Arctic. Deep Sea Res. Part II 52, 3175–3198.

Pickart, R.S., Moore, G.W.K., Torres, D.J., Fratantoni, P.S., Goldsmith, R.A., Yang, J., 2009. Upwelling on the continental slope of the Alaskan Beaufort Sea: storms, ice, and oceanographic response. J. Geophys. Res. 114, C00A13. http://dx.doi.org/10.1029/2208JC005009.

Pickart, R.S., Spall, M.A., Aagaard, K., Shimada, K., 2011. Upwelling in the Alaskan Beaufort Sea: atmospheric forcing and local versus non-local response. Prog. Oceanogr. 88, 78–100.

Polyakov, I.V., Alexeev, V.A., Belchansky, G.I., Dmitrenko, I.A., Ivanov, V.V., Kirillov, S.A., Korablev, A.A., Steele, M., Timokhov, L.A., Yashayaev, I., 2008. Arctic Ocean freshwater changes over the past 100 years and their causes. J. Clim. 21 (2), 364–384.

Polyakov, I.V., Timokhov, L.A., Alexeev, V.A., Bacon, S., Dmitrenko, I.A., Fortier, L., Frolov, I.E., Gascard, J.C., Hansen, E., Ivanov, V.V., Laxon, S., Mauritzen, C., Perovich, D., Shimada, K., Simmons, H.L., Sokolov, V.T., Steele, M., Toole, J., 2010. Arctic Ocean warming contributes to reduced polar ice cap. J. Phys. Oceanogr. 40, 2743–2756. http://dx.doi.org/10.1175/2010JPO4339.1.

Polyakov, I.V., Alexeev, V.A., Ashik, I.M., Bacon, S., Beszczynska-Möller, A., Carmack, E.C., Dmitrenko, I.A., Fortier, L., Gascard, J.-C., Hansen, E., Hölemann, J., Ivanov, V.V., Kikuchi, T., Kirillov, S., Lenn, Y.-D., McLaughlin, F., Piechura, J., Repina, I., Timokhov, L.A., Walczowski, W., Woodgate, R., 2011. Fate of early 2000s Arctic warm water pulse. Bull. Am. Meteorol. Soc. 92, 561–566. http://dx.doi.org/10.1175/2010BAMS2921.1.

Polyakov, I.V., Pnyushkov, A.V., Timokhov, L.A., 2012. Warming of the Intermediate Atlantic Water of the Arctic Ocean in the 2000s. J. Clim. 25 (23), 8362–8370.

Prinsenberg, S., Hamilton, J., Peterson, I., Pettipas, R., 2009. Observing and interpreting the seasonal variability of the oceanographic fluxes passing through Lancaster Sound of the Canadian Arctic Archipelago. Influence of Climate Change on the Changing Arctic and Sub-Arctic Conditions. Springer, Netherlands, pp. 125–143.

Proshutinsky, A.Y., Johnson, M.A., 1997. Two circulation regimes of the wind-driven Arctic Ocean. J. Geophys. Res. 102 (C6), 12493–12514.

Proshutinsky, A., Bourke, R.H., McLaughlin, F.A., 2002. The role of the Beaufort Gyre in Arctic climate variability: seasonal to decadal climate scales. Geophys. Res. Lett. 29 (23), 2100.

Quadfasel, D., Rudels, B., Kurz, K., 1988. Outflow of dense water from a Svalbard fjord into the Fram Strait. Deep Sea Res. 35, 1143–1150.

Quadfasel, D., Sy, A., Wells, D., Tunik, A., 1991. Warming in the Arctic. Nature 350, 385.

Rabe, B., Münchow, A., Johnson, H.L., Melling, H., 2010. Nares Strait hydrography and salinity field from a 3-year moored array. J. Geophys. Res. 115, C07010. http://dx.doi.org/10.1029/2009JC005966.

Rabe, B., Karcher, M., Schauer, U., Toole, J.M., Krishfield, R.A., Pisarev, S., Kauker, F., Gerdes, R., Kikuchi, T., 2011. An assessment of Arctic Ocean freshwater content changes from the 1990s to the 2006–2008 period. Deep Sea Res. Part I 58 (2), 173–185. http://dx.doi.org/10.1016/j.dsr.2010.12.002.

Rainville, L., Winsor, P., 2008. Mixing across the Arctic Ocean: microstructure observations during the Beringia 2005 expedition. Geophys. Res. Lett. 35, L08606. http://dx.doi.org/10.1029/2008GL033532.

Rainville, L., Woodgate, R.A., 2009. Observations of internal wave generation in the seasonally ice-free Arctic. Geophys. Res. Lett. 36, L23604. http://dx.doi.org/10.1029/2009GL041291.

Richter, K., Furevik, T., Orvik, K.A., 2009. Effect of wintertime low-pressure systems on the Atlantic inflow to the Nordic seas. J. Geophys. Res. 114, C09006. http://dx.doi.org/10.1029/2009JC005392.

Richter-Menge, J., Jeffries, M. (Eds.), 2011. The Arctic. In "State of the Climate in 2010", In: Bull. Am. Meteorol. Soc., 92, pp. S161–S1716.

Rigor, I.G., Wallace, J.M., 2004. Variations in the age of Arctic sea-ice and summer sea-ice extent. Geophys. Res. Lett. 31 (9), L09401.

Riis Carstensen, E., 1936. The "Godthaab" expedition 1928. The hydrographic work and material. Medd. Grönland 78 (3), 101.

Roach, A., Aagaard, K., Pease, C., Salo, S., Weingartner, T., Pavlov, V., Kulakov, M., 1995. Direct measurements of transport and water properties through the Bering Strait. J. Geophys. Res. 100 (C9), 18443–18457.

Ross, C.K., 1984. Temperature-salinity characteristics of the "overflow" water in Denmark Strait during "Overflow 73". Rapp. Proc. Verb. Reun. Explor. Mer. 185, 111–119.

Rossby, T., Prater, M.D., Søiland, H., 2009. Pathways of inflow and dispersion of warm waters in the Nordic seas. Geophys. Res. Lett. 114, C04011. http://dx.doi.org/10.1029/2008JC005073.

Rudels, B., 1986. The Θ-S relations in the northern seas: implications for the deep circulation. Polar Res. 4, 133–159.

Rudels, B., 2012. Arctic Ocean circulation and variability-advection and external forcing encounter constraints and local processes. Ocean Sci. 8, 261–286.

Rudels, B., Jones, E.P., Anderson, L.G., Kattner, G., 1994. On the intermediate depth waters of the Arctic Ocean. In: Johannessen, O.M., Muench, R.D., Overland, J.E. (Eds.), The Role of the Polar Oceans in Shaping the Global Climate. American Geophysical Union, Washington, DC, pp. 33–46.

Rudels, B., Anderson, L.G., Jones, E.P., 1996. Formation and evolution of the surface mixed layer and halocline of the Arctic Ocean. J. Geophys. Res. Oceans 101 (C4), 8807–8821. http://dx.doi.org/10.1029/96JC00143.

Rudels, B., Friedrich, H.J., Quadfasel, D., 1999. The Arctic Circumpolar Boundary Current. Deep Sea Res. Part II 46, 1023–1062.

Rudels, B., Meyer, R., Fahrbach, E., Ivanov, V.V., Østerhus, S., Quadfasel, D., Schauer, U., Tverberg, V., Woodgate, R.A., 2000a. Water mass distribution in Fram Strait and over the Yermak Plateau in summer 1997. Ann. Geophys. 18, 687–705.

Rudels, B., Muench, R.D., Gunn, J., Schauer, U., Friedrich, H.J., 2000b. Evolution of the Arctic Ocean boundary current north of the Siberian shelves. J. Mar. Syst. 25, 77–99.

Rudels, B., Jones, E., Schauer, U., Eriksson, P., 2004. Atlantic sources of the Arctic Ocean surface and halocline waters. Polar Res. 23, 181–208. http://dx.doi.org/10.1111/j.1751-8369.2004.tb00007.x.

Rudels, B., Björk, G., Nilsson, J., Winsor, P., Lake, I., Nohr, C., 2005. The interactions between waters from the Arctic Ocean and the Nordic Seas north of Fram Strait and along the East Greenland Current: results from the Arctic Ocean-02 Oden expedition. J. Mar. Syst. 55, 1–30. http://dx.doi.org/10.1016/j.jmarsys.2004.06.008.

Rudels, B., Marnela, M., Eriksson, P., 2008. Constraints on estimating mass, heat and freshwater transport in the Arctic Ocean: an exercise.

In: Dickson, R.R., Meincke, J., Rhines, P. (Eds.), Arctic-Subarctic Ocean Fluxes. Springer, Dordrecht, pp. 315–341.

Rudels, B., Anderson, L., Eriksson, P., Fahrbach, E., Jakobsson, M., Jones, E.P., Melling, H., Prinsenberg, S., Schauer, U., Yao, T., 2012a. Observations in the ocean. Arctic Clim. Change 43, 117–198.

Rudels, B., Korhonen, M., Budéus, G., Beszczynska-Möller, A., Schauer, U., Nummelin, A., Quadfasel, D., Valdimarsson, H., 2012b. The East Greenland Current and its impacts on the Nordic Seas: observed trends in the past decade. ICES J. Mar. Sci. 69 (5), 841–851.

Rudels, B., Schauer, U., Björk, G., Korhonen, M., Pisarev, S., Rabe, B., Wisotski, A., 2013. Observations of water masses and circulation with focus on the Eurasian basin of the Arctic Ocean from the 1990s to the late 2000s. Ocean Sci. 9, 147–169. http://dx.doi.org/10.5194/os-9-147-2013.

Sadler, H.E., 1976. Water, heat, and salt transports through Nares Strait, Ellesmere Island. J. Fish. Res. Board Can. 33, 2286–2295.

Schauer, U., Beszczynska-Möller, A., 2009. Problems with estimation and interpretation of oceanic heat transport–conceptual remarks for the case of Fram Strait in the Arctic Ocean. Ocean Sci. 5 (4), 487–494.

Schauer, U., Muench, R.D., Rudels, B., Timokhov, L., 1997. Impact of eastern Arctic shelf waters on the Nansen Basin intermediate layers. J. Geophys. Res. 102, 3371–3382.

Schauer, U., Beszczynska-Möller, A., Walczowski, W., Fahrbach, E., Piechura, J., Hansen, E., 2008. Variation of measured heat flux through the Fram Strait between 1997 and 2006. In: Dickson, R.R., Meincke, J., Rhines, P. (Eds.), Arctic-Subarctic Ocean Fluxes. Springer, Dordrecht, pp. 65–85.

Schlitzer, R., 2013. Ocean Data View. http://odv.awi.de.

Sciascia, R., Straneo, F., Cenedese, C., Heimbach, P., 2013. Seasonal variability of submarine melt rate and circulation in an east Greenland Fjord. J. Geophys. Res. 118 (5), 2492–2506. http://dx.doi.org/10.1002/jgrc.20142.

Serreze, M.C., Francis, J.A., 2006. The Arctic amplification debate. Clim. Change 76 (3), 241–264.

Serreze, M.C., Barrett, A.P., Slater, A.G., Woodgate, R.A., Aagaard, K., Lammers, R.B., Steele, M., Moritz, R., Meredith, M., Lee, C.M., 2006. The large-scale freshwater cycle of the Arctic. J. Geophys. Res. 111, C11010. http://dx.doi.org/10.1029/2005JC003424.

Shiklomanov, A.I., Lammers, R.B., 2009. Record Russian river discharge in 2007 and the limits of analysis. Environ. Res. Lett. 4, 045015. http://dx.doi.org/10.1088/1748-9326/4/4/045015.

Shimada, K., McLaughlin, F., Carmack, E., Proshutinsky, A., Nishino, S., Motoyo, I., 2004. Penetration of the 1990s warm temperature anomaly of Atlantic Water in the Canada Basin. Geophys. Res. Lett. 31 (20), L20301. http://dx.doi.org/10.1029/2004GL020860.

Shimada, K., Itoh, M., Nishino, S., McLaughlin, F., Carmack, E., Proshutinsky, A., 2005. Halocline structure in the Canada Basin of the Arctic Ocean. Geophys. Res. Lett. 32 (3), L03605.

Shimada, K., Kamoshida, T., Itoh, M., Nishino, S., Carmack, E., McLaughlin, F., Zimmermann, S., Proshutinsky, A., 2006. Pacific Ocean inflow: influence on catastrophic reduction of sea ice cover in the Arctic Ocean. Geophys. Res. Lett. 33, L08605. http://dx.doi.org/10.1029/2005GL025624.

Shtokman, V.B., 1957. Influence of wind on currents in the Bering Strait and causes of their high velocities and predominant northern direction. Trans. Inst. Okeanol. Akad. Nauk SSSR 25, 171–197.

Skagseth, Ø., Orvik, K.A., Furevik, T., 2004. Coherent variability of the Norwegian Atlantic Slope Current derived from TOPEX/ERS altimeter data. Geophys. Res. Lett. 31, L14304. http://dx.doi.org/10.1029/2004GL020057.

Skagseth, Ø., Furevik, T., Ingvaldsen, R., Loeng, H., Mork, K.A., Orvik, K.A., Ozhigin, V., 2008. Volume and heat transports to the Arctic Ocean via the Norwegian and Barents Seas. Arctic-Subarctic Ocean Fluxes 45–64.

Skagseth, Ø., Drinkwater, K.F., Terrile, E., 2011. Wind- and buoyancy-induced transport of the Norwegian Coastal Current in the Barents Sea. J. Geophys. Res. 116, C08007. http://dx.doi.org/10.1029/2011JC006996.

Smethie, W.M., Schlosser, P., Bonisch, G., Hopkins, T.S., 2000. Renewal and circulation of intermediate waters in the Canadian Basin observed on the SCICEX 96 cruise. J. Geophys. Res. Oceans 105 (C1), 1105–1121. http://dx.doi.org/10.1029/1999JC900233.

Smith, J.N., Ellis, K.M., Boyd, T., 1999. Circulation features in the central Arctic Ocean revealed by nuclear fuel reprocessing tracers from Scientific Ice Expeditions 1995 and 1996. J. Geophys. Res. Oceans 104 (C12), 29663–29677. http://dx.doi.org/10.1029/1999JC900244.

Spall, M.A., Pickart, R.S., Fratantoni, P.S., Plueddemann, A.J., 2008. Western Arctic Shelfbreak Eddies: formation and transport. J. Phys. Oceanogr. 38, 1644–1668.

Stanton, T., Shaw, W., Hutchings, J., 2012. Observational study of relationships between incoming radiation, open water fraction, and ocean-to-ice heat flux in the transpolar drift: 2002-2010. J. Geophys. Res. 117, C07005. http://dx.doi.org/10.1029/2011JC007871.

Steele, M., Morley, R., Ermold, W., 2001. PHC: a global ocean hydrography with a high-quality Arctic Ocean. J. Clim. 14, 2079–2087.

Steele, M., Morison, J., Ermold, W., Rigor, I., Ortmeyer, M., Shimada, K., 2004. Circulation of summer Pacific halocline water in the Arctic Ocean. J. Geophys. Res. 109, C02027. http://dx.doi.org/10.1029/2003JC002009.

Steele, M., Zhang, J., Ermold, W., 2010. Mechanisms of summertime upper Arctic Ocean warming and the effect on sea ice melt. J. Geophys. Res. 115, C11004. http://dx.doi.org/10.1029/2009-JC005849.

Stern, M.E., 1967. Lateral mixing of water masses. Deep Sea Res. Oceanogr. Abstr. 14 (6), 747–753.

Stigebrandt, A., 1984. The North Pacific: a global estuary. J. Phys. Oceanogr. 14, 464–470.

Stott, P.A., Sutton, R.T., Smith, D.M., 2008. Detection and attribution of Atlantic salinity changes. Geophys. Res. Lett. 35, L21702. http://dx.doi.org/10.1029/2008GL035874.

Straneo, F., Sutherland, D.A., Holland, D., Gladish, C., Hamilton, G., Johnson, H., Rignot, E., Xu, Y., Koppes, M., 2012. Characteristics of ocean waters reaching Greenland's glaciers. Ann. Glaciol. 53 (60), 202–210.

Sundby, S., Drinkwater, K., 2007. On the mechanisms behind salinity anomaly signals of the northern North Atlantic. Prog. Oceanogr. 73 (2), 190–202. http://dx.doi.org/10.1016/j.pocean.2007.02.002.

Sundby, S., Nakken, O., 2008. Spatial shifts in spawning habitats of Arcto-Norwegian cod related to multidecadal climate oscillations and climate change. ICES J. Mar. Sci. 65, 953–962.

Sutherland, D.A., Straneo, F., 2012. Estimating ocean heat transport and submarine melt rate in Sermilik Fjord, Greenland, using lowered ADCP velocity profiles. Ann. Glaciol. 53 (60), 50–58.

Swift, J.H., Jones, E.P., Aagaard, K., Carmack, E.C., Hingston, M., MacDonald, R.W., McLaughlin, F.A., Perkin, R.G., 1997. Waters of the Makarov and Canada basins. Deep Sea Res. Part II 0967-064544 (8), 1503–1529. http://dx.doi.org/10.1016/S0967-0645(97)00055-6.

Tang, C.C.L., Ross, C.-K., Yao, T., Petrie, B., DeTracey, B.M., Dunlap, E., 2004. The circulation, water masses and sea-ice of Baffin Bay. Prog. Oceanogr. 63, 183–228.

Timmermans, M.L., Garrett, C., 2006. Evolution of the deep water in the Canadian Basin in the Arctic Ocean. J. Phys. Oceanogr. 36 (5), 866–874.

Timmermans, M.-L., Toole, J., Krishfield, R., Winsor, P., 2008. Ice-Tethered Profiler observations of the double-diffusive staircase in the Canada Basin thermocline. J. Geophys. Res. 113, C00A02. http://dx.doi.org/10.1029/2008JC004829.

Timmermans, M.L., Cole, S., Toole, J., 2012. Horizontal density structure and restratification of the Arctic Ocean surface layer. J. Phys. Oceanogr. 42 (4), 659–668.

Toole, J.M., Georgi, D.T., 1981. On the dynamics and effects of double-diffusively driven intrusions. Prog. Oceanogr. 10 (2), 123–145.

Toole, J.M., Timmermans, M.-L., Perovich, D.K., Krishfield, R.A., Proshutinsky, A., Richter-Menge, J.A., 2010. Influences of the ocean surface mixed layer and thermohaline stratification on Arctic Sea Ice in the Central Canada Basin. J. Geophys. Res. 115, C10018. http://dx.doi.org/10.1029/2009JC005660.

Tschudi, M., Fowler, C., Maslanik, J., Stroeve, J., 2010. Tracking the movement and changing surface characteristics of Arctic sea ice. IEEE J. Sel. Top. Appl. Earth Obs. Remote Sens. 3 (4), 536–540.

Tsubouchi, T., Bacon, S., Naveira Garabato, A.C., Aksenov, Y., Laxon, S.W., Fahrbach, E., Beszczynska-Möller, A., Hansen, E., Lee, C.M., Ingvaldsen, R.B., 2012. The Arctic Ocean in summer: a quasi-synoptic inverse estimate of boundary fluxes and water mass transformation. J. Geophys. Res. 117, C01024. http://dx.doi.org/10.1029/2011JC007174.

Turner, J.S., 2010. The melting of ice in the Arctic Ocean: the influence of double-diffusive transport of heat from below. J. Phys. Oceanogr. 40, 249–256.

Walczowski, W., Piechura, J., Osinski, R., Wieczorek, P., 2005. The West Spitsbergen Current volume and heat transport from synoptic observations in summer. Deep Sea Res. Part I 52, 1374–1391.

Walsh, D., Carmack, E., 2003. The nested structure of Arctic thermohaline intrusions. Ocean Model. 5 (3), 267–289.

Walsh, D., Polyakov, I., Timokhov, L., Carmack, E., 2007. Thermohaline structure and variability in the eastern Nansen Basin as seen from historical data. J. Mar. Res. 65, 685–714.

Watanabe, E., 2011. Beaufort shelf break eddies and shelf-basin exchange of Pacific summer water in the western Arctic Ocean detected by satellite and modeling analyses. J. Geophys. Res. 116, C08034. http://dx.doi.org/10.1029/2010JC006259.

Weingartner, T.J., Cavalieri, D.J., Aagaard, K., Sasaki, Y., 1998. Circulation, dense water formation, and outflow on the northeast Chukchi shelf. J. Geophys. Res. 103, 7647–7661.

Wijffels, S.E., Schmitt, R.W., Bryden, H.L., Stigebrandt, A., 1992. Transport of freshwater by the oceans. J. Phys. Oceanogr. 22, 155–162.

Woodgate, R.A., Aagaard, K., 2005. Revising the Bering Strait freshwater flux into the Arctic Ocean. Geophys. Res. Lett. 32 (2), L02602.

Woodgate, R.A., Aagaard, K., Weingartner, T.J., 2005a. Monthly temperature, salinity, and transport variability of the Bering Strait through flow. Geophys. Res. Lett. 32 (4), L04601.

Woodgate, R.A., Aagaard, K., Weingartner, T.J., 2005b. A year in the physical oceanography of the Chukchi Sea: moored measurements from autumn 1990-1991. Deep Sea Res. Part II 52, 3116–3149. http://dx.doi.org/10.1016/j.dsr2.2005.10.016.

Woodgate, R.A., Aagaard, K., Weingartner, T.J., 2006. Interannual changes in the Bering Strait fluxes of volume, heat and freshwater between 1991 and 2004. Geophys. Res. Lett. 33, L15609. http://dx.doi.org/10.1029/2006GL026931.

Woodgate, R.A., Weingartner, T., Lindsay, R., 2010. The 2007 Bering Strait oceanic heat flux and anomalous Arctic sea-ice retreat. Geophys. Res. Lett. 37, L01602. http://dx.doi.org/10.1029/2009GL041621.

Yang, J., 2006. The seasonal variability of the Arctic Ocean Ekman transport and its role in the mixed layer heat and salt fluxes. J. Clim. 19, 5366–5387. http://dx.doi.org/10.1175/JCLI3892.1.

Yin, J.H., 2005. A consistent poleward shift of the storm tracks in simulations of 21st century climate. Geophys. Res. Lett. 32, L18701. http://dx.doi.org/10.1029/2005GL023684.

Chapter 18

Dynamics of the Southern Ocean Circulation

Stephen R. Rintoul* and Alberto C. Naveira Garabato†

*CSIRO Marine and Atmospheric Research, Antarctic Climate and Ecosystems Cooperative Research Centre, University of Tasmania, Centre for Australian Weather and Climate Research, Hobart, Tasmania, Australia

†National Oceanography Centre, University of Southampton, Southampton, United Kingdom

Chapter Outline

1. INTRODUCTION

The existence of a circumpolar oceanic channel in the latitude band of Drake Passage has profound implications for the global ocean circulation and climate. The Drake Passage gap permits the existence of the Antarctic Circumpolar Current (ACC), a system of ocean currents that flows from west to east along a roughly 25,000-km-long path circling Antarctica (Figure 18.1). The strong eastward flow of the ACC has several important implications for the global ocean circulation and its influence on regional and global climate. It promotes exchange between the three major ocean basins, allowing transmission of climate signals and smoothing out zonal differences in water properties, yet inhibits north–south exchange and isolates Antarctica from the warm waters to the north. The interbasin connection (see Chapter 19) allows the establishment of a global-scale overturning circulation, which transports heat, moisture, and carbon dioxide around the globe and strongly influences the Earth's climate. The eastward geostrophic flow of the ACC is associated with steeply sloping density surfaces, which shoal to the south across the current and expose dense waters to the surface in the high-latitude Southern Ocean (Figure 18.2). Where the dense waters outcrop at the sea surface, they exchange heat, moisture, and gases like oxygen and carbon dioxide with the atmosphere. In this sense, the Southern Ocean acts as a valve regulating exchange between the deep and shallow layers of the global ocean. Upwelling of deep water returns nutrients and carbon to the upper ocean, while sinking of waters from the surface transfers oxygen and anthropogenic carbon dioxide (CO_2) into the ocean interior. As a result of the

Ocean Circulation and Climate, Vol. 103. http://dx.doi.org/10.1016/B978-0-12-391851-2.00018-0

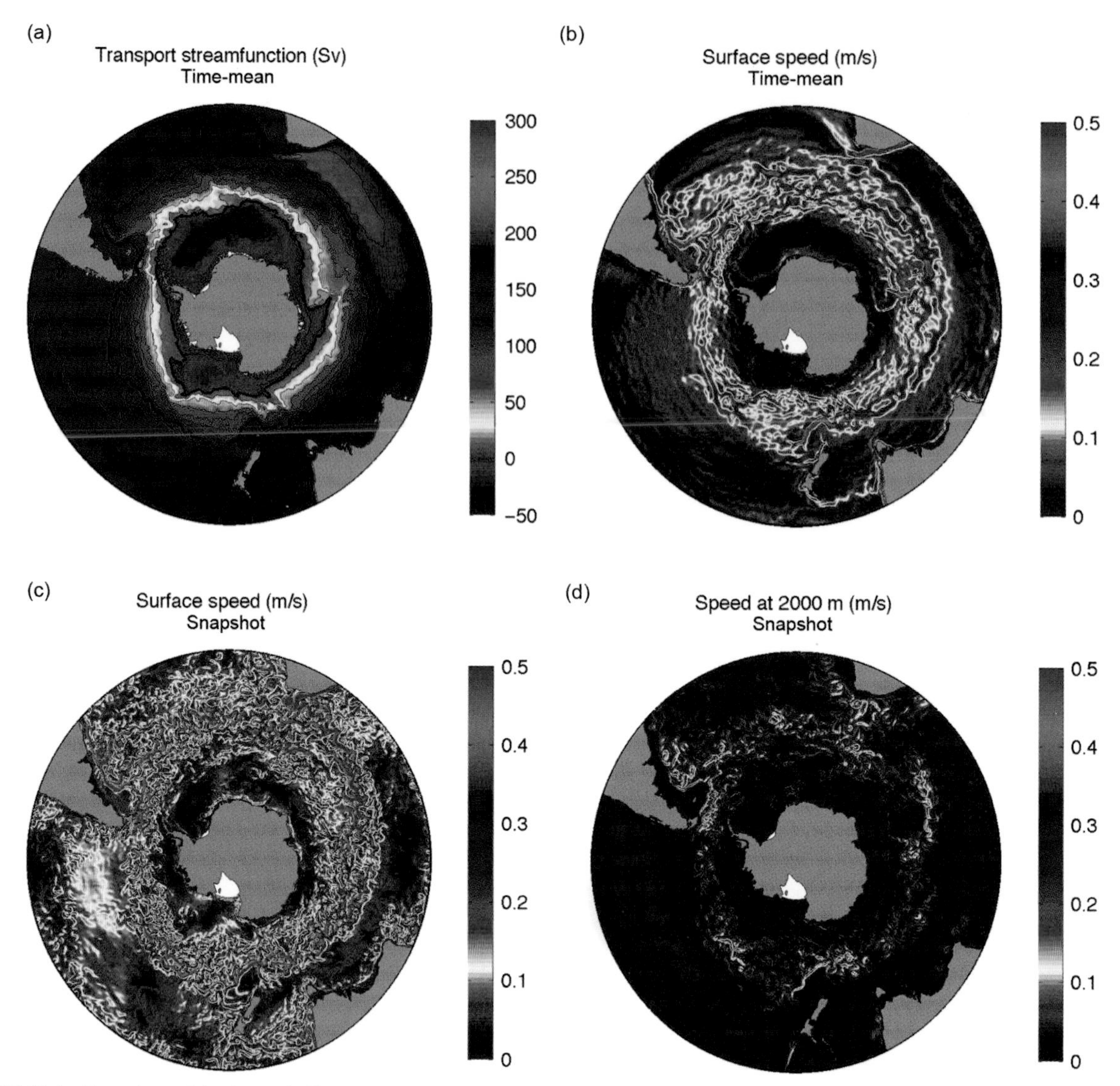

FIGURE 18.1 Four views of the Antarctic Circumpolar Current based on the Southern Ocean State Estimate of Mazloff et al. (2010). (a) Five-year mean (2005–2009) vertically integrated transport streamfunction (in Sv); bold contours indicate the northern and southernmost streamlines to pass through Drake Passage. (b) Five-year mean surface velocity (m s^{-1}). (c) Five-day mean snapshot of surface velocity in mid-January 2005. (d) Five-day mean velocity at 2000 m depth. *Figure courtesy of M. Mazloff and B. Peña Molino.*

vigorous overturning circulation, the Southern Ocean stores large amounts of heat and anthropogenic CO_2 (see Chapter 30). For example, 40% of the total ocean inventory of anthropogenic CO_2 entered the ocean south of 40°S (Khatiwala et al., 2009, 2013).

Understanding of the Southern Ocean circulation has been slowed by a scarcity of observations. The measurements collected during the World Ocean Circulation Experiment (WOCE) were a significant achievement. The WOCE hydrographic program, for example, provided the first circumpolar survey of the Southern Ocean since the epic voyages of the *Eltanin* in the late 1960s and early 1970s. But over the last decade or so, even more remarkable advances have been made in our capacity to observe the Southern Ocean (Chapter 3). More than 150,000 temperature and salinity profiles to 2000 m depth have been collected south of 40°S since the Argo profiling float program began in 2003. Oceanographic sensors deployed on elephant seals for biology studies have provided more than 50,000 temperature and salinity profiles, many of them from the sea ice zone in winter where few measurements have been made with traditional techniques. The satellite altimeter record now exceeds two decades in length, allowing an unprecedented year-round, circumpolar view of the Southern Ocean circulation (Chapter 4). Repeated hydrographic sections have quantified the evolution of a wide range of ocean properties throughout the full ocean depth. Direct observations of diapycnal mixing are now possible (Chapter 7). Taken together, these advances have revolutionized our ability to observe and understand the Southern Ocean.

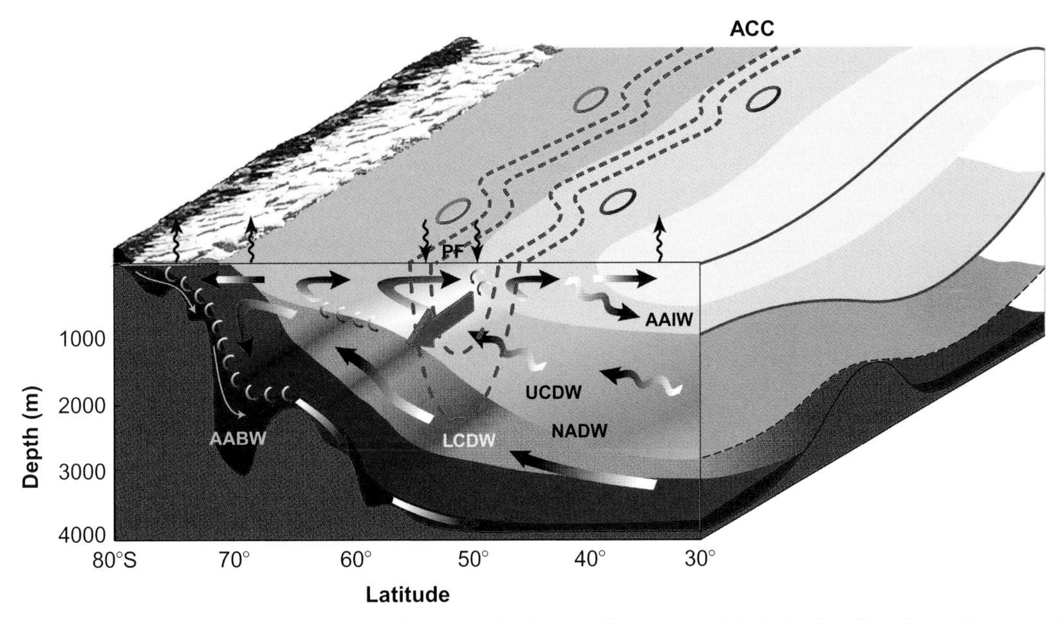

FIGURE 18.2 A schematic view of the Southern Ocean circulation. The heavy yellow arrow and dashed yellow lines denote the eastward flow of the Antarctic Circumpolar Current. The overturning circulation is indicated by dark arrows, with wavy arrows intended to represent transport by eddies. The meridional transport is largely along layers of constant density, represented by the colored surfaces. Vertical curly arrows at the sea surface indicate air–sea buoyancy exchange (upward arrows mean a buoyancy loss by the ocean). Small light arrows show diapycnal mixing. The east–west section on the right face of the diagram illustrates the surface and isopycnal tilts in relation to bottom topography (brown) associated with interfacial and bottom form stress. *From Olbers and Visbeck (2005), adapted from Speer et al. (2000).*

Progress in modeling the Southern Ocean has been equally dramatic. Near the beginning of WOCE, the Fine Resolution Antarctic Model (FRAM) was completed, a regional model of the Southern Ocean with (1/4° latitude) × (1/2° longitude) resolution (an achievement so remarkable at that time that a large atlas was published with results from a snapshot of a single model run (Webb et al., 1991)). The FRAM model, which took 3 years of integration time to complete 6 years of simulation, was the first numerical representation of the Southern Ocean to include mesoscale features like eddies, fronts, and meanders. Global models are now being run for decades with a spatial resolution of 1/10° (Chapter 20). Even more importantly for improving understanding, advances in computing power have allowed many runs to be completed at high resolution, allowing testing of the sensitivity of the circulation to changes in forcing and to model parameters. Eddies play a particularly important role in the Southern Ocean, and the development of high-resolution models has allowed the dynamical significance of eddies to be explored in detail for the first time.

Several contemporary reviews have done a comprehensive job of summarizing progress in understanding the Southern Ocean circulation (e.g., Olbers et al., 2004; Meredith et al., 2011; Marshall and Speer, 2012; chapter 16 of Olbers et al., 2012). Here we aim to complement these studies by summarizing recent advances in the understanding of Southern Ocean dynamics gained from observations, theory, and modeling. We begin with a brief review of what had been learned about Southern Ocean dynamics during WOCE, as described in the first edition of the "WOCE book" (Rintoul et al., 2001) (Section 2). In Section 3, advances in understanding the structure and dynamics of the ACC are summarized. A common element of these recent developments is an increasing recognition of the central role played by small- and mesoscale processes, often localized by topography, in the dynamics and climatic sensitivity of the Southern Ocean circulation. Recent progress in understanding the dynamics of the Southern Ocean overturning circulation is reviewed in Section 4. New observations have provided time series of sufficient length to provide insight into the nature and causes of Southern Ocean change, as discussed in Section 5. In Section 6, the main conclusions of this chapter are presented, and a brief outlook on future challenges put forward.

2. PROGRESS IN UNDERSTANDING SOUTHERN OCEAN DYNAMICS DURING WOCE (1990–2002)

Rintoul et al. (2001) reviewed progress made during the WOCE decade in understanding the ACC system. New observations and advances in theory and numerical modeling during WOCE allowed the synthesis of a "dynamical recipe" for the circulation of the Southern Ocean. The key ingredients include wind and buoyancy forcing, eddies, stratification, and topography. Both wind and buoyancy forcing (including ocean–ice interaction) act to tilt isopycnals so they shoal to the south, producing the strong eastward geostrophic flow of the ACC. When the tilt is sufficiently large, baroclinic instability taps the available potential energy stored in the sloping isopycnals to produce an energetic eddy field. Eddies play a number of critical roles in the ACC system. They carry heat poleward, to balance oceanic heat loss to the atmosphere at high latitude. Poleward eddy heat flux is associated with vertical transport of horizontal momentum. The vertical momentum transport (or interfacial form stress) establishes deep geostrophic flows balanced by pressure gradients across bathymetric features, resulting in bottom form stress that opposes the wind stress at the sea surface. Divergence of the eddy-induced interfacial form stress supports meridional transport in layers not blocked by topography, where no time-mean meridional geostrophic flow is possible. The dynamics of the ACC and the meridional overturning circulation are therefore fundamentally linked, as illustrated schematically in Figure 18.2.

Interaction of the flow with topography is also central to the vorticity balance, with the dominant balance in the barotropic vorticity equation being between bottom pressure torque and advection of planetary vorticity (in contrast to the flat-bottom Sverdrup balance between wind stress curl and advection of planetary vorticity that holds in most other parts of the ocean) (Hughes and de Cuevas, 2001). Bottom topography has a greater influence in the Southern Ocean because of the relatively weak stratification and deep-reaching nature of the flow. The stratification is weak, but not zero, and partially shields the flow from the effects of the bathymetry: barotropic models with realistic topography, forcing, and friction typically have very weak circumpolar flow (e.g., Olbers et al., 2004).

The Southern Ocean overturning circulation consists of two cells (Figure 18.2). Circumpolar Deep Water (CDW) spreads poleward and shoals, ultimately outcropping at the sea surface. The relatively dense Lower CDW upwells to the south of the ACC, where interactions with the atmosphere and sea ice convert the upwelled water to dense Antarctic Bottom Water (AABW), forming the lower cell of the overturning circulation. Relatively light Upper CDW outcrops within the ACC belt, where it is driven equatorward in the Ekman layer and converted to less dense intermediate waters by heat and freshwater gain from the atmosphere, forming the upper cell. In this way, the Southern Ocean connects the upper and lower layers of the ocean and allows a global-scale overturning circulation to exist even in the limit of very weak diapycnal mixing in the ocean interior (e.g., Speer et al., 2000).

While the observations and theoretical advances made during the WOCE era led to much deeper understanding of the Southern Ocean circulation, understanding of the interactions between the dynamical ingredients identified above remained incomplete. In particular, the sensitivity of the ACC and overturning circulation to changes in wind and buoyancy forcing was poorly understood. Observational evidence for change in the Southern Ocean was sparse due to the lack of long records. In the following, we review progress made in the last decade in understanding the dynamics of the Southern Ocean circulation, much of it built on the foundation laid during WOCE.

3. THE ANTARCTIC CIRCUMPOLAR CURRENT (ACC)

3.1. Structure of the ACC

The eastward flow of the ACC is concentrated in narrow fronts or jets, as first recognized by Sir George Deacon in the 1930s. For example, Orsi et al. (1995) identified three circumpolar fronts: from north to south, the Subantarctic Front (SAF), Polar Front (PF), and southern ACC front, as well as a fourth feature they called the "southern boundary of the ACC," coincident with the southern limit of the flow through Drake Passage and not always associated with a current core. Each of the fronts was found to coincide with a water mass boundary (and hence could be reliably identified in hydrographic data), as well as a maximum in current speed (and hence associated with enhanced horizontal gradients in density and sea surface height).

This classical view of the ACC fronts is not easily recognizable in Figure 18.1, which shows four perspectives on the structure of the ACC derived from the Southern Ocean State Estimate (SOSE; Mazloff et al., 2010). SOSE assimilates a large volume of observations (e.g., altimetry, moorings, and temperature and salinity profiles collected by floats, ships, and seals) into an eddy-permitting (1/6° resolution) general circulation model (Chapter 21 discusses ocean state estimates). The resulting flow field can be thought of as a dynamically consistent interpolation of sparse observations. The circumpolar flow of the ACC clearly dominates the vertically integrated circulation south of 25°S. While the zonal continuity of the ACC is the most striking feature in Figure 18.1a, it is also clear that the path and width of the ACC varies significantly around the circumpolar belt. Cyclonic gyres south of the ACC are evident in the Ross and Weddell Seas.

Growing evidence from observations and models suggests that the frontal structure of the ACC is richer than suggested by earlier studies. The 3-year mean surface velocity field in SOSE shows that the ACC consists of multiple narrow filaments (Figure 18.1b). The jet structure is even more complicated in a 5-day snapshot (Figure 18.1c), as expected, but it is notable how much of this structure remains in the 3-year mean field. The multiple jets are a robust characteristic of the ACC, both at the surface and at depth (Figure 18.1d). Multiple ACC jets have now been observed in a number of data sets, including repeat hydrographic sections (Sokolov and Rintoul, 2007), repeat acoustic Doppler current profiler (ADCP) transects (Firing et al., 2011), sea surface temperature data

(Hughes and Ash, 2001), dynamic height fields derived from surface drifters (Niiler et al., 2003; Hughes, 2005), and sea surface height measurements (Sokolov and Rintoul, 2007, 2009a,b; Hughes et al., 2010). The current branches are observed to merge and diverge along the circumpolar path and to vary in time (see also Chapter 12). Numerical simulations with sufficient spatial resolution produce a filamented ACC similar to observations (e.g., Hallberg and Gnanadesikan, 2006; Mazloff et al., 2010; Thompson et al., 2010). Recognition of the existence of multiple jets in the ACC has provided a new and useful perspective from which to view the eddy–mean flow and flow–topography interactions that are central to ACC dynamics (e.g., Hughes, 2005; Thompson et al., 2010; Thompson and Sallée, 2012). The jets result from eddy-mean flow interactions and, in turn, the structure of the jets determines the stability of the flow and the nature of the eddy field. Interaction of the jets with topography influences eddy generation and mixing and establishes the bottom stresses and pressure torques that enable closure of the momentum and vorticity budgets, respectively.

3.2. Transport of the ACC

The ACC is the primary means of interbasin exchange and substantial effort has been devoted to measuring the ACC transport. Observing the mean absolute transport of the ACC remains a difficult challenge. The geostrophic transport relative to a level of no motion (the baroclinic transport) can be determined in a straightforward way from measurements of temperature and salinity (although few sections are repeated with sufficiently high frequency to avoid aliasing unresolved variability). The velocity at the level of no motion (often taken to lie near the sea floor in the Southern Ocean), and hence the barotropic transport, needs to be determined by long-duration, closely spaced direct velocity measurements (e.g., current meter moorings) or inferred indirectly (e.g., by inverse methods). The difficulty of directly observing the barotropic transport over large space and timescales means that studies have often focused on the baroclinic contribution to the total transport. When discussing variations in transport, it can be important to distinguish between net interbasin exchange (i.e., the total flow between Antarctic and the southern hemisphere continents) and the transport of the ACC itself, as noted by Hughes et al. (1999). The former may be sensitive to phenomena on the boundaries (e.g., the southern mode, discussed below, or boundary waves) and to extensions of the subtropical current systems crossing the northern end of sections spanning the Southern Ocean (e.g., south of Africa and Australia).

Meredith et al. (2011) provide a summary of historical and recent measurements of the ACC transport in Drake Passage, the best-measured region of the Southern Ocean. The mean baroclinic volume transport for 16 sections occupied between 1993 and 2009 was 136.7 Sv, with a standard deviation of 6.9 Sv, and no indication of a trend. Four years of inverted echo sounder measurements spanning Drake Passage gave a mean full-depth baroclinic transport of 126.7 ± 1.5 Sv, with root mean square (rms) fluctuations of 9.5 Sv (Chidichimo et al., 2013). The mean absolute transport above 1000 m depth in Drake Passage was 95 Sv, with a standard error of 2 Sv, based on the average of 53 repeat ADCP sections between 2005 and 2009 (Firing et al., 2011). The mean transport through Drake Passage in SOSE, illustrated in Figure 18.1, is 153 ± 5 Sv (Mazloff et al., 2010). Measurements of the ACC transport at other locations (e.g., south of Africa (Swart et al., 2008) and south of Australia (Rintoul et al., 2002)) are largely consistent with those made at Drake Passage, once the additional transport south of Australia needed to balance the Indonesian Throughflow is taken into account.

As found in the studies cited earlier, estimates of the ACC transport from observations and models consistently show that the variability in ACC transport on timescales of years to decades is small relative to the transport's mean value (e.g., Rintoul et al., 2002; Cunningham et al., 2003; Meredith et al., 2004; Hallberg and Gnanadesikan, 2006; Böning et al., 2008). The relative stability of the ACC transport has motivated new efforts aimed at understanding the response of the ACC to climatic perturbations in wind and buoyancy forcing.

3.3. Response of the ACC to Wind and Buoyancy Forcing

Both wind and buoyancy forcing can alter the slope of isopycnals across the Southern Ocean, and therefore the transport of the ACC. Variability and trends in wind and buoyancy forcing may arise from internal variability of the climate system, including teleconnections to low-latitude phenomena such as the El Niño–Southern Oscillation, and from anthropogenic drivers including global atmospheric warming and loss of stratospheric ozone. The response of the ACC depends on the frequency of the forcing, with high-frequency forcing favoring a barotropic response and baroclinic mechanisms becoming more important at lower frequencies.

3.3.1. Response to High-Frequency Wind Forcing: The Southern Mode

For wind-forcing perturbations on timescales shorter than O (1 year), the variability in Antarctic circumpolar transport is dominated by a barotropic mode that follows f/H contours (where f is the Coriolis parameter and H is the ocean depth), and which is therefore focused along the southern edge of the Southern Ocean. This was termed the "southern mode" by

Hughes et al. (1999). Propagation of wind-forced barotropic topographic Rossby waves along *f*/*H* contours governs the dynamics of this mode. The characteristic propagation speed of the waves is several meters per second, yielding a circumpolar adjustment timescale of a few days (Webb and de Cuevas, 2007). The mode has been shown to explain a substantial proportion of the daily-to-interannual variance in sea level and bottom pressure near the Antarctic continent, and in the transport through Drake Passage, in both observations and eddy-permitting general circulation models (Aoki, 2002; Hughes et al., 2003; Zika et al., 2013a).

3.3.2. Response to Low-Frequency Wind Forcing: Eddy Saturation

On timescales longer than O(1 year), the response of the ACC to forcing perturbations is distinct in both spatial structure and dynamical character compared with the shorter timescale response. In mid-latitudes, variations in wind forcing excite baroclinic Rossby waves that propagate westward and are instrumental in the adjustment of the basin-scale gyres to forcing perturbations, confining the response to the upper ocean, and establishing western boundary currents (Anderson and Killworth, 1977; Hughes and de Cuevas, 2001). The dynamical balance of the ACC is distinct from that of mid-latitude basins in two subtle yet profound aspects. First, the eastward flow of the ACC exceeds the westward phase speed of baroclinic Rossby waves (i.e., the flow is supercritical) and baroclinic disturbances propagate eastward in the jets of the ACC (Hughes, 1996). As baroclinic waves are unable to propagate westward in the ACC, the response to wind forcing is not limited to the upper ocean there and the ACC extends to greater depth than the mid-latitude gyres. Second, owing to the absence of continental boundaries at the latitude band of Drake Passage, variations in wind forcing must be balanced by topographic form drag on submarine obstacles in the path of the ACC, rather than pressure gradients along the boundaries of the basin (e.g., Hughes and de Cuevas, 2001). Therefore, in order for the ACC to reach a stationary state, the horizontal momentum imparted by wind forcing must be transported vertically to great depth before it can be removed by submarine topography. This interfacial form stress is effected by mesoscale eddies (e.g., Olbers et al., 2004). In the following, we will argue that these two distinctive features of the dynamics of the ACC and the details of their interaction are key to understanding the response of the current to perturbations in forcing.

Models and observations suggest that the ACC responds to an increase in wind forcing primarily by increasing eddy kinetic energy (EKE) rather than increasing transport, for realistically strong wind forcing (e.g., Tansley and Marshall, 2001; Cunningham et al., 2003; Hallberg and Gnanadesikan, 2006; Meredith and Hogg, 2006; Böning et al., 2008; Firing et al., 2011; Munday et al., 2013). The limit in which the time-mean ACC transport is insensitive to changes in wind forcing, first discussed by Straub (1993), has come to be known as the "eddy saturation" regime. The models indicate that stronger wind forcing increases the baroclinicity of the ACC; baroclinic instability then releases available potential energy (reduces isopycnal tilt) and increases EKE, reestablishing the pre-perturbation baroclinic state. Eddy saturation therefore describes the stationary-state response to the change in forcing, rather than implying completely invariant ACC transport. The characteristic time lag between the wind perturbation and the change in EKE is typically a few years in models. An analysis of altimeter data suggested a similar delay in the EKE response to a strong wind perturbation in 1998 (Meredith and Hogg, 2006), while a subsequent investigation of a longer period of altimetric measurements found that the EKE response varied strongly with location and with the nature of the wind perturbation (Morrow et al., 2010). Models and observations indicate that the ACC approaches, but does not reach, the eddy saturation limit, with most models showing a transport increase of less than 10% to a doubling of the wind stress (e.g., Hallberg and Gnanadesikan, 2006; Farneti et al., 2010; Dufour et al., 2012).

A complete theory of eddy saturation does not yet exist. Straub (1993) used a simple theoretical argument to show that once the ACC reached the threshold for baroclinic instability, the transport of the current would be independent of the wind forcing. However, the details of how interactions between the mean flow, the standing and transient eddy fields, and topography establish the eddy-saturated response to changes in forcing are only beginning to emerge. A common thread is the significance of local dynamics. The standing and transient eddy fields, and hence the eddy fluxes of momentum and vorticity, vary along the path of the current, and these variations are likely to be important to the dynamics of the ACC as a whole. For example, a transient strengthening of the ACC in response to an increase in wind forcing might alter the standing meander pattern and hence the vertical momentum transfer by standing and transient eddies (e.g., Dufour et al., 2012; Thompson and Naveira Garabato, 2013). The change in interfacial form stress can then accelerate the deep flow of the ACC and produce additional topographic form drag to balance the wind forcing and restore the pre-perturbation conditions. The response of the ACC to changes in wind forcing also depends on the spatial structure of the wind perturbation (e.g., Morrow et al., 2010; Mazloff, 2012), again underscoring the importance of local dynamics. Yet another example is the likely importance of internal waves radiated by small-lateral-scale (O(1–10 km)) topography in arresting the ACC flow (e.g., Naveira Garabato et al., 2013a) and in dissipating the Southern Ocean mesoscale eddy field (e.g., Nikurashin et al., 2013; Waterman et al., 2013).

3.3.3. *Response to Buoyancy and Remote Forcing*

Any local or remote forcing that alters the meridional density gradient across the ACC can change the transport. In addition to the wind effects discussed above, buoyancy forcing and processes that influence the depth of the pycnocline on the northern flank of the Southern Ocean can also shape the baroclinicity of the ACC. The sensitivity of ACC transport to buoyancy forcing has been discussed by a number of authors (e.g., Gent et al., 2001; Borowski et al., 2002; Hogg, 2010). Cooling in the south and warming in the north will tend to increase the tilt of isopycnals and therefore the baroclinic transport. Buoyancy forcing may also alter potential vorticity (PV) gradients in the upper ocean and therefore the "slumping" of isopycnals by eddy PV fluxes (e.g., Marshall and Radko, 2003).

The baroclinicity of the ACC is closely related to the depth and structure of the global pycnocline. Processes that affect the meridional slope of isopycnals in the Southern Ocean influence the global pycnocline (e.g., Wolfe and Cessi, 2010; Chapter 11), and processes outside the Southern Ocean that affect the global pycnocline influence the meridional slope of isopycnals in the ACC. The latter set of processes includes deep water formation in the North Atlantic (e.g., Gnanadesikan 1999; Hallberg and Gnanadesikan, 2001), interior diapycnal mixing at mid- and low latitudes (e.g., Nikurashin and Vallis, 2011), and wind forcing to the north of the ACC (Allison et al., 2011). Anomalies in mid- and low-latitude pycnocline depth induced by forcing both local to and remote from the ACC trigger a global wave adjustment lasting between many decades and several centuries, where the long timescale relates to the large volume of water resident in the global pycnocline that must be displaced vertically for the adjustment to occur (e.g., Allison et al., 2011; Jones et al., 2011). Evidence of the ACC's baroclinic adjustment to low-latitude wind forcing is provided by Close and Naveira Garabato (2012), who show that isopycnals in the upper kilometer of northern Drake Passage have deepened markedly between the early 1990s and 2010 in the wake of poleward-propagating boundary waves generated by changes in tropical Pacific winds. The baroclinic adjustment of the ACC therefore occurs on multiple timescales: an interannual response to local forcing, and a multidecadal-to-centennial response set by the timescale for equilibration of the ACC's baroclinic structure with the stratification of the ocean basins to the north. The existence of this longer timescale provides an alternative explanation for the observed resilience of the ACC transport, though wind perturbation experiments with idealized eddy-resolving models run to equilibrium (Munday et al., 2013) suggest that eddy saturation ideas may apply on these long timescales as well.

4. SOUTHERN OCEAN OVERTURNING CIRCULATION

The existence of a double-celled overturning circulation in the Southern Ocean (Figure 18.2) was inferred in the early days of oceanography from observed hydrographic distributions. However, the structure and transport of the overturning have been difficult to quantify. In this section, we first discuss the water mass transformations at the heart of the overturning circulation and review recent progress toward a quantitative representation of the overturning. We then provide an overview of the insights gained from residual-mean theory and introduce the concept of (partial) eddy compensation. Eddy stirring of PV provides a useful framework to consider interactions between eddies, the mean flow, and topography and exposes potential limitations of the zonally averaged view of the Southern Ocean circulation.

4.1. Water Mass Transformations and Southern Ocean Overturning

The role of the Southern Ocean in the global overturning circulation can be appreciated by following a water parcel along the global circuit, starting in the North Atlantic. North Atlantic Deep Water (NADW) flows southward in the Atlantic, largely along isopycnals, to reach the Southern Ocean. Some of the NADW flows poleward across the Southern Ocean, where it forms a high-salinity layer known as Lower Circumpolar Deep Water (LCDW). The LCDW outcrops near Antarctica, where cooling and brine rejected during sea ice formation transform LCDW to denser AABW. The poleward spreading of LCDW and equatorward return flow of AABW form the lower cell of the Southern Ocean overturning circulation (Figure 18.2).

LCDW also enters the deep Indian and Pacific Oceans, where enhanced diapycnal mixing at depth converts the inflowing water to slightly less dense Upper Circumpolar Deep Water (UCDW), which is returned to the Southern Ocean (e.g., Lumpkin and Speer, 2007). The UCDW is associated with an oxygen minimum layer, reflecting the relatively long isolation from the atmosphere during its transit of the deep ocean basins. UCDW spreads poleward, above the LCDW, and outcrops within the ACC belt. Warming and freshening of the surface layer converts UCDW into less dense Antarctic Intermediate Water (AAIW) and Subantarctic Mode Water (SAMW), which flow equatorward to ventilate the pycnocline of the Southern Hemisphere oceans. The poleward flow of UCDW and equatorward flow of AAIW and SAMW form the upper cell of the Southern Ocean overturning circulation (Figure 18.2). A third cell, sometimes called the subtropical cell, involves poleward transport of subtropical waters that are transformed to cooler, denser SAMW before returning equatorward, and will not be discussed here (see Chapter 9).

The classical view of the global overturning circulation emphasized a diffusive return path: sinking of dense water in the polar regions was balanced by uniform upwelling and vertical mixing in the ocean interior (e.g., Munk, 1966). Direct measurements later showed the rate of interior mixing to be up to an order of magnitude too small to support the rate of upwelling required for mass balance (e.g., Ledwell et al., 1993; Chapter 7). These observations helped motivate studies of overturning in the weak mixing limit (e.g., Toggweiler and Samuels, 1998). In this limit, water mass transformations driven by buoyancy forcing in the Southern Ocean, rather than diapycnal mixing in the ocean interior, act to connect the upper and lower limbs of the global overturning circulation, while the interior flow is largely adiabatic.

Recent observations, theory and numerical simulations suggest that both adiabatic and diabatic processes are important. While the overturning circulation is largely adiabatic in the ocean interior, diapycnal mixing drives water mass transformations that are central to the global overturning. The conversion of LCDW to UCDW in the deep Indian and Pacific oceans is one example. Efforts to quantify the water mass transformations driven by breaking of internal waves are just beginning. An illustration is provided by Figure 18.3, where estimates of the energy input to internal waves by tides and geostrophic motions and a turbulent mixing parameterization are used to quantify the diapycnal water mass transformations driven by diapycnal mixing (Nikurashin and Ferrari, 2013). Mixing is largest in the deep ocean and decays with height above the bottom. While such estimates are uncertain due to numerous assumptions and poorly constrained parameters, they suggest that mixing by breaking internal waves makes a significant contribution to water mass transformation in the abyssal ocean. Observations show that diapycnal mixing is enhanced in the deep Southern Ocean where the ACC jets interact with rough bathymetry (e.g., Naveira Garabato et al., 2004; Waterman et al., 2013).

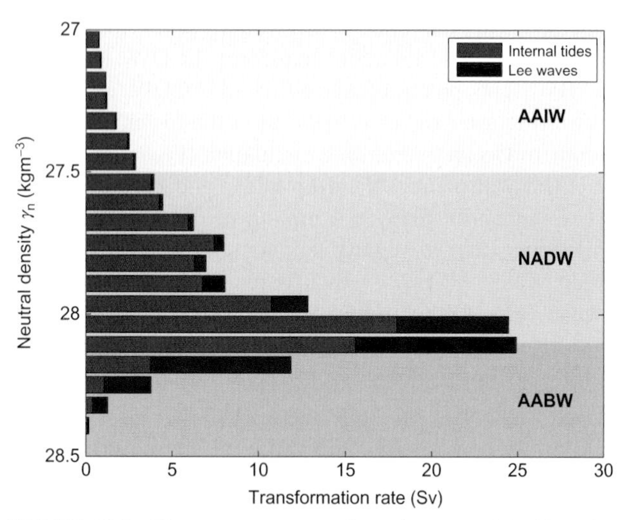

FIGURE 18.3 The water mass transformation rate (Sv) driven by breaking of internal waves, as a function of neutral density, with the density range of different water masses indicated. Both internal tides (blue) and deep geostrophic flows (red) contribute to generation of internal waves and diapycnal mixing; the latter is particularly important in the ACC. While there is growing evidence to support the shape of the curve, with elevated mixing at depth and weak mixing in the upper ocean, the magnitude of the implied water mass transformations is sensitive to a number of assumptions and poorly known parameters. The density range of Antarctic Bottom Water (AABW), North Atlantic Deep Water (NADW), and Antarctic Intermediate Water (AAIW) are shown. *From Nikurashin and Ferrari (2013).*

Diapycnal mixing in the thermocline and upper ocean is generally weak, even in the Southern Ocean (e.g., Naveira Garabato et al., 2004; Nikurashin and Ferrari, 2013). The conversion of UCDW to intermediate waters required to close the global overturning circulation is instead accomplished by air–sea buoyancy fluxes and diapycnal mixing by horizontal eddy motions in the surface mixed layer (Speer et al., 2000; Marshall and Speer, 2012).

In summary, the Southern Ocean plays several key roles in the global overturning circulation. The circumpolar flow of the ACC provides the interbasin connection that permits a global-scale overturning to exist. The steeply sloping isopycnals of the ACC allow upwelling of deep water to the sea surface, where air–sea buoyancy forcing and enhanced diapycnal mixing in the surface layer act to transform deep water to intermediate waters, thereby connecting the deep and shallow limbs of the global overturning. Intermediate waters (AAIW + SAMW) exported from the Southern Ocean ultimately supply the northward transport in the upper layers of the Atlantic required to balance the export of NADW (e.g., Sloyan and Rintoul, 2001; Marshall and Speer, 2012). Note, however, that NADW/LCDW is too dense to outcrop in regions of net ocean buoyancy gain in the ACC belt, as is often assumed in conceptual models of the zonally averaged overturning circulation. LCDW must first be transformed to less dense UCDW by diapycnal mixing in the deep Indian and Pacific oceans. Vigorous diapycnal mixing by internal wave breaking in the deep Southern Ocean helps close the lower cell of the overturning by converting AABW to CDW. The Southern Ocean overturning therefore lies somewhere between the adiabatic and diabatic limits: while much of the flow associated with the overturning cells is adiabatic, and mixing is weak in the pycnocline, water mass transformations driven by diapycnal mixing in the deep ocean, by air–sea buoyancy exchange, and by diapycnal mixing in the surface mixed layer are fundamental to the overturning.

4.2. Estimates of the Rate of Southern Ocean Overturning

Quantifying the strength of the Southern Ocean overturning circulation has long been a challenge. Small biases in the observed or inferred absolute horizontal velocity can accumulate to large errors on long sections, while vertical motions are not observable with present technology. In addition, the meridional flow associated with the overturning is a small residual of larger wind- and eddy-induced transports and it depends on a variety of small- and mesoscale processes that are difficult to measure (e.g., turbulent diapycnal mixing and eddy stirring). Nonetheless, some progress has been made in quantifying the net overturning by combining observations and dynamical constraints with a variety of inverse methods. Recent estimates suggest that the strength of the upper cell across 32°S is about 12–21 Sv (with largely overlapping uncertainties; Ganachaud, 2003; Lumpkin and Speer, 2007; Mazloff et al., 2010; Naveira Garabato et al., 2013b), consistent with estimates of 15 Sv based on calculations of water mass transformations driven by climatological surface forcing (e.g., Olbers et al., 2004).

There is less agreement on the partitioning of the zonally averaged upper-cell flow among the three major ocean basins. While most inverse studies find about 20 Sv of water in the CDW density class enter the Southern Ocean from the subtropical Atlantic (e.g., Sloyan and Rintoul, 2001; Ganachaud, 2003; Lumpkin and Speer, 2007; Mazloff et al., 2010; Naveira Garabato et al., 2013b), there are substantial discrepancies in the Indian and Pacific oceans. Most recent studies estimate that about 10 Sv of deep water flows poleward in each of these two basins, but the distribution of the flow with respect to density differs, affecting the degree of cancellation in the zonal integral. Finally, investigations of the vertical circulation on subbasin scales suggest that zonally averaged estimates conceal important spatial variations in the strength and sense of overturning. For example, the subduction of fluid across the base of the winter mixed layer to supply the SAMW and AAIW layers occurs in highly localized regions, as set by the combined influence of wind, eddies, and lateral induction resulting from along-stream variations in winter mixed layer depth (Figure 18.4; Sallée et al., 2010a; Herraiz-Borreguero and Rintoul, 2011).

In contrast to studies of the upper cell, estimates of the strength of the lower cell still vary over a wide range. Inverse studies explicitly representing the circulation of the subpolar Southern Ocean (e.g., Lumpkin and Speer, 2007; Mazloff et al., 2010) tend to estimate a relatively weak lower cell of ca. 10 Sv across nominally 60°S, in reasonably good alignment with transient tracer-based estimates of AABW ventilation (Orsi et al., 2002). Naveira Garabato et al. (2013b) estimate a somewhat larger 15 Sv. Entrainment tends to increase the transport of deep and bottom waters flowing northward in the lower cell and the deep overturning circulation tends to be stronger near 30°S than at 60°S. Recent estimates of the deep overturning at 30°S (e.g., 18.6±0.9 Sv (Naveira Garabato et al., 2013b), 20.9±6.7 Sv (Lumpkin & Speer, 2007), 29 Sv (Talley, 2013)) tend to be somewhat smaller than the O(50 Sv) estimated earlier by Schmitz (1996) and Sloyan and Rintoul (2001). Part of the difficulty in quantifying the strength of the lower cell arises from the fact that the northward and southward limbs of the cell are connected by poorly known diapycnal mixing processes in the deep ocean. Direct measurements of the equatorward export of AABW at selected outflow sites have helped quantify the pathways of the lower cell. For example, northward AABW transports of ca. 6 Sv through the Orkney Passage (north of the Weddell Sea) and ca. 12 Sv in the deep boundary current east of the Kerguelen Plateau have been measured by Naveira Garabato et al. (2002) and Fukamachi et al. (2010), respectively. However, comparison of transport estimates near the sources of AABW with estimates in the export pathways is not straightforward, as entrainment and recirculation may increase the transport downstream of the sources.

4.3. Residual-Mean Circulation

Despite the intrinsically three-dimensional nature of the Southern Ocean overturning circulation, much theoretical progress has been made by considering a zonally averaged view of the circulation. Residual-mean theory considers the time-mean, steady, zonal- (or streamwise-) average buoyancy and zonal momentum budgets of the ACC to assess how the interplay between wind and surface buoyancy forcing, diapycnal mixing, and mesoscale ocean dynamics shapes both the zonal transport and the Lagrangian-mean overturning across the region (Marshall and Radko, 2003; Olbers et al., 2004 and references therein). It is in making explicit the dynamical connection between zonal and meridional (overturning) flow that the residual-mean approach is particularly insightful. In short, meridional flow is induced by the vertical divergence in a zonal stress, which is initially imparted on the ocean by the wind, transmitted downward by transient eddies (as an interfacial form stress), and ultimately opposed by topographic form drag at the sea floor.

An equivalent description of the residual-mean overturning circulation is as the sum of the wind-driven and eddy-induced circulations. This description is illustrated in Figure 18.5, which presents results from an eddy-resolving channel model of the circumpolar current (Abernathey et al., 2011; Marshall and Speer, 2012). The wind-driven (Eulerian mean) overturning cell consists of northward flow in the surface Ekman layer, a southward return flow at depth, and upwelling (downwelling) to the south (north) of the current's axis. It thus acts to tilt isopycnals. Release of the potential energy stored in the sloping isopycnals by baroclinic instability generates transient eddies. The eddy-induced overturning opposes the Ekman cell, driving a slumping of isopycnals, a tendency referred to as "eddy compensation". The sum of the wind-driven and eddy-induced circulations yields the residual-mean circulation, which is responsible for tracer transports as illustrated by the simulated "dye release" in Figure 18.5.

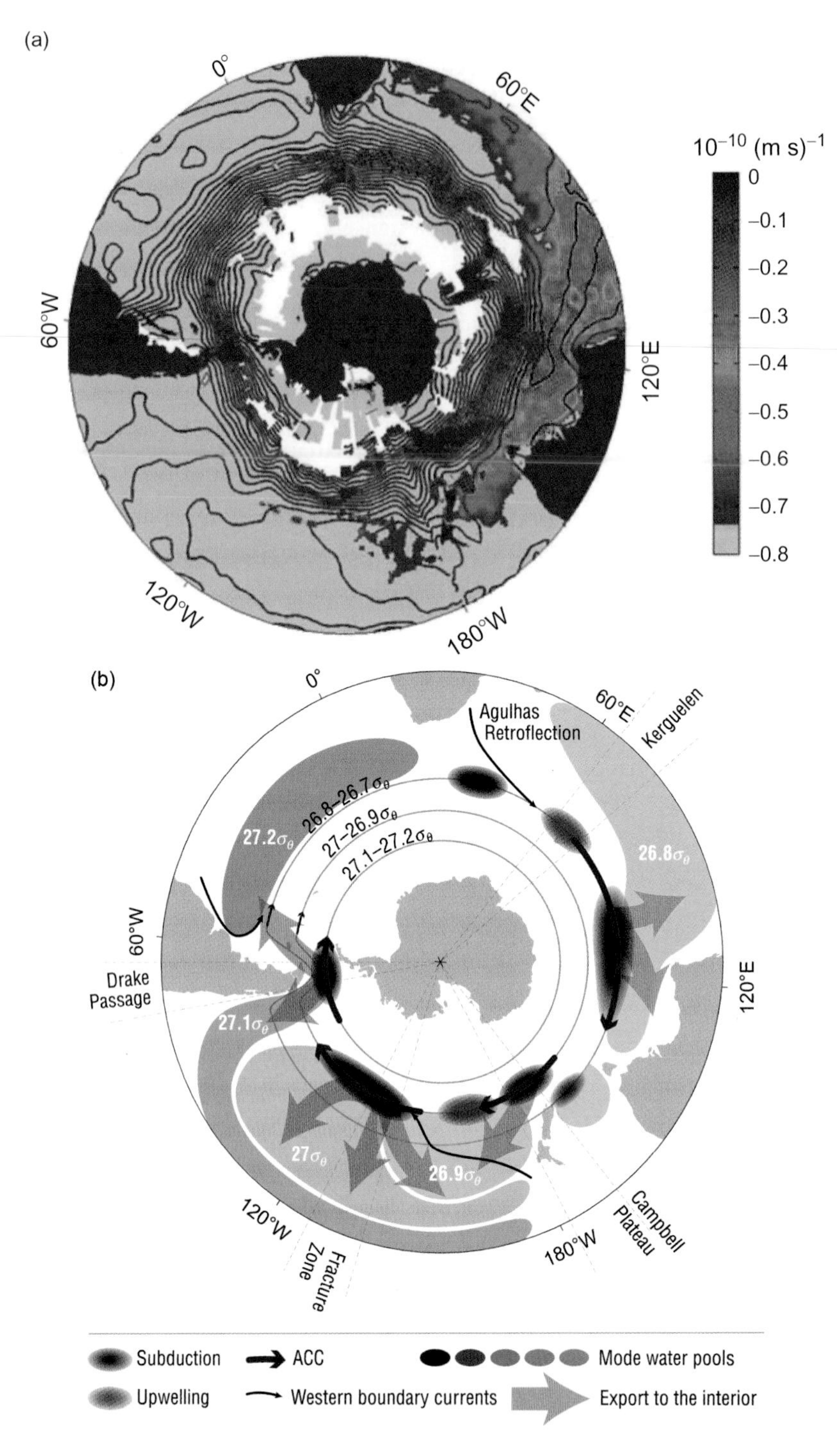

FIGURE 18.4 Subduction hot spots in the Southern Ocean inferred from Argo floats, satellite altimetry, and wind forcing (see Sallée et al., 2010a for details). (a) Subduction on the 26.8σ_θ surface (red dots), overlaid on the Montgomery streamlines indicating the flow on this density surface (black lines). The potential vorticity on the isopycnal is shown in color. A tongue of low-potential vorticity spreads along streamlines from the subduction hot spot in the southeast Indian Ocean into the interior (anthropogenic CO_2 follows similar pathways; Sallée et al., 2012). (b) A schematic view of subduction in the Southern Ocean. Each density surface is ventilated through a narrow window, with denser surfaces ventilated progressively further east from the Indian Ocean. Subduction in some locations is ineffective, because subducted water is upwelled again further downstream (e.g., in the central Indian and western Pacific basins, regions of net upwelling shown in orange). Subduction and upwelling are defined as transfer across the base of the winter mixed layer. *From Sallée et al. (2010a).*

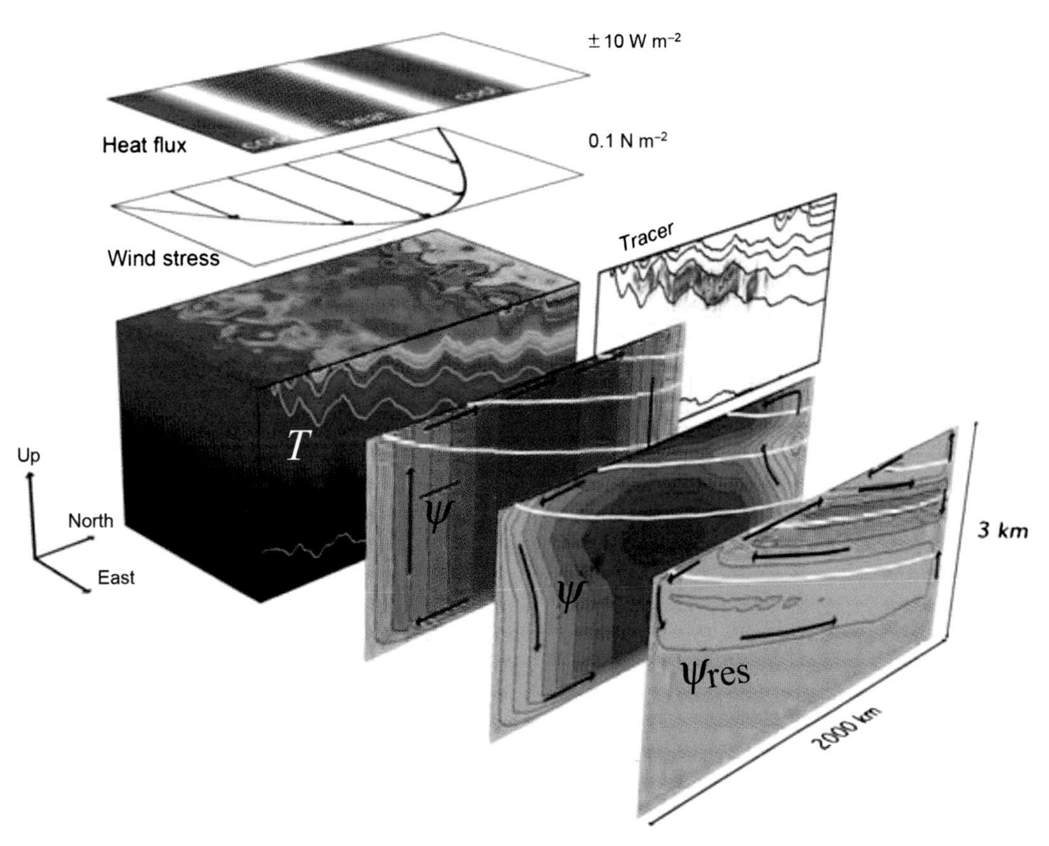

FIGURE 18.5 An illustration of the residual-mean circulation in the Southern Ocean, derived from an idealized but eddy-resolving ocean circulation model (Abernathey et al., 2011; see http://vimeo.com/17822148 for a movie of the spin-up of the circulation). The reentrant channel is forced by wind and heat fluxes at the sea surface. From left to right, the lower panels show a snapshot of the temperature (T) field; the time-mean overturning associated with the wind-driven Ekman transport (Ψ); the time-mean counter-rotating overturning associated with eddy fluxes (Ψ^*); and their sum, the residual-mean overturning streamfunction (Ψ_{res}). The panel labeled "Tracer" illustrates the poleward, along-isopycnal transport of tracer by the residual-mean flow. *From Marshall and Speer (2012), redrawn from Abernathey et al. (2011).*

In the residual-mean framework, the strength of the net overturning is set by the buoyancy forcing, which transforms upwelled deep water to less dense intermediate water. Both air–sea exchange and diapycnal mixing in the surface mixed layer can contribute to the net buoyancy forcing.

The tendency for the eddy-induced circulation to compensate the wind-driven circulation is a common feature of models that resolve or effectively parameterize the effect of eddies. (Coarse resolution models tend to underestimate the eddy-driven circulation and therefore overestimate the sensitivity of the overturning to changes in wind stress (e.g., Farneti et al., 2010).) For example, increasing the wind stress in a reentrant sector model of the Southern Ocean results in an increase in EKE, a small increase in ACC transport, and a larger increase in overturning (Figure 18.6; Morrison and Hogg, 2013). However, the degree of compensation varies widely between models. For a doubling of wind stress, ACC transport increases by about 25% and overturning increases by about 50–70% in Morrison and Hogg's (2013) model. Many other models are closer to the eddy saturation and eddy compensation limits, showing an increase in ACC transport of less than 10% and an increase in overturning of less than 25% for a doubling of wind stress (e.g., Hallberg and Gnanadesikan, 2006; Farneti et al., 2010; Abernathey et al., 2011; Dufour et al. 2012). Detailed comparisons are difficult to make because the model simulations differ in many respects, including geographic domain, model resolution and complexity, and the nature of the imposed wind-forcing perturbation. Perhaps most importantly, the effective buoyancy forcing applied is likely to have a strong impact on the sensitivity of the overturning response. We note that the coupled model study of Farneti et al. (2010), in which the buoyancy flux adjusts to changes in the ocean and atmosphere, is the closest to the eddy saturation and eddy compensation limits.

Despite quantitative disagreement between results from different models, dynamical arguments (Meredith et al., 2012) and the modeling studies cited above provide a consistent picture of the response of the Southern Ocean circulation to changes in wind forcing. Both the ACC transport and overturning circulation are less sensitive to wind-forcing perturbations than would be expected in a non-eddying ocean, and the ACC transport is less sensitive than the overturning circulation to wind changes. While the

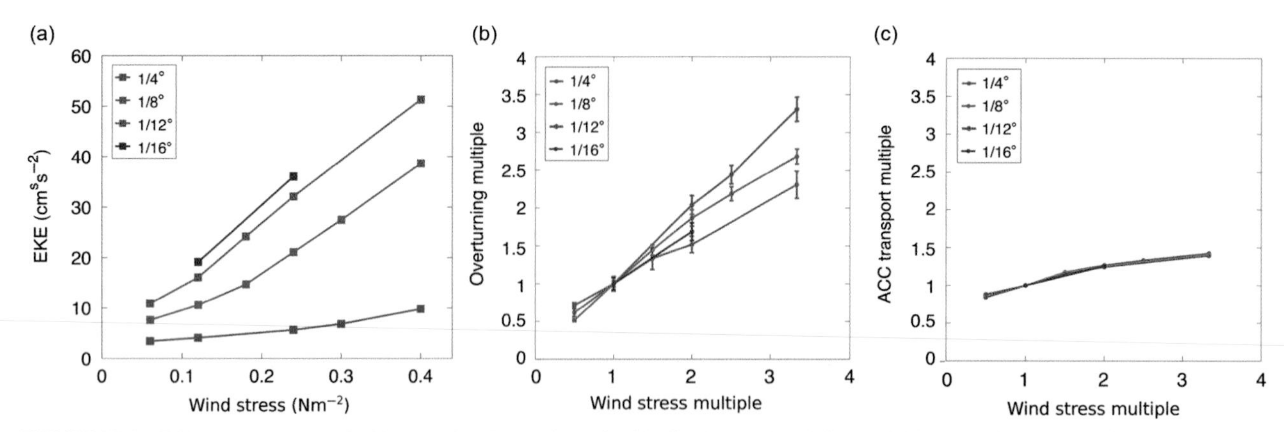

FIGURE 18.6 Eddy compensation and eddy saturation. Integrations of an idealized ocean circulation model illustrate the concepts of eddy saturation and eddy compensation, and their dependence on the extent to which eddies are resolved. (a) As wind forcing increases, eddy kinetic energy (EKE) increases, with a tendency for greater increase in EKE with increasing model resolution. (b) Sensitivity of the overturning circulation at 30°S to wind forcing (transport and wind stress have been scaled relative to a control case). The overturning strength increases with wind stress, but more slowly in models that better resolve eddies, illustrating that eddy-driven overturning partially compensates wind-driven changes. Zero eddy compensation would follow a diagonal (1:1) line; perfect eddy compensation would correspond to a horizontal line. (c) Sensitivity of ACC transport to wind forcing. A linear increase with wind would follow a diagonal (1:1) line; perfect eddy saturation would fall on a horizontal line. *Redrawn from Morrison and Hogg (2013), figure supplied by A. Morrison.*

degree of eddy saturation and eddy compensation tends to increase when refining model resolution, the simulations do not reach the eddy saturation or eddy compensation limits, even with very fine resolution. Unless surface buoyancy forcing is fixed, neither the magnitude nor the shape of the vertical profile of eddy-induced stress divergence will necessarily match that of the Ekman component, resulting in incomplete compensation.

4.4. Eddy Stirring of PV

While model results suggest that eddies tend to partially compensate the wind-driven circulation response to a change in wind forcing, the amount of compensation remains uncertain. Progress requires a better understanding of the eddy-induced circulation. This circulation arises from transient eddy fluctuations on scales of O(10–100 km) that are difficult to quantify in observations and to describe theoretically. A common approach is to relate the stress divergence (or eddy-induced meridional flow) to a downgradient flux of PV resulting from eddy stirring of a background isopycnal PV gradient, where the efficiency of the eddy stirring process is characterized by an isopycnal diffusivity (e.g., Rhines and Young, 1982). This approach reduces the overturning problem to an investigation of the physical controls of the isopycnal diffusivity and its spatial covariability with the background isopycnal PV gradient. The simplification is artificial in that the background PV distribution is itself shaped by the action of eddies. However, the substantial resilience of the mean PV gradient to changes in forcing in an eddy-saturated regime (Meredith et al., 2012) suggests that much may be learned about the dynamical controls of the Southern Ocean overturning by focusing on the physics of eddy stirring.

Recognition of the importance of eddy stirring in the Southern Ocean overturning circulation has motivated a recent flurry of investigations on the subject (e.g., Marshall et al., 2006; Abernathey et al., 2010; Ferrari and Nikurashin, 2010; Naveira Garabato et al., 2011; Thompson and Sallée, 2012). Two general findings have emerged from that body of work. First, the intensity of eddy stirring is not solely governed by EKE, contrary to the predictions of classical quasi-homogeneous turbulence theories. Rather, it is commonly reduced at the upper-ocean core of the ACC frontal jets (where EKE is often highest) relative to surrounding regions, in conjunction with large isopycnal PV gradients (e.g., Thompson et al., 2010; Naveira Garabato et al., 2011). Second, jets are found to be leaky or permeable in the proximity of prominent topographic obstacles, where the mean flow adopts a nonzonal configuration (Figure 18.7). There, vigorous eddy stirring may occur in association with large isopycnal PV gradients, suggesting that those sites are major contributors to the circumpolar integral of the cross-stream PV flux, vertical divergence in along-stream stress, and ultimately cross-stream overturning, as found by, for example, Zika et al. (2013b) in an eddy-permitting GCM. The localization of eddy fluxes by topography provides another example of how the zonally averaged view might conceal important aspects of Southern Ocean dynamics.

The reduction of isopycnal diffusivity in the upper layers of the ACC frontal jets may be interpreted in at least two alternative, partially compatible ways. Some authors (e.g., Marshall et al., 2006; Abernathey et al., 2010), treating mesoscale eddies as linear Rossby waves, suggest that the intensification of eddy stirring on the flanks of jets is associated with a Rossby wave critical layer, as in the atmospheric literature on eddy stirring. Others (Ferrari and Nikurashin, 2010; Naveira Garabato et al., 2011) note

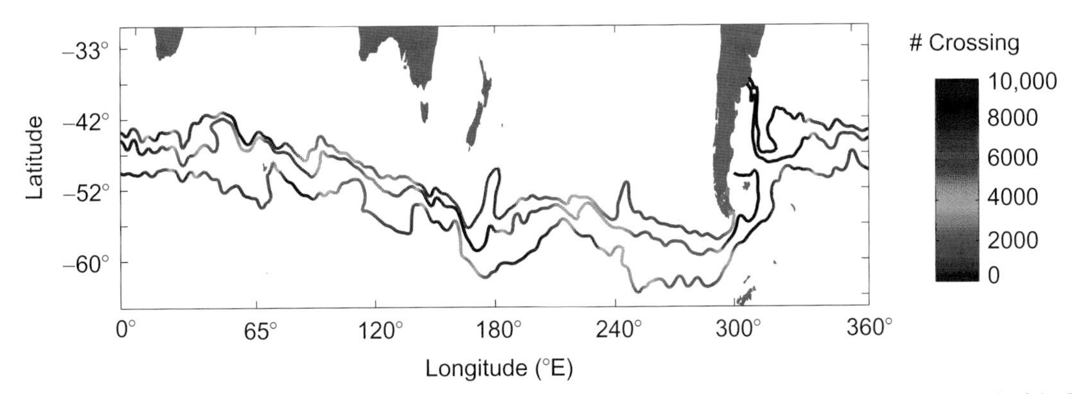

FIGURE 18.7 Number of particle crossings (a metric of the rate of eddy stirring) across (from north to south) the northern branch of the Subantarctic Front, the main branch of the Subantarctic Front, and the Polar Front, from a synthetic release of Lagrangian particles in the altimetric velocity field by Thompson and Sallée (2012). Note the localization of particle crossings to nonzonal segments of the fronts, which correspond to regions where the ACC fronts are steered by topographic obstacles. *From Thompson and Sallée (2012).*

that oceanic mesoscale eddies possess a significant degree of nonlinearity, which results in kinematic suppression of the eddy mixing length (the typical cross-stream displacement of water parcels induced by stirring) by eddy–mean flow interaction. By contrast, there is as yet no consensus on how to interpret the permeability of ACC jets in nonzonal segments of the current close to topography. Using ray-tracing arguments and evidence from idealized studies (e.g., Thompson, 2010), Naveira Garabato et al. (2011) conjecture that nonparallel structure in the mean flow, such as that associated with standing meanders, leads to a breakdown of the conditions required for suppression of eddy stirring: eddies are locally strained, amplified, and eventually "torn" by narrow and twisted jets rather than advected as coherent entities by broader, parallel jets.

The preceding discussion suggests that in the limit of an invariant PV distribution, both the wind-driven Ekman and eddy-induced components of the overturning are expected to respond positively to wind-forcing perturbations. The eddy-induced overturning increases because isopycnal eddy diffusivity depends on EKE, and hence on wind stress (Meredith et al., 2012). As the vertical profiles of the stress divergence associated with the Ekman and eddy contributions to the overturning are generally different, eddy compensation will not be complete and the net (residual) Southern Ocean overturning will scale with changes in wind forcing.

Nonetheless, the extent to which the limit of invariable PV applies in the Southern Ocean is uncertain. The background PV field can be modified by wind (e.g., cross-frontal advection in the Ekman layer) or buoyancy forcing (e.g., convection) (Thomas, 2005), or by eddy buoyancy fluxes in the surface mixed layer (Marshall and Radko, 2003). Any modification of the upper-ocean PV field upon which eddy stirring acts will change the eddy flux of PV along isopycnals, the vertical profile of stress divergence, and the eddy-induced overturning. Various mechanisms can alter the PV distribution in the interior as well, including the wave-mediated adjustment of low-latitude stratification to local and remote forcing (e.g., Allison et al., 2011; Munday et al., 2013) and the modification of stratification by changes in diapycnal mixing (e.g., Ito and Marshall, 2008; Nikurashin and Vallis, 2011). Diapycnal mixing in the deep Southern Ocean may depend on wind forcing, either through wind-forced barotropic fluctuations in the intensity of deep boundary currents (Polzin and Naveira Garabato, 2013) or by changes in internal wave generation and breaking produced by wind-driven changes in the ACC eddy field interacting with rough topography (Nikurashin et al., 2013; Waterman et al., 2013).

5. SOUTHERN OCEAN CHANGE

Over the last decade, evidence has grown of strong variability and trends in water properties in the Southern Ocean. Given the influence of Southern Ocean processes on the global ocean circulation, climate, and biogeochemical cycles, change in the circulation of the Southern Ocean might have widespread consequences. Observations of the response to changes in natural and anthropogenic forcing can also provide useful insights into the dynamics of the system. Here we briefly summarize recent evidence of changes in the Southern Ocean. The observational record is short and in most cases insufficient to distinguish long-term trends from low-frequency variability.

5.1. Warming and Freshening of the Southern Ocean

The upper 2000 m of the Southern Ocean has warmed at a faster rate and to greater depth than the global ocean average in recent decades. Gille (2008) demonstrated that the upper 1000 m of the Southern Ocean has warmed since at least the 1930s, with the largest warming concentrated in the ACC. Böning et al. (2008) compared Argo observations from 2003 to 2008 to a historical climatology along mean streamlines of the ACC and showed that the Southern Ocean has both warmed and freshened to depths of at least 2000 m (Figure 18.8; see also Helm et al., 2010 and Durack and Wijffels, 2010). The zonally integrated linear trend in 0–2000 m ocean heat content (1955–1959 to 2006–2010) reaches a maximum between 40°S and 50°S (Levitus et al., 2012). Warming of the Southern Ocean below 1000 m accounts for about 10% of the increase in global ocean 0–700 m heat content between 1993 and 2003, or about 1 mm year^{-1} of local sea level rise, indicating both

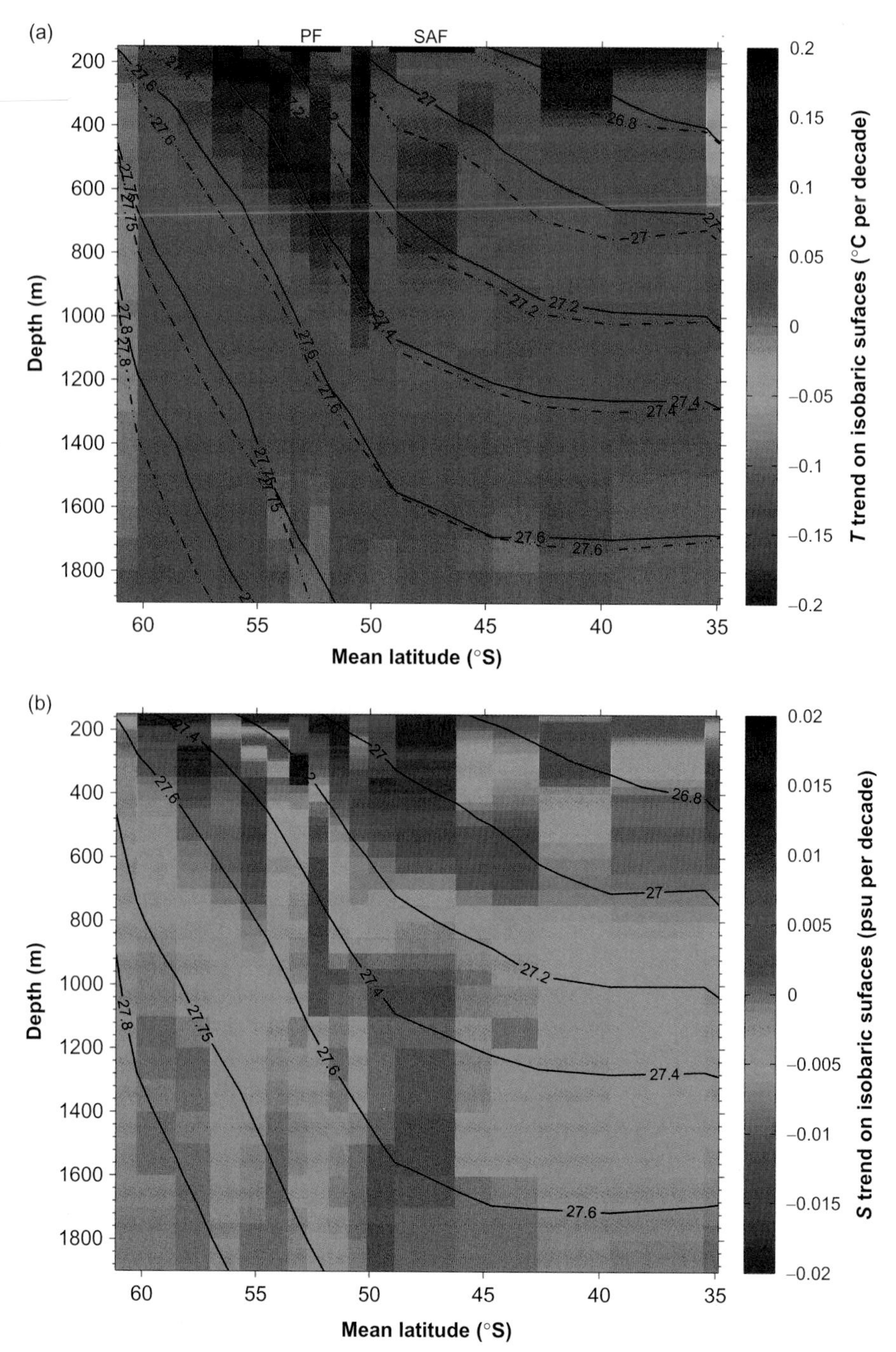

FIGURE 18.8 Temperature and salinity changes across the ACC, estimated by taking the difference between recent Argo measurements and a long-term climatology along mean streamlines. Mean decadal trends on pressure surfaces of potential temperature (a) and salinity (b). Black contours in (a) illustrate the migration of isopycnal surfaces during the past four decades: continuous (dashed) curves represent potential densities obtained by subtracting (adding) the linear trends over two decades from (to) the climatological mean density field. The position of the Polar Front (PF) and Subantarctic Front (SAF) are indicated in (a). *From Böning et al. (2008).*

the deep-reaching nature of temperature changes in the Southern Ocean and their importance to global ocean energy budgets (Purkey and Johnson, 2010).

The warming of the Southern Ocean has been linked to a southward shift of the ACC, likely caused by a southward shift of the westerly winds (e.g., Böning et al., 2008; Gille, 2008). The wind changes have, in turn, been associated with a decadal-scale trend in the Southern Annular Mode (SAM), the dominant mode of variability in the southern hemisphere atmospheric circulation (e.g., Thompson et al., 2011). An increasing trend in sea surface height measured by satellite altimetry since 1993 is also consistent with an overall southward migration of the ACC (Figure 18.9). SSH contours corresponding to each of the ACC fronts migrated southward by about 0.6° of latitude between 1992 and 2007, but with significant regional variations (Sokolov and Rintoul, 2009a,b). Meijers et al. (2011) demonstrated that both adiabatic shifts of the ACC fronts and diabatic water mass changes driven by surface warming and an increase in precipitation–evaporation have contributed to the observed changes in temperature and salinity in the Southern Ocean. The inference of strong freshening driven by increased precipitation–evaporation in the ACC belt is consistent with the global analyses of salinity changes by Helm et al. (2010) and Durack and Wijffels (2010), who concluded that freshening of the Southern Ocean was the regional expression of a global amplification of the water cycle induced by the increased water vapor content of a warmer atmosphere.

5.2. Changes in the Southern Ocean Inventory of Dissolved Gases

The Southern Ocean inventory of carbon, oxygen, and chlorofluorocarbons (CFCs) has also changed in recent decades. As a result of the vigorous and deep-reaching overturning circulation, the Southern Ocean makes an important contribution to the global carbon cycle (e.g., Sabine et al., 2004). More than 40% of the ocean inventory of anthropogenic CO_2 entered the ocean south of 40°S (Khatiwala et al., 2013), mostly via the upper cell of the overturning circulation. More specifically, anthropogenic CO_2 is carried into the ocean interior through localized subduction windows, where Ekman pumping, eddy flux convergence, and the correlation between changes in mixed layer depth and the mean flow (lateral induction) conspire to transfer fluid across the base of the winter mixed layer (Sallée et al., 2012).

The Southern Ocean is the only region of the ocean where deep waters upwell to the sea surface and can exchange properties, including CO_2, with the atmosphere. While orbital changes determine the timing of glacial–interglacial cycles, carbon dioxide released to the atmosphere by upwelling of carbon-rich deep waters in the Southern Ocean provides the amplifier needed to complete the transition from glacial to interglacial conditions (e.g., Sigman et al., 2010). The effectiveness with which the Southern Ocean stores carbon dioxide depends on the response of both the natural and anthropogenic carbon cycles to changes in forcing. Le Quéré et al. (2007) argued on the basis of atmospheric measurements and an ocean circulation model that the Southern Ocean sink for CO_2 had saturated, due to increased outgassing of natural carbon resulting from a wind-driven increase in the overturning circulation in recent decades. The hypothesis sparked active debate (e.g., Law et al., 2007; Böning et al., 2008; Zickfeld et al., 2008) that helped focus attention on the question of how the overturning circulation responds to changes in wind forcing, or in other words, to what extent do eddies compensate wind-driven increases in Ekman transport? (See Section 4.)

Observational evidence of an increase in strength of the Southern Ocean overturning circulation has been difficult

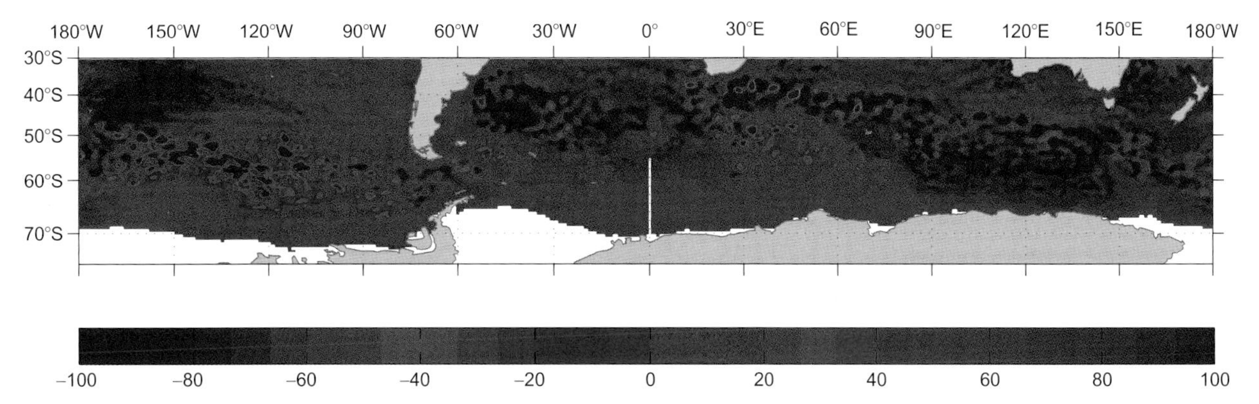

FIGURE 18.9 Change in sea surface height (SSH, in mm) between 1992 and 2012 based on satellite altimetry data (Collecte, Localization, Satellites (CLS)/Archiving, Validation, and Interpretation of Satellite Oceanographic Data (AVISO) "Mean Sea Level Anomaly" (MSLA), Le Traon et al. (1998)). SSH has increased over much (but not all) of the Southern Ocean and the southern flank of the subtropical gyres (yellow to red colors), consistent with an overall poleward shift of the ACC and poleward expansion and intensification of the subtropical gyres. *Figure courtesy of Serguei Sokolov, CSIRO.*

to obtain. A few recent studies provide some support for the hypothesis that the overturning circulation has increased in recent decades. While data are sparse, Takahashi et al. (2012) show that the pCO_2 of Southern Ocean winter surface waters is increasing more rapidly than it is in the atmosphere, consistent with stronger upwelling of carbon from below. Changes in the inventory of CFCs along repeat hydrographic sections also suggest that the upper cell of the Southern Ocean overturning has intensified, with more rapid upwelling of deep water and greater formation and subduction of intermediate waters (Figure 18.10; Waugh et al., 2013). These observational studies are thus at least broadly consistent with the modeling and theoretical results discussed in Section 4, which indicate that eddies likely compensate some, but not all, of the larger Ekman transport in response to strengthening winds. On the other hand, Helm et al. (2011) interpret a decline in oxygen between the 1970s and 1990s in AAIW and SAMW as evidence for reduced ventilation as a result of enhanced stratification.

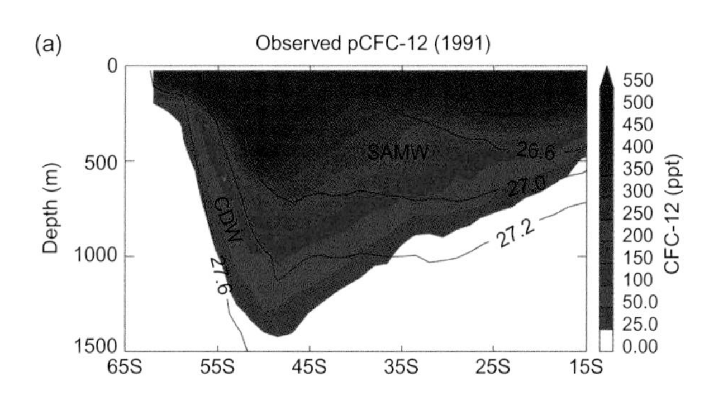

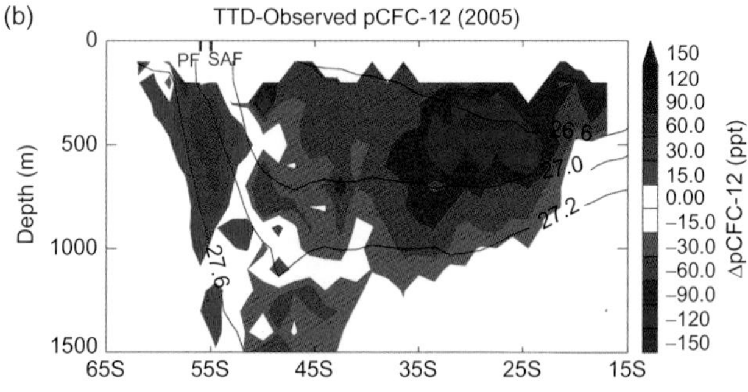

FIGURE 18.10 (a) Observed chlorofluorocarbon-12 concentration (pCFC-12) in 1991 along a north–south transect at 150°W in the Pacific. (b) Difference between modeled and observed pCFC-12 in 2005, where the modeled value is based on a transit time distribution approach assuming steady transport. Positive (negative) values indicate modeled pCFC-12 values are higher (lower) than observed. The pattern of differences is consistent with an increase in the overturning circulation, with stronger sinking of well ventilated intermediate waters north of the ACC and stronger upwelling of poorly ventilated deep water south of the ACC. Black lines show the potential density distribution in 1991. CDW, Circumpolar Deep Water; SAMW, Subantarctic Mode Water; PF, Polar Front; SAF, Subantarctic Front. *From Waugh et al. (2013).*

5.3. Changes in Southern Ocean Water Masses

As water mass properties are largely set at the sea surface through air–sea interaction, and then transferred into the ocean interior by subduction, analysis of changes in water mass properties can provide clues about changes in surface climate. Water masses also influence the interior stratification and distribution of PV, so water mass changes can have dynamical implications.

5.3.1. Changes in Mode and Intermediate Waters

Coherent changes in the properties of the water masses involved in the upper cell (SAMW and AAIW) have been observed in recent decades. SAMW has generally cooled and freshened (e.g., Böning et al., 2008; Helm et al., 2010). The AAIW salinity minimum core shoaled and warmed between 1970 and 2009, with regionally varying trends in salinity (Schmidtko and Johnson, 2012). Both an increase in precipitation–evaporation and a warming-induced poleward migration of isopycnal outcrops (i.e., into regions of higher or lower surface salinity) have likely contributed to the observed trends in intermediate water properties (Durack and Wijffels, 2010; Helm et al., 2010). AAIW changes in particular locations have been linked to other processes, including exchange between the Indian and Atlantic basins (McCarthy et al., 2011) and changes in surface forcing related to modes of climate variability (Naveira Garabato et al., 2009).

5.3.2. Changes in AABW

Perhaps the most dramatic water mass changes observed in the Southern Ocean are those of the AABW. AABW has warmed, freshened, decreased in density, and contracted over the last 40 years (e.g., Jacobs et al., 2002; Rintoul, 2007; Jacobs and Giulivi, 2010; Purkey and Johnson, 2010, 2012, 2013; Fahrbach et al., 2011; Shimada et al., 2012; van Wijk and Rintoul, 2013; see also Chapter 19). Basin-average warming is largest near Antarctica and decreases with distance along the AABW export pathways (Purkey and Johnson, 2010). Globally, the volume of water cooler than 0 °C (a common definition of AABW) has decreased by 8.2 ± 2.6 Sv over the last two decades (Purkey and Johnson, 2012); in the Australian Antarctic Basin, the layer of AABW denser than the 28.3 kg m^{-3} neutral density surface has thinned by more than 100 m decade^{-1} and lost more than half its volume since 1970 (van Wijk and Rintoul, 2013). The strongest freshening signal is found in the Indian and Pacific basins, where both the Ross Sea and Adélie Land sources of bottom water have freshened (Jacobs et al., 2002; Aoki et al., 2005; Rintoul, 2007;

Jacobs and Giulivi, 2010; Purkey and Johnson, 2013). Freshening of Ross Sea shelf waters has been attributed to inflow of glacial ice melt in the Amundsen–Bellingshausen Seas, although changes in sea ice formation and melt or precipitation may also contribute (Jacobs and Giulivi, 2010). van Wijk and Rintoul (2013) argue that freshening of the dense source waters is the primary cause of the observed changes in AABW, based on the evolution of water mass properties along the boundary currents exporting AABW to the north in the Australian Antarctic Basin. Near the sources, strong freshening is observed, with little change in temperature. Salinity changes decrease downstream, while temperature changes increase, consistent with gradual warming by enhanced mixing along the export pathway due to freshening-induced changes in abyssal stratification (e.g., Shimada et al., 2012). The oxygen content of the densest waters remained constant within measurement error between the mid-1990s and 2012, suggesting that the contraction of the AABW layer results from formation of slightly less dense varieties of AABW rather than a reduction in the formation rate.

5.4. Ocean–Ice Shelf Interaction

Enhanced basal melt of floating ice shelves has been identified as a likely source of additional freshwater input to the Southern Ocean (Shepherd et al., 2004; Rignot et al., 2011). Thinning of the Pine Island Glacier in the eastern Pacific sector of the Southern Ocean has been particularly dramatic. Recent measurements of the sub-ice shelf cavity using the autonomous underwater vehicle Autosub have revealed that water with temperatures 4 °C above the *in situ* freezing point reaches the grounding line, driving rapid melt (Jenkins et al., 2010; Jacobs et al., 2011). Pritchard et al. (2012) concluded from satellite measurements that 20 of Antarctica's 54 ice shelves and glacier tongues thinned during the short period from 2003 to 2008 as a result of enhanced basal melt. Because ice shelves buttress the interior ice sheet, thinning or collapse of the floating ice shelves can cause more rapid loss of grounded ice and hence more rapid rise in sea level.

The oceanic heat flux toward the Antarctic margin is therefore a key factor controlling the mass balance of the Antarctic ice sheet. Changes in either temperature or circulation could alter the ocean heat flux to the base of the ice shelf. More specifically, warm ocean waters need to cross a number of barriers to reach the base of the floating ice: the shelf break, the continental shelf, the ice front, and the boundary layer beneath the ice. The dynamics controlling the net ocean heat flux to the base of the ice are therefore complex and likely to depend on large-scale forcing (e.g., wind stress, sea ice, buoyancy forcing) and circulation (e.g., proximity to the ACC), as well as small-scale (e.g., turbulent mixing rates) and local factors (e.g., topography of the sea floor and the ice shelf base). Measurements from beneath the Pine Island Glacier underscore this point: basal melt rates likely increased when the glacier thinned enough to float free of a submarine ridge, allowing stronger inflow of relatively warm ocean waters (Jenkins et al., 2010). This example highlights the difficulty of assessing the sensitivity of the Antarctic ice sheet mass balance to warming of the Southern Ocean. The situation is made even more challenging by the fact that very few observations of temperature and velocity have been made on the Antarctic continental shelf, and modeling studies are hampered by lack of knowledge of even such basic features as the bathymetry.

5.5. Changes in Southern Ocean Sea Ice

Satellite data suggest that there has been a small increase in the extent of Southern Ocean sea ice (by about 1% per decade between 1978 and 2010; Parkinson and Cavalieri, 2012), in stark contrast to the Arctic. The small overall increase is the sum of strong trends of opposite sign in the Bellingshausen–Amundsen and Ross Seas. The retreat of sea ice in the Bellingshausen–Amundsen seas has extended the summer ice-free period by more than 3 months since the late 1970s, a change comparable to that observed in the Arctic; expansion of sea ice in the Ross Sea has reduced the ice-free period by 2 months over the same time period (e.g., Stammerjohn et al., 2012). These regional changes in the extent and duration of the Southern Ocean sea ice can largely be explained by trends in surface winds, through both sea ice advection and changes in ice formation and melt rates (Holland and Kwok, 2012). Input of glacial meltwater may have contributed to the expansion of Southern Ocean sea ice by enhancing near-surface ocean stratification and restricting heat input from below (Hellmer, 2004; Bintanja et al., 2013). The changing extent and seasonality of sea ice over broad regions of the Southern Ocean is likely to alter the surface stress and buoyancy fluxes that drive the subpolar circulation, but these effects have yet to be quantified.

5.6. Causes of Recent Southern Ocean Change

As in many other parts of the ocean, changes in recent decades are likely to be associated, directly or indirectly, with changes in wind forcing. Many studies have linked observed changes in the Southern Ocean to the trend in the westerly winds associated with the SAM (e.g., Hall and Visbeck, 2002; Thompson et al., 2011), including the southward shift of the ACC and resulting changes in temperature and salinity (e.g., Gille, 2008), warming and freshening of shelf waters (e.g., Jacobs and Giulivi, 2010), and changes in ocean carbon uptake (e.g., Le Quéré et al., 2007), among many others. The schematic ocean response to an intensification of winds (e.g., Hall and Visbeck, 2002) is straightforward to understand: stronger westerly winds mean more equatorward Ekman transport, with enhanced upwelling to the south and enhanced downwelling to the north; a southward shift of the westerlies will drive a southward shift of the ACC; and stronger northward Ekman transport might drive an expansion of the Southern Ocean sea ice. However, in many cases a quantitative understanding of the link between changes in winds (i.e., the SAM) and ocean change is lacking, in part due to a scarcity of observations. The oceanographic literature has not always taken the seasonal (strongest trend in summer) and longer term variations of the SAM (the time history is not a simple monotonic trend in time) into account. A key question, once again, is the response of the overturning circulation to a change in wind forcing. Schematics like that of Hall and Visbeck (2002) are based on the assumption that eddies do not compensate Ekman transport. In addition, while most studies have focused on the effect of changes in the westerlies, changes in meridional winds over the Southern Ocean are likely to be important for many oceanographic phenomena such as air–sea heat flux and the depth of the surface mixed layer (e.g., Sallée et al., 2010b), and sea ice extent (e.g., Turner et al., 2009; Holland and Kwok, 2012). Warming of the surface layer and an increase in precipitation and ice melt have likely changed the buoyancy forcing of the Southern Ocean, with likely impacts on the circulation, but these effects have not yet been quantified.

6. SUMMARY AND OUTSTANDING CHALLENGES

A common theme to emerge from the observational, theoretical, and numerical studies reviewed here is the significance of departures from zonal symmetry. Topography steers the mean flow and localizes many of the dynamical interactions that govern the circulation of the Southern Ocean and its sensitivity to changes in climatic forcing, including ACC instability, EKE, eddy stirring and meridional exchange, interfacial and bottom form stress, and diapycnal mixing through the generation and breaking of internal waves. Another example of the localization of important dynamical processes is the generation of narrow subduction windows by changes in mixed layer depth along mean streamlines (Figure 18.4). Thus, while the key ingredients in the dynamical recipe for the Southern Ocean circulation have arguably been identified, it is likely that further progress will require disentangling the details of the dynamics of local regions that dominate the circumpolar budgets of momentum, vorticity, and other dynamical quantities. In particular, the climatic sensitivity of the overturning circulation is intrinsically linked to eddy–mean flow–topography interactions in specific areas, which influence the response of upper-ocean diapycnal mixing to wind and eddy effects and the resilience of the Southern Ocean PV structure.

Changes in the Southern Ocean overturning circulation may feed back on climate by altering the capacity of the ocean to take up heat and carbon dioxide or by affecting the stability of the Antarctic cryosphere. Therefore, improved understanding of the response of the Southern Ocean overturning to wind and buoyancy forcing perturbations remains a high priority. While a consensus seems to be emerging from models and theory that the Southern Ocean circulation approaches an eddy-saturated and partially eddy-compensated state, the details of how this state is maintained are not yet clear.

The observational record is often too short to provide clear evidence to support or refute specific hypotheses concerning the response of the Southern Ocean circulation to changes in forcing. However, evidence is growing that changes are underway in the region that are of potential significance to climate, biogeochemical cycles, and the mass balance of the Antarctic ice sheet. Observations are particularly sparse in the deep ocean and within the sea ice zone. The need to better understand the nature, causes, and consequences of Southern Ocean change provides strong motivation for enhancements to the Southern Ocean observing system.

ACKNOWLEDGMENTS

The authors gratefully acknowledge the contributions from Andy Hogg, five anonymous reviewers, and the editors, whose comments have helped improve the clarity of the text. We also thank Adele Morrison, John Marshall, Darryn Waugh, Serguei Sokolov, Beatriz Peña Molino, Jean-Baptiste Sallée, Matt Mazloff, Dirk Olbers, and Max Nikurashin for assistance with figures. This work was supported in part by the Australian Government's Cooperative Research Centres Program, through the Antarctic Climate and Ecosystems Cooperative Research Centre (ACE CRC), and by the Department of Climate Change and Energy Efficiency through the Australian Climate Change Science Program. A. C. N. G. acknowledges the support of a Philip Leverhulme Prize.

REFERENCES

Abernathey, R., Marshall, J., Mazloff, M., Shuckburgh, E., 2010. Critical layer enhancement of mesoscale eddy stirring in the Southern Ocean. J. Phys. Oceanogr. 40, 170–184. http://dx.doi.org/10.1175/2009JPO4201.1.

Abernathey, R., Marshall, J., Ferreira, D., 2011. The dependence of Southern Ocean meridional overturning on wind stress. J. Phys. Oceanogr. 41, 2261–2278. http://dx.doi.org/10.1175/JPO-D-11-023.1.

Allison, L.C., Johnson, H.L., Marshall, D.P., 2011. Spin-up and adjustment of the Antarctic Circumpolar Current and global pycnocline. J. Mar. Res. 69, 167–189.

Anderson, D.L.T., Killworth, P.D., 1977. Spin-up of a stratified ocean, with topography. Deep Sea Res. 24, 709–732.

Aoki, S., 2002. Coherent sea level response to the Antarctic Oscillation. Geophys. Res. Lett. 29, 1950. http://dx.doi.org/10.1029/2002GL015733.

Aoki, S., Rintoul, S.R., Ushio, S., Watanabe, S., Bindoff, N.L., 2005. Freshening of the Adélie Land Bottom Water near 140°E. Geophys. Res. Lett. 32, L23601. http://dx.doi.org/10.1029/2005FL024246.

Bintanja, R., van Oldenborgh, G.J., Drijfout, S.S., Wouters, B., Katsman, C.A., 2013. Important role for ocean warming and increased ice-shelf melt in Antarctic sea-ice expansion. Nat. Geosci. 6, 376–379. http://dx.doi.org/10.1038/NGEO1767.

Böning, C.W., Dispert, A., Visbeck, M., Rintoul, S.R., Schwarzkopf, F.U., 2008. The response of the Antarctic Circumpolar Current to recent climate change. Nat. Geosci. 1, 864–869.

Borowski, D., Borowski, R., Olbers, D., 2002. Thermohaline and wind forcing of a circumpolar channel with blocked geostrophic contours. J. Phys. Oceanogr. 32, 2520–2538.

Chidichimo, M.P., Donohue, K.A., Watts, D.R., Tracey, K.L., 2013. Baroclinic transport time series of the Antarctic Circumpolar Current in Drake Passage. J. Phys. Oceanogr., submitted.

Close, S.E., Naveira Garabato, A.C., 2012. Baroclinic adjustment in Drake Passage driven by tropical Pacific forcing. Geophys. Res. Lett. 39, L19610. http://dx.doi.org/10.1029/2012GL053402.

Cunningham, S.A., Alderson, S.G., King, B.A., Brandon, M.A., 2003. Transport and variability of the Antarctic Circumpolar Current in Drake Passage. J. Geophys. Res. 108, 8084. http://dx.doi.org/10.1029/2001JC001147.

Dufour, C.O., Le Sommer, J., Zika, J., Gehlen, M., Orr, J.C., Mathiot, P., Barnier, B., 2012. Standing and transient eddies in the response of the Southern Ocean meridional overturning to the Southern Annular Mode. J. Clim. 25, 6958–6974.

Durack, P.J., Wijffels, S.E., 2010. Fifty-year trends in global ocean salinities and their relationship to broad-scale warming. J. Clim. 23, 4342–4362.

Fahrbach, E., Hoppema, M., Rohardt, G., Boebel, O., Klatt, O., Wisotzki, A., 2011. Warming of deep and abyssal water masses along the Greenwich meridian on decadal time scales: the Weddell gyre as a heat buffer. Deep Sea Res. Part II 58, 2509–2523.

Farneti, R., Delworth, T.L., Rosati, A.J., Griffies, S.M., Zeng, F., 2010. The role of mesoscale eddies in the rectification of the Southern Ocean response to climate change. J. Phys. Oceanogr. 40, 1539–1557.

Ferrari, R., Nikurashin, M., 2010. Suppression of eddy diffusivity across jets in the Southern Ocean. J. Phys. Oceanogr. 40, 1501–1519.

Firing, Y.L., Chereskin, T.K., Mazloff, M.R., 2011. Vertical structure and transport of the Antarctic Circumpolar Current in Drake Passage from direct velocity observations. J. Geophys. Res. 116, C08015. http://dx.doi.org/10.1029/2011JC006999.

Fukamachi, Y., Rintoul, S.R., Church, J.A., Aoki, S., Sokolov, S., Rosenberg, M.A., Wakatsuchi, M., 2010. Strong export of Antarctic Bottom Water east of the Kerguelen plateau. Nat. Geosci. 3, 327–331.

Ganachaud, A., 2003. Large scale mass transports, water mass formation and diffusivities estimated from the WOCE hydrographic data. J. Geophys. Res. 108, 3213. http://dx.doi.org/10.1029/2002JC001565.

Gent, P.R., Large, W.G., Bryan, F.O., 2001. What sets the mean transport through Drake Passage? J. Geophys. Res. 106, 2693–2712.

Gille, S.T., 2008. Decadal-scale temperature trends in the Southern Hemisphere ocean. J. Clim. 21, 4749–4765.

Gnanadesikan, A., 1999. A simple predictive model for the structure of the oceanic pycnocline. Science 283, 2077–2079.

Hall, A., Visbeck, M., 2002. Synchronous variability in the southern hemisphere atmosphere, sea ice, and ocean resulting from the annular mode. J. Clim. 15, 3043–3057.

Hallberg, R., Gnanadesikan, A., 2001. An exploration of the role of transient eddies in determining the transport of a zonally reentrant current. J. Phys. Oceanogr. 31, 3312–3330. http://dx.doi.org/10.1175/1520-0485(2001)031<3312:AEOTRO>2.0.CO;2.

Hallberg, R.W., Gnanadesikan, A., 2006. The role of eddies in determining the structure and response of the wind-driven Southern Hemisphere overturning: results from the modeling eddies in the Southern Ocean (MESO) project. J. Phys. Oceanogr. 36, 2232–2252.

Hellmer, H.H., 2004. Impact of Antarctic ice shelf basal melting on sea ice and deep ocean properties. Geophys. Res. Lett. 31, L10307. http://dx.doi.org/10.1029/2004GL019506.

Helm, K.P., Bindoff, N.L., Church, J.A., 2010. Changes in the global hydrological-cycle inferred from ocean salinity. Geophys. Res. Lett. 37, L18701. http://dx.doi.org/10.1029/2010GL044222.

Helm, K.P., Bindoff, N.L., Church, J.A., 2011. Observed decreases in oxygen content of the global ocean. Geophys. Res. Lett. 38, L23602. http://dx.doi.org/10.1029/2011GL049513.

Herraiz-Borreguero, L., Rintoul, S.R., 2011. Subantarctic mode water: distribution and circulation. Ocean Dyn. 61 (1), 103–126.

Hogg, A.M.C., 2010. An Antarctic Circumpolar Current driven by surface buoyancy forcing. Geophys. Res. Lett. 37, L23601. http://dx.doi.org/10.1029/2010GL044777.

Holland, P.R., Kwok, R., 2012. Wind-driven trends in Antarctic sea-ice drift. Nat. Geosci. 5, 872–875.

Hughes, C.W., 1996. The Antarctic Circumpolar Current as a waveguide for Rossby waves. J. Phys. Oceanogr. 26, 1375–1387.

Hughes, C.W., 2005. Nonlinear vorticity balance of the Antarctic Circumpolar Current. J. Geophys. Res. 110, C11008. http://dx.doi.org/10.1029/2004JC002753.

Hughes, C.W., Ash, E.R., 2001. Eddy forcing of the mean flow in the Southern Ocean. J. Geophys. Res. 106, 2713–2722.

Hughes, C.W., de Cuevas, B.A., 2001. Why western boundary currents in realistic oceans are inviscid: a link between form stress and bottom pressure torques. J. Phys. Oceanogr. 31, 2871–2885.

Hughes, C.W., Meredith, M.P., Heywood, K.J., 1999. Wind-driven transport fluctuations through Drake Passage: a southern mode. J. Phys. Oceanogr. 29, 1971–1992.

Hughes, C.W., Woodworth, P.L., Meredith, M.P., Stepanov, V., Whitworth, T., Pyne, A.R., 2003. Coherence of Antarctic sea levels, southern hemisphere annular mode, and flow through Drake Passage. Geophys. Res. Lett. 30, 1464. http://dx.doi.org/10.1029/2003GL017240.

Hughes, C.W., Thompson, A.F., Wilson, C., 2010. Identification of jets and mixing barriers from sea level and vorticity measurements using simple statistics. Ocean Model. 32, 44–57. http://dx.doi.org/10.1016/j.ocemod.2009.10.004.

Ito, T., Marshall, J., 2008. Control of lower-limb overturning circulation in the Southern Ocean by diapycnal mixing and mesoscale eddy transfer. J. Phys. Oceanogr. 38, 2832–2845.

Jacobs, S.S., Giulivi, C.F., 2010. Large multidecadal salinity trends near the Pacific–Antarctic continental margin. J. Clim. 23, 4508–4524.

Jacobs, S.S., Giulivi, C.F., Mele, P.A., 2002. Freshening of the Ross Sea during the late 20th century. Science 297, 386–389.

Jacobs, S.S., Jenkins, A., Giulivi, C.F., Dutrieux, P., 2011. Stronger ocean circulation and increased melting under Pine Island Glacier ice shelf. Nat. Geosci. 4, 519–523.

Jenkins, A., Dutrieux, P., Jacobs, S.S., McPhail, S.D., Perrett, J.R., Webb, A.T., White, D., 2010. Observations beneath Pine Island Glacier in West Antarctica and implications for its retreat. Nat. Geosci. 3, 468–472.

Jones, D.C., Ito, T., Lovenduski, N.S., 2011. The transient response of the Southern Ocean pycnocline to changing atmospheric winds. Geophys. Res. Lett. 38, L15604. http://dx.doi.org/10.1029/2011GL048145.

Khatiwala, S., Primeau, F., Hall, T., 2009. Reconstruction of the history of anthropogenic CO_2 concentrations in the ocean. Nature 462, 346–349.

Khatiwala, S., Tanhua, T., Fletcher, S., Gerber, M., Doney, S.C., Graven, H.D., Gruber, N., McKinley, G.A., Murata, A., Rios, A.F., Sabine, C.L., 2013. Global ocean storage of anthropogenic carbon. Biogeosciences 10, 2169–2191. http://dx.doi.org/10.5194/bg-10-2169-2013.

Law, R.M., Matear, R.J., Francey, R.J., 2007. Comment on "Saturation of the Southern Ocean CO_2 sink due to recent climate change" Science 319, 570.

Le Quéré, C., Rodenbeck, C., Buitenhuis, E.T., Conway, T.J., Langenfelds, R., Gomez, A., Labuschagne, C., Ramonet, M., Nakazawa, T., Metzl, N., Gillett, N., Heimann, M., 2007. Saturation of the southern ocean CO_2 sink due to recent climate change. Science 316, 1735–1738. http://dx.doi.org/10.1126/science.1136188.

Le Traon, P., Nadal, F., Ducet, N., 1998. An improved mapping method of multisatellite altimeter data. J. Atmos. Oceanic Technol. 15, 522–534. http://dx.doi.org/10.1175/1520-0426,(1998)015<0522:AIMMOM>2.0.CO;2.

Ledwell, J.R., Watson, A.J., Law, C.S., 1993. Evidence for slow mixing across the pycynocline from an open-ocean tracer-release experiment. Nature 364, 701–703. http://dx.doi.org/10.1038/364701a0.

Levitus, S., Antonov, J.I., Boyer, T.P., Baranova, O.K., Garcia, H.E., Locarnini, R.A., Mishonov, A.V., Reagan, J.R., Seidov, D., Yarosh, E.S., Zweng, M.M., 2012. World ocean heat content and thermosteric sea level change (0-2000). Geophys. Res. Lett. 39, L10603. http://dx.doi.org/10.1029/2012GL051106.

Lumpkin, R., Speer, K., 2007. Global ocean meridional overturning. J. Phys. Oceanogr. 37, 2550–2562.

Marshall, J., Radko, T., 2003. Residual-mean solutions for the Antarctic Circumpolar Current and its associated overturning circulation. J. Phys. Oceanogr. 33, 2341–2354.

Marshall, J., Speer, K., 2012. Closure of the meridional overturning circulation through Southern Ocean upwelling. Nat. Geosci. 5, 171–180.

Marshall, J., Shuckburgh, E., Jones, H., Hill, C., 2006. Estimates and implications of surface eddy diffusivity in the Southern Ocean derived from tracer transport. J. Phys. Oceanogr. 36, 1806–1821.

Mazloff, M.R., 2012. On the sensitivity of the Drake Passage transport to air–sea momentum flux. J. Clim. 25, 2279–2290.

Mazloff, M.R., Heimbach, P., Wunsch, C., 2010. An eddy-permitting Southern Ocean State Estimate. J. Phys. Oceanogr. 40, 880–899.

McCarthy, G., McDonagh, E., King, B., 2011. Decadal variability of thermocline and intermediate waters at 24°s in the South Atlantic. J. Phys. Oceanogr. 41, 157–165.

Meijers, A.J.S., Bindoff, N.L., Rintoul, S.R., 2011. Frontal movements and property fluxes; contributions to heat and freshwater trends in the Southern Ocean. J. Geophys. Res. 116, C08024. http://dx.doi.org/10.1029/2010JC006832.

Meredith, M.M., Hogg, A.M., 2006. Circumpolar response of Southern Ocean eddy activity to a change in the Southern Annular Mode. Geophys. Res. Lett. 33, L16608. http://dx.doi.org/10.1029/2006GL026499.

Meredith, M.P., Woodworth, P.L., Hughes, C.W., Stepanov, V., 2004. Changes in the ocean transport through Drake Passage during the 1980s and 1990s, forced by changes in the Southern Annular Mode. Geophys. Res. Lett. 31, L21305. http://dx.doi.org/10.1029/2004GL021169.

Meredith, M.P., Woodworth, P.L., Chereskin, T.K., Marshall, D.P., Allison, L.C., Bigg, G.R., Donohue, K., Heywood, K.J., Hughes, C.W., Hibbert, A., Hogg, A.M.C., Johnson, H.L., Jullion, L., King, B.A., Leach, H., Lenn, Y.-D., Morales Maqueda, M.A., Munday, D.R., Naveira Garabato, A.C., Provost, C., Sallée, J.-B., Sprintall, J., 2011. Sustained monitoring of the Southern Ocean at Drake Passage: past achievements and future priorities. Rev. Geophys. 49, RG4005. http://dx.doi.org/10.1029/2010RG000348.

Meredith, M.P., Naveira Garabato, A.C., Hogg, A.M.C., Farneti, R., 2012. Sensitivity of the overturning circulation in the Southern Ocean to decadal changes in wind forcing. J. Clim. 25, 99–110.

Morrison, A.K., Hogg, A.M.C., 2013. On the relationship between Southern Ocean overturning and ACC transport. J. Phys. Oceanogr. 43, 140–148.

Morrow, R., Ward, M.L., Hogg, A.M., Pasquet, S., 2010. Eddy response to Southern Ocean climate modes. J. Geophys. Res. Oceans 115, C10030. http://dx.doi.org/10.1029/2009JC005894.

Munday, D.R., Johnson, H.L., Marshall, D.P., 2013. Eddy saturation of equilibrated circumpolar currents. J. Phys. Oceanogr. 43, 507–532.

Munk, W., 1966. Abyssal recipes. Deep Sea Res. Oceanogr. Abstr. 13, 707–730.

Naveira Garabato, A.C., McDonagh, E.L., Stevens, D.P., Heywood, K.J., Sanders, R.J., 2002. On the export of Antarctic Bottom Water from the Weddell Sea. Deep Sea Res. Part II 49, 4715–4742.

Naveira Garabato, A.C., Polzin, K.L., King, B.A., Heywood, K.J., Visbeck, M., 2004. Widespread intense turbulent mixing in the Southern Ocean. Science 303, 210–213.

Naveira Garabato, A.C., Jullion, L., Stevens, D.P., Heywood, K.J., King, B.A., 2009. Variability of subantarctic mode water and Antarctic Intermediate Water in the Drake Passage during the late-twentieth and early-twenty-first centuries. J. Clim. 22, 3661–3688.

Naveira Garabato, A.C., Ferrari, R., Polzin, K.L., 2011. Eddy stirring in the Southern Ocean. J. Geophys. Res. 116, C09019. http://dx.doi.org/10.1029/2010JC006818.

Naveira Garabato, A.C., Nurser, A.G.J., Scott, R.B., Goff, J.A., 2013a. The impact of small-scale topography on the dynamical balance of the ocean. J. Phys. Oceanogr. 43, 647–668.

Naveira Garabato, A.C., Williams, A.P., Bacon, S., 2013b. The three-dimensional overturning circulation of the Southern Ocean during the WOCE era. Prog. Oceanogr., in press.

Niiler, P.P., Maximenko, N.A., McWilliams, J.C., 2003. Dynamically balanced absolute sea level of the global ocean derived from near-surface velocity observations. Geophys. Res. Lett. 30, 2164. http://dx.doi.org/10.1029/2003GL018628.

Nikurashin, M., Ferrari, R., 2013. Overturning circulation driven by breaking internal waves in the deep ocean. Geophys. Res. Lett. 40, 1–5. http://dx.doi.org/10.1002/grl.50542.

Nikurashin, M., Vallis, G., 2011. A theory of deep stratification and overturning circulation in the ocean. J. Phys. Oceanogr. 41, 485–502.

Nikurashin, M., Vallis, G., Adcroft, A., 2013. Routes to energy dissipation for geostrophic flows in the Southern Ocean. Nat. Geosci. 6, 48–51.

Olbers, D., Borowski, D., Völker, C., Wölff, J.-O., 2004. The dynamical balance, transport and circulation of the Antarctic Circumpolar Current. Antarct. Sci. 16, 439–470.

Olbers, D., Visbeck, M., 2005. A model of the zonally averaged stratification and overturning in the Southern Ocean. J Phys. Oceanogr. 35, 1190–1205. http://dx.doi.org/10.1175/JPO2750.1.

Olbers, D., Willebrand, J., Eden, C., 2012. Ocean Dynamics. Springer, Heidelberg, 704 pp.

Orsi, A.H., Whitworth III., T., Nowlin Jr., W.D., 1995. On the meridonal extent and fronts of the Antarctic Circumpolar Current. Deep Sea Res. Part I 42, 641–673.

Orsi, A.H., Smethie Jr., W.M., Bullister, J.L., 2002. On the total input of Antarctic waters to the deep ocean: a preliminary estimate from chlorofluorocarbon measurements. J. Geophys. Res. 107, 3122. http://dx.doi.org/10.1029/2001JC000976.

Parkinson, C.L., Cavalieri, D.J., 2012. Antarctic sea ice variability and trends, 1979–2010. The Cryosphere 6, 871–880. http://dx.doi.org/10.5194/tc-6-871-2012.

Polzin, K.L., Naveira Garabato, A.C., 2013. Boundary mixing in the Orkney Passage outflow. J. Geophys. Res., submitted.

Pritchard, H.D., Ligtenberg, S.R.M., Fricker, H.A., Vaughan, D.G., Van den Broeke, M.R., Padman, L., 2012. Antarctic ice-sheet loss driven by basal melting of ice shelves. Nature 484, 502–505. http://dx.doi.org/10.1038/nature10968.

Purkey, S.G., Johnson, G.C., 2010. Warming of global abyssal and deep southern ocean waters between the 1990s and 2000s: contributions to global heat and sea level rise budgets. J. Clim. 23, 6336–6351. http://dx.doi.org/10.1175/2010jcli3682.1.

Purkey, S.G., Johnson, G.C., 2012. Global contraction of Antarctic Bottom Water between the 1980s and 2000s. J. Clim. 25, 5830–5844. http://dx.doi.org/10.1175/JCLI-D-11-00612.1.

Purkey, S.G., Johnson, G.C., 2013. Antarctic Bottom Water warming and freshening: contributions to sea level rise, ocean freshwater budgets, and global heat gain. J. Clim. 26, 6105–6122. http://dx.doi.org/10.1175/JCLI-D-12-00834.1, in press.

Rhines, P.B., Young, W.R., 1982. Homogenizaton of potential vorticity in planetary gyres. J. Fluid Mech. 122, 347–367.

Rignot, E., Velicogna, I., Van den Broeke, M.R., Monaghan, A., Lenaerts, J., 2011. Acceleration of the contribution of the Greenland and Antarctic ice sheets to sea level rise. Geophys. Res. Lett. 38, L05503.

Rintoul, S.R., 2007. Rapid freshening of Antarctic Bottom Water formed in the Indian and Pacific Oceans. Geophys. Res. Lett. 34, L06606. http://dx.doi.org/10.1029/2006GL028550.

Rintoul, S.R., Hughes, C., Olbers, D., 2001. The Antarctic Circumpolar Current system. In: Siedler, G., Church, J., Gould, J. (Eds.), Ocean Circulation and Climate. Academic Press, London, pp. 271–302.

Rintoul, S.R., Sokolov, S., Church, J., 2002. A six year record of baroclinic transport variability of the Antarctic Circumpolar Current at 140°E, derived from XBT and altimeter measurements. J. Geophys. Res. Oceans 107, 3155. http://dx.doi.org/10.1029/2001JC000787.

Sabine, C.L., Feely, R.A., Gruber, N., Key, R.M., Lee, K., Bullister, J.L., Wanninkhof, R., Wong, C.S., Wallace, D.W.R., Tilbrook, B., Millero, F.J., Peng, T.H., Kozyr, A., Ono, T., Rios, A.F., 2004. The oceanic sink for anthropogenic CO_2. Science 305, 367–371.

Sallée, J.-B., Speer, K., Rintoul, S.R., Wijffels, S., 2010a. Southern Ocean thermocline ventilation. J. Phys. Oceanogr. 40, 509–529. http://dx.doi.org/10.1175/2009JPO4291.1.

Sallée, J.-B., Speer, K., Rintoul, S.R., 2010b. Zonally asymmetric response of the Southern Ocean mixed-layer depth to climate variability. Nat. Geosci. 3, 273–279. http://dx.doi.org/10.1038/ngeo812.

Sallée, J.B., Matear, R., Rintoul, S.R., Lenton, A., 2012. Surface to interior pathways of anthropogenic CO_2 in the southern hemisphere oceans. Nat. Geosci. 5, 579–584. http://dx.doi.org/10.1038/ngeo1523.

Schmidtko, S., Johnson, G.C., 2012. Multi-decadal warming and shoaling of Antarctic Intermediate Water. J. Clim. 25, 201–221. http://dx.doi.org/10.1175/JCLI-D-11-00021.1.

Schmitz, W.J., 1996. On the world ocean circulation: volume I. Some global features/North Atlantic circulation. Woods Hole Oceanographic Institution Technical Report. WHOI-96-03, Woods Hole, MA, 141 pp.

Shepherd, A., Wingham, D., Rignot, E., 2004. Warm ocean is eroding West Antarctic ice sheet. Geophys. Res. Lett. 31, L23402. http://dx.doi.org/10.1029/2004gl021106.

Shimada, K., Aoki, S., Oshima, K.I., Rintoul, S.R., 2012. Influence of the Ross Sea Bottom Water changes on the warming and freshening of the Antarctic Bottom Water in the Australian–Antarctic Basin. Ocean Sci. 8 (419–432), 2012.

Sigman, D.M., Hain, M.P., Haug, G.H., 2010. The polar ocean and glacial cycles in atmospheric CO_2 concentration. Nature 466, 47–55. http://dx.doi.org/10.1038/nature09149.

Sloyan, B.M., Rintoul, S.R., 2001. The Southern Ocean limb of the global deep overturning circulation. J. Phys. Oceanogr. 31, 143–173.

Sokolov, S., Rintoul, S.R., 2007. Multiple jets of the Antarctic Circumpolar Current south of Australia. J. Phys. Oceanogr. 37, 1394–1412.

Sokolov, S., Rintoul, S.R., 2009a. The circumpolar structure and distribution of the Antarctic Circumpolar Current fronts. Part 1: mean circumpolar paths. J. Geophys. Res. Oceans 114, C11018. http://dx.doi.org/10.1029/2008JC005108.

Sokolov, S., Rintoul, S.R., 2009b. The circumpolar structure and distribution of the Antarctic Circumpolar Current fronts. Part 2: variability and relationship to sea surface height. J. Geophys. Res. Oceans 114, C11019. http://dx.doi.org/10.1029/2008JC005248.

Speer, K., Rintoul, S.R., Sloyan, B., 2000. The diabatic Deacon cell. J. Phys. Oceanogr. 30, 3212–3222.

Stammerjohn, S., Massom, R., Rind, D., Martinson, D., 2012. Regions of rapid sea ice change: an inter-hemispheric seasonal comparison. Geophys. Res. Lett. 39, L06501. http://dx.doi.org/10.1029/2012GL050874.

Straub, D.N., 1993. On the transport and angular momentum balance of channel models of the Antarctic Circumpolar Current. J. Phys. Oceanogr. 23, 776–782. http://dx.doi.org/10.1175/1520-0485.

Swart, S., Speich, S., Ansorge, I.J., Goni, G.J., Gladyshev, S., Lutjeharms, J.R.E., 2008. Transport and variability of the Antarctic Circumpolar Current south of Africa. J. Geophys. Res. Oceans 113, C09014. http://dx.doi.org/10.1029/2007JC004223.

Takahashi, T., Sweeney, C., Hales, B., Chipman, D.W., Newberger, T., Goddard, J.G., Iannuzzi, R.A., Sutherland, S.C., 2012. The changing carbon cycle in the Southern Ocean. Oceanography 25 (3), 26–37. http://dx.doi.org/10.5670/oceanog.2012.71.

Talley, L.D., 2013. Closure of the global overturning circulation through the Indian, Pacific, and Southern Oceans: schematics and transports. Oceanography 26 (1), 80–97. http://dx.doi.org/10.5670/oceanog.2013.07.

Tansley, C.E., Marshall, D.P., 2001. On the dynamics of wind-driven circumpolar currents. J. Phys. Oceanogr. 31, 3258–3273. http://dx.doi.org/10.1175/1520-0485(2001).

Thomas, L.N., 2005. Destruction of potential vorticity by winds. J. Phys. Oceanogr. 35, 2457–2466.

Thompson, A.F., 2010. Jet formation and evolution in baroclinic turbulence with simple topography. J. Phys. Oceanogr. 40, 257–278.

Thompson, A.F., Naveira Garabato, A.C., 2013. Equilibration of the Antarctic Circumpolar Current by standing meanders. J. Phys. Oceanogr, submitted.

Thompson, A.F., Sallée, J.B., 2012. Jets and topography: jet transitions and the impact on transport in the Antarctic Circumpolar Current. J. Phys. Oceanogr. 42, 956–972.

Thompson, A.F., Haynes, P.H., Wilson, C., Richards, K.J., 2010. Rapid Southern Ocean front transitions in an eddy-resolving ocean GCM. Geophys. Res. Lett. 37, L2360237.

Thompson, D.W.J., Solomon, S., Kushner, P.J., England, M.H., Grise, K.M., Karoly, D.J., 2011. Signatures of the Antarctic ozone hole in Southern Hemisphere surface climate change. Nat. Geosci. 4, 741–749. http://dx.doi.org/10.1038/ngeo1296.

Toggweiler, J.R., Samuels, B., 1998. On the ocean's large-scale circulation near the limit of no vertical mixing. J. Phys. Oceanogr. 28, 1832–1852. http://dx.doi.org/10.1175/1520-0485(1998)028<1832:OTOSLS>2.0.CO;2.

Turner, J., Comiso, J.C., Marshall, G.J., Lachlan-Cope, T.A., Bracegirdle, T.J., Maksym, T., Meredith, M.P., Wang, Z., Orr, A., 2009. Non-annular atmospheric circulation change induced by stratospheric ozone depletion and its role in the recent increase of Antarctic sea ice extent. Geophys. Res. Lett. 36, L08502. http://dx.doi.org/10.1029/2009GL037524.

van Wijk, E.M., Rintoul, S.R., 2013. Freshening drives contraction of Antarctic Bottom Water in the Australian Antarctic Basin. Geophys. Res. Lett, submitted.

Waterman, S.N., Naveira Garabato, A.C., Polzin, K.L., 2013. Internal waves and turbulence in the Antarctic Circumpolar Current. J. Phys. Oceanogr. 43, 259–282.

Waugh, D.W., Primeau, F., DeVries, T., Holzer, M., 2013. Recent changes in the ventilation of the southern oceans. Science 339, 568–570. http://dx.doi.org/10.1126/science.1225411.

Webb, D.J., de Cuevas, B.A., 2007. On the fast response of the Southern Ocean to changes in the zonal wind. Ocean Sci. 3, 417–427.

Webb, D.J., Killworth, P.D., Coward, A.C., Thompson, S.R., 1991. The FRAM Atlas of the Southern Ocean. Natural Environment Research Council, Swindon, 67 pp.

Wolfe, C.L., Cessi, P., 2010. What sets the strength of the mid-depth stratification and overturning circulation in eddying ocean models? J. Phys. Oceanogr. 40, 1520–1538.

Zickfeld, K., Fyfe, J.C., Eby, M., Weaver, A.J., 2008. Comment on "Saturation of the Southern Ocean CO_2 sink due to recent climate change" Science 319, 570b.

Zika, J.D., Le Sommer, J., Dufour, C.O., Naveira Garabato, A.C., Blaker, A., 2013a. Acceleration of the Antarctic Circumpolar Current by wind stress along the coast of Antarctica. J. Phys. Oceanogr, in press.

Zika, J.D., Le Sommer, J., Dufour, C.O., Molines, J.-M., Barnier, B., Brasseur, P., Dussin, R., Iudicone, D., Lenton, A., Madec, G., Mathiot, P., Orr, J., Penduff, T., Shuckburgh, E., Vivier, F., 2013b. Vertical eddy fluxes in the Southern Ocean. J. Phys. Oceanogr. 43, 941–955. http://dx.doi.org/10.1175/JPO-D-12-0178.1.

Chapter 19

Interocean and Interbasin Exchanges

Janet Sprintall*, Gerold Siedler† and Herlé Mercier‡
*Scripps Institution of Oceanography, U.C. San Diego, La Jolla, California, USA
†Helmholtz Centre for Ocean Research Kiel (GEOMAR), Kiel, Germany
‡CNRS, Laboratoire de Physique des Océans, Plouzané, France

Chapter Outline

1. INTRODUCTION

The meridional overturning circulation (MOC) is characterized by a complex system of currents that facilitate interbasin exchanges within and between the global oceans. Dense water overflows across the sills of the Greenland–Scotland ridge and mixes with salty surface waters that have cooled to sink in the North Atlantic and form North Atlantic Deep Water (NADW) that penetrates southward along the western boundary and crosses the equator. Part of this deep water mixes with Antarctic Bottom Water (AABW) that sinks at specific sites around the continental margin of Antarctica. The mixing can occur either directly at depth or nearer the surface with NADW that has upwelled within the Antarctic divergence zone. The resulting bottom water spreads northward over deep passages to fill the abyssal basins of the Atlantic (via the Romanche Fracture Zone, Vema, and Hunter Channels), the Pacific (via Samoan and Wake Island Passages), and the Indian (via the Southwest Indian Ridge and Amirante Passage) Oceans and forms the lower limb of the overturning circulation. The remainder of the southward flowing deep water shoals through diapycnal mixing in the Southern Ocean and is converted by air–sea fluxes into intermediate and mode waters that also spread northward into the major ocean basins to form the upper overturning cell. In the Pacific, this upper limb eventually exits as a warm, relatively fresh current through the Indonesian seas into the Indian Ocean and then returns via the Agulhas leakage south of Africa back to the North Atlantic sinking regions.

Marginal seas play an important role in the MOC, either as regions where water mass formation directly forces the circulation like the Nordic Sea north of the Greenland–Scotland Ridge and the Labrador Sea, or as regions where water masses are formed which modify and mark the water masses in the MOC. These latter regions include the Mediterranean and Caribbean Seas adjacent to the North Atlantic, along with several seas that border the eastern boundary of Asia, of which the Okhotsk Sea is particularly important for water mass formation in the North Pacific. Because all these adjacent seas are semienclosed and mostly of relatively small size, they are frequently sites of intense air–sea exchange. Intra-basin exchanges with marginal seas, therefore, provide contrasting thermohaline fluxes that can potentially affect the strength, stability, and meridional distribution of the overturning circulation.

While the meridional circulation is thought to be the primary mechanism for the spreading of heat, salt, carbon, and other biogeochemical properties (e.g., Lumpkin et al., 2008), at subtropical and subpolar latitudes the distribution

Ocean Circulation and Climate, Vol. 103. http://dx.doi.org/10.1016/B978-0-12-391851-2.00019-2

is further enhanced by the basin-wide horizontal gyre circulations and their strong western boundary currents. In the Southern Ocean, the Antarctic Circumpolar Current (ACC) plays a critical role in the transportation of oceanic properties between the major ocean basins. Together, the meridional and vertical currents within the overturning circulation and the strong horizontal flows of the gyres, the ACC, and the interbasin exchanges provide a unifying concept of the three-dimensional, interconnected global ocean circulation.

The multiple streams and components, along with the complex physical processes that actively transform water masses and their properties along the various pathways, can make it difficult to quantitatively measure and understand the variability within the global circulation system. While there has been a recent concerted effort to establish trans-basin arrays to monitor the MOC, particularly in the North Atlantic (Rayner et al., 2011), the vast zonal width of the major ocean basins has historically made this logistically challenging. For this reason, narrower "choke points" at key locations, where there is interocean and/or interbasin exchange (Figure 19.1), have provided natural geographical constraints for more convenient monitoring of climate variability. In the Southern Ocean, Drake Passage and the Agulhas leakage around southern Africa are particularly attractive locations, since they fully constrain the meridional width of the ACC and provide direct linkages for the interocean exchange of the Pacific and Indian Oceans, respectively, with the Atlantic Ocean. Furthermore, the Drake Passage lies in the Southern Ocean along the "cold-water route" for the overturning pathway to the Atlantic, while the "warm-water route" follows the tropical Indonesian Throughflow (ITF) and returns via the Agulhas leakage (Gordon, 1986; Rintoul, 1991; Schmitz, 1995; Richardson, 2008). However, the relative contribution of each route, their variability, and distribution with depth has not as yet been adequately quantitatively observed and described. In the tropical oceans, the Indonesian seas provide the only low-latitude pathway for surface to intermediate waters to pass from the Pacific to Indian Oceans. Although the Indonesian archipelago is the largest in the world consisting of islands, sills, and basins of various widths and depths, recent observations suggest that much of the ITF variability in mass, heat, and freshwater can be captured by measurements within the two main inflow passages (Gordon et al., 2008; van Aken et al., 2009) and three relatively narrow passages that provide the main exit pathways into the southeast Indian Ocean (Sprintall et al., 2009). Semi-enclosed marginal seas similarly encompass a highly complex geometry where water masses are exchanged with the open ocean through a number of interconnected straits and sills (e.g., Okhotsk Sea) or through a single restricted channel or strait (e.g., Mediterranean Sea). In the lower limb of the overturning circulation, it is thought that roughly equivalent volumes of dense water sink as deep water in the North Atlantic and as bottom water around Antarctica (Orsi et al., 1999, 2002; Johnson et al., 2008a; Lumpkin et al., 2008).

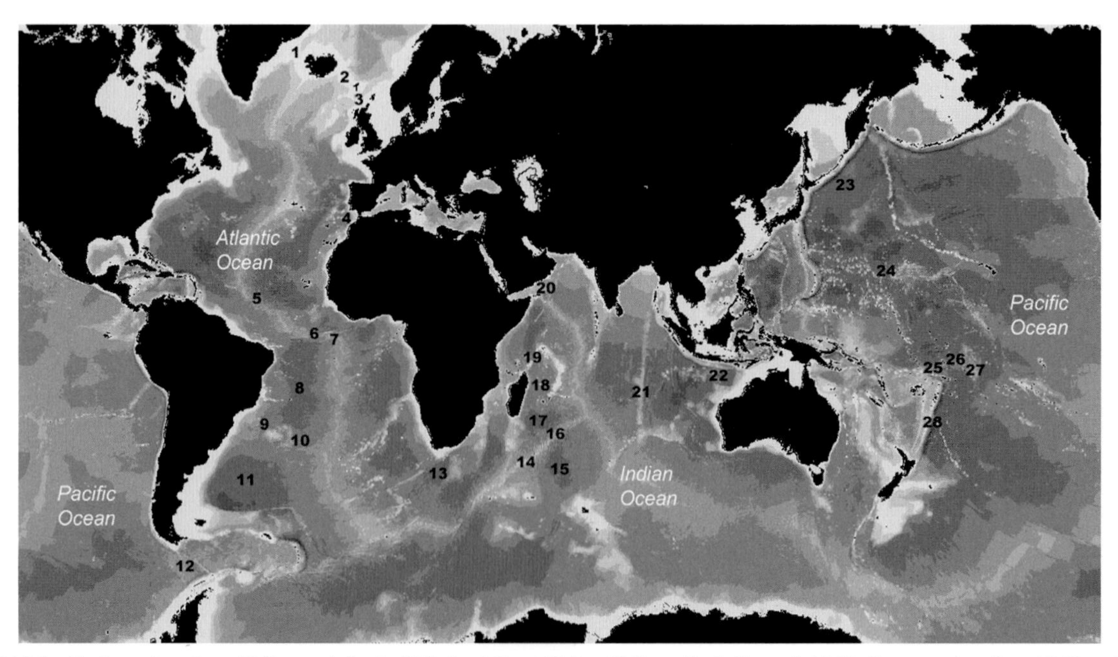

FIGURE 19.1 Exchange locations: (1) Denmark Strait, (2) Iceland–Faroe Ridge, (3) Faroe Bank Channel, (4) Mediterranean outflow, (5) Vema Fracture Zone, (6) Romanche Fracture Zone, (7) Chain Fracture Zone, (8) Brazil Basin, (9) Vema Channel, (10) Hunter Channel, (11) Argentine Basin, (12) Drake Passage, (13) Agulhas Retroflection, (14) Southwest Indian Ridge, (15) Crozet Basin, (16) Atlantis II Fracture Zone, (17) Madagascar Basin, (18) Mascarene Basin, (19) Amirante Passage, (20) Red Sea outflow, (21) Ninety East Ridge, (22) Indonesian Throughflow, (23) Okhotsk Sea outflow, (24) Wake Island Passage, (25) Robbie Ridge, (26), Samoan Passage, (27) Manihiki Plateau, (28) Tonga–Kermadec Ridge. Shallow to deep topography (IOC, IHO, and BODC, 2003) is indicated by light to dark spaces.

However, since only limited long-term direct velocity measurements exist in a few select deep passage choke points, the fate of the deep flows is poorly resolved once they circulate into the basin interiors.

The complex bathymetry and many unique dynamical processes associated with interocean and interbasin exchanges at the passages and choke points provide challenges for observations and models alike. A complicated system of currents may be driven by winds, buoyancy forcing through air–sea exchange, and by mixing due to winds, tides, and eddies. The pressure gradient force setup within the water column between two interconnected basins can drive the water above sill depth out through the passage into another basin. Deep interbasin exchanges occur through passages in ridges that are often narrow and shallow, which are favorable conditions for a hydraulic control of the flow. Whitehead (1998) identified some of the complex mechanisms that control the throughflow in the deep passages using a simple model of rotating critical flow. This model required knowledge of the bathymetry (sill depth) and hydrographic conditions (height of the upstream reservoir and density gradient between the deep controlled layer and the layer above) to predict throughflow transports under the assumption that the passage sills are control points. The flow transport was estimated below the bifurcation point, that is, the depth at which the density profiles of the neighboring basins diverge. The ratio of predicted to observed transports showed good agreement, ranging from 1 to 2.7. This suggests that control dynamics may play a role in the flow through deep passages (Whitehead, 1998). Several studies concur (e.g., Mercier and Morin, 1997 and Polzin et al., 1996 for the Romanche Fracture Zone; Hansen and Østerhus, 2007 for the Faroe Bank Channel; Dickson et al., 2008 for the Denmark Strait) and also show evidence for the critical role of the narrows (e.g., Hogg, 1983).

This chapter examines our present knowledge from observational evidence of interocean exchanges at choke points (Section 2), interbasin exchanges with marginal seas (Section 3), and overflows in deep passages (Section 4). As in the "Interocean" chapter of the first edition of this volume (Gordon, 2001), the three major interocean choke points are included: the Drake Passage (Section 2.1) and the Agulhas system (Section 2.2) at high latitudes representing the possible cold-water and warm-water routes for the thermohaline circulation, and the Indonesian Sea at tropical latitudes (Section 2.3). Further details of the Southern Ocean circulation can be found in Chapter 18 and of the role of the Agulhas Current as a western boundary current in Chapter 13. The cold, low-salinity interocean exchange between the Pacific Ocean and the Arctic through the Bering Strait is found in Chapter 17.

Among the range of adjacent seas, the Nordic and Mediterranean Seas are presented for the Atlantic Ocean (Sections 3.1 and 3.2). The western boundary flow through the Caribbean Sea and Gulf of Mexico is part of Chapter 13. The Red Sea outflow (Section 3.3) and the Okhotsk Sea (Section 3.4) are selected as important examples of interbasin exchange sites in the Indian and Pacific Oceans, respectively. Finally, deep passage overflows are discussed in the Atlantic (Section 4.1), the Pacific (Section 4.2), and the Indian (Section 4.3) Oceans.

2. INTEROCEAN EXCHANGES AT CHOKE POINTS

2.1. Drake Passage

Drake Passage is the narrowest constriction (~800 km) through which the ACC must pass on its global journey around the Southern Ocean (Figure 19.2). Dynamically the width of the Drake Passage gap is important as, in the absence of continental barriers and submerged topography, a zonal pressure gradient cannot be supported in the upper ocean at these latitudes. Thus the zonally integrated meridional geostrophic flow must vanish above the mid-depth topographic obstructions. This leads to a Southern Ocean overturning circulation that extends deep into the water column. The overturning cell is largely maintained by the strong westerly winds and buoyancy gradient across the ACC, and these mechanisms also play key roles in the formation and transformation of many global water masses. At present, determining the strength and intensity of the Southern Ocean meridional overturning circulation from observations can only be achieved through inverse techniques (e.g., Sloyan and Rintoul, 2001; Lumpkin and Speer, 2007). However, the width of Drake Passage also means that it has historically provided an ideal location for monitoring the ACC transport, water masses, and properties that participate within the meridional cells and the interbasin exchange of the global circulation. Because of the logistical convenience of access, many international bases are maintained on the Antarctic Peninsula, and the frequent visits by supply ships have long fostered many opportunities for oceanographic surveys. Drake Passage remains the only feasible location for the high-resolution, picket-fence, coherent moored array needed to capture the full width and depth of the ACC in the Southern Ocean. Such an array was implemented during the 1970s in the International Southern Ocean Studies (ISOS) program (Whitworth and Peterson, 1985) and more recently as part of DRAKE (Provost et al., 2011) and cDrake (Chereskin et al., 2009, 2012). This chapter briefly details recent science highlights from the various monitoring programs that have been sustained over the past decade(s) within Drake Passage. This long history of measurements makes Drake Passage the most observed region in the Southern Ocean and a recent study of these observations can be found in Meredith et al. (2011). Collectively the observations have provided much insight as to the mechanisms and impacts of

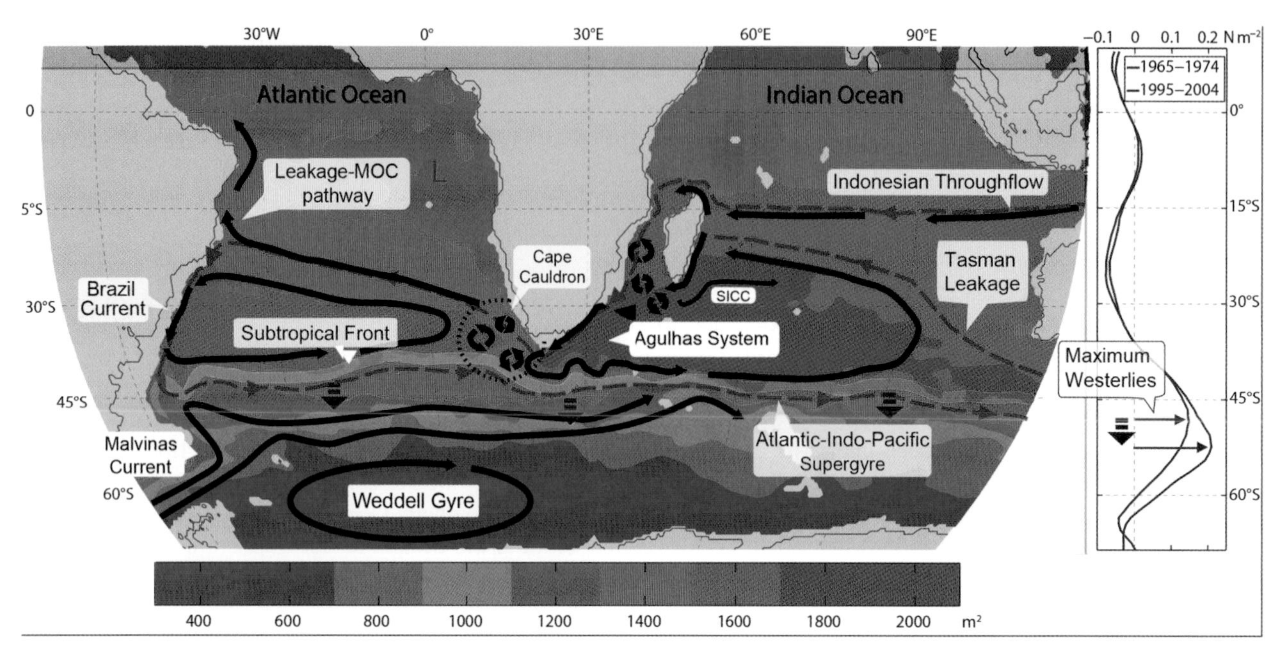

FIGURE 19.2 Schematic of Drake Passage inflow and the greater Agulhas system embedded in the Southern Hemisphere supergyre (after Beal et al., 2011). Background colors show the climatological dynamic height integrated between the surface and 2000 dbar. Black arrows illustrate the main features of the flow. The Cape Cauldron is the key Agulhas Current leakage area. The flow from this region to the northwest provides the upper part of the Meridional Overturning Circulation, with some return transport by the Brazil Current. The Southern Hemisphere supergyre is indicated by the gray-dashed line. The plot on the right shows the southward expansion of the Southern Hemisphere westerlies over a 30-year period, with the expected corresponding southward shift of the subtropical front being illustrated by red-dashed arrows. MOC, Meridional Overturning Circulation; SICC, South Indian Ocean Countercurrent.

the observed changes in both cells of the overturning circulation.

The upper limb of the overturning circulation in the Southern Ocean consists of the Subantarctic Mode Waters (SAMW) and Antarctic Intermediate Waters (AAIW) that act to ventilate the pycnocline of the global oceans (Gnanadesikan, 1999). The AAIW and SAMW originate from the cold, fresh surface waters of the Winter Water layer that lie south of the Polar Front. These surface waters are transported northward in the wind-driven Ekman layer and are then subducted at the Polar Front (AAIW) or by winter overturning on the equatorward flank of the ACC (SAMW). Although the smaller Ekman currents are notoriously difficult to distinguish from the background geostrophic circulation, high-resolution multi-year repeat measurements of the upper ocean velocity (Lenn et al., 2007) and temperature (Sprintall, 2003) profiles within Drake Passage enabled the first directly observed Ekman transport estimate in the Southern Ocean (Lenn and Chereskin, 2009). Of interest, the mean observed equatorward transport over the upper 100 m depth of $0.81 \pm 0.17\ m^2\ s^{-1}$ agreed within 95% confidence limits with the indirect Ekman transport inferred from the wind, while the mean temperature of the Ekman layer was no different from the surface temperature: this suggests that remotely sensed winds and SSTs might be used to provide mean estimates of the Ekman heat flux contribution of the shallow upper cell of the overturning circulation. In Drake Passage, both SAMW and AAIW enter primarily from the major source region associated with the deep winter mixed layers in the southeast Pacific Ocean (Dong et al., 2008; Hartin et al., 2011), although recent evidence suggests that both water masses may also be formed locally within the passage itself (Naveira Garabato et al., 2009). Properties and volumes of the SAMW and AAIW layers are dependent on variations in the winter-time air–sea turbulent heat fluxes and the freshwater fluxes (evaporation and sea-ice melt) within the Winter Water layer. As observed in other regions of the Southern Ocean (e.g., Downes et al., 2009), a recent freshening and warming has occurred in the AAIW and SAMW within Drake Passage (Naveira Garabato et al., 2009). Significant modulation can also occur on interannual to interdecadal timescales in response to the large-scale climate modes such as ENSO, the Southern Annular Mode, and the Interdecadal Pacific Oscillation (Gille, 2008; Sprintall, 2008; Naveira Garabato et al., 2009).

The lower limb of the overturning circulation in Drake Passage is dominated by the presence of AABW that floods the deep basins of the global abyssal oceans. AABW forms from the transformation of Warm Deep Water and Weddell Sea Deep Water in the Weddell Sea, and thus enters Drake Passage from the east (Gordon et al., 2001). Significant property variations in all these deep water masses have been observed on seasonal to interannual to decadal timescales from mooring and hydrographic data within the Weddell Sea (Naveira Garabato et al., 2002; Robertson et al.,

2002; Fahrbach et al., 2004) with recent trends pointing to a warming and salinification (Fahrbach et al., 2011). Within Drake Passage, discrete annual full-depth hydrographic surveys over the past 15 years (nominally along WOCE-designated transect SR-1) are not sufficiently temporally resolved to determine trends, although they point to stronger wind forcing in the northern Weddell gyre, leading to warmer AABW within Drake Passage on a relatively short timescale of 5 months (Jullion et al., 2010).

Understanding the dynamical structure of the Southern Ocean overturning circulation and the ACC momentum balance critically depends upon understanding the horizontal and vertical distribution of the eddy heat and momentum fluxes. Since there can be no mean meridional transport above bounding topography at the latitude band of Drake Passage, eddies are thought to play a dual role: first, the eddies carry the poleward heat flux needed to balance the heat lost to the atmosphere at higher latitudes with the Ekman heat transport carried equatorward; and second, the eddies furnish the downward energy flux that balances the wind-driven eastward momentum with the interfacial form stress (Bryden, 1979; de Szoeke and Levine, 1981; Johnson and Bryden, 1989, Hughes, 2005). Observations have provided conflicting evidence within this framework, and eddy heat and momentum fluxes vary widely with depth and location around the Southern Ocean (Sciremammano, 1979; Johnson and Bryden, 1989; Morrow et al., 1994; Phillips and Rintoul, 2000). Part of the challenge is that the flux calculations require observations of a long duration that are densely sampled in space and time in order to obtain statistically significant results. Using 7 years of high-resolution temperature and velocity profiles in the upper 250 m of Drake Passage, Lenn et al. (2011) found statistically significant stream-averaged cross-stream eddy momentum and heat fluxes. The eddies exchanged momentum with the mean Subantarctic Front (SAF) and Polar Front (PF), strengthening and sharpening the fronts, and acting to decelerate the flow in between. Correspondingly, elevated eddy kinetic energy (EKE) is also evident in this interfrontal zone in Drake Passage, with values around 600–800 $cm^2 s^{-2}$ in the surface layer (Lenn et al., 2007) and $\sim$200 $cm^2 s^{-2}$ at depth (Chereskin et al., 2009). The intensified EKE indicates significant eddy stirring in this zone, and since EKE is much smaller than the kinetic energy in the jets themselves (Firing et al., 2011), the intensified EKE suggests that the PF and SAF might present barriers to mixing (Ferrari and Nikurashin, 2010). The near-surface-observed eddy heat fluxes (Lenn et al., 2011) were poleward, surface intensified and large (–290 kW m^{-2})–an order of magnitude larger than the poleward eddy heat fluxes determined from ISOS moored observations deeper in the water column (Johnson and Bryden, 1989). In the momentum balance, Lenn et al. (2011) showed that the interfacial form stress, derived from the eddy heat fluxes in the SAF, had little depth dependence and balanced the wind stress. Furthermore, the vertical divergence of the eddy heat flux was only a magnitude more than the eddy momentum forcing, suggesting that this latter effect is not entirely negligible in the balance as historically assumed (Bryden and Heath, 1985; Morrow et al., 2004; Stammer and Theiss, 2004).

The substantial field effort associated with the 1-year ISOS deployment of current meter moorings during the 1970s in Drake Passage derived a canonical value for the ACC volume transport of $\sim$134 Sv (1 Sv = 10^6 $m^3 s^{-1}$) with a standard deviation of $\sim$10 Sv (Whitworth and Peterson, 1985). The veracity of this estimate has been confirmed with the mean transport estimated from 15 annually occupied full-depth hydrographic surveys of SR-1 in Drake Passage (Cunningham et al., 2003; Meredith et al., 2011). Firing et al. (2011) report an observed mean transport of 95 ± 2 Sv in the upper 1000 m (i.e., 71% of the full-depth 134 Sv estimate) from a 4.5-year record of shipboard ADCP velocity data collected on 105 year-round crossings of Drake Passage. The mean observed current speeds in the Drake Passage ACC jets vary slowly with depth and can be quite strong: velocities of 10–20 cm s^{-1} were observed at 1000 m depth (Firing et al., 2011) and also in current meter records only 50 m above the bottom (Chereskin et al., 2009). The strength of the deep currents has significance for the ACC momentum balance, suggesting that vorticity dissipation through bottom friction may play a more significant role than considered previously. Both the direct velocity measurements (Firing et al., 2011) and the reanalysis of the ISOS transports (Cunningham et al., 2003) find a significant fraction (45–50%) of the total variability of the ACC transport to be baroclinic, with perhaps increasing importance on interannual timescales (Meredith et al., 2004). Firing et al. (2011) also showed that the mean vertical structure of the observed ACC in Drake Passage was consistent with an equivalent-barotropic form (i.e., streamlines that are parallel with depth), although the vertical length scales varied across the passage. The equivalent-barotropic structure has important implications for both theories (e.g., Smith and Marshall, 2009) and numerical models (e.g., Killworth and Hughes, 2002) of ACC dynamics, and from an observational perspective, it also allows for the extrapolation of the remotely sensed surface dynamic topography to be mapped to the full water column depth to improve our understanding of the ACC vorticity balance (e.g., Hughes, 2005).

2.2. Agulhas System

The Agulhas Current is the strongest western boundary current in the southern hemisphere. Beal and Bryden (1999) obtained 73 Sv for the mean Agulhas Current transport at 32°S from direct observations. In comparison, volume transports for the other two western boundary currents in the southern hemisphere are only 22 Sv for the East Australian Current at 30°S (Mata et al., 2000) and range from 10 Sv at 23°S to 24 Sv at 36°S for the Brazil Current (Garzoli, 1993).

The greater Agulhas system (Lutjeharms, 2006) consists of the western boundary Agulhas Current, the retroflection region to the south of the Agulhas Bank, the eastward Agulhas Return Current between about 36° and 42°S, and the Agulhas leakage region southwest of the Cape of Good Hope (Figure 19.2). The westward leakage of eddies that spawn from the Agulhas Current provides the main source for the warm and salty water return flow from the South to the North Atlantic as part of the MOC (Gordon, 1985, 1986; Donners and Drijfhout, 2004). The total northward flow is contributed by both the inflow from the Indian Ocean via the "warm-water route," and the modification of the eastward flow through Drake Passage via the "cold-water route" that ultimately arrives in the Benguela region (Figure 19.2). Observations from subsurface floats and surface drifters give transports of 14–17 Sv for this northward flow (Richardson, 2007), resulting mostly from the Agulhas leakage. A part of this flow will, however, recirculate to the south in the South Atlantic subtropical gyre. Long-term changes in the Agulhas leakage can be expected to influence the MOC and cause corresponding variations in the global climate (e.g., Biastoch et al., 2008a,c). Identifying the possible cause of changes in the warm-water route requires an understanding of the western boundary Agulhas Current. This is discussed in Chapter 13.

The water masses in the Agulhas region have their sources both in the Indian and in the Atlantic Oceans (see Beal and Bryden, 1999). Indian Ocean water masses are found in the surface layer (Tropical Surface Water and the Subtropical Surface Water) and the main thermocline (South Indian Central). Below this there is interleaving of AAIW from the Atlantic Ocean and Red Sea Outflow Water (RSOW) from the Indian Ocean. At depth, the NADW flows from the Atlantic into the Indian Ocean. Gordon (1986) described the essential roles of thermocline water versus NADW in his pioneering publication on interocean exchange, and Beal (2009) provided details of the NADW flow.

The Agulhas Current is part of the South Indian Ocean subtropical gyre and driven by the annual wind stress curl regime (Matano et al., 1999). However, the gyre is also embedded in a Southern Hemisphere supergyre (Figure 19.2) that connects the South Atlantic, Indian, and Pacific Ocean subtropical gyres (Speich et al., 2002; Ridgway and Dunn, 2007). Observational studies of the Agulhas Current were summarized by Lutjeharms (2006, 2007). The Agulhas Current is fed by contributions from three sources: through the Mozambique Channel, the East Madagascar Current, and by recirculation in the South Indian Ocean subtropical gyre (Figure 19.2). Anticyclonic deep-reaching eddies migrating southward dominate the transport of mass and heat in the Mozambique Channel (Schouten et al., 2003; Harlander et al., 2009; Swart et al., 2010), and observations give a mean transport of 16.7 ± 3.1 Sv (Ridderinkhof et al., 2010). The lack of a persistent boundary current in the Mozambique Channel is due to topographic effects of the large island of Madagascar (Penven et al., 2006). A more persistent East Madagascar Current provides another portion of warm water originating from the equatorial region, but also serving as a source of the South Indian Ocean Countercurrent (Siedler et al., 2006; Palastanga et al., 2007). The eddy-dominated regime to the southwest of Madagascar and the Mozambique Channel (Quartly and Srokosz, 2004; Siedler et al., 2009) is the source region for the Agulhas Current.

At its western boundary, the South Indian Ocean subtropical gyre is controlled by the African continent that forces the flow southward. The continental shelf boundary ends at ~37°S, and so does not reach the zone of maximum westerly winds where the wind stress curl becomes zero (climatologically between 46 and 50°S). South of the southern tip of Africa, the current would tend to flow to the west if it were not for its inertia and effects of bottom topography forcing it to the south (de Ruijter et al., 1999). There, most of the flow turns eastward when approaching the latitude of zero wind stress curl. Increasing South Indian Ocean wind stress curl will cause a stronger Agulhas Current with higher inertia and lower leakage contributions to the Atlantic (van Sebille et al., 2009). The opposite effect will occur when the meridional extension of the Agulhas regime is increased by a displacement of the latitude of zero wind stress curl to the south on decadal or longer timescales, making it more difficult for the inertia-driven Agulhas Current to reach that latitude.

Ultimately, only a small part of the Agulhas Current reaches the Atlantic Ocean (Richardson, 2007). Processes in the retroflection region are most essential for the contribution of Indian Ocean Water to the Atlantic. When going through the retroflection, the Agulhas Current spawns large rings and generates eddies and meanders, with most moving into the South Atlantic. The dependence of the ring generation on eddies and the Natal pulse is still an unresolved problem (Biastoch et al., 2008b). In their study of the impact of wind-induced intensification of the South Equatorial Current from 1993 to 2009, Backeberg et al. (2012) showed that there was a resulting increase of eddy motion in the Mozambique Channel and south of Madagascar, that is, in the source region of the Agulhas Current. However, it was not possible to prove that the upstream changes had a direct impact on the retroflection and leakage. The Agulhas rings provide a key source of warm upper ocean and intermediate waters, which contribute to the northward return flow of the MOC in the South Atlantic. While the retroflection loop is usually located west of 20°E, an early retroflection sometimes occurs south of Port Elizabeth (Lutjeharms and van Ballegooyen, 1988), causing a shortcut to the Agulhas Return Current during a period of 2–3 months. Only minor seasonal variability was usually assumed in the Agulhas system. However, Dencausse et al. (2010) found a well-defined seasonal signal in the retroflection longitude from 1992 to 1997 during the first 5 years of a 13-year time series. In the Agulhas Current further upstream, Krug and Tournadre (2012), by using a method for isolating the core of the geostrophic current, were able to identify annual changes with the seasonal cycle accounting for up to 30% of the observed current variance.

Combining Lagrangian measurements with altimeter sea surface height data, Boebel et al. (2003) found a field of common anticyclonic rings with typical diameters of 200 km, and an energetic field of smaller anticyclonic and cyclonic eddies in the Cape Basin region (the "Cape Cauldron," Figure 19.2). Giulivi and Gordon (2006) show that there is a 2:1 anticyclone/cyclone ratio in this region. The cyclonic eddies are partly formed by vortex shedding when the Agulhas Current overshoots the western boundary of the Agulhas Bank (Penven et al., 2001). About half of the rings crossed 5°E to the northwest, while the others were entrained back into the Agulhas retroflection. The high eddy energy, particularly at intermediate depths, suggests intensive mixing processes in the Cape Basin, including a contribution of Drake Passage water arriving within the South Atlantic Current (Stramma and England, 1999; You, 2002).

Attempts have been made to estimate the volume transport of surface and intermediate Indian Ocean waters to the South Atlantic from observations and models (see de Ruijter et al., 1999). Hydrographic and tracer data lead to a wide range of transports between 2 and 10 Sv. Typically six rings per year each contribute 0.5–1.5 Sv. Coarse-resolution models, such as used for climate studies, do not allow for the highly nonlinear processes in the Agulhas retroflection and leakage region and usually give transports that appear too large (Biastoch et al., 2008c). Transports in better agreement with the observations are obtained using regional high-resolution eddy-resolving models and a Lagrangian diagnostic that seeds particles into the Agulhas Current. Doglioli et al. (2006) obtained an exchange transport of about 14 Sv with a regional model, while Biastoch et al. (2008b) found about 12 Sv when using a high-resolution model nested in a coarse-resolution global model. This value increased up to 17 Sv toward the 2000s concurrent with a southward migration of the westerlies (Biastoch et al., 2009). These numbers are within the range of observational results (see Richardson, 2007) and correspond to approximately one fifth of the Agulhas Current transport at 32°S (Beal and Bryden, 1999). While these transports suggest a dominance of the Agulhas warm water over the Drake Passage cold-water route, a final answer to this partitioning question still remains (Dong et al., 2011).

Is there any evidence that Agulhas leakage variability will have a distinct influence on the strength of the MOC? Strong variability of the MOC on decadal scales has been observed (Latif et al., 2006) and was related to changes in convection in the subarctic North Atlantic. Biastoch et al. (2008a,c) used the combination of coarse-resolution global models and an embedded high-resolution nested regional model to compare decadal MOC variability due to subarctic convection and that due to changes in the Agulhas leakage. They found that the MOC decadal variability is dominated by subarctic origin in the northern hemisphere and by Agulhas origin in the southern hemisphere. However, both effects are of the same order of magnitude at tropical/subtropical latitudes in the North Atlantic. It follows then that both effects need be considered when interpreting decadal-scale MOC variability in the subtropical North Atlantic.

There is evidence for a poleward shift and intensification of the westerly wind maximum south of Africa (see Figure 19.2, right side), and a simultaneous warming of the Agulhas region since the 1980s that is related to an increase in Agulhas Current transport (Rouault et al., 2009). High-resolution nested model results show a simultaneous enlargement in the Agulhas leakage with increased salt transport to the South Atlantic (Biastoch et al., 2009). Model studies demonstrated that changes in the wind field also induced a southward intensification of the supergyre (Cai, 2006). The effect of increasing salinization of the South Atlantic may potentially counteract the observed dilution of the North Atlantic over the past decades, which is considered as an indication of a slowdown in the MOC due to greenhouse warming (Curry and Mauritzen, 2005). On yet longer timescales, paleoceanographic records from deep-sea cores provide information of the near-surface water masses for more than the past 500,000 years (e.g., Peeters et al., 2004). Increased Agulhas leakage usually occurred when the glacial ice coverage was at a maximum, although still increasing after the last ice age (approx. 18,000 years ago), and it was suggested that the Agulhas leakage was critical for the resumption of the MOC after such glacial ice periods (Weijer et al., 2002).

2.3. Indonesian Throughflow

The ITF is the leakage of water from the western Pacific into the eastern tropical Indian Ocean through the Indonesian archipelago, driven by the pressure head built up in the western Pacific by the easterly trade winds (Wyrtki, 1961, 1987). The relationship between the large-scale Pacific winds and the ITF can also be theoretically explained through a simple linear formula, known as the Island Rule (Godfrey, 1989). The ITF is the only place in the global ocean circulation where warm tropical water is transferred between basins, and this water has to return to the other ocean basins at a cooler temperature. This warmer and fresher water from the Pacific is mixed and transformed into the unique Indonesian tropical stratification within the internal seas. After exiting into the Indian Ocean, the ITF is recognizable as a distinct low-salinity tongue of surface to thermocline water (Gordon et al., 1997) and a separate salinity minimum/silica maximum core at intermediate depths (Talley and Sprintall, 2005) across the entire Indian Ocean basin. Models show that the interocean exchange of properties substantially impacts the heat and freshwater budgets along with the atmosphere–ocean coupling, not just in the connected tropical Indian and Pacific Oceans, but on a global scale (e.g., McCreary and Lu, 2001; Lee et al., 2002, Vranes et al., 2002, Song et al., 2004; Potemra and Schneider, 2007a; Song et al., 2007).

Within the Indonesian seas, the thousands of islands and numerous passages connect a series of large, deep basins, so that the ITF actually consists of several filaments of flow that occupy different depth levels (Figure 19.3). The primary source of the surface to thermocline waters in the ITF is from the North Pacific entering via Makassar Strait, while the lower thermocline and intermediate water masses are drawn from the South Pacific entering across the deeper Lifamatola Passage (Gordon, 1986). Secondary ITF portals enter the Indonesian seas via the western Pacific marginal seas, such as the South China Sea, and although relatively shallow, these portals can provide a significant source of freshwater that influences the ITF stratification (Fang et al., 2005, 2010; Tozuka et al., 2007) and results in a subsurface ITF maximum during winter in Makassar Strait (Gordon et al., 2003). During transit, the Pacific temperature and salinity stratification are mixed and modified by the strong air–sea fluxes, monsoonal wind-induced upwelling, and extremely large tidal forces in the internal seas (e.g., Ffield and Gordon, 1996; Gordon and Susanto, 2001; Koch-Larrouy et al., 2007, 2008, 2010). The enhanced mixing not only impacts the ITF water mass properties, but also the SST distribution that in turn may modulate air–sea interaction and the monsoonal response (McBride et al., 2003; Jochum and Potemra, 2008; Kida and Wijffels, 2012), as well as upwell nutrients that influence regional primary productivity (Moore et al., 2003). To date, the spatial and temporal patterns of turbulent mixing within the internal Indonesian seas are still poorly understood, yet a quantitative grasp of the associated small-scale processes is needed to properly model the regional circulation, and its role in the climate and marine ecosystems.

Historical estimates of the mean ITF mass transport were obtained at different times and in different channels and over different phases of the ENSO cycle, and were thus wide ranging (Figure 19.3). The recent International Nusantara Stratification and Transport (INSTANT) program provided the first simultaneous measurements within both the major inflow and outflow passages (Sprintall et al., 2004; Gordon et al., 2010). INSTANT observations infer a three-year (2004–2006) mean ITF transport via the inflow passages of 11.6 Sv for Makassar Strait (Gordon et al., 2008) and ~1 Sv for Lifamatola Passage that was only resolved from ~200 m depth to the bottom (van Aken et al., 2009). The outflow transports within the three primary passages along

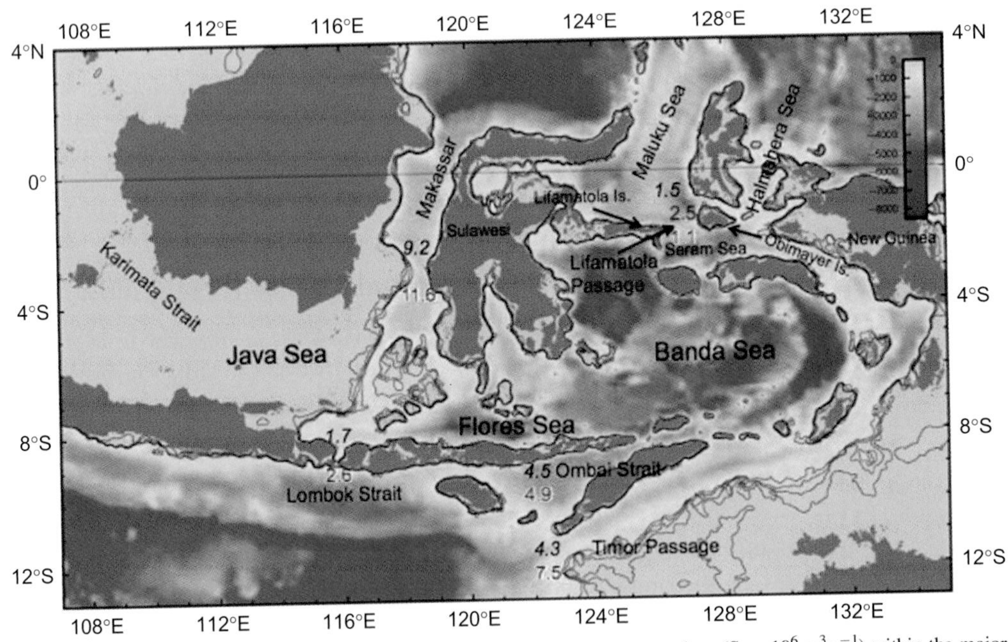

FIGURE 19.3 Major seas and straits of the Indonesian archipelago. Also shown are transport values (Sv $= 10^6$ m^3 s^{-1}) within the major passages of the Indonesian Throughflow: Makassar, Lifamatola, Ombai, Timor, and Lombok Straits. Black transport values are based on data collected prior to the INSTANT program (2003) for full-depth estimates in Makassar (Gordon et al., 1999), Lombok (Murray and Arief, 1988), Timor (Molcard et al., 1996), and Ombai (Molcard et al., 2001) Straits. Lifamatola Passage represents the overflow of dense water at depths greater than 1500 m based on 3.5 months of current meter measurement in early 1985 (van Aken et al., 1988). The red numbers are the 2004–2006 3-year mean transports measured by INSTANT (Gordon et al., 2008; Sprintall et al., 2009). In Lifamatola Passage, the green number is the INSTANT overflow transport >1250 m depth, representing the overflow into the deep Seram and Banda Seas, and the red number is the total transport measured by INSTANT below 200 m (van Aken et al., 2009). *After Gordon et al. (2010).*

the Nusa Tengarra archipelago were 2.6 Sv for Lombok Strait, 4.9 Sv for Ombai Strait, and 7.5 Sv for Timor Passage (Sprintall et al., 2009). The difference between the measured inflow and outflow over the 3-year period is attributed to the missing near-surface layer flow through Lifamatola Passage and unresolved flow through other channels, such as from the South China Sea via Karimata Strait (Gordon et al., 2010). Interestingly, the mean ITF agrees extremely well with the 16 Sv ITF calculated using the Island Rule by Godfrey (1989), which for many years was thought to be an overestimate.

The INSTANT measurements revealed a number of major new features of the ITF that had previously been undocumented. The transport in Timor Passage is nearly twice as large as historical estimates and represents half of the 15 Sv full-depth ITF transport that enters the Indian Ocean, which itself is about 25–30% greater than the historical nonsynoptic measurements (Sprintall et al., 2009). The semiannual signal in ITF transport is much more dominant than previously thought (relative to the annual signal). Maximum ITF occurs during the longer southeast monsoon of July–September (Gordon et al., 2008; Sprintall et al., 2009). During the northwest monsoon of November–January, the surface transport relaxes in Timor Strait and weakly reverses in Ombai, Lombok, and Makassar Strait, so the main core of the ITF is subsurface. Below the thermocline, semiannual reversals occur during the monsoon transitions in response to the passage of equatorial Indian Ocean wind-forced Kelvin waves (Wyrtki, 1973). These Kelvin waves can also occur on intraseasonal timescales in response to Madden Julian Oscillation wind fluctuations and correspondingly drive transport, thermocline, and temperature anomalies in the ITF exit passages (Drushka et al., 2010) and up into Makassar Strait (Pujiana et al., 2010). The transport changes in response to Kelvin waves occur over different depth levels in each passage reflecting the influence of different sill depths along the coastal waveguide. The seasonal cycle of depth-integrated transports in Lombok and Ombai Straits is strongly out of phase with Timor Passage, suggesting that the subthermocline flow is largely gated by these Kelvin waves (Sprintall et al., 2009).

It is largely the vertical distribution of transport through each passage that sets the heat flux carried by the ITF. Earlier estimates of the heat transport from inverse models and climatology ranged from 0.5 to 1.0 PW, as the ITF was assumed to be surface intensified and reflect the relatively warm temperatures found in the western Pacific source waters. Since full-depth temperature measurements are as yet unavailable in the ITF passages, heat fluxes are frequently represented as the effective temperature of the total throughflow, that is, the transport-weighted temperature (TWT). In Makassar Strait, earlier moored measurements from 1996 to 1998 (Gordon et al., 1999) led to a TWT of 14.7 °C to 16.3 °C (Vranes et al., 2002), while during INSTANT, the Makassar Strait TWT was 15.6 °C (Gordon et al., 2008). In the outflow passages, Lombok carries the warmest ITF (21.5 °C) due to its shallow sill (~300 m), Timor the next warmest (17.8 °C) because it is more surface intensified, and Ombai the coldest (15.2 °C) because the maximum ITF occurs at depth (Sprintall et al., 2009). A 20-year mean TWT of 21.2 °C was determined in the upper 750 m along a frequently repeated XBT transect near where the ITF enters into the southeast Indian Ocean (Wijffels et al., 2008). The warmer temperatures in this region can be reconciled with the much cooler Makassar Strait estimates by accounting for an unmeasured barotropic and deeper component from the Indonesian Intermediate Water (a significant contributor in Ombai Strait), and local surface air–sea heat fluxes that warm the ITF during its passage through the Banda Sea (Wijffels et al., 2008; Gordon et al., 2010). Because of the lack of subsurface salinity measurements, the freshwater flux from the ITF to the Indian Ocean is even less well constrained than the heat flux. However, since the Pacific waters entering the Indonesian seas are fresher than the inflow northward from the Southern Ocean into the Pacific, the ITF probably represents a net transport of freshwater into the Indian Ocean, where the salinity gradually increases before export into the Atlantic Ocean via the Agulhas system (Talley, 2008).

Many fundamental questions about the structure and variability of the ITF on climate timescales remain unanswered. Equatorial and coastal waveguides in the tropical Pacific and Indian Ocean and the Indonesian Archipelago allow the ITF to be modulated by climate variability from both the Pacific and Indian Ocean (e.g., Meyers, 1996; Masumoto, 2002; Wijffels and Meyers, 2004; Cai et al., 2005; McClean et al., 2005; Potemra and Schneider, 2007b). During El Niño, when the tropical easterly Pacific winds relax, the thermocline cooling and shoaling response in the western Pacific extend into the Indonesian seas, and the ITF is thought to cool and weaken (Ffield et al., 2000; Tillinger and Gordon, 2010). The opposite conditions prevail during La Niña. The Indian Ocean Dipole (IOD) also appears to influence the ITF transport in both the inflow and outflow straits through the passage of Kelvin waves driven by IOD-related wind anomalies in the equatorial Indian Ocean. A strong positive phase of the IOD in late 2006 appeared to strengthen the ITF transport through the outflow passages (Sprintall et al., 2009), hence modulating the expected influence from the concurrent 2006 El Niño. In contrast, the ITF transport was reduced in the surface layer but enhanced at depth during the 2004 El Niño, when the IOD was near zero. There is also evidence that ITF transport can vary by a few Sv on decadal timescales. For example, the weakening easterly trade winds associated with the Pacific "climate shift" in 1976/1977 (Vecchi et al., 2006) caused a 23% decrease of 2.5 Sv in ITF volume transport, a surface warming, and subsurface cooling in the main thermocline measured along the XBT line between Indonesia and Australia (Wainwright et al., 2008). Longer time series are needed to understand the links of the vertical ITF transport profile to the IOD and ENSO phenomena and decadal variability, and their implications for climate variations within the Australasian region.

3. INTERBASIN EXCHANGES

3.1. Nordic Seas—Atlantic Ocean

The circulation over the Greenland–Scotland Ridge permits mass, heat, and freshwater exchange between the northern North Atlantic and the Nordic (Greenland–Iceland–Norwegian) Seas. This circulation is composed of an inflow of warm and salty subtropical water carried through the North Atlantic into the Nordic Seas and an export of dense water by the Denmark Strait and Iceland–Scotland overflow (see Chapter 10). As noted above, it is a key component of the Atlantic MOC (Lumpkin and Speer, 2007).

The North Atlantic Current crosses the Greenland–Scotland Ridge in three main branches (Figure 19.4): the Shetland and Faroe Currents to the east of Iceland and the Icelandic Current to the west of Iceland (Hansen and Østerhus, 2000). The total transport of these North Atlantic Current branches was estimated at 8.5 Sv and is thought to be relatively stable at interannual to decadal timescales (Olsen et al., 2008). The North Atlantic Current branches northward through the Nordic Seas and the Arctic Ocean where the Atlantic water is cooled by intense heat loss to the atmosphere (see Chapter 17). This transformed Atlantic Water flows back into the southward East Greenland Current as Return Atlantic Water (Mauritzen, 1996; Hansen and Østerhus, 2000). The East Greenland Current entrains Greenland Sea Water such that both the Return Atlantic Water and the Greenland Sea Water are constituents of the Denmark Strait overflow (Köhl et al., 2007; Eldevik et al., 2009). The part of the Return Atlantic Water that does not cross the Denmark Strait flows to the Norwegian Sea, where it entrains

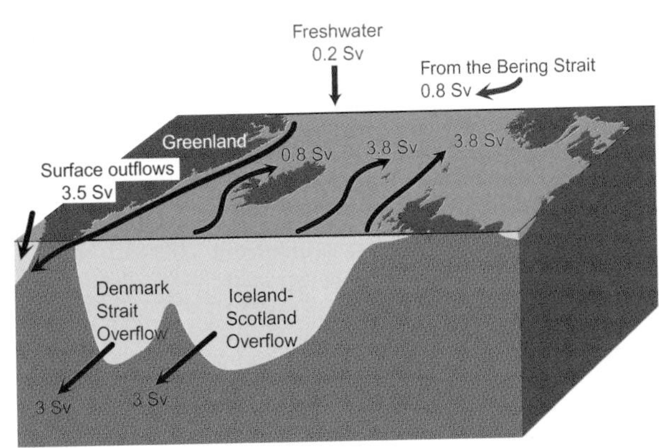

FIGURE 19.4 Volume exchanges across the Greenland–Scotland Ridge. *From Hansen et al. (2008).*

Norwegian North Atlantic Water before leaving the Nordic Seas through the Faroe–Scotland Channel and the Faroe Bank Channel (Hansen and Østerhus, 2000). Additional minor contributions to the dense water flow over the Iceland–Scotland Ridge were also reported over the Iceland–Faroe Ridge (Hansen and Østerhus 2000) and over the Wyville–Thomson Ridge from the Faroe–Shetland Channel (Hansen and Østerhus, 2000; Shervin et al., 2008). The total transport of the overflows is estimated at about 6 Sv with no significant decadal trend (Dickson et al., 2008; Østerhus et al., 2008). Collectively, taking into account the ~3.5 Sv flowing southward on the East Greenland Shelf and through the Davis Strait west of Greenland, most of the 8.5 Sv of the inflow from the Atlantic flows back to the Atlantic after being cooled in the Nordic Seas.

The overflow waters at the Greenland–Scotland sills are cold ($\theta_{min} < 0$ °C) and relatively fresh ($S_{min} \sim 34.9$) and are most often defined as waters denser than 27.80 σ_θ. In the past few decades, the overflow waters showed significant property variability and Dickson et al. (2002) reported a general decrease in the salinity of the overflow water since the early 1970s. Since then, the properties and transports of the overflows at the sills have been the subject of much attention due to the possibility of an abrupt change in the Atlantic MOC in response to the increased freshwater flux into the Nordic Seas driven by global warming (Rahmstorf, 1996). Curry and Mauritzen (2005) noted that the overflow freshening was concomitant with freshening of the Atlantic subpolar gyre and most likely related to the shift between a negative and a positive phase of the North Atlantic Oscillation that prevailed in the 1970s until the mid-1990s. In the early 2000s, when the North Atlantic Oscillation was in a negative/neutral phase, the freshening of the overflows stopped and an increase in the salinity of the overflows occurred concomitant with a salinification of the subpolar gyre (Hátún et al., 2005; Eldevik et al., 2009; Sarafanov et al., 2010).

The overflow water at the Denmark Strait sill, monitored between 1999 and 2002 by three bottom-mounted Acoustic Doppler Current Profilers, showed a transport decrease from 3.7 to 3.1 Sv (Macrander et al., 2005). In their analysis, the thickness of the overflow layer was computed by estimating the upper boundary of the overflow as the depth of maximum current shear (the "kinematic overflow"). The overflow bottom temperature exhibited large variability, with a temporary increase of 0.5 °C from the beginning of the record in 1999 until mid-2002, followed by a decrease of similar amplitude (Macrander et al., 2005).

The main pathway for the Iceland–Scotland Overflow is through the Faroe Bank Channel. A mean transport was estimated at 2.1 ± 0.2 Sv from direct current measurements over 1995–2005. Seasonal and interannual variability represented 10% of the mean value and there was no evidence of a significant trend over the deployment period (Hansen and Østerhus, 2007). This transport value, corresponding to the total southward flow (i.e., the "kinematic overflow"), reduced to 1.9 ± 0.3 Sv if only water denser than 27.80 σ_θ was considered. In addition, ~1 Sv flowed over the Iceland–Faroe Ridge (Perkins et al., 1998) and ~0.3 Sv across the Wyville–Thomson Ridge (Shervin et al., 2008).

The interannual variability of the Greenland–Scotland overflow transports was further discussed by Dickson et al. (2008) and Olsen et al. (2008) using numerical simulations that satisfactorily reproduced the observations. The simulations showed that the Faroe Bank Channel overflow transport as well as the total overflow transport between Greenland and Scotland did not decrease significantly over the 1948–2005 period (Olsen et al., 2008).

Downstream of the sills, the cascading overflows entrain Atlantic waters, producing the Iceland Scotland Overflow Water (ISOW) and the Denmark Strait Overflow Water (DSOW). As a consequence of the entrainment, the transports below $\sigma_\theta = 27.80$ increase. The ISOW flows along the eastern flank of the Reykjanes Ridge, crosses the Mid-Atlantic Ridge at the Charlie Gibbs Fracture Zone, and joins the DSOW in the Deep Western Boundary Current (DWBC) that flows east of Greenland in the Irminger Sea. The absolute DWBC ($\sigma_\theta > 27.80$) transport at Cape Farewell (south of Greenland) was estimated at ~9 Sv in the early 2000s from hydrographic measurements by Lherminier et al. (2007), and at 7.8 Sv by Bacon and Saunders (2010) from a 9-month current meter array deployed in 2005. The baroclinic component of the DWBC transport at Cape Farewell exhibits a large decadal variability (Bacon, 1998; Sarafanov et al., 2009). Sarafanov et al. (2009) diagnosed an increasing DWBC transport of 2.1 ± 0.7 Sv between 1994 and 2007 and noted that the decadal variability was (1) not related to that of the overflow transports and (2) anticorrelated to the thickness of Labrador Sea Water. This issue clearly deserves further attention.

The DWBC flows around the Labrador Sea and in so doing, incorporates Labrador Sea Water. The Labrador Sea Water and the Overflow Waters form the two main components of the NADW that is transported southward toward the equator, the South Atlantic, and the ACC before being distributed throughout the world ocean. Several interior pathways have been found for the NADW throughout the Atlantic: at Newfoundland (Bower et al., 2011), at the equator (Mercier and Morin, 1997), and at ~20°S (Speer et al., 1995). At the equator, the flow of the lower component of the NADW is through the Romanche and Chain Fracture Zone. At 20°S, the route for the NADW toward the Indian Ocean is through the fracture zones of the Mid-Atlantic Ridge and the Namib Col, a gap in the Walvis Ridge (Speer et al., 1995), and the eastern South Atlantic (Arhan et al., 2003).

3.2. Mediterranean Sea—Atlantic Ocean

The Mediterranean is an arid type of sea adjacent to the Atlantic, with the loss of water due to evaporation exceeding the combined influences of precipitation, river input, and the inflow of less saline Black Sea water. The long-term balances of water volume and salt content therefore require a typical flow pattern in the Strait of Gibraltar of a near-surface inflow of low-salinity Atlantic water and a somewhat smaller outflow of high-salinity Mediterranean water near the bottom. The details of Mediterranean Water formation in different subbasins were described by Candela (2001) in the first edition of this book and will not be discussed again here. For recent studies on the Mediterranean circulation, we refer the reader to Roether et al. (2007) and Bergamasco and Malanotte-Rizzoli (2010). Here we will examine changes in the Mediterranean Outflow Water (MOW; see Figure 19.1) that may influence the Atlantic intermediate and deep waters (Baringer and Price, 1999), and identify possible impacts on the deep water masses in the MOC.

Components from different water masses in the subbasins of the Mediterranean Sea contribute to the outflow. Hydrographic data from 1959 to 1997 in the western basin near the Strait of Gibraltar showed a trend of increasing temperature and salinity of approximately 0.03 °C and 0.01 per decade, respectively, below 800 m depth (Béthoux et al., 1998). However, observations at the Camarinal Sill in the Strait of Gibraltar indicated a much higher increase over the two decades up till the early 2000s, of about 0.15 °C and 0.03 per decade (Millot et al., 2006). The larger change is explained as a stronger contribution of waters from the eastern Mediterranean, providing a mixture of Levantine Intermediate Water and deeper water masses from the Adriatic and Aegean Seas that have higher temperature and salinity. However, Vargas-Yanez et al. (2010) argued, on the basis of an upper-ocean time series in the western Mediterranean, that part of the trend in the outflow water could also be caused by changes in the inflow water that in turn affect the Levantine Intermediate Water in the eastern Mediterranean.

After passing the Strait of Gibraltar, with a sill depth close to 300 m, the MOW cascades down the slope, leveling off at its local density level between 700 and 1400 m between the North Atlantic Central Water and the NADW. The Gulf of Cadiz is the first major site for entrainment and mixing of the MOW with the surrounding Atlantic water masses and thus provides the source region for the spreading of MOW into the North Atlantic. The topographic control of the outflow in this region, the branching toward the north and west off Cape St. Vincent, and Meddy formation have been the subject of numerous studies (e.g., Jungclaus and Mellor, 2000; Bower et al., 2002; Ambar et al., 2008). Fusco et al. (2008) determined trends at

1200 m depth in the Gulf of Cadiz and found that monthly mean temperature increased by 0.16 °C and salinity by 0.05 per decade. These values are quite similar to those identified above the Camarinal Sill by Millot et al. (2006) and support the argument for Mediterranean outflow changes as being the main cause.

Do these observed property changes have any influence on the large-scale property distributions in the Atlantic or impact the MOC? Although the outflow transport is only about 0.8 Sv (Sánchez-Román et al., 2009), the high temperatures and salinities in the MOW could make a difference with regard to Atlantic Basin property distributions. There are two possible processes to be considered in the source region of the Gulf of Cadiz. The first process is the entrainment of fresher North Atlantic Central Water (Baringer and Price, 1997) that acts to dilute the MOW and increase its transport by a factor of three. The second process is detrainment (Mauritzen et al., 2001) that leads to a cross-isopycnal flux of MOW to the Central Water. This flux causes higher salinities in the Central Water in addition to the effect of winter cooling in the North Atlantic that increases salinity at specific density levels (Pollard et al., 1996).

The MOW can be identified in the central Atlantic by its salinity maximum at approximately 1200 m depth. It is correlated with a temperature maximum and is largely density compensated. Arbic and Owens (2001) and earlier authors cited therein studied the interdecadal temperature variability in hydrographic sections and found the largest change occurring in the 1000–2000 db layer. They identified warming of about 0.5 °C per century south of 36°N for the period from the 1950s to the 1980s, with the warming trend being recognizable to 32°S in the South Atlantic. Levitus et al. (2000) had earlier identified the largest warming at intermediate depths and concluded that changes along isopycnals were more important than changes due to vertical displacement (see also Levitus et al., 2005). Potter and Lozier (2004) used hydrographic data to determine trends in the MOW layer in the eastern North Atlantic and found increases of about 0.1 °C and 0.03 per decade for temperature and salinity, respectively, for the period 1955–1993 (Figure 19.5). The corresponding heat gain in the MOW in the North Atlantic was about 2×10^{6} J m^{-3}, which is larger than the average gain in heat content of the North Atlantic Basin over the latter half of the twentieth century.

While all these results seem to indicate a correspondence of long-term changes in the Mediterranean outflow and changes in the North Atlantic intermediate waters, the study of Leadbetter et al. (2007) is seemingly contradictory. They compared three hydrographic sections at 36°N from 1959, 1981, and 2005 and found a cooling of intermediate waters across the whole North Atlantic averaged over the period 1959–2005 that is contrary to the warming for the period 1959–1981 (Roemmich and Wunsch, 1984). However, the eastern North Atlantic does not show this general cooling over the longer time interval, and the cooling signal could be primarily caused by changes in the Labrador Sea Water in the western North Atlantic.

There exist different ideas about the pathways of MOW in the North Atlantic, away from the source region in the Gulf of Cadiz. Reid (1979) suggested that the MOW flows northward and contributes to the waters of the Norwegian Current. However, three different models in the study of New et al. (2001) all found that the MOW does not upwell and contribute to the flow into the Nordic Seas. Lozier and Stewart (2008) were able to explain these seemingly conflicting results through temporal variability in the northward transport of the MOW. When the North Atlantic Oscillation Index (NAO) is low and the subpolar gyre is withdrawing to the west, there is evidence of MOW following the northward pathway. Conversely, when the NAO is high, the signal on the northward pathway disappears. Jia et al. (2007) also provided a model with a MOW inflow and flow patterns in the North Atlantic. However, although their model represents many features in correspondence with observations, unfortunately the source water is somewhat too warm and the MOW core is about 200 m shallower than observed, and hence the results have to be considered with caution. The Jia et al. (2007) model MOW shows three pathways into the North Atlantic: the well-known routes to the west into the open Atlantic and to the north along the eastern boundary, but also a northward route toward the North Atlantic Current. Along all these pathways the water becomes part of the NADW and exits the region to the south.

Two questions need to be addressed. Are the temperature and salinity changes observed at the Mediterranean source large enough to generate significant decadal variability in the intermediate waters of the North Atlantic, and can changes in the salt or heat content of the MOW have a significant impact on changes to the MOC? Lozier and Sindlinger (2009) used a box model of water mass modification to study the relation between long-term changes of water masses in the Mediterranean Sea and the North Atlantic at intermediate depths. They conclude that changes in the Mediterranean source waters need to be much larger in order to cause the observed magnitude of decadal variability in the open North Atlantic. Rahmstorf (1998) simulated a salinity increase in the Gibraltar outflow that corresponds with an assumed cutoff to the Nile River. He concluded that the MOW outflow has a noticeable effect on the circulation, warming the North Atlantic by a few tenths of a degree and enhancing the transport of NADW by 1–2 Sv. Changes of the magnitude detected in the observed trends, however, are too small to compete with the effects from changing precipitation patterns over the North Atlantic induced by climate change. Van Sebille

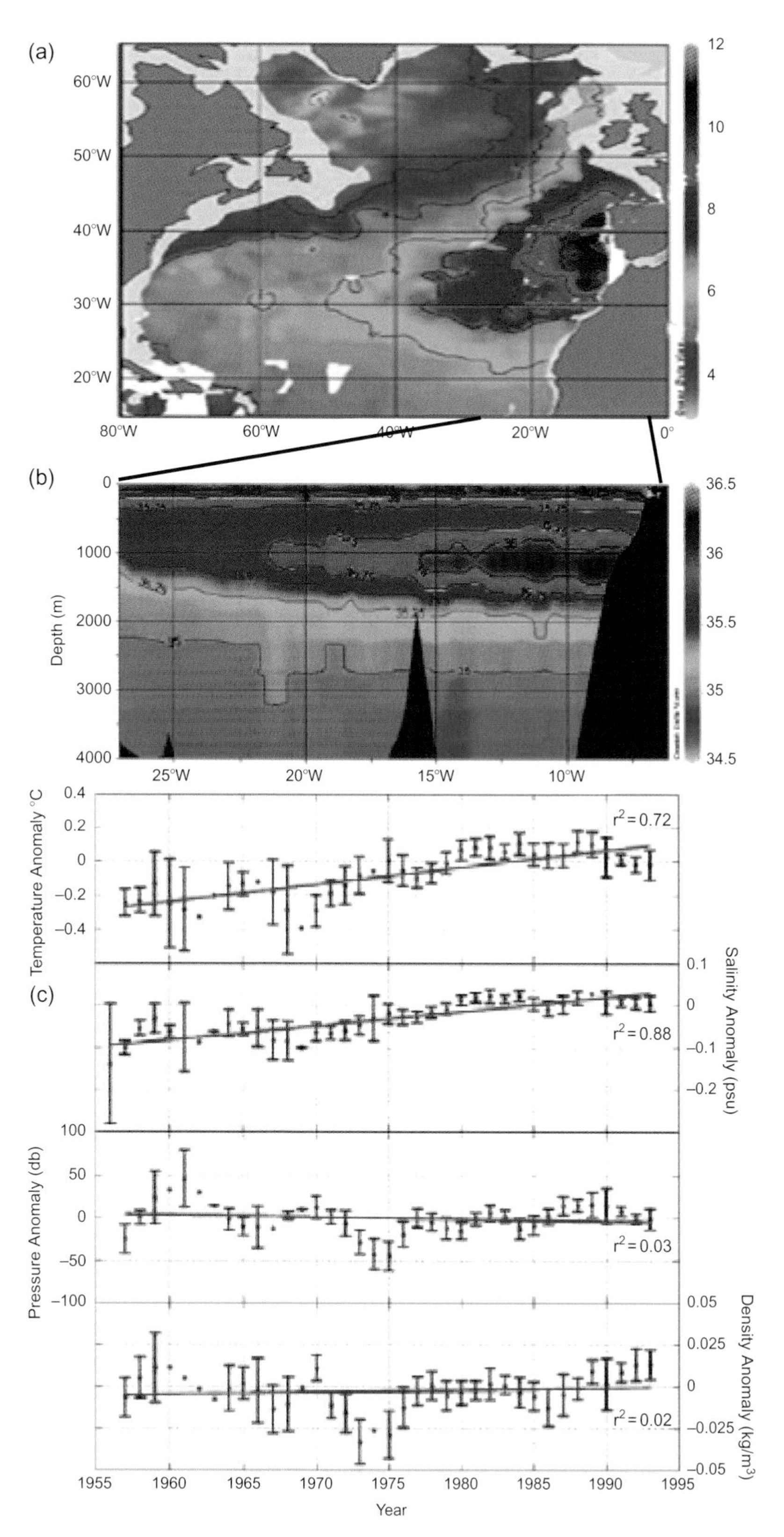

FIGURE 19.5 Distribution and temporal variation of Mediterranean Overflow Water (MOW) in the North Atlantic: (a) potential temperature (°C) at 1200 m; (b) 36°N cross section of the climatological salinity field with MOW salinity maximum; (c) time series for potential temperature, salinity, pressure, and potential density anomalies of the core MOW. Spatial values are averaged in one-year time bins from 1955 to 1993 and smoothed. Error bars show standard errors for annual means, points without error bars were interpolated. Green lines give linear regression fit and correlation for each fit is given as the r^2 value *After Potter and Lozier (2004).*

et al. (2011) reached a similar conclusion in their model. Although the MOW contributes to the western boundary flow of Labrador Sea Water by 20–40%, in the end the variability of the MOW properties is negligible in comparison to the effect of changes in the Labrador Sea Water.

3.3. Red Sea—Indian Ocean

The Red Sea, like the Mediterranean, is an arid sea adjacent to a larger ocean. It is connected to the Indian Ocean by the shallow strait of Bab el Mandeb, with the 154 m Hanish Sill at its northern entrance. Because evaporation far exceeds precipitation, the Red Sea has the highest salinities found in the world oceans, up to more than 40. Temperatures are also high at more than 30 °C. Recent estimates of annual mean net freshwater loss and heat gain were about 1.7–2 m year^{-1} and 8–11 W m^{-12}, respectively (Tragou et al., 1999; Sofianos et al., 2002). Episodic convection due to evaporation and heat loss probably occurs every 4–7 years in the Gulf of Suez, with the generated water mass forming a plume that descends down the slope, while entraining ambient water (Woelk and Quadfasel, 1996). This process accounts for most of the deep water renewal, although convection in the Gulf of Aqaba and open ocean convection may also play a role. The overall thermohaline circulation in the Red Sea is mainly forced by evaporation and less by thermal effects (Cember, 1988). Internal upwelling throughout the Red Sea leads to a sharp thermocline at a depth level close to the sill depth at Bab el Mandeb. The total deep water renewal time is estimated at 40–90 years (Woelk and Quadfasel, 1996).

The monsoon wind system produces mostly southward winds throughout the year in the northern region and a seasonally changing pattern in the southern region, with southward winds in summer and northward winds in winter. As a result, the water exchange through Bab el Mandeb has a seasonally changing pattern: a two-layer structure with a bottom outflow and near-surface inflow in winter, and a three-layer structure with an additional surface outflow in summer. An 18-month time series of observations (Murray and Johns, 1997; Sofianos et al., 2002) resolved this seasonal cycle and found a maximum bottom outflow in February of over 0.6 Sv, with a minimum of 0.05 Sv in August. The three-layer structure in summer had maximum inflow of Gulf of Aden Intermediate Water of about 0.3 Sv in August. There was little variation of temperature and salinity with mean values of 23 °C and 39.7, respectively, over the deployment period.

Although the annual mean transport through Bab el Mandeb can be small (<0.4 Sv), the RSOW produces a strong signal in the Indian Ocean because of its high salinity. A better understanding of RSOW flow through the Gulf of Aden (see Figure 19.1) has been obtained in recent years (Bower et al., 2000; Özgökmen et al., 2003; Peters and Johns, 2005, 2006; Peters et al., 2005). With

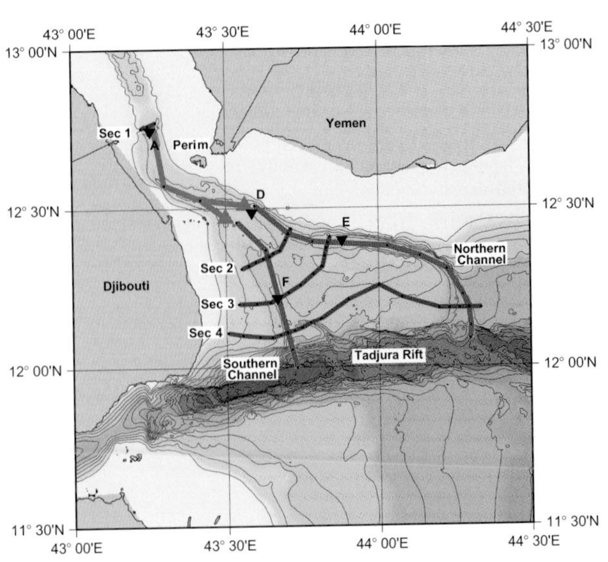

FIGURE 19.6 The Red Sea Water outflow region (Matt and Johns, 2007): The plume splits into two channels (yellow), with the topography shown from shallow (light blue) to deep (dark blue). Section hydrographic stations (black dots) and mooring positions (triangles) during the projects REDSOX-I and -II are indicated.

the outflow occurring at low latitude and of limited width and layer thickness, Ekman numbers are of order one. This implies frictional density currents modified by the earth rotation and entrainment processes play an important role. The RSOW descends from the shallow sill of the strait down the continental slope and branches into two topographic channels: the southern channel along the African slope and the deeper northern channel in the center of the Gulf of Aden (Figure 19.6). In addition to the narrow boundary currents in these channels, an intrusion layer with outflow water components occurs at shallower depth. The overall dilution rate was estimated at about 2–2.5. The different pathways generate different mixing histories for the corresponding water masses and, together with the seasonality of the outflow, are probably responsible for the multilayer structure and patchiness observed in the RSOW that spreads through the Indian Ocean.

Beyond the Gulf of Aden, the core of high-salinity water is found at 500–800 m depth. It spreads along the African continental slope, with an eastward detour at 5–10°S in the region of the South Equatorial Current (Beal et al., 2000). It then flows along the Mozambique Channel (Donohue and Toole, 2003) and becomes part of the Agulhas Current. The flux of salt across 32°S is similar to the flux into the Gulf of Aden, implying that all RSOW is exported at the western boundary and provides a dominant component of the salt budget at intermediate depths (Beal et al., 2000). The notable patchiness of the high-salinity water mass is also found in the Agulhas Current proper and south to the retroflection region, with 5–20% of this water probably contributed by the RSOW (Roman and Lutjeharms, 2007). Once it becomes part of the leakage in Agulhas rings, the RSOW provides salt to the Atlantic MOC.

3.4. Okhotsk Sea—Pacific Ocean

The North Pacific Intermediate Water (NPIW) is the main salinity minimum water mass with high oxygen content in the North Pacific subtropical gyre (Figure 19.7). NPIW is predominantly formed in the northwestern subtropical gyre in a mixing region northeast of Japan, between the fronts associated with the Oyashio and Kuroshio (Talley, 1993, 1997), and does not outcrop in the open ocean North Pacific. It was Wüst (1930) who first suggested that low-salinity surface water sinking in the Okhotsk Sea might provide a possible source for the formation of NPIW. The Okhotsk Sea is connected to the North Pacific mainly through Bussol Strait (sill depth 2300 m) and Kruzenshtern Strait (sill depth 1900 m) in the Kuril Ridge (Figure 19.7). The idea of possible water formation in the region came into focus again when a low-salinity anomaly was identified south of Bussol Strait (Talley, 1991), hinting at an origin of NPIW in the Okhotsk Sea (see Figures 19.1 and 19.7).

Direct observations of the water formation process in the Okhotsk Sea later confirmed Wüst's hypothesis (Shcherbina et al., 2003). On the northern and western shelves, that are less than 100 m depth, coastal polynas support effective salt brine rejection when new ice is being formed. The density of the underlying water increases with the enhanced salt content, and surface water is transferred to deeper layers because of the loss of stability. Shelf water production estimates range from 0.2 to 0.5 Sv, depending on cooling changes from year to year. With its near-freezing temperature, the convected water is relatively fresh in comparison with surrounding water masses of equal density.

The Dense Shelf Water (DSW) flows along the east coast of the island of Sakhalin to the south (Fukamachi et al., 2004) as part of the cyclonic circulation in the Okhotsk Sea. According to Itoh et al. (2003), the intermediate water that contributes to the Kuroshio–Oyashio interfrontal mixing zone has three sources: Western Subarctic Water (WSAW) enters the Okhotsk Sea through the Kuril Straits and arrives at the brine rejection regions in the cyclonic Okhotsk Sea flow, the above-mentioned DSW, and to a smaller extent a contribution of Soya Current Water (SCW) that enters through Soya Strait (sill depth about 50 m) from the Japan Sea. Approximate isopycnic mixing ratios of DSW:WSAW:SCW were estimated at 1:1:0.1, with a flow rate of 1.4 Sv of the mixed water masses (Itoh et al., 2003). Tidal current mixing near the Kuril Ridge

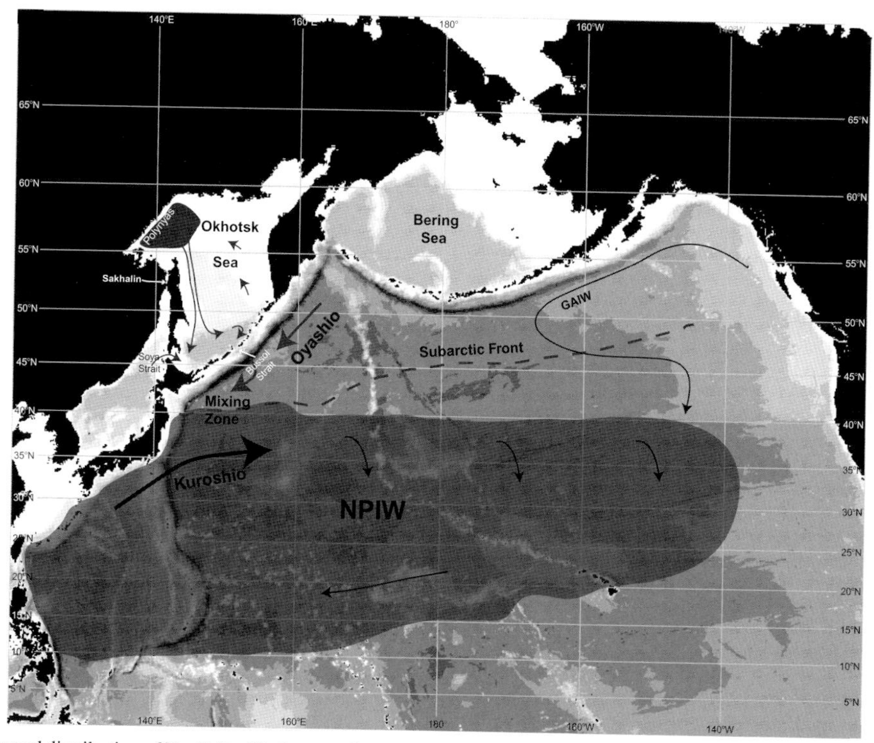

FIGURE 19.7 Sources and distribution of North Pacific Intermediate Water (NPIW): the polynya region with convection on the shelf in the Okhotsk Sea, dense water flow along Sakhalin Island, with some addition from Soya Strait, then outflow through Bussol Strait, transport with Oyashio toward the interfrontal mixing zone between the Kuroshio and Oyashio, then spreading as NPIW throughout the subtropical North Pacific, with possible addition of Gulf of Alaska Intermediate Water (GAIW). The schematic is based on You et al. (2000), Shcherbina et al. (2003), Minobe and Nakamura (2004), and Masujima and Yasuda (2009). Topography is based on IOC, IHO, and BODC (2003).

plays an important role in the inflow region of WSAW and also in the modification of water masses in the outflow region (Kowalik and Polyakov, 1998; Nakamura et al., 2006). With additional entrainment and mixing expected in the interfrontal zone where the NPIW is formed, the Okhotsk Sea water could well be the dominant factor contributing approximately 3 Sv to sustain the NPIW (Talley, 2003). However, You et al. (2003) concluded in an analysis of WOCE sections that the contribution from the Okhotsk Sea is only 1.5 Sv when accounting for aged NPIW that has recirculated within the subtropical gyre. You et al. (2003) further suggested some input from Gulf of Alaska Intermediate Water in the Northeast Pacific.

How constant are the water properties within the Okhotsk Sea? Itoh (2007) and Nakanowatari et al. (2007) studied time series of temperature and salinity in the Okhotsk Sea intermediate water from 1955 to 2003 or 2004, respectively, and found an increase of about 0.4–0.6 °C over the period while the increasing trend in salinity was small (about 0.02). This temperature increase is much larger than the temperature trend of about 0.1 °C identified by Levitus et al. (2005) in the 0–700 m layer of the North Pacific over the same period. One can speculate that the cause for increasing temperatures could be a reduction of sea-ice formation in the Okhotsk Sea. Considering the major influence of Okhotsk Sea water on NPIW formation, such changes could be an important factor for the long-term variability of North Pacific water masses.

4. DEEP PASSAGES

4.1. Atlantic Ocean: Romanche Fracture Zone, Vema, and Hunter Channels

The formation and sinking of AABW around Antarctica and the subsequent northward spreading and upwelling in the three oceans at depths below the crests of the main ocean ridges constitute the deep MOC (Lumpkin and Speer, 2007). Since the AABW formation rate is of comparable amplitude to that of the NADW (Orsi et al., 1999), the deep MOC has an amplitude similar to that of the NADW overturning cell (Lumpkin and Speer, 2007).

The spreading of the AABW, as inferred from the potential temperature distribution at the ocean bottom, shows an accentuated increase in bottom temperature at deep passages due to blocking and intense mixing (Mantyla and Reid, 1983). In the eastern South Atlantic Ocean (Figure 19.1), the northward spreading of AABW from the Weddell Sea is blocked east of the Mid-Atlantic Ridge by the Walvis Ridge, and consequently only small leakages of AABW into the Angola Basin at the Walvis Passage have been reported (Shannon and Chapman, 1991). To the west of the Mid-Atlantic Ridge, the Rio Grande Rise that separates the Argentine Basin from the Brazil Basin is cut by both the Vema Channel and the Hunter Channel, providing pathways into the Brazil Basin for the AABW. The AABW exits the Brazil Basin through an equatorial passage to the west and the Romanche and Chain Fracture Zones to the east. An extensive study of the circulation of AABW in the Brazil Basin was conducted during WOCE, and the flows through the four passages were estimated (see Hogg et al., 1996 for a summary of the Deep Basin Experiment (DBE) program).

The AABW is defined hereinafter as water with potential temperature less than 2 °C and therefore its upper part is lower circumpolar water (Hogg et al., 1982). The total flow of AABW into the Brazil Basin was estimated from a combination of direct velocity measurements and geostrophic estimates gathered between 1991 and 1996 (Hogg et al., 1999). The total transport was 6.9 Sv, with ~4 Sv flowing through the Vema Channel and the rest flowing through the Hunter Channel (Hogg et al., 1999). Mercier and Speer (1998) reported a total flow through the Romanche and Chain Fracture Zone of 1.2±0.25 Sv from 2 years of direct current measurements. The flow through the equatorial channel at 36°W was estimated at 2 Sv from a moored array deployment of 600 days (Hall et al., 1997).

Morris et al. (2001) computed mass and heat budgets for AABW in the Brazil Basin from the DBE measurements. A net upwelling of 3.7 Sv of AABW in the Brazil Basin across the 2 °C isotherm was diagnosed. This produced a balance between the vertical advection of heat and the vertical diffusion of heat, leading to estimates of the vertical mixing diffusivities of between 3 and $5\times10^{-4}\,m^2\,s^{-1}$ (Morris et al., 2001). Although the spatial distribution of the vertical diffusivity within the Brazil Basin was found to be nonuniform and intensified over the rough bathymetry of the Mid-Atlantic Ridge (Figure 19.8; Polzin et al., 1996; Ledwell et al., 2000), the estimates based on the large-scale heat budget of Morris et al. (2001) were consistent with estimates based on a relation between mixing intensity and topography roughness reported by St. Laurent et al. (2001).

Property modifications of AABW through the Romanche and Chain Fracture Zones were documented by Mercier et al. (1994) and Mercier and Morin (1997), showing that the densest AABW warms by ~0.7 °C and becomes saltier by ~0.07 due to intense vertical mixing as established by Ferron et al. (1998) and Polzin et al. (1996). Interestingly, Ferron et al. (1998) showed that the diffusive heat flux through the 1.4 °C isotherm in the Romanche Fracture Zone was half that across the 1.8 °C isotherm in the Sierra Leone and Guinea Abyssal Plains downstream, emphasizing the important role of the deep passages in the mixing of the bottom water masses.

Recent observations reviewed by Purkey and Johnson (2010) showed that AABW has warmed. Among these observations, the analysis of the time series of high-precision CTD casts acquired at the Vema sill Zenk and

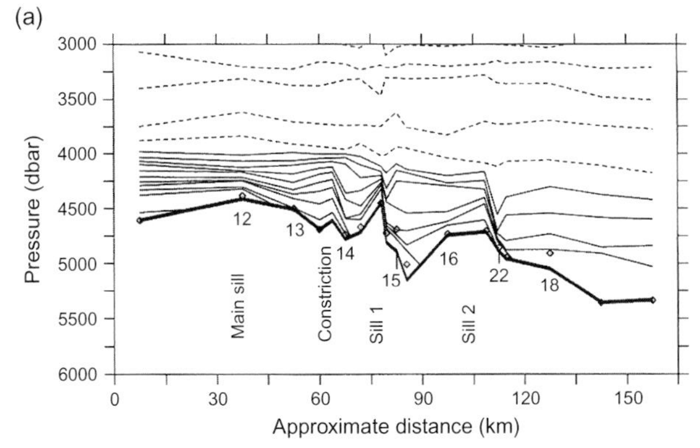

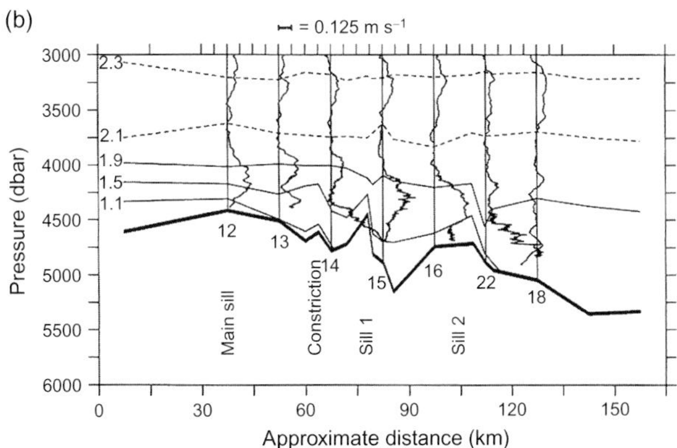

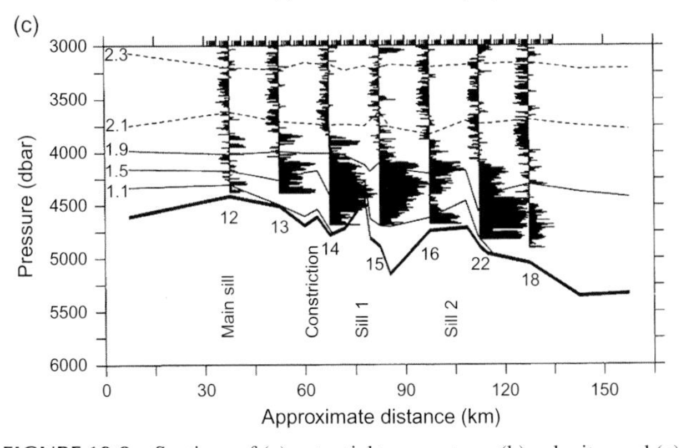

FIGURE 19.8 Sections of (a) potential temperature, (b) velocity, and (c) turbulent dissipation rate along the Romanche Fracture Zone from Polzin et al. (1996). The contour interval is 0.1 °C for the potential temperature section (a). The velocity profiles are referenced to the 2.1 °C isotherm (b). Note the logarithmic axis for the turbulent dissipation rate shown with a reference value of 10^{-10} W kg^{-1} (c).

Morozov (2007) showed that, after a period of remarkable stability between the early 1970s and the early 1990s when bottom potential temperature at the Vema Fracture Zone sill was stable at -0.184 ± 0.006 °C, the AABW was measured at -0.123 °C in November 2002, indicating a 0.061 °C warming. Temperature tendency was estimated at 2–3 mK year^{-1} over the time period with an associated salinity decrease of –0.3 to -1.2×10^{-4} year^{-1}. This warming signal propagated and was evident in the Equatorial Channel (Andrié et al., 2003) and in the western North Atlantic along 52 and 66°W (Johnson et al., 2008b).

4.2. Pacific Ocean: Samoan Passage, Wake Island Passage

Unlike the Atlantic and Indian Oceans, there are no deep passages that allow the spreading of the densest AABW formed in the Pacific sector of the Antarctic Ocean toward the Southwest Pacific Basin (Mantyla and Reid, 1983; Orsi et al., 1999; Purkey and Johnson, 2010). Following the literature, AABW in the Pacific Ocean will be referred hereinafter as Lower Circumpolar Deep Water (LCDW). In the Southwest Pacific Basin (Figure 19.1), LCDW flows northward in a deep western boundary current along the continental rise east of New Zealand and the Tonga–Kermadec Ridge (e.g., Warren, 1973; Whitworth et al., 1999). The Samoan Passage near 10°S, 169°W is thought to be the main conduit for the LCDW toward the Central Pacific Basin (Freeland, 2001). A gap in Robbie Ridge and a flow along the Manihiki Plateau located to the west and east of the Samoan Passage, respectively, contained less transport (Roemmich et al., 1996; Freeland, 2001). Analyzing direct velocity measurements obtained between 1992 and 1994 at the entrance of the Samoan Passage, Rudnick (1997) reported a variable transport with a 17-month mean of 6.0 Sv oriented northward for the flow below 4000 m (4.8 Sv for flow colder than 0.8 °C) with a standard deviation of 1.5 Sv, and minimum and maximum transports of 1.1 and 10.7 Sv, respectively. Rudnick (1997) argued that the destination of the water passing through the array is certainly toward the Central Pacific Basin. After crossing the 4200 m deep sill of the Samoan Passage, LCDW gradually sinks and undergoes strong property modifications. Roemmich et al. (1996) estimated the flow across the coldest isotherms and reported high rates of mixing with heating of the abyssal layer up to 20 W m^{-2}, corresponding to diffusivities of up to 10^{-1} m^2 s^{-1}. In the bottom temperature maps of Mantyla and Reid (1983), the LCDW exiting the Samoan Passage spreads northward in the Central Pacific Basin. Kawabe et al. (2003) showed that the coldest LCDW ($\theta < 0.98$ °C) flows northward into the Central Pacific Basin and passes the Mid-Pacific Seamounts between 162°10′E and 170°10′E at 18°20′N through the Wake Island Passage and other passages to the west. Kawabe et al. (2003) estimated a mean transport of 3.6 ± 1.3 Sv northward for the flow ($\theta < 1.2$ °C) through the Wake Island Passage from the deployment of two moorings in 1999–2000, with minimum and maximum transports of –5.3 and 14.8 Sv, respectively. The northward flowing LCDW upwells across isopycnals and returns southward as Pacific Deep Water that can be traced toward the Drake Passage (Reid, 1997). This deep overturning cell has an amplitude of about 10 Sv at 30°S (Talley, 2003; Macdonald et al., 2009).

4.3. Indian Ocean: Southwest Indian Ridge, Amirante Passage

In the Indian Ocean, the MOC consists of northward flowing deep water originating from the North Atlantic and bottom water from the Weddell Sea, with a southward return flow at intermediate levels (Ganachaud and Wunsch, 2000; Ganachaud et al., 2000). Whereas in the Atlantic and Pacific Oceans dense bottom water from Antarctica spreads northward in the western basins as Deep Western Boundary Currents, in the Indian Ocean (Figure 19.1) the bottom waters flow into both the western and eastern basins from the Weddell–Enderby Basin and the Australian–Antarctica Basin, respectively (Mantyla and Reid, 1995).

Hydrographic measurements suggested that equatorward spreading of AABW from the Weddell Sea through a gap in the Southwest Indian Ridge at about 50°S, 30°E, fills the dead-end Agulhas, Cape, Natal, and Mozambique Basins, and passes through the Southwest Indian Ridge at about 30°S, 60°E where the Ridge is heavily fractured (Mantyla and Reid, 1995). CFC measurements in the AABW confirmed the flow at 30°E and also identified a weaker flow through the Prince Edward Fracture Zone (Boswell and Smythe-Wright, 2002). East of the Madagascar Ridge, the Indomed, Atlantis II (main conduit), and Melville Fracture Zones were identified as pathways for bottom water (Warren, 1978; Toole and Warren, 1993; Donohue and Toole, 2003). MacKinnon et al. (2008) provided the first evidence of a strong bottom water flow through the Atlantis II Fracture Zone, with transports (for $\gamma^n \geq 28.11$) estimated from direct velocity measurements at 2.2 and 1.9 Sv at the entrance and the main sill of the Fracture Zone, respectively. High mixing was evident within the Atlantis II Fracture Zone with mixing estimates increasing from 10^{-4} to 10^{-2} m^2 s^{-1} from 2500-m to 4000-m depth. Equatorward of the Southwest Indian Ridge, in the Mascarene and Madagascar Basins, the AABW flows northward along the eastern boundary of the Madagascar Ridge and its transport was estimated at 3 Sv for $\sigma_4 > 45.89$ ($\gamma^n \geq 28.11$) by Donohue and Toole (2003). This is in agreement with Johnson et al. (1998) who reported 2.5–3.8 Sv northward flow in the Mascarene Basin for potential temperature less than 1.1 °C. The Amirante Passage at 9°S, 53°E is the sill through which the AABW spreads into the Somali Basin from the Mascarene Basin (Johnson et al., 1998). Johnson et al. (1998) reported a northward transport of bottom water from 1 to 1.7 Sv, with the range reflecting the variation for the geostrophic transport computation below a zero velocity reference level that varies between 1 and 1.1 °C.

AABW enters the eastern basin of the Indian Ocean through the Australian–Antarctica Discordance Zone at about 50°S, 125°E (Mantyla and Reid, 1995). A gap near 33°S, 105°E allows the flow of AABW into the western Australian Basin, while the spreading of AABW in the central Indian Basin occurs through gaps in the Ninetyeast Ridge at 10°S (Warren, 1982) and 28°S (Toole and Warren, 1993).

5. DISCUSSION

The choke points formed at high latitudes by the separation of the Southern Hemisphere continents with Antarctica, and in the tropics via the Indonesian archipelago, provide natural locations for observing and monitoring interocean exchanges. Indeed, there have been many concerted efforts to sustain long-term monitoring programs in these regions, although in the past the geographical (Southern Ocean) and logistical (Indonesia) isolation of these extreme locations has made this difficult. Nonetheless, in the more recent decades, the many ongoing observational programs along with remotely sensed measurements have successfully provided valuable information on the variability over a range of timescales of interocean exchange through these choke points and their importance to the global climate system.

In the Southern Ocean, the fast-flowing ACC provides an efficient equalizer of interocean properties acting to reduce the contrasts between each of the major ocean basins in the Southern Hemisphere. Nonetheless, energetic eddies in the Agulhas region counteract the strong eastward flow of the ACC and inject salty Indian Ocean water that can be traced across the South Atlantic and potentially influence the MOC. While the strong air–sea interaction, tidal and wind-induced mixing within the Indonesian seas significantly alters the Pacific source water masses that comprise the ITF thermohaline profile that enters the Indian Ocean, its signature appears to be largely overwhelmed possibly by the saltier RSOW by the time the Indonesian waters reach the Indian Ocean western boundary.

As with interocean exchanges, the exchanges between the oceans and their adjacent seas carry different weights with respect to their relevance for variations of the global MOC and climate. The Arctic Sea and the Labrador Sea are the most significant members, with a strong influence on changes in the MOC. The freshwater injection from these subarctic marginal seas will counter against the contribution of saltier waters from the Agulhas system as well as the Mediterranean, and subsequently have a competing influence on the stabilization of the MOC of the North Atlantic. Other marginal seas influence the mean oceanic circulation, but apparently not its variations. However they provide important markers in water mass properties that can be used to identify changes in the transfers between atmosphere and ocean, and in the budgets of heat, salt, carbon, nutrients, and other properties. Notwithstanding the open question of MOW pathways, it is clear that the Mediterranean outflow results in strong property signals in the North Atlantic and in part of the South Atlantic, above or as part of the NADW. Similarly, the outflow from the Red Sea and the Okhotsk Sea has a strong influence on water mass properties at intermediate depths in the Indian and North Pacific Ocean, respectively. However, on decadal timescales, it was shown that source property changes in MOW were too small to have a significant effect on the open Atlantic. This may also prove to be the case for decadal changes in the Red Sea and Okhotsk Sea waters, impacting their adjoining basins. Nevertheless, all these marginal sea inflows can be thought of as indicators of climatic change affecting larger regions.

Deep ocean passages between neighboring oceanic basins permit throughflows of deep and bottom water from one basin to the next. Deep passages are also choke points that, due to their limited extent, potentially provide a relatively easy monitoring site for the amplitude and property variability of the deep branch of the MOC. We focused our discussion on the deep passages that control the spreading of the NADW in the Atlantic Ocean and AABW in the world oceans, and we reviewed the characteristics of these flows. These deep passages are of considerable interest since they are the location of high levels of turbulence, strong water mass modification, and impact the dynamics of the upstream basin (Whitehead, 1998). Mixing is intense ($\sim 10^{-2}$ m^2 s^{-1}) in deep passages due to the unstable nature of the strongly sheared flows. Downstream of a critical point, the flow becomes supercritical and mixing may be even more intense (values as high as 10^{-1} m^2 s^{-1} were reported by Ferron et al., 1998 for a hydraulic jump region in the Romanche Fracture Zone). This enhanced mixing strongly affects the deep and bottom water properties of the downstream basins. Accurately modeling these regions of intense mixing in general circulation models (GCMs) remains a challenge (Legg et al., 2009).

As documented in other parts of the world oceans, there have been recent significant measurable changes both in the properties and fluxes at interocean and interbasin exchange sites. It is remarkable that all of the examples discussed in this chapter, with the exception of the Red Sea outflow, indicate increasing temperatures over recent decades, thereby strongly suggesting a response of the oceans to global warming. Long-term changes in the Pacific tropical tradewinds have resulted in changes to the ITF transport (Wainwright et al., 2008). While model studies suggest that the poleward shift and intensification of the Southern Ocean westerlies have led to an increased Agulhas leakage (Biastoch et al., 2009), the impact of these wind changes on ACC transport itself remains less clear. Although there is much recent evidence that property changes have occurred in the deep ocean (e.g., Fukasawa et al., 2004; Johnson and Doney, 2006; Kawano et al., 2006; Johnson and Gruber, 2007; Rintoul, 2007; Zenk and Morozov, 2007; McKee et al., 2011), unfortunately no long-term transport measurements in deep passages are presently available. The extreme complexity of abyssal topography along with the technological challenges of making long-term observations of the relatively small signals at low temperatures under immense pressure in remote locations complicates our ability to maintain an optimal sampling array in the deep ocean. Garzoli et al. (2010) recommended the setup of sustainable measurements in the deep passages that are not yet instrumented (e.g., Vema Chanel, Romanche Fracture Zone, Samoan, and Amirante Passages). Indeed, the observed changes highlight the need for long-term monitoring in all interbasin choke points that ultimately connects the MOC system. Such measurements are critical for climate monitoring and GCM validation.

ACKNOWLEDGMENTS

The authors would like to thank Gary Meyers and an anonymous reviewer for their valuable comments to improve this chapter. We also appreciate the contribution of accompanying figures from our colleagues: 19.2 Lisa Beal; 19.3 Arnold Gordon; 19.4 Bogi Hansen; 19.5 Susan Lozier; 19.6 Silvia Matt; 19.8 Kurt Polzin.

REFERENCES

Ambar, I., Serra, N., Neves, F., Ferreira, T., 2008. Observations of the Mediterranean Undercurrent and eddies in the Gulf of Cadiz during 2001. J. Mar. Syst. 71 (1–2), 195–220.

Andrié, C., Gouriou, Y., Bourlès, B., Braga, E.S., Ternon, J.F., Morin, P., Oudot, C., 2003. Variability of the Antarctic Bottom Water properties in the equatorial channel at 35°W. Geophys. Res. Lett. 30 (5), 8007. http://dx.doi.org/10.1029/2002GL015766.

Arbic, B.K., Owens, W.B., 2001. Climatic warming of Atlantic Intermediate Waters. J. Clim. 14, 4091–4108.

Arhan, M., Mercier, H., Park, Y.H., 2003. On the deep water circulation of the eastern South Atlantic Ocean. Deep Sea Res. Part I 50, 889–916.

Backeberg, B.C., Penven, P., Rouault, M., 2012. Impact of intensified Indian Ocean winds on the mesoscale variability of the Agulhas system. Nat. Clim. Chang 2, 608–612. http://dx.doi.org/10.1038/nclimate1587.

Bacon, S., 1998. Decadal variability in the outflow from the Nordic seas to the deep Atlantic Ocean. Nature 394, 871–874.

Bacon, S., Saunders, P., 2010. The Deep Western Boundary Current at Cape Farewell: results from a moored current meter array. J. Phys. Oceanogr. 40, 815–829. http://dx.doi.org/10.1175/2009JPO4091.1.

Baringer, M.O., Price, J.F., 1997. Mixing and spreading of the Mediterranean Outflow. J. Phys. Oceanogr. 27 (8), 1654–1677.

Baringer, M.O., Price, J.F., 1999. A review of the physical oceanography of the Mediterranean outflow. Mar. Geol. 155 (1–2), 63–82.

Beal, L.M., 2009. A time series of Agulhas Undercurrent transport. J. Phys. Oceanogr. 39, 2436–2450.

Beal, L.M., Bryden, H.L., 1999. The velocity and vorticity structure of the Agulhas Current at 32°S. J. Geophys. Res. 104, 5151–5176.

Beal, L.M., Ffield, A., Gordon, A.L., 2000. Spreading of Red Sea overflow waters. J. Geophys. Res. 105 (C4), 8549–8564.

Beal, L.M., de Ruijter, W.P.M., Biastoch, A., Zahn, R., members of SCOR/WCRP/IAPSO Working Group 136, 2011. On the role of the Agulhas system in global climate. Nature 472, 429–436.

Bergamasco, A., Malanotte-Rizzoli, P., 2010. The circulation of the Mediterranean Sea: a historical review of experimental investigations. Adv. Oceanogr. Limnol. 1 (1), 11–28.

Béthoux, J.P., Gentili, B., Tailliez, D., 1998. Warming and freshwater budget change in the Mediterranean since the 1940s, their possible relation to the greenhouse effect. Geophys. Res. Lett. 25 (7), 1023–1026.

Biastoch, A., Böning, C.W., Getzlaff, J., Molines, J.-M., Madec, G., 2008a. Mechanisms of interannual-decadal variability in the meridional overturning circulation of the mid-latitude North Atlantic Ocean. J. Clim. 21, 6599–6615.

Biastoch, A., Lutjeharms, J.R.E., Böning, C.W., Scheinert, M., 2008b. Mesoscale perturbations control inter-ocean exchange south of Africa. Geophys. Res. Lett. 35, L20602.

Biastoch, A., Böning, C.W., Lutjeharms, J.R.E., 2008c. Agulhas leakage dynamics affects decadal variability in Atlantic overturning circulation. Nature 456, 489–492.

Biastoch, A., Böning, C.W., Schwarzkopf, F.U., Lutjeharms, J.R.E., 2009. Increase in Agulhas leakage due to poleward shift of the Southern Hemisphere westerlies. Nature 462, 495–498.

Boebel, O., Lutjeharms, J.R.E., Schmid, C., Zenk, W., Rossby, T., Barron, C., 2003. The Cape Cauldron: a regime of turbulent inter-ocean exchange. Deep Sea Res. Part II 50 (1), 57–86.

Boswell, S.M., Smythe-Wright, D., 2002. The tracer signature of Antarctic Bottom Water and its spread in the Southwest Indian Ocean: part IFCFC-derived translation rate and topographic control around the Southwest Indian Ridge and the Conrad Rise. Deep Sea Res. Part I 49, 555–573.

Bower, A.S., Hunt, H.D., Price, F., 2000. Character and dynamics of the Red Sea and Persian Gulf outflows. J. Geophys. Res. 105 (C3), 6387–6414.

Bower, A.S., Serra, N., Ambar, I., 2002. Structure of the Mediterranean Undercurrent and Mediterranean Water spreading around the Southwestern Iberian Peninsula. J. Geophys. Res. 107, 3161.

Bower, A., Lozier, S., Stephan, G., 2011. Export of Labrador Sea Water from the subpolar North Atlantic: a Lagrangian perspective. Deep Sea Res. Part II 58, 1798–1818. http://dx.doi.org/10.1016/j.dsr2.2010.10.060.

Bryden, H.L., 1979. Poleward heat flux and conversion of available kinetic energy in Drake Passage. J. Mar. Res. 37, 1–22.

Bryden, H.L., Heath, R., 1985. Energetic eddies at the northern edge of the Antarctic Circumpolar Current in the southwest Pacific. Prog. Oceanogr. 14, 65–87.

Cai, W., 2006. Antarctic ozone depletion causes an intensification of the Southern Ocean super-gyre circulation. Geophys. Res. Lett. 33, L03712.

Cai, W., Meyers, G., Shi, G., 2005. Transmission of ENSO signal to the Indian Ocean. Geophys. Res. Lett. 32, L05616. http://dx.doi.org/10.1029/2004GL021736.

Candela, J., 2001. Mediterranean water and global circulation. In: Siedler, G. et al., (Eds.), Ocean Circulation and Climate. Academic Press, London UK, pp. 419–429.

Cember, R.P., 1988. On the sources, formation, and circulation of Red Sea Deep Water. J. Geophys. Res. 93 (C7), 8175–8191.

Chereskin, T.K., Donohue, K.A., Watts, D.R., 2012. cDrake: Dynamics and transport of the Antarctic Circumpolar Current in Drake Passage. Oceanography 25 (3), 134–135.

Chereskin, T.K., Donohue, K.A., Watts, D.R., Tracey, K.L., Firing, Y.L., Cutting, A.L., 2009. Strong bottom currents and cyclogenisis in Drake Passage. Geophys. Res. Lett. 36, L23602.

Cunningham, S.A., Alderson, S.G., King, B.A., Brandon, M.A., 2003. Transport and variability of the Antarctic Circumpolar Current in Drake Passage. J. Geophys. Res. 108. 8084–8101. http://dx.doi.org/10.1029/2001JC001147.

Curry, R., Mauritzen, C., 2005. Dilution of the northern North Atlantic in the recent decades. Science 308, 1772–1774.

Dencausse, G., Arhan, M., Speich, S., 2010. Spatio-temporal characteristics of the Agulhas Current retroflection. Deep Sea Res. Part I 57, 1392–1405.

de Ruijter, W.P.M., Biastoch, A., Drijfhout, S.S., Lutjeharms, J.R.E., Matano, R., Pichevin, T., van Leeuwen, P.J., Weijer, W., 1999. Indian-Atlantic inter-ocean exchange: dynamics, estimation and impact. J. Geophys. Res. 104, 20, 885–20, 910.

de Szoeke, R.A., Levine, M.D., 1981. The advective flux of heat by mean geostrophic motions in the Southern Ocean. Deep Sea Res. 28A, 1057–1085.

Dickson, B., Yashayaev, I., Meincke, J., Turrell, B., 2002. Rapid freshening of the deep North Atlantic Ocean over the past four decades. Nature 416, 832–836.

Dickson, R.R., Dye, S., J´onsson, S., K¨ohl, A., Macrander, A., et al., 2008. The overflow flux west of Iceland: variability, origins and forcing. In: Dickson, R.R., Meincke, J., Rhines, P. (Eds.), Arctic–Subarctic Ocean Fluxes. Springer, Dordrecht, the Netherlands, pp. 443–474.

Doglioli, A.M., Veneziani, M., Blanke, B., Speich, S., Griffa, A., 2006. A Lagrangian analysis of the Indian-Atlantic interocean exchange in a regional model. Geophys. Res. Lett. 33, L14611.

Dong, S., Sprintall, J., Gille, S.T., Talley, L., 2008. Southern Ocean mixed-layer depth from Argo float profiles. J. Geophys. Res. 113, C06013. http://dx.doi.org/10.1029/2006JC004051.

Dong, S., Garzoli, S., Baringer, M., 2011. The role of interocean exchanges on decadal variations of the meridional heat transport in the south Atlantic. J. Phys. Oceanogr. 41, 1498–1511.

Donners, J., Drijfhout, S.S., 2004. The Lagrangian view of South Atlantic interocean exchange in a global ocean model compared with inverse model results. J. Phys. Oceanogr. 34, 1019–1035.

Donohue, K.A., Toole, J.M., 2003. A near-synoptic survey of the Southwest Indian Ocean. Deep Sea Res. Part II 50, 1893–1931.

Downes, S.M., Bindoff, N.L., Rintoul, S.R., 2009. Impacts of climate change on the subduction of Mode and Intermediate Water Masses in the Southern Ocean. J. Clim. 22 (12), 3289–3330.

Drushka, K., Sprintall, J., Gille, S.T., Brodjonegoro, I., 2010. Vertical structure of Kelvin waves in the Indonesian Throughflow exit passages. J. Phys. Oceanogr. 40 (9), 1965–1987. http://dx.doi.org/10.1175/2010JPO4380.1.

Eldevik, T., Nilsen, J.E.Ø., Iovino, D., Olsson, K.A., Sandø, A.B., et al., 2009. Observed sources and variability of Nordic Seas overflow. Nat. Geosci. 2, 406–410.

Fahrbach, E., Hoppema, M., Rohardt, G., Schroder, M., Wisotzki, A., 2004. Decadal scale variations of water ass properties in the deep Weddell Sea. Ocean Dyn. 54, 77–91.

Fahrbach, E., Hoppema, M., Rohardt, G., Boebel, O., Klatt, O., Wisotzki, A., 2011. Warming of deep and abyssal water masses along the Greenwich meridian on decadal time scales: the Weddell gyre as a heat buffer. Deep Sea Res. 58, 2509–2523. http://dx.doi.org/10.1016/j.dsr2.2011.06.007.

Fang, G., Susanto, R.D., Soesilo, I., Zheng, Q., Qiao, F., Wei, Z., 2005. A note on the South China Sea shallow interocean circulation. Adv. Atmos. Sci. 22, 945–954.

Fang, G., Susanto, R.D., Wirasantosa, S., Qiao, F., Supangat, A., Fan, B., Wei, Z., Sulistiyo, B., Li, S., 2010. Volume, heat, and freshwater transports from the South China Sea to the Indonesian seas in the boreal winter of 2007–2008. J. Geophys. Res. 115, C12020. http://dx.doi.org/10.1029/2010JC006225.

Ferrari, R., Nikurashin, M., 2010. Suppression of eddy mixing across jets in the Southern Ocean. J. Phys. Oceanogr. 40, 1501–1519.

Ferron, B., Mercier, H., Speer, K.G., Gargett, A., Polzin, K., 1998. Mixing in the Romanche Fracture Zone. J. Phys. Oceanogr. 28, 1929–1945.

Ffield, A., Gordon, A.L., 1996. Tidal mixing signatures in the Indonesian seas. J. Phys. Oceanogr. 26, 1924–1937.

Ffield, A., Vranes, K., Gordon, A.L., Susanto, R.D., Garzoli, S.L., 2000. Temperature variability within Makassar Strait. Geophys. Res. Lett. 27, 237–240.

Firing, Y.L., Chereskin, T.K., Mazloff, M.R., 2011. Vertical structure and transport of the Antarctic Circumpolar Current in Drake Passage from direct velocity observations. J. Geophys. Res. 116, C08015. http://dx.doi.org/10.1029/2011JC006999.

Freeland, H., 2001. Observations of flow of abyssal water through the Samoa Passage. J. Phys. Oceanogr. 31, 2273–2279.

Fukamachi, Y., Mizuta, G., Ohshima, K.I., Talley, L.D., Riser, S.C., Wakatsuchi, M., 2004. Transport and modification processes of dense shelf water revealed by long-term moorings off Sakhalin in the Sea of Okhotsk. J. Geophys. Res. 109 (C9), 16.

Fukasawa, M., Freeland, H., Perkin, R., Watanabe, T., Uchida, H., Nishina, A., 2004. Bottom water warming in the North Pacific Ocean. Nature 427, 825–827.

Fusco, G., Artale, V., Cotroneo, Y., Sannino, G., 2008. Thermohaline variability of Mediterranean Water in the Gulf of Cadiz, 1948-1999. Deep Sea Res. 55 (12), 1624–1638.

Ganachaud, A., Wunsch, C., 2000. Improved estimates of global ocean circulation, heat transport and mixing from hydrographic data. Nature 408, 453–457.

Ganachaud, A., Wunsch, C., Marotzke, J., Toole, J., 2000. Meridional overturning and large-scale circulation of the Indian Ocean. J. Geophys. Res. 105, 26117–26134.

Garzoli, S.L., 1993. Geostrophic velocity and transport variability in the Brazil-Malvinas Confluence. Deep Sea Res. Part I 40 (7), 1379–1403.

Garzoli, S.L., et al., 2010. Progressing towards global sustained deep ocean observations. In: Hall, J., Harrison, D.E., Stammer, D. (Eds.), Proceedings of OceanObs'09: Sustained Ocean Observations and Information for Society, Venice, Italy, 21–25 September 2009, vol. 2. http://dx.doi.org/10.5270/OceanObs09.cwp.34, ESA Publication WPP-306.

Gille, S.T., 2008. Decadal scale temperature trends in the Southern Hemisphere ocean. J. Clim. 21 (18), 4749–4764.

Giulivi, C.F., Gordon, A.L., 2006. Isopycnal displacements within the Cape Basin thermocline as revealed by the Hydrographic Data Archive. Science 53 (8), 1285–1300.

Gnanadesikan, A., 1999. A simple predictive model for the structure of the oceanic pycnocline. Science 238, 2077–2079.

Godfrey, J.S., 1989. A Sverdrup model of the depth-integrated flow for the world ocean allowing for island circulations. Geophys. Astrophys. Dyn. 45, 89–112.

Gordon, A.L., 1985. Indian-Atlantic transfer of thermocline water at the Agulhas retroflection. Science 227, 1030–1033.

Gordon, A.L., 1986. Interocean exchange of thermocline water. J. Geophys. Res. 91, 5037–5046.

Gordon, A.L., 2001. Interocean exchange. In: Siedler, G. et al., (Eds.), Ocean Circulation and Climate. Academic Press, London, UK, pp. 305–314.

Gordon, A.L., Susanto, R.D., 2001. Banda Sea surface-layer divergence. Ocean Dyn. 52, 2–10.

Gordon, A.L., Ma, S., Olson, D.B., Hacker, P., Ffield, A., Talley, L.D., Wilson, D., Baringer, M., 1997. Advection and diffusion of Indonesian Throughflow water within the Indian Ocean South Equatorial Current. Geophys. Res. Lett. 24, 2573–2576.

Gordon, A.L., Susanto, R.D., Ffield, A., 1999. Throughflow within Makassar Strait. Geophys. Res. Lett. 26, 3325–3328.

Gordon, A.L., Visbeck, M., Huber, B., 2001. Export of Weddell Sea deep and Bottom Water. J. Geophys. Res. 106 (C5), 9005–9017.

Gordon, A.L., Susanto, R.D., Vranes, K., 2003. Cool Indonesian Throughflow as a consequence of restricted surface layer flow. Nature 425, 824–828.

Gordon, A.L., Susanto, R.D., Ffield, A., Huber, B.A., Pranowo, W., Wirasantosa, S., 2008. Makassar Strait Throughflow, 2004 to 2006. Geophys. Res. Lett. 35, L24605. http://dx.doi.org/10.1029/2008GL036372.

Gordon, A.L., Sprintall, J., van Aken, H.M., Susanto, R.D., Wijffels, S., Molcard, R., Ffield, A., Pranowo, W., Wirasantosa, S., 2010. The Indonesian Throughflow during 2004-2006 as observed by the INSTANT program. Dyn. Atmos. Oceans 50 (2), 115–128.

Hall, M.M., McCartney, M., Whitehead, J.A., 1997. Antarctic Bottom Water flux in the equatorial western Atlantic. J. Phys. Oceanogr. 27, 1903–1927.

Hansen, B., Østerhus, S., 2000. North Atlantic–Nordic Seas exchanges. Prog. Oceanogr. 45, 109–208.

Hansen, B., Østerhus, S., 2007. Faroe Bank overflow 1995–2005. Prog. Oceanogr. 52, 817–856. http://dx.doi.org/10.1016/j.pocean.2007.09.004.

Hansen, B., Østerhus, S., Turrell, W.R., Jónsson, S., Valdimarsson, H., Hátún, H., Malskær Olsen, S., 2008. The inflow of Atlantic water, heat, and salt to the Nordic seas across the Greenland–Scotland ridge. In: Dickson, D., Meincke, J., Rhines, P. (Eds.), Arctic–Subarctic Ocean Fluxes: Defining the Role of the Northern Seas in Climate, pp. 15–44.

Harlander, U., Ridderinkhof, H., Schouten, M.W., de Ruijter, W.P.M., 2009. Long-term observations of transport, eddies, and Rossby waves in the Mozambique Channel. J. Geophys. Res. 114, C02003.

Hartin, C.A., Fine, R.A., Sloyan, B.M., Talley, L.D., Chereskin, T.K., Happell, J., 2011. Formation rates of Subantarctic mode water and Antarctic intermediate water within the South Pacific. Deep Sea Res. 58, 524–534. http://dx.doi.org/10.1016/j.dsr.2011.02.010.

Hátún, H., Sando, A.B., Drange, H., 2005. Influence of the Atlantic subpolar gyre on the thermohaline circulation. Science 309, 1841–1844. http://dx.doi.org/10.1126/science.1114777.

Hogg, N.G., 1983. Hydraulic control and flow separation in a multi-layered fluid with applications to the Vema Channel. J. Phys. Oceanogr. 13, 695–708.

Hogg, N.G., Biscaye, P., Gardner, W., Schmitz Jr., W.J., 1982. On the transport and modification of Antarctic Bottom Water in the Vema Channel. J. Mar. Res. 40 (Suppl.), 231–263.

Hogg, N.G., Owens, W., Siedler, G., Zenk, W., 1996. Circulation in the Deep Brazil Basin. In: Wefer, G. et al., (Eds.), The South Atlantic, Present and Past Circulation. Springer-Verlag, Berlin, Heidelberg, pp. 249–260.

Hogg, N.G., Siedler, G., Zenk, W., 1999. Circulation and variability at the southern boundary of the Brazil Basin. J. Phys. Oceanogr. 29, 145–157.

Hughes, C.W., 2005. Nonlinear vorticity balance of the Antarctic Circumpolar Current. J. Geophys. Res. 110, C11008. http://dx.doi.org/10.1029/2004JC002753.

IOC, IHO, and BODC, 2003. Centenary Edition of the GEBCO Digital Atlas. Published on CD-ROM on behalf of the IOC and the IHO as part of the General Bathymetric Chart of the Oceans, British Oceanographic Center, Liverpool.

Itoh, M., 2007. Warming of intermediate water in the sea of Okhotsk since the 1950s. J. Oceanogr. 63, 637–641.

Itoh, M., Ohshima, K.I., Wakatsuchi, M., 2003. Distribution and formation of Okhotsk Sea Intermediate Water: an analysis of isopycnal climatological data. J. Geophys. Res. 108 (C8), 18.

Jia, Y., Coward, A.C., de Cuevas, B.A., Webb, D.J., Drijfhout, S.S., 2007. A model analysis of the behavior of the Mediterranean Water in the North Atlantic. J. Phys. Oceanogr. 37, 764–786.

Jochum, M., Potemra, J.T., 2008. Sensitivity of tropical rainfall to Banda Sea diffusivity in the Community Climate System Model. J. Clim. 21, 6445–6454.

Johnson, G.C., Bryden, H.L., 1989. On the size of the Antarctic Circumpolar Current. Deep Sea Res. 36, 39–53.

Johnson, G.C., Doney, S.C., 2006. Recent western South Atlantic bottom water warming. Geophys. Res. Lett. 33, L14614. http://dx.doi.org/10.1029/2006GL026769.

Johnson, G.C., Gruber, N., 2007. Decadal water mass variations along 20°W in the northeastern Atlantic Ocean. Prog. Oceanogr. 73, 277–295. http://dx.doi.org/10.1016/j.pocean.2006.03.022.

Johnson, G.C., Musgrave, D.L., Warren, B.A., Ffield, A., Olson, D.B., 1998. Flow of deep and bottom water in the Amirante Passage and Mascarene Basin. J. Geophys. Res. 103, 30, 973–30, 984.

Johnson, G.C., Purkey, S.G., Bullister, J.L., 2008a. Warming and freshening in the abyssal southeastern Indian Ocean. J. Clim. 21, 5351–5363. http://dx.doi.org/10.1175/2008JCLI2384.1.

Johnson, G.C., Purkey, S.G., Toole, J.L., 2008b. Reduced Antarctic meridional overturning circulation reaches the North Atlantic Ocean. Geophys. Res. Lett. 35, L222601. http://dx.doi.org/10.1029/2008GL035619.

Jullion, L., Jones, S.C., Naveira Garabato, A.C., Meredith, M.P., 2010. Wind controlled export of Antarctic Bottom water from the Weddell Sea. Geophys. Res. Lett. 37, L09609.

Jungclaus, J.H., Mellor, G.L., 2000. A three-dimensional model study of the Mediterranean outflow. J. Mar. Syst. 24, 41–66.

Kawabe, M., Fujio, S., Yanagimoto, D., 2003. Deep-water circulation at low latitudes in the western North Pacific. Deep Sea Res. Part I 50, 631–656.

Kawano, T., Fukusawa, M., Kouketsu, S., Ushida, H., Doi, T., et al., 2006. Bottom water warming along the pathway of Lower Circumpolar Deep Water in the Pacific Ocean. Geophys. Res. Lett. 33, L23613. http://dx.doi.org/10.1029/2006GL027933.

Kida, S., Wijffels, S., 2012. The impact of the Indonesian Throughflow and tidal mixing on the summertime Sea Surface Temperature in the western Indonesian Seas. J. Geophys. Res. 117, C09007. http://dx.doi.org/10.1029/2012JC008162.

Killworth, P.D., Hughes, C.W., 2002. The Antarctic Circumpolar Current as a free equivalent barotropic jet. J. Mar. Res. 60, 19–45.

Koch-Larrouy, A., Madec, G., Bouret-Aubertot, P., Gerkema, T., Bessieres, L., Molcard, R., 2007. On the transformation of Pacific water into Indonesian Throughflow Water by internal tidal mixing. Geophys. Res. Lett. 34, L04604. http://dx.doi.org/10.1029/2006GL028405.

Koch-Larrouy, A., Madec, G., Iudicone, D., Atmadipoera, A., Molcard, R., 2008. Physical processes contributing to the water mass transformation of the Indonesian Throughflow. Ocean Dyn. 58, 275–288. http://dx.doi.org/10.1007/s10236-008-0154-5.

Koch-Larrouy, A., Lengaigne, M., Terray, P., et al., 2010. Tidal mixing in the Indonesian seas and its effect on the tropical climate system. Clim. Dyn. 34, 891–904.

Köhl, A., Käse, R.H., Stammer, D., Serra, N., 2007. Causes of changes in the Denmark Strait Overflow. J. Phys. Oceanogr. 37, 1678–1696.

Kowalik, Z., Polyakov, I., 1998. Tides in the Sea of Okhotsk. J. Phys. Oceanogr. 28 (7), 1389–1409.

Krug, M., Tournadre, J., 2012. Satellite observations of an annual cycle in the Agulhas Current. Geophys. Res. Lett. 39, L15607.

Latif, M., Böning, C., Willebrand, J., Biastoch, A., Dengg, J., Keenlyside, N., Schweckendiek, U., Madec, G., 2006. Is the thermohaline circulation changing? J. Clim. 19, 4631–4637.

Leadbetter, S.J., Williams, R.G., McDonagh, E.L., King, B.A., 2007. A twenty year reversal in water mass trends in the subtropical North Atlantic. Geophys. Res. Lett. 34, L12608.

Ledwell, J., Montgomery, E., Polzin, K., St. Laurent, L., Schmitt, R., Toole, J., 2000. Evidence for enhanced mixing over rough topography in the abyssal ocean. Nature 403, 179–182.

Lee, T., Fukumori, I., Menemenlis, D., Xing, Z., Fu, L.-L., 2002. Effects of the Indonesian Throughflow on the Pacific and Indian Ocean. J. Phys. Oceanogr. 32, 1404–1429.

Legg, S., Briegleb, B., Chang, Y., Chassignet, E.P., Danabasoglu, G., Ezer, T., Gordon, A.L., Griffies, S., Hallberg, R., Jackson, L., Large, W., Özgökmen, T.M., Peters, H., Price, J., Riemenschneider, U., Wu, W., Xu, W., Yang, J., 2009. Improving oceanic overflow representation in climate models: the gravity current entrainment climate process team. Bull. Am. Meteorol. Soc. 90, 657–670.

Lenn, Y.-D., Chereskin, T.K., 2009. Observations of Ekman Currents in the Southern Ocean. J. Phys. Oceanogr. 39, 768–779.

Lenn, Y.-D., Chereskin, T.K., Sprintall, J., Firing, E., 2007. Mean Jets, mesoscale variability and eddy momentum fluxes in the surface layer of the Antarctic Circumpolar Current in Drake Passage. J. Mar. Res. 65, 27–58.

Lenn, Y.-D., Chereskin, T.K., Sprintall, J., McLean, J., 2011. Near-surface eddy heat and momentum fluxes in the Antarctic Circumpolar Current in Drake Passage. J. Phys. Oceanogr. 41, 1385–1407. http://dx.doi.org/10.1175/JPO-D-10-05017.1.

Levitus, S., Antonov, J.I., Boyer, T.P., Stephens, C., 2000. Warming of the world ocean. Science 287 (5461), 2225–2229.

Levitus, S., Antonov, J.I., Boyer, T.P., 2005. Warming of the world ocean, 1955–2003. Geophys. Res. Lett. 32, L02604.

Lherminier, P., Mercier, H., Gourcuff, C., Alvarez, M., Bacon, S., Kermabon, C., 2007. Transports across the 2002 Greenland–Portugal Ovide section and comparison with 1997. J. Geophys. Res. 112, C07003. http://dx.doi.org/10.1029/2006JC003716.

Lozier, M.S., Sindlinger, L., 2009. On the source of Mediterranean overflow water property changes. J. Phys. Oceanogr. 39, 1800–1817.

Lozier, M., Stewart, N., 2008. On the temporally varying northward penetration of Mediterranean Overflow Water and eastward penetration of Labrador Sea Water. J. Phys. Oceanogr. 38, 2097–2103.

Lumpkin, R., Speer, K., 2007. Global ocean meridional overturning. J. Phys. Oceanogr. 37, 2550–2562.

Lumpkin, R., Speer, K., Koltermann, K.P., 2008. Transport across 48°N in the Atlantic Ocean. J. Phys. Oceanogr. 38 (4), 733–752.

Lutjeharms, J.R.E., 2006. The Agulhas Current. Springer, Berlin, Heidelberg, New York, 329 pp.

Lutjeharms, J.R.E., 2007. Three decades of research on the greater Agulhas Current. Ocean Sci. 3 (1), 129–147.

Lutjeharms, J.R.E., van Ballegooyen, R.C., 1988. Anomalous upstream retroflection in the Agulhas Current. Science 240 (4860), 1770.

Macdonald, A.M., Mecking, S., Robbins, P.E., Toole, J.M., Johnson, G.C., Talley, L., Cook, M., Wijffels, S.E., 2009. The WOCE-era 3-D Pacific Ocean circulation and heat budget. Prog. Oceanogr. 82, 281–325. http://dx.doi.org/10.1016/j.pocean.2009.08.002.

MacKinnon, J.A., Johnston, M.S., Pinkel, R., 2008. Strong transport and mixing of the deep water through the southwest Indian Ridge. Nat. Geosci. 1, 766–768. http://dx.doi.org/10.1038/ngeo340.

Macrander, A., Send, U., Valdimarsson, H., J´onsson, S., K¨ase, R.H., 2005. Interannual changes in the overflow from the Nordic Seas into the Atlantic Ocean through Denmark Strait. Geophys. Res. Lett. 32, L06606.

Mantyla, A.W., Reid, J.L., 1983. Abyssal characteristics of the World Ocean waters. Deep Sea Res. 30, 805–833.

Mantyla, A.W., Reid, J.L., 1995. On the origins of deep and bottom waters of the Indian Ocean. J. Geophys. Res. 100, 2417–2439.

Masujima, M., Yasuda, I., 2009. Distribution and modification of North Pacific Intermediate Water around the Subarctic Frontal Zone east of 150°E. J. Phys. Oceanogr. 39 (6), 1462–1474.

Masumoto, Y., 2002. Effects of interannual variability in the eastern Indian Ocean on the Indonesian Throughflow. J. Oceanogr. Jpn. 58, 175–182.

Mata, M.M., Tomczak, M., Wijffels, S.E., Church, J., 2000. East Australian Current volume transports at 30°S: estimates from the World Ocean Circulation Experiment hydrographic sections PR11/P6 and the PCM3 current meter array. J. Geophys. Res. 105 (C12), 28,509–28,526.

Matano, R.P., Simonato, C.G., Strub, P.T., 1999. Modelling the wind driven variability of the South Indian Ocean. J. Phys. Oceanogr. 29, 217–230.

Matt, S., Johns, W., 2007. Transport and entrainment in the Red Sea outflow plume. J. Phys. Oceanogr. 37, 819–836.

Mauritzen, C., 1996. Production of dense overflow waters feeding the North Atlantic across the Greenland-Scotland Ridge. 1. Evidence for a revised circulation scheme. Deep Sea Res. Part I 43, 769–806.

Mauritzen, C., Morel, Y., Paillet, J., 2001. On the influence of Mediterranean Water on the Central Waters of the North Atlantic Ocean. Deep Sea Res. 48 (2), 347–381.

McBride, J.L., Haylock, M.R., Nichols, N., 2003. Relationships between the maritime continent heat source and the El Nino-Southern Oscillation phenomenon. J. Clim. 16 (756), 2905–2914.

McClean, J.L., Ivanova, D.P., Sprintall, J., 2005. Remote origins of interannual variability in the Indonesian Throughflow region from data and a global Parallel Ocean Program simulation. J. Geophys. Res. 110, C10013. http://dx.doi.org/10.1029/2004JC002477.

McCreary, J.P., Lu, P., 2001. Influence of the Indonesian Throughflow on the circulation of intermediate water in the Pacific Ocean. J. Phys. Oceanogr. 31, 932–942.

McKee, D.C., Yuan, X., Gordon, A.L., Huber, B., Dong, Z., 2011. Climate impact on interannual variability of Weddell Sea Bottom Water. J. Geophys. Res. 116. http://dx.doi.org/10.1029/2010JC00648, C05020.

Mercier, H., Morin, P., 1997. Hydrography of the Romanche and Chain Fracture Zones. J. Geophys. Res. 102, 10373–10389.

Mercier, H., Speer, K.G., 1998. The transport of Bottom Water through the Romanche Fracture Zone and the Chain Fracture Zone. J. Phys. Oceanogr. 28, 779–790.

Mercier, H., Speer, K.G., Honnorez, J., 1994. Flow pathways of Bottom Water through the Romanche and Chain Fracture Zones. Deep Sea Res. Part I 41, 1457–1477.

Meredith, M.P., Woodworth, P.L., Hughes, C.W., Stepanov, V., 2004. Changes in the ocean transport through Drake Passage during the 1980s and 1990s, forced by changes in the Southern Annular Mode. Geophys. Res. Lett. 31, L21305. http://dx.doi.org/10.1029/2004GL021169.

Meredith, M., Woodworth, P., Chereskin, T.K., Marshall, D., Allison, L., Bigg, G.R., et al., 2011. Sustained monitoring of the Southern Ocean at Drake Passage: past achievements and future priorities. Rev. Geophys. 49, RG4005. http://dx.doi.org/10.1029/2010RG000348.

Meyers, G., 1996. Variation of Indonesian throughflow and the El Niño-Southern Oscillation. J. Geophys. Res. 101, 12,255–12,263.

Millot, C., Candela, J., Fuda, J.-L., Tber, Y., 2006. Large warming and salinification of the Mediterranean outflow due to changes in its composition. Deep Sea Res. Part I 53 (4), 656–666.

Minobe, S., Nakamura, M., 2004. Interannual to decadal variability in the southern Okhotsk Sea based on a new gridded upper water temperature data set. J. Geophys. Res. 109, C09S05.

Molcard, R.M., Fieux, M., Ilahude, A.G., 1996. The Indo-Pacific throughflow in the Timor Passage. J. Geophys. Res. 101, 12411–12420.

Molcard, R.M., Fieux, M., Syamsudin, F., 2001. The throughflow within Ombai Strait. Deep Sea Res. 48, 1237–1253.

Moore, T.S., Marra, J., Alkatiri, A., 2003. Response of the Banda Sea to the southeast monsoon. Mar. Ecol. Prog. Ser. 261, 41–49.

Morris, M.Y., Hall, M.M., St. Laurent, L.C., Hogg, N.G., 2001. Abyssal mixing in the Brazil Basin. J. Phys. Oceanogr. 31, 3331–3348.

Morrow, R.A., Coleman, R., Church, J.A., Chelton, D.B., 1994. Surface eddy momentum flux and velocity variances in the Southern Ocean from Geosat altimetry. J. Phys. Oceanogr. 24, 2050–2071.

Morrow, R.A., Birol, F., Griffin, D., Sudre, J., 2004. Divergent pathways of cyclonic and anti-cyclonic ocean eddies. Geophys. Res. Lett. 31, L24311. http://dx.doi.org/10.1029/2004GL020974.

Murray, S.P., Arief, D., 1988. Throughflow into the Indian Ocean through the Lombok Strait, January 1985-January 1986. Nature 333, 444–447.

Murray, S.P., Johns, W., 1997. Direct observations of seasonal exchange through the Bab el Mandab Strait. Geophys. Res. Lett. 24 (21), 2557–2560.

Nakamura, T., Toyoda, T., Ishikawa, Y., Awaji, T., 2006. Enhanced ventilation in the Okhotsk Sea through tidal mixing at the Kuril Straits. Deep Sea Res. 53 (3), 425–448.

Nakanowatari, T., Ohshima, K., Wakatsuchi, M., 2007. Warming and oxygen decrease of intermediate water in the northwestern North Pacific, originating from the Sea of Okhotsk. 1955-2004. Geophys. Res. Lett. 34 (4), L04602.

Naveira Garabato, A.C., McDonagh, E.L., Stevens, D.P., Heywood, K.J., Sanders, R.J., 2002. On the export of Antarctic Bottom Water from the Weddell Sea. Deep Sea Res. Part II 49 (21), 4715–4742.

Naveira Garabato, A.C., Jullion, L., Stevens, D.P., Heywood, K.J., King, B.A., 2009. Variability of Subantarctic Mode Water and Antarctic Intermediate Water in the Drake Passage during the late-twentieth and early-twenty-first centuries. J. Clim. 22 (7), 3661–3688.

New, A.L., Barnard, S., Herrmann, P., Molines, J.-M., 2001. On the origin and pathway of the saline inflow to the Nordic Seas: insights from models. Prog. Oceanogr. 48 (2–3), 255–287.

Olsen, S.M., Hansen, B., Quadfasel, D., Østerhus, S., 2008. Observed and modelled stability of overflow across the Greenland-Scotland Ridge. Nature 455, 519–523. http://dx.doi.org/10.1038/nature07302.

Orsi, A.H., Johnson, G.C., Bullister, J.L., 1999. Circulation, mixing, and production of Antarctic Bottom Water. Prog. Oceanogr. 43, 55–109.

Orsi, A.H., Smethie Jr., W.M., Bullister, J.L., 2002. On the total input of Antarctic waters to the deep ocean: a preliminary estimate from chlorofluorocarbon measurements. J. Geophys. Res. 107, 3122. http://dx.doi.org/10.1029/2001JC000976.

Østerhus, S., Sherwin, T., Quadfasel, D., Hansen, B., 2008. The overflow transport east of Iceland. In: Dickson, R.R., Meincke, J., Rhines, P. (Eds.), Arctic–Subarctic Ocean Fluxes. Springer, Dordrecht, the Netherlands, pp. 427–441.

Özgökmen, T.M., Johns, W.E., Peters, H., Matt, S., 2003. Turbulent mixing in the Red Sea outflow plume from a high-resolution nonhydrostatic model. J. Phys. Oceanogr. 33 (8), 1846–1869.

Palastanga, V., van Leeuwen, P.J., Schouten, M.W., De Ruijter, W.P.M., 2007. Flow structure and variability in the subtropical Indian Ocean: instability of the South Indian Ocean Countercurrent. J. Geophys. Res. 112, C01001, 11 pp.

Peeters, F.J.C., Acheson, R., Brummer, G.-J.A., de Ruijter, W.P.M., Schneider, R.R., 2004. Vigorous exchange between the Indian and Atlantic Oceans at the end of the past five glacial periods. Nature 430, 661–665.

Penven, P., Lutjeharms, J.R.E., Marchesiello, P., Roy, C., Weeks, S.J., 2001. Generation of cyclonic eddies by the Agulhas Current in the lee of the Agulhas Bank. Geophys. Res. Lett. 27, 1055–1058.

Penven, P., Lutjeharms, J.R.E., Florenchie, P., 2006. Madagascar: a pacemaker for the Agulhas Current system? Geophys. Res. Lett. 33, L17609.

Perkins, H., Hopkins, T.S., Malmberg, S.-A., Poulain, P.-M., Warn-Varnas, A., 1998. Oceanic conditions east of Iceland. J. Geophys. Res. 03, 21, 531–21, 542.

Peters, H., Johns, W.E., 2005. Mixing and entrainment in the Red Sea outflow plume. Part II: turbulence characteristics. J. Phys. Oceanogr. 35 (5), 584–600.

Peters, H., Johns, W.E., 2006. Bottom layer turbulence in the Red Sea outflow plume. J. Phys. Oceanogr. 36, 1763–1785.

Peters, H., Johns, W.E., Bower, A.S., Fratantoni, D.M., 2005. Mixing and entrainment in the Red Sea outflow plume. Part I: plume structure. J. Phys. Oceanogr. 35 (5), 569–583.

Phillips, H.E., Rintoul, S.R., 2000. Eddy variability and energetics from direct current measurements in the Antarctic Circumpolar Current south of Australia. J. Phys. Oceanogr. 30, 3050–3076.

Pollard, R., Griffiths, M., Cunningham, S., Read, J., Pérez, F., Rios, A., 1996. Vivaldi 1991—a study of the formation, circulation and ventilation of Eastern North Atlantic Central Water. Prog. Oceanogr. 37, 167–192.

Polzin, K.L., Speer, K.G., Toole, J.M., Schmitt, R., 1996. Intense mixing of Antarctic Bottom Water in the equatorial Atlantic Ocean. Nature 380, 54–57.

Potemra, J.T., Schneider, N., 2007a. Influence of low-frequency Indonesian throughflow transport on temperatures in the Indian Ocean in a coupled model. J. Clim. 20, 1439–1452.

Potemra, J.T., Schneider, N., 2007b. Interannual variations of the Indonesian throughflow. J. Geophys. Res. 112, C05035. http://dx.doi.org/10.1029/2006JC003808.

Potter, R.A., Lozier, M.S., 2004. On the warming and salinification of the Mediterranean outflow waters in the North Atlantic. Geophys. Res. Lett. 31.

Provost, C., Renault, A., Barre, N., Sennechael, N., Garcon, V., Sudre, J., Huhn, O., 2011. Two repeat crossings of Drake Passage in austral summer 2006: short term variations and evidence for considerable ventilation of intermediate and deep waters. Deep Sea Res. 58, 2555–2571. http://dx.doi.org/10.1016/j.dsr2/2011.06.009.

Pujiana, K., Gordon, A.L., Sprintall, J., Susanto, R.D., 2010. Intraseasonal variability in the Makassar Strait thermocline. J. Mar. Res. 67 (6), 757–777.

Purkey, S.G., Johnson, G.C., 2010. Warming of global abyssal and deep Southern Ocean waters between the 1990s and 2000s: contributions to global heat and sea level rise budgets. J. Clim. 23, 6336–6351.

Quartly, G.D., Srokosz, M.A., 2004. Eddies in the southern Mozambique Channel. Deep Sea Res. Part II 51 (1–3), 69–83.

Rahmstorf, S., 1996. On the freshwater forcing and transport of the Atlantic thermohaline circulation. Clim. Dyn. 12, 799–811.

Rahmstorf, S., 1998. Influence of Mediterranean outflow on climate. Eos 79 (24), 281.

Rayner, D., Hirschi, J.-M., Kanzow, T., Johns, W.E., Wright, P.G., Frajka-Williams, E., et al., 2011. Monitoring the Atlantic meridional overturning circulation. Deep Sea Res. 58, 1744–1753. http://dx.doi.org/10.1016/j.dsr2.2010.10.056.

Reid, J.L., 1979. On the contribution of the Mediterranean Sea outflow to the Norwegian-Greenland Sea. Deep Sea Res. A 26 (11), 1199–1223.

Reid, J.L., 1997. On the total geostrophic circulation of the Pacific Ocean: flow patterns, tracers, and transports. Prog. Oceanogr. 39, 263–352. http://dx.doi.org/10.1016/S0079-6611(97)00012-8.

Richardson, P.L., 2007. Agulhas leakage into the Atlantic estimated with subsurface floats and surface drifters. Deep Sea Res. Part I 54 (8), 1361–1389.

Richardson, P.L., 2008. On the history of meridional overturning circulation schematic diagrams. Prog. Oceanogr. 76, 466–486.

Ridderinkhof, H., van der Werf, P.M., Ullgren, J.E., van Aken, H.M., van Leeuwen, P.J., de Ruijter, W.P.M., 2010. Seasonal and interannual variability in the Mozambique Channel from moored current observations. J. Geophys. Res. 115, C06010.

Ridgway, K.R., Dunn, J.R., 2007. Observational evidence for a southern hemisphere oceanic supergyre. Geophys. Res. Lett. 34, L13612. http://dx.doi.org/10.1029/2007GL030392.

Rintoul, S.R., 1991. South Atlantic interbasin exchange. J. Geophys. Res. 96, 2675–2692.

Rintoul, S.R., 2007. Rapid freshening of Antarctic Bottom water formed in the Indian and Pacific Oceans. Geophys. Res. Lett. 34, L06606. http://dx.doi.org/10.1029/2006GL028550.

Robertson, R., Visbeck, M., Gordon, A.L., Fahrbach, E., 2002. Long-term temperature trends in the deep waters of the Weddell Sea. Deep Sea Res. 49, 4791–4806.

Roemmich, D., Wunsch, C., 1984. Apparent changes in the climatic state of the deep North Atlantic Ocean. Nature 307 (5950), 447–450.

Roemmich, D., Hautala, S., Rudnick, D., 1996. Northward abyssal transport through the Samoan Passage and adjacent regions. J. Geophys. Res. 101, 14039–14055.

Roether, W., Klein, B., Manca, B.B., Theocharis, A., Kioroglou, S., 2007. Transient Eastern Mediterranean deep waters in response to the massive dense-water output of the Aegean Sea in the 1990s. Prog. Oceanogr. 74 (4), 540–571.

Roman, R., Lutjeharms, J.R.E., 2007. Red Sea Intermediate Water at the Agulhas Current termination. Deep Sea Res. 54 (8), 1329–1340.

Rouault, M., Penven, P., Pohl, B., 2009. Warming in the Agulhas Current system since the 1980s. Geophys. Res. Lett. 36, L12602.

Rudnick, D.L., 1997. Direct velocity measurements in the Samoan Passage. J. Geophys. Res. 102, 3293–3302.

Sánchez-Román, A., Sannino, G., García-Lafuente, J., Carillo, A., Criado-Aldeanueva, F., 2009. Transport estimates at the western section of the Strait of Gibraltar: a combined experimental and numerical modeling study. J. Geophys. Res. 114 (C6), C06002.

Sarafanov, A., Falina, A., Mercier, H., Lherminier, P., Sokov, A., 2009. Recent changes in the Greenland-Scotland overflow-derived water transport inferred from hydrographic observations in the southern Irminger Sea. Geophys. Res. Lett. 36, L13707. http://dx.doi.org/10.1029/2009GL038041.

Sarafanov, A., Mercier, H., Falina, A., Lherminier, P., Sokov, A., 2010. Cessation and partial reversal of deep water freshening in the northern North Atlantic: observation-based estimates and attribution. Tellus 62, 80–90.

Schmitz Jr., W.J., 1995. On the interbasin-scale thermohaline circulation. Rev. Geophys. 33 (2), 151–173.

Schouten, M.W., de Ruijter, W.P.M., van Leeuwen, P.J., Ridderinkhof, H., 2003. Eddies and variability in the Mozambique Channel. Deep Sea Res. Part II 50, 1987–2004.

Sciremammano, F., 1979. Observations of Antarctic Polar Front motions in a deep water expression. J. Phys. Oceanogr. 9 (1), 221–226.

Shannon, L.V., Chapman, P., 1991. Evidence of Antarctic Bottom Water in the Angola Basin at 32°S. Deep Sea Res. Part A 38, 1299–1304. http://dx.doi.org/10.1016/0198-0149(91)90028-E.

Shcherbina, A.Y., Talley, L.D., Rudnick, D.L., 2003. Direct observations of North Pacific ventilation: brine rejection in the Okhotsk Sea. Science 302 (5652), 1952–1955.

Shervin, T.J., Griffiths, C.R., Inall, M.E., Turrell, W.R., 2008. Quantifying the overflow across the Wyville Thomson Ridge into the Rockall Trough. Deep Sea Res. Part I 55, 396–404.

Siedler, G., Rouault, M., Lutjeharms, J.R.E., 2006. Structure and origin of the subtropical South Indian Ocean Countercurrent. J. Geophys. Res. 33, L24609, 5 pp.

Siedler, G., Rouault, M., Biastoch, A., Backeberg, B., Reason, C.J.C., Lutjeharms, J.R.E., 2009. Modes of the southern extension of the East Madagascar Current. J. Geophys. Res. 114, C01005.

Sloyan, B.M., Rintoul, S.R., 2001. The Southern Ocean limb of the global deep overturning circulation. J. Phys. Oceanogr. 31 (1), 143–173.

Smith, K.S., Marshall, J., 2009. Evidence for enhanced eddy mixing at mid-depth in the Southern Ocean. J. Phys. Oceanogr. 39, 50–69.

Sofianos, S., Johns, W., Murray, S., 2002. Heat and freshwater budgets in the Red Sea from direct observations at Bab el Mandeb. Deep Sea Res. Part II 49 (7–8), 1323–1340.

Song, Q., Gordon, A.L., Visbeck, M., 2004. Spreading of the Indonesian throughflow in the Indian Ocean. J. Phys. Oceanogr. 34, 772–779.

Song, Q., Vecchi, G.A., Rosati, A.J., 2007. The role of the Indonesian Throughflow in the Indo-Pacific climate variability in the GFDL Coupled Climate Model. J. Clim. 20, 2434–2451.

Speer, K.G., Siedler, G., Talley, L.D., 1995. The Namib Col Current. Deep Sea Res. 42, 1933–1950.

Speich, S., Blanke, B., de Vries, P., Doos, K., Drijfhout, S., Ganachaud, A., Marsh, R., 2002. Tasman leakage: a new route for the global conveyor belt. Geophys. Res. Lett. 29 (10), 1416. http://dx.doi.org/10.1029/2001GL014586.

Sprintall, J., 2003. Seasonal to interannual upper-ocean variability in the Drake Passage. J. Mar. Res. 61, 1–31.

Sprintall, J., 2008. Long-term trends and interannual variability of temperature in Drake Passage. Prog. Oceanogr. 77, 316–330.

Sprintall, J., Wijffels, S.E., Gordon, A.L., et al., 2004. INSTANT: a new international array to measure the Indonesian Throughflow. Eos 85 (39), 369.

Sprintall, J., Wijffels, S.E., Molcard, R., Jaya, I., 2009. Direct estimates of the Indonesian throughflow entering the Indian Ocean: 2004–2006. J. Geophys. Res. 114, C7. http://dx.doi.org/10.1029/2008JC005257.

St. Laurent, L.C., Toole, J.M., Schmitt, R.W., 2001. Buoyancy forcing by turbulence over the rough topography in the Brazil Basin. J. Phys. Oceanogr. 31, 3476–3495.

Stammer, D., Theiss, J., 2004. Velocity statistics inferred from the TOPEX/Poseidon-Jason-1 tandem mission data. Mar. Geod. 27, 551–575.

Stramma, L., England, M., 1999. On the water masses and mean circulation of the South Atlantic Ocean. J. Geophys. Res. 104 (C9), 20, 863–20, 883.

Swart, N.C., Lutjeharms, J.R.E., Ridderinkhof, H., de Ruijter, W.P.M., 2010. Observed characteristics of Mozambique Channel eddies. J. Geophys. Res. 115, C09006/1–C09006/14.

Talley, L.D., 1991. An Okhotsk Sea water anomaly: implications for ventilation in the North Pacific. Deep Sea Res. Part A 38 (1), S171–S190.

Talley, L.D., 1993. Distribution and formation of North Pacific Intermediate Water. J. Phys. Oceanogr. 23, 517–537.

Talley, L.D., 1997. North Pacific Intermediate Water transports in the mixed water region. J. Phys. Oceanogr. 27 (8), 1795–1803.

Talley, L.D., 2003. Shallow, intermediate, and deep overturning components of the global heat budget. J. Phys. Oceanogr. 33 (3), 530–560.

Talley, L.D., 2008. Freshwater transport estimates and the global overturning circulation: shallow, deep and throughflow components. Prog. Oceanogr. 78, 257–303. http://dx.doi.org/10.1016/j.pocean.2008.05.001.

Talley, L.D., Sprintall, J., 2005. Deep expression of the Indonesian Throughflow: Indonesian Intermediate Water in the South Equatorial Current. J. Geophys. Res. 110, C10009. http://dx.doi.org/10.1029/2004JC002826.

Tillinger, D., Gordon, A.L., 2010. Transport weighted temperature and internal energy transport of the Indonesian Throughflow. Dyn. Atmos. Oceans 50 (2), 224–232.

Toole, J., Warren, B.A., 1993. A hydrographic section across the South Indian Ocean. Deep Sea Res. Part I 40, 1973–2019.

Tozuka, T., Qu, T., Yamagata, T., 2007. Dramatic impact of the South China Sea on the Indonesian Throughflow. Geophys. Res. Lett. 34, L12612. http://dx.doi.org/10.1029/2007GL030420.

Tragou, E., Garrett, C., Outerbridge, R., Gilman, C., 1999. The heat and freshwater budgets of the Red Sea. J. Phys. Oceanogr. 29 (10), 2504–2522.

van Aken, H.M., Punjanan, J., Saimima, S., 1988. Physical aspects of the flushing of the East Indonesian basins. Netherlands J. Sea Res. 22, 315–339.

van Aken, H.M., Brodjonegoro, I.S., Jaya, I., 2009. The deep water motion through Lifamatola Passage and its contribution to the Indonesian throughflow. Deep Sea Res. 56, 1203–1216.

van Sebille, E., Biastoch, A., van Leeuwen, P.J., de Ruijter, W.P.M., 2009. A weaker Agulhas Current leads to more Agulhas leakage. Geophys. Res. Lett. 36 (3), L03601.

van Sebille, E., Baringer, M.O., Johns, W.E., Meinen, C.S., Beal, L., de Jong, M.M.F., van Aken, H.M., 2011. Propagation pathways of classical Labrador Sea water from its source region to 26°N. J. Geophys. Res. 116, C12027.

Vargas-Yanez, M., Moya, F., Garcia-Martinez, M.C., Tel, E., Zunino, P., Plaza, F., Salat, J., Pascual, J., Lopez-Jurado, J.L., Serra, M., 2010. Climate change in the Western Mediterranean Sea 1900–2008. J. Mar. Syst. 82 (3), 171–176.

Vecchi, G.A., Soden, B.J., Wittenberg, A.T., Isaac, H.M., Leetmaa, A., Harrison, M.J., 2006. Weakening of the tropical Pacific atmospheric circulation due to anthropogenic forcing. Nature 441, 73–76. http://dx.doi.org/10.1038/nature04744.

Vranes, K., Gordon, A.L., Field, A., 2002. The heat transport of the Indonesian throughflow and implications for the Indian Ocean heat budget. Deep Sea Res. Part II 49, 1410–1491.

Wainwright, L., Meyers, G., Wijffels, S.E., 2008. Change in the Indonesian Throughflow with the climatic shift of 1976/77. Geophys. Res. Lett. 35, L03604. http://dx.doi.org/10.1029/2007GL031911.

Warren, B.A., 1973. TransPacific hydrographic sections at 43°S and 28°S: the SCORPIO Expedition—II: deep water. Deep Sea Res. 20, 9–38.

Warren, B.A., 1978. Bottom water transport through the Southwest Indian Ridge. Deep Sea Res. 25, 315–321.

Warren, B.A., 1982. The deep water of the Central Indian Basin. J. Mar. Res. 40 (Suppl.), 823–860.

Weijer, W., De Ruijter, W.P.M., Sterl, A., Drijfhout, S.S., 2002. Response of the Atlantic overturning circulation to South Atlantic sources of buoyancy. Glob. Planet. Change 34 (3–4), 293–311.

Whitehead, J.A., 1998. Topographic control of oceanic flows in deep passages and straits. Rev. Geophys. 36, 423–440.

Whitworth, T., Peterson, R.G., 1985. Volume transport of the Antarctic Circumpolar Current from bottom pressure measurements. J. Phys. Oceanogr. 15 (6), 810–816.

Whitworth, T., Warren, B.A., Nowlin, W.D., Rutz, S.B., Pillsbury, R.D., Moore, M.I., 1999. On the deep western boundary current in the Southwest Pacific Basin. Prog. Oceanogr. 43, 1–54.

Wijffels, S., Meyers, G., 2004. An intersection of oceanic waveguides: variability in the Indonesian Throughflow region. J. Phys. Oceanogr. 34, 1232–1253.

Wijffels, S., Meyers, G., Godfrey, J.S., 2008. A twenty year average of the Indonesian Throughflow: regional currents and the interbasin exchange. J. Phys. Oceanogr. 38, 1965–1978.

Woelk, S., Quadfasel, D., 1996. Renewal of deep water in the Red Sea during 1982-1987. J. Geophys. Res. 101, 18155–18165.

Wüst, G., 1930. Meridionale Schichtung und Tiefenzirkulation in den Westhälften der drei Ozeane. Cons. Intl. Explor. Mer. 5 (1), 21 pp.

Wyrtki, K., 1961. Physical oceanography of the Southeast Asian waters. Naga Report 2. Scripps Institution of Oceanography, 195 pp.

Wyrtki, K., 1973. An equatorial jet in the Indian Ocean. Science 181, 262–264.

Wyrtki, K., 1987. Indonesian Throughflow and the associated pressure gradient. J. Geophys. Res. 92, 12, 941–12, 946.

You, Y., 2002. Quantitative estimate of Antarctic Intermediate Water contributions from the Drake Passage and the southwest Indian Ocean to the South Atlantic. J. Geophys. Res. 107, 3031.

You, Y., Suginohara, N., Fukasawa, M., Yasuda, I., Kaneko, I., Yoritaka, H., Kawamiya, M., 2000. Roles of the Okhotsk Sea and Gulf of Alaska in forming the North Pacific Intermediate Water. J. Geophys. Res. 105, 3253–3280.

You, Y., Suginohara, N., Fukasawa, M., Yoritaka, H., Mizuno, K., Kashino, Y., Hartoyo, D., 2003. Transport of North Pacific Intermediate Water across Japanese WOCE sections. J. Geophys. Res. 108 (C6), 23.

Zenk, W., Morozov, E., 2007. Decadal warming of the coldest Antarctic Bottom Water flow through the Vema Channel. Geophys. Res. Lett. 34, L14607. http://dx.doi.org/10.1029/2007GL030340.

Part V

Modeling the Ocean Climate System

Ocean models are a repository for mechanistic theories of how the ocean works, with numerical methods used to transform mathematical expressions of the theories into a computational tool for scientific investigations. Physical theories and numerical methods provide the foundation for numerical *ocean models*, whereas *ocean modeling* uses numerical model simulations as an experimental tool to help deduce mechanisms for emergent space–time patterns of ocean phenomena. These two *pillars* of numerical oceanography are supported by, and in turn support, observational oceanography and laboratory studies.

The increasing realism of numerical simulations, particularly since the first edition of this book in 2001, has greatly expanded the role of ocean models in oceanography and climate science. Notably, ocean models are now used, almost routinely, to help synthesize and mechanistically interpret observational-based analyses. Observational-based analyses are in turn essential for evaluating the fidelity of simulations. The increasingly vigorous interaction between numerical simulations and observational-based analyses stimulate synergistic advances to oceanography and climate science well beyond what could be achieved in isolation. In this Part V of the book, we encounter examples of this synergy that span a broad range of phenomena.

Chapters in This Part of the Book

As detailed in Chapter 20, ocean models marry theories of ocean fluid mechanics and thermodynamics to computational physics, whereas the exercise of ocean modeling uses numerical models as an experimental tool to address questions about the ocean; as an operational method to make predictions; or as part of an ocean state estimation system to mechanistically understand the present and past. The exercise of ocean modeling thus tests the theories and methods by comparing to the observed ocean, and in so doing closes the loop so that models can in turn improve both in their functionality and scientific fidelity.

Numerical ocean models enable a suite of scientific and operational applications. In particular, numerical models are used for hypothesis-driven experimental and theoretical studies that make them an essential tool in the hierarchy of methods used by scientists to better understand the earth climate system. When initialized through the assimilation of observations, ocean and coupled climate models allow for meaningful predictions of future states of the ocean, such as weekly forecasts of ocean eddies (Chapter 22), forecasts of seasonal to interannual fluctuations such as El Niño (Chapter 24), and potentially for decadal to multidecadal fluctuations of the large-scale overturning circulation (Chapter 25). Conversely, models provide the means for mechanistically interpreting the present and past history of the ocean through state estimation and retrospective or historical simulations (Chapters 21 and 22).

Finally, ocean models are an essential component of earth system models that aim to project future states of the global climate system, including biogeochemical cycles (Chapter 26), with questions related to human-induced climate change framing nearly all such studies (Chapter 23).

The Need for Ongoing Efforts from Future Generations

Although they are becoming a ubiquitous tool in oceanography and climate science, numerical ocean models have important limitations that require significant research to overcome. One limitation, largely outside the hands of oceanographers, is based on computer technology, in which ocean modelers have a proven ability to reach the boundaries of computational speed and archive space. Even though computer technology continues to evolve, these boundaries will inevitably be reached, given the wide range of space–time scales relevant for global and regional ocean circulation. The boundaries in turn necessitate difficult compromises on grid spacing, extent of the simulation time, number of prognostic tracers such as for simulating biogeochemical cycles, and the amount of diagnostic output to be archived.

A more basic limitation of models, which is central to the questions of oceanography and climate science, concerns our incomplete understanding of ocean processes, including interior mixing and stirring as well as boundary fluxes of momentum and buoyancy. Limited understanding of physical processes impacts the accuracy of parameterizations used for unresolved physics; these topics are the subject of Chapters 7 and 8. Limitations imposed by inaccurate parameterizations motivate the ubiquitous push for increasingly refined resolution models. The underlying assumption is that more accurately resolving a process leads to a greater simulation fidelity than parameterizing the process. This assumption is partially supported through improvements in flow representation realized by global climate models that admit a vigorous mesoscale eddy field.

Nonetheless, the range of space and time scales of importance for global, regional, and coastal ocean circulation are incredibly broad, thus removing any pretense of fully resolving all important features. Furthermore, there are large uncertainties associated with interactions between the ocean and other components of the climate system (Chapters 5 and 16), which in turn lead to uncertainties in the mechanical and buoyancy fluxes acting to force the ocean circulation. Many details impacting these interactions are at the fine scale (meters) and microscale (down to millimeters), leaving large-scale modelers no choice but to seek faithful parameterizations of these processes as well as more accurate surface flux datasets to evaluate the simulations. Hence, although progress continues to help better understand the ocean, great uncertainties remain, and they in turn lead to uncertainties in simulations.

It is largely for these reasons that the development of numerical ocean models, and the use of these models for ocean and climate modeling, is an active area of intense research. It is notable that this research spans many disciplines, from nearly all facets of ocean science, to various areas of physics, chemistry, biology, applied mathematics, engineering, and computational sciences. In general, there are many important fundamental and applied questions that remain to be answered by scientists and engineers. We therefore expect that numerical oceanography will remain both exciting and central to the evolution of oceanography and climate science for the foreseeable future.

Chapter 20

Ocean Circulation Models and Modeling

Stephen M. Griffies* and Anne Marie Treguier†
*NOAA Geophysical Fluid Dynamics Laboratory, Princeton, New Jersey, USA
†Laboratoire de Physique des Océans, LPO, Brest, France

Chapter Outline

1. SCOPE OF THIS CHAPTER

We focus in this chapter on numerical models, used to understand and predict large-scale ocean circulation, such as the circulation comprising basin and global scales. It is organized according to two themes, which we consider the "pillars" of numerical oceanography. The first addresses physical and numerical topics forming a foundation for ocean models. We focus here on the science of ocean models, in which we ask questions about fundamental processes and develop the mathematical equations for ocean thermo-hydrodynamics. We also touch upon various methods used to represent the continuum ocean fluid with a discrete computer model, raising such topics as the finite volume formulation of the ocean equations; the choice for vertical coordinate; the complementary issues related to horizontal gridding; and the pervasive questions of subgrid scale parameterizations. The second theme of this chapter

Ocean Circulation and Climate. Vol. 103. http://dx.doi.org/10.1016/B978-0-12-391851-2.00020-9

concerns the applications of ocean models, in particular, how to design an experiment and how to analyze results. This material forms the basis for ocean modeling, with the aim being to mechanistically describe, interpret, understand, and predict emergent features of the simulated, and ultimately the observed, ocean.

2. PHYSICAL AND NUMERICAL BASIS FOR OCEAN MODELS

As depicted in Figure 20.1, the ocean experiences a wide variety of boundary interactions and possesses numerous internal physical processes. Kinematic constraints on the fluid motion are set by the geometry of the ocean domain, and by assuming each fluid parcel conserves mass, save for the introduction of mass across the ocean surface (i.e., precipitation, evaporation, river runoff), or bottom (e.g., crustal vents). Dynamical interactions are described by Newton's Laws, in which the acceleration of a continuum fluid parcel is set by forces acting on the parcel. The dominant forces in the ocean interior are associated with pressure, the Coriolis force, gravity, and, to a lesser degree, friction. Boundary forces arise from interactions with the atmosphere, cryosphere, and solid earth, with each interaction generally involving buoyancy and momentum exchanges. Material budgets for tracers, such as salt and biogeochemical species, as well as thermodynamic tracers such as heat or enthalpy, are affected by circulation, mixing from turbulent processes, surface and bottom boundary fluxes, and internal sources and sinks, especially for biogeochemical tracers (see Chapter 26).

2.1. Scales of Motion

The ocean's horizontal gyre and overturning circulations occupy nearly the full extent of ocean basins (10^3 to

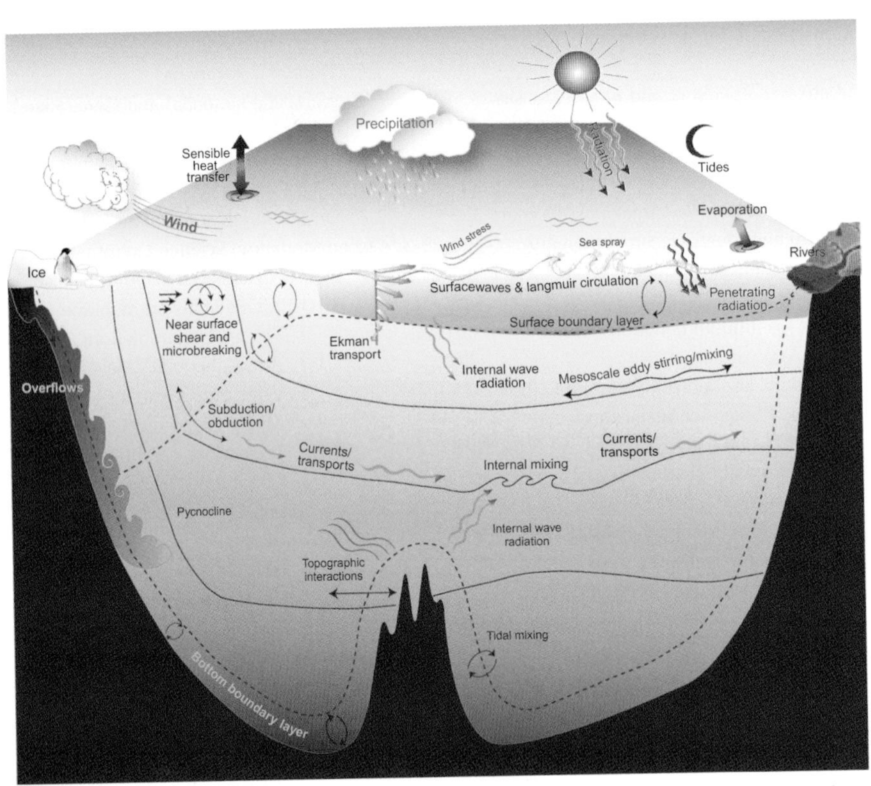

FIGURE 20.1 Understanding and quantifying the ocean's role in the earth system, including coastal, regional, and global phenomena, involves a variety of questions related to how physical processes impact the movement of tracers (e.g., heat, salt, carbon, nutrients) and momentum across the ocean boundaries and within the ocean interior. The ocean interacts with the variety of earth system components, including the atmosphere, sea ice, land ice shelves, rivers, and the solid earth lower boundary. Ocean processes transport material between the ventilated surface boundary layer and the ocean interior. When in the interior, it is useful to characterize processes according to whether they transport material across density surface (dianeutrally) or along neutral directions (epineutrally). In this figure, we illustrate the turbulent air–sea exchanges and upper ocean wave motions (including wave breaking and Langmuir circulations); subduction/obduction which exchanges material between the boundary layer and interior; gyre-scale, mesoscale, and submesoscale transport that largely occurs along neutral directions; high latitude convective and downslope exchange; and mixing induced by breaking internal gravity waves energized by winds and tides. Missing from this schematic are mixing due to double diffusive processes (Schmitt, 1994) and nonlinear equation of state effects (Chapter 6). Nearly all such processes are subgrid scale for present day global ocean climate simulations. The formulation of sensible parameterizations, including schemes that remain relevant under a changing climate (e.g., modifications to stratification and boundary forcing), remains a key focus of oceanographic research efforts, with Chapters 7 and 8 detailing many issues.

10^4 km in horizontal extent and roughly 4 km in depth on average), with typical recirculation times for the horizontal gyres of decadal, and overturning time scales of millennial. The ocean microscale is on the order of 10^{-3} m, and it is here that mechanical energy is transferred to internal energy through Joule heating. The microscale is set by the *Kolmogorov length*

$$L_{\mathrm{Kol}} = \left(\nu^3/\varepsilon\right)^{1/4}, \tag{20.1}$$

where $\nu \approx 10^{-6}\ \mathrm{m}^2\ \mathrm{s}^{-1}$ is the molecular kinematic viscosity for water, and ε is the energy dissipation rate. In turn, molecular viscosity and the Kolmogorov length imply a time scale $T = L^2/\nu \approx 1$ s.

Consider a direct numerical simulation (DNS) of ocean climate, where all space and time scales between the Kolmogorov scale and the global scale are explicitly resolved by the simulation. One second temporal resolution over a millennial time scale climate problem requires more than 3×10^{10} time steps of the model equations. Resolving space into cubes of dimension 10^{-3} m for an ocean with volume roughly $1.3 \times 10^{18}\ \mathrm{m}^3$ requires 1.3×10^{27} discrete grid cells, which is roughly 10^4 larger than Avogadro's Number. These numbers far exceed the capacity of any computer, thus necessitating approximated or truncated descriptions for practical ocean simulations, further promoting the central importance of subgrid scale parameterizations.

2.2. Thermo-Hydrodynamic Equations for a Fluid Parcel

As a starting point for developing ocean model equations, we consider the thermo-hydrodynamic equations for an infinitesimal seawater parcel. Some of this material is standard from geophysical fluid dynamics as applied to the ocean (e.g., see books such as Gill, 1982; Pedlosky, 1987; Vallis, 2006; Olbers et al., 2012), so the presentation here will be focused on setting the stage for later discussions.

2.2.1. Mass Conservation for Seawater and Trace Constituents

When formulating the tracer and dynamical equations for seawater, it is convenient to focus on a fluid parcel whose mass is constant. Writing the mass as $M = \rho\,\mathrm{d}V$, with $\mathrm{d}V$ the parcel's infinitesimal volume and ρ the *in situ* density, parcel mass conservation $\mathrm{d}M/\mathrm{d}t = 0$ yields the continuity equation

$$\frac{\mathrm{d}\rho}{\mathrm{d}t} = -\rho \nabla\cdot\mathbf{v}. \tag{20.2}$$

The three-dimensional velocity of the parcel is the time derivative of its position, $\mathbf{v} = \mathrm{d}\mathbf{x}/\mathrm{d}t$, and the horizontal and vertical components are written as $\mathbf{v} = (\mathbf{u}, w)$. Transforming this parcel or material Lagrangian expression into a fixed space or Eulerian perspective leads to the equivalent form

$$\frac{\partial\rho}{\partial t} = -\nabla\cdot(\rho\mathbf{v}), \tag{20.3}$$

where we related the material time derivative to the Eulerian time derivative through

$$\frac{\mathrm{d}}{\mathrm{d}t} = \partial_t + \mathbf{v}\cdot\nabla. \tag{20.4}$$

Seawater is comprised of freshwater along with a suite of constituents such as salt, nutrients, and biogeochemical elements and compounds. The tracer concentration, C, which is the mass of trace matter within a seawater parcel per mass of the parcel, is affected through the convergence of a tracer flux plus a potentially nonzero source/sink term $\mathcal{S}^{(C)}$ (sources and sinks are especially important for describing biogeochemical tracers; Chapter 26)

$$\rho\frac{\mathrm{d}C}{\mathrm{d}t} = -\nabla\cdot\mathbf{J}^{(C)} + \rho\mathcal{S}^{(C)}. \tag{20.5}$$

The canonical form of the tracer flux is associated with isotropic downgradient molecular diffusion

$$\mathbf{J}^{(C)}_{\mathrm{molecular}} = -\rho\kappa\nabla C, \tag{20.6}$$

where $\kappa > 0$ is a kinematic molecular diffusivity with units of length times a velocity, and $\rho\kappa$ is the corresponding dynamic diffusivity. For large-scale ocean models, the tracer flux $\mathbf{J}^{(C)}$ is modified according to the parameterization of various unresolved physical processes (see Chapters 7 and 8).

The Eulerian perspective converts the material time derivative into a local Eulerian time derivative plus advection $\rho(\partial_t + \mathbf{v}\cdot\nabla)C = -\nabla\cdot\mathbf{J}^{(C)} + \rho\mathcal{S}^{(C)}$. Combining this advective-form tracer equation with the seawater mass equation (20.3) leads to the Eulerian flux-form of the tracer equation

$$\partial_t(\rho C) = -\nabla\cdot\left(\rho\mathbf{v}C + \mathbf{J}^{(C)}\right) + \rho\mathcal{S}^{(C)}. \tag{20.7}$$

Setting the tracer concentration to a uniform constant in the tracer equation (20.7) recovers the mass continuity equation (20.3), where we assume there is no seawater mass source, and the tracer flux $\mathbf{J}^{(C)}$ vanishes with the concentration constant (e.g., see Section II.2 of DeGroot and Mazur, 1984; Section 8.4 of Chaikin and Lubensky, 1995, or Section 3.3 of Müller, 2006). This connection between the tracer equation and the seawater mass continuity equation is sometimes referred to as a *compatibility condition* (see Griffies et al., 2001 or Chapter 12 of Griffies, 2004). Equivalently, requiring that the tracer equation maintain a uniform tracer unchanged in the absence of boundary fluxes is sometimes referred to as *local* tracer conservation, which is a property required for conservative numerical algorithms. The flux-form in

Equation (20.7) is used in Section 2.4 as the basis for developing finite volume equations for a region of seawater.

2.2.2. *Conservative Temperature and* In Situ *Density*

As detailed by McDougall (2003), potential enthalpy provides a useful measure of heat in a seawater parcel (see also Chapter 6). Conservative temperature, Θ, is the potential enthalpy divided by a constant heat capacity. According to the First Law of Thermodynamics, it satisfies, to an extremely good approximation, a scalar conservation equation directly analogous to material tracers

$$\rho \frac{d\Theta}{dt} = -\nabla \cdot \mathbf{J}^{(\Theta)}. \tag{20.8}$$

This equation, or its Eulerian form, is termed "conservative" since the net heat content in a region is impacted only through fluxes passing across the boundary of that region (see Chapter 26 for more discussion of conservative and nonconservative tracers). In fact, there are actually nonzero source terms that are neglected in Equation (20.8), so that conservative temperature is not precisely "conservative." However, McDougall (2003) noted that these omitted source terms are negligible, as they are about 100 times smaller than those source terms omitted when considering potential temperature, θ, to be a conservative scalar. It is for this reason that IOC et al. (2010) recommend the use of conservative temperature, Θ, as a means to measure the heat of a seawater parcel.

The equation of state, which provides an empirical expression for the *in situ* density ρ, is written as a function of conservative temperature, salinity, and pressure

$$\rho = \rho(\Theta, S, p). \tag{20.9}$$

Note that the equation of state as derived in IOC et al. (2010) is written in terms of the Gibb's thermodynamic potential, thus making it self-consistent with other thermodynamic properties of seawater. Based on this connection, efforts are underway to update ocean model codes and analysis methods toward the recommendations of IOC et al. (2010).

2.2.3. *Momentum Equation*

Newton's Second Law of Motion applied to a continuum fluid in a rotating frame of reference leads to the equation describing the evolution of linear momentum per volume of a fluid parcel

$$\rho\left(\frac{d}{dt} + 2\boldsymbol{\Omega}\wedge\right)\mathbf{v} = -\rho\nabla\Phi + \nabla\cdot(\boldsymbol{\tau} - \mathbf{I}p). \tag{20.10}$$

The momentum equation (20.10) encapsulates nearly all the phenomena of ocean and atmospheric fluid mechanics. Such wide applicability is a testament to the power of classical mechanics to describe observed natural phenomena. The terms in the equation are the following.

Acceleration: When considering fluid dynamics on a flat space, the acceleration times density, $\rho d\mathbf{v}/dt$, takes the following Eulerian flux-form

$$\rho \frac{d\mathbf{v}}{dt} = \frac{\partial(\rho \mathbf{v})}{\partial t} + \nabla\cdot(\rho \mathbf{v}\mathbf{v}) \quad \text{flat space,} \tag{20.11}$$

which is directly analogous to the flux-form tracer equation (20.7). However, for fluid dynamics on a curved surface such as a sphere, the acceleration picks up an extra source-like term that is associated with curvature of the surface. When using locally orthogonal coordinates to describe the motion, acceleration takes the form (see Section 4.4.1 of Griffies, 2004)

$$\rho \frac{d\mathbf{v}}{dt} = \frac{\partial(\rho \mathbf{v})}{\partial t} + \nabla\cdot(\rho \mathbf{v}\mathbf{v}) + \mathcal{M}(\hat{\mathbf{z}}\wedge\rho\mathbf{v}) \quad \text{sphere.} \tag{20.12}$$

For spherical coordinates, $\mathcal{M} = (u/r)\tan\phi$, with ϕ the latitude and r the radial position. At latitude $\phi = 45°$ with $r \approx 6.37 \times 10^6$ m, and for a zonal current of $u = 1$ m s^{-1}, $\mathcal{M} \approx 10^{-3} f$, where

$$f = 2\Omega \sin\phi, \tag{20.13}$$

is the Coriolis parameter (see below). Hence, $\mathcal{M}$ is generally far smaller than the inertial frequency, f, determined by the Earth's rotation, except near the equator where f vanishes.

The nonlinear self-advective transport term $\rho\mathbf{v}\mathbf{v}$ contributing to the acceleration (see Equation 20.12) accounts for the rich variety of nonlinear and cross-scale turbulent processes that pervade the ocean. At the small scales (hundreds of meters and smaller), such processes increase three-dimensional gradients of tracer and velocity through straining and filamentation effects, and in so doing increase diffusive fluxes. In turn, tracer variance and kinetic energy cascade to the small scales through the effects of three-dimensional turbulence (*direct cascade*), and are dissipated at the microscale (millimeters) by molecular viscosity and diffusivity. At the larger scales where vertical stratification and quasi-geostrophic dynamics dominate (Chapter 11), kinetic energy preferentially cascades to the large scales (*inverse cascade*) as in two-dimensional fluid dynamics, whereas tracer variance continues to preferentially cascade to the small scales. Such cascade processes are fundamental to how energy and tracer variance are transferred across the many space-time scales within the ocean fluid.

Coriolis force: Angular rotation of the earth about the polar axis, measured by $\boldsymbol{\Omega}$, leads to the Coriolis force per volume, $2\rho\boldsymbol{\Omega}\wedge\mathbf{v}$. The locally horizontal component to the rotation vector, $f^* = 2\Omega\cos\phi$, can induce *tilted convection* that causes convecting plumes to deflect laterally (Denbo and Skyllingstad, 1996; Wirth and Barnier, 2006, 2008). Another effect was noted by Stewart and Dellar (2011), who argue for the importance of f^* in cross-equatorial flow of abyssal currents. However, hydrostatic primitive equation ocean models, which are the most common basis for large-scale models of the ocean, retain only the local vertical component of the earth's rotation, and thus approximate the Coriolis force according to

$$2\rho\mathbf{\Omega}\wedge\mathbf{v}\approx\hat{\mathbf{z}}f\wedge(\rho\mathbf{v}), \tag{20.14}$$

where f (Equation 20.13) is termed the Coriolis parameter. Marshall et al. (1997) provide a discussion of this approximation and its connection to hydrostatic balance. It is this form of the Coriolis force that gives rise to many of the characteristic features of geophysical fluid motions, such as Rossby waves, Kelvin waves, western boundary currents, and other large-scale features (Chapter 11).

Gravitational force: The gravitational potential, Φ, is commonly approximated in global circulation models as a constant gravitational acceleration, g, times the displacement, z, from resting sea level or the surface ocean geopotential (geoid),

$$\Phi\approx gz. \tag{20.15}$$

However, the geopotential must be considered in its more general form when including astronomical tide forcing and/or changes to the geoid due to rearrangements of mass; for example, melting land ice such as in the studies of Mitrovica et al. (2001) and Kopp et al. (2010).

Frictional stresses and pressure: The symmetric second order deviatoric stress tensor, $\boldsymbol{\tau}$, accounts for the transfer of momentum between fluid parcels due to shears, whereas p is the pressure force acting normal to the boundary of the parcel, with $\mathbf{I}$ the unit second order tensor. At the microscale, frictional stresses are parameterized by molecular diffusive fluxes in the same way as for tracers in Equation (20.6), with this parameterization based on analogy with the kinetic theory of gases (e.g., Section 12.3 of Reif, 1965). Vertical stresses in the ocean interior are thought to be reasonably well parameterized in this manner for large-scale ocean models, with the eddy viscosity far larger than molecular viscosity due to momentum mixing by unresolved eddy processes. In contrast, there is no consensus on how to represent lateral frictional stress in large-scale ocean models, with modelers choosing lateral friction based on empirical (i.e., "tuning") perspectives (Part 5 in Griffies, 2004, as well as Jochum et al., 2008 and Fox-Kemper and Menemenlis, 2008 for further discussion). In Section 2.6, we have more to say about certain issues involved with setting lateral friction in models.

2.2.4. Comments on Parcel Equations

The mass conservation equation (20.2), tracer equation (20.5), conservative temperature equation (20.8), equation of state (20.9), momentum equation (20.10), and boundary conditions (Section 2.4), are the basic building blocks for a mathematical physics description of ocean thermo-hydrodynamics. However, these equations alone do not provide an algorithm for numerical simulations. Indeed, we know of no algorithm, much less a working numerical code, based on a realistic nonlinear equation of state for a mass conserving and nonhydrostatic ocean. Instead, various approximations are made, either together or separately, that have proved useful for developing numerical ocean model algorithms.

2.3. Approximation Methods

Three general approaches to approximation, or truncation, are employed in computational fluid dynamics, and we outline here these approaches as used for ocean models.

2.3.1. Coarse Grid and Realistic Large-Scale Domain

One approach is to coarsen the space and time resolution used by the discrete grid forming the basis for the numerical simulation. By removing scales smaller than the grid, the truncated system carries less information than the continuum. Determining how the resolved scales are affected by the unresolved scales is fundamental to the science of ocean models: this is the parameterization problem (Section 2.5).

2.3.2. Refined Grid and Idealized Small Domain

A complementary approach is to configure a small space–time domain so as to maintain the very fine space and time resolution set by either molecular viscosity and diffusivity (DNS), or somewhat larger eddy viscosity and diffusivity (large eddy simulation (LES)). These simulations are necessarily idealized both because of their small domain and the associated need to include idealized boundary conditions. Both DNS and LES are important for process studies aimed at understanding the mechanisms active in fine scale features of the ocean. Insights gained via DNS and LES have direct application to the development of subgrid scale parameterizations used in large-scale models. Large-scale simulations that represent a wide range of mesoscale and submesoscale eddies (e.g., finer than 1-km grid spacing) share much in common with LES (Fox-Kemper and Menemenlis, 2008). Such simulations will conceivably be more common for global climate scales within the next one or two decades, as computational power increases.

2.3.3. Filtering the Continuum Equations: Hydrostatic Approximation

A third truncation method filters the continuum equations by truncating the fundamental modes of motion admitted by the equations. This approach reduces the admitted motions and reduces the space–time scales required to simulate the system.

The hydrostatic approximation is a prime example of mode filtering used in large-scale modeling. Here, the admitted vertical motions possess far less kinetic energy than horizontal motions, thus rendering a simplified vertical momentum balance where the weight of fluid above a point in the ocean determines the pressure at that point

$$\frac{\partial p}{\partial z} = -\rho \frac{\partial \Phi}{\partial z} \quad \text{hydrostatic balance.} \tag{20.16}$$

Since vertical convective motion involves fundamentally nonhydrostatic dynamics (Marshall and Schott, 1999), hydrostatic primitive equation models must parameterize these effects (Klinger et al., 1996). Although the hydrostatic approximation is ubiquitous in large scale ocean modeling (for scales larger than roughly 1 km), there are many process studies that retain nonhydrostatic dynamics, with the MIT general circulation model a common publicly available code used for such studies (Marshall et al., 1997).

2.3.4. Filtering the Continuum Equations: Oceanic Boussinesq Approximation

In situ density in the large-scale ocean varies by a relatively small amount, with a 5% variation over the full ocean column mostly due to compressibility. Further, the dynamically relevant horizontal density variations are on the order of 0.1%. These observations motivate the *oceanic Boussinesq approximation*.

As detailed in Section 9.3 of Griffies and Adcroft (2008), the first step to the oceanic Boussinesq approximation applies a linearization to the momentum equation by removing the nonlinear product of density times velocity, in which the product $\rho \mathbf{v}$ is replaced by $\rho_o \mathbf{v}$, where ρ_o is a constant Boussinesq reference density. However, one retains the *in situ* density dependence of the gravitational potential energy, and correspondingly it is retained for computing pressure. The second step considers the mass continuity equation (20.3), where the three-dimensional flow is incompressible to leading order

$$\nabla \cdot \mathbf{v} = 0 \quad \text{volume conserving Boussinesq approximation.} \tag{20.17}$$

This step filters acoustic modes (i.e., sound waves), if they are not already filtered, by making the hydrostatic approximation.

As revealed by the mass conservation equation (20.2), a nontrivial material evolution of *in situ* density requires a divergent velocity field. However, a divergent velocity field is unresolved in oceanic Boussinesq models. Not resolving the divergent velocity field does not imply this velocity vanishes. Indeed, the oceanic Boussinesq approximation retains the dependence of density on pressure (or depth), temperature, and salinity (Equation 20.9), thus avoiding any assumption regarding the fluid properties. In turn, such models allow for a consistent material evolution of *in situ* density, with this evolution critical for representing the thermohaline-induced variations in density (and hence pressure) that are key drivers of the large scale ocean circulation (Chapter 11).

An element missing from Boussinesq ocean models concerns the calculation of global mean sea level. Greatbatch (1994) noted that the accumulation of seawater compressibility effects over an ocean column leads to meaningful systematic changes in global sea level when, for example, the ocean is heated. These global steric effects must therefore be added *a posteriori* to a Boussinesq simulation of sea level to provide a meaningful measure of global sea-level changes associated with buoyancy forcing (see also the sea-level discussion in Chapter 27). Griffies and Greatbatch (2012) build on the work of Greatbatch (1994) by detailing how physical processes impact global mean sea level in ocean models.

2.4. Thermo-Hydrodynamic Equations for a Finite Region

Our next step in developing the equations of an ocean model involves integrating the continuum parcel equations over a finite region, with the region boundaries generally moving and permeable. The resulting budget equations form the basis for a finite volume discretization of the ocean equations. They may also be used to develop basin-wide budgets for purposes of large-scale analysis (Section 3.2). The finite volume approach serves our pedagogical aims, and it forms the basis for most ocean models in use today for large-scale studies. We make reference to the schematic shown in Figure 20.2 relevant for a numerical model.

2.4.1. Finite Volume Budget for Scalars and Momentum

Consider a volume of fluid, V, with a moving and permeable boundary $\mathcal{S}$. The tracer mass budget within this region satisfies

$$\frac{\partial}{\partial t}\left(\int_V C\rho \, \mathrm{d}V\right) = -\int_{\mathcal{S}} \hat{\mathbf{n}} \cdot \left[\left(\mathbf{v} - \mathbf{v}^{\mathcal{S}}\right)\rho C + \mathbf{J}\right] \mathrm{d}\mathcal{S}, \tag{20.18}$$

where we ignored tracer source/sink terms for brevity, dropped the superscript (C) on the subgrid scale tracer flux $\mathbf{J}$, and wrote $\hat{\mathbf{n}}$ for the outward normal to the boundary. Tracer mass within a region (left hand side) changes due to the passage of tracer through the boundary, either from advective transport or subgrid scale transport (right hand side).

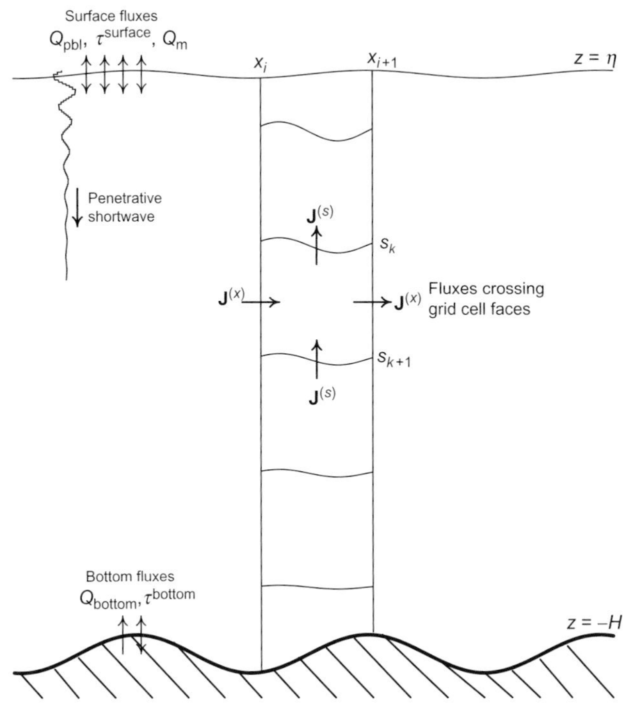

FIGURE 20.2 A longitudinal–vertical slice of ocean fluid from the surface at $z = \eta(x, y, t)$ to bottom at $z = -H(x, y)$, along with a representative column of discrete grid cells (a latitudinal–vertical slice is analogous). Most ocean models used for large-scale climate studies assume the horizontal boundaries of a grid cell at x_i and x_{i+1} are static, whereas the vertical extent, defined by surfaces of constant generalized vertical coordinate s_k and s_{k+1}, can be time dependent. The tracer flux **J** is decomposed into horizontal and dia-surface components, with the convergence of these fluxes onto a grid cell determining the evolution of tracer content within the cell. Similar decomposition occurs for momentum fluxes. Additional terms contributing to the evolution of tracer include source terms, and momentum evolution also includes body forces (Coriolis and gravity). Among the fluxes crossing the ocean surface, the shortwave flux penetrates into the ocean column as a function of the optical properties of seawater (e.g., Manizza et al., 2005).

Advective transport is measured according to the normal projection of the fluid velocity in a frame moving with the surface, $\mathbf{v} - \mathbf{v}^{\mathcal{S}}$. The subgrid scale tracer transport must likewise be measured relative to the moving surface. The finite volume budget for seawater mass is obtained by setting the tracer concentration to a constant in the tracer budget (20.18)

$$\frac{\partial}{\partial t}\left(\int_V \rho \, \mathrm{d}V\right) = -\int_{\mathcal{S}} \hat{\mathbf{n}} \cdot \left(\mathbf{v} - \mathbf{v}^{\mathcal{S}}\right) \rho \, \mathrm{d}\mathcal{S}. \tag{20.19}$$

The relation between the mass budget (20.19) and tracer budget (20.18) is a manifestation of the compatibility condition discussed following the continuum tracer equation (20.7). An analogous finite volume budget follows for the hydrostatic primitive equations, in which we consider the horizontal momentum over a finite region with the Coriolis force in its simplified form (20.14)

$$\begin{aligned}\partial_t\left(\int_V \mathbf{u}\rho \, \mathrm{d}V\right) = &-\int_V [g\hat{\mathbf{z}} + (f + \mathcal{M})\hat{\mathbf{z}} \wedge \mathbf{u}]\rho \, \mathrm{d}V \\ &- \int_{\mathcal{S}} \left[\hat{\mathbf{n}} \cdot \left(\mathbf{v} - \mathbf{v}^{\mathcal{S}}\right)\right] \mathbf{u}\rho \, \mathrm{d}\mathcal{S} + \int_{\mathcal{S}} \hat{\mathbf{n}} \cdot (\boldsymbol{\tau} - \mathbf{I}p) \, \mathrm{d}\mathcal{S}.\end{aligned} \tag{20.20}$$

The volume integral on the right hand side arises from the gravitational and Coriolis body forces, whereas the surface integrals arise from both advective transport and contact forces associated with stress and pressure.

Some domain boundaries are static, such as the lateral boundaries for a model grid cell or the solid earth boundaries of an ocean basin (Figure 20.2). However, vertical boundaries are quite often moving, with the ocean-free surface

$$z = \eta(x, y, t), \tag{20.21}$$

a canonical example. In this case, the projection of the boundary velocity onto the normal direction is directly proportional to the time tendency of the free surface

$$\hat{\mathbf{n}} \cdot \mathbf{v}^{\mathcal{S}} = \left(\frac{\partial \eta}{\partial t}\right) |\nabla(z - \eta)|^{-1}. \tag{20.22}$$

Iso-surfaces of a generalized vertical coordinate

$$s = s(x, y, z, t), \tag{20.23}$$

are generally space and time dependent. For example, the grid cell top and bottom may be bounded by surfaces of constant pressure, potential density, or another moving surface. Here, the normal component of the surface velocity is proportional to the tendency of the generalized vertical coordinate

$$\hat{\mathbf{n}} \cdot \mathbf{v}^{\mathcal{S}} = -\left(\frac{\partial s}{\partial t}\right) |\nabla s|^{-1}. \tag{20.24}$$

2.4.2. Generalized Vertical Coordinates and Dia-Surface Transport

To make use of a finite volume budget for layers defined by generalized vertical coordinates requires that the vertical coordinate be monotonically stacked in the vertical, so that there is a one-to-one relation between the geopotential coordinate, z, and the generalized vertical coordinate. Mathematically, this constraint means that the *specific thickness* $\partial s/\partial z$ never vanishes, and thus remains of one sign throughout the domain so there are no inversions in the generalized vertical coordinate iso-surfaces. An important case is where $\partial s/\partial z = 0$ occurs for isopycnal models in regions of zero vertical density stratification. Handling such regions necessitates either a transformation to a stably stratified vertical coordinate such as pressure, as in the Hybrid Ocean Model (HYCOM) code of Bleck

(2002), or appending a bulk mixed layer (Hallberg, 2003) to the interior isopycnal layers as in the Miami Isopycnal Coordinate Ocean Model code of Bleck (1998), or the General Ocean Layer Dynamics (GOLD) code used in Adcroft et al. (2010).

The monotonic assumption (i.e., $\partial s/\partial z$ remains single signed) allows us to measure the advective transport across the constant s surfaces according to the dia-surface velocity component (Section 2.2 of Griffies and Adcroft, 2008)

$$\rho w^{(s)} \equiv \frac{(\text{Mass/time}) \text{ of fluid through surface}}{\text{Area of horizontal projection of surface}} \quad (20.25a)$$

$$= \frac{\hat{\mathbf{n}}\cdot(\mathbf{v}-\mathbf{v}^S)\rho\, \mathrm{d}\mathcal{S}}{\mathrm{d}A}, \quad (20.25b)$$

where $\mathrm{d}A$ is the horizontal projection of the surface area $\mathrm{d}\mathcal{S}$. Questions of how to measure dia-surface mass transport arise in many areas of ocean model formulation as well as construction of budgets for ocean domains. We present here two equivalent expressions

$$w^{(s)} = \left(\frac{\partial z}{\partial s}\right)\frac{\mathrm{d}s}{\mathrm{d}t} \quad (20.26a)$$

$$= w - (\partial_t + \mathbf{u}\cdot\nabla_s)z, \quad (20.26b)$$

in which $\nabla_s z = -(\partial z/\partial s)\nabla_z s$ is the slope of the s surface as projected onto the horizontal plane (Chapter 6 of Griffies, 2004). Equation (20.26a) indicates that if the vertical coordinate has zero material time derivative, then there is zero dia-surface mass transport. Equation (20.26b) is commonly encountered when studying subduction of water from the mixed layer to the ocean interior, in which the generalized vertical coordinate is typically an isopycnal or isotherm (e.g., Marshall et al., 1999). A final example of dia-surface transport arises from motion across the ocean-free surface at $z=\eta(x, y, t)$, in which case

$$Q_\mathrm{m}\,\mathrm{d}A \equiv (\text{Mass/time}) \text{ of fluid through free surface} \quad (20.27a)$$

$$= -\mathrm{d}A(w - \mathbf{u}\cdot\nabla\eta - \partial_t\eta)\rho, \quad (20.27b)$$

with $Q_\mathrm{m} > 0$ if mass enters the ocean. Rearrangement leads to the surface kinematic boundary condition

$$\rho(\partial_t + \mathbf{u}\cdot\nabla)\eta = w + Q_\mathrm{m} \quad \text{at } z=\eta. \quad (20.28)$$

2.4.3. *Surface and Bottom Boundary Conditions*

The tracer flux leaving the ocean through the free surface is given by (see Equation 20.18)

$$\int_{z=\eta} \hat{\mathbf{n}}\cdot\left[(\mathbf{v}-\mathbf{v}^S)\rho C + \mathbf{J}\right]\mathrm{d}\mathcal{S} = \int_{z=\eta}\left(-Q_\mathrm{m}C + J^{(s)}\right)\mathrm{d}A, \quad (20.29)$$

where

$$\mathrm{d}A J^{(S)} = \mathrm{d}\mathcal{S}\hat{\mathbf{n}}\cdot\mathbf{J}, \quad (20.30)$$

is the dia-surface tracer transport associated with subgrid scale processes and/or parameterized turbulent boundary fluxes. Boundary fluxes are often given in terms of bulk formula (see, e.g., Taylor, 2000; Appendix C of Griffies et al., 2009; Section 3.1; Chapter 5), allowing for the boundary flux to be written in the form

$$-\int_{z=\eta} \hat{\mathbf{n}}\cdot\left[(\mathbf{v}-\mathbf{v}^S)\rho C + \mathbf{J}\right]\mathrm{d}\mathcal{S} = \int_{z=\eta}(Q_\mathrm{m}C_\mathrm{m} + Q_\mathrm{pbl})\mathrm{d}A, \quad (20.31)$$

where C_m is the tracer concentration within the incoming mass flux Q_m. The first term on the right hand side of Equation (20.31) represents the advective transport of tracer through the surface with the water (i.e., ice melt, rivers, precipitation, evaporation). The term Q_pbl arises from parameterized turbulence and/or radiative fluxes within the surface planetary boundary layer, such as sensible, latent, shortwave, and longwave heating as occurs for the temperature equation, with $Q_\mathrm{pbl} > 0$ signaling tracer entering the ocean through its surface. A similar expression to (20.31) holds at the ocean bottom $z=-H(x, y)$, though it is common in climate modeling to only consider geothermal heating (Adcroft et al., 2001; Emile-Geay and Madec, 2009) with zero mass flux.

The force acting on the bottom surface of the ocean is given by

$$\mathbf{F}_\text{bottom} = -\int_{z=-H}[\nabla(z+H)\cdot\boldsymbol{\tau} - p\nabla(z+H)]\mathrm{d}A. \quad (20.32)$$

In the presence of a nonzero topography gradient, $\nabla H \neq 0$, the term $-p\nabla H$ at the ocean bottom gives rise to a topographic *form stress* that affects horizontal momentum. Such stress is especially important for strong flows that reach to the ocean bottom, such as in the Southern Ocean (Chapter 18). Parameterization of this stress is particularly important for models that only resolve a coarse-grained representation of topography. In addition to form stress, we assume that a boundary layer model, typically in the form of a drag law, provides information so that we can parameterize the bottom vector stress

$$\boldsymbol{\tau}^\text{bottom} \equiv \nabla(z+H)\cdot\boldsymbol{\tau} \quad \text{at } z=-H, \quad (20.33)$$

associated with bottom boundary layer momentum exchange. This parameterization of bottom stress necessarily incorporates interactions between the ocean fluid with small scale topography variations, so that there is a nonzero vector stress $\boldsymbol{\tau}^{\text{bottom}}$ even if the large-scale topography resolved by a numerical model is flat. Additional considerations for the interactions between unresolved mesoscale eddies with topography lead to the Neptune parameterization of Holloway (1986, 1989, 1992).

Momentum transfer through the ocean surface is given by

$$\mathbf{F}_{\text{surface}} = \int_{z=\eta} \left[\boldsymbol{\tau}^{\text{surface}} - p_{\text{a}} \nabla(z-\eta) + Q_{\text{m}} \mathbf{u}_{\text{m}}\right] \text{d}A. \quad (20.34)$$

In this equation, $\mathbf{u}_{\text{m}}$ is the horizontal velocity of the mass transferred across the ocean boundary. This velocity is typically taken equal to the velocity of the ocean currents in the top cell of the ocean model, but such is not necessarily the case when considering the different velocities of, say, river water and precipitation. The vector stress

$$\boldsymbol{\tau}^{\text{surface}} \equiv \nabla(z-\eta) \cdot \boldsymbol{\tau} \ \text{ at } z=\eta, \quad (20.35)$$

arises from the wind, as well as interactions between the ocean and ice. As for the bottom stress parameterization (20.33), a boundary layer model determining the surface vector stress, $\boldsymbol{\tau}^{\text{surface}}$, must consider subgrid-scale fluctuations of the sea surface, such as nonlinear effects associated with surface waves (Sullivan and McWilliams, 2010; Belcher et al., 2012; Cavaleri et al., 2012). Finally, we take the applied pressure at $z=\eta$ to equal the pressure p_{a} from the media sitting above the ocean; namely, the atmosphere and ice. As for the bottom force, there is generally a nonzero horizontal projection of the applied pressure acting on the curved free surface, $p_{\text{a}} \nabla \eta$, thus contributing to an applied surface pressure form stress on the ocean.

2.5. Physical Considerations for Transport

Working with a discrete rather than continuous fluid presents many fundamental and practical issues. One involves the introduction of unphysical *computational modes* whose presence can corrupt the simulation; for example, dispersion arising from discrete advection operators can lead to spurious mixing (Griffies et al., 2000b; Ilicak et al., 2012). Another issue involves the finite grid size, Δ, or more generally the finite degrees of freedom available to simulate a continuum fluid. The grid scale is generally many orders larger than the Kolmogorov scale (Equation 20.1)

$$\Delta \gg L_{\text{Kol}}, \quad (20.36)$$

and Δ determines the degree to which an oceanic flow feature can be resolved by a simulation.

There are two reasons to parameterize a physical process impacting the ocean. The first is if the process is filtered from the continuum equations forming the basis for the model, such as the hydrostatic approximation (20.16). The second concerns the finite grid scale. To understand how the grid introduces a closure or parameterization problem, consider a *Reynolds decomposition* of an advective flux

$$\overline{u\Psi} = \bar{u}\bar{\Psi} + \overline{u'\Psi'}, \quad (20.37)$$

where $u=\bar{u}+u'$ expresses a velocity component as the sum of a mean and fluctuation, and the average of a fluctuating field is assumed to vanish, $\bar{u'}=0$. The same decomposition is assumed for the field being transported, Ψ, which could be a tracer concentration or velocity component. The discrete grid represents the product of the averaged fields, $\bar{u}\bar{\Psi}$, through a numerical advection operator. Computing this resolved transport using numerical methods is the *representation problem*, which involves specification of a numerical advection operator. The correlation term, $\overline{u'\Psi'}$, is not explicitly represented on the grid, with its specification constituting the *subgrid scale parameterization problem*. The correlation term is referred to as a *Reynolds stress* if Ψ is a velocity component, and an *eddy flux* if Ψ is a tracer. To deduce information about the second order correlation $\overline{u'\Psi'}$ requires third order correlations, which are functions of fourth order correlations, etc., thus forming the *turbulence closure problem*. Each process depicted in Figure 20.1 contributes to fluctuations, so they each engender a closure problem if unresolved.

The theory required to produce *mean field* or averaged fluid equations is extensive and nontrivial. A common aim is to render the resulting subgrid scale correlations in a form subject to physical insight and sensible parameterization. The variety of averaging methods amount to different mathematical approaches that are appropriate under differing physical regimes and are functions of the vertical coordinates used to describe the fluid. A nonexhaustive list of examples specific to the ocean include the following (see also Olbers et al., 2012 for further discussion of even more averaging methods).

- The microscale or infra-grid averaging of DeSzoeke and Bennett (1993), Davis (1994a,b), and DeSzoeke (2009) focuses on scales smaller than a few tens of meters.
- The density weighted averaging of Hesselberg (1926) (see also McDougall et al., 2002 and Chapter 8 of Griffies, 2004), provides a framework to account for the mass conserving character of the non-Boussinesq ocean equations, either hydrostatic or nonhydrostatic.

- The isopycnal thickness weighted methods of DeSzoeke and Bennett (1993), McDougall and McIntosh (2001), DeSzoeke (2009), and Young (2012) (see also Chapter 9 of Griffies, 2004) provide a framework to develop parameterizations of mesoscale eddy motions in the stratified ocean interior; see also the combined density and thickness weighted methods of Greatbatch and McDougall (2003). Eden et al. (2007) propose an alternative that averages over the same mesoscale phenomena, but maintains an Eulerian perspective rather than moving to isopycnal space.

There are few robust, and even fewer first principle, approaches to parameterization, with simulations often quite sensitive to the theoretical formulation as well as specific details of the numerical implementation. One may choose to ignore the topic of parameterizations, invoking an *implicit large eddy simulation* philosophy (Shchepetkin and McWilliams, 1998; Margolin et al., 2006; Grinstein et al., 2007), whereby the responsibility for closing the transport terms rests on the numerical methods used to represent advection. For large-scale modeling, especially with applications to climate, this approach is not common since the models are far from resolving many of the known important dynamical scales, such as the mesoscale. However, it is useful to test this approach to expose simulation features where the absence of a parameterization leads to obvious biases. Delworth et al. (2012) provide one such example, in which the ocean model component of a coupled climate model permits, but does not resolve, mesoscale eddies, and yet there is no parameterization of the unresolved portion of the mesoscale eddies. Determining methods of mesoscale parameterization for use in mesoscale eddy permitting models is an active research area. In general, simulations extending over decadal to longer times must confront an ocean whose circulation and associated water masses are fundamentally impacted by the zoo of physical processes depicted in Figure 20.1, most of which are unresolved and have nontrivial impacts on the simulation.

2.5.1. Parameterizing Transport in a Stratified Ocean

In an ideal ocean without mixing, tracer concentration is reversibly stirred by the resolved velocity field (Eckart, 1948). That is, tracer concentration is materially constant (Equation 20.5 with zero right hand side), and all tracer iso-surfaces are impenetrable to the resolved fluid flow. Mixing changes this picture, with molecular diffusion the ultimate cause of mixing and irreversibility. Upon averaging the equations according to the grid scale of a numerical model of a stratified ocean, subgrid eddy tracer fluxes associated with mesoscale eddies are generally parameterized by downgradient diffusion, oriented according to neutral directions (Solomon, 1971; Redi, 1982), with this parameterization termed *neutral*, *epineutral*, or *isoneutral* diffusion. As noted by Gent and McWilliams (1990), there is an additional eddy advective flux (see also Gent et al., 1995; Griffies, 1998). Over the past decade, the use of such *neutral physics* parameterizations has become ubiquitous in ocean climate models since they generally improve simulations of water masses (Chapter 8).

Dianeutral processes mix material across neutral directions (Chapter 7). These processes arise from enhanced mixing in upper and lower boundary layers (e.g., Large et al., 1994; Legg et al., 2009), as well as regions above rough topography (Kunze and Sanford, 1996; Polzin et al., 1997; Toole et al., 1997; Naveira-Garabato et al., 2004; Kunze et al., 2006; MacKinnon et al., 2010). Dianeutral mixing in the ocean interior away from rough topography is far smaller (Ledwell et al., 1993, 2011). Additionally, double diffusive processes (salt fingering and diffusive convection) arise from the differing rates for heat and salt diffusion (Schmitt, 1994). Finally, cabbeling and thermobaricity (Chapter 6) may play an important role in dianeutral transport within the ocean interior, especially in the Southern Ocean (Marsh, 2000; Iudicone et al., 2008; Klocker and McDougall, 2010). Cabbeling and thermobaricity arise from epineutral mixing of temperature and salinity in the presence of the nonlinear equation of state for seawater (McDougall, 1987).

Although vigorous in parts of the ocean, dianeutral transport is extremely small in other parts in comparison to the far larger epineutral transport. Indeed, ocean measurements indicate that the ratio of dianeutral to epineutral transport is roughly 10^{-8} in many regions away from boundaries and above relatively smooth bottom topography (Ledwell et al., 1993, 2011), and it can become even smaller at the equator (Gregg et al., 2003). Although tiny by comparison for much of the ocean, dianeutral transport in the ocean interior is in fact an important process involved with modifying vertical stratification. Consequently, it impacts fundamentally on the ocean's role in climate. In ocean climate models, the parameterization of interior dianeutral mixing has evolved from a prescribed and static vertical diffusivity proposed by Bryan and Lewis (1979), to a collection of subgrid scale processes largely associated with breaking internal gravity waves and other sources of enhanced vertical shear (e.g., Large et al., 1994; Simmons et al., 2004; Jackson et al., 2008; Melet et al., 2013) (Chapter 6).

2.5.2. Two Emerging Ideas for Parameterization

We mention two emerging approaches to account for subgrid scale processes that may impact on ocean climate modeling in the near future. Although much work remains to determine whether either will become practical, there are compelling physical and numerical reasons to give these proposals serious investigation.

Stochastic closure: Hasselmann (1976) noted that certain components of the climate system can be considered a stochastic, or noise, forcing that contributes to the variability of other components. The canonical example is an ocean that transfers the largely white noise fluctuations from the atmospheric weather patterns into a red noise response (i.e., increased power at the low frequencies) (Frankignoul and Hasselmann, 1977; Hall and Manabe, 1997). More recently, elements of the atmospheric and climate communities have considered a stochastic term in the numerical model equations used for weather forecasting and climate projections, with particular emphasis on the utility for tropical convection; for example, see Williams (2005) and Palmer and Williams (2008) for pedagogical discussions. This noise term is meant to parameterize elements of unresolved fluctuations as they feedback onto the resolved fields.

Depending on the phenomena, there are cases where subgrid scale ocean fields are indeed fluctuating chaotically. Further, the averaging operation applied to the nonlinear terms does not generally satisfy the Reynolds assumption of zero average for the fluctuating terms (i.e., $\overline{u'} \neq 0$) (Davis, 1994b; DeSzoeke, 2009). So along with the compelling results from atmospheric models, there are reasons to consider introducing a stochastic element to the subgrid scale terms used in an ocean model (Berlov, 2005; Brankart, 2013; Kitsios et al., 2013).

Super-parameterization: In an effort to improve the impact of atmospheric convective processes on the large-scale, Grabowski (2001) embedded a two-dimensional nonhydrostatic model into a three-dimensional large-scale hydrostatic primitive equation model. The nonhydrostatic model feeds information to the hydrostatic model about convective processes, and the hydrostatic model in turn provides information about the large-scale to the nonhydrostatic model. Khairoutdinov et al. (2008) further examined this *super-parameterization* approach and showed some promising results. Campin et al. (2011) in turn have applied the approach to oceanic convection (see Figure 20.3). Some processes are perhaps not parameterizable, and so must be explicitly represented. Additionally, some processes are not represented or parameterized well using a particular modeling framework. Both these cases may lend themselves to super-parameterizations.

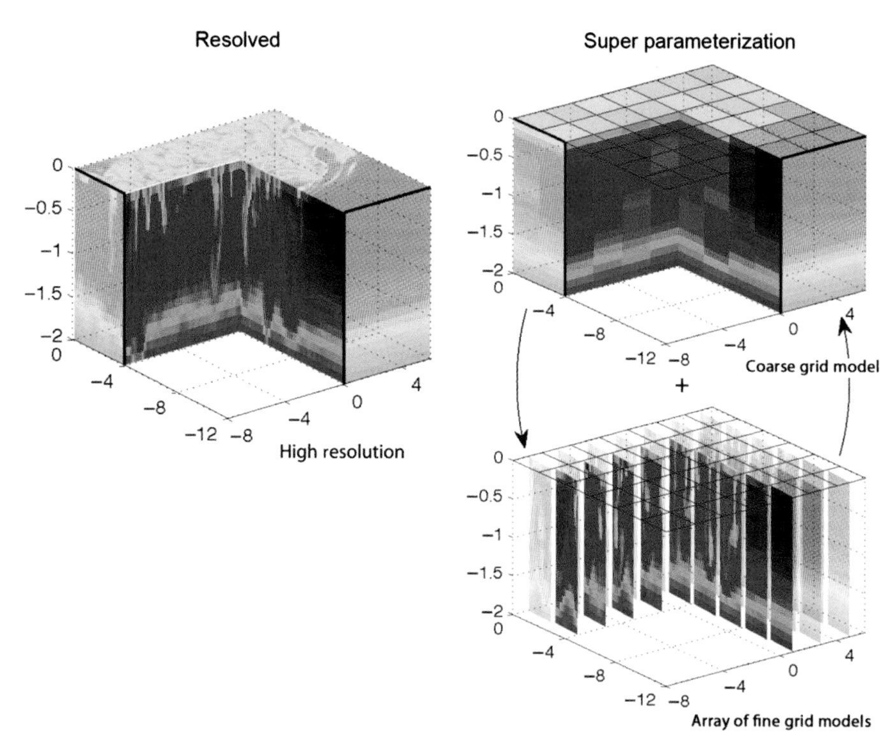

FIGURE 20.3 Three-dimensional view of the temperature field (red is warm, blue is cold) in two simulations of chimney convection similar to Jones and Marshall (1993). Left side: from a high-resolution simulation which resolves small scale plume processes. Right side: from a super-parameterized model in which a coarse-grain (CG) large-scale model (top right panel) representing balanced motion is integrated forward with embedded fine-grained (FG) (bottom right panel) running at each column of the large-scale grid. The FG is nonhydrostatic and attempts to resolve the small-scale processes. The FGs and the CG are integrated forward together and exchange information following the algorithm set out in Campin et al. (2011). *This figure is based on figure 1 of Campin et al. (2011).*

We consider a super-parameterization to be the use of a submodel (or child model) that is two-way embedded into the main or parent-model, with the submodel focused on representing certain processes that the parent-model either cannot resolve, or does a poor job of representing, due to limitations of its numerical methods. In this regard, super-parameterization ideas share features with two-way nesting approaches (Debreu and Blayo, 2008), in which a nested fine grid region resolves processes that the coarse grid parent-model cannot (we have more to say on nesting in Section 3.1). The approach of Bates et al. (2012a,b) is another example, in which a dynamic and interactive three-dimensional Lagrangian submodel is embedded in an Eulerian model.

2.5.3. *Where We Stand with Physical Parameterizations*

Many of the same questions regarding parameterizations raised in the review of Griffies et al. (2000a) remain topical in the research community today. This longevity is both a reflection of the difficulty of the associated theoretical and numerical issues and the importance of developing robust parameterizations suitable for a growing suite of applications. We offer the following assessment regarding the parameterization question:

A necessary condition for the evaluation of a physical process parameterization in global ocean climate simulations is to examine companion climate simulations that fully resolve the process.

That is, we will not know the physical integrity of a parameterization until the parameterized process is fully resolved. This assessment does not mean that comparisons between models and field observations, laboratory studies, or process studies, are irrelevant to the parameterization question. It does, however, summarize the situation with regard to certain phenomena such as the mesoscale, as supported with recent experience studying the Southern Ocean response to wind stress changes. As shown by Farneti et al. (2010), mesoscale eddying models respond in a manner closer to the observational analysis from Böning et al. (2008) than certain coarse resolution noneddying models using parameterizations. Prompted by this study, numerous authors have made compelling suggestions for improving the mesoscale eddy parameterizations (e.g., Farneti and Gent, 2011; Gent and Danabasoglu, 2011; Hofmann and Morales-Maqueda, 2011).

The above assessment does not undermine the ongoing quest to understand processes, such as mesoscale eddy transport, and to develop parameterizations for use in coarse grid models. However, it does lend a degree of humility to those arguing for the validity of their favorite parameterization. It also supports the use of ensembles of model simulations whose members differ by perturbing the physical parameterizations and numerical methods in sensible manners to more fully test the large space of unknown parameters.

2.6. Numerical Considerations for Transport

We propose the following as an operational definition for resolution of a flow feature.

A flow feature is resolved only so far as there are no less than 2π grid points spanning the feature.

This definition is based on resolving a linear wave with a discrete, nonspectral, representation so that any admitted wave has length no smaller than $2\pi\Delta$. We consider this a sensible operational definition even when representing nonlinear and turbulent motion. Decomposing the flow into vertical baroclinic modes, one is then led to considering the baroclinic flow as resolved only so far as there are 2π grid points for each baroclinic wave whose contribution to the flow energy is nontrivial. Traditionally we write the Rossby radius as $R=1/\kappa$, with the wavenumber $\kappa=2\pi/\lambda$ and λ the baroclinic wavelength. Hence, if the grid spacing is less than the Rossby radius, $\Delta\leq R$, then the grid indeed resolves the corresponding baroclinic wave since $2\pi\Delta\leq\lambda$. As the first baroclinic mode dominates much of the mid-latitude ocean (Wunsch and Stammer, 1995; Wunsch, 1997; Stammer, 1998; Smith and Vallis, 2001; Smith, 2007), modelers generally look to the first baroclinic Rossby radius as setting the scale whereby baroclinic flow is resolved (Smith et al., 2000). Higher baroclinic modes, submesoscale modes (Boccaletti et al., 2007; Fox-Kemper et al., 2008; Klein and Lapeyre, 2009), internal gravity wave modes (Arbic et al., 2010), and other filamentary features require even finer resolution.

2.6.1. *Ensuring that Admitted Flow Features Are Resolved*

Nonlinear eddying flows contain waves with many characteristic lengths, and turbulent flows experience an energy and variance cascade between scales (see Section 2.2 or Vallis, 2006). Further, in the presence of a strongly nonlinear flow, certain discretizations of the nonlinear self-advection term $\rho\mathbf{v}\mathbf{v}$ (see Equation 20.12) can introduce grid scale energy even when the eddying flow is geostrophic and thus subject to the inverse cascade. So quite generally, resolving all flow features admitted in a simulation requires one to minimize the energy and variance contained at scales smaller than $2\pi\Delta$.

There are two general means to dissipate energy and variance of unresolved flow features. The implicit LES approach places responsibility for dissipation with the numerical advection operators acting on momentum and tracer. When coupled to a highly accurate underlying discretization of the advection fluxes, and with a monotonicity constraint that retains physically sensible values for the transport field, such numerical transport operators can be constructed to ensure that only well-resolved flow features are admitted. The Regional Ocean Model System (ROMS) (Shchepetkin and McWilliams, 2005) has incorporated elements of this approach, in which lateral friction or diffusion operators are not required for numerical purposes.

The second approach to dissipating unresolved flow features is to incorporate a friction operator into the momentum equation, and diffusion operator into the tracer equation, each using a transport coefficient that is far larger than molecular. There are straightforward ways to do so, yet a naive use of these methods can lead to over-dissipation of the simulation (Griffies and Hallberg, 2000; Large et al., 2001; Smith and McWilliams, 2003; Jochum, 2009), and/or spurious dianeutral mixing associated with diffusion across density fronts (Veronis, 1975; Roberts and Marshall, 1998). Consequently, more sophisticated dissipation operators are typically considered, with their design based on a mix between physical and numerical needs (e.g., see Chapter 14 of Griffies, 2004).

2.6.2. *Representing Transport in a Numerical Ocean*

The extreme anisotropy between dianeutral and epineutral transport in the ocean interior has motivated the development of ocean models based on potential density as the vertical coordinate (see Section 2.7). Respecting the epineutral/dianeutral anisotropy in nonisopycnal models is a nontrivial problem in three-dimensional numerical transport. Level models or terrain following models must achieve small levels of spurious dianeutral mixing through a combination of highly accurate tracer advection schemes, and properly chosen momentum and tracer closure schemes, all in the presence of hundreds to thousands of mesoscale eddy turnover times and a nonlinear equation of state.

As noted by Griffies et al. (2000a), resolving all flow features (see Section 2.6) is difficult for mesoscale eddying simulations, since eddies pump tracer variance to the grid scale and thus increase tracer gradients. At some point, a tracer advection scheme will either produce dispersive errors, and so introduce spurious extrema and thus expose the simulation to spurious convection, or add dissipation via a mixing operator or low order upwind biased advection operator in order to preserve monotonicity.

2.6.3. *Methods for Reducing Spurious Dianeutral Transport*

Mechanical energy cascades to the large scale in a geostrophically turbulent flow. However, grid scale energy can appear as the nonlinear advection of momentum becomes more dominant with eddies, thus stressing the numerical methods used to transport momentum. This issue is directly connected to the spurious dianeutral tracer transport problem, since even very accurate tracer advection schemes, such as the increasingly popular scheme from Prather (1986) (see Maqueda and Holloway, 2006; Tatebe and Hasumi, 2010; Hill et al., 2012 for ocean model examples) will be exposed to unphysically large spurious transport and/or dispersion error (which produce tracer extrema) if the velocity field contains too much energy (i.e., noise) near the grid scale. Hence, the integrity of momentum transport, and the associated momentum closure, becomes critical for maintaining physically sensible tracer transport, particularly with an eddying flow or any flow where momentum advection is important (Ilicak et al., 2012).

Results from Griffies et al. (2000a), Jochum et al. (2008), and Ilicak et al. (2012) emphasize the need to balance the quest for more kinetic energy, which generally pushes the model closer to observed energy levels seen in satellites (see, e.g., Figure 20.4 discussed in Section 3, or Chapter 4), with the need to retain a negligible spurious potential energy source whose impact accumulates over decades and longer. Following Ilicak et al. (2012), we suggest that maintaining a grid Reynolds number so that

$$\mathrm{Re}_\Delta = \frac{U\Delta}{\nu} < 2, \tag{20.38}$$

ensures unresolved flow features are adequately filtered. In this equation, U is the velocity scale of currents admitted in the simulation, Δ is the grid scale, and ν is the generally nonconstant Laplacian eddy viscosity used to dissipate mechanical energy.

The constraint (20.38) has multiple origins. One is associated with the balance between advection and diffusion in a second order discretization (see Bryan et al., 1975 or Section 18.1.1 of Griffies, 2004), in which $\mathrm{Re}_\Delta < 2$ eliminates an unphysical mode. More recently, Ilicak et al. (2012) identified this constraint as necessary to ensure that spurious dianeutral mixing is minimized. ROMS (Shchepetkin and McWilliams, 2005) has this constraint built into the advection of momentum, whereas most other codes require specification of a friction operator. Selective use of a flow-dependent viscosity, such as from a Laplacian or biharmonic Smagorinsky scheme (see Smagorinsky, 1993; Griffies and Hallberg, 2000; or Section 18.3 of Griffies, 2004); or the scheme of Leith (1996) discussed by Fox-Kemper and Menemenlis (2008), assists in maintaining the constraint (20.38) while aiming to avoid over-dissipating kinetic energy in the larger scales.

2.7. Vertical Coordinates

There are three traditional approaches to choosing vertical coordinates: geopotential, terrain-following, and potential density (isopycnal). Work continues within each model class to expand its regimes of applicability, with significant progress occurring in many important areas. The review by Griffies et al. (2010) provides an assessment of recent efforts, which we now summarize.

We start this discussion by noting that all vertical coordinates found to be useful in ocean modeling remain "vertical" in the sense they retain a simple expression for the hydrostatic balance (20.16), thus allowing for a hydrostatic balance to be trivially maintained in a simulation. This constraint is a central reason ocean and atmospheric modelers favor the projected nonorthogonal coordinates first introduced by Starr (1945), rather than locally orthogonal coordinates whose form of hydrostatic balance is generally far more complex.

2.7.1. Geopotential and Generalized Level Models

Geopotential z-coordinate models have found widespread use in global climate applications for several reasons, such as their simplicity and straightforward nature of parameterizing the surface boundary layer and associated air–sea interaction. For example, of the 25 coupled climate models contributing to the CMIP3 archive used for the IPCC AR4 (Meehl et al., 2007), 22 employ geopotential ocean models, one is terrain following, one is isopycnal, and one is hybrid pressure–isopycnal–terrain.

There are two key shortcomings ascribed to z-coordinate ocean models.

- *Spurious mixing*: This issue has been discussed in Section 2.6.
- *Overflows*: Downslope flows (Legg, 2012) in z-models tend to possess excessive entrainment (Roberts and Wood, 1997; Winton et al., 1998; Legg et al., 2006, 2008; Treguier et al., 2012), and this behavior compromises simulations of deep watermasses derived from dense overflows. Despite much effort and progress in understanding both the physics and numerics (Dietrich et al., 1987; Beckmann and Döscher, 1997; Beckmann, 1998; Price and Yang, 1998; Campin and Goosse, 1999; Killworth and Edwards, 1999; Nakano and Suginohara, 2002; Wu et al., 2007; Danabasoglu et al., 2010; Laanaia et al., 2010), the representation/parameterization of overflows remains difficult at horizontal grid spacing coarser than a few kilometers (Legg et al., 2006).

A shortcoming related to the traditional representation of topography (e.g., Cox, 1984) has largely been overcome by partial cells now commonly used in level models (Adcroft et al., 1997; Pacanowski and Gnanadesikan, 1998; Barnier et al., 2006). It is further reduced by the use of a momentum advection scheme conserving both energy and enstrophy, and by reducing near-bottom sidewall friction (Penduff et al., 2007; Le Sommer et al., 2009). A complementary problem arises from the use of free surface geopotential coordinate models, whereby they can lose their surface grid cell in the presence of refined vertical spacing. Generalizations of geopotential coordinates, such as the stretched geopotential coordinate, z^*, introduced by Stacey et al. (1995) and Adcroft and Campin (2004), overcome this problem (see Griffies et al., 2011; Dunne et al., 2012 for recent global model applications). Leclair and Madec (2011) introduce an extension to z^* that aims to reduce spurious dianeutral mixing. Additional efforts toward mass conserving non-Boussinesq models have been proposed by Huang et al. (2001), DeSzoeke and Samelson (2002), and Marshall et al. (2004), with one motivation being the direct simulation of the global steric effect required for sea-level studies (Greatbatch, 1994; Griffies and Greatbatch, 2012). What has emerged from the geopotential model community is a movement toward such *generalized level* coordinates that provide enhanced functionality while maintaining essentially the physical parameterizations developed for geopotential models. We thus hypothesize that the decades of experience and continued improvements with numerical methods, parameterizations, and applications suggest that generalized level methods will remain in use for ocean climate studies during the next decade and likely much longer.

2.7.2. Isopycnal Layered and Hybrid Models

Isopycnal models generally perform well in the ocean interior, where flow is dominated by quasi-adiabatic dynamics, as well as in the representation/parameterization of dense overflows (Legg et al., 2006). Their key liability is that resolution is limited in weakly stratified water columns. For ocean climate simulations, isopycnal models attach a nonisopycnal surface layer to describe the surface boundary layer. Progress has been made with such *bulk mixed layer* schemes, so that Ekman-driven restratification and diurnal cycling are now well simulated (Hallberg, 2003). Additionally, when parameterizing lateral mixing along constant potential density surfaces rather than neutral directions, the isopycnal models fail to incorporate dianeutral mixing associated with thermobaricity (McDougall, 1987) (see Section A.27 of IOC et al., 2010). Iudicone et al. (2008) and Klocker and McDougall (2010) suggest that thermobaricity contributes more to water mass transformation in the Southern Ocean than from breaking internal gravity waves.

Hybrid models offer an alternative means to eliminate liabilities of the various traditional vertical coordinate classes. The HYCOM code of Bleck (2002) exploits elements of the hybrid approach, making use of the Arbitrary Lagrangian–Eulerian (ALE) method for vertical remapping (Donea et al., 2004). As noted by Griffies et al. (2010), progress is being made to address issues related to the use of isopcynal layered models, or their hybrid brethren, thus providing a venue for the use of such models for a variety of applications, including global climate (Megann et al., 2010; Dunne et al., 2012).

A physical system of growing importance for sea-level and climate studies concerns the coupling of ocean circulation to ice shelves whose grounding lines can evolve. Required of such models is a land–ocean boundary that evolves, in which case ocean models require a wetting and drying method. We have in mind the growing importance of studies of coupled ice-shelf ocean processes with evolving grounding lines (Goldberg et al., 2012) (see Chapter 16). Though not uncommon for coastal modeling applications, wetting and drying for ocean climate model codes remain rare, with the study of Goldberg et al. (2012) using the GOLD isopycnal code of Adcroft et al. (2010) the first to our knowledge. It is notable that climate applications require exact conservation of mass and tracer to remain viable for long-term (decadal and longer) simulations, whereas certain of the wetting and drying methods used for coastal applications fail to meet this constraint. We conjecture that isopycnal models, or their generalizations using ALE methods, will be very useful for handling the evolving coastlines required for such studies.

2.7.3. Terrain-Following Vertical Coordinate Models

Terrain-following coordinate models (TFCM) have found extensive use for coastal and regional applications, where bottom boundary layers and topography are well resolved. As with geopotential models, TFCMs generally suffer from spurious dianeutral mixing due to problems with numerical advection (Marchesiello et al., 2009). Also, the formulation of neutral diffusion (Redi, 1982) and eddy-induced advection (Gent and McWilliams, 1990) have until recently not been considered for TFCMs. However, recent studies by Lemarié et al. (2012a,b) have proposed new methods to address both these issues.

A well-known problem with TFCMs is calculation of the horizontal pressure gradient, with errors leading to potentially nontrivial spurious flows. Errors are a function of topographic slope and near-bottom stratification (Haney, 1991; Deleersnijder and Beckers, 1992; Beckmann and Haidvogel, 1993; Mellor et al., 1998; Shchepetkin and McWilliams, 2002). The pressure-gradient problem has typically meant that TFCMs are not useful for global-scale climate studies with realistic topography, at least until horizontal grid spacing is very fine (order of 10 km or finer). However, Lemarié et al. (2012b), following from Mellor et al. (1998), identify an intriguing connection between pressure-gradient errors and the treatment of lateral diffusive transport. Namely, the use of neutral diffusion rather than terrain-following diffusion in grids of order of 50 km with the ROMS (Shchepetkin and McWilliams, 2005) significantly reduces the sensitivity of the simulation to the level of topographic smoothing. This result suggests that it is not just the horizontal pressure-gradient error that has plagued terrain-following models, but the additional interaction between numerically induced mixing of active tracers and the pressure gradient.

2.7.4. Where We Stand with Vertical Coordinates

Table 20.1 provides a list of open-source codes maintaining an active development process, providing updated and thorough documentation, and supporting an international user community. There are fewer codes listed in Table 20.1 than in the Griffies et al. (2000a) review written at the close of the WOCE era. It is inevitable that certain codes will not continue to be widely supported. There has also been a notable merger of efforts, such as in Europe where the majority of the larger modeling projects utilize the NEMO ocean component, and in the regional/shelf modeling community that focuses development on ROMS.

Numerical methods utilized for many of the community ocean codes have greatly improved during the past decade through intense development and a growing suite of applications. We are thus motivated to offer the following hypothesis.

Physical parameterizations, more so than vertical coordinates, determine the physical integrity of a global ocean climate simulation.

This hypothesis was untenable at the end of the WOCE era, which was the reason that Griffies et al. (2000a) emphasized vertical coordinates as the central defining feature of a model simulation. However, during the past decade, great strides in numerical methods have removed many of the "features" that distinguish large-scale simulations with different vertical coordinates. Hence, so long as the model configuration resolves flow features admitted by the simulation, there are fewer compelling reasons today than in the year 2000 to choose one vertical coordinate over another.

2.8. Unstructured Horizontal Grid Meshes

Within the past decade, there has been a growing focus on unstructured horizontal meshes, based on finite volume or finite element methods. These approaches are very distinct from the structured Arakawa grids (Arakawa, 1966; Arakawa and Lamb, 1981) used since the 1960s in both the atmosphere and ocean. The main motivation for generalization is to economically capture multiple scales seen in the ocean geometry (i.e., land–sea boundaries) and various scales of oceanic flow (i.e., boundary currents, coastal and shelf processes, active mesoscale eddy regimes). Griffies

TABLE 20.1 Open-Source Ocean Model Codes with Structured Horizontal Grids Applicable for a Variety of Studies Including Large-Scale Circulation

Model	Vertical Coordinate	Web Site
HYCOM	Hybrid $\sigma - \rho - p$	http://hycom.org/ocean-prediction
MIT	General level	http://mitgcm.org/
MOM	General level	http://mom-ocean.org
NEMO	General level	http://nemo-ocean.eu/
POM	Terrain following	http://aos.princeton.edu/WWWPUBLIC/PROFS/NewPOMPage.html
POP	Geopotential	http://climate.lanl.gov/Models/POP/
ROMS	Terrain following	http://myroms.org/

These codes are currently undergoing active development (i.e., updated algorithms, parameterizations, diagnostics, applications), possess thorough documentation, and maintain widespread community support and use. Listed are the model names, vertical coordinate features, and Web sites where code and documentation are available. We failed to find other model codes that satisfy these criteria.
Each model is coded in Fortran with generalized orthogonal horizontal coordinates. MOM and POP use an Arakawa B-grid layout of the discrete momentum equations, whereas others use an Arakawa C-grid (see Griffies et al., 2000a for a summary of B and C grids). General-level models are based on the traditional z-coordinate approach, but may be generalized to include other vertical coordinates such as pressure or terrain-following. HYCOM's vertical coordinate algorithm is based on vertical remapping to return coordinate surfaces at each time step to their predefined targets. In contrast, general-level models diagnose the dia-surface velocity component through the continuity equation (Adcroft and Hallberg, 2006), which is the fundamental distinction from general layered or quasi-Lagrangian models such as HYCOM.
Acronyms are the following: HYCOM, Hybrid Coordinate Ocean Model; MIT, Massachusetts Institute of Technology; MOM, Modular Ocean Model; NEMO, Nucleus for European Modeling of the Ocean; POM, Princeton Ocean Model; POP, Parallel Ocean Program; ROMS, Regional Ocean Modeling System.

TABLE 20.2 A Nonexhaustive List of Ongoing Development Efforts Utilizing the Flexibility of Unstructured Horizontal Meshes

Model/Institute	Web Site
FESOM/AWI	http://www.awi.de/en
ICOM/Imperial	http://amcg.ese.ic.ac.uk/index.php?title=ICOM
ICON/MPI	http://www.mpimet.mpg.de/en/science/models/icon.html
MPAS-ocean/LANL	public.lanl.gov/ringler/ringler.html
SLIM/Louvain	http://sites-final.uclouvain.be/slim/

These efforts remain immature for large-scale climate applications, though there are some showing promise (e.g., Timmermann et al., 2009; Ringler et al., 2013). Furthermore, many efforts are not yet supporting open-source public use due to their immaturity. Acronyms are the following: FESOM, Finite Element Sea-ice Ocean circulation Model, AWI, Alfred Wegener Institute for Polar and Marine Research in Germany; MPI, Max Planck Institute für Meteorologie in Germany; ICOM, Imperial College Ocean Model in the UK; MPAS, Model for Prediction Across Scales; LANL, Los Alamos National Laboratory in the United States; SLIM, Second-generation Louvain-la-Neuve Ice–ocean Model; Louvain, Louvain-la-Neuve in Belgium.

et al. (2010), Danilov (2013), and Ringler et al. (2013) review recent efforts with applications to the large-scale circulation.

There are many challenges facing finite element and unstructured finite volume methods. Even if the many technical issues listed in Section 4 of Griffies et al. (2010) are overcome, it remains unclear if these approaches will be computationally competitive with structured meshes. That is, more generality in grid meshing comes with a cost in added computational requirements. Nonetheless, the methods are sufficiently compelling to have motivated a new wave in efforts and to have entrained many smart minds towards seeing the ideas fully tested. Table 20.2 lists nascent efforts focused on aspects of this approach, with applications in the ocean. We anticipate that within 5–10 years, realistic coupled climate model simulations using unstructured ocean meshes will be realized.

3. OCEAN MODELING: SCIENCE EMERGING FROM SIMULATIONS

We now move from the reductionist theme focused on formulating a physically sound and numerically robust ocean model tool, to the needs of those aiming to use this tool for exploring wholistic questions of ocean circulation and climate. The basis for this exercise in *ocean modeling* is that the model tool has been formulated with sufficient respect to the fundamental physics so that simulated patterns and responses are physically meaningful. A successful ocean modeling activity thus requires a high-fidelity numerical tool, a carefully designed experiment, and a variety of analysis methods helping to unravel a mechanistic storyline.

We present a selection of topics relevant to the formulation of a numerical experiment aimed at understanding aspects of the global ocean circulation. Foremost is the issue of how to force an ocean or ocean–ice model. We rely for this discussion on the more thorough treatment given of global ocean–ice modeling in Griffies et al. (2009). In particular, we do not address the extremely difficult and ambiguous issues of model initialization and spinup, leaving such matters to the Griffies et al. (2009) paper for ocean–ice models and Chapter 23 for coupled climate models. Other important issues, such as boundary conditions for regional models and community model experiment strategies, are introduced very briefly. Additionally, we do not consider here the issues of fully coupled climate models (see Chapters 23–25).

3.1. Design Considerations for Ocean–Ice Simulations

The ocean is a forced and dissipative system. Forcing occurs at the upper boundary from interactions with the atmosphere, rivers, sea ice, and land ice shelves, and at its lower boundary from the solid earth (see Figure 20.1). Forcing also occurs from astronomical effects of the sun and moon to produce tidal motions.[1] Important atmospheric forcing occurs over basin scales, with time scales set by the diurnal cycle, synoptic weather variability (days), the seasonal cycle, and interannual fluctuations such as the North Atlantic Oscillation and even longer time scales. Atmospheric momentum and buoyancy fluxes are predominantly responsible for driving the ocean's large scale horizontal and overturning circulations (e.g., Kuhlbrodt et al., 2007). Additional influences include forcing at continental boundaries from river inflow and calving glaciers, as well as in polar regions where sea ice dynamics greatly affect the surface buoyancy fluxes.

Since the successes at reproducing the El Niño-Southern Oscillation phenomenon with linear ocean models in the early 1980s (Philander, 1990), a large number of forced ocean models have demonstrated skill in reproducing the main modes of tropical variability without assimilation of *in situ* ocean data, in part because of the linear character of the tropical ocean response to the winds (e.g., Illig et al., 2004). Further, studies from the past decade show that forced ocean models can, to some extent, reproduce interannual ocean variability in mid-latitudes (e.g., regional patterns of decadal sea-level trends, Lombard et al., 2009). Hence, a critical issue for the fidelity of an ocean and/or coupled ocean–ice simulation is the forcing methodology.

In the following, we introduce issues associated with how ocean models are forced through boundary fluxes. There is a spectrum of methods that go from the fully coupled climate models detailed in Chapter 23, to highly simplified boundary conditions such as damping of surface tracers to an "observed" dataset. Our focus is with ocean and ocean–ice models that are not coupled to an interactive atmosphere. Use of uncoupled ocean models allows one to remove biases inherent in the coupled climate models associated with the prognostic atmosphere component. Yet there is a price to pay when removing feedbacks. We outline these issues in the following.

3.1.1. Air–Sea Flux Formulation for Coupled Ocean–Ice Simulations

Ice–ocean fluxes are not observed, and as a result ocean–ice coupled models are more commonly used than ocean-only models for investigations of the basin to global-scale forced ocean circulation. Coupled ocean–ice models require surface momentum, heat, and hydrological fluxes to drive the simulated ocean and ice fields. When decoupling the ocean and sea ice models from the atmosphere and land, one must introduce a method to generate these fluxes. One approach is to damp sea surface temperature (SST) and salinity (SSS) to prescribed values. This approach for SST is sensible because SST anomalies experience a local negative feedback (Haney, 1971), whereby they are damped by interactions with the atmosphere. Yet the same is not true for salinity. Further, the associated buoyancy fluxes generated by SST and SSS restoring can be unrealistic (Large et al., 1997; Killworth et al., 2000). Barnier et al. (1995) introduced another method by combining prescribed fluxes and restoring. However, fluxes from observations and/or reanalysis products have large uncertainties (Taylor, 2000; Large and Yeager, 2004, 2009; Chapter 5), which can lead to unacceptable model drift (Rosati and Miyakoda, 1988).

Another forcing method prognostically computes turbulent fluxes for heat, moisture, and momentum from a planetary boundary layer scheme (Parkinson and Washington, 1979; Barnier, 1998), in addition to applying radiative heating, precipitation and river runoff. Turbulent fluxes are computed from bulk formulae as a function of the ocean surface state (SST and surface currents) and a prescribed atmospheric state (air temperature, humidity, sea-level pressure, and wind velocity or wind speed). It is this approach that has been recommended by the CLIVAR Working Group for Ocean Model Development for running

1. Climate modelers tend to ignore tidal forcing, but we may soon reach the limitations of assuming tidal motions merely add linearly to the low frequency solution (Schiller and Fiedler, 2007; Arbic et al., 2010).

Coordinated Ocean-ice Reference Experiments (COREs) (Griffies et al., 2009). Although motivated from its connection to fully coupled climate models, a fundamental limitation of this method relates to the use of a prescribed and nonresponsive atmospheric state that effectively has an infinite heat capacity, moisture capacity, and inertia.

The first attempts to define a forcing protocol for COREs have shown that a restoring to observed SSS is necessary to prevent multidecadal drift in the ocean–ice simulations, even though such restoring has no physical basis (see Chapter 28 as well as Rivin and Tziperman, 1997). It is thus desirable to use a weak restoring that does not prevent variability in the surface salinity and deep circulation. Unfortunately, when the restoring timescale for SSS is much longer than the effective SST restoring timescale, the thermohaline fluxes move into a regime commonly known as *mixed boundary conditions* (Bryan, 1987), with rather unphysical sensitivities to buoyancy fluxes present in such regimes (Griffies et al., 2009). Further, Griffies et al. (2009) have demonstrated that model solutions are very dependent on the arbitrary strength of the salinity restoring. Artificial salinity restoring may become unnecessary for short-term simulations (a few years maximum), if the fidelity of ocean models and the observations of precipitation and runoff improve. For long-term simulations, some way of parameterizing the missing feedback between evaporation and precipitation through atmospheric moisture transport is needed.

Another drawback of using a prescribed atmosphere to force an ocean–ice model is the absence of atmospheric response as the ice edge moves. Windy, cold, and dry air is often found near the sea ice edge in nature. Interaction of this air with the ocean leads to large fluxes of latent and sensible heat which cool the surface ocean, as well as evaporation, which increases salinity. This huge buoyancy loss increases surface density, which provides a critical element in the downward branch of the thermohaline circulation (e.g., Marshall and Schott, 1999). When the atmospheric state is prescribed, where the simulated sea ice cover increases relative to the observed, the air–sea fluxes are spuriously shut down in the ocean–ice simulation.

3.1.2. Atmospheric Datasets and Continental Runoff

In order to be widely applicable in global ocean–ice modeling, an atmospheric dataset from which to derive surface boundary fluxes should produce near zero global mean heat and freshwater fluxes when used in combination with observed SSTs. These criteria preclude the direct use of atmospheric reanalysis products (see Chapter 5). As discussed in Taylor (2000), a combination of reanalysis and remote sensing products provides a reasonable choice to force global ocean–ice models. Further, it is desirable for many research purposes to provide both a repeating "normal" year forcing as well as an interannually varying forcing. The dataset compiled by Large and Yeager (2004, 2009) satisfies these criteria.

The Large and Yeager (2004, 2009) atmospheric state has been chosen for COREs. The most recent version of the dataset is available from

```
http://data1.gfdl.noaa.gov/nomads/forms/core/
COREv2.html,
```

and it covers the period 1948 to 2009. It is based on NCEP–NCAR reanalysis temperature, wind, and humidity, and satellite observations of radiation and precipitation (a climatology is used when satellite products are not available). Similar datasets have been developed by Röske (2006), and more recently by Brodeau et al. (2010), both of which are based on ECMWF products instead of NCEP. The Brodeau et al. (2010) dataset is used in the framework of the European Drakkar project (Drakkar Group, 2007). The availability of multiple forcing datasets is useful in light of large uncertainties of air–sea fluxes. In addition, short term (i.e., interannual) or regional simulations can take advantage of other forcing data, such as scatterometer wind measurements, which have been shown to improve ocean simulations locally (Jiang et al., 2008).

For the multidecadal global problem, further efforts are needed to improve the datasets used to force ocean models. For example, in the CORE simulations considered by Griffies et al. (2009), interannual variability of river runoff and continental ice melt are not taken into account. However, recent efforts have incorporated both a seasonal cycle and interannually varying climatology into the river runoff, based on the Dai et al. (2009) analysis. Further, the interpretation of trends in the forcing datasets is a matter of debate. For example, the increase of Southern Ocean winds between the early 1970s and the late 1990s is probably exaggerated in the atmospheric reanalyses due to the lack of Southern Ocean observations before 1979. This wind increase is retained in Large and Yeager (2009), used for COREs, whereas Brodeau et al. (2010) attempt to remove it for the Drakkar Forcing. These different choices lead, inevitably, to different decadal trends in the ocean simulations.

Considering the key role of polar regions and their high sensitivity to climate change, ocean–ice simulations will need improved forcings near the polar continents. One issue is taking into account the discharge of icebergs, which can provide a source of freshwater far from the continent, especially in the Antarctic (Jongma et al., 2009; Martin and Adcroft, 2010). The ice–ocean exchanges that occur due to the ocean circulation underneath the ice shelves is an additional complex process that needs to be taken into account, both for the purpose of modeling water mass properties near ice shelves and for the purpose of modeling the flow and stability of continental ice sheets (Chapter 16).

3.1.3. Wind Stress, Surface Waves, and Surface Mixed Layer

Mechanical work done by atmospheric winds provides a source of available potential energy that in turn drives much of the ocean circulation. A successful ocean simulation thus requires an accurate mechanical forcing. This task is far from trivial, not only because of wind uncertainties (reanalysis or scatterometer measurements) but also because of uncertainties in the transfer function between 10 m wind vector and the air–sea wind stress. During the WOCE years, the wind stress was generally prescribed to force ocean models. However, with the generalization of the bulk approach led by Large and Yeager (2004, 2009), modelers began to use a bulk formula to compute the wind stress, with some choosing to do so as a function of the difference between the 10 m wind speed and the ocean velocity (Pacanowski, 1987). The use of such *relative winds* in the stress calculation has a significant damping effect on the surface eddy kinetic energy, up to 50% in the tropical Atlantic and about 10% in mid-latitudes (Xu and Scott, 2008; Eden and Dietze, 2009). Relative winds are clearly what the real system uses to exchange momentum between the ocean and atmosphere, so it is sensible to use such for coupled climate models where the atmosphere responds to the exchange of momentum with the ocean. However, we question the physical relevance of relative winds for the computation of stress in ocean–ice models, where the atmosphere is prescribed.

In general, the classical bulk formulae used to compute the wind stress are being questioned, given the complex processes relating surface wind, surface waves, ocean currents, and high-frequency coupling with fine resolution atmosphere and ocean simulations (McWilliams and Sullivan, 2001; Sullivan et al., 2007; Sullivan and McWilliams, 2010). It is potentially important to take into account surface waves and swell, not only in the wind stress formulation but also in the parameterization of vertical mixing in the surface boundary layer (Belcher et al., 2012).

3.1.4. Boundary Conditions for Regional Domains

In order to set up a numerical experiment in a regional domain, one needs to represent the lateral exchanges with the rest of the global ocean, at the "open" boundaries of the region of interest. When knowledge of the solution outside the simulated region is limited, an approach similar to the one advocated for ROMS is often used (Marchesiello et al., 2001). This method combines relaxation to a prescribed solution outside the domain with a radiation condition aimed at avoiding spurious reflection or trapping of perturbations at the open boundary. Treguier et al. (2001) have noted that in a realistic primitive equation model where Rossby waves, internal waves and turbulent eddies are present, the phase velocities calculated from the radiation condition have no relationship to the physical processes occurring at the boundary. Despite this fact, radiation appears to have a positive effect on the model solution, perhaps because it introduces stochastic noise in an otherwise over-determined problem. When the solution outside the domain is considered reliable, a "sponge" layer with relaxation to the outside solution is often preferred. Blayo and Debreu (2005) and Herzfeld et al. (2011) provide a review of various methods.

For the purpose of achieving regional simulations of good fidelity, the main progress accomplished in the past decade has come less from improved theory or numerics, and more from the availability of improved global model output that can be used to constrain the boundaries of regional models. These global datasets include operational products, ocean state estimates (Chapters 21 and 22) and prognostic global simulations (Maltrud and McClean, 2005; Barnier et al., 2006).

The quality of a regional model depends critically on the consistency between the solution outside and inside the domain. Consistency can be ensured by using the same numerical code for the global and regional solution; by using the same (or similar) atmospheric forcing; or by using strategies of grid refinement and nesting. Nesting can be one-way or two-way. For two-way, the large scale or global model is modified at each time step to fit the regional fine-scale solution. Although complex, two-way grid nesting is a promising strategy (Debreu and Blayo, 2008), with impressive applications documenting the role of Agulhas eddies in the variability of the Atlantic meridional overturning (Biastoch et al., 2008). Further considerations are being given to nesting a number of fine resolution regions within a global model.

3.1.5. Community Model Experiments

In Chapter 7.2 of the first edition of this book, Willebrand and Haidvogel wrote:

> *One therefore can argue that the principal limitation for model development arises from the limited manpower in the field, and that having an overly large model diversity may not be the most efficient use of human resources. A more efficient way is the construction of community models that can be used by many different groups.*

This statement seems prescient with regard to model codes, as noted by the reduced number of codes listed in Table 20.1 relative to the Griffies et al. (2000a) review. Additionally, it applies to the coordination of large simulation efforts. Indeed, WOCE has motivated the first Community Modeling Experiment (CME). This pioneering eddy-permitting simulation of the Atlantic circulation (Bryan et al., 1995) and its companion sensitivity experiments have engaged a wide community of oceanographers. The results gave insights into the origin of mesoscale eddies (Beckmann and Haidvogel, 1994), the mechanical energy balance (Treguier, 1992), and mechanisms driving the Atlantic meridional overturning circulation (Redler and Böning, 1997).

As ocean model simulations refine their grid spacing over longer time periods, such community strategies become more useful, whereby simulations are performed in a coordinated fashion by a small group of scientists and distributed to a wider user community. An example of such a strategy is carried out within the European DYNAMO project using regional models (Willebrand et al., 1997), and the more recent Drakkar project (Drakkar Group, 2007) that focuses on global ocean–ice models. Global hind-cast simulations of the past 50 years have been performed using the NEMO modeling framework for Drakkar (see Table 20.1), at different spatial resolutions from 2° to 1/12°, with different forcings and model parameters. A few examples illustrate the usefulness of this approach.

- Analyses of a hierarchy of global simulations with differing resolutions have revealed the role of mesoscale eddies in generating large scale, low frequency variability of sea surface height (SSH) (Penduff et al., 2010). Figure 20.4 shows that a significant part of the SSH variability observed at periods longer than 18 months is not captured by the coarse resolution version of the model, but is reproduced in an eddy-permitting version, especially in western boundary currents and in the Southern Ocean.
- Using experiments with different strategies for salinity restoring helped assess the robustness of modeled freshwater transports from the Arctic to the Atlantic (Lique et al., 2009).
- A long experiment (obtained by cycling twice over the 50 years of forcing) with a 1/4° global model has been used to estimate the respective role of ocean heat transport and surface heat fluxes in variability of the Atlantic ocean heat content (Grist et al., 2010). The same simulation helped sort out the influence of model drift on the simulated response of the Antarctic Circumpolar Current to the recent increase in Southern Ocean winds (Treguier et al., 2010).

3.2. Analysis of Simulations

As models grow more realistic, they become tools of discovery. Important features of the ocean circulation have been discovered in models before being observed in nature. We highlight here two such discoveries.

- *Zapiola anticyclone*: The Zapiola anticyclone is a large barotropic circulation ($\sim$100 Sv) in the Argentine basin south of the Brazil–Malvinas confluence zone. It is a prominent feature in satellite maps of SSH variability (Figure 20.5), causing a minimum of eddy activity located near 45°S, 45°W inside a characteristic "C"-shaped maximum. The satellite record is now long enough to allow a detailed analysis of its variability (Volkov and Fu, 2008). This region is thus a key location for the evaluation of eddy processes represented in ocean circulation models.

  The Zapiola anticyclone initially appeared in a terrain-following ocean model of the South Atlantic (B. Barnier, personal communication). Yet the circulation was considered a model artifact until observations confirmed its existence (Saunders and King, 1995). As facilitated through studies with ocean models, the Zapiola anticyclone arises from eddy-topography interactions (De Miranda et al., 1999). More precisely, it results from a balance between eddy vorticity fluxes and dissipation, mainly due to bottom drag. For this reason, different models or different numerical choices lead to different simulated strengths of this circulation (Figure 20.5).

  The Zapiola Drift rises 1100 m above the bottom of the Argentine Abyssal plain. In model simulations that truncate the bottom to be no deeper than 5500 m, the topographic seamount rises only 500 m above the maximum model depth, whereas models with a maximum depth of 6000 m render a far more realistic representation. Merryfield and Scott (2007) argue that the strength of the simulated anticyclone can be dependent on the maximum depth in the model, with shallower representations reducing the strength of the anticyclone.
- *Zonal jets in the Southwest Pacific*: Another model-driven discovery is the existence of zonal jets in the Southwest Pacific, between 30°S and 10°S in the region northeast of Australia. These jets, constrained by topography of the islands, were first documented by the OCCAM eddy-permitting model (Webb, 2000). Their existence in the real ocean was later confirmed by satellite altimetry (Hughes, 2002).

Whereas the science of ocean model development consists of the construction of a comprehensive tool, the analysis of ocean simulations mechanistically deconstructs and simplifies the output of the simulation to aid interpretation and to make connections to observations and theory. Analysis methods are prompted by the aims of the research. For example, one may aim to develop a reduced or simplified description, with dominant pieces of the physics

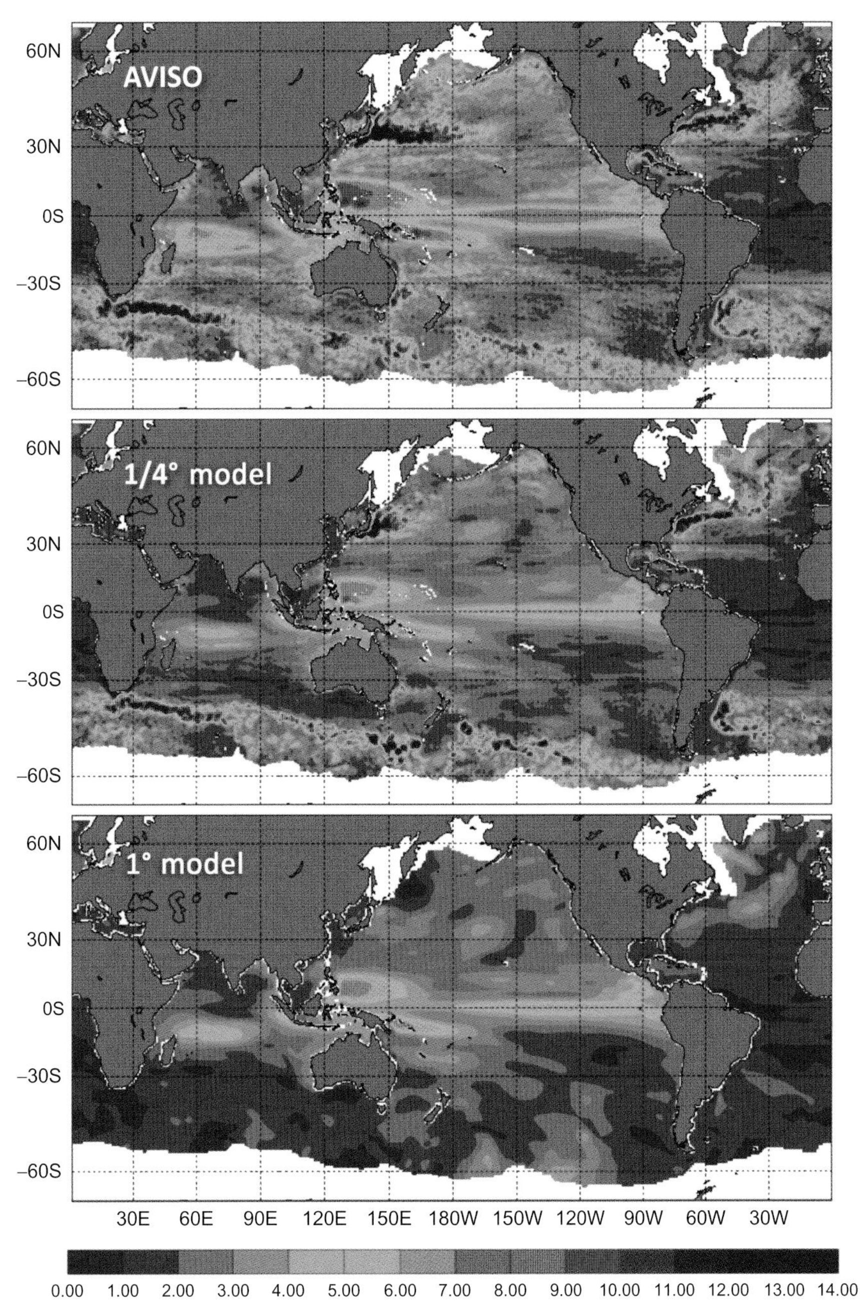

FIGURE 20.4 Variability of the sea surface height for periods longer than 18 months. Top: AVISO altimetric observations (Archiving, Validation, and Interpolation of Satellite Oceanographic; Le Traon et al. (1998); Ducet et al., 2000); bottom panels: Drakkar model simulations at 1/4° and 1° horizontal grid spacing. Note the absence of much variability in the 1° simulation. Note the enhanced intrinsic ocean variability in the 1/4° model, in contrast to the one-degree model. See Penduff et al. (2010) for details of the models and the temporal filtering.

identified to aid understanding and provoke further hypotheses, predictions, and theories. By doing so, understanding may arise concerning how the phenomena emerges from the underlying physical laws, making simplifications where appropriate, to remove less critical details and to isolate essential mechanisms. The following material represents a nonexhaustive selection of physically based analysis methods used in ocean modeling. It is notable that options for analyses are enriched, and correspondingly more complex and computationally burdensome, as the model resolution is refined to expand the admitted space and time scales, especially when turbulent elements of the ocean mesoscale and finer scales are included.

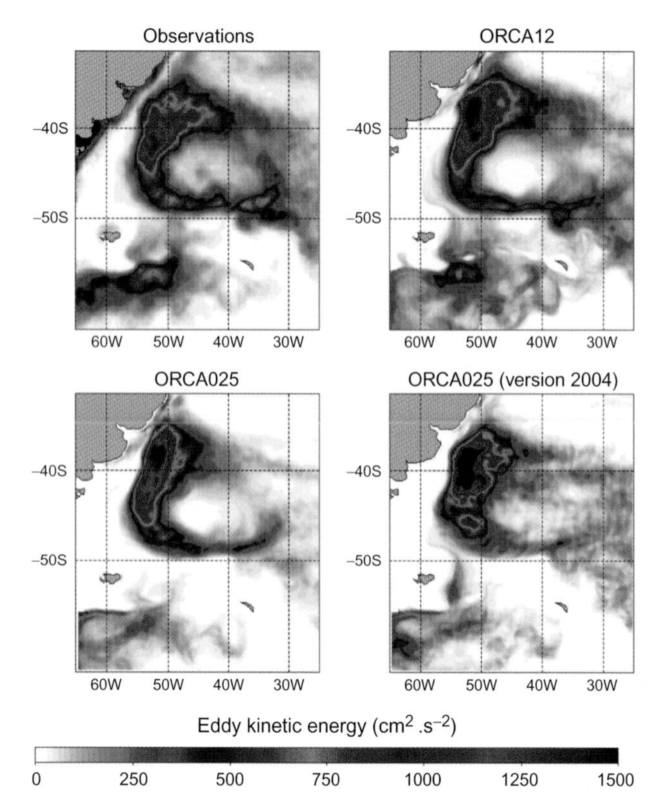

FIGURE 20.5 Variability of surface eddy kinetic energy (EKE) (units of $cm^2\ s^{-2}$) in the Argentine basin of the South Atlantic. (a) EKE of geostrophic currents calculated from altimetric observations (AVISO; Le Traon et al., 1998; Ducet et al., 2000), based on 10-year mean (October 1992 until February 2002), (b) Drakkar ORCA12 1/12° global model, with simulated data taken from a 10-year mean (1998–2007: past 10 years of a multidecade run), (c) recent version of the Drakkar ORCA025 1/4° global model, with simulated data taken from years 2000–2009 from a multidecadal run, (d) same model as (c), but using an older model version with full step bathymetry and a different momentum advection scheme (referenced as ORCA025 G04 in Barnier et al., 2006), with simulated data taken from 3-year mean (0008–0010: past 3 years from a climatological 10-year run). Note the good agreement with satellite measurements for both the ORCA12 and more recent ORCA025 simulations.

Our focus in the following concerns methods used to unravel elements of a particular simulation. To complement these methods, modelers often make use of perturbation approaches whereby elements of the simulation are altered relative to a control case. We have in mind those simulations that alter the boundary fluxes (e.g., remove buoyancy and/or mechanical forcing, swap one forcing dataset for another, modify fluxes over selected geographical regions); alter elements of the model's prognostic equations (e.g., modify subgrid scale parameterizations, remove nonlinear terms in the momentum equation); and refine the horizontal and/or vertical grid spacing. When combined with analysis methods such as those discussed below, these experimental approaches are fundamental to why numerical models are useful for understanding the ocean.

3.2.1. Budget Analysis

Identifying dominant terms in the tracer, momentum, and/or vorticity budgets assists in the quest to develop a reduced description, which in turn isolates the physical processes that are essential. The straightforward means for doing so consists of a budget analysis, which generally occurs within the framework of the model equations associated with the finite volume budgets as developed in Section 2.4.

As one example, the mechanical energy cycle of the ocean has been the subject of interest since a series of papers pointed out the potential role of dianeutral mixing as a key energy source for the overturning circulation (Wunsch and Ferrari, 2004; Kuhlbrodt et al., 2007). This work has motivated model-based studies aimed at understanding the energy cycle of the ocean. For example, Gnanadesikan et al. (2005) demonstrated that the link between mechanical mixing and meridional heat transport is rather weak in a climate model with parameterized ocean mesoscale eddies. No unifying view has emerged, but the approach is promising, and will gain momentum when results can be confirmed in more refined eddying global models.

Other examples include budgets of heat or salt in key regions, such as the surface mixed layer (Vialard and Delecluse, 1998) or the subtropical waters (McWilliams et al., 1996). Griffies and Greatbatch (2012) and Palter et al. (2013) present detailed budget analyses focusing on the role of buoyancy at the global and regional sea levels. Following the pioneering studies of the 1990s, a large number of model-based analyses have considered such tracer budgets in various parts of the ocean. The Argo observing network now makes possible similar analyses that can be partially compared with model results (e.g., de Boisseson et al., 2010). The confrontation of model-based and observation-based tracer budgets will undoubtedly help improve the representation of mixing processes in models.

3.2.2. Isopycnal Watermass Analysis

How much seawater or tracer transport passes through an isopycnal layer is a common question asked of model analysts. Relatedly, isopycnal mass analysis as per methods of Walin (1982) have proved useful for inferring the amount of watermass transformation associated with surface boundary fluxes (e.g., Tziperman, 1986; Speer and Tziperman, 1992; Williams et al., 1995; Marshall et al., 1999; Large and Nurser, 2001; Maze et al., 2009; Downes et al., 2011). Numerical models allow one to go beyond an analysis based solely on surface fluxes, so that interior transformation processes can be directly deduced. For example, the effect of mesoscale eddies on the subduction from the surface mixed layer into the ocean interior has been quantified in the North Atlantic (Costa et al., 2005). By performing a full three-dimensional analysis in a neutral density framework, Iudicone et al. (2008) discovered the essential importance of light penetration on the formation of tropical water masses.

3.2.3. Lagrangian Analysis

The Lagrangian parcel perspective often provides useful complementary information relative to the more commonly used Eulerian (fixed point) perspective. One method of Lagrangian analysis proposed by Blanke et al. (1999), as well as Vries and Döös (2001) and van Sebille et al. (2009), uses mass conservation (or volume conservation in Boussinesq models) to decompose mass transport into a large number of "particles", each carrying a tiny fraction of the transport. By following these particles using a Lagrangian algorithm, one can recover the transport of water masses and diagnose their transformation.

Applications of such Lagrangian analyses are numerous. Examples include the tropical Atlantic study of Blanke et al. (1999); the first quantification of the contribution of the Tasman leakage to the global conveyor belt (Speich et al., 2002); the Lagrangian view of the meridional circulation in the Southern Ocean (Döös et al., 2008; Iudicone et al., 2011); and quantification of how water masses are transferred between different regions (Rodgers et al., 2003; Koch-Larrouy et al., 2008; Melet et al., 2011). These Lagrangian methods have been applied to models absent mesoscale eddies, or only partially admitting such eddies, where a significant part of the dispersion of water masses is parameterized rather than explicitly resolved. The application to eddying models requires large computer resources, and thus have to date only been applied in regional models (Melet et al., 2011). More classical Lagrangian analysis, following arbitrary parcels without relation to the mass transports, have also been applied to eddying models, with a focus on statistical analyses of dispersion (Veneziani et al., 2005).

3.2.4. Passive Tracer Methods

Many of the ocean's trace constituents have a negligible impact on ocean density, in which case these tracers are dynamically passive (Chapter 26). England and Maier-Reimer (2001) review how chemical tracers, such as CFCs and radioactive isotopes, can be used to help understand both the observed and simulated ocean circulation, largely by providing means of tracking parcel motions as well as diagnosing mixing processes. Purposefully released tracers have provided benchmarks for measurements of mixing across the ocean thermocline and abyss (Ledwell et al., 1993, 2011; Ledwell and Watson, 1998). Ocean modelers have used similar tracer methods to assess physical and spurious numerical mixing (Section 2.6). Tracers can also provide estimates for the time it takes water to move from one region to another, with such timescale or generalized age methods exemplified by the many articles in Deleersnijder et al. (2010).

4. SUMMARY REMARKS

The evolution of numerical methods, physical parameterizations, and ocean climate applications has been substantial since the first edition of this book in 2001. Today, we better understand the requirements of, for example, maintaining a realistic tropical thermocline essential for simulations of El Niño fluctuations (Meehl et al., 2001), whereas earlier models routinely suffered from an overly diffuse thermocline. We understand far more about the importance of and sensitivity to various physical parameterizations, such as mixing induced by breaking internal waves (Chapter 7) and lateral mixing/stirring from mesoscale and submesoscale eddies (Chapter 8). Nonetheless, many of the key questions from the first edition remain with us today, in part because the ocean "zoo" (Figure 20.1) is so diverse and difficult to tame.

Questions about resolution of physical processes and/or their parameterization sit at the foundation of nearly all compelling questions of ocean models and modeling. What does it mean to fully resolve a physical process? What sorts of numerical methods and/or vertical coordinates are appropriate? Are the multiscale methods offered by unstructured meshes an optimal means for representing and parameterizing (using scale aware schemes) the multiscales of ocean fluid dynamics and fractal structure of the land–sea geometry? How well does a parameterization support high-fidelity simulations? How do we parameterize a process that is partially resolved, without suppressing and/or double-counting those elements of the process that are resolved? Relatedly, how do subgrid scale parameterizations impact on an *effective* resolution? What are the climate impacts from a particular physical process? Are these impacts robust to whether the process is unresolved and parameterized, partially resolved and partially parameterized, or fully resolved? We suggested potential avenues in pursuit of answers to these questions, while noting that robust answers will perhaps only be available after global climate models routinely resolve processes to determine their role in a holistic context.

Among the most important transitions to have occurred during the past decade is the growing presence of mesoscale eddying global ocean climate simulations. Changes may appear in air–sea fluxes in coupled simulations due to refined representation of frontal-scale features (Bryan et al., 2010); circulation can be modified through eddy-mean flow interactions (Holland and Rhines, 1980); stochastic features are introduced through eddy fluctuations; and currents interact with a refined representation of bathymetry. Relative to their more laminar predecessors, eddying simulations necessitate enhanced fidelity from numerical methods and require a wide suite of analysis methods to unravel mechanisms. There is progress, but more is required before mesoscale eddying simulations achieve the trust and familiarity required to make them a robust scientific tool for numerical oceanography and climate. In particular, we need a deeper understanding of the generation and decay of mesoscale eddies, both to ensure their proper representation in eddying simulations, and to parameterize in coarse models. We must also address the difficulties associated with managing the huge amounts of simulated data generated by global eddying simulations.

No sound understanding exists of what is required from both grid spacing and numerical methods to fully resolve the mesoscale in global models. The work of Smith et al. (2000) suggests that the mesoscale is resolved so long as the grid spacing is finer than the first baroclinic Rossby radius. This is a sensible hypothesis, given that the mesoscale eddy scales are proportional to the Rossby radius, and given that much of the mid-latitude ocean energy is contained in the barotropic and first baroclinic modes. However, this criterion was proposed without a rigorous examination of how important higher modes may be; how sensitive this criteria is to specifics of numerical methods and subgrid scale parameterizations; or whether the criteria is supported by a thorough resolution study. We propose that a solid understanding of the mesoscale eddy resolution question will greatly assist in answering many of the questions regarding the role of the ocean in climate.

A related question concerns the relation between the numerical modeling of mesoscale eddies and dianeutral mixing. Namely, is it sensible to consider mesoscale eddying climate simulations using a model that includes unphysically large spurious dianeutral mixing? Are isopycnal models, or their generalizations to ALE methods, the optimal means for ensuring spurious numerical mixing is sufficiently small to accurately capture physical mixing processes, even in the presence of realistic stirring from mesoscale eddies? Or will the traditional-level model approaches be enhanced sufficiently to make the modeler's choice based on convenience rather than fundamentals? We conjecture that an answer will be clear within a decade.

As evidenced by the increasing "operational" questions being asked by oceanographers, spanning the spectrum from real time ocean forecasting (Chapter 22) to interannual to longer term climate projections (Chapters 23–26) as well as reanalysis and state estimation (Chapter 21), numerical oceanography is being increasingly asked to address applied questions that have an impact on decisions reaching outside of science. As with the atmospheric sciences, the added responsibility, and the associated increased visibility, arising from applications bring great opportunities for enhancing ocean science. The increased functionality and applications of ocean models must in turn be strongly coupled to a continued focus on the physics and numerics forming their foundation.

ACKNOWLEDGMENTS

We thank WCRP/CLIVAR for sponsoring the Working Group for Ocean Model Development, where the authors have participated since 1999. The presentation in this chapter was greatly assisted by comments from John A. Church, Carolina Dufour, John Gould, Trevor McDougall, Angélique Melet, Maxim Nikurashin, Gerold Siedler, and an anonymous reviewer. Bernard Barnier kindly provided the Drakkar/ORCA simulation results for Figure 20.5. Cathy Raphael kindly drafted Figure 20.1.

REFERENCES

Adcroft, A., Campin, J.-M., 2004. Rescaled height coordinates for accurate representation of free-surface flows in ocean circulation models. Ocean Model. 7, 269–284.

Adcroft, A., Hallberg, R.W., 2006. On methods for solving the oceanic equations of motion in generalized vertical coordinates. Ocean Model. 11, 224–233.

Adcroft, A., Hill, C., Marshall, J., 1997. Representation of topography by shaved cells in a height coordinate ocean model. Mon. Weather Rev. 125, 2293–2315.

Adcroft, A., Scott, J.R., Marotzke, J., 2001. Impact of geothermal heating on the global ocean circulation. Geophys. Res. Lett. 28, 1735–1738.

Adcroft, A., Hallberg, R., Dunne, J., Samuels, B., Galt, J., Barker, C., Payton, D., 2010. Simulations of underwater plumes of dissolved oil in the Gulf of Mexico. Geophys. Res. Lett. 37. http://dx.doi.org/10.1029/2010GL044689.

Arakawa, A., 1966. Computational design for long-term numerical integration of the equations of fluid motion: two-dimensional incompressible flow. Part 1. J. Comput. Phys. 1, 119–143.

Arakawa, A., Lamb, V., 1981. A potential enstrophy and energy conserving scheme for the shallow water equations. Mon. Weather Rev. 109, 18–36.

Arbic, B., Wallcraft, A., Metzger, E., 2010. Concurrent simulation of the eddying general circulation and tides in a global ocean model. Ocean Model. 32, 175–187.

Barnier, B., 1998. Forcing the ocean. In: Chassignet, E.P., Verron, J. (Eds.), Ocean Modeling and Parameterization. NATO ASI Mathematical and Physical Sciences Series, vol. 516. Kluwer Academic Publishers, pp. 45–80.

Barnier, B., Siefridt, L., Marchesiello, P., 1995. Thermal forcing for a global ocean circulation model using a three-year climatology of ECMWF analyses. J. Mar. Res. 6, 363–380.

Barnier, B., Madec, G., Penduff, T., Molines, J., Treguier, A., Sommer, J.L., Beckmann, A., Biastoch, A., Böning, C.W., Dengg, J., Derval, C., Durand, E., Gulev, S., Remy, E., Talandier, C., Theetten, S., Maltrud, M., McClean, J., Cuevas, B.D.,

2006. Impact of partial steps and momentum advection schemes in a global ocean circulation model at eddy permitting resolution. Ocean Dyn. 56, 543–567.

Bates, M., Griffies, S.M., England, M., 2012a. A dynamic, embedded Lagrangian model for ocean climate models, Part I: theory and implementation. Ocean Model. 59–60, 41–59.

Bates, M., Griffies, S.M., England, M., 2012b. A dynamic, embedded Lagrangian model for ocean climate models, Part II: idealised overflow tests. Ocean Model. 59–60, 60–76.

Beckmann, A., 1998. The representation of bottom boundary layer processes in numerical ocean circulation models. In: Chassignet, E.P., Verron, J. (Eds.), Ocean Modeling and Parameterization. NATO ASI Mathematical and Physical Sciences Series, 516, Kluwer Academic Publishers, pp. 135–154.

Beckmann, A., Döscher, R., 1997. A method for improved representation of dense water spreading over topography in geopotential-coordinate models. J. Phys. Oceanogr. 27, 581–591.

Beckmann, A., Haidvogel, D., 1993. Numerical simulation of flow around a tall isolated seamount. Part I: problem formulation and model accuracy. J. Phys. Oceanogr. 23, 1736–1753.

Beckmann, A., Haidvogel, D., 1994. On the generation and role of eddy variability in the central North Atlantic Ocean. J. Geophys. Res. 99, 20381–20391.

Belcher, S., Grant, A., Hanley, K., Fox-Kemper, B., Van Roekel, L., Sullivan, P., Large, W., Brown, A., Hines, A., Calvert, D., Rutgersson, A., Petterson, H., Bidlot, J., Janssen, P., Polton, J.A., 2012. A global perspective on Langmuir turbulence in the ocean surface boundary layer. Geophys. Res. Lett. 39, L18605. http://dx.doi.org/10.1029/2012GL052932.

Berlov, P., 2005. Random-forcing model of the mesoscale oceanic eddies. J. Fluid Mech. 529, 71–95.

Biastoch, A., Böning, C., Lutjeharms, J., 2008. Agulhas leakage dynamics affects decadal variability in Atlantic overturning circulation. Nature 456 (7221), 489–492.

Blanke, B., Arhan, M., Madec, G., Roche, S., 1999. Warm water paths in the equatorial Atlantic as diagnosed with a general circulation model. J. Phys. Oceanogr. 29, 2753–2768.

Blayo, E., Debreu, L., 2005. Revisiting open boundary conditions from the point of view of characteristic variables. Ocean Model. 9, 231–252.

Bleck, R., 1998. Ocean modeling in isopycnic coordinates. In: Chassignet, E.P., Verron, J. (Eds.), Ocean Modeling and Parameterization. NATO ASI Mathematical and Physical Sciences Series, 516, Kluwer Academic Publishers, pp. 423–448.

Bleck, R., 2002. An oceanic general circulation model framed in hybrid isopycnic-cartesian coordinates. Ocean Model. 4, 55–88.

Boccaletti, G., Ferrari, R., Fox-Kemper, B., 2007. Mixed layer instabilities and restratification. J. Phys. Oceanogr. 35, 1263–1278.

Böning, C.W., Dispert, A., Visbeck, M., Rintoul, S., Schwarzkopf, F., 2008. The response of the Antarctic Circumpolar Current to recent climate change. Nat. Geosci. 1, 864–869.

Brankart, J.-M., 2013. Impact of uncertainties in the horizontal density gradient upon low resolution global ocean modelling. Ocean Model. 66, 64–76.

Brodeau, L., Barnier, B., Treguier, A., Penduff, T., Gulev, S., 2010. An ERA40-based atmospheric forcing for global ocean circulation models. Ocean Model. 31, 88–104.

Bryan, F., 1987. Parameter sensitivity of primitive equation ocean general circulation models. J. Phys. Oceanogr. 17, 970–985.

Bryan, K., Lewis, L.J., 1979. A water mass model of the world ocean. J. Geophys. Res. 84, 2503–2517.

Bryan, K., Manabe, S., Pacanowski, R.C., 1975. A global ocean-atmosphere climate model. Part II. The oceanic circulation. J. Phys. Oceanogr. 5, 30–46.

Bryan, F., Böning, C., Holland, W., 1995. On the mid-latitude circulation in a high resolution model of the North Atlantic. J. Phys. Oceanogr. 25, 289–305.

Bryan, F., Thomas, R., Dennis, J., Chelton, D., Loeb, N., McClean, J., 2010. Frontal scale air-sea interaction in high-resolution coupled climate models. J. Clim. 23, 6277–6291.

Campin, J.-M., Goosse, H., 1999. Parameterization of density-driven downsloping flow for a coarse-resolution ocean model in z-coordinate. Tellus 51A, 412–430.

Campin, J.-M., Marshall, J., Ferreira, D., 2011. Super-parameterization in ocean modeling: application to deep convection. Ocean Model. 36, 90–101.

Cavaleri, L., Fox-Kemper, B., Hemer, M., 2012. Wind waves in the coupled climate system. Bull. Am. Meteorol. Soc. 93, 1651–1661.

Chaikin, P.M., Lubensky, T.C., 1995. Principles of Condensed Matter Physics. Cambridge University Press, Cambridge, United Kingdom.

Costa, M.D., Mercier, H., Treguier, A.-M., 2005. Effects of the mixed layer time variability on kinematic subduction rate diagnostics. J. Phys. Oceanogr. 35, 427–443.

Cox, M.D., 1984. A Primitive Equation, 3-Dimensional Model of the Ocean. NOAA/Geophysical Fluid Dynamics Laboratory, Princeton, USA.

Dai, A., Qian, T., Trenberth, K., Milliman, J., 2009. Changes in continental freshwater discharge from 1948-2004. J. Clim. 22, 2773–2791.

Danabasoglu, G., Large, W., Briegleb, B., 2010. Climate impacts of parameterized nordic sea overflows. J. Geophys. Res. 115, C11005. http://dx.doi.org/10.1029/2010JC006243.

Danilov, S.D., 2013. Ocean modeling on unstructured meshes. Ocean Model. 69, 195–210. http://dx.doi.org/10.1016/j.ocemod.2013.05.005.

Davis, R.E., 1994a. Diapycnal mixing in the ocean: equations for large-scale budgets. J. Phys. Oceanogr. 24, 777–800.

Davis, R.E., 1994b. Diapycnal mixing in the ocean: the Osborn-Cox model. J. Phys. Oceanogr. 24, 2560–2576.

de Boisseson, E., Thierry, V., Mercier, H., Caniaux, G., 2010. Mixed layer heat budget in the iceland basin from argo. J. Geophys. Res. 115, C10055.

De Miranda, A., Barnier, B., Dewar, W., 1999. On the dynamics of the Zapiola Anticyclone. J. Geophys. Res. 104, 21137–21149.

Debreu, L., Blayo, E., 2008. Two-way embedding algorithms: a review. Ocean Dyn. 58, 415–428.

DeGroot, S.R., Mazur, P., 1984. Non-Equilibrium Thermodynamics. Dover Publications, New York, p. 510.

Deleersnijder, E., Beckers, J.-M., 1992. On the use of the σ-coordinate system in regions of large bathymetric variations. J. Mar. Syst. 3, 381–390.

Deleersnijder, E., Cornaton, F., Haine, T., Vanclooster, M., Waugh, D.W., 2010. Tracer and timescale methods for understanding complex geophysical and environmental fluid flows. Environ. Fluid Mech. 10, http://dx.doi.org/10.1007/s10652-009-9164-1.

Delworth, T.L., Rosati, A., Anderson, W., Adcroft, A.J., Balaji, V., Benson, R., Dixon, K., Griffies, S.M., Lee, H.-C., Pacanowski, R.C., Vecchi, G.A., Wittenberg, A.T., Zeng, F., Zhang, R., 2012. Simulated climate and climate change in the GFDL CM2.5 high-resolution coupled climate model. J. Clim. 25, 2755–2781.

Denbo, D., Skyllingstad, E., 1996. An ocean large-eddy simulation model with application to deep convection in the Greenland Sea. J. Geophys. Res. 101, 1095–1110.

DeSzoeke, R.A., 2009. Isentropic averaging. J. Mar. Res. 67, 533–567.

DeSzoeke, R.A., Bennett, A.F., 1993. Microstructure fluxes across density surfaces. J. Phys. Oceanogr. 23, 2254–2264.

DeSzoeke, R.A., Samelson, R.M., 2002. The duality between the Boussinesq and non-Boussinesq hydrostatic equations of motion. J. Phys. Oceanogr. 32, 2194–2203.

Dietrich, D., Marietta, M., Roache, P., 1987. An ocean modeling system with turbulent boundary layers and topography: part 1. Numerical studies of small island wakes in the ocean. Int. J. Numer. Methods Fluids 7, 833–855.

Donea, J., Huerta, A., Ponthot, J.-P., Rodríguez-Ferran, A., 2004. Arbitrary Lagrangian-Eulerian methods. In: Stein, E., de Borst, R., Hughes, T.J.R. (Eds.), Encyclopedia of Computational Mechanics. John Wiley and Sons, Chapter 14. http://dx.doi.org/10.1002/0470091355.

Döös, K., Nycander, J., Coward, A., 2008. Lagrangian decomposition of the Deacon Cell. J. Geophys. Res. Oceans 113. http://dx.doi.org/10.1029/2007JC004351.

Downes, S.M., Gnanadesikan, A., Griffies, S.M., Sarmiento, J., 2011. Water mass exchange in the Southern Ocean in coupled climate models. J. Phys. Oceanogr. 41, 1756–1771.

Drakkar Group, 2007. Eddy-permitting ocean circulation hindcasts of past decades. CLIVAR Exchanges 42, 8–10.

Ducet, N., Le Traon, P.-Y., Reverdin, G., 2000. Global high-resolution mapping of ocean circulation from TOPEX/Poseidon and ERS-1 and -2. J. Geophys. Res. 105, 19477–19498.

Dunne, J.P., John, J.G., Hallberg, R.W., Griffies, S.M., Shevliakova, E.N., Stouffer, R.J., Krasting, J.P., Sentman, L.A., Milly, P.C.D., Malyshev, S.L., Adcroft, A.J., Cooke, W., Dunne, K.A., Harrison, M.J., Levy, H., Samuels, B.L., Spelman, M., Winton, M., Wittenberg, A.T., Phillips, P.J., Zadeh, N., 2012. GFDLs ESM2 global coupled climate-carbon Earth System Models Part I: physical formulation and baseline simulation characteristics. J. Clim. 25, 6646–6665.

Eckart, C., 1948. An analysis of the stirring and mixing processes in incompressible fluids. J. Mar. Res. 7, 265–275.

Eden, C., Dietze, H., 2009. Effects of mesoscale eddy/wind interactions on biological new production and eddy kinetic energy. J. Geophys. Res. 114, C05023.

Eden, C., Greatbatch, R., Olbers, D., 2007. Interpreting eddy fluxes. J. Phys. Oceanogr. 37, 1282–1296.

Emile-Geay, J., Madec, G., 2009. Geothermal heating, diapycnal mixing and the abyssal circulation. Ocean Sci. 5, 203–217.

England, M.H., Maier-Reimer, E., 2001. Using chemical tracers to assess ocean models. Rev. Geophys. 39, 29–70.

Farneti, R., Gent, P., 2011. The effects of the eddy-induced advection coefficient in a coarse-resolution coupled climate model. Ocean Model. 39, 135–145.

Farneti, R., Delworth, T., Rosati, A., Griffies, S.M., Zeng, F., 2010. The role of mesoscale eddies in the rectification of the Southern Ocean response to climate change. J. Phys. Oceanogr. 40, 1539–1557.

Fox-Kemper, B., Menemenlis, D., 2008. Can large eddy simulation techniques improve mesoscale rich ocean models? In: Hecht, M., Hasumi, H. (Eds.), Ocean Modeling in an Eddying Regime. Geophysical Monograph, 177, American Geophysical Union, Washington, DC, pp. 319–338.

Fox-Kemper, B., Ferrari, R., Hallberg, R., 2008. Parameterization of mixed layer eddies. I: theory and diagnosis. J. Phys. Oceanogr. 38, 1145–1165.

Frankignoul, C., Hasselmann, K., 1977. Stochastic climate models. Part II: application to sea-surface temperature variability and thermocline variability. Tellus 29, 284–305.

Gent, P., Danabasoglu, G., 2011. Response of increasing Southern Hemisphere winds in CCSM4. J. Clim. 24, 4992–4998.

Gent, P.R., McWilliams, J.C., 1990. Isopycnal mixing in ocean circulation models. J. Phys. Oceanogr. 20, 150–155.

Gent, P.R., Willebrand, J., McDougall, T.J., McWilliams, J.C., 1995. Parameterizing eddy-induced tracer transports in ocean circulation models. J. Phys. Oceanogr. 25, 463–474.

Gill, A., 1982. Atmosphere-Ocean Dynamics. International Geophysics Series, vol. 30. Academic Press, London, 662+xv pp.

Gnanadesikan, A., Slater, R.D., Swathi, P.S., Vallis, G.K., 2005. The energetics of ocean heat transport. J. Clim. 17, 2604–2616.

Goldberg, D., Little, C., Sergienko, O., Gnanadesikan, A., Hallberg, R., Oppenheimer, M., 2012. Investigation of land ice-ocean interaction with a fully coupled ice-ocean model: 1. Model description and behavior. J. Geophys. Res. 117, F02037. http://dx.doi.org/10.1029/2011JF002246.

Grabowski, W., 2001. Coupling cloud processes with the large-scale dynamics using the Cloud-Resolving Convection Parameterization (CRCP). J. Atmos. Sci. 58, 978–997.

Greatbatch, R.J., 1994. A note on the representation of steric sea level in models that conserve volume rather than mass. J. Geophys. Res. 99, 12767–12771.

Greatbatch, R.J., McDougall, T.J., 2003. The non-Boussinesq temporal-residual-mean. J. Phys. Oceanogr. 33, 1231–1239.

Gregg, M., Sanford, T., Winkel, D., 2003. Reduced mixing from the breaking of internal waves in equatorial waters. Nature 422, 513–515.

Griffies, S.M., 1998. The Gent-McWilliams skew-flux. J. Phys. Oceanogr. 28, 831–841.

Griffies, S.M., 2004. Fundamentals of Ocean Climate Models. Princeton University Press, Princeton, USA, 518+xxxiv pp.

Griffies, S.M., Adcroft, A.J., 2008. Formulating the equations for ocean models. In: Hecht, M., Hasumi, H. (Eds.), Eddy Resolving Ocean Models. Geophysical Monograph, vol. 177. American Geophysical Union, pp. 281–317.

Griffies, S.M., Greatbatch, R.J., 2012. Physical processes that impact the evolution of global mean sea level in ocean climate models. Ocean Model. 51, 37–72.

Griffies, S.M., Hallberg, R.W., 2000. Biharmonic friction with a Smagorinsky viscosity for use in large-scale eddy-permitting ocean models. Mon. Weather Rev. 128, 2935–2946.

Griffies, S.M., Böning, C.W., Bryan, F.O., Chassignet, E.P., Gerdes, R., Hasumi, H., Hirst, A., Treguier, A.-M., Webb, D., 2000a. Developments in ocean climate modelling. Ocean Model. 2, 123–192.

Griffies, S.M., Pacanowski, R.C., Hallberg, R.W., 2000b. Spurious diapycnal mixing associated with advection in a *z*-coordinate ocean model. Mon. Weather Rev. 128, 538–564.

Griffies, S.M., Pacanowski, R., Schmidt, M., Balaji, V., 2001. Tracer conservation with an explicit free surface method for *z*-coordinate ocean models. Mon. Weather Rev. 129, 1081–1098.

Griffies, S.M., Biastoch, A., Böning, C.W., Bryan, F., Danabasoglu, G., Chassignet, E., England, M.H., Gerdes, R., Haak, H., Hallberg, R.W., Hazeleger, W., Jungclaus, J., Large, W.G., Madec, G., Pirani, A., Samuels, B.L., Scheinert, M., Gupta, A.S., Severijns, C.A.,

Simmons, H.L., Treguier, A.M., Winton, M., Yeager, S., Yin, J., 2009. Coordinated Ocean-ice Reference Experiments (COREs). Ocean Model. 26, 1–46.

Griffies, S.M., Adcroft, A.J., Banks, H., Böoning, C.W., Chassignet, E.P., Danabasoglu, G., Danilov, S., Deleersnijder, E., Drange, H., England, M., Fox-Kemper, B., Gerdes, R., Gnanadesikan, A., Greatbatch, R.J., Hallberg, R., Hanert, E., Harrison, M.J., Legg, S.A., Little, C.M., Madec, G., Marsland, S., Nikurashin, M., Pirani, A., Simmons, H.L., Schröooter, J., Samuels, B.L., Treguier, A.-M., Toggweiler, J.R., Tsujino, H., Vallis, G.K., White, L., 2010. Problems and prospects in large-scale ocean circulation models. In: Hall, J., Harrison, D., Stammer, D. (Eds.), Proceedings of the OceanObs09 Conference: Sustained Ocean Observations and Information for Society, Venice, Italy, 21–25 September 2009, vol. 2. ESA Publication WPP-306.

Griffies, S.M., Winton, M., Donner, L.J., Downes, S.M., Farneti, R., Gnanadesikan, A., Horowitz, L.W., Hurlin, W.J., Lee, H.-C., Liang, Z., Palter, J.B., Samuels, B.L., Wittenberg, A.T., Wyman, B.L., Yin, J., Zadeh, N.T., 2011. GFDL's CM3 coupled climate model: characteristics of the ocean and sea ice simulations. J. Clim. 24, 3520–3544.

Grinstein, F.F., Margolin, L.G., Rider, W.J., 2007. A rationale for implicit LES. In: Grinstein, F., Margolin, L., Rider, W. (Eds.), Implicit Large Eddy Simulation: Computing Turbulent Fluid Dynamics. Cambridge University Press, Cambridge, United Kingdom.

Grist, J.P., Josey, S.A., Marsh, R., Good, S.A., Coward, A., de Cuevas, B., Alderson, S., New, A., Madec, G., 2010. The roles of surface heat flux and ocean heat transport convergence in determining Atlantic Ocean temperature variability. Ocean Dyn. 60, 771–790.

Hall, A., Manabe, S., 1997. Can local, linear stochastic theory explain sea surface temperature and salinity variability? Clim. Dyn. 13, 167–180.

Hallberg, R.W., 2003. The suitability of large-scale ocean models for adapting parameterizations of boundary mixing and a description of a refined bulk mixed layer model. In: Müller, P., Garrett, C. (Eds.), Near-Boundary Processes and Their Parameterization. Proceedings of the 13th Aha Huliko'a Hawaiian Winter Workshop. University of Hawaii, Manoa, pp. 187–203.

Haney, R.L., 1971. Surface thermal boundary conditions for ocean circulation models. J. Phys. Oceanogr. 1, 241–248.

Haney, R.L., 1991. On the pressure gradient force over steep topography in sigma-coordinate ocean models. J. Phys. Oceanogr. 21, 610–619.

Hasselmann, K., 1976. Stochastic climate models. Part I: theory. Tellus 28, 473–485.

Herzfeld, M., Schmidt, M., Griffies, S.M., Liang, Z., 2011. Realistic test cases for limited area ocean modelling. Ocean Model. 37, 1–34.

Hesselberg, T., 1926. Die Gesetze der ausgeglichenen atmosphaerischen Bewegungen. Beitr. Phys. Atmos. 12, 141–160.

Hill, C., Ferreira, D., Campin, J.-M., Marshall, J., Abernathey, R., Barrier, N., 2012. Controlling spurious diapyncal mixing in eddy-resolving height-coordinate ocean models—insights from virtual deliberate tracer release experiments. Ocean Model. 45–46, 14–26.

Hofmann, M., Morales-Maqueda, M., 2011. The response of Southern Ocean eddies to increased midlatitude westerlies: a non-eddy resolving model study. Geophys. Res. Lett. 38, L03605. http://dx.doi.org/10.1029/2010GL045972.

Holland, W.R., Rhines, P.B., 1980. An example of eddy-induced ocean circulation. J. Phys. Oceanogr. 10, 1010–1031.

Holloway, G., 1986. Eddies, waves, circulation, and mixing: statistical geofluid mechanics. Annu. Rev. Fluid Mech. 18, 91–147.

Holloway, G., 1989. Subgridscale representation. In: Anderson, D.L., Willebrand, J. (Eds.), Oceanic Circulation Models: Combining Data and Dynamics. NATO ASI Series C, vol. 284. Kluwer Academic Publishers, pp. 513–593.

Holloway, G., 1992. Representing topographic stress for large-scale ocean models. J. Phys. Oceanogr. 22, 1033–1046.

Huang, R.X., Jin, X., Zhang, X., 2001. An oceanic general circulation model in pressure coordinates. Adv. Atmos. Phys. 18, 1–22.

Hughes, C., 2002. Zonal jets in and near the Coral Sea seen by satellite altimetry. Geophys. Res. Lett. 29, 1330.

Ilicak, M., Adcroft, A.J., Griffies, S.M., Hallberg, R.W., 2012. Spurious dianeutral mixing and the role of momentum dissipation. Ocean Model. 45–46, 37–58.

Illig, S., Dewitte, B., Ayoub, N., du Penhoat, Y., De Mey, P., Reverdin, G., Bonjean, F., Lagerloef, G.S.E., 2004. Interannual long equatorial waves in the tropical Atlantic from a high-resolution ocean general circulation model experiment in 1981–2000. J. Geophys. Res. 109, C02022. http://dx.doi.org/10.1029/2003JC001771.

IOC, SCOR, IAPSO, 2010. The international thermodynamic equation of seawater-2010: calculation and use of thermodynamic properties. Intergovernmental Oceanographic Commission, Manuals and Guides No. 56. UNESCO, p. 196, Available from http://www.TEOS-10.org.

Iudicone, D., Madec, G., McDougall, T.J., 2008. Water-mass transformations in a neutral density framework and the key role of light penetration. J. Phys. Oceanogr. 38, 1357–1376.

Iudicone, D., Rodgers, K., Stendardo, I., Aumont, O., Madec, G., Bopp, L., Mangoin, O., d'Alcala, M.R., 2011. Water masses as a unifying framework for understanding the Southern Ocean Carbon Cycle. Biogeosciences 8, 1031–1052.

Jackson, L., Hallberg, R., Legg, S., 2008. A parameterization of shear-driven turbulence for ocean climate models. J. Phys. Oceanogr. 38, 1033–1053.

Jiang, C., Thompson, L., Kelly, K., 2008. Equatorial influence of QuikSCAT winds in an isopycnal ocean model compared to NCEP2 winds. Ocean Model. 24, 65–71.

Jochum, M., 2009. Impact of latitudinal variations in vertical diffusivity on climate simulations. J. Geophys. Res. 114, C01010. http://dx.doi.org/10.1029/2008JC005030.

Jochum, M., Danabasoglu, G., Holland, M., Kwon, Y.-O., Large, W., 2008. Ocean viscosity and climate. J. Geophys. Res. 114, C06017. http://dx.doi.org/10.1029/2007JC004515.

Jones, H., Marshall, J., 1993. Convection with rotation in a neutral ocean: a study of open-ocean deep convection. J. Phys. Oceanogr. 23, 1009–1039.

Jongma, J., Driesschaert, E., Fichefet, T., Goosse, H., Renssen, H., 2009. The effect of dynamic-thermodynamic icebergs on the Southern Ocean climate in a three-dimensional model. Ocean Model. 26, 104–113.

Khairoutdinov, M., DeMott, C., Randall, D., 2008. Evaluation of the simulated interannual and subseasonal variability in an AMIP-style simulation using the CSU Multiscale Modeling Framework. J. Clim. 21, 413–431.

Killworth, P.D., Edwards, N., 1999. A turbulent bottom boundary layer code for use in numerical ocean models. J. Phys. Oceanogr. 29, 1221–1238.

Killworth, P.D., Smeed, D., Nurser, A., 2000. The effects on ocean models of relaxation toward observations at the surface. J. Phys. Oceanogr. 30, 160–174.

Kitsios, V., Frederiksen, J.S., Zidikheri, M.J., 2013. Scaling laws for parameterisations of subgrid eddy-eddy interactions in simulations of oceanic circulations. Ocean Model. http://dx.doi.org/10.1016/j.ocemod.2013.05.001.

Klein, P., Lapeyre, G., 2009. The oceanic vertical pump induced by mesoscale and submesoscale turbulence. Ann. Rev. Mar. Sci. 1, 351–375.

Klinger, B.A., Marshall, J., Send, U., 1996. Representation of convective plumes by vertical adjustment. J. Geophys. Res. 101, 18175–18182.

Klocker, A., McDougall, T.J., 2010. Influence of the nonlinear equation of state on global estimates of dianeutral advection and diffusion. J. Phys. Oceanogr. 40, 1690–1709.

Koch-Larrouy, A., Madec, G., Blanke, B., Molcard, R., 2008. Water mass transformation along the Indonesian throughflow in an OGCM. J. Phys. Oceanogr. 58, 289–309.

Kopp, R.E., Mitrovica, J.X., Griffies, S.M., Yin, J., Hay, C.C., Stouffer, R.J., 2010. The impact of Greenland melt on regional sea level: a preliminary comparison of dynamic and static equilibrium effects. Clim. Change Lett. 103, 619–625.

Kuhlbrodt, T., Griesel, A., Montoya, M., Levermann, A., Hofmann, M., Rahmstorf, S., 2007. On the driving processes of the Atlantic meridional overturning circulation. Rev. Geophys. 45. http://dx.doi.org/10.1029/2004RG000166.

Kunze, E., Sanford, T.B., 1996. Abyssal mixing: where it is not. J. Phys. Oceanogr. 26, 2286–2296.

Kunze, E., Firing, E., Hummon, J.M., Chereskin, T.K., Thurnherr, A.M., 2006. Global abyssal mixing inferred from lowered ADCP shear and CTD strain profiles. J. Phys. Oceanogr. 36, 1553–1576.

Laanaia, N., Wirth, A., Molines, J., Barnier, B., Verron, J., 2010. On the numerical resolution of the bottom layer in simulations of oceanic gravity currents. Ocean Sci. 6, 563–572.

Large, W.B., Nurser, A.G., 2001. Ocean surface water mass transformation. In: Siedler, G., Church, J., Gould, J. (Eds.), Ocean Circulation and Climate. International Geophysics Series, vol. 77. Academic Press, San Diego, pp. 317–336.

Large, W., Yeager, S., 2004. Diurnal to decadal global forcing for ocean and sea-ice models: the data sets and flux climatologies. NCAR Technical Note: NCAR/TN-460+STR. CGD Division of the National Center for Atmospheric Research.

Large, W.G., Yeager, S., 2009. The global climatology of an interannually varying air-sea flux data set. Clim. Dyn. 33, 341–364.

Large, W., McWilliams, J., Doney, S., 1994. Oceanic vertical mixing: a review and a model with a nonlocal boundary layer parameterization. Rev. Geophys. 32, 363–403.

Large, W.G., Danabasoglu, G., Doney, S.C., McWilliams, J.C., 1997. Sensitivity to surface forcing and boundary layer mixing in a global ocean model: annual-mean climatology. J. Phys. Oceanogr. 27, 2418–2447.

Large, W.G., Danabasoglu, G., McWilliams, J.C., Gent, P.R., Bryan, F.O., 2001. Equatorial circulation of a global ocean climate model with anisotropic horizontal viscosity. J. Phys. Oceanogr. 31, 518–536.

Le Sommer, J., Penduff, T., Theetten, S., Madec, G., Barnier, B., 2009. How momentum advection schemes influence current-topography interactions at eddy permitting resolution. Ocean Model. 29, 1–14.

Le Traon, P.-Y., Nadal, F., Ducet, N., 1998. An improved mapping method of multi-satellite altimeter data. J. Atmos. Oceanic Technol. 15, 522–534.

Leclair, M., Madec, G., 2011. $\tilde{z}$-coordinate, an arbitrary Lagrangian Eulerian coordinate separating high and low frequency motions. Ocean Model. 37, 139–152.

Ledwell, J.R., Watson, A.J., 1998. Mixing of a tracer in the pycnocline. J. Geophys. Res. 103, 21499–21529.

Ledwell, J.R., Watson, A.J., Law, C.S., 1993. Evidence for slow mixing across the pycnocline from an open-ocean tracer-release experiment. Nature 364, 701–703.

Ledwell, J.R., St. Laurent, L., Girton, J., Toole, J., 2011. Diapycnal mixing in the Antarctic Circumpolar Current. J. Phys. Oceanogr. 41, 241–246.

Legg, S., 2012. Overflows and convectively driven flows. In: Chassignet, E., Cenedese, C., Verron, J. (Eds.), Buoyancy-Driven Flows. Cambridge University Press, Cambridge, UK, pp. 203–239.

Legg, S., Hallberg, R., Girton, J., 2006. Comparison of entrainment in overflows simulated by z-coordinate, isopycnal and non-hydrostatic models. Ocean Model. 11, 69–97.

Legg, S., Jackson, L., Hallberg, R., 2008. Eddy-resolving modeling of overflows. In: Hecht, M., Hasumi, H. (Eds.), Eddy Resolving Ocean Models. Geophysical Monograph, 177, American Geophysical Union, Washington, DC, pp. 63–82.

Legg, S., Briegleb, B., Chang, Y., Chassignet, E.P., Danabasoglu, G., Ezer, T., Gordon, A.L., Gries, S.M., Hallberg, R.W., Jackson, L., Large, W., Özgökmen, T.M., Peters, H., Price, J., Riemenschneider, U., Wu, W., Xu, X., Yang, J., 2009. Improving oceanic overow representation in climate models: the Gravity Current Entrainment Climate Process Team. Bull. Am. Meteorol. Soc. 90, 657–670.

Leith, C.E., 1996. Stochastic models of chaotic systems. Phys. D 98, 481–491.

Lemarié, F., Debreu, L., Shchepetkin, A.F., McWilliams, J.C., 2012a. On the stability and accuracy of the harmonic and biharmonic isoneutral mixing operators in ocean models. Ocean Model. 52–53, 9–35.

Lemarié, F., Kurian, J., Shchepetkin, A.F., Molemaker, M.J., Colas, F., McWilliams, J.C., 2012b. Are there inescapable issues prohibiting the use of terrain-following coordinates in climate models? Ocean Model. 42, 57–79.

Lique, C., Treguier, A.-M., Scheinert, M., Penduff, T., 2009. A model-based study of ice and freshwater transport variability along both sides of Greenland. Clim. Dyn. 33, 685–705.

Lombard, A., Garric, G., Penduff, T., 2009. Regional patterns of observed sea level change: insights from a 1/4° global ocean/sea-ice hindcast. Ocean Dyn. 59, 433–449.

MacKinnon, J., Alford, M., Bouruet-Aubertot, P., Bindoff, N., Elipot, S., Gille, S., Girton, J., Gregg, M., Kunze, E., Naveira Garabato, A., Phillips, H., Pinkel, R., Polzin, K., Sanford, T., Simmons, H., Speer, K., 2010. Using global arrays to investigate internal-waves and mixing. In: Hall, J., Harrison, D., Stammer, D. (Eds.), Proceedings of the Ocean. Obs09 Conference: Sustained Ocean Observations and Information for Society, Venice, Italy, 21–25 September 2009. vol. 2. ESA Publication WPP-306.

Maltrud, M., McClean, J., 2005. An eddy resolving global 1/10° ocean simulation. Ocean Model. 8, 31–54.

Manizza, M., Le Quere, C., Watson, A., Buitenhuis, E., 2005. Bio-optical feedbacks among phytoplankton, upper ocean physics and sea-ice in a global model. Geophys. Res. Lett. 32. http://dx.doi.org/10.1029/2004GL020778.

Maqueda, M.M., Holloway, G., 2006. Second-order moment advection scheme applied to Arctic Ocean simulation. Ocean Model. 14, 197–221.

Marchesiello, P., McWilliams, J., Shchepetkin, A., 2001. Open boundary conditions for long-term integration of regional oceanic models. Ocean Model. 3, 1–20.

Marchesiello, J.M.P., Debreu, L., Couvelard, X., 2009. Spurious diapycnal mixing in terrain-following coordinate models: the problem and a solution. Ocean Model. 26, 156–169.

Margolin, L., Rider, W., Grinstein, F., 2006. Modeling turbulent flow with implicit LES. J. Turbul. 7, 1–27.

Marsh, R., 2000. Cabbeling due to isopycnal mixing in isopycnic coordinate models. J. Phys. Oceanogr. 30, 1757–1775.

Marshall, J., Schott, F., 1999. Open-ocean convection: observations, theory, and models. Rev. Geophys. 37, 1–64.

Marshall, J., Hill, C., Perelman, L., Adcroft, A., 1997. Hydrostatic, quasi-hydrostatic, and nonhydrostatic ocean modeling. J. Geophys. Res. 102, 5733–5752.

Marshall, J., Jamous, D., Nilsson, J., 1999. Reconciling thermodynamic and dynamic methods of computation of water-mass transformation rates. Deep Sea Res. Part I 46, 545–572.

Marshall, J., Adcroft, A., Campin, J.-M., Hill, C., White, A., 2004. Atmosphere-ocean modeling exploiting fluid isomorphisms. Mon. Weather Rev. 132, 2882–2894.

Martin, T., Adcroft, A., 2010. Parameterizing the fresh-water flux from land ice to ocean with interactive icebergs in a coupled climate model. Ocean Model. 34, 111–124.

Maze, G., Forget, G., Buckley, M., Marshall, J., Cerovecki, I., 2009. Using transformation and formation maps to study the role of air-sea heat fluxes in the North Atlantic eighteen degree water formation. J. Phys. Oceanogr. 39, 1818–1835.

McDougall, T.J., 1987. Thermobaricity, cabbeling, and water-mass conversion. J. Geophys. Res. 92, 5448–5464.

McDougall, T.J., 2003. Potential enthalpy: a conservative oceanic variable for evaluating heat content and heat fluxes. J. Phys. Oceanogr. 33, 945–963.

McDougall, T.J., McIntosh, P.C., 2001. The temporal-residual-mean velocity. Part II: isopycnal interpretation and the tracer and momentum equations. J. Phys. Oceanogr. 31, 1222–1246.

McDougall, T.J., Greatbatch, R., Lu, Y., 2002. On conservation equations in oceanography: how accurate are Boussinesq ocean models? J. Phys. Oceanogr. 32, 1574–1584.

McWilliams, J.C., Sullivan, P., 2001. Vertical mixing by Langmuir circulations. Spill Sci. Technol. Bull. 6, 225–237.

McWilliams, J.C., Danabasoglu, G., Gent, P.R., 1996. Tracer budgets in the warm water sphere. Tellus A 48, 179–192.

Meehl, G., Gent, P.R., Arblaster, J., Otto-Bliesner, B., Brady, E., Craig, A., 2001. Factors that affect the amplitude of El Niño in global coupled climate models. Clim. Dyn. 17, 515–526.

Meehl, G., Covey, C., Delworth, T., Latif, M., McAvaney, B., Mitchell, J., Stouffer, R., Taylor, K., 2007. The WCRP CMIP3 multimodel dataset: a new era in climate change research. Bull. Am. Meteorol. Soc. 88, 1383–1394.

Megann, A., New, A., Blaker, A., Sinha, B., 2010. The sensitivity of a coupled climate model to its ocean component. J. Clim. 23, 5126–5150.

Melet, A., Verron, J., Gourdeau, L., Koch-Larrouy, A., 2011. Solomon Sea water masses pathways to the equator and their modifications. J. Phys. Oceanogr. 41, 810–826.

Melet, A., Hallberg, R., Legg, S., Polzin, K., 2013. Sensitivity of the Pacific Ocean state to the vertical distribution of internal-tide driven mixing. J. Phys. Oceanogr. 43, 602–615. http://dx.doi.org/10.1175/JPO-D-12-055.1.

Mellor, G.L., Oey, L.-Y., Ezer, T., 1998. Sigma coordinate pressure gradient errors and the seamount problem. J. Atmos. Oceanic Technol. 15, 1122–1131.

Merryfield, W.J., Scott, R., 2007. Bathymetric influence on mean currents in two high resolution near-global ocean models. Ocean Model. 16, 76–94.

Mitrovica, J.X., Tamisiea, M.E., Davis, J.L., Milne, G.A., 2001. Recent mass balance of polar ice sheets inferred from patterns of global sea-level change. Nature 409, 1026–1029.

Müller, P., 2006. The Equations of Oceanic Motions, first ed. Cambridge University Press, Cambridge, p. 302.

Nakano, H., Suginohara, N., 2002. Effects of bottom boundary layer parameterization on reproducing deep and bottom waters in a world ocean model. J. Phys. Oceanogr. 32, 1209–1227.

Naveira-Garabato, A., Polzin, K., King, B., Heywood, K., Visbeck, M., 2004. Widespread intense turbulent mixing in the Southern Ocean. Science 303, 210–213.

Olbers, D.J., Willebrand, J., Eden, C., 2012. Ocean Dynamics, first ed. Springer, Berlin, Germany, p. 704.

Pacanowski, R.C., 1987. Effect of equatorial currents on surface stress. J. Phys. Oceanogr. 17, 833–838.

Pacanowski, R.C., Gnanadesikan, A., 1998. Transient response in a z-level ocean model that resolves topography with partial-cells. Mon. Weather Rev. 126, 3248–3270.

Palmer, T., Williams, P., 2008. Stochastic physics and climate modelling. Philos. Trans. R. Soc. A 366, 2421–2427.

Palter, J.B., Griffies, S.M., Galbraith, E.D., Gnanadesikan, A., Samuels, B.L., Klocker, A., 2013. The deep ocean buoyancy budget and its temporal variability. J. Clim. http://dx.doi.org/10.1175/JCLI-D-13-00016.1.

Parkinson, C., Washington, W., 1979. A large-scale numerical model of sea ice. J. Geophys. Res. 84, 311–337.

Pedlosky, J., 1987. Geophysical Fluid Dynamics, second ed. Springer-Verlag, Berlin Heidelberg New York, 710+xv pp.

Penduff, T., Sommer, J.L., Barnier, B., Treguier, A.-M., Molines, J.-M., Madec, G., 2007. Influence of numerical schemes on current-topography interactions in 1/4° global ocean simulations. Ocean Sci. 3, 509–524.

Penduff, T., Juza, M., Brodeau, L., Smith, G.C., Barnier, B., Molines, J.M., Treguier, A.M., Madec, G., 2010. Impact of global ocean model resolution on sea-level variability with emphasis on interannual time scales. Ocean Sci. 6, 269–284.

Philander, S.G., 1990. El Niño, La Niña, and the Southern Oscillation. Academic Press, San Diego.

Polzin, K.L., Toole, J.M., Ledwell, J.R., Schmitt, R.W., 1997. Spatial variability of turbulent mixing in the abyssal ocean. Science 276, 93–96.

Prather, M., 1986. Numerical advection by conservation of second-order moments. J. Geophys. Res. 91, 6671–6681.

Price, J., Yang, J., 1998. Marginal sea overows for climate simulations. In: Chassignet, E.P., Verron, J. (Eds.), Ocean Modeling and Parameterization. NATO ASI Mathematical and Physical Sciences Series, vol. 516. Kluwer Academic Publishers, pp. 155–170.

Redi, M.H., 1982. Oceanic isopycnal mixing by coordinate rotation. J. Phys. Oceanogr. 12, 1154–1158.

Redler, R., Böning, C.W., 1997. Effect of the overflows on the circulation in the subpolar north Atlantic: a regional model study. J. Geophys. Res. 102, 18529–18552.

Reif, F., 1965. Fundamentals of Statistical and Thermal Physics. McGraw-Hill, New York.

Ringler, T., Petersen, M., Higdon, R.L., Jacobsen, D., Jones, P.W., Maltrud, M., 2013. A multi-resolution approach to global ocean modeling. Ocean Model. 69, 211–232. http://dx.doi.org/10.1016/j.ocemod.2013.04.010.

Rivin, I., Tziperman, E., 1997. Sensitivity of air-sea fluxes to SST perturbations. J. Clim. 11, 2431–2446.

Roberts, M.J., Marshall, D., 1998. Do we require adiabatic dissipation schemes in eddy-resolving ocean models? J. Phys. Oceanogr. 28, 2050–2063.

Roberts, M.J., Wood, R., 1997. Topographic sensitivity studies with a Bryan-Cox-type ocean model. J. Phys. Oceanogr. 27, 823–836.

Rodgers, K., Blanke, B., Madec, G., Aumont, O., Ciais, P., Dutay, J.-C., 2003. Extratropical sources of Equatorial Pacific upwelling in an OGCM. Geophys. Res. Lett. 30. http://dx.doi.org/10.1029/2002GL016003.

Rosati, A., Miyakoda, K., 1988. A general circulation model for upper ocean simulation. J. Phys. Oceanogr. 18, 1601–1626.

Röske, F., 2006. A global heat and freshwater forcing dataset for ocean models. Ocean Model. 11, 235–297.

Saunders, P., King, B., 1995. Bottom currents derived from a shipborne ADCP on WOCE cruise A11 in the South Atlantic. J. Phys. Oceanogr. 25, 329–347.

Schiller, A., Fiedler, R., 2007. Explicit tidal forcing in an ocean general circulation model. Geophys. Res. Lett. 34. http://dx.doi.org/10.1029/2006GL028363.

Schmitt, R.W., 1994. Double diffusion in oceanography. Annu. Rev. Fluid Mech. 26, 255–285.

Shchepetkin, A.F., McWilliams, J.C., 1998. Quasi-monotone advection schemes based on explicit locally adaptive dissipation. Mon. Weather Rev. 126, 1541–1580.

Shchepetkin, A., McWilliams, J., 2002. A method for computing horizontal pressure-gradient force in an ocean model with a non-aligned vertical coordinate. J. Geophys. Res. 108, 35.1–35.34.

Shchepetkin, A., McWilliams, J., 2005. The regional oceanic modeling system (ROMS): a split-explicit, free-surface, topography-following-coordinate oceanic model. Ocean Model. 9, 347–404.

Simmons, H.L., Jayne, S.R., St. Laurent, L.C., Weaver, A.J., 2004. Tidally driven mixing in a numerical model of the ocean general circulation. Ocean Model. 6, 245–263.

Smagorinsky, J., 1993. Some historical remarks on the use of nonlinear viscosities. In: Galperin, B., Orszag, S.A. (Eds.), Large Eddy Simulation of Complex Engineering and Geophysical Flows. Cambridge University Press, Cambridge, United Kingdom, pp. 3–36.

Smith, K.S., 2007. The geography of linear baroclinic instability in earth's oceans. J. Mar. Res. 65, 655–683.

Smith, R.D., McWilliams, J.C., 2003. Anisotropic horizontal viscosity for ocean models. Ocean Model. 5, 129–156.

Smith, K.S., Vallis, G.K., 2001. The scales and equilibration of midocean eddies: freely evolving flow. J. Phys. Oceanogr. 31, 554–570.

Smith, R., Maltrud, M., Bryan, F., Hecht, M., 2000. Numerical simulation of the North Atlantic at 1/10°. J. Phys. Oceanogr. 30, 1532–1561.

Solomon, H., 1971. On the representation of isentropic mixing in ocean models. J. Phys. Oceanogr. 1, 233–234.

Speer, K., Tziperman, E., 1992. Rates of water mass formation in the North Atlantic Ocean. J. Phys. Oceanogr. 22, 2444–2457.

Speich, S., Blanke, B., de Vries, P., Drijfhout, S., Döös, K., Ganachaud, A., Marsh, R., 2002. Tasman leakage: a new route in the global ocean conveyor belt. Geophys. Res. Lett. 29.

Stacey, M.W., Pond, S., Nowak, Z.P., 1995. A numerical model of the circulation in Knight Inlet, British Columbia, Canada. J. Phys. Oceanogr. 25, 1037–1062.

Stammer, D., 1998. On eddy characteristics, eddy transports, and mean flow properties. J. Phys. Oceanogr. 28, 727–739.

Starr, V.P., 1945. A quasi-Lagrangian system of hydrodynamical equations. J. Meteorol. 2, 227–237.

Stewart, A., Dellar, P., 2011. The role of the complete Coriolis force in cross-equatorial flow of abyssal ocean currents. Ocean Model. 38, 187–202.

Sullivan, P.P., McWilliams, J.C., 2010. Dynamics of winds and currents coupled to surface waves. Annu. Rev. Fluid Mech. 42, 19–42.

Sullivan, P.P., McWilliams, J.C., Melville, W.K., 2007. Surface gravity wave effects in the oceanic boundary layer: large-eddy simulation with vortex force and stochastic breakers. J. Fluid Mech. 593, 405–452.

Tatebe, H., Hasumi, H., 2010. Formation mechanism of the Pacific equatorial thermocline revealed by a general circulation model with a high accuracy tracer advection scheme. Ocean Model. 35, 245–252.

Taylor, P., 2000. Final report of the joint WCRP/SCOR working group on air-sea fluxes: intercomparison and validation of ocean-atmosphere energy flux fields. World Climate Research Programme, p. 303, WCRP-112, WMO/TD-No.1036.

Timmermann, R., Danilov, S., Schröter, J., Böning, C., Sidorenko, D., Rollenhagen, K., 2009. Ocean circulation and sea ice distribution in a finite element global sea ice-ocean model. Ocean Model. 27, 114–129.

Toole, J.M., Schmitt, R.W., Polzin, K.L., 1997. Near-boundary mixing above the flanks of a mid-latitude seamount. J. Geophys. Res. 102, 947–959.

Treguier, A.M., 1992. Kinetic energy analysis of an eddy resolving, primitive equation model of the North Atlantic. J. Geophys. Res. 97, 687–701.

Treguier, A., Barnier, B., Miranda, A., Molines, J., Grima, N., Imbard, M., Madec, G., Messager, C., Reynaud, T., Michel, S., 2001. An eddy-permitting model of the Atlantic circulation: evaluating open boundary conditions. J. Geophys. Res. Oceans 106, 22115–22129.

Treguier, A.M., Le Sommer, J., Molines, J.M., de Cuevas, B., 2010. Response of the Southern Ocean to the Southern Annular Mode: interannual variability and multidecadal trend. J. Phys. Oceanogr. 40, 1659–1668.

Treguier, A., Ferron, B., Dussin, R., 2012. Buoyancy-driven currents in eddying ocean models. In: Chassignet, E., Cenedese, C., Verron, J. (Eds.), Buoyancy-Driven Flows. Cambridge University Press, Cambridge, UK, pp. 281–311.

Tziperman, E., 1986. On the role of interior mixing and air-sea fluxes in determining the stratification and circulation in the oceans. J. Phys. Oceanogr. 16, 680–693.

Vallis, G.K., 2006. Atmospheric and Oceanic Fluid Dynamics: Fundamentals and Large-scale Circulation. first ed. Cambridge University Press, Cambridge, p. 745+xxv pp.

van Sebille, E., Jan van Leeuwen, P., Biastoch, A., Barron, C.N., de Ruijter, W.P.M., 2009. Lagrangian validation of numerical drifter trajectories using drifting buoys: application to the Agulhas system. Ocean Model. 29, 269–276.

Veneziani, M., Griffa, A., Garraffo, Z., Chassignet, E., 2005. Lagrangian spin parameter and coherent structures from trajectories released in a high-resolution ocean model. J. Mar. Res. 63, 753–788.

Veronis, G., 1975. The role of models in tracer studies. In: Numerical Models of Ocean Circulation. National Academy of Sciences, Washington, DC, pp. 133–146.

Vialard, J., Delecluse, P., 1998. An OGCM study for the TOGA decade. Part I: role of salinity in the physics of the Western Pacific Fresh Pool. J. Phys. Oceanogr. 28, 1071–1088.

Volkov, D., Fu, L.-L., 2008. The role of vorticity fluxes in the dynamics of the Zapiola Anticyclone. J. Geophys. Res. 113, C11015.

Vries, P., Döös, K., 2001. Calculating Lagrangian trajectories using time-dependent velocity fields. J. Atmos. Oceanic Technol. 18, 1092–1101.

Walin, G., 1982. On the relation between sea-surface heat flow and thermal circulation in the ocean. Tellus 34, 187–195.

Webb, D., 2000. Evidence for shallow zonal jets in the South Equatorial Current region of the Southwest Pacific. J. Phys. Oceanogr. 20, 706–720.

Willebrand, J., Barnard, S., Barnier, B., Beckmann, A., Böning, C., Coulibaly, M., deCuevas, B., Dengg, J., Dieterich, C., Ernst, U., Herrmann, P., Jia, Y., Killworth, P., Kröger, J., Lee, M.-M., Provost, C., Molines, J., New, A., Oschlies, A., Reynaud, T., West, L., 1997. DYNAMO: dynamics of North Atlantic models: simulation and assimilation with high resolution models. DYNAMO Scientific Report No 3. Available from hdl:10013/epic.10757.

Williams, P., 2005. Modelling climate change: the role of unresolved processes. Philos. Trans. R. Soc. A 363, 2931–2946.

Williams, R., Spall, M., Marshall, J., 1995. Does Stommel's mixed layer 'demon' work? J. Phys. Oceanogr. 25, 3089–3102.

Winton, M., Hallberg, R., Gnanadesikan, A., 1998. Simulation of density-driven frictional downslope flow in *z*-coordinate ocean models. J. Phys. Oceanogr. 28, 2163–2174.

Wirth, A., Barnier, B., 2006. Tilted convective plumes in numerical experiments. Ocean Model. 12, 101–111.

Wirth, A., Barnier, B., 2008. Mean circulation and structures of tilted ocean deep convection. J. Phys. Oceanogr. 38, 803–816.

Wu, W., Danabasoglu, G., Large, W., 2007. On the effects of parameterized Mediterranean overflow on North Atlantic ocean circulation and climate. Ocean Model. 19, 31–52.

Wunsch, C., 1997. The vertical partition of oceanic horizontal kinetic energy and the spectrum of global variability. J. Phys. Oceanogr. 27, 1770–1794.

Wunsch, C., Ferrari, R., 2004. Vertical mixing, energy, and the general circulation of the ocean. Annu. Rev. Fluid Mech. 36, 281–314.

Wunsch, C., Stammer, D., 1995. The global frequency-wavenumber spectrum of oceanic variability estimated from TOPEX/POSEIDON altimetric measurements. J. Geophys. Res. 100, 24895–24910.

Xu, Y., Scott, R.B., 2008. Subtleties in forcing eddy resolving ocean models with satellite wind data. Ocean Model. 20, 240–251.

Young, W.R., 2012. An exact thickness-weighted average formulation of the Boussinesq equations. J. Phys. Oceanogr. 42, 692–707.

Chapter 21

Dynamically and Kinematically Consistent Global Ocean Circulation and Ice State Estimates

Carl Wunsch and Patrick Heimbach

Department of Earth, Atmospheric and Planetary Sciences, Massachusetts Institute of Technology, Cambridge, Massachusetts, USA

Chapter Outline

1. INTRODUCTION

The goal of what we call "state estimates" of the oceans arose directly out of the plans for the World Ocean Circulation Experiment (WOCE). That program, out of necessity, employed in a pragmatic way observational tools of a very wide diversity of type—including classical hydrography, current meters, tracers, satellite altimeters, floats, and drifters. The designers of WOCE realized that to obtain a coherent picture of the global ocean circulation approaching a timescale of a decade, they would require some form of synthesis method: one capable of combining very disparate observational types, but also having greatly differing space–time sampling, and geographical coverage.

Numerical weather forecasting, in the form of what had become labeled "data assimilation" (DA), was a known analogue of what was required: a collection of tools for combining the best available global numerical model representation of the ocean with any and all data, suitably weighted to account for both model and data errors (e.g., Talagrand, 1997; Kalnay, 2003; Evensen, 2009). Several major, and sometimes ignored, obstacles existed in employing meteorological methods for the oceanic problem. These included the large infrastructure used to carry out DA within the national weather forecast centers—organizations for which no oceanographic equivalent existed or exists. DA had developed for the purposes of *forecasting* over timescales of hours to a few days, whereas the climate goals of WOCE were directed at timescales of years to decades, with a goal of understanding and *not* forecasting. Another, more subtle, difficulty was the WOCE need for state estimates capable of being used for global-scale energy, heat, and water cycle budgets. Closed global budgets are of little concern to a weather forecaster, as their violation has no impact on short-range prediction skill, but they are crucial to the understanding of climate change. Construction of closed budgets is also rendered physically impossible by the forecasting goal: solutions "jump" toward the data at every analysis time, usually

Ocean Circulation and Climate, Vol. 103. http://dx.doi.org/10.1016/B978-0-12-391851-2.00021-0

every 6 h, introducing spurious sources and sinks of basic properties.

Because of these concerns, the widespread misunderstanding of what DA usually does, and what oceanographers actually require, the first part of this essay is devoted to a sketch of the basic principles of DA and the contrast with methods required in practice for use for climate-relevant state estimates. More elaborate accounts can be found in Wunsch (2006) and Wunsch and Heimbach (2007) among others. Within the atmospheric sciences literature itself, numerous publications exist (e.g., Trenberth et al., 1995, 2001; Bengtsson et al., 2004; Bromwich and Fogt, 2004; Bromwich et al., 2004, 2007, 2011; Thorne, 2008; Nicolas and Bromwich, 2011), warning against the use of DA and the associated "reanalyses" for the study of climate change. These warnings have been widely disregarded.

A theme of this chapter is that both DA and state estimation can be understood from elementary principles, ones not going beyond beginning calculus. Those concepts must be distinguished from the far more difficult numerical engineering problem of finding practical methods capable of coping with large volumes of data, large model state dimensions, and a variety of computer architectures. But one can understand and use an automobile without being an expert in the manufacture of an internal combustion engine or of the chemistry of tire production.

At the time of the writing of the first WOCE volume, (Siedler et al., 2001), two types of large-scale synthesis existed: (1) the time-mean global inverse results of Macdonald (1998) based upon the pre-WOCE hydrography and that of Ganachaud (2003b) using the WOCE hydrographic sections. (2) Preliminary results from the first ECCO (Estimating the Circulation and Climate of the Ocean) synthesis (Stammer et al., 2002) were based upon a few years data and comparatively coarse resolution models. Talley et al. (2001) summarized these estimates, but little time had been available for their digestion.

In the intervening years, Lumpkin and Speer (2007) produced a revision of the Ganachaud results using somewhat different assumptions, but with similar results, and a handful of other static global estimates (e.g., Schlitzer, 2007) appeared. The ECCO project greatly extended its capabilities and duration for time-dependent estimates. A number of regional, assumed steady-state, box inversions also exist (e.g., Macdonald et al., 2009).

As part of his box inversions, Ganachaud (2003a) had shown that the dominant errors in trans-oceanic property transports of volume (mass), heat (enthalpy), salt, etc., arose from the temporal variability. Direct confirmation of that inference can be seen in the ECCO-based time-varying solutions and from *in situ* measurements (Rayner et al., 2011). So-called synoptic sections spanning ocean basins, which had been the basis for most global circulation pictures, at best produce "blurred" snapshots of transport properties. We are now well past the time in which they can be labeled and interpreted as being the time-average. A major result of WOCE was to confirm the conviction that *the ocean must be observed and treated as a fundamentally time-varying system*, especially for any property involving the flow field. Gross scalar properties such as the temperature or nitrogen concentrations have long been known to be stable on the largest scales: that their distributions are nonetheless often dominated by intense temporal fluctuations, sometimes involving very high wavenumbers, represents a major change in the understanding of classical ocean properties. That understanding inevitably drives one toward state estimation methods.

2. DEFINITION

Consider any model of a physical system satisfying known equations, written generically in discrete time as,

$$\mathbf{x}(t) = L(\mathbf{x}(t-\Delta t), \mathbf{q}(t-\Delta t), \mathbf{u}(t-\Delta t)), \quad 1 \le t \le t_f = M\Delta t, \tag{21.1}$$

where $\mathbf{x}(t)$ is the "state" at time t, discrete at intervals Δt, and includes those prognostic or dependent variables usually computed by a model, such as temperature or salinity in an advection–diffusion equation or a stream function in a flow problem. $\mathbf{q}(t)$ denotes known forcings, sources, sinks, boundary and initial conditions, and internal model parameters, and $\mathbf{u}(t)$ are any such elements that are regarded as only partly or wholly unknown, hence subject to adjustment and termed independent or control variables (or simply "controls"). Model errors of many types are also represented by $\mathbf{u}(t)$. L is an operator and can involve a large range of calculations, including derivatives, or integrals, or any other mathematically defined function. In practise, it is usually a computer code working on arrays of numbers. (Notation is approximately that of Wunsch, 2006.) Time, $t = m\Delta t$, is assumed to be discrete, with $m = 0, \ldots, M$, as that is almost always true of models run on computers.[1] Note that the steady-state situation is a special case, in which one writes an additional relationship, $\mathbf{x}(t) = \mathbf{x}(t-\Delta t)$ and $\mathbf{q}$, $\mathbf{u}$ are then time-independent. For computational efficiency, steady models are normally rewritten so that time does not appear at all, but that step is not necessary. Thus the static box inverse methods and their relatives such as the beta-spiral are special cases of the ocean estimation problem (Wunsch, 2006).

1. An interesting mathematical literature surrounds state estimation carried out in continuous time and space in formally infinite dimensional spaces. Most of it proves irrelevant for calculations on computers, which are always finite dimensional. Digression into functional analysis can be needlessly distracting.

Useful observations at time t are all functions of the state and, in almost all practical situations, are a linear combination of one or more state vector elements,

$$\mathbf{y}(t) = \mathbf{E}(t)\mathbf{x}(t) + \mathbf{n}(t), \quad 0 \le t \le t_f, \qquad (21.2)$$

where $\mathbf{n}(t)$ is the inevitable noise in the observations. $\mathbf{y}(t)$ is a vector of whatever observations of whatever diverse type are available at t. (Uncertain initial conditions are included here at $t=0$, representing them as noisy observations.) Standard matrix–vector notation is being used. In a steady-state formulation, parameter t would be suppressed. On rare occasions, data are a nonlinear combination of the state vector: an example would be a speed measurement in terms of two components of the velocity, or a frequency spectrum for some variable. Methods exist (not discussed here) for dealing with such observations. Observations relating to the control vector may exist and one easy approach to using them is to redefine elements of $\mathbf{u}(t)$ as being part of the state vector. The "state estimation problem"[2] is defined as determining $\tilde{\mathbf{x}}(t)$, $0 \le t \le t_f$, $\tilde{\mathbf{u}}(t)$, $0 \le t \le t_f - \Delta t$, exactly satisfying both Equations (21.1) and (21.2). Tildes here denote estimates to distinguish them from the true values.

Important note: "exact" satisfaction of Equation (21.1) must be understood as meaning the model *after adjustment* by $\tilde{\mathbf{u}}(t)$. Because $\mathbf{u}(t)$ can represent, if necessary, very complex, nonlinear, and large changes to the original model, which is usually defined with $\mathbf{u}(t)=\mathbf{0}$, the adjusted model can be very different from the initial version. *But the adjusted model is known, fully specified, and exactly satisfied*, and is what is used for discussion of the physics or chemistry. It thus differs in a fundamental way from other types of estimates rendered discontinuous by "data injection," or forcing to data, during the final forward calculation.

Typically, one must also have some knowledge of the statistics of the controls, $\mathbf{u}(t)$, and observation noise, $\mathbf{n}(t)$, commonly as the first and second-order moments,

$$\left.\begin{array}{l} <\mathbf{u}(t)>=0, \\ \left\langle \mathbf{u}(t)\mathbf{u}(t')^T \right\rangle = \mathbf{Q}(t)\delta_{tt'}, \end{array}\right\} \quad 0 \le t \le t_f - \Delta t = (M-1)\Delta t, \qquad (21.3a)$$

$$\left.\begin{array}{l} <\mathbf{n}(t)>=0, \\ \left\langle \mathbf{n}(t)\mathbf{n}(t')^T \right\rangle = \mathbf{R}(t)\delta_{tt'}, \end{array}\right\} \quad 0 \le t \le t_f = M\Delta t \qquad (21.3b)$$

The brackets denote expected values and superscript T is the vector or matrix transpose.

In generic terms, the problem is one of *constrained estimation/optimization*, in which, usually, one seeks to minimize both the normalized quadratic model-data differences,

$$\left\langle (\mathbf{y}(t) - \mathbf{E}(t)\mathbf{x}(t))^{\mathrm{T}} \mathbf{R}^{-1}(t)(\mathbf{y}(t) - \mathbf{E}(t)\mathbf{x}(t)) \right\rangle \qquad (21.4)$$

and the normalized independent variables (controls),

$$\left\langle \mathbf{u}(t)^{\mathrm{T}} \mathbf{Q}^{-1}(t)\mathbf{u}(t) \right\rangle \qquad (21.5)$$

—subject to the exact satisfaction of the *adjusted model* in Equation (21.1).

For data sets and controls that are Gaussian or nearly so, the problem as stated is equivalent to weighted least-squares minimization of the scalar,

$$J = \sum_{m=0}^{M} (\mathbf{y}(t) - \mathbf{E}(t)\tilde{\mathbf{x}}(t))^{\mathrm{T}} \mathbf{R}^{-1}(t)(\mathbf{y}(t) - \mathbf{E}(t)\tilde{\mathbf{x}}(t)) + \sum_{m=0}^{M-1} \tilde{\mathbf{u}}(t)^{\mathrm{T}} \mathbf{Q}^{-1}(t)\tilde{\mathbf{u}}(t), \quad t = m\Delta t, \qquad (21.6)$$

subject to Equation (21.1). It is a *least-squares problem constrained by partial differential equations*, and nonlinear if the model or its connection with observations are nonlinear. The uncertain initial conditions, contained implicitly in Equation (21.6), are readily written out separately if desired.

In comparing the solutions to DA, note that *the latter problem is different*. It seeks to minimize,

$$\mathrm{diag}\left\langle (\tilde{\mathbf{x}}(t_{\mathrm{now}} + \tau) - \mathbf{x}(t_{\mathrm{now}} + \tau))(\tilde{\mathbf{x}}(t_{\mathrm{now}} + \tau) - \mathbf{x}(t_{\mathrm{now}} + \tau))^{\mathrm{T}} \right\rangle, \qquad (21.7)$$

that is the variance of the state about the true value at some time *future* to t_{now}. Brackets again denote the expected value. The role of the model is to make the forecast, by setting $\mathbf{u}(t)=0$, $t_{\mathrm{now}} + \Delta t \le t \le t_{\mathrm{now}} + \tau$, because it is unknown, and starting with the most recent estimate $\tilde{\mathbf{x}}(t_{\mathrm{now}})$ at t_{now}. Equation (21.7) is itself equivalent to a requirement of minimum square deviation at $t_{\mathrm{now}} + \tau$. A bit more will be said about this relationship later.

Model error deserves an extended discussion by itself. A consequence of exact satisfaction of the model equations is that we assume the discretized version of Equation (21.1) to be error-free, but only after determination of $\mathbf{u}(t)$. Model errors come in roughly three flavors: (a) the equations are incomplete or an approximated form of the real system; (b) errors are incurred in their discretization (e.g., numerical diffusion); and (c) sub-grid scale parameterizations are incomplete, and/or their parameter choices sub-optimal. Methods exist to quantify these errors in an estimation framework. As in any multi-parameter optimization problem, data sets are commonly inadequate to distinguish completely between errors in the model structure, including resolution, and in other components such as the initial and boundary conditions. Errors in one element can show up (be compensated) by incorrect adjustments made to other elements. The current approach in ECCO is to introduce explicit adjustments to the most important interior parameters such as mixing coefficients (e.g., Ferreira et al.,

2. A terminology borrowed from control theory (e.g., Gelb, 1974).

2005; Stammer, 2005) as likely representing the dominant model inadequacies. Other errors, including those arising from inadequate resolution in regions of higher order dynamics, nonetheless inevitably distort some elements of the best-estimate solution.

Most of the fundamental principles of practical state estimation and of DA can be understood from the common school problem of the least-squares fitting of lines and curves to data in one dimension. The central point is that the *concepts* of state estimation and DA are very simple; but it is equally simple to surround them with an aura of mystery and complexity that is needless for anyone who wishes primarily to understand the meaning of the results.

3. DATA ASSIMILATION AND THE REANALYSES

Despite the technical complexities of the numerical engineering practice, DA and what are called "reanalyses" should be understood as approximate methods for obtaining a solution of a least-squares problem. Using the same notation as in Equation (21.7), consider again an analysis time, $t_{\text{now}} = t_{\text{ana}} + \tau$, when data have become available, and where t_{ana} is the previous analysis time, $\tau > 0$, typically 6 h earlier. The weather forecaster's model has been run forward to make a prediction, $\tilde{\mathbf{x}}(t_{\text{now}}, -)$, with the minus sign denoting that newer observations have *not yet* been used. The new observations are $\mathbf{E}(t_{\text{now}})\ \mathbf{x}(t_{\text{now}}) + \mathbf{n}(t_{\text{now}}) = \mathbf{y}(t_{\text{now}})$. With some understanding of the quality of the forecast, expressed in the form of an uncertainty matrix (2nd moments about the truth) called $\mathbf{P}(t_{\text{now}}, -)$, and of the covariance matrix of the observational noise, $\mathbf{R}(t_{\text{now}})$, the best combination in the L_2-norm of the information of the model and the data is the minimum of,

$$\begin{aligned} J_1 = & (\tilde{\mathbf{x}}(t_{\text{now}}) - \tilde{\mathbf{x}}(t_{\text{now}}, -))^T \mathbf{P}(t_{\text{now}}, -)^{-1} \\ & (\tilde{\mathbf{x}}(t_{\text{now}}) - \tilde{\mathbf{x}}(t_{\text{now}}, -)) \\ & + (\mathbf{y}(t_{\text{now}}) - \mathbf{E}(t_{\text{now}})\mathbf{x}(t_{\text{now}}))^T \mathbf{R}(t_{\text{now}})^{-1} \\ & (\mathbf{y}(t_{\text{now}}) - \mathbf{E}(t_{\text{now}})\mathbf{x}(t_{\text{now}})), \end{aligned} \tag{21.8}$$

and whose least-squares minimum for a linear model is given rigorously by the Kalman (1960) filter. In DA practise, only some very rough approximation of that minimum is sought and obtained. True Kalman filters are never used for prediction in real geophysical fluid flow problems as they are computationally overwhelming (for more detail, see e.g., Wunsch, 2006). *Notice that J_1 assumes that a summation of errors is appropriate, even in the presence of strong nonlinearities.*

A brief excursion into meteorological "reanalyses" is worthwhile here for several reasons: (1) they are often used as an atmospheric "truth" to drive ocean, ice, chemical, and biological models. (2) A number of ocean circulation estimates have followed their numerical engineering methodology. (3) With the long history of the atmospheric DA effort, many have been unwilling to believe that any alternatives exist.

Note that the "analysis" consists of an operational weather model run in conventional prediction mode, analogous to the simple form described in the previous section, adjusted, and thus displaying discontinuities at the analysis times, by attempts to approximately minimize J_1. Because of the operational/real-time requirements, only a fraction of the global operational meteorological observations are relayed and quality-controlled in time to be available at the time of analysis. Furthermore, because models have changed so much over the years, the stored analyses are inhomogeneous in the underlying physics[3] and model codes. Oceanographers have no such products at this time; global "*analyses*" *in the meteorological sense do not exist*, and thus the term "reanalysis" for ocean state estimates is inappropriate.

Meteorological reanalysis is the recomputation, using the same prediction methodology as previously used in the analysis, but with the differences that (1) the model code and combination methodology are held fixed over the complete time duration of the calculation (e.g., over 50 years) thus eliminating artificial changes in the state from model or method improvements and, (2) including many data that arrived too late to be incorporated into the real-time analysis (see Kalnay, 2003; Evensen, 2009).

Estimated states still have the same discontinuities at the analysis times when the model is forced toward the data. Of even greater significance for oceanographic and climatic studies are the temporal shifts induced in the estimates by the major changes that have taken place in the observational system over several decades—most notably, but not solely, the appearance of meteorological satellites. Finally, no use is made of the information content in the observations of the *future* evolution of the state.

Although as already noted above, clear warnings have appeared in the literature—that spurious trends and values are artifacts of changing observation systems (see, e.g., Elliott and Gaffen, 1991; Marshall et al., 2002; Thompson et al., 2008)—the reanalyses are rarely used appropriately, meaning with the recognition that they are subject to large errors. In Figure 21.1, for example, the jump in precipitation minus evaporation ($P - E$) with the advent of the polar orbiting satellites implies either that the unspecified error estimates prior to that time must, at a minimum, encompass the jump, and/or that computation has been erroneous, or that a remarkable coincidence has occurred. But even the smaller transitions in $P - E$, for example, over the more recent period of 1992 onward, are likely too large to be physical; see Table 21.1.

3. We employ "physics" in its conventional meaning as encompassing all of dynamics and thermodynamics.

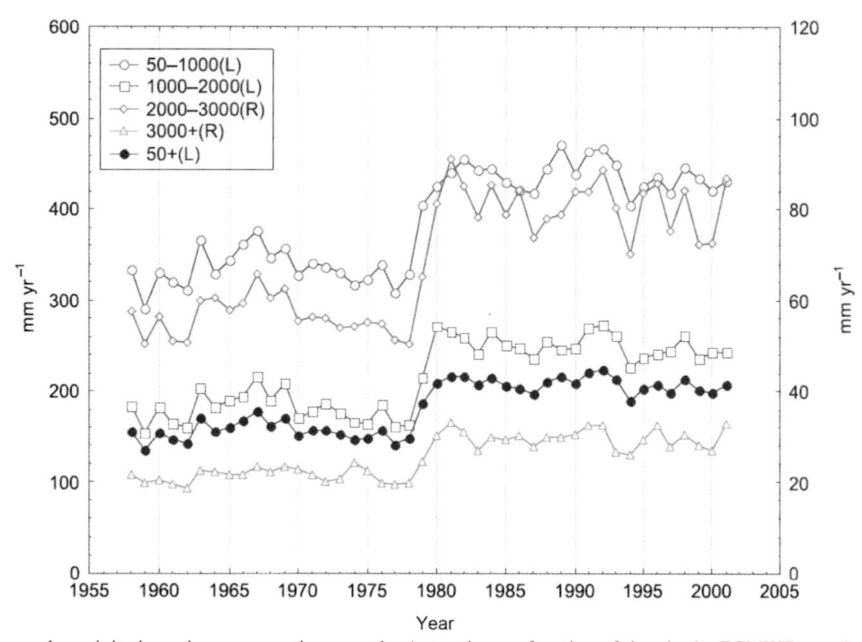

FIGURE 21.1 Mean annual precipitation minus evaporation over the Antarctic as a function of time in the ECMWF reanalysis ERA-40 showing the impact of new observations, in this case, the arrival of the polar orbiting satellites. Different curves are for different elevations. The only simple inference is that the uncertainties must exceed the size of the rapid transition seen in the late 1970s. L and R identify whether the left or right axis is to be used for that curve. *From Bromwich et al. (2007).*

TABLE 21.1 Negative Heat Fluxes refer to Oceanic Heating, Positive Freshwater Imbalances to Evaporation

	Net Freshwater Imbalance (mm/year)		Net Heat Flux Imbalance (W/m^2)	
Reanalysis Product	Ocean-Only	Global	Ocean-Only	Global
NCEP/NCAR-I (1992–2010)	159	62	−0.7	−2.2
NCEP/DOE-II (1992–2004)	740	–	−10	–
ERA-Interim (1992–2010)	199	53	−8.5	−6.4
JRA-25 (1992–2009)	202	70	15.3	10.1
CORE-II (1992–2007)	143	58		

Figure 21.2 and other, similar ones, are further disquieting, showing that reanalyses using essentially the same data, and models that have been intercompared over decades, have significant qualitative disagreements on climate timescales. Differences in the reanalyses in the northern hemisphere are not so large and are generally agreed to be the result of a much greater data density. They remain, nevertheless, significant, as evidenced in the discussion of analysis increments over the Arctic by Cullather and Bosilovich (2012). *Evidently, considerations of data density and types and their handling dominate the reanalyses, with the models being of secondary importance.*

For climate studies, another major concern is the failure of the reanalyses to satisfy basic global conservation requirements. So, for example, Table 21.1 shows the global imbalances on a per year basis of several reanalysis products in apparent heating of the oceans and in the net freshwater flux from the atmosphere. Such imbalances can arise either because global constraints are not implied by the model equations, and/or because biased data have not been properly handled, or most likely, some combination of these effects is present. Trenberth and Solomon (1994), for example, note that the NCEP/NCAR reanalysis implies a meridional heat transport within continental land masses. "User beware" is the best advice we can give.

State estimation as defined in the ECCO context is a much more robust and tractable problem than is, for

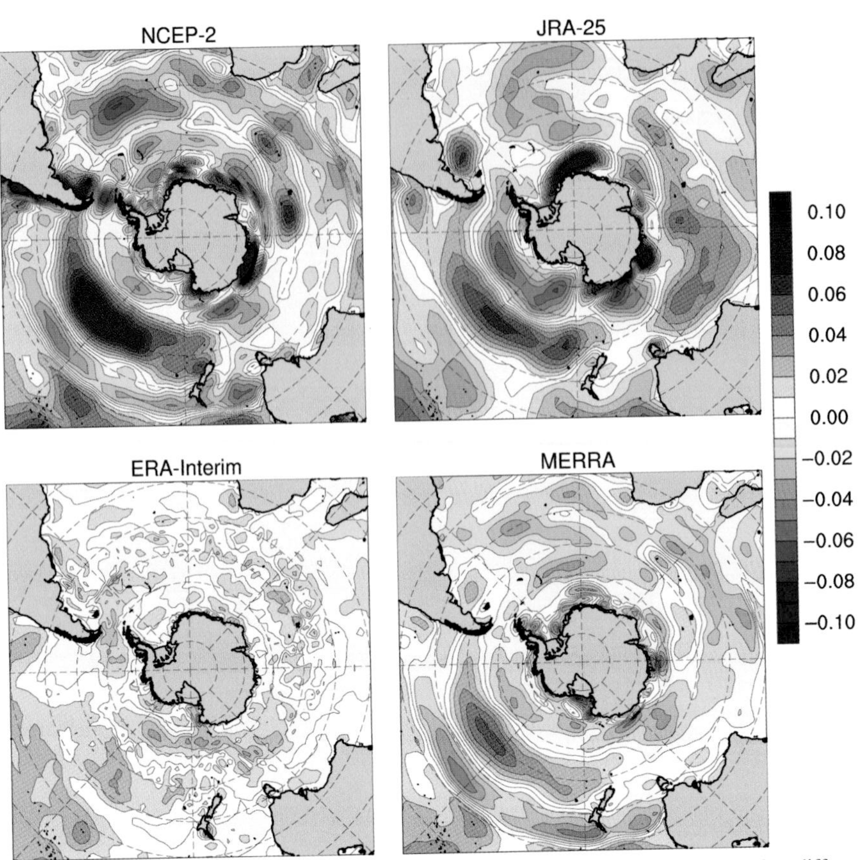

FIGURE 21.2 Calculated trends (meters/second/year) in the 10-m zonal wind fields at high southern latitudes from four different atmospheric reanalyses (D. Bromwich and J. P. Nicolas, of Ohio State University, private communication, 2010). Note particularly the different patterns in the Indian Ocean and the generally discrepant amplitudes. Because of the commonality of data sets, forecast models, and methodologies, the differences here must be lower bounds on the true uncertainties of trends. See Bromwich et al. (2011) for a description of the four different estimates. Acronyms denote National Center for Environmental Prediction; Japanese Reanalysis; European Centre for Medium Range Forecasts Reanalysis; Modern Era-Retrospective Analysis for Research and Applications (NASA).

example, prediction of future climate states. As is well known even to beginning scientists, extrapolation of very simple models can be extremely unstable, with interpolation[4], or curve-fitting, remaining robust. (A classical example is the use of a cubic polynomial to fit some noisy data, and which can be very effective. But one is advised *never* to use such a fit to extrapolate the curve; see Figure 21.3). The ECCO process is effectively a temporal curve-fit of the WOCE-era data sets by a model and which, with some care to avoid data blunders, produces a robust result. It is the interpolating (smoothing) character, coupled with the expectation of thermal wind balance over most of the domain, that produces confidence in the basic system products. As is well-known, least-squares methods tend to generate meaningless structures in unconstrained parts of the domain. Some regions of spatial extrapolation do exist here, depending upon the time-varying distribution of observations, and although they tend to be limited in both space and time, detailed values there should be regarded skeptically.

Terminological note: the observational community has lost control of the word "data," which has come to be used, confusingly, for the output of models, rather than having any direct relationship to instrumental values. In the context of reanalyses and state estimates involving both measurements and computer codes, the word generally no longer conveys any information. For the purposes of this essay, "data" always represents instrumental values of some sort, and anything coming out of a general circulation model (GCM) is a "model-value" or "model-datum," or has a similar label. We recognize that models are involved in all real observations, even in such familiar values as those coming from, for example, a mercury thermometer, in which a measured length is converted to a temperature.

4. The commonplace term "interpolation" is used in numerical analysis to imply that fitted curves pass exactly through data points—an inappropriate requirement here.

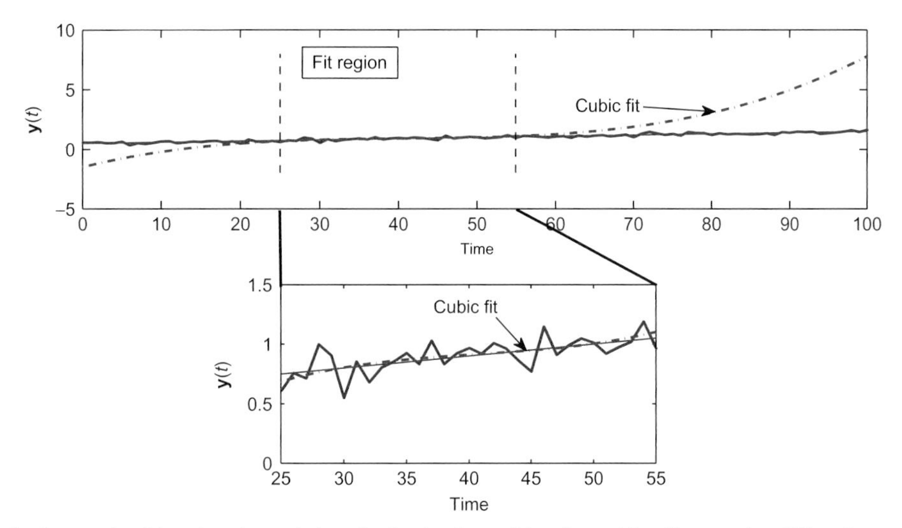

FIGURE 21.3 A textbook example of the robust interpolation of noisy data by a cubic polynomial and its gross instability when used to extrapolate. This analogue is a very simplified representation of the problem of extrapolating a GCM state into unobserved time-spans.

Even so, most readers can recognize the qualitative difference between conventional observations and the output of a 100,000+line computer code.

4. OCEAN STATE ESTIMATES

The rest of this chapter is primarily devoted to a summary description and discussion of some results of the ECCO groups which, beginning with Stammer et al. (2002), were directed at decadal and longer state estimates satisfying known equations and using as much of the WOCE-era-and-beyond data as possible. No claim is made that these estimates are definitive, nor that the discussion is comprehensive. A number of other, superficially similar, estimates exist (Carton et al., 2000; Martin et al., 2007; Hurlburt et al., 2009), but these generally have had different goals, for example, a fast approximate estimate primarily of the upper ocean, or prediction of the mesoscales over ocean basins. Some weather forecasting centers have undertaken "operational oceanography" products closely resembling atmospheric weather forecasts. To our knowledge, however, the ECCO estimates are today the only ones specifically directed at physically continuous, dynamically consistent, top-to-bottom estimates from a comprehensive data set.

A number of review papers exist that attempt to compare various such solutions (e.g., Carton and Santorelli, 2008; Lee et al., 2010) as though they were equivalent. But as the above discussion tries to make clear, estimates are not equally reliable for all purposes and comparisons make no sense unless their individual purposes are well understood. Although one could compare a crop-dusting airplane to a jet fighter, and both have their uses, few would regard that effort as helpful, except as a vehicle for discussion of the highly diverse applications of aero-physics. Thus a numerical scheme directed primarily at mesoscale prediction, and using a model not conserving energy, may well be a useful tool for forecasting the trajectory of the Gulf Stream over a few weeks, but it would be unsuited to a discussion of global ocean heat transports—a useful model of which is, in turn, unsuitable for mesoscale interests. These other applications are discussed in this volume by Schiller et al. (2013).

Originally, ECCO was meant primarily to be a demonstration of the practicality of its approach to finding the oceanic state. When the first ECCO estimates did become available (Stammer et al., 2002) they proved sufficiently useful even with that short duration and coarse resolution, that a decision was made to continue with a gradually improving data set and computer power. This review summarizes mainly what has been published thus far, but as optimization is an asymptotic process, the reader should be aware that newer, and likely better, solutions are being prepared continuously and the specific results here will have been refined in the intervals between writing, publishing, and reading.

4.1. Basic Notions

As described above, most state estimation problems in practice are generically those of constrained least-squares, in which one seeks to minimize objective or cost or misfit functions similar to Equation (21.6) subject to the solution (including both the estimated state $\mathbf{x}(t)$, and the controls, $\mathbf{u}(t)$) of the model-time stepping equations.[5] One approach,

5. Advantages exist to using norms other than L_2 including those such as one and infinity norms commonly regarded as robust. These norms are not normally used in ocean and atmosphere state estimation or data assimilation systems because software development for parallel computers permitting computation at super-large dimensions has not yet occurred.

among many, to solving such problems is the method of Lagrange multipliers dating back 200 years. This method is discussed at length in Wunsch (2006) and the references there. In a very brief summary, one "adjoins" the model equations using vectors of Lagrange multipliers, $\boldsymbol{\mu}(t)$, to produce a new objective function,

$$J' = \sum_{m=0}^{M}(\mathbf{y}(t)-\mathbf{E}(t)\mathbf{x}(t))^{\mathrm{T}}\mathbf{R}(t)^{-1}(\mathbf{y}(t)-\mathbf{E}(t)\mathbf{x}(t)) + \sum_{m=0}^{M-1}\mathbf{u}(t)^{\mathrm{T}}\mathbf{Q}(t)^{-1}\mathbf{u}(t) - 2\sum_{m=1}^{M-1}\mu(t)^{\mathrm{T}}[\mathbf{x}(t)-\mathbf{L}\mathbf{x}(t-\Delta t),\mathbf{B}\mathbf{q}(t-\Delta t),\Gamma\mathbf{u}(t-\Delta t)],$$

$$t = m\Delta t,\ m = 0,\ldots,M \tag{21.9}$$

Textbooks explain that the problem can now be treated as a conventional, unconstrained least-squares problem in which the $\boldsymbol{\mu}(t)$ are part of the solution. In principle, one simply does vector differentiation with respect to all of $\mathbf{x}(t)$, $\mathbf{u}(t)$, $\boldsymbol{\mu}(t)$, sets the results to zero, and solves the resulting "normal equations" (they are written out in Wunsch, 2006). J and J' are very general, and one easily adds, for example, internal model parameters such as mixing coefficients, water depths, etc., as further parameters to be calculated, thus rendering the problem one of combined state and parameter estimation.

The entire problem of state estimation thus reduces to finding the stationary values of J'. The large literature on what is commonly called the "adjoint method" ("4DVAR" in weather forecasting, where it is used only incrementally over short time-spans) reduces to coping with a very large set of simultaneous equations (and some are nonlinear). But as an even larger literature deals with solving linear and nonlinear simultaneous equations by many methods, ranging from direct solution, to downhill search, to Monte Carlo, etc., most of the discussion of adjoint methods reduces to technical details, many of which are complex, but which are primarily of interest to computer-code constructors (Heimbach et al., 2005). Within the normal equations, the time-evolution of the Lagrange multipliers is readily shown to satisfy a set of equations usually known as the "adjoint" or "dual" model. This dual model can be manipulated into a form having time run "backward," although that interpretation is unnecessary; see the references.

A very interesting complication is worth noting: the description in the last two paragraphs assumes one can actually differentiate J and J'. In oceanographic practise, that implies differentiating the computer code which does everything. The "trick" that has made this method practical for GCMs is the so-called automatic differentiation (or AD), in which a software tool can be used to produce the partial derivatives and their transposed values—in the form of another software code (see, e.g., Giering and Kaminski, 1998; Griewank and Walther, 2008; Utke et al., 2008). This somewhat bland statement hides a complex set of practical issues; see, for example, Heimbach et al. (2005) for discussion in the context of the MIT general circulation model (MITgcm). Most of the difficult problems are of no particular concern to someone mainly interested in the results.[6]

As discussed in more detail by Wunsch and Heimbach (2007), the central ECCO estimates are based upon this Lagrange multiplier method, with the state estimates obtained from the adjusted, but then *freely running*, MITgcm, as is required in our definition of state estimation. At the time of this writing, most of the estimates have restricted the control variables (the adjustable parameters) to the initial conditions and the meteorological forcing, although following exploratory studies by Ferreira et al. (2005), Stammer (2005), and Liu et al. (2012), state estimates are becoming available that also adjust internal model parameters, such as isopycnal, thickness, or vertical diffusion.

A full modern oceanic general circulation model (GCM or OGCM) such as that of Marshall et al. (1997) as modified over subsequent years (e.g., Adcroft et al., 2004; Campin et al., 2004), is a complex machine consisting of hundreds of thousands of lines of code encompassing the Navier–Stokes equations, the relevant thermodynamics, sea ice and mixed-layer sub-codes, various schemes to represent motions below the model resolution (whatever it may be), and further subsidiary codes for overflow entrainment, etc. Understanding such a model is a difficult proposition, in part because different elements were written by different people over many years, sometimes without full understanding of the potential interactions of the existing or future subcomponents. Further, various studies have shown the inevitability of coding errors (e.g., Basili et al., 1992) and unlike the situation with the real ocean, one is faced with determining if some interesting or unusual behavior is real or an artifact of interacting, possibly very subtle, errors. (Nature presumably never solves the incorrect equations; but observational systems do have their own mysteries that must be understood: recent examples include the discovery of systematic errors in fall rates to infer the depth of XBT data, e.g., by Wijffels et al., 2008, and calibration errors of pressure sensors onboard some of the Argo floats Barker et al., 2011).

By recognizing that most algorithms can be regarded as directed at the approximate solution of a least-squares

6. The situation is little different from that in ordinary ocean GCM studies. Technical details of time-stepping, storage versus recomputation, re-starts, etc., are very important and sometimes very difficult, but not often of consequence to most readers, except where the author necessarily calls attention to them.

problem, one can exploit the 200-year history of methodologies that have emerged (e.g., Björck, 1996), substituting differing numerical algorithms where necessary. For example, Köhl and Willebrand (2002) and Lea et al. (2002) suggested that the Lagrange multiplier method would fail when applied at high resolution to oceanic systems that had become chaotic. Although such behavior has been avoided in oceanographic practice (Gebbie et al., 2006; Hoteit et al., 2006; Mazloff et al., 2010), one needs to separate the *possible* failure of a particular numerical algorithm to find a constrained minimum from the inference that no minimum exists. If local gradient descent methods are not feasible in truly chaotic systems, one can fall back on variations of Monte Carlo or other more global methods. Obvious failure of search methods using local derivatives has had limited importance in oceanographic practice. This immunity is likely a consequence of the observed finite time interval in the state estimation problem, in which structures such as bifurcations are tracked adequately by the formally future data, providing adequate estimates of the algorithmic descent directions. Systematic failure to achieve an acceptable fit to the observations can lead to accepting the hypothesis that the model should be rejected as an inadequate representation. Potential model falsification is part of the estimation problem, and is the pathway to model improvement.

Modern physical oceanography is largely based upon inferences from the thermal wind, or geostrophic–hydrostatic, equations. Scale analyses of the primitive equations (e.g., Pedlosky, 1987; Vallis, 2006; Huang, 2010) all demonstrate that apart from some very exceptional regions of small area and volume, deviations from geostrophic balance are slight. This feature is simultaneously an advantage and a liability. It is an advantage because any model, be it analytical or numerical must, to a first approximation, satisfy the linear thermal wind equations. It is a liability because it is only the deviations which define the governing physics of the flow maintenance and evolution, and which are both difficult to observe and compute with adequate accuracy. In the present context, one anticipates that over the majority of the oceanic volume, any plausible model fit to the data sets must be, to a good approximation, a rendering of the ocean circulation in geostrophic, hydrostatic, balance, with Ekman forcing, and volume or mass conservation imposed regionally and globally as an automatic consequence of the model configuration. The most visible ageostrophic physics are the variability, seen as slow accumulating deviations from an initial state.

4.2. The Observations

Data sets used for many (not all) of the ECCO family of solutions are displayed in Table 21.2. As noted in Section 1, they are of very diverse types, geographical and temporal distribution, and with very different accuracies and precisions.

As is true of any least-squares solution, no matter how it is obtained, the results are directly dependent upon the weights or error variances assigned to the data sets. An over-estimate of the error corresponds to the suppression of useful information; an under-estimate to imposing erroneous values and structures. Although an unglamorous and not well-rewarded activity, a quantitative description of the errors is essential and is often where oceanographic expertise is most central. Partial discussions are provided by Stammer et al. (2007), Ponte et al. (2007), Forget and Wunsch (2007), and Ablain et al. (2009). Little is known about the space–time covariances of these errors, information, which if it were known, could improve the solutions (see Weaver and Courtier, 2001, for a useful direction now being used in representing spatial covariances). Model errors, which dictate how well estimates should fit to hypothetical perfect data, are extremely poorly known and are generally added to the true data error—as in linear problems the two types of error are algebraically indistinguishable.

5. GLOBAL-SCALE SOLUTIONS

Solutions of this type were first described by Stammer et al. (2001, 2002, 2003) and were computed on a $2^\circ \times 2^\circ$ grid with 22 vertical levels. As the computing power increased, a shift was made to a $1^\circ \times 1^\circ$, 23-level solution and that, until very recently, has remained the central vehicle for the global ECCO calculations. Although some discrepancies continue to exist in the ability to fit certain data types, these solutions (Wunsch and Heimbach, 2007) based as they are on geostrophic, hydrostatic balance over most of the domain, were and are judged adequate for the calculation of large-scale transport and variability properties. The limited resolution does mean that systematic misfits were expected, and are observed, in special regions such as the western boundary currents. Often the assumed error structures of the data are themselves of doubtful accuracy.

As noted above, Ganachaud (2003a) inferred that the dominant error in trans-oceanic transport calculations of properties arose from the temporal variability. Perhaps the most important lesson of the past decade has been the growing recognition of the extent to which temporal aliasing is a serious problem in calculating the oceanic state. For example, Figures 21.4 and 21.5 display the global meridional heat and freshwater transport as a function of latitude along with their standard errors computed from the monthly fluctuations. The figures suggest that errors inferred from hydrography are under-estimated and error estimates of the non-eddy resolving ECCO estimates are themselves lower bounds of the noise encountered in the real ocean. The classical oceanographic notion that semi-synoptic sections are accurate renderings of the time-average properties,

TABLE 21.2 Data Used in the ECCO Global 1° Resolution State Estimates Until About 2011

Observation	Instrument	Product/Source	Area	Period	dT
Mean dynamic topography (MDT)	• GRACE SM004-GRACE3	CLS/GFZ (A.M. Rio)	Global	Time-mean	Mean
	• EGM2008/DNSC07	N. Pavlis/Andersen & Knudsen	Global		
Sea level anomaly (SLA)	• TOPEX/POSEIDON	NOAA/RADS & PO.DAAC	65°N/S	1993–2005	Daily
	• Jason	NOAA/RADS & PO.DAAC	82°N/S	2001–2011	Daily
	• ERS, ENVISAT	NOAA/RADS & PO.DAAC	65°N/S	1992–2011	Daily
	• GFO	NOAA/RADS & PO.DAAC	65°N/S	2001–2008	Daily
SST	• Blended, AVHRR (O/I)	Reynolds & Smith	Global	1992–2011	Monthly
	• TRMM/TMI	GHRSST	40°N/S	1998–2004	Daily
	• AMSR-E (MODIS/Aqua)	GHRSST	Global	2001–2011	Daily
SSS	Various *in situ*	WOA09 surface	Global	Climatology	Monthly
In situ T, S	• Argo, P-Alace	Ifremer	"Global"	1992–2011	Daily
	• XBT	D. Behringer (NCEP)	"Global"	1992–2011	Daily
	• CTD	Various	Sections	1992–2011	Daily
	• SEaOS	SMRU & BAS (UK)	SO	2004–2010	Daily
	• TOGA/TAO, Pirata	PMEL/NOAA	Tropics	1992–2011	Daily
Mooring velocities	• TOGA/TAO, Pirata	PMEL/NOAA	Tropics	1992–2006	Daily
	• Florida Straits	NOAA/AOML	North Atlantic	1992–2011	Daily
Average T, S	• WOA09	WOA09	"Global"	1950–2000	Mean
	• OCCA	Forget (2010)	"Global"	2004–2006	Mean
Sea ice cover	• Satellite passive microwave radiometry	NSIDC (bootstrap)	Arctic, SO	1992–2011	Daily
Wind stress	QuickScat	• NASA (Bourassa)	Global	1999–2009	Daily
		• SCOW (Risien & Chelton)		Climatology	Monthly
Tide gauge SSH	Tide gauges	NBDC/NOAA	Sparse	1992–2006	Monthly
Flux constraints	From ERA-Interim, JRA-25, NCEP, CORE-2 variances	Various	Global	1992–2011	2-day to 14-day
Balance constraints			Global	1992–2011	Mean
Bathymetry		Smith & Sandwell, ETOPO5	Global	–	–

An estimated 22×10^8 individual values have been used in Equation (21.6), of which about 4×10^8 are assigned to the control terms.

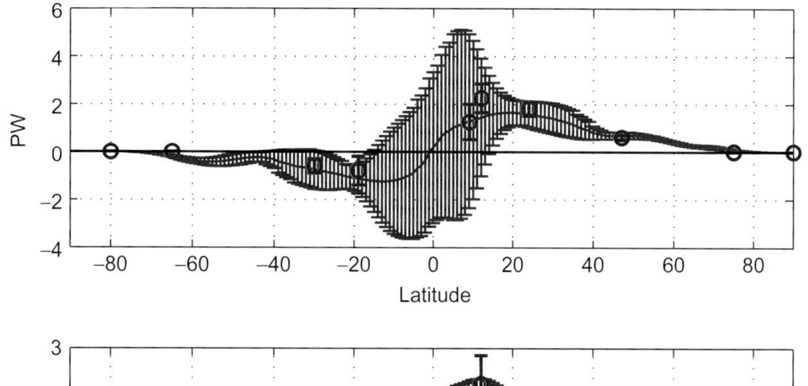

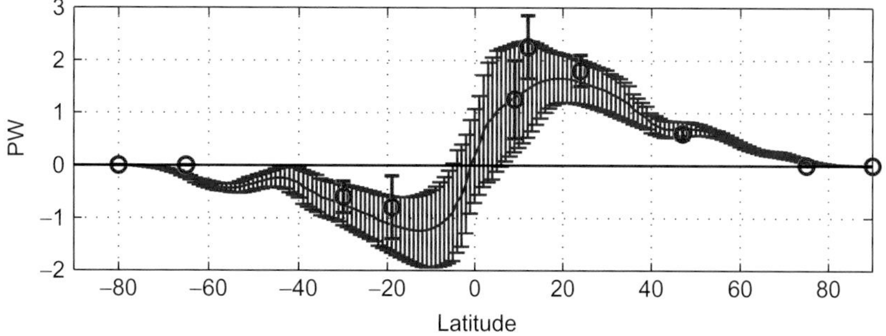

FIGURE 21.4 Global meridional heat transport in the ocean from ECCO-Production version 4 (G. Forget, private communication, 2011). Upper panel shows the standard error including the annual cycle and the lower one, with the annual cycle removed—as being largely predictable. Possible systematic errors are not included. Red dots with error bars are estimates from Ganachaud and Wunsch (2002). Note that the WOCE-era hydrographic survey failed to capture the southern hemisphere extreme near 10°S, thus giving an exaggerated picture of the oceanic heat transport asymmetry about the equator.

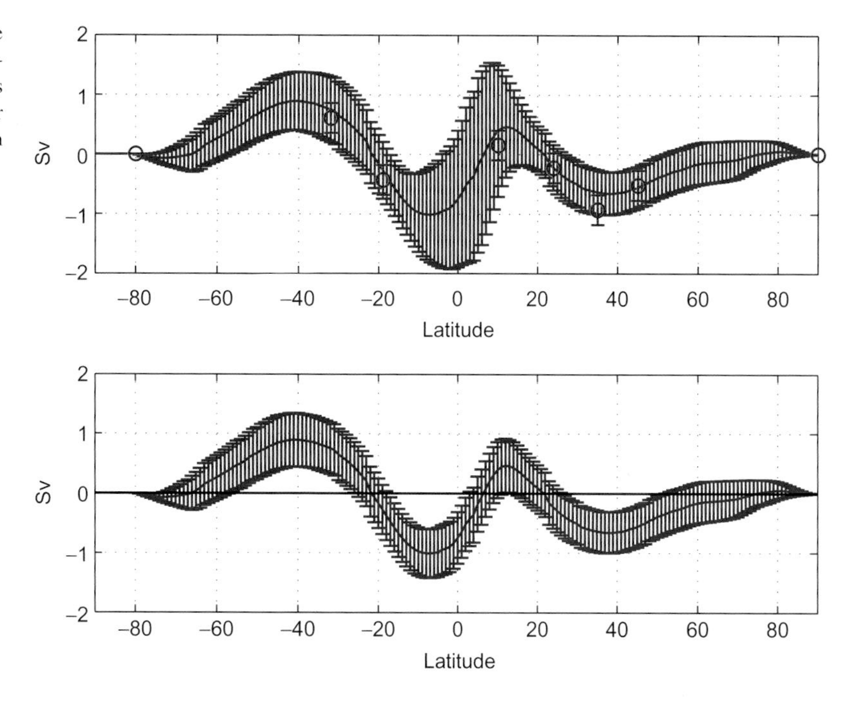

FIGURE 21.5 Same as Figure 21.4 except for the freshwater transport (G. Forget, private communication, 2011). Upper panel shows standard errors that include the seasonal cycle, and the lower without the seasonal cycle. Red dots are again from Ganachaud and Wunsch (2002).

while having some qualitative utility, has now to be painfully abandoned—an essential step if the subject is to be a quantitative one. Temporal effects are most conspicuous at low latitudes, but in many ways, the difficulty is greatest at high latitudes: the long timescales governing behavior there mean that the hydrographic structure is very slowly changing, requiring far longer times to produce an accurate time-mean. In other words, a 10-year average at 10° N will be a more accurate estimate of the longer term mean than one at 50° N. Even this comment begs the question of whether a stable long-term mean exists, or whether the system drifts over hundreds and thousands of years. This latter is a question concerning the frequency spectrum of oceanic variability and which is very poorly known at periods beyond a few years.

For the 19+ years now available in the global state estimates, most of the large-scale properties, including the time variations, are stable from one particular set of assumptions to others, probably as a consequence of the dominance of overall geostrophic balance and the comparatively well-sampled hydrography and altimetric slopes. They are thus worth analyzing in detail. The intricacies of the global, time-varying ocean circulation are a serious challenge to the summarizing capabilities of authors. A full description of the global state estimates becomes a discussion of the complete three-dimensional time-varying ocean circulation, a subject requiring a book, if not an entire library, encompassing distinctions among time and space scales, geographical position, depth, season, trends, and the forcing functions (controls). No such synthesis is attempted here! Instead we can only give a bit of the flavor of what can and has been done with the estimates along with enough references for the reader to penetrate the wider literature.

Note too, as discussed, for example, by Heimbach et al. (2011) and other, earlier efforts, the Lagrange multipliers are the solution to the dual model. As such, they are complete solutions in three spatial dimensions and time, and convey the sensitivity of the forward model to essentially any parameter or boundary or initial condition in the system. The information content of the dual solutions is very large—representing not only the sensitivities of the solution to the data and model parameters and boundary and initial conditions, but also the flow of information through the system. Analyzing the dual solution does, however, require the same three-dimensional time-dependent representations of any full GCM, and these elements of the state estimates remain greatly under-exploited at the present time.

5.1. Summary of Major, Large-Scale Results

None of the results obtained so far can be regarded as the final state estimate: obtaining fully consistent misfits by the model to the observations has never been achieved (see the residual misfit figures in the references). Misfits linger for a variety of reasons, including the sometimes premature termination of the descent algorithms before full optimization, mis-representation of the true model or data errors, or selection of a local rather than a global minimum in the major nonlinear components of the model. As with all very large nonlinear optimization problems, approach to the "best" solution is asymptotic. With these caveats, we describe some of the more salient oceanographic features of the recent solutions, with no claim to being comprehensive. Note that results from a variety of ECCO-family estimates are used, largely dictated by the particular problem that was the focus of the calculation.

5.1.1. *Volume, Enthalpy, Freshwater Transports and their Variability*

The most basic elements describing the ocean circulation and its large-scale variability are usually the mass (or volume, which is nearly identical) transports. Stammer et al. (2001, 2002, 2003) depicted the basic global-scale elements of the mass transport as averaged over the duration of their estimates. A longer duration estimate (v3.73) has been used (Figure 21.6) to compute the vertically integrated volume stream function. We reiterate that diagrams such as this one are finite duration averages whose relationship to hypothetical hundred year or longer climatologies remains uncertain.

Figure 21.7 shows the zonally integrated and vertically accumulated meridional transport as a function of depth and ocean. The very large degree of temporal variability can be seen in Figure 21.4 from a new, fully global solution, which is about to become available online at the time of writing (ECCO-Production version 4; see Table 21.3) with error bars derived from the temporal variances. These time averages have been a historically important goal of physical oceanography, albeit estimates derived from unaveraged data were commonly assumed without basis to accurately

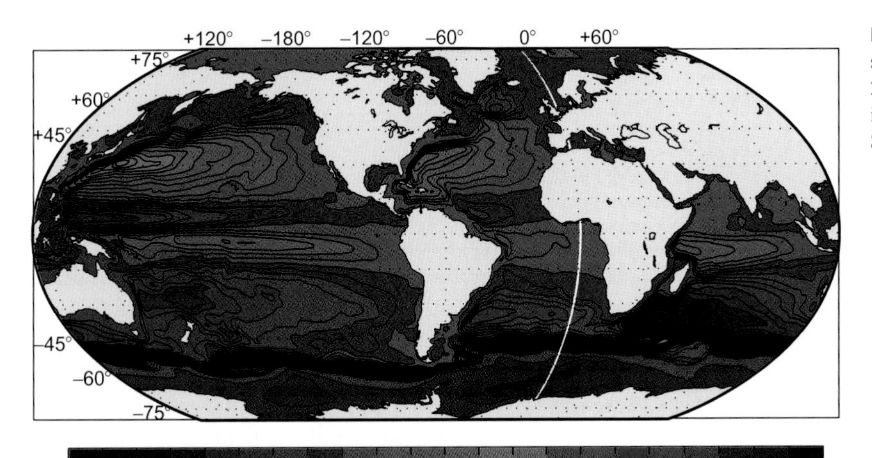

FIGURE 21.6 The top-to-bottom transport stream function from ECCO v3.73 (Wunsch, 2011). Qualitatively, the wind-driven gyres dominate the result, with the intense transports in the Southern Ocean particularly conspicuous.

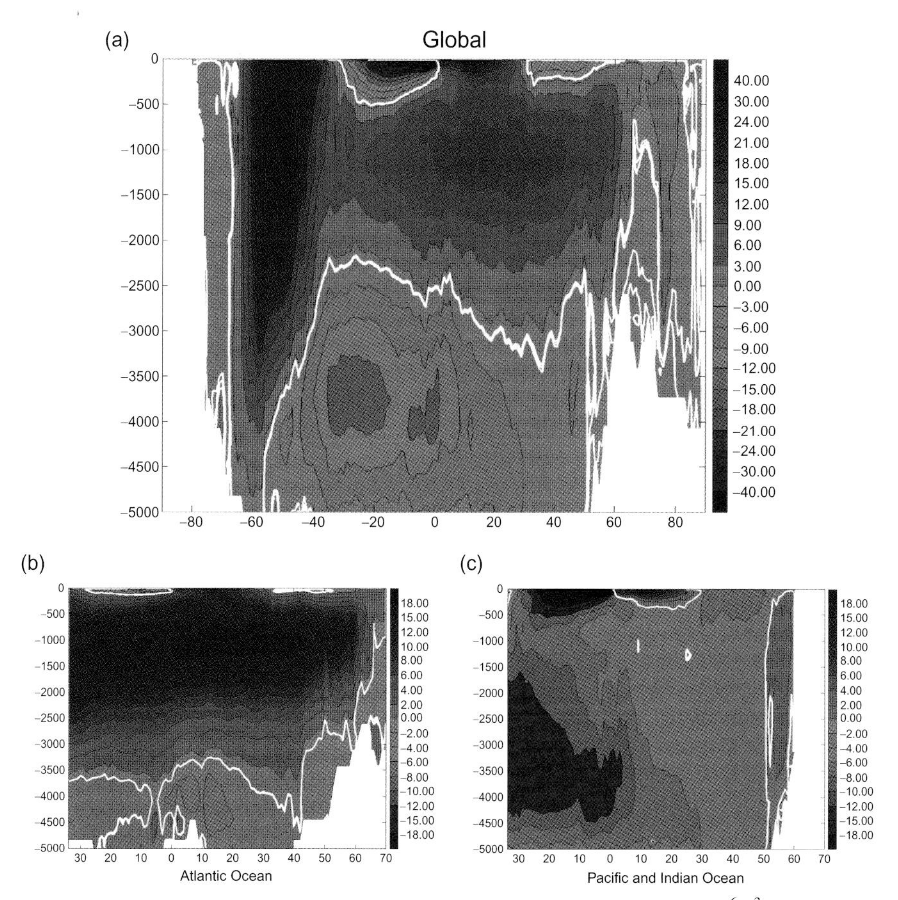

FIGURE 21.7 Mean (1992–2010) of the meridional volume transport stream function in Sverdrups (Sv-$10^6 m^3/s$) from ECCO-Production version 4 (Wunsch and Heimbach, 2013; Forget et al., in preparation, 2013). Panel (a) is the global result; panels (b and c) are the Atlantic, and the combined Indo-Pacific, respectively. Note the complex equatorial structure, and that this representation integrates out a myriad of radically different dynamical sub-regimes. In the Southern Ocean, interpretation of zonally integrated Eulerian means requires particular care owing to the complex topography and relatively important eddy transport field.

depict the true time-average. Perhaps the most important utility of the existing state estimates has been the ability, at last, to estimate the extent of the time-variability taking place in the oceans (Wunsch and Heimbach, 2007, 2009, 2013). Withheld, direct *in situ* observations in a few isolated regions (Kanzow et al., 2009; Baehr, 2010) are consistent with the inference that even volume transports integrated across entire ocean basins have a large and qualitative temporal variability. More generally, mooring data and the now almost 20-year high-resolution high-accuracy altimetric records all show the intense variations that exist everywhere. With ECCO-like systems, syntheses of these data sets are now possible.

5.1.2. The Annual Cycle

The annual cycle of oceanic response is of interest in part because the ultimate forcing function (movement of the sun through the year) is very large and with very accurately known structure. In practice, that forcing is mediated through the very complex atmospheric annual changes, and understanding how and why the ocean shifts seasonally on a global basis is a difficult problem. Using the ECCO state estimates, Vinogradov et al. (2008) mapped the amplitude and relative contributions for salt and heat of the annual cycle in sea level (Figure 21.8). The importance of the annual cycle, more generally, is visible in Figures 21.4 and 21.5 as the large contribution to the standard errors.

5.1.3. Sea-Level Change

The sea surface height is simultaneously a boundary condition on the oceanic general circulation and a consequence of that circulation. Because of the intense interest in possible large-scale changes in its height, the potential shifts in vulnerability to storm surges, and associated issues such as ecosystem and freshwater reservoir declines, the ECCO state estimation system has been used to estimate the shifts taking place in the era since 1992 (Wunsch et al., 2007). A summary of a complex subject is that sea level change is dominated by regional variations more than an order

TABLE 21.3 Published ECCO Family State Estimates, Divided Roughly into Categories

Label and Version	Hor./Ver. Grid	Domain	Duration	Scope	References
ECCO-Production Sustained production of decadal climate state estimates (former ECCO-GODAE)					
ver.0 (ECCO-MIT)	2°/22	80°N/S	1992–1997	First ECCO product—proof of feasibility	Stammer et al. (2002, 2004)
ver.1 (ECCO-SIO)	1°/23	80°N/S	1992–2002	Begin of 1° sustained production	Köhl et al. (2007)
ver.2 (ECCO-GODAE)	1°/23	80°N/S	1992–2004	Air–sea flux constraints for sea level studies	Wunsch and Heimbach (2006, 2007)
(OCCA)	1°/50	80°N/S	2004/2005/2006/2007	Atlas from 1-year "synoptic snapshots"	Forget (2010)
(GECCO)	1°/23	80°N/S	1951–2000	50-year solution covering NCEP/NCAR period	Köhl and Stammer (2008a,b)
ver.3 (ECCO-GODAE)	1°/23	80°N/S	1992–2007	Switch to atmospheric state controls and sea ice	Wunsch and Heimbach (2009)
ver. 4 (ECCO-Production)	1°/50	Global	1992–2010	First full-global estimate including Arctic	Forget et al. (in preparation, 2013)
ECCO-ICES Ocean–ice interactions in Earth system models (former ECCO2)					
ver.1 (CS510 GF)	18 km/50	Global	1992–2002/2010	Green's function optimization, of eddying model	Menemenlis et al. (2005a,b)
ECCO-JPL near real-time filter and reduced-space smoother					
ver.1 (KF)	1°/46	80°N/S	1992–present	Near-real-time Kalman Filter (KF) assimilation	Fukumori et al. (1999)
ver.2 (RTS)	1°/46	80°N/S	1992–present	Smoother update of KF solution	Fukumori (2002)
Regional efforts					
Southern Ocean (SOSE)[a]	1/6°/42	25°–80°S	2005–2009	Eddy-permitting SO state estimate	Mazloff et al. (2010)
ECCO2 Arctic and ASTE[a]	18 and 4 km/50	Arctic and SPG	1992–2009	Arctic/subpolar gyre ocean–sea ice estimate	Nguyen et al. (2011, 2012)
North Atlantic	1°/23	25°–80°N	1993	Experimental 2° versus 1° nesting	Ayoub (2006)
Subtropical Atlantic	1/6°/42	–	1992/1993	Experimental 1° versus 1/6° nesting	Gebbie et al. (2006)
Tropical Pacific	–	–		Experimental 1° versus 1/3° nesting	Hoteit et al. (2006, 2010)
Labrador Sea and Baffin Bay	–	–	1996/1997	First full coupled ocean–sea ice estimate	Fenty and Heimbach (2013a,b)

The global decade+estimates are labeled as "ECCO-Production," while others are either regional or experimental.
[a]*Denotes ongoing efforts.*

of magnitude larger than the putative global average, and arising primarily from wind field shifts. Varying spatial contributions from competing exchanges of freshwater and heat with the atmosphere and the extremely inhomogeneous (space and time) *in situ* data sets render the global mean and its underlying causes far more uncertain than some authors have claimed.

At the levels of accuracy appearing to be required, very careful attention must now be paid to modeling issues such as water self-attraction and load (Kuhlmann et al., 2011; Vinogradova et al., 2011) not normally accounted for in OGCMs. Conventional approximations to the moving free-surface boundary conditions generate systematic errors no longer tolerable (e.g., Huang, 1993; Wunsch et al., 2007). Usefully accurate sea level estimation over multiple decades may be the most demanding requirement on both models and data sets now facing oceanographers (Griffies and Greatbatch, 2012). The global means are claimed by some to have

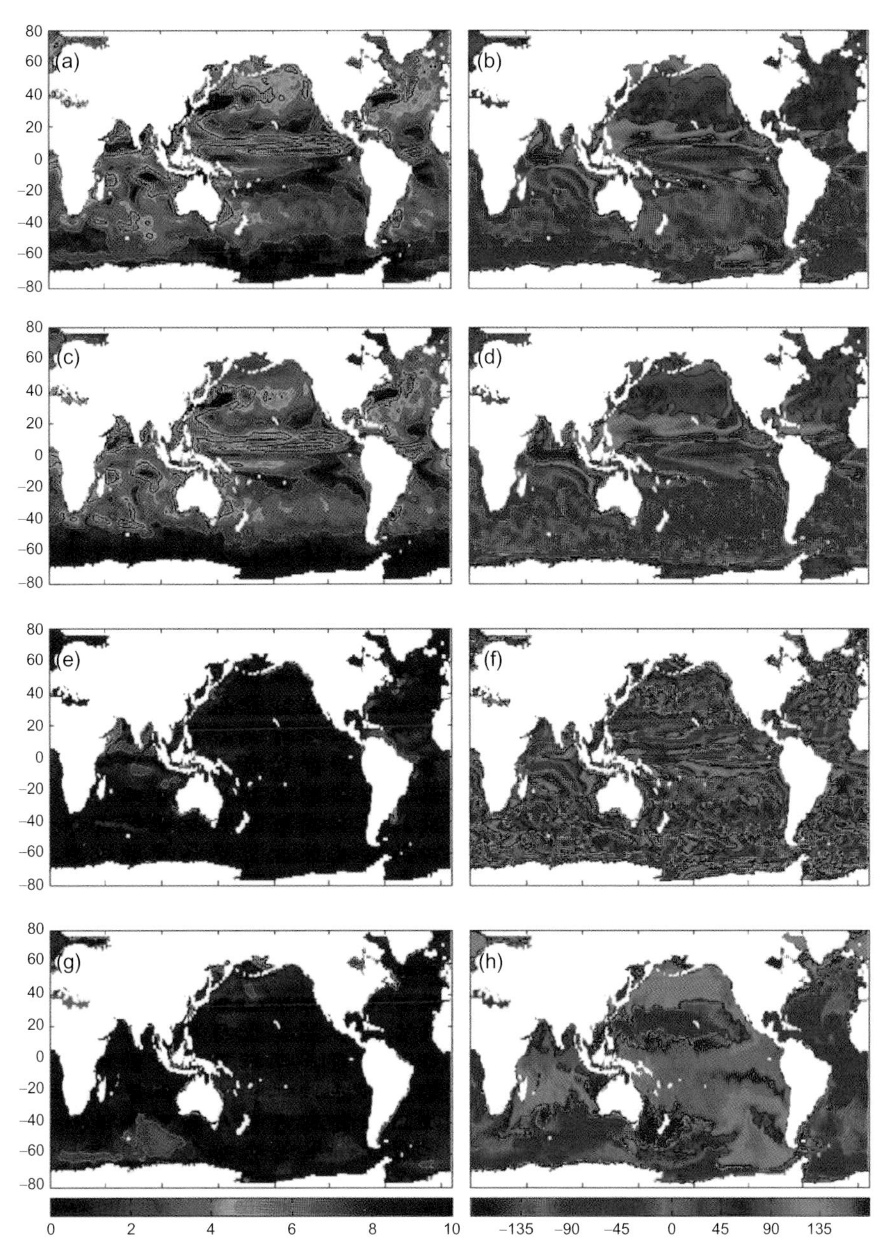

FIGURE 21.8 Showing the annual cycle in sea level from ECCO Climate State v2.177. Left column is the amplitude in cm and the right column the phase. From top-to-bottom, they are the surface elevation (a and b), the thermosteric component (c and d), the halosteric component (e and f), and at bottom, the bottom pressure (g and h). Phases, ϕ are in degrees relative to a January time origin as $cos(\omega t + \phi)$. *From Vinogradov et al. (2008).*

accuracies approaching a few tenths of a millimeter per year—a historically extraordinary requirement on any ocean estimate. Despite widely publicized claims to the contrary (e.g., Cazenave and Remy, 2011; Church et al., 2011), state estimate results suggest that at the present time, the global observing system is insufficient to provide robust partitioning amongst heat content changes, land and ice sheet runoff, and large-scale shifts in circulation patterns (the recent paper by King et al., 2013 discusses an example of the remaining uncertainty in current ice sheet mass loss estimates, with implications for sea level budgets). A particular difficulty pertains to the deep ocean, below depths measured by the Argo array, where the distinction between apparent changes occurring (Kouketsu et al., 2011; Purkey and Johnson, 2010) and the significant deep eddy variability (Ponte, 2012) remains obscure due to poor observational coverage. Claims for closed budget elements involve accuracies much coarser than are stated for the total value.[7]

7. We have omitted here the distinction between absolute sea level with respect to the geoid, and relative sea level measured by tide gauges, and ignored processes associated with the unloading of the solid Earth from ice sheet shrinkage (e.g., Munk, 2002; Milne et al., 2009; Mitrovica et al., 2011). Only recently have these phenomena begun to appear in climate models: Kopp et al. (2010) and Slangen et al. (2012).

5.1.4. Biogeochemical Balances

From the adjoint of the tracer concentration sub-model of the ECCO system, Dutkiewicz et al. (2006) calculated the sensitivity of the nutrient production in the system to iron enrichment. This work is representative of the use of dual solutions to probe large complex models in any scientific field. They found a strong dependence upon the available light, and that the tropical ocean had the greatest sensitivity to iron limitation. Among other considerations, these inferences are important in the erstwhile debate over whether iron fertilization makes any sense for control of atmospheric CO_2.

Woloszyn et al. (2011) used the ECCO higher resolution Southern Ocean State Estimates (SOSE) of Mazloff et al. (2010) to demonstrate the great importance of adequate resolution in calculating carbon exchange between the atmosphere and ocean. The same configuration was adopted by Ito et al. (2010) to describe the Ekman layer contribution to the movement of carbon dioxide.

The emerging field of microbial oceanography seeks a zeroth-order understanding of the biogeography and diversity of marine microbes. Coupling between ocean physics and ecology is being explored through the use of ECCO state estimates, which drive models of marine ecology (e.g., Follows et al., 2007; Follows and Dutkiewicz, 2011). Crucial requirements of the estimates are (1) to be in sufficiently close agreement with the observed physical ocean state such as to reduce uncertainties in the coupled models from the physical component, and (2) to furnish an evolution of the physical state in agreement with conservation laws.

5.1.5. Sea Ice

The importance of sea ice to both the ocean circulation and climate more generally has become much more conspicuous in recent years. Sea ice models have been developed within the state estimation framework as fully coupled sub-systems influenced by and influencing the ocean circulation (Menemenlis et al., 2005a,b; Losch et al., 2010). By way of example, Figure 21.9, taken from Losch et al. (2010) depicts 1992–2002 mean March and September effective ice thickness distributions representing the months of maximum and minimum ice cover in both

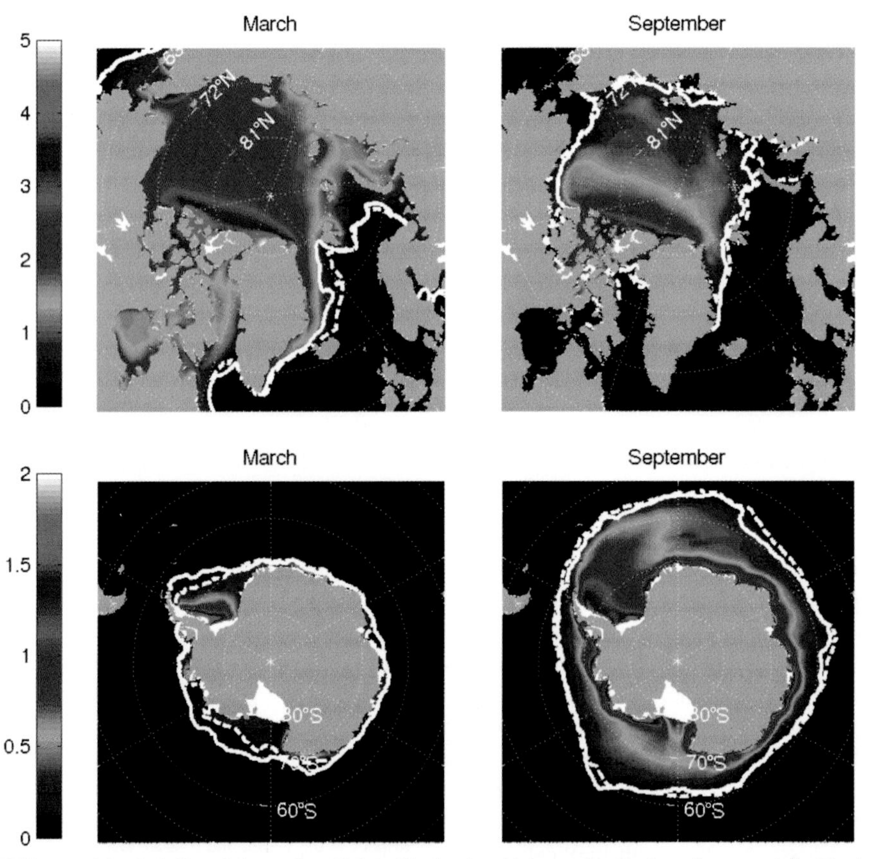

FIGURE 21.9 1992–2002 mean March (left) and September (right) effective ice thickness distributions (in meters) for Northern (top) and Southern (bottom) hemispheres. Obtained from a global eddy-permitting ECCO2 simulation, for which a set of global parameters has been adjusted. Also indicated are the ice edge (15% ice concentration isoline) inferred from the model (dashed line) and from satellite-retrieved passive microwave radiometry (solid line). *From Losch et al. (2010).*

hemispheres. Also shown are the modeled and observed ice edges, represented as 15% isolines of the fractional sea ice concentrations (0–100%). The results were obtained from an early version of the alternative ECCO2 eddy-permitting optimization using Green functions (Menemenlis et al., 2005a,b) on the cubed-sphere grid at 18 km horizontal resolution (see Table 21.3). More detailed studies focusing on the Arctic were carried out with similar and higher resolution (4 km) configurations (Nguyen et al., 2011, 2012), but with a very limited control space available for parameter adjustment via the Green function approach.

A comprehensive step toward full coupled ocean–sea ice estimation, in which both ocean and sea ice observation were synthesized, was made by Fenty and Heimbach (2013a,b) for the limited region of the Labrador Sea and Baffin Bay. Figure 21.10a shows an annual cycle of total sea ice area in the domain from observations, the state estimate, and the unconstrained model solution. Also shown are the remaining misfits, as evidence of the random nature of the residuals, as required by theory, Equations (21.2) and (21.3b). An important result of that study is the demonstration that adjustment well within their prior uncertainties in the high-dimensional space of uncertain surface atmospheric forcing, patterns can achieve an acceptable fit between model and observation, placing stringent requirements on process studies that aim to discriminate between model errors and forcing deficiencies.

As in the discussion of biogeochemical balances above, the adjoint or dual solution of the coupled ocean–sea ice model can provide detailed sensitivity analyses. Heimbach et al. (2010) used the dual solution to study sensitivities of sea ice export through the Canadian Arctic Archipelago to changes in atmospheric forcing patterns in the domain. Kauker et al. (2009) investigated the causes of the 2007 September minimum in Arctic sea ice cover in terms of sensitivities to atmospheric forcing over the preceding months. A similar sensitivity study on longer timescales is shown in Figure 21.11 of solid (sea ice and snow) freshwater export through Fram Strait for two study periods, January 1989 to September 1993, and January 2003 to September 2007 (unpublished work). The objective function was chosen to be the annual sea ice export between October 1992 and September 1993, and October 2006 and September 2007. Export sensitivities to changes in effective sea ice thickness, 24 months prior to September 1993 and 2007, respectively, are shown. The dominant patterns are positive sensitivities upstream of Fram Strait, for which an increase in ice thickness will increase ice export at Fram Strait 24 months later. (Spurious patterns south of Svalbard have been attributed to masking errors in the sea ice adjoint model and were corrected in Fenty and Heimbach, 2013a.) Sensitivities are linearized around their respective states, and depend on the state trajectory. The extended domain of influence for the 2007 case compared to 1993 suggests

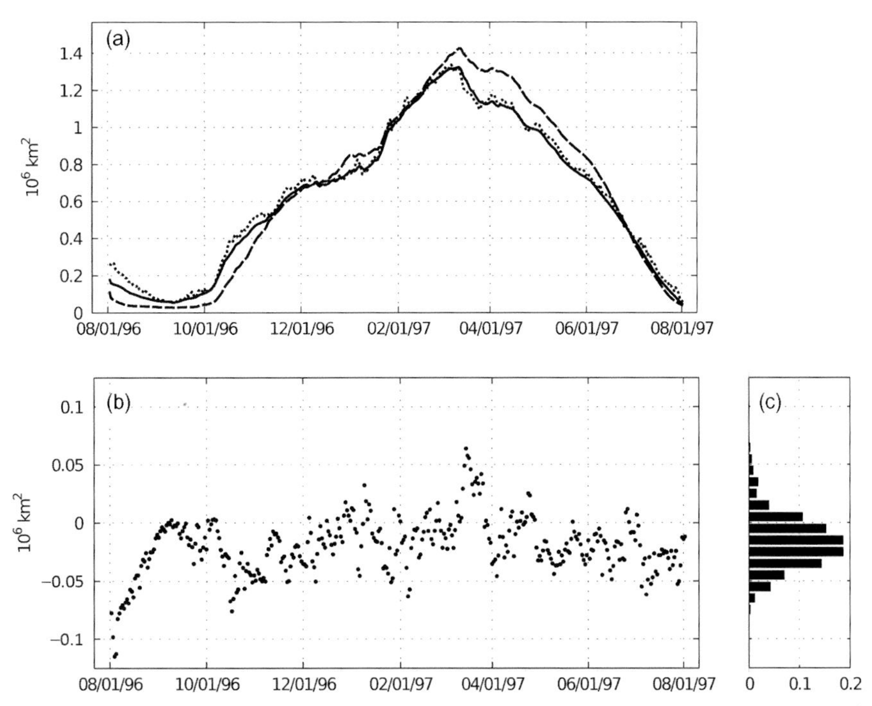

FIGURE 21.10 (Top): Annual cycle from August 1996 to July 1997 of daily mean total sea ice area in the Labrador Sea and Baffin Bay from observations (dotted), a regional state estimate (black), and the unconstrained model solution (dashed). (Bottom): Residual misfits between estimated and observed sea ice area and its frequency of occurrence histogram (right panel). *Taken from Fenty and Heimbach (2013a).*

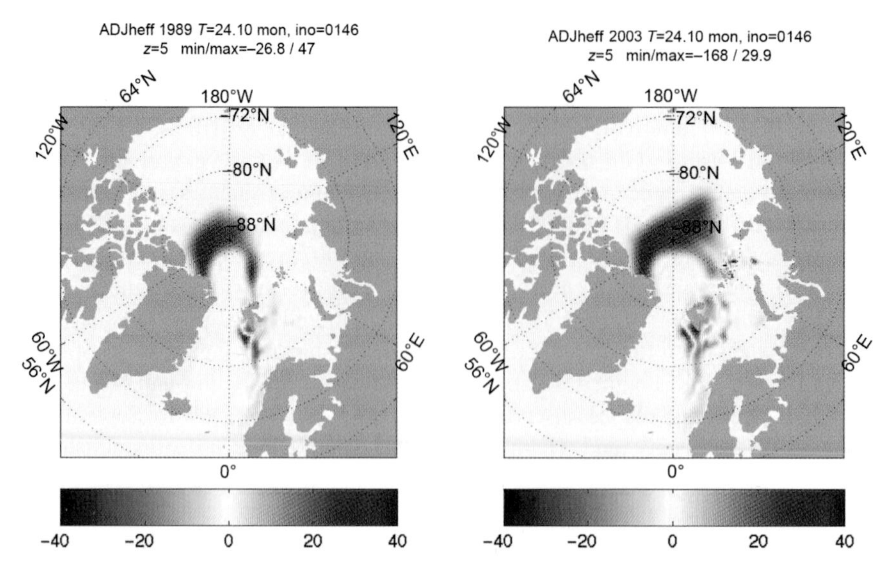

FIGURE 21.11 Sensitivity of sea ice export through Fram Strait to changes in effective sea ice thickness 24 months back in time. Two integration periods were considered, January 1989 to September 1993 (left) and January 2003 to September 2007 (right). The objective function is annual sea ice export between October 1992 and September 1993 (left), and October 2006 and September 2007 (right).

more swift transport conditions in the central Arctic, possibly due to favorable atmospheric conditions, or to weaker sea ice, or both.

5.1.6. Ice Sheet–Ocean Interactions

The intense interest in sea level change and the observed acceleration of outlet glaciers spilling into narrow deep fjords in Greenland and ice streams feeding vast ice shelves in Antarctica (e.g., Payne et al., 2004; Alley et al., 2005; Shepherd and Wingham, 2007; Pritchard et al., 2009; Rignot et al., 2011; Straneo et al., 2013) has led to inferences that much of the ice response may be due to regional oceanic warming at the glacial grounding lines, an area termed by Munk (2011) "*this last piece of unknown ocean*." One such region is the Amundsen Sea Embayment in West Antarctica (Figure 21.12, taken from Schodlok et al., 2012), where the ocean is in contact with several large shelves, among which Pine Island Ice Shelf (PIIS) and Pine Island Glacier exhibit one of the largest changes in terms of ice sheet acceleration, thinning, and mass loss. Recent, and as yet incomplete, model developments have been directed at determining the interactions of changing ocean temperatures and ice sheet response, and for the purpose of inclusion into the coupled state estimation system (Losch, 2008). Simulated melt rates under PIIS are depicted in Figure 21.13, but cannot be easily measured directly (Dansereau et al., 2013). A first step toward their estimation in terms of measured hydrography has been undertaken by Heimbach and Losch (2012) who developed an adjoint model complementing the sub-ice shelf melt rate parameterization. By way of example, Figure 21.14 depicts transient sensitivities of integrated melt rates (Figure 21.13) to changes in ocean temperatures. The spatial inhomogeneous patterns have implications for the interpretation of isolated measurements and optimal observing design.

The critical dependence of sub-ice shelf cavity circulation and melt rates to details of the bathymetry and grounding line position noted by Schodlok et al. (2012) revives the issue of bottom topography as a dominant control on ocean circulation and the necessity for its inclusion into formal estimation systems (Losch and Wunsch, 2003; Losch and Heimbach, 2007).

5.1.7. Air–Sea Transfers and Property Budgets

By definition, state estimates permit calculations up to numerical accuracies of global budgets of energy, enthalpy, etc. Many of these budgets are of interest for the insight they provide into the forces powering the ocean circulation. Josey et al. (2013) discuss estimates of the air–sea property transfers using the ECCO estimates. As an example, Figure 21.15 is an estimate by Stammer et al. (2004) of the net air–sea transfers of freshwater. That paper compares this estimate to other more ad hoc calculations and evaluates its relative accuracy.

As examples of more specific studies using the state estimates, we note only Piecuch and Ponte (2011, 2012), who examined the role of transport fluctuations on the regional sea level and oceanic heat content distribution, and Roquet et al. (2011), who used them to depict the regions in which mechanical forcing by the atmosphere enters into the interior geostrophic circulation. Many more such studies are expected in the future.

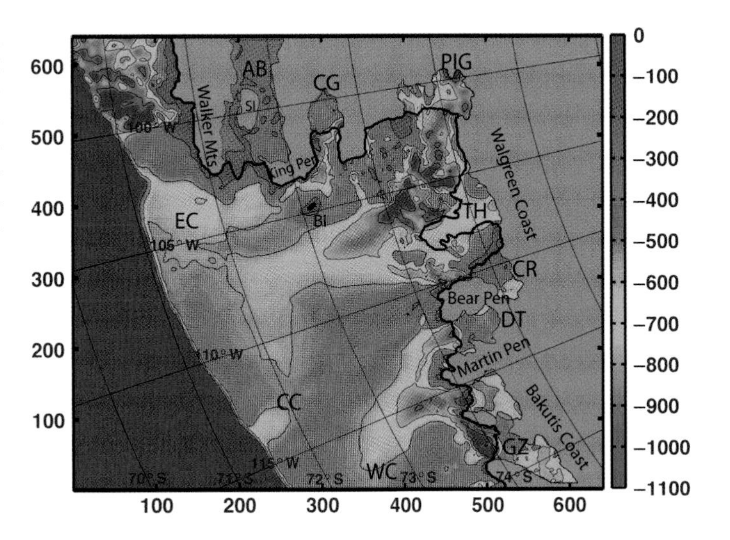

FIGURE 21.12 Bottom topography (in meters) of the Amundsen Sea Embayment, West Antarctica, with thick black lines delineating the edge of several large ice shelves which buttress the following glaciers grounded deep below sea level: Abbot (AB), Cosgrove (CG), Pine Island Glacier (PIG), Thwaites (TH), Crosson (CR), Dotson (DT), and Getz (GZ). Also indicated are prominent topographic features, such as Sherman Island (SI), Burke Island (BI), Eastern Channel (EC), Central Channel (CC), and Western Channel (WC). *From Schodlok et al. (2012).*

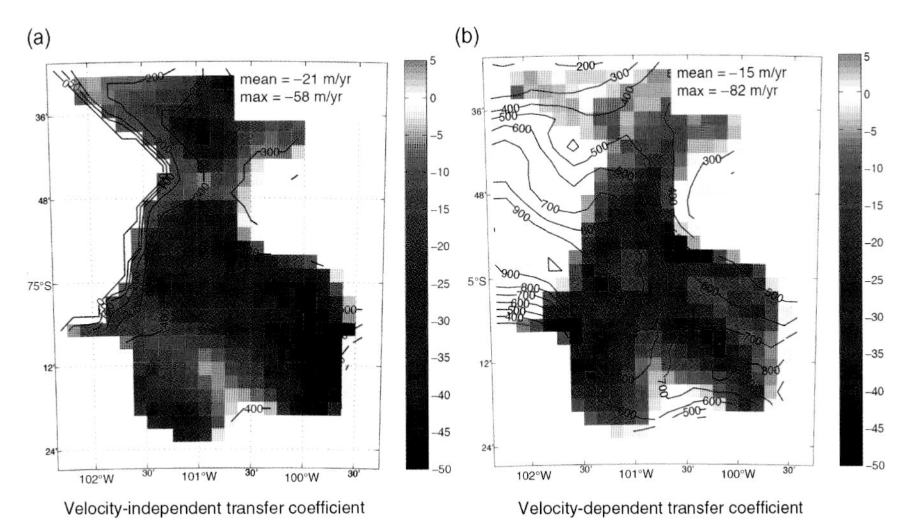

FIGURE 21.13 Simulated melt rates (colors, in meters/year) under Pine Island Ice Shelf (PIIS) derived from variants of the Holland and Jenkins (1999) melt rate parameterization, using either velocity-independent (a) or velocity-dependent transfer coefficients. Large melt rates correspond to either locations deep inside the cavity where the ice shelf is in contact with the warmest Circumpolar Deep Waters, or to locations of highest flow at the ice shelf-ocean interface. Direct measurement of melt rates is challenging, making robust inferences difficult. *From Dansereau et al. (2013).*

5.2. Longer Duration Estimates

Although the original ECCO estimates were confined to the period beginning in the early 1990s with the improved observational coverage that became available in association with WOCE, the intense interest in decadal scale climate change has led to some estimates of the ocean state emulating the meteorological reanalyses, extending 50 years and longer into the past. Some of these estimates are based essentially on the reanalysis methods already described (e.g., Rosati et al., 1995; Hurlburt et al., 2009), and having all of their known limitations.

Köhl and Stammer (2008) and Wang et al. (2010) have pioneered the application of the ECCO least-squares methods to an oceanic state estimate extending back to 1960. Their estimates have the same virtue as the wider ECCO family of solutions, in satisfying known model equations of motion and dynamics and with known misfits to all data types. The major problem is the extreme paucity of data in the ocean preceding the WOCE-era; see, for example, figures 1 and 2 of Forget and Wunsch (2007), and the accompanying very limited meteorological forcing observations in the early days. Note that polar orbiting meteorological satellites did not exist prior to 1979—see Figure 21.2 and Bromwich and Fogt (2004). Useful altimetry appears only at the end of 1992. "Whole domain" methods such as smoothers or Lagrange multipliers do carry information backward in time, and in the estimates for the underconstrained decades prior to about 1992, the gross properties of the ocean circulation are better determined because of

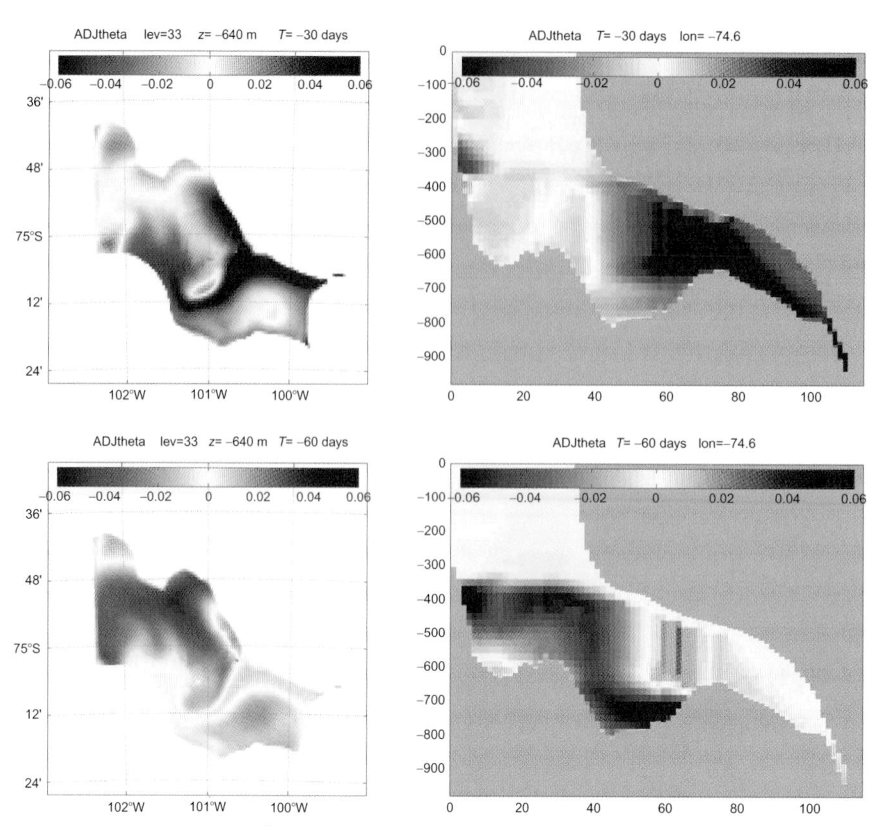

FIGURE 21.14 Transient sensitivities, $\delta^* T=(\partial J/\partial T)^{\mathrm{T}}$, of integrated melt rates J under PIIS (from Figure 21.11b) to changes in temperature T at times $t=\tau_f-30$ days (upper row) and -60 days (lower row) prior to computing J. Left panels are horizontal slices at 640 m depth, right panels are two vertical slices taken across the domain. Units are in $\mathrm{m}^3\,\mathrm{s}^{-1}\,\mathrm{K}^{-1}$, where $0.1\ \mathrm{m}^3\,\mathrm{s}^{-1}\,\mathrm{K}^{-1} \approx 3\ \mathrm{Mt\,a}^{-1}\,\mathrm{K}^{-1} \approx 3\ \mathrm{mm\,a}^{-1}\,\mathrm{K}^{-1}$.

the later, denser, data sets. But the memory of the upper ocean, which is most prominent, for example, in climate forecasting attempts, appears to be restricted to a few years, and one expects considerable near-surface uncertainty to occur even as recently as the 1980s.

A preliminary step of assessing the impact of observational assets in constraining the ECCO solutions has been taken through observing system experiments in the context of short-duration optimizations during the Argo array period (Heimbach et al., 2009; Zhang et al., 2010). Results suggest that the impact of altimetry and Argo floats in constraining, for example, the MOC is drastic, compared to the pre-WOCE period when only hydrographic sections were available.

The published solutions for the interval prior to about 1992 are best regarded as physically possible, but whose uncertainty estimates, were they known, would surely be very much greater than they are in the later times, but diminishing as the WOCE-era is approached. These long-duration estimates, decades into the past, thus present a paradox: if they are quantitatively useful—other than as examples of *possible* solutions—then the relatively large investment in observation systems the community has made since the early 1990s was unnecessary. If that investment has been necessary, then one cannot readily quantitatively interpret the early estimates. We leave the subject here as one awaiting the necessary time-dependent uncertainty estimates.

5.3. Short-Duration Estimates

Finding a least-squares fit over 19+ years is computationally very demanding and for some purposes, estimates over shorter time intervals can be useful. In particular, Forget (2010) used the same model and methodology as that of the ECCO Climate State 1° system (Wunsch and Heimbach, 2007), but limited the calculation to three overlapping 18-month periods in the years 2004–2006. In his estimate, the model-data misfit is considerably reduced compared to that in the 19+year solution. The reasons for that better fit are easy to understand from the underlying least-squares methodology: the number of adjustable parameters (the control vector) has the same number of degrees-of-freedom in the initial condition elements as does the decade+calculations, but with many fewer data to fit, and with little time to evolve away from the opening state. (Meteorological elements change over the same timescales in both calculations.) It is much more demanding of a model and its initial condition controls to produce fits to a 19-year evolution than to an 18-month one. Although both calculations have short timescales compared to oceanic equilibrium times of hundreds to thousands of years, in an 18-month interval little coupling exists between the meteorological controls and the deep data sets—which are then easily fit by the estimated initial state.

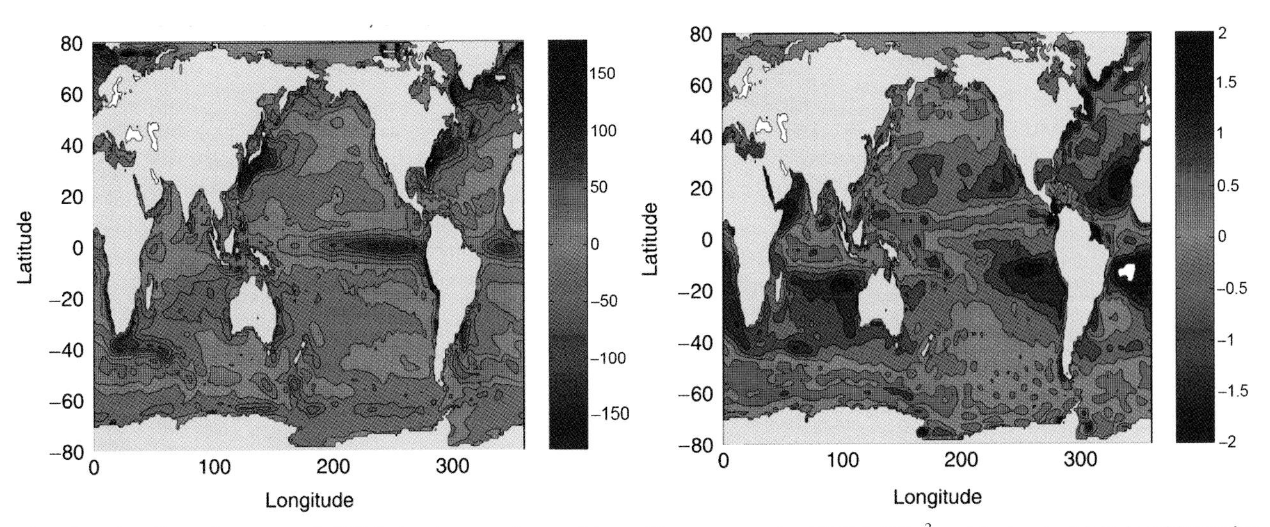

FIGURE 21.15 From Stammer et al. (2004) showing an estimate of the multi-year average heat (left, in W/m^2) and freshwater (right, in m/y) transfers between ocean and atmosphere.

Solutions of this type are very useful, particularly for upper ocean and regional oceanographic estimates (see the water mass formation rate application in Maze et al., 2009). An important caveat, however, is that one must resist the temptation to regard them as climatologies. They *do* bring us much closer to the ancient oceanographic goal of obtaining a synoptic "snapshot."

5.4. Global High-Resolution Solutions

Ocean modelers have been pursuing ever-higher resolution from the very beginning of ocean modeling and the effort continues. In classical computational fluid dynamics, one sought "numerical convergence": the demonstration that further improvements in resolution did not qualitatively change the solutions, and preferably that they reproduced known analytical values. Such demonstrations with GCMs are almost non-existent, and thus a very large literature has emerged attempting to demonstrate the utility of "parameterizations"—constructs intended to mimic the behavior of motions smaller than the resolution capability of any particular model. A recent review is by Ringler and Gent (2011). Absent fully resolved solutions with which to compare the newer parameterizations, the question of their quantitative utility remains open. They do represent clear improvements on older schemes.

Despite the parameterization efforts, considerable evidence exists (e.g., Hecht and Smith, 2008; Lévy et al., 2010) that qualitative changes take place in GCM solutions when the first baroclinic deformation radius, at least, is fully resolved. From the state estimation point of view, one seeks as much skill as possible in the model—which is meant to represent the fullest possible statement of physical understanding. On the other hand, state estimation, as a curve-fitting procedure, is relatively immune to many of the problems of prediction. In particular, because of the dominant geostrophic balance, its mass transport properties are insensitive to unresolved spatial scales—bottom topographic interference being an exception. In data dense regions, away from boundary currents, one anticipates robust results even at modest resolution.

Ultimately, however, the boundary current regions particularly must be resolved (no parameterizations exist for unresolved boundary currents) so as to accurately compute transport properties for quantities such as heat or carbon that depend upon the rendering of the second moments, $\langle C\mathbf{v}\rangle$, where C is any scalar property, and $\mathbf{v}$ is the velocity. Thus a major effort has been devoted to producing global or near-global state estimates from higher resolution models (Menemenlis et al., 2005a,b). The same methodologies used at coarser resolution are also appropriate at high resolution—as has been demonstrated in the regional estimates taken up next, but the computational load rapidly escalates with the state and control vector dimensions. Thus available globally constrained models have used reduced data sets, and have been calculated only over comparatively short time intervals (see Table 21.3).

Because of the short-duration, much of the interest in these high-resolution models lies with the behavior of the eddy field rather than in the large-scale circulation (e.g., Wortham, 2012). As with ordinary forward modeling, how best to adjust the eddy flux parameterizations when parts of the eddy field have been resolved, is a major unknown.

5.5. Regional Solutions

Because the computational load of high-resolution global models is so great, efforts have been made to produce regional estimates, typically embedded in a coarser resolution global system. Embedding, with appropriate open boundary conditions, is essential because so much of the ocean state in any finite region is directly dependent upon the boundary values. Implementing open boundary conditions is technically challenging, particularly where the velocity field is directly involved—with slight barotropic imbalances producing large volume imbalances (Ayoub, 2006).

Gebbie et al. (2006) discussed estimates in a small region of the North Atlantic, and their results were used to calculate (Gebbie, 2007) the eddy contribution to near-surface subduction processes. In a much-larger region, the Mazloff et al. (2010) SOSE, was computed initially over the restricted time interval 2005–2006 (now being extended) at 1/6° horizontal and 42 vertical-level resolution.

6. THE UNCERTAINTY PROBLEM

From the earliest days of least-squares as used by Gauss and Lagrange, it was recognized that an important advantage of the methods is their ability to produce uncertainty estimates for the solutions, generally as covariances about the expected solution or the underlying true solution. The art of calculating those errors in historically large systems (especially in geodesy and orbit estimation—the fields where the method originated) is highly refined. Unhappily, large as those systems are, their dimensions pale in comparison to the state and control vector sizes encountered in the oceanographic problem. This dimensionality issue renders impractical any of the conventional means that are useful at small and medium sizes. Numerous methods have been proposed, including direct calculation of the coefficients of the normal equations (the matrix $\mathbf{A}$, defining any system of simultaneous equations) and inversion or pseudo inversion, of $\mathbf{A}^T\mathbf{A}$ (the Hessian); the indirect calculation of the lowest eigenvalues and eigenvectors of the inverse Hessian from algorithmic differentiation (AD) tools; to solutions for the probability density through the Fokker–Planck equation; to the generation of ensembles of solutions. Mostly they have been applied to "toy" problems—somewhat similar to designing a bridge to span the Strait of Gibraltar, and then pointing at a local highway bridge as a demonstration of its practicality. Serious efforts, more generally, to calculate the uncertainties of any large model solution are continuing, but when a useful outcome will emerge is unknown at this time.

In the interim, we generally have only so-called standard errors, representing the temporal variances about the mean of the estimate (Figures 21.4 and 21.5). These are useful and helpful. Sensitivities, derived from the adjoint solutions (e.g., Heimbach et al., 2011; and see Figures 21.11 and 21.16), are computationally feasible and need to be

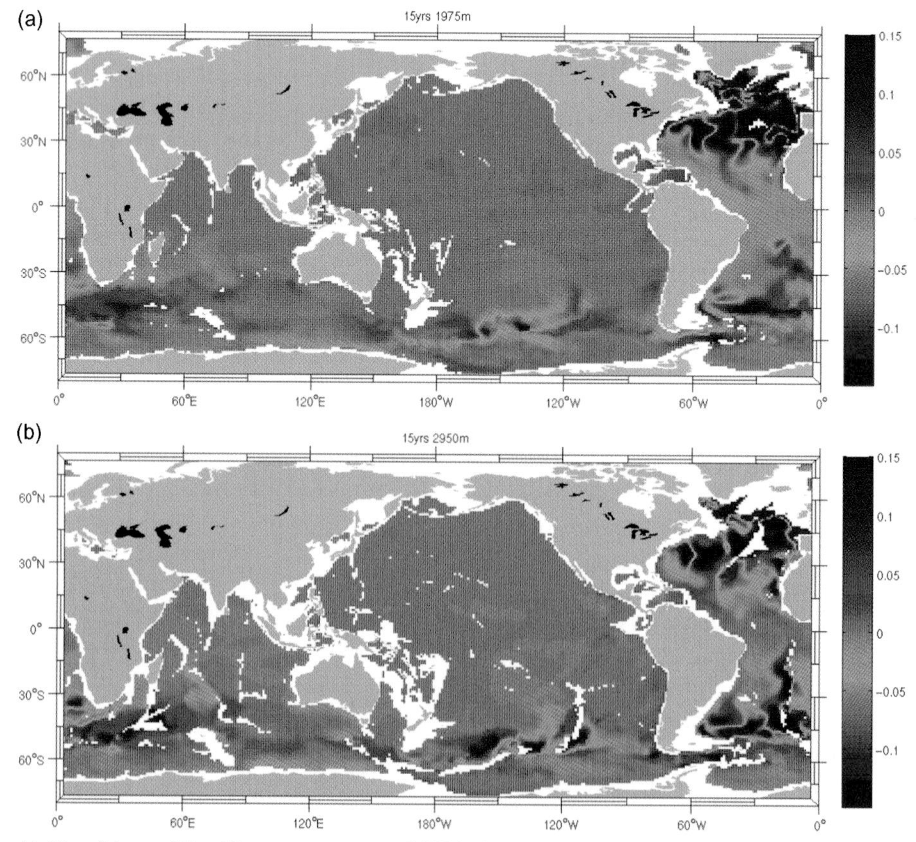

FIGURE 21.16 Sensitivities of the meridional heat transport across 26°N in the North Atlantic from temperature perturbations at two depths, *15 years earlier*. Top panel is for 1875 m, and lower panel is for 2960 m. *From Heimbach et al. (2011).*

more widely used. In the meantime, the quest of ocean and climate modelers and for the state estimation community more specifically, for useful understanding of reliability, remains a central, essential, goal. One should note that conventional ocean GCMs or coupled climate models, run without state estimation, are almost never accompanied by uncertainty estimates—a serious lack—particularly in an era in which "prediction skill" is being claimed.

Some authors compare their solutions to those inferred from more conventional methods, for example, transport calculations from box inversions of hydrographic sections. These comparisons are worthwhile but it is a major error to treat the hydrographic solutions as if they were true time-averages or climatologies. It is now possible to compare a state estimate from data obtained over a short interval (e.g., March 2003) with a state estimate for that time, sampled in the same way. Differences will appear in the objective function, J. Inevitable discrepancies raise all the fundamental questions of allocating errors amongst the data, the model, and external controls. In the decadal+ prediction problem (not discussed here), by definition there are no data, and measures of error and skill are far more difficult to obtain. Divergence of IPCC (2007) models over time (e.g., Schmittner et al., 2005; Stroeve et al., 2007), even where fitted to the historical observations, is a strong indicator of the fundamental difficulties involved in extrapolating even systems that appear to give an apparently good fit to historical data, and they are reminiscent of the parable above of fitting cubic polynomials to data.

7. DISCUSSION

The history of fluid dynamics generally, and of complex model use in many fields, supports the inference that models unconstrained by data can and do often go wildly wrong (in the wider sense, see, e.g., May, 2004; Post and Votta, 2005). Readers will recognize the strong point of view taken by the present authors: that models unaccompanied by detailed, direct, comparisons with and constraints by data are best regarded as a kind of science novel with a mixture of truth and fiction.

As we go forward collectively, the need to develop methods describing GCM and state estimate uncertainties is compelling: how else can one combine the quantitative understanding of oceanographic, meteorological and cryospheric physics with the diverse sets of system observations? Such syntheses are the overarching goal of any truly scientific field. Existing state estimates have many known limitations, some of which will be overcome by waiting for the outcome of Moore's Law over the coming years. Other problems, including the perennial and difficult problem of oceanic mixing and dissipation (Munk and Wunsch, 1998; Wunsch and Ferrari, 2004) are unlikely to simply vanish with any forseeable improvement in computer power. Further insight is required.

Lack of long-duration, large-scale, observations generates a fundamental knowledge gap. Without the establishment and maintenance of a comprehensive global ocean observing system, which satisfies the stringent requirements for climate research and monitoring, progress over the coming decades will remain limited (Baker et al., 2007; Wunsch et al., 2013).

Oceanographers now also directly confront the limits of knowledge of atmospheric processes. Until about 20 years ago, meteorological understanding so greatly exceeded that of the ocean circulation that estimated state errors for the atmosphere were of little concern. The situation has changed emphatically with the global observations starting in WOCE, along with the development of oceanic state estimates.[8] These estimation systems are better suited for the purposes of climate research than those developed for numerical weather prediction.

For climate change purposes, what is needed are useful state estimates for the coupled Earth system such that property evolution within and exchanges across its components are fully accounted for by closed cycles including heat, freshwater, energy and momentum. The coupled system must ultimately include oceanic, atmospheric, terrestrial, and cryospheric physics, as well as associated property transports (e.g., representing the carbon cycle), and the entirety of the properly understood relevant observations in those fields. Thus, atmospheric precipitation and evaporation pattern changes can be constrained tightly by changes in the oceanic state. ECCO and related programs have demonstrated how to carry out such recipes. Conventional weather forecast methods are not appropriate, and implementation of a fully coupled state estimation system that will be ongoing is a challenge to governments, universities, and research organizations alike. Bengtsson et al. (2007) proposed a limited step in this direction. Sugiura et al. (2008) and Mochizuki et al. (2009) have made some tentative starts. Surely we must have the capability.

ACKNOWLEDGMENTS

Support was provided by the US National Ocean Partnership Program with contributions from the National Aeronautics and Space Administration, the National Science Foundation, and the National Oceanographic and Atmospheric Administration. The collaboration of our many ECCO partners is gratefully acknowledged. G. Forget was particularly helpful with the calculations involving version 4. CW also thanks AOPP and Balliol College, Oxford, for support and hospitality through the George Eastman Visiting Professorship. Detailed comments by R. Ponte and the anonymous reviewers were very helpful.

8. The authors have been asked repeatedly at meetings "Why don't oceanographers adopt the sophisticated methods used by meteorologists?" The shoe, however, is now firmly on the other foot.

REFERENCES

Ablain, M., Cazenave, A., Valladeau, G., Guinehut, S., 2009. A new assessment of the error budget of global mean sea level rate estimated by satellite altimetry over 1993-2008. Ocean Sci. 5, 193–201.

Adcroft, A., Hill, C., Campin, J.-M., Marshall, J., Heimbach, P., 2004. Overview of the formulation and numerics of the MIT GCM. In: Proceedings of the ECMWF Seminar on Recent Developments in Numerical Methods for Atmospheric and Ocean Modelling, 6–10 September 2004, Shinfield Park, Reading, UK, pp. 139–150.

Alley, R.B., Clark, P.U., Huybrechts, P., Joughin, I., 2005. Ice-sheet and sea-level changes. Science 310, 456–460.

Ayoub, N., 2006. Estimation of boundary values in a North Atlantic circulation model using an adjoint method. Ocean Model. 12 (3–4), 319–347.

Baehr, J., 2010. Influence of the RAPID-MOCHA array and Florida current cable observations on the ECCO-GODAE state estimate. J. Phys. Oceanogr. 40, 865–879.

Baker, D.J., Schmitt, R.W., Wunsch, C., 2007. Endowments and new institutions for long term observations. Oceanography 20 (4), 10–14.

Barker, P.M., Dunn, J.R., Domingues, C.M., Wijffels, S.E., 2011. Pressure sensor drifts in Argo and their impacts. J. Atmos. Oceanic Technol. 28, 1036–1049.

Basili, V., Caldiera, G., McGarry, F., Pajerski, R., Page, G., Waligora, S., 1992. The software engineering laboratory: an operational software experience factory. In: Proceedings of the 14th International Conference on Software Engineering, pp. 370–381.

Bengtsson, L., Hagemann, S., Hodges, K.I., 2004. Can climate trends be calculated from reanalysis data? J. Geophys. Res. 109, D11111.

Bengtsson, L., et al., 2007. The need for a dynamical climate reanalysis. Bull. Am. Meteorol. Soc. 88, 495–501.

Björck, A., 1996. Numerical Methods for Least Squares Problems. Society for Industrial Mathematics, Philadelphia, 408 pp.

Bromwich, D.H., Fogt, R.L., 2004. Strong trends in the skill of the ERA-40 and NCEP-NCAR reanalyses in the high and midlatitudes of the southern hemisphere, 1958-2001. J. Clim. 17, 4603–4619.

Bromwich, D.H., Guo, Z.C., Bai, L.S., Chen, Q.S., 2004. Modeled Antarctic precipitation. Part I: spatial and temporal variability. J. Clim. 17, 427–447.

Bromwich, D.H., Fogt, R.L., Hodges, K.I., Walsh, J.E., 2007. A tropospheric assessment of the ERA-40, NCEP, and JRA-25 global reanalyses in the polar regions. J. Geophys. Res. 112, D10.

Bromwich, D.H., Nicolas, J.P., Monaghan, A.J., 2011. An assessment of precipitation changes over Antarctica and the southern ocean since 1989 in contemporary global reanalyses. J. Clim. 24, 4189–4209.

Campin, J.-M., Adcroft, A., Hill, C., Marshall, J., 2004. Conservation of properties in a free surface model. Ocean Model. 6, 221–244.

Carton, J.A., Santorelli, A., 2008. Global decadal upper-ocean heat content as viewed in nine analyses. J. Clim. 21, 6015–6035.

Carton, J.A., Chepurin, G., Cao, X.H., Giese, B., 2000. A simple ocean data assimilation analysis of the global upper ocean 1950-95. Part I: methodology. J. Phys. Oceanogr. 30, 294–309.

Cazenave, A., Remy, F., 2011. Sea level and climate: measurements and causes of changes. Wiley Interdiscip. Rev. Clim. Change 2, 647–662.

Church, J.A., White, N.J., Konikow, L.F., Domingues, C.M., Cogley, J.G., Rignot, E., Gregory, J.M., van den Broeke, M.R., Monaghan, A.J., Velicogna, I., 2011. Revisiting the Earth's sea-level and energy budgets from 1961 to 2008. Geophys. Res. Lett. 38, L18601.

Cullather, R.I., Bosilovich, M., 2012. The energy budget of the polar atmosphere in MERRA. J. Clim. 25, 5–24.

Dansereau, V., Heimbach, P., Losch, M., 2013. Simulation of sub-ice shelf melt rates in a general circulation model: velocity-dependent transfer and the role of friction. J. Geophys. Res.

Dutkiewicz, S., Follows, M., Heimbach, P., Marshall, J., 2006. Controls on ocean productivity and air-sea carbon flux: an adjoint model sensitivity study. Geophys. Res. Lett. 33, L02603. http://dx.doi.org/10.1029/2005GL024987.

Elliott, W.P., Gaffen, D.J., 1991. On the utility of radiosonde humidity archives for climate studies. Bull. Am. Meteorol. Soc. 72, 1507–1520.

Evensen, G., 2009. Data Assimilation: The Ensemble Kalman Filter. Springer Verlag, Berlin.

Fenty, I.G., Heimbach, P., 2013a. Coupled Sea Ice-Ocean State estimation in the Labrador Sea and Baffin Bay. J. Phys. Oceanogr. 43 (6), 884–904. http://dx.doi.org/10.1175/JPO-D-12-065.1.

Fenty, I.G., Heimbach, P., 2013b. Hydrographic preconditioning for seasonal sea ice anomalies in the Labrador Sea. J. Phys. Oceanogr. 43 (6), 863–883. http://dx.doi.org/10.1175/JPO-D-12-064.1.

Ferreira, D., Marshall, J., Heimbach, P., 2005. Estimating eddy stresses by fitting dynamics to observations using a residual-mean ocean circulation model and its adjoint. J. Phys. Oceanogr. 35, 1891–1910.

Follows, M.J., Dutkiewicz, S., 2011. Modeling diverse communities of marine microbes. Ann. Rev. Mar. Sci. 3, 427–451.

Follows, M.J., Dutkiewicz, S., Grant, S., Chisholm, S.W., 2007. Emergent biogeography of microbial communities in a model ocean. Science 315, 1843–1846.

Forget, G., 2010. Mapping ocean observations in a dynamical framework: a 2004-06 ocean atlas. J. Phys. Oceanogr. 40, 1201–1221.

Forget, G., Wunsch, C., 2007. Estimated global hydrographic variability. J. Phys. Oceanogr. 37, 1997–2008.

Forget, G., Heimbach, P., Ponte, R., Wunsch, C., Campin, J.M., Hill, C., 2013. A new-generation global ocean state estimate, ECCO version 4: System formulation and basic characteristics. Unpublished Report.

Fukumori, I., Raghunath, R., Fu, L., Chao, Y., 1999. Assimilation of TOPEX/POSEIDON data into a global ocean circulation model: how good are the results? J. Geophys. Res. 104, 25647–25665.

Fukumori, I., 2002. A partitioned Kalman filter and smoother. Mon. Weather Rev. 130, 1370–1383.

Ganachaud, A., 2003a. Error budget of inverse box models: the North Atlantic. J. Atmos. Oceanic Technol. 20, 1641–1655.

Ganachaud, A., 2003b. Large-scale mass transports, water mass formation, and diffusivities estimated from World Ocean Circulation Experiment (WOCE) hydrographic data. J. Geophys. Res. 108, 3213.

Ganachaud, A., Wunsch, C., 2002. Large-scale ocean heat and freshwater transports during the World Ocean Circulation Experiment. J. Clim. 16, 696–705.

Gebbie, G., 2007. Does eddy subduction matter in the Northeast Atlantic Ocean? J. Geophys. Res. 112, C06007.

Gebbie, G., Heimbach, P., Wunsch, C., 2006. Strategies for nested and eddy-permitting state estimation. J. Geophys. Res. Oceans 111, C10073.

Gelb, A. (Ed.), 1974. Applied Optimal Estimation. MIT Press, Cambridge, MA, 382 pp.

Giering, R., Kaminski, T., 1998. Recipes for adjoint code construction. ACM Trans. Math. Softw. 24, 437–474.

Griewank, A., Walther, A., 2008. Evaluating Derivatives. Principles and Techniques of Algorithmic Differentiation. SIAM, Philadelphia, 442 pp.

Griffies, S.M., Greatbatch, R.J., 2012. Physical processes that impact the evolution of global mean sea level in ocean climate models. Ocean Model. 51(C), 37–72. http://dx.doi.org/10.1016/j.ocemod.2012.04.003.

Hecht, M.W., Smith, R.D., 2008. Towards a physical understanding of the North Atlantic: a review of model studies. In: Hecht, M.W., Hasumi, H. (Eds.), Ocean Modeling in an Eddying Regime. AGU Geophysical Monograph, vol. 177. American Geophysical Union, Washington, DC, pp. 213–240.

Heimbach, P., Losch, M., 2012. Adjoint sensitivities of sub-ice shelf melt rates to ocean circulation under Pine Island Ice Shelf, West Antarctica. Ann. Glaciol. 53 (60), 59–69. http://dx.doi.org/10.3189/2012/AoG60A025.

Heimbach, P., Hill, C., Giering, R., 2005. An efficient exact adjoint of the parallel MIT General Circulation Model, generated via automatic differentiation. Future Gener. Comput. Syst. 21, 1356–1371.

Heimbach, P., Forget, G., Ponte, R., Wunsch, C., 2009. Observational requirements for global-scale ocean climate analysis: lessons from ocean state estimation. Community White Paper. In: Hall, J., Harrison, D.E., Stammer, D. (Eds.), 2010: Proceedings of OceanObs'09: Sustained Ocean Observations and Information for Society, Venice, Italy, 21-25 September 2009, vol. 2. ESA Publication WPP-306. ESA, Frascati, Italy. http://dx.doi:10.5270/OceanObs09.cwp.42.

Heimbach, P., Menemenlis, D., Losch, M., Campin, J.M., Hill, C., 2010. On the formulation of sea-ice models. Part 2: lessons from multi-year adjoint sea ice export sensitivities through the Canadian Arctic Archipelago. Ocean Model. 33 (1–2), 145–158.

Heimbach, P., Wunsch, C., Ponte, R.M., Forget, G., Hill, C., Utke, J., 2011. Timescales and regions of the sensitivity of Atlantic meridional volume and heat transport magnitudes: toward observing system design. Deep Sea Res. Part II 58, 1858–1879.

Hoteit, I., Cornuelle, B., Köhl, A., Stammer, D., 2006. Treating strong adjoint sensitivities in tropical eddy-permitting variational data assimilation. Q. J. R. Meteorol. Soc. 131 (613), 3659–3682.

Hoteit, I., Cornuelle, B., Heimbach, P., 2010. An eddy-permitting, dynamically consistent adjoint-based assimilation system for the Tropical Pacific: hindcast experiments in 2000. J. Geophys. Res. 115, C03001. http://dx.doi.org/10.1029/2009JC005437.

Huang, R.X., 1993. Real freshwater flux as a natural boundary condition for the salinity balance and thermohaline circulation forced by evaporation and precipitation. J. Phys. Oceanogr. 23, 2428–2446.

Huang, R.X., 2010. Ocean Circulation: Wind-Driven and Thermohaline Processes. vol. xiii. Cambridge University Press, Cambridge, 791 pp.

Hurlburt, H.E., et al., 2009. High-resolution global and basin-scale ocean analyses and forecasts. Oceanography 22, 110–127.

IPCC Intergovernmental Panel on Climate Change, 2007. Climate Change 2007—The Physical Science Basis. Cambridge University Press, Cambridge, 1009 pp.

Ito, T., Woloszyn, M., Mazloff, M., 2010. Anthropogenic carbon dioxide transport in the Southern Ocean driven by Ekman flow. Nature 463, 80–83.

Josey, S.A., Gulev, S., Yu, L., 2013. Exchanges through the ocean surface. Chapter 5, this volume.

Kalman, R.E., 1960. A new approach to linear filtering and prediction problems. J. Basic Eng. 82, 35–45.

Kalnay, E., 2003. Atmospheric Modeling, Data Assimilation, and Predictability. xxii. Cambridge University Press, Cambridge, pp. 341–344.

Kanzow, T., et al., 2009. Basinwide integrated volume transports in an eddy-filled ocean. J. Phys. Oceanogr. 39, 3091–3110.

Kauker, F., et al., 2009. Adjoint analysis of the 2007 all time Arctic sea-ice minimum. Geophys. Res. Lett. 36, L03707.

King, M.A., Bingham, R.J., Moore, P., Whitehouse, P.L., Bentley, M.J., Milne, G.A., 2013. Lower satellite-gravimetry estimates of Antarctic sea-level contribution. Nature 491 (7425), 586–589. http://dx.doi.org/10.1038/nature11621.

Köhl, A., Cornuelle, B., Stammer, D., 2007. Interannual to decadal changes in the ECCO global synthesis. J. Phys. Oceanogr. 37, 313–337. http://dx.doi.org/10.1175/JPO3014.1.

Köhl, A., Stammer, D., 2008. Decadal sea level changes in the 50-year GECCO ocean synthesis. J. Clim. 21, 1876–1890.

Köhl, A., Willebrand, J., 2002. An adjoint method for the assimilation of statistical characteristics into eddy-resolving models. Tellus 54A, 406–425.

Kopp, R.E., Mitrovica, J.X., Griffies, S.M., Yin, J., Hay, C.C., Stouffer, R.J., 2010. The impact of Greenland melt on local sea levels: a partially coupled analysis of dynamic and static equilibrium effects in idealized water-hosing experiments. Climatic Change 103 (3–4), 619–625. http://dx.doi.org/10.1007/s10584-010-9935-1.

Kouketsu, S., et al., 2011. Deep ocean heat content changes estimated from observation and reanalysis product and their influence on sea level change. J. Geophys. Res. 116, C03012. http://dx.doi.org/10.1029/2010JC006464.

Kuhlmann, J., Dobslaw, H., Thomas, M., 2011. Improved modeling of sea level patterns by incorporating self-attraction and loading. J. Geophys. Res. 116, C11036.

Lea, D.J., Haine, T.W.N., Allen, M.R., Hansen, J.A., 2002. Sensitivity analysis of the climate of a chaotic ocean circulation model. Q. J. R. Meteorol. Soc. 128, 2587–2605.

Lee, T., et al., 2010. Consistency and fidelity of Indonesian-throughflow total volume transport estimated by 14 ocean data assimilation products. Dyn. Atmos. Oceans 50, 201–223.

Lévy, M., Klein, P., Treguier, A.M., Iovino, D., Madec, G., Masson, S., Takahashi, K., 2010. Modifications of gyre circulation by submesoscale physics. Ocean Model. 34, 1–15.

Liu, C.Y., Köhl, A., Stammer, D., 2012. Adjoint-based estimation of eddy-induced tracer mixing parameters in the global ocean. J. Phys. Oceanogr. 42, 1186–1206.

Losch, M., 2008. Modeling ice shelf cavities in a z-coordinate ocean general circulation model. J. Geophys. Res. 113, C08043.

Losch, M., Heimbach, P., 2007. Adjoint sensitivity of an ocean general circulation model to bottom topography. J. Phys. Oceanogr. 37 (2), 377–393.

Losch, M., Wunsch, C., 2003. Bottom topography as a control variable in an ocean model. J. Atmos. Oceanic Technol. 20, 1685–1696.

Losch, M., Menemenlis, D., Campin, J.M., Heimbach, P., Hill, C., 2010. On the formulation of sea-ice models. Part 1: effects of different solver implementations and parameterizations. Ocean Model. 33 (1–2), 129–144.

Lumpkin, R., Speer, K., 2007. Global ocean meridional overturning. J. Phys. Oceanogr. 37, 2550–2562.

Macdonald, A.M., 1998. The global ocean circulation: a hydrographic estimate and regional analysis. Prog. Oceanogr. 41, 281–382.

Macdonald, A.M., et al., 2009. The WOCE-era 3-D Pacific Ocean circulation and heat budget. Prog. Oceanogr. 82, 281–325.

Marshall, J., Adcroft, A., Hill, C., Perelman, L., Heisey, C., 1997. A finite-volume, incompressible Navier Stokes model for studies of the ocean on parallel computers. J. Geophys. Res. Oceans 102, 5753–5766.

Marshall, G.J., Lagun, V., Lachlan-Cope, T.A., 2002. Changes in Antarctic Peninsula tropospheric temperatures from 1956 to 1999: a synthesis of observations and reanalysis data. Int. J. Climatol. 22, 291–310.

Martin, A.J., Hines, A., Bell, M.J., 2007. Data assimilation in the FOAM operational short-range ocean forecasting system: a description of the scheme and its impact. Q. J. R. Meteorol. Soc. 133, 981–995.

May, R., 2004. Uses and abuses of mathematics in biology. Science 303 (5659), 790–793.

Maze, G., Forget, G., Buckley, M., Marshall, J., Cerovecki, I., 2009. Using transformation and formation maps to study the role of air-sea heat fluxes in North Atlantic eighteen degree water formation. J. Phys. Oceanogr. 39, 1818–1835.

Mazloff, M.R., Heimbach, P., Wunsch, C., 2010. An eddy-permitting southern ocean state estimate. J. Phys. Oceanogr. 40, 880–899.

Menemenlis, D., Fukumori, I., Lee, T., 2005a. Using Green's functions to calibrate an ocean general circulation model. Mon. Weather Rev. 133, 1224–1240.

Menemenlis, D., et al., 2005b. NASA supercomputer improves prospects for ocean climate research. Eos 86 (9), 89.

Milne, G.A., Gehrels, W.R., Hughes, C.W., Tamisiea, M.E., 2009. Identifying the causes of sea-level change. Nat. Geosci. 2 (7), 471–478. http://dx.doi.org/10.1038/ngeo544.

Mitrovica, J.X., Gomez, N., Morrow, E., Hay, C., Latychev, K., Tamisiea, M.E., 2011. On the robustness of predictions of sea level fingerprints. Geophys. J. Int. 187 (2), 729–742. http://dx.doi.org/10.1111/j.1365-246X.2011.05090.x.

Mochizuki, T., Sugiura, N., Awaji, T., Toyoda, T., 2009. Seasonal climate modeling over the Indian Ocean by employing a 4D-VAR coupled data assimilation approach. J. Geophys. Res. 114, C11003.

Munk, W., 2002. Twentieth century sea level: an enigma. Proc. Natl. Acad. Sci. U.S.A. 99 (10), 6550–6555. http://dx.doi.org/10.1073/pnas.092704599.

Munk, W., 2011. The sound of climate change. Tellus 63A, 190–197.

Munk, W., Wunsch, C., 1998. Abyssal recipes II: energetics of tidal and wind mixing. Deep Sea Res. Part I 45, 1977–2010.

Nguyen, A.T., Kwok, R., Menemenlis, D., 2011. Arctic ice-ocean simulation with optimized model parameters: approach and assessment. J. Geophys. Res. 116, C04025. http://dx.doi.org/10.1029/2010JC006573.

Nguyen, A.T., Kwok, R., Menemenlis, D., 2012. Source and pathway of the Western arctic upper halocline in a data-constrained coupled ocean and sea ice model. J. Phys. Oceanogr. 42 (5), 802–823. http://dx.doi.org/10.1175/JPO-D-11-040.1.

Nicolas, J.P., Bromwich, D.H., 2011. Precipitation changes in high Southern latitudes from global reanalyses: a cautionary tale. Surv. Geophys. 32 (4–5), 475–494.

Payne, A.J., Vieli, A., Shepherd, A.P., Wingham, D.J., Rignot, E., 2004. Recent dramatic thinning of largest West Antarctic ice stream triggered by oceans. Geophys. Res. Lett. 31, L23401.

Pedlosky, J., 1987. Geophysical Fluid Dynamics. second ed. vol. xiv. Springer-Verlag, New York, 710 pp.

Piecuch, C.G., Ponte, R.M., 2011. Mechanisms of interannual steric sea level variability. Geophys. Res. Lett. 38, L15605.

Piecuch, C.G., Ponte, R.M., 2012. Importance of circulation changes to Atlantic heat storage rates on seasonal and interannual time scales. J. Clim. 25, 350–362.

Ponte, R.M., 2012. An assessment of deep steric height variability over the global ocean. Geophys. Res. Lett. 39, L04601. http://dx.doi.org/10.1029/2011GL050681.

Ponte, R.M., Wunsch, C., Stammer, D., 2007. Spatial mapping of time-variable errors in TOPEX/POSEIDON and Jason-1 sea surface height measurements. J. Atmos. Oceanic Technol. 24, 1078–1085.

Post, D.E., Votta, L.G., 2005. Computational science demands a new paradigm. Phys. Today 58, 35–41.

Pritchard, H.D., Arthen, R.J., Vaughan, D.G., Edwards, L.A., 2009. Extensive dynamic thinning on the margins of the Greenland and Antarctic ice sheets. Nature 461, 971–975.

Purkey, S.G., Johnson, G.C., 2010. Warming of Global Abyssal and Deep Southern Ocean Waters between the 1990s and 2000s: Contributions to Global Heat and Sea Level Rise Budgets. J. Clim. 23 (23), 6336–6351. http://dx.doi.org/10.1175/2010JCLI3682.1.

Rayner, D., et al., 2011. Monitoring the Atlantic meridional overturning circulation. Deep Sea Res. Part II 58, 1744–1753.

Rignot, E., Velicogna, I., van den Broeke, M.R., Monaghan, A., Lenaerts, J., 2011. Acceleration of the contribution of the Greenland and Antarctic ice sheets to sea level rise. Geophys. Res. Lett. 38, L05503.

Ringler, T., Gent, P., 2011. An eddy closure for potential vorticity. Ocean Model. 39, 125–134.

Roquet, F., Wunsch, C., Madec, G., 2011. On the patterns of wind-power input to the ocean circulation. J. Phys. Oceanogr. 41, 2328–2342.

Rosati, A., Gudgel, R., Miyakoda, K., 1995. Decadal analysis produced from an ocean data assimilation system. Mon. Weather Rev. 123, 2206–2228.

Schiller, A., Lee, T., Masuda, S., 2013. Methods and applications of ocean synthesis in climate research. Chapter 22, this volume.

Schlitzer, R., 2007. Assimilation of radiocarbon and chlorofluorocarbon data to constrain deep and bottom water transports in the world ocean. J. Phys. Oceanogr. 37, 259–276.

Schmittner, A., Latif, M., Schneider, B., 2005. Model projections of the North Atlantic thermohaline circulation for the 21st century assessed by observations. Geophys. Res. Lett. 32, L23710. http://dx.doi.org/10.1029/2005GL024368.

Schodlok, M.P., Menemenlis, D., Rignot, E., Studinger, M., 2012. Sensitivity of the ice shelf ocean system to the sub-ice shelf cavity shape measured by NASA Ice Bridge in Pine Island Glacier West Antarctica. Ann. Glaciol. 53 (60), 156–162.

Shepherd, A., Wingham, D., 2007. Recent sea-level contributions of the Antarctic and Greenland ice sheets. Science 315, 1529–1532.

Siedler, G., Church, J., Gould, W.J. (Eds.), 2001. Ocean Circulation and Climate: Observing and Modeling the Global Ocean. Academic, San Diego, 715 pp.

Slangen, A.B.A., Katsman, C.A., Wal, R.S.W., Vermeersen, L.L.A., Riva, R.E.M., 2012. Towards regional projections of twenty-first century sea-level change based on IPCC SRES scenarios. Clim. Dyn. 38 (5–6), 1191–1209. http://dx.doi.org/10.1007/s00382-011-1057-6.

Stammer, D., 2005. Adjusting internal model errors through ocean state estimation. J. Phys. Oceanogr. 35, 1143–1153.

Stammer, D., et al., 2001. Transport processes of the global ocean circulation between 1992 and 1997 Estimated from global altimeter data, sst fields, daily NCEP surface fluxes, the levitus climatology and a general circulation model. In: Fifth Symposium on Integrated Observing Systems, pp. 41–44.

Stammer, D., et al., 2002. Global ocean circulation during 1992-1997, estimated from ocean observations and a general circulation model. J. Geophys. Res. Oceans 107 (C9), 3118. http://dx.doi.org/10.1029/2001JC000888.

Stammer, D., et al., 2003. Volume, heat, and freshwater transports of the global ocean circulation 1993-2000, estimated from a general circulation model constrained by World Ocean Circulation Experiment (WOCE) data. J. Geophys. Res. Oceans 108 (C1), 3007. http://dx.doi.org/10.1029/2001JC001115.

Stammer, D., Ueyoshi, K., Köhl, A., Large, W.B., Josey, S., Wunsch, C., 2004. Estimating air-sea fluxes of heat, freshwater and momentum through global ocean data assimilation. J. Geophys. Res. 109, C05023. http://dx.doi.org/10.1029/2003JC002082.

Stammer, D., Köhl, A., Wunsch, C., 2007. Impact of accurate geoid fields on estimates of the ocean circulation. J. Atmos. Oceanic Technol. 24, 1464–1478.

Straneo, F., Heimbach, P., Sergienko, O., et al., 2013. Challenges to understanding the dynamic response of Greenlands marine terminating glaciers to oceanic and atmospheric forcing. Bull. Am. Meteorol. Soc. 94, 1131–1144. http://dx.doi.org/10.1175/BAMS-D-12-00100.

Stroeve, J., Holland, M.M., Meier, W., Scambos, T., Serreze, M., 2007. Arctic sea ice decline: faster than forecast. Geophys. Res. Lett. 34, L09501. http://dx.doi.org/10.1029/2007GL029703.

Sugiura, N., Awaji, T., Masuda, S., Mochizuki, T., Toyoda, T., Miyama, T., Igarashi, H., Ishikawa, Y., 2008. Development of a four-dimensional variational coupled data assimilation system for enhanced analysis and prediction of seasonal to interannual climate variations. J. Geophys. Res. 113 (C10), C10017.

Talagrand, O., 1997. Assimilation of observations, an introduction. J. Meteorol. Soc. Jpn. 75, 191–209.

Talley, L.D., Stammer, D., Fukumori, I., 2001. Towards a WOCE synthesis. In: Siedler, G., Church, J., Gould, W.J. (Eds.), Ocean Circulation and Climate: Observing and Modelling the Global Ocean. Int. Geophys. Ser., 77. Academic, San Diego, pp. 525–546.

Thompson, D.W.J., Kennedy, J.J., Wallace, J.M., Jones, P.D., 2008. A large discontinuity in the mid-twentieth century in observed global-mean surface temperature. Nature 453, 646–649.

Thorne, P.W., 2008. Arctic tropospheric warming amplification? Nature 455, E1–E2. http://dx.doi.org/10.1038/nature07256.

Trenberth, K.E., Solomon, A., 1994. The global heat-balance—heat transports in the atmosphere and ocean. Clim. Dyn. 10, 107–134.

Trenberth, K.E., Hurrell, J.W., Solomon, A., 1995. Conservation of mass in 3-dimensions in global analyses. J. Clim. 8, 692–708.

Trenberth, K.E., Stepaniak, D.P., Hurrell, J.W., Fiorino, M., 2001. Quality of reanalyses in the tropics. J. Clim. 14, 1499–1510.

Utke, J., Naumann, U., Fagan, M., Thallent, N., Strout, M., Heimbach, P., Hill, C., Wunsch, C., 2008. OpenAD/F: a modular, open-source tool for automatic differentiation of Fortran codes. ACM Trans. Math. Softw. 34 (4), 18. http://dx.doi.org/10.1145/1377596.1377598.

Vallis, G.K., 2006. In: Atmospheric and Oceanic Fluid Dynamics: Fundamentals and Large-Scale Circulation. vol. xxv. Cambridge University Press, Cambridge, UK, 745 pp.

Vinogradov, S.V., Ponte, R.M., Heimbach, P., Wunsch, C., 2008. The mean seasonal cycle in sea level estimated from a data-constrained general circulation model. J. Geophys. Res. Oceans 113, C03032.

Vinogradova, N.T., Ponte, R.M., Tamisiea, M.E., Quinn, K.J., Hill, E.M., Davis, J.L., 2011. Self-attraction and loading effects on ocean mass redistribution at monthly and longer time scales. J. Geophys. Res. 116, C08041. http://dx.doi.org/10.1029/2011JC007037.

Wang, W.Q., Kohl, A., Stammer, D., 2010. Estimates of global ocean volume transports during 1960 through 2001. Geophys. Res. Lett. 37, L15601.

Weaver, A., Courtier, P., 2001. Correlation modelling on the sphere using a generalized diffusion equation. Q. J. R. Meteorol. Soc. 127, 1815–1846.

Wijffels, S.E., Willis, J., Domingues, C.M., Barker, P., White, N.J., Gronell, A., Ridgway, K., Church, J.A., 2008. Changing expendable bathythermograph fall rates and their impact on estimates of thermosteric sea level rise. J. Clim. 21, 5657–5672.

Woloszyn, M., Mazloff, M., Ito, T., 2011. Testing an eddy-permitting model of the Southern Ocean carbon cycle against observations. Ocean Model. 39, 170–182.

Wortham IV, C.J.L., 2012. A multi-dimensional spectral description of ocean variability with applications. PhD Thesis, MIT and WHOI, 184 pp.

Wunsch, C., 2006. Discrete Inverse and State Estimation Problems: With Geophysical Fluid Applications. vol. xi. Cambridge University Press, Cambridge, 371 pp.

Wunsch, C., 2011. The decadal mean ocean circulation and Sverdrup balance. J. Mar. Res. 69, 417–434.

Wunsch, C., Ferrari, R., 2004. Vertical mixing, energy, and the general circulation of the oceans. Annu. Rev. Fluid Mech. 36 (1), 281–314. http://dx.doi.org/10.1146/annurev.fluid.36.050802.122121.

Wunsch, C., Heimbach, P., 2006. Estimated decadal changes in the North Atlantic meridional overturning circulation and heat flux 1993-2004. J. Phys. Oceanogr. 36, 2012–2024.

Wunsch, C., Heimbach, P., 2007. Practical global oceanic state estimation. Phys. D 230, 197–208.

Wunsch, C., Heimbach, P., 2009. The global zonally integrated ocean circulation, 1992-2006: seasonal and decadal variability. J. Phys. Oceanogr. 39, 351–368.

Wunsch, C., Heimbach, P., 2013. Two decades of the Atlantic Meridional Overturning Circulation: anatomy, variations, extremes, prediction, and overcoming its limitations. J. Clim. 26, 7167–7186. http://dx.doi.org/10.1175/JCLI-D-12- 00478.1.

Wunsch, C., Ponte, R.M., Heimbach, P., 2007. Decadal trends in sea level patterns: 1993-2004. J. Clim. 20, 5889–5911.

Wunsch, C., Schmitt, R.W., Baker, D.J., 2013. Climate change as an intergenerational problem. Proc. Natl. Acad. Sci. U.S.A. 110 (12), 4435–4436. http://dx.doi.org/10.1073/pnas.1302536110.

Zhang, S., Rosati, A., Delworth, T., 2010. The adequacy of observing systems in monitoring the Atlantic meridional overturning circulation and North Atlantic climate. J. Clim. 23 (19), 5311–5324. http://dx.doi.org/10.1175/2010JCLI3677.1.

Chapter 22

Methods and Applications of Ocean Synthesis in Climate Research

Andreas Schiller*, Tong Lee[†] and Shuhei Masuda[‡]

*Centre for Australian Weather and Climate Research, CSIRO Wealth from Oceans Flagship, Hobart, Australia

[†]Jet Propulsion Laboratory, California Institute of Technology, Pasadena, California, USA

[‡]Research Institute for Global Change, Japan Agency for Marine-Earth Science and Technology (JAMSTEC), Yokohama, Japan

Chapter Outline

1. INTRODUCTION

1.1. Definitions

Model–observation synthesis is a concept encompassing any method for combining observations of variables like temperature or velocity components into numerical models like the ones used to project climate or predict weather. There are many benefits from model–data syntheses, the details depend on the models and synthesis tools chosen but two of the most common benefits are: a reduction in uncertainties and biases associated with the use of numerical models and the ability of numerical models to dynamically interpolate information into observation-void regions.

Most applications are used to estimate the ocean state or model parameters or the forcing state through a unified process known in applied mathematics and engineering as control theory (see, e.g., Wunsch, 1988 and references therein). It is widely recognized that combining data with advanced numerical models yields the "best"—and in case of ocean state estimation dynamically consistent—estimates of ocean fields, of errors of these fields, and of certain model parameters such as mixing coefficients and sub-grid scale closure. Here, "best" is used as a technical term as in statistical estimation, corresponding to some optimum principle.

The fundamental approach common to all variants of ocean syntheses is the use of a model estimate and an observational estimate—the ocean state of the model, the

Ocean Circulation and Climate, Vol. 103. http://dx.doi.org/10.1016/B978-0-12-391851-2.00022-2

external forcing, a model parameter, or the observable—and a minimization of a cost function according to constraints and model and observational error statistics. The implementations may differ according to the way temporal variations are handled (e.g., continuous vs. intermittent), the length of the assimilation window, or different implementations of constraints according to models and observations. Both, ocean state estimation and data assimilation, have been used for analyses of climate variability in the ocean and both have been used for initialization of weather, ocean mesoscale and climate predictions.

Ocean state estimation and data assimilation can be interpreted as two related components of ocean synthesis. Ocean state estimation aims to improve our understanding and modeling of the global ocean circulation, the role of the oceans in the Earth's climate, and, ultimately, to increase skill in climate projections and predictions. Conversely, assimilation of observations into meteorological and ocean prediction systems focuses on forecasting timescales of days to weeks (but also contributes to our understanding of the climate system). Due to requirements for operational efficiency, these types of data assimilation are usually not dynamically fully consistent (as they obtain additional source/sink terms and because of the asymmetry of the observing system for real-time applications) but, if run in reanalysis mode, provide useful insights into some aspects of the ocean circulation on longer timescales. The term "reanalyses" is used here to denote "retrospective analyses" by freezing the model and data assimilation codes for the duration of a reanalysis such as the BLUElink reanalysis (e.g., Schiller et al., 2008). As our observing system capability improves over time it is often useful to repeat reanalyses with enhanced observations, thus providing improved model-based estimates of the ocean circulation. Especially high-resolution ocean reanalyses based on such systems provide information about the ocean mesoscale and its link to climate variability.

More specifically, least-squares Lagrange multipliers/adjoint-type or smoother-type methods are subsequently referred to as "state estimation" and sequential methods are referred to as "data assimilation." Nonsequential and sequential approaches are major categories, which differ according to whether observations are treated all at once or sequentially (e.g., Anderson and Moore, 1979). An essential commonality is that for all model–data synthesis problems, both nonsequential and sequential, uncertainties in the observations are as important as observations themselves and have a comparable role in determining the outcome. In this context, data assimilation in ocean prediction systems and associated reanalyses should be understood as approximate methods for obtaining a solution of a least-squares problem. Similarly, we demonstrate below that virtually all state estimation techniques also apply approximations to obtain a solution. For instance, the ocean circulation is strongly nonlinear but most state estimation methods apply tangent-linear models and as such are based on an approximate ocean. Therefore even an optimal state estimation approach, though self-consistent with a model, is still an approximation to the true ocean state. Thus, regarding dynamical consistency, an approximate state estimation technique is comparable to an approximation using sequential data assimilation and there can be no absolute distinction between them. However, for reasons of consistency with Chapter 21, we adopt their terminology here.

It is obvious from above comments that a comprehensive overview of all methods and applications in model–observation synthesis would be a large undertaking which is beyond the scope of this chapter, and we refer the reader to the extensive literature in this field. Instead, we are aiming to explore methods and applications of ocean syntheses in the dual approaches of ocean state estimation and ocean reanalyses, and to highlight scientific advances in these fields in the last decade.

1.2. Ocean Climate Models

The development of complex ocean climate models and their application to simulate climate phenomena either in uncoupled or coupled mode with, for example, the atmosphere or sea-ice (Griffies et al., 2010), has made significant progress since the 1960s (e.g., Bryan and Cox, 1967; Chapter 20). Although simplified versions of ocean models existed prior to the 1960s, it was not until the advent of large computers in the 1960s that scientists could simulate the large-scale global ocean circulation. By then, atmospheric modeling had already made major advances since its early days in the 1920s (Richardson, 1922). Because ocean and atmospheric circulations follow the same dynamical principles (Müller and Willebrand, 1989) the progress in atmospheric modeling benefited the development of ocean models.

The fundamental challenge for early ocean climate models was to resolve the large-scale geostrophic circulation (Rossby number < 0.1) to allow the evaluation of ocean responses to large-scale atmospheric climate forcing. There was also a need for "realistic" ocean simulations, which required the development of a new class of models entitled "primitive equation" models (Bryan, 1969). In these early days of ocean modeling (with very limited large-scale ocean observations), the use of observations was usually restricted to providing guidance on the initial conditions of the potential temperature and salinity fields in the ocean model and the use of some kind of mapped surface temperature and salinity fields onto the model grid to constrain the surface boundary of the model in the form of a Newtonian nudging (e.g., Haney, 1971). Model evaluation with observations, especially for global models, was hard to achieve due to the lack of coherent three-dimensional global-scale ocean observations.

Based on significant advances in supercomputing technologies, the 1990s saw the emergence of the first large-scale mesoscale-eddying models (e.g., Semtner and Chervin, 1992) and the first realistic ocean–atmosphere coupled climate change projections (e.g., IPCC First Assessment Report, 1990). This period coincided with the first multi-year satellite altimetry missions (see Section 1.3), thus for the first time offering the opportunity to constrain eddying models by global mesoscale observations (Hurlburt, 1984).

1.3. The Global Ocean Observing System

The ocean observing system (Chapter 3) consists of a multitude of *in situ* and satellite-based measurement platforms, communication components and data analysis centers. Observational datasets released by these centers have typically undergone some basic level of quality control (QC) and serve numerous applications such as monitoring of climate change and variability, near-real-time weather and ocean forecasting and, more recently, also contribute to the improved management of marine and coastal ecosystems and resources (e.g., http://www.ioc-goos.org/).

The *Tropical-Ocean-Global-Atmosphere* (TOGA) program was launched in 1985 to investigate the global atmospheric response to the coupled ocean–atmosphere forcing from the tropical regions. This 10-year program was one of the first large-scale programs that addressed the predictability of the coupled tropical oceans and global atmosphere by drawing on observations and by recognizing the key role of numerical models for understanding tropical air–sea interactions as a prerequisite for launching successful climate predictions into the future (also see Chapter 24). Related to TOGA, the Pacific TAO/TRITON oceanographic buoy array was established in the 1980s to allow real-time monitoring of low-latitude variability in the Pacific Ocean (http://www.pmel.noaa.gov/tao/proj_over/taohis.html). Enhancements to this *in situ* observing system, augmented by satellite observations, and the development of the first El Niño-Southern Oscillation (ENSO) forecast model led to the first successful ENSO prediction (Zebiak and Cane, 1987).

The *World Ocean Circulation Experiment* (WOCE, 1990–2001) was a component of the international World Climate Research Program, and aimed to establish the role of the world ocean in the Earth's climate system. The goals of WOCE were

- To develop models useful for predicting climate change and to collect the data necessary to test them.
- To determine the representativeness of the specific WOCE datasets for the long-term behavior of the ocean and to find methods for determining long-term changes in the ocean circulation (WOCE IPO, 1997; also see Chapter 3).

The WOCE field phase ran between 1990 and 1998, and was followed by an analysis and model–data synthesis phase that lasted until 2002. The results were summarized in *"Ocean Circulation and Climate: Observing and Modelling the Global Ocean"* (Siedler et al., 2001).

Until the 1980s, when satellites became more commonly available, oceanographers were "data poor." Since then, significant technological and scientific advances in satellite remote sensing have provided regular, global observations of the oceans. Sea surface temperature, surface wind vectors, surface wave height, and surface topography can all be measured in near-real time with reliability from space. Although regional forecasting systems have been developed earlier (e.g., Robinson et al., 1986), the advent of satellite altimetry observations have, for the first time, enabled global ocean forecasting applications (Fu and Cazenave, 2001; Chapter 4).

The realization of the network of 3000 Argo profiling floats reporting temperature and salinity profiles to 2000 m depth in a timely fashion has transformed the *in situ* ocean measurement network in the new millennium. This allows, for the first time, continuous monitoring of the temperature and salinity of the upper ocean on basin-scales, with all data being relayed and made publicly available within hours after collection. Another emerging technology is the new satellite missions which monitor sea surface salinity (e.g., Aquarius mission, 2011; Klemas, 2011).

Despite the significant progress made in the last decade, significant gaps in the observing system remain (such as the lack of large-scale and sustained observations in the deep ocean below 2000 m, including associated current observations). Another issue that is relevant to climate analyses (and, as such, to ocean syntheses) is the challenge of evaluating climate variability and change from an observing system and forcing that change markedly over time.

The OceanObs'99 Conference was a major event that recognized the need for coordinating and implementing an integrated sustained global ocean observing system to be able to serve a broad range of requirements (Smith and Koblinsky, 2001). Robust El Niño forecasts, research on climate variability and change, and ocean and marine forecasts were prominent in the rationale. While the focus of the Conference was on measurement networks, all participants recognized the fundamental importance of models and state estimation to the progress of ocean and climate sciences based on prosperity and evolution of the observing system. The new paradigm that emerged from this Conference was fashioned around the use of models to interpret and exploit observations and to develop products that expand the values of observing systems.

OceanObs'09 recognized the progress made in ocean observations in the last decade but also acknowledged that achieving a sustained ocean observing system remains a challenge to meet the needs of societies (Fischer et al., 2010).

In addition to the calls to implement the initial physical and carbon global ocean observing system envisioned at OceanObs'99 the participants also called on all governments to commit to the implementation and international coordination of systematic global biogeochemical and biological observations. For further details on the status of the global ocean observing system the reader is referred to Chapters 3 and 4).

1.4. Ocean Syntheses

The 1970s and 1980s saw the development and first applications of so-called inverse methods to oceanographic data analyses (e.g., Menke, 1984; Bennett, 1992; Wunsch, 1996). An inverse problem is a general framework that is used to convert observed measurements into information about a physical object or system that one is interested in. For example, if we have measurements of the ocean's temperature and salinity, one might ask the question: "given the data that we have available, what inference can we make from the density distribution about the reference velocity and the circulation in that area of the ocean?" The solution to this problem is useful because it generally contains information about a physical parameter that one did not directly observe (e.g., reference velocity in thermal wind equation, mixing coefficients). These methods, often derived from numerical weather prediction applications, were applied to steady-state ocean models and their applications paved the way for the development of time-dependent state estimation and model initialization used nowadays in ocean analyses. Although inverse methods provided new insights into the dynamical behavior of the ocean, they were mostly applied to steady-state models where the time-rate-of-change is assumed zero despite the time-dependent nature of ocean circulation. This can introduce systematic errors in the estimated quantities due to an underestimation of (unresolved) features such as seasonal variability. Time-dependent ocean state estimation and, to some extent, ocean data assimilation overcome this problem.

Wunsch and Heimbach (Chapter 21; also see Bennett, 2002) provide a comprehensive account of the basic principles that underpin ocean state estimation and data assimilation. In brief, all major assimilation schemes in use today calculate in one form or another a cost function J, which seeks to minimize deviations of the state of the model from observations. Many schemes also allow for the simultaneous optimization of model parameters such as mixing coefficients. The schemes vary in the degree to which they attempt to accurately find stationary values of J, in whether minimizing J is treated as a constrained or unconstrained optimization problem; in their ability to produce robust estimates of the final solution; in the extent to which they assimilate available observations; and in regard to the level of sophistication and accuracy of *a priori* error estimates in observations and model. Application of ocean state estimation and data assimilation in climate research range from early methods such as "nudging" to statistically and dynamically rigorous methods such as the sequential Kalman filter (KF) methods, and the adjoint iterative approach (Stammer et al., 2000).

A set of state estimation and data assimilation methods of varying complexity and their applications to ocean climate research is outlined in the subsequent sections of this chapter. Section 2 outlines the basic principles and methods used today, covering sequential and smoother methods and their applications. Section 3 focuses on recent applications of ocean state estimation and data assimilation products in climate research, addressing ocean circulation, heat budgets, water mass pathways, and initialization of climate prediction. Ocean state estimation and data assimilation methods also provide guidance as to where and when to extend the observing system, which is discussed in Section 4 (observing system evaluation). Section "Conclusion and future challenges" provides an outlook about emerging science trends and future applications in ocean, biogeochemical and coupled state estimation and ocean forecasting.

2. METHODS WITH A FOCUS ON DEVELOPMENTS IN THE LAST DECADE

What progress has been made in ocean data assimilation and state estimation in the last decade? After the end of the WOCE, when the first edition of this book was published, data synthesis had been shown to be technically feasible, and some global ocean state estimates had been initiated. Fukumori (2002) provided a useful historical perspective on ocean data assimilation in a book edited by Fu and Cazenave (2001). Hólm (2003) summarized the methodologies clearly. Subsequently ocean data assimilation and state estimation techniques were further developed. These activities are facilitating the discovery of new information about climate change and enhancing the accuracy of seasonal–interannual (S–I) forecasts by providing substantial information on subsurface oceanic physics, which play an important role in medium- to long-term climate variability. The aim of this discussion is to introduce methodological developments in data assimilation and state estimation during the last decade, with special emphasis on valuable contributions from some recent work. Rigorous descriptions of the formalism are given in the references.

2.1. Sequential Methods

Sequential data assimilation consists of sequentially estimating the state of a system at a fixed point in time based on observations around that time, the estimate then being available for use in real-time assimilation systems.

We focus here on three major approaches to the sequential method, optimal interpolation (OI), the three-dimensional variational (3D-VAR) approach, and the KF; and we discuss the development or extension of these methods for practical application to climate research. Comprehensive mathematical expositions of the original approaches are summarized by Bouttier and Courtier (1999).

2.1.1. *Static Approaches*

"Static" here means that background noise statistics is assumed to be constant in time as opposed to being "dynamic" or time-dependent. Static sequential approaches are usually characterized by a requirement for modest computational resources compared to other approaches discussed later.

OI is one of the simplest forms of optimal least-squares estimators, a minimum variance estimator with prescribed weights (e.g., Gandin, 1963). For each observation innovation is defined as the difference between observation and the corresponding initial guess, which is the model value in this context. Interpolated values are then calculated from a linear combination of the innovations weighted by an ad hoc error covariance, which minimizes the error variance of the solution.

In general, OI provides optimal instantaneous estimates for the particular set of weights. The solution is suboptimal over the entire measurement period because a time dimension is absent from the problem it solves (e.g., Fukumori, 2002). To enhance the accuracy of the solution for improved state estimation or data analyses, an advanced OI method was developed by considering the dimension of time in the definition of the ad hoc error covariance (the so-called 4D-OI). Kuragano and Kamachi (2000) applied this method to the analysis of satellite altimeter data and successfully integrated the acquired sequence of spatially and temporally sporadic data in accordance with the orbital track.

Another extension of the OI method is ensemble optimal interpolation (EnOI), which was first described by Oke et al. (2002) and Evensen (2003); EnOI involves the use of a stationary ensemble of anomalies, or modes, to approximate background-error covariances. EnOI has many attractive characteristics: it is quasi-dynamically consistent and the covariances are inhomogeneous and anisotropic (Oke et al., 2010). Applications of EnOI are introduced in Sections 3 and "Conclusion and future challenges."

Extended applications of the OI method include the Simple Ocean Data Assimilation (SODA), the operational Ocean Reanalysis System 3 (Balmaseda et al., 2007b) of the European Centre for Medium-Range Weather Forecasts (ECMWF), and the Ocean Data Assimilation System of NASA's Global Modeling and Assimilation Office (GMAO) (Yang et al., 2010), the products of which are widely used for climate research (see Section 3).

The 3D-VAR approach is another least-squares estimator, a maximum likelihood estimator. A cost function is defined as a "performance index" through the addition of several summation terms that involve the square of the difference between the observations and first guesses; the general form of the cost function for the least-squares estimator is

$$J = \sum_{m=1}^{M} (\mathbf{y}(t) - \mathbf{E}(t)\bar{\mathbf{x}}(t))^{\mathrm{T}} \mathbf{R}^{-1}(t)(\mathbf{y}(t) - \mathbf{E}(t)\bar{\mathbf{x}}(t)) + \sum_{\mathrm{m}=0}^{M-1} \mathbf{u}(t)^{T} \mathbf{Q}^{-1}(t)\mathbf{u}(t), \quad t = m\Delta t, \tag{22.1}$$

where the vector $\mathbf{x}(t)$ represents the state of the ocean at time t, $\bar{\mathbf{x}}(t)$ is the vector to be analyzed, $\mathbf{u}(t)$ is a control variable to be adjusted, and $\mathbf{y}(t) = \mathbf{E}(t)\mathbf{x}(t) + \mathbf{n}(t)$ is a set of observational data. The vector $\mathbf{n}(t)$ is the observational noise related to the state, and $\mathbf{E}(t)$ is a matrix that relates observations $\mathbf{y}(t)$ to the state of the ocean $\mathbf{x}(t)$, the so-called observation operator. $\mathbf{R}$ and $\mathbf{Q}$ represent the error covariance matrices of the data and the model, respectively. This equation is identical to Equation (21.6) in Chapter 21.

In the conventional 3D-VAR formalism, the elements in the cost function are treated separately along the dimension of time by assuming $\mathbf{u}(t) = \bar{\mathbf{x}}(t) - \mathbf{x}(t)$. Through use of the cost function, the method seeks to find an approximate solution to the equivalent minimization problem (e.g., Derber and Rosati, 1989; Bourlès et al., 1992; Parrish and Derber, 1992; Courtier et al., 1998). The implementation requires the observation operators and the adjoints of the observation operators. The solution is often sought iteratively by performing several evaluations of the cost function. The 3D-VAR method bears a close similarity to OI methods based on the common assumption that errors follow a normal distribution. Bennett (1990), Wunsch (1996), and Malanotte-Rizzoli (1996) provide comprehensive discussions of alternatives.

The advantage of the so-called variational method inclusive of 3D-VAR is the fact that complex (including nonlinear) observation operators, and additional terms in the cost function can be easily introduced. (Note that the conventional OI method assumes linear observation operators.) A great deal of effort was put into their development to improve synthesis products. Some of these studies are summarized below.

Fujii and Kamachi (2003) introduced a nonlinear observation operator that included calculation of sea surface dynamic heights from gridded temperature and salinity data and interpolation. They successfully improved the estimation of the salinity field by using Topex/Poseidon altimetry data with temperature observations with their operator.

Several researchers successfully incorporated dynamic constraints by adding some terms to the cost function.

Xie et al. (2002) added geostrophic balance, Pegion et al. (2000) smoothness, and Fujii et al. (2005) the condition for stability of the density field. Other possible constraints can be proposed and applied to enhance the quality of state estimation or save computational resources. The MOVE/MRI.COM (e.g., Usui et al., 2006) is one of the extended applications of the 3D-VAR method (see Sections 3 and "Conclusion and future challenges").

To conclude the discussion of the above two static methods, we introduce one approach called First Guess at Appropriate Time (e.g., Fisher and Andersson, 2001; Ricci et al., 2005), which is useful for synthesizing observations into a model by OI or 3D-VAR. In this formulation, an observation operator is used to compute the difference between background fields and observations closest to the observation time (e.g., Storto et al., 2010). A tangent-linear model is applied around the relevant observation time to perform this calculation. This approach propagates forward in time the background state along with the general circulation model (GCM) in order to compare this state to the observations at the observation time (Massart et al., 2010).

2.1.2. *Dynamical Approaches*

The KF (Kalman, 1960) is a kind of minimum variance estimator that is algorithmically similar to OI. The advantage of the KF is the fact that the analysis includes a forecast error covariance matrix that evolves in time according to the underlying model and the statistics of the observations that have been assimilated. The adoption of a time-dependent forecast error covariance is the reason for the categorization of the KF as a "dynamical" approach. This is in contrast to the OI, which uses ad hoc error covariances. The original KF is formalized for a linear process. The extended Kalman filters (EKFs; Gelb, 1974; Anderson and Moore, 1979) can be applied to weakly nonlinear problems under the tangent-linear approximation. Evensen (1994) proposed an assimilation method for strongly nonlinear problems called the ensemble Kalman filter (EnKF). An adaptive filter is a simple extension of the KF to estimate temporally constant parameters.

From a practical standpoint, the large computational requirement associated with error evaluation made the KF approach problematic for use with GCMs. A number of studies have explored approximating the errors of the model state to reduce the computational requirements of the KF. We here take note of recent work aimed directly at reducing the large computational requirement of the KF by suitable approximation of the errors.

Fukumori (2002) advanced a new approach to approximate Kalman filtering (and optimal smoothing, which is discussed later) that is suitable for oceanic and atmospheric data assimilation. The method solves the larger estimation problem by partitioning it into a series of smaller calculations. Errors with small correlation distances are derived by regional approximations, and errors associated with independent processes, which can be regarded as a separate reduced state approximation of the model state, are evaluated separately from one another. The smaller dimensionality of each separate element renders application of Kalman filtering (and smoothing) to the larger problem much more practical than would otherwise be the case. In particular, the approximation makes high-resolution global eddy-permitting data assimilation computationally viable. The KF (and smoother) based on such partitioning has been termed the partitioned KF (and partitioned smoother).

There are various promising filters that will be available for application to ocean synthesis in the near future (e.g., Manda et al., 2003; Van Leeuwen, 2003; Hunt et al., 2004; Ott et al., 2004).

2.2. Smoother Methods

A smoothing approach provides the optimal solution to the assimilation of data distributed in space and time into model, where observations from the future can be used, for instance, in a retrospective ocean state estimation. Smoothing methods can provide a system state close to the best time-trajectory fit to the observations (e.g., Chapter 21). In this section, we discuss developmental approaches for oceanographic applications of two of the major smoothing methods: the optimal smoother and the adjoint method (a kind of four-dimensional variational, 4D-VAR method). Comprehensive mathematical expositions of the original approaches are summarized again by Bouttier and Courtier (1999) for the 4D-VAR method, and by Anderson and Moore (1979) and Kitagawa and Gersch (1996) for smoothers. Note that these methods differ with respect to algorithms but are equivalent to each other as long as assumptions about data and model dynamic constraint errors are the same (e.g., Bennett, 2002; Lee et al., 2009).

2.2.1. *Optimal Smoother*

Optimal smoothers, which are minimum variance estimators, are related to the KF. They smooth the filtered results by reducing the temporal discontinuities that result from the sequential input of data by the filters in case of a fixed interval smoother. The use of observations from the future leads to expected uncertainties that are smaller than those associated with filtered results (e.g., Fukumori, 2002). The smoother is a recursive algorithm to seek estimates of the state vector and associated uncertainty at each point in time based on all observations both before and after (e.g., Cohn et al., 1994; Tingley, 2010). Most recently, some studies have applied this method to ocean GCMs (e.g., Freychet et al., 2012).

Cosme et al. (2010) implemented a square-root smoother algorithm (e.g., Anderson and Moore, 1979), an extension of the modified EKF (Pham et al., 1997), for ocean circulation models with an idealized basin. The authors described a modified algorithm that implemented a particular parameterization of model error. Based on twin experiments with some noise-corrupted system states, they concluded that the smoother was efficient in reducing errors, particularly in regions poorly covered by the observations at the time of filter analysis.

A variety of promising smoothers are potentially available for application to retrospective state estimation. One of them is a new smoother algorithm based on ensemble statistics and was presented as an Ensemble Kalman Smoother by Evensen and van Leeuwen (2000). As mentioned above in the discussion of the KF, the large computational requirement associated with error evaluation is still the central issue that must be resolved to enable routine use of smoothers with GCMs.

2.2.2. *Adjoint Method*

The 4D-VAR is a generalization of the 3D-VAR for observations that are distributed in time (e.g., Sasaki, 1970; Talagrand and Courtier, 1987), meaning that the elements in Equation 22.1 include a time dimension in contrast to 3D-VAR. The adjoint method is based on "strong constraint" formalism in which model equations are assumed to hold exactly. The approach consists of performing iterative procedures to converge on a solution. Note that care must be taken when choosing the criterion for convergence or termination of the iterative procedure because a suitable state is obtained when the iterations converge on a solution.

An ocean GCM is defined as an algorithm for deriving state variables from its independent variables (e.g., initial conditions and boundary conditions). Optimization procedures control the independent variables to minimize the cost function (e.g., Marotzke et al., 1999). The result is hence a dynamically and kinematically consistent estimate with no artificial internal sources or sinks for the temperature and salinity fields (Stammer et al., 2002a; Chapter 21). This characteristic is suitable for an ocean state estimate. After a feasibility study appeared in the first edition of this book, applications of the adjoint method in ocean state estimation have increased in the last decade.

The ECCO project (Estimating the Circulation and Climate of the Ocean, http://www.ecco-group.org), funded by the U.S. National Oceanographic Partnership Program, is a collaboration between the Massachusetts Institute of Technology, the Jet Propulsion Laboratory, and the Scripps Institution of Oceanography. Under the ECCO Consortium, Stammer et al. (2002b) estimated a long-term oceanic state for the global ocean by using the adjoint method. Their efforts provided results capable of assessing the dynamical state of the global ocean. The K7 consortium in Japan (http://www.jamstec.go.jp/e/medid/dias/kadai/clm/clm_or.html) has also made an effort to improve global ocean state estimates by using an adjoint method (e.g., Awaji et al., 2003; Masuda et al., 2003). Wenzel and Schröter (2007) tried state estimation by using the Hamburg Large Scale Geostrophic Model. Such estimates are useful for a variety of new scientific applications (see Section 3).

Most recently, ECCO, its German partner G-ECCO, and K7 have updated their models and provided more sophisticated ocean state estimates (e.g., Köhl and Stammer, 2008a,b; Hoteit et al., 2010; Masuda et al., 2010; Chapter 21). Although the basic architecture of the adjoint formalism in their system is conventional, more complicated cost functions and error covariance structures were assessed, for example, for the deep ocean in the K7 system.

In some cases it is possible for the adjoint method based on "strong constraint" formalism to lead to an unrealistic "optimal" state because of model deficiencies that have not been accounted for by appropriate control variables. Furthermore, the direct application of the adjoint method might not be appropriate for systems with turbulence closure parameterizations. Chua and Bennett (2001) reviewed variational assimilation systems and illustrated the efficiency of a "weak constraint" formalism that allowed departures from model dynamics while obtaining an optimal state estimation. This more generalized approach represents a larger computational demand to strong constraint formulations. The representer method is often applied to seek the solution effectively in observational space (e.g., Bennett, 2002). Apte et al. (2008) summarize the mathematical and statistical perspectives for 3D-VAR, 4D-VAR, and weak constraint 4D-VAR and discuss their interrelation.

Note: Model Green's functions can be utilized as a kind of 4D-VAR approach without an adjoint (e.g., Stammer and Wunsch, 1996). Green's functions provide a convenient and effective method to test and to calibrate model parameterizations, to study and quantify model and data errors, to correct model biases and trends, and to blend estimates from different solutions and data products. Menemenlis et al. (2005) applied this method to an OGCM; the result was a substantial improvement of the solution relative to observations when compared to prior estimates.

2.3. Improved Procedure for Data Assimilation

It is important to develop an appropriate procedure by which all information that can possibly be derived from limited observational data can be deduced (e.g., Bennett, 2002). In this section, we introduce technical developments in the various aspects of data assimilation and state estimation. These can be applied without regard to the choice of the synthesis method.

We agree with Fukumori's suggestion (2002) that suitable specification of weights is essential to obtaining sensible solutions and is the most fundamental issue in data assimilation. Weaver and Courtier (2001) applied a statistical model to represent the correlations of background errors. Application of their algorithm involves a numerical integration of a generalized diffusion-type equation (GDE). The GDE is formed by replacing the Laplacian operator in the classical diffusion equation by a polynomial in the Laplacian. The integral solution of the GDE defines a correlation operator whose kernel is an isotropic correlation function; this is a useful technique for obtaining idealized correlations of background errors for practical use. Fujii and Kamachi (2003) also developed a practical method to obtain an effective background covariance matrix by adopting a vertically coupled T–S empirical orthogonal function decomposition.

The choice of the covariance relationship for data assimilation elements is also an essential factor to obtain reasonable solutions. Weaver et al. (2005) implemented a multivariate balance operator for ocean data assimilation, which they used to implicitly specify the error covariances of the model background state via a transformation from model space, where variables were highly correlated, to a control space, where variables could be considered to be approximately uncorrelated. They showed that the balance operator could explain a significant percentage of background-error variance. Haines et al. (2006) proposed a new method for assimilation of salinity data. From a historical ocean database, they showed that over large regions of the globe (mainly mid-latitudes and lower latitudes) the variance of salinity on an isotherm $S(T)$ was often less than the variance measured at a particular depth $S(z)$. They also showed that the dominant temporal variations in $S(T)$ were associated with a lower frequency than the dominant variations in $S(z)$. Based on this background, Haines et al. (2006) showed that assimilating salinity data onto isotherms led to better use of scarce salinity observations because longer space and timescales could be used for the $S(T)$ assimilations. Their paper identified salinity increments coming separately from temperature and salinity data and demonstrated the independence of these increments.

System bias is a serious obstacle to the reliable representation of climate variability because, when a bias is present, a time-dependent observing system can induce spurious time variability in the analysis (e.g., Segschneider et al., 2000; Vialard et al., 2005). Balmaseda et al. (2007a) implemented a generalized algorithm for treatment of bias in sequential data assimilation. The scheme allows the control variables of the bias to be treated separately from those of the state vector. They illustrate, for example, that in some cases the direct correction of the bias in temperature leads to a significant increase in the error in the velocity field. It is not the case when correcting the bias in the pressure field. Some insight into the nature of the error is needed for the formulation of bias error statistics.

QC of the assimilated elements is also an important factor in the synthesis procedure. A conventional method to exclude outliers, for example, is to ignore data that lie outside ± 3 standard deviations of the long-term mean. However, such ignored observations might be truly associated, for example, with large El Niño events, and their exclusion would lead to a poor representation of the analysis. Application of variational QC (e.g., Andersson and Järvinen, 1999) is an efficient way to improve the accuracy of the analysis without discarding anomalous data. Fujii et al. (2005), for example, were able to retain reasonable information among their observations by simply adopting a clipped covariance matrix of observational errors. A detailed discussion of QC was recently published by Cummings (2011).

To conclude this section, we introduce an example of careful input of synthesis products into a model to avoid "initialization shock," which is caused by the input of artificial data, in the process of which model equations do not exactly hold (e.g., Balmaseda and Anderson, 2009). COMPASS-K adopts a time-retrospective nudging method that prevents the delay of model response associated with the standard nudging method and leads to more consistent initial states. A characteristic of the time-retrospective nudging method is that the timescale of the delay is proportional to the inverse of the coefficient of the nudging term. In the study of Kamachi et al. (2004), observed data were inserted into the model beginning 3 days before an observation time in the time-retrospective nudging method to suppress the possibility of initialization shock. Sandery et al. (2011) developed a similar technique, adaptive nonlinear dynamical initialization, which they effectively applied in BLUElink (see Section "Conclusion and future challenges").

3. APPLICATIONS FOR CLIMATE RESEARCH

3.1. Significant Progress in the Past Decade

Few ocean synthesis products were publically available in the 1990s. Since then, a suite of global and regional synthesis systems have been developed to synthesize various observational datasets and extend in time period on a regular basis. These products have been widely disseminated for a broad scope of applications related to climate research, for instance, meridional overturning circulation (MOC) and heat transport (Section 3.2), the upper-ocean heat budget (Section 3.3), water-pass pathways (Section 3.4), and initialization of climate prediction

(Section 3.5). There has been a substantial increase in the number of applications of ocean synthesis products for various research topics from the 1990s to the 2000s (by at least several times as shown by literature search). The sustained measurements from some observing platform (e.g., nearly two decades of satellite altimetry) and new observing system over much of the global upper ocean (e.g., Argo floats since the mid 2000s) have significantly enhanced the applications of ocean synthesis products to investigate various subjects that were not addressed in the 1990s (by ocean synthesis products), such as many topics related to decadal variability and subsurface structure of the ocean. The readers are referred to Chapter 21 for additional description and perspectives of the progress in the past decade in ocean synthesis. In terms of initialization of climate prediction, the limited efforts in the 1990s have primarily used ocean synthesis products to initialize S–I forecasts. Important progress in the past few years has been in the emerging efforts of coupled assimilation to obtain initial states that are more consistent with coupled model dynamics to alleviate the so-called initial shocks (Section 3.5). Due to limited space, however, here we can only highlight a small number of examples for ocean and climate studies in an attempt to make a relatively focused and coherent discussion within the limited space.

Ocean synthesis products combined various observations that are incomplete in space and time with model dynamics to perform space–time interpolation and dynamical inference with the aim to provide a more complete description of the time-varying, three-dimensional ocean circulation. Note that some of the processes described by ocean synthesis products are not necessarily very different from those inferred from the underlying models without synthesizing the data. However, the observational constraints in ocean synthesis improve many aspects of the estimated state of the ocean, thus often provide a better confidence and more accurate description of the ocean. However, understanding the uncertainty of ocean synthesis products and the consistency of these products among one another and with observations is an important challenge that the ocean synthesis community is tackling (Section "Understanding consistency and uncertainty").

Some of the synthesis methods render the corresponding synthesis products more amendable to certain applications. For example, the applications of ocean synthesis products to study mixed-layer heat budget analysis and water mass pathway analysis are mostly based on ocean synthesis products that use estimation methods that do not produce internal sources and sinks of heat and salt (e.g., the adjoint method), thus facilitating the closure of budgets. For initialization of climate prediction, the examples given are mostly based on synthesis products that use sequential filter methods because the smoother-type methods typically require the estimation of the entire temporal trajectory of the model state (often through iterative optimization) and thus not practical for applications of initializing climate models on a routine basis.

3.2. Ocean Circulation

MOC plays an important role in climate variability and change by regulating meridional heat transport (MHT) in the ocean and the subsequent air–sea interaction. Direct measurement or inference of the MOC and MHT with observations is difficult. Apart from the RAPID-MOCHA array near 26°N of the Atlantic Ocean (Cunningham et al., 2007; Johns et al., 2011), there is no transoceanic array that monitors the MOC at other latitudes of the Atlantic or in other oceans. Ocean reanalysis products provide an important tool to characterize and quantify the MOC and related MHT and to understand the mechanisms of their variations. For instance, estimates of volume, heat, and freshwater transports of the global ocean were studied by fitting an OGCM to WOCE data (Stammer et al., 2003), which facilitated the analysis of interannual-to-decadal changes of these transports in the ocean (e.g., Köhl et al., 2007). Many ocean state estimate products have been applied to studies of MOCs in different basins and their relations to heat transport and heat content changes.

Ocean state estimation has been applied to the studies of interannual and decadal variability of the Atlantic MOC (AMOC). For example, Cabanes et al. (2008) used an ECCO product to study the mechanism of interannual variability of MOC in the subtropical North Atlantic. The estimated MOC variability in the subtropics was found to be well correlated with the east–west difference in pycnocline depth, which reflects the zonal pressure gradient that drives the meridional flow associated with the MOC. They found that variability in the east–west difference of pycnocline depth was primarily associated with fluctuations of the pycnocline depth near the western boundary due to a substantially larger interannual variability of Ekman pumping in the western part of the basin. The mechanism identified by this study would be helpful to the interpretation of dominant interannual variability captured by *in situ* monitoring systems such as the RAPID array.

On decadal timescales, the processes that affect AMOC variability are more complex because of remote forcing. For example, Köhl and Stammer (2008a) found that decadal changes in the estimated AMOC strength at 25°N in the G-ECCO product were strongly influenced by the southward communication of density anomalies along the western boundary originating from the subpolar North Atlantic. These density anomalies were related to changes in the Denmark Strait overflow but were only marginally influenced by water mass formation in the Labrador Sea. The influence of density anomalies propagating along the southern edge of the subtropical gyre associated with baroclinically unstable Rossby waves was found to be equally important. They also suggested that wind driven processes such as local Ekman transport explained a smaller fraction of the variability on those long timescales (compared to interannual timescales).

Decadal change of AMOC structure is also complex. Bryden et al. (2005), using the data from several synoptic hydrographic sections occupied in different decades, suggested a substantial slowdown of the AMOC at 25°N. Based on estimates by an ECCO-GODAE product, Wunsch and Heimbach (2006) found a much weaker and barely significant decrease in the strength of the AMOC at 26°N as described by the northward volume transport above 1200 m (Figure 22.1). However, it is accompanied by a strengthening of the deeper meridional circulations, that is, the southward outflow of North Atlantic Deep Water and northward inflow of abyssal water. The decrease of MHT was not statistically significant. It should be noted that the estimated AMOC decadal trend and associated error estimate were inferred from a product not containing mesoscale eddies. Moreover, the abyssal circulation was poorly constrained by local direct observations.

An analysis of ECMWF ORA-S3 operational ocean reanalysis (Balmaseda et al., 2007b) showed that the assimilation significantly improves the time-mean estimates of AMOC strength. The resultant estimate of AMOC strength at 26°N showed only a weak decreasing trend. No decrease in heat transport was found because the enhanced vertical temperature gradient due to near-surface warming counteracted the effect due to the decrease in AMOC strength. Both Wunsch and Heimbach (2006) and Balmaseda et al. (2007b) discussed the large high-frequency fluctuations in estimated MOC strength and the potential for these fluctuations to be aliased into low frequencies changes if the sampling is infrequent. These studies indicate the need to enhance the scope of the observations of AMOC (or ocean circulation in general) both in space and time in order to (1) capture low-frequency signals without being aliased by high-frequency variability, (2) to have a more complete picture of the AMOC changes at different depths and latitudes, and (3) to accurately estimate climatically important quantities such as MHT.

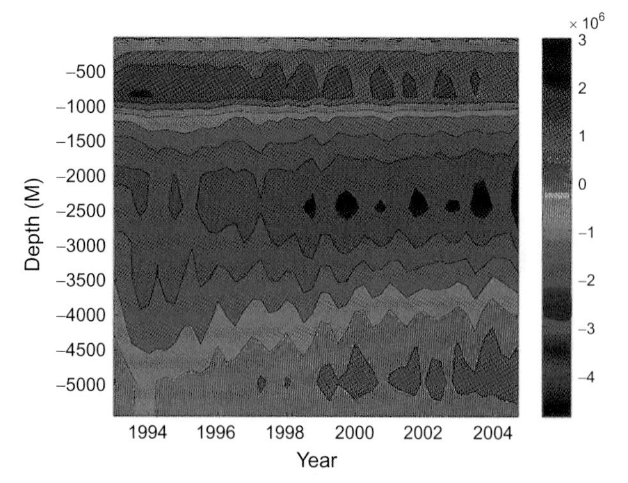

FIGURE 22.1 Seasonal averages (3 months) of volume transport ($m^3\ s^{-1}$) in different layers through time as a function of depth at 26°N of the Atlantic Ocean inferred from an ECCO-GODAE ocean state estimation product. *After Wunsch and Heimbach (2006).*

ECCO and G-ECCO ocean state estimation products (http://www.ecco-group.org) and a SODA assimilation product (http://www.atmos.umd.edu/~ocean/) have been used to analyze the structure and variability of the shallow MOCs that connect the tropical and subtropical oceans in different basins, namely, the subtropical cell or STC (Lee and Fukumori, 2003; Schoenefeldt and Schott, 2006; Schott et al., 2007, 2008; Rabe et al., 2008). In particular, Lee and Fukumori (2003) and Schott et al. (2007) identified a horizontal structure in the interannual and decadal variations of the lower limb of the STC (pycnocline flow) in the Pacific Ocean: the anomalies of the pycnocline volume transports of the low-latitude western boundary currents (the Mindanao Current and the New Guinea Coastal Undercurrent) were found to be anti-correlated with that of the interior flow, with the latter being more dominant. Lee and Fukumori (2003) attributed such a partial compensation of the meridional pycnocline transport to fluctuations in Ekman pumping in the western Pacific associated with the variability of the ITCZ and SPCZ. This process has important implications to the understanding of ENSO and its decadal modulation in terms of how tropical upper-ocean heat content is "charged" and "discharged." It also underscores the need to have sustained *in situ* measurements near low-latitude western boundary, a region not well resolved by existing observing systems. The readers are referred to the article by Lee et al. (2009) for additional description of the applications of ocean state estimation products for ocean and climate research related to MOCs and other topics.

3.3. Upper-Ocean Heat Budget

The knowledge about upper-ocean heat balance is critical to the understanding of ocean–atmosphere interaction associated with climate variability and change. Because of their incompleteness in space, time, and observed parameters, observations are usually inadequate to estimate the complete budget terms associated with three-dimensional advection and diffusion of heat (e.g., vertical advection and diffusion). Ocean synthesis products, constrained by the incomplete observations, provide a potentially important tool to perform spatial, temporal, and dynamical interpolation. Satisfying the conservation of heat, ECCO state estimation products have been used to demonstrate the potential to advance our understanding of upper-ocean heat budget (e.g., Kim et al., 2004, 2007; Halkides and Lee, 2009, 2011; Halkides et al., 2011a,b). For example, Kim et al. (2004) investigated the relative contributions of ocean dynamics and surface heat flux in causing an abrupt warming in the north-central Pacific Ocean in the late 1990s. They found that the ocean played an equally important role in causing that warming even though that part of the ocean is away from the strong western boundary current region that is known to be significantly affected by heat advection. Kim et al. (2007) and Halkides et al. (2011b) found that interannual and intraseasonal mixed-layer heat budgets at an individual TAO mooring location inferred from ECCO state estimation products were very similar to those inferred from *in situ* observations based on TAO mooring data (Wang and McPhaden, 2000). However, the point-wise budgets were very different from those averaged over regions that span a few degrees across the equator. This is because meridional Ekman advection and tropical instability waves have larger influence off the equator than at the equator, and equatorial Kelvin waves affect the equator more than the off-equatorial regions. These examples illustrate the complementarity between ocean state estimation and *in situ* observations in understanding the spatial inhomogeneity of upper-ocean dynamics and thermodynamics.

3.4. Water Mass Pathways

The adjoint models used by several ocean state estimation systems (e.g., ECCO and K-7) offer powerful tools to study the origin, destination, and pathway of water masses. Fukumori et al. (2004) used an ECCO product and the adjoint tool to quantify the origin of water in the NINO3 region in the eastern equatorial Pacific (i.e., the relative contributions of source waters from various parts of the North and South Pacific). They also examined the destination of the NINO3 waters and found that, while NINO3 waters primarily came from the eastern part of the subtropical gyre (the subduction region), the NINO3 waters carried by the surface flow mostly end up in the recirculation region in the western subtropical oceans. Thus, where the NINO3 waters came from and where they are going form an "open circuit" on advective timescales. Using the same technique, Wang et al. (2004) investigate the cause for the interannual co-variation of salinity and temperature in the Pacific cold-tongue region (higher salinity coinciding with higher temperature during El Nino, vice versa). They found that the anomalous advection of higher salinity waters from the South Pacific (off the equator) was the cause for the salinity variation, which is very different from the mechanism causing interannual variation of cold-tongue temperature (the latter is dominated by interannual change in local upwelling and vertical mixing).

Masuda et al. (2006) used the ocean state estimate obtained by Japan's K-7 system to study the interannual variability of the temperature inversion that characterizes the upper layer of the subarctic North Pacific. They found that the depth of the maximum temperature associated with that layer exhibited significant interannual variability that was closely related to ENSO. Using an adjoint method similar to that of Fukumori et al. (2004), they found that when the depth of the maximum temperature was shallow, the source waters came from the Gulf of Alaska and the Kuroshio Extension region. When the depth of the maximum temperature was greater, the source water came only from the Kuroshio Extension region. They concluded that the depth of maximum temperature was an important indication for the origin of water mass property in the subarctic Pacific on interannual timescales. Toyoda et al. (2011) used the product and tool of Japan's K-7 system to study the processes responsible for interannual variation of the eastern part of the subtropical mode water in the North Pacific and identified three mechanisms: (1) salinity convergence by Ekman flow in the preconditioning phase during summer and autumn, (2) solar insolation affected by the amount of stratocumulus in the preconditioning phase, and (3) wintertime surface cooling. They also suggested that the counterpart in the South Pacific exhibited a similar interannual variability and could be explained by the mechanisms described above.

3.5. Initialization for Climate Prediction

One of the important applications of ocean data assimilation and state estimation is to improve initialization of climate prediction (also see Chapter 24). The use of ocean data assimilation to initialize seasonal climate forecasts has become an important routine practice in several operational centers around the world. This includes the National Center for Environmental Prediction (NCEP) (Ji et al., 1998; Behringer, 2007), ECMWF (Balmaseda et al., 2008), Meteorological Research Institute (MRI) of Japan's Meteorological Agency (Usui et al., 2006), Australian Bureau of Meteorology (Yin et al., 2011), U.K. Met Office (Martin et al., 2007). NASA's GMAO also use GMAO's ocean data assimilation product to initialize S–I forecast by a coupled model (Yang et al., 2010). In most cases, positive impacts on coupled model seasonal hindcasts and forecasts due to the use of ocean data assimilation products for the initialization have been reported, an example of which is shown in Figure 22.2.

Despite the positive impacts, significant challenges remain in using ocean synthesis products to improve S–I prediction. The merit of using ocean data assimilation products for initializing climate prediction is often shadowed by errors in coupled models and "initialization shock" arising from the lack of dynamical consistency

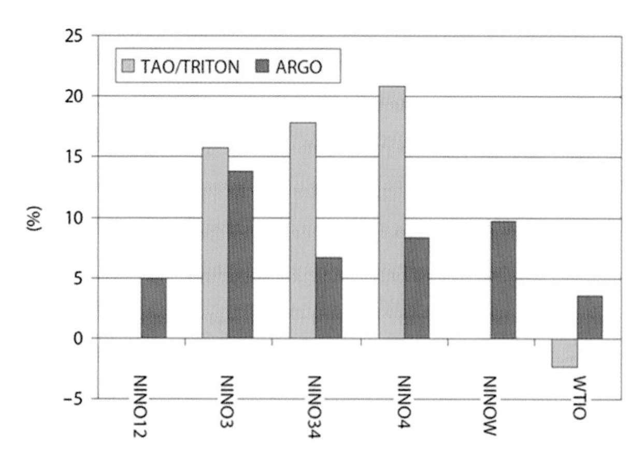

FIGURE 22.2 Impact of Tropical Atmosphere Ocean/Triangle Trans-Ocean Buoy Network (TAO/TRITON) and Argo data on seasonal forecast skill. The bars show the relative increase in root mean square errors of 1–7-month forecasts of monthly sea surface temperature resulting from withholding TAO/TRITON and Argo data from the initialization of seasonal forecasts, as reported by Fujii et al. (2008). The regions are defined as the following: NINO12 (10°–0°S, 90–80°W), NINO3 (5°S–5°N, 150°–90°W), NINO34 (5°S–5°N, 170°–120°W), NINO4 (5°S–5°N, 160°E–150°W), NINOW (0°–15°N, 130°–150°E), WTIO (10°S–10°N, 50°–70°W). *After Balmaseda et al. (2009b).*

of the ocean initial state either among different variables of the initial state or between the initial ocean state and the coupled model dynamics. There have been various experimental efforts attempting to obtain initial state that is more dynamically consistent among ocean variables (often referred to as "balanced" initial state). For example, there have been efforts to use dynamically consistent ocean state to initialize coupled models (e.g., Dommenget and Stammer, 2004; Cazes-Boezio et al., 2008), The inconsistency of the background or mean state of the estimated ocean with coupled model dynamics has been considered an important factor that contributes to the initial shocks in forecast. Some efforts used the so-called anomaly initialization method to alleviate this problem (in which only the estimated anomaly of the ocean initial state referenced to the background state is used in the initialization).

Obtaining an initial state that is consistent with the coupled model physics can potentially minimize the initial shock. An early attempt was made by Lee et al. (2000) to assimilate sea surface temperature, sea level anomaly, and ocean surface wind data into an intermediate coupled model by optimizing the initial conditions as well as coupled model parameters that control growth and decay of interannual anomalies. The parameter optimization helped reduce the error of the coupled model and allowed a better fit of the coupled model to oceanic and atmospheric observations during the initialization procedure, which led to an improvement in hindcast skill for SST across the equatorial Pacific. A much more sophisticated optimization of a coupled model was accomplished by Sugiura et al. (2008), who assimilated *in situ* and satellite observations into a fully coupled GCM (the K-7 system). In their procedure, the optimization of the spatially and temporally dependent transfer coefficients for the bulk formula air–sea fluxes allowed the coupled model to capture the active ocean–atmosphere coupling associated with MJO events that often precede the development of El Nino events and thus a better representation of the state of the ocean and atmosphere. This led to a better hindcast of the 1997–1998 El Nino event. There are other ongoing efforts to develop coupled ocean–atmosphere data assimilation systems that aim to improve seasonal forecasts, for example, NCEP's Climate Forecast Reanalysis effort (Saha et al., 2010; http://cfs.ncep.noaa.gov/), NOAA Geophysical Dynamics Laboratory's coupled assimilation effort (Zhang et al., 2007). The readers are referred to the paper by Balmaseda et al. (2009b) for an overview of the efforts by operational centers on the utility of ocean data assimilation for initializing seasonal prediction. Ocean data assimilation products are also being used to test the impacts on decadal climate prediction (e.g., Pohlmann et al., 2009). Such type of efforts, being in their infancy, are strongly encouraged by the World Climate Research Program (also see Section "Conclusion and future challenges"; Chapter 25). Ocean synthesis or coupled synthesis products also have the potential to initialize climate projection using IPCC models. This type of effort is yet to be initiated.

4. ASSESSMENTS OF THE IMPACT OF NEW AND FUTURE CLIMATE OBSERVING SYSTEMS

Ocean data assimilation and state estimation provide tools to bridge the gap between modeling and observation communities and to assess the skill and deficiencies in both observations and models. The same tools can also be used for defining observational requirements for a climate monitoring and prediction system and for evaluating the impact of different observing types and space–time distributions on climate diagnostics. A central aim of observing strategies targeting climate timescales and diagnostics is to improve the understanding of the origin, propagation, and growth of climate signals (Heimbach et al., 2010a). Focusing on global datasets that measure ocean climate variables, the existing observing system includes, among others, satellite altimetry, scatterometer wind stresses, sea surface temperature, Argo profiles, expendable bathythermographs, and surface drifters.

Various approaches exist for identifying and quantifying the importance of single observations or whole observing systems. For example, Observing System

Simulation Experiments (OSSEs) are intended to guide the design of specific observing systems and the configuration of multi-sensor observing networks, to provide an assessment of the potential for future observing systems and innovative uses of existing systems to achieve major improvements in forecast skill, to test advanced data assimilation methods, and to assess the relative role of observations and forecasting methods in improving the utility of forecasts (Oke and Schiller, 2007a).

Adjoint and singular vector methods are powerful tools to detect and track regions of high climate sensitivity and those that lead to optimal growth of climate signals (via specified climate norms). However, assessing the effectiveness of model sensitivities over climate timescales in the presence of highly nonlinear physics and the robustness across different models remains an outstanding task for all types of OSSEs (Heimbach et al., 2010a).

Whether the results from OSSEs such as the examples shown below are reliable and applicable to real observing arrays depends as much on the technical design (e.g., choice of model, optimization technique, objective function) as on its subsequent physical interpretation. Users of information based on OSSEs need to be aware of the potential caveats when using these still maturing tools as the results obtained from these techniques are highly system dependent, are sensitive to their assumptions, and need careful interpretation.

The Science Working Group of the "Pilot Research Moored Array in the Tropical Atlantic" adopted the results of an observing system study by Hackert et al. (1998) to optimize the exact mooring locations prior to the deployment of the tropical Atlantic mooring array. To the best of our knowledge, this was the first time an OSSE based on dynamical models contributed to the design of a large-scale *in situ* ocean observing system.

4.1. Indian Ocean Observing System

One of the first coordinated international efforts to design a basin-scale ocean observing system by performing OSSEs was undertaken in 2003/2004. The CLIVAR Indian Ocean Panel initiated OSSE efforts to guide the deployment of a mooring system in the Indian Ocean to complement other components of the observing system (IOP, 2005). More specifically, the international efforts examined the efficiency of the proposed Indian Ocean Observing System over a range of timescales from intraseasonal variability to decadal and longer changes. This coordinated international effort also used a variety of approaches, from assessing how well the initial planned observing system performs to determining what set of observations is optimal.

A moored array is an essential part of the Indian Ocean Observing System to capture subseasonal and near-equatorial variability. Through a series of OSSEs the international group concluded that while the proposed empirical array did a reasonable job of resolving seasonal to interannual variabilities, it might not adequately resolve intraseasonal variability. It was suggested that the proposed array probably oversampled the region within 3° of the equator in the western tropical Indian Ocean and undersampled the region south of 8°S where seasonal Rossby waves are prevalent (Oke and Schiller, 2007b). Similarly, Ballabrera-Poy et al. (2007)—based on results from a series of OSSEs using a reduced-order KF—and Vecchi and Harrison (2007) also suggested that the proposed array might oversample the region within a few degrees of the equator. They argued that Argo observations should give good coverage poleward of 5° latitudes but that equatorward of 5° moored buoys and XBT lines are essential for completing the integrated observing system in the Indian Ocean.

The OSSEs provided valuable insights that led to modifications of the initial plan. As indicated above, the results were qualitatively robust for different methods and for different models. The OSSEs were a useful complement to the usual observationalist approach to designing fieldwork. In particular, the cumulative results of all OSSEs led to improved approaches to observe critical regions where upwelling occurs and remote forcing of the thermocline affects SST (e.g., Java/Sumatra upwelling and the South Equatorial Counter Current ridge); they also shed light on what observations are critical for observing change in the cross-equatorial heat transport (IOP, 2005).

Figure 22.3 is an example of one of the Indian Ocean OSSEs performed by Vecchi and Harrison (2007). It shows the correlation of monthly mean 50-m temperature anomaly from various sampling strategies and that from an OGCM. Figure 22.3 (top left) indicates the standard deviation of the OGCM monthly 50-m temperature anomaly (i.e., the "truth" in this sampling study). The strongest variability is along the southern Indian Ocean thermocline ridge, along the Java/Sumatra upwelling zone, and in the southeast Arabian Sea. There is also significant variability all along the equatorial waveguide, the eastern boundary of the Bay of Bengal, and the western Arabian Sea. Figure 22.3b and c shows the correlation with two XBT sampling strategies, indicating that much of the Indian Ocean is not adequately sampled by XBTs alone. Existing XBT lines provide key information, however, along the coast of Java, the Somali upwelling zone, and the southeast Indian Ocean; they also provide sampling along the thermocline ridge. Addition of a tropical and subtropical moored buoy array to the XBT lines (Figure 22.3d) improves the basin-wide description of the interannual 50-m temperature variability in this OSSE. The impact of an array of Argo profiling floats at two sampling intervals is shown in Figure 22.3e and f. The

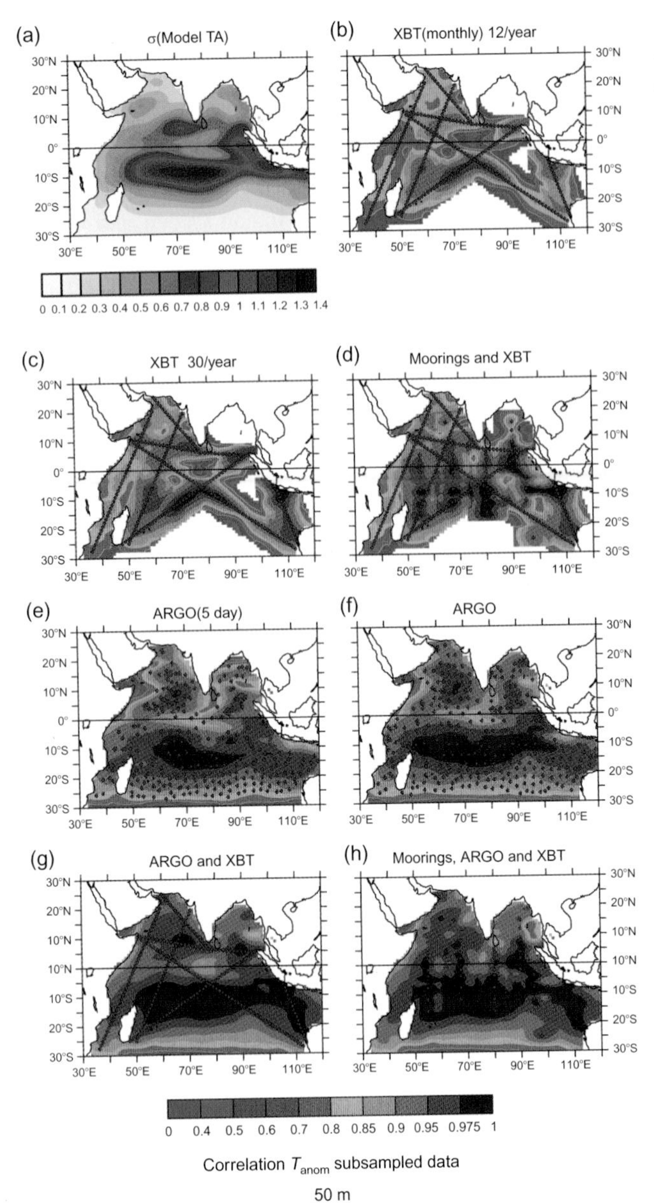

FIGURE 22.3 Computed over the period 1986–2002: (a) standard deviation of monthly 50-m temperature anomaly from an ECMWF-forced OGCM, units: °C. Correlation of monthly 50-m temperature anomaly from the ECMWF-forced OGCM and that subsampled by (b) XBT lines done 12 times per year; (c) XBT lines performed 30 times per year except for IX-01 (52 times per year); (d) moored buoy array and 30 times per year XBTs; (e) 5-day interval Argo array; (f) 10-day interval Argo array; (g) 10-day interval Argo array and 30 times per year XBTs; and (h) Argo, moorings, and 30 times per year XBTs. *After Vecchi and Harrison (2007).*

enhanced divergence of Argo floats from regions of surface divergence results in a reduced effectiveness of sampling; the degradation of sampling is more evident from the 5-day sampling strategy. The addition of frequent XBT lines brings the correlations along the coast of Java above 0.97 and in the Somali upwelling zone above 0.9 (Figure 22.3g). The full observing system comprising a 10-day sampling Argo array, a moored buoy array, and frequent XBT lines gives an extremely accurate description of the 50-m temperature anomaly field: correlations exceed 0.95 throughout much of the basin.

4.2 North Atlantic Meridional Overturning

As discussed in Section 3.2 monitoring of the North AMOC has recently attracted increased attention. Focus areas are the strength and variability of the MOC with a particular emphasis on interannual-to-decadal variability (Cunningham et al., 2010). Both Wunsch and Heimbach (2006) and Balmaseda et al. (2007b) showed that the trend in the MHT was smaller than that of the MOC strength because surface warming partially counteracted the weakening upper MOC. Therefore, an observing system that is capable of inferring changes in the volume transport alone may not be adequate to monitor the heat transport. Heimbach et al. (2011) discussed the complexity of monitoring the AMOC in space and time. These findings suggest that a systematic measurement network for the AMOC and heat transport at different latitudes and different depths beyond the traditional synoptic hydrographic survey is needed. The deployment of such an observing system as the *Rapid Climate Change—Meridional Overturning Circulation and Heatflux Array* (*RAPID-MOCHA*) at 26.5° N in 2004 is a step toward that direction (Lee et al., 2009; Johns et al, 2011).

Baehr et al. (2008) analyze various model-based methods for the design of an ocean observing system to monitor the MOC in the North Atlantic. Zhang et al. (2010) use output from OSSEs with the GFDL CM2.0 coupled climate model (IPCC AR4) to assess the adequacy of the observing system in monitoring the MOC. Both studies indicate that presently neither observations nor modeling provide a complete picture of the North AMOC in time and space, and that our knowledge of the basic features of the MOC remains limited. A key result of the observing system experiments by Zhang et al. is the need for a coherent combination of atmosphere (temperature and winds) and ocean observations (temperature and salinity from XBT and Argo) to monitor the MOC accurately.

This is illustrated in Figure 22.4 (from Zhang et al., 2010), which illustrates how the structure of the AMOC mean state is constrained by various observing systems. The figure compares the vertical structure of the time-mean errors of the estimated AMOC streamfunction from the control simulation (Figure 22.4a, no use of simulated observations) and five simulated assimilations of observations. Compared to the constraint of an ocean observing system (O_{XBT} or O_{Argo}), the constraint of surface forcings (O_{SST}^{Atm}) produces a more consistent AMOC mean state

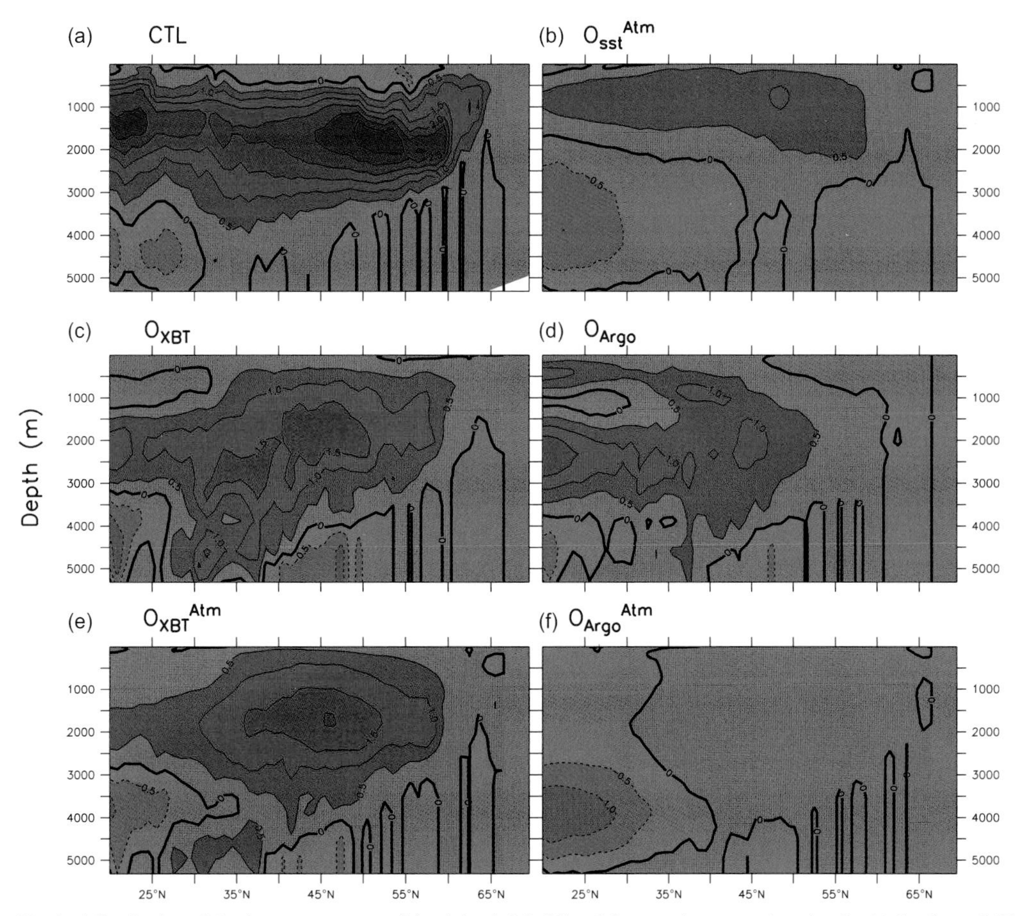

FIGURE 22.4 Vertical distribution of the time-mean errors of the Atlantic Meridional Overturning streamfunction for (a) the Control (CTL) and assimilations of (b) O_{SST}^{Atm}, (c) O_{XBT}, (d) O_{Argo}, (e) O_{XBT}^{Atm}, and (f) O_{Argo}^{Atm}. Color and contour interval is 0.5 Sv (1 Sv = 10^6 m^3/s). *After Zhang et al. (2010).*

(Figure 22.4b), although the accuracy of the estimated decadal variability and regime transition is lower. Furthermore, Figure 22.4c and d clearly show that, consistent with the variability estimate for the AMOC, the time-mean transport constrained by O_{Argo} is much better than the one constrained by O_{XBT}, suggesting that Argo's direct salinity observations play a critical role in accurately reproducing the AMOC. While the addition of the surface forcing constraint to O_{XBT} marginally increases the AMOC error over the middle to upper ocean it reduces the error at the bottom ocean (Figure 22.4e), presumably because of inconsistencies in the simulated observations chosen for assimilation. The combination of the atmospheric and Argo constraints consistently reduces the AMOC error dramatically, except for the subtropical deep ocean (Figure 22.4f).

In summary, we note that OSSEs can yield valuable information on the design of future observing systems; but, given their immaturity, they are no substitute for an in-depth understanding of the methodological limitations and physical mechanisms behind a proposed monitoring array. A long-term goal of observation impact assessments remains the consideration of feedbacks between various components of the Earth system through coupled observing system assessments.

5. CONCLUSION AND FUTURE CHALLENGES

Advance in ocean synthesis has led to a better understanding of ocean circulation and climate variability (Section 3) and may contribute to development of improved ocean observing systems (Section 4) in the next decade. Recent *in situ* and satellite ocean observations provide high-quality data from a wide variety of sensors. Advances in computer science have made possible the execution of huge numerical calculations that were previously impossible. Considering these improvements in data and processing, Rienecker et al. (2010) discussed the advantages of "Integrated Earth System Analyses" at OceanObs'09. A physical–biogeochemical ocean model that couples the atmosphere, sea-ice, and land surface is a promising prognostic tool for climate research in the future. Novel data assimilation techniques suitable for such coupled systems are expected to be developed in the near future. Another challenge is the need for higher resolution ocean data assimilation techniques, including methods to reduce the computational cost, applicable not only to the mesoscale-eddying estimation of ocean state but also to research on the coastal environment, which is directly connected with human activities such as fisheries, marine transportation, coastal security, and marine leisure.

5.1. Coupled Data Assimilation

5.1.1. Ocean–Atmosphere Coupling

Although not a focus of this book, international work for innovation in ocean climate research is well underway in the form of coupled Earth system models. Coupled Earth system models seamlessly link together models of the oceans, atmosphere, sea-ice, land surface, ocean waves, tides, the global carbon cycle and chemistry, aerosols, and so on, to simulate changes in the Earth's climate systems with ever-increasing precision (WMO, 2009). These models enable scientists to not only project major long-term changes in the Earth's climate, but also make short- and medium-term atmospheric weather forecasts and seasonal predictions for particular regions. Skill and reliability on one timescale engender confidence for other timescales (such as those for climate change), and vice versa. These more complex models increasingly inform socio-economic simulations of human behavior in the context of climate. Furthermore, recent advances in Earth system modeling have been accompanied by progress in coupled data assimilation (CDA) (WMO, 2010). CDA uses observations in more than one component of a coupled model (e.g., atmosphere and ocean) so that the whole coupled model is optimized simultaneously, and observations in one subcomponent can influence the estimated state in another component.

Since major S–I climate phenomena, such as the Asian monsoon, the ENSO, and the Indian Ocean Dipole Mode, are strongly influenced by interactions between the atmosphere, ocean, and land surface, a comprehensive coupled model is required to further understand and predict realistic S–I climate fluctuations (e.g., Mason et al., 1999; Alves et al., 2004; Dombrowsky et al., 2009; Schiller and Brassington, 2011). Data assimilation by use of a coupled ocean–atmosphere model (hereafter CDA) has gathered increasing attention as a possible breakthrough in attempts to better understand climate variability by providing a consistent estimation of the state of the ocean–atmosphere system that would enable us to directly investigate such interactions. Also, for accurate initialization of S–I forecasts, Rienecker (2003) has shown that using the analysis fields obtained from CDA is better than using those derived in earlier studies from an atmospheric or oceanic GCM alone.

A fully coupled GCM sometimes generates rapidly growing modes, particularly in the atmospheric component, which make optimization of the simulated state of the atmosphere difficult. The actual coupled phenomena are thought to include a controllable dynamical nature in S–I processes because they should contain low-frequency modes generated and controlled by oceanic processes (Palmer et al., 2005). It is therefore possible that use of data assimilation could produce the optimal state in a coupled system.

CDA experiments have been initiated with intermediate models (e.g., Lee et al., 2000), intermediate complexity anomaly models (e.g., Chen et al., 2000), or hybrid models (e.g., Galanti et al., 2003) as the platforms, primarily because of the huge computational burden that data assimilation experiments with fully coupled models would require. Zhang et al. (2007) have recently applied an EnKF approach to an ocean–atmosphere CDA system with a fully coupled GCM. The assimilation successfully reconstructs the twentieth century ocean heat content variability and trends in most locations. The critical advance applied by Zhang et al. to produce their model was a "super"-parallelization technique for ensemble integrations.

Sugiura et al. (2008) took up the challenge of developing a sophisticated CDA system with a fully coupled GCM. Using the adjoint data assimilation method, they corrected the coupled fields by adjusting both the oceanic initial conditions and the drag (coupling) coefficients associated with mass, momentum, and heat exchange between the different spheres. One of the most fascinating essential elements of their approach is filtering out chaotic fluctuations that take place on the timescales of weather modes by operating an averaging procedure in order to highlight the representation and forecast of S–I variations. Their products provide dynamically self-consistent coupled fields that are suitable for the initial states in S–I prediction experiments.

The huge computational costs of fully CDA experiments sometimes limit their operational use; thus, a quasi-CDA approach is one solution. For example, Fujii et al. (2009) used only ocean observations for assimilation elements and consistently estimated coupled fields with a fully coupled model. They demonstrated that incorporation of modeled SST–precipitation interactions with this approach could improve estimation of the precipitation field. Most recently, a loosely coupled ocean–atmosphere data assimilation system has been implemented for a coupled global NCEP reanalysis for the period 1979–2011 (e.g., Saha et al., 2010); in this system, the data assimilation for the atmosphere and ocean proceeded independently. The ocean–atmosphere CDA approach is technically feasible and is expected to become more sophisticated and evolve markedly in the coming decade through continuing efforts by researchers.

5.1.2. Ocean–Sea Ice Data Assimilation

Sea-ice concentrations and thickness are important climate variables and foci of research because they not only constrain the freshwater budget but also, along with sea surface temperature, are lower boundary conditions for the atmosphere. The ability of a coupled GCM to represent interannual to interdecadal changes in climate sometimes depends on the performance of the sea-ice model (e.g., Lindsay and Zhang, 2006). Observations in polar regions are increasing over time. In particular, remote sensing by active radar and satellites (e.g., CryoSat-2) is providing increasingly higher quality sea-ice information that data assimilation techniques may use to better estimate sea-ice concentrations and thickness by dynamical interpolation of observations. Improved estimation leads to better understanding of the climate system by making possible more accurate forcing of atmospheric models and better estimation of ocean state. However, the solid/liquid phase transition of water makes implementation of a sea-ice data assimilation system difficult.

Lisæter et al. (2003) implemented the EnKF with a coupled ice–ocean model. They assimilated sea-ice concentration data to successfully estimate the distribution of the ice area. Recently, Caya et al. (2010) have developed a 3D-VAR ocean–ice data assimilation system with covariance parameters estimated from EnKF. Their system has significantly improved short-term sea-ice forecasts. Many others have attempted global sea-ice data assimilation (e.g., Stark et al., 2008; Scott et al., 2012).

Application of the adjoint method to the coupled ocean/sea-ice system is also being developed as part of the ECCO and ECCO2 (http://www.ecco2.org) efforts (Heimbach et al., 2010a,b). Sea-ice data assimilation for new observational elements (e.g., thickness, drift speed, and ocean state beneath the sea-ice) is another challenge (e.g., Lydersen et al., 2002; Lisæter et al., 2007; Rollenhagen et al., 2009). Further work on the adjoint method and new observational elements will lead to better understanding and more reliable forecasts of long-term climate changes that originate in the polar regions.

5.1.3. Biogeochemical Data Assimilation

The development of data assimilation methods and their application to a wider range of problems will likely proceed as a result of their proven utility in ocean climate research. Data assimilation for oceanic biogeochemical and ecological modeling is of interest because of the possible application of such models to sustainable management of marine resources. However, there are many issues to be addressed in data assimilation for such a complex system. In particular, there remains uncertainty in a variety of oceanic biogeochemical and ecosystem model parameters. Furthermore, a dynamically and kinematically consistent ocean state, deduced from some smoother approach or equivalent methods (e.g., Daley, 1991), may be required to ensure realistic performance of the biogeochemical and ecosystem models, because three-dimensional advection is one of the key processes determining the distributions of nutrients and plankton (e.g., Anderson and Robinson, 2001). At present, ocean color measurement from satellites is one of the most effective methods to constrain biogeochemical and ecosystem models.

Pioneering assimilation research has been done in recent years. Nerger and Gregg (2007) employed a global synthesis of chlorophyll-*a* data (SeaWiFS data derived from satellite measurements) with an ocean biogeochemical model to successfully enhance the representation of chlorophyll distributions. Ourmières et al. (2009) assimilated physical and nutrient data into a coupled physical–biogeochemical model and showed that nutrient data played a key role in representing phytoplankton patterns in the North Atlantic.

Recently, subsurface observation systems based on Argo floats have been under development for use in biogeochemical and ecosystem studies (e.g., Claustre et al., 2010). Although the spatio-temporal coverage of subsurface biogeochemical data is sparse at present, the number of *in situ* measurements and input from some multi-function floats are growing (e.g., Suga et al., 2011).

Brasseur et al. (2009) summarized recent progress in integrating biogeochemistry and ecology into ocean data assimilation. Improved oceanic environmental datasets will lead to new scientific findings, for example, with respect to marine biodiversity (e.g., Barton et al., 2010) and may also contribute to understanding of fish stocks and fisheries management. "During the next decade coupled physical–biogeochemical assimilation can be expected to mature, providing new insights not only to ocean biological variations and the marine carbon cycle but also into the feedbacks within the physical climate system" (Rienecker et al., 2010).

5.2. High-Resolution Data Assimilation and Climate Research

Because of recent progress in data assimilation technology and advances in computing power, a number of national centers have developed ocean analysis and forecasting systems that operate on regional and global scales (e.g., Brasseur et al., 2005). The Global Ocean Data Assimilation

Experiment (GODAE) and now GODAE OceanView (www.godae-oceanview.org) have fostered the development and improvement of operational ocean analysis and forecast systems worldwide. These systems include, inter alia, BLUElink in Australia (e.g., Schiller and Smith, 2006; Oke et al., 2008; Brassington et al. 2012); C-NOOFS in Canada; HYCOM/NCODA (e.g., Crosnier and Le Provost, 2006; Chassignet et al., 2007) and RTOFS in the United States; FOAM in the United Kingdom (e.g., Martin et al., 2007); INDOFOS in India; NMEFC in China; MERCATOR in France (e.g., Lagarde et al., 2001; De Mey and Benkiran, 2002; Drillet et al., 2008); MFS in Italy (e.g., Pinardi et al., 2003; Tonani et al., 2009); MOVE/MRI.COM in Japan (e.g., Usui et al., 2006); REMO in Brazil; and TOPAZ in Norway (Bertino and Lisæter, 2008; Oddo et al., 2009). Most of these systems assimilate real-time observations; more than half of them provide daily short-term forecasts. Could operational products from such systems contribute to better understanding of ocean dynamics or lead to new scientific findings in climate research?

Short-term ocean analysis and forecasting as championed by GODAE and GODAE OceanView involve timescales of days, whereas climate-oriented state estimation and assimilation fall in CLIVAR's realm and involve intraseasonal to decadal timescales. The interplay between these two types of efforts is developing with time as ocean analysis and forecasting extend the scope to reanalysis and as the resolution of climate-oriented ocean estimation and assimilation systems improves. Some short-term ocean forecast and decadal assimilation systems now share much of the data assimilation methodology and infrastructure. For example, the post-1991 eddy-resolving assimilation product for the Asian–Australian region (Schiller et al., 2008) is being created with a methodology similar to that of the BLUElink ocean forecasting system. The ongoing Global Ocean Reanalysis and Simulation project adopted the assimilation scheme developed for the MERCATOR ocean forecasting system. The high-resolution nature of the ocean analysis and forecasting systems benefits studies of regional ocean dynamics and climate (including regional sea level change). Some examples are described below.

The goals of BLUElink include the development of eddy-permitting, data-assimilating, ocean forecast systems. The BLUElink ocean data assimilation system is an EnOI system that utilizes one of the extended OI methods (Oke et al., 2008). BLUElink uses an ensemble of intraseasonal anomalies from simulated values to estimate background-error covariances. Ensemble-based background-error covariances can reflect the major features of mesoscale oceanic processes; BLUElink has been shown to realistically reproduce the mesoscale circulation around Australia. Accurate representation of mesoscale eddy and circulation behavior provides important information needed to realistically estimate mass and heat transport and to elucidate processes associated with water mass formation in conjunction with climate variability. For instance, Schiller et al. (2010) demonstrated the utility of BLUElink assimilation product BRAN2.1 to represent the observations collected by the INSTANT program and to study the dynamics of intraseasonal variability associated with the complicated pathways of the Indonesian throughflow (ITF) (Figure 22.5). Hill et al. (2011) reported that the 1/10° BLUElink assimilation product was more accurate in representing the regional circulation of the Tasman Front and East Australian Current extension than lower-resolution state estimation and assimilation products; such an accurate representation is important to the study of regional climate. Most recently, Divakaran and Brassington (2010) have discovered ocean zonal mean currents in the southeast Indian Ocean by using similar BLUElink products.

The Kuroshio is one of the main engines of horizontal mass and heat transport in the Northern Hemisphere and also plays an important role in various aspects of coastal oceanography around the western boundary of the North Pacific. By diagnosing the distribution of potential vorticity (PV) with an analytical dataset in the operational system MOVE/MRI.COM with a variable resolution of 1/6° to 1/10°, Usui et al. (2008) have studied the mechanism responsible for generating the large Kuroshio meander south of Japan: the process was associated with formation of a small meander (cyclonic eddy) in the upstream region, which triggered the subsequent large meander. The analysis revealed that a large frontal wave could contribute to the generation of a fully developed trigger meander by transporting a large amount of higher-PV water to the upstream region.

Even though the high-resolution systems described above were originally developed to improve the analytical and forecasting capabilities for shorter timescales (days to weeks), the results of studies based on these systems improve our understanding of regional dynamics that is also important in climate research. Note that the ocean is turbulent and dominated by mesoscale variability (Chelton et al., 2011). Hence, high-resolution ocean reanalyses can provide important first-order insights into basin-scale ocean current systems (e.g., Maximenko et al., 2008; Divakaran and Brassington, 2010).

An important future challenge is the development of seamless systems that will enable scientists to fully investigate multi-scale interactions (i.e., between short- and long-term and between small- and large-scale phenomena). This development is important because high-frequency and small-scale features may rectify low-frequency and large-scale phenomena, and large-scale climate signals may compound with synoptic variability (e.g., storm surge) to affect regional changes (e.g., for regional sea level).

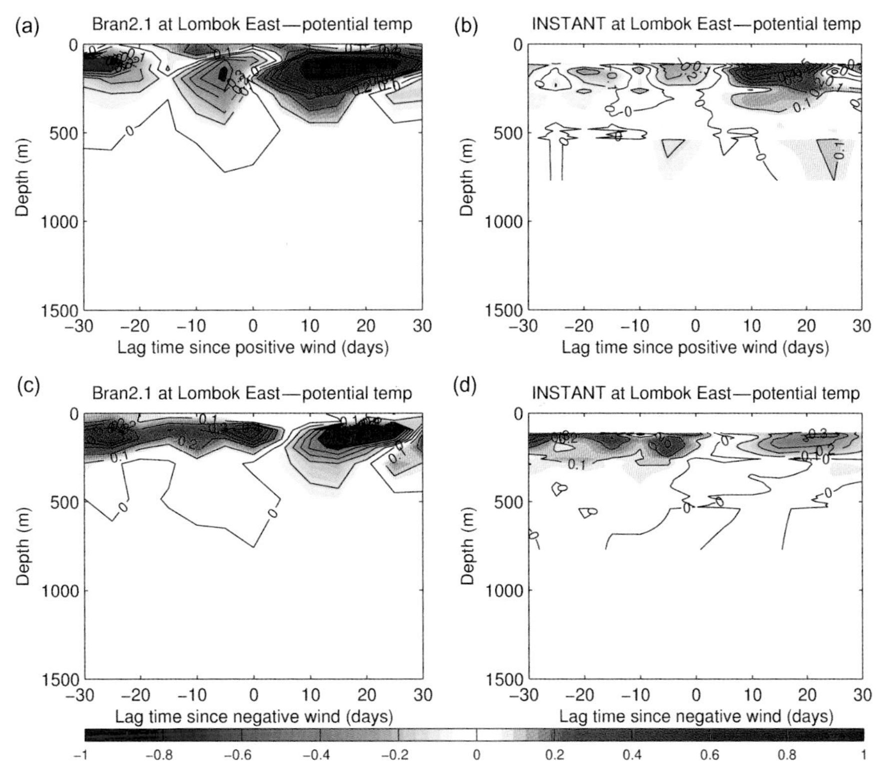

FIGURE 22.5 Time–depth plots of composites of potential temperature anomalies at mooring Lombok East for (a) and (c) BRAN2.1 reanalysis and (b) and (d) for INSTANT observations. Day 0: peak of negative and positive wind stress anomalies in equatorial eastern Indian Ocean. The color scheme saturates at ±1.0 °C; C.I.: 0.1 °C. Note missing near-surface data in observations. *Reprinted from Schiller et al. (2010).*

5.2.1. *Applications to Coastal Oceanography*

It will be important to provide analyses of the coastal zone to better understand land–ocean exchange processes that are relevant to climate change, for instance, in conjunction with the distribution and fluxes of freshwater. Highly accurate estimates of coastal sea conditions have been eagerly awaited. However, application of the data assimilation approach to coastal oceanography involves many complications (e.g., De Mey et al., 2009).

High-resolution models are required to represent near-shore phenomena on relatively fine temporal and spatial scales. Such models often produce strong currents that reduce controllability during the assimilation procedure because of inherent nonlinearities (Köhl and Willebrand, 2002). For example, adjoint sensitivities grow exponentially when the fields are highly nonlinear because of inconsistencies with a linearized adjoint model (Zhu et al., 2002). Previous studies have attempted to overcome such problems by, for example, using smoothed forward results (Köhl and Willebrand, 2002) or a larger diffusivity and viscosity in the adjoint model than in the forward model (Hoteit et al., 2005). Some other improved methods to estimate highly nonlinear fields have been developed based on established statistical methodologies or original techniques (e.g., Ishikawa et al., 2009; Abarbanel et al., 2010). Many such methods will be of practical use in coastal oceanographic applications.

Most researchers studying coastal ocean data assimilation have used regional models (e.g., Veneziani et al., 2009; Hirose, 2011). The boundary conditions imposed at the open boundary are important factors that constrain model results. An advanced downscaling technique that can provide a perfectly consistent interchange between outer and inner models is also a vital factor. Further development of the data assimilation technique and improved model implementation will inevitably require sustained observations of the finer structure of water properties and precise topographic information to improve model representations of near-shore phenomena.

5.3. Understanding Consistency and Uncertainty

A suite of ocean synthesis products have been produced in the past decade for various purposes. Few ocean synthesis products provide uncertainty estimates for inferred quantities (e.g., global ocean heat content and sea level change). There is an increasing need to understand the consistency and uncertainty of these products. This is a very challenging task because of the large number of factors that can contribute to the differences among these products. Among these factors are the differences in model (including the configuration, parameterization, resolution, etc.), in forcing, in assimilation or estimation method (including the way they are implemented such as the treatment of error estimates for model and data), and in the observational data being assimilated (e.g., data types, data sources). Decadal and longer variability and temporal inhomogeneity of observations could also contribute to the differences among different products. This challenge is not unique to ocean products, but for atmospheric analysis and reanalysis products as well. However, it is arguably more challenging for ocean synthesis products because the atmospheric products are not as diverse in terms of model, assimilation method, and even the purpose of the assimilation (originated from the need to initialize weather forecast). In practice it is very difficult to isolate the effect of one particular factor.

In the past few years, CLIVAR and GODAE OceanView have been coordinating an evaluation effort for global ocean synthesis products. Comparison of various products for many diagnostics has been conducted. But understanding the consistency of the products among one another and the consistency with observations (i.e., fidelity) are ongoing tasks. The spread among the ensemble of ocean synthesis products for a diagnostic quantity (e.g., global sea level or heat content change) can potentially be used as a measure of the uncertainty estimate (e.g., Corre et al., 2012). However, it is important to determine if ocean synthesis products have common bias or limitation that would give the appearance of relatively good consistency. The following are two examples showing that a good consistency among ocean synthesis products does not necessarily reflect a good fidelity.

Global ocean heat content and sea level are important indicators of the climate change. Carton and Santorelli (2008) compared estimates of global upper-ocean (0–700 m) heat content for the period of 1960–2002 from seven ocean synthesis products and two observation-based estimates (Levitus et al., 2005; Ishii et al., 2006). Six of the seven synthesis products are based on sequential assimilation methods that force the model state quite closely to temperature profiles from XBT data. As a result, their heat content estimates are relatively close to the two observational estimates. In particular, all have a "bump" in the 1970s to mid 1980s (with an increase in the early to mid 1970s and decrease in the early to mid 1980s) (Figure 22.6a). The GECCO product, based on the adjoint method (or 4D-VAR, a smoother method) rather than sequential methods, did not show that bump. This is presumably because the system was unable to adjust the surface forcing within the assumed uncertainty range to produce the bump in the heat content. One may interpret GECCO as incorrect as it does not capture that bump in heat content. However, a subsequent study (Domingues et al., 2008) showed that the heat content bump was actually due to the bias in XBT fall rate that had not been corrected in the XBT data at the time and the uncorrected XBT data were all used by the two observational analysis and the synthesis products. The revised estimate of heat content that accounted for XBT fall rate bias no longer shows that bump (Figure 22.6b) (Domingues et al., 2008; Ishii and Kimoto, 2009; Levitus et al., 2009). Domingues et al. (2008) showed that the correction of biases in *in situ* temperature data also had an effect on the estimate of global sea level change.

In another evaluation effort, Lee et al. (2010a,b) compared the estimates of ITF volume transport derived from 14 ocean synthesis products with the estimate based on *in situ* observations of the INSTANT program (Sprintall et al., 2009). On seasonal timescales, the ITF transports from 13 of the 14 synthesis products have a dominant annual signal. Only 1 of the 14 products shows a dominant semiannual signal like the estimate based on INSTANT data. This difference turns out to be caused by the difference in resolution. The 13 products that have a dominant annual signal all have relatively coarse resolution ($(1/4)^\circ$ or coarser). The one that has a dominant semiannual signal has a 18-km resolution (approximately $(1/6)^\circ$). The variability of ITF transport is affected by signals propagated from the Pacific and Indian Oceans. The signals from the Pacific Ocean have a dominant annual cycle and come into the ITF channels at shallower depths (thermocline and above, or shallower than 150 m) whereas the signals from the Indian Ocean have a dominant semiannual cycle and reach great depths (down to 1800 m) where the widths of the channels become very narrow. The coarser models capture the shallow signals from the Pacific well (that have a strong annual cycle) but failed to adequately capture the deep semiannual signals from the Indian Ocean.

In summary, understanding the consistency and uncertainty of ocean synthesis products requires an international coordination among ocean synthesis groups such as the ongoing evaluation effort coordinated by CLIVAR and GODAE OceanView. There are many factors that can affect the uncertainty of and consistency among ocean synthesis products, including the errors in surface forcing, observational data, and errors that are internal to the underlying dynamical models. The latter include many sources such as missing or inadequate physics (e.g., those related to mixing), poor representation of topography, and limited resolution. While ocean synthesis products attempt to correct model errors, understanding the sources or nature of model errors is crucial to the improvement of ocean synthesis (e.g., Fukumori, 2006). Therefore, a close collaboration among the ocean synthesis, modeling, and observational communities becomes increasingly important. Moreover, the ocean synthesis and atmospheric reanalysis communities need to work together to tackle over-arching issues such as the estimation of air–sea fluxes.

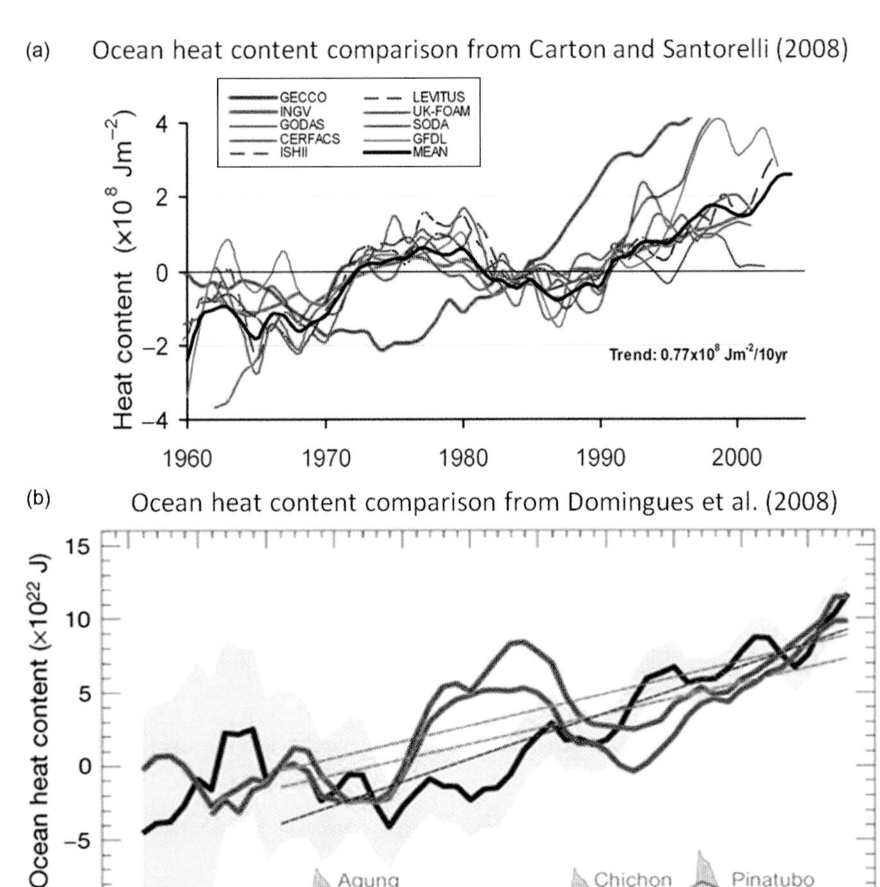

FIGURE 22.6 (a) Estimates of global upper-ocean (0–700 m) heat content from two observational analyses (Levitus et al., 2005; Ishii et al., 2006) and seven ocean data assimilation and estimation products. GECCO (using a smoother method) did not show the dump in 1970–1980. Used with permission. (b) Global upper-ocean heat content estimates based on XBT data corrected for fall rate bias (black curve) and those from Levitus et al (2005) and Ishii et al. (2006) without correcting for such the bias. The bump in the 1970s is gone when the bias in XBT data was corrected for the fall rate bias.

ACKNOWLEDGMENTS

Six anonymous reviewers provided valuable comments that helped to significantly improve the chapter. They deserve our gratitude. We greatly benefited from reading an early draft of Chapter 21. We also thank our editor Stephen M. Griffies for guiding us through the review process. A. S. received support from BLUElink, a partnership project between the Royal Australian Navy, CSIRO, and the Bureau of Meteorology. Funding for his work was provided by CSIRO and through the Royal Australian Navy.

REFERENCES

Abarbanel, H.D., Kostuk, M., Whartenby, W., 2010. Data assimilation with regularized nonlinear instabilities. Q. J. R. Meteorol. Soc. 136, 769–783. http://dx.doi.org/10.1002/qj.600.

Alves, O., Balmaseda, M.A., Anderson, D., Stockdale, T., 2004. Sensitivity of dynamical seasonal forecasts to ocean initial conditions. Q. J. R. Meteorol. Soc. 130, 647–667.

Anderson, B.D.O., Moore, J.B., 1979. Optimal Filtering. Prentice Hall, Upper Saddle River, New Jersey, 357 pp.

Anderson, L.A., Robinson, A.R., 2001. Physical and biological modeling in the Gulf Stream region. Part II. Physical and biological processes. Deep Sea Res. 48, 1139–1168.

Andersson, E., Järvinen, H., 1999. Variational quality control. Q. J. R. Meteorol. Soc. 125, 697–722.

Apte, A., Jones, C.K.R.T., Stuart, A.M., Voss, J., 2008. Data assimilation: mathematical and statistical perspectives. Int. J. Numer. Methods Fluids 56 (8), 1033–1046.

Aquarius/SAC-D Mission brochure, 2011. Sea surface salinity from space. NASA. http://aquarius.nasa.gov/index.html.

Awaji, T., Masuda, S., Ishikawa, Y., Sugiura, N., Toyoda, T., Nakamura, T., 2003. State estimation of the North Pacific Ocean by a four-dimensional variational data assimilation experiment. J. Oceanogr. 59, 931–943.

Baehr, J., McInerny, D., Keller, K., Marotzke, J., 2008. Optimization of an observing system design for the North Atlantic meridional overturning circulation. J. Atmos. Oceanic Technol. 25, 625–634. http://dx.doi.org/10.1175/2007JTECHO535.1.

Ballabrera-Poy, J., Hackert, E., Murtugudde, R., Busalacchi, A.J., 2007. An observing system simulation experiment for an optimal moored instrument array in the tropical Indian Ocean. J. Clim. 20, 3284–3299.

Balmaseda, M.A., Anderson, D., 2009. Impact of initialization strategies and observations on seasonal forecast skill. Geophys. Res. Lett. 36, L01701. http://dx.doi.org/10.1029/2008GL035561.

Balmaseda, M., Dee, D., Vidard, A., Anderson, D.L.T., 2007a. A multivariate treatment of bias for sequential data assimilation: application to the tropical oceans. Q. J. R. Meteorol. Soc. 133, 167–179.

Balmaseda, M.A., Smith, G.C., Haines, K., et al., 2007b. Historical reconstruction of the Atlantic Meridional Overturning Circulation from the ECMWF operational ocean reanalysis. Geophys. Res. Lett. 34, L23615. http://dx.doi.org/10.1029/2007GL031645.

Balmaseda, M.A., Vidard, A., Anderson, D., 2008. The ECMWF ORA-S3 ocean analysis system. Mon. Weather Rev. 136, 3018–3034.

Balmaseda, M.A., Alves, O.J., Arribas, A., Awaji, T., Behringer, D.W., Ferry, N., Fujii, Y., Lee, T., Rienecker, M., Rosati, T., Stammer, D., 2009. Ocean initialization for seasonal forecasts. Oceanography 22, 154–159.

Barton, A., Dutkiewicz, S., Flierl, G., Bragg, J., Follows, M., 2010. Patterns of diversity in marine phytoplankton. Science 327, 1509–1511. http://dx.doi.org/10.1126/science.1184961.

Behringer, D.W., 2007. The global ocean data assimilation system at NCEP. Paper Presented at the 11th Symposium on Integrated Observing and Assimilation Systems for Atmosphere, Oceans, and Land Surface, AMS 87th Annual Meeting, San Antonio, Texas.

Bennett, A.F., 1990. Inverse methods for assessing ship-of-opportunity networks and estimating cir- culation and winds from tropical expendable bathythermograph data. J. Geophys. Res. 95, 16111–16148.

Bennett, A.F., 1992. Inverse Methods in Physical Oceanography. Cambridge University Press, Cambridge.

Bennett, A.F., 2002. Inverse Modeling of the Ocean and Atmosphere. Cambridge University Press, Cambridge, 256 pp.

Bertino, L., Lisæter, K.A., 2008. The TOPAZ monitoring and pre- diction system for the Atlantic and Arctic Oceans. J. Oper. Oceanogr. 2, 15–18.

Bourlès, B., Arnault, S., Provost, C., 1992. Toward altimetric data assimilation in a Tropical Atlantic model. J. Geophys. Res. 97 (C12), 20/271–20/283.

Bouttier, F., Courtier, P., 1999. Data assimilation concepts and methods. http://www.ecmwf.int/newsevents/training/rcourse_notes/DATA_ASSIMILATION/ASSIM_CONCEPTS/Assim_concepts8.html.

Brasseur, P., Bahurel, P., Bertino, L., Birol, F., Brankart, J.M., Ferry, N., Losa, S., et al., 2005. Data assimilation for marine monitoring and prediction: the MERCATOR operational assimilation systems and the MERSEA developments. Q. J. R. Meteorol. Soc. 131 (613), 3561–3582. http://dx.doi.org/10.1256/qj.05.142.

Brasseur, P., Gruber, N., Barciela, R., Brander, K., Doron, M., El Moussaoui, A., Hobday, A.J., Huret, M., Kremeur, A.-S., Lehodey, P., Matear, R., Moulin, C., Murtugudde, R., Senina, I., Svendsen, E., 2009. Integrating biogeochemistry and ecology into ocean data assimilation systems. Oceanography 22 (3), 206–215. http://dx.doi.org/10.5670/oceanog.2009.80.

Brassington, G.B., Freeman, J., Huang, X., Pugh, T., Oke, P.R., Sandery, P. A., Taylor, A., Andreu-Burillo, I., Schiller, A., Griffin, D.A., Fiedler, R., Mansbridge, J., Beggs, H., Spillman, C.M., 2012. Ocean model, analysis and prediction system (OceanMAPS): version 2, CAWCR Technical Report No. 052, 110 pp.

Bryan, K., 1969. Climate and ocean circulation. III. The ocean model. Mon. Weather Rev. 97, 806–827.

Bryan, K., Cox, M.D., 1967. A numerical investigation of the oceanic general circulation. Tellus 19, 54–80.

Bryden, H.L., Longworth, H.R., Cunningham, S.A., 2005. Slowing of the Atlantic meridional overturning circulation at 25 degrees N. Nature 438, 655–657.

Cabanes, C., Lee, T., Fu, L.-L., 2008. Mechanisms of interannual variations of the meridional overturning circulation of the North Atlantic Ocean. J. Phys. Oceanogr. 38, 467–480.

Carton, J.A., Santorelli, A., 2008. Global decadal upper-ocean heat content as viewed in nine analyses. J. Clim. 21, 6015–6035.

Caya, A., Buehner, M., Carrieres, T., 2010. Analysis and forecasting of sea ice conditions with three dimensional variational data assimilation and a coupled ice-ocean model. J. Atmos. Oceanic Tech. 27, 353–369.

Cazes-Boezio, G., Menemenlis, D., Mechoso, C.R., 2008. Impact of ECCO ocean-state estimates on the initialization of seasonal climate forecasts. J. Clim. 21, 1929–1947.

Chassignet, E.P., Hurlburt, H.E., Smedstad, O.M., Halliwell, G.R., Hogan, P.J., Wallcraft, A.J., Baraille, R., Bleck, R., 2007. The HYCOM (HYbrid Coordinate Ocean Model) data assimilative system. J. Mar. Syst. 65, 60–83.

Chelton, D.B., Schlax, M.G., Samelson, R.M., 2011. Global observations of nonlinear mesoscale eddies. Prog. Oceanogr. 91, 167–216.

Chen, D., Cane, M.A., Zebiak, S.E., Canizales, R., Kaplan, A., 2000. Bias correction of an ocean-atmosphere coupled model. Geophys. Res. Lett. 27 (16), 2585–2588.

Chua, B., Bennett, A.F., 2001. An inverse ocean modeling system. Ocean Model. 3, 137–165.

Claustre, H., et al., 2010. Bio-Optical profiling floats as new observational tools for biogeochemical and ecosystem studies: potential synergies with ocean color remote sensing. In: Hall, J., Harrison, D.E., Stammer, D. (Eds.), Proceedings of OceanObs'09: Sustained Ocean Observations and Information for Society (vol. 2), Venice, Italy, 21–25 September 2009. ESA Publication WPP-306, New York http://dx.doi.org/10.5270/OceanObs09.cwp.17.

Cohn, S.E., Sivakumaran, N.S., Todling, R., 1994. A fixed-lag Kalman smoother for retrospective data assimilation. Mon. Weather Rev. 122, 2838–2867.

Corre, L., Terray, L., Balmaseda, M., et al., 2012. Can oceanic reanalyses be used to assess recent anthropogenic changes and low-frequency internal variability of upper ocean temperature? Clim. Dyn. 38, 877–896. http://dx.doi.org/10.1007/s00382-010-0950-8.

Cosme, E., Brankart, J.M., Verron, J., Brasseur, P., Krysta, M., 2010. Implementation of a reduced rank square-root smoother for high resolution ocean data assimilation. Ocean Model. 33 (1–2), 87–100.

Courtier, P., Andersson, E., Heckley, W., Pailleux, J., Vasiljevic, D., Hamrud, M., Hollingsworth, A., Rabier, F., Fisher, M., 1998. The ECMWF implementation of three-dimensional variational assimilation (3D-Var). Part 1: formulation. Q. J. R. Meteorol. Soc. 124, 1783–1807.

Crosnier, L., Le Provost, Ch., 2006. Internal metrics definition for operational forecast systems intercomparison: example in the North Atlantic and Mediterranean Sea. In: Chassignet, E., Verron, J.

(Eds.), Ocean Weather Forecasting: An Integrated View of Oceanography. Springer, Dordrecht, The Netherlands, pp. 455–465.

Cummings, J.A., 2011. Ocean data quality control. In: Schiller, A., Brassington, G. (Eds.), Operational Oceanography in the 21st Century. Springer, The Netherlands, pp. 91–121.

Cunningham, S.A., Kanzow, T., Rayner, D., Baringer, M.O., Johns, W.E., Marotzke, J., Longworth, H.R., Grant, E.M., Hirschi, J.J.-M., Beal, L.M., et al., 2007. Temporal variability of the Atlantic meridional overturning circulation at 26.5°N. Science 317, 935–938. http://dx.doi.org/10.1126/science.1141304.

Cunningham, S., Baringer, M., Johns, B., Toole, J., Østerhus, S., Fisher, J., Piola, A., McDonagah, E., Lozier, S., Send, U., Kanzow, T., Marotzke, J., Rhein, M., Garzoli, S., Rintoul, S., Sloyan, B., Speich, S., Talley, L., Baehr, J., Meinen, C., Treguier, A., Lherminier, P., 2010. The present and future system for measuring the Atlantic meridional overturning circulation and heat transport. In: Hall, J., Harrison, D.E., Stammer, D. (Eds.), Proceedings of OceanObs'09: Sustained Ocean Observations and Information for Society (vol. 2), Venice, Italy, 21–25 September 2009. ESA Publication WPP-306, New York. http://dx.doi.org/10.5270/OceanObs09.cwp.21.

Daley, R., 1991. Atmospheric Data Analysis. Cambridge University Press, Cambridge.

De Mey, P., Benkiran, M., 2002. A multivariate reduced-order optimal interpolation method and its application to the Mediterranean basin-scale circulation. In: Pinardi, N., Woods, J.D. (Eds.), Ocean Forecasting, Conceptual Basis and Applications. Springer-Verlag, Berlin, Heidelberg, New York, 472 pp.

De Mey, P., Craig, P., Davidson, F., Edwards, C.A., Ishikawa, Y., Kindle, J.C., Proctor, R., Thompson, K.R., Zhu, J., The GODAE Coastal and Shelf Seas Working Group (CSSWG) community, 2009. Applications in coastal modelling and forecasting. Oceanography 22 (3), 198–205.

Derber, J., Rosati, A., 1989. A global oceanic data assimilation system. J. Phys. Oceanogr. 19, 1333–1347, Oceanography 22, 3, 198–205.

Divakaran, P., Brassington, G.B., 2010. Arterial ocean circulation of the Southeast Indian Ocean. Geophys. Res. Lett. 38, http://dx.doi.org/10.1029/2010GL045574.

Dombrowsky, E., Bertino, L., Brassington, G.B., Chassignet, E.P., Davidson, F., Hurlburt, H.E., Kamachi, M., Lee, T., Martin, M.J., Mei, S., Tonani, M., 2009. GODAE systems in operation. Oceanography 22 (3), 80–95.

Domingues, C.M., Church, J.A., White, N.J., et al., 2008. Improved estimates of upper-ocean warming and multi-decadal sea level rise. Nature 453, 1090–1093. http://dx.doi.org/10.1038/nature07080.

Dommenget, D., Stammer, D., 2004. Assessing ENSO simulations and predictions using adjoint ocean state estimation. J. Clim. 17, 4301–4315. http://dx.doi.org/10.1175/3211.1.

Drillet, Y., Bricaud, C., Bourdallé-Badie, R., Derval, C., Le Galloudec, O., Garric, G., Testut, C.E., Tranchant, B., 2008. The Mercator Ocean global 1/12° operational system: demonstration phase in the MERSEA context, The Mercator Ocean Newsletter #29, April 2008.

Evensen, G., 1994. Sequential data assimilation with nonlinear quasi-geostrophic model using Monte Carlo methods to forecast error statistics. J. Geophys. Res. 99 (C5), 143–162.

Evensen, G., 2003. The Ensemble Kalman Filter: theoretical formulation and practical implementation. Ocean Dyn. 53, 343–367.

Evensen, G., van Leeuwen, P.J., 2000. An ensemble kalman smoother for nonlinear dynamics. Mon. Weather Rev. 128, 1852–1867.

Fischer, A. S., Hall, J., Harrison, D.E., Stammer D., 2010. Ocean information for society: sustaining the benefits, realizing the potential. OceanObs'09 Conference Summary. http://www.oceanobs09.net/documents/OceanObs09-Conference_Summary-draft_24OCT10.pdf. Unpublished manuscript.

Fisher, M., Andersson, E., 2001. Developments in 4D-Var and Kalman filtering. ECMWF Tech. Memo. 347, 36 pp.

Freychet, N., Cosme, E., Brasseur, P., Brankart, J.-M., Kpemlie, E., 2012. Obstacles and benefits of the implementation of a reduced rank smoother with a high resolution model of the Atlantic ocean. Ocean Sci. Discuss. 9, 1187–1229. http://dx.doi.org/10.5194/osd-9-1187-2012.

Fu, L.L., Cazenave, A., 2001. Satellite altimetry and earth sciences. A Handbook of Techniques and Applications. Academic Press, San Diego.

Fujii, Y., Ishizaki, S., Kamachi, M., 2005. Application of nonlinear constraints in a three-dimensional variational ocean analysis. J. Oceanogr. 61, 655–662.

Fujii Y., M. Kamachi, 2003. Three dimensional analysis of temperature and salinity in the equatorial Pacific using a variational method with vertical coupled temperature-Salinity empirical orthogonal function modes. J. Geophys. Res. 108(C9): 3297, http://dx.doi.org/10.1029/2002JC001745.

Fujii, Y., Yasuda, T., Matsumoto, S., Kamachi, M., Ando, K., 2008. Observing system evaluation (OSE) using the El Niño forecasting system in Japan Meteorological Agency. Paper Presented at the Fall Meeting of the Oceanographic Society of Japan (in Japanese).

Fujii, Y., Nakaegawa, T., Matsumoto, S., Yasuda, T., Yamanaka, G., Kamachi, M., 2009. Coupled climate simulation by constraining ocean fields in a coupled model with ocean data. J. Clim. 22, 5541–5557.

Fukumori, I., 2002. A partitioned kalman filter and smoother. Mon. Weather Rev. 130, 1370–1383. http://dx.doi.org/10.1175/1520-0493(2002).

Fukumori, I., 2006. What is data assimilation really solving, and how is the calculation done? In: Chasignet, E.P., Verron, J. (Eds.), Ocean Weather Forecasting: An integrated view of oceanography. Springer, Dordrecht. 578 pp.

Fukumori, I., Lee, T., Cheng, B., Menemenlis, D., 2004. The origin, pathway, and destination of NINO3 water estimated by a simulated passive tracer and its adjoint. J. Phys. Oceanogr. 34, 582–604.

Galanti, E., Tziperman, E., Harrison, M., Rosati, A., Sirkes, Z., 2003. A study of ENSO prediction using a hybrid coupled model and the adjoint method for data assimilation. Mon. Weather Rev. 131, 2748–2764.

Gandin, L.S., 1963. Objective Analysis of Meteorological Fields. Hydromet Press, Leningrad.

Gelb, A., 1974. Applied Optimal Estimation. MIT press, Cambridge Mass, 374 pp.

Griffies, S., Adcroft, A., Banks, H., Boening, C., Chassignet, E., Danabasoglu, G., Danilov, S., Deleersnijder, E., Drange, H., England, M., Fox-Kemper, B., Gerdes, R., Gnanadesikan, A., Greatbatch, R., Hallberg, R., Hanert, E., Harrison, M., Legg, S., Little, C., Madec, G., Marsland, S., Nikurashin, M., Pirani, A., Simmons, H., Schroter, J., Samuels, B., Treguier, A., Toggweiler, J., Tsujino, H., Vallis, G., White, L., 2010. Problems and prospects in large-scale ocean circulation models. In: Hall, J., Harrison, D.E., Stammer, D. (Eds.), Proceedings of OceanObs'09: Sustained Ocean Observations and Information for Society (Vol. 2),

Venice, Italy, 21–25 September 2009. ESA Publication WPP-306, New York. http://dx.doi.org/10.5270/OceanObs09.cwp.38.

Hackert, E.C., Miller, R.N., Busalacchi, A.J., 1998. An optimized design for a moored instrument array in the tropical Atlantic Ocean. J. Geophys. Res. 103, 7491–7509.

Haines, K., Blower, J.D., Drecourt, J.P., Liu, C., Vidard, A., Astin, I., Zhou, X., 2006. Salinity assimilation using S(T): covariance relationships. Mon. Weather Rev. 134 (3), 759–771.

Halkides, D., Lee, T., 2009. Mechanisms controlling seasonal-to-interannual mixed-layer temperature variability in the southeastern tropical Indian Ocean. J. Geophys. Res. 114, C02012. http://dx.doi.org/10.1029/2008JC004949.

Halkides, D., Lee, T., 2011. Mechanisms controlling seasonal mixed layer temperature and salinity in the southwestern tropical Indian Ocean. Dyn. Atmos. Oceans 51, 77–93. http://dx.doi.org/10.1016/j.dynatmoce.2011.03.002.

Halkides, D., Lee, T., Kida, S., 2011a. Mechanisms controlling seasonal mixed layer temperature and salinity of the Indonesian Seas. Ocean Dyn. 6 (4), 481. http://dx.doi.org/10.1007/s10236-010-0374-3.

Halkides, D., Lucas, L.E., Waliser, D.E., Lee, T., Murtugudde, R., 2011b. Mechanisms controlling mixed layer temperature variability in the eastern tropical Pacific on the intraseasonal timescale. Geophys. Res. Lett. 38. http://dx.doi.org/10.1029/2011GL048545.

Haney, R.L., 1971. Surface thermal boundary condition for ocean circulation models. J. Phys. Oceanogr. 1, 241–248.

Heimbach, P., Forget, G., Ponte, R., Wunsch, C., Balmaseda, M., Awaji, T., Baehr, J., Behringer, D., Carton, J., Ferry, N., Fischer, A., Fukumori, I., Giese, B., Haines, K., Harrison, E., Hernandez, F., Kamachi, M., Keppenne, C., Köhl, A., Lee, T., Menemenlis, D., Oke, P., Remy, E., Rienecker, M., Rosati, A., Smith, D., Speer, K., Stammer, D., Weaver, A., 2010a. Observational requirements for global-scale ocean climate analysis: lessons from ocean state estimation. In: Hall, J., Harrison, D.E., Stammer, D. (Eds.), Proceedings of OceanObs'09: Sustained Ocean Observations and Information for Society (vol. 2), Venice, Italy, 21–25 September 2009. ESA Publication WPP-306, New York. http://dx.doi.org/10.5270/OceanObs09.cwp.42.

Heimbach, P., Menemenlis, D., Losch, M., Campin, J., Hill, C., 2010b. On the formulation of sea-ice models. Part 2: lessons from multi-year adjoint sea ice export sensitivities through the Canadian Arctic Archipelago. Ocean Model. 33, 145–158.

Heimbach, P., Wunsch, C., Ponte, R.M., Forget, G., Hill, C., Utke, J., 2011. Timescales and regions of the sensitivity of Atlantic meridional volume and heat transport: Toward observing system design. Deep Sea Res. II 58, 1858–1879.

Hill, K.L., Rintoul, S.R., Ridgeway, K.R., Oke, P.R., 2011. Decadal changes in the South Pacific western boundary current system revealed in observations and ocean state estimates. J. Geophys. Res. 116. http://dx.doi.org/10.1029/2009JC005926.

Hirose, N., 2011. Inverse estimation of empirical parameters used in a regional ocean circulation model. J. Oceanogr. 67, 323–336. http://dx.doi.org/10.1007/s10872-011-0041-4.

Hoteit, I., Cornuelle, B., Köhl, A., Stammer, D., 2005. Treating strong adjoint sensitivities in tropical eddy-permitting variational data assimilation. Q. J. R. Meteorol. Soc. 131, 3659–3682.

Hoteit, I., Cornuelle, B., Heimbach, P., 2010. An eddy-permitting, dynamically consistent adjoint-based assimilation system for the tropical Pacific: hindcast experiments in 2000. J. Geophys. Res. 115, C03001.

Hunt, B.R., Kalnay, E., Kostelich, E.J., Ott, E., Patil, D.J., Sauer, T., Szunyogh, I., Yorke, J.A., Zimin, A.V., 2004. Four-dimensional ensemble kalman filtering. Tellus 56A, 273–277.

Hurlburt, H.E., 1984. The potential for ocean prediction and the role of altimeter data. Mar. Geod. (1–4), 17–66.

Hólm, E.V., 2003. Lecture notes on assimilation algorithms, http://www.ecmwf.int/newsevents/training/rcourse_notes/pdf_files/Assim_algorithms.pdf.

IOP (CLIVAR-GOOS Indian Ocean Panel), 2005. Report of the second Indian ocean panel meeting. International CLIVAR Project Office. CLIVAR Publication Series No. 92, March/April 2005, 22 pp, unpublished manuscript.

IPCC First Assessment Report, 1990. Scientific Assessment of Climate Change—Report of Working Group I. In: Houghton, J.T., Jenkins, G.J., Ephraums, J.J. (Eds.), Cambridge University Press, UK, 365 pp.

Ishii, M., Kimoto, M., 2009. Reevaluation of historical ocean heat content variations with time-varying xbt and mbt depth bias. J. Oceanogr. 65, 287–299. http://dx.doi.org/10.1007/s10872-009-0027-7.

Ishii, M., Kimoto, M., Sakamoto, K., Wasaki, S.-I., 2006. Steric sea level changes estimated from historical ocean subsurface temperature and salinity analyses. J. Oceanogr. 62, 155–170.

Ishikawa, Y., Awaji, T., Toyoda, T., Nishina, I.K., Nakayama, T., Shima, S., Masuda, S., 2009. High-resolution synthetic monitoring by a 4-dimensional variational data assimilation system in the northwestern North Pacific. J. Geophys. Res. 114, C11003. http://dx.doi.org/10.1029/2008JC005208.

Ji, M., Behringer, D.W., Leetma, A., 1998. An improved coupled model for ENSO prediction and implications for ocean initialization. Part II: the coupled model. Mon. Weather Rev. 4, 1022–1034. http://dx.doi.org/10.1175/1520-0493.

Johns, W.E., Baringer, M.O., Beal, L.M., Cunningham, S.A., Kanzow, T., Bryden, H.L., Hirschi, J.J.M., Marotzke, J., Meinen, C.S., Shaw, B., Curry, R., 2011. Continuous, array-based estimates of Atlantic ocean heat transport at 26.5 degrees N. J. Clim. 24, 2429–2449. http://dx.doi.org/10.1175/2010JCLI3997.1.

Kalman, R.E., 1960. A new approach to linear filtering and prediction problems. J. Basic Eng. 82 (1), 35–45.

Kamachi, M., Kuragano, T., Ichikawa, H., Nakamura, H., Nishina, A., Isobe, A., Ambe, D., Arai, M., Gohda, N., Sugimoto, S., Yoshita, K., Sakurai, T., Uboldi, F., 2004. Operational data assimilation system for the Kuroshio south of Japan: reanalysis and validation. J. Oceanogr. 60, 303–312.

Kim, S.-B., Lee, T., Fukumori, I., 2004. The 1997–99 abrupt change of the upper ocean temperature in the north central Pacific. Geophys. Res. Lett. 31, L22304. http://dx.doi.org/10.1029/2004GL021142.

Kim, S.-B., Lee, T., Fukumori, I., 2007. Mechanisms controlling the interannual variation of mixed layer temperature averaged over the NINO3 region. J. Clim. 20, 3822–3843.

Kitagawa, G., Gersch, W., 1996. Lecture Notes in Statistics 116: Smoothness Priors Analysis of Time Series. Springer-Verlag, New York.

Klemas, V., 2011. Remote sensing techniques for studying coastal ecosystems: an overview. J. Coast. Res. 27, 2–17.

Köhl, A., Stammer, D., 2008a. Variability of the meridional overturning in the North Atlantic from 50-year GECCO state estimation. J. Phys. Oceanogr. 38, 1913–1930.

Köhl, A., Stammer, D., 2008b. Decadal sea level changes in the 50-year GECCO ocean synthesis. J. Clim. 21, 1866–1890.

Köhl, A., Willebrand, J., 2002. An adjoint method for the assimilation of statistical characteristics into eddy-resolving ocean models. Tellus A 54, 406–425.

Köhl, A., Stammer, D., Cornulle, B., 2007. Interannual to decadal changes in the ECCO global synthesis. J. Phys. Oceanogr. 37, 313–337.

Kuragano, T., Kamachi, M., 2000. Global statistical spacetime scales of oceanic variability estimated from the TOPEX/POSEIDON altimeter data. J. Geophys. Res. 105, 955–974.

Lagarde, T., Piacentini, A., Thual, O., 2001. A new representation of data-assimilation methods: the PALM flow-charting approach. Q. J. R. Meteorol. Soc. 127 (571), 189–207.

Lee, T., Fukumori, I., 2003. Interannual to decadal variation of tropical-subtropical exchange in the Pacific Ocean: boundary versus interior pycnocline transports. J. Clim. 16, 4022–4042.

Lee, T., Boulanger, J.-P., Foo, A., Fu, L.-L., Giering, R., 2000. Data assimilation by an intermediate coupled ocean–atmosphere model: application to the 1997–1998 El Nino. J. Geophys. Res. 105, 26063–26087.

Lee, T., Awaji, T., Balmaseda, M., Greiner, E., Stammer, D., 2009. Ocean state estimation for climate research. Oceanography 22, 160–167.

Lee, T., Awaji, T., Balmaseda, M., et al., 2010a. Consistency and fidelity of Indonesian-throughflow total volume transport estimated by 14 ocean data assimilation products. Dyn. Atmos. Oceans. http://dx.doi.org/10.1016/j.dynatmoce.2009.12.004.

Lee, T., Stammer, D., Awaji, T., Balmaseda, M., Behringer, D., Carton, J., Ferry, N., Fischer, A., Fukumori, I., Giese, B., Haines, K., Harrison, E., Heimbach, P., Kamachi, M., Keppenne, C., Köhl, A., Masina, S., Menemenlis, D., Ponte, R., Remy, E., Rienecker, M., Rosati, A., Schroeter, J., Smith, D., Weaver, A., Wunsch, C., Xue, Y., 2010b. Ocean state estimation for climate research. In: Hall, J., Harrison, D.E., Stammer, D. (Eds.), Proceedings of OceanObs'09: Sustained Ocean Observations and Information for Society (vol. 2), Venice, Italy, 21–25 September 2009. ESA Publication WPP-306, New York. http://dx.doi.org/10.5270/OceanObs09.cwp.55.

Levitus, S., Antonov, J.I., Boyer, T., 2005. Warming of the world ocean, 1955–2003. Geophys. Res. Lett. 32, L02604. http://dx.doi.org/10.1029/2004GL021592.

Levitus, S., Antonov, J.I., Boyer, T.P., et al., 2009. Global ocean heat content 1955–2008 in light of recently revealed instrumentation problems. Geophys. Res. Lett. 36, L07608. http://dx.doi.org/10.1029/2008GL037155.

Lindsay, R.W., Zhang, J., 2006. Assimilation of ice concentration in an ice-ocean model. J. Atmos. Oceanic Tech. 23, 742–749.

Lisæter, K.A., Rosanova, J., Evensen, G., 2003. Assimilation of ice concentration in a coupled ice-ocean model using the ensemble kalman filter. Ocean Dyn. 53, 368–388.

Lisæter, K.A., Evensen, G., Laxon, S., 2007. Assimilating synthetic CryoSat sea ice thickness in a coupled ice-ocean model. J. Geophys. Res. 112, C07023. http://dx.doi.org/10.1029/2006JC003786.

Lydersen, C., Nøst, O.A., Lovell, P., McConnell, B.J., Gammelsrød, T., Hunter, C., Fedak, M.A., Kovacs, K.M., 2002. Salinity and temperature structure of a freezing Arctic fjord—monitored by white whales (Delphinapterus leucas). Geophys. Res. Lett. 29 (23), 2119. http://dx.doi.org/10.1029/2002GL015462.

Malanotte-Rizzoli, P., 1996. Modern Approaches to Data Assimilation in Ocean Modeling. Elsevier, Amsterdam, 455 pp.

Manda, A., Hirose, N., Yanagi, T., 2003. Application of a nonlinear and non- Gaussian sequential estimation method for an ocean mixed layer model. Eng. Sci. Rep. Kyushu Univ. 25, 285–289.

Marotzke, J., Giering, R., Zhang, K.Q., Stammer, D., Hill, C., Lee, T., 1999. Construction of the adjoint MIT ocean general circulation model and application to Atlantic heat transport sensitivity. J. Geophys. Res. 104, 29529–29547.

Martin, M.J., Hines, A., Bell, M.J., 2007. Data assimilation in the FOAM operational short-range ocean forecasting system: a description of the scheme and its impact. Q. J. R. Meteorol. Soc. 133, 981–995.

Mason, S.J., Goddard, L., Graham, N.E., Yulaeva, E., Sun, L., Arkin, P.A., 1999. The IRI seasonal climate prediction system and the 1997/98 El Niño event. Bull. Am. Meteorol. Soc. 80, 1853–1873.

Massart, S., Pajot, B., Piacentini, A., Pannekoucke, O., 2010. On the merits of using a 3D-FGAT assimilation scheme with an outer loop for atmospheric situations governed by transport. Mon. Weather Rev. 138, 4509–4522. http://dx.doi.org/10.1175/2010MWR3237.1.

Masuda, S., Awaji, T., Sugiura, N., Ishikawa, Y., Baba, K., Horiuchi, K., Komori, N., 2003. Improved estimates of the dynamical state of the North Pacific Ocean from a 4 dimensional variational data assimilation. Geophys. Res. Lett. 30 (16), 1868. http://dx.doi.org/10.1029/2003GL017604.

Masuda, S., Awaji, T., Sugiura, N., Toyoda, T., Ishikawa, Y., Horiuchi, K., 2006. Interannual variability of temperature inversions in the subarctic North Pacific. Geophys. Res. Lett. 33, L24610. http://dx.doi.org/10.1029/2006GL027865.

Masuda, S., Awaji, T., Sugiura, N., Matthews, J.P., Toyoda, T., Kawai, Y., Doi, T., Kouketsu, S., Igarashi, H., Katsumata, K., Uchida, H., Kawano, T., Fukasawa, M., 2010. Simulated rapid warming of abyssal north Pacific waters. Science 329, 319–322. http://dx.doi.org/10.1126/science.1188703.

Maximenko, N.A., Melnichenko, O.V., Niiler, P.P., Sasaki, H., 2008. Stationary mesoscale jet-like features in the ocean. Geophys. Res. Lett. 35, L08603. http://dx.doi.org/10.1029/2008GL033267.

Menemenlis, D., Fukumori, I., Lee, T., 2005. Using Green's functions to calibrate an ocean general circulation model. Mon. Weather Rev. 133, 1224–1240.

Menke, W., 1984. Geophysical Data Analysis: Discrete Inverse Theory. Academic Press, Inc., New York.

Müller, P., Willebrand, J., 1989. Equations for oceanic motions. In: Sündermann, J. (Ed.), Oceanography, Landolt-Börnstein—Numerical Data and Functional Relationships in Science and Technology, Group V Geophysics. http://dx.doi.org/10.1007/10312310_1. 1-14.

Nerger, L., Gregg, W.W., 2007. Assimilation of SeaWiFS data into a global ocean-biogeochemical model using a Local SEIK filter. J. Mar. Syst. 68, 237–254.

Oddo, P., Adani, M., Pinardi, N., Fratianni, C., Tonani, M., Pettenuzzo, D., 2009. A nested Atlantic-Mediterranean sea general circulation model for operational forecasting. Ocean Sci. 5, 461–473.

Oke, P.R., Brassington, G.B., Griffin, D.A., Schiller, A., 2010. Ocean data assimilation: a case for ensemble optimal interpolation. Aust. Meteorol. Oceanogr. J. 59, 67–76.

Oke, P.R., Schiller, A., 2007a. Impact of Argo, SST and altimetry on an eddy-resolving ocean reanalysis. Geophys. Res. Lett. 34, L19601. http://dx.doi.org/10.1029/2007GL031549.

Oke, P., Schiller, A., 2007b. A model-based assessment and design of a tropical Indian Ocean mooring array. J. Clim. 20, 3269–3283.

Oke, P.R., Allen, J.S., Miller, R.N., Egbert, G.D., Kosro, P.M., 2002. Assimilation of surface velocity data into a primitive equation coastal ocean model. J. Geophys. Res. 107, 3122. http://dx.doi.org/10.1029/2000JC000511.

Oke, P.R., Brassington, G.B., Griffin, D.A., Schiller, A., 2008. The Bluelink Ocean Data Assimilation System (BODAS). Ocean Model. 20, 46–70. http://dx.doi.org/10.1016/j.ocemod.2007.11.002.

Ott, E., Hunt, B.R., Szunyogh, I., Zimin, A.V., Kostelich, E.J., Kostelich, M., Corazza, M., Sauer, T., Kalnay, E., Patil, D.J., Yorke, J.A., 2004. A local ensemble kalman filter for atmospheric data assimilation. Tellus 56A, 415–428.

Ourmières, Y., Brasseur, P., Lévy, M., Brankart, J.-M., Verron, J., 2009. On the key role of nutrient data to constrain a coupled physical-biogeochemical assimilative model of the North Atlantic ocean. J. Mar. Syst. 75, 100–115. http://dx.doi.org/10.1016/j.jmarsys.2008.08.003.

Palmer, T.N., Doblas-Reyes, F.J., Hagedorn, R., Weisheimer, A., 2005. Probabilistic prediction of climate using multi-model ensembles: from basics to applications. Philos. Trans. R. Soc. Lond. B 360, 1991–1998. http://dx.doi.org/10.1098/rstb.2005.1750.

Parrish, D., Derber, J., 1992. The National Meteorological Center's spectral statistical-interpolation analysis system. Mon. Weather Rev. 120, 1747–1763.

Pegion, P.J., Bourassa, M.A., Legler, D.M., O'Brien, J.J., 2000. Objectively derived daily "winds" from satellite scatterometer data. Mon. Weather Rev. 128, 3150–3168. http://dx.doi.org/10.1175/1520-0493(2000).

Pham, D., Verron, J., Roubaud, M., 1997. Singular evolutive Kalman filter with EOF initialization for data assimilation in oceanography. J. Mar. Syst. 16, 323–340.

Pinardi, N., Allen, I., Demirov, E., De Mey, P., Korres, G., Lascaratos, A., Le Traon, P.-Y., Maillard, C., Manzella, G., Tziavos, C., 2003. The Mediterranean ocean forecasting system: first phase of implementation (1998–2001). Ann. Geophys. 21, 3–20.

Pohlmann, H., Jungcaus, J.H., Köhl, A., Stammer, D., Marotzke, J., 2009. Initializing decadal climate predictions with the GECCO oceanic synthesis: effects on the North Atlantic. J. Clim. 22, 3926–3938. http://dx.doi.org/10.1175/2009JCLI2535.1.

Rabe, B., Schott, F.A., Köhl, A., 2008. Mean circulation and variability of the tropical Atlantic during 1952–2001 in the GECCO assimilation fields. J. Phys. Oceanogr. 38, 177–192.

Ricci, S., Weaver, A.T., Vialard, J., Rogel, P., 2005. Incorporating state-dependent temperature-salinity constraints in the background-error covariance of variational ocean data assimilation. Mon. Weather Rev. 133, 317–338.

Richardson, L.F., 1922. Weather Prediction by Numerical Process. Cambridge University Press, Cambridge, 248 pp.

Rienecker, M.M., 2003. Report of the coupled data assimilation workshop (NOAA/OGP), http://www.amath.unc.edu/Faculty/nburrell/samsi/CoupledDAreptfinal.pdf,online.

Rienecker, M., Awaji, T., Balmaseda, M., Barnier, B., Behringer, D., Bell, M., Bourassa, M., Brasseur, P., Brevik, L.-A., Carton, J., 2010. Synthesis and assimilation systems—essential adjuncts to the global ocean observing system. In: Proceedings of OceanObs'09: Sustained Ocean Observations and Information for Society (vol. 1), Venice, Italy, 21–25.

Robinson, A.R., Carton, J.A., Pinardi, N., Mooers, C.N.K., 1986. Dynamical forecasting and dynamical interpolation: an experiment in the California current. J. Phys. Oceanogr. 16, 1561–1579.

Rollenhagen, K., Timmermann, R., Janjic, T., Schröter, J., Danilov, S., 2009. Assimilation of sea ice motion in a finite element sea ice model. J. Geophys. Res. 114, C05007. http://dx.doi.org/10.1029/2008JC005067.

Saha, S., et al., 2010. The NCEP climate forecast system reanalysis. Bull. Am. Meteorol. Soc. 91, 1015–1057.

Sandery, P.A., Brassington, G.B., Freeman, J., 2011. Adaptive nonlinear dynamical initialization. J. Geophys. Res. 116, C01021. http://dx.doi.org/10.1029/2010JC006260.

Sasaki, Y., 1970. Some basic formulations in numerical variational analysis. Mon. Weather Rev. 98, 875–883.

Schiller, A., Brassington, G.B. (Eds.), 2011. Operational Oceanography in the 21st Century. Springer Science+Business Media B.V., Dordrecht. http://dx.doi.org/10.1007/978-94-007-0332-2_18.

Schiller, A., Smith, N.R., 2006. BLUElink: large-to-coastal scale operational oceanography in the southern hemisphere. In: Chassignet, E., Verron, J. (Eds.), Ocean Weather Forecasting: An Integrated View of Oceanography. Springer, Dordrecht, The Netherlands, pp. 427–439.

Schiller, A., Oke, P.R., Brassington, G., Entel, M., Fiedler, R., Griffin, D.A., Mansbridge, J.V., 2008. Eddy-resolving ocean circulation in the Asian-Australian region inferred from an ocean reanalysis effort. Prog. Oceanogr. 76 (3), 334–365. http://dx.doi.org/10.1016/j.pocean.2008.01.003.

Schiller, A., Wiffels, S.E., Sprintall, J., Molcard, R., Oke, P.R., 2010. Pathways of intraseasonal variability in the Indonesian throughflow. Dyn. Atmos. Oceans 50 (2), 174–200. http://dx.doi.org/10.1016/j.dynatmoce.2010.02.003.

Schoenefeldt, R., Schott, F.A., 2006. Decadal variability of the Indian Ocean cross-equatorial exchange in SODA. Geophys. Res. Lett. 33, L08602. http://dx.doi.org/10.1029/2006GL025891.

Schott, F.A., Wang, W.-Q., Stammer, D., 2007. Variability of Pacific subtropical cells in the 50-year ECCO assimilation. Geophys. Res. Lett. 34, L05604. http://dx.doi.org/10.1029/2006GL028478.

Schott, F.A., Stramma, L., Wang, W., et al., 2008. Pacific subtropical cell variability in the SODA 2.0.2/3 assimilation. Geophys. Res. Lett. 35, L10607. http://dx.doi.org/10.1029/2008GL033757.

Scott, K.A., Buehner, M., Caya, A., Carrieres, T., 2012. Direct assimilation of AMSR-E brightness temperatures for estimating sea ice concentration. Mon. Weather Rev. 140, 997–1013. http://dx.doi.org/10.1175/MWR-D-11-00014.1.

Segschneider, J., Balmaseda, M.A., Anderson, D.L.T., 2000. Anomalous temperature and salinity variations in the tropical Atlantic: possible causes and implications for the use of altimeter data. Geophys. Res. Lett. 17 (15), 2281. http://dx.doi.org/10.1029/1999GL011310.

Semtner, A.J., Chervin, R.M., 1992. Ocean general circulation from a global eddy resolving model. J. Geophys. Res. 97, 5493–5550.

Siedler, G., Church, J., Gould, J. (Eds.), 2001. Ocean Circulation and Climate: Observing and Modelling the Global Ocean. Academic Press, San Diego.

Smith, N.R., Koblinsky, C.J. (Eds.), 2001. Observing the Oceans in the 21st Century. GODAE Project Office, Bureau of Meteorology, Melbourne, 604 pp.

Sprintall, J., Wijffels, S.E., Molcard, R., Jaya, I., 2009. Direct estimates of the Indonesian throughflow entering the Indian Ocean: 2004–2006. J. Geophys. Res. 114, C7. http://dx.doi.org/10.1029/2008JC005257.

Stammer, D., Wunsch, C., 1996. The determination of the large-scale circulation of the Pacific Ocean from satellite altimetry using model Green's functions. J. Geophys. Res. 101 (C8), 18409–18432.

Stammer, D., ECCO Consortium, Chassignet, E., HYCOM Consortium, 2000. Ocean state estimation and prediction in support of oceanographic research. Oceanography 13, 51–56.

Stammer, D., Wunsch, C., Giering, R., et al., 2002a. Global ocean circulation during 1992–1997, estimated from ocean observations and a general circulation model. J. Geophys. Res. 107, 3118–3127. http://dx.doi.org/10.1029/2001JC000888.

Stammer, D., Wunsch, C., Fukumori, I., Marshall, J., 2002b. State estimation in modern oceanographic research. Eos 83 (27), 289, 294–295.

Stammer, D., Wunsch, C., Giering, R., et al., 2003. Volume, heat, and freshwater transports of the global ocean circulation 1993–2000, estimated from a general circulation model constrained by World Ocean Circulation Experiment (WOCE) data. J. Geophys. Res. Oceans 108, http://dx.doi.org/10.1029/2001JC001115.

Stark, J.D., Ridley, J., Martin, M., Hines, A., 2008. Sea ice concentration and motion assimilation in a sea ice-ocean model. J. Geophys. Res. 113, C05S91.

Storto, A., Dobricic, S., Masina, S., Pietro, P.D., 2010. Global oceanographic variational data assimilation of in-situ observations and space-borne altimeter data for reanalysis applications, extended abstract, 5th WMO Symposium on Data Assimilation, World Meteorological Organisation, World Weather Watch Programme, WWRP 2010-5, 78 pp.

Suga, T., Sukigara, C., Saino, T., Toyama, K., Yanagimoto, D., Hanawa, K., Shikama, N., 2011. Physical-biogeochemical study using a profiling float: subsurface primary production in the subtropical North Pacific. In: Hall, J., Harrison, D.E., Stammer, D. (Eds.), Proceedings of the OceanObs'09: Sustained Ocean Observations and Information for Society Conference (Annex), Venice, Italy, 21–25 September 2009. ESA Publication WPP-306, New York.

Sugiura, N., Awaji, T., Masuda, S., Mochizuki, T., Toyoda, T., Miyama, T., Igarashi, H., Ishikawa, Y., 2008. Development of a four-dimensional variational coupled data assimilation system for enhanced analysis and prediction of seasonal to interannual climate variations. J. Geophys. Res. 113, C10017. http://dx.doi.org/10.1029/2008JC004741.

Talagrand, O., Courtier, P., 1987. Variational assimilation of meteorological observations with the adjoint vorticity equation. I: theory. Q. J. R. Meteorol. Soc. 113, 1311–1328.

Tingley, M., 2010. A comparison of the Kalman smoother and BARCAST, note and miscellany, www.people.fas.harvard.edu/tingley/Kalman_vs_BARSCAST.pdf.

Tonani, M., Pinardi, N., Fratianni, C., Pistoia, J., Dobricic, S., Pensieri, S., de Alfonso, M., Nittis, K., 2009. Mediterranean forecasting system: forecast and analysis assessment through skill scores. Ocean Sci. 5, 649–660.

Toyoda, T., Awaji, T., Masuda, S., Sugiura, N., Igarashi, H., Mochizuki, T., Ishikawa, Y., 2011. Interannual variability of North Pacific eastern subtropical mode water formation in the 1990s derived from a 4-dimensional variational ocean data assimilation experiment. Dynam. Atmos. Oceans 51, 1–25. http://dx.doi.org/10.1016/j.dynatmoce.2010.09.001.

Usui, N., Ishizaki, S., Fujii, Y., Tsujino, H., Yasuda, T., Kamachi, M., 2006. Meteorological research institute multivariate ocean variational estimation (MOVE) system: some early results. Adv. Space Res. 37, 806–822.

Usui, N., Tsujino, H., Fujii, Y., Kamachi, M., 2008. Generation of a trigger meander for the 2004 Kuroshio large meander. J. Geophys. Res. 113, C01012. http://dx.doi.org/10.1029/2007JC004266.

Van Leeuwen, P.J., 2003. A variance-minimizing filter for large-scale applications. Mon. Weather Rev. 131, 2071–2084.

Vecchi, G.A., Harrison, M.J., 2007. An observing system simulation experiment for the Indian Ocean. J. Clim. 20, 3300–3319.

Veneziani, M., Edwards, C.A., Moore, A.M., 2009. A central California coastal ocean modeling study: 2. Adjoint sensitivities to local and remote forcing mechanisms. J. Geophys. Res. 114, C04020. http://dx.doi.org/10.1029/2008JC004775.

Vialard, J., Vitart, F., Balmaseda, M.A., Stockdale, T.N., Anderson, D.L.T., 2005. An ensemble generation method for seasonal forecasting with an ocean–atmosphere coupled model. Mon. Weather Rev. 131, 1379–1395.

Wang, W., McPhaden, M.J., 2000. The surface-layer heat balance in the equatorial Pacific ocean. Part II: interannual variability. J. Phys. Oceanogr. 30, 2989–3008. http://dx.doi.org/10.1175/1520-0485.

Wang, O., Fukumori, I., Lee, T., Cheng, B., 2004. On the cause of eastern equatorial Pacific Ocean T-S variations associated with El Nino. Geophys. Res. Lett. 31, L15310. http://dx.doi.org/10.1029/2004GL02472.

Weaver, A., Courtier, P., 2001. Correlation modeling on the sphere using a generalized diffusion equation. Q. J. R. Meteorol. Soc. 127, 1815–1846.

Weaver, A.T., Deltel, C., Machu, E., Ricci, S., Daget, N., 2005. A multivariate balance operator for variational ocean data assimilation. Q. J. R. Meteorol. Soc. 131, 3605–3625. http://dx.doi.org/10.1256/qj.05.119.

Wenzel, M., Schröter, J., 2007. The global ocean mass budget in 1993–2003 estimated from sea level. J. Phys. Oceanogr. 37 (2), 203–213. http://dx.doi.org/10.1175/JPO3007.1.

WOCE International Project Office, 1997. Ocean circulation and climate, WOCE Report 154/97, 13p.

World Meteorological Organisation (WMO), 2010. 5th WMO Symposium on Data Assimilation. Melbourne, Australia, 5–9 October 2009. WWRP 2010-5, QMO/TD No. 1549.

World Meteorological Organisation (WMO), 2009. World modelling summit for climate prediction. Reading, UK, 6–9 May 2008. WCRP No. 131, WMO/TD No. 1468.

Wunsch, C., 1988. Transient tracers as a problem in Conrol Theory. J. Geophys. Res. Oceans 92, 8099–8110.

Wunsch, C., 1996. The ocean circulation inverse problem. Cambridge University Press, Cambridge, 437 pp.

Wunsch, C., Heimbach, P., 2006. Estimated decadal changes in the North Atlantic meridional overturning circulation and heat flux 1993–2004. J. Phys. Oceanogr. 36 (11), 2012–2024.

Xie, Y., Chungu L., Browning, G. L., 2002. Impact of formulation of cost function and constraints on three-dimensional variational data assimilation. Mon. Weather Rev. 130, 2433–2447. http://dx.doi.org/10.1175/1520-0493(2002).

Yang, S.-C., Rienecker, M., Keppenne, C., 2010. The impact of ocean data assimilation on seasonal-to-interannual forecasts: a case study of the

2006 El Nino event. J. Clim. 23, 4080–4095. http://dx.doi.org/10.1175/2010JCLI3319.

Yin, Y., Alves, O., Oke, P.R., 2011. An ensemble ocean data assimilation system for seasonal prediction. Mon. Weather Rev. 139, 786–806.

Zebiak, S.E., Cane, M.A., 1987. A model of El Nino—southern oscillation. Mon. Weather Rev. 115, 2262–2278.

Zhang, S., Harrison, M.J., Rosati, A., Wittenberg, A., 2007. System design and evaluation of coupled ensemble data assimilation for global oceanic studies. Mon. Weather Rev. 135, 3541–3564.

Zhang, S., Rosati, A., Delworth, T., 2010. The adequacy of observing systems in monitoring the Atlantic meridional overturning circulation and North Atlantic Climate. J. Clim. 23, 5311–5324. http://dx.doi.org/10.1175/2010JCLI3677.1.

Zhu, J., Kamachi, M., Hui, W., 2002. The improvement made by a modified TLM in 4D-Var with a geophysical boundary layer model. Adv. Atmos. Sci. 19, 563–582.

Chapter 23

Coupled Models and Climate Projections

Peter R. Gent

National Center for Atmospheric Research, Boulder, Colorado, USA

Chapter Outline

1. FORMULATION OF COUPLED MODELS

Coupled climate models consist of atmosphere, land, and sea ice components, as well as the ocean component. The atmosphere component is similar to the ocean in that it solves the primitive equations of motion, usually with hydrostatic balance, the continuity equation and predictive equations for temperature, and specific humidity. The vertical coordinate is usually pressure, or terrain following, or a combination of both. A comprehensive review of atmosphere components can be found in Randall et al. (2007). The land surface component usually solves equations in the vertical direction for heat and water that include many complex interactions at the land surface and soil layers below. It also calculates water runoff, which is then routed by a realistic horizontal pattern of river basins to create the river runoff field that is sent to the ocean component. The sea ice component solves a complicated equation for the ice rheology and thermodynamic equations for the heat balance of the sea ice. When the ice melts, freshwater is sent to the ocean component, and the ocean receives a brine rejection flux when the sea ice is formed.

The formulation of the ocean component of coupled models has been described in detail earlier in Chapter 20. In that chapter it is conjectured, and I concur, that ocean model simulations depend much more strongly on the parameterizations of mixing and the effects of unresolved scales than on details of the numerical discretization of the equations (Chassignet et al., 1996). The parameterization of vertical mixing has been discussed earlier in Chapter 7, and the parameterization of unresolved lateral transport in Chapter 8.

The first climate model that used realistic geometry was assembled by Syukuro Manabe, Kirk Bryan, and coworkers at the Geophysical Fluid Dynamics Laboratory (GFDL), and the results were published in two landmark papers (Bryan et al., 1975; Manabe et al., 1975). The horizontal grid spacing was $5° \times 5°$, and there were nine vertical levels in the atmosphere component and five levels in the ocean component. Since then, the components have become much more sophisticated and complex, and the horizontal resolution has increased to about $1° \times 1°$ or finer in present-day climate models that use about 30–60 vertical levels in the atmosphere and ocean components.

2. FLUX ADJUSTMENTS

Through most of the 1990s, present-day control runs of all climate models would quite quickly drift away from the realistic initial conditions with which they were initialized. This drift was usually corrected by the use of flux adjustments described in Sausen et al. (1988), which could be calculated in two different ways. The heat and freshwater fluxes between the atmosphere and ocean components were diagnosed when the atmosphere component was run using observed sea surface temperatures (SSTs) and the ocean component was run using observed wind stresses and atmosphere surface variables. These flux diagnoses were quite different, and their differences were the flux adjustments. Alternatively, the coupled model was run with strong

Ocean Circulation and Climate, Vol. 103. http://dx.doi.org/10.1016/B978-0-12-391851-2.00023-4

relaxation back to observations of SST and sea surface salinity, and these relaxation terms were used as the flux adjustments. The flux adjustments were added to the heat and freshwater fluxes between the atmosphere and ocean in a present-day coupled control run at each time step. This method corrected most of the model drift, but was very unsatisfactory because flux adjustments are completely unphysical, mask deficiencies in the atmosphere and ocean components that require their use, and likely give an unrealistic response to large perturbations of the climate system.

The first model that could maintain the present-day temperature climate in a control run without the use of flux adjustments was the Community Climate System Model, version 1 (CCSM1). A 300-year control simulation that showed very little drift in surface temperature was run during the second half of 1996, and documented in Boville and Gent (1998). The reason for this success was refinements of the atmosphere, and especially the ocean component, so that the surface heat flux produced by the two components in stand-alone runs was close enough to be compatible. However, the ocean salinity did drift because CCSM1 did not have a realistic river runoff component. A history of this success is contained in a recent review by Gent (2011). Quite quickly, the climate centers in Australia and the United Kingdom implemented two of the new ocean parameterizations from the CCSM1, and were also able to run their models with much reduced, or without, flux adjustments (Hirst et al., 1996; Gordon et al., 2000).

Now, almost all climate models are run without using flux adjustments. A consequence is that current climate models cannot reproduce the present-day atmosphere surface temperature and SSTs quite as well as when flux adjustments were used. However, this small detraction is far outweighed by the fact that climate models no longer use the unphysical flux adjustments. For example, in some models the flux adjustment transported nearly as much heat northward in the North Atlantic Ocean as did the ocean component. This can be seen in Figure 2.3.1 of northward heat transport in the Hadley Centre models with and without flux adjustments shown in Section 2.3.3 of the first edition of this book by Wood and Bryan (2001).

3. CONTROL RUNS

A control run is an integration of a climate model in which the solar input at the top of the atmosphere component has a repeating annual cycle, and all the other forcings have a repeating annual cycle or are kept constant in time. The forcings can include the atmospheric concentrations of carbon dioxide (CO_2) and other greenhouse gases, and the distributions of several aerosols that affect the atmosphere radiation balance. Quite often a control run is made using forcings from present-day conditions. This has the advantage that the simulation can be compared to a whole range of observations that have been made in the recent past. However, it is well known that the present-day climate is not in equilibrium because important forcings, namely, the greenhouse gas concentrations and aerosol emissions, have been increasing quite rapidly. Therefore, an alternative control run is made for "preindustrial" conditions, when it is assumed that the climate was in equilibrium. Preindustrial is often chosen to be 1850, which is not truly before the industrial revolution. However, 1850 is chosen as a compromise between only a very small increase in CO_2 over truly preindustrial levels, and reducing the length of climate model runs between preindustrial times and the present day.

Choosing the initial conditions to start an 1850 control run is not straightforward because the ocean, sea ice, and land states for this period are unknown. The atmosphere initial condition is not as important because it is forgotten within a few weeks of the start of the run, so a state from an atmosphere model run forced by observed SSTs is usually chosen. Present-day states for the sea ice are usually chosen, although the 1850 sea ice volume is thought to be larger, so that sea ice should grow during the control run. It is important not to put far too much sea ice into the initial condition, because when it melts quickly, it will suppress convection in the ocean component, which can take a few hundred years to overcome. It is also important that the water content of the soil in the land component is initialized realistically, because the timescale to change the water content of the deepest soil level can also be a few hundred years. The traditional way to initialize the ocean component is to use a climatology from the late twentieth century (Levitus et al., 1998; Steele et al., 2001). This initial condition has a larger heat content than in 1850 so that the climate model would be expected to lose heat initially from the ocean component and the whole system, although this does not happen in all climate models as some have the ocean gaining heat in 1850 control runs (e.g., Griffies et al., 2011).

The SST acts as a negative feedback on the atmosphere-to-ocean heat flux, so it would be expected that the atmosphere-to-ocean heat flux, and consequently the top of atmosphere heat budget, would converge toward zero after a few decades of an 1850 control run. However, in practice this seldom happens because the ocean heat loss or gain occurs in the deeper ocean and not in the surface layers. The climate model heat loss or gain is often almost constant in time and is difficult to get very close to zero. Most climate modeling groups aim to get the globally averaged heat imbalance less than 0.1 W/m^2, usually by tuning a cloud parameter or two in the atmosphere component (e.g., Gent et al., 2011). Even so, this rate of ocean heat loss or gain sustained over an integration of 1000 years means the ocean volume average temperature decreases or increases by 0.2 °C compared to the 3.7 °C value in the Levitus et al. (1998) data used to initialize ocean components. Many well-constructed climate models conserve water quite well, even though not to machine accuracy. The volume average ocean salinity can vary because of changes in the volume of sea ice and soil water content over the course of the integration. However, the volume average ocean salinity changes only marginally even over a 1000 year control run if water is almost conserved.

One disadvantage of a preindustrial control run compared to a present-day control run is that there are very few observations relevant to that period. However, an 1850 control run does reveal the internal variability of the climate model given repeating annual cycle forcing, which is important to document. Variability occurs on all timescales from the diurnal to centennial, and this variability can be compared to late twentieth century observations. For example, the largest interannual signal in the climate system is the El Nino/Southern Oscillation (ENSO) phenomena. This has been a difficult signal for climate models to simulate well. However, there has been significant progress in recent years. For example, Wittenberg (2009) looks at ENSO in a 2000 year control run of the GFDL CM2.1 climate model and shows there is decadal and centennial variability in both the frequency and amplitude of the signal. Figure 23.1 shows the frequency of NINO3 area (90°W–150°W, 5°S–5°N) monthly SST anomalies from observations and the CM2.1 when the run is divided into 20, 100, and 500 year segments. The amplitude has a lot of variability over the 20 year segments, and still significant variability over the 100 year segments of the control run. A period in the run can be found where the model ENSO matches quite well the observed record from 1956 to 2010. The ENSO signal in the recently completed CCSM4 model is significantly improved over earlier model versions (Neale et al., 2008), and the above comments also apply to ENSO in the 1300 year long CCSM4 1850 control run (Deser et al., 2012). The other purpose of an 1850 control run is to provide initial conditions for an ensemble of integrations of the model from 1850 to the present day, which are often called twentieth century runs.

4. TWENTIETH CENTURY RUNS

Time series over 1850 to the present day of four or five types of input variables are needed to force twentieth century runs. They are the atmospheric concentrations of CO_2 and other greenhouse gases, the level of solar output, the levels of natural and man-made aerosols, the amount of atmospheric aerosols from volcanic eruptions, and some models use historical changes in land use in the land component. The volcanic aerosol level is determined from the observed levels of aerosols from recent eruptions, such as El Chicon in 1982 and Pinatubo in 1991, and then scaled by the size of eruptions earlier in the period, such as the large Krakatoa eruption in 1883. Usually, an ensemble of these twentieth century runs is made with each climate model, where the initial conditions are taken from different times in the 1850 control run. These initial conditions are chosen after a few hundred years of the control run, so that the modeled climate system, including the upper kilometer or so of the ocean, has had time to come into equilibrium. If a climate model is to be useful, then its twentieth century runs must reproduce well many of the observed changes in the Earth's climate over the last 150 years. Most of these comparisons with observations will use the last 50 years of these runs, which is when virtually all of the observations were made.

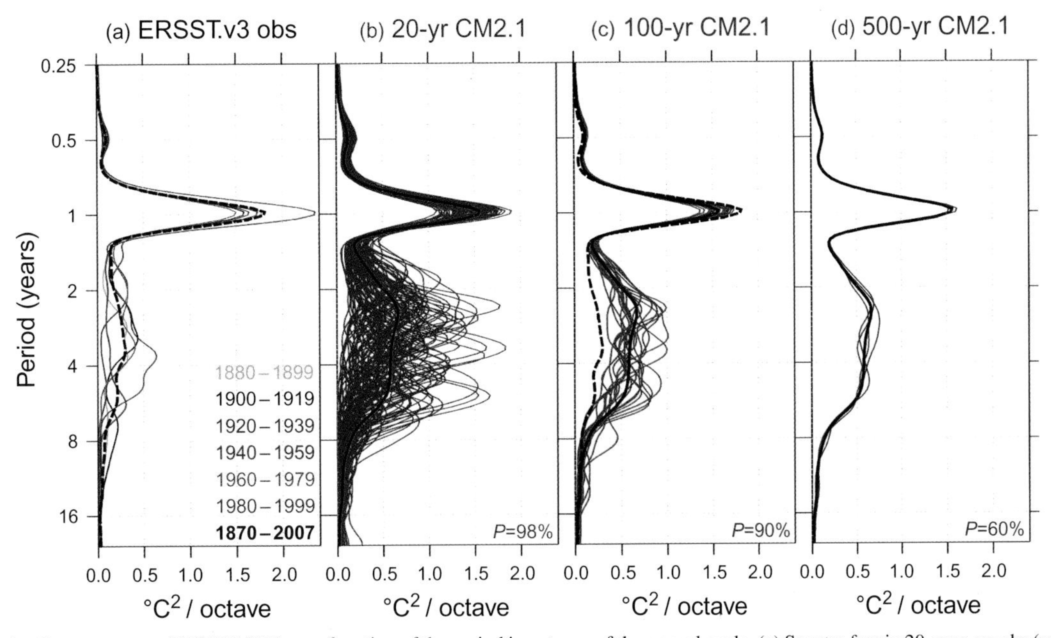

FIGURE 23.1 Power spectra of NINO3 SSTs as a function of the period in octaves of the annual cycle. (a) Spectra for six 20-year epochs (solid) and one 138-year epoch (dashed and repeated in (c)) from the ERSST observations, and spectra from the CM2.1 preindustrial control run for (b) 20-year epochs, (c) 100-year epochs, and (d) 500-year epochs, where the thick black solid line is the spectrum from the full 2000-year run. *From Wittenberg (2009).*

The best measured quantities going back to 1850 are surface temperature over land and SST. These can be combined into a globally averaged surface temperature with only a small uncertainty, so that it is well known that the Earth's surface temperature has increased by about 0.7 °C over the time 1850–2000. Figure 23.2 shows both the observations and the ensemble of five twentieth century simulations using the GFDL CM2.1 model. The CM2.1 ensemble mean reproduces the observations rather well. There are an enormous number of other quantities that can be compared to observations, and choices have to be made about which are compared because they are most important. For example, satellite observations of Arctic sea ice between 1979 and 2012 have shown a very significant decrease in the ice area during September, which is the month when the ice area is a minimum. If the climate model is to be used to project the future state of Arctic sea ice, including when the Arctic Ocean might be virtually ice free in September, then the model must reproduce the observations well within the ensemble of twentieth century runs. Figure 23.3 shows

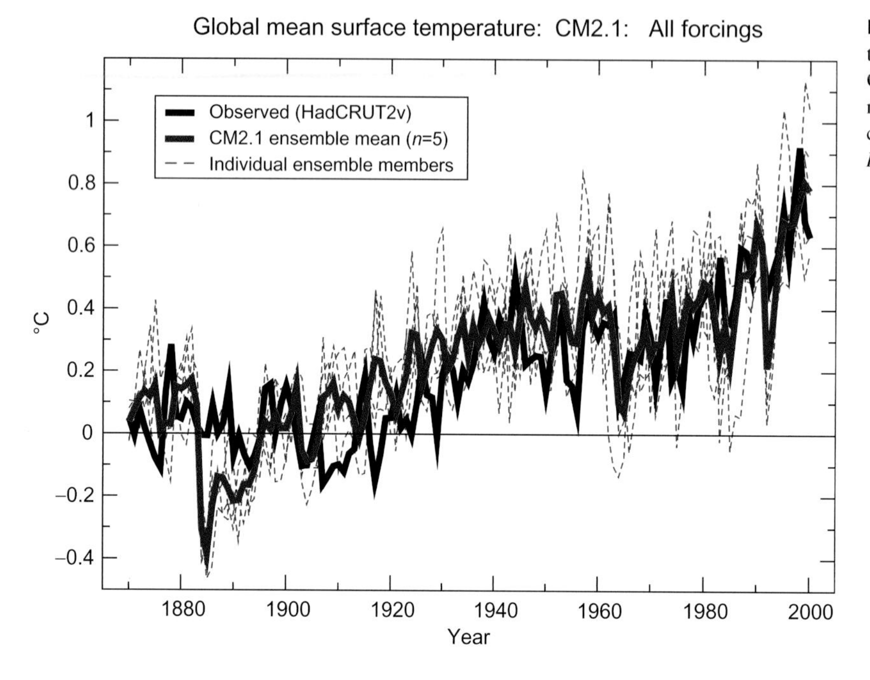

FIGURE 23.2 Globally averaged surface air temperature over 1870–2000 from the HadCRUT2v observations and both the individual members and the ensemble mean from twentieth century runs using the GFDL CM2.1 model. *From Knutson et al. (2006).*

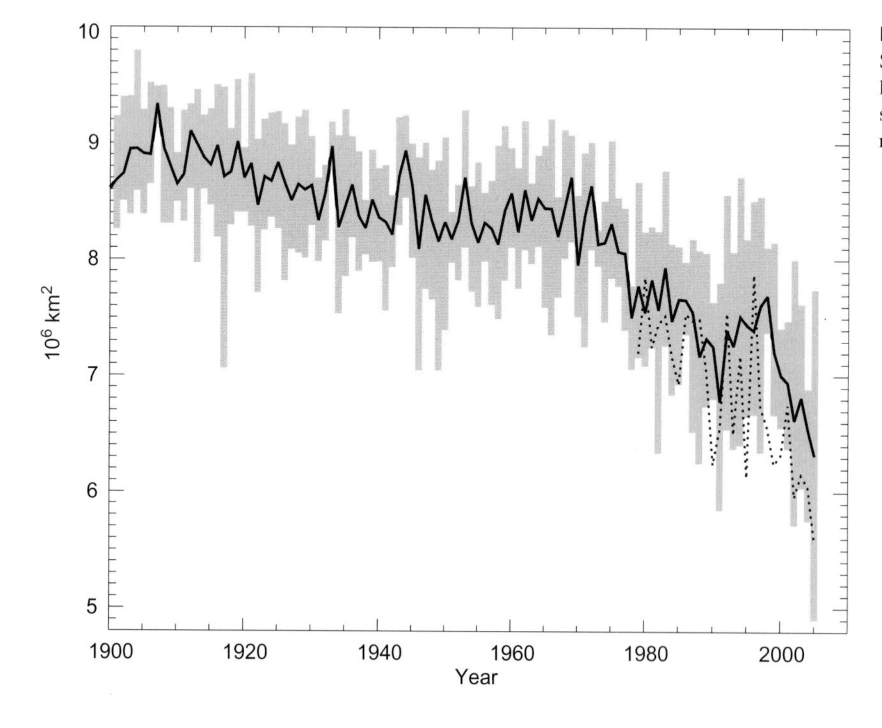

FIGURE 23.3 Area of Arctic sea ice in September from satellite observations (dotted line) and the ensemble mean (solid line) and spread (gray shading) from six twentieth century runs using the CCSM4. *From Gent et al. (2011).*

the September Arctic sea ice area from satellite observations (dotted line), and the ensemble mean and spread of twentieth century runs using the CCSM version 4 (solid line and shading) through 2005, which is when the twentieth century runs end. The CCSM4 reproduces the observed decline of September sea ice quite well.

Another use of twentieth century runs is to determine the cause of the Earth's warming that has increased markedly since the mid 1970s. Additional runs are made where all the natural forcings are retained and the levels of the greenhouse gases and man-made aerosols are kept at their preindustrial values, and the converse of this run where just the anthropogenic forcings are allowed to vary. Figure 23.4 shows the globally averaged surface temperature from ensembles of these two runs using the GFDL CM2.1 model. The conclusion is that the all natural forcings runs cannot explain the accelerated warming since 1975, whereas the anthropogenic forcings runs explain the observed temperature rise within the internal variability of the model. The confidence with which this result can be stated depends on the amplitude of the model internal variability and the length of time since the model surface temperature started

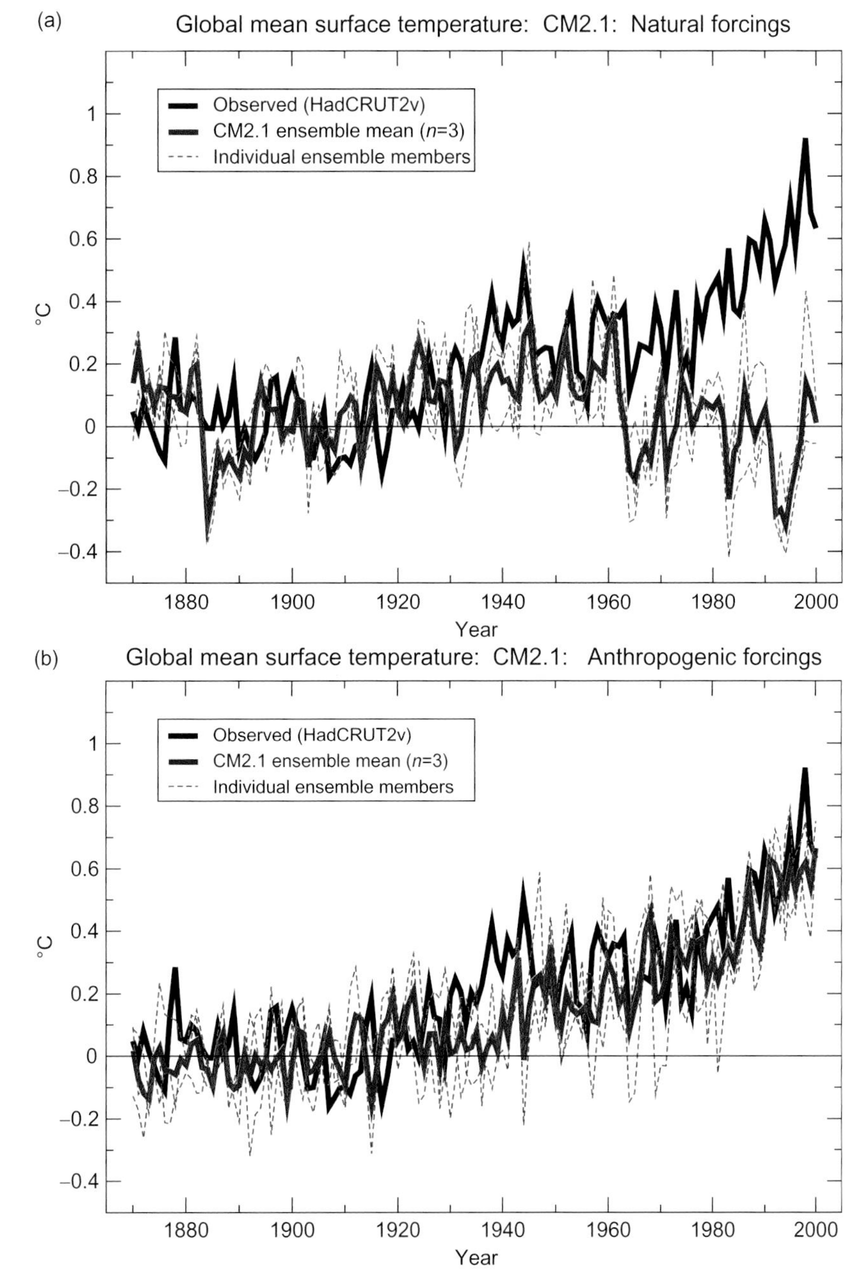

FIGURE 23.4 Globally averaged surface temperature 1870–2000 from observations, and ensembles of (a) all natural forcing runs and (b) anthropogenic forcings runs using the GFDL CM2.1. *From Knutson et al. (2006).*

to increase markedly. The increase in the model surface temperature over the last 35 years is now above two standard deviations of most models' internal variability. The 4th Assessment Report of the Intergovernmental Panel on Climate Change (IPCC) (Solomon et al., 2007) concluded that it is now "very likely" that the recent warming is due to anthropogenic causes. The level of certainty about this conclusion has increased in successive IPCC reports as, over time, the size of the observed temperature increase has become considerably larger compared to climate models' internal variability. It is important to note here that there is considerable uncertainty about the parameterization of clouds and the interaction of aerosols and clouds in the atmosphere component of climate models. A detailed discussion of these topics can be found in Chapters 1 and 2 of Solomon et al. (2007).

5. FUTURE PROJECTIONS

In order for a climate model to run projections, the future levels need to be estimated of the same four or five types of input variables that were required for twentieth century runs. The future level of solar output is usually chosen to be constant, although a very small 11 year cycle is sometimes imposed. Future volcanic eruptions are unknown, so a constant background level of volcanic aerosols is usually used. The future levels of natural aerosols are often chosen to be constant, along with a gradual reduction of man-made aerosols later in the twenty-first century. Finally, a scenario for the future atmosphere concentrations of CO_2 and other greenhouse gases has to be chosen. There is much debate and uncertainty about whether future emissions of CO_2 will remain at their present levels, accelerate over the next few decades, or reduce in the second half of the twenty-first century as the world turns to alternate energy sources. Because of this uncertainty, climate models have been run with a wide variety of scenarios for the future concentrations of CO_2. These concentrations are usually based on a fixed percentage of the CO_2 emissions remaining in the atmosphere, or are the output of integrated assessment models that use CO_2 emissions as input. This approach ignores the possible feedbacks from the land and ocean components, which may take up less or more of the emitted CO_2 in the future compared to the past. If a climate model is to predict the future CO_2 concentrations given the emissions, then it must have an interactive carbon cycle that can predict the future levels of CO_2 uptake by the land and ocean components (see Chapter 30).

An informative future projection is obtained if the concentrations of CO_2 and other greenhouse gases are kept constant at their present-day values. This is called a "commitment" run, because it shows the future changes that we have committed to by raising the atmosphere CO_2 concentration to its present level. Assuming no future changes in the forcings, most climate models simulate an increase in the globally averaged surface temperature of between 0.2 °C and 0.4 °C over the next 100 years (see Figure 23.5b). This increase depends on the model's equilibrium climate sensitivity, sensitivity to aerosols, and the

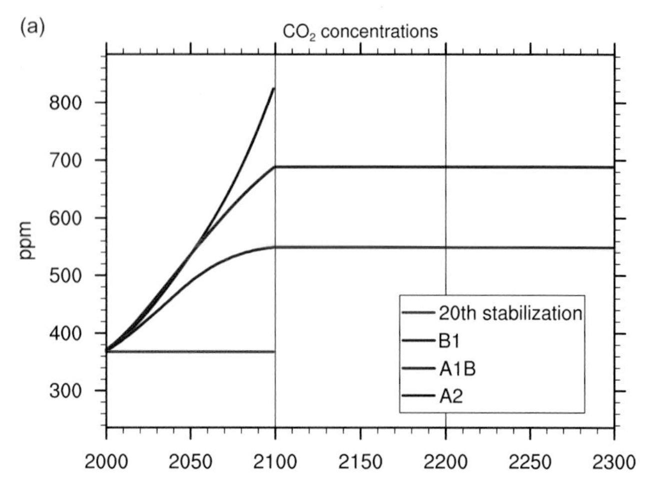

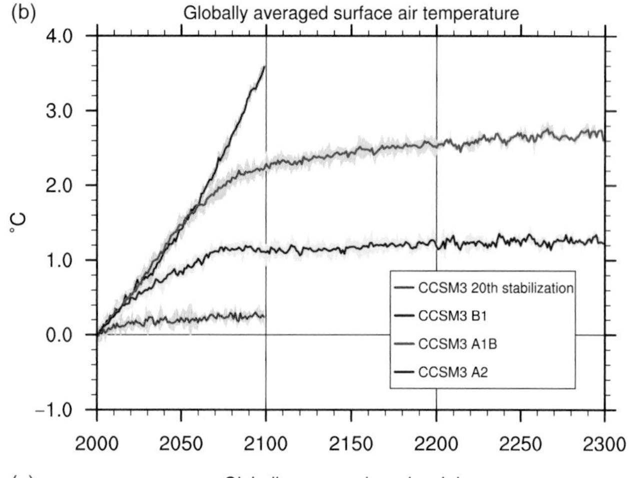

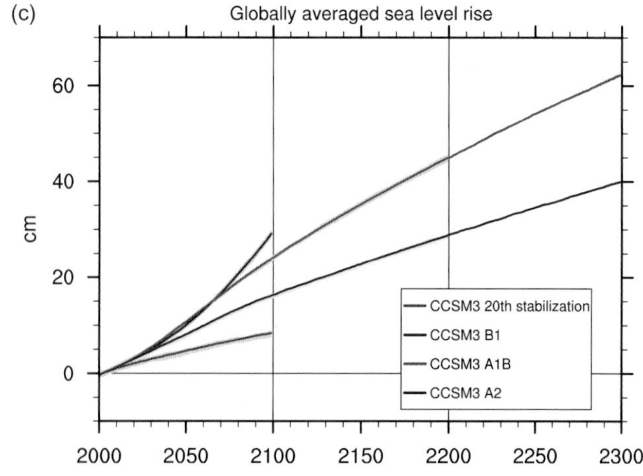

FIGURE 23.5 (a) CO_2 concentrations from four future scenarios, (b) the resulting globally averaged surface temperature increase, and (c) the globally averaged sea-level rise due to thermal expansion from runs using the CCSM3. *From Meehl et al. (2006).*

rate of ocean heat uptake. The equilibrium climate sensitivity is defined as the surface temperature increase due to a doubling of the CO_2 concentration when the ocean component is a simple mixed layer model. Using a mixed layer model ensures an equilibrium response after about 30 years, rather than the order 3000 years if the full depth ocean component is used (Stouffer, 2004; Danabasoglu and Gent, 2009). It is interesting to note that the future temperature rise is very small after about 50 years of a commitment run. This is in sharp contrast to the future heat uptake by the ocean component and the resulting sea-level rise due to thermal expansion, which increases almost linearly with time over the entire duration of commitment runs (see Figure 23.5c). As stated above, the time for the full ocean to reach equilibrium is on the order of 3000 years, so that the ocean heat content will not start to asymptote away from the linear increase for well over 1000 years. A similar, extremely long timescale is also relevant for glaciers and ice caps to come into equilibrium with the temperature increase that we have committed to. The conclusion is that, even if future temperature rises are stabilized by reductions in CO_2 emissions, the sea-level rise we have committed to will continue for the next 1000 years. This extremely long timescale for sea-level rise is frequently not appreciated; more discussions are in Chapter 27.

Figure 23.5a shows the future CO_2 concentrations from three different scenarios and Figure 23.5b shows the resulting globally averaged surface temperatures. As is to be expected, the faster the rise in greenhouse gases, the faster the rise in surface temperature. The rate of temperature increase varies across climate models depending upon their equilibrium climate sensitivity, sensitivity to aerosols, on the rate of heat uptake by their ocean components, and several other factors. Some future projections have been run where the CO_2 concentration reaches a maximum and is then kept constant. The temperature increases quickly, but then only increases slowly when CO_2 is constant, and so do other important quantities such as the rate of decrease of Arctic sea ice. However, as shown in Figure 23.5c, ocean heat uptake and the resulting rise in sea level due to thermal expansion do not level off; the rate of heat uptake only decreases slightly once the CO_2 concentrations are kept constant because of the very long ocean adjustment timescale.

6. NORTH ATLANTIC MERIDIONAL OVERTURNING CIRCULATION

The Gulf Stream transports warm water northward in the North Atlantic. Most of this water returns to the south in the gyre circulation in the upper ocean, but some is returned southward in a vertical circulation. Deepwater is formed in the Labrador and Greenland/Iceland/Norwegian Seas and is then returned south in the deep western boundary current along the west side of the North Atlantic. This meridional overturning circulation (MOC) is important because it carries a majority of the heat that is transported toward the Arctic by all the oceans in the Northern Hemisphere. This heat transport modulates the climate of Western Europe compared to other land masses at the same latitude. An equivalent, large-scale vertical overturning circulation does not occur in the North Pacific Ocean. An important question is how future increases in CO_2 will affect modes of climate variability, such as the North Atlantic MOC?

In models, the MOC is characterized by the streamfunction formed by the meridional and vertical velocities when the continuity equation is integrated zonally across the ocean basins. A typical Atlantic Ocean MOC streamfunction from the CCSM4 is shown in Figure 23.6. It shows water moving northward in the upper kilometer, sinking down to between 2 and 4 km between 60°N and 65°N, and then returning southward. Note that a large fraction of this return flow crosses the equator into the Southern Hemisphere. Below this deep return flow, the negative contours show a weak northward flow near the ocean floor, which is the model representation of Antarctic Bottom Water flowing northward. The magnitude of the North Atlantic MOC is virtually always taken to be the maximum value of the overturning streamfunction in Sverdrups (Sv), which is usually at a depth of about 1 km. Note that this maximum value cannot be directly measured from observations, although estimates can be made both from observations at particular latitudes (Cunningham et al., 2007) and using assimilation models of ocean circulation (Wunsch and Heimbach, 2006).

Figure 23.7 shows the maximum North Atlantic MOC values from the preindustrial control runs of the climate models CCSM3 and CCSM4 (from the box shown in Figure 23.6). It shows that the maximum value takes on the order of 400–500 years to adjust to the model climatology from the initial condition. This is one reason why the initial conditions for twentieth century runs described earlier should be taken after 500 years or longer of the control runs. Figure 23.7 also shows a large difference in the internal variability of the North Atlantic MOC between the two versions of the model. CCSM3 has a fairly regular, large amplitude oscillation with a period of about 20 years, although its amplitude decreases after 500 years, whereas CCSM4 shows much smaller variability with no regular oscillation. This change is mostly caused by the implementation of a new overflow parameterization in CCSM4, which is described in Danabasoglu et al. (2010). This parameterization improves the representation of the Denmark Strait and Greenland/Iceland/Scotland overflows, which results in North Atlantic deep water reaching down to 4 km near 60°N, and is therefore more realistic. This highlights the fact that the representation of the North Atlantic MOC in the ocean components of climate models is rather

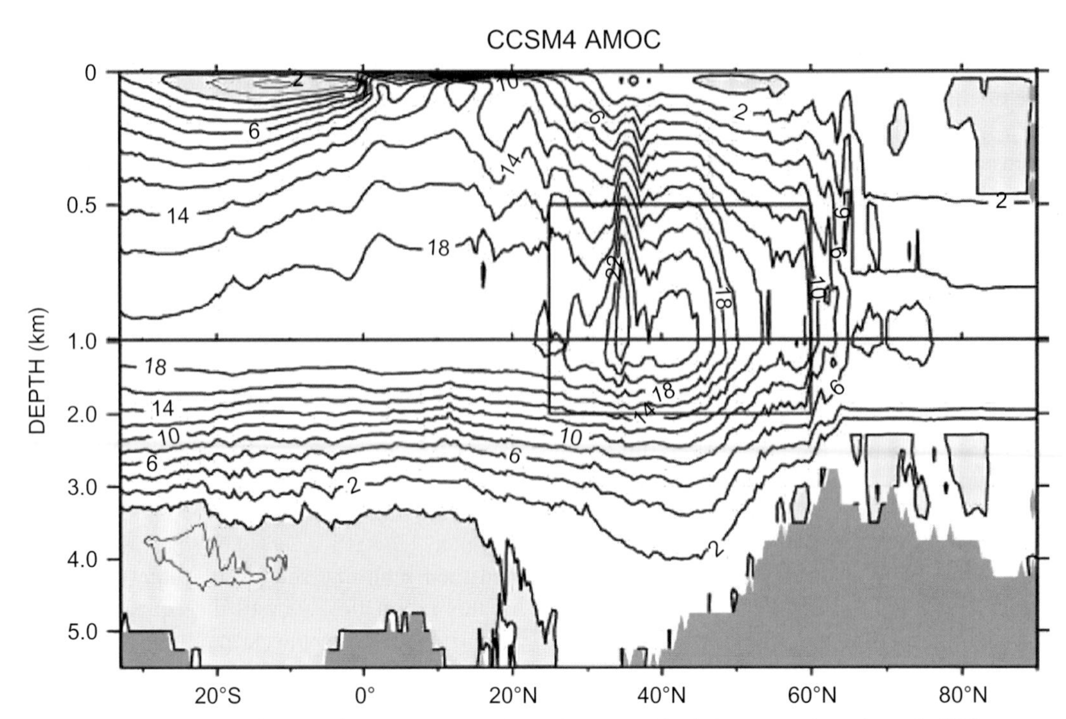

FIGURE 23.6 North Atlantic meridional overturning stream function in Sv from the CCSM4. *From Danabasoglu et al. (2012).*

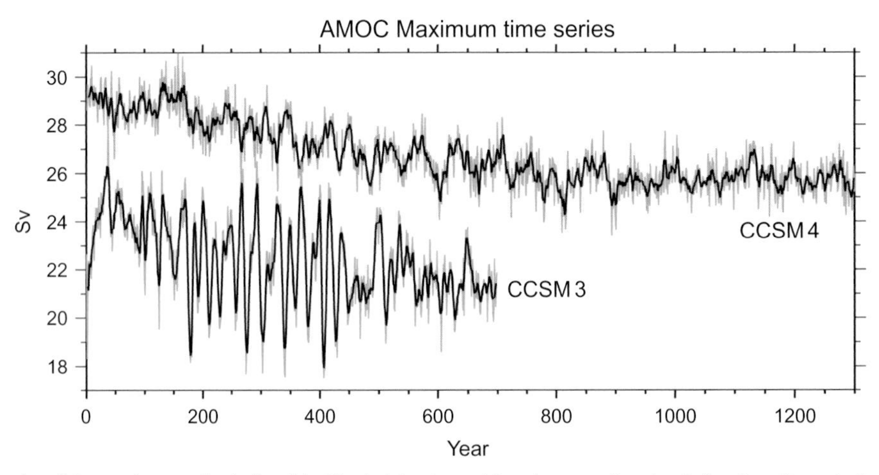

FIGURE 23.7 Time series of the maximum value in Sv of the North Atlantic meridional overturning circulation from the preindustrial control runs of the CCSM3 and CCSM4. *From Danabasoglu et al. (2012).*

sensitive to several of the parameterizations used. This results in both the mean maximum value and the internal variability being quite different across all the climate models used to make future climate projections. Much more discussion of the North Atlantic MOC is in Chapter 25.

It has long been thought that future climate change could reduce the magnitude of the North Atlantic MOC (Manabe et al., 1991). Both heating of the upper ocean and an increase in surface freshwater from melting sea ice and land ice, especially around Greenland, would reduce the density of the near-surface water. This would increase the stability of the upper ocean compared to the deeper ocean, and hence has the potential to reduce the rate of deepwater formation in the high-latitude North Atlantic. There has also been much speculation that, if the MOC does reduce significantly, then Europe's climate could become colder in the future, rather than becoming warmer. The IPCC 4th Assessment Report covered the most recent work on this subject in Chapter 10.3.4 of Solomon et al. (2007). The results show that the MOC does get weaker in the future in all 19 climate models assessed, and the rate of weakening increases as the future rate of CO_2 rise increases in the forcing scenarios used. In addition, there is a range of weakening rates in the MOC across climate models that are forced with the same scenario CO_2 increase. The range of MOC decreases are bounded by no change at all in the commitment run where the CO_2 remains constant at the present-day value, and a decrease of about 50% in the most sensitive models using a scenario where the CO_2 concentration has doubled from its 2000 value by the end of the twenty-first century. Some of these future projections have been run beyond 2100, where the CO_2 concentration is then kept constant at the 2100 value. In all "modern" models, and for all the scenarios used for CO_2 concentration at 2100, the MOC begins to strengthen once the CO_2 concentration is kept constant (Meehl et al., 2006; Solomon et al., 2007). Again, the rate of recovery depends on the particular model and the future scenarios used. "Modern" here means full climate models using a horizontal resolution of 1° or finer, no flux adjustments, and modern physics in the ocean component including the much more physically realistic diffusion of heat and salt along sloping isopycnal surfaces rather than along horizontal surfaces. In addition, for all climate models and all future CO_2 scenarios used in the IPCC 4th Assessment Report, the radiative warming effect of the increased greenhouse gases over Europe is larger than the cooling effect of a reduction in the MOC. So, current climate models do not support a future cooling of Europe's climate (Solomon et al., 2007).

It has frequently been suggested that the North Atlantic MOC could "collapse" as a result of future increases in CO_2 concentration, and then remain at a very small magnitude for a long time into the future. Manabe and Stouffer (1993) showed that when the CO_2 concentration was quadrupled after 140 years and then held constant, the North Atlantic MOC did become very small, and then stayed very small for the duration of the 500 year integration. However, a later paper (Stouffer and Manabe, 2003) shows that the MOC did recover after about 1500 years, and then stayed near to its initial value before CO_2 was increased for the remainder of the 5000 year run. The reason was that the warming near the surface diffused down slowly over the 1500 years, so that the upper 2–3 km of ocean became less stratified, and deepwater formation started again. Thus, the MOC recovery time was set by the diffusive timescale

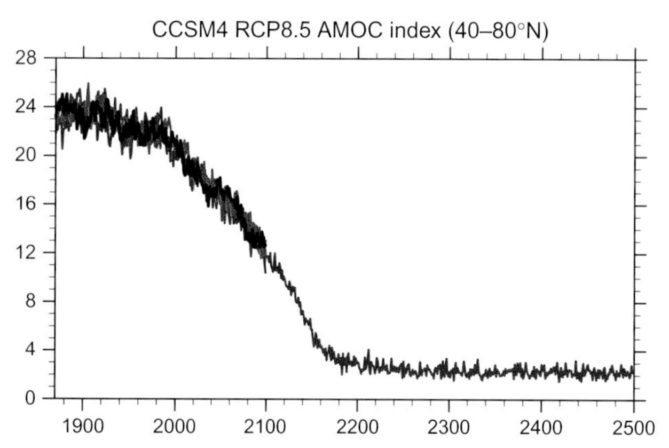

FIGURE 23.8 Maximum value of the North Atlantic MOC streamfunction north of 40°N in an ensemble of CCSM4 RCP8.5 runs, and in one extension from 2100 to 2500. The CO_2 value is constant after 2300. *Courtesy of A. Hu.*

for heat to reach the ocean mid-depths. When the same run up to $4\times CO_2$ was made with more "modern" climate models, then the MOC did start to recover soon after the CO_2 was held constant in both the HadCM3 (Wood et al., 2003) and the CCSM3 (Bryan et al., 2006b). Therefore, I believe that the Manabe and Stouffer (1993) result is influenced by the very coarse ocean resolution of about 4° in the horizontal and 12 vertical levels, horizontal mixing of heat and salt, and the large flux adjustments of heat and freshwater. Mikolajewicz et al. (2007) also showed that the North Atlantic MOC can become small and stay small when the increase in CO_2 is a factor of 5, and sometimes when it is a factor of 3. However, the horizontal resolution of their model is very coarse at 5.6^o, and it has a very large flux adjustment of freshwater in the North Atlantic. The question then arises: Is there a final CO_2 concentration where the MOC becomes very small and subsequently does not recover in "modern" climate models? Very recently, the Representative Concentration Pathway 8.5 (RCP8.5) has been used to force the CCSM4 between 2005 and 2300 from the end of a twentieth century run from 1850 to 2005. The CO_2 starts at 285 ppm in 1850, is 385 ppm in 2000, and rises to near 1962 ppm by 2250 before leveling off. This run has already been discussed in Chapter 1. The globally averaged surface temperature and the North Atlantic MOC index from Meehl et al. (2012) are shown in Figure 1.13. The MOC index starts at $\sim$25 Sv in 1850, reduces to $\sim$8 Sv in 2250, and remains near this value as the CO_2 forcing becomes almost constant. The average surface temperature in the Arctic Ocean has risen by more than 20° C by 2300 in this RCP8.5 run, and there is no sea ice in the Arctic all year round. The deepwater formation in the Labrador Sea, Greenland–Iceland–Norwegian Seas, and Arctic Ocean is completely shut off by 2200 because of the large rise in SSTs and rapid freshening in surface salinity caused by the sea ice melt. The maximum of the MOC streamfunction north of 40^oN, shown in Figure 23.8, has fallen to $\sim$3 Sv by 2200, so that the index of $\sim$8 Sv shown in Figure 1.13 is associated with the subtropical gyre near 20^oN at 500 m depth, rather than the MOC itself. This run has been continued out to 2500 with constant CO_2 forcing, and the maximum MOC value north of 40^oN remains $<$3 Sv throughout to 2500. If this run were to be continued further, I think that the MOC would remain small for several hundred years, and a recovery would probably be on the long diffusive timescale of heat to reach the ocean mid-depths, as in the 5000 year run of Stouffer and Manabe (2003). Thus, the CCSM4 does show that its MOC can be switched off quasi-permanently when it is forced by a CO_2 rise of a factor of almost 7 between 1850 and 2250. I suspect that other "modern" climate models display similar behavior when the forcing is so strong that all the Arctic sea ice melts, and deepwater formation completely shuts off.

Stommel (1961) showed that in a simple box model the MOC could possess multiple equilibria, which is two different stable solutions with the MOC either strong or weak, given identical forcing of the model. In addition, Marotzke and Willebrand (1991), Hughes and Weaver (1994), and many more recent studies have found multiple equilibria of the MOC in global ocean models that use very coarse resolution of about 4°, diffusion of heat and salt along horizontal surfaces, and are forced by mixed boundary conditions of restoring to an atmospheric temperature, but an imposed flux of freshwater. More recently, Marsh et al. (2004) and Sijp et al. (2012) have found MOC multiple equilibria in Intermediate Complexity coupled models that use a full ocean component with coarse resolution, but a considerably simplified atmosphere component. The last of these results is despite the fact that, using the same model, Sijp et al. (2006) find that the stability of North Atlantic deepwater formation to imposed freshwater inputs is significantly increased when diffusion of heat and salt along horizontal surfaces is replaced by diffusion along sloping isopycnal surfaces. The reason is that vertical exchange can then be accomplished by the diffusion, whereas it has to be accomplished by convection when the diffusion is along horizontal surfaces. Manabe and Stouffer (1988, 1999) did find multiple equilibria of the North Atlantic MOC in a full climate model. However, I believe this result is a consequence of the very coarse ocean resolution, horizontal mixing of heat and salt, and the very large flux adjustment of freshwater. I know of no evidence of multiple equilibria of the North Atlantic MOC in full, "modern" climate models, so there are two possible explanations. Either multiple equilibria of the MOC exist in full climate models, but have not yet been found because of computational time constraints, or they do not exist when all the ocean–atmosphere feedbacks are working, which is not the case in Intermediate Complexity climate models. I favor the second explanation, but am willing to be proved wrong.

7. EL NINO/SOUTHERN OSCILLATION

ENSO is the largest and best observed interannual signal in the Earth's climate system. However, it has proved rather difficult to simulate well in climate models. Most models have difficulty reproducing the mean precipitation pattern in the tropical Pacific Ocean, because they have too much rain south of the equator in the western ocean that reaches too far into the central Pacific. In addition, the frequency of model ENSO variability in several models is too short and is not the broadband maximum between 3 and 7 years seen in observations. Much work over recent years has resulted in good ENSO simulations in a small number of climate models, such as the Hadley Centre HadCM3 (Collins et al., 2001), the GFDL CM2.1 (Wittenberg, 2009), and the CCSM4 (Deser et al., 2012). These improvements have resulted mainly from refinements to the deep convection parameterizations in the atmosphere components of these models, and from increased resolution in both the atmosphere and ocean (Guilyardi et al., 2009). Figure 23.9 shows the correlation of monthly mean NINO3 SST anomalies with global SST anomalies from observations, and the CCSM versions 3 and 4. The improvements in the CCSM4 are the much wider region of positive correlation in the eastern Pacific Ocean that reaches into the subtropics, and the horseshoe pattern of negative correlation in the west Pacific Ocean that extends into the central Pacific Ocean at midlatitudes. Degradations are that the negative correlation is now much stronger than observations in the western Pacific Ocean, and it reaches into the east tropical Indian Ocean, probably because the ENSO amplitude is too large. These improvements result from changes to the atmosphere component deep convection scheme (see Neale et al., 2008).

The models that now simulate ENSO well show large very low-frequency variability in its amplitude on decadal and centennial timescales. Wittenberg (2009) shows the Nino3 SST anomaly from a 2000-year control run using 1860 conditions from the GFDL CM2.1 model (see Figure 23.1). The very low-frequency variability is large, and is reminiscent of the ENSO simulations from the earlier, much simpler coupled model of Zebiak and Cane (1987). A frequently asked question is how ENSO will change in the future. However, Wittenberg (2009) and Stevenson et al. (2012) show that the very low-frequency ENSO variability in the CM2.1 and CCSM4 control runs is so large that runs of at least 500 years would be required in order to detect a statistically significant change. Therefore, these models suggest that ENSO variability in the twenty-first century will be large, but will not be statistically different than the intrinsic variability of ENSO if the climate were stationary and in equilibrium. Much more discussion of ENSO and its prediction is in Chapter 24.

8. USES OF CLIMATE MODELS

Sections 3–5 describe the use of climate models to simulate a control run when the forcing is stationary, the Earth's climate from 1850 to the present time, and to make future climate projections out to 2100 and beyond. In addition to the variables and phenomena described earlier, these runs make future projections of the sea-level rise due to the warming of the oceans and the regional changes in sea level due to changes in ocean circulation. There are two other factors that change sea level; the first is the sea-level rise due to melting of ice on land, including mountain glaciers and the Greenland and Antarctic ice sheets (see Chapter 27). There is strong evidence from GRACE satellite data of increasing ice melt off Greenland, Antarctica, and other land glaciers over the last decade (Velicogna,

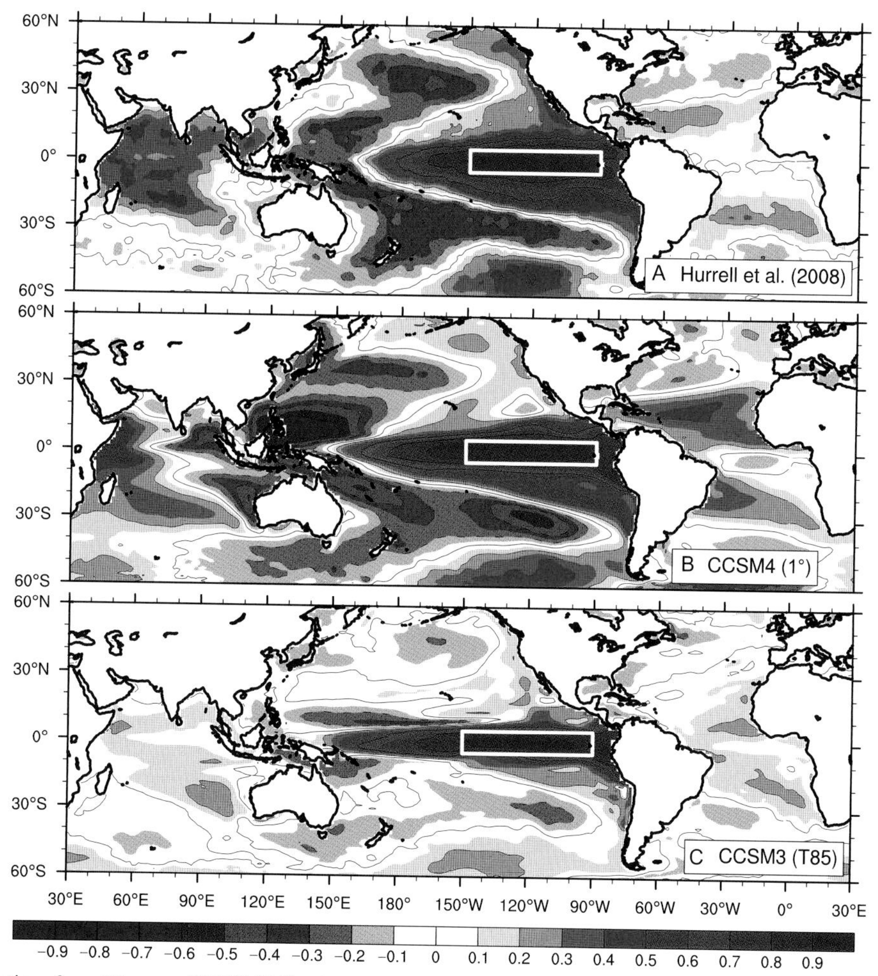

FIGURE 23.9 Correlation of monthly mean NINO3 (defined by the white box) SST anomalies with global SST anomalies from: (a) observations, (b) the CCSM version 4, and (c) the CCSM version 3. *From Gent et al. (2011).*

2009; Rignot et al., 2011), and during this time, it is estimated that this freshwater source has contributed an equal amount to sea-level rise as the thermosteric rise due to warming of the oceans. This contribution to sea-level rise can be included in a climate model, if it has components that represent land glaciers and the two large ice caps. Several climate models are now in the process of including these two components. The final contributor to sea-level changes is the height of the Earth's crust, which is affected by melting of the Greenland ice sheet, changes in the Earth's geoid, and glacial isostatic adjustment (Douglas et al., 2001). It is very unlikely these effects will be included in climate models; but they can be estimated using glacial rebound models (Kopp et al., 2010).

If a climate model has an interactive carbon cycle, then it must have a biogeochemistry module in its ocean component. Several climate models now do have an interactive carbon cycle, which means they predict the concentration of CO_2 in the atmosphere given a scenario of emissions, rather than taking the atmospheric concentration as an input. In this case, the model will make future projections of the oceanic carbon cycle, which can be used as a projection for future ocean acidification, and for how this affects the future state of marine ecosystems (see Chapters 30 and 31).

Another very important use of the control and twentieth century climate model runs is to explore the mechanisms and processes of climate variability on all timescales. For timescales from the diurnal, through the seasonal cycle and out to the several year timescale of ENSO, the model mechanisms can be compared to our knowledge from ocean observations. However, ocean variability on decadal and longer timescales is not well documented by observations, so our knowledge of the mechanisms driving ocean decadal variability comes mostly from models. Another use of climate models is to document how ocean variability, such as the North Atlantic MOC described in Section 6, affects the atmosphere and sea ice. In general, climate models are used to assess how processes in one component affect all the other components of the climate system, which cannot be determined by runs of the individual components alone.

Another use of climate models is to run simulations for many different paleoclimates of the past. These range from the last glacial maximum 21,000 years ago when the orbital parameters were different (e.g., Otto-Bliesner et al., 2006), to deep time experiments many million years ago, when the locations of the continents were very different (e.g., Kiehl and Shields, 2005).

Over the past 10 years or more, climate models have been used to make ENSO and seasonal forecasts out to 6 months or a year. The difference between a forecast and a future projection is that the climate system needs to be initialized to the current climate state in order to make a forecast, whereas it is not initialized to observations when making a projection. Making a seasonal forecast is similar to making a weather forecast for the next few days, except that in the climate system on seasonal timescales it is just as important to initialize the upper ocean state as well as the atmospheric state. For an ENSO forecast, it is only important to initialize the upper part of the tropical Pacific Ocean (Rosati et al., 1997), and seasonal forecasts require the upper ocean to be initialized globally. More information on these ENSO and seasonal forecasts is in Chapter 24. A very recent development is the use of climate models to make decadal forecasts of the future climate. For a decadal forecast, the deeper parts of all the oceans need to be initialized, not just the upper ocean, and it may also be important to initialize the current state of sea ice in the Arctic and Antarctic, and the levels of soil moisture in the land component, although this is as yet unproven. Initializing the full depth ocean component is a major new challenge because it is a new aspect of ocean science, and much more information and details of decadal forecasts are in Chapter 25.

Obtaining an estimate of the current ocean state entails assimilating ocean data into a run of an ocean model that is driven by the best estimate of atmosphere forcing over the past few years. This has been done much more frequently with atmosphere models to produce "reanalyses" of the past state of the atmosphere. It is now done routinely by several groups using ocean models, and some ocean "reanalyses" have been produced recently (see Chapter 22). The accuracy of these "reanalyses" has increased over the past few years (Carton and Santorelli, 2008), not only because of the experience obtained, but mostly because since about 2003 there are many more observations of temperature and salinity down to 2 km depth that have been obtained using Argo floats (Roemmich et al., 2009). However, whether these "reanalyses" are accurate enough to initialize the ocean component so that climate models can make useful decadal forecasts is a research question still to be explored in the future.

9. LIMITATIONS OF CLIMATE MODELS

The standard future projections of climate change made with the current models mostly use a horizontal resolution of around 1° or a little finer in all the components, and quite coarse resolution in the vertical. Almost certainly, the largest limitation of 1° resolution is that several very important processes must be parameterized because they cannot be resolved. Probably the best known of these processes is clouds in the atmosphere component. It has been known since the first climate models were assembled that the radiative properties, including the greenhouse effect, depend strongly on the various cloud parameterizations that have been used in different models (Cess et al., 1989). The various cloud schemes, and the parameterization of the interaction between aerosols and clouds, also dictate to a large degree the model equilibrium climate sensitivity, which is the globally averaged surface temperature increase in response to a doubling of the carbon dioxide concentration. Despite almost 30 years of research, the range of model equilibrium climate sensitivities remains about a factor of two.

In the ocean component, a resolution of 1° means that the viscosity required has to be much higher than desired or used in higher resolution models so that the intrinsic variability is much smaller than in the real ocean. Also, a resolution of 1° only begins to resolve the first Rossby radius of deformation near the equator, which is helped by the finer meridional resolution which is often used around the equator. Therefore, the effects of mesoscale eddies have to be parameterized. Most climate models use the Gent and McWilliams (1990) (GM) eddy parameterization, but there are significant differences in how it is implemented near the ocean surface, in the value of the coefficient chosen, and whether the coefficient is a constant or varies with position. It has been shown that a resolution of 1/10° is necessary to resolve the eddy effects so that a model's sea surface height variability matches that from satellite observations (Bryan et al., 2006a). Thus, it will still be over a decade before eddy-resolving ocean components will be used for standard climate projections. However, some climate model control runs with the ocean and sea ice components using 1/10° resolution have recently been completed (McClean et al., 2011; Kirtman et al., 2012). In general, these high-resolution integrations have a worse climate in many respects compared to the coarse resolution runs because we have less experience in how to choose the best parameter values. Chapter 8 shows how these high-resolution integrations can be used to evaluate the GM parameterization.

Another important limitation of climate models is that they lack components that may be important for simulating certain potentially large future changes in the climate system. Good examples have already been mentioned, such as an active carbon cycle and a component for the

Greenland and Antarctic ice sheets. Without these components, a climate model cannot address the possibility that the ocean will take up a smaller fraction of the CO_2 output in the future, which will leave a larger fraction in the atmosphere, and the future sea-level rise due to water melting from the ice sheets. In addition, there are several more possibilities of severe, possibly abrupt, climate changes with a very small probability, but with very large consequences. Examples are the release of large amounts of methane from ocean clathrates (Archer, 2007) and the possible fast breakup of the West Antarctic Ice Sheet (Bamber et al., 2009). These are difficult to simulate accurately and to assess quantitatively the possibilities that they will occur.

10. CUTTING EDGE ISSUES

The numerical discretization of the depth coordinate ocean components that are used in most climate models is quite old. They are based on latitude and longitude grids, although nearly all models now transpose the North Pole into a nearby land mass. There are also grids that have two poles in northern land masses so that the resolution of the Arctic Ocean can be more comparable to the resolution in the rest of the global oceans. However, over the past 10 years, there has been much work on new grids that are much more uniform over the globe than latitude and longitude grids. These have mostly been developed with the atmosphere component in mind and have been designed to work efficiently on the tens of thousands of processors that make up modern supercomputers (Staniforth and Thuburn, 2012). These grids can be made to vary in resolution from one part of the globe to another quite routinely. Therefore, the prospect is that grids with variable resolution will soon be available for global ocean components. Where to enhance the resolution and where to keep it coarse will then have to be decided, but enhanced resolution at the equator, along coasts and to resolve important narrow passages, is now in prospect. The hope is that a variable grid would have many fewer grid points than a global $1/10°$ resolution grid but would give comparable results.

Also over the past 10 years, it has been well documented that depth coordinate models contain some cross-isopycnal mixing due to the numerics (Griffies et al., 2000; Ilıcak et al., 2012). This is a problem if the numerical mixing is comparable to the very small level of cross-isopycnal mixing imposed in the deeper ocean. One way to avoid this is to use an isopycnal model where the vertical coordinate is potential density (Hallberg, 2000; Bleck, 2002). This type of model can be run stably with no cross-isopycnal mixing at all. Some climate models use an isopycnal ocean component (Furevik et al., 2003; Sun and Bleck, 2006; Dunne et al., 2012), but still the large majority of climate models use a depth coordinate ocean component. One reason is familiarity, but benefits of depth coordinates are the ease of simulating the upper mixed layer and that they keep uniform vertical resolution throughout the global ocean. Maintaining realistic deep overflows has also been a traditional problem using depth coordinates, but this can be improved by incorporating an explicit overflow parameterization (Danabasoglu et al., 2010). Isopycnal coordinate models represent deep overflows quite well, but their traditional drawbacks are the difficulty of representing the mixed layer, where vertical density gradients are small, and poor resolution in the high-latitude oceans, where the top-to-bottom density gradient is very small. I think that both types of models should continue to be developed so that the positives and negatives of the two coordinate systems can be further compared and contrasted. A third vertical coordinate system used in coastal ocean models is sigma, or terrain-following, coordinates, which automatically have very thin grid cells in the shallow regions of the ocean. This is an advantage when studying coastal currents and processes, but is a real disadvantage in the climate context. The reason is that the allowable time step due to vertical advection in the shallow ocean is much smaller than that using the other vertical coordinates so that the ocean model takes much more computer time to run.

As important as upgrading the numerical aspects of ocean climate components is further development of all the parameterizations used in the model. The vertical mixing scheme in the momentum and tracer equations not only sets the mixed layer and thermocline depths but also dictates the amount of cross-isopycnal mixing in the deeper ocean. It should include contributions from many sources such as internal wave breaking, tidal mixing, possibly mesoscale eddies and other processes (see Chapter 7). It also needs to work well across a range of vertical resolutions, because the vertical grid varies considerably with depth in most models. How to specify the coefficient in the GM eddy parameterization as a function of position is the subject of ongoing work (Eden et al., 2009), as well as how to transition the GM scheme into horizontal tracer mixing in the mixed layer (Danabasoglu et al., 2008; Ferrari et al., 2010; see Chapter 8). The prospect of variable resolution grids discussed above leads to the requirement that ocean component parameterizations work well over a large range of scales. A good example is the GM parameterization for the effects of eddies on the mean flow. How should the coefficient be specified in a grid that varies between $1°$ and $1/10°$? The same question needs to be asked of the horizontal viscosity scheme.

A final cutting edge issue is which processes and interactions are missing from current ocean components that could be important in future climate change? One example is the interaction of the ocean with ice shelves. This is very important because the breakup of small ice shelves has already been observed in Greenland and the Antarctic Peninsula. There is some potential for much larger Antarctic ice shelves to break up into the ocean. Current ocean components do not include the coastal and estuarine environments. These are areas of strong interactions between physical ocean properties and the carbon cycle. It is also where nutrients are injected from river outflows and most ocean biology occurs. These are potentially important aspects of the climate system, which are discussed in Chapters 30–31.

ACKNOWLEDGMENTS

I would like to thank my NCAR colleagues: Gokhan Danabasoglu for Figures 23.6 and 23.7, Marika Holland for Figure 23.3, Jerry Meehl for Figure 23.5, and Rich Neale for Figure 23.9. Aixue Hu ran the continuation of the CCSM4 RCP8.5 run out to 2500, and provided Figure 23.8. Thanks is also due to several colleagues at GFDL: Andrew Wittenberg for Figure 23.1, and Tom Knutson for Figures 23.2 and 23.4. Steve Griffies, Ron Stouffer, and two other reviewers provided very thorough reviews of this chapter, which contributed to a much improved final version. Thomas Stocker provided some comments, and his views on AMOC bifurcations and the possibility of multiple equilibria in full climate models.

REFERENCES

Archer, D., 2007. Methane hydrate stability and anthropogenic climate change. Biogeosciences 4, 521–544.

Bamber, G.L., Riva, R.E., Vermeersen, B.L., LeBrocq, A.M., 2009. Reassessment of the potential sea-level rise from a collapse of the West Antarctic ice sheet. Science 324, 901–903.

Bleck, R., 2002. An oceanic general circulation model framed in hybrid isopycnic-Cartesian coordinates. Ocean Model. 37, 55–88.

Boville, B.A., Gent, P.R., 1998. The NCAR Climate System Model, version one. J. Clim. 11, 1115–1130.

Bryan, F.O., Hecht, M.W., Smith, R.D., 2006a. Resolution convergence and sensitivity studies with North Atlantic circulation models. Part I: the western boundary current system. Ocean Model. 16, 141–159.

Bryan, F.O., Danabasoglu, G., Nakashiki, N., Yoshida, Y., Kim, D.-H., Tsutsui, J., Doney, S.C., 2006b. Response of the North Atlantic thermohaline circulation and ventilation to increasing carbon dioxide in CCSM3. J. Clim. 19, 2382–2397.

Bryan, K., Manabe, S., Pacanowski, R.C., 1975. A global ocean-atmosphere climate model. Part II. The oceanic circulation. J. Phys. Oceanogr. 5, 30–46.

Carton, J.A., Santorelli, A., 2008. Global decadal upper-ocean heat content as viewed in nine analyses. J. Clim. 21, 6015–6035.

Cess, R.D., et al., 1989. Interpretation of cloud-climate feedback as produced by 14 atmospheric general circulation models. Science 245, 513–516.

Chassignet, E.P., Smith, L.T., Bleck, R., Bryan, F.O., 1996. A model comparison: numerical simulations of the north and equatorial Atlantic Oceanic circulation in depth and isopycnic coordinates. J. Phys. Oceanogr. 26, 1849–1867.

Collins, M., Tett, S.F., Cooper, C., 2001. The internal climate variability of HadCM3, a version of the Hadley Centre coupled model without flux adjustments. Clim. Dyn. 17, 61–81.

Cunningham, S.A., et al., 2007. Temporal variability of the Atlantic meridional overturning circulation at 26.5°N. Science 317, 935–938.

Danabasoglu, G., Gent, P.R., 2009. Equilibrium climate sensitivity: is it accurate to use a slab ocean model? J. Clim. 22, 2494–2499.

Danabasoglu, G., Ferrari, R., McWilliams, J.C., 2008. Sensitivity of an ocean general circulation model to a parameterization of near-surface eddy fluxes. J. Clim. 21, 1192–1208.

Danabasoglu, G., Large, W.G., Briegleb, B.P., 2010. Climate impacts of parameterized Nordic Sea overflows. J. Geophys. Res. 115, C11005. http://dx.doi.org/10.1029/2010JC006243.

Danabasoglu, G., Yeager, S.G., Kwon, Y.O., Tribbia, J.J., Phillips, A.S., Hurrell, J.W., 2012. Variability of the Atlantic meridional overturning circulation in CCSM4. J. Clim. 25, 5153–5172.

Deser, C., Phillips, A.S., Tomas, R.A., Okumura, Y., Alexander, M.A., Capotondi, A., Scott, J.D., Kwon, Y.O., Ohba, M., 2012. ENSO and Pacific decadal variability in Community Climate System Model version 4. J. Clim. 25, 2622–2651.

Douglas, B.C., Kearney, M.S., Leatherman, S.P. (Eds.), 2001. Sea Level Rise: History and Consequences. In: International Geophysics Series, vol. 75. Academic Press.

Dunne, J.P., et al., 2012. GFDL's ESM2 global coupled climate-carbon Earth system model. Part I: physical formulation and baseline simulation characteristics. J. Clim. 25, 6646–6665.

Eden, C., Jochum, M., Danabasoglu, G., 2009. Effect of different closures for thickness diffusivity. Ocean Model. 26, 47–59.

Ferrari, R., Griffies, S.M., Nurser, A.J., Vallis, G.K., 2010. A boundary-value problem for the parameterized mesoscale eddy transport. Ocean Model. 32, 143–156.

Furevik, T., Bentsen, M., Drange, H., Kindem, I.K., Kvamstø, N.G., Sorteberg, A., 2003. Description and evaluation of the Bergen climate model: ARPEGE coupled with MICOM. Clim. Dyn. 21, 27–51.

Gent, P.R., 2011. The Gent-McWilliams parameterization: 20/20 hindsight. Ocean Model. 39, 2–9.

Gent, P.R., McWilliams, J.C., 1990. Isopycnal mixing in ocean circulation models. J. Phys. Oceanogr. 20, 150–155.

Gent, P.R., et al., 2011. The Community Climate Model version 4. J. Clim. 24, 4973–4991.

Gordon, C., Cooper, C., Senior, C.A., Banks, H., Gregory, J.M., Johns, T.C., Mitchell, J.F., Wood, R.A., 2000. The simulation of SST, sea ice extents and ocean heat transports in a version of the Hadley Centre coupled model without flux adjustments. Clim. Dyn. 16, 147–168.

Griffies, S.M., Pacanowski, R.C., Hallberg, R.W., 2000. Spurious diapycnal mixing associated with advection in a z-coordinate ocean model. Mon. Weather Rev. 128, 538–564.

Griffies, S.M., et al., 2011. The GFDL CM3 coupled climate model: characteristics of the ocean and sea ice simulations. J. Clim. 24, 3520–3544.

Guilyardi, E., et al., 2009. Understanding El Nino in ocean-atmosphere general circulation models: progress and challenges. Bull. Am. Meteorol. Soc. 90, 325–340.

Hallberg, R.W., 2000. Time integration of diapycnal diffusion and Richardson number-dependent mixing in isopycnal coordinate ocean models. Mon. Weather Rev. 128, 1402–1419.

Hirst, A.C., Gordon, H.B., O'Farrell, S.P., 1996. Global warming in a coupled climate model including oceanic eddy-induced advection. Geophys. Res. Lett. 23, 3361–3364.

Hughes, T.M.C., Weaver, A.J., 1994. Multiple equilibria of an asymmetric two-basin ocean model. J. Phys. Oceanogr. 24, 619–637.

Hurrell, J.W., Hack, J.J., Shea, D., Caron, J.M., Rosinski, J., 2008. A new sea surface temperature and sea ice boundary dataset for the Community Atmosphere Model. J. Clim. 21, 5145–5153.

Ilıcak, M., Adcroft, A.J., Griffies, S.M., Hallberg, R.W., 2012. Spurious dianeutral mixing and the role of momentum closure. Ocean Model. 45, 37–58.

Kiehl, J.T., Shields, C.A., 2005. Climate simulation of the latest Permian: implications for mass extinction. Geology 33, 757–760.

Kirtman, B.P., et al., 2012. Impact of ocean model resolution on CCSM climate simulations. Clim. Dyn. 39, 1303–1328.

Knutson, T.R., et al., 2006. Assessment of twentieth-century regional surface temperature trends using the GFDL CM2 coupled models. J. Clim. 19, 1624–1651.

Kopp, R.E., Mitrovica, J.X., Griffies, S.M., Yin, J., Hay, C.C., Stouffer, R.J., 2010. The impact of Greenland melt on local sea levels: a partially coupled analysis of dynamic and static equilibrium effects in idealized water-hosing experiments. Clim. Change 103, 619–625.

Levitus, S., Boyer, T., Conkright, M., Johnson, D., O'Brien, T., Antonov, J., Stephens, C., Gelfeld, R., 1998. Introduction. In: World Ocean Database 1998, vol. 1. NOAA Atlas NESDIS 18.

Manabe, S., Stouffer, R.J., 1988. Two stable equilibria of a coupled ocean-atmosphere model. J. Clim. 1, 841–866.

Manabe, S., Stouffer, R., 1993. Century-scale effects of increased atmospheric CO_2 on the ocean-atmosphere system. Nature 364, 215–218.

Manabe, S., Stouffer, R.J., 1999. Are two modes of thermohaline circulation stable? Tellus A 51, 400–411.

Manabe, S., Bryan, K., Spelman, M.J., 1975. A global ocean-atmosphere climate model. Part I. The atmospheric circulation. J. Phys. Oceanogr. 5, 3–29.

Manabe, S., Stouffer, R.J., Spelman, M.J., Bryan, K., 1991. Transient responses of a coupled ocean-atmosphere model to gradual changes of atmospheric CO_2. Part I: annual mean response. J. Clim. 4, 785–818.

Marotzke, J., Willebrand, J., 1991. Multiple equilibria of the global thermohaline circulation. J. Phys. Oceanogr. 21, 1372–1385.

Marsh, R., Yool, A., Lenton, T.M., Gulamali, M.Y., Edwards, N.R., Shepherd, J.G., Krznaric, M., Newhouse, S., Cox, S.J., 2004. Bistability of the thermohaline circulation identified through comprehensive 2-parameter sweeps of an efficient climate model. Clim. Dyn. 23, 761–777.

McClean, J.L., et al., 2011. A prototype two-decade fully-coupled fine-resolution CCSM simulation. Ocean Model. 39, 10–30.

Meehl, G.A., et al., 2006. Climate change projections for the twenty-first century and climate change commitment in the CCSM3. J. Clim. 19, 2597–2616.

Meehl, G.A., et al., 2012. Climate system response to external forcings and climate change projections in CCSM4. J. Clim. 25, 3661–3683.

Mikolajewicz, U., Groger, M., Maier-Reimer, E., Schurgers, G., Vizcaino, M., Winguth, A., 2007. Long-term effects of anthropogenic CO_2 emissions simulated with a complex earth system model. Clim. Dyn. 28, 599–633.

Neale, R.B., Richter, J.H., Jochum, M., 2008. The impact of convection on ENSO: from a delayed oscillator to a series of events. J. Clim. 21, 5904–5924.

Otto-Bliesner, B.L., Brady, E.C., Clauzet, G., Tomas, R., Levis, S., Kothavala, Z., 2006. Last glacial maximum and holocene climate in CCSM3. J. Clim. 19, 2526–2544.

Randall, D.A., et al., 2007. Climate models and their evaluation. In: Climate Change 2007: The Physical Science Basis. Contribution of Working Group I to the fourth Assessment Report of the IPCC. Cambridge University Press.

Rignot, E., Velicogna, I., Van den Broeke, M.R., Monaghan, A., Lenaerts, J., 2011. Acceleration of the contribution of the Greenland and Antarctic ice sheets to sea level rise. Geophys. Res. Lett. 38, L05503. http://dx.doi.org/10.1029/2011GL046583.

Roemmich, D., Johnson, G.C., Riser, S., Davis, R., Gilson, J., Owens, W.B., Owens, S.L.G., Schmid, C., Ignaszewski, M., 2009. The Argo Program: observing the global oceans with profiling floats. Oceanography 22, 24–33.

Rosati, A., Miyakoda, K., Gudgel, R., 1997. The impact of ocean initial conditions on ENSO forecasting with a coupled model. Mon. Weather Rev. 125, 754–772.

Sausen, R., Barthels, R.K., Hasselmann, K., 1988. Coupled ocean-atmosphere models with flux correction. Clim. Dyn. 2, 154–163.

Sijp, W.P., Bates, M., England, M.H., 2006. Can isopycnal mixing control the stability of the thermohaline circulation in ocean climate models? J. Clim. 19, 5637–5651.

Sijp, W.P., England, M.H., Gregory, J.M., 2012. Precise calculations of the existence of multiple AMOC equilibria in coupled climate models. Part I: equilibrium states. J. Clim. 25, 282–298.

Solomon, S., Qin, D., Manning, M., Chen, Z., Marquis, M., Averyt, K.B., Tignor, M., Miller, H.L., 2007. Climate Change 2007: The Physical Science Basis. Contribution of Working Group I to the Fourth Assessment Report of the IPCC. Cambridge University Press.

Staniforth, A., Thuburn, J., 2012. Horizontal grids for global weather and climate prediction models: a review. Q. J. R. Meteorol. Soc. 138, 1–26.

Steele, M., Morley, R., Ermold, W., 2001. PHC: a global ocean hydrography with a high-quality Arctic Ocean. J. Clim. 14, 2079–2087.

Stevenson, S., Fox-Kemper, B., Jochum, M., Neale, R., Deser, C., Meehl, G., 2012. Will there be a significant change to El Nino in the 21st century? J. Clim. 25, 2129–2145.

Stommel, H., 1961. Thermohaline convection with two stable regimes of flow. Tellus 13, 224–230.

Stouffer, R.J., 2004. Time scales of climate response. J. Clim. 17, 209–217.

Stouffer, R.J., Manabe, S., 2003. Equilibrium response of thermohaline circulation to large changes in atmospheric CO_2 concentration. Clim. Dyn. 20, 759–773.

Sun, S., Bleck, R., 2006. Multi-century simulations with the coupled GISS-HYCOM climate model: control experiments. Clim. Dyn. 26, 407–428.

Velicogna, I., 2009. Increasing rates of ice mass loss from the Greenland and Antarctic ice sheets revealed by GRACE. Geophys. Res. Lett. 36, L19503, http://dx.doi.org/10.1029/2009GL040222.

Wittenberg, A.T., 2009. Are historical records sufficient to constrain ENSO simulations? Geophys. Res. Lett. 36, L12702. http://dx.doi.org/10.1029/2009GL038710.

Wood, R.A., Bryan, F.O., 2001. Coupled ocean-atmosphere models. In: Siedler, G., Church, J., Gould, J. (Eds.), Ocean Circulation and Climate. International Geophysics Series, vol. 77. Academic Press.

Wood, R.A., Vellinga, M., Thorpe, R., 2003. Global warming and thermohaline circulation stability. Philos. Trans. R. Soc. Lond. A 361, 1961–1975.

Wunsch, C., Heimbach, P., 2006. Estimated decadal changes in the North Atlantic meridional overturning circulation and heat flux 1993-2004. J. Phys. Oceanogr. 36, 2012–2024.

Zebiak, S.E., Cane, M.A., 1987. A model El-Nino-Southern Oscillation. Mon. Weather Rev. 115, 2262–2278.

Chapter 24

The Ocean's Role in Modeling and Predicting Seasonal-to-Interannual Climate Variations

Ben P. Kirtman*, Tim Stockdale† and Robert Burgman‡
*University of Miami—RSMAS, Miami, Florida, USA
†European Centre for Medium-Range Weather Forecasts, Shinfield Park, United Kingdom
‡Florida International University, Miami, Florida, USA

Chapter Outline

1. INTRODUCTION

The feasibility of seasonal prediction (see Table 24.1 for definitions of prediction, forecast, and predictability) largely rests on the existence of slow, and predictable, variations in the Earth's boundary conditions of soil moisture, snow cover, sea ice, and ocean surface temperature (e.g., Shukla and Kinter, 2005) and how the atmosphere interacts with and is affected by these boundary conditions. For example, a warm sea surface temperature anomaly (SSTA) in, for example, the tropical Pacific Ocean will lead to increased heat flux from the ocean to the atmosphere. This increased flux, if sufficiently large in magnitude and spatial scale, will alter the atmospheric boundary layer and ultimately change the structure of the rainfall and the release of latent heat in the free troposphere. The change in tropospheric latent heat release, in turn, will perturb the circulation leading to climatic anomalies in remote regions of the globe. Of course the atmospheric response itself interacts and affects the SST, thus the prediction problem is coupled. Similar effects are expected with tropical Atlantic and Indian Ocean SSTA. However, the response to extratropical SSTA is less well understood, and the direct coupling of the atmosphere and the ocean is known to be more complex.

For the purposes of this chapter, the emphasis is on the slowly evolving boundary condition of SSTA. So, the question then becomes what is the source or mechanism for the slow evolution of the SSTA, and whether it is predictable. Often this is referred to as ocean memory. The slow evolution of the SSTA could potentially come from thermal inertia associated with a finite depth ocean-mixed layer, which may in fact be the case in the midlatitude North Pacific, for example (e.g., Jin and Kirtman, 2010). Alternatively, ocean dynamics associated with thermocline variations may be the source of ocean memory or predictability. Here, we plot in Figure 24.1 a diagnostic demonstrating a connection between local thermocline variations and local

Ocean Circulation and Climate, Vol. 103. http://dx.doi.org/10.1016/B978-0-12-391851-2.00024-6

TABLE 24.1 Definitions

Definitions	
Predictability	The common procedure for investigating predictability consists of two or more numerical simulations with slightly perturbed initial states. The rate at which the solutions diverge as time progresses is used to estimate predictability, so that initial error, error growth rate, and error saturation value all influence estimates of the limit of predictability (see Lorenz, 1965). Since seasonal-to-interannual predictability is well beyond the limit of deterministic weather predictability, the predictable features are limited to certain statistics (e.g., time mean or spatial mean values, or higher moments). Predictability is estimated with a model and thus is neither a lower bound nor an upper bound (NRC, 2010)
Prediction	Information on future climate (deterministic or probabilistic) from a specific tool (statistical or dynamical) (see NRC, 2010)
Forecast	Issued guidance on future climate, can be quantitative or qualitative, is usually based on multiple tools, and can include subjective input (see NRC, 2010)
Skill	Statistical evaluation of accuracy, expressed quantitatively in terms of a specific metric (see NRC, 2010)
Quality	Broad assessment of forecast performance encompassing a range of metrics related to the fidelity of physical processes (see Kirtman and Pirani, 2009)

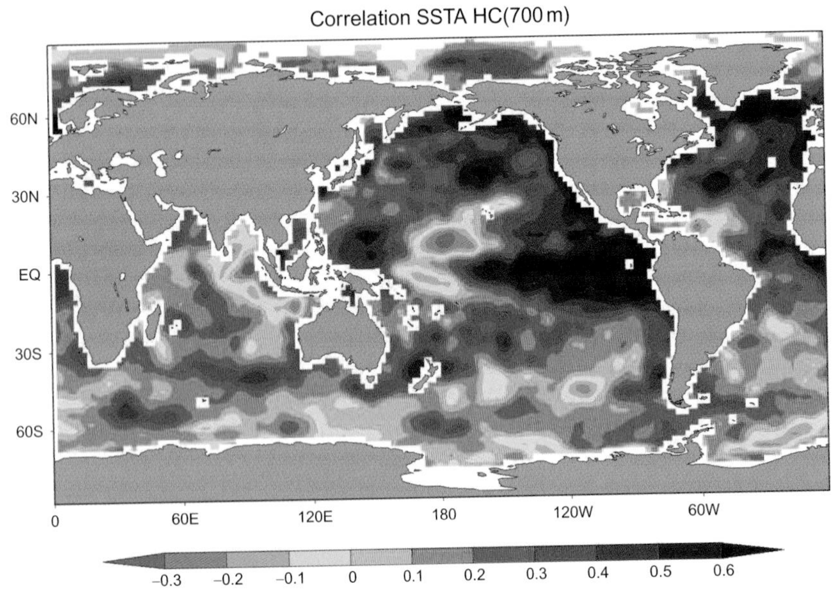

FIGURE 24.1 Correlation between SSTA and upper 700 m heat content anomalies from observational estimates 1955–2010. Observational estimates are from Levitus et al. (2012).

SSTA. In this example, heat content from the surface to 700 m is used as an estimate of thermocline variations, and we correlate observed estimates of heat content with observed estimates of SSTA. Large positive correlations suggest the slowly evolving thermocline variations are connected to SSTA variations, and are a potential source of predictability. For example, we see relatively large positive correlation in the tropical eastern Pacific and Atlantic. There are also relatively large positive correlations in the North Atlantic and to a smaller degree the North Pacific. This diagnostic has limitations—in particular 700 m is unlikely to be the best choice to represent thermocline variations in all regions, and more importantly, the correlation is local in space and time so that any propagation characteristics are not represented. Nevertheless, the correlation does identify many regions where ocean dynamics is playing an important role in seasonal-to-interannual predictability.

As in the example discussed earlier, the emphasis placed in this chapter is on the role of the oceans and there is little doubt that the largest source of seasonal predictability and the most studied is the El Niño–Southern Oscillation (ENSO, e.g., Chang et al., 2006) phenomenon. Considerable progress has been made in understanding and predicting ENSO during the last two decades, and this chapter seeks to summarize some of this progress and identify critical outstanding problems. However, seasonal predictability is not limited to ENSO, and we also seek to describe other key phenomena (e.g., tropical Atlantic variability, Indian Ocean zonal mode, Meridional Modes in the Atlantic, Pacific, and Indian Oceans) as potential sources of predictability.

2. THE SCIENTIFIC BASIS FOR SEASONAL-TO-INTERANNUAL PREDICTION

2.1. El Niño and the Southern Oscillation

The variability in the tropical Pacific on seasonal-to-interannual timescales is dominated by ENSO (e.g., Philander, 1990). During the warm phase of ENSO, ocean–atmosphere interactions lead to a positive SSTA in the eastern equatorial Pacific. At the same time, precipitation is displaced eastward from the warm pool region toward the dateline, and the normally easterly trade winds weaken. Conversely, during a cold event the eastern tropical Pacific SSTA is negative, the trade winds are anomalously strong, and the precipitation is tightly confined to the warm pool region. ENSO events last approximately 9–15 months and the time between events is around 2–7 years, although ENSO also has power on much lower frequency timescales. The large spatial shifts in tropical Pacific rainfall associated with ENSO lead to large-scale changes in global circulation and precipitation. It is these changes in circulation and precipitation that make ENSO a primary source for predictability in remote regions.

The importance of coupled atmosphere–ocean interactions for ENSO can be traced back to the pioneering work of Bjerknes (1966, 1969) who argued for a coupling between the atmospheric Southern Oscillation and oceanic warming in the eastern Pacific (El Niño) through interactions between the sea surface temperature and the surface winds. Ultimately, Bjerknes's work was the first to identify the physical basis and a conceptual framework for understanding the mechanisms for ENSO (Neelin et al., 1998; Burgers et al., 2005). As described in Kirtman (1997), the coupled instability mechanism or Bjerknes feedback works as follows: If a positive eastern equatorial SSTA exists, the temperature gradient between the eastern Pacific and the western Pacific is reduced, which then produces a weakening of the easterly trade winds, augmenting the warming in the eastern Pacific. The additional warming in the east further weakens the trade winds, leading to a coupled ocean–atmosphere positive feedback. A reversal of the argument explains the growth of a cold event. This positive feedback between the SSTA and the atmospheric wind anomaly leads to a continually growing anomaly, although it does not provide an explanation for what causes the transition from one extreme state to the other.

The importance of the Tropical Ocean-Global Atmosphere (TOGA) program in establishing a quantitative understanding of the Bjerknes feedback (e.g., Philander et al., 1984), theories for the transition between warm and cold ENSO events (e.g., Suarez and Schopf, 1988; Neelin, 1991) and the data necessary to support prediction and predictability research (McPhaden et al., 1998) cannot be overstated. For example, Philander et al. (1984) and Hirst (1986) demonstrated how coupled air–sea feedbacks similar to those first postulated by Bjerknes could destabilize equatorial oceanic Kelvin and Rossby waves depending on the details of the physical processes determining the SSTA. If surface current advective processes dominated the SSTA, Rossby waves were destabilized and if upwelling was the primary controlling factor, Kelvin waves were found to be unstable. Moreover, these ocean wave-stability arguments were also used to explain why there are no ENSO events in the extratropical Oceans. Similar to Bjerknes' original mechanism, the destabilized equatorial ocean waves did not provide an adequate explanation for the transition between warm and cold event or an accepted explanation for the timescale of ENSO.

The understanding of equatorial ocean wave dynamics was further refined to provide theories for the transitions between warm and cold events (McCreary, 1983; Battisti, 1988; Suarez and Schopf, 1988; Battisti and Hirst, 1989; Schopf and Suarez, 1990) and the timescale of ENSO (Cane et al., 1990). These theories rely heavily on equatorial wave dynamics and western boundary reflections for the transitions between events and for setting the timescale for ENSO, and are conventionally referred to as delayed oscillator theories. For example, Kirtman (1997) summarized the relevant ocean processes emphasizing the completion between time delay effects associated with Rossby and Kelvin wave propagation and the local growth in the eastern Pacific due to Bjerknes instability (see also, e.g., Philander et al., 1984).

Clearly, ocean dynamics play a central role in the transition between warm and cold events and the ocean memory that is the source of predictability. Nevertheless, there has been a vigorous debate regarding the specific role of ocean *wave* dynamics. For example, Neelin (1991), Neelin and Jin (1993), Jin and Neelin (1993a,b), and Hao et al. (1993) suggest an alternative mechanism for the slow ENSO timescales, where ocean wave dynamics play a minimal role and oceanic surface layer processes provide mechanisms for transitions between warm and cold and the predictive memory. In this so-called slow SST-fast wave limit the slow adjustment time of the SST is a result of local air–sea coupling and advective processes as opposed to Kelvin and Rossby wave propagation and western boundary reflections associated with the delayed oscillator coined by Suarez and Schopf (1988).

It is worth noting two key elements for motivation that led to development of the slow SST-fast wave theory for ENSO discussed above—indeed, this motivation highlights some of the conceptual difficulties in understanding or accepting the delayed oscillator theory for ENSO. First, if the dominant westward propagating signal is carried by a first meridional mode Rossby wave (the gravest mode), then twice the transit time for the Kelvin and Rossby waves is substantially shorter than the observed ENSO period.

Second, the western boundary is considerably more complicated than formulated in the delayed oscillator theory fostering skepticism regarding the central role of western boundary reflections.

Regarding the first issue Cane et al. (1990), using a linear delayed oscillator equation, argued that the oscillation period is longer than twice the delay because the signal in the east is growing due to local instability and, as a consequence of this local growth, the upwelling Kelvin wave that is the result of the western boundary reflections does not completely eliminate the initial downwelling. Additional Rossby wave propagation into the western boundary and the associated reflected Kelvin waves are required to reverse the oscillation. Schneider et al. (1995), using an ocean general circulation model (GCM), and Kirtman (1997) using a simple coupled model, argue that in contrast to Cane et al. (1990) that off equatorial Rossby waves forced by off equatorial wind anomalies via their slower propagation speeds explain the slower timescale. Wang and An (2001) use similar arguments to explain variations in the dominant timescale for ENSO.

In response to the second issue, Jin (1997a,b) proposed a recharge/discharge oscillator for ENSO where the slow timescale is associated with near equatorial Sverdrup balance. Wave dynamics remain important in terms of setting up or leading to near equatorial Sverdrup balance, but the details of western boundary reflections are of secondary importance. A major advantage of this approach is that the essential mechanism, interior Sverdrup flow, is more amenable to measurement compared to western boundary reflections, and monitoring of the low-frequency convergence of the tropical thermocline can be undertaken (Johnson and McPhaden, 1999; McPhaden and Zhang, 2002, 2004). The challenge to such monitoring, however, occurs because the hard-to-measure western boundary current transports may counteract the interior mass convergence.

Ultimately, it is now recognized that the predictability of ENSO is largely determined by the life cycle of individual events, which depends on the memory or inertia associated with upper ocean heat content and coupled ocean–atmosphere interactions (Latif et al., 1998). Understanding the importance of upper ocean heat content for the predictability of ENSO is based on both observations and models that show that equatorial Pacific SSTA and wind stress anomalies are largely in equilibrium with each other (i.e., no substantial memory in either field) and are dominated by a standing oscillation, although the results of Penland and Sardeshmukh (1995) suggest that SSTA has a predictive capability. On the other hand, the winds stress anomaly and the heat content anomaly are not in equilibrium (e.g., Zebiak and Cane, 1987; Schneider et al., 1995; Jin, 1997a,b; Kirtman, 1997; Meinen and McPhaden, 2000) and it is this disequilibrium that is the primary mechanism for ocean memory or predictability. Figure 24.2, for example, shows a time series of SSTA, heat content anomaly, and wind stress anomaly along the equator in the Pacific from 1980–2005. The standing nature of the SSTA can be detected by the fact that there is very little east–west propagation east of the dateline for the large events (e.g., 1998 warm and 1989 cold events), and the disequilibrium between heat content anomaly and the wind stress can best be seen in the western Pacific where the heat content leads central Pacific wind stress anomalies by several months. For example, cold heat content anomalies during early 1998 in the western Pacific appear to lead the easterly wind stress anomalies in the central Pacific in late 1998 and 1999. These leading heat content anomalies are often viewed as the western Pacific "preconditioning" required for the coming ENSO event. It should be noted that all of the theories for ENSO referenced above including the slow-SST fast wave limit can capture the equilibrium between wind stress anomaly and SSTA and the disequilibrium between wind stress anomaly and heat content anomaly.

The western Pacific preconditioning noted above is viewed as a necessary condition for ENSO prediction, but is not necessarily a sufficient condition. Indeed, atmospheric variability associated with westerly wind bursts (WWBs) (e.g., McPhaden, 2004; Lopez et al., 2012), Pacific meridional modes (e.g., Vimont et al., 2003; Chang et al., 2007), Madden–Julian Oscillation (e.g., Zavala-Garay et al., 2005 among many others) or even Indian Summer Monsoon variability (Kirtman and Shukla, 2000) may all serve to modify the amplitude or as potential triggers of ENSO events. These primarily atmospheric triggers, of course, have a distinct ocean forced response that initiates the development of an ENSO event. The possible role of atmospheric forcing can also be detected in Figure 24.2. This is particularly evident for the 1997/98 warm event, but can also be detected during the 2002/03 warm event as enhanced westerly wind anomalies in the far western equatorial Pacific. Indeed, the precise role of atmospheric stochastic forcing of ENSO remains the subject of some debate (see Kirtman et al., 2005 for discussion).

The mechanisms for the loss of predictability are also a major research question. There are currently two main hypotheses that describe the loss of ENSO predictability: (i) ENSO is intrinsically chaotic because of the nonlinear dynamics of the coupled system (Zebiak and Cane, 1987; Zebiak, 1989; Münnich et al., 1991; Chang et al., 1994; Jin et al., 1994; Tziperman et al., 1994, 1995; Wang et al., 1999) and (ii) stochastic forcing (e.g., Penland and Matrosova, 1994; Penland and Sardeshmukh, 1995; Flügel and Chang, 1996; Moore and Kleeman, 1996; Blanke et al., 1997; Chen et al., 1997; Eckert and Latif, 1997; Kleeman and Moore, 1997; Xue et al., 1997; Kirtman and Schopf, 1998; Moore and Kleeman, 1999a,b;

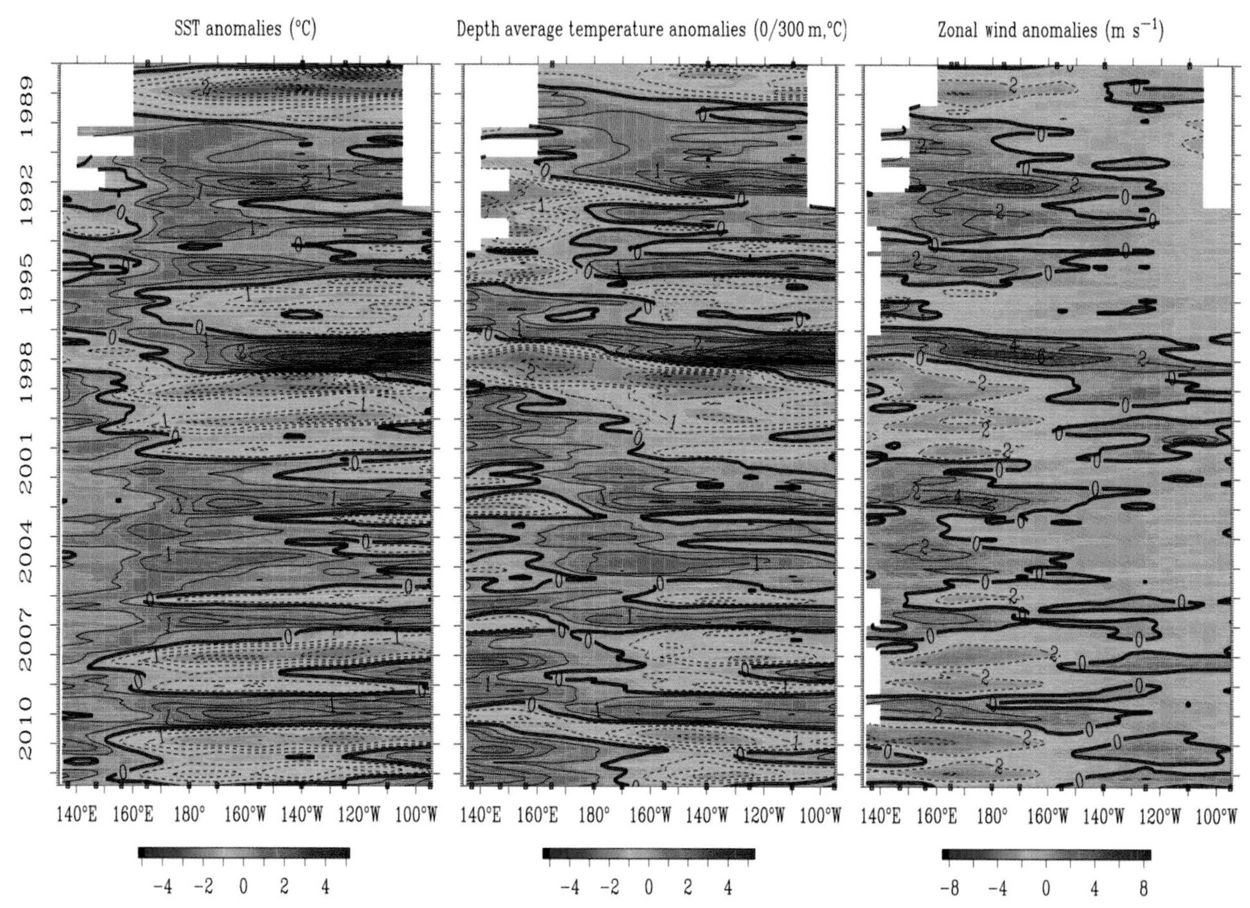

FIGURE 24.2 Time-longitude sections of SSTA (left), depth averaged (upper 300 m) temperature (center), and zonal wind stress anomalies (right) average 2S–2N in the Pacific Ocean. The data come from www.pmel.noaa.gov/tao and are 5 days means.

Thompson and Battisti, 2001; Kleeman et al., 2003; Zavala-Garay et al., 2003, 2004, 2005, 2008; Flügel et al., 2004). At first glance this may appear as an academic argument, but in fact it has been shown that which hypothesis dominates affects ENSO predictability (e.g., Kirtman and Schopf, 1998; Thompson and Battisti, 2001; Chen et al., 2004).

The precise mechanism for ENSO's change of phase (i.e., onset and termination) is also the subject of debate, particularly in terms of the role of ocean dynamics versus atmospheric forcing and response. This debate arises because observations of onset and termination show asymmetries that are not predicted by our basic delayed or recharge oscillator theories. Ultimately, heat content anomalies both on and off the equator in the western Pacific are the necessary precursors for onset and termination (as noted earlier), but precisely how this heat content accumulates during onset and termination is an open question. For example, Guilyardi et al. (2003) argue that heat content buildup in the western equatorial North Pacific leading to ENSO warm phase onset is due to local Ekman pumping (i.e., local atmospheric forcing) as opposed the free Rossby waves as predicted by theory. Termination, however, appears to be consistent with the basic theory. Harrison and Vecchi (1999) and Vecchi and Harrison (2003) emphasize the importance of seasonality (i.e., late in the year southward shift of wind stress anomalies) in the termination. This asymmetry between onset and termination suggests that an event is not simply due to the previous opposite phased event as suggested by the conceptual theories. Indeed, this asymmetry may be the cause for relatively poor forecast quality of ENSO onset, and relatively good forecast quality of ENSO termination.

An important characteristic of ENSO, which is also not fully understood, is the tendency to peak toward the end of the calendar year (Rasmusson and Carpenter, 1982; Wang, 1995). The suggested mechanisms include the seasonal migration of the intertropical convergence zones (ITCZs, Philander et al., 1984); associated seasonal variations in the strength of the coupled atmosphere–ocean instability (Battisti and Hirst, 1989; Li, 1997); seasonal amplification of Kelvin and Rossby waves (Tziperman et al., 1998; Galanti and Tziperman, 2000); Indian Monsoon forcing (Kirtman and Shukla, 2000); and seasonal preference for

WWBs (Eisenman et al., 2005), which effectively enhance coupling strength in the latter part of the year.

2.2. Tropical Atlantic Variability

On seasonal-to-interannual timescales, tropical Atlantic SST variability is typically separated into two patterns of variability—the gradient pattern and the equatorial pattern (Kushnir et al., 2006). The gradient pattern is characterized as a north–south dipole centered at the equator with the largest signals in the subtropics (see figure 15 of Kushnir et al., 2006), and is typically associated with variability in the southernmost position of the ITCZ. The instantaneous correlation between the northern and southern hemispheric components is small suggesting that the variability may in fact be independent; but this is the subject of some debate. Huang (2004) and Bates (2010), for example, separate the gradient pattern into a north tropical Atlantic (NTA) and a southern subtropical Atlantic (SSA) pattern (see figure 1 of Bates, 2008, 2010). The equatorial pattern is sometimes referred to as the zonal mode (e.g., Chang et al., 2006), or the "Atlantic Niño" because of its structural similarities to the ENSO pattern in the Pacific, although the phase locking with the annual cycle is quite different and the air–sea feedbacks are weaker leading to a more clearly damped mode of variability (e.g., Nobre et al., 2003).

The gradient pattern is linked to large rainfall variability over South America and the northeast region (Nordeste) of Brazil, in particular during the boreal spring (Moura and Shukla, 1981; Nobre and Shukla, 1996). The positive gradient pattern (i.e., warm SSTA to the north of the equator) is associated with a failure of the ITCZ to shift its southern most location during boreal spring. This leads to large-scale drought in much of Brazil and coastal equatorial Africa and affects Sahel rainfall. The equatorial pattern in the positive phase is linked to increased maritime rainfall just south of the climatological position of the boreal summer ITCZ. The associated terrestrial rainfall anomalies are typically relatively small.

In terms of mechanisms for predictability, both the gradient and zonal modes require some "external" excitation (Zebiak, 1993; Chang et al., 2001) from either ENSO (e.g., Huang et al., 2002; Chang et al., 2003) or the North Atlantic Oscillation (NAO; e.g., Melice and Servain, 2003). For the gradient mode surface layer, thermodynamic processes appear to be the dominant physics (i.e., the variability is captured reasonably well with thermodynamic mixed-layer models), but ocean dynamics may provide a negative feedback, which is particularly important during the decay phase.

A wind-evaporation-SST (WES) feedback is primarily responsible for the gradient mode. The WES feedback works as follows: the positive phase of the gradient mode is associated with cross-equatorial pressure gradient leading to meridional wind anomalies from south to north. These anomalous winds through evaporation lead to a positive feedback on the SSTA.

Early predictability studies (Penland and Matrosova, 1998) suggest that the NTA component of the gradient mode (and variability in the Caribbean) can be predicted one to two seasons in advance largely due to the "disruptive" or excitation influence from the Indo-Pacific SSTA, but this does not suggest that local coupled processes in the NTA region are unimportant (e.g., Nobre et al., 2003). The NAO can also be an external excitation mechanism, but again local processes remain important for the life cycle of the variability. The predictability of the SSA component of the gradient mode has not been well established, and is largely viewed as independent from ENSO (Huang et al., 2002).

As noted earlier the zonal mode is structurally and mechanistically similar to ENSO in the sense that its maintenance involves a Bjerknes like feedback, that is, weaker winds leading to weaker thermocline slope and anomalously warmer eastern basin SST. The dominant timescale appears to be about 30 months (e.g., Latif and Grötzner, 2000), which is shorter than ENSO presumably due to the smaller basin size (Jin, 1996). Contemporaneous SSTA correlations seem to suggest that this zonal mode is independent of ENSO, but both observation and modeling results suggest a 6-month lagged response to ENSO. The lagged response is related to the timescale of tropical Atlantic thermocline adjustment forced by local wind anomalies that are teleconnected to ENSO. It is well accepted that the feedbacks in the Atlantic are too weak to maintain a self-sustained mode, whereas this remains the subject of some debate in the Pacific (e.g., Kirtman et al., 2005). Attempts to predict the zonal mode have met with modest success (Nobre et al., 2003) due, in part, to the large systematic errors in the Atlantic in the coupled (GCMs).

2.3. Tropical Indian Ocean Variability

There are three dominant patterns of variability in the tropical Indian Ocean that affect remote seasonal-to-interannual rainfall variability over land: (i) a basin wide pattern that is remotely forced by ENSO (e.g., Krishnamurthy and Kirtman, 2003); (ii) the so-called Indian Ocean Dipole/Zonal Mode (IOD for simplicity) that can be excited by ENSO, but can also develop independently of ENSO (e.g., Saji et al., 1999; Webster et al., 1999; Huang and Kinter, 2002); and (iii) a gradient pattern or meridional mode similar to the Atlantic and Pacific that is prevalent during boreal spring (Wu et al., 2008). The basin wide pattern is slave to ENSO and thus its predictability is determined by the predictability of ENSO. The IOD plays an important role in the Indian Ocean sector response to ENSO and contributes to regional rainfall anomalies that are independent of ENSO. Idealized predictability studies suggest that the IOD should be predictable up to about 6 months (Wajsowicz, 2007; Zhao and Hendon, 2009), but prediction experiments are less optimistic (e.g., Zhao and Hendon, 2009). The predictability of the Indian Ocean meridional mode has not been investigated to date.

Mechanistically, the basin wide mode is captured in thermodynamic slab mixed-layer models, suggesting that ocean dynamics is of secondary importance and that the pattern is due to an "atmospheric bridge" associated with ENSO (e.g., Lau and Nath, 1996; Klein et al., 1999). The IOD, on the other hand, depends on coupled air–sea interactions and ocean dynamics. For example, Saji et al. (1999) noted that the IOD was associated with east–west shifts in rainfall and substantial wind anomalies. Huang and Kinter (2002) argued for well-defined (although not as well defined as for ENSO) interannual oscillations where thermocline variations due to asymmetric equatorial Rossby waves play an integral role in the evolution of the IOD. The importance of thermocline variations is that they are a potential source of ocean memory and hence predictability. The development and decay of the meridional mode is largely driven by local thermodynamic cloud and wind feedbacks induced by either ENSO or the IOD, whereas thermocline variations do not seem to be important Wu et al. (2008).

2.4. Extratropical SST Predictability

The discussion to this point has emphasized the role of the tropical Oceans in seasonal-to-interannual predictability. Here, we discuss some aspect of seasonal-to-interannual predictability associated with midlatitude ocean processes, but we note that the largest signals occur and much of the literature focus on longer (e.g., multiannual or decadal) timescales. For example, Smith et al. (2012) and references cited show (see their Figures 24.5 and 24.6) signals associated with Pacific decadal variability and Atlantic multidecadal variability. While we acknowledge that careful consideration of the superposition of decadal and season-to-interannual variability could lead to improved seasonal-to-interannual predictability, here we focus on just the seasonal-to-interannual timescale. The decadal timescale is a focus of Chapter 25.

A fundamental problem in understanding the mechanisms that cause low-frequency variability of SST in mid-latitudes has been separating the relative roles of "one-way" versus "two-way" air–sea interactions (e.g., Latif, 1998). In the common explanation of "one-way" interactions, stochastic forcing due to internal atmospheric dynamics drives low-frequency variability of the ocean. The atmosphere does not respond to the resulting ocean variability. Conversely, in two-way interactions, coupled feedbacks between the ocean and atmosphere are important elements driving the variability. The feedbacks may be unstable leading to self-sustained oscillations, or they can be damped oscillations that occur at a preferred timescale due to, for example, ocean dynamics, and for which stochastic forcing is required. In the midlatitudes, the feedbacks (if they exist) are most likely damped; however, we emphasize that even damped local two-way interactions have been shown to effectively increase atmospheric persistence and predictability on seasonal timescales (Barsugli and Battisti, 1998). The problem is further complicated by how much of the midlatitude SST variability is remotely forced from the tropics (e.g., ENSO). Nevertheless, put simply, if the mechanisms for midlatitude SST variability are "one-way," then any impact on terrestrial climate predictability is likely to be low. On the other hand, "two-way" interactions allow for increased persistence and potential predictability of terrestrial climate.

2.4.1. *North Pacific SSTA*

Namias (1959) was among the first to note that phase relationships between midlatitude SSTA and atmospheric circulation patterns suggest that there are "two-way" interactions and that these feedbacks could lead to potential seasonal-to-interannual predictability of terrestrial climate. Deser and Blackmon (1995), Barlow et al. (2001), and Yeh and Kirtman (2004a) identified North Pacific variability in observational estimates that is distinctly interannual and that appears to be associated with ENSO. Lau and Nath (2001) argued that seasonality complicates the North Pacific teleconnection to the tropics, and Jin and Kirtman (2010) noted similar complications with the Southern Hemisphere. Yeh and Kirtman (2004b) used coupled GCM experiments to argue that seasonal-to-interannual SST variability in the North Pacific is mostly due to stochastic forcing with little "two-way" interactions detected, and that the local atmospheric stochastic forcing largely masks the remote tropical forcing. Lau and Nath (2001) used the so-called Tropical Ocean-Global Atmosphere-Mixed-Layer (TOGA-ML) experimental frame work to show that two-way interactions in the extratropics enhance the persistence of ENSO-related atmospheric anomalies. Jin and Kirtman (2010) used similar TOGA-ML experiments to show that the strength of the ENSO teleconnections in the North Pacific depends on dominant ENSO periodicity. In summary, the balance of evidence suggests that two-way interactions are important, but this has not been clearly demonstrated in actual prediction experiments. Finally, we note that the correlation between predicted and observed North Pacific SST (i.e., a measure of forecast quality) is lower than for the tropical Pacific, but is on the order of 0.6 in a relatively large region (see Figure 24.4 below).

2.4.2. *North Atlantic SSTA*

The NAO is the dominant mode of seasonal variability in the North Atlantic sector and is linked to boreal winter precipitation, temperature and storminess throughout the entire region (e.g., Hurrell, 1995). Seasonal prediction studies with prescribed SST indicate modest midlatitude predictability (Rodwell et al., 1999; Mehta et al. 2000), but this could be due to model deficiencies or due to the lack of two-way interactions (i.e., Barsugli and Battisti, 1998). Indeed, many observational studies, which implicitly include two-way interactions, suggest considerably more seasonal predictability (e.g., Saunders and Qian, 2002). Recent experience at the UK Met Office with their seasonal prediction system also indicates improving seasonal NAO forecast quality (A. A. Scaife, personal communication). Figure 24.4 shows an estimate of the SSTA forecast quality in current seasonal prediction systems—again, like the North Pacific there has not been a clean set of prediction experiments demonstrating some terrestrial predictability or prediction quality arises exclusive from North Atlantic processes.

3. DEVELOPMENT OF SEASONAL-TO-INTERANNUAL PREDICTION SYSTEMS

3.1. Historical Review

Once the connections between ENSO became understood, it was an obvious step to consider whether it might be possible to predict ENSO-related variations in advance. Much basic work was undertaken in the 1970s and 1980s to observe and better understand the equatorial Pacific and its interactions with the atmosphere. This culminated in the afore-mentioned TOGA programme sponsored by World Climate Research Program during the decade 1985–1995 (Latif et al., 1998). This programme was predicated on the expectation that variations in the tropical oceans could be predicted, and the knowledge that these variations have a global impact via the response of the atmosphere.

Together with observational and theoretical work, computer modeling studies in the 1980s began to make progress in both simulating and predicting the ENSO cycle. One approach was to develop "intermediate" models of the ocean–atmosphere system, such as Anderson and McCreary (1985) and Zebiak and Cane (1987). The "Cane-Zebiak" model became famous as the first dynamical forecast model to predict the onset of an El Niño event in 1986 (Cane et al., 1986). Implicit in this approach was the idea that if an El Niño or La Niña event could be predicted, something useful could be said about expected weather patterns based on correlations established from previous events. Indeed, work also progressed using statistical methods to predict atmospheric behavior on seasonal timescales directly from predictors such as SST. If only a "short-range" seasonal forecast was needed (such as a zero-lead forecast for the next 3 months) then just persisting the initial observed SST anomalies already gives a fair degree of skill relative to climatology, so practical work could be done even when dynamical prediction of SST was still in its infancy. One notable area of success was the relatively strong relationship between Atlantic SSTs and March–May rainfall in the Nordeste region of Brazil (Ward and Folland, 1991; Hastenrath and Greischar, 1993). Even today, statistical methods still find wide use, both for predicting the atmospheric response to ocean conditions, and for giving baseline predictions of aspects of the ocean state such as ENSO.

A second approach to ENSO and seasonal prediction was not to use statistical methods or intermediate models, but to directly use coupled ocean–atmosphere GCMs. In principle, these represent our knowledge about ocean and atmosphere dynamics in the most complete form possible for a global scale model. In practice, computer resources limited the resolution that could be used, and our knowledge and experience of parameterizing subgrid scale processes were less developed than today, so even if it was viewed as the correct long-term strategy, early success was by no means assured. In particular, much of the physics of ENSO is sensitive to the ocean mean state; so intermediate models that could specify a mean state had an immediate head start over complex models that might struggle to produce a mean state close to that observed.

Initial work with coupled GCMs focused on simulating the tropical Pacific, both the mean state and the interannual variability. This was a prerequisite to being able to use such models for prediction. Delecluse et al. (1998) summarize the progress that was made during the TOGA decade of 1985–1995, during which time a variety of coupled GCMs were developed and studied by research groups around the world. Early work to predict ENSO using coupled GCMs used low-resolution models in the atmosphere—the first such published study (Latif et al., 1993) used a T21 spectral atmosphere model with a 5.6° grid, although ocean resolution was higher to enable a representation of the equatorial Kelvin and Rossby waves. Other systems developed in the research community included those of Kirtman et al. (1997) and Rosati et al. (1997), the latter of which included an ocean data assimilation scheme to prepare initial conditions. Work was also taking place at some of the major operational numerical weather prediction (NWP) centers. NCEP was the first to routinely run and issue experimental coupled GCM forecasts for ENSO (Ji et al., 1994), using a Pacific Basin ocean model coupled to a T40 AGCM. The European Centre for Medium Range Weather Forecasts (ECMWF) started routine forecasts in 1997 (Stockdale et al., 1998a), using a T63 AGCM coupled to a global ocean model.

It was apparent at this stage (and perhaps before) that the forecast models had significant imperfections, which were only slowly being reduced by the process of model development and improvement. One way to deal with this was to work toward multimodel systems, which can sample and partially average out some of the model-induced errors. This was already implicit in the ENSO forecasting practices of the community, where many different forecasts would be compared before conclusions were drawn. A major research project in Europe, called DEMETER (Palmer et al., 2004), took a systematic approach and used seven coupled GCM forecasting systems to show that a multimodel approach was indeed powerful in improving the reliability of model based seasonal prediction systems (e.g., Hagedorn et al. 2005). Multimodel ensemble methods are also being used for real-time seasonal prediction in the United States (see http://origin.cpc.ncep.noaa.gov/products/people/wd51yf/NMME/index.html) and for the Asian Pacific Climate Center.

Seasonal prediction with coupled GCMs is now firmly established operational practice. At present there are 12 WMO-designated "Global Producing Centres" established at NWP centers around the world, supplemented by additional predictions run elsewhere. There is also an established and growing user interface community, both in the public/humanitarian sector (such as the International Research Institute for Climate and Society), and in the commercial sector. Seasonal prediction is also a natural bridge by which society can learn to cope with a fluctuating and changing climate, and as such is a key part of the "Climate Services" being developed internationally to assist human adaptation to climate change.

We now look in more detail at the role of the ocean in the practice of seasonal prediction.

3.2. Ocean Data Assimilation for Initializing Forecasts and for Assessing Models

Any form of dynamical forecast needs initial conditions. The paradigm for seasonal forecasting in the 1980s and 1990s was that it was the ocean initial state that mattered—indeed, in many intermediate models the atmosphere was represented as a purely diagnostic response to the instantaneous SST field. In the early days of seasonal prediction, there was very little direct information on the interior state of the ocean. Instead, the initial state of the ocean was usually obtained by relying on the history of the wind field to determine the position of the model thermocline. Information on the winds came from hand-made analyses of ship reports, most famously in the "FSU winds" dataset produced by J. J. O'Brien and coworkers at Florida State University (FSU; see also Legler et al., 1997). This provided maps of monthly mean pseudo-stress over the tropical Pacific. Given the production method, these were inevitably noisy, but at the time were the best data going to initialize forecast models. Bourassa et al. (2005) provide a comparison between the original FSU dataset and a later objective *in situ* wind analysis. Although using only boundary forcing to estimate the interior state of the ocean may seem a surprising strategy, in principle it can work very well for the ENSO prediction problem. That is because the oceanic side of ENSO is a large-scale, upper ocean phenomenon—the eddy scale, which is determined by internal ocean dynamics rather than forcing, does not need initializing. Modeling studies show that the large-scale behavior of the upper tropical ocean is dominated by the history of the surface forcing, although internal variability is not entirely absent. Modeling studies have shown that the equatorial upper ocean variability is mostly wind driven (Stockdale et al., 1998b).

The wind forcing largely constrains the position of the thermocline, assuming the overall water mass distributions are correctly represented in the model. An additional important input for a seasonal forecasting system is the initial state of the surface mixed layer, particularly its temperature. This is because it is via the SST that the ocean influences the atmosphere, and the thermal inertia of the mixed layer means that it must be accurately initialized to allow a good forecast. Since the SST is generally a fairly well observed field, simple relaxation techniques allow a reasonable initialization of the mixed-layer temperature in an ocean model, at least if the mixed-layer physics is reasonably represented.

The above wind/SST forcing strategy can provide baseline ocean initial states good enough to use in a forecasting system, in particular for the tropical Pacific where wind-forced thermocline variations are hugely dominant. It is worth noting that over the last 20 years the quality of available wind forcing has improved markedly, due to better *in situ* and satellite data and better analysis/reanalysis capabilities in the NWP community, so purely wind forced ocean initial conditions are substantially better now than in the past. Despite this improvement, studies have repeatedly shown that assimilation of *in situ* and other data gives improved ocean analyses and improved ENSO forecasts (Ji and Leetmaa, 1997; Balmaseda and Anderson, 2009).

Today's dynamical seasonal forecasting systems typically rely on global ocean data assimilation systems to prepare the necessary ocean initial conditions. The most critical subsurface *in situ* data are temperature and salinity data from moorings (see http://www.pmel.noaa.gov/tao/proj_over/proj_over.html and http://www.pmel.noaa.gov/pirata/) and from Argo floats (see http://www.argo.ucsd.edu/About_Argo.html). SST is also crucial, and can be analyzed together with the rest of the ocean (e.g., Saha et al., 2010), or else introduced from a separate analysis (e.g., ECMWF). Velocity data are typically used only for validation. Near the equator, the large-scale velocity field holds significant energy (e.g., an equatorial Kelvin wave has half its energy in kinetic form), and so whether or not velocity data are available, it is important that the velocity field is initialized. This can be implemented in multivariate schemes by ensuring adjustments of the T and S field near the equator are balanced by corresponding velocity adjustments (Burgers et al., 2002). Present day ocean analyses are also highly dependent on remotely sensed satellite data. Altimetry is the most "obvious" oceanographic data, but scatterometer data for surface winds and infrared and microwave radiance data for SST are also crucial; indeed, the quality of the atmospheric forcing of the ocean depends on the full panoply of satellite data, which ensures the quality of NWP analyses.

Seasonal forecast models are not perfect, and their biases must be assessed by running them over a past set of years. It is also important for users to have estimates of the skill of forecast systems, and this is even more demanding in the number of past years for which model runs are required. Thus, when creating an ocean analysis system for use in seasonal forecasting, it is important that attention is paid to the quality of reanalyses for past dates as well as the quality of realtime analyses, and to the extent possible, the consistency of the analyses for different periods. This is a real challenge, because of the very large changes in the ocean observing system over recent decades. The major change points are roughly 1981 (start of satellite SST data, from Advanced Very High Resolution Radiometer; http://noaasis.noaa.gov/NOAASIS/ml/avhrr.html); 1993 (start of Tropical Atmosphere Ocean mooring array giving well sampled *in situ* data in the equatorial Pacific, and start of altimetry); and 2002–2004 (period when global Argo array was implemented). Current data availability allows production of high quality ocean analyses of the larger scales in the ocean, even if substantial work and care is needed to use all of the data well. Further back in time, as the data become much sparser, the solution is likely to be more sensitive to the forcing fields used, the quality of the assimilating model, and the use of effective and appropriate techniques to extract the most information from a small number of temperature data, in the presence of almost no corresponding salinity data. More information on ocean data assimilation techniques is given in Chapters 21 and 22.

3.3. Current Forecast Quality

Seasonal forecasting is a challenging subject, still very much in its main development phase. The initial emphasis was on predicting ENSO-related SST variations and their impacts around the world, but today's systems are much more comprehensive in their concerns. Anomalies in soil moisture and snow cover, the role of the stratosphere and various forcings acting on it (including ozone, volcanic aerosols, and solar variation), variations and trends in sea ice, tropospheric aerosol variations and changes in atmospheric composition are all processes that have relevance for seasonal prediction. Producing credible and accurate ensemble forecast distributions of atmospheric behavior in response to the full spectrum of forcings remains a real challenge. Much of the work to improve our forecasting systems is undertaken at major NWP and climate modeling centers, and involves continuous processes of reducing errors, increasing model accuracy, and increasing forecast quality.

It remains true, however, that on a global scale the majority of achievable forecast skill on seasonal timescales comes via prediction of the SST field. It is thus helpful to consider what we presently achieve for SST prediction, what the prospects are for improvement, and what obstacles still lie in the way.

Forecasting of ENSO, and more generally of tropical Pacific SST variations, is now thoroughly established. In terms of quantifying forecast skill, it is important to note that skill scores are quite sensitive to the period over which forecasts are verified—ENSO variations are rather irregular, and the amplitude, number and predictability of ENSO events varies from decade to decade. This was realized early on (Balmaseda et al., 1995), and means that progress in the quality of our forecasting systems does not necessarily lead to the most recent periods of operational forecasts having the highest score (Barnston et al., 2012). It is also important to remember that GCM-based forecasts are ensemble in nature, and thus predict a range of values for SST, with the range growing with lead time. Verification is often applied to the ensemble mean, and this can be a useful way to compare forecasting systems, but probabilistic verification is in fact more appropriate. Because "chance" plays a part in the outcome of each ENSO event, our assessment of the quality of a forecasting system is necessarily approximate, and distinguishing between two good systems becomes hard except perhaps over a large number of cases. Poor performance is easier to detect. These difficulties are due to the fact that forecast systems are typically verified over approximately a 30-year period, and that a larger number of years (i.e., independent samples) are needed.

Analysis of recent operational forecasts suggest that today's GCM-based forecast systems generally beat statistical models, and is also suggestive that better developed GCM systems tend to do best (Barnston et al., 2012). This fits with the development experience at operational centers, where testing of successive new systems over common hindcast periods has shown significant long-term progress in ENSO forecast skill (Stockdale et al., 2011). Progress has needed much work to improve both the preparation of ocean initial conditions and the realism of the models. GCMs remain imperfect, and are particularly prone to

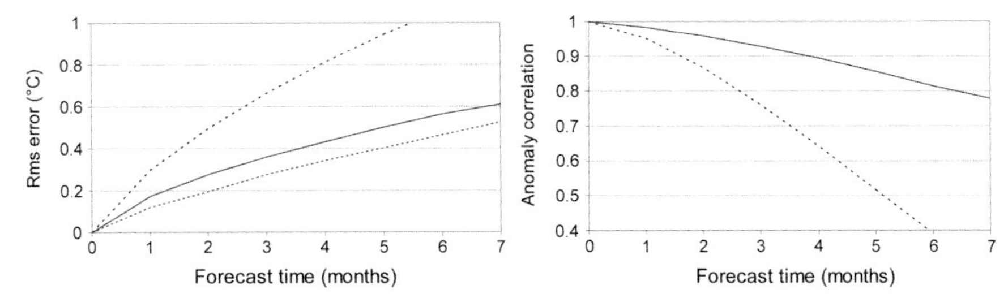

FIGURE 24.3 RMS error (left) and ACC (right) for predictions of NINO3.4 SST anomalies taken from the ECMWF seasonal forecasting System 4. The statistics are for forecasts initialized in the 30.5-year period 1981–mid 2011. Ensemble mean (15 members) forecast skill (red), anomaly persistence (dashed black), and ensemble spread (dashed red, RMSE plot).

pattern-type errors, for example, the amplitude and location of the maximum SST variations are often misplaced in a systematic way for a given model. Errors in the best models are smaller than they used to be, but the far east Pacific and the cold tongue/warm pool boundary remain easy to get wrong. Some systematic errors can be accounted for before forecast products are issued, although where the errors are nonlinear (as they sometimes are) the scope for this is limited. Large multi-model ensembles have given very good ENSO skill scores in hindcast experiments such as those performed in ENSEMBLES (Weisheimer et al., 2009). Individual models can also do well as illustrated by Figure 24.3, taken from the latest ECMWF operational forecast system (System 4). This shows a high skill for forecasts up to 7 months, although the gap between the RMS error and the ensemble spread indicates that there is still scope for improvement—a perfect system has identical RMS error and ensemble spread. Ensemble mean forecasts will never be perfect predictors of reality—the best that we can achieve is to have errors that are on average no larger than an ensemble spread that properly represents the irreducible uncertainty in the forecast.

Outside of the tropical Pacific, interannual variations are generally less predictable, and the weaker "signals" have been more easily lost to initial condition errors and model inaccuracies. For example, SST predictions in the equatorial Atlantic have often been well below predictability limits (Stockdale et al., 2006), and it is only recently that some coupled GCM systems have started to become more credible in this region. Preparing high quality initial conditions has proved a challenge, especially given the much less extensive array of moorings in the Atlantic compared to the Pacific.

As noted earlier, multimodel ensembles are a practical (but imperfect) methodology of sampling forecast uncertainty due to errors in model formulation. Here, we show one result highlighting multimodel forecast quality for global SSTA taken from the US multimodel effort noted earlier. Figure 24.4 shows the ensemble mean SSTA correlation coefficient for hindcasts initialized in August 1982–2010 verifying 6 month later in February. In this example, the multimodel combination produces 83 ensemble members for each initial condition and each ensemble member has equal weight in calculating the ensemble mean. The correlation is relatively large throughout most of the equatorial Pacific and Indian Oceans and in much of the tropical Atlantic. The correlation drops off rapidly in the extratropics particularly in regions of western boundary currents.

3.4. Biases and the Need to Improve Models—Resolved Eddies

Many biases in coupled GCM forecast systems originate in the atmosphere model. The exact structure and variability of the tropical winds can be difficult to get right, even if the SST is correct (Richter and Xie, 2008). Nonetheless, there is plenty of evidence that some of the errors originate in the ocean models. Indeed, one of the most glaring systematic errors in coupled GCM today is the fact the zonal SST gradient along the equator in the Atlantic Ocean is reversed compared to observational estimates. It is likely that this severely limits the coupled model's ability to predict the zonal mode in the Atlantic (Stockdale et al., 2006).

Ocean models are typically simpler than atmosphere models, in that the range of physical processes that must be treated by parameterization is smaller (though not necessarily easier). Despite this, there is scope for continued development of numerical methods, physical parameterization and code organization of ocean models. Mixing processes are particularly important in accurate representation of the equatorial undercurrent systems and the surface mixed layer, which are both important for SST predictions. However, an unavoidable fact is that the ocean has a small Rossby radius of deformation, and that a high resolution is needed to resolve the eddy field. In common parlance, *eddy-resolving* models have horizontal resolution finer than 1/6°, in contrast to *eddy-parameterized* models with 1° or greater grid spacing or *eddy-permitting* ocean models whose resolution lies between 1/6° and 1°. These bounds are also a function of latitude with finer resolution required in the middle and high latitudes, and less resolution in the tropics (e.g., 1/3°). Ocean models used for seasonal prediction so

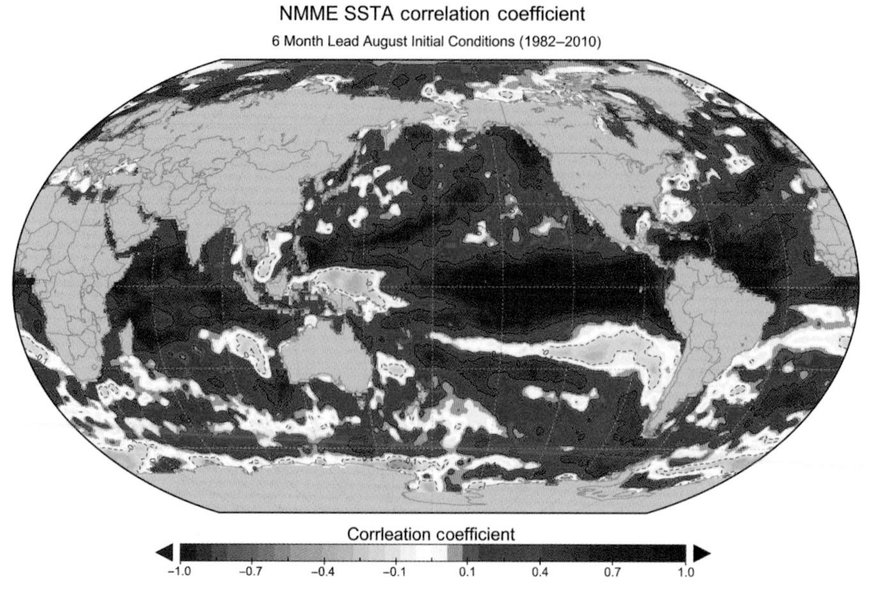

FIGURE 24.4 The ensemble mean SSTA correlation coefficient for hindcasts from the US National Multimodel Ensemble prediction effort (http://origin.cpc.ncep.noaa.gov/products/people/wd51yf/NMME/index.html). The hindcasts are initialized in each August from 1982 to 2010, verifying 6 months later in February. Eighty three ensemble members covering seven different models are used to calculate the ensemble mean. Each ensemble member from each model has equal weight in calculating the ensemble mean.

far are not eddy resolving, but are eddy permitting—how important and how feasible is it to change this?

Ocean modelers have repeatedly reported benefits from running ocean models at higher resolution (Smith et al., 2000; Oschlies, 2002; Hurlburt et al., 2008). Unfortunately, it seems that the benefits do not asymptote until a very high resolution is reached—for example, Hurlburt et al. consider that a resolution of 1/25° in tropics and 1/32° in midlatitudes to be the point at which the mesoscale eddy field is sufficiently well resolved to allow high quality simulations of the ocean circulation.

While this might just be affordable for operational ocean forecast systems making forecasts of a few days, it is substantially beyond what will be feasible for seasonal prediction systems in the near to midterm future. To understand the computational requirements, consider that seasonal forecast systems need to calculate the evolution of the ocean state many times—each forecast is an ensemble, and each forecast needs a set of "calibration" hindcasts covering a set of past years, each of which is again an ensemble. For example, the ECMWF seasonal forecast system introduced in 2011 calculates 51 forecasts of 7 months each month (about 30 years of model integration), and the hindcast set totals more than 3000 years of model integration.

Although fully eddy-resolving models are likely to be unaffordable for a while, resolutions such as 1/4 by 1/4 degree are becoming affordable; this resolution is planned for use in the UK Met Office seasonal forecast system, for example. Experiments have shown benefits of even modest eddy resolution in the tropical performance of coupled GCMs (e.g., Roberts et al., 2009), and as computer resources continue to increase over the next decade and more, it should be possible to incrementally improve the eddy resolution and performance of ocean models in seasonal forecast systems. Key benefits that are expected are improvements in the simulation of equatorial SST, and big improvements in the biases seen in the Kuroshio extension and Gulf Stream regions.

As an example of the potential benefit of ocean eddy-resolving models (in the extratropics), we show the time mean rainfall distribution from a long simulation of two coupled model simulations in comparison to observational estimates of rainfall. The two simulations use identical atmospheric models and resolutions (approximately 50 km), but two different ocean resolutions. In the first case, the ocean resolution is approximately 100 km and in the second case the ocean resolution is eddy resolving (10 km). The ocean models are identical in all other aspects. Here, we show the time mean rainfall (Figure 24.5) in the Gulf Stream region. Significant structure changes with ocean resolution are noted. In particular, the orientation of the rainfall maximum in the eddy-resolving simulation agrees with the observational estimates, whereas in the eddy-permitting simulation the rainfall is erroneously oriented along the coast of North America. The details of the experimental design and results are described in Kirtman et al. (2012).

The need for improving or increasing resolution is not limited to the horizontal. There is compelling evidence that

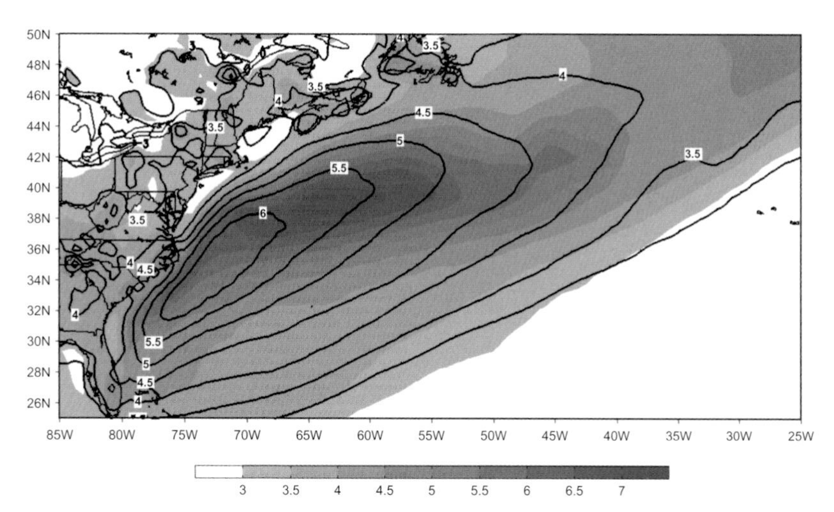

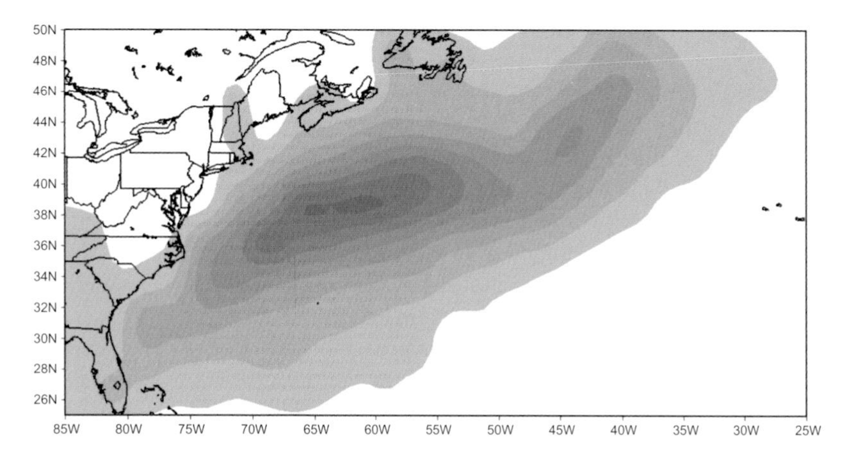

FIGURE 24.5 The top panel shows the time mean precipitation from the eddy-permitting simulation in contour and the eddy-resolving simulation in shaded. The bottom panel shows the observational estimates from Xie and Arkin (1997). See Kirtman et al. (2012) for detailed discussion of these results.

diurnal SST variations are intimately connected to capturing the correct air–sea coupling associated with the Madden–Julian Oscillation, and that vertical ocean resolution on the order of 1 m in the near surface may be required (Bernie et al., 2007; Bernie et al., 2008). Most current seasonal prediction systems have on the order of 5 m vertical resolution in the ocean.

Seasonal forecast systems need multidecadal ocean reanalyses, and assuming these are run sequentially, parallel performance must be sufficient to allow a reasonable wall-clock time (in contrast, running large numbers of relatively short ensemble forecasts is a much more naturally parallel problem). High-resolution ocean models are also notorious for the volumes of data they can generate, a problem that will only be made worse if ensemble assimilation techniques are used to sample the uncertainties in the ocean initial state, as is already done in some operational systems. With all seasonal forecast systems, there is a trade-off between the costs and the value addition of various possible enhancements—more ensemble members, a larger set of hindcast dates, a higher resolution atmosphere, and a higher resolution ocean. Thus, the degree to which improving computer resources will actually be directed toward improving ocean resolution is hard to predict. If large ensemble sizes are needed to sample midlatitude noise (the unpredictable part of the atmospheric flows), and a high ocean resolution is needed to maintain a realistic mean state in the midlatitude boundary currents, the necessity of running all ensemble members with the full midlatitude ocean resolution might conceivably be possible to sidestep. Adequate resolution to properly represent the equatorial oceans, with enough ensemble members to properly sample the uncertainty in ENSO development, is likely to remain the highest priority for ocean models in seasonal forecast systems.

4. CLOSING REMARKS: CHALLENGES FOR THE FUTURE RESEARCH

In a qualitative sense many of the sources of seasonal-to-interannual predictability associated with the tropical Oceans are well known, for example, ENSO, gradient modes, and zonal modes. We also have a fairly detailed understanding of the basic physical mechanisms underpinning these sources of predictability. However, quantifying the limit of predictability and the physical processes that limit predictability remains elusive. For example, it is unclear whether atmospheric "noise" or internal nonlinear dynamics (i.e., chaos) is the primary limiter of ENSO predictability. Is the so-called spring predictability barrier a fundamental property of the coupled system or a profound modeling problem? One merely has to consider the large range of ENSO responses in climate change projections to recognize that serious gaps in our understanding and our ability to model seasonal-to-interannual variability remain.

Certainly, from a practical point of view we can at least quantify the lower bound of predictability based on our current ability to "skillfully" predict seasonal-to-interannual variability. But is this the best we can do? As an example of where we stand today, we show Figure 24.6 captured from the International Research Institute for Climate and Society (IRI; http://portal.iri.columbia.edu/portal/server.pt) web page. This figure shows ENSO forecasts (measured by Nino3.4 SSTA) from a large number of realtime dynamical prediction systems initialized during March 2010–December 2011. The observational estimates are given by the black curve throughout the period in question. On the positive side, the range or the forecast plume encompasses the observational estimate. Put simply, the forecasts did not "bust" in that the observed evolution was predicted with a nonzero probability. On the negative side, one merely has to consider the SON2011 forecast, which ranges from −1.25 to 1.25 °C with the greatest probability of well above zero, whereas the observational estimate is about −1 °C. Current capabilities give us no concrete assessment of whether this arguably poor forecast was just because of an unpredictable state and that the observed evolution was a statistical outlier or whether our prediction systems continue to suffer from serious flaws. Our intuition based on our experience with seasonal-to-interannual prediction systems suggests that this poor forecast is a due to the models' inability to adequately capture the observed physical processes and that the forecast can be improved.

Improving forecast quality will require resolving several points of tension. Here, we highlight some of these key tension points in need of hard-nosed research:

i. As noted above, there is compelling evidence that resolved ocean eddies reduce systematic biases in the tropics and extratropics, but to make predictions with ocean eddy-resolving resolutions will require smaller ensemble sizes. (This reduced ensemble size is due to limits in computational resources.) In other words, we need to quantitatively determine the benefit from increased ocean (and atmosphere) resolution versus ensemble size. How much effort and resources need to be invested in uncertainty quantification (i.e., larger ensembles) versus taking your best shot (i.e., higher resolution). Indeed, we do not even have a clear understanding of how ensemble size and resolution might be related. In relation to computing costs the assimilation system also depends strongly on the model resolution and often this is one of the main limiting factors, particularly for ensemble based or 4Dvar assimilation systems. The resolution question is particularly challenging since there may be physical phenomena that are important for predictability that are unresolved

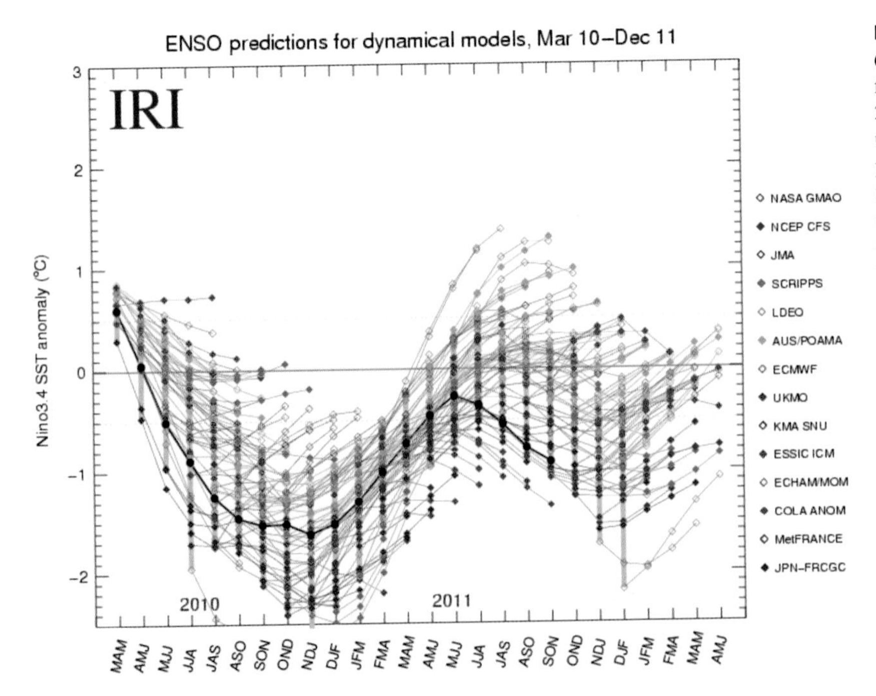

FIGURE 24.6 International Research Institute (IRI) for Climate and Society compiled model forecasts issued from March 2010 to December 2011. The observations are also shown in black up to the most recently completed 3-month period. The plots allow comparison of plumes from the previous start times, or examination of the forecast behavior of a given model over time. See http://portal.iri.columbia.edu/portal/server.pt for details.

and unrepresented in the models. Do we need to have models that resolve tropical cyclones to predict seasonal changes in tropical cyclone frequency? Are forecasts for changes in tropical cyclone frequency at all *credible* when our models fail to resolve tropical cyclones? Nevertheless, seasonal predictions of tropical cycle frequencies are made and have significant skill when measured against observed frequencies. Are air–sea interactions associated with tropical instability waves (TIWs) critical for ENSO predictability? Both tropical cyclones and TIWs are at best poorly resolved in current seasonal-to-interannual prediction systems.

ii. There are studies that show ocean biology (e.g., turbidity) can affect the systematic errors in ocean models (e.g., Murtugudde et al., 2002). This raises challenging question of how to deploy resources into model complexity versus resolution versus ensemble size. In fact, it is arguable that we do not even know how to answer this question today nor do we know which biogeochemical processes need to be included in our prediction systems.

iii. There is significant tension in multimodel methodologies. The approach is largely ad hoc—the collection of models is based on those that happen to be available. The multimodel ensemble is an ensemble of opportunity. We do not know if the collection of models is optimal, and we have, to date, been unable to identify models that are of no benefit to the forecast. We do not know if the available models are sufficiently different and how many models are needed to resolve uncertainty in model formulation. It is also unclear how this model diversity detracts from ensuring that there is a critical mass of scientists working to improve a given model. More purposeful multi-model methodologies are being explored that also attempt to explicitly include model uncertainty in the evolution of the prediction (i.e., perturbed physics ensembles and stochastic physics ensembles), but much research needs to be done in this regard, particularly in terms of including uncertainty in ocean model formulation.

iv. There is no clear path or process for model improvement. There are many examples of how confronting model parameterizations with detailed observations of physical processes leads to improved parameterizations or even new parameterizations, and this should and will continue. Nevertheless, questions remain about how to ensure that this process-by-process approach, which is typically done within the context of one model, is transferable to other models. We also do not have a mechanism for identifying the physical processes that, if improved or incorporated into the model, will also improve seasonal-to-interannual prediction quality. Indeed, there are even questions whether the concept of local parameterized physical processes based on the larger scale environment is a well-posed problem, and whether nonlocal strategies (e.g., stochastic physics) must be emphasized (Berner et al., 2008).

Addressing the points of tension noted above represents exciting and demanding research problems that encompass basic theoretical question regarding how to quantify predictability, understanding mechanisms for predictability and processes that limit predictability, how to optimally use multiple prediction systems to produce the best forecast probability distribution that accounts for all sources of uncertainty, and how to improve models and their representation of the physical phenomenon. These are daunting research challenges, and yet even modest progress has the potential for significant forecast quality improvements and societal benefit.

ACKNOWLEDGMENTS

The authors thank the anonymous reviewers and the editor (S. Griffies) for their comments and suggestion to improve this contribution.

REFERENCES

Anderson, D.L.T., McCreary, J.P., 1985. Slowly propagating disturbances in a coupled ocean atmosphere model. J. Atmos. Sci. 42, 615–629.

Balmaseda, M., Anderson, D., 2009. Impact of initialization strategies and observations on seasonal forecast skill. Geophys. Res. Lett. 36, L01701. http://dx.doi.org/10.1029/2008GL035561.

Balmaseda, M.A., Davey, M.K., Anderson, D.L.T., 1995. Decadal and seasonal dependence of Enso prediction skill. J. Clim. 8, 2705–2715.

Barlow, M., Nigam, S., Berbery, H.E., 2001. ENSO, pacific decadal variability, and U.S. summertime precipitation, drought, and stream flow. J. Clim. 14, 2105–2128.

Barnston, A.G., Tippett, M.K., L'Heureux, M., Li, G., Dewitt, D., 2012. Skill of real-time seasonal ENSO model predictions during 2002–2011: is our capability increasing? Bull. Am. Meteorol. Soc. 93, 631–651.

Barsugli, J., Battisti, D.S., 1998. The basic effects of atmosphere–ocean thermal coupling on midlatitude variability. J. Atmos. Sci. 55, 473–493.

Bates, S.C., 2008. Coupled ocean–atmosphere interaction and variability in the tropical Atlantic ocean with and without an annual cycle. J. Clim. 21, 5501–5523.

Bates, S.C., 2010. Seasonal influences on coupled ocean–atmosphere variability in the tropical Atlantic ocean. J. Clim. 23, 582–604.

Battisti, D.S., 1988. Dynamics and thermodynamics of a warming event in a coupled tropical atmosphere ocean model. J. Atmos. Sci. 45, 2889–2919.

Battisti, D.S., Hirst, A.C., 1989. Interannual variability in a tropical atmosphere ocean model—influence of the basic state, ocean geometry and nonlinearity. J. Atmos. Sci. 46, 1687–1712.

Berner, J., et al., 2008. Impact of a quasi-stochastic cellular automaton backscatter scheme on the systematic error and seasonal prediction skill of a global climate model. Philos. Trans. R. Soc. A 366, 2561–2579. http://dx.doi.org/10.1098/rsta.2008.0033.

Bernie, D.J., Guilyardi, E., Madec, G., Woolnough, S.J., Slingo, J.M., 2007. Impact of resolving the diurnal cycle in an ocean–atmosphere GCM. Part 1: diurnally forced OGCM. Clim. Dyn. 29, 575–590.

Bernie, D.J., Guilyardi, E., Madec, G., Slingo, J.M., Woolnough, S.J., Cole, J., 2008. Impact of resolving the diurnal cycle in an ocean–atmosphere GCM. Part 2: a diurnally coupled CGCM. Clim. Dyn. 31, 909–925.

Bjerknes, J., 1966. A possible response of atmospheric Hadley circulation to equatorial anomalies of ocean temperature. Tellus 18, 820–829.

Bjerknes, J., 1969. Atmospheric teleconnections from equatorial pacific. Mon. Weather Rev. 97, 163–172.

Blanke, B., Neelin, J.D., Gutzler, D., 1997. Estimating the effect of stochastic wind stress forcing on ENSO irregularity. J. Clim. 10, 1473–1487.

Bourassa, M.A., Romero, R., Smith, S.R., O'Brien, J.J., 2005. A new FSU winds climatology. J. Clim. 18, 3686–3698.

Burgers, G., Balmaseda, M.A., Vossepoel, F.C., van Oldenborgh, G.J., van Leeuwen, P.J., 2002. Balanced ocean-data assimilation near the equator. J. Phys. Oceanogr. 32, 2509–2519. http://dx.doi.org/10.1175/1520-0485-32.9.2509.

Burgers, G., Jin, F.-F., van Oldenborgh, G.J., 2005. The simplest ENSO recharge oscillator. Geophys. Res. Lett. 32, L13706. http://dx.doi.org/10.1029/2005GL022951.

Cane, M.A., Zebiak, S.E., Dolan, S.C., 1986. Experimental forecasts of El-Nino. Nature 321, 827–832.

Cane, M.A., Munnich, M., Zebiak, S.E., 1990. A study of self-excited oscillations of the tropical ocean–atmosphere system. 1. Linear-analysis. J. Atmos. Sci. 47, 1562–1577.

Chang, P., Wang, B., Li, T., Ji, Li., 1994. Interactions between the seasonal cycle and the Southern Oscillation-frequency entrainment and chaos In an intermediate coupled ocean–atmosphere model. Geophys. Res. Lett. 21, 2817–2820.

Chang, P., Ji, L., Saravanan, R., 2001. A hybrid coupled model study of tropical Atlantic variability. J. Clim. 14, 361–390.

Chang, P., Saravanan, R., Ji, L., 2003. Tropical Atlantic seasonal predictability: The roles of El Niño remote influence and thermodynamic air-sea feedback. Geophys. Res. Lett. 30, 1501. http://dx.doi.org/10.1029/2002GL016119, 10.

Chang, P., et al., 2006. Climate fluctuations of tropical coupled systems—the role of ocean dynamics. J. Clim. 19, 5122–5174.

Chang, P., Zhang, L., Saravanan, R., Vimont, D.J., Chiang, J.C.H., Ji, L., Seidel, H., Tippett, M.K., 2007. Pacific meridional mode and El Niño—Southern Oscillation. Geophys. Res. Lett. 34, L16608. http://dx.doi.org/10.1029/2007GL030302.

Chen, Y.-Q., Battisti, D.S., Palmer, T.N., Barsugli, J., Sarachik, E.S., 1997. A study of the predictability of tropical Pacific SST in a coupled atmosphere–ocean model using singular vector analysis: the role of the annual cycle and the ENSO cycle. Mon. Weather Rev. 125, 831–845.

Chen, D., Cane, M.A., Kaplan, A., Zebiak, S., Huang, D., 2004. Predictability of El Nino over the past 148 years. Nature 428, 733–736.

Delecluse, P., Davey, M.K., Kitamura, Y., Philander, S.G.H., Suarez, M., Bengtsson, L., 1998. Coupled general circulation modeling of the tropical Pacific. J. Geophys. Res. Oceans 103, 14357–14373.

Deser, C., Blackmon, M.L., 1995. On the relationship between tropical and North Pacific sea surface temperature variations. J. Clim. 8, 1677–1680.

Eckert, C., Latif, M., 1997. Predictability of a stochastically forced hybrid coupled model of El Niño. J. Clim. 10, 1488–1504.

Eisenman, I., Yu, L., Tziperman, E., 2005. Westerly wind bursts: ENSO's tail rather than the dog? J. Clim. 18, 5224–5238.

Flügel, M., Chang, P., 1996. Impact of dynamical and stochastic processes on the predictability of ENSO. Geophys. Res. Lett. 23, 2089–2092.

Flügel, M., Chang, P., Penland, C., 2004. The role of stochastic forcing in modulating ENSO predictability. J. Clim. 17, 3125–3140.

Galanti, E., Tziperman, E., 2000. ENSO's Phase locking to the seasonal cycle in the fast-SST, fast-wave, and mixed-mode regimes. J. Atmos. Sci. 57, 2936–2950.

Guilyardi, E., Delecluse, P., Gualdi, S., Navarra, A., 2003. Mechanisms for ENSO phase change in a coupled GCM. J. Clim. 16, 1141–1158.

Hagedorn, R., Doblas-Reyes, F.J., Palmer, T.N., 2005. The rationale behind the success of multi-model ensembles in seasonal forecasting – I. Basic concept. Tellus A 57, 219–233. http://dx.doi.org/10.1111/j.1600-0870.2005.00103.x.

Hao, Z., Neelin, J.D., Jin, F.F., 1993. Nonlinear tropical Air-Sea interaction in the fast-wave limit. J. Clim. 6, 1523–1544.

Harrison, D.E., Vecchi, G.A., 1999. On the termination of El Niño. Geophys. Res. Lett. 26, 1593–1596.

Hastenrath, S., Greischar, L., 1993. Further work on the prediction of northeast Brazil rainfall anomalies. J. Clim. 6, 743–758.

Hirst, A.C., 1986. Unstable and damped equatorial modes in simple coupled ocean atmosphere models. J. Atmos. Sci. 43, 606–630.

Huang, B., 2004. Remotely forced variability in the tropical Atlantic Ocean. Clim. Dyn. 23, 133–152.

Huang, B., Kinter III, J.L., 2002. Interannual variability in the tropical Indian Ocean. J. Geophys. Res. 107 (C11), 3199. http://dx.doi.org/10.1029/2001JC001278.

Huang, B., Schopf, P.S., Pan, Z., 2002. The ENSO effect on the tropical Atlantic variability: A regionally coupled model study. Geophys. Res. Lett. 29 (21), 2039. http://dx.doi.org/10.1029/2002GL014872.

Hurlburt, H.E., et al., 2008. Ocean modeling in an eddying regime. In: Hecht, M.W., Hasumi, H. (Eds.), Eddy Resolving Ocean Prediction. American Geophysical Union, Washington, DC.

Hurrell, J.W., 1995. Decadal trends in the north Atlantic oscillation: regional temperature and precipitation. Science 269, 676–679.

Jin, F.F., 1996. Tropical Ocean-Atmosphere Interactions, the Pacific Cold Tongue, and the El Nino-Southern Oscillation. Science 274, 76–78. http://dx.doi.org/10.1126/science.274.5284.76.

Ji, M., Leetmaa, A., 1997. Impact of data assimilation on ocean initialization and El Nino prediction. Mon. Weather Rev. 125, 742–753.

Ji, M., Kumar, A., Leetmaa, A., 1994. A multiseason climate forecast system at the national-meteorological-center. Bull. Am. Meteorol. Soc. 75, 569–577.

Jin, F.F., 1997a. An equatorial ocean recharge paradigm for ENSO.1. Conceptual model. J. Atmos. Sci. 54, 811–829.

Jin, F.F., 1997b. An equatorial ocean recharge paradigm for ENSO.2. A stripped-down coupled model. J. Atmos. Sci. 54, 830–847.

Jin, D., Kirtman, B., 2010. Impact of ENSO periodicity on North Pacific SST variability. Clim. Dyn. 34, 1015–1039. http://dx.doi.org/10.1007/s00382-009-0619-3.

Jin, F.F., Neelin, J.D., 1993a. Modes of interannual tropical ocean–atmosphere interaction—a unified view. 3. Analytical results in fully coupled cases. J. Atmos. Sci. 50, 3523–3540.

Jin, F.F., Neelin, J.D., 1993b. Modes of interannual tropical ocean–atmosphere interaction—a unified view. 1. Numerical results. J. Atmos. Sci. 50, 3477–3503.

Jin, F.-F., Neelin, J.D., Ghil, M., 1994. El Niño on the devil's staircase: annual subharmonic steps to chaos. Science 264, 70–72.

Johnson, G.C., McPhaden, M.J., 1999. Interior pycnocline flow from the subtropical to the equatorial Pacific Ocean. J. Phys. Oceanogr. 29, 3073–3089.

Kirtman, B.P., 1997. Oceanic Rossby wave dynamics and the ENSO period in a coupled model. J. Clim. 10, 1690–1704.

Kirtman, B., Pirani, A., 2009. The state of the art of seasonal prediction outcomes and recommendations from the first world climate research program workshop on seasonal prediction. Bull. Am. Meteorol. Soc. 90, 455–458.

Kirtman, B.P., Schopf, P.S., 1998. Decadal variability in ENSO predictability and prediction. J. Clim. 11, 2804–2822.

Kirtman, B.P., Shukla, J., 2000. Influence of the Indian summer monsoon on ENSO. Q. J. R. Meteorol. Soc. 126, 213–239.

Kirtman, B.P., Shukla, J., Huang, B.H., Zhu, Z.X., Schneider, E.K., 1997. Multiseasonal predictions with a coupled tropical ocean-global atmosphere system. Mon. Weather Rev. 125, 789–808.

Kirtman, B.P., Pegion, K., Kinter, S.M., 2005. Internal atmospheric dynamics and tropical indo-Pacific climate variability. J. Atmos. Sci. 62, 2220–2233.

Kirtman, B.P., et al., 2012. Impact of ocean model resolution on CCSM climate simulations. Clim. Dyn. 39, 1303–1328. http://dx.doi.org/10.1007/s00382-012-1500-3.

Kleeman, R., Moore, A.M., 1997. A theory for the limitation of ENSO predictability due to stochastic atmospheric transients. J. Atmos. Sci. 54, 753–767.

Kleeman, R., Tang, Y., Moore, A.M., 2003. The calculation of climatically relevant singular vectors in the presence of weather noise as applied to the ENSO problem. J. Atmos. Sci. 60, 2856–2868.

Klein, S.A., Soden, B.J., Lau, N.C., 1999. Remote sea surface temperature variations during ENSO: evidence for a tropical atmospheric bridge. J. Clim. 12, 917–932.

Krishnamurthy, V., Kirtman, B.P., 2003. Variability of the Indian Ocean: relation to monsoon and ENSO. Q. J. R. Meteorol. Soc. 129, 1623–1646.

Kushnir, Y., Robinson, W.A., Chang, P., Robertson, A.W., 2006. The physical basis for predicting Atlantic sector seasonal-to-interannual climate variability. J. Clim. 19, 5949–5970.

Latif, M., 1998. Dynamics of interdecadal variability in coupled ocean-atmosphere models. J. Clim. 11, 602–624.

Latif, M., Grotzner, A., 2000. The equatorial Atlantic oscillation and its response to ENSO. Clim. Dyn. 16, 213–218.

Latif, M., Sterl, A., Maierreimer, E., Junge, M.M., 1993. Structure and predictability of the El-Nino southern oscillation phenomenon in a coupled ocean atmosphere general-circulation model. J. Clim. 6, 700–708.

Latif, M., et al., 1998. A review of the predictability and prediction of ENSO. J. Geophys. Res. Oceans 103, 14375–14393.

Lau, N.C., Nath, M.J., 1996. The role of the "atmospheric bridge" in linking tropical Pacific ENSO events to extratropical SST anomalies. J. Clim. 9, 2036–2057.

Lau, N.-C., Nath, M.J., 2001. Impact of ENSO on SST variability in the North Pacific and North Atlantic: seasonal dependence and role of extratropical Sea–Air coupling. J. Clim. 14, 2846–2866.

Legler, D.M., Stricherz, J.N., O'Brien, J.J., 1997. Indian Ocean. TOGA Pseudo-Stress Atlas 1985–1994, vol. 3, COAPS Rep. 97–3, 175 pp. Available from COAPS, The Florida State University, Tallahassee, FL 32306.

Levitus, S., et al., 2012. World ocean heat content and thermosteric sea level change (0–2000 m), 1955–2010. Geophys. Res. Lett. 39, L10603. http://dx.doi.org/10.1029/2012GL051106.

Li, T., 1997. Phase transition of the El Niño–Southern Oscillation: a stationary SST mode. J. Atmos. Sci. 54, 2872–2887.

Lopez, H., Kirtman, B.P., Tziperman, E., Gebbe, G., 2012. Impact of interactive westerly wind bursts on CCSM3. Dyn. Atmos. Oceans. http://dx.doi.org/10.1016/j.jynamtmoce.2012.11.001.

Lorenz, E.N., 1965. A study of the predictability of a 28-variable atmospheric model. Tellus 17, 321–333.

McCreary, J.P., 1983. A model of tropical ocean atmosphere interaction. Mon. Weather Rev. 111, 370–387.

McPhaden, M.J., 2004. Evolution of the 2002/03 El Nino. Bull. Am. Meteorol. Soc. 85, 677–695.

McPhaden, M.J., Zhang, D.X., 2002. Slowdown of the meridional overturning circulation in the upper Pacific Ocean. Nature 415, 603–608.

McPhaden, M.J., Zhang, D., 2004. Pacific Ocean circulation rebounds. Geophys. Res. Lett. 31, L18301. http://dx.doi.org/10.1029/2004GL020727.

McPhaden, M.J., et al., 1998. The tropical ocean global atmosphere observing system: a decade of progress. J. Geophys. Res. Oceans 103, 14169–14240.

Mehta, V.M., Suarez, M.J., Manganello, J.V., Delworth, T.L., 2000. Oceanic influence on the North Atlantic Oscillation and the associated northern hemisphere climate variations: 1959–1993. Geophys. Res. Lett. 27, 121–124.

Meinen, C.S., McPhaden, M.J., 2000. Observations of warm water volume changes in the equatorial Pacific and their relationship to El Nino and La Nina. J. Clim. 13, 3551–3559.

Melice, J.L., Servain, J., 2003. The tropical Atlantic meridional SST gradient index and its relationships with the SOI, NAO and Southern Ocean. Clim. Dyn. 20, 447–464.

Moore, A.M., Kleeman, R., 1996. The dynamics of error growth and predictability in a coupled model of ENSO. Q. J. R. Meteorol. Soc. 122, 1405–1446.

Moore, A.M., Kleeman, R., 1999a. The nonnormal nature of El Niño and intraseasonal variability. J. Clim. 12, 2965–2982.

Moore, A.M., Kleeman, R., 1999b. Stochastic forcing of ENSO by the intraseasonal oscillation. J. Clim. 12, 1199–1220.

Moura, A.D., Shukla, J., 1981. On the dynamics of droughts in Northeast Brazil—observations, theory and numerical experiments with a general-circulation model. J. Atmos. Sci. 38, 2653–2675.

Münnich, M., Cane, M.A., Zebiak, S.E., 1991. A study of self-excited oscillations in a tropical ocean–atmosphere system. Part II: nonlinear cases. J. Atmos. Sci. 48, 1238–1248.

Murtugudde, R., Beauchamp, J., McClain, C.R., Lewis, M., Busalacchi, A.J., 2002. Effects of penetrative radiation on the upper tropical ocean circulation. J. Clim. 15, 470–486. http://dx.doi.org/10.1175/1520-0442(2002)015<0470:EOPROT>2.0.CO;2.

Namias, J., 1959. Recent seasonal interactions between North Pacific waters and the overlying atmospheric circulation. J. Geophys. Res. 64 (6), 631. http://dx.doi.org/10.1029/JZ064i006p00631.

National Research Council, National Acadamies, 2010. Assessment of Intraseasonal to Interannual Climate Prediction and Predictability. The National Academies Press, 192 pp. (NRC2010).

Neelin, J.D., 1991. The slow Sea-surface temperature mode and the fast-wave limit—analytic theory for tropical interannual oscillations and experiments in a hybrid coupled model. J. Atmos. Sci. 48, 584–606.

Neelin, J.D., Jin, F.F., 1993. Modes of interannual tropical ocean–atmosphere interaction—a unified view. 2. Analytical results in the weak-coupling limit. J. Atmos. Sci. 50, 3504–3522.

Neelin, J.D., Battisti, D.S., Hirst, A.C., Jin, F.F., Wakata, Y., Yamagata, T., Zebiak, S.E., 1998. Enso theory. J. Geophys. Res. Oceans 103, 14261–14290.

Nobre, P., Shukla, J., 1996. Variations of sea surface temperature, wind stress, and rainfall over the tropical Atlantic and South America. J. Clim. 9, 2464–2479.

Nobre, P., Zebiak, S.E., Kirtman, B.P., 2003. Local and remote sources of tropical atlantic variability as inferred from the results of a hybrid ocean-atmosphere coupled model. Geophys. Res. Lett. 30, 8008. http://dx.doi.org/10.1029/2002GL015785, 5.

Oschlies, A., 2002. Improved representation of upper-ocean dynamics and mixed layer depths in a model of the North Atlantic on switching from eddy-permitting to eddy-resolving grid resolution. J. Phys. Oceanogr. 32, 2277–2298.

Palmer, T.N., et al., 2004. Development of a European multimodel ensemble system for seasonal-to-interannual prediction (DEMETER). Bull. Am. Meteorol. Soc. 85, 853–872.

Penland, C., Matrosova, L., 1994. A balance condition for stochastic numerical models with application to the El Niño–Southern Oscillation. J. Clim. 7, 1352–1372.

Penland, C., Matrosova, L., 1998. Prediction of tropical Atlantic sea surface temperatures using linear inverse modeling. J. Clim. 11, 483–496.

Penland, C., Sardeshmukh, P.D., 1995. The optimal growth of tropical sea surface temperature anomalies. J. Clim. 8, 1999–2024.

Philander, S.G.H., 1990. El Nino, La Nina, and the Southern Oscillation. Academic Press, San Diego, California, 293 pp.

Philander, S.G.H., Yamagata, T., Pacanowski, R.C., 1984. Unstable Air Sea interactions in the tropics. J. Atmos. Sci. 41, 604–613.

Rasmusson, E.M., Carpenter, T.H., 1982. Variations in tropical sea surface temperature and surface wind fields associated with the Southern Oscillation/El Niño. Mon. Weather Rev. 110, 354–384.

Richter, I., Xie, X.-P., 2008. On the origin of equatorial Atlantic biases in coupled general circulation models. Clim. Dyn. 31 (5), 587–598. http://dx.doi.org/10.1007/s00382-008-0364-z.

Roberts, M.J., et al., 2009. Impact of resolution on the tropical pacific circulation in a matrix of coupled models. J. Clim. 22, 2541–2556.

Rodwell, M.J., Rowell, D.P., Folland, C.K., 1999. Oceanic forcing of the wintertime North Atlantic Oscillation and European climate. Nature 398, 320–323.

Rosati, A., Miyakoda, K., Gudgel, R., 1997. The impact of ocean initial conditions on ENSO forecasting with a coupled model. Mon. Weather Rev. 125, 754–772.

Saha, S., et al., 2010. The NCEP climate forecast system reanalysis. Bull. Am. Meteorol. Soc. 91, 1015–1057.http://dx.doi.org/10.1175/2010BAMS3001.1.

Saji, N.H., Goswami, B.N., Vinayachandran, P.N., Yamagata, T., 1999. A dipole mode in the tropical Indian Ocean. Nature 401, 360–363.

Saunders, S.A., Qian, B., 2002. Seasonal predictability of the winter NAO from north Atlantic sea surface temperatures. Geophys. Res. Lett. 29, 2049. http://dx.doi.org/10.1029/2002GL014952, 4 pp.

Schneider, E.K., Huang, B., Shukla, J., 1995. Ocean wave dynamics and El-Nino. J. Clim. 8, 2415–2439.

Schopf, P.S., Suarez, M.J., 1990. Ocean wave dynamics and the time scale of ENSO. J. Phys. Oceanogr. 20, 629–645.

Shukla, J., Kinter III, J.L., 2005. Predictability of seasonal climate variations: A pedagogical review. In: Palmer, T., Hagedorn, R. (Eds.), Predictability of Weather and Climate. Cambridge University Press, Cambridge, UK, 702 pp, 306–341.

Smith, R.D., Maltrud, M.E., Bryan, F.O., Hecht, M.W., 2000. Numerical simulation of the North Atlantic Ocean at 1/10 degrees. J. Phys. Oceanogr. 30, 1532–1561.

Smith, D.M., Scaife, A.A., Kirtman, B., 2012. What is the current state of scientific knowledge with regard to seasonal and decadal forecasting. Environ. Res. Lett. 7, 015602. http://dx.doi.org/10.1088/1748-9326/7/1/015602.

Stockdale, T.N., Anderson, D.L.T., Alves, J.O.S., Balmaseda, M.A., 1998a. Global seasonal rainfall forecasts using a coupled ocean–atmosphere model. Nature 392, 370–373.

Stockdale, T.N., Busalacchi, A.J., Harrison, D.E., Seager, R., 1998b. Ocean modeling for ENSO. J. Geophys. Res. 103, 14,325–14,355.

Stockdale, T.N., Balmaseda, M.A., Vidard, A., 2006. Tropical Atlantic SST prediction with coupled ocean–atmosphere GCMs. J. Clim. 19, 6047–6061.

Stockdale, T.N., et al., 2011. ECMWF seasonal forecast system 3 and its prediction of sea surface temperature. Clim. Dyn. 37, 455–471.

Suarez, M.J., Schopf, P.S., 1988. A delayed action oscillator for ENSO. J. Atmos. Sci. 45, 3283–3287.

Thompson, C.J., Battisti, D.S., 2001. A linear stochastic dynamical model of ENSO. Part II: analysis. J. Clim. 14, 445–466.

Tziperman, E., Stone, L., Cane, M., Jarosh, H., 1994. El Nino chaos: overlapping of resonances between the seasonal cycle and the Pacific ocean–atmosphere oscillator. Science 264, 72–74.

Tziperman, E., Cane, M.A., Zebiak, S.E., 1995. Irregularity and locking to the seasonal cycle in an ENSO prediction model as explained by the quasi-periodicity route to chaos. J. Atmos. Sci. 52, 293–306.

Tziperman, E., Cane, M.A., Zebiak, S., Xue, Y., Blumenthal, B., 1998. Locking of El Nino's peak time to the end of the calendar year in the delayed oscillator picture of ENSO. J. Clim. 11, 2191–2199.

Vecchi, G.A., Harrison, D.E., 2003. On the termination of the 2002–03 El Niño event. Geophys. Res. Lett. 30, 1964. http://dx.doi.org/10.1029/2003GL017564.

Vimont, D.J., Battisti, D.S., Hirst, A.C., 2003. The seasonal footprinting mechanism in the CSIRO general circulation models. J. Clim. 16, 2653–2667.

Wajsowicz, R.C., 2007. Seasonal-to-interannual forecasting of tropical Indian Ocean sea surface temperature anomalies: potential predictability and barriers. J. Clim. 20, 3320–3343.

Wang, B., 1995. Interdecadal changes in El Niño onset in the last four decades. J. Clim. 8, 267, 258.

Wang, B., An, S.I., 2001. Why the properties of El Nino changed during the late 1970s. Geophys. Res. Lett. 28, 3709–3712.

Wang, B., Barcilon, A., Fang, Z., 1999. Stochastic dynamics of El Niño–Southern Oscillation. J. Atmos. Sci. 56, 5–23.

Ward, M.N., Folland, C.K., 1991. Prediction of seasonal rainfall in the north Nordeste of Brazil using eigenvectors of Sea-surface temperature. Int. J. Climatol. 11, 711–743.

Webster, P.J., Moore, A.M., Loschnigg, J.P., Leben, R.R., 1999. Coupled ocean–atmosphere dynamics in the Indian Ocean during 1997–98. Nature 401, 356–360.

Weisheimer, A., Doblas-Reyes, F.J., Palmer, T.N., Alessandri, A., Arribas, A., Déqué, M., Keenlyside, N., MacVean, M., Navarra, A., Rogel, P., 2009. ENSEMBLES: A new multi-model ensemble for seasonal-to-annual predictions—Skill and progress beyond DEMETER in forecasting tropical Pacific SSTs. Geophys. Res. Lett. 36, L21711. http://dx.doi.org/10.1029/2009GL040896.

Wu, R., Kirtman, B.P., Krishnamurthy, V., 2008. An asymmetric mode of tropical Indian Ocean rainfall variability in boreal spring. J. Geophys. Res. 113, D05104. http://dx.doi.org/10.1029/2007JD009316.

Xie, P., Arkin, P.A., 1997. Global precipitation: a 17-Year monthly analysis based on gauge observations, satellite estimates, and numerical model outputs. Bull. Amer. Meteor. Soc. 78, 2539–2558.

Xue, Y., Cane, M.A., Zebiak, S.E., 1997. Predictability of a coupled model of ENSO using singular vector analysis. Part I: optimal growth in seasonal background and ENSO cycles. Mon. Weather Rev. 125, 2043–2056.

Yeh, S.-W., Kirtman, B.P., 2004a. The impact of internal atmospheric dynamics for the North Pacific SST variability. Clim. Dyn. 22, 721–732. http://dx.doi.org/10.1007/s00382-004-0399-8.

Yeh, S.-W., Kirtman, B.P., 2004b. The North Pacific Oscillation–ENSO and internal atmospheric variability. Geophys. Res. Lett. 31, L13206. http://dx.doi.org/10.1029/2004GL019983.

Zavala-Garay, J., Moore, A.M., Perez, C.L., Kleeman, R., 2003. The response of a coupled model of ENSO to observed estimates of stochastic forcing. J. Clim. 16, 2827–2842.

Zavala-Garay, J., Moore, A.M., Kleeman, R., 2004. Influence of stochastic forcing on ENSO prediction. J. Geophys. Res. 109, C11007. http://dx.doi.org/10.1029/2004JC002406.

Zavala-Garay, J., Zhang, C., Moore, A.M., Kleeman, R., 2005. The linear response of ENSO to the Madden-Julian oscillation. J. Clim. 18, 2441–2459.

Zavala-Garay, J., et al., 2008. Sensitivity of hybrid ENSO models to unresolved atmospheric variability. J. Clim. 21, 3704–3721. http://dx.doi.org/10.1175/2007JCLI1188.1.

Zebiak, S.E., 1989. On the 30–60 Day oscillation and the prediction of El Niño. J. Clim. 2, 1381–1387.

Zebiak, S.E., 1993. Air-Sea interaction in the equatorial Atlantic region. J. Clim. 6, 1567–1568.

Zebiak, S.E., Cane, M.A., 1987. A model El-Nino southern oscillation. Mon. Weather Rev. 115, 2262–2278.

Zhao, M., Hendon, H.H., 2009. Representation and prediction of the Indian Ocean dipole in the POAMA seasonal forecast model. Q. J. R. Meteorol. Soc. 135, 337–352.

Chapter 25

The Ocean's Role in Modeling and Predicting Decadal Climate Variations

Mojib Latif*,†
*Helmholtz Centre for Ocean Research Kiel (GEOMAR), Kiel, Germany
†Cluster of Excellence "The Future Ocean," Kiel University, Kiel, Germany

Chapter Outline

Abbreviations

ACC Antarctic Circumpolar Current
ACW Antarctic circumpolar wave
AGCM atmospheric general circulation model
AMOC Atlantic meridional overturning circulation
AMV Atlantic multidecadal variability
AR4 fourth assessment report (of IPCC)
AR5 fifth assessment report (of IPCC)
CGCM coupled general circulation model
CM column model
ENSO El Niño/Southern Oscillation
EOF empirical orthoganal function
IPCC Intergovernmental Panel on Climate Change
ITCZ intertropical convergence zone
KE Kuroshio extension
NAO North Atlantic Oscillation
NPGO North Pacific Gyre Oscillation
NPO North Pacific Oscillation
OGCM oceanic general circulation model
PDO Pacific Decadal Oscillation
PDV Pacific decadal variability
SAM southern annular mode
SAT surface air temperature
SOCV Southern Ocean centennial variability
SSH sea surface height
SSS sea surface salinity
SST sea surface temperature
SSTA sea surface temperature anomaly
STC subtropical cell
T2m 2 m air temperature
TC tropical cell
WES wind-evaporation-SST (feedback)

1. INTRODUCTION

The ocean circulation and its variability is a key factor controlling natural climate fluctuations. It is also critical to anthropogenic climate change and its regional manifestations, biogeochemical cycling, marine ecosystems, or food supply. Reliable future projections of anthropogenic change thus require an advanced understanding of the physical processes and phenomena in the ocean, which are important to the climatology and variability of the coupled ocean–atmosphere–sea ice system. The outstanding role of the

Ocean Circulation and Climate, Vol. 103. http://dx.doi.org/10.1016/B978-0-12-391851-2.00025-8

oceans in shaping the mean climate and the seasonal cycle and in driving climate variability on a wide range of timescales is due to their large heat storage capacity and dynamical inertia and the ability of the ocean currents to carry vast amounts of heat and substances over large distances. The focus of this review is on the possible mechanisms of decadal ocean circulation variability, its theoretical predictability, and attempts to forecast decadal ocean change and its impacts.

Natural climate fluctuations and anthropogenic climate change go side by side. Global average surface air temperature (SAT) during the twentieth century (Figure 25.1), for instance, displays a gradual warming on the order of 0.7 °C and superimposed shorter-term fluctuations on interannual to multidecadal timescales. The long-term warming trend contains the climate response to enhanced atmospheric greenhouse gas levels, in particular those of carbon dioxide (Figure 25.1), which have gradually increased from its preindustrial value of about 280 ppmv (parts per million by volume) to its present concentration of well above 390 ppmv. The temperature trend also contains an unknown component from natural centennial variability. It is commonly believed, however, that the natural contribution to the twentieth century global warming is considerably smaller than the anthropogenic share (IPCC, 2007).

The temperature ups and downs around the long-term warming trend largely reflect shorter-term natural variability, and it is the variability of the ocean circulation, which plays a key role in that. The El Niño/Southern Oscillation (ENSO) phenomenon (see Chapter 23 in this book) originating in the Equatorial Pacific is the strongest interannual climate signal in the tropics. The positive phase of the quasiperiodic cycle is termed El Niño. Such events are typified by strong warming of the order of a few degrees of the eastern and central tropical Pacific Ocean with

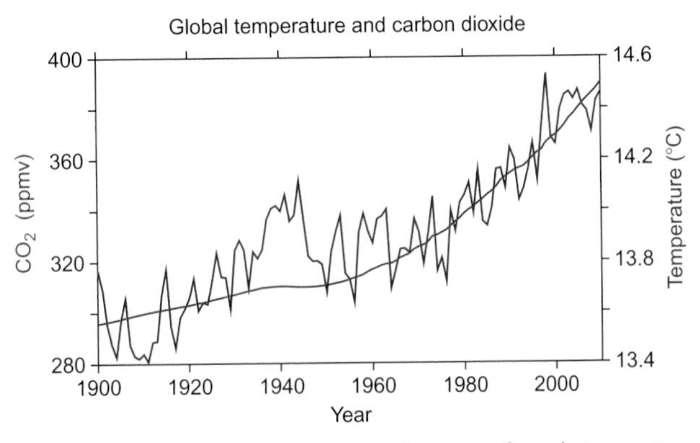

FIGURE 25.1 Globally averaged annual mean surface air temperature (°C, HadCRUT3, blue curve) and the smoothed atmospheric CO_2 concentration (ppmv, red curve) 1900–2010. HadCRUT3 data were obtained from http://www.cru.uea.ac.uk/cru/data/temperature. CO_2 data are a merged product of Law Dome reconstruction and Mauna Loa observation obtained from http://chartsgraphs.wordpress.com/climate-data-links/.

cooling over portions of the subtropics and midlatitudes and the tropical western Pacific. This pattern of sea surface temperature (SST) anomalies drives anomalous diabatic heating in the upper atmosphere, thereby perturbing the global atmospheric circulation and generating global teleconnections, which have diverse impacts on the regional climate and ecology in many regions with impacts on the economies of several countries. ENSO is even seen in global average SAT. The major El Niño warming of winter 1997/1998, for instance, "helped" to make the year 1998 one of the warmest years on record globally since instrumental measurements had started in 1850 (Figure 25.1).

ENSO-like SST variability also exists at decadal timescales (e.g., Trenberth and Hurrell, 1994; Zhang et al., 1997), and like ENSO itself the ENSO-like decadal SST variability perturbs the global climate. The decadal tropical Pacific SST fluctuations drive, for instance, decade-long changes of the Aleutian low (Trenberth and Hurrell, 1994) and therefore have major impacts on the climate of North America. The leading mode of decadal SST variability in the Pacific basin is often referred to as the Pacific Decadal Oscillation (PDO). The latter switched its phase in the early 1990s. As a consequence, large parts of the tropical Pacific exhibited a cooling trend during the last decade (Figure 25.2), which retarded global warming (Figure 25.1). The PDO has been linked in many studies to upper ocean processes. The timescale at which surface density-driven deeper ocean circulation changes become important in the Pacific is controversial and requires further investigation.

Another region of strong decadal variability is the Atlantic Ocean (Folland et al., 1986; Latif et al., 2004; Knight et al., 2005). The decadal to multidecadal SST variability in the Atlantic is often termed the Atlantic multidecadal oscillation (AMO). Here surface density perturbations in the high latitudes are thought to be important in driving decadal variability of the ocean circulation. In many climate models, decadal to centennial SST variability in the Atlantic Ocean is linked to variations of the Atlantic meridional overturning circulation (AMOC), a major current system characterized by a northward flow of warm and salty water in the upper layers of the Atlantic, and a southward flow of colder water in the deep Atlantic (e.g., Rintoul et al., 2011).

Natural climate variability such as that associated with the PDO or the AMO introduces an element of irregularity as global warming develops, making the detection of anthropogenic trends a challenge. Climate variations can be categorized into two types: internal and external. Internal variability is produced by the climate system itself due to its chaotic nature. It is the oceans that are a major agent in producing the internal variability, especially on decadal and longer timescales. External fluctuations need a forcing, in mathematical terms a change in the boundary conditions.

Volcanic eruptions and fluctuations in solar output are examples of natural external drivers of climate. The eruption of the Philippine volcano Mt. Pinatubo in 1991, for instance, caused a short-lived drop in global SAT of a few tenths of a degree Celsius in 1992 (Figure 25.1; IPCC, 2007). An increase of the solar radiation reaching the Earth contributed to the early twentieth century warming during 1920–1940 (Figure 25.1; e.g., Hegerl et al., 1997). Zhou and Tung (2010) described a weak but consistent response of global average SST to decadal solar variations during the instrumental period. Ottera et al. (2010) and Zanchettin et al. (2010) conducting forced climate model simulations highlighted the role of volcanic forcing for the climate of the last millennium. The anthropogenic influence on climate is also considered as external, but not natural, within this framework. Here we largely concentrate on the internal decadal variability for which changes in the ocean circulation is so important.

Figure 25.2 (taken from Keenlyside and Ba, 2010) in its top panel repeats a version of the global average SAT (solid line) and depicts also the mean temperature evolution simulated by an ensemble of climate models driven with estimated observed external forcing taken from the CMIP3 database (http://www-pcmdi.llnl.gov/ipcc/about_ipcc.php). The average over all models (dashed line), sometimes misleadingly referred to as the "consensus," is a measure of the externally driven climate change, provided both the models and the forcing are realistic. The long-term upward trend in the multimodel average is largely consistent with the observed SAT trend and has been attributed primarily to the increase in the concentrations of long-lived atmospheric greenhouse gases (e.g., Hegerl et al., 1997; IPCC, 2007). The deviations of the observations from the multimodel mean are a measure of the internal variability in the data. Within such a framework, the early twentieth century warming and the subsequent decadal cooling, for instance, would reflect internal variability. In contrast, almost all of the warming during the recent decades would be externally forced, since the multimodel mean closely follows the observed SAT, a picture which has been challenged in some studies speculating about a larger share of internal variability during the recent decades (e.g., Semenov et al., 2010).

The next two panels of Figure 25.2 depict the observed North Atlantic (0–60 °N) and Eastern tropical Pacific (150–90 °W and 20 °S–20 °N) average SST, again together with the corresponding multimodel mean. Differences between observed and multimodel mean SST are clearly enhanced at regional scales in comparison to global average SAT, possibly indicating a more prominent role of internal variability. North Atlantic SST exhibited, superimposed on a long-term warming trend, pronounced multidecadal variability coherent with global SAT. The causal link between the two indices, however, remains unclear. Eastern Pacific

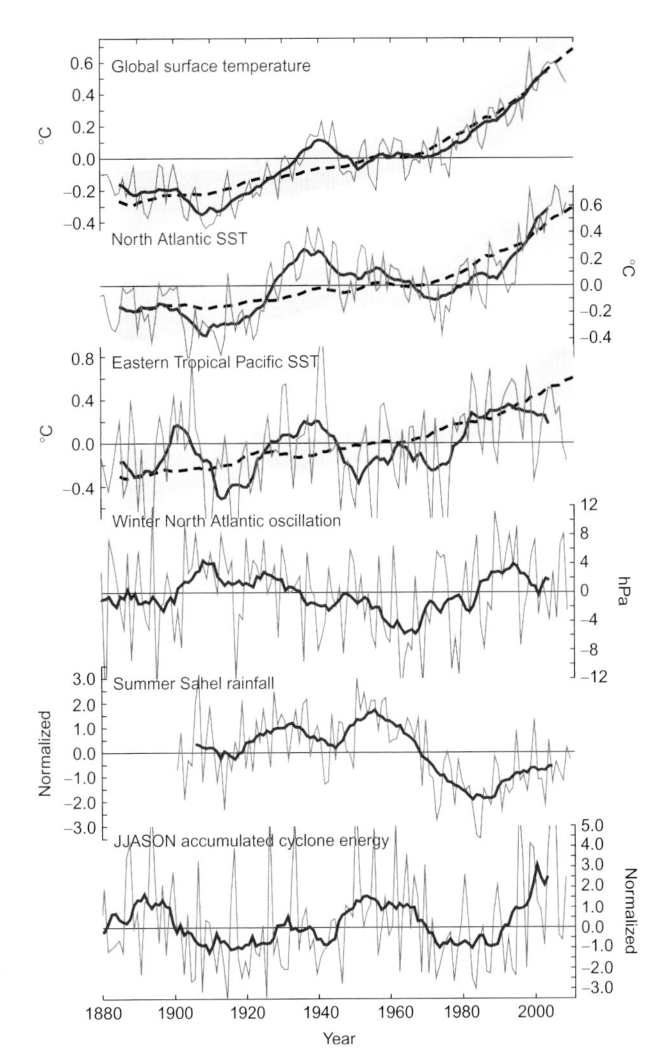

FIGURE 25.2 Observed (blue) indices of climate variability and the ensemble mean of 21 climate model simulations (black dashed) that are driven with all (known) external forcing agents. Shown are global, North Atlantic (0–60 °N), and Eastern Tropical Pacific (150–90 °W and 20 °S–20 °N) average surface temperature; December–February NAO index; June–October precipitation averaged over the Sahel; and June–November accumulated cyclone energy (ACE) index of Atlantic hurricane activity. Thick (thin) lines show 11-year running (annual/seasonal) means, and gray shading the 90% confidence interval computed from model spread. Model data are from CMIP3 database used for the Intergovernmental Panel on Climate Change (IPCC) fourth assessment report. *Figure from Keenlyside and Ba (2010).*

SST also shows decadal-scale changes and some correspondence to global changes, but has featured decadal variability with a somewhat shorter timescale.

The North Atlantic Oscillation (NAO; Hurrell, 1995, 2003), which is a vacillation in sea-level pressure between Iceland and the subtropical North Atlantic and the most important mode of internal atmospheric variability in the North Atlantic sector, has also featured pronounced multidecadal variations (Figure 25.2, fourth panel). These were associated with strong changes in wintertime storminess over the North Atlantic, and European and North American surface temperature and precipitation, and thus had major economic impacts. A significant long-term trend possibly indicating an anthropogenic influence cannot be identified in the NAO index. Large multidecadal fluctuations were also seen in summertime Sahel rainfall (Figure 25.2, fifth panel), with profound consequences for people living in the region. For example, the drought of the 1970s–1980s caused the death of at least 100,000 people, and displaced many more. Again, a clear trend during the twentieth century is hard to detect, stressing the importance of natural variability. Finally, the twentieth century Atlantic hurricane activity (Figure 25.2, bottom panel) also exhibited strong multidecadal variability and no clear trend, hindering the detection of a possible global warming signal.

The detection of anthropogenic climate change against the natural background variability is obviously a complicated issue. To some extent, we need to "ignore" the natural fluctuations, if we want to "see" the human influence on climate during the twentieth century. Of particular importance in this context is the decadal variability, by which one can be easily "fooled." Had forecasters extrapolated the early-twentieth-century warming during 1920–1940 (Figure 25.1) into the future, they would have predicted far more warming than actually occurred. Likewise, the subsequent cooling trend during 1950–1970 if used as the basis for a long-range forecast could have erroneously supported the idea of a rapidly approaching ice age. The simplest approach to filter out the short-term climate fluctuations is that of just fitting a smooth function to the data without taking into account any physical reasoning. A physically based way is to make use of the ensemble mean of the climate models to estimate the external contribution to twentieth century climate change (e.g., Ting et al., 2009; Figure 25.2, upper three panels). More sophisticated detection and attribution methods have been recently reviewed by Stott et al. (2010) and Hegerl and Zwiers (2011).

Natural variability is also an important source of uncertainty in climate change projections for the twenty-first century (e.g., Hawkins and Sutton, 2009). Considering global average SAT and time horizons of a few decades, the dominant sources of uncertainty are model (response) uncertainty and internal variability. In general, the importance of internal variability increases at shorter time and smaller space scales. Internal decadal variability was shown to be predictable to some extent (e.g., Collins et al., 2006; Latif et al., 2006a), and near-term climate projections could be potentially improved by initializing the models with an estimate of the current climate state. The fourth assessment report (AR4) of the Intergovernmental Panel on Climate Change (IPCC) published only climate projections based on certain scenarios with no attempt to take account of the likely evolution of the internal variability. The next-generation climate change projections to be presented in the upcoming fifth assessment report (AR5) of the IPCC will be initialized. The initialization of the ocean models is most important in this respect.

Section 2 provides some aspects of tropical Pacific and tropical Atlantic decadal variability, while Section 3 describes decadal variability featuring strong signals in the extratropics. The null hypothesis for the generation of internal variability, the stochastic climate model scenario, is introduced in Section 4 by describing some theoretical concepts and results from extended-range integrations of a variety of climate models, which either were explicitly driven with stochastic forcing or contained the latter as an integral part. Decadal predictability and examples of decadal forecasts are discussed in Section 5. This chapter is concluded in Section 6 with a summary of the current knowledge about decadal variability and a discussion of the future needs to obtain a better understanding of decadal variability and to enhance decadal forecasts.

2. TROPICAL PACIFIC AND TROPICAL ATLANTIC DECADAL VARIABILITY

Many studies have documented tropical Pacific decadal variability (e.g., Jacobs et al., 1994; Trenberth and Hurrell, 1994; Gu and Philander, 1997; Latif et al., 1997; Zhang et al., 1997; McPhaden and Zhang, 2002; Rodgers et al., 2004; Lohmann and Latif, 2005; Merryfield and Boer, 2005; Zhang and McPhaden, 2006; Lee and McPhaden, 2008; Meehl et al., 2009). Lohmann and Latif (2005) described the two leading modes of the observed SST variability in the western-central Equatorial Pacific during the instrumental period. The leading mode is decadal with a period of about 10–20 years, and the second most energetic mode is ENSO with a dominant period of about 4 years. The SST anomaly (SSTA) pattern of the decadal mode was shown to project on ENSO, but the decadal mode explains most variance in the western Equatorial Pacific and off the Equator, in contrast to ENSO itself explaining most variance in the east and at the Equator. The meridional scale of the decadal mode is larger than that of ENSO, as previously shown by Zhang et al. (1997).

Forced ocean general circulation model (OGCM) and coupled general circulation model (CGCM) simulations revealed that the variability of the shallow subtropical cells (STCs) and tropical overturning cells (TCs) is an important factor in driving the decadal Equatorial Pacific SST variability (e.g., Merryfield and Boer, 2005), which gives rise to some multiyear predictability of Equatorial Pacific SST. STC strength anomalies derived from a forced OGCM integration lead observed Equatorial SST anomalies by 1–2 years in the study of Lohmann and Latif (2005). It was argued in that same study that the El Niños of 1982–1983 and 1997–1998 had become record events, simply because the decadal mode was also in its positive phase;

the amplitude of the ENSO mode did not depict any exceptional decadal change during that time.

Another mechanism proposed for tropical Pacific decadal variability is based on the asymmetry of ENSO. The spatial SSTA pattern of warm events (El Niños) and that of cold events (La Niñas) differ from each other on average. Rodgers et al. (2004) hypothesized that the tropical Pacific decadal variability is simply due to this asymmetry, with the asymmetry reflecting a nonlinearity of ENSO. As a result, the residual (i.e., the sum) of the composite El Niño and La Niña patterns exhibits a nonzero dipole structure across the Equatorial Pacific, with positive deviations in the east and negative values in the west. During periods when ENSO variability is strong, the difference between the two extreme phase patterns manifests itself as a rectified change in the mean state.

Decadal variability is observed not only in the mean state of the tropical Pacific but also in the statistics of the interannual variability. Stability analysis of the relevant dynamical equations suggests that decadal changes in the background mean state can drive changes in the statistics of ENSO such as spatial expression, variance, or skewness. Stochastic forcing and nonlinear processes were also proposed to explain decadal variability of ENSO (see Neelin et al., 1998 for a review on ENSO theory). The ENSO anomaly patterns, for instance, change on decadal and longer timescales. Two studies (Yeh et al., 2009; Lee and McPhaden, 2010) showed that the canonical El Niño has become less frequent and that a different kind of El Niño became more common during the late twentieth century, in which warm SST anomalies in the central Pacific are flanked on the east and west by cooler anomalies. This type of El Niño, termed the central Pacific El Niño, differs from the canonical eastern Pacific El Niño in both the location of maximum SST anomalies and atmospheric teleconnection patterns. The origin of the change in the El Niño pattern is controversial. It has been speculated by Yeh et al. (2009) that the change in the El Niño pattern may be a consequence of global warming. However, Yeh et al. (2011) and McPhaden et al. (2011) find evidence in support of that the character of El Niño could have varied naturally.

Tropical Atlantic quasidecadal SST variability was described from observations and coupled models by, for example, Chang et al. (1997) and Ruiz-Barradas et al. (2000). The tropical Atlantic SST is important in several respects. It impacts, for instance, Atlantic hurricane activity via a series of mechanisms: one mechanism is through changes in easterly wave propagation. Another mechanism is a local effect through changes in moist static stability and vertical wind shear. The SST in the tropical Atlantic has also a significant impact on the rainfall in northeast South America (Moura and Shukla, 1981) and western Africa (Rowell et al., 1995). Tropical Atlantic SST variations are also a factor in producing drought conditions over portions of North America, although tropical Pacific SST variations appear to play a more dominant role (e.g., Schubert et al., 2004; Seager et al., 2005; Seager, 2007).

Climate variability in the tropical Atlantic Sector is often associated with relative changes in SST between the hemispheres, a phenomenon referred to as the tropical Atlantic SST dipole, meridional mode, or gradient mode. SST anomalies manifest their most pronounced features around 10–15 degrees of latitude off of the Equator. Chang et al. (1997) hypothesized the existence of a thermodynamic coupled air–sea mode. The dominant physical processes are a positive feedback (WES) between the surface wind, surface heat flux through evaporation and SST, and a delayed negative feedback due to the advection of heat by the steady upper ocean circulation. Imagine that the Northern Hemisphere is warmer than the Southern Hemisphere. The anomalous meridional SST gradient drives cross-equatorial surface wind anomalies, which affect the surface heat fluxes in the trade wind zones in a way that reinforces the SST anomalies in the two hemispheres. The mean ocean circulation will transport anomalously cold water from the Southern to the Northern Hemisphere. This northward advection of cold water acts against the atmospheric warming in the Northern Hemisphere, thereby enabling oscillatory behavior.

While Xie (1999) basically confirmed the mechanism, Kushnir et al. (2002) challenged it. In the coupled model study of the latter, the ocean plays only a damping role and cannot advect anomalies across the Equator. The long timescales up to decadal are merely the result of a weak positive heat flux–SST feedback that "leaks" into the trade wind regions and interferes with the prescribed random forcing there. Remote forcing by the NAO needs also to be considered when discussing tropical Atlantic decadal variability, as the NAO has significant power at decadal timescales (Figure 25.2) and strongly affects both the trade winds over the tropical Atlantic and the tropical Atlantic Ocean circulation (e.g., Visbeck et al., 1998). It should be mentioned that the tropical Atlantic is also impacted by the basin-scale AMO which is discussed in Section 3.

3. DESCRIPTION OF EXTRATROPICAL DECADAL VARIABILITY FROM OBSERVATIONS

The focus of this section is on the PDO and the AMO, the two leading modes of extratropical decadal SST variability in the North Pacific and North Atlantic. The Southern Ocean centennial variability (SOCV) is also described in this section. Instrumental records (Figure 25.3) are not long enough to determine whether the three phenomena are robust and have a well-defined period rather than a simpler character, such as red noise. Biondi et al. (2001), by using a

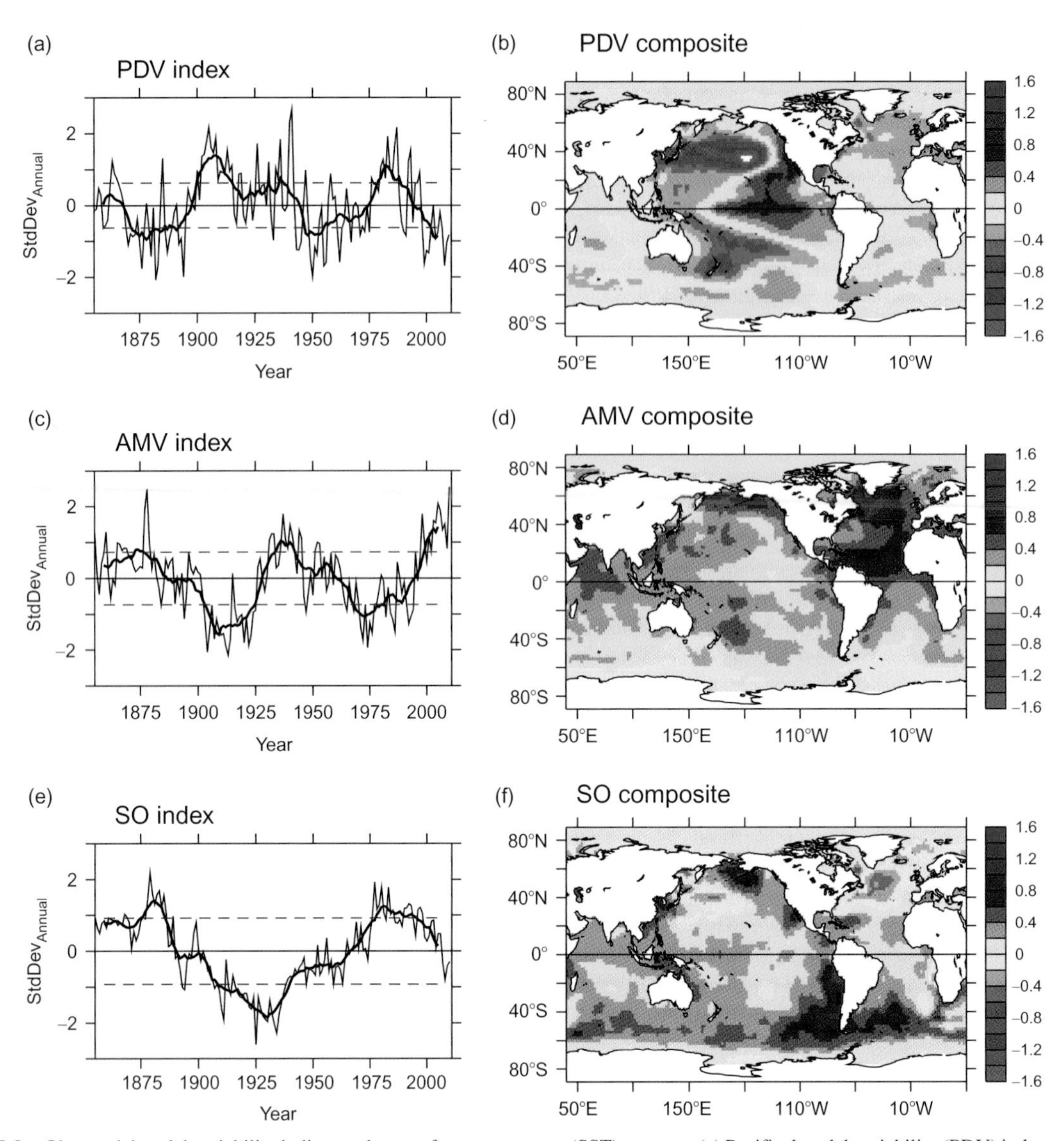

FIGURE 25.3 Observed decadal variability indices and sea surface temperature (SST) patterns. (a) Pacific decadal variability (PDV) index as defined by the first EOF of North Pacific (20–60 °N) SST and (b) SST pattern computed as the composite difference between positive and negative phases (the dashed lines denote the thresholds). (c) Atlantic multidecadal variability (AMV) index as defined as North Atlantic (0–60 °N) average SST and (d) SST pattern (computed as in b). (e) Southern Ocean SST index (SO-index) as defined as the average over the region 50–70 °S, and (f) SST pattern (computed as in b). Thick (thin) lines indicate 11-year running (annual) mean. Please note that the scale of the panels on the left is given in terms of the standard deviation of the annual mean time series. The right panels are in (°C). The linear trend at each point was removed prior to the analysis. *From Park and Latif (2011).*

network of tree ring chronologies for Southern and Baja California, extended the instrumental record back to AD 1661 and revealed a robust PDO with a dominant bidecadal period. The robustness of the AMO signal has been addressed by Delworth and Mann (2000), and similar fluctuations have been documented from a multiproxy dataset through the last four centuries. The SOCV also appears to be robust and is clearly seen in the spectrum of the Tasmanian summer temperatures of the last 3600 years, which were reconstructed from tree rings (Cook et al., 2000).

It is commonly assumed that PDO, AMO, and SOCV largely reflect modes of internal climate variability, although contributions from external forcing, especially during the twentieth century, cannot be ruled out (e.g., Stott et al., 2000; Rotstayn and Lohmann, 2002). Booth et al. (2012) even claim that aerosol forcing, natural and anthropogenic, can explain most of the AMO signal during the instrumental period. However, the past emissions of and the climate response to aerosols are subject to large uncertainties. Decadal to centennial SST variability during the instrumental period may also contain some contribution from solar forcing, which features variability with periods of about 70–100 years and about 200 years known as the Gleissberg and Suess cycle, respectively. The internal nature of PDO, AMO, and SOCV is supported by many climate models, which simulate similar variability without applying external forcing. It is the ocean that provides the long-term memory in these models, as extensively described in Section 3.1.

Are PDO, AMO, and SOCV related to each other? We cannot satisfactorily answer this question at this stage of research. The time series and spatial patterns presented in Figure 25.3 seem to be sufficiently different to think of them as independent modes. d'Orgeville and Peltier (2007) reported, however, that PDO and AMO are significantly lag correlated during the instrumental record. Yet the last decades displayed a more synchronous evolution of the two phenomena. Modeling studies contradict each other. Timmermann et al. (1998) found that PDO and AMO are phase locked to each other in a multicentury control integration of a climate model and suggested coupling through an atmospheric bridge. In contrast, Park and Latif (2010) by investigating the variability in a multimillennial control run with another climate model reported that PDO and AMO can be regarded to first order as independent phenomena with slightly different periods. The SOCV seems to be independent of both PDO and AMO, but not much research on this topic has been conducted to date.

3.1. Pacific

Following Mantua et al. (1997), an index of PDO is defined as the time series of the first empirical orthogonal function (EOF) of the SST anomalies north of 20 °N. According to this definition, the PDO has a fixed pattern (Figure 25.3b) and describes only part of the North Pacific decadal SST variability. The PDO index features both a strong bidecadal and a multidecadal component (Figure 25.3a). The multidecadal signal was more prominent during the twentieth century; but tree ring reconstructions suggest the bidecadal component was the dominant component during the previous centuries (Biondi et al., 2001). The composite pattern associated to the PDO index exhibits a v-shaped pattern of SST anomalies in the tropical Pacific and opposite-signed anomalies in the western and central extratropical North and South Pacific (Figure 25.3b). Overall, there is a remarkable symmetry between the North and South Pacific.

Phase changes of the PDO are associated with pronounced changes in temperature and rainfall patterns across North and South America, Asia, and Australia, as well as important ecological consequences, including major shifts in distribution and abundance of zooplankton and important commercial species of fish (Hare and Mantua, 2000). Furthermore, ENSO teleconnections on interannual timescales around the Pacific basin are significantly modified by the PDO, as reported by Gershunov and Barnett (1998). They show that the PDO exerts a modulating effect on ENSO teleconnections over the United States. As a consequence, seasonal climate anomalies over North America exhibit rather large variability between years characterized by the same ENSO phase. This lack of consistency reduces potential statistically based ENSO-related climate predictability.

We lack a good understanding of the dynamics of the decadal variability in the extratropical Pacific. For instance, how does decadal variability in the tropics and extratropics interact? Different competing proposals were made (see for a review on the interactions between the tropics and the extratropics Liu and Alexander, 2007). First, decadal variability in the Equatorial Pacific is internal to that region, and either a linear stochastic scenario applies (e.g., Eckert and Latif, 1997; Neelin et al., 1998), or nonlinear processes (e.g., Jin et al., 1994; Rodgers et al., 2004) produce the decadal variability. Second, Equatorial Pacific decadal variability is driven by processes in the subtropics and/or midlatitudes (e.g., Pierce et al., 2000; Lohmann and Latif, 2005; Merryfield and Boer, 2005; Dommenget and Latif, 2008). It should be noted here that the Equatorial Pacific decadal SST variability, independent of its origin, can impact the extratropical Pacific through atmospheric teleconnections. Third, the decadal variability in the North Pacific originates in the North Pacific itself (e.g., Latif and Barnett, 1994; Wu et al., 2003; Ceballos et al., 2009). And finally, fourth, the decadal-scale variability in the Pacific basin constitutes a coupled air–sea mode, involving the tropics and the extratropics, a fast bridge through the atmosphere and a slow bridge through the ocean which sets the timescale (e.g., Gu and Philander, 1997).

Many different processes are likely to operate simultaneously to produce the decadal variability in the Pacific. They are associated with rather different timescales, ranging from interannual to decadal and even longer, which leads to the multitimescale character of the PDO index (Figure 25.3a). Schneider and Cornuelle (2005) showed that the PDO can be recovered from a reconstruction of North Pacific SST anomalies based on a first-order autoregressive model and forcing by variability of the Aleutian low, remote forcing by the tropical Pacific, and Rossby wave induced oceanic zonal advection anomalies in the Kuroshio Oyashio Extension (KOE) region. This supports the hypothesis that the PDO is not a dynamical mode, but arises from the superposition of SST fluctuations with different dynamical origins.

Newman (2007) argued that the North Pacific decadal variability is the sum of several phenomena, each of which with its own spatial pattern and time evolution. The variability of the Aleutian low accounting for most of the sea-level pressure variability in the North Pacific Sector is a prime driver of the PDO. Persistence of SST anomalies for several years in an elongated band near 40 °N is achieved through the subduction/reemergence mechanism (e.g., Alexander et al., 1999). The North Pacific Oscillation (NPO) is the second most energetic mode of sea-level pressure variability in the North Pacific Sector and an important driver of the North Pacific Gyre Oscillation (NPGO; see e.g., Ceballos et al., 2009). The NPGO has a prominent quasidecadal component and features a dipolar anomaly pattern in SST and sea surface height (SSH). Changes in the strength of the central and eastern branches of the subpolar and subtropical gyres are an important element of the NPGO. Rossby waves propagate the wind-forced signal from the central North Pacific into the KOE region, and trigger changes in the strength of the KOE with a lag of 2–3 years.

3.2. Atlantic

The AMO is the leading mode of decadal SST variability in the North Atlantic (e.g., Alvarez-Garcia et al., 2008). An AMO index is defined as the linearly detrended North Atlantic (0–60 °N) average anomalous SST. The index exhibits enhanced variability at periods of about 60–80 year during the instrumental record (Figures 25.2 and 25.3c). Proxy-based reconstructions of surface temperatures during the past 330 years support such a periodicity (Delworth and Mann, 2000). The spatial pattern associated to the AMO index is characterized by monopolar SST anomalies covering the whole North Atlantic and anomalies of opposite sign in large parts of the South Atlantic. This characteristic SSTA pattern was originally described by Folland et al. (1986) and is commonly referred to as the interhemispheric SST contrast or dipole. It should not be confused with the decadal tropical Atlantic SST dipole described earlier, as the interhemispheric dipole encompasses the whole Atlantic. The AMO pattern shown here (Figure 25.3d; Park and Latif, 2011) depicts a strong global warming signature and features only a relative SST contrast between the North Atlantic and the South Atlantic. This demonstrates a large sensitivity to the way of preprocessing the data before computing indices of internal modes such as the AMO index (Trenberth and Shea, 2006). It should be mentioned in this context that the AMO variability projects on global average SAT in climate models (Semenov et al., 2010).

The slow changes in the Atlantic SSTs as expressed by the AMO index (Figure 25.3c) have affected regional climate trends over parts of North America and Europe (e.g., Sutton and Hodson, 2005), Northern Hemisphere surface temperature anomalies (e.g., Zhang et al., 2007), Arctic sea ice (e. g., Deser et al., 2000; Venegas and Mysak, 2000), Arctic Ocean conditions (Bengtsson et al., 2004), Atlantic hurricane activity (e.g., Goldenberg et al., 2001; Zhang and Delworth, 2006), and fisheries production in the northern North Atlantic and North Sea (e.g., Beaugrand, 2004; Drinkwater, 2006). In addition, tropical Atlantic SST anomalies associated with AMO have contributed to shifts in the intertropical convergence zone (ITCZ) and to rainfall anomalies over the Caribbean and the Nordeste region of Brazil and severe multiyear droughts over parts of Africa including the Sahel (see Figure 25.2; e.g., Folland et al., 1986; Bader and Latif, 2003; Giannini et al., 2003; Lu and Delworth, 2005).

Bjerknes (1964) concluded from his early analysis of the observations in the midlatitudinal Atlantic Region that the atmosphere drives SST at interannual timescales, while, at longer timescales, it is the ocean dynamics that produce SST variability. He suggested that if the top-of-the-atmosphere fluxes and the oceanic heat storage did not vary too much, then the total energy transport by the climate system would not vary much either. This implies that any large anomalies of oceanic and atmospheric energy transport should be equal and opposite. This scenario has become known as Bjerknes compensation. Shaffrey and Sutton (2006) found evidence for this picture in a climate model, in which the decadal variability of the energy transport was linked to AMOC. Additional observational evidence for the Bjerknes compensation was recently provided by Gulev et al. (2013). They analyzed a new, 130-year-long, turbulent surface heat flux dataset and found a cutoff period of about 10 years, above which oceanic forcing dominates North Atlantic SST.

Latif et al. (2006b) suggested that there have occurred considerable multidecadal changes in AMOC strength during the last century. AMOC variations were indirectly determined and reconstructed from the history of observed Atlantic SST. Since AMOC variations are associated in climate models with variations in the meridional heat transport, a fingerprint of relative AMOC strength can be defined as the difference between the North and South Atlantic SST making use of the interhemispheric SST contrast characterizing the AMO. The observed changes in the interhemispheric SST contrast are argued to be driven by the low-frequency variations of the NAO through changes in air–sea heat exchange over the Labrador Sea (Latif and Keenlyside, 2011). A persistent positive NAO phase, for instance, would enhance Labrador Sea convection (Figure 25.4), which in turn would spin up the AMOC. Accordingly, multidecadal variations in the interhemispheric SST contrast follow those of the NAO index with a time delay of about a decade (Figure 25.4).

OGCMs forced by observations at the surface are useful tools to understand observed variability. Alvarez-Garcia et al. (2008) analyzed such a forced simulation. Snapshots of the ocean model's overturning stream function 5 years apart from each other (Figure 25.5), as reconstructed from the leading (multidecadal) mode, show clearly how negative overturning anomalies develop in the 1960s and subsequently slowly propagate southward. During 1970–1980, the height of the cold AMO phase in the North Atlantic (Figure 25.3c), the tendency in the overturning is reversed and the negative anomalies start to weaken, until they are replaced by positive overturning anomalies in the mid-1980s in the north. The positive overturning anomalies expand southward and initiate the subsequent warm phase in the North Atlantic during the 1990s corresponding to a positive AMO phase. The latter is characterized by an anomalously strong overturning circulation in the model and anomalously high SSTs, which persist up to the present (Figure 25.3c). Consistent with Figure 25.4, the evolution of the overturning can be regarded as the model's AMOC response to the multidecadal NAO forcing. OGCM studies by Eden and Jung (2001) and Eden and Willebrand (2001) have also suggested this mechanism.

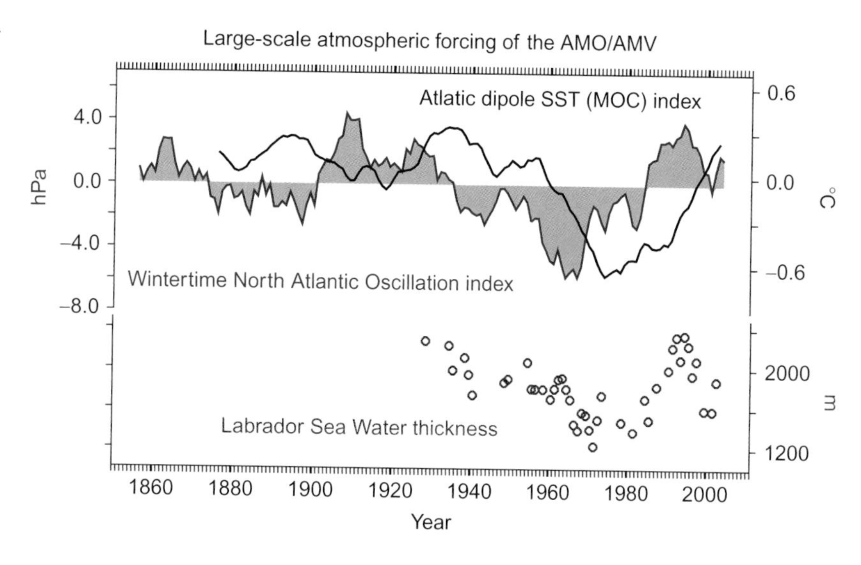

FIGURE 25.4 Time series of the winter (December–March (DJFM)) NAO index (hPa, shaded curve), a measure of the strength of the westerlies and heat fluxes over the North Atlantic and the Atlantic dipole SST anomaly index (°C, black curve), a measure of the strength of the AMOC. The NAO index is smoothed with an 11-year running mean; the dipole index is also smoothed with an 11-year running mean filter. Multidecadal changes of the AMOC as indicated by the dipolar SST index lag those of the NAO by about a decade, supporting the notion that a significant fraction of the low-frequency variability of the AMOC is driven by that of the NAO. Shown in red are annual data of LSW thickness (m), a measure of convection in the Labrador Sea, at ocean weather ship Bravo, defined between isopycnals σ 1.5=34.72−34.62, following Curry et al. (1998). *From Latif and Keenlyside (2011).*

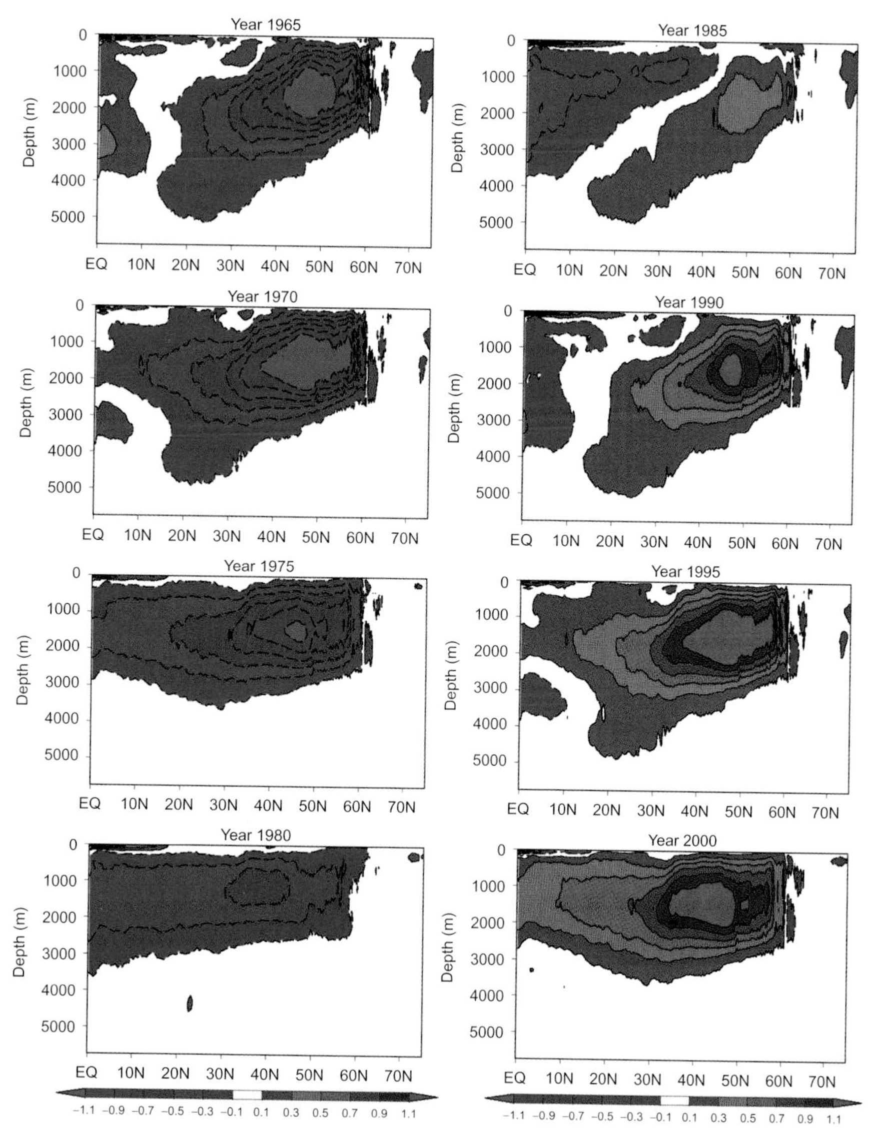

FIGURE 25.5 Snapshots (annual means) of the meridional overturning stream function (Sv) 5 years apart from each other, as reconstructed from the multidecadal SSA mode obtained from the results of a forced simulation with an OGCM using observed boundary forcing. The transition from an anomalously weak overturning during the 1970s to an anomalously strong overturning during the 1990s is clearly seen. *From Alvarez-Garcia et al. (2008).*

Quasidecadal SST variability in the North Atlantic was reported by Deser and Blackmon (1993). The corresponding pattern is different to the AMO pattern and depicts anomalies of one sign east of Newfoundland, and anomalies of the opposite polarity off the southeast coast of the United States. There is also clear evidence of decadal variability in the heat and freshwater content of the North Atlantic Ocean (Curry and Mauritzen, 2005; Lozier et al., 2008). Moreover, decadal subpolar gyre circulation changes have been indirectly inferred from satellite observations of North Atlantic SSH (Häkkinen and Rhines, 2004). The relationship between the changes in the subpolar gyre region of the North Atlantic with the AMOC and the AMO, however, is still controversial and likely to be dependent on the timescale.

3.3. Southern Ocean Centennial Variability

An analysis of Southern Ocean SST variability similar to the analysis of the PDO and the AMO (Figure 25.3) was performed. The instrumental database, however, is poor. An index was defined as the zonally averaged SST over the region 50–70 °S (Figure 25.3e) by Latif et al. (2013), a region where both the PDO and the AMO have only weak projections on the zonal mean SST anomalies. The index, hereafter termed SOCV index, exhibits a rather strong centennial variability with a maximum during the mid-1870s, a minimum during the mid-1920s and another maximum during the mid-1970s. The SOCV index considerably dropped thereafter. As mentioned earlier, the tree ring reconstruction of Tasmanian summer temperatures of the last 3600 years confirms the existence of enhanced variability in the Southern Ocean Sector at centennial timescales well above the red background spectrum.

The composite SSTA pattern associated with the SOCV index (Figure 25.3f) depicts some interesting features. First, there is a pronounced north–south asymmetry, with a relatively strong warming in the Southern Hemisphere. Second, the strongest warming signals are seen off the west coast of South America and in the midlatitudinal South Atlantic. And third, both the Kuroshio and Gulfstream/North Atlantic Current Regions are characterized by negative SST anomalies. Statistical significance cannot be assigned to these results, as the instrumental data cover only one realization of the centennial variability.

4. THE STOCHASTIC CLIMATE MODEL: THE NULL HYPOTHESIS FOR CLIMATE VARIABILITY

The climate system is comprised of components with very different internal timescales. Weather phenomena, for instance, have typical lifetimes of hours or days, while the deep ocean needs many centuries to millennia to adjust to changes in surface boundary conditions. Hasselmann (1976) introduced an approach to modeling the effect of the fast variables on the slow in analogy to Brownian motion. He suggested treating the former not as deterministic but as stochastic variables, so that the slow variables evolve following dynamical equations with stochastic forcing. The chaotic components of the system often have well-defined statistical properties and these can be built into approximate stochastic representations of the high-frequency variability. The resulting models for the slow variables are referred to collectively as stochastic climate models, although the precise timescale considered slow may vary greatly from model to model. The stochastic climate model can be considered as *the* null hypothesis for the generation of natural climate variability.

4.1. The Zero-Order Stochastic Climate Model

In the following, the atmosphere is the fast (chaotic) and the ocean–sea ice system the slow component. The simplest model is linear and local in which the atmospheric forcing at one location drives only changes in the ocean–sea ice system at this very point; neither the atmosphere nor the ocean–sea ice system exhibit spatial coherence. In this model, the ocean mixed layer simply integrates the random surface heat flux (or freshwater) fluctuations. To avoid a singularity at zero frequency, a damping was introduced by Hasselmann (1976). If the atmospheric forcing is white in time, which means that the variability is equally distributed over all frequencies, the power of the SST or sea surface salinity (SSS) increases with the inverse of the squared frequency (ω^{-2}-shape). The spectrum flattens at a frequency that depends on the damping timescale. Frankignoul and Hasselmann (1977) have shown that observed SST variability is consistent with such a local model in parts of the midlatitudes, away from coasts and fronts, whereas the simple stochastic model fails in regions where mesoscale eddies or advection are important. Hall and Manabe (1997) analyzed data from four ocean weather stations in midlatitudes and explained differences between SST and SSS spectra by the simple model in terms of the different damping timescales and reported that a complex climate model did reproduce this behavior.

Barsugli and Battisti (1998) by extending the Hasselmann (1976) model constructed a simple stochastically forced, one-dimensional and linear, coupled energy balance model and obtained important insight into the nature of coupled interactions in the midlatitudes. The physical mechanism that this model presents is that ocean–atmosphere coupling reduces the internal damping of temperature anomalies due to surface heat fluxes. Reduced damping in the presence of a constant level of stochastic forcing in the atmosphere leads to greater thermal variance of the coupled atmosphere and the coupled ocean compared to the uncoupled atmosphere and to the uncoupled "slave" ocean. It was suggested that the experimental design of an atmospheric general circulation model (AGCM) coupled to a mixed layer ocean model (more sophisticated than a slab model) would provide a reasonable null hypothesis against which to test for the presence of distinctive decadal variability.

4.2. Hyper Mode

Following the proposal of Barsugli and Battisti (1998), Dommenget and Latif (2008) developed a coupled ocean–atmosphere model in which the atmosphere is represented by a state-of-the-art AGCM, while the ocean is represented by a vertical column model (CM) in which the individual levels communicate only by vertical diffusion (see also Alexander and Penland, 1996). A similar setup was used by Kwon et al. (2010) to investigate the winter response of the coupled atmosphere–ocean mixed layer system to anomalous geostrophic ocean heat flux convergence in the Kuroshio Extension (KE). In the model of Dommenget and Latif (2008), stochastic surface heat flux variability associated with regional atmospheric variability patterns (e.g., fluctuations of the Aleutian low) is integrated by the large heat capacity of the extratropical oceans, leading to a continuous increase of SST variance toward longer timescales. Atmospheric teleconnections spread the extratropical signal to the tropical regions. Once tropical SST anomalies have developed, global atmospheric teleconnections spread the signal around the world, creating a global climate mode termed hyper mode. Such a simplified coupled model reproduces reasonably well the SSTA structure of the PDO (Figure 25.3b).

4.3. Stochastic Models with Mean Advection and Spatial Coherence

Atmospheric variability on timescales of a month or longer is dominated by a small number of large-scale spatial patterns, whose time evolution has a significant stochastic component (Davis, 1976). The atmospheric patterns play an important role in ocean–sea ice–atmosphere interactions, and ocean advection can play a role in this coupling (e.g., Saravanan and McWilliams, 1997, 1998). The ocean, for instance, can feature preferred timescales, although there is no underlying oscillatory mechanism, neither in the ocean nor in the atmosphere. Furthermore, the existence of the preferred timescale in the ocean does not depend on the existence of an atmospheric response to SST anomalies. The preferred timescale is determined by the advective velocity scale in the upper ocean and the length scale associated with the low-frequency atmospheric variability. This mechanism is often referred to as "spatial resonance" or "optimal forcing."

For the extratropical North Atlantic basin, this timescale would be of the order of a decade. Deser and Blackmon (1993), Sutton and Allen (1997), Czaja and Marshall (2001), and Alvarez-Garcia et al. (2008) find such a quasidecadal timescale in surface observations of the North Atlantic. However, the findings of these studies substantially differ in several aspects and concerning the derived propagation characteristics of the SST anomalies. It therefore remains unclear whether the stochastic–advective mechanism applies to the North Atlantic quasidecadal variability. The mechanism was also proposed as an explanation for the Antarctic circumpolar wave (ACW; White and Peterson, 1996) by Weisse et al. (1999) investigating a stochastically forced ocean model integration. The ACW is associated with anomalies in SLP, wind stress, SST, and sea ice extent. These anomalies propagate eastward with the Antarctic Circumpolar Current (ACC), with a period of 4–5 years and taking 8–10 years to encircle the pole.

4.4. Stochastic Wind Stress Forcing of a Dynamical Ocean

Frankignoul et al. (1997) used a linear model to estimate the dynamical response of the extratropical ocean to spatially coherent stochastic wind stress forcing with a white frequency spectrum. For forcing without zonal variation, the response propagates westward at twice the baroclinic Rossby wave phase speed. The baroclinic response is spread over a continuum of frequencies, with a dominant timescale determined by the time it takes a long Rossby wave to propagate across the basin and thus increasing with the basin width. The baroclinic predictions for a white wind stress curl spectrum were shown to be broadly consistent with the frequency spectrum of sea-level changes and temperature fluctuations in the thermocline observed near Bermuda.

Taguchi et al. (2007) analyzed low-frequency variability of the KE in a multidecadal hindcast by an eddy-permitting OGCM. The simulated North Pacific SSH variability in this complex model was shown to be consistent with satellite observations. The following scenario was suggested for decadal KE variability: basin-scale wind variability excites broad-scale Rossby waves, which propagate westward, triggering intrinsic modes of the KE jet and reorganizing SSH variability in space. Schneider et al. (2002) previously found evidence in data that Rossby waves carry thermocline depth perturbations to the west, which then results in a shift of the KE and subsequently in SST anomalies.

4.5. Stochastically Driven AMOC Variability

Competing mechanisms were proposed to explain the decadal-scale AMOC variability. One idea is that low-frequency AMOC variability, consistent with a simple stochastic scenario, is driven by the low-frequency portion of the spectrum of atmospheric forcing. This hypothesis was first tested in uncoupled mode under mixed boundary conditions (i.e., surface temperature is relaxed to a specified value and salinity flux is prescribed) by Mikolajewicz and Maier-Reimer (1990). They performed a multimillennial integration of a global OGCM forced by spatially correlated white-noise freshwater flux anomalies. In addition to the expected red-noise character of the oceanic response, enhanced AMOC variability in a frequency band around 320 years was reported, which was shown by Pierce et al. (1995) to originate in the Southern Ocean. The oscillations involve movement between two states: one characterized by strong Weddell Sea convection and an active overturning circulation in the Southern Ocean and the other with a halocline around Antarctica capping off the water column, thus preventing convection. The physical mechanism that forces the model from the quiescent state to an actively convecting one is subsurface heating around Antarctica, which destabilizes the water column; the ultimate source of this heat is advected North Atlantic Deep Water.

Is the variability simulated in ocean-only integrations as that in Mikolajewicz and Maier-Reimer (1990) too vigorous? Especially the use of mixed boundary conditions in ocean-only experiments has been criticized (e.g., Weaver and Sarachik, 1991). Self-sustained oscillations were found under mixed boundary conditions in simulations with flat bottom ocean models (e.g., Greatbatch and Zhang, 1995; Yin and Sarachik, 1995), but Winton (1997) showed that coastal topography has a considerable damping influence upon internal decadal oscillations of the thermohaline circulation. Dijkstra and Neelin (1999) pointed out that care should be used in drawing conclusions on the existence of multiple equilibria and the stability of the thermohaline circulation when mixed boundary conditions are used. The results of Mikolajewicz and Maier-Reimer (1990) obtained with mixed boundary conditions were basically reproduced with a fully coupled ocean–atmosphere–sea ice general circulation model (Park and Latif, 2008), suggesting that ocean model integrations with mixed boundary conditions can be useful to obtain some insight into the dynamics of ocean circulation variability.

CGCMs employ realistic bottom topography to the extent it can be resolved and also retain full ocean–atmosphere–sea ice coupling. Delworth et al. (1993) were the first to report pronounced multidecadal variability of the AMOC in a multicentury control run with a coupled ocean–atmosphere general circulation model. The simulated SST variability in the North Atlantic is consistent with observations concerning both spatial structure and timescale. Delworth and Greatbatch (2000) showed that the multidecadal AMOC fluctuations described in Delworth et al. (1993) are damped and primarily driven by a spatial pattern of surface heat flux variations that bear a strong resemblance to the NAO. No conclusive evidence was found that the AMOC variability is part of a dynamically coupled atmosphere-ocean mode in this particular model. Griffies and Tziperman (1995) interpreted the multidecadal variability in terms of a stochastically forced four-box model of the AMOC. Jungclaus et al. (2005) investigated the multidecadal AMOC variability in another CGCM and suggested a mechanism in which the Arctic plays a major role. The multidecadal variability in that model is sustained by the interplay between the storage of freshwater in the central Arctic and its release into the subpolar North Atlantic and circulation changes in the Nordic Seas caused by variations in the Atlantic heat and salt transport into the Arctic.

Recently, Delworth and Zeng (2012) reported pronounced unforced, that is, internal, centennial variability of the AMOC in a climate model, which had a significant impact on Northern Hemisphere average SAT. The centennial variability in the model fits the stochastic framework, although the precise mechanism is still under investigation. Pronounced stochastically driven centennial variability was also reported from another climate model by Park and Latif (2008). The studies of Mikolajewicz and Maier-Reimer (1990), Park and Latif (2008), and Delworth and Zeng (2012) support the applicability of the stochastic concept also at centennial timescales, a result expected given the long timescales associated with the deep ocean circulation. This has potentially significant implications for understanding past climate variations.

4.6. Stochastic Coupled Variability Involving the AMOC

Coupled modes in which the feedback from the ocean to the atmosphere is necessary to produce enhanced variability above the background red spectrum at a distinctive timescale (as it is the case for ENSO at interannual timescales) were also proposed. The coupled modes have also to be considered within a stochastic framework, as we expect them to be damped and not self sustained. It should be mentioned in this context that AMOC indices obtained from CGCMs generally exhibit spectra that substantially deviate from the ω^{-2}-shape predicted by the simplest version of the stochastic climate model (Section 4.1): the increase of the AMOC variance toward lower frequencies is considerably steeper than in typical spectra of midlatitudinal SST or SSS anomalies. Timmermann et al. (1998) described coupled variability with a 35-year period in a multicentury integration of a climate model, in which AMOC played an important role. The coupled feedback involves SST-forced freshwater flux anomalies, which drive SSS anomalies in the North Atlantic. The salinity anomalies are advected into the sinking region in the subpolar North Atlantic, thereby providing a delayed negative feedback.

Eden and Greatbatch (2003) analyzed results from a simple stochastic atmospheric feedback model coupled to an OGCM of the North Atlantic. The coupled model simulates a damped quasidecadal oscillation for sufficiently strong coupling, in which the overturning component of the northward heat transport provides the delayed negative feedback. A fast positive feedback arising from the wind-driven (gyre) component of the northward heat transport turns out to be necessary to distinguish the coupled oscillation from that in a model without any feedback from the ocean to the atmosphere.

Vellinga and Wu (2004) described a coupled mode with a centennial timescale from a climate model. The mechanism relies on interactions between the extratropics and the tropics. Suppose the AMOC is anomalously strong. The ITCZ both strengthens and moves northward in this situation, leading to an increased freshwater flux into the northern tropical Atlantic. The resulting negative salinity anomaly is gradually advected northward by the mean upper ocean circulation into the subpolar region on a timescale of a few decades. A negative salinity anomaly in the subpolar region reduces the density and weakens deep convection there, providing a delayed negative feedback on the AMOC. A similar mechanism was derived from two other climate models by Menary et al. (2012), but the strength of the coupled mode strongly varies from model to model.

4.7. Stochastically Forced Southern Ocean Variability

Hall and Visbeck (2002) investigated the decadal to centennial Southern Ocean variability in a coarse-resolution CGCM. Overall the variability was found to be consistent with the stochastic climate model scenario. The simulated southern annular mode (SAM) index is basically white in that model, while ocean spectra are red. The fluctuations associated with the SAM constitute a significant fraction of the simulated ocean variability south of 30 °S in the coupled model. Martin et al. (2012) and Latif et al. (2012) investigating the results of another coarse-resolution climate model describe a quasioscillatory multicentennial mode of open ocean deep convection in the Weddell Sea with global impacts. The deep convection is stimulated by a built-up of heat at mid-depth provided by the lower branch of the AMOC, similar to the mechanism proposed by Pierce et al. (1995). The heat is released to the atmosphere during periods of deep convection. When the heat reservoir is depleted, a coincidental strong freshening event at the sea surface stalls the convection.

Böning et al. (2008) by analyzing the Argo network of profiling floats and historical oceanographic data challenged the applicability of coarse-resolution ocean models to study Southern Ocean climate change and variability. This pertains especially to the response of the Southern Ocean circulation to decadally varying westerlies, as coarse-resolution models do not resolve the stabilizing feedback by mesoscale eddies on the large-scale flow. Böning et al. (2008) concluded that the transport in the ACC and meridional overturning in the Southern Ocean have been insensitive to decadal changes in wind stress. On the other hand, Meredith et al. (2012)—using novel theoretical considerations and fine-resolution ocean models to develop a new scaling for the sensitivity of eddy-induced mixing to changes in winds—demonstrated that changes in Southern Ocean overturning in response to recent and future changes in wind stress forcing are likely to be substantial, even in the presence of a decadally varying eddy field.

How the parameterized mesoscale diffusivities are prescribed in noneddy resolving ocean models makes a big difference in the response to wind stress, and the incorporation of spatially (horizontally and vertically) dependent coefficients in the eddy parameterization appears to be of fundamental importance (e.g., Farneti and Gent, 2011). Sallee and Rintoul (2011) showed that the use of spatially varying coefficients is very important to realistically simulate the regional pattern of the eddy-induced subduction in the Southern Ocean, a process that counters the Ekman flow which is only simulated by coarse-resolution models.

5. DECADAL PREDICTABILITY

Decadal prediction is a relatively new emerging field in climate research. However, what class of problem is decadal prediction: an initial value or boundary value problem? The prediction of variations associated with the internal modes would pose an initial value problem. On the other hand, projections of future climate indicate a rise in global mean temperature of between 2 ° and 4 °C by 2100, dependant on emission scenario and model. This translates to an average rise in global mean temperature of the order of 0.3 °C per decade. This is large compared, for instance, with the observed increase of about 0.7 °C during the last century (Figure 25.1), and argues that decadal prediction is also a boundary value problem, because external forcing cannot be neglected. Thus the prediction of the climate over the next few decades poses a joint initial/boundary value problem.

While the predictability of internal fluctuations on seasonal timescales, specifically ENSO, has been intensively studied for more than 20 years (see Chapter 23 in this book), decadal predictability has been systematically investigated for only a few years. Lack of understanding of predictable dynamics at decadal timescales and shortness of observational records are two main reasons that prevent us from studying decadal predictability in a systematic way.

Another reason for this is the much longer timescale which requires rather long model integrations and therefore the availability of large computer resources. Moreover, realtime predictions such as the decadal forecasts by Smith et al. (2007) and Keenlyside et al. (2008) need a long time until we can verify them.

Conceptually, one distinguishes between potential (diagnostic) and classical (prognostic) predictability studies. Potential decadal predictability (see Boer and Lambert, 2008 and references therein) is simply defined as the ratio of the variance on the decadal timescales to the total variance. It does not discriminate among variability arising from a zero-order stochastic model (red-noise process) or higher-order models. Fitted linear inverse models or constructed analogues provide more discriminative estimates of diagnostic predictability (e.g., Hawkins et al., 2011; Teng and Branstator, 2011; Branstator et al., 2012).

Classical predictability studies consist of performing ensemble experiments with a single coupled model perturbing the initial conditions (Griffies and Bryan, 1997a,b; Grötzner et al., 1999; Collins, 2002; Collins and Sinha, 2003; Pohlmann et al., 2004). The predictability of a variable is given by the ratio of the actual signal variance to the ensemble variance. This method provides in most cases an upper limit of decadal predictability since it assumes a perfect model and, very often, near-perfect oceanic initial conditions. A third method compares the variability simulated in coupled models with and without active ocean–sea ice dynamics. Those regions in which ocean–sea ice dynamics are important in generating decadal-scale variability are believed to be regions of relatively high decadal predictability potential (Park and Latif, 2005).

All three types of studies yield similar patterns of decadal predictability (e.g., Latif et al., 2006a). In contrast to seasonal to interannual predictability, potential decadal predictability is found predominately over the mid to high-latitude oceans (e.g., Boer and Lambert, 2008). The potential decadal predictability decreases with increasing timescale, but appreciable values exist up to multidecadal timescales, especially for the North Atlantic and the Southern Ocean (Figure 25.6). The result for the North

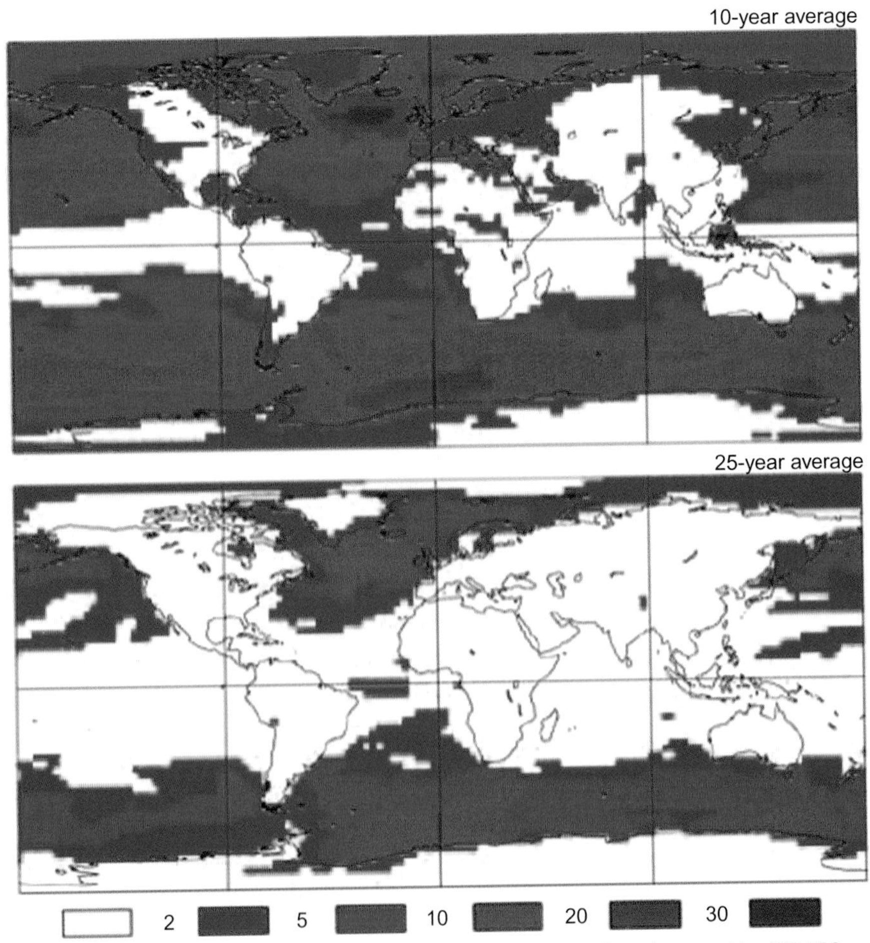

FIGURE 25.6 Potential decadal predictability (%) of SAT for 10- and 25-year means obtained from the ensemble of CMIP3 models. See text for details. *From Boer and Lambert (2008).*

Atlantic could have been anticipated on the basis of the discussion around Figures 25.4 and 25.5 in Section 4.

The decadal predictability of internal variability was investigated through decadal hindcasts (retrospective forecasts) by Yang et al. (2013) using an average predictability time analysis. Comparison of initialized with uninitialized historical forcing simulations using the same climate model allowed the identification of internal multidecadal patterns for SST and 2 m air temperature (T2m). The most predictable pattern of T2m (Figure 25.7a) depicts a general bipolar seesaw, with warm anomalies centered in the North Atlantic and the Arctic, and cold anomalies centered in Antarctica. The hindcast skill of the initialized system, verified against independent observations (Figure 25.7b and c), indicates that both the SST and T2m pattern may be predictable up to a decade. The decadal predictive skill primarily originates from initialization of multidecadal variations of northward oceanic heat transport in the North Atlantic, which in turn is related to AMOC.

Although atmospheric decadal predictability is found to be rather limited relative to oceanic decadal predictability (for instance, that of the NAO in comparison to that of the AMO), there are some exceptions. Multidecadal changes in the Atlantic SST drive changes in Atlantic hurricane variability and have impacts in the Sahel, India, Brazil, Central America, and the Arctic. Smith et al. (2010) presented skilful multiyear predictions of Atlantic hurricane frequency (Figure 25.8a). The initialized hindcasts (red circles) are significantly more skilful than both the uninitialized (blue squares) and persistence forecasts made by persisting anomalous storm counts from previous years (green diamonds). The skill is robust to ensemble size, tracking algorithm choices and geographical region. The hindcast period 1960–2005 was largely dominated by an increasing trend in storm frequency (Figure 25.8b), which gives rise to increasing skill at longer forecast periods for all forecasts (Figure 25.8a). These low-frequency variations are not caused by internal variability alone in the model, but are partly externally forced by a combination of anthropogenic changes in greenhouse gas, ozone and aerosol concentrations, and natural variations in solar irradiance and volcanic aerosol.

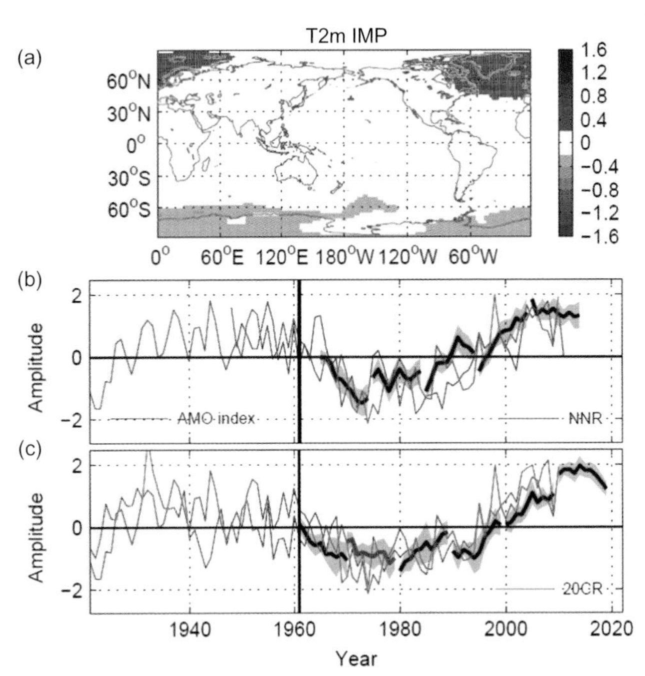

FIGURE 25.7 (a) The spatial structure of internal multidecadal pattern (IMP) for the 2-m temperature in the decadal hindcasts. (b) The ensemble mean (black solid) and spread (gray shading) time series of IMP as a function of forecast lead time for the decadal hindcasts initialized on 1 January every 10 years from 1965 to 2005, the time series for projecting the NCEP–NCAR Reanalysis (NNR) data onto IMP (red solid) and the normalized AMO index (blue solid) from 1920 to 2010. (c) Same as (b) but for hindcasts initialized on 1 January 1961 and every 10 years from 1970 to 2010. The green line denotes the projected time series of the twentieth century reanalysis (20CR) data onto IMP. *From Yang et al. (2013).*

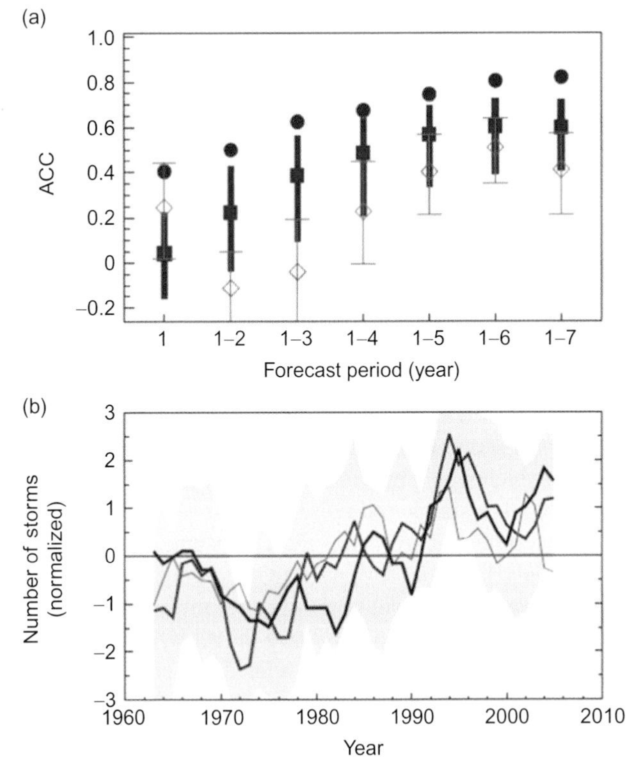

FIGURE 25.8 (a) Anomaly correlation coefficient for predictions of the number of Atlantic tropical storms for increasing forecast periods. Forecast period "1" is the first hurricane season (months 8–13 from November hindcasts) and "1–7" is the average of years 1–7 inclusive. Initialized predictions (red circles) are compared with externally forced ones (blue squares) and persistence (green diamonds), with the blue/green bars indicating the 5–95% confidence interval in which differences in skill from the initialized hindcasts are not significant. (b) The same as for Figure 25.1 but for 5-year rolling means. The blue curve shows the uninitialized hindcasts. *From Smith et al. (2010).*

The origin of the decadal predictability during the twenty-first century may be timescale dependent, because different factors could become important at different lead times (Hawkins and Sutton, 2009). AMOC-related decadal SST variations are strong in the North Atlantic and appear to be predictable out to about a decade. In contrast, the expected anthropogenic weakening of the AMOC in response to enhanced greenhouse gas concentrations and the corresponding North Atlantic SST changes will probably develop over several decades. So, the nature of decadal predictability will critically depend on the lead time. On short lead times of a decade, the internal variability may contribute most to the predictability and the initial conditions strongly matter. On longer lead times of several decades, the AMOC weakening in response to external forcing will give rise to the predictability.

6. SUMMARY AND DISCUSSION

Many coupled general models simulate decadal variability that is consistent with the available observations. Yet, the mechanisms differ strongly from model to model, and the poor observational database does not allow a distinction between "realistic" and "unrealistic" simulations. In the Pacific, decadal variability seems to originate mostly from two types of interactions. First, there is a strong red-noise component, that is, the ocean just integrates the spatially coherent surface heat flux fluctuations associated with the internal (chaotic) atmospheric variability. Global atmospheric teleconnections and local coupled (surface layer) feedbacks can spread regional signals to produce basin-scale decadal variability. This type of decadal variability would not exhibit enhanced predictability relative to persistence. Second, the slow adjustment of the upper ocean circulations to low-frequency wind stress variations, both in the tropics and extratropics, can lead to (damped) decadal modes, which would provide some enhanced predictability compared to a simple red-noise process.

Concerning the Atlantic, there is some consensus in the models that the AMOC is a key player in driving decadal to multidecadal variability in surface quantities. However, different processes force the AMOC in different models. The mechanisms can be related to thermal or haline processes. They can be tropical or extratropical, coupled or uncoupled. Furthermore, although enhanced variability relative to the red background is found in many models, the frequency of the spectral peak varies greatly from model to model, which clearly has implications for predictability. Finally, Southern Ocean decadal to centennial variability is prominent in some models, but not much is known about the mechanisms from observational-based data.

The decadal predictability potential appears to be rather large in the North Atlantic sector. This is consistent with the notion that the AMOC exhibits a large dynamical inertia and possibly even damped eigenmodes at decadal timescales that are excited by the chaotic atmospheric variability. The NAO plays an important role in this respect, as it has the potential to efficiently force AMOC by changing the surface fluxes over the subpolar North Atlantic. We expect that the climate of the next few decades will remain under a strong influence of decadal variability, although the effects of anthropogenic climate change are likely to introduce trends. The NAO index, for instance, exhibited a downward trend from the early 1990s to the present, with a record low value in 2010 (http://www.cgd.ucar.edu/cas/jhurrell/nao.stat.winter.html). Will this negative trend in the NAO-index slow down the AMOC during the next years, as expected from the above discussion? Several impacts of decadal AMOC variations on the atmosphere have been demonstrated in a number of model studies, so that a successful decadal AMOC prediction for the next decade or two would be of potentially large economic benefit. Tropical Atlantic SST, for instance, could cool temporarily if AMOC weakens, which could possibly retard Atlantic hurricane activity. However, unpredictable external forcing through explosive volcanic eruptions and/or anomalous solar radiation originating from internal solar dynamics may offset the internal variations and introduce an additional source of uncertainty.

An attempt should be made to identify key regions for long-term intensive ocean observations that will eventually help to understand the fundamental mechanisms of decadal variability in the real world. Furthermore, key indices should be defined, which can be reconstructed from paleoclimatic data to extend the record backward in time as much as possible. We also need to define and deploy a "suitable" climate-observing system to initialize our climate models for decadal predictions. For the past, not many subsurface ocean observations were available, which hindered initialization and verification of decadal hindcasts. Dunstone and Smith (2010) concluded that the current Argo array of profiling floats provides a good basis for predicting decadal AMOC variations. However, much more research is needed to define what a suitable *global* observing system really means for decadal prediction.

Finally, we need to improve our models. Experience gained from numerical weather and seasonal prediction shows that reduction of systematic model error helps to considerably improve forecast skill. Biases are still large in state-of-the-art climate models. Typical errors in SAT, for instance, can amount up to 10 °C in certain regions in individual models. Hotspots in this respect are, for example, the eastern tropical and subtropical oceans exhibiting a large warm bias, and the North Atlantic and North Pacific generally suffering from a large cold bias. Moreover, oceanic deep convection is often misplaced in climate models. An example is the Labrador Sea convection, a key region for decadal to multidecadal AMOC variability. This model bias can result, for instance, in unrealistic links between the NAO and the AMOC and can lead to mechanisms for AMOC variability that may not operate in the real world.

Likewise significant discrepancies to observations exist concerning the simulation of interannual and decadal variability. Many models still fail to simulate a realistic ENSO, which is also relevant to the simulation of decadal variability. Moreover, several climate models simulate an AMO with too short a period. Thus it cannot be assumed that current climate models are well suited to realistically simulate decadal variability and to realize the full decadal predictability potential. Some models have used flux corrections or anomaly coupling techniques to cope with the biases. However, it is largely unknown how such techniques affect internal variability and its predictability.

A better representation of the physics is an obvious way to improve models. Much higher resolution, horizontal and vertical, in both the ocean and atmosphere models is certainly an important step to improve models, as has been shown in numerous studies. Up until the late 1960s, oceanographers thought of the ocean circulation as consisting of slowly moving interior gyres and fast-moving boundary currents. However, the ocean circulation also contains vigorous eddies at spatial scales from roughly 100 km and smaller, evolving over timescales from weeks to months. These eddies are important in establishing the ocean's large-scale circulation, they determine the ocean circulation response to external forcing and tracer properties. Most ocean components of climate models still parameterize eddies. The explicit simulation of ocean eddies and the resolution of other small-scale structures such as fronts or vertically propagating internal waves requires a significant increase in the computing capacity available to the world's weather and climate centers in order to accelerate progress in improving models and eventually predictions. The World Modeling Summit for Climate Prediction in 2008 (WCRP, 2009) recommended computing systems dedicated to climate research at least a thousand times more powerful than those currently available.

Although many problems remain, both with respect to the observing system and the climate models, decadal climate forecasting is worth pursuing, as discussed earlier, and presently conducted at about a dozen institutions worldwide. An example is shown in Figure 25.9 depicting a forecast of global average SAT until 2021 issued in 2011 by the British Meteorological Office (Hadley Centre). The predictions account for natural variability and anthropogenic climate change, the latter by assuming a scenario for future greenhouse and aerosol concentrations. Model verification is based on a posteriori predictions. These retrospective forecasts, also called hindcasts, have been made from numerous dates in the past. Some of these are shown in

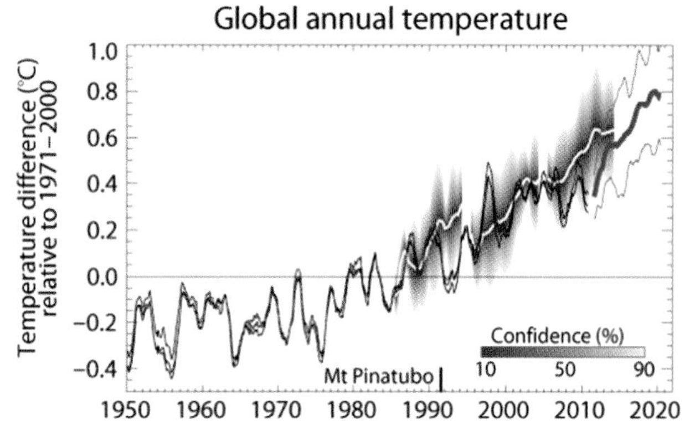

FIGURE 25.9 Observed (black, from Hadley Centre, GISS and NCDC) and predicted global average annual surface temperature difference relative to 1971–2000. Previous predictions starting from June 1985, 1995, and 2005 are shown as white curves, with red shading representing their probable range, such that the observations are expected to lie within the shading 90% of the time. The most recent forecast (thick blue curve with thin blue curves showing range) starts from September 2011. All data are rolling annual mean values. The gap between the black and blue curves arises because the last observed value represents the period November 2010–October 2011, whereas the first forecast period is September 2011–August 2012. *From http://www.metoffice.gov.uk/research/climate/seasonal-to-decadal/long-range/decadal-fc.*

Figure 25.9 (white curves and red uncertainty regions from 1985, 1995, and 2005).

Generally, the forecasts predict rises in temperatures similar to those observed (black curve). Many facets of the variability, such as the record warming caused by the large 1997–1998 El Niño event and the cooling caused by the 2008 La Niña event, are within the range of the predictions (red shading), which were made in ensemble mode starting with different initial states. An exception is the short cool spell after the large volcanic eruption of Mount Pinatubo in 1991, which could not have been forecast years in advance. Furthermore, the global average surface temperature did not warm further after 1998, in contrast to the hindcasts that generally depict a warming trend, especially since the mid-2000s. A likely candidate for this decade-long "halt" is the negative phase of the PDO during that time, which switched sign in 2005 and is known to exert a cooling effect on global average SAT. Another climate model predicted this decadal change in the Pacific with more success (Keenlyside et al., 2008).

Global average temperature is expected to rise to between 0.36 ° and 0.72 °C (90% confidence range) above the long-term (1971–2000) average during the period 2012–2016, with values most likely to be about 0.54 °C higher than average (see thick blue curve in Figure 25.8). From 2017 to 2021, global average temperature is forecast to rise further to between 0.54 ° and 0.97 °C, with most likely values of about 0.76 °C above average. The warmest year in the 160-year Met Office Hadley Centre global temperature record is 1998, with a temperature of 0.40 °C above the long-term average. 2009 had a temperature of 0.32 °C above average. The forecast trend of further global warming is largely driven by increasing levels of greenhouse gases. The large forecast spread (thin blue lines) reflects the internal variability. Decadal forecasts are still experimental, so at this early stage of development skill levels vary from place to place, for different variables, and from model to model. As a result, the decadal forecast shown in Figure 25.8 should be taken with caution and expert advice is needed to assess the reliability of such predictions.

ACKNOWLEDGMENTS

The author would like to thank the four anonymous referees for their helpful comments and Dr. Wonsun Park for providing some of the figures. The author's work was supported by the "Nordatlantik" project of BMBF and the European Union's "THOR" project.

REFERENCES

Alexander, M., Penland, C., 1996. Variability in a mixed layer ocean model driven by stochastic atmospheric forcing. J. Clim. 9, 2424.

Alexander, M.A., Deser, C., Timlin, M.S., 1999. The re-emergence of SST anomalies in the North Pacific Ocean. J. Clim. 12, 2419–2431.

Alvarez-Garcia, F., Latif, M., Biastoch, A., 2008. On multidecadal and quasi-decadal North Atlantic variability. J. Clim. 21, 3433–3452.

Bader, J., Latif, M., 2003. North Atlantic Oscillation response to anomalous Indian Ocean SST in a coupled GCM. J. Clim. 18, 5382–5389.

Barsugli, J.J., Battisti, D.S., 1998. The basic effects of atmosphere-ocean thermal coupling on midlatitude variability. J. Atmos. Sci. 55, 477–493.

Beaugrand, G., 2004. The North Sea regime shift: evidence, causes, mechanisms and consequences. Prog. Oceanogr. 60, 245–262.

Bengtsson, L., Semenov, V.A., Johannessen, O.M., 2004. The early twentieth century warming in the Arctic—a possible mechanism. J. Clim. 17, 4045–4057.

Biondi, F., Gershunov, A., Cayan, D., 2001. North Pacific decadal climate variability since AD 1661. J. Clim. 14, 5–10.

Bjerknes, J., 1964. Atlantic air-sea interaction. Advances in Geophysics, 10 Academic Press, London, 1–82.

Boer, G.J., Lambert, S.J., 2008. Multi-model decadal potential predictability of precipitation and temperature. Geophys. Res. Lett. 35, http://dx.doi.org/10.1029/2008GL033234, L05706.

Böning, C.W., Dispert, A., Visbeck, M., Rintoul, S., Schwarzkopf, F.U., 2008. The response of the Antarctic circumpolar current to recent climate change. Nat. Geosci. 1, 864–869. http://dx.doi.org/10.1038/ngeo362.

Booth, B.B., Dunstone, N.J., Halloran, P.R., Andrews, T., Bellouin, N., 2012. Aerosols implicated as a prime driver of twentieth-century North Atlantic climate variability. Nature 484, 228–232. http://dx.doi.org/10.1038/nature10946.

Branstator, G., Teng, H., Meehl, G.A., Kimoto, M., Knight, J.R., Latif, M., Rosati, A., 2012. Systematic estimates of initial value decadal predictability for six AOGCMs. J. Clim. 25, 1827–1846.

Ceballos, L.I., Di Lorenzo, E., Hoyos, C.D., Schneider, N., Taguchi, B., 2009. North Pacific gyre oscillation synchronizes climate fluctuations in the eastern and western boundary systems. J. Clim. 22, 5163–5174.

Chang, P., Ji, L., Li, H., 1997. A decadal climate variation in the tropical Atlantic Ocean from thermodynamic air-sea interactions. Nature 385, 516–518.

Collins, M., Sinha, B., 2003. Predictable decadal variations in the thermohaline circulation and climate. Geophys. Res. Lett. 30, 1306. http://dx.doi.org/10.1029/2002GLO16504.

Collins, M., 2002. Climate predictability on interannual to decadal time scales: The initial value problem. Clim. Dyn. 19, 671–692.

Collins, M., Botzet, M., Carril, A., Drange, H., Jouzeau, A., Latif, M., Ottera, O.H., Masina, S., Pohlmann, H., Sorteberg, A., Sutton, R., Terray, L., 2006. Inter-annual to decadal climate predictability in the North Atlantic: a multimodel-ensemble study. J. Clim. 19, 1195–1203.

Cook, E.R., Buckley, B.M., D'Arrigo, R.D., Peterson, M.J., 2000. Warm-season temperatures since 1600 BC reconstructed from Tasmanian tree rings and their relationship to large-scale sea surface temperature anomalies. Clim. Dyn. 16 (2), 79–91. http://dx.doi.org/10.1007/s003820050006.

Curry, R., Mauritzen, C., 2005. Dilution of the northern North Atlantic in recent decades. Science 308, 1772–1774.

Curry, R.G., McCartney, M.S., Joyce, T.M., 1998. Oceanic transport of subpolar climate signals to mid-depth subtropical waters. Nature 391, 575–577.

Czaja, A., Marshall, J., 2001. Observations of atmosphere-ocean coupling in the North Atlantic. Q. J. R. Meteorol. Soc. 127, 1893–1916.

d'Orgeville, M., Peltier, W.R., 2007. On the Pacific decadal oscillation and the Atlantic multidecadal oscillation: might they be related? Geophys. Res. Lett. 34, L23705. http://dx.doi.org/10.1029/2007GL031584.

Davis, R.E., 1976. Predictability of sea surface temperature and sea level pressure anomalies over the North Pacific Ocean. J. Phys. Oceanogr. 6, 249–266.

Delworth, T.L., Greatbatch, R.J., 2000. Multidecadal thermohaline circulation variability driven by atmospheric surface flux forcing. J. Clim. 13, 1481–1495.

Delworth, T.L., Mann, M.E., 2000. Observed and simulated multidecadal variability in the Northern Hemisphere. Clim. Dyn. 16, 661–676. http://dx.doi.org/10.1007/s003820000075.

Delworth, T.L., Zeng, F., 2012. Multicentennial variability of the Atlantic meridional overturning circulation and its climatic influence in a 4000 year simulation of the GFDL CM2.1 climate model. Geophys. Res. Lett. 39, L13702. http://dx.doi.org/10.1029/2012GL052107.

Delworth, T., Manabe, S., Stouffer, R.J., 1993. Interdecadal variations of the thermohaline circulation in a coupled ocean-atmosphere model. J. Clim. 6, 1993–2011.

Deser, C., Blackmon, M.L., 1993. Surface climate variations over the North Atlantic Ocean during winter: 1900–1989. J. Clim. 6, 1743–1753.

Deser, C., Walsh, J., Timlin, M., 2000. Arctic sea ice variability in the context of recent atmospheric circulation trends. J. Clim. 13, 617–633.

Dijkstra, H.A., Neelin, J.D., 1999. Imperfections of the thermohaline circulation: multiple equilibria and flux correction. J. Clim. 12, 1382–1392.

Dommenget, D., Latif, M., 2008. Generation of hyper climate mode. Geophys. Res. Lett. 35, L02706. http://dx.doi.org/10.1029/2007GL031087.

Drinkwater, K.F., 2006. The regime shift of the 1920s and 1930s in the North Atlantic. Prog. Oceanogr. 68, 134–151.

Dunstone, N.J., Smith, D.M., 2010. Impact of atmosphere and sub-surface ocean data on decadal climate prediction. Geophys. Res. Lett. 37, L02709. http://dx.doi.org/10.1029/2009GL041609.

Eckert, C., Latif, M., 1997. Predictability limits of ENSO: the role of stochastic forcing. J. Clim. 10, 1488–1504.

Eden, C., Greatbatch, R.J., 2003. A damped decadal oscillation in the North Atlantic Ocean climate system. J. Clim. 16, 4043–4060.

Eden, C., Jung, T., 2001. North Atlantic interdecadal variability: oceanic response to the North Atlantic oscillation (1865-1997). J. Clim. 14, 676–691.

Eden, C., Willebrand, J., 2001. Mechanism of inter-annual to decadal variability of the North Atlantic circulation. J. Clim. 14, 2266–2280.

Farneti, R., Gent, P.R., 2011. The effects of the eddy-induced advection coefficient in a coarse-resolution coupled climate model. Ocean Model. 39, 135–145.

Folland, C.K., Palmer, T.N., Parker, D.E., 1986. Sahel rainfall and worldwide sea temperatures, 1901-85. Nature 320, 602–607.

Frankignoul, C., Hasselmann, K., 1977. Stochastic climate models. Part II: application to sea surface temperature anomalies and thermocline variability. Tellus 29, 284–305.

Frankignoul, C., Müller, P., Zorita, E., 1997. A simple model of the decadal response of the ocean to stochastic wind forcing. J. Phys. Oceanogr. 27, 1533–1546.

Gershunov, A., Barnett, T., 1998. Inter-decadal modulation of ENSO teleconnections. Bull. Am. Meteorol. Soc. 79, 2715–2725.

Giannini, A., Saravanan, R., Chang, P., 2003. Oceanic forcing of Sahel rainfall on inter-annual to interdecadal timescales. Science 302, 1027–1030.

Goldenberg, S.B., Landsea, C.W., Mestas-Nuñez, A.M., Gray, W.M., 2001. The recent increase in Atlantic hurricane activity: causes and implications. Science 293, 474–479.

Greatbatch, R.J., Zhang, S., 1995. An interdecadal oscillation in an idealized ocean basin forced by constant heat flux. J. Clim. 8, 81–91.

Griffies, S.M., Bryan, K., 1997a. Predictability of North Atlantic multidecadal climate variability. Science 275, 181–184.

Griffies, S.M., Bryan, K., 1997b. A predictability of simulated North Atlantic multidecadal variability. Clim. Dyn. 13, 459–487.

Griffies, S.M., Tziperman, E., 1995. A linear thermohaline oscillator driven by stochastic atmospheric forcing. J. Clim. 8, 2440–2453.

Grötzner, A., Latif, M., Timmermann, A., Voss, R., 1999. Interannual to decadal predictability in a coupled ocean-atmosphere general circulation model. J. Clim. 12, 2607–2624.

Gu, D., Philander, S.G.H., 1997. Interdecadal climate fluctuations that depend on exchanges between the tropics and extratropics. Science 275, 805–807. http://dx.doi.org/10.1126/science.275.5301.805.

Gulev, S.K., Latif, M., Keenlyside, N.S., Koltermann, K.P., 2013. North Atlantic Ocean control on surface heat flux at multidecadal timescales. Nature 499, 464–467.

Häkkinen, S., Rhines, P.B., 2004. Decline of subpolar North Atlantic circulation during the 1990s. Science 304, 555–559.

Hall, A., Manabe, S., 1997. Can local, linear stochastic theory explain sea surface temperature and salinity variability? Clim. Dyn. 13, 167–180.

Hall, A., Visbeck, M., 2002. Synchronous variability in the southern hemisphere atmosphere, sea ice, and ocean resulting from the annular mode. J. Clim. 13, 3043–3057.

Hare, S.R., Mantua, N.J., 2000. Empirical evidence for North Pacific regime shifts in 1977 and 1989. Prog. Oceanogr. 47 (2–4), 103–145.

Hasselmann, K., 1976. Stochastic climate models. Part I: theory. Tellus 28, 473–485.

Hawkins, E., Sutton, R., 2009. The potential to narrow uncertainty in regional climate predictions. Bull. Am. Meteorol. Soc. 90, 1095–1107.

Hawkins, E., Robson, J., Sutton, R., Smith, D., Keenlyside, N.S., 2011. Evaluating the potential for statistical decadal predictions of sea surface temperatures with a perfect model approach. Clim. Dyn. 37, 2495–2509.

Hegerl, G.C., Zwiers, F.W., 2011. Use of models in detection and attribution of climate change. WIREs Clim. Change 2, 570–591.

Hegerl, G.C., Hasselmann, K., Cubasch, U., Mitchell, J.F.B., Roeckner, E., Voss, R., Waszkewitz, J., 1997. Multi-fingerprint detection and attribution analysis of greenhouse gas, greenhouse gas-plus-aerosol and solar forced climate change. Clim. Dyn. 13, 613–634.

Hurrell, J.W., 1995. Decadal trends in the North Atlantic oscillation: regional temperatures and precipitation. Science 269, 676–679.

Hurrell, J.W., Kushnir, Y., Visbeck, M., Ottersen, G., 2003. An overview of the North Atlantic oscillation. In: Hurrell, J.W., Kushnir, Y., Ottersen, G., Visbeck, M. (Eds.), The North Atlantic Oscillation: Climate Significance and Environmental Impact, Geophysical Monograph Series, vol. 134. pp. 1–35.

IPCC, 2007. Climate change 2007: the physical science basis. In: Solomon, S. et al., (Eds.), Contribution of Working Group I to the Fourth Assessment Report of the Intergovernmental Panel on Climate Change. Cambridge University Press, Cambridge, United Kingdom and New York, NY, USA.

Jacobs, G.A., Hurlburt, H.E., Kindle, J.C., Metzger, E.J., Mitchell, J.L., Teague, W.J., Wallcraft, A.J., 1994. Decade-scale trans-Pacific propagation and warming effects of an El Niño anomaly. Nature 370, 360–363.

Jin, F.-F., Neelin, J.D., Ghil, M., 1994. El Niño on the devil's staircase: annual subharmonic steps to chaos. Science 264, 70–72.

Jungclaus, J.H., Haak, H., Latif, M., Mikolajewicz, U., 2005. Arctic–North Atlantic interactions and multidecadal variability of the meridional overturning circulation. J. Clim. 18, 4013–4031.

Keenlyside, N.S., Ba, J., 2010. Prospects for decadal climate prediction. Wiley Interdiscip. Rev. Clim. Chang. 1, 627–635. http://dx.doi.org/10.1002/wcc.69.

Keenlyside, N.S., Latif, M., Jungclaus, J., Kornblueh, L., Roeckner, E., 2008. Advancing decadal-scale climate prediction in the North Atlantic sector. Nature 453, 84–88. http://dx.doi.org/10.1038/nature06921.

Knight, J.R., Allan, R.J., Folland, C.K., Vellinga, M., Mann, M.E., 2005. A signature of persistent natural thermohaline circulation cycles in observed climate. Geophys. Res. Lett. 32, L20708. http://dx.doi.org/10.1029/2005GL024233.

Kushnir, Y., Seager, R., Miller, J., Chiang, J.C.H., 2002. A simple coupled model of tropical Atlantic decadal climate variability. Geophys. Res. Lett. 29, http://dx.doi.org/10.1029/2002GL015874.

Kwon, Y.-O., Deser, C., Cassou, C., 2010. Coupled atmosphere-mixed layer ocean response to ocean heat flux convergence along the Kuroshio Current Extension. Clim. Dyn. http://dx.doi.org/10.1007/s00382-010-0764-8.

Latif, M., Barnett, T.P., 1994. Causes of decadal climate variability over the North Pacific and North America. Science 266, 634–637.

Latif, M., Keenlyside, N.S., 2011. A perspective on decadal climate variability and predictability. Deep Sea Res. 58, 1880–1894. http://dx.doi.org/10.1016/j.dsr2.2010.10.066.

Latif, M., Kleeman, R., Eckert, C., 1997. Greenhouse warming, decadal variability, or El Niño: an attempt to understand the anomalous 1990's. J. Clim. 10, 2221–2239.

Latif, M., Roeckner, E., Botzet, M., Esch, M., Haak, H., Hagemann, S., Jungclaus, J., Legutke, S., Marsland, S., Mikolajewicz, U., Mitchell, J., 2004. Reconstructing, monitoring, and predicting multidecadal-scale changes in the North Atlantic thermohaline circulation with sea surface temperature. J. Clim. 17, 1605–1614.

Latif, M., Collins, M., Pohlmann, H., Keenlyside, N., 2006a. A review of predictability studies of the Atlantic sector climate on decadal time scales. J. Clim. 19, 5971–5987.

Latif, M., Böning, C., Willebrand, J., Biastoch, A., Dengg, J., Keenlyside, N., Schweckendiek, U., Madec, G., 2006b. Is the thermohaline circulation changing? J. Clim. 19, 4631–4637.

Latif, M., Martin, T., Park, W., 2013. Southern Ocean sector centennial climate variability: dynamics and implications for recent decadal trends. J. Clim. http://dx.doi.org/10.1175/JCLI-D-12-00281.1.

Lee, T., McPhaden, M.J., 2008. Decadal phase change in large-scale sea level and winds in the Indo-Pacific region at the end of the 20th century. Geophys. Res. Lett. 35, L01605. http://dx.doi.org/10.1029/2007GL032419.

Lee, T., McPhaden, M.J., 2010. Increasing intensity of El Niño in the central-equatorial Pacific. Geophy. Res. Lett. 37, L14603. http://dx.doi.org/10.1029/2010GL044007.

Liu, Z., Alexander, M., 2007. Atmospheric bridge, oceanic tunnel, and global climatic teleconnections. Rev. Geophys. 45, RG2005. http://dx.doi.org/10.1029/2005RG000172.

Lohmann, K., Latif, M., 2005. Tropical Pacific decadal variability and the subtropical–tropical cells. J. Clim. 18, 5163–5178.

Lozier, M.S., Leadbetter, S.J., Williams, R.G., Roussenov, V., Reed, M.S.C., Moore, N.J., 2008. The spatial pattern and mechanisms of heat content change in the North Atlantic. Science 319, 800–803.

Lu, J., Delworth, T., 2005. Oceanic forcing of the late 20th century Sahel drought. Geophys. Res. Lett. 32, L22706. http://dx.doi.org/10.1029/2005GL023316.

Mantua, N.J., Hare, S.R., Zhang, Y., Wallace, J.M., Francis, R.C., 1997. A Pacific decadal climate oscillation with impacts on salmon. Bull. Am. Meteorol. Soc. 78, 1069–1079.

Martin, T., Park, W., Latif, M., 2012. Multi-centennial variability controlled by Southern ocean convection. Clim. Dyn. 40, 2005–2022.

McPhaden, M.J., Zhang, D., 2002. Slowdown of the meridional overturning circulation in the upper Pacific Ocean. Nature 415, 603–608.

McPhaden, M.J., Lee, T., McClurg, D., 2011. El Niño and its relationship to changing background conditions in the tropical Pacific Ocean. Geophys. Res. Lett. 38, L15709. http://dx.doi.org/10.1029/2011GL048275.

Meehl, G.A., Arblaster, J.M., Matthes, K., Sassi, F., van Loon, H., 2009. Amplifying the Pacific climate system response to a small 11-year solar cycle forcing. Science 2009 (325), 1114–1118.

Menary, M.B., Park, W., Lohmann, K., Vellinga, M., Palmer, M., Latif, M., Jungclaus, J., 2012. A multimodel comparison of centennial Atlantic meridional overturning circulation variability. Clim. Dyn. 38, 2377–2388. http://dx.doi.org/10.1007/s00382-011-1172-4.

Meredith, M.P., Naveira Garabato, A.C., Hogg, A.McC., Farneti, R., 2012. Sensitivity of the overturning circulation in the Southern Ocean to decadal changes in wind forcing. J. Clim. 25, 99–110. http://dx.doi.org/10.1175/2011JCLI4204.1.

Merryfield, W.J., Boer, G.J., 2005. Variability of upper Pacific Ocean overturning in a coupled climate model. J. Clim. 18, 666–683.

Mikolajewicz, U., Maier-Reimer, E., 1990. Internal secular variability in an ocean general circulation model. Clim. Dyn. 4, 145–156.

Moura, A.D., Shukla, J., 1981. On the dynamics of droughts in northeast Brazil: observation, theory, and numerical experiments with a general circulation model. J. Atmos. Sci. 38, 2653–2675.

Neelin, J.D., Battisti, D.S., Hirst, A.C., Jin, F.-F., Wakata, Y., Yamagata, T., Zebiak, S.E., 1998. ENSO theory. J. Geophys. Res. 103 (C7), 14261–14290. http://dx.doi.org/10.1029/97JC03424.

Newman, M., 2007. Interannual to decadal predictability of tropical and North Pacific sea surface temperatures. J. Clim. 20, 2333–2356.

Ottera, O.H., Bentsen, M., Drange, H., Suo, L., 2010. External forcing as a metronome for Atlantic multidecadal variability. Nat. Geosci. 3, 688–694. http://dx.doi.org/10.1038/ngeo955.

Park, W., Latif, M., 2005. Ocean dynamics and the nature of air–sea interactions over the North Atlantic at decadal time scales. J. Clim. 18, 982–995.

Park, W., Latif, M., 2008. Multidecadal and multicentennial variability of the meridional overturning circulation. Geophys. Res. Lett. 35, L22703. http://dx.doi.org/10.1029/2008GL035779.

Park, W., Latif, M., 2010. Pacific and Atlantic multidecadal variability in the Kiel climate model. Geophys. Res. Lett. 37, L24702. http://dx.doi.org/10.1029/2010GL045560.

Park, W., Latif, M., 2011. Atlantic meridional overturning circulation response to idealized external forcing. Clim. Dyn. 39, 1709–1726. http://dx.doi.org/10.1007/s00382-011-1212-0.

Pierce, D.W., Barnett, T.P., Mikolajewicz, U., 1995. Competing roles of heat and fresh water fluxes in forcing thermohaline oscillations. J. Phys. Oceanogr. 25, 2046–2064.

Pierce, D.W., Barnett, T.P., Latif, M., 2000. Connections between the Pacific Ocean tropics and midlatitudes on decadal timescales. J. Clim. 13, 1173–1194.

Pohlmann, H., Botzet, M., Latif, M., Roesch, A., Wild, M., Tschuck, P., 2004. Estimating the decadal predictability of a coupled AOGCM. J. Clim. 17, 4463–4472.

Rintoul, S.R., Balmeseda, M., Cunningham, S., Dushaw, B., Garzoli, S., Gordon, A., Heimbach, P., Hood, M., Johnson, G., Latif, M., Send, U., Shum, C., Speich, S., Stammer, D., 2011. Deep circulation and meridional overturning: recent progress and a strategy for sustained observations. In: Hall, J., Harrison, D.E., Stammer, D. (Eds.), Proceedings of the "OceanObs'09: Sustained Ocean Observations and Information for Society" Conference, vol. 1. ESA Publication WPP-306, Venice, Italy, 21–25 September 2009.

Rodgers, K.B., Friederichs, P., Latif, M., 2004. Tropical Pacific decadal variability and its relation to decadal modulations of ENSO. J. Clim. 17, 3761–3774.

Rotstayn, L.D., Lohmann, U., 2002. Tropical rainfall trends and the indirect aerosol effect. J. Clim. 15, 2103–2116.

Rowell, D.P., Folland, C.K., Maskell, K., Ward, M., 1995. Variability of summer rainfall over tropical North Africa (1906–92): observations and modelling. Q. J. R. Meteorol. Soc. 121, 669–704.

Ruiz-Barradas, A., Carton, J.A., Nigam, S., 2000. Structure of interannual-to-decadal climate variability in the tropical Atlantic sector. J. Clim. 13, 3285–3297.

Sallee, J.-B., Rintoul, S.R., 2011. Parameterization of eddy-induced subduction in the Southern Ocean surface-layer. Ocean Model. 39, 146–153.

Saravanan, R., McWilliams, J.C., 1997. Stochasticity and spatial resonance in interdecadal climate fluctuations. J. Clim. 10, 2299–2320.

Saravanan, R., McWilliams, J.C., 1998. Advective ocean–atmosphere interaction: an analytical stochastic model with implications for decadal variability. J. Clim. 11, 165–188.

Schneider, N., Cornuelle, B.D., 2005. The forcing of the Pacific decadal oscillation. J. Clim. 18, 4355–4373. http://dx.doi.org/10.1175/JCLI3527.1.

Schneider, N., Miller, A.J., Pierce, D.W., 2002. Anatomy of North Pacific decadal variability. J. Clim. 15, 586–605.

Schubert, S.D., Suarez, M.J., Pegion, P.J., Koster, R.D., Bacmeister, J.T., 2004. Causes of long-term drought in the United States Great Plains. J. Clim. 17, 485–503.

Seager, R., 2007. The turn of the Century drought across North America: global context, dynamics and past analogues. J. Clim. 20, 5527–5552.

Seager, R., Kushnir, Y., Herweijer, C., Naik, N., Velez, J., 2005. Modeling of tropical forcing of persistent droughts and pluvials over western North America: 1856–2000. J. Clim. 18 (19), 4065–4088.

Semenov, V.A., Latif, M., Dommenget, D., Keenlyside, N.S., Strehz, A., Martin, T., Park, W., 2010. The impact of North Atlantic–Arctic multidecadal variability on Northern Hemisphere surface air temperature. J. Clim. 23, 5668–5677. http://dx.doi.org/10.1175/2010JCLI3347.1.

Shaffrey, L., Sutton, R., 2006. Bjerknes compensation and the decadal variability of the energy transports in a coupled climate model. J. Clim. 19, 1167–1181.

Smith, D.M., Cusack, S., Colman, A.W., Folland, C.K., Harris, G.R., Murphy, J.M., 2007. Improved surface temperature prediction for the coming decade from a global climate model. Science 317, 796–799.

Smith, D.M., Eade, R., Dunstone, N.J., Fereday, D., Murphy, J.M., Pohlmann, H., Scaife, A.A., 2010. Skilful multi-year predictions of Atlantic hurricane frequency. Nat. Geosci. 3, 846–849.

Stott, P.A., Tett, S.F.B., Jones, G.S., Allen, M.R., Mitchell, J.F.B., Jenkins, G.J., 2000. External control of 20th century temperature by natural and anthropogenic forcings. Science 290, 2133–2137.

Stott, P.A., Gillett, N.P., Hegerl, G.C., Karoly, D.J., Stone, D.A., Zhang, X., Zwiers, F., 2010. Detection and attribution of climate change: a regional perspective. WIREs Clim. Change 1, 192–211. http://dx.doi.org/10.1002/wcc.34.

Sutton, R.T., Allen, M.R., 1997. Decadal predictability of North Atlantic sea surface temperature and climate. Nature 388, 563–567.

Sutton, R.T., Hodson, D.L.R., 2005. Atlantic Ocean forcing of North American and European summer climate. Science 309, 115–118.

Taguchi, B., Xie, S.-P., Schneider, N., Nonaka, M., Sasaki, H., Sasai, Y., 2007. Decadal variability of the Kuroshio Extension: observations and an eddy-resolving model hindcast. J. Clim. 20 (11), 2357–2377.

Teng, H., Branstator, G., 2011. Initial-value predictability of prominent modes of North Pacific subsurface temperature in a CGCM. Clim. Dyn. 36, 1813–1834. http://dx.doi.org/10.1007/s00382-010-0749-7.

Timmermann, A., Latif, M., Voss, R., Grötzner, A., 1998. Northern Hemisphere interdecadal variability: a coupled air-sea mode. J. Clim. 11, 1906–1931.

Ting, M.F., Kushnir, Y., Seager, R., Li, C.H., 2009. Forced and internal twentieth-century SST trends in the North Atlantic. J. Clim. 22, 1469–1481. http://dx.doi.org/10.1175/2008jcli2561.1.

Trenberth, K.E., Hurrell, J.W., 1994. Decadal atmosphere–ocean variations in the Pacific. Clim. Dyn. 9, 303–319.

Trenberth, K.E., Shea, D.J., 2006. Atlantic hurricanes and natural variability in 2005. Geophys. Res. Lett. 33, L12704. http://dx.doi.org/10.1029/2006GL026894.

Vellinga, M., Wu, P., 2004. Low-latitude freshwater influence on centennial variability of the Atlantic thermohaline circulation. J. Clim. 17, 4498–4511.

Venegas, S.A., Mysak, L.A., 2000. Is there a dominant timescale of natural climate variability in the Arctic? J. Clim. 13, 3412–3434.

Visbeck, M., Cullen, H., Krahmann, G., Naik, N., 1998. An ocean model's response to North Atlantic Oscillation-like wind forcing. Geophys. Res. Lett. 25, 4521–4525.

WCRP, 2009. Workshop report: world modelling summit for climate prediction, reading, UK, 6-9 May 2008. WCRP No. 13, WMO/TD No. 1468.

Weaver, A.J., Sarachik, E.S., 1991. The role of mixed boundary conditions in numerical models of the ocean's climate. J. Phys. Oceanogr. 21, 1470–1493.

Weisse, R., Mikolajewicz, U., Sterl, A., Drijfhout, S.S., 1999. Stochastically forced variability in the Antarctic Circumpolar Current. J. Geophys. Res. 104 (C5), 11049–11064.

White, W.B., Peterson, R., 1996. An Antarctic circumpolar wave in surface pressure, wind, temperature, and sea ice extent. Nature 380, 699–702.

Winton, M., 1997. The damping effect of bottom topography on internal decadal-scale oscillations of the thermohaline circulation. J. Phys. Oceanogr. 27, 203–208.

Wu, L., Liu, Z., Gallimore, R., Jacob, R., Lee, D., Zhong, Y., 2003. A coupled modeling study of Pacific decadal variability: the Tropical mode and the North Pacific mode. J. Clim. 16, 1101–1120.

Xie, S.-P., 1999. A dynamic ocean–atmosphere model of the tropical Atlantic decadal variability. J. Clim. 12, 64–70.

Yang, X., Rosati, A., Zhang, S., Deworth, T.L., Gudgel, R.G., Zhang, R., Vecchi, G., Anderson, W., Chang, Y.-S., DelSole, T., Dixon, K., Msadek, R., Stern, W.F., Wittenberg, A., Zheng, F., 2013. A predictable AMO-like pattern in the GFDL fully coupled ensemble initialization and decadal forecasting system. J. Clim. 26, 650–661. http://dx.doi.org/10.1175/JCLI-D-12-00231.1.

Yeh, S.-W., Kug, J.-S., Dewitte, B., Kwon, M.-H., Kirtman, B.P., Jin, F.-F., 2009. El Niño in a changing climate. Nature 461, 511–514.

Yeh, S.-W., Kirtman, B.P., Kug, J.-S., Park, W., Latif, M., 2011. Natural variability of the central Pacific El Niño event on multi-centennial timescales. Geophys. Res. Lett. 38, L02704. http://dx.doi.org/10.1029/2010GL045886.

Yin, F.L., Sarachik, E.S., 1995. Interdecadal thermohaline oscillations in a sector ocean general circulation model: advective and convective processes. J. Phys. Oceanogr. 25, 2465–2484.

Zanchettin, D., Rubino, A., Jungclaus, J.H., 2010. Intermittent multidecadal to centennial fluctuations dominate global temperature evolution over the last millennium. Geophys. Res. Lett. 37, L14702. http://dx.doi.org/10.1029/2010GL043717.

Zhang, R., Delworth, T.L., 2006. Impact of Atlantic multidecadal oscillations on India/Sahel rainfall and Atlantic hurricanes. Geophys. Res. Lett. 33, L17712. http://dx.doi.org/10.1029/2006GL026267.

Zhang, D., McPhaden, M.J., 2006. Decadal variability of the shallow Pacific meridional overturning circulation: relation to tropical sea surface temperatures in observations and climate change models. Ocean Model. 15, 250–273. http://dx.doi.org/10.1016/j.ocemod.2005.12.005.

Zhang, Y., Wallace, J.M., Battisti, D.S., 1997. ENSO-like interdecadal variability: 1900–93. J. Clim. 10, 1004–1020.

Zhang, R., Delworth, T.L., Held, I.M., 2007. Can the Atlantic Ocean drive the observed multidecadal variability in Northern Hemisphere mean temperature? Geophys. Res. Lett. 34, L02709. http://dx.doi.org/10.1029/2006GL028683.

Zhou, J., Tung, K.-K., 2010. Solar cycles in 150 years of global sea surface temperature data. J. Clim. 23, 3234–3324.

Chapter 26

Modeling Ocean Biogeochemical Processes and the Resulting Tracer Distributions

Christoph Heinze*,†,‡ and Marion Gehlen§

*Geophysical Institute, University of Bergen, Bergen, Norway

†Bjerknes Centre for Climate Research, Bergen, Norway

‡Uni Klima, Uni Research, Bergen, Norway

§Laboratoire des Sciences du Climat et de l'Environnement (LSCE), UMR, CEA-CNRS, Gif-sur-Yvette, France

Chapter Outline

1. GOALS OF OCEAN BIOGEOCHEMICAL MODELING WITHIN CLIMATE RESEARCH

Geophysical fluid dynamics and physical climate sciences have made key contributions to the understanding of ocean circulation and its role in climate. The last 20 years have brought increasing evidence for the contribution of biogeochemical processes in shaping the mean state of the earth's climate and driving its variability.

What is biogeochemistry? It is the science of the processes that govern the chemical structure and composition of the natural environment in the earth's surface reservoirs and the transport of matter between them. These reservoirs are the biosphere, the hydrosphere (oceans, lakes, rivers, groundwater), the atmosphere, the pedosphere (soils), the cryosphere (ice on land and on the ocean), and the surface of the lithosphere (sediments, different rock types). Biogeochemistry deals in particular with the cycles of chemical

Ocean Circulation and Climate, Vol. 103. http://dx.doi.org/10.1016/B978-0-12-391851-2.00026-X

elements and related transports of matter as well as energy that involve interaction with biota, such as the cycles of carbon, nitrogen, and phosphorus. Biogeochemistry is a part of the more comprehensive discipline of geochemistry, which also deals with other planets. A more elaborate, but nevertheless concise, introduction to what biogeochemistry is can be found in Schlesinger (2004).

Climate science and, in particular, climate modeling, have started to include biogeochemical processes in order to understand the climate system, to quantify its variability, and to provide realistic future scenarios under given boundary conditions for natural and human-induced forcing. There are two major reasons for combining climate physics and biogeochemistry under the more general term *Earth system science* (Bretherton, 1985; NASA Advisory Council, 1986):

- Ocean circulation and climate physics control, in combination with the carbon cycle and related biogeochemical cycles, the radiation budget of the atmosphere through greenhouse gases (Denman et al., 2007).
- Biogeochemical tracers provide key information about the physical climate system, and especially ocean circulation, which cannot be provided by the physical state variables *per se* (Broecker and Peng, 1982; England and Maier-Reimer, 2001).

Advanced coupled climate models (Earth system models) include a biogeochemical ocean module as one component. While these marine biogeochemical components are of varying complexity, they always depend on the input of variables from a dynamical physical ocean model, in particular, the three-dimensional velocity field (vector **v**), subgrid scale stirring and mixing processes (see Chapters 7 and 8), seawater temperature and salinity, sea ice cover (concentration, thickness), and solar radiation. The ocean biogeochemical model interacts with the physical model components of the Earth system, namely atmosphere, cryosphere, and ocean circulation, through the exchange of radiatively active gases, for example, carbon dioxide—CO_2—and aerosols (especially through biogenic dimethylsulfide production—DMS—Charlson et al., 1987; Kloster et al., 2007; Quinn and Bates, 2011). Marine biogeochemistry feeds back on ocean physics by modifying sea surface temperature through bio-optical heating (Manizza et al., 2005; Wetzel et al., 2006; Lengaigne et al., 2007; Gnanadesikan and Anderson, 2009; Gnanadesikan et al., 2010), ocean bottom *albedo* influenced by biota (Warrior and Carder, 2007), and bio-mixing (Lin et al., 2011).

Biogeochemistry studies cover a broad range of timescales from the earliest traces of life processes 3.7 billion years ago (e.g., Schidlowski et al., 1979), over the evolution from a CO_2- to an O_2- dominated atmosphere (e.g., Berner and Canfield, 1989), the role of higher plants and mountain lifting on weathering and respective matter transports (e.g., Raymo, 1994; Berner, 2006), the glacial–interglacial cycles (e.g., Brovkin et al., 2007), the holocene (e.g., Joos et al., 2004), the ongoing anthropocene (Crutzen, 2002a,b), and also the future including the next few decades to centuries (Denman et al., 2007), or even "deep time" in the future (e.g., Berger et al., 2003; Archer, 2005).

In this chapter, the state-of-the-art in biogeochemical ocean modeling is synthesized with emphasis on two aspects:

- How can we quantify and predict the marine contribution to global carbon cycling? The focus here is primarily on the modern and future ocean and the problem of ocean acidification is briefly touched upon. Ocean acidification is the decrease in pH value caused by anthropogenic CO_2 invasion of the ocean; the pH value is the negative decadic logarithm of the effective H^+ ion concentration in a solution and hence a measure of acidity—for details, see Zeebe and Wolf-Gladrow (2001).
- How can we evaluate ocean circulation models through implementation of tracers that reflect ocean circulation? Here, a complete review on available tracers is not provided, but key examples are mentioned, especially of those tracers that can efficiently complement measurements of physical ocean variables.

In a more general sense, marine biogeochemical modeling focuses on the quantification of geochemical and biogeochemical cycles involving all relevant Earth system reservoirs (interaction with the atmosphere and continental sources, surface ocean processes including primary production and the first levels of the marine food chain, cycling of matter throughout the water column, and interaction with the marine sediments and hence the lithosphere). We will not focus on ecosystems in detail (see Chapter 31), but will include a discussion of ecosystem representations insofar as they are essential for the simulation of biogeochemical cycles. The chapter is structured into three major sections apart from this brief introduction: We will first describe the concepts of biogeochemical ocean models and the related methodology, with particular emphasis on the marine carbon cycle. We will then discuss some model results of biogeochemical tracer distributions and how these can be combined with tracer observations in a useful way. Finally, we will give an overview of major new results from biogeochemical ocean simulations from the last decade.

2. CONCEPTS AND METHODS OF BIOGEOCHEMICAL OCEAN MODELING

2.1. Tracer Conservation and Classification of Tracers

The primary goal of biogeochemical ocean models is to quantify the time- and space-dependent distributions of tracers—substances or other seawater state variables of physical-, biological-, or chemical-type—in the ocean as well as exchange fluxes with the other components of the Earth system that are land, atmosphere, and the marine sea floor. The archetype (time averaged over short-term turbulent fluctuations) differential prognostic tracer conservation equation relates the temporal change in a tracer concentration at specific locations within a parcel of seawater with the respective controlling processes:

$$\frac{\partial C}{\partial t} = \mathbf{v}\cdot\nabla C + \nabla\cdot(\mathbf{D}\cdot\nabla C) + \mathrm{Sources}(C) - \mathrm{Sinks}(C) \quad (26.1)$$

with C being the tracer concentration; $\mathbf{v}$, the three-dimensional velocity field (with components u, v, w again time averaged over short-term turbulent fluctuations), and $\mathbf{D} = \mathbf{D}_{\mathbf{eddy}} + \mathbf{D}_{\mathbf{molec}}$, a subgrid scale transport tensor including the sum of eddy and molecular diffusivities as diagonal components. This tensor includes both downgradient diffusive processes as well as skew diffusive or quasi-Stokes processes, the latter of which could equally be added to the advection velocity (cf. Chapter 8). The biogeochemistry is inferred through source and sink terms. The velocity vector $\mathbf{v}$ is impacted by the eddy diffusion of momentum. The tracer equation adds a further eddy diffusion term for tracer concentrations (in order to account for time-averaged turbulent tracer transport terms such as $\overline{C'\cdot u'}$). Equation (26.1) clearly states that any prediction of oceanic tracer distributions depends critically on the quality of the underlying velocities. In addition, Equation (26.1) indicates nicely that the tracer concentration $C(x, y, z, t)$ reflects the ocean circulation and allows its quantification, if the tracer sources and sinks are well known.

In general, tracer conservation is the underlying fundamental principle of all simulations of biogeochemical cycles. Numerical algorithms for process descriptions have to conserve tracers exactly in order to arrive at quasi-steady states for simulations of biogeochemical cycles. The requirements for tracer conservation are very demanding, especially for simulations of the carbon cycle. On account of the buffer factor, that is, the sensitivity of the CO_2 partial pressure for changes in the total dissolved inorganic carbon concentration, small errors in the dissolved inorganic carbon would translate into about a ten times larger error in the CO_2 partial pressure.

A (nonexhaustive) classification of oceanic tracers relevant in the context of ocean circulation is given in Table 26.1. Most emphasis in marine biogeochemistry is placed on the minor constituents of seawater, that is, those substances that occur in such small quantities in seawater that the variation of their concentrations does not influence the density field and hence does not contribute to actively inducing oceanic motion. These passive tracers are transported with the flow field without changing it, that is, they have a negligible influence on the ocean density field (see Chapter 6). In contrast, the active tracers temperature and salinity, where salinity includes all major constituents of seawater, such as the chlorine ion concentration, which precisely covaries with salinity itself, do influence oceanic motion through their impact on the density field. However, the boundary between active and passive tracers can become blurred on specific occasions. For example, phytoplankton blooms might lead to increases in suspended matter large enough to modify the absorption of incoming solar radiation, thus inducing temperature changes in the ocean surface waters through bio-optical heating. Seawater layers including large amounts of suspended and particulate matter can become denser than water in neighboring layers and effectively sink along the ocean bottom as turbidity currents.

2.2. Classification of Models

2.2.1. Kinematic Box Models

A model is always a simplification of a real process or system maintaining the essential parts only and discarding unnecessary elements. The degree of simplification is determined by the question to be answered. For many purposes of very general nature, kinematic box models (or batch reactor models) provide an appropriate tool where the ocean is divided in a number of large-scale boxes. The mixing between these boxes and any uni-directional transport rates from one box to another are prescribed so that the mean tracer concentrations in the boxes come as close as possible to those of the observed counterparts. Thus, mixing and transport rates are usually not based on results from a dynamical circulation model. Some prominent representatives of this model class are, for example, the HILDA model (Shaffer and Sarmiento, 1995), the PANDORA model (Broecker and Peng, 1986), and the TOTEM model (Mackenzie et al., 1998). At least three boxes are needed for a reasonable simultaneous representation of marine nutrient and oxygen cycles (a two-box model does not result in plausible results for oxygen as demonstrated by, e.g., Sarmiento and Gruber, 2006). When carefully calibrated with age tracers, simple box models can give useful results for problems such as the approximate global oceanic anthropogenic CO_2 uptake rate under the assumption of a constant ocean circulation (Siegenthaler and Oeschger, 1978; Peng, 1986), conceptual studies on potential mechanisms of the glacial atmospheric CO_2 draw down by the oceans (Knox and McElroy, 1984; Sarmiento and Toggweiler, 1984; Siegenthaler and Wenk, 1984), and the quantification of ocean water renewal through thermocline ventilation (Sarmiento, 1983) and deep-water production (Peterson and Rooth, 1976; Smethie et al., 1986; Heinze et al., 1990).

2.2.2. Three-Dimensional Biogeochemical Ocean Circulation Models

For any detailed, time-dependent, and prognostic simulations of biogeochemical cycling, ocean biogeochemical

TABLE 26.1 Overview of the Different Classes of Tracers and Their Specific Features

Tracer Class with Respect to	Feature	Examples
Influence on density field and circulation		
Active tracers	They alter the density and hence the velocity field of the ocean	Temperature, salinity
Passive tracers	They do not affect the oceanic velocity field	Minor constituents of seawater, drifting buoys, etc.
Conservation properties		
Conservative tracers	Their concentration depends only on physics (circulation and mixing) (Touratier and Goyet, 2004) and on additions/removal of water (Libes, 1992); these tracers can have variable delivery functions at the boundaries of the water volume considered	Potential temperature, most CFCs
Nonconservative tracers	Their concentrations depend on physics, water additions/removal, and also on local transformations (such as chemical and biological processes as well as radioactive decay)	PO_4^{3-}, NO_3^-, O_2, $Si(OH)_4$
Biogeochemical properties		
Dissolved gases	Tracers that exchange with the atmosphere across the air–water interface	O_2, CFCs, CO_2 or H_2CO_3, ^3He
Dissolved tracers	Tracers in solution in the water column or in the sediment porewaters; substances for which dissolution is a source and precipitation (biological or chemical) is a sink	HCO_3^-, CO_3^{2-}, Ca^{2+}, $Si(OH)_4$, PO_4^{3-}, NO_3^-, O_2
Solid tracers	Tracers occurring in particulate matter within the water column or as solid tracers within the sediment; substances for which dissolution is a sink and precipitation from dissolved substances is a source	$CaCO_3$, biogenic silica/opal, particulate organic carbon clay minerals, quartz
Biolimiting elements/tracers	Substances upon which biota depend to survive and develop and which are strongly depleted in the euphotic zone and limited in availability	PO_4^{3-}, NO_3^-, Fe
Biointermediate elements/tracers	Substances upon which biota depend to survive and develop and which are always available to organisms	Total inorganic carbon, Se
Biounlimiting elements/tracers	Substances not necessary for life in the ocean	^3He, ^4He, CFCs
Involvement of time		
Transient tracers	Substances with a time-dependent source or sink in the ocean	^3H, CFCs, ^{85}Kr, ^{137}Cs, bomb ^{14}C
Radioactive tracers	Tracers that decay exponentially, with a given half-life after having been produced	^{14}C, ^{230}Th, ^3H, ^{39}Ar, ^{90}Sr, ^{137}Cs
Age tracers	Tracers that allow estimating travel times of water masses, particles, or solid sediment tracers, from a certain start point on	^{14}C age, tritium/freon ages
Ideal age tracers	A tracer whose age is set to 0 at the surface and whose age increases at the rate of 1 year per year in subsurface waters	(No real counterpart)

models must include a realistic velocity field computed through a physical dynamical ocean general circulation model (cf. Chapter 20). Velocity fields can be used either in "frozen form," when time-averaged fields are read in from a fixed data set (previously computed through a physical dynamical ocean model) by the biogeochemical model (e.g., Maier-Reimer and Hasselmann, 1987; Sarmiento and Orr, 1991), or through direct coupling of a biogeochemical model with a physical ocean general circulation model, where the velocity field for the tracer transport is also regularly updated (e.g., Maier-Reimer et al., 1996; Le Quéré et al., 2000). Within state-of-the-art Earth system models (e.g., Community Earth System Model based on NCAR CCSM4, Gent et al., 2011; BCM-C—Tjiputra et al., 2010a; GFDL—Dunne et al., 2012, 2013; see overview in Friedlingstein et al., 2006; cf. also Chapter 23), ocean circulation and ocean biogeochemistry are always interactively coupled, as are the ocean and atmosphere through exchanges of gases, heat, and moisture. Prognostic biogeochemical models, which use the velocity field of a physical dynamical ocean model are also called biogeochemical ocean general circulation models—BOGCMs. In this chapter, we will focus on BOGCMs with global coverage. This is done for two reasons: (1) The focus of this book is on observations and model simulations of the global ocean. (2) Ocean biogeochemistry is of climatic relevance because of its influence on atmospheric greenhouse gas concentrations.

BOGCMs need initial and boundary conditions to be specified. Initial conditions are a description of the marine physical as well as biogeochemical state—such as tracer concentrations—at the start of a model run. Boundary conditions are values that need to be specified continuously during a model's integration forward in time at the spatial boundaries of the model domain. Typical boundary conditions for biogeochemical models include the increase in the atmospheric CO_2 concentration from human activities, the delivery of matter through riverine input or through dust deposition from the atmosphere. As initial conditions for global models, homogeneous tracer concentrations, average profiles, or interpolated gridded products from observational data sets can be used.

Limited area high-resolution biogeochemical ocean models are useful for regional applications on short timescales, and tracer boundary conditions as well as detailed initial tracer fields from observations are needed. In addition, for regional models, tracer concentrations have to be imposed along the open boundary of the model domain. They are either derived from a global model or specified from observations. Long timescales of model integration result in the spreading of conditions imposed at the open boundary into the domain of the regional model. This, in addition to the limited availability of observations, might result in artificial redistributions of tracers including changes in air–sea fluxes. BOGCMs, in general, render as output time- and space-dependent distributions of tracer concentrations in the ocean water column and, if specified, in surface sediments (e.g., Archer and Maier-Reimer, 1994; Heinze et al., 1999).

2.2.3. *Inverse Models*

Fully inverse ocean models are a special class of models. This class of models uses oceanic tracer distributions from observations as input variables and provides values for process parameters as model output. So far, this approach has been applied mostly to steady state situations. It relies on the minimization of cost functions of varying complexity to produce the closest fit between modeled and observed tracer distributions by adjusting selected model parameters (e.g., Bolin et al., 1987; Wunsch, 1989; Schlitzer, 2000; Ganachaud and Wunsch, 2002; Kwon and Primeau, 2006; see also Chapters 21 and 22). In addition to providing optimal parameter values, these diagnostic models are extremely useful for a principal understanding of ocean circulation and biogeochemical cycles. However, they cannot be used to predict changes in ocean circulation and tracer cycling, as may result, for example, from human-induced climate change. Prognostic BOGCMs are appropriate simulation tools whenever time-varying ocean circulation phenomena are important for the model results, such as mesoscale eddies, short-term climate variability (as expressed by many climatic variability modes such as El Niño Southern Oscillation and North Atlantic Oscillation - two of the major large-scale patterns of climatic change on an interannual timescale), explosive volcanic eruptions, and human-induced longer term climate change. The combination of prognostic BOGCMs, observational data, and advanced data assimilation methods can in principle also be used for interpolation of observational data and parameter optimization where "ocean circulation" (i.e., the three-dimensional velocity field) would be one of these "parameters." However, in many cases the computational effort as well as the engineering effort is still immense (see Chapter 21), so these studies are still fairly rare (e.g., Winguth et al., 1999; Schlitzer, 2007; also Chapter 21).

2.3. Biogeochemical Cycles and Processes Included in BOGCMs

Which tracers and which biogeochemical cycles are the most important ones for interacting with ocean circulation and climate? The answer to this question depends strongly on the timescales of the variability involved and the specific research question. For paleoclimate studies and analysis of the oceanic sediment core record, a largely different set of tracers is considered than for studies of the modern and future ocean. Here, we focus on the modern and future ocean. We first discuss a number of tracers whose implementation in circulation models provides a good quality check for the simulation of the velocity field. We then describe the simulation of the abiotic and biotic parts of the ocean carbon cycle and their link to nutrient and oxygen cycles. Finally, we describe the processes at the air–water as well as water column-sediment boundaries.

For diagnosing the pattern of ocean circulation, deep-water renewal, and thermocline ventilation on decadal to centennial timescales, radioactive tracers and transient tracers are particularly useful. These tracers interact, at best, in a negligible way with biological or chemical processes. Among these tracers are the following:

- Natural as well as bomb radiocarbon $\Delta^{14}C$ (e.g., Cooper, 1956; Schlosser et al., 1994a,b, 1995; Holzer et al., 2010),
- chlorofluorocarbons (e.g., $CFCL_3$, CF_2Cl_2; e.g., Bullister and Weiss, 1983; Steinfeldt et al., 2007; Holzer et al., 2010),
- the noble gas ^{39}Ar (e.g., Schlosser et al., 1994a,b, 1995),

- bomb tritium ^3H (as ideal water tracer HTO instead of H_2O; e.g., Jenkins, 1980; Schlosser et al., 1995; Jean-Baptiste et al., 2004), and
- tritiugenic ^3He produced via the decay of bomb tritium (e.g., Schlosser et al., 1994a,b, 1995) though the latter has to be discriminated from naturally occurring ^3He (e.g., Bianchi et al., 2010).

However, there are tracers involving biogeochemical processes that also allow direct qualitative and quantitative assessments of ocean circulation, especially the following tracers:

- dissolved phosphate (PO_4^{3-}, Broecker et al., 1985),
- silicic acid (dissolved silica, $Si(OH)_4$, Broecker et al., 1995; Heinze, 2002a), and
- dissolved oxygen (Wüst, 1935; Wattenberg, 1939).

The carbon cycle—through its involvement of the most important anthropogenic excess greenhouse gas, CO_2, for the alteration of the atmospheric radiation budget (Forster et al., 2007)—is one of the most important biogeochemical cycles within the climate system to be modeled by ocean biogeochemists. The marine carbon cycle is studied for two reasons, among others: (1) The marine carbon cycle interacts with the climate system by influencing the atmospheric pCO_2 at glacial–interglacial timescales. (2) The ocean is a major sink for anthropogenic carbon. The present day ocean sink amounts to roughly 1/4 of the annual emissions (Peters et al., 2012), but the ocean will take up over 90% of the total carbon emitted by mankind once the Earth system has attained quasi-equilibrium tens of thousands of years after the anthropogenic perturbation (Bolin and Eriksson, 1959; Archer, 2005). We have to distinguish between the "natural" carbon cycle and the anthropogenic perturbation. The latter is a small amount of additional dissolved inorganic carbon (defined in Section 2.3.1) against a large natural background. Here, we focus on the assessment of the marine carbon cycle from the preindustrial to the present time, the anthropocene (Crutzen, 2002a,b). The ocean carbon cycle is governed by three carbon pumps: the solubility pump, the organic carbon pump, and the carbonate counter-pump (Volk and Hoffert, 1985). The concept of "pump" refers to a group of processes acting to create vertical gradients in tracer distributions. In a fully coupled Earth system model, these three pumps (Figure 26.1) are combined with oceanic circulation and interact via exchange with the other reservoirs (see Figures 26.1 and 26.2).

2.3.1. *Simulation of the Inorganic Carbon System (Solubility Pump)*

The solubility of gases, thus that of CO_2 as well, increases with decreasing temperature. It is greater in high-latitude cold waters such as those found in areas of deep-water formation. These water masses, which are rich in CO_2 taken up from the atmosphere, are mixed downward away from the air–sea interface and transported equatorward with the large-scale ocean circulation. The solubility pump results from the combination of the thermodynamic properties of the CO_2 and ocean circulation.

The ocean exchanges CO_2 with the atmosphere across the sea–air interface through gas exchange fluxes. The flux of CO_2 across the interface follows Fick's first law,

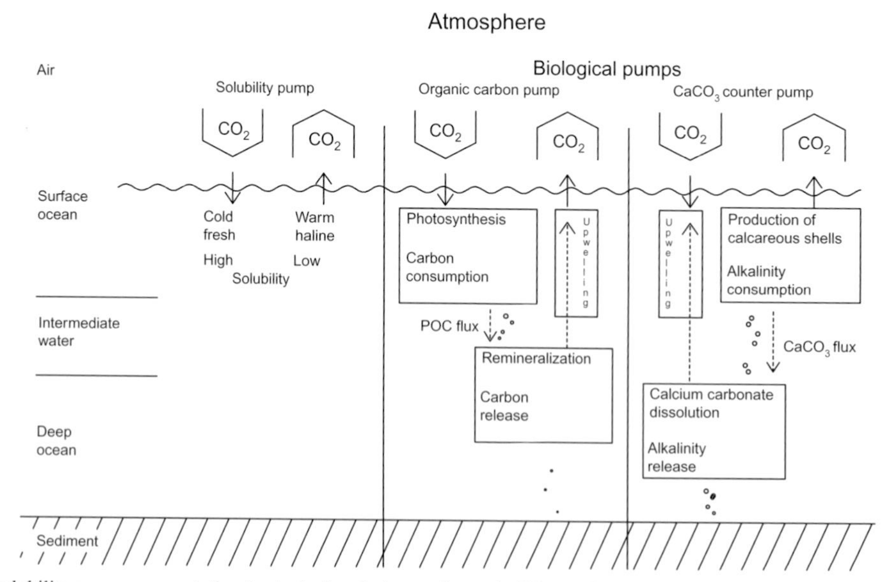

FIGURE 26.1 The solubility pump represents the physical–chemical part of oceanic CO_2 uptake and release. The two biological carbon pumps work in an opposing way: While organic carbon fixation by plankton tends to decrease the surface ocean CO_2 partial pressure, biogenic $CaCO_3$ production increases the pCO_2 (opposite effect for upwelling water, which carries a lot of remineralized carbon or redissolved carbonate ions). *Redrawn and modified after Heinze et al. (1991).*

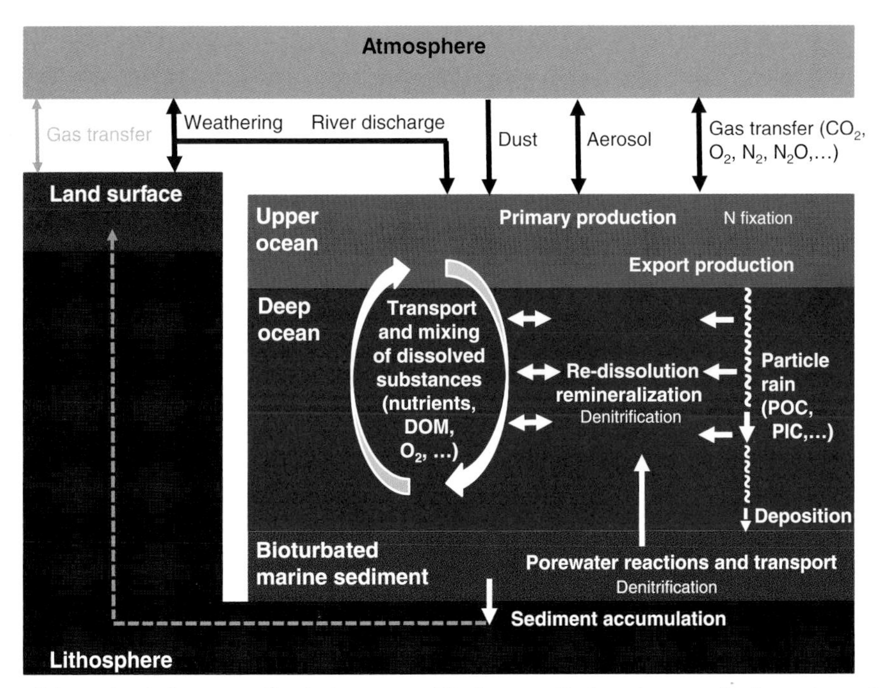

FIGURE 26.2 Scheme of biogeochemical matter cycling in the ocean and Earth system. During primary production, carbon and nutrients are consumed to form organic matter while oxygen is released. The biogenic particles sinking vertically out of the ocean surface layer provide the export production. During remineralization of organic particulate matter (including particulate organic carbon POC), carbon and nutrients are transferred to their inorganic dissolved forms while dissolved oxygen is consumed. Biogenic inorganic shell material (including calcium carbonate—particulate inorganic carbon PIC) sinks through the water column and gets partly redissolved. Particles that survive the sinking through the water column become deposited on the top sediment. Chemical reactions in the pore waters of the upper ca. 20 cm of sediment and pore water diffusion bring respectively remineralized and dissolved constituents, respectively, back to the water column. Some marine biota can convert molecular nitrogen to the nutrient ammonium, while others are able to use nitrate for remineralization of organic matter. These processes are indicated through the terms N fixation and denitrification. The inorganically dissolved constituents as well as dissolved organic matter (DOM) are transported with the oceanic flow field.

$$F(CO_2) = k \cdot \alpha \cdot \left(pCO_2^{sea} - pCO_2^{air}\right), \qquad (26.2)$$

where k is the gas transfer velocity (Wanninkhof, 1992; Jähne and Haußecker, 1998; Nightingale et al., 2000); α is the solubility of CO_2, a function of temperature (T), salinity (S), and pressure (p); and pCO_2^{sea} and pCO_2^{air} stand for the partial pressures of CO_2 in the ocean and in the atmosphere, respectively. A positive flux stands for outgassing (i.e., carbon flux from ocean to atmosphere). The CO_2 concentration $[CO_2]_{equil}$ (in general, the square brackets denote concentrations as moles of tracer per unit of seawater mass, e.g., μmol kg^{-1}) in sea water for solubility equilibrium with the CO_2 partial pressure of the atmosphere pCO_2^{air} is computed from Henry's law:

$$[CO_2]_{equil} = K_0 \cdot pCO_2^{air} \qquad (26.3)$$

The chemical reactions of CO_2 in seawater are well understood (Zeebe and Wolf-Gladrow, 2001). CO_2 reacts with water to form carbonic acid, H_2CO_3, which in turn dissociates into bicarbonate and carbonate ions:

$$\begin{aligned} H_2O + CO_2 &\rightleftharpoons H_2CO_3 \\ &\rightleftharpoons HCO_3^- + H^+; \ K_1 = \frac{[HCO_3^-] \cdot [H^+]}{[H_2CO_3]}; \end{aligned} \qquad (26.4)$$

$$HCO_3^- \rightleftharpoons CO_3^{2-} + H^+; \ K_2 = \frac{[CO_3^{2-}] \cdot [H^+]}{[HCO_3^-]}; \qquad (26.5)$$

where K_1 and K_2 are the first and second dissociation constants of carbonic acid, and depend on temperature, salinity, and pressure. The sum of the dissolved forms is referred to as dissolved inorganic carbon or DIC:

$$DIC = [CO_2^*] + [HCO_3^-] + [CO_3^{2-}]; \qquad (26.6)$$

where $[CO_2^*]$ stands for the sum of true carbonic acid H_2CO_3 and aqueous CO_2, which cannot be chemically separated. Under typical present day surface ocean conditions of temperature, pressure, salinity, and pH, bicarbonate (HCO_3^-) is by far the dominant species (about 90%), followed by carbonate (CO_3^{2-}, about 9%) and CO_2^* (about 1%). Seawater is a buffer system, which is quite resilient to additions of acids or bases. At average seawater pH, the marine buffer system is dominated by the dissociation of carbonic acid and boric acid. Alkalinity (Rakestraw, 1949; Dickson, 1992; Wolf-Gladrow et al., 2007), denoted by Alk, quantifies how well weak acids such as CO_2 (or its hydrated form carbonic acid H_2CO_3) dissociate in water. Alkalinity describes the acid–base distribution in seawater and links the inorganic carbon cycle reactions with all other relevant elements in seawater. It is defined by:

$$\mathrm{Alk} = 2[CO_3^{2-}] + [HCO_3^-] + [B(OH)_4^-] + [OH^-] - [H^+] + [\text{minor contributions}]; \quad (26.7)$$

where $[B(OH)_4^-]$ is the borate ion, which comes from the dissociation of boric acid, and $[OH^-]$ plus $[H^+]$ come from the auto-protolysis of water:

$$H_3BO_3 + H_2O_2 \rightleftharpoons B(OH)_4^- + H^+; K_B = \frac{[B(OH)_4^-]\cdot[H^+]}{[H_3BO_3]}; \quad (26.8)$$

$$H_2O \rightleftharpoons H^+ + OH^-; K_w = [H^+]\cdot[OH^-]; \quad (26.9)$$

The total borate concentration can be determined as a function of salinity:

$$TB = [H_3BO_3] + [B(OH)_4^-]. \quad (26.10)$$

Among the minor contributors to alkalinity would be phosphate, sulfate, and ammonia. According to the definition by Dickson (1981), total alkalinity can be regarded as a measure of the proton deficit of the solution relative to an (arbitrarily) defined zero level of protons. In order to enable unambiguous incorporation of any particular acid–base system into a respective definition of alkalinity, Dickson (1981) proposed to consider acids with a dissociation constant $K > 10^{-4.5}$ (at 25 °C and zero ionic strength) as proton donors and to consider the bases formed from weak acids with $K \leq 10^{-4.5}$ as proton acceptors (for details, see Dickson, 1981).

The marine carbonate system with its six unknowns (DIC, Alk, $[CO_2^*]$, $[HCO_3^-]$, $[CO_3^{2-}]$, $[H^+]$) is fully characterized by the two equilibrium conditions (Equations 26.4 and 26.5), the mass balance for total inorganic carbon (Equation 26.6), and the charge balance (Equation 26.7) (see Zeebe and Wolf-Gladrow, 2001), where the remaining variables $[OH^-]$, $[B(OH)_4^-]$, and $[H_3BO_3]$ are derived from Equations 26.8–26.10. Standard analytical procedures allow the routine determination of pCO_2, pH, DIC, and Alk. Because of their conservative property of being independent of pressure, temperature, and salinity, DIC and Alk serve usually as "master variables" for computations of the marine inorganic carbon system. In BOGCMs, DIC and Alk are used for advection with the models' velocity fields.

The carbonate system buffers seawater against large changes in DIC in response to uptake of CO_2 from the atmosphere. The buffer capacity, or Revelle factor, is defined as the change in dissolved CO_2 relative to that of DIC assuming seawater in equilibrium with atmospheric CO_2 and under constant alkalinity (Zeebe and Wolf-Gladrow, 2001). Water masses with a high Revelle factor have a low capacity in taking up CO_2 from the atmosphere (such as water masses in high latitudes). The dissociation of carbonic acid produces hydrogen ions, some of which react with carbonate ions to produce bicarbonate ions (Equations 26.4 and 26.5). This suite of reactions has two important consequences: (1) it consumes carbonate ions and (2) produces hydrogen ions.

As a first approximation, the uptake of anthropogenic carbon by the oceans in response to the increase in atmospheric CO_2 caused by fossil fuel burning and land use (Keeling et al., 1976; Houghton, 1999; Hurtt et al., 2006; Raupach et al., 2007; Andres et al., 2011) can be simulated as a perturbation of the natural carbon distributions in the ocean, that is, by considering only the solubility pump in connection with the ocean circulation. In this case, the ocean models do not need to include biological processes or ecosystem representations (e.g., Maier-Reimer and Hasselmann, 1987, including free prognostic atmospheric pCO_2; Sarmiento et al., 1992, prescribed atmospheric pCO_2 with diagnosed uptake fluxes).

In purely inorganic carbon cycle models (i.e., those that neglect the influence of biota), initial constant background values for DIC and generally fixed values for Alk are chosen. The perturbation of the carbon system through solution and dissociation of CO_2 additions from the atmosphere via gas exchange is then predicted. If the prescribed initial Alk and DIC values are not too far away from real values (i.e., which would result when the biogenic part of the carbon cycle was also taken into account), and if the underlying velocity field for the respective model is realistic, first order quantifications of inorganic carbon uptake by the oceans can be carried out. Even the decreasing buffer effect (see Zeebe and Wolf-Gladrow, 2001) with rising CO_2 concentrations can be accounted for. However, these studies are limited by two key disadvantages:

1. The approach relies on the assumption of an unchanged organic carbon cycle. Modifications of the organic carbon cycle through climate change, rising CO_2 concentrations, changes in the seawater pH value ($pH = -\log_{10}([H^+])$), nutrient supply, and interactions with the sediment and marine calcium carbonate are thus ignored.
2. Models including only the inorganic carbon cycle show an unrealistic oceanic distribution of DIC and Alk and hence cannot be verified through field observations. The reason is that the gradients of DIC and Alk in the ocean interior are caused mainly by biological and biogeochemical processes, which are not accounted for in purely inorganic carbon models.

The fraction of DIC in excess of the natural preindustrial background, called anthropogenic carbon or C_{ant} could, in principle, be compared to measurements. However, analytical methods give access to total DIC and do not allow us to discriminate between the natural and anthropogenic fraction. Observation-based reconstructions of the distribution of anthropogenic carbon rely mostly on the combination of ocean tracer measurements and statistical models. While different algorithms might agree on the cumulated uptake since the onset of industrialization, differences remain significant on the regional scale as, for example, in the Southern Ocean domain (e.g., Vázquez-Rodríguez et al., 2009; Sabine and Tanhua, 2010).

2.3.2. Simulation of the Coupled Organic–Inorganic Carbon System (Biological Pumps)

For the improved representation of the marine carbon cycle in BOGCMs, biological processes and related biogeochemical fluxes need to be taken into account. The biogeochemical cycling of carbon between its dissolved and particulate pools, both as organic and inorganic constituents, is directly coupled to the cycles of nutrients (nitrate NO_3^-, ammonium NH_4^+, phosphate PO_4^{3-}, silicic acid, and micronutrients such as iron) and oxygen through processes of organic matter production and remineralization (see Figure 26.2). BOGCMs used for coupled climate carbon cycle applications are by necessity simplified representations of the complex network of biological interactions. Most BOGCMs take advantage of the observation made 50 years ago by Redfield and coworkers (1963) of a remarkably constant mean elemental ratio of marine organic matter of C:N:P = 106:16:1. It reflects the average composition of marine phytoplankton and their degradation products. Since 1963, several studies revisited the stoichiometry of marine organic matter. They yielded slightly different sets of mean elemental ratios depending on the individual data sets used and the analytical methods applied (e.g., Takahashi et al., 1985; Anderson, 1995). In general, BOGCMs used in coupled climate carbon cycle studies are based on fixed stoichiometry. They are referred to as "Redfieldian models" in the literature and use a single element as the basic currency (C, N, or P). The other elements are derived from the stoichiometric relationship, thereby limiting the number of computationally costly state variables. A limited number of models allow for variable stoichiometry by letting the ratios between major elements vary as a function of environmental conditions and physiological requirements (e.g., Vichi et al., 2007a, b; Tagliabue et al., 2011).

The biological carbon pump starts with primary production (photosynthesis) in sunlit surface ocean waters sustained by the supply of nutrients and modulated by light and temperature. The physical environment acts on primary production through turbulent motion and mixed layer dynamics. For instance, the seasonal deepening of the mixed layer at high latitudes fuels the upper ocean with nutrients. However, a deep winter mixed layer such as typical for the North Atlantic entrains cells out of their optimal light regime, and optimal conditions for growth are reached with the onset of restratification in spring. Turbulence and nutrient availability favor specific phytoplankton groups, giving rise to typical succession patterns over a seasonal cycle. Products of primary production are either recycled in the mixed layer or exported to the ocean interior; the latter fraction is called "export production (EP)." The relative contribution of recycling and export changes over the seasonal cycle as does the composition of the phytoplankton community. Recycling and export can be conceptualized as "functions" within the biogeochemical cycle of carbon, which are assigned to specific plankton groups, lumped together into "Plankton Functional Types" or PFTs. The concept of functional types was first inferred by the terrestrial vegetation modeling community in order to manage the complexity of plants through a simplifying structuring into plant functional types or groups based on their specific function in the ecosystem or their specific resource requirements (Smith et al., 2001).

Most BOGCMs rely on PFTs as a means to deal with the complexity of first levels of the marine ecosystem. Typical PFTs include phytoplankton, which produces $CaCO_3$ shells (e.g., coccolithophores) or siliceous shells (e.g., diatoms), nitrogen fixers (e.g., diazotrophs), and small and large zooplankton (Moore et al., 2002a; Le Quéré et al., 2005; Aumont and Bopp, 2006). The size of organisms is an additional means to differentiate between PFTs. Size controls nutrient uptake by phytoplankton cells with small cells favored under nutrient-poor conditions, such as those prevailing in oligotrophic (i.e., having low levels of nutrient concentrations) ocean gyres or during summer stratification at high latitudes. To the contrary, nutrient-rich conditions, such as typical for the end of winter mixing at high latitudes, favor large cells. Large cells are in turn less tightly controlled by grazing, allowing for a large build-up of phytoplankton biomass (bloom).

In the model world, phytoplankton growth is commonly a function of the external level of nutrients, with one or more limiting nutrient modulated by light and temperature. Grazing by zooplankton, respiration, excretion, and mortality are common sink terms. The production of either $CaCO_3$ or biogenic silica (BSi) shell material is determined by the availability of, among others, silicic acid, which tends to be consumed quicker than phosphate or nitrate (Chipman et al., 1993), leading to a sequence of diatom and coccolithophorid production during phytoplankton blooms. As long as sufficient silicic acid is available, diatoms can outcompete $CaCO_3$ producers, so that the availability of silicic acid also indirectly determines the degree of $CaCO_3$ production.

The complexity of biogeochemical models has evolved over the past. Early models included EP fuelled by only one nutrient (e.g., Maier-Reimer, 1993). Present state-of-the-art BOGCMs are of the more complex NPZD model-type (Fasham et al., 1990), where N stands for nutrient, P for phytoplankton, Z for zooplankton, and D for detritus. These models are three-dimensional representations of the first levels of the marine ecosystems and distinguish, in general, between several limiting nutrients and PFTs (e.g., Moore et al., 2002a; Le Quéré et al., 2005; Maier-Reimer et al., 2005; Aumont and Bopp, 2006). These models also include a formulation, though as yet rudimentary, of the microbial loop (Azam et al., 1983) in the ocean surface layer, which enables them to simulate the seasonal cycle of biogeochemical interactions in the surface ocean. Thus, while biogeochemical models are mostly based on empirical parameterizations, their conceptual frameworks are often similar. However, recent years have seen the emergence of a new generation of models that have started to take into account aspects of evolutionary self-organization of ecosystems (Follows et al., 2007; Barton et al., 2010).

Production of particulate organic carbon (POC) in the sunlit surface ocean, and its export and gravitational settling across the water column, along with the subduction of DOC are constitutive of the biological carbon pump. Equation (26.11) provides an example of the production/remineralization reaction for organic matter assuming stoichiometric ratios after Sarmiento and Gruber (2006):

$$106CO_2 + 16HNO_3 + H_3PO_4 + 78H_2O \overbrace{\rightleftharpoons}^{\text{production}} \underbrace{\rightleftharpoons}_{\text{degradation}} C_{106}H_{175}O_{42}N_{16}P|_{\text{particulate/suspended}} + 150O_2 . \quad (26.11)$$

Taken as a whole, the succession of process depletes surface waters in nutrients and DIC and enriches deeper water masses as DIC and nutrients get released by remineralization (see Figure 26.2). Carbon exported to depth is removed from exchange with the atmosphere on timescales directly related to the depth of remineralization. The production of PIC (or $CaCO_3$) and its dissolution at depth are called the carbonate counter-pump. The denomination of "counter-pump" refers to the effect of PIC production and dissolution on pCO_2. While the production of POC lowers the pCO_2 of surface waters and its remineralization at depth increases it, the production and dissolution of PIC have the opposite effect. For the regulation of the atmospheric CO_2 concentration by the biological carbon pumps, the part of the biological production that leaves the euphotic zone as "EP" is the most relevant.

The attenuation of the vertical particle flux through the water column is described by process parameterizations of varying degrees of complexity. Simple approaches fold the combination of the sinking velocity and the degradation rate constant into one governing constant parameter assuming an exponential or hyperbolic decrease in particulate matter with depth (Suess, 1980; Martin et al., 1987). More complex schemes enable three-dimensionally varying sinking velocity and degradation rates, two size classes of particles (Moore et al., 2002b; Aumont et al., 2003), or even an entire spectrum of particle-size classes and representations for coagulation and disaggregation of particles (Kriest and Evans, 1999; Gehlen et al., 2006). The degree of complexity needed for the simulation of the export flux of particulate matter is not yet fully clarified. There are indications that increasing model complexity does not necessarily lead to a better reproduction of measurements (Kriest et al., 2010). However, the models of different complexity discussed in the paper of Kriest et al. (2010) have not all been optimized with respect to observations. At first order, simple models work well, while more complex food web models are needed to predict the export flux and its efficiency under a larger variety of climatic, biogeochemical, and environmental conditions (Steinacher et al., 2010).

2.4. Links Between the Water Column and Other Reservoirs

The distribution of biogeochemical tracers is influenced by external forcings at the boundaries of the ocean domain: atmosphere, land-ocean transition, and marine sediments. The following section presents an overview of forcings to be considered in the set-up of a BOGCM.

2.4.1. *Forcing by the Atmosphere*

At the air–sea interface, the physical forcing factors, wind speed and wind stress, directly control the gas transfer velocity and, thus, fluxes of, for example, CO_2, O_2, and N_2O. Moreover, wind speed and stress, solar radiation, heat, and fresh water fluxes participate in controlling mixed layer dynamics and, hence, biological production (see Section 2.3.2). Next to these physical forcings, biogeochemical boundary conditions need to be specified, namely dust deposition and the concentration of gases in the atmosphere. The gases specified depend on the scope of the model simulation. For BOGCMs, they are, in general, CO_2 and O_2. Mobilization of particles from deserts is by far the major source of aeolian dust (e.g., Mahowald et al., 1999). Their biogeochemical relevance stems from the fact that dust particles release the micronutrient Fe upon deposition and dissolution in seawater (e.g., Jickells et al., 2005). More recently, volcanic ashes have been recognized as potential external sources of nutrients (e.g., Hamme et al., 2010). For most applications, the simplifying assumption is made that the composition of dust equals the mean composition of the continental crust (Al:Si:Fe:P). For simulations under preindustrial conditions, dust fluxes as well as the concentrations of atmospheric gases are held constant. For coupled climate–BOGCMs, both physical and biogeochemical forcings evolve in response to global climate change. The most prominent change in biogeochemical forcing is the increasing atmospheric CO_2 concentration caused by human-produced emissions (Keeling et al., 1976; Houghton, 1999; Hurtt et al., 2006; Raupach et al., 2007; Andres et al., 2011). They induce a net uptake flux of CO_2 by the ocean and a decrease in buffer capacity as well as increasing ocean acidification.

2.4.2. *Forcing by Lateral Sources and Seabed Sources*

Substantial amounts of matter are brought into the oceans through rivers, groundwater seeping, mobilization from sediments, gas hydrate destabilization, and hydrothermal activity. Associated tracers can provide important insight into the functioning of the ocean and ocean circulation. Rivers are important sources of carbon and nutrients, in both inorganic and organic form (e.g., Seitzinger et al., 2010). Groundwater seeping can also provide a significant input of nutrients and other tracers to the coastal ocean (e.g., Slomp and Van Cappellen, 2004). While most BOGCMs take into account riverine input of nutrients and carbon, groundwater discharge is usually neglected. Both—direct river input and groundwater seeping—have changed considerably with time as a result of anthropogenic action.

In recent days, the potential for a destabilization of methane from gas hydrate (gas hydrate is a solid material comparable to ice, which can contain large amounts of methane) has been receiving increased attention (e.g., Milkov, 2004; Archer et al., 2009; Biastoch et al., 2011). Methane from gas hydrates could contribute to inorganic and organic carbon pools through microbial conversion to CO_2 (Hoehler et al., 1994; Pohlman et al., 2010). The increase in DIC along with methane release would further aggravate ocean acidification.

Outgassing, from the earth's interior and through hydrothermal vents, releases tracer plumes. Among the geochemical tracers released, mantle ^3He was used with success in model studies as a constraint on deep-water circulation (Dutay et al., 2004). Trace metals, such as Fe, are strongly enriched in hydrothermal plumes. The addition of hydrothermal Fe to a BOGCM suggests that this Fe source, which is relatively constant over long timescales, has the potential to buffer the dissolved Fe inventory against short-term fluctuations in dust input (Tagliabue et al., 2010).

2.4.3. *Output Through Sediments*

Particles exported from the surface ocean, along with lateral inputs from continental margins and rivers accumulate at the seafloor. POC is a small yet important contribution to the total mass flux as it provides the main energy source for many deep-sea communities (Bishop, 2009). Remineralization products accumulate in porewaters and are exchanged with overlying water masses making sediments a source of nutrients such as nitrate, dissolved Si, phosphate, and Fe. Most BOGCMs do not include a sediment module and recycle the flux, which normally would be deposited on the ocean floor to the last wet model box over the bottom. Models including the Fe cycle consider inputs of dissolved Fe from sediments (e.g., Moore et al., 2002b; Aumont and Bopp, 2006).

A limited number of BOGCMs include an interactive sediment module coupled to the water column. These modules allow for a process-based description of the fate of matter that has been deposited onto the model ocean floor. They distinguish between redissolution/reoxidation and burial (sediment accumulation) below the chemically reactive and bioturbated sediment mixed layer (Archer and Maier-Reimer, 1994; Heinze et al., 1999; Maier-Reimer et al., 2005; Gehlen et al., 2008). Temporal imbalances between input of matter from continental sources and output of matter through sediment accumulation can cause substantial changes in the geochemical inventory of the ocean and associated changes in atmospheric CO_2 concentrations (Archer and Maier-Reimer, 1994; Heinze et al., 1999; Tschumi et al., 2011). Such imbalances usually develop and smooth out over a long timescale of several 10,000 years. For the uptake and compensation of anthropogenic CO_2 by the ocean, the dissolution of $CaCO_3$ from the sea floor plays a crucial role. It represents a key negative feedback on the 10,000 to 100,000 year timescale and accounts for significant percentages (about 50% in case of an instantaneous CO_2 release to the atmosphere of 5000 GtC, Archer, 2005) of the ultimate oceanic buffer capacity for human-produced CO_2 (Bolin and Eriksson, 1959; Broecker and Takahashi, 1977; Archer et al., 1998; Archer, 2005).

Though sediment processes significantly influence the simulated tracer distributions on longer timescales, consideration of the oceanic sediment in climate models is also worthwhile and important for shorter timescales. One reason is the threat to benthic deep-sea ecosystems from ocean acidification, where $CaCO_3$ sediments close to deep-water production areas may be already starting to dissolve (Gehlen et al., 2008; Olafsson et al., 2009). The other reason is that a proper sediment simulation is an excellent test for adequate simulations of other biogeochemical processes such as surface production, marine particle fluxes, and the attenuation of these fluxes with depth. BOGCMs that can correctly simulate the marine sediment coverage for carbon ($CaCO_3$ and organic carbon), BSi, and clay, may thus be more reliable in their overall results in the sense that the best sediment trap is the sediment itself. Though POC deposition to the sea floor is relatively small for deep-sea environments, the resulting sediment distribution and the associated oxygen fluxes are good indicators for appropriate simulations of carbon and nutrient cycles (Najjar et al., 1992; Jahnke, 1996; Heinze, 2002b; Gehlen et al., 2006). As upwelling regimes are tightly coupled with the EP of BSi, simulations of the BSi (or opal, mainly shell material of diatoms and radiolarians) sediment are an efficient reality check for BOGCMs with respect to the ocean flow field (e.g., Ragueneau et al., 2000).

2.5. Model Coupling, Model Resolution, and Model Complexity

Ocean biogeochemical cycles involve a series of processes on a wide spectrum of timescales: Biological particle production and particle fluxes can occur on a timescale of days; the ventilation of the thermocline can occur on timescales of 10–100 years; the flushing time for the ocean is on the order of 1000–1500 years; and sediment build-up timescales are on the order of 10,000 to 100,000 years. Also, residence times of elements occur on a wide range of timescales (e.g., Broecker and Peng, 1982; Sarmiento and Gruber, 2006). Therefore, ocean models have to be tailored to the purpose of the study in question. For climatic ocean circulation studies involving biogeochemical cycling contributing to assessment reports of the "Intergovernmental Panel on Climate Change" (IPCC, www.ipcc.ch), typical timescales are of a few hundred years, with spin-ups of at least 1000 years and experiment durations of around 500 years. Within this time frame, in the year 2011, models must still have a fairly coarse resolution (order of 1° horizontally) so that water masses can build up prognostically and the tracer distributions as well as their inventories have reached quasi-equilibrium. Acceleration methods have been proposed (DEGINT—Aumont et al., 1998; "transport matrix" method—Kwon and Primeau, 2006; Khatiwala, 2007) that are powerful so as to see whether model quasi-equilibria will be very incorrect or acceptable, but as these methods do not render the exact quasi-equilibrium of the full BOGCMs, a longer spin-up with the full model (higher resolution and/or full nonlinearity) can often not be avoided.

Generally, BOGCMs can be run as pure ocean models (stand-alone models forced by input data) or interactively coupled to other models of the Earth system. BOGCMs as stand-alone versions are forced by atmospheric climatological data (e.g., Hellerman and Rosenstein, 1983; COADS—Woodruff et al., 1987) and reanalyses (e.g., Kalnay et al., 1996; Simmons et al., 2004; Large and Yeager, 2009) and to some degree also by surface ocean *in situ* data (as for salinity restoring). Stand-alone ocean models are considerably more economical to run than coupled models at the same resolution. When forced by atmospheric reanalysis products, stand-alone studies allow producing hindcasts of the past 50 years—essentially the timespan covered by atmospheric reanalysis—presenting realistic modes of interannual variability. This is an important prerequisite for a sound time-dependent comparison with observations and for attributing changes in carbon fluxes to changes in forcing (e.g., Wetzel et al., 2005; Assmann et al., 2010). Recently, ocean circulation fields from assimilation products (ocean reanalysis, see Lee et al., 2009 for a review) have been used to force biogeochemical models (e.g., see Brasseur et al., 2009; Ito et al., 2010). While this is an emerging field of research, first studies suggest the advantages of physical ocean state estimates over free ocean circulation models (that is without data assimilation) for specific biogeochemical studies (Mikaloff Fletcher et al., 2006, 2007).

The significant increase in computing resources over the last years, has allowed the development of eddy-permitting BOGCMs (e.g., Lachkar et al., 2009; Woloszyn et al., 2011). These models resolve mesoscale eddies at least in the tropical ocean (where the baroclinic Rossby radius of deformation is relatively large). The influence of meso- and sub-mesoscale activity on biogeochemical processes is an area of active research (Lévy et al., 2001; Mahadevan et al., 2004; Lévy, 2008; Ito et al., 2010). The main disadvantage of these models is their high computation costs, which preclude a long spin-up followed by a multidecadal simulation. Most studies published to date present results obtained after, at most, one decade of simulation. While mean surface distributions of properties and their seasonal cycles might be stable, tracer distributions still display a fairly strong drift at depth. Deep tracer distributions will, moreover, largely reflect initial conditions prescribed by climatological biogeochemical data sets. Or, to put it in a different way, the mismatch between model output and climatological data will largely reflect model adjustment rather than providing insight into model skill. Eddy-resolving BOGCMs, nevertheless, have an important field of application, namely, developing and testing new parameterizations for future high-resolution models, highlighting missing processes or inappropriate process parameterizations in coarse resolution models, and operational forecasting (Brasseur et al., 2009).

In contrast to forced ocean-only models stand coupled Earth system models, where the ultimate driving force is almost exclusively solar insolation (next to this, greenhouse gas concentrations and volcanic eruptions also play a role, of course). In these Earth system models, the model ocean interacts with the model atmosphere and, at best, the statistics of real ocean variability are reproduced (not single events, however, in the respective calendar years). Nevertheless, complex Earth system models are, at present, the best tool for future projections of climate including the interaction with the carbon cycle (e.g., Friedlingstein et al., 2006; Roy et al., 2011). For longer term studies (timescales of several 1000 years up to millions of years), the so-called Earth system models of intermediate complexity are employed. They include a drastically simplified atmospheric module (the limiting factor for long integrations) but often retain the ocean in a reasonable though still coarse resolution (e.g., Montoya et al., 2005; Tachirii et al., 2010). These models have proved useful, especially for studies of glacial–interglacial cycles (e.g., Brovkin et al., 2007), though coarse full Earth system models are emerging (Vizcaíno et al., 2008).

3. MODEL RESULTS, EVALUATION, SKILL, AND LIMITS, AND MODEL DATA FUSION/DATA ASSIMILATION

3.1. Ability of BOGCMs to Match Natural Tracer Distributions to First Order

Present-day ocean circulation models are able to adequately reproduce large-scale features of key natural tracer distributions. This has been illustrated nicely for radiocarbon in bottom waters for an inverse ocean biogeochemical circulation model (Schlitzer, 2007). Likewise, NPZD-type BOGCMs simulate the distributions of nutrients, oxygen, and the carbon cycle in line with the climatological mean state. However, model data fit is at best as good as the underlying flow field. Numerical algorithms for advection and diffusion, as well as the discretization in space and time result in somewhat smoothed tracer distributions with smaller gradients than observed (Figures 26.3 and 26.4). The simulation of dissolved phosphate (PO_4^{3-}) (Figure 26.3) is a particularly good test for ocean circulation models, as the external input is small and the phosphorus cycling (on short timescales) is less complicated than the nitrogen and carbon cycles, and where the nitrogen cycle is, in addition, influenced by nitrogen fixation as well as denitrification (e.g., Zehr and Ward, 2002) and the carbon cycle is influenced by air–sea gas exchange. For studies aiming at the quantification of ocean carbon uptake and ocean acidification, the simulation of the carbonate ion distribution is key because it provides the most important influence on alkalinity and is essential for the saturation

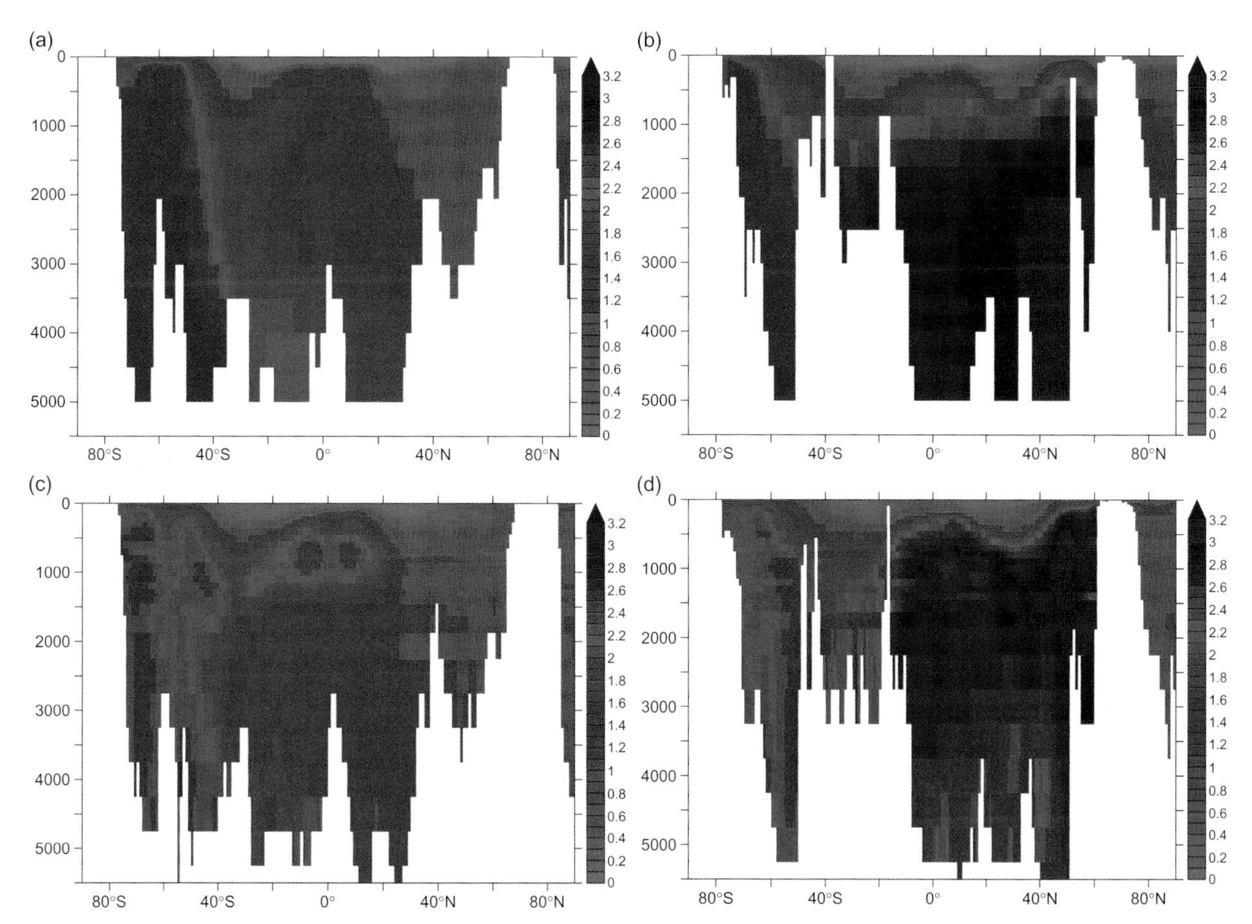

FIGURE 26.3 Meridional cross-sections of dissolved phosphate in the ocean (Gehlen, personal communication). The simulated values are the result of a run with the coupled IPSL-CM5-LR Earth system model as used in Séférian et al. (2013). Shown are model averages over the years 1990–2000. (a) Model, Atlantic along 30°W. (b) Model, western Pacific along 180°W. (c) Observations, western Atlantic along 30°W. (d) Observations, western Pacific along 180°W. All values are in [micromol/kg]. *The observational data are from Conkright et al. (1994).*

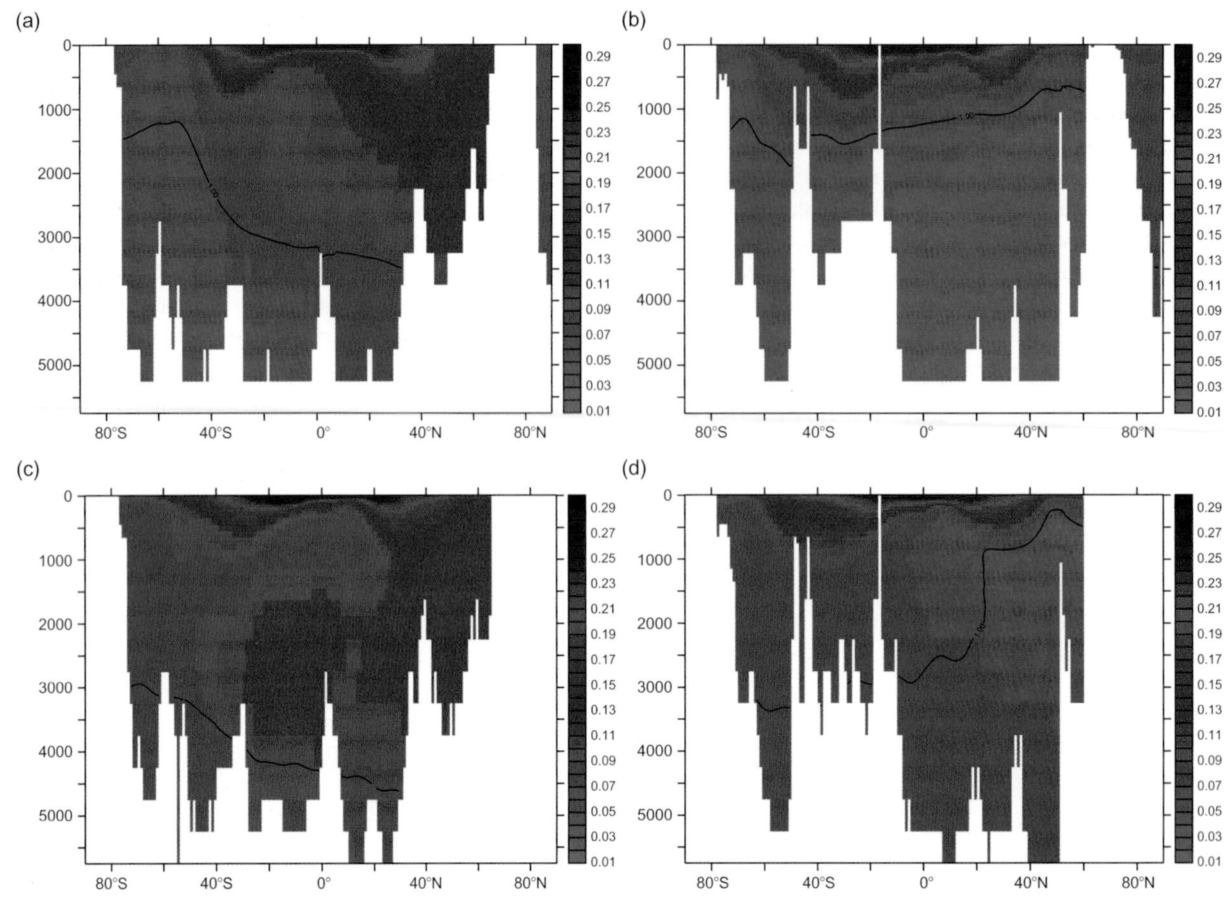

FIGURE 26.4 Meridional cross-sections of the marine carbonate ion concentration (Gehlen, personal communication). The simulated values are the result of a run with the coupled IPSL-CM5-LR Earth system model as used in Séférian et al. (2013). Shown are model averages over the years 1990–2000. The 100% carbonate saturation level is indicated through the black line. (a) Model, Atlantic along 30°W. (b) Model, western Pacific along 180°W. (c) Observations, western Atlantic along 30°W. (d) Observations, western Pacific along 180°W. All values are in [millimol/kg]. *The observational data are from Key et al. (2004).*

of seawater with respect to $CaCO_3$. Figure 26.4 reveals that the CO_3^{2-} concentration including the depth level of the calcite saturation horizon—representing the boundary between waters that preserve $CaCO_3$ shell material and waters that start to corrode $CaCO_3$—is well simulated compared to observations. Both the PO_4^{3-} and the CO_3^{2-} distributions show the interaction of the biological pumps with ocean circulation. It can be clearly seen that the PO_4^{3-} concentration increases from the deep-water source areas in the North Atlantic to the large-scale upwelling in the North Pacific while the CO_3^{2-} concentration decreases with water mass age. PO_4^{3-} concentrations are low at the ocean surface where phytoplankton grows and increase in deeper layers where nutrients and carbon are released through organic matter degradation. CO_3^{2-} shows the highest values at the sea surface because of the low prevailing CO_2 partial pressure (as a consequence of plankton growth) though carbonate ions are consumed for the building of $CaCO_3$ shell material. Deep waters get enriched with dissolved inorganic carbon and as a result of high pressure and a low pH value, the local *in situ* equilibrium for the three inorganic carbon species free CO_2, bicarbonate HCO_3^-, and carbonate CO_3^{3-}, the speciation is shifted away from the carbonate ion toward bicarbonate and free CO_2. The $CaCO_3$ saturation horizon shallows from the Atlantic to the Pacific (Figure 26.4).

3.2. Optimization of BOGCMs

In spite of the demonstrated capability of state-of-the-art BOGCMs to simulate the climatological mean state of marine tracer distributions, including the carbon and coupled nutrient and oxygen cycles, there exist considerable discrepancies between modeled and observational fields. This holds in particular when it comes to temporal and spatial variability. On one hand, our representation of biogeochemical processes is likely to be incomplete with processes still to be discovered and on the other hand, known ones remain poorly quantified in a detailed process-based sense. Further, the simulated tracer patterns depend strongly on the quality of the simulated flow field. It cannot be excluded that part of the resulting appropriate-appearing tracer distributions may be "right for the wrong reasons." Because of the sometimes limited knowledge of the exact values of adjustable parameters in process parameterizations, a firm interest exists in optimizing these free parameters through fits of modeled tracer distributions to observed values. In the following section, we discuss the uncertainties associated with the observations, which have to be considered when optimizing models with respect to these observations. Then we present a brief overview on data assimilation methods applied to biogeochemical models.

3.2.1. Uncertainties Associated with the Observations

Several factors limit the straightforward optimization or even evaluation of simulated tracer data with observations. The data coverage for geochemical and carbon cycle tracers is still limited as compared to temperature and salinity. Aliasing because of low data coverage (aliasing through eddy motion, for example) is also a considerable problem for geochemical tracers. The surface pCO_2 observations using VOS lines (VOS = voluntary observing ships) have largely improved our understanding of air–sea carbon fluxes (SOCAT http://www.socat.info/; Pfeil et al., 2013; Sabine et al., 2013; for a surface ocean pCO_2, see also Takahashi et al., 2009), but these VOS lines are mostly focused on regular shipping lines leaving vast areas of the ocean, especially in the Pacific sector of the Southern Ocean, not well observed. Full seasonal cycle coverage is achieved only in the densely measured VOS line domains and at selected Eulerian time series stations (such as BATS, Lohrenz et al., 1992; HOTS, Keeling et al., 2004; and ESTOC, Santana-Casiano et al., 2007). Repeat hydrography measurements (CARINA, Key et al., 2010; and other emerging three-dimensional data syntheses such as PACIFICA, http://www.pmel.noaa.gov/co2/story/PACIFICA) merge observations covering several years and all seasons into one big data set. While the number of individual observations is impressive and provides a satisfying basis for deriving gridded products representing the mean state, data coverage is too limited to address temporal variability at seasonal to interannual timescales. Mostly, water column data of dissolved tracers are used for model evaluation as these can be most accurately measured (Sarmiento and Gruber, 2006).

Sediment trap data measuring the flux of sinking particulate matter are still associated with considerable potential systematic errors (Buesseler et al., 1994). The temporal pattern of fluxes can nevertheless help to constrain simulations of the seasonal cycle and plankton blooms in models.

Remotely sensed data have helped to constrain model predictions of marine phytoplankton primary production through ocean color data (Antoine et al., 1996; Carr et al., 2006; Saba et al., 2011) and the respective algorithms, which convert ocean color data into chlorophyll concentrations from which primary production may be estimated. The overall uncertainty of the chlorophyll estimates is considerable (relative error of 35% for chlorophyll-*a* concentrations, Hooker et al., 1992).

For both the amount of anthropogenic carbon stored in the ocean and the impact of progressive ocean acidification on marine biota, the observational data-derived estimates are associated with considerable uncertainties. Elevated levels of dissolved inorganic carbon in the oceans caused by uptake of human-produced carbon from the atmosphere can be almost perfectly analyzed in the model worlds of BOGCMs as the preindustrial distributions within these worlds are known . It is more difficult to arrive at the concentrations of the respective excess carbon from observations as the preindustrial dissolved inorganic distribution in the ocean has not been measured and the human perturbation in dissolved inorganic carbon in seawater occurs on a huge background value. The different reconstruction methods for marine anthropogenic carbon are not always fully consistent (Sabine and Tanhua, 2010). Though trends for progressive ocean acidification were inferred unequivocally from both Eulerian time series stations (e.g., Santana-Casiano et al., 2007) and repeat hydrography (e.g., Olafsson et al., 2009), the exact process description of the large-scale impact of ocean acidification is still in its infancy. In order to at least "get an idea" of how ecosystem and biogeochemistry may react to high ambient CO_2 levels, laboratory experiments as well as mesocosm experiments have been carried out (e.g., Zondervan et al., 2001; Riebesell et al., 2007). (Mesocosms are experimental devices that bring a part of the natural marine ecosystem into controllable conditions, for example, with specialized large "plastic bags" that are lowered into the surface ocean.) It is still unclear how such experiments can be generalized for other regions and time periods, and what part of the experiments may be dominated by local conditions and the experimental set-ups themselves.

3.2.2. Data Assimilation in Biogeochemical Modeling

In order to systematically combine observations and biogeochemical models, different optimization and data assimilation procedures have been attempted. The goal of these procedures on the one hand is to arrive at optimally interpolated fields for the biogeochemical data sets (of comparatively low spatio-temporal coverage) and, on the other hand, to derive optimal choices of adjustable parameters for which the discrepancies between modeled and observed data would be minimized. A further approach is to "tie" model results to observations. A good introduction to the topic is found in Brasseur et al. (2009).

Let us start with the last approach first. In order to provide quantitatively realistic surface ocean biological production rates even with nonideal physical circulation input fields, oceanographers had restored surface ocean nutrient concentrations to interpolated (gridded) observational data sets. The difference between the annual upward transport of nutrients by the respective model's flow field and the prevailing surface nutrient concentration is then—if positive—converted into an EP rate of organic carbon and nutrients. Such methods have been used in global ocean modeling studies including the coupled inorganic and organic carbon cycle (e.g., Najjar et al., 1992) the silicon cycle (Gnanadesikan, 1999), as well as the parallel simulation of the silicon, organic carbon, and $CaCO_3$ cycling (Jin et al., 2006). Though such a "restoring approach"—similar to restoring to surface temperature and surface salinity using Neumann boundary conditions in physical oceanography—renders in many cases reasonable global particle EP rates of the respective particle species under consideration, incompatibilities between the simulated ocean circulation and the surface nutrient fields can result in locally unrealistic production rates. Further, such approaches cannot be used in truly prognostic computations, where the surface nutrient concentration must also be allowed to vary (see, e.g., Maier-Reimer, 1993, for fully prognostic simulations without restoring).

The task of interpolating between observations is important for upscaling individual pCO_2 data from VOS lines to basin-scale air–sea fluxes. Neural network methods have been used successfully for achieving integrated air–sea CO_2 fluxes for basin-wide areas such as the North Atlantic (Telszewski et al., 2009), covering specific time intervals. In the respective self-organizing maps (Kohonen, 2001), surface ocean pCO_2 is estimated through correlations with other variables (e.g., sea surface temperature, mixed layer depth, and chlorophyll, as well as time and geographical position).

Inverse approaches have been pursued in order to quantify preindustrial as well as anthropogenically influenced carbon fluxes across the air–sea interface as well as in the interior of the ocean (e.g., Gloor et al., 2001, 2003; Mikaloff Fletcher et al., 2006, 2007).

For time-dependent data assimilation, both sequential and variational methods have been applied to systematically improve BOGCMs. Through sequential assimilation methods, the model state is adjusted step-wise in time with respect to the observed data (e.g., Ensemble Kalman Filtering; Evensen, 2009). This approach has been used, for example, by Gerber et al. (2009) to determine regional air–sea fluxes of CO_2 as well as transport and storage of anthropogenic carbon. In contrast to the sequential approaches, variational data assimilation methods allow the fitting of entire model trajectories to observations, while directly obtaining the quantitative link between model parameter adjustments and the model fit to the observations. Variational methods, such as the adjoint method (see also Chapter 21), however, are more difficult to implement than sequential methods. Combined biogeochemical state estimates and biogeochemical model parameter estimations have been carried out through adjoint approaches (e.g., Tjiputra et al., 2007; Gregg, 2008; Kane et al., 2011). An overview on data assimilation of satellite-derived chlorophyll data into ocean models using a broad spectrum of different methods is given in Gregg (2008).

4. MAJOR MARINE CARBON MODELING FINDINGS OF THE RECENT DECADE

There has been a large increase in manuscripts addressing ocean carbon cycle modeling in the last decade. The citation index ISI lists 1443 new peer-reviewed publications for the combined search key words "ocean" and "carbon cycle" and "model" between the beginning of 2001 and October 2011 (ISI Web of knowledge citation index). This is almost two times the number of papers (1857) published in the 50 years' time span 1945–2001. The key impetus for the large number of recent publications has been the need to include the carbon cycle interactively in models used for future climate projections and the awareness of progressive ocean acidification.

4.1. Future Biogeochemical Climate Projections Including Oceanic Carbon Cycle Feedback

While a series of ocean modeling studies had dealt with estimating the ocean uptake of anthropogenic carbon from the atmosphere (Maier-Reimer and Hasselmann, 1987; Sarmiento et al., 1992; Orr et al., 2001), the interactive coupling of ocean (as well as land) carbon cycle modules into complex climate models started only about 10 years ago (Cox et al., 2000). After this first Earth system model based on general circulation models, a suite of further interactive carbon cycle climate models followed (e.g., Doney et al., 2006; Friedlingstein et al., 2006; Crueger et al., 2008; Tjiputra et al., 2010a). In general, in all Earth system models, the ocean acts as a sink for excess atmospheric carbon dioxide. However, the sink strength varies differently over time in the various models (Friedlingstein et al., 2006; Frölicher and Joos, 2010; Roy et al., 2011). The differences can be attributed to between-model discrepancies in simulated surface ocean properties, such as temperature, salinity, and sea surface pCO_2, and in the ocean circulation field, which itself has an impact on nutrient supply and biological particle production.

Generally, the ocean carbon sink weakens with rising atmospheric CO_2 because of the pCO_2 dependency of the oceanic CO_2 buffer system (e.g., Zeebe and Wolf-Gladrow, 2001; high CO_2 waters can take up further CO_2 additions to the atmosphere less well). A weakening of ocean circulation as predicted for rising global temperatures by state-of-the-art climate models (Meehl et al., 2007) also acts as a net positive feedback to climate change because surface ocean waters loaded with high CO_2 from the atmosphere can less quickly be mixed downward and get replenished with waters not yet carrying a high anthropogenic carbon signature (e.g., Plattner et al., 2001). Both forcings, the rising CO_2 concentration in the atmosphere and the physical climate change, individually and in conjunction lead to a weaker ocean carbon uptake rate and hence a positive feedback to climate change. This holds for both the global ocean (Figure 26.5; Friedlingstein et al., 2006) and regional sub-domains (Figure 26.6; Roy et al., 2011). According to the models used in Roy et al. (2011), next to the equatorial ocean, the largest regional sink until the end of this century is expected to be the Southern Ocean. The Southern Ocean will develop into a major lateral export region for excess CO_2, while the North Atlantic will receive laterally inflowing water with already substantial amounts of anthropogenic CO_2, which will limit local CO_2 uptake from the atmosphere in this area (Sarmiento

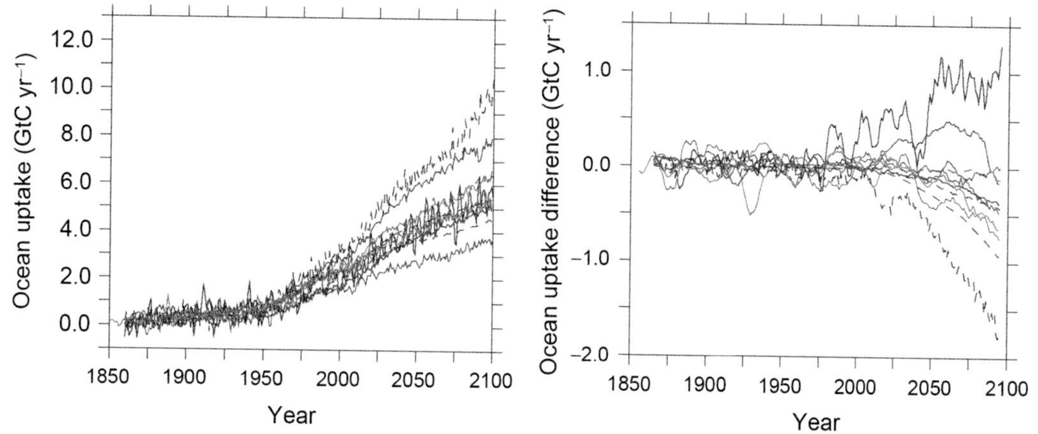

FIGURE 26.5 Varying global oceanic carbon uptake rates for different interactive carbon cycle climate models (Friedlingstein et al., 2006) for an IPCC SRES A2 CO_2 emission scenario. Models: HadCM3LC (solid black), IPSL-CM2C (solid red), IPSL-CM4-LOOP (solid yellow), CSM-1 (solid green), MPI (solid dark blue), LLNL (solid light blue), FRCGC (solid purple), UMD (dash black), UVic-2.7 (dash red), CLIMBER (dash green), and BERN-CC (dash blue). Left: Air–sea carbon fluxes for the Earth system model runs with fully interactive carbon cycle (GtC yr^{-1}). Right: Differences in air–sea carbon fluxes (GtC yr^{-1}) between runs using a fully interactive carbon cycle and runs where the CO_2 emissions have no influence on the radiative budget (only a biogeochemical effect on the ocean and land carbon modules). The HadCM3LC model shows an increase in the oceanic sink with time, which is attributed to the extremely large land carbon source in this specific model that is compensated for partially through a higher oceanic carbon uptake. *Source: Friedlingstein et al. (2006). © American Meteorological Society. Used with permission.*

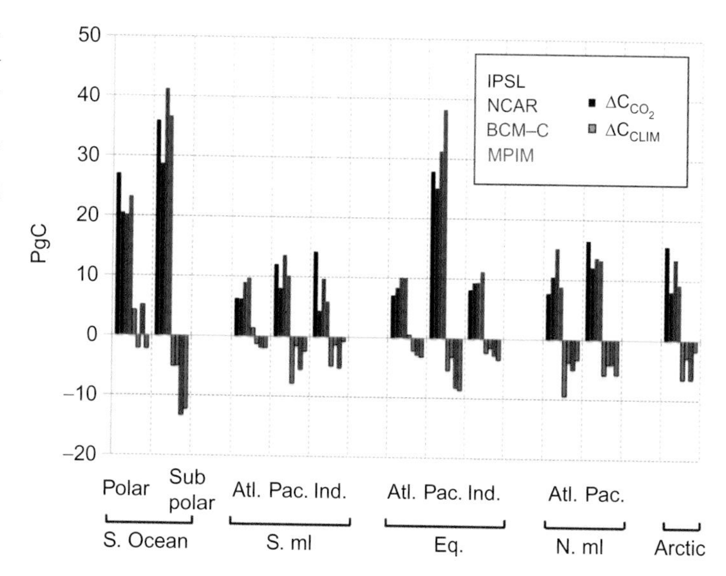

FIGURE 26.6 Integrated cumulative CO_2 uptake (PgC) over different regions of the world ocean to the end of this century (2010–2100) in four different carbon cycle climate models (Roy et al., 2011). The carbon flux changes because of CO_2 forcing, ΔC_{CO_2} (GtC), are shown through color bars without black lines as frame, and the carbon flux changes caused by climate forcing, ΔC_{CO_2} (GtC), are shown through color bars surrounded by black solid lines. The coordinates of the regional domains are as follows: Southern Ocean (S. Ocean; Polar, >58°S; Subpolar, 44°–58°S), Southern mid-latitudes (S. ml; 18°–44°S), equatorial (Eq.; 18°S–18°N), Northern mid-latitudes (N. ml; 18°–49°N), and Arctic (>49°N). *Source: Roy et al. (2011). © American Meteorological Society. Used with permission.*

et al., 1992; Orr et al., 2001; Mikaloff Fletcher et al., 2006; Tjiputra et al., 2010b).

The development of the oceanic biogeochemical modules for prognostic Earth system models for climate projections has to a substantial degree been supported by comprehensive model intercomparison projects. In ocean biogeochemistry, among others, the OCMIP (phases I and II) (Ocean Carbon-Cycle Model Intercomparison Project; e.g., Orr, 2002), the C[4]MIP (Coupled Climate–Carbon-Cycle Model Intercomparison Project; e.g., Friedlingstein et al., 2006), and the ongoing Marine Ecosystem Model Intercomparison Project as well as Coupled Model Intercomparison Project Phase 5 are examples of such projects.

4.2. Modeling the Interaction of Ocean Circulation with Greenhouse Gas Fluxes and Biological Production

BOGCMs, in stand-alone mode and coupled to full-fledged Earth system models, have been successfully used to quantify the integrated seawater column burdens of anthropogenic carbon (Figure 26.7; e.g., Assmann et al., 2010; Tjiputra et al., 2010a). The North Atlantic and the southern hemisphere oceans between 30° and 40° S stand out as regional maxima for storage of anthropogenic carbon. These areas are adjacent to the formation regions of deep and mode waters. From these areas, water with high loadings of anthropogenic carbon is transported further equatorward at depth. The North Atlantic water column storage of excess carbon is the highest per unit area. This confirms the studies with coupled Earth system models as described in the previous paragraph.

Time-dependent ocean carbon uptake studies with models confirm the decreasing trend in uptake efficiency of the "classical" carbon storage areas as given in Figure 26.7. Using BOGCMs, Le Quéré et al. (2007) and Lenton et al. (2009) confirm evidence from observations on a decrease in Southern Ocean carbon uptake efficiency and attribute this to transient changes in climatic forcing. However, this carbon uptake efficiency change toward a positive carbon cycle climate feedback can change to an overall negative feedback with a further rise in atmospheric CO_2 concentrations as pointed out by Matear and Lenton (2007). A decrease in CO_2 sink efficiency can similarly be simulated for the North Atlantic (Wetzel et al., 2005) using climatic forcing from reanalysis data though the amplitude in uptake flux (for an observation-derived estimate see, e.g., Watson et al., 2009) seems as yet to be underestimated by models. McKinley et al. (2011) suggested—using a limited area model and observations—that over longer time intervals, fluctuations in regional sink strength for anthropogenic carbon could possibly compensate for each other.

Modeling studies have also helped to elucidate the role of the biological pumps with progressing climatic change. Warming of the surface ocean and the associated increase in stratification is anticipated to result in a more sluggish ocean circulation. The delivery of nutrients to the productive surface ocean by upwelling or injection into the mixed layer will likely decrease. The reduced availability of nutrients will translate into a globally less productive ocean and a decrease in EP (Schmittner, 2005; Steinacher et al., 2010). This can provide a weak negative feedback (e.g., Sarmiento et al., 1998; Plattner et al., 2001) to rising CO_2. Though the integrated EP rate may decrease, the efficiency of the downward pumping carbon mechanism may increase slightly. The negative feedback is about one order of magnitude smaller than the positive feedback through reduced downward mixing of high CO_2 from the surface (bottleneck effect) (see, e.g., Plattner et al., 2001). A potential restructuring of ecosystems accompanying a reduction in primary production could turn into a weakly positive feedback to climate change (Bopp et al., 2005). A reduction in primary production could lead to a decrease in standing stock biomass in the oceans and may have consequences for food production through fisheries (Sherman et al., 2009). Laboratory and mesocosm experiments indicate the possibility of an increase in phytoplankton growth rates under elevated ambient CO_2 (see Zondervan et al., 2001; Riebesell et al., 2007; cf. also Chapter 31). Depending on the depth at which POC will

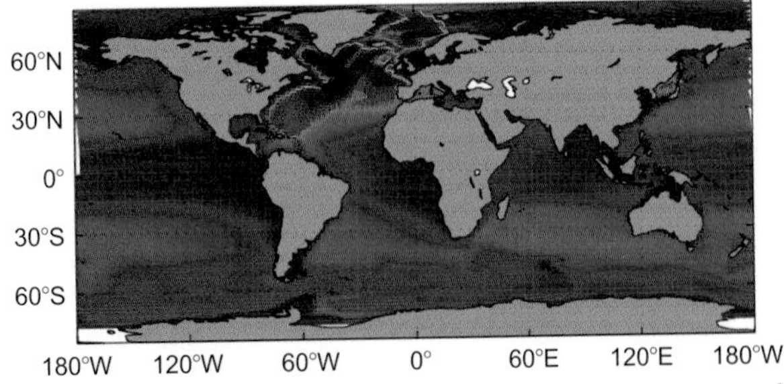

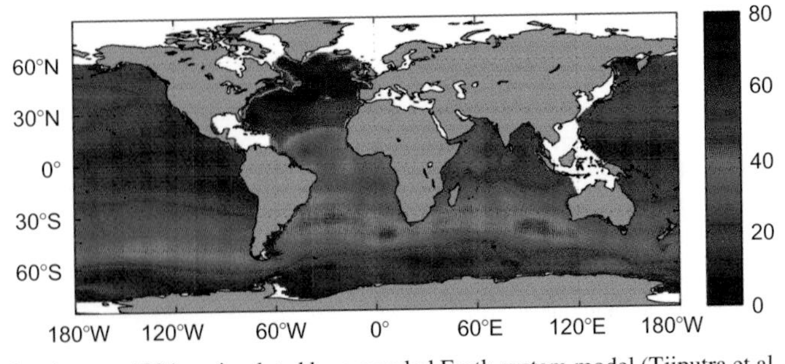

FIGURE 26.7 Anthropogenic CO_2 water column inventory (mol m^{-2}) for the year 1994 as simulated by a coupled Earth system model (Tjiputra et al., 2010a) (left). The simulation compares well with observation-based estimates of Sabine et al. (2004) (right). The units are [mol C m^{-2}]. *Source: Tjiputra, J.F., K. Assmann, M. Bentsen, I. Bethke, O.H. Ottera, C. Sturm, and C. Heinze, (2010a), Bergen earth system model (BCM-C): Model description and regional climate-carbon cycle feedbacks assessment, Geoscientific Model Development, 3, 123–141.*

be remineralized, this could lead to either an increase in biological production or—after some time—a further vertical fractionation of nutrients in the water column and a reduction in surface productivity.

4.3. Model Assessment and Detection Limits of Ocean Acidification

In the last decade, ocean acidification—the decrease in carbonate saturation and pH value caused by anthropogenic CO_2 invasion into the ocean (for a more complete definition see Field et al., 2011)—has emerged as a key topic in marine biogeochemical research (Caldeira and Wickett, 2003; Orr et al., 2005; Raven et al., 2005; Gattuso and Hansson, 2011). The uptake of excess CO_2 from the atmosphere by the ocean (involving the reactions of Equations 26.4 and 26.5) leads to a decrease in carbonate ion concentration $[CO_3^{2-}]$ while concentrations of bicarbonate $[HCO_3^-]$, carbonic acid, gaseous CO_2, and protons increase. Dissolution of calcium carbonate shell material

$$CaCO_3 \rightleftharpoons Ca^{2+} + CO_3^{2-} \quad (26.12)$$

depends on the solubility product (function of pressure, temperature, and salinity) for $CaCO_3$:

$$K_{sp}^*(S,T,p) = \left[Ca^{2+}\right]_{saturation} \cdot \left[CO_3^{2-}\right]_{saturation}. \quad (26.13)$$

As the calcium ion concentration does not vary much in seawater, the critical value determining the saturation state of seawater with respect to $CaCO_3$, Ω, is governed by the *in situ* $[CO_3^{2-}]$ itself:

$$\Omega = \left(\left[Ca^{2+}\right] \cdot \left[CO_3^{2-}\right]\right) / K_{sp}^*. \quad (26.14)$$

Ω values larger than 1 indicate oversaturation. For $\Omega < 1$, dissolution of $CaCO_3$ occurs. The solubility product of the various $CaCO_3$ polymorphs (e.g., aragonite, calcite) differs and is larger for the less stable aragonite than for calcite. Should oceanic uptake of anthropogenic CO_2 lead to a decrease in biological $CaCO_3$ production, a somewhat higher alkalinity background can be expected as less carbonate ions are being taken out of the system for shell material production. Biogeochemical ocean models have indicated so far that the climatic feedback caused by a change in biological $CaCO_3$ production is most probably a minor one in relation to the ongoing and expected anthropogenic CO_2 emissions (Heinze, 2004; Gehlen et al., 2007; Ridgwell et al., 2007). Open questions remain, such as a potential shift in Redfield ratios (carbon overconsumption) in high CO_2 waters as indicated in mesocosm experiments (Riebesell et al., 2007, 2009).

The timing of progressive ocean acidification and the associated shoaling of the saturation horizons for the different $CaCO_3$ polymorphs has been simulated by BOGCMs for prescribed CO_2 emission scenarios (Orr et al., 2005; Steinacher et al., 2009; Frölicher and Joos, 2010; Gangstø et al., 2011). Because of the nonlinear inorganic carbon chemistry relationships in seawater, cold (and thus high-latitude) waters are most vulnerable to ocean acidification and local impacts on biota may occur in the Southern Ocean and Arctic Ocean soon. Steinacher et al. (2009) estimate that by year 2016, parts of the Arctic Ocean surface waters will start to be undersaturated with respect to aragonite (Figure 26.8). That ocean acidification on human time-scales also affects deep and abyssal waters has been shown by Gehlen et al. (2008) who used a biogeochemical ocean model with an interactive sediment module.

Models are also useful in identifying expected detection thresholds for large-scale impacts of ocean acidification. Ilyina et al. (2009) showed that a large-scale decrease in $CaCO_3$ shell material production can be

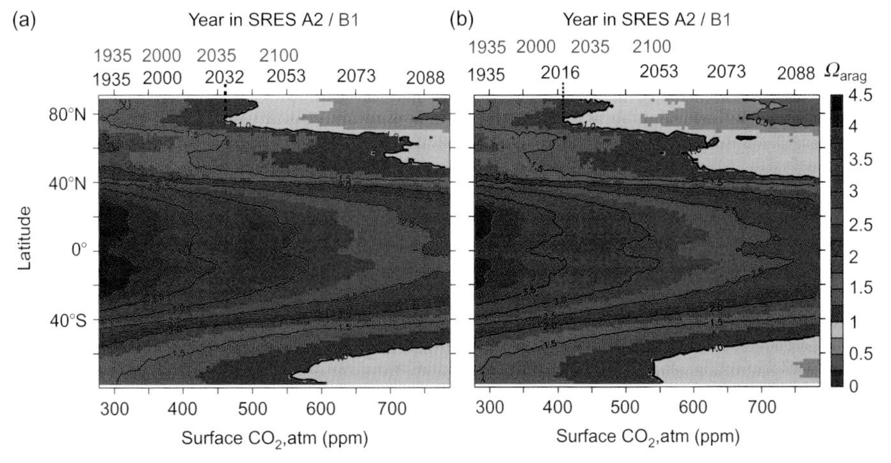

FIGURE 26.8 Timing of projected ocean acidification (Steinacher et al., 2009). Shown are (a) annual-mean and (b) lowest monthly mean zonally averaged aragonite saturation for a model scenario using IPCC SRES A2 and B1 forcing. The evolution of aragonite saturation is given in relation to the mean annual atmospheric CO_2 mixing ratio (almost identical to the numerical values of the pCO_2) at the sea surface. For comparison to the temporal evolution for the A2 scenario (lower row of year numbers), years for a more moderate B1 scenario (upper row of year numbers) are listed on the top of the figures as well. The dotted line marks the boundary between aragonite oversaturation and undersaturation at 77°N. *Source: Steinacher et al., (2009), Imminent ocean acidification in the Arctic projected with the NCAR global coupled carbon cycle-climate model, Biogeosciences, 6, 515–533.*

unambiguously identified from *in situ* alkalinity measurements only in about 20–30 years from now given the high-quality measurements through the last decade. Though the pH change will be largest in high-latitude waters, the alkalinity effect is likely to be seen first in the equatorial Pacific Ocean because of the overall higher biological production rate there.

5. CONCLUSION

Biogeochemical ocean models have become an integral part of climate modeling, either as stand-alone models for purely oceanographic studies or as modules of comprehensive Earth system models. Typical fields in which these models have been and are successfully being used are as follows: Future climate projections including the carbon cycle climate feedback, studies of the extent and timing as well as variability of anthropogenic CO_2 uptake by the oceans, studies on the progress and impact of ocean acidification, and studies on the natural cycling of carbon and related biogeochemical cycles. In most modeling exercises, ecosystem models of the NPZD-type have been used. With a few exceptions, riverine inflow and sediment processes have not yet been included. In most global applications, horizontal model resolution is on the order of 1°.

Though a larger number of BOGCMs have become conceptually quite similar, different models still yield considerably different results. This can be mainly attributed to the different flow fields from the physical dynamical ocean models that drive the biogeochemical modules. A considerable number of process parameterizations are still based, at best, on semi-empirical evidence and the choice of specific adjustable model parameters is associated with large uncertainties. This applies to air–sea interaction processes and details of biological production and ecosystem dynamics, as well as to details of marine particle fluxes.

A suite of innovative initiatives has started which in the coming years could alter BOGCMs substantially and will lead to more process-based and detailed simulations of ocean tracers and the carbon cycle. First of all, more emphasis will be placed on the coupled cycling of carbon, nitrogen, phosphate, silicon, iron (and further micronutrients), and oxygen, as well as radionuclides. Global BOGCMs that are able to simulate a large number of tracers simultaneously will provide a solid ground for reliable predictions. Ecosystem models will include a larger number of functional groups and processes through better defined process parameterizations as functions of physical and biogeochemical ambient conditions (Le Quéré et al., 2005; Follows et al., 2007). Also, the representation of microbial organisms (bacteria, viruses) as well as higher trophic levels (especially important for impact studies) is expected to improve. A higher spatial resolution and increased computer power will allow a more realistic inclusion of the effects of smaller scale motion (e.g., Gruber et al., 2011).

In general, detailed anaerobic processes are as yet not included in climate-type pelagic ocean biogeochemisty models. For example, denitrification at low or zero dissolved oxygen conditions is currently accounted for in only a few state-of-the-art BOGCMs. Such anaerobic processes will have to be prescribed more comprehensively in forthcoming BOGCMs for climate applications including the cycles of N_2O and CH_4. Higher resolution models will allow for an improved simulation of the transition from rivers/estuaries, shelf seas, and the open ocean including anthropogenically induced perturbations. These perturbations include increased nutrient inflow into the ocean—also through aeolian deposition—rising temperatures, and decreasing pH values. The problem of widespread oxygen drawdown in the oceans will have to be included in the projections of greenhouse gas budgets. Advanced carbon cycle data assimilation systems are currently under development, which will in the near future enable a consistent best possible estimate of marine greenhouse gas budgets within the Earth system. These studies will also pinpoint specific issues to be improved in current models and observation systems. As marine biogeochemistry is still a relatively young discipline, exciting new findings about the ocean as a chemical plant driven by circulation as a key factor are to be expected.

ACKNOWLEDGMENTS

We are grateful for the thorough reading and constructive comments by Friederike Hoffmann. Six anonymous reviewers provided excellent comments and suggestions for improving the original chapter. Thanks are due to Stephen M. Griffies, Carolina Dufour, and Gerold Siedler for their support in improving the chapter. This study was cosponsored by the EU FP7 large-scale integrating project CARBOCHANGE (grant agreement no. 264879), EU FP7 project COMBINE (grant agreement no. 226520), and the EU FP7 coordination action COCOS (grant agreement no. 212196). C. H. is grateful for sabbatical funds from the University of Bergen (faculty of mathematics and natural sciences) in 2011 where parts of this chapter have been prepared. This is a contribution to the Center for Climate Dynamics (SKD), core project BIOFEEDBACK, at the Bjerknes Centre for Climate Research in Bergen. This is publication nr. A409 of the Bjerknes Centre for Climate Research.

REFERENCES

Anderson, L.A., 1995. On the hydrogen and oxygen content of marine phytoplankton. Deep Sea Res. Part I 42, 1675–1680.

Andres, J., Gregg, J.S., Losey, L., Marland, G., Boden, T.A., 2011. Monthly, global emissions of carbon dioxide from fossil fuel consumption. Tellus 63B, 309–327.

Antoine, D., André, J.M., Morel, A., 1996. Oceanic primary production: 2. Estimation at global scale from satellite (Coastal Zone Color Scanner) chlorophyll. Global Biogeochem. Cycles 10 (1), 57–69.

Archer, D., 2005. Fate of fossil fuel CO_2 in geologic time. J. Geophys. Res. Oceans 110 (C9), C09S05.

Archer, D., Maier-Reimer, E., 1994. The effect of deep-sea sedimentary calcite preservation on atmospheric CO_2 concentration. Nature 367 (6460), 260–263.

Archer, D., Kheshgi, H., Maier-Reimer, E., 1998. Dynamics of fossil fuel CO_2 neutralization by marine $CaCO_3$. Global Biogeochem. Cycles 12 (2), 259–276.

Archer, D., Buffett, B., Brovkin, V., 2009. Ocean methane hydrates as a slow tipping point in the global carbon cycle. Proc. Natl. Acad. Sci. U.S.A. 106 (49), 20596–20601.

Assmann, K.M., Bentsen, M., Segschneider, J., Heinze, C., 2010. An isopycnic ocean carbon cycle model. Geosci. Model Dev. 3, 143–167.

Aumont, O., Bopp, L., 2006. Globalizing results from ocean in situ iron fertilization studies. Global Biogeochem. Cycles 20, GB2017.

Aumont, O., Orr, J.C., Jamous, D., Monfray, P., Marti, O., Madec, G., 1998. A degradation approach to accelerate simulations to steady-state in a 3-D tracer transport model of the global ocean. Clim. Dyn. 14, 101–116.

Aumont, O., Maier-Reimer, E., Blain, S., Monfray, P., 2003. An ecosystem model of the global ocean including Fe, Si, P colimitations. Global Biogeochem. Cycles 17 (2), 1060.

Azam, F.T., Fenchel, J.G., Gray, L.A. Meyer-Reil, Thingstad, F., 1983. The ecological role of water-column microbes in the sea. Mar. Ecol. Prog. Ser. 10, 257–263.

Barton, A.D., Dutkiewicz, S., Flierl, G., Bragg, J., Follows, M.J., 2010. Patterns of diversity in marine phytoplankton. Science 327, 1509–1511.

Berger, A., Loutre, M.F., Crucifix, M., 2003. The Earth's climate in the next hundred thousand years (100 kyr). Surv. Geophys. 24 (2), 117–138.

Berner, R.A., 2006. GEOCARBSULF: a combined model for Phanerozoic atmospheric O_2 and CO_2. Geochim. Cosmochim. Acta 70, 5653–5664.

Berner, R.A., Canfield, D.E., 1989. A new model for atmospheric oxygen over Phanerozoic time. Am. J. Sci. 289 (4), 333–361.

Bianchi, D., Sarmiento, J.L., Gnanadesikan, A., 2010. Low helium flux from the mantle inferred from simulations of oceanic helium isotope data. Earth Planet. Sci. Lett. 297 (3–4), 379–386.

Biastoch, A.T., Treude, L.H., Rüpke, U., Riebesell, C., Roth, E.B., Burwicz, W., Park, M., Latif, C.W., Böning, G. Madec, Wallmann, K., 2011. Rising Arctic Ocean temperatures cause gas hydrate destabilization and ocean acidification. Geophys. Res. Lett. 38, L08602.

Bishop, J.K.B., 2009. Autonomous observations of the ocean biological carbon pump. Oceanography 22 (2), 182–193.

Bolin, B., Eriksson, E., 1959. Changes in the carbon dioxide content of the atmosphere and sea due to fossil fuel combustion. In: Bolin, B. (Ed.), The Atmosphere and the Sea in Motion, Rossby Memorial Volume. Rockefeller Inst., New York, pp. 130–142.

Bolin, B., Björkström, A., Holmén, K., Moore, B., 1987. On inverse methods for combining chemical and physical oceanographic data: a state-state analysis of the Atlantic ocean, Report CM-71, August 1987, UDC 551.464:551.465, ISSN 0280-445X, University of Stockholm Sweden, 188 pp. plus appendices.

Bopp, L., Aumont, O., Cadule, P., Alvain, S., Gehlen, M., 2005. Response of diatoms distribution to global warming and potential implications: a global model study. Geophys. Res. Lett. 32, L19606.

Brasseur, P., Gruber, N., Barciela, R., Brander, K., Doron, M., El Moussaoui, A., Hobday, A.J., Huret, M., Kremeur, A.-S., Lehodey, P., Matear, R., Moulin, C., Murtugudde, R., Senina, I., Svendsen, E., 2009. Integrating biogeochemistry and ecology into ocean data assimilation systems. Oceanography 22 (3), 206–215.

Bretherton, F.P., 1985. Earth system science and remote sensing. Proc. IEEE 73 (6), 1118–1127.

Broecker, W.S., Peng, T.-H., 1982. Tracers in the sea, ELDGIO Press, Lamont-Doherty Geological Observatory, Columbia University, Palisades, New York, 690 pp.

Broecker, W.S., Peng, T.-H., 1986. Carbon-cycle—1985 glacial to interglacial changes in the operation of the global carbon-cycle. Radiocarbon 28 (2A), 309–327.

Broecker, W.S., Takahashi, T., 1977. Neutralization of fossil fuel CO_2 by marine calcium carbonate. In: Andersen, N.R., Malahoff, A. (Eds.), The Fate of Fossil Fuel CO_2 in the Oceans. Plenum Press, New York, pp. 213–241.

Broecker, W.S., Takahashi, T., Takahashi, T., 1985. Sources and flow patterns of deep-ocean waters as deduced from potential temperature, salinity, and initial phosphate concentration. J. Geophys. Res. 90 (C4), 6925–6939.

Broecker, W.S., Sutherland, S., Smethie, W., Peng, T.-H., Ostlund, G., 1995. Oceanic radiocarbon: separation of the natural and bomb components. Global Biogeochem. Cycles 9 (2), 263–288.

Brovkin, V., Ganopolski, A., Archer, D., Rahmstorf, S., 2007. Lowering of glacial atmospheric CO_2 in response to changes in oceanic circulation and marine biogeochemistry. Paleoceanography 22, 14, PA4202.

Buesseler, K.O., Michaelis, A.F., Siegel, D.A., Knap, A.H., 1994. A 3-dimensional time-dependent approach to calibrating sediment trap fluxes. Global Biogeochem. Cycles 8 (2), 179–193.

Bullister, J.L., Weiss, R.F., 1983. Anthropogenic chlorofluoromethanes in the Greenland and Norwegian Seas. Science 221, 265–268.

Caldeira, K., Wickett, M.E., 2003. Anthropogenic carbon and ocean pH. Nature 425, 365.

Carr, M.-E., Friedrichs, M.A.M., Schmeltz, M., Noguchi Aita, M., Antoine, D., Arrigo, K.R., Asanuma, I., Aumont, O., Barber, R., Behrenfeld, M., Bidigare, R., Buitenhuis, E.T., Campbell, J., Ciotti, A., Dierssen, H., Dowell, M., Dunne, J., Esaias, W., Gentili, B., Gregg, W., Groom, S., Hoepffner, N., Ishizaka, J., Kameda, T., Le Quéré, C., Lohrenz, S., Marra, J., Melino, F., Moore, K., Morel, A., Redd, T.E., Ryan, J., Scardi, M., Smyth, T., Turpie, K., Tilstone, G., Waters, K., Yamanaka, Y., 2006. A comparison of global estimates of marine primary production from ocean color. Deep Sea Res. Part II 53, 741–770.

Charlson, R.J., Lovelock, J.E., Andreae, M.O., Warren, S.G., 1987. Oceanic phytoplankton, atmospheric sulphur, cloud albedo and climate. Nature 326, 655–661.

Chipman, D.W., Marra, J., Takahashi, T., 1993. Primary production at 47°N and 20°W in the North Atlantic Ocean: a comparison between the 14C incubation method and the mixed layer carbon budget. Deep Sea Res. Part II 40 (1–2), 151–169.

Conkright, M.E., Levitus, S., Boyer, T., 1994. NOAA Atlas NESDIS 1, World Ocean Atlas 1994, vol. 1: Nutrients, U.S. Department of Commerce, Washington D.C, 150 pp.

Cooper, L.H.N., 1956. On assessing the age of deep oceanic water by carbon-14. J. Mar. Biol. Assoc. U.K. 35, 341–354.

Cox, P.M., Betts, R.A., Jones, C.D., Spall, S.A., Totterdell, I.J., 2000. Acceleration of global warming due to carbon-cycle feedbacks in a coupled climate model. Nature 408, 184–187.

Crueger, T., Roeckner, E., Raddatz, T., Schnur, R., Wetzel, P., 2008. Ocean dynamics determine the response of oceanic CO_2 uptake to climate change. Clim. Dyn. 31, 151–168.

Crutzen, P.J., 2002a. Geology of mankind. Nature 415, 23.

Crutzen, P.J., 2002b. The "anthropocene" J. Phys. IV 12 (PR10), 1–5.

Denman, K.L., Brasseur, G., Chidthaisong, A., Ciais, P., Cox, P.M., Dickinson, R.E., Hauglustaine, D., Heinze, C., Holland, E., Jacob, D., Lohmann, U., Ramachandran, S., da Silva Dias, P.L., Wofsy, S.C., Zhang, X., 2007. Couplings between changes in the climate system and biogeochemistry. In: Solomon, S., Qin, D., Manning, M., Chen, Z., Marquis, M., Averyt, K.B., Tignor, M., Miller, H.L. (Eds.), Climate Change 2007: The Physical Science Basis. Contribution of Working Group I to the Fourth Assessment Report of the Intergovernmental Panel on Climate Change. Cambridge University Press, Cambridge, United Kingdom, and NY, USA, pp. 499–587.

Dickson, A.G., 1981. An exact definition of total alkalinity and a procedure for the estimation of alkalinity and total inorganic carbon from titration data. Deep Sea Res. 28A (6), 609–623.

Dickson, A.G., 1992. The development of the alkalinity concept in marine chemistry. Mar. Chem. 40, 49–63.

Doney, S.C., Lindsay, K., Fung, I., John, J., 2006. Natural variability in a stable, 1000-yr global coupled climate-carbon cycle simulation. J. Clim. 19 (13), 3033–3054.

Dunne, J.P., John, J.G., Adcroft, A.J., Griffies, S.M., Hallberg, R.W., Shevliakova, E., Stouffer, R.J., Cooke, W., Dunne, K.A., Harrison, M.J., Krasting, J.P., Malyshev, S.L., Milly, P.C.D., Phillipps, P.J., Sentman, L.T., Samuels, B.L., Spelman, M.J., Winton, M., Wittenberg, A.T., Zadeh, N., 2012. GFDL's ESM2 global coupled climate-carbon Earth System Models Part I: physical formulation and baseline simulation characteristics. J. Clim. 25 (19), 6646–6665.

Dunne, J.P., John, J.G., Shevliakova, E., Stouffer, R.J., Krasting, J.P., Malyshev, S.L., Milly, P.C.D., Sentman, L.T., Adcroft, A.J., Cooke, W., Dunne, K.A., Griffies, S.M., Hallberg, R.W., Harrison, M.J., Levy, H., Wittenberg, A.T., Phillips, P.J., Zadeh, N., 2013. GFDLs ESM2 global coupled climate-carbon Earth System Models Part II: carbon system formulation and baseline simulation characteristics. J. Clim. 26 (7), 2247–2267.

Dutay, J.-C., Jean-Baptiste, P., Campin, J.-M., Ishida, A., Maier-Reimer, E., Matear, R.J., Mouchet, A., Totterdell, I.J., Yamanaka, Y., Rodgers, K., Madec, G., Orr, J.C., 2004. Evaluation of OCMIP-2 ocean models deep circulation with mantle helium-3. J. Mar. Syst. 48, 1536.

England, M., Maier-Reimer, E., 2001. Using chemical tracers to assess ocean models. Rev. Geophys. 39 (1), 29–70.

Evensen, G., 2009. The Ensemble Kalman Filter for combined state and parameter estimation. IEEE Control Syst. Mag. 83–104.

Fasham, M.J.R., Ducklow, H.W., McKelvie, S.M., 1990. A nitrogen-based model of plankton dynamics in the oceanic mixed layer. J. Mar. Res. 48, 591–639.

Field, C.B., Barros, V., Stocker, T.F., Dahe, Q., Mach, K.J., Plattner, G.-K., Mastrandrea, M., Tignor, M., Ebi, K.L. (Eds.), 2011. IPCC workshop on impacts of ocean acidification on marine biology and ecosystems, Bankoku Shinryokan, Okinawa, Japan, 17th–19th January 2011, Workshop Report, 174 pp.

Follows, M.J., Dutkiewicz, S., Grant, S., Chisholm, S.W., 2007. Emergent biogeography of microbial communities in a model ocean. Science 315 (5820), 1843–1846.

Forster, P., Ramaswamy, V., Artaxo, P., Berntsen, T., Betts, R., Fahey, D.W., Haywood, J., Lean, J., Lowe, D.C., Myhre, G., Nganga, J., Prinn, R., Raga, G., Schulz, M., Van Dorland, R., 2007. Changes in atmospheric constituents and in radiative forcing. In: Solomon, S., Qin, D., Manning, M., Chen, Z., Marquis, M., Averyt, K.B., Tignor, M., Miller, H.L. (Eds.), Climate Change 2007: The Physical Science Basis. Contribution of Working Group I to the Fourth Assessment Report of the Intergovernmental Panel on Climate Change. Cambridge University Press, Cambridge, United Kingdom, and NY, USA.

Friedlingstein, P., Cox, P., Betts, R., Bopp, L., van Bloh, W., Brovkin, V., Cadule, P., Doney, S., Eby, M., Fung, I., Bala, G., John, J., Jones, C., Joos, F., Kato, T., Kawamiya, M., Knorr, W., Lindsay, K., Matthews, D., raddatz, T., Tayner, P., Reick, C., Roeckner, E., Schnitzler, K.-G., Schnur, R., Strassmann, K., Weaver, A.J., Yoshikawa, C., Zeng, N., 2006. Climate carbon cycle feedback analysis: results from the C^4MIP model intercomparison. J. Clim. 19, 3337–3353.

Frölicher, T.I., Joos, F., 2010. Reversible and irreversible impacts of greenhouse gas emissions in multi-century projections with the NCAR global coupled carbon cycle—climate model. Clim. Dyn. 35, 1439–1459.

Ganachaud, A., Wunsch, C., 2002. Oceanic nutrient and oxygen transports and bounds on export production during the World Ocean Circulation Experiment. Global Biogeochem. Cycles 16 (4), Article Number: 1057, pages 5-1 to 5-14 (14 pages).

Gangstø, R., Joos, F., Gehlen, M., 2011. Sensitivity of pelagic calcification to ocean acidification. Biogeosciences 8, 433–458.

Gattuso, J.-P., Hansson, L. (Eds.), 2011. Ocean Acidification. Oxford University Press, Oxford, New York, 326 pp.

Gehlen, M., Bopp, L., Ernprin, N., Aumont, O., Heinze, C., Raguencau, O., 2006. Reconciling surface ocean productivity, export fluxes and sediment composition in a global biogeochemical ocean model. Biogeosciences 3 (4), 521–537.

Gehlen, M., Gangstø, R., Schneider, B., Bopp, L., Aumont, O., Ethe, C., 2007. The fate of pelagic $CaCO_3$ production in a high CO_2 ocean: a model study. Biogeosciences 4, 505–519.

Gehlen, M., Bopp, L., Aumont, O., 2008. Short-term dissolution response of pelagic carbonate sediments to the invasion of anthropogenic CO_2: a model study. Geochem. Geophys. Geosyst. 9, 11, Q02012.

Gent, P.R., Danabasoglu, G., Donner, L.J., Holland, M.M., Hunke, E.C., Jayne, S.R., Lawrence, D.M., Neale, R.B., Rasch, P.J., Vertenstein, M., Worley, P.H., Yang, Z.-L., Zhang, M., 2011. The community climate system model version 4. J. Clim. 24, 4973–4991.

Gerber, M., Joos, F., Vázquez-Rodríguez, M., Touratier, F., Goyet, C., 2009. Regional air-sea fluxes of anthropogenic carbon inferred with an Ensemble Kalman Filter. Global Biogeochem. Cycles 23, 15, GB1013.

Gloor, M., Gruber, N., Hughes, T.M.C., Sarmiento, J.L., 2001. Estimating net air-sea fluxes from ocean bulk data: methodology and application to the heat cycle. Global Biogeochem. Cycles 15 (4), 767–782.

Gloor, M., Gruber, N., Sarmiento, J., Sabine, C.L., Feely, R.A., Rödenbeck, C., 2003. A first estimate of present and preindustrial air-sea CO_2 flux patterns based on ocean interior carbon measurements and models. Geophys. Res. Lett. 30 (1), 1010.

Gnanadesikan, A., 1999. A global model of silicon cycling: sensitivity to eddy parameterization and dissolution. Global Biogeochem. Cycles 13 (1), 199–230.

Gnanadesikan, A., Anderson, W.G., 2009. Ocean water clarity and the ocean general circulation in a coupled climate model. J. Phys. Oceanogr. 39 (2), 314–332.

Gnanadesikan, A., Emanuel, K.A., Vecchi, G.A., Anderson, W.G., Hallberg, R.W., 2010. How ocean color can steer Pacific tropical cyclones. Geophys. Res. Lett. 37, L18802.

Gregg, W.W., 2008. Assimilation of SeaWiFS ocean chlorophyll data into a three-dimensional global ocean model. J. Mar. Syst. 69, 205–225.

Gruber, N., Lachkar, Z., Frenzel, H., Marchesiello, P., Münnich, M., McWilliams, J.C., Nagai, T., Plattner, G.-K., 2011. Eddy-induced reduction of biological production in eastern boundary upwelling systems. Nat. Geosci. 4 (11), 787–792.

Hamme, R.C., Webley, P.W., Crawford, W.R., Whitney, F.A., DeGrandpre, M.D., Emerson, S.R., Eriksen, C.C., Giesbrecht, K.E., Gower, J.F.R., Kavanaugh, M.T., Peña, M.A., Sabine, C.L., Batten, S.D., Coogan, L.A., Grundle, D.S., Lockwood, D., 2010. Volcanic ash fuels anomalous plankton bloom in subarctic northeast Pacific. Geophys. Res. Lett. 37, L19604.

Heinze, C., 2002a. Modelling of the global marine Si cycle. achievements and missing links. Oceanis 28 (3–4), 557–566.

Heinze, C., 2002b. Das marine Sediment als Klimazeuge und Komponente des Klimasystems—eine Modellstudie, Habilitationsschrift (habilitation thesis), Fachbereich Geowissenschaften, Universitt Hamburg, GCA-Verlag, Serie Forschen und Wissen—Physik, ISBN 3-89863-084-6, Herdecke, Germany, 124 pp.

Heinze, C., 2004. Simulating oceanic $CaCO_3$ export production in the greenhouse. Geophys. Res. Lett. 31, L16308.

Heinze, C., Schlosser, P., Koltermann, K.P., Meincke, J., 1990. A tracer study of the deep water renewal in the European Polar Seas. Deep Sea Res. 37 (a), 1425–1453.

Heinze, C., Maier-Reimer, E., Winn, K., 1991. Glacial pCO_2 reduction by the World Ocean—experiments with the Hamburg Carbon Cycle Model. Paleoceanography 6, 395–430.

Heinze, C., Maier-Reimer, E., Winguth, A.M.E., Archer, D., 1999. A global oceanic sediment model for long-term climate studies. Global Biogeochem. Cycles 13, 221–250.

Hellerman, S., Rosenstein, M., 1983. Normal monthly wind stress over the world ocean with error estimates. J. Phys. Oceanogr. 13, 1093–1104.

Hoehler, T.M., Alperin, M.J., Albert, D.B., Martens, C.S., 1994. Field and laboratory studies of methane oxidation in an anoxic marine sediment evidence for a methanogensulfate reducer consortium. Global Biogeochem. Cycles 8, 451–463.

Holzer, M., Primeau, F.W., Smethie Jr., W.M., Khatiwala, S., 2010. Where and how long ago was water in the western North Atlantic ventilated? Maximum entropy inversions of bottle data from WOCE line A20. J. Geophys. Res. 115, 26, C07005.

Hooker, S.B., Esaias, W.E., Feldman, G.C., Gregg, W.W., McClain, C.R., 1992. An overview of SeaWiFS and ocean color, NASA Tech. Memo. vol. 104566, National Aeronautics and Space Administration, Goddard Space Flight Center, Greenbelt, MD.

Houghton, R.A., 1999. The annual net flux of carbon to the atmosphere from changes in land use 18501990. Tellus 51B, 298–313.

Hurtt, G.C., Frolking, S., Fearon, M.G., Moore, B., Shevliakova, E., Malyshev, S., Pacala, S.W., Houghton, R.A., 2006. The underpinnings of land-use history: three centuries of global gridded land-use transitions, wood-harvest activity, and resulting secondary lands. Glob. Chang. Biol. 12 (7), 1208–1229.

Ilyina, T., Zeebe, R.E., Maier-Reimer, E., Heinze, C., 2009. Early detection of ocean acidification effects on marine calcification. Global Biogeochem. Cycles 23, 11, GB1008.

Ito, T., Woloszyn, M., Mazloff, M., 2010. Anthropogenic carbon dioxide transport in the Southern Ocean driven by Ekman flow. Nature 463, 80–83.

Jähne, B., Haußecker, H., 1998. Air-water gas exchange. Annu. Rev. Fluid Mech. 30, 443–468.

Jahnke, R.A., 1996. The global ocean flux of particulate organic carbon: areal distribution and magnitude. Global Biogeochem. Cycles 10 (1), 71–88.

Jean-Baptiste, P., Jenkins, W.J., Dutay, J.C., Fourre, E., Leboucher, V., Fieux, M., 2004. Temporally integrated estimate of the Indonesian throughflow using tritium. Geophys. Res. Lett. 31, 4, L21301.

Jenkins, W.J., 1980. Tritium and He-3 in the Sargasso Sea. J. Mar. Res. 38 (3), 533–569.

Jickells, T.D., An, Z.S., Andersen, K.K., Baker, A.R., Bergametti, G., Brooks, N., Cao, J.J., Boyd, P.W., Duce, R.A., Hunter, K.A., Kawahata, H., Kubilay, N., laRoche, J., Liss, P.S., Mahowald, N., Prospero, J.M., Ridgwell, A.J., Tegen, I., Torres, R., 2005. Global iron connections between desert dust, ocean biogeochemistry, and climate. Science 308 (5718), 67–71.

Jin, X., Gruber, N., Dunne, J.P., Sarmiento, J.L., Armstrong, R.A., 2006. Diagnosing the contribution of phytoplankton functional groups to the production and export of particulate organic carbon, $CaCO_3$, and opal from global nutrient and alkalinity distributions. Global Biogeochem. Cycles 20, GB2015.

Joos, F., Gerber, S., Prentice, I.C., Otto-Bliesner, B.L., Valdes, P.J., 2004. Transient simulations of Holocene atmospheric carbon dioxide and terrestrial carbon since the last glacial maximum. Global Biogeochem. Cycles 18 (2), GB2002.

Kalnay, E., Kanamitsu, M., Kistler, R., Collins, W., Deaven, D., Gandin, L., Iredell, M., Saha, S., White, G., Woollen, J., Zhu, Y., Chelliah, M., Ebisuzaki, W., Higgins, W., Janowiak, J., Mo, K.C., Ropelewski, C., Wang, J., Leetmaa, A., Reynolds, R., Jenne, R., Joseph, D., 1996. The NCEP/NCAR 40-year reanalysis project. Bull. Am. Meteorol. Soc. 77, 437–470.

Kane, A., Moulin, C., Thiria, S., Bopp, L., Berrada, M., Tagliabue, A., Crepon, M., Aumont, O., Badran, F., 2011. Improving the parameters of a global ocean biogeochemical model via variational assimilation of in situ data at five time series stations. J. Geophys. Res. Oceans 116, 14, C06011.

Keeling, C.D., Bacastow, R.B., Bainbridge, A.E., Ekdahl, C.A., Guenther, P.R., Watermann, L.S., Chin, F.J.S., 1976. Atmospheric carbon-dioxide variations at Mauna-Loa Observatory, Hawaii. Tellus 28 (6), 538–551.

Keeling, C.D., Brix, H., Gruber, N., 2004. Seasonal and long-term dynamics of the upper ocean carbon cycle at Station ALOHA near Hawaii. Global Biogeochem. Cycles 18, 18, GB4006.

Key, R.M., Kozyr, A., Sabine, C.L., Lee, K., Wanninkhof, R., Bullister, J.L., Feely, R.A., Millero, F.J., Mordy, C., Peng, T.-H., 2004. A global ocean carbon climatology: results from Global Data Analysis Project (GLODAP). Global Biogeochem. Cycles 18, 23, GB4031.

Key, R.M., Tanhua, T., Olsen, A., Hoppema, M., Jutterström, S., Schirnick, C., van Heuven, S., Kozyr, A., Lin, X., Velo, A., Wallace, D.W.R., Mintrop, L., 2010. The CARINA data synthesis project: introduction and overview. Earth Syst. Sci. Data 2, 105–121.

Khatiwala, S., 2007. A computational framework for simulation of biogeochemical tracers in the ocean. Global Biogeochem. Cycles 21, 14, GB3001.

Kloster, S., Six, K.D., Feichter, J., Maier-Reimer, E., Roeckner, E., Wetzel, P., Stier, P., Esch, M., 2007. Response of dimethylsulfide (DMS) in the ocean and atmosphere to global warming. J. Geophys. Res. 112, G03005.

Knox, F., McElroy, M.B., 1984. Changes in atmospheric CO_2—influence of the marine biota at high-latitude. J. Geophys. Res. 89, 4629–4637, ND3.

Kohonen, T., 2001. Self-Organizing Maps, third ed. Springer-Verlag, Berlin, Heidelberg, New York, 501 pp.

Kriest, I., Evans, G.T., 1999. Representing phytoplankton aggregates in biogeochemical models. Deep Sea Res. Part I 46, 1841–1859.

Kriest, I., Khatiwala, S., Oschlies, A., 2010. Towards an assessment of simple global marine biogeochemical models of different complexity. Prog. Oceanogr. 86, 337–360.

Kwon, E.Y., Primeau, F., 2006. Optimization and sensitivity study of a biogeochemistry ocean model using an implicit solver and in situ phosphate data. Global Biogeochem. Cycles 20, GB4009.

Lachkar, Z., Orr, J.C., Dutay, J.-C., 2009. Seasonal and mesoscale variability of oceanic transport of anthropogenic CO_2. Biogeosciences 6, 2509–2523.

Large, W.G., Yeager, S.G., 2009. The global climatology of an interannually varying air-sea flux data set. Clim. Dyn. 33, 341–364.

Le Quéré, C., Orr, J.C., Monfray, P., Aumont, O., Madec, G., 2000. Interannual variability of the oceanic sink of CO_2 from 1979 through 1997. Global Biogeochem. Cycles 14 (4), 1247–1265.

Le Quéré, C., Harrison, S.P., Prentice, I.C., Buitenhuis, E.T., Aumont, O., Bopp, L., Claustre, H., Da Cunha, L.C., Geider, R., Giraud, X., Klaas, C., Kohfeld, K.E., Legendre, L., Manizza, M., Platt, T., Rivkin, R.B., Sathyendranath, S., Uitz, J., Watson, A.J., Wolf-Gladrow, D., 2005. Ecosystem dynamics based on plankton functional types for global ocean biogeochemistry models. Glob. Chang. Biol. 11 (11), 2016–2040.

Le Quéré, C., Rödenbeck, E.T., Buitenhuis, T.J., Conway, R., Langenfelds, A., Gomez, C., Labuschagne, M., Ramonet, T., Nakazawa, N., Metzl, N. Gillett, Heimann, M., 2007. Saturation of the Southern Ocean CO_2 sink due to recent climate change. Science 316 (5832), 1735–1738.

Lee, T., Awaji, T., Balmaseda, M.A., Greiner, E., Stammer, D., 2009. Ocean state estimation for climate research. Oceanography 22 (3), 160–167.

Lengaigne, M., Menkes, C., Aumont, O., Gorgues, T., Bopp, L., André, J.-M., Madec, G., 2007. Influence of the oceanic biology on the tropical Pacific climate in a coupled general circulation model. Clim. Dyn. 28, 503–516.

Lenton, A., Codron, F., Bopp, L., Metzl, N., Cadule, P., Tagliabue, A., Le Sommer, J., 2009. Stratospheric ozone depletion reduces ocean carbon uptake and enhances ocean acidification. Geophys. Res. Lett. 36, 5, L12606.

Lévy, M., 2008. The modulation of biological production by oceanic mesoscale turbulence. In: Weiss, J.B., Provenzale, A. (Eds.), Transport and Mixing in Geophysical Flows. Lecture Notes in Physics, vol. 744. Springer Verlag, Berlin, Heidelberg, pp. 219–261.

Lévy, M., Klein, P., Treguier, A.-M., 2001. Impact of sub-mesoscale physics on production and subduction of phytoplankton in an oligotrophic regime. J. Mar. Res. 59, 535–565.

Libes, S.M., 1992. An Introduction to Marine Biogeochemistry. John Wiley and sons, New York, 734 pp.

Lin, Z., Thiffeault, J.-L., Childers, S., 2011. Stirring by squirmers. J. Fluid Mech. 669, 167–177. http://dx.doi.org/10.1017/S002211201000563.

Lohrenz, S.E., Knauer, G.A., Asper, V.L., Tuel, M., Michaels, A.F., Knap, A.H., 1992. Seasonal variability in primary production and particle flux in the northwestern Sargasso Sea: U.S. JGOFS Bermuda Atlantic time-series study. Deep Sea Res. Part A 39 (7–8), 1373–1391.

Mackenzie, F.T., Lerman, A., Ver, L.M.B., 1998. Role of the continental margin in the global carbon balance during the past three centuries. Geology 26 (5), 423–426.

Mahadevan, A., Lévy, M., Mémery, L., 2004. Mesoscale variability of sea surface pCO_2: what does it respond to? Global Biogeochem. Cycles 18, GB1017.

Mahowald, N., Kohfeld, K., Hansson, M., Balkanski, Y., Harrison, Sandy P., Prentice, I.C., Schulz, M., Rodhe, H., 1999. Dust sources and deposition during the last glacial maximum and current climate: a comparison of model results with paleodata from ice cores and marine sediments. J. Geophys. Res. 104 (D13), 15895–15916.

Maier-Reimer, E., 1993. Geochemical tracers in an ocean general circulation model. Preindustrial tracer distributions. Global Biogeochem. Cycles 7 (3), 645–677.

Maier-Reimer, E., Hasselmann, K., 1987. Transport and storage of CO_2 in the ocean—an inorganic ocean-circulation carbon cycle model. Clim. Dyn. 2, 63–90.

Maier-Reimer, E., Mikolajewicz, U., Winguth, A., 1996. Future ocean uptake of CO_2: interaction between ocean circulation and biology. Clim. Dyn. 12 (10), 711–721.

Maier-Reimer, E., Kriest, I., Segschneider, J., Wetzel, P., 2005. The HAMburg Ocean Carbon Cycle Model HAMOCC 5.1. Technical Description Release 1.1, Berichte zur Erdsystemforschung, 14, 1614–1199, Max Planck Institute for Meteorology, Hamburg, Germany, 50 pp.

Manizza, M., Le Quéré, C., Watson, A.J., Buitenhuis, E.T., 2005. Bio-optical feedbacks among phytoplankton, upper ocean physics and sea-ice in a global model. Geophys. Res. Lett. 32, 4. http://dx.doi.org/10.1029/2004GL020778, L05603.

Martin, J.H., Knauer, G.A., Karl, D.M., Broenkow, W.W., 1987. VERTEX: carbon cycling in the northeast Pacific. Deep Sea Res. 34, 267–285.

Matear, R.J., Lenton, A., 2007. Impact of historical climate change on the Southern Ocean carbon cycle. J. Clim. 21, 5820–5834.

McKinley, G.A., Fay, A.R., Takahashi, T., Metzl, N., 2011. Convergence of atmospheric and North Atlantic carbon dioxide trends on multidecadal timescales. Nat. Geosci. 4 (9), 606–610.

Meehl, G.A., Stocker, T.F., Collins, W.D., Friedlingstein, P., Gaye, A.T., Gregory, J.M., Kitoh, A., Knutti, R., Murphy, J.M., Noda, A., Raper, S.C.B., Watterson, I.G., Weaver, A.J., Zhao, Z.-C., 2007. Global climate projections. In: Solomon, S., Qin, D., Manning, M., Chen, Z., Marquis, M., Averyt, K.B., Tignor, M., Miller, H.L. (Eds.), Climate Change 2007: The Physical Science Basis. Contribution of Working Group I to the Fourth Assessment Report of the Intergovernmental Panel on Climate Change. Cambridge University Press, Cambridge, United Kingdom, and NY, USA.

Mikaloff Fletcher, S.E., Gruber, N., Jacobson, A.R., Doney, S.C., Dutkiewicz, S., Gerber, M., Follows, M., Joos, F., Lindsay, K., Menemenlis, D., Mouchet, A., Müller, S.A., Sarmiento, J.L., 2006. Inverse estimates of anthropogenic CO_2 uptake, transport, and storage by the ocean. Global Biogeochem. Cycles 20, GB2002.

Mikaloff Fletcher, S.E., Gruber, N., Jacobson, A.R., Gloor, M., Doney, S.C., Dutkiewicz, S., Gerber, M., Follows, M., Joos, F., Lindsay, K., Menemenlis, D., Mouchet, A., Müller, S.A., Sarmiento, J.L., 2007. Inverse estimates of the oceanic sources and sinks of natural CO_2 and the implied oceanic carbon transport. Global Biogeochem. Cycles 21, GB1010.

Milkov, A.V., 2004. Global estimates of hydrate-bound gas in marine sediments: how much is really out there? Earth Sci. Rev. 66, 183–197.

Montoya, M., Griesel, A., Levermann, A., Mignot, J., Hofmann, M., Ganopolski, A., Rahmstorf, S., 2005. The earth system model of

intermediate complexity CLIMBER3a. Part I: description and performance for present-day conditions. Clim. Dyn. 25, 237–263.
Moore, J.K., Doney, S.C., Glover, D.M., Fung, I.Y., 2002a. Iron cycling and nutrient-limitation patterns in surface waters of the World Ocean. Deep Sea Res. Part II 49, 463–507.
Moore, J.K., Doney, S.C., Kleypas, J.A., Glover, D.M., Fung, I.Y., 2002b. An intermediate complexity marine ecosystem model for the global domain. Deep Sea Res. Part II 49, 403–462.
Najjar, R.G., Sarmiento, J.L., Toggweiler, J.R., 1992. Downward transport and fate of organic matter in the ocean: simulations with a general circulation model. Global Biogeochem. Cycles 6, 45–76.
NASA Advisory Council, 1986. Earth System Sciences Committee, United States, National Aeronautics and Space Administration, National Academies, 48 pp.
Nightingale, P.D., Malin, G., Law, C.S., Watson, A.J., Liss, P.S., Liddicoat, M.I., Boutin, J., Upstill-Goddard, R.C., 2000. In situ evaluation of air-sea gas exchange parameterizations using novel conservative and volatile tracers. Global Biogeochem. Cycles 14 (1), 373–387.
Olafsson, J., Olafsdottir, S.R., Benoit-Cattin, A., Danielsen, M., Arnarson, T.S., Takahashi, T., 2009. Rate of Iceland Sea acidification from time series measurements. Biogeosciences 6, 2661–2668.
Orr, J.C., 2002. Global ocean of anthropogenic carbon (GOSAC), EC Environment and Climate Programme, Final Report, IPSL/SNRS, France, 117 pp.
Orr, J.C., Maier-Reimer, E., Mikolajewicz, U., Monfray, P., Sarmiento, J.L., Toggweiler, J.R., Taylor, N.K., Palmer, J., Gruber, N., Sabine, C.L., Le Quéré, C., Key, R.M., Boutin, J., 2001. Estimates of anthropogenic carbon uptake from four three-dimensional global ocean models. Global Biogeochem. Cycles 15 (1), 43–60.
Orr, J.C., Fabry, V.J., Aumont, O., Bopp, L., Doney, S.C., Feely, R.A., Gnanadesikan, A., Gruber, N., Ishida, A., Joos, F., Key, R.M., Lindsay, K., Maier-Reimer, E., Matear, R., Monfray, P., Mouchet, A., Najjar, R.G., Plattner, G.-K., Rodgers, K.B., Sabine, C.L., Sarmiento, J.L., Schlitzer, R., Slater, R.D., Totterdell, I.J., Weirig, M.-F., Yamanaka, Y., Yool, A., 2005. Anthropogenic ocean acidification over the twenty-first century and its impact on calcifying organisms. Nature 437, 681–686.
Peng, T.-H., 1986. Uptake of anthropogenic CO_2 by lateral transport models of the ocean based on the distribution of bomb-produced ^{14}C. Radiocarbon 28 (2A), 363–375.
Peters, G.P., Marland, G., Le Quéré, C., Boden, T., Canadell, J.G., Raupach, M.R., 2012. Rapid growth in CO_2 emissions after the 2008-2009 global financial crisis. Nat. Clim. Chang. 2, 2–4.
Peterson, W.H., Rooth, C.G., 1976. Formation and exchange of the deep water in the Greenland and Norwegian Seas. Deep Sea Res. 23, 273–283.
Pfeil, B., Olsen, A., Bakker, D.C.E., Hankin, S., Koyuk, H., Kozyr, A., Malczyk, J., Manke, A., Metzl, N., Sabine, C.L., Akl, J., Alin, S.R., Bellerby, R.G.J., Borges, A., Boutin, J., Brown, P.J., Cai, W.-J., Chavez, F.P., Chen, A., Cosca, C., Fassbender, A.J., Feely, R.A., Gonzlez-Dvila, M., Goyet, C., Hardman-Mountford, N., Heinze, C., Hood, M., Hoppema, M., Hunt, C.W., Hydes, D., Ishii, M., Johannessen, T., Jones, S.D., Key, R.M., Krtzinger, A., Landschtzer, P., Lauvset, S.K., Lefvre, N., Lenton, A., Lourantou, A., Merlivat, L., Midorikawa, T., Mintrop, L., Miyazaki, C., Murata, A., Nakadate, A., Nakano, Y., Nakaoka, S., Nojiri, Y., Omar, A.M., Padin, X.A., Park, G.-H., Paterson, K., Perez, F.F., Pierrot, D., Poisson, A., Ríos, A.F., Salisbury, J., Santana-Casiano, J.M., Sarma, V.V.S.S., Schlitzer, R., Schneider, B., Schuster, U., Sieger, R., Skjelvan, I., Steinhoff, T., Suzuki, T., Takahashi, T., Tedesco, K., Telszewski, M., Thomas, H., Tilbrook, B., Tjiputra, J., Vandemark, D., Veness, T., Wanninkhof, R., Watson, A.J., Weiss, R., Wong, C.S., Yoshikawa-Inoue, H., 2013. A uniform, quality controlled surface ocean CO_2 atlas (SOCAT). Earth Syst. Sci. Data 5, 125–143.
Plattner, G.-K., Joos, F., Stocker, T.F., Marchal, O., 2001. Feedback mechanisms and sensitivities of ocean carbon uptake under global warming. Tellus 53B, 564–592.
Pohlman, J.W., Bauer, J.E., Waite, W.F., Osburn, C.L., Chapman, N.R., 2010. Methane hydrate-bearing seeps as a source of aged dissolved organic carbon to the oceans. Nat. Geosci. 4, 37–41.
Quinn, P.K., Bates, T.S., 2011. The case against climate regulation via oceanic phytoplankton sulphur emissions. Nature 480, 51–56.
Ragueneau, O., Tréguer, P., Leynaert, A., Anderson, R.F., Brzezinski, M.A., DeMaster, D.J., Dugdale, R.C., Dymond, J., Fischer, G., Francois, R., Heinze, C., Maier-Reimer, E., Martin-Jeézéquel, V., Nelson, D.M., Quéguiner, B., 2000. A review of the Si cycle in the modern ocean: recent progress and missing gaps in the application of biogenic opal as a paleoproductivity tracer. Glob. Planet. Change 26, 317–365.
Rakestraw, N.W., 1949. The conception of alkalinity or excess base of sea water. J. Mar. Res. 8 (1), 14–20.
Raupach, M.R., Marland, G., Ciais, P., Le Quéré, C., Canadell, J.G., Klepper, G., Field, C.B., 2007. Global and regional drivers of accelerating CO_2 emissions. Proc. Natl. Acad. Sci. U.S.A. 104 (24), 10288–10293.
Raven, J., Caldeira, K., Elderfield, H., Hoegh-Guldberg, O., Liss, P., Riebesell, U., Shepherd, J., Turley, C., Watson, A., Heap, R., Banes, R., Quinn, R., 2005. Ocean Acidification Due to Increasing Atmospheric Carbon Dioxide POLICY Document 12/05. The Royal Society, London, ISBN 0 85403 617 2, 60 pp.
Raymo, M.E., 1994. The Himalayas, organic carbon burial, and climate in the Miocene. Paleoceanography 9 (3), 399–404.
Redfield, A.C., Ketchum, H., Richards, F.A., 1963. The influence of organisms on the composition of seawater. In: Hill, M.N. (Ed.), The Sea, vol. 2. Wiley Interscience, New York, pp. 26–77.
Ridgwell, A., Zondervan, I., Hargreaves, J.C., Bijma, J., Lenton, T.M., 2007. Assessing the potential long-term increase of oceanic fossil fuel CO_2 uptake due to CO_2-calcification feedback. Biogeosciences 4, 481–492.
Riebesell, U., Schulz, K.G., Bellerby, R.G.J., Botros, M., Fritsche, P., Meyerhöfer, M., Neill, C., Nondal, G., Oschlies, A., Wohlers, J., Zöllner, E., 2007. Enhanced biological carbon consumption in a high CO_2 ocean. Nature 450 (7169), 545–548.
Riebesell, U., Körtzinger, A., Oschlies, A., 2009. Sensitivities of marine carbon fluxes to ocean change. Proc. Natl. Acad. Sci. U.S.A. 106 (49), 20602–20609.
Roy, T., Bopp, L., Gehlen, M., Schneider, B., Cadule, P., Frölicher, T.L., Segschneider, J., Tjiputra, J., Heinze, C., Joos, F., 2011. Regional impacts of climate change and atmospheric CO_2 on future ocean carbon uptake: a multimodel linear feedback analysis. J. Clim. 24, 2300–2318.
Saba, V.S., Friedrichs, M.A.M., Antoine, D., Armstrong, R.A., Asanuma, I., Behrenfeld, M.J., Ciotti, A.M., Dowell, M., Hoepffner, N., Hyde, K.J.W., Ishizaka, J., Kameda, T., Marra, J., Mélin, F., Morel, A., OReilly, J., Scardi, M., Smith Jr., W.O.,

Smyth, T.J., Tang, S., Uitz, J., Waters, K., Westberry, T.K., 2011. An evaluation of ocean color model estimates of marine primary productivity in coastal and pelagic regions across the globe. Biogeosciences 8, 489–503.

Sabine, C.L., Tanhua, T., 2010. Estimation of anthropogenic CO_2 inventories in the ocean. Ann. Rev. Mar. Sci. 2, 175–198.

Sabine, C.L., Feely, R.A., Gruber, N., Kee, R.M., Lee, K., Bullister, J.L., Wanninkhof, R., Wong, C.S., Wallace, D.W.R., Tilbrook, B., Millero, F.J., Peng, T.-H., Kozyr, A., Ono, T., Ríos, A., 2004. The oceanic sink for anthropogenic CO_2. Science 305, 367–371.

Sabine, C.L., Hankin, S., Koyuk, H., Bakker, D.C.E., Pfeil, B., Olsen, A., Metzl, N., Kozyr, A., Fassbender, A., Manke, A., Malczyk, J., Akl, J., Alin, S.R., Bellerby, R.G.J., Borges, A., Boutin, J., Brown, P.J., Cai, W.-J., Chavez, F.P., Chen, A., Cosca, C., Feely, R.A., Gonzlez-Dvila, M., Goyet, C., Hardman-Mountford, N., Heinze, C., Hoppema, M., Hunt, C.W., Hydes, D., Ishii, M., Johannessen, T., Key, R.M., Krtzinger, A., Landschtzer, P., Lauvset, S.K., Lefvre, N., Lenton, A., Lourantou, A., Merlivat, L., Midorikawa, T., Mintrop, L., Miyazaki, C., Murata, A., Nakadate, A., Nakano, Y., Nakaoka, S., Nojiri, Y., Omar, A.M., Padin, X.A., Park, G.-H., Paterson, K., Perez, F.F., Pierrot, D., Poisson, A., Ríos, A.F., Salisbury, J., Santana-Casiano, J.M., Sarma, V.V.S.S., Schlitzer, R., Schneider, B., Schuster, U., Sieger, R., Skjelvan, I., Steinhoff, T., Suzuki, T., Takahashi, T., Tedesco, K., Telszewski, M., Thomas, H., Tilbrook, B., Vandemark, D., Veness, T., Watson, A.J., Weiss, R., Wong, C.S., Yoshikawa-Inoue, H., 2013. Surface ocean CO_2 atlas (SOCAT) gridded data products. Earth Syst. Sci. Data 5, 145–153.

Santana-Casiano, J.M., González-Dávila, M., Llinás, O., González-Dávila, O., Rueda Maria-Jose, M.-J., 2007. The interannual variability of oceanic CO(2) parameters in the northeast Atlantic subtropical gyre at the ESTOC site. Global Biogeochem. Cycles 21 (1), 16, GB1015.

Sarmiento, J.L., 1983. A tritium box model of the North-Atlantic Thermocline. J. Phys. Oceanogr. 13 (7), 1269–1274.

Sarmiento, J.L., Gruber, N., 2006. Ocean Biogeochemical Dynamics. Princeton University Press, USA, 503 pp.

Sarmiento, J.L., Orr, J.C., 1991. Three-dimensional simulations of the impact of Southern Ocean nutrient depletion on atmospheric CO_2 and ocean chemistry. Limnol. Oceanogr. 36 (8), 1928–1950.

Sarmiento, J.L., Toggweiler, J.R., 1984. A new model for the role of the oceans in determining atmospheric CO_2. Nature 308 (5960), 621–624.

Sarmiento, J.L., Orr, J.C., Siegenthaler, U., 1992. A perturbation simulation of CO_2 uptake in an ocean general circulation model. J. Geophys. Res. 97 (C3), 3621–3645.

Sarmiento, J.L., Hughes, T.M.C., Stouffer, R.J., Manabe, S., 1998. Simulated response of the ocean carbon cycleto anthropogenic climate warming. Nature 393, 245–249.

Schidlowski, M., Appel, P.W.U., Eichmann, R., Junge, C.E., 1979. Carbon isotope geochemistry of the 3.7×10^9-yr-old Isua sediments, West Greenland: implications for the Archean carbon and oxygen cycles. Geochim. Cosmochim. Acta 43, 189–199.

Schlesinger, W.H., 2004. Volume editor's introduction. In: Schesinger, W.H., Holland, H.D., Turekian, K.K. (Eds.), The Treatise of Geochemistry, first ed. Biogeochemistry, vol. 8, pp. xv–xvii.

Schlitzer, R., 2000. Applying the adjoint method for global biogeochemical modeling. In: Kasibhatla, P., Heimann, M., Rayner, P., Mahowald, N., Prinn, E.G., Hartley, D.E. (Eds.), Inverse Methods in Global Biogeochemical Cycles. Geophys. Monogr., vol. 114. American Geophysical Union, Washington D.C., pp. 107–124.

Schlitzer, R., 2007. Assimilation of radiocarbon and chlorofluorocarbon data to constrain deep and bottom water transports in the world ocean. J. Phys. Oceanogr. 37, 259–276.

Schlosser, P., Kromer, B., Östlund, G., Ekwurzel, B., Bönisch, G., Loosli, H.H., Purtschert, R., 1994a. On the ^{14}C and ^{39}Ar distribution in the central Arctic Ocean: implications for deep water formation. Radiocarbon 36 (3), 327–343.

Schlosser, P., Kromer, B., Weppernig, R., Loosli, H.H., Bayer, R., Bonani, G., Suter, M., 1994b. The distribution of ^{14}C and ^{39}Ar in the Weddell Sea. J. Geophys. Res. 99 (C5), 10275–10287.

Schlosser, P., Bönisch, G., Kromer, B., Loosli, H.H., Bühler, R., Bayer, R., Bonani, G., Koltermann, K.P., 1995. Mid 1980s distribution of tritium, ^3He, ^{14}C and ^{39}Ar in the Greenland/Norwegian Seas and the Nansen Basin of the Arctic Ocean. Prog. Oceanogr. 35, 1–28.

Schmittner, A., 2005. Decline of the marine ecosystem caused by a reduction in the Atlantic overturning circulation. Nature 443, 628–633.

Séférian, R., Bopp, L., Gehlen, M., Orr, J., Ethé, C., Cadule, P., Aumont, O., Salas y Mélia, D., Voldoire, A., Madec, G., 2013. Skill assessment of three earth system models with common marine biogeochemistry. Clim. Dyn., 9–10, 2549–2573. http://dx.doi.org/10.1007/s00382-012-1362-8.

Seitzinger, S.P., Mayorga, E., Bouwman, A.F., Kroeze, C., Beusen, A.H.W., Billen, G., Van Drecht, G., Dumont, E., Fekete, B.M., Garnier, J., Harrison, J.A., 2010. Global river nutrient export: a scenario analysis of past and future trends. Global Biogeochem. Cycles 24, GB0A08.

Shaffer, G., Sarmiento, J.L., 1995. Biogeochemical cycling in the global ocean 1. A new, analytical model with continuous vertical resolution and high-latitude dynamics. J. Geophys. Res. 100 (C2), 2659–2672.

Sherman, K., Belkin, I.M., Friedland, K.D., O'Reilly, J., Hyde, K., 2009. Accelerated warming and emergent trends in fisheries biomass yields of the world's large marine ecosystems. Ambio 38 (4), 215–224.

Siegenthaler, U., Oeschger, H., 1978. Predicting future atmospheric carbon dioxide levels. Science 199, 388–395, No. 4327.

Siegenthaler, U., Wenk, T., 1984. Rapid atmospheric CO_2 variations and ocean circulation. Nature 308 (5960), 624–626.

Simmons, A.J., Jones, P.D., da Costa Bechtold, V., Beljaars, A.C.M., Kallberg, P.W., Saarinen, S., Uppala, S.M., Viterbo, P., Wedi, N., 2004. Comparison of trends and low-frequency variability in CRU, ERA-40, and NCEP/NCAR analyses of surface air temperature. J. Geophys. Res. 109, 18, D24115.

Slomp, C.P., Van Cappellen, P., 2004. Nutrient inputs to the coastal ocean through submarine groundwater discharge: controls and potential impact. J. Hydrol. 295, 64–86.

Smethie Jr., W.M., Östlund, H.G., Loosli, H.H., 1986. Ventilation of the deep Greenland and Norwegian seas: evidence from krypton-85, tritium, carbon-14 and argon-39. Deep Sea Res. Part A 33 (5), 675–703.

Smith, B., Prentice, I.C., Sykes, M.T., 2001. Representation of vegetation dynamics in the modelling of terrestrial ecosystems: comparing two contrasting approaches within European climate space. Glob. Ecol. Biogeogr. 10 (6), 621–637.

Steinacher, M., Joos, F., Frölicher, T.L., Plattner, G.-K., Doney, S.C., 2009. Imminent ocean acidification in the Arctic projected with the NCAR global coupled carbon cycle-climate model. Biogeosciences 6, 515–533.

Steinacher, M., Joos, F., Frölicher, T.L., Bopp, L., Cadule, P., Cocco, V., Doney, S.C., Gehlen, M., Lindsay, K., Moore, J.K., Schneider, B., Segschneider, J., 2010. Projected 21st century decrease in marine productivity: a multi-model analysis. Biogeosciences 7, 979–1005.

Steinfeldt, R., Rhein, M., Walter, M., 2007. NADW transformation at the western boundary between 66°W/26°N and 606°W/106°N. Deep Sea Res. Part I 54, 835–855.

Suess, E., 1980. Particulate organic carbon flux in the oceans—surface productivity and oxygen utilization. Nature 288, 260–263.

Tachirii, K., Hargreaves, J.C., Annan, J.D., Oka, A., Abe-Ouchi, A., Kawamiya, M., 2010. Development of a system emulating the global carbon cycle in Earth system models. Geosci. Model Dev. 3, 365–376.

Tagliabue, A., Bopp, L., Dutay, J.-C., Bowie, A.R., Chever, F., Jean-Baptiste, P., Bucciarelli, E., Lannuzel, D., Remenyi, T., Sarthou, G., Aumont, O., Gehlen, M., Jeandel, C., 2010. Hydrothermal contribution to the oceanic dissolved iron inventory. Nat. Geosci. 3, 252–256.

Tagliabue, A., Bopp, L., Gehlen, M., 2011. The response of marine carbon and nutrient cycles to ocean acidification: large uncertainties related to phytoplankton physiological assumptions. Global Biogeochem. Cycles 25, GB3017.

Takahashi, T., Broecker, W.S., Langer, S., 1985. Redfield ratio based on chemical data from isopycnic surfaces. J. Geophys. Res. Oceans 90 (NC4), 6907–6924.

Takahashi, T., Sutherland, S.C., Wanninkhof, R., Sweeney, C., Feely, R.A., Chipman, D., Hales, B., Friederich, G., Chavez, F., Sabine, C., Watson, A., Bakker, D.C.E., Schuster, U., Metzl, N., Yoshikawa-Inoue, H., Ishii, M., Midorikawa, T., Nojiri, Y., Körtzinger, A., Steinhoff, T., Hoppema, M., Olafsson, J., Arnarson, T.S., Tilbrook, B., Johannessen, T., Olsen, A., Bellerby, R., Wong, C.S., Delille, B., Bates, N.R., de Baar, H.J.W., 2009. Climatological mean and decadal changes in surface ocean pCO_2, and net sea-air CO_2 flux over the global oceans, SOCOVV Symposium Volume. Deep Sea Res. Part II 810, 554–577.

Telszewski, M., Chazottes, A., Schuster, U., Watson, A.J., Moulin, C., Bakker, D.C.E., González-Dávila, M., Johannessen, T., Körtzinger, A., Lüger, H., Olsen, A., Omar, A., Padin, X.A., Ríos, A.F., Steinhoff, T., Santana-Casiano, M., Wallace, D.W.R., Wanninkhof, R., 2009. Estimating the monthly pCO_2 distribution in the North Atlantic using a self-organizing neural network. Biogeosciences 6, 1405–1421.

Tjiputra, J.F., Polzin, D., Winguth, A.M.E., 2007. Assimilation of seasonal chlorophyll and nutrient data into an adjoint three-dimensional ocean carbon cycle model: sensitivity analysis and ecosystem parameter optimization. Global Biogeochem. Cycles 21 (1), GB1001.

Tjiputra, J.F., Assmann, K., Bentsen, M., Bethke, I., Ottera, O.H., Sturm, C., Heinze, C., 2010a. Bergen earth system model (BCM-C): Model description and regional climate-carbon cycle feedbacks assessment. Geosci. Model Dev. 3, 123–141.

Tjiputra, J.F., Assmann, K., Heinze, C., 2010b. Anthropogenic carbon dynamics in the changing ocean. Ocean Sci. 6 (3), 605–614.

Touratier, F., Goyet, C., 2004. Definition, properties, and Atlantic Ocean distribution of the new tracer TrOCA. J. Mar. Syst. 46, 169–179.

Tschumi, T., Joos, F., Gehlen, M., Heinze, C., 2011. Deep ocean ventilation, carbon isotopes, marine sedimentation and the deglacial CO_2 rise. Clim. Past 7, 771–800.

Vázquez-Rodríguez, M., Touratier, F., Lo Monaco, C., Waugh, D.W., Padin, X.A., Bellerby, R.G.J., Goyet, C., Metzl, N., Ríos, A.F., Pérez, F.F., 2009. Anthropogenic carbon distributions in the Atlantic Ocean: data-based estimates from the Arctic to the Antarctic. Biogeosciences 6 (3), 439–451.

Vichi, M., Pinardi, N., Masina, S., 2007a. A generalized model of pelagic biogeochemistry for the global ocean ecosystem. Part I: theory. J. Mar. Syst. 64 (1–4), 89–109.

Vichi, M., Masina, S., Navarra, A., 2007b. A generalized model of pelagic biogeochemistry for the global ocean ecosystem. Part II: numerical simulations. J. Mar. Syst. 64 (1–4), 110–134.

Vizcaíno, M., Mikolajewicz, U., Gröger, M., Maier-Reimer, E., Schurgers, G., Winguth, A.M.E., 2008. Long-term ice sheet climate interactions under anthropogenic greenhouse forcing simulated with a complex Earth System Model. Clim. Dyn. 31, 665–690.

Volk, T., Hoffert, M.I., 1985. Ocean carbon pumps: analysis of relative strengths and efficiencies in ocean-driven atmospheric CO_2 changes. In: Sundquist, E.T., Broecker, W.S. (Eds.), The Carbon Cycle and Atmospheric CO_2: Natural Variations Archean to Present. Geophysical Monograph, 32, American Geophysical Union, Washington, D.C., pp. 99–110.

Wanninkhof, R., 1992. Relationship between gas exchange and wind speed over the ocean. J. Geophys. Res. 97, 7373–7381.

Warrior, H., Carder, K., 2007. An optical model for heat and salt budget estimation for shallow seas. J. Geophys. Res. Oceans 112, C12021.

Watson, A.J., Schuster, U., Bakker, D.C.E., Bates, N.R., Corbière, A., González-Dávila, M., Friedrich, T., Hauck, J., Heinze, C., Johannessen, T., Körtzinger, A., Metzl, N., Olafsson, J., Olsen, A., Oschlies, A., Padin, X.A., Pfeil, B., Santana-Casiano, J.M., Steinhoff, T., Telszewski, M., Ríos, A.F., Wallace, D.W.R., Wanninkhof, R., 2009. Tracking the variable north Atlantic sink for atmospheric CO_2. Science 326, 1391–1393.

Wattenberg, H., 1939. Atlas zu: Die Verteilung des Sauerstoffs im Atlantischen Ozean, Verlag von Walter de Gruyter & Co., Berlin und Leipzig, 1939. Translated from German as: Atlas of the dissolved oxygen content of the Atlantic Ocean, Scientific Results of the German Atlantic Expedition of the Research Vessel 'Meteor', 1925–1927, vol. 9. Atlas, published for the Division of Ocean Sciences, National Science Foundation, Washington, D.C. Amerind Publishing Co. Pvt. Ltd., New Delhi, 1992, OCE/NSF 85-00-002.

Wetzel, P., Winguth, A., Maier-Reimer, E., 2005. Sea-to-air CO_2 flux from 1948 to 2003: a model study. Global Biogeochem. Cycles 19, 28, GB2005.

Wetzel, P., Maier-Reimer, E., Botzet, M., Jungclaus, J., Keenlyside, N., Latif, M., 2006. Effects of ocean biology on the penetrative radiation in a coupled climate model. J. Clim. 19 (16), 3973–3987.

Winguth, A.M.E., Archer, D., Duplessy, J.-C., Maier-Reimer, E., Mikolajewicz, U., 1999. Sensitivity of paleonutrient tracer distributions and deep-sea circulation to glacial boundary conditions. Paleoceanography 14 (3), 304–323.

Wolf-Gladrow, D.A., Zeebe, R.E., Klaas, Christine, Körtzinger, Arne, Dickson, A.G., 2007. Total alkalinity: the explicit conservative expression and its application to biogeochemical processes. Mar. Chem. 106 (1–2), 287–300.

Woloszyn, M., Mazloff, M., Ito, T., 2011. Testing an eddy-permitting model of the Southern Ocean carbon cycle against observations. Ocean Model. 39 (1–2), 170–182.

Woodruff, S.D., Slurz, R.J., Jenne, R.L., Steurer, P.M., 1987. A comprehensive ocean-atmosphere data set. Bull. Am. Meteorol. Soc. 68, 1239–1250.

Wunsch, C., 1989. Tracer inverse problems. In: Anderson, D.L.T., Willebrand, J. (Eds.), Ocean Circulation Models: Combining Data and Dynamics. Mathematical and Physical Sciences, 284, Kluwer Academic Publishers, Dordrecht, pp. 1–77, NATO ASI Series, Series C.

Wüst, G., 1935. Schichtung und Zirkulation des Atlantischen Ozeans. Die Stratosphäre. Wissenschaftliche Ergebnisse der Deutschen Atlantischen Expedition auf dem Forschungsund Vermessungsschiff 'Meteor' 1925–1927. English translation by W.J. Emery, 1978, of vol. VI. Section 1. The stratosphere of the Atlantic Ocean. Amerind Publishing Co., New Delhi, 112 pp.

Zeebe, R.E., Wolf-Gladrow, D., 2001. CO_2 in seawater: equilibrium, kinetics, isotopes. In: Elsevier Oceanography Series, 65, Elsevier Science B.V., Amsterdam, 346 pp.

Zehr, J.P., Ward, B.B., 2002. Nitrogen cycling in the ocean: new perspectives on processes and paradigms. Appl. Environ. Microbiol. 68 (3), 1015–1024.

Zondervan, I., Zeebe, R.E., Rost, B., Riebesell, U., 2001. Decreasing marine biogenic calcification: a negative feedback on rising atmospheric pCO_2. Global Biogeochem. Cycles 15 (2), 507–516.

Part VI

The Changing Ocean

Earlier parts of this book introduced the ocean, its role in the climate system, and ocean and atmosphere–ocean general circulation models (AOGCMs). This Part VI builds on these to address five major scientific and societally relevant challenges of The Changing Ocean: namely the oceanic uptake of heat and rising sea levels, changes in the global hydrological cycle, the ocean's transport of heat, the ocean's uptake of carbon, and the impact of these changes for the ocean ecosystem.

Understanding the energy balance of the earth is fundamental to understanding the Earth's climate. The ocean has stored more than 90% of the total increase in energy of the Earth's warming climate system over recent decades (Chapter 27). Much of this heat is in the upper ocean but temperatures have also increased in the abyssal waters. Accurate estimates of this heat-content increase are fundamental to the evaluation of climate models and to determining the sensitivity of the climate system to greenhouse gas forcing. This warming causes the ocean to expand, contributing to sea-level rise.

In low-latitude regions, more energy is received by the Earth as short-wave radiation from the sun than is emitted by the Earth as long-wave radiation, with the reverse occurring in high-latitude regions. Poleward transport of energy by the combined circulations of the atmosphere and the ocean (Chapter 29) close the energy balance. The ocean transports heat by vertical overturning and horizontal circulation. The transport can be estimated by direct ocean observations, from models of ocean temperatures and circulation and by indirect means, involving the integral of air–sea heat fluxes or the difference between radiative balance of the Earth and the atmospheric energy transport. Poleward ocean heat transport is not stationary but we are only beginning to unravel its variability.

The ocean contains 97% of the available water in the world, is central to the global hydrological cycle (Chapter 28), and is the ultimate source of rainfall that is fundamental to life. The distribution of salinity in the ocean is determined by air–sea exchange (precipitation and evaporation), ocean circulation and mixing, and river and ice-sheet runoff. Observations of the changing salinity pattern over recent decades provide evidence of an accelerating global hydrological cycle. Estimates of the inferred changes in the hydrological cycle are a valuable means of evaluating AOGCMs.

Atmospheric carbon dioxide (CO_2) is an important regulator of Earth's climate. As explained in Chapter 30, the ocean contains 60 times more CO_2 than the atmosphere and a large fraction of emitted anthropogenic carbon has been and will be absorbed by the oceans, thus slowing the rate of climate change. This ocean storage has implications for sea-water chemistry, ocean acidity, and for the oceans' biological communities. The storage of anthropogenic carbon is highest in the North Atlantic Ocean and in a quasi-zonal band in the mid-latitude Southern Ocean.

Ocean climate variability and change play a central role in the large-scale patterns and functioning of ocean ecosystems (Chapter 31). Primary production by phytoplankton forms the base of the pelagic-ocean food web and is modulated by temperature, nutrient supply, light, and mixed layer depth. Historical observations already indicate a biological response to anthropogenic ocean warming and sea-ice melt. Future climate change impacts are expected to increase in magnitude with further warming, ocean acidification, deoxygenation, and coastal nutrient eutrophication.

Chapter 27

Sea-Level and Ocean Heat-Content Change

John A. Church*,†, Neil J. White*,†, Catia M. Domingues‡, Didier P. Monselesan*,† and Elaine R. Miles¶

*Centre for Australian Weather and Climate Research, GPO Box 1538, Hobart, Tasmania, Australia

†CSIRO Marine and Atmospheric Research and Wealth from Oceans Flagship, GPO Box 1538, Hobart, Tasmania, Australia

‡Antarctic Climate and Ecosystems Cooperative Research Centre, University of Tasmania, Private Bag 80, Hobart, Tasmania, Australia

¶Bureau of Meteorology, GPO Box 1289, Melbourne, Victoria, Australia

Chapter Outline

1. INTRODUCTION

The ocean has the largest heat capacity in the climate system; the upper 3 m of the ocean has about the same heat capacity as the entire atmosphere. As a result, the ocean plays a critical role in the climate system through its ability to store vast amounts of heat, transport it thousands of kilometers, and release it back to the atmosphere. Oceanic heat storage is thus a critical contribution to natural climate variability and anthropogenic climate change. The sequestering of heat in the ocean slows the rate of increase in surface atmospheric temperature and influences the regional distribution of the effects of climate change. However, the ocean's large heat capacity means that it is out of equilibrium with the atmosphere and to return to equilibrium it will continue to warm for centuries even if greenhouse gas concentrations were to be stabilized in the atmosphere. Accurately estimating changes in ocean heat content is thus one of the most important and fundamental challenges in understanding climate variability and change and it is important in assessing the sensitivity of the climate system to changes in radiative forcing.

The ocean expands as it warms and thermosteric sea-level change is a major component of sea-level variability and rise. Other contributions to sea-level change come from changes in the mass of the ocean, particularly through exchanges with glaciers and ice sheets. Large sea-level changes have occurred in the past (time scales of thousands to millions of years) in response to climate changes and will occur in the future as a result of natural and anthropogenic climate change. The multitude of factors contributing to sea-level change makes understanding it a considerable scientific challenge and a fundamental measure of climate change.

In addition to its importance as an indicator of climate change, sea level has a clear socioeconomic impact. At the coast, where sea level impacts society directly, it is affected by changes on the shelf, particularly from local and remote wind forcing, and in deepwater and changes in the mass of water stored on land. With the order of

Ocean Circulation and Climate, Vol. 103. http://dx.doi.org/10.1016/B978-0-12-391851-2.00027-1

150 million people living within 1 m of the current day mean high tide level (Anthoff et al., 2006), sea-level change impacts modern society through flooding of coastal developments and natural ecosystems, coastal erosion, and salt water intrusion into coastal waterways and aquifers (Nicholls and Cazenave, 2010). These impacts are felt most acutely through extreme sea-level events and it is thus important to determine if there are, and understand the reasons for, any changes in the frequency and/or intensity of such events. Surface waves are also a critical component of these coastal impacts.

Church et al. (2010) provide a comprehensive report on the major factors affecting mean sea-level change and its impact on society. Cazenave and Llovel (2010) and Stammer et al. (2013) provide recent reviews. Here, we summarize past sea-level changes (Section 3) and changes in ocean heat content and thermal expansion (Section 4). Glaciers and ice sheets are critically important for understanding observed and future sea-level change (Section 5), but they receive only passing attention in this ocean-focused book, although ocean-ice interactions are covered in Chapter 16. Seasonal sea-level predictions, centennial sea-level projections, and future challenges are discussed briefly (Sections 6 and 7). See Church et al. (2013a) for a more in-depth discussion of projections for glaciers and ice sheets and global averaged and regional sea level.

2. FUNDAMENTAL CONCEPTS OF SEA-LEVEL CHANGE

Sea level varies on time scales from seconds to millennia and longer. On the shortest time scales of seconds, surface waves forced by local and distant winds are important and these directly impact coastal erosion and flooding. The impact of surface waves is strongly modulated by the height of the tide (semidiurnal to interannual time scales) and the presence of nontidal sea-level variations, particularly storm surges driven by severe weather events (winds and also low atmospheric pressures) as well as longer-period changes in mean sea level, one of the foci of this chapter.

Globally, sea level changes because the density of ocean water changes as it warms or cools (like the liquid in a thermometer) and because of changes in the mass (and hence volume) of the ocean (see Figure 27.1 for a schematic of processes important for sea-level change and Figure 27.2 for the thermal expansion and haline contraction coefficients as a function of depth and latitude). The sea-level change resulting from changes in ocean mass (termed barystatic sea-level change, Gregory et al., 2013b) may come from exchange with glaciers (Arendt et al., 2012, including the peripheral glaciers surrounding the ice sheets), the Greenland and Antarctic ice sheets and water stored in the terrestrial environment (Church et al. 2010). The latter can be through natural fluctuations of the amount of water stored in soils, lakes, and rivers or through anthropogenic activities such as the impoundment of water in reservoirs and the depletion of ground water.

Changes in ocean density, ocean currents, and sea level (with respect to the geoid, the equipotential surface that best approximates sea level and is the level the ocean surface would take if the ocean was at rest) are all closely linked and are directly related to surface fluxes of momentum (surface winds), heat and freshwater, and changes in atmospheric pressure. As a result, changes at one location are felt globally. For example, warming of the deep ocean results in a thermosteric sea-level rise which also results in an increase in bottom pressure on the shelf and a sea-level rise at the coast (Landerer et al., 2007). Similarly, the addition of mass to the ocean, for example, from the loss of ice-sheet mass, is communicated rapidly around the globe through fast external gravity (barotropic) waves such that all areas experience a sea-level rise within days (Lorbacher et al., 2012).

The movement of ocean eddies, interannual to decadal and longer term climate variability, such as the El Niño-Southern Oscillation (ENSO) phenomenon with

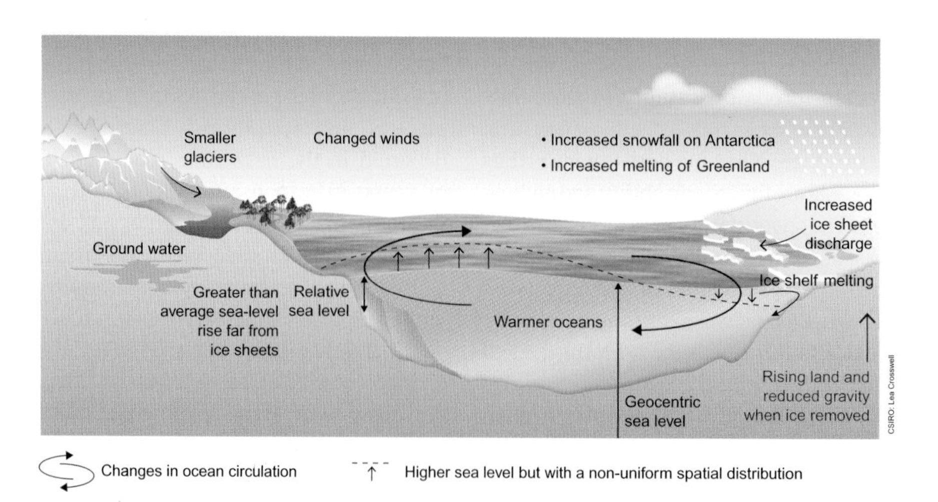

FIGURE 27.1 Processes affecting global mean and regional sea-level change. *Adapted from Climate Change 2013: The Physical Science Basis. Working Group I Contribution to the Fifth Assessment Report of the Intergovernmental Panel on Climate Change, Figure 13.1, Cambridge University Press (in press).*

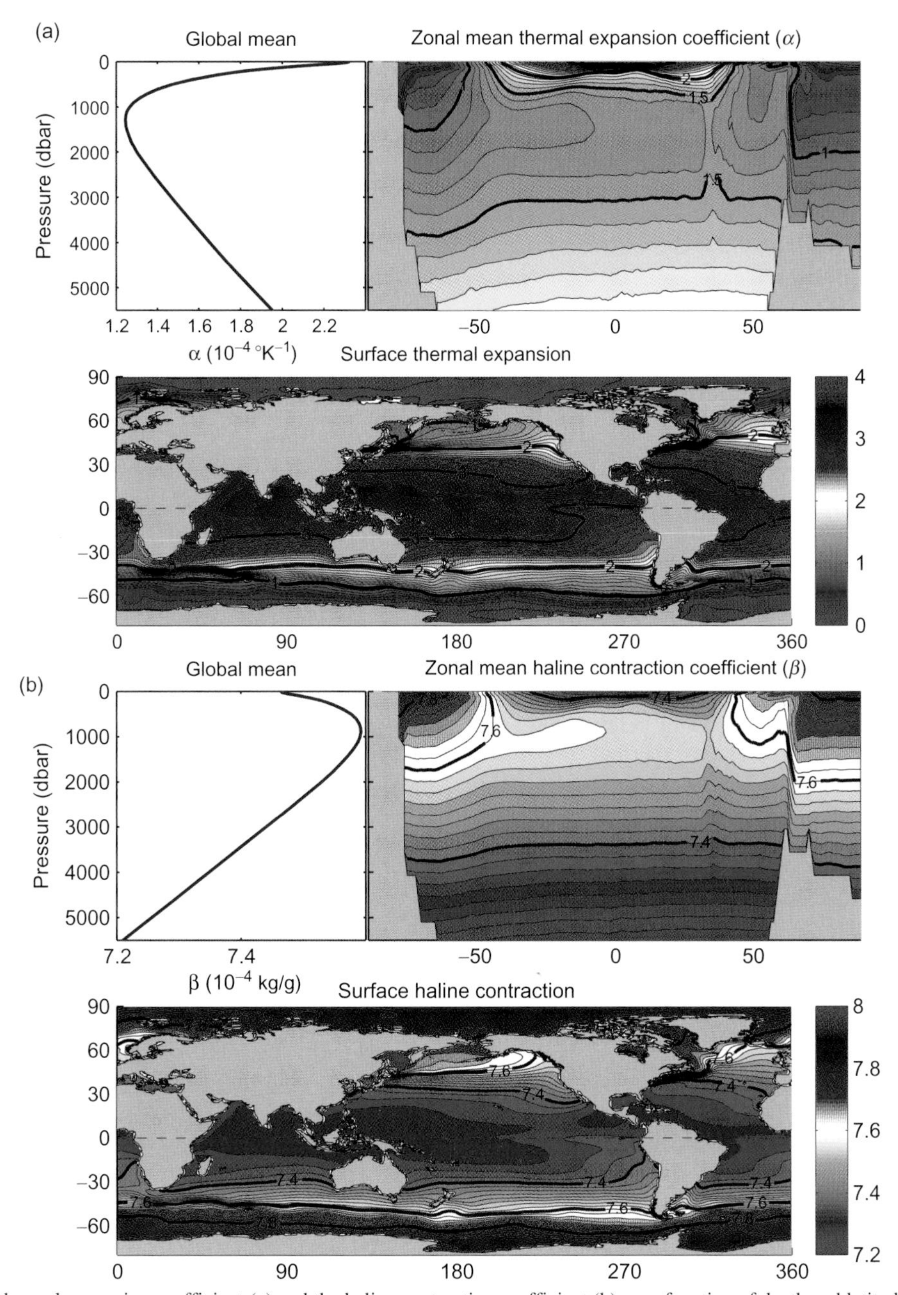

FIGURE 27.2 The thermal expansion coefficient (a) and the haline contraction coefficient (b), as a function of depth and latitude and longitude at the surface. Both coefficients are derived from mean World Ocean Atlas 2009 *in situ* temperature and practical salinity using the Gibbs-Sea Water (GSW) Oceanographic Toolbox based on Thermodynamic Equation of Seawater 2010 (TEOS-10), http://www.teos-10.org/index.htm. (See Chapter 6 for a discussion of the equation of state of sea water.)

the associated changes in ocean circulation and anthropogenic climate change, all cause changes in ocean density. Much of the regional variability is associated with the heave of the thermocline, as part of natural variability in the ocean (e.g., Roemmich and Gilson, 2011) and in response to surface wind-stress changes (e.g., Zhang and Church, 2012).

The ocean's density is dependent on temperature (water expands with increasing temperature, termed the thermosteric effect or thermal expansion), on salinity (density increases with salinity, haline contraction or halosteric effect), and pressure (see Chapter 6 for a discussion of the equation of the state of sea water). As the total salt content of the ocean is effectively constant over time scales relevant to society and the haline contraction coefficient changes by only a few percent over the ocean (Figure 27.2b), the global mean halosteric contribution is an indication of a change in the ocean mass through the exchange of water with other components of the climate system (Munk, 2003). This ocean-mass change is substantially larger than the global mean halosteric effect. Although both thermosteric and halosteric effects contribute to regional sea-level change, only the thermal expansion (thermosteric) component is important for global mean sea-level change (Gregory and Lowe, 2000). Note that estimating the mass change of the ocean to a useful accuracy from observations of salinity would require improvements in the current measurement system by an order of magnitude.

The amount of ocean thermal expansion is dependent on temperature (greater expansion at higher temperatures). At the ocean surface (Griffies and Greatbatch, 2012; Figure 27.2), there is an order of magnitude decrease in the thermal expansion coefficient from the warm equatorial region to the colder high latitudes (Figure 27.2a). The coefficient also decreases with decreasing temperature over the upper 1000 m of the water column. As a result, the warming of the main thermocline, the ocean area warming most rapidly, has the largest impact on global averaged thermal expansion. The thermal-expansion coefficient is also affected by pressure (Figure 27.2), increasing toward the bottom of the ocean. The ocean absorbs heat in low latitudes where the thermal expansion coefficient is large and loses heat at higher latitudes where the thermal expansion coefficient is smaller. This net buoyancy flux to the ocean is balanced by a poleward eddy heat transport and ocean mixing, transferring heat from regions with a large thermal-expansion coefficient to regions with a smaller expansion coefficient. See Griffies and Greatbatch (2012) for a more detailed discussion of these issues.

With the waxing and waning of ice sheets over ice-age cycles, there were large movements of mass between the ice sheets and the ocean, thus changing the surface loading of the solid Earth. This has resulted in an ongoing and significant viscous response of the solid Earth and a resultant smooth and large-scale change in the geoid (Glacial Isostatic Adjustment, GIA; see, e.g., Mitrovica et al., 2010; Tamisiea, 2011 for a discussion of the issues). The GIA results in a fall in relative sea level (sea level relative to the land; RSL) as measured by coastal tide gauges in regions of former ice sheets (the land is rising as a result of removal of the surface load at the rate of several mm year^{-1}), rising RSL in the area immediately adjacent to the former ice sheets as the mantle material moves back toward the regions of former ice sheets (order of mm year^{-1}), and a small fall in RSL in the region distant from the former ice sheets (generally of the order of a few tenths mm year^{-1}). As a result, these GIA signals (and other vertical land motions) need to be removed from sea-level records to estimate changes in ocean volume, that are unrelated to the GIA.

In addition to past mass changes, present day changes in mass distribution and surface loading cause changes in the gravitational field and in the elastic rotational response of the Earth (Mitrovica et al., 2001; Figure 27.1). As a result, the effect of an ice sheet (or any other body of stored water) losing mass is to cause a fall in the local RSL (less gravitational attraction and a rising crust) whereas in the far field, relative sea level can be up to 30% larger than the global-mean sea-level (GMSL) rise from the mass loss (Mitrovica et al., 2001, 2009). The current generation of ocean models does not directly include these temporal and spatial variations of the geoid and in practice these "fingerprints" are added to the results from climate models (Kopp et al., 2010; Church et al., 2011a; Slangen et al., 2012).

Satellite altimeters measure sea level in the International Terrestrial Reference Frame (ITRF, Blewitt et al., 2010) describing the overall shape of the Earth and its center of mass. Geocentric sea level (sea level measured with respect to the ITRF) is also affected by the ongoing GIA through changes of the ocean floor and thus the altimeter record needs to be adjusted for the ongoing GIA to estimate changes in ocean water volume. For the global average, this correction results in the equivalent ocean water volume change of the order of 0.3 mm year^{-1} larger than the direct altimeter measurements of global-averaged geocentric sea-level rise (Tamisiea, 2011).

Other changes in RSL can occur as a result of tectonic motion and sediment compaction (especially if there is ground water withdrawal from the sediments). Sediment compaction can be large (several mm year^{-1} or more; see, e.g., Nicholls, 2010) but tends to be confined to smaller areas and these effects have to be estimated by direct local measurements and modeling.

3. OBSERVATIONS OF SEA-LEVEL CHANGE

3.1. Sea-Level Change on Multimillenial Time Scales

As indicated in Chapter 2, the main control on sea level over time scales of millennia and longer is the ice sheets. Over the last 450,000 years, ice sheets have waxed and waned with periods of about 100,000 years, resulting in sea levels ranging from 130 m below present day sea level to meters higher than today's value (Lambeck et al., 2002; Rohling et al., 2009; Austermann et al., 2013; figure 2.13). At the last interglacial, 129,000–116,000 years ago, peak sea level was between 5 and 9 m higher than present day values (Kopp et al., 2009, 2013; Dutton and Lambeck, 2012). A combination of observations and ice-sheet models indicate that the Greenland Ice Sheet contributed between 1.4 and 4.3 m of sea-level equivalent (SLE) (NEEM Community Members, 2013). The ocean thermal expansion contribution is estimated to have been 0.4 ± 0.3 m (McKay et al., 2011) and glaciers could not have contributed more than their present mass, equivalent to about 0.41 ± 0.11 m of sea level

(Vaughan et al., 2013). These estimates imply a contribution from Antarctica for the higher sea-level estimates for the last interglacial. However, at present there is no clear observational or modeling evidence to provide a direct estimate of the Antarctic Ice Sheet contribution.

Sea level then fell to about 130 m below the present day value before the decay of the ice sheets commenced about 20,000 years ago (Lambeck et al., 2002; Clark et al., 2009; Austermann et al., 2013; figures 2.7 and 2.14). Over the next 13,000 years, sea level rose at an average of about 1 m/century (Lambeck et al., 2002; Stanford et al., 2011), but with possible peak rates of order 4–5 m/century (Clark et al., 2002; Deschamps et al., 2012). From about 7000 to 3000 years ago, the sea level is estimated to have risen about 2–3 m (Lambeck et al., 2010; figure 2.15). Salt marsh sediment cores from both northern and southern hemisphere sites have been used to estimate coastal sea level for as long as the last 2000 years (Donnelly et al., 2004; Gehrels et al., 2006, 2008, 2011, 2012; Kemp et al., 2009, 2011; figures 2.16 and 2.17). A number of these cores overlap with the tide-gauge records, confirming the veracity of the salt-marsh inferred RSLs. These salt-marsh records indicate an increase in the rate of sea-level rise during the late nineteenth to early twentieth century from the late Holocene rate, which locally is dominated by the ongoing GIA, to the twentieth century rates of the order of mm year^{-1} higher than the preindustrial rate. Gehrels and Woodworth (2013) infer that the start of this period when the rate of RSL rise exceeded the background Holocene rate was within a 40-year period centered on 1925 (Figure 27.3).

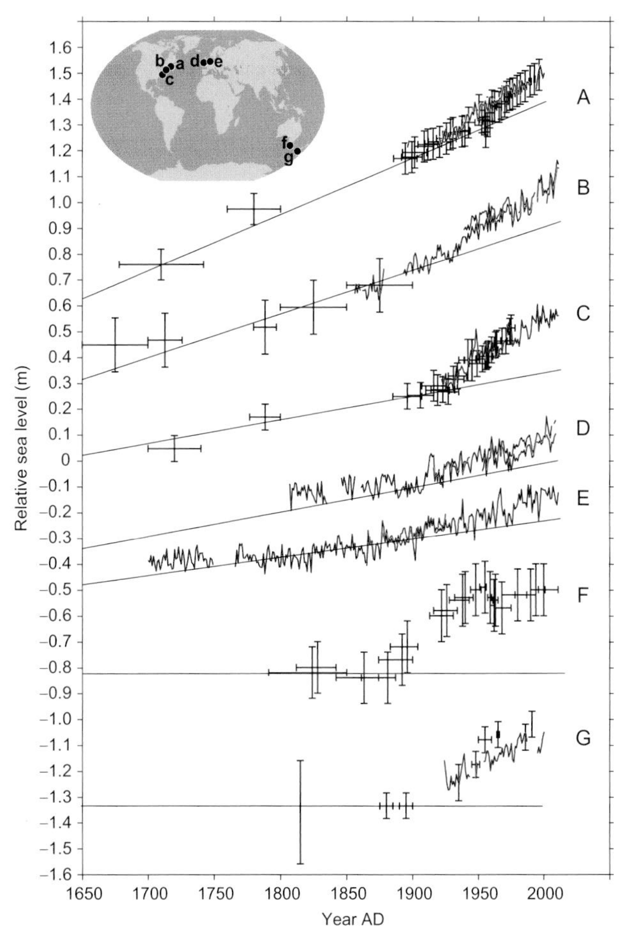

FIGURE 27.3 Sea-level changes compared with late Holocene background trend of sea-level change. Red and blue lines are tide-gauge records. Sea-level index points from salt marsh sea-level estimates (see the map for locations) are shown as crosses reflecting age and altitudinal uncertainties. (a) Chezzetcook, Nova Scotia (Gehrels et al., 2005), with tide-gauge record from Halifax. (b) Barn Island, Connecticut, USA (Donnelly et al., 2004), with tide-gauge records from New York City (red line) and New London (blue line). (c) Sand Point, North Carolina, USA (Kemp et al., 2009, 2011), with tide-gauge record from Charleston, South Carolina. (d) Tide-gauge records from Brest (red line) and Newlyn (blue line) compared with late Holocene trend of relative sea-level change at Thurlestone, Devon, United Kingdom (Gehrels et al., 2011). (e) Instrumental sea-level record from Amsterdam (red line) and tide-gauge record from Den Helder (blue line), compared with late Holocene trend of relative sea-level change at Schokland, the Netherlands (van de Plassche et al., 2005). (f) Little Swanport, Tasmania, Australia (Gehrels et al., 2012). (g) Pounawea south-eastern New Zealand (Gehrels et al., 2008), with tide-gauge record from Lyttelton. *From Gehrels and Woodworth (2013).*

3.2. Instrumental Observations of Sea-Level Change

The first continuous measurements of RSL come from tide gauges installed in the eighteenth century in Europe to provide information for shipping purposes rather than to provide climate records (Woodworth, 1999). By the late 1860s, 10 records, all in the northern hemisphere, were available. It was not until the late nineteenth century that records from the southern hemisphere were available, with coverage increasing significantly during the twentieth century, particularly between 1940 and 1960 (Figure 27.4). The decrease in the number of available records in the past few years is mostly because of the late submission of data to PSMSL but in some cases may indicate a real decline in the network. However, even today, global coverage is incomplete. Monthly average tide-gauge data are available from the PSMSL (Woodworth and Player, 2003; see figures 3.11 and 27.4 for the locations where data were available for the Church and White, 2011 study) with high-frequency near real-time data available from the University of Hawaii Sea-Level Center as well as PSMSL.

The longest records are mostly from northern Europe, but some long records are also available from North America. There have been many studies of the PSMSL data set. For example, Douglas (1991) estimated a twentieth century rate of sea-level rise of 1.8 mm year^{-1} by combining sea-level data from carefully selected tide-gauge

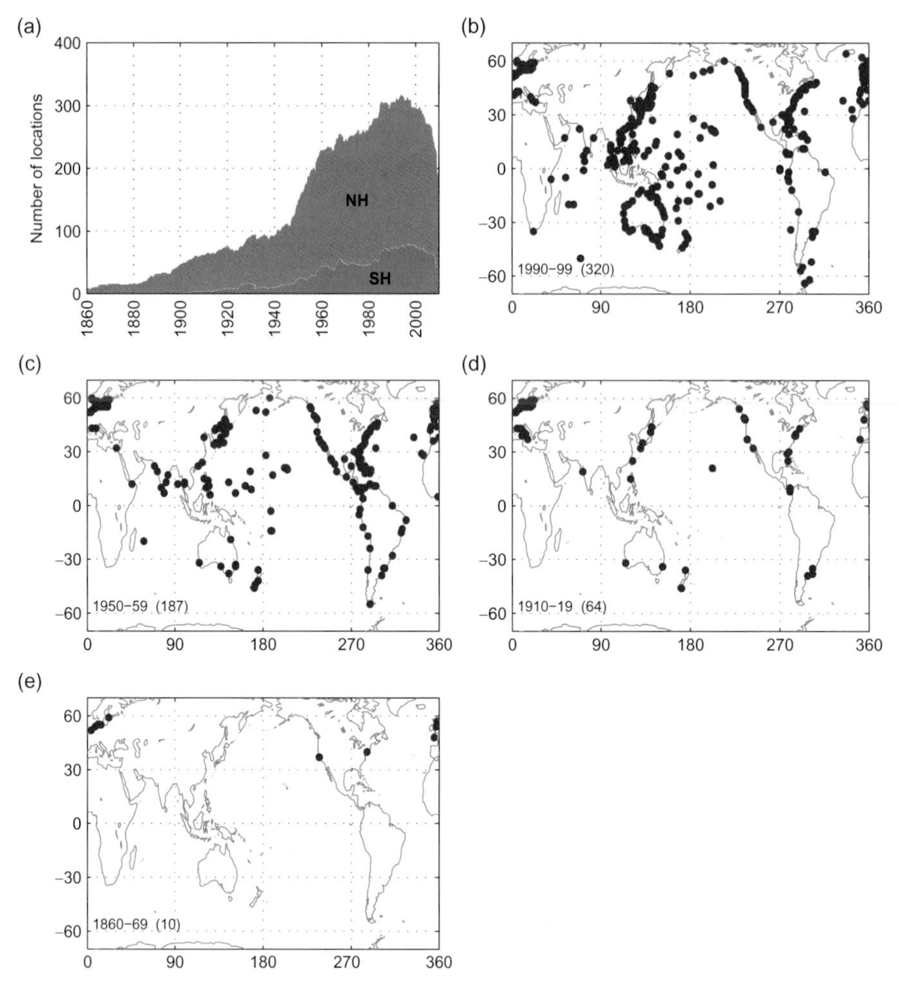

FIGURE 27.4 The number and distribution of sea-level records available for the reconstruction of Church and White (2011a). (a) The number of locations for the globe and the northern and southern hemispheres. (b–e) indicate the distribution of gauges in the 1990s, 1950s, 1910s and 1860s. The locations indicated have at least 60 months of data in the decade and the number of records is indicated in brackets. *Modified from Church and White (2011).*

sites (to avoid tectonic motions) with model estimates of GIA. Similar to Woodworth (1990), Douglas (1992) was unable to detect any significant acceleration in the rate of rise during the twentieth century. However, the longest records starting prior to 1850 all indicate an increase in the rate of rise over a multicentury period (Woodworth, 1999), in agreement with the salt-marsh data. There have been a number of recent studies of regional mean sea-level changes, for example, in the British Isles (Woodworth et al., 2009), the English Channel (Haigh et al., 2009), the German Bight (Wahl et al., 2011), the Norwegian and Russian coasts (Henry et al., 2012), the Mediterranean (Calafat and Jordà, 2011; Tsimplis et al., 2011), the United States (Snay et al., 2007; Sallenger et al., 2012; Ezer et al., 2013), New Zealand (Hannah and Bell, 2012), and Australia (Burgette et al., 2013). Although RSL is falling in some regions (e.g., the high northern latitudes in regions of former ice sheets as a result of a large GIA signal), these studies substantiate a GMSL rise even though there are regional differences in RSL rates.

The highest quality near-global (66°S–66°N) satellite altimeter observations of sea level cover 93.9% of the ocean surface and are available since 1993, following the launch of the TOPEX/Poseidon satellite in 1992, and a series of follow-on missions (Jason-1 and OSTM/Jason-2, see Chapter 4). These satellites measure sea level in a geocentric reference frame and many corrections are necessary to allow for environmental effects. (For a description of the data sets and the various corrections, see Benada (1997) for TOPEX/Poseidon, AVISO (2003) for Jason-1, and CNES (2009) for OSTM/Jason-2.) The altimeter sea-level measurements are compared with *in situ* observations at three sites (White et al., 1994; Bonnefond et al., 2003, 2010; Haines et al., 2003, 2010; Watson et al., 2003, 2011) located within the geocentric ITRF to give a geocentric verification of the sea-level measurements and a network of tide gauges

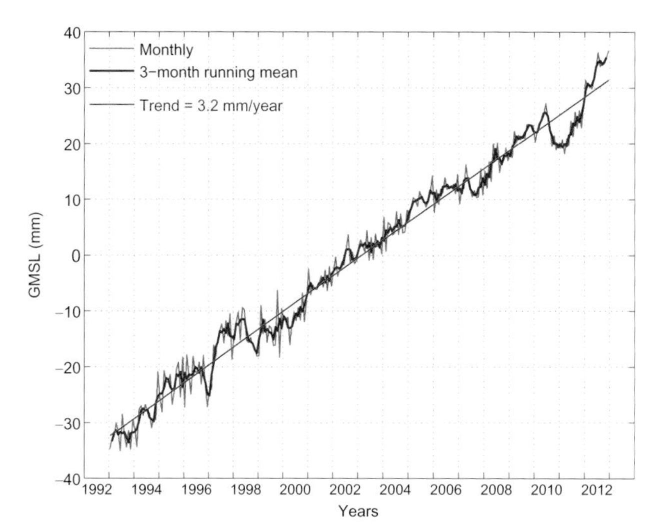

FIGURE 27.5 Global mean sea level from satellite altimeter data, following the processing techniques in Church and White (2011). The red line is the least squares trend line. The altimeter data have had the seasonal signal removed, and the inverse barometer correction and the GIA correction applied.

to estimate time-dependent biases (Mitchum, 1998, 2000). To date, these measurements have (mostly) been used as a check on the overall quality of the altimeter data rather than as a correction to the data.

A number of different approaches have been used to estimate GMSL change from satellite altimeter data and all agree that the global average rate of rise since 1993 is 3.2 ± 0.4 mm year^{-1} after allowing for GIA corrections (Masters et al., 2012; Figure 27.5). However, over this period, the rate of rise has not been uniform over the globe, with the rate of rises in sea levels of over three times the global average in the western equatorial Pacific and falls in sea level in the eastern Pacific Ocean (figure 4.3). Linear trends in these regional rates of rise are not a good representation of the local time series. Rather, variability from interannual and decadal climate variability is responsible for much of the regional pattern (e.g., Roemmich and Gilson, 2011; Zhang and Church, 2012). The high rates of rise in the western Pacific are also present in tide-gauge records and are a response to strengthening Pacific trade winds (Feng et al., 2010, 2011; Timmermann et al., 2010; Merrifield, 2011; Merrifield and Maltrud, 2011) resulting in ocean circulation changes and therefore steric sea-level changes. Merrifield et al. (2012) suggests that the west-east gradient of the DSL change reflects the negative phase of the Pacific Decadal Oscillation, rather than a trend induced by external climate forcing as originally proposed by Merrifield (2011) and Merrifield and Maltrud (2011). The departures from the linear trend, particularly the fall in GMSL in 2010 and its recovery in 2011 were associated with ocean-land exchanges (as measured by the GRACE gravity mission, see Chapter 4; Section 5), in which a large mass of freshwater from the ocean was transferred to land, causing major floods in Australia, Asia, and South America (Boening et al., 2012), and later returned to the ocean.

3.3. Reconstructions of Global Mean Sea Level

The satellite altimeter record provides measurements of 93.9% of the global ocean surface, thus providing a good estimate of GMSL change. However, the record is only 20 years long compared with the much longer tide-gauge sea-level observations (Figure 27.4). Chambers et al. (2002) used tide-gauge data and information on large-scale spatial correlation patterns, as represented in empirical orthogonal functions (EOFs), from the altimeter data to study sea-level variability. Church et al. (2004) and Church and White (2006, 2011) extended this approach to reconstruct regional sea levels since 1950 and global mean sea levels (because of seasonal sea-ice coverage, 90% of the ocean is covered in this reconstruction) since the late nineteenth century. They used a reduced space optimal interpolation method developed for sea-surface temperature and pressure (Kaplan et al., 1998, 2000) to reconstruct sea level $H^r(x,y,t)$

$$H^r(x,y,t) = U^r(x,y)\beta(t) + \varepsilon, \qquad (27.1)$$

where $U^r(x,y)$ is a matrix of the leading EOFs calculated from monthly satellite altimeter data (after removing the mean and the global mean trend), ε is the uncertainty, x and y are latitude and longitude, and t is time. This matrix is augmented by an additional "mode" that is constant in space and used to represent any global average sea-level change. In the reduced space optimal interpolation, the amplitudes of the constant mode and the leading EOFs, $\beta(t)$, are calculated by minimizing the difference between the reconstructed sea levels and the observed coastal and island sea levels. Berge-Nguyen et al. (2008) and Llovel et al. (2009) used modeled sea level to calculate the EOFs in an attempt to capture longer period variability than present in the altimeter record.

The resultant GMSL (without the annual cycle, Figure 27.6) shows a total rise of about 210 mm from 1870 to 2010, a twentieth century trend of 1.7 ± 0.2 mm year^{-1} and an acceleration of 0.009 ± 0.003 mm year^{-2} (one sigma) over the period, and agrees, within uncertainties, with the satellite altimeter record since 1993. Jevrejeva et al. (2006, 2008) estimated global mean sea level by averaging coastal tide-gauge records locally, then regionally, and then globally. The Jevrejeva et al. GMSL series is in broad agreement with that shown in Figure 27.6 (see Church and White, 2011 for a comparison), has a similar acceleration of 0.01 mm year^{-2}, but contains significantly higher frequency variability. Ray and Douglas (2011) used a similar approach to that of Church et al. (2004) but with a smaller number of sea-level records. They found the GMSL record was not sensitive to the details

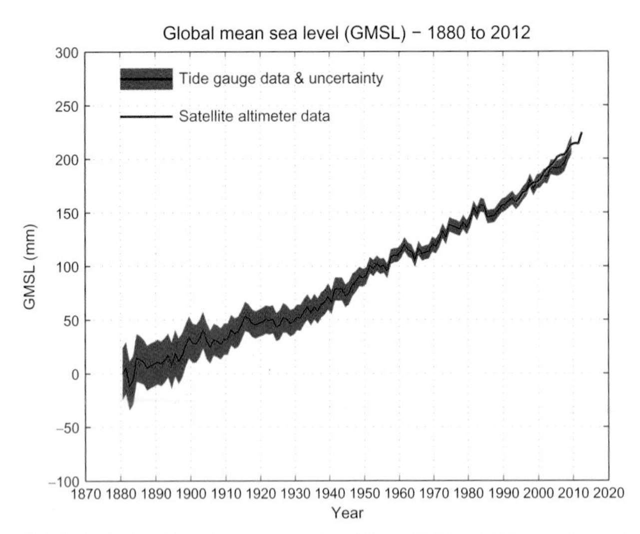

FIGURE 27.6 Global average sea level from 1880 to 2009 as estimated from the coastal and island sea-level data (blue). The one standard deviation uncertainty estimates plotted about the low passed sea level are indicated by the shading. The satellite altimeter data since 1993 is also shown in red. *Modified and updated from Church and White (2011).*

of the reconstruction but the regional patterns were more sensitive and seemed unrealistic prior to 1950. All three studies find a faster rate of rise since 1993, in agreement with the altimeter data, and a similarly high rate between about 1930 and 1960. Ray and Douglas, however, did not find a significant acceleration for the twentieth century as a whole. Meyssignac et al. (2012) used a similar reconstruction method (but with EOFs from an ocean model) to show that the pattern of sea-level rise present in the altimeter data was not representative of a long-term trend but instead was part of decadal variability, similar to the finding of White et al. (2005) and Zhang and Church (2012).

4. OBSERVATIONS OF OCEAN HEAT-CONTENT AND STERIC SEA-LEVEL CHANGE

4.1. Historical Observations

As discussed in Chapter 3, systematic observations of the physical distribution of ocean properties began with the Challenger Expedition of 1872–1876, which provided the first essentially global (although very sparse) measurements of subsurface ocean temperatures (Roemmich et al., 2012). This was followed in the 1920s–1950s by a series of voyages of the Meteor, Discovery, Discovery II, and Atlantis in the Atlantic and Southern Oceans (see Arbic and Owens, 2001). These early observations were made by a few dedicated research expeditions using maximum/minimum and reversing thermometers. Despite these efforts, it was not until about the 1950s, and perhaps the 1970s, that enough data began to become available to begin mapping the three dimensional structure of the ocean and to begin to explore its temporal variability (Levitus, 1982, 1987). The availability of ocean temperatures, although of lower accuracy than with a mercury thermometer, increased considerably following the development of the Mechanical BathyThermograph (in the late 1930s) and the eXpendable BathyThermograph (XBT, in the late 1960s), both primarily developed for naval purposes (Abraham et al., 2013). In addition to their use by the research community, XBTs, in particular, were deployed from naval and fisheries vessels and more recently from merchant marine vessels, coordinated by and in support of research activities.

The development of the World Climate Research Programme (WCRP) and its research projects, beginning in the early 1980s, was and remains critically important in the coordination and further development of observational activities. The WCRP Tropical Ocean Global Atmosphere project (TOGA, 1985–1994) focused on the coupled atmosphere–ocean El Niño–Southern Oscillation (ENSO) phenomenon with observations in the equatorial Pacific Ocean, leading to the development of the TOGA Tropical Atmosphere Ocean (TAO) mooring array in the Pacific and a coherent XBT program (Chapter 3). The TOGA array was slowly extended to the tropical Atlantic and Indian oceans as the TAO/TRITON/PIRATA/RAMA array during the CLIVAR Project. The World Ocean Circulation Experiment (WOCE, Siedler et al., 2001) designed a global observational system, incorporating research ship observations (figure 3.4), merchant ship XBT transects (figure 3.3), moored observations, satellite altimeter missions (TOPEX/Poseidon, Fu and Cazenave, 2001), and the development of new instrumentation. A major outcome of the instrumental development encouraged by WOCE was the profiling floats that now provide global temperature and salinity profiles over the (almost) global ocean from the surface to 2000 dbar as part of the Argo project (Gould et al., 2004; Chapter 3). Efforts to extend Argo to give full global coverage (including under sea ice) and to the full ocean depth are continuing. Importantly, the Argo project has developed increasingly robust quality control mechanisms (Wong et al., 2013). Although Argo data have much better coverage and accuracy than most previous data sets, care is still required as several problems have been found and corrected to the extent possible (Willis et al., 2007; Barker et al., 2011). There is also likely to be a bias between XBT data and Argo data, and as a result, there may be spurious variability and trends over periods spanning the transition from XBT observations (early 2000s) to the period of full Argo implementation (about 2007). Ongoing development and implementation of these ocean observational programs is shared between the WCRP's CLIVAR Project and the Ocean Observing Panel for Climate of the Global Climate Observing System. For a more complete description of these historical developments, see Chapter 3 and Abraham et al. (2013).

4.2. Global Ocean Heat-Content and Thermal-Expansion Estimates

There are two major challenges in accurately determining the global and regional distribution of ocean heat-content and steric sea-level variability and change. First, ocean observations are sparse (figures 3.12–3.14). In the 1950s, observations were concentrated in the northern hemisphere with very few observations in the Southern Ocean (south of 30°S), and fewer observations at high latitudes in winter. The WOCE project resulted in a global network of full-depth ocean observations (figure 3.4). These high quality observations were predominantly one-off observations and the variability was measured by a few repeat research ship observations and the global (but with significant gaps, particularly in the Southern Ocean) upper-ocean XBT network (figure 3.3). Objective mapping schemes that are used to make global estimates of ocean heat and steric sea level (e.g., Ishii et al., 2003, 2006; Levitus et al., 2005) assume zero anomalies (relative to the poorly known climatological average that can contain biases as a result of nonuniform spatial and temporal sampling) where there are no data. It is likely that the resultant regional and global heat-content estimates are biased, particularly in the data sparse Southern Ocean (Gille, 2008). Using output from global climate models, Gregory et al. (2004) demonstrated that the way in which gaps in the observational data are filled can significantly affect estimates of regional and global averaged temperatures. Since the start of 2006, the Argo Array has achieved almost uniform global coverage of temperatures and salinities in the upper 2000 dbar of the ocean, at a relatively high-frequency sampling rate (every ~10 days).

The second major issue is instrumental biases. Gouretski and Koltermann (2007) demonstrated that XBTs have time-dependent instrumental biases. Since XBTs dominate the historical data base, Gouretski and Koltermann postulated that the large decadal variability of ocean-heat content (Levitus et al., 2005; Ishii et al., 2006), with anomalously warm temperatures in the 1970s, was primarily a result of XBT biases (Figure 27.7). Climate models were unable to reproduce this decadal variability (AchutaRao et al., 2006, 2007). Wijffels et al. (2008) argued these instrumental biases were predominantly a result of errors in the depth of the observations (calculated using an estimate of the fall rate of the XBT probes through the water column).

For the upper 700 m of the ocean, Domingues et al. (2008) addressed the spatial mapping issue using a reduced space optimal interpolation scheme (see Section 3) and the XBT biases using the Wijffels et al. (2008) fall-rate corrections. The resultant time series (Figure 27.7) no longer had the large peak in ocean heat content in the 1970s as in earlier estimates (primarily because the XBT temperature biases were minimized). They did find falls in ocean heat content following the major volcanic eruptions of Mt. Agung in Indonesia in 1963, El Chichon in Mexico in 1982 and Mt. Pinatubo in

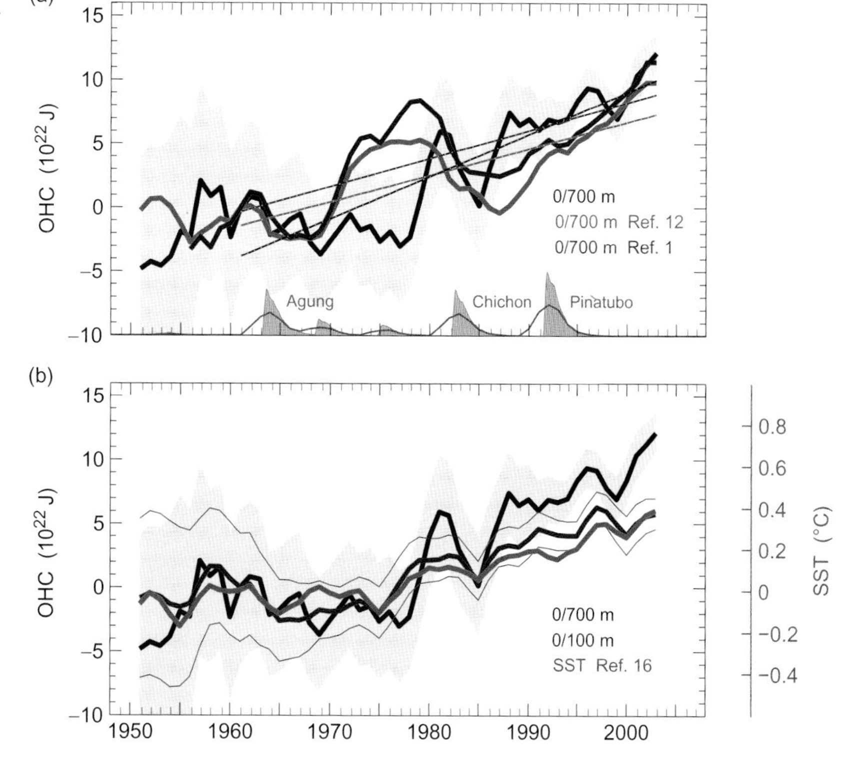

FIGURE 27.7 Estimates of ocean heat content and sea-surface temperature. (a) Comparison of our upper-ocean heat content (black; gray shading indicates an estimate of one standard deviation error) with previous estimates (red, Levitus et al., 2005, and blue, Ishii et al., 2006) for the upper 700 m. The straight lines are linear fits to the estimates. The global mean stratospheric optical depth (beige, arbitrary scale; Ammann et al., 2003) at the bottom indicates the timing of major volcanic eruptions. The brown curve is a 3-year running average of these values, included for comparison with the smoothed observations. (b) Comparison of the 700-m (thick black line, as in (a) and 100-m (thick red line; thin red lines indicate estimates of one standard deviation error) results with sea-surface temperature (blue; right-hand scale; Rayner et al. 2003). All time series were smoothed with a 3-year running average and are relative to 1961. *From Domingues et al. (2008).*

the Philippines in 1991, of magnitude similar to those present in climate model simulations that included volcanic forcing, and a larger trend since the mid 1970s to the end of the record. Subsequent analyses by Levitus et al. (2009) and Ishii and Kimoto (2009) produced similar results, but significant differences between the analyses remained.

Since 2008, much more work has been done on understanding and providing corrections to reduce the XBT biases (Ishii and Kimoto, 2009; Levitus et al., 2009, 2012; Cheng et al., 2010; Good, 2010; Gouretski and Reseghetti, 2010; Gouretski, 2012; Cowley et al., 2013; see Abraham et al., 2013). Although it is clear that the corrections developed (for both temperature offsets and depth errors) have improved estimates of ocean heat content, further work is required to come to an agreement on the best set of corrections for the historical data.

Estimates of ocean heat content (Figure 27.8) are impacted by mapping procedures (see the earlier comments on filling data gaps), climatological reference fields (particularly given the lack of high latitude observations in winter; see Chapter 3, Section 4), and quality control procedures (Palmer et al., 2010). Levitus et al. (2005, 2009, 2012), Ishii et al. (2006), and Ishii and Kimoto (2009) used objective mapping where the mapped field relaxes back to zero anomaly in areas of data gaps. In contrast, Lyman et al. (2010) and Johnson et al. (2012, 2013) used the global averaged mean anomaly to fill data gaps. For the period since 1993, Boening et al. (2012), Willis et al. (2004), Guinehut et al. (2004, 2006), Lombard et al. (2006), and Johnson et al. (2012, 2013) used spatially variable linear regressions between ocean temperatures and satellite altimeter measurements of sea-surface height anomalies to fill data gaps. Area-weighted box averages have also been used (Palmer et al., 2007; von Schuckmann and Le Traon, 2011; Gouretski et al., 2012; von Schuckmann et al., 2013). Palmer et al. (2007) estimated heat content above a particular isotherm in an attempt to minimize the impact of thermocline heave associated with variability and changes in ocean circulation rather than changes in net global-ocean heat content. Ocean reanalysis is being used increasingly for interpolating between the sparse ocean data. These analyses range from procedures where the model is strongly constrained by the data (Carton et al., 2005; Carton and Giese, 2008; Balmaseda et al., 2013a; see Chapter 22) to more sophisticated adjoint methods (Wunsch et al., 2007; Kohl and Stammer, 2008 for the deep ocean Kouketsu et al., 2011; see Chapter 21). Even for the most sophisticated schemes, the heat content estimates are strongly dependent on the quantity and quality of the available data and their uncertainties.

Recent estimates of upper ocean (top 400, 700, and 900 m) heat content changes (Abraham et al., 2013; Figure 27.8) all show multidecadal warming, with in excess of about 150×10^{21} J of heat stored between 1970 and 2012. However, there are significant differences between the various estimates. A number of these estimates also have falls in ocean heat content following major volcanic eruptions, similar to those estimated with climate model simulations with some (but not all) indicating a tendency toward reduced heat uptake since 2002. Over the globe (5.1×10^{14} m^2), these ocean heat storages are equivalent to a global energy imbalance of about 0.3 W m^{-2} for 1970 and from 1993 to 2012. (Note that the substantially larger trend computed by Lyman et al. (2010), and close to the "robust" average presented by them, for 1993–2008 has been superseded by their updated estimate with a significantly smaller trend; Johnson et al., 2012, 2013.)

The majority of observational analyses have focused on the upper 700 m of the ocean because of the greater data availability. Although the observations indicate that the largest increases have occurred in the upper ocean, heat storage and ocean thermal expansion occurs throughout the water column. In climate models, integrating ocean heat content over increasing depth provides an increasingly good estimate of the top-of-the-atmosphere radiation imbalances (Palmer et al., 2011; they agree to within 0.05 W m^{-2} for the full ocean depth). Observations also indicate the importance of the deeper ocean. Whereas Levitus et al. (2012) indicate a leveling off of heat content increases in the upper 700 m after 2004, analyses to 2000 m indicate continuing ocean heat uptake, emphasizing the importance of warming between 700 and 2000 m. Their updated ocean heat-content trend to 2000 m is for a net increase of $240 \pm 20 \times 10^{21}$ J from 1955–2010, corresponding to a global energy imbalance of 0.27 W m^{-2}. Whether or not the ocean heat uptake between 700 and 2000 m has been underestimated prior to the Argo observations remains an open question. Argo observations (von Schuckmann and Le Traon, 2011; von Schuckmann et al., 2013) confirm considerable ocean heat uptake between 700 and 2000 m. For the 2005–2012 period, von Schuckmann et al. (2013) find an ocean heat uptake of 0.5 ± 0.1 W m^{-2}, equivalent to a global energy imbalance of 0.35 ± 0.1 W m^{-2}. With the freely available Argo data set, there are now several different gridded products available for estimating ocean heat content and steric level (see http://www.argo.ucsd.edu/Gridded_fields.html).

Below 2000 m, repeat hydrographic sections reveal warming of deep ocean and bottom waters (e.g., Fukasawa et al., 2004; Johnson and Doney, 2006; Johnson et al., 2007, 2008; Kawano et al., 2010). Purkey and Johnson (2010) analyzed the available repeat sections, finding that the maximum abyssal and deep warming occurred in the Southern Ocean, weakening northward in the central Pacific, western Atlantic, and eastern Indian Ocean, with cooling in the western Indian Ocean and the eastern Atlantic Ocean. The global ocean heat content change below 3000 m in the

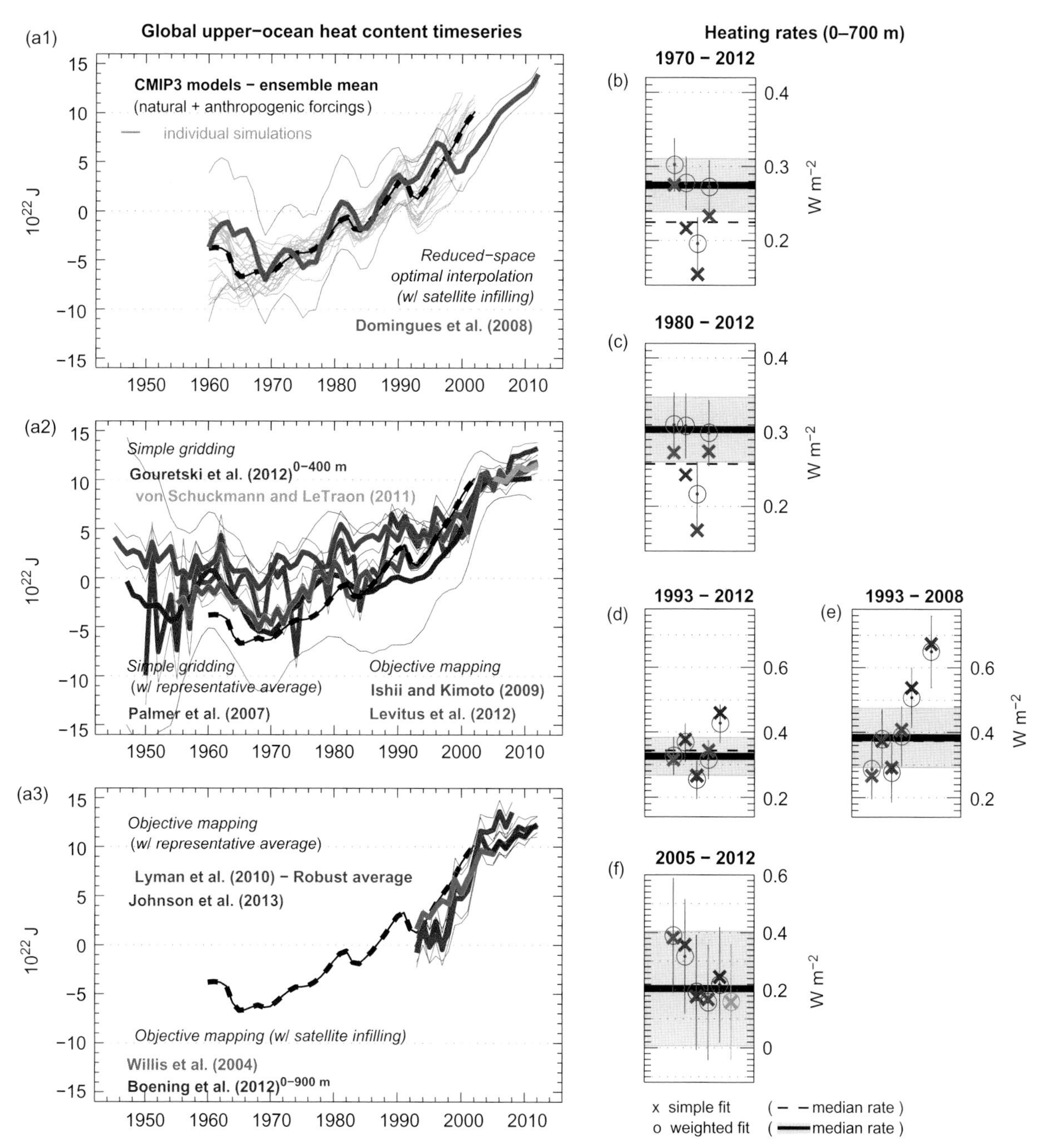

FIGURE 27.8 Time series of observed global upper ocean heat content (Panels a1–a3) for the upper 400 m (Gouretski et al., 2012), upper 900 m (Boening et al., 2012), and 700 m (others) and heating rates for the upper 700 m (Panels b–f). Results from CMIP3 model simulations (a1) with the most complete set of natural (e.g., solar and volcanic) and anthropogenic (e.g., greenhouse gases and aerosols) forcings are shown for individual runs (thin gray lines) and as a multimodel ensemble mean (black bold dashed line), with the latter repeated in panels (a2) and (a3) for reference. Updated observational time series (colored bold lines) and their error bars when available (colored thin lines) provided by the originators. Time series stopping before 2011 were originally included in Palmer et al. (2010) or Lyman et al. (2010). All observational time series are relative to 2005–2007 (except the robust time series which was referenced to the 1993–2001 period from Johnson et al. (2013)) whereas the model time series are relative to 1960–1999 plus an offset for plotting purposes. In the right panels (b–f), the simple least squares (SLS) linear trends are shown as crosses and the weighted least squares (WLS) linear trends are shown as circles. The median rates of the trends for the upper 700 m are indicated by horizontal lines, solid for WLS (accompanied by the gray shading uncertainties), and dashed for SLS. Only the estimates shown in panels (b–f) were used to calculate the median rates, except for the robust average estimate from Lyman et al. (2010). The OHC time series from Domingues et al. (2008) has been smoothed by a 3-year running mean to reduce noise. *Adapted from Abraham et al. (2013).*

1990s and 2000s is equivalent to a heat flux of 0.05 ± 0.03 W m^{-2} over the surface of the earth and is equivalent to a ocean averaged thermal expansion (3.61×10^{14} m^2) of 0.10 ± 0.06 mm year^{-1}. Kouketsu et al. (2011) obtained similar results from an analysis of the observations and the results of an adjoint data assimilation system.

The total ocean heat content and global-averaged thermal expansion (evaluated as in Rhein et al., 2013 from the sum of the contributions for the upper 700 m from the Domingues et al. (2008) updated time series, the Levitus et al. (2012) estimate from 700 to 2000 m, and the deep-ocean estimate from Purkey and Johnson (2010)) indicate an ocean heat uptake of 255×10^{21} J from 1971 to 2010 and a global averaged thermal expansion of 31 mm (Figure 27.9). The ocean heat uptake over this period represents 93% of the energy stored in the total climate system (Rhein et al., 2013; similar to Church et al., 2011a, 2013a). The ocean heat content (and thermal expansion) temporarily decreased (relative to the trend) following the major volcanic eruption of El Chichon in 1982 and Mt. Pinatubo in 1991 and during the major El Niño event of 1998. In the recent reanalysis of Balmaseda et al. (2013b), volcanic eruptions and the 1997–1998 El Niño event resulted in a loss of heat from the ocean. Their estimate of the total heat uptake is slightly lower at about 240×10^{21} J over 1971–2010. Note that although satellite observations of the top of the atmosphere radiation can be used to estimate variability in the rate of energy storage in the climate system, ocean heat-content estimates are necessary to determine the absolute storage rate (Loeb et al., 2012).

Warming has occurred in essentially all the ocean basins (see figure 3.1 from Rhein et al., 2013 and figure S6 from Levitus et al., 2012), including the Southern Ocean where it has occurred over many decades and to considerable depth (Böning et al., 2008; Gille, 2008). The warming is largest above the main thermocline, but there are also maxima at the poleward boundaries of the subtropical gyres. In the south-west Pacific, these are associated with a spin-up of the subtropical gyre in response to increased wind-stress curl (Roemmich et al., 2007), that is a redistribution of heat rather than a local air-sea flux of heat.

4.3. The Global Energy Balance

The oceans dominate the increase in energy stored in the climate system, being responsible for 93% of the total energy increase in the climate system from 1972 to 2008 and 1993 to 2008 (Church et al., 2011a). The storage of energy in the Earth system N is given by (Gregory and Forster, 2008)

$$N = F - \alpha\Delta T \tag{27.2}$$

where F is the radiative forcing and $\alpha\Delta T$ is the long-wave radiative response of a warming Earth (i.e., greater long-wave emission from the Earth as it warms). Using the historical radiative forcings F of Myhre et al. (2013) and the

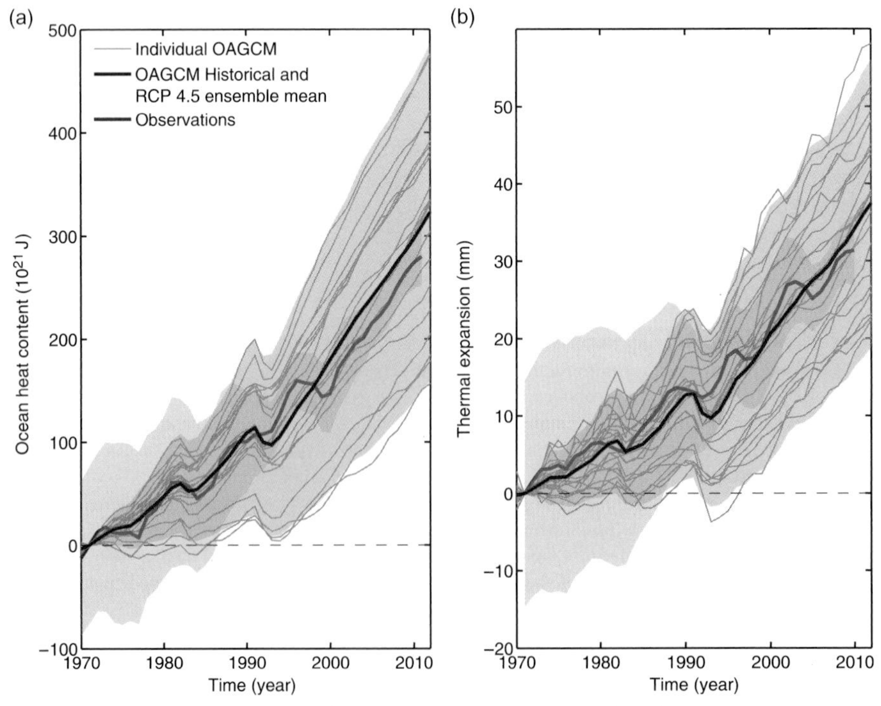

FIGURE 27.9 Model and observed estimates of (a) full-depth ocean heat-content increases and (b) global averaged thermal expansion. The light gray lines are individual CMIP-5 AOGCM simulations and the blue lines are the model averages with blue shading indicating the 90% (5–95%) confidence range. The black lines are observational estimates as described in the text, with the shading indicating the 90% (5–95%) confidence range.

warming of the Earth system N of Rhein et al. (2013), Church et al. (2013a) tested the consistency of the Earth's energy balance for values of the climate feedback factor α of 0.82 W m^{-2} °C^{-1} (equivalent to an equilibrium climate sensitivity (ECS) of 4.5 °C for an effective radiative forcing F_{2x} of 3.7±0.4 W m^{-2} for a doubling of CO_2 concentration), 1.23 W m^{-2} °C^{-1} (climate sensitivity of 3.0 °C), and 1.85 W m^{-2} °C^{-1} (climate sensitivity of 2.0 °C). [Note that Gregory and Forster (2008) called α both the climate sensitivity parameter and the climate feedback parameter.] The consistency of the energy storage and surface warming with the estimated effective radiative forcing for the range of climate sensitivities (Figure 27.10) provides strong support for our understanding of anthropogenic climate change. The change in the energy stored in the ocean is also a powerful observation for the detection and attribution of climate change (Gleckler et al., 2012; Pierce et al., 2012).

The Earth's energy storage is also important in observational estimates of the climate sensitivity. With $\alpha = F_{2x}/\text{ECS}$, the energy balance equation can be written as

$$\Delta T = 1/\alpha(F - N) = \text{ECS}(F - N)/F_{2x} \quad (27.3)$$

Otto et al. (2013) regressed surface temperature increase against $(F - N)$, estimating the ECS to be 2.2 °C (with a 5–95% range of 1.4–4.0 °C) for an effective forcing for doubling CO_2 concentration of 3.7±0.7 W m^{-2} (2.0 °C for 3.4 W m^{-2}).

5. UNDERSTANDING OBSERVED SEA-LEVEL CHANGE

Explaining the observed sea-level rise has been a conundrum for several decades. Munk (2002) termed our inability to explain the observed twentieth century sea-level rise as a sum of thermal expansion and loss of mass from the cryosphere, and at the same time satisfying constraints imposed by observed changes in the rotation of the Earth, as the "sea-level enigma". This riddle has now been solved. First, Mitrovica et al. (2006) recognized that responses within the mantle could relax the constraints imposed by Earth rotation changes. Second, we can now close the sea-level budget using observations, particularly from the past two decades and also from the twentieth century as a whole, using a combination of models and observations.

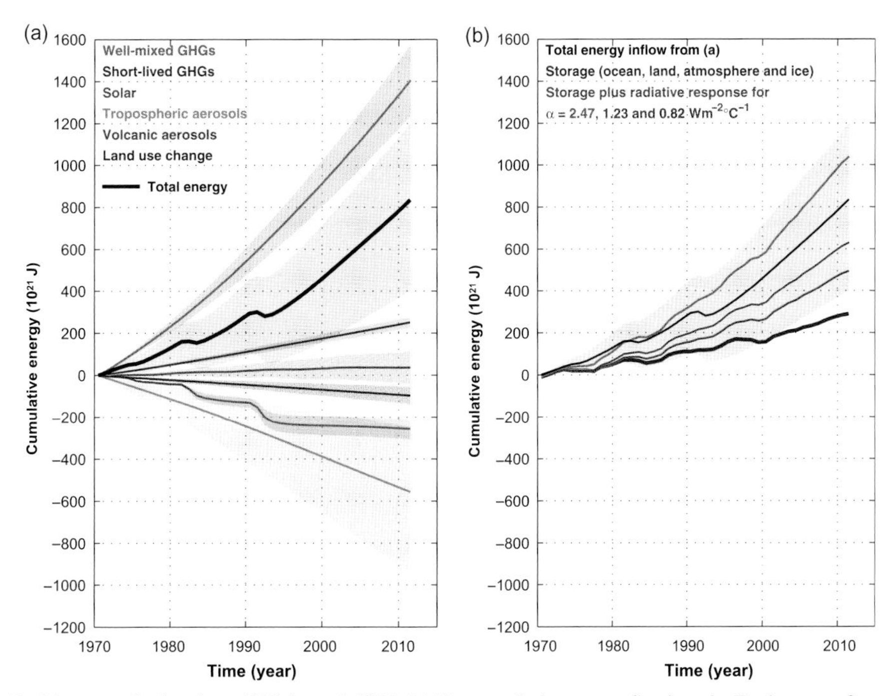

FIGURE 27.10 The Earth's energy budget from 1970 through 2011. (a) The cumulative energy flux into the Earth system from changes in well-mixed and short-lived greenhouse gases, solar forcing, changes in tropospheric aerosol forcing, volcanic forcing, and surface albedo, (relative to 1860–1879; from Myhre et al., 2013) are shown by the colored lines and these are added to give the cumulative energy inflow (black; including black carbon on snow and combined contrails and contrail induced cirrus, not shown separately). (b) The cumulative total energy inflow from (a, black) is balanced by the sum of the warming of the Earth system (blue; energy absorbed in warming the ocean, the atmosphere and the land and in the melting of ice; from Rhein et al., 2013) and an increase in outgoing radiation inferred from changes in the global averaged surface temperature. The sum of these two terms is given for a climate feedback parameter α of 0.82, 1.23, and 2.47 W m^{-2} °C^{-1}, (corresponding to an equilibrium climate sensitivity of 4.5, 3.0, and 1.5 °C). The energy budget would be closed for a particular value of α if that line coincided with the total energy inflow. For clarity, all uncertainties (shading) shown are for a likely (5–95%) range. *Adapted from Climate Change 2013: The Physical Science Basis. Working Group I Contribution to the Fifth Assessment Report of the Intergovernmental Panel on Climate Change, Box 13.1, Figure 1, Cambridge University Press (in press).*

The largest single contribution during these periods was ocean thermal expansion, as discussed in Section 4. However, the increase in ocean mass, as a result of loss of mass from glaciers, ice sheets, and land water storage, comprises more than 50% of the observed sea-level rise.

5.1. Changes in Ocean Mass

After thermal expansion, the next largest contribution comes from losses in glacier mass around the world (Arendt et al., 2012). Cogley (2009) used a combination of glaciological (the surface mass balance (SMB) of glaciers) and geodetic (the height of the glaciers) observations to estimate glacier mass loss every 5 years since 1950. There are limited observations of the more than 160,000 glaciers and the estimate (23 mm from 1960 to 2008, Figure 27.11) required upscaling of the observations to all areas of the globe. Recent comprehensive observations using *in situ* and satellite observations (Gardner et al., 2013) indicate that for 2003–2010 the Cogley estimates were biased high, by perhaps 25%. This is thought to relate to the individual point locations of the historical measurement and it is not clear at this time how to correct the historical record. Other estimates come from measurements of glacier length since the nineteenth century (Leclercq et al., 2011) and use of a glacier model with observed climate data for the twentieth century (Marzeion et al., 2012). Both of these agree with the Cogley (2009) estimate for the period since 1950, although there is now increased uncertainty about the accuracy of the Cogley (2009) estimate.

The storage of water on land makes a significant contribution to sea-level change (see Milly et al., 2010). Using hydrological models, Milly et al. (2003) and Ngo-Duc et al. (2005) found little long-term contribution related to natural climate variability. However, Chao et al. (2008) estimated a significant (22 mm) storage of water in reservoirs built during the twentieth century, mostly since 1960, thus offsetting other contributions to sea-level rise. The rate of dam building and water impoundment has decreased significantly since the peak in the 1970s. Konikow (2011) estimated that ground water depletion had contributed 11 mm to sea-level rise since 1960 and the rate of depletion continues to increase. The net contribution to sea-level change from these two terms over the full twentieth century was negative but small (Figure 27.11).

The contributions from the ice sheets of Greenland and particularly Antarctica are the most uncertain. For more detailed discussion of ocean-ice sheet interactions, see Chapter 16.

Sophisticated high-resolution models of SMB for Greenland (e.g., Fettweis et al., 2011, 2013; van Angelen et al., 2012) agree with the observational estimates of climatological surface mass balance (within uncertainties) and show little trend in SMB from the 1960s to the 1980s.

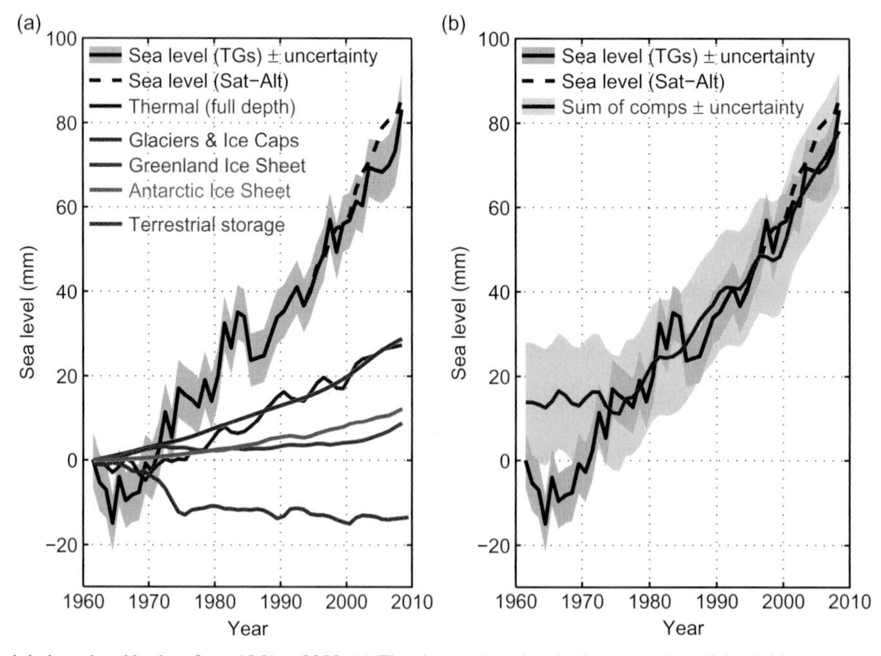

FIGURE 27.11 The global sea-level budget from 1961 to 2008. (a) The observed sea level using coastal and island tide gauges (solid black line with gray shading indicating the estimated uncertainty) and using TOPEX/Poseidon/Jason-1&2 satellite altimeter data (dashed black line). The two estimates have been matched at the start of the altimeter record in 1993. Also shown are the various components. (b) The observed sea level and the sum of components. The estimated uncertainties are indicated by the shading. The two time series are plotted such that they have the same average over 1972–2008. *From Church et al. (2011a).*

For the first half of the twentieth century, mass balance estimates depend on reconstructions of surface climate variables and there is a greater disparity of results between studies. For example, Box and Colgan (2013) estimated a small positive contribution to sea-level rise whereas Hanna et al. (2011) estimated a small negative contribution, and Wake et al.'s (2012) estimate is approximately zero. A careful intercomparison of several methods (altimetry, gravity measurements, surface mass-balance modeling, and velocity of glacier discharge) indicates that Greenland lost 8 mm of SLE between 1993 and 2010, with the rate of loss increasing during the period (Shepherd et al., 2012). About equal amounts come from a decrease in SMB, as a result of an increase in melting and runoff, and an increase in discharge from the Greenland outlet glaciers (van den Broeke et al., 2009; Rignot et al., 2011; Sasgen et al., 2012). Observations (Holland et al., 2008) and simulations of the increase in discharge (Nick et al., 2009, 2013) indicate that ocean warming at the glacier front leads to a rapid adjustment of the calving glacier that propagates landwards. The few available observations (Straneo et al., 2010, 2011) indicate complex ocean dynamics related to a variety of external parameters, likely including offshore ocean conditions. Further observational and modeling studies are required to understand the complex linkages and the implications for future Greenland Ice Sheet discharge (Straneo et al., 2013). For the twenty-first century, surface melting at lower elevations is projected to increase more rapidly than increased snowfall leading to an ongoing and likely accelerated mass loss from Greenland (e.g., Fettweis et al., 2013). The outlet glaciers are likely to retreat from the coast thus ending ocean-ice interaction.

For Antarctica, there is little surface melting; so in steady state snowfall must be balanced by solid ice discharge to the floating ice shelves. There is agreement that the West Antarctic Ice Sheet is losing mass through increased discharge (Rignot et al., 2008a; Pritchard et al., 2009; Velicogna, 2009; Velicogna and Wahr, 2013; figure 4.8). This is thought to be a result of penetration of warm ocean water beneath the ice shelves (Jacobs et al., 2011), melting them at their base and allowing a dynamic response of the ice streams, as modeled for the Pine Island Glacier (Joughin et al., 2012). A similar effect was observed with a significant increased discharge of glaciers following the rapid collape of the Larsen-B ice shelf on the eastern side of the Antarctic Peninsula (Rignot et al., 2004; Scambos et al., 2004). Rignot et al. (2013) estimate that more than half the ice-shelf mass loss occurs from bottom melting with the remainder from iceberg calving. Thus adequately understanding the current (e.g., Jacobs et al., 2011) and future warming (Yin et al., 2011; Galton-Fenzi et al., 2012; Hellmer et al., 2012; Kusahara and Hasumi, 2013) of the Southern Ocean adjacent to and under ice sheets is critically important and one of the major challenges in projecting sea-level rise. There is less agreement for East Antarctica with some studies showing a mass gain and others a mass loss. Shepherd et al. (2012) estimated a 4-mm sea-level contribution from 1992 to 2011 for Antarctica as a whole. There is little information on the Antarctic mass balance prior to the 1990s, although ice-sheet modeling studies (Huybrechts et al., 2011) suggest a small ongoing contribution (of order 0.2 mm year^{-1} SLE) in response to changes since the last glacial maximum. Increasing atmospheric temperatures during the twenty-first century are expected to lead to greater accumulation partially offsetting other contributions to sea-level rise. However, the West Antarctic Ice Sheet is grounded below sea level on bathymetry that deepens inland and thus could be unstable with small initial losses of mass leading to a growing discharge (e.g., Hindmarsh, 1993; Schoof, 2007), although lateral stresses (Gudmundsson, 2013) and isostatic and gravitational changes (Gomez et al., 2010) are potentially stabilizing mechanisms.

5.2. The Sea-Level Budget

Closing the sea-level budget, whether with models or observations or some combination of both, relies on many different contributions. As a result, any budget closure study must be updated as new estimates of individual contributions are completed.

Church et al. (2011b) used the observational estimates for ocean thermal expansion discussed earlier, the Cogley (2009) glacier estimate, the Greenland estimate from Rignot et al. (2008b; based on a number of estimates of glacier outflow and an estimate of surface mass balance). For Antarctica, they used the accelerated ice loss estimated by Rignot et al. (2008a) since 1990 and assumed a small ongoing contribution of 0.1 mm year^{-1} prior to 1990 (see Figure 27.11a). The sum of these terms satisfactorily explained the observed rise since the early 1970s (Figure 27.11b). However, they were unable to explain the fall in sea level in the early 1960s. They suggested this may be a result of lack of ocean observations following the eruption of Mt. Agung in 1963 resulting in an underestimate of the decrease in ocean heat uptake, and possibly increased glacier mass. Moore et al. (2011) and Hanna et al. (2013) also provided closures of the sea-level budget for recent decades.

For the twentieth century as a whole, there are insufficient ocean observations to robustly estimate thermal expansion prior to 1970. Gregory et al. (2013a) completed a thorough analysis of the twentieth century sea-level budget using all available observed and modeled time series. For ocean thermal expansion, they used results from the Coupled Model Intercomparison Project Phase 3 (CMIP3) forced by historical changes in radiative forcing (greenhouse gases, aerosols, natural forcing from volcanoes and solar changes). However, many of these coupled ocean atmosphere general circulation models (Chapter 23) have been spun up ignoring volcanic eruptions, which in the time average have a negative radiative forcing. As a result, the historical simulations start out too warm and have an artificial negative trend in ocean thermal expansion when the volcanic forcing is suddenly applied (Gregory, 2010; Gregory et al., 2013b). To minimise this error the model results have to be corrected. Gregory et al. (2013a) used the estimates of glacier mass evaluated from observations of glacier length (Leclercq et al., 2011) and estimated by driving a glacier model, tuned to the Cogley (2009) observations, with observations of twentieth century climate (Marzeion et al., 2012). Both of these estimates have a period of greater glacier mass loss in the first half compared with the second half of the twentieth century, with the Marzeion et al. estimate being significantly larger. For Greenland, they used three different estimates of surface mass balance (Fettweis et al., 2008; Hanna et al., 2011; Wake et al., 2012) and one estimate of total Greenland mass balance (Box and Colgan, 2013). For land water storage they used an update of the terrestrial reservoir retention of Chao et al. (2008), the Konikow (2011) aquifer depletion estimate, and a new model estimate for aquifer depletion from Wada et al. (2012). Lacking information on Antarctica, they assumed a small linear trend, estimated from the difference between the sum of the time series of the contributions and the observations of sea level produced by Church and White (2011), Ray and Douglas (2011), Jevrejeva et al. (2008), and Wenzel and Schröter (2010). This small Antarctic trend is assumed to be the result of an ongoing ice-sheet imbalance with its magnitude constrained by paleo sea-level observations. With the range of different possible estimates for each term, a total of 144 solutions were possible, with a number satisfying the constraints imposed. An example of one of the solutions is given in Figure 27.12.

5.3. Recent Direct Observations of Ocean-Mass Changes

GRACE satellite gravity measurements (Chapter 4) can be used to infer changes in the mass of the ocean (e.g., Chambers et al., 2010; Llovel et al., 2010) and the ice sheets (e.g., Chen et al., 2009, 2011; Velicogna, 2009; Velicogna and Wahr, 2013). Together with accurate satellite altimeter observations to measure ocean volume changes and the implementation of Argo to measure changes in upper-ocean density, GRACE ocean-mass observations offers a new opportunity to understand sea-level change and to examine the sea-level budget (Chapter 4). Leuliette and Miller (2009) showed that the sum of measured ocean thermal expansion and changes in ocean mass explained the observed annual cycle in global mean sea level. For 2005–2012, von Schuckmann et al. (2013) found that the trend from the sum of ocean thermal expansion and GRACE ocean-mass changes were equal to the global mean sea-level change but regionally there remained significant differences, indicating some inconsistencies in the observations.

The ENSO cycle also results in significant perturbations to the hydrological cycle (see, e.g., Llovel et al., 2010, 2011). During the 2010–2011 La Niña event, global mean sea level fell by 5 mm as a result of large transfers of mass from the ocean to the land (Boening et al., 2012; Figure 27.13) with resultant flooding in Australia, Southeast Asia, and northern South America. This water has since returned to the ocean and sea level has returned to the long-term rising trend (Figures 27.5 and 27.13).

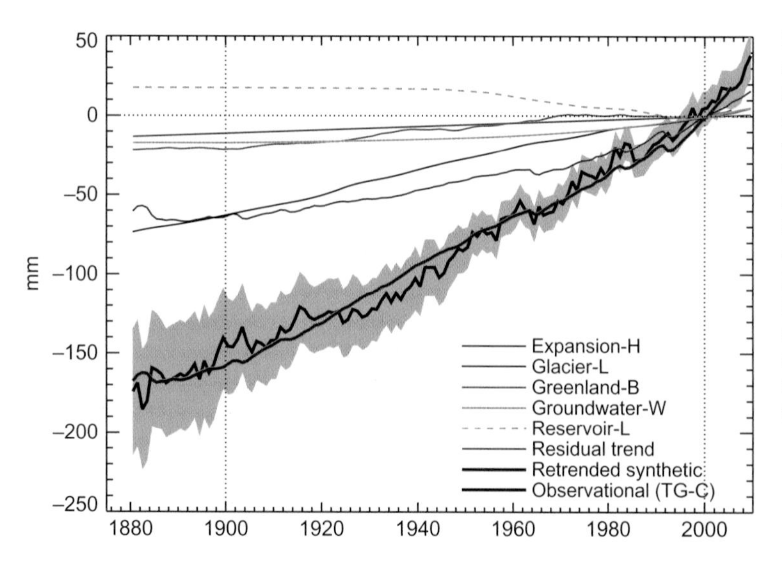

FIGURE 27.12 An example comparing observational and retrended synthetic time series of global mean sea-level rise for the twentieth century (thick lines, with 5–95% observational uncertainty shaded), also showing the contributions to the latter (thin lines), identified by the time series initials in the key (see Gregory et al., 2013a,b for a detailed explanation of the terms). The thin blue line is the residual trend to minimize the misfit between the sum of terms and the observed sea level. The observational and retrended (i.e., included a linear term for an ongoing Antarctic contribution) synthetic time series have the same time-mean during 1986–2005, and the latter and its components are all plotted relative to zero in 2000. The horizontal dotted line indicates zero; the vertical dotted lines delimit the twentieth century. *From Gregory et al. (2013).*

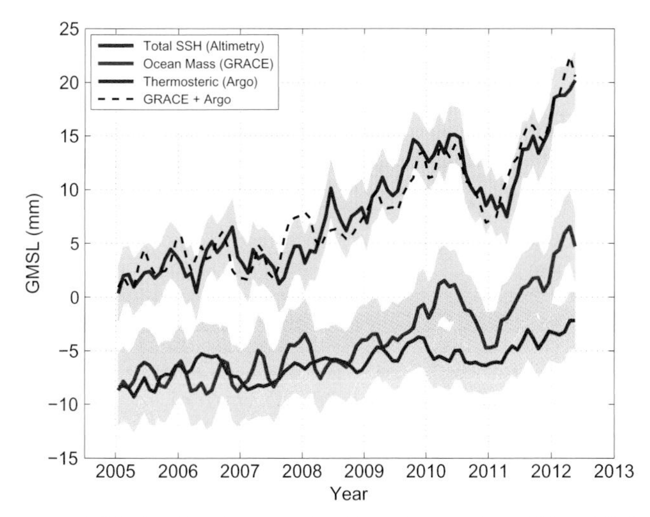

FIGURE 27.13 Global mean sea level from altimetry from 2005 to 2012 (blue line). Ocean mass changes are shown in green (as measured by GRACE) and thermosteric sea-level changes (as measured by Argo) are shown in red. The dashed black line shows the sum of the ocean mass and thermosteric contributions. *Adapted from Boening et al. (2012).*

5.4. Modeling Global Mean Sea-Level Change

Ocean general circulation models and coupled atmosphere-ocean general circulation models, which are at the heart of projections of sea-level change, are discussed in Chapters 20 and 23. Griffies and Greatbatch (2012) extensively discussed the ocean modeling aspects of sea-level change. Here, we focus on the ability of models to represent the twentieth century sea-level change.

Church et al. (2013b) used the CMIP5 simulations that have been forced with the best estimates of historical radiative forcings up to 2005 and then radiative forcing from the RCP4.5 scenario (Moss et al., 2010; Taylor et al., 2011) until 2010. The CMIP5 model mean estimates of ocean heat content and global mean thermal expansion are in good agreement with the observations for the period since 1971 when the observational estimates are more robust (Figure 27.9). As discussed above, the models depict the oceanic response to volcanic forcing (El Chichon in 1982 and Mt Pinatubo in 1991) more clearly than the observations; which is probably a result of inadequate ocean observations and hence a critical evaluation of the model response to volcanic forcing is not yet possible. We now have an observing system (Argo, altimeter and GRACE) in place that, if maintained, could track the impact of a major volcanic eruption if it happens and this could be used to evaluate/improve model simulations. The warming and thermal expansion in the first half of the twentieth century (Gouretski et al., 2012) is consistent with the Hobbs and Willis (2013) comparison of the *HMS Challenger* observations (Roemmich et al., 2012) with more recent observations and CMIP5 model results. Kuhlbrodt and Gregory (2012) found that the models were in general too weakly stratified. They also found that the ocean heat-uptake efficiency was negatively correlated with the climatological mean stratification, the stratification increase in a changing climate and the diffusivity used in the models' mesoscale eddy parameterization.

For the twentieth century as a whole (Figure 27.14a), Church et al. (2013b) argued that in addition to the impact of the Mt. Agung eruption in 1963, the low rate of sea-level rise from 1950 to 1975 (Figure 27.15d) may also be related to the increase in sulfur dioxide emissions by more than a factor of two (Smith et al., 2011) and the resultant increased negative aerosol forcing. From 1970, the rate of expansion increases significantly, with the fastest rate occurring from 1993.

The Marzeion et al. (2012) glacier estimate (Figure 27.14b and d) is larger in the first half than the second half of the twentieth century, and even larger when the observed climate rather than AOGCM results are used to force their glacier model. Church et al. (2013b) suggested this might be a result of decadal variability (Delworth and Knutson, 2000), consistent with the warmer conditions over Greenland in the 1930s (Chylek et al., 2004). Estimates of loss of glacier mass from the glacier length time series (Leclercq et al., 2011) also shows a larger contribution during the first half of the twentieth century, but with a slightly different timing and magnitude.

The net contribution from the land water storage terms over the twentieth century is small (Figure 27.14c and d).

When these terms are summed, particularly when using the larger modeled glacier contribution estimated from the observed (rather than the modeled) climate, they are not significantly different from the observed sea level for the three periods since 1900, 1970, and 1993 (Church et al., 2013b; Figure 27.15). The inclusion of the observed ice-sheet contributions (Shepherd et al., 2012) slightly improves the agreement between the observed and modeled rise, particularly for the period since 1993. Both the observed and modeled rate of rise have a peak in the first half of the twentieth century with the largest rate of rise occurring at the end of the record. This latter rate is similar to the rate estimated from the satellite altimeter record and is related to increases in natural and anthropogenic radiative forcing (Church et al., 2013b).

The ability to balance the sea-level budget using observations (although with some uncertainty as time series of individual components are updated) and the demonstrated improved ability to simulate the observed rise is reason for increased confidence in projections of twenty-first century sea-level rise. However, the Greenland and Antarctic ice sheets appear to have made a significant contribution only over the last two decades and hence the closure of the sea-level budget is not a test of the ice-sheet models' ability to reliably estimate future sea-level rise. Nevertheless, there is significant agreement between observations and models of surface

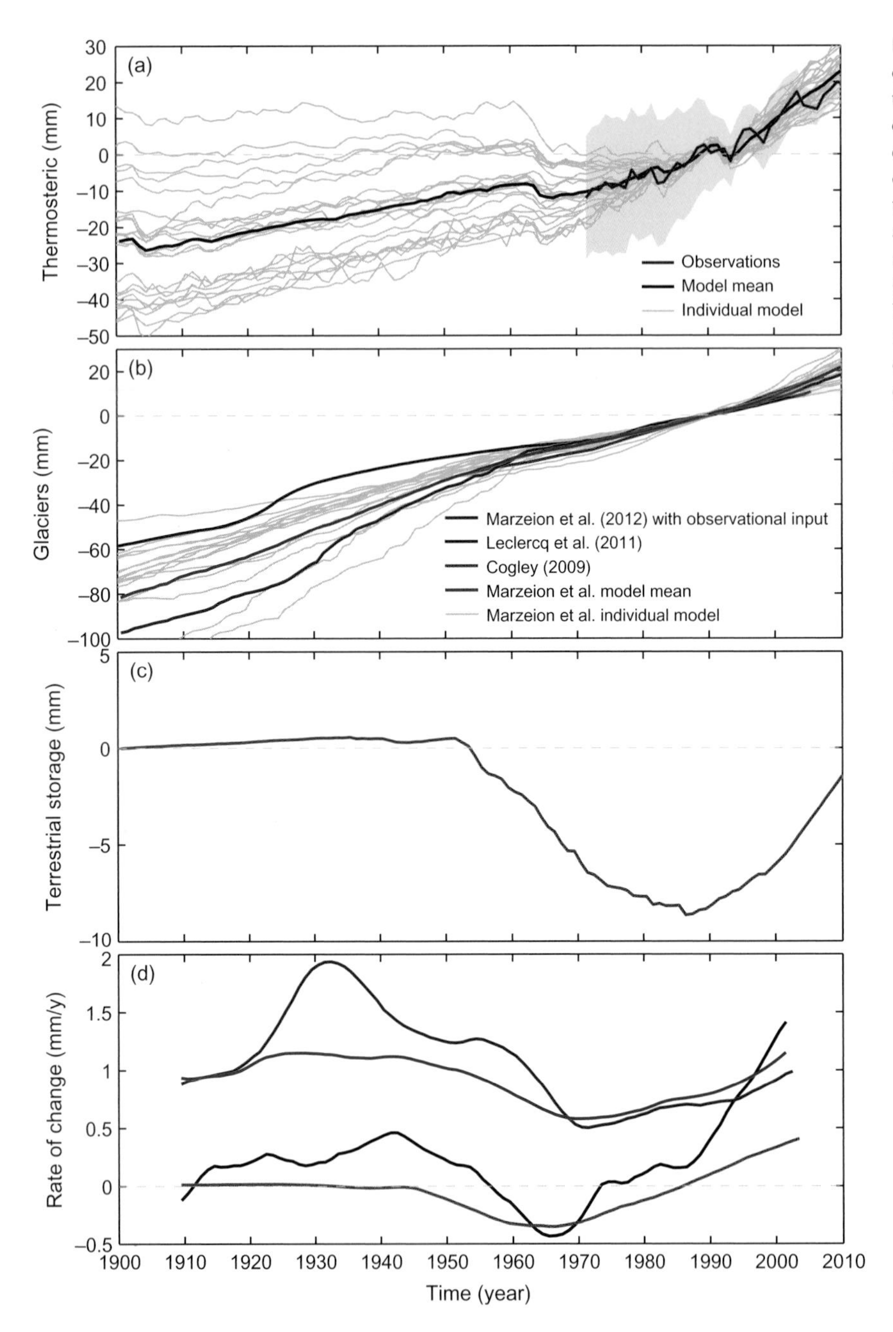

FIGURE 27.14 Comparisons of modeled and observed (a) ocean thermal expansion (observations in blue), (b) glacier contributions, (c) changes in terrestrial storage (the sum of aquifer depletion and reservoir storage), and (d) the rate of change (10 year centered average) for the terms in (a)–(c). Individual model simulations are shown by gray lines with the model average shown in black (thermal expansion) and purple (glaciers). The glacier contributions estimated by Cogley (2009, green), Leclercq et al. (2011, red) and using the model of Marzeion et al. (2012, dark blue) forced by observed climate are also shown in (b). All curves in (a) and (b) are normalized over the period 1980–1999 and the colors in (d) are matched to earlier panels. *From Church et al. (2013b).*

mass balance for Greenland and Antarctica, and models of ice-sheet dynamical responses are improving rapidly.

6. PREDICTION AND PROJECTIONS OF FUTURE SEA-LEVEL CHANGE

6.1. Interannual Sea-Level Predictions

Sea-level change research has been pursued for several decades. However, perhaps surprisingly, seasonal sea-level predictions from ENSO prediction models have not been utilized, even though sea-level change is possibly one of the more robust parameters in these models. Recognizing this, a study to test the skill of seasonal sea-level prediction has been initiated. Early results (Miles et al., 2013; Figure 27.16) indicate skill (considerably exceeding persistence) at period of 6 months in the low latitude Pacific and to a lesser extent Indian Oceans. Sea-level prediction experiments are likely to be initiated by additional groups in the near future.

6.2. Sea-Level Projections

Projections of sea-level change are a central element in IPCC assessments of climate change. The improved understanding of the twentieth century sea-level budget and the understanding of the contributions from climate models

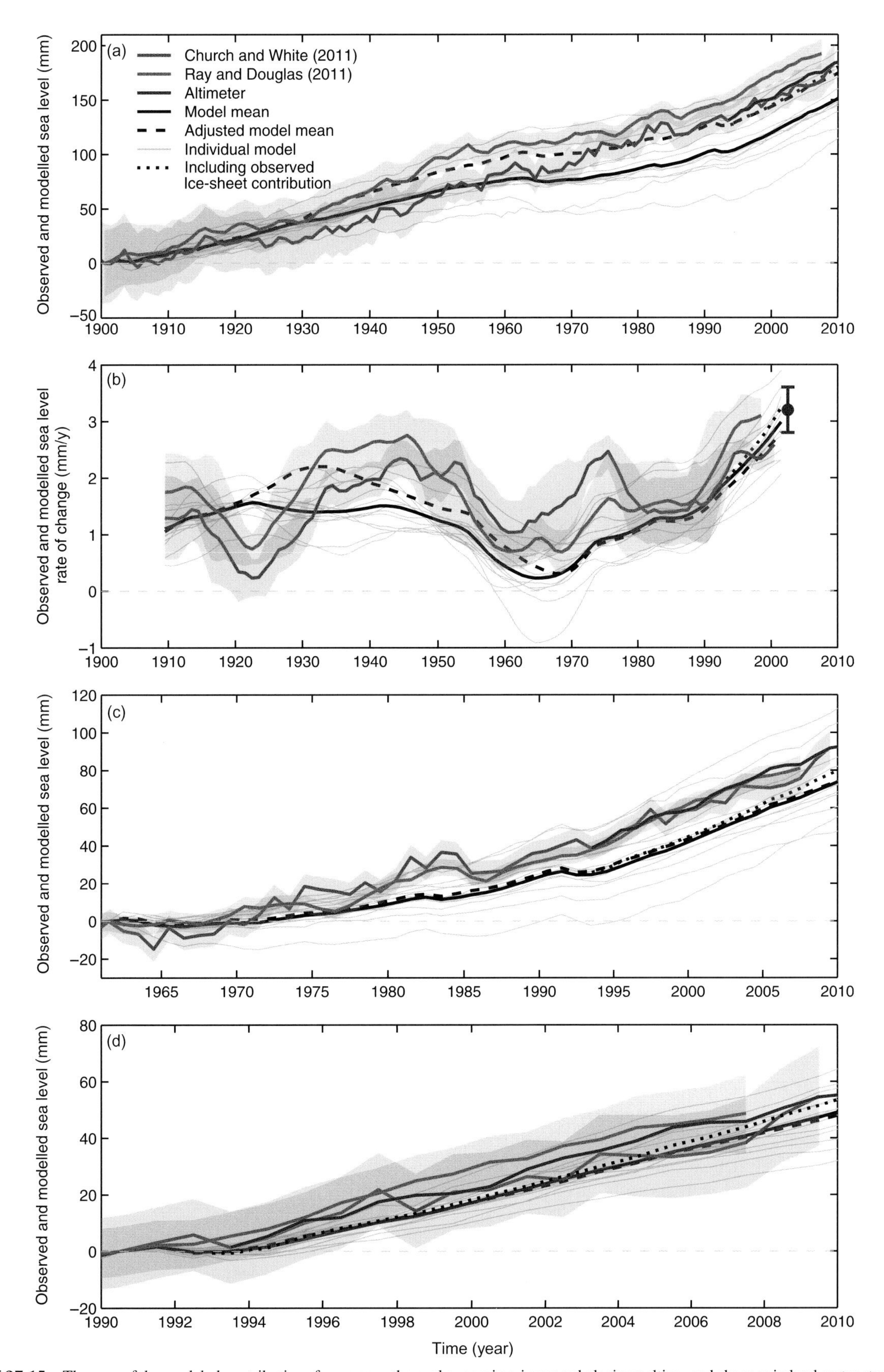

FIGURE 27.15 The sum of the modeled contributions from ocean thermal expansion, increased glacier melting, and changes in land-water storage. The light gray lines are individual models with the black line, the model mean. The twentieth century estimates of GMSL are indicated by the blue (CW11) and green (RD) lines with the shading indicating the uncertainty estimates (90% (5–95%) confidence range). The satellite altimeter data since 1993 is shown in red. The adjusted model mean (dashed black line) is the model mean after an allowance for the impact of natural variability on glacier contributions and a potential long-term ice sheet contribution are included. The results are given (a) for the period 1900–2010, (b) the rates of sea-level change for the same period, (c) for 1961–2010, and (d) for 1990–2010. The dotted black line is after inclusion of the Shepherd et al. (2012) ice sheet observational estimates but excluding the peripheral glacier contribution (to avoid double counting). The red dot (with uncertainty estimates) is the average rate from the altimeter record. *From Church et al. (2013b).*

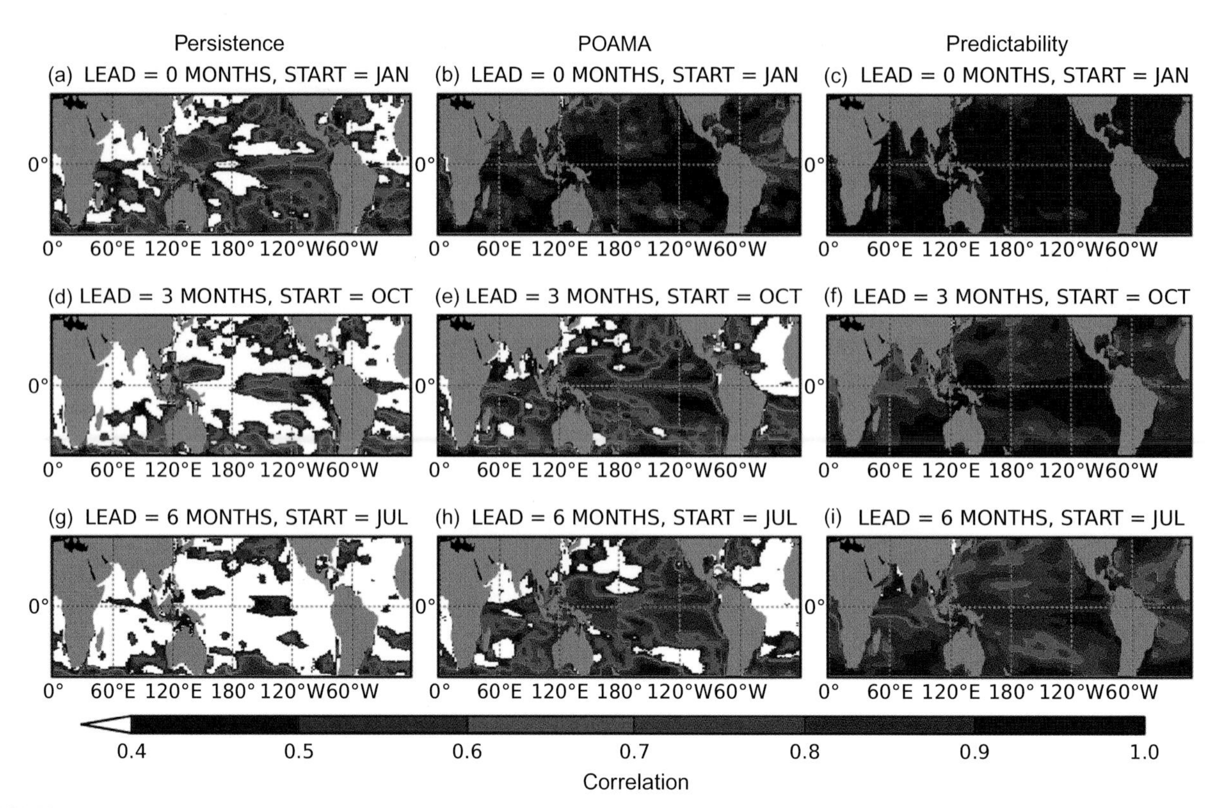

FIGURE 27.16 Correlations of seasonal forecasts for sea-level anomalies for the target season January–February–March from 1981 to 2010 for (left column) persistence and (center column) POAMA with the reanalyses for 0, 3, and 6 months lead-times. The right column is potential model predictability. Significant correlations are shaded ($|r|>0.361$ is significant at the 95% confidence level; two-tailed t-test, $n=30$, degrees of freedom determined by auto-correlation). *From Miles et al. (submitted).*

are reasons for increased confidence in projections of GMSL for the twenty-first century. These projections (Church et al., 2013a) combined with the instrumental observations of the twentieth century and the paleo observations indicate a sea-level rise of order of a meter for the highest (RCP8.5) emission scenario by 2100 compared to preindustrial conditions and with rates by 2100 approaching the average rate for the last deglaciation (1 m century^{-1} of sea-level rise), and substantially less for the lowest (RCP2.6) emission scenario (Figure 27.17).

Because of the large inertia of the oceans, there is little difference between the projections of GMSL for different scenarios in the first half of the twenty-first century. However, the projections subsequently diverge to and beyond 2100. Simulations of Greenland SMB indicate that the higher emission scenarios result in crossing the threshold where melting exceeds accumulation during the latter half of the twenty-first century (Fettweis et al., 2013), whereas for the lowest emission scenarios this threshold can be potentially avoided. Crossing this threshold would likely lead to an ongoing and irretrievable loss of much of the Greenland ice sheet leading to a sea-level rise of several meters from this source alone over the next millennia or so (Gregory and Huybrechts, 2006; Fettweis et al., 2013; note that with a crossing of this

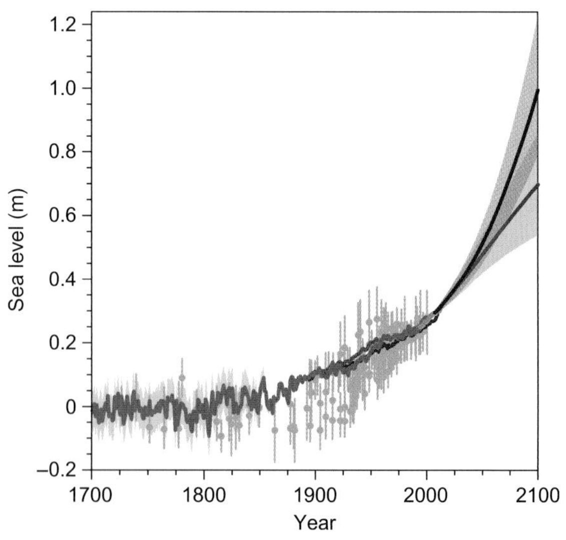

FIGURE 27.17 Compilation of paleo sea-level data (dots with uncertainties), GMSL estimates from tide-gauge, and altimeter data (continuous lines until 2010), and central estimates and likely ranges for projections of global-mean sea-level rise for RCP2.6 (blue) and RCP8.5 (red) scenarios, all relative to preindustrial values. *Adapted from Climate Change 2013: The Physical Science Basis. Working Group I Contribution to the Fifth Assessment Report of the Intergovernmental Panel on Climate Change, Figure 13.27, Cambridge University Press (in press).*

threshold for a limited time there can be a significant and possibly irreversible sea-level contribution, Ridley et al., 2010). Ice-sheet dynamics (Robinson et al., 2012) may lead to further lowering of the threshold for loss of the Greenland Ice Sheet. To follow the lower sea-level projections and to avoid this melting of the Greenland Ice Sheet will require significant and sustained mitigation of greenhouse gas emissions. The multicentury contribution from Antarctica remains uncertain but it is potentially large. The ice grounded below sea level, and therefore potentially unstable to ocean warming, in East Antarctica is equivalent to about 3.3 m of sea-level rise (Bamber et al., 2009). Paleo observations (Naish et al., 2009; Levermann et al., 2013) and modeling (Pollard and DeConto, 2009) indicate significant sea-level contributions from past changes in the Antarctic Ice Sheet.

6.3. The Regional Distribution of Sea-Level Change

Although progress has been made on understanding GMSL rise, the regional distribution of sea-level rise remains a challenge, with many poorly explored factors affecting both the past and future regional distribution of sea level. It is clear that natural variability is the dominant contribution to the regional pattern of sea-level change (at least in the low latitude Pacific Ocean) during the satellite era (Meyssignac et al., 2012; Zhang and Church, 2012). Feng et al. (2010, 2011), Timmermann et al. (2010), Merrifield (2011), Merrifield and Maltrud (2011) and Merrifield et al. (2012) have shown that the larger rate of rise in the western equatorial Pacific is related to a strengthening of the south-east trades (both observed over recent decades and in the CMIP3 simulations for the twenty-first century). In the Indian Ocean, Han et al. (2010) find that winds over the Indian Ocean are responsible for lower sea levels in the southern tropics. In contrast, Schwarzkopf and Böning (2011) argue that this feature is the result of sea-level anomalies generated by changes of the winds in the western equatorial Pacific Ocean as part of the Pacific Decadal Oscillation. The sea-level anomalies then propagate through the Indonesian Archipelago, southward along the eastern boundary and then westward as baroclininc Rossby waves. Couttes et al. (2012) has shown that for the Southern Ocean changes in the strength and latitude of the westerly winds are an important driver of projected regional sea-level change. In the Atlantic, there are observations of an acceleration on the north American east coast (Sallenger et al., 2012; Ezer et al., 2013), and projections (Yin et al., 2009) of a greater than global averaged rise in this region. The projected regional rise is associated with a slowing of the Atlantic Meridional Overturning Circulation. Although there are some common features in projections of future sea-level rise (e.g., Pardaens et al., 2011), there are also significant unexplained regional differences between the projected regional patterns (e.g., Yin et al., 2010; Yin, 2012). Better historical reconstructions of the regional distribution of sea-level change, detection and attribution studies and further analysis of model simulations are all required to understand the reasons for the remaining differences and whether they can be attributed to a particular cause.

In addition to these coupled climate issues, the mass loss from the glaciers or ice sheets results in a regional fingerprint of sea-level change (Mitrovica et al., 2001, 2009). These fingerprints result in a fall in sea level in the region near the ice sheet and a greater than average rise in the far field. Kopp et al. (2010) estimated that during the twenty-first century these fingerprints are likely to become detectable despite the presence of natural climate variability. Hay et al. (2013) outline an approach for detecting theses signals in the available sea-level observations.

The total projected sea-level rise pattern can be estimated by adding the dynamical ocean contribution to the fingerprints of the mass contributions and ongoing GIA (Church et al., 2011b; Slangen et al., 2012; Figure 27.18), on the assumption that interactions between these terms are small.

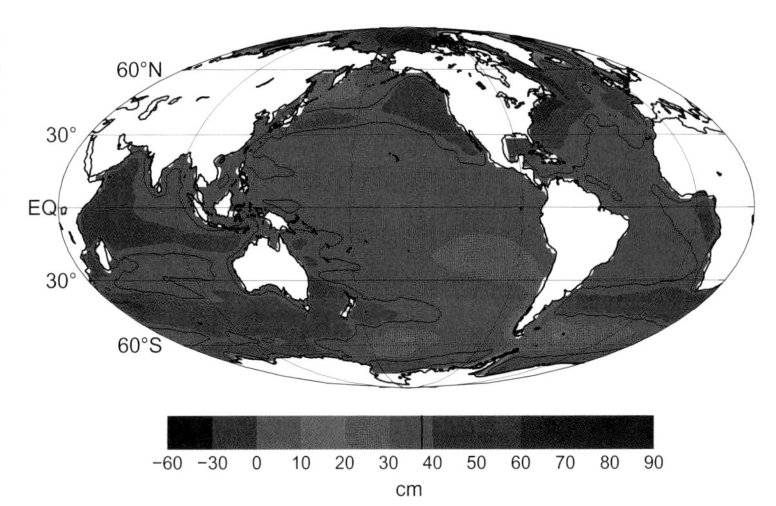

FIGURE 27.18 The regional distribution of the projections of sea-level change for 2090 compared to 1990, combining global average sea-level projections, dynamic ocean departure from the global average, and regional changes associated with the changing mass distribution in the cryosphere, based on the results in Meehl et al. (2007). The black contour is the "average" value at 2090 of 38 cm, dividing those regions with above and below-average sea-level rise. *From Church et al. (2011a).*

Any potential feedback of changes in the gravity field and the geoid (including from changes in ocean-mass distribution) on sea level and ocean circulation have not been explored.

The addition of freshwater to ocean-only models, from ice sheets or glaciers, results in a rapid rise in GMSL associated with barotropic waves propagating around the globe (Lorbacher et al., 2012) and regional changes that propagate slowly through the ocean as baroclinic coastal and equatorial Kelvin waves and westward propagating Rossby waves (Stammer, 2008). It takes decades for these baroclinic signals to propagate throughout the oceans. In addition, in coupled models the changes in surface conditions lead to air-sea interactions and teleconnections through the atmosphere forcing sea-level changes in the global ocean only a year after commencement of adding freshwater to the ocean (Stammer et al., 2011). Exploration of these issues is in its early stages.

The interested reader is referred to Church et al. (2013a) for a more detailed discussion of the GMSL and regional projections and their importance for the changes in extreme events.

7. FUTURE OUTLOOK

The significant progress over the last decade in our understanding and projections of future sea-level rise has largely been the result of improvements in global *in situ* and satellite observing systems, coupled AOGCMs, and the design of model intercomparison projects. However, the current uncertainties are still large and significant challenges remain. These will only be satisfactorily addressed with further improvements in our fundamental understanding. Priority action items include:

- Maintaining the current observing system, particularly the GLOSS sea-level observations, high-quality satellite altimetry of the oceans, glaciers, and the ice sheets, improved higher-resolution, time-variable satellite gravity measurements, Argo for measuring steric sea-level change, and inSAR for measuring ice-sheet motions. Additions required are the completion and improvement of the GLOSS program (particularly at island sites and in regions with few observations, and more sites with measurements of vertical land motion), extending Argo-like observations to the full-depth ocean, under ice and into marginal seas, and more complete glacier and ice-sheet observations. Reconstructing past sea levels will be of growing importance.
- A focus on observing and understanding ocean-ice shelf/ice-sheet interactions and the implications for a dynamic ice-sheet and glacier response and determining the magnitude of ice-sheet contributions, requiring improved GIA models. Detection of the spatial sea-level fingerprint of ice-sheet changes potentially offers an additional perspective.
- Understanding the factors that control the regional distribution of sea-level change. Issues to address include better characterization and understanding of natural variability, the impact of different climatic forcings, detection and attribution studies, and the impact of freshwater additions for the regional changes in ocean circulation.
- Improvements in our understanding of basic climatic parameters including the uptake of heat by the ocean, climate sensitivity, and better representation of climatic processes in models. Detection and attribution studies would likely lead to sea-level projections constrained by observations.
- Bringing together all elements of sea-level change thus allowing critical evaluation of model simulations of global and regional sea levels, including an improved distribution and density of paleo observations for longer periods and different climate states. The opportunity of seasonal to interannual and decadal prediction of sea levels should be pursued.

The comprehensive 2006 WRCP workshop on sea-level variability and change brought together what was then a disparate community to review progress, agree on future priorities, and produce a comprehensive and authoritative requirements document. A second such multidisciplinary meeting 10 years after the initial meeting would be appropriate.

ACKNOWLEDGMENTS

We thank several referees (W. John Gould, Stephen M. Griffies, Jianjun Yin, John Hunter, and Peter U. Clark) for excellent suggestions leading to improvements from an earlier version of this chapter. We also thank colleagues who generously gave permission for us to reproduce figures. This publication is supported by the Australian Government Department of Industry, Innovation, Climate Change, Science, Research and Tertiary Education, the Bureau of Meteorology and CSIRO through the Australian Climate Change Science Program.

REFERENCES

Abraham, J.P., Baringer, M., Bindoff, N.L., Boyer, T., Cheng, L.J., Church, J.A., Conroy, J.L., Domingues, C.M., Fasullo, J.T., Gilson, J., Goni, G., Good, S.A., Gorman, J.M., Gouretski, V., Ishii, M., Johnson, G.C., Kizu, S., Lyman, J.M., Macdonald, A.M., Minkowycz, W.J., Moffitt, S.E., Palmer, M.D., Piola, A., Reseghetti, F., von Schuckmann, K., Trenberth, K.E., Velicogna, I., Willis, J.K., 2013. A review of global ocean temperature observations: Implications for ocean heat content estimates and climate change. Rev. Geophys., 51, http://dx.doi.org/10.1002/rog.20022.

AchutaRao, K.M., Santer, B.D., Gleckler, P.J., Taylor, K.E., Pierce, D.W., Barnett, T.P., Wigley, T.M.L., 2006. Variability of ocean heat uptake: reconciling observations and models. J. Geophys. Res. 111, C05019. http://dx.doi.org/10.1029/2005JC003136.

AchutaRao, K.M., Ishii, M., Santer, B.D., Gleckler, P.J., Taylor, K.E., Barnett, T.P., Pierce, D.W., Stouffer, R.J., Wigley, T.M.L., 2007.

Simulated and observed variability in ocean temperature and heat content. Proc. Natl. Acad. Sci. U.S.A. 204 (26), 10768–10773.

Ammann, C.M., Meehl, G.A., Washington, W.M., 2003. A monthly and lattitudinally varying volcanic forcing dataset in simulations of the 20th century climate. Geophys. Res. Lett. 30 (12), 16257. http://dx.doi.org/10.1029/12003GLO16875.

Anthoff, D., Nicholls, R.J., Tol, R.S.J., Vafeidis, A.T., 2006. Global and regional exposure to large rises in sea-level: a sensitivity analysis. Tyndall Centre for Climate Change Research. Working paper 96.

Arbic, B.K., Owens, W.B., 2001. Climatic warming of Atlantic Intermediate Waters. J. Clim. 14 (20), 4091–4108.

Arendt, A., et al., 2012. Randolph Glacier Inventory [v2.0]: A Dataset of Global Glacier Outlines. Global Land Ice Measurements from Space, Boulder Colorado, USA. Digital Media.

Austermann, J., Mitrovica, J.X., Latychev, K., Milne, G.A., 2013. Barbados-based estimate of ice volume at Last Glacial Maximum affected by subducted plate. Nat. Geosci. 6 (7), 553–557.

AVISO, 2003. AVISO and PODAAC user handbook—IGDR and GDR Jason products. Edition 2.0. SMMMU-M5-OP-13184-CN.

Balmaseda, M.A., Mogensen, K., Weaver, A.T., 2013a. Evaluation of the ECMWF ocean reanalysis system ORAS4. Q. J. R. Meteorol. Soc. 139 (674), 1132–1161.

Balmaseda, M.A., Trenberth, K.E., Källén, E., 2013b. Distinctive climate signals in reanalysis of global ocean heat content. Geophys. Res. Lett. 40 (9), 1754–1759.

Bamber, J.L., Riva, R.E.M., Vermeersen, B.L.A., LeBrocq, A.M., 2009. Reassessment of the potential sea-level rise from a collapse of the West Antarctic Ice Sheet. Science 324 (5929), 901–903.

Barker, P.M., Dunn, J.R., Domingues, C.M., Wijffels, S.E., 2011. Pressure sensor drifts in Argo and their impacts. J. Atmos. Oceanic Technol. 28 (8), 1036–1049.

Benada, J.R., 1997. PO.DAAC Merged GDR (TOPEX/POSEIDON) Generation B user's Handbook, version 2.0, JPL D-11007.

Berge-Nguyen, M., Cazenave, A., Lombard, A., Llovel, W., Viarre, J., Cretaux, J.F., 2008. Reconstruction of past decades sea level using thermosteric sea level, tide gauge, satellite altimetry and ocean reanalysis data. Glob. Planet. Change 62 (1–2), 1–13.

Blewitt, G., et al., 2010. Geodetic observations and global reference frame contributions to understanding sea-level rise and variability. In: Church, J.A., Woodworth, P.L., Aarup, T., Wilson, W.S. (Eds.), Understanding Sea-level Rise and Variability. Wiley-Blackwell Publishing, Chichester, UK, pp. 256–284.

Boening, C., Willis, J.K., Landerer, F.W., Nerem, R.S., Fasullo, J., 2012. The 2011 La Niña: so strong, the oceans fell. Geophys. Res. Lett. 39 (19), L19602.

Böning, C.W., Dispert, A., Visbeck, M., Rintoul, S.R., Schwarzkopf, F.U., 2008. The response of the Antarctic Circumpolar Current to recent climate change. Nat. Geosci. 1 (12), 864–869.

Bonnefond, P., Exertier, P., Laurain, O., Ménard, Y., Orsoni, A., Jan, G., Jeansou, E., 2003. Absolute calibration of Jason-1 and TOPEX/Poseidon altimeters in Corsica special issue: Jason-1 calibration/validation. Mar. Geod. 26 (3–4), 261–284.

Bonnefond, P., Exertier, P., Laurain, O., Jan, G., 2010. Absolute Calibration of Jason-1 and Jason-2 Altimeters in Corsica during the Formation Flight Phase. Mar. Geod. 33 (Suppl. 1), 80–90.

Bouttes, N., Gregory, J.M., Lowe, J.A., 2012. The reversibility of sea level rise. J. Clim. 26 (8), 2502–2513.

Box, J.E., Colgan, W., 2013. Greenland ice sheet mass balance reconstruction. Part III: marine ice loss and total mass balance (1840–2010). J. Clim. 26, 6990–7002. http://dx.doi.org/10.1175/jcli-d-12-00546.1.

Burgette, R.J., Watson, C.S., Church, J.A., White, N.J., Tregoning, P., Coleman, R., 2013. Characterizing and minimizing the effects of noise in tide gauge time series: relative and geocentric sea level rise around Australia. Geophys. J. Int. 194, 719–736. http://dx.doi.org/10.1093/gji/ggt131.

Calafat, F.M., Jordà, G., 2011. A Mediterranean sea level reconstruction (1950–2008) with error budget estimates. Glob. Planet. Change 79 (1–2), 118–133.

Carton, J.A., Giese, B.S., 2008. A reanalysis of ocean climate using Simple Ocean Data Assimilation (SODA). Mon. Weather Rev. 136 (August), 2999–3017.

Carton, J.A., Giese, B.S., Grodsky, S.A., 2005. Sea level rise in the SODA ocean reanlaysis and the warming of the oceans. J. Geophys. Res. 110, C09006. http://dx.doi.org/10.01029/02004JC002817.

Cazenave, A., Llovel, W., 2010. Contemporary sea level rise. Ann. Rev. Mar. Sci. 2, 145–173.

Chambers, D.P., Melhaff, C.A., Urban, T.J., Fuji, D., Nerem, R.S., 2002. Low-frequency variations in global mean sea level: 1950-2000. J. Geophys. Res. 107 (C4), 3026. http://dx.doi.org/10.129/2001JC001089.

Chambers, D.P., Wahr, J., Tamisiea, M.E., Nerem, R.S., 2010. Ocean mass from GRACE and glacial isostatic adjustment. J. Geophys. Res. Solid Earth 115 (B11), B11415.

Chao, B.F., Wu, Y.H., Li, Y.S., 2008. Impact of artificial reservoir water impoundment on global sea level. Science 320 (5873), 212–214.

Chen, J.L., Wilson, C.R., Blankenship, D., Tapley, B.D., 2009. Accelerated Antarctic ice loss from satellite gravity measurements. Nat. Geosci. 2 (12), 859–862.

Chen, J.L., Wilson, C.R., Tapley, B.D., 2011. Interannual variability of Greenland ice losses from satellite gravimetry. J. Geophys. Res. 116 (B7), B07406.

Cheng, L., Zhu, J., Reseghetti, F., Liu, Q., 2010. A new method to estimate the systematical biases of expendable bathythermograph. J. Atmos. Oceanic Technol. 28 (2), 244–265.

Church, J.A., White, N.J., 2006. A 20th century acceleration in global sea-level rise. Geophys. Res. Lett. 33, L10602. http://dx.doi.org/10.11029/12005GL024826.

Church, J., White, N., 2011. Sea-level rise from the late 19th to the early 21st century. Surv. Geophys. 32 (4), 585–602.

Church, J.A., White, N.J., Coleman, R., Lambeck, K., Mitrovica, J.X., 2004. Estimates of the regional distribution of sea level rise over the 1950-2000 period. J. Clim. 17 (13), 2609–2625.

Church, J.A., Woodworth, P.L., Aarup, T., Stanley Wilson, W. (Eds.), 2010. Understanding Sea-level Rise and Variability. Wiley-Blackwell Publishing, Chichester, UK, 427 pp.

Church, J.A., Gregory, J.M., White, N.J., Platten, S.M., Mitrovica, J.X., 2011a. Understanding and projecting sea level change. Oceanography 24 (2), 130–143. http://dx.doi.org/10.5670/oceanog.2011.33.

Church, J.A., White, N.J., Konikow, L.F., Domingues, C.M., Cogley, J. G., Rignot, E., Gregory, J.M., van den Broeke, M.R., Monaghan, A. J., Velicogna, I., 2011b. Revisiting the Earth's sea-level and energy budgets from 1961 to 2008. Geophys. Res. Lett. 38(18), L18601. And Correction Church, J.A. White, N.J., Konikow, L.F., Domingues, C.M., Cogley, J.G., Rignot E., Gregory, J.M., 2013. Correction to Revisiting the Earth's sea-level and energy budgets for 1961 to 2008. Geophys. Res. Lett. 40, 4066.

Church, J.A., Clark, P.U., Cazenave, A., Gregory, J.M., Jevrejeva, S., Levermann, A., Merrifield, M.A., Milne, G.A., Nerem, R.S.,

Nunn, P.D., Payne, A.J., Pfeffer, W.T., Stammer, D., Unnikrishnan, A.S., 2013a. Sea Level Change. In: Stocker, T.F. et al., (Ed.), Climate Change 2013: The Physical Science Basis. Contribution of Working Group I to the Fifth 1376 Assessment Report of the Intergovernmental Panel on Climate Change. Cambridge University Press, Cambridge, published online September 30, 2013.

Church, J.A., Monselesan, D., Gregory, J.M., Marzeion, B., 2013b. Evaluating the ability of process based models to project sea-level change. Environ. Res. Lett. 8 (1), 014051.

Chylek, P., Box, J., Lesins, G., 2004. Global warming and the greenland ice sheet. Clim. Change 63 (1–2), 201–221.

Clark, P.U., Mitrovica, J.X., Milne, G.A., Tamisiea, M.E., 2002. Sea-level fingerprinting as a direct test for the source of global meltwater pulse IA. Science 295 (5564), 2438–2441.

Clark, P., Dyke, A., Shakun, J., Carlson, A., Clark, J., Wohlfarth, B., Mitrovica, J., Hostetler, S., McCabe, A., 2009. The last glacial maximum. Science 325, 710–714.

CNES, 2009. OSTM/Jason-2 Products Handbook. SALP-MU-M-OP-15815-CN.

Cogley, J.G., 2009. Geodetic and direct mass-balance measurements: comparison and joint analysis. Ann. Glaciol. 50, 96–100.

Cowley, R., Wijffels, S., Cheng, L., Boyer, T., Kizu, S., 2013. Biases in expendable bathythermograph data: a new view based on historical side-by-side comparisons. J. Atmos. Oceanic Technol. 30 (6), 1195–1225.

Delworth, T.L., Knutson, T.R., 2000. Simulation of early 20th century global warming. Science 287 (5461), 2246–2250.

Deschamps, P., Durand, N., Bard, E., Hamelin, B., Camoin, G., Thomas, A.L., Henderson, G.M., Okuno, J.I., Yokoyama, Y., 2012. Ice-sheet collapse and sea-level rise at the Bolling warming 14,600 years ago. Nature 483 (7391), 559–564.

Domingues, C.M., Church, J.A., White, N.J., Gleckler, P.J., Wijffels, S.E., Barker, P.M., Dunn, J.R., 2008. Improved estimates of upper-ocean warming and multi-decadal sea-level rise. Nature 453 (7198), 1090–1093.

Donnelly, J.P., Cleary, P., Newby, P., Ettinger, R., 2004. Coupling instrumental and geological records of sea-level change: evidence from southern New England of an increase in the rate of sea-level rise in the late 19th century. Geophys. Res. Lett. 31, L05203, http://dx.doi.org/10.01029/02003GL018933.

Douglas, B.C., 1991. Global sea-level rise. J. Geophys. Res. Oceans 96 (C4), 6981–6992.

Douglas, B.C., 1992. Global sea level acceleration. J. Geophys. Res. 97, 12,699–612,706.

Dutton, A., Lambeck, K., 2012. Ice volume and sea level during the last interglacial. Science 337 (6091), 216–219.

Ezer, T., Atkinson, L.P., Corlett, W.B., Blanco, J.L., 2013. Gulf Stream's induced sea level rise and variability along the U.S. mid-Atlantic coast. J. Geophys. Res. Oceans 118 (2), 685–697.

Feng, M., McPhaden, M.J., Lee, T., 2010. Decadal variability of the Pacific subtropical cells and their influence on the southeast Indian Ocean. Geophys. Res. Lett. 37 (9), L09606.

Feng, M., Böning, C., Biastoch, A., Behrens, E., Weller, E., Masumoto, Y., 2011. The reversal of the multi-decadal trends of the equatorial Pacific easterly winds, and the Indonesian Throughflow and Leeuwin Current transports. Geophys. Res. Lett. 38 (11), L11604.

Fettweis, X., Hanna, E., Gallée, H., Huybrechts, P., Erpicum, M., 2008. Estimation of the Greenland ice sheet surface mass balance for the 20th and 21st centuries. The Cryosphere 2 (2), 117–129.

Fettweis, X., Mabille, G., Erpicum, M., Nicolay, S., den Broeke, M., 2011. The 1958–2009 Greenland ice sheet surface melt and the mid-tropospheric atmospheric circulation. Clim. Dyn. 36 (1), 139–159.

Fettweis, X., Franco, B., Tedesco, M., van Angelen, J.H., Lenaerts, J.T.M., van den Broeke, M.R., Gallée, H., 2013. Estimating the Greenland ice sheet surface mass balance contribution to future sea level rise using the regional atmospheric climate model MAR. The Cryosphere 7 (2), 469–489.

Fu, L.-L., Cazeenave, A., 2001. Satellite altimetry and earth sciences. In: International Geophysics Series, vol. 69. Academic Press, San Diego.

Fukasawa, M., Freeland, H., Perkin, R., Watanabe, T., Uchida, H., Nishina, A., 2004. Bottom water warming in the North Pacific Ocean. Nature 427 (26 (February)), 825–827.

Galton-Fenzi, B.K., Hunter, J.R., Coleman, R., Marsland, S.J., Warner, R.C., 2012. Modeling the basal melting and marine ice accretion of the Amery Ice Shelf. J. Geophys. Res. Oceans 117 (C9), C09031.

Gardner, A.S., et al., 2013. A reconciled estimate of glacier contributions to sea level rise: 2003 to 2009. Science 340 (6134), 852–857.

Gehrels, W.R., Woodworth, P.L., 2013. When did modern rates of sea-level rise start? Glob. Planet. Change 100, 263–277.

Gehrels, W.R., Kirby, J.R., Prokoph, A., Newnham, R.M., Achterberg, E.P., Evans, H., Black, S., Scott, D., 2005. Onset of recent rapid sea-level rise in the western Atlantic Ocean. Quat. Sci. Rev. 24, 2083–2100.

Gehrels, W.R., Marshall, W.A., Gehrels, M.J., Larsen, G., Kirby, J.R., Eiriksson, J., Heinemeier, J., Shimmield, T., 2006. Rapid sea-level rise in the North Atlantic Ocean since the first half of the nineteenth century. Holocene 16 (7), 949–965.

Gehrels, W.R., Hayward, B., Newnham, R.M., Southall, K.E., 2008. A 20th century acceleration of sea-level rise in New Zealand. Geophys. Res. Lett. 35 (2), L02717.

Gehrels, W.R., Dawson, D.A., Shaw, J., Marshall, W.A., 2011. Using Holocene relative sea-level data to inform future sea-level predictions: an example from southwest England. Glob. Planet. Change 78 (3–4), 116–126.

Gehrels, W.R., Callard, S.L., Moss, P.T., Marshall, W.A., Blaauw, M., Hunter, J., Milton, J.A., Garnett, M.H., 2012. Nineteenth and twentieth century sea-level changes in Tasmania and New Zealand. Earth Planet. Sci. Lett. 315–316, 94–102.

Gille, S.T., 2008. Decadal-scale temperature trends in the southern hemisphere ocean. J. Clim. 21, 4749–4765.

Gleckler, P.J., et al., 2012. Human-induced global ocean warming on multidecadal timescales. Nat. Clim. Chang. 2 (7), 524–529.

Gomez, N., Mitrovica, J.X., Huybers, P., Clark, P.U., 2010. Sea level as a stabilizing factor for marine-ice-sheet grounding lines. Nat. Geosci. 3, 850–853.

Good, S.A., 2010. Depth biases in XBT data diagnosed using bathymetry data. J. Atmos. Oceanic Technol. 28 (2), 287–300.

Gould, J., The Argo Science Team, 2004. Argo Profiling Floats Bring New Era of In Situ Ocean Observations, Eos 85(19, 11 May).

Gouretski, V., 2012. Using GEBCO digital bathymetry to infer depth biases in the XBT data. Deep Sea Res. Part I 62, 40–52.

Gouretski, V., Koltermann, K., 2007. How much is the ocean really warming? Geophys. Res. Lett. 34 (1), L01610.

Gouretski, V., Reseghetti, F., 2010. On depth and temperature biases in bathythermograph data: development of a new correction scheme based on analysis of a global ocean database. Deep Sea Res. Part I 57 (6), 812–833.

Gouretski, V., Kennedy, J., Boyer, T., Kohl, A., 2012. Consistent near-surface ocean warming since 1900 in two largely independent observing networks. Geophys. Res. Lett. 39 (19), L19606.

Gregory, J.M., 2010. Long-term effect of volcanic forcing on ocean heat content. Geophys. Res. Lett. 37 (22), L22701.

Gregory, J.M., Forster, P.M., 2008. Transient climate response estimated from radiative forcing and observed temperature change. J. Geophys. Res. 113 (D23), D23105.

Gregory, J.M., Lowe, J.A., 2000. Predictions of global and regional sea-level rise using AOGCMs with and without flux adjustment. Geophys. Res. Lett. 27, 3060–3072.

Gregory, J.M., Banks, H.T., Stott, P.A., Lowe, J.A., Palmer, M.D., 2004. Simulated and observed decadal variability in ocean heat content. Geophys. Res. Lett. 31, L15312. http://dx.doi.org/10.11029/12004GL02058.

Gregory, J.M., Huybrechts, P., 2006. Ice-sheet contributions to future sea-level change. Philos. Trans. R. Soc. A 1709–1731.

Gregory, J.M., et al., 2013a. Twentieth-century global-mean sea level rise: is the whole greater than the sum of the parts? J. Clim. 26 (13), 4476–4499.

Gregory, J.M., Bi, D., Collier, M.A., Dix, M.R., Hirst, A.C., Hu, A., Huber, M., Knutti, R., Marsland, S.J., Meinshausen, M., Rashid, H.A., Rotstayn, L.D., Schurer, A., Church, J.A., 2013b. Climate models without preindustrial volcanic forcing underestimate historical ocean thermal expansion. Geophys. Res. Lett. 40 (8), 1600–1604.

Griffies, S.M., Greatbatch, R.J., 2012. Physical processes that impact the evolution of global mean sea level in ocean climate models. Ocean Model. 51, 37–72.

Gudmundsson, G.H., 2013. Ice-shelf buttressing and the stability of marine ice sheets. The Cryosphere 7 (2), 647–655.

Guinehut, S., Traon, P.Y., Larnicol, G., Philipps, S., 2004. Combining Argo and remote-sensing data to estimate the ocean three-dimensional temperature fields—a first approach based on simulated observations. J. Mar. Syst. 46, 85–98.

Guinehut, S., Traon, P.-Y.l., Larnicol, G., 2006. What can we learn from Global Altimetry/Hydrography comparisons? Geophys. Res. Lett. 33, L10604, 10.11029/GL025551.

Haigh, I., Nicholls, R., Wells, N., 2009. Mean sea level trends around the English Channel over the 20th century and their wider context. Con. Shelf Res. 29 (17), 2083–2098.

Haines, B.J., Dong, D., Born, G.H., Gill, S.K., 2003. The harvest experiment: monitoring Jason-1 and TOPEX/POSEIDON from a California Offshore Platform Special Issue: Jason-1 calibration/validation. Mar. Geod. 26 (3–4), 239–259.

Haines, B.J., Desai, S.D., Born, G.H., 2010. The harvest experiment: calibration of the climate data record from TOPEX/Poseidon, Jason-1 and the ocean surface topography mission. Mar. Geod. 33 (Suppl. 1), 91–113.

Han, W.Q., et al., 2010. Patterns of Indian Ocean sea-level change in a warming climate. Nat. Geosci. 3 (8), 546–550.

Hanna, E., et al., 2011. Greenland Ice Sheet surface mass balance 1870 to 2010 based on Twentieth Century Reanalysis, and links with global climate forcing. J. Geophys. Res. 116 (D24), D24121.

Hanna, E., et al., 2013. Ice-sheet mass balance and climate change. Nature 498 (7452), 51–59.

Hannah, J., Bell, R.G., 2012. Regional sea level trends in New Zealand. J. Geophys. Res. 117 (C1), C01004.

Hay, C.C., Morrow, E., Kopp, R.E., Mitrovica, J.X., 2013. Estimating the sources of global sea level rise with data assimilation techniques. Proc. Natl. Acad. Sci. U.S.A. 110 (Suppl. 1), 3692–3699.

Hellmer, H.H., Kauker, F., Timmermann, R., Determann, J., Rae, J., 2012. Twenty-first-century warming of a large Antarctic ice-shelf cavity by a redirected coastal current. Nature 485 (7397), 225–228.

Henry, O., Prandi, P., Llovel, W., Cazenave, A., Jevrejeva, S., Stammer, D., Meyssignac, B., Koldunov, N., 2012. Tide gauge-based sea level variations since 1950 along the Norwegian and Russian coasts of the Arctic Ocean: contribution of the steric and mass components. J. Geophys. Res. 117 (C6), C06023.

Hindmarsh, R.C.A., 1993. Qualitative dynamics of marine ice sheet. In: Peltier, W.R. (Ed.), Ice in the Climate System. Springer-Verlag, Berlin, pp. 67–99.

Hobbs, W.R., Willis, J.K., 2013. Detection of an observed 135-year ocean temperature change from limited data. Geophys. Res. Lett. 40 (10), 2252–2258.

Holland, D.M., Thomas, R.H., De Young, B., Ribergaard, M.H., Lyberth, B., 2008. Acceleration of Jakobshavn Isbrae triggered by warm subsurface ocean waters. Nat. Geosci. 1 (10), 659–664.

Huybrechts, P., Goelzer, H., Janssens, I., Driesschaert, E., Fichefet, T., Goosse, H., Loutre, M.F., 2011. Response of the Greenland and Antarctic ice sheets to multi-millennial greenhouse warming in the Earth system model of intermediate complexity LOVECLIM. Surv. Geophys. 32 (4–5), 397–416.

Ishii, M., Kimoto, M., 2009. Reevaluation of historical ocean heat content variations with time-varying XBT and MBT depth bias corrections. J. Oceanogr. 65 (3), 287–299.

Ishii, M., Kimoto, M., Kachi, M., 2003. Historical ocean subsurface temperature analysis with error estimates. Mon. Weather Rev. 131 (January), 51–73.

Ishii, M., Kimoto, M., Sakamoto, K., Iwasaki, S.-I., 2006. Steric sea level changes estimated from historical ocean subsurface temperature and salinity analysis. J. Oceanogr. 62, 155–170.

Jacobs, S.S., Jenkins, A., Giulivi, C.F., Dutrieux, P., 2011. Stronger ocean circulation and increased melting under Pine Island Glacier ice shelf. Nat. Geosci. 4 (8), 519–523.

Jevrejeva, S., Grinsted, A., Moore, J.C., Holgate, S., 2006. Nonlinear trends and multiyear cycles in sea level records. J. Geophys. Res. 111 (C9), C09012.

Jevrejeva, S., Moore, J.C., Grinsted, A., Woodworth, P.L., 2008. Recent global sea level acceleration started over 200 years ago. Geophys. Res. Lett. 35, L08715. http://dx.doi.org/10.01029/02008GL033611.

Johnson, G.C., Doney, S.C., 2006. Recent western South Atlantic bottom water warming. Geophys. Res. Lett. 33, L14614. http://dx.doi.org/10.11029/12006GL026769.

Johnson, G., Mecking, S., Sloyan, B., Wijffels, S., 2007. Recent bottom water warming in the Pacific Ocean. J. Clim. 20 (21), 5365–5375.

Johnson, G.C., Purkey, S.G., Bullister, J.L., 2008. Warming and freshening in the abyssal southeastern Indian Ocean. J. Clim. 21 (15 October), 5351–5363.

Johnson, G.C., Lyman, J.M., Willis, J.K., Levitus, S., Boyer, T., Antonov, J., Good, S.A., 2012. Global oceans: ocean heat content. In: Blunden, J., Arndt, D.S. (Eds.), State of the Climate in 2011, Bulletin of the American Meteorological Society, vol. 93. pp. S62–S65. http://dx.doi.org/10.1175/2012BAMSStateoftheClimate.

Johnson, G.C., Lyman, J.M., Willis, J.K., Levitus, S., Boyer, T., Antonov, J., Good, S.A., Domingues, C.M., Wijffels, S., Bindoff, N., 2013. Ocean heat content. In: Blunden, J., Arndt, D.S.

(Eds.), State of the Climate in 2012, Bulletin of the American Meteorological Society, 94, pp. S50–S53.

Joughin, I., Alley, R.B., Holland, D.M., 2012. Ice-sheet response to oceanic forcing. Science 338 (6111), 1172–1176.

Kaplan, A., Cane, M.A., Kushnir, Y., Clement, A.C., 1998. Analyses of global sea surface temperatures 1856-1991. J. Geophys. Res. 103 (C9), 18567–18589.

Kaplan, A., Kushnir, Y., Cane, M.A., 2000. Reduced Space optimal interpolation of historical marine sea level pressure. J. Clim. 13 (16), 2987–3002.

Kawano, T., Doi, T., Uchida, H., Kouketsu, S., Fukasawa, M., Kawai, Y., Katsumata, K., 2010. Heat Content Change in the Pacific Ocean between the 1990s and 2000s. Deep Sea Res. Part II 57 (13–14), 1141–1151.

Kemp, A., Horton, B., Culver, S., Corbett, D., van de Plassche, O., Gehrels, W., Douglas, B., Parnell, A., 2009. Timing and magnitude of recent accelerated sea-level rise (North Carolina, United States). Geology 37, 1035–1038.

Kemp, A.C., Horton, B.P., Donnelly, J.P., Mann, M.E., Vermeer, M., Rahmstorf, S., 2011. Climate related sea-level variations over the past two millennia. Proc. Natl. Acad. Sci. U.S.A. 108 (27), 11017–11022.

Kohl, A., Stammer, D., 2008. Decadal sea level changes in the 50-year GECCO ocean synthesis. J. Clim. 21 (9), 1876–1890.

Konikow, L.F., 2011. Contribution of global groundwater depletion since 1900 to sea-level rise. Geophys. Res. Lett. 38 (17), L17401.

Kopp, R.E., Simons, F.J., Mitrovica, J.X., Maloof, A.C., Oppenheimer, M., 2009. Probabilistic assessment of sea level during the last interglacial stage. Nature 462 (7275), 863–867.

Kopp, R., Mitrovica, J., Griffies, S., Yin, J., Hay, C., Stouffer, R., 2010. The impact of Greenland melt on local sea levels: a partially coupled analysis of dynamic and static equilibrium effects in idealized water-hosing experiments. Clim. Change 103 (3), 619–625.

Kopp, R.E., Simons, F.J., Mitrovica, J.X., Maloof, A.C., Oppenheimer, M., 2013. A probabilistic assessment of sea level variations within the last interglacial stage. Geophys. J. Int. 193, 711–716.

Kouketsu, S., et al., 2011. Deep ocean heat content changes estimated from observation and reanalysis product and their influence on sea level change. J. Geophys. Res. Oceans 116, C03012.

Kuhlbrodt, T., Gregory, J.M., 2012. Ocean heat uptake and its consequences for the magnitude of sea level rise and climate change. Geophys. Res. Lett. 39 (18), L18608.

Kusahara, K., Hasumi, H., 2013. Modeling Antarctic ice shelf responses to future climate changes and impacts on the ocean. J. Geophys. Res. Oceans 118 (5), 2454–2475.

Lambeck, K., Esat, T., Potter, E., 2002. Links between climate and sea levels for the past three million years. Nature 419, 199–206.

Lambeck, K., Woodroffe, C.D., Antonioli, F., Anzidei, M., Gehrels, W.R., Laborel, J., Wright, A.J., 2010. Paleoenvironmental records, geophysical modeling, and reconstruction of sea-level trends and variability on centennial and longer timescales. In: Church, J.A., Woodworth, P.L., Aarup, T., Wilson, W.S. (Eds.), Understanding Sea-level Rise and Variability. Wiley-Blackwell Publishing, Chichester, UK, pp. 61–121.

Landerer, F.W., Jungclaus, J.H., Marotzke, J., 2007. Regional dynamic and steric sea level change in response to the IPCC-A1B scenario. J. Phys. Oceanogr. 37 (2), 296–312.

Leclercq, P.W., Oerlemans, J., Cogley, J.G., 2011. Estimating the glacier contribution to sea-level rise for the period 1800-2005. Surv. Geophys. 32 (4–5), 519–535.

Leuliette, E.W., Miller, L., 2009. Closing the sea level rise budget with altimetry, Argo, and GRACE. Geophys. Res. Lett. 36, L04608.

Levermann, A., Clark, P.U., Marzeion, B., Milne, G.A., Pollard, D., Radic, V., Robinson, A., 2013. The multimillennial sea-level commitment of global warming. Proc. Natl. Acad. Sci. U.S.A. 110, 13745–13750. http://dx.doi.org/10.1073/pnas.1219414110.

Levitus, S., 1982. Climatological atlas of the world. Rep. No. 13, 173pp.

Levitus, S., 1987. Rate of change of heat storage in the world ocean. J. Phys. Oceanogr. 17 (4), 518–528.

Levitus, A., Antonov, J.I., Boyer, T.P., Garcia, H.E., Locarnini, R.A., 2005. Linear trends of zonally averaged thermosteric, halosteric, and total steric sea level for individual ocean basins and the world ocean, (1955-1959)-(1994-1998). Geophys. Res. Lett. 32, L16601. http://dx.doi.org/10.11029/12005GL023761.

Levitus, S., Antonov, J.I., Boyer, T.P., Locarnini, R.A., Garcia, H.E., Mishonov, A.V., 2009. Global ocean heat content 1955-2008 in light of recently revealed instrumentation problems. Geophys. Res. Lett. 36, L07608. http://dx.doi.org/10.01029/02008GL037155.

Levitus, S., et al., 2012. World ocean heat content and thermosteric sea level change (0-2000 m), 1955-2010. Geophys. Res. Lett. 39 (10), L10603.

Llovel, W., Cazenave, A., Rogel, P., Lombard, A., Nguyen, M.B., 2009. Two-dimensional reconstruction of past sea level (1950-2003) from tide gauge data and an Ocean General Circulation Model. Clim. Past 5 (2), 217–227.

Llovel, W., Guinehut, S., Cazenave, A., 2010. Regional and interannual variability in sea level over 2002–2009 based on satellite altimetry, Argo float data and GRACE ocean mass. Ocean Dyn. 60 (5), 1193–1204.

Llovel, W., Becker, M., Cazenave, A., Jevrejeva, S., Alkama, R., Decharme, B., Douville, H., Ablain, M., Beckley, B., 2011. Terrestrial waters and sea level variations on interannual time scale. Glob. Planet. Change 75 (1–2), 76–82.

Loeb, N.G., Lyman, J.M., Johnson, G.C., Allan, R.P., Doelling, D.R., Wong, T., Soden, B.J., Stephens, G.L., 2012. Observed changes in top-of-the-atmosphere radiation and upper-ocean heating consistent within uncertainty. Nat. Geosci. 5 (2), 110–113.

Lombard, A., Cazenave, A., Traon, P., Guinehut, S., Cabanes, C., 2006. Perspectives on present-day sea level change: a tribute to Christian le Provost. Ocean Dyn. 56 (5–6), 445–451.

Lorbacher, K., Marsland, S.J., Church, J.A., Griffies, S.M., Stammer, D., 2012. Rapid barotropic sea level rise from ice sheet melting. J. Geophys. Res. 117 (C6), C06003.

Lyman, J.M., Good, S.A., Gouretski, V.V., Ishii, M., Johnson, G.C., Palmer, M.D., Smith, D.M., Willis, J.K., 2010. Robust warming of the global upper ocean. Nature 465 (7296), 334–337.

Marzeion, B., Jarosch, A.H., Hofer, M., 2012. Past and future sea-level change from the surface mass balance of glaciers. The Cryosphere 6 (6), 1295–1322.

Masters, D., Nerem, R.S., Choe, C., Leuliette, E., Beckley, B., White, N., Ablain, M., 2012. Comparison of global mean sea level time series from TOPEX/Poseidon, Jason-1, and Jason-2. Mar. Geod. 35 (Supp.1), 20–41.

McKay, N.P., Overpeck, J.T., Otto-Bliesner, B.L., 2011. The role of ocean thermal expansion in Last Interglacial sea level rise. Geophys. Res. Lett. 38 (14), L14605.

Meehl, G.A., Stocker, T.F., Collins, W.D., Friedlingstein, P., Gaye, A.T., Gregory, J.M., Kitoh, A., Knutti, R., Murphy, J.M., Noda, A., Raper, S.C.B., Watterson, I.G., Weaver, A.J., Zhao, Z.-C., 2007. Global climate projections. In: Solomon, D.Q.S., Manning, M.,

Marquis, M., Averyt, K., Tignor, M.M.B., Miller Jr., H.L., Chen, Z. (Eds.), Climate Change 2007: The Physical Science Basis. Contribution of Working Group 1 to the Fourth Assessment Report of the Intergovernmental Panel on Climate Change. Cambridge University Press, Cambridge, UK/ New York, USA, pp. 747–845.

Merrifield, M.A., 2011. A shift in Western Tropical Pacific sea level trends during the 1990s. J. Clim. 24 (15), 4126–4138.

Merrifield, M.A., Maltrud, M.E., 2011. Regional sea level trends due to a Pacific trade wind intensification. Geophys. Res. Lett. 38 (21), L21605.

Merrifield, M., Thompson, P.R., Lander, M., 2012. Multidecadal sea level anomalies and trends in the western tropical Pacific. Geophys. Res. Lett. 39, L13602.

Meyssignac, B., Salas y Melia, D., Becker, M., Llovel, W., Cazenave, A., 2012. Tropical Pacific spatial trend patterns in observed sea level: internal variability and/or anthropogenic signature? Clim. Past 8 (2), 787–802.

Miles, E.R., Spillman, C.M., Church, J.A., McIntosh, P.C. Seasonal prediction of global sea level anomalies using an ocean-atmosphere dynamical model. Clim. Dyn. submitted.

Milly, P.C.D., Cazenave, A., Gennero, M.C., 2003. Contribution of climate-driven change in continental water storage to recent sea-level rise. Proc. Natl. Acad. Sci. U.S.A. 100 (23), 13158–13161.

Milly, P.C.D., Cazenave, A., Famiglietti, J.S., Gornitz, V., Laval, K., Lettenmaier, D.P., Sahagian, D.L., Wahr, J.M., Wilson, C.R., 2010. Terrestrial water-storage contributions to sea-level rise and variability. In: Church, J.A., Woodworth, P.L., Aarup, T., Wilson, W.S. (Eds.), Understanding Sea-level Rise and Variability. Wiley-Blackwell Publishing, Chichester, UK, pp. 226–255.

Mitchum, G.T., 1998. Monitoring the stability of satellite altimeters with tide gauges. J. Atmos. Oceanic Technol. 15 (June), 721–730.

Mitchum, G.T., 2000. An improved calibration of satellite altimetric height using tide gauge sea levels with adjustment for land motion. Mar. Geod. 23, 145–166.

Mitrovica, J.X., Tamisiea, M.E., Davis, J.L., Milne, G.A., 2001. Recent mass balance of polar ice sheets inferred from patterns of global sea-level change. Nature 409 (6823), 1026–1029.

Mitrovica, J.X., Wahr, J., Matsuyama, I., Paulson, A., Tamisiea, M.E., 2006. Reanalysis of ancient eclipse, astronomic and geodetic data: a possible route to resolving the enigma of global sea-level rise. Earth Planet. Sci. Lett. 243 (3–4), 390–399.

Mitrovica, J.X., Gomez, N., Clark, P.U., 2009. The sea-level fingerprint of West Antarctic collapse. Science 323 (5915), 753.

Mitrovica, J.X., Tamisiea, M.E., Ivins, E.R., Vermeersen, L.L.A., Milne, G.A., Lambeck, K., 2010. Surface mass loading on a dynamic earth: complexity and contamination in the geodetic analysis of global sea-level trends. In: Church, J.A., Woodworth, P.L., Aarup, T., Stanley Wilson, W. (Eds.), Understanding Sea-level Rise and Variability. Wiley-Blackwell Publishing, Chichester, UK, pp. 285–325.

Moore, J.C., Jevrejeva, S., Grinsted, A., 2011. The historical global sea level budget. Ann. Glaciol. 52 (59), 8–14.

Moss, R.H., et al., 2010. The next generation of scenarios for climate change research and assessment. Nature 463 (7282), 747–756.

Munk, W., 2002. Twentieth century sea level: an enigma. Proc. Natl. Acad. Sci. U.S.A. 99 (10), 6550–6555.

Munk, W., 2003. Ocean freshening, sea level rising. Science 300 (27 June), 2041–2043.

Myhre, G., Shindell, D., Bréon, F.-M., Collins, W., Fuglestvedt, J., Huang, J., Koch, D., Lamarque, J.-F., Lee, D., Mendoza, B., Nakajima, T., Robock, A., Stephens, G., Takemura, T., Zhang, H., 2013. Anthropogenic and natural radiative forcing. In: Stocker, T.F. et al., (Ed.), Climate Change 2013: The Physical Science Basis. Contribution of Working Group I to the Fifth 1376 Assessment Report of the Intergovernmental Panel on Climate Change. Cambridge University Press, Cambridge, Published online 30 September 2013.

Naish, T., et al., 2009. Obliquity-paced Pliocene West Antarctic ice sheet oscillations. Nature 458 (7236), 322–328.

NEEM Community Members, 2013. Eemian interglacial reconstructed from a Greenland folded ice core. Nature 493 (7433), 489–494.

Ngo-Duc, T., Laval, K., Polcher, J., Lombard, A., Cazenave, A., 2005. Effects of land water storage on global mean sea level over the past half century. Geophys. Res. Lett. 32 (9), 9704–9707.

Nicholls, R.J., 2010. Impacts of and Responses to Sea-Level Rise. In: Church, John A., Woodworth, Philip L., Aarup, Thorkild, Stanley Wilson, W. (Eds.), Understanding Sea-level Rise and Variability. Wiley-Blackwell Publishing, Chichester, UK.

Nicholls, R.J., Cazenave, A., 2010. Sea-level rise and its impact on coastal zones. Science 328 (5985), 1517–1520.

Nick, F.M., Vieli, A., Howat, I.M., Joughin, I., 2009. Large-scale changes in Greenland outlet glacier dynamics triggered at the terminus. Nat. Geosci. 2 (2 February), 110–114.

Nick, F.M., Vieli, A., Andersen, M.L., Joughin, I., Payne, A., Edwards, T.L., Pattyn, F., van de Wal, R.S.W., 2013. Future sea-level rise from Greenland's main outlet glaciers in a warming climate. Nature 497 (7448), 235–238.

Otto, A., et al., 2013. Energy budget constraints on climate response. Nat. Geosci. 6 (6), 415–416.

Palmer, M.D., Haines, K., Tett, S.F.B., Ansell, T.J., 2007. Isolating the signal of ocean global warming. Geophys. Res. Lett. 34, L23610. http://dx.doi.org/10.1029/2007GL031712.

Palmer, M.D., Antonov, J., Barker, P., Bindoff, N.L., Boyer, T., Carson, M., Domingues, C.M., Gille, S., Gleckler, P.J., Good, S.A., Gouretski, V., Guinehut, S., Haines, K., Harrison, D.E., Ishii, M., Johnson, G.C., Levitus, S., Lozier, M.S., Lyman, J.M., Meijers, A.J., von Schuckmann, K., Smith, D., Wijffels, S.E., Willis, J.K., 2010. Future Observations for Monitoring Global Ocean Heat Content, Proceedings of OceanObs'09: Sustained Ocean Observations and Information for Society, 21–25 September 2009, Venice, pp. 1–13. http://dx.doi.org/10.5270/OceanObs09.cwp.68.

Palmer, M.D., McNeall, D.J., Dunstone, N.J., 2011. Importance of the deep ocean for estimating decadal changes in Earth's radiation balance. Geophys. Res. Lett. 38 (13), L13707.

Pardaens, A., Gregory, J., Lowe, J., 2011. A model study of factors influencing projected changes in regional sea level over the twenty-first century. Clim. Dyn. 36 (9), 2015–2033.

Pierce, D.W., Gleckler, P.J., Barnett, T.P., Santer, B.D., Durack, P.J., 2012. The fingerprint of human-induced changes in the ocean's salinity and temperature fields. Geophys. Res. Lett. 39 (21), L21704.

Pollard, D., DeConto, R., 2009. Modelling West Antarctic ice sheet growth and collapse through the past five million years. Nature 458, 329–332.

Pritchard, H.D., Arthern, R.J., Vaughan, D.G., Edwards, L.A., 2009. Extensive dynamic thinning on the margins of the Greenland and Antarctic ice sheets. Nature 461 (7266), 971–975.

Purkey, S.G., Johnson, G.C., 2010. Warming of global abyssal and deep Southern Ocean waters between the 1990s and 2000s: contributions to global heat and sea level rise budgets. J. Clim. 23 (23), 6336–6351.

Ray, R.D., Douglas, B.C., 2011. Experiments in reconstructing twentieth-century sea levels. Prog. Oceanogr. 91 (4), 496–515.

Rayner, N.A., Parker, D.E., Horton, E.B., Folland, C.K., Alexander, L.V., Rowell, D.P., Kent, E.C., Kaplan, A., 2003. Global analyses of sea surface temperature, sea ice, and night marine air temperature since the late nineteenth century. J. Geophys. Res. 108 (D14), 4407. http://dx.doi.org/10.1029/2002JD002670.

Rhein, M., Rintoul, S.R., Aoki, S., Campos, E., Chambers, D., Feely, R.A., Gulev, S., Johnson, G.C., Josey, S.A., Kostianoy, A., Mauritzen, C., Roemmich, D., Talley, L., Wang, F., 2013. Observations: Ocean. In: Stocker, T.F. et al., (Ed.), Climate Change 2013: The Physical Science Basis. Contribution of Working Group I to the Fifth 1376 Assessment Report of the Intergovernmental Panel on Climate Change. Cambridge University Press, Cambridge published online September 30, 2013.

Ridley, J., Gregory, J.M., Huybrechts, P., Lowe, P., 2010. Thresholds for irreversible decline of the Greenland ice sheet. Clim. Dyn. 35 (6), 1065–1073.

Rignot, E., Casassa, G., Gogineni, P., Krabill, W., Rivera, A., Thomas, R., 2004. Accelerated ice discharge from the Antarctic Peninsula following the collapse of Larsen B ice shelf. Geophys. Res. Lett. 31 (18), L18401.

Rignot, E., Bamber, J.L., Van Den Broeke, M.R., Davis, C., Li, Y.H., Van De Berg, W.J., Van Meijgaard, E., 2008a. Recent Antarctic ice mass loss from radar interferometry and regional climate modelling. Nat. Geosci. 1 (2), 106–110.

Rignot, E., Box, J.E., Burgess, E., Hanna, E., 2008b. Mass balance of the Greenland Ice Sheet from 1958 to 2007. Geophys. Res. Lett. 35, L20502. http://dx.doi.org/10.21029/22008GL035417.

Rignot, E., Velicogna, I., van den Broeke, M.R., Monaghan, A., Lenaerts, J., 2011. Acceleration of the contribution of the Greenland and Antarctic ice sheets to sea level rise. Geophys. Res. Lett. 38 (5), L05503.

Rignot, E., Jacobs, S., Mouginot, J., Scheuchl, B., 2013. Ice shelf melting around Antarctica. Science 341, 266–270.

Robinson, A., Calov, R., Ganopolski, A., 2012. Multistability and critical thresholds of the Greenland ice sheet. Nat. Clim. Chang. 2 (6), 429–432.

Roemmich, D., Gilson, J., 2011. The global ocean imprint of ENSO. Geophys. Res. Lett. 38, L13606. http://dx.doi.org/10.1029/2011GL047992.

Roemmich, D., Gilson, J., Davis, R., Sutton, P., Wijffels, S., Riser, S., 2007. Decadal spinup of the South Pacific Subtropical Gyre. J. Phys. Oceanogr. 37, 162–173.

Roemmich, D., John Gould, W., Gilson, J., 2012. 135 years of global ocean warming between the challenger expedition and the Argo Programme. Nat. Clim. Chang. 2 (6), 425–428.

Rohling, E.J., Grant, K., Bolshaw, M., Roberts, A.P., Siddall, M., Hemleben, C., Kucera, M., 2009. Antarctic temperature and global sea level closely coupled over the past five glacial cycles. Nat. Geosci. 2 (7), 500–504.

Sallenger, A.H., Doran, K.S., Howd, P.A., 2012. Hotspot of accelerated sea-level rise on the Atlantic coast of North America. Nat. Clim. Chang. 2, 884–888.

Sasgen, I., van den Broeke, M., Bamber, J.L., Rignot, E., Sørensen, L.S., Wouters, B., Martinec, Z., Velicogna, I., Simonsen, S.B., 2012. Timing and origin of recent regional ice-mass loss in Greenland. Earth Planet. Sci. Lett. 333–334, 293–303.

Scambos, T.A., Bohlander, J.A., Shuman, C.A., Skvarca, P., 2004. Glacier acceleration and thinning after ice shelf collapse in the Larsen B embayment, Antarctica. Geophys. Res. Lett. 31, L18402. http://dx.doi.org/10.11029/12004GL020670.

Schoof, C., 2007. Ice sheet grounding line dynamics: steady states, stability, and hysteresis. J. Geophys. Res. Earth Surf. 112 (F3), F03S28.

Schwarzkopf, F.U., Böning, C.W., 2011. Contribution of Pacific wind stress to multi-decadal variations in upper-ocean heat content and sea level in the tropical south Indian Ocean. Geophys. Res. Lett. 38 (12), L12602.

Shepherd, A., et al., 2012. A reconciled estimate of ice-sheet mass balance. Science 338 (6111), 1183–1189.

Siedler, G., Church, J., Gould, J. (Eds.), 2001. Ocean Circulation and Climate, Observing and Modelling the Global Ocean. In: International Geophysics Series, vol. 77. Academic Press, San Diego, p. 715.

Slangen, A., Katsman, C., van de Wal, R., Vermeersen, L., Riva, R., 2012. Towards regional projections of twenty-first century sea-level change based on IPCC SRES scenarios. Clim. Dyn. 38 (5), 1191–1209.

Smith, S.J., van Aardenne, J., Klimont, Z., Andres, R.J., Volke, A., Delgado Arias, S., 2011. Anthropogenic sulfur dioxide emissions: 1850–2005. Atmos. Chem. Phys. 11 (3), 1101–1116.

Snay, R., Cline, M., Dillinger, W., Foote, R., Hilla, S., Kass, W., Ray, J., Rohde, J., Sella, G., Soler, T., 2007. Using global positioning system-derived crustal velocities to estimate rates of absolute sea level change from North American tide gauge records. J. Geophys. Res. 112, B04409. http://dx.doi.org/10.01029/02006JB003606.

Stammer, D., 2008. Response of the global ocean to Greenland and Antarctic ice melting. J. Geophys. Res. 113, C06022.

Stammer, D., Agarwal, N., Herrmann, P., Kohl, A., Mechoso, C.R., 2011. Response of a coupled ocean-atmosphere model to Greenland ice melting. Surv. Geophys. 32 (4–5), 621–642.

Stammer, D., Cazenave, A., Ponte, R., Tamisiea, M., 2013. Contemporary regional sea level changes. Ann. Rev. Mar. Sci. 5, 21–46. http://dx.doi.org/10.1146/annurev-marine-121211-172406.

Stanford, J.D., Hemingway, R., Rohling, E.J., Challenor, P.G., Medina-Elizalde, M., Lester, A.J., 2011. Sea-level probability for the last deglaciation: a statistical analysis of far-field records. Glob. Planet. Change 79 (3–4), 193–203.

Straneo, F., Hamilton, G.S., Sutherland, D.A., Stearns, L.A., Davidson, F., Hammill, M.O., Stenson, G.B., Rosing-Asvid, A., 2010. Rapid circulation of warm subtropical waters in a major glacial fjord in East Greenland. Nat. Geosci. 3 (3), 182–186.

Straneo, F., Curry, R.G., Sutherland, D.A., Hamilton, G.S., Cenedese, C., Vage, K., Stearns, L.A., 2011. Impact of fjord dynamics and glacial runoff on the circulation near Helheim Glacier. Nat. Geosci. 4 (5), 322–327.

Straneo, F., et al., 2013. Challenges to understand the dynamic response of Greenland's marine terminating glaciers to oceanic and atmospheric forcing. Bull. Am. Meteorol. Soc. 94, 1131–1144. http://dx.doi.org/10.1175/bams-d-12-00100.

Tamisiea, M.E., 2011. Ongoing glacial isostatic contributions to observations of sea level change. Geophys. J. Int. 186 (3), 1036–1044.

Taylor, K.E., Stouffer, R.J., Meehl, G.A., 2011. An overview of CMIP5 and the experiment design. Bull. Am. Meteorol. Soc. 93 (4), 485–498.

Timmermann, A., McGregor, S., Jin, F.-F., 2010. Wind effects on past and future regional sea level trends in the Southern Indo-Pacific. J. Clim. 23 (16), 4429–4437.

Tsimplis, M., Spada, G., Marcos, M., Flemming, N., 2011. Multi-decadal sea level trends and land movements in the Mediterranean Sea with estimates of factors perturbing tide gauge data and cumulative uncertainties. Glob. Planet. Change 76 (1–2), 63–76.

van Angelen, J.H., Lenaerts, J.T.M., Lhermitte, S., Fettweis, X., Kuipers Munneke, P., van den Broeke, M.R., van Meijgaard, E., Smeets, C.J.P.P., 2012. Sensitivity of Greenland Ice Sheet surface mass balance to surface albedo parameterization: a study with a regional climate model. The Cryosphere 6 (5), 1175–1186.

van de Plassche, O., Bohncke, S.J.P., Makaske, B., van der Plicht, J., 2005. Water-level changes in the Flevo area, central Netherlands (5300–1500 BC): implications for relative mean sea-level rise in the Western Netherlands. Quat. Int. 133–134, 77–93.

van den Broeke, M., Bamber, J., Ettema, J., Rignot, E., Schrama, E., van de Berg, W.J., van Meijgaard, E., Velicogna, I., Wouters, B., 2009. Partitioning recent Greenland mass loss. Science 326 (5955), 984–986.

Vaughan, D.G., Comiso, J.C., Allison, I., Carrasco, J., Kaser, G., Kwok, R., Mote, P., Murray, T., Paul, F., Ren, J., Rignot, Eric, Solomina, O., Steffen, K., Zhang, T., 2013. Observations: cryosphere. In: Stocker, T.F. et al., (Ed.), Climate Change 2013: The Physical Science Basis. Contribution of Working Group I to the Fifth 1376 Assessment Report of the Intergovernmental Panel on Climate Change. Cambridge University Press, Cambridge, published online 30 September, 2013.

Velicogna, I., 2009. Increasing rates of ice mass loss from the Greenland and Antarctic ice sheets revealed by GRACE. Geophys. Res. Lett. 36, L19503.

Velicogna, I., Wahr, J., 2013. Time-variable gravity observations of ice sheet mass balance: precision and limitations of the GRACE satellite data. Geophys. Res. Lett. 40 (12), 3055–3063.

von Schuckmann, K., Le Traon, P.Y., 2011. How well can we derive Global Ocean indicators from Argo data? Ocean Sci. 7 (6), 783–791.

von Schuckmann, K., Sallée, J.B., Chambers, D., Le Traon, P.Y., Cabanes, C., Gaillard, F., Speich, S., Hamon, M., 2013. Monitoring ocean heat content from the current generation of global ocean observing systems. Ocean Sci. Discuss. 10 (3), 923–949.

Wada, Y., van Beek, L.P.H., Sperna Weiland, F.C., Chao, B.F., Wu, Y.-H., Bierkens, M.F.P., 2012. Past and future contribution of global groundwater depletion to sea-level rise. Geophys. Res. Lett. 39 (9), L09402.

Wahl, T., Jensen, J., Frank, T., Haigh, I., 2011. Improved estimates of mean sea level changes in the German Bight over the last 166 years. Ocean Dyn. 61 (5), 701–715.

Wake, L.M., Milne, G.A., Long, A.J., Woodroffe, S.A., Simpson, M.J.R., Huybrechts, P., 2012. Century-scale relative sea-level changes in West Greenland—a plausibility study to assess contributions from the cryosphere and the ocean. Earth Planet. Sci. Lett. 315–316, 86–93.

Watson, C., Coleman, R., White, N., Church, J., Govind, R., 2003. Absolute calibration of TOPEX/Poseidon and Jason-1 Using GPS Buoys in Bass Strait, Australia Special Issue: Jason-1 Calibration/Validation. Mar. Geod. 26 (3–4), 285–304.

Watson, C., White, N., Church, J., Burgette, R., Tregoning, P., Coleman, R., 2011. Absolute calibration in Bass Strait, Australia: TOPEX, Jason-1 and OSTM/Jason-2. Mar. Geod. 34 (3–4), 242–260.

Wenzel, M., Schröter, J., 2010. Reconstruction of regional mean sea level anomalies from tide gauges using neural networks. J. Geophys. Res. 115 (C8), C08013.

White, N.J., Coleman, R., Church, J.A., Morgan, P.J., Walker, S.J., 1994. A southern hemisphere verification for the Topex/Poseidon satellite altimeter mission. J. Geophys. Res. 99 (24), 505–24516.

White, N.J., Church, J.A., Gregory, J.M., 2005. Coastal and global averaged sea-level rise for 1950 to 2000. Geophys. Res. Lett. 32 (1), L01601. http://dx.doi.org/10.1029/2004GL021391.

Wijffels, S., Willis, J., Domingues, C., Barker, P., White, N., Gronell, A., Ridgway, K., Church, J., 2008. Changing expendable bathythermograph fall rates and their impact on estimates of thermosteric sea level rise. J. Clim. 21 (21), 5657–5672.

Willis, J., Roemmich, D., Cornuelle, B., 2004. Interannual variability in upper-ocean heat content, temperature and thermosteric expansion on global scales. J. Geophys. Res. 109 (C12), C12037. http://dx.doi.org/10.10292003JC10002260.

Willis, J.K., Lyman, J.M., Johnson, G.C., Gilson, J., 2007. Correction to "Recent cooling of the upper ocean". Geophys. Res. Lett. 34, L16601, 10.11029/12007GL030323.

Wong, A., Keeley, R., Carval, T., and the Argo Data Management Team, 2013. Argo quality control manual, version 2.8. Available at: http://www.argodatamgt.org/content/download/15699/102401/file/argo-quality-control-manual-version2.8.pdf.

Woodworth, P.L., 1990. A search for accelerations in records of European mean sea level. Int. J. Climatol. 10, 129–143.

Woodworth, P.L., 1999. High waters at Liverpool since 1768: the UK's longest sea level record. Geophys. Res. Lett. 26 (11), 1589–1592.

Woodworth, P.L., Player, R., 2003. The permanent service for mean sea level: an update to the 21st century. J. Coast. Res. 19, 287–295.

Woodworth, P.L., Teferle, F.N., Bingley, R.M., Shennan, I., Williams, S.D.P., 2009. Trends in UK mean sea level revisited. Geophys. J. Int. 176 (1), 19–30.

Wunsch, C., Ponte, R.M., Heimbach, P., 2007. Decadal trends in sea level patterns: 1993–2004. J. Clim. 20 (24), 5889–5911.

Yin, J., 2012. Century to multi-century sea level rise projections from CMIP5 models. Geophys. Res. Lett. 39 (17), L17709.

Yin, J., Schlesinger, M.E., Stouffer, R.J., 2009. Model projections of rapid sea-level rise on the northeast coast of the United States. Nat. Geosci. 2 (4), 262–266.

Yin, J.J., Griffies, S.M., Stouffer, R.J., 2010. Spatial variability of sea level rise in twenty-first century projections. J. Clim. 23 (17), 4585–4607.

Yin, J., Overpeck, J.T., Griffies, S.M., Hu, A., Russell, J.L., Stouffer, R.J., 2011. Different magnitudes of projected subsurface ocean warming around Greenland and Antarctica. Nat. Geosci. 4 (8), 524–528.

Zhang, X., Church, J.A., 2012. Sea level trends, interannual and decadal variability in the Pacific Ocean. Geophys. Res. Lett. 39 (21), L21701.

Chapter 28

Long-term Salinity Changes and Implications for the Global Water Cycle

Paul J. Durack[*,†], **Susan E. Wijffels**[†] **and Tim P. Boyer**[‡]

[*] *Program for Climate Model Diagnosis and Intercomparison (PCMDI), Lawrence Livermore National Laboratory (LLNL), Livermore, California, USA*

[†] *CSIRO Marine and Atmospheric Research, Hobart, Australia*

[‡] *National Oceanographic Data Center, NOAA, Silver Spring, Maryland, USA*

Chapter Outline

1. INTRODUCTION

The global ocean covers 71% of the Earth's surface and contains 97% of all free water when considered by storage (Figure 28.1); it's where the water is. If we consider the total freshwater flux at the surface of the Earth, the global ocean dominates totals; 85% of surface evaporation and 77% of surface precipitation occurs at the ocean–atmosphere interface (Trenberth et al., 2007; Schanze et al., 2010; Figure 28.1). With the oceans dominating global water cycle totals by every measure, they provide a unique insight into the global water cycle operation, and by considering changes to ocean properties linked to the water cycle, to its long-term change.

The ocean's surface freshwater fluxes of evaporation (E) and precipitation (P) are poorly measured. Large global mean discrepancies of order 25 Wm^{-2} are apparent between independent current generation surface heat flux climatologies, suggesting similar issues exist with current climatological evaporation minus precipitation (E–P) estimates (see Chapter 5). Estimates of P are further complicated by its episodic nature. Sea-surface salinity (SSS) provides an integrated and smoothed field from which the difficult to measure E–P field (Chapter 5)—which set the spatial pattern of ocean SSS—can be inferred over the long term (Section 6.2). The spatial structure of the ocean's subsurface salinity field is maintained by ocean circulation and mixing, which are driven by the global ocean density gradients and surface winds acting at the ocean–atmosphere interface (Chapter 11). The relationship between E–P and salinity motivates the ocean "rain gauge" concept (e.g., Schmitt, 2008), with the hope that continued and improving observation of ocean salinity can lead to better estimates of E–P over the ocean, and as a consequence, an enhanced understanding of water cycle operation, variability, and change. Additionally, global mean ocean salinity and its spatial patterns (along with other chemical tracers) are considered stable over the period of direct observational coverage, with large deviations from present day values having occurred during geologic (~ 100,000-year),

Ocean Circulation and Climate, Vol. 103. http://dx.doi.org/10.1016/B978-0-12-391851-2.00028-3

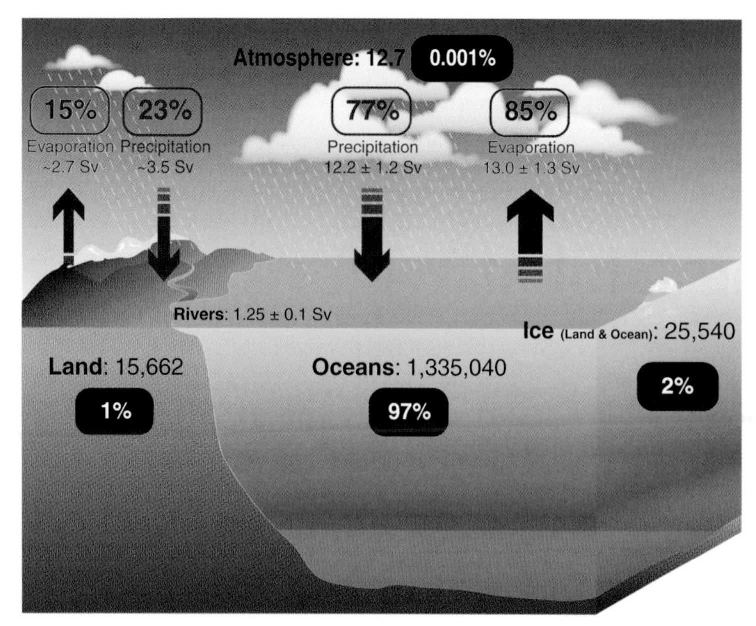

Reservoirs represented by solid boxes: 10^3 km^3, fluxes represented by arrows: Sverdrups (10^6 m^3 s^{-1})
Sources: Baumgartner & Reichel, 1975; Schmitt, 1995; Trenberth et al., 2007; Schanze et al., 2010; Steffen et al., 2010

FIGURE 28.1 A schematic representation of the global water cycle, which represents the dominant role of the ocean. Reservoir estimates represent storages in 10^3 km^3 (unboxed numbers), flux estimates represent transports in Sverdrups (10^6 m^3 s^{-1}) and values within boxes represent the approximate percentage of total storages (black filled boxes) or flux estimates (rainfall: blue, evaporation: red) for the global surface. Total ice volume expressed is dominated by terrestrial sources, with sea-ice comprising approximately 0.1% of the total ice storage. Following Trenberth et al. (2007) ice volumes are scaled by a 0.917 density factor to convert to liquid water equivalent. *Reproduced from Durack et al. (2012).*

rather than climate (~100-year), history (e.g., Rubey, 1951; Holland, 1972). The relative temporal stability of ocean salinity over the observational period (Section 2) underscores its usefulness, which allows small perturbations to be assessed as a marker of oceanic and water cycle changes.

When we think of the global water cycle, regional rainfall is generally the key focus. The dependence of humankind on rainfall cannot be understated, as we are reliant upon rainfall to supply available water, enabling societal operation, agriculture and the built communities that are a feature of modern life. The nature of rainfall, with its sporadic temporal and spatial patterns, along with the sparse terrestrial rain gauge observing network, makes it very difficult to accurately observe, let alone estimate, small changes over the observational record.

Since the 1980s rainfall has been measured using satellites. In practice satellite observations are merged with the sparse *in situ* gauge measurements (e.g., Huffman et al., 1997) or in addition with reanalysis products (e.g., Xie and Arkin, 1997) to form monthly mean climatological data products. However, changes between satellite missions have led to issues in accurately validating and calibrating between the resulting data. Such issues lead to inconsistencies between differing data streams, the resulting temporal (and spatial) observations, and estimates of their changes (Dai et al., 1997; Yin et al., 2004; Lau and Wu, 2007; Quartly et al., 2007; Wentz et al., 2007; Tian et al., 2009; Arkin et al., 2010). Accurate absolute rainfall measurement from satellites are confounded by the temporal coverage of such satellite missions, with satellite overpasses occurring at most twice during a daily cycle, whereas rainfall could occur anytime during a 24-h observing period. As a consequence, satellite observations may have only captured part of the *in situ* temporal total (see Huffman et al., 1997 for a more detailed discussion).

These issues become obvious when directly comparing current observational precipitation products. Differing signs of regional and global (1979–near present) trends highlight the uncertainty with long-term water cycle change estimated from precipitation measurements (Figure 28.2). For these reasons, it is necessary to consider other climate observations which provide insights into the global water cycle, and most importantly, its long-term change.

It has long been noted that the climatological mean SSS spatial pattern is highly correlated with the climatological mean E–P spatial pattern (Section 5; Figure 28.8; Wüst, 1936). This correlation reflects the long-term balance between ocean advection and mixing processes and E–P fluxes acting at the ocean surface which maintain salinity gradients. The E–P field is a more accurate measure of the complete water cycle than P alone, as both these terms considered together (along with smaller runoff term, R; Schanze et al., 2010; Figure 28.1), provide the water sources and sinks that comprise the complete global freshwater budget for the ocean. The ocean is the ultimate source of all terrestrial rainfall (Gimeno et al., 2010; van der Ent and Savenije, 2013) and it is assumed that broad-scale pattern changes occurring over 71% of the Earth covered by oceans will also broadly reflect changes over corresponding terrestrial zones. However, we note that freshwater cycling and recycling over land is complex (van der Ent et al., 2010) and is not addressed in this chapter.

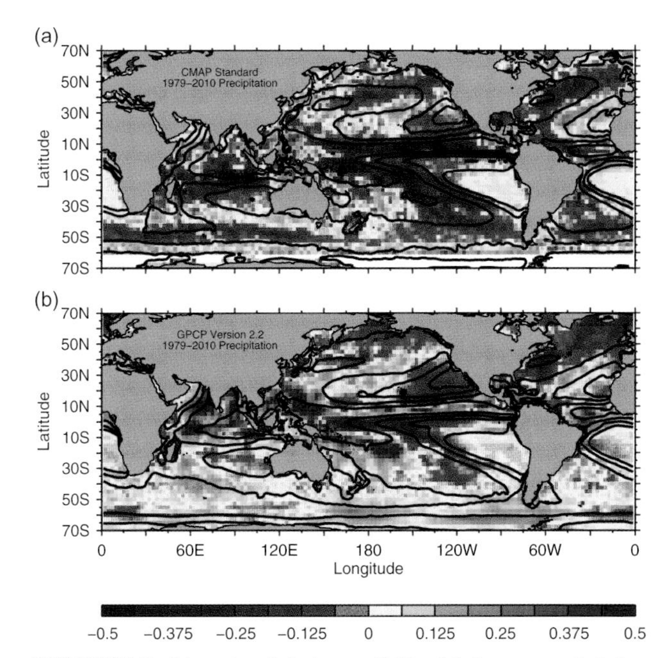

FIGURE 28.2 Linear trends in two available global ocean precipitation products (a) CMAP 1979–2010 and (b) GPCP v2.2 1979–2010 obtained from annual-mean climatological values. Units are absolute changes scaled to represent m 50-year^{-1}. Black contours express the associated mean precipitation for the analysis period for each climatology. Even though these independent climatology products are generated from similar input data sources, their long-term trends differ markedly in spatial patterns and magnitudes.

2. SALINITY OBSERVATIONS IN THE GLOBAL OCEANS

Ocean salinity was traditionally measured from seawater samples obtained at a small number of discrete ocean depths using specifically designed "Nansen" sampling bottles. Prior to electronics, these were mechanically closed by "messenger" weights sent down along the line supporting the bottles beneath the research vessel. Such measurements are now obtained from a "rosette" sampler containing an array of sampling bottles. These modern bottles are closed by an electronic signal from the ship above and also provide samples at discrete depths. Seawater samples are routinely analyzed on board within a day of collection to minimize evaporation and ensure the most accurate salinity estimates. In addition, *in situ* estimates of salinity at meter-scale depth intervals are obtained from temperature and electrical conductivity measurements from "profiling" sensors installed on the "rosette" or used independently on a stand-alone system. Once the cruise is complete, careful calibration tests are undertaken in land-based laboratories to further validate the underway analyses.

The major historical deep-sea oceanographic research expeditions which measured ocean salinity were undertaken between 1872 and 1960 and were documented by Wüst (1964). This work described four eras of oceanographic observation during the period prior to 1960, and we shall summarize some of the key expeditions and those subsequent to 1960 below.

Modern ocean observing began with the Challenger Expedition. The HMS *Challenger* was the first true oceanographic research vessel specifically designed to investigate physical, biological, geological, and chemical properties of the Atlantic, Pacific, and Indian Oceans, and seafloor. A 4-year long research cruise was undertaken on this vessel from 1872–1876. Since this time, oceanography has undergone many observational, platform, and data-precision revolutions. A more complete coverage of the history and future of ocean observations are detailed in an earlier chapter (see Chapter 3).

Hydrographic (salinity, temperature, and select chemical tracers) sampling first came to prominence in an early Atlantic Ocean survey undertaken in 1925–1927 on the RV *Meteor*. This expedition used discrete (Nansen) bottle measurements obtained at ocean depths and then analyzed for salinity and chemical properties, along with temperature from reversing thermometers. Since this early twentieth century expedition, there have been a number of major efforts to explore the full-depth properties of the global and regional oceans. Key expeditions include: the International Geophysical Year (IGY; 1956–1960) which provided Atlantic Ocean coverage with a systematic, high quality, top-to-bottom, continent-to-continent grid of hydrographic stations; the Geochemical Ocean Sections Study (GEOSECS; 1972–1978) which provided a global survey of chemical, isotopic, and radiochemical tracers in the ocean for the Atlantic (1972–1973), Pacific (1973–1974), and Indian Oceans (1977–1978); and the Transient Tracers in the Ocean (TTO; 1981–1983) which considered the North Atlantic (1981) and tropical Atlantic (1983) hydrography.

By the 1970s the significant variability associated with the ocean mesoscale had been recognized, and in the 1990s resulted in the World Ocean Circulation Experiment (WOCE) Hydrographic Programme (WHP). The WOCE program (1990–1998) undertook surveys along densely-sampled sections for all the three major ocean basins. These measurements were complemented by regional programs in the Nordic Seas, the Arctic Ocean, the Mediterranean, the Southern Ocean, and numerous continental shelf regions to produce an unprecedented, high quality, global dataset of physical and chemical oceanographic observations. The observed sections were sampled during the experiment and for some sections, repeated in an attempt to capture the interannual variability of the ocean.

A new era of automated ocean observations began in 1999 with the development and implementation of the Argo Program (Gould et al., 2004). The program was specifically designed to continuously track the oceans broad-scale structure, addressing the issues associated with discontinuous global hydrographic observations and the resultant poor spatial and temporal data coverage. A particular emphasis of the program was to improve coverage in the poorly sampled Southern Hemisphere oceans. The clear improvement in the observational, spatial, and temporal coverage is expressed in Figure 3.12, with near complete global coverage achieved around 2005.

Argo floats provide unprecedented observational data coverage from the near surface to 2000 dbar, and record quality-controlled temperature and salinity measurements which approach ship-based data accuracy (Janzen et al., 2008). For the first time, near-global, upper (0–2000 dbar) ocean observation was achieved, providing complete seasonal data coverage. In particular the Argo Program is currently providing much higher temporal and spatial coverage than data previously obtained from research vessels, and is enabling a better understanding of ocean variability and structure. With its continued operation, the Argo Program will provide a much-needed baseline from which a quantitative assessment of long-term ocean climate change can be made. Active Argo floats number 3639 as of October 2013, and the current database includes profiles from over 8600 floats, which collect more than 120,000 profiles every year (Figure 28.3). Over one million individual profiles from 1999 to the present have been obtained, with this milestone reached in November 2012. This new data stream now accounts for well over half the profiles stored in the high-quality historical hydrographic database (e.g., Durack and Wijffels, 2010). Regionally, Argo provides more than half the entire austral winter profile coverage south of 30°S, and two and a half times the Southern Hemisphere coverage with just over 13 years of data (Figure 28.3), compared to the historical database which spans 137 years (Figure 3.12). When combined, the complete historical database totals 9.4 million temperature and 4.2 million salinity observations from the various profiling platforms (these data comprise the World Ocean Database 2009 updated to 2012; Boyer et al., 2009; Figure 28.3). If data from *in situ* moorings, gliders, ice-tethered buoys, and pinniped-mounted sensors are included these numbers increase to 12.2 million temperature and 5.5 million salinity observations.

Complementing *in situ* observations, remotely sensed salinity observations using satellites have been underway since 2009. Two platforms are presently providing data for use in oceanographic research, with both platforms still undergoing calibration and validation to ensure accuracy of their data. A further discussion of new insights from satellites is found in Section 7.

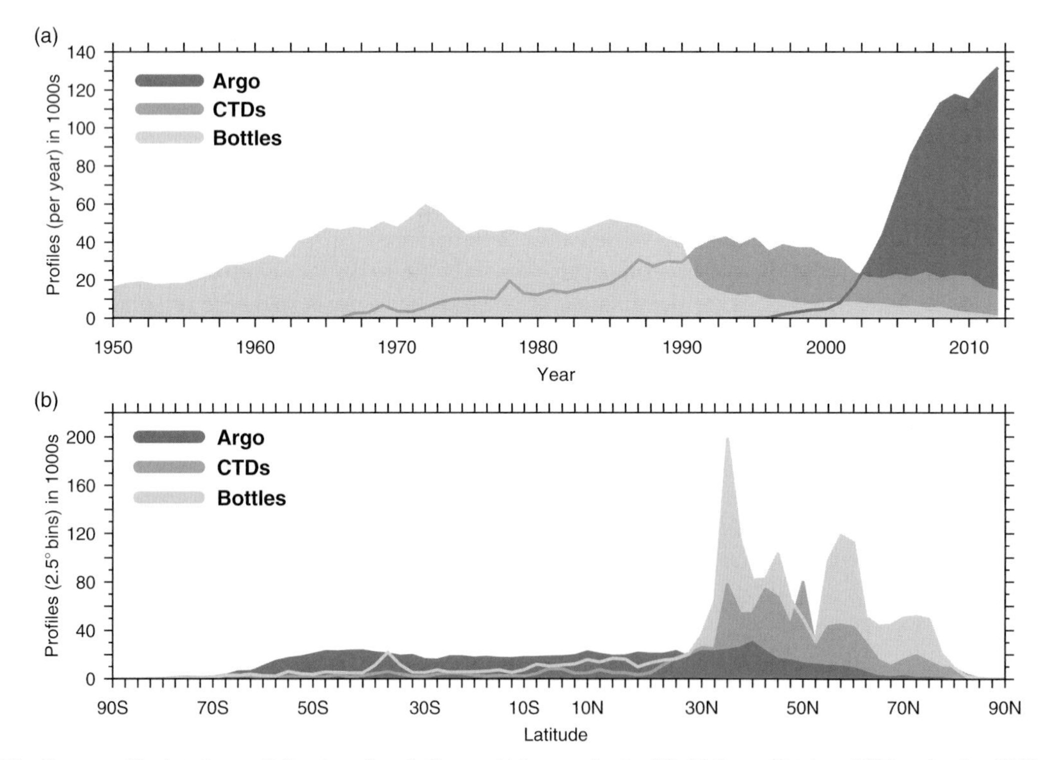

FIGURE 28.3 Ocean profile data from salinity observing platforms which comprise the World Ocean Database 2009 updated to 2012 (Boyer et al., 2009). Platform type for (a) per year (1950–2012) and (b) in 2.5° zonal (latitude) bins for the period 1950 to 2012. The global nature of the Argo program is evident in the fairly even distribution of profiles across both hemispheres, with the clear Northern Hemisphere bias evident in the historical archive comprised of bottles and CTDs.

2.1. Measuring Ocean Salinity

In parallel to observational expeditions, new and evolving platforms ensured that data quality was increasingly more accurate, and easier to obtain. Nansen bottles (or insulated metal cylinders) were designed in the 1890s during Fridtjof Nansen's famous *Fram* expedition in the Arctic Ocean and were commercially produced by 1900 (Pettersson-Nansen insulating water-bottle; Mill, 1900). This new technology allowed deep seawater samples to be retrieved and for the first time provided an efficient method to record *in situ* salinity alongside temperature measurements from reversing thermometers.

Due to the considerable growth in hydrographic investigations made possible by the new technology, the oceanographic community recognized a need for standardization of these measurements, to allow effective intercomparisons between independent studies. In 1899, participating European countries convened a conference in Stockholm to establish the International Council for the Exploration of the Sea (ICES). The ICES Chair Martin Knudsen proposed the use of Standard Seawater (SSW; which historically was also called "Copenhagen water") to introduce greater consistency between independent oceanographic assessments by providing a single standard for salinity (or chlorinity) analysis. Knudsen felt this was necessary as the numerous methods of measuring salinity at the time yielded unreliable results (e.g., Culkin and Smed, 1979). Accordingly there was a need for a better method to determine total dissolved salts (or salinity), based on the premise of constancy of ionic ratios in seawater (e.g., Marcet, 1819; Forchhammer, 1862; Dittmar, 1884). The commission defined "chlorinity," a property that could be determined by a simple volumetric titration using silver nitrate, to be used as the standard measure of salinity. Knudsen and colleagues tested seawater from different regions of the World Ocean, and based on comparison of nine determinations (from the Baltic, Mediterranean, and Red seas, along with the North Atlantic Ocean) of salinity and chlorinity, defined a formula to estimate salinity from the chloride content (grams) in one kilogram of seawater (Knudsen, 1901).

To date, SSW has been collected at the surface of the North Atlantic. It is then pumped through filters into a storage tank and circulated through filters for 2–3 weeks to achieve thorough mixing and remove organic material. During this period, the seawater is gradually diluted with distilled water to give a final salinity near 35 (Practical Salinity Scale, 1978; PSS-78) and the temperature is raised to 26 °C so that when it is transferred to glass ampules the seawater is slightly under saturated with dissolved air (Culkin and Smed, 1979). Following the introduction of SSW by Knudsen, the responsibility was transferred from the ICES Central Laboratory in 1908 to what later became known as the Standard Seawater Service, which was based in Copenhagen. In 1948, Knudsen proposed that the Association International d'Oceanographie Physique (AIOP, now IAPSO) take over the responsibility for preparing SSW, and production was maintained in a series of locations throughout Denmark. In 1974, it was transferred to Wormley, England at the Institute of Oceanographic Sciences (IOS; which subsequently became the National Oceanography Centre NOC). In 1989, production moved to Ocean Scientific International Ltd (OSIL) a privatized laboratory located in Hampshire, England, where SSW production continues today. SSW is still in use today and rigorous standards are maintained by intercomparisons of SSW samples across the available batches (e.g., Park, 1964; Mantyla, 1980; Culkin, 1986; Mantyla, 1987; Culkin and Ridout, 1998; Aoyama et al., 2002; Kawano et al., 2006; Bacon et al., 2007).

The Knudsen methodology served oceanographers for the next 65 years. In his proposal, Knudsen stressed the importance of salinity measurement of seawater in physical, climatological, and biological investigations and maintained that the measurement could be carried out using titration with an accuracy of 0.04‰; however, such accuracy was not generally achieved by the methods in use at the time. Usually, a few titrations were carried out by weighing and all subsequent volumetric titrations were then referred to these. Knudsen pointed out that titration by weighing was a fairly difficult operation at the time, and was almost impossible to carry out often enough to obtain the desired accuracy. He noted that errors in salinity determinations up to 0.15‰ were common. From Knudsen's own investigations on the *Ingolf* Expeditions of 1895 and 1896, he knew that some Atlantic water types differed by only 0.1‰ to 0.25‰, therefore greater accuracy in salinity determination was required.

Advancing technology in the twentieth century allowed for the development of salinometers (an instrument that determines the concentration of salt by measuring electrical conductivity; e.g., Wenner et al., 1930) and was first used in the U.S. Coast Guard International Ice Patrol Vessels in 1934. The new technology improved oceanographic measurements by leveraging the more consistent conductivity–density ratio when compared to the chlorinity–density ratio (obtained through titration techniques). The advantage of the new technique was an insensitivity to subtle ionic differences between differing seawater samples, a problem with the earlier titration methods (Cox et al., 1962). This apparatus provided measurements at least as precise as the standard chemical titration method but was simpler and quicker to use (Cox, 1963). However, it is worth noting that conductivity is foremost a function of temperature and secondarily salinity, and therefore introduced an additional source of error if temperature was not accurately accounted for during analysis. Additionally, precise conductivity methods are usually comparative rather than absolute and therefore require the use of SSW for calibration.

Prior to the late 1950s, salinity was typically determined at laboratory benches by silver nitrate titration of seawater chloride concentration and the Knudsen methodology, with shipboard titration salinities resolved only to two decimal places (Wooster and Taft, 1958). However, three decimal precision is required for good density calculations and to map subtle deep salinity features. The instruments of choice between the late 1950s and early 1970s were conductivity salinometers (e.g., Schleicher and Bradshaw, 1956; Paquette, 1959) which replaced titration methods, and data from those instruments were accurate to ~0.005‰ (Mantyla, 1994). The development of bench salinometers continued through to the 1960s, with Brown and Hamon (1961) developing the first portable unit (15 kg) which was subsequently mass produced, and incorporated inherently better sensitivity than the older conductivity instruments, along with improved usability over the original Wenner-Smith-Soule version which weighed around 200 kg. These units were widely used in the 1970s and their use extended into the 1990s. The instruments were operated at ambient temperatures and were subject to nonlinear drifts, consequently the results were often no better than from the earlier salinometers. More recent conductive salinometers (Guildline), in use since the 1980s, feature a double sensitivity configuration, operating in a temperature controlled water bath, with high stability and repeatable sensitivities to 0.001‰ (Dauphinee and Klein, 1975; Knapp and Stalcup, 1987). More recently, a number of companies have introduced equally precise bench salinometers with different designs (RBR; Optimare).

In situ methods for measuring salinity profiles through the water column were developed using salinity–temperature–depth (STD; Hamon, 1955) and conductivity–temperature–depth (CTD; Brown, 1974) platforms in the mid-1950s and for the first time provided an efficient method to record salinity alongside temperature measurements from thermistors (thermally sensitive resistors used in oceanography; e.g., Becker et al., 1946). This approach further improved the precision of salinity estimates, as the issues related to temperature drifts in salinometers were removed due to the coincident temperature measurement. Although a great improvement over earlier methods, the conductivity sensors are far less stable than thermistors as they are open cells and changes in the cell geometry (e.g., by fouling) affects the calibration. Therefore, measurement accuracy at the highest precision is still dependent upon calibration and field checks run on salinometers standardized with SSW. Owing to the technical improvements in the measurement of seawater conductivity, a dramatic increase in global ocean observations and greater vertical resolution to full depth was achieved, along with an improvement in salinity precision as a result of the concurrent temperature measurement (see Table 28.1).

The temporal and zonal coverage from these various platform types is presented in Figure 28.3. The dominance of Northern Hemisphere historical observations is clear, with particularly good spatial and temporal coverage found in the Atlantic and the North Pacific basins (Figure 3.12; Chapter 3). The transition to trusted electronic platforms is apparent, with CTDs overtaking bottle measurements during the WOCE Program after 1990 and the Argo floats dominating total annual profile counts with their onboard CTDs in 2002 (Figure 28.3).

2.2. Definitions of Ocean Salinity

Although "salinity" can be intuitively understood as the amount of salts in solution in seawater, an exact definition is difficult to develop because seawater is not created by dissolving a simple salt in water. Traditionally, numerical values have been approximately linked to the mass of residue that remains after evaporation from a defined volume of seawater. However, the latest definition of salinity (Absolute

TABLE 28.1 Representative Salinity Precision Over Time (Standard Deviation) from Selected Research Cruises in the Northeast Atlantic Basin

Year	Cruise Detail	Latitude	Platform/Method	Salinity Precision (‰)
1927	Meteor	18–19°N	Titration	±0.010
1957	Crawford 10	16–24°N	Schleicher & Bradshaw conductivity bridge	±0.005
1973	GEOSECS	21–28°N	U.W. conductivity bridge	±0.0013
1983	TTO/TAS	20–27°N	Double conductivity ratio Autosal salinometer	±0.0016
1988	McTT	17–23°N	Double conductivity ratio Autosal salinometer	±0.0015
1988	McTT	16–24°N	NBIS CTD	±0.0006

Reproduced from Mantyla (1994).

Salinity, S_A) is linked to the mass of all dissolved matter. The two definitions differ numerically because the process of evaporation removes some of the chemicals dissolved in seawater additional to the water itself, and also changes the chemical composition of the solid mass remaining.

In addition, the ongoing development of observing technology along with an improving knowledge of ocean chemistry has led to three major variants of observed salinity populating global hydrographic databases to date [alongside salinity techniques used to measure temperature have also undergone changes, with seven definitions (including amendments) dating back to the early 1900s (Preston-Thomas, 1990), not discussed here]. For most data collected during the years from 1901 to about 1966, Knudsen Salinity (S_K; Knudsen, 1901; Thomas et al., 1934; Jacobsen and Knudsen, 1940; Millero et al., 2008) based on titration was recorded. From 1967 to 1977, a variety of measurement techniques were used, including conductivity-based Cox Salinity (S_C; Cox et al., 1967; Lyman, 1969; Wooster et al., 1969). Finally, in 1978 the conductivity-based Practical Salinity Scale of 1978 (PSS-78; S_P; Lewis, 1980) was established as a standard that is in use through to the present day. All three of these standards result in approximately identical numerical values for the same seawater sample. However, the new Thermodynamic Equation of Seawater 2010 (TEOS-10 Chapter 6) relegates Practical Salinity to database archival use only, with the recommendation that all scientific observational analysis use the new numerically different variable called Absolute Salinity (S_A; IOC, SCOR, and IAPSO, 2010).

However, within these major variants, there are also a long list of smaller changes in the definition of salinity. To further complicate measurement, by the 1930s changes in atomic weight values (for Na^+ and Cl^- ions) led to a difference between computed estimates of Knudsen Salinity and the SSW working standard of 0.009‰ (Reeburgh, 1976). In response to this change, Jacobsen and Knudsen (1940) redefined chlorinity so that computed estimates were identical to the original chlorinity definition (Knudsen, 1901). Also during the 1930s comprehensive investigations of the chlorinity and temperature dependence of electrical conductivity were performed (Thomas et al., 1934). Their chlorinities, while accurate in 1934, were no longer following changes to relevant atomic weights which led to the redefinition of chlorinity in 1937 (Jacobsen and Knudsen, 1940). These measurements then became chlorine-equivalent, with a division by 1.00045 required to convert these to true chlorinities, and obtain accurate salinity estimates which were consistent with the methodology of Knudsen (1901) and Jacobsen and Knudsen (1940).

By the 1960s more problems began to arise in the Knudsen methodology. The first problem was the salinity–chlorinity relation. The Knudsen (1901) formula, found by fitting to simultaneous measurements of evaporation and titration-based chlorinity:

$$S_K = 1.805\,\mathrm{Cl‰} + 0.030 \quad (28.1)$$

although conservative for chlorinity, was not for salinity because of the offset term. Further, it was found that Baltic Sea waters (on which (28.1) was dependent) were responsible for this offset, as they contain a significant river contribution. Baltic Sea water has a different chlorine ion ratio than water from the open Atlantic Ocean diluted to the same mass of salt per kilogram of solution. This meant that salinities calculated using (28.1) were not truly representative of the World Ocean (Cox, 1963; Lewis and Perkin, 1978). These issues led to the development of Cox Salinity (S_C; Cox et al., 1967), which was obtained through joint chlorinity and conductivity estimations using salinometers.

We shall now consider the apparent accuracy of salinity observing methods, by reviewing some key and useful studies. Such accuracies are particularly relevant when considering the resolved long-term changes (Section 4).

In more modern surveys, Knudsen Salinity yielded repeatable accuracies for multiple instruments and laboratories of ±0.02‰ for Mediterranean waters (~38‰) and ±0.05‰ for Baltic waters (~8‰) (Grasshoff and Hermann, 1976). These estimates were obtained from an intercomparison of separate experiments undertaken for the ICES program, and much exceed the smaller errors attainable by careful use of equipment in a single laboratory (Lewis and Perkin, 1978). A salinity intercomparison undertaken as part of the GARP Atlantic Tropical Experiment (GATE) suggested that titration-based differences of ±0.01‰ were commonplace (Farland, 1976). Cox (1963) reviewed salinity precisions obtained from ten separate cruises, which included assessments in the South Atlantic (RV *Meteor*) as well as the Southern Ocean (RRS *Discovery II*). He noted that accuracies of ±0.01‰ or a little better were possible by a good analyst; however, values of ±0.02‰ were noted in Scripps Institution of Oceanography reports, about average for titrated results, and worst case titration standard deviations of up to ±0.06‰ were also found. The GATE intercomparison (Farland, 1976) suggested that titration for chlorinity was a highly skilled routine. They reported an average operator yielded a less precise estimation of salinity using this method when compared to those derived from electrical conductivity, which at this time was becoming the standard observing technique through the use of salinometers.

Although one of the earliest STD platforms provided a precision of ±0.05‰ (Hamon, 1955), precision improved over time, and with later estimates, the standard deviation was around the same as titration-based estimates. However, if outlier observations were discarded the precision increased to ±0.005‰ (Cox, 1963). By the end of the IGY expeditions in the early 1960s, accuracy improved to

±0.003‰ (Cox, 1963). However, salinometers, which did not account for the sample temperature (nonthermostated), could still yield errors exceeding ±0.01‰ (Lewis and Perkin, 1978). The ongoing development of the CTD platform continued to improve measurement accuracies. The precision CTD microprofiler developed at Woods Hole Oceanographic Institution in the 1970s (Brown, 1974) yielded accuracies within ±0.003‰ for 75% of samples taken from the surface to 4500 m depth (Lewis and Perkin, 1978).

SSW, originally developed to standardize titrations, was also being used to standardize conductivity-based measurements. However, although the chlorinity of SSW was strictly defined, its conductivity was not, with different batches varying in conductivity-based salinities by greater than 0.01‰ from titration-based salinities (Lewis and Perkin, 1978). Also as *in situ* instruments were measuring profiles of conductivity in the deep ocean with *in situ* temperatures colder than 10 °C, and the available conversion tables only listed values greater than 10 °C, various other equations (six variants are documented by Lewis and Perkin, 1978; Lewis and Perkin, 1981) began to be used to correct for the temperature effect. By 1975, it was clear that something had to be done, a new standard was needed that would be (a) reproducible, (b) conservative, and (c) would allow density to be computed to acceptable accuracy limits.

With the development of the Practical Salinity Scale 1978 (PSS-78; S_P; Lewis and Perkin, 1978), many issues were resolved. Its first outcome was to define a reproducible primary standard for conductivity, by this time the most common method of salinity measurement. Equations were developed which were valid for seawater over a wide range of salinities (2 to 42‰) and temperatures (−2 to 35 °C), with this equation developed so that $S_P = S_C$ at 35‰, the mean salinity value for the global ocean. This also ensured that for the bulk of open ocean waters (33 to 37 PSS-78) measurement accuracies were around ±0.01 (PSS-78; Lewis and Perkin, 1978).

PSS-78 envisaged a measurement accuracy of ±0.02 (PSS-78) at best (Lewis and Perkin, 1978), and at this level a number of issues related to the changing chemical composition of seawater regionally can be ignored. However, the development of the new standard spurred advances in measurement technology. By the WOCE period hydrographic observations were providing very high quality *in situ* salinity (±0.002 PSS-78) and temperature (±0.002 °C) measurements (Joyce, 1991). Table 28.1 contains some representative precision estimates from regionally consistent oceanographic research cruises from 1927 to 1988. However, at this precision, PSS-78 salinity is more usefully considered a temperature and pressure-corrected electrical conductivity measurement, rather than an estimate of the dissolved matter in seawater.

Along with the development of the new thermodynamic equation of seawater 2010 (TEOS-10; IOC, SCOR, and IAPSO, 2010; see Chapter 6) the use of a new variable Absolute Salinity (S_A) is now the recommended practice for observational oceanographic analyses. Absolute Salinity is numerically different from historical salinity measurements because it provides an estimate of the total dissolved matter. In addition, it accounts for the varying ionic composition (constituents) of regional seawater samples which effects the conductivity-based salinity assessments measured by research vessels and most recently by the Argo floats. In practice this is achieved by first obtaining a Reference Salinity (S_R; Millero et al., 2008), which is calculated by scaling a Practical Salinity (S_P) estimate, which is itself based on conductivity, temperature, and pressure. A small correction factor δS_A is then added to S_R, which accounts for the spatial variations in World Ocean seawater composition to give S_A. To maintain database continuity, S_A will only be utilized in oceanographic analysis whereas conductivity-based S_P will continue to be the variable recorded by ship-based and Argo float observations.

When considering long-term changes, accuracy uncertainties as detailed earlier add to the larger issue of the sparse temporal and spatial coverage provided by ocean salinity profiles. These together lead to challenges in obtaining estimates of long-term property changes in the ocean. It would be useful to consider the effects of data accuracy in addition to the poor temporal and spatial coverage on resolved long-term changes to salinity.

3. OBSERVED SALINITY VARIABILITY

In addition to the lack of comprehensive observational records (see previous sections; Chapter 3), natural (unforced) climate variability is another effect which confounds detection and attribution of long-term change in response to increasing anthropogenic greenhouse gases. This manifests in largely cyclical climate modes on global and regional-scales. Such climate modes include: the El Niño Southern Oscillation (ENSO; Bjerknes, 1969; Philander, 1990), Pacific Decadal Oscillation (PDO; Mantua et al., 1997), North Atlantic Oscillation (NAO; Barnston and Livezey, 1987), Atlantic Multidecadal Oscillation (AMO; Schlesinger and Ramankutty, 1994), and the Southern Annular Mode (SAM; Limpasuvan and Hartmann, 1999), amongst others. These modes have regional, and in some cases, broad-scale influence over many coupled ocean–atmosphere climate variables. Their influence can range from 3 to 7 years in the case of ENSO and 20–30 years in the case of the PDO to 65–70 years for the AMO. These phenomena manifest in fluctuations to broad-scale surface ocean temperature and rainfall patterns along with undocumented effects on ocean salinity. The climate variability "envelope," determined from observed secular amplitudes, needs to be considered explicitly when attempting to ascertain long-term changes. The much longer timescales of the ocean when compared to the atmosphere also need to be considered when assessing long-term change, and our understanding of long-term variability of the global ocean is significantly limited by the poor historical coverage of full-depth ocean observations. With a clear focus on atmospheric variables the World Meteorological Organisation defines mean climate over a 30-year averaging interval, an attempt to average out inherent variability, which provides a "baseline" over which changes can be computed. Changes determined from timeseries <30 years therefore can be problematic (such as those obtained from satellite precipitation estimates), as biases due to the effects of variability may skew the resolved changes, a relevant issue particularly when considering present ocean observational coverage. This long-term trend versus climate variability "envelope" is a key concept, and provides a framework through which to consider estimates of change.

Even though the limitation of poor observational coverage impedes analyses of salinity variability, a number of studies have attempted to investigate variability either globally or regionally. Global analyses using observed data and focusing on SSS have highlighted ocean regions with high-salinity variability (>0.3 PSS-78) throughout the annual cycle. These regions include the intertropical convergence zones (ITCZ) in the Pacific and Atlantic Oceans, the South Pacific convergence zone (SPCZ), and both the Arabian Sea and Bay of Bengal in the northern Indian Ocean, with the remainder of the ocean having a comparatively lower annual amplitude (<0.3 PSS-78; Boyer and Levitus, 2002; Delcroix et al., 2005; Bingham et al., 2012). An analysis which compared the annual variability of SSS between the World Ocean Atlas 2001 (Boyer et al., 2002) and the Argo Period (2004–2008) suggested that variability was largely in agreement between the different analyses; however, Argo expressed lower amplitudes during the period (Roemmich and Gilson, 2009). The effect of ENSO on tropical (30°S–30°N) ocean salinity was investigated by Delcroix et al. (2005) who found high variability (standard deviations of 0.4–1.4 PSS-78) was associated with the positive phase, with the negative phase (La Niña) leading to western Pacific Warm/Fresh Pool variability reducing and becoming more like the central Pacific. More recent studies using the well-sampled Argo period have also elucidated the influence of ENSO variability on ocean properties (2004–2011; Roemmich and Gilson, 2011).

A number of global studies investigating the covariability over an annual cycle of E–P and terrestrial runoff with SSS (Boyer and Levitus, 2002; Bingham et al., 2012), along with mixed-layer salinity (MLS; Yu, 2011) have also been undertaken. The dominant effect of P on the seasonal cycle was highlighted by Bingham et al. (2012). They noted that seasonal cycles were bimodal in their distribution with the Northern Hemisphere SSS amplitude peaking in March/April and Southern Hemisphere in September/October, approximately 1–2 months after the peak in E–P (dominated by the annual P maxima). The analysis of Yu (2011) suggested that salinity variability is governed by complex upper-ocean dynamics, which include E–P, Ekman and geostrophic advection, vertical entrainment and horizontal diffusion. Regions where E–P controls seasonal MLS were identified as the ITCZ (a P-dominant regime, therefore fresh), and the western north Pacific and Atlantic (an E-dominant regime, therefore salty) with 40–70% of MLS variance accounted for by E–P variability, and peak correlations occurring at a 2- to 4-month lead time. Outside of the tropics, Yu (2011) reported MLS variations are governed predominantly by Ekman advection, and then vertical entrainment. A similar analysis using a model-only approach was undertaken by Vinogradova and Ponte (2013); however, they concluded that E–P variability from MLS using linear models was only possible in subtropical gyres, and only on monthly to annual timescales.

A number of studies have also considered regional salinity variability, with a focus on the Pacific Ocean (e.g., Delcroix et al., 1996; Delcroix, 1998; Delcroix and Picaut, 1998; Gouriou and Delcroix, 2001; Maes et al., 2005; Delcroix et al., 2007; Bosc et al., 2009; Cravatte et al., 2009; Hasson et al., 2013). Many of these studies investigated the role of ENSO on salinity variability and highlighted the role of anomalous P during the positive ENSO phase driving fresh salinity anomalies along with meridional migration (extension) of the western Pacific warm/fresh pool. Delcroix et al. (1996) noted two dominant modes, with the ITCZ expressing a seasonally dominant mode and an ENSO mode dominating the central and SPCZ, with Gouriou and Delcroix (2001) suggesting this ENSO mode has twice the amplitude as the seasonal mode for the SPCZ region. Maes et al. (2005) suggested that vertical salinity stratification and barrier layers are important for preconditioning an ENSO event, with this feature affecting ocean dynamics, facilitating anomalous heat buildup in the western equatorial Pacific. The relationship between barrier layers (fresh anomalies) and high SSTs (>28 °C) was also highlighted by Bosc et al. (2009). They suggested such features are an effective proxy, marking regions of maximum atmospheric convection (P). This tropical region of maximum interannual variability was supported by the modeling study of Hasson et al. (2013), who found variability in the warm pool and both convergence zones to be driven by ENSO. This study suggested that all terms have to be considered to close the salinity budget on monthly to interannual timescales. A PDO-like influence on SSS was reported by Delcroix et al. (2007). They suggested the Pacific warm pool, the SPCZ, and the Equatorial Cold Tongue experienced positive anomalies from the mid-1970s to the 1990s and negative anomalies before and after this period. They indicated that the timing of the anomalies qualitatively agreed with E–P and circulation changes apparent during the coincident period.

Regional assessments over the historically well-sampled North Atlantic (e.g., Dickson et al., 1988; Belkin et al., 1998, Belkin, 2004; Curry and Mauritzen, 2005; Wadley and Bigg, 2006; Boyer et al., 2007; Holliday et al., 2008; Reverdin, 2010) and tropical Atlantic (e.g., Grodsky et al., 2006; Foltz and McPhaden, 2008; Grodsky et al., 2012; Tzortzi et al., 2013) have also been undertaken. Due to the temporal coverage of profile data for this region, long-term observations provide a better insight into the true variability envelope of the observed ocean. One of these assessments considered the 1895–2009 period, with such extensive temporal data coverage only available for the northeastern part of the North Atlantic subpolar gyre (Reverdin, 2010). In this assessment the response of both surface temperature and salinity to the regional NAO mode was considered. The study found large deviations over time in surface temperature and salinity properties over this record, with a peak to peak range of 2 °C and 0.2 (PSS-78) over the 114-year observed record for temperature and salinity, respectively. Such magnitudes of salinity (and temperature) variability have also been discussed in terms of the "Great Salinity Anomaly(ies)" (Dickson et al., 1988; Belkin et al., 1998; Belkin, 2004; Grodsky et al., 2006). Such anomalies have been suggested as a response to anomalous advective events in the North Atlantic (Belkin, 2004); however, there is some contention as to the source of these, with anomalous currents or surface fluxes suggested by modeling studies to be the source (Wadley and Bigg, 2006). Nonlocal teleconnections were suggested by Grodsky et al. (2006), who concluded that year-to-year changes in salinity are related to P; however, decadal changes, at least in part were attributed to low frequency changes in winds in the deep tropics and their role in altering upwelling (and possibly evaporation rates). Recent studies have focused on the tropical North Atlantic, and suggest a complex balance exists between the western and eastern tropical basin, with zonal and meridional salinity advection balancing a positive E–P forcing and leading to a weak seasonal cycle in the E-dominated north-central basin (Foltz and McPhaden, 2008). They suggested that the largest seasonal amplitude was apparent for 5–15°N, a region with the largest seasonal variations in P, and that a vertical entrainment term is necessary to balance the seasonal budget for the region. Using the latest available data from the SMOS satellite, Tzortzi et al. (2013) also suggest the tropical Atlantic is a region of high variability with an eastern and western dipole apparent. They suggest that very large seasonal ranges of 6.5 (PSS-78) occur in the region out of phase by 6 months, which essentially compensate each other over an annual cycle.

Regional salinity variability in the Indian Ocean has also been assessed in a number of studies (e.g., Han and McCreary, 2001; Perigaud et al., 2003; Rao and Sivakumar, 2003; Illig and Perigaud, 2007; Vinayachandran et al., 2013). An idealized study using a simplified dynamical model was undertaken by Han and McCreary (2001) to elucidate the effects of five separate forcing terms. They suggested that the E–P forcing alone allowed their model to qualitatively reproduce the observed SSS climatology, however with high biases evident through the entire basin. These high biases were rectified with the addition of river runoff into the Bay of Bengal and the Indonesian Through flow as forcing terms. To most effectively reproduce the three-dimensional structure, along with an increase to Arabian Sea salinity, fluxes from the Persian Gulf and Red Sea had to be included. The role of E–P and wind forcing on salinity was investigated by Perigaud et al. (2003) and Illig and Perigaud (2007). These studies suggested that regional salinity is maintained by complex dynamics which include E–P and wind forced advection, however they confirmed that SSS responds directly to P, and these coupled variations play an important role in the coupling of the ocean to the overlying atmosphere. Focusing on the north Indian Ocean, Rao and Sivakumar (2003) investigated the seasonal cycle of salinity in the Bay of Bengal and Arabian Sea. They showed the Bay of Bengal to be P dominated, both directly and through large runoff events with a clear and coherent freshening evident through the annual cycle. They also suggested that over the 1982–1995 period, coherent SSS variability in phase with ENSO was evident. Also focusing on the Bay of Bengal, Vinayachandran et al. (2013) investigated the role of higher salinity import water which balances the large freshwater inputs to the region. They suggest that large intraseasonal Ekman pumping favored upwelling events that are responsible for mixing saltier subsurface water sourced from the Arabian Sea and maintaining the higher salinity of the Bay.

Of the studies that considered three-dimensional salinity variability beneath the mixed layer, Forget and Wunsch (2007) concluded that temperature and salinity variability are usually strongly correlated except in the mixed layer (upper 100 m). Salinity contributions to dynamic height anomalies are found to follow those of the temperature in the tropics and subtropics, but these salinity anomalies become the leading source of variability in the high latitudes. The analysis of von Schuckmann et al. (2009) suggested that over the Argo period (2003–2008) the largest and deepest variability estimates were found in the North Atlantic. They also concluded that the spatial scales over which this variability manifests was larger in the Atlantic compared to the Pacific, for example. However, such a conclusion could be linked to the higher spatial (and temporal) sampling provided by the Argo program in the Atlantic when compared to the Pacific basins and hence, caution in interpreting these results is required.

It is likely that salinity variability is also linked to other described modes of large-scale climate variability, such as the AMO and the SAM, additional to the as yet undocumented modes of natural climate variability. Therefore when ascertaining long-term changes, it is preferable to consider the longest temporal period available (>30-years) which best accounts for observed modes of variability.

4. OBSERVED LONG-TERM CHANGES TO OCEAN SALINITY

Patterns of long-term changes to SSS are available, based on both trends fitted to ocean data (e.g., Freeland et al., 1997; Wong et al., 1999; Curry et al., 2003; Boyer et al., 2005; Delcroix et al., 2007; Gordon and Giulivi, 2008; Cravatte et al., 2009; Durack and Wijffels, 2010; Chen et al., 2012) and comparisons of an Argo era (2003–present) modern- to historical-ocean climatologies (e.g., Johnson and Lyman, 2007; Hosoda et al., 2009; Roemmich and Gilson, 2009; von Schuckmann et al., 2009; Helm et al., 2010). Although these studies suggest that long-term, broad-scale, and coherent changes have occurred over the observed record, it must be noted that because of the long timescales of ocean variability (Section 3) elucidating what is a forced change (in response to increasing temperatures) and what is longer timescale variability remain the focus of continuing research.

Of these independent studies, four (Boyer et al., 2005; Hosoda et al., 2009; Durack and Wijffels, 2010; Helm et al., 2010) have considered changes globally with the explicit objective of estimating long-term trends, with three providing near-surface changes for comparison. Each of these three studies used different temporal periods for their analysis. The objective averaging methodology of Boyer et al. (2005) generated pentads over 1955–1998, which excludes the dominant modern footprint of Argo data, and was fundamental in the other global studies (Hosoda et al., 2009; Durack and Wijffels, 2010). A methodology which compared two temporal climatologies was the approach of Hosoda et al. (2009), with a historical climatology centered around 1974 (pre-Argo) and another using just the modern Argo data (2003–2007), centered around 2005. The approach of Durack and Wijffels (2010) was to directly calculate the 1950–2008 changes, by using a combination of the historical hydrographic as well as the Argo data. The analysis also took care to attempt to minimize the aliasing associated with seasonal and ENSO variability. A disadvantage of this study was a spatially varying footprint because of the methodology having to increase its spatial search radius in order to sample across decadal "bins" from 1950–2010, and the assumption that long-term trends would be zonally homogenous across regions where large spatial footprints due to data sparsity were used. In comparison, the Hosoda et al. (2009) analysis has a fixed spatial footprint, but varied temporal "epochs," which leads to the resolved trend map representing differing time periods in response to the observational data sparsity. (rather than spatially varying as with Durack and Wijffels, 2010). Temporally averaged pentads are a feature of the Boyer et al. (2005) analysis, which ensures a sound temporal analysis. However, their analysis returns a zero estimate in regions of data sparsity and leads to patchy, noisy, and difficult to interpret spatial patterns. At the time of their analysis, the Argo platform was in its infancy (and their analysis omitted the Argo period, concluding in 1998). In the subsequent studies, Argo has provided a solid baseline for near complete seasonal coverage of the global ocean, which is particularly evident in the Southern Hemisphere, where sampling during the Austral winter was very sparse. Additionally, the analyses of Boyer et al. (2005) and Hosoda et al. (2009) did not attempt to resolve the mean spatial structure, or resolve and de-bias the data for seasonal coverage (particularly an issue in the Southern Ocean) and ENSO cycles, with noisier and less spatially coherent patterns the result. For these reasons, although spatially varying, it is considered that the Durack and Wijffels (2010) analysis provides a more robust measure of the absolute temporal trend.

4.1. Observed Surface Salinity Changes

The surface patterns of multidecadal salinity change from each of these analyses shows remarkable similarities with the mean salinity field and also the mean E–P field (Section 5; Figures 28.4 and 28.8). Rainfall (P) dominated regions such as the Western Pacific Warm (and fresh) Pool, for example, have undergone a long-term freshening and subtropical regions in the evaporation (E) dominated "desert latitudes" have generally increased in salinity in each basin (Figure 28.4). In the broad-scale, these changes suggest that the existing gradients of SSS have intensified, with increased values in the salinity maxima and decreased values in the salinity minima zones. This global tendency toward an enhanced SSS pattern is in broad-scale agreement with the regional studies of Cravatte et al. (2009), Curry et al. (2003), and numerous global analyses of SSS change (e.g., Boyer et al., 2005; Hosoda et al., 2009; Roemmich and Gilson, 2009; Durack and Wijffels, 2010). The SSS changes demonstrate that wet regions get fresher and dry regions saltier, following the expected response of an amplified water cycle (Allen and Ingram, 2002; Held and Soden, 2006).

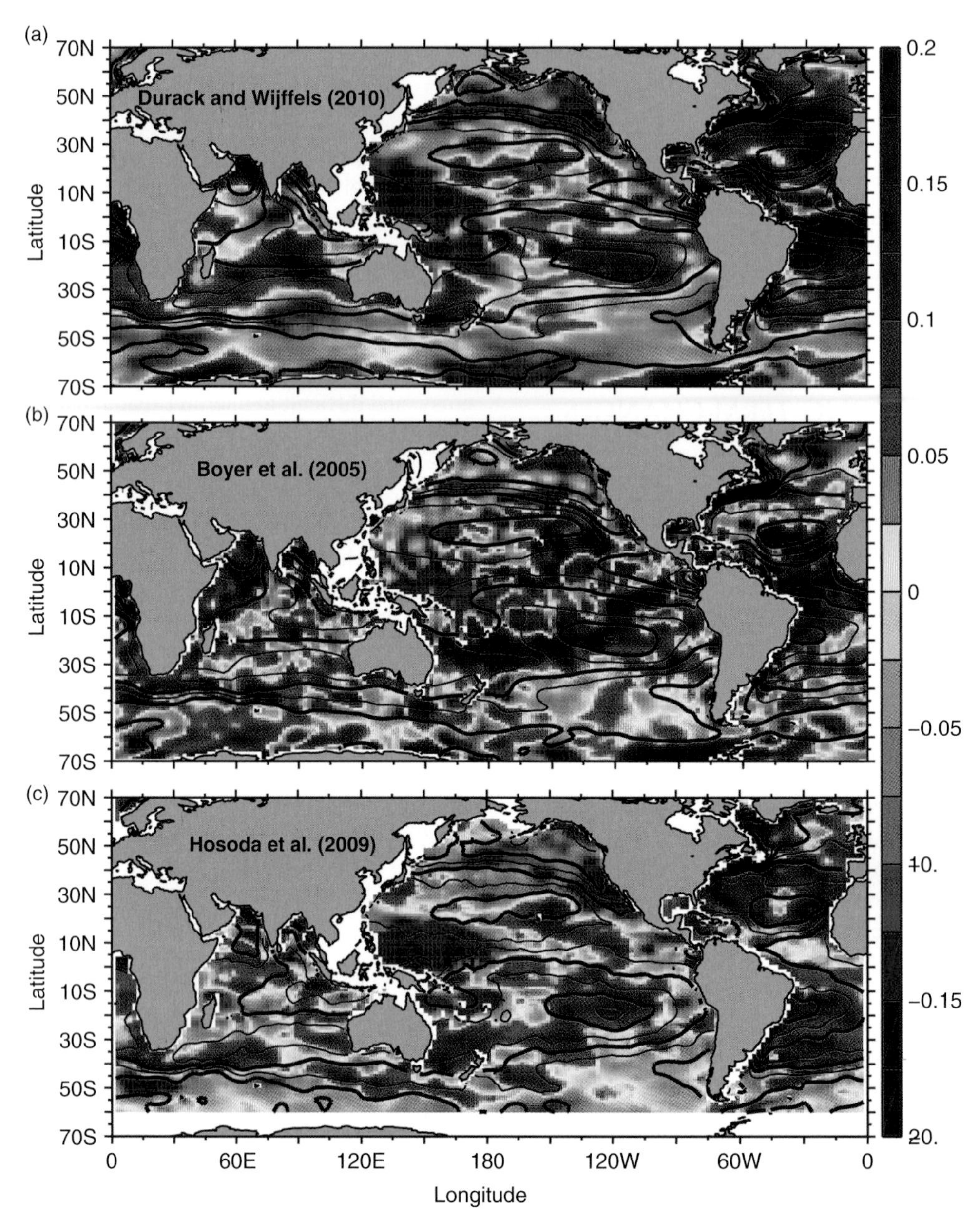

FIGURE 28.4 Three long-term estimates of global sea surface salinity (SSS) change after (a) Durack and Wijffels (2010; Analysis period 1950–2008), (b) Boyer et al. (2005; Analysis period 1955–1998), and (c) Hosoda et al. (2009; Analysis period 1975–2005) all scaled to represent equivalent magnitude changes over a 50-year period (PSS-78 50-year^{-1}). Black contours express the associated mean SSS for the analysis period. Broad-scale similarities exist between each independent estimate of long-term change, suggesting an increase in the spatial gradients of salinity have occurred over the observed record. However, regional scale differences are due to differing data sources and analysis methodologies.

4.2. Observed Zonal-Mean SSS Changes

A synthesis plot expressing the basin zonal-mean SSS changes for each basin and each of the three key global studies (Boyer et al., 2005; Hosoda et al., 2009; Durack and Wijffels, 2010) is presented in Figure 28.5. While providing a heavily smoothed synthesis of changes, the broad-scale zonal patterns for each basin are captured well by this analysis, and such a smoothing technique facilitates comparison to the coarse resolution global climate models presented later (Section 6.2).

In this synthesis, the enhanced salinity in the Atlantic Basin shows clearly the largest magnitude of change for any of the basins, and is present in each of the available analyses. The freshening of the Pacific Basin is most pronounced in the Northern Hemisphere, with the complex

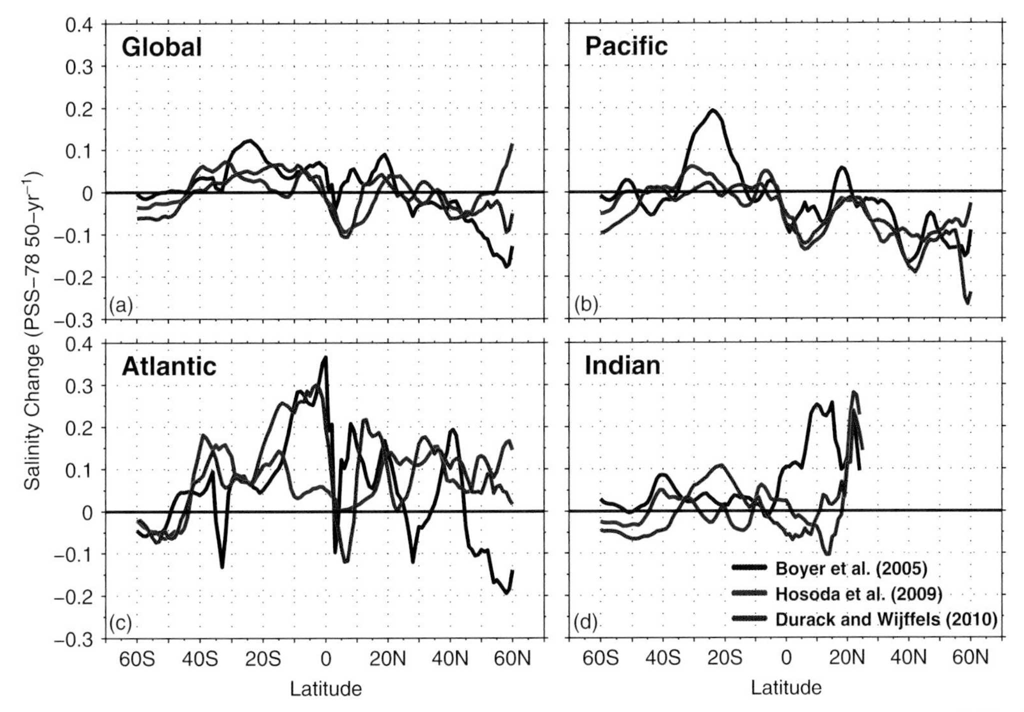

FIGURE 28.5 Three long-term estimates of global (a) and basin zonal-mean (b–d) SSS change. Included studies are Boyer et al. (2005; Black; Analysis period 1955–1998) and Hosoda et al. (2009; Blue; Analysis period 1975–2005) Durack and Wijffels (2010; Red; Analysis period 1950–2008). All analyses are scaled to represent equivalent magnitude changes over a 50-year period (PSS-78 50-year^{-1}).

counteracting patterns on each side of the southern basins from the west to the east (Figure 28.5) leading to a near zero integrated zonal-mean anomaly for most of the Southern Hemisphere. In the Indian basin, the strong regional response on the Arabian Sea dominates the Northern Hemisphere enhanced salinity patterns, and there is less agreement in the Southern Hemisphere because of data sparsity.

4.3. Observed Subsurface Salinity Changes

Patterns of long-term subsurface salinity changes on pressure surfaces also largely follow an enhancement of the existing mean pattern (Figure 28.6). The interbasin contrast between the Atlantic (salty) and Pacific (fresh) intensifies over the observed record (Figure 28.6; e.g., Boyer et al., 2005; Roemmich and Gilson, 2009; von Schuckmann et al., 2009; Durack and Wijffels, 2010). The penetration of salinity anomalies into the upper 1000 m suggest that persistent surface changes from past decades have propagated into the ocean interior, with a clear enhancement to the high-salinity subtropical waters, and freshening of the high-latitude waters. The salty subtropical gyre bowls in each basin, particularly those of the Southern Hemisphere Atlantic and Indian Oceans, show positive anomalies in the salinity maxima in depths shallower than 300 m. The patterns found in the Northern Hemisphere are less homogenous across basins. The North Pacific shows a strong high-latitude freshening in the upper 100 m and muted patterns of deeper changes. A strong salinity increase is found in the North Atlantic subtropical gyre bowl, with this feature extending to 500 m depth. A suggestion of weak high-latitude freshening is also found, consistent in sign but weaker than those found by Curry and Mauritzen (2005) and Boyer et al. (2007). Such regional changes have been further investigated for strong multidecadal oscillations, from a 30-year freshening trend (1960s to 1990s) to enhanced salinities in the more modern period (e.g., Holliday et al., 2008; Reverdin, 2010). In the North Indian Ocean surface and subsurface salinity increase. A shallow anomaly maximum (~50 m depth) is found in both the Arabian Sea and the Bay of Bengal, whereas a deeper maximum is found primarily in the Arabian Sea, and is likely linked to high-salinity outflows from the Red Sea and Persian Gulf (Durack and Wijffels, 2010). A particularly strong and coherent freshening expressed in the Antarctic Intermediate Water subduction pathway centered around 50°S has also been detected (Johnson and Orsi, 1997; Wong et al., 1999; Bindoff and McDougall, 2000; Antonov et al., 2002; Curry et al., 2003; Boyer et al., 2005; Roemmich and Gilson, 2009; Durack and Wijffels, 2010; Helm et al., 2010) and is visible in each of the separately sampled ocean basins.

Studies have also reported long-term and coherent salinity changes on subsurface density horizons (e.g., Wong

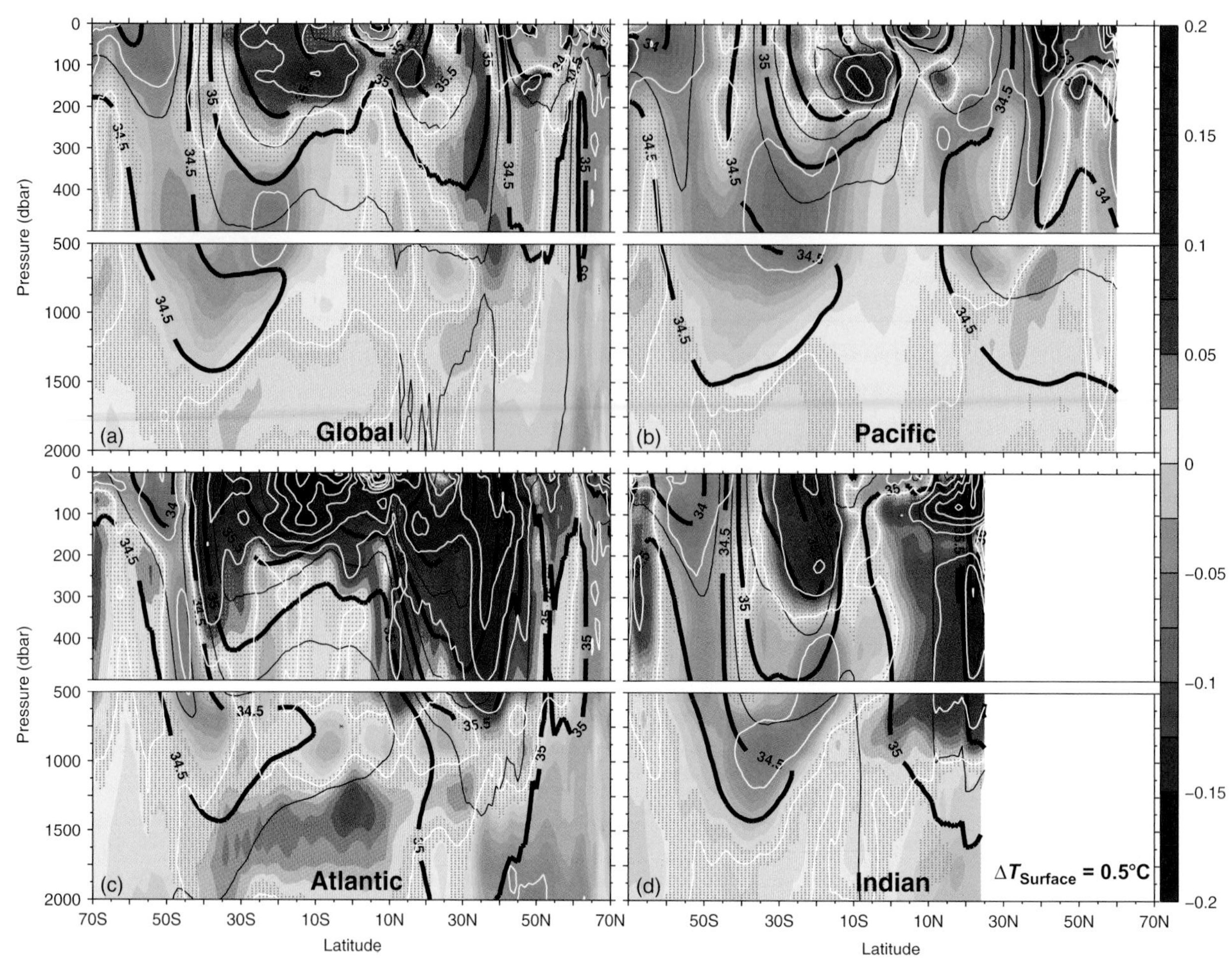

FIGURE 28.6 Observed subsurface salinity changes for the global zonal mean (a) and each of the global basins (b–d) (Durack and Wijffels, 2010; PSS-78 50-year^{-1}).

et al., 1999; Curry et al., 2003; Helm et al., 2010). In a density framework, Durack and Wijffels (2010) show that many salinity changes are dominated by subduction into the deep ocean driven by a broad-scale surface warming. A temperature-driven lateral shift of an isopycnal outcrop (generally poleward) can lead to a migration through ocean surface climate zones (e.g., from a low salinity, P-dominated region to a high-salinity region of E-dominance) and result in a subsurface salinity anomaly unrelated to surface E–P changes. Such changes are particularly relevant in regions which have large SSS gradients. According to these results, subsurface salinity changes are less useful in directly reflecting E–P changes as suggested by other studies (e.g., Helm et al., 2010), as warming plays a large role. As discussed later (Section 5) this is an active area of research and further quantitative assessments using idealized model simulations are required to understand and attribute ocean surface and subsurface property changes to their respective causes.

4.4. Quantifying Rates of SSS Change

To quantify salinity changes over the observed record Durack et al. (2012) developed a technique to quantify and compare the strength of broad-scale salinity pattern amplification (PA) between observed estimates and models. To achieve this, they developed zonal ocean basin (Pacific, Atlantic, and Indian) averages for both the 50-year (1950–2000) climatological mean SSS and its corresponding change (Figure 28.4a). A linear regression was then undertaken using the basin zonal anomalies from the climatological global SSS mean (x-axis) against the corresponding basin zonal change pattern (y-axis). The resulting scatterplot then expresses regions with respect to the climatology that are fresher than the global mean (negative x values) and saltier than the global mean (positive x values), and their corresponding regional change either freshening (negative y values) or expressing a salinity enhancement (positive y values). The resulting slope of this relationship

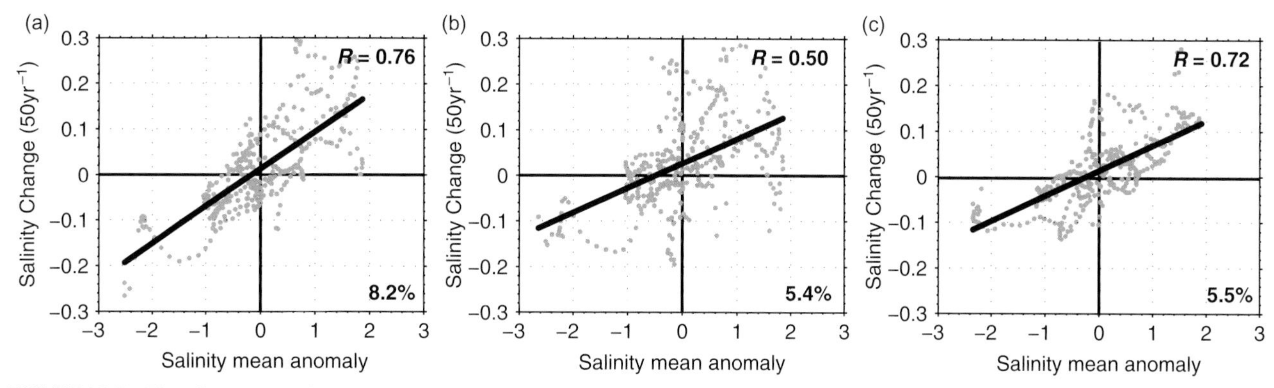

FIGURE 28.7 Three long-term estimates of basin zonal SSS pattern amplification (PA, %) and pattern correlation (PC, R) 1950–2000 after (a) Durack and Wijffels (2010), (b) Boyer et al. (2005), and (c) Hosoda et al. (2009) all scaled to represent equivalent magnitude changes over a 50-year period (% PA 50-year^{-1}). PA and PC values for 28.7a were generated from the spatially smoothed map presented in Figure 28.4a and consequently slightly differ from values published in Durack et al. (2012). *Adapted from Durack et al. (2012).*

was defined as the PA and the corresponding correlation coefficient (R) as the pattern correlation (PC). Using this technique, the results from each of the three observed global near-surface estimates (Boyer et al., 2005; Hosoda et al., 2009; Durack and Wijffels, 2010), scaled to represent 50-year changes are presented in Figure 28.7.

An enhancement to the spatial surface salinity gradients, a PA (% values) is apparent in each of the independent analyses, with fresh regions becoming fresher and salty regions saltier. This feature dominates each of the results, with the bulk of data points residing in the upper right (salty getting saltier) or lower left quadrants (fresh getting fresher). The pattern correlations (PC; R values) also suggest a strong spatial coherence, and are stronger in the two analyses which benefit from the modern data coverage provided by Argo (Durack and Wijffels (2010); Figure 28.7a and Hosoda et al. (2009); Figure 28.7c).

While providing the largest estimate of basin zonal-mean SSS change from the three independent estimates (Figure 28.7), the result of Durack and Wijffels (2010; Figure 28.7a) does not repeatedly return the largest SSS PA values. Their result yields the median global zonal-mean PA of 5.2% (PC = 0.69) when compared to 6.6% (PC = 0.71) for Boyer et al. (2005) and 2.8% (PC = 0.41) for Hosoda et al. (2009). A global zonal-mean analysis is a less rigorous test, which averages out the strong zonal salinity gradients that occur in each basin at differing latitudes, heavily smoothing the analysis. The Southern Hemisphere zonal (latitudinal) salinity maximum in the Indian Basin is found at 30°S, whereas this maximum is located at 20°S in the Pacific and Atlantic Basins (Figure 28.8a and b). The basin zonal-mean analysis depends upon these independent basin features and its relationship to the basin pattern of change to obtain the PA estimate. It was clear from the analysis of Durack et al. (2012) that basin or global zonal smoothing enhances the PC values in both models and observations.

In summary, several recent studies employing different analysis techniques find a clear multidecadal ocean SSS change. Broad-scale changes can be characterized as an amplification of the climatological salinity pattern, a tendency also found in the subsurface. The consensus view of coherent salinity change arises, even though many different analysis techniques and ocean salinity observing platforms have been used—reflecting the robustness of the signal. To first order this suggests that broad zonal changes to E–P have changed SSS, and changes are propagating into the subsurface ocean following the mean circulation pathways. As noted by Durack and Wijffels (2010) the effect of a concurrent warming of the ocean cannot be discounted as a driver of subsurface salinity changes analyzed on density surfaces. However, irrespective of their cause, the enhancement to mean salinity patterns and basin contrasts are expressed in the observational record.

5. OCEAN SALINITY—RELATIONSHIP TO THE GLOBAL WATER CYCLE

Although oceanographers are interested in properties of the global ocean and their variability and changes over time, the most challenging responses to a changing climate are those of alterations to terrestrial water availability and extreme events—particularly modifications to regional rainfall patterns and the associated changes to regional water availability.

5.1. Linking Evaporation and Precipitation Fluxes to Salinity

Broad-scale patterns of climatological mean ocean salinity are set by the local evaporation (E) minus precipitation (P; E–P) fluxes, with regions dominated by P in the annual mean expressing fresher surface salinities than

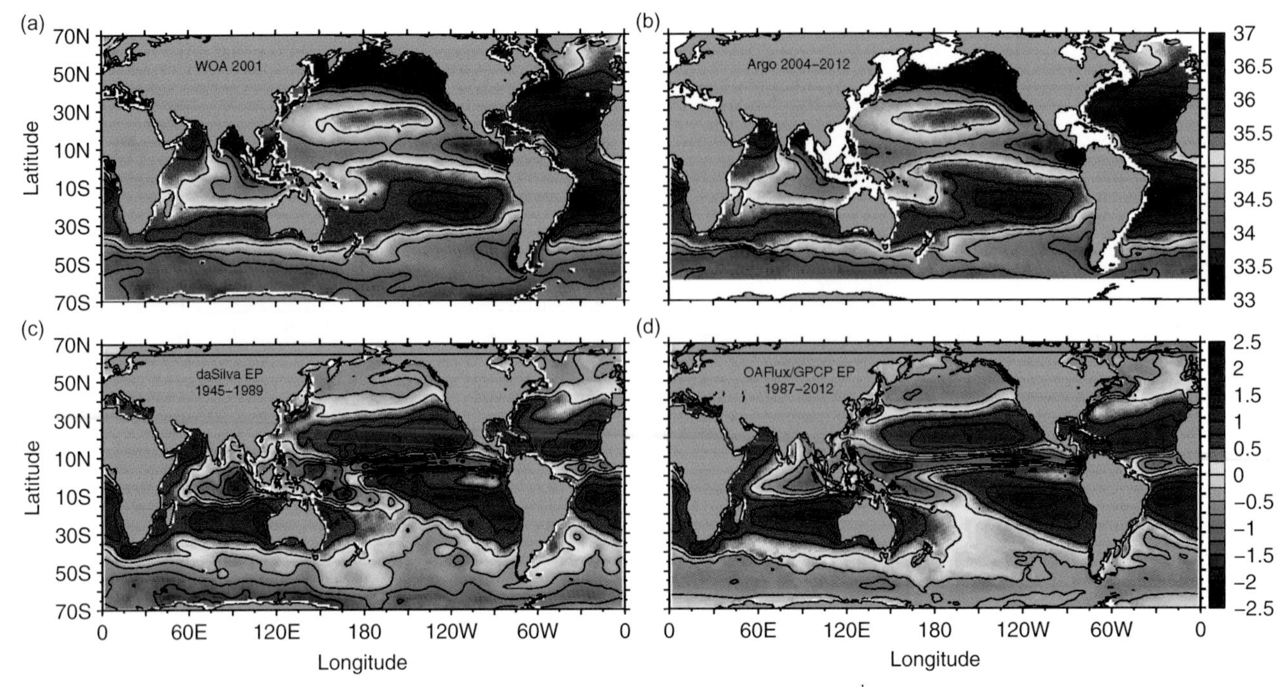

FIGURE 28.8 (c and a) Presatellite annual mean estimates of ocean surface E–P (1945–1989; m yr^{-1}; da Silva et al., 1994) and pre-Argo (SSS; World Ocean Atlas 2001; PSS-78; Boyer et al., 2002), respectively. (d and b) Satellite annual mean estimates of ocean surface E–P (1987–2012; m yr^{-1}; OAFlux v3 and GPCP v2.2) and the Argo-period SSS (2004–2012; PSS-78; Roemmich and Gilson, 2009). Black contours express the associated mean SSS or E–P for the analysis period.

the global mean (35 PSS-78), and regions dominated by E expressing saltier values (Figure 28.8; Schmitt, 2008). As detailed by Yu (2011) in an assessment of the drivers of climatological mean seasonal SSS, the local salinity is controlled by complex upper-ocean dynamics; local E–P, Ekman and geostrophic advection, vertical entrainment and horizontal diffusion. These dynamic effects ensure that there is a spatial offset with regards to the surface regions of high/low salinity, and the corresponding source regions of E or P maxima, which set the local salinity pattern. The E–P field sets the broad-scale patterns of salinity and ocean circulation (and mixing) then acts to distribute these fresh and salty anomalies throughout the global ocean in balancing oceanic freshwater transports (e.g., Ganachaud and Wunsch, 2003; Talley, 2008; also see Chapters 19 and 29) that mirror atmospheric water transports in the long-term mean. However, in the broad-scale this pattern-lock between the spatial patterns of E–P and SSS is telling. This relationship provides oceanographers with an independent tool to investigate changes to the poorly observed E–P field (Chapter 5) and water cycle changes corresponding to the period of observational salinity coverage over the long-term.

The earliest estimates of observational climatological annual mean E–P values (Figure 28.8c) provided an accurate spatial pattern and amplitude assessment, when compared to the most recent, spatially complete merged satellite estimates (Figure 28.8d). Schanze et al. (2010) found that the ocean freshwater flux (E–P–R) is in balance within measurable uncertainties when E (OAFlux) and P (GPCP) from recent satellite E–P estimates are used for the 1987–2006 period (Figure 28.8d; OAFlux/GPCP E-P presented for the 1987-2012 period). However, they found that the oceanic freshwater budget is far from closed when considering current generation reanalysis products (e.g., ERA-40, MERRA, NCEP-1, and NCEP-2).

Similarly, long-term climatological annual mean SSS has also been accurately estimated from the sparse hydrographic observations, which contributed to the global hydrographic database pre-Argo (Figure 28.8a). These feature very similar spatial distributions and a hint of larger gradients (fresher salinity minima, and saltier salinity maxima) in the Argo-only annual mean climatology for the modern period (2004–2012; Figure 28.8b; Roemmich and Gilson, 2009); however, the effect of the changing observing system must also be considered when assessing such differences between climatologies.

Clearly there is a strong link between the E–P patterns and SSS in the climatological mean. However, a more difficult question is how variability and change in these fields are related, particularly over the long-term. Due to the sparsity of observational data, both for E–P and ocean salinity, the only method currently available to investigate such long-term relationships are those undertaken through climate model simulations.

5.2. Idealized Ocean Responses to E–P Forcing Experiments

To investigate long-term salinity changes, idealized ocean-only simulations were undertaken by Durack et al. (2012). This study concentrated on E–P perturbations (5%, 10%, 15%, and 20% trends over the 50-year simulations; so linear trends which result in the target % magnitude being achieved in the final year of simulation) over time and assessed the response of the ocean salinity field to this forcing. To simplify the analysis, the effects of atmospheric, land-surface, or ice (cryospheric) changes were not considered. The idealized E–P forcing experiments were motivated by the results of Allen and Ingram (2002) and Held and Soden (2006), which concluded that in future coupled climate model simulations using the Coupled Model Intercomparison Project phase 2 and 3 (CMIP2/3) model suites, broad-scale patterns of wet regions becoming wetter and dry regions becoming drier were apparent, with an amplification of the mean patterns of E–P driving these regional changes.

To minimize interannual variability, a single year of daily surface ocean forcing was obtained from the NCEP/NCAR reanalysis (1948; Kalnay et al., 1996). The results presented from this study for the 10% E–P case (Figures 28.9 and 28.10) tend to qualitatively agree with observational estimates of long-term broad-scale salinity changes in the global ocean (Section 4). The idealized E–P change of 10% over a 50-year period approximately aligns with a global surface warming of ~1.5 °C, following the Clausius–Clapeyron relation, which suggests a ~7% °C^{-1} response at temperatures typical of the lower troposphere (Held and Soden, 2006). As expressed in Figure 28.9a, the imposed idealized forcing leads to an increase in the spatial gradients of E–P, with more intense regions of E and P being the result.

Considering the surface changes first (Figure 28.9b) it is clear that, particularly in the Southern Hemisphere, enhanced salinities are found in the subtropical regions dominated by increasing E (Figure 28.9a), with E being the dominant term in the annual mean E–P budget. A broad-scale freshening pattern is found in high latitudes and in the equatorial and Northern Pacific basin. The salinity contrasts between a broad-scale freshening Pacific and a saltier Atlantic are also clear, agreeing well with observed patterns of global SSS changes (Boyer et al., 2005; Hosoda et al., 2009; Durack and Wijffels, 2010). These simulations suggest that the spatial gradients of SSS are intensifying, resulting in saltier salinity maxima and fresher minima in response to amplified E–P. This result is hinted at by comparisons of the modern (Argo) versus pre-Argo climatologies (e.g., WOA01; Figure 28.8a vs. b; Roemmich and Gilson, 2009).

When subsurface changes are assessed (Figure 28.10), this same pattern of mean climatological salinity gradient intensification is evident, with the salty subtropical gyre bowls increasing their salinities and high latitude and tropical salinity minima becoming fresher. These results again qualitatively reproduce those patterns of change found in the observed record (Section 4; Boyer et al., 2005; Durack and Wijffels, 2010).

A remarkable feature in both the observed and simulated analyses is the apparent spatial pattern lock in each of the separate ocean basins. Durack et al. (2012) undertook an additional simulation that included a further forcing term, which simulates the observed warming (additional to the imposed E–P change) over the 1950–2000 period. This subsequent analysis provided results which simulated the

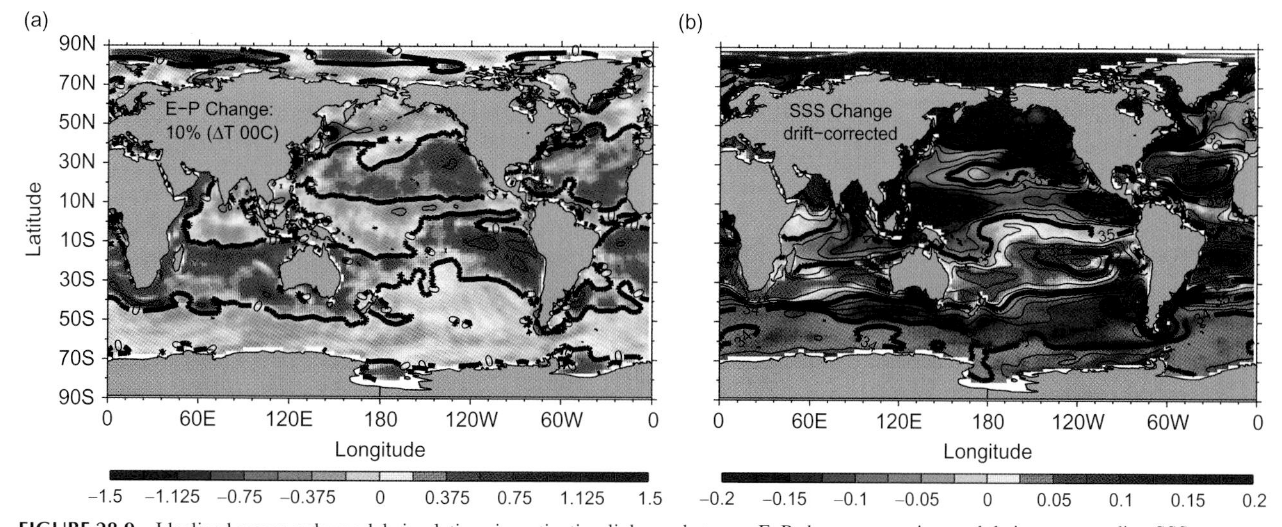

FIGURE 28.9 Idealized ocean-only model simulations investigating linkages between E–P changes over time and their corresponding SSS response. (a) A 50-year trend in ocean surface E–P as obtained from year 1948 of the NCEP R1 reanalysis (Kalnay et al., 1996; m 50-year^{-1}) and (b) the 50-year trend response for SSS as simulated by the MOM3 (Pacanowski and Griffies, 1999) ocean-only model (Durack et al., 2012; PSS-78 50-year^{-1}).

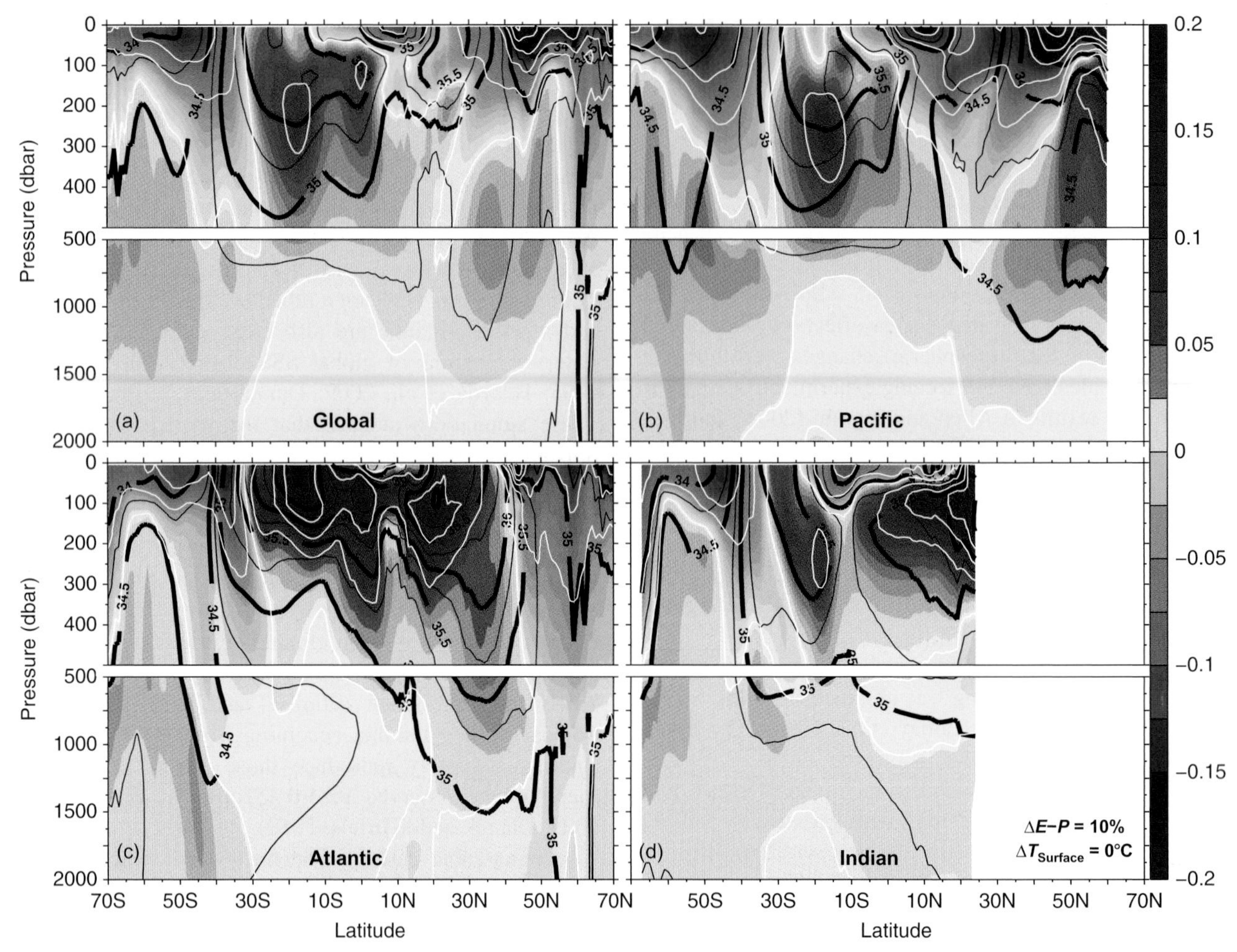

FIGURE 28.10 Idealized subsurface salinity changes for the global zonal-mean (a) and each of the global basins (b–d) as simulated by the MOM3 (Pacanowski and Griffies, 1999) ocean-only model with a 10% E–P forcing over the 50-year simulation (Durack et al., 2012; PSS-78 50-year^{-1}).

spatial patterns and distribution of subsurface salinity anomalies more closely than the idealized simulations which only utilized E–P forcing (Figures 28.9 and 28.10). However, it must be noted that these simulations are idealized in their construction, and only qualitative comparisons can be made as a large salinity drift was associated with these simulations because of no SSS damping or relaxation being applied during the simulations (Section 6; Chapter 20).

The relationship presented between changing E–P and resulting changes to ocean salinity in these idealized simulations suggests that ocean changes provide a quantitative insight into global water cycle changes. However, further quantitative assessments of surface and subsurface salinity changes are needed to improve our understanding of their relationship to long-term E–P changes. An improved understanding will determine how we can use long-term salinity change estimates to best quantify the coincident long-term changes to the global water cycle.

6. MODELING OCEAN SALINITY VARIABILITY AND CHANGE

Until recently ocean salinity has been infrequently evaluated in state of the art coupled general circulation models (see Chapter 23), primarily owing to the lack of quantitative and coherent estimates of long-term salinity changes by which to compare model fields. An additional issue is the apparent inability for models from the Coupled Model Intercomparison Project phase 3 (CMIP3; Meehl et al., 2007) and phase 5 (CMIP5; Taylor et al., 2012) to accurately reproduce the spatial patterns of climatological mean salinity, both at the ocean surface (e.g., Durack et al., 2012) and through depth (e.g., Flato et al., 2013). It is highly likely that well documented atmospheric circulation biases in the CMIP3 models (Lin, 2007; Belluci et al., 2010) and associated biases in the atmosphere–ocean surface freshwater fluxes, drive some of the errors in the modeled ocean salinity pattern. An additional source of difficulty in modeling realistic salinity is the lack of a strong atmosphere–ocean feedback between E–P and salinity. A negative feedback provides a self-limiting constraint for sea-surface temperature that is apparent in both coupled and ocean-only simulations (Haney, 1971; Griffies et al., 2009), and the absence of a feedback allows small unchecked errors in the E–P fluxes to integrate into large drifts from the observed state.

6.1. Considerations when Analyzing Modeled Salinity

A detailed ocean-only model intercomparison is provided by Griffies et al. (2009), which considered seven independent ocean models and their responses to idealized forcing from the Coordinated Ocean-ice Reference Experiments (COREs) atmospheric dataset (Large and Yeager, 2009). We shall review and expand upon their discussion with respect to salinity analyses in ocean-only and coupled climate modeling systems.

Particular issues that need to be considered when assessing the temporal change fields from CMIP3/5 ocean models are briefly discussed. First, model drift (e.g., Rahmstorf, 1995a; Covey et al., 2006; Sen Gupta et al., 2012; Sen Gupta et al., 2013) manifests when the long-term model salinity (and temperature) fields drifts away from the initialized climatological observed values, generally with longer term and larger drift magnitudes in the deeper ocean. There appears to be improvements between CMIP3 and CMIP5 modeled ocean fields, with a reduction in drift in the newer simulations (Sen Gupta et al., 2013); however, pronounced drifts in the three-dimensional salinity field continue to be an issue in the latest CMIP5 models (e.g., Voldoire et al., 2012). Model drift can locally either magnify or suppress spatial patterns of long-term forced change, and consequently needs to be considered when comparing model responses to forcing experiments with observed changes. At the surface, drift is unlikely to strongly influence modeled trends in salinity that are due to anthropogenic forcing (e.g., Durack et al., 2012), whereas at depth the combined effect of strong salinity and temperature drifts can reverse the sign of depth-integrated global mean steric sea-level rise estimates (Sen Gupta et al., 2012). Possible causes are linked with spurious local E–P fluxes from the atmosphere that integrate over time, the realism of the modeled discretization of the global ocean (not fully-resolving features which play an important role in maintaining regional balances, that is, boundary currents and mesoscale eddies) and deficiencies in the parameterisations which are required to represent the finer-scale processes, which occur at resolutions not explicitly resolved by the model grid.

In atmosphere–ocean coupled models, drift is generally less when compared to ocean-only simulations that do not use unphysical restoring fluxes. Furthermore, thanks to ongoing model improvement, unphysical freshwater flux corrections, which have been used to minimize drift, were largely absent in the CMIP3 model suite when compared to preceding versions (CMIP2+; e.g., Covey et al., 2006). However, of the 23 models contributing ocean data in CMIP3, four used freshwater and heat flux corrections to maintain a realistic base climatology, and unsurprisingly when comparing the annual mean climatological SSS, these models were closest to observed estimates (Durack et al., 2012). The drift reduction in fully coupled models can be linked to the feedback between evolving SSTs and the interactive atmospheric state, which provides a nontrivial space-time dependent damping, and is better resolved in more modern modeling systems. This SST damping acts to reduce model drift as evaporation and its associated turbulent latent heat flux link the thermal and hydrological fluxes (Griffies et al., 2009). Since the surface heat flux/SST feedback does not damp the SSS anomalies but generally damps existing SST anomalies, SSS anomalies have a longer characteristic timescale. As a result they are more influenced by the mean currents and the geostrophic variability, which dominate SSS changes at low frequency as shown in a study by Mignot and Frankignoul (2003) for the Atlantic Ocean.

There is no significant direct feedback between SSS and E–P (aside from a negligible lowering of adjacent atmospheric vapor pressure). As a consequence SSS is fundamentally distinct when compared to SST. This absence of a direct local feedback is the key reason why the simulation of salinity is difficult in both coupled climate and ocean-only modeling systems, and deviations from the observed mean state are apparent (e.g., Durack et al., 2012). In a modeling study, Rivin and Tziperman (1997) found that the local E–P flux strongly depends on the SST perturbation amplitude, and in the case of midlatitudes, that evaporation and both large-scale and local convective rainfall depend linearly on SST perturbation. They found that the sensitivity of evaporation to SST (dE/dT) is around three times greater than the sensitivity of total rainfall. This implies that most of the evaporation due to the local SST anomaly is not precipitated locally, and instead is carried by the atmospheric circulation to convergence (precipitation) zones. This is a key reason why coupled models tend to experience less surface drift than their ocean-only counterparts, the interaction between the ocean and atmospheric state minimizes and stabilizes the spurious fluxes which are imposed on ocean-only simulations lacking this feedback. They noted however that tropical behaviors are strongly nonlinear, linked to the dominance of the highly variable large-scale and convective rainfall term, whereas midlatitudes are dominated by the more spatially homogenous evaporative term. Getting this rainfall term correct along with the other freshwater flux sources of evaporation, runoff, and ice melt (along with wind and ocean circulation) in both coupled and ocean-only models would provide an improvement to salinity simulation when compared to observations. Over much longer timescales an indirect feedback driven by salinity anomalies could occur. Large positive (negative) salinity anomalies at high latitudes in the Atlantic, for example, could potentially reinforce (dampen) the thermohaline overturning circulation. In response to enhanced (decreased) overturning a global change in sea-surface temperatures could result, and the resulting changes to evaporation would in turn affect salinity (e.g., Stommel, 1961; Rooth, 1982; Bryan, 1986; Manabe and Stouffer, 1988; Rahmstorf, 1995b; Rahmstorf, 1996; Rahmstorf, 2000).

To suppress drift in ocean-only model simulations, in practice a relaxation or damping term (freshwater flux correction) is applied to SSS. Modern model configurations ensure that SST damping occurs in a more physically realistic way, through heat fluxes specified by bulk formulae. In idealized simulations or simulations using prescribed atmospheric fluxes, direct heat flux corrections may be applied to resolve the issue of a noninteractive atmosphere. This damping term then nudges fields towards prescribed climatological values to maintain model realism when compared to observations (e.g., Griffies et al., 2009). In practice, if the SSS is larger than the salinity from the restoring (observed) data, there is a positive net input of freshwater into the ocean, thus damping the SSS back toward the data. The strength of this restoration forcing is highly dependent on the ocean model configuration, and is generally undertaken on a monthly time step or similar (Griffies et al., 2009). There is no *a priori* reason that this freshwater flux will integrate to zero over the globe. Hence, its use will generally lead to a drift in ocean volume (in modern free-surface models), unless a global adjustment is made at each ocean model time step (Griffies, 2004; Griffies et al., 2009). However, such corrections are unphysical and can lead to issues in the simulation, particularly with salinity, and as a consequence in the modeled thermohaline circulation (Large et al., 1997; Killworth et al., 2000; see Chapter 20). Thermohaline damping is typically associated with rather short damping time scales (i.e., monthly or more frequently), which additionally can suppress potentially interesting internal modes of variability such as mesoscale eddies (Griffies et al., 2009).

With respect to atmospheric circulation (E–P), ocean salinity is a passive tracer which responds to the surface E–P fluxes, whereas in the ocean it is highly dynamically active (particularly in the cold high latitudes) due to its influence on density through the haline contraction coefficient. High-latitude, cold-water ocean densities are weakly affected by temperature, with salinity being the dominant driver of density changes. As a consequence, modeled salinity errors in these high-latitude regions strongly impact the realism of the model circulation. In coupled modeling systems accumulated deficiencies in E–P over land impact river runoff, and local salinity as a consequence.

An additional source of drift can be attributed to inconsistencies with the observational climatological mean fields from which the models are initialized, with models attempting to reach a dynamical equilibrium and during this process "drift" from the imposed unbalanced climatological mean state. Such issues are particularly prevalent when considering the relatively short ocean reanalysis or decadal experiments which assimilate or are initialised by estimated ocean states (e.g., Falkovich et al., 2005; Pohlmann et al., 2009; Zhang and Rosati, 2010).

A further consideration when analyzing salinity fields from ocean models is the use of a virtual salt flux (Huang, 1993), rather than a freshwater flux when coupling an ocean, atmosphere, and land-surface model together. Around a third of current generation CMIP5 models presently use this scheme rather than the more physical freshwater flux. This assumption was introduced as a result of the rigid lid approximation, which suppressed fast external oceanic gravity waves to allow longer time steps but led to a fixed volume for a modeled ocean (Bryan, 1969). Under the rigid-lid approximation, the freshening and salinification effects of E–P are parameterized as a salt extraction or input, and the oceanic water cycle is felt by the ocean model as a spurious salt-, rather than a water-cycle (Yin et al., 2010a). As a consequence, a model using the virtual salt flux assumption excludes the Goldsborough–Stommel circulation induced by the large-scale E–P fluxes (Huang and Schmitt, 1993). Most CMIP3/5 models impose freshwater fluxes, rather than the unphysical flux of salt to simulate either a freshwater loss (add salt) or freshwater gain (lose salt). However, an investigation into the sensitivities of this unphysical virtual salt flux assumption, compared to a more modern free-surface model which imposed real freshwater fluxes was undertaken by Yin et al. (2010a). They concluded that although unphysical, the model configured to use the virtual salt flux assumption did not respond differently when considering current and near-term climate projections. They cautioned however, that for strong freshwater hosing perturbation experiments, the use of the virtual salt flux assumption would likely lead to spurious results and thus should be avoided if practical.

Two final considerations when comparing observed and modeled salinities are the tendency for modelers to equate temperature and salinity in the upper model grid cell with the sea-surface temperature and salinity, respectively. This equality is not precise as the model grid values represent a grid volume average and do not precisely reflect observed surface skin values as measured from satellites, for example (see Robinson, 2006 for a more detailed discussion). The determination of the mixed-layer depth, particularly in the high latitudes can also be sensitive to the diagnostic method used by the model, whether this is determined using temperature alone or density (and therefore sensitive to salinity, particularly in the high-latitude, colder oceans) threshold (Griffies et al., 2009; Griffies et al., in preparation) along with the model vertical resolution. In most models, the upper level vertical resolution is of order 5 m and consequently the presence of barrier layers (Lukas and Lindstrom, 1991; Sprintall and Tomczak, 1992) because of salinity stratification are often neglected (Griffies et al., 2009). Additionally, while the implementation of TEOS-10 and the new salinity variables (IOC, SCOR, and IAPSO, 2010; also see Chapter 6) will allow for the compositional information (Absolute Salinity and Reference Salinity; Millero et al., 2008) to be accounted for in observational assessments, models do not currently resolve such compositional differences.

6.2. Modeled Water Cycle Changes Assessed from Ocean Salinity

Regardless of the apparent issues in modeling ocean salinity (Section 6.1), a number of analyses have used the output of CMIP3/5 ocean models to assess changes to salinity due to anthropogenic influence (Stott et al., 2008; Helm et al., 2010; Durack et al., 2012; Pierce et al., 2012; Terray et al., 2012). These analyses have used historical and future coupled climate simulations, which incorporate anthropogenic greenhouse gas and aerosol forcing, and compared the response of the modeled ocean salinity changes to observed regional (Stott et al., 2008; Terray et al., 2012) or global change patterns (Helm et al., 2010; Durack et al., 2012; Pierce et al., 2012). These studies found that the changes captured by the CMIP3/5 model suite simulations qualitatively agree with the broad-scale patterns of change expressed in observations. In the case of Stott et al. (2008), Pierce et al. (2012), and Terray et al. (2012), using a detection and attribution approach (Hasselmann, 1979), they found that the patterns of salinity change were attributable to anthropogenic influence, and were not present in model simulations which excluded anthropogenic greenhouse gas forcing.

6.2.1. Modeled Water Cycle Changes Assessed from Surface Ocean Salinity

The analysis of Durack et al. (2012) used a different approach to compare SSS changes in the CMIP3 simulations and observations. In this analysis, the link between E–P changes and SSS changes was investigated using both idealized model simulations (Section 5.2) and the results from the CMIP3 model suite. The analysis focused on relative changes per degree of global surface temperature change, which is the natural framework driving thermodynamic water cycle amplification in the atmospheric response (e.g., Allen and Ingram, 2002; Held and Soden, 2006; Durack et al., 2012). This study concluded that there is a strong relationship in the long-term SSS PA (Section 4), and that the most representative magnitudes of PA when compared to observations were found in CMIP3 simulations of future twenty-first century climate. A summary of these CMIP3 results are expressed in Figure 28.11. The model results presented are those of the model

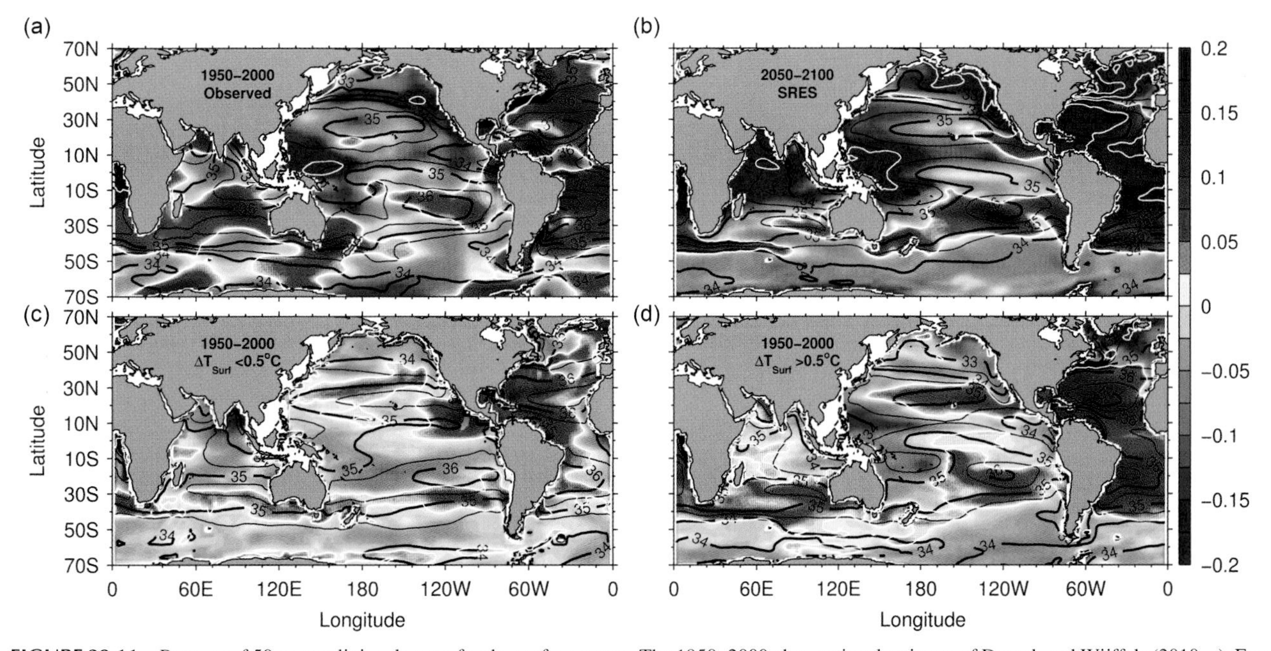

FIGURE 28.11 Patterns of 50-year salinity changes for the surface ocean. The 1950–2000 observational estimate of Durack and Wijffels (2010; a). For an ensemble mean from 2050 to 2100 for the CMIP3 SRES scenarios (b). For an ensemble mean from 1950 to 2000 for the CMIP3 twentieth century simulations that warm <0.5 °C (24 simulations; c) and for an ensemble mean from 1950 to 2000 for the CMIP3 twentieth century simulations that warm >0.5 °C (26 simulations; d). Changes in color are expressed as PSS-78 50-year^{-1}, and for each panel the corresponding mean SSS is contoured in black with thick lines every 1 PSS-78 and thin lines every 0.5. *Adapted from Durack et al. (2012).*

ensemble mean, so are not directly quantitatively comparable to the observed estimates.

It is clear that magnitudes and spatial patterns of change are most closely replicated by Figure 28.11b–the ensemble mean change pattern for all available CMIP3 SRES realizations–expressing future changes from 2050–2099. These simulations show a strongly enhanced Atlantic basin salinity with larger magnitudes than observed (contrast Figure 28.11a vs. b), whereas the western Pacific Warm Pool undergoes a strong freshening. These increasing basin contrasts, and mean climatological SSS PA follow those suggested by many independent observed studies (Section 4) and those of the idealized E–P forcing simulations (Section 5.2). The lower panels show simulated changes for the twentieth century, with the time period corresponding directly with the observations (Figure 28.11a) for 1950–2000. In CMIP3 the twentieth century simulation magnitudes are realistic, but only for the strongly warming twentieth century realizations (>0.5 °C; Figure 28.11c vs. d). Those simulations with less than the observed warming of 0.5 °C over 1950–2000 often incorporate aerosol effects that act to reduce warming and thus underpredict the subsequent water cycle amplification as expressed in SSS (Durack et al., 2012).

In their analysis Durack et al. (2012) considered each of the available 93 CMIP3 simulations of salinity, using both twentieth century historical (1950–2000) and twenty-first century future projected simulations. A synthesis plot from their analysis is presented below (Figure 28.12). The issue of surface drift in the models (Section 6.1) was explicitly accounted for in this analysis.

For all simulations Durack et al. (2012) found a strong relationship between the rate of average surface temperature increase, and the salinity PA and PC strength (Figure 28.12a). The strength and coherence (high PC values) of this relationship was strongest in the twenty-first century simulations, which are strongly forced with projected CO_2 emissions that drive a strong warming pattern. The model ensemble sensitivity indicated a value of 8% PA per degree, whereas the observed estimates yield 16% per degree (Figure 28.12a) suggesting that the models are conservative in their change estimates. A similar, though less clear relationship was found in the modeled E–P fields (Figure 28.12c), which suggested a weaker sensitivity of E–P to warming around 4% PA per degree. By analyzing the modeled internal relationship for each available simulation and comparing the resolved SSS and E–P PA values, they found a strong relationship between the rates of property changes, suggesting that salinity PA responds to warming at twice the rate of E–P. By using this modeled relationship Durack et al. (2012) then inferred a 1950–2000 observed E–P PA change of 4% from their salinity PA estimates (8% 1950 to 2000), concurrent to the 0.5 °C observed warming experienced during the period.

6.2.2. *Modeled Water Cycle Changes Assessed from Subsurface Ocean Salinity*

Considering changes to the depth-integrated ocean provides a more rigorous test for models, as surface and mixed-layer variability is not apparent in the deeper ocean (>500 m). As a consequence, deeper changes are likely more representative of persistent long-term changes, and reflect anomalies subducted from their surface source regions over decadal timescales. If large and coherent subsurface salinity changes are apparent, this provides further evidence which suggests water cycle intensification has occurred.

Seawater density (volume) is a fundamental physical property which decreases (increases) when heat is added (removed). Similarly, changing salinity also drives density change, with an expansion (contraction) occurring in response to decreasing (increasing) salinity concentrations for a contained water parcel. These steric (density) components are termed the thermosteric (heat) and halosteric (salt) properties of seawater (see Chapter 27 for a more detailed discussion of historical thermosteric sea-level rise estimates;

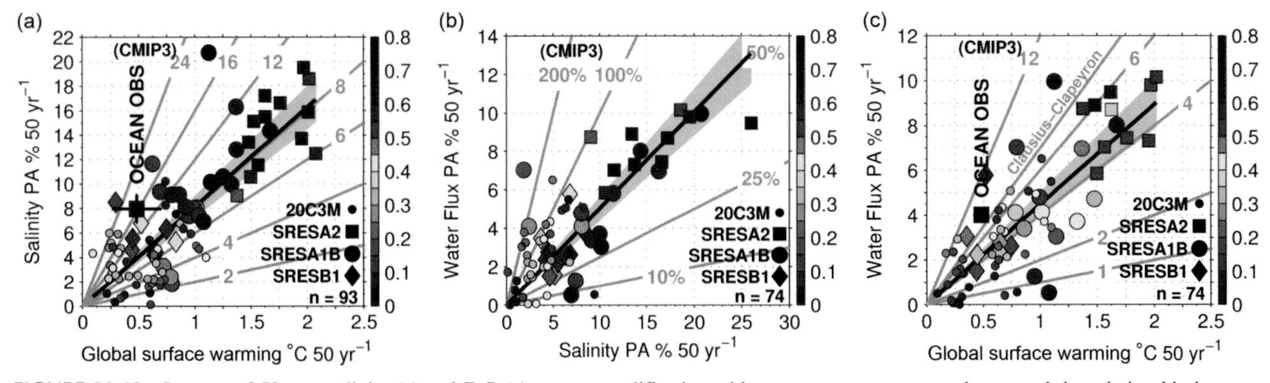

FIGURE 28.12 Patterns of 50-year salinity (a) and E–P (c) pattern amplification with respect to temperature change and the relationship between E–P and salinity pattern amplification (b) for individual CMIP3 twentieth century and future twenty-first century simulations. *Adapted from Durack et al. (2012).*

see Griffies and Greatbatch, 2012 for a detailed description of physical processes driving the evolution of global mean sea level). When integrated over the ocean depth, steric changes impact sea levels both globally and locally.

While there have been numerous investigations into observed changes to ocean steric properties, there have been relatively few which focus on long-term salinity changes (e.g., Antonov et al., 2002; Levitus et al., 2005; Ishii et al., 2006). Additionally, analyses using future twenty-first century projected changes simulated by coupled models have also been undertaken, with these studies considering model simulations forced by strong greenhouse gas emissions (Landerer et al., 2007; Yin et al., 2010b; Pardaens et al., 2011). Comparing halosteric changes in both observed and model-simulated estimates of twentieth century change then allows an investigation into whether models are correctly capturing the processes that have driven observed changes. As the timescales for anomalies to propagate into the deeper ocean are much longer than those of surface and mixed-layer changes, such a comparison provides a more difficult test for the modeling systems.

Similar to the surface change results (Figure 28.11) the depth-integrated results (Figure 28.13) are not directly quantitatively comparable to the observed estimates. However, the spatial patterns of change do appear to share broad-scale similarities to those of observations.

The results presented in Figure 28.13 capture the depth-averaged salinity from the surface to 1800 m for 1950–2000 observations (Figure 28.13a; Durack and Wijffels, 2010) and twentieth century model ensemble mean (Figure 28.13b). The agreement between the climatological mean estimates is encouraging, with the key features of a salty Atlantic Basin, particularly the North Atlantic, and a comparatively fresh Pacific Basin matching very well. Surprisingly, even regional salinity contrasts between the Arabian Sea (salty) and the Bay of Bengal (fresh) are captured in the model ensemble mean (Figure 28.13b) replicating the observed estimate (Figure 28.13a). So in the climatological mean, this model suite is performing well in its broad-scale replication of key salinity features. The more difficult test, however, is the comparison of the long-term depth-integrated halosteric change fields between observations and models (Figure 28.13c and d). These changes indicate regions where the depth-integrated water column has contracted (blue) because of enhanced depth-integrated salinity, or expanded (orange/red) because of a corresponding freshening. Although the models share less similarity compared with the mean climatology (top row), the broad-scale similarity of a halosteric contraction in the Atlantic Basin, with largest magnitudes in the North Atlantic, and a tendency for depth-integrated freshening in the Pacific suggests that the model ensemble mean is capturing the processes which are driving the observed changes well. As noted earlier, these results cannot be quantitatively compared, as the process of developing the model

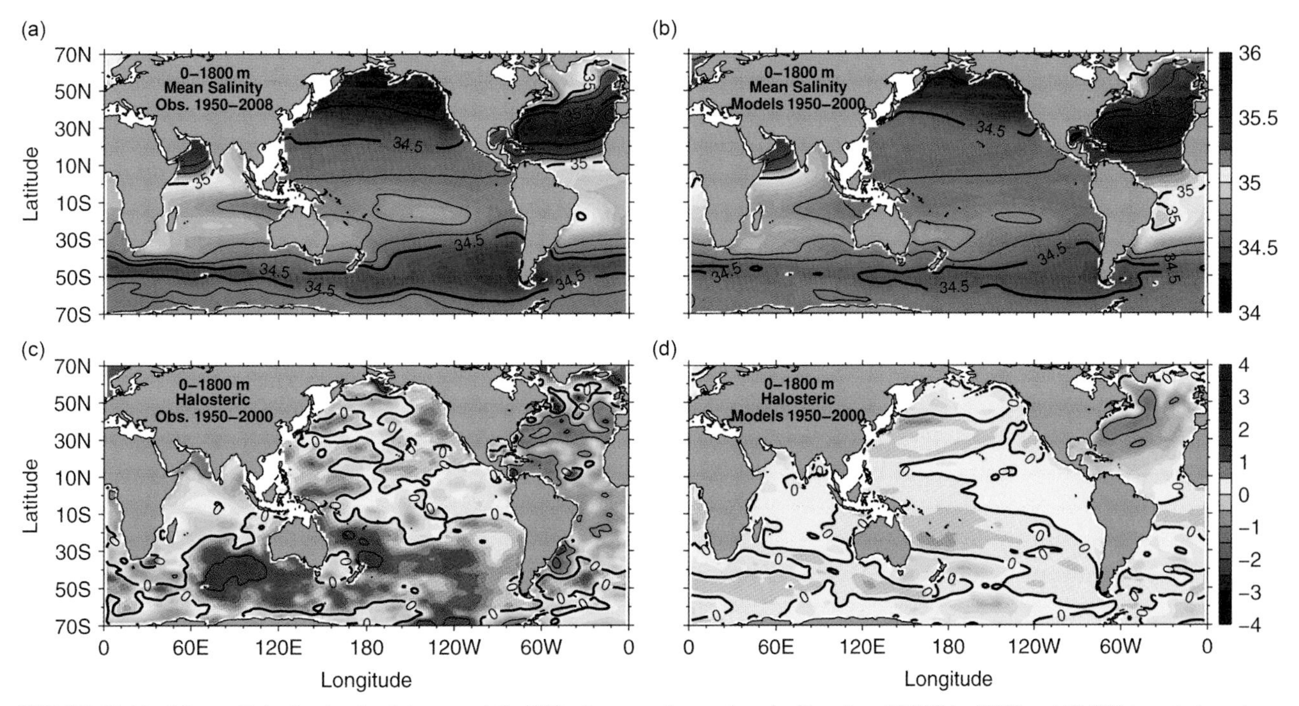

FIGURE 28.13 Mean salinity for the depth-integrated (0–1800 m) ocean, observations (a; Durack and Wijffels, 2010) and CMIP3 twentieth century models (b); PSS-78. Calculated 1950–2000 depth-integrated (0–1800 m) halosteric (salinity-driven) sea-level change for observations (c) and CMIP3 twentieth century models (d); mm 50-year^{-1}.

ensemble mean change patterns tends to dissipate and smooth the results from individual realizations; however, their broad-scale correspondence is encouraging.

7. SUMMARY AND OUTLOOK

An improved understanding of the water cycle through salinity observation looks bright in the early twenty-first century because of the development of a series of new observing technologies and platforms. Along with the existing Argo array, these platforms will provide oceanographers with an unprecedented insight into global ocean variability and change. By monitoring salinity we will dramatically enhance our capability to observe water cycle operation, variability, and change. There are no such observational breakthroughs anticipated in the observation of oceanic rainfall, so these big forward steps in global water cycle observation constitute a truly exciting period for climate science.

Two new satellite platforms have begun to provide salinity estimates of the global ocean, with both the Soil Moisture and Ocean Salinity (SMOS; Berger et al., 2002) satellite mission launched in November 2009, and the dedicated salinity satellite AQUARIUS (Lagerloef et al., 2008) launched in June 2011. Both platforms aim to provide near real-time SSS measurements for the global ocean, with the dedicated AQUARIUS mission aiming to resolve salinity to 0.2 (PSS-78), global coverage every 7 days and a resolved horizontal resolution near 150 km. Even though AQUARIUS data is still undergoing calibration and validation against other observing and forecasting systems to ensure data accuracy, available preliminary data are providing a good representation of the annual mean SSS field (Figure 28.14) when compared to other estimates (Figure 28.8a and b). This satellite has already captured high-frequency salinity anomalies (Lagerloef et al., 2012), associated with high Amazon River runoff (e.g., Grodsky et al., 2012), and the early data has begun to provide an insight into tropical SSS variability and spatial structure and its coupling to tropical rainfall patterns.

The objective of these new platforms is to facilitate oceanographic research by tracking high-frequency salinity changes to (1) improve seasonal to interannual ENSO predictability by better resolving the presence of barrier layers and improving the mixed-layer heat budget in the tropical oceans; (2) improve rainfall estimates and estimates of the global water cycle budgets; This is facilitated by using the "ocean rain gauge" concept that salinity on short-timescales provides an integrator of the turbulent, sporadic freshwater fluxes at the surface; and (3) monitor large-scale salinity change events, such as ice melt, major river runoff events, and monsoons. Additionally, more effective observation of high-latitude oceans will provide an insight into oceanic convection and poleward heat transport. Early studies using this data (e.g., Lee et al., 2012) are already providing new insights into high-frequency salinity variability in the tropical Pacific. New details are also becoming apparent at the finer spatial scales that Argo does not resolve because of outflow events from the Amazon and Orinoco Rivers into the Atlantic Ocean (e.g., Grodsky et al., 2012). Better-defined and sharper gradients are shown around the ITCZ in the Pacific and Atlantic basins, and it is expected that AQUARIUS will resolve the transient, highly stratified, and surface-intensified fresh pools from precipitation and river outflows found in low-wind conditions (Lagerloef et al., 2012). Monitoring global ocean salinity at the high temporal and spatial resolution provided by AQUARIUS will provide an insight into a large part of the currently under sampled and little understood water cycle. AQUARIUS is observing regions with very low Argo

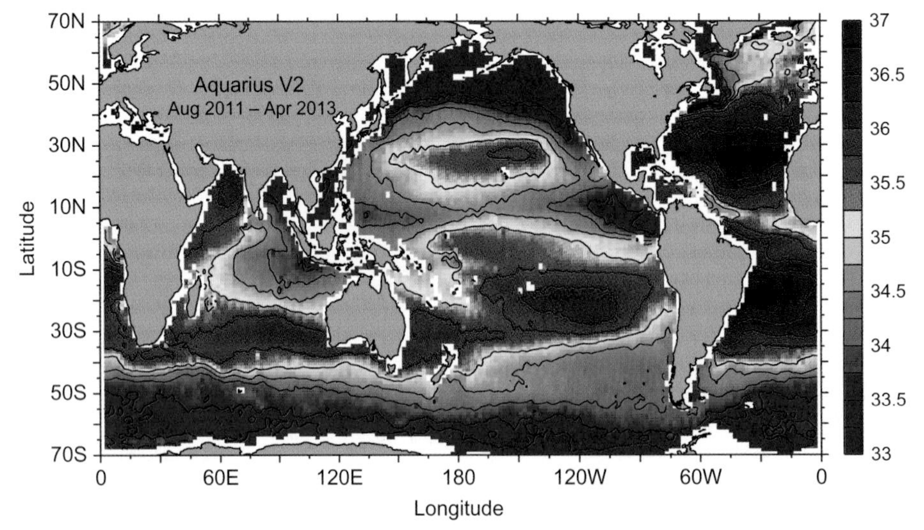

FIGURE 28.14 Global surface mean salinity as captured by the AQUARIUS (version 2) satellite over the period August 2011 to April 2013.

coverage, the salinity responses from regions of currently ungauged terrestrial outflow, and areas of high seasonal precipitation (and in response, salinity) variability. Such new data coverage will fill a large gap in our understanding of global water cycle variability and change. With validated data released in February 2013, this platform is expected to provide a key new insight into the oceanic water cycle and improve our understanding of the marine freshwater budget and water cycle variability over the global ocean. However, because of its relatively poor precision when compared to other observing platforms (0.2 vs. 0.002 PSS-78) it is uncertain how useful such observations will be to elucidate small changes to SSS on climate timescales.

Alongside the new AQUARIUS measurements, a new process-based oceanographic experiment was undertaken in the North Atlantic during September 2012 and March 2013. The Salinity Processes in the Upper Ocean Regional Study (SPURS) field experiment aims to assess the salinity budget for a small region located in the SSS maxima zone (25N, 38W) and will utilize all available observational platforms to attempt to resolve the temperature changes, current, turbulent effects, evaporation, and precipitation events which drive the ocean salinity budget and temporal changes. Plans to follow this field campaign with a future experiment, which will assess a regional salinity minima zone, are underway.

In addition to these new observing platforms and experiments, much new information about the global ocean variability and change has been uncovered since the early 2000s thanks to the Argo Program. The implementation of this 3000+ float array has strongly improved the spatial and temporal coverage of ocean observation and for the first time provided a consistent and coherent observing baseline for the modern ocean. Without the development of the Argo fleet, many recent assessments of ocean variability and long-term change would not have been possible. It is imperative that this relatively new observing platform, along with the continued array of repeat observations obtained from research vessels, be maintained and supplemented into the future. Such ongoing observations will allow us to better quantify and understand how ocean heat content is changing in response to climate change (e.g., Church et al., 2011; Chapter 27) and allow us to better resolve global and regional freshwater changes, through observing the global ocean salinity field.

Alongside developments in ocean observing, the latest version of the CMIP5 model suite is further helping the progress of scientific investigation in our understanding of ocean climate variability and change. When compared to CMIP3 (Meehl et al., 2007), CMIP5 (Taylor et al., 2012) has a number of notable improvements in ocean simulation. These include increased model resolution, both in their resolved horizontal and vertical grids, along with better eddy parameterisations focusing on eddy fluxes and diffusivity (Flato et al., 2013). Additionally, the total number of models contributing to CMIP5 is 49, over double the contribution to CMIP3 (23 models). As the CMIP5 data archive is still growing, this project in combination with observed data will underpin climate research in the coming decade.

As discussed in Section 2, to date there has not been a systematic investigation into the accuracy of salinity data associated with changes in the hydrographic database over time. Issues such as those found with the expendable bathythermographs are not applicable to the bulk of salinity observations, as most historical salinity observations were derived from dedicated research cruises by agencies that routinely calibrate and validate their cruise measurements against international seawater standards. Consequently, this has limited error propagation into the hydrographic database. If such errors were discovered, they would be particularly relevant in light of resolved trend estimates (Section 4) and it would be useful to understand the effect of such errors on resolved long-term trends. When considering previous intercomparisons (Lewis and Perkin, 1978) the effect of errors on the order of ± 0.01 (PSS-78) is uncertain, although such errors are certainly smaller than the resolved long-term changes that have been documented in numerous assessments (Boyer et al., 2005; Hosoda et al., 2009; Durack and Wijffels, 2010; Helm et al., 2010).

As presented in Section 6.2, on average, CMIP3 models underestimate the long-term SSS changes when compared to observations over the 1950 to 2000 period (16% vs. 8% $°C^{-1}$; Durack et al., 2012). The observed salinity estimates indicate a larger rate of change than models, which is consistent with independent estimates of water cycle changes obtained from both precipitation and evaporation change estimates over the satellite record (Wentz et al., 2007; Allan et al., 2010). With the availability of new ocean salinity observing platforms, there is hope that the concept of the "ocean rain gauge" will provide an insight into water cycle variability and change. With the issues highlighted with the available precipitation climatologies (Section 1), we still do not have an unambiguous view from terrestrial or satellite observations on whether the Earth's water cycle has intensified over the observed record—it is hoped that continued quantitative investigations using ocean salinity observations will provide a definitive view on how the global water cycle has changed and will continue to change into the future.

ACKNOWLEDGMENTS

P. J. D. was supported by Lawrence Livermore National Laboratory, which is funded by the U.S. Department of Energy under contract DE-AC52-07NA27344. S. E. W. was supported by the Australian Climate Change Science Program, funded jointly by the Department of Climate Change and Energy Efficiency, the Bureau of Meteorology and CSIRO. T. P. B. was supported by NOAA's Office of Climate Observations. The authors would like to thank the Editors, along with Juliana Durack (UC Berkeley, USA), Anne-Marie Treguier (CNRS, France), Rich Pawlowicz (UBC, Canada), Ray Schmitt (WHOI, USA) and a number of anonymous reviewers for their comments which greatly improved the chapter. We acknowledge the modeling groups, the Program for Climate Model Diagnosis and Intercomparison (PCMDI) and the WCRP's Working Group on Coupled Modeling (WGCM), for their roles in making available the WCRP, CMIP3, and CMIP5 multimodel dataset.

REFERENCES

Allan, R.P., Soden, B.J., John, V.O., Ingram, W., Good, P., 2010. Current changes in tropical precipitation. Environ. Res. Lett. 5 (2), 025205. http://dx.doi.org/10.1088/1748-9326/5/2/025205.

Allen, M.R., Ingram, W.J., 2002. Constraints on future changes in climate and the hydrologic cycle. Nature 419, 224–232. http://dx.doi.org/10.1038/nature01092.

Antonov, J.I., Levitus, S., Boyer, T.P., 2002. Steric sea level variations during 1957-1994: importance of salinity. J. Geophys. Res. 107 (C12). http://dx.doi.org/10.1029/2001JC000964.

Aoyama, M., Joyce, T.M., Kawano, T., Takatsuki, Y., 2002. Standard seawater comparison up to P129. Deep Sea Res. Part I 49 (6), 1103–1114. http://dx.doi.org/10.1016/S0967-0637(02)00018-3.

Arkin, P.A., Smith, T.M., Sapiano, M.R.P., Janowiak, J., 2010. The observed sensitivity of the global hydrological cycle to changes in surface temperature. Environ. Res. Lett. 5, 035201. http://dx.doi.org/10.1088/1748-9326/5/3/035201.

Bacon, S., Culkin, F., Higgs, N., Ridout, P., 2007. IAPSO standard seawater: definition of the uncertainty in the calibration procedure, and stability of recent batches. J. Atmos. Oceanic Technol. 24 (10), 1785–1799. http://dx.doi.org/10.1175/JTECH2081.1.

Barnston, A.G., Livezey, R.E., 1987. Classification, seasonality and persistence of low-frequency atmospheric circulation patterns. Mon. Weather Rev. 115, 1083–1126. http://dx.doi.org/10.1175/1520-0493(1987)115<1083:CSAPOL>2.0.CO;2.

Baumgartner, A., Reichel, E., 1975. World Water Balance: Mean Annual Global, Continental and Maritime Precipitation, Evaporation and Runoff. Elsevier Science Ltd., Amsterdam, 179 pp.

Becker, J.A., Green, C.B., Pearson, G.L., 1946. Properties and uses of thermistors—thermally sensitive resistors. Trans. Am. Inst. Electr. Eng. 65 (11), 711–725. http://dx.doi.org/10.1109/T-AIEE.1946.5059235.

Belkin, I.M., 2004. Propagation of the "Great Salinity Anomaly" of the 1990s around the northern North Atlantic. Geophys. Res. Lett. 31, L08306. http://dx.doi.org/10.1029/2003GL019334.

Belkin, I.M., Levitus, S., Antonov, J., Malmberg, S.-A., 1998. "Great Salinity Anomalies" in the North Atlantic. Prog. Oceanogr. 41, 1–68. http://dx.doi.org/10.1016/S0079-6611(98)00015-9.

Belluci, A., Gualdi, S., Navarra, A., 2010. The double-ITCZ syndrome in coupled general circulation models: the role of large-scale vertical circulation regimes. J. Clim. 23, 1127–1145. http://dx.doi.org/10.1175/2009JCLI3002.1.

Berger, M., Camps, A., Font, J., Berr, Y., Miller, J., Johannessen, J., Boutin, J., Drinkwater, M.R., Skou, N., Floury, N., Rast, M., Rebhan, H., Attema, E., 2002. Measuring ocean salinity with ESA's SMOS mission. ESA Bull. 111, 113–121. Available online, http://esamultimedia.esa.int/docs/Cryosat/ESABulletin111-SMOSOSactivities.pdf.

Bindoff, N.L., McDougall, T.J., 2000. Decadal changes along an Indian Ocean section at 32°S and their interpretation. J. Phys. Oceanogr. 30, 1207–1222. http://dx.doi.org/10.1175/1520-0485(2000)030<1207:DCAAIO>2.0.CO;2.

Bingham, F.M., Foltz, G.R., McPhaden, M.J., 2012. Characteristics of the seasonal cycle of surface layer salinity in the global ocean. Ocean Sci. 8, 915–929. http://dx.doi.org/10.5194/os-8-915-2012.

Bjerknes, J., 1969. Atmospheric teleconnections from the Equatorial Pacific. Mon. Weather Rev. 97 (3), 163–172. http://dx.doi.org/10.1175/1520-0493(1969)097<0163:ATFTEP>2.3.CO;2.

Bosc, C., Delcroix, T., Maes, C., 2009. Barrier layer variability in the western Pacific warm pool from 2000 to 2007. J. Geophys. Res. 114, C06023. http://dx.doi.org/10.1029/2008JC005187.

Boyer, T.P., Levitus, S., 2002. Harmonic analysis of climatological sea surface salinity. J. Geophys. Res. 107 (C12), 8006. http://dx.doi.org/10.1029/2001JC000829.

Boyer, T.P., Stephens, C., Antonov, J.I., Conkright, M.E., Locarnini, R.A., O'Brien, T.D., Garcia, H.E., 2002. World Ocean Atlas 2001, Volume 2: Salinity. In: Levitus, S. (Ed.), NOAA Atlas NESDIS 50. U.S. Gov. Printing Office, Washington, DC165 pp.

Boyer, T.P., Levitus, S., Antonov, J.I., Locarnini, R.A., Garcia, H.E., 2005. Linear trends in salinity for the World Ocean, 1955–1998. Geophys. Res. Lett. 32, L01604. http://dx.doi.org/10.1029/2004GL021791.

Boyer, T., Levitus, S., Antonov, J., Locarnini, R., Mishonov, A., Garcia, H., Josey, S.A., 2007. Changes in freshwater content in the North Atlantic Ocean 1955–2006. Geophys. Res. Lett. 34, L16603. http://dx.doi.org/10.1029/2007GL030126.

Boyer, T.P., Antonov, J.I., Baranova, O.K., Garcia, H.E., Johnson, D.R., Locarnini, R.A., Mishonov, A.V., O'Brien, T.D., Seidov, D., Smolyar, I.V., Zweng, M.M., 2009. In: Levitus, S. (Ed.), World Ocean Database 2009. NOAA Atlas NESDIS 66, U.S. Gov. Printing Office, Washington, DC, p. 216.

Brown, N.L., 1974. A Precision CTD Microprofiler. In: International Conference on Engineering in the Ocean Environment, Ocean '74–IEEE2, pp. 270–278. http://dx.doi.org/10.1109/OCEANS.1974.1161443.

Brown, N.L., Hamon, B.V., 1961. An inductive salinometer. Deep Sea Res. 8 (1), 65–75. http://dx.doi.org/10.1016/0146-6313(61)90015-6.

Bryan, K., 1969. A numerical method for the study of the circulation of the World Ocean. J. Comput. Phys. 4, 347–376. http://dx.doi.org/10.1016/0021-9991(69)90004-7.

Bryan, F., 1986. High-latitude salinity effects and interhemispheric thermohaline circulations. Nature 323, 301–304. http://dx.doi.org/10.1038/323301a0.

Chen, J., Zhang, R., Wang, H., An, Y., Peng, P., Zhang, W., 2012. Isolation of sea surface salinity maps on various timescales in the tropical Pacific Ocean. J. Oceanogr. 68 (5), 687–701. http://dx.doi.org/10.1007/s10872-012-0126-8.

Church, J.A., White, N.J., Konikow, L.F., Domingues, C.M., Cogley, J.G., Rignot, E., Gregory, J.M., van den Broeke, M.R., Monaghan, A.J., Velicogna, I., 2011. Revisiting the Earth's sea-level and energy budgets from 1961 to 2008. Geophys. Res. Lett. 38, L18601. http://dx.doi.org/10.1029/2011GL048794.

Covey, C., Gleckler, P.J., Phillips, T.J., Bader, D.C., 2006. Secular trends and climate drift in coupled ocean-atmosphere general circulation models. J. Geophys. Res. 111, D03107. http://dx.doi.org/10.1029/2005JD006009.

Cox, R.A., 1963. The salinity problem. Prog. Oceanogr. 1, 243–261. http://dx.doi.org/10.1016/0079-6611(63)90006-5.

Cox, R.A., Culkin, F., Greenhalgh, R., Riley, J.P., 1962. Chlorinity, conductivity and density of sea-water. Nature 193, 518–520. http://dx.doi.org/10.1038/193518a0.

Cox, R.A., Culkin, F., Riley, J.P., 1967. The electrical conductivity/chlorinity relationship in natural sea water. Deep Sea Res. Oceanogr. Abstr. 14 (2), 203–220. http://dx.doi.org/10.1016/0011-7471(67)90006-X.

Cravatte, S., Delcroix, T., Zhang, D., McPhaden, M., LeLoup, J., 2009. Observed freshening and warming of the western Pacific Warm Pool. Clim. Dyn. 33, 565–589. http://dx.doi.org/10.1007/s00382-009-0526-7.

Culkin, F., 1986. Calibration of standard seawater in electrical conductivity. Sci. Total Environ. 49, 1–7. http://dx.doi.org/10.1016/0048-9697(86)90230-5.

Culkin, F., Ridout, P.S., 1998. Stability of IAPSO standard seawater. J. Atmos. Oceanic Technol. 15 (4), 1072–1075. http://dx.doi.org/10.1175/1520-0426(1998)015<1072:SOISS>2.0.CO;2.

Culkin, F., Smed, J., 1979. The history of standard seawater. Oceanol. Acta 2 (3), 355–364., Available online, http://archimer.ifremer.fr/doc/00122/23351/21178.pdf.

Curry, R., Mauritzen, C., 2005. Dilution of the northern North Atlantic Ocean in recent decades. Science 308, 1772–1774. http://dx.doi.org/10.1126/science.1109477.

Curry, R., Dickson, B., Yashayaev, I., 2003. A changes in the freshwater balance of the Atlantic Ocean over the past four decades. Nature 426, 826–829. http://dx.doi.org/10.1038/nature02206.

da Silva, A.M., Young, C.C., Levitus, S., 1994. In: Atlas of Marine Surface Data 1994, Volume 4: Anomalies of Fresh Water Fluxes. NOAA Atlas Series, vol. 4. National Oceanic and Atmospheric Administration, National Environmental Satellite, Data and Information Service, Silver Spring, MD, p. 308, 20910-3292.

Dai, A., Fung, I.Y., Del Genio, A.D., 1997. Surface observed global land precipitation variations during 1900-88. J. Clim. 10, 2943–2962. http://dx.doi.org/10.1175/1520-0442(1997)010<2943:SOGLPV>2.0.CO;2.

Dauphinee, T.M., Klein, H.P., 1975. A new automated laboratory salinometer. Sea Technol. 16 (3), 23–25.

Delcroix, T., 1998. Observed surface oceanic and atmospheric variability in the tropical Pacific at seasonal and ENSO timescales: a tentative overview. J. Geophys. Res. 103 (C9), 18611–18633. http://dx.doi.org/10.1029/98JC00814.

Delcroix, T., Picaut, J., 1998. Zonal displacement of the western equatorial Pacific "fresh pool" J. Geophys. Res. 103 (C1), 1087–1098. http://dx.doi.org/10.1029/97JC01912.

Delcroix, T., Henin, C., Porte, V., Arkin, P., 1996. Precipitation and sea-surface salinity in the tropical Pacific Ocean. Deep-Sea Res. I Oceanogr. Res. Pap. 43 (7), 1123–1141. http://dx.doi.org/10.1016/0967-0637(96)00048-9.

Delcroix, T., McPhaden, M.J., Dessier, A., Gouriou, Y., 2005. Time and space scales for sea surface salinity in the tropical oceans. Deep-Sea Res. I Oceanogr. Res. Pap. 52 (5), 787–813. http://dx.doi.org/10.1016/j.dsr.2004.11.012.

Delcroix, T., Cravatte, S., McPhaden, M.J., 2007. Decadal variations and trends in tropical Pacific sea surface salinity since 1970. J. Geophys. Res. 112, C03012. http://dx.doi.org/10.1029/2006JC003801.

Dickson, R.R., Meincke, J., Malmberg, S.-A., Lee, A.J., 1988. The "Great Salinity Anomaly" in the Northern North Atlantic 1968–1982. Prog. Oceanogr. 20, 103–151. http://dx.doi.org/10.1016/0079-6611(88)90049-3.

Dittmar, W., 1884. Thomson, C.W., Murray, J. (Eds.), Report on Researches in the Composition of Ocean-Water, collected by H.M.S. Challenger, during the years 1873-1876. The Voyage of H.M.S. Challenger, Physics and Chemistry, 1, Composition of Ocean, London, pp. 1–257. Part 1, Available online: http://www.escholarship.org/uc/item/38p2q583.

Durack, P.J., Wijffels, S.E., 2010. Fifty-year trends in global ocean salinities and their relationship to broad-scale warming. J. Clim. 23, 4342–4362. http://dx.doi.org/10.1175/2010JCLI3377.1.

Durack, P.J., Wijffels, S.E., Matear, R.J., 2012. Ocean salinities reveal strong global water cycle intensification during 1950–2000. Science 336 (6080), 455–458. http://dx.doi.org/10.1126/science.1212222.

Falkovich, A., Ginis, I., Lord, S., 2005. Ocean data assimilation and initialization procedure for the Coupled GFDL/URI Hurricane Prediction System. J. Atmos. Oceanic Technol. 22 (12), 1918–1932. http://dx.doi.org/10.1175/JTECH1810.1.

Farland, R.J., 1976. Salinity intercomparison report, the oceanographic subprogramme for the GARP Atlantic tropical experiment (GATE). National Oceanographic Instrument Center, Washington, DC.

Flato, G., et al., 2013. Evaluation of climate models. In: Climate Change 2013: The Physical Science Basis. Contribution to Working Group I to the Fifth Assessment Report of the Intergovernmental Panel on Climate Change, (Chapter 9).

Foltz, G.R., McPhaden, M.J., 2008. Seasonal mixed layer salinity balance of the Tropical North Atlantic Ocean. J. Geophys. Res. 113, C02013. http://dx.doi.org/10.1029/2007JC004178.

Forchhammer, G., 1862. On the constitution of sea-water, at different depths, and in different latitudes. Proc. R. Soc. Lond. 12, 129–132. http://dx.doi.org/10.1098/rspl.1862.0021.

Forget, G., Wunsch, C., 2007. Estimated global hydrographic variability. J. Phys. Oceanogr. 37, 1997–2008. http://dx.doi.org/10.1175/JPO3072.1.

Freeland, H., Denman, K., Wong, C.S., Whitney, F., Jacques, R., 1997. Evidence of change in the winter mixed layer in the Northeast Pacific Ocean. Deep-Sea Res. I Oceanogr. Res. Pap. 44 (12), 2117–2129. http://dx.doi.org/10.1016/S0967-0637(97)00083-6.

Ganachaud, A., Wunsch, C., 2003. Large-scale ocean heat and freshwater transports during the World Ocean Circulation experiment. J. Clim. 16, 696–705. http://dx.doi.org/10.1175/1520-0442(2003)016<0696:LSOHAF>2.0.CO;2.

Gimeno, L., Drumon, A., Nieto, R., Trigo, R.M., Stohl, A., 2010. On the origin of continental precipitation. Geophys. Res. Lett. 37, L13804. http://dx.doi.org/10.1029/2010GL043712.

Gordon, A.L., Giulivi, C.F., 2008. Sea surface salinity trends over fifty years within the Subtropical North Atlantic. Oceanography 21 (1), 20–29. http://dx.doi.org/10.5670/oceanog.2008.64.

Gould, J., Roemmich, D., Wijffels, S., Freeland, H., Ignaszewsky, N., Jianping, X., Pouliquen, S., Desaubies, Y., Send, U., Radhakrishnan, K., Takeuchi, K., Kim, K., Danchenkov, M., Sutton, P., King, B., Owens, B., Riser, S., 2004. Argo profiling floats bring new era of in situ ocean observations. Eos 85 (19), 185. http://dx.doi.org/10.1029/2004EO190002.

Gouriou, Y., Delcroix, T., 2001. Seasonal and ENSO variations of sea surface salinity and temperature in the South Pacific Convergence

Zone during 1976–2000. J. Geophys. Res. 107 (C12), 8011. http://dx.doi.org/10.1029/2001JC000830.

Grasshoff, K., Hermann, F., 1976. Salinometer Intercalibration Experiment. In: Seventh report of the joint panel on oceanographic tables and standards. Grenoble, France, 2–5 September 1975pp. 19–23, UNESCO technical papers in marine science 24.

Griffies, S.M., 2004. Fundamentals of Ocean Models. Princeton University Press, Princeton, New Jersey, USA, 463pp.

Griffies, S.M., Greatbatch, R.J., 2012. Physical processes that impact the evolution of global mean sea level in ocean climate models. Ocean Model. 51, 37–72. http://dx.doi.org/10.1016/j.ocemod.2012.04.003.

Griffies, S.M., Biastoch, A., Böning, C., Bryan, F., Danabasoglu, G., Chassignet, E.P., England, M.H., Gerdes, R., Haak, H., Hallberg, R.W., Hazeleger, W., Jungclaus, J., Large, W.G., Madec, G., Pirani, A., Samuels, B.L., Scheinert, M., Sen Gupta, A., Severijns, C.A., Simmons, H.L., Treguier, A.M., Winton, M., Yeager, S., Yin, J., 2009. Coordinated Ocean-ice Reference Experiments (COREs). Ocean Model. 26, 1–46. http://dx.doi.org/10.1016/j.ocemod.2008.08.007.

Griffies, S.M., Yin, J., Durack, P.J., Goddard, P., Bates, S.C., Behrens, E., Bentsen, M., Bi, D., Biastoch, A., Boning, C., Bozec, A., Cassou, C., Chassignet, E., Danabasoglu, G., Danilov, S., Domingues, C., Drange, H., Farnetti, R., Fernandez, E., Greatbatch, R.J., Ilicak, M., Lu, J., Marsland, S.J., Mishra, A., Lorbacher, K., Nurser, A.J.G., Salas y Melia, D., Palter, J.B., Samuels, B.L., Schroter, J., Schwarzkopf, F.U., Sidorenko, D., Treguier, A.-M., Tseng, Y., Tsujino, H., Uotila, P., Valcke, S., Voldoire, A., Wang, Q., Winton, M., Zhang, X., 2013. Global and regional sea level in a suite of CORE-forced global ocean-ice simulations. Ocean Model. (in preparation).

Grodsky, S.A., Carton, J.A., Bingham, F.M., 2006. Low frequency variation of sea surface salinity in the tropical Atlantic. Geophys. Res. Lett. 33, L14604. http://dx.doi.org/10.1029/2006GL026426.

Grodsky, S.A., Reul, N., Lagerloef, G., Reverdin, G., Carton, J.A., Chapron, B., Quilfen, Y., Kudryavtsev, V.N., Kao, H.-Y., 2012. Haline hurricane wake in the Amazon/Orinco plume: AQUARIUS/SACD and SMOS observations. Geophys. Res. Lett. 39, L14604. http://dx.doi.org/10.1029/2012GL053335.

Hamon, B.V., 1955. A temperature-salinity-depth recorder. J. Conseil 21, 72–73. http://dx.doi.org/10.1093/icesjms/21.1.72.

Han, W., McCreary, J.P., 2001. Modelling salinity distributions in the Indian Ocean. J. Geophys. Res. 106 (C1), 859–877. http://dx.doi.org/10.1029/2000JC000316.

Haney, R.L., 1971. Surface thermal boundary condition for Ocean Circulation Models. J. Phys. Oceanogr. 1, 241–248. http://dx.doi.org/10.1175/1520-0485(1971)001<0241:STBCFO>2.0.CO;2.

Hasselmann, K., 1979. On the signal-to-noise problem in atmospheric response studies. In: Shaw, D.B. (Ed.), Meteorology over the Tropical Oceans. Royal Meteorological Society, Reading, UK, pp. 251–259.

Hasson, A.E.A., Delcroix, T., Dussin, R., 2013. An assessment of the mixed layer salinity budget in the tropical Pacific Ocean. Observations and modelling (1990–2009). Ocean Dyn. 63 (2–3), 179–194. http://dx.doi.org/10.1007/s10236-013-0596-2.

Held, I.M., Soden, B.J., 2006. Robust responses of the hydrological cycle to global warming. J. Clim. 19, 5686–5699. http://dx.doi.org/10.1175/JCLI3990.1.

Helm, K.P., Bindoff, N.L., Church, J.A., 2010. Changes in the global hydrological-cycle inferred from ocean salinity. Geophys. Res. Lett. 37, L18701. http://dx.doi.org/10.1029/2010GL044222.

Holland, H.D., 1972. The geologic history of sea water—an attempt to solve the problem. Geochim. Cosmochim. Acta 36 (6), 637–651. http://dx.doi.org/10.1016/0016-7037(72)90108-1.

Holliday, N.P., Hughes, S.L., Bacon, S., Beszczynska-Möller, A., Hansen, B., Lavín, A., Loeng, H., Mork, K.A., Østerhus, S., Sherwin, T., Walczowski, W., 2008. Reversal of the 1960s to 1990s freshening trend in the northeast North Atlantic and Nordic Seas. Geophys. Res. Lett. 35, L03614. http://dx.doi.org/10.1029/2007GL032675.

Hosoda, S., Suga, T., Shikama, N., Mizuno, K., 2009. Global surface layer salinity change detected by Argo and its implication for hydrological cycle intensification. J. Oceanogr. 65, 579–596. http://dx.doi.org/10.1007/s10872-009-0049-1.

Huang, R.X., 1993. Real freshwater flux as a natural boundary condition for the salinity balance and thermohaline circulation forced by evaporation and precipitation. J. Phys. Oceanogr. 23, 2428–2446. http://dx.doi.org/10.1175/1520-0485(1993)023<2428:RFFAAN>2.0.CO;2.

Huang, R.X., Schmitt, R.W., 1993. The Goldsbrough-Stommel circulation of the World Oceans. J. Phys. Oceanogr. 23, 1277–1284. http://dx.doi.org/10.1175/1520-0485(1993)023<1277:TGCOTW>2.0.CO;2.

Huffman, G.J., Adler, R.F., Arkin, P., Chang, A., Ferraro, R., Gruber, A., Janowiak, J., McNab, A., Rudolf, B., Schneider, U., 1997. The Global Precipitation Climatology Project (GPCP) Combined Precipitation Dataset. Bull. Am. Meteorol. Soc. 78, 5–20. http://dx.doi.org/10.1175/1520-0477(1997)078<0005:TGPCPG>2.0.CO;2.

Illig, S., Perigaud, C., 2007. Yearly impact of submonthly rain fluctuations on the Indian Ocean salinity. Geophys. Res. Lett. 34, L12609. http://dx.doi.org/10.1029/2007GL029655.

IOC, SCOR, IAPSO, 2010. The international thermodynamic equation of seawater—2010: calculation and use of thermodynamic properties. Intergovernmental Oceanographic Commission, Manuals and Guides No. 56, UNESCO (English), 196 pp.

Ishii, M., Kimoto, M., Sakamoto, K., Iwasaki, S.-I., 2006. Steric sea level changes estimated from historical ocean subsurface temperature and salinity analyses. J. Oceanogr. 62 (2), 155–170. http://dx.doi.org/10.1007/s10872-006-0041-y.

Jacobsen, J.P., Knudsen, M., 1940. Urnormal 1937 or primary standard sea-water 1937. Int. Assoc. Phys. Sci. Oceans (IAPSO) 7, 38. Available online, http://iapso.iugg.org/images/stories/pdf/IAPSO_publications/Pub_Sci_No_7.pdf.

Janzen, C., Larson, N., Beed, R., Anson, K., 2008. Accuracy and Stability of Argo SBE 41 and SBE 41CP CTD Conductivity and Temperature Sensors. *Seabird Electronics Inc*. Seabird Technical Papers (unpublished). Available online: http://www.seabird.com/technical_references/LongtermTSstabilityAGUDec08Handout2Pages.pdf.

Johnson, G.C., Lyman, J.M., 2007. Global Oceans: sea surface salinity. In: Arguez, A. (Ed.), State of the Climate in 2006, 88, Bulletin of the American Meteorological Society, pp. s34–s35. http://dx.doi.org/10.1175/BAMS-88-6-StateoftheClimate.

Johnson, G.C., Orsi, A.H., 1997. Southwest Pacific Ocean Water-Mass Changes between 1968/69 and 1990/91. J. Clim. 10, 306–316. http://dx.doi.org/10.1175/1520-0442(1997)010<0306:SPOWMC>2.0.CO;2.

Joyce, T.M., 1991. Introduction To The Collection of Expert Reports Compiled for the WHP Programme. WOCE Hydrographic Programme Operations and Methods. WOCE Operations Manual. WHP Office Report WHPO-91-1, WOCE Report No. 68/91. 4 pp. Available online: http://woce.nodc.noaa.gov/woce_v3/wocedata_1/whp/manuals.htm.

Kalnay, E., Kanamitsu, M., Kistler, R., Collins, W., Deaven, D., Gandin, L., Iredell, M., Saha, S., White, G., Woollen, J., Zhu, Y.,

Chelliah, M., Ebisuzaki, W., Higgins, W., Janowiak, J., Mo, K.C., Ropelewski, C., Wang, J., Leetmaa, A., Reynolds, R., Jenne, R., Joseph, D., 1996. The NCEP/NCAR 40-year reanalysis project. Bull. Am. Meteorol. Soc. 77, 437–471. http://dx.doi.org/10.1175/1520-0477(1996)077<0437:TNTRP>2.0.CO;2.

Kawano, T., Aoyama, M., Joyce, T., Uchida, H., Takatsuki, Y., Fukasawa, M., 2006. The latest batch-to-batch difference table of standard seawater and its application to the WOCE onetime sections. J. Oceanogr. 62 (6), 777–792. http://dx.doi.org/10.1007/s10872-006-0097-8.

Killworth, P.D., Smeed, D., Nurser, A., 2000. The effects on ocean models of relaxation toward observations at the surface. J. Phys. Oceanogr. 30, 160–174. http://dx.doi.org/10.1175/1520-0485(2000)030<0160:TEOOMO>2.0.CO;2.

Knapp, G.P., Stalcup, M.C., 1987. Progress in the measurement of salinity and oxygen at the Woods Hole Oceanographic Institution. WHOI Technical Report, WHOI-87-4, Woods Hole Oceanographic Institution, 27 pp. Available online: http://hdl.handle.net/1912/5870.

Knudsen, M., 1901. Hydrographical Tables According to the Measurings of Carl Forch, P. Jacobsen, Martin Knudsen and S. P. L. Sørensen. In: Knudsen, M. (Ed.), G. E. C. Gad, Copenhagen, p. 63., Available online: http://hdl.handle.net/2027/uc1.31822000615591.

Lagerloef, G., Colomb, F.R., Le Vine, D., Wentz, F., Yueh, S., Ruf, C., Lilly, J., Gunn, J., Chao, Y., deCharon, A., Feldman, G., Swift, C., 2008. The Aquarius/SAC-D Mission: designed to meet the salinity remote-sensing challenge. Oceanography 21 (1), 68–81. http://dx.doi.org/10.5670/oceanog.2008.68.

Lagerloef, G., Wentz, F., Yueh, S., Kao, H.-Y., Johnson, G.C., Lyman, J.M., 2012. Global Oceans] Aquarius Satellite Mission provides new detailed view of sea surface salinity. [In "State of the Climate 2011"]. Bull. Am. Meteorol. Soc. 97 (7), S70–S71. http://dx.doi.org/10.1175/2012BAMSStateoftheClimate.1.

Landerer, F.W., Jungclaus, J.H., Marotzke, J., 2007. Regional dynamic and steric sea-level change in response to the IPCC-A1B scenario. J. Clim. 37, 296–312. http://dx.doi.org/10.1175/JPO3013.1.

Large, W.G., Yeager, S.G., 2009. The global climatology of an interannually varying air-sea flux data set. Clim. Dyn. 33, 341–364. http://dx.doi.org/10.1007/s00382-008-0441-3.

Large, W.G., Danabasoglu, G., Doney, S.C., McWilliams, J.C., 1997. Sensitivity to surface forcing and boundary layer mixing in a global ocean model: annual-mean climatology. J. Phys. Oceanogr. 27, 2418–2447. http://dx.doi.org/10.1175/1520-0485(1997)027<2418:STSFAB>2.0.CO;2.

Lau, K.-M., Wu, H.-T., 2007. Detecting trends in tropical rainfall characteristics, 1979–2003. Int. J. Climatol. 27, 979–988. http://dx.doi.org/10.1002/joc.1454.

Lee, T., Lagerloef, G., Gierach, M.M., Kao, H.-Y., Yueh, S., Dohan, K., 2012. Aquarius reveals salinity structure of tropical instability waves. Geophys. Res. Lett. 39, L12610. http://dx.doi.org/10.1029/2012GL052232.

Levitus, S., Antonov, J.I., Boyer, T.P., Garcia, H.E., Locarnini, R.A., 2005. Linear trends of zonally averaged thermosteric, halosteric and total steric sea-level for individual ocean basins and the world ocean, (1955-1959)-(1994-1998). Geophys. Res. Lett. 32, L16601. http://dx.doi.org/10.1029/2005GL023761.

Lewis, E., 1980. The practical salinity scale 1978 and its antecedents. IEEE J. Oceanic Eng. 5 (1), 3–8. http://dx.doi.org/10.1109/JOE.1980.1145448.

Lewis, E.L., Perkin, R.G., 1978. Salinity: its definition and calculation. J. Geophys. Res. 83 (C1), 466–478. http://dx.doi.org/10.1029/JC083iC01p00466.

Lewis, E.L., Perkin, R.G., 1981. The Practical Salinity Scale 1978: conversion of existing data. Deep-Sea Res. 28A (4), 307–328. http://dx.doi.org/10.1016/0198-0149(81)90002-9.

Limpasuvan, V., Hartmann, D.L., 1999. Eddies and the annular modes of climate variability. Geophys. Res. Lett. 26 (20), 3133–3136. http://dx.doi.org/10.1029/1999GL010478.

Lin, J.-L., 2007. The double-ITCZ problem in IPCC AR4 coupled GCMs: ocean-atmosphere feedback analysis. J. Clim. 20, 4497–4525. http://dx.doi.org/10.1175/JCLI4272.1.

Lukas, R., Lindstrom, E., 1991. The mixed layer of the western equatorial Pacific Ocean. J. Geophys. Res. 96 (S01), 3343–3357. http://dx.doi.org/10.1029/90JC01951.

Lyman, J., 1969. Redefinition of salinity and chlorinity. Limnol. Oceanogr. 14 (6), 928–929. http://dx.doi.org/10.4319/lo.1969.14.6.0928.

Maes, C., Picaut, J., Belamari, S., 2005. Importance of the salinity barrier layer for the buildup of El Niño. J. Clim. 18 (1), 104–118. http://dx.doi.org/10.1175/JCLI-3214.1.

Manabe, S., Stouffer, R.J., 1988. Two stable equilibria of a coupled ocean-atmosphere model. J. Clim. 1 (9), 841–866. http://dx.doi.org/10.1175/1520-0442(1988)001<0841:TSEOAC>2.0.CO;2.

Mantua, N.J., Hare, S.R., Zhang, Y., Wallace, J.M., Francis, R.C., 1997. Pacific interdecadal climate oscillation with impacts on Salmon Production. Bull. Am. Meteorol. Soc. 78, 1069–1079. http://dx.doi.org/10.1175/1520-0477(1997)078<1069:APICOW>2.0.CO;2.

Mantyla, A.W., 1980. Electrical conductivity comparisons of Standard Seawater batches P29 to P84. Deep-Sea Res. 27 (10), 837–846. http://dx.doi.org/10.1016/0198-0149(80)90047-3.

Mantyla, A.W., 1987. Standard seawater comparisons updated. J. Phys. Oceanogr. 17 (4), 543–548. http://dx.doi.org/10.1175/1520-0485(1987)017<0543:SSCU>2.0.CO;2.

Mantyla, A.W., 1994. The treatment of inconsistencies in Atlantic deep water salinity data. Deep-Sea Res. I Oceanogr. Res. Pap. 41 (9), 1387–1405. http://dx.doi.org/10.1016/0967-0637(94)90104-X.

Marcet, A., 1819. On the specific gravity, and temperature of sea waters, in different parts of the ocean, and in particular seas; With some account of their Haline contents. Philos. Trans. R. Soc. Lond. 109, 161–208. http://dx.doi.org/10.1087/rstl.1819.0014.

Meehl, G.A., Covey, C., Delworth, T., Latif, M., McAvaney, B., Mitchell, J.F.B., Stouffer, R.J., Taylor, K.E., 2007. The WCRP CMIP3 Multimodel Dataset. Bull. Am. Meteorol. Soc. 88, 1383–1394. http://dx.doi.org/10.1175/BAMS-88-9-1383.

Mignot, J., Frankignoul, C., 2003. On the interannual variability of surface salinity in the Atlantic. Clim. Dyn. 20 (6), 555–565. http://dx.doi.org/10.1007/s00382-002-0294-0.

Mill, H.R., 1900. The Pettersson-Nansen insulating water-bottle. Geogr. J. 16 (4), 469–471. http://dx.doi.org/10.2307/1774328.

Millero, F.J., Feistel, R., Wright, D.G., McDougall, T.J., 2008. The composition of standard seawater and the definition of the reference-composition salinity scale. Deep-Sea Res. I Oceanogr. Res. Pap. 55, 50–72. http://dx.doi.org/10.1016/j.dsr.2007.10.001.

Pacanowski, R.C., Griffies, S.M., 1999. The MOM3 ManualGFDL Ocean Group Technical Report No. 4. NOAA/Geophysical Fluid Dynamics Laboratory, Princeton, USA., 680 pp. Available online, http://www.gfdl.noaa.gov/cms-filesystem-action/model_development/ocean/mom3_manual.pdf.

Paquette, R.G., 1959. A modification of the Wenner-Smith-Soule salinity bridge for the determination of salinity in sea waterTechnical Report 61. University of Washington, Department of Oceanography, 61 pp.

Pardaens, A.K., Gregory, J.M., Lowe, J.A., 2011. A model study of factors influencing projected changes in regional sea-level over the twenty-first century. Clim. Dyn. 36, 2015–2033. http://dx.doi.org/10.1007/s00382-009-0738-x.

Park, K., 1964. Reliability of standard seawater as a conductivity standard. Deep Sea Res. Oceanogr. Abstr. 11 (1), 85–87. http://dx.doi.org/10.1016/0011-7471(64)91084-8.

Perigaud, C., McCreary, J.P., Zhang, K.Q., 2003. Impact of interannual rainfall anomalies on Indian Ocean salinity and temperature variability. J. Geophys. Res. 108 (C10), 3319. http://dx.doi.org/10.1029/2002JC001699.

Philander, S.G., 1990. El Niño, La Niña, and the Southern Oscillation. International Geophysics, vol. 46. Academic Press, Elsevier Science Ltd., p. 293. http://dx.doi.org/10.1016/S0074-6142(08)60168-0.

Pierce, D.W., Gleckler, P.J., Barnett, T.P., Santer, B.D., Durack, P.J., 2012. The fingerprint of human-induced changes in the ocean's salinity and temperature fields. Geophys. Res. Lett. 39, L21704. http://dx.doi.org/10.1029/2012GL053389.

Pohlmann, H., Jungclaus, J.H., Köhl, A., Stammer, D., Marotzke, J., 2009. Initializing decadal climate predictions with the GECCO oceanic synthesis: effect on the North Atlantic. J. Clim. 22 (14), 3926–3928. http://dx.doi.org/10.1175/2009JCLI2535.1.

Preston-Thomas, H., 1990. The International Temperature Scale of 1990 (ITS-90). Metrologia 27 (1), 3–10. http://dx.doi.org/10.1088/0026-1394/27/1/002, (2) p 107.

Quartly, G.D., Kyte, E.A., Srokosz, M.A., Tsimplis, M.N., 2007. An intercomparison of global oceanic precipitation climatologies. J. Geophys. Res. 112, D10121. http://dx.doi.org/10.1029/2006JD007810.

Rahmstorf, S., 1995a. Climate drift in an ocean model coupled to a simple, perfectly matched atmosphere. Clim. Dyn. 11 (8), 447–458. http://dx.doi.org/10.1007/BF00207194.

Rahmstorf, S., 1995b. Bifurcations of the Atlantic thermohaline circulation in responses to changes in the hydrological cycle. Nature 378, 145–149. http://dx.doi.org/10.1038/378145a0.

Rahmstorf, S., 1996. On the freshwater forcing and transport of the Atlantic thermohaline circulation. Clim. Dyn. 12 (12), 799–811. http://dx.doi.org/10.1007/s003820050144.

Rahmstorf, S., 2000. The thermohaline ocean circulation: a system with dangerous thresholds. Clim. Change 46 (3), 247–256. http://dx.doi.org/10.1023/A:1005648404783.

Rao, R.R., Sivakumar, R., 2003. Seasonal variability of sea surface salinity and salt budget of the mixed layer of the north Indian Ocean. J. Geophys. Res. 108 (C1), 3009. http://dx.doi.org/10.1029/2001JC000907.

Reeburgh, W.S., 1976. Some implications for the 1940 redefinition of chlorinity. Deep Sea Res. Oceanogr. Abstr. 13 (5), 975–976. http://dx.doi.org/10.1016/0011-7471(76)90916-5.

Reverdin, G., 2010. North Atlantic Subpolar Gyre Surface Variability (1895-2009). J. Clim. 23, 4571–4584. http://dx.doi.org/10.1175/2010JCLI3493.1.

Rivin, I., Tziperman, E., 1997. Sensitivity of air-sea fluxes to SST perturbations. J. Clim. 10, 2431–2446. http://dx.doi.org/10.1175/1520-0442(1997)010<2431:SOASFT>2.0.CO;2.

Robinson, I., 2006. Satellite measurements for operational ocean models. In: Chassignet, E.P., Verron, J. (Eds.), Ocean Weather Forecasting: An Integrated View of Oceanography. Springer, The Netherlands, pp. 147–189. http://dx.doi.org/10.1007/1-4020-4028-8_6.

Roemmich, D., Gilson, J., 2009. The 2004-2008 mean and annual cycle of temperature, salinity, and steric height in the global ocean from the Argo Program. Prog. Oceanogr. 82, 81–100. http://dx.doi.org/10.1016/j.pocean.2009.03.004.

Roemmich, D., Gilson, J., 2011. The global ocean imprint of ENSO. Geophys. Res. Lett. 38, L13606. http://dx.doi.org/10.1029/2011GL047992.

Rooth, C., 1982. Hydrology and ocean circulation. Prog. Oceanogr. 11 (2), 131–149. http://dx.doi.org/10.1016/0079-6611(82)90006-4.

Rubey, W.W., 1951. Geologic history of sea water: an attempt to state the problem. Bull. Geol. Soc. Am. 62 (9), 1111–1148. http://dx.doi.org/10.1130/0016-7606(1951)62[1111:GHOSW]2.0.CO;2.

Schanze, J.J., Schmitt, R.W., Yu, L.L., 2010. The global oceanic freshwater cycle: a state-of-the-art quantification. J. Mar. Res. 68, 569–595. http://dx.doi.org/10.1357/002224010794657164.

Schleicher, K.E., Bradshaw, A., 1956. A conductivity bridge for measurement of the salinity of sea water. J. Cons. Permanent Int. pour l'Explor. Mer 22 (1), 9–20. http://dx.doi.org/10.1093/icesjms/22.1.9.

Schlesinger, M.E., Ramankutty, N., 1994. An oscillation in the global climate system of period 65-70 years. Nature 367, 723–726. http://dx.doi.org/10.1038/367723a0.

Schmitt, R.W., 1995. The ocean component of the global water cycle. U.S. National Report to the International Union of Geodesy and Geophysics, 1991-1994. Rev. Geophys. 33 (Suppl.), 1395–1409.

Schmitt, R.W., 2008. Salinity and the global water cycle. Oceanography 21 (1), 12–19. http://dx.doi.org/10.5670/oceanog.2008.63.

Sen Gupta, A., Muir, L.C., Brown, J.N., Phipps, S.J., Durack, P.J., Monselesan, D., Wijffels, S.E., 2012. Climate drift in the CMIP3 models. J. Clim. 25, 4621–4640. http://dx.doi.org/10.1175/JCLI-D-11-00312.1.

Sen Gupta, A., Jourdain, N.C., Brown, J.N., Monselesan, D., 2013. Climate Drift in the CMIP5 models. J. Clim. http://dx.doi.org/10.1175/JCLI-D-12-00521.1, in press.

Sprintall, J., Tomczak, M., 1992. Evidence of the barrier layer in the surface layer of the Tropics. J. Geophys. Res. 97 (C5), 7305–7316. http://dx.doi.org/10.1029/92JC00407.

Steffen, K., Thomas, R.H., Rignot, E., Cogley, J.G., Dyurgerov, M.B., Raper, S.C.B., Huybrechts, P., Hanna, E., 2010. Cryospheric contributions to sea-level rise and variability. In: Church, J.A., Woodworth, P.L., Aarup, T., Wilson, W.S. (Eds.), Understanding Sea-Level Rise and Variability. Wiley-Blackwell, Oxford, UK, pp. 178–225. http://dx.doi.org/10.1002/9781444323276.

Stommel, H., 1961. Thermohaline convection with two stable regimes of flow. Tellus A 13 (2), 224–230. http://dx.doi.org/10.1111/j.2153-3490.1961.tb00079.x.

Stott, P.A., Sutton, R.A., Smith, D.M., 2008. Detection and attribution of Atlantic salinity changes. Geophys. Res. Lett. 35, L21702, http://dx.doi.org/10.129/2008GL035874.

Talley, L.D., 2008. Freshwater transport estimates and the global overturning circulation: shallow, deep and throughflow components. Prog. Oceanogr. 78, 257–303. http://dx.doi.org/10.1016/j.pocean. 2008.05.001.

Taylor, K.E., Stouffer, R.J., Meehl, G.A., 2012. An overview of CMIP5 and the experimental design. Bull. Am. Meteorol. Soc. 93, 485–498. http://dx.doi.org/10.1175/BAMS-D-11-00094.1.

Terray, L., Corre, L., Cravatte, S., Delcroix, T., Reverdin, G., Ribes, A., 2012. Near-surface salinity as nature's rain gauge to detect human influence on the tropical water cycle. J. Clim. 25, 958–977. http://dx.doi.org/10.1175/JCLI-D-10-05025.1.

Thomas, B.D., Thompson, T.G., Utterback, C.L., 1934. The electrical conductivity of sea water. J. Conseil 9, 28–34. http://dx.doi.org/10.1093/icesjms/9.1.28.

Tian, Y., Peters-Lidard, C.D., Eylander, J.N., Joyce, R.J., Huffman, G.J., Adler, R.F., Hsu, K.-L., Turk, F.J., Garcia, M., Zeng, J., 2009. Component analysis of errors in satellite-based precipitation estimates. J. Geophys. Res. 114, D24101. http://dx.doi.org/10.1029/2009JD011949.

Trenberth, K.E., Smith, L., Qian, T., Dai, A., Fasullo, J., 2007. Estimates of the global water budget and its annual cycle using observational and model data. J. Hydrometeorol. 8, 758–769. http://dx.doi.org/10.1175/JHM600.1.

Tzortzi, E., Josey, S.A., Srokosz, M., Gommenginger, C., 2013. Tropical Atlantic salinity variability: new insights from SMOS. Geophys. Res. Lett. 40 (2143–2147). http://dx.doi.org/10.1002/grl.50225.

van der Ent, R.J., Savenije, H.H., 2013. Oceanic sources of continental precipitation and the correlation with sea surface temperature. Water Resour. Res. 49 (7), 3993–4004. http://dx.doi.org/10.1002/wrcr.20296.

van der Ent, R.J., Savenije, H.H.G., Schaefli, B., Steele-Dunne, S.C., 2010. Origin and fate of atmospheric moisture over continents. Water Resour. Res. 46, W09525. http://dx.doi.org/10.1029/2010WR009127.

Vinayachandran, P.N., Shankar, D., Vernekar, S., Sandeep, K.K., Prakash, A., Neema, C.P., Chatterjee, A., 2013. A summer monsoon pump to keep the Bay of Bengal salty. Geophys. Res. Lett. 40 (1–6) http://dx.doi.org/10.1002/grl.50274.

Vinogradova, N.T., Ponte, R.M., 2013. Clarifying the link between surface salinity and freshwater fluxes on monthly to inter-annual time-scales. J. Geophys. Res. 118 (6), 3190–3201. http://dx.doi.org/10.1002/jgrc.20200.

Voldoire, A., Sanchez-Gomez, E., Salas y Mélia, D., Decharne, B., Cassou, C., Sénési, S., Valcke, S., Beau, I., Alias, A., Chevallier, M., Déqué, M., Deshayes, J., Douville, H., Fernandez, E., Madec, G., Maisonnave, E., Moine, M.-P., Planton, S., Saint-Martin, D., Szopa, S., Tyteca, S., Alkama, R., Belamari, S., Braun, A., Coquart, L., Chauvin, G., 2012. The CNRM-CM5.1 global climate model: description and basic evaluation. Clim. Dyn. 40 (9-10), 2091–2121. http://dx.doi.org/10.1007/s00382-011-1259-y.

von Schuckmann, K., Gaillard, F., Le Traon, P.-Y., 2009. Global hydrographic variability patterns during 2003-2008. J. Geophys. Res. 114, C09007. http://dx.doi.org/10.1029/2008JC005237.

Wadley, M.R., Bigg, G.R., 2006. Are "Great Salinity Anomalies" Advective? J. Clim. 19, 1080–1088. http://dx.doi.org/10.1175/JCLI3647.1.

Wenner, F., Smith, E.H., Soule, F.M., 1930. Apparatus for the determination aboard ship of the salinity of sea water by the electrical conductivity method. Bur. Stand. J. Res. 5 (3), 711–732., Available online, http://cdm16009.contentdm.oclc.org/cdm/ref/collection/p13011coll6/id/145548.

Wentz, F.J., Ricciardulli, L., Hilburn, K., Mears, C., 2007. How much more rain will global warming bring? Science 317 (5835), 233–235. http://dx.doi.org/10.1126/science.1140746.

Wong, A.P.S., Bindoff, N.L., Church, J.A., 1999. Large-scale freshening of intermediate waters in the Pacific and Indian oceans. Nature 400 (6743), 440–443. http://dx.doi.org/10.1038/22733.

Wooster, W.S., Taft, B.A., 1958. On the reliability of field measurements of temperature and salinity in the ocean. J. Mar. Res. 17, 552–566.

Wooster, W.S., Lee, A.J., Dietrich, G., 1969. Redefinition of salinity. Limnol. Oceanogr. 14 (3), 437–438. http://dx.doi.org/10.4319/lo.1969.14.3.0437.

Wüst, G., 1936. Surface Salinity, Evaporation and Precipitation over the World's Ocean. In: Länderkundliche Forschung. Festschrift Norbert Krebs, Stuttgart, Germany, pp. 347–359.

Wüst, G., 1964. The major deep-sea expeditions and research vessels 1873-1960: a contribution to the history of oceanography. Prog. Oceanogr. 2, 1–52. http://dx.doi.org/10.1016/0079-6611(64)90002-3.

Xie, P., Arkin, P.A., 1997. Global precipitation: a 17-year monthly analysis based on gauge observations, satellite estimates, and numerical model outputs. Bull. Am. Meteorol. Soc. 78, 2539–2558. http://dx.doi.org/10.1175/1520-0477(1997)078<2539:GPAYMA>2.0.CO;2.

Yin, X., Gruber, A., Arkin, P., 2004. Comparison of the GPCP and CMAP merged gauge-satellite monthly precipitation products for the period 1979-2001. J. Hydrometeorol. 5, 1207–1222. http://dx.doi.org/10.1175/JHM-392.1.

Yin, J., Stouffer, R.J., Spelman, M.J., Griffies, S.M., 2010a. Evaluating the uncertainty induced by the virtual salt flux assumption in climate simulations and future projections. J. Clim. 23, 80–96. http://dx.doi.org/10.1175/2009JCLI3084.1.

Yin, J., Griffies, S.M., Stouffer, R.J., 2010b. Spatial variability of sea-level rise in 21st century projections. J. Clim. 23, 4585–4607. http://dx.doi.org/10.1175/2010JCLI3533.1.

Yu, L., 2011. A global relationship between the ocean water cycle and near-surface salinity. J. Geophys. Res. 116, C10025. http://dx.doi.org/10.1029/2010JC006937.

Zhang, S., Rosati, A., 2010. An inflated ensemble filter for ocean data assimilation with a biased coupled GCM. Mon. Weather Rev. 138 (10), 3905–3931. http://dx.doi.org/10.1175/2010MWR3326.1.

Chapter 29

Ocean Heat Transport

Alison M. Macdonald* and Molly O. Baringer†

*Woods Hole Oceanographic Institution, Woods Hole, Massachusetts, USA

†NOAA/Atlantic Oceanographic and Meteorological Laboratory, Miami, Florida, USA

Chapter Outline

1. BACKGROUND

At the top of the atmosphere, the global energy balance is solely determined as the difference between the amount of solar energy that is absorbed by the earth and the amount reflected and reemitted back into space. At the surface, however, it is the distribution of the incoming and outgoing components that determines the balance and defines the climate of the environments in which we live. Radiant solar energy drives movement within the atmosphere and ocean, and the resulting circulation patterns within both systems provide feedbacks that continuously define and shape the global energy budget. This intertwined relationship between the atmosphere and ocean implies that oceanic transports of mass, and heat in particular, but more generally all properties, are intrinsically related to one another. They are affected by, and in some cases can affect, atmospheric circulation and conditions. Although both the atmosphere and ocean behave according to the laws of thermal and fluid dynamics, their differences allow them to play contrasting roles in the global energy balance. The ocean has a much greater capacity for energy storage and, being comparatively slow, has a long thermal memory. The atmosphere, moving more quickly, is an efficient mixer. Understanding the flow of energy both within and between the two systems is fundamental to our understanding and prediction of both natural climate variability and anthropogenically induced climate change.

In this chapter, as well as elsewhere in the literature, ocean heat transport is discussed with the recognition that although temperature is a property of water, heat is not. Here, the term *heat transport* is used synonymously with the more correct term *energy transport* (see Bryan (1962) and Warren (1999) for more complete discussions), while we reserve the term *temperature transport* for enthalpy, that is, for the specific instance where mass is not balanced (see Section 2.2). Several acronyms are used throughout the text. OHT refers to ocean heat transport. The term MHT refers specifically to meridional heat transport and is used here only in reference to the ocean. MOC is meridional overturning circulation.

In this chapter, we provide some historical background on the various methods used to calculate OHT and the resulting estimates; discussion of the techniques that have been used to decompose these estimates so as to better understand the physical mechanisms involved; and a discussion of OHT variability and its effects on OHT estimates. The reader is referred to earlier chapters such as 7

Ocean Circulation and Climate, Vol. 103. http://dx.doi.org/10.1016/B978-0-12-391851-2.00029-5

and 8 (diapycnal and lateral transport, respectively), 9 and 10 (overturning circulation), 11 (wind-driven circulation), and 13 and 14 (western and eastern boundary currents, respectively) for more in-depth consideration of the underlying circulation.

1.1. Energy Balance in the Atmosphere

At the top of the atmosphere, the global energy balance is between total incoming radiation from the sun and total outgoing radiation. The latter includes reflected solar radiation and emitted blackbody radiation. Total solar irradiance at the top of the atmosphere, also known as the solar constant (*S*), has a canonical mean value of 1365.4 ± 0.7 Wm^{-2} (Lee et al., 1995) and varies with the 11-year sunspot cycle by 1–2 Wm^{-2} (Lee et al., 1995; Willson and Hudson, 1991). The total solar radiation that the earth receives at the top of the atmosphere is a function of the earth's cross-section (πr^2) and because it rotates, its surface area ($4\pi r^2$). Therefore, the net incoming solar radiation is one-fourth the solar constant, *S*/4.

Estimates of the earth's radiation budget were made as early as 1837, but it was not until the early 1900s that a ground-based monitoring observatory allowed for observed estimates of incoming shortwave and outgoing longwave radiative components to be made (Abbot and Fowle, 1908 from Hunt et al., 1986). A balance was obtained using a global annual average incident solar flux of 364.4 Wm^{-2} based on the local observatory measurements. A decade later, Dines (1917) (from Hunt et al., 1986) improved upon this calculation, including terms representing downward longwave radiation from the atmosphere and upward nonradiative fluxes from the surface. These early estimates were not unreasonable, but they contained much detail based on broad assumptions made necessary by the dearth of observations. By the 1950s, ground-based observations had developed to the point that truly global estimates were possible. Hunt et al. (1986) attribute the first global study to London (1957).

The first satellite capable of observing meteorology was launched in 1959, but it was the Nimbus satellite program of the 1970s that flew instruments specifically designed to observe the earth's net energy budget at the top of the atmosphere by directly measuring solar irradiance, reflected shortwave radiation and emitted longwave radiation. It provided the first global continuous time series of these radiative components (House et al., 1986). Today's estimates of these components of the global energy budget (Figure 29.1) have been further refined by satellite monitoring of the top of the atmosphere, as well as by radiative models that estimate the partitioning of energy for different atmospheric conditions (e.g., high clouds, cloud-free, etc.). Together, these estimates suggest that about 23% of incoming solar radiation is reflected back out into space by the atmosphere. The atmosphere itself absorbs approximately the same amount, and yet it emits a great deal more (~60% of the incoming value) due to the complex energy exchange between the atmosphere and the surface of the earth. The earth as a whole, including the ocean, absorbs approximately half the incoming solar radiation (Figure 29.1), which is then returned to the atmosphere through remitted blackbody

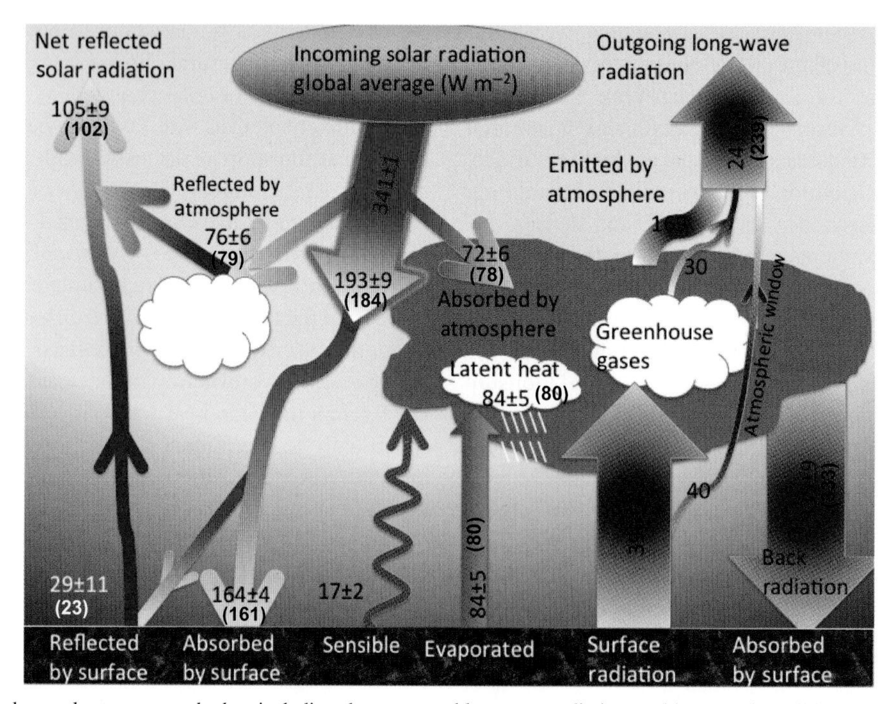

FIGURE 29.1 Global annual mean energy budget including shortwave and longwave radiation, and latent and sensible heat fluxes based on CERES (Clouds and the Earth's Radiant Energy System) data for the period March 2000 to May 2004. Values are the means and standard deviations of the four products compared in Trenberth et al. (2009). Values in parenthesis are the Trenberth et al. estimates where they differ from the mean (Units Wm^{-2}).

longwave radiation, evaporation (latent heat of vaporization), and conduction (sensible heat transfer).

The orientation of the earth to the sun causes the magnitude of net incoming shortwave solar radiation to be greatest near the equator (>300 Wm^{-2}) where it dominates the radiation budget at the top of the atmosphere, and least near the poles (<100 Wm^{-2}) where net upward longwave radiation dominates (Figure 29.2a). This configuration results in a tropical surplus and polar deficit in energy. The equator-to-pole difference in incoming radiation exceeds the equator-to-pole difference in upward radiation. This asymmetry is largely due to differences in absorbed radiation that are caused by the equator-to-pole gradient in albedo (Enderton and Marshall, 2009). According to climate models, there is maximum poleward heat transport (5.6 ± 0.8 PW in the northern hemisphere and -5.3 ± 1.1 PW in the southern hemisphere, $1\ PW = 1 \times 10^{15}$ W) at approximately 35° latitude in each hemisphere (Donohoe and Battisti, 2012). Observational estimates are consistent, if a little higher, at 5.9 ± 0.3 PW in the northern hemisphere and -5.9 ± 0.5 PW in the southern hemisphere (Fasullo and Trenberth, 2008). As a consequence, equatorward of this latitude there would be a net warming, and poleward, a net cooling, if the atmosphere and ocean did not provide the poleward transport of energy necessary to offset the imbalance (Figure 29.2d). Both regimes contribute, but here we focus on the ocean's role in the process.

1.2. Energy Balance at the Ocean Surface

The energy budget at the sea surface is a combination of the four components: net short and longwave radiation (e.g., Figure 29.2b), and net latent and sensible heat flux (e.g., Figure 29.2c). The term shortwave radiation is often used to describe the ultraviolet, visible, and infrared wavelengths (0.2–5 μ) of incoming solar energy, whereas longwave radiation is often associated with the infrared (5–200 μ) emitted by the cooler earth. However, it should be kept in mind that upward and downward components of both longwave and shortwave radiation exist within the energy budget below the top of the atmosphere (Figure 29.1).

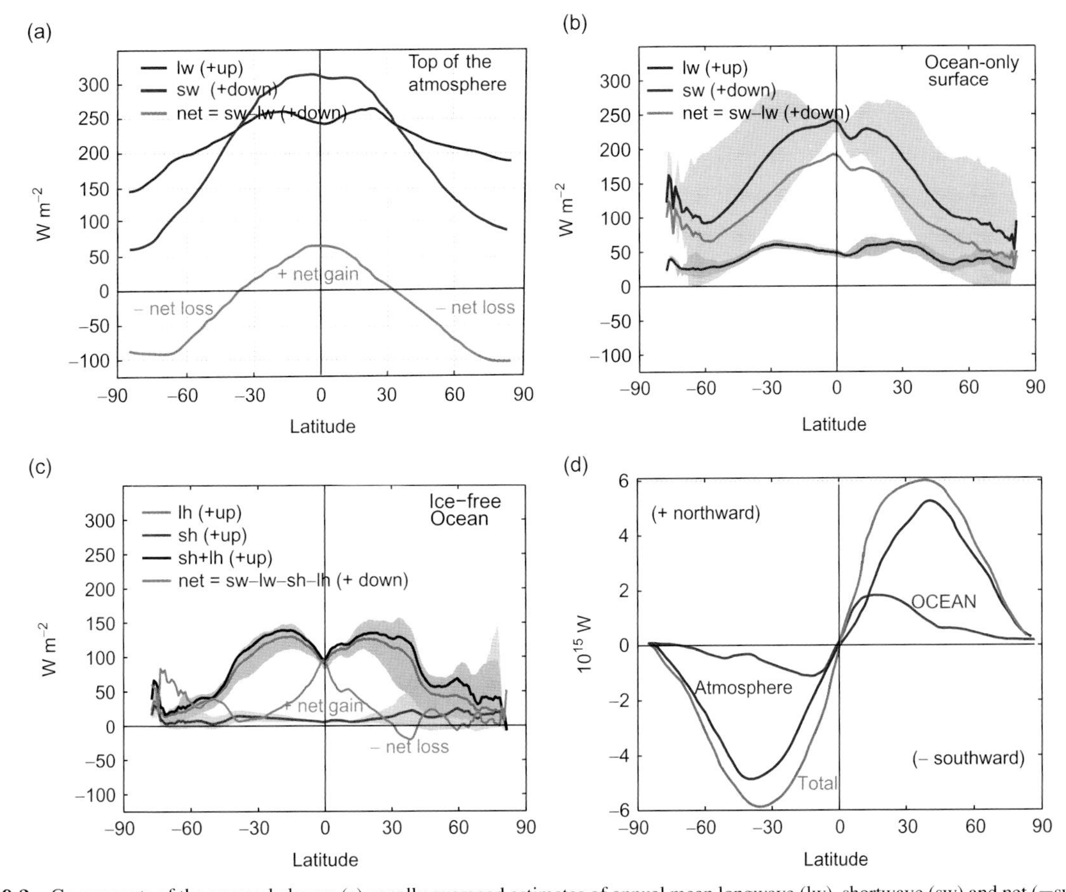

FIGURE 29.2 Components of the energy balance: (a) zonally averaged estimates of annual mean longwave (lw), shortwave (sw) and net (=sw – lw) radiation at (a) the top-of-the-atmosphere (estimates from Hatzianastassiou et al., 2004) and (b) the ocean surface (ISCCP data 1984–2007); (c) zonally averaged annual mean estimates of latent (lh), sensible (sh), latent + sensible and net (=sw – lw – lh – sh, postive down) heat flux. Gray shading illustrates range in monthly average values (OAFlux data from 1958 to 2007. Note this product does not balance globally—see text. Ocean heat flux estimates courtesy of L. Yu) (Units Wm^{-2}); (d) annual mean energy transport by the atmosphere (blue), ocean (green), and atmosphere + ocean (purple). Atmospheric and net values are based on the National Centers for Environmental Prediction-National Center for Atmospheric Research reanalysis, the Earth Radiation Budget Experiment and Global Ocean Data Assimilation System. The ocean curve is a residual estimate (Units PW). *Adapted from Fasullo and Trenberth (2008).*

At the surface of the ocean (Figure 29.2b), the shape of the curve describing the net incoming shortwave radiation is similar to the top-of-the-atmosphere global average (Figure 29.2a). Both imply greater warming at low latitudes than at high latitudes. Latent and sensible heat fluxes from the ocean are large over WBC regions (e.g., Yu et al., 2004). While the influence of subtropical trade winds is clear in the zonally averaged latent fluxes (Figure 29.2c, magenta curve), it is less obvious in the sensible fluxes which show maximum loss in the northern high latitudes (Figure 29.2c, green curve). Loss of energy from the ocean through sensible and latent heat fluxes reduces the magnitude of the net radiation (Figure 29.2c, cyan curve).

Without oceanic transport of energy from the equator to the poles, the global energy balance could not be maintained. Although this unidirectional transport (Figure 29.2d) is clearly implied by the top-of-the-atmosphere radiation distribution (e.g., Figure 29.2a and d), poleward movement of energy in the ocean is by no means ubiquitous (Figure 29.3). Restricted by land and topography, OHT maintains a complex relationship of feedbacks with the atmosphere that influences both the temperature and dynamics within and between the basins. The details of how energy is moved within the individual ocean basins and how it is exchanged between basins are therefore fundamental to our understanding of how ocean energy balances are maintained and how they might be vulnerable to change.

1.3. Heat Transport in the Ocean

The ocean transfers heat from one location to another by transporting waters of different temperatures in opposite directions. Through the dedicated efforts of the ocean-going research community to collect *in situ* full-depth observations in the late twentieth century, estimates of mean basin-scale circulation and MHT have been much improved. By way of introduction to these estimates, we begin here with a brief description of geographical patterns of MHT.

There is a prevailing pattern (Figure 29.3d) of northward MHT throughout the Atlantic Ocean. In the southern hemisphere, this heat transport is composed of northward flow of warm upper and intermediate layer waters entering the basin from the Agulhas region and Drake Passage (Beal et al., 2011; Garzoli and Matano, 2011; Rintoul, 1991; Weijer et al., 2002) (see Chapter 19) and southward flow of North Atlantic Deep Water (NADW) (see Chapter 10). In the North Atlantic, poleward MHT is carried by the relatively warm shallow western boundary current (WBC) and cooler southward return (NADW and abyssal overflow waters) both in the deep western boundary current (DWBC) and the interior. It is the loss of heat from the northward flowing surface waters to the atmosphere along their path throughout the Atlantic that preconditions the deep overturn in the sub-Arctic seas and southward return of cooler waters (Talley, 2003) (see Chapter 13).

In the Indian and Pacific basins, MHT is poleward. In the southern Indo-Pacific (Figure 29.3c), it is dominated by southward transport in the Indian Ocean (Figure 29.3b) supported mainly by a warm upper layer (but, not necessarily surface) export from the tropics of both basins and cooler inflow of Lower Circumpolar Deep Water (LCDW) from the Southern Ocean. Unlike Atlantic MHT, which is predominantly supported by thermohaline overturn, North Pacific MHT (Figure 29.3a), is dominated by the relatively shallow wind-driven gyre circulation and the temperature difference between the northward flowing WBC and the interior return (Bryden et al., 1991; Wijffels et al., 1996).

The simplicity of the above descriptions belies the complexity of the system and the many uncertainties that remain in our understanding of OHT. These uncertainties will be discussed in the context of the estimates themselves in the next two sections. Section 2 provides a basic description of indirect (Section 2.1) and direct (Section 2.2) techniques used to calculate MHT. Section 3 is a more detailed presentation of the resulting estimates that were briefly described earlier. There is an intended focus on the direct, *in situ*, non-numerical model-based estimates. For more discussion on indirect estimates, the reader is directed to Chapter 5 and detailed discussion of state estimates and reanalyses are provided in Chapters 21 and 22.

2. CALCULATION OF OCEAN HEAT TRANSPORT

While a variety of different techniques have been used to assess the role oceans play in the global energy balance, historically speaking, estimates of OHT fall into one of three categories: bulk formula calculations, residual calculations, and direct calculations. The first two methods produce what are considered indirect OHT estimates, as they do not require ocean water column observations. The third method is considered direct because it uses observations of both water temperature and velocity to estimate OHT.

2.1. Indirect Ocean Heat Transport Estimation

The main advantages of indirect methods are that they are capable of providing both global coverage and long, high-resolution, timeseries estimates of OHT. Their disadvantage is that they tend to contain strong regional biases that are not easily removed and can result in large uncertainties in OHT.

- Bulk formulae estimates use observations of radiation, air–sea freshwater and gas exchange, air and sea surface

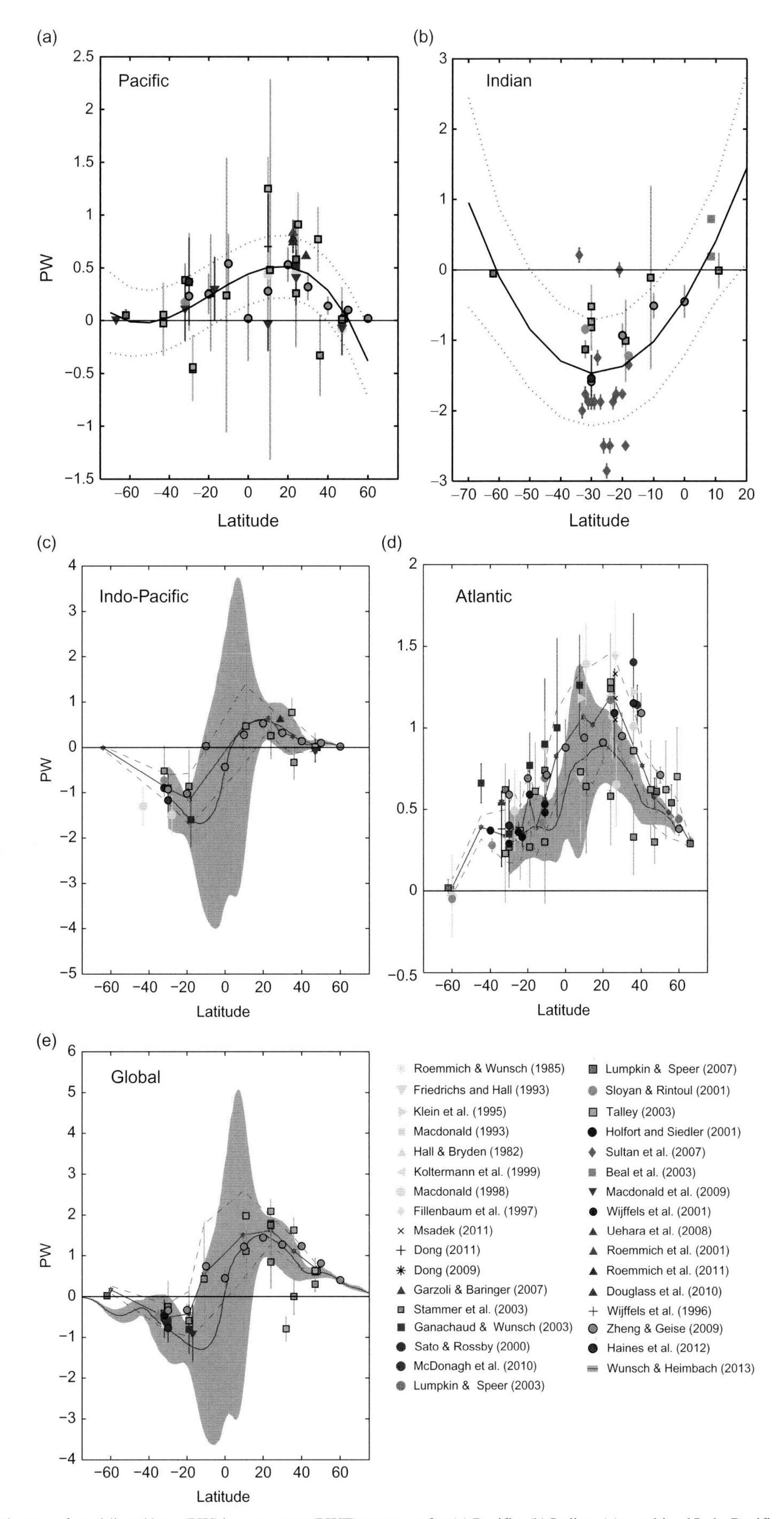

FIGURE 29.3 Estimates of meridional heat (PW)/temperature (PWT) transport for (a) Pacific, (b) Indian, (c) combined Indo-Pacific, (d) Atlantic, and (e) global ocean. Light gray and color markers indicate estimates from publications prior to and later than 2000, respectively. Values in (a) and (b) affected by ITF have been normalized to a 10 Sv throughflow. In (a–b), the thin solid and dashed lines represent a polynomial fit through the data points and the standard deviation of uncertainties. In (c–e), solid and dashed blue lines indicate the mean and standard deviation from Table 29.5. An illustration of the range in seasonal cycles is provided by the dark gray shading representing the standard deviation of the monthly ECCO-Production version 4 (ECCO-GODAE follow-on), global bi-decadal (1992–2010) state estimate (Wunsch and Heimbach, 2013).

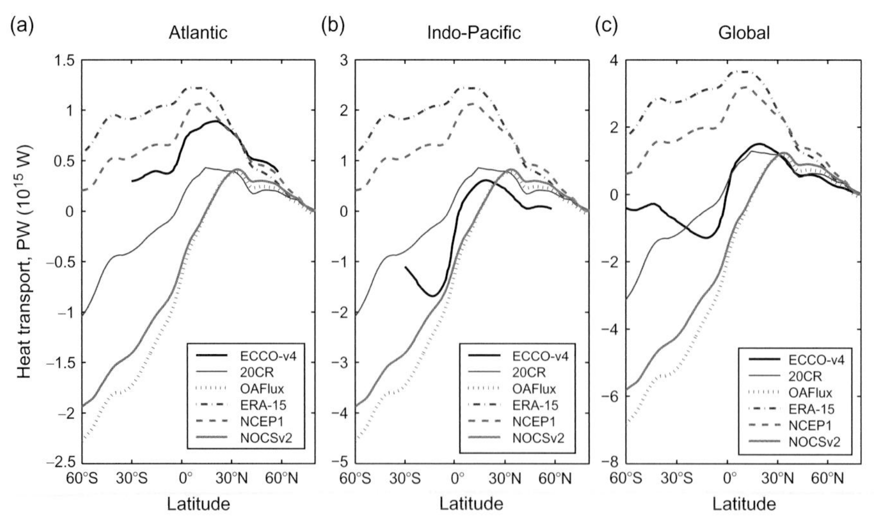

FIGURE 29.4 Meridional heat flux based on different flux products computed by southward integration for (a) the Atlantic, (b) the Indo-Pacific, and (c) the global ocean. The 20CR, OAFlux-ISCCP, ERA-15 (Gibson et al., 1997) and NOCSv2 products and references are listed in Table 29.1. NCEP1 is the original 40-year reanalysis climatology (Kalnay et al., 1996). The large spread away from 0 PW at the southern end of the integration is indicative of the various model biases. We have opted not to remove values representing the global bias from these curves under the assumption that had the individual authors understood these biases well enough to remove them, they would have done so. Please note the different vertical axis limits in each panel.

temperatures, wind speed, etc., along with equations that include empirical coefficients to calculate estimates of each of the four components of the surface energy budget (short and longwave radiation and latent and sensible heat fluxes). Air–sea heat fluxes computed globally are balanced by ocean heat storage and divergence of OHT. Assuming that there is no regional storage of heat within the ocean, the component fluxes can be integrated from north to south to obtain estimates of the OHT necessary to balance the loss or gain through air–sea exchange.

- Residual estimates use atmospheric energy transport estimates calculated from meteorological observations subtracted from the observed top-of-the-atmosphere energy budget to obtain the ocean transport component necessary to provide a balanced budget. The disadvantage of this technique is that any noise in the atmospheric energy budgets is propagated to the OHT estimates.

Bulk formulae parameterizations have steadily improved since the 1990s, from the 30–40 Wm^{-2} global mean uncertainties (da Silva et al., 1994; Josey et al., 1998; Large et al., 1997) that integrated to produce MHT uncertainties of 0.5 PW (1 PW $= 1 \times 10^{15}$ W) or more. Major steps forward included the use of global satellite observations and application of inverse methods that allowed regional parameterizations. Yu et al. (2004) looked to improve estimates using a variational objective analysis, and although they still found a large (30 Wm^{-2}) global imbalance, regional estimates agreed well with direct estimates outside the tropics (Macdonald et al., 2009; Yu et al., 2008). The implied MHT for several commonly available net surface flux products is shown in Figure 29.4. Improvements have also been derived from satellite time series that now span more than two decades. With adjustments, Large and Yeager (2009) find a climatological, global mean air–sea heat flux of 2 Wm^{-2}. Nevertheless, regional uncertainties remain as large as 40 Wm^{-2}. For more detail on where these uncertainties arise in the currently available products, see Chapter 5. Despite the issues with uncertainties, an important advantage to bulk formulae calculations of OHT is that they can provide times series for investigation of interannual variability (Section 5). It is also a technique that can be tuned to specific data sets, observations, or regional dynamics.

Residual methods have also been greatly improved by the global observations afforded by satellite technology including radiation and atmospheric temperature and moisture datasets. Other improvements include separation of land and sea domains, global models that have improved the accuracy and detail of estimates of nonradiative flux components, and improved understanding of the annual and interannual variations in the atmospheric energy budget. Nevertheless, satellite observations still find imbalances in the global mean top-of-the-atmosphere energy budget (6.4 Wm^{-2} from CERES (Trenberth et al., 2009)) that suggest warming greater than that expected from the anthropogenic increase in greenhouse aerosols alone (1.6 ± 0.3 Wm^{-2} (Hansen et al., 2011)). Likewise, models do not generally provide a radiative balance at the top of the atmosphere. Most predict radiative heating (Hatzianastassiou et al., 2004; Loeb et al., 2002; Wang et al., 2011). Further, the variety of observations now available are not easily synthesized into internally compatible data sets (Bourras, 2006; Brodeau et al., 2010; Garnier et al., 2000), which are needed to drive simulations or validate model results.

Looking at the range in estimates (Bourras, 2006; Liu et al., 2011b; Smith et al., 2011; Trenberth et al., 2009, 2011) for the various top-of-the-atmosphere and surface components of the energy budget from the extensive set of currently available flux products (Table 29.1) provides some measure of present-day uncertainty in the individual components. However, Trenberth et al. (2009) point out that the spread of estimates exaggerates the actual uncertainties, since many of the weaknesses in the individual products are understood (Berry and Kent, 2009; Moore and Refrew, 2002; Siqueira and Nobre, 2006; Smith et al., 2011; Uppala et al., 2005). For this reason, there is no one flux product that is best suited for all applications (Berry and Kent, 2009; Smith et al., 2011; Tomita et al., 2010). According to Trenberth et al. (2009), global top-of-the-atmosphere components are known to within about 3%, surface reflected solar radiation, latent heat of evaporation and net longwave radiation are good to about 10%, and other surface components to about 5% (see Figure 29.1 to convert percentages to approximate values in Wm^{-2}). They consider the largest remaining uncertainty to be due to clouds and the biased estimates of downward longwave radiation (back radiation in Figure 29.1). Adjustments of flux products have been able to reduce regional rms differences with *in situ* surface estimates to about 1–2 Wm^{-2} (Grist and Josey, 2003). However, global biases in the individual components of radiation remain on the order of 5–30 Wm^{-2} (Bourras, 2006; Large and Yeager, 2009; Liu et al., 2011a; Trenberth and Smith, 2009; Wang et al., 2011). These biases are particularly large in the Southern Ocean due to the sparsity in both time and space of directly observed surface meteorological data. The lack of Southern Ocean data is largely due to an absence of ship tracks. The result is a large spread in indirectly estimated MHT (Figure 29.4), particularly in the Southern Hemisphere.

In summary, over the past 10 years, there has been great improvement in flux products. Time series, multiple decades in length, are becoming the norm, making climate analyses feasible. To understand and reduce uncertainties, there remains a need for metrics and comparison, both to observations and among products. When it comes to estimating OHT, lack of direct observations in the Southern Ocean for comparison and constraint is likely the most difficult issue we face in coming years. Further details on surface fluxes and flux products can be found in Chapter 5.

2.2. Direct Ocean Heat Transport Estimation

Direct estimates of OHT are based on observations of ocean water column temperature and on either direct or inferred velocity estimates that are also derived from observations. The general equation for the calculation of heat transport perpendicular to a hydrographic transect (Bryan, 1962) is:

$$\iint \rho(x,z)C_p(x,z)\Theta(x,z)v(x,z)\mathrm{d}x\mathrm{d}z, \qquad (29.1)$$

where ρ represents density, C_p the specific heat of water at constant pressure, which is a function of temperature (T) and salinity (S), Θ the potential temperature, v the velocity perpendicular to the transect, and x and z the distance and depth coordinates. Density, C_p, Θ, and v are all functions of temperature, salinity and pressure and v is typically computed as a geostrophic balance with corrections for ageostrophic motion like Ekman transport included. The integral is taken vertically and horizontally across a slice of the ocean where a mass balance,

$$\iint \rho(x,z)v(x,z)\mathrm{d}x\mathrm{d}z = 0 \qquad (29.2)$$

exists.

It is understood that to compare estimates of heat transport there must be an underlying mass balance, otherwise a change in units such as from °C to °K would create a change in the magnitude of the heat transport estimate. To make the distinction, where mass balance is either not possible or not sensible, heat transport is referred to as *temperature transport*. It is referenced to °C and is accompanied by a net mass or volume transport estimate. Following Talley (2003), temperature transport is reported in units of PWT (PW with the T reminding the reader that the value is a nonmass-balanced temperature transport). Positive values indicate northward (and eastward) flow.

Estimates using Equations (29.1) and (29.2) are typically constructed from temperature, salinity, and pressure observations from hydrographic transects using a variety of techniques described in basic physical oceanography texts. One specific goal of the 1990s World Ocean Circulation Experiment (WOCE) was to obtain a quantitative description of the circulation of heat throughout the global ocean. Previously, the historical hydrographic record contained few basin-scale transects, and even fewer transects which conformed to any particular standard of measurement quality. The WOCE program expanded ocean observations to include multiple high-quality hydrographic zonal sections in each of the ocean basins, with well-defined best practices and standards. Supplemented by transect data from other programs that occurred from the late 1980s through the WOCE-period, a 10-year "synoptic" observational data base was obtained, which could be used to quantify the ocean component of MHT. The WOCE program included observations of direct velocity and temperature in strongly temporally varying boundary current regions, as well as interior absolute velocity measurements on some of the transects from Acoustic Doppler Current

TABLE 29.1 Some Currently Available Global Flux Products that Can be Used to Estimate OHT

Product	References	Comment
SOC/NOC1.1	Grist and Josey (2003)	COADS air–sea flux VOS climatology constrained by hydrography
	http://iridl.ldeo.columbia.edu/SOURCES/.SOC/.GASC97/.dataset_documentation.html	
ERA-15	Bromwich and Wang (2005)	ECMWF 15-year reanalysis, daily/monthly (1978–1994)
	http://www.ecmwf.int/research/era/ERA-15/	
ERA-40	Uppala et al. (2005)	ECMWF 40-year reanalysis, daily/monthly (1957–2002)
	http://www.ecmwf.int/products/data/archive/descriptions/e4/index.html	
JRA-25	Onogi et al. (2007)	Japanese 25-year reanalysis, 6 hourly (1979–2004)
	http://jra.kishou.go.jp/JRA-25/AboutJRA25_en.html	
OAFlux	Yu et al. (2008)	Objective analysis, near real-time reanalysis, daily/monthly (1958 on)
	http://oaflux.whoi.edu/index.html	
CORE v2	Large and Yeager (2009)	Air–sea flux components from atmospheric state fields (1949 on)
	http://rda.ucar.edu/datasets/ds260.2/	
NOCS v2	Berry and Kent (2009)	Air–sea flux from ICOADS (1973–2009)
	http://www.noc.soton.ac.uk/noc_flux/noc2.php	
ERA-I	Dee et al. (2011)	ECMWF near real-time reanalysis, daily/monthly (1979 on)
	http://www.ecmwf.int/products/data/archive/descriptions/ei/index.html	
MERRA	Bosilovich et al. (2011)	Satellite era daily reanalysis of GEOS-5 (1979 on)
	http://gmao.gsfc.nasa.gov/merra/	
CFSR	Saha et al. (2010) and Wang et al. (2011)	NCEP daily reanalysis including trace gases (1979 on)
	http://cfs.ncep.noaa.gov/	
20CR	Compo et al. (2011)	NOAA Sub-daily/daily/monthly (1870–2010)
	https://climatedataguide.ucar.edu/category/data-set-variables/reanalysis/noaa-20cr	
HOAPS 3.2	Fennig et al. (2012)	Satellite-only based climatology, 6 hourly/monthly (1987–2008)
	http://www.hoaps.zmaw.de/	
J-OFURO	Dee et al. (2011)	Satellite based climatology, daily/monthly (1988–2006)
	http://dtsv.scc.u-tokai.ac.jp/j-ofuro/dataset_information.html	
GSSTF 3	Shie et al. (2012)	Satellite based climatology, daily/monthly (1998–2008)
	http://disc.sci.gsfc.nasa.gov/datareleases/gsstf-version	
HOAPS 3.2	Fennig et al. (2012)	Satellite-only based climatology, 6 hourly/monthly (1987–2008)
	http://www.hoaps.zmaw.de/	

HOAPS, Hamburg Ocean Atmosphere Parameters and Fluxes from Satellite Data; SOC, Southampton Oceanography Centre; NOC, National Oceanography Centre; COADS, Comprehensive Ocean-Atmosphere Data Set; VOS, volunteer ship of opportunity; ECMWF, European Centre for Medium Range Weather Forecasts; CORE, Common Ocean Reference Experiment; MERRA, Modern-Era Retrospective Analysis for Research and Applications; GEOS, Goddard Earth Observing System data assimilation; CFSR, Climate Forcing System Reanalysis; OFURO, Japanese Ocean data Flux sets with Use of Remote sensing Observations; GSSTF, Goddard Satellite-based Surface Turbulent Fluxes. For further details and data, see the links provided. Also details on comparative features can be found in the climate reanalysis data guide provided by the National Center for Atmospheric Research: http://climatedataguide.ucar.edu/.

TABLE 29.2 Comparison of the BBH and SOV Decompositions from Talley (2003) Provided as Percentages of the Total Heat Transport According to Each Method Across 5 Zonal Sections Representing the Northern and Southern Hemisphere Subtropical Gyres

		SOV Decomposition (% of Total)			BBH Decomposition (% of Total)		
Section	Total Heat Transport (PW)	ITF/BS Throughflow (%)	SOV (%)	Int./Deep Overturn (%)	Barotropic (%)	Horizontal (%)	Baroclinic (%)
Atlantic 24°N	1.28	−2	32	**70**	−3	9	**94**
Pacific 24°N	0.81	17	**68**	15	4	**60**	37
Atlantic 32°S	0.23	−10	−52	**162**	−12	−108	**220**
Pacific 28°S	−0.43	−14	57	57	**52**	36	11
Indian 32°S	−0.59	12	**47**	41	−43	45	**98**

Positive percentages indicate transport in the direction of the total. Bold face values indicate the dominant components at individual latitudes according to each of the two methods.

Profilers (SADCP and LADCP, ship and lowered, respectively), and volunteer ship of opportunity repeat temperature profiles (Festa and Molinari, 1992). Over the last decade, the WOCE data set has been expanded to include repeats of many of these transects through the international efforts of the hydrographic component of the Climate Variability and Predictability (CLIVAR) program, which is now being coordinated through Global Ocean Ship-based Hydrographic Investigations Program (GO-SHIP) (see Chapter 3). Along with estimates at individual latitudes, many of these observations have been used in data syntheses that can compute or correct estimates of both heat transport and air–sea heat fluxes such as the inverse box model solutions of Ganachaud and Wunsch (2003), Lumpkin and Speer (2007), and Macdonald et al. (2009).

A major roadblock to calculating meridional heat transports from long-line hydrographic transects lies in calculating reliable estimates of the absolute velocity field. A variety of methods are used to estimate the unknown barotropic velocities inherent in velocity fields based on hydrography and these methods vary considerably by region and investigator. They will not be discussed here as they are described in most introductory texts. A major step forward was taken when Wunsch (1977) introduced the physical oceanography community to the inverse technique (Wiggins, 1972) for determining least squares (Lawson and Hanson, 1974) best-estimate solutions for these unknown velocity corrections. Initially used on single transects, it quickly became popular as a technique for combining observations from multiple transects and different data sets (Wunsch, 1996). The two main advantages of this method are that it encourages users to state all assumptions clearly, and when possible, produces solutions consistent with the stated assumptions. Describing the inverse method simply, observations are used to write a set of constraint equations (i.e., physical assumptions about dynamics, conservation, and uncertainties), which are formulated as an eigenvalue problem that can then be solved in a variety of ways. It was this type of formulism that eventually led to state-estimates (Chapters 21 and 22).

In the next section, we take a basin-by-basin look at some of the OHT estimates arising from the now diverse set of observations and calculation techniques.

3. OBSERVATION-BASED ESTIMATES OF OCEAN HEAT TRANSPORT

Using a hypothetical-closed overturning circulation, Jung (1952) made the remarkably accurate calculation that at 30°N the Atlantic Ocean could support a MHT of 3×10^{14} gm-cal s^{-1} (about 1.3 PW in modern units). He confirmed the result with observations from the Wüst and Defant (1936) Meteor Atlas. Half a century later, Bryden and Imawaki (2001) reported on a single global synthesis that estimated MHT across 15 zonal pre-WOCE transects (Macdonald, 1998) and was able to include 6 WOCE-era estimates in the Atlantic (Bacon, 1997; Holfort and Siedler 2001; Klein et al., 1995; Lavin et al., 1998; Saunders and King, 1995; Speer et al., 1996). In the intervening decade, the WOCE and CLIVAR programs, and volunteer ship of opportunity XBT programs have provided a relative plethora of ocean observations with which to investigate OHT (Figures 29.3 and 29.5, Tables 29.3–29.5). These new observations and analyses have provided further insights into our understanding of ocean heat transports, but have not yet answered many of the open questions. In the following subsections, we discuss direct steady-state OHT estimates in light of these estimates. Here and throughout this chapter, results are considered to be inconsistent (i.e., significantly different) when the stated values

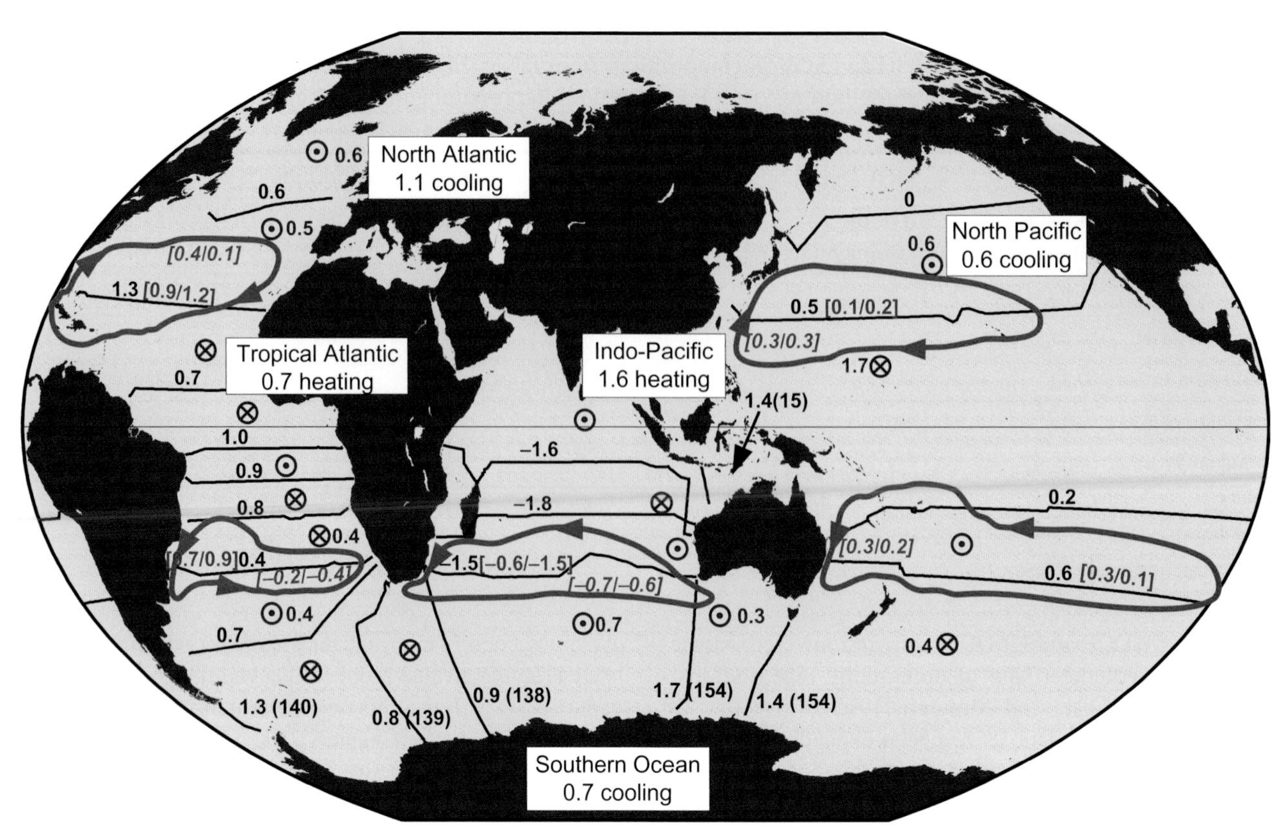

FIGURE 29.5 Estimates of ocean heat transport (black) with significant net volume transport in parentheses, convergence (gain of heat from the atmosphere) and divergence (loss of heat to the atmosphere) from Ganachaud and Wunsch (2000) (GW00). Red arrow tails and blue heads indicate convergence and divergence, respectively and estimated magnitudes are shown where they differ significantly from zero. Full basin convergence estimates are provided in boxes. Black lines indicate the location of hydrographic sections used by GW00 in their WOCE-era global inversion. Estimated uncertainties are provided in Tables 29.3 and 29.4. Values delineated by square brackets in cyan and blue are the estimated components of GW00's MHT at 24°N and 30°S due to gyre and overturn, respectively, based on Table 29.3. These components, separated by a "/," are the SOV shallow subducting overturn/BBH horizontal gyre components (cyan), and the SOV deep overturn/BBH baroclinic components (blue). The associated anticyclonic wind-driven gyre circulation is shown schematically in light blue with directional arrows according to T03 based on the predominantly pre-WOCE adjusted geostrophic fields of Reid (1994, 1997). Positive values indicate northward and eastward flow. Units (PW, PWT in ACC). *Adapted from both GW00 and T03.*

plus/minus the associated uncertainty reported by the authors do not overlap.

3.1. Atlantic

The Atlantic Ocean, the first place where MHT was estimated (Jung, 1952; Sverdrup, 1957), has arguably become the best-observed basin in terms of both mass and heat transport (Figure 29.3d). Throughout much of the Atlantic, the vertically integrated overturn and the temperature difference between surface/upper layer waters and the southward flowing NADW are primarily responsible for the northward transport of heat in both hemispheres (Figures 29.3a and 29.5, and see detailed discussion of particular estimates in Bryden and Imawaki, 2001).

Beginning our discussion in the north, Lumpkin and Speer (2007) estimate a MHT across 66°N of 0.29 ± 0.02 PW. They used five different surface air–sea flux products (da Silva et al., 1994; Grist and Josey, 2003; Josey et al., 1998; NCEP/NCAR and ERA-15) in their ocean inverse box model. They found that none of these products were capable of supplying sufficient heat loss to the atmosphere to allow formation of observed rates of Denmark Strait and Faroe Bank overflow and Antarctic Bottom Water (AABW) transport across 62°S. They reasoned that the models were unable to resolve the strong air–sea exchange that occurs at small scales (e.g., in narrow boundary currents, leads, and polynas that are common) at high latitudes. This observation may explain the disagreement between their estimate of MHT across 66°N and that of various flux products (Figure 29.4).

At 45–50°N, there are numerous MHT estimates. They consistently suggest a northward heat transport of about 0.6 PW. The weighted mean of the estimates in Table 29.3 (including only those estimates with uncertainties), and the rms uncertainty is 0.58 ± 0.24 PW (column 13). The average is lowered by the assimilation estimate of Stammer et al. (2003) who noted that their ocean MHT estimates generally tended to fall below those made directly from the observations, hypothesizing that the 2° horizontal resolution ECCO model used at that time did not well resolve either WBCs or the temperature difference

TABLE 29.3 Atlantic Ocean MHT Estimates, Predominantly Based on Observations (Lower Case Superscripts) with Indirect and Model Estimates (Upper Case Superscripts) Included Where Calculated by the Same Authors

Atlantic Meridional Heat Transport Estimates (PW)

Latitude	WH13	ZG09	LS07	T03	GW03	SE03	Other Estimates					Mean
66°N			0.29 ± 0.02					0.30 ± 0.03[j]				0.29 ± 0.03
59–60°N		0.38 ± 0.04		0.7				0.44 ± 0.03[j]				0.43 ± 0.18
53–56°N	0.43 ± 0.05		0.54 ± 0.11	0.62								0.48 ± 0.19
45–48°N	0.50 ± 0.09	0.71 ± 0.05	0.61 ± 0.13	0.62	0.60 ± 0.09	0.30 ± 0.13	0.65 ± 0.25[hh]	0.60 ± 0.04[j]	0.27 ± 0.15[a]	0.62 ± 0.11[a]	0.53 ± 0.12[a]	0.58 ± 0.24
40–41°N		1.09 ± 0.12					0.50 ± 0.10[ii]					0.77 ± 0.11
36–38°N	0.62 ± 0.17			0.86		0.33 ± 0.23	1.01 ± 0.26[hh]	1.14 ± 0.12[i]	0.47 ± 0.24[a]	1.29 ± 0.17[a]	0.70 ± 0.15[a]	0.88 ± 0.22
							0.8[n]	1.4 ± 0.3[o]	1.15[ll]			
24–26°N	0.84 ± 0.21	0.95 ± 0.14	1.24 ± 0.25	1.28	1.27 ± 0.15	0.58 ± 0.30	1.07 ± 0.26[hh]	1.17 ± 0.08[j]	1.38 ± 0.29[a]	1.48 ± 0.20[a]	1.54 ± 0.19[a]	1.20 ± 0.27
							1.33[b]	1.05–1.18[B]	1.44 ± 0.33[e]	1.2[dd]	1.21 ± 0.34[ff]	
							1.2 ± 0.3[p]	1.27 ± 0.26[q]	1.2 ± 0.26[q]	1.33 ± 0.26[q]	1.51 ± 0.39[bb]	
							1.33 ± 0.40[cc]	1.09 ± 0.27[ll]				
15°N	0.87 ± 0.31						1.22 ± 0.42[x]					0.39 ± 0.21
8–11°N	0.82 ± 0.47	0.94 ± 0.15		0.73	1.26 ± 0.31	0.64 ± 0.41	1.39 ± 0.25[hh]	1.1 ± 0.2[r]	1.18 ± 0.52[x]	1.28 ± 0.17[ee]		1.07 ± 0.33
0–5°S	0.40 ± 0.23	0.88 ± 0.17			1.00 ± 0.55			1.00 ± 0.14[ee]				0.83 ± 0.32
11–12°S	0.38 ± 0.20	0.71 ± 0.11	0.74 ± 0.36		0.9 ± 0.4	0.30 ± 0.38	0.89 ± 0.25[hh]	0.69 ± 0.10[j]	0.42 ± 0.05[k]	0.53 ± 0.08[s]	0.48 ± 0.09[s]	0.56 ± 0.26
15–16°S	0.40 ± 0.19			0.16			0.4 ± 0.02[y]					0.39 ± 0.21
19–20°S	0.38 ± 0.18	0.69 ± 0.08			0.77 ± 0.20	0.27 ± 0.25		0.59 ± 0.07[s]				0.58 ± 0.17
23–25°S	0.32 ± 0.17			0.37			0.33 ± 0.25[hh]	0.36 ± 0.06[s]	0.33 ± 0.06[s]			0.34 ± 0.19
27°S	0.31 ± 0.18						0.49 ± 0.25[hh]					0.39 ± 0.18
30–34°S	0.30 ± 0.20	0.59 ± 0.07	0.62 ± 0.16	0.23	0.35 ± 0.15	0.27 ± 0.25	0.55 ± 0.13[d]	0.38 ± 0.23[C]	0.29 ± 0.05[s]	0.22 ± 0.08[z]		0.34 ± 0.18
							0.25 ± 0.12[t]	0.3 ± 0.02[y]	0.40 ± 0.28[ll]			
45°S					0.66 ± 0.12		0.28 ± 0.04[k]	0.43 ± 0.08[z]	0.5 ± 0.1[aa]	0.37 ± 0.02[s]		0.39 ± 0.08
57–62°S			0.02 ± 0.05				−0.03 ± 0.25[hh]	−0.05 ± 0.01[k]				−0.03 ± 0.01

Columns 2–7: Global estimates from five publications, WH13—Wunsch and Heimbach (2013) uncertainties are a measure of the seasonal signal; ZG09—Zheng and Giese (2009) SODA Reanalysis (1958–2004) standard deviation calculated with annual cycle was removed; LS07—Lumpkin and Speer (2007); T03—Talley (2003) uncertainties 0.2–0.3 PW; GW03—Ganachaud and Wunsch (2003); SE03—Stammer et al. (2003). Columns 8–11: Section, basin, or hemisphere specific and pre-2000 estimates. Superscripts: [a]Koltermann et al. (1999) 1957–59, 1981–82, 1992–93 data from left to right, respectively; [b]Msadek (2011); [c]Dong et al. (2011); [d]Garzoli and Baringer (2007); [e]Fillenbaum et al. (1997); [f]Roemmich et al. (2001); [g]Roemmich and Gilson (2011); [h]Uehara et al. (2008); [i]McDonagh et al. (2010); [j]Lumpkin and Speer (2003); [k]Sloyan and Rintoul (2001) (ITF = 7.4–10.3); [m]Beal et al. (2003) Just the Arabian Sea; [n]Macdonald et al. (2009); [o]Sato and Rossby (2000); [p]Hall and Bryden (1982); [q]Lavín et al. (1998) 1957, 1981, 1992 data from left to right; [r]Friedrichs and Hall (1993); [s]Holfort and Siedler (2001); errors do not include ~0.24 PW unresolved uncertainty associated with wind stress and temporal variability. [t]Rintoul (1991); [u]Roemmich and McCallister (1989); [v]Bryden et al. (1991); [w]Wijffels et al. (2001); [x]Klein et al. (1995); [y]Macdonald (1993)(ITF = 10); [z]McDonagh and King (2005); [aa]Saunders and King (1995); [bb]Lavín et al. (2003); [cc]Johns et al. (2011); [dd]Roemmich and Wunsch (1985) data from 1957 and 1981; [ee]Marin (1998); [ff]Molinari et al. (1990); [gg]Wijffels et al. (1996); [hh]Macdonald (1998)(ITF = 8.1); [ii]Hobbs and Willis (2012) Argo; [ll]Haines et al. (2012) UR025.3 Reanalysis (1993–2009). Column 13: Weighted mean and rms uncertainty of OHT values in the row.

TABLE 29.4 Same as Table 29.3, but for Indo-Pacific

Indo-Pacific Meridional Heat Transport Estimates (PW)

	WH13	ZG09	LS07	T03	GW03	SE03				
Latitude	ITF = 16.4 ± 3.7	ITF = 15*	ITF = 13.2 ± 1.8	ITF = 8	ITF = 15 ± 3	ITF = 11	Other Estimates			Mean
47–48°N	0.09 ± 0.09	0.10 ± 0.02[+]	0.00 ± 0.07	0.01	0.00 ± 0.05	0.00 ± 0.13	−0.08 ± 0.25[hh]	−0.09[u]	−0.07 ± 0.1[n]	0.04 ± 0.16
35–36°N	0.24 ± 0.39	0.14 ± 0.39[++]		0.77		−0.33 ± 0.38		−0.16[u]		0.24 ± 0.37
29–30°N	0.47 ± 0.52	0.32 ± 0.12						0.61[h]		0.35 ± 0.38
22–24°N	0.55 ± 0.43	0.53 ± 0.16[+++]	0.58 ± 0.35	0.81	0.52 ± 0.20	0.26 ± 0.51	0.45 ± 0.26[hh]	0.83 ± 0.12[f]	0.77 ± 0.12[g]	0.64 ± 0.29
							0.74 ± 0.10[h]	0.41 ± 0.26[n]	0.75[u] ± 0.76[v]	
10–11°N	0.45 ± 2.55	0.28 ± 0.36		1.25		0.47 ± 1.8	0.44 ± 0.25[hh]	0.03 ± 0.26[n]	0.7 ± 0.5[gg]	0.51 ± 1.22
9°N	0.37 ± 3.19						0.19–0.72[m]			
17–20°S	−1.56 ± 0.59	−1.02 ± 0.18			−1.6 ± 0.6	−0.86 ± 0.80	−0.9 ± 0.7[n]			−1.15 ± 0.61
28–32°S[#]	−1.10 ± 0.68	−0.93 ± 0.12	−1.00 ± 0.19	−1.02	−0.9 ± 0.3	−0.52 ± 0.56	−1.34 ± 0.30[hh]	−0.74 ± 0.07[k]	−1.0 ± 0.1[y]	−0.91 ± 0.36
							−1.17 ± 0.37[jj]	−0.90[w]	−0.6 ± 0.4[n]	
62°S			0.00 ± 0.05							

Although some individual basin values have been summed to obtain total Indo-Pacific estimates, no ITF normalization has been performed, for the reason that ITF values are provided in the headers of columns 2–6 and in the caption of Table 29.1 for columns 7–10. *ZG09 average ITF transport is from Carton and Giese (2008). [+]Estimate is for 50°N. [++]Estimate is for 40°N. [+++]Estimate is for 20°N. [#]Pre-WOCE estimates of net Indo-Pacific heat transport across 32°S usually used the 1967 Scorpio 28°S observations in the Pacific Ocean (ITF units $10^6 ms^{-1}$).

TABLE 29.5 Same as Table 29.3 and 29.4, Zonally Integrated at Each Latitude Across all Basins

Global Ocean Meridional Heat Transport Estimates (PW)

Latitude	WH13	ZG09	LS07	T03	GW03	SE03	Other Estimates				Mean
47–48°N	0.58 ± 0.15	[+]0.81 ± 0.04	0.62 ± 0.08	0.63	0.6 ± 0.1	0.60 ± 0.10					0.68 ± 0.15
36°N	0.90 ± 0.50	[++]1.23 ± 0.10		1.63		0.00 ± 0.44					1.11 ± 0.37
24°N	1.42 ± 0.54	[+++]1.44 ± 0.15	1.75 ± 0.19	2.09	1.8 ± 0.3	0.84 ± 0.64					1.62 ± 0.40
8–11°N	1.27 ± 2.94	1.22 ± 0.28		1.98		1.11 ± 0.80					1.50 ± 1.54
10–11°S	−1.28 ± 2.08	0.74 ± 0.20				0.43 ± 1.40					0.55 ± 1.45
19–20°S	−1.18 ± 0.62	−0.33 ± 0.15			−0.8 ± 0.6	0.59 ± 0.84					−0.43 ± 0.61
30–32°S	−0.72 ± 0.77	−0.34 ± 0.11	0.41 ± 0.18	−0.79	−0.6 ± 0.3	−0.25 ± 0.62	−0.7 ± 0.1[y]	−0.46 ± 0.38[w]	−0.46 ± 0.08[k]	−0.77 ± 0.34[jj]	−0.51 ± 0.39
60–62°S	−0.40 ± 0.11		0.02 ± 0.07								0.17 ± 0.09

between WBCs and the interior. The mean estimate is also reduced by the 0.27 ± 0.15 PW estimate from Koltermann et al. (1999). This value was based on the IGY 1957 observations which, as Koltermann et al. point out, had much lower resolution O(100 nm) than more modern estimates. Using data from 1957, 1981, and 1993, they concluded that the differences in their MHT estimates (0.27–0.62 PW) were due to variations in the geostrophic circulation and, in particular, differences in MOC (estimated as 9.2 ± 2.0 Sv (1957), 18.9 ± 1.3 Sv (1981), and 14.7 ± 1.2 Sv (1993), where $1\ \text{Sv} = 1 \times 10^6\ \text{m}^3\ \text{s}^{-1}$). How much these particular MOC estimates are influenced by the details of the observations (resolution, seasonality, etc.) and physical assumptions as opposed to actual decadal changes is not obvious. The wide range of estimates, however, points to the tremendous influence of circulation on MHT and likely explains a great deal about the range in estimates at all latitudes.

At 36°N, the ocean MHT is about 1 PW (weighted mean $= 0.88 \pm 0.22$). Again estimates from the assimilation and the IGY dataset lie at the lower end of the spectrum. Koltermann et al. (1999) would suggest that the latter is for the same reason as at 48°N. Their MOC estimates at this latitude are 7.5 ± 2.2 Sv (spring 1959), 20.1 ± 1.4 Sv (summer 1981), and 12.3 ± 1.2 Sv (fall 1993). Similarly, Sato and Rossby (2000) found that WBC temperature flux at this latitude is dependent upon station density. Using historical hydrography (boundary current data from the early 1930s to the late 1980s), they find a seasonal cycle in heat transport at this latitude with a peak-to-peak range of 0.6 ± 0.1 PW, suggesting the spread seen by Koltermann et al. may also be attributable to a seasonal signal. It should be noted that the global inverse solution of Ganachaud and Wunsch (2000) did not include the 1993 data set used by Koltermann et al. at this latitude because it was not thought to meet WOCE standards. Combining hydrography and ADCP data from the May–June 2005 CLIVAR dataset, McDonagh et al. (2010) find 1.14 ± 0.12 PW crossing 36°N, an estimate higher than, but consistent with, Koltermann's summer 1981 estimate, and inconsistent with both the spring 1957 and fall 1993 estimates. McDonagh et al. determined that as expected the heat transport here is dominated by the overturn: 0.75 PW associated with overturn and a 10 °C temperature difference between the upper 1000 m and the water below and 0.39 PW associated with horizontal transport predominantly in the upper 800 m.

Across the line that lies nominally between 24.5 and 27°N, a diverse assortment of measurement techniques have been applied to the direct estimation of both the MOC and MHT, including repeat hydrography, moorings, XBTs, and most recently, the integrated MOCHA/RAPID array (see Section 5 for time series estimates). As at latitudes to the north, the estimate of WBC transport has been found to be important to MHT estimates, as has the resolution of the particular sections used. The weighted mean of the values in Table 29.3 is 1.20 ± 0.27 PW, which is consistent with the mean estimate from 3.5 years of time series data from the MOCHA/RAPID array (1.33 ± 0.40 PW (Johns et al., 2011)). The range in nonassimilation-based estimates (0.5 PW) is a large fraction of the mean. A portion of the spread is due to the choice by some to use *in situ* cruise surface temperatures and winds, rather than annual means, in calculating the Ekman component of MHT. Lavín et al. (2003) attribute 0.14 PW to what they term "seasonal sampling," referring to MHT calculated using the mean Florida Strait temperature from their 1992 cruise compared to that using an annual mean. The half petawatt spread in values is similar to the 0.6 PW annual range reported by Johns et al. (2011) for the RAPID time series.

At 15°N, the latitude at which indirect estimates would suggest there is a maximum in poleward MHT (Figures 29.2d and 29.4), there is only a single direct estimate (Klein et al., 1995) based on the combination of two pre-WOCE hydrographic lines. The main challenges encountered in directly measuring MHT with hydrography in the tropics are high variability in winds and in the WBC. The geometry near 15°N in the Atlantic further complicates matters because the unique Lesser Antilles Island chain enclosing the Caribbean Sea makes designing a hydrographic section to cleanly measure WBC transport extremely difficult. Klein et al. (1995) report that of the 1.18 ± 0.52 PW and 1.22 ± 0.42 PW at 8 and 14.5°N, respectively, most is associated with the annual mean Ekman transport. Due to the strong seasonal variation in the winds, using seasonal (i.e., *in situ*) estimates of the Ekman components increases these estimates to 1.67 and 1.37 PW, respectively. This result is consistent with Friedrichs and Hall (1993) who found their MHT estimate based on *in situ* winds at 11°N was 0.8 PW, less than the 1.1 ± 0.2 PW estimated as an annual average. Nevertheless, it appears that large seasonal variations in wind-induced heat transport do not necessarily invalidate "steady-state" hydrographic estimates as long as annual average winds are used. This is because while time-mean Ekman mass transports are balanced at relatively shallow depths (explaining their importance to the total heat transport in regions where they are strong), seasonal variations in Ekman transport are compensated by depth-independent flow (Jayne and Marotzke, 2001) that is not measured by hydrography.

Estimates of Atlantic Ocean MHT between 8°N and 5°S have average values of 1.07 ± 0.33 PW and 0.83 ± 0.32 PW northward, respectively, indicating a small O(0.2 PW) cross-hemispheric gradient that is not significantly different from zero. Only the somewhat lower Talley (2003) estimates would indicate otherwise. She predicts a strong significant divergence of $\sim 0.57 \pm 0.3$ PW. It should be noted that the Reid (1994) data set as interpreted by Talley (2003) offers our only MHT estimate at 8°S. It is southward (−0.22 PW) as is the accompanying estimate at 11°S (−0.38 PW). However, as there was concern that these values may have been tainted by suspect geostrophic velocities (Talley, 2003) and since the 11°S value differs significantly from nearly all other MHT estimates at this latitude, they were not included in the tables or figures.

Apart from these two particular values, throughout the South Atlantic, direct MHT estimates suggest northward transport with little gradient with latitude. Displaying remarkable consistency through data sets and methods, the weighted average of all the northward MHT estimates in Table 29.3 between 11 and 45°S that include estimates of uncertainty is 0.4 ± 0.2 PW. Note however, that the air–sea flux products disagree about the magnitude of the gradient in MHT in the South Atlantic (Figure 29.4). While the earlier NCEP1 and ERA-40 show little gradient, agreeing with the ECCO state estimation and direct observations, other and notably newer flux products show a fairly strong gradient in the South Atlantic. This disagreement is likely attributable to the regional biases mentioned in Section 2.1 and the fact that the older flux products contained specific flux corrections that are no longer used.

In summary, observation-based MHT estimates in the Atlantic reveal little gradient to the south of 15°S (about 0.4 PW); an increase to about 0.8 PW at 5°S; a cross-equatorial gradient that cannot be statistically distinguished from zero due to the large uncertainty arising from highly variable tropical winds and too few observations; and a maximum MHT of about 1.3 PW in the northern hemisphere subtropics (25°N), which decreases with increasing latitude (0.9 PW at 36°N, 0.6 PW at 47°N, 0.3 PW at 66°N). Issues affecting the estimates include technical details such as the low resolution of earlier sections and assimilative models, the choice of wind and air–sea flux products used as well as choices in how they are applied, and changes in the circulation itself, such as variable winds, seasonal changes in upper water column temperature, and differing MOC estimates.

3.2. Indo-Pacific

Like the Atlantic, the meridional transport of heat in the Pacific Ocean is principally poleward in the northern hemisphere. Recall that in the North Atlantic, the wind-driven gyre and the MOC are closely related and northward MHT arises from poleward transport of warm western boundary surface waters and southward return of waters that are on average colder and much deeper. In contrast, the Pacific wind-driven upper layer and deepwater overturning circulations exist in virtual isolation (Bryden et al., 1991; Wijffels et al., 1996). Pacific vertical overturn does not entail deepwater formation from ventilated surface waters but rather a deep overturning cell with a much smaller temperature difference between northward and southward components than in the Atlantic. The bulk of the northward MHT is a result of the wind-driven gyre circulation with poleward transport of warm surface waters and an only slightly cooler return flow within the thermocline (Bryden et al., 1991; Talley, 2003; Macdonald et al., 2009). Hence, in the North Pacific, poleward MHT occurs more as a result of the zonal relationship between temperature and velocity (wind-driven gyre) than the vertical relationship (overturn).

Reviewing the observations from north to south, at 47°N in the Pacific, none of the available estimates suggest any significant heat transport, but that is not to say that there are not significant north/south flows. Most of the strong flows appear to occur in the west where, according to Macdonald et al. (2009), the WBC carries 20.2×10^9 kg s^{-1} northward at a mean temperature of 3.7 °C (0.3 ± 0.1 PWT), which is compensated by a balancing return transport to the west of 160°E.

At 36°N, results are less conclusive as there is little consistency among the few available MHT estimates. Roemmich and McCallister (1989) suggest that their −0.16 PW is possibly underestimated due to poor sampling in the region of the WBC and note that a 10 Sv increase in Kuroshio transport at 15 °C would increase estimated MHT to +0.2 PW. If this is indeed the case, then the even greater southward estimate of Stammer et al. (2003) likely also suffers from the poor horizontal resolution of the underlying GCM. On the other hand, Wang and Carton's (2002) assimilation estimate using 2.5° horizontal resolution model finds a northward transport of ~0.35 PW, and the Talley (2003) estimate based on several cruises between 1965 and 1991, also northward, is twice as large again as this at 0.77 PW. This spread in estimates, that may have its origin in the spatial and temporal resolution of the observations and models used, is an indication that understanding the role Kuroshio plays in the region in both transporting energy northward and releasing it to the atmosphere is vital to our interpretation of the resulting estimates of MHT and its variability.

Though still presenting a fairly large range in possible values, MHT estimates near 24°N in the Pacific are more consistent than those at 36°N. This may be because the Kuroshio transport at this latitude is weaker, 24–33 Sv (Bingham and Talley, 1991; Ichikawa and Beardsley, 1993; Macdonald et al., 2009; Roemmich and McCallister, 1989) and more stable than to the north, 57–90 Sv (Book et al., 2002; Imawaki et al., 2001; Jayne et al., 2009; Macdonald et al., 2009). However, unlike 24°N in the Atlantic where a myriad of data sets are available, the mean Pacific MHT of 0.66 ± 0.28 PW comes from nine calculations, nearly all of which have been based on the same 1985 transect. The Uehara et al. (2008) estimate of 0.61 PW estimate stands out as it is based on XBT sections conducted between 1998 and 2002 between Honolulu and the east and west coast at ~22°N and 40°N, respectively.

The uncertainties in their calculation involved the estimated salinities and dearth of observations below the depth of the XBTs (about 800 m) and, as with all sections, sampling and Ekman transport estimation; the latter particularly because the section reaches as far south as 15°N in the tropics where the wind stress is more variable.

At 10°N, the weighted mean of the available estimates is 0.56 ± 0.86 PW with a spread of 1.2 PW. All these estimates are also surprisingly based on the same 1989 transect. It appears that the strong trade winds (capable of producing 25–40 Sv of Ekman transport across this latitude) and the strong currents that are known to be subject to both seasonal and interannual variations (Johnson et al., 2002) have made consistent estimation of a mean heat transport in the region difficult and highly susceptible to the details in choices of reference level, wind product, and so on.

Consideration of ocean heat or temperature transport closer to the equator is complicated by the highly variable, and therefore difficult to observe, flow through the Indonesian Archipelago (abbreviated as the Indonesian Throughflow or ITF; see Chapter 19). On a technical level, comparison of heat transport estimates is made more difficult because assumptions about where within the water column the compensation is expected to occur differ from study to study. Investigations within the Archipelago itself are further complicated by both geography and politics. The result is varying estimates such as the Indian Ocean hydrographic estimate of 1.4 PWT for a 15 Sv ITF (Ganachaud and Wunsch, 2000), a mooring estimate in the Makassar Strait of 0.55 PWT for a 10 Sv ITF (Vranes et al., 2002), Ekman transport estimates of ~0.5 PW for the southeast monsoon and ~1 PWT for the northwest monsoon timeframes (Sprintall and Liu, 2005), and a high-resolution modeling result of 1.1 PWT for 10.5 Sv (Pandey and Pandey, 2006).

In the South Pacific, the MHT is more properly called temperature transport because there is a net transport supporting the ITF. Because much of the water in the throughflow comes from the north, the temperature transport in the South Pacific is weak (see Table 29.4 references). Southern Indian Ocean temperature transports are strong (Figure 29.5), dominate the Indo-Pacific heat budget, and according to Vranes et al. (2002) lose negligible energy to the north of 30°S. The observational synthesis of Sultan et al. (2007) links southern Indian MHT intensity to the Agulhas Current strength and to the vertical mixing within the basin.

To summarize, the Indo-Pacific as a whole supports poleward heat transport in both hemispheres. There is little net meridional temperature transport in the South Pacific and strong southward temperature flux in the South Indian Ocean, supported partially by input from the North Pacific. As in the Atlantic, there is a maximum northern hemisphere MHT in the subtropics with lower values at higher latitudes. Many of the issues are the same as found in the Atlantic. In particular, more repeat observations are needed as strong, variable tropical winds present an even more serious problem to MHT estimates in the wide Pacific, while strong seasonality in the Indian Ocean requires more observations than presently available to gain better statistics.

3.3. Southern Ocean

In the Southern Ocean (SO), eddies play an increasingly important role in setting the MHT (Bryden, 1979; de Szoeke and Levine, 1981; Gille, 2003). Lagrangian float observations have determined an implied mean poleward eddy-driven MHT across the ACC of 0.3 ± 0.1 to 0.6 ± 0.1 PW. It is dominated by cross-stream fluxes in the Agulhas Retroflection region and supported by poleward fluxes in the ACC core (Gille, 2003). Satellite observations suggest that on decadal scales, SO wind-stress forcing is increasing (e.g., Dong et al., 2011; Thompson and Solomon, 2002). Increased winds contribute to both a stronger ACC and a richer eddy field. An eddy-resolving modeling study finds that such an increase first leads to a transient SO cooling due to increased northward Ekman transport, followed on longer timescales by a surface warming induced by the wind-enhanced eddy field (Hogg et al., 2008).

However, the SO is not solely dependent on eddies for MHT. Supported by an import of heat from the Brazil and Agulhas Currents, the ACC carries energy zonally from basin to basin within the Southern Ocean (Figure 29.5). Through its meridional excursions, it is warmed on its northward path in the Atlantic and Indian Basins and cooled on its way southward through the Pacific. In this way, the ACC directly contributes to the overall meridional energy budget with a mean baroclinic flow relative to 3000 dbar carrying -0.14 ± 0.01 PW, 0.095 ± 0.009 PW, and 0.082 ± 0.008 PW across 56°S, 58°S, and 60°S, respectively (Sun and Watts, 2002). Note that the flow below 3000 m is excluded from the Sun and Watts study. The ECCO2 synthesis of Volkov et al. (2010) concludes that transient features contribute to the gyre component, but not the overturning component. They show that the OHT in the SO is largely confined to the richly energetic fields of the Agulhas Retroflection and the Brazil-Malvinas Confluence. They also determine that the Indian Ocean sector is the dominant pathway for MHT due to the relatively large temperature contrast associated with this Warm Water Pathway of the MOC. The general northward MHT in the SO is opposed by deep AABW overturn. An OCCAM high-resolution model designed to better resolve AABW export finds that although transients contribute to the MHT, deep overturn is responsible in the annual mean for a southward MHT of −0.033 PW, and on shorter timescales for as much as −0.061 PW (Heywood and Stevens, 2007).

In summary, there are few hydrographically based estimates of MHT in the SO. This lack is due partially to the difficulty of the calculation in a region with minimal boundaries, where so many isopycnals outcrop and winds are so strong. It is also due to a severe lack of data. SO observations are notoriously difficult to obtain and hard to analyze. For this reason, much of what is known comes from other sources such as remote instrumentation (e.g., floats and satellites) and models. With strong variability and strong surface fluxes, the SO exemplifies the need to understand the interaction between the ocean and the atmosphere to better interpret OHT estimates and to predict seasonal and climate-scale changes.

4. UNDERSTANDING MECHANISMS

As described in Section 3, there is a large and growing database of observations that can provide MHT and OHT values to be compared and used to improve our understanding of how the ocean moves energy from one location to another. A variety of techniques have been developed to investigate the physical mechanisms of these energy transfers. In particular, to better understand how the underlying ocean circulation transports energy (i.e., heat) from one place to another, a number of different decomposition techniques have been employed. Here we describe two such methods.

4.1. Barotropic-Baroclinic-Horizontal Decomposition (BBH)

Hall and Bryden (1982) decomposed MHT in their analysis at 25°N in the Atlantic into a depth-averaged transport (barotropic) and a depth dependent transport (baroclinic). Bryan (1962), in his analysis of model results, suggested a similar decomposition that included a third, smaller term representing horizontal diffusion. Recognizing the unique contribution of both horizontal and vertical overturn to ocean MHT, Bryden and Imawaki (2001) put forth a variation on the barotropic/baroclinic (BB) decomposition. Reminiscent of the Bryan (1962) method, this decomposition more directly discriminates between the two components. It separates the velocity and temperature terms in Equation (29.1) into a barotropic component based on the section averaged velocity $[v]$, temperature $[\Theta]$ and density $[\rho]$:

$$[\rho]\ [\mathrm{Cp}]\ [v]\ [\Theta]\iint \mathrm{d}x\mathrm{d}z,$$

a baroclinic component based on the horizontal average values $\langle v\rangle\ (z)$ and $\langle\Theta\rangle\ (z)$:

$$\int \rho \mathrm{Cp}\langle v\rangle(z)\ \langle\Theta\rangle(z)\int \mathrm{d}x\mathrm{d}z,$$

and a horizontal (gyre) component based on the horizontal anomalies: $v'(x, z)$ and $\Theta'(x, z)$

$$\int \mathrm{d}z\int \rho \mathrm{Cp}v'(x,z)\ \Theta'(x,z)\mathrm{d}x,$$

where, $v=[v]+\langle v\rangle(z)+v'(x,z)$ and $\Theta=[\Theta]+\langle\Theta\rangle(z)+\Theta'(x,z)$. We refer to this technique as the BBH—barotropic, baroclinic, and horizontal decomposition.

In their investigation comparing the 2009 and 1983 meridional heat and freshwater transports across 24°S in the Atlantic, Bryden et al. (2011) used this decomposition to illustrate the strong heat transport associated with overturning circulation (0.76 PW in 2009 and 0.53 PW in 1983) and the weaker (−0.07 PW in 2009 and −0.14 PW in 1983) horizontal component dominated by the shallow southward flowing western boundary Brazil Current. The removal of the section-average barotropic term in this decomposition forces a zero net mass transport for both the baroclinic and the horizontal terms. This allows for a comparison of these terms even when a net mass transport is unavoidable as is the case for the ACC, Pacific-Indian Throughflow, and indeed any zonal transect affected by Bering Strait throughflow and/or freshwater transport. The utility of intercomparing different sections using this decomposition was shown by Lavín et al.'s (2003) analysis of hydrographic lines spanning three decades. Here, rather than using an estimate of the absolute velocity field, they simplified their analysis by combining the wind-driven component with the barotropic term, thus making the other two terms more easily comparable.

4.2. Shallow Subducting Overturn Decomposition (SOV)

A different decomposition technique (Roemmich and Wunsch, 1985) has been applied globally by Talley (2003) to separate the heat transport associated with the shallow, subducting component of subtropical gyres from that accompanying deep meridional overturn. This decomposition is based on the premise that whether or not deep overturn exists or is dominant, shallow gyre overturn will occur in the presence of a warm poleward-oriented WBC, whose waters are cooled along its path and subducted (i.e., overturned) before returning equatorward. This process is typical of wind-driven gyres subject to air–sea buoyancy loss and should be distinguished from intergyre/interhemispheric large-scale overturn. As mentioned earlier, this shallow subducting overturn (SOV) dominates in the Pacific (Bryden et al., 1991). Talley (2003) hypothesizes that it is present to a greater or lesser degree in all subtropical gyres, and the goal of the decomposition is to estimate the relative influence of intermediate, deep, and bottom flows on the MHT.

The technique which is referred to here as the SOV method is outlined as follows: (a) The base of the SOV is defined as the maximum wintertime density along the zero-Sverdrup transport line between the subtropical and subpolar gyres. This definition is chosen to separate the shallow heat transport arising from cooling of the near-surface wind-driven gyre circulation from that supported by deeper overturn as it represents the densest water that could be subducted within the gyre. The density and depth of the SOV vary considerably from basin to basin. For example (Talley, 2003), the SOV base taken as $\sigma_0 = 27.3$ lies between ~700 and 800 m at 24°N in the Atlantic compared to the $\sigma_0 = 26.2$ SOV base lying between ~100 and 600 m at the same latitude in the Pacific; (b) Above the SOV base, along a transect within the subtropical gyre, interior geostrophic transports are summed with the Ekman transport perpendicular to the section and that portion of the WBC transport that provides either mass balance or a known net mass transport. Talley suggests two methods of distributing the WBC portion to yield upper and lower estimates of poleward heat transport. Net transports across the sections are also apportioned so as to provide extrema for the estimated heat transport values; (c) The estimated SOV mass transports are subtracted from the total transports in the SOV layers to provide an estimate of the transport that is assumed to be transformed into water denser than that associated with the SOV, and here Talley assumes that this water is transformed into the nearest available density class; (d) Then, heat and temperature transports are calculated by zonal integration within density classes.

Although subjective in nature, with judicious selection of extrema, the SOV technique provides a unique view into the circulation of energy within the ocean. It improves understanding and quantification of the water mass transformations associated with the subduction process, which otherwise have a tendency to be lumped into a single "horizontal" or "gyre," nonoverturn value. Using this technique, Talley (2003) found that there is a measureable SOV within each of the subtropical gyres. This SOV water is transformed into an intermediate density class (500–2000 m) in the northern-hemisphere gyres, or into deep density classes in the southern-hemisphere gyres and that portion of the North Atlantic influenced by deep overflows. Further, she determined that separated from the SOV, Indo-Pacific temperature transport associated with ITF is modest ($\leq$0.1 PWT) and the bulk of the Indo-Pacific MHT is carried in the shallow "supergyre" overturn that includes the ITF. Formation rates of NADW and AABW are similar. However, because temperature differences between NADW/LCDW and AABW at 30°S are relatively small, the heat transport associated with Antarctic dense water formation is similarly small.

In summary, both the BBH and SOV techniques look to separate the throughflow, overturning, and horizontal transports. According to all the decomposition methods, deep overturn is the dominant mechanism by which heat is transported northward (88% according to Johns et al. (2011) BB method; 67% Talley (2003) SOV method; 62% Lavin et al. (2003) BBH method) in the subtropical North Atlantic. Nevertheless, by most accounts, the horizontal/gyre component is nontrivial unless, as Talley (2003) points out, the warm WBC water is not strongly surface-intensified. If this is the case, the portion of the total heat transport for which the SOV is responsible is reduced due to a reduction in temperature contrast between the WBC and the subtropical gyre return flow. In the North Pacific, all studies find that MHT is dominated by shallow gyre scale overturn (Table 29.2, Figure 29.5). However, Talley (2003) concluded that due to differing underlying assumptions, the BBH and SOV decompositions are not directly comparable. Regardless of the decomposition used, the user must still make choices regarding the Ekman component, which wind product to use, and how to apply it. Neither method includes a time varying component although there is no mathematical reason why they should not. Variability within the ocean has been recognized for a long time, but acquiring the necessary observations for estimating variations in OHT has only recently been initiated.

5. OCEAN HEAT TRANSPORT VARIABILITY

Ocean heat transport variability estimates have been derived from observations using several techniques, including repeat full water column hydrography (providing exceptional spatial resolution and extremely poor temporal resolution); *in situ* partial water column (surface to 750 or 2000 m) XBT and Argo data (providing better temporal resolution and for the latter, improved spatial coverage); and moored instruments spanning the breadth of a basin providing a spatially subsampled data set with excellent temporal resolution. Global ocean general circulation model and coupled model assimilations (see Chapters 21 and 22) provide a means to combine multiple and varied observations and generate time series linking the more sparse observations. The expectation is that model results will better represent reality when they are consistent with observations, and interpretation of diverse observations will be improved by imposing conservation equations. Originally including satellite altimetry, historical hydrography, and atmospheric estimates of surface forcing (Carton et al., 2000a,b; Stammer et al., 2002), they now assimilate most of the types of available water column observations (e.g., repeat hydrography, profiling float, expendable bathythermograph (XBT) and mooring time series data, etc.) as constraints on temporally varying state estimates of ocean circulation (Davis et al., 2011; Heimbach et al., 2011; Wunsch and Heimbach, 2009).

The discussion here focuses on observations in the Atlantic, and in particular, at 24–27°N because this is where most time series observations are presently available and the most research has been done. Results from other basins are provided where they are available. We include discussion of historical hydrography because, although not ideal as a time series, at most locations these are the only data available for temporal comparison of full water column MOC/MHT.

5.1. Repeat Hydrography

Initial variability estimates based on repeat hydrography along 48°N in the Atlantic suggest conflicting conclusions relative to the state of the thermohaline circulation and MHT. In particular, Koltermann et al. (1999) conclude that the strong MOC/MHT variability they saw was related to the strength of Labrador Sea Water (LSW) production, with larger (smaller) MOC/MHT corresponding to less (more) LSW export. The same data (1993–2000), reanalyzed to formally test whether MOC circulation is steady, concluded that a steady MOC could not be ruled out due to the uncertainty in determining the barotropic circulation, and the observed range in MHT variability (0.51–0.55 PW) could not be characterized as a trend (Lumpkin et al., 2008).

A similar full analysis, at 24°N in the subtropical Atlantic, concluded that the thermohaline circulation had declined over the previous 50 years by 30% (Bryden et al., 2005), a much higher rate of change than predicted in coupled climate model simulations (e.g., Schmittner et al., 2005). A comparable study using 24°N hydrography collected through 1998 showed similar MOC variability, but no decadal-scale trend in MHT could be verified due to strong seasonal variability (Baringer and Molinari, 1999). As more complete time series became available, these differences between hydrographic estimates were recognized as aliased seasonal and interannual variations (see Section 5.2). Hernández-Guerra et al. (2009) used Argo float data to reduce the eddy signal seen in the 24–26°N repeat hydrographic sections and showed that no significant change in MOC had occurred since the earliest 1957 transect. Longworth et al. (2011) examined 26°N MOC variability using a database of end-point measurements from the past 50 years. This improvement to the Bryden et al. (2005) statistics verified that no significant trend in the MOC had been observed. Nevertheless, the indirectly estimated NCEP record does suggest some longer time-scale trends both in MHT and the latitude of the MHT (Figure 29.6).

5.2. Transition to Timeseries Arrays

Since 1982, variations in the upper limb of the MOC at 27°N have been monitored by measuring the Florida Current transport using a submarine cable across the Straits of Florida in combination with repeat hydrography (Baringer and Larsen, 2001; Meinen et al., 2010). In 1984, monitoring was expanded to the deep limb of the MOC through regular hydrographic cruises (more than 30 to date) across the DWBC east of the Bahamas (van Sebille et al., 2011), and in 2001 the program commenced time series monitoring of the DWBC by adding moored inverted echo sounders and bottom pressure gauges east of Abaco Island in the Bahamas (Meinen et al., 2006). In 2004, the effort was substantially enhanced and became international, expanding the MOC monitoring array to span

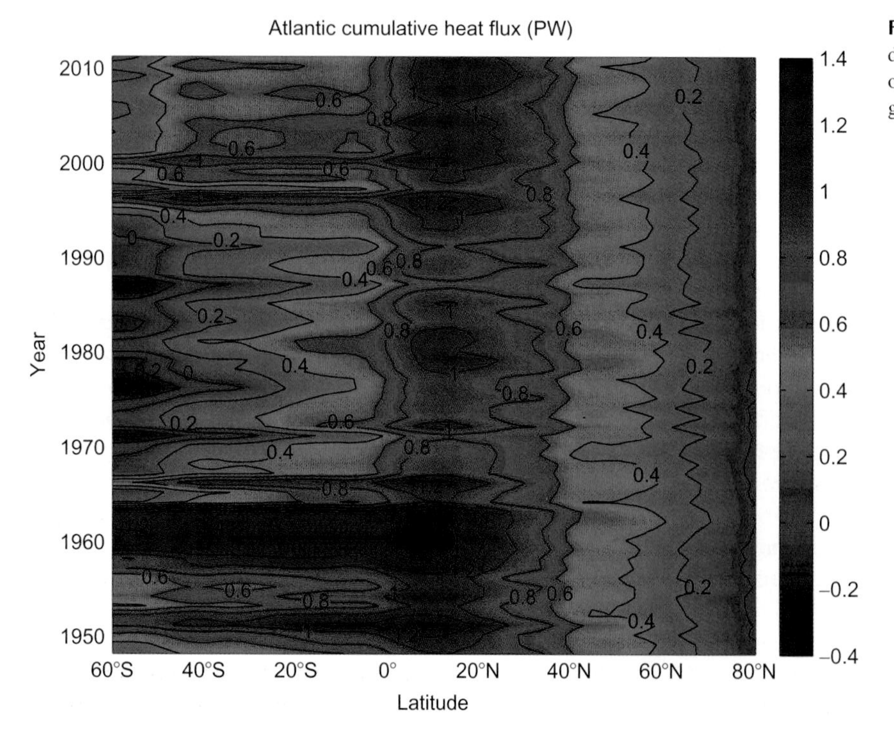

FIGURE 29.6 Illustration of annual to decadal variability. NCEP CORE-v2 estimates of meridional heat transport in the Atlantic integrated from north to south.

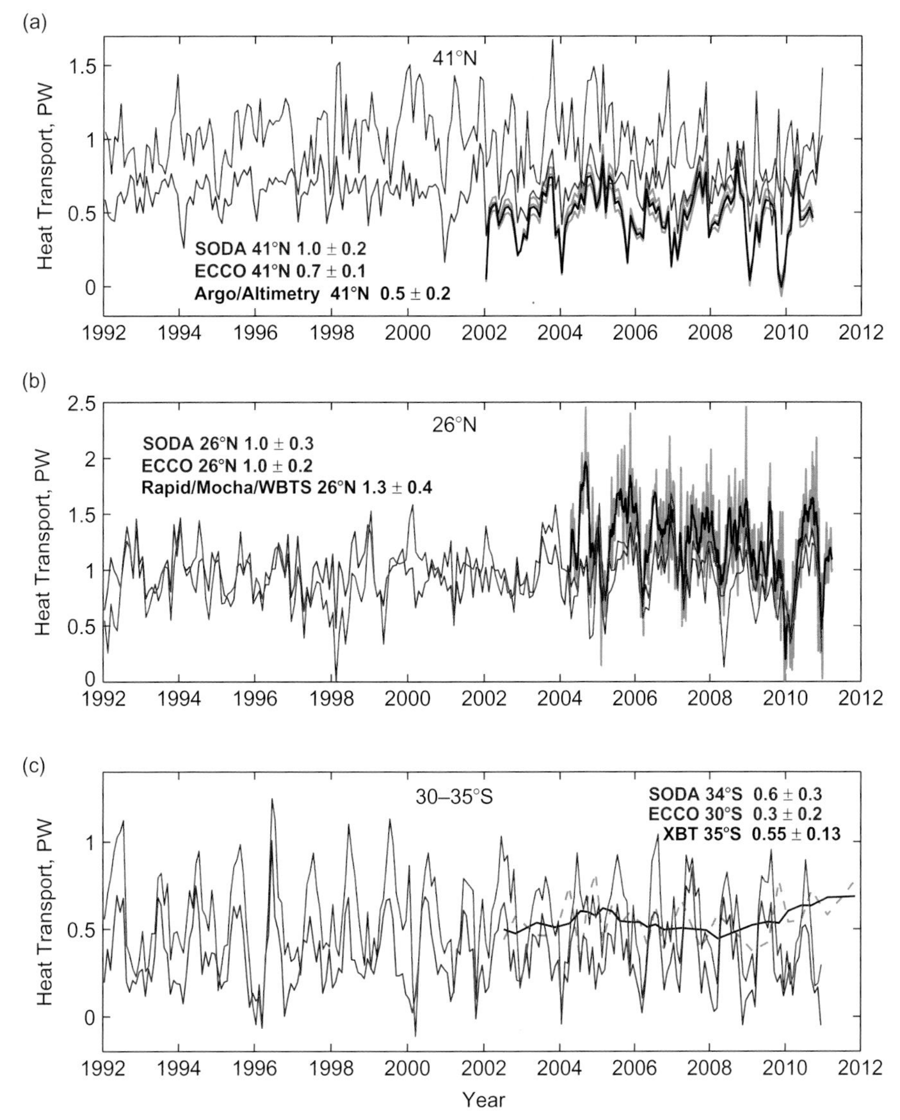

FIGURE 29.7 Assimilation timeseries of Atlantic MHT from SODA V2.2.4 (Giese and Ray, 2011) and ECCO-PROD version 4 (Wunsch and Heimbach, 2013) compared to: (a) the profiling float time series (Hobbs and Willis, 2012) at 41°N; (b) the RAPID/MOCHA mooring/hydrography time series (Johns et al., 2011) at 26°N; and (c) the repeat XBT ship-of-opportunity time series at ~30°S (updated from Garzoli et al., 2012). Values in the legends indicate means and standard deviations over the illustrated time periods. Please note the limits of the vertical axes in the panels are different (units in PW), positive is northward.

the entire Atlantic Basin at 26°N using a variety of measurement systems (Rayner et al., 2011). Called the RAPID Climate Change Program by the UK contributors and the Meridional Overturning Circulation Heat-Transport Array (MOCHA) by the US contributors, this expanded project seeks to develop a cost-effective basin-wide MOC monitoring system that will lead to a much greater degree of certainty in the magnitude of the variations in the integrated, basin-wide MOC and the time scales of variation (Kanzow et al., 2007, 2008).

The results from this array quickly indicated significant variability in all components of the MOC, and the entire range of MOC values suggested by the repeat hydrography was encountered within the first year of the time series (Figure 29.7b; Cunningham et al., 2007). The time series continues to indicate strong variability at a range of time scales, including the emergence of an annual cycle (Kanzow et al., 2010) that looks similar to the annual cycle postulated on the basis of the repeat hydrography at 24°N (Baringer and Molinari, 1999). Of note is that all the upper ocean transport values estimated by Bryden et al. (2005) from five repeat CTD sections can be found within the seasonal range of the interior transport time series (Baringer et al., 2009; Kanzow et al., 2010). This array is now providing an unprecedented continuous time series of MHT in the subtropical North Atlantic (Johns et al., 2011).

Results have provided unique quantification of the seasonal amplitude in heat transport (0.6 PW annual range), confirmation of the large variability found at short time scales (MHT values ranging from 0.2 to 2.5 PW), and a change in our understanding about what controls the variability in MHT versus the mean MHT. As noted earlier, the mean MHT in the subtropical North Atlantic is largely determined by the mean overturn strength (carrying 88% of the total MHT). Based on 3.5 years of data, the Ekman layer is responsible for 50% of the variance in MHT, while it is only responsible for a small portion of the mean MHT.

5.3. Float Program

The Argo float program has provided upper water column temperature measurements throughout the global ocean for nearly a decade. With approximately 3800 floats presently in the water, Argo is producing 100,000–135,000 profiles each year (P.E. Robbins, personal communication, 2013) in a data set that improves our global coverage with unprecedented spatial resolution and little seasonal bias. These profiles can be combined with velocity estimates and assumptions about the circulation below 2000 m to estimate, among other things, heat transport. Time-varying estimates of Argo-based Atlantic MHT suggest a 7-year mean at 41°N of 0.50 ± 0.1 PW with strong seasonal variability (Figure 29.7a, Hobbs and Willis, 2012). Strong, seasonal variability is a characteristic of the Atlantic ECCO solution (Figure 29.7) as well as indirect and model estimates (see Chapter 24). The latter can, with a longer time-series, see evidence of MHT variability on longer timescales even away from the tropics (Figure 29.6 and see Chapter 25). As RAPID/MOCHA concluded at 24°N, Hobbs and Willis (2012) reported, based on the float record, a nonsignificant decreasing trend of -0.05 PW $decade^{-1}$.

Hobbs and Willis (2012) consider their uncertainty estimate to be a lower limit because it does not include decadal-scale variability. Going to longer time-scales requires the use of models. Zheng and Giese (2009) found a positive trend of 0.036 PW $decade^{-1}$ significant at >95% at 40°N using data assimilation over a 50-year period. In fact, all the significant trends they reported for the Atlantic suggested an increase in northward heat transport, which they believe is associated with an increasing trend in mid-latitude overturn. This strengthening of the overturn appears to be related to the northward shift of the atmospheric circulation discussed in the AR4 Report.

5.4. Ships of Opportunity and Expendable Instrumentation

Volunteer ships of opportunity and expendable instrumentation have allowed repeat measurements of upper ocean hydrography along specific shipping lanes. Note that in the following discussion estimates are calculated for flow across these lanes, which are not always directly comparable to the basin transects occupied by research vessels. Roemmich et al. (2001) used high-resolution XBT data and XCTD observations to determine MHT variability from 1992 to 1999 in the North Pacific, where the upper layers that carry most of the heat transport (Bryden et al., 1991) are also subject to strong temporal variability. Using the repeat transects (~four 17-day, 10 km resolution sections between 15 and 35°N each year over the period of the study), they found a mean 0.83 ± 0.12 PW, slightly higher relative to other midlatitude estimates (Table 29.4) with the variability in Ekman transport (northward) balanced by variability in the interior transport (southward). The interannual variability O(0.4 PW) represents a substantial fraction of their estimated mean and appears to correspond to variation in the North Pacific Index. Roemmich et al. also showed that interannual variability in eddy heat fluxes can be quite large O(0.15 PW) and concluded that closure of the net heat budget was attainable with improved MHT time series estimates, air–sea flux products, and ocean heat storage estimates.

The South Atlantic has been historically one of the least sampled basins. As a consequence, model estimates in this region have been relatively unconstrained. In the past decade, however, the picture has been changed by multiple, high-density XBT lines (AX18 at 35°S from Brazil to South Africa, AX22 across Drake Passage, and AX25 across the Agulhas retroflection; Garzoli and Baringer, 2007; Stephenson et al., 2012; Swart et al., 2008), and broad scale temperature and salinity profiles collected by the Argo program (Roemmich and Owens, 2000). AX18 observations from 2002 onward have been used to study MOC variability and its effect on the net northward South Atlantic MHT (Garzoli et al., 2012). This analysis shows that the South Atlantic is responsible for a (2002–2011) mean northward MHT of 0.55 ± 0.13 PW (slightly higher than section estimates, Table 29.3), and in agreement with the 50-year SODA assimilation (Zheng and Giese, 2009) with no significant trend (Figure 29.7c; Dong et al., 2011). Although geostrophic transport dominates the time-mean MOC/MHT, both the geostrophic and Ekman components are important in explaining the variability as they display out of phase annual cycles that result in weak seasonal variability (Baringer and Garzoli, 2007; Dong et al., 2009). Examination of the contributions from boundary currents and the interior indicates that transport variability of all three regions is comparable. MHT variability is significantly correlated with the MOC variability, where a 1 Sv increase in the strength of MOC would yield a 0.05 ± 0.01 PW increase in the MHT. Thus a 5–6% MOC change results in a 10% change in the MHT, clearly illustrating the climatic importance of accurately monitoring the MOC.

Until recently, models have suggested a strong annual cycle for the total MHT and MOC in the South Atlantic that depends mainly on the seasonal cycle of the Ekman component and very little on the geostrophic component of the fluxes (Böning et al., 2001; Dong et al., 2009; Jayne and Marotzke, 2001). Now, new results suggest that climate models may be underestimating oceanic geostrophic variability (e.g., Sarojini et al., 2011).

5.5. Summary of Observational Approaches

A variety of measurement and analysis techniques are currently being used to better understand heat transport variability on differing temporal scales. Hydrography is presently the only method available for obtaining full water-column land-to-land observations at multiple latitudes. Repeat transects have suggested signals (e.g., annual cycle) that more complete time series have measured or could measure, but seasonal variability reduces its effectiveness as a means for determining temporal trends in transport. Use of ships of opportunity and expendable instruments has allowed improved temporal and spatial coverage at some cost to choice of path and vertical coverage, while the now prodigious profiling float record allows for greater horizontal coverage.

The RAPID/MOCHA array in the subtropical North Atlantic is providing an enviably complete OHT time series at a single latitude and has been improved over time to include bottom and mid-ocean measurements. This time series continues to indicate strong variability at a range of time scales, including the emergence of a possible annual cycle and substantial short-term variability such as the abrupt weakening of the MOC in the winter of 2009–2010 (McCarthy et al., 2012). Given another 5 years of observations, examination of interannual cycles should be feasible (Cunningham et al., 2007). Nevertheless, it is unlikely that this type of array can provide the large-scale latitudinal coverage that is needed. In some regions, time-series that focus on particular aspects of the circulation (e.g., WBC or shallow overturn) are necessary for improved statistics. For other regions, continued repeat hydrography, VOS/XBT, float programs, and improved data assimilating models represent our best opportunities for obtaining the spatial and temporal coverage necessary for observing changes in OHT, as well as long-term trends in temperature and salinity that could directly or indirectly affect OHT.

6. SYNTHESIS AND SUMMARY

Transport of heat by the oceans is one of the dynamic components of the earth's energy budget. In providing a brief background on the global energy balance, we have touched upon the similarities and differences in the roles played by the atmosphere (an efficient homogenizer) and ocean (the long term memory) in maintaining the balance, as well as the importance of the interactions between them. There are spatial asymmetries in the energy budget resulting from the earth's orientation to the sun and the spatial variation in absorbed radiation. These asymmetries support a transfer of energy from the tropics toward the poles. Observational and model estimates of maximum total poleward heat transport are consistent and range between 5.3 and 5.9 PW depending upon hemisphere and method of calculation. As at the top of the atmosphere, incoming shortwave solar radiation at the surface of the ocean supplies stronger warming at the equator than at the poles. However, there are spatial variations in the loss of heat by the ocean through sensible and latent heat fluxes, as well as differences in ocean basin geometry and current systems. These complexities support a pattern of ocean heat transport that is not strictly from lower to higher latitudes.

Three methods for calculating OHT have been discussed. Two of these, the bulk formula method and the residual method, are considered "indirect" as they do not require observations of the ocean. The advantage of these estimation techniques is that they can provide complete global coverage and long, high-resolution timeseries OHT estimates. They also directly connect OHT to atmospheric processes through air–sea interactions. They afford an excellent basis from which to investigate variations in OHT on seasonal and longer time-scales, as well as feedbacks and changes in feedbacks that could result in changes in OHT. Their disadvantage is that they tend to contain strong regional biases that are likely due to missing or inadequately represented physics and spatial inhomogeneity of the underlying data (in both amount and type of data) that cannot be easily removed. The third method for calculating OHT is considered "direct" because it is based on observations of ocean properties. This technique, which requires estimates of ocean temperature, salinity, and velocity fields, was originally used with long-line hydrography, but is now being used with float profiles, data from volunteer ship of opportunity expendable instrumentation, trans-basin time series mooring-array data, and altimetry.

On average, MHT estimates support a global poleward heat flux in the Atlantic and Indo-Pacific with maxima in the subtropics. The mean maximum global MHT from the values presented here is 1.6 ± 0.4 PW at 24°N (slightly less than 1/3 of total). In the Atlantic, the maximum lies in the northern hemisphere and is sustained by meridional overturn. In the Indo-Pacific, the southern hemisphere MHT maximum dominates. Supported mainly by Indian Ocean outflow, it represents a temperature flux maintained by the warm ITF. Mean estimates of South Atlantic and North Pacific subtropical MHT maxima are about half the magnitude of their counterparts (in the North Atlantic and southern Indo-Pacific, respectively) and are associated with strong gyre components of the circulation.

Although uncertainties are greatest in the tropics mainly due to strong, variable winds, there remains a relatively large range in estimates in the subtropical North Pacific. Although not presented, the same is true in the South Pacific. The range appears to be largely due to how shallow overturn is represented, as shallower overturn produces smaller MHTs. The XBT record has also pointed to a strong MHT/MOC connection in the South Atlantic.

Uncertainties in OHT estimates have been attributed to insufficient spatial resolution of both observations and models that lead to uncertainties in WBC and MOC mass transports. OHT estimates are also affected by strong unresolved seasonal signals in the winds and circulation. Some specific challenges for the future include clarifying the effects of WBC variability (e.g., Garzoli and Matano, 2011); the role of eddies, particularly in the South Atlantic (e.g., Biastoch et al., 2008; Duncombe Rae et al., 1996) and Southern Ocean (e.g., Gille, 2003) where their effects need to be separated from changes in wind-forcing (Meredith et al., 2012); the effect of ITF variability (e.g., Corell et al., 2009); the influence of strong seasonal changes circulation and in the magnitude of the southern deep inflow in the Indian Ocean (Schott et al., 2009; Sultan et al., 2007).

To improve statistics in these regions, as well as in the observation-poor Southern Ocean, more complete or longer *in situ* time series are necessary. Such observations can be used on their own or to confirm and/or constrain model estimates and predictions. At present, observing systems capable of quantifying changes in MHT are at fledgling stages. At 26.5°N in the Atlantic, the RAPID/MOCHA program (Johns et al., 2011; Kanzow et al., 2008) has for the first time provided the elements of a continuous timeseries of meridional overturn and its associated heat transport across a latitude line. However, such data sets need not necessarily be always cross-basin and full-depth. For instance, in the North and South Pacific, observations that provide a clearer picture of the depth, magnitude, and variability in shallow overturn would go a long way toward improving our understanding of Pacific OHT. In the Indian Ocean, observations that focus on deep overturn, mixing, and seasonal changes would likewise benefit our interpretation of OHT estimates. And in the Atlantic, observations such as the MOVE array monitoring the deep circulation (Send et al., 2011) and the float time series (Willis, 2010) will become increasingly valuable as we seek to understand the meridional coherence of MOC and MHT fluctuations. Throughout the global ocean, the long-transect hydrographic survey that began as WOCE in the 1990s and was extended as CLIVAR in the 2000s, now continues into the new century as GO-SHIP. Still temporally sparse, at present it provides the only high-quality, global, full-depth, cross-basin timeseries available for studying OHT.

REFERENCES

Abbot, C.G., Fowle, F.E., 1908. Determination of the intensity of the solar radiation outside the earth's atmosphere, otherwise termed "the Solar Constantant of Radiation" Annals of Astrophysical Observatory of the Smithsonian Institution. 2(1), Smithsonian Institution, Washington, DC, pp. 1–237.

Bacon, S., 1997. Circulation and fluxes in the North Atlantic between Greenland and Ireland. J. Phys. Oceanogr. 27, 1420–1435.

Baringer, M.O., Molinari, R., 1999. Atlantic Ocean baroclinic heat flux at 24° to 26°N. Geophys. Res. Lett. 26, 353–356.

Baringer, M.O., Larsen, J., 2001. Sixteen years of Florida Current transport at 27 N. Geophys. Res. Lett. 28 (16), 3179–3182.

Baringer, M.O., Garzoli, S.L., 2007. Meridional heat transport determined with expendable bathythermographs, Part I: error estimates from model and hydrographic data. Deep Sea Res. Part I 54 (8), 1390–1401.

Baringer, M.O., Meinen, C.S., Johnson, G.C., Kanzow, T.O., Cunningham, S.A., Johns, W.E., Beal, L.M., Hirschi, J.J.-M., Rayner, D., Longworth, H.R., Bryden, H.L., Marotzke, J., 2009. The meridional overturning circulation. In: Peterson, T.C., Baringer, M.O. (Eds.), State of the Climate in 2008, Bulletin of the American Meteorological Society, vol. 90(8). pp. S59–S62.

Beal, L.M., Chereskin, T.K., Bryden, H., Ffield, A., 2003. Variability of water properties, heat and salt fluxes in the Arabian Sea, between the onset and wane of the 1995 southwest monsoon. Deep Sea Res. Part II 50, 2049–2075.

Beal, L.M., De Ruijter, W.P.M., Biastoch, A., Zahn, R., SCOR/WCRP/IAPSO Working Group 136, 2011. On the role of the Agulhas system in ocean circulation and climate. Nature 472, 429–436.

Berry, D.I., Kent, E.C., 2009. A new air–sea interaction dataset from ICOADS with uncertainty estimates. Bull. Am. Meteorol. Soc. 90 (5), 645–656.

Biastoch, A., Böning, C.W., Lutjeharms, J.R.E., 2008. Agulhas leakage dynamics affects decadal variability in the Atlantic overturning circulation. Nature 456, 489–492.

Bingham, F.M., Talley, L.D., 1991. Estimates of Kuroshio transport using an inverse technique. Deep Sea Res. 38 (Suppl. 1A), 21–43.

Böning, C.W., Dieterich, C., Barnier, B., Jia, Y., 2001. Seasonal cycle of meridional heat transport in the subtropical North Atlantic: a model intercomparison in relation to observations near 25°N. Prog. Oceanogr. 48, 231–253.

Book, J.A., Wimbush, M., Imawaki, S., Ichikawa, H., Uchida, H., Kinoshita, H., 2002. Kuroshio temporal and spatial variations south of Japan determined from inverted echo sounder measurements. J. Geophys. Res. 107 (C9), 3121. http://dx.doi.org/10.1029/2001JC000795.

Bosilovich, M.G., Robertson, F.R., Chen, J., 2011. Global energy and water budgets in MERRA. J. Clim. 24, 5721–5739.

Bourras, D., 2006. Comparison of five satellite-derived latent heat flux products to moored buoy data. J. Clim. 19, 6291–6313.

Brodeau, L., Barnier, B., Treguier, A.-M., Penduff, T., Gulev, S., 2010. An ERA40-based atmospheric forcing for global ocean circulation models. Ocean Model. 31, 88–104.

Bromwich, D.H., Wang, S.-H., 2005. Evaluation of the NCEP–NCAR and ECMWF 15- and 40-Yr reanalyses using Rawinsonde Data from two independent arctic field experiments. Mon. Weather Rev. Special Section 133, 3562–3577.

Bryan, K., 1962. Measurements of meridional heat transport by currents. J. Geophys. Res. 67 (9), 3403–3414.

Bryden, H.L., 1979. Poleward heat flux and conversion of available potential energy in Drake Passage. J. Mar. Res. 37, 1–22.

Bryden, H.L., Imawaki, S., 2001. Ocean heat transport. In: Siedler, G., Church, J., Gould, J. (Eds.), Ocean Circulation and Climate. Academic Press, London, pp. 455–474, Chapter 6.2.

Bryden, H.L., King, B.A., McCarthy, G.D., 2011. South Atlantic overturning circulation at 24°S. J. Mar. Res. 69 (1), 38–55.

Bryden, H.L., Longworth, R., Cunningham, S., 2005. Slowing of the Atlantic meridional overturning circulation at 25°N. Nature 438, 655–657.

Bryden, H.L., Roemmich, D.H., Church, J.A., 1991. Ocean heat transport across 24°N in the Pacific. Deep Sea Res. 38, 297–324.

Carton, J.A., Chepurin, G., Cao, X., Giese, B.S., 2000a. A Simple Ocean Data Assimilation analysis of the global upper ocean 1950–1995, Part 1: methodology. J. Phys. Oceanogr. 30, 294–309.

Carton, J.A., Chepurin, G., Cao, X., 2000b. A Simple Ocean Data Assimilation analysis of the global upper ocean 1950–1995 Part 2: results. J. Phys. Oceanogr. 30, 311–326.

Carton, J.A., Giese, B.S., 2008. A reanalysis of ocean climate using Simple Ocean Data Assimilation (SODA). Mon. Weather Rev. 136, 2999–3017.

Compo, G.P., Whitaker, J.S., Sardeshmukh, P.D., Matsui, N., Allan, R.J., Yin, X., Gleason, B.E., Vose, R.S., Rutledge, G., Bessemoulin, P., Brönnimann, S., Brunet, M., Crouthamel, R.I., Grant, A.N., Groisman, P.Y., Jones, P.D., Kruk, M., Kruger, A.C., Marshall, G.J., Maugeri, M., Mok, H.Y., Nordli, Ø., Ross, T.F., Trigo, R.M., Wang, X.L., Woodruff, S.D., Worley, S.J., 2011. The Twentieth Century Reanalysis Project. Q. J. R. Meteorol. Soc. 137, 1–28. http://dx.doi.org/10.1002/qj.776.

Corell, H., Nilsson, J., Döös, K., Broström, G., 2009. Wind sensitivity of the inter-ocean heat exchange. Tellus 61A, 635–653.

Cunningham, S.A., Kanzow, T., Rayner, D., Baringer, M.O., Johns, W.E., Marotzke, J., Longworth, H.R., Grant, E.M., Hirschi, J.J.-M., Beal, L.M., Meinen, C.S., Bryden, H.L., 2007. Temporal variability of the Atlantic meridional overturning circulation at 26.5°N. Science 312, 335–938. http://dx.doi.org/10.1126/science.1141304.

da Silva, A., Young, A.C., Levitus, S., 1994. Atlas of surface marine data 1994. Technical Report 6, vol. 1 U.S. Department of Commerce, NOAA, NESDIS, 83 pp.

Davis, X.J., Rothstein, L.M., Dewar, W.K., Menemenlist, D., 2011. Numerical investigations of seasonal and interannual variability of North Pacific. J. Clim. 24, 2648–2665.

Dee, D., Uppala, S.M., Simmons, A.J., Berrisford, P., Poli, P., Kobayashi, S., Andrae, U., Balmaseda, M.A., Balsamo, G., Bauer, P., Bechtold, P., Beljaars, A.C.M., van de Berg, L., Bidlot, J., Bormann, N., Delsol, C., Dragani, R., Fuentes, M., Geer, A.J., Haimberger, L., Healy, S.B., Hersbach, H., Hólm, E.V., Isaksen, L., Kållberg, P., Köhler, M., Matricardi, M., McNally, A.P., Monge-Sanz, B.M., Morcrette, J.-J., Park, B.-K., Peubey, C., de Rosnay, P., Tavolato, C., Thépaut, J.-N., Vitart, F., 2011. The ERA-Interim reanalysis: configuration and performance of the data assimilation system. Q. J. R. Meteorol. Soc. 137, 553–597.

de Szoeke, R.A., Levine, M.D., 1981. The advective flux of heat by mean geostrophic motions in the Southern Ocean. Deep Sea Res. Part A 28, 1057–1085.

Dines, W.H., 1917. The heat balance of the atmosphere. J. R. Meteorol. Soc. 43, 151–158.

Dong, S., Garzoli, S., Baringer, M., 2011. The role of interocean exchanges on decadal variations of the meridional transport of heat in the South Atlantic. J. Phys. Oceanogr. 41, 1498–1511.

Dong, S., Garzoli, S., Baringer, M., Meinen, C., Goni, G., 2009. Interannual variations in the Atlantic meridional overturning circulation and its relationship with the net northward heat transport in the South Atlantic. Geophys. Res. Lett. 36. http://dx.doi.org/10.1029/2009GL039356, L20606.

Donohoe, A., Battisti, D.S., 2012. What determines meridional heat transport in climate models? J. Clim. 25, 3832–3850. http://dx.doi.org/10.1175/JCLI-D-11-00257.1.

Douglass, E., Roemmich, D., Stammer, D., 2010. Interannual variability in North Pacific heat and freshwater budgets. Deep Sea Res. Part II 57 (13–14), 1127–1140.

Duncombe Rae, C.M., Garzoli, S.L., Gordon, A.L., 1996. The eddy field of the southeast Atlantic Ocean: a statistical census from the Benguela Sources and Transports Project. J. Geophys. Res. 101 (C5), 11949–11964.

Enderton, D., Marshall, J., 2009. Controls on the total dynamical heat transport of the atmosphere and oceans. J. Atmos. Sci. 66, 1593–1611.

Fasullo, J.T., Trenberth, K.E., 2008. The annual cycle of the energy budget: part 2. Meridional structures and poleward transports. J. Clim. 21, 2313–2325.

Fennig, K., Andersson, A., Bakan, S., Klepp, C., Schroeder, M., 2012. Hamburg Ocean Atmosphere Parameters and Fluxes from Satellite Data—HOAPS 3.2—Monthly Means/6-Hourly Composites. Satellite Application Facility on Climate Monitoring. http://dx.doi.org/10.5676/EUM_SAF_CM/HOAPS/V001.

Festa, J.F., Molinari, R.L., 1992. An evaluation of the WOCE volunteer observing ship–XBT network in the Atlantic. J. Atmos. Oceanic Technol. 9, 305–317.

Fillenbaum, E.R., Lee, T.N., Johns, W.E., Zantopp, R.J., 1997. Meridional heat transport at 26.5°N in North Atlantic. J. Phys. Oceanogr. 27, 153–174.

Friedrichs, M.A.M., Hall, M.M., 1993. Deep circulation in the tropical Atlantic. J. Mar. Res. 51, 697–736.

Ganachaud, A., Wunsch, C., 2000. The oceanic meridional overturning circulation, mixing, bottom water formation, and heat transport. Nature 408, 453–457.

Ganachaud, A., Wunsch, C., 2003. Large-scale ocean heat and freshwater transports during the world ocean circulation experiment. J. Clim. 16, 696–705.

Garnier, E., Barnier, B., Siefridt, L., Béranger, K., 2000. Investigating the 15 years air–sea flux climatology from the ECMWF re-analysis project as a surface boundary condition for ocean models. Int. J. Climatol. 20 (14), 1653–1673.

Garzoli, S.L., Baringer, M.O., 2007. Meridional heat transport determined with expendable bathythermographs. Part II: South Atlantic transport. Deep Sea Res. Part I 54, 1402–1420.

Garzoli, S.L., Baringer, M., Dong, S., Perez, R., Yao, Q., 2012. South Atlantic meridional fluxes. Deep Sea Res. Part I 71, 21–32. http://dx.doi.org/10.1016/j.dsr.2012.09.003.

Garzoli, S.L., Matano, R., 2011. The South Atlantic and the Atlantic meridional overturning circulation. Deep Sea Res. Part II 58, 1837–1847.

Gibson, J.K., Kallberg, P., Uppala, S., Noumura, A., Hernandez, A., Serrano, E., 1997. ERA Description. ECMWF Re-Analysis Project Report Series, vol. 1. ECMWF, Reading, UK, 77 pp.

Giese, B.S., Ray, S., 2011. El Niño variability in simple ocean data assimilation (SODA), 1871–2008. J. Geophys. Res. 116, C02024. http://dx.doi.org/10.1029/2010JC006695.

Gille, S.T., 2003. Float observations of the Southern Ocean. Part II: eddy fluxes. J. Phys. Oceanogr. 33, 1182–1196.

Grist, J.P., Josey, S.A., 2003. Inverse analysis adjustment of the SOC air–sea flux climatology using ocean heat transport constraints. J. Clim. 20, 3274–3295.

Haines, K., Valdivieso, M., Zuo, H., Stepanov, V.N., 2012. Transports and budgets in a ¼° global ocean reanalysis 1989–2010. Ocean Sci. 8, 333–344.

Hall, M.M., Bryden, H.L., 1982. Direct estimates and mechanisms of ocean heat transport. Deep Sea Res. 29 (3A), 339–359.

Hansen, J., Sato, M., Kharecha, P., 2011. Earth's energy imbalance and implications. Atmos. Chem. Phys. 11, 13421–13449. http://dx.doi.org/10.5194/acp-11-13421-2011.

Hatzianastassiou, N., Matsoukas, C., Hatzidimitriou, D., Pavlakis, C., Drakakis, M., Vardavas, I., 2004. Ten year radiation budget of the Earth: 1984–93. Int. J. Climatol. 24 (14), 1785. http://dx.doi.org/10.1002/joc.1110.

Heimbach, P., Wunsch, C., Ponte, R.M., Forget, G., Hill, C., Utke, J., 2011. Timescales and regions of the sensitivity of the Atlantic meridional volume and heat transport: toward observing system design. Deep Sea Res. Part II 58, 1858–1879.

Hernández-Guerra, A., Joyce, T.M., Fraile-Nuez, E., Vélez Belchí, P., 2009. Using Argo data to investigate the meridional overturning circulation in the North Atlantic. Deep Sea Res. Part I 57, 29–36.

Heywood, K.J., Stevens, D.P., 2007. Meridional heat transport across the Antarctic Circumpolar Current by the Antarctic Bottom Water overturning cell. Geophys. Res. Lett. 34, L11610. http://dx.doi.org/10.1029/2007GL030130.

Hobbs, W., Willis, J.K., 2012. Estimates of North Atlantic heat transport from satellite and drifter data. J. Geophys. Res. 17 (C01008). http://dx.doi.org/10.1029/2011JC007039.

Hogg, A. McC, Meredith, M.P., Blundell, J.R., Wilson, C., 2008. Eddy heat flux in the Southern Ocean: response to variable wind forcing. J. Clim. 21 (4), 608–620.

Holfort, J., Siedler, G., 2001. The Meridional Oceanic transports of heat and nutrients in the South Atlantic. J. Phys. Oceanogr. 31, 5–29.

House, F.B., Gruber, A., Hunt, G.E., Mecherikunnel, A.T., 1986. History of satellite missions and measurements of the earth radiation budget (1957-1984). Rev. Geophys. 24, 357–377. http://dx.doi.org/10.1029/RG024i002p00357.

Hunt, G.E., Kandel, R., Mecherikunnel, A.T., 1986. A history of presatellite investigations of the earth's radiation budget. Rev. Geophys. 24, 351–356. http://dx.doi.org/10.1029/RG024i002p00351.

Ichikawa, H., Beardsley, R.C., 1993. Temporal and spatial variability of volume transport of the Kuroshio in the East China Sea. Deep Sea Res. 40, 583–605.

Imawaki, S., Uchida, H., Ichikawa, H., Fukasawa, M., Umatani, S., the ASUKA Group, 2001. Satellite altimeter monitoring the Kuroshio transport south of Japan. Geophys. Res. Lett. 24 (1), 17–20.

Jayne, S.R., Marotzke, J., 2001. The dynamics of ocean heat transport variability. Rev. Geophys. 39, 385–412.

Jayne, S.R., Hogg, N.G., Waterman, S.N., Rainville, L., Donohue, K.A., Watts, D.R., Tracey, K.L., McClean, J.L., Maltrud, M.E., Qiu, B., Chen, S., Hacker, P., 2009. The Kuroshio Extension and its recirculation gyres. Deep Sea Res. Part I 56, 2088–2099.

Johns, W.E., Baringer, M.O., Beal, L.M., Cunningham, S.A., Kanzow, T., Bryden, H.L., Hirschi, J.J.M., Marotzke, J., Meinen, C.S., Shaw, B., Curry, R., 2011. Continuous array-based estimates of Atlantic Ocean heat transport at 26.5°N. J. Clim. 24, 2429–2449.

Johnson, G.C., Sloyan, B.M., Kessler, W.S., McTaggart, K.E., 2002. Direct measurements of upper ocean currents and water properties across the tropical Pacific during the 1990s. Prog. Oceanogr. 52, 31–61.

Josey, S., Kent, E.C., Taylor, P.K., 1998. The Southampton Oceanography Centre (SOC) Ocean-Atmosphere, Heat, Momentum and Freshwater Flux Atlas. Report 6. Southampton Oceanography Centre, UK, 30 pp.

Jung, G.H., 1952. Note on the meridional transport of energy by the ocean. J. Mar. Res. 11, 139–146.

Kalnay, E., Kanamitsu, M., Kistler, R., Collins, W., Deaven, D., Gandin, L., Iredell, M., Saha, S., White, G., Woollen, J., Zhu, Y., Leetmaa, A., Reynolds, R., Chelliah, M., Ebisuzaki, W., Higgins, W., Janowiak, J., Mo, K.C., Ropelewski, C., Wang, J., Jenne, R., Joseph, D., 1996. The NCEP/NCAR 40-year reanalysis project. Bull. Am. Meteorol. Soc. 77, 437–470. http://dx.doi.org/10.1175/1520-0477, 077 < 0437:TNYRP > 2.0.CO;2.

Kanzow, T., Cunningham, S.A., Johns, W.E., Hirschi, J.J.-M., Marotzke, J., Baringer, M.O., Meinen, C.S., Chidichimo, M.P., Atkinson, C., Beal, L.M., Bryden, H.L., Collins, J., 2010. Seasonal variability of the Atlantic meridional overturning circulation at 26.5°N. J. Clim. 23. http://dx.doi.org/10.1175/2010JCLI3389.1171.

Kanzow, T., Cunningham, S.A., Rayner, D., Hirschi, J.J.-M., Johns, W.E., Baringer, M.O., Bryden, H.L., Beal, L.M., Meinen, C.S., Marotzke, J., 2007. Observed flow compensation associated with the meridional overturning circulation near 26.5°N in the Atlantic. Science 317, 938–941. http://dx.doi.org/10.1126/science.1141293.

Kanzow, T., Hirschi, J.J.-M., Meinen, C.S., Rayner, D., Cunningham, S.A., Marotzke, J., Johns, W.E., Bryden, H.L., Beal, L.M., Baringer, M.O., 2008. A prototype system of observing the Atlantic Meridional Overturning Circulation: scientific basis, measurement and risk mitigation strategies, and first results. J. Oper. Oceanogr. 1, 19–28.

Klein, B., Molinari, R.L., Müller, T.J., Siedler, G., 1995. A transatlantic section at 24.5°N: meridional volume and heat fluxes. J. Mar. Res. 53, 929–957.

Koltermann, K.P., Sokov, A.V., Tereschenkov, V.P., Dobroliubov, S.A., Lorbacher, K., Sy, A., 1999. Decadal changes in the thermohaline circulation of the North Atlantic. Deep Sea Res. Part II 46, 109–138.

Large, W.G., Danabasoglu, G., Doney, S.C., McWilliams, J.C., 1997. Sensitivity to surface forcing and boundary layer mixing in a global ocean model: annual-mean climatology. J. Phys. Oceanogr. 27, 2418–2447.

Large, W.G., Yeager, S.G., 2009. The global climatology of an interannually varying air–sea flux data set. Clim. Dyn. 33, 341–364.

Lavín, A., Bryden, H.L., Parrilla, G., 1998. Meridional transport and heat flux variations in the subtropical North Atlantic. Glob. Atmos. Ocean Syst. 6, 269–293.

Lavín, A., Bryden, H.L., Parrilla, G., 2003. Mechanisms of heat, freshwater, oxygen and nutrient transports and budgets at 24.5°N in the subtropical North Atlantic. Deep Sea Res. Part I 50 (9), 1099–1128.

Lawson, C.L., Hanson, D.J., 1974. Solving Least Squares Problems. Prentice-Hall, Englewood Cliffs, NJ.

Lee III, R.B.., Gibson, M.A., Wilson, R.S., Thomas, S., 1995. Long-term total solar irradiance variability during sunspot cycle 22. J. Geophys. Res. 100 (A2), 1667–1675.

Liu, H., Liu, X., Zhang, M., Lin, W., 2011a. A critical evaluation of the upper ocean heat budget in the Climate Forecast System Reanalysis data for the south central equatorial Pacific. Environ. Res. Lett. 6, 034022. http://dx.doi.org/10.1088/1748-9326/6/3/034022.

Liu, J., Xiao, T., Chen, L., 2011b. Intercomparisons of air–sea heat fluxes over the Southern Ocean. J. Clim. 24, 1198–1211.

Loeb, N.G., Kato, S., Wielicki, B.A., 2002. Defining top-of-the-atmosphere flux reference level for earth radiation budget studies. J. Clim. 15, 3301–3309.

London, J., 1957. A study of the atmospheric heat balance. New York, USA, Final report, contract USAF8S1911221165, New York University.

Longworth, H.R., Bryden, H.L., Baringer, M.O., 2011. Historical variability in Atlantic meridional baroclinic transport at 26.5°N from boundary dynamic height observations. Deep Sea Res. Part II 58 (17–18), 1754–1767. http://dx.doi.org/10.1016/j.dsr2.2010.10.057.

Lumpkin, R., Speer, K., 2003. Large-scale vertical and horizontal circulation in the North Atlantic Ocean. J. Phys. Oceanogr. 33, 1902–1920.

Lumpkin, R., Speer, K., 2007. Global Ocean meridional overturning. J. Phys. Oceanogr. 37, 2550–2562.

Lumpkin, R., Speer, K.G., Koltermann, K.P., 2008. Transport across 48°N in the Atlantic Ocean. J. Phys. Oceanogr. 38, 733–752.

Macdonald, A.M., 1993. Property fluxes at 30°S and their implications for the Pacific–Indian throughflow and the global heat budget. J. Geophys. Res. 98 (C4), 6851–6868.

Macdonald, A.M., 1998. The global ocean circulation: a hydrographic estimate and regional analysis. Prog. Oceanogr. 41, 281–382.

Macdonald, A.M., Mecking, S., Robbins, P.E., Toole, J.M., Johnson, G.C., Talley, L., Cook, M., Wijffels, S.E., 2009. The WOCE-era 3-D Pacific Ocean circulation and heat budget. Prog. Oceanogr. 82, 281–325.

Marin, F., 1998. Heat flux estimates across A6 and A7 WOCE sections. International WOCE Newsletter 31, 28–31.

McCarthy, G., Frajka-Williams, E., Johns, W.E., Baringer, M.O., Meinen, C.S., Bryden, H.L., Rayner, D., Duchez, A., Roberts, C., Cunningham, S.A., 2012. Observed interannual variability of the Atlantic meridional overturning circulation at 26.5°N. Geophys. Res. Lett. 39, L19609. http://dx.doi.org/10.1029/2012GL052933.

McDonagh, E.L., King, B.A., 2005. Oceanic fluxes in the South Atlantic. J. Phys. Oceanogr. 35, 109–122.

McDonagh, E.L., McLeod, P., King, B.A., Bryden, H.L., Valdés, V.T., 2010. Circulation, heat, and freshwater transport at 36°N in the Atlantic. J. Phys. Oceanogr. 40, 2661–2678.

Meinen, C.S., Baringer, M.O., Garcia, R.F., 2010. Florida Current transport variability: an analysis of annual and longer-periods. Deep Sea Res. Part I 57 (7), 835–846. http://dx.doi.org/10.1016/j.dsr.2010.04.001.

Meinen, C.S., Baringer, M.O., Garzoli, S.L., 2006. Variability in deep western boundary current transports: preliminary results from 26.5°N in the Atlantic. Geophys. Res. Lett. 33, L17610. http://dx.doi.org/10.1029/2006GL026965.

Meredith, M.P., Naveira Garabato, A.C., Hogg, A. McC, Farneti, R., 2012. Sensitivity of the overturning circulation in the Southern Ocean to decadal changes in wind forcing. J. Clim. 25, 99–110.

Molinari, R.L., Johns, E., Festa, J.F., 1990. The annual cycle of meridional heat flux in the Atlantic Ocean at 26.5°N. J. Phys. Oceanogr. 20, 476–482.

Moore, G.W.K., Refrew, I.A., 2002. An assessment of the surface turbulent heat fluxes from the NCEP–NCAR reanalysis over the Western Boundary currents. J. Clim. 15, 2020–2037.

Msadek, R., 2011. Comparing the meridional heat transport at 26.5°N and its relationship with the MOC in two CMIP5. WCRP Open Science Conference, Denver, USA, October 24–28, 2011.

Onogi, K., Tsutsui, J., Koide, H., Sakamoto, M., Kobayashi, S., Hatsushika, H., Matsumoto, T., Yamazaki, N., Kamahori, H., Takahashi, K., Kadokura, S., Wada, K., Kato, K., Oyama, R., Ose, T., Mannoji, N., Taira, R., 2007. The JRA-25 reanalysis. J. Meteorol. Soc. Jpn. 85, 369–432.

Pandey, V.K., Pandey, A.C., 2006. Heat transport through Indonesian throughflow. J. Indian Geophys. Union 10 (4), 273–277.

Rayner, D., Hirschi, J.J.-M., Kanzow, T., Johns, W.E., Cuningham, S.A., Wright, P.G., Frajka-Williams, E., Bryden, H.L., Meinen, C.S., Baringer, M.O., Marotzke, J., Beal, L.M., 2011. Monitoring the Atlantic meridional overturning circulation. Deep Sea Res. Part II 58, 1744–1753. http://dx.doi.org/10.1016/j.dsr2.2010.10.056.

Reid, J.L., 1994. On the total geostrophic circulation of the North Atlantic Ocean: flow patterns, tracers, and transports. Prog. Oceanogr. 33, 1–92.

Rintoul, S., 1991. South Atlantic interbasin exchange. J. Geophys. Res. 96, 2675–2692.

Roemmich, D., Gilson, J., 2011. Ocean circulation and the mass and heat budgets of large ocean regions. First XBT Science Workshop July 7-8, 2011, Melbourne, Australia.

Roemmich, D., Gilson, J., Cornuelle, B., Weller, R., 2001. Mean and time-varying transport of heat at the tropical/subtropical boundary of the North Pacific Ocean. J. Geophys. Res. 106 (C5), 8957–8970.

Roemmich, D., McCallister, T., 1989. Large scale circulation of the North Pacific Ocean. Prog. Oceanogr. 22, 171–204.

Roemmich, D., Owens, W.B., 2000. The Argo Project: global ocean observations for understanding and prediction of climate variability. Oceanography 13, 45–50.

Roemmich, D., Wunsch, C., 1985. Two transatlantic sections: meridional circulation and heat flux in the subtropical North Atlantic Ocean. J. Phys. Oceanogr. 32, 619–664.

Saha, S., Moorthi, S., Pan, H.-L., Wu, X., Wang, J., Nadiga, S., Tripp, P., Kistler, R., Woollen, J., Behringer, D., Liu, H., Stokes, D., Grumbine, R., Gayno, G., Wang, J., Hou, Y.-T., Chuang, H.-Y., Juang, H.-M.H., Sela, J., Iredell, M., Treadon, R., Kleist, D., Van Delst, P., Keyser, D., Derber, J., Ek, M., Meng, J., Wei, H., Yang, R., Lord, S., Van Den Dool, H., Kumar, A., Wang, W., Long, C., Chelliah, M., Xue, Y., Huang, B., Schemm, K.-K., Ebisuzaki, W., Lin, R., Xie, P., Chen, M., Zhou, S., Higgins, W., Zou, C.Z., Liu, Q., Chen, Y., Han, Y., Cucurull, L., Reynolds, R.W., Rutledge, G., Goldberg, M., 2010. The NCEP Climate Forecast System Reanalysis. Bull. Am. Meteorol. Soc. 91, 1015–1057. http://dx.doi.org/10.1175/2010BAMS3001.1.

Sarojini, B.B., Gregory, J.M., Tailleux, R., Bigg, G., Blaker, A., Cameron, D., Edwards, N., Megann, A., Shaffrey, L., Sinha, B., 2011. High frequency variability of the Atlantic meridional overturning circulation. Ocean Sci. 7, 471–486. http://dx.doi.org/10.5194/os-7-471-2011.

Sato, O.T., Rossby, T., 2000. Seasonal and low-frequency variability of the meridional heat flux at 36°N in the North Atlantic. J. Phys. Oceanogr. 30, 606–621.

Saunders, P.M., King, B.A., 1995. Oceanic fluxes on the WOCE A11 section. J. Phys. Oceanogr. 25, 1942–1958.

Schmittner, A., Latif, M., Schneider, B., 2005. Model projections of the North Atlantic thermohaline circulation for the 21st century assessed by observations. Geophys. Res. Lett. 32. http://dx.doi.org/10.1029/2005GL024368, L23710.

Schott, F.A., Xie, S.-P., McCreary Jr., J.P., 2009. Indian Ocean circulation and climate variability. Rev. Geophys. 47. http://dx.doi.org/10.1029/2007RG000245, RG1002.

Send, U., Lankhorst, M., Kanzow, T., 2011. Observation of decadal change in the Atlantic meridional overturning circulation using 10 years of continuous transport data. Geophys. Res. Lett. 38, L24606. http://dx.doi.org/10.1029/2011GL049801.

Shie, C.-L., Hilburn, K., Chiu, L. S., Adler, R., Lin, I.-I., Nelkin, E., Ardizzone, J., Gao, S., 2012. Goddard Satellite-Based Surface Turbulent Fluxes, Daily Grid, version 3. In: Savtchenko, A. (Ed.), Goddard Earth Science Data and Information Services Center (GES DISC), Greenbelt, MD, USA.. Accessed at http://dx.doi.org/10.5067/MEASURES/GSSTF/DATA301.

Siqueira, L., Nobre, P., 2006. Tropical Atlantic sea surface temperature and heat flux simulations in a coupled GCM. Geophys. Res. Lett. 33, L15708. http://dx.doi.org/10.1029/2006GL026528.

Sloyan, B.M., Rintoul, S.R., 2001. The Southern Ocean limb of the global deep overturning circulation. J. Phys. Oceanogr. 31, 143–173.

Smith, S.R., Hughes, P.J., Bourassa, M.A., 2011. A comparison of nine air-sea flux products. Int. J. Climatol. 31, 1002–1027.

Speer, K., Holfort, J., Reynaud, T., Siedler, G., 1996. South Atlantic heat transport at 11°S. In: Wefer, G. et al., (Eds.), The South Atlantic: Present and Past Circulation. Springer-Verlag, Berlin, pp. 105–120.

Sprintall, J., Liu, W.T., 2005. Ekman mass and heat transport in the Indonesian seas. Oceanography 18 (4), 88–97.

Stammer, D., Wunsch, C., Giering, R., Ekert, C., Heimbach, P., Marotzke, J., Adcroft, A., Hill, C., Marshall, J., 2002. The global ocean circulation during 1992–1997, estimated from ocean observations and a general circulation model. J. Geophys. Res. 107 (C9), 3118. http://dx.doi.org/10.1029/2001JC000888.

Stammer, D., Wunsch, C., Giering, R., Ekert, C., Heimbach, P., Marotzke, J., Adcroft, A., Hill, C., Marshall, J., 2003. Volume, heat, and freshwater transports of the global ocean circulation 1993–2000, estimated from a general circulation model constrained by World Ocean Circulation Experiment (WOCE) data. J. Geophys. Res. 108 (C1), 3007. http://dx.doi.org/10.1029/2001JC001115.

Stephenson Jr., G.R., Gille, S.T., Sprintall, J., 2012. Seasonal variability of upper ocean heat content in Drake Passage. J. Geophys. Res. 117, C04019. http://dx.doi.org/10.1029/2011JC007772.

Sultan, E., Mercier, H., Pollard, R.T., 2007. An inverse model of the large scale circulation in the South Indian Ocean. Prog. Phys. Oceanogr. 74, 71–94.

Sun, C., Watts, D.R., 2002. Heat flux carried by the Antarctic Circumpolar Current mean flow. J. Geophys. Res. 107 (C9), 3119. http://dx.doi.org/10.1029/ 2001JC001187.

Sverdrup, H.U., 1957. Oceanography. In: Bartels, J. (Ed.), Handbuch der Physik. 48. Springer-Verlag, Berlin, pp. 608–670.

Swart, S., Speich, S., Ansorge, I.J., Goni, G.J., Gladyshev, S., Lutjeharms, J.R.E., 2008. Transport and variability of the Antarctic Circumpolar Current south of Africa. J. Geophys. Res. 113 (C09014). http://dx.doi.org/10.1029/2007JC004223.

Talley, L.D., 2003. Shallow, intermediate and deep overturning components of the global heat budget. J. Phys. Oceanogr. 33, 530–560.

Thompson, D.W.J., Solomon, S., 2002. Interpretation of recent Southern Hemisphere climate change. Science 296, 895–899.

Tomita, H., Kubota, M., Cronin, M.F., Iwasaki, S., Konda, M., Ichikawa, H., 2010. An assessment of surface heat fluxes from J-OFURO2 at the KEO and JKEO sites. J. Geophys. Res. 115, C03018. http://dx.doi.org/10.1029/2009JC005545.

Trenberth, K.E., Fasullo, J.T., Kiehl, J., 2009. Earth's global energy budget. Bull. Am. Meteorol. Soc. 90 (3), 311–323.

Trenberth, K.E., Fasullo, J.T., Mackard, J., 2011. Atmospheric moisture transports from ocean to land and global energy flows in reanalyses. J. Clim. 24, 4907–4924.

Trenberth, K.E., Smith, L., 2009. The three dimensional structure of the atmospheric energy budget: methodology and evaluation. Clim. Dyn. 32, 1065–1079.

Uehara, H., Kizu, S., Hanawa, K., Yoshikawa, Y., Roemmich, D., 2008. Estimation of heat and freshwater transports in the North Pacific using high-resolution expendable bathythermograph data. J. Geophys. Res. 113, C02014. http://dx.doi.org/10.1029/2007JC00.

Uppala, S.M., Kållberg, P.W., Simmons, A.J., Andrae, U., da Costa Bechtold, V., Fiorino, M., Gibson, J.K., Haseler, J., Hernandez, A., Kelly, G.A., Li, X., Onogi, K., Saarinen, S., Sokka, N., Allan, R.P., Andersson, E., Arpe, K., Balmaseda, M.A., Beljaars, A.C.M., van de Berg, L., Bidlot, J., Bormann, N., Caires, S., Chevallier, F., Dethof, A., Dragosavac, M., Fisher, M., Fuentes, M., Hagemann, S., Holm, E., Hoskins, B.J., Isaksen, L., Janssen, P.A.E.M., Jenne, R., McNally, A.P., Mahfouf, J.-F., Morcrette, J.-J., Rayner, N.A., Saunders, R.W., Simon, P., Sterl, A., Trenberth, K.E., Untch, A., Vasiljevic, D., Viterbo, P., Woollen, J., 2005. The ERA-40 re-analysis. Q. J. R. Meteorol. Soc. 131, 2961–3012.

van Sebille, E., Baringer, M.O., Johns, W.E., Meinen, C.S., Beal, L.M., Femke de Jong, M., van Aken, H.M., 2011. Propagation pathways of classical Labrador Sea Water from its source region to 26°N. J. Geophys. Res. 116, C12027. http://dx.doi.org/10.1029/2011JC007171.

Vranes, K., Gordon, A.L., Ffield, A., 2002. The heat transport of the Indonesian throughflow and implications for the Indian Ocean heat budget. Deep Sea Res. 49, 1391–1410.

Volkov, D.L., Fu, L.-L., Lee, T., 2010. Mechanisms of the meridional heat transport in the Southern Ocean. Ocean Dyn. 60, 791–801.

Wang, J., Carton, J.A., 2002. Seasonal Heat Budgets of the North Pacific and North Atlantic Oceans. J. Phys. Oceanogr. 32, 3474–3489.

Wang, W., Xie, P., Yoo, S.-H., Xue, Y., Kumar, A., Wu, X., 2011. An assessment of the surface climate in the NCEP climate forecast system reanalysis. Clim. Dyn. 37, 1601–1620.

Warren, B.A., 1999. Approximating the energy transport across oceanic sections. J. Geophys. Res. 104, 7915–7979.

Weijer, W., De Ruijter, W.P.M., Sterl, A., Drijfhout, S.S., 2002. Response of the Atlantic overturning circulation to South Atlantic sources of buoyancy. Glob. Planet. Change 34, 293–311.

Wiggins, R.A., 1972. The general linear inverse problem: implication of surface waves and free oscillations for earth structure. Rev. Geophys. Space Phys. 10, 251–283.

Wijffels, S.E., Toole, J.M., Bryden, H.L., Fine, R.A., Jenkins, W.J., Bullister, J.L., 1996. The water masses and circulation at 10°N in the Pacific. Deep Sea Res. 43, 501–544.

Wijffels, S.E., Toole, J.M., Davis, R., 2001. Revisiting the South Pacific subtropical circulation: a synthesis of World Ocean Circulation Experiment observations along 32°S. J. Geophys. Res. 106, 19481–19513.

Willis, J.K., 2010. Can *in situ* floats and satellite altimeters detect long-term changes in Atlantic Ocean overturning? Geophys. Res. Lett. 37, L06602. http://dx.doi.org/10.1029/ 2010GL042372.

Willson, R.C., Hudson, H.S., 1991. The sun's luminosity over a complete lunar cycle. Nature 351, 41–44.

Wunsch, C., 1977. Determining the general circulation of the oceans: a preliminary discussion. Science 196, 871–875.

Wunsch, C., 1996. The Ocean Circulation Inverse Problem. Cambridge University Press, 442 pp.

Wunsch, C., Heimbach, P., 2009. The global zonally integrated ocean circulation, 1992–2006: seasonal and decadal variability. J. Phys. Oceanogr. 39, 351–368.

Wunsch, C.,Heimbach, P., 2013. Dynamically and kinematically consistent global ocean circulation and ice state estimates.

Wüst, G., Defant, A., 1936. Atlas zur Zirculation des Atlantischen Ozeans, VI. Walter de Gruyter and Co., Berlin/Leipzig, Germany, 103 pp.

Yu, L.S., Jin, X., Weller, R.A., 2008. Multidecade Global Flux Datasets from the Objectively Analyzed Air–sea Fluxes (OAFlux) Project: Latent and Sensible Heat Fluxes, Ocean Evaporation, and Related Surface Meteorological Variables. OAFlux Project Technical Report (OA-2008-01), Woods Hole, Woods Hole Oceanographic Institution, USA.

Yu, L.S., Weller, R.A., Sun, B., 2004. Improving latent and sensible heat flux estimates for the Atlantic Ocean (1988–1999) by a synthesis approach. J. Clim. 17, 373–393.

Zheng, Y., Giese, B.S., 2009. Ocean heat transport in simple ocean data assimilation: structure and mechanisms. J. Geophys. Res. 114. http://dx.doi.org/10.1029/2008JC005190.

Chapter 30

The Marine Carbon Cycle and Ocean Carbon Inventories

Toste Tanhua*, Nicholas R. Bates[†] and Arne Körtzinger*
*Helmholtz Centre for Ocean Research Kiel (GEOMAR), Kiel, Germany
[†]Bermuda Institute of Ocean Sciences, Ferry Reach, Bermuda

Chapter Outline

1. INTRODUCTION AND BACKGROUND TO THE MARINE CARBON CYCLE

The marine carbon cycle is a key elemental cycle of the ocean and, by virtue of its interaction with the atmospheric carbon dioxide (CO_2), of fundamental importance for the Earth's climate. If we ignore the comparatively inert (on timescales relevant to the Anthropocene) lithospheric carbon pools, the ocean constitutes by far the largest carbon reservoir of the atmosphere–terrestrial biosphere–ocean system. In fact, the preindustrial ocean contained more than 60 times as much carbon as the atmosphere. This large reservoir with its complex internal dynamics is the major determining factor for the atmospheric CO_2 content over time scales of centuries and longer. Therefore even small changes in the natural components of the marine carbon cycle—as may be expected under the evolving anthropogenic forcing through changes known as ocean warming, ocean eutrophication, ocean acidification, and ocean deoxygenation—bear the potential to significantly feedback to the Earth's climate system. On the other hand, the CO_2 buffering property of the marine carbon reservoir provides a large uptake capacity for additional CO_2, that is, anthropogenic CO_2.

The absorption of carbon dioxide by seawater and subsequent equilibrium reactions within this medium give rise to a complex chemical system, often referred to as the seawater CO_2–carbonate system (or alternatively referred to as the marine CO_2 system or marine carbonate system). The seawater CO_2–carbonate system is responsible for about 95% of the acid–base buffering capacity of seawater and hence essentially controls the pH of ocean waters. This complex chemical system is influenced by physical, chemical, biological and geological processes in the ocean such as primary production and the organic carbon cycle, and in turn, influences, for example, chemical rates and speciation of trace elements, compounds and isotopes in the ocean. It also has a major impact on key processes such as the precipitation and dissolution of calcium carbonate ($CaCO_3$) predominantly in the form of particulate inorganic carbon in the ocean.

Ocean Circulation and Climate, Vol. 103. http://dx.doi.org/10.1016/B978-0-12-391851-2.00030-1

The chemical basis of the seawater CO_2–carbonate system is the reaction of CO_2 with water to form carbonic acid (H_2CO_3), that is, $CO_{2(gas)} \rightleftharpoons CO_{2(aq)} + H_2O \rightleftharpoons H_2CO_3 \rightleftharpoons HCO_3^- + H^+ \rightleftharpoons CO_3^{2-} + 2H^+$. The hydration reaction (i.e., the second step in the sequence above) is slow and most of the CO_2 in seawater remains in physically dissolved state rather than in the combined form of true carbonic acid. As a diprotic acid, H_2CO_3 can dissociate in two steps that lead to the formation of two ionic species, bicarbonate (HCO_3^-) and carbonate (CO_3^{2-}). The seawater CO_2–carbonate system is thus comprised of four inorganic carbon species—CO_2, H_2CO_3, HCO_3^-, and CO_3^{2-}—which are connected through chemical equilibria that have been thermodynamically characterized to rather high accuracy (e.g., refs in Dickson et al., 2007). The chemical speciation within the inorganic carbon acid–base system is a function of pH. Over the rather narrow pH range of seawater (approximately 7.4–8.2), HCO_3^- is by far the dominant species (~90%), followed by CO_3^{2-} (~10%) and finally CO_2/H_2CO_3 (<1%). The chemical characterization of the seawater CO_2–carbonate system is thus achieved through knowledge of these acid–base equilibria and kinetics (e.g., Stumm and Morgan, 1981; Butler, 1982; Zeebe and Wolf-Gladrow, 2002) and by measurement of two or more of four commonly observable carbonate parameters such as dissolved inorganic carbon (DIC), total alkalinity (TA), pH (i.e., $-\log_{10}[H^+]$), and partial pressure or fugacity of CO_2 (i.e., pCO_2 or fCO_2). DIC reflects the sum of the concentrations of CO_2, H_2CO_3, HCO_3^-, and CO_3^{2-} species, while TA is a measureable parameter that reflects acid binding capacity of seawater, both independent of temperature and pressure. Seawater pCO_2 or fCO_2 is a measureable parameter related to the only volatile species, CO_2, and hence relevant for air–sea CO_2 gas exchange assessments, and pH, a parameter describing the acidity of seawater. Both are dependent on temperature, salinity, and pressure (Dickson et al., 2007).

The seawater CO_2–carbonate system is influenced by complex interactions between physical, biological, and chemical processes that are highly variable across time and space, which, while outside the scope of this chapter, are described in other chapters of this book. We do however provide a brief and simplified overview of key processes that are relevant to understanding the marine carbon cycle. More information about the dynamics of the marine carbon system and on significant processes can be found in several text books, for instance, by Sarmiento and Gruber (2003) or Williams and Follows (2011).

The kinetics and equilibria of the seawater CO_2–carbonate system discussed above are well characterized and are mainly functions of salinity and temperature, but highly influenced by physico-biogeochemical processes as well. The ocean is constantly exchanging CO_2 (and other gases) with the atmosphere across the air–sea interface, and the chemical reactions in seawater act toward establishing equilibrium between the atmospheric and surface ocean partial pressure of CO_2 (see Section 4 for a discussion of spatial variability of air–sea fluxes). The air–sea equilibrium of CO_2 is however seldom completely reached, if ever, across the global ocean due to the influence of other processes. The first such process is the equilibration of the inorganic carbon chemical reactions in seawater. Even though the air–sea exchange is a relatively fast process, the buffering capacity of the seawater CO_2–carbonate system is such that roughly only one molecule of CO_2 out of 20 that enter the ocean from the atmosphere remains as CO_2, see reactions above. The time scale required to reach chemical equilibrium between ocean and atmosphere is thus significantly larger for CO_2 than for nonreactive gases such as oxygen. The exact relationship and balance of chemically driven transfer processes is a complex interplay where the equilibrium constants of the chemical reactions above are all functions of both salinity and temperature, and where the ratio of DIC to alkalinity is an important factor. The flux of CO_2 across the air–sea interface is also influenced by turbulent processes such as waves, bubbles, and gas mixing in seawater, and barriers to gas exchange such as sea–ice cover or surface microlayers.

Within the ocean, multiple physico-biogeochemical processes act upon the marine carbon cycle. For example, the transport of carbon between the surface and deep ocean is highly influenced by the global thermohaline circulation (see Chapter 11); deep, intermediate, and mode water formation (see Chapters 9 and 10); and horizontal transport of carbon between hemispheres and ocean basins as well as between open and coastal/shelf ocean regions (see Chapter 19). Several internal biological processes also affect the concentrations of the chemical species of the seawater CO_2–carbonate system and observable parameters such as DIC, TA, pCO_2, and pH. The process of photosynthesis uses CO_2 and water to form organic matter, so that the DIC and pCO_2 decrease (TA is influenced by the concurrent uptake or regeneration of nitrate; while pH increases). Photosynthesis is, obviously, limited to the photic zone in the ocean, but its role of reducing DIC and pCO_2 tends to drive an air-to-sea flux of CO_2 (i.e., ocean sink of CO_2). This process is variably influenced by the diversity of the biological community present (across all trophic levels), which in turn is reinforced by variability and heterogeneity of the biological pump of carbon from surface to deep by the vertical export of organic matter. The remineralization (i.e., respiration) of organic matter also takes place in the deep (and dark) ocean through sinking organic particles (and by dissolved organic carbon transported downward by physical processes) and adds CO_2 and DIC to the water column, so that the deep ocean tends to be rich in DIC compared to the surface ocean.

The seawater CO_2–carbonate system is also highly influenced by the formation and dissolution of $CaCO_3$ (mostly in the form of calcite or aragonite) minerals used by many marine animals and plants (such as hermatypic and ahermatypic coral reefs, shelly fauna, and pelagic coccolithophores, foraminifera, and pteropods). $CaCO_3$ shells, tests, or skeletons are mostly (but not always) formed in the upper ocean, and as the particles sink they create a net downward transport of DIC. The overall reaction can be written as $Ca^{2+} + 2\,HCO_3^- \rightleftharpoons CaCO_3 + CO_2 + H_2O$ from which it can be seen that formation of $CaCO_3$ removes both TA and DIC (in a ratio of 2:1 to DIC) from the surface ocean. Unlike photosynthesis and respiration, $CaCO_3$ formation leads to the generation of CO_2, and depending on the calcifying taxa/species and the ratio of organic carbon to $CaCO_3$ production, has implications for the net effect on air–sea exchange of CO_2. Furthermore, the vertical sinking or export of $CaCO_3$ particles from the surface ocean tends to also increase the downward flux of organic particles. In the mid-depth and deep ocean, sinking $CaCO_3$ particles dissolve in the deep ocean, below the saturation horizon of $CaCO_3$, with some fraction accumulating in seafloor sediments. Variability of the saturation depths of $CaCO_3$ minerals such as calcite and aragonite (e.g., the depth at which these minerals become undersaturated at the lysocline and experience net dissolution at the carbonate compensation depth) and the process of burial and redissolution of carbonates from the sediment are important regulators of the seawater CO_2–carbonate system but more relevant over centennial to millennial to geological time scales. The flux of carbon from the surface to the deep ocean by biological transformation is highly influenced by temporal and spatial variability and heterogeneity in marine primary production, export of carbon, and the remineralization of organic matter at depth, and $CaCO_3$ production and dissolution.

As briefly described above, there is a global heterogeneity of the seawater CO_2–carbonate system between the deep open ocean and the shallow coastal ocean. Indeed, while we cannot comprehensively describe all the influences on the marine carbon cycle, it is worthwhile remembering that there are many factors that need to be considered. The impact of seawater CO_2–carbonate system processes on anthropogenic CO_2 uptake is discussed further in Section 3. Temporal variability of physico-biogeochemical processes, occurring on diurnal to seasonal to multidecadal to centennial timescales, are all important drivers for changes of the seawater CO_2–carbonate system in the Anthropocene. Here, the Anthropocene refers to the emerging geological epoch that encapsulates the relatively short-term impact of humans on global biogeochemical cycles and processes.

From our knowledge of the seawater CO_2–carbonate system and the timescale of ocean ventilation, it can be predicted that, millennia from now, in a future steady state, this capacity will account for an oceanic storage of 80% of the cumulative emissions of anthropogenic CO_2, a fraction that will be further enhanced toward 95% when equilibration with carbonate sediments is also taken into account (Archer, 2005). In contrast, our best estimate of the ocean's current fossil fuel share amounts to 155 Pg C in 2010, which represents only about 45% of the cumulative fossil fuel emissions of 350 Pg C in 2009 (Khatiwala et al., 2013). The situation is therefore characteristic of a transient signal, the propagation and implication of which requires detailed understanding and hence observation of the properties of inorganic carbon dynamics in the ocean.

2. HISTORY OF OBSERVATIONS AND CAPACITY TO COLLECT MARINE CARBON CYCLE MEASUREMENTS

The history of ocean carbon cycle research has been marked by continual reassessment of appropriate needs for scientific investigation with several common themes throughout, including improvement in analytical methodologies, the introduction of reference materials, improvements in sampling strategies, and inception of global surveys, fixed and sustained ocean time series, and process studies for understanding important feedbacks between the marine carbon cycle and physico-biogeochemical processes, and the evolution of observational networks for the open and coastal oceans.

2.1. Measurable Parameters of the Seawater CO_2–Carbonate System

Accurate knowledge of the properties of seawater CO_2–carbonate system is a prerequisite for understanding the chemical forcing and consequences of key biogeochemical processes such as biological production, respiration, biocalcification, or uptake of anthropogenic carbon (C_{ant}). Particularly challenging is the quantification of the ocean interior storage of C_{ant} since the anthropogenic signal is very small compared to the large natural background, so that accurate measurements of the seawater CO_2–carbonate system are required. This includes measurements of DIC, TA, $p$$CO_2$, and pH in the ocean.

On the other hand, in order to better observe and understand natural variability and feedback potential of the marine carbon cycle, the severe undersampling of classical hydrography both in space and time has to be addressed. Seawater CO_2–carbonate system measurements of the highest precision and accuracy (Table 30.1) require experienced operators to collect discrete samples during dedicated cruises on research vessels. The collection of surface $p$$CO_2$ observations from a global and international network of "Voluntary Observing Ships" is one example of how spatio-temporal coverage can be extended significantly (Monteiro et al., 2010). In this section we give an overview of analytical options for the marine carbonate system and then move on to discuss data quality control and synthesis.

TABLE 30.1 Estimates of the Analytical Precision and Accuracy Achievable with At-Sea Measurements of DIC, TA, pCO_2, and pH (Millero, 2007) Assuming the Use of Certified Reference Materials (CRM)

Parameter	Method	Precision	Accuracy
DIC	CO_2 extraction with coulometric detection	±1 μmol kg^{-1}	±2 μmol kg^{-1}
TA	Potentiometric titration	±1 μmol kg^{-1}	±3 μmol kg^{-1}
pCO_2	Air–seawater equilibration with infrared detection	±0.5 μatm	±2 μatm
pH	Spectrophotometry with pH indicator dye	±0.0004	±0.002

For all of the measurable variables, state-of-the-art methods are available that routinely achieve the required very high precision and accuracy requirements even at sea (Table 30.1). For full access to the seawater CO_2–carbonate system, at least two of these four parameters need to be measured. In principle, any combination of two of the parameters is possible but there are a few less favorable choices where unfortunate error propagation leads to a significant loss of accuracy. The most extreme example is the combination of pH–pCO_2. For measurements in the interior ocean, the pairing of DIC and TA is by far the most commonly used combination. This has mostly practical reasons since the sampling is straightforward, both measurements can be done on the same sample, and DIC and TA are both state variables facilitating the reporting of the values. Our current understanding of the marine carbon cycle in the interior of the global ocean and its anthropogenic perturbations is therefore largely based on measurements of these two parameters.

2.2. History and Coordination of Global-Scale Marine Carbon Cycle Measurements

The assessment of the global ocean carbon cycle is clearly a very large task that cannot be undertaken by one group of investigators, but where coordination and cooperation between many groups in many different countries is necessary. The history and coordination of ocean carbon observations were recently summarized by Sabine et al. (2010). The early efforts in the 1960s and 1970s, and those that have followed, have been critically important for supporting the collection of high-quality seawater CO_2-carbonate data over the last four decades, which have allowed quantitative assessments of trends in marine carbon cycle parameters (such as DIC, pCO_2, pH), determination of the rate and inventories of anthropogenic CO_2 in the global ocean and, more recently, assessment of ocean chemical changes associated with ocean acidification.

The first comprehensive survey and collection of inorganic carbon in the open ocean occurred as part of the Geochemical Ocean Sections Study (GEOSECS) program. The program was initiated in 1969 with an intercalibration exercise at sea to test sampling and analytical methodologies (e.g., Craig and Weiss, 1970; Takahashi et al., 1982) and was followed by a full global ocean survey that took place between 1971 and 1978. The GEOSECS project laid the foundation for subsequent scientific expeditions, with the Transient Tracers in the Ocean (TTO) following in the early 1980s (Brewer et al., 1985). Both GEOSECS and TTO seawater CO_2–carbonate chemistry data, corrected for comparison to modern measurements (e.g., Tanhua and Wallace, 2005) and incorporated into global data synthesis products (e.g., Key et al., 2004), have been widely used for data-model syntheses and for determining long-term trends in changes in DIC and anthropogenic CO_2 (e.g., Olsen et al., 2006; Tanhua et al., 2006; Peng and Wanninkhof, 2010; Pérez et al., 2010; Bates et al., 2012).

In the early and mid-1980s, planning for the World Ocean Circulation Experiment (WOCE) and Joint Global Ocean Flux Study (JGOFS; SCOR, 1987) began, with expeditions initiated in the late 1980s. These scientific programs spanned the next decade, and together completed a global ocean survey by the end of the 1990s. As part of the WOCE hydrographic program, studies of sections of seawater CO_2–carbonate chemistry (e.g., DIC and TA) were completed in the major ocean basins. In contrast, the JGOFS program focused on physico-biogeochemical process studies such as the North Atlantic Bloom Experiment (NABE), and the Arabian Sea, Equatorial Pacific Ocean and Ross Sea projects. As part of U.S. JGOFS efforts, two ocean time-series sites, ALOHA (A Long-term Ocean Habitat Assessment, also known as Hawaii Ocean Time-series, HOT) near Hawaii, and BATS (Bermuda Atlantic Time-series Study) near Bermuda (e.g., Karl et al., 2003) were initiated—continuing to the present—to complement other time-series efforts; for example, Line P (Wong et al., 2010), Icelandic Sea (Olafsson et al., 2010), with other time series following thereafter (e.g., CARIACO, CArbon Retention In A Colored Ocean, off Venezuela; ESTOC, European Station for Time Series in the OCean, near the Canary Islands; KERFIX, KERguelen FIXed station, in the Southern Ocean off the Kerguelen Islands; DYFAMED, DYnamique des Flux Atmosphériques en MEDiterranée, in the Mediterranean Sea).

Since the WOCE and JGOFS programs ended in the 1990s, inorganic carbon measurements along repeat hydrography sections have continued primarily with the CLIVAR-CO_2 (Climate Variability program) in the wider context of GO-SHIP (the Global Ocean Ship-based Hydrographic Investigations Program), but also as part of GEOTRACES (a study focused on large-scale distributions of trace elements and their isotopes in the marine environment) and several other projects.

The rather stringent quality requirements of oceanic CO_2 measurements had already been recognized at the time of the WOCE/JGOFS CO_2 surveys that led to the development of standard operating procedures. The resulting first *Handbook of Methods for the Analysis of the Various Parameters of the Carbon Dioxide System in Seawater* (DOE, 1994) received great worldwide attention in the scientific community and after several revisions is still in use in its most updated version (Dickson et al., 2007). This early emphasis on high quality is reflected not only in the development of the analytical techniques themselves but also in the implementation of quality control and assessment measures. The most important such measure was the development of a certified reference material (CRM), which started in 1989 and achieved certification for DIC in 1991 and TA in 1996 and has since then been distributed in large numbers to scientists worldwide (Dickson, 2010). The CRM-based approach of the marine CO_2 community is a success story that is being implemented also in other areas of chemical oceanography (i.e., nutrients, dissolved organic carbon).

Figure 30.1 shows the frequency of interior ocean measurements of DIC from open ocean samples. The rapid increase of the measurements in the early WOCE period is a dominant feature in this figure. After a brief lull in the frequency of cruises and number of observations after the end of the WOCE survey, the observational effort increased again during the mid-2000s, that is, as the repeat hydrography program started. The decline in observations toward the end of the time series is likely a reflection of the slowness for data to be reported to data-centers rather than a real decline in the observational effort. A large internal effort has to be made, however, to keep the repeat hydrography program at its current density.

Similarly, Figure 30.2 shows a dramatic increase in the number of surface ocean pCO_2 observations per year reported to the data collection SOCAT version 1.5 (Pfeil et al., 2013). This partly reflects improvements in the capability to conduct surface ocean pCO_2 measurements (through improved and more uniform analytical techniques, and calibration and increasing capacity to conduct shipboard observations. This version of SOCAT holds 6.3 million data points of quality controlled pCO_2 measurements. Even though the community is now logging approximately 1 million observations per year, the ocean remains vastly undersampled both in space and time. This is partly because of the large temporal and spatial variability of pCO_2 resulting in relatively short decorrelation length scales, and partly due to large holes in the observational network, that is, there are regions where the shipping lines hardly ever go to, such as the southeast Pacific (e.g., Pfeil et al., 2013).

2.3. Data Synthesis Products and Quality Control Procedures

Although sampling and measurements of carbon parameters in the ocean have a long history, it was only by the advancement of data synthesis products that a global

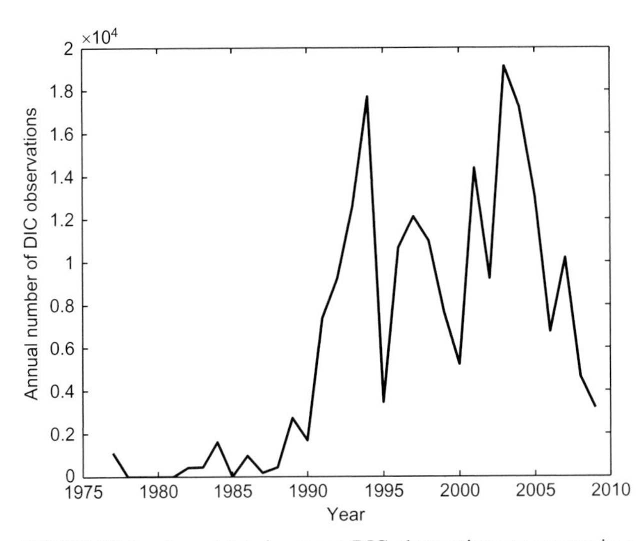

FIGURE 30.1 Annual interior ocean DIC observations as present in a preliminary version of the GLODAPv2 data collection as of April 2013. *This figure is an update of the figure in Sabine et al. (2010).*

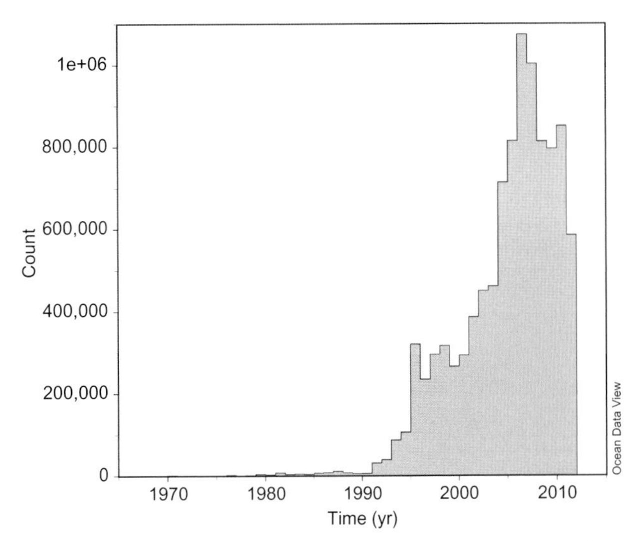

FIGURE 30.2 Annual number of surface ocean pCO_2 observations from the SOCATv1.5 data product. *Figure updated from Pfeil et al. (2013), thanks to Benjamin Pfeil.*

estimate of the oceanic storage of anthropogenic carbon was made possible. The most important aspects of the interior ocean carbon data products are that they consist of carefully quality controlled, internally consistent, data available in a common format. The Global Ocean Data Analysis Project (GLODAP) provided a data set from the global CO_2 survey of the 1990s (Key et al., 2004; Sabine et al., 2005), including significant historic cruises. It was, however, soon apparent that large areas of the ocean were heavily undersampled and underrepresented in GLODAP, particularly the North Atlantic Ocean and the not-represented Arctic Ocean. Therefore, a data collection for these areas was published in 2009, called CARINA (Carbon IN the Atlantic) (Key et al., 2010 and additional articles in the special issue). Recently a data product covering the Pacific Ocean including data from 213 cruises (additionally 59 data sets from line P, and 34 WOCE cruises) was published as PACIFCA (PACIFic ocean Interior CArbon) (Suzuki et al., 2013). In addition, a current effort known as GLODAPv2 aims at merging those three products and additional data not included in any of those (see Table 30.2 and Figure 30.3).

Surface ocean pCO_2 data products are also available from two different but complementary sources: the LDEO pCO_2 data product (V2012) contains 6.7 million surface pCO_2 data points (Takahashi and Sutherland, 2013) from which climatological fields have been constructed (Takahashi et al., 2009); The SOCAT v2 data products contain 10 million data points (Pfeil et al., 2013) and a gridded product (Sabine et al., 2013).

The interior ocean data products consist of two or three main components; individual cruise files, merged data products, and (for GLODAP) gridded products (Key et al., 2004). The individual cruise files are all reported in a common format with standardized units and quality flags, and were in all instances scrutinized and quality controlled (first level of QC). As discussed in Tanhua et al. (2010); "Primary QC is a process in which data are studied in order to identify outliers and obvious errors." Only data flagged as "good" are included in the merged products. The primary QC is designed to find outliers but is insensitive to systematic biases; those can be assessed by the so-called secondary quality control (secondary QC), which is a process in which the data are "objectively studied in order to quantify systematic biases in the reported values" (Tanhua et al., 2010). Biases in the reported data are often due to incorrectly quantified standard concentrations, blank problems or other analytical difficulties that are very difficult to assess in the field. The use of internationally agreed on CRMs tends to greatly reduce intercruise biases. Note that secondary QC only addresses the accuracy of the data, not the precision.

The most important tool in the secondary QC process is the so-called cross-over analysis. This is based on the assumption that the deep ocean is only changing slowly so that one can compare values obtained at reasonably close station locations by multiple investigators making measurements from different ships at different times. If systematic differences in the reported values from the investigators are found, these are likely to be due to systematic biases in, at least, one of the measurements. Offsets between available cruises for an ocean basin are calculated and compared to each other by the use of least square models (e.g., Wunsch, 1996), typically using the inversion scheme described by Johnson et al. (2001). The inversion then makes suggestions on how individual cruises should be adjusted to produce the most internally consistent data collection possible. As described by Tanhua et al. (2010), lower bounds were applied for actually making an adjustment to the seawater CO_2–carbonate data. These adjustments were, respectively, as follows: for DIC—4 $\mu mol\ kg^{-1}$, for TA—6 $\mu mol\ kg^{-1}$, and for pH to 0.005 units (Tanhua et al., 2010).

The gridded products are valuable components of the data products, particularly for easy comparison with model results and for calculating inventories of properties such as C_{ant}. Due to sparseness of data, significant interpolation and

TABLE 30.2 Summary of Data Products of Marine Carbon System Data

Data Product	Time Period	Region	Repository
GLODAP	1972–1998	Global	http://cdiac.ornl.gov/oceans/glodap/
CARINA	1977–2005	Atlantic Arctic Southern Ocean	http://cdiac.ornl.gov/oceans/CARINA/
PACIFICA		Pacific Ocean	http://cdiac.ornl.gov/oceans/PACIFICA/
GLODAPv2	1972–2012	Global	http://cdiac.ornl.gov/oceans/
LDEO pCO_2 database	1957–2012	Global	http://cdiac.ornl.gov/oceans/LDEO_Underway_Database/
SOCAT	1968–2011	Global	http://www.socat.info/

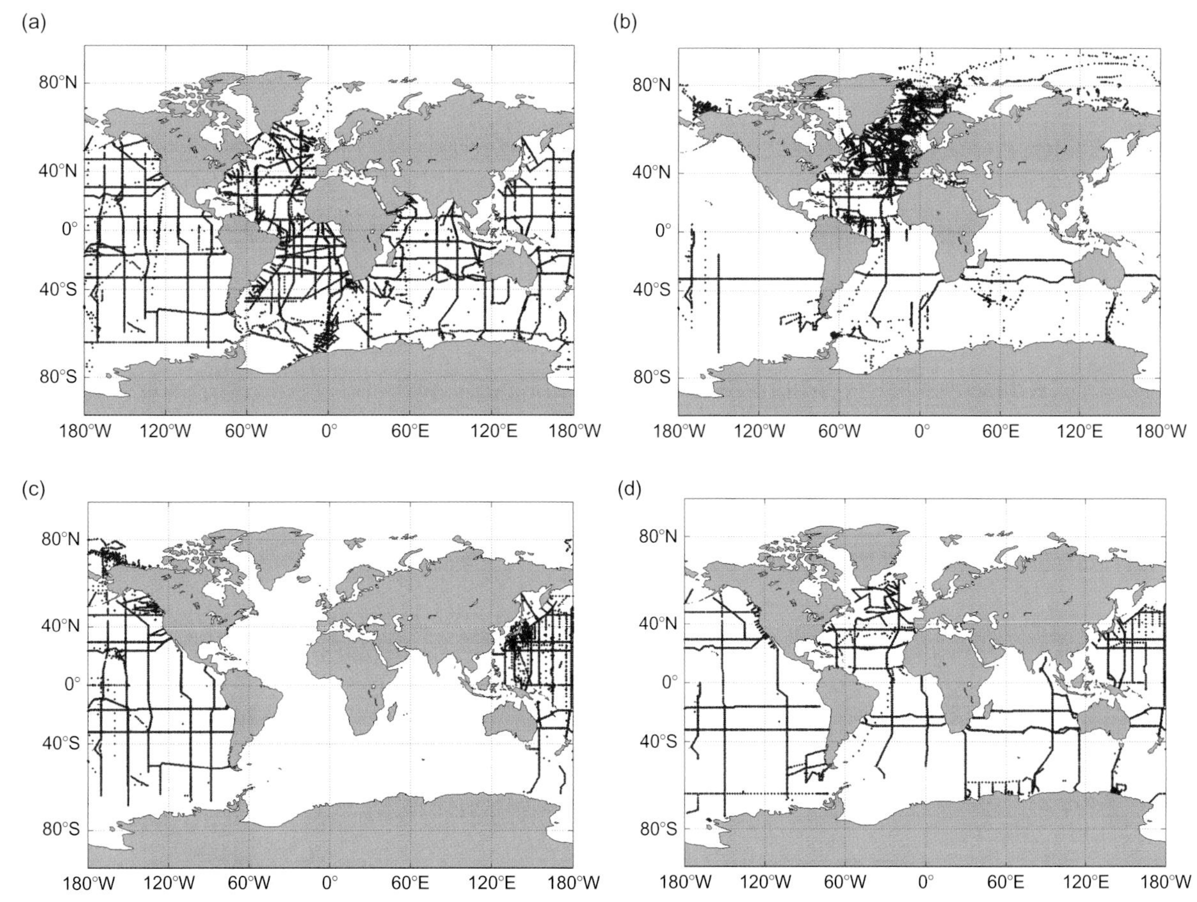

FIGURE 30.3 Cruises available in the GLODAP (a), CARINA (b), and PACIFICA (c) data collections. Panel d shows additional cruises not available in any of those, but that will be included in GLODAPv2 (i.e., CLIVAR). *Figure from Are Olsen (personal communication).*

extrapolation errors can be expected on local scales (e.g., Schneider et al., 2012).

3. THE ANTHROPOGENIC PERTURBATION OF THE MARINE CARBONATE SYSTEM

Quantifying the uptake and storage of anthropogenic CO_2 in the interior ocean requires that a distinction or estimate be made between "natural" contributions (i.e., $C_{natural}$) and the "anthropogenic" contribution to changes in DIC over time, that is, C_{ant}. With the anthropogenic part of the inorganic carbon, we normally understand the "extra" DIC present in the ocean as a direct consequence of the higher CO_2 concentration in the atmosphere compared to "preindustrial" times, that is, about year 1750, and the shift in air–sea gas exchange this has led to. However, higher atmospheric CO_2 concentration is not the only "anthropogenic" driver of changes in the seawater CO_2–carbonate system in the ocean. Other anthropogenic drivers that can directly influence the DIC content are, for instance, warming (outgassing) and increased stratification (slower exchange between surface and the interior ocean). Even though the definition of C_{ant} might be slightly fuzzy, it is practical to be able to quantify the amount of C_{ant} in the ocean, and put this number in relation to other quantities of (anthropogenic) carbon pools in the global system.

Uptake of anthropogenic or "excess" CO_2 (C_{ant}) into the global ocean causes shifts in the balance of the seawater CO_2–carbonate system, which affects all species and all but one measurable parameter such that the concentration of CO_2, HCO_3^-, H^+, DIC, and pCO_2 increases, CO_3^{2-} and pH decreases while TA remains unchanged. Although this "excess" CO_2, that is C_{ant}, is not distinguishable from "natural" CO_2 ($C_{natural}$) and cannot be measured directly (except for its isotopic signature), historically the two have been treated as independent quantities. The motivation for treating C_{ant} as an independent transient tracer is its usefulness since this quantity can be compared to or inferred from other transient tracers and simulated even in ocean circulation models that lack a fully characterized marine carbonate system. Doing so, however, requires the assumption that all components of the natural carbon cycle remain unaffected by the accumulating C_{ant}. Since carbon is not a bio-limiting element in the ocean, the first order assumption of no effect on the biological carbon pumps appeared reasonable (i.e., the ocean remains in steady state). We have long since recognized, however, that a whole suite of direct and indirect sensitivities and potential feedback mechanisms of C_{ant} (Riebesell et al., 2008; Sabine and Tanhua, 2010) will make this assumption less and less justifiable as evidence grows that the natural ocean carbon is not in steady state any more under direct and indirect anthropogenic forcing, and this will eventually render the entire C_{ant} concept less useful, see Section 3.3. This is even more the case, as the various methods in use to estimate C_{ant} are biased to a variable, and mostly unknown, degree by changes in the natural carbon background that will make C_{ant} increasingly poorly defined and incomparable between the different estimation techniques. It is therefore necessary for these estimations that the quantity of prime interest should be DIC that allows for keeping track of changes in the ocean's inorganic carbon storage, regardless of their cause and source/sink. In the end, it is this quantity that determines atmospheric CO_2 levels.

3.1. Review of Recent Estimates of Global C_{ant} Storage

Several methods have therefore been developed to calculate the C_{ant} concentration through the water column, and ultimately the integrated inventory of C_{ant}, in the ocean. Fundamentally, these calculations are approached either by (1) back-calculation methods based on measurements of the seawater CO_2–carbonate system, together with biogeochemical variables such as oxygen and nutrients, or (2) based on measurements of transient tracers that provide information about the ventilation characteristics of the interior ocean. Recently published works provide useful overviews of the currently available methods to quantify C_{ant} concentrations (e.g., Sabine and Tanhua, 2010; Khatiwala et al., 2013). Back-calculation methods include the ΔC^* method (Gruber et al., 1996), and variations thereof such as the LM05 technique (Lo Monaco et al., 2005) and φC_T° method (Vázquez-Rodríguez et al., 2009) and TrOCA (tracer combining oxygen, inorganic carbon and TA) (Touratier et al., 2007) and depend on estimates of preindustrial DIC concentrations from other biogeochemical parameters. Transient tracer-based methods treat C_{ant} as a passive tracer with well-known input function that can be calculated with knowledge about ocean ventilation. Commonly used approaches include the TTD (transit time tistribution) method (e.g., Hall et al., 2002, 2004; Waugh et al., 2004) or the more complex variation thereof (i.e., the "Green Function (GF) method"; Khatiwala et al., 2009).

To date, all observationally based estimates of the global ocean C_{ant} inventory are based on the GLODAP data collection. These global estimates range from 106 ± 17 Pg C (Pg = Petagram = 10^{15} g carbon) based on the ΔC^* method (Sabine et al., 2004a), 94–112 Pg C based on the TTD method (Waugh et al., 2006), and 114 ± 22 Pg C based on the GF method (Khatiwala et al., 2009) for the year 1994. It should be noted that all of these estimates are based on the region covered by GLODAP such that the world ocean estimates exclude marginal seas such as the Mediterranean Sea and the Arctic Ocean (see Sections 4.4 and 4.5). However, regional estimates of the C_{ant} inventories based on more recent data have been published (e.g., Steinfeldt et al., 2009; Schneider et al., 2010; Ríos et al., 2012; Schneider et al., 2012; Perez et al., 2013, see Section 4). The GLODAP data collection covers data collected, roughly, until the late 1990s (i.e., to the end of the WOCE global survey), so that changes in ocean circulation and biogeochemistry during the last 15 years are not captured by these estimates (assuming that the C_{ant} inference methods are able to compensate for a non-steady-state ocean, see Section 3.3). In a recent compilation, Khatiwala et al. (2013) scaled GLODAP-based data to 2010 inventories, assuming steady-state ocean circulation and biology, and compared those to estimates from ocean inversions and model-based estimates. These estimates vary from 106 to 152 Pg C for the GLODAP region. Khatiwala et al. (2013) have also added an estimate of 9–14 Pg C to the global estimates to account for those ocean regions not covered by the GLODAP region (based on model estimates, upper limit; and scaled observational estimates, lower limit). The combination of these estimates, albeit not entirely independent, led the authors to suggest a range of the global C_{ant} inventory to be 152–157 Pg C (155 Pg C as the "best estimate") in 2010, with an uncertainty of $\pm 20\%$. A caveat for these estimates—as for most previous estimates—is that they assume a time invariant ocean circulation and biogeochemistry to scale 1994-based data to year 2010 estimates.

The current state of the marine carbon cycle (Figure 30.4) is reasonably well constrained. The marine carbon reservoir receives carbon in various forms (particulate–dissolved, organic–inorganic) from the terrestrial system and exports both particulate organic and inorganic carbon into shelf and deep-sea sediments. The estimated input–output balance of 0.6 Pg C yr^{-1} is assumed to drive a natural sea-to-air flux of CO_2 (i.e., natural source of CO_2 to the atmosphere from the ocean) both in the preindustrial and current ocean.

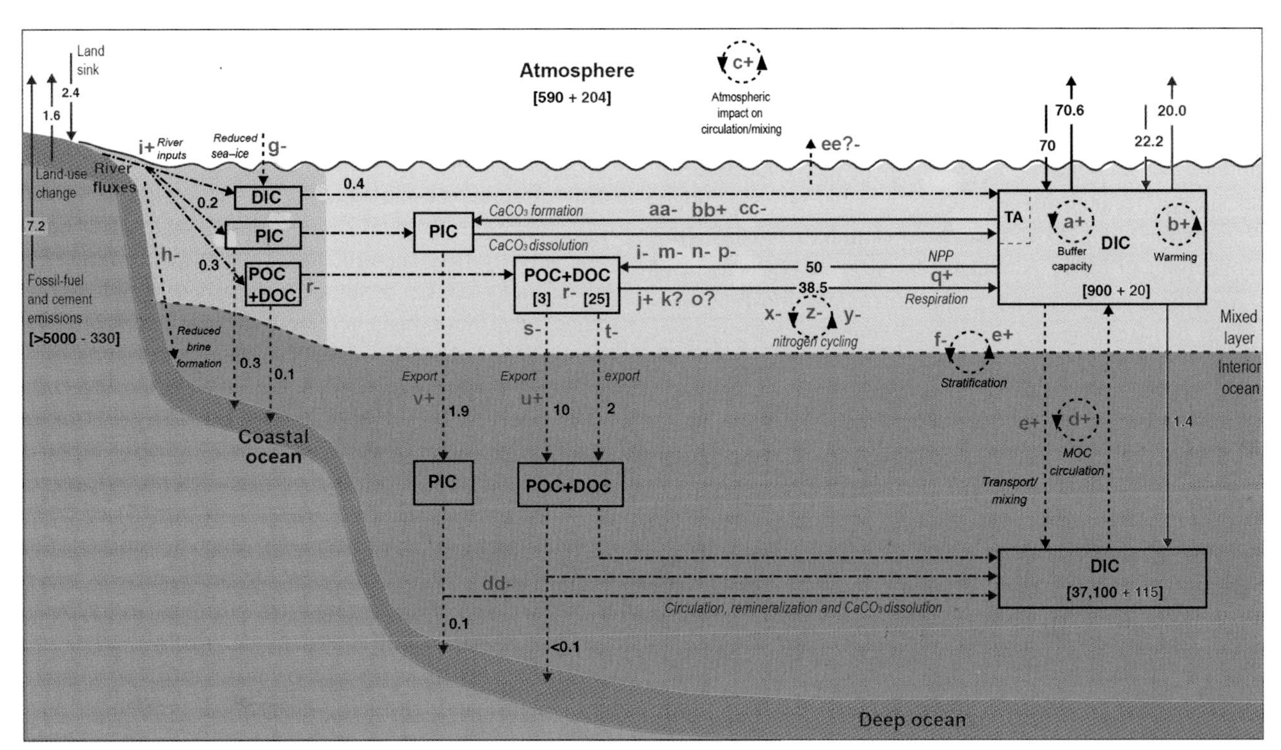

FIGURE 30.4 Schematic figure of physical and biological feedbacks impacting the ocean carbon cycle and anthropogenic inventories and uptake in the open ocean. Bracketed values in black are approximate reservoir size (Pg C) with anthropogenic perturbation listed in red. Un-bracketed values accompanying arrows denote fluxes in Pg C yr^{-1}. The dashed arrows denote physical transport and circulation primarily, while the solid and dashed green arrows denote biologically driven processes. The green arrows denote biogeochemical processes. Letters in purple denote feedback mechanisms listed in Table 30.3. Please note that many of these feedback mechanisms are also relevant in the coastal ocean but have not been identified here in order to reduce the complexity of the figure. Abbreviations included: PIC, particulate inorganic carbon; POC, particulate organic carbon; DOC, dissolved organic carbon; and DIC, dissolved inorganic carbon. *This figure is modified from Sabine and Tanhua (2010) with additional inputs from Gehlen et al. (2011), Riebesell and Tortell (2011), and Andersson et al. (2011).*

The anthropogenic CO_2 uptake in the global ocean, estimated at about 2.2 Pg C yr^{-1} at present, more than offsets the natural sea-to-air flux of CO_2. The resulting net air-to-sea flux of CO_2 of 1.6 ± 0.9 Pg C yr^{-1} is in good agreement with an independent flux estimate based on a global surface ocean pCO_2 climatology that includes more than 3 million pCO_2 measurements from almost four decades (Takahashi et al., 2009). The closure between inventory-based and flux-based estimates of anthropogenic CO_2 uptake is rather reassuring although the uncertainties in these estimates remain large.

3.2. Monitoring Decadal Change of DIC and C_{ant}

Assessments of decadal changes of DIC and C_{ant} (and related biogeochemical variables like oxygen and nutrients) have been made through analyses of repeat hydrographic sections across the major basins and over time at the few ocean time-series sites. With repeat sections, changes in DIC and C_{ant} between two periods of observation (typically interannual to multidecadal intervals between observations) can be used to directly assess the rate of DIC change and C_{ant} uptake over regional scales. Such analyses are based on direct measurements of DIC and do not, in principal, require knowledge of preformed preindustrial concentrations as with back-calculation methods such as using ΔC^* (Gruber et al., 1996). In Section 4, we summarize direct assessments of changes in C_{ant} and/or DIC using repeat section for estimating C_{ant} inventories and/or uptake rates in all major ocean basins. Similarly, semicontinuous observations of DIC and oxygen at ocean time series can be used to assess C_{ant} uptake rates over time (Section 5).

One complication to the conceptually simple method of determining C_{ant} storage rates from repeated ocean sections is the sensitivity to biases due to ocean interior variability such as eddies and moving fronts. One way to compensate for such variability is to use multiple linear regression (MLR) techniques (Wallace, 1995) or "extended MLR" (eMLR), a variation subsequently suggested by Friis et al. (2005). These techniques can be used to assess the rate of DIC and C_{ant} changes along isopycnal surfaces (e.g., Tanhua et al., 2007). The shortcomings and biases in the eMLR/MLR methodologies have been discussed in the literature and are commonly associated with the choice of predictor variables, biases in measurements, regional issues, etc. (e.g., Tanhua et al., 2007; Levine et al., 2008; Wanninkhof et al., 2010; Plancherel et al., 2012). Different applications of MLR techniques on the same dataset, such as with respect to the choice of predictors to construct the MLRs, can result in significant differences in the decadal storage of C_{ant}. This complicates the comparison of estimates from different studies (see Tables 30.5, 30.7, and 30.8).

Another caveat with using repeat observations to monitor direct changes in DIC or C_{ant} is the necessity to assume that the ocean is in steady state. As discussed in Section 3.3, there are complex positive and negative feedbacks between ocean physics and biology that can impact C_{ant} over multidecadal and longer time scales. This also includes the impact of natural modes of climate variability such as El Niño-Southern Oscillation (ENSO), Atlantic Multidecadal Oscillation, and North Atlantic Oscillation (NAO) on the seawater CO_2–carbonate system. For instance, Goodkin et al. (2011) found that analysis of repeat section data with more than a couple of decades between them caused significant biases in the eMLR technique due to changing climate and ocean chemistry. Another disadvantage of using this approach is that there remains a limited number of repeat section datasets and few ocean time series of sufficient length and with sufficiently high-quality data to assess relatively small changes in ocean chemistry over time.

A different approach was taken by Tanhua and Keeling (2012), who used a large dataset to calculate the decadal difference of water-column inventories of DIC in the Atlantic Ocean. This approach reduces biasing effects resulting from the need to separate "anthropogenic" and "natural" carbon, as well as reduces the influence of temporal changes in water mass abundance. The result is the net change in DIC in the water column, however with larger uncertainties.

3.3. Feedbacks and the Non-Steady-State Ocean

Studies of the cumulative inventories of C_{ant}, the present day ocean uptake of C_{ant}, and prediction/invocation of future scenarios assume that the ocean carbon cycle is in steady state. This assumes, for example, that the DIC fields in the preindustrial ocean can be approximated by the current ocean state, and that the only difference is the rising surface ocean pCO_2 concentration with time. There is mounting evidence that the concept of a steady-state ocean is not valid (e.g., Levitus et al., 2012; Waugh et al., 2013) and that this influences the ocean carbon cycle through complex sensitivities and feedbacks in the physics, chemistry, and biology of the ocean. For instance, recent work has shown a slowdown of the storage rates in the North Atlantic Ocean (Steinfeldt et al., 2009; Pérez et al., 2010; Perez et al., 2013) during the first decade of the twenty-first century, likely associated with variations of the NAO. On the other hand, higher than expected storage rates have been reported for the South Atlantic Ocean (Wanninkhof et al., 2010) that could potentially have a compensating effect. Similar variations in the uptake rate of CO_2, or the interior ocean carbon storage rates, has been reported for other parts of the ocean as well, so that it is unclear if the scaling of the GLODAP data to 2010 is completely justified. A more direct measure of the non-steady state of the ocean biogeochemistry over the last 60 years or so is evident from the fact that most ocean basins have shown decreasing oxygen concentrations with time (e.g., Ono et al., 2001; Keeling et al., 2010; Stramma et al., 2010; Helm et al., 2011; Stendardo and Gruber, 2012), with deoxygenation rates of $\sim$2–3 μmol kg^{-1} per decade, implying a non-steady-state ocean impacting the anthropogenic carbon as well.

In addition to climate variability, such as the NAO, long-time trends will affect the uptake and storage of C_{ant}. There are a number of feedback mechanisms, ultimately caused by anthropogenic influences, that have the potential to either reduce or enhance the storage of carbon in the ocean. Some of these feedback mechanisms are reviewed in Sabine and Tanhua (2010), which is updated and summarized in Figure 30.4. Embedded within these feedbacks are processes that have either direct or indirect impacts, or both, with the additional caveat that most of these feedback mechanisms and their quantitative impact on C_{ant} uptake is poorly known or quantified. We cannot fully address this topic here and instead refer the reader to other reviews that have addressed this issue and those related to the impacts of ocean acidification (Riebesell et al., 2009; Sabine and Tanhua, 2010; Andersson et al., 2011; Gehlen et al., 2011; Pörtner et al., 2011; Riebesell and Tortell, 2011; Weinbauer et al., 2011). Rather, here, we summarize current concepts that relate to feedback mechanisms in Table 30.3 and Figure 30.4.

These hard to quantify, but potentially significant, changes of the oceanic carbon pool due to ocean variability ought to be included in the definition of anthropogenic carbon, in addition to the atmospheric perturbation of the CO_2 concentrations (e.g., Wang et al., 2012). An alternative would be to quantify changes in the total pool of DIC in the ocean, as suggested by Tanhua and Keeling (2012). Although conceptually simple, the undersampled nature of the seawater CO_2–carbonate system renders this method vulnerable to large uncertainties due to the relative paucity of data. However, Tanhua and Keeling (2012) were able to show statistically significant temporal changes in DIC in the Atlantic Ocean, and that these are often significantly different from estimates of decadal changes in "anthropogenic carbon." Such analyses underscore problem of separating "natural" and "anthropogenic" carbon pools in the ocean. The concept of C_{ant} determination in a non-steady-state ocean has recently been formalized by McNeil and Matear (2013) where they claim that the non-steady-state signal is rapidly increasing and is currently up to 18% of the steady-state signal. They also suggest that by using a combination of methods to determine the C_{ant} inventories in the ocean, the non-steady-state signal can be quantified. However, Wang et al. (2012) found in a model experiment (in a coupled carbon cycle ocean circulation model, CCSM3.1) that the change in C_{ant} inventories between a constant-climate run and a run where the circulation was allowed to vary was less than 1%.

TABLE 30.3 Selected Physical and Biological Feedbacks on Anthropogenic CO_2

Process	Causality	Sign of Feedback	Magnitude	Level of Understanding	Figure Label
A. Solubility pump (including changes in ocean properties and processes such as circulation, stratification and temperature, and salinity)					
Seawater CO_2 chemistry	Change in Revelle Factor (e.g., reduced capacity for ocean uptake)	Positive	Large	High	a a*
Warming	Higher pCO_2; reduced CO_2 uptake (e.g., increased seawater pCO_2; reduced capacity for ocean uptake)	Positive	Small	High	b b*
Atmospheric circulation	Reduced CO_2 uptake (e.g., changing wind patterns, Ekman transport, intensification of Southern Ocean winds)	Positive/ negative	Medium	Low	c k*
Reduced MOC	Reduced uptake of CO_2 (e.g., reduced solubility pump through reduced deep and mode water formation)	Positive	Large	Medium	d l/m*
Stratification	Reduction in net primary production (NPP) (e.g., shoaling of mixed layer depth; increased water column stability	Positive	Medium	Medium	e j*
	Reduced upward flux of CO_2	Negative	Medium	Medium	f
Increased sea–ice melt	Reduced sea ice barrier to gas exchange (e.g., more open waters in polar regions; reduced barrier to gas exchange)	Negative	Small/ medium	Low	g
Reduced sea–ice brine formation	Reduced transport of carbon with brine (e.g., less winter sea ice formed)	Positive	Small	Low	h
Changed riverine inputs	Reduced DOC input (e.g., less input of terrestrial organic carbon and its respiration in surface waters)	Negative?	Small	Low	i
B. Biological processes (including direct impacts on ocean biology and physiology)					
Microbial loop processes	Increased respiration/remineralization (e.g., includes prokaryotes, TEP production, zooplankton grazing	Positive	Small/ medium	Low	j j*
	Increased viral activity	?	Small	Low	k
Phytoplankton physiology	Increased biomass (e.g., includes diatom, coccolithophore, dinoflagellate, and cyanobacteria under higher CO_2 conditions)	Negative	Small/ medium	Low/medium	l
	Diatom, coccolithophore, dinoflagellate, cyanobacteria Growth (e.g., higher growth rates at higher pCO_2 except for coccolithophores)	Negative	Small/ medium	Low/medium	m
Phytoplankton calcification	Reduced calcification by calcifying taxa (e.g., includes coccolithophores, foraminifera)	Negative	Medium	Low/medium	n
Zooplankton activity	Zooplankton respiration	Uncertain	Small/ medium	Low/medium	o

Continued

TABLE 30.3 Selected Physical and Biological Feedbacks on Anthropogenic CO_2—cont'd

Process	Causality	Sign of Feedback	Magnitude	Level of Understanding	Figure Label
Cellular stoichiometry	Enhanced C:N and C:P ratios of phytoplankton (e.g., higher phytoplankton C:N and C:P rations under higher CO_2 conditions; generally for most ecotypes but not all)	Negative	Small/ medium	Low/medium	p
C. Biogeochemical processes (including direct and indirect impacts on ocean biology and biogeochemical processes)					
Balance of NPP versus respiration	Enhanced respiration	Positive	Small/ medium	Low/medium	q
Change in POC/ DOC production	Enhanced dissolved organic carbon production	Positive	Small/ medium	Low/medium	r e*
Carbon overconsumption	Enhanced organic matter export	Negative	Small/ medium	Low/medium	s d*
Changes in DOC export	Increased remineralization depth	Negative	Small/ medium	Low/medium	t
Increased POC export	Increased OMZ, denitrification (e.g., reduced $CaCO_3$ ballasting effect)	Positive	Small/ medium	Low/medium	u
Reduced $CaCO_3$ ballasting	Less remineralization at depth	Positive	Small/ medium	Low/medium	v l*
Phytoplankton community	Phytoplankton community shifts (e.g., ecotype shift from diatoms to coccolithophore or picoplankton)	Positive	small/ medium	Low/medium	w
Increased denitrification	Increased NPP	Negative	Medium	Low	x f*
Increased nitrogen fixation	Higher N_2 fixation under higher CO_2 conditions	Negative	Medium	Low	y g*
Nitrification reduction	Favor small picoplankton, less N_2O	Negative	Small	Low/medium	z
Reduced calcification	Less CO_2 released during calcification	Negative	Medium	Low	aa h*
Changes in calcification	Positive coccolithophore-CO_2 feedback (e.g., balance shifts to calcification from NPP)	Positive	Small/ medium	Low/medium	bb
	Negative coccolithophore-CO_2 feedback (e.g., reduced calcification relative to NPP)	Negative	Small/ medium	Low/medium	cc
Increased $CaCO_3$ dissolution	Uptake of CO_2 and release of alkalinity	Negative	Small (long term)	Low/medium	dd n*
DMS production	Reduced DMS production from diatoms	?	Uncertain	Uncertain	ee

Notes: Modified from Sabine and Tanhua (2010), Gehlen et al. (2011), Riebesell and Tortell (2011), and Andersson et al. (2011). In the last column, a* to n* refers to processes illustrated in Figure 5 of Sabine and Tanhua (2010). A positive feedback implies that a higher fraction of the CO_2 signal remains in the atmosphere.

4. OCEAN INVENTORIES, STORAGE RATES, AND UPTAKE OF CO_2 AND C_{ant}

In this section, we review patterns of anthropogenic carbon storage in the ocean interior and fluxes of natural and anthropogenic CO_2 across the air–sea interface for the major ocean basins. We also provide an overview of the significant horizontal transport patterns for each basin. We refer to relevant literature and have compiled tables of published estimates of C_{ant} inventories and the decadal change in C_{ant} inventories (or concentrations) of the interior ocean. Initially, an assessment of global anthropogenic carbon storage is made, followed by discussion of individual ocean basins in more detail.

The storage of C_{ant} in the global ocean is highest in the North Atlantic Ocean, where deep water is formed from surface water that is well saturated with C_{ant}. Similarly, quasizonal bands of high C_{ant} inventories are found in the

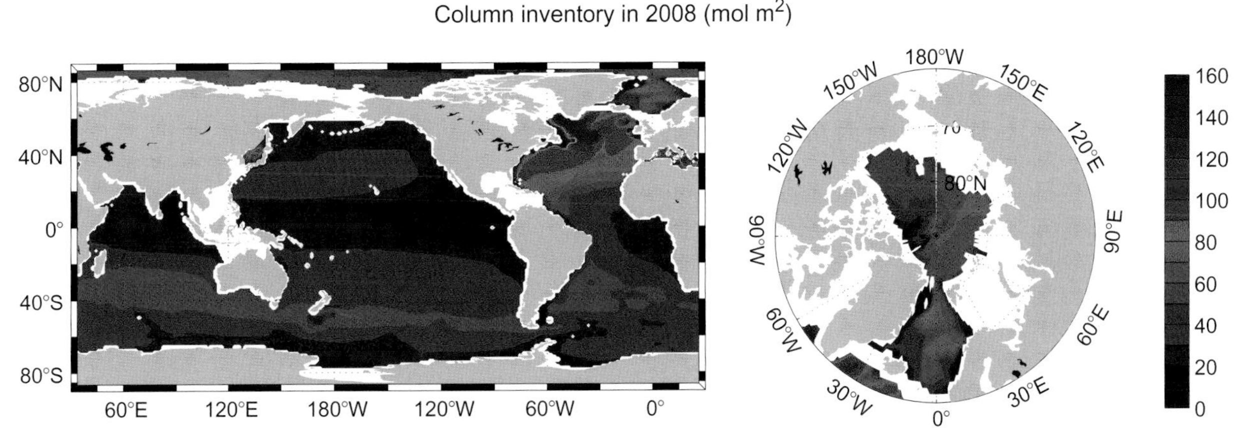

FIGURE 30.5 Column inventories of interior ocean C_{ant}, including the main marginal seas where C_{ant} estimates are available, for the reference year 2008. *Figure from Khatiwala et al. (2013) based on data from Park et al. (2006), Khatiwala et al. (2009), Tanhua et al. (2009), Olsen et al. (2010), and Schneider et al. (2010).*

mid-latitude Southern Ocean. Figure 30.5 shows a recent compilation by (Khatiwala et al., 2013) based on the GF method for the reference year 2008. This map also includes estimates from marginal seas that tend to have a high water column C_{ant} inventory, albeit small in volume compared to the global ocean.

The global CO_2 fluxes across the air–sea interface are based on the difference in seawater and atmospheric pCO_2 (i.e., ΔpCO_2). These ΔpCO_2-based estimates can be separated into a natural and an anthropogenic component where the natural flux equals the (estimated) preindustrial, and the anthropogenic flux is the difference between preindustrial and contemporary (measured) fluxes. In the following, we refer to contemporary fluxes unless stated otherwise. The sign convention for the air–sea flux is negative for a flux directed from the atmosphere to the ocean.

In general terms, the ocean tends to be a net sink of atmospheric CO_2 in mid and high latitudes of the northern hemisphere (Takahashi et al., 2009). There are characteristic zonal bands of outgassing from the ocean in areas of upwelling, such as in the equator regions and in the Southern Ocean around 60°S (Figure 30.6) This is due to the upwelled water being rich in CO_2 due to interior ocean respiration.

4.1. Indian Ocean

The Indian Ocean north of 50°S is a contemporary net sink for atmospheric CO_2. The magnitude of this sink (−0.32 Pg C yr^{-1}, Takahashi et al., 2009) represents about 23% of the global oceanic net CO_2 sink (−1.42 Pg C yr^{-1}, Takahashi et al., 2009), which is about 40% higher than the areal fraction of the Indian Ocean (16%). It is driven by a solid sink pattern in the zonal band 14–50°S which is somewhat more pronounced than in the other basins. In stark contrast to the Pacific Ocean, this sink is only moderately offset by the tropical CO_2 source band (14°S–14°N, 0.10 Pg C yr^{-1}, Takahashi et al., 2009). The weakness of this CO_2 outgassing is due to the complete lack of equatorial upwelling in the eastern Indian Ocean.

A regional "hotspot" CO_2 source in the Indian Ocean, however, is associated with coastal upwelling in the northwestern Arabian Sea during the southwest monsoon period. The CO_2 flux rates observed here during peak upwelling (Körtzinger et al., 1997) are among the largest found in the open ocean and largely responsible for the net CO_2 source in 14°S–14°N band (0.02 Pg C yr^{-1}, Takahashi et al., 2009), and in the northern Indian Ocean. The pCO_2 climatology also does not show the "hotspot" CO_2 source that occurs in the Bay of Bengal during the Southwest Monsoon due to remineralization of river organic carbon (Bates et al., 2006). Estimates of net air–sea CO_2 fluxes based on inversions of interior ocean carbon observations (Gruber et al., 2009) yield fluxes that are quantitatively consistent with the above flux estimates based on the global ΔpCO_2 climatology (Takahashi et al., 2009).

It is interesting to separate the net CO_2 fluxes into natural and anthropogenic components. This has been done with data-based inversion techniques (Gruber et al., 2009) and this approach shows strong outgassing of natural CO_2 in the Indian Ocean south of 44°S, which however is counteracted by an even stronger uptake of anthropogenic CO_2; the net result shows that this region is a net CO_2 sink. Both fluxes are a result of the exposure of comparatively old Circumpolar Deep Water that has high concentrations of CO_2, to the atmosphere in this region. In the temperate region of the southern Indian Ocean, the net CO_2 sink is composed of roughly equal contributions of natural and anthropogenic CO_2. The tropical band shows a moderate anthropogenic CO_2 sink superimposed upon a somewhat larger natural CO_2 source which reflects a significant

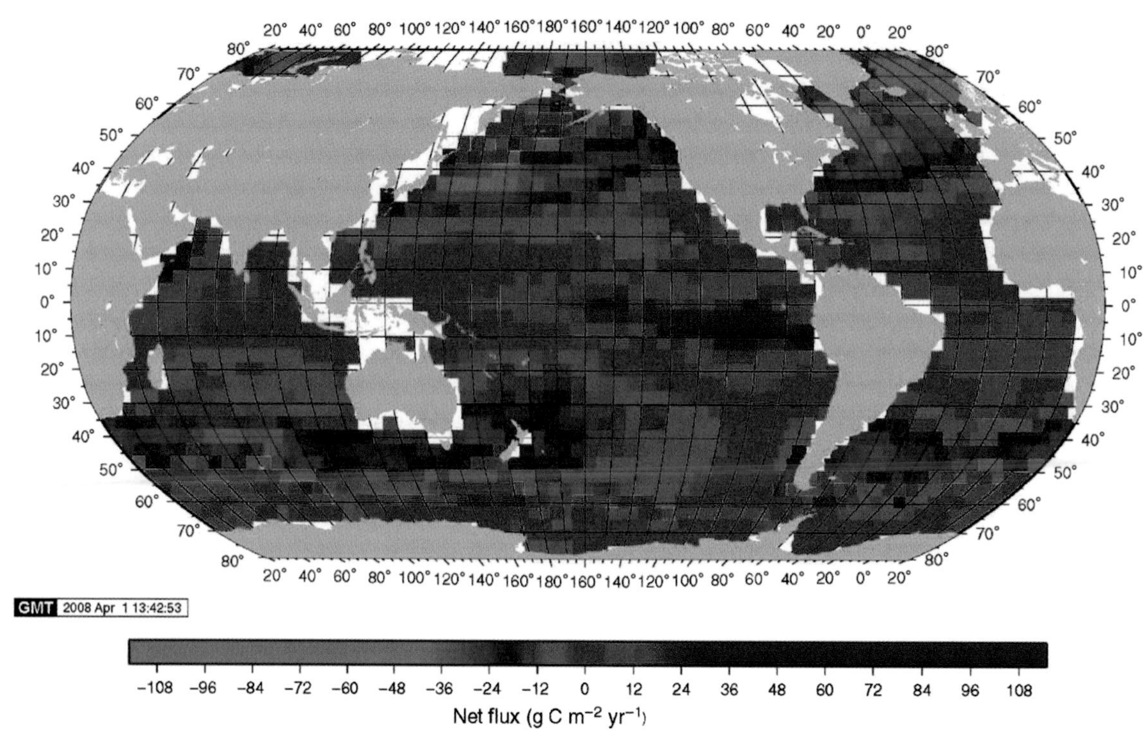

FIGURE 30.6 Climatological mean annual air–sea flux for the reference year 2000 (non-El Niño conditions). *From Takahashi et al. (2009).*

riverine component and remineralization signature (Gruber et al., 2009). Estimates of the anthropogenic CO_2 column inventory yield highest values (30–40 mol m^{-2}) between 15°S and 45°S (Sabine and Tanhua, 2010) which is largely supported by other methods (Alvarez et al., 2009). The inventories of C_{ant} in the Indian Ocean ranged from ~14.3 to 22 Pg C (Table 30.4). The band of elevated anthropogenic CO_2 storage, which is also found in the other basins, can be explained by the formation and northward propagation of Antarctic Intermediate Water as well as Subantarctic Mode Water. Globally, this is the single most important pathway for anthropogenic CO_2 to enter the ocean, followed by the deep-water formation pathway in the North Atlantic Ocean. There is some indication for a recent enhancement of anthropogenic CO_2 storage by changes in circulation and enhanced ventilation (Murata et al., 2010; Álvarez et al., 2011).

Interior ocean transports are characterized by a small northward net carbon transport in the Indian Ocean, which is the balance of significant northward transport of natural CO_2 and a slightly smaller southward transport of anthropogenic CO_2. Much of the latter can be traced to the Indonesian Through-Flow (ITF) that appears to be an important exit pathway for anthropogenic CO_2 from the Pacific Ocean. Estimates of the total anthropogenic CO_2 inventory for the Indian Ocean range from 14.3–20.5 Pg C for the year 2000 (Hall et al., 2004) to 22 Pg C for the year 1994 (Sabine et al., 2004a).

4.2. Pacific Ocean

The Pacific Ocean, north of 50°S, is also an overall net CO_2 sink (−0.46 Pg C yr^{-1}, Takahashi et al., 2009) and represents 32% of the net global ocean CO_2 sink. With a surface area of ~47% of the world ocean, the total flux is substantial compared to global fluxes, but, if flux is averaged to area, this region contributes the smallest per area CO_2 sink of all ocean basins. The meridional flux pattern is nearly symmetric with strongest sinks around 40°N (50–14°N, −0.50 Pg C yr^{-1}, Takahashi et al., 2009) and 40°S (14–50°S, −0.541 Pg C yr^{-1}, Takahashi et al., 2009). The CO_2 source from the tropical band (14°N–14°S, 0.48 Pg C yr^{-1}, Takahashi et al., 2009) is far larger than in the other basins owing to the strong and spatially extensive equatorial upwelling in the eastern Pacific Ocean. Those regions that are poleward of 50° latitudes in both hemispheres contribute rather weakly to the overall CO_2 air–sea balance. Again a reasonable agreement of these ΔpCO_2-based estimates with those from ocean inversions is found (Gruber et al., 2009).

Natural CO_2 outgasses strongly in the tropical band of the Pacific Ocean and from the Southern Ocean sector,

TABLE 30.4 Pacific and Indian Ocean Inventories of Anthropogenic CO_2

Region	Zone	Time	Inventory (Pg C)	Source
Pacific Ocean				
Pacific	>50°S–65°N	1994	44	Sabine et al. (2004a)
Pacific	70°S–65°N	1994	44.5±4.5	Sabine et al. (2002)
Indian Ocean				
Indian	North of 0°N	1994	4	Sabine et al. (2004a) (Split 14°S–14°N)
Indian	South of 0°N	1994	18	Sabine et al. (2004a) (Split 14°S–14°N)
Indian	North of 35°S	1995	13.6±2	Sabine et al. (1999)
Indian	South of 35°S	1995	6.7±1	Sabine et al. (1999)
Indian		2000	14.3–20.5	Hall et al. (2004)
North Pacific				
North Pacific	0–65°N	1994	16	Sabine et al. (2004a) (Split 14°S–14°N)
North Pacific	0–65°N	1994	16.5	Sabine et al. (2002)
South Pacific				
South Pacific	>50–0°S	1994	28	Sabine et al. (2004a) (Split 14°S14°N)
South Pacific	0–70°S	1994	28	Sabine et al. (2002)
Regional seas				
East Sea/Sea of Japan		1999	0.40±0.06	Park et al. (2006)
South China Sea		1994	0.6±0.15	Chen et al. (2006)

while both regions also show the strongest uptake of anthropogenic CO_2. These patterns can be explained as a result of the ventilation of older subsurface waters that are rich in respiratory CO_2 but, due to the relatively old age and "isolation" from the anthropogenic CO_2 transient in the atmosphere, contain little anthropogenic CO_2. The observed residual CO_2 fluxes are thus the balance of two rather large fluxes that act in opposite directions. Another way of phrasing this is that the tropical Pacific outgassing of CO_2 is less today than during preindustrial times, and that reduced outgassing has effectively increased the uptake of CO_2 by the oceans.

In total, the Pacific Ocean contains about 44 Pg C or 41% of the global ocean C_{ant} inventory (Table 30.4). The anthropogenic CO_2 column inventories are generally smallest in the Pacific Ocean, with C_{ant} uptake values generally lower than the Indian Ocean (Table 30.4). In the absence of deep-water formation, intermediate water formation represents the dominant entry pathway for C_{ant} in the interior of the Pacific Ocean. This is more pronounced and of comparable magnitude to other basins in the southern hemisphere. Globally, the tropical band of the Pacific Ocean exhibits the second lowest anthropogenic CO_2 column inventory. For waters bordering the Antarctic continent, the picture is less clear, with different methods producing either very small (Sabine et al., 2004a) or moderate column C_{ant} inventories (Waugh et al., 2006). A recent study on decadal carbon trends in the Pacific Ocean revealed relatively high increases in C_{ant} inventories for the subtropical regions of both hemispheres (>0.7 mol m^{-2} yr^{-1}, Kouketsu et al., 2013) while almost no storage was found north of 40°N (except in the western subarctic Pacific Ocean). This follows the general distribution pattern of C_{ant} and leads to a total anthropogenic CO_2 inventory increase of 8.4±0.5 Pg C for the decade following the mid-1990s (Table 30.5).

Contemporary net carbon transports exhibit a strong convergence of carbon in the equatorial Pacific. In the South Pacific, natural and anthropogenic CO_2 fluxes enhance each other, resulting in large northward fluxes (Gruber et al., 2009). In the North Pacific Ocean, the situation is complex but the anthropogenic component is generally small and both compensate for and enhance natural carbon fluxes. A sizeable fraction of anthropogenic CO_2 leaves the equatorial region via the ITF toward the Indian Ocean as mentioned above.

TABLE 30.5 Pacific and Indian Ocean Rates of Anthropogenic CO_2 Uptake

Region	Zone	Period	Uptake (Pg C yr^{-1})	Uptake (mol m^{-2} yr^{-1})	Source
Pacific Ocean					
South Pacific	32°S	1992–2003	n/a	1.0±0.4	Murata et al. (2007)
West Pacific	149°E	1993–2005	n/a	0.5±0.1	Murata et al. (2009)
Central Pacific	152°W	1991/2–2005/6	n/a	0.25	Sabine et al. (2008)
North Pacific	30°N	1994–2004	n/a	0.43	Sabine et al. (2008)
Northeast Pacific		1973–1991	n/a	1.3±0.5	Peng et al. (2003)
South-central Pacific		1973–1991	n/a	0.9±0.3	Peng et al. (2003)
Northwest Pacific	45°N, 160°E	1992–2008	n/a	0.40±0.08	Wakita et al. (2010)
Southern Ocean	40–70°S	1968–1996	n/a	0.4±0.2	Matear and McNeil (2003)
Pacific	100°W	1994–2007/9	n/a	0.46±0.2	Waters et al. (2011)
South Pacific	32°S, E of 170°W	1992–2009/10	n/a	0.87±0.1	Waters et al. (2011)
South Pacific	32°S, W of 170°W	1992–2009/10	n/a	0.66±0.1	Waters et al. (2011)
Pacific Ocean	N of 50°S	1990s	8.4±0.5	n/a	Kouketsu et al., 2013
Indian Ocean					
Indian	North of 35°S	1995	0.23±0.05		Sabine et al. (1999)
Indian		2000	0.26–0.36		Hall et al. (2004)
South Indian	20°S	1995–2003/4	n/a	1.1±0.1	Murata et al. (2010)
Tropical Indian	20–10°S	1978–1995	0.075	0.65	Peng et al. (1998)
Tropical Indian	10°S–5°N	1978–1995	0.015	0.1	Peng et al. (1998)

4.3. Atlantic Ocean

The Atlantic Ocean shows a rather symmetric CO_2 source/sink pattern of contemporary atmospheric CO_2 with sizeable ocean CO_2 sinks in the 14°–50° latitude band of the northern (−0.20 Pg C yr^{-1}, Takahashi et al., 2009) and southern hemisphere (−0.21 Pg C yr^{-1}, Takahashi et al., 2009). Both ocean CO_2 sink regions are dominated by the latitude band 35°–50° since the subtropical regions (14°–35°) are neutral or weak sources. The tropical band (14°N–14°S) acts as a source of CO_2 to the atmosphere (0.10 Pg C yr^{-1}, Takahashi et al., 2009) comparable in size to the Indian Ocean but far weaker than estimated for the Pacific Ocean. The Southern Ocean sector of the Atlantic Ocean behaves very much like the Southern Ocean of the other two basins, the region north of 50°N represents a significant sink despite its comparatively small area (−0.27 Pg C yr^{-1}, Takahashi et al., 2009) due to very high CO_2 flux rates. Overall, the Atlantic Ocean constitutes an oceanic sink for CO_2 (−0.58 Pg C yr^{-1}, Takahashi et al., 2009) that represents 41% of the global ocean CO_2 sink, which is far in excess of its areal fraction (23%). It should also be noted that the Atlantic Ocean was a net sink of CO_2 in preindustrial times (e.g., Gruber et al., 2009).

For the entire Atlantic Ocean basin (North and South Atlantic, excluding marginal seas such as the Mediterranean Sea) published estimates of C_{ant} inventories range from 40 to 55 Pg C (Table 30.6; Sabine et al., 2004a; Touratier and Goyet, 2004; Vázquez-Rodríguez et al., 2009; Velo et al., 2010), representing approximately 35–40% of the global ocean estimates (e.g., Sabine et al., 2004a; Khatiwala et al., 2013). In the North Atlantic Ocean, the anthropogenic CO_2 inventories range from 20 to 36 Pg C (Table 30.6; Gruber et al., 1996; Gruber, 1998; Lee et al., 2003; Sabine et al., 2004a; Steinfeldt et al., 2009), or about 50–60% of the total Atlantic inventory, with variability in the estimates imparted in part from the different latitudinal zones used. Estimates for the South Atlantic Ocean are more limited and range from 16.5 to 18.5 Pg C (Table 30.6; Gruber, 1998; Lee et al., 2003; Sabine et al., 2004a).

TABLE 30.6 Atlantic Ocean Inventories of Anthropogenic CO_2

Region	Zone	Time	Inventory (Pg C)	Source
Atlantic Ocean				
Atlantic	>50°S–65°N	1994	40	Sabine et al. (2004a)
Atlantic	50°S–80°N	1980s	40	Touratier and Goyet (2004)
Atlantic	50°S–80°N	1990s	45	Touratier and Goyet (2004)
Atlantic	40°S–60°N	1990s	54 ± 8	Vázquez-Rodríguez et al. (2009)
Atlantic			55.1	Velo et al. (2010)
North Atlantic				
North Atlantic	10°–80°N	1989	20 ± 4	Gruber et al. (1996)
North Atlantic	10°–70°N	1989	22 ± 5	Gruber (1998)
North Atlantic	0°–70°N	1990–1998	28.4 ± 4.7	Lee et al. (2003)
North Atlantic +	20°S–65°N	1997	32.5 ± 9.5	Steinfeldt et al. (2009)
North Atlantic +	20°S–65°N	2003	36 ± 10.5	Steinfeldt et al. (2009)
North Atlantic +	14°S–65°N	1994	27	Sabine et al. (2004a)
North Atlantic	0–65°N	1994	23.5	Sabine et al. (2004a) (Split 14°S–14°N)
South Atlantic				
South Atlantic	10°–70°S	1989	18.0 ± 4	Gruber (1998)
South Atlantic	>50°S–0°N	1994	16.5	Sabine et al. (2004a) (Split 14°S–14°N)
South Atlantic	0°–70°S	1990–1998	18.5 ± 3.9	Lee et al. (2003)
Regional seas				
Nordic Seas		2002	1.15 ± 0.25	Olsen et al. (2010)
Nordic Seas		2002	1.2	Jutterström et al. (2008)
Labrador Sea		2001	1.0	Terenzi et al. (2007)
Mediterranean Sea		2001	1.7 ± 0.25	Schneider et al. (2010)
Equatorial Atlantic	4.5°S–4.5°N	2007	1.5 ± 0.3	Schneider et al. (2012)
Guinea Dome Atlantic	5–15°N	2007	1.0 (0.9–1.5)	Schneider et al. (2012)
Arctic Ocean		1991	1.35 ± 0.6	Anderson et al. (1998)
Arctic Ocean		2005	2.9 ± 0.4	Tanhua et al. (2009)

The anthropogenic CO_2 uptake rates reported in the literature are highly variable. For example, for the entire Atlantic Ocean basin, the C_{ant} uptake rates ranged from about 0.27 to 0.66 Pg C yr^{-1} (see Table 30.7; also given in mol C m^{-2} yr^{-1}). In the North and South Atlantic, anthropogenic CO_2 uptake rates ranged from 0.32 to 0.58 Pg C yr^{-1}, and 0.29–0.44 Pg C yr^{-1}, respectively (see Table 30.7). A contributor to this variability relates to the approaches used, the time period for the calculation and the area considered in each analysis.

In the North Atlantic Ocean, the lateral transport of anthropogenic CO_2 northward across 25°N appears to dominate the transport terms (Álvarez and Gourcuff, 2010; Perez et al., 2013). The transport of anthropogenic CO_2 into the subtropical gyre of the North Atlantic Ocean was much greater at 0.25 Pg C yr^{-1} and higher than transports between the subtropical and subpolar gyres (0.092 Pg C yr^{-1}), and northward into the Nordic Sea (0.063 Pg C yr^{-1}), and into the Arctic Ocean (0.039 Pg C yr^{-1}).

4.4. Arctic Ocean

There is relatively little information about the inventory and uptake rate of anthropogenic CO_2 in the Arctic

TABLE 30.7 Atlantic Ocean Rates of Anthropogenic CO_2 Uptake

Region	Zone	Period	Uptake (Pg C yr^{-1})	Uptake (mol m^{-2} yr^{-1})	Source
Atlantic Ocean					
Atlantic	50°S–80°N	1980s–1990s	0.66	0.65[a]	Touratier and Goyet (2004)
Atlantic	56°S–63°N	1988–2005	0.5 ± 0.1	0.53 ± 0.05	Wanninkhof et al. (2010)
Atlantic	50°S–65°N	2000s	0.49 ± 0.05	0.48[a]	Schuster et al. (2013)
Atlantic		2000s		0.86 ± 0.14	Tanhua and Keeling (2012)[c]
North Atlantic					
North Atlantic	15–65°N	1988–2005	n/a	0.57	Wanninkhof et al. (2010)
North Atlantic	0–65°N	1990s	0.32 ± 0.08	0.63 ± 0.16	Quay et al. (2007)
North Atlantic	20°S–65°N	1997	0.58	1.16[a]	Steinfeldt et al. (2009)
North Atlantic	0°S–65°N		0.34[a]	0.70 ± 0.53	Peng and Wanninkhof (2010)
North Atlantic	40–65°N	1981–1997	0.27	2.2 ± 0.7	Friis et al. (2005)
North Atlantic	20–40°N	1981–2004	n/a	1.2 ± 0.3	Tanhua et al. (2007)
North Atlantic	25°N—Bering Strait	2004	0.386 ± 0.012		Perez et al. (2013)
Subpolar NA		1997	0.083 ± 0.008		Perez et al. (2013)
Subpolar NA		2004	0.026 ± 0.004		Perez et al. (2013)
Subtropical NA		2004	0.28 ± 0.011		Perez et al. (2013)
Nordic Seas		2004	0.018 ± 0.004		Perez et al. (2013)
Arctic Ocean		2004	0.043 ± 0.004		Perez et al. (2013)
South Atlantic					
South Atlantic	54–15°S		n/a	0.76	Wanninkhof et al. (2010)
South Atlantic	55°S–0°N		0.29[a]	0.6 ± 0.1	Murata et al. (2008)
South Atlantic	55°S–10°N		0.44[a]	0.92 ± 0.13	Ríos et al. (2012)
South Atlantic	80°S–10°N	1975–2005	0.39 ± 0.08[a]	0.81 ± 0.17	Ríos et al. (2012)
Regional					
Azores			n/a	1.32 ± 0.11	Perez et al. (2010)
Irminger Sea		1981–2006	n/a	1.1 ± 0.1	Perez et al. (2008)
Irminger Sea		1981–1991	n/a	0.55 ± 0.39	Perez et al. (2008)
Irminger Sea		1991–1996	n/a	2.3 ± 0.6	Perez et al. (2008)
Irminger Sea		1997–2006	n/a	0.75 ± 01.6	Perez et al. (2008)
Iceland Basin		1991–1998	n/a	1.88 ± 0.45	Pérez et al. (2010)
Iceland Basin		1997–2006	n/a	0.30 ± 0.2	Pérez et al. (2010)
Eastern North Atlantic		1981–2006	n/a	0.72 ± 0.03	Pérez et al. (2010)
Labrador Sea		1996–2001	0.02	1.98[a]	Terenzi et al. (2007)
Arctic Ocean		2000s	0.12 ± 0.06	0.64[a]	Schuster et al. (2013)[b]
Nordic Seas		1981–2002	0.12	0.9 ± 0.2	Olsen et al. (2010)
Weddell Gyre		1973–2008	0.012 ± 0.006	n/a	van Heuven et al. (2011)

[a]*Calculated from area and mean depth: (i) Atlantic Ocean: 85,133,000 km^2, 3646 m depth; (ii) North Atlantic: 41,490,000 km^2, 3519 m depth; (iii) South Atlantic: 40,270,000 km^2, 3973 m depth; (iv) Labrador Sea: 841,000 km^2, 1898 m depth; (v) Arctic Ocean: 15,558,000 km^2, 1205 m depth.*
[b]*Both estimates from Schuster et al. (2013) for the North Atlantic Ocean and Arctic Ocean are total changes, including natural and anthropogenic; BS, Bering Strait.*
[c]*Upper 2000 m, ΔDIC.*

Ocean. This has been primarily due to the difficulties of conducting shipboard carbon cycle sampling and a paucity of seawater carbonate chemistry data from which to determine anthropogenic carbon inventories and uptake rates. As such the Arctic Ocean has not typically been considered in the global estimates of anthropogenic CO_2 (e.g., Sabine et al., 2004a; Waugh et al., 2006; Khatiwala et al., 2013). The earliest estimate of anthropogenic CO_2 inventory was 1.35 ± 0.12 Pg C (Anderson et al., 1998) based on model interpretation of limited data. More recently, Tanhua et al. (2009) estimated that the anthropogenic CO_2 inventory was higher at 2.9 ± 0.4 Pg C based on the TTD approach, salinity–alkalinity relationships from other oceans and equilibrium concentrations of DIC estimated for present day and preindustrial concentration. Complicating any such assessment of anthropogenic CO_2 inventory in the Arctic Ocean is the presence of wintertime sea–ice that acts as a barrier to air–sea gas exchange, rapidly changing air–sea disequilibrium due to sea–ice loss and warming (Bates et al., 2006; Bates and Mathis, 2009), and changes in the biology of the Arctic (e.g., rates of primary production; Arrigo et al., 2008). The Arctic Ocean also has a complex circulation and a variety of water mass constituents on the different Arctic shelves and in the deep basin that reflect the varying contributions of Atlantic and Pacific Ocean waters that flow into the region, and the influences of sea–ice meltwater and river freshwater (which have a variety of DIC and TA contents and variety of nonconservative and conservative relationships with physical properties such as salinity). Although the Arctic Ocean is only 1% of the global ocean volume, it receives 10% of the global freshwater river input (primarily from Siberian rivers, but also the Mackenzie and Yukon rivers in North America). Preformed DIC and TA concentrations, integral to back-calculation methods (e.g., C^*, TrOCA) to calculate anthropogenic CO_2 inventories, are difficult to estimate due to the complexities of end-members mixing in the Arctic Ocean. As yet, there is insufficient data to allow direct decadal change assessments (e.g., Sabine and Tanhua, 2010) to be made.

Lateral transport estimates of anthropogenic CO_2 into and out of the Arctic are also in their infancy due to lack of data, and instead have relied on model and inferential assessments (Macdonald et al., 2003; Mikaloff Fletcher et al., 2006). Recently, Perez et al. (2013) estimated a transport of anthropogenic carbon from the Arctic Ocean to the Nordic Seas at a rate of 0.039 Pg C yr^{-1}, and an anthropogenic carbon input of 0.008 Pg C yr^{-1} with Pacific Ocean water transported northward through Bering Strait into the Arctic. This latter estimate appears to be underestimated by a factor of two, given recent DIC sampling across the entire Bering Strait (N.R. Bates, unpublished data). The flux of C_{ant} to the Nordic Sea is reported by (Jeansson et al., 2011) who calculated C_{ant} concentrations from transient tracer data. They found a neutral net flux through the Fram Strait (1 ± 17 Tg C yr^{-1}), but a significant northward flux of C_{ant} through the Barents Sea (41 ± 8 Tg C yr^{-1}). Although there is a significant southward flux of C_{ant} through the Denmark Strait (58 ± 3 Tg C yr^{-1}), the northward flux east of Iceland associated with Atlantic Water (109 ± 121 Tg C yr^{-1}) leads to a close to neutral transport from the Nordic Seas to the Atlantic Ocean of C_{ant}.

The saturation horizon of $CaCO_3$ minerals in the Arctic Ocean is relatively shallow compared to other ocean basins. In the deep basin of the Arctic (Canada and Eurasian Basin), undersaturated waters with respect to aragonite and calcite are found in the upper thermocline layer that lies beneath the polar mixed layer at ~50–100 m depth (Jutterström and Anderson, 2005; Anderson et al., 2010). On the Arctic shelves, bottom waters (~30–70 m) are seasonally undersaturated with respect to aragonite and calcite during the sea–ice free summertime on shallow shelves such as the Chukchi Sea (Bates et al., 2009, 2013) with Siberian shelves also likely to be undersaturated, given high subsurface pCO_2 contents (Semiletov et al., 2007; Anderson et al., 2011).

4.5. Marginal Seas

The inventory and fluxes of C_{ant} in marginal seas and on continental shelves were initially estimated based on the global fields since these regions were not included in the GLODAP data collection (Sabine et al., 2004a). However, several reports document significant fluxes and inventories; these are summarized in a review by Lee et al. (2011). Several marginal seas have active overturning circulation and store proportionally more C_{ant} than the open ocean. However, due to the relatively small size of marginal seas, the total C_{ant} inventories are small compared to the total ocean C_{ant} storage (e.g., Labrador Sea, 1 Pg C, Terenzi et al., 2007). Based on the estimates review by Lee et al. (2011), model results and inversion estimates, Khatiwala et al. (2013) estimate the combined global storage of marginal seas (i.e., the part of the ocean not covered by the GLODAP data set) to be 9–14 Pg C in 2010. This corresponds to 6–9% of the global ocean C_{ant} storage, hence not a negligible fraction. The storage is particularly high in the Arctic Ocean (see above) and in the Mediterranean Sea, where an active overturning circulation combined with high temperatures and alkalinity are particularly favorable for C_{ant} storage (Schneider et al., 2010). As pointed out by Lee et al. (2011), the overturning circulation in marginal seas is potentially more sensitive to climate perturbations than the global ocean. For most marginal seas and continental margins, C_{ant} sequestered there tends to be exported to the global ocean so that variability in the marginal seas could have larger implications on the global C_{ant} storage. We refer to Lee et al. (2011) for a listing of published estimates of marginal seas C_{ant} storage and transport (to major ocean basins).

Although the total storage of C_{ant} is low on continental shelves and in coastal estuaries due to the low volume, there are intense air–sea fluxes of CO_2 over these areas. In a recent review, Chen et al. (2013) estimate a CO_2 flux to the atmosphere of 0.10 Pg C yr^{-1} from all estuaries, and a CO_2 flux from the atmosphere to the ocean of 0.4 Pg C yr^{-1} from the world continental shelves. The estuarine flux from Chen et al. (2013) is significantly lower than previously published estimates (see Chen et al., 2013 for a summary).

5. OCEAN TIME-SERIES VALIDATION OF TRENDS IN DIC/pCO_2/C_{ant}

There are several indirect methods to determine the rates of uptake and cumulative inventories of anthropogenic CO_2 uptake in the global ocean. These rate and inventory estimates have significant uncertainties and caveats associated with the methodologies, but these approaches can be validated by direct observations of changes in inorganic carbon over time (either DIC, DIC $= CO_2 + HCO_3^- + CO_3^{2-}$ or partial pressure of CO_2, pCO_2) in the open ocean, regional seas, or coastal shelf waters. Supporting data of a second carbonate parameter (e.g., alkalinity) and physical data are important for attribution of observed temporal variability/trends.

Direct observations of inorganic carbon changes in the ocean over time have been conducted with relatively high frequency repeated monthly or seasonal sampling at a fixed ocean time-series location (e.g., Steinberg et al., 2001; Karl et al., 2003; Feely et al., 2013), annual reoccupations of a fixed location or ocean section (e.g., Olafsson et al., 2010), long-term opportunistic surface sampling, irregular in time and space but sufficient to establish long-term trends in ocean regions, for example, surface pCO_2 (Takahashi et al., 2002, 2009; McKinley et al., 2011; Lenton et al., 2012), or lower frequency sampling along an ocean section reoccupied through projects like WOCE and CLIVAR/CO_2 repeat hydrography (see Section 3.2). In this chapter, we expand the "time-series" term to include repeat occupations at ocean sections. However, since trends estimated from two or more repeats only are more likely to be subject to variation and uncertainty imparted by interannual variability, depending on timing of sampling and length of time between repeats, we treat "time series" and repeat hydrographic effort separately.

The longest of these time series extends back in time about 30 years, sufficient to record and establish interannual and multidecadal trends in anthropogenic CO_2 changes. These observations reflect changes in both the natural carbon cycle and the uptake of anthropogenic CO_2 from the atmosphere. Thus these direct observations allow examination of the influence of natural variability on the ocean carbon cycle imparted by climate modes, such as ENSO, Pacific Decadal Oscillation (PDO), and NAO (e.g., Hurrell, 1995; Hurrell et al., 2001; Hurrell and Deser, 2009). Given the relatively short period of direct observations of inorganic carbon changes made in the ocean, it remains difficult to examine the influence of longer-term variability such as the Atlantic Multidecadal Variability (AMV) on the ocean carbon cycle (McKinley et al., 2004, 2011; Ullman et al., 2009; Metzl et al., 2010).

In Table 30.8, we list trend analyses for seawater DIC and pCO_2 at individual ocean time-series sites, annual reoccupation and repeat hydrographic-CO_2 sections. Note that

TABLE 30.8 Trends in Seawater pCO_2 and DIC Concentrations

Location	Period	Water Mass	DIC_{time} (μmol kg^{-1} yr^{-1})	pCO_{2time} (μatm yr^{-1})	Source	Notes
A. Surface trends (high-frequency ¶, annual #, and low-frequency occupation §)						
Atlantic Ocean						
¶BATS, Bermuda	1983–2011	Surface	1.53±0.12	1.62±0.12	Bates et al. (2012)	1
¶ESTOC, Canary Islands	1995–2004	Surface	0.94±0.50	n/a	Gonzalez-Davila et al. (2010)	2
¶ESTOC, Canary Islands	1995–2008	Surface	n/a	1.7±0.7	Gonzalez-Davila et al. (2010)	3
#Icelandic Sea	1985–2008	Surface	1.48±0.23	2.15±0.16	Olafsson et al. (2009)	4
#Norwegian Sea	2001–2006	Surface	1.3±0.7	2.6±1.2	Skjelvan et al. (2008)	5
§Rockall Trough (52–56°N; 11–21°W)	1991–2010	Surface	~1.0±0.3	N/a	McGrath et al. (2012)	6

Continued

TABLE 30.8 Trends in Seawater pCO_2 and DIC Concentrations—cont'd

Location	Period	Water Mass	DIC_{time} (μmol kg^{-1} yr^{-1})	pCO_{2time} (μatm yr^{-1})	Source	Notes
Pacific Ocean						
¶ALOHA, Hawaii	1988–2007	Surface	0.85 ± 0.08	1.88 ± 0.16	Dore et al. (2009)	7
#Japan; (137°E; Kuroshio)	1994–2008	Surface	1.23 ± 0.40	1.54 ± 0.33	Ishii et al. (2011)	8
#Line P; North Pacific;	1973–2005	Surface	n/a	1.36 ± 0.16	Wong et al. (2010)	9
§N. Pacific (137°E; 3–33°N)	1983–2007	Surface	0.96 ± 0.26	1.58 ± 0.12	Midorikawa et al. (2012)	10
#N. Pacific (137°E; 3–30°N)	1983–2003	Surface	n/a	1.3–1.7	Midorikawa et al. (2005)	11
§KNOT (44°N, 155°E)	1992–2002	Surface	1.0–1.3	0.5–2.5	Sabine et al. (2004b)	12
§KNOT (44°N, 155°E)	1992–2001	Surface	1.3–2.3	1.9 ± 0.7	Wakita et al. (2005)	13
§KNOT (44°N, 155°E)	1999–2006	Upper	0.86 ± 0.12	n/a	Watanabe et al. (2011)	14
§KNOT and K2 (44°N, 155°E)	1992–2009	Surface	1.3 ± 0.13	n/a	Wakita et al. (2010)	15
§150*W, 30°N	1991–2006	Surface	1.5–2.0	1.0	Sabine et al. (2008)	16
§Eq. Pacific (144°E–160°W, 5°S–5°N)	1985–2004	Surface	n/a	1.5 ± 0.4	Ishii et al. (2009)	17
§149*E	1993–2005	Surface	0.9–1.1	n/a	Murata et al. (2007)	18
#Equatorial Pacific	1992–2001	Surface	n/a	1.5	Takahashi et al. (2003)	19
§Subarctic Pacific	1992–2008	Surface	1.3–1.5	0.7 ± 0.5	Wakita et al. (2010)	15
Indian Ocean						
§S. Indian Ocean	1991–2007	Surface	n/a	2.11 ± 0.07	Metzl (2009)	19
§Indian Ocean (Kerfix)	1990–1995	Surface	n/a	2.4	Louanchi et al. (1999)	20
B. Deeper water mass trends						
¶BATS	1983–2011	STMW	1.51 ± 0.08	2.13 ± 0.16	Bates et al. (2012)	1
¶ESTOC	1995–2004	300 m	0.81 ± 0.50	n/a	Gonzalez-Davila et al. (2010)	2
#Norwegian Sea	2001–2006	Deep	0.57 ± 0.24	n/a	Skjelvan et al. (2008)	5
#Dyfamed	1993–2005	500 m	0.42	n/a	Touratier and Goyet (2009)	21
¶ALOHA	1988–2007	250 m	1.86 ± 0.30	1.88 ± 0.16	Dore et al. (2009)	7
§Indian Ocean (20°S)	1995–2003	100–300 m	~1.0		Murata et al. (2010)	22

Notes: (1) Ocean time-series data with seasonally detrended trends determined from monthly observed DIC and TA collected semi-continuously at Hydrostation S (1983–1989) and BATS (1988–present). pCO_2 derived from DIC, TA, temperature, and salinity data (Bates et al., 2012). (2) Ocean time-series data with trends determined from monthly observed pH and TA. DIC derived from pH, TA, temperature, and salinity data (Gonzalez-Davila et al., 2010). Long-term trends for deeper waters are also given. (3) Ocean time-series data with Gonzalez-Davila et al. (2010) giving trends for measured pCO_2 for the years 1995–2008. (4) Ocean time-series data with multivariate linear regression methods for trend analysis (Olafsson et al., 2009) and based on annual winter observations. (5) Ocean time-series data with trends determined from monthly to seasonal observation of DIC. (6) Repeat section data with trends determined from observed DIC and TA data (7) Ocean time-series data with trends from observed salinity normalized DIC data and pCO_2 derived from DIC and TA data. (8) Ocean time-series data with trends determined from annual and biannual winter observations of DIC and pCO_2. (9) Ocean time-series data with trends determined from observed salinity normalized DIC and TA and pCO_2 derived from DIC and TA data. (10) Repeat section data with trends determined from observed DIC and pH data. (11) Ocean time-series data with trends determined from seasonally observed DIC and pH and pCO_2 derived from DIC and pH data. (12) Repeat section data with trends determined from observed DIC and TA data. (13) Ocean time-series data with trends determined from seasonally observed DIC and TA and pCO_2 derived from DIC and TA data. (14) Repeat section data with trends determined from observed DIC and TA data. (15) Ocean time-series data with trends determined from seasonally observed DIC and TA and pCO_2 derived from DIC and TA data. (16) Repeat section data with trends determined from observed DIC and TA data. (17) Ocean time-series data with trends determined from seasonally observed DIC and pCO_2 data. 144°E–160°W, 5°N–5°S (Ishii et al., 2009)—time-series observations in the coast of western N. Pacific, with the seasonal cycle removed. (18) Repeat section data with trends determined from observed DIC and TA data. (19) Ocean time-series data with trends determined from seasonally observed pCO_2 data. (20) Repeat section data with trends determined from observed DIC and TA data. (21) Ocean time-series data with trends determined from observed DIC and TA data. (22) Repeat section data with trends determined from observed DIC and TA data.

results from repeat hydrography are also reported in Tables 30.5 and 30.7, although as inventories or column inventories, whereas Table 30.8 is reporting trends in concentrations. This list is not fully comprehensive, and sources given in the table provide trends for the longest time period reported in the literature, and the frequencies of observations are indicated in the table. The data listed in the table are visualized in Figure 30.7, and illustrate how the trend of δpCO_{2time} and δDIC_{time} tends to converge to the values that can be expected from the increased atmospheric CO_2 concentration as the length of the time-series record increases.

In the Atlantic Ocean, the concentration in surface/mixed layer DIC has increased by 0.85–1.53 μmol kg^{-1} yr^{-1}, Table 30.8. The longest time series (>20 years) that began in the 1980s have trends at the higher end of the range of DIC increase (e.g., Sargasso Sea, Icelandic Sea; Table 30.8) with the DIC increase similar to that expected due to ocean uptake of anthropogenic CO_2. The range of change of seawater pCO_2 was 1.62–2.60 μatm yr^{-1} (Table 30.8), with rates of seawater pCO_2 increase similar to atmosphere pCO_2 at longest time-series sites (Table 30.8), or longest spatiotemporal records of seawater pCO_2 (e.g., Takahashi et al., 2009; McKinley et al., 2011). These findings indicate that, over multidecadal time periods, surface seawater DIC and pCO_2 have kept pace with contemporaneous atmospheric pCO_2 increase, and that one of the two primary driving forces for air-sea CO_2 gas exchange (i.e., ΔpCO_2; the difference in pCO_2 between ocean and atmosphere) has remained fairly constant. Over the last thirty years, surface seawater DIC and pCO_2 have increased by ~1.5–2% and ~12–18%, respectively, in the North Atlantic Ocean, with deeper thermocline waters such as the subtropical mode water (STMW; Bates et al., 2012) and eastern subtropical gyre waters (Gonzalez-Davila et al., 2010) increasing at similar rates to the surface mixed layer (Table 30.8).

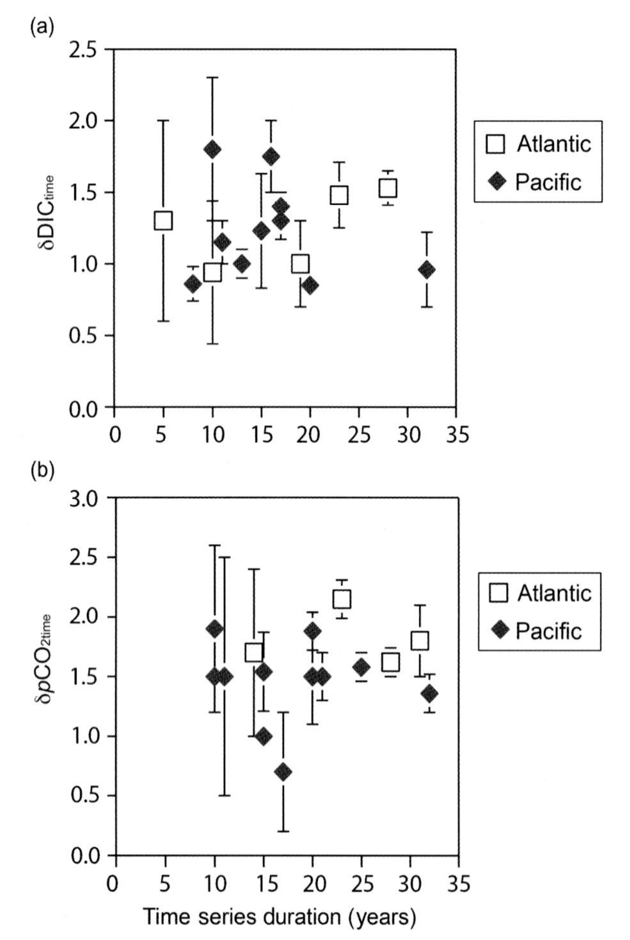

FIGURE 30.7 Annual trends in DIC and pCO_2 (μmol kg^{-1} yr^{-1} and μatm yr^{-1}, respectively) from ocean time series and repeat sections. This data is taken from Table 30.7, and shows trends against period of observation.

In contrast to the longer time-series records, shorter time-series of seawater DIC and pCO_2 in the North Atlantic Ocean have increased at rates either above or below that expected for equilibration with atmospheric pCO_2 increase. For example, this includes lower rates of DIC and pCO_2 increase at the ESTOC site near the Canary Islands and DYFAMED site, and higher rates of DIC and pCO_2 increase in the Norwegian Sea (Table 30.8). Such variability reflects interannual changes in the North Atlantic Ocean, associated with climate phenomena such as the North Atlantic Oscillation and AMV, or the Eastern Mediterranean Transit for the DYFAMED site (Bates, 2001, 2007; Gruber et al., 2002; McKinley et al., 2004, 2011; Ullman et al., 2009; Metzl et al., 2010), with ΔpCO_2 more variable over the short term and the sink for CO_2 decelerating/accelerating for select periods over the last couple of decades (e.g., Olsen et al., 2003; Omar et al., 2007; Schuster and Watson, 2007; Watson et al., 2009; Tjiputra et al., 2012).

In the Pacific Ocean, the increase in surface/mixed layer DIC was more variable, ranging from 0.85 to 2.3 μmol kg^{-1} yr^{-1}, and the rate of change of seawater pCO_2 ranged from 0.5 to 2.6 μatm yr^{-1} (Table 30.8). Similar changes in DIC and pCO_2 have also been observed in the Indian Ocean (Table 30.8). As in the North Atlantic Ocean, the longer-term rate of increase in DIC and pCO_2 in the Pacific and Indian Oceans was similar to atmospheric pCO_2 changes (e.g., Wong et al., 2010) but more variable over short observation periods. This reflects short-term variability in the carbon cycle (e.g., Midorikawa et al., 2005, 2012; Ono et al., 2005; Currie et al., 2011) that relates to physical changes in the ocean basin, some of which are in turn related to changes associated with the El Niño-Southern Oscillation (e.g., Feely et al., 1999, 2004, 2006; Takahashi et al., 2009) and the PDO (e.g., Takahashi et al., 2003; McKinley et al.,

2006). Thus, over longer time periods, changes in inorganic carbon due to anthropogenic CO_2 become increasingly prominent relative to changes imparted by natural, physical, and biological variability.

6. CONCLUSION AND OUTLOOK

Major progress in the understanding of the marine carbon cycle has been reached during the last few decades. An important component consists of accurate and precise observation of the global ocean carbonate system, albeit with large temporal and spatial gaps. Particularly important are the global survey during WOCE/JGOFS in the 1990s and the repeat hydrography during the 2000s, but numerous regional surveys as well as process-oriented studies have been pivotal for advancing the observationally constrained knowledge. A major research focus is the ocean uptake and storage of anthropogenic carbon, C_{ant}. Several independent methods to calculate are converging with regard to the storage of C_{ant} and for the annual oceanic uptake of C_{ant}, although the error bars are somewhat larger on the latter due to larger temporal and spatial variability. Several studies on the decadal storage rate of C_{ant} are reporting on deviations from the expected trajectory based on the atmospheric history of CO_2. The reasons for this are poorly understood and more work and data is needed to verify and understand these trends and the implications for the development of the atmospheric CO_2 concentrations. During the last decade, the focus has increasingly been on understanding feedback mechanisms in the ocean and how this impacts uptake and storage of carbon.

Regarding the future trajectory of the marine carbon cycle, we note that the chemical environment can be predicted with reasonable certainty with assumptions about future atmospheric trajectories. In contrast to the chemistry, our knowledge of potential feedbacks from its natural components is rather immature. In some cases, we are not even sure about the sign of a feedback, let alone its magnitude and longevity. The biological responses to these are extremely hard to assess. While evidence from manipulative experiments both with cultures and natural seawater enclosures (mesocosms) can be used to identify and quantify these responses, we are far from being able to extrapolate with confidence these findings on small space and time scales to the scale of the global ocean and the decades to come.

Measurements of the seawater CO_2–carbonate system made on discrete water samples remains the only method to obtain accurate and precise data on the changes in inorganic carbon within the interior of the global ocean. Therefore shipboard measurements are a key element of the observational effort to monitor the seawater CO_2–carbonate system, and inventories of C_{ant} in the ocean. Sustaining the ship-based repeat hydrography program is therefore a prerequisite for obtaining decadal assessments of the state of the marine carbon cycle (Hood et al., 2010). Ship-based hydrography will probably remain the only method for obtaining highest-quality measurements of a suite of physical, chemical, and biological parameters over the full water column.

However, recent promising technological advances in terms of biogeochemical sensors open a possibility to increase the spatial and temporal data coverage in the oceans, and to better understand biogeochemical ocean dynamics from observations. The significant progress made for oxygen and bio-optical measurements from floats shows the potential of the float-based approach (Johnson et al., 2009) and other emerging technologies (e.g., gliders and AUVs) for observing the marine carbon cycle in the future. First examples of new systems/sensors for autonomous platforms are available for pH (Martz et al., 2010; Assmann et al., 2011) and pCO_2 (Fiedler et al., 2013). For instance, the Argo float observatory—a great success story of physical oceanography—needs to be embraced by the CO_2 community. However, it is not expected that the newly developed sensors will be accurate enough (within the next few years, at least) to be able to monitor decadal scale changes in carbon inventories in the ocean interior, underlining the continuous need for shipboard carbon observations complementary to observations by sensors on autonomous platforms.

There is greater hope for observations of surface ocean pCO_2 measurements where a range of sensors and autonomous instrumentation has been developed for deployment on "ships of opportunity" and other platforms, such as surface drifters and gliders. Here the challenge is to support an observational system with sufficient temporal and spatial resolution to facilitate observations of variability in the air–sea exchange of carbon. In the future observation system of the ocean carbon system, it is likely that a combination of autonomous vehicles equipped with "carbon sensors" will complement more accurate, ship-based, measurements, thus providing a useful mix of long-term trend estimates and the understanding of variability needed to understand the trends. Ocean time series, which provide long-term data at high temporal resolution until today, remain the most incontrovertible evidence of the changing marine carbon cycle (Bates, 2007; Dore et al., 2009). There is a need to establish additional time series in both the tropical and high latitude oceans.

ACKNOWLEDGMENTS

T. T. acknowledge funding from the EU FP7 project CARBOCHANGE "Changes in carbon uptake and emissions by oceans in a changing climate" which received funding from the European Community's Seventh Framework Program under grant agreement no. 264879. N. R. B. acknowledges support from NSF award OCE-0752366.

REFERENCES

Álvarez, M., Gourcuff, C., 2010. Uncoupled transport of chlorofluorocarbons and anthropogenic carbon in the subpolar North Atlantic. Deep Sea Res. Part I 57, 860–868.

Álvarez, M., Lo Monaco, C., Tanhua, T., Yool, A., Oschlies, A., Bullister, J.L., Goyet, C., Metzl, N., Touratier, F., McDonagh, E., Bryden, H.L., 2009. Estimating the storage of anthropogenic carbon in the subtropical Indian Ocean: a comparison of five different approaches. Biogeosciences 6, 681–703.

Álvarez, M., Tanhua, T., Brix, H., Lo Monaco, C., Metzl, N., McDonagh, E.L., Bryden, H.L., 2011. Decadal biogeochemical changes in the subtropical Indian Ocean associated with Subantarctic Mode Water. J. Geophys. Res. 116, C09016. http://dx.doi.org/10.1029/2010jc006475.

Anderson, L.G., Olsson, K., Jones, E.P., Chierici, M., Fransson, A., 1998. Anthropogenic carbon dioxide in the Arctic Ocean: inventory and sinks. J. Geophys. Res. 103, 27707–27716.

Anderson, L.G., Tanhua, T., Björk, G., Hjalmarsson, S., Jones, E.P., Jutterström, S., Rudels, B., Swift, J.H., Wåhlstöm, I., 2010. Arctic ocean shelf-basin interaction: an active continental shelf CO_2 pump and its impact on the degree of calcium carbonate solubility. Deep Sea Res. Part I 57, 869–879. http://dx.doi.org/10.1016/j.dsr.2010.03.012.

Anderson, L.G., Bjork, G., Jutterstrom, S., Pipko, I., Shakhova, N., Semiletov, I., Wahlstrom, I., 2011. East Siberian Sea, an Arctic region of very high biogeochemical activity. Biogeosciences 8, 1745–1754. http://dx.doi.org/10.5194/bg-8-1745-2011.

Andersson, A.J., Mackenzie, F.T., Gattuso, J.P., 2011. Effects of ocean acidification on benthic processes, organisms and ecosystems. In: Gattuso, J.P., Hansson, L. (Eds.), Ocean Acidification. Oxford University Press, New York, pp. 122–153.

Archer, D., 2005. Fate of fossil fuel CO_2 in geologic time. J. Geophys. Res. 110, C09S05. http://dx.doi.org/10.1029/2004JC002625.

Arrigo, K.R., van Dijken, G., Pabi, S., 2008. Impact of a shrinking Arctic ice cover on marine primary production. Geophys. Res. Lett. 35, L19603. http://dx.doi.org/10.1029/2008GL035028.

Assmann, S., Frank, C., Kortzinger, A., 2011. Spectrophotometric high-precision seawater pH determination for use in underway measuring systems. Ocean Sci. 7, 597–607. http://dx.doi.org/10.5194/os-7-597-2011.

Bates, N.R., 2001. Interannual variability of oceanic CO_2 and biogeochemical properties in the Western North Atlantic subtropical gyre. Deep Sea Res. Part II 48, 1507–1528. http://dx.doi.org/10.1016/S0967-0645(00)00151-X.

Bates, N.R., 2007. Interannual variability of the oceanic CO_2 sink in the subtropical gyre of the North Atlantic Ocean over the last 2 decades. J. Geophys. Res. 112, C09013. http://dx.doi.org/10.1029/2006JC003759.

Bates, N.R., Mathis, J.T., 2009. The Arctic Ocean marine carbon cycle: evaluation of air-sea CO_2 exchanges, ocean acidification impacts and potential feedbacks. Biogeosciences 6, 2433–2459.

Bates, N.R., Moran, S.B., Hansell, D.A., Mathis, J.T., 2006. An increasing CO_2 sink in the Arctic Ocean due to sea-ice loss. Geophys. Res. Lett. 33, L23609. http://dx.doi.org/10.1029/2006GL027028.

Bates, N.R., Mathis, J.T., Cooper, L.W., 2009. Ocean acidification and biologically induced seasonality of carbonate mineral saturation states in the western Arctic Ocean. J. Geophys. Res. 114, C11007. http://dx.doi.org/10.1029/2008JC004862.

Bates, N.R., Best, M.H., Neely, K., Garley, R., Dickson, A.G., Johnson, R.J., 2012. Indicators of anthropogenic carbon dioxide uptake and ocean acidification in the North Atlantic Ocean. Biogeosciences 9, 2509–2522. http://dx.doi.org/10.5194/bg-9-2509-2012.

Bates, N.R., Orchowska, M.I., Garley, R., Mathis, J.T., 2013. Summertime calcium carbonate undersaturation in shelf waters of the western Arctic ocean – how biological processes exacerbate the impact of ocean acidification. Biogeosciences 10, 5281–5309. http://dx.doi.org/10:105194/bg/bg-10-5281-2013.

Brewer, P.G., Sarmiento, J.L., Smethie Jr., W.M., 1985. The Transient Tracers in the Ocean (TTO) program: The North Atlantic Study, 1981; The Tropical Atlantic Study, 1983. J. Geophys. Res. 90, 6903–6905. http://dx.doi.org/10.1029/JC090iC04p06903.

Butler, J.N., 1982. Carbon Dioxide Equilibria and Their Applications. Addison-Wesley, Reading, MA.

Chen, C.T.A., Wang, S.L., Chou, W.C., Sheu, D.D., 2006. Carbonate chemistry and projected future changes in pH and $CaCO_3$ saturation state of the South China Sea. Mar. Chem. 101, 277–305. http://dx.doi.org/10.1016/j.marchem.2006.01.007.

Chen, C.T.A., Huang, T.H., Chen, Y.C., Bai, Y., He, X., Kang, Y., 2013. Review article "Air-sea exchanges of CO_2 in world's coastal seas" Biogeosci. Discuss. 10, 5041–5105. http://dx.doi.org/10.5194/bgd-10-5041-2013.

Craig, H., Weiss, R.F., 1970. The Geosecs 1969 Intercalibration Station: introduction, hydrographic features, and total CO_2-O_2 relationships. J. Geophys. Res. 75, 7641–7647. http://dx.doi.org/10.1029/JC075i036p07641.

Currie, K.I., Reid, M.R., Hunter, K.A., 2011. Interannual variability of carbon dioxide drawdown by subantarctic surface water near New Zealand. Biogeochemistry 104, 3–34. http://dx.doi.org/10.1007/s10533-009-9355-3.

Dickson, A.G., 2010. Standards for ocean measurements. Oceanography 23, 34–47. http://dx.doi.org/10.5670/oceanog.2010.22.

Dickson, A.G., Sabine, C.L., Christian, J.R., 2007. Guide to Best Practices for Ocean CO_2 Measurements, Carbon Dioxide Information and Analysis center (CDIAC), USA, 191 pp.

DOE, 1994. In: Dickson, A.G., Goyet, C. (Eds.), Handbook of Methods for the Analysis of the Various Parameters of the Carbon Dioxide System in Sea Water; Version 2. Department of Energy, ORNL/CDIAC-74.

Dore, J.E., Lukas, R., Sadler, D.W., Church, M.J., Karl, D.M., 2009. Physical and biogeochemical modulation of ocean acidification in the central North Pacific. Proc. Natl. Acad. Sci. U.S.A. 106, 12235–12240. http://dx.doi.org/10.1073/pnas.0906044106.

Feely, R.A., Wanninkhof, R., Takahashi, T., Tans, P., 1999. Influence of El Nino on the equatorial Pacific contribution to atmospheric CO_2 accumulation. Nature 398, 597–601.

Feely, R.A., Wanninkhof, R., McGillis, W., Carr, M.E., Cosca, C.E., 2004. Effects of wind speed and gas exchange parameterizations on the air-sea CO_2 fluxes in the equatorial Pacific Ocean. J. Geophys. Res. 109, C08S03.

Feely, R.A., Takahashi, T., Wanninkhof, R., McPhaden, M.J., Cosca, C.E., Sutherland, S.C., Carr, M.E., 2006. Decadal variability of the air-sea CO_2 fluxes in the equatorial Pacific Ocean. J. Geophys. Res. 111, C08S90. http://dx.doi.org/10.1029/2005JC003129.

Feely, R.A., Aoyama, M., Bates, N.R., Byrne, R.H., González-Dávila, M., Dore, J., Gruber, N., Hydes, D., Karl, D.M., Khatiwala, S., Kleypas, J., Lee, K., Mordy, C., Olafsson, J., Orr, J., Stramma, L., Takahashi, T., Tanhua, T., Wanninkhof, R., 2013. IPCC AR5 Chapter 3 Ocean Observations Subsection 3.7 (submitted for inclusion to the forthcoming IPCC 5th Assessment).

Fiedler, B., Fietzek, P., Vieira, N., Silva, P., Bittig, H.C., Kortzinger, A., 2013. In situ CO_2 and O_2 measurements on a profiling float. J. Atmos. Oceanic Technol. 30, 112–126. http://dx.doi.org/10.1175/Jtech-D-12-00043.1.

Friis, K., Körtzinger, A., Pätsch, J., Wallace, D.W.R., 2005. On the temporal increase of anthropogenic CO_2 in the subpolar North Atlantic. Deep Sea Res. Part I 52, 681–698. http://dx.doi.org/10.1016/j.dsr.2004.11.017.

Gehlen, M., Gruber, N., Gangstø, R., Bopp, L., Oschlies, A., 2011. Biogeochemical consequences of ocean acidification and feedbacks to the earth system. In: Gattuso, J.P., Hansson, L. (Eds.), Ocean Acidification. Oxford University Press, New York, pp. 210–229.

Gonzalez-Davila, M., Santana-Casiano, J.M., Rueda, M.J., Llinas, O., 2010. The water column distribution of carbonate system variables at the ESTOC site from 1995 to 2004. Biogeosciences 7, 3067–3081. http://dx.doi.org/10.5194/bg-7-3067-2010.

Goodkin, N.F., Levine, N.M., Doney, S.C., Wanninkhof, R., 2011. Impacts of temporal CO_2 and climate trends on the detection of ocean anthropogenic CO_2 accumulation. Global Biogeochem. Cycles 25, GB3023. http://dx.doi.org/10.1029/2010GB004009.

Gruber, N., 1998. Anthropogenic CO_2 in the Atlantic Ocean. Global Biogeochem. Cycles 12, 165–191.

Gruber, N., Sarmiento, J.L., Stocker, T.F., 1996. An improved method for detecting anthropogenic CO_2 in the Oceans. Global Biogeochem. Cycles 10, 809–837.

Gruber, N., Keeling, C.D., Bates, N.R., 2002. Interannual variability in the North Atlantic Ocean carbon sink. Science 298, 2374–2378. http://dx.doi.org/10.1126/science.1077077.

Gruber, N., Gloor, M., Mikaloff Fletcher, S.E., Doney, C.S., Dutkiewicz, S., Follows, M., Greber, M., Jacobson, A.R., Joos, F., Lindsay, K., Menemenlis, D., Mouchet, A., Müller, S.A., Sarmiento, J., Takahashi, K., 2009. Ocean sources, sinks, and transport of atmospheric CO_2. Global Biogeochem. Cycles 23, GB1005.

Hall, T.M., Haine, T.N., Waugh, D.W., 2002. Inferring the concentration of anthropogenic carbon in ocean from tracers. Global Biogeochem. Cycles 16, 1131. http://dx.doi.org/10.1029/2001GB001835.

Hall, T.M., Waugh, D.W., Haine, T.W.N., Robbins, P.E., Khatiwala, S., 2004. Estimates of anthropogenic carbon in the Indian Ocean with allowance for mixing and time-varying air-sea CO_2 disequilibrium. Global Biogeochem. Cycles 18, GB1031. http://dx.doi.org/10.1029/2003GB02120.

Helm, K.P., Bindoff, N.L., Church, J.A., 2011. Observed decreases in oxygen content of the global ocean. Geophys. Res. Lett. 38, L23602. http://dx.doi.org/10.1029/2011gl049513.

Hood, M., Fukasawa, M., Gruber, N., Johnson, C.G., Körtzinger, A., Sabine, C.L., Sloyan, B., Stansfield, K., Tanhua, T., 2010. Ship-based repeat hydrography: a strategy for a sustained global program. In: Hall, J., Harrison, D.E., Stammer, D. (Eds.), Proceedings of OceanObs'09: Sustained Ocean Observations and Information for Society. vol. 2. ESA Publication, Venice, Italy.

Hurrell, J.W., 1995. Decadal trends in the North-Atlantic oscillation—regional temperatures and precipitation. Science 269, 676–679. http://dx.doi.org/10.1126/science.269.5224.676.

Hurrell, J.W., Deser, C., 2009. North Atlantic climate variability: the role of the North Atlantic oscillation. J. Mar. Syst. 78, 28–41. http://dx.doi.org/10.1016/j.jmarsys.2008.11.026.

Hurrell, J.W., Kushnir, Y., Visbeck, M., 2001. Climate—the North Atlantic oscillation. Science 291, 603. http://dx.doi.org/10.1126/science.1058761.

Ishii, M., Inoue, H.Y., Midorikawa, T., Saito, S., Tokieda, T., Sasano, D., Nakadate, A., Nemoto, K., Metzl, N., Wong, C.S., Feely, R.A., 2009. Spatial variability and decadal trend of the oceanic CO_2 in the western equatorial Pacific warm/fresh water. Deep Sea Res. Part II 56, 591–606. http://dx.doi.org/10.1016/j.dsr2.2009.01.002.

Ishii, M., Kosugi, N., Sasano, D., Saito, S., Midorikawa, T., Inoue, H.Y., 2011. Ocean acidification off the south coast of Japan: a result from time series observations of CO_2 parameters from 1994 to 2008. J. Geophys. Res. 116, C06022. http://dx.doi.org/10.1029/2010jc006831.

Jeansson, E., Olsen, A., Eldevik, T., Skjelvan, I., Omar, A.M., Lauvset, S.K., Nilsen, J.E.Ö., Bellerby, R.G.J., Johannessen, T., Falck, E., 2011. The Nordic Seas carbon budget: sources, sinks, and uncertainties. Global Biogeochem. Cycles 25, GB4010. http://dx.doi.org/10.1029/2010GB003961.

Johnson, G.C., Robbins, P.E., Hufford, G.E., 2001. Systematic adjustments of hydrographic sections for internal consistency. J. Atmos. Oceanic Technol. 18, 1234–1244.

Johnson, K.S., Berelson, W.M., Boss, E.S., Chase, Z., Claustre, H., Emerson, S.R., Gruber, N., Kortzinger, A., Perry, M.J., Riser, S.C., 2009. Observing biogeochemical cycles at global scales with profiling floats and gliders prospects for a global array. Oceanography 22, 216–225.

Jutterström, S., Anderson, L.G., 2005. The saturation of calcite and aragonite in the Arctic Ocean. Mar. Chem. 94, 101–110. http://dx.doi.org/10.1016/j.marchem.2004.08.010.

Jutterström, S., Jeansson, E., Anderson, L.G., Bellerby, R., Jones, E.P., Smethie, W.M., Swift, J.H., 2008. Evaluation of anthropogenic carbon in the Nordic Seas using observed relationships of N, P and C versus CFCs. Prog. Oceanogr. 78, 78–84.

Karl, D.M., Bates, N.R., Emerson, S., Harrison, P.J., Jeandal, C., Llinas, O., Liu, K.K., Matry, J.-C., Michaels, A.F., Miquel, J.C., Neuer, S., Nojiri, Y., Wong, C.S., 2003. Temporal studies of biogeochemical processes in the world's oceans during the JGOFS era. In: Fasham, M.R. (Ed.), Ocean Biogeochemistry: The Role of the Ocean Carbon Cycle in Global Climate Change. Springer, Berlin, pp. 239–265.

Keeling, R.F., Körtzinger, A., Gruber, N., 2010. Ocean deoxygenation in a warming World. Ann. Rev. Mar. Sci. 2, 199–229. http://dx.doi.org/10.1146/annurev.marine.010908.163855.

Key, R.M., Kozyr, A., Sabine, C.L., Lee, K., Wanninkhof, R., Bullister, J.L., Feely, R.A., Millero, F.J., Mordy, C., Peng, T.H., 2004. A global ocean carbon climatology: results from Global Data Analysis Project (GLODAP). Global Biogeochem. Cycles 18, GB4031. http://dx.doi.org/10.1029/2004GB002247.

Key, R.M., Tanhua, T., Olsen, A., Hoppema, M., Jutterström, S., Schirnick, C., van Heuven, S., Kozyr, A., Lin, X., Velo, A., Wallace, D.W.R., Mintrop, L., 2010. The CARINA data synthesis project: introduction and overview. Earth Syst. Sci. Data 2, 105–121. http://dx.doi.org/10.5194/essd-2-105-2010.

Khatiwala, S., Primeau, F., Hall, T., 2009. Reconstruction of the history of anthropogenic CO_2 concentrations in the ocean. Nature 462. http://dx.doi.org/10.1038/nature08526, 346-U110.

Khatiwala, S., Tanhua, T., Mikaloff Fletcher, S., Gerber, M., Doney, C.S., Graven, H.D., Gruber, N., McKinley, G.A., Murata, A., Sabine, C., 2013. Global storage of anthropogenic carbon. Biogeosciences 10, 2169–2191. http://dx.doi.org/10.5194/bg-10-2169-2013.

Körtzinger, A., Duinker, J.C., Mintrop, L., 1997. Strong CO_2 emissions from the Arabian Sea during South-West Monsoon. Geophys. Res. Lett. 24, 1763–1766.

Kouketsu, S., Murata, A., Doi, T., 2013. Decadal changes in dissolved inorganic carbon in the Pacific Ocean. Global Biogeochem. Cycles 27, 1–12. http://dx.doi.org/10.1029/2012GB004413.

Lee, K., Choi, S.-D., Park, G.-H., Wanninkhof, R., Peng, T.H., Key, R.M., Sabine, C.L., Feely, R.A., Bullister, J.L., Millero, F.J., Kozyr, A., 2003. An updated anthropogenic CO_2 inventory in the Atlantic Ocean. Global Biogeochem. Cycles 17, 1116. http://dx.doi.org/10.1029/2003GB002067.

Lee, K., Sabine, C.L., Tanhua, T., Kim, T.W., Feely, R.A., Kim, H.C., 2011. Roles of marginal seas in absorbing and storing fossil fuel CO_2. Energy Environ. Sci. 4, 1133–1146.

Lenton, A., Metzl, N., Takahashi, T., Kuchinke, M., Matear, R.J., Roy, T., Sutherland, S.C., Sweeney, C., Tilbrook, B., 2012. The observed evolution of oceanic pCO_2 and its drivers over the last two decades. Global Biogeochem. Cycles 26, GB2012. http://dx.doi.org/10.1029/2011gb004095.

Levine, N.M., Doney, S.C., Wanninkhof, R., Lindsay, K., Fung, I.Y., 2008. Impact of ocean carbon system variability on the detection of temporal increases in anthropogenic CO_2. J. Geophys. Res. 113, C03019.

Levitus, S., Antonov, J.I., Boyer, T.P., Baranova, O.K., Garcia, H.E., Locarnini, R.A., Mishonov, A.V., Reagan, J.R., Seidov, D., Yarosh, E.S., Zweng, M.M., 2012. World ocean heat content and thermosteric sea level change (0–2000 m), 1955–2010. Geophys. Res. Lett. 39, L10603. http://dx.doi.org/10.1029/2012gl051106.

Lo Monaco, C., Goyet, C., Metzl, N., Poisson, A., Tourtier, F., 2005. Distribution and inventory of anthropogenic CO_2 in the Southern Ocean: comparison of three data-based methods. J. Geophys. Res. 110. C09S02. http://dx.doi.org/10.1029/2004JC002571.

Louanchi, F., Ruiz-Pino, D.P., Poisson, A., 1999. Temporal variations of mixed-layer oceanic CO_2 at JGOFS-KERFIX time-series station: physical versus biogeochemical processes. J. Mar. Res. 57, 165–187. http://dx.doi.org/10.1357/002224099765038607.

Macdonald, A.M., Baringer, M.O., Wanninkhof, R., Lee, K., Wallace, D.W.R., 2003. A 1998–1992 comparison of inorganic carbon and its transport across 24.5°N in the Atlantic. Deep Sea Res. Part II 50, 3041–3064. http://dx.doi.org/10.1016/j.dsr2.2003.07.009.

Martz, T.R., Connery, J.G., Johnson, K.S., 2010. Testing the Honeywell Durafet (R) for seawater pH applications. Limnol. Oceanogr. Methods 8, 172–184. http://dx.doi.org/10.4319/lom.2010.8.172.

Matear, R.J., McNeil, B.I., 2003. Decadal accumulation of anthropogenic CO_2 in the Southern Ocean: a comparison of CFC-age derived estimates to multiple-linear regression estimates. Global Biogeochem. Cycles 17, 24.

McGrath, T., Kivimäe, C., Tanhua, T., Cave, R.R., McGovern, E., 2012. Inorganic carbon and pH levels in the Rockall Trough 1991–2010. Deep Sea Res. Part I 68, 79–91. http://dx.doi.org/10.1016/j.dsr.2012.05.011.

McKinley, G.A., Follows, M.J., Marshall, J., 2004. Mechanisms of air-sea CO_2 flux variability in the equatorial Pacific and the North Atlantic. Global Biogeochem. Cycles 18, GB2011. http://dx.doi.org/10.1029/2003gb002179.

McKinley, G.A., Takahashi, T., Buitenhuis, E., Chai, F., Christian, J.R., Doney, S.C., Jiang, M.S., Lindsay, K., Moore, J.K., Le Quere, C., Lima, I., Murtugudde, R., Shi, L., Wetzel, P., 2006. North Pacific carbon cycle response to climate variability on seasonal to decadal timescales. J. Geophys. Res. 111, C07506. http://dx.doi.org/10.1029/2005JC003173.

McKinley, G.A., Fay, A.R., Takahashi, T., Metzl, N., 2011. Convergence of atmospheric and North Atlantic carbon dioxide trends on multidecadal timescales. Nat. Geosci. 4, 606–610. http://dx.doi.org/10.1038/Ngeo1193.

McNeil, B.I., Matear, R., 2013. The non-steady state oceanic CO_2 signal: its importance, magnitude and a novel way to use it. Biogeosciences 10, 2219–2228. http://dx.doi.org/10.5194/bg-10-2219-2013.

Metzl, N., 2009. Decadal increase of oceanic carbon dioxide in Southern Indian Ocean surface waters (1991–2007). Deep Sea Res. Part II 56, 607–619. http://dx.doi.org/10.1016/j.dsr2.2008.12.007.

Metzl, N., Corbière, A., Reverdin, G., Lenton, A., Takahashi, T., Olsen, A., Johannessen, T., Pierrot, D., Wanninkhof, R., Ólafsdóttir, S.R., Olafsson, J., Ramonet, M., 2010. Recent acceleration of the sea surface fCO_2 growth rate in the North Atlantic subpolar gyre (1993–2008) revealed by winter observations. Global Biogeochem. Cycles 24, GB4004. http://dx.doi.org/10.1029/2009GB003658.

Midorikawa, T., Nemoto, K., Kamiya, H., Ishii, M., Inoue, H.Y., 2005. Persistently strong oceanic CO_2 sink in the western subtropical North Pacific. Geophys. Res. Lett. 32, L2601. http://dx.doi.org/10.1029/2004gl021952.

Midorikawa, T., Ishii, M., Kosugi, N., Sasano, D., Nakano, T., Saito, S., Sakamoto, N., Nakano, H., Inoue, H.Y., 2012. Recent deceleration of oceanic pCO_2 increase in the western North Pacific in winter. Geophys. Res. Lett. 39, L12601. http://dx.doi.org/10.1029/2012gl051665.

Mikaloff Fletcher, S.E., Gruber, N., Jacobson, A.R., Doney, S.C., Dutkiewicz, S., Gerber, M., Follows, M., Joos, F., Lindsay, K., Menemenlis, D., Mouchet, A., Muller, S.A., Sarmiento, J.L., 2006. Inverse estimates of anthropogenic CO_2 uptake, transport, and storage by the ocean. Global Biogeochem. Cycles 20, GB2002.

Millero, F.J., 2007. The marine inorganic carbon cycle. Chem. Rev. 107, 308–341. http://dx.doi.org/10.1021/cr0503557.

Monteiro, P.M.S., Schuster, U., Hood, M., Lenton, A., Metzl, N., Olsen, A., Rogers, K., Sabine, C.L., Takahashi, T., Tilbrook, B., Yoder, J., Wanninkhof, R., Watson, A.J., 2010. A global sea surface carbon observing system: assessment of changing sea surface CO_2 and air-sea CO_2 fluxes. In: Hall, J., Harrison, D.E., Stammer, D. (Eds.), Proceedings of OceanObs'09: Sustained Ocean Observations and Information for Society. vol. 2. ESA Publication, Venice, Italy.

Murata, A., Kumamoto, Y., Watanabe, S., Fukasawa, M., 2007. Decadal increases of anthropogenic CO_2 in the South Pacific subtropical ocean along 32 degrees S. J. Geophys. Res. 112, C05033. http://dx.doi.org/10.1029/2005JC003405.

Murata, A., Kumamoto, Y., Sasaki, K., Watanabe, S., Fukasawa, M., 2008. Decadal increases of anthropogenic CO_2 in the subtropical South Atlantic Ocean along 30 degrees S. J. Geophys. Res. 113, C06007. http://dx.doi.org/10.1029/2007JC004424.

Murata, A., Kumamoto, Y., Sasaki, K., Watanabe, S., Fukasawa, M., 2009. Decadal increases of anthropogenic CO_2 along 149°E in the western North Pacific. J. Geophys. Res. 114, C04018.

Murata, A., Kumamoto, Y., Sasaki, K., Watanabe, S., Fukasawa, M., 2010. Decadal increases in anthropogenic CO_2 along 20°S in the South Indian Ocean. J. Geophys. Res. 115, C12055. http://dx.doi.org/10.1029/2010JC006250.

Olafsson, J., Olafsdottir, S.R., Benoit-Cattin, A., Danielsen, M., Arnarson, T.S., Takahashi, T., 2009. Rate of Iceland Sea acidification from time series measurements. Biogeosciences 6, 2661–2668.

Olafsson, J., Olafsdottir, S.R., Benoit-Cattin, A., Takahashi, T., 2010. The Irminger Sea and the Iceland Sea time series measurements of sea water carbon and nutrient chemistry 1983–2006. Earth Syst. Sci. Data 2, 99–104. www.earth-syst-sci-data.net/2/99/2010/.

Olsen, A., Bellerby, R.G.J., Johannessen, T., Omar, A.M., Skjelvan, I., 2003. Interannual variability in the wintertime air-sea flux of carbon dioxide in the northern North Atlantic, 1981–2001. Deep Sea Res. Part I 50, 1323–1338. http://dx.doi.org/10.1016/S0967-0637(03)00144-4.

Olsen, A., Omar, A.M., Bellerby, R.G.J., Johannessen, T., Ninnemann, U., Brown, K.R., Olsson, K.A., Olafsson, J., Nondal, G., Kivimae, C., Kringstad, S., Neill, C., Olafsdottir, S., 2006. Magnitude and origin of the anthropogenic CO_2 increase and C-13 Suess effect in the Nordic seas since 1981. Global Biogeochem. Cycles 20, GB3027. http://dx.doi.org/10.1029/2005GB002669.

Olsen, A., Omar, A.M., Jeansson, E., Anderson, L.G., Bellerby, R.G.J., 2010. Nordic seas transit time distributions and anthropogenic CO_2. J. Geophys. Res. 115, C05005.

Omar, A.M., Johannessen, T., Olsen, A., Kaltin, S., Rey, F., 2007. Seasonal and interannual variability of the air-sea CO_2 flux in the Atlantic sector of the Barents Sea. Mar. Chem. 104, 203–213. http://dx.doi.org/10.1016/j.marchem.2006.11.002.

Ono, T., Midorikawa, T., Watanabe, Y.W., Tadokoro, K., Saino, T., 2001. Temporal increases of phosphate and apparent oxygen utilization in the subsurface waters of western subarctic Pacific from 1968 to 1998. Geophys. Res. Lett. 28, 3285–3288. http://dx.doi.org/10.1029/2001gl012948.

Ono, T., Kasai, H., Midorikawa, T., Takatani, Y., Saito, K., Ishii, M., Watanabe, Y.W., Sasaki, K., 2005. Seasonal and interannual variation of DIC in surface mixed layer in the Oyashio region: a climatological view. J. Oceanogr. 61, 1075–1087. http://dx.doi.org/10.1007/s10872-006-0023-0.

Park, G.H., Lee, K., Tishchenko, P., Min, D.H., Warner, M.J., Talley, L.D., Kang, D.J., Kim, K.R., 2006. Large accumulation of anthropogenic CO_2 in the East (Japan) Sea and its significant impact on carbonate chemistry. Global Biogeochem. Cycles 20, GB4013.

Peng, T.-H., Wanninkhof, R., 2010. Increase in anthropogenic CO_2 in the Atlantic Ocean in the last two decades. Deep Sea Res. Part I 57, 755–770. http://dx.doi.org/10.1016/j.dsr.2010.03.008.

Peng, T.-H., Wanninkhof, R., Bullister, J.L., Feely, R.A., Takahashi, T., 1998. Quantification of decadal anthropogenic CO_2 uptake in the ocean based on dissolved inorganic carbon measurements. Nature 396, 560–563.

Peng, T.-H., Wanninkhof, R., Feely, R.A., 2003. Increase of anthropogenic CO_2 in the Pacific Ocean over the last two decades. Deep Sea Res. Part A 50, 3065–3082.

Perez, F.F., Vázquez-Rodríguez, M., Louarn, E., Padín, X.A., Mercier, H., Ríos, A.F., 2008. Temporal variability of the anthropogenicv CO_2 storage in the Irminger Sea. Biogeosciences 5, 1669–1679. http://dx.doi.org/10.5194/bg-5-1669-2008.

Perez, F.F., Aristegui, J., Vazquez-Rodriguez, M., Rios, A.F., 2010. Anthropogenic CO_2 in the Azores region. Sci. Mar. 74, 11–19. http://dx.doi.org/10.3989/scimar.2010.74s1011.

Pérez, F.F., Vázquez-Rodríguez, M., Mercier, H., Velo, A., Lherminier, P., Ríos, A.F., 2010. Trends of anthropogenic CO_2 storage in North Atlantic water masses. Biogeosciences 7, 1789–1807. http://dx.doi.org/10.5194/bg-7-1789-2010.

Perez, F.F., Mercier, H., Vazquez-Rodriguez, M., Lherminier, P., Velo, A., Pardo, P.C., Roson, G., Rios, A.F., 2013. Atlantic Ocean CO_2 uptake reduced by weakening of the meridional overturning circulation. Nat. Geosci. 6, 146–152. http://dx.doi.org/10.1038/NGEO1680.

Pfeil, B., Olsen, A., Bakker, D.C.E., Hankin, S., Koyuk, H., Kozyr, A., Malczyk, J., Manke, A., Metzl, N., Sabine, C.L., Akl, J., Alin, S.R., Bates, N., Bellerby, R.G.J., Borges, A., Boutin, J., Brown, P.J., Cai, W.J., Chavez, F.P., Chen, A., Cosca, C., Fassbender, A.J., Feely, R.A., González-Dávila, M., Goyet, C., Hales, B., Hardman-Mountford, N., Heinze, C., Hood, M., Hoppema, M., Hunt, C.W., Hydes, D., Ishii, M., Johannessen, T., Jones, S.D., Key, R.M., Körtzinger, A., Landschützer, P., Lauvset, S.K., Lefèvre, N., Lenton, A., Lourantou, A., Merlivat, L., Midorikawa, T., Mintrop, L., Miyazaki, C., Murata, A., Nakadate, A., Nakano, Y., Nakaoka, S., Nojiri, Y., Omar, A.M., Padin, X.A., Park, G.H., Paterson, K., Perez, F.F., Pierrot, D., Poisson, A., Ríos, A.F., Santana-Casiano, J.M., Salisbury, J., Sarma, V.V.S.S., Schlitzer, R., Schneider, B., Schuster, U., Sieger, R., Skjelvan, I., Steinhoff, T., Suzuki, T., Takahashi, T., Tedesco, K., Telszewski, M., Thomas, H., Tilbrook, B., Tjiputra, J., Vandemark, D., Veness, T., Wanninkhof, R., Watson, A.J., Weiss, R., Wong, C.S., Yoshikawa-Inoue, H., 2013. A uniform, quality controlled Surface Ocean CO_2 Atlas (SOCAT). Earth Syst. Sci. Data 5, 125–143. http://dx.doi.org/10.5194/essd-5-125-2013.

Plancherel, Y., Rodgers, K.B., Key, R.M., Jacobson, A.R., Sarmiento, J.L., 2012. Role of regression model selection and station distribution on the estimation of oceanic anthropogenic carbon change by eMLR. Biogeosci. Discuss. 9, 14589–14638. http://dx.doi.org/10.5194/bgd-9-14589-2012.

Pörtner, H.-O., Gutowska, M., Ishimatsu, A., Lucassen, M., Melzner, F., Seibel, B., 2011. Effects of ocean acidification on nektonic organisms. In: Gattuso, J.P., Hansson, L. (Eds.), Ocean Acidification. Oxford University Press, New York, pp. 154–175.

Quay, P., Sonnerup, R., Stutsman, J., Maurer, J., Körtzinger, A., Padin, X.A., Robinson, C., 2007. Anthropogenic CO_2 accumulation rates in the North Atlantic Ocean from changes in the 13C/12C of dissolved inorganic carbon. Global Biogeochem. Cycles 21, GB1009. http://dx.doi.org/10.1029/2006GB002761.

Riebesell, U., Tortell, P.D., 2011. Effects of ocean acidification on pelagic organisms and ecosystems. In: Gattuso, J.P., Hansson, L. (Eds.), Ocean Acidification. Oxford University Press, New York, pp. 99–121.

Riebesell, U., Bellerby, R.G.J., Engel, A., Fabry, V.J., Hutchins, D.A., Reusch, T.B.H., Schulz, K.G., Morel, F.M.M., 2008. Comment on "Phytoplankton Calcification in a High-CO_2 World. Science 322, 1466.

Riebesell, U., Körtzinger, A., Oschlies, A., 2009. Sensitivities of marine carbon fluxes to ocean change. Proc. Natl. Acad. Sci. U.S.A. 106, 20602–20609. http://dx.doi.org/10.1073/pnas.0813291106.

Ríos, A.F., Velo, A., Pardo, P.C., Hoppema, M., Pérez, F.F., 2012. An update of anthropogenic CO_2 storage rates in the western South Atlantic basin and the role of Antarctic Bottom Water. J. Mar. Syst. 94, 197–203. http://dx.doi.org/10.1016/j.jmarsys.2011.11.023.

Sabine, C.L., Tanhua, T., 2010. Estimation of anthropogenic CO_2 inventories in the ocean. Ann. Rev. Mar. Sci. 2, 175–198. http://dx.doi.org/10.1146/annurev-marine-120308-080947.

Sabine, C.L., Key, R.M., Johnson, K.M., Millero, F.J., Poisson, A., Sarmiento, J.L., Wallace, D.W.R., Winn, C.D., 1999. Anthropogenic CO_2 inventory of the Indian Ocean. Global Biogeochem. Cycles 13, 179–198.

Sabine, C.L., Feely, R.A., Key, R.M., Bullister, J.L., Millero, F.J., Lee, K., Peng, T.H., Tilbrook, B., Ono, T., Wong, C.S., 2002. Distribution of anthropogenic CO_2 in the Pacific Ocean. Global Biogeochem. Cycles (16).

Sabine, C.L., Feely, R.A., Gruber, N., Key, R.M., Lee, K., Bullister, J.L., Wanninkhof, R., Wong, C.S., Wallace, D.W.R., Tilbrook, B.,

Millero, F.J., Peng, T.-H., Kozyr, A., Ono, T., Rios, A.F., 2004a. The oceanic sink for anthropogenic CO_2. Science 305, 367–371.

Sabine, C.L., Feely, R.A., Watanabe, Y.W., Lamb, M., 2004b. Temporal evolution of the North Pacific CO_2 uptake rate. J. Oceanogr. 60, 5–15. http://dx.doi.org/10.1023/B:Joce.0000038315.23875.Ae.

Sabine, C.L., Key, R.M., Kozyr, A., Feely, R.A., Wanninkhof, R., Millero, F., Peng, T.H., Bullister, J., Lee, K., 2005. Global Ocean Data Analysis Project (GLODAP): Results and Data. ORNL/CDIAC, Oak Ridge, TN, USA, NDP-083.

Sabine, C.L., Feely, R.A., Millero, F., Dickson, A.G., Langdon, C., Mecking, S., Greeley, D., 2008. Decadal changes in Pacific Carbon. J. Geophys. Res. 113, C07021. http://dx.doi.org/10.1029/2007JC004577.

Sabine, C.L., Ducklow, H., Hood, M., 2010. International carbon coordination: Roger Revelle's legacy in the Intergovernmental Oceanographic Commission. Oceanography 23, 48–61. http://dx.doi.org/10.5670/oceanog.2010.23.

Sabine, C.L., Hankin, S., Koyuk, H., Bakker, D.C.E., Pfeil, B., Olsen, A., Metzl, N., Kozyr, A., Fassbender, A., Manke, A., Malczyk, J., Akl, J., Alin, S.R., Bellerby, R.G.J., Borges, A., Boutin, J., Brown, P.J., Cai, W.J., Chavez, F.P., Chen, A., Cosca, C., Feely, R.A., González-Dávila, M., Goyet, C., Hardman-Mountford, N., Heinze, C., Hoppema, M., Hunt, C.W., Hydes, D., Ishii, M., Johannessen, T., Key, R.M., Körtzinger, A., Landschützer, P., Lauvset, S.K., Lefèvre, N., Lenton, A., Lourantou, A., Merlivat, L., Midorikawa, T., Mintrop, L., Miyazaki, C., Murata, A., Nakadate, A., Nakano, Y., Nakaoka, S., Nojiri, Y., Omar, A.M., Padin, X.A., Park, G.H., Paterson, K., Perez, F.F., Pierrot, D., Poisson, A., Ríos, A.F., Salisbury, J., Santana-Casiano, J.M., Sarma, V.V.S.S., Schlitzer, R., Schneider, B., Schuster, U., Sieger, R., Skjelvan, I., Steinhoff, T., Suzuki, T., Takahashi, T., Tedesco, K., Telszewski, M., Thomas, H., Tilbrook, B., Vandemark, D., Veness, T., Watson, A.J., Weiss, R., Wong, C.S., Yoshikawa-Inoue, H., 2013. Surface Ocean CO_2 Atlas (SOCAT) gridded data products. Earth Syst. Sci. Data 5, 145–153. http://dx.doi.org/10.5194/essd-5-145-2013.

Sarmiento, J.L., Gruber, N., 2003. Ocean Biogeochemical Dynamics. Princeton University Press, Princeton.

Schneider, A., Tanhua, T., Körtzinger, A., Wallace, D.W.R., 2010. High anthropogenic carbon content in the eastern Mediterranean. J. Geophys. Res. 115, C12050. http://dx.doi.org/10.1029/2010JC006171.

Schneider, A., Tanhua, T., Körtzinger, A., Wallace, D.W.R., 2012. An evaluation of tracer fields and anthropogenic carbon in the equatorial and the tropical North Atlantic. Deep Sea Res. Part I 67, 85–97. http://dx.doi.org/10.1016/j.dsr.2012.05.007.

Schuster, U., Watson, A.J., 2007. A variable and decreasing sink for atmospheric CO_2 in the North Atlantic. J. Geophys. Res. 112, C11006. http://dx.doi.org/10.1029/2006JC003941.

Schuster, U., McKinley, G.A., Bates, N., Chevallier, F., Doney, S.C., Fay, A.R., González-Dávila, M., Gruber, N., Jones, S., Krijnen, J., Landschützer, P., Lefèvre, N., Manizza, M., Mathis, J., Metzl, N., Olsen, A., Rios, A.F., Rödenbeck, C., Santana-Casiano, J.M., Takahashi, T., Wanninkhof, R., Watson, A.J., 2013. An assessment of the Atlantic and Arctic sea–air CO_2 fluxes, 1990–2009. Biogeosciences 10, 607–627. http://dx.doi.org/10.5194/bg-10-607-2013.

Semiletov, I.P., Pipko, I.I., Repina, I., Shakhova, N.E., 2007. Carbonate chemistry dynamics and carbon dioxide fluxes across the atmosphere-ice-water interfaces in the Arctic Ocean: Pacific sector of the Arctic. J. Mar. Syst. 66, 204–226. http://dx.doi.org/10.1016/j.jmarsys.2006.05.012.

Skjelvan, I., Falck, E., Rey, F., Kringstad, S.B., 2008. Inorganic carbon time series at Ocean Weather Station M in the Norwegian Sea. Biogeosciences 5, 549–560.

Steinberg, D.K., Carlson, C.A., Bates, N.R., Johnson, R.J., Michaels, A.F., Knap, A.H., 2001. Overview of the US JGOFS Bermuda Atlantic Time-series Study (BATS): a decade-scale look at ocean biology and biogeochemistry. Deep Sea Res. Part II 48, 1405–1447. http://dx.doi.org/10.1016/S0967-0645(00)00148-X.

Steinfeldt, R., Rhein, M., Bullister, J.L., Tanhua, T., 2009. Inventory changes in anthropogenic carbon from 1997–2003 in the Atlantic Ocean between 20 degrees S and 65 degrees N. Global Biogeochem. Cycles 23, GB3010. http://dx.doi.org/10.1029/2008GB003311.

Stendardo, I., Gruber, N., 2012. Oxygen trends over five decades in the North Atlantic. J. Geophys. Res. 117, C11004. http://dx.doi.org/10.1029/2012jc007909.

Stramma, L., Schmidtko, S., Levin, L.A., Johnson, G.C., 2010. Ocean oxygen minima expansions and their biological impacts. Deep Sea Res. Part I 57, 587–595. http://dx.doi.org/10.1016/j.dsr.2010.01.005.

Stumm, W., Morgan, J.J., 1981. Aquatic Chemistry. An Introduction Emphasizing Chemical Equilibria in Natural Waters. Wiley-Interscience, New York.

Suzuki, T., Ishii, M., Aoyama, M., Christian, J.R., Enyo, K., Kawano, T., Key, R.M., Kosugi, N., Kozyr, A., Miller, L.A., Murata, A., Nakano, T., Ono, T., Saino, T., Sasaki, K., Sasano, D., Takatani, Y., Wakita, M., Sabine, C., 2013. PACIFICA Data Synthesis Project, Carbon Dioxide Information Analysis Center, Oak Ridge National Laboratory, U.S. Department of Energy, Oak Ridge, Tennessee.

Takahashi, T., Williams, R.T., Bos, D.L., 1982. Carbonate chemistry. In: Broecker, W.S., Spencer, D.S., Craig, H. (Eds.), GEOSECS Pacific Expedition, Hydrographic Data 1973–1974, vol. 3. National Science Foundation, Washington, DC, pp. 77–83.

Takahashi, T., Sutherland, S.C., Sweeney, C., Poisson, A., Metzl, N., Tilbrook, B., Bates, N., Wanninkhof, R., Feely, R.A., Sabine, C., Olafsson, J., Nojiri, Y., 2002. Global sea-air CO_2 flux based on climatological surface ocean pCO_2, and seasonal biological and temperature effects. Deep Sea Res. Part II 49, 1601–1622.

Takahashi, T., Sutherland, S.C., Feely, R.A., Cosca, C.E., 2003. Decadal variation of the surface water PCO_2 in the western and central equatorial Pacific. Science 302, 852–856. http://dx.doi.org/10.1126/science.1088570.

Takahashi, T., Sutherland, S.C., Wanninkhof, R., Sweeney, C., Feely, R.A., Chipman, D.W., Hales, B., Friederich, G., Chavez, F., Sabine, C., Watson, A., Bakker, D.C.E., Schuster, U., Metzl, N., Yoshikawa-Inoue, H., Ishii, M., Midorikawa, T., Nojiri, Y., Kortzinger, A., Steinhoff, T., Hoppema, M., Olafsson, J., Arnarson, T.S., Tilbrook, B., Johannessen, T., Olsen, A., Bellerby, R., Wong, C.S., Delille, B., Bates, N.R., de Baar, H.J.W., 2009. Climatological mean and decadal change in surface ocean pCO_2, and net sea-air CO_2 flux over the global oceans. Deep Sea Res. Part I 56, 2075–2076. http://dx.doi.org/10.1016/j.dsr2.2008.12.009.

Takahashi, T., Sutherland, S.C., Kozyr, A., 2013. Global Ocean Surface Water Partial Pressure of CO_2 Database: Measurements Performed During 1957–2012 (Version 2012).

Tanhua, T., Keeling, R.F., 2012. Changes in column inventories of carbon and oxygen in the Atlantic Ocean. Biogeosciences 9, 4819–4833. http://dx.doi.org/10.5194/bg-9-4819-2012.

Tanhua, T., Wallace, D.W.R., 2005. Consistency of TTO-NAS Inorganic Carbon Data with modern measurements. Geophys. Res. Lett. 32, L14618. http://dx.doi.org/10.1029/2005GL023248.

Tanhua, T., Biastoch, A., Körtzinger, A., Lüger, H., Böning, C., Wallace, D.W.R., 2006. Changes of anthropogenic CO_2 and CFCs in the North Atlantic between 1981 and 2004. Global Biogeochem. Cycles 20, GB4017. http://dx.doi.org/10.1029/2006GB002695.

Tanhua, T., Körtzinger, A., Friis, K., Waugh, D.W., Wallace, D.W.R., 2007. An estimate of anthropogenic CO_2 inventory from decadal changes in ocean carbon content. Proc. Natl. Acad. Sci. U.S.A. 104, 3037–3042. http://dx.doi.org/10.1073/pnas.0606574104.

Tanhua, T., Jones, E.P., Jeansson, E., Jutterstrom, S., Smethie, W.M., Wallace, D.W.R., Anderson, L.G., 2009. Ventilation of the Arctic Ocean: mean ages and inventories of anthropogenic CO_2 and CFC-11. J. Geophys. Res. 114, C01002. http://dx.doi.org/10.1029/2008JC004868.

Tanhua, T., van Heuven, S., Key, R.M., Velo, A., Olsen, A., Schirnick, C., 2010. Quality control procedures and methods of the CARINA database. Earth Syst. Sci. Data 2, 35–49.

Terenzi, F., Hall, T.M., Khatiwala, S., Rodehacke, B., LeBel, D.A., 2007. Uptake of natural and anthropogenic carbon by the Labrador Sea. Geophys. Res. Lett. 34, L06608. http://dx.doi.org/10.1029/2006GL028543.

Tjiputra, J.F., Olsen, A., Assmann, K., Pfeil, B., Heinze, C., 2012. A model study of the seasonal and long-term North Atlantic surface pCO_2 variability. Biogeosciences 9, 907–923. http://dx.doi.org/10.5194/bg-9-907-2012.

Touratier, F., Goyet, C., 2004. Applying the new TrOCA approach to assess the distribution of anthropogenic CO_2 in the Atlantic Ocean. J. Mar. Syst. 46, 181–197.

Touratier, F., Goyet, C., 2009. Decadal evolution of anthropogenic CO_2 in the northwestern Mediterranean Sea from the mid-1990s to the mid-2000s. Deep Sea Res. Part I 56, 1708–1716. http://dx.doi.org/10.1016/j.dsr.2009.05.015.

Touratier, F., Azouzi, L., Goyet, C., 2007. CFC-11, Delta C-14 and H-3 tracers as a means to assess anthropogenic CO_2 concentrations in the ocean. Tellus B Chem. Phys. Meteorol. 59, 318–325.

Ullman, D.J., McKinley, G.A., Bennington, V., Dutkiewicz, S., 2009. Trends in the North Atlantic carbon sink: 1992–2006. Global Biogeochem. Cycles 23, Gb4011. http://dx.doi.org/10.1029/2008gb003383.

van Heuven, S.M.A.C., Hoppema, M., Huhn, O., Slagter, H.A., de Baar, H.J.W., 2011. Direct observation of increasing CO_2 in the Weddell Gyre along the Prime Meridian during 1973–2008. Deep Sea Res. Part II 58, 2613–2635. http://dx.doi.org/10.1016/j.dsr2.2011.08.007.

Vázquez-Rodríguez, M., Touratier, F., Lo Monaco, C., Waugh, D.W., Padin, X.A., Bellerby, R., Goyet, C., Metzl, N., Ríos, A.F., Pérez, F.F., 2009. Anthropopgenic carbon distributions in the Atlantic Ocean: data-based estimates from the Arctic to the Antarctic. Biogeosciences 6, 439–451. http://dx.doi.org/10.5194/bg-6-439-2009.

Velo, A., Vazquez-Rodriguez, M., Padin, X.A., Gilcoto, M., Rios, A.F., Perez, F.F., 2010. A multiparametric method of interpolation using WOA05 applied to anthropogenic CO_2 in the Atlantic. Sci. Mar. 74, 21–32. http://dx.doi.org/10.3989/scimar.2010.74s1021.

Wakita, M., Watanabe, S., Watanabe, Y.W., Ono, T., Tsurushima, N., Tsunogai, S., 2005. Temporal change of dissolved inorganic carbon in the subsurface water at Station KNOT (44 degrees N, 155 degrees E) in the western North Pacific subpolar region. J. Oceanogr. 61, 129–139. http://dx.doi.org/10.1007/s10872-005-0026-2.

Wakita, M., Watanabe, S., Murata, A., Tsurushima, N., Honda, M., 2010. Decadal change of dissolved inorganic carbon in the subarctic western North Pacific Ocean. Tellus B Chem. Phys. Meteorol. 62, 608–620. http://dx.doi.org/10.1111/j.1600-0889.2010.00476.x.

Wallace, D.W.R., 1995. Monitoring global ocean carbon inventories, ocean observing system development panel background. Texas A&M Univeristy, College Station, Texas, 54 pp.

Wang, S., Moore, J.K., Primeau, F.W., Khatiwala, S., 2012. Simulation of anthropogenic CO_2 uptake in the CCSM3.1 ocean circulation-biogeochemical model: comparison with data-based estimates. Biogeosciences 9, 1321–1336. http://dx.doi.org/10.5194/bg-9-1321-2012.

Wanninkhof, R., Doney, S.C., Bullister, J.L., Levine, N.M., Warner, M., Gruber, N., 2010. Detecting anthropogenic CO_2 changes in the interior Atlantic Ocean between 1989 and 2005. J. Geophys. Res. 115, C11028. http://dx.doi.org/10.1029/2010JC006251.

Watanabe, Y.W., Chiba, T., Tanaka, T., 2011. Recent change in the oceanic uptake rate of anthropogenic carbon in the North Pacific subpolar region determined by using a carbon-13 time series. J. Geophys. Res. 116, C02006. http://dx.doi.org/10.1029/2010jc006199.

Waters, J.F., Millero, F.J., Sabine, C.L., 2011. Changes in South Pacific anthropogenic carbon. Global Biogeochem. Cycles 25, GB4011. http://dx.doi.org/10.1029/2010GB003988.

Watson, A.J., Schuster, U., Bakker, D.C.E., Bates, N.R., Corbiere, A., Gonzalez-Davila, M., Friedrich, T., Hauck, J., Heinze, C., Johannessen, T., Kortzinger, A., Metzl, N., Olafsson, J., Olsen, A., Oschlies, A., Padin, X.A., Pfeil, B., Santana-Casiano, J.M., Steinhoff, T., Telszewski, M., Rios, A.F., Wallace, D.W.R., Wanninkhof, R., 2009. Tracking the Variable North Atlantic Sink for Atmospheric CO_2. Science 326, 1391–1393. http://dx.doi.org/10.1126/science.1177394.

Waugh, D.W., Hall, T.M., Haine, T.W.N., 2004. Transport times and anthropogenic carbon in the Subpolar North Atlantic Ocean. Deep Sea Res. Part I 51, 1471–1491. http://dx.doi.org/10.1016/j.dsr.2004.06.011.

Waugh, D.W., Hall, T.M., McNeil, B.I., Key, R., Matear, R.J., 2006. Anthropogenic CO_2 in the Oceans estimated using transit-time distributions. Tellus 58B, 376–389. http://dx.doi.org/10.1111/j.1600-0889.2006.00222.x.

Waugh, D.W., Primeau, F., DeVries, T., Holzer, M., 2013. Recent changes in the ventilation of the southern oceans. Science 339, 568. http://dx.doi.org/10.1126/science.1225411.

Weinbauer, M.G., Mari, M., Gattuso, J.-P., 2011. Effects of ocean acidificiation on the diversity and activity of heterotrophic marine microorganisms. In: Gattuso, J.P., Hansson, L. (Eds.), Ocean Acidification. Oxford University Press, New York, pp. 83–98.

Williams, R.G., Follows, M.J., 2011. Ocean Dynamics and the Carbon Cycle: Principles and Mechanisms. Cambridge University Press, Cambridge.

Wong, C.S., Christian, J.R., Wong, S.K.E., Page, J., Xie, L.S., Johannessen, S., 2010. Carbon dioxide in surface seawater of the eastern North Pacific Ocean (Line P), 1973–2005. Deep Sea Res. Part I 57, 687–695. http://dx.doi.org/10.1016/j.dsr.2010.02.003.

Wunsch, C., 1996. The Ocean circulation inverse problem. Cambridge University Press, Cambridge, 442 pp.

Zeebe, R.E., Wolf-Gladrow, D., 2002. CO_2 in Seawater: Equilibrium, Kinetics, Isotopes: Equilibrium, Kinetics. Elsevier Science, Amsterdam.

Chapter 31

Marine Ecosystems, Biogeochemistry, and Climate

Scott C. Doney
Woods Hole Oceanographic Institution, Woods Hole, Massachusetts, USA

Chapter Outline

1. INTRODUCTION

This chapter introduces key basic concepts on marine ecosystem dynamics, discusses how physical variability impacts ocean biota on large scales (i.e., gyre to basin, interannual to centennial), and touches on how ocean biogeochemical processes can modify physical climate (Sarmiento and Gruber, 2006; Miller and Wheeler, 2012). An ecosystem consists of a complete set of biotic and abiotic components of a system or location—all the living organisms, nutrients and detrital materials, and the physical environment—as well as the interactions among all of these components. For the pelagic ocean, key organism groups span from microscopic phytoplankton and bacteria through zooplankton all the way up the trophic ladder to fish, marine mammals, and seabirds (Figure 31.1). Relevant biological interactions include inter-species competition, predator–prey relationships, disease, and parasitism. Important physical processes involve seawater chemistry, temperature and light, vertical and horizontal turbulent mixing, and ocean circulation that helps govern nutrient supply and the dispersal of organisms. The spatial extent of an ecosystem is defined more by the strength of the interactions rather than by spatial homogeneity. Of course, no marine ecosystem is fully self-contained, and constraining physical and biological transport (Williams and Follows, 2011) often is essential to understand the functioning of the ecosystem.

Given the focus here on ecosystem–climate scale interactions, by necessity, the chapter neglects the many fascinating smaller-scale biological–physical phenomenon—for example, microscale diffusion and molecular viscosity; local-scale Langmuir cells and internal waves; and mesoscale fronts and eddies (Mann and Lazier, 2005)—that are critical for maintaining the base state upon which climate variability acts. Similarly, one cannot hope to capture in a single chapter, a detailed discussion on the hierarchy of biological scales from individual organisms and populations of distinct species up through the interacting communities of different species that compose the core of an ecosystem. Nor, in many cases, do we have sufficient information to document climate–biological interactions across all biological scales, particularly at the community and ecosystem level. Rather, current understanding depends heavily on theory and laboratory experimental results at the organism level to help explain observed spatial patterns and historical trends in aggregated measures such as phytoplankton primary production or the abundance and spatial range of particular species. The emphasis in the chapter falls particularly on understanding the biological response to observed physical variability and trends over the twentieth and early twenty-first centuries as well as projecting the potential impacts on marine ecosystems due to further anthropogenic climate change over the next several decades to centuries.

Marine ecosystems are already experiencing large-scale trends in physical climate, ocean chemistry, and other human environmental perturbations (Figure 31.2) (Doney,

Ocean Circulation and Climate, Vol. 103. http://dx.doi.org/10.1016/B978-0-12-391851-2.00031-3

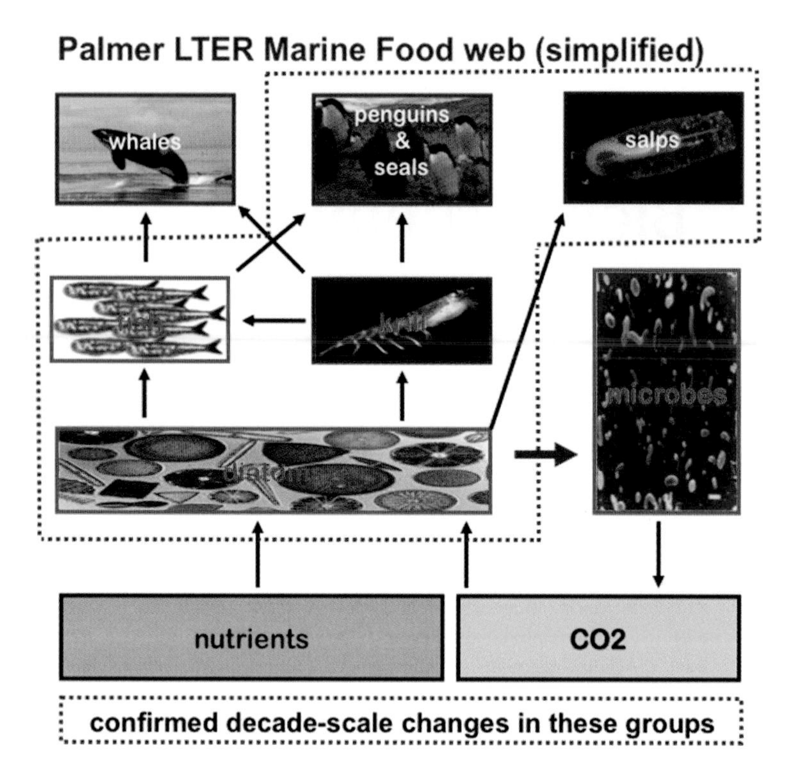

FIGURE 31.1 Food-web schematic for the coastal waters near Palmer Station on the West Antarctica Peninsula. Significant regime shifts are occurring in the marine ecosystem in this region linked to warming and sea-ice decline, and decade time-scale shifts in population levels are confirmed from field observations for the biological groups indicated by the red dashed box (Ducklow et al., 2007; Schofield et al., 2010). *Figure courtesy of Hugh Ducklow.*

2010; Gruber, 2011; Doney et al., 2012). Documented physical climate changes relevant to marine biota include rising sea-surface temperature (SST), upper-ocean warming, sea-level rise, altered precipitation patterns and river runoff rates, and sea-ice retreat in the Arctic and west Antarctic Peninsula (Figure 31.2) (Bindoff et al., 2007). Reduced stratospheric ozone over Antarctica appears to be causing a major shift in atmospheric pressure (more positive Southern Annular Mode conditions), which strengthens and displaces poleward the westerly winds in the Southern Ocean and which also may be increasing ocean vertical upwelling. Future climate projections indicate continuation and, in many cases, acceleration of these trends as well as other changes such as more intense Atlantic hurricanes (Bender et al., 2010), an ice-free summer in the Arctic (Stroeve et al., 2012), and a very likely reduction in the strength of the Atlantic deepwater formation (Bryan et al., 2006).

Relevant chemical trends include rising seawater CO_2 levels (leading to ocean acidification) (Gattuso and Hansson, 2011), reduced dissolved oxygen (O_2) concentrations reflecting warming and altered circulation (deoxygenation) (Keeling et al., 2010), and growing coastal nutrient levels leading to eutrophication, and expanding coastal and estuarine hypoxia (very low dissolved O_2) (Rabalais et al., 2010). These chemical trends are caused by the same global human pressures driving climate change, namely, fossil-fuel burning, deforestation, and industrial-scale agriculture (Le Quéré et al., 2009). Climate change also may exacerbate the ecosystem impacts of other human pressures and stressors such as coastal habitat loss, coastal urbanization, and overfishing, which have increased in magnitude dramatically over the past several decades (bottom panel Figure 31.2; Doney et al., 2012). Therefore, organisms (and ecosystems) will experience simultaneously multiple physical and chemical stressors that may exceed their capability to acclimate or adapt (Boyd et al., 2008).

The physical environment directly influences organism physiology through multiple pathways including temperature, salinity, O_2, CO_2, pH, etc. (Somero, 2012). Temperature variations and thermal stress are perhaps the most straightforward to understand because most marine plants and animals are ectothermic and cannot regulate their internal body temperature. Therefore, seawater temperature plays a central role modulating almost all biological rates, including growth and reproduction as well as microbial processes that dominate ocean biogeochemical cycling. Metabolic rates tend to rise exponentially with temperature up to some threshold temperature, above which thermal stress kicks in and biological rates drop sharply (Pörtner and Farrell, 2008). The exponential relationship with temperature can be captured by a Q_{10} value, that is, the rate increase resulting from a 10 °C rise in temperature. For example, Eppley (1972) reported a Q_{10} of ~1.9 for the upper envelope of growth rates among ~130 species and clones of phytoplankton; a 2 °C warming would, therefore, yield a 37% increase in growth rate.

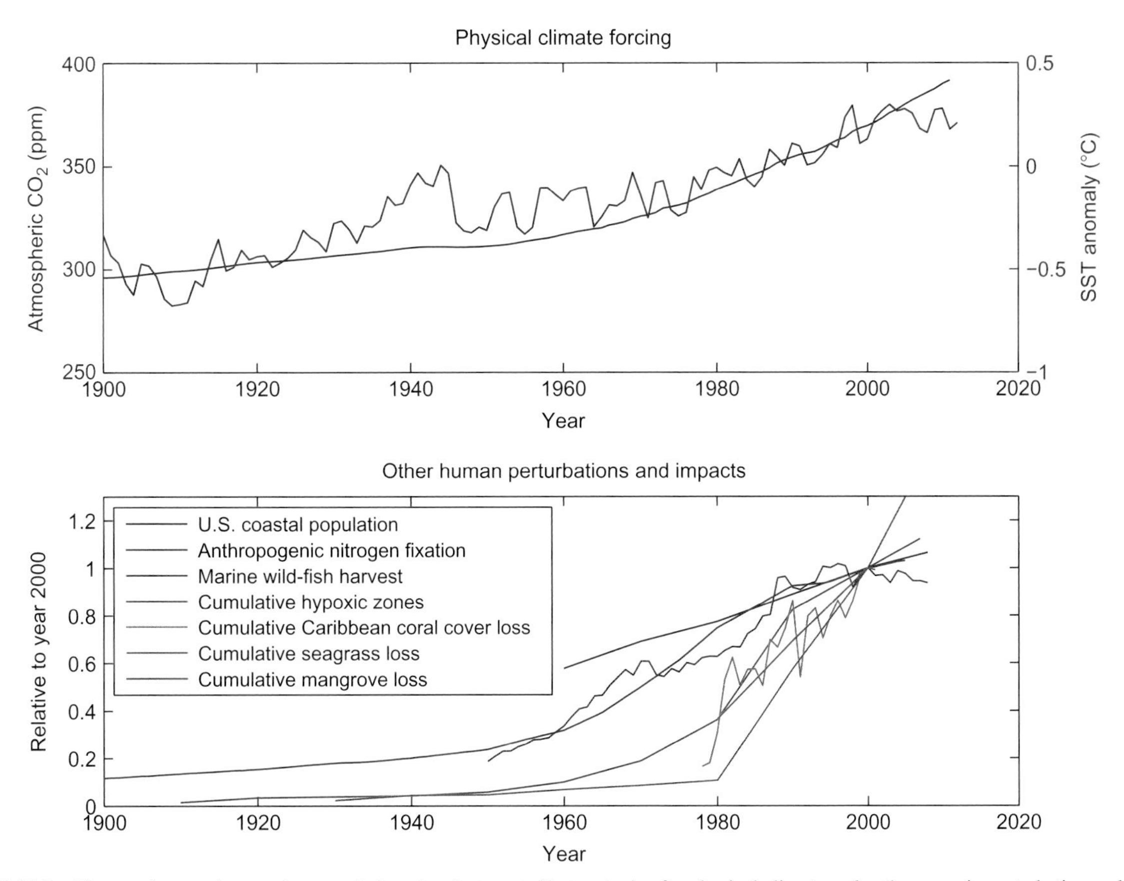

FIGURE 31.2 Time-series trends over the twentieth and early twenty-first centuries for physical climate and anthropogenic perturbations relevant to marine ecosystem dynamics. Top panel: annual-average atmospheric CO_2 from ice cores prior to 1959 (MacFarling Meure et al., 2006) and Mauna Loa instrumental record from 1959 to present (Tans and Keeling, 2012); and global-mean SST anomalies (ERSST data referenced to 1971–2000 climatology) (Smith et al., 2008). Lower panel: U.S. coastal population (Wilson and Fischetti, 2010), anthropogenic nitrogen fixation (Davidson, 2009), global marine wild-fish harvest (Food Agric. Org. U.N., 2010), cumulative global hypoxic zones (Diaz and Rosenberg, 2008), cumulative seagrass loss (Waycott et al., 2009), cumulative Caribbean coral cover loss (Gardner et al., 2003), cumulative mangrove loss (Food Agric. Org. U.N., 2007). All time series in lower panel are normalized to 2000 levels. *Adapted from Doney et al. (2012).*

On this basis, it might be expected that primary production, as well as the growth rates of ectothermic animals and pathogens, will increase in a warmer ocean. However, nutritional status, thermal tolerance, O_2 availability, environmental chemistry, food availability, or other factors may limit growth and production or other biological processes, regardless of metabolic rate. Further, most organisms inhabit a geographic range often bounded by upper and lower temperature limits, which may be further constrained by biological interactions with prey, competitors, predators, parasites, and diseases. As climate warms, species's geographic ranges may shift poleward to maintain a similar thermal niche, all other factors remaining favorable.

Over evolutionary time scales, organism life histories adapt to the physical climate and biological community in which the species population is embedded. Rapid environmental variability on short time scales can disrupt key biological relationships that underpin an organism's food supply or reproductive success. Many species exhibit seasonal variations in the timing or phenology of major life events such as reproduction. Changes in the spatial pattern, abundance or timing of prey blooms, for example, could result in dramatic indirect climate impacts on a predator. Climate variations and trends can create mismatches in time or space due to differential responses of species, potentially leading to cascading effects through a food web (Edwards and Richardson, 2004; Parmesan, 2006). For example, the seasonal match/mismatch in the timing of fish larval production to planktonic food supply has been suggested as an important factor driving year to year variability in fish recruitment (e.g., Cushing, 1990). This could translate into substantial and nonlinear biological responses to climate change from shifts in phytoplankton and zooplankton phenology (Stenseth and Mysterud, 2002). Organisms attempt to cope with such disruptions through physiological acclimation, behavior modifications, and eventually evolutionary adaption. The cumulative direct and indirect climate responses of individual organisms and species populations alter aggregated properties of an ecosystem such as primary production, energy and mass flow, community structure, and biodiversity.

The remainder of the chapter is organized as follows. Section 2 discusses the influence of physics and climate variability on phytoplankton distributions and primary production. This is followed by a survey of climate impacts on higher trophic levels, focusing primarily on thermal effects that are relatively more well documented in the literature compared to most other stressors (Section 3). Section 4 touches on the seawater chemistry changes associated with rising atmospheric CO_2 as well as the biological responses to the resulting ocean acidification. Section 5 highlights the effects of climate and nutrient eutrophication on ocean O_2 distributions. Section 6 talks about the coupling between marine biogeochemistry and global climate in terms of ocean CO_2 storage and O_2 distributions as well as climate-active trace gases. The chapter concludes with a brief discussion on future observational and research directions (Section 7). The chapter draws on and builds from several recent review articles on carbon cycle-climate coupling (Doney and Schimel, 2007), ocean acidification (Doney et al., 2009), ocean biogeochemistry (Doney, 2010), and climate change impacts on ocean ecosystems (Doney et al., 2012; Griffis and Howard, 2012).

2. PHYTOPLANKTON, PRIMARY PRODUCTION, AND CLIMATE

In the upper-ocean, small floating photosynthetic microbes and plants, collectively called autotrophic phytoplankton, use sunlight to convert inorganic CO_2 into organic matter and O_2 via the simplified net overall equation for photosynthesis:

$$CO_2 + H_2O \Rightarrow CH_2O + O_2 \quad (31.1)$$

where CH_2O is a generic carbohydrate. Associated metabolic processes, such as synthesis of proteins and enzymes, DNA and RNA, and lipids, also require bioavailable forms of nitrogen, phosphorus, and trace elements, most notably iron (Geider et al., 1997). Phytoplankton growth rates are governed "bottom-up" by temperature, light, and limiting macro and micronutrients. Typically, growth rates increase linearly with light or nutrients at low illumination and nutrient levels, eventually saturating at a temperature-dependent maximum growth rate. Diatoms also require silicon to build their shells, whereas coccolithophores need carbonate ions (CO_3^{2-}) to build calcium carbonate ($CaCO_3$) shells. The local time rate of change in phytoplankton biomass (P) depends on physical advection, mixing, sinking, and the net balance of biological growth and loss terms:

$$\frac{\partial P}{\partial t} + \nabla\cdot(\vec{u}\,P) - \nabla\cdot(K\nabla P) = \mathrm{RHS_{bio}} \quad (31.2)$$

where $\vec{u}$ is velocity and K is turbulent diffusivity. The biological right-hand-side terms, $\mathrm{RHS_{bio}}$, can be expressed as the net specific growth rate μ:

$$\frac{1}{P}\frac{\mathrm{d}P}{\mathrm{d}t} = \mu = \text{photosynthesis} - \text{grazing} - \text{other loss terms} \quad (31.3)$$

where "top-down" losses are dominated by zooplankton grazing as well as other, less well quantified, processes such as viral lysis, cell death, and phytoplankton aggregation that leads to gravitational sinking out of the well-lit upper ocean.

The stored chemical energy from phytoplankton primary production supports rich pelagic food webs in both the coastal and open-ocean, including deep sea and benthic ecosystems (Figure 31.3). Recent estimates for globally integrated marine net primary production are in the range of 60–70 PgC year^{-1} (where 1 Pg $= 10^{15}$ g) (Behrenfeld et al., 2005). Most of the organic carbon produced by phytoplankton is converted back to CO_2 in the upper ocean through respiration (the reverse of Equation 31.1). The primary loss mechanisms are via cycling through the heterotrophic bacterial loop or grazing by zooplankton. Transfer of organic carbon to higher trophic levels—that is fish, marine mammals, etc.—is inefficient, and marine biogeochemical mass and energy cycling are dominated by the activity of microbes and plankton. A small and relatively uncertain fraction of primary production is exported into the subsurface ocean, roughly 5–12 PgC year^{-1} (Dunne et al., 2007; Henson et al., 2011), where respiration releases CO_2 and nutrients and consumes O_2. Export flux is modulated by phytoplankton size structure with a greater fraction of production export from regions with larger cells and especially diatoms with siliceous shells. As a result of the net fixation of organic carbon in the euphotic zone and respiration in deeper waters, surface waters tend to have lower dissolved inorganic carbon (DIC) levels, whereas thermocline and deep waters are marked by higher DIC and nutrient concentrations and lower O_2, even after accounting for variations in temperature-dependent solubility (cold water holds more gases than warm water).

Relative to terrestrial systems, the elemental stoichiometry of marine plankton and sinking particles is relatively uniform, a fact first noted in a series of seminal papers (Redfield, 1958; Redfield et al., 1963) and codified in the so-called Redfield ratios relating the molar ratio of P:N:C:O_2 during net production and remineralization (1:16:117: −170) (Anderson and Sarmiento, 1994). Redfield ratios provide conversion factors for interrelating different types of ocean biogeochemical measurements of new production, net community production, and export flux that often are constructed from different elemental currencies. Although fixed Redfield ratios are a good guide, recent work indicates systematic spatial and temporal variations in plankton elemental composition in response to

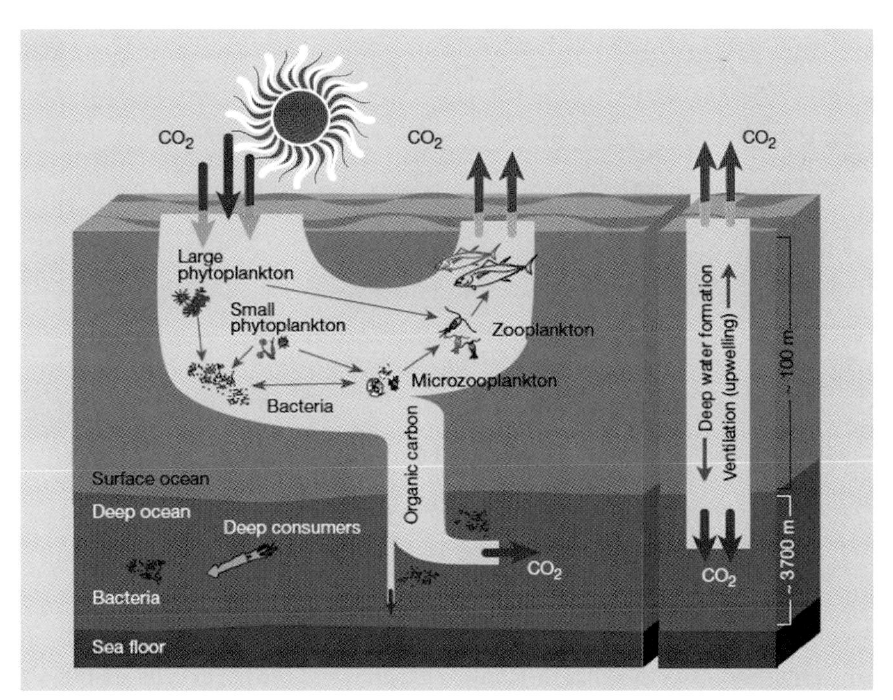

FIGURE 31.3 Schematic of the flow of organic carbon through a generic open–ocean pelagic food web, the so-called biological carbon pump that transports carbon from the surface ocean to the deep sea and increases natural ocean carbon storage (left panel). A schematic of the physiochemical "solubility" carbon pump driven by CO_2 solubility and ocean circulation is also shown (right panel). The thicknesses of the light blue bands in the left panel indicate that most of the organic carbon produced by phytoplankton primary production is respired in the upper ocean by bacteria, zooplankton, and, to a smaller degree, higher tropic levels. Export of organic carbon to the deep sea typically is a small fraction of primary production. *Figure from Chisholm (2000).*

variations in community structure, nutrient stress, and other biological factors (Geider and LaRoche, 2002; Deutsch and Weber, 2012; Martiny et al., 2013).

The large-scale patterns of phytoplankton biomass, mapped from satellite remote sensing in terms of concentration of the photosynthetic pigment chlorophyll (ocean color), broadly reflect the patterns of the wind-driven gyre circulation and coastal upwelling (McClain, 2009; Siegel et al., 2013; Figure 31.4). Export of sinking organic matter strips surface waters of vital nutrients over time, in effect-transporting nutrients diapycnally from light surface waters to dense thermocline and deep waters. Biomass levels, therefore, are modulated by physical processes that result in upward fluxes of nutrient-rich subsurface waters—large-scale upwelling, seasonal convection, and mesoscale eddies (McGillicuddy et al., 2003). Surface chlorophyll levels are low in nutrient-poor, subtropical gyres characterized by downwelling and deep thermoclines. In contrast, surface chlorophyll levels are more than an order of magnitude higher in nutrient-rich subpolar waters marked by large-scale upwelling of cold, nutrient-rich water and shallow thermoclines. Chlorophyll also can be considerably greater on continental shelves and upwelling eastern boundary current systems such as the California Current off the west coast of North America and the Benguela Current off the west coast of southern Africa. The ultimate source of thermocline nutrients and, thus, low-latitude productivity involves global-scale circulation; wind-driven upwelling in the Southern Ocean forms nutrient-rich mode and intermediate waters that then flow northward at mid-depth indirectly supplying the nutrients that feed productivity over much of the globe (Sarmiento et al., 2004a).

Some surface ocean regions have abundant macronutrients but relatively low chlorophyll, and phytoplankton growth appears limited by surface iron levels (Martin and Fitzwater, 1988; Boyd et al., 2007). Sources of iron to the surface ocean include atmospheric dust deposition, continental shelf sediments, and upwelling of recycled iron (Jickells et al., 2005). Inputs of bioavailable iron are relatively low in the subpolar North Pacific, eastern Equatorial Pacific, and the Southern Ocean, resulting in High-Nitrate, Low Chlorophyll (HNLC) conditions, under which iron limitation slows phytoplankton growth rates, particularly for larger cells, and reduces export fluxes.

Spatial and temporal variations in surface solar radiation and mixed layer depth also affect phytoplankton production. A little less than half of total solar irradiance falls in the bands classified as photosynthetically available radiation (PAR) (~400–700 nm), and PAR levels drop approximately exponentially with depth away from the ocean surface. Deeper mixed layers, therefore, reduce the average light level seen by the phytoplankton community over the mixed layer and, thus, lower the average photosynthesis rate. Large seasonal phytoplankton blooms occur

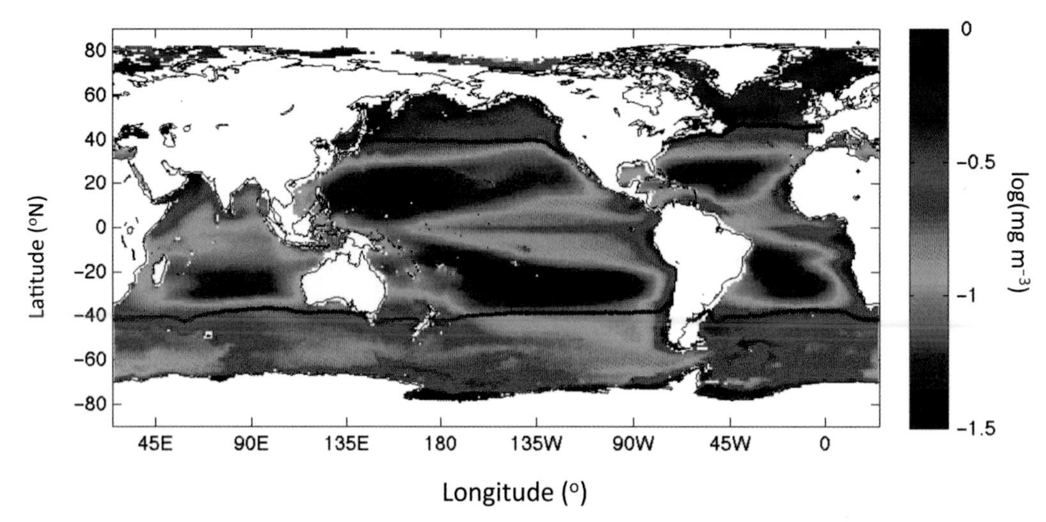

FIGURE 31.4 Satellite estimated ocean surface chlorophyll concentration from the Sea-viewing Wide Field-of-view Sensor (SeaWiFS) using the OC4v6 band-ratio chlorophyll algorithm. Mean values over the length of the SeaWiFS mission (1997–2010) are calculated for 1° bins in latitude and longitude over the global ocean. Units are $\log_{10}(\text{mg m}^{-3})$. The mean SST = 15 °C isotherm is shown as the black lines. *Figure from Siegel et al. (2013).*

in late-winter through spring and early summer in many temperate waters; the traditional explanation in terms of the Sverdrup's Critical Depth Hypothesis suggests that blooms are triggered by the relief of community light limitation due to increasing surface irradiance and shoaling mixed layers (Siegel et al., 2002). Alternatively, bloom initiation can be traced back to mixed layer deepening earlier in the season, which decouples the zooplankton–phytoplankton grazing relationship that normally keeps net phytoplankton growth μ near zero (Evans and Parslow, 1985; Behrenfeld, 2010). Reconciliation of these two hypotheses may lie in recognizing that community respiration varies over time and that grazing and light-driven variations in productivity more strongly influence bloom dynamics during different stages of the seasonal cycle.

Physical and biological factors influence plankton community composition as well as primary production. For example, low nutrients in the subtropical oligotrophic ocean favor species with smaller cells, prokaryotic and small eukaryotic picoplankton, of the order of $O(1)\,\mu$m in diameter, and nanoplankton of the order of $O(10)\,\mu$m in diameter. Cells adapt to severe phosphorus limitation by replacing phosphorus-based lipids in cell membranes with unique sulfur and nitrogen-based lipids (Van Mooy et al., 2009). Nitrogen-fixing diazotrophic organisms, which can create bioavailable nitrogen from otherwise inert N_2 gas, also arise in low-nitrate oligotrophic waters (Karl et al., 1997). Warm, well-stratified oligotrophic waters, typically, are characterized by a microbial food web with low biomass, rapid recycling of organic matter by small microzooplankton, long complex food chains, reduced export flux and elevated production of dissolved organic matter and bacterial activity. At the other extreme, nutrient-rich, productive, coastal, and polar waters typically contain both small and large phytoplankton cells and exhibit higher export rates. The larger cells are often dominated by bloom forming diatoms (10–200 μm in diameter) grazed by larger meso- and macro-zooplankton that, in turn, support relatively shorter, more direct food chains to higher trophic level predators such as fish.

Associated with El Niño-Southern Oscillation (ENSO) and other climate modes, marine phytoplankton and primary production exhibit substantial climate-driven interannual variability (Henson et al., 2009), estimated, on a global scale from satellite remote sensing, to be roughly ± 2PgC year^{-1} (Chavez et al., 2011), that is, a few percent of the total; regional variations can be substantially larger in a fractional sense. Satellite ocean color records for the past couple of decades indicate a robust anticorrelation of tropical and subtropical surface chlorophyll to SST, upper-ocean heat content, and thermocline depth (Figure 31.5; Behrenfeld et al., 2006). For example, one of the largest interannual signals in satellite ocean color and in situ chlorophyll involves a shift from low to high surface chlorophyll in the tropics and subtropics linked to the transition from warm El Niño to cold La Niña conditions in 1998–1999 in the tropical Pacific (Chavez et al., 1999; McClain, 2009). These findings are consistent with arguments that increased vertical stratification limits nutrient supply in stratified waters. However, recent work (Siegel et al., 2013) suggests a more subtle interpretation of the satellite ocean color record that most of the chlorophyll variability in the tropics and subtropics reflects physiological adjustments in intracellular chlorophyll concentrations rather than biomass variations; nutrient supply could still be the distal cause of the observed chlorophyll variations, but the signal is then more one of change in cell health rather than abundance. Ocean color variations in temperate and high latitudes appear to be driven more by changes in phytoplankton biomass; the northern hemisphere waters

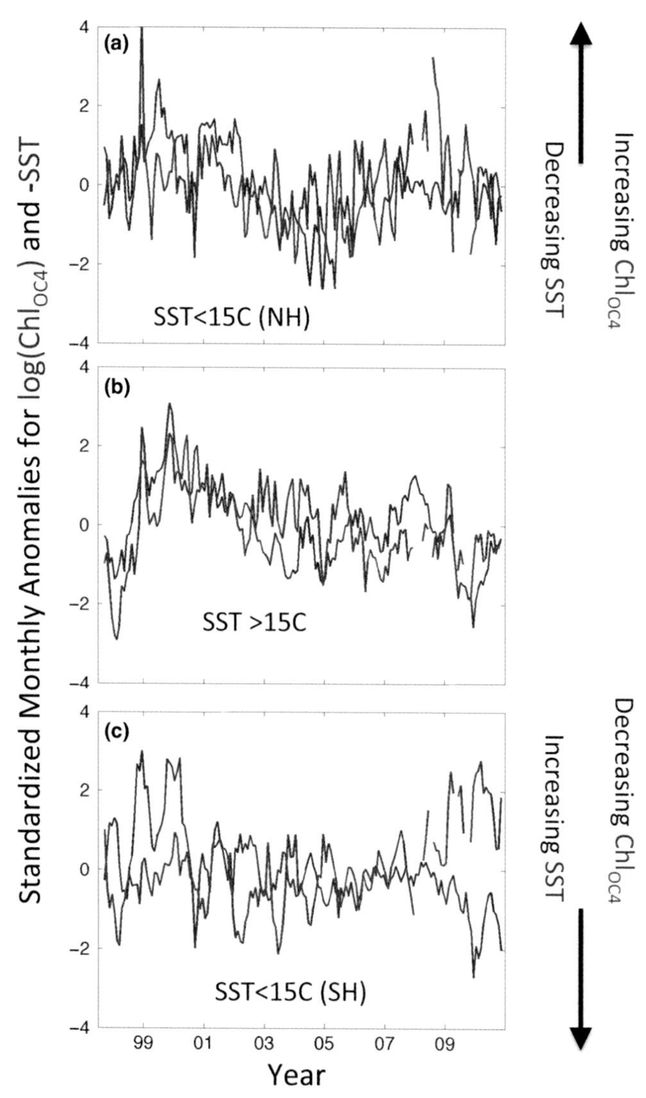

FIGURE 31.5 Time series of SeaWiFS surface ocean chlorophyll (Chl) (Figure 31.4) and sea-surface temperature (SST) monthly standardized anomalies (z-scores) for three global regions delineated by the mean SST isotherm for (a) the cool (mean SST < 15 °C) northern hemisphere (NH) aggregate, (b) the warm, permanently stratified ocean aggregate (mean SST > 15 °C) and (c) the cool southern hemisphere (SH) (mean SST < 15 °C). Anomalies are constructed by first removing the monthly mean value for each 1° bin of each property and then aggregating the regional, monthly anomalies into global aggregates. *Figure from Siegel et al. (2013).*

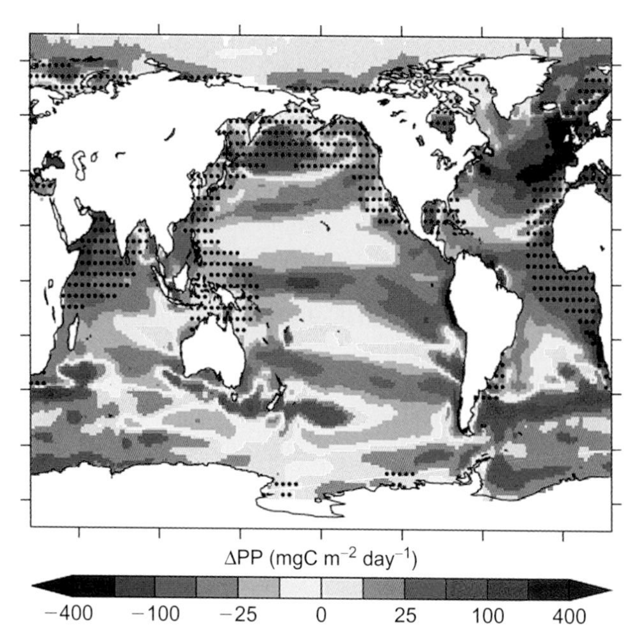

FIGURE 31.6 Projected climate-driven changes in vertically integrated, annual mean net primary production by the end of the twenty-first century (difference between 2090–2099 and 1860–1869 decadal means). Multi-model means are weighted by model skill under contemporary conditions, and dotted areas indicate regions where all of the models have low skill scores. *Figure from Steinacher et al. (2010).*

also exhibit an anticorrelation between chlorophyll and SST, with a less clear relationship in the southern hemisphere. Climate-driven variability in surface chlorophyll and primary production is also abundantly evident in longer, multidecade *in situ* time series such as BATS, HOT and more coastal stations (Chavez et al., 2011).

Discerning decadal and longer-term trends in phytoplankton and primary production from existing *in situ* and satellite observations is more difficult because of the short time series available and the presence of substantial natural interannual variability (Saba et al., 2010; Chavez et al., 2011). Modeling studies suggest some caution in the interpretation of historical data, arguing that it may require several decades or more of observations to clearly discern any anthropogenic signal (Henson et al., 2010; Beaulieu et al., 2013). Boyce et al. (2010) published a provocative result indicating an ~1% year^{-1} decline in global median chlorophyll over the past century based on ocean transparency and *in situ* chlorophyll data; this would indicate a wholesale change in ocean circulation and marine ecosystems. Other researchers, however, argue that the decline is an artifact of temporal sampling bias or merging of different data types (e.g., Rykaczewski and Dunne, 2011). Bridging satellite ocean color data between the Coastal Zone Color Scanner (CZCS; 1979–1986) and initial SeaWiFS data (1998–2002), Antoine et al. (2005) found a 22% increase in global average chlorophyll. Using SeaWiFS data, Polovina et al. (2008) observed a 15% increase over 1998–2006 in the spatial extent of the most oligotrophic surface waters ($\leq$0.07 mg Chl m^{-3}), but this time-slice includes the large 1997–1998 El Niño event and subsequent La Niña (Figure 31.5).

Anthropogenic climate change impacts on phytoplankton are expected to grow with time over the twenty-first century in response to further upper-ocean warming and increased vertical stratification. The resulting decline in nutrient supply into subtropical surface waters is projected to reduce primary production (Figure 31.6; Sarmiento et al., 2004b; Doney, 2006; Steinacher et al., 2010) and increase nitrogen fixation (Boyd and Doney,

2002). The situation is less clear in temperate and polar waters though there is a tendency in most models for increased production due to warming, reduced vertical mixing, and reduced sea-ice cover (Bopp et al., 2013). The spatial extent of biomes may expand or contract with the potential for the emergence of new regions with combined biotic and abiotic conditions that have not been observed before. For example, Polovina et al. (2011) forecast an ~30% expansion in the spatial extent of the North Pacific subtropical biome by 2100 due to stratification and a poleward shift of the mid-latitude westerly winds. Also, a novel thermal habitat is created in the area of very warm tropical and subtropical surface waters, mean annual SST exceeding 31 °C, growing in the simulation from a negligible amount to over 25 million km^2. Higher SSTs likely will cause poleward migration of phytoplankton thermal niches and may sharply reduce tropical phytoplankton diversity (Thomas et al., 2012). Warming also may cause the fraction of small phytoplankton (i.e., picophytoplankton) to increase, reducing the energy flow to higher trophic levels (Moran et al., 2010; Marinov et al., 2010).

Marine phytoplankton also can influence ocean physics and climate. In the upper ocean, the vertical attenuation of solar radiation depends strongly on the abundance of chlorophyll as well as biologically derived detritus and colored dissolved organic matter (Morel and Antoine, 1994). In more productive regions, solar radiation is absorbed closer to the surface, resulting in shallower mixed layers, warmer SSTs, and altered air–sea heat and freshwater fluxes, particularly in the tropics and subtropics (Ohlmann et al., 1996). Nonlocal and often nonintuitive effects can arise because altered upper-ocean stratification affects ocean currents and heat transport (Sweeney et al., 2005), and modeling studies indicate that ocean chlorophyll may substantially influence equatorial Pacific Ocean thermal structure (Murtugudde et al., 2002), modify ENSO interannual variability (Jochum et al., 2010), and steer tropical cyclones (Gnanadesikan et al., 2010).

3. CLIMATE IMPACTS ON HIGHER TROPHIC LEVELS

Climate-driven variations in primary production introduce bottom-up effects on ocean food webs that couple with direct impacts on higher trophic level organisms. These include physiological intolerance to changing physical and chemical environments, altered dispersal and migration patterns, and shifts in species interactions such as predation and competition. Population-level changes may occur in abundance, spatial range, and seasonal timing of major species's life history events or phenology. Populations of different organisms interacting through predator–prey, competition, and parasite–host relationships constitute a biological community. Together with local climate-driven invasion and local (and perhaps global-scale) extinction, climate processes may result in altered community structure and diversity, including possible emergence of novel ecosystems. Biological variability at a particular trophic level can be classified as controlled resources (bottom-up) or predation (top-down); in wasp-waist ecosystems, population variations in a crucial intermediate trophic level can generate both top-down influences on lower trophic levels and bottom-up influences on higher predators (Cury et al., 2000).

Ocean biological populations exhibit substantial interannual to decadal variability, and in many cases, a substantial component of the variability can be correlated with global-scale climate modes such as ENSO, the Pacific Decadal Oscillation (PDO), or the North Atlantic Oscillation (NAO) (Stenseth et al., 2003). Climate-related variability has been identified across a wide range of biological taxa from zooplankton (Brodeur and Ware, 1992; Mackas and Beaugrand, 2010) to commercial fish species such as North Pacific salmon and groundfish (Mantua et al., 1997; Hollowed et al., 2001).

Zooplankton cover a wide range of taxonomic groups and play a pivotal role in marine food webs because they feed directly on phytoplankton, bacteria, and often other smaller zooplankton. Zooplankton, in turn, serve as prey for many fish, marine mammals, and seabirds. Long records of zooplankton abundance and community composition are available for the eastern North Atlantic from the Continuous Plankton Recorder (CPR) survey; the CPR data exhibit climate-related trends and variability in, for example, calanoid copepod (zooplankton crustaceans) biodiversity and biogeographic distributions (Beaugrand et al., 2002) associated with warming and shifts in the NAO.

At subtropical open-ocean time series sites off Hawaii and Bermuda, mesozooplankton (>200 μm) dry-weight biomass is increasing with time and is positively correlated with SST at Bermuda (Steinberg et al., 2012). These sites exhibit positive or neutral primary productivity trends (Saba et al., 2010), despite surface warming that is expected to decrease both primary and secondary production. Possible explanations for the zooplankton trends include a combination of transient responses, increased nitrogen fixation supporting great primary production, alteration in top-down controls, and northward expansion of tropical species' ranges.

Other interesting zooplankton–climate relationships arise from the California Cooperative Oceanic Fisheries Investigations (CALCOFI) data in the California Current System. The CALCOFI data exhibit, among other features, increases in cold-water krill species on interannual scales in association with strong coastal upwelling and La Nina events as well as decadal variations in the abundance of warm water, coastal krill reflecting changes in the strength of poleward transport (Brinton and Townsend, 2003;

Bestelmeyer et al., 2011). The CALCOFI data also document long-term warming and declining zooplankton biomass trends (Roemmich and McGowan, 1995), primarily reflecting decreasing abundance of pelagic tunicates (Lavaniegos and Ohman, 2003), possibly because warm-water taxa (small subtropical copepods) were favored by a shift toward greater relative abundance of small prey (picoplankton). Climate-driven fluctuations in individual zooplankton populations in the California Current system also shift planktonic food-web structure and dynamics (Francis et al., 2012).

Boundary current systems exhibit striking decadal-scale oscillations in the abundance of small pelagic fish (Kawasaki, 1992). In the Pacific, oscillations between anchovies and sardines are approximately synchronous in the California Current and Peru/Chile (Humboldt Current) upwelling systems as well as in the Kuroshio, with anchovies dominating during cold, negative phases of the PDO, replaced by sardines during warm phases (Chavez et al., 2003). The large-scale synchrony must reflect atmospheric teleconnections, though the exact mechanisms are not fully resolved (Alheit and Bakun, 2010). Rykaczewski and Checkley (2008), for example, suggest that the abundance of Pacific sardines covaries with wind-stress curl-driven upwelling that has relatively low vertical velocity compared to near-shore coastal upwelling caused by along-shore winds; they hypothesize that sardines are favored because of the smaller-sized plankton assemblage under those conditions.

Marine ecosystem structure as a whole also appears to vary at decadal time scales with some of the best explored examples involving the response of the North Pacific ecosystem to the PDO. Hare and Mantua (2000) and others argue that the system undergoes periodic regime shifts, that is, relatively abrupt, synchronous transitions from one quasi-stable state to another on decadal time scales in conjunction with changes between cold and warm phases of the PDO. Ocean dynamics acts to integrate high-frequency atmospheric weather resulting in low-frequency (red-spectrum) variance of ocean properties that could appear at first glance like a regime shift, but which, in actuality, is simply a linear response to external forcing (Doney and Sailley, 2013). Thus, it is often difficult to determine whether decadal ecological trends simply track external physical forcing or reflect more complicated, nonlinear responses that may arise from internal biological interactions (Overland et al., 2010). Careful analysis suggests that, at least to some degree, the North Pacific regime shift is real and reflects a combination of forced linear responses and internal nonlinear biological dynamics (Hsieh et al., 2005). Nonlinear regime shifts have been detected in several other marine population time series suggesting the potential for more climate-driven ecological surprises (Bestelmeyer et al., 2011).

Turning to anthropogenic climate change, polar marine ecosystems appear likely to be particularly sensitive to climate-driven sea-ice variations, which can substantially restructure food-web pathways from plankton through to higher trophic levels (Ducklow et al., 2007; Grebmeier, 2012). For example, along the western side of the Antarctic Peninsula seasonal sea-ice duration has declined by nearly 90 days since the beginning of satellite-based measurements in 1978, reflecting both warming of the ocean-atmosphere system and strengthening of wind patterns that drive sea ice off-shore (Schofield et al., 2010). The reduced sea-ice duration lengthens the growing season for water-column phytoplankton, which are shaded from sunlight by sea-ice cover for much of the year. Further, sea-ice melting during the spring releases freshwater, acting to cap the water column, shallow the ocean mixed layer depth, and promote surface plankton blooms by trapping plankton in a well-lit surface layer. Time series satellite ocean color records for the northern regions of the west Antarctic Peninsula indicate that phytoplankton stocks have declined by over 80% because there is less sea ice to stabilize the upper water column and because of increased wind-driven turbulence (Montes-Hugo et al., 2009). In contrast, phytoplankton blooms have increased in the south in the present-day seasonal marginal ice zone because more light is penetrating the ocean as the sea ice declines.

The decline of sea ice and warmer, more maritime weather conditions may be the cause of the major ecosystem regime shifts now being observed around the Antarctic Peninsula (Figure 31.1). Impacts can be especially strong for organisms adapted to polar conditions that require the presence of sea ice for part of their life history. Ice-dependent species include krill, relatively large planktonic crustaceans that are quite abundant in Antarctic waters and a critical prey for seabirds and marine mammals. Small juvenile krill over-winter under the ice pack, hiding from predators and grazing on ice algae. Krill have declined by an order of magnitude in the Atlantic sector of the Southern Ocean since 1950 with a corresponding increase in the abundance of salps, a gelatinous tunicate filter feeder (Atkinson et al., 2004). Salps are not a very palatable food source for most marine predators, and therefore, a switch from krill to salps could lead to important consequences for higher trophic levels.

Sea ice also provides an important habitat for many polar seabirds and mammals (e.g., penguins, polar bears, walruses, seals) that use the ice as a foraging platform or breeding habitat, suggesting that these species will face problems with warming. Under warming conditions, the spatial range of the polar species will likely contract toward the pole, whereas for subpolar species adapted to warmer conditions, this may allow migration into newly available habitats. Temporal mismatch between prey availability and predator food demands, which are often constrained

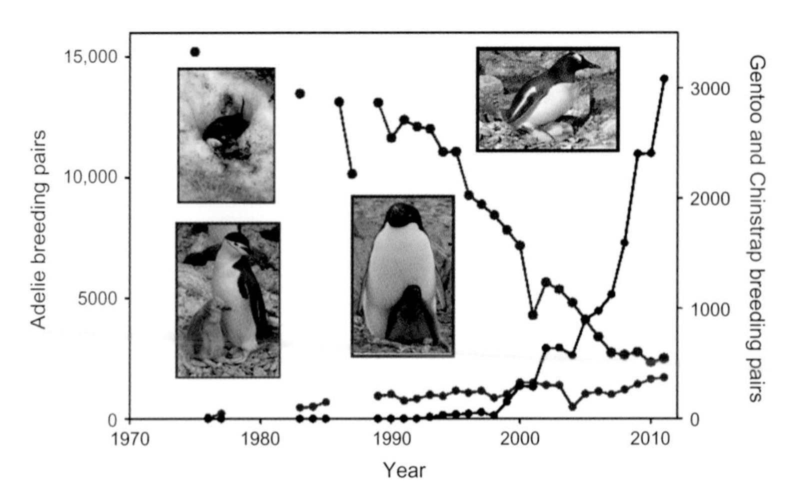

FIGURE 31.7 Time series of penguin populations in the local area around Palmer Station, Antarctica. The population is estimated from the number of breeding pairs measured during the summer breeding season. Three species are shown, polar ice-dependent Adélie (red; left-hand axis) and subpolar Chinstrap (blue) and Gentoo (black) (both on right-hand axis). Over time, in response to sea-ice decline and increased snow, the polar Adélie population has declined by about 80%, replaced by an influx from the north of Chinstrap and Gentoo penguins that are ice intolerant. *Data and images courtesy of William Fraser and Hugh Ducklow.*

by aspects of their life history, can be especially problematic for polar and migratory seabirds that depend upon an ample and reliable food supply of krill and small fish during the limited summer breeding season. Along the Antarctic Peninsula, the population of Adélie penguins has declined by 80% in the Palmer Station region because of cascading responses to sea-ice loss, reduced food availability, and elevated late-spring snowfalls (Figure 31.7; Schofield et al., 2010). Similar regional declines have been observed in crab-eater seals, another ice-dependent species. Conversely, in a case of opening niche space and species expansion, ice-intolerant gentoo and chinstrap penguins, as well as southern fur seals, perhaps, are now migrating into the region and establishing new breeding colonies. In another example of a climate-related invasive event with broad ecological impacts, warming of bottom waters along the continental slope and shelf appears to be allowing the recent and ongoing colonization of the Antarctic Peninsula by king crab, shell-crushing predators that have been absent from the Antarctic food web for about 25 million years (Fox, 2012; Smith et al., 2012).

In the Arctic and adjacent marginal seas, emerging biological–climate signals include: marine species range shifts; changes in abundance, growth, condition, behavior, and phenology of some species; and community and regime shifts (Wassmann et al., 2011). An unusually large fraction of primary production in the seasonally ice-covered northern Bering shelf ecosystem sinks to the seafloor and supports a large and diverse benthic community; pelagic fish predation is limited by cold-water temperatures and ice cover, allowing diving seabirds, bearded seals, walrus, and gray whales to harvest the high benthic production (Grebmeier et al., 2006). Warming and variability in sea-ice retreat coincide with declines in clam populations, which in turn co-occur with dramatic declines in diving sea ducks, more northerly migrations of large vertebrate predators (walrus and gray whales), and potentially poleward-expanding ranges for pelagic fishes (Grebmeier et al., 2010).

In temperate oceans, poleward range shifts are evident for many fish species based on long time series from commercial fish stock surveys (Perry et al., 2005). Nye et al. (2009), for example, analyzed temporal records of fish species' distributions in the Northeast United States continental shelf ecosystem. The southern stocks tended to move northward in time in response to warming, appearing to maintain an approximately constant thermal habitat (Figure 31.8). In the more geographically restricted Gulf of Maine, northern stock data indicated no significant latitudinal migration but rather movement toward deeper and colder waters.

The concept that species' ranges are bounded within a fixed range of environmental properties is the basis for bioclimate envelope models that can be used to project fish species responses to climate change including local extinction and species invasion (Cheung et al., 2009). Habitat suitability, defined using present-day biogeographic distributions and environmental conditions, is folded into dynamic population models that capture growth, larval dispersal, and adult migration. Cheung et al. (2009) forecast substantial reorganization of global fishery catch potential with large declines in the tropics and increases in high-latitude regions, overlain with large regional variations. Refinements to this class of models and other methods are being used to examine likely future changes in fish body size with warming, in interactions with ocean acidification and declining O_2, in interactions of climate effects with overfishing, and in impacts on the economics of fisheries (Brander, 2007, 2010; Sumaila et al., 2011).

However, shifts in species' geographic ranges depend on much more than simply temperature, and good forecasting likely will require an understanding of the life history of particular species. For example, while warming may open up new habitat in polar Arctic regions for cod, herring, and pollock (Loeng et al., 2005), the continued presence of cold bottom-water temperatures on the shelf could limit northward migration into the northern Bering

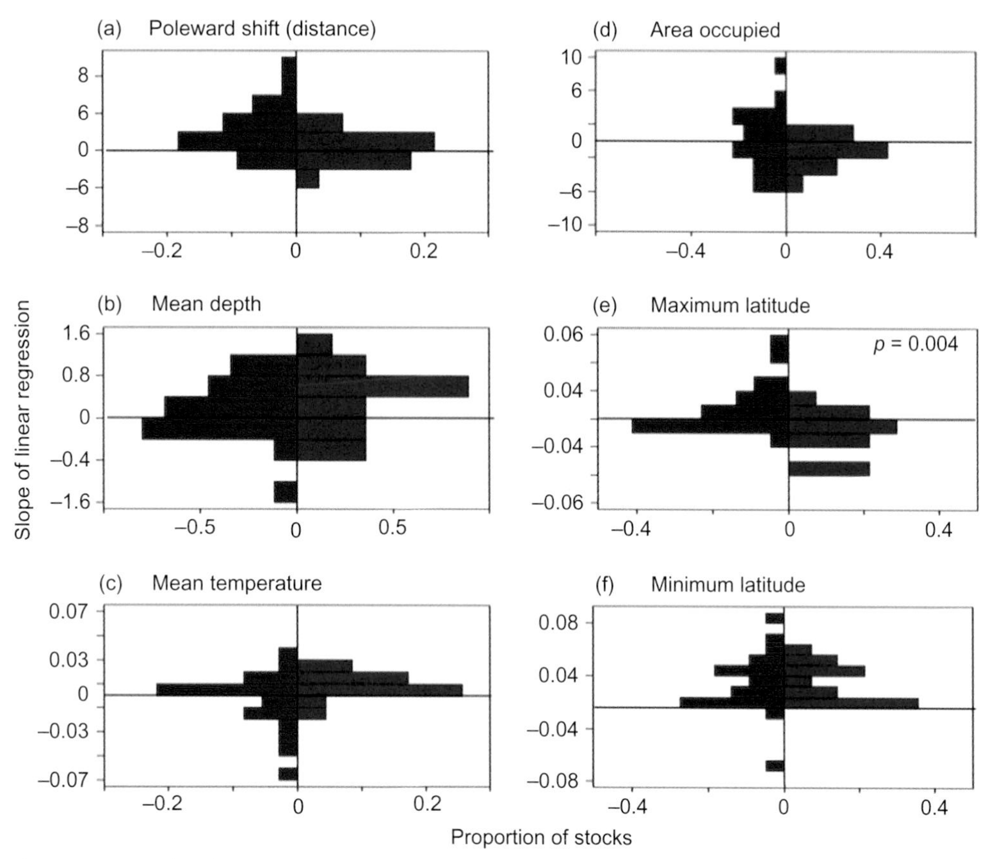

FIGURE 31.8 Histograms comparing distributional responses for southern (red) and northern (blue) stocks in (a) distance shifted poleward, (b) mean depth, (c) mean temperature, (d) area occupied, (e) maximum latitude, and (f) minimum latitude. Significant differences detected with a Mann–Whitney *U*-test between species found in the two ecoregions are indicated by *p* values. *From Nye et al. (2009).*

Sea and Chukchi Sea (Sigler et al., 2011). In addition, warming may cause reductions in the abundances of some species, such as pollock, over their current ranges in the Bering Sea (Mueter et al., 2011), linked to changes in overall food-web dynamics and bottom-up food resources, not just direct thermal effects (Hunt et al., 2011). Climate change information needs to be incorporated into management strategies for specific fisheries (Ianelli et al., 2011) and assessments of the vulnerabilities of keystone, sentinel, iconic, and endangered marine species to local population collapse. Wolf et al. (2010) illustrate such an approach integrating climate trends into a demographic model for Cassin's auklet, a seabird that feeds on plankton and a sentinel species in the California Current System. More broadly, climate model projections are increasingly being used in a variety of ways to evaluate climate impacts on living marine resources (Stock et al., 2011).

Climate variations and climate change also influence the spread and impact of marine diseases and parasites (Harvell et al., 2002). Marine disease appears to be on the rise with time, and higher SSTs have been linked with higher intensity and increased spatial ranges of diseases that attack corals, abalones, oysters, fishes, and marine mammals (Ward and Lafferty, 2004). Climate warming acts through several different pathogen-specific mechanisms. Warming can increase pathogen over-wintering survival, tied to the northward spread of Dermo disease, an oyster parasite, up the U.S. east coast (Cook et al., 1998) and the growth of coral disease lesions (Weil et al., 2009). Higher seasonal temperatures may cause an expansion of *Vibrio* species, pathogenic bacteria that infect oysters, and may cause human illness (Baker-Austin et al., 2013). Warming can also increase pathogen susceptibility by intensifying thermal stress because of the elevated size and duration of positive temperature anomalies. Record warm tropical SSTs have caused widespread coral disease outbreaks (Miller et al., 2009) and coral bleaching (Eakin et al., 2010). Bleaching occurs in response to environmental stress when the naturally colorless coral polyps expel their zooxanthellae, the colored symbiotic dinoflagellates whose photosynthesis fuels the growth of their coral hosts (Figure 31.9).

In fact, coral reefs are some of the most susceptible ecosystems to climate warming because of the sensitivity of coral-algal symbiosis to minor increases in maximum seasonal temperature; warming of as little as 1 °C can cause coral bleaching (Hoegh-Guldberg et al., 2007; Donner,

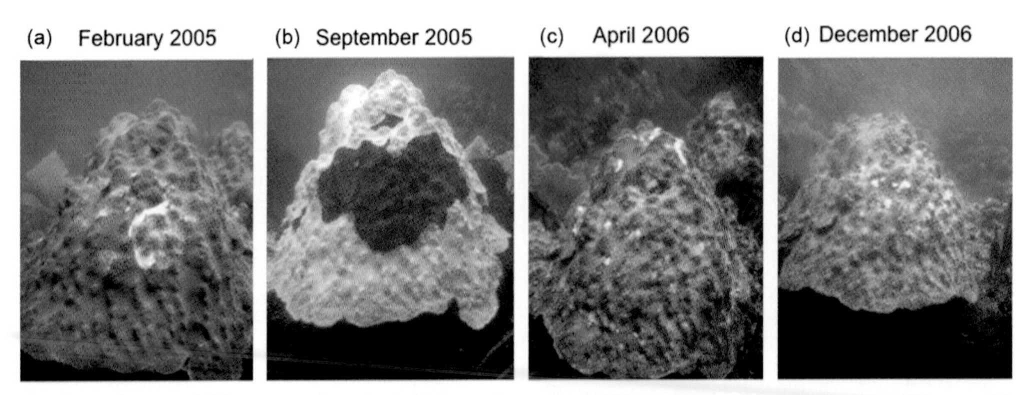

FIGURE 31.9 A colony of star coral (*Montastraea faveolata*) off the southwest coast of Puerto Rico, estimated to be about 500 years old, exemplifies the effect of rising water temperatures. Increasing diseases due to warming waters (a) were followed by such high temperatures that bleaching or loss of symbiotic microalgae from coral occurred (b), followed by more disease (c) that finally killed the colony (d). *Figure courtesy of Ernesto Weil.*

2009). Warm-tolerant zooxanthellae may become more predominant on reefs in the future, allowing some corals to survive moderate temperature increases, but this may have negative impacts on other aspects of coral health such as growth (Jones and Berkelmans 2010). Reefs are also threatened by a variety of local human pressures including pollution, overfishing, and dredging, and the degradation of reefs affects not only corals themselves but also the rich biodiversity of other organisms living within the structural complexity of the coral reef seascape. Coral reefs are important for human societies, often supporting locally essential artisanal reef fisheries, and reefs are some of the most valuable marine ecosystems because of tourism and recreation income and coastal protection (Cooley et al., 2009). Ocean acidification due to rising atmospheric CO_2 is an additional threat to corals and other important reef calcifying organisms such as crustose coralline algae that build reef frameworks (Anthony et al., 2008).

4. OCEAN ACIDIFICATION

Climate change is not the only important effect of rising atmospheric CO_2, which also causes direct changes in seawater acid–base and inorganic carbon chemistry, termed ocean acidification that can impact marine organisms and ecosystems. Compared to most freshwater systems, the acid–base chemistry of seawater is relatively stable because the inorganic carbon system and large alkalinity levels buffer seawater pH. Many marine organisms appear to be adapted to relatively constant local acid–base conditions and are sensitive to relatively small variations in pH and the concentrations of various inorganic carbon species (Doney et al., 2009; Gattuso and Hansson, 2011). The ocean uptake of anthropogenic CO_2 is causing global-scale shifts in upper-ocean chemistry that are rapid compared to variations in the geological past; for example, the surface ocean pH change caused by the $\sim$30% rise in atmospheric CO_2 associated with the last deglaciation was roughly two orders of magnitude slower than the current rate driven largely by fossil-fuel burning (Hönisch et al., 2012). The chemistry of ocean acidification is relatively well understood (Feely et al., 2009); biological implications are slowly becoming clearer at the level of individual species, but substantial uncertainties remain particularly at the ecosystem level (Gattuso et al., 2011).

CO_2 acts as a weak acid when added to water at seawater pH levels:

$$CO_2 + H_2O \Leftrightarrow H^+ + HCO_3^- \qquad (31.4)$$

The forward reaction releases hydrogen (H^+) and bicarbonate (HCO_3^-) ions and lowers pH, defined as $pH = -\log_{10}\{H^+\}$. Most of the extra H^+ ions react with and lower CO_3^{2-} concentrations:

$$H^+ + CO_3^{2-} \Leftrightarrow HCO_3^- \qquad (31.5)$$

CO_2 input also increases aqueous $CO_{2(aq)}$ and DIC, $[DIC] = [CO_{2(aq)}] + [HCO_3^-] + [CO_3^{2-}]$. Another important reaction is the dissolution or precipitation of solid $CaCO_3$ used by many marine plants and animals to form shells and hard body parts:

$$CaCO_3(s) \Leftrightarrow Ca^{2+} + CO_3^{2-} \qquad (31.6)$$

$CaCO_3$ becomes more soluble as CO_2 rises and CO_3^{2-} declines, represented mathematically by a lowering of the $CaCO_3$(s) saturation state, $\Omega = [Ca^{2+}][CO_3^{2-}]/K_{sp}$, where K_{sp} is the thermodynamic solubility product that varies with temperature, pressure, and mineral form. Present-day ocean surface waters are currently supersaturated ($\Omega > 1$) for the two major forms used by marine organisms, the more soluble form, aragonite (corals, many mollusks), and the less soluble form, calcite (coccolithophores, foraminifera, and some mollusks). Other relatively soluble forms include amorphous $CaCO_3$ and $CaCO_3$ containing various amounts of magnesium.

Natural physical and biological processes influence seawater acid–base chemistry, leading to large-scale spatial gradients and seasonal variability in pH, Ω, and inorganic carbon speciation (Feely et al., 2009). High-frequency temporal variability on diurnal to weekly timescales is also observed in coastal, estuarine, and coral reef systems (Hofmann et al., 2011). In general, surface waters tend to have lower CO_2 and DIC and, therefore, slightly higher pH because of phytoplankton uptake of inorganic carbon as part of photosynthesis. The opposite pattern and low O_2 values are found in the thermocline because of the respiration of sinking organic matter and downward transported dissolved organic matter. $CaCO_3$ saturation state decreases with depth because of organic matter respiration and pressure effects on $CaCO_3$ solubility. Aragonite and calcite often become undersaturated ($\Omega < 1$) below some depth in the water column, at which point, unprotected shells and skeletons begin to dissolve; the saturation depth horizon is particularly shallow in the Pacific that has a high burden of metabolic CO_2 (Feely et al., 2009). Because of increased CO_2 solubility and temperature effects on the thermodynamic equations, cold polar surface waters exhibit lower CO_3^{2-} ion concentrations and Ω values.

Long-term trends in pH and inorganic carbon chemistry are clearly evident over the past several decades in ocean time series and hydrographic surveys (Dore et al., 2009; Byrne et al., 2010; Bates et al., 2012). From preindustrial levels, contemporary surface ocean pH is estimated to have dropped on average by about 0.1 pH units (a 26% increase in $[H^+]$), and further decreases of 0.2 and 0.3 pH units will occur over this century unless anthropogenic CO_2 emissions are curtailed dramatically (Orr et al., 2005). Surface ocean $CaCO_3$ saturation state is declining everywhere, and model simulations indicate that polar surface waters will become undersaturated for aragonite when atmospheric CO_2 reaches 400–450 ppm for the Arctic and 550–600 ppm for the Antarctic (Orr et al., 2005; Steinacher et al., 2009). Because of the larger natural background CO_2 levels, subsurface waters have a lower buffer capacity and exhibit a larger pH drop per amount of CO_2 added; this increases the susceptibility to acidification of O_2 minimum zones (Brewer and Peltzer, 2009), coastal waters that are already experiencing nutrient eutrophication and hypoxia (Feely et al., 2010; Cai et al., 2011), and eastern boundary current upwelling systems (Feely et al., 2008; Gruber et al., 2012).

Numerous biological effects have been measured in response to ocean acidification for both pelagic (Riebesell and Tortell, 2011) and benthic (Andersson et al., 2011) organisms (see also Fabry et al., 2008; Doney et al., 2009). Most biological impacts have been inferred from short-term manipulation experiments at the organism level to step-increases in CO_2, for example, lower calcification rates in corals and mollusks, higher photosynthesis rates for seagrasses and some phytoplankton groups, and increased nitrogen fixation by cyanobacteria. For example, Figure 31.10 shows a summary plot for tropical corals of the relative decrease in calcification rate as a function of declining $CaCO_3$ saturation state.

Organism responses may vary with life history stage, with juveniles often more susceptible than adults, and some organisms may be able to accommodate elevated CO_2 but at an additional energetic cost with consequences for development, reproduction, and fitness. Different organism groups may be sensitive to different aspects of seawater chemical

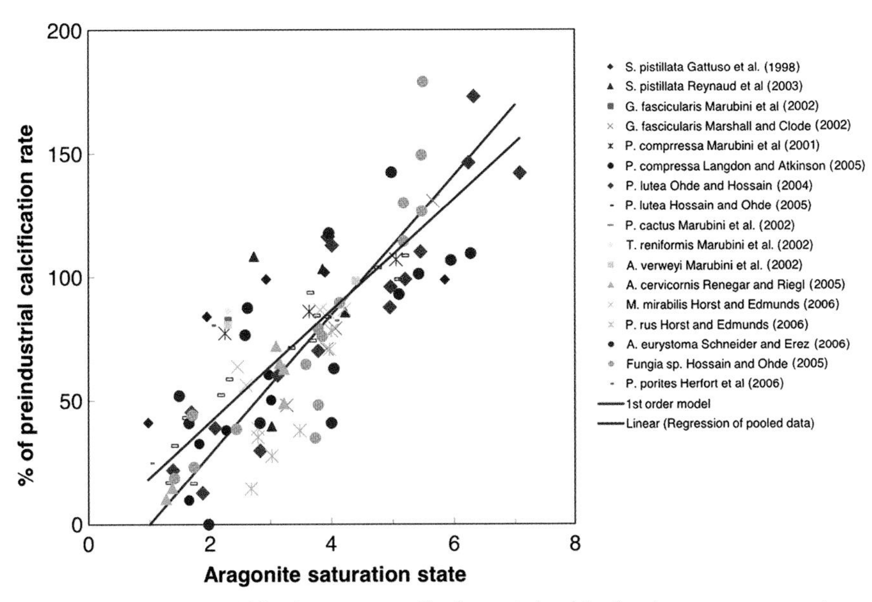

FIGURE 31.10 Summary plot of variations in calcification rate, normalized to preindustrial values in percent, to aragonite saturation state for tropical corals. *Figure courtesy of Chris Langdon.*

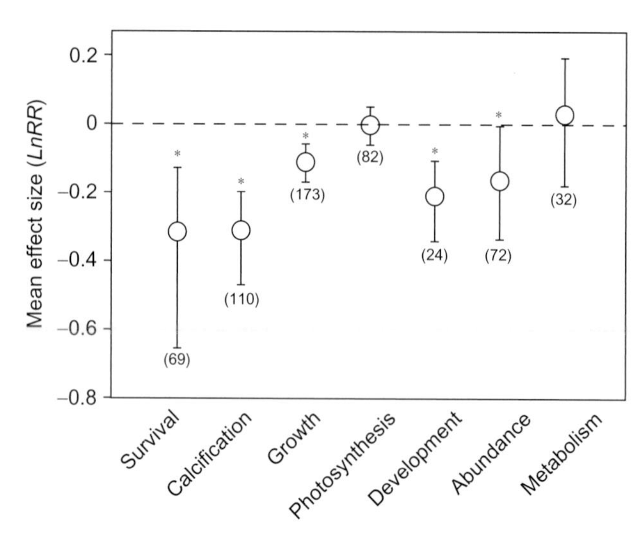

FIGURE 31.11 Summary results for a meta-analysis of biological impacts of ocean acidification reported in the literature. Key physiological responses are aggregated across all taxa. The effect size is the ratio of the mean effect in the acidification treatment to the mean effect in a control group and is scaled to ~0.5 unit reduction in pH. Error bars represent 95% confidence intervals. Asterisks denote statistically significant effects, and the number of studies is shown in parentheses. *Figure from Kroeker et al. (2013).*

trends: for calcifiers declining CO_3^{2-} levels; for autotrophs, increased aqueous CO_2; and for adult fish and cephalopods, acid–base regulation and CO_2/O_2 transport and gas exchange. In a recent meta-analysis of available literature studies, Kroeker et al. (2010, 2013) show that statistically significant declines are observed for survival, calcification, growth, development, and abundance, with substantial variations across taxonomic groups (Figure 31.11).

Effects on natural populations and communities so far have been more difficult to detect outside of a limited number of pelagic mesocosm experiments and some studies in isolated high-CO_2 environments such as shallow volcanic vents that tend to support laboratory findings (Hall-Spencer et al., 2008; Fabricius et al., 2011). Shellfish hatcheries along the Oregon–Washington coast have experienced dramatic declines in oyster harvest (Barton et al., 2012) in response to the upwelling onto the shelf of strongly acidified waters with low $CaCO_3$ saturation state (Feely et al., 2008). In general though, the nature and magnitude of the responses within natural populations and the ability of organisms to acclimate or adapt to gradual CO_2 trends are still mostly unknown.

5. DEOXYGENATION AND HYPOXIA

Marine biota can also be influenced by ocean oxygen distributions, particularly at low O_2 values in oxygen minimum zones and hypoxic coastal environments (Levin et al., 2009; Keeling et al., 2010). Dissolved O_2 gas is required for aerobic respiration, and below certain organism-specific thresholds, low O_2 begins to affect metabolic rates and behavior. Low O_2 leads to marine habitat degradation and, in extreme cases, extensive fish and invertebrate mortality, and larger mobile animals often move out of low oxygen environments, resulting in so-called dead-zones where many macrofauna are nearly absent (Diaz and Rosenberg, 2008; Rabalais et al., 2010). Thresholds for hypoxia vary by organism but are typically ~60 μmol kg^{-1} or about 30% of surface saturation. Under suboxic conditions (<5 μmol kg^{-1}), microbes begin to utilize nitrate (NO_3^-) rather than O_2 as the terminal electron acceptor for organic matter respiration (leading to denitrification), resulting in reactive nitrogen loss and N_2O production. By simultaneously removing O_2 and adding CO_2, organic matter respiration can induce multiple stressors on organism physiology, and O_2 stress can also reduce organism thermal tolerances (Pörtner and Farrell, 2008; Pörtner et al., 2011).

Hypoxic conditions occur naturally in open-ocean and coastal subsurface waters from a combination of weak ventilation, warming, and organic matter degradation—features that are all exacerbated by climate warming and coastal nutrient eutrophication. Coastal hypoxic systems are widespread globally with more than 400 instances covering an area >245,000 km^2 (Diaz and Rosenberg, 2008). An expansion in the duration, intensity, and extent of coastal hypoxia over the past several decades is attributed to growing coastal urbanization, land runoff of excess nutrients from fertilizers and sewage, and atmospheric nitrogen deposition from fossil-fuel combustion. About half the global riverine nitrogen input is anthropogenic in origin (Seitzinger et al., 2010), and coastal nutrient eutrophication is also associated with increased frequency of harmful algal blooms (Anderson et al., 2002). The emergence of hypoxia in other coastal regions may be related to variations or trends in ocean-atmospheric physics. Increased wind-driven upwelling is linked to the first appearance of hypoxia and even anoxia on the inner-shelf off Oregon–Washington coast after five decades of hypoxia-free observations (Chan et al., 2008).

Oxygen minimum zones occur naturally in the open-ocean in the tropics and subtropics, with the lowest O_2 suboxic waters restricted to the Arabian Sea, the Bay of Bengal, and the eastern tropical Pacific (Paulmier and Ruiz-Pino, 2009; Bianchi et al., 2012). In most subtropical thermocline regions, oxygen levels tend to be relatively high because the anticyclonic gyre circulation transports oxygen-rich water relatively rapidly and directly from surface isopycnal outcrops into the interior. In the tropics, low oxygen regions typically occur equatorward of these directly ventilated subtropical waters in regions more cut off from the atmosphere where ventilation occurs more indirectly through diffusive processes, so-called shadow zones of the wind-driven circulation. Oxygen minimum zones often underlie biologically productive regions such as the eastern boundary current upwelling systems (e.g., Peru, Benguela). The lowest water-column O_2 values are observed in the upper and mid-thermocline where biological oxygen consumption rates are high. A number of studies indicate that subsurface oxygen values are declining with time in the midlatitudes (Whitney et al., 2007; Helm et al., 2011) and that oxygen minimum zones are expanding spatially in both the vertical and horizontal directions (Stramma et al., 2008), which could limit the habitat range for some fish species and alter diel vertical migration patterns (Stramma et al., 2012). Only a fraction of the oxygen loss is directly related to warming and decreased O_2 solubility, and, therefore, most of the deoxygenation must reflect alterations and slowdowns of thermocline ventilation. In response to climate warming over the twenty-first century, model projections indicate further reductions in the global oxygen inventory and expansions of open-ocean oxygen minimum zones (Bopp et al., 2002; Frölicher et al., 2009).

6. MARINE BIOGEOCHEMICAL CYCLES–CLIMATE INTERACTIONS

Marine biota and especially microbes can affect the composition of the atmosphere by producing and consuming a number of trace gases that influence climate and atmospheric chemistry (Denman et al., 2007). Ocean sources and sinks of radiatively active trace gases such as CO_2, N_2O, methane, and dimethylsulfide (DMS) are of particular interest because they can mediate biological–climate feedbacks and the amplification or damping of external climate perturbations (Boyd and Doney, 2003; Doney and Schimel, 2007).

Methane, CO_2, and N_2O are powerful "greenhouse" or heat-trapping gases in the atmosphere, absorbing far-infrared or longwave radiation emitted by the land and ocean surface; an increase in atmospheric greenhouse gases thus leads to surface heating and climate warming (Kiehl and Trenberth, 1997; Solomon et al., 2007). Positive feedbacks with other components of the physical climate system (e.g., sea ice and snow cover) tend to amplify the heating caused by the initial biogeochemical radiative perturbations, leading to further warming; key ocean climate feedbacks include warming SSTs that increase evaporation and elevate atmospheric water vapor, another powerful greenhouse gas, and retreating bright, high-albedo sea ice that exposes dark, low-albedo waters that absorb more solar radiation.

Atmospheric CO_2 and climate have coevolved with the biosphere over the history of the Earth, with periods of elevated CO_2 generally reflected in warmer global climate conditions (Siegenthaler et al., 2005; Doney and Schimel, 2007; Royer et al., 2012). The net balance of sources and sinks that determines atmospheric CO_2 is sensitive to climate variations, and climate warming may both result from and cause higher atmospheric CO_2 levels. Ocean circulation and biogeochemistry are major factors governing atmospheric CO_2 (Denman et al., 2007; see also Chapter 30 on the carbon cycle; Tanhua et al. 2013). The large ocean inventory of inorganic carbon ($\sim$37,100 PgC) is roughly 50 times the CO_2 inventory in the preindustrial atmosphere ($\sim$590 PgC), and the ocean carbon pool is the largest mobile reservoir on the planet on timescales of decades to millennia (Sarmiento and Gruber, 2002).

Volk and Hoffert (1985) described a simple conceptual model for how solubility and biological processes or "pumps" affect the vertical redistribution of inorganic carbon within the ocean and thus total ocean carbon storage. CO_2 solubility is temperature dependent; warm water holds less DIC, and climate warming will therefore reduce ocean carbon storage. The biological pump consists of two components, the sinking fluxes of organic matter and inorganic $CaCO_3$ that transport carbon from the surface ocean to depth. Organic carbon production lowers surface water CO_2, driving a net CO_2 uptake from the atmosphere; in contrast, $CaCO_3$ shell formation reduces surface water alkalinity more than it reduces DIC and, thus, effectively causes a net efflux of CO_2 to the atmosphere.

As recorded in Antarctic ice cores, atmospheric CO_2 levels underwent large variations over glacial–interglacial cycles with low values of $\sim$180 ppm during cold glacial maxima and high values of $\sim$280 ppm during warm interglacial periods (EPICA community members, 2004). The large atmospheric CO_2 variations must reflect reorganizations of ocean circulation and biogeochemistry, leading to substantial changes in ocean carbon storage (Sigman and Boyle, 2000). During the last deglaciation, paleo-evidence indicates that initial northern hemisphere warming led to a reduction in Atlantic meridional overturning circulation that, in turn, triggered Southern Ocean CO_2 release and global warming (Shakun et al., 2012a,b). Numerous hypotheses have been proposed to explain glacial–interglacial ocean CO_2 variations associated with circulation, biological productivity, and sea-ice effects; recent work argues that CO_2 degassing was related to increases in Southern Ocean upwelling (Anderson et al., 2009). Atmospheric CO_2 variations have been considerably smaller during the recent warm Holocene (the past $\sim$11,000 years) with a rise of only $\sim$20 ppm over the past 7000 years attributed to a mix of terrestrial and oceanic processes including shallow-water carbonate deposition (coral reefs) and slow adjustment of deepwater chemistry and sediments (Menviel and Joos, 2012) (Figure 31.12).

Since the preindustrial era (i.e., since $\sim$1800 CE), atmospheric CO_2 levels have increased by more than 40% from a preindustrial level of approximately 280 ppm to 395 ppm at Mauna Loa observatory by the end of 2012 (Tans and Keeling, 2012). Based on isotopic composition and detailed carbon budgets, this CO_2 rise can be tied definitively to

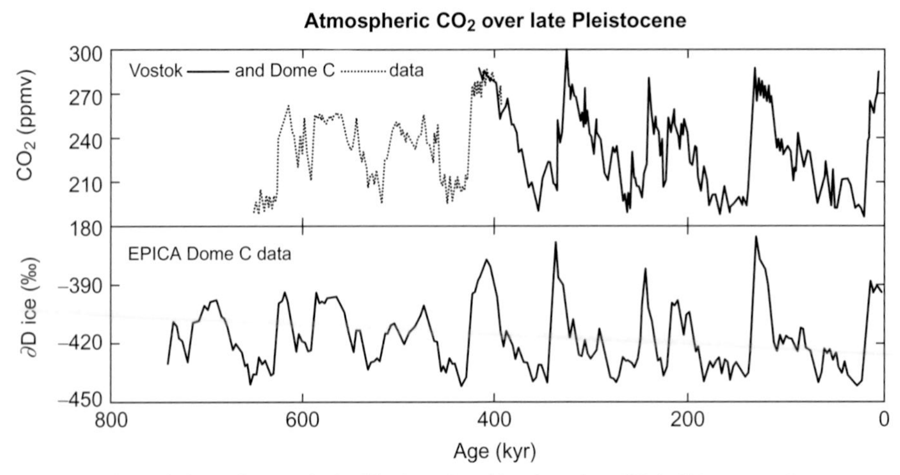

FIGURE 31.12 Glacial–interglacial variations of atmospheric CO_2 (ppmv) and ice deuterium (δD in ‰), a proxy for temperature (higher δD reflects warmer conditions), from Antarctic ice cores for the past 650,000 years. *Figure from Doney and Schimel (2007).*

human activities, in particular deforestation, fossil-fuel combustion, and cement manufacture. The ocean has played a critical climate service by removing some of the excess or anthropogenic CO_2 from the atmosphere (Sabine et al., 2004; Sabine and Tanhua, 2010). Estimated ocean carbon uptake in 2008 was 2.3 ± 0.4 PgC year^{-1} compared to a fossil-fuel combustion release to the atmosphere of 8.7 ± 0.5 PgC year^{-1} (Le Quéré et al., 2009). Cumulative ocean carbon uptake since the beginning of the industrial age is equivalent to about 25–30% of total human CO_2 emissions (Sabine and Tanhua, 2010). The global ocean uptake rate is governed primarily by the atmospheric CO_2 excess and trend and the rate of ocean circulation that exchanges surface waters equilibrated with elevated CO_2 levels with subsurface waters that have not yet been exposed to the anthropogenic CO_2 transient (Sarmiento et al., 1992; Khatiwala et al., 2009).

Ocean carbon storage is enhanced when more of subsurface nutrient inventory is biologically released rather than "preformed," the latter component referring to nutrients that are advected into the ocean interior from nutrient-rich surface waters (Ito and Follows, 2005). The largest reservoir of unused surface macronutrients resides in the Southern Ocean, and modeling studies suggest that ocean carbon storage is especially sensitive to Southern Ocean deepwater formation (Marinov et al., 2006). A number of other biogeochemical factors can enhance ocean carbon storage, and many of these mechanisms are sensitive to physical climate (Boyd and Doney, 2003; Denman et al., 2007). Warmer stratified conditions or increased iron inputs via dust deposition could promote subtropical nitrogen fixation, increasing the pool of bioavailable nitrogen (Boyd and Doney, 2002; Moore et al., 2006). Atmospheric iron deposition could also increase the extent of surface nutrient utilization in HNLC areas (Martin, 1990). Ocean acidification and climate-driven shifts in community composition may increase the carbon to nutrient and organic carbon to $CaCO_3$ stoichiometry of sinking export material (Oschlies et al., 2008; Gehlen et al., 2011). Plankton community shifts could also increase (or perhaps decrease) the vertical length scale over which sinking organic matter is regenerated (Kwon et al., 2009).

Climate warming is projected to reduce ocean uptake of anthropogenic CO_2 due to decreased solubility, increased vertical stratification, and slowing of intermediate and deepwater formation in the North Atlantic (Sarmiento and Le Quéré, 1996; Friedlingstein et al., 2006). The ocean also becomes less efficient with time at removing further atmospheric CO_2 because ocean acidification lowers seawater buffer capacity. Climate-governed changes in ocean vertical exchange also slow the nutrient supply to the surface resulting in reduced biological carbon export. In principle this should lower ocean CO_2 uptake even further; however this effect is counteracted by the reduced upward flux of metabolic CO_2, and the net climate effect on the biological pump is a modest increase in the effective carbon sink (Fung et al., 2005). Strengthening of the westerly winds in the Southern Ocean may be increasing vertical upwelling of CO_2-rich Circumpolar Deep Water, which would increase the ocean efflux of natural CO_2, reducing the global net anthropogenic CO_2 uptake (Le Quéré et al., 2007; Lovenduski et al., 2008). There is some evidence that anthropogenic climate change is already slowing ocean CO_2 uptake (Le Quéré et al., 2010). However, detecting a climate signal trend is difficult given the large seasonal and spatial variations in ocean CO_2 uptake and release (Takahashi et al., 2009) and the substantial interannual to interdecadal variability in air–sea CO_2 flux driven by natural climate modes, in particular ENSO (Park et al., 2010).

Ocean acidification may alter the ocean carbon cycle via impacts on the export flux and subsurface remineralization for either $CaCO_3$ or organic matter. The net effect on ocean carbon storage varies with both positive and negative feedbacks but is relatively small in current models. Increased carbon to nutrient ratios in sinking organic matter seen in some mesocosm experiments exposed to high CO_2 could expand subsurface low oxygen zones and increase N_2O production (Oschlies et al., 2008).

N_2O is produced by marine microbes as a minor byproduct of two biogeochemical pathways involved with organic matter respiration, nitrification ($NH_4^+ \rightarrow NO_3^-$) throughout most of the water column, and denitrification ($NO_3^- \rightarrow N_2$) restricted to hypoxic and suboxic waters (Codispoti, 2010). The traditional paradigm suggested that the two processes contribute about equal amounts to global N_2O production, though there is growing evidence that nitrification dominates (Dore et al., 1998; Freing et al., 2012). Subsurface ocean N_2O distributions are correlated with apparent oxygen utilization ($AOU = [O_2]_{saturation} - [O_2]_{measured}$), consistent with observations that the nitrification yield of N_2O per mole NO_3^- produced increases with declining O_2 (Nevison et al., 2003). Under climate warming scenarios, areal expansion of suboxic waters will likely increase marine denitrification but may not substantially alter ocean N_2O production (Bianchi et al., 2012), particularly if denitrification is a minor global source. On the other hand, broad-scale deoxygenation in the ocean thermocline could increase N_2O yield from nitrification. This may be partially countered by ocean acidification, which has been shown to slow microbial nitrification (Beman et al., 2011). The net effect is as yet uncertain.

Surface ocean DMS levels and air–sea fluxes are governed by a complex set of food-web interactions: phytoplankton production of nongaseous organic sulfur precursors; biological cleavage to DMS; bacterial and photochemical DMS destruction (Toole et al., 2008). In temperate waters, elevated DMS is associated with increased primary production and particular organosulfur-rich phytoplankton species; in subtropical and tropical regions, high DMS production is associated with ultraviolet and low-nutrient stress, both of which could result from warming and increased vertical stratification (Toole and Siegel, 2004). Modeling studies suggest climate warming may increase Southern Ocean DMS fluxes to the atmosphere because of sea-ice retreat and changes in phytoplankton composition (Cameron-Smith et al., 2011) though current model DMS parameterizations may be insufficient to make robust climate change projections (Halloran et al., 2010).

When released to the atmosphere, biologically produced DMS can form aerosols and cloud condensation nuclei in the remote marine atmosphere. In a seminal paper, Charlson et al. (1987) argued for a biological-climate regulation mechanism, by which warming would increase DMS flux to the atmosphere, leading to more marine stratus cloud cover and surface cooling. The so-called CLAW hypothesis, named after the authors of Charlson et al., involves a complex suite of biological and chemical steps (Vogt and Liss, 2009), and some researchers argue that, after more than two decades, there is little evidence to support biological control over cloud condensation nuclei levels and that the CLAW hypothesis should be abandoned (Quinn and Bates, 2011).

7. OBSERVATIONAL AND RESEARCH DIRECTIONS

Studying climate-ecosystem dynamics in the ocean is quite challenging because of the long timescales and large space scales involved as well as the complexity of marine food webs that span from viruses and bacteria to apex predators. In many cases, we know considerably more about the direct responses of a particular species to short-term physical variations and have to infer longer time-scale effects. Or we may have time series data for ecosystem parameters or species abundance but do not fully understand the underlying mechanisms or the indirect feedbacks on other species through altered food webs. Individual research techniques each have their own strengths and weaknesses, and a mix of different, complementary approaches is required combining observations, experiments, theory, and modeling. The time scales required to resolve climate–ecosystem interactions are inherently multiannual to multidecadal, and most available ocean field datasets are of insufficient duration. The situation is improving with time, but establishing and maintaining long-term observational records should be a top priority for the ocean research community. In many cases, observing the ecological response to interannual and decadal variability on a regional scale may inform estimates of future climate-driven biological trends that are more difficult or impossible to address through other approaches (Boyd and Doney, 2002).

Many key insights have been derived by co-opting time-series data that were originated for other purposes such as surveying commercial fishery stocks or addressing specific process-oriented questions. Problems may arise related to standardization of measurements, data quality, and data continuity when data were not collected with climate-scale analysis in mind. Even with such caveats, the availability of long-term ecological data is invaluable. Commonly used records include local and regional datasets such as CALCOFI, Joint Global Ocean Flux Study time series, the Atlantic Meridional Transect line, Long-Term Ecological Research (LTER) sites, and the CPR (Ducklow et al., 2009). For the upper-ocean, satellite data records of ocean color only began with the CZCS in the late 1970s, and the Sea-viewing Wide Field-of-view Sensor (SeaWiFS) provided more than a decade of nearly continuous global coverage of surface chlorophyll and primary productivity from late 1997 through 2010 (McClain, 2009; Siegel et al., 2013). Efforts are also underway to deploy more extensive global *in situ* observing systems that will inform marine ecology and biogeochemistry. Major advances have been made on integrating bio-optical and chemical sensors on new and autonomous observing platforms including: moorings, profiling floats, drifters, subsurface gliders, wave gliders, and AUVs (Johnson et al., 2009), and observational plans have been developed, for example, to add O_2, pH, and bio-optical sensors to the Argo float array (Gruber et al., 2010) or to track ocean acidification and subsequent biological impacts in open-ocean and coastal regions (Iglesias-Rodriguez et al., 2010).

Numerous process studies have been conducted on laboratory cultures and field samples to access the sensitivity of organisms and biological communities to temperature, CO_2, pH, O_2, nutrients, trace metals, and other environmental factors (Somero, 2012). However, some care must be taken in the interpretation of these results and their extrapolation to longer-term and ecosystem-level impacts, especially for short-duration perturbation experiments that simulate climate change from the acute biological responses to large, abrupt environmental changes. Process-based studies will remain an essential tool for providing a mechanistic framework to explain the phenomenological signals and trends in field observations, and moving forward, more emphasis is needed on the synergistic effect of multiple stressors (warming, lower pH, etc.) that will occur contemporaneously in the future (Boyd et al., 2008). Also, more emphasis is needed on the ecological resilience of marine systems (Bernhardt and Leslie, 2013) and microevolution in response to climate change, especially for planktonic organisms with relatively short generation time scales (Dam, 2013). Considerable insights have been derived from mesocosm experiments on planktonic communities in which various environmental parameters are manipulated either individually or in a factorial fashion. Open-ocean iron fertilization experiments have been a great success (Boyd et al., 2007), and emerging technologies, such as wave pumps (White et al., 2010) and free-ocean CO_2 release may allow for other types of ocean manipulation experiments (Kline et al., 2012).

A deeper mechanistic understanding is also desirable for deciphering the past and predicting the future using numerical models. Although statistical relationships can be deduced relating biological response to climate forcing, these relationships may not hold outside the bounds of present-day conditions. Over the twenty-first century, anthropogenic climate change may be substantial enough to create no-analogue or novel ocean ecosystem states, where, because of spatial range shifts and changes in abundance, the biological community does not match well any present-day system. Prognostic modeling can be a powerful tool for projecting into an uncertain future but only if the model dynamics is adequately known and the model tested thoroughly against available data (Glover et al., 2011). Marine ecology and biogeochemistry are increasingly being incorporated into basin and global ocean general circulation models as well as coupled ocean-atmosphere climate models, and substantial progress has been made from only a decade ago (Doney, 1999; see also Chapter 26 on modeling ocean biogeochemistry, Heinze and Gehlen, 2013). The skill of biological impact models, however, needs to be tested and improved, and in many cases, we may run up against the problem of predictability of complex biological systems, especially as stakeholders ask more focused questions related to individual species and specific locations.

Finally, climate change and other human activities, especially fishing and coastal habitat degradation, are negatively impacting the marine resources and fisheries upon which humans depend for food, personal security, and livelihoods (Allison et al., 2009; Cooley and Doney, 2009; Halpern et al., 2012). More research is needed to evaluate the efficacy of potential adaptation strategies and to better understand the consequences for human, social, and economic systems (Ruckelshaus et al., 2013).

ACKNOWLEDGMENTS

The author gratefully acknowledges support from the U.S. National Science Foundation through the Palmer LTER project (http://pal.lternet.edu/) (NSF OPP-0823101) and the Center for Microbial Oceanography Research and Education (C-MORE, http://cmore.soest.hawaii.edu/) (NSF EF-0424599). The author thanks H. Ducklow, K. Kroeker, C. Langdon, and D. Siegel for providing figures as well as H. Ducklow, W. Gould, S. Sailley, O. Schofield, and J. Shepherd for constructive comments.

REFERENCES

Alheit, J., Bakun, A., 2010. Population synchronies within and between ocean basins: apparent teleconnections and implications as to physical–biological linkage mechanisms. J. Mar. Syst. 79, 267–285.

Allison, E.H., Perry, A.L., Badjeck, M.-C., Adger, W.N., Brown, K., Conway, D., Halls, A.S., Pilling, G.M., Reynolds, J.D., Andrew, N.L., Dulvy, N.K., 2009. Vulnerability of national economies to the impacts of climate change on fisheries. Fish Fish. 10, 173–196.

Anderson, L.A., Sarmiento, J.L., 1994. Redfield ratios of remineralization determined by nutrient data analysis. Global Biogeochem. Cycles 8, 65–80.

Anderson, D.M., Glibert, P.M., Burkholder, J.M., 2002. Harmful algal blooms and eutrophication: nutrient sources, composition, and consequences. Estuaries 25, 704–726.

Anderson, R.F., Ali, S., Bradtmiller, L.I., Nielsen, S.H.H., Fleisher, M.Q., Anderson, B.E., Burckle, L.H., 2009. Wind-driven upwelling in the Southern Ocean and the deglacial rise in atmospheric CO_2. Science 323, 1443–1448.

Andersson, A.J., Mackenzie, F.T., Gattuso, J.-P., 2011. Effects of ocean acidification on pelagic organism and ecosystems. In: Gattuso, J.P.,

Hansson, L. (Eds.), Ocean Acidification. Oxford University Press, Oxford, UK, pp. 122–153.

Anthony, K.R.N., Kline, D.I., Diaz-Pulido, G., Dove, S., Hoegh-Guldberg, O., 2008. Ocean acidification causes bleaching and productivity loss in coral reef builders. Proc. Natl. Acad. Sci. U.S.A. 105, 17442–17446.

Antoine, D., Morel, A., Gordon, H.R., Banzon, V.F., Evans, R.H., 2005. Bridging ocean color observations of the 1980s and 2000s in search of long-term trends. J. Geophys. Res. 110, C06009. http://dx.doi.org/10.1029/2004JC002620.

Atkinson, A., Siegel, V., Pakhomov, E., Rothery, P., 2004. Long-term decline in krill stock and increase in salps within the Southern Ocean. Nature 432, 100–103.

Baker-Austin, C., Trinanes, J.A., Taylor, N.G.H., Hartnell, R., Siitonen, A., Martinez-Urtaza, J., 2013. Emerging *Vibrio* risk at high latitudes in response to ocean warming. Nat. Clim. Chang. 3, 73–77.

Barton, A., Hales, B., Waldbusser, G.G., Langdon, C., Feely, R., 2012. The Pacific oyster, *Crassostrea gigas*, shows negative correlation to naturally elevated carbon dioxide levels: implications for near-term ocean acidification effects. Limnol. Oceanogr. 57, 698–710.

Bates, N.R., Best, M.H.P., Neely, K., Garley, R., Dickson, A.G., Johnson, R.J., 2012. Detecting anthropogenic carbon dioxide uptake and ocean acidification in the North Atlantic Ocean. Biogeosciences 9, 2509–2522.

Beaugrand, G., Reid, P.C., Ibanez, F., Lindley, J.A., Edwards, M., 2002. Reorganization of North Atlantic marine copepod biodiversity and climate. Science 296, 1692–1694.

Beaulieu, C., Henson, S.A., Sarmiento, J.L., Dunne, J.P., Doney, S.C., Rykaczewski, R.R., Bopp, L., 2013. Factors challenging our ability to detect long-term trends in ocean chlorophyll. Biogeosciences 10, 2711–2724.

Behrenfeld, M.J., 2010. Abandoning Sverdrup's critical depth hypothesis on phytoplankton blooms. Ecology 91, 977–989.

Behrenfeld, M.J., Boss, E., Siegel, D.A., Shea, D.M., 2005. Carbon based ocean productivity and phytoplankton physiology from space. Global Biogeochem. Cycles 19, GB1006. http://dx.doi.org/10.1029/2004GB002299.

Behrenfeld, M.J., O'Malley, R.T., Siegel, D.A., McClain, C.R., Sarmiento, J.L., Feldman, G.C., Milligan, A.J., Falkowski, P.G., Letelier, R.M., Boss, E.S., 2006. Climate-driven trends in contemporary ocean productivity. Nature 444, 752–755.

Beman, J.M., Chow, C.-E., King, A.L., Feng, Y., Fuhrman, J.A., Andersson, A., Bates, N.R., Popp, B.N., Hutchins, D.A., 2011. Global declines in oceanic nitrification rates as consequence of ocean acidification. Proc. Natl. Acad. Sci. U.S.A. 108, 208–213.

Bender, M.A., Knutson, T.R., Tuleya, R.E., Sirutis, J.J., Vecchi, G.A., Garner, S.T., Held, I.M., 2010. Modeled impact of anthropogenic warming on the frequency of intense Atlantic hurricanes. Science 327, 454–458.

Bernhardt, J.R., Leslie, H.M., 2013. Resilience to climate change in coastal marine ecosystems. Ann. Rev. Mar. Sci. 5, 371–392.

Bestelmeyer, B.T., Ellison, A.M., Fraser, W.R., Gorman, K.B., Holbrook, S.J., Laney, C.M., Ohman, M.D., Peters, D.P.C., Pillsbury, F.C., Rassweiler, A., Schmitt, R.J., Sharma, S., 2011. Analysis of abrupt transitions in ecological systems. Ecosphere 2 (12), 1–26, Article 129.

Bianchi, D., Dunne, J.P., Sarmiento, J.L., Galbraith, E.D., 2012. Data-based estimates of suboxia, denitrification and N_2O production in the ocean and their sensitivities to dissolved oxygen. Global Biogeochem. Cycles 26, GB2009. http://dx.doi.org/10.1029/2011GB004209.

Bindoff, N.L., Willebrand, J., Artale, V., Cazenave, A., Gregory, J., Gulev, S., Hanawa, K., Le Quéré, C., Levitus, S., Nojiri, Y., Shum, C.K., Talley, L.D., Unnikrishnan, A., 2007. Observations: oceanic climate change and sea level. In: Solomon, S., Qin, D., Manning, M., Chen, Z., Marquis, M., Averyt, K.B., Tignor, M., Miller, H.L. (Eds.), Climate Change 2007: The Physical Science Basis, Contribution of Working Group I to the Fourth Assessment Report of the Intergovernmental Panel on Climate Change. Cambridge University Press, Cambridge, United Kingdom and New York, NY, USA, pp. 385–432.

Bopp, L., Le Quéré, C., Heimann, M., Manning, A.C., Monfray, P., 2002. Climate induced oceanic oxygen fluxes: implications for the contemporary carbon budget. Global Biogeochem. Cycles 16, 1022. http://dx.doi.org/10.1029/2001GB001445.

Bopp, L., Resplandy, L., Orr, J.C., Doney, S.C., Dunne, J.P., Gehlen, M., Halloran, P., Heinze, C., Ilyina, T., Seferian, R., Tjiputra, J., Vichi, M., 2013. Multiple stressors of ocean ecosystems in the 21st century: projections with CMIP5 models. Biogeosci. Discuss. 10, 3627–3676.

Boyce, D.G., Lewis, M.R., Worm, B., 2010. Global phytoplankton decline over the past century. Nature 466, 591–596.

Boyd, P.W., Doney, S.C., 2002. Modelling regional responses by marine pelagic ecosystems to global climate change. Geophys. Res. Lett. 29 (16), 1806. http://dx.doi.org/10.1029/2001GL014130.

Boyd, P., Doney, S.C., 2003. The impact of climate change and feedback process on the ocean carbon cycle. In: Fasham, M. (Ed.), Ocean Biogeochemistry. Springer, New York, NY, USA, pp. 157–193.

Boyd, P.W., Jickells, T., Law, C.S., Blain, S., Boyle, E.A., Buesseler, K.O., Coale, K.H., Cullen, J.J., de Baar, H.J.W., Follows, M., Harvey, M., Lancelot, C., Levasseur, M., Owens, N.P.J., Pollard, R., Rivkin, R.B., Sarmiento, J., Schoemann, V., Smetacek, V., Takeda, S., Tsuda, A., Turner, S., Watson, A.J., 2007. Mesoscale iron enrichment experiments 1993–2005: synthesis and future directions. Science 315, 612–627.

Boyd, P.W., Doney, S.C., Strzepek, R., Dusenberry, J., Lindsay, K., Fung, I., 2008. Climate-mediated changes to mixed-layer properties in the Southern Ocean: assessing the phytoplankton response. Biogeosciences 5, 847–864.

Brander, K.M., 2007. Global fish production and climate change. Proc. Natl. Acad. Sci. U.S.A. 104, 19709–19714.

Brander, K., 2010. Impacts of climate change on fisheries. J. Mar. Syst. 79, 389–402.

Brewer, P.G., Peltzer, E.T., 2009. Limits to marine life. Science 324, 347–348.

Brinton, E., Townsend, A., 2003. Decadal variability in abundances of the dominant euphausiid species in southern sectors of the California Current. Deep Sea Res. Part II 50, 2469–2492.

Brodeur, R.D., Ware, D.M., 1992. Interannual and interdecadal changes in zooplankton biomass in the subarctic Pacific Ocean. Fish. Oceanogr. 1, 32–38.

Bryan, F.O., Danabasoglu, G., Nakashiki, N., Yoshida, Y., Kim, D.H., Tsutsui, J., Doney, S.C., 2006. Response of North Atlantic thermohaline circulation and ventilation to increasing carbon dioxide in CCSM3. J. Clim. 19, 2382–2397.

Byrne, R.H., Mecking, S., Feely, R.A., Liu, X.W., 2010. Direct observations of basin-wide acidification of the North Pacific Ocean. Geophys. Res. Lett. 37, L02601. http://dx.doi.org/10.1029/2009GL040999.

Cai, W.-J., Xinping Hu, X., Huang, W.-J., Murrell, M.C., Lehrter, J.C., Lohrenz, S.E., Chou, W.-C., Zhai, W., Hollibaugh, J.T., Wang, Y., Zhao, P., Guo, X., Gundersen, K., Dai, M., Gong, G.-C., 2011. Acidification of subsurface coastal waters enhanced by eutrophication. Nat. Geosci. 4, 766–770.

Cameron-Smith, P., Elliott, S., Maltrud, M., Erickson, D., Wingenter, O., 2011. Changes in dimethyl sulfide oceanic distribution due to climate change. Geophys. Res. Lett. 38, L07704. http://dx.doi.org/10.1029/2011GL047069.

Chan, F., Barth, J.A., Lubchenco, J., Kirincich, A., Weeks, H., Peterson, W.T., Menge, B.A., 2008. Emergence of anoxia in the California Current large marine ecosystem. Science 319, 920.

Charlson, R.J., Lovelock, J.E., Andreae, M.O., Warren, S.G., 1987. Oceanic phytoplankton, atmospheric sulphur, cloud albedo and climate. Nature 326, 655–661.

Chavez, F.P., Strutton, P.G., Friederich, C.E., Feely, R.A., Feldman, G.C., Foley, D.C., McPhaden, M.J., 1999. Biological and chemical response of the equatorial Pacific Ocean to the 1997–98 El Nino. Science 286, 2126–2131.

Chavez, F.P., Ryan, J., Lluch-Cota, S.E., Niquen, M., 2003. From anchovies to sardines and back: multidecadal change in the Pacific Ocean. Science 299, 217–221.

Chavez, F.P., Messié, M., Pennington, J.T., 2011. Marine primary production in relation to climate variability and change. Ann. Rev. Mar. Sci. 3, 227–260.

Cheung, W.W.L., Lam, V.W.Y., Sarmiento, J.L., Kearney, K., Watson, R., Pauly, D., 2009. Projecting global marine biodiversity impacts under climate change scenarios. Fish Fish. 10, 235–251.

Chisholm, S.W., 2000. Oceanography: stirring time in the Southern Ocean. Nature 407, 685–687.

Codispoti, L.A., 2010. Interesting tines for marine N_2O. Science 327, 1339–1340.

Cook, T., Folli, M., Klinck, J., Ford, S., Miller, J., 1998. The relationship between increasing sea-surface temperature and the northward spread of *Perkinsus marinus* (Dermo) disease epizootics in oysters. Estuar. Coast. Shelf Sci. 46, 587–597.

Cooley, S.R., Doney, S.C., 2009. Anticipating ocean acidification's economic consequences for commercial fisheries. Environ. Res. Lett. 4, 024007. http://dx.doi.org/10.1088/1748-9326/4/2/024007.

Cooley, S., Kite-Powell, H.L., Doney, S.C., 2009. Ocean acidification's potential to alter global marine ecosystem services. Oceanography 22 (4), 172–180.

Cury, P., Bakun, A., Crawford, R.J.M., Jarre, A., Quiñones, R.A., Shannon, L.J., Verheye, H.M., 2000. Small pelagics in upwelling systems: patterns of interaction and structural changes in "wasp-waist" ecosystems. ICES J. Mar. Sci. 57, 603–618.

Cushing, D.H., 1990. Plankton production and year-class strength in fish populations: an update of the match/mismatch hypothesis. Adv. Mar. Biol. 26, 249–293.

Dam, H.G., 2013. Evolutionary adaptation of marine zooplankton to global change. Ann. Rev. Mar. Sci. 5, 349–370.

Davidson, E., 2009. The contribution of manure and fertilizer nitrogen to atmospheric nitrous oxide since 1860. Nat. Geosci. 2, 659–662.

Denman, K.L., Brasseur, G., Chidthaisong, A., Ciais, P., Cox, P.M., Dickinson, R.E., Hauglustaine, D., Heinze, C., Holland, E., Jacob, D., Lohmann, U., Ramachandran, S., da Silva Dias, P.L., Wofsy, S.C., Zhang, X., 2007. Couplings between changes in the climate system and biogeochemistry. In: Solomon, S., Qin, D., Manning, M., Chen, Z., Marquis, M., Averyt, K.B., Tignor, M., Miller, H.L. (Eds.), Climate Change 2007: The Physical Science Basis, Contribution of Working Group I to the Fourth Assessment Report of the Intergovernmental Panel on Climate Change. Cambridge University Press, Cambridge, United Kingdom and New York, NY, USA, pp. 499–587.

Deutsch, C., Weber, T., 2012. Nutrient ratios as a tracer and driver of ocean biogeochemistry. Ann. Rev. Mar. Sci. 4, 113–141.

Diaz, R.J., Rosenberg, R., 2008. Spreading dead zones and consequences for marine ecosystems. Science 321, 926–929.

Doney, S.C., 1999. Major challenges confronting marine biogeochemical modeling. Global Biogeochem. Cycles 13, 705–714.

Doney, S.C., 2006. Oceanography: plankton in a warmer world. Nature 444, 695–696.

Doney, S.C., 2010. The growing human footprint on coastal and open-ocean biogeochemistry. Science 328, 1512–1516.

Doney, S.C., Sailley, S.F., 2013. When an ecological regime shift is really just stochastic noise. Proc. Natl. Acad. Sci. U.S.A. 110 (7), 2438–2439.

Doney, S.C., Schimel, D.S., 2007. Carbon and climate system coupling on timescales from the Precambrian to the Anthropocene. Ann. Rev. Environ. Resour. 32, 31–66.

Doney, S.C., Fabry, V.J., Feely, R.A., Kleypas, J.A., 2009. Ocean acidification: the other CO_2 problem. Ann. Rev. Mar. Sci. 1, 169–192.

Doney, S.C., Ruckelshaus, M., Duffy, J.E., Barry, J.P., Chan, F., English, C.A., Galindo, H.M., Grebmeier, J.M., Hollowed, A.B., Knowlton, N., Polovina, J., Rabalais, N.N., Sydeman, W.J., Talley, L.D., 2012. Climate change impacts on marine ecosystems. Ann. Rev. Mar. Sci. 4, 11–37.

Donner, S.D., 2009. Coping with commitment: projected thermal stress on coral reefs under different future scenarios. PLoS One 4, e5712.

Dore, J.E., Popp, B.N., Karl, D.M., Sansone, F.J., 1998. A large source of atmospheric nitrous oxide from subtropical North Pacific surface waters. Nature 396, 63–66.

Dore, J.E., Lukas, R., Sadler, D.W., Church, M.J., Karl, D.M., 2009. Physical and biogeochemical modulation of ocean acidification in the central North Pacific. Proc. Natl. Acad. Sci. U.S.A. 106, 12235–12240.

Ducklow, H.W., Baker, K., Martinson, D.G., Quetin, L.B., Ross, R.M., Smith, R.C., Stammerjohn, S.E., Vernet, M., Fraser, W., 2007. Marine pelagic ecosystems: the West Antarctic Peninsula. Philos. Trans. R. Soc. B Biol. Sci. 362, 67–94.

Ducklow, H., Doney, S.C., Steinberg, D.K., 2009. Contributions of long-term research and time-series observations to marine ecology and biogeochemistry. Ann. Rev. Mar. Sci. 1, 279–302.

Dunne, J.P., Sarmiento, J.L., Gnanadesikan, A., 2007. A synthesis of global particle export from the surface ocean and cycling through the ocean interior and on the seafloor. Global Biogeochem. Cycles 21, GB4006. http://dx.doi.org/10.1029/2006GB002907.

Eakin, C.M., Morgan, J.A., Heron, S.F., Smith, T.B., Liu, G., Alvarez-Filip, L., Baca, B., Bartels, E., Bastidas, C., Bouchon, C., 2010. Caribbean corals in crisis: record thermal stress, bleaching, and mortality in 2005. PLoS One 5, e13969.

Edwards, M., Richardson, A.J., 2004. Impact of climate change on marine pelagic phenology and trophic mismatch. Nature 430, 881–884.

EPICA community members, 2004. Eight glacial cycles from an Antarctic ice core. Nature 429, 623–628.

Eppley, R., 1972. Temperature and phytoplankton growth in the sea. Fish. Bull. 70, 1063–1085.

Evans, G.T., Parslow, J.S., 1985. A model of annual plankton cycles. Biol. Oceanogr. 3, 327–347.

Fabricius, K.E., Langdon, C., Uthicke, S., Humphrey, C., Noonan, S., De'ath, G., Okazaki, R., Muehllehner, N., Glas, M.S., Lough, J.M., 2011. Losers and winners in coral reefs acclimatized to elevated carbon dioxide concentrations. Nat. Clim. Chang. 1, 165–169.

Fabry, V.J., Seibel, B.A., Feely, R.A., Orr, J.C., 2008. Impacts of ocean acidification on marine fauna and ecosystem processes. ICES J. Mar. Sci. 65, 414–432.

Feely, R.A., Sabine, C.L., Hernandez-Ayon, J.M., Ianson, D., Hales, B., 2008. Evidence for upwelling of corrosive "acidified" water onto the continental shelf. Science 320, 1490–1492.

Feely, R.A., Doney, S.C., Cooley, S.R., 2009. Ocean acidification: present conditions and future changes in a high-CO_2 world. Oceanography 22 (4), 36–47.

Feely, R.A., Alin, S.R., Newton, J., Sabine, C.L., Warner, M., Devol, A., Krembs, C., Maloy, C., 2010. The combined effects of ocean acidification, mixing, and respiration on pH and carbonate saturation in an urbanized estuary. Estuar. Coast. Shelf Sci. 88, 442–449.

Food and Agricultural Organization of the United Nations, 2007. The world's mangroves 1980–2005. FAO Forestry Paper 153. Food and Agricultural Organization of the United Nations, Rome, Italy, 77 pp.

Food and Agricultural Organization of the United Nations, 2010. Fishery and Aquaculture Statistics. FAO Yearbook 2008. Food and Agricultural Organization of the United Nations, Rome, Italy, 72 pp.

Fox, D., 2012. Trouble bares its claws. Nature 492, 170–172.

Francis, T.B., Scheuerell, M.D., Brodeur, R.D., Levin, P.S., Ruzicka, J.J., Tolimieri, N., Peterson, W.T., 2012. Climate shifts the interaction web of a marine plankton community. Glob. Chang. Biol 18, 2498–2508.

Freing, A., Wallace, D.W.R., Bange, H.W., 2012. Global oceanic production of nitrous oxide. Philos. Trans. R. Soc. B 367, 1245–1255.

Friedlingstein, P., Cox, P., Betts, R., Bopp, L., von Bloh, W., Brovkin, V., Cadule, P., Doney, S., Eby, M., Fung, I., Bala, G., John, J., Jones, C., Joos, F., Kato, T., Kawamiya, M., Knorr, W., Lindsay, K., Matthews, H.D., Raddatz, T., Rayner, P., Reick, C., Roeckner, E., Schnitzler, K.-G., Schnur, R., Strassmann, K., Weaver, A.J., Yoshikawa, C., Zeng, N., 2006. Climate–carbon cycle feedback analysis: results from the C^4MIP model intercomparison. J. Clim. 19 (14), 3337–3353.

Frölicher, T.L., Joos, F., Plattner, G.-K., Steinacher, M., Doney, S.C., 2009. Natural variability and anthropogenic trends in oceanic oxygen in a coupled carbon cycle-climate model ensemble. Global Biogeochem. Cycles 23, GB1003. http://dx.doi.org/10.1029/2008GB003316.

Fung, I., Doney, S.C., Lindsay, K., John, J., 2005. Evolution of carbon sinks in a changing climate. Proc. Natl. Acad. Sci. U.S.A. 102, 11201–11206.

Gardner, T.A., Côté, I.M., Gill, J.A., Grant, A., Watkinson, A.R., 2003. Long-term region-wide declines in Caribbean corals. Science 301, 958–960.

Gattuso, J.P., Hansson, L. (Eds.), 2011. Ocean Acidification. Oxford University Press, Oxford, UK, 326 pp.

Gattuso, J.P., Bijma, J., Gehlen, M., Riebesell, U., Turley, C., 2011. Ocean acidification: knowns, unknowns, and perspectives. In: Gattuso, J.P., Hansson, L. (Eds.), Ocean Acidification. Oxford University Press, Oxford, UK, pp. 291–311.

Gehlen, M., Gruber, N., Gangstø, R., Bopp, L., Oschlies, A., 2011. Biogeochemical consequences of ocean acidification and feedbacks to the earth system. In: Gattuso, J.P., Hansson, L. (Eds.), Ocean Acidification. Oxford University Press, Oxford, UK, pp. 230–248.

Geider, R.J., LaRoche, J., 2002. Redfield revisited: variability of C:N:P in marine microalgae and its biochemical basis. Eur. J. Phycol. 37, 1–17.

Geider, R.J., MacIntyre, H.L., Kana, T.M., 1997. Dynamic model of phytoplankton growth and acclimation: responses of the balanced growth rate and the chlorophyll a:carbon ratio to light, nutrient-limitation and temperature. Mar. Ecol. Prog. Ser. 148, 187–200.

Glover, D.M., Jenkins, W.J., Doney, S.C., 2011. Modeling Methods for Marine Science. Cambridge University Press, Cambridge, UK, 592 pp.

Gnanadesikan, A., Emanuel, K., Vecchi, G.A., Anderson, W.G., Hallberg, R., 2010. How ocean color can steer Pacific tropical cyclones. Geophys. Res. Lett. 37, L18802. http://dx.doi.org/10.1029/2010GL044514.

Grebmeier, J.M., 2012. Shifting patterns of life in the pacific Arctic and sub-Arctic seas. Ann. Rev. Mar. Sci. 4, 63–78.

Grebmeier, J.M., Cooper, L.W., Feder, H.M., Sirenko, B.I., 2006. Ecosystem dynamics of the Pacific-influenced Northern Bering and Chukchi Seas in the Amerasian Arctic. Prog. Oceanogr. 71, 331–361.

Grebmeier, J.M., Moore, S.E., Overland, J.E., Frey, K.E., Gradinger, R., 2010. Biological response to recent pacific Arctic sea ice retreats. Eos 91, 161–163.

Griffis, R., Howard J. (Eds.), 2012. Oceans and Marine Resources in a Changing Climate: Technical Input to the 2013 National Climate Assessment. United States Global Change Research Project, Washington, DC, 313 pp. Available from: http://downloads.usgcrp.gov/NCA/technicalinputreports/Griffis_Howard_Ocean_Marine_Resources.pdf

Gruber, N., 2011. Warming up, turning sour, losing breath: ocean biogeochemistry under global change. Philos. Trans. R. Soc. A 369, 1980–1996.

Gruber, N., Körtzinger, A., Borges, A., Claustre, H., Doney, S.C., Feely, R.A., Hood, M., Ishii, M., Kozyr, A., Monteiro, P., Nojiri, Y., Sabine, C.L., Schuster, U., Wallace, D.W.R., Wanninkhof, R., 2010. Toward an integrated observing system for ocean carbon and biogeochemistry at a time of change. In: Hall, J., Harrison, D.E., Stammer, D. (Eds.), Proceedings of OceanObs'09: Sustained Ocean Observations and Information for Society (Vol. 1). ESA Publication WPP-306. http://dx.doi.org/10.5270/OceanObs09.pp.18.

Gruber, N., Hauri, C., Lachkar, Z., Loher, D., Frölicher, T.L., Plattner, G.-K., 2012. Rapid progression of ocean acidification in the California current system. Science 337, 220–223.

Halloran, P.R., Bell, T.G., Totterdell, I.J., 2010. Can we trust empirical marine DMS parameterisations within projections of future climate? Biogeosciences 7, 1645–1656.

Hall-Spencer, J.M., Rodolfo-Metalpa, R., Martin, S., Ransome, E., Fine, M., Turner, S.M., Rowley, S.J., Tedesco, D., Buia, M.-C., 2008. Volcanic carbon dioxide vents show ecosystem effects of ocean acidification. Nature 454, 96–99.

Halpern, B.S., Longo, C., Hardy, D., McLeod, K.L., Samhouri, J.F., Katona, S.K., Kleisner, K., Lester, S.E., O'Leary, J., Ranelletti, M., Rosenberg, A.A., Scarborough, C., Selig, E.R., Best, B.D., Brumbaugh, D.R., Chapin, F.S., Crowder, L.B., Daly, K.L., Doney, S.C., Elfes, C., Fogarty, M.J., Gaines, S.D., Jacobsen, K.I., Karrer, L.B., Leslie, H.M., Neeley, E., Pauly, D., Polasky, S., Ris, B., St. Martin, K., Stone, G.S., Sumaila, U.R., Zeller, D., 2012.

An index to assess the health and benefits of the global ocean. Nature 488, 615–620.

Hare, S.R., Mantua, N.J., 2000. Empirical evidence for North Pacific regime shifts in 1977 and 1989. Prog. Oceanogr. 47, 103–145.

Harvell, C.D., Mitchell, C.E., Ward, J.R., Altizer, S., Dobson, A.P., Ostfeld, R.S., Samuel, M.D., 2002. Climate warming and disease risks for terrestrial and marine biota. Science 296, 2158–2162.

Heinze, C., Gehlen, M., 2013. Modeling ocean biogeochemical processes and resulting tracer distributions. In: Siedler, G., Church, J., Gould, J., Griffies, S. (Eds.), Ocean Circulation and Climate. second ed. Academic Press (Elsevier), Waltham, MA, USA.

Helm, K.P., Bindoff, N.L., Church, J.A., 2011. Observed decreases in oxygen content of the global ocean. Geophys. Res. Lett. 38, L23602. http://dx.doi.org/10.1029/2011GL049513.

Henson, S.A., Dunne, J.P., Sarmiento, J.L., 2009. Decadal variability in North Atlantic phytoplankton blooms. J. Geophys. Res. 114, C04013. http://dx.doi.org/10.1029/2008JC005139.

Henson, S.A., Sarmiento, J.L., Dunne, J.P., Bopp, L., Lima, I., Doney, S.C., John, J., Beaulieu, C., 2010. Detection of anthropogenic climate change in satellite records of ocean chlorophyll and productivity. Biogeosciences 7, 621–640.

Henson, S.A., Sanders, R., Madsen, E., Morris, P.J., Le Moigne, F., Quartly, G.D., 2011. A reduced estimate of the strength of the ocean's biological carbon pump. Geophys. Res. Lett. 38, L04606. http://dx.doi.org/10.1029/2011GL046735.

Hoegh-Guldberg, O., Mumby, P.J., Hooten, A.J., Steneck, R.S., Greenfield, P., Gomez, E., Harvell, C.D., Sale, P.F., Edwards, A.J., Caldeira, K., Knowlton, N., Eakin, C.M., Iglesias-Prieto, R., Muthiga, N., Bradbury, R.H., Dubi, A., Hatziolos, M.E., 2007. Coral reefs under rapid climate change and ocean acidification. Science 318, 1737–1742.

Hofmann, G.E., Smith, J.E., Johnson, K.S., Send, U., Levin, L.A., Micheli, F., Paytan, A., Price, N.M., Peterson, B., Takeshita, Y., Matson, P.G., Crook, E.D., Kroeker, K.J., Gambi, M.C., Rivest, E.B., Frieder, C.A., Yu, P.C., Martz, T.R., 2011. High-frequency dynamics of ocean pH: a multi-ecosystem comparison. PLoS One 6 (12), e28983.

Hollowed, A.B., Hare, S.R., Wooster, W.S., 2001. Pacific basin climate variability and patterns of Northeast Pacific marine fish production. Prog. Oceanogr. 49, 257–282.

Hönisch, B., Ridgwell, A., Schmidt, D.N., Thomas, E., Gibbs, S.J., Sluijs, A., Zeebe, R., Kump, L., Martindale, R.C., Greene, S.E., Kiessling, W., Ries, J., Zachos, J.C., Royer, D.L., Barker, S., Marchitto Jr., T.M., Moyer, R., Pelejero, C., Ziveri, P., Foster, G.L., Williams, B., 2012. The geological record of ocean acidification. Science 335, 1058–1063.

Hsieh, C.H., Glaser, S.M., Lucas, A.J., Sugihara, G., 2005. Distinguishing random environmental fluctuations from ecological catastrophes for the North Pacific Ocean. Nature 435, 336–340.

Hunt Jr., G.L., Coyle, K.O., Eisner, L.B., Farley, E.V., Heintz, R.A., Mueter, F., Napp, J.M., Overland, J.E., Ressler, P.H., Salo, S., Stabeno, P.J., 2011. Climate impacts on eastern Bering Sea foodwebs: a synthesis of new data and an assessment of the Oscillating Control Hypothesis. ICES J. Mar. Sci. 68, 1230.

Ianelli, J.N., Hollowed, A.B., Haynie, A.C., Mueter, F.J., Bond, N.A., 2011. Evaluating management strategies for eastern Bering Sea walleye pollock (*Theragra chalcogramma*) in a changing environment. ICES J. Mar. Sci. 68, 1297–1304.

Iglesias-Rodriguez, M.D., Fabry, V.J., Dickson, A., Gattuso, J.-P., Bijma, J., Riebesell, U., Doney, S., Turley, C., Saino, T., Lee, K., Anthony, K., Kleypas, J., 2010. Developing a global ocean acidification observation network. In: Hall, J., Harrison, D.E., Stammer, D. (Eds.), Proceedings of OceanObs'09: Sustained Ocean Observations and Information for Society. vol. 1. ESA Publication WPP-306. http://dx.doi.org/10.5270/OceanObs09.pp.24.

Ito, T., Follows, M.J., 2005. Preformed phosphate, soft tissue pump and atmospheric CO_2. J. Mar. Res. 63 (4), 813–839.

Jickells, T.D., An, Z.S., Andersen, K.K., Baker, A.R., Bergametti, G., Brooks, N., Cao, J.J., Boyd, P.W., Duce, R.A., Hunter, K.A., Kawahata, H., Kubilay, N., LaRoche, J., Liss, P.S., Mahowald, N., Prospero, J.M., Ridgwell, A.J., Tegen, I., Torres, R., 2005. Global iron connections between desert dust, ocean biogeochemistry, and climate. Science 308, 67–71.

Jochum, M., Yeager, S., Lindsay, K., Moore, K., Murtugudde, R., 2010. Quantification of the feedback between phytoplankton and ENSO in the Community Climate System Model. J. Clim. 11, 2916–2925.

Johnson, K.S., Berelson, W.M., Boss, E.S., Chase, Z., Claustre, H., Emerson, S.R., Gruber, N., Körtzinger, A., Perry, M.J., Riser, S.C., 2009. Observing biogeochemical cycles at global scales with profiling floats and gliders: prospects for a global array. Oceanography 22 (3), 216–225.

Jones, A., Berkelmans, R., 2010. Potential costs of acclimatization to a warmer climate: growth of a reef coral with heat tolerant versus sensitive symbiont types. PLoS One 5, e10437.

Karl, D., Letelier, R., Tupas, L., Dore, J., Christian, J., Hebel, D., 1997. The role of nitrogen fixation in biogeochemical cycling in the subtropical North Pacific Ocean. Nature 388, 533–538.

Kawasaki, T., 1992. Mechanisms governing fluctuations in pelagic fish populations. S. Afr. J. Mar. Sci. 12, 873–879.

Keeling, R.F., Körtzinger, A., Gruber, N., 2010. Ocean deoxygenation in a warming world. Ann. Rev. Mar. Sci. 2, 199–229.

Khatiwala, S., Primeau, F., Hall, T., 2009. Reconstruction of the history of anthropogenic CO_2 concentrations in the ocean. Nature 462, 346–349.

Kiehl, J.T., Trenberth, K.E., 1997. Earth's annual global mean energy budget. Bull. Am. Meteorol. Soc. 78 (2), 197–208.

Kline, D.I., Teneva, L., Schneider, K., Miard, T., Chai, A., Marker, M., Headley, K., Opdyke, B., Nash, M., Valetich, M., Caves, J.K., Russell, B.D., Connell, S.D., Kirkwood, B.J., Brewer, P., Peltzer, E., Silverman, J., Caldeira, K., Dunbar, R.B., Koseff, J.R., Monismith, S.G., Mitchell, B.G., Dove, S., Hoegh-Guldberg, O., 2012. A short-term in situ CO_2 enrichment experiment on Heron Island (GBR). Sci. Rep. 2, 413. http://dx.doi.org/10.1038/srep00413.

Kroeker, K.J., Kordas, R.L., Crim, R.N., Singh, G.G., 2010. Meta-analysis reveals negative yet variable effects of ocean acidification on marine organisms. Ecol. Lett. 13, 1419–1434.

Kroeker, K.J., Kordas, R.L., Crim, R.N., Hendriks, I.E., Ramajo, L., Singh, G.G., Duarte, C., Gattuso, J.P., 2013. Impacts of ocean acidification on marine organisms: quantifying sensitivities and interaction with warming. Glob. Chang. Biol 19, 1884–1896. http://dx.doi.org/10.1111/gcb.12179.

Kwon, E.Y., Primeau, F., Sarmiento, J.L., 2009. The impact of remineralization depth on the air-sea carbon balance. Nat. Geosci. 2, 630–635.

Lavaniegos, B., Ohman, M., 2003. Long-term changes in pelagic tunicates of the California Current. Deep Sea Res. Part II 50, 2473–2498.

Le Quéré, C., Rödenbeck, C., Buitenhuis, E.T., Conway, T.J., Langenfelds, R., Gomez, A., Labuschagne, C., Ramonet, M.,

Nakazawa, T., Metzl, N., Gillett, N., Heimann, M., 2007. Saturation of the southern ocean CO_2 sink due to recent climate change. Science 316, 1735–1738.

Le Quéré, C., Raupach, M.R., Canadell, J.G., Marland, G., Bopp, L., Ciais, P., Conway, T.J., Doney, S.C., Feely, R.A., Foster, P., Friedlingstein, P., Gurney, K., Houghton, R.A., House, J.I., Huntingford, C., Levy, P.E., Lomas, M.R., Majkut, J., Metzl, N., Ometto, J.P., Peters, G.P., Prentice, I.C., Randerson, J.T., Running, S.W., Sarmiento, J.L., Schuster, U., Sitch, S., Takahashi, T., Viovy, N., van der Werf, G.R., Woodward, F.I., 2009. Trends in the sources and sinks carbon dioxide. Nat. Geosci. 2, 831–836.

Le Quéré, C., Takahashi, T., Buitenhuis, E.T., Rödenbeck, C., Sutherland, S.C., 2010. Impact of climate change and variability on the global oceanic sink of CO_2. Global Biogeochem. Cycles 24, GB4007. http://dx.doi.org/10.1029/2009GB003599.

Levin, L.A., Ekau, W., Gooday, A.J., Jorissen, F., Middelburg, J.J., Naqvi, S.W.A., Neira, C., Rabalais, N.N., Zhang, J., 2009. Effects of natural and human-induced hypoxia on coastal benthos. Biogeosciences 6, 2063–2098.

Loeng, H., Brander, K., Carmack, E., Denisenko, S., Drinkwater, K., Hansen, B., Kovacs, K., Livingston, P., McLaughlin, F., Saksburg, E., 2005. Marine systems. In: Symon, C., Arris, L., Heal, B. (Eds.), Arctic Climate Impact Assessment. Cambridge University Press, Cambridge, UK, pp. 453–538.

Lovenduski, N., Gruber, N., Doney, S.C., 2008. Towards a mechanistic understanding of the decadal trends in the Southern Ocean carbon sink. Global Biogeochem. Cycles 22, GB3016. http://dx.doi.org/10.1029/2007GB003139.

MacFarling Meure, C., Etheridge, D., Trudinger, C., Steele, P., Langenfelds, R., van Ommen, T., Smith, A., Elkins, J., 2006. The Law Dome CO_2, CH_4 and N_2O ice core records extended to 2000 years BP. Geophys. Res. Lett. 33, L14810. http://dx.doi.org/10.1029/2006GL026152.

Mackas, D.L., Beaugrand, G., 2010. Comparisons of zooplankton time series. J. Mar. Syst. 79, 286–304.

Mann, K.H., Lazier, J.R., 2005. Dynamics of Marine Ecosystems: Biological–Physical Interactions in the Oceans. Wiley & Blackwell, Malden, Massachusetts, 512 pp.

Mantua, N., Hare, S., Zhang, Y., Wallace, J., Francis, R., 1997. A Pacific interdecadal climate oscillation with impacts on salmon production. Bull. Am. Meteorol. Soc. 78, 1069–1079.

Marinov, I., Gnanadesikan, A., Toggweiler, J.R., Sarmiento, J.L., 2006. The southern ocean biogeochemical divide. Nature 441, 964–967.

Marinov, I., Doney, S.C., Lima, I.D., 2010. Response of ocean phytoplankton community structure to climate change over the 21st century: partitioning the effects of nutrients, temperature and light. Biogeosciences 7, 3941–3959.

Martin, J.H., 1990. Glacial-interglacial CO_2 change: the iron hypothesis. Paleoceanography 5, 1–13.

Martin, J.H., Fitzwater, S.E., 1988. Iron deficiency limits phytoplankton growth in the north-east Pacific subarctic. Nature 331, 341–343.

Martiny, A.C., Pham, C.T.A., Primeau, F.W., Vrugt, J.A., Moore, J.K., Levin, S.A., Lomas, M.W., 2013. Strong latitudinal patterns in the elemental ratios of marine plankton and organic matter. Nat. Geosci. 6, 279–283.

McClain, C.R., 2009. A decade of satellite ocean color observations. Ann. Rev. Mar. Sci. 1, 19–42.

McGillicuddy Jr., D.J., Anderson, L.A., Doney, S.C., Maltrud, M.E., 2003. Eddy-driven sources and sinks of nutrients in the upper ocean: results from a 0.1 resolution model of the North Atlantic. Global Biogeochem. Cycles 17 (2), 1035. http://dx.doi.org/10.1029/2002GB001987.

Menviel, L., Joos, F., 2012. Toward explaining the Holocene carbon dioxide and carbon isotope records: results from transient ocean carbon cycle-climate simulations. Paleoceanography 27, PA1207. http://dx.doi.org/10.1029/2011PA002224.

Miller, C.B., Wheeler, P.A., 2012. Biological Oceanography, second ed. John Wiley & Sons, Oxford, UK, 480 pp.

Miller, J., Muller, E., Rogers, C.S., Waara, R., Atkinson, A., Whelan, K.R.T., Patterson, M., Witcher, B., 2009. Coral disease following massive bleaching in 2005 causes 60% decline in coral cover on reefs in the US Virgin Islands. Coral Reefs 28, 925–937.

Montes-Hugo, M., Doney, S.C., Ducklow, H.W., Fraser, W., Martinson, D., Stammerjohn, S.E., Schofield, O., 2009. Recent changes in phytoplankton communities associated with rapid regional climate change along the western Antarctic Peninsula. Science 323, 1470–1473.

Moore, J.K., Doney, S.C., Lindsay, K., Mahowald, N., Michaels, A.F., 2006. Nitrogen fixation amplifies the ocean biogeochemical response to decadal timescale variations in mineral dust deposition. Tellus 58B, 560–572.

Moran, X.A.G., Lopez-Urrutia, A., Calvo-Diaz, A., Li, W.K.W., 2010. Increasing importance of small phytoplankton in a warmer ocean. Glob. Chang. Biol 16 (3), 1137–1144.

Morel, A., Antoine, D., 1994. Heating rate within the upper ocean in relation to its biooptical state. J. Phys. Oceanogr. 24, 1652–1665.

Mueter, F.J., Bond, N.A., Ianelli, J.N., Hollowed, A.B., 2011. Expected declines in recruitment of walleye pollock (*Theragra chalcogramma*) in the eastern Bering Sea under future climate change. ICES J. Mar. Sci. 68, 1284–1296.

Murtugudde, R., Beauchamp, J., McClain, C.R., Lewis, M., Busalacchi, A.J., 2002. Effects of penetrative radiation on the upper tropical ocean circulation. J. Clim. 15, 470–486.

Nevison, C., Butler, J.H., Elkins, J.W., 2003. Global distribution of N_2O and ΔN_2O–AOU yield in the subsurface ocean. Global Biogeochem. Cycles 17, 1119, http://dx.doi.org/10.029/2003GB002068.

Nye, J.A., Link, J.S., Hare, J.A., Overholtz, W.J., 2009. Changing spatial distribution of fish stocks in relation to climate and population size on the Northeast United States continental shelf. Mar. Ecol. Prog. Ser. 393, 111–129.

Ohlmann, J.C., Siegel, D.A., Gautier, C., 1996. Ocean mixed layer radiant heating and solar penetration: a global analysis. J. Clim. 9, 2265–2280.

Orr, J.C., Fabry, V.J., Aumont, O., Bopp, L., Doney, S.C., Feely, R.A., Gnanadesikan, A., Gruber, N., Ishida, A., Joos, F., Key, R.M., Lindsay, K., Maier-Reimer, E., Matear, R., Monfray, P., Mouchet, A., Najjar, R.G., Plattner, G.-K., Rodgers, K.B., Sabine, C.L., Sarmiento, J.L., Schlitzer, R., Slater, R.D., Totterdell, I.J., Weirig, M.-F., Yamanaka, Y., Yool, A., 2005. Anthropogenic ocean acidification over the twenty-first century and its impact on marine calcifying organisms. Nature 437, 681–686.

Oschlies, A., Shulz, K.G., Riebesell, U., Schmittner, A., 2008. Simulated 21st century's increase in oceanic suboxia by CO_2-enhanced biotic carbon export. Global Biogeochem. Cycles 22, GB4008. http://dx.doi.org/10.1029/2007GB003147.

Overland, J.E., Alheit, J., Bakun, A., Hurrell, J.W., Mackas, D.L., Miller, A.J., 2010. Climate controls on marine ecosystems and fish populations. J. Mar. Syst. 79, 305–315.

Park, G.-H., Wanninkhof, R., Doney, S.C., Takahashi, T., Lee, K., Feely, R.A., Sabine, C.L., Triñanes, J., Lima, I.D., 2010. Variability of global net air-sea CO_2 fluxes over the last three decades using empirical relationships. Tellus 62B, 352–368.

Parmesan, C., 2006. Ecological and evolutionary responses to recent climate change. Ann. Rev. Ecol. Evol. Syst. 37, 637–669.

Paulmier, A., Ruiz-Pino, D., 2009. Oxygen minimum zones (OMZs) in the modern ocean. Prog. Oceanogr. 80, 113–128.

Perry, A.L., Low, P.J., Ellis, J.R., Reynolds, J.D., 2005. Climate change and distribution shifts in marine fishes. Science 308, 1912–1915.

Polovina, J.J., Howell, E.A., Abecassis, M., 2008. Ocean's least productive waters are expanding. Geophys. Res. Lett. 35, L03618. http://dx.doi.org/10.1029/2007GL031745.

Polovina, J.J., Dunne, J.P., Woodworth, P.A., Howell, E.A., 2011. Projected expansion of the subtropical biome and contraction of the temperate and equatorial upwelling biomes in the North Pacific under global warming. ICES J. Mar. Sci. 68, 986–995.

Pörtner, H.O., Farrell, A.P., 2008. Ecology: physiology and climate change. Science 322, 690–692.

Pörtner, A.P., Gutowska, M., Ishimatsu, A., Lucassen, M., Melzner, F., Seibel, B., 2011. Effects of ocean acidification on nektonic organisms. In: Gattuso, J.P., Hansson, L. (Eds.), Ocean Acidification. Oxford University Press, Oxford, UK, pp. 154–175.

Quinn, P.K., Bates, T.S., 2011. The case against climate regulation via oceanic phytoplankton sulphur emissions. Nature 480, 51–56.

Rabalais, N.N., Diaz, R.J., Levin, L.A., Turner, R.E., Gilbert, D., Zhang, J., 2010. Dynamics and distribution of natural and human-caused hypoxia. Biogeosciences 7, 585–619.

Redfield, A.C., 1958. The biological control of chemical factors in the environment. Am. Sci. 46, 205–221.

Redfield, A.C., Ketchum, B.H., Richards, F.A., 1963. The influence of organisms on the composition of seawater. In: Hill, M.N. (Ed.), The Sea, 2. Interscience, New York, USA, pp. 26–77.

Riebesell, U., Tortell, P.D., 2011. Effects of ocean acidification on pelagic organism and ecosystems. In: Gattuso, J.P., Hansson, L. (Eds.), Ocean Acidification. Oxford University Press, Oxford, UK, pp. 99–121.

Roemmich, D., McGowan, J., 1995. Climatic warming and the decline of zooplankton in the California Current. Science 267, 1324–1326.

Royer, D.L., Pagani, M., Beerling, D.J., 2012. Geobiological constraints on Earth system sensitivity to CO_2 during the Cretaceous and Cenozoic. Geobiology 10, 298–310.

Ruckelshaus, M., Doney, S.C., Galindo, H.M., Barry, J.P., Chan, F., Duffy, J.E., English, C.A., Gaines, S.D., Grebmeier, J.M., Hollowed, A.B., Knowlton, N., Polovina, J., Rabalais, N.N., Sydeman, W.J., Talley, L.D., 2013. Securing ocean benefits for society in the face of climate change. Mar. Policy, 40, 154–159.

Rykaczewski, R.R., Checkley Jr., D.M., 2008. Influence of ocean winds on the pelagic ecosystem in upwelling regions. Proc. Natl. Acad. Sci. U.S.A. 105, 1965–1970.

Rykaczewski, R.R., Dunne, J.P., 2011. A measured look at ocean chlorophyll trends. Nature 472, E5–E6. http://dx.doi.org/10.1038/nature09952.

Saba, V.S., Friedrichs, M.A.M., Carr, M.-E., Antoine, D., Armstrong, R.A., Asanuma, I., Aumont, O., Bates, N.R., Behrenfeld, M.J., Bennington, V., Bopp, L., Bruggeman, J., Buitenhuis, E.T., Church, M.J., Ciotti, A.M., Doney, S.C., Dowell, M., Dunne, J., Dutkiewicz, S., Gregg, W., Hoepffner, N., Hyde, K.J.W., Ishizaka, J., Kameda, T., Karl, D.M., Lima, I., Lomas, M.W., Marra, J., McKinley, G.A., Mélin, F., Moore, J.K., Morel, A., O'Reilly, J., Salihoglu, B., Scardi, M., Smyth, T.J., Tang, S., Tjiputra, J., Uitz, J., Vichi, M., Waters, K., Westberry, T.K., Yool, A., 2010. Challenges of modeling depth-integrated marine primary productivity over multiple decades: a case study at BATS and HOT. Global Biogeochem. Cycles 24, GB3020. http://dx.doi.org/10.1029/2009GB003655.

Sabine, C.L., Toste Tanhua, T., 2010. Estimation of anthropogenic CO_2 inventories in the ocean. Ann. Rev. Mar. Sci. 2, 175–198.

Sabine, C.L., Feely, R.A., Gruber, N., Key, R.M., Lee, K., Gruber, Nicolas, Bullister, J.L., Wanninkhof, R., Wong, C.S., Wallace, D.W.R., Tilbrook, B., Millero, F.J., Peng, T.-H., Kozyr, A., Ono, T., Rios, A.F., 2004. The oceanic sink for anthropogenic CO_2. Science 305, 367–371.

Sarmiento, J.L., Gruber, N., 2002. Sinks for anthropogenic carbon. Phys. Today 55, 30–36.

Sarmiento, J.L., Gruber, N., 2006. Ocean Biogeochemical Dynamics. Princeton University Press, Princeton, NJ, 503 pp.

Sarmiento, J.L., Orr, J.C., Siegenthaler, U., 1992. A perturbation simulation of CO2 uptake in an ocean general circulation model. J. Geophys. Res. 97, 3621–3645.

Sarmiento, J.L., Le Quéré, C., 1996. Oceanic carbon dioxide uptake in a model of century-scale global warming. Science 274, 1346–1350.

Sarmiento, J.L., Gruber, N., Brzezinski, M.A., Dunne, J.P., 2004a. High-latitude controls of thermocline nutrients and low latitude biological productivity. Nature 427, 56–60.

Sarmiento, J., Slater, R., Barber, R., Bopp, L., Doney, S.C., Hirst, A.C., Kleypas, J., Matear, R., Mikolajewicz, U., Monfray, P., Soldatov, V., Spall, S., Slater, R., Stouffer, R., 2004b. Response of ocean ecosystems to climate warming. Global Biogeochem. Cycles 18, GB3003. http://dx.doi.org/10.1029/2003GB002134.

Schofield, O., Ducklow, H.W., Martinson, D.G., Meredith, M.P., Moline, M.A., Fraser, W.R., 2010. How do polar marine ecosystems respond to rapid climate change? Science 328, 1520–1523.

Seitzinger, S.P., Mayorga, E., Bouwman, A.F., Kroeze, C., Beusen, A.H.W., Billen, G., Van Drecht, G., Dumont, E., Fekete, B.M., Garnier, J., Harrison, J.A., 2010. Global river nutrient export: a scenario analysis of past and future trends. Global Biogeochem. Cycles 24, GB0A08. http://dx.doi.org/10.1029/2009GB003587.

Shakun, J.D., Clark, P.U., He, F., Marcott, S.A., Mix, A.C., Liu, Z., Otto-Bliesner, B., Schmittner, A., Bard, E., 2012a. Global warming preceded by increasing carbon dioxide concentrations during the last deglaciation. Nature 484, 49–54.

Shakun, J.D., Clark, P.U., He, F., Marcott, S.A., Mix, A.C., Liu, Z., Otto-Bliesner, B., Schmittner, A., Bard, E., 2012b. Global warming preceded by increasing carbon dioxide concentrations during the last deglaciation. Nature 484, 49–54.

Siegel, D.A., Doney, S.C., Yoder, J.A., 2002. The North Atlantic spring phytoplankton bloom and Sverdrup's critical depth hypothesis. Science 296, 730–733.

Siegel, D.A., Behrenfeld, M.J., Maritorena, S., McClain, C.R., Antoine, D., Bailey, S.W., Bontempi, P.S., Boss, E.S., Dierssen, H.M., Doney, S.C., Eplee Jr., R.E., Evans, R.H., Feldman, G.C., Fields, E., Franz, B.A., Kuring, N.A., Mengelt, C., Nelson, N.B., Patt, F.S., Robinson, W.D., Sarmiento, J.L., Swan, C.S., Werdell, P.J., Westberry, T.K., Wilding, J.G., Yoder, J.A., 2013. Regional to global assessments of phytoplankton dynamics from the SeaWiFS mission. Remote Sens. Environ. 135, 77–91.

Siegenthaler, U., Stocker, T.F., Monnin, E., Lüthi, D., Schwander, J., Stauffer, B., Raynaud, D., Barnola, J.-M., Fischer, H., Masson-Delmotte, V., Jouzel, J., 2005. Stable carbon cycle-climate relationship during the late Pleistocene. Science 310, 1313–1317.

Sigler, M.F., Renner, M., Danielson, S.L., Eisner, L.B., Lauth, R.R., Kuletz, K.J., Logerwell, E.A., Hunt Jr., G.L., 2011. Fluxes, fins, and feathers: relationships among the Bering, Chukchi, and Beaufort Seas in a time of climate change. Oceanography 24 (3), 250–265.

Sigman, D.M., Boyle, E.A., 2000. Glacial/interglacial variations in atmospheric carbon dioxide. Nature 407, 859–869.

Smith, T.M., Reynolds, R.W., Peterson, T.C., Lawrimore, J., 2008. Improvements to NOAA's historical merged land-ocean surface temperature analysis (1880–2006). J. Clim. 21, 2283–2296.

Smith, C.R., Grange, L.J., Honig, D.L., Naudts, L., Huber, B., Guidi, L., Domack, E., 2012. A large population of king crabs in the Palmer Deep on the west Antarctic Peninsula shelf and potential invasive impacts. Proc. R. Soc. B 279, 1017–1026.

Solomon, S., Qin, D., Manning, M., Alley, R.B., Berntsen, T., Bindoff, N.L., Chen, Z., Chidthaisong, A., Gregory, J.M., Hegerl, G.C., Heimann, M., Hewitson, B., Hoskins, B.J., Joos, F., Jouzel, J., Kattsov, V., Lohmann, U., Matsuno, T., Molina, M., Nicholls, N., Overpeck, J., Raga, G., Ramaswamy, V., Ren, J., Rusticucci, M., Somerville, R., Stocker, T.F., Whetton, P., Wood, R.A., Wratt, D., 2007. Technical summary. In: Solomon, S., Qin, D., Manning, M., Chen, Z., Marquis, M., Averyt, K.B., Tignor, M., Miller, H.L. (Eds.), Climate Change 2007: The Physical Science Basis, Contribution of Working Group I to the Fourth Assessment Report of the Intergovernmental Panel on Climate Change. Cambridge University Press, Cambridge, United Kingdom and New York, NY, USA, pp. 19–91.

Somero, G.N., 2012. The physiology of global change: linking patterns to mechanisms. Ann. Rev. Mar. Sci. 4, 39–61.

Steinacher, M., Joos, F., Frölicher, T.L., Plattner, G.-K., Doney, S.C., 2009. Imminent ocean acidification in the Arctic projected with the NCAR global coupled carbon cycle-climate model. Biogeosciences 6, 515–533.

Steinacher, M., Joos, F., Frölicher, T.L., Bopp, L., Cadule, P., Cocco, V., Doney, S.C., Gehlen, M., Lindsay, K., Moore, J.K., Schneider, B., Segschneider, J., 2010. Projected 21st century decrease in marine productivity: a multi-model analysis. Biogeosciences 7, 979–1005.

Steinberg, D.K., Lomas, M.W., Cope, J.S., 2012. Long-term increase in mesozooplankton biomass in the Sargasso Sea: linkage to climate and implications for food web dynamics and biogeochemical cycling. Global Biogeochem. Cycles 26, GB1004. http://dx.doi.org/10.1029/2010GB004026.

Stenseth, N.C., Mysterud, A., 2002. Climate, changing phenology, and other life history traits: nonlinearity and match–mismatch to the environment. Proc. Natl. Acad. Sci. U.S.A. 99 (21), 13379–13381.

Stenseth, N.C., Ottersen, G., Hurrell, J.W., Mysterud, A., Lima, M., Chan, K.-S., Yoccoz, N.G., Adlandsvik, B., 2003. Studying climate effects on ecology through the use of climate indices: the North Atlantic Oscillation, El Nino Southern Oscillation and beyond. Proc. R. Soc. B Biol. Sci. 270 (1529), 2087–2096.

Stock, C.A., Alexander, M.A., Bond, N.A., Brander, K.M., Cheung, W.W.L., Curchitser, E.N., Delworth, T.L., Dunne, J.P., Griffies, S.M., Haltuch, M.A., Hare, J.A., Hollowed, A.B., Lehodey, P., Levin, S.A., Link, J.S., Rose, K.A., Rykaczewski, R.R., Sarmiento, J.L., Stouffer, R.J., Schwing, F.B., Vecchi, G.A., Werner, F.E., 2011. On the use of IPCC-class models to assess the impact of climate on living marine resources. Prog. Oceanogr. 88, 1–27.

Stramma, L., Johnson, G.C., Sprintall, J., Mohrholz, V., 2008. Expanding oxygen-minimum zones in the tropical oceans. Science 320, 655–658.

Stramma, L., Prince, E.D., Schmidtko, S., Jiangang Luo, J., Hoolihan, J.P., Visbeck, M., Wallace, D.W.R., Brandt, P., Körtzinger, A., 2012. Expansion of oxygen minimum zones may reduce available habitat for tropical pelagic fishes. Nat. Clim. Chang. 2, 33–37.

Stroeve, J.C., Kattsov, V., Barrett, A., Serreze, M., Pavlova, T., Holland, M., Meier, W.N., 2012. Trends in Arctic sea ice extent from CMIP5, CMIP3 and observations. Geophys. Res. Lett. 39, L16502. http://dx.doi.org/10.1029/2012GL052676.

Sumaila, U.R., Cheung, W.W.L., Lam, V.W.Y., Pauly, D., Herrick, S., 2011. Climate change impacts on the biophysics and economics of world fisheries. Nat. Clim. Chang. 1, 449–456.

Sweeney, C., Gnanadesikan, A., Griffies, S.M., Harrison, M.J., Rosati, A.J., Samuels, B.L., 2005. Impacts of shortwave penetration depth on large-scale ocean circulation and heat transport. J. Phys. Oceanogr. 35, 1103–1119.

Takahashi, T., Sutherland, S.C., Wanninkhof, R., Sweeney, C., Feely, R.A., Chipman, D.W., Hales, B., Friederich, G., Chavez, F., Sabine, C., Watson, A., Bakker, D.C.E., Schuster, U., Metzl, N., Inoue, H.Y., Ishii, M., Midorikawa, T., Nojiri, Y., Koertzinger, A., Steinhoff, T., Hoppema, M., Olafsson, J., Arnarson, T.S., Tilbrook, B., Johannessen, T., Olsen, A., Bellerby, R., Wong, C.S., Delille, B., Bates, N.R., de Baar, H.J.W., 2009. Climatological mean and decadal change in surface ocean pCO_2, and net sea-air CO_2 flux over the global oceans. Deep Sea Res. Part II 56, 554–577.

Tanhua, et al., 2013. The carbon cycle and the ocean. In: Siedler, G., Church, J., Gould, J., Griffies, S. (Eds.), Ocean Circulation and Climate. second ed. Academic Press (Elsevier), Waltham, MA, USA.

Tans, P., Keeling, R., 2012. Trends in atmospheric carbon dioxide, full Mauna Loa CO_2 record. http://www.esrl.noaa.gov/gmd/ccgg/trends/, Dr. Pieter Tans, NOAA/ESRL (www.esrl.noaa.gov/gmd/ccgg/trends/) and Dr. Ralph Keeling, Scripps Institution of Oceanography (scrippsco2.ucsd.edu/) (Accessed Dec. 2012).

Thomas, M.K., Kremer, C.T., Klausmeier, C.A., Litchman, E., 2012. A global pattern of thermal adaptation in marine phytoplankton. Science 338, 1085–1088.

Toole, D.A., Siegel, D.A., 2004. Light-driven cycling of dimethylsulfide (DMS) in the Sargasso Sea: closing the loop. Geophys. Res. Lett. 31, L09308. http://dx.doi.org/10.1029/2004GL019581.

Toole, D.A., Siegel, D.A., Doney, S.C., 2008. A light-driven, one-dimensional dimethylsulfide biogeochemical cycling model for the Sargasso Sea. J. Geophys. Res. 113, G02009. http://dx.doi.org/10.1029/2007JG000426.

Van Mooy, B.A.S., Fredricks, H.F., Pedler, B.E., Dyhrman, S.T., Karl, D.M., Koblížek, M., Lomas, M.W., Mincer, T.J., Moore, L.R., Moutin, T., Rappé, M.S., Webb, E.A., 2009. Phytoplankton in the ocean use non-phosphorus lipids in response to phosphorus scarcity. Nature 458, 69–72.

Vogt, M., Liss, P.S., 2009. Dimethylsulfide and climate. In: Le Quéré, C., Saltzman, E.S. (Eds.), Surface Ocean—Lower Atmosphere Processes. Geophysical Monograph Series, vol. 187. American Geophysical Union, Washington, DC, pp. 197–232. http://dx.doi.org/10.1029/2008GM000790.

Volk, T., Hoffert, M.I., 1985. Ocean carbon pumps: analysis of relative strengths and efficiencies in ocean-driven atmospheric CO_2 changes.

In: The Carbon Cycle and Atmospheric CO_2: Natural Variations Archean to Present. Geophysical Monograph Series, vol. 32. American Geophysical Union, Washington, DC, USA, pp. 99–110.

Ward, J.R., Lafferty, K.D., 2004. The elusive baseline of marine disease: are diseases in ocean ecosystems increasing? PLoS Biol. 2, e120.

Wassmann, P., Duarte, C.M., Agusti, S., Sejr, M.K., 2011. Footprints of climate change in the Arctic marine ecosystem. Glob. Chang. Biol 17, 1235–1249.

Waycott, M., Duarte, C.M., Carruthers, T.J.B., Orth, R.J., Dennison, W.C., Olyarnik, S., Calladine, A., Fourqurean, J.W., Heck Jr., K.L., Hughes, A.R., Kendrick, G.A., Kenworthy, W.J., Short, F.T., Williams, S.L., 2009. Accelerating loss of seagrasses across the globe threatens coastal ecosystems. Proc. Natl. Acad. Sci. U.S.A. 106, 12377–12381.

Weil, E., Croquer, A., Urreiztieta, I., 2009. Temporal variability and consequences of coral diseases and bleaching in La Parguera, Puerto Rico from 2003–2007. Caribb. J. Sci. 45, 221–246.

White, A.E., Björkman, K., Grabowski, E., Letelier, R.M., Poulos, S., Watkins, B., Karl, D.M., 2010. An open ocean trial of controlled upwelling using wave pump technology. J. Atmos. Oceanic Technol. 27, 385–396.

Whitney, F.A., Freeland, H.J., Robert, M., 2007. Persistently declining oxygen levels in the interior waters of the eastern subarctic Pacific. Prog. Oceanogr. 75, 179–199.

Williams, R.G., Follows, M.J., 2011. Ocean Dynamics and the Carbon Cycle. Cambridge University Press, Cambridge, UK, 404 pp.

Wilson, S., Fischetti, T., 2010. Coastline population trends in the United States: 1960 to 2008. In: Current Population Reports: Estimates and projections, P25-1139. US Census Bureau, Washington DC, 27 pp.

Wolf, S.G., Snyder, M.A., Sydeman, W.J., Doak, D.F., 2010. Predicting population consequences of ocean climate change for an ecosystem sentinel, the seabird Cassin's auklet. Glob. Chang. Biol 16, 1923–1935.

Index

Note: Page numbers followed by '*f*' indicate figures and '*t*' indicate tables.

A

B

C

D

E

F

H

I

J

K

L

M

O

P

Q

R

S

T

U

V

W

X

Y

Z